FOURTH EDITION

CALCULUS
EARLY TRANSCENDENTALS

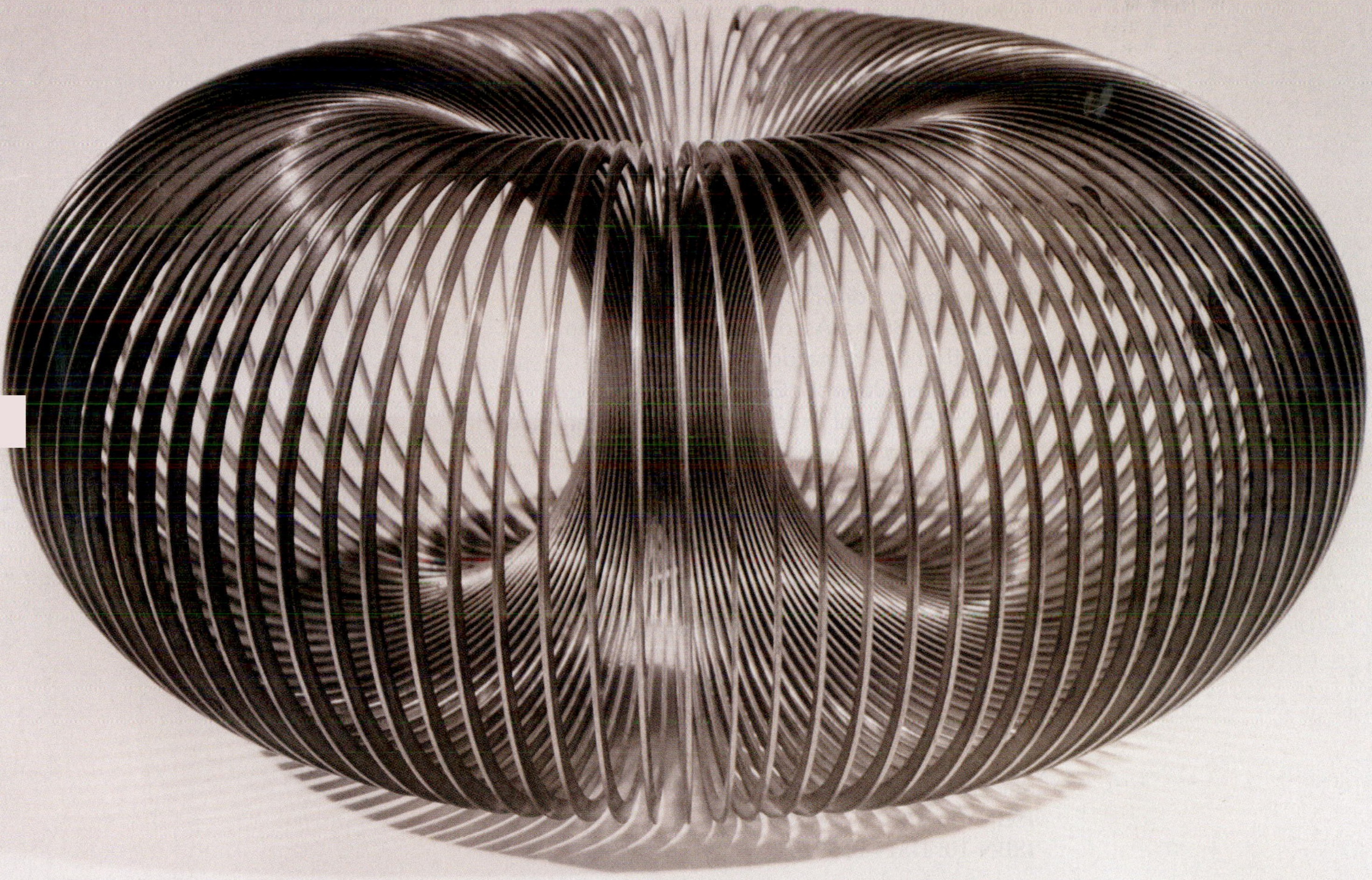

Jon Rogawski · Colin Adams · Robert Franzosa

University of California, Los Angeles   Williams College   The University of Maine

w.h.freeman
Macmillan Learning

New York

TO JULIE –Jon
TO ALEXA AND COLTON –Colin
TO MY FAMILY –Bob

Vice President, STEM: Daryl Fox
Program Director: Andy Dunaway
Program Manager: Nikki Miller Dworsky
Senior Marketing Manager: Nancy Bradshaw
Marketing Assistant: Madeleine Inskeep
Executive Development Editor: Katrina Mangold
Development Editor: Tony Palermino
Executive Media Editor: Catriona Kaplan
Associate Editor: Andy Newton
Editorial Assistant: Justin Jones
Director, Content Management Enhancement: Tracey Kuehn
Senior Managing Editor: Lisa Kinne
Senior Content Project Manager: Kerry O'Shaughnessy
Senior Workflow Project Manager: Paul Rohloff
Director of Design, Content Management: Diana Blume
Design Services Manager: Natasha Wolfe
Cover Design Manager: John Callahan
Interior & Cover Design: Lumina Datamatics, Inc.
Senior Photo Editor: Sheena Goldstein
Rights and Billing Associate: Alexis Gargin
Illustration Coordinator: Janice Donnola
Illustrations: Network Graphics
Director of Digital Production: Keri deManigold
Senior Media Project Manager: Alison Lorber
Media Project Manager: Hanna Squire
Composition: Lumina Datamatics, Inc.
Printing and Binding: LSC Communications
Cover Photo: Bettmann/Getty Images

Library of Congress Control Number: 2018959764

Student Edition Hardcover:
ISBN-13: 978-1-319-05074-0
ISBN-10: 1-319-05074-3

Student Edition Loose-leaf:
ISBN-13: 978-1-319-05591-2
ISBN-10: 1-319-05591-5

Instructor Complimentary Copy:
ISBN-13: 978-1-319-05581-3
ISBN-10: 1-319-05581-8

Printed in the United States of America

1 2 3 4 5 6   23 22 21 20 19 18

W. H. Freeman and Company
One New York Plaza
Suite 4500
New York, NY 10004-1562
www.macmillanlearning.com

ABOUT THE AUTHORS

Jon Rogawski

As a successful teacher for more than 30 years, Jon Rogawski listened and learned much from his own students. These valuable lessons made an impact on his thinking, his writing, and his shaping of a calculus text.

Jon Rogawski received his undergraduate and master's degrees in mathematics simultaneously from Yale University, and he earned his PhD in mathematics from Princeton University, where he studied under Robert Langlands. Before joining the Department of Mathematics at UCLA in 1986, where he was a full professor, he held teaching and visiting positions at the Institute for Advanced Study, the University of Bonn, and the University of Paris at Jussieu and Orsay.

Jon's areas of interest were number theory, automorphic forms, and harmonic analysis on semisimple groups. He published numerous research articles in leading mathematics journals, including the research monograph *Automorphic Representations of Unitary Groups in Three Variables* (Princeton University Press). He was the recipient of a Sloan Fellowship and an editor of the *Pacific Journal of Mathematics* and the *Transactions of the AMS*.

Sadly, Jon Rogawski passed away in September 2011. Jon's commitment to presenting the beauty of calculus and the important role it plays in students' understanding of the wider world is the legacy that lives on in each new edition of *Calculus*.

Colin Adams

Colin Adams is the Thomas T. Read professor of Mathematics at Williams College, where he has taught since 1985. Colin received his undergraduate degree from MIT and his PhD from the University of Wisconsin. His research is in the area of knot theory and low-dimensional topology. He has held various grants to support his research and written numerous research articles.

Colin is the author or co-author of *The Knot Book, How to Ace Calculus: The Streetwise Guide, How to Ace the Rest of Calculus: The Streetwise Guide, Riot at the Calc Exam and Other Mathematically Bent Stories, Why Knot?, Introduction to Topology: Pure and Applied,* and *Zombies & Calculus.* He co-wrote and appears in the videos "The Great Pi vs. e Debate" and "Derivative vs. Integral: the Final Smackdown."

He is a recipient of the Haimo National Distinguished Teaching Award from the Mathematical Association of America (MAA) in 1998, an MAA Polya Lecturer for 1998-2000, a Sigma Xi Distinguished Lecturer for 2000-2002, and the recipient of the Robert Foster Cherry Teaching Award in 2003.

Colin has two children and one slightly crazy dog, who is great at providing the entertainment.

Robert Franzosa

Robert (Bob) Franzosa is a professor of mathematics at the University of Maine where he has been on the faculty since 1983. Bob received a BS in mathematics from MIT in 1977 and a PhD in mathematics from the University of Wisconsin in 1984. His research has been in dynamical systems and in applications of topology in geographic information systems. He has been involved in mathematics education outreach in the state of Maine for most of his career.

Bob is a co-author of *Introduction to Topology: Pure and Applied* and *Algebraic Models in Our World.* He was awarded the University of Maine's Presidential Outstanding Teaching award in 2003.

Bob is married, has two children, three step-children, and one grandson.

SECTION 2.1 was largely rewritten so that its focus is on motivating the need for limits via the concepts of velocity and the tangent line. The content on rate of change that did not treat velocity was moved elsewhere.

SECTION 4.1 was rewritten and reorganized to clarify the relationship between the different types of linear approximation. In particular, we wanted to reinforce the understanding that the various types of linear approximation are all based on the idea that the tangent line approximates the curve close to the point of tangency.

SECTION 8.1 The section on probability was moved from Section 7.8 to 8.1 so that it appears in the chapter on applications of the integral rather than the chapter techniques of integration.

THE PREVIOUS EDITION'S SECTION 5.9 was eliminated because it appeared in the main introductory chapter on integration yet had little content involving the integral. The main topics from the section were placed elsewhere where they fit better. For example, the material on differential equations and on exponential growth and decay was moved to the chapter on differential equations.

SECTION 9.1 was rewritten to provide a more-straightforward introduction to differential equations and methods of solving them. Furthermore we wrote a few new examples that replaced a rather technical derivation to provide for a wider variety of simpler, more accessible application examples.

SECTION 14.4 The development of the concept of differentiability in Section 14.4 was rewritten to provide a clearer pathway from the basic idea of the existence of partial derivatives to the more-technical notion of differentiability. We dropped the concept of local linearity introduced in previous editions because it is redundant and adds an extra layer of technical detail that can be avoided.

CALCULUS: EARLY TRANSCENDENTALS

CONTENTS

Additional content can be accessed online at www.macmillanlearning.com/calculuset4e:

Additional Proofs:

L'Hôpital's Rule
Error Bounds for Numerical Integration
Comparison Test for Improper Integrals

Additional Content:

Second-Order Differential Equations
Complex Numbers

SECTION 10.7 We have chosen a somewhat traditional location for the section on Taylor polynomials, placing it directly before the section on Taylor series in Chapter 10. We feel that this placement is an improvement over the previous edition where the section was isolated in a chapter that primarily was about applications of the integral. The subject matter in the Taylor polynomials section works well as an initial step toward the important topic of Taylor series representations of specific functions. The Taylor polynomials section can serve as a follow-up to linear approximation in Section 4.1. Consequently, Taylor polynomials (except for Taylor's Theorem at the end of the section, which involves integration) can be covered at any point after Section 4.1.

CALCULUS: EARLY TRANSCENDENTALS, FOURTH EDITION

On Teaching Mathematics

We consider ourselves very lucky to have careers as teachers and researchers of mathematics. Through many years (over 30 each) teaching and learning mathematics we have developed many ideas about how best to present mathematical concepts and to engage students working with and exploring them. We see teaching mathematics as a form of storytelling, both when we present in a classroom and when we write materials for exploration and learning. The goal is to explain to students in a captivating manner, at the right pace, and in as clear a way as possible, how mathematics works and what it can do for them. We find mathematics to be intriguing and immensely beautiful. We want students to feel that way, too.

On Writing a Calculus Text

It has been an exciting challenge to author the recent editions of Jon Rogawski's calculus book. We both had experience with the early editions of the text and had a lot of respect for Jon's approach to them. Jon's vision of what a calculus book could be fits very closely with our own. Jon believed that as math teachers, how we present material is as important as what we present. Although he insisted on rigor at all times, he also wanted a book that was clearly written, that could be read by a calculus student and would motivate them to engage in the material and learn more. Moreover, Jon strived to create a text in which exposition, graphics, and layout would work together to enhance all facets of a student's calculus experience. Jon paid special attention to certain aspects of the text:

1. Clear, accessible exposition that anticipates and addresses student difficulties.
2. Layout and figures that communicate the flow of ideas.
3. Highlighted features that emphasize concepts and mathematical reasoning including Conceptual Insight, Graphical Insight, Assumptions Matter, Reminder, and Historical Perspective.
4. A rich collection of examples and exercises of graduated difficulty that teach basic skills as well as problem-solving techniques, reinforce conceptual understanding, and motivate calculus through interesting applications. Each section also contains exercises that develop additional insights and challenge students to further develop their skills.

Our approach to writing the recent editions has been to take the strong foundation that Jon provided and strengthen it in two ways:

- To fine-tune it, while keeping with the book's original philosophy, by enhancing presentations, clarifying concepts, and emphasizing major points where we felt such adjustments would benefit the reader.
- To expand it slightly, both in the mathematics presented and the applications covered. The expansion in mathematics content has largely been guided by input from users and reviewers who had good suggestions for valuable additions (for example, a section on how to decide which technique to employ on an integration problem). The original editions of the text had very strong coverage of applications in physics and engineering; consequently, we have chosen to add examples that provide applications in the life and climate sciences.

We hope our experience as mathematicians and teachers enables us to make positive contributions to the continued development of this calculus book. As mathematicians, we want to ensure that the theorems, proofs, arguments, and derivations are correct and are presented with an appropriate level of rigor. As teachers, we want the material to be accessible and written at the level of a student who is new to the subject matter. Working from the strong foundation that Jon set, we have strived to maintain the level of quality of the previous editions while making the changes that we believe will bring the book to a new level.

What's New in the Fourth Edition

In this edition we have continued the themes introduced in the third edition and have implemented a number of new changes.

A Focus on Concepts

We have continued to emphasize conceptual understanding over the memorization of formulas. Memorization can never be completely avoided, but it should play a minor role in the process of learning calculus. Students will remember how to apply a procedure or technique if they see the logical progression of the steps in the proof that generates it. And they then understand the underlying concepts rather than seeing the topic as a black box. To further support conceptual understanding of calculus, we have added a number of new Graphical and Conceptual Insights through the book. These include insights that discuss:

- The differences between the expressions "undefined," "does not exist," and "indeterminate" in Section 2.5 on indeterminate forms,
- How measuring angles in radians is preferred in calculus over measuring in degrees because the resulting derivative formulas are simpler (in Section 3.6 on derivative rules of trigonometric functions),
- How the Fundamental Theorem of Calculus (Part II) guarantees the existence of an antiderivative for continuous functions (in Section 5.5 on the Fundamental Theorem of Calculus, Part II),
- How the volume-of-revolution formulas in Section 6.3 are special cases of the main volume-by-slices approach in Section 6.2,
- The relationships between a curve, parametrizations of it, and arc length computed from a parametrization (in Section 11.2 on arc length and speed),
- The relationship between linear approximation in multivariable calculus (in Section 14.4) and linear approximation for a function of one variable in Section 4.1.

Simplified Derivations

We simplified a number of derivations of important calculus formulas. These include:

- The derivative rule for the exponential function in Section 3.2,
- The formula for the area of a surface of revolution in Section 8.2,
- The vector-based formulas for lines and planes in 3-space in Sections 12.2 and 12.5.

New Examples in the Life and Climate Sciences

Expanding on the strong collection of applications in physics and engineering that were already in the book, we added a number of applications from other disciplines, particularly in the life and climate sciences. These include:

- The rate of change of day length in Section 3.7
- The log-wind profile in Section 3.9

- A grid-connected energy system in Section 5.2
- A glacier height differential-equations model in Section 9.1
- A predator-prey interaction in Section 11.1
- Geostrophic wind flow in Section 14.5
- Gulf Stream heat flow in Section 15.1

An Introduction to Calculus

In previous editions of the text, the first mathematics material that the reader encountered was a review of precalculus. We felt that a brief introduction to calculus would be a more meaningful start to this important body of mathematics. We hope that it provides the reader with a motivating glimpse ahead and a perspective on why a review of precalculus is a beneficial way to begin.

Additional Historical Content

Historical Perspectives and margin notes have been a well-received feature of previous editions. We added to the historical content by including a few new margin notes about past and contemporary mathematicians throughout the book. For example, we added a margin note in Section 3.1 about the contributions of Sir Isaac Newton and Gottfried Wilhelm Leibniz to the development of calculus in the seventeenth century, and a margin note in Section 12.2 about recent Field's medalist Maryam Mirzakhani.

New Examples, Figures, and Exercises

Numerous examples and accompanying figures have been added to expand on the variety of applications and to clarify concepts. Figures marked with a **DF** icon have been made dynamic and can be accessed via WebAssign Premium. A selection of these figures also includes brief tutorial videos explaining the concepts at work.

A variety of exercises have also been added throughout the text, particularly following up on new examples in the sections. The comprehensive section exercise sets are closely coordinated with the text. These exercises vary in difficulty from routine to moderate as well as more challenging. Specialized exercises are identified by icons. For example, indicates problems that require the student to give a written response. There also are icons for problems that require the use of either graphing-calculator technology **GU** or more advanced software such as a computer algebra system **CAS**.

CALCULUS: EARLY TRANSCENDENTALS, FOURTH EDITION

offers an ideal balance of formal precision and dedicated conceptual focus, helping students build strong computational skills while continually reinforcing the relevance of calculus to their future studies and their lives.

FOCUS ON CONCEPTS

CONCEPTUAL INSIGHTS encourage students to develop a conceptual understanding of calculus by explaining important ideas clearly but informally.

> **CONCEPTUAL INSIGHT** In our work with functions and limits so far, we have encountered three expressions that are similar but have different meanings: *undefined*, *does not exist*, and *indeterminate*. It is important to understand the meanings of these expressions so that you can use them correctly to describe functions and limits.
>
> - The word "undefined" is used for a mathematical expression that is not defined, such as 2/0 or ln 0.
> - The phrase "does not exist" means $\lim_{x \to c} f(x)$ does not exist, that is, $f(x)$ does not approach a particular numerical value as x approaches c.
> - The term "indeterminate" is used when, upon substitution, a function or limit has one of the indeterminate forms.

GRAPHICAL INSIGHTS enhance students' visual understanding by making the crucial connections between graphical properties and the underlying concepts.

> **GRAPHICAL INSIGHT** The formula $(\sin x)' = \cos x$ seems reasonable when we compare the graphs in Figure 1. The tangent lines to the graph of $y = \sin x$ have positive slope on the interval $\left(-\frac{\pi}{2}, \frac{\pi}{2}\right)$, and on this interval, the derivative $y' = \cos x$ is positive. The tangent lines have negative slope on the interval $\left(\frac{\pi}{2}, \frac{3\pi}{2}\right)$, where $y' = \cos x$ is negative. The tangent lines are horizontal at $x = -\frac{\pi}{2}, \frac{\pi}{2}, \frac{3\pi}{2}$, where $\cos x = 0$.

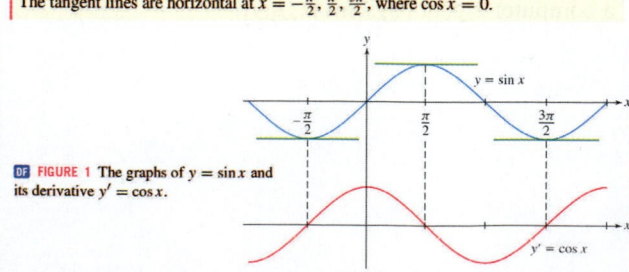

DF FIGURE 1 The graphs of $y = \sin x$ and its derivative $y' = \cos x$.

HISTORICAL PERSPECTIVES are brief vignettes that place key discoveries and conceptual advances in their historical context. They give students a glimpse into some of the accomplishments of great mathematicians and an appreciation for their significance.

> **HISTORICAL PERSPECTIVE**
>
>
>
> *Mechanics Magazine, London, 1824*
>
> Geometric series were used as early as the third century BCE by Archimedes in a brilliant argument for determining the area S of a "parabolic segment" (shaded region in Figure 3). Given two points A and C on a parabola, there is a point B between A and C where the tangent line is parallel to $\overline{AC}$ (apparently, Archimedes was aware of the Mean Value Theorem more than 2000 years before the invention of calculus). Let T be the area of triangle $\triangle ABC$. Archimedes proved that if D is chosen in a similar fashion relative to $\overline{AB}$ and E is chosen relative to $\overline{BC}$, then
>
> $$\frac{1}{4}T = \text{area}(\triangle ADB) + \text{area}(\triangle BEC) \quad \boxed{5}$$
>
> This construction of triangles can be continued. The next step would be to construct the four triangles on the segments $\overline{AD}, \overline{DB}, \overline{BE}, \overline{EC}$, of total area $\left(\frac{1}{4}\right)^2 T$. Then construct eight triangles of total area $\left(\frac{1}{4}\right)^3 T$, and so on. In this way, we obtain infinitely many triangles that completely fill up the parabolic segment. By the formula for the sum of a geometric series, we get
>
> $$S = T + \frac{1}{4}T + \frac{1}{16}T + \cdots = T \sum_{n=0}^{\infty} \frac{1}{4^n} = \frac{4}{3}T$$
>
> For this and many other achievements, Archimedes is ranked together with Newton and Gauss as one of the greatest scientists of all time.
>
> The modern study of infinite series began in the seventeenth century with Newton, Leibniz, and their contemporaries. The divergence of $\sum_{n=1}^{\infty} 1/n$ (called the **harmonic series**) was known to the medieval scholar Nicole d'Oresme (1323–1382), but his proof was lost for centuries, and the result was rediscovered on more than one occasion. It was also known that the sum of the reciprocal squares $\sum_{n=1}^{\infty} 1/n^2$ converges, and in the 1640s, the Italian Pietro Mengoli put forward the challenge of finding its sum. Despite the efforts of the best mathematicians of the day, including Leibniz and the Bernoulli brothers Jakob and Johann, the problem resisted solution for nearly a century. In 1735, the great master Leonhard Euler (at the time, 28 years old) astonished his contemporaries by proving that
>
> $$\frac{1}{1^2} + \frac{1}{2^2} + \frac{1}{3^2} + \frac{1}{4^2} + \frac{1}{5^2} + \frac{1}{6^2} + \cdots = \frac{\pi^2}{6}$$
>
> We examine the convergence of this series in Exercises 85 and 91 in Section 10.3.

FOCUS ON CLEAR, ACCESSIBLE EXPOSITION that anticipates and addresses student difficulties

REMINDERS are margin notes that link the current discussion to important concepts introduced earlier in the text to give students a quick review and make connections with related ideas.

⟵ **REMINDER** *Useful identities:*

$$\sin^2 x = \frac{1}{2}(1 - \cos 2x)$$

$$\cos^2 x = \frac{1}{2}(1 + \cos 2x)$$

$$\sin 2x = 2 \sin x \cos x$$

$$\cos 2x = \cos^2 x - \sin^2 x$$

Using the trigonometric identities in the margin, we can also integrate $\cos^2 x$, obtaining the following:

$$\int \sin^2 x \, dx = \frac{x}{2} - \frac{\sin 2x}{4} + C = \frac{x}{2} - \frac{1}{2}\sin x \cos x + C \qquad \boxed{1}$$

$$\int \cos^2 x \, dx = \frac{x}{2} + \frac{\sin 2x}{4} + C = \frac{x}{2} + \frac{1}{2}\sin x \cos x + C \qquad \boxed{2}$$

CAUTION NOTES warn students of common pitfalls they may encounter in understanding the material.

CAUTION *When using L'Hôpital's Rule, be sure to take the derivative of the numerator and denominator separately:*

$$\lim_{x \to a} \frac{f(x)}{g(x)} = \lim_{x \to a} \frac{f'(x)}{g'(x)}$$

Do not take the derivative of the function $y = f(x)/g(x)$ as a quotient, for example using the Quotient Rule.

EXAMPLE 1 Use L'Hôpital's Rule to evaluate $\displaystyle\lim_{x \to 2} \frac{x^3 - 8}{x^4 + 2x - 20}$.

Solution Let $f(x) = x^3 - 8$ and $g(x) = x^4 + 2x - 20$. Both f and g are differentiable and $f(x)/g(x)$ is indeterminate of type $0/0$ at $a = 2$ because $f(2) = g(2) = 0$.

ASSUMPTIONS MATTER uses short explanations and well-chosen counterexamples to help students appreciate why hypotheses are needed in theorems.

EXAMPLE 3 **Assumptions Matter** Show that the Product Law cannot be applied to $\displaystyle\lim_{x \to 0} f(x)g(x)$ if $f(x) = x$ and $g(x) = x^{-1}$.

Solution For all $x \neq 0$, we have $f(x)g(x) = x \cdot x^{-1} = 1$, so the limit of the product exists:

$$\lim_{x \to 0} f(x)g(x) = \lim_{x \to 0} 1 = 1$$

However, there is an issue with the product of the limits because $\displaystyle\lim_{x \to 0} x^{-1}$ does not exist (since $g(x) = x^{-1}$ becomes infinite as $x \to 0$). Therefore, the Product Law cannot be applied and its conclusion does not hold even though the limit of the products does exist. Specifically, $\displaystyle\lim_{x \to 0} f(x)g(x) = 1$, but the product of the limits is not defined:

$$1 \neq \left(\lim_{x \to 0} f(x)\right)\left(\lim_{x \to 0} g(x)\right) = \left(\lim_{x \to 0} x\right)\underbrace{\left(\lim_{x \to 0} x^{-1}\right)}_{\text{Does not exist}}$$ ∎

SECTION SUMMARIES summarize a section's key points in a concise and useful way and emphasize for students what is most important in each section.

FOCUS ON EXERCISES AND EXAMPLES

SECTION EXERCISE SETS offer a comprehensive set of exercises closely coordinated with the text. These exercises vary in difficulty from routine, to moderate, to more challenging. Also included are icons indicating problems that require the student to give a written response or require the use of technology.

Each section offers **PRELIMINARY QUESTIONS** that test student understanding.

Preliminary Questions

1. Assume that

$$\lim_{x \to \infty} f(x) = L \quad \text{and} \quad \lim_{x \to L} g(x) = \infty$$

Which of the following statements are correct?

(a) $x = L$ is a vertical asymptote of g.

(b) $y = L$ is a horizontal asymptote of g.

(c) $x = L$ is a vertical asymptote of f.

(d) $y = L$ is a horizontal asymptote of f.

2. What are the following limits?

(a) $\lim_{x \to \infty} x^3$ (b) $\lim_{x \to -\infty} x^3$ (c) $\lim_{x \to -\infty} x^4$

3. Sketch the graph of a function that approaches a limit as $x \to \infty$ but does not approach a limit (either finite or infinite) as $x \to -\infty$.

4. What is the sign of a if $f(x) = ax^3 + x + 1$ satisfies $\lim_{x \to -\infty} f(x) = \infty$?

5. What is the sign of the coefficient multiplying x^7 if f is a polynomial of degree 7 such that $\lim_{x \to -\infty} f(x) = \infty$?

6. Explain why $\lim_{x \to \infty} \sin \frac{1}{x}$ exists but $\lim_{x \to 0} \sin \frac{1}{x}$ does not exist. What is $\lim_{x \to \infty} \sin \frac{1}{x}$?

A main set of **EXERCISES** teaches basic skills as well as problem-solving techniques, reinforces conceptual understanding, and motivates calculus through interesting applications.

Exercises

1. What are the horizontal asymptotes of the function in Figure 7?

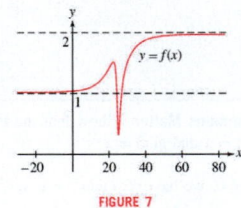

FIGURE 7

2. Sketch the graph of a function f that has both $y = -1$ and $y = 5$ as horizontal asymptotes.

3. Sketch the graph of a function f with a single horizontal asymptote $y = 3$.

21. $f(t) = \dfrac{e^t}{1 + e^{-t}}$ 22. $f(t) = \dfrac{t^{1/3}}{(64t^2 + 9)^{1/6}}$

23. $g(t) = \dfrac{10}{1 + 3^{-t}}$ 24. $p(t) = e^{-t^2}$

The following statement is incorrect: "If f has a horizontal asymptote $y = L$ at ∞, then the graph of f approaches the line $y = L$ as x gets greater and greater, but never touches it." In Exercises 25 and 26, determine $\lim_{x \to \infty} f(x)$ and indicate how f demonstrates that the statement is incorrect.

25. $f(x) = \dfrac{2x + |x|}{x}$

26. $f(x) = \dfrac{\sin x}{x}$

In Exercises 27–34, evaluate the limit.

27. $\lim_{x \to \infty} \dfrac{\sqrt{9x^4 + 3x + 2}}{4x^3 + 1}$ 28. $\lim_{x \to \infty} \dfrac{\sqrt{x^3 + 20x}}{10x - 2}$

FURTHER INSIGHTS & CHALLENGES develop additional insights and challenge students to further develop their skills.

Further Insights and Challenges

40. Show that if both $\lim_{x \to c} f(x) g(x)$ and $\lim_{x \to c} g(x)$ exist and $\lim_{x \to c} g(x) \neq 0$, then $\lim_{x \to c} f(x)$ exists. *Hint:* Write $f(x) = \dfrac{f(x) g(x)}{g(x)}$.

41. Suppose that $\lim_{t \to 3} t g(t) = 12$. Show that $\lim_{t \to 3} g(t)$ exists and equals 4.

42. Prove that if $\lim_{t \to 3} \dfrac{h(t)}{t} = 5$, then $\lim_{t \to 3} h(t) = 15$.

43. Assuming that $\lim_{x \to 0} \dfrac{f(x)}{x} = 1$, which of the following statements is necessarily true? Why?

(a) $f(0) = 0$ (b) $\lim_{x \to 0} f(x) = 0$

44. Prove that if $\lim_{x \to c} f(x) = L \neq 0$ and $\lim_{x \to c} g(x) = 0$, then the limit $\lim_{x \to c} \dfrac{f(x)}{g(x)}$ does not exist.

45. Suppose that $\lim_{h \to 0} g(h) = L$.

(a) Explain why $\lim_{h \to 0} g(ah) = L$ for any constant $a \neq 0$.

(b) If we assume instead that $\lim_{h \to 1} g(h) = L$, is it still necessarily true that $\lim_{h \to 1} g(ah) = L$?

(c) Illustrate (a) and (b) with the function $f(x) = x^2$.

46. Assume that $L(a) = \lim_{x \to 0} \dfrac{a^x - 1}{x}$ exists for all $a > 0$. Assume also that $\lim_{x \to 0} a^x = 1$.

(a) Prove that $L(ab) = L(a) + L(b)$ for $a, b > 0$. *Hint:* $(ab)^x - 1 = a^x b^x - a^x + a^x - 1 = a^x(b^x - 1) + (a^x - 1)$. [This shows that $L(a)$ behaves like a logarithm, in the sense that $\ln(ab) = \ln a + \ln b$. In fact, it can be shown that $L(a) = \ln a$.]

(b) Verify numerically that $L(12) = L(3) + L(4)$.

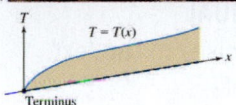

FIGURE 3 The glacier's thickness T is modeled as a function of distance x from the terminus.

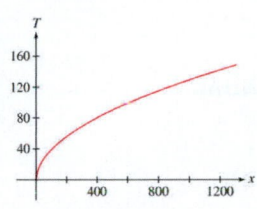

FIGURE 4 $T(x) = \sqrt{16.7x}$.

EXAMPLE 3 A Glacial Thickness Model Let $\rho = 917$ kg/m^3, $g = 9.8$ m/s^2, and $\tau = 75{,}000$ N/m^2 in Eq. (2). Use $T(0) = 0$ for an initial condition, and solve for $T(x)$. Then use $T(x)$ to determine the thickness of the glacier 1 km from its terminus.

Solution The differential equation that we need to solve is

$$T\frac{dT}{dx} = \frac{75{,}000}{(917)(9.8)}$$

It is a separable differential equation. We use the approximate value of 8.35 for the right-hand side, and proceed as follows:

$$\int T\, dT = \int 8.35\, dx$$

$$\frac{1}{2}T^2 = 8.35x + C$$

$$T(x) = \sqrt{16.7x + C}$$

Since $T(0) = 0$, we obtain $T(x) = \sqrt{16.7x}$ (Figure 4).

At a distance of 1 km from the terminus, the thickness is $T(1000) = \sqrt{16{,}700} \approx 129$ m. ∎

RICH APPLICATIONS such as this exercise on smart phone growth (below) and this example discussing glacier thickness (left) reinforce the relevance of calculus to students' lives and demonstrate the importance of calculus in scientific research.

37. In 2009, 2012, and 2015, the number (in millions) of smart phones sold in the world was 172.4, 680.1, and 1423.9, respectively.

(a) ⬚CAS⬚ Let t represent time in years since 2009, and let S represent the number of smart phones sold in millions. Determine M, A, and k for a logistic model, $S(t) = \dfrac{M}{1 + Ae^{-kt}}$, that fits the given data points.

(b) What is the long-term expected maximum number of smart phones sold annually? That is, what is $\lim\limits_{t \to \infty} S(t)$?

(c) In what year does the model predict that smart-phone sales will reach 98% of the expected maximum?

CHAPTER REVIEW EXERCISES

1. The position of a particle at time t (s) is $s(t) = \sqrt{t^2 + 1}$ m. Compute its average velocity over $[2, 5]$ and estimate its instantaneous velocity at $t = 2$.

2. A rock dropped from a state of rest at time $t = 0$ on the planet Ginormon travels a distance $s(t) = 15.2t^2$ m in t seconds. Estimate the instantaneous velocity at $t = 5$.

3. For $f(x) = \sqrt{2x}$ compute the slopes of the secant lines from 16 to each of 16 ± 0.01, 16 ± 0.001, 16 ± 0.0001 and use those values to estimate the slope of the tangent line at $x = 16$.

4. Show that the slope of the secant line for $f(x) = x^3 - 2x$ over $[5, x]$ is equal to $x^2 + 5x + 23$. Use this to estimate the slope of the tangent line at $x = 5$.

21. $\lim\limits_{x \to -1+} \dfrac{1}{x+1}$

22. $\lim\limits_{y \to \frac{1}{3}} \dfrac{3y^2 + 5y - 2}{6y^2 - 5y + 1}$

23. $\lim\limits_{x \to 1} \dfrac{x^3 - 2x}{x - 1}$

24. $\lim\limits_{a \to b} \dfrac{a^2 - 3ab + 2b^2}{a - b}$

25. $\lim\limits_{x \to 0} \dfrac{e^{3x} - e^x}{e^x - 1}$

26. $\lim\limits_{\theta \to 0} \dfrac{\sin 5\theta}{\theta}$

27. $\lim\limits_{x \to 1.5} \left\lfloor \dfrac{1}{x} \right\rfloor$

28. $\lim\limits_{\theta \to \frac{\pi}{4}} \sec \theta$

29. $\lim\limits_{z \to -3} \dfrac{z + 3}{z^2 + 4z + 3}$

30. $\lim\limits_{x \to 1} \dfrac{x^3 - ax^2 + ax - 1}{x - 1}$

CHAPTER REVIEW EXERCISES offer a comprehensive set of exercises closely coordinated with the chapter material to provide additional problems for self-study or assignments.

FOCUS ON MEDIA AND RESOURCES

WebAssign Premium

https://www.webassign.net/whfreeman

WebAssign Premium offers course and assignment customization to extend and enhance the classroom experience for instructors and students.

The fully customizable WebAssign Premium for *Calculus: Early Transcendentals* integrates an interactive e-book, a powerful answer evaluator, algorithmically generated homework and quizzes, and Macmillan's acclaimed CalcTools:

DYNAMIC FIGURES—Interactive versions of 165 text figures. Tutorial videos explain how to use select figures.

CALCCLIPS—Step-by-step whiteboard tutorials explain key concepts from exercises in the book.

E-BOOK—Easy to navigate, with highlighting and note-taking features.

LEARNINGCURVE—A powerful, self-paced assessment tool provides instant feedback tied to specific sections of the e-book. Difficulty level and topic selection adapt based on each student's performance.

TUTORIAL QUESTIONS—This feature reviews difficult questions one segment at a time.

A PERSONAL STUDY PLAN (PSP)—Lets each student use chapter and section assessments to gauge their mastery of the material and generate an individualized study plan.

FOR INSTRUCTORS

available at https://www.webassign.net/whfreeman

OVER 7,000 EXERCISES from the text, with detailed solutions available to students at your discretion.

READY-TO-USE COURSE PACK ASSIGNMENTS drawn from the exercise bank to save you time.

A SUITE OF INSTRUCTOR RESOURCES, including iClicker questions, Instructor's Manuals, PowerPoint lecture slides, a printable test bank, and more.

ADDITIONAL SUPPLEMENTS

FOR INSTRUCTORS

INSTRUCTOR'S SOLUTIONS MANUAL
Worked-out solutions to all exercises in the text.
ISBN: (SV) 978-1-319-25215-1;
(MV) 978-1-319-25217-5

TEST BANK
ISBN: 978-1-319-24292-3

INSTRUCTOR'S RESOURCE MANUAL
ISBN: 978-1-319-22135-5

LECTURE SLIDES (customizable)

IMAGE SLIDES (all text figures and tables)

iClicker

iClicker's two-way radio-frequency classroom response solution was developed by educators for educators. To learn more about packaging iClicker with this textbook, please contact your local sales representative or visit www.iclicker.com.

FOR STUDENTS

STUDENT SOLUTIONS MANUAL
Worked-out solutions to all odd-numbered exercises.
ISBN: (SV) 978-1-319-25442-1;
(MV) 978-1-319-25441-4

MAPLE™ MANUAL
ISBN: 978-1-319-22151-5

MATHEMATICA® MANUAL
ISBN: 978-1-319-22137-9

 WeBWorK

WEBWORK.MAA.ORG
Macmillan Learning offers thousands of algorithmically generated questions (with full solutions) from this book through WeBWorK's free, open-source online homework system.

ACKNOWLEDGMENTS

We are grateful to the many instructors from across the United States and Canada who have offered comments that assisted in the development and refinement of this book. These contributions included class testing, manuscript reviewing, exercise reviewing, and participating in surveys about the book and general course needs.

ALABAMA Tammy Potter, *Gadsden State Community College*; David Dempsey, *Jacksonville State University*; Edwin Smith, *Jacksonville State University*; Jeff Dodd, *Jacksonville State University*; Douglas Bailer, *Northeast Alabama Community College*; Michael Hicks, *Shelton State Community College*; Patricia C. Eiland, *Troy University, Montgomery Campus*; Chadia Affane Aji, *Tuskegee University*; James L. Wang, *The University of Alabama*; Stephen Brick, *University of South Alabama*; Joerg Feldvoss, *University of South Alabama*; Ulrich Albrecht, *Auburn University* **ALASKA** Mark A. Fitch, *University of Alaska Anchorage*; Kamal Narang, *University of Alaska Anchorage*; Alexei Rybkin, *University of Alaska Fairbanks*; Martin Getz, *University of Alaska Fairbanks* **ARIZONA** Stefania Tracogna, *Arizona State University*; Bruno Welfert, *Arizona State University*; Light Bryant, *Arizona Western College*; Daniel Russow, *Arizona Western College*; Jennifer Jameson, *Coconino College*; George Cole, *Mesa Community College*; David Schultz, *Mesa Community College*; Michael Bezusko, *Pima Community College, Desert Vista Campus*; Garry Carpenter, *Pima Community College, Northwest Campus*; Paul Flasch, *Pima County Community College*; Jessica Knapp, *Pima Community College, Northwest Campus*; Roger Werbylo, *Pima County Community College*; Katie Louchart, *Northern Arizona University*; Janet McShane, *Northern Arizona University*; Donna M. Krawczyk, *The University of Arizona* **ARKANSAS** Deborah Parker, *Arkansas Northeastern College*; J. Michael Hall, *Arkansas State University*; Kevin Cornelius, *Ouachita Baptist University*; Hyungkoo Mark Park, *Southern Arkansas University*; Katherine Pinzon, *University of Arkansas at Fort Smith*; Denise LeGrand, *University of Arkansas at Little Rock*; John Annulis, *University of Arkansas at Monticello*; Erin Haller, *University of Arkansas, Fayetteville*; Shannon Dingman, *University of Arkansas, Fayetteville*; Daniel J. Arrigo, *University of Central Arkansas* **CALIFORNIA** Michael S. Gagliardo, *California Lutheran University*; Harvey Greenwald, *California Polytechnic State University, San Luis Obispo*; Charles Hale, *California Polytechnic State University*; John Hagen, *California Polytechnic State University, San Luis Obispo*; Donald Hartig, *California Polytechnic State University, San Luis Obispo*; Colleen Margarita Kirk, *California Polytechnic State University, San Luis Obispo*; Lawrence Sze, *California Polytechnic State University, San Luis Obispo*; Raymond Terry, *California Polytechnic State University, San Luis Obispo*; James R. McKinney, *California State Polytechnic University, Pomona*; Robin Wilson, *California State Polytechnic University, Pomona*; Charles Lam, *California State University, Bakersfield*; David McKay, *California State University, Long Beach*; Melvin Lax, *California State University, Long Beach*; Wallace A. Etterbeek, *California State University, Sacramento*; Mohamed Allali, *Chapman University*; George Rhys, *College of the Canyons*; Janice Hector, *DeAnza College*; Isabelle Saber, *Glendale Community College*; Peter Stathis, *Glendale Community College*; Douglas B. Lloyd, *Golden West College*; Thomas Scardina, *Golden West College*; Kristin Hartford, *Long Beach City College*; Eduardo Arismendi-Pardi, *Orange Coast College*; Mitchell Alves, *Orange Coast College*; Yenkanh Vu, *Orange Coast College*; Yan Tian, *Palomar College*; Donna E. Nordstrom, *Pasadena City College*; Don L. Hancock, *Pepperdine University*; Kevin Iga, *Pepperdine University*; Adolfo J. Rumbos, *Pomona College*; Virginia May, *Sacramento City College*; Carlos de la Lama, *San Diego City College*; Matthias Beck, *San Francisco State University*; Arek Goetz, *San Francisco State University*; Nick Bykov, *San Joaquin Delta College*; Eleanor Lang Kendrick, *San Jose City College*; Elizabeth Hodes, *Santa Barbara City College*; William Konya, *Santa Monica College*; John Kennedy, *Santa Monica College*; Peter Lee, *Santa Monica College*; Richard Salome, *Scotts Valley High School*; Norman Feldman, *Sonoma State University*; Elaine McDonald, *Sonoma State University*; John D. Eggers, *University of California, San Diego*; Adam Bowers, *University of California, San Diego*; Bruno Nachtergaele, *University of California, Davis*; Boumediene Hamzi, *University of California, Davis*; Olga Radko, *University of California, Los Angeles*; Richard Leborne, *University of California, San Diego*; Peter Stevenhagen, *University of California, San Diego*; Jeffrey Stopple, *University of California, Santa Barbara*; Guofang Wei, *University of California, Santa Barbara*; Rick A. Simon, *University of La Verne*; Alexander E. Koonce, *University of Redlands*; Mohamad A. Alwash, *West Los Angeles College*; Calder Daenzer, *University of California, Berkeley*; Jude Thaddeus Socrates, *Pasadena City College*; Cheuk Ying Lam, *California State University Bakersfield*; Borislava Gutarts, *California State University, Los Angeles*; Daniel Rogalski, *University of California, San Diego*; Don Hartig, *California Polytechnic State University*; Anne Voth, *Palomar College*; Jay Wiestling,

Palomar College; Lindsey Bramlett-Smith, *Santa Barbara City College*; Dennis Morrow, *College of the Canyons*; Sydney Shanks, *College of the Canyons*; Bob-Tolar, *College of the Canyons*; Gene W. Majors, *Fullerton College*; Robert Diaz, *Fullerton College*; Gregory Nguyen, *Fullerton College*; Paul Sjoberg, *Fullerton College*; Deborah Ritchie, *Moorpark College*; Maya Rahnamaie, *Moorpark College*; Kathy Fink, *Moorpark College*; Christine Cole, *Moorpark College*; K. Di Passero, *Moorpark College*; Sid Kolpas, *Glendale Community College*; Miriam Castrconde, *Irvine Valley College*; Ilkner Erbas-White, *Irvine Valley College*; Corey Manchester, *Grossmont College*; Donald Murray, *Santa Monica College*; Barbara McGee, *Cuesta College*; Marie Larsen, *Cuesta College*; Joe Vasta, *Cuesta College*; Mike Kinter, *Cuesta College*; Mark Turner, *Cuesta College*; G. Lewis, *Cuesta College*; Daniel Kleinfelter, *College of the Desert*; Esmeralda Medrano, *Citrus College*; James Swatzel, *Citrus College*; Mark Littrell, *Rio Hondo College*; Rich Zucker, *Irvine Valley College*; Cindy Torigison, *Palomar College*; Craig Chamberline, *Palomar College*; Lindsey Lang, *Diablo Valley College*; Sam Needham, *Diablo Valley College*; Dan Bach, *Diablo Valley College*; Ted Nirgiotis, *Diablo Valley College*; Monte Collazo, *Diablo Valley College*; Tina Levy, *Diablo Valley College*; Mona Panchal, *East Los Angeles College*; Ron Sandvick, *San Diego Mesa College*; Larry Handa, *West Valley College*; Frederick Utter, *Santa Rosa Junior College*; Farshod Mosh, *DeAnza College*; Doli Bambhania, *DeAnza College*; Charles Klein, *DeAnza College*; Tammi Marshall, *Cauyamaca College*; Inwon Leu, *Cauyamaca College*; Michael Moretti, *Bakersfield College*; Janet Tarjan, *Bakersfield College*; Hoat Le, *San Diego City College*; Richard Fielding, *Southwestern College*; Shannon Gracey, *Southwestern College*; Janet Mazzarella, *Southwestern College*; Christina Soderlund, *California Lutheran University*; Rudy Gonzalez, *Citrus College*; Robert Crise, *Crafton Hills College*; Joseph Kazimir, *East Los Angeles College*; Randall Rogers, *Fullerton College*; Peter Bouzar, *Golden West College*; Linda Ternes, *Golden West College*; Hsiao-Ling Liu, *Los Angeles Trade Tech Community College*; Yu-Chung Chang-Hou, *Pasadena City College*; Guillermo Alvarez, *San Diego City College*; Ken Kuniyuki, *San Diego Mesa College*; Laleh Howard, *San Diego Mesa College*; Sharareh Masooman, *Santa Barbara City College*; Jared Hersh, *Santa Barbara City College*; Betty Wong, *Santa Monica College*; Brian Rodas, *Santa Monica College*; Veasna Chiek, *Riverside City College*; Kenn Huber, *University of California, Irvine*; Berit Givens, *California State Polytechnic University, Pomona*; Will Murray, *California State University, Long Beach*; Alain Bourget, *California State University, Fullerton* **COLORADO** Tony Weathers, *Adams State College*; Erica Johnson, *Arapahoe Community College*; Karen Walters, *Arapahoe Community College*; Joshua D. Laison, *Colorado College*; G. Gustave Greivel, *Colorado School of Mines*; Holly Eklund, *Colorado School of the Mines*; Mike Nicholas, *Colorado School of the Mines*; Jim Thomas, *Colorado State University*; Eleanor Storey, *Front Range Community College*; Larry Johnson, *Metropolitan State College of Denver*; Carol Kuper, *Morgan Community College*; Larry A. Pontaski, *Pueblo Community College*; Terry Chen Reeves, *Red Rocks Community College*; Debra S. Carney, *Colorado School of the Mines*; Louis A. Talman, *Metropolitan State College of Denver*; Mary A. Nelson, *University of Colorado at Boulder*; J. Kyle Pula, *University of Denver*; Jon Von Stroh, *University of Denver*; Sharon Butz, *University of Denver*; Daniel Daly, *University of Denver*; Tracy Lawrence, *Arapahoe Community College*; Shawna Mahan, *University of Colorado Denver*; Adam Norris, *University of Colorado at Boulder*; Anca Radulescu, *University of Colorado at Boulder*; MikeKawai, *University of Colorado Denver*; Janet Barnett, *Colorado State University–Pueblo*; Byron Hurley, *Colorado State University–Pueblo*; Jonathan Portiz, *Colorado State University–Pueblo*; Bill Emerson, *Metropolitan State College of Denver*; Suzanne Caulk, *Regis University*; Anton Dzhamay, *University of Northern Colorado*; Stephen Pankavich, *Colorado School of Mines*; Murray Cox, *University of Colorado at Boulder*; Anton Betten, *Colorado State University* **CONNECTICUT** Jeffrey McGowan, *Central Connecticut State University*; Ivan Gotchev, *Central Connecticut State University*; Charles Waiveris, *Central Connecticut State University*; Christopher Hammond, *Connecticut College*; Anthony Y. Aidoo, *Eastern Connecticut State University*; Kim Ward, *Eastern Connecticut State University*; Joan W. Weiss, *Fairfield University*; Theresa M. Sandifer, *Southern Connecticut State University*; Cristian Rios, *Trinity College*; Melanie Stein, *Trinity College*; Steven Orszag, *Yale University* **DELAWARE** Patrick F. Mwerinde, *University of Delaware* **DISTRICT OF COLUMBIA** Jeffrey Hakim, *American University*; Joshua M. Lansky, *American University*; James A. Nickerson,

Gallaudet University **FLORIDA** Gregory Spradlin, *Embry-Riddle University at Daytona Beach*; Daniela Popova, *Florida Atlantic University*; Abbas Zadegan, *Florida International University*; Gerardo Aladro, *Florida International University*; Gregory Henderson, *Hillsborough Community College*; Pam Crawford, *Jacksonville University*; Penny Morris, *Polk Community College*; George Schultz, *St. Petersburg College*; Jimmy Chang, *St. Petersburg College*; Carolyn Kistner, *St. Petersburg College*; Aida Kadic-Galeb, *The University of Tampa*; Constance Schober, *University of Central Florida*; S. Roy Choudhury, *University of Central Florida*; Kurt Overhiser, *Valencia Community College*; Jiongmin Yong, *University of Central Florida*; Giray Okten, *The Florida State University*; Frederick Hoffman, *Florida Atlantic University*; Thomas Beatty, *Florida Gulf Coast University*; Witny Librun, *Palm Beach Community College North*; Joe Castillo, *Broward County College*; Joann Lewin, *Edison College*; Donald Ransford, *Edison College*; Scott Berthiaume, *Edison College*; Alexander Ambrioso, *Hillsborough Community College*; Jane Golden, *Hillsborough Community College*; Susan Hiatt, *Polk Community College–Lakeland Campus*; Li Zhou, *Polk Community College–Winter Haven Campus*; Heather Edwards, *Seminole Community College*; Benjamin Landon, *Daytona State College*; Tony Malaret, *Seminole Community College*; Lane Vosbury, *Seminole Community College*; William Rickman, *Seminole Community College*; Cheryl Cantwell, *Seminole Community College*; Michael Schramm, *Indian River State College*; Janette Campbell, *Palm Beach Community College–Lake Worth*; KwaiLee Chui, *University of Florida*; Shu-Jen Huang, *University of Florida*; Sidra Van De Car, *Valencia College* **GEORGIA** Christian Barrientos, *Clayton State University*; Thomas T. Morley, *Georgia Institute of Technology*; Doron Lubinsky, *Georgia Institute of Technology*; Ralph Wildy, *Georgia Military College*; Shahram Nazari, *Georgia Perimeter College*; Alice Eiko Pierce, *Georgia Perimeter College, Clarkson Campus*; Susan Nelson, *Georgia Perimeter College, Clarkson Campus*; Laurene Fausett, *Georgia Southern University*; Scott N. Kersey, *Georgia Southern University*; Jimmy L. Solomon, *Georgia Southern University*; Allen G. Fuller, *Gordon College*; Marwan Zabdawi, *Gordon College*; Carolyn A. Yackel, *Mercer University*; Blane Hollingsworth, *Middle Georgia State College*; Shahryar Heydari, *Piedmont College*; Dan Kannan, *The University of Georgia*; June Jones, *Middle Georgia State College*; Abdelkrim Brania, *Morehouse College*; Ying Wang, *Augusta State University*; James M. Benedict, *Augusta State University*; Kouong Law, *Georgia Perimeter College*; Rob Williams, *Georgia Perimeter College*; Alvina Atkinson, *Georgia Gwinnett College*; Amy Erickson, *Georgia Gwinnett College* **HAWAII** Shuguang Li, *University of Hawaii at Hilo*; Raina B. Ivanova, *University of Hawaii at Hilo* **IDAHO** Uwe Kaiser, *Boise State University*; Charles Kerr, *Boise State University*; Zach Teitler, *Boise State University*; Otis Kenny, *Boise State University*; Alex Feldman, *Boise State University*; Doug Bullock, *Boise State University*; Brian Dietel, *Lewis-Clark State College*; Ed Korntved, *Northwest Nazarene University*; Cynthia Piez, *University of Idaho* **ILLINOIS** Chris Morin, *Blackburn College*; Alberto L. Delgado, *Bradley University*; John Haverhals, *Bradley University*; Herbert E. Kasube, *Bradley University*; Marvin Doubet, *Lake Forest College*; Marvin A. Gordon, *Lake Forest Graduate School of Management*; Richard J. Maher, *Loyola University Chicago*; Joseph H. Mayne, *Loyola University Chicago*; Marian Gidea, *Northeastern Illinois University*; John M. Alongi, *Northwestern University*; Miguel Angel Lerma, *Northwestern University*; Mehmet Dik, *Rockford College*; Tammy Voepel, *Southern Illinois University Edwardsville*; Rahim G. Karimpour, *Southern Illinois University*; Thomas Smith, *University of Chicago*; Laura DeMarco, *University of Illinois*; Evangelos Kobotis, *University of Illinois at Chicago*; Jennifer Mc-Neilly, *University of Illinois at Urbana-Champaign*; Timur Oikhberg, *University of Illinois at Urbana-Champaign*; Manouchehr Azad, *Harper College*; Minhua Liu, *Harper College*; Mary Hill, *College of DuPage*; Arthur N. DiVito, *Harold Washington College* **INDIANA** Vania Mascioni, *Ball State University*; Julie A. Killingbeck, *Ball State University*; Kathie Freed, *Butler University*; Zhixin Wu, *DePauw University*; John P. Boardman, *Franklin College*; Robert N. Talbert, *Franklin College*; Robin Symonds, *Indiana University Kokomo*; Henry L. Wyzinski, *Indiana University Northwest*; Melvin Royer, *Indiana Wesleyan University*; Gail P. Greene, *Indiana Wesleyan University*; David L. Finn, *Rose-Hulman Institute of Technology*; Chong Keat Arthur Lim, *University of Notre Dame* **IOWA** Nasser Dastrange, *Buena Vista University*; Mark A. Mills, *Central College*; Karen Ernst, *Hawkeye Community College*; Richard Mason, *Indian Hills Community College*; Robert S. Keller, *Loras College*; Eric Robert Westlund, *Luther College*; Weimin Han, *The University of Iowa*; Man Basnet, *Iowa State University* **KANSAS** Timothy W. Flood, *Pittsburg State University*; Sarah Cook, *Washburn University*; Kevin E. Charlwood, *Washburn University*; Conrad Uwe, *Cowley County Community College*; David N. Yetter, *Kansas State University*; Matthew Johnson, *University of Kansas* **KENTUCKY** Alex M. Mc-Allister, *Center College*; Sandy Spears, *Jefferson Community & Technical College*; Leanne Faulkner, *Kentucky Wesleyan College*; Donald O. Clayton, *Madisonville Community College*; Thomas Riedel, *University of Louisville*; Manabendra Das, *University of Louisville*; Lee Larson, *University of Louisville*; Jens E. Harlander, *Western Kentucky University*; Philip Mc-Cartney, *Northern Kentucky University*; Andy Long, *Northern Kentucky University*; Omer Yayenie, *Murray State University*; Donald Krug, *Northern Kentucky University*; David Royster, *University of Kentucky* **LOUISIANA** William Forrest, *Baton Rouge Community College*; Paul Wayne Britt, *Louisiana State University*; Galen Turner, *Louisiana Tech University*; Randall Wills, *Southeastern Louisiana University*; Kent Neuerburg, *Southeastern Louisiana University*; Guoli Ding, *Louisiana State University*; Julia Ledet, *Louisiana State University*; Brent Strunk, *University of Louisiana at Monroe*; Michael Tom, *Louisiana State University* **MAINE** Andrew Knightly, *The University of Maine*; Sergey Lvin, *The University of Maine*; Joel W. Irish, *University of Southern Maine*; Laurie Woodman, *University of Southern Maine*; David M. Bradley, *The University of Maine*; William O. Bray, *The University of Maine* **MARYLAND** Leonid Stern, *Towson University*; Jacob Kogan, *University of Maryland Baltimore County*; Mark E. Williams, *University of Maryland Eastern Shore*; Austin A. Lobo, *Washington College*; Supawan Lertskrai, *Harford Community College*; Fary Sami, *Harford Community College*; Andrew Bulleri, *Howard Community College* **MASSACHUSETTS** Sean McGrath, *Algonquin Regional High School*; Norton Starr, *Amherst College*; Renato Mirollo, *Boston College*; Emma Previato, *Boston University*; Laura K Gross, *Bridgewater State University*; Richard H. Stout, *Gordon College*; Matthew P. Leingang, *Harvard University*; Suellen Robinson, *North Shore Community College*; Walter Stone, *North Shore Community College*; Barbara Loud, *Regis College*; Andrew B. Perry, *Springfield College*; Tawanda Gwena, *Tufts University*; Gary Simundza, *Wentworth Institute of Technology*; Mikhail Chkhenkeli, *Western New England College*; David Daniels, *Western New England College*; Alan Gorfin, *Western New England College*; Saeed Ghahramani, *Western New England College*; Julian Fleron, *Westfield State College*; Maria Fung, *Worchester State University*; Brigitte Servatius, *Worcester Polytechnic Institute*; John Goulet, *Worcester Polytechnic Institute*; Alexander Martsinkovsky, *Northeastern University*; Marie Clote, *Boston College*; Alexander Kastner, *Williams College*; Margaret Peard, *Williams College*; Mihai Stoiciu, *Williams College*; Maciej Szczesny, *Boston University* **MICHIGAN** Mark E. Bollman, *Albion College*; Jim Chesla, *Grand Rapids Community College*; Jeanne Wald, *Michigan State University*; Allan A. Struthers, *Michigan Technological University*; Debra Pharo, *Northwestern Michigan College*; Anna Maria Spagnuolo, *Oakland University*; Diana Faoro, *Romeo Senior High School*; Andrew Strowe, *University of Michigan–Dearborn*; Daniel Stephen Drucker, *Wayne State University*; Christopher Cartwright, *Lawrence Technological University*; Jay Treiman, *Western Michigan University* **MINNESOTA** Bruce Bordwell, *Anoka-Ramsey Community College*; Robert Dobrow, *Carleton College*; Jessie K. Lenarz, *Concordia College–Moorhead Minnesota*; Bill Tomhave, *Concordia College*; David L. Frank, *University of Minnesota*; Steven I. Sperber, *University of Minnesota*; Jeffrey T. Mc-Lean, *University of St. Thomas*; Chehrzad Shakiban, *University of St. Thomas*; Melissa Loe, *University of St. Thomas*; Nick Christopher Fiala, *St. Cloud State University*; Victor Padron, *Normandale Community College*; Mark Ahrens, *Normandale Community College*; Gerry Naughton, *Century Community College*; Carrie Naughton, *Inver Hills Community College* **MISSISSIPPI** Vivien G. Miller, *Mississippi State University*; Ted Dobson, *Mississippi State University*; Len Miller, *Mississippi State University*; Tristan Denley, *The University of Mississippi* **MISSOURI** Robert Robertson, *Drury University*; Gregory A. Mitchell, *Metropolitan Community College–Penn Valley*; Charles N. Curtis, *Missouri Southern State University*; Vivek Narayanan, *Moberly Area Community College*; Russell Blyth, *Saint Louis University*; Julianne Rainbolt, *Saint Louis University*; Blake Thornton, *Saint Louis University*; Kevin W. Hopkins, *Southwest Baptist University*; Joe Howe, *St. Charles Community College*; Wanda Long, *St. Charles Community College*; Andrew Stephan, *St. Charles Community College* **MONTANA** Kelly Cline, *Carroll College*; Veronica Baker, *Montana State University, Bozeman*; Richard C. Swanson, *Montana State University*; Thomas Hayes-McGoff, *Montana State University*; Nikolaus Vonessen, *The University of Montana*; Corinne Casolara, *Montana State University, Bozeman* **NEBRASKA** Edward G. Reinke Jr., *Concordia University*; Judith Downey, *University of Nebraska at Omaha* **NEVADA** Jennifer Gorman, *College of Southern Nevada*; Jonathan Pearsall, *College of Southern Nevada*; Rohan Dalpatadu, *University of Nevada, Las Vegas*; Paul Aizley, *University of Nevada, Las Vegas*; Charlie Nazemian, *University of Nevada, Reno* **NEW HAMPSHIRE** Richard Jardine, *Keene State College*; Michael Cullinane, *Keene State College*; Roberta Kieronski, *University of New Hampshire at Manchester*; Erik Van Erp, *Dartmouth College* **NEW JERSEY** Paul S. Rossi, *College of Saint Elizabeth*; Mark Galit, *Essex County College*; Katarzyna Potocka, *Ramapo College of New Jersey*; Nora S. Thornber, *Raritan Valley Community College*; Abdulkadir Hassen, *Rowan University*; Olcay Ilicasu, *Rowan University*; Avraham Soffer, *Rutgers, The State University of New Jersey*; Chengwen Wang, *Rutgers, The State University of New Jersey*; Shabnam Beheshti, *Rutgers University, The State University of New Jersey*; Stephen J. Greenfield, *Rutgers, The State University of New Jersey*; John T. Saccoman, *Seton Hall University*; Lawrence E. Levine, *Stevens Institute of Technology*; Jana Gevertz, *The College of New Jersey*; Barry Burd, *Drew University*; Penny Luczak, *Camden County College*; John Climent, *Cecil Community College*; Kristyanna Erickson, *Cecil Community College*; Eric Compton, *Brookdale Community College*; John Atsu-Swanzy, *Atlantic Cape Community College*; Paul Laumakis, *Rowan University*; Norman Beil, *Rowan University* **NEW MEXICO** Kevin Leith, *Central New Mexico Community College*; David Blankenbaker, *Central New Mexico Community College*; Joseph Lakey, *New Mexico State University*; Kees Onneweer, *University of New Mexico*; Jurg Bolli, *The University of New Mexico*; Amal

Mostafa, *New Mexico State University*; Christopher Stuart, *New Mexico State University* **NEW YORK** Robert C. Williams, *Alfred University*; Timmy G. Bremer, *Broome Community College State University of New York*; Joaquin O. Carbonara, *Buffalo State College*; Robin Sue Sanders, *Buffalo State College*; Daniel Cunningham, *Buffalo State College*; Rose Marie Castner, *Canisius College*; Sharon L. Sullivan, *Catawba College*; Fabio Nironi, *Columbia University*; Camil Muscalu, *Cornell University*; Maria S. Terrell, *Cornell University*; Margaret Mulligan, *Dominican College of Blauvelt*; Robert Andersen, *Farmingdale State University of New York*; Leonard Nissim, *Fordham University*; Jennifer Roche, *Hobart and William Smith Colleges*; James E. Carpenter, *Iona College*; Peter Shenkin, *John Jay College of Criminal Justice/CUNY*; Gordon Crandall, *LaGuardia Community College/CUNY*; Gilbert Traub, *Maritime College, State University of New York*; Paul E. Seeburger, *Monroe Community College Brighton Campus*; Abraham S. Mantell, *Nassau Community College*; Daniel D. Birmajer, *Nazareth College*; Sybil G. Shaver, *Pace University*; Margaret Kiehl, *Rensselaer Polytechnic Institute*; Carl V. Lutzer, *Rochester Institute of Technology*; Michael A. Radin, *Rochester Institute of Technology*; Hossein Shahmohamad, *Rochester Institute of Technology*; Thomas Rousseau, *Siena College*; Jason Hofstein, *Siena College*; Leon E. Gerber, *St. Johns University*; Christopher Bishop, *Stony Brook University*; James Fulton, *Suffolk County Community College*; John G. Michaels, *SUNY Brockport*; Howard J. Skogman, *SUNY Brockport*; Cristina Bacuta, *SUNY Cortland*; Jean Harper, *SUNY Fredonia*; David Hobby, *SUNY New Paltz*; Kelly Black, *Union College*; Thomas W. Cusick, *University at Buffalo/The State University of New York*; Gino Biondini, *University at Buffalo/The State University of New York*; Robert Koehler, *University at Buffalo/The State University of New York*; Donald Larson, *University of Rochester*; Robert Thompson, *Hunter College*; Ed Grossman, *The City College of New York*; David Hemmer, *University at Buffalo/The State University of New York* **NORTH CAROLINA** Jeffrey Clark, *Elon University*; William L. Burgin, *Gaston College*; Manouchehr H. Misaghian, *Johnson C. Smith University*; Legunchim L. Emmanwori, *North Carolina A&T State University*; Drew Pasteur, *North Carolina State University*; Demetrio Labate, *North Carolina State University*; Mohammad Kazemi, *The University of North Carolina at Charlotte*; Richard Carmichael, *Wake Forest University*; Gretchen Wilke Whipple, *Warren Wilson College*; John Russell Taylor, *University of North Carolina at Charlotte*; Mark Ellis, *Piedmont Community College* **NORTH DAKOTA** Jim Coykendall, *North Dakota State University*; Anthony J. Bevelacqua, *The University of North Dakota*; Richard P. Millspaugh, *The University of North Dakota*; Thomas Gilsdorf, *The University of North Dakota*; Michele Iiams, *The University of North Dakota*; Mohammad Khavanin, *University of North Dakota*; Jessica Striker, *North Dakota State University*; Benton Duncan, *North Dakota State University* **OHIO** Christopher Butler, *Case Western Reserve University*; Pamela Pierce, *The College of Wooster*; Barbara H. Margolius, *Cleveland State University*; Tzu-Yi Alan Yang, *Columbus State Community College*; Greg S. Goodhart, *Columbus State Community College*; Kelly C. Stady, *Cuyahoga Community College*; Brian T. Van Pelt, *Cuyahoga Community College*; David Robert Ericson, *Miami University*; Frederick S. Gass, *Miami University*; Thomas Stacklin, *Ohio Dominican University*; Vitaly Bergelson, *The Ohio State University*; Robert Knight, *Ohio University*; John R. Pather, *Ohio University, Eastern Campus*; Teresa Contenza, *Otterbein College*; Ali Hajjafar, *The University of Akron*; Jianping Zhu, *The University of Akron*; Ian Clough, *University of Cincinnati Clermont College*; Atif Abueida, *University of Dayton*; Judith McCrory, *The University at Findlay*; Thomas Smotzer, *Youngstown State University*; Angela Spalsbury, *Youngstown State University*; James Osterburg, *The University of Cincinnati*; Mihaela A. Poplicher, *University of Cincinnati*; Frederick Thulin, *University of Illinois at Chicago*; Weimin Han, *The Ohio State University*; Crichton Ogle, *The Ohio State University*; Jackie Miller, *The Ohio State University*; Walter Mackey, *Owens Community College*; Jonathan Baker, *Columbus State Community College*; Vincent Graziano, *Case Western Reserve University*; Sailai Sally, *Cleveland State University* **OKLAHOMA** Christopher Francisco, *Oklahoma State University*; Michael McClendon, *University of Central Oklahoma*; Teri Jo Murphy, *The University of Oklahoma*; Kimberly Adams, *University of Tulsa*; Shirley Pomeranz, *University of Tulsa* **OREGON** Lorna TenEyck, *Chemeketa Community College*; Angela Martinek, *Linn-Benton Community College*; Filix Maisch, *Oregon State University*; Tevian Dray, *Oregon State University*; Mark Ferguson, *Chemekata Community College*; Andrew Flight, *Portland State University*; Austina Fong, *Portland State University*; Jeanette R. Palmiter, *Portland State University*; Jean Nganou, *University of Oregon*; Juan Restrepo, *Oregon State University* **PENNSYLVANIA** John B. Polhill, *Bloomsburg University of Pennsylvania*; Russell C. Walker, *Carnegie Mellon University*; Jon A. Beal, *Clarion University of Pennsylvania*; Kathleen Kane, *Community College of Allegheny County*; David A. Santos, *Community College of Philadelphia*; David S. Richeson, *Dickinson College*; Christine Marie Cedzo, *Gannon University*; Monica Pierri-Galvao, *Gannon University*; John H. Ellison, *Grove City College*; Gary L. Thompson, *Grove City College*; Dale McIntyre, *Grove City College*; Dennis Benchoff, *Harrisburg Area Community College*; William A. Drumin, *King's College*; Denise Reboli, *King's College*; Chawne Kimber, *Lafayette College*; Elizabeth McMahon, *Lafayette College*; Lorenzo Traldi, *Lafayette College*; David L. Johnson, *Lehigh University*; Matthew Hyatt, *Lehigh University*; Zia Uddin, *Lock Haven University of Pennsylvania*; Donna A. Dietz, *Mansfield University of Pennsylvania*; Samuel Wilcock, *Messiah College*; Richard R. Kern, *Montgomery County Community College*; Michael Fraboni, *Moravian College*; Neena T. Chopra, *The Pennsylvania State University*; Boris A. Datskovsky, *Temple University*; Dennis M. DeTurck, *University of Pennsylvania*; Jacob Burbea, *University of Pittsburgh*; Mohammed Yahdi, *Ursinus College*; Timothy Feeman, *Villanova University*; Douglas Norton, *Villanova University*; Robert Styer, *Villanova University*; Michael J. Fisher, *West Chester University of Pennsylvania*; Peter Brooksbank, *Bucknell University*; Larry Friesen, *Butler County Community College*; LisaAngelo, *Bucks County College*; Elaine Fitt, *Bucks County College*; Pauline Chow, *Harrisburg Area Community College*; Diane Benner, *Harrisburg Area Community College*; Erica Chauvet, *Waynesburg University*; Mark McKibben, *West Chester University*; Constance Ziemian, *Bucknell University*; Jeffrey Wheeler, *University of Pittsburgh*; Jason Aran, *Drexel University*; Nakia Rimmer, *University of Pennsylvania*; Nathan Ryan, *Bucknell University*; Bharath Narayanan, *Pennsylvania State University* **RHODE ISLAND** Thomas F. Banchoff, *Brown University*; Yajni Warnapala-Yehiya, *Roger Williams University*; Carol Gibbons, *Salve Regina University*; Joe Allen, *Community College of Rhode Island*; Michael Latina, *Community College of Rhode Island* **SOUTH CAROLINA** Stanley O. Perrine, *Charleston Southern University*; Joan Hoffacker, *Clemson University*; Constance C. Edwards, *Coastal Carolina University*; Thomas L. Fitzkee, *Francis Marion University*; Richard West, *Francis Marion University*; John Harris, *Furman University*; Douglas B. Meade, *University of South Carolina*; GeorgeAndroulakis, *University of South Carolina*; Art Mark, *University of South Carolina Aiken*; Sherry Biggers, *Clemson University*; Mary Zachary Krohn, *Clemson University*; Andrew Incognito, *Coastal Carolina University*; Deanna Caveny, *College of Charleston* **SOUTH DAKOTA** Dan Kemp, *South Dakota State University* **TENNESSEE** Andrew Miller, *Belmont University*; Arthur A. Yanushka, *Christian Brothers University*; Laurie Plunk Dishman, *Cumberland University*; Maria Siopsis, *Maryville College*; Beth Long, *Pellissippi State Technical Community College*; Judith Fethe, *Pellissippi State Technical Community College*; Andrzej Gutek, *Tennessee Technological University*; Sabine Le Borne, *Tennessee Technological University*; Richard Le Borne, *Tennessee Technological University*; Maria F. Bothelho, *University of Memphis*; Roberto Triggiani, *University of Memphis*; Jim Conant, *The University of Tennessee*; Pavlos Tzermias, *The University of Tennessee*; Luis Renato Abib Finotti, *University of Tennessee, Knoxville*; Jennifer Fowler, *University of Tennessee, Knoxville*; Jo Ann W. Staples, *Vanderbilt University*; Dave Vinson, *Pellissippi State Community College*; Jonathan Lamb, *Pellissippi State Community College*; Stella Thistlewaite, *University of Tennessee, Knoxville* **TEXAS** Sally Haas, *Angelina College*; Karl Havlak, *Angelo State University*; Michael Huff, *Austin Community College*; John M. Davis, *Baylor University*; Scott Wilde, *Baylor University and The University of Texas at Arlington*; Rob Eby, *Blinn College*; Tim Sever, *Houston Community College–Central*; Ernest Lowery, *Houston Community College–Northwest*; Brian Loft, *Sam Houston State University*; Jianzhong Wang, *Sam Houston State University*; Shirley Davis, *South Plains College*; Todd M. Steckler, *South Texas College*; Mary E. Wagner-Krankel, *St. Mary's University*; Elise Z. Price, *Tarrant County College, Southeast Campus*; David Price, *Tarrant County College, Southeast Campus*; Runchang Lin, *Texas A&M University*; Michael Stecher, *Texas A&M University*; Philip B. Yasskin, *Texas A&M University*; Brock Williams, *Texas Tech University*; I. Wayne Lewis, *Texas Tech University*; Robert E. Byerly, *Texas Tech University*; Ellina Grigorieva, *Texas Woman's University*; Abraham Haje, *Tomball College*; Scott Chapman, *Trinity University*; Elias Y. Deeba, *University of Houston Downtown*; Jianping Zhu, *The University of Texas at Arlington*; Tuncay Aktosun, *The University of Texas at Arlington*; John E. Gilbert, *The University of Texas at Austin*; Jorge R. Viramontes-Olivias, *The University of Texas at El Paso*; Fengxin Chen, *University of Texas at San Antonio*; Melanie Ledwig, *The Victoria College*; Gary L. Walls, *West Texas A&M University*; William Heierman, *Wharton County Junior College*; Lisa Rezac, *University of St. Thomas*; Raymond J. Cannon, *Baylor University*; Kathryn Flores, *McMurry University*; Jacqueline A. Jensen, *Sam Houston State University*; James Galloway, *Collin County College*; Raja Khoury, *Collin County College*; Annette Benbow, *Tarrant County College–Northwest*; Greta Harland, *Tarrant County College–Northeast*; Doug Smith, *Tarrant County College–Northeast*; Marcus McGuff, *Austin Community College*; Clarence McGuff, *Austin Community College*; Steve Rodi, *Austin Community College*; Vicki Payne, *Austin Community College*; Anne Pradera, *Austin Community College*; Christy Babu, *Laredo Community College*; Deborah Hewitt, *McLennan Community College*; W. Duncan, *McLennan Community College*; Hugh Griffith, *Mt. San Antonio College*; Qin Sheng, *Baylor University*, My Linh Nguyen, *University of Texas at Dallas*; Lorenzo Sadun, *University of Texas at Austin* **UTAH** Ruth Trygstad, *Salt Lake City Community College* **VIRGINIA** Verne E. Leininger, *Bridgewater College*; Brian Bradie, *Christopher Newport University*; Hongwei Chen, *Christopher Newport University*; John J. Avioli, *Christopher Newport University*; James H. Martin, *Christopher Newport University*; David Walnut, *George Mason University*; Mike Shirazi, *Germanna Community College*; Julie Clark, *Hollins University*; Ramon A. Mata-Toledo, *James Madison University*; Adrian Riskin, *Mary Baldwin College*; Josephine Letts, *Ocean Lakes High School*; Przemyslaw Bogacki, *Old Dominion University*; Deborah Denvir, *Randolph-Macon Woman's College*; Linda Powers, *Virginia Tech*; Gregory Dresden, *Washington and Lee University*; Jacob A. Siehler, *Washington and*

Lee University; Yuan-Jen Chiang, *University of Mary Washington*; Nicholas Hamblet, *University of Virginia*; Bernard Fulgham, *University of Virginia*; Manouchehr "Mike" Mohajeri, *University of Virginia*; Lester Frank Caudill, *University of Richmond* **VERMONT** David Dorman, *Middlebury College*; Rachel Repstad, *Vermont Technical College* **WASHINGTON** Jennifer *Laveglia, Bellevue Community College*; David Whittaker, *Cascadia Community College*; Sharon Saxton, *Cascadia Community College*; Aaron Montgomery, *Central Washington University*; Patrick Averbeck, *Edmonds Community College*; Tana Knudson, *Heritage University*; Kelly Brooks, *Pierce College*; Shana P. Calaway, *Shoreline Community College*; Abel Gage, *Skagit Valley College*; Scott MacDonald, *Tacoma Community College*; Jason Preszler, *University of Puget Sound*; Martha A. Gady, *Whitworth College*; Wayne L. Neidhardt, *Edmonds Community College*; Simrat Ghuman, *Bellevue College*; Jeff Eldridge, *Edmonds Community College*; Kris Kissel, *Green River Community College*; Laura Moore-Mueller, *Green River Community College*; David Stacy, *Bellevue College*; Eric Schultz, *Walla Walla Community College*; Julianne Sachs, *Walla Walla Community College* **WEST VIRGINIA** David Cusick, *Marshall University*; Ralph Oberste-Vorth, *Marshall University*; Suda Kunyosying, *Shepard University*; Nicholas Martin, *Shepherd University*; Rajeev Rajaram, *Shepherd University*; Xiaohong Zhang, *West Virginia State University*; Sam B. Nadler, *West Virginia University* **WYOMING** Claudia Stewart, *Casper College*; Pete Wildman, *Casper College*; Charles Newberg, *Western Wyoming Community College*; Lynne Ipina, *University of Wyoming*; John Spitler, *University of Wyoming* **WISCONSIN** Erik R. Tou, *Carthage College*; Paul Bankston, *Marquette University*; Jane Nichols, *Milwaukee School of Engineering*; Yvonne Yaz, *Milwaukee School of Engineering*; Simei Tong, *University of Wisconsin–Eau Claire*; Terry Nyman, *University of Wisconsin–Fox Valley*; Robert L. Wilson, *University of Wisconsin–Madison*; Dietrich A. Uhlenbrock, *University of Wisconsin–Madison*; Paul Milewski, *University of Wisconsin–Madison*; Donald Solomon, *University of Wisconsin–Milwaukee*; Kandasamy Muthuvel, *University of Wisconsin–Oshkosh*; Sheryl Wills, *University of Wisconsin–Platteville*; Kathy A. Tomlinson, *University of Wisconsin–River Falls*; Cynthia L. McCabe, *University of Wisconsin–Stevens Point*; Matthew Welz, *University of Wisconsin–Stevens Point*; Joy Becker, *University of Wisconsin-Stout*; Jeganathan Sriskandarajah, *Madison Area Tech College*; Wayne Sigelko, *Madison Area Tech College*; James Walker, *University of Wisconsin–Eau Claire* **CANADA** Don St. Jean, *George Brown College*; Robert Dawson, *St. Mary's University*; Len Bos, *University of Calgary*; Tony Ware, *University of Calgary*; Peter David Papez, *University of Calgary*; John O'Conner, *Grant MacEwan University*; Michael P. Lamoureux, *University of Calgary*; Yousry Elsabrouty, *University of Calgary*; Darja Kalajdzievska, *University of Manitoba*; Andrew Skelton, *University of Guelph*; Douglas Farenick, *University of Regina*; Daniela Silvesan, *Memorial University of Newfoundland*; Beth Ann Austin, *Memorial University*; Brenda Davison, *Simon Fraser University*; Robert Steacy, *University of Victoria*; Dan Kucerovsky, *University of New Brunswick*; Bernardo GalvaoSousa, *University of Toronto*; Hadi Zibaeenejad, *University of Waterloo*

The creation of this fourth edition could not have happened without the help of many people. First, we want to thank the primary individuals with whom we have worked over the course of the project. Katrina Mangold, Michele Mangelli, and Nikki Miller Dworsky have been our main contacts managing the flow of the work, doing all that they could to keep everything coming together within a reasonable schedule, and efficiently arranging the various contributions of review input that helped keep the project well informed. Their work was excellent, and that excellence in project management helped greatly in bringing this new edition of the book together. Tony Palermino has provided expert editorial help throughout the process. Tony's experience with the book since its beginning helped to keep the writing and focus consistent with the original structure and vision of the book. His eye for detail and knowledge of the subject matter helped to focus the writing to deliver its message as clearly and effectively as possible.

Kerry O'Shaughnessy kept the production process moving forward in a timely manner. Thanks to Ron Weickart at Network Graphics for his skilled and creative execution of the art program. Sarah Wales-McGrath (copyeditor) and Christine Sabooni (proofreader) both provided expert feedback. Our thanks are also due to Macmillan Learning's superb production team: Janice Donnola, Sheena Goldstein, Alexis Gargin, and Paul Rohloff.

Many faculty gave critical feedback on the third edition and drafts of the fourth, and their names appear above. We are very grateful to them. We want to particularly thank all of the advisory board members who gave very valuable input on very specific questions about the approach to and the presentation of many important topics: John Davis (Baylor University), Judy Fethe (Pellissippi State Community College), Chris Francisco (Oklahoma State University), and Berit Givens (California State Polytechnic University, Pomona). The accuracy reviewers at Math Made Visible helped to bring the final version into the form in which it now appears.

We also want to thank our colleagues in the departments where we work. We are fortunate to work in departments that are energized by mathematics, where many interesting projects take place, and clever pedagogical ideas are employed and debated. We would also like to thank our students who, over many years, have provided the energy, interest, and enthusiasm that help make teaching rewarding.

Colin would like to thank his two children, Alexa and Colton. Bob would like to thank his family. They are the ones who keep us well grounded in the real world, especially when mathematics tries to steer us otherwise. This book is dedicated to them.

Bob and Colin

INTRODUCTION TO CALCULUS

Maria Gaetana Agnesi (1718–1799), an Italian mathematician and theologian, is credited with writing one of the first books about calculus, *Instituzioni analitiche ad uso della gioventù italiana*. It was self-published and was written as a textbook for her brothers, who she was tutoring.

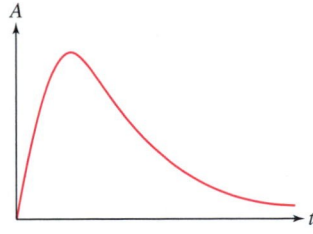

FIGURE 1

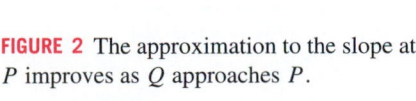

FIGURE 2 The approximation to the slope at *P* improves as *Q* approaches *P*.

We begin with a brief introduction to some key ideas in calculus. It is not an exaggeration to say that calculus is one of the great intellectual achievements of humankind. Sending spacecraft to other planets, building computer systems for forecasting the weather, explaining the interactions between plants, insects, and animals, and understanding the structure of atoms are some of the countless scientific and technological advances that could not have been achieved without calculus. Moreover, calculus is a foundational part of the mathematical theory of analysis, a field that is under continuous development.

The primary formulation of calculus dates back to independent theories of Sir Isaac Newton and Gottfried Wilhelm Liebnitz in the 1600s. However, their work only remotely resembles the topics presented in this book. Through a few centuries of development and expansion, calculus has grown into the theory we present here. Newton and Liebnitz would likely be quite impressed that their calculus has evolved into a theory that many thousands of students around the world study each year.

There are two central concepts in calculus: the derivative and the integral. We introduce them next.

The Derivative   The derivative of a function is simply the slope of its graph; it represents the rate of change of the function. For a linear function $y = 2.3x - 8.1$, the slope 2.3 indicates that y changes by 2.3 for each one-unit change in x. How do we find the slope of a graph of a function that is not linear, such as the one in Figure 1?

Imagine that this function represents the amount A of a drug in the bloodstream as a function of time t. Clearly, this situation is more complex than the linear case. The slope varies as we move along the curve. Initially positive because the amount of the drug in the bloodstream is increasing, the slope becomes negative as the drug is absorbed. Having an expression for the slope would enable us to know the time when the amount of the drug is a maximum (when the slope turns from positive to negative) or the time when the drug is leaving the bloodstream the fastest (a time to administer another dose).

To define the slope for a function that is not linear, we adapt the notion of slope for linear relationships. Specifically, to estimate the slope at point P in Figure 2, we select a point Q on the curve and draw a line between P and Q. We can use the slope of this line to approximate the slope at P. To improve this approximation, we move Q closer to P and calculate the slope of the new line. As Q moves closer to P, this approximation gets more precise. Although we cannot allow P and Q to be the same point (because we could no longer compute a slope), we instead "take the limit" of these slopes. We develop the concept of the limit in Chapter 2. Then in Chapter 3, we show that the limiting value may be defined as the exact slope at P.

The Definite Integral   The definite integral, another key calculus topic, can be thought of as adding up infinitely many infinitesimally small pieces of a whole. It too is obtained through a limiting process. More precisely, it is a limit of sums over a domain that is divided into progressively more and more pieces. To explore this idea, consider a solid

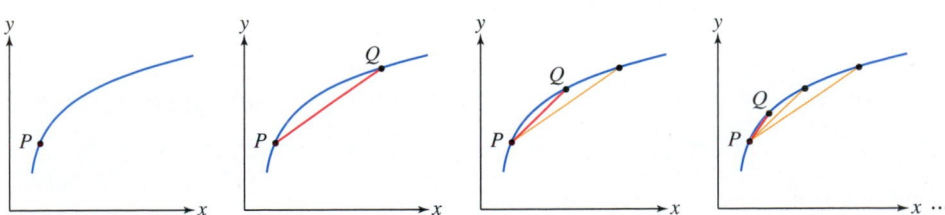

FIGURE 3 For a ball of uniform density, mass is the product of density and volume. For a nonuniform ball, a limiting process needs to be used to determine the mass.

The irregular density of the moon presented a navigational challenge for spacecraft orbiting it. The first group of spacecraft (unmanned!) that circled the moon exhibited unexpected orbits. Space scientists realized that the density of the moon varied considerably and that the gravitational attraction of concentrations of mass (referred to as mascons) deflected the path of the spacecraft from the planned trajectory.

ball of volume 2 cm³ whose density (mass per unit volume) throughout is 1.5 g/cm³. The mass of this ball is the product of density and volume, (1.5)(2) = 3 grams.

If the density is not the same throughout the ball (Figure 3), we can approximate its mass as follows:

- Chop the ball into a number of pieces,
- Assume the density is uniform on each piece and approximate the mass of each piece by multiplying density by volume,
- Add the approximate masses of the pieces to estimate the total mass of the ball.

We continually improve this approximation by chopping the ball into ever smaller pieces (Figure 4). Ultimately, an exact value is obtained by taking a limit of the approximate masses.

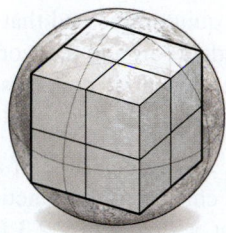

FIGURE 4

In Chapter 5, we define the definite integral in exactly this way; it is a limit of sums over an interval that is divided into progressively smaller subintervals.

The Fundamental Theorem of Calculus   Although the derivative and the definite integral are very different concepts, it turns out they are related through an important theorem called the Fundamental Theorem of Calculus presented in Chapter 5. This theorem demonstrates that the derivative and the definite integral are, to some extent, inverses of each other, a relationship that we will find beneficial in many ways.

Oscillatory phenomena, such as the extreme tides of the Bay of Fundy in Atlantic Canada, are modeled by the sine function from trigonometry.

1 PRECALCULUS REVIEW

Calculus builds on the foundation of algebra, analytic geometry, and trigonometry. In this chapter, therefore, we review some concepts, facts, and formulas from precalculus that are used throughout the text. In the last section, we discuss ways in which technology can be used to enhance your visual understanding of functions and their properties.

1.1 Real Numbers, Functions, and Graphs

We begin with a short discussion of real numbers. This gives us the opportunity to recall some basic properties and standard notation.

A **real number** is a number represented by a decimal or "decimal expansion." There are three types of decimal expansions: finite, infinite repeating, and infinite but non-repeating. For example,

$$\frac{3}{8} = 0.375, \qquad \frac{1}{7} = 0.142857142857\ldots = 0.\overline{142857} \qquad \pi = 3.141592653589793\ldots$$

The number $\frac{3}{8}$ is represented by a finite decimal, whereas $\frac{1}{7}$ is represented by an infinite repeating decimal. The bar over 142857 indicates that this sequence repeats indefinitely. The decimal expansion of π is infinite but nonrepeating.

The set of all real numbers is denoted by a boldface **R**. When there is no risk of confusion, we refer to a real number simply as a *number*. We also use the standard symbol $\in$ for the phrase "belongs to." Thus,

$$a \in \mathbf{R} \qquad \text{reads} \qquad \text{"}a \text{ belongs to } \mathbf{R}\text{"}$$

The set of integers is commonly denoted by the letter **Z** (this choice comes from the German word *Zahl*, meaning "number"). Thus, $\mathbf{Z} = \{\ldots, -2, -1, 0, 1, 2, \ldots\}$. A **whole number** is a nonnegative integer—that is, one of the numbers $0, 1, 2, \ldots$.

A real number is called **rational** if it can be represented by a fraction p/q, where p and q are integers with $q \neq 0$. The set of rational numbers is denoted **Q** (for "quotient"). Numbers that are not rational, such as π and $\sqrt{2}$, are called **irrational**.

We can tell whether a number is rational from its decimal expansion: Rational numbers have finite or infinite repeating decimal expansions, and irrational numbers have infinite, nonrepeating decimal expansions. Furthermore, the decimal expansion of a number is unique, apart from the following exception: Every finite decimal is equal to an infinite decimal in which the digit 9 repeats. For example, $1/5 = 0.5 = 0.499999\ldots$

Two algebraic properties of the real numbers are the commutative property of addition, $a + b = b + a$, and the distributive property of multiplication over addition, $a(b + c) = ab + ac$. A list of further properties can be found in Appendix B. Next, we present some properties of exponents that are used regularly when we work with exponential expressions and functions.

	Rule	Example
Exponent zero	$b^0 = 1$	$5^0 = 1$
Products	$b^x b^y = b^{x+y}$	$2^5 \cdot 2^3 = 2^{5+3} = 2^8$
Quotients	$\dfrac{b^x}{b^y} = b^{x-y}$	$\dfrac{4^7}{4^2} = 4^{7-2} = 4^5$
Negative exponents	$b^{-x} = \dfrac{1}{b^x}$	$3^{-4} = \dfrac{1}{3^4}$
Power to a power	$\left(b^x\right)^y = b^{xy}$	$\left(3^2\right)^4 = 3^{2(4)} = 3^8$
Roots	$b^{1/n} = \sqrt[n]{b}$	$5^{1/2} = \sqrt{5}$

EXAMPLE 1 Rewrite as a whole number or fraction:

(a) $16^{-1/2}$ **(b)** $27^{2/3}$ **(c)** $4^{16} \cdot 4^{-18}$ **(d)** $\dfrac{9^3}{3^7}$

Solution

(a) $16^{-1/2} = \dfrac{1}{16^{1/2}} = \dfrac{1}{\sqrt{16}} = \dfrac{1}{4}$ **(b)** $27^{2/3} = \left(27^{1/3}\right)^2 = 3^2 = 9$

(c) $4^{16} \cdot 4^{-18} = 4^{-2} = \dfrac{1}{4^2} = \dfrac{1}{16}$ **(d)** $\dfrac{9^3}{3^7} = \dfrac{\left(3^2\right)^3}{3^7} = \dfrac{3^6}{3^7} = 3^{-1} = \dfrac{1}{3}$ ■

Another important algebraic relationship is the binomial expansion of $(a + b)^n$. It is proved in Appendix C and is needed in the proof of the power law for derivatives in Section 3.2.

Expanding $(a + b)^n$ for $n = 2, 3, 4$, we obtain

- $(a + b)^2 = (a + b)(a + b) = a^2 + 2ab + b^2$
- $(a + b)^3 = (a + b)(a + b)^2 = (a + b)(a^2 + 2ab + b^2) = a^3 + 3a^2b + 3ab^2 + b^3$
- $(a + b)^4 = (a + b)(a + b)^3 = (a + b)(a^3 + 3a^2b + 3ab^2 + b^3) = a^4 + 4a^3b + 6a^2b^2 + 4ab^3 + a^4$

Notice there are some patterns emerging here. In each case, the first and second terms are a^n and $na^{n-1}b$, while the last two terms are nab^{n-1} and b^n. There is a general formula for the expansion, called the **binomial expansion formula**. It is expressed using summation notation as

$$(a + b)^n = \sum_{p=0}^{n} \frac{n!}{(n - p)! \, p!} a^{n-p} b^p$$

We introduce summation notation in Section 5.1. For now, you can understand the formula as saying that $(a + b)^n$ is a sum of terms $\frac{n!}{(n-p)! \, p!} a^{n-p} b^p$, with a term for each p going from 0 to n. So, for example, in $(a + b)^8$, the first four terms are: $\frac{8!}{8!0!} a^8 = a^8$, $\frac{8!}{7!1!} a^7 b = 8a^7 b$, $\frac{8!}{6!2!} a^6 b^2 = 28a^6 b^2$, and $\frac{8!}{5!3!} a^5 b^3 = 56a^5 b^3$. Working out the rest of the terms, we find that:

$$(a + b)^8 = a^8 + 8a^7b + 28a^6b^2 + 56a^5b^3 + 70a^4b^4 + 56a^3b^5 + 28a^2b^6 + 8ab^7 + a^8$$

We visualize real numbers as points on a line (Figure 1), and we refer to that line as the **real line**. For this reason, real numbers are often called **points**. The point corresponding to 0 is called the **origin**.

The real numbers are ordered, and we can view that ordering in terms of position on the real line: p is **greater than** q, written $p > q$, if p is to the right of q on the number line. p is **less than** q, written $p < q$, if p is to the left of q on the number line.

A real number x is said to be **positive** if $x > 0$, **negative** if $x < 0$, **nonpositive** if $x \leq 0$, and **nonnegative** if $x \geq 0$.

Two other important terms we use, related to position on the real line, are "large" and "small." We say that p is **large** if p is distant from the origin, and p is **small** if p is close to the origin. While these definitions are somewhat vague, the meaning should be clear in the contexts in which they are used.

The **absolute value** of a real number a, denoted $|a|$, is defined by (Figure 2):

$$|a| = \text{distance from the origin} = \begin{cases} a & \text{if } a \geq 0 \\ -a & \text{if } a < 0 \end{cases}$$

← REMINDER n-factorial is the number

$$n! = n(n - 1)(n - 2) \cdots (2)(1)$$

Thus,

$$1! = 1, \quad 2! = (2)(1) = 2$$
$$3! = (3)(2)(1) = 6$$

By convention, we set $0! = 1$.

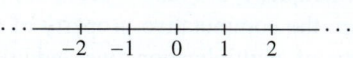

FIGURE 1 The set of real numbers represented as a line.

In some texts, "larger than" is used synonymously with "greater than." We will avoid that usage in this text.

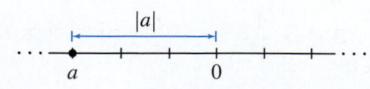

FIGURE 2 $|a|$ is the distance from a to the origin.

For example, $|1.2| = 1.2$ and $|-8.35| = -(-8.35) = 8.35$. The absolute value satisfies

$$|a| = |-a|, \qquad |ab| = |a|\,|b|$$

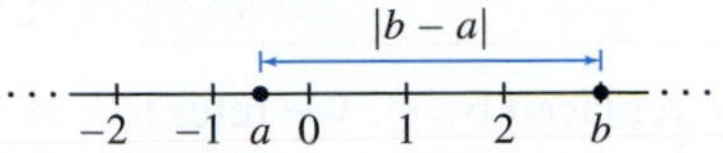

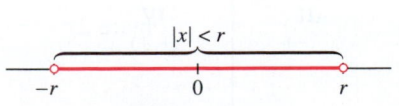

$|b - a|$

FIGURE 3 The distance between a and b is $|b - a|$.

The **distance** between two real numbers a and b is $|b - a|$, which is the length of the line segment joining a and b (Figure 3).

Two real numbers a and b are close to each other if $|b - a|$ is small, and this is the case if their decimal expansions agree to many places. More precisely, *if the decimal expansions of a and b agree to k places (to the right of the decimal point), then the distance $|b - a|$ is at most 10^{-k}.* Thus, the distance between $a = 3.1415$ and $b = 3.1478$ is at most 10^{-2} because a and b agree to two places. In fact, the distance is exactly $|3.1478 - 3.1415| = 0.0063$.

Beware that $|a + b|$ is not equal to $|a| + |b|$ unless a and b have the same sign or at least one of a and b is zero. If they have opposite signs, cancellation occurs in the sum $a + b$, and $|a + b| < |a| + |b|$. For example, $|2 + 5| = |2| + |5|$ but $|-2 + 5| = 3$, which is less than $|-2| + |5| = 7$. In any case, $|a + b|$ is never greater than $|a| + |b|$, and this gives us the simple but important **triangle inequality**:

$$|a + b| \le |a| + |b| \qquad \boxed{1}$$

We use standard notation for intervals. Given real numbers $a < b$, there are four intervals with endpoints a and b (Figure 4). They all have length $b - a$ but differ according to which endpoints are included.

FIGURE 4 The four intervals with endpoints a and b.

| Closed interval $[a, b]$ (endpoints included) | Open interval (a, b) (endpoints excluded) | Half-open interval $[a, b)$ | Half-open interval $(a, b]$ |

The **closed interval** $[a, b]$ is the set of all real numbers x such that $a \le x \le b$:

$$[a, b] = \{x \in \mathbf{R} : a \le x \le b\}$$

We usually write this more simply as $\{x : a \le x \le b\}$, it being understood that x belongs to $\mathbf{R}$. The **open** and **half-open intervals** are the sets

$$\underbrace{(a, b) = \{x : a < x < b\}}_{\text{Open interval (endpoints excluded)}}, \qquad \underbrace{[a, b) = \{x : a \le x < b\},}_{\text{Half-open interval}} \qquad \underbrace{(a, b] = \{x : a < x \le b\}}_{\text{Half-open interval}}$$

The infinite interval $(-\infty, \infty)$ is the entire real line $\mathbf{R}$. A half-infinite interval is closed if it contains its finite endpoint and is open otherwise (Figure 5):

$$[a, \infty) = \{x : x \ge a\}, \qquad (-\infty, b] = \{x : x \le b\}$$

$[a, \infty)$ $\qquad$ $(-\infty, b]$

FIGURE 5 Closed half-infinite intervals.

Open and closed intervals may be described by absolute-value inequalities. For example, the interval $(-r, r)$ is described by the inequality $|x| < r$ (Figure 6):

$$|x| < r \quad \Leftrightarrow \quad -r < x < r \quad \Leftrightarrow \quad x \in (-r, r) \qquad \boxed{2}$$

$|x| < r$

FIGURE 6 The interval $(-r, r) = \{x : |x| < r\}$.

The symbol ⇔ is read as "is equivalent to," and the symbol ⇒, that we will also use, is read as "implies."

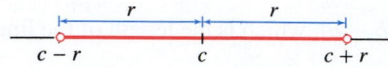

FIGURE 7 $(a, b) = (c - r, c + r)$.

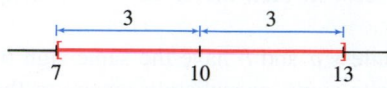

FIGURE 8 The interval $[7, 13]$ is described by $|x - 10| \leq 3$.

In Example 3, we use the notation ∪ to denote "union": The union $A \cup B$ of sets A and B consists of all elements that belong to either A or B (or to both).

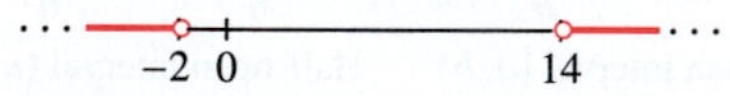

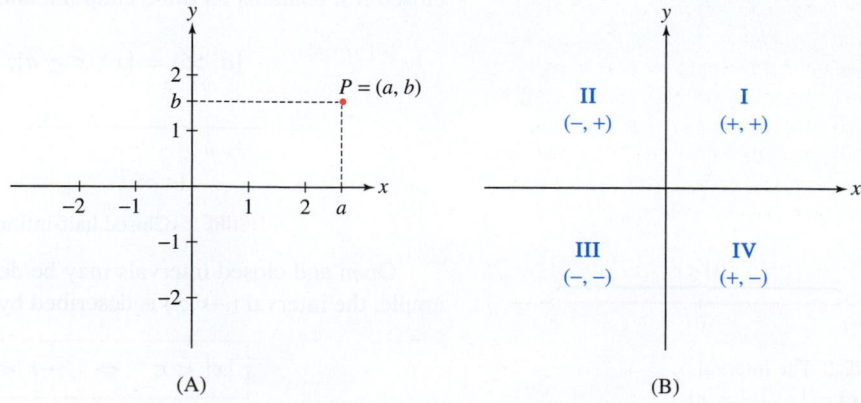

FIGURE 9 The set $S = \left\{ x : \left| \frac{1}{2}x - 3 \right| > 4 \right\}$.

More generally, for an interval symmetric about the value c (Figure 7):

$$|x - c| < r \quad \Leftrightarrow \quad c - r < x < c + r \quad \Leftrightarrow \quad x \in (c - r, c + r) \qquad \boxed{3}$$

Closed intervals can be represented similarly, with $<$ replaced by $\leq$. We refer to r as the **radius** and c as the **midpoint** or **center** of the intervals $(c - r, c + r)$ and $[c - r, c + r]$. The intervals (a, b) and $[a, b]$ have midpoint $c = \frac{1}{2}(a + b)$ and radius $r = \frac{1}{2}(b - a)$ (Figure 7).

EXAMPLE 2 Describe $[7, 13]$ using an absolute-value inequality.

Solution The midpoint of the interval $[7, 13]$ is $c = \frac{1}{2}(7 + 13) = 10$, and its radius is $r = \frac{1}{2}(13 - 7) = 3$ (Figure 8). Therefore,

$$[7, 13] = \left\{ x \in \mathbf{R} : |x - 10| \leq 3 \right\} \qquad \blacksquare$$

EXAMPLE 3 Describe the set $S = \left\{ x : \left| \frac{1}{2}x - 3 \right| > 4 \right\}$ in terms of intervals.

Solution It is easier to consider the opposite inequality $\left| \frac{1}{2}x - 3 \right| \leq 4$ first. By (2):

$$\left| \frac{1}{2}x - 3 \right| \leq 4 \quad \Leftrightarrow \quad -4 \leq \frac{1}{2}x - 3 \leq 4$$

$$-1 \leq \frac{1}{2}x \leq 7 \qquad \text{(add 3)}$$

$$-2 \leq x \leq 14 \qquad \text{(multiply by 2)}$$

Thus, $\left| \frac{1}{2}x - 3 \right| \leq 4$ is satisfied when x belongs to $[-2, 14]$. The set S is the *complement*, consisting of all numbers x *not in* $[-2, 14]$. We can describe S as the union of two intervals: $S = (-\infty, -2) \cup (14, \infty)$ (Figure 9). $\qquad \blacksquare$

Graphing

Graphing is a basic tool in calculus, as it is in algebra and trigonometry. Recall that rectangular (or Cartesian) coordinates in the plane are defined by choosing two perpendicular axes, the x-axis and the y-axis. To a pair of numbers (a, b) we associate the point P located at the intersection of the line perpendicular to the x-axis at a and the line perpendicular to the y-axis at b [Figure 10(A)]. The numbers a and b are the x- and y-**coordinates** of P. The x-coordinate is sometimes called the **abscissa** and the y-coordinate the **ordinate**. The **origin** is the point with coordinates $(0, 0)$.

The term "Cartesian" refers to the French philosopher and mathematician René Descartes (1596–1650), whose Latin name was Cartesius. He is credited (along with Pierre de Fermat) with the invention of analytic geometry. In his great work La Géométrie, Descartes used the letters x, y, z for unknowns and a, b, c for constants, a convention that has been followed ever since.

The notation (a, b) could mean the open interval that is equal to the set of points $\{x : a < x < b\}$ or it could mean the point in the xy-plane with $x = a$ and $y = b$. In general, the meaning will be apparent from the context.

(A)

(B)

FIGURE 10 The rectangular coordinate system (A) and the four quadrants (B).

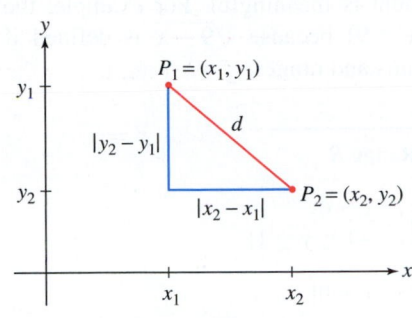

FIGURE 11 Distance d is given by the distance formula.

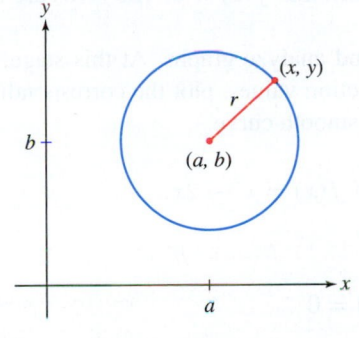

FIGURE 12 Circle with equation $(x-a)^2 + (y-b)^2 = r^2$.

The axes divide the plane into four quadrants labeled I–IV, determined by the signs of the coordinates [Figure 10(B)]. For example, quadrant III consists of points (x, y) such that $x < 0$ and $y < 0$.

The distance d between two points $P_1 = (x_1, y_1)$ and $P_2 = (x_2, y_2)$ is computed using the Pythagorean Theorem. In Figure 11, we see that $\overline{P_1 P_2}$ is the hypotenuse of a right triangle with sides $a = |x_2 - x_1|$ and $b = |y_2 - y_1|$. Therefore,

$$d^2 = a^2 + b^2 = (x_2 - x_1)^2 + (y_2 - y_1)^2$$

We obtain the distance formula by taking square roots.

Distance Formula   The distance between $P_1 = (x_1, y_1)$ and $P_2 = (x_2, y_2)$ is

$$d = \sqrt{(x_2 - x_1)^2 + (y_2 - y_1)^2}$$

Once we have the distance formula, we can derive the equation of a circle of radius r and center (a, b) (Figure 12). A point (x, y) lies on this circle if the distance from (x, y) to (a, b) is r:

$$\sqrt{(x - a)^2 + (y - b)^2} = r$$

Squaring both sides, we obtain the standard equation of the circle of radius r centered at (a, b):

$$(x - a)^2 + (y - b)^2 = r^2$$

We now review some definitions and notation concerning functions.

DEFINITION   A **function** f from a set D to a set Y is a rule that assigns, to each element x in D, a unique element $y = f(x)$ in Y. We write

$$f : D \rightarrow Y$$

The set D, called the **domain** of f, is the set of "allowable inputs." For $x \in D$, $f(x)$ is called the **value** of f at x (Figure 13). The **range** R of f is the subset of Y consisting of all values $f(x)$:

$$R = \{y \in Y : f(x) = y \text{ for some } x \in D\}$$

Informally, we think of f as a "machine" that produces an output y for every input x in the domain D (Figure 14).

A function $f : D \rightarrow Y$ is also called a **map**. The sets D and Y can be arbitrary. For example, we can define a map from the set of living people to the set of whole numbers by mapping each person to his or her year of birth. The range of this map is the set of years in which a living person was born. In multivariable calculus, the domain might be a set of points in the two-dimensional plane and the range a set of numbers, points, or vectors.

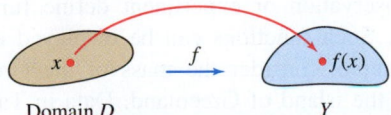

Domain D Y

FIGURE 13 A function assigns an element $f(x)$ in Y to each $x \in D$.

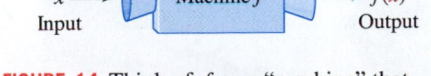

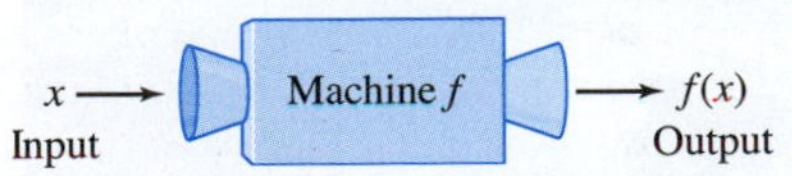

Input   Output

FIGURE 14 Think of f as a "machine" that takes the input x and produces the output $f(x)$.

Writing $y = f(x)$ for a function f, we refer to x as the **independent variable** and y as the **dependent variable** (because its value depends on the choice of x).

The first part of this text deals with functions f, where both the domain and the range are sets of real numbers. When f is defined by a formula, its natural domain is the set of real numbers x for which the formula is meaningful. For example, the function $f(x) = \sqrt{9 - x}$ has domain $D = \{x : x \le 9\}$ because $\sqrt{9 - x}$ is defined if $9 - x \ge 0$. Here are some other examples of domains and ranges:

$f(x)$	Domain D	Range R
x^2	$\mathbf{R}$	$\{y : y \ge 0\}$
$\cos x$	$\mathbf{R}$	$\{y : -1 \le y \le 1\}$
$\dfrac{1}{x+1}$	$\{x : x \ne -1\}$	$\{y : y \ne 0\}$

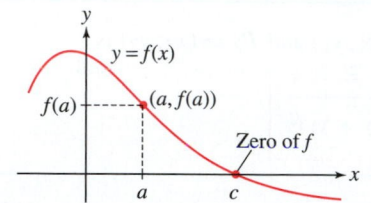

$y = f(x)$

$f(a)$ — $(a, f(a))$

Zero of f

a c

FIGURE 15

The **graph** of a function $y = f(x)$ is obtained by plotting the points $(a, f(a))$ for a in the domain D (Figure 15). If you start at $x = a$ on the x-axis, and move up to the graph and then over to the y-axis, you arrive at the value $f(a)$.

A **zero** or **root** of a function f is a number c such that $f(c) = 0$. The zeros are the values of x where the graph intersects the x-axis.

In Chapter 4, we will use calculus to sketch and analyze graphs. At this stage, to sketch a graph by hand, we can make a table of function values, plot the corresponding points (including any zeros), and connect them by a smooth curve.

EXAMPLE 4 Find the roots and sketch the graph of $f(x) = x^3 - 2x$.

Solution First, we solve

$$x^3 - 2x = x(x^2 - 2) = 0$$

The roots of f are $x = 0$ and $x = \pm\sqrt{2}$. To sketch the graph, we plot the roots and a few values listed in Table 1 and join them by a curve (Figure 16). ∎

TABLE 1

x	$x^3 - 2x$
-2	-4
-1	1
0	0
1	-1
2	4

FIGURE 16 Graph of $f(x) = x^3 - 2x$.

Functions arising in applications are not always given by formulas. Data collected from observation or experiment define functions for which there may be no exact formula. Such functions can be displayed either graphically or by a table of values. For example, consider the mass of the Greenland ice sheet (Figure 17) that covers most of the island of Greenland. Data in Table 2 and Figure 18 collected by NASA's GRACE (Global Recovery and Climate Experiment) satellite show the change in the mass of the ice, C, as a function of time, t, since the beginning of 2012. (Note, for example, $t = 1.46$ means 0.46 years into 2013.) To plot this function, we plot the data points in the table and connect the points with a smooth curve. We will see that many of the tools of calculus can be applied to functions constructed from data in this way.

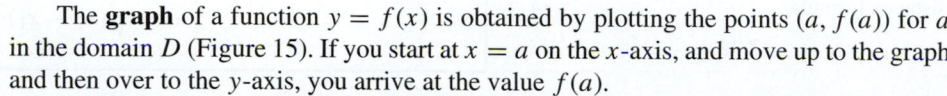

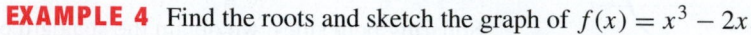

Harvepino/Shutterstock

FIGURE 17 The Greenland ice sheet.

TABLE 2

Time (years since Jan. 1, 2012)	Change in Mass from Jan. 1, 2012 (in gigatonnes)	Time (years since Jan. 1, 2012)	Change in Mass from Jan. 1, 2012 (in gigatonnes)
0	0	2.79	−794.51
0.21	138.53	3.12	−623.4
0.54	−139.14	3.32	−624.41
0.89	−487.05	3.70	−960.08
1.12	−386.78	4.12	−899.86
1.46	−355.26	4.46	−869.46
1.87	−518.52	4.91	−1153.08
2.21	−475.14	5.25	−1110.29
2.45	−474.96	5.44	−1115.94

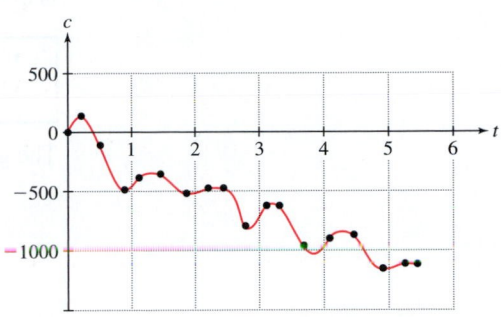

FIGURE 18 Change in mass of the Greenland ice sheet.

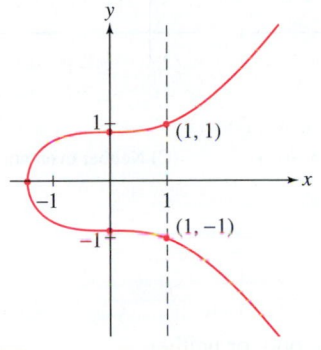

FIGURE 19 Graph of $4y^2 - x^3 = 3$. This graph fails the Vertical Line Test, so it is not the graph of a function.

We can graph not just functions but, more generally, any equation relating y and x. Figure 19 shows the graph of the equation $4y^2 - x^3 = 3$; it consists of all pairs (x, y) satisfying the equation. This curve is not the graph of a function of x because some x-values are associated with two y-values. For example, $x = 1$ is associated with both $y = 1$ and $y = -1$. A curve is the graph of a function of x if and only if it passes the **Vertical Line Test**; that is, every vertical line $x = a$ intersects the curve in at most one point.

We are often interested in whether a function is increasing or decreasing. Roughly speaking, a function f is increasing if its graph goes up as we move to the right and is decreasing if its graph goes down [Figures 20(A) and (B)]. More precisely, we define the notion of increase/decrease on an open interval.

> A function f is
>
> - **Increasing** on (a, b) if $f(x_1) < f(x_2)$ for all $x_1, x_2 \in (a, b)$ such that $x_1 < x_2$
> - **Decreasing** on (a, b) if $f(x_1) > f(x_2)$ for all $x_1, x_2 \in (a, b)$ such that $x_1 < x_2$

We say that f is **monotonic** if it is either increasing or decreasing. In Figure 20(C), the function is not monotonic because, while it is increasing for some intervals of x and decreasing for others, it is neither increasing nor decreasing for all x.

A function f is called **nondecreasing** if $f(x_1) \leq f(x_2)$ for $x_1 < x_2$ (defined by $\leq$ rather than a strict inequality $<$). **Nonincreasing** functions are defined similarly. Function (D) in Figure 20 is nondecreasing, but it is not increasing on the intervals where the graph is horizontal. Function (E) is increasing everywhere, even though it levels off momentarily.

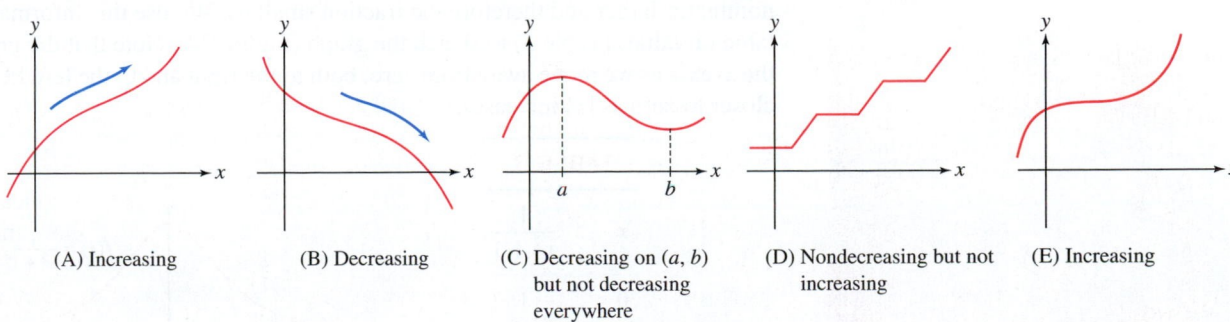

(A) Increasing   (B) Decreasing   (C) Decreasing on (a, b) but not decreasing everywhere   (D) Nondecreasing but not increasing   (E) Increasing

FIGURE 20

Another important property of functions is **parity**, which refers to whether a function is even or odd:

> - f is **even** if $f(-x) = f(x)$.
> - f is **odd** if $f(-x) = -f(x)$.

The graphs of functions with even or odd parity have a special symmetry:

- **Even function:** The graph is symmetric about the y-axis. This means that if $P = (a, b)$ lies on the graph, then so does $Q = (-a, b)$ [Figure 21(A)].
- **Odd function:** The graph is symmetric with respect to the origin. This means that if $P = (a, b)$ lies on the graph, then so does $Q = (-a, -b)$ [Figure 21(B)].

Many functions are neither even nor odd [Figure 21(C)].

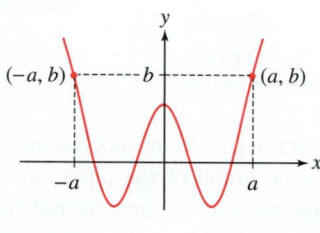
(A) Even function: $f(-x) = f(x)$
Graph is symmetric
about the y-axis.

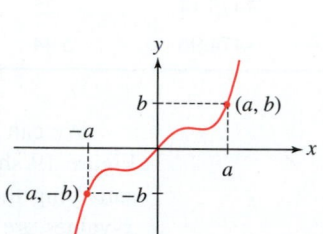
(B) Odd function: $f(-x) = -f(x)$
Graph is symmetric
about the origin.

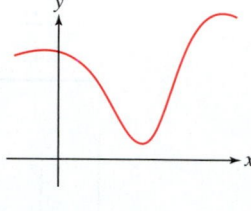
(C) Neither even nor odd

FIGURE 21

EXAMPLE 5 Determine whether the function is even, odd, or neither.

(a) $f(x) = x^4$ **(b)** $g(x) = x^{-1}$ **(c)** $h(x) = x^2 + x$

Solution

(a) $f(-x) = (-x)^4 = x^4$. Thus, $f(x) = f(-x)$, and f is even.

(b) $g(-x) = (-x)^{-1} = -x^{-1}$. Thus, $g(-x) = -g(x)$, and g is odd.

(c) $h(-x) = (-x)^2 + (-x) = x^2 - x$. We see that $h(-x)$ is not equal to $h(x)$ or to $-h(x) = -x^2 - x$. Therefore, h is neither even nor odd. ∎

EXAMPLE 6 **Using Symmetry** Sketch the graph of $f(x) = \dfrac{1}{x^2 + 1}$.

Solution The function f is positive [$f(x) > 0$] and even [$f(-x) = f(x)$]. Therefore, the graph lies above the x-axis and is symmetric with respect to the y-axis.

Furthermore, f is decreasing for $x \geq 0$ (because a larger value of x makes the denominator larger and therefore the fraction smaller). We use this information and a short table of values (Table 3) to sketch the graph (Figure 22). Note that the graph approaches the x-axis as we move away from zero, both to the right and to the left, because $f(x)$ gets closer to zero as $|x|$ increases. ∎

TABLE 3

x	$\dfrac{1}{x^2 + 1}$
0	1
± 1	$\frac{1}{2}$
± 2	$\frac{1}{5}$

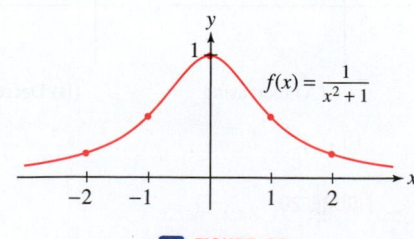

DF **FIGURE 22**

Two important ways of modifying a graph are **translation** (or **shifting**) and **scaling**. Translation consists of moving the graph horizontally or vertically:

Remember that $f(x) + c$ and $f(x + c)$ are different. The graph of $y = f(x) + c$ is a vertical translation and $y = f(x + c)$ is a horizontal translation of the graph of $y = f(x)$.

> **DEFINITION Translation (Shifting)**
> - **Vertical Translation** $y = f(x) + c$: Shifts the graph of f by $|c|$ units *vertically*, upward if $c > 0$ and downward if $c < 0$.
> - **Horizontal Translation** $y = f(x + c)$: Shifts the graph of f by $|c|$ units *horizontally*, to the right if $c < 0$ and to the left if $c > 0$.

Figure 23 shows the effect of translating the graph of $f(x) = 1/(x^2 + 1)$ vertically and horizontally.

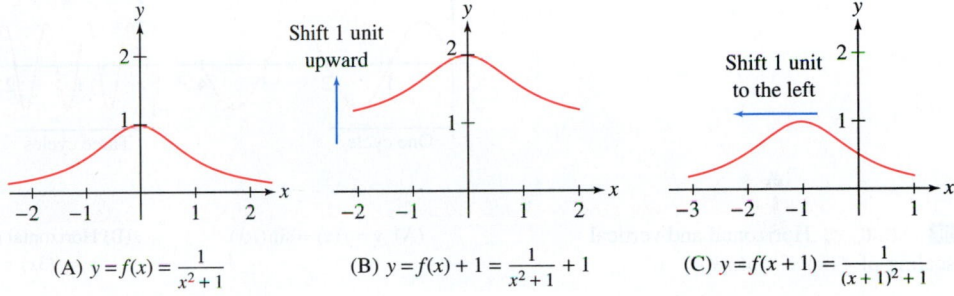

(A) $y = f(x) = \dfrac{1}{x^2 + 1}$ (B) $y = f(x) + 1 = \dfrac{1}{x^2 + 1} + 1$ (C) $y = f(x + 1) = \dfrac{1}{(x + 1)^2 + 1}$

FIGURE 23

EXAMPLE 7 Figure 24(A) is the graph of $f(x) = x^2$, and Figure 24(B) is a horizontal and vertical shift of (A). What is the equation of graph (B)?

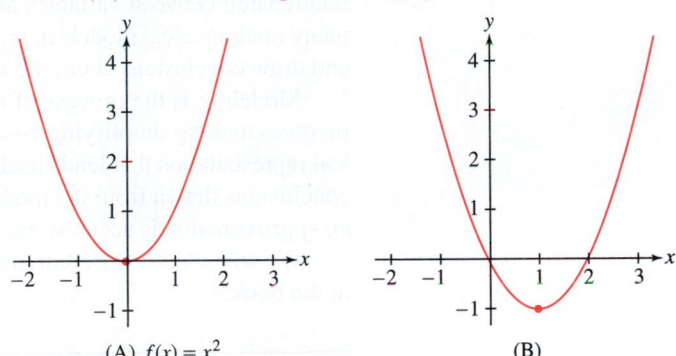

(A) $f(x) = x^2$ (B)

DF **FIGURE 24**

Solution Graph (B) is obtained by shifting graph (A) 1 unit to the right and 1 unit down. We can see this by observing that the point $(0, 0)$ on the graph of f is shifted to $(1, -1)$. Therefore, (B) is the graph of $g(x) = (x - 1)^2 - 1$.   ■

Scaling (also called **dilation**) consists of compressing or expanding the graph in the vertical or horizontal directions:

> **DEFINITION Scaling**
> - **Vertical scaling** $y = kf(x)$: If $|k| > 1$, the graph of f is expanded vertically by the factor $|k|$. If $0 < |k| < 1$, the graph of f is compressed vertically by the factor $|k|$. If $k < 0$, then the graph is also reflected across the x-axis (Figure 25).
> - **Horizontal scaling** $y = f(kx)$: If $|k| > 1$, the graph of f is compressed horizontally by the factor $|k|$. If $0 < |k| < 1$, the graph of f is expanded horizontally by the factor $|k|$. If $k < 0$, then the graph is also reflected across the y-axis.

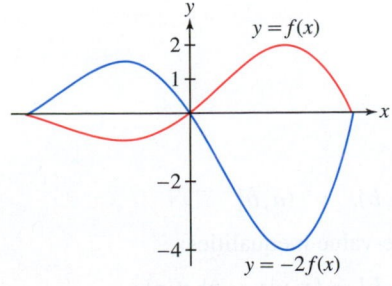

FIGURE 25 Negative vertical scale factor $k = -2$.

EXAMPLE 8 Sketch the graphs of $f(x) = \sin(\pi x)$ and its dilates $f(3x)$ and $3f(x)$.

Solution The graph of $f(x) = \sin(\pi x)$ is a sine curve with period 2. It completes one cycle over every interval of length 2—see Figure 26(A).

- The graph of $f(3x) = \sin(3\pi x)$ is a compressed version of $y = f(x)$, completing three cycles instead of one over intervals of length 2 [Figure 26(B)].
- The graph of $y = 3f(x) = 3\sin(\pi x)$ is obtained from $y = f(x)$ by expanding in the vertical direction by a factor of 3 [Figure 26(C)]. ∎

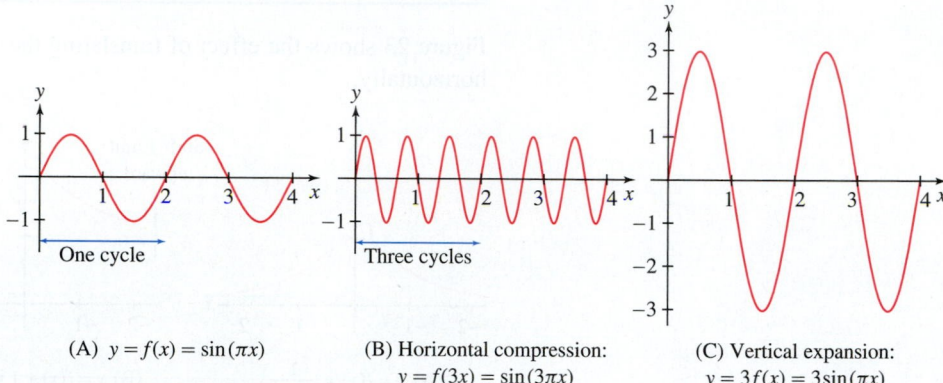

DF **FIGURE 26** Horizontal and vertical scaling of $f(x) = \sin(\pi x)$.

(A) $y = f(x) = \sin(\pi x)$

(B) Horizontal compression: $y = f(3x) = \sin(3\pi x)$

(C) Vertical expansion: $y = 3f(x) = 3\sin(\pi x)$

Mathematical Models

A mathematical model is a representation of a real-world phenomenon using mathematical concepts. Functions are often used as models; they provide a simple way to express a relationship between variables associated with a real-world situation. We will introduce many mathematical models in this book. Using the tools of calculus we will study models and draw conclusions about the situations they describe.

Modeling is the process of developing a mathematical model. The process usually involves making simplifying assumptions about a system in order to develop a mathematical representation that lends itself well to analysis. When such assumptions are made, the conclusions drawn from the model only approximate the real-world system. Ideally such an approximation is accurate enough to make useful predictions.

We will address different important aspects of the modeling process at various points in the book.

1.1 SUMMARY

- Important exponent laws:

  (i) $b^x b^y = b^{x+y}$ **(ii)** $\frac{b^x}{b^y} = b^{x-y}$ **(iii)** $b^{-x} = \frac{1}{b^x}$ **(iv)** $(b^x)^y = b^{xy}$

- Binomial expansion formula: $(a+b)^n$ is a sum of terms $\frac{n!}{(n-p)!\,p!} a^{n-p} b^p$, with a term for each p going from 0 to n.

- Absolute value: $|a| = \begin{cases} a & \text{if } a \geq 0 \\ -a & \text{if } a < 0 \end{cases}$

- Triangle inequality: $|a + b| \leq |a| + |b|$
- Four intervals with endpoints a and b:

$$(a, b), \qquad [a, b], \qquad [a, b), \qquad (a, b]$$

- Writing open and closed intervals using absolute-value inequalities:

$$(a, b) = \{x : |x - c| < r\}, \qquad [a, b] = \{x : |x - c| \leq r\}$$

where $c = \frac{1}{2}(a + b)$ is the midpoint and $r = \frac{1}{2}(b - a)$ is the radius.

- Distance d between (x_1, y_1) and (x_2, y_2):

$$d = \sqrt{(x_2 - x_1)^2 + (y_2 - y_1)^2}$$

- Equation of circle of radius r with center (a, b):

$$(x - a)^2 + (y - b)^2 = r^2$$

- A *zero* or *root* of a function f is a number c such that $f(c) = 0$.
- Vertical Line Test: A curve in the plane is the graph of a function of x if and only if each vertical line $x = a$ intersects the curve in at most one point.

Increasing:	$f(x_1) < f(x_2)$ if $x_1 < x_2$
Nondecreasing:	$f(x_1) \leq f(x_2)$ if $x_1 < x_2$
Decreasing:	$f(x_1) > f(x_2)$ if $x_1 < x_2$
Nonincreasing:	$f(x_1) \geq f(x_2)$ if $x_1 < x_2$

- Even function: $f(-x) = f(x)$ (graph is symmetric about the y-axis)
- Odd function: $f(-x) = -f(x)$ (graph is symmetric about the origin)
- Four ways to transform the graph of f:

$f(x) + c$	Shifts graph vertically $|c|$ units (upward if $c > 0$, downward if $c < 0$)
$f(x + c)$	Shifts graph horizontally $|c|$ units (to the right if $c < 0$, to the left if $c > 0$)
$kf(x)$	Scales graph vertically by factor $|k|$, stretching if $|k| > 1$, compressing if $0 < |k| < 1$; if $k < 0$, graph is reflected across x-axis
$f(kx)$	Scales graph horizontally by factor $|k|$, compressing if $|k| > 1$, stretching if $0 < |k| < 1$; if $k < 0$, graph is reflected across y-axis

1.1 EXERCISES

Preliminary Questions

1. Give an example of numbers a and b such that $a < b$ and $|a| > |b|$.

2. Which numbers satisfy $|a| = a$? Which satisfy $|a| = -a$? What about $|-a| = a$?

3. Give an example of numbers a and b such that $|a + b| < |a| + |b|$.

4. Are there numbers a and b such that $|a + b| > |a| + |b|$?

5. What are the coordinates of the point lying at the intersection of the lines $x = 9$ and $y = -4$?

6. In which quadrant do the following points lie?
(a) $(1, 4)$ (b) $(-3, 2)$ (c) $(4, -3)$ (d) $(-4, -1)$

7. What is the radius of the circle with equation $(x - 7)^2 + (y - 8)^2 = 9$?

8. The equation $f(x) = 5$ has a solution if (choose one):
(a) 5 belongs to the domain of f.
(b) 5 belongs to the range of f.

9. What kind of symmetry does the graph have if $f(-x) = -f(x)$?

10. Is there a function that is both even and odd?

Exercises

1. Which of the following equations is incorrect?
(a) $3^2 \cdot 3^5 = 3^7$ (b) $(\sqrt{5})^{4/3} = 5^{2/3}$
(c) $3^2 \cdot 2^3 = 1$ (d) $(2^{-2})^{-2} = 16$

2. Rewrite as a whole number (without using a calculator):
(a) 7^0 (b) $10^2(2^{-2} + 5^{-2})$
(c) $\dfrac{(4^3)^5}{(4^5)^3}$ (d) $27^{4/3}$
(e) $8^{-1/3} \cdot 8^{5/3}$ (f) $3 \cdot 4^{1/4} - 12 \cdot 2^{-3/2}$

3. Use the binomial expansion formula to expand $(2 - x)^7$.

4. Use the binomial expansion formula to expand $(x + 1)^9$.

5. Which of (a)–(d) are true for $a = 4$ and $b = -5$?
(a) $-2a < -2b$ (b) $|a| < -|b|$ (c) $ab < 0$
(d) $\dfrac{1}{a} < \dfrac{1}{b}$

6. Which of (a)–(d) are true for $a = -3$ and $b = 2$?
(a) $a < b$ (b) $|a| < |b|$ (c) $ab > 0$
(d) $3a < 3b$

In Exercises 7–12, express the interval in terms of an inequality involving absolute value.

7. $[-2, 2]$ 8. $(-4, 4)$ 9. $(0, 4)$

10. $[-4, 0]$ 11. $[-1, 8]$ 12. $(-2.4, 1.9)$

In Exercises 13–16, write the inequality in the form $a < x < b$.

13. $|x| < 8$

14. $|x - 12| < 8$

15. $|2x + 1| < 5$

16. $|3x - 4| < 2$

In Exercises 17–22, express the set of numbers x satisfying the given condition as an interval.

17. $|x| < 4$

18. $|x| \le 9$

19. $|x - 4| < 2$

20. $|x + 7| < 2$

21. $|4x - 1| \le 8$

22. $|3x + 5| < 1$

In Exercises 23–26, describe the set as a union of finite or infinite intervals.

23. $\{x : |x - 4| > 2\}$

24. $\{x : |2x + 4| > 3\}$

25. $\{x : |x^2 - 1| > 2\}$

26. $\{x : |x^2 + 2x| > 2\}$

27. Match (a)–(f) with (i)–(vi).

(a) $a > 3$

(b) $|a - 5| < \frac{1}{3}$

(c) $\left|a - \frac{1}{3}\right| < 5$

(d) $|a| > 5$

(e) $|a - 4| < 3$

(f) $1 \le a \le 5$

   (i) a lies to the right of 3.

  (ii) a lies between 1 and 7.

 (iii) The distance from a to 5 is less than $\frac{1}{3}$.

 (iv) The distance from a to 3 is at most 2.

  (v) a is less than 5 units from $\frac{1}{3}$.

 (vi) a lies either to the left of -5 or to the right of 5.

28. Describe $\left\{x : \dfrac{x}{x + 1} < 0\right\}$ as an interval. *Hint:* Consider the sign of x and $x + 1$ individually.

29. Describe $\{x : x^2 + 2x < 3\}$ as an interval. *Hint:* Consider the graph of $y = x^2 + 2x - 3$.

30. Describe the set of real numbers satisfying $|x - 3| = |x - 2| + 1$ as a half-infinite interval.

31. Show that if $a > b$, and $a, b \ne 0$, then $b^{-1} > a^{-1}$, provided that a and b have the same sign. What happens if $a > 0$ and $b < 0$?

32. Which x satisfies both $|x - 3| < 2$ and $|x - 5| < 1$?

33. Show that if $|a - 5| < \frac{1}{2}$ and $|b - 8| < \frac{1}{2}$, then we can conclude that $|(a + b) - 13| < 1$. *Hint:* Use the triangle inequality ($|a + b| \le |a| + |b|$).

34. Suppose that $|x - 4| \le 1$.

(a) What is the maximum possible value of $|x + 4|$?

(b) Show that $|x^2 - 16| \le 9$.

35. Suppose that $|a - 6| \le 2$ and $|b| \le 3$.

(a) What is the largest possible value of $|a + b|$?

(b) What is the smallest possible value of $|a + b|$?

36. Prove that $|x| - |y| \le |x - y|$. *Hint:* Apply the triangle inequality to y and $x - y$.

37. Express $r_1 = 0.\overline{27}$ as a fraction. *Hint:* $100r_1 - r_1$ is an integer. Then express $r_2 = 0.2666\ldots$ as a fraction.

38. Represent $1/7$ and $4/27$ as infinite repeating decimals.

39. Plot each pair of points and compute the distance between them:

(a) $(1, 4)$ and $(3, 2)$

(b) $(2, 1)$ and $(2, 4)$

40. Plot each pair of points and compute the distance between them:

(a) $(0, 0)$ and $(-2, 3)$

(b) $(-3, -3)$ and $(-2, 3)$

41. Find the equation of the circle with center $(2, 4)$:

(a) With radius $r = 3$

(b) That passes through $(1, -1)$

42. Find all points in the xy-plane with integer coordinates located at a distance 5 from the origin. Then find all points with integer coordinates located at a distance 5 from $(2, 3)$.

43. Determine the domain and range of the function

$$f : \{r, s, t, u\} \to \{A, B, C, D, E\}$$

defined by $f(r) = A$, $f(s) = B$, $f(t) = B$, $f(u) = E$.

44. Give an example of a function whose domain D has three elements and whose range R has two elements. Does a function exist whose domain D has two elements and whose range R has three elements?

In Exercises 45–52, find the domain and range of the function.

45. $f(x) = -x$

46. $g(t) = t^4$

47. $f(x) = x^3$

48. $g(t) = \sqrt{2 - t}$

49. $f(x) = |x|$

50. $h(s) = \dfrac{1}{s}$

51. $f(x) = \dfrac{1}{x^2}$

52. $g(t) = \dfrac{1}{\sqrt{1 - t}}$

In Exercises 53–56, determine where f is increasing.

53. $f(x) = |x + 1|$

54. $f(x) = x^3$

55. $f(x) = x^4$

56. $f(x) = \dfrac{1}{x^4 + x^2 + 1}$

In Exercises 57–62, find the zeros of f and sketch its graph by plotting points. Use symmetry and increase/decrease information where appropriate.

57. $f(x) = x^2 - 4$

58. $f(x) = 2x^2 - 4$

59. $f(x) = x^3 - 4x$

60. $f(x) = x^3$

61. $f(x) = 2 - x^3$

62. $f(x) = \dfrac{1}{(x - 1)^2 + 1}$

63. Which of the curves in Figure 27 is the graph of a function of x?

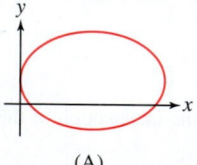

(A)

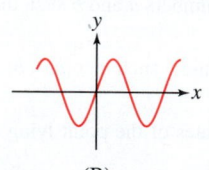

(B)

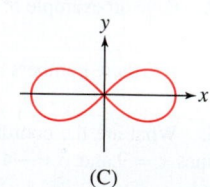

(C)

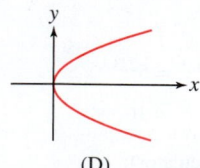

(D)

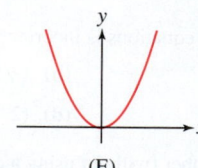

(E)

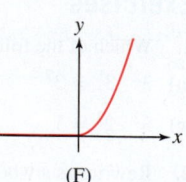

(F)

FIGURE 27

64. Of the curves in Figure 27 that are graphs of functions, which is the graph of an odd function? Of an even function?

65. Determine whether the function is even, odd, or neither.

(a) $f(x) = x^5$

(b) $g(t) = t^3 - t^2$

(c) $F(t) = \dfrac{1}{t^4 + t^2}$

66. Determine whether the function is even, odd, or neither.

(a) $f(x) = 2x - x^2$

(b) $k(w) = (1 - w)^3 + (1 + w)^3$

(c) $f(t) = \dfrac{1}{t^4 + t + 1} - \dfrac{1}{t^4 - t + 1}$

(d) $g(t) = 2^t - 2^{-t}$

67. Write $f(x) = 2x^4 - 5x^3 + 12x^2 - 3x + 4$ as the sum of an even and an odd function.

68. Assume that p is a function that is defined for all x.

(a) Prove that if f is defined by $f(x) = p(x) + p(-x)$ then f is even.

(b) Prove that if g is defined by $g(x) = p(x) - p(-x)$ then g is odd.

69. Assume that p is a function that is defined for $x > 0$ and satisfies $p(a/b) = p(b) - p(a)$. Prove that $f(x) = p\left(\dfrac{2 - x}{2 + x}\right)$ is an odd function.

70. State whether the function is increasing, decreasing, or neither.

(a) Surface area of a sphere as a function of its radius

(b) Temperature at a point on the equator as a function of time

(c) Price of an airline ticket as a function of the price of oil

(d) Pressure of the gas in a piston as a function of volume

In Exercises 71–76, let f be the function shown in Figure 28.

71. Find the domain and range of f.

72. Sketch the graphs of $y = f(x + 2)$ and $y = f(x) + 2$.

73. Sketch the graphs of $y = f(2x)$, $y = f\left(\tfrac{1}{2}x\right)$, and $y = 2f(x)$.

74. Sketch the graphs of $y = f(-x)$ and $y = -f(-x)$.

75. Extend the graph of f to $[-4, 4]$ so that it is an even function.

76. Extend the graph of f to $[-4, 4]$ so that it is an odd function.

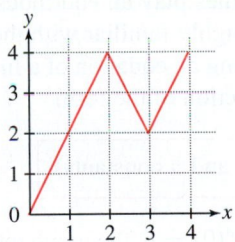

FIGURE 28

77. Suppose that f has domain $[4, 8]$ and range $[2, 6]$. Find the domain and range of:

(a) $y = f(x) + 3$

(b) $y = f(x + 3)$

(c) $y = f(3x)$

(d) $y = 3f(x)$

78. Let $f(x) = x^2$. Sketch the graph over $[-2, 2]$ of:

(a) $y = f(x + 1)$

(b) $y = f(x) + 1$

(c) $y = f(5x)$

(d) $y = 5f(x)$

79. Suppose that the graph of $f(x) = x^4 - x^2$ is compressed horizontally by a factor of 2 and then shifted 5 units to the right.

(a) What is the equation for the new graph?

(b) What is the equation if you first shift by 5 and then compress by 2?

(c) [GU] Verify your answers by plotting your equations.

80. Figure 29 shows the graph of $f(x) = |x| + 1$. Match the functions (a)–(e) with their graphs (i)–(v).

(a) $y = f(x - 1)$

(b) $y = -f(x)$

(c) $y = -f(x) + 2$

(d) $y = f(x - 1) - 2$

(e) $y = f(x + 1)$

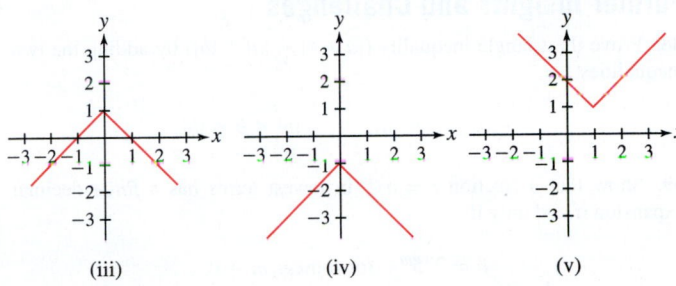

$y = f(x) = |x| + 1$

FIGURE 29

81. Sketch the graph of $y = f(2x)$ and $y = f\left(\tfrac{1}{2}x\right)$, where $f(x) = |x| + 1$ (Figure 29).

82. Find the function f whose graph is obtained by shifting the parabola $y = x^2$ by 3 units to the right and 4 units down, as in Figure 30.

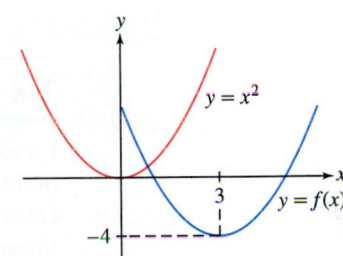

FIGURE 30

83. Define $f(x)$ to be the larger of x and $2 - x$. Sketch the graph of f. What are its domain and range? Express $f(x)$ in terms of the absolute value function.

84. For each curve in Figure 31, state whether it is symmetric with respect to the y-axis, the origin, both, or neither.

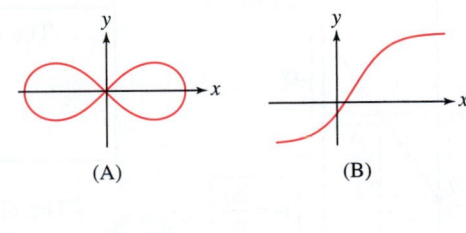

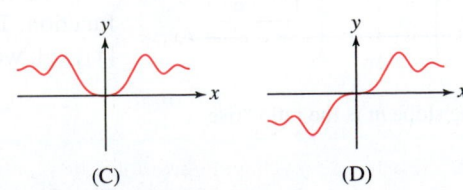

FIGURE 31

85. Show that the sum of two even functions is even and the sum of two odd functions is odd.

86. Suppose that f and g are both odd. Which of the following functions are even? Which are odd?

(a) $y = f(x)g(x)$

(b) $y = f(x)^3$

(c) $y = f(x) - g(x)$

(d) $y = \dfrac{f(x)}{g(x)}$

87. Prove that the only function whose graph is symmetric with respect to both the y-axis and the origin is the function $f(x) = 0$.

Further Insights and Challenges

88. Prove the triangle inequality ($|a + b| \le |a| + |b|$) by adding the two inequalities:

$$-|a| \le a \le |a|, \qquad -|b| \le b \le |b|$$

89. Show that a fraction $r = a/b$ in lowest terms has a *finite* decimal expansion if and only if

$$b = 2^n 5^m \quad \text{for some } n, m \ge 0$$

Hint: Observe that r has a finite decimal expansion when $10^N r$ is an integer for some $N \ge 0$ (and hence b divides 10^N).

90. Let $p = p_1 \dots p_s$ be an integer with digits $p_1, \dots, p_s$. Show that

$$\frac{p}{10^s - 1} = 0.\overline{p_1 \dots p_s}$$

Use this to find the decimal expansion of $r = \frac{2}{11}$. Note that

$$r = \frac{2}{11} = \frac{18}{10^2 - 1}$$

91. 🖊 A function f is symmetric with respect to the vertical line $x = a$ if $f(a - x) = f(a + x)$.

(a) Draw the graph of a function that is symmetric with respect to $x = 2$.

(b) Show that if f is symmetric with respect to $x = a$, then $g(x) = f(x + a)$ is even.

92. 🖊 Formulate a condition for f to be symmetric with respect to the point $(a, 0)$ on the x-axis.

1.2 Linear and Quadratic Functions

Linear functions are the simplest of all functions, and their graphs (lines) are the simplest of all curves. However, linear functions and lines play an enormously important role in calculus. For this reason, you should be thoroughly familiar with the basic properties of linear functions and the different ways of writing an equation of a line.

Let's recall that a **linear function** is a function of the form

$$f(x) = mx + b \quad (m \text{ and } b \text{ constants})$$

The graph of f is a line of slope m, and since $f(0) = b$, the graph intersects the y-axis at the point $(0, b)$ (Figure 1). The number b is called the y-intercept.

> The **slope-intercept form** of the line with slope m and y-intercept b is given by
>
> $$y = mx + b$$

The Greek letter Δ (delta) is commonly used to denote the change in a variable or function. Thus, letting Δx and Δy denote the change in x and $y = f(x)$ over an interval $[x_1, x_2]$, we have

$$\Delta x = x_2 - x_1, \qquad \Delta y = y_2 - y_1 = f(x_2) - f(x_1)$$

The slope m of a line (Figure 1) is equal to the ratio

$$m = \frac{\Delta y}{\Delta x} = \frac{\text{vertical change}}{\text{horizontal change}} = \frac{\text{rise}}{\text{run}}$$

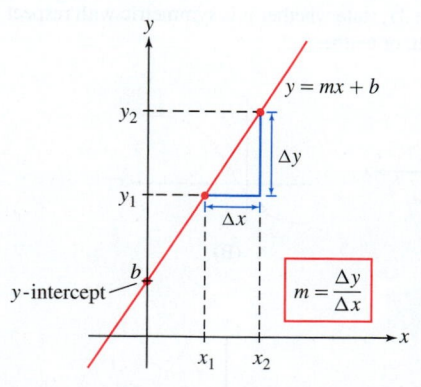

FIGURE 1 The slope m is the ratio "rise over run."

This follows from the formula $y = mx + b$:

$$\frac{\Delta y}{\Delta x} = \frac{y_2 - y_1}{x_2 - x_1} = \frac{(mx_2 + b) - (mx_1 + b)}{x_2 - x_1} = \frac{m(x_2 - x_1)}{x_2 - x_1} = m$$

The slope m measures the *rate of change* of y with respect to x. In fact, by writing

$$\Delta y = m \Delta x$$

we see that a 1-unit increase in x (i.e., $\Delta x = 1$) produces an m-unit change Δy in y. For example, if $m = 5$, then y increases by 5 units per unit increase in x. The rate-of-change interpretation of the slope is fundamental in calculus.

Graphically, the slope m measures the steepness of the line $y = mx + b$. Figure 2(A) shows lines through a point of varying slope m. Note the following properties:

- **Steepness:** The larger the absolute value $|m|$, the steeper the line.
- **Positive slope:** If $m > 0$, the line slants upward from left to right.
- **Negative slope:** If $m < 0$, the line slants downward from left to right.
- $f(x) = mx + b$ is increasing if $m > 0$ and decreasing if $m < 0$.
- The **horizontal line** $y = b$ has slope $m = 0$ [Figure 2(B)].
- A **vertical line** has equation $x = c$, where c is a constant. The slope of a vertical line is undefined. It is not possible to write the equation of a vertical line in slope-intercept form $y = mx + b$. A vertical line is not the graph of a function [Figure 2(B)].

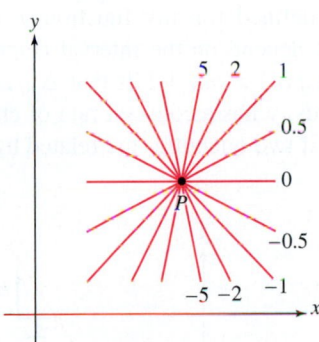
(A) Lines of varying slopes through P

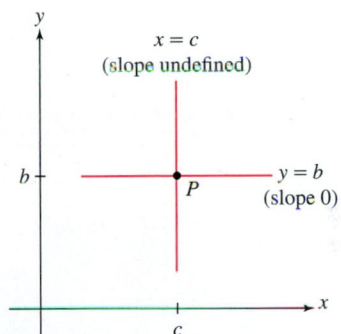
(B) Horizontal and vertical lines through P

FIGURE 2

Scale is especially important in applications because the steepness of a graph depends on the choice of units for the x- and y-axes. We can create very different *subjective* impressions by changing the scale. Figure 3 shows the growth of company profits over a 4-year period. The two plots convey the same information, but the left-hand plot makes the growth look more dramatic.

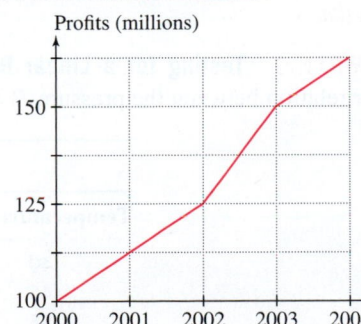

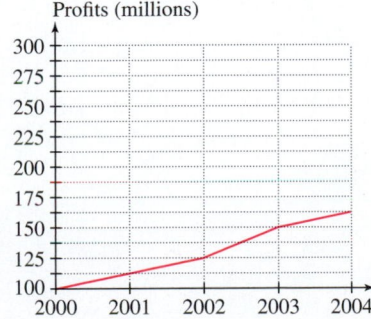

FIGURE 3

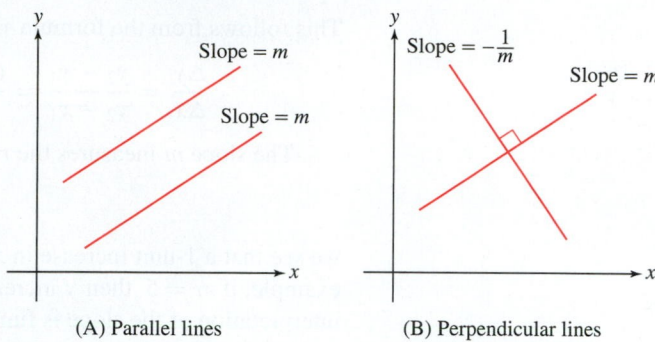

FIGURE 4 Parallel and perpendicular lines.

(A) Parallel lines   (B) Perpendicular lines

Next, we recall the relation between the slopes of parallel and perpendicular lines that are not vertical (Figure 4):

- Lines of slopes m_1 and m_2 are **parallel** if and only if $m_1 = m_2$.
- Lines of slopes m_1 and m_2 are **perpendicular** if and only if

$$m_1 = -\frac{1}{m_2} \quad (\text{or } m_1 m_2 = -1)$$

CONCEPTUAL INSIGHT   The changes over an interval $[x_1, x_2]$

$$\Delta x = x_2 - x_1, \qquad \Delta y = f(x_2) - f(x_1)$$

are defined for any function f (linear or not), but the rise-over-run ratio $\Delta y / \Delta x$ may depend on the interval (Figure 5). The characteristic property of a linear function $f(x) = mx + b$ is that $\Delta y / \Delta x$ has the same value m for every interval. In other words, y has a constant rate of change with respect to x. We can use this property to test if two quantities are related by a linear equation.

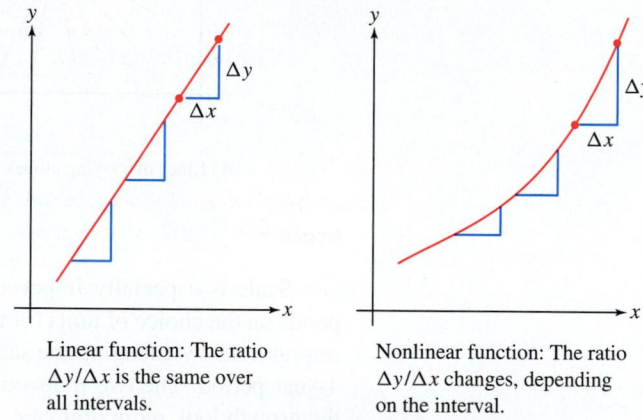

Linear function: The ratio $\Delta y / \Delta x$ is the same over all intervals.

Nonlinear function: The ratio $\Delta y / \Delta x$ changes, depending on the interval.

DF **FIGURE 5**

EXAMPLE 1   Testing for a Linear Relationship   Do the data in Table 1 suggest a linear relation between the pressure P and temperature T of a gas?

TABLE 1

Temperature (°C)	Pressure (kPa)
40	1365.80
45	1385.40
55	1424.60
70	1483.40
80	1522.60

Real experimental data generally do not display perfect linearity. To model a data set, the statistical tool called linear regression is used to find the linear function that best approximates the data.

Solution We calculate $\Delta P/\Delta T$ at successive data points and check whether this ratio is constant:

(T_1, P_1)	(T_2, P_2)	$\dfrac{\Delta P}{\Delta T}$
$(40, 1365.80)$	$(45, 1385.40)$	$\dfrac{1385.40 - 1365.80}{45 - 40} = 3.92$
$(45, 1385.40)$	$(55, 1424.60)$	$\dfrac{1424.60 - 1385.40}{55 - 45} = 3.92$
$(55, 1424.60)$	$(70, 1483.40)$	$\dfrac{1483.40 - 1424.60}{70 - 55} = 3.92$
$(70, 1483.40)$	$(80, 1522.60)$	$\dfrac{1522.60 - 1483.40}{80 - 70} = 3.92$

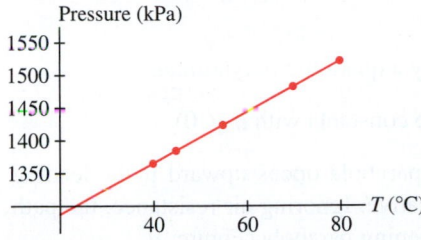

FIGURE 6 Line through pressure–temperature data points.

Because $\Delta P/\Delta T$ has the constant value 3.92, the data points lie on a line with slope $m = 3.92$. This is confirmed in the plot in Figure 6. ∎

As mentioned above, it is important to be familiar with the standard ways of writing the equation of a line. The general **linear equation** is

$$ax + by = c \qquad \boxed{1}$$

where a and b are not *both* zero. For $b = 0$, we obtain the vertical line $ax = c$. For $a = 0$, we obtain the horizontal line $by = c$. When $b \neq 0$, we can rewrite Eq. (1) in slope-intercept form. For example, $-6x + 2y = 3$ can be rewritten as $y = 3x + \frac{3}{2}$.

Another important form for an equation of a line is the **point-slope** form. Given the slope of the line and a point on it (Figure 7), we can use this form to obtain an equation for the line.

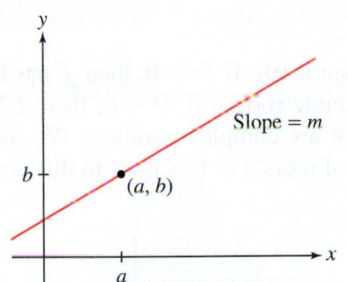

FIGURE 7 Line with slope m through (a, b).

> The **point-slope form** of the line through $P = (a, b)$ with slope m is
>
> $$y - b = m(x - a)$$

EXAMPLE 2 **Line of Given Slope Through a Given Point** Find the slope-intercept equation of the line through $(9, 2)$ with slope $-\frac{2}{3}$.

Solution In point-slope form:

$$y - 2 = -\frac{2}{3}(x - 9)$$

In slope-intercept form: $y = -\frac{2}{3}(x - 9) + 2$ or $y = -\frac{2}{3}x + 8$. See Figure 8. ∎

EXAMPLE 3 **Line Through Two Points** Find an equation of the line through $(2, 1)$ and $(9, 5)$.

Solution The line has slope

$$m = \frac{5 - 1}{9 - 2} = \frac{4}{7}$$

Because $(2, 1)$ lies on the line, its equation in point-slope form is $y - 1 = \frac{4}{7}(x - 2)$. ∎

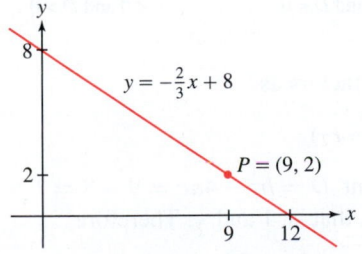

FIGURE 8 Line through $P = (9, 2)$ with slope $m = -\frac{2}{3}$.

Recall that the slope m of a line measures the rate of change of the dependent variable y with respect to the independent variable x, that is $m = \frac{\Delta y}{\Delta x}$. In applications, the slope has units:

$$\text{units of slope} = \frac{\text{units of dependent variable}}{\text{units of independent variable}}$$

For example:

- Let $T = -0.1h + 52$ represent the temperature (in °C) as a function of height (in m) measured by a weather balloon as it rose from the ground to 1000 meters. The slope is $-0.1\frac{°C}{m}$, indicating that the temperature dropped by one-tenth of a degree for every meter the balloon rose.
- For the time period 1900 to 1920, $P = 33.3t + 57.7$ models the population of Saskatchewan (in thousands) t years after 1900. The slope is 33.3 $\frac{\text{thousand people}}{\text{year}}$, indicating the population rose by 33.3 thousand people per year during the time period.

A **quadratic function** is a function defined by a quadratic polynomial:

$$f(x) = ax^2 + bx + c \quad (a, b, c \text{ are constants with } a \neq 0)$$

The graph of f is a **parabola** (Figure 10). The parabola opens upward if the leading coefficient a is positive and downward if a is negative. Ignoring air resistance, the path of a struck baseball is modeled by a downward-opening parabola (Figure 9).

The **discriminant** of $f(x) = ax^2 + bx + c$ is the quantity

$$D = b^2 - 4ac$$

The roots of f are given by the **quadratic formula** (see Exercise 60):

$$\boxed{\text{roots of } f = \frac{-b \pm \sqrt{b^2 - 4ac}}{2a} = \frac{-b \pm \sqrt{D}}{2a}}$$

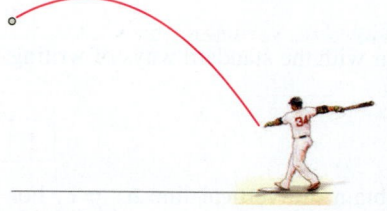

FIGURE 9 A parabola models the path of the baseball.

The sign of D determines the nature of the roots (Figure 10). If $D > 0$, then f has two real roots, and if $D = 0$, it has one real root (a "double root"). If $D < 0$, then f has no roots that are real numbers, but has two roots that are complex numbers. We focus primarily on real numbers and real-number roots ("real roots") of functions in this text.

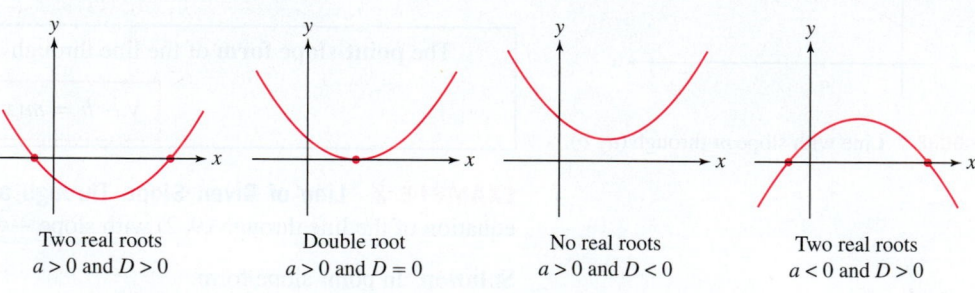

FIGURE 10 Graphs of quadratic functions $f(x) = ax^2 + bx + c$.

Two real roots
$a > 0$ and $D > 0$

Double root
$a > 0$ and $D = 0$

No real roots
$a > 0$ and $D < 0$

Two real roots
$a < 0$ and $D > 0$

When f has two real roots r_1 and r_2, then $f(x)$ factors as

$$f(x) = a(x - r_1)(x - r_2)$$

For example, $f(x) = 2x^2 - 3x + 1$ has discriminant $D = b^2 - 4ac = 9 - 8 = 1 > 0$, and by the quadratic formula, its roots are $(3 \pm 1)/4$, that is, 1 and $\frac{1}{2}$. Therefore,

$$f(x) = 2x^2 - 3x + 1 = 2(x - 1)\left(x - \frac{1}{2}\right)$$

The technique of **completing the square** consists of writing a quadratic polynomial as a multiple of a square plus a constant. For $x^2 + bx + c$, add and subtract the square of half the coefficient of the x-term so that a square term $\left(x + \frac{b}{2}\right)^2$ can be made:

$$x^2 + bx + c = x^2 + bx + \left(\frac{b}{2}\right)^2 - \left(\frac{b}{2}\right)^2 + c = \left(x + \frac{b}{2}\right)^2 - \left(\frac{b}{2}\right)^2 + c$$

If the x^2 term has a coefficient a, we factor that out first, as demonstrated in the following example.

Cuneiform texts written on clay tablets show that the method of completing the square was known to ancient Babylonian mathematicians who lived some 4000 years ago.

EXAMPLE 4   Completing the Square   Complete the square for the quadratic polynomial $f(x) = 4x^2 - 12x + 3$.

Solution First factor out the leading coefficient:

$$4x^2 - 12x + 3 = 4\left(x^2 - 3x + \frac{3}{4}\right)$$

Then complete the square for the term $x^2 - 3x$:

$$x^2 - 3x = x^2 - 3x + \left(\frac{3}{2}\right)^2 - \left(\frac{3}{2}\right)^2 = \left(x - \frac{3}{2}\right)^2 - \frac{9}{4}$$

Therefore,

$$4x^2 - 12x + 3 = 4\left(\left(x - \frac{3}{2}\right)^2 - \frac{9}{4} + \frac{3}{4}\right) = 4\left(x - \frac{3}{2}\right)^2 - 6 \qquad \blacksquare$$

The method of completing the square can be used to find the minimum or maximum value of a quadratic function, as we do in the next example.

EXAMPLE 5   Finding the Maximum of a Quadratic Function   Complete the square and find the maximum value of $f(x) = -x^2 + 4x + 1$.

Solution Since the x^2 term has a coefficient of -1, we factor it out:

$$f(x) = -(x^2 - 4x - 1) = -(x^2 - 4x + 4 - 4 - 1) = -((x-2)^2 - 5) = -(x-2)^2 + 5$$

Now, $f(2) = 5$, and $-(x-2)^2 + 5 < 5$ for all other x. Thus, the maximum value of f is 5 occurring at $x = 2$ (Figure 11). $\qquad \blacksquare$

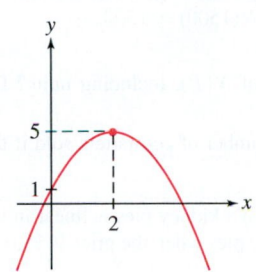

DF FIGURE 11 Graph of $f(x) = -x^2 + 4x + 1$.

1.2 SUMMARY

- A linear function is a function of the form $f(x) = mx + b$.
- The general equation of a line is $ax + by = c$. The line $y = c$ is horizontal, and $x = c$ is vertical.
- Two convenient ways of writing the equation of a nonvertical line:

  - Slope-intercept form: $y = mx + b$ (slope m and y-intercept b)
  - Point-slope form: $y - b = m(x - a)$ [slope m, passes through (a, b)]

- Two lines of slopes m_1 and m_2 are parallel if and only if $m_1 = m_2$, and they are perpendicular if and only if $m_1 = -1/m_2$.
- Quadratic function: $f(x) = ax^2 + bx + c$. The roots are $x = (-b \pm \sqrt{D})/2a$, where $D = b^2 - 4ac$ is the discriminant. The roots are real and distinct if $D > 0$, there is a double real root if $D = 0$, and there are no real roots if $D < 0$.
- Completing the square consists of writing a quadratic function $f(x) = ax^2 + bx + c$ as a multiple of a square plus a constant; that is, as $f(x) = a(x + p)^2 + q$.

1.2 EXERCISES

Preliminary Questions

1. What is the slope of the line $y = -4x - 9$?

2. Are the lines $y = 2x + 1$ and $y = -2x - 4$ perpendicular?

3. When is the line $ax + by = c$ parallel to the y-axis? To the x-axis?

4. Suppose $y = 3x + 2$. What is Δy if x increases by 3?

5. What is the minimum of $f(x) = (x + 3)^2 - 4$?

6. What is the result of completing the square for $f(x) = x^2 + 1$?

7. ✏️ Describe how the parabolas $y = ax^2 - 1$ change as a changes from $-\infty$ to ∞.

8. ✏️ Describe how the parabolas $y = x^2 + bx$ change as b changes from $-\infty$ to ∞.

Exercises

In Exercises 1–4, find the slope, the y-intercept, and the x-intercept of the line with the given equation.

1. $y = 3x + 12$

2. $y = 4 - x$

3. $4x + 9y = 3$

4. $y - 3 = \frac{1}{2}(x - 6)$

In Exercises 5–8, find the slope of the line.

5. $y = 3x + 2$

6. $y = 3(x - 9) + 2$

7. $3x + 4y = 12$

8. $3x + 4y = -8$

In Exercises 9–20, find the equation of the line with the given description.

9. Slope 3, y-intercept 8

10. Slope -2, y-intercept 3

11. Slope 3, passes through $(7, 9)$

12. Slope -5, passes through $(0, 0)$

13. Horizontal, passes through $(0, -2)$

14. Passes through $(-1, 4)$ and $(2, 7)$

15. Parallel to $y = 3x - 4$, passes through $(1, 1)$

16. Passes through $(1, 4)$ and $(12, -3)$

17. Perpendicular to $3x + 5y = 9$, passes through $(2, 3)$

18. Vertical, passes through $(-4, 9)$

19. Horizontal, passes through $(8, 4)$

20. Slope 3, x-intercept 6

21. Find the equation of the perpendicular bisector of the segment joining $(1, 2)$ and $(5, 4)$ (Figure 12). *Hint:* The midpoint Q of the segment joining (a, b) and (c, d) is $\left(\dfrac{a + c}{2}, \dfrac{b + d}{2} \right)$.

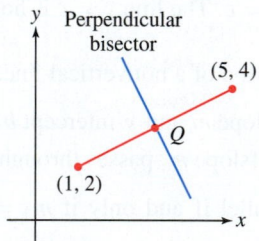

FIGURE 12

22. Intercept–Intercept Form   Show that if $a, b \neq 0$, then the line with x-intercept $x = a$ and y-intercept $y = b$ has equation (Figure 13)

$$\frac{x}{a} + \frac{y}{b} = 1$$

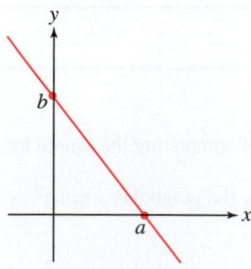

FIGURE 13

23. Find an equation of the line with x-intercept $x = 4$ and y-intercept $y = 3$.

24. Find y such that $(3, y)$ lies on the line of slope $m = 2$ through $(1, 4)$.

25. Determine whether there exists a constant c such that the line $x + cy = 1$:

(a) Has slope 4

(b) Passes through $(3, 1)$

(c) Is horizontal

(d) Is vertical

26. Determine whether there exists a constant c such that the line $cx - 2y = 4$:

(a) Has slope 4

(b) Passes through $(1, -4)$

(c) Is horizontal

(d) Is vertical

27. Suppose that the number of Bob's Bits computers that can be sold when the computer's price is P (in dollars) is given by a linear function $N(P)$, where $N(1000) = 10{,}000$ and $N(1500) = 7500$.

(a) Determine $N(P)$.

(b) What is the slope of the graph of $N(P)$, including units? Describe what the slope represents.

(c) What is the change ΔN in the number of computers sold if the price is increased by $\Delta P = \$100$?

28. Suppose that the demand for Colin's kidney pies is linear in the price P. Further, assume that he can sell 100 pies when the price is \$5.00 and 40 pies when the price is \$10.00.

(a) Determine the demand N (number of pies sold) as a function of the price P (in dollars).

(b) What is the slope of the graph of $N(P)$, including units? Describe what the slope represents.

(c) Determine the revenue $R = N \times P$ for prices $P = 5, 6, 7, 8, 9, 10$ and then choose a price to maximize the revenue.

29. In each case, identify the slope and give its meaning with the appropriate units.

(a) The function $N = -70t + 5000$ models the enrollment at Maple Grove College during the fall of 2018, where N represents the number of students and t represents the time in weeks since the start of the semester.

(b) The function $C = 3.5n + 700$ represents the cost (in dollars) to rent the Shakedown Street Dance Hall for an evening if n people attend the dance.

30. In each case, identify the slope and give its meaning with the appropriate units.

(a) The function $N = 3.9T - 178.8$ models the the number of times, N, that a cricket chirps in a minute when the temperature is $T°$ Celsius.

(b) The function $V = 47{,}500d$ gives the volume (V, in gallons) of molasses in the storage tank in relation to the depth (d, in feet) of the molasses.

31. Materials expand when heated. Consider a metal rod of length L_0 at temperature T_0. If the temperature is changed by an amount ΔT, then the rod's length approximately changes by $\Delta L = \alpha L_0 \Delta T$, where α is the thermal expansion coefficient and ΔT is not an extreme temperature change. For steel, $\alpha = 1.24 \times 10^{-5} \, °C^{-1}$.

(a) A steel rod has length $L_0 = 40$ cm at $T_0 = 40°C$. Find its length at $T = 90°C$.

(b) Find its length at $T = 50°C$ if its length at $T_0 = 100°C$ is 65 cm.

(c) Express length L as a function of T if $L_0 = 65$ cm at $T_0 = 100°C$.

32. Do the points $(0.5, 1)$, $(1, 1.2)$, $(2, 2)$ lie on a line?

33. Find b such that $(2, -1)$, $(3, 2)$, and $(b, 5)$ lie on a line.

34. Find an expression for the velocity v as a linear function of t that matches the following data:

t (s)	0	2	4	6
v (m/s)	39.2	58.6	78	97.4

35. The period T of a pendulum is measured for pendulums of several different lengths L. Based on the following data, does T appear to be a linear function of L?

L (cm)	20	30	40	50
T (s)	0.9	1.1	1.27	1.42

36. Show that f is linear of slope m if and only if

$$f(x+h) - f(x) = mh \quad \text{(for all } x \text{ and } h)$$

That is to say, prove the following two statements:

(a) f is linear of slope m implies that $f(x+h) - f(x) = mh$ (for all x and h).

(b) $f(x+h) - f(x) = mh$ (for all x and h) implies that f is linear of slope m.

37. Find the roots of the quadratic polynomials:

(a) $f(x) = 4x^2 - 3x - 1$ **(b)** $f(x) = x^2 - 2x - 1$

In Exercises 38–45, complete the square and find the minimum or maximum value of the quadratic function.

38. $y = x^2 + 2x + 5$ **39.** $y = x^2 - 6x + 9$

40. $y = -9x^2 + x$ **41.** $y = x^2 + 6x + 2$

42. $y = 2x^2 - 4x - 7$ **43.** $y = -4x^2 + 3x + 8$

44. $y = 3x^2 + 12x - 5$ **45.** $y = 4x - 12x^2$

46. Sketch the graph of $y = x^2 - 6x + 8$ by plotting the roots and the minimum point.

47. Sketch the graph of $y = x^2 + 4x + 6$ by plotting the minimum point, the y-intercept, and one other point.

48. If the alleles A and B of the cystic fibrosis gene occur in a population with frequencies p and $1 - p$ (where p is between 0 and 1), then the frequency of heterozygous carriers (carriers with both alleles) is $2p(1 - p)$. Which value of p gives the largest frequency of heterozygous carriers?

49. For which values of c does $f(x) = x^2 + cx + 1$ have a double root? No real roots?

50. Let $f(x) = x^2 + x - 1$.

(a) Show that the lines $y = x + 3$, $y = x - 1$, and $y = x - 3$ intersect the graph of f in two, one, and zero points, respectively.

(b) Sketch the graph of f and the three lines from (a).

(c) 🖊 Describe the relationship between the graph of f and the lines $y = x + c$ as c changes from $-\infty$ to ∞.

51. Let $f(x) = x^2 + 2x - 21$.

(a) Show that the lines $y = 3x - 25$, $y = 6x - 25$, and $y = 9x - 25$ intersect the graph of f in zero, one, and two points, respectively.

(b) Sketch the graph of f and the three lines from (a).

(c) 🖊 Describe the relationship between the graph of f and the lines $y = cx - 25$ as c changes from 0 to ∞.

52. Let $a, b > 0$. Show that the *geometric mean* $\sqrt{ab}$ is not larger than the *arithmetic mean* $(a + b)/2$. *Hint:* Consider $(a^{1/2} - b^{1/2})^2$.

53. If objects of weights x and w_1 are suspended from the balance in Figure 14(A), the cross-beam is horizontal if $bx = aw_1$. If the lengths a and b are known, we may use this equation to determine an unknown weight x by selecting w_1 such that the cross-beam is horizontal. If a and b are not known precisely, we might proceed as follows. First balance x by w_1 on the left, as in (A). Then switch places and balance x by w_2 on the right, as in (B). The average $\bar{x} = \frac{1}{2}(w_1 + w_2)$ gives an estimate for x. Show that $\bar{x}$ is greater than or equal to the true weight x.

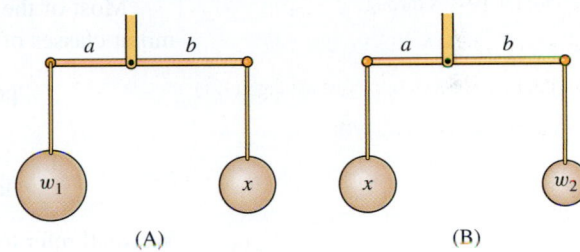

(A)   (B)

FIGURE 14

54. Find numbers x and y with sum 10 and product 24. *Hint:* Find a quadratic polynomial satisfied by x.

55. Find a pair of numbers whose sum and product are both equal to 8.

56. Show that the parabola $y = x^2$ consists of all points P such that $d_1 = d_2$, where d_1 is the distance from P to $\left(0, \frac{1}{4}\right)$ and d_2 is the distance from P to the line $y = -\frac{1}{4}$ (Figure 15).

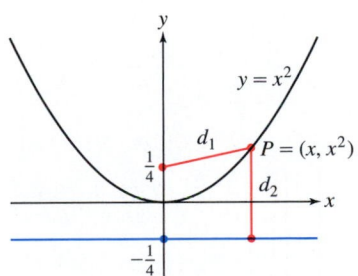

FIGURE 15

Further Insights and Challenges

57. Show that if f and g are linear, then so is $f + g$. Is the same true of fg?

58. Show that if f and g are linear functions such that $f(0) = g(0)$ and $f(1) = g(1)$, then $f = g$.

59. Show that $\Delta y / \Delta x$ for the function $f(x) = x^2$ over the interval $[x_1, x_2]$ is not a constant, but depends on the interval. Determine the exact dependence of $\Delta y / \Delta x$ on x_1 and x_2.

60. Complete the square and use the result to derive the quadratic formula for the roots of $ax^2 + bx + c = 0$.

61. Let $a, c \neq 0$. Show that the roots of

$$ax^2 + bx + c = 0 \quad \text{and} \quad cx^2 + bx + a = 0$$

are reciprocals of each other.

62. Show, by completing the square, that the parabola

$$y = ax^2 + bx + c$$

can be obtained from $y = ax^2$ by a vertical and horizontal translation.

63. Prove **Viète's Formulas**: The quadratic polynomial with α and β as roots is $x^2 + bx + c$, where $b = -\alpha - \beta$ and $c = \alpha\beta$.

1.3 The Basic Classes of Functions

The primary condition on a function f is that it assigns to each element x of its domain a unique element $f(x)$ in its range. There are no other restrictions on how that relation is defined. Usually we describe the relationship by a formula for the function, sometimes by a table or a graph, but the association could be quite complicated, not lending itself to any simple description. The possibilities for functions are endless. In calculus we make no attempt to deal with all possible functions. The techniques of calculus, powerful and general as they are, apply only to functions that are sufficiently "well-behaved" (we will see what well-behaved means when we study the derivative in Chapter 3). Fortunately, such functions are adequate for a vast range of applications.

Most of the functions considered in this text are constructed from the following familiar classes of well-behaved functions:

polynomials   rational functions   algebraic functions

exponential functions   trigonometric functions

logarithmic functions   inverse trigonometric functions

We shall refer to these as the **basic functions**.

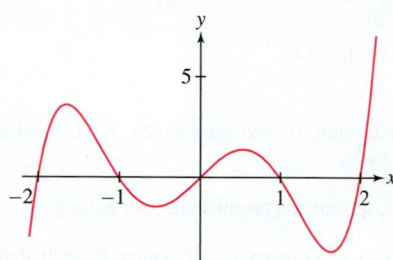

FIGURE 1   The polynomial function $f(x) = x^5 - 5x^3 + 4x$.

- **Polynomials:** For any real number m, $f(x) = x^m$ is called the **power function** with exponent m. Power functions include $f(x) = x^3$, $f(x) = x^{-7}$, and $f(x) = x^\pi$. The base is the variable, and the exponent is a constant. For now, we are interested in power functions with exponents that are positive integers. A **polynomial** is a sum of multiples of power functions with exponents that are positive integers or zero (Figure 1):

$$f(x) = x^5 - 5x^3 + 4x, \qquad g(t) = 7t^6 + t^3 - 3t - 1, \qquad h(x) = x^9$$

Thus, the function $f(x) = x + x^{-1}$ is not a polynomial because it includes a term x^{-1} with a negative exponent. The general polynomial P in the variable x may be written

$$P(x) = a_n x^n + a_{n-1} x^{n-1} + \cdots + a_1 x + a_0$$

- The numbers $a_0, a_1, \ldots, a_n$ are called **coefficients**.
- The **degree** of P is n (assuming that $a_n \neq 0$).
- The coefficient a_n is called the **leading coefficient**.
- The domain of P is **R**.

- A **rational function** is a *quotient* of two polynomials (Figure 2):

$$f(x) = \frac{P(x)}{Q(x)} \qquad [P(x) \text{ and } Q(x) \text{ polynomials}]$$

The domain of f is the set of numbers x such that $Q(x) \neq 0$. For example,

$$f(x) = \frac{1}{x^2} \qquad\qquad \text{domain } \{x : x \neq 0\}$$

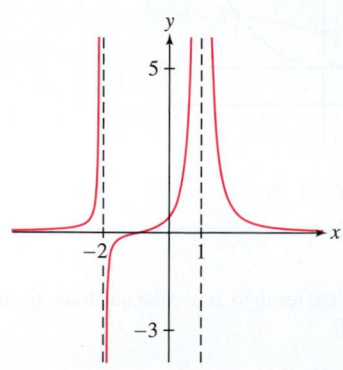

FIGURE 2   The rational function $f(x) = \dfrac{x+1}{x^3 - 3x + 2}$.

$$h(t) = \frac{7t^6 + t^3 - 3t - 1}{t^2 - 1} \qquad\qquad \text{domain } \{t : t \neq \pm 1\}$$

Every polynomial is also a rational function [with $Q(x) = 1$].

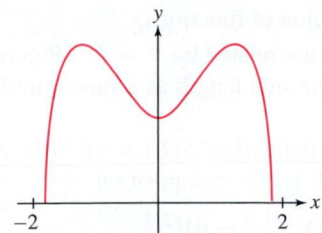

FIGURE 3 The algebraic function
$f(x) = \sqrt{1 + 3x^2 - x^4}$.

Any function that is not algebraic is called **transcendental**. *Exponential and trigonometric functions are examples, as are the Bessel and gamma functions that appear in engineering and statistics. The term "transcendental" goes back to the 1670s, when it was used by Gottfried Wilhelm Leibniz (1646–1716) to describe functions of this type.*

- An **algebraic function** is produced by taking sums, products, and quotients of *roots* of polynomials and rational functions (Figure 3):

$$f(x) = \sqrt{1 + 3x^2 - x^4}, \qquad g(t) = (\sqrt{t} - 2)^{-2}, \qquad h(z) = \frac{z + z^{-5/3}}{5z^3 - \sqrt{z}}$$

  A number x belongs to the domain of f if each term in the formula is defined and the result does not involve division by zero. For example, $g(t)$ is defined if $t \geq 0$ and $\sqrt{t} \neq 2$, so the domain of g is $D = \{t : t \geq 0 \text{ and } t \neq 4\}$.
- **Exponential functions:** The function $f(x) = b^x$, where $b > 0$ and $b \neq 1$, is called the exponential function with base b. Some examples are

$$f(x) = 2^x, \qquad g(t) = 10^t, \qquad h(x) = \left(\frac{1}{3}\right)^x, \qquad p(t) = (\sqrt{5})^t$$

  Exponential functions and their *inverses*, the **logarithmic functions**, are treated in greater detail in Section 1.6.
- **Trigonometric functions** are functions built from $\sin x$ and $\cos x$. These functions and their inverses are discussed in the next two sections.

Constructing New Functions

Given functions f and g, we can construct new functions by forming the sum, difference, product, and quotient functions:

$$(f + g)(x) = f(x) + g(x), \qquad (f - g)(x) = f(x) - g(x)$$

$$(fg)(x) = f(x)\,g(x), \qquad \left(\frac{f}{g}\right)(x) = \frac{f(x)}{g(x)} \quad \text{(where } g(x) \neq 0\text{)}$$

For example, if $f(x) = x^2$ and $g(x) = \sin x$, then

$$(f + g)(x) = x^2 + \sin x, \qquad (f - g)(x) = x^2 - \sin x$$

$$(fg)(x) = x^2 \sin x, \qquad \left(\frac{f}{g}\right)(x) = \frac{x^2}{\sin x}$$

We can also multiply functions by constants. A function of the form

$$h(x) = c_1\, f(x) + c_2\, g(x) \quad (c_1, c_2 \text{ constants})$$

is called a **linear combination** of f and g.

   Composition is another important way of constructing new functions. The composition of f and g is the function $f \circ g$ defined by $(f \circ g)(x) = f(g(x))$. The domain of $f \circ g$ is the set of values of x in the domain of g such that $g(x)$ lies in the domain of f.

EXAMPLE 1 Compute the composite functions $f \circ g$ and $g \circ f$ and discuss their domains, where

$$f(x) = \sqrt{x}, \qquad g(x) = 1 - x$$

Example 1 shows that the composition of functions is not commutative: The functions $f \circ g$ and $g \circ f$ may be (and usually are) different.

Solution We have

$$(f \circ g)(x) = f(g(x)) = f(1 - x) = \sqrt{1 - x}$$

The square root $\sqrt{1 - x}$ is defined if $1 - x \geq 0$, that is, for $x \leq 1$. Therefore, the domain of $f \circ g$ is $\{x : x \leq 1\}$. On the other hand,

$$(g \circ f)(x) = g(f(x)) = g(\sqrt{x}) = 1 - \sqrt{x}$$

The domain of $g \circ f$ is $\{x : x \geq 0\}$. ■

EXAMPLE 2 **Surface Area and Volume** Express the surface area S of a cube as a function of its volume V.

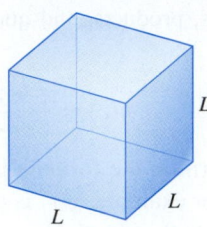

FIGURE 4 A cube of side length L.

A **power law** is a relationship in the form $y = kx^{\alpha}$ for constants k and α. They are quite common in biology and ecology. We will introduce a number of examples in the text.

Solution We will derive a relationship via a composition of functions.

The volume V and the length L of a side of a cube are related by $V = L^3$ (Figure 4). Therefore, $L = V^{1/3}$. Thus, $L(V) = V^{1/3}$ expresses the side length as a function of the volume.

The surface area S is a function of the side length defined by $S(L) = 6L^2$ (the cube has six sides, each with area L^2). Thus, S depends on V by the composition

$$S \circ L(V) = S(L(V)) = S(V^{1/3}) = 6(V^{1/3})^2 = 6V^{2/3}$$

It follows that we can express surface area as a function of volume by $S(V) = 6V^{2/3}$. ■

The simple geometric relationship derived in the previous example is the basis for a variety of theoretical power laws in biology and ecology in which an attribute proportional to an animal's surface area is related to an attribute proportional to its volume. For example, in a particular species, the mass M of an individual is proportional to its volume, and the mass F of its fur might be proportional to its surface area. Thus, the relationship between fur mass and animal mass could be modeled by a power law $F = kM^{2/3}$. Typically, scientists collect data to check the proposed relationship that either confirms this model or suggests adjustments or other factors that must be considered.

Elementary Functions

As noted above, we can produce new functions by applying the operations of addition, subtraction, multiplication, division, and composition. It is convenient to refer to a function constructed in this way from the basic functions listed above as an **elementary function**. The following functions are elementary:

$$f(x) = \sqrt{2x + \sin x}, \qquad f(x) = 10^{\sqrt{x}}, \qquad f(x) = \frac{1 + x^{-1}}{1 + \cos x}$$

Piecewise-Defined Functions

We can also create new functions by piecing together functions defined over limited domains, obtaining **piecewise-defined functions**. One example we have already seen is the absolute value function defined by

$$|x| = \begin{cases} -x & \text{when } x < 0 \\ x & \text{when } x \geq 0 \end{cases}$$

EXAMPLE 3 Given the function f, determine its domain, range, and intervals where it is increasing or decreasing.

$$f(x) = \begin{cases} 1 & \text{when } x < 0 \\ x + 1 & \text{when } x \geq 0 \end{cases}$$

Solution The graph of f appears in Figure 5. The function is defined for all values of x, so the domain is all real numbers. Now, for all $x < 0$, the output of f is just the single value 1, and for $x \geq 0$, the output covers all values greater than or equal to 1. Hence, the range of the function is $\{y : y \geq 1\}$. The function is neither increasing nor decreasing for $x < 0$; however, the function is increasing for $x \geq 0$. ■

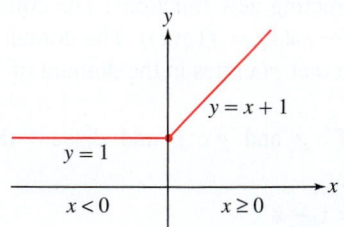

FIGURE 5 A function defined piecewise.

1.3 SUMMARY

- For m a real number, $f(x) = x^m$ is called the *power function* with exponent m. A polynomial P is a sum of multiples of x^m, where m is a whole number:

$$P(x) = a_n x^n + a_{n-1} x^{n-1} + \cdots + a_1 x + a_0$$

This polynomial has degree n (assuming that $a_n \neq 0$), and a_n is called the leading coefficient.

- A rational function is a quotient P/Q of two polynomials [defined when $Q(x) \neq 0$].
- An algebraic function is produced by taking sums, products, and quotients of roots of polynomials and rational functions.
- Exponential function: $f(x) = b^x$, where $b > 0$ and $b \neq 1$ (b is called the base).
- The composite function $f \circ g$ is defined by $(f \circ g)(x) = f(g(x))$. The domain of $f \circ g$ is the set of x in the domain of g such that $g(x)$ belongs to the domain of f.
- The elementary functions are obtained by taking products, sums, differences, quotients, and compositions of the basic functions, which include polynomials, rational functions, algebraic functions, exponential functions, trigonometric functions, logarithmic functions, and inverse trigonometric functions.
- A piecewise-defined function is obtained by defining a function over two or more distinct domains.

1.3 EXERCISES

Preliminary Questions

1. Explain why both $f(x) = x^3 + 1$ and $g(x) = \frac{1}{x^3 + 1}$ are rational functions.

2. Is $y = |x|$ a polynomial function? What about $y = |x^2 + 1|$?

3. What is unusual about the domain of the composite function $f \circ g$ for the functions $f(x) = x^{1/2}$ and $g(x) = -1 - |x|$?

4. Is $f(x) = \left(\frac{1}{2}\right)^x$ increasing or decreasing?

5. Explain why both $f(x) = \frac{x}{1-x^4}$ and $g(x) = \frac{x}{\sqrt{1-x^4}}$ are algebraic functions.

6. We have $f(x) = (x+1)^{1/2}$, $g(x) = x^{-2} + 1$, $h(x) = 2^x$, and $k(x) = x^2 + 1$. Identify which of the functions may be described by each of the following.

(a) Transcendental

(b) Polynomial

(c) Rational but not polynomial

(d) Algebraic but not rational

Exercises

In Exercises 1–12, determine the domain of the function.

1. $f(x) = x^{1/4}$

2. $g(t) = t^{2/3}$

3. $f(x) = x^3 + 3x - 4$

4. $h(z) = z^3 + z^{-3}$

5. $g(t) = \dfrac{1}{t+2}$

6. $f(x) = \dfrac{1}{x^2 + 4}$

7. $G(u) = \dfrac{1}{u^2 - 4}$

8. $f(x) = \dfrac{\sqrt{x}}{x^2 - 9}$

9. $f(x) = x^{-4} + (x-1)^{-3}$

10. $F(s) = \sin\left(\dfrac{s}{s+1}\right)$

11. $g(y) = 10^{\sqrt{y} + y^{-1}}$

12. $f(x) = \dfrac{x + x^{-1}}{(x-3)(x+4)}$

In Exercises 13–24, identify each of the following functions as polynomial, rational, algebraic, or transcendental.

13. $f(x) = 4x^3 + 9x^2 - 8$

14. $f(x) = x^{-4}$

15. $f(x) = \sqrt{x}$

16. $f(x) = \sqrt{1 - x^2}$

17. $f(x) = \dfrac{x^2}{x + \sin x}$

18. $f(x) = 2^x$

19. $f(x) = \dfrac{2x^3 + 3x}{9 - 7x^2}$

20. $f(x) = \dfrac{3x - 9x^{-1/2}}{9 - 7x^2}$

21. $f(x) = \sin(x^2)$

22. $f(x) = \dfrac{x}{\sqrt{x} + 1}$

23. $f(x) = x^2 + 3x^{-1}$

24. $f(x) = \sin(3^x)$

25. Is $f(x) = 2^{x^2}$ a transcendental function?

26. Show that $f(x) = x^2 + 3x^{-1}$ and $g(x) = 3x^3 - 9x + x^{-2}$ are rational functions—that is, quotients of polynomials.

In Exercises 27–34, calculate the composite functions $f \circ g$ and $g \circ f$, and determine their domains.

27. $f(x) = \sqrt{x}$, $g(x) = x + 1$

28. $f(x) = \dfrac{1}{x}$, $g(x) = x^{-4}$

29. $f(x) = 2^x$, $g(x) = x^2$

30. $f(x) = |x|$, $g(\theta) = \sin\theta$

31. $f(\theta) = \cos\theta$, $g(x) = x^3 + x^2$

32. $f(x) = \dfrac{1}{x^2 + 1}$, $g(x) = x^{-2}$

33. $f(t) = \dfrac{1}{\sqrt{t}}, \quad g(t) = -t^2$

34. $f(t) = \sqrt{t}, \quad g(t) = 1 - t^3$

35. The volume V and surface area of a sphere [Figure 6(A)] are expressed in terms of radius r by $V(r) = \frac{4}{3}\pi r^3$ and $S(r) = 4\pi r^2$, respectively. Determine $r(V)$, the radius as a function of volume. Then determine $S(V)$, the surface area as a function of volume, by computing the composite $S \circ r(V)$.

36. A tetrahedron is a polyhedron with four equilateral triangles as its faces [Figure 6(B)]. The volume V and surface area of a tetrahedron are expressed in terms of the side-length L of the triangles by $V(L) = \frac{\sqrt{2}L^3}{12}$ and $S(L) = \sqrt{3}L^2$, respectively. Determine $L(V)$, the side length as a function of volume. Then determine $S(V)$, the surface area as a function of volume, by computing the composite $S \circ L(V)$.

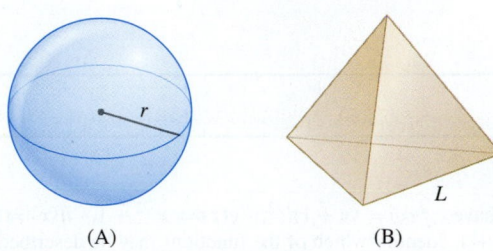

(A)   (B)

FIGURE 6 A sphere (A) and tetrahedron (B).

In Exercises 37–40, draw the graphs of each of the piecewise-defined functions.

37. $f(x) = \begin{cases} 3 & \text{when } x < 0 \\ x^2 + 3 & \text{when } x \ge 0 \end{cases}$

38. $f(x) = \begin{cases} x + 1 & \text{when } x < 0 \\ 1 - x & \text{when } x \ge 0 \end{cases}$

39. $f(x) = \begin{cases} x^2 & \text{when } x < 0 \\ -x^2 & \text{when } x \ge 0 \end{cases}$

40. $f(x) = \begin{cases} 2x - 2 & \text{when } x < 0 \\ x & \text{when } x \ge 0 \end{cases}$

41. Let $f(x) = \frac{x}{|x|}$.

(a) What are the domain and range of f?

(b) Sketch the graph of f.

(c) Express f as a piecewise-defined function where each of the "pieces" is a constant.

42. The **Heaviside function** (named after Oliver Heaviside, 1850–1925) is defined by:

$$H(x) = \begin{cases} 0 & \text{when } x < 0 \\ 1 & \text{when } x \ge 0 \end{cases}$$

The Heaviside function can be used to "turn on" another function at a specific value in the domain, as seen in the four examples here. For each of the following, sketch the graph of f.

(a) $f(x) = H(x)x^2$

(b) $f(x) = H(x)(1 - x^2)$

(c) $f(x) = H(x - 1)x$

(d) $f(x) = H(x + 2)x^2$

43. The population (in millions) of Calcedonia as a function of time t (years) is $P(t) = 30 \cdot 2^{0.1t}$. Show that the population doubles every 10 years. Show more generally that for any positive constants a and k, the function $g(t) = a2^{kt}$ doubles after $1/k$ years.

44. Find all values of c such that $f(x) = \dfrac{x + 1}{x^2 + 2cx + 4}$ has domain **R**.

Further Insights and Challenges

In Exercises 45–51, we define the first difference δf *of a function* f *by* $\delta f(x) = f(x + 1) - f(x)$.

45. Show that if $f(x) = x^2$, then $\delta f(x) = 2x + 1$. Calculate δf for $f(x) = x$ and $f(x) = x^3$.

46. Show that $\delta(10^x) = 9 \cdot 10^x$ and, more generally, that $\delta(b^x) = (b - 1)b^x$.

47. Show that for any two functions f and g, $\delta(f + g) = \delta f + \delta g$ and $\delta(cf) = c\delta(f)$, where c is any constant.

48. Suppose we can find a function P such that $\delta P(x) = (x + 1)^k$ and $P(0) = 0$. Prove that $P(1) = 1^k$, $P(2) = 1^k + 2^k$, and, more generally, for every whole number n,

$$P(n) = 1^k + 2^k + \cdots + n^k \qquad \boxed{1}$$

49. Show that if

$$P(x) = \frac{x(x + 1)}{2}$$

then $\delta P = (x + 1)$. Then apply Exercise 48 to conclude that

$$1 + 2 + 3 + \cdots + n = \frac{n(n + 1)}{2}$$

50. Calculate $\delta(x^3)$, $\delta(x^2)$, and $\delta(x)$. Then find a polynomial P of degree 3 such that $\delta P = (x + 1)^2$ and $P(0) = 0$. Conclude that $P(n) = 1^2 + 2^2 + \cdots + n^2$.

51. This exercise combined with Exercise 48 shows that for all whole numbers k, there exists a polynomial P satisfying Eq. (1). The solution requires the Binomial Theorem and proof by induction (see Appendix C).

(a) Show that $\delta(x^{k+1}) = (k + 1)x^k + \cdots$, where the dots indicate terms involving smaller powers of x.

(b) Show by induction that there exists a polynomial of degree $k + 1$ with leading coefficient $1/(k + 1)$:

$$P(x) = \frac{1}{k + 1}x^{k+1} + \cdots$$

such that $\delta P = (x + 1)^k$ and $P(0) = 0$.

1.4 Trigonometric Functions

We begin our trigonometric review by recalling the two systems of angle measurement: **radians** and **degrees**. They are best described using the relationship between angles and rotation. As is customary, we often use the lowercase Greek letter θ (theta) to denote angles and rotations.

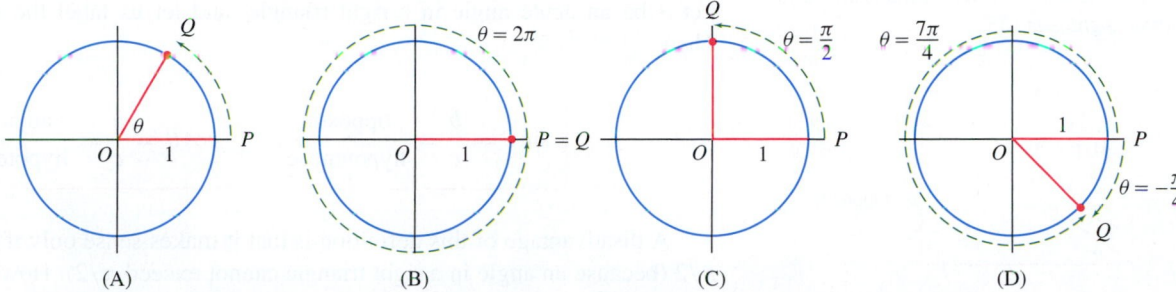

FIGURE 1 The radian measure θ of a counterclockwise rotation is the length along the unit circle of the arc traversed by P as it rotates into Q.

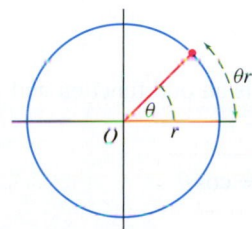

FIGURE 2 On a circle of radius r, the arc traversed by a counterclockwise rotation of θ radians has length θr.

TABLE 1

Rotation through	Radian measure
Two full circles	4π
Full circle	2π
Half circle	π
Quarter circle	$2\pi/4 = \pi/2$
One-sixth circle	$2\pi/6 = \pi/3$

Radians	Degrees
0	$0°$
$\dfrac{\pi}{6}$	$30°$
$\dfrac{\pi}{4}$	$45°$
$\dfrac{\pi}{3}$	$60°$
$\dfrac{\pi}{2}$	$90°$

Figure 1(A) shows a unit circle with radius $\overline{OP}$ rotating counterclockwise into radius $\overline{OQ}$. *The radian measure of this rotation is the length θ of the circular arc traversed by P as it rotates into Q.* On a circle of radius r, the arc traversed by a counterclockwise rotation of θ radians has length θr (Figure 2).

The unit circle has circumference 2π. Therefore, a rotation through a full circle has radian measure $\theta = 2\pi$ [Figure 1(B)]. The radian measure of a rotation through one-quarter of a circle is $\theta = 2\pi/4 = \pi/2$ [Figure 1(C)] and, in general, the rotation through one-nth of a circle has radian measure $2\pi/n$ (Table 1). A negative rotation (with $\theta < 0$) is a rotation in the *clockwise* direction [Figure 1(D)].

The radian measure of an angle such as $\angle POQ$ in Figure 1(A) is defined as the radian measure of a rotation that carries $\overline{OP}$ to $\overline{OQ}$. Notice, however, that the radian measure of an angle is not unique. The rotations through θ and $\theta + 2\pi$ both carry $\overline{OP}$ to $\overline{OQ}$. Therefore, θ and $\theta + 2\pi$ represent the same angle, even though the rotation through $\theta + 2\pi$ takes an extra trip around the circle. In general, *two radian measures represent the same angle if the corresponding rotations differ by an integer multiple of 2π.* For example, $\pi/4$, $9\pi/4$, and $-15\pi/4$ all represent the same angle because they differ by multiples of 2π:

$$\frac{\pi}{4} = \frac{9\pi}{4} - 2\pi = -\frac{15\pi}{4} + 4\pi$$

Every angle has a unique radian measure satisfying $0 \le \theta < 2\pi$. With this choice, the angle θ subtends an arc of length θr on a circle of radius r (Figure 2).

Degrees are defined by dividing the circle (not necessarily the unit circle) into 360 equal parts. A degree is $\frac{1}{360}$ of a circle. A rotation through θ degrees (denoted $\theta°$) is a rotation through the fraction $\theta/360$ of the complete circle. For example, a rotation through $90°$ is a rotation through the fraction $\frac{90}{360}$, or $\frac{1}{4}$, of a circle.

As with radians, the degree measure of an angle is not unique. Two degree measures represent that same angle if they differ by an integer multiple of 360. For example, the angles $-45°$ and $675°$ coincide because $675 = -45 + 2(360)$. Every angle has a unique degree measure θ with $0 \le \theta < 360$.

To convert between radians and degrees, remember that 2π radians is equal to $360°$. Therefore, 1 radian equals $360/2\pi$ or $180/\pi$ degrees.

- To convert from radians to degrees, multiply by $180/\pi$.
- To convert from degrees to radians, multiply by $\pi/180$.

Radian measurement is usually the better choice for mathematical purposes, but there are good practical reasons for using degrees. The number 360 has many divisors ($360 = 8 \cdot 9 \cdot 5$), and consequently, many fractional parts of the circle can be expressed as an integer number of degrees. For example, one-fifth of the circle is $72°$, two-ninths is $80°$, and three-eighths is $135°$.

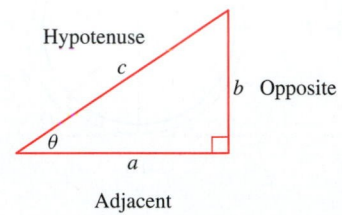

FIGURE 3

EXAMPLE 1 Convert **(a)** $55°$ to radians and **(b)** 0.5 radians to degrees.

Solution

(a) $55° \times \dfrac{\pi}{180°} \approx 0.9599$ radians

(b) 0.5 radians $\times \dfrac{180°}{\pi} \approx 28.648°$ ∎

Convention *Unless otherwise stated, we always measure angles in radians.*

The trigonometric functions sine and cosine can be defined in terms of right triangles. Let θ be an acute angle in a right triangle, and let us label the sides as in Figure 3. Then

$$\sin\theta = \frac{b}{c} = \frac{\text{opposite}}{\text{hypotenuse}}, \qquad \cos\theta = \frac{a}{c} = \frac{\text{adjacent}}{\text{hypotenuse}}$$

A disadvantage of this definition is that it makes sense only if θ lies between 0 and $\pi/2$ (because an angle in a right triangle cannot exceed $\pi/2$). However, sine and cosine can be defined for all angles in terms of the unit circle. Let $P = (x, y)$ be the point on the unit circle corresponding to the angle θ, as in Figures 4(A) and (B), and define

$$\cos\theta = x\text{-coordinate of } P, \qquad \sin\theta = y\text{-coordinate of } P$$

This agrees with the right-triangle definition when $0 < \theta < \frac{\pi}{2}$. On the circle of radius r (centered at the origin), the point corresponding to the angle θ has coordinates

$$(r\cos\theta, r\sin\theta)$$

Furthermore, we see from Figure 4(C) that $f(\theta) = \sin\theta$ is an odd function and $f(\theta) = \cos\theta$ is an even function:

$$\sin(-\theta) = -\sin\theta, \qquad \cos(-\theta) = \cos\theta$$

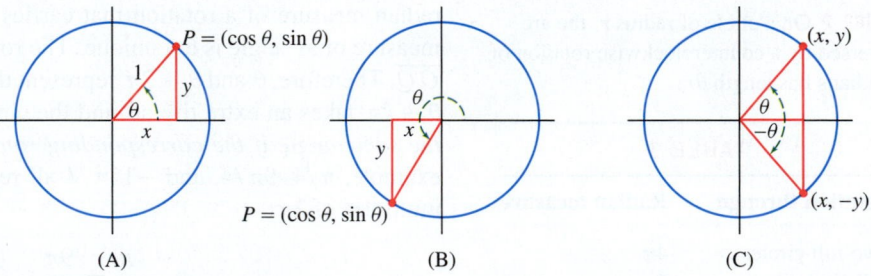

FIGURE 4 The unit circle definition of sine and cosine is valid for all angles θ.

(A) (B) (C)

Although we can use a calculator to evaluate sine and cosine for general angles, the standard values listed in Figure 5 and Table 2 appear often and are worth knowing.

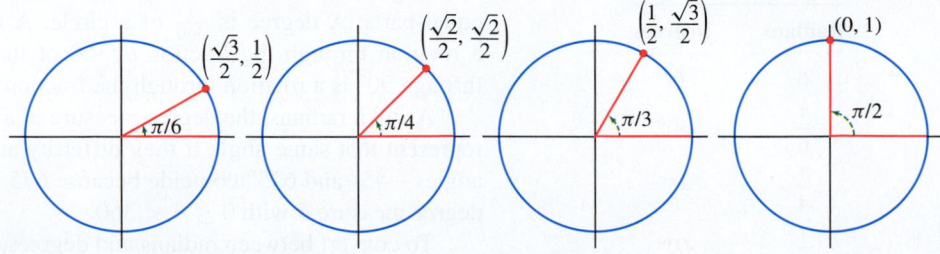

FIGURE 5 Four standard angles: The x- and y-coordinates of the points are $\cos\theta$ and $\sin\theta$.

The graph of $y = \sin\theta$ is the familiar "sine wave" shown in Figure 6. Observe how the graph is generated by the y-coordinate of the point $P = (\cos\theta, \sin\theta)$ moving around the unit circle.

TABLE 2

θ	0	$\dfrac{\pi}{6}$	$\dfrac{\pi}{4}$	$\dfrac{\pi}{3}$	$\dfrac{\pi}{2}$	$\dfrac{2\pi}{3}$	$\dfrac{3\pi}{4}$	$\dfrac{5\pi}{6}$	π
$\sin\theta$	0	$\dfrac{1}{2}$	$\dfrac{\sqrt{2}}{2}$	$\dfrac{\sqrt{3}}{2}$	1	$\dfrac{\sqrt{3}}{2}$	$\dfrac{\sqrt{2}}{2}$	$\dfrac{1}{2}$	0
$\cos\theta$	1	$\dfrac{\sqrt{3}}{2}$	$\dfrac{\sqrt{2}}{2}$	$\dfrac{1}{2}$	0	$-\dfrac{1}{2}$	$-\dfrac{\sqrt{2}}{2}$	$-\dfrac{\sqrt{3}}{2}$	-1

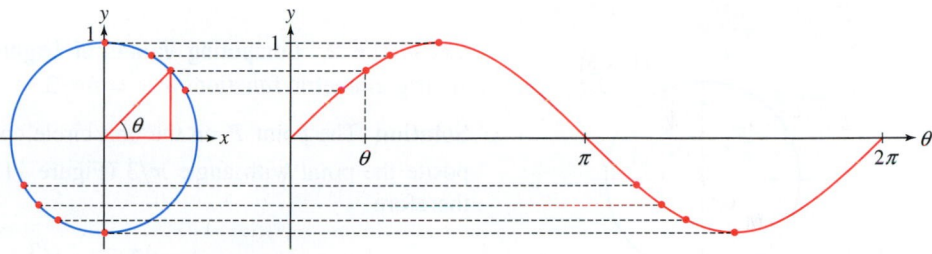

DF **FIGURE 6** The graph of $y = \sin\theta$ is generated as the point $P = (\cos\theta, \sin\theta)$ moves around the unit circle.

The graph of $y = \cos\theta$ has the same shape but is shifted to the left $\pi/2$ units (Figure 7). The signs of $\sin\theta$ and $\cos\theta$ vary as $P = (\cos\theta, \sin\theta)$ changes quadrant.

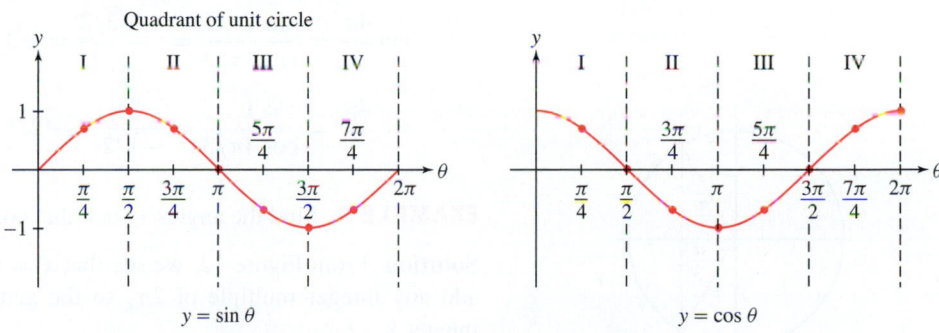

FIGURE 7 Graphs of $y = \sin\theta$ and $y = \cos\theta$ over $[0, 2\pi]$.

$y = \sin\theta$ $y = \cos\theta$

A function f is called **periodic** with period T if $f(x + T) = f(x)$ (for all x) and T is the smallest positive number with this property. The sine and cosine functions are periodic with period $T = 2\pi$ (Figure 8) because the radian measures x and $x + 2\pi k$ correspond to the same point on the unit circle for any integer k:

We often write $\sin x$ and $\cos x$, using x instead of θ. Depending on the application, we may think of x as an angle or simply as a real number.

$$\sin x = \sin(x + 2\pi k), \qquad \cos x = \cos(x + 2\pi k)$$

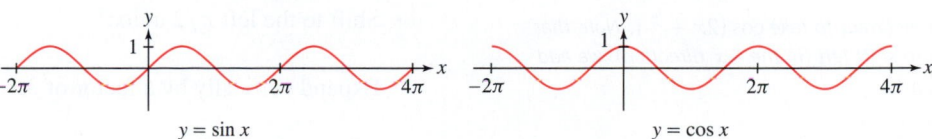

FIGURE 8 Sine and cosine have period 2π.

$y = \sin x$ $y = \cos x$

There are four other standard trigonometric functions, each defined in terms of $\sin x$ and $\cos x$ or as ratios of sides in a right triangle (Figure 9):

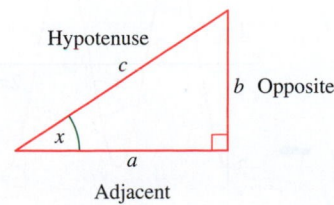

Hypotenuse
c
b Opposite
x
a
Adjacent

FIGURE 9

Tangent:	$\tan x = \dfrac{\sin x}{\cos x} = \dfrac{b}{a},$	Cotangent:	$\cot x = \dfrac{\cos x}{\sin x} = \dfrac{a}{b}$
Secant:	$\sec x = \dfrac{1}{\cos x} = \dfrac{c}{a},$	Cosecant:	$\csc x = \dfrac{1}{\sin x} = \dfrac{c}{b}$

These functions are periodic (Figure 10): $y = \tan x$ and $y = \cot x$ have period π, and $y = \sec x$ and $y = \csc x$ have period 2π (see Exercise 57).

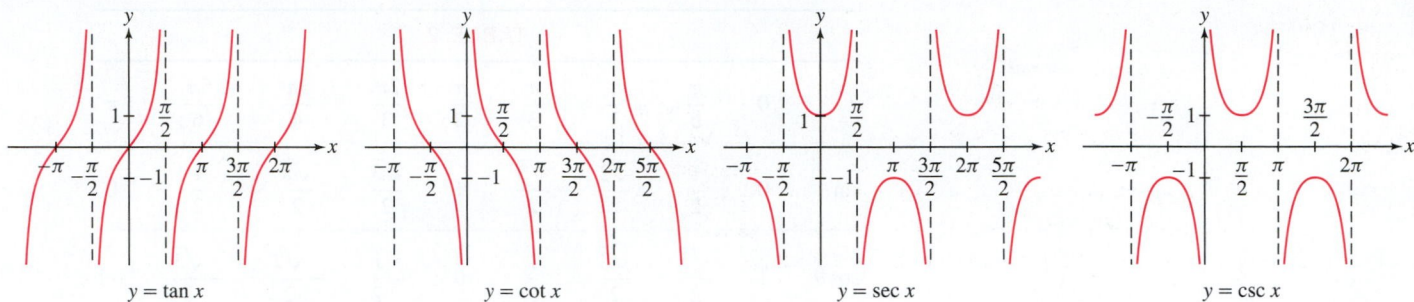

FIGURE 10 Graphs of standard trigonometric functions.

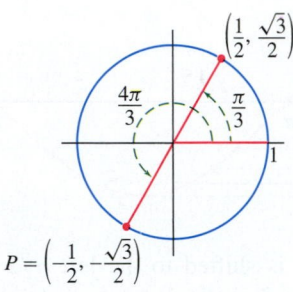

FIGURE 11

EXAMPLE 2 **Computing Values of Trigonometric Functions** Find the values of the six trigonometric functions at $x = 4\pi/3$.

Solution The point P on the unit circle corresponding to the angle $x = 4\pi/3$ lies opposite the point with angle $\pi/3$ (Figure 11). It follows that $P = (-1/2, -\sqrt{3}/2)$, and therefore

$$\sin\frac{4\pi}{3} = -\frac{\sqrt{3}}{2}, \qquad \cos\frac{4\pi}{3} = -\frac{1}{2}$$

The remaining values are

$$\tan\frac{4\pi}{3} = \frac{\sin 4\pi/3}{\cos 4\pi/3} = \frac{-\sqrt{3}/2}{-1/2} = \sqrt{3}, \quad \cot\frac{4\pi}{3} = \frac{\cos 4\pi/3}{\sin 4\pi/3} = \frac{\sqrt{3}}{3}$$

$$\sec\frac{4\pi}{3} = \frac{1}{\cos 4\pi/3} = \frac{1}{-1/2} = -2, \quad \csc\frac{4\pi}{3} = \frac{1}{\sin 4\pi/3} = \frac{-2\sqrt{3}}{3}$$

■

EXAMPLE 3 Find the angles x such that $\cos x = \frac{1}{2}$.

Solution From Figure 12, we see that $x = \pi/3$ and $x = -\pi/3$ are solutions. We may add any integer multiple of 2π, so the general solution is $x = \pm\pi/3 + 2\pi k$ for any integer k.

■

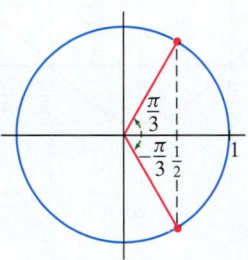

FIGURE 12 $\cos x = \frac{1}{2}$ for $x = \pm\frac{\pi}{3}$

CAUTION To shift the graph of $y = \cos 2x$ to the left $\pi/2$ units, we must replace x by $x + \frac{\pi}{2}$ to obtain $\cos\left(2\left(x + \frac{\pi}{2}\right)\right)$. It is incorrect to take $\cos\left(2x + \frac{\pi}{2}\right)$. Note that to shift left (in the $-x$ direction), we add $\pi/2$.

EXAMPLE 4 Sketch the graph of $f(x) = 3\cos\left(2\left(x + \frac{\pi}{2}\right)\right)$ over $[0, 2\pi]$.

Solution The graph is obtained by scaling and shifting the graph of $y = \cos x$ in three steps (Figure 13):

• Compress horizontally by a factor of 2: $\quad y = \cos 2x$

• Shift to the left $\pi/2$ units: $\quad y = \cos\left(2\left(x + \frac{\pi}{2}\right)\right)$

• Expand vertically by a factor of 3: $\quad y = 3\cos\left(2\left(x + \frac{\pi}{2}\right)\right)$

■

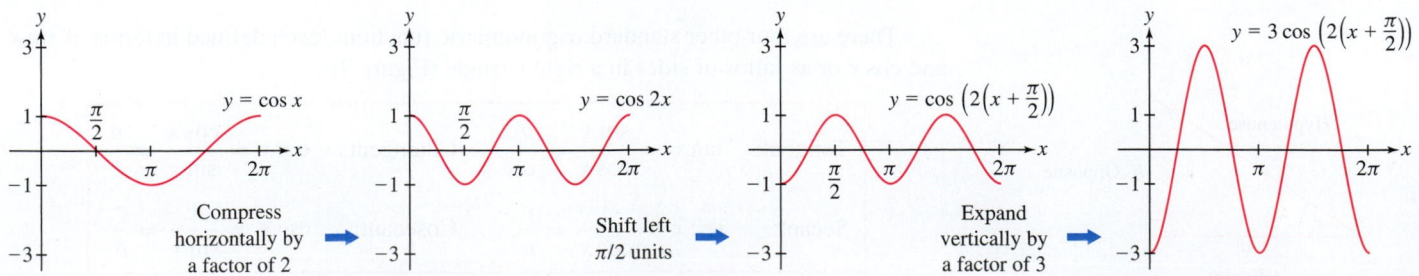

DF **FIGURE 13**

In the previous example the coefficient 3 of the cosine term is referred to as the **amplitude** of the oscillation. The idea is that the function oscillates by 3 above and below a central value of 0.

The sine function lends itself well to modeling oscillatory phenomena. By scaling and translating in both the vertical and horizontal directions, we can fit a sine curve to data representing many different relationships. In the next example, we use the sine function to model the varying day length throughout the year.

EXAMPLE 5 **A Day-Length Model** The function $L(t) = 12 + 3.1 \sin(\frac{2\pi}{365}t)$ approximates the length of a day, in hours from sunrise to sunset, in Orange City, Iowa, where t represents the day in the year assuming $t = 0$ is the spring equinox on March 21 (Figure 14). What are the lengths of the longest day and the shortest day? What are the day lengths on May 1, August 1, and November 1?

Solution The function L oscillates with amplitude 3.1 on either side of a central value of 12. According to the model, the longest day is approximately 15.1 hours long, and the shortest is approximately 8.9 hours long.

May 1, August 1, and November 1 correspond to $t = 41$, $t = 133$, and $t = 225$, respectively. Evaluating $L(t)$ at each of these values of t, we find that the day lengths are approximately 14.0 hours on May 1, 14.3 hours on August 1, and 10.0 hours on November 1. ∎

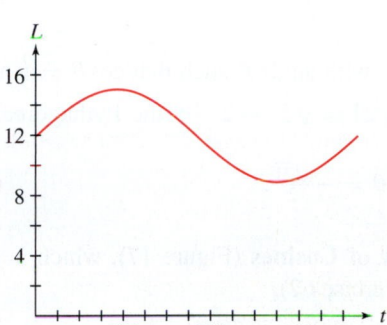

FIGURE 14 Day length in Orange City, Iowa.

This model $L(t)$, with amplitude 3.1, works for locations along the same latitude as Orange City, Iowa, 42 degrees north. At other latitudes, different amplitudes would need to be used. In more northern latitudes the days are longer in the summer and shorter in the winter, and a greater amplitude would be used. At more southern latitudes (in the northern hemisphere), smaller amplitudes would be used (see Exercises 41 and 42).

Trigonometric Identities

A key feature of trigonometric functions is that they satisfy a large number of identities. First and foremost, sine and cosine satisfy a fundamental identity, which is equivalent to the Pythagorean Theorem:

$$\sin^2 x + \cos^2 x = 1 \qquad \boxed{1}$$

Equivalent versions are obtained by dividing Eq. (1) by $\cos^2 x$ or $\sin^2 x$:

$$\tan^2 x + 1 = \sec^2 x, \qquad 1 + \cot^2 x = \csc^2 x \qquad \boxed{2}$$

Here is a list of some other commonly used identities. The identities for complementary angles are justified by Figure 15.

> The expression $(\sin x)^k$ is usually denoted $\sin^k x$. For example, $\sin^2 x$ is the square of $\sin x$. We use similar notation for the other trigonometric functions. However, we reserve $\sin^{-1} x$ for the inverse sine function discussed in the next section, rather than for $\frac{1}{\sin x}$.

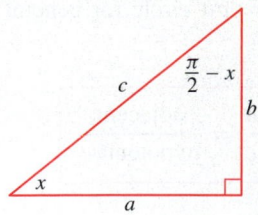

FIGURE 15 For complementary angles, the sine of one angle is equal to the cosine of the complementary angle.

Basic Trigonometric Identities

Complementary angles: $\sin\left(\frac{\pi}{2} - x\right) = \cos x, \qquad \cos\left(\frac{\pi}{2} - x\right) = \sin x$

Addition formulas: $\sin(x + y) = \sin x \cos y + \cos x \sin y$

$\cos(x + y) = \cos x \cos y - \sin x \sin y$

Double-angle formulas: $\sin^2 x = \frac{1}{2}(1 - \cos 2x), \qquad \cos^2 x = \frac{1}{2}(1 + \cos 2x)$

$\cos 2x = \cos^2 x - \sin^2 x, \qquad \sin 2x = 2 \sin x \cos x$

Shift formulas: $\sin\left(x + \frac{\pi}{2}\right) = \cos x, \qquad \cos\left(x + \frac{\pi}{2}\right) = -\sin x$

EXAMPLE 6 For θ between 0 and 2π, the equation $\cos\theta = \frac{2}{5}$ has a solution in $(0, \frac{\pi}{2})$ and a solution in $(\frac{3\pi}{2}, 2\pi)$. Calculate $\tan\theta$ in each case.

Solution First, using the identity $\cos^2\theta + \sin^2\theta = 1$, we obtain

$$\sin\theta = \pm\sqrt{1 - \cos^2\theta} = \pm\sqrt{1 - \frac{4}{25}} = \pm\frac{\sqrt{21}}{5}$$

If $0 < \theta < \frac{\pi}{2}$, then $\sin\theta$ is positive and we take the positive square root:

$$\tan\theta = \frac{\sin\theta}{\cos\theta} = \frac{\sqrt{21}/5}{2/5} = \frac{\sqrt{21}}{2}$$

To visualize this computation, draw a right triangle with angle θ such that $\cos\theta = \frac{2}{5}$ as in Figure 16. The opposite side then has length $\sqrt{21} = \sqrt{5^2 - 2^2}$ by the Pythagorean Theorem.

If $\frac{3\pi}{2} < \theta < 2\pi$, then $\sin\theta$ is negative and $\tan\theta = -\frac{\sqrt{21}}{2}$. ∎

We conclude this section by quoting the **Law of Cosines** (Figure 17), which is a generalization of the Pythagorean Theorem (see Exercise 62).

> **THEOREM 1 Law of Cosines** If a triangle has sides a, b, and c, and θ is the angle opposite side c, then
>
> $$c^2 = a^2 + b^2 - 2ab\cos\theta$$

If $\theta = \pi/2$, then $\cos\theta = 0$ and the Law of Cosines reduces to the Pythagorean Theorem.

Hypotenuse 5
Opposite $\sqrt{21}$
θ
Adjacent 2

FIGURE 16

c
b
θ
a

FIGURE 17

1.4 SUMMARY

• An angle of θ radians subtends an arc of length θr on a circle of radius r.
• To convert from radians to degrees, multiply by $180/\pi$.
• To convert from degrees to radians, multiply by $\pi/180$.
• Unless otherwise stated, all angles in this text are given in radians.
• The functions $f(\theta) = \cos\theta$ and $f(\theta) = \sin\theta$ are defined in terms of right triangles for acute angles and as coordinates of a point on the unit circle for general angles (Figure 18):

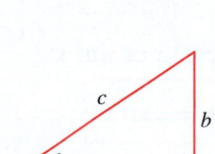

c
b
θ
a

$(\cos\theta, \sin\theta)$
θ
1

FIGURE 18

$$\sin\theta = \frac{b}{c} = \frac{\text{opposite}}{\text{hypotenuse}}, \qquad \cos\theta = \frac{a}{c} = \frac{\text{adjacent}}{\text{hypotenuse}}$$

• Basic properties of sine and cosine:

– Periodicity: $\sin(\theta + 2\pi) = \sin\theta$, $\cos(\theta + 2\pi) = \cos\theta$
– Parity: $\sin(-\theta) = -\sin\theta$, $\cos(-\theta) = \cos\theta$
– Basic identity: $\sin^2\theta + \cos^2\theta = 1$

• The four additional trigonometric functions:

$$\tan\theta = \frac{\sin\theta}{\cos\theta}, \qquad \cot\theta = \frac{\cos\theta}{\sin\theta}, \qquad \sec\theta = \frac{1}{\cos\theta}, \qquad \csc\theta = \frac{1}{\sin\theta}$$

1.4 EXERCISES

Preliminary Questions

1. How is it possible for two different rotations to define the same angle?

2. Give two different positive rotations that define the angle $\pi/4$.

3. Give a negative rotation that defines the angle $\pi/3$.

4. The definition of $\cos\theta$ using right triangles applies when (choose the correct answer):

(a) $0 < \theta < \dfrac{\pi}{2}$ (b) $0 < \theta < \pi$ (c) $0 < \theta < 2\pi$

5. What is the unit circle definition of $\sin\theta$?

6. How does the periodicity of $f(\theta) = \sin\theta$ and $f(\theta) = \cos\theta$ follow from the unit circle definition?

Exercises

1. Find the angle between 0 and 2π equivalent to $13\pi/4$.

2. Describe $\theta = \pi/6$ by an angle of negative radian measure.

3. Convert from radians to degrees:

(a) 1   (b) $\dfrac{\pi}{3}$ (c) $\dfrac{5}{12}$ (d) $-\dfrac{3\pi}{4}$

4. Convert from degrees to radians:

(a) $1°$ (b) $30°$ (c) $25°$ (d) $120°$

5. Find the lengths of the arcs subtended by the angles θ and ϕ radians in Figure 19.

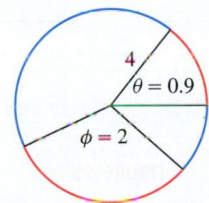

FIGURE 19 Circle of radius 4.

6. Calculate the values of the six standard trigonometric functions for the angle θ in Figure 20.

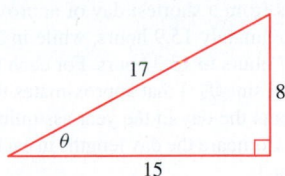

FIGURE 20

7. Fill in the remaining values of $(\cos\theta, \sin\theta)$ for the points in Figure 21.

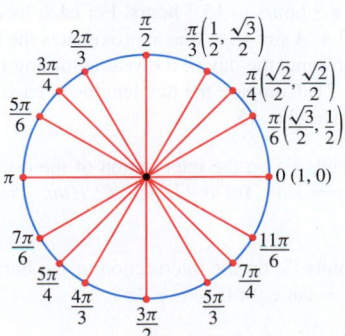

FIGURE 21

8. Find the values of the six standard trigonometric functions at $\theta = 11\pi/6$.

In Exercises 9–14, use Figure 21 to find all angles between 0 and 2π satisfying the given condition.

9. $\cos\theta = \dfrac{1}{2}$

10. $\tan\theta = 1$

11. $\tan\theta = -1$

12. $\csc\theta = 2$

13. $\sin x = \dfrac{\sqrt{3}}{2}$

14. $\sec t = 2$

15. Fill in the following table of values:

θ	$\dfrac{\pi}{6}$	$\dfrac{\pi}{4}$	$\dfrac{\pi}{3}$	$\dfrac{\pi}{2}$	$\dfrac{2\pi}{3}$	$\dfrac{3\pi}{4}$	$\dfrac{5\pi}{6}$
$\tan\theta$							
$\sec\theta$							

16. Complete the following table of signs:

θ	$\sin\theta$	$\cos\theta$	$\tan\theta$	$\cot\theta$	$\sec\theta$	$\csc\theta$
$0 < \theta < \dfrac{\pi}{2}$	+	+				
$\dfrac{\pi}{2} < \theta < \pi$						
$\pi < \theta < \dfrac{3\pi}{2}$						
$\dfrac{3\pi}{2} < \theta < 2\pi$						

17. Show that if $\tan\theta = c$ and $0 \le \theta < \pi/2$, then $\cos\theta = 1/\sqrt{1+c^2}$. *Hint:* Draw a right triangle whose opposite and adjacent sides have lengths c and 1.

18. Suppose that $\cos\theta = \frac{1}{3}$.

(a) Show that if $0 \le \theta < \pi/2$, then $\sin\theta = 2\sqrt{2}/3$ and $\tan\theta = 2\sqrt{2}$.

(b) Find $\sin\theta$ and $\tan\theta$ if $3\pi/2 \le \theta < 2\pi$.

In Exercises 19–24, assume that $0 \le \theta < \pi/2$.

19. Find $\sin\theta$ and $\tan\theta$ if $\cos\theta = \frac{5}{13}$.

20. Find $\cos \theta$ and $\tan \theta$ if $\sin \theta = \frac{3}{5}$.

21. Find $\sin \theta$, $\sec \theta$, and $\cot \theta$ if $\tan \theta = \frac{2}{7}$.

22. Find $\sin \theta$, $\cos \theta$, and $\sec \theta$ if $\cot \theta = 4$.

23. Find $\cos 2\theta$ if $\sin \theta = \frac{1}{5}$.

24. Find $\sin 2\theta$ and $\cos 2\theta$ if $\tan \theta = \sqrt{2}$.

25. Find $\cos \theta$ and $\tan \theta$ if $\sin \theta = 0.4$ and $\pi/2 \le \theta < \pi$.

26. Find $\cos \theta$ and $\sin \theta$ if $\tan \theta = 4$ and $\pi \le \theta < 3\pi/2$.

27. Find $\cos \theta$ if $\cot \theta = \frac{4}{3}$ and $\sin \theta < 0$.

28. Find $\tan \theta$ if $\sec \theta = \sqrt{5}$ and $\sin \theta < 0$.

29. Find the values of $\sin \theta$, $\cos \theta$, and $\tan \theta$ for the angles corresponding to the eight points on the unit circles in Figure 22(A) and (B).

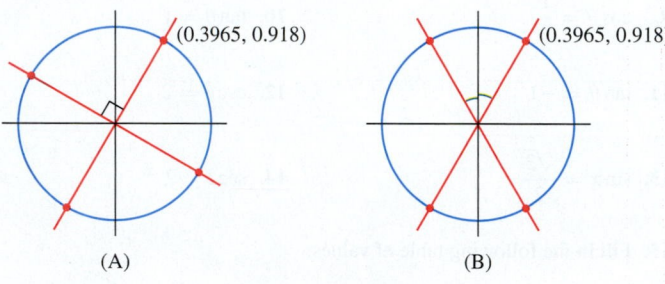

(A)   (B)

FIGURE 22

30. Refer to Figure 23(A). Express the functions $\sin \theta$, $\tan \theta$, and $\csc \theta$ in terms of c.

31. Refer to Figure 23(B). Compute $\cos \psi$, $\sin \psi$, $\cot \psi$, and $\csc \psi$.

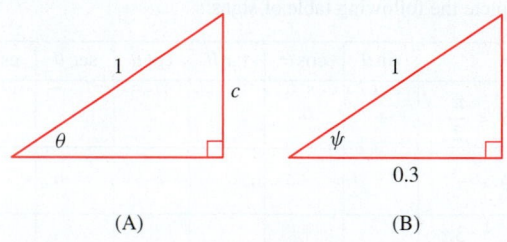

(A)   (B)

FIGURE 23

32. Express $\cos \left(\theta + \frac{\pi}{2} \right)$ and $\sin \left(\theta + \frac{\pi}{2} \right)$ in terms of $\cos \theta$ and $\sin \theta$. *Hint:* Find the relation between the coordinates (a, b) and (c, d) in Figure 24.

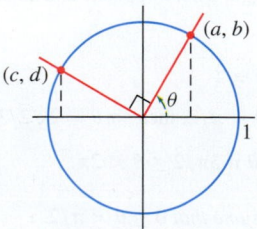

FIGURE 24

33. Use addition formulas and the values of $\sin \theta$ and $\cos \theta$ for $\theta = \frac{\pi}{3}, \frac{\pi}{4}$ to compute $\sin \frac{7\pi}{12}$ and $\cos \frac{7\pi}{12}$ exactly.

34. Use addition formulas and the values of $\sin \theta$ and $\cos \theta$ for $\theta = \frac{\pi}{3}, \frac{\pi}{4}$ to compute $\sin \frac{\pi}{12}$ and $\cos \frac{\pi}{12}$ exactly.

In Exercises 35–38, sketch the graph over $[0, 2\pi]$.

35. $f(\theta) = 2 \sin 4\theta$

36. $f(\theta) = \cos \left(2 \left(\theta - \frac{\pi}{2} \right) \right)$

37. $f(\theta) = \cos \left(2\theta - \frac{\pi}{2} \right)$

38. $f(\theta) = \sin \left(2 \left(\theta - \frac{\pi}{2} \right) + \pi \right) + 2$

39. Determine a function that would have a graph as in Figure 25(A), stating the period and amplitude.

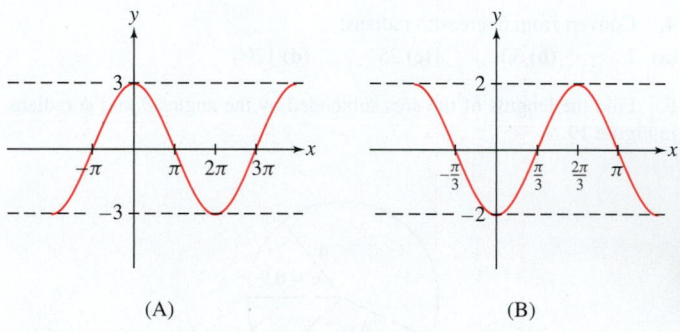

(A)   (B)

FIGURE 25

40. Determine a function that would have a graph as in Figure 25(B), stating the period and amplitude.

41. During a year, the length of a day, from sunrise to sunset, in Wolf Point, Montana, varies from a shortest day of approximately 8.1 hours to a longest day of approximately 15.9 hours, while in Mexico City, the day lengths vary from 10.7 hours to 13.3 hours. For each location, determine a function $L(t) = 12 + A \sin(\frac{2\pi}{365} t)$ that approximates the length of a day, in hours, where t represents the day in the year assuming $t = 0$ is the spring equinox on March 21. Compare the day lengths in each location on April 1, July 15, and November 1.

42. During a year, the length of a day, from sunrise to sunset, in Taloga, Oklahoma, varies from a shortest day of approximately 9.6 hours to a longest day of approximately 14.4 hours, while in Montreal, the day lengths vary from 8.3 hours to 15.7 hours. For each location, determine a function $L(t) = 12 + A \sin(\frac{2\pi}{365} t)$ that approximates the length of a day, in hours, where t represents the day in the year assuming $t = 0$ is the spring equinox on March 21. Compare the day lengths in each location on April 15, July 30, and November 15.

43. How many points lie on the intersection of the horizontal line $y = c$ and the graph of $y = \sin x$ for $0 \le x < 2\pi$? *Hint:* The answer depends on c.

44. How many points lie on the intersection of the horizontal line $y = c$ and the graph of $y = \tan x$ for $0 \le x < 2\pi$?

In Exercises 45–46, solve for $0 \le \theta < 2\pi$.

45. $\sin \theta = \sin 2\theta$ *Hint:* Use the double angle formula for sine.

46. $\sin\theta = \cos 2\theta$ *Hint:* Use appropriate identities to express $\cos 2\theta$ in terms of the sine function.

In Exercises 47–56, derive the identity using the identities listed in this section.

47. $\cos 2\theta = 2\cos^2\theta - 1$

48. $\cos^2\dfrac{\theta}{2} = \dfrac{1+\cos\theta}{2}$

49. $\sin^2\dfrac{\theta}{2} = \dfrac{1-\cos\theta}{2}$

50. $\sin(\theta + \pi) = -\sin\theta$

51. $\cos(\theta + \pi) = -\cos\theta$

52. $\tan x = \cot\left(\dfrac{\pi}{2} - x\right)$

53. $\tan(\pi - \theta) = -\tan\theta$

54. $\tan 2x = \dfrac{2\tan x}{1 - \tan^2 x}$

55. $\tan x = \dfrac{\sin 2x}{1 + \cos 2x}$

56. $\sin^2 x \cos^2 x = \dfrac{1 - \cos 4x}{8}$

57. Use Exercises 50 and 51 to show that $\tan\theta$ and $\cot\theta$ are periodic with period π.

58. Use the double-angle formulas to show that $\sin^2\theta$ and $\cos^2\theta$ are periodic with period π.

59. Use the identity of Exercise 48 to show that $\cos\frac{\pi}{8}$ is equal to
$$\sqrt{\dfrac{1}{2} + \dfrac{\sqrt{2}}{4}}.$$

60. Use Exercise 55 to compute $\tan\frac{\pi}{8}$.

61. Use the Law of Cosines to find the distance from P to Q in Figure 26.

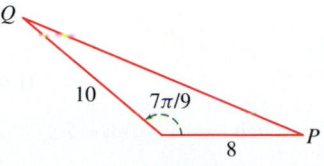

FIGURE 26

Further Insights and Challenges

62. Use Figure 27 to derive the Law of Cosines from the Pythagorean Theorem.

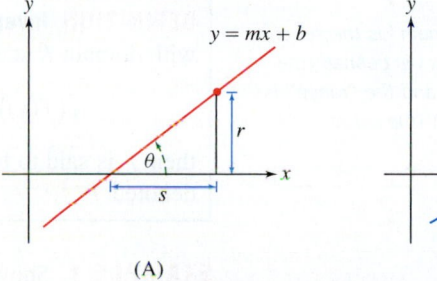

FIGURE 27

(A)   (B)

FIGURE 28

63. Use the addition formula to prove
$$\cos 3\theta = 4\cos^3\theta - 3\cos\theta$$

64. Use the addition formulas for sine and cosine to prove
$$\tan(a+b) = \dfrac{\tan a + \tan b}{1 - \tan a\tan b}$$
$$\cot(a-b) = \dfrac{\cot a\cot b + 1}{\cot b - \cot a}$$

65. Let θ be the angle between the line $y = mx + b$ and the x-axis [Figure 28(A)]. Prove that $m = \tan\theta$.

66. Let L_1 and L_2 be the lines of slope m_1 and m_2 [Figure 28(B)]. Show that the angle θ between L_1 and L_2 satisfies $\cot\theta = \dfrac{m_2 m_1 + 1}{m_2 - m_1}$.

67. Perpendicular Lines Use Exercise 66 to prove that two lines with nonzero slopes m_1 and m_2 are perpendicular if and only if $m_2 = -1/m_1$.

68. Apply the double-angle formula to prove:

(a) $\cos\dfrac{\pi}{8} = \dfrac{1}{2}\sqrt{2 + \sqrt{2}}$

(b) $\cos\dfrac{\pi}{16} = \dfrac{1}{2}\sqrt{2 + \sqrt{2 + \sqrt{2}}}$

Guess the values of $\cos\dfrac{\pi}{32}$ and of $\cos\dfrac{\pi}{2^n}$ for all n.

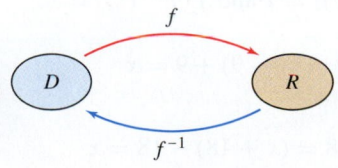

FIGURE 1 A function and its inverse.

1.5 Inverse Functions

Many important functions, such as logarithms, roots, and the arcsine, are defined as inverse functions. In this section, we review inverse functions and their graphs, and we discuss the inverse trigonometric functions.

The inverse of f, denoted f^{-1}, is the function that *reverses* the effect of f (Figure 1). For example, the inverse of $f(x) = x^3$ is the cube root function $f^{-1}(x) = x^{1/3}$. Given

In general, $f^{-1}(x) \neq \frac{1}{f(x)}$. The expression $f^{-1}(x)$ is simply a notation for the inverse function, and the -1 does not represent an exponent.

a table of function values for f, we obtain a table for f^{-1} by interchanging the x and y columns, assuming the resulting f^{-1} is a function:

Function			Inverse	
x	$f(x) = x^3$		x	$f^{-1}(x) = x^{1/3}$
-2	-8	(Interchange columns) $\Longrightarrow$	-8	-2
-1	-1		-1	-1
0	0		0	0
1	1		1	1
2	8		8	2
3	27		27	3

If we apply both f and f^{-1} to a number x in either order, we get back x. For instance,

Apply f and then f^{-1}: $\quad 2 \overset{\text{(apply } x^3\text{)}}{\longrightarrow} 8 \overset{\text{(apply } x^{1/3}\text{)}}{\longrightarrow} 2$

Apply f^{-1} and then f: $\quad 8 \overset{\text{(apply } x^{1/3}\text{)}}{\longrightarrow} 2 \overset{\text{(apply } x^3\text{)}}{\longrightarrow} 8$

This property is used in the formal definition of the inverse function:

← **REMINDER** The "domain" is the set of numbers x such that $f(x)$ is defined (the set of allowable inputs), and the "range" is the set of all values $f(x)$ (the set of outputs).

DEFINITION Inverse Let f have domain D and range R. If there is a function g with domain R such that

$$g\big(f(x)\big) = x \quad \text{for } x \in D \qquad \text{and} \qquad f\big(g(x)\big) = x \quad \text{for } x \in R$$

then f is said to be **invertible**. The function g is called the **inverse function** and is denoted f^{-1}.

EXAMPLE 1 Show that $f(x) = 2x - 18$ is invertible. What are the domain and range of f^{-1}?

Solution We show that f is invertible by computing the inverse function in two steps.

Step 1. Solve the equation $y = f(x)$ for x in terms of y.

$$y = 2x - 18$$
$$y + 18 = 2x$$
$$x = \frac{1}{2}y + 9$$

This gives us the inverse as a function of the variable y: $f^{-1}(y) = \frac{1}{2}y + 9$.

The variable y in $f^{-1}(y) = \frac{1}{2}y + 9$ is called a **dummy variable**. It is "local" to the equation, which means that changing y to a different symbol does not change the meaning of the relationship, as well as the mathematics that precedes and follows the relationship. We could write $f^{-1}(A) = \frac{1}{2}A + 9$ or $f^{-1}(DOG) = \frac{1}{2}(DOG) + 9$ and not affect the mathematics in Example 1. Generally, x is preferred as the independent variable when writing expressions for functions.

Step 2. Interchange variables.

We usually prefer to write the inverse as a function of x, so we interchange the roles of x and y:

$$f^{-1}(x) = \frac{1}{2}x + 9$$

Graphs of f and f^{-1} are shown in Figure 2.

To check our calculation, let's verify that $f^{-1}(f(x)) = x$ and $f(f^{-1}(x)) = x$:

$$f^{-1}\big(f(x)\big) = f^{-1}(2x - 18) = \frac{1}{2}(2x - 18) + 9 = (x - 9) + 9 = x$$

$$f\big(f^{-1}(x)\big) = f\left(\frac{1}{2}x + 9\right) = 2\left(\frac{1}{2}x + 9\right) - 18 = (x + 18) - 18 = x$$

Because f^{-1} is a linear function, its domain and range are **R**.

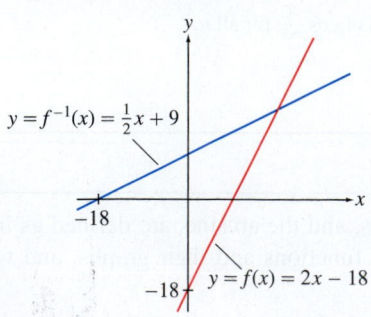

$y = f^{-1}(x) = \frac{1}{2}x + 9$

$y = f(x) = 2x - 18$

FIGURE 2

The inverse function, if it exists, is unique. However, some functions do not have an inverse. Consider $f(x) = x^2$. When we interchange the columns in a table of values (which should give us a table of values for f^{-1}), the resulting table does not define a function:

Function			Inverse (?)	
x	$f(x) = x^2$		x	$f^{-1}(x)$
-2	4		4	-2
-1	1		1	-1
0	0		0	0
1	1		1	1
2	4		4	2

(Interchange columns) $\Longrightarrow$

$f^{-1}(1)$ has two values: 1 and -1.

The problem is that every positive number occurs twice as an output of $f(x) = x^2$. For example, 1 occurs twice as an *output* in the first table and therefore occurs twice as an *input* in the second table. So the second table gives us two possible values for $f^{-1}(1)$, namely $f^{-1}(1) = 1$ and $f^{-1}(1) = -1$. Neither value satisfies the inverse property. For instance, if we set $f^{-1}(1) = 1$, then $f^{-1}(f(-1)) = f^{-1}(1) = 1$, but an inverse would have to satisfy $f^{-1}(f(-1)) = -1$.

So when does a function f have an inverse? The answer is: if f is **one-to-one**, which means that f takes on each value at most once (Figure 3). Here is the formal definition:

Another standard term for one-to-one is **injective**.

DEFINITION One-to-One Function   A function f is one-to-one on a domain D if, for every value c, the equation $f(x) = c$ has at most one solution for $x \in D$. Or, equivalently, if for all $a, b \in D$, if $a \neq b$, then $f(a) \neq f(b)$

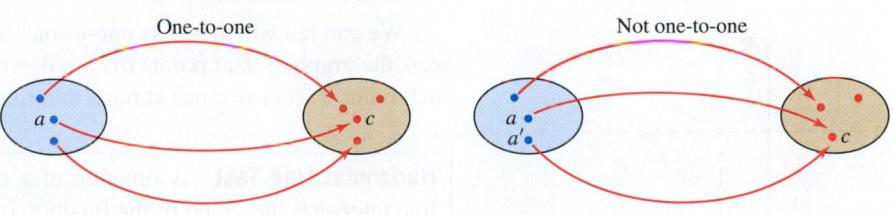

One-to-one   Not one-to-one

$f(x) = c$ has at most one solution for all c. $f(x) = c$ has two solutions: $x = a$ and $x = a'$.

FIGURE 3 A one-to-one function takes on each value at most once.

Think of a function as a device for "labeling" members of the range by members of the domain. When f is one-to-one, this labeling is unique and f^{-1} maps each number in the range back to its label.

When f is one-to-one on its domain D, the inverse function f^{-1} exists and its domain is equal to the range R of f (Figure 4). Indeed, for every $c \in R$, there is precisely one element $a \in D$ such that $f(a) = c$ and we may define $f^{-1}(c) = a$. With this definition, $f(f^{-1}(c)) = f(a) = c$ and $f^{-1}(f(a)) = f^{-1}(c) = a$. This proves the following theorem:

THEOREM 1 Existence of Inverses   The inverse function f^{-1} exists if and only if f is one-to-one on its domain D. Furthermore,

- Domain of f = range of f^{-1}
- Range of f = domain of f^{-1}

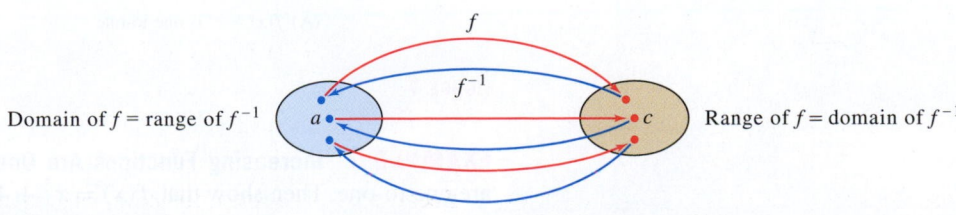

Domain of f = range of f^{-1} Range of f = domain of f^{-1}

FIGURE 4 In passing from f to f^{-1}, the domain and range are interchanged.

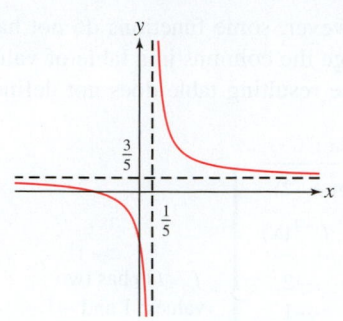

FIGURE 5 Graph of $f(x) = \dfrac{3x+2}{5x-1}$.

EXAMPLE 2 Show that $f(x) = \dfrac{3x+2}{5x-1}$ is invertible. Determine the domain and range of f and f^{-1}.

Solution The domain of f is $D = \left\{x : x \neq \tfrac{1}{5}\right\}$ (Figure 5). Assume that $x \in D$, and let's solve $y = f(x)$ for x in terms of y:

$$y = \frac{3x+2}{5x-1}$$

$$y(5x-1) = 3x+2$$

$$5xy - y = 3x+2$$

$$5xy - 3x = y + 2 \qquad \text{(gather terms involving } x\text{)}$$

$$x(5y-3) = y + 2 \qquad \text{(factor out } x \text{ in order to solve for } x\text{)} \qquad \boxed{1}$$

$$x = \frac{y+2}{5y-3} \qquad \text{(divide by } 5y-3\text{)} \qquad \boxed{2}$$

Often, it is impossible to find a formula for the inverse because we cannot solve for x explicitly in the equation $y = f(x)$. For example, the function $f(x) = x + \sin x$ has an inverse, but we must make do without an explicit formula for it.

The last step is valid if $5y - 3 \neq 0$—that is, if $y \neq \tfrac{3}{5}$. But note that $y = \tfrac{3}{5}$ is not in the range of f. For if it were, Eq. (1) would yield the false equation $0 = \tfrac{3}{5} + 2$. Now Eq. (2) shows that for all $y \neq \tfrac{3}{5}$, there is a unique value x such that $f(x) = y$. Therefore, f is one-to-one on its domain. By Theorem 1, f is invertible. The range of f is $R = \left\{x : x \neq \tfrac{3}{5}\right\}$ and

$$f^{-1}(x) = \frac{x+2}{5x-3}$$

The inverse function has domain R and range D. ∎

We can tell whether f is one-to-one from its graph. The horizontal line $y = c$ intersects the graph of f at points $(a, f(a))$, where $f(a) = c$ (Figure 6). There is at most one such point if $f(x) = c$ has at most one solution. This gives us the following:

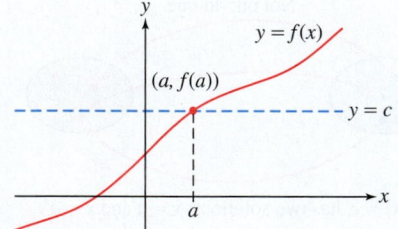

DF **FIGURE 6** The line $y = c$ intersects the graph at points where $f(a) = c$.

> **Horizontal Line Test** A function of x is one-to-one if and only if every horizontal line intersects the graph of the function in at most one point.

In Figure 7, we see that $f(x) = x^3$ passes the Horizontal Line Test and therefore is one-to-one, whereas $f(x) = x^2$ fails the test and is not one-to-one.

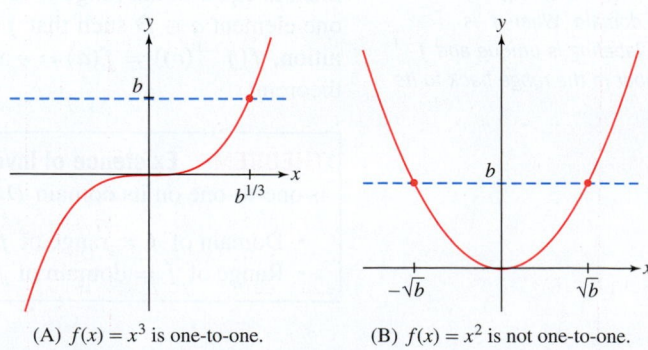

(A) $f(x) = x^3$ is one-to-one.

(B) $f(x) = x^2$ is not one-to-one.

FIGURE 7

EXAMPLE 3 **Increasing Functions Are One-to-One** Show that increasing functions are one-to-one. Then show that $f(x) = x^5 + 4x + 3$ is one-to-one.

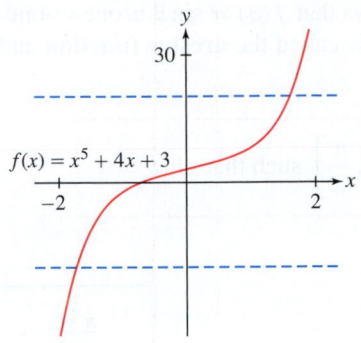

FIGURE 8 The increasing function $f(x) = x^5 + 4x + 3$ satisfies the Horizontal Line Test.

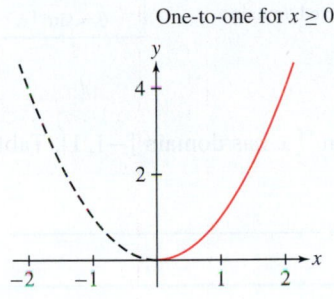

FIGURE 9 $f(x) = x^2$ satisfies the Horizontal Line Test on the domain $\{x : x \geq 0\}$.

Solution An increasing function satisfies $f(a) < f(b)$ if $a < b$. Therefore, f cannot take on any value more than once, and thus f is one-to-one.

Now observe that

- If n is odd and $c > 0$, then $f(x) = cx^n$ is increasing.
- A sum of increasing functions is increasing.

Thus, $g(x) = x^5$, $h(x) = 4x$, and hence the sum $k(x) = x^5 + 4x$ are increasing. It follows that the function $f(x) = x^5 + 4x + 3$ is increasing and therefore one-to-one (Figure 8). However, determining an explicit formula for its inverse would be difficult. ∎

Note that using an argument like the one in the previous example, we can prove that decreasing functions are also one-to-one and therefore have inverses.

We can make a function one-to-one by restricting its domain suitably.

EXAMPLE 4 **Restricting the Domain** Find a domain on which $f(x) = x^2$ is one-to-one and determine its inverse on this domain.

Solution The function $f(x) = x^2$ is one-to-one on the domain $D = \{x : x \geq 0\}$, for if $a^2 = b^2$, where a and b are both nonnegative, then $a = b$ (Figure 9). The inverse of f on D is the positive square root $f^{-1}(x) = \sqrt{x}$. Alternatively, we may restrict f to the domain $\{x : x \leq 0\}$, on which the inverse function is $f^{-1}(x) = -\sqrt{x}$. ∎

Next, we describe the graph of the inverse function. The **reflection** of a point (a, b) through the line $y = x$ is defined to be the point (b, a) (Figure 10). Note that if the x- and y-axes are drawn to the same scale, then (a, b) and (b, a) are equidistant from the line $y = x$ and the segment joining them is perpendicular to $y = x$.

The graph of f^{-1} is the reflection of the graph of f through $y = x$ (Figure 11). To check this, note that (a, b) lies on the graph of f if $f(a) = b$. But $f(a) = b$ if and only if $f^{-1}(b) = a$, and in this case, (b, a) lies on the graph of f^{-1}.

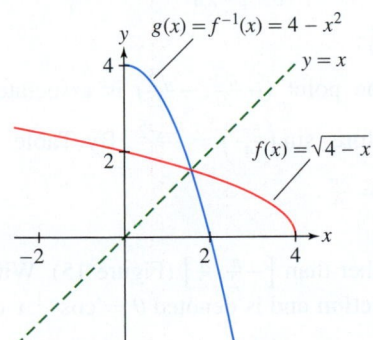

FIGURE 12 Graph of the inverse g of $f(x) = \sqrt{4 - x}$.

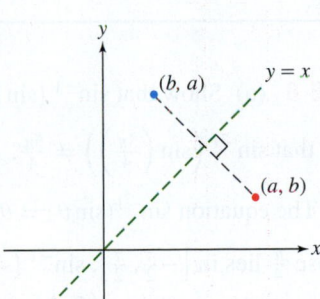

FIGURE 10 The reflection (a, b) through the line $y = x$ is the point (b, a).

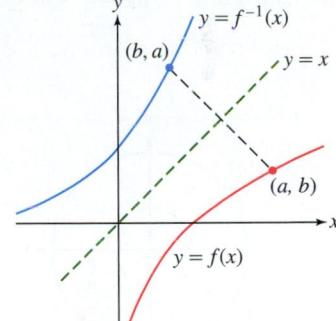

FIGURE 11 The graph of f^{-1} is the reflection of the graph of f through the line $y = x$.

EXAMPLE 5 **Sketching the Graph of the Inverse** Sketch the graph of the inverse of $f(x) = \sqrt{4 - x}$.

Solution Let $g(x) = f^{-1}(x)$. Observe that the domain of f is $\{x : x \leq 4\}$, and the range of f is $\{x : x \geq 0\}$. We do not need a formula for $g(x)$ to draw its graph. We simply reflect the graph of f through the line $y = x$, as in Figure 12. If desired, however, we can easily solve $y = \sqrt{4 - x}$ to obtain $x = 4 - y^2$ and thus $g(x) = 4 - x^2$ with domain $\{x : x \geq 0\}$. ∎

Inverse Trigonometric Functions

We have seen that the inverse function f^{-1} exists if and only if f is one-to-one on its domain. Because the trigonometric functions are not one-to-one, we must restrict their domains to define their inverses.

Do not confuse the inverse $\sin^{-1} x$ *with the* reciprocal

$$(\sin x)^{-1} = \frac{1}{\sin x} = \csc x$$

The inverse functions $\sin^{-1} x$, $\cos^{-1} x$, ... *are often denoted* $\arcsin x$, $\arccos x$, *etc.*

First, consider the sine function. Figure 13 shows that $f(\theta) = \sin\theta$ is one-to-one on $\left[-\frac{\pi}{2}, \frac{\pi}{2}\right]$. With this interval as domain, the inverse is called the **arcsine function** and is denoted $\theta = \sin^{-1} x$ or $\theta = \arcsin x$. By definition,

> $\theta = \sin^{-1} x$ is the unique angle in $\left[-\frac{\pi}{2}, \frac{\pi}{2}\right]$ such that $\sin\theta = x$

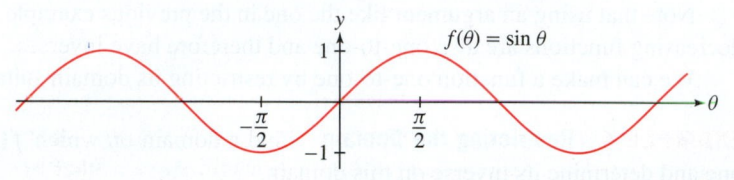

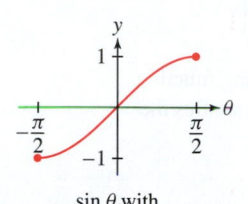
sin θ with restricted domain

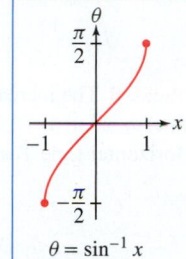
$\theta = \sin^{-1} x$

FIGURE 13

The range of $f(x) = \sin x$ is $[-1, 1]$, so $f^{-1}(x) = \sin^{-1} x$ has domain $[-1, 1]$. Table 1 gives some values of $\theta = \sin^{-1} x$.

Summary of inverse relation between the sine and arcsine functions:

$\sin(\sin^{-1} x) = x$ for $-1 \le x \le 1$

$\sin^{-1}(\sin\theta) = \theta$ for $-\frac{\pi}{2} \le \theta \le \frac{\pi}{2}$

TABLE 1

x	-1	$-\frac{\sqrt{3}}{2}$	$-\frac{\sqrt{2}}{2}$	$-\frac{1}{2}$	0	$\frac{1}{2}$	$\frac{\sqrt{2}}{2}$	$\frac{\sqrt{3}}{2}$	1
$\theta = \sin^{-1} x$	$-\frac{\pi}{2}$	$-\frac{\pi}{3}$	$-\frac{\pi}{4}$	$-\frac{\pi}{6}$	0	$\frac{\pi}{6}$	$\frac{\pi}{4}$	$\frac{\pi}{3}$	$\frac{\pi}{2}$

EXAMPLE 6   **(a)** Show that $\sin^{-1}\left(\sin\left(\frac{\pi}{4}\right)\right) = \frac{\pi}{4}$.

(b) Show that $\sin^{-1}\left(\sin\left(\frac{5\pi}{4}\right)\right) \ne \frac{5\pi}{4}$.

Solution The equation $\sin^{-1}(\sin\theta) = \theta$ is valid if θ lies in $\left[-\frac{\pi}{2}, \frac{\pi}{2}\right]$.

(a) Because $\frac{\pi}{4}$ lies in $\left[-\frac{\pi}{2}, \frac{\pi}{2}\right]$, $\sin^{-1}\left(\sin\left(\frac{\pi}{4}\right)\right) = \frac{\pi}{4}$.

(b) Now consider $\sin^{-1}\left(\sin\left(\frac{5\pi}{4}\right)\right)$. Note that the point $\left(-\frac{\sqrt{2}}{2}, -\frac{\sqrt{2}}{2}\right)$ is associated with the angle $\frac{5\pi}{4}$, as shown in Figure 14. Therefore, $\sin\left(\frac{5\pi}{4}\right) = \frac{-\sqrt{2}}{2}$. By Table 1, $\sin^{-1}\left(\frac{-\sqrt{2}}{2}\right) = -\frac{\pi}{4}$. So, $\sin^{-1}\left(\sin\left(\frac{5\pi}{4}\right)\right) = -\frac{\pi}{4} \ne \frac{5\pi}{4}$. ∎

FIGURE 14 $\sin\left(\frac{5\pi}{4}\right) = -\frac{\sqrt{2}}{2}$.

Summary of inverse relation between the cosine and arccosine functions:

$\cos(\cos^{-1} x) = x$ for $-1 \le x \le 1$

$\cos^{-1}(\cos\theta) = \theta$ for $0 \le \theta \le \pi$

The cosine function is one-to-one on $[0, \pi]$ rather than $\left[-\frac{\pi}{2}, \frac{\pi}{2}\right]$ (Figure 15). With this domain, the inverse is called the **arccosine function** and is denoted $\theta = \cos^{-1} x$ or $\theta = \arccos x$. It has domain $[-1, 1]$. By definition,

> $\theta = \cos^{-1} x$ is the unique angle in $[0, \pi]$ such that $\cos\theta = x$

When we study the calculus of inverse trigonometric functions in Section 3.8, we will need to simplify composite expressions such as $\cos(\sin^{-1} x)$ and $\tan(\sin^{-1} x)$. This can be done in two ways: by referring to the appropriate right triangle or by using trigonometric identities.

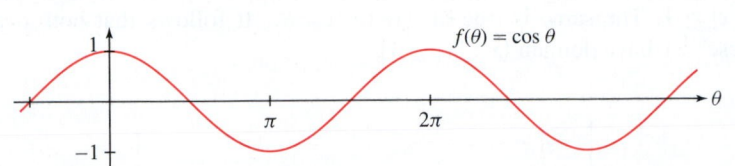

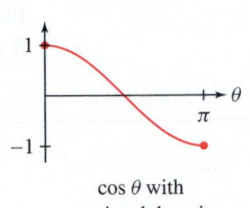
cos θ with
restricted domain

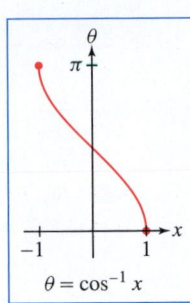
$\theta = \cos^{-1} x$

FIGURE 15

EXAMPLE 7 Find an alternative form in terms of x for each of $\cos(\sin^{-1} x)$ and $\tan(\sin^{-1} x)$.

Solution This problem asks for the values of $\cos\theta$ and $\tan\theta$ at the angle $\theta = \sin^{-1} x$. Consider a right triangle with hypotenuse of length 1 and angle θ such that $\sin\theta = x$, as in Figure 16. By the Pythagorean Theorem, the adjacent side has length $\sqrt{1 - x^2}$. Now we can read off the values from Figure 16:

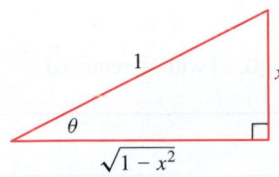

FIGURE 16 Right triangle constructed such that $\sin\theta = x$.

$$\cos(\sin^{-1} x) = \cos\theta = \frac{\text{adjacent}}{\text{hypotenuse}} = \sqrt{1 - x^2}$$

$$\tan(\sin^{-1} x) = \tan\theta = \frac{\text{opposite}}{\text{adjacent}} = \frac{x}{\sqrt{1 - x^2}}$$

Alternatively, we may argue using trigonometric identities. Because $\sin\theta = x$,

$$\cos(\sin^{-1} x) = \cos\theta = \sqrt{1 - \sin^2\theta} = \sqrt{1 - x^2}$$

We are justified in taking the positive square root in either approach because $\theta = \sin^{-1} x$ lies in $\left[-\frac{\pi}{2}, \frac{\pi}{2}\right]$ and $\cos\theta$ is positive in this interval. ∎

We now address the remaining trigonometric functions. The function $f(\theta) = \tan\theta$ is one-to-one on $\left(-\frac{\pi}{2}, \frac{\pi}{2}\right)$, and $f(\theta) = \cot\theta$ is one-to-one on $(0, \pi)$ (see Figure 10 in Section 1.4). We define their inverses by restricting them to these domains:

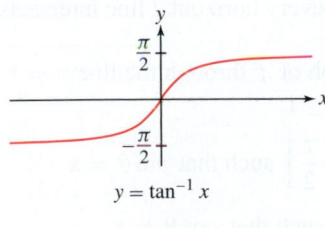
$y = \tan^{-1} x$

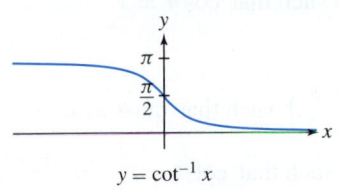
$y = \cot^{-1} x$

FIGURE 17

> $\theta = \tan^{-1} x$ is the unique angle in $\left(-\frac{\pi}{2}, \frac{\pi}{2}\right)$ such that $\tan\theta = x$

> $\theta = \cot^{-1} x$ is the unique angle in $(0, \pi)$ such that $\cot\theta = x$

The range of both $f(\theta) = \tan\theta$ and $f(\theta) = \cot\theta$ is the set of all real numbers **R**. Therefore, $\theta = \tan^{-1} x$ and $\theta = \cot^{-1} x$ have domain **R** (Figure 17).

The function $f(\theta) = \sec\theta$ is not defined at $\theta = \frac{\pi}{2}$, but we see in Figure 18 that it is one-to-one on $\left[0, \frac{\pi}{2}\right) \cup \left(\frac{\pi}{2}, \pi\right]$. Similarly, $f(\theta) = \csc\theta$ is not defined at $\theta = 0$, but it is one-to-one on $\left[-\frac{\pi}{2}, 0\right) \cup \left(0, \frac{\pi}{2}\right]$. We define the inverse functions as follows:

> $\theta = \sec^{-1} x$ is the unique angle in $\left[0, \frac{\pi}{2}\right) \cup \left(\frac{\pi}{2}, \pi\right]$ such that $\sec\theta = x$

> $\theta = \csc^{-1} x$ is the unique angle in $\left[-\frac{\pi}{2}, 0\right) \cup \left(0, \frac{\pi}{2}\right]$ such that $\csc\theta = x$

Figure 18 shows that the range of $f(\theta) = \sec\theta$ is the set of real numbers x such that $|x| \geq 1$. The same is true of $f(\theta) = \csc\theta$. It follows that both $\theta = \sec^{-1} x$ and $\theta = \csc^{-1} x$ have domain $\{x : |x| \geq 1\}$.

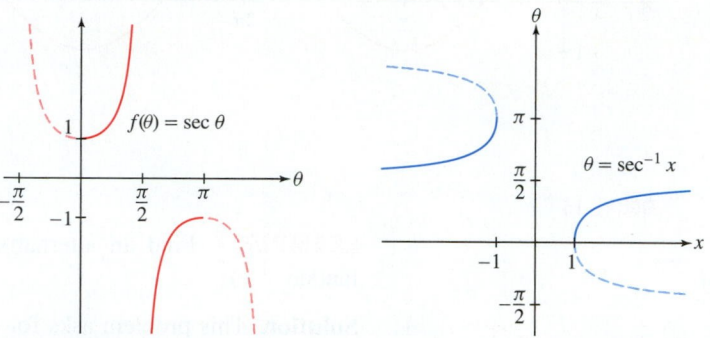

FIGURE 18 $f(\theta) = \sec\theta$ is one-to-one on the interval $[0, \pi]$ with $\frac{\pi}{2}$ removed.

1.5 SUMMARY

- A function f is *one-to-one* on a domain D if for every value c, the equation $f(x) = c$ has at most one solution for $x \in D$, or, equivalently, if for all $a, b \in D$, if $a \neq b$, then $f(a) \neq f(b)$.
- Let f have domain D and range R. The *inverse* f^{-1} (if it exists) is the unique function with domain R and range D satisfying $f(f^{-1}(x)) = x$ and $f^{-1}(f(x)) = x$.
- The inverse of f exists if and only if f is one-to-one on its domain.
- To find the inverse function, solve $y = f(x)$ for x in terms of y to obtain $x = g(y)$. The inverse is the function g.
- *Horizontal Line Test*: f is one-to-one if and only if every horizontal line intersects the graph of f in at most one point.
- The graph of f^{-1} is obtained by reflecting the graph of f through the line $y = x$.
- The *arcsine* and *arccosine* are defined for $-1 \leq x \leq 1$:

$$\theta = \sin^{-1} x \text{ is the unique angle in } \left[-\frac{\pi}{2}, \frac{\pi}{2}\right] \text{ such that } \sin\theta = x$$

$$\theta = \cos^{-1} x \text{ is the unique angle in } [0, \pi] \text{ such that } \cos\theta = x$$

- $\tan^{-1} x$ and $\cot^{-1} x$ are defined for all x:

$$\theta = \tan^{-1} x \text{ is the unique angle in } \left(-\frac{\pi}{2}, \frac{\pi}{2}\right) \text{ such that } \tan\theta = x$$

$$\theta = \cot^{-1} x \text{ is the unique angle in } (0, \pi) \text{ such that } \cot\theta = x$$

- $\sec^{-1} x$ and $\csc^{-1} x$ are defined for $|x| \geq 1$:

$$\theta = \sec^{-1} x \text{ is the unique angle in } \left[0, \frac{\pi}{2}\right) \cup \left(\frac{\pi}{2}, \pi\right] \text{ such that } \sec\theta = x$$

$$\theta = \csc^{-1} x \text{ is the unique angle in } \left[-\frac{\pi}{2}, 0\right) \cup \left(0, \frac{\pi}{2}\right] \text{ such that } \csc\theta = x$$

1.5 EXERCISES

Preliminary Questions

1. Which of the following satisfy $f^{-1}(x) = f(x)$?

(a) $f(x) = x$

(b) $f(x) = 1 - x$

(c) $f(x) = 1$

(d) $f(x) = \sqrt{x}$

(e) $f(x) = |x|$

(f) $f(x) = x^{-1}$

2. The function f maps teenagers in the United States to their last names. Explain why the inverse function f^{-1} does not exist.

3. The following fragment of a train schedule for the New Jersey Transit System defines a function f from towns to times. Is f one-to-one? What is $f^{-1}(6{:}27)$?

Trenton	6:21
Hamilton Township	6:27
Princeton Junction	6:34
New Brunswick	6:38

4. A homework problem asks for a sketch of the graph of the *inverse* of $f(x) = x + \cos x$. Frank, after trying but failing to find a formula for $f^{-1}(x)$, says it's impossible to graph the inverse. Bianca hands in an accurate sketch without solving for f^{-1}. How did Bianca complete the problem?

5. Which of the following quantities is undefined?

(a) $\sin^{-1}\left(-\frac{1}{2}\right)$

(b) $\cos^{-1}(2)$

(c) $\csc^{-1}\left(\frac{1}{2}\right)$

(d) $\csc^{-1}(2)$

6. Give an example of an angle θ such that $\cos^{-1}(\cos\theta) \neq \theta$. Does this contradict the definition of inverse function?

Exercises

1. Show that $f(x) = 7x - 4$ is invertible and find its inverse.

2. Is $f(x) = x^2 + 2$ one-to-one? If not, describe a domain on which it is one-to-one.

3. What is the largest interval containing zero on which $f(x) = \sin x$ is one-to-one?

4. Show that $f(x) = \dfrac{x-2}{x+3}$ is invertible and find its inverse.

(a) What is the domain of f? The range of f^{-1}?

(b) What is the domain of f^{-1}? The range of f?

5. Verify that $f(x) = x^3 + 3$ and $g(x) = (x-3)^{1/3}$ are inverses by showing that $f(g(x)) = x$ and $g(f(x)) = x$.

6. Repeat Exercise 5 for $f(t) = \dfrac{t+1}{t-1}$ and $g(t) = \dfrac{t+1}{t-1}$.

7. The escape velocity from a planet of radius R is $v(R) = \sqrt{\dfrac{2GM}{R}}$, where G is the universal gravitational constant and M is the mass. Find the inverse of $v(R)$ expressing R in terms of v.

8. Show that the power law relationship $P(Q) = kQ^r$, for $Q \geq 0$ and $r \neq 0$, has an inverse that is also a power law, $Q(P) = mP^s$, where $m = k^{-1/r}$ and $s = 1/r$.

9. The volume V of a cone that has height equal to its radius r is given by $V(r) = \frac{1}{3}\pi r^3$. Find the inverse of $V(r)$, expressing r as a function of V.

10. The surface area S of a sphere of radius r is given by $S(r) = 4\pi r^2$. Explain why, in the given context, $S(r)$ has an inverse function. Find the inverse of $S(r)$, expressing r as a function of S.

In Exercises 11–17, find a domain on which f is one-to-one and a formula for the inverse of f restricted to this domain. Sketch the graphs of f and f^{-1}.

11. $f(x) = 4 - x$

12. $f(x) = \dfrac{1}{x+1}$

13. $f(x) = \dfrac{1}{7x-3}$

14. $f(s) = \dfrac{1}{s^2}$

15. $f(x) = \dfrac{1}{\sqrt{x^2+1}}$

16. $f(z) = z^3$

17. $f(x) = \sqrt{x^3 + 9}$

18. For each function shown in Figure 19, sketch the graph of the inverse (restrict the function's domain if necessary).

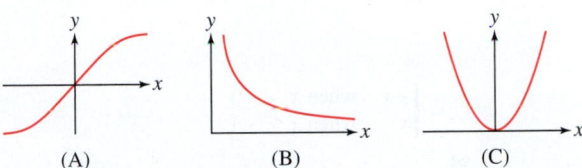

(A)   (B)   (C)

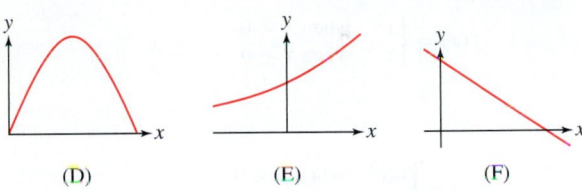

(D)   (E)   (F)

FIGURE 19

19. Which of the graphs in Figure 20 is the graph of a function satisfying $f^{-1} = f$?

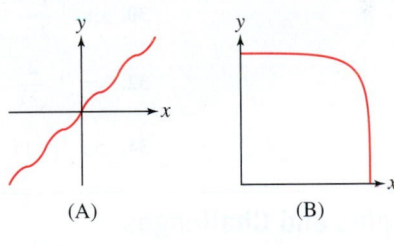

(A)   (B)

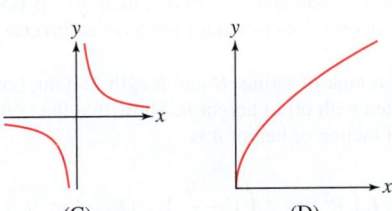

(C)   (D)

FIGURE 20

20. Let n be a nonzero integer. Find a domain on which the function $f(x) = (1 - x^n)^{1/n}$ coincides with its inverse. *Hint:* The answer depends on whether n is even or odd.

21. Let $f(x) = x^7 + x + 1$.

(a) Show that f^{-1} exists (but do not attempt to find it). *Hint:* Show that f is increasing.

(b) What is the domain of f^{-1}?

(c) Find $f^{-1}(3)$.

22. Show that $f(x) = (x^2 + 1)^{-1}$ is one-to-one on $(-\infty, 0]$, and find a formula for f^{-1} for this domain of f.

23. Let $f(x) = x^2 - 2x$. Determine a domain on which f^{-1} exists, and find a formula for f^{-1} for this domain of f.

24. Show that $f(x) = x + x^{-1}$ is one-to-one on $[1, \infty)$, and find the corresponding inverse f^{-1}. What is the domain of f^{-1}?

For each of the piecewise-defined functions in Exercises 25–28, determine whether or not the function is one-to-one, and if it is, determine its inverse function.

25.
$$f(x) = \begin{cases} x & \text{when } x < 0 \\ 2x & \text{when } x \geq 0 \end{cases}$$

26.
$$g(x) = \begin{cases} -x & \text{when } x < -1 \\ x & \text{when } x \geq -1 \end{cases}$$

27.
$$f(x) = \begin{cases} x^2 & \text{when } x < 0 \\ x & \text{when } x \geq 0 \end{cases}$$

28.
$$g(x) = \begin{cases} -x^2 & \text{when } x < 0 \\ x^2 & \text{when } x \geq 0 \end{cases}$$

In Exercises 29–34, evaluate without using a calculator.

29. $\cos^{-1} 1$

30. $\sin^{-1} \dfrac{1}{2}$

31. $\cot^{-1} 1$

32. $\sec^{-1} \dfrac{2}{\sqrt{3}}$

33. $\tan^{-1} \sqrt{3}$

34. $\sin^{-1}(-1)$

In Exercises 35–44, compute without using a calculator.

35. $\sin^{-1}\left(\sin \dfrac{\pi}{3}\right)$

36. $\sin^{-1}\left(\sin \dfrac{4\pi}{3}\right)$

37. $\cos^{-1}\left(\cos \dfrac{3\pi}{2}\right)$

38. $\sin^{-1}\left(\sin\left(-\dfrac{5\pi}{6}\right)\right)$

39. $\tan^{-1}\left(\tan \dfrac{3\pi}{4}\right)$

40. $\tan^{-1}(\tan \pi)$

41. $\sec^{-1}(\sec 3\pi)$

42. $\sec^{-1}\left(\sec \dfrac{3\pi}{2}\right)$

43. $\csc^{-1}\left(\csc(-\pi)\right)$

44. $\cot^{-1}\left(\cot\left(-\dfrac{\pi}{4}\right)\right)$

In Exercises 45–48, simplify by referring to the appropriate triangle or trigonometric identity.

45. $\tan(\cos^{-1} x)$

46. $\cos(\tan^{-1} x)$

47. $\cot(\sec^{-1} x)$

48. $\cot(\sin^{-1} x)$

In Exercises 49–56, refer to the appropriate triangle or trigonometric identity to compute the given value.

49. $\cos\left(\sin^{-1} \frac{2}{3}\right)$

50. $\tan\left(\cos^{-1} \frac{2}{3}\right)$

51. $\tan\left(\sin^{-1} 0.8\right)$

52. $\cos\left(\cot^{-1} 1\right)$

53. $\cot\left(\csc^{-1} 2\right)$

54. $\tan\left(\sec^{-1}(-2)\right)$

55. $\cot\left(\tan^{-1} 20\right)$

56. $\sin\left(\csc^{-1} 20\right)$

57. **GU** Let $f(x) = \frac{x^2}{x^2-1}$.

(a) From a graph of f, explain why f is not invertible. Furthermore, explain why f is invertible if we restrict the domain to $[0, \infty)$.

(b) Find the inverse function for f restricted to $[0, \infty)$.

58. **GU** Let $f(x) = \frac{x}{1-x^2}$.

(a) From a graph of f, explain why f is not invertible. Furthermore, explain why f is invertible if we restrict the domain to $(-1, 1)$.

(b) Let h be the inverse function for f restricted to $(-1, 1)$. What are $h(-10), h(-2), h(0), h(3), h(9)$?

(c) Use the quadratic formula to find an expression for $h(x)$.

Further Insights and Challenges

59. Show that if f is odd and f^{-1} exists, then f^{-1} is odd. Show, on the other hand, that an even function does not have an inverse.

60. A cylindrical tank of radius R and length L lying horizontally, as in Figure 21, is filled with oil to height h. Show that the volume $V(h)$ of oil in the tank as a function of height h is

$$V(h) = L\left(R^2 \cos^{-1}\left(1 - \frac{h}{R}\right) - (R - h)\sqrt{2hR - h^2}\right)$$

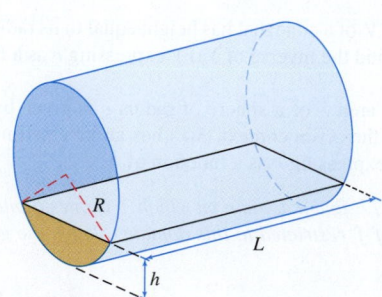

FIGURE 21 Oil in the tank has level h.

1.6 Exponential and Logarithmic Functions

An **exponential function** is a function of the form $f(x) = b^x$, where $b > 0$ and $b \neq 1$. The number b is called the **base**. Some examples are $f(x) = 2^x$, $g(x) = (1.4)^x$, and $h(x) = 10^x$. The case $b = 1$ is excluded because $f(x) = 1^x$ is a constant function. Calculators give good decimal approximations to values of exponential functions:

$$2^4 = 16, \quad 2^{-3} = 0.125, \quad (1.4)^{0.8} \approx 1.309, \quad 10^{4.6} \approx 39{,}810.717$$

Three properties of exponential functions should be singled out from the start (see Figure 1 for the case $b = 2$):

- *Exponential functions are positive*: $b^x > 0$ for all x.
- The *range* of $f(x) = b^x$ is the set of all positive real numbers.
- $f(x) = b^x$ is increasing if $b > 1$ and decreasing if $0 < b < 1$.

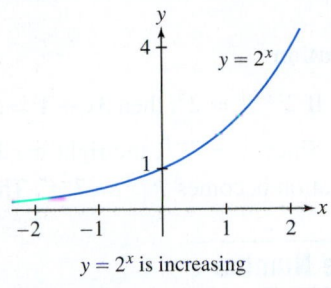

$y = 2^x$ is increasing

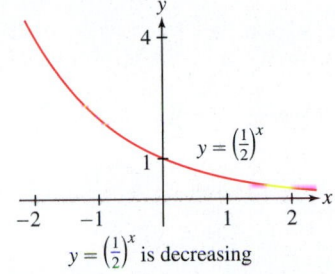

$y = \left(\frac{1}{2}\right)^x$ is decreasing

DF **FIGURE 1**

If $b > 1$, the exponential function $f(x) = b^x$ is not merely increasing but is, in a certain sense, rapidly increasing. Although the term "rapid increase" is perhaps subjective, the following precise statement is true: For all n, if x is positive and large enough, then $f(x) = b^x$ increases more rapidly than the power function $g(x) = x^n$ (we will prove this in Section 4.5). For example, Figure 2 shows that $f(x) = 3^x$ eventually overtakes and increases faster than the power functions $g(x) = x^3$, $g(x) = x^4$, and $g(x) = x^5$. Table 1 compares $f(x) = 3^x$ and $g(x) = x^5$.

	TABLE 1	
x	x^5	3^x
1	1	3
5	3125	243
10	100,000	59,049
15	759,375	14,348,907
25	9,765,625	847,288,609,443

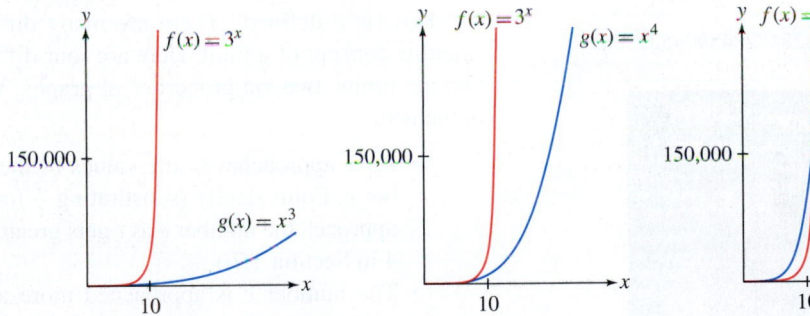

DF **FIGURE 2** Comparison of $f(x) = 3^x$ and power functions.

CONCEPTUAL INSIGHT The expressions "exponential growth" and "increases exponentially" (and related expressions) describe the growth in or modeled by an exponential function. A function whose graph increases more and more steeply does not necessarily increase exponentially. What distinguishes true exponential growth from other forms of growth is the precise way the function values increase. In exponential growth, an increase in x by 1 causes an increase in the function value by the same fixed percent, regardless of the value of x involved. In contrast, power functions do not exhibit such increase.

For example, Table 2 demonstrates how $f(x) = 3(2^x)$ increases by 100% with each increase by 1 in x, but $g(x) = 4x^2$ does not exhibit growth by a fixed percent.

	TABLE 2			
x	$f(x) = 3(2^x)$	% increase	$g(x) = 4x^2$	% increase
1	$3(2^1) = 6$		$4(1^2) = 4$	
2	$3(2^2) = 12$	100	$4(2^2) = 16$	300
3	$3(2^3) = 24$	100	$4(3^2) = 36$	125
4	$3(2^4) = 48$	100	$4(4^2) = 64$	78
5	$3(2^5) = 96$	100	$4(5^2) = 100$	56

In the next example, to solve exponential equations, we use the fact that $f(x) = b^x$ is one-to-one. In other words, if $b^x = b^y$, then $x = y$.

EXAMPLE 1 Solve for the unknown:

(a) $2^{3x+1} = 2^5$ **(b)** $7^{t+1} = \left(\dfrac{1}{7}\right)^{2t}$

Solution

(a) If $2^{3x+1} = 2^5$, then $3x + 1 = 5$ and thus $x = \frac{4}{3}$.

(b) Since $\frac{1}{7} = 7^{-1}$, the right-hand side of the equation is $\left(\frac{1}{7}\right)^{2t} = (7^{-1})^{2t} = 7^{-2t}$. The equation becomes $7^{t+1} = 7^{-2t}$. Therefore, $t + 1 = -2t$, or $t = -\frac{1}{3}$. ∎

The Number e

In Chapter 3, we will use calculus to study exponential functions. One of the surprising insights of calculus is that the most convenient or "natural" base for an exponential function is not $b = 10$ or $b = 2$, as one might think at first, but rather a certain irrational number, denoted by e, whose value is approximately $e \approx 2.718$. A calculator is used to evaluate specific values of $f(x) = e^x$. For example,

$$e^3 \approx 20.0855, \qquad e^{-1/4} \approx 0.7788$$

In calculus, when we speak of *the* exponential function, it is understood that the base is e. Another common notation for e^x is $\exp(x)$.

How is e defined? There are many different definitions, but they all rely on the calculus concept of a limit. Here are four different ways of characterizing or defining e, two via limits, two via properties of graphs. We will investigate each at different points in the text.

- As x approaches 0, the values of the expression $(1 + x)^{1/x}$ approach the number e. Equivalently (substituting $\frac{1}{t}$ for x), we can say that the values of $(1 + \frac{1}{t})^t$ approach the number e as t gets greater and greater without bound. (See Example 4 in Section 1.7.)
- The number e is approached more and more closely as we progressively add terms to the sum $1 + 1 + \frac{1}{2} + \frac{1}{6} + \frac{1}{24} + \frac{1}{120} + \frac{1}{720} + \cdots + \frac{1}{n}$. (With infinitely many terms in the sum, the expression is known as an infinite series. We study infinite series in Chapter 10.)
- Using Figure 3(A): Among all exponential functions $y = b^x$, $b = e$ is the unique base for which the slope of the tangent line to the graph at $(0, 1)$ is equal to 1.
- Using Figure 3(B): The number e is the unique number such that the area of the region under the hyperbola $y = 1/x$ for $1 \le x \le e$ is equal to 1.

Although written references to the number π go back more than 4000 years, mathematicians first became aware of the special role played by e in the seventeenth century. The notation e was introduced by Leonhard Euler, who discovered many fundamental properties of this important number. The number e has been computed to an accuracy of more than 100 billion decimal places. Here's a start:

$$e = 2.71828182845904523536\ldots$$

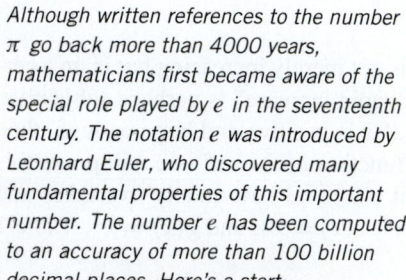

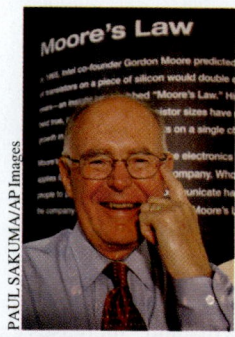

PAUL SAKUMA/AP Images

Gordon Moore (1929–). Moore, who later became chairman of Intel Corporation, predicted that following 1965, the number of transistors per integrated circuit would grow "exponentially." This prediction has held up for over five decades and may well continue into the future. Moore said, "Moore's Law is a term that got applied to a curve I plotted in the mid-sixties showing the increase in complexity of integrated circuits versus time. It's been expanded to include a lot more than that, and I'm happy to take credit for all of it."

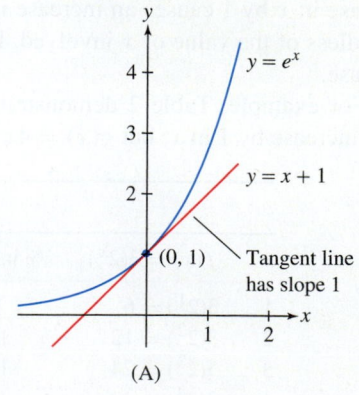

(A)

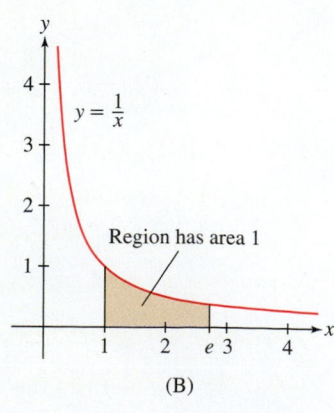

(B)

FIGURE 3

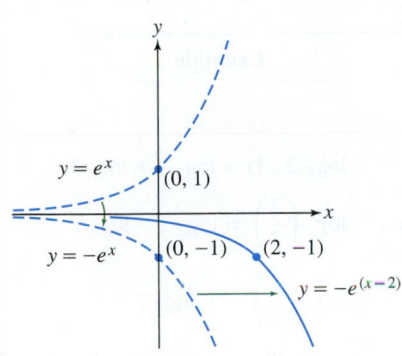

FIGURE 4 Graphing $y = -e^{(x-2)}$.

From these descriptions, it is not clear why e is important. As we will learn, however, the exponential function $f(x) = e^x$ plays a fundamental role because it behaves in a particularly simple way with respect to the basic operations of calculus.

EXAMPLE 2 Draw the graph of $y = -e^{(x-2)}$.

Solution Figure 3(A) shows the graph of $y = e^x$. The graph of $y = -e^x$ is simply the reflection of this graph over the x-axis, making all values negative instead of positive, as in Figure 4. The graph of $y = -e^{(x-2)}$ is obtained by translating this graph 2 units to the right. ∎

Logarithms

Logarithmic functions are inverses of exponential functions. More precisely, if $b > 0$ and $b \neq 1$, then the *logarithm to the base b*, denoted $\log_b x$, is the inverse of $f(x) = b^x$. By definition, $y = \log_b x$ if $b^y = x$, so we have

$$b^{\log_b x} = x \qquad \text{and} \qquad \log_b(b^x) = x$$

In other words, $\log_b x$ is the number to which b must be raised in order to get x. For example,

$$\log_2(8) = 3 \quad \text{because} \quad 2^3 = 8$$

$$\log_{10}(1) = 0 \quad \text{because} \quad 10^0 = 1$$

$$\log_3\left(\frac{1}{9}\right) = -2 \quad \text{because} \quad 3^{-2} = \frac{1}{3^2} = \frac{1}{9}$$

The logarithm to the base e, denoted $\ln x$, plays a special role and is called the **natural logarithm**.

$$\ln x = \log_e x$$

We use a calculator to evaluate logarithms numerically. For example,

$$\ln 17 \approx 2.83321 \quad \text{because} \quad e^{2.83321} \approx 17$$

As in Figure 5, $f(x) = \ln x$ and $g(x) = e^x$ are inverse functions, so we have

$$e^{\ln x} = x \qquad \text{and} \qquad \ln(e^x) = x$$

Recall that the domain of $f(x) = b^x$ is **R**, and its range is the set of positive real numbers $\{x : x > 0\}$. Since the domain and range are reversed in the inverse function,

- The *domain* of $f(x) = \log_b x$ is $\{x : x > 0\}$.
- The *range* of $f(x) = \log_b x$ is the set of all real numbers **R**.

If $b > 1$, then $\log_b x$ is positive for $x > 1$ and negative for $0 < x < 1$. Figure 5 illustrates these facts for the base $b = e$. Keep in mind that the logarithm of a negative number does not exist. For example, $\log_{10}(-2)$ does not exist because $10^y = -2$ has no solution.

For each law of exponents (see Section 1.1), there is a corresponding law for logarithms. They are listed in the following table.

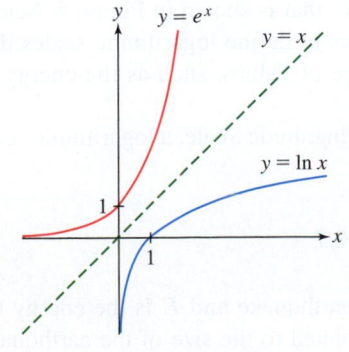

FIGURE 5 $y = \ln x$ is the inverse of $y = e^x$.

Laws of Logarithms

	Law	Example
Log of 1	$\log_b(1) = 0$	
Log of b	$\log_b(b) = 1$	
Products	$\log_b(xy) = \log_b x + \log_b y$	$\log_5(2 \cdot 3) = \log_5 2 + \log_5 3$
Quotients	$\log_b\left(\dfrac{x}{y}\right) = \log_b x - \log_b y$	$\log_2\left(\dfrac{3}{7}\right) = \log_2 3 - \log_2 7$
Reciprocals	$\log_b\left(\dfrac{1}{x}\right) = -\log_b x$	$\log_2\left(\dfrac{1}{7}\right) = -\log_2 7$
Powers (any n)	$\log_b(x^n) = n\log_b x$	$\log_{10}(8^2) = 2 \cdot \log_{10} 8$

EXAMPLE 3   Using the Logarithm Laws   Evaluate:

(a) $\log_6 9 + \log_6 4$ **(b)** $\ln\left(\dfrac{1}{\sqrt{e}}\right)$ **(c)** $10\log_b(b^3) - 4\log_b(\sqrt{b})$

Solution

(a) $\log_6 9 + \log_6 4 = \log_6(9 \cdot 4) = \log_6(36) = \log_6(6^2) = 2$

(b) $\ln\left(\dfrac{1}{\sqrt{e}}\right) = \ln(e^{-1/2}) = -\dfrac{1}{2}\ln(e) = -\dfrac{1}{2}$

(c) $10\log_b(b^3) - 4\log_b(\sqrt{b}) = 10(3) - 4\log_b(b^{1/2}) = 30 - 4\left(\dfrac{1}{2}\right) = 28$ ■

We can change the base in exponential functions and in logarithmic functions using the following **change-of-base** formulas that can be proven via the laws of exponents and the laws of logarithms (see Exercises 49 and 50):

$$b^x = a^{x\log_a b}, \qquad b^x = e^{x\ln b}, \qquad \log_b x = \frac{\log_a x}{\log_a b}, \qquad \log_b x = \frac{\ln x}{\ln b} \qquad \boxed{1}$$

CONCEPTUAL INSIGHT   By these change-of-base formulas, we can easily convert between bases in exponentials and logarithms. Therefore, from the perspective of algebra, no base is preferred over any other. Once we start working with calculus, we will see that base e provides for simpler calculations of important calculus operations, and therefore in the calculus setting, $f(x) = e^x$ and $g(x) = \ln x$ are preferred over exponentials and logarithms with other bases.

Since logarithmic functions are inverses of exponential functions (which grow rapidly) logarithmic functions grow slowly, a property that is shown in Figure 5. Scientists exploit this slow growth of logarithmic functions to define logarithmic scales that measure phenomena that can have a very large range of values, such as the energy in earthquakes.

One scale for earthquakes is called the Moment Magnitude Scale, a logarithmic scale defined by

The Moment Magnitude Scale was developed to improve on some shortcomings in the more familiar Richter Scale. It was defined to be relatively consistent with Richter-scale values for medium-to-strong earthquakes and is the scale currently used by the U.S. Geological Survey when reporting earthquake magnitudes.

$$M_w = \frac{2}{3}\log_{10} E - 10.7$$

where M_w is the unitless moment magnitude of an earthquake and E is the energy (in ergs) released by the earthquake, which is directly related to the size of the earthquake fault and the distance the fault moved.

EXAMPLE 4 **Earthquake Measurement** The 2011 Tohoku earthquake in Japan, one of the strongest ever recorded, released 3.9×10^{29} ergs of energy, about 10 billion times as much energy as a mild earthquake that rattles the windows in a house. What are the magnitudes of the mild (10^{21} ergs of energy) and Tohoku earthquakes?

Solution If $E = 10^{21}$, then

$$M_w = \frac{2}{3}\log_{10} 10^{21} - 10.7 = \frac{2}{3}(21) - 10.7 = 3.3$$

For the Tohoku earthquake, $E = 3.9 \times 10^{29}$, so

$$M_w = \frac{2}{3}\log_{10} 3.9 \times 10^{29} - 10.7 = \frac{2}{3}(\log_{10} 3.9 + 29) - 10.7 \approx 9.0 \quad\blacksquare$$

EXAMPLE 5 **Solving an Exponential Equation** The bacteria population in a bottle at time t (in hours) has size $P(t) = 1000e^{0.35t}$. After how many hours will there be 5000 bacteria?

Solution We must solve $P(t) = 1000e^{0.35t} = 5000$ for t (Figure 6):

$$e^{0.35t} = \frac{5000}{1000} = 5$$

$$\ln(e^{0.35t}) = \ln 5 \qquad \text{(take logarithm of both sides)}$$

$$0.35t = \ln 5 \approx 1.609 \qquad [\text{because } \ln(e^a) = a]$$

$$t \approx \frac{1.609}{0.35} \approx 4.6 \text{ hours} \quad\blacksquare$$

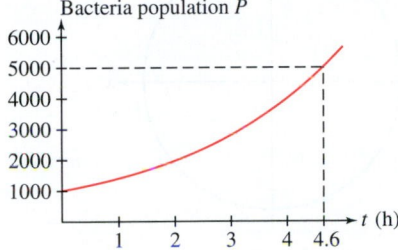

Bacteria population P

DF **FIGURE 6** Bacteria population as a function of time.

Hyperbolic Functions

The hyperbolic functions are certain special combinations of e^x and e^{-x} that play a role in engineering and physics (see Figure 7 for a real-life example). The hyperbolic sine and cosine, pronounced "cinch" and "cosh," are shown in Figure 8 and defined as follows:

$$\sinh x = \frac{e^x - e^{-x}}{2}, \qquad \cosh x = \frac{e^x + e^{-x}}{2}$$

© Corbis

FIGURE 7 The St. Louis Arch has the shape of an inverted hyperbolic cosine.

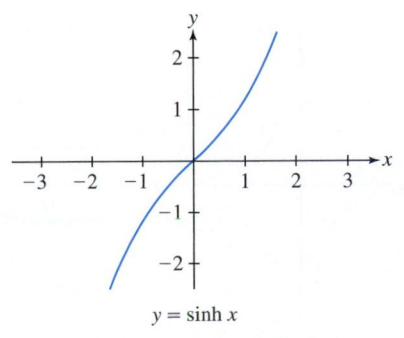

$y = \sinh x$

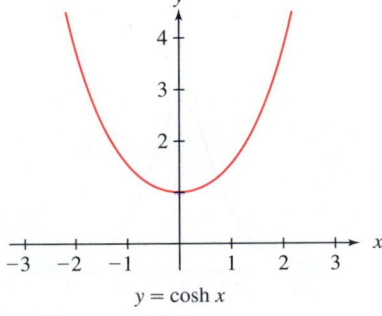

$y = \cosh x$

FIGURE 8 The hyperbolic sine and cosine functions.

As the terminology suggests, there are similarities between the hyperbolic and trigonometric functions. Here are some examples:

- **Parity:** The trigonometric functions and their hyperbolic analogs have the same parity. Thus, $f(x) = \sin x$ and $f(x) = \sinh x$ are both odd, and $f(x) = \cos x$ and $f(x) = \cosh x$ are both even (Figure 8):

$$\sinh(-x) = -\sinh x, \qquad \cosh(-x) = \cosh x$$

- **Identities:** The basic trigonometric identity $\sin^2 x + \cos^2 x = 1$ has a hyperbolic analog:

$$\cosh^2 x - \sinh^2 x = 1 \qquad \boxed{2}$$

The addition formulas satisfied by $\sin x$ and $\cos x$ also have hyperbolic analogs:

$$\sinh(x + y) = \sinh x \cosh y + \cosh x \sinh y$$
$$\cosh(x + y) = \cosh x \cosh y + \sinh x \sinh y$$

- **Hyperbola instead of the circle:** Because of the identity $\cosh^2 t - \sinh^2 t = 1$, the point $(\cosh t, \sinh t)$ lies on the hyperbola $x^2 - y^2 = 1$, just as $(\cos t, \sin t)$ lies on the unit circle $x^2 + y^2 = 1$ (Figure 9).

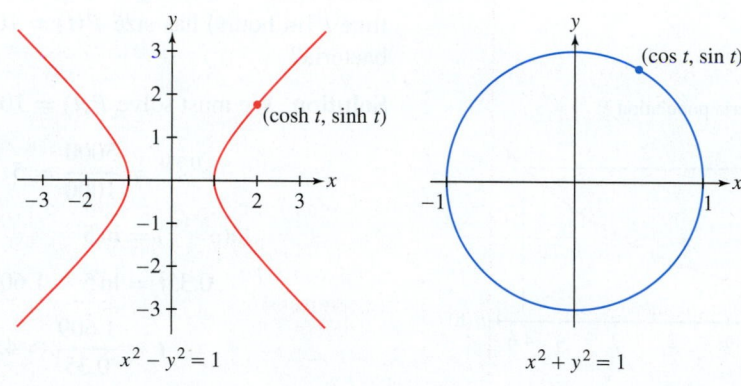

$$x^2 - y^2 = 1 \qquad\qquad x^2 + y^2 = 1$$

FIGURE 9

- **Other hyperbolic functions:** The hyperbolic tangent, cotangent, secant, and cosecant functions (see Figure 10) are defined like their trigonometric counterparts:

$$\tanh x = \frac{\sinh x}{\cosh x} = \frac{e^x - e^{-x}}{e^x + e^{-x}}, \qquad \operatorname{sech} x = \frac{1}{\cosh x} = \frac{2}{e^x + e^{-x}}$$

$$\coth x = \frac{\cosh x}{\sinh x} = \frac{e^x + e^{-x}}{e^x - e^{-x}}, \qquad \operatorname{csch} x = \frac{1}{\sinh x} = \frac{2}{e^x - e^{-x}}$$

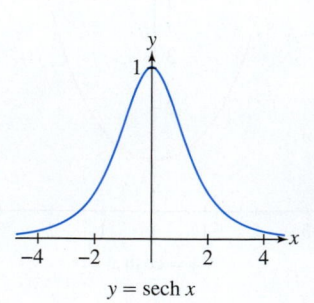

$y = \operatorname{sech} x$

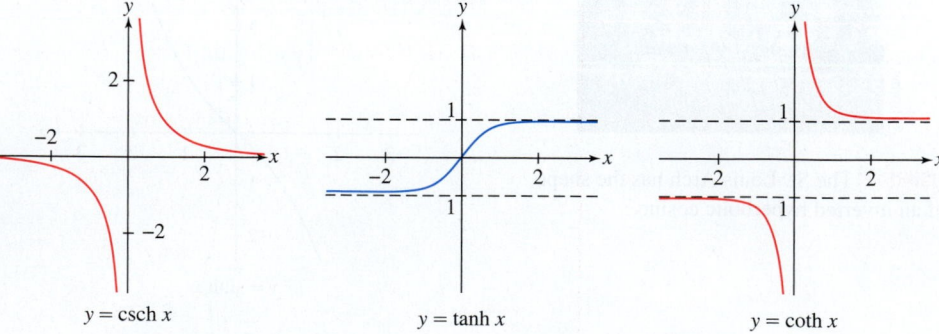

$y = \operatorname{csch} x$ $y = \tanh x$ $y = \coth x$

FIGURE 10

EXAMPLE 6 **Verifying the Basic Identity** Verify Eq. (2): $\cosh^2 x - \sinh^2 x = 1$.

Solution Because $\cosh x = \frac{1}{2}(e^x + e^{-x})$ and $\sinh x = \frac{1}{2}(e^x - e^{-x})$, we have

$$\cosh x + \sinh x = e^x, \qquad \cosh x - \sinh x = e^{-x}$$

We obtain Eq. (2) by multiplying these two equations together:

$$\cosh^2 x - \sinh^2 x = (\cosh x + \sinh x)(\cosh x - \sinh x) = e^x \cdot e^{-x} = 1 \qquad \blacksquare$$

Inverse hyperbolic functions

Function	Domain
$y = \sinh^{-1} x$	all x
$y = \cosh^{-1} x$	$x \geq 1$
$y = \tanh^{-1} x$	$|x| < 1$
$y = \coth^{-1} x$	$|x| > 1$
$y = \text{sech}^{-1} x$	$0 < x \leq 1$
$y = \text{csch}^{-1} x$	$x \neq 0$

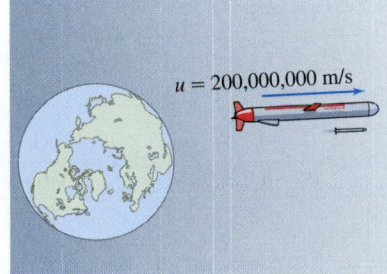

$u = 200{,}000{,}000$ m/s

FIGURE 11 What is the missile's velocity relative to the earth?

Inverse Hyperbolic Functions

Each of the hyperbolic functions, except $y = \cosh x$ and $y = \text{sech} \, x$, is one-to-one on its domain and therefore has a well-defined inverse. The functions $y = \cosh x$ and $y = \text{sech} \, x$ are one-to-one on the restricted domain $\{x : x \geq 0\}$. We let $y = \cosh^{-1} x$ and $y = \text{sech}^{-1} x$ denote the corresponding inverses.

Einstein's Law of Velocity Addition

The inverse hyperbolic tangent plays a role in the Special Theory of Relativity, developed by Albert Einstein in 1905. One consequence of this theory is that no object can travel faster than the speed of light, $c \approx 3 \times 10^8$ m/s. Einstein realized that this contradicts a law stated by Galileo more than 250 years earlier, namely that *velocities add*. Imagine a train traveling at $u = 50$ m/s and a man walking down the aisle in the train at $v = 2$ m/s. According to Galileo, the man's velocity relative to the ground is $u + v = 52$ m/s. This agrees with our everyday experience. But now imagine an (unrealistic) rocket traveling away from the earth at $u = 2 \times 10^8$ m/s, and suppose that the rocket fires a missile with velocity $v = 1.5 \times 10^8$ m/s (relative to the rocket). If Galileo's Law were correct, the velocity of the missile relative to the earth would be $u + v = 3.5 \times 10^8$ m/s, which exceeds Einstein's maximum speed limit of $c \approx 3 \times 10^8$ m/s.

However, Einstein's theory replaces Galileo's Law with a new law stating that the *inverse hyperbolic tangents of velocities add*. More precisely, if u is the rocket's velocity relative to the earth and v is the missile's velocity relative to the rocket, then the velocity of the missile relative to the earth (Figure 11) is w, where

$$\tanh^{-1}\left(\frac{w}{c}\right) = \tanh^{-1}\left(\frac{u}{c}\right) + \tanh^{-1}\left(\frac{v}{c}\right) \qquad \boxed{3}$$

EXAMPLE 7 A rocket travels away from the earth at a velocity of 2×10^8 m/s. A missile is fired from the rocket at a velocity of 1.5×10^8 m/s (relative to the rocket) away from the earth. Use Einstein's Law to find the velocity w of the missile relative to the earth.

Solution According to Eq. (3),

$$\tanh^{-1}\left(\frac{w}{c}\right) = \tanh^{-1}\left(\frac{2 \times 10^8}{3 \times 10^8}\right) + \tanh^{-1}\left(\frac{1.5 \times 10^8}{3 \times 10^8}\right) \approx 0.805 + 0.549 \approx 1.354$$

Therefore, $w/c \approx \tanh(1.354) \approx 0.875$, and $w \approx 0.875c \approx 2.6 \times 10^8$ m/s. This value obeys the Einstein speed limit of 3×10^8 m/s. ∎

Einstein's Law of Velocity Addition [Eq. (3)] reduces to Galileo's Law, $w = u + v$, when u and v are small relative to the velocity of light c. See Exercise 52 for another way of expressing Eq. (3).

EXAMPLE 8 **Low Velocities** A plane traveling at 300 m/s fires a missile forward at a velocity of 200 m/s. Calculate the missile's velocity w relative to the earth using both Einstein's Law and Galileo's Law.

Solution According to Einstein's Law,

$$\tanh^{-1}\left(\frac{w}{c}\right) = \tanh^{-1}\left(\frac{300}{c}\right) + \tanh^{-1}\left(\frac{200}{c}\right)$$

$$w = c \cdot \tanh\left(\tanh^{-1}\left(\frac{300}{c}\right) + \tanh^{-1}\left(\frac{200}{c}\right)\right) \approx 499.99999999967 \text{ m/s}$$

This is practically indistinguishable from the value $w = 300 + 200 = 500$ m/s obtained using Galileo's Law. ∎

1.6 SUMMARY

- $f(x) = b^x$ is the *exponential function* with base b (where $b > 0$ and $b \neq 1$).
- $f(x) = b^x$ is increasing if $b > 1$ and decreasing if $b < 1$.
- The number $e \approx 2.718$.

- For $b > 0$ with $b \neq 1$, the *logarithmic function* $f(x) = \log_b x$ is the inverse of $f(x) = b^x$:

$$y = \log_b x \quad \Leftrightarrow \quad x = b^y$$

- The *natural logarithm* is the logarithm with base e and is denoted $\ln x$.
- $e^{\ln x} = x$ for $x > 0$ and $\ln(e^x) = x$ for all x.
- Important logarithm laws:

(i) $\log_b(xy) = \log_b x + \log_b y$ (ii) $\log_b\left(\dfrac{x}{y}\right) = \log_b x - \log_b y$

(iii) $\log_b(x^n) = n \log_b x$ (iv) $\log_b 1 = 0$ and $\log_b b = 1$

- Change of base formulas: $b^x = a^{x \log_a b}$, $b^x = e^{x \ln b}$, $\log_b x = \dfrac{\log_a x}{\log_a b}$, $\log_b x = \dfrac{\ln x}{\ln b}$
- The *hyperbolic sine and cosine*:

$$\sinh x = \frac{e^x - e^{-x}}{2} \quad \text{(odd function)}, \qquad \cosh x = \frac{e^x + e^{-x}}{2} \quad \text{(even function)}$$

The remaining hyperbolic functions:

$$\tanh x = \frac{\sinh x}{\cosh x}, \qquad \coth x = \frac{\cosh x}{\sinh x}, \qquad \operatorname{sech} x = \frac{1}{\cosh x}, \qquad \operatorname{csch} x = \frac{1}{\sinh x}$$

- Basic identity: $\cosh^2 x - \sinh^2 x = 1$
- The inverse hyperbolic functions and their domains:

$$f(x) = \sinh^{-1} x, \text{ for all } x \qquad f(x) = \coth^{-1} x, \text{ for } |x| > 1$$

$$f(x) = \cosh^{-1} x, \text{ for } x \geq 1 \qquad f(x) = \operatorname{sech}^{-1} x, \text{ for } 0 < x \leq 1$$

$$f(x) = \tanh^{-1} x, \text{ for } |x| < 1 \qquad f(x) = \operatorname{csch}^{-1} x, \text{ for } x \neq 0$$

1.6 EXERCISES

Preliminary Questions

1. When is $\ln x$ negative?

2. What is $\ln(-3)$? Explain.

3. Explain the phrase "The logarithm converts multiplication into addition."

4. What are the domain and range of $f(x) = \ln x$?

5. Explain why $f(x) = x^3$ does not grow exponentially.

6. Compute $\log_{b^2}(b^4)$.

7. Which hyperbolic functions take on only positive values?

8. Which hyperbolic functions are increasing on their domains?

9. Describe three properties of hyperbolic functions that have trigonometric analogs.

Exercises

In Exercises 1–8, solve for the unknown variable.

1. $e^{2x} = e^{x+1}$

2. $e^{t^2} = e^{4t-3}$

3. $3^x = \left(\frac{1}{3}\right)^{x+1}$

4. $(\sqrt{5})^x = 125$

5. $4^{-x} = 2^{x+1}$

6. $b^4 = 10^{12}$

7. $k^{3/2} = 27$

8. $(b^2)^{x+1} = b^{-6}$

In Exercises 9–24, calculate without using a calculator.

9. $\log_3 27$

10. $\log_5 \frac{1}{25}$

11. $\ln 1$

12. $\log_5(5^4)$

13. $\log_2(2^{5/3})$

14. $\log_2(8^{5/3})$

15. $\log_{64} 4$

16. $\log_7(49^2)$

17. $\log_8 2 + \log_4 2$

18. $\log_{25} 30 + \log_{25} \frac{5}{6}$

19. $\log_4 48 - \log_4 12$

20. $\ln(\sqrt{e} \cdot e^{7/5})$

21. $\ln(e^3) + \ln(e^4)$

22. $\log_2 \frac{4}{3} + \log_2 24$

23. $7^{\log_7(29)}$

24. $8^{3\log_8(2)}$

25. Write as the natural log of a single expression:
(a) $2 \ln 5 + 3 \ln 4$
(b) $5 \ln(x^{1/2}) + \ln(9x)$

26. Solve for x: $\ln(x^2 + 1) - 3 \ln x = \ln(2)$.

In Exercises 27–32, solve for the unknown.

27. $7e^{5t} = 100$

28. $6e^{-4t} = 2$

29. $2^{x^2 - 2x} = 8$

30. $e^{2t+1} = 9e^{1-t}$

31. $\ln(x^4) - \ln(x^2) = 2$

32. $\log_3 y + 3\log_3(y^2) = 14$

33. Find the inverse of $y = e^{2x-3}$.

34. Find the inverse of $y = \ln(x^2 - 2)$ for $x > \sqrt{2}$.

35. Use a calculator to compute $\sinh x$ and $\cosh x$ for $x = -3, 0, 5$.

36. Compute $\sinh(\ln 5)$ and $\tanh(3 \ln 5)$ without using a calculator.

37. Show, by producing a counterexample, that $\ln(ab)$ is not equal to $(\ln a)(\ln b)$.

38. For which values of x are $y = \sinh x$ and $y = \cosh x$ increasing and decreasing?

39. Show that $y = \tanh x$ is an odd function.

40. The population of Integraton (in millions) at time t (years) is $P(t) = 2.4e^{0.06t}$, where $t = 0$ is the year 2000.
(a) What is the population at time $t = 0$?
(b) When will the population double from its size at $t = 0$?

41. The decibel level for the intensity of a sound is a logarithmic scale defined by $D = 10 \log_{10} I + 120$, where I is the intensity of the sound in watts per square meter.
(a) Express I as a function of D.
(b) Show that when D increases by 20, the intensity increases by a factor of 100.

42. Consider the equation $M_w = \frac{2}{3} \log_{10} E - 10.7$ relating the moment magnitude of an earthquake and the energy E (in ergs) released by it.

(a) Express E as a function of M_w.

(b) Show that when M_w increases by 1, the energy increases by a factor of approximately 31.6.

43. [✎] Refer to the graphs to explain why the equation $\sinh x = t$ has a unique solution for every t and why $\cosh x = t$ has two solutions for every $t > 1$.

44. Compute $\cosh x$ and $\tanh x$, assuming that $\sinh x = 0.8$.

45. Prove the addition formula for $\cosh x$ given by

$$\cosh(x + y) = \cosh x \cosh y + \sinh x \sinh y.$$

46. Use the addition formulas to prove

$$\sinh(2x) = 2 \cosh x \sinh x$$

$$\cosh(2x) = \cosh^2 x + \sinh^2 x$$

47. A train moves along a track at velocity v. Bionica walks down the aisle of the train with velocity u in the direction of the train's motion. Compute the velocity w of Bionica relative to the ground using the laws of both Galileo and Einstein in the following cases:

(a) $v = 500$ m/s and $u = 10$ m/s. Is your calculator accurate enough to detect the difference between the two laws?

(b) $v = 10^7$ m/s and $u = 10^6$ m/s.

Further Insights and Challenges

48. Show that $\log_a b \log_b a = 1$ for all $a, b > 0$ such that $a \neq 1$ and $b \neq 1$.

49. Verify that for all x, the change-of-base formula holds:
$$\log_b x = \frac{\log_a x}{\log_a b} \text{ for } a, b > 0 \text{ such that } a \neq 1, b \neq 1.$$

50. Verify that for all x, the change-of-base formula holds: $b^x = a^{x \log_a b}$ for $a, b > 0$ such that $a \neq 1, b \neq 1$.

51. Use the addition formulas for $\sinh x$ and $\cosh x$ to prove

$$\tanh(u + v) = \frac{\tanh u + \tanh v}{1 + \tanh u \tanh v}$$

52. Use Exercise 51 to show that Einstein's Law of Velocity Addition [Eq. (3)] is equivalent to

$$w = \frac{u + v}{1 + \dfrac{uv}{c^2}}$$

53. Prove that every function f can be written as a sum in the form $f(x) = f_+(x) + f_-(x)$ where $f_+(x)$ is an even function and $f_-(x)$ is an odd function. Express $f(x) = 5e^x + 8e^{-x}$ in terms of $\cosh x$ and $\sinh x$.
Hint: $y = f(x) + f(-x)$ is an even function, and $y = f(x) - f(-x)$ is an odd function.

1.7 Technology: Calculators and Computers

Computer technology has vastly extended our ability to calculate and visualize mathematical relationships. In applied settings, computers are indispensable for solving complex systems of equations and analyzing data, as in weather prediction and medical imaging. Mathematicians use computers to study complex structures and relationships such as the geometry and symmetry of cubes of dimensions higher than 3 (Figure 1). We take advantage of this technology to explore the ideas of calculus visually and numerically.

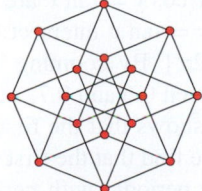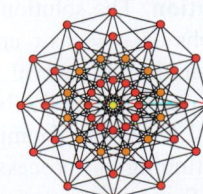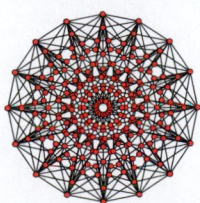

FIGURE 1 Representations of cubes of dimension 4, 6, 8, and 10.

When we plot a function with a graphing calculator or computer algebra system, the graph is contained within a **viewing rectangle**, the region determined by the range of x- and y-values in the plot. We write $[a, b] \times [c, d]$ to denote the rectangle for which $a \leq x \leq b$ and $c \leq y \leq d$.

The appearance of the graph depends heavily on the choice of viewing rectangle. Different choices may convey very different impressions that are sometimes misleading. Compare the three viewing rectangles for the graph of $f(x) = 12 - x - x^2$ in Figure 2. Only (A) successfully displays the shape of the graph as a parabola. In (B), the graph is cut off, and no graph at all appears in (C). Keep in mind that the scales along the axes may change with the viewing rectangle. For example, the unit increment along the y-axis is larger in (B) than in (A), so the graph in (B) is steeper.

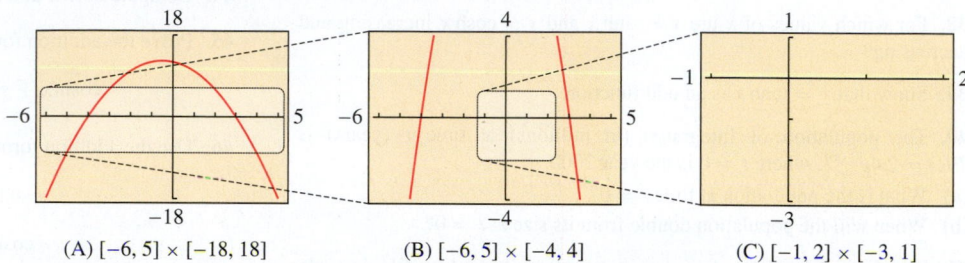

FIGURE 2 Viewing rectangles for the graph of $f(x) = 12 - x - x^2$.

(A) $[-6, 5] \times [-18, 18]$ (B) $[-6, 5] \times [-4, 4]$ (C) $[-1, 2] \times [-3, 1]$

There is no single "correct" viewing rectangle. The goal is to select the viewing rectangle that displays the properties you wish to investigate. This usually requires experimentation.

Technology is indispensable but also has its limitations. When shown the computer-generated results of a complex calculation, the Nobel prize–winning physicist Eugene Wigner (1902–1995) is reported to have said, "It is nice to know that the computer understands the problem, but I would like to understand it too."

EXAMPLE 1 How Many Roots and Where? How many real roots does the function $f(x) = x^9 - 20x + 1$ have? Find their approximate locations.

Solution We experiment with several viewing rectangles (Figure 3). Our first attempt (A) displays a cut-off graph, so we try a viewing rectangle that includes a larger range of y-values. Plot (B) shows that the roots of f probably lie somewhere in the interval $[-3, 3]$, but it does not reveal how many real roots there are. Therefore, we try the viewing rectangle in (C). Now we can see clearly that f has three roots. A further zoom in (D) shows that these roots are located near $-1.5, 0.1$, and 1.5. Further zooming would provide their locations with greater accuracy. ∎

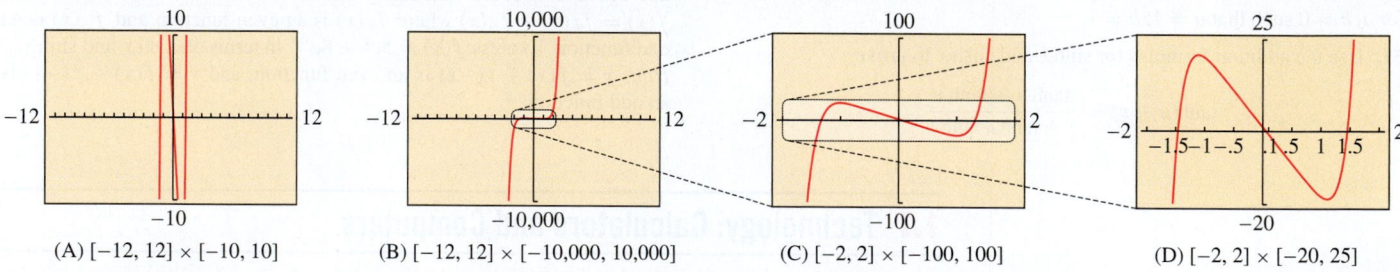

(A) $[-12, 12] \times [-10, 10]$ (B) $[-12, 12] \times [-10{,}000, 10{,}000]$ (C) $[-2, 2] \times [-100, 100]$ (D) $[-2, 2] \times [-20, 25]$

FIGURE 3 Graphs of $f(x) = x^9 - 20x + 1$.

EXAMPLE 2 Does a Solution Exist? Does $\cos x = \tan x$ have a solution? Describe the set of all solutions.

Solution The solutions of $\cos x = \tan x$ are the x-coordinates of the points where the graphs of $y = \cos x$ and $y = \tan x$ intersect. Figure 4(A) shows that there are two solutions in the interval $[0, 2\pi]$. By zooming in on the graph as in (B), we see that the first positive root lies between 0.6 and 0.7, and the second positive root lies between 2.4 and 2.5. Further zooming shows that the first root is approximately 0.67 [Figure 4(C)]. Continuing this process, we find that the first two roots are $x \approx 0.666$ and $x \approx 2.475$.

Since $f(x) = \cos x$ is periodic with period 2π, and $f(x) = \tan x$ is periodic with period π, the picture repeats itself with period 2π. All solutions are obtained by adding multiples of 2π to the two solutions in $[0, 2\pi]$:

$$x \approx 0.666 + 2\pi k \qquad \text{and} \qquad x \approx 2.475 + 2\pi k \quad \text{(for any integer } k\text{)} \qquad \blacksquare$$

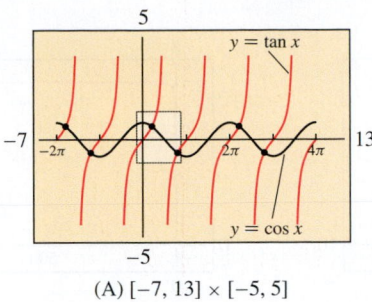

(A) [−7, 13] × [−5, 5]

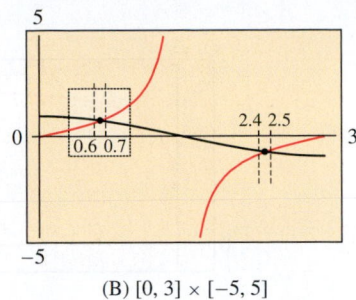

(B) [0, 3] × [−5, 5]

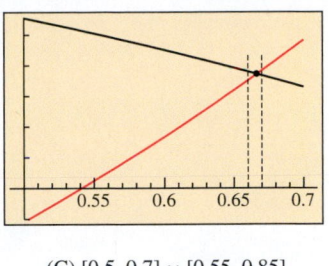

(C) [0.5, 0.7] × [0.55, 0.85]

FIGURE 4 Graphs of $y = \cos x$ and $y = \tan x$.

CAUTION When considering the graph of a function such as $y = \ln x$ (Figure 5), it may appear to approach a horizontal asymptote, but in fact, it does not. For any given horizontal line, the graph eventually rises above it. In this respect, graphing calculators and computer graphing systems must be used judiciously.

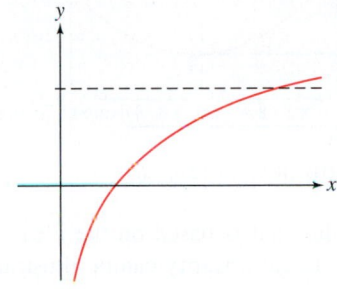

FIGURE 5 $y = \ln x$.

EXAMPLE 3 **Functions with Asymptotes** Plot the function $f(x) = \dfrac{1 - 3x}{x - 2}$ and describe its asymptotic behavior.

Solution First, we plot f in the viewing rectangle $[-10, 20] \times [-5, 5]$, as in Figure 6(A). The vertical line $x = 2$ is called a **vertical asymptote**. Many graphing calculators display this line, but it is *not* part of the graph (and it can usually be eliminated by choosing a smaller range of y-values). We see that $f(x)$ tends to ∞ as x approaches 2 from the left, and to $-\infty$ as x approaches 2 from the right. To display the horizontal asymptotic behavior of f, we use the viewing rectangle $[-10, 20] \times [-10, 5]$ [Figure 6(B)]. Here, we see that the graph approaches the horizontal line $y = -3$, called a **horizontal asymptote** (which we have added as a dashed horizontal line in the figure). ∎

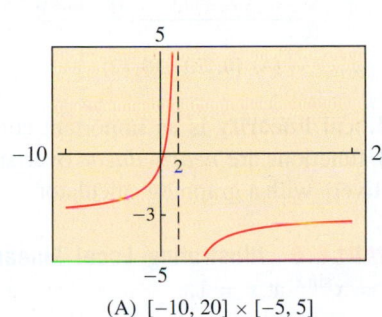

(A) [−10, 20] × [−5, 5]

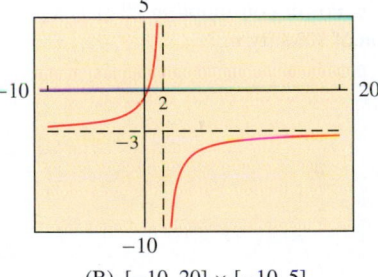

(B) [−10, 20] × [−10, 5]

FIGURE 6 Graphs of $f(x) = \dfrac{1 - 3x}{x - 2}$.

Calculators and computer algebra systems give us the freedom to experiment numerically. For instance, we can explore the behavior of a function by constructing a table of values. In the next example, we investigate a function related to the exponential function.

EXAMPLE 4 **Investigating the Behavior of a Function** How does $f(n) = (1 + 1/n)^n$ behave for large positive whole-number values of n?

Solution First, we make a table of values of $f(n)$ for larger and larger positive values of n. Table 1 suggests that $f(n)$ appears to get closer to some value near 2.718. In fact, as we indicated in Section 1.6, the values of $f(n)$ approach e as n gets greater and greater. This is an example of limiting behavior that we will discuss in Chapter 2. Next, replace n by the variable x and plot the function $f(x) = (1 + 1/x)^x$. The graphs in Figure 7 also suggest that $f(x)$ approaches a limiting value near 2.7. ∎

TABLE 1

n	$\left(1 + \dfrac{1}{n}\right)^n$
10	2.59374
10^2	2.70481
10^3	2.71692
10^4	2.71815
10^5	2.71827
10^6	2.71828

EXAMPLE 5 **Bird Flight: Finding a Minimum Graphically** According to one model of bird flight, the power consumed by a pigeon flying at velocity v (in meters per second) is $P(v) = 17v^{-1} + 10^{-3}v^3$ (in joules per second). Use a graph of P to find the velocity that minimizes power consumption.

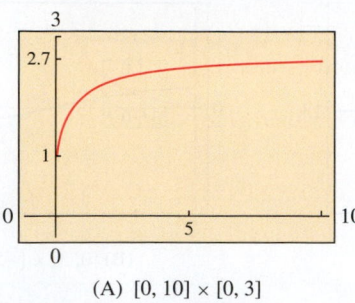

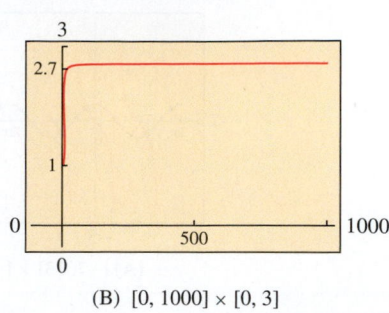

(A) $[0, 10] \times [0, 3]$

(B) $[0, 1000] \times [0, 3]$

FIGURE 7 Graphs of $f(x) = \left(1 + \dfrac{1}{x}\right)^x$.

Solution The velocity that minimizes power consumption corresponds to the lowest point on the graph of P. We plot P first in a large viewing rectangle (Figure 8). This figure reveals the general shape of the graph and shows that P takes on a minimum value for v somewhere between $v = 8$ and $v = 9$. In the viewing rectangle $[8, 9.2] \times [2.6, 2.65]$, we see that the minimum occurs at approximately $v = 8.65$ m/s. ∎

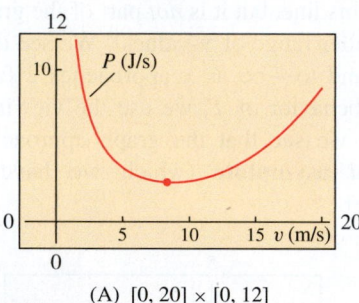

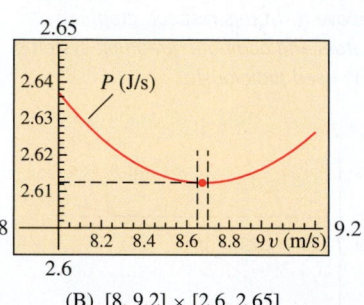

FIGURE 8 Power consumption $P(v)$ as a function of velocity v.

(A) $[0, 20] \times [0, 12]$

(B) $[8, 9.2] \times [2.6, 2.65]$

Local linearity is an important concept in calculus that is based on the idea that many functions are *nearly linear* over small intervals. Local linearity can be illustrated effectively with a graphing calculator.

EXAMPLE 6 **Illustrating Local Linearity** Illustrate local linearity for the function $f(x) = x^{\sin x}$ at $x = 1$.

Solution First, we plot $f(x) = x^{\sin x}$ in the viewing window of Figure 9(A). The graph moves up and down and appears very wavy. However, as we zoom in, the graph straightens out. Figures (B)–(D) show the result of zooming in on the point $(1, f(1))$. When viewed up close, the graph looks like a straight line. This illustrates the local linearity of f at $x = 1$. ∎

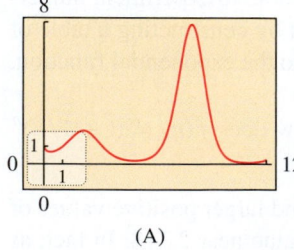

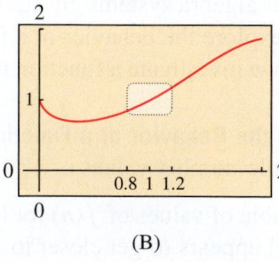

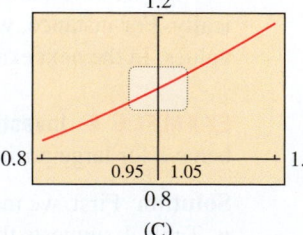

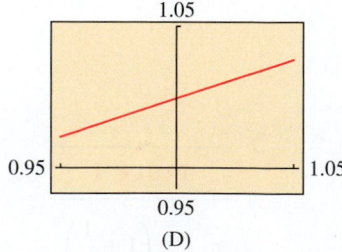

(A)

(B)

(C)

(D)

FIGURE 9 Zooming in on the graph of $f(x) = x^{\sin x}$ near $x = 1$.

1.7 SUMMARY

- The appearance of a graph on a graphing calculator depends on the choice of viewing rectangle. Experiment with different viewing rectangles until you find one that displays the information you want. Keep in mind that the scales along the axes may change as you vary the viewing rectangle.

• The following are some ways in which graphing calculators and computer algebra systems can be used in calculus:

  – Visualizing the behavior of a function
  – Finding solutions graphically or numerically
  – Conducting numerical or graphical experiments
  – Illustrating theoretical ideas (such as local linearity)

1.7 EXERCISES

Preliminary Questions

1. Is there a definite way of choosing the optimal viewing rectangle, or is it best to experiment until you find a viewing rectangle appropriate to the problem at hand?

2. Describe the calculator screen produced when the function $y = 3 + x^2$ is plotted with a viewing rectangle:
(a) $[-1, 1] \times [0, 2]$ (b) $[0, 1] \times [0, 4]$

3. According to the evidence in Example 4, it appears that $f(n) = (1 + 1/n)^n$ never takes on a value greater than 3 for $n > 0$. Does this evidence *prove* that $f(n) \leq 3$ for $n > 0$?

4. How can a graphing calculator be used to find the minimum value of a function?

Exercises

The exercises in this section should be done using a graphing calculator or computer algebra system.

1. Plot $f(x) = 2x^4 + 3x^3 - 14x^2 - 9x + 18$ in the appropriate viewing rectangles and determine its roots.

2. How many solutions does $x^3 - 4x + 8 = 0$ have?

3. How many *positive* solutions does $x^3 - 12x + 8 = 0$ have?

4. Does $\cos x + x = 0$ have a solution? A positive solution?

5. Find all the solutions of $\sin x = \sqrt{x}$ for $x > 0$.

6. How many solutions does $\cos x = x^2$ have?

7. Let $f(x) = (x - 100)^2 + 1000$. What will the display show if you graph f in the viewing rectangle $[-10, 10]$ by $[-10, 10]$? Find an appropriate viewing rectangle.

8. Plot $f(x) = \dfrac{8x + 1}{8x - 4}$ in an appropriate viewing rectangle. What are the vertical and horizontal asymptotes?

9. Plot the graph of $f(x) = x/(4 - x)$ in a viewing rectangle that clearly displays the vertical and horizontal asymptotes.

10. Illustrate local linearity for $f(x) = x^2$ by zooming in on the graph at $x = 0.5$ (see Example 6).

11. Plot $f(x) = \cos(x^2)\sin x$ for $0 \leq x \leq 2\pi$. Then illustrate local linearity at $x = 3.8$ by choosing appropriate viewing rectangles.

12. By zooming in on the graph of $f(x) = \sqrt{x}$ at $x = 0$ examine the local linearity. How does the resulting "line" appear?

13. By examining the graph of $f(x) = 2x^2 - x^3 - 3x^4$ in appropriate viewing rectangles, approximate the maximum value of $f(x)$ and the value of x at which it occurs.

14. (a) Plot the graph of $f(x) = \frac{2x^2 + x}{1 - x^2}$ in a viewing rectangle that clearly shows the two vertical asymptotes.

(b) By examining the graph of f in appropriate viewing rectangles, approximate the minimum value of $f(x)$ between the vertical asymptotes, and approximate the value of x at which the minimum occurs.

15. If \$500 is deposited in a bank account paying 3.5% interest compounded monthly, then the account has value $V(N) = 500\left(1 + \frac{0.035}{12}\right)^N$ dollars after N months. By examining the graph of $V(N)$ find, to the nearest integer N, the number of months it takes for the account value to reach \$800.

16. If \$1000 is deposited in a bank account paying 5% interest compounded monthly, then the account has value $V(N) = 1000\left(1 + \frac{0.05}{12}\right)^N$ dollars after N months. By examining the graph of $V(N)$ find, to the nearest integer N, the number of months it takes for the account value to double.

In Exercises 17–22, investigate the behavior of the function as n or x grows large by making a table of function values and plotting a graph (see Example 4). Describe the behavior in words.

17. $f(n) = n^{1/n}$

18. $f(n) = \dfrac{4n + 1}{6n - 5}$

19. $f(n) = \left(1 + \dfrac{1}{n}\right)^{n^2}$

20. $f(x) = \left(\dfrac{x + 6}{x - 4}\right)^x$

21. $f(x) = \left(x \tan \dfrac{1}{x}\right)^x$

22. $f(x) = \left(x \tan \dfrac{1}{x}\right)^{x^2}$

23. The graph of $f(\theta) = A\cos\theta + B\sin\theta$ is a sinusoidal wave for any constants A and B. Confirm this for $(A, B) = (1, 1)$, $(1, 2)$, and $(3, 4)$ by plotting f.

24. Find the maximum value of f for the graphs produced in Exercise 23. Can you guess the formula for the maximum value in terms of A and B?

25. Find the intervals on which $f(x) = x(x + 2)(x - 3)$ is positive by plotting a graph.

26. Find the set of solutions to the inequality $(x^2 - 4)(x^2 - 1) < 0$ by plotting a graph.

Further Insights and Challenges

27. [CAS] Let $f_1(x) = x$ and define a sequence of functions by $f_{n+1}(x) = \frac{1}{2}(f_n(x) + x/f_n(x))$. For example, $f_2(x) = \frac{1}{2}(x + 1)$. Use a computer algebra system to compute $f_n(x)$ for $n = 3$, 4, 5 and plot $y = f_n(x)$ together with $y = \sqrt{x}$ for $x \geq 0$. What do you notice?

28. Set $P_0(x) = 1$ and $P_1(x) = x$. The **Chebyshev polynomials** (useful in approximation theory) are defined inductively by the formula $P_{n+1}(x) = 2x P_n(x) - P_{n-1}(x)$.

(a) Show that $P_2(x) = 2x^2 - 1$.

(b) Compute $P_n(x)$ for $3 \leq n \leq 6$ using a computer algebra system or by hand, and plot $y = P_n(x)$ over $[-1, 1]$.

(c) Check that your plots confirm the following important properties of the Chebyshev polynomials. (A) $y = P_n(x)$ has n real roots in $[-1, 1]$, and (B) for $x \in [-1, 1]$, $P_n(x)$ lies between -1 and 1.

CHAPTER REVIEW EXERCISES

1. Match each quantity (a)–(d) with (i), (ii), or (iii) if possible, or state that no match exists.

(a) $2^a 3^b$

(b) $\dfrac{2^a}{3^b}$

(c) $(2^a)^b$

(d) $2^{a-b} 3^{b-a}$

(i) 2^{ab}

(ii) 6^{a+b}

(iii) $\left(\frac{2}{3}\right)^{a-b}$

2. Indicate which of the following are correct, and correct the ones that are not.

(a) $5^2 \cdot 5^{\frac{1}{2}} = 5$

(b) $(\sqrt{8})^{4/3} = 8^{2/3}$

(c) $\dfrac{3^8}{3^4} = 3^2$

(d) $(2^4)^{-2} = 2^2$

3. Express $(4, 10)$ as a set $\{x : |x - a| < c\}$ for suitable a and c.

4. Express as an interval:

(a) $\{x : |x - 5| < 4\}$

(b) $\{x : |5x + 3| \leq 2\}$

5. Express $\{x : 2 \leq |x - 1| \leq 6\}$ as a union of two intervals.

6. Give an example of numbers x, y such that $|x| + |y| = x - y$.

7. Describe the pairs of numbers x, y such that $|x + y| = x - y$.

8. Sketch the graph of $y = f(x + 2) - 1$, where $f(x) = x^2$ for $-2 \leq x \leq 2$.

In Exercises 9–12, let f be the function whose graph is shown in Figure 1.

9. Sketch the graphs of $y = f(x) + 2$ and $y = f(x + 2)$.

10. Sketch the graphs of $y = \frac{1}{2}f(x)$ and $y = f\left(\frac{1}{2}x\right)$.

11. Continue the graph of f to the interval $[-4, 4]$ as an even function.

12. Continue the graph of f to the interval $[-4, 4]$ as an odd function.

In Exercises 13–16, find the domain and range of the function.

13. $f(x) = \sqrt{x + 1}$

14. $f(x) = \dfrac{4}{x^4 + 1}$

15. $f(x) = \dfrac{2}{3 - x}$

16. $f(x) = \sqrt{x^2 - x + 5}$

17. Determine whether the function is increasing, decreasing, or neither:

(a) $f(x) = 3^{-x}$

(b) $f(x) = \dfrac{1}{x^2 + 1}$

(c) $g(t) = t^2 + t$

(d) $g(t) = t^3 + t$

18. Determine whether the function is even, odd, or neither:

(a) $f(x) = x^4 - 3x^2$

(b) $g(x) = \sin(x + 1)$

(c) $f(x) = 2^{-x^2}$

In Exercises 19–26, find the equation of the line.

19. Line passing through $(-1, 4)$ and $(2, 6)$

20. Line passing through $(-1, 4)$ and $(-1, 6)$

21. Line of slope 6 through $(9, 1)$

22. Line of slope $-\frac{3}{2}$ through $(4, -12)$

23. Line through $(2, 1)$ perpendicular to the line given by $y = 3x + 7$

24. Line through $(3, 4)$ perpendicular to the line given by $y = 4x - 2$

25. Line through $(2, 3)$ parallel to $y = 4 - x$

26. Horizontal line through $(-3, 5)$

27. Does the following table of market data suggest a linear relationship between price and number of homes sold during a one-year period? Explain.

Price (thousands of \$)	180	195	220	240
No. of homes sold	127	118	103	91

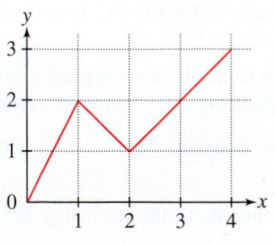

FIGURE 1

28. Does the following table of revenue data for a computer manufacturer suggest a linear relation between revenue and time? Explain.

Year	2005	2009	2011	2014
Revenue (billions of $)	13	18	15	11

29. Suppose that a cell phone plan that is offered at a price of P dollars per month attracts C customers, where $C(P)$ is a linear demand function for $\$100 \le P \le \$500$. Assume $C(100) = 1{,}000{,}000$ and $C(500) = 100{,}000$.

(a) Determine the demand function $C(P)$.

(b) What is the slope of the graph of $C(P)$? Describe what the slope represents.

(c) What is the decrease in the number of customers for each increase of $100 in the price?

30. Suppose that Internet domain names are sold at a price of $\$P$ per month for $\$2 \le P \le \$100$. The number of customers C who buy the domain names is a linear function of the price. Assume that 10,000 customers buy a domain name when the price is $2 per month, and 1000 customers buy when the price is $100 per month.

(a) Determine the demand function $C(P)$.

(b) What is the slope of the graph of $C(P)$? Describe what the slope represents.

31. Find the roots of $f(x) = x^4 - 4x^2$ and sketch its graph. On which intervals is f decreasing?

32. Let $h(z) = -2z^2 + 12z + 3$. Complete the square and find the maximum value of h.

33. Let $f(x)$ be the square of the distance from the point $(2, 1)$ to a point $(x, 3x + 2)$ on the line $y = 3x + 2$. Show that f is a quadratic function, and find its minimum value by completing the square.

34. Prove that $x^2 + 3x + 3 \ge 0$ for all x.

In Exercises 35–40, sketch the graph by hand.

35. $y = t^4$

36. $y = t^5$

37. $y = \sin \dfrac{\theta}{2}$

38. $y = 10^{-x}$

39. $y = x^{1/3}$

40. $y = \dfrac{1}{x^2}$

41. Show that the graph of $y = f\left(\frac{1}{3}x - b\right)$ is obtained by shifting the graph of $y = f\left(\frac{1}{3}x\right)$ to the right $3b$ units. Use this observation to sketch the graph of $y = \left|\frac{1}{3}x - 4\right|$.

42. Let $h(x) = \cos x$ and $g(x) = x^{-1}$. Compute the composite functions $h \circ g$ and $g \circ h$, and find their domains.

43. Find functions f and g such that the function

$$f(g(t)) = (12t + 9)^4$$

44. Sketch the points on the unit circle corresponding to the following three angles, and find the values of the six standard trigonometric functions at each angle:

(a) $\dfrac{2\pi}{3}$

(b) $\dfrac{7\pi}{4}$

(c) $\dfrac{7\pi}{6}$

45. What are the periods of these functions?

(a) $y = \sin 2\theta$

(b) $y = \sin \dfrac{\theta}{2}$

(c) $y = \sin 2\theta + \sin \dfrac{\theta}{2}$

46. Determine A, B, and C so that $f(x) = A\cos(Bx) + C$ cycles once from 8 to -2 and back to 8 as x goes from 0 to 2.

47. $H(t) = A\sin(Bt) + C$ models the height (in meters) of the tide in Happy Harbor at time t (hours since midnight) in a day. Determine A, B, and C if the high tide of 18 m occurs at 6:00 AM and the subsequent low tide of 15 m occurs at 6:00 PM.

48. Assume that $\sin \theta = \frac{4}{5}$, where $\pi/2 < \theta < \pi$. Find:

(a) $\tan \theta$

(b) $\sin 2\theta$

(c) $\csc \dfrac{\theta}{2}$

49. Give an example of values a, b such that

(a) $\cos(a + b) \ne \cos a + \cos b$

(b) $\cos \dfrac{a}{2} \ne \dfrac{\cos a}{2}$

50. Let $f(x) = \cos x$. Sketch the graph of $y = 2f\left(\frac{1}{3}x - \frac{\pi}{4}\right)$ for $0 \le x \le 6\pi$.

51. Solve $\sin 2x + \cos x = 0$ for $0 \le x < 2\pi$.

52. How does $h(n) = n^2/2^n$ behave for large whole-number values of n? Does $h(n)$ tend to infinity?

53. [GU] Use a graphing calculator to determine whether the equation $\cos x = 5x^2 - 8x^4$ has any solutions.

54. [GU] Using a graphing calculator, find the number of real roots and estimate the largest root to two decimal places:

(a) $f(x) = 1.8x^4 - x^5 - x$

(b) $g(x) = 1.7x^4 - x^5 - x$

55. Match each quantity (a)–(d) with (i), (ii), or (iii) if possible, or state that no match exists.

(a) $\ln\left(\dfrac{a}{b}\right)$

(b) $\dfrac{\ln a}{\ln b}$

(c) $e^{\ln a - \ln b}$

(d) $(\ln a)(\ln b)$

(i) $\ln a + \ln b$

(ii) $\ln a - \ln b$

(iii) $\dfrac{a}{b}$

56. Indicate which of the following are correct, and correct the ones that are not.

(a) $\ln e^x = x$

(b) $e^3 e^{1/3} = e$

(c) $\dfrac{\ln 6}{\ln 3} = \ln 2$

(d) $\ln 2 + \ln 4 = \ln 8$

57. The decibel level for the intensity of a sound is a logarithmic scale defined by $D = 10\log_{10} I + 120$, where I is the intensity of the sound in watts per square meter. If the intensity of one sound is 5000 times greater than the intensity of another, how much greater is the decibel level of the more intense sound?

58. Consider the equation $M_w = \frac{2}{3}\log_{10} E - 10.7$ relating the moment magnitude of an earthquake and the energy E (in ergs) released by it. If M_w increases by 2, by what factor does the energy increase?

59. Find the inverse of $f(x) = \sqrt{x^3 - 8}$ and determine its domain and range.

60. Find the inverse of $f(x) = \dfrac{x - 2}{x - 1}$ and determine its domain and range.

61. Find a domain on which $h(t) = (t - 3)^2$ is one-to-one and determine the inverse on this domain.

62. Show that $g(x) = \dfrac{x}{x - 1}$ is equal to its inverse on the domain $\{x : x \neq 1\}$.

63. Let

$$f(x) = \begin{cases} -x^2 & \text{when } x < 0 \\ x & \text{when } x \geq 0 \end{cases}$$

(a) Is f increasing?

(b) Does f have an inverse? If so, what is it?

64. Let

$$f(x) = \begin{cases} x - 1 & \text{when } x < 1 \\ \ln x & \text{when } x \geq 1 \end{cases}$$

(a) Is f increasing?

(b) Does f have an inverse? If so, what is it?

65. Suppose that g is the inverse of f. Match the functions (a)–(d) with their inverses (i)–(iv).

(a) $f(x) + 1$ **(b)** $f(x + 1)$ **(c)** $4f(x)$ **(d)** $f(4x)$

(i) $g(x)/4$ **(ii)** $g(x/4)$ **(iii)** $g(x - 1)$ **(iv)** $g(x) - 1$

66. **GU** Plot $f(x) = xe^{-x}$ and use the zoom feature to find two solutions of $f(x) = 0.3$.

2 LIMITS

Moustapha Temsah/EyeEm/Getty Images

Many observed phenomena can be understood as a long-term stable limit of interacting processes. For example, the moon once turned rapidly on its axis, but its rotation gradually slowed. Under the influence of the earth's gravity the moon's rotational period (the time for one complete turn about its axis) eventually became equal to its orbital period (the time for one complete revolution around the earth) so that now the same side of the moon always faces earth.

In motion along a line, velocity may be positive (forward motion), negative (backward motion), or zero. Speed, by definition, is the absolute value of velocity and is always nonnegative.

Calculus is usually divided into two branches, differential and integral, partly for historical reasons. The subject grew out of efforts in the seventeenth century to solve two important geometric problems: finding tangent lines to curves (differential calculus) and computing areas under curves (integral calculus). However, calculus is a broad subject with no clear boundaries. It includes other topics, such as the theory of infinite series, and it has an extraordinarily wide range of applications. What makes these methods and applications part of calculus is that they all rely on the concept of a limit. We will see throughout the text how limits allow us to make computations and solve problems that cannot be solved using algebra alone.

This chapter introduces the limit concept and sets the stage for our study of the derivative in Chapter 3. The first section, intended as motivation, discusses how limits arise in the study of instantaneous velocity and tangent lines.

2.1 The Limit Idea: Instantaneous Velocity and Tangent Lines

A limit describes how the values $f(x)$, of a function f, behave as x approaches a number a. Does $f(x)$ get closer and closer to a number L? If so, we say L is the limit of $f(x)$ as x approaches a. If not, we say the limit does not exist.

Limits play a key role throughout calculus. In this section, we motivate the need for the limit concept by examining two closely related questions that are central to differential calculus:

- How do we define and compute instantaneous velocity at a particular time?
- How do we define and compute the slope of the line tangent to a graph at a particular point?

When we speak of velocity, we usually mean *instantaneous* velocity, which indicates the speed and direction of an object at a particular moment. The idea of instantaneous velocity makes intuitive sense, but care is required to define it precisely. We do that via the concept of a limit.

Consider an object traveling along a line (linear motion). The **average velocity** over a given time interval has a straightforward definition as the ratio

$$\text{average velocity} = \frac{\text{change in position}}{\text{change in time}}$$

Thus, average velocity is the rate of change of position with respect to time over a particular interval. For example, if an automobile travels forward 200 km in 4 h, then its average velocity during this 4-h period is $\frac{200}{4} = 50$ km/h. At any given moment, the automobile may be going faster or slower than the average.

Like average velocity, instantaneous velocity is the rate of change of position with respect to time, but at a particular instant (thus the term "instantaneous") rather than over an interval. We cannot define instantaneous velocity as a ratio, as above, because we would have to divide by a change in time equal to zero. However, we can estimate instantaneous velocity by computing average velocity over a small time interval. And we can continually improve the estimate by using smaller and smaller time intervals. Ultimately, we can obtain an exact value of the instantaneous velocity as a number, which is called the limit, that the estimates approach more and more closely as the time interval shrinks to zero.

We investigate the relationship between average velocity and instantaneous velocity for the motion of an object falling to earth under the influence of gravity (assuming no air resistance).

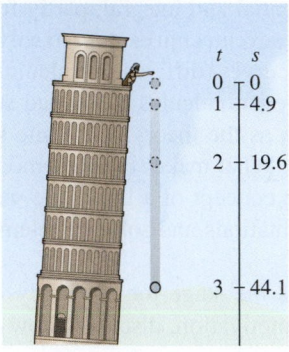

FIGURE 1 Distance traveled by a falling object dropped from rest after t seconds is approximately $s(t) = 4.9t^2$ meters.

Galileo discovered that if the object is released at time $t = 0$ from a state of rest (Figure 1), then the distance traveled (in meters) after t seconds is approximately given by the formula

$$s(t) = 4.9t^2$$

To compute average velocity over a time interval $[t_0, t_1]$, we set

$$\Delta s = s(t_1) - s(t_0) = \text{change in position}$$

$$\Delta t = \quad t_1 - t_0 \quad = \text{change in time (length of time interval)}$$

The change in position Δs is also called the **displacement**, or **net change** in position. For $t_1 \neq t_0$,

$$\text{average velocity over } [t_0, t_1] = \frac{\Delta s}{\Delta t} = \frac{s(t_1) - s(t_0)}{t_1 - t_0}$$

EXAMPLE 1 A ball is dropped at time $t = 0$. What is the average velocity over the time interval from $t_0 = 0$ to $t_1 = 0.8$ s? Estimate the instantaneous velocity at $t = 0.8$ s.

Solution We use Galileo's formula $s(t) = 4.9t^2$ to compute the average velocity over the interval $[t_0, t_1] = [0, 0.8]$:

$$\Delta s = s(0.8) - s(0) = 4.9(0.8)^2 - 4.9(0)^2 = 3.136 \text{ m}$$

$$\Delta t = 0.8 - 0 = 0.8 \text{ s}$$

The average velocity over $[0, 0.8]$ is the ratio

$$\frac{\Delta s}{\Delta t} = \frac{3.136}{0.8} = 3.92 \text{ m/s}$$

To estimate the instantaneous velocity at $t = 0.8$ s, we examine the average velocity over the five short time intervals listed in Table 1. Consider the first interval $[t_0, t_1] = [0.8, 0.81]$:

$$\Delta s = s(0.81) - s(0.8) = 4.9(0.81)^2 - 4.9(0.8)^2 \approx 3.2149 - 3.1360 = 0.07889 \text{ m}$$

$$\Delta t = 0.81 - 0.8 = 0.01 \text{ s}$$

The average velocity over $[0.8, 0.81]$ is the ratio

$$\frac{\Delta s}{\Delta t} = \frac{s(0.81) - s(0.8)}{0.81 - 0.8} = \frac{0.07889}{0.01} = 7.889 \text{ m/s}$$

TABLE 1

Time intervals (s)	Average velocity (m/s)
[0.8, 0.81]	7.889
[0.8, 0.805]	7.8645
[0.8, 0.8001]	7.8405
[0.8, 0.80005]	7.84024
[0.8, 0.800001]	7.840005

There is nothing special about the particular time intervals in Table 1. We are looking for a trend. In fact, we could have chosen any intervals $[0.8, t]$ for values of t approaching 0.8. We could also have chosen intervals $[t, 0.8]$ for $t < 0.8$.

Table 1 shows the results of similar calculations for intervals of successively shorter lengths. It looks like these average velocities are getting closer to 7.84 m/s as the length of the time interval shrinks:

$$7.889, \quad 7.8645, \quad 7.8405, \quad 7.84024, \quad 7.840005$$

This suggests that 7.84 m/s is a good estimate for the instantaneous velocity at $t = 0.8$. ■

Our estimate of the instantaneous velocity is a guess at what happens to the average velocities as we shrink the time interval to 0. Formally, the limit of the average velocities is the instantaneous velocity:

Instantaneous velocity is the limit of average velocity as the length of the time interval shrinks to zero. That is,

$$\text{instantaneous velocity} = \lim_{\Delta t \to 0} (\text{average velocity})$$

We read the right side of the equality as, "the limit as delta t goes to 0 of the average velocity."

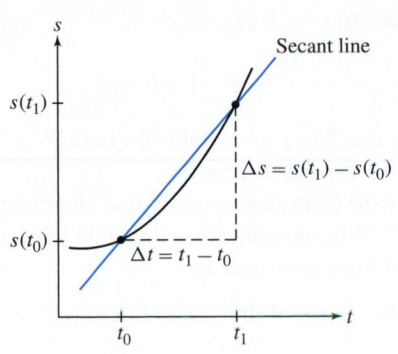

FIGURE 2 The average velocity over $[t_0, t_1]$ is equal to the slope of the secant line.

To employ the concept of the limit properly in calculus, we need to define the limit, establish its properties, and develop rules for computing limits. Those are the goals of the remaining sections in this chapter.

The notion that the limit of average velocities is the instantaneous velocity is vividly revealed using graphs. Notice, first, that the ratio defining the average velocity over $[t_0, t_1]$ is the slope of the line, called the **secant line**, through the points $(t_0, s(t_0))$ and $(t_1, s(t_1))$ on the graph of s (Figure 2).

$$\text{average velocity} = \text{slope of secant line} = \frac{\Delta s}{\Delta t} = \frac{s(t_1) - s(t_0)}{t_1 - t_0} \qquad \boxed{1}$$

By interpreting average velocity as a slope, we can visualize what happens as the time interval gets smaller. Figure 3 shows a closeup of the graph of position for the falling ball in Example 1. As the time interval shrinks, the secant lines get closer to—*and seem to rotate into*—a line that appears to be tangent to the graph at $t = 0.8$.

Time interval (s)	Average velocity (m/s)
[0.8, 0.805]	7.8645
[0.8, 0.8001]	7.8405
[0.8, 0.80005]	7.8402

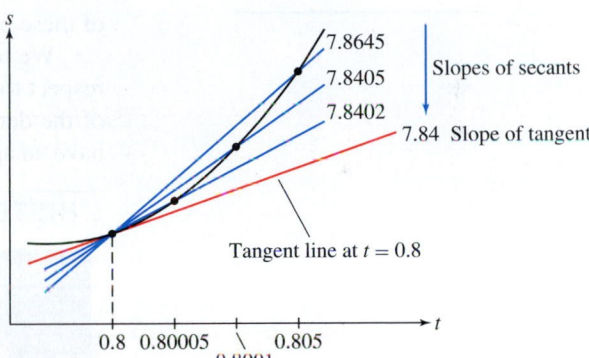

FIGURE 3 The secant lines "rotate into" the tangent line as the time interval shrinks. *Note:* The graph is not drawn to scale.

And since the secant lines approach the tangent line, the slopes of the secant lines get closer and closer to the slope of the tangent line. In other words,

> The slope of the tangent line is the limit of the slopes of the secant lines as Δt shrinks to zero. That is,
>
> $$\text{slope of tangent line} = \lim_{\Delta t \to 0} (\text{slopes of secant lines})$$

Now, putting together the relationships between instantaneous velocity, average velocity, slope of tangent, and slope of secant, it follows that

$$\text{slope of tangent line} = \text{instantaneous velocity}$$

The concepts of secant lines and tangent lines carry over to graphs of any function f, not just functions representing position changing in time.

EXAMPLE 2 The graph of $f(x) = x^3$ is shown in Figure 4.

(a) Compute the slope of the secant line from $x = 2$ to $x = 3$.

(b) Compute the slope of the secant line from $x = 2$ to $x = P$, and then investigate the slope of the tangent line at $x = 2$ by considering what happens to the slopes of the secant lines as we let the length of the interval from 2 to P shrink to 0.

Solution

(a) Using Eq. (1) for $f(x)$ [rather than $s(t)$], we have that the slope of the secant line from $x = 2$ to $x = 3$ is

$$\frac{f(3) - f(2)}{3 - 2} = \frac{27 - 8}{1} = 19$$

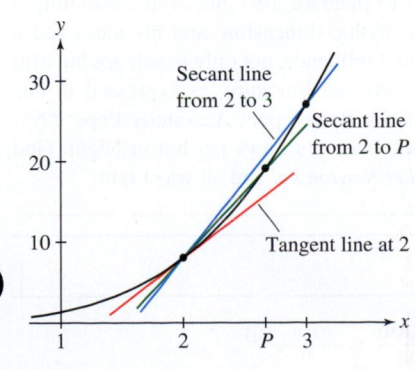

FIGURE 4 $f(x) = x^3$.

We are using the difference-of-cubes factoring formula:
$$a^3 - b^3 = (a - b)(a^2 + ab + b^2)$$

TABLE 2

P	$P^2 + 2P + 4$
1.9	11.41
1.99	11.9401
1.999	11.994001
1.9999	11.99940001
2.1	12.61
2.01	12.0601
2.001	12.006001
2.0001	12.00060001

(b) Similarly, the slope of the secant line from $x = 2$ to $x = P$ is

$$\frac{f(P) - f(2)}{P - 2} = \frac{P^3 - 8}{P - 2} = \frac{(P - 2)(P^2 + 2P + 4)}{P - 2} = P^2 + 2P + 4$$

This expression for the slope of the secant line from $x = 2$ to $x = P$ holds for both $P > 2$ and $P < 2$. In Table 2 we investigate these slopes for various P near 2.

The slope of the tangent line at $x = 2$ is obtained from these secant-line slopes by shrinking to 0 the length of the interval from 2 to P. It is apparent from the table that, in the limit as P approaches 2, the slopes of the secant lines approach 12. ∎

In the previous example we can also see that the slopes of the secant lines approach 12 by directly examining the expression for the secant slopes. That is, as P approaches 2, the expression $P^2 + 2P + 4$ approaches $2^2 + 2(2) + 4 = 12$. This reasoning for computing limits is justified by limit "laws" introduced in Section 2.3 and essentially amounts to substituting 2 for P in the expression for the secant slope. Sometimes taking a limit can be this simple. Other times a substitution might result in an undefined expression, in which case special care needs to be taken to understand what that means for the limit. All of these points are addressed in our investigation of limits in this chapter.

We return to the study of tangent lines, velocity (the rate of change of position with respect to time), and other rates of change in Chapter 3 when we begin our development of the derivative. For now, we need to build a good understanding of limits so that we have an appropriate foundation on which we can build the theory of calculus.

HISTORICAL PERSPECTIVE

NASA/JPL–Caltech/STScI

Philosophy is written in this grand book—I mean the universe—which stands continually open to our gaze, but it cannot be understood unless one first learns to comprehend the language... in which it is written. It is written in the language of mathematics...
—GALILEO GALILEI, 1623

The scientific revolution of the sixteenth and seventeenth centuries reached its high point in the work of Isaac Newton (1643–1727), who was the first scientist to show that the physical world, despite its complexity and diversity, is governed by a small number of universal laws. One of Newton's great insights was that the universal laws are dynamic, describing how the world changes over time in response to forces, rather than how the world actually is at any given moment in time. These laws are expressed best in the language of calculus, which is the mathematics of change.

More than 50 years before the work of Newton, the astronomer Johannes Kepler (1571–1630) discovered his three laws of planetary motion, the most famous of which states that the path of a planet around the sun is an ellipse. Kepler arrived at these laws through a painstaking analysis of astronomical data, but he could not explain why they were true. According to Newton, the motion of any object—planet or pebble—is determined by the forces acting on it. The planets, if left undisturbed, would travel in straight lines. Since their paths are elliptical, some force—in this case, the gravitational force of the sun—must be acting to make them change direction continuously. In his magnum opus *Principia Mathematica*, published in 1687, Newton proved that Kepler's laws follow from Newton's own universal laws of motion and gravity.

For these discoveries, Newton gained widespread fame in his lifetime. His fame continued to increase after his death, assuming a nearly mythic dimension, and his ideas had a profound influence, not only in science but also in the arts and literature, as expressed in this epitaph by British poet Alexander Pope: "Nature and Nature's Laws lay hid in Night. God said, *Let Newton be!* and all was Light."

2.1 SUMMARY

- Average velocity over $[t_0, t_1] = \dfrac{\text{change in position}}{\text{change in time}} = \dfrac{s(t_1) - s(t_0)}{t_1 - t_0}$.
- Instantaneous Velocity $= \lim_{\Delta t \to 0}$ (average velocity)

- The slope of the secant line through the points $(t_0, s(t_0))$ and $(t_1, s(t_1))$ on the graph of $s(t)$ is $\dfrac{s(t_1) - s(t_0)}{t_1 - t_0}$.
- Slope of the tangent line = $\lim_{\Delta t \to 0}$ (slopes of the secant lines)
- Average velocity over an interval $[t_0, t_1]$ is the slope of the secant line through the points $(t_0, s(t_0))$ and $(t_1, s(t_1))$ on the graph of $s(t)$.
- Instantaneous velocity at t_0 is the slope of the tangent line at t_0.
- To estimate the instantaneous velocity or tangent-line slope at t_0, compute the average velocity or secant-line slope over several intervals $[t_0, t_1]$ (or $[t_1, t_0]$) for t_1 close to t_0 and estimate from those values.

2.1 EXERCISES

Preliminary Questions

1. Average velocity is equal to the slope of a secant line through two points on a graph. Which graph?

2. Can instantaneous velocity be defined as a ratio? If not, how is instantaneous velocity computed?

3. With t in hours, at $t = 0$ Dale entered Highway 1. At $t = 2$ he was 126 miles down the highway, on the side of the road with a flat tire. At $t = 3$ he was still on the side of the road, waiting for road assistance. What was Dale's average velocity over each of the time intervals:

(a) From $t = 0$ to $t = 2$

(b) From $t = 0$ to $t = 3$

(c) From $t = 2$ to $t = 3$

4. What is the graphical interpretation of instantaneous velocity at a specific time $t = t_0$?

Exercises

1. A ball dropped from a state of rest at time $t = 0$ travels a distance $s(t) = 4.9t^2$ m in t seconds.

(a) How far does the ball travel during the time interval $[2, 2.5]$?

(b) Compute the average velocity over $[2, 2.5]$.

(c) Compute the average velocity for the time intervals in the table and estimate the ball's instantaneous velocity at $t = 2$.

Interval	$[2, 2.01]$	$[2, 2.005]$	$[2, 2.001]$	$[2, 2.00001]$
Average velocity				

2. A wrench dropped from a state of rest at time $t = 0$ travels a distance $s(t) = 4.9t^2$ m in t seconds. Estimate the instantaneous velocity at $t = 3$.

3. On her bicycle ride Fabiana's position (in km) as a function of time (in hours) is $s(t) = 22t + 17$. What was her average velocity between $t = 2$ and $t = 3$? What was her instantaneous velocity at $t = 2.5$?

4. Compute $\Delta y / \Delta x$ for the interval $[2, 5]$, where $y = 4x - 9$. What is the slope of the tangent line at $x = 2$?

In Exercises 5–6, a ball is dropped on Mars where the distance traveled is $s(t) = 1.9t^2$ meters in t seconds.

5. Compute the ball's average velocity over the time interval $[3, 6]$ and estimate the instantaneous velocity at $t = 3$.

6. Compute the ball's average velocity over the time interval $[5, 9]$ and estimate the instantaneous velocity at $t = 5$.

In Exercises 7–8, a stone is tossed vertically into the air from ground level with an initial velocity of 15 m/s. Its height at time t is $h(t) = 15t - 4.9t^2$ m.

7. Compute the stone's average velocity over the time interval $[0.5, 2.5]$ and indicate the corresponding secant line on a sketch of the graph of h.

8. Compute the stone's average velocity over the time intervals $[1, 1.01]$, $[1, 1.001]$, $[1, 1.0001]$ and $[0.99, 1]$, $[0.999, 1]$, $[0.9999, 1]$, and then estimate the instantaneous velocity at $t = 1$.

9. The position of a particle at time t is $s(t) = 2t^3$. Compute the average velocity over the time interval $[2, 4]$ and estimate the instantaneous velocity at $t = 2$.

10. The position of a particle at time t is $s(t) = t^3 + t$. Compute the average velocity over the time interval $[1, 4]$ and estimate the instantaneous velocity at $t = 1$.

In Exercises 11–20, estimate the slope of the tangent line at the point indicated.

11. $f(x) = x^2 + x$; $\quad x = 0$

12. $P(x) = 3x^2 - 5$; $\quad x = 2$

13. $f(t) = 12t - 7$; $\quad t = -4$

14. $y(x) = \dfrac{1}{x + 2}$; $\quad x = 2$

15. $y(t) = \sqrt{3t + 1}$; $\quad t = 1$

16. $f(x) = e^x$; $\quad x = 0$

17. $f(x) = e^x$; $\quad x = e$

18. $f(x) = \ln x$; $\quad x = 3$

19. $f(x) = \tan x$; $\quad x = \dfrac{\pi}{4}$

20. $f(x) = \tan x$; $\quad x = 0$

21. The height (in centimeters) at time t (in seconds) of a small mass oscillating at the end of a spring is $h(t) = 3\sin(2\pi t)$. Estimate its instantaneous velocity at $t = 4$.

22. The height (in centimeters) at time t (in seconds) of a small mass oscillating at the end of a spring is $h(t) = 8\cos(12\pi t)$.

(a) Calculate the mass's average velocity over the time intervals $[0, 0.1]$ and $[3, 3.5]$.

(b) Estimate its instantaneous velocity at $t = 3$.

23. Consider the function $f(x) = \sqrt{x}$.

(a) Compute the slope of the secant lines from $(0,0)$ to $(x, f(x))$ for $x = 1, 0.1, 0.01, 0.001, 0.0001$.

(b) Discuss what the secant-line slopes in (a) suggest happens to the tangent line at 0.

(c) ⬛GU⬛ Plot the graph of f near $x = 0$ and verify your observation from (b).

24. Consider the function $f(x) = (x - 1)^{1/3}$.

(a) Compute the slope of the secant lines between $(1, 0)$ and $(x, f(x))$ for $x = 0.9, 0.99, 0.9999$ and for $x = 1.1, 1.01, 1.0001$.

(b) Discuss what the secant-line slopes in (a) suggest happens to the tangent line at 1.

(c) ⬛GU⬛ Plot the graph of f near $x = 1$ and verify your observation from (b).

25. 🖊 If an object in linear motion (but with changing velocity) covers Δs meters in Δt seconds, then its average velocity is $v_0 = \Delta s/\Delta t$ m/s. Show that it would cover the same distance if it traveled at constant velocity v_0 over the same time interval. This justifies our calling $\Delta s/\Delta t$ the *average velocity*.

26. 🖊 Sketch the graph of $f(x) = x(1 - x)$ over $[0, 1]$. Refer to the graph and, without making any computations, find:

(a) The slope of the secant line over $[0, 1]$

(b) The slope of the tangent line at $x = \frac{1}{2}$

(c) The values of x at which the slope of the tangent line is positive

27. 🖊 Which graph in Figure 5 has the following property: For all x, the slope of the secant line over $[0, x]$ is greater than the slope of the tangent line at x. Explain.

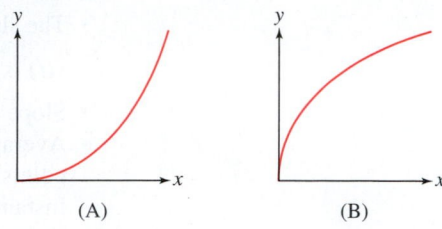

(A)   (B)

FIGURE 5

28. The height of a projectile fired in the air vertically with initial velocity 25 m/s is

$$h(t) = 25t - 4.9t^2 \text{ m}$$

(a) Compute $h(1)$. Show that $h(t) - h(1)$ can be factored with $(t - 1)$ as a factor.

(b) Using part (a), show that the average velocity over the interval $[1, t]$ is $20.1 - 4.9t$.

(c) Use this formula to estimate the instantaneous velocity at time $t = 1$.

29. Let $Q(t) = t^2$. Find a formula for the slope of the secant line over the interval $[1, t]$ and use it to estimate the slope of the tangent line at $t = 1$. Repeat for the interval $[2, t]$ and for the slope of the tangent line at $t = 2$.

30. For $f(x) = x^3$, show that the slope of the secant line over $[1, x]$ is $x^2 + x + 1$, and use this to estimate the slope of the tangent line at $x = 1$.

31. For $f(x) = x^3$, show that the slope of the secant line over $[-3, x]$ is $x^2 - 3x + 9$, and use this to estimate the slope of the tangent line at $x = -3$.

Further Insights and Challenges

The next two exercises involve limit estimates related to the definite integral, an important topic introduced in Chapter 5.

32. (a) Figure 6(A) shows two rectangles whose combined area is an over-estimate of the area A under the graph of $y = x^2$ from $x = 0$ to $x = 1$. Compute the combined area of the rectangles.

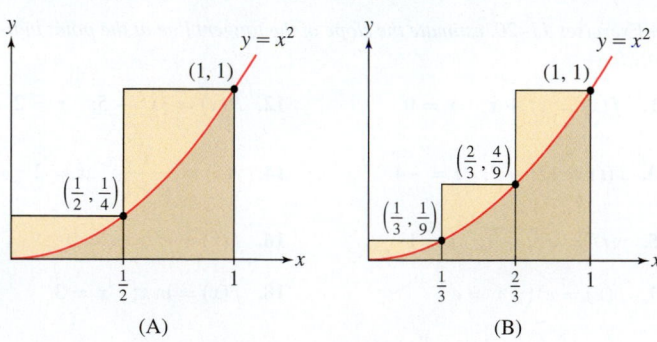

(A)   (B)

FIGURE 6

(b) We can improve the estimate by using three rectangles obtained by dividing $[0, 1]$ into thirds, as shown in Figure 6(B). Compute the combined areas of the three rectangles.

(c) Now divide $[0, 1]$ into subintervals of width $1/5$, and, on a graph of f, sketch the corresponding five rectangles obtained similar to those in (a) and (b). Compute the combined area of the five rectangles to estimate the area A.

(d) Improve your area estimate by dividing $[0, 1]$ into 10 subintervals of width $1/10$ and computing the combined area of the 10 resulting rectangles.

By dividing $[0, 1]$ into more and more subintervals, you can improve your estimate. You can use technology to carry out these computations for large numbers of rectangles. The exact value of the area is the limit of the estimates as the number of subintervals gets larger and larger.

Alternatively, for this example there is a formula (that we show how to derive in Section 5.1) that gives the total area $A(n)$ of the rectangles formed when $[0, 1]$ is divided into n subintervals of equal width:

$$A(n) = \frac{(n + 1)(2n + 1)}{6n^2}$$

(e) Compute $A(n)$ for $n = 2, 3, 5, 10$ to verify your results from (a)–(d).

(f) Compute $A(n)$ for $n = 100, 1000$, and $10,000$. Use your results to conjecture what the area A equals.

33. Let A represent the area under the graph of $y = x^3$ between $x = 0$ and $x = 1$. In this problem, we will follow the process in Exercise 32 to approximate A.

(a) As in (a)–(d) in Exercise 32, separately divide $[0, 1]$ into 2, 3, 5, and 10 equal-width subintervals, and in each case compute an overestimate of A using rectangles on each subinterval whose height is the value of x^3 at the right end of the subinterval.

In this case, it can be shown that if we use n equal-width subintervals, then the total area $A(n)$ of the n rectangles is:

$$A(n) = \frac{(n + 1)^2}{4n^2}$$

(b) Compute $A(n)$ for $n = 2, 3, 5, 10$ to verify your results from (a).

(c) Compute $A(n)$ for $n = 100, 1000$, and $10,000$. Use your results to conjecture what the area A equals.

2.2 Investigating Limits

The goal in this section is to define limits and study them. Here we primarily use numerical and graphical techniques, but in upcoming sections we will develop other means for investigating and evaluating limits. We begin with the following question: *How do the values $f(x)$ of a function f behave when x approaches a number c, whether or not $f(c)$ is defined?*

To explore this question, we'll experiment with the function

$$f(x) = \frac{\sin x}{x} \quad (x \text{ in radians})$$

Notice that $f(0)$ is not defined. In fact, when we set $x = 0$ in

$$f(x) = \frac{\sin x}{x}$$

we obtain the undefined expression $(\sin 0)/0$. Nevertheless, we can compute $f(x)$ for values of x *close* to 0. When we do this, a clear trend emerges.

To describe the trend, we use the phrase "x approaches 0" or "x tends to 0" to indicate that x takes on values (both positive and negative) that get closer and closer to 0. The notation for this is $x \to 0$, and more specifically we write

- $x \to 0^+$ if x approaches 0 from the right (on the number line).
- $x \to 0^-$ if x approaches 0 from the left (on the number line).

Now consider the values listed in Table 1. The table gives the unmistakable impression that $f(x)$ gets closer and closer to 1 as $x \to 0^+$ and as $x \to 0^-$.

This conclusion is supported by the graph of f in Figure 1. The point $(0, 1)$ is missing from the graph because $f(x)$ is not defined at $x = 0$, but the graph approaches this missing point as x approaches 0 from the left and right. We say that the *limit* of $f(x)$ as $x \to 0$ is equal to 1, and we write

$$\lim_{x \to 0} f(x) = 1$$

We also say that $f(x)$ *approaches* or *converges to* 1 as $x \to 0$.

TABLE 1

x	$\dfrac{\sin x}{x}$	x	$\dfrac{\sin x}{x}$
1	0.841470985	−1	0.841470985
0.5	0.958851077	−0.5	0.958851077
0.1	0.998334166	−0.1	0.998334166
0.05	0.999583385	−0.05	0.999583385
0.01	0.999983333	−0.01	0.999983333
0.005	0.999995833	−0.005	0.999995833
0.001	0.999999833	−0.001	0.999999833
$x \to 0^+$	$f(x) \to 1$	$x \to 0^-$	$f(x) \to 1$

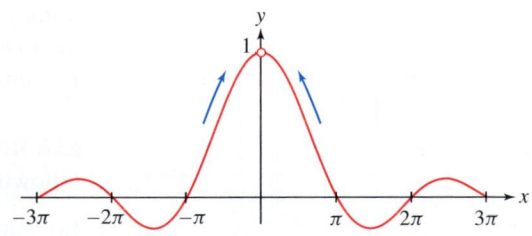

DF **FIGURE 1** Graph of $f(x) = \dfrac{\sin x}{x}$.

In Section 2.6 we will verify the numerical and graphical evidence seen here and prove that $\lim\limits_{x \to 0} \frac{\sin x}{x} = 1$. Note that, if instead we use degrees for x rather than radians, then this limit is $\frac{\pi}{180} \approx 0.01745$ (since π radians = 180 degrees). Exploring $\frac{\sin x}{x}$ numerically, with x in degrees, you can verify that the values approach 0.01745 as x approaches 0 (Exercise 6). Because this limit is 1 when x is measured in radians, we primarily work with radians when using the trigonometric functions in calculus. Important calculus formulas are much simpler as a result (see Section 3.6).

CONCEPTUAL INSIGHT Could we arrive at $\lim\limits_{x\to 0}\frac{\sin x}{x}$ by simply substituting 0 for x and saying that $\frac{\sin 0}{0} = \frac{0}{0} = 1$? The answer is no. *We cannot divide by 0 under any circumstances*, and it is not correct to say that the undefined expression $0/0$ equals 1 or any other number.

This example shows that a function g may approach a limit as $x \to c$ even if the formula for $g(c)$ produces an undefined expression. In this example, the limit turns out to be 1. We will encounter examples of other functions g where the formula for $g(c)$ produces the undefined expression $0/0$ but the limit is a number other than 1 (or the limit does not exist).

Definition of a Limit

The formal definition of a limit is somewhat technical, and we will wait until Section 2.9 to present it. Here we present a definition that is more conceptually oriented. It will serve our purposes for the time being. To begin, let us recall that the distance between two numbers a and b is the absolute value $|a - b|$. So we can express the idea that $f(x)$ is close to L by saying that $|f(x) - L|$ is small.

In a more precise version of the limit definition "$|f(x) - L|$ can be made arbitrarily small" is expressed as "$|f(x) - L|$ can be made less than s for any small number $s > 0$," and "by taking x sufficiently close to c" is expressed as "by finding a small number $r > 0$ and taking $|x - c| < r$."

> **DEFINITION** Limit Assume that $f(x)$ is defined for all x in an open interval containing c, but not necessarily at c itself. We say that
>
> $$\text{the limit of } f(x) \text{ as } x \text{ approaches } c \text{ is equal to the number } L$$
>
> if $|f(x) - L|$ can be made arbitrarily small by taking x sufficiently close (but not equal) to c. In this case, we write
>
> $$\lim_{x\to c} f(x) = L$$
>
> We also say that $f(x)$ *approaches* or *converges to* L as $x \to c$ [and we write $f(x) \to L$].

In other words, as x approaches c, $f(x)$ approaches L. See Figure 2 for the graphical interpretation. If the values of $f(x)$ do not converge to any number L as $x \to c$, we say that $\lim\limits_{x\to c} f(x)$ *does not exist*. It is important to note that the value $f(c)$ itself, which may or may not be defined, plays no role in the limit. All that matters are the values of $f(x)$ for x close to c. Furthermore, if $f(x)$ approaches a limit as $x \to c$, then the limiting value L is unique.

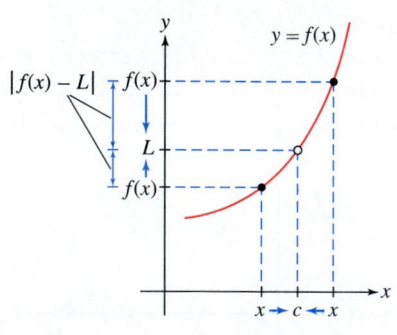

FIGURE 2 As $x \to c$, $f(x) \to L$.

EXAMPLE 1 Let $f(x) = 5$ and $g(x) = 3x + 1$. Use the definition above to verify the following limits:

(a) $\lim\limits_{x\to 7} f(x) = 5$ **(b)** $\lim\limits_{x\to 4} g(x) = 13$

Solution

(a) To show that $\lim\limits_{x\to 7} f(x) = 5$, we must show that $|f(x) - 5|$ becomes arbitrarily small when x is sufficiently close (but not equal) to 7. But note that $|f(x) - 5| = |5 - 5| = 0$ *for all x*, so what we are required to show is automatic.

(b) To show that $\lim\limits_{x\to 4} g(x) = 13$, we must show that $|g(x) - 13|$ becomes arbitrarily small when x is sufficiently close (but not equal) to 4. We have

$$|g(x) - 13| = |(3x + 1) - 13| = |3x - 12| = 3|x - 4|$$

Utkin/Sputnik/The Image Works

Olga Ladyzhenskaya (1922–2004) was a Soviet-Russian mathematician who worked in the field of partial differential equations. A major contribution of hers was a proof of a theorem verifying that a numerical approximation method converged in the limit to solutions to the Navier–Stokes equations, an important model for fluid motion. Such verification is valuable because it ensures the accuracy of estimates made via the approximation method.

Keep in mind that graphical and numerical investigations provide evidence for a limit, but they do not prove that the limit exists or has a given value. This is done using the Limit Laws established in the following sections.

Because $|g(x) - 13|$ is a multiple of $|x - 4|$, we can make $|g(x) - 13|$ arbitrarily small by taking x sufficiently close to 4. ∎

Reasoning as in Example 1 but with arbitrary constants, we obtain the following simple but important results:

> **THEOREM 1**   For any constants k and c,   (a) $\lim\limits_{x \to c} k = k$,   (b) $\lim\limits_{x \to c} x = c$.

To deal with more complicated limits and, especially, to provide mathematically rigorous proofs, the precise version of the above limit definition, presented in Section 2.9, is needed. There inequalities are used to pin down the exact meaning of the phrases "arbitrarily small" and "sufficiently close."

Graphical and Numerical Investigation

Our goal in the rest of this section is to develop a better intuitive understanding of limits by investigating them graphically and numerically.

Graphical Investigation   Use a graphing utility to produce a graph of f. The graph should give a visual impression of whether or not a limit exists. It can often be used to estimate the value of the limit.

Numerical Investigation   We write $x \to c^-$ to indicate that x approaches c through values less than c (i.e., from the left), and we write $x \to c^+$ to indicate that x approaches c through values greater than c (i.e., from the right). To investigate $\lim\limits_{x \to c} f(x)$,

(i) Make a table of values of $f(x)$ for x close to but less than c—that is, as $x \to c^-$.

(ii) Make a second table of values of $f(x)$ for x close to but greater than c—that is, as $x \to c^+$.

(iii) If both tables indicate convergence to the same number L, we take L to be an estimate for the limit.

The tables should contain enough values to reveal a clear trend of convergence to a value L. If $f(x)$ approaches a limit, the successive values of $f(x)$ will generally agree to more and more decimal places as x is taken closer to c. If no pattern emerges, then the limit may not exist.

EXAMPLE 2   Investigate $\lim\limits_{x \to 9} \dfrac{x - 9}{\sqrt{x} - 3}$ graphically and numerically.

Solution   The function $f(x) = \dfrac{x - 9}{\sqrt{x} - 3}$ is undefined at $x = 9$ because the formula for $f(9)$ leads to the undefined expression $0/0$. Therefore, the graph in Figure 3 has a gap at $x = 9$. However, the graph suggests that $f(x)$ approaches 6 as x approaches 9.

For numerical evidence, we consider a table of values of $f(x)$ for x approaching 9 from both the left and the right. Table 2 supports our impression that

$$\lim_{x \to 9} \frac{x - 9}{\sqrt{x} - 3} = 6$$

In Section 2.5, we will revisit this limit and show how we can use algebraic simplification to prove that this limit is, in fact, 6. ∎

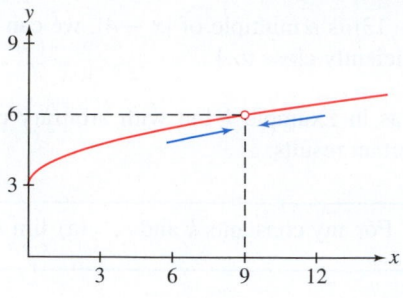

DF FIGURE 3 Graph of $f(x) = \dfrac{x-9}{\sqrt{x}-3}$.

TABLE 2			
$x \to 9^-$	$\dfrac{x-9}{\sqrt{x}-3}$	$x \to 9^+$	$\dfrac{x-9}{\sqrt{x}-3}$
8.9	5.98329	9.1	6.01662
8.99	5.99833	9.01	6.001666
8.999	5.99983	9.001	6.000167
8.9999	5.9999833	9.0001	6.0000167

EXAMPLE 3   **Limit Equals Value of the Function**   Investigate $\lim\limits_{x \to 4} x^2$.

Solution   Figure 4 and Table 3 both suggest that $\lim\limits_{x \to 4} x^2 = 16$. Furthermore, note that $f(x) = x^2$ is defined at $x = 4$ and $f(4) = 16$, so in this case, *the limit is equal to the function value*. This pleasant conclusion is valid whenever f is a *continuous* function, a concept treated in Section 2.4.   ∎

FIGURE 4 Graph of $f(x) = x^2$. The limit is equal to the value of the function $f(4) = 16$.

TABLE 3			
$x \to 4^-$	x^2	$x \to 4^+$	x^2
3.9	15.21	4.1	16.81
3.99	15.9201	4.01	16.0801
3.999	15.992001	4.001	16.008001
3.9999	15.99920001	4.0001	16.00080001

EXAMPLE 4   **A Defining Condition of e**   Investigate $\lim\limits_{x \to 0} (1 + x)^{1/x}$ numerically and graphically.

Solution   The function $f(x) = (1 + x)^{1/x}$ is undefined at $x = 0$, but both Figure 5 and Table 4 suggest that a limit exists. As we indicated in Section 1.6, one approach to defining e is via this limit; that is, $e = \lim\limits_{x \to 0} (1 + x)^{1/x}$. From our numerical investigation, we see that $e \approx 2.71828$ accurate to 5 decimal places.   ∎

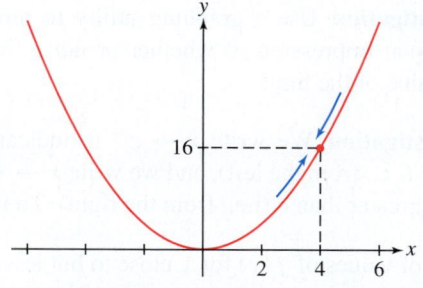

FIGURE 5 $f(0)$ is undefined but $\lim\limits_{x \to 0} f(x)$ exists.

TABLE 4			
$x \to 0^-$	$(1+x)^{1/x}$	$x \to 0^+$	$(1+x)^{1/x}$
-0.01	2.731999	0.01	2.704814
-0.001	2.719642	0.001	2.716924
-0.0001	2.718418	0.0001	2.718146
-0.00001	2.718295	0.00001	2.718268
-0.000001	2.718283	0.000001	2.718280

CAUTION *Numerical investigations are often suggestive, but may be misleading in some cases. If, in Example 5, we had chosen to evaluate $f(x) = \sin\dfrac{\pi}{x}$ at the values $x = 0.1, 0.01, 0.001, \ldots$, we might have concluded incorrectly that $f(x)$ approaches the limit 0 as $x \to 0$. The problem is that $f(10^{-n}) = \sin(10^n \pi) = 0$ for every whole number n, but $f(x)$ itself does not approach any limit.*

EXAMPLE 5   **A Limit That Does Not Exist**   Investigate $\lim\limits_{x \to 0} \sin\dfrac{\pi}{x}$ graphically and numerically.

Solution   The function $f(x) = \sin\dfrac{\pi}{x}$ is not defined at $x = 0$, but Figure 6 suggests that it oscillates between $+1$ and -1 infinitely often as $x \to 0$. It appears, therefore, that $\lim\limits_{x \to 0} \sin\dfrac{\pi}{x}$ does not exist. This impression is confirmed by Table 5, which shows that the values of $f(x)$ bounce around and do not tend toward any limit L as $x \to 0$.   ∎

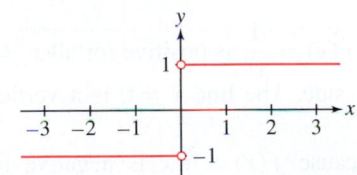

FIGURE 6 Graph of $f(x) = \sin \frac{\pi}{x}$.

TABLE 5	The Function $f(x) = \sin \frac{\pi}{x}$ Does Not Approach a Limit as $x \to 0$		
$x \to 0^-$	$\sin \dfrac{\pi}{x}$	$x \to 0^+$	$\sin \dfrac{\pi}{x}$
-0.1	0	0.1	0
-0.03	0.866	0.03	-0.866
-0.007	-0.434	0.007	0.434
-0.0009	0.342	0.0009	-0.342
-0.00065	-0.935	0.00065	0.935

One-Sided Limits

The limits discussed so far are *two-sided*. To show that $\lim\limits_{x \to c} f(x) = L$, it is necessary to check that $f(x)$ converges to L as x approaches c through values both greater than and less than c. In some instances, $f(x)$ may approach L from one side of c without necessarily approaching it from the other side, or $f(x)$ may be defined on only one side of c. For this reason, we define the one-sided limits

$$\lim_{x \to c^-} f(x) \quad \text{(left-hand limit)}, \qquad \lim_{x \to c^+} f(x) \quad \text{(right-hand limit)}$$

> The limit itself exists if and only if both one-sided limits exist and are equal. Otherwise the limit does not exist.

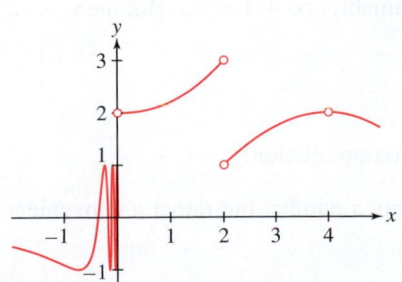

FIGURE 7 Graph of $f(x) = \dfrac{x}{|x|}$.

EXAMPLE 6 **Left- and Right-Hand Limits Not Equal** Investigate the one-sided limits of $f(x) = \dfrac{x}{|x|}$ as $x \to 0$. Does $\lim\limits_{x \to 0} f(x)$ exist?

Solution Figure 7 shows what is going on. For $x < 0$,

$$f(x) = \frac{x}{|x|} = \frac{x}{-x} = -1$$

Therefore, the left-hand limit is $\lim\limits_{x \to 0^-} f(x) = -1$. But for $x > 0$,

$$f(x) = \frac{x}{|x|} = \frac{x}{x} = 1$$

Therefore, $\lim\limits_{x \to 0^+} f(x) = 1$. These one-sided limits are not equal, so $\lim\limits_{x \to 0} f(x)$ does not exist. ■

EXAMPLE 7 The function f in Figure 8 is not defined at $c = 0, 2, 4$. Investigate the one- and two-sided limits at these points.

Solution

- $c = 0$: The left-hand limit $\lim\limits_{x \to 0^-} f(x)$ does not seem to exist because $f(x)$ appears to oscillate infinitely often to the left of $x = 0$. On the other hand, $\lim\limits_{x \to 0^+} f(x) = 2$.
- $c = 2$: The one-sided limits exist but are not equal:

$$\lim_{x \to 2^-} f(x) = 3 \quad \text{and} \quad \lim_{x \to 2^+} f(x) = 1$$

Therefore, $\lim\limits_{x \to 2} f(x)$ does not exist.
- $c = 4$: The one-sided limits exist and both have the value 2. Therefore, the two-sided limit exists and $\lim\limits_{x \to 4} f(x) = 2$. ■

FIGURE 8

← *REMINDER* Recall that "large" refers to distance from zero. So being negative and becoming arbitrarily large means becoming arbitrarily far from zero to the left of zero.

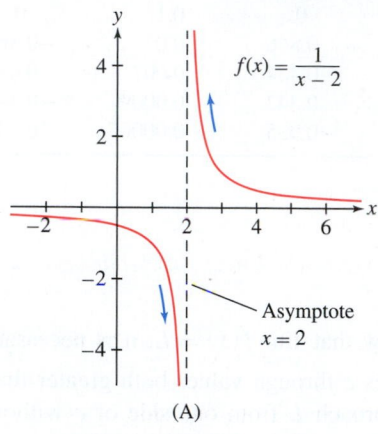

(A)

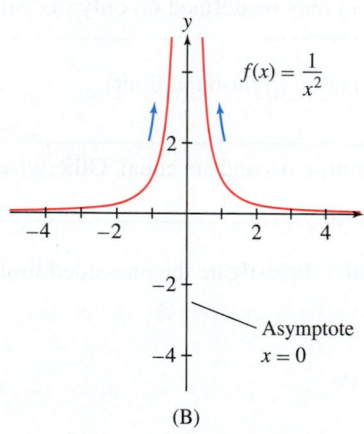

(B)

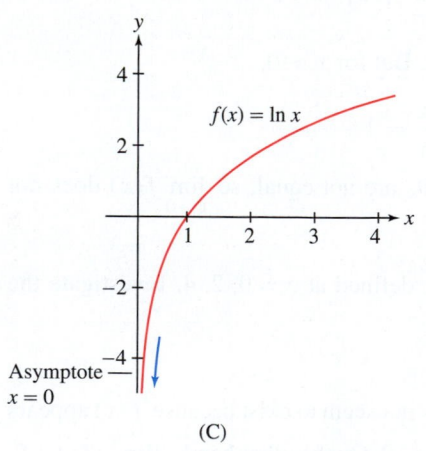

(C)

DF FIGURE 9

Infinite Limits

For some functions, $f(x)$ tends to ∞ or $-\infty$ as x approaches a value c. It is important to understand that ∞ and $-\infty$ are not numbers, and therefore $\lim\limits_{x \to c} f(x)$ does not exist. However, we say that $f(x)$ has an *infinite limit*. More precisely, we write

- $\lim\limits_{x \to c} f(x) = \infty$ if $f(x)$ is positive and becomes arbitrarily large as $x \to c$.
- $\lim\limits_{x \to c} f(x) = -\infty$ if $f(x)$ is negative and becomes arbitrarily large as $x \to c$.

One-sided infinite limits are defined similarly.

When $f(x)$ approaches ∞ or $-\infty$ as x approaches c from one or both sides, the line $x = c$ is called a **vertical asymptote**. In Figure 9, the line $x = 2$ is a vertical asymptote in (A), and $x = 0$ is a vertical asymptote in both (B) and (C).

In the next example, the notation $x \to c^{\pm}$ is used to indicate that the left- and right-hand limits are to be considered separately.

EXAMPLE 8 Investigate the one-sided limits graphically:

(a) $\lim\limits_{x \to 2^{\pm}} \dfrac{1}{x - 2}$ (b) $\lim\limits_{x \to 0^{\pm}} \dfrac{1}{x^2}$ (c) $\lim\limits_{x \to 0^+} \ln x$

Solution

(a) Figure 9(A) suggests that

$$\lim_{x \to 2^-} \frac{1}{x - 2} = -\infty, \qquad \lim_{x \to 2^+} \frac{1}{x - 2} = \infty$$

The line $x = 2$ is a vertical asymptote. Why are the one-sided limits different? Because $f(x) = \dfrac{1}{x - 2}$ is negative for $x < 2$ (so the limit from the left is $-\infty$) and $f(x)$ is positive for $x > 2$ (so the limit from the right is ∞).

(b) Figure 9(B) suggests that $\lim\limits_{x \to 0} \dfrac{1}{x^2} = \infty$. Indeed, $f(x) = \dfrac{1}{x^2}$ is positive for all $x \neq 0$ and becomes arbitrarily large as $x \to 0$ from either side. The line $x = 0$ is a vertical asymptote.

(c) Figure 9(C) suggests that $\lim\limits_{x \to 0^+} \ln x = -\infty$ because $f(x) = \ln x$ is negative for $0 < x < 1$ and tends to $-\infty$ as $x \to 0^+$. The line $x = 0$ is a vertical asymptote. ∎

CONCEPTUAL INSIGHT You should not think of an infinite limit as a true limit. The notation $\lim\limits_{x \to c} f(x) = \infty$ is merely a shorthand way of saying that $f(x)$ is positive and arbitrarily large as x approaches c. The limit itself does not exist. We must be careful when using this notation because ∞ and $-\infty$ *are not numbers*, and contradictions can arise if we try to manipulate them as numbers. For example, if ∞ were a number, it would be larger than any finite number, and presumably, $\infty + 1 = \infty$. But then

$$\infty + 1 = \infty$$

$$(\infty + 1) - \infty = \infty - \infty$$

$$1 = 0 \qquad \text{(contradiction!)}$$

To avoid errors like this, keep in mind the ∞ is not a number but rather a convenient shorthand notation.

2.2 SUMMARY

- By definition, $\lim\limits_{x \to c} f(x) = L$ if $|f(x) - L|$ can be made arbitrarily small by taking x sufficiently close (but not equal) to c. We say that

- *The limit of $f(x)$ as x approaches c is L, or*
  - *$f(x)$ approaches (or converges to) L as x approaches c.*
- If $f(x)$ approaches a limit as $x \to c$, then the value of the limit L is unique.
- If $f(x)$ does not approach a limit as $x \to c$, we say that $\lim_{x \to c} f(x)$ does not exist.
- The limit may exist even if $f(c)$ is not defined.
- *One-sided limits*:
  - $\lim_{x \to c^-} f(x) = L$ if $f(x)$ converges to L as x approaches c through values less than c.
  - $\lim_{x \to c^+} f(x) = L$ if $f(x)$ converges to L as x approaches c through values greater than c.
- The limit exists if and only if both one-sided limits exist and are equal.
- *Infinite limits*: $\lim_{x \to c} f(x) = \infty$ if $f(x)$ is positive and becomes arbitrarily large as x approaches c, and $\lim_{x \to c} f(x) = -\infty$ if $f(x)$ is negative and becomes arbitrarily large as x approaches c.
- In the case of a one- or two-sided infinite limit at c, the vertical line $x = c$ is called a *vertical asymptote*.

2.2 EXERCISES

Preliminary Questions

1. What is the limit of $f(x) = 1$ as $x \to \pi$?

2. What is the limit of $g(t) = t$ as $t \to \pi$?

3. Is $\lim_{x \to 10} 20$ equal to 10 or 20?

4. Can $f(x)$ approach a limit as $x \to c$ if $f(c)$ is undefined? If so, give an example.

5. What does the following table suggest about $\lim_{x \to 1^-} f(x)$ and $\lim_{x \to 1^+} f(x)$?

x	0.9	0.99	0.999	1.001	1.01	1.1
$f(x)$	7	25	4317	3.00011	3.0047	3.0126

6. Can you tell whether $\lim_{x \to 5} f(x)$ exists from a plot of f for $x > 5$? Explain.

7. If you know in advance that $\lim_{x \to 5} f(x)$ exists, can you determine its value from a plot of f for all $x > 5$?

Exercises

In Exercises 1–5, fill in the table and guess the value of the limit.

1. $\lim_{x \to 1} f(x)$, where $f(x) = \dfrac{x^3 - 1}{x^2 - 1}$.

x	$f(x)$	x	$f(x)$
1.002		0.998	
1.001		0.999	
1.0005		0.9995	
1.00001		0.99999	

2. $\lim_{t \to 0} h(t)$, where $h(t) = \dfrac{\cos t - 1}{t^2}$. Note that h is even; that is, $h(t) = h(-t)$.

t	± 0.002	± 0.0001	± 0.00005	± 0.00001
$h(t)$				

3. $\lim_{y \to 2} f(y)$, where $f(y) = \dfrac{y^2 - y - 2}{y^2 + y - 6}$.

y	$f(y)$	y	$f(y)$
2.002		1.998	
2.001		1.999	
2.0001		1.9999	

4. $\lim_{x \to 0^+} f(x)$, where $f(x) = x \ln x$.

x	1	0.5	0.1	0.05	0.01	0.005	0.001
$f(x)$							

5. $\lim_{t \to 0} f(t)$, where $f(t) = \dfrac{1 - \cos 2t}{t}$.

t	$f(t)$	t	$f(t)$
0.002		-0.002	
0.001		-0.001	
0.0005		-0.0005	
0.00001		-0.00001	

6. Numerically investigate $\lim_{x \to 0} \frac{\sin x}{x}$, computing the values of $\sin x$ with x in degrees. Make an estimate of the limit accurate to 5 decimal places.

7. Determine $\lim_{x \to 0.5} f(x)$ for f as in Figure 10.

8. Determine $\lim_{x \to 0.5} g(x)$ for g as in Figure 11.

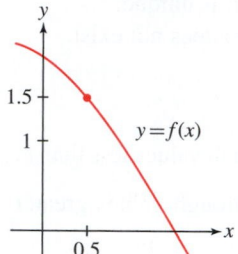

FIGURE 10

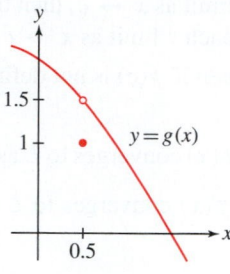

FIGURE 11

In Exercises 9–10, evaluate the limit.

9. $\lim_{x \to 21} x$

10. $\lim_{x \to 4.2} \sqrt{3}$

11. Show, via illustration, that the limits $\lim_{x \to a} x$ and $\lim_{x \to a} a$ are equal but the functions in each limit are different.

12. Give examples of functions f and g such that $\lim_{x \to 0} f(x) = \lim_{x \to 0} g(x)$, but $f(x) \neq g(x)$ for all x, including 0.

In Exercises 13–20, verify each limit using the limit definition. For example, in Exercise 13, show that $|3x - 12|$ can be made as small as desired by taking x close to 4.

13. $\lim_{x \to 4} 3x = 12$

14. $\lim_{x \to 5} 3 = 3$

15. $\lim_{x \to 3} (5x + 2) = 17$

16. $\lim_{x \to 2} (7x - 4) = 10$

17. $\lim_{x \to 0} x^2 = 0$

18. $\lim_{x \to 0} (3x^2 - 9) = -9$

19. $\lim_{x \to 0} (4x^2 + 2x + 5) = 5$

20. $\lim_{x \to 0} (x^3 + 12) = 12$

In Exercises 21–44, estimate the limit numerically or state that the limit does not exist. If infinite, state whether the one-sided limits are ∞ or $-\infty$.

21. $\lim_{x \to 1} \dfrac{\sqrt{x} - 1}{x - 1}$

22. $\lim_{x \to -4} \dfrac{2x^2 - 32}{x + 4}$

23. $\lim_{x \to 2} \dfrac{x^2 + x - 6}{x^2 - x - 2}$

24. $\lim_{x \to 3} \dfrac{x^3 - 2x^2 - 9}{x^2 - 2x - 3}$

25. $\lim_{x \to 0} \dfrac{\sin 2x}{x}$

26. $\lim_{x \to 0} \dfrac{\sin 5x}{x}$

27. $\lim_{x \to 0} \dfrac{\sin 3x}{3x}$

28. $\lim_{x \to 0} \dfrac{\cos x}{3x}$

29. $\lim_{\theta \to 0} \dfrac{\cos \theta - 1}{\theta}$

30. $\lim_{x \to 0} \dfrac{\sin x}{x^2}$

31. $\lim_{x \to 4} \dfrac{1}{(x - 4)^3}$

32. $\lim_{x \to 1^-} \dfrac{3 - x}{x - 1}$

33. $\lim_{x \to -3} \dfrac{x + 3}{x^2 + x - 6}$

34. $\lim_{x \to -2^-} \dfrac{x + 1}{x + 2}$

35. $\lim_{x \to 3^+} \dfrac{x - 4}{x^2 - 9}$

36. $\lim_{h \to 0} \dfrac{3^h - 1}{h}$

37. $\lim_{h \to 0} \sin h \cos \dfrac{1}{h}$

38. $\lim_{h \to 0} \cos \dfrac{1}{h}$

39. $\lim_{x \to 0} |x|^x$

40. $\lim_{x \to 1^+} \dfrac{\sec^{-1} x}{\sqrt{x - 1}}$

41. $\lim_{t \to e} \dfrac{t - e}{\ln t - 1}$

42. $\lim_{r \to 0} (1 + 2r)^{1/r}$

43. $\lim_{x \to 1^-} \dfrac{\tan^{-1} x}{\cos^{-1} x}$

44. $\lim_{x \to 0} \dfrac{\tan^{-1} x - x}{\sin^{-1} x - x}$

45. The **greatest integer function**, also known as the **floor function**, is defined by $\lfloor x \rfloor = n$, where n is the unique integer such that $n \leq x < n + 1$. Sketch the graph of $y = \lfloor x \rfloor$. Calculate for c an integer:

(a) $\lim_{x \to c^-} \lfloor x \rfloor$ **(b)** $\lim_{x \to c^+} \lfloor x \rfloor$ **(c)** $\lim_{x \to 2.6} \lfloor x \rfloor$

46. Determine the one-sided limits at $c = 1$, 2, and 4 of the function g shown in Figure 12, and state whether the limit exists at these points.

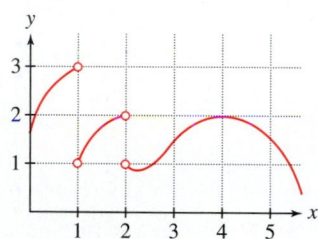

FIGURE 12

In Exercises 47–54, determine the one-sided limits numerically or graphically. If infinite, state whether the one-sided limits are ∞ or $-\infty$, and describe the corresponding vertical asymptote. In Exercise 54, $f(x) = \lfloor x \rfloor$ is the greatest integer function defined in Exercise 45.

47. $\lim_{x \to 0^\pm} \dfrac{\sin x}{|x|}$

48. $\lim_{x \to 0^\pm} |x|^{1/x}$

49. $\lim_{x \to 0^\pm} \dfrac{x - \sin |x|}{x^3}$

50. $\lim_{x \to 4^\pm} \dfrac{x + 1}{x - 4}$

51. $\lim_{x \to -2^\pm} \dfrac{4x^2 + 7}{x^3 + 8}$

52. $\lim_{x \to -3^\pm} \dfrac{x^2}{x^2 - 9}$

53. $\lim_{x \to 1^\pm} \dfrac{x^5 + x - 2}{x^2 + x - 2}$

54. $\lim_{x \to 2^\pm} \cos \left(\dfrac{\pi}{2} (x - \lfloor x \rfloor) \right)$

55. Determine the one-sided limits at $c = 2$ and $c = 4$ of the function f in Figure 13. What are the vertical asymptotes of f?

56. Determine the infinite one- and two-sided limits in Figure 14.

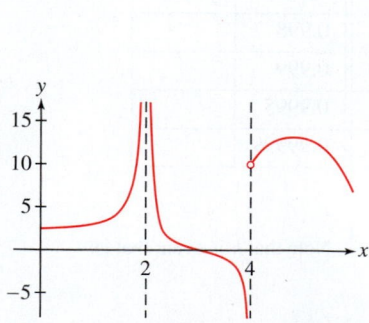

FIGURE 13

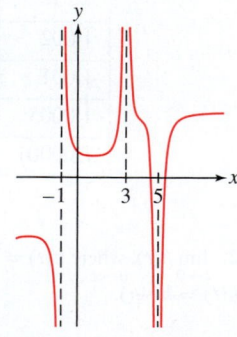

FIGURE 14

In Exercises 57–60, sketch the graph of a function with the given limits.

57. $\lim_{x \to 1} f(x) = 2$, $\lim_{x \to 3^-} f(x) = 0$, $\lim_{x \to 3^+} f(x) = 4$

58. $\lim_{x \to 1} f(x) = \infty$, $\lim_{x \to 3^-} f(x) = 0$, $\lim_{x \to 3^+} f(x) = -\infty$

59. $\lim\limits_{x \to 2^{+}} f(x) = f(2) = 3, \quad \lim\limits_{x \to 2^{-}} f(x) = -1, \quad \lim\limits_{x \to 4} f(x) = 2 \neq f(4)$

60. $\lim\limits_{x \to 1^{+}} f(x) = \infty, \quad \lim\limits_{x \to 1^{-}} f(x) = 3, \quad \lim\limits_{x \to 4} f(x) = -\infty$

61. Determine the one-sided limits of the function f in Figure 15, at the points $c = 1, 3, 5, 6$.

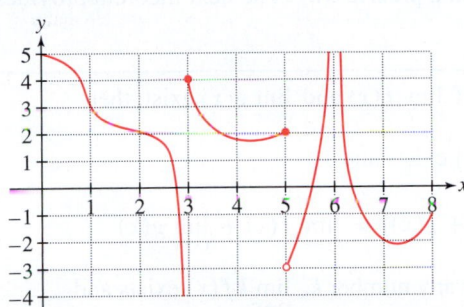

FIGURE 15 Graph of f.

62. Does either of the two oscillating functions in Figure 16 appear to approach a limit as $x \to 0$?

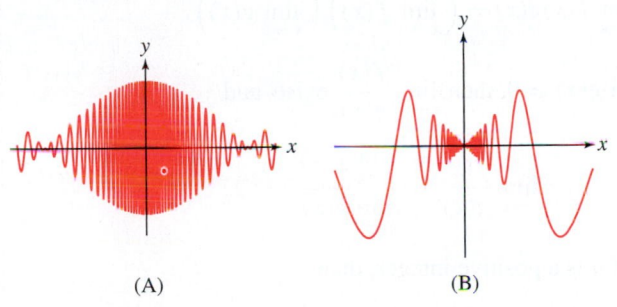

(A)   (B)

FIGURE 16

GU In Exercises 63–68, plot the function and use the graph to estimate the value of the limit.

63. $\lim\limits_{\theta \to 0} \dfrac{\sin 5\theta}{\sin 2\theta}$

64. $\lim\limits_{x \to 0} \dfrac{12^{x} - 1}{4^{x} - 1}$

65. $\lim\limits_{x \to 0} \dfrac{2^{x} - \cos x}{x}$

66. $\lim\limits_{\theta \to 0} \dfrac{\sin^{2} 4\theta}{\cos \theta - 1}$

67. $\lim\limits_{\theta \to 0} \dfrac{\cos 7\theta - \cos 5\theta}{\theta^{2}}$

68. $\lim\limits_{\theta \to 0} \dfrac{\sin^{2} 2\theta - \theta \sin 4\theta}{\theta^{4}}$

69. Let n be a positive integer. For which n are the two infinite one-sided limits $\lim\limits_{x \to 0^{\pm}} 1/x^{n}$ equal?

70. Let $L(n) = \lim\limits_{x \to 1} \left(\dfrac{n}{1 - x^{n}} - \dfrac{1}{1 - x} \right)$ for n a positive integer. Investigate $L(n)$ numerically for several values of n, and then guess the value of $L(n)$ in general.

71. GU In some cases, numerical investigations can be misleading. Plot $f(x) = \cos \dfrac{\pi}{x}$.

(a) Does $\lim\limits_{x \to 0} f(x)$ exist?

(b) Show, by evaluating $f(x)$ at $x = \pm\dfrac{1}{2}, \pm\dfrac{1}{4}, \pm\dfrac{1}{6}, \ldots$, that you might be able to trick your friends into believing that the limit exists and is equal to $L = 1$.

(c) Which sequence of evaluations might trick them into believing that the limit is $L = -1$?

Further Insights and Challenges

72. Light waves of frequency λ passing through a slit of width a produce a **Fraunhofer diffraction pattern** of light and dark fringes (Figure 17). The intensity as a function of the angle θ is

$$I(\theta) = I_{m} \left(\dfrac{\sin(R \sin \theta)}{R \sin \theta} \right)^{2}$$

where $R = \pi a / \lambda$ and I_{m} is a constant. Show that the intensity function is not defined at $\theta = 0$. Then choose any two values for R and check numerically that $I(\theta)$ approaches I_{m} as $\theta \to 0$.

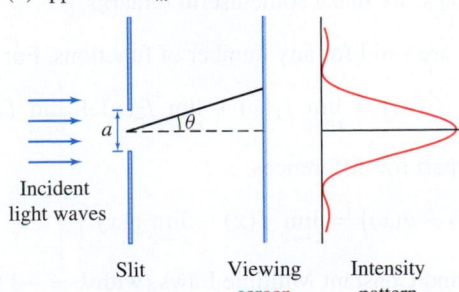

FIGURE 17 Fraunhofer diffraction pattern.

73. Investigate $\lim\limits_{\theta \to 0} \dfrac{\sin n\theta}{\theta}$ numerically for several positive integer values of n. Then guess the value in general.

74. Show numerically that $\lim\limits_{x \to 0} \dfrac{b^{x} - 1}{x}$ is less than 2 with $b = 7$ and is greater than 2 with $b = 8$. Experiment with values of b to find an approximate value of b for which the limit is 2.

75. Investigate $\lim\limits_{x \to 1} \dfrac{x^{n} - 1}{x^{m} - 1}$ for (m, n) equal to $(2, 1)$, $(1, 2)$, $(2, 3)$, and $(3, 2)$. Then guess the value of the limit in general and check your guess for two additional pairs.

76. Find by numerical experimentation the positive integers k such that $\lim\limits_{x \to 0} \dfrac{\sin(\sin^{2} x)}{x^{k}}$ exists.

77. ✎ GU Plot the graph of $f(x) = \dfrac{2^{x} - 8}{x - 3}$.

(a) Zoom in on the graph to estimate $L = \lim\limits_{x \to 3} f(x)$.

(b) Explain why

$$f(2.99999) \leq L \leq f(3.00001)$$

Use this to determine L to three decimal places.

78. GU The function $f(x) = \dfrac{2^{1/x} - 2^{-1/x}}{2^{1/x} + 2^{-1/x}}$ is defined for $x \neq 0$.

(a) Investigate $\lim\limits_{x \to 0^{+}} f(x)$ and $\lim\limits_{x \to 0^{-}} f(x)$ numerically.

(b) Plot the graph of f and describe its behavior near $x = 0$.

2.3 Basic Limit Laws

In Section 2.2, we relied on graphical and numerical approaches to investigate limits and estimate their values. In the next four sections, we go beyond this intuitive approach and develop tools for computing limits in a precise way. The next theorem provides our first set of tools.

The proof of Theorem 1 is discussed in Section 2.9 and Appendix D. To illustrate the underlying idea, consider two numbers such as 2.99 and 5.001. Observe that 2.99 is close to 3 and 5.0001 is close to 5, so certainly the sum 2.99 + 5.0001 is close to 3 + 5 and the product 2.99 × 5.0001 is close to 3 × 5. In the same way, if $f(x)$ approaches L and $g(x)$ approaches M as $x \to c$, then $f(x) + g(x)$ approaches the sum $L + M$, and $f(x)g(x)$ approaches the product LM. The other laws are similar.

THEOREM 1 Basic Limit Laws   If $\lim\limits_{x \to c} f(x)$ and $\lim\limits_{x \to c} g(x)$ exist, then

(i) Sum Law: $\lim\limits_{x \to c} \big(f(x) + g(x)\big)$ exists and

$$\lim_{x \to c} \big(f(x) + g(x)\big) = \lim_{x \to c} f(x) + \lim_{x \to c} g(x)$$

(ii) Constant Multiple Law: For any number k, $\lim\limits_{x \to c} kf(x)$ exists and

$$\lim_{x \to c} kf(x) = k \lim_{x \to c} f(x)$$

(iii) Product Law: $\lim\limits_{x \to c} f(x)g(x)$ exists and

$$\lim_{x \to c} f(x)g(x) = \Big(\lim_{x \to c} f(x)\Big)\Big(\lim_{x \to c} g(x)\Big)$$

(iv) Quotient Law: If $\lim\limits_{x \to c} g(x) \neq 0$, then $\lim\limits_{x \to c} \dfrac{f(x)}{g(x)}$ exists and

$$\lim_{x \to c} \frac{f(x)}{g(x)} = \frac{\lim\limits_{x \to c} f(x)}{\lim\limits_{x \to c} g(x)}$$

(v) Powers and Roots: If n is a positive integer, then

$$\lim_{x \to c} [f(x)]^n = \Big(\lim_{x \to c} f(x)\Big)^n, \qquad \lim_{x \to c} \sqrt[n]{f(x)} = \sqrt[n]{\lim_{x \to c} f(x)}$$

In the second limit, assume that $\lim\limits_{x \to c} f(x) \geq 0$ if n is even.

If p, q are integers with $q \neq 0$, then $\lim\limits_{x \to c} [f(x)]^{p/q}$ exists and

$$\lim_{x \to c} [f(x)]^{p/q} = \Big(\lim_{x \to c} f(x)\Big)^{p/q}$$

Assume that $\lim\limits_{x \to c} f(x) \geq 0$ if q is even, and that $\lim\limits_{x \to c} f(x) \neq 0$ if $p/q < 0$.

Before proceeding to the examples, we make some useful remarks.

- The Sum and Product Laws are valid for any number of functions. For example,

$$\lim_{x \to c} \big(f_1(x) + f_2(x) + f_3(x)\big) = \lim_{x \to c} f_1(x) + \lim_{x \to c} f_2(x) + \lim_{x \to c} f_3(x)$$

- The Sum Law has a counterpart for differences:

$$\lim_{x \to c} \big(f(x) - g(x)\big) = \lim_{x \to c} f(x) - \lim_{x \to c} g(x)$$

This follows from the Sum and Constant Multiple Laws (with $k = -1$):

$$\lim_{x \to c} \big(f(x) - g(x)\big) = \lim_{x \to c} f(x) + \lim_{x \to c} \big(-g(x)\big) = \lim_{x \to c} f(x) - \lim_{x \to c} g(x)$$

- Recall two basic limits from Theorem 1 in Section 2.2:

$$\lim_{x \to c} k = k, \qquad \lim_{x \to c} x = c$$

Applying Law (v) to $f(x) = x$, we obtain

$$\boxed{\lim_{x \to c} x^{p/q} = c^{p/q}} \qquad \boxed{1}$$

for integers p and q such that $q \neq 0$. Note, in Eq. (1) we need to assume that $c \geq 0$ if q is even and that $c \neq 0$ if $p/q < 0$.

EXAMPLE 1 Use the Basic Limit Laws to evaluate:

(a) $\displaystyle\lim_{x \to 2} x^3$ **(b)** $\displaystyle\lim_{x \to 2} (x^3 + 5x + 7)$ **(c)** $\displaystyle\lim_{x \to 2} \sqrt{x^3 + 5x + 7}$

Solution

(a) By Eq. (1), $\displaystyle\lim_{x \to 2} x^3 = 2^3 = 8$.

(b)
$$\lim_{x \to 2} (x^3 + 5x + 7) = \lim_{x \to 2} x^3 + \lim_{x \to 2} 5x + \lim_{x \to 2} 7 \qquad \text{(Sum Law)}$$

$$= \lim_{x \to 2} x^3 + 5 \lim_{x \to 2} x + \lim_{x \to 2} 7 \qquad \text{(Constant Multiple Law)}$$

$$= 8 + 5(2) + 7 = 25$$

(c) By Law (v) for roots and (b),

$$\lim_{x \to 2} \sqrt{x^3 + 5x + 7} = \sqrt{\lim_{x \to 2} (x^3 + 5x + 7)} = \sqrt{25} = 5 \qquad \blacksquare$$

EXAMPLE 2 Evaluate   **(a)** $\displaystyle\lim_{t \to -1} \frac{t + 6}{2t^4}$ and   **(b)** $\displaystyle\lim_{t \to 3} t^{-1/4}(t + 5)^{1/3}$.

Solution

(a) Use the Quotient, Sum, and Constant Multiple Laws:

$$\lim_{t \to -1} \frac{t + 6}{2t^4} = \frac{\lim\limits_{t \to -1} (t + 6)}{\lim\limits_{t \to -1} 2t^4} = \frac{\lim\limits_{t \to -1} t + \lim\limits_{t \to -1} 6}{2 \lim\limits_{t \to -1} t^4} = \frac{-1 + 6}{2(-1)^4} = \frac{5}{2}$$

> You may have noticed that each of the limits in Examples 1 and 2 could have been evaluated by a simple substitution. For example, set $t = -1$ to evaluate
>
> $$\lim_{t \to -1} \frac{t + 6}{2t^4} = \frac{-1 + 6}{2(-1)^4} = \frac{5}{2}$$
>
> Substitution is valid when the function is **continuous**, a concept we shall study in the next section.

(b) Use the Product, Powers, and Sum Laws:

$$\lim_{t \to 3} t^{-1/4}(t + 5)^{1/3} = \left(\lim_{t \to 3} t^{-1/4}\right)\left(\lim_{t \to 3} \sqrt[3]{t + 5}\right) = \left(3^{-1/4}\right)\left(\sqrt[3]{\lim_{t \to 3} t + 5}\right)$$

$$= 3^{-1/4}\sqrt[3]{3 + 5} = 3^{-1/4}(2) = \frac{2}{3^{1/4}} \qquad \blacksquare$$

The next example reminds us that the Basic Limit Laws apply only when the limits of both $f(x)$ and $g(x)$ exist.

EXAMPLE 3 **Assumptions Matter** Show that the Product Law cannot be applied to $\displaystyle\lim_{x \to 0} f(x)g(x)$ if $f(x) = x$ and $g(x) = x^{-1}$.

Solution For all $x \neq 0$, we have $f(x)g(x) = x \cdot x^{-1} = 1$, so the limit of the product exists:

$$\lim_{x \to 0} f(x)g(x) = \lim_{x \to 0} 1 = 1$$

However, there is an issue with the product of the limits because $\displaystyle\lim_{x \to 0} x^{-1}$ does not exist (since $g(x) = x^{-1}$ becomes infinite as $x \to 0$). Therefore, the Product Law cannot be applied and its conclusion does not hold even though the limit of the products does exist. Specifically, $\lim\limits_{x\to 0} f(x)g(x) = 1$, but the product of the limits is not defined:

$$1 \neq \left(\lim_{x\to 0} f(x) \right)\left(\lim_{x\to 0} g(x) \right) = \left(\lim_{x\to 0} x \right)\underbrace{\left(\lim_{x\to 0} x^{-1} \right)}_{\text{Does not exist}}$$

∎

2.3 SUMMARY

- The Basic Limit Laws: If $\lim\limits_{x\to c} f(x)$ and $\lim\limits_{x\to c} g(x)$ both exist, then

  (i) $\lim\limits_{x\to c} \big(f(x) + g(x)\big) = \lim\limits_{x\to c} f(x) + \lim\limits_{x\to c} g(x)$

  (ii) $\lim\limits_{x\to c} kf(x) = k \lim\limits_{x\to c} f(x)$

  (iii) $\lim\limits_{x\to c} f(x) g(x) = \left(\lim\limits_{x\to c} f(x) \right)\left(\lim\limits_{x\to c} g(x) \right)$

  (iv) If $\lim\limits_{x\to c} g(x) \neq 0$, then $\lim\limits_{x\to c} \dfrac{f(x)}{g(x)} = \dfrac{\lim\limits_{x\to c} f(x)}{\lim\limits_{x\to c} g(x)}$

  (v) If p, q are integers with $q \neq 0$,

  $$\lim_{x\to c}[f(x)]^{p/q} = \left(\lim_{x\to c} f(x) \right)^{p/q}$$

  For n a positive integer,

  $$\lim_{x\to c}[f(x)]^n = \left(\lim_{x\to c} f(x) \right)^n, \qquad \lim_{x\to c} \sqrt[n]{f(x)} = \sqrt[n]{\lim_{x\to c} f(x)}$$

- If $\lim\limits_{x\to c} f(x)$ or $\lim\limits_{x\to c} g(x)$ does not exist, then the Basic Limit Laws cannot be applied.

2.3 EXERCISES

Preliminary Questions

1. State the Sum Law and Quotient Law.

2. Which of the following is a verbal version of the Product Law (assuming the limits exist)?

(a) The product of two functions has a limit.

(b) The limit of the product is the product of the limits.

(c) The product of a limit is a product of functions.

(d) A limit produces a product of functions.

3. Which statement is correct? The Quotient Law does not hold if

(a) The limit of the denominator is zero

(b) The limit of the numerator is zero

Exercises

In Exercises 1–26, evaluate the limit using the Basic Limit Laws and the limits $\lim\limits_{x\to c} x^{p/q} = c^{p/q}$ *and* $\lim\limits_{x\to c} k = k$.

1. $\lim\limits_{x\to 9} x$

2. $\lim\limits_{x\to -3} 14$

3. $\lim\limits_{x\to \frac{1}{2}} x^4$

4. $\lim\limits_{z\to 27} z^{2/3}$

5. $\lim\limits_{t\to 2} t^{-1}$

6. $\lim\limits_{x\to 5} x^{-2}$

7. $\lim\limits_{x\to 0.2} (3x + 4)$

8. $\lim\limits_{x\to \frac{1}{3}} (3x^3 + 2x^2)$

9. $\lim\limits_{x\to -1} (3x^4 - 2x^3 + 4x)$

10. $\lim\limits_{x\to 8} (3x^{2/3} - 16x^{-1})$

11. $\lim\limits_{x\to 2} (x + 1)(3x^2 - 9)$

12. $\lim\limits_{x\to \frac{1}{2}} (4x + 1)(6x - 1)$

13. $\lim\limits_{t\to 4} \dfrac{1}{t + 4}$

14. $\lim\limits_{z\to 0} \dfrac{3}{z - 1}$

15. $\lim\limits_{t\to 4} \dfrac{3t - 14}{t + 1}$

16. $\lim\limits_{z\to 9} \dfrac{\sqrt{z}}{z - 2}$

17. $\lim\limits_{y\to \frac{1}{4}} (16y + 1)(2y^{1/2} + 1)$

18. $\lim\limits_{x\to 2} x(x + 1)(x + 2)$

19. $\lim\limits_{y\to 4} \dfrac{1}{\sqrt{6y + 1}}$

20. $\lim\limits_{w\to 7} \dfrac{\sqrt{w + 2} + 1}{\sqrt{w - 3} - 1}$

21. $\lim\limits_{x\to -1} \dfrac{x}{x^3 + 4x}$

22. $\lim\limits_{t\to -1} \dfrac{t^2 + 1}{(t^3 + 2)(t^4 + 1)}$

23. $\displaystyle\lim_{t\to 25} \frac{3\sqrt{t} - \frac{1}{5}t}{(t-20)^2}$

24. $\displaystyle\lim_{y\to\frac{1}{3}} (18y^2 - 4)^4$

25. $\displaystyle\lim_{t\to\frac{3}{2}} (4t^2 + 8t - 5)^{3/2}$

26. $\displaystyle\lim_{t\to 7} \frac{(t+2)^{1/2}}{(t+1)^{2/3}}$

27. Use the Quotient Law to prove that if $\displaystyle\lim_{x\to c} f(x)$ exists and is nonzero, then

$$\lim_{x\to c} \frac{1}{f(x)} = \frac{1}{\displaystyle\lim_{x\to c} f(x)}$$

28. Assuming that $\displaystyle\lim_{x\to 6} f(x) = 4$, compute:

(a) $\displaystyle\lim_{x\to 6} f(x)^2$

(b) $\displaystyle\lim_{x\to 6} \frac{1}{f(x)}$

(c) $\displaystyle\lim_{x\to 6} x\sqrt{f(x)}$

In Exercises 29–32, evaluate the limit assuming that $\displaystyle\lim_{x\to -4} f(x) = 3$ *and* $\displaystyle\lim_{x\to -4} g(x) = 1$.

29. $\displaystyle\lim_{x\to -4} f(x)g(x)$

30. $\displaystyle\lim_{x\to -4} (2f(x) + 3g(x))$

31. $\displaystyle\lim_{x\to -4} \frac{g(x)}{x^2}$

32. $\displaystyle\lim_{x\to -4} \frac{f(x)+1}{3g(x)-9}$

33. 📝 Can the Quotient Law be applied to evaluate $\displaystyle\lim_{x\to 0} \frac{\sin x}{x}$? Explain.

34. Show that the Product Law cannot be used to evaluate the limit $\displaystyle\lim_{x\to\pi/2} \left(x - \frac{\pi}{2}\right) \tan x$.

35. Assume that if $\displaystyle\lim_{x\to a} f(x) = L$, then $\displaystyle\lim_{x\to a} \sin f(x) = \sin L$. In each case evaluate the limit or indicate that the limit does not exist.

(a) $\displaystyle\lim_{x\to 0} \sin\left(\frac{x}{x-1}\right)$

(b) $\displaystyle\lim_{x\to\pi/2} \frac{\sin x}{x}$

(c) $\displaystyle\lim_{x\to 1} \frac{3x}{\sin(1-x)}$

(d) $\displaystyle\lim_{x\to 1} x^2 \sin(\pi x^2)$

36. Assume that if $\displaystyle\lim_{x\to a} f(x) = L$, then $\displaystyle\lim_{x\to a} \cos f(x) = \cos L$. In each case evaluate the limit or indicate that the limit does not exist.

(a) $\displaystyle\lim_{x\to 0} \cos\left(\frac{2x}{1-2x}\right)$

(b) $\displaystyle\lim_{x\to\pi/2} \frac{\cos x}{x}$

(c) $\displaystyle\lim_{x\to 1} x^3 \cos(1-x)$

(d) $\displaystyle\lim_{x\to 0} \frac{1-x^2}{1-\cos(x^2)}$

37. Give an example where $\displaystyle\lim_{x\to 0}(f(x) + g(x))$ exists but neither $\displaystyle\lim_{x\to 0} f(x)$ nor $\displaystyle\lim_{x\to 0} g(x)$ exists.

38. Give an example where $\displaystyle\lim_{x\to 0}(f(x) \cdot g(x))$ exists but neither $\displaystyle\lim_{x\to 0} f(x)$ nor $\displaystyle\lim_{x\to 0} g(x)$ exists.

39. Give an example where $\displaystyle\lim_{x\to 0} \frac{f(x)}{g(x)}$ exists but neither $\displaystyle\lim_{x\to 0} f(x)$ nor $\displaystyle\lim_{x\to 0} g(x)$ exists.

Further Insights and Challenges

40. Show that if both $\displaystyle\lim_{x\to c} f(x)\,g(x)$ and $\displaystyle\lim_{x\to c} g(x)$ exist and $\displaystyle\lim_{x\to c} g(x) \neq 0$, then $\displaystyle\lim_{x\to c} f(x)$ exists. *Hint:* Write $f(x) = \dfrac{f(x)\,g(x)}{g(x)}$.

41. Suppose that $\displaystyle\lim_{t\to 3} t g(t) = 12$. Show that $\displaystyle\lim_{t\to 3} g(t)$ exists and equals 4.

42. Prove that if $\displaystyle\lim_{t\to 3} \frac{h(t)}{t} = 5$, then $\displaystyle\lim_{t\to 3} h(t) = 15$.

43. 📝 Assuming that $\displaystyle\lim_{x\to 0} \frac{f(x)}{x} = 1$, which of the following statements is necessarily true? Why?

(a) $f(0) = 0$

(b) $\displaystyle\lim_{x\to 0} f(x) = 0$

44. Prove that if $\displaystyle\lim_{x\to c} f(x) = L \neq 0$ and $\displaystyle\lim_{x\to c} g(x) = 0$, then the limit $\displaystyle\lim_{x\to c} \frac{f(x)}{g(x)}$ does not exist.

45. 📝 Suppose that $\displaystyle\lim_{h\to 0} g(h) = L$.

(a) Explain why $\displaystyle\lim_{h\to 0} g(ah) = L$ for any constant $a \neq 0$.

(b) If we assume instead that $\displaystyle\lim_{h\to 1} g(h) = L$, is it still necessarily true that $\displaystyle\lim_{h\to 1} g(ah) = L$?

(c) Illustrate (a) and (b) with the function $f(x) = x^2$.

46. Assume that $L(a) = \displaystyle\lim_{x\to 0} \frac{a^x - 1}{x}$ exists for all $a > 0$. Assume also that $\displaystyle\lim_{x\to 0} a^x = 1$.

(a) Prove that $L(ab) = L(a) + L(b)$ for $a, b > 0$.
Hint: $(ab)^x - 1 = a^x b^x - a^x + a^x - 1 = a^x(b^x - 1) + (a^x - 1)$.
[This shows that $L(a)$ behaves like a logarithm, in the sense that $\ln(ab) = \ln a + \ln b$. In fact, it can be shown that $L(a) = \ln a$.]

(b) Verify numerically that $L(12) = L(3) + L(4)$.

2.4 Limits and Continuity

In everyday speech, the word "continuous" means having no breaks or interruptions. In calculus, continuity is used to describe functions whose graphs have no breaks. If we imagine the graph of a function f as a wavy metal wire, then f is continuous if its graph consists of a single piece of wire as in Figure 1.

Many physical phenomena can be considered as continuous. Our position and velocity vary continuously with time. Barometric pressure varies continuously with altitude above the earth. The current in a simple circuit varies continuously with the voltage applied to it. Ultimately, when we determine the rate of change of a function as we do in the next chapter, we will need the function to be continuous for the mathematics to work properly.

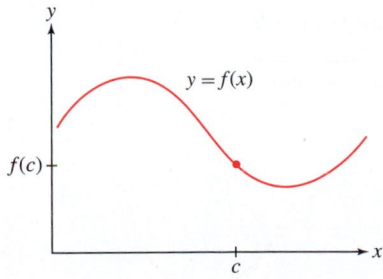

FIGURE 1 f is continuous at $x = c$.

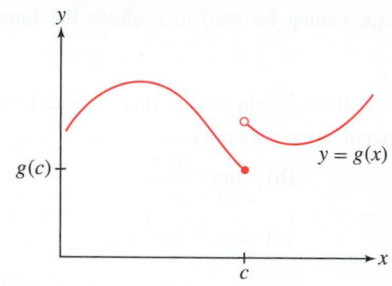

DF **FIGURE 2** Discontinuity at $x = c$: The left- and right-hand limits as $x \to c$ are not equal.

A break in the wire as in Figure 2 is called a **discontinuity**. Observe in Figure 2 that the break in the graph occurs because the left- and right-hand limits as x approaches c are not equal and thus $\lim_{x \to c} g(x)$ does not exist. By contrast, in Figure 1, $\lim_{x \to c} f(x)$ exists and is equal to the function value $f(c)$. This suggests the following definition of continuity in terms of limits.

> **DEFINITION   Continuity at a Point**   Assume that $f(x)$ is defined on an open interval containing $x = c$. Then f is **continuous** at $x = c$ if
>
> $$\lim_{x \to c} f(x) = f(c)$$
>
> If the limit does not exist, or if it exists but is not equal to $f(c)$, we say that f has a **discontinuity** (or is **discontinuous**) at $x = c$.

Note that for f to be continuous at c, three conditions must hold:

1. $f(c)$ is defined.   **2.** $\lim_{x \to c} f(x)$ exists.   **3.** They are equal.

A function f may be continuous at some points and discontinuous at others. If f is continuous at all points in its domain, then f is simply called continuous.

EXAMPLE 1   Show that the following functions are continuous:

(a) $f(x) = k$ (k any constant)   **(b)** $g(x) = x^n$ (n a whole number)

Solution

(a) We have $\lim_{x \to c} f(x) = \lim_{x \to c} k = k$ and $f(c) = k$. The limit exists and is equal to the function value for all c, so f is continuous (Figure 3).

(b) By Eq. (1) in Section 2.3, $\lim_{x \to c} g(x) = \lim_{x \to c} x^n = c^n$ for all c. Also $g(c) = c^n$, so again, the limit exists and is equal to the function value. Therefore, g is continuous. (Figure 4 illustrates the case $n = 1$.) ■

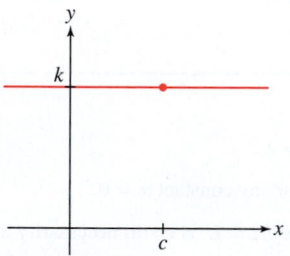

FIGURE 3 The function $f(x) = k$ is continuous.

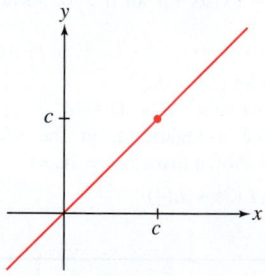

FIGURE 4 The function $g(x) = x$ is continuous.

Examples of Discontinuities

To understand continuity better, let's consider some ways in which a function can fail to be continuous. Keep in mind that continuity at a point $x = c$ requires that:

1. $f(c)$ is defined.   **2.** $\lim_{x \to c} f(x)$ exists.   **3.** They are equal.

If $\lim_{x \to c} f(x)$ exists, but either the limit is not equal to $f(c)$, or $f(c)$ is not defined, then we say that f has a **removable discontinuity** at $x = c$. The function in Figure 5(A) has a removable discontinuity at $c = 2$ because

$$f(2) = 10 \quad \text{but} \quad \underbrace{\lim_{x \to 2} f(x) = 5}$$

Limit exists but is not equal to function value

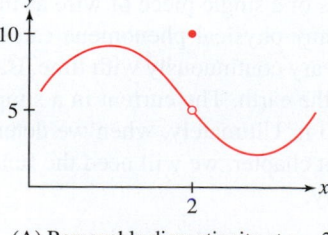

(A) Removable discontinuity at $x = 2$

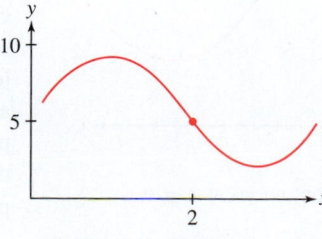

(B) Function redefined at $x = 2$

FIGURE 5 Removable discontinuity: The discontinuity can be removed by redefining $f(2)$.

Removable discontinuities are mild in the following sense: We can make f continuous at $x = c$ by redefining $f(c)$ [in the case $\lim\limits_{x \to c} f(x) \neq f(c)$] or defining $f(c)$ [in the case $f(c)$ is not defined] so that $f(c) = \lim\limits_{x \to c} f(x)$. In Figure 5(B), $f(2)$ has been redefined as $f(2) = 5$, and this makes f continuous at $x = 2$.

*A removable discontinuity at $x = c$ that occurs because $f(c)$ is not defined is sometimes referred to as a **removable singularity**.*

EXAMPLE 2   Show that $g(x) = \dfrac{x^3 - 8}{x - 2}$ has a removable discontinuity at $x = 2$. How should $g(2)$ be defined so that g is continuous at $x = 2$?

Solution   First note that g is not defined at $x = 2$ since evaluating g at 2 involves division by 0. Also,

$$\lim_{x \to 2} g(x) = \lim_{x \to 2} \frac{x^3 - 8}{x - 2} = \lim_{x \to 2} \frac{(x - 2)(x^2 + 2x + 4)}{x - 2} = \lim_{x \to 2} (x^2 + 2x + 4) = 12$$

where the Basic Limit Laws are used to determine the value of the limit. Since $\lim\limits_{x \to 2} g(x)$ exists, but $g(2)$ is not defined, g has a removable discontinuity at $x = 2$. If we define $g(2) = 12$, then g would be continuous at $x = 2$. ∎

A worse type of discontinuity is a **jump discontinuity**, which occurs if the one-sided limits $\lim\limits_{x \to c^-} f(x)$ and $\lim\limits_{x \to c^+} f(x)$ exist but are not equal. In this case f is not continuous at c because $\lim\limits_{x \to c} f(x)$ does not exist. Figure 6 shows two functions with jump discontinuities at $c = 2$. Unlike the removable case, we cannot make f continuous simply by redefining f at the single point c.

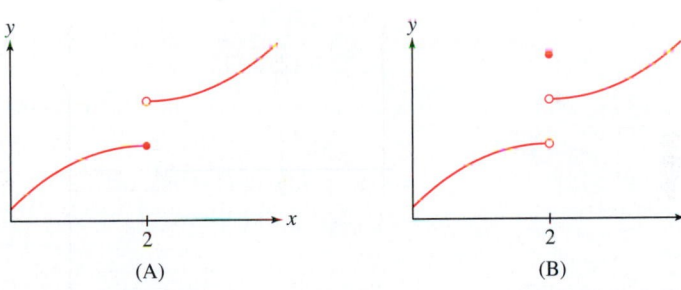

FIGURE 6 These functions have jump discontinuities at $x = 2$.

In connection with jump discontinuities, it is convenient to define *one-sided continuity*.

DEFINITION   One-Sided Continuity   A function f is called

- **Left-continuous** at $x = c$ if $\lim\limits_{x \to c^-} f(x) = f(c)$
- **Right-continuous** at $x = c$ if $\lim\limits_{x \to c^+} f(x) = f(c)$

In Figure 6 above, the function in (A) is left-continuous at $x = 2$ but the function in (B) is neither left- nor right-continuous at $x = 2$.

Many theorems in calculus apply to functions that are continuous on an interval, a concept that is defined as follows:

DEFINITION   Continuity on an Interval I   Assume that I is an interval in the form (a, b), $[a, b)$, $(a, b]$, or $[a, b]$. Then f is **continuous on I** if f is continuous at each point in (a, b), f is right-continuous at a (if a is in I), and f is left-continuous at b (if b is in I).

The next example explores one-sided continuity using a piecewise-defined function—that is, a function defined by different formulas on different intervals.

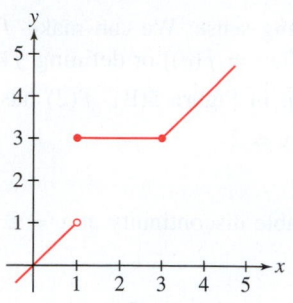

FIGURE 7 Piecewise-defined function F in Example 3.

CAUTION *Piecewise-defined functions may or may not be continuous at points where they are pieced together.*

EXAMPLE 3   Piecewise-Defined Function   Discuss the continuity of

$$F(x) = \begin{cases} x & \text{for } x < 1 \\ 3 & \text{for } 1 \le x \le 3 \\ x & \text{for } x > 3 \end{cases}$$

Solution   The functions $f(x) = x$ and $g(x) = 3$ are continuous, so F is also continuous, except possibly at the transition points $x = 1$ and $x = 3$, where the formula for $F(x)$ changes (Figure 7).

- At $x = 1$, the one-sided limits exist but are not equal:

$$\lim_{x \to 1^-} F(x) = \lim_{x \to 1^-} x = 1, \qquad \lim_{x \to 1^+} F(x) = \lim_{x \to 1^+} 3 = 3$$

Thus, F has a jump discontinuity at $x = 1$. However, the right-hand limit is equal to the function value $F(1) = 3$, so F is *right-continuous* at $x = 1$.

- At $x = 3$, the left- and right-hand limits exist and both are equal to $F(3)$, so F is *continuous* at $x = 3$:

$$\lim_{x \to 3^-} F(x) = \lim_{x \to 3^-} 3 = 3, \qquad \lim_{x \to 3^+} F(x) = \lim_{x \to 3^+} x = 3 \qquad \blacksquare$$

We say that f has an **infinite discontinuity** at $x = c$ if one or both of the one-sided limits are infinite [even if $f(x)$ itself is not defined at $x = c$]. Like with a jump discontinuity, in this case f is not continuous at c because $\lim_{x \to c} f(x)$ does not exist. Figure 8 illustrates three types of infinite discontinuities occurring at $x = 2$. Notice that $x = 2$ does not belong to the domain of the function in cases (A) and (B).

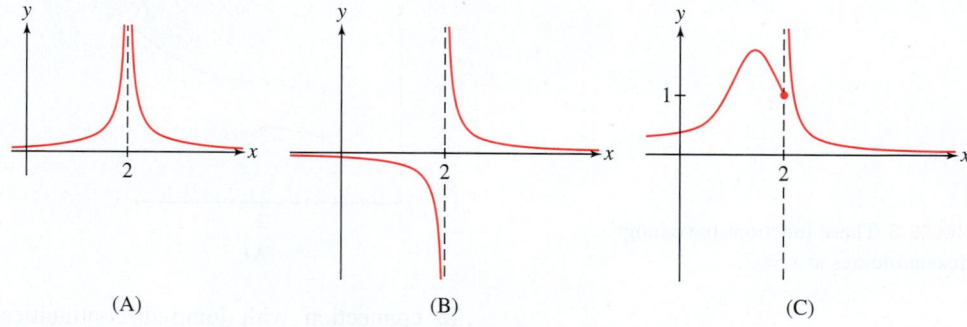

(A)   (B)   (C)

FIGURE 8 Functions with an infinite discontinuity at $x = 2$.

EXAMPLE 4   The Intensity of a Light Source   A standard model for the intensity I of a light source at varying distances d from the light is an inverse-square law, $I(d) = k/d^2$ for a constant $k > 0$ depending on the light source (Figure 9). Show that I has an infinite discontinuity at $d = 0$.

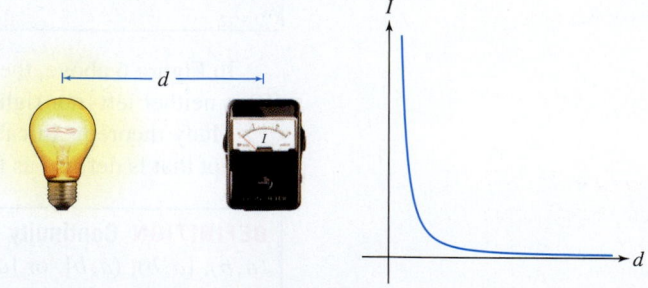

FIGURE 9 Light intensity is inversely proportional to d^2.

Solution   Regardless of the value of $k > 0$, as d approaches 0 from the right, the values of $I(d) = k/d^2$ are positive and become arbitrarily large. Therefore $\lim_{d \to 0^+} I(d) = \infty$, and it follows that I has an infinite discontinuity at 0.

Note that this does not mean that the intensity of the light is actually unbounded as we get closer and closer to it. The relationship $I(d) = k/d^2$ is a *model* for the true behavior. The model does, however, properly reflect the fact that the intensity rises rapidly as we approach the source. ∎

Finally, we note that some functions have more severe types of discontinuities than those discussed above. For example, $f(x) = \sin \frac{1}{x}$ oscillates infinitely often between $+1$ and -1 as $x \to 0$ (Figure 10). Neither the left- nor the right-hand limit exists at $x = 0$, so this discontinuity is not a jump discontinuity. See Exercises 96 and 97 for even stranger examples.

DF **FIGURE 10** Graph of $y = \sin \frac{1}{x}$. The discontinuity at $x = 0$ is not a jump, removable, or infinite discontinuity.

Building Continuous Functions

Having studied some examples of discontinuities, we focus again on continuous functions. How can we show that a function is continuous? One way is to use the **Laws of Continuity**, which state, roughly speaking, that a function is continuous if it is built out of functions that are known to be continuous.

THEOREM 1 **Basic Laws of Continuity** If f and g are continuous at $x = c$, then the following functions are also continuous at $x = c$:

  (i) $f + g$ and $f - g$ **(iii)** fg

  (ii) kf for any constant k **(iv)** f/g if $g(c) \neq 0$

Proof These laws follow directly from the corresponding Basic Limit Laws (Theorem 1, Section 2.3). We illustrate by proving the first part of (i) in detail. The remaining laws are proved similarly. By definition, we must show that $\lim\limits_{x \to c}(f(x) + g(x)) = f(c) + g(c)$. Because f and g are both continuous at $x = c$, we have

$$\lim_{x \to c} f(x) = f(c), \qquad \lim_{x \to c} g(x) = g(c)$$

The Sum Law for limits yields the desired result:

$$\lim_{x \to c}(f(x) + g(x)) = \lim_{x \to c} f(x) + \lim_{x \to c} g(x) = f(c) + g(c) \qquad ∎$$

In Section 2.3, we noted that the Basic Limit Laws for Sums and Products are valid for an arbitrary number of functions. The same is true for continuity; that is, if $f_1, \ldots, f_n$ are continuous, then so are the functions

$$f_1 + f_2 + \cdots + f_n, \qquad f_1 f_2 \cdots f_n$$

When a function f is defined and continuous for all values of x, we say that f is continuous on the real line.

The next two theorems assert that the basic functions are continuous on their domains. Recall (Section 1.3) that the term "basic function" refers to polynomials, rational functions, nth-root and algebraic functions, trigonometric functions and their inverses, and exponential and logarithmic functions.

*◀ **REMINDER** A rational function is a quotient of two polynomials P/Q.*

THEOREM 2 **Continuity of Polynomial and Rational Functions** Let P and Q be polynomials. Then:

- P and Q are continuous on the real line.
- P/Q is continuous on its domain [at all values $x = c$ such that $Q(c) \neq 0$].

Proof The function $f(x) = x^m$ is continuous for all whole numbers m by Example 1. By Continuity Law (ii), $f(x) = ax^m$ is continuous for every constant a. A polynomial

$$P(x) = a_n x^n + a_{n-1} x^{n-1} + \cdots + a_1 x + a_0$$

is a sum of continuous functions, so it too is continuous. By Continuity Law (iv), a quotient function P/Q is continuous at $x = c$, provided that $Q(c) \neq 0$. ∎

This result shows, for example, that $f(x) = 3x^4 - 2x^3 + 8x$ is continuous for all x, and that

$$g(x) = \frac{x + 3}{x^2 - 1}$$

is continuous for $x \neq \pm 1$. Note that if n is a positive integer, then $f(x) = x^{-n}$ is continuous for $x \neq 0$ because $f(x) = x^{-n} = 1/x^n$ is a rational function.

The continuity of the nth-root, sine, cosine, exponential, and logarithmic functions should not be surprising because their graphs have no visible breaks (Figure 11). However, complete proofs of continuity are somewhat technical and are omitted.

← REMINDER The domain of $y = x^{1/n}$ is the real line if n is odd and the half-line $[0, \infty)$ if n is even.

THEOREM 3 Continuity of Some Basic Functions

- $y = x^{1/n}$ is continuous on its domain for n a natural number.
- $y = \sin x$ and $y = \cos x$ are continuous on the real line.
- $y = b^x$ is continuous on the real line (for $b > 0$, $b \neq 1$).
- $y = \log_b x$ is continuous for $x > 0$ (for $b > 0$, $b \neq 1$).

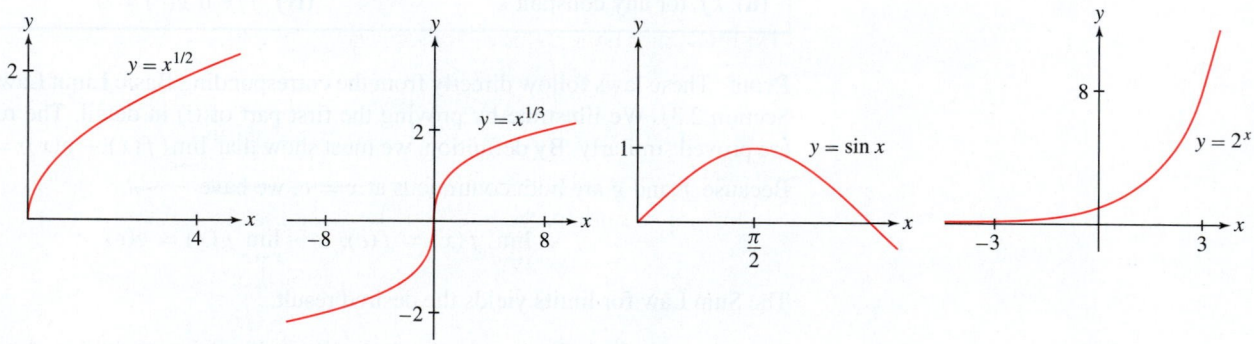

FIGURE 11 As the graphs suggest, these functions are continuous on their domains.

Because $f(x) = \sin x$ and $f(x) = \cos x$ are continuous, the Continuity Law (iv) for Quotients implies that the other standard trigonometric functions are continuous on their domains, consisting of the values of x where the denominators, in the following quotient expressions for them, are nonzero:

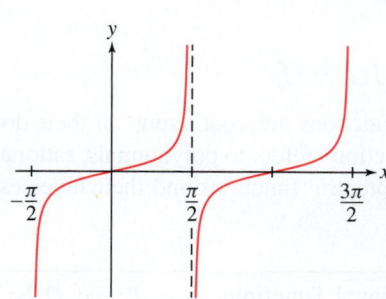

FIGURE 12 Graph of $y = \tan x$.

$$\tan x = \frac{\sin x}{\cos x}, \qquad \cot x = \frac{\cos x}{\sin x}, \qquad \sec x = \frac{1}{\cos x}, \qquad \csc x = \frac{1}{\sin x}$$

They have infinite discontinuities at points where the denominators are zero. For example, as illustrated in Figure 12, $f(x) = \tan x$ has infinite discontinuities at the points

$$x = \pm\frac{\pi}{2}, \quad \pm\frac{3\pi}{2}, \quad \pm\frac{5\pi}{2}, \ldots$$

The next theorem states that the inverse f^{-1} of a continuous function f is continuous. This is to be expected because the graph of f^{-1} is the reflection of the graph of f through the line $y = x$. If the graph of f has "no breaks," the same ought to be true of the graph of f^{-1}.

THEOREM 4 Continuity of the Inverse Function   If f is continuous on an interval I with range R, and if f^{-1} exists, then f^{-1} is continuous with domain R.

One consequence of this theorem is that the logarithms, nth roots, and inverse trigonometric functions $[f(x) = \sin^{-1} x, \ f(x) = \cos^{-1} x, \ f(x) = \tan^{-1} x,$ and so on$]$ are all continuous on their domains.

Finally, it is important to know that a composition of continuous functions is again continuous. The following theorem is proved in Appendix D.

THEOREM 5 Continuity of Composite Functions   If g is continuous at $x = c$, and f is continuous at $x = g(c)$, then the composite function $F(x) = f(g(x))$ is continuous at $x = c$.

For example, $F(x) = (x^2 + 9)^{1/3}$ is continuous because it is the composite of the continuous functions $f(x) = x^{1/3}$ and $g(x) = x^2 + 9$. Similarly, $F(x) = \cos(x^{-1})$ is continuous for all $x \neq 0$, and $F(x) = 2^{\sin x}$ is continuous for all x.

More generally, an **elementary function** is a function that is constructed out of basic functions using the operations of addition, subtraction, multiplication, division, and composition. Since the basic functions are continuous (on their domains), an elementary function is also continuous on its domain by the Laws of Continuity. An example of an elementary function is

$$F(x) = \tan^{-1}\left(\frac{x^2 + \cos(2^x + 9)}{x - 8}\right)$$

This function is continuous on its domain $\{x : x \neq 8\}$.

Substitution: Evaluating Limits Using Continuity

It is easy to evaluate a limit when the function in question is known to be continuous. In this case, by definition, the limit is equal to the function value:

$$\lim_{x \to c} f(x) = f(c)$$

We call this the **Substitution Method** because the limit is evaluated by substituting $x = c$ in $f(x)$.

EXAMPLE 5   Evaluate   **(a)** $\displaystyle\lim_{y \to \frac{\pi}{3}} \sin y$ and   **(b)** $\displaystyle\lim_{x \to -1} \frac{3^x}{\sqrt{x + 5}}$.

Solution

(a) We can use substitution because $f(y) = \sin y$ is continuous.

$$\lim_{y \to \frac{\pi}{3}} \sin y = \sin\frac{\pi}{3} = \frac{\sqrt{3}}{2}$$

(b) The function $f(x) = 3^x/\sqrt{x + 5}$ is continuous at $x = -1$ because the numerator and denominator are both continuous at $x = -1$ and the denominator $\sqrt{x + 5}$ is nonzero at $x = -1$. Therefore, we can use substitution:

$$\lim_{x \to -1} \frac{3^x}{\sqrt{x + 5}} = \frac{3^{-1}}{\sqrt{-1 + 5}} = \frac{1}{6}$$ ∎

The **greatest integer function** $f(x) = \lfloor x \rfloor$ is the function defined by $\lfloor x \rfloor = n$, where n is the unique integer such that $n \leq x < n + 1$ (Figure 13). For example, $\lfloor 4.7 \rfloor = 4$ and $\lfloor -2.3 \rfloor = -3$. This function is called the greatest integer function because $\lfloor x \rfloor = n$ represents the greatest integer less than or equal to x.

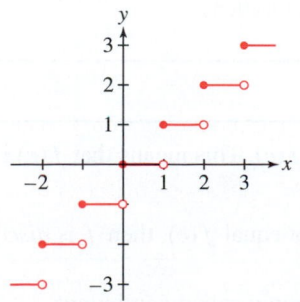

FIGURE 13 Graph of $f(x) = \lfloor x \rfloor$.

EXAMPLE 6 **Assumptions Matter**   Can we evaluate $\lim_{x \to 2} \lfloor x \rfloor$ using substitution?

Solution Substitution cannot be applied because $f(x) = \lfloor x \rfloor$ is not continuous at $x = 2$. In fact, $\lim_{x \to 2} \lfloor x \rfloor$ does not exist, and that follows because the one-sided limits are not equal:

$$\lim_{x \to 2^+} \lfloor x \rfloor = 2 \quad \text{and} \quad \lim_{x \to 2^-} \lfloor x \rfloor = 1 \qquad \blacksquare$$

CONCEPTUAL INSIGHT Real-World Modeling by Continuous Functions   Continuous functions are often used to model relationships between physical quantities such as position and time, temperature and altitude, and voltage and resistance. This reflects our everyday experience that change in the physical world tends to occur continuously rather than through abrupt transitions. However, mathematical models are approximations to reality and are based on assumptions about the phenomenon being studied. It is always important to be aware of a model's assumptions and the limitations they impose.

In Figure 14, atmospheric temperature is represented as a continuous function of altitude. At such a large scale the assumption of continuity is reasonable because it is consistent with our experience. However, at smaller, less familiar scales the situation can be different. In fact, in 2002 scientists McGaughey and Ward reported observing temperature discontinuities at the surface of evaporating water droplets, suggesting that it may not be appropriate to assume temperature is continuous at such small scales.

The size of a population is often treated as a continuous function of time. Strictly speaking, population size is a whole number that changes by ±1 when an individual is born or dies, so it really is not continuous. At the scale of the size of your family, it does not make sense to consider the number of people as a continuous variable. However, if a population is large, the effect of an individual birth or death is small, and at such a scale it is both reasonable and convenient to treat population as a continuous function of time. Ultimately, the test of a model is how well it enables us to understand and predict the behavior of the actual system. When it fails to do so, the assumptions need to be examined, and possibly adjusted, to try to find a better fit between model and reality.

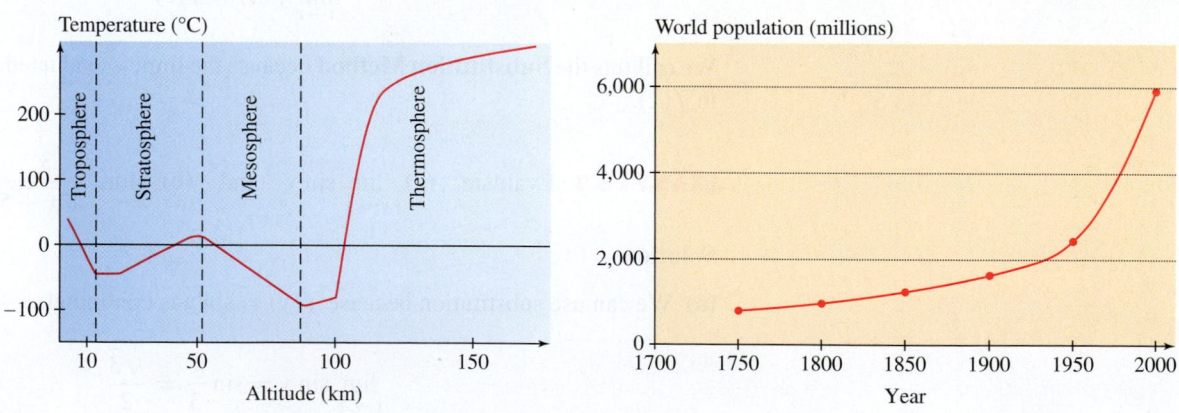

FIGURE 14 Atmospheric temperature and world population are modeled by continuous functions.

2.4 SUMMARY

- Definition: f is *continuous* at $x = c$ if $\lim_{x \to c} f(x) = f(c)$. This means that $f(c)$ exists, $\lim_{x \to c} f(x)$ exists, and they are equal.
- If $\lim_{x \to c} f(x)$ does not exist, or if it exists but does not equal $f(c)$, then f is *discontinuous* at $x = c$.
- If f is continuous at all points in its domain, f is simply called *continuous*.
- *Right-continuous* at $x = c$: $\lim_{x \to c^+} f(x) = f(c)$.

• *Left-continuous* at $x = c$: $\lim\limits_{x \to c^-} f(x) = f(c)$.
• Three common types of discontinuities:

 – *Removable discontinuity:* $\lim\limits_{x \to c} f(x)$ exists, but either the limit does not equal $f(c)$ or $f(c)$ is not defined.

 – *Jump discontinuity:* The one-sided limits both exist but are not equal.

 – *Infinite discontinuity:* The limit is infinite as x approaches c from one or both sides.

• Laws of Continuity: Sums, products, multiples, inverses, and composites of continuous functions are continuous. The same holds for a quotient f/g at points where $g(x) \neq 0$.

• The basic functions are continuous on their domains where the basic functions are polynomials, rational functions, nth-root and algebraic functions, trigonometric functions and their inverses, and exponential and logarithmic functions.

• Substitution Method: If f is known to be continuous at $x = c$, then the value of the limit $\lim\limits_{x \to c} f(x)$ is $f(c)$.

2.4 EXERCISES

Preliminary Questions

1. Which property of $f(x) = x^3$ allows us to conclude that $\lim\limits_{x \to 2} x^3 = 8$?

2. What can be said about $f(3)$ if f is continuous and $\lim\limits_{x \to 3} f(x) = \frac{1}{2}$?

3. Suppose that $f(x) < 0$ if x is positive and $f(x) > 1$ if x is negative. Can f be continuous at $x = 0$?

4. Is it possible to determine $f(7)$ if $f(x) = 3$ for all $x < 7$ and f is right-continuous at $x = 7$? What if f is left-continuous?

5. Are the following true or false? If false, then draw or give a counterexample, and state a correct version.

(a) f is continuous at $x = a$ if the left- and right-hand limits of $f(x)$ as $x \to a$ exist and are equal.

(b) f is continuous at $x = a$ if the left- and right-hand limits of $f(x)$ as $x \to a$ exist and equal $f(a)$.

(c) If the left- and right-hand limits of $f(x)$ as $x \to a$ exist, then f has a removable discontinuity at $x = a$.

(d) If f and g are continuous at $x = a$, then $f + g$ is continuous at $x = a$.

(e) If f and g are continuous at $x = a$, then f/g is continuous at $x = a$.

Exercises

1. Referring to Figure 15, state whether f is left- or right-continuous (or neither) at each point of discontinuity. Does f have any removable discontinuities?

Exercises 2–4 refer to the function g whose graph appears in Figure 16.

2. State whether g is left- or right-continuous (or neither) at each of its points of discontinuity.

3. At which point c does g have a removable discontinuity? How should $g(c)$ be redefined to make g continuous at $x = c$?

4. Find the point c_1 at which g has a jump discontinuity but is left-continuous. How should $g(c_1)$ be redefined to make g right-continuous at $x = c_1$?

5. In Figure 17, determine the one-sided limits at the points of discontinuity. Which discontinuity is removable and how should f be redefined to make it continuous at this point?

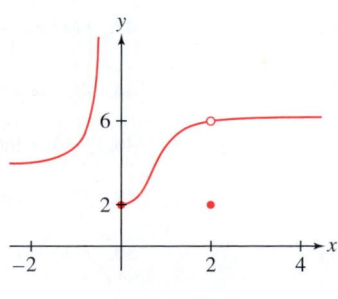

FIGURE 17

6. Suppose that $f(x) = 2$ for $x < 3$ and $f(x) = -4$ for $x > 3$.
(a) What is $f(3)$ if f is left-continuous at $x = 3$?
(b) What is $f(3)$ if f is right-continuous at $x = 3$?

In Exercises 7–16, use Theorems 1–5 to show that the function is continuous.

7. $f(x) = x + \sin x$ **8.** $f(x) = x \sin x$

9. $f(x) = 3x + 4 \sin x$ **10.** $f(x) = 3x^3 + 8x^2 - 20x$

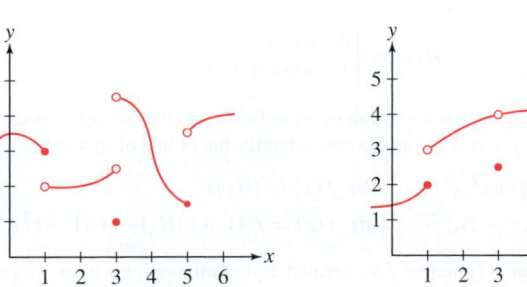

FIGURE 15 Graph of $y = f(x)$.

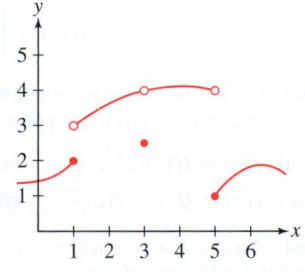

FIGURE 16 Graph of $y = g(x)$.

11. $f(x) = \dfrac{1}{x^2 + 1}$

12. $f(x) = \dfrac{x^2 - \cos x}{3 + \cos x}$

13. $f(x) = \cos(x^2)$

14. $f(x) = \tan^{-1}(4^x)$

15. $f(x) = 3^x \cos 3x$

16. $f(x) = \log_2(x^2 + 1)$

In Exercises 17–38, determine the points of discontinuity. State the type of discontinuity (removable, jump, infinite, or none of these) and whether the function is left- or right-continuous.

17. $f(x) = \dfrac{1}{x}$

18. $f(x) = |x|$

19. $f(x) = \dfrac{x - 2}{|x - 1|}$

20. $f(x) = \lfloor x \rfloor$

21. $f(x) = \left\lfloor \dfrac{x}{2} \right\rfloor$

22. $g(t) = \dfrac{1}{t^2 - 1}$

23. $h(x) = \dfrac{1}{2 - |x|}$

24. $k(x) = \dfrac{x - 2}{|2 - x|}$

25. $f(x) = \dfrac{x + 1}{4x - 2}$

26. $h(z) = \dfrac{1 - 2z}{z^2 - z - 6}$

27. $f(x) = 3x^{2/3} - 9x^3$

28. $g(t) = 3t^{-2/3} - 9t^3$

29. $f(x) = \begin{cases} \dfrac{x - 2}{|x - 2|} & x \neq 2 \\ -1 & x = 2 \end{cases}$

30. $f(x) = \begin{cases} \cos \dfrac{1}{x} & x \neq 0 \\ 1 & x = 0 \end{cases}$

31. $f(x) = \dfrac{2x^2 - 50}{x + 5}$

32. $w(t) = \dfrac{t + 1}{t^2 - 1}$

33. $g(t) = \tan 2t$

34. $f(x) = \csc(x^2)$

35. $f(x) = \tan(\sin x)$

36. $f(x) = \cos(\pi \lfloor x \rfloor)$

37. $f(x) = \dfrac{1}{e^x - e^{-x}}$

38. $f(x) = \ln|x - 4|$

In Exercises 39–52, determine the domain of the function and prove that it is continuous on its domain using Theorems 1–5.

39. $f(x) = 2\sin x + 3\cos x$

40. $f(x) = \sqrt{x^2 + 9}$

41. $f(x) = \sqrt{x} \sin x$

42. $f(x) = \dfrac{x^2}{x + x^{1/4}}$

43. $f(x) = x^{2/3} 2^x$

44. $f(x) = x^{1/3} + x^{3/4}$

45. $f(x) = x^{-4/3}$

46. $f(x) = \ln(9 - x^2)$

47. $f(x) = \tan^2 x$

48. $f(x) = \cos(2^x)$

49. $f(x) = (x^4 + 1)^{3/2}$

50. $f(x) = e^{-x^2}$

51. $f(x) = \dfrac{\cos(x^2)}{x^2 - 1}$

52. $f(x) = 9^{\tan x}$

53. The graph of the following function is shown in Figure 18.

$$f(x) = \begin{cases} x^2 + 3 & \text{for } x < 1 \\ 10 - x & \text{for } 1 \le x \le 2 \\ 6x - x^2 & \text{for } x > 2 \end{cases}$$

Show that f is continuous for $x \neq 1, 2$. Then compute the right- and left-hand limits at $x = 1, 2$, and determine whether f is left-continuous, right-continuous, or continuous at these points.

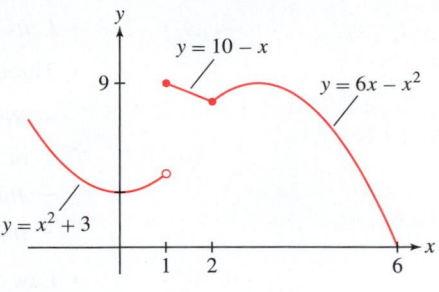

FIGURE 18

54. Sawtooth Function Draw the graph of $f(x) = x - \lfloor x \rfloor$. At which points is f discontinuous? Is it left- or right-continuous at those points?

*In Exercises 55–56, $\lceil x \rceil$ refers to the **least integer function**. It is defined by $\lceil x \rceil = n$, where n is the unique integer such that $n - 1 < x \le n$. In each case, provide the graph of f, indicate the points of discontinuity and type of each (removable, jump, infinite, or none of these), and indicate whether f is left- or right-continuous.*

55. $f(x) = \lceil x \rceil$

56. $f(x) = \lceil x \rceil - \lfloor x \rfloor$

In Exercises 57–60, sketch the graph of f. At each point of discontinuity, state whether f is left- or right-continuous.

57. $f(x) = \begin{cases} x^2 & \text{for } x \le 1 \\ 2 - x & \text{for } x > 1 \end{cases}$

58. $f(x) = \begin{cases} x + 1 & \text{for } x < 1 \\ \dfrac{1}{x} & \text{for } x \ge 1 \end{cases}$

59. $f(x) = \begin{cases} \dfrac{x^2 - 3x + 2}{|x - 2|} & x \neq 2 \\ 0 & x = 2 \end{cases}$

60. $f(x) = \begin{cases} x^3 + 1 & \text{for } -\infty < x \le 0 \\ -x + 1 & \text{for } 0 < x < 2 \\ -x^2 + 10x - 15 & \text{for } x \ge 2 \end{cases}$

61. Show that the function

$$f(x) = \begin{cases} \dfrac{x^2 - 16}{x - 4} & x \neq 4 \\ 10 & x = 4 \end{cases}$$

has a removable discontinuity at $x = 4$.

62. **GU** Define $f(x) = x \sin \dfrac{1}{x} + 2$ for $x \neq 0$. Plot f. How should $f(0)$ be defined so that f is continuous at $x = 0$?

In Exercises 63–64, H is the Heaviside function, defined by

$$H(x) = \begin{cases} 0 & \text{when } x < 0 \\ 1 & \text{when } x \ge 0 \end{cases}$$

63. In each case, sketch the graph of f, indicate whether or not f is continuous, and—if f is not continuous—identify the points of discontinuity.

(a) $f(x) = H(x)(x^2 + 1)$ **(b)** $f(x) = H(x)x$

(c) $f(x) = H(x - 2)\sqrt{x}$ **(d)** $f(x) = H(1 + x)H(1 - x)(1 - x^2)$

64. Assume that a function f is defined and continuous for all x. Under what condition on f are we assured that the function g, defined by $g(x) = H(x - a)f(x)$, is continuous?

In Exercises 65–67, find the value of the constant (a, b, or c) that makes the function continuous.

65. $f(x) = \begin{cases} x^2 - c & \text{for } x < 5 \\ 4x + 2c & \text{for } x \geq 5 \end{cases}$

66. $f(x) = \begin{cases} 2x + 9x^{-1} & \text{for } x \leq 3 \\ -4x + c & \text{for } x > 3 \end{cases}$

67. $f(x) = \begin{cases} x^{-1} & \text{for } x < -1 \\ ax + b & \text{for } -1 \leq x \leq \frac{1}{2} \\ x^{-1} & \text{for } x > \frac{1}{2} \end{cases}$

68. Define

$$g(x) = \begin{cases} x + 3 & \text{for } x < -1 \\ cx & \text{for } -1 \leq x \leq 2 \\ x + 2 & \text{for } x > 2 \end{cases}$$

Find a value of c such that g is

(a) left-continuous   **(b)** right-continuous

In each case, sketch the graph of g.

69. Define $g(t) = \tan^{-1}\left(\dfrac{1}{t-1}\right)$ for $t \neq 1$. Answer the following questions, using a plot if necessary.

(a) Can $g(1)$ be defined so that g is continuous at $t = 1$?

(b) How should $g(1)$ be defined so that g is left-continuous at $t = 1$?

70. Each of the following statements is *false*. For each statement, sketch the graph of a function that provides a counterexample.

(a) If $\lim\limits_{x \to a} f(x)$ exists, then f is continuous at $x = a$.

(b) If f has a jump discontinuity at $x = a$, then $f(a)$ is equal to either $\lim\limits_{x \to a^-} f(x)$ or $\lim\limits_{x \to a^+} f(x)$.

In Exercises 71–74, draw the graph of a function on $[0, 5]$ with the given properties.

71. f is not continuous at $x = 1$, but $\lim\limits_{x \to 1^+} f(x)$ and $\lim\limits_{x \to 1^-} f(x)$ exist and are equal.

72. f is left-continuous but not continuous at $x = 2$, and right-continuous but not continuous at $x = 3$.

73. f has a removable discontinuity at $x = 1$, a jump discontinuity at $x = 2$, and

$$\lim\limits_{x \to 3^-} f(x) = -\infty, \qquad \lim\limits_{x \to 3^+} f(x) = 2$$

74. f is right- but not left-continuous at $x = 1$, left- but not right-continuous at $x = 2$, and neither left- nor right-continuous at $x = 3$.

In Exercises 75–88, evaluate using substitution.

75. $\lim\limits_{x \to -1} (2x^3 - 4)$

76. $\lim\limits_{x \to 2} (5x - 12x^{-2})$

77. $\lim\limits_{x \to 3} \dfrac{x + 2}{x^2 + 2x}$

78. $\lim\limits_{x \to \pi} \sin\left(\dfrac{x}{2} - \pi\right)$

79. $\lim\limits_{x \to \frac{\pi}{4}} \tan(3x)$

80. $\lim\limits_{x \to \pi} \dfrac{1}{\cos x}$

81. $\lim\limits_{x \to 4} x^{-5/2}$

82. $\lim\limits_{x \to 2} \sqrt{x^3 + 4x}$

83. $\lim\limits_{x \to -1} (1 - 8x^3)^{3/2}$

84. $\lim\limits_{x \to 2} \left(\dfrac{7x + 2}{4 - x}\right)^{2/3}$

85. $\lim\limits_{x \to 3} 10^{x^2 - 2x}$

86. $\lim\limits_{x \to -\frac{\pi}{2}} 3^{\sin x}$

87. $\lim\limits_{x \to 4^-} \sin^{-1}\left(\dfrac{x}{4}\right)$

88. $\lim\limits_{x \to 0} \tan^{-1}(e^x)$

89. Suppose that f and g are discontinuous at $x = c$. Does it follow that $f + g$ is discontinuous at $x = c$? If not, give a counterexample. Does this contradict Theorem 1(i)?

90. Prove that $f(x) = |x|$ is continuous for all x. *Hint:* To prove continuity at $x = 0$, consider the one-sided limits.

91. Use the result of Exercise 90 to prove that if g is continuous, then $f(x) = |g(x)|$ is also continuous.

92. Which of the following quantities would be represented by continuous functions of time and which would have one or more discontinuities?

(a) Velocity of an airplane during a flight

(b) Temperature in a room under ordinary conditions

(c) Value of a bank account with interest paid yearly

(d) Salary of a teacher

(e) Population of the world

93. In 2017, the federal income tax T on income of x dollars (up to $91,900) was determined by the formula

$$T(x) = \begin{cases} 0.10x & \text{for } 0 \leq x < 9325 \\ 0.15x - 466.25 & \text{for } 9325 \leq x < 37{,}950 \\ 0.25x - 4261.25 & \text{for } 37{,}950 \leq x \leq 91{,}900 \end{cases}$$

Sketch the graph of T. Does T have any discontinuities? Explain why, if T had a jump discontinuity, it might be advantageous in some situations to earn *less* money.

Further Insights and Challenges

94. If f has a removable discontinuity at $x = c$, then it is possible to redefine $f(c)$ so that f is continuous at $x = c$. Can this be done in more than one way? Explain.

95. Give an example of functions f and g such that $f(g(x))$ is continuous but g has at least one discontinuity.

96. Continuous at Only One Point Show that the following function is continuous only at $x = 0$:

$$f(x) = \begin{cases} x & \text{for } x \text{ rational} \\ -x & \text{for } x \text{ irrational} \end{cases}$$

97. Show that f is a discontinuous function for all x, where $f(x)$ is defined as follows:

$$f(x) = \begin{cases} 1 & \text{for } x \text{ rational} \\ -1 & \text{for } x \text{ irrational} \end{cases}$$

Show that f^2 is continuous for all x.

2.5 Indeterminate Forms

Substitution can be used to evaluate limits when the function in question is known to be continuous. For example, $f(x) = x^{-2}$ is continuous at $x = 3$, and therefore,

$$\lim_{x \to 3} x^{-2} = 3^{-2} = \frac{1}{9}$$

When we study derivatives in Chapter 3, we will be faced with limits $\lim_{x \to c} f(x)$, where $f(c)$ is not defined. In such cases, substitution cannot be used directly. However, many of these limits can be evaluated if we use algebra to rewrite the formula for $f(x)$.

To illustrate, consider this limit (Figure 1):

$$\lim_{x \to 4} \frac{x^2 - 16}{x - 4}$$

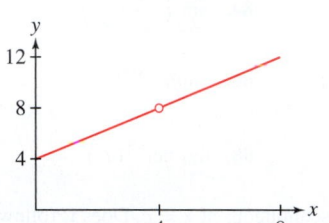

The function $f(x) = \dfrac{x^2 - 16}{x - 4}$ is not defined at $x = 4$ because the formula for $f(4)$ produces the undefined expression $0/0$. However, the numerator of $f(x)$ factors:

$$\frac{x^2 - 16}{x - 4} = \frac{(x + 4)(x - 4)}{x - 4} = x + 4 \quad \text{(valid for } x \neq 4\text{)}$$

FIGURE 1 Graph of $f(x) = \dfrac{x^2 - 16}{x - 4}$. This function is undefined at $x = 4$, but the limit as $x \to 4$ exists.

This shows that f coincides with the *continuous* function $y = x + 4$ for all $x \neq 4$. Since the limit depends only on the values of $f(x)$ for $x \neq 4$, we have

$$\lim_{x \to 4} \frac{x^2 - 16}{x - 4} = \underbrace{\lim_{x \to 4} (x + 4) = 8}_{\text{Evaluate by substitution}}$$

> Given a function f, if the formula for $f(c)$ yields an undefined expression in one of the forms $\frac{0}{0}$, $\frac{\infty}{\infty}$, $\infty \cdot 0$, or $\infty - \infty$, then we say that $f(x)$ has an **indeterminate form** (or is **indeterminate**) at $x = c$.

Some other indeterminate forms are 1^∞, ∞^0, and 0^0. These are treated in Section 4.5.

An indeterminate form, as the name suggests, indicates that the limit cannot be determined from the form. It *does not* mean that the limit does not exist. Instead, we think of it as a warning sign that tells us more work needs to be done to evaluate the limit. One strategy, when $f(x)$ has an indeterminate form at $x = c$, is to transform $f(x)$ algebraically, if possible, into a new expression that is defined and continuous at $x = c$, and then evaluate the limit by substitution. As you study the following examples, notice that the critical step is to cancel a common factor from the numerator and denominator at the appropriate moment, thereby removing the indeterminacy.

EXAMPLE 1 Calculate $\lim_{x \to 3} \dfrac{x^2 - 4x + 3}{x^2 + x - 12}$.

Solution The function has the indeterminate form $0/0$ at $x = 3$ because

Numerator at $x = 3$: $x^2 - 4x + 3 = 3^2 - 4(3) + 3 = 0$

Denominator at $x = 3$: $x^2 + x - 12 = 3^2 + 3 - 12 = 0$

Step 1. **Transform algebraically and cancel.**

$$\frac{x^2 - 4x + 3}{x^2 + x - 12} = \underbrace{\frac{(x - 3)(x - 1)}{(x - 3)(x + 4)}}_{\text{Cancel common factor}} = \underbrace{\frac{x - 1}{x + 4}}_{\text{Continuous at } x = 3} \quad \text{(if } x \neq 3\text{)} \qquad \boxed{1}$$

***Step 2.* Substitute (evaluate using continuity).**

Because the expression on the right in Eq. (1) is *continuous* at $x = 3$,

$$\lim_{x \to 3} \frac{x^2 - 4x + 3}{x^2 + x - 12} = \underbrace{\lim_{x \to 3} \frac{x - 1}{x + 4} = \frac{2}{7}}_{\text{Evaluate by substitution}}$$

■

The next example illustrates the algebraic technique of multiplying by the conjugate, which can be used to treat some indeterminate forms involving square roots.

EXAMPLE 2 **Multiplying by the Conjugate** Evaluate $\displaystyle\lim_{x \to 9} \frac{x - 9}{\sqrt{x} - 3}$.

Solution We check that $f(x) = \dfrac{x - 9}{\sqrt{x} - 3}$ has the indeterminate form $0/0$ at $x = 9$:

Numerator at $x = 9$: $x - 9 = 9 - 9 = 0$

Denominator at $x = 9$: $\sqrt{x} - 3 = \sqrt{9} - 3 = 0$

***Step 1.* Multiply by the conjugate and cancel.**

> Note, in Step 1, that the conjugate of $\sqrt{x} - 3$ is $\sqrt{x} + 3$, so $(\sqrt{x} - 3)(\sqrt{x} + 3) = x - 9$.

$$\left(\frac{x - 9}{\sqrt{x} - 3} \right) \left(\frac{\sqrt{x} + 3}{\sqrt{x} + 3} \right) = \frac{(x - 9)(\sqrt{x} + 3)}{x - 9} = \sqrt{x} + 3 \quad (\text{if } x \neq 9)$$

***Step 2.* Substitute (evaluate using continuity).**

Because $g(x) = \sqrt{x} + 3$ is continuous at $x = 9$, we can now evaluate the limit by substitution:

$$\lim_{x \to 9} \frac{x - 9}{\sqrt{x} - 3} = \lim_{x \to 9} (\sqrt{x} + 3) = 6$$

■

EXAMPLE 3 Calculate $\displaystyle\lim_{x \to 1} \frac{x^3 - 1}{(x - 1)^3}$.

Solution The function has the indeterminate form $0/0$ at $x = 1$ because

Numerator at $x = 1$: $x^3 - 1 = 1^3 - 1 = 0$

Denominator at $x = 1$: $(x - 1)^3 = (1 - 1)^3 = 0$

***Step 1.* Transform algebraically and cancel.**

$$\frac{x^3 - 1}{(x - 1)^3} = \underbrace{\frac{(x - 1)(x^2 + x + 1)}{(x - 1)^3}}_{\text{Cancel common factor}} = \frac{x^2 + x + 1}{(x - 1)^2} \quad (\text{if } x \neq 1) \qquad \boxed{2}$$

***Step 2.* Determine the limit.**

Because $x^2 + x + 1$ approaches 3 as $x \to 1$ and $(x - 1)^2$ is positive and approaches 0 as $x \to 1$, it follows that $\dfrac{x^2 + x + 1}{(x - 1)^2}$ is positive and becomes arbitrarily large as $x \to 1$. Therefore, $\displaystyle\lim_{x \to 1} \frac{x^3 - 1}{(x - 1)^3}$ does not exist, but we can say $\displaystyle\lim_{x \to 1} \frac{x^3 - 1}{(x - 1)^3} = \infty$.

■

Note that in this example, we obtained the undefined expression $3/0$ for the limit form after simplifying. This is not an indeterminate form, and from it we were able to evaluate the limit (determining that it does not exist). A limit form $a/0$ with nonzero a is *not* an indeterminate form. If $\displaystyle\lim_{x \to c} f(x)$ is in the form $a/0$ with $a \neq 0$, then f takes on arbitrarily large values arbitrarily close to c, and we can conclude that the limit does not exist.

The previous three examples involve limits in the indeterminate form 0/0. In the first case the limit is 2/7, in the second it is 6, and in the third the limit does not exist. This underscores the meaning of indeterminate. We do not know *yet* whether or not the limit exists. Indeterminate means further work is needed to evaluate the limit.

EXAMPLE 4   **The Form** $\dfrac{\infty}{\infty}$ Calculate $\displaystyle\lim_{x \to \frac{\pi}{2}} \frac{\tan x}{\sec x}$.

Solution As we see in Figure 2, both $f(x) = \tan x$ and $f(x) = \sec x$ have infinite discontinuities at $x = \frac{\pi}{2}$, so this limit has the indeterminate form ∞/∞ at $x = \frac{\pi}{2}$.

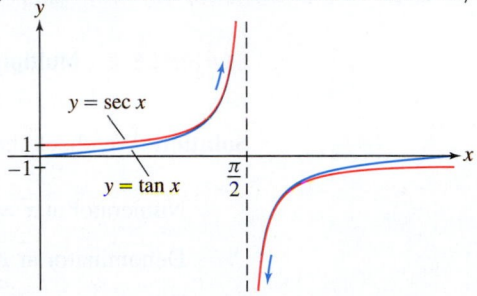

DF **FIGURE 2**

Step 1. **Transform algebraically and cancel.**

$$\frac{\tan x}{\sec x} = \frac{(\sin x)\left(\dfrac{1}{\cos x}\right)}{\dfrac{1}{\cos x}} = \sin x \quad (\text{if } \cos x \neq 0)$$

Step 2. **Substitute (evaluate using continuity).**
Because $f(x) = \sin x$ is continuous,

$$\lim_{x \to \frac{\pi}{2}} \frac{\tan x}{\sec x} = \lim_{x \to \frac{\pi}{2}} \sin x = \sin \frac{\pi}{2} = 1 \qquad \blacksquare$$

EXAMPLE 5   **The Form** $\infty - \infty$ Calculate $\displaystyle\lim_{x \to 1} \left(\frac{1}{x - 1} - \frac{2}{x^2 - 1} \right)$.

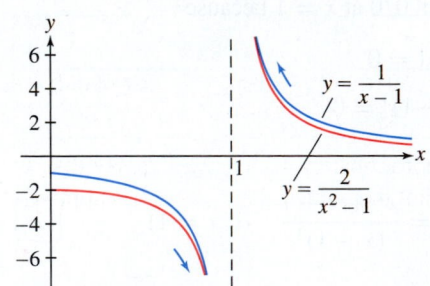

FIGURE 3

Solution As we see in Figure 3, $y = \dfrac{1}{x - 1}$ and $y = \dfrac{2}{x^2 - 1}$ both have infinite discontinuities at $x = 1$. This limit has the indeterminate form $\infty - \infty$.

Step 1. **Transform algebraically and cancel.**
Combine the fractions and simplify (for $x \neq 1$):

$$\frac{1}{x - 1} - \frac{2}{x^2 - 1} = \frac{x + 1}{x^2 - 1} - \frac{2}{x^2 - 1} = \frac{x - 1}{x^2 - 1} = \frac{x - 1}{(x - 1)(x + 1)} = \frac{1}{x + 1}$$

Step 2. **Substitute (evaluate using continuity).**

$$\lim_{x \to 1} \left(\frac{1}{x - 1} - \frac{2}{x^2 - 1} \right) = \lim_{x \to 1} \frac{1}{x + 1} = \frac{1}{1 + 1} = \frac{1}{2} \qquad \blacksquare$$

As preparation for work we will do computing derivatives in Chapter 3, we evaluate a limit involving a symbolic constant.

EXAMPLE 6   **Symbolic Constant**   Calculate $\displaystyle\lim_{h \to 0} \frac{(h + a)^2 - a^2}{h}$, where a is a constant.

Solution We have the indeterminate form 0/0 at $h = 0$ because

Numerator at $h = 0$: $(h + a)^2 - a^2 = (0 + a)^2 - a^2 = 0$
Denominator at $h = 0$: $h = 0$

Expand the numerator and simplify (for $h \neq 0$):

$$\frac{(h+a)^2 - a^2}{h} = \frac{(h^2 + 2ah + a^2) - a^2}{h} = \frac{h^2 + 2ah}{h} = \frac{h(h+2a)}{h} = h + 2a$$

The function $f(h) = h + 2a$ is continuous (for any constant a), so

$$\lim_{h \to 0} \frac{(h+a)^2 - a^2}{h} = \lim_{h \to 0}(h + 2a) = 2a \qquad \blacksquare$$

Some expressions that yield an indeterminate form cannot be simplified algebraically to determine the limit. In the next two examples we consider limits that cannot be found with the rules we have developed so far. Nevertheless, we can investigate the limits numerically to find estimates for them. In the next section and in Section 4.5 we introduce important theorems that provide other tools for evaluating limits in indeterminate forms.

EXAMPLE 7   Maximum Height Under Air Resistance   A one-kilogram bocce ball is launched straight upward at 30 m/sec (Figure 4) and is acted on by gravity and an air resistance force. We assume that the latter force is in the form $-kv(t)$, where $v(t)$ is the ball's upward velocity and k is a positive constant reflecting the strength of the air resistance (the stronger the air resistance, the greater the value of k). It can be shown (see Section 9.2) that the maximum height (in meters) that the ball reaches depends on the strength of the air resistance and is given by

$$H(k) = \frac{30k - 9.8 \ln\left(\frac{150k}{49} + 1\right)}{k^2}$$

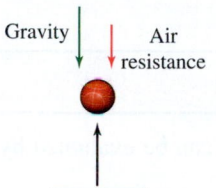

FIGURE 4 A ball acted on by gravity and air resistance.

Values of k between 0 and 1 are physically reasonable. Intuitively, it makes sense that the stronger the air resistance, the lower the maximum height attained by the ball (Figure 5). What is $\lim_{k \to 0} H(k)$? In other words, what happens to the maximum height with less and less air resistance, and ultimately with none at all?

Solution   This limit has the indeterminate form 0/0. We cannot evaluate it by algebraic simplification, so we will estimate it by examining it numerically:

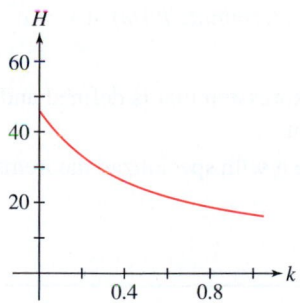

FIGURE 5 The maximum height decreases as the air resistance (and k) increases.

TABLE 1

k	0.1	0.01	0.001	0.0001	0.00001	0.000001	0.0000001
$H(k)$	38.2785	45.0023	45.8249	45.9090	45.9174	45.9173	45.9184

The values in Table 1 suggest the limit is approximately 45.92. We can say that as the air resistance vanishes, the maximum height that the ball attains approaches approximately 45.92 m. $\qquad \blacksquare$

CONCEPTUAL INSIGHT   In our work with functions and limits so far, we have encountered three expressions that are similar but have different meanings: *undefined*, *does not exist*, and *indeterminate*. It is important to understand the meanings of these expressions so that you can use them correctly to describe functions and limits.

- The word "undefined" is used for a mathematical expression that is not defined, such as $2/0$ or $\ln 0$.
- The phrase "does not exist" means $\lim_{x \to c} f(x)$ does not exist, that is, $f(x)$ does not approach a particular numerical value as x approaches c.
- The term "indeterminate" is used when, upon substitution, a function or limit has one of the indeterminate forms.

The next limit will be important to us in the next chapter when determining the derivative of the function $f(x) = e^x$. It is in the indeterminate form 0/0, but we cannot simplify it algebraically.

EXAMPLE 8 Verify graphically and numerically that $\lim\limits_{h \to 0} \dfrac{e^h - 1}{h} = 1$.

Solution The graph in Figure 6 and the results in Table 2 suggest that $\lim\limits_{h \to 0} \dfrac{e^h - 1}{h} = 1$.

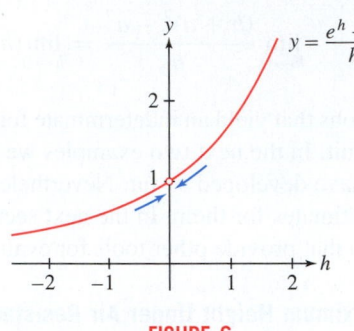

	TABLE 2		
$h \to 0^-$	$\dfrac{e^h - 1}{h}$	$h \to 0^+$	$\dfrac{e^h - 1}{h}$
-0.02	0.990	0.02	1.0101
-0.005	0.99750	0.005	1.00250
-0.001	0.999500	0.001	1.000500
-0.0001	0.99995000	0.0001	1.00005000

FIGURE 6

2.5 SUMMARY

- When f is known to be continuous at $x = c$, the limit can be evaluated by substitution: $\lim\limits_{x \to c} f(x) = f(c)$.
- If the formula for $f(c)$ yields an undefined expression of the type

$$\frac{0}{0}, \quad \frac{\infty}{\infty}, \quad \infty \cdot 0, \quad \infty - \infty$$

then we say that $f(x)$ is *indeterminate* (or has an *indeterminate form*) at $x = c$.
- If $f(x)$ is indeterminate at $x = c$:

  - Try to transform $f(x)$ algebraically into a new expression that is defined and continuous at $x = c$, and then evaluate by substitution.
  - Examine it numerically or graphically, or evaluate it with specialized theorems (see Sections 2.6 and 4.5).

2.5 EXERCISES

Preliminary Questions

1. Which of the following is indeterminate at $x = 1$?

$$\frac{x^2 + 1}{x - 1}, \qquad \frac{x^2 - 1}{x + 2}, \qquad \frac{x^2 - 1}{\sqrt{x + 3} - 2}, \qquad \frac{x^2 + 1}{\sqrt{x + 3} - 2}$$

2. Give counterexamples to show that these statements are false:

(a) If $f(c)$ is indeterminate, then the right- and left-hand limits as $x \to c$ are not equal.

(b) If $\lim\limits_{x \to c} f(x)$ exists, then $f(c)$ is not indeterminate.

(c) If $f(x)$ is undefined at $x = c$, then $f(x)$ has an indeterminate form at $x = c$.

3. The method for evaluating limits discussed in this section is sometimes called simplify and substitute. Explain how it actually relies on the property of continuity.

Exercises

In Exercises 1–4, show that the limit leads to an indeterminate form. Then carry out the two-step procedure: Transform the function algebraically and evaluate using continuity.

1. $\lim\limits_{x \to 6} \dfrac{x^2 - 36}{x - 6}$

2. $\lim\limits_{h \to 3} \dfrac{9 - h^2}{h - 3}$

3. $\lim\limits_{x \to -1} \dfrac{x^2 + 2x + 1}{x + 1}$

4. $\lim\limits_{t \to 9} \dfrac{2t - 18}{5t - 45}$

In Exercises 5–34, evaluate the limit, if it exists. If not, determine whether the one-sided limits exist (finite or infinite).

5. $\lim\limits_{x \to 7} \dfrac{x - 7}{x^2 - 49}$

6. $\lim\limits_{x \to 8} \dfrac{x^2 - 64}{x - 9}$

7. $\lim\limits_{x \to -2} \dfrac{x^2 + 3x + 2}{x + 2}$

8. $\lim\limits_{x \to 8} \dfrac{x^3 - 64x}{x - 8}$

9. $\displaystyle\lim_{x\to 5}\frac{2x^2-9x-5}{x^2-25}$

10. $\displaystyle\lim_{h\to 0}\frac{(1+h)^3-1}{h}$

11. $\displaystyle\lim_{x\to -\frac{1}{2}}\frac{2x+1}{2x^2+3x+1}$

12. $\displaystyle\lim_{x\to 3}\frac{x^2-x}{x^2-9}$

13. $\displaystyle\lim_{x\to 2}\frac{3x^2-4x-4}{2x^2-8}$

14. $\displaystyle\lim_{h\to 0}\frac{(3+h)^3-27}{h}$

15. $\displaystyle\lim_{t\to 0}\frac{4^{2t}-1}{4^t-1}$

16. $\displaystyle\lim_{h\to 4}\frac{(h+2)^2-9h}{h-4}$

17. $\displaystyle\lim_{x\to 16}\frac{\sqrt{x}-4}{x-16}$

18. $\displaystyle\lim_{t\to -2}\frac{2t+4}{12-3t^2}$

19. $\displaystyle\lim_{h\to 0}\frac{\frac{1}{(h+2)^2}-\frac{1}{4}}{h}$

20. $\displaystyle\lim_{y\to 3}\frac{y^2+y-12}{y^3-10y+3}$

21. $\displaystyle\lim_{h\to 0}\frac{\sqrt{2+h}-2}{h}$

22. $\displaystyle\lim_{x\to 8}\frac{\sqrt{x-4}-2}{x-8}$

23. $\displaystyle\lim_{x\to 4}\frac{x-4}{\sqrt{x}-\sqrt{8-x}}$

24. $\displaystyle\lim_{x\to 4}\frac{\sqrt{5-x}-1}{2-\sqrt{x}}$

25. $\displaystyle\lim_{x\to 4}\left(\frac{1}{\sqrt{x}-2}-\frac{4}{x-4}\right)$

26. $\displaystyle\lim_{x\to 0^+}\left(\frac{1}{\sqrt{x}}-\frac{1}{\sqrt{x^2+x}}\right)$

27. $\displaystyle\lim_{x\to 0}\frac{\cot x}{\csc x}$

28. $\displaystyle\lim_{\theta\to \frac{\pi}{2}}\frac{\cot\theta}{\csc\theta}$

29. $\displaystyle\lim_{x\to 1}\left(\frac{1}{1-x}-\frac{2}{1-x^2}\right)$

30. $\displaystyle\lim_{x\to \frac{\pi}{4}}\frac{\sin x-\cos x}{\tan x-1}$

31. $\displaystyle\lim_{t\to 2}\frac{2^{2t}+2^t-20}{2^t-4}$

32. $\displaystyle\lim_{\theta\to \frac{\pi}{2}}\left(\sec\theta-\tan\theta\right)$

33. $\displaystyle\lim_{\theta\to \frac{\pi}{4}}\left(\frac{1}{\tan\theta-1}-\frac{2}{\tan^2\theta-1}\right)$

34. $\displaystyle\lim_{x\to \frac{\pi}{3}}\frac{2\cos^2 x+3\cos x-2}{2\cos x-1}$

35. The following limits all have the indeterminate form $0/0$. One of the limits does not exist, one is equal to 0, and one is a nonzero limit. Evaluate each limit algebraically if you can or investigate it numerically if you cannot.

(a) $\displaystyle\lim_{x\to -2}\frac{x^2+3x+2}{x+2}$

(b) $\displaystyle\lim_{x\to 1}\frac{1-x^{-1}}{x-2+x^{-1}}$

(c) $\displaystyle\lim_{x\to 0}\frac{x^2}{1-e^x}$

36. The following limits all have the indeterminate form ∞/∞. One of the limits does not exist, one is equal to 0, and one is a nonzero limit. Evaluate each limit algebraically if you can or investigate it numerically if you cannot.

(a) $\displaystyle\lim_{x\to 0}\frac{x^{-4}}{4+x^{-1}}$

(b) $\displaystyle\lim_{x\to 0}\frac{3\cot x}{\csc x}$

(c) $\displaystyle\lim_{x\to 0}\frac{1+\frac{1}{x^2}}{1+\frac{1}{x^4}}$

In Exercises 37 and 38, a ball is launched straight up in the air and is acted on by air resistance and gravity as in Example 7. The function H gives the maximum height that the ball attains as a function of the air-resistance parameter k.

37. If the mass of the ball is one kilogram and it is launched upward with an initial velocity of 60 m/sec, then

$$H(k)=\frac{60k-9.8\ln\left(\frac{300k}{49}+1\right)}{k^2}$$

Estimate the maximum height without air resistance by investigating $\displaystyle\lim_{k\to 0}H(k)$ numerically.

38. If the mass of the ball is 500 grams and it is launched upward with an initial velocity of 30 m/sec, then

$$H(k)=\frac{15k-2.45\ln\left(\frac{300k}{49}+1\right)}{k^2}$$

Estimate the maximum height without air resistance by investigating $\displaystyle\lim_{k\to 0}H(k)$ numerically.

39. GU Use a plot of $f(x)=\dfrac{x-4}{\sqrt{x}-\sqrt{8-x}}$ to estimate $\displaystyle\lim_{x\to 4}f(x)$ to two decimal places. Compare with the answer obtained algebraically in Exercise 23.

40. GU Use a plot of $f(x)=\dfrac{1}{\sqrt{x}-2}-\dfrac{4}{x-4}$ to estimate $\displaystyle\lim_{x\to 4}f(x)$ numerically. Compare with the answer obtained algebraically in Exercise 25.

41. Show numerically that for $b=3$ and $b=5$, $\displaystyle\lim_{x\to 0}\frac{b^x-1}{x}$ appears to equal $\ln 3$ and $\ln 5$, respectively.

42. Show numerically that for $b=2$ and $b=4$, $\displaystyle\lim_{x\to 0}\frac{b^x-1}{x}$ appears to equal $\ln 2$ and $\ln 4$, respectively.

In Exercises 43–48, evaluate using the identity

$$a^3-b^3=(a-b)(a^2+ab+b^2)$$

43. $\displaystyle\lim_{x\to 2}\frac{x^3-8}{x-2}$

44. $\displaystyle\lim_{x\to 3}\frac{x^3-27}{x^2-9}$

45. $\displaystyle\lim_{x\to 1}\frac{x^2-5x+4}{x^3-1}$

46. $\displaystyle\lim_{x\to -2}\frac{x^3+8}{x^2+6x+8}$

47. $\displaystyle\lim_{x\to 1}\frac{x^4-1}{x^3-1}$

48. $\displaystyle\lim_{x\to 27}\frac{x-27}{x^{1/3}-3}$

In Exercises 49–56, evaluate in terms of the constant a.

49. $\displaystyle\lim_{x\to 0}(2a+x)$

50. $\displaystyle\lim_{h\to -2}(4ah+7a)$

51. $\displaystyle\lim_{t\to -1}(4t-2at+3a)$

52. $\displaystyle\lim_{x\to a}\frac{(x+a)^2-4x^2}{x-a}$

53. $\displaystyle\lim_{x\to a}\frac{\sqrt{x}-\sqrt{a}}{x-a}$

54. $\displaystyle\lim_{h\to 0}\frac{\sqrt{a+2h}-\sqrt{a}}{h}$

55. $\displaystyle\lim_{x\to 0}\frac{(x+a)^3-a^3}{x}$

56. $\displaystyle\lim_{h\to a}\frac{\frac{1}{h}-\frac{1}{a}}{h-a}$

57. Evaluate $\displaystyle\lim_{h\to 0}\frac{\sqrt[4]{1+h}-1}{h}$. *Hint:* Set $x=\sqrt[4]{1+h}$, express h as a function of x, and rewrite as a limit as $x\to 1$.

58. Evaluate $\displaystyle\lim_{h\to 0}\frac{\sqrt[3]{1+h}-1}{\sqrt[2]{1+h}-1}$. *Hint:* Set $x=\sqrt[6]{1+h}$, express h as a function of x, and rewrite as a limit as $x\to 1$.

Further Insights and Challenges

In Exercises 59–62, find all values of c such that the limit exists.

59. $\displaystyle \lim_{x \to c} \frac{x^2 - 5x - 6}{x - c}$

60. $\displaystyle \lim_{x \to 1} \frac{x^2 + 3x + c}{x - 1}$

61. $\displaystyle \lim_{x \to 1} \left(\frac{1}{x - 1} - \frac{c}{x^3 - 1} \right)$

62. $\displaystyle \lim_{x \to 0} \frac{1 + cx^2 - \sqrt{1 + x^2}}{x^4}$

63. For which sign, $+$ or $-$, does the following limit exist?

$$\lim_{x \to 0} \left(\frac{1}{x} \pm \frac{1}{x(x - 1)} \right)$$

2.6 The Squeeze Theorem and Trigonometric Limits

In our study of the derivative, we will need to evaluate certain limits involving transcendental functions such as sine and cosine. The algebraic techniques of the previous section are often ineffective for such functions, and other tools are required. In this section, we discuss one such tool—the Squeeze Theorem—and use it to evaluate the trigonometric limits needed in Section 3.6.

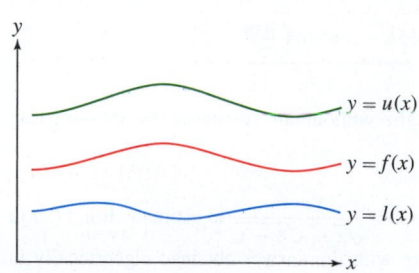

FIGURE 1 f is trapped between l and u.

The Squeeze Theorem

Consider a function f that is "trapped" between two functions l, for lower bound, and u, for upper bound, on an interval I. In other words,

$$l(x) \le f(x) \le u(x) \quad \text{for all } x \in I$$

Thus, the graph of f lies between the graphs of l and u (Figure 1).

The Squeeze Theorem applies when f is not just trapped but **squeezed** at a point $x = c$ (Figure 2). By this we mean that for all $x \ne c$ in some open interval containing c,

$$l(x) \le f(x) \le u(x) \qquad \text{and} \qquad \lim_{x \to c} l(x) = \lim_{x \to c} u(x) = L$$

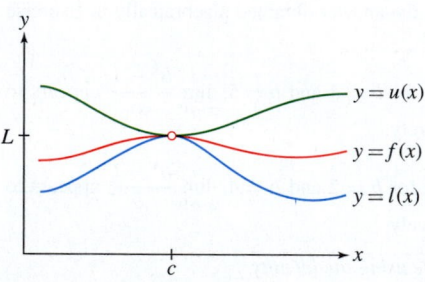

DF **FIGURE 2** f is squeezed by l and u at $x = c$.

We do not require that $f(x)$ be defined at $x = c$, but it is clear graphically that $f(x)$ must approach the limit L, as stated in the next theorem. See Appendix D for a proof.

> **THEOREM 1 Squeeze Theorem** Assume that for $x \ne c$ (in some open interval containing c),
>
> $$l(x) \le f(x) \le u(x) \qquad \text{and} \qquad \lim_{x \to c} l(x) = \lim_{x \to c} u(x) = L$$
>
> Then $\displaystyle\lim_{x \to c} f(x)$ exists and $\displaystyle\lim_{x \to c} f(x) = L$.

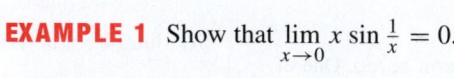

EXAMPLE 1 Show that $\displaystyle \lim_{x \to 0} x \sin \frac{1}{x} = 0$.

Solution Although $f(x) = x \sin \frac{1}{x}$ is a product of two functions, we cannot use the Product Law because $\displaystyle \lim_{x \to 0} \sin \frac{1}{x}$ does not exist. However, the sine function takes on values between 1 and -1, and therefore $\left| \sin \frac{1}{x} \right| \le 1$ for all $x \ne 0$. Multiplying by $|x|$, we obtain $\left| x \sin \frac{1}{x} \right| \le |x|$ and conclude that (Figure 3)

$$-|x| \le x \sin \frac{1}{x} \le |x|$$

Furthermore, we have

$$\lim_{x \to 0} |x| = 0 \qquad \text{and} \qquad \lim_{x \to 0} (-|x|) = 0$$

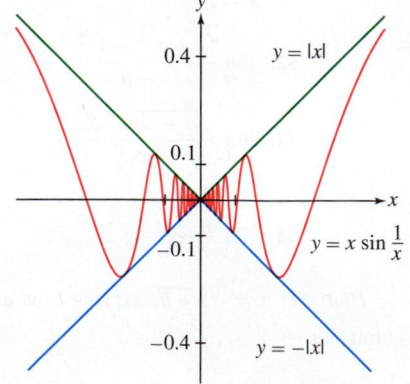

DF **FIGURE 3**

and therefore it follows that $f(x) = x \sin \frac{1}{x}$ is squeezed between $l(x) = -|x|$ and $u(x) = |x|$ at $x = 0$. The Squeeze Theorem now applies, and we can conclude that
$$\lim_{x \to 0} x \sin \frac{1}{x} = 0.$$ ■

In Section 2.2, we found numerical and graphical evidence suggesting that the limit $\lim_{\theta \to 0} \dfrac{\sin \theta}{\theta}$ is equal to 1. The Squeeze Theorem will allow us to prove this fact.

> **THEOREM 2 Important Trigonometric Limits**
>
> $$\lim_{\theta \to 0} \frac{\sin \theta}{\theta} = 1 \qquad \text{and} \qquad \lim_{\theta \to 0} \frac{1 - \cos \theta}{\theta} = 0$$

Note that both $\frac{\sin \theta}{\theta}$ and $\frac{\cos \theta - 1}{\theta}$ are indeterminate at $\theta = 0$, so Theorem 2 cannot be proved by substitution.

To apply the Squeeze Theorem to prove that $\lim_{\theta \to 0} \dfrac{\sin \theta}{\theta} = 1$, we must find functions that squeeze $\dfrac{\sin \theta}{\theta}$ at $\theta = 0$. These are illustrated in Figure 4 and provided by the next theorem.

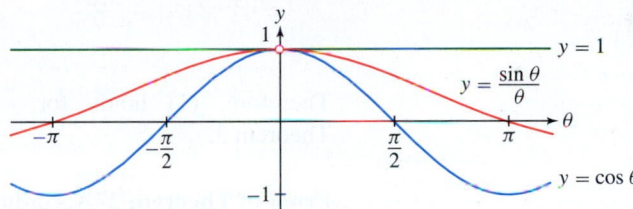

DF **FIGURE 4** Graph illustrating the inequalities of Theorem 3.

> **THEOREM 3**
>
> $$\cos \theta \le \frac{\sin \theta}{\theta} \le 1 \qquad \text{for} \qquad -\frac{\pi}{2} < \theta < \frac{\pi}{2}, \qquad \theta \ne 0$$ **1**

Proof Assume first that $0 < \theta < \frac{\pi}{2}$. Our proof is based on the following relation between the areas in Figure 5:

$$\text{area of } \triangle OAB < \text{area of sector } BOA < \text{area of } \triangle OAC$$ **2**

Let's determine these three areas. First, $\triangle OAB$ has base 1 and height $\sin \theta$, so its area is $\frac{1}{2} \sin \theta$. Next, recall that a sector of angle θ has area $\frac{1}{2}\theta$. Finally, to compute the area of $\triangle OAC$, we observe that

*◀— **REMINDER** Let's recall why a sector of angle θ in a circle of radius r has area $\frac{1}{2}r^2\theta$. A sector of angle θ represents a fraction $\frac{\theta}{2\pi}$ of the entire circle. The circle has area πr^2, so the sector has area $\left(\frac{\theta}{2\pi}\right)\pi r^2 = \frac{1}{2}r^2\theta$. In the unit circle $(r = 1)$, the sector has area $\frac{1}{2}\theta$.*

$$\tan \theta = \frac{\text{opposite side}}{\text{adjacent side}} = \frac{AC}{OA} = \frac{AC}{1} = AC$$

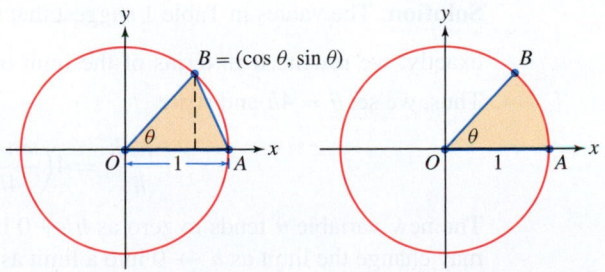

Area of triangle = $\frac{1}{2}\sin \theta$ Area of sector = $\frac{1}{2}\theta$ Area of triangle = $\frac{1}{2}\tan \theta$

FIGURE 5

Thus, $\triangle OAC$ has base 1, height $\tan\theta$, and area $\frac{1}{2}\tan\theta$. We have shown, therefore, that

$$\underbrace{\frac{1}{2}\sin\theta}_{\text{Area }\triangle OAB} \;\leq\; \underbrace{\frac{1}{2}\theta}_{\text{Area of sector}} \;\leq\; \underbrace{\frac{1}{2}\frac{\sin\theta}{\cos\theta}}_{\text{Area }\triangle OAC} \qquad \boxed{3}$$

The first inequality yields $\sin\theta \leq \theta$, and because $\theta > 0$, we obtain

$$\frac{\sin\theta}{\theta} \leq 1 \qquad \boxed{4}$$

Next, multiply the second inequality in (3) by $\dfrac{2\cos\theta}{\theta}$ to obtain

$$\cos\theta \leq \frac{\sin\theta}{\theta} \qquad \boxed{5}$$

The combination of (4) and (5) gives us (1) when $0 < \theta < \frac{\pi}{2}$. However, the functions in (1) do not change value when θ is replaced by $-\theta$ because both $f(x) = \cos\theta$ and $f(x) = \dfrac{\sin\theta}{\theta}$ are even functions. Indeed, $\cos(-\theta) = \cos\theta$ and

$$\frac{\sin(-\theta)}{-\theta} = \frac{-\sin\theta}{-\theta} = \frac{\sin\theta}{\theta}$$

Therefore, (1) holds for $-\frac{\pi}{2} < \theta < 0$ as well. This completes the proof of Theorem 3. ∎

Proof of Theorem 2 According to Theorem 3,

$$\cos\theta \leq \frac{\sin\theta}{\theta} \leq 1$$

Since $\lim\limits_{\theta\to 0}\cos\theta = \cos 0 = 1$ and $\lim\limits_{\theta\to 0} 1 = 1$, the Squeeze Theorem yields $\lim\limits_{\theta\to 0}\dfrac{\sin\theta}{\theta} = 1$, as required. It then follows that

$$\lim_{\theta\to 0}\frac{1-\cos\theta}{\theta} = \lim_{\theta\to 0}\left(\frac{1+\cos\theta}{1+\cos\theta}\right)\frac{1-\cos\theta}{\theta} = \lim_{\theta\to 0}\frac{1-\cos^2\theta}{(1+\cos\theta)\theta}$$

$$= \lim_{\theta\to 0}\left(\frac{1}{1+\cos\theta}\right)\frac{\sin^2\theta}{\theta} = \lim_{x\to 0}\left(\frac{\sin\theta}{1+\cos\theta}\right)\frac{\sin\theta}{\theta} = \frac{0}{2}\cdot 1 = 0 \quad \blacksquare$$

In the next example, we evaluate another trigonometric limit. The key idea is to rewrite the function of h in terms of the new variable $\theta = 4h$.

EXAMPLE 2 Evaluating a Limit by Changing Variables Investigate $\lim\limits_{h\to 0}\dfrac{\sin 4h}{h}$ numerically and then evaluate it exactly.

Solution The values in Table 1 suggest that the limit is equal to 4. To evaluate the limit exactly, we rewrite it in terms of the limit of $\dfrac{\sin\theta}{\theta}$ so that Theorem 2 can be applied. Thus, we set $\theta = 4h$ and write

$$\frac{\sin 4h}{h} = 4\left(\frac{\sin 4h}{4h}\right) = 4\frac{\sin\theta}{\theta}$$

The new variable θ tends to zero as $h \to 0$ because θ is a multiple of h. Therefore, we may change the limit as $h \to 0$ into a limit as $\theta \to 0$ to obtain

$$\lim_{h\to 0}\frac{\sin 4h}{h} = \lim_{\theta\to 0}4\frac{\sin\theta}{\theta} = 4\left(\lim_{\theta\to 0}\frac{\sin\theta}{\theta}\right) = 4(1) = 4 \qquad \blacksquare$$

TABLE 1

h	$\dfrac{\sin 4h}{h}$
± 1.0	-0.75680
± 0.5	1.81859
± 0.2	3.58678
± 0.1	3.89418
± 0.05	3.97339
± 0.01	3.99893
± 0.005	3.99973

Note that the change of variables $\theta = kx$ demonstrates that $\dfrac{\sin kx}{kx}$ approaches 1 as $x \to 0$. We use this limit to our advantage in the next example.

EXAMPLE 3 Find $\lim\limits_{x \to 0} \dfrac{\tan 3x}{\tan 2x}$.

Solution

$$\lim_{x \to 0} \frac{\tan 3x}{\tan 2x} = \lim_{x \to 0} \frac{\sin 3x}{\cos 3x} \cdot \frac{\cos 2x}{\sin 2x} = \lim_{x \to 0} \frac{\sin 3x}{\cos 3x} \cdot \frac{\cos 2x}{\sin 2x} \cdot \frac{x}{x}$$

$$= \lim_{x \to 0} \frac{3}{2} \left(\frac{\sin 3x}{3x} \right) \left(\frac{2x}{\sin 2x} \right) \frac{\cos 2x}{\cos 3x} = \frac{3}{2} \cdot 1 \cdot 1 \cdot \frac{1}{1} = \frac{3}{2} \qquad \blacksquare$$

2.6 SUMMARY

- We say that a function f is *squeezed* at $x = c$ if there exist functions l and u such that $l(x) \le f(x) \le u(x)$ for all $x \ne c$ in an open interval I containing c, and

$$\lim_{x \to c} l(x) = \lim_{x \to c} u(x) = L$$

The Squeeze Theorem states that in this case, $\lim\limits_{x \to c} f(x) = L$.

- Two important trigonometric limits:

$$\lim_{\theta \to 0} \frac{\sin \theta}{\theta} = 1 \qquad \text{and} \qquad \lim_{\theta \to 0} \frac{1 - \cos \theta}{\theta} = 0$$

2.6 EXERCISES

Preliminary Questions

1. Assume that $-x^4 \le f(x) \le x^2$. What is $\lim\limits_{x \to 0} f(x)$? Is there enough information to evaluate $\lim\limits_{x \to \frac{1}{2}} f(x)$? Explain.

2. State the Squeeze Theorem carefully.

3. If you want to evaluate $\lim\limits_{h \to 0} \dfrac{\sin 5h}{3h}$, it is a good idea to rewrite the limit in terms of the variable (choose one):

(a) $\theta = 5h$ **(b)** $\theta = 3h$ **(c)** $\theta = \dfrac{5h}{3}$

Exercises

In Exercises 1–10, evaluate using the Squeeze Theorem.

1. $\lim\limits_{x \to 0} x^2 \cos \dfrac{1}{x}$

2. $\lim\limits_{x \to 0} x \sin \dfrac{1}{x^2}$

3. $\lim\limits_{x \to 1} (x - 1) \sin \dfrac{\pi}{x - 1}$

4. $\lim\limits_{x \to 3} (x^2 - 9) \dfrac{x - 3}{|x - 3|}$

5. $\lim\limits_{t \to 0} (2^t - 1) \cos \dfrac{1}{t}$

6. $\lim\limits_{x \to 0^+} \sqrt{x}\, e^{\cos(\pi/x)}$

7. $\lim\limits_{t \to 2} (t^2 - 4) \cos \dfrac{1}{t - 2}$

8. $\lim\limits_{x \to 0} \tan x \cos \left(\sin \dfrac{1}{x} \right)$

9. $\lim\limits_{\theta \to \frac{\pi}{2}} \cos \theta \cos(\tan \theta)$

10. $\lim\limits_{t \to 0^+} \sin t \tan^{-1}(\ln t)$

11. State precisely the hypothesis and conclusions of the Squeeze Theorem for the situation in Figure 6.

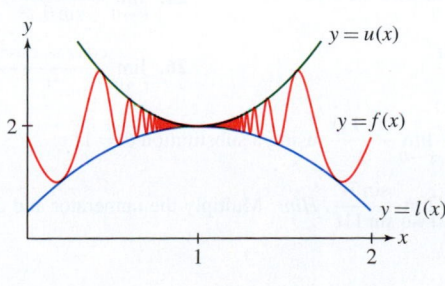

FIGURE 6

12. In Figure 7, is f squeezed by u and l at $x = 3$? At $x = 2$?

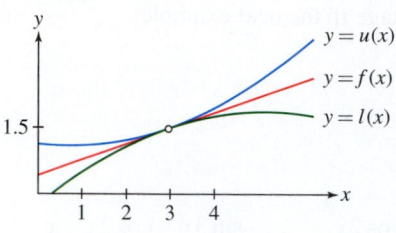

FIGURE 7

13. What does the Squeeze Theorem say about $\lim_{x \to 7} f(x)$ if the limits $\lim_{x \to 7} l(x) = \lim_{x \to 7} u(x) = 6$ and f, u, and l are related as in Figure 8? The inequality $f(x) \le u(x)$ is not satisfied for all x. Does this affect the validity of your conclusion?

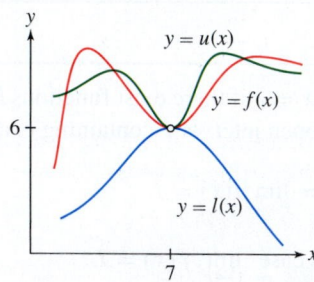

FIGURE 8

14. Determine $\lim_{x \to 0} f(x)$ assuming that $\cos x \le f(x) \le 1$.

15. State whether the inequality provides sufficient information to determine $\lim_{x \to 1} f(x)$, and if so, find the limit.

(a) $4x - 5 \le f(x) \le x^2$
(b) $2x - 1 \le f(x) \le x^2$
(c) $4x - x^2 \le f(x) \le x^2 + 2$

16. [GU] Plot the graphs of $u(x) = 1 + |x - \frac{\pi}{2}|$ and $l(x) = \sin x$ on the same set of axes. What can you say about $\lim_{x \to \frac{\pi}{2}} f(x)$ if f is squeezed by l and u at $x = \frac{\pi}{2}$?

In Exercises 17–26, evaluate using Theorem 2 as necessary.

17. $\lim_{x \to 0} \dfrac{\tan x}{x}$

18. $\lim_{x \to 0} \dfrac{\sin x \sec x}{x}$

19. $\lim_{t \to 0} \dfrac{\sqrt{t^3 + 9} \sin t}{t}$

20. $\lim_{t \to 0} \dfrac{\sin^2 t}{t}$

21. $\lim_{x \to 0} \dfrac{x^2}{\sin^2 x}$

22. $\lim_{t \to \frac{\pi}{2}} \dfrac{1 - \cos t}{t}$

23. $\lim_{\theta \to 0} \dfrac{\sec \theta - 1}{\theta}$

24. $\lim_{\theta \to 0} \dfrac{1 - \cos \theta}{\sin \theta}$

25. $\lim_{t \to \frac{\pi}{4}} \dfrac{\sin t}{t}$

26. $\lim_{t \to 0} \dfrac{\cos t - \cos^2 t}{t}$

27. Evaluate $\lim_{x \to 0} \dfrac{\sin 11x}{x}$ using a substitution $\theta = 11x$.

28. Evaluate $\lim_{t \to 0} \dfrac{\sin 7t}{\sin 11t}$. *Hint:* Multiply the numerator and denominator by $(7)(11)t$.

In Exercises 29–48, evaluate the limit.

29. $\lim_{h \to 0} \dfrac{\sin 9h}{h}$

30. $\lim_{h \to 0} \dfrac{\sin 4h}{4h}$

31. $\lim_{h \to 0} \dfrac{\sin h}{5h}$

32. $\lim_{x \to \frac{\pi}{6}} \dfrac{x}{\sin 3x}$

33. $\lim_{\theta \to 0} \dfrac{\sin 7\theta}{\sin 3\theta}$

34. $\lim_{x \to 0} \dfrac{\tan 4x}{9x}$

35. $\lim_{x \to 0} x \csc 25x$

36. $\lim_{t \to 0} \dfrac{\tan 4t}{t \sec t}$

37. $\lim_{h \to 0} \dfrac{\sin 2h \sin 3h}{h^2}$

38. $\lim_{z \to 0} \dfrac{\sin(z/3)}{\sin z}$

39. $\lim_{\theta \to 0} \dfrac{\sin(-3\theta)}{\sin 4\theta}$

40. $\lim_{x \to 0} \dfrac{\tan 4x}{\tan 9x}$

41. $\lim_{t \to 0} \dfrac{\csc 8t}{\csc 4t}$

42. $\lim_{x \to 0} \dfrac{\sin 5x \sin 2x}{\sin 3x \sin 5x}$

43. $\lim_{x \to 0} \dfrac{\sin 3x \sin 2x}{x \sin 5x}$

44. $\lim_{h \to 0} \dfrac{1 - \cos 2h}{h}$

45. $\lim_{h \to 0} \dfrac{\sin(2h)(1 - \cos h)}{h^2}$

46. $\lim_{t \to 0} \dfrac{1 - \cos 2t}{\sin^2 3t}$

47. $\lim_{\theta \to 0} \dfrac{\cos 2\theta - \cos \theta}{\theta}$

48. $\lim_{h \to \frac{\pi}{2}} \dfrac{1 - \cos 3h}{h}$

49. Use the identity $\sin 2\theta = 2 \sin \theta \cos \theta$ to evaluate $\lim_{\theta \to 0} \dfrac{\sin 2\theta - 2 \sin \theta}{\theta^2}$.

50. Use the identity $\sin 3\theta = 3 \sin \theta - 4 \sin^3 \theta$ to evaluate the limit, $\lim_{\theta \to 0} \dfrac{\sin 3\theta - 3 \sin \theta}{\theta^3}$.

51. Explain why $\lim_{\theta \to 0} (\csc \theta - \cot \theta)$ involves an indeterminate form, and then prove that the limit equals 0.

52. Explain why $\lim_{\theta \to \frac{\pi}{2}} (2 \tan \theta - \sec \theta)$ involves an indeterminate form, and then evaluate the limit.

53. [GU] Investigate $\lim_{h \to 0} \dfrac{1 - \cos 2h}{h^2}$ numerically or graphically. Then evaluate the limit using the double angle formula $\cos 2h = 1 - 2 \sin^2 h$.

54. [GU] Investigate $\lim_{h \to 0} \dfrac{1 - \cos h}{h^2}$ numerically or graphically. Then prove that the limit is equal to $\frac{1}{2}$. *Hint:* See the proof of Theorem 2.

In Exercises 55–57, evaluate using the result of Exercise 54.

55. $\lim_{h \to 0} \dfrac{\cos 3h - 1}{h^2}$

56. $\lim_{h \to 0} \dfrac{\cos 3h - 1}{\cos 2h - 1}$

57. $\lim_{t \to 0} \dfrac{\sqrt{1 - \cos t}}{t}$

58. Use the Squeeze Theorem to prove that if $\lim_{x \to c} |f(x)| = 0$, then $\lim_{x \to c} f(x) = 0$.

Further Insights and Challenges

59. Use the result of Exercise 54 to prove that for $m \neq 0$,

$$\lim_{x \to 0} \frac{\cos mx - 1}{x^2} = -\frac{m^2}{2}.$$

60. Using a diagram of the unit circle and the Pythagorean Theorem, show that

$$\sin^2 \theta \leq (1 - \cos \theta)^2 + \sin^2 \theta \leq \theta^2$$

Conclude that $\sin^2 \theta \leq 2(1 - \cos \theta) \leq \theta^2$ and use this to give an alternative proof that the limit in Exercise 51 equals 0. Then give an alternative proof of the result in Exercise 54.

61. (a) Investigate the limit $\lim\limits_{x \to c} \dfrac{\sin x - \sin c}{x - c}$ numerically for the five values $c = 0, \frac{\pi}{6}, \frac{\pi}{4}, \frac{\pi}{3}, \frac{\pi}{2}$.

(b) Can you guess the answer for general c?

(c) Check numerically that your answer to (b) works for two other values of c.

2.7 Limits at Infinity

So far we have considered limits as x approaches a number c. It is also important to consider limits where x approaches ∞ or $-\infty$, which we refer to as **limits at infinity**. In applications, limits at infinity arise naturally when we describe the "long-term" behavior of a system as in Figure 1.

The notation $x \to \infty$ indicates that x increases without bound, and $x \to -\infty$ indicates that x decreases (through negative values) without bound. We write

- $\lim\limits_{x \to \infty} f(x) = L$ if $f(x)$ gets closer and closer to L as $x \to \infty$
- $\lim\limits_{x \to -\infty} f(x) = L$ if $f(x)$ gets closer and closer to L as $x \to -\infty$

As before, "closer and closer" means that $|f(x) - L|$ becomes arbitrarily small. In either case, the line $y = L$ is called a **horizontal asymptote**. We use the notation $x \to \pm\infty$ to indicate that we are considering both infinite limits, as $x \to \infty$ and as $x \to -\infty$.

Infinite limits describe the **asymptotic behavior** of a function, which is determined by the behavior of the graph as we move out indefinitely to the right or the left.

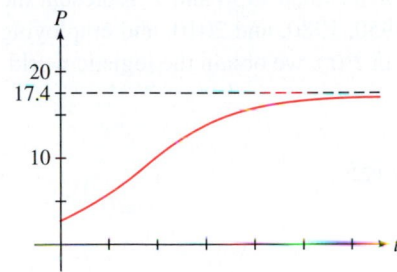

FIGURE 1 The graphed model $P(t)$ fits recent world population data and is derived with the assumption that the population eventually levels off (see Example 2). In the model $\lim\limits_{t \to \infty} P(t) = 17.4$, suggesting that the human carrying capacity of the earth is about 17.4 billion people.

EXAMPLE 1 Discuss the asymptotic behavior in Figure 2.

Solution The function g approaches $L = 7$ as we move to the right and it approaches $L = 3$ as we move to left, so

$$\lim_{x \to \infty} g(x) = 7 \qquad \text{and} \qquad \lim_{x \to -\infty} g(x) = 3$$

Accordingly, the lines $y = 7$ and $y = 3$ are horizontal asymptotes of g. ∎

A function may approach an infinite limit as $x \to \pm\infty$. We write

$$\lim_{x \to \infty} f(x) = \infty \qquad \text{or} \qquad \lim_{x \to -\infty} f(x) = \infty$$

if $f(x)$ is positive and becomes arbitrarily large as $x \to \infty$ or $-\infty$. Similar notation is used if $f(x)$ approaches $-\infty$ as $x \to \pm\infty$. For example, we see in Figure 3(A) that

$$\lim_{x \to \infty} e^x = \infty \qquad \text{and} \qquad \lim_{x \to -\infty} e^x = 0$$

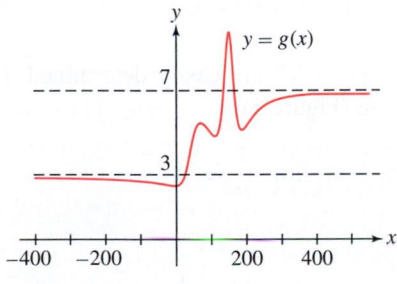

FIGURE 2 The lines $y = 7$ and $y = 3$ are horizontal asymptotes of g.

However, limits at infinity do not always exist. For example, $f(x) = \sin x$ oscillates indefinitely [Figure 3(B)], so the following limits do not exist:

$$\lim_{x \to \infty} \sin x \qquad \text{and} \qquad \lim_{x \to -\infty} \sin x$$

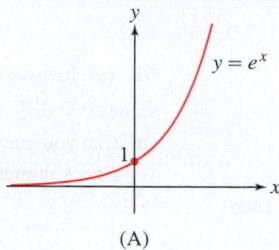

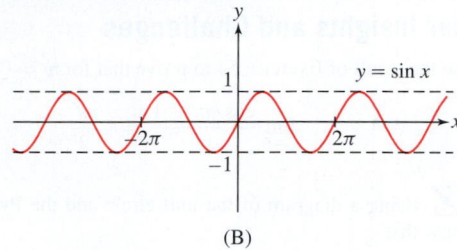

(A)   (B)

DF FIGURE 3

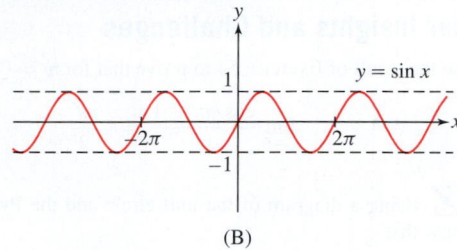

FIGURE 4 The logistic function
$P(t) = \frac{M}{1+Ae^{-kt}}$.

EXAMPLE 2 **Logistic Functions and the Carrying Capacity of the Earth** The function $P(t) = \frac{M}{1+Ae^{-kt}}$, with M, A, k all positive, is known as a *logistic function* (Figure 4). Such functions are used to model phenomena that have an initial rapid increase but then level off toward some finite value.

From 1950 to 1980 to 2010 the human population of the earth grew from 2.65 to 4.45 to 6.90 billion people. Let t represent time in years since 1950 and P represent the population in billions. Using the (t, P) values for 1950, 1980, and 2010, and employing a computer algebra system to solve for M, A, and k in $P(t)$, we obtain the logistic world-population model:

$$P(t) = \frac{17.4}{1 + 5.56e^{-0.022t}}$$

Find $\lim_{t\to\infty} P(t)$.

Solution Since $\lim_{t\to\infty} e^{-0.022t} = 0$, then

$$\lim_{t\to\infty} P(t) = \lim_{t\to\infty} \frac{17.4}{1 + 5.56e^{-0.022t}} = \frac{17.4}{1+0} = 17.4$$

We can interpret this limiting value of 17.4 billion as a theoretical carrying capacity of the earth. Opinions vary on the earth's actual human carrying capacity, and many approaches have been taken to develop estimates. This simply derived estimate is consistent with the carrying capacity models that are presented in a United Nations Environment Program 2012 review. ∎

The limits at infinity of the power functions $f(x) = x^n$ are easily determined. If $n > 0$, then x^n increases without bound as $x \to \infty$, so (Figure 5)

$$\lim_{x\to\infty} x^n = \infty \qquad \text{and} \qquad \lim_{x\to\infty} x^{-n} = \lim_{x\to\infty} \frac{1}{x^n} = 0$$

FIGURE 5

(A) n even: $\lim_{x\to\infty} x^n = \lim_{x\to-\infty} x^n = \infty$ (B) n odd: $\lim_{x\to\infty} x^n = \infty$, $\lim_{x\to-\infty} x^n = -\infty$ (C) $\lim_{x\to\infty} \frac{1}{x} = \lim_{x\to-\infty} \frac{1}{x} = 0$

CAUTION $\lim_{x\to-\infty} x^{1/2}$ does not exist, since the square root of a negative number is not a real number.

To describe the limits as $x \to -\infty$, assume that n is a whole number so that x^n is defined for $x < 0$. If n is even, then x^n becomes large and positive as $x \to -\infty$, and if n is odd, it becomes large and negative. We summarize these limits in the following theorem:

> **THEOREM 1** For all $n > 0$,
>
> $$\lim_{x \to \infty} x^n = \infty \qquad \text{and} \qquad \lim_{x \to \infty} x^{-n} = \lim_{x \to \infty} \frac{1}{x^n} = 0$$
>
> If n is a positive whole number,
>
> $$\lim_{x \to -\infty} x^n = \begin{cases} \infty & \text{if } n \text{ is even} \\ -\infty & \text{if } n \text{ is odd} \end{cases} \qquad \text{and} \qquad \lim_{x \to -\infty} x^{-n} = \lim_{x \to -\infty} \frac{1}{x^n} = 0$$

Note also that if p and q are positive integers and q is odd, then $\lim\limits_{x \to -\infty} x^{p/q} = \infty$ if p is even and $\lim\limits_{x \to -\infty} x^{p/q} = -\infty$ if p is odd. In the case that q is even, $x^{p/q}$ is not defined for negative x, so it does not make sense to address the limit as $x \to -\infty$.

The Basic Limit Laws (Theorem 1 in Section 2.3) are valid for limits at infinity. For example, the Sum and Constant Multiple Laws yield

$$\lim_{x \to \infty} \left(3 - 4x^{-3} + 5x^{-5} \right) = \lim_{x \to \infty} 3 - 4 \lim_{x \to \infty} x^{-3} + 5 \lim_{x \to \infty} x^{-5}$$

$$= 3 + 0 + 0 = 3$$

EXAMPLE 3 Calculate $\lim\limits_{x \to \infty} \dfrac{20x^2 - 3x}{3x^5 - 4x^2 + 5}$.

Solution It would be nice if we could apply the Quotient Law directly, but this law is valid only if the denominator has a finite, nonzero limit. Our limit has the indeterminate form ∞/∞ because

$$\lim_{x \to \infty} (20x^2 - 3x) = \infty \qquad \text{and} \qquad \lim_{x \to \infty} (3x^5 - 4x^2 + 5) = \infty$$

The way around this difficulty is to divide the numerator and denominator by x^5 (the highest power of x in the denominator):

$$\frac{20x^2 - 3x}{3x^5 - 4x^2 + 5} = \frac{x^{-5}(20x^2 - 3x)}{x^{-5}(3x^5 - 4x^2 + 5)} = \frac{20x^{-3} - 3x^{-4}}{3 - 4x^{-3} + 5x^{-5}}$$

Now we can use the Quotient Law:

$$\lim_{x \to \infty} \frac{20x^2 - 3x}{3x^5 - 4x^2 + 5} = \frac{\lim\limits_{x \to \infty} \left(20x^{-3} - 3x^{-4} \right)}{\lim\limits_{x \to \infty} \left(3 - 4x^{-3} + 5x^{-5} \right)} = \frac{0}{3} = 0 \qquad \blacksquare$$

In general, if

$$f(x) = \frac{a_n x^n + a_{n-1} x^{n-1} + \cdots + a_0}{b_m x^m + b_{m-1} x^{m-1} + \cdots + b_0}$$

where $a_n \neq 0$ and $b_m \neq 0$, divide the numerator and denominator by x^m:

$$f(x) = \frac{a_n x^{n-m} + a_{n-1} x^{n-1-m} + \cdots + a_0 x^{-m}}{b_m + b_{m-1} x^{-1} + \cdots + b_0 x^{-m}}$$

$$= x^{n-m} \left(\frac{a_n + a_{n-1} x^{-1} + \cdots + a_0 x^{-n}}{b_m + b_{m-1} x^{-1} + \cdots + b_0 x^{-m}} \right)$$

The quotient in parentheses approaches the finite limit a_n/b_m because

$$\lim_{x \to \infty} (a_n + a_{n-1} x^{-1} + \cdots + a_0 x^{-n}) = a_n$$

$$\lim_{x \to \infty} (b_m + b_{m-1} x^{-1} + \cdots + b_0 x^{-m}) = b_m$$

This also holds true for $x \to -\infty$, and therefore,

$$\lim_{x \to \pm\infty} f(x) = \lim_{x \to \pm\infty} x^{n-m} \lim_{x \to \pm\infty} \frac{a_n + a_{n-1}x^{-1} + \cdots + a_0 x^{-n}}{b_m + b_{m-1}x^{-1} + \cdots + b_0 x^{-m}} = \frac{a_n}{b_m} \lim_{x \to \pm\infty} x^{n-m}$$

THEOREM 2 Limits at Infinity of a Rational Function The asymptotic behavior of a rational function depends only on the leading terms of its numerator and denominator. If $a_n, b_m \neq 0$, then

$$\lim_{x \to \pm\infty} \frac{a_n x^n + a_{n-1}x^{n-1} + \cdots + a_0}{b_m x^m + b_{m-1}x^{m-1} + \cdots + b_0} = \frac{a_n}{b_m} \lim_{x \to \pm\infty} x^{n-m}$$

Here are some examples:

- $n = m$:

$$\lim_{x \to \infty} \frac{3x^4 - 7x + 9}{7x^4 - 4} = \frac{3}{7} \lim_{x \to \infty} x^0 = \frac{3}{7}$$

- $n < m$:

$$\lim_{x \to \infty} \frac{3x^3 - 7x + 9}{7x^4 - 4} = \frac{3}{7} \lim_{x \to \infty} x^{-1} = 0$$

- $n > m$, $n - m$ odd:

$$\lim_{x \to -\infty} \frac{3x^8 - 7x + 9}{7x^3 - 4} = \frac{3}{7} \lim_{x \to -\infty} x^5 = -\infty$$

- $n > m$, $n - m$ even:

$$\lim_{x \to -\infty} \frac{3x^7 - 7x + 9}{7x^3 - 4} = \frac{3}{7} \lim_{x \to -\infty} x^4 = \infty$$

Our method can be adapted to noninteger exponents and algebraic functions.

EXAMPLE 4 Calculate the limits   **(a)** $\displaystyle\lim_{x \to \infty} \frac{3x^{7/2} + 7x^{-1/2}}{x^2 - x^{1/2}}$ **(b)** $\displaystyle\lim_{x \to \pm\infty} \frac{4x}{\sqrt{x^2 + 1}}$.

Solution

> *The Quotient Law is valid if $\lim\limits_{x \to c} f(x) = \infty$ and $\lim\limits_{x \to c} g(x) = L$, where $L \neq 0$:*
>
> $$\lim_{x \to c} \frac{f(x)}{g(x)} = \frac{\lim\limits_{x \to c} f(x)}{\lim\limits_{x \to c} g(x)} = \begin{cases} \infty & \text{if } L > 0 \\ -\infty & \text{if } L < 0 \end{cases}$$
>
> *A similar result holds when $\lim\limits_{x \to c} f(x) = -\infty$.*

(a) As before, divide the numerator and denominator by x^2, which is the highest power of x occurring in the denominator (this means multiply by x^{-2}):

$$\frac{3x^{7/2} + 7x^{-1/2}}{x^2 - x^{1/2}} = \left(\frac{x^{-2}}{x^{-2}}\right) \frac{3x^{7/2} + 7x^{-1/2}}{x^2 - x^{1/2}} = \frac{3x^{3/2} + 7x^{-5/2}}{1 - x^{-3/2}}$$

$$\lim_{x \to \infty} \frac{3x^{7/2} + 7x^{-1/2}}{x^2 - x^{1/2}} = \frac{\lim\limits_{x \to \infty}(3x^{3/2} + 7x^{-5/2})}{\lim\limits_{x \to \infty}(1 - x^{-3/2})} = \frac{\infty}{1} = \infty$$

(b) First, consider $x \to \infty$. The key is to observe that the denominator of $\dfrac{4x}{\sqrt{x^2 + 1}}$ behaves like $x^1 = x$:

$$\sqrt{x^2 + 1} = \sqrt{x^2(1 + x^{-2})} = x\sqrt{1 + x^{-2}} \qquad (\text{for } x > 0)$$

This suggests that we divide the numerator and denominator by x:

$$\frac{4x}{\sqrt{x^2 + 1}} = \frac{\frac{4x}{x}}{\frac{\sqrt{x^2+1}}{x}} = \frac{4x}{x\sqrt{1 + x^{-2}}} = \frac{4}{\sqrt{1 + x^{-2}}}$$

Then apply the Quotient Law:

$$\lim_{x \to \infty} \frac{4x}{\sqrt{x^2 + 1}} = \lim_{x \to \infty} \frac{4}{\sqrt{1 + x^{-2}}} = \frac{\lim\limits_{x \to \infty} 4}{\lim\limits_{x \to \infty} \sqrt{1 + x^{-2}}} = \frac{4}{1} = 4$$

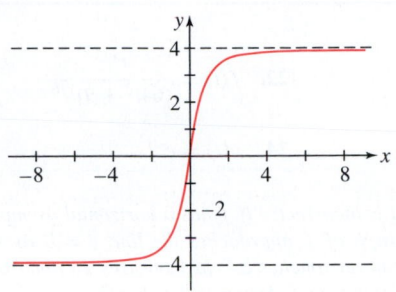

FIGURE 6 There are horizontal asymptotes at $y = \pm 4$.

For the limit as $x \to -\infty$, one approach is to replace x with $-t$. Since $x = -t$ and $x \to -\infty$, then $t \to \infty$. So we have

$$\lim_{x \to -\infty} \frac{4x}{\sqrt{x^2 + 1}} = \lim_{t \to \infty} \frac{4(-t)}{\sqrt{(-t)^2 + 1}} = \lim_{t \to \infty} \frac{-4t}{\sqrt{t^2 + 1}} = -\lim_{t \to \infty} \frac{4t}{\sqrt{t^2 + 1}} = -4$$

where the last equality holds by the limit we previously calculated.

The limits in (b) indicate that the graph of $f(x) = \frac{4x}{\sqrt{x^2+1}}$ has horizontal asymptotes at $y = 4$ and $y = -4$, which is confirmed in Figure 6. ∎

2.7 SUMMARY

- *Limits at infinity*:

  - $\lim\limits_{x \to \infty} f(x) = L$ if $|f(x) - L|$ becomes arbitrarily small as x increases without bound.
  - $\lim\limits_{x \to -\infty} f(x) = L$ if $|f(x) - L|$ becomes arbitrarily small as x decreases without bound.
  - $\lim\limits_{x \to \infty} e^x = \infty$ and $\lim\limits_{x \to -\infty} e^x = 0$

- A horizontal line $y = L$ is a *horizontal asymptote* of f if

$$\lim_{x \to \infty} f(x) = L \qquad \text{or} \qquad \lim_{x \to -\infty} f(x) = L$$

  A function can have 0, 1 or 2 horizontal asymptotes.
- If $n > 0$, then $\lim\limits_{x \to \pm\infty} x^{-n} = 0$.
- If $n > 0$ is a whole number, then

$$\lim_{x \to \infty} x^n = \infty \quad \text{and} \quad \lim_{x \to -\infty} x^n = \begin{cases} \infty & \text{if } n \text{ is even} \\ -\infty & \text{if } n \text{ is odd} \end{cases}$$

- If $f(x) = \dfrac{a_n x^n + a_{n-1} x^{n-1} + \cdots + a_0}{b_m x^m + b_{m-1} x^{m-1} + \cdots + b_0}$ with $a_n, b_m \neq 0$, then

$$\lim_{x \to \pm\infty} f(x) = \frac{a_n}{b_m} \lim_{x \to \pm\infty} x^{n-m}$$

2.7 EXERCISES

Preliminary Questions

1. Assume that

$$\lim_{x \to \infty} f(x) = L \quad \text{and} \quad \lim_{x \to L} g(x) = \infty$$

Which of the following statements are correct?

(a) $x = L$ is a vertical asymptote of g.

(b) $y = L$ is a horizontal asymptote of g.

(c) $x = L$ is a vertical asymptote of f.

(d) $y = L$ is a horizontal asymptote of f.

2. What are the following limits?

(a) $\lim\limits_{x \to \infty} x^3$ **(b)** $\lim\limits_{x \to -\infty} x^3$ **(c)** $\lim\limits_{x \to -\infty} x^4$

3. Sketch the graph of a function that approaches a limit as $x \to \infty$ but does not approach a limit (either finite or infinite) as $x \to -\infty$.

4. What is the sign of a if $f(x) = ax^3 + x + 1$ satisfies $\lim\limits_{x \to -\infty} f(x) = \infty$?

5. What is the sign of the coefficient multiplying x^7 if f is a polynomial of degree 7 such that $\lim\limits_{x \to -\infty} f(x) = \infty$?

6. Explain why $\lim\limits_{x \to \infty} \sin \frac{1}{x}$ exists but $\lim\limits_{x \to 0} \sin \frac{1}{x}$ does not exist. What is $\lim\limits_{x \to \infty} \sin \frac{1}{x}$?

Exercises

1. What are the horizontal asymptotes of the function in Figure 7?

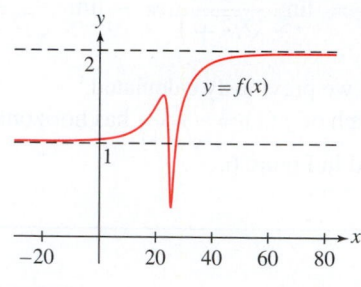

FIGURE 7

2. Sketch the graph of a function f that has both $y = -1$ and $y = 5$ as horizontal asymptotes.

3. Sketch the graph of a function f with a single horizontal asymptote $y = 3$.

4. Sketch the graphs of functions f and g that have both $y = -2$ and $y = 4$ as horizontal asymptotes but $\lim_{x\to\infty} f(x) \neq \lim_{x\to\infty} g(x)$.

5. [GU] Investigate the asymptotic behavior of $f(x) = \dfrac{x^2}{x^2+1}$ numerically and graphically:

(a) Make a table of values of $f(x)$ for $x = \pm 50$, ± 100, ± 500, ± 1000.

(b) Plot the graph of f.

(c) What are the horizontal asymptotes of f?

6. [GU] Investigate $\lim_{x\to\pm\infty} \dfrac{12x+1}{\sqrt{4x^2+9}}$ numerically and graphically:

(a) Make a table of values of $f(x) = \dfrac{12x+1}{\sqrt{4x^2+9}}$ for the following: $x = \pm 100$, ± 500, ± 1000, $\pm 10,000$.

(b) Plot the graph of f.

(c) What are the horizontal asymptotes of f?

In Exercises 7–16, evaluate the limit.

7. $\lim_{x\to\infty} \dfrac{x}{x+9}$

8. $\lim_{x\to\infty} \dfrac{3x^2+20x}{4x^2+9}$

9. $\lim_{x\to\infty} \dfrac{3x^2+20x}{2x^4+3x^3-29}$

10. $\lim_{x\to\infty} \dfrac{4}{x+5}$

11. $\lim_{x\to\infty} \dfrac{7x-9}{4x+3}$

12. $\lim_{x\to\infty} \dfrac{9x^2-2}{6-29x}$

13. $\lim_{x\to-\infty} \dfrac{7x^2-9}{4x+3}$

14. $\lim_{x\to-\infty} \dfrac{5x-9}{4x^3+2x+7}$

15. $\lim_{x\to-\infty} \dfrac{3x^3-10}{x+4}$

16. $\lim_{x\to-\infty} \dfrac{2x^5+3x^4-31x}{8x^4-31x^2+12}$

In Exercises 17–24, find the horizontal asymptotes.

17. $f(x) = \dfrac{2x^2-3x}{8x^2+8}$

18. $f(x) = \dfrac{8x^3-x^2}{7+11x-4x^4}$

19. $f(x) = \dfrac{\sqrt{36x^2+7}}{9x+4}$

20. $f(x) = \dfrac{\sqrt{36x^4+7}}{9x^2+4}$

21. $f(t) = \dfrac{e^t}{1+e^{-t}}$

22. $f(t) = \dfrac{t^{1/3}}{(64t^2+9)^{1/6}}$

23. $g(t) = \dfrac{10}{1+3^{-t}}$

24. $p(t) = e^{-t^2}$

📝 *The following statement is incorrect: "If f has a horizontal asymptote $y = L$ at ∞, then the graph of f approaches the line $y = L$ as x gets greater and greater, but never touches it." In Exercises 25 and 26, determine $\lim_{x\to\infty} f(x)$ and indicate how f demonstrates that the statement is incorrect.*

25. $f(x) = \dfrac{2x+|x|}{x}$

26. $f(x) = \dfrac{\sin x}{x}$

In Exercises 27–34, evaluate the limit.

27. $\lim_{x\to\infty} \dfrac{\sqrt{9x^4+3x+2}}{4x^3+1}$

28. $\lim_{x\to\infty} \dfrac{\sqrt{x^3+20x}}{10x-2}$

29. $\lim_{x\to-\infty} \dfrac{8x^2+7x^{1/3}}{\sqrt{16x^4+6}}$

30. $\lim_{x\to-\infty} \dfrac{4x-3}{\sqrt{25x^2+4x}}$

31. $\lim_{t\to\infty} \dfrac{t^{4/3}+t^{1/3}}{(4t^{2/3}+1)^2}$

32. $\lim_{t\to\infty} \dfrac{t^{4/3}-9t^{1/3}}{(8t^4+2)^{1/3}}$

33. $\lim_{x\to-\infty} \dfrac{|x|+x}{x+1}$

34. $\lim_{t\to-\infty} \dfrac{4+6e^{2t}}{5-9e^{3t}}$

35. 📝 Determine $\lim_{x\to\infty} \tan^{-1} x$. Explain geometrically.

36. Show that $\lim_{x\to\infty} (\sqrt{x^2+1}-x) = 0$. *Hint:* Observe that

$$\sqrt{x^2+1}-x = \dfrac{1}{\sqrt{x^2+1}+x}$$

37. In 2009, 2012, and 2015, the number (in millions) of smart phones sold in the world was 172.4, 680.1, and 1423.9, respectively.

(a) [CAS] Let t represent time in years since 2009, and let S represent the number of smart phones sold in millions. Determine M, A, and k for a logistic model, $S(t) = \dfrac{M}{1+Ae^{-kt}}$, that fits the given data points.

(b) What is the long-term expected maximum number of smart phones sold annually? That is, what is $\lim_{t\to\infty} S(t)$?

(c) In what year does the model predict that smart-phone sales will reach 98% of the expected maximum?

38. Sam was 28 inches tall on her first birthday, 50 inches tall on her 8th, and 62 inches tall on her 14th.

(a) [CAS] Let t represent Sam's age in years, and let h represent her height in inches. Determine the values of M, A, and k for a logistic model, $h(t) = \dfrac{M}{1+Ae^{-kt}}$, that fits the given height data.

(b) What is Sam's theoretical long-term expected height? That is, what is $\lim_{t\to\infty} h(t)$?

(c) At what age does the model predict that Sam will reach 95% of her expected maximum height?

In Exercises 39–46, calculate the limit.

39. $\lim_{x\to\infty} (\sqrt{4x^4+9x}-2x^2)$

40. $\lim_{x\to\infty} (\sqrt{9x^3+x}-x^{3/2})$

41. $\lim_{x\to\infty} (2\sqrt{x}-\sqrt{x+2})$

42. $\lim_{x\to\infty} \left(\dfrac{1}{x}-\dfrac{1}{x+2}\right)$

43. $\displaystyle\lim_{x\to\infty}\,(\ln(3x+1)-\ln(2x+1))$

44. $\displaystyle\lim_{x\to\infty}\left(\ln(\sqrt{5x^2+2})-\ln x\right)$

45. $\displaystyle\lim_{x\to\infty}\tan^{-1}\left(\frac{x^2+9}{9-x}\right)$ **46.** $\displaystyle\lim_{x\to\infty}\tan^{-1}\left(\frac{1+x}{1-x}\right)$

47. Let $P(n)$ be the perimeter of an n-gon inscribed in a unit circle (Figure 8).

(a) Explain, intuitively, why $P(n)$ approaches 2π as $n\to\infty$.

(b) Show that $P(n)=2n\sin\left(\frac{\pi}{n}\right)$.

(c) Combine (a) and (b) to conclude that $\displaystyle\lim_{n\to\infty}\frac{n}{\pi}\sin\left(\frac{\pi}{n}\right)=1$.

(d) Use this to give another argument that $\displaystyle\lim_{\theta\to 0}\frac{\sin\theta}{\theta}=1$.

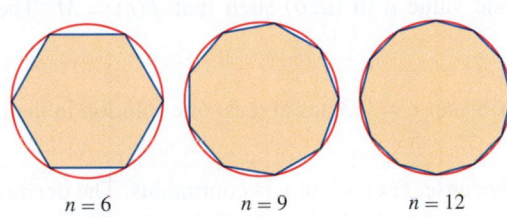

$n=6$ $n=9$ $n=12$

FIGURE 8

48. Physicists have observed that Einstein's theory of **special relativity** reduces to Newtonian mechanics in the limit as $c\to\infty$, where c is the speed of light. This is illustrated by a stone tossed up vertically from ground level so that it returns to Earth 1 s later. Using Newton's Laws, we find that the stone's maximum height is $h=g/8$ m ($g=9.8$ m/s^2). According to special relativity, the stone's mass depends on its velocity divided by c, and the maximum height is

$$h(c)=c\sqrt{c^2/g^2+1/4}-c^2/g$$

Prove that $\displaystyle\lim_{c\to\infty}h(c)=g/8$.

49. According to the **Michaelis–Menten equation**, when an enzyme is combined with a substrate of concentration s (in millimolars), the reaction rate (in micromolars/min) is

$$R(s)=\frac{As}{K+s}\qquad(A,\ K\ \text{constants})$$

(a) Show, by computing $\displaystyle\lim_{s\to\infty}R(s)$, that A is the limiting reaction rate as the concentration s approaches ∞.

(b) Show that the reaction rate $R(s)$ attains one-half of the limiting value A when $s=K$.

(c) For a certain reaction, $K=1.25$ mM and $A=0.1$. For which concentration s is $R(s)$ equal to 75% of its limiting value?

Further Insights and Challenges

50. Every limit as $x\to\infty$ can be expressed alternatively as a one-sided limit as $t\to 0^+$, where $t=x^{-1}$. Setting $g(t)=f(t^{-1})$, we have

$$\lim_{x\to\infty}f(x)=\lim_{t\to 0^+}g(t)$$

Show that $\displaystyle\lim_{x\to\infty}\frac{3x^2-x}{2x^2+5}=\lim_{t\to 0^+}\frac{3-t}{2+5t^2}$, and evaluate using the Quotient Law.

51. Rewrite the following as one-sided limits as in Exercise 50 and evaluate.

(a) $\displaystyle\lim_{x\to\infty}\frac{3-12x^3}{4x^3+3x+1}$ **(b)** $\displaystyle\lim_{x\to\infty}e^{1/x}$

(c) $\displaystyle\lim_{x\to\infty}x\sin\frac{1}{x}$ **(d)** $\displaystyle\lim_{x\to\infty}\ln\left(\frac{x+1}{x-1}\right)$

52. Let $G(b)=\displaystyle\lim_{x\to\infty}(1+b^x)^{1/x}$ for $b\geq 0$. Investigate $G(b)$ numerically and graphically for $b=0.2,\ 0.8,\ 2,\ 3,\ 5$ (and additional values if necessary). Then make a conjecture for the value of $G(b)$ as a function of b. Draw a graph of $y=G(b)$. Does G appear to be continuous? We will evaluate $G(b)$ using L'Hôpital's Rule in Section 4.5 (see Exercise 73 there).

2.8 The Intermediate Value Theorem

The **Intermediate Value Theorem (IVT)** says, roughly speaking, that *a continuous function cannot skip values*. Consider a plane that takes off and climbs from 0 to 10,000 m in 20 min. The plane must reach every altitude between 0 and 10,000 m during this 20-min interval. Thus, at some moment, the plane's altitude must have been exactly 8371 m. Of course, this assumes that the plane's motion is continuous, so its altitude cannot jump abruptly from, say, 8000 to 9000 m.

To state this conclusion formally, let $A(t)$ be the plane's altitude at time t. The IVT asserts that for every altitude M between 0 and 10,000, there is a time t_0 between 0 and 20 min such that $A(t_0)=M$. In other words, the graph of A must intersect the horizontal line $y=M$ [Figure 1(A)].

By contrast, a discontinuous function can skip values. The greatest integer function $f(x)=\lfloor x\rfloor$ in Figure 1(B) satisfies $\lfloor 1\rfloor=1$ and $\lfloor 2\rfloor=2$, but it does not take on the value 1.5 (or any other value between 1 and 2).

THEOREM 1   Intermediate Value Theorem   If f is continuous on a closed interval $[a,b]$, then for every value M, strictly between $f(a)$ and $f(b)$, there exists at least one value $c\in(a,b)$ such that $f(c)=M$.

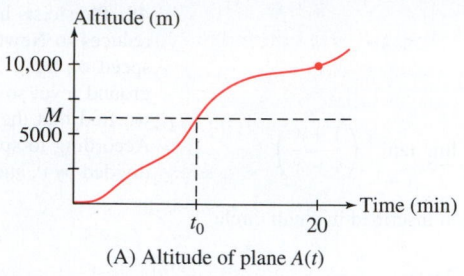

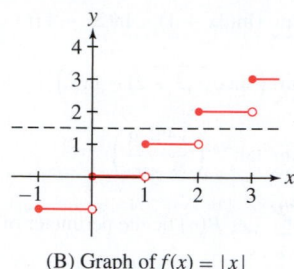

(A) Altitude of plane $A(t)$ (B) Graph of $f(x) = \lfloor x \rfloor$

DF FIGURE 1

Graphically, as in Figure 2, the result appears obvious. For a continuous function, every horizontal line at height M between $f(a)$ and $f(b)$ is forced to hit the graph, and therefore there must be at least one value c in (a, b) such that $f(c) = M$. The proof appears in Appendix B.

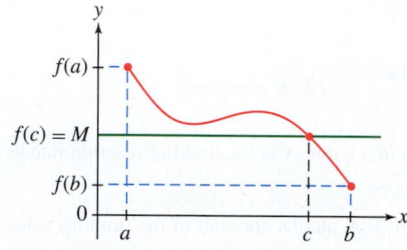

FIGURE 2 For every M between $f(a)$ and $f(b)$, there is a c between a and b such that $f(c) = M$.

EXAMPLE 1 Prove that the equation $\sin x = 0.3$ has at least one solution in the interval $\left(0, \frac{\pi}{2}\right)$.

Solution We may apply the IVT because $f(x) = \sin x$ is continuous. The desired value 0.3 lies between the values of the function at the endpoints of the interval:

$$\sin 0 = 0 \qquad \text{and} \qquad \sin \frac{\pi}{2} = 1$$

as illustrated in (Figure 3). The IVT tells us that $\sin x = 0.3$ has at least one solution in $\left(0, \frac{\pi}{2}\right)$. ∎

The IVT can be used to show the existence of zeros of functions. If f is continuous and takes on both nonpositive and nonnegative values (say, $f(a) \le 0$ and $f(b) \ge 0$) then the IVT guarantees that $f(c) = 0$ for some c in $[a, b]$. This is extremely useful when we cannot explicitly solve for the zero but would like to know that there is one in the interval.

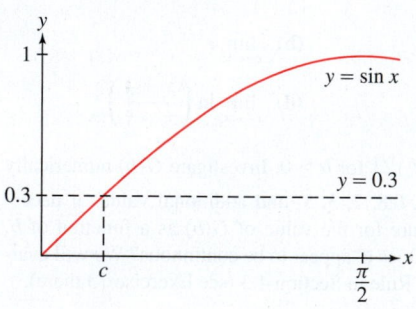

FIGURE 3

> **REMINDER** A zero or root of a function is a value c such that $f(c) = 0$.

COROLLARY 2 Existence of Zeros If f is continuous on $[a, b]$, and if one of $f(a)$ and $f(b)$ is nonnegative and the other is nonpositive, then f has a zero in $[a, b]$.

We can locate zeros of functions to arbitrary accuracy using the **Bisection Method**. The idea is to find an interval $[a, b]$ such that the function has opposite signs at the endpoints. Then Corollary 2 tells us that there is a zero on this interval. To find its location more precisely, we cut the interval into two equal subintervals. Then, check the signs at the endpoints of each of these intervals to see which one Corollary 2 tells us has a zero. (But keep in mind that there may be more than one zero, so both could contain a zero). Next, we repeat the process on this smaller interval. Eventually, we narrow down on a zero. This is illustrated in the next example.

EXAMPLE 2   The Bisection Method Show that $f(x) = \cos^2 x - 2 \sin \frac{x}{4}$ has a zero in $(0, 2)$. Then, using the Bisection Method, find a subinterval of $(0, 2)$ of length $1/8$ that contains a zero of f.

Solution To begin, we note that f is continuous on $[0, 2]$. Calculating $f(0)$ and $f(2)$, we find that they have opposite signs:

$$f(0) = 1 > 0, \qquad f(2) \approx -0.786 < 0$$

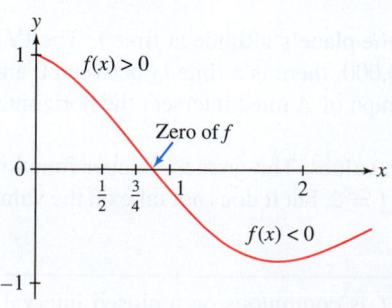

FIGURE 4 Graph of $f(x) = \cos^2 x - 2 \sin \frac{x}{4}$.

Corollary 2 guarantees that $f(x) = 0$ has a solution in $(0, 2)$ (Figure 4).

To locate a zero more accurately, divide $[0, 2]$ into two intervals $[0, 1]$ and $[1, 2]$. At least one of these intervals must contain a zero of f. To determine which, we evaluate f at the midpoint $m = 1$, obtaining $f(1) \approx -0.203 < 0$. Since $f(0) = 1$, we see that

$$f(x) \text{ takes on opposite signs at the endpoints of } [0, 1]$$

Therefore, $(0, 1)$ must contain a zero. Note that from the function values $f(1)$ and $f(2)$ alone, we cannot conclude that f *does not* have a zero in the interval $[1, 2]$, but it is clear from the graph in Figure 4 that it does not.

The Bisection Method consists of continuing this process until we narrow down the location of a zero to any desired accuracy. In the following table, the process is carried out three times to find an interval of length $1/8$ containing a zero of f.

Interval	Midpoint of interval	Function values		Conclusion
$[0, 1]$	$\frac{1}{2}$	$f\left(\frac{1}{2}\right) \approx 0.521$	$f(1) \approx -0.203$	A zero lies in $\left(\frac{1}{2}, 1\right)$.
$\left[\frac{1}{2}, 1\right]$	$\frac{3}{4}$	$f\left(\frac{3}{4}\right) \approx 0.163$	$f(1) \approx -0.203$	A zero lies in $\left(\frac{3}{4}, 1\right)$.
$\left[\frac{3}{4}, 1\right]$	$\frac{7}{8}$	$f\left(\frac{7}{8}\right) \approx -0.0231$	$f\left(\frac{3}{4}\right) \approx 0.163$	A zero lies in $\left(\frac{3}{4}, \frac{7}{8}\right)$.

We conclude that f has a zero c satisfying $0.75 < c < 0.875$. ∎

EXAMPLE 3 **A Meteorological Consequence of the IVT** Take a map and draw a circle anywhere on it such as in Figure 5. With the IVT we can show that there must be a pair of points that have the same temperature and lie opposite each other through the center of the circle.

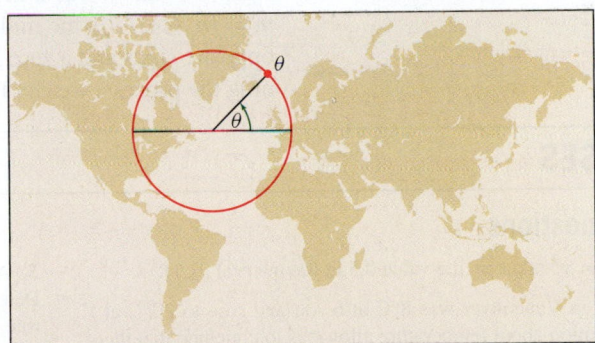

FIGURE 5

Let $T(\theta)$ be the temperature of the location on the circle that is at angle θ from the horizontal (as shown in Figure 5). We assume that T is a continuous function. Define another function f by $f(\theta) = T(\theta) - T(\theta + \pi)$. Thus, $f(\theta)$ is the difference between the temperature at location θ and the temperature at the location opposite it on the circle. Since T is continuous, it follows that f is as well.

Show that $f(\pi) = -f(0)$, and use Corollary 2 and the relationship between f and T to argue that there are opposite points on the circle with the same temperature.

Solution For f we have

$$f(0) = T(0) - T(\pi)$$

$$f(\pi) = T(\pi) - T(\pi + \pi) = T(\pi) - T(2\pi) = T(\pi) - T(0) = -f(0) \qquad \boxed{1}$$

Computer algebra systems have built-in commands for finding roots of a function or solving an equation numerically. These systems use a variety of methods, including versions of the Bisection Method and Newton's Method, a process that we introduce in Section 4.8.

Since $f(\pi) = -f(0)$ by Eq. (1), it follows that one of $f(0)$ and $f(\pi)$ is nonnegative and the other is nonpositive. Corollary 2 then implies that there is c between 0 and π such that $f(c) = 0$. So $T(c) - T(c + \pi) = 0$, implying that c and $c + \pi$ are opposite points with the same temperature. ∎

Note that the argument in Example 3 works with any continuous variable defined over the map. Temperature could be replaced with barometric pressure, relative humidity, elevation, and so on, and the same type of conclusion could be drawn.

CONCEPTUAL INSIGHT   The IVT seems to state the obvious, namely that a continuous function cannot skip values. Yet its proof (given in Appendix B) is subtle because it depends on the *completeness property* of real numbers. To highlight the subtlety, observe that the IVT is *false* for functions defined only on the *rational numbers*. For example, $f(x) = x^2$ is continuous, but the Intermediate Value Theorem does not apply if we restrict its domain to the rational numbers. Indeed, $f(0) = 0$, $f(2) = 4$, and 3 is between 0 and 4 but $f(c) = 3$ *has no solution* for c rational. The solution $c = \sqrt{3}$ is missing from the set of rational numbers because it is irrational. No doubt the IVT was always regarded as obvious, but it was not possible to give a correct proof until the completeness property was clarified in the second half of the nineteenth century.

2.8 SUMMARY

- The Intermediate Value Theorem (IVT) says that a continuous function cannot *skip* values.
- More precisely, if f is continuous on $[a, b]$ with $f(a) \neq f(b)$, and if M is a number strictly between $f(a)$ and $f(b)$, then $f(c) = M$ for some $c \in (a, b)$.
- Existence of zeros: If f is continuous on $[a, b]$, and if one of $f(a)$ and $f(b)$ is nonnegative and the other is nonpositive, then f has a zero in $[a, b]$.
- Bisection Method: Assume f is continuous and that $f(a)$ and $f(b)$ have opposite signs, so that f has a zero in (a, b). Then f has a zero in $[a, m]$ or $[m, b]$, where $m = (a + b)/2$ is the midpoint of $[a, b]$. A zero lies in (a, m) if $f(a)$ and $f(m)$ have opposite signs and a zero lies in (m, b) if $f(m)$ and $f(b)$ have opposite signs. Continuing the process, we can locate zeros with arbitrary accuracy.

2.8 EXERCISES

Preliminary Questions

1. Prove that $f(x) = x^2$ takes on the value 0.5 in the interval $[0, 1]$.

2. The temperature in Vancouver was 8°C at 6 AM and rose to 20°C at noon. Which assumption about temperature allows us to conclude that the temperature was 15°C at some moment of time between 6 AM and noon?

3. What is the graphical interpretation of the IVT?

4. Show that the following statement is false by drawing a graph that provides a counterexample:

If f is continuous and has a root in $[a, b]$, then $f(a)$ and $f(b)$ have opposite signs.

5. Assume that f is continuous on $[1, 5]$ and that $f(1) = 20$, $f(5) = 100$. Determine whether each of the following statements is always true, never true, or sometimes true.

(a) $f(c) = 3$ has a solution with $c \in [1, 5]$.

(b) $f(c) = 75$ has a solution with $c \in [1, 5]$.

(c) $f(c) = 50$ has no solution with $c \in [1, 5]$.

(d) $f(c) = 30$ has exactly one solution with $c \in [1, 5]$.

Exercises

1. Use the IVT to show that $f(x) = x^3 + x$ takes on the value 9 for some x in $[1, 2]$.

2. Show that $g(t) = \dfrac{t}{t + 1}$ takes on the value 0.499 for some t in $[0, 1]$.

3. Show that $g(t) = t^2 \tan t$ takes on the value $\frac{1}{2}$ for some t in $\left[0, \frac{\pi}{4}\right]$.

4. Show that $f(x) = \dfrac{x^2}{x^7 + 1}$ takes on the value 0.4.

5. Show that $\cos x = x$ has a solution in the interval $[0, 1]$. *Hint:* Show that $f(x) = x - \cos x$ has a zero in $[0, 1]$.

6. Use the IVT to find an interval of length $\frac{1}{2}$ containing a root of $f(x) = x^3 + 2x + 1$.

In Exercises 7–16, prove using the IVT.

7. $\sqrt{c} + \sqrt{c+2} = 3$ has a solution.

8. For all integers n, $\sin nx = \cos x$ for some $x \in [0, \pi]$.

9. $\sqrt{2}$ exists. *Hint:* Consider $f(x) = x^2$.

10. A positive number c has an nth root for all positive integers n.

11. For all positive integers k, $\cos x = x^k$ has a solution.

12. $2^x = bx$ has a solution if $b > 2$.

13. $2^x + 3^x = 4^x$ has a solution.

14. $\cos x = \cos^{-1} x$ has a solution in $(0, 1)$.

15. $e^x + \ln x = 0$ has a solution.

16. $\tan^{-1} x = \cos^{-1} x$ has a solution.

17. Use the Intermediate Value Theorem to show that the equation $x^6 - 8x^4 + 10x^2 - 1 = 0$ has at least six distinct solutions.

In Exercises 18–20, determine whether or not the IVT applies to show that the given function takes on all values between $f(a)$ and $f(b)$ for $x \in [a, b]$. If it does not apply, determine any values between $f(a)$ and $f(b)$ that the function does not take on for $x \in [a, b]$.

18.
$$f(x) = \begin{cases} x & \text{for } x < 0 \\ x^2 & \text{for } x \ge 0 \end{cases} \quad \text{for the interval } [-1, 1].$$

19.
$$f(x) = \begin{cases} -x & \text{for } x < 0 \\ x^3 + 1 & \text{for } x \ge 0 \end{cases} \quad \text{for the interval } [-1, 1].$$

20.
$$f(x) = \begin{cases} -x^2 & \text{for } x < 0 \\ 1 & \text{for } x = 0 \\ x & \text{for } x > 0 \end{cases} \quad \text{for the interval } [-2, 2].$$

21. Carry out three steps of the Bisection Method for $f(x) = 2^x - x^3$ as follows:

(a) Show that f has a zero in $[1, 1.5]$.

(b) Show that f has a zero in $[1.25, 1.5]$.

(c) Determine whether $[1.25, 1.375]$ or $[1.375, 1.5]$ contains a zero.

22. Figure 6 shows that $f(x) = x^3 - 8x - 1$ has a root in the interval $[2.75, 3]$. Apply the Bisection Method twice to find an interval of length $\frac{1}{16}$ containing this root.

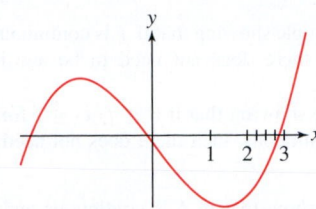

FIGURE 6 Graph of $y = x^3 - 8x - 1$.

23. Find an interval of length $\frac{1}{4}$ in $[1, 2]$ containing a root of the equation $x^7 + 3x - 10 = 0$.

24. Show that $\tan^3 \theta - 8 \tan^2 \theta + 17 \tan \theta - 8 = 0$ has a root in $[0.5, 0.6]$. Apply the Bisection Method twice to find an interval of length 0.025 containing this root.

In Exercises 25–28, draw the graph of a function f on $[0, 4]$ with the given property.

25. Jump discontinuity at $x = 2$ and does not satisfy the conclusion of the IVT

26. Jump discontinuity at $x = 2$ and satisfies the conclusion of the IVT on $[0, 4]$

27. Infinite one-sided limits at $x = 2$ and does not satisfy the conclusion of the IVT

28. Infinite one-sided limits at $x = 2$ and satisfies the conclusion of the IVT on $[0, 4]$

29. 🖉 Can Corollary 2 be applied to $f(x) = x^{-1}$ on $[-1, 1]$? Does f have any roots?

30. (a) Assume that g and h are continuous on $[a, b]$. Use Corollary 2 to show that if $g(a) < h(a)$ and $h(b) < g(b)$, then there exists $c \in [a, b]$ such that $g(c) = h(c)$.

(b) Interpret the result of (a) in terms of the graphs of g and h, and show, by a graphical example, that the conclusion in (a) need not hold if one of g or h is not continuous.

31. At 1:00 PM, Jacqueline began to climb up Waterpail Hill from the bottom. At the same time Giles began to climb down from the top. Giles reached the bottom at 2:20 PM, when Jacqueline was 85% of the way up. Jacqueline reached the top at 2:50. Use the result in Exercise 30 to prove that there was a time when they were at the same elevation on the hill.

32. 🖉 On Wednesday at noon the weather was fair in Boston with a barometric pressure of 1018 mb. At the same time, a low-pressure storm system was passing by Buffalo, where the pressure was 996 mb. At noon Thursday the storm was approaching Boston, where the pressure was 1002 mb, while the weather was clearing in Buffalo and the pressure there had risen to 1014 mb. Use the result in Exercise 30 to prove that there was a time between noon Wednesday and noon Thursday when Boston and Buffalo had the same barometric pressure.

Further Insights and Challenges

Exercises 33 and 34, address the 1-Dimensional Brouwer Fixed Point Theorem. It indicates that every continuous function f mapping the closed interval $[0, 1]$ to itself must have a fixed point; that is, a point c such that $f(c) = c$.

33. 🖉 Show that if f is continuous and $0 \le f(x) \le 1$ for $0 \le x \le 1$, then $f(c) = c$ for some c in $[0, 1]$ (Figure 7).

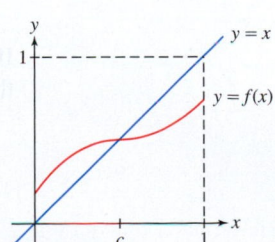

FIGURE 7 If $0 \le f(x) \le 1$ for $0 \le x \le 1$, then f has a fixed point c.

34. (a) Give an example showing that if f is continuous and $0 < f(x) < 1$ for $0 < x < 1$, then there does not need to be a c in $(0,1)$ such that $f(c) = c$.
(b) Give an example showing that if $0 \leq f(x) \leq 1$ for $0 \leq x \leq 1$, but f is not necessarily continuous, then there does not need to be a c in $(0,1)$ such that $f(c) = c$.

35. Use the IVT to show that if f is continuous and one-to-one on an interval $[a,b]$, then f is either an increasing or a decreasing function.

36. ✏️ **Ham Sandwich Theorem** Figure 8(A) shows a slice of ham. Prove that for any angle θ ($0 \leq \theta \leq \pi$), it is possible to cut the slice in half with a cut of incline θ. *Hint:* The lines of inclination θ are given by the equations $y = (\tan\theta)x + b$, where b varies from $-\infty$ to ∞. Each such line divides the slice into two pieces (one of which may be empty). Let $A(b)$ be the amount of ham to the left of the line minus the amount to the right, and let A be the total area of the ham. Show that $A(b) = -A$ if b is sufficiently large and $A(b) = A$ if b is sufficiently negative. Then use the IVT. This works if $\theta \neq 0$ or $\frac{\pi}{2}$. If $\theta = 0$, define $A(b)$ as the amount of ham above the line $y = b$ minus the amount below. How can you modify the argument to work when $\theta = \frac{\pi}{2}$ (in which case $\tan\theta = \infty$)?

37. ✏️ Figure 8(B) shows a slice of ham on a piece of bread. Prove that it is possible to slice this open-faced sandwich so that each part has equal amounts of ham and bread. *Hint:* By Exercise 36, for all $0 \leq \theta \leq \pi$ there

is a line $L(\theta)$ of incline θ (which we assume is unique) that divides the ham into two equal pieces. Let $B(\theta)$ denote the amount of bread to the left of (or above) $L(\theta)$ minus the amount to the right (or below). Notice that $L(\pi)$ and $L(0)$ are the same line, but $B(\pi) = -B(0)$ since left and right get interchanged as the angle moves from 0 to π. Assume that B is continuous and apply the IVT. (By a further extension of this argument, one can prove the full Ham Sandwich Theorem, which states that if you allow the knife to cut at a slant, then it is possible to cut a sandwich consisting of a slice of ham and two slices of bread so that all three layers are divided in half.)

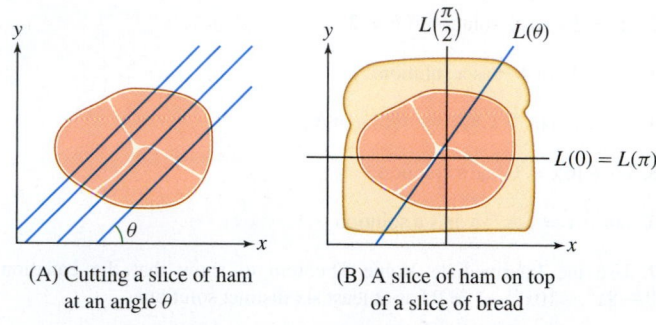

(A) Cutting a slice of ham at an angle θ

(B) A slice of ham on top of a slice of bread

FIGURE 8

2.9 The Formal Definition of a Limit

In this section, we reexamine the definition of a limit in order to state it in a more rigorous and precise fashion. Why is this necessary? In Section 2.2, we defined limits by saying that $\lim\limits_{x \to c} f(x) = L$ if $|f(x) - L|$ becomes arbitrarily small when x is sufficiently close (but not equal) to c. The problem with this definition lies in the phrases "arbitrarily small" and "sufficiently close." We must find a way to specify just how close is sufficiently close.

The Size of the Gap

A rigorous proof in mathematics is an argument based on a complete chain of logic where each step follows unambiguously from what proceeds it. The formal definition of a limit is a key ingredient of rigorous proofs in calculus. A few such proofs are included in Appendix D. More complete developments can be found in textbooks on the branch of mathematics called analysis.

Recall that the distance from $f(x)$ to L is $|f(x) - L|$. It is convenient to refer to the quantity $|f(x) - L|$ as the *gap* between the value $f(x)$ and the limit L.

Let's reexamine the trigonometric limit

$$\lim_{x \to 0} \frac{\sin x}{x} = 1 \qquad \boxed{1}$$

In this example, $f(x) = \dfrac{\sin x}{x}$ and $L = 1$, so Eq. (1) tells us that the gap $|f(x) - 1|$ gets arbitrarily small when x is sufficiently close, but not equal, to 0 [Figure 1(A)].

Suppose we want the gap $|f(x) - 1|$ to be less than 0.2. How close to 0 must x be? Figure 1(B) shows that $f(x)$ lies within 0.2 of $L = 1$ for all values of x in the interval $[-1, 1]$. In other words, the following statement is true:

$$\text{If } 0 < |x| < 1, \quad \text{then} \quad \left| \frac{\sin x}{x} - 1 \right| < 0.2$$

If we insist instead that the gap be smaller than 0.004, we can check by zooming in on the graph, as in Figure 1(C), that

$$\text{If } 0 < |x| < 0.15, \quad \text{then} \quad \left| \frac{\sin x}{x} - 1 \right| < 0.004$$

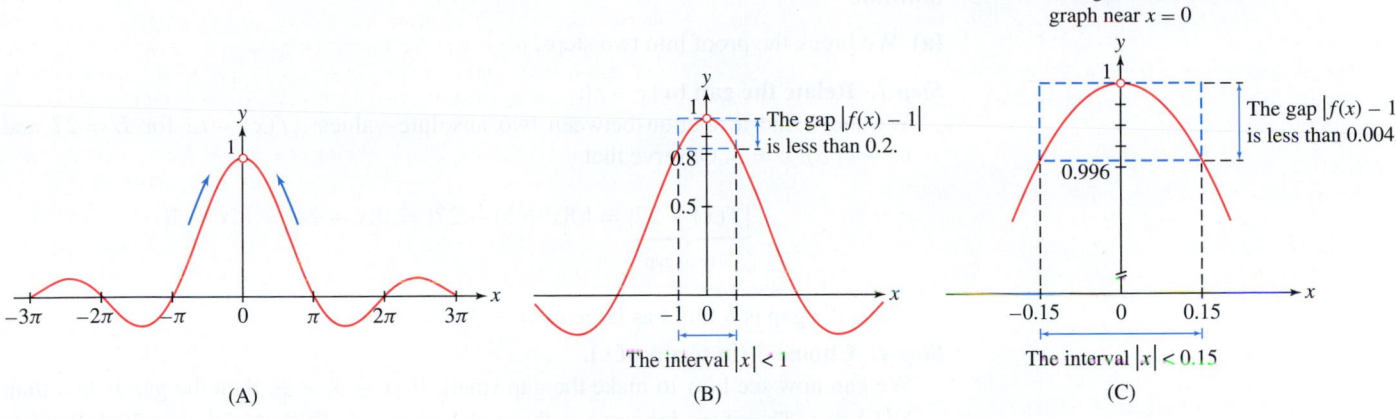

Enlarged view of
graph near $x = 0$

The gap $|f(x) - 1|$
is less than 0.2.

The gap $|f(x) - 1|$
is less than 0.004.

The interval $|x| < 1$

The interval $|x| < 0.15$

(A)   (B)   (C)

DF **FIGURE 1** Graphs of $y = \dfrac{\sin x}{x}$. To shrink the gap from 0.2 to 0.004, we require that x lie within 0.15 of 0.

It would seem that this process can be continued: Given any positive number, no matter how small, by zooming in on the graph we can find a small interval around $c = 0$ where the gap $|f(x) - 1|$ is smaller than the given number.

To express this in a precise fashion, we follow time-honored tradition of using the Greek letters ϵ (epsilon) and δ (delta) to denote small numbers specifying the sizes of the gap and the quantity $|x - c|$, respectively. In our case, $c = 0$ and $|x - c| = |x|$. The precise meaning of Eq. (1) is that for every choice of $\epsilon > 0$, there exists some δ (depending on ϵ) such that

$$\text{If } 0 < |x| < \delta, \quad \text{then} \quad \left| \frac{\sin x}{x} - 1 \right| < \epsilon$$

The number δ pins down just how close is "sufficiently close" for a given ϵ. With this motivation, we are ready to state the formal definition of the limit.

The formal definition of a limit is often called the ϵ-δ definition. The tradition of using the symbols ϵ and δ originated in the writings of Augustin-Louis Cauchy on calculus and analysis in the 1820s.

FORMAL DEFINITION OF A LIMIT   Suppose that $f(x)$ is defined for all x in an open interval containing c (but not necessarily at $x = c$). Then

$$\lim_{x \to c} f(x) = L$$

if for all $\epsilon > 0$, there exists $\delta > 0$ such that

$$\text{if } 0 < |x - c| < \delta, \quad \text{then} \quad |f(x) - L| < \epsilon$$

The condition $0 < |x - c| < \delta$ in this definition excludes $x = c$. In other words, the limit depends only on values of $f(x)$ near c but not on $f(c)$ itself. As we have seen in previous sections, the limit may exist even when $f(c)$ is not defined.

EXAMPLE 1   Let $f(x) = 8x + 3$.

(a) Prove that $\lim_{x \to 3} f(x) = 27$ using the formal definition of the limit.

(b) Find values of δ that work for $\epsilon = 0.2$ and 0.001.

Solution

(a) We break the proof into two steps.

Step 1. **Relate the gap to $|x - c|$.**

We must find a relation between two absolute values: $|f(x) - L|$ for $L = 27$ and $|x - c|$ for $c = 3$. Observe that

$$\underbrace{|f(x) - 27|}_{\text{Size of gap}} = |(8x + 3) - 27| = |8x - 24| = 8|x - 3|$$

Thus, the gap is 8 times as large as $|x - 3|$.

Step 2. **Choose δ (in terms of ϵ).**

We can now see how to make the gap small: If $|x - 3| < \frac{\epsilon}{8}$, then the gap is less than $8\left(\frac{\epsilon}{8}\right) = \epsilon$. Therefore, for any $\epsilon > 0$, we choose $\delta = \frac{\epsilon}{8}$. With this choice, the following statement holds:

$$\boxed{\text{If } 0 < |x - 3| < \delta, \quad \text{then } |f(x) - 27| < \epsilon, \quad \text{where } \delta = \frac{\epsilon}{8}}$$

Since we have specified δ for all $\epsilon > 0$, we have fulfilled the requirements of the formal definition, thus proving rigorously that $\lim_{x \to 3}(8x + 3) = 27$.

(b) For the particular choice $\epsilon = 0.2$, we may take $\delta = \frac{\epsilon}{8} = \frac{0.2}{8} = 0.025$:

$$\text{If } 0 < |x - 3| < 0.025, \quad \text{then } |f(x) - 27| < 0.2$$

This statement is illustrated in Figure 2. But note that *any positive δ smaller than 0.025 will also work*. For example, the following statement is also true, although it places an unnecessary restriction on x:

$$\text{If } 0 < |x - 3| < 0.019, \quad \text{then } |f(x) - 27| < 0.2$$

Similarly, to make the gap less than $\epsilon = 0.001$, we may take

$$\delta = \frac{\epsilon}{8} = \frac{0.001}{8} = 0.000125 \qquad\blacksquare$$

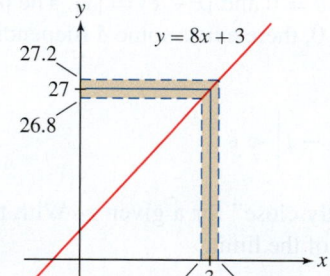

FIGURE 2 To make the gap less than 0.2, we may take $\delta = 0.025$ (not drawn to scale).

The difficulty in applying the limit definition lies in trying to relate $|f(x) - L|$ to $|x - c|$. The next two examples illustrate how this can be done in special cases.

EXAMPLE 2 Prove that $\lim_{x \to 2} x^2 = 4$.

Solution Let $f(x) = x^2$.

Step 1. **Relate the gap to $|x - c|$.**

In this case, we must relate the gap $|f(x) - 4| = |x^2 - 4|$ to the quantity $|x - 2|$ (Figure 3). This is more difficult than in the previous example because the gap is not a constant multiple of $|x - 2|$. To proceed, consider the factorization

$$|x^2 - 4| = |x + 2|\,|x - 2|$$

Because we are going to require that $|x - 2|$ be small, we may as well assume from the outset that $|x - 2| < 1$, which means that $1 < x < 3$. In this case, $|x + 2|$ is less than 5 and the gap satisfies

$$\text{If } |x - 2| < 1, \quad \text{then } |x^2 - 4| = |x + 2|\,|x - 2| < 5\,|x - 2| \qquad \boxed{2}$$

DF **FIGURE 3** Graph of $f(x) = x^2$. We may choose δ so that $f(x)$ lies within ϵ of 4 for all x in $[2 - \delta, 2 + \delta]$.

Step 2. **Choose δ (in terms of ϵ).**

We see from Eq. (2) that if $|x - 2|$ is smaller than both $\frac{\epsilon}{5}$ and 1, then the gap satisfies

$$|x^2 - 4| < 5|x - 2| < 5\left(\frac{\epsilon}{5}\right) = \epsilon$$

Therefore, the following statement holds for all $\epsilon > 0$:

> If $0 < |x - 2| < \delta$, then $|x^2 - 4| < \epsilon$, where δ is the smaller of $\frac{\epsilon}{5}$ and 1

We have specified δ for all $\epsilon > 0$, so we have fulfilled the requirements of the formal limit definition, thus proving that $\lim\limits_{x \to 2} x^2 = 4$. ∎

EXAMPLE 3 Prove that $\lim\limits_{x \to 3} \dfrac{1}{x} = \dfrac{1}{3}$.

Solution

Step 1. **Relate the gap to $|x - c|$.**

The gap is equal to

$$\left|\frac{1}{x} - \frac{1}{3}\right| = \left|\frac{3 - x}{3x}\right| = |x - 3|\left|\frac{1}{3x}\right|$$

← **REMINDER** If $a > b > 0$, then $\frac{1}{a} < \frac{1}{b}$. Thus, if $3x > 6$, then $\frac{1}{3x} < \frac{1}{6}$.

Because we are going to require that $|x - 3|$ be small, we may as well assume from the outset that $|x - 3| < 1$, or equivalently, $2 < x < 4$. Now observe that if $x > 2$, then $3x > 6$ and $\frac{1}{3x} < \frac{1}{6}$, so the following inequality is valid if $|x - 3| < 1$:

$$\left|f(x) - \frac{1}{3}\right| = \left|\frac{3 - x}{3x}\right| = \left|\frac{1}{3x}\right| |x - 3| < \frac{1}{6}|x - 3| \qquad \boxed{3}$$

Step 2. **Choose δ (in terms of ϵ).**

By Eq. (3), if $|x - 3| < 1$ and $|x - 3| < 6\epsilon$, then

$$\left|\frac{1}{x} - \frac{1}{3}\right| < \frac{1}{6}|x - 3| < \frac{1}{6}(6\epsilon) = \epsilon$$

Therefore, given any $\epsilon > 0$, we let δ be the smaller of the numbers 6ϵ and 1. Then we have

> If $0 < |x - 3| < \delta$, then $\left|\frac{1}{x} - \frac{1}{3}\right| < \epsilon$, where δ is the smaller of 6ϵ and 1

Again, we have fulfilled the requirements of the formal limit definition, thus proving rigorously that $\lim\limits_{x \to 3} \frac{1}{x} = \frac{1}{3}$. ∎

GRAPHICAL INSIGHT   Keep the graphical interpretation of limits in mind. In Figure 4(A), $f(x)$ approaches L as $x \to c$ because for any $\epsilon > 0$, we can make the gap less than ϵ by taking δ sufficiently small. By contrast, consider the function g in Figure 4(B). It has a jump discontinuity at $x = c$. Because of the jump, the gap at c cannot be made less than half the distance between a and b. Therefore if $\epsilon < \frac{b-a}{2}$, then no matter how small we choose δ, the gap corresponding to $(c - \delta, c + \delta)$ cannot be made smaller than ϵ. Thus, the limit as $x \to c$ of $g(x)$ does not exist.

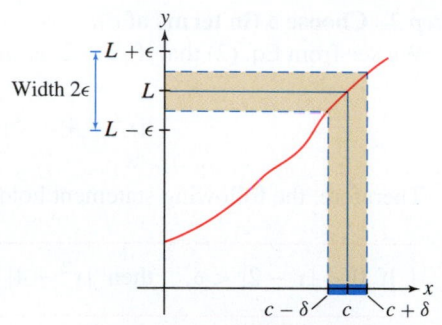

(A) The function f is continuous at $x = c$.
By taking δ sufficiently small, we
can make the gap smaller than ϵ.

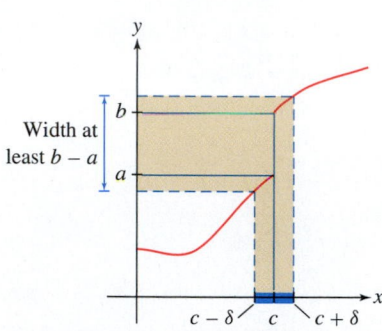

(B) The function g is not continuous at $x = c$.
The gap is always larger than $(b - a)/2$,
no matter how small δ is.

FIGURE 4

Proving Limit Theorems

In practice, the formal definition of the limit is rarely used to evaluate limits. Most limits are evaluated using the Basic Limit Laws or other techniques such as the Squeeze Theorem. However, the formal definition allows us to prove these laws in a rigorous fashion and thereby ensure that calculus is built on a solid foundation. We illustrate by proving the Sum Law. Other proofs are given in Appendix D.

Proof of the Sum Law   Assume that

$$\lim_{x \to c} f(x) = L \qquad \text{and} \qquad \lim_{x \to c} g(x) = M$$

We must prove that $\lim_{x \to c}(f(x) + g(x)) = L + M$.

Apply the Triangle Inequality (see margin) with $a = f(x) - L$ and $b = g(x) - M$:

$$|(f(x) + g(x)) - (L + M)| \leq |f(x) - L| + |g(x) - M| \qquad \boxed{4}$$

> **REMINDER** *The Triangle Inequality [Eq. (1) in Section 1.1] states*
>
> $$|a + b| \leq |a| + |b|$$
>
> *for all a and b.*

Each term on the right in (4) can be made small by the limit definition. More precisely, given $\epsilon > 0$, we can choose δ such that if $0 < |x - c| < \delta$, then $|f(x) - L| < \frac{\epsilon}{2}$ and $|g(x) - M| < \frac{\epsilon}{2}$ (in principle, we might choose different δ's for f and g, but we may then use the smaller of the two δ's). Thus, Eq. (4) gives

$$\text{If } 0 < |x - c| < \delta, \quad \text{then } |f(x) + g(x) - (L + M)| < \frac{\epsilon}{2} + \frac{\epsilon}{2} = \epsilon \qquad \boxed{5}$$

This proves that

$$\lim_{x \to c}\big(f(x) + g(x)\big) = L + M = \lim_{x \to c} f(x) + \lim_{x \to c} g(x) \qquad \blacksquare$$

2.9 SUMMARY

- Informally speaking, the statement $\lim_{x \to c} f(x) = L$ means that the gap $|f(x) - L|$ tends to 0 as x approaches c.
- The *formal definition* (called the ϵ-δ definition): $\lim_{x \to c} f(x) = L$ if, for all $\epsilon > 0$, there exists a $\delta > 0$ such that

$$\text{if } 0 < |x - c| < \delta, \quad \text{then } |f(x) - L| < \epsilon$$

2.9 EXERCISES

Preliminary Questions

1. Given that $\lim_{x \to 0} \cos x = 1$, which of the following statements is true?

(a) If $|\cos x - 1|$ is very small, then x is close to 0.
(b) There is an $\epsilon > 0$ such that if $0 < |\cos x - 1| < \epsilon$, then $|x| < 10^{-5}$.
(c) There is a $\delta > 0$ such that if $0 < |x| < \delta$, then $|\cos x - 1| < 10^{-5}$.
(d) There is a $\delta > 0$ such that if $0 < |x - 1| < \delta$, then $|\cos x| < 10^{-5}$.

2. Suppose it is known that for a given ϵ and δ, if $0 < |x - 3| < \delta$, then $|f(x) - 2| < \epsilon$. Which of the following statements must also be true?

(a) If $0 < |x - 3| < 2\delta$, then $|f(x) - 2| < \epsilon$.
(b) If $0 < |x - 3| < \delta$, then $|f(x) - 2| < 2\epsilon$.
(c) If $0 < |x - 3| < \dfrac{\delta}{2}$, then $|f(x) - 2| < \dfrac{\epsilon}{2}$.
(d) If $0 < |x - 3| < \dfrac{\delta}{2}$, then $|f(x) - 2| < \epsilon$.

Exercises

1. Based on the information conveyed in Figure 5(A), find values of L, ϵ, and $\delta > 0$ such that the following statement holds: If $|x| < \delta$, then $|f(x) - L| < \epsilon$.

2. Based on the information conveyed in Figure 5(B), find values of c, L, ϵ, and $\delta > 0$ such that the following statement holds: If $0 < |x - c| < \delta$, then $|f(x) - L| < \epsilon$.

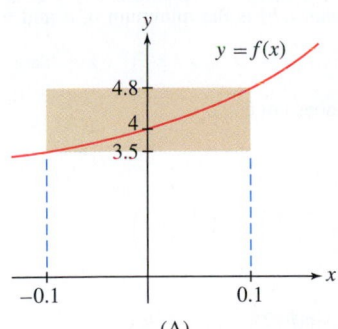

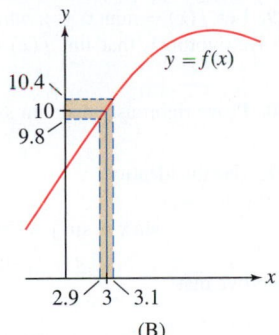

(A) (B)

FIGURE 5

3. Make a sketch illustrating the following statement: To prove $\lim_{x \to a} x = a$, given $\epsilon > 0$, we can take $\delta = \epsilon$ to have the gap be small enough.

4. Make a sketch illustrating the following statement: To prove $\lim_{x \to c} a = a$, given $\epsilon > 0$, we can choose any $\delta > 0$ to have the gap be small enough.

5. Consider $\lim_{x \to 4} f(x)$, where $f(x) = 8x + 3$.

(a) Show that $|f(x) - 35| = 8|x - 4|$.
(b) Show that for any $\epsilon > 0$, if $0 < |x - 4| < \delta$, then $|f(x) - 35| < \epsilon$, where $\delta = \frac{\epsilon}{8}$. Explain how this proves rigorously that $\lim_{x \to 4} f(x) = 35$.

6. Consider $\lim_{x \to 2} f(x)$, where $f(x) = 4x - 1$.

(a) Show that if $0 < |x - 2| < \delta$, then $|f(x) - 7| < 4\delta$.
(b) Find a δ such that

$$\text{If } 0 < |x - 2| < \delta, \quad \text{then } |f(x) - 7| < 0.01$$

(c) Prove rigorously that $\lim_{x \to 2} f(x) = 7$.

7. Consider $\lim_{x \to 2} x^2 = 4$ (refer to Example 2).

(a) Show that if $0 < |x - 2| < 0.01$, then $|x^2 - 4| < 0.05$.
(b) Show that if $0 < |x - 2| < 0.0002$, then $|x^2 - 4| < 0.0009$.
(c) Find a value of δ such that if $0 < |x - 2| < \delta$, then $|x^2 - 4|$ is less than 10^{-4}.

8. Consider the limit $\lim_{x \to 5} x^2 = 25$.

(a) Show that if $4 < x < 6$, then $|x^2 - 25| < 11|x - 5|$. *Hint:* Write $|x^2 - 25| = |x + 5| \cdot |x - 5|$.
(b) Find a δ such that if $0 < |x - 5| < \delta$, then $|x^2 - 25| < 10^{-3}$.
(c) Give a rigorous proof of the limit by showing that if $0 < |x - 5| < \delta$, then $|x^2 - 25| < \epsilon$, where δ is the smaller of $\frac{\epsilon}{11}$ and 1.

9. Refer to Example 3 to find a value of $\delta > 0$ such that

$$\text{If } 0 < |x - 3| < \delta, \quad \text{then } \left|\frac{1}{x} - \frac{1}{3}\right| < 10^{-4}$$

10. Use Figure 6 to find a value of $\delta > 0$ such that the following statement holds: If $0 < |x - 2| < \delta$, then $\left|\frac{1}{x^2} - \frac{1}{4}\right| < \epsilon$ for $\epsilon = 0.03$. Then find a value of δ that works for $\epsilon = 0.01$.

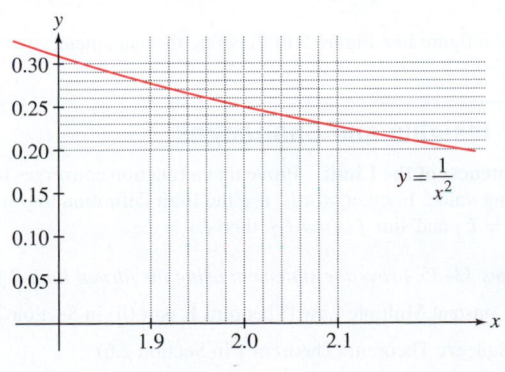

FIGURE 6

11. **GU** Plot the function $f(x) = \sqrt{2x-1}$ together with the horizontal lines $y = 2.9$ and $y = 3.1$. Use this plot to find a value of $\delta > 0$ such that if $0 < |x - 5| < \delta$, then $|\sqrt{2x-1} - 3| < 0.1$.

12. **GU** Plot the function $f(x) = \tan x$ together with the horizontal lines $y = 0.99$ and $y = 1.01$. Use this plot to find a value of $\delta > 0$ such that if $0 < \left|x - \frac{\pi}{4}\right| < \delta$, then $|\tan x - 1| < 0.01$.

13. **GU** The number e has the following property: $\lim\limits_{x \to 0} \dfrac{e^x - 1}{x} = 1$.

Use a plot of the function $f(x) = \dfrac{e^x - 1}{x}$ to find a value of $\delta > 0$ such that if $0 < |x| < \delta$, then $|f(x) - 1| < 0.01$.

14. **GU** Let $f(x) = \dfrac{4}{x^2 + 1}$ and $\epsilon = 0.5$. Using a plot of f, find a value of $\delta > 0$ such that if $0 < \left|x - \frac{1}{2}\right| < \delta$, then $\left|f(x) - \frac{16}{5}\right| < \epsilon$. Repeat for $\epsilon = 0.2$ and 0.1.

15. Consider $\lim\limits_{x \to 2} \dfrac{1}{x}$.

(a) Show that if $|x - 2| < 1$, then
$$\left|\frac{1}{x} - \frac{1}{2}\right| < \frac{1}{2}|x - 2|$$

(b) Find a $\delta > 0$ such that if $0 < |x - 2| < \delta$, then $\left|\frac{1}{x} - \frac{1}{2}\right| < 0.01$.

(c) Let δ be the smaller of 1 and 2ϵ. Prove the following:

$$\text{If } 0 < |x - 2| < \delta, \qquad \text{then} \quad \left|\frac{1}{x} - \frac{1}{2}\right| < \epsilon$$

Then explain why this proves that $\lim\limits_{x \to 2} \dfrac{1}{x} = \dfrac{1}{2}$.

16. Consider $\lim\limits_{x \to 1} \sqrt{x + 3}$.

(a) Show that if $|x - 1| < 4$, then $|\sqrt{x+3} - 2| < \frac{1}{2}|x - 1|$. *Hint:* Multiply the inequality by $|\sqrt{x+3} + 2|$ and observe that $|\sqrt{x+3} + 2| > 2$.

(b) Find $\delta > 0$ such that if $0 < |x - 1| < \delta$, then $|\sqrt{x+3} - 2| < 10^{-4}$.

(c) Prove rigorously that the limit is equal to 2.

17. 📝 Let $f(x) = \sin x$. Using a calculator, we find

$$f\left(\frac{\pi}{4} - 0.1\right) \approx 0.633, \quad f\left(\frac{\pi}{4}\right) \approx 0.707, \quad f\left(\frac{\pi}{4} + 0.1\right) \approx 0.774$$

Use these values and the fact that f is increasing on $\left[0, \frac{\pi}{2}\right]$ to justify the statement

$$\text{If } 0 < \left|x - \frac{\pi}{4}\right| < 0.1, \qquad \text{then} \quad \left|f(x) - f\left(\frac{\pi}{4}\right)\right| < 0.08$$

Then draw a figure like Figure 3 to illustrate this statement.

18. Adapt the argument in Example 1 to prove rigorously that $\lim\limits_{x \to c}(ax + b) = ac + b$, where a, b, c are arbitrary.

19. Adapt the argument in Example 2 to prove rigorously that $\lim\limits_{x \to c} x^2 = c^2$ for all c.

20. Adapt the argument in Example 3 to prove rigorously that $\lim\limits_{x \to c} x^{-1} = \frac{1}{c}$ for all $c \neq 0$.

In Exercises 21–26, use the formal definition of the limit to prove the statement rigorously.

21. $\lim\limits_{x \to 4} \sqrt{x} = 2$

22. $\lim\limits_{x \to 1}(3x^2 + x) = 4$

23. $\lim\limits_{x \to 1} x^3 = 1$

24. $\lim\limits_{x \to 0}(x^2 + x^3) = 0$

25. $\lim\limits_{x \to 2} x^{-2} = \dfrac{1}{4}$

26. $\lim\limits_{x \to 0} x \sin \dfrac{1}{x} = 0$

27. Let $f(x) = \dfrac{x}{|x|}$. Prove rigorously that $\lim\limits_{x \to 0} f(x)$ does not exist. *Hint:* Show that for any L, there always exists some x such that $|x| < \delta$ but $|f(x) - L| \geq \frac{1}{2}$, no matter how small δ is taken.

28. Prove rigorously that $\lim\limits_{x \to 0} |x| = 0$.

29. Let $f(x) = \min(x, x^2)$, where $\min(a, b)$ is the minimum of a and b. Prove rigorously that $\lim\limits_{x \to 1} f(x) = 1$.

30. Prove rigorously that $\lim\limits_{x \to 0} \sin \frac{1}{x}$ does not exist.

31. Use the identity

$$\sin x + \sin y = 2 \sin\left(\frac{x+y}{2}\right) \cos\left(\frac{x-y}{2}\right)$$

to prove that

$$\sin(a + h) - \sin a = h \frac{\sin(h/2)}{h/2} \cos\left(a + \frac{h}{2}\right) \qquad \boxed{6}$$

Then use the inequality $\left|\dfrac{\sin x}{x}\right| \leq 1$ for $x \neq 0$ to show that $|\sin(a + h) - \sin a| < |h|$ for all a. Finally, prove rigorously that $\lim\limits_{x \to a} \sin x = \sin a$.

Further Insights and Challenges

32. Uniqueness of the Limit Prove that a function converges to at most one limiting value. In other words, use the limit definition to prove that if $\lim\limits_{x \to c} f(x) = L_1$ and $\lim\limits_{x \to c} f(x) = L_2$, then $L_1 = L_2$.

In Exercises 33–35, prove the statement using the formal limit definition.

33. The Constant Multiple Law [Theorem 1, part (ii) in Section 2.3]

34. The Squeeze Theorem (Theorem 1 in Section 2.6)

35. The Product Law [Theorem 1, part (iii) in Section 2.3]. *Hint:* Use the identity.

$$f(x)g(x) - LM = (f(x) - L)g(x) + L(g(x) - M)$$

36. Let $f(x) = 1$ if x is rational and $f(x) = 0$ if x is irrational. Prove that $\lim\limits_{x \to c} f(x)$ does not exist for any c. *Hint:* There exist rational and irrational numbers arbitrarily close to any c.

37. ✏️ Here is a function with strange continuity properties:

$$f(x) = \begin{cases} \dfrac{1}{q} & \text{if } x \text{ is the rational number } p/q \text{ in lowest terms} \\ 0 & \text{if } x \text{ is an irrational number} \end{cases}$$

(a) Show that f is discontinuous at c if c is rational. *Hint:* There exist irrational numbers arbitrarily close to c.
(b) Show that f is continuous at c if c is irrational. *Hint:* Let I be the interval $\{x : |x - c| < 1\}$. Show that for any $Q > 0$, I contains at most finitely many fractions p/q with $q < Q$. Conclude that there is a δ such that all fractions in $\{x : |x - c| < \delta\}$ have a denominator larger than Q.

38. ✏️ Write a formal definition of the following:
$$\lim_{x \to \infty} f(x) = L$$

39. ✏️ Write a formal definition of the following:
$$\lim_{x \to a} f(x) = \infty$$

CHAPTER REVIEW EXERCISES

1. The position of a particle at time t (s) is $s(t) = \sqrt{t^2 + 1}$ m. Compute its average velocity over $[2, 5]$ and estimate its instantaneous velocity at $t = 2$.

2. A rock dropped from a state of rest at time $t = 0$ on the planet Ginormon travels a distance $s(t) = 15.2t^2$ m in t seconds. Estimate the instantaneous velocity at $t = 5$.

3. For $f(x) = \sqrt{2x}$ compute the slopes of the secant lines from 16 to each of 16 ± 0.01, 16 ± 0.001, 16 ± 0.0001 and use those values to estimate the slope of the tangent line at $x = 16$.

4. Show that the slope of the secant line for $f(x) = x^3 - 2x$ over $[5, x]$ is equal to $x^2 + 5x + 23$. Use this to estimate the slope of the tangent line at $x = 5$.

In Exercises 5–10, estimate the limit numerically to two decimal places or state that the limit does not exist.

5. $\displaystyle\lim_{x \to 0} \frac{1 - \cos^3(x)}{x^2}$

6. $\displaystyle\lim_{x \to 1} x^{1/(x-1)}$

7. $\displaystyle\lim_{x \to 2} \frac{x^x - 4}{x^2 - 4}$

8. $\displaystyle\lim_{x \to 2} \frac{x - 2}{\ln(3x - 5)}$

9. $\displaystyle\lim_{x \to 1} \left(\frac{7}{1 - x^7} - \frac{3}{1 - x^3} \right)$

10. $\displaystyle\lim_{x \to 2} \frac{3^x - 9}{5^x - 25}$

In Exercises 11–50, evaluate the limit if it exists. If not, determine whether the one-sided limits exist. For limits that don't exist indicate whether they can be expressed as $= -\infty$ or $= \infty$.

11. $\displaystyle\lim_{x \to 4} (3 + x^{1/2})$

12. $\displaystyle\lim_{x \to 1} \frac{5 - x^2}{4x + 7}$

13. $\displaystyle\lim_{x \to -2} \frac{4}{x^3}$

14. $\displaystyle\lim_{x \to -1} \frac{3x^2 + 4x + 1}{x + 1}$

15. $\displaystyle\lim_{t \to 9} \frac{\sqrt{t} - 3}{t - 9}$

16. $\displaystyle\lim_{x \to 3} \frac{\sqrt{x + 1} - 2}{x - 3}$

17. $\displaystyle\lim_{x \to 1} \frac{x^3 - x}{x - 1}$

18. $\displaystyle\lim_{h \to 0} \frac{2(a + h)^2 - 2a^2}{h}$

19. $\displaystyle\lim_{t \to 9} \frac{t - 6}{\sqrt{t} - 3}$

20. $\displaystyle\lim_{s \to 0} \frac{1 - \sqrt{s^2 + 1}}{s^2}$

21. $\displaystyle\lim_{x \to -1^+} \frac{1}{x + 1}$

22. $\displaystyle\lim_{y \to \frac{1}{3}} \frac{3y^2 + 5y - 2}{6y^2 - 5y + 1}$

23. $\displaystyle\lim_{x \to 1} \frac{x^3 - 2x}{x - 1}$

24. $\displaystyle\lim_{a \to b} \frac{a^2 - 3ab + 2b^2}{a - b}$

25. $\displaystyle\lim_{x \to 0} \frac{e^{3x} - e^x}{e^x - 1}$

26. $\displaystyle\lim_{\theta \to 0} \frac{\sin 5\theta}{\theta}$

27. $\displaystyle\lim_{x \to 1.5} \left\lfloor \frac{1}{x} \right\rfloor$

28. $\displaystyle\lim_{\theta \to \frac{\pi}{4}} \sec \theta$

29. $\displaystyle\lim_{z \to -3} \frac{z + 3}{z^2 + 4z + 3}$

30. $\displaystyle\lim_{x \to 1} \frac{x^3 - ax^2 + ax - 1}{x - 1}$

31. $\displaystyle\lim_{x \to b} \frac{x^3 - b^3}{x - b}$

32. $\displaystyle\lim_{x \to 0} \frac{\sin 4x}{\sin 3x}$

33. $\displaystyle\lim_{x \to 0} \left(\frac{1}{3x} - \frac{1}{x(x + 3)} \right)$

34. $\displaystyle\lim_{\theta \to \frac{1}{4}} 3^{\tan(\pi\theta)}$

35. $\displaystyle\lim_{x \to 0^-} \frac{\lfloor x \rfloor}{x}$

36. $\displaystyle\lim_{x \to 0^+} \frac{\lfloor x \rfloor}{x}$

37. $\displaystyle\lim_{\theta \to \frac{\pi}{2}} \theta \sec \theta$

38. $\displaystyle\lim_{y \to 2} \ln \left(\sin \frac{\pi}{y} \right)$

39. $\displaystyle\lim_{\theta \to 0} \frac{\cos \theta - 2}{\theta}$

40. $\displaystyle\lim_{x \to 4.3} \frac{1}{x - \lfloor x \rfloor}$

41. $\displaystyle\lim_{x \to 2^-} \frac{x - 3}{x - 2}$

42. $\displaystyle\lim_{t \to 0} \frac{\sin^2 t}{t^3}$

43. $\displaystyle\lim_{x \to 1^+} \left(\frac{1}{\sqrt{x - 1}} - \frac{1}{\sqrt{x^2 - 1}} \right)$

44. $\displaystyle\lim_{t \to e} \sqrt{t}(\ln t - 1)$

45. $\displaystyle\lim_{x \to \frac{\pi}{2}} \tan x$

46. $\displaystyle\lim_{t \to 0} \cos \frac{1}{t}$

47. $\displaystyle\lim_{t \to 0^+} \sqrt{t} \cos \frac{1}{t}$

48. $\displaystyle\lim_{x \to 5^+} \frac{x^2 - 24}{x^2 - 25}$

49. $\displaystyle\lim_{x \to 0} \frac{\cos x - 1}{\sin x}$

50. $\displaystyle\lim_{\theta \to 0} \frac{\tan \theta - \sin \theta}{\sin^3 \theta}$

51. Find the left- and right-hand limits of the function f in Figure 1 at $x = 0, 2, 4$. State whether f is left- or right-continuous (or both) at these points.

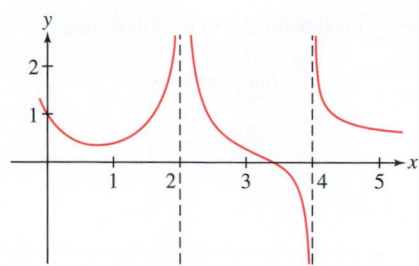

FIGURE 1

52. Sketch the graph of a function f such that

(a) $\lim\limits_{x \to 2^-} f(x) = 1, \qquad \lim\limits_{x \to 2^+} f(x) = 3$

(b) $\lim\limits_{x \to 4} f(x)$ exists but does not equal $f(4)$

53. Graph h and describe the discontinuity:

$$h(x) = \begin{cases} e^x & \text{for } x \le 0 \\ \ln x & \text{for } x > 0 \end{cases}$$

Is h left- or right-continuous?

54. Sketch the graph of a function g such that

$$\lim\limits_{x \to -3^-} g(x) = \infty, \qquad \lim\limits_{x \to -3^+} g(x) = -\infty, \qquad \lim\limits_{x \to 4} g(x) = \infty$$

55. Find the points of discontinuity of

$$g(x) = \begin{cases} \cos\left(\dfrac{\pi x}{2}\right) & \text{for } |x| < 1 \\ |x - 1| & \text{for } |x| \ge 1 \end{cases}$$

Determine the type of discontinuity and whether g is left- or right-continuous.

56. Show that $f(x) = x e^{\sin x}$ is continuous on its domain.

57. Find a constant b such that h is continuous at $x = 2$, where

$$h(x) = \begin{cases} x + 1 & \text{for } |x| < 2 \\ b - x^2 & \text{for } |x| \ge 2 \end{cases}$$

With this choice of b, find all points of discontinuity.

In Exercises 58–65, find the horizontal asymptotes of the function by computing the limits at infinity.

58. $f(x) = \dfrac{9x^2 - 4}{2x^2 - x}$

59. $f(x) = \dfrac{x^2 - 3x^4}{x - 1}$

60. $f(u) = \dfrac{8u - 3}{\sqrt{16u^2 + 6}}$

61. $f(u) = \dfrac{2u^2 - 1}{\sqrt{6 + u^4}}$

62. $f(x) = \dfrac{3x^{2/3} + 9x^{3/7}}{7x^{4/5} - 4x^{-1/3}}$

63. $f(t) = \dfrac{t^{1/3} - t^{-1/3}}{(t - t^{-1})^{1/3}}$

64. $f(t) = \dfrac{17}{1 + 2^t}$

65. $g(x) = \pi + 2\tan^{-1} x$

66. $\boxed{\text{CAS}}$ Determine values of M, A, and k for a logistic function $f(t) = \dfrac{M}{1 + Ae^{-kt}}$ satisfying $f(0) = 1$, $f(1) = 8$, and $f(2) = 14$. What are the horizontal asymptotes of f?

67. $\boxed{\text{CAS}}$ Determine values of M, A, and k for a logistic function $p(t) = \dfrac{M}{1 + Ae^{-kt}}$ satisfying $p(0) = 10$, $p(4) = 35$, and $p(10) = 60$. What are the horizontal asymptotes of p?

68. Calculate (a)–(d), assuming that

$$\lim\limits_{x \to 3} f(x) = 6, \qquad \lim\limits_{x \to 3} g(x) = 4$$

(a) $\lim\limits_{x \to 3} (f(x) - 2g(x))$

(b) $\lim\limits_{x \to 3} x^2 f(x)$

(c) $\lim\limits_{x \to 3} \dfrac{f(x)}{g(x) + x}$

(d) $\lim\limits_{x \to 3} (2g(x)^3 - g(x)^{3/2})$

69. Assume that the following limits exist:

$$A = \lim\limits_{x \to a} f(x), \qquad B = \lim\limits_{x \to a} g(x), \qquad L = \lim\limits_{x \to a} \dfrac{f(x)}{g(x)}$$

Prove that if $L = 1$, then $A = B$. *Hint:* You cannot use the Quotient Law if $B = 0$, so apply the Product Law to L and B instead.

70. $\boxed{\text{GU}}$ Define $g(t) = (1 + 2^{1/t})^{-1}$ for $t \ne 0$. How should $g(0)$ be defined to make g left-continuous at $t = 0$?

71. $\boxed{\text{✐}}$ In the notation of Exercise 69, give an example where L exists but neither A nor B exists.

72. True or false?

(a) If $\lim\limits_{x \to 3} f(x)$ exists, then $\lim\limits_{x \to 3} f(x) = f(3)$.

(b) If $\lim\limits_{x \to 0} \dfrac{f(x)}{x} = 1$, then $f(0) = 0$.

(c) If $\lim\limits_{x \to -7} f(x) = 8$, then $\lim\limits_{x \to -7} \dfrac{1}{f(x)} = \dfrac{1}{8}$.

(d) If $\lim\limits_{x \to 5^+} f(x) = 4$ and $\lim\limits_{x \to 5^-} f(x) = 8$, then $\lim\limits_{x \to 5} f(x) = 6$.

(e) If $\lim\limits_{x \to 0} \dfrac{f(x)}{x} = 1$, then $\lim\limits_{x \to 0} f(x) = 0$.

(f) If $\lim\limits_{x \to 5} f(x) = 2$, then $\lim\limits_{x \to 5} f(x)^3 = 8$.

73. $\boxed{\text{✐}}$ Let $f(x) = \left\lfloor \dfrac{1}{x} \right\rfloor$, where $\lfloor x \rfloor$ is the greatest integer function. Show that for $x \ne 0$,

$$\dfrac{1}{x} - 1 < \left\lfloor \dfrac{1}{x} \right\rfloor \le \dfrac{1}{x}$$

Then use the Squeeze Theorem to prove that

$$\lim\limits_{x \to 0} x \left\lfloor \dfrac{1}{x} \right\rfloor = 1$$

Hint: Treat the one-sided limits separately.

74. Let r_1 and r_2 be the roots of $f(x) = ax^2 - 2x + 20$. Observe that f "approaches" the linear function $L(x) = -2x + 20$ as $a \to 0$. Because $r = 10$ is the unique root of L, we might expect one of the roots of f to approach 10 as $a \to 0$ (Figure 2). Prove that the roots can be labeled so that $\lim_{a \to 0} r_1 = 10$ and $\lim_{a \to 0} r_2 = \infty$.

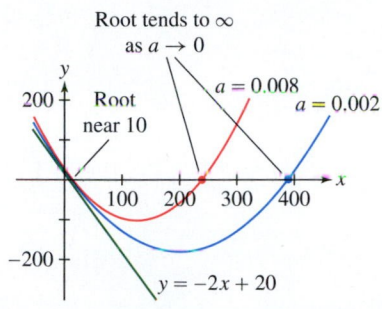

FIGURE 2 Graphs of $f(x) = ax^2 - 2x + 20$.

75. Use the IVT to prove that the curves $y = x^2$ and $y = \cos x$ intersect.

76. Use the IVT to prove that $f(x) = x^3 - \dfrac{x^2 + 2}{\cos x + 2}$ has a root in the interval $[0, 2]$.

77. Use the IVT to show that $e^{-x^2} = x$ has a solution on $(0, 1)$.

78. Use the Bisection Method to locate a solution of $x^2 - 7 = 0$ to two decimal places.

79. Give an example of a (discontinuous) function that does not satisfy the conclusion of the IVT on $[-1, 1]$. Then show that the function

$$f(x) = \begin{cases} \sin \dfrac{1}{x} & x \neq 0 \\ 0 & x = 0 \end{cases}$$

satisfies the conclusion of the IVT on every interval $[-a, a]$.

80. Let $f(x) = \dfrac{1}{x + 2}$.

(a) Show that if $|x - 2| < 1$, then $\left| f(x) - \frac{1}{4} \right| < \dfrac{|x - 2|}{12}$. *Hint:* Observe that if $|x - 2| < 1$, then $|4(x + 2)| > 12$.

(b) Find $\delta > 0$ such that if $|x - 2| < \delta$, then $\left| f(x) - \frac{1}{4} \right| < 0.01$.

(c) Prove rigorously that $\lim_{x \to 2} f(x) = \frac{1}{4}$.

81. **GU** Plot the function $f(x) = x^{1/3}$. Use the zoom feature to find a $\delta > 0$ such that if $|x - 8| < \delta$, then $|x^{1/3} - 2| < 0.05$.

82. Use the fact that $f(x) = 2^x$ is increasing to find a value of δ such that $|2^x - 8| < 0.001$ if $|x - 2| < \delta$. *Hint:* Find c_1 and c_2 such that $7.999 < f(c_1) < f(c_2) < 8.001$.

83. Prove rigorously that $\lim_{x \to -1} (4 + 8x) = -4$.

84. Prove rigorously that $\lim_{x \to 3} (x^2 - x) = 6$.

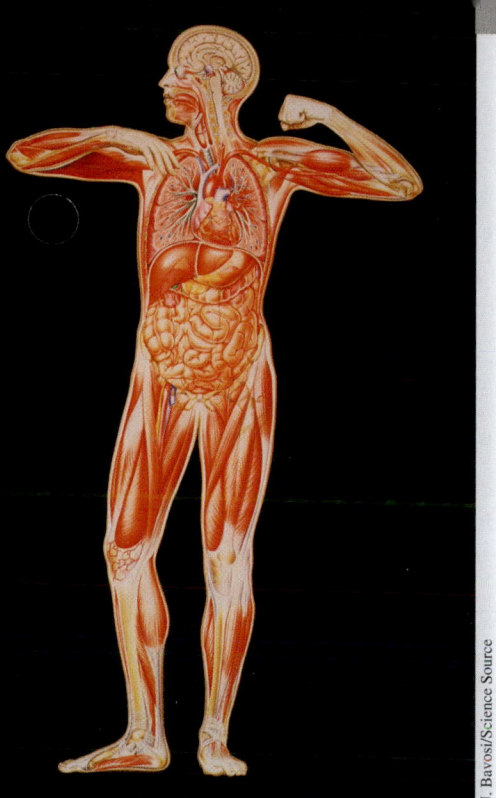

J. Bavosi/Science Source

Our bodies are complex systems of moving parts, evolving structures, flowing masses, and electrical currents. Using the derivative we can study the rates at which these changes occur, and we can analyze the impact of these changes on body systems. Following are some typical rates of change for human body functions:

Neuron impulse speed: 120 m/second
Blood flow rate out of the aorta: 220 mL/s
Food speed through the esophagus: 2 cm/s
Sneeze particle speed: 40 m/s
Skin cell regeneration rate: 1500 cells/hour
Air intake rate: 7 L/minute
Eyelid blink rate: 10 cycles/min

FIGURE 1 The secant line has slope $\Delta f / \Delta x$. Our goal is to compute the slope of the tangent line at $(a, f(a))$.

← **REMINDER** A **secant line** is any line through two points on a curve or graph.

3 DIFFERENTIATION

Differential calculus is the study of the derivative, and differentiation is the process of computing derivatives. What is a derivative? There are three equally important answers: A derivative is a rate of change, it is the slope of a tangent line, and (more formally) it is the limit of a difference quotient, as we will explain shortly. In this chapter, we explore all three facets of the derivative and develop the basic rules of differentiation. When you master these techniques, you will possess one of the most useful and flexible tools that mathematics has to offer.

3.1 Definition of the Derivative

We begin with two questions: What is the precise definition of a tangent line? And how can we compute its slope? To answer these questions, let's return to the relationship between tangent and secant lines first mentioned in Section 2.1.

The secant line through distinct points $P = (a, f(a))$ and $Q = (x, f(x))$ on the graph of a function f has slope [Figure 1(A)]

$$\frac{\Delta f}{\Delta x} = \frac{f(x) - f(a)}{x - a}$$

where

$$\Delta f = f(x) - f(a) \qquad \text{and} \qquad \Delta x = x - a$$

The expression $\dfrac{f(x) - f(a)}{x - a}$ is called the **difference quotient**. We can think of the secant line through P and Q as a rough approximation to the tangent line at P [Figure 1(B)].

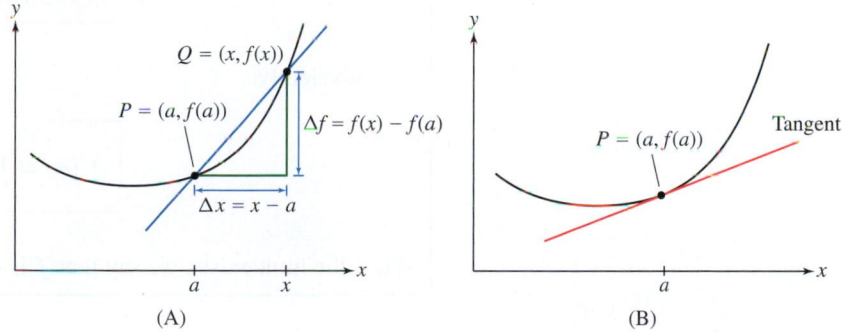

(A)   (B)

We can improve the secant-line approximation to the tangent line by moving point Q closer to point P, equivalently by moving x closer to a [Figure 2(A–C)]. As Q approaches P, the secant lines get progressively closer to the tangent line as in Figure 2(D). Therefore, we may expect the slopes of the secant lines to approach the slope of the tangent line; that is, we expect that as x approaches a, the limit of the secant-line slopes is equal to the slope of the tangent line.

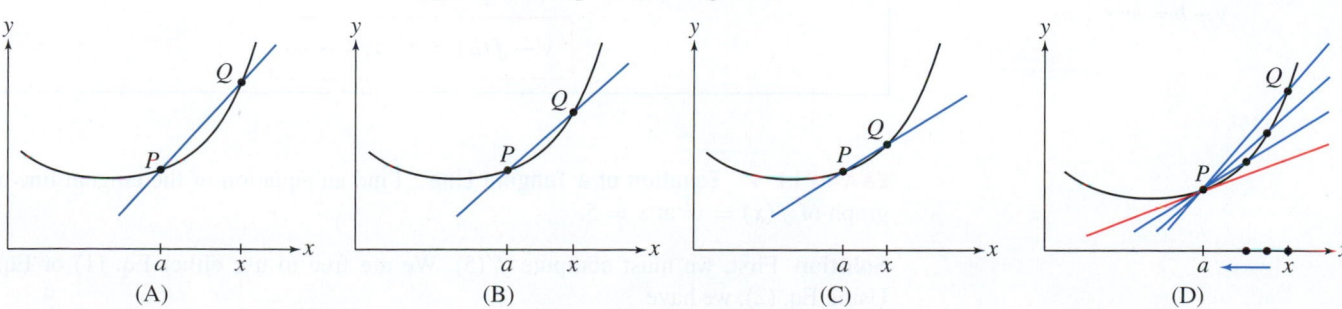

(A)   (B)   (C)   (D)

DF **FIGURE 2** The secant lines approach the tangent line as Q approaches P.

Based on this intuition, we define the **derivative** $f'(a)$ (which is read "f prime of a") as the limit

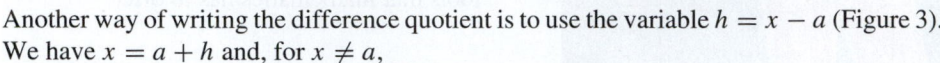

$$\underbrace{f'(a)}_{\substack{\text{Slope of the}\\ \text{tangent line}}} = \underbrace{\lim_{x \to a} \frac{f(x) - f(a)}{x - a}}_{\text{Limit of slopes of secant lines}}$$

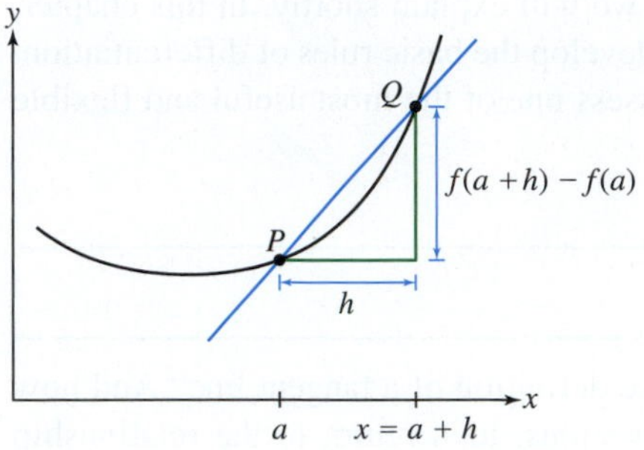

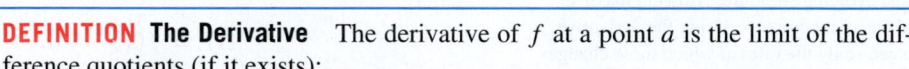

Another way of writing the difference quotient is to use the variable $h = x - a$ (Figure 3). We have $x = a + h$ and, for $x \neq a$,

$$\frac{f(x) - f(a)}{x - a} = \frac{f(a + h) - f(a)}{h}$$

The variable h approaches 0 as $x \to a$, so we can rewrite the derivative as

$$f'(a) = \lim_{h \to 0} \frac{f(a + h) - f(a)}{h}$$

Each way of writing the derivative is useful. In examples and proofs in this and the upcoming sections, we will use the formula for the derivative that is most appropriate for the situation.

FIGURE 3 In terms of h the difference quotient is $\frac{f(a+h) - f(a)}{h}$.

> **DEFINITION** **The Derivative** The derivative of f at a point a is the limit of the difference quotients (if it exists):
>
> $$f'(a) = \lim_{h \to 0} \frac{f(a + h) - f(a)}{h} \qquad \boxed{1}$$
>
> or equivalently:
>
> $$f'(a) = \lim_{x \to a} \frac{f(x) - f(a)}{x - a} \qquad \boxed{2}$$
>
> When the limit exists, we say that f is **differentiable** at a.

We can now define the tangent line in a precise way, as the line of slope $f'(a)$ through $P = (a, f(a))$.

REMINDER The equation of the line through $P = (a, b)$ of slope m in point-slope form:

$$y - b = m(x - a)$$

> **DEFINITION** **Tangent Line** Assume that f is differentiable at a. The tangent line to the graph of $y = f(x)$ at $P = (a, f(a))$ is the line through P of slope $f'(a)$. The equation of the tangent line in point-slope form is
>
> $$y - f(a) = f'(a)(x - a) \qquad \boxed{3}$$

EXAMPLE 1 **Equation of a Tangent Line** Find an equation of the tangent line to the graph of $f(x) = x^2$ at $x = 5$.

Solution First, we must compute $f'(5)$. We are free to use either Eq. (1) or Eq. (2). Using Eq. (2), we have

$$f'(5) = \lim_{x \to 5} \frac{f(x) - f(5)}{x - 5} = \lim_{x \to 5} \frac{x^2 - 25}{x - 5}$$

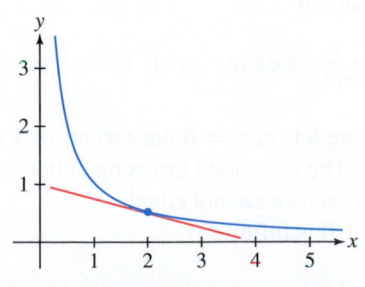

DF **FIGURE 4** Tangent line to $y = x^2$ at $x = 5$.

This limit is in the indeterminate form $0/0$. We can simplify and then evaluate by substitution:

$$f'(5) = \lim_{x \to 5} \frac{(x-5)(x+5)}{x-5} = \lim_{x \to 5}(x+5) = 10$$

Next, we apply Eq. (3) with $a = 5$. Because $f(5) = 25$, an equation of the tangent line is $y - 25 = 10(x - 5)$, or in slope-intercept form, $y = 10x - 25$ (Figure 4). ∎

CONCEPTUAL INSIGHT In the previous example we encountered an indeterminate limit in the form $0/0$. In general, if f is continuous at a, then the difference-quotient limits in Eq. (1) and Eq. (2) are guaranteed to be in the form $0/0$.

Although we do not always indicate that we have an indeterminate form when we compute derivatives via the limit definition, it is important to realize that the approach we take in evaluating these limits is usually motivated by the fact that we are working with an indeterminate form.

In the next two examples, we perform the differentiation (the process of computing the derivative) using Eq. (1). For clarity, we break up the computations into three steps.

EXAMPLE 2 Compute $f'(3)$, where $f(x) = x^2 - 8x$.

Solution Using Eq. (1), we write the difference quotient at $a = 3$ as

$$\frac{f(a+h) - f(a)}{h} = \frac{f(3+h) - f(3)}{h} \quad (h \neq 0)$$

Step 1. Write out the numerator of the difference quotient.

$$f(3+h) - f(3) = \big((3+h)^2 - 8(3+h)\big) - \big(3^2 - 8(3)\big)$$
$$= \big((9 + 6h + h^2) - (24 + 8h)\big) - (9 - 24)$$
$$= h^2 - 2h$$

Step 2. Divide by h and simplify.

$$\frac{f(3+h) - f(3)}{h} = \frac{h^2 - 2h}{h} = \underbrace{\frac{h(h-2)}{h}}_{\text{Cancel } h} = h - 2$$

Step 3. Compute the limit.

$$f'(3) = \lim_{h \to 0} \frac{f(3+h) - f(3)}{h} = \lim_{h \to 0}(h - 2) = -2$$ ∎

EXAMPLE 3 Sketch the graph of $f(x) = \dfrac{1}{x}$ and the tangent line at $x = 2$.

(a) Based on the sketch, do you expect $f'(2)$ to be positive or negative?
(b) Find $f'(2)$.

Solution The graph and tangent line at $x = 2$ are shown in Figure 5.

(a) We see that the tangent line has negative slope, so $f'(2)$ must be negative.
(b) We compute $f'(2)$ in three steps as before.

FIGURE 5 The graph of $f(x) = \frac{1}{x}$ and the tangent line at $x = 2$.

Step 1. Write out the numerator of the difference quotient.

$$f(2+h) - f(2) = \frac{1}{2+h} - \frac{1}{2} = \frac{2}{2(2+h)} - \frac{2+h}{2(2+h)} = -\frac{h}{2(2+h)}$$

Step 2. Divide by h and simplify.

$$\frac{f(2+h) - f(2)}{h} = \frac{1}{h} \cdot \left(-\frac{h}{2(2+h)} \right) = -\frac{1}{2(2+h)}$$

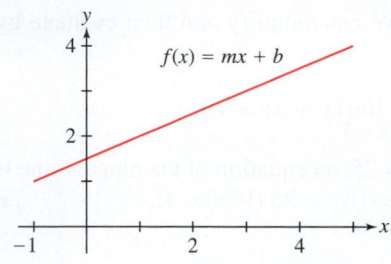

FIGURE 6 The derivative of $f(x) = mx + b$ is $f'(a) = m$ for all a.

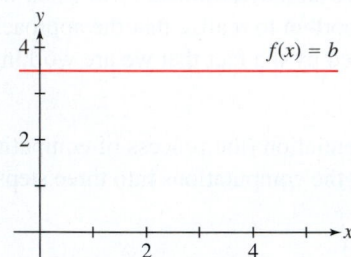

FIGURE 7 The derivative of a constant function $f(x) = b$ is $f'(a) = 0$ for all a.

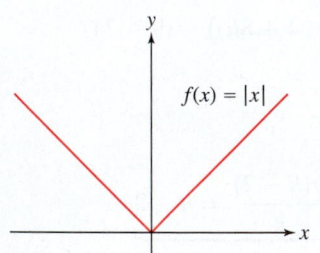

FIGURE 8 f is not differentiable at $x = 0$.

In general, if $\lim\limits_{h \to 0^-} \frac{f(x+h) - f(x)}{h}$ and $\lim\limits_{h \to 0^+} \frac{f(x+h) - f(x)}{h}$ exist but are not equal, then f is not differentiable at x, and the graph of f is said to have a **corner** at x (see Exercises 47 and 48).

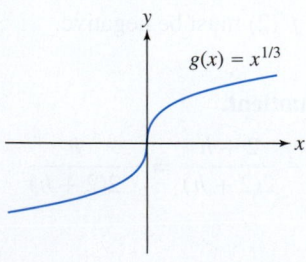

FIGURE 9 g is not differentiable at $x = 0$.

Step 3. **Compute the limit.**

$$f'(2) = \lim_{h \to 0} \frac{f(2+h) - f(2)}{h} = \lim_{h \to 0} \frac{-1}{2(2+h)} = -\frac{1}{4} \quad \blacksquare$$

The graph of a **linear function** $f(x) = mx + b$ (where m and b are constants) is a line of slope m. The tangent line at any point coincides with the line itself (Figure 6), so we should expect that $f'(a) = m$ for all a. Let's check this by computing the derivative:

$$f'(a) = \lim_{h \to 0} \frac{f(a+h) - f(a)}{h} = \lim_{h \to 0} \frac{(m(a+h) + b) - (ma + b)}{h}$$

$$= \lim_{h \to 0} \frac{mh}{h} = \lim_{h \to 0} m = m$$

If $m = 0$, then $f(x) = b$ is constant and $f'(a) = 0$ (Figure 7). In summary:

> **THEOREM 1 Derivative of Linear and Constant Functions**
> - If $f(x) = mx + b$ is a linear function, then $f'(a) = m$ for all a.
> - If $f(x) = b$ is a constant function, then $f'(a) = 0$ for all a.

EXAMPLE 4 Find the derivative of $f(x) = 9x - 5$ at $x = 2$ and $x = 5$.

Solution We have $f'(a) = 9$ for all a. Hence, $f'(2) = f'(5) = 9$. $\quad \blacksquare$

EXAMPLE 5 Failure to be Differentiable Show that the functions $f(x) = |x|$ and $g(x) = x^{1/3}$ are not differentiable at $x = 0$.

Solution First, note that

$$\lim_{h \to 0} \frac{f(0+h) - f(0)}{h} = \lim_{h \to 0} \frac{|0+h| - |0|}{h} = \lim_{h \to 0} \frac{|h|}{h}$$

Since

$$\frac{|h|}{h} = \begin{cases} 1 & \text{if } h > 0 \\ -1 & \text{if } h < 0 \end{cases}$$

we have the one-sided limits

$$\lim_{h \to 0^+} \frac{|h|}{h} = 1 \quad \text{and} \quad \lim_{h \to 0^-} \frac{|h|}{h} = -1$$

These one-sided limits are not equal; therefore, $\lim\limits_{h \to 0} \frac{|h|}{h}$ does not exist. Thus, f is not differentiable at $x = 0$.

Figure 8 reveals the problem we have here. To the left of $x = 0$ the secant lines all have slope -1, and to the right they all have slope 1. The one-sided limits on either side of 0 equal the corresponding secant-line slopes and therefore are not equal.

Now consider $g(x) = x^{1/3}$. The limit defining $g'(0)$ is infinite:

$$\lim_{h \to 0} \frac{g(h) - g(0)}{h} = \lim_{h \to 0} \frac{h^{1/3} - 0}{h} = \lim_{h \to 0} \frac{h^{1/3}}{h} = \lim_{h \to 0} \frac{1}{h^{2/3}} = \infty$$

Therefore, $g'(0)$ does not exist, and it follows that g is not differentiable at $x = 0$.

In this case, the problem is that the secant lines on either side of $x = 0$ become infinitely steep as h approaches 0. The tangent line is therefore vertical, having an undefined slope (see Figure 9). $\quad \blacksquare$

Estimating the Derivative

Approximations to the derivative are useful in situations where we cannot evaluate $f'(a)$ exactly. Since the derivative is the limit of difference quotients, the difference quotient should give a good numerical approximation when h is sufficiently small:

$$f'(a) \approx \frac{f(a+h) - f(a)}{h} \qquad \text{if } h \text{ is small}$$

We refer to this estimate as the **difference quotient approximation**. Graphically, this says that for small h, the slope of the secant line is nearly equal to the slope of the tangent line (Figure 10).

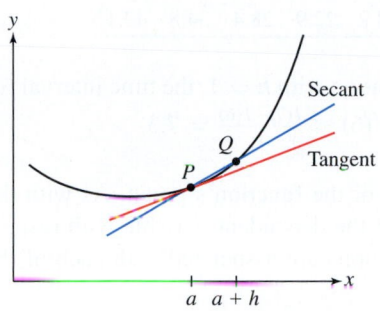

FIGURE 10 When h is small, the secant line has nearly the same slope as the tangent line.

EXAMPLE 6 Estimate the derivative of $f(x) = \sin x$ at $x = \frac{\pi}{6}$.

Solution We calculate the difference quotient for several small values of h:

$$\frac{\sin\left(\frac{\pi}{6} + h\right) - \sin\frac{\pi}{6}}{h} = \frac{\sin\left(\frac{\pi}{6} + h\right) - 0.5}{h}$$

This difference quotient represents the slope of a secant line through the graph at $\frac{\pi}{6}$. The resulting secant-line slopes are shown in Table 1. Note that Figure 11 indicates that the slope of the tangent line at $\frac{\pi}{6}$ lies between the slopes of the secant lines for $h > 0$ and those for $h < 0$. Thus, we can infer that

$$0.8660229 < f'\left(\frac{\pi}{6}\right) < 0.8660279$$

It follows that $f'\left(\frac{\pi}{6}\right) \approx 0.8660$, accurate to four decimal places. ∎

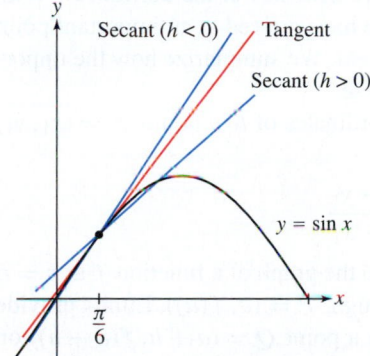

FIGURE 11 The tangent line is squeezed *in between* the secant lines with $h > 0$ and $h < 0$.

TABLE 1 Values of the Difference Quotient for Small h

$h > 0$	$\dfrac{\sin\left(\frac{\pi}{6} + h\right) - 0.5}{h}$	$h < 0$	$\dfrac{\sin\left(\frac{\pi}{6} + h\right) - 0.5}{h}$
0.01	0.863511	−0.01	0.868511
0.001	0.865775	−0.001	0.866275
0.0001	0.866000	−0.0001	0.866050
0.00001	0.8660229	−0.00001	0.8660279

In Section 3.6 we develop a derivative formula for $f(x) = \sin x$ and will see that $f'\left(\frac{\pi}{6}\right) = \cos\left(\frac{\pi}{6}\right) = \frac{\sqrt{3}}{2} \approx 0.8660254$.

In the previous example, because we had a specific formula for $f(x)$, we were able to choose h as small as we like to obtain an accurate approximation to $f'(x)$. In applied situations, as in the following example, we might have only a data set of values available for approximating a derivative. We can use the difference quotient approximation, but accuracy is limited because we cannot choose h to be arbitrarily small.

We can also approximate $f'(x)$ with the slope of the secant line between $(x - h, f(x - h))$ and $(x + h, f(x + h))$. This is called the **symmetric difference quotient approximation**:

$$f'(x) \approx \frac{f(x+h) - f(x-h)}{2h}$$

We investigate this approximation approach further in the exercises.

EXAMPLE 7 The Internet of Things The Internet of things (IoT) refers to the collection of devices that are connected to the Internet, such as computers, cell phones, smart watches, home security systems, and so on. Table 2 gives estimates of the size of the IoT for the years 2012–2019. Assume that $I(t)$ represents the size of the IoT (in billions of devices) t years after January 1, 2012, and that the data in the table represents the size of the IoT at the start of the indicated year. Approximate $I'(3)$ and $I'(6)$.

TABLE 2

t (years after 2012)	0	1	2	3	4	5	6	7
I (billions of devices)	8.7	11.2	14.4	18.2	22.9	28.4	34.8	42.1

Solution We use the difference quotient approximation with $h = 1$, the time interval for the data set. We have $I'(3) \approx \frac{I(4)-I(3)}{1} = 4.7$ and $I'(6) \approx \frac{I(7)-I(6)}{1} = 7.3$ ■

The derivative of a function defines the slope of the function's graph. As with the slope of a line, the derivative is a rate of change of the dependent variable with respect to the independent variable. In applications, when units are associated with each of the variables, the derivative has units

$$\text{units of derivative} = \frac{\text{units of dependent variable}}{\text{units of independent variable}}$$

When we include the units in the previous example, the derivatives are the rates of change $I'(3) \approx 4.7$ billion devices per year and $I'(6) \approx 7.3$ billion devices per year. These rates of change represent the rate of growth of the Internet of things at the start of 2015 and 2018, respectively.

While some early contributions to mathematics introduced ideas similar to those in calculus, it was Sir Isaac Newton (left) and Gottfried Wilhelm Leibniz (right) who independently initiated the development of mathematical concepts that eventually became the modern theory of calculus.

Sir Isaac Newton (1643–1727) was an English mathematician, scientist, and theologian. Beginning in the mid-1660s, he developed many of the foundational ideas of calculus. He used the term "fluxion" (from the Latin word for flow) for the concept that was eventually to become known as the derivative.

Gottfried Wilhelm Leibniz (1646–1716) was a German mathematician and philosopher. In the mid-1670s he began publishing works on the theory of "calculus differentialis" where he also formulated the main concepts of calculus.

Because Newton did not immediately publish his work, a long-running dispute arose over who should be credited as the inventor of calculus. The controversy had the effect of creating a nationalistic division among mathematicians. Over time, the contributions of both mathematicians were generally acknowledged, and today they are considered independent developers of the theory of calculus.

CONCEPTUAL INSIGHT The Significance of Limits in the Definition of the Derivative With the introduction of the derivative in this section, we have arrived at an important point in the development of the calculus concepts in this text. We summarize how the important ideas of slopes of lines and limits brought us here.

The slope of a line can be computed if the coordinates of *two* points $P = (x_1, y_1)$ and $Q = (x_2, y_2)$ on the line are known:

$$\text{slope of line} = \frac{y_2 - y_1}{x_2 - x_1}$$

This formula cannot be applied to the tangent line to the graph of a function f at $x = a$ because we know only one point that it passes through, $P = (a, f(a))$. Limits provide an ingenious way around this obstacle. We choose a point $Q = (a + h, f(a + h))$ on the graph near P and form the secant line. The slope of this secant line is just an approximation to the slope of the tangent line:

$$\text{slope of secant line} = \frac{f(a+h) - f(a)}{h} \approx \text{slope of tangent line}$$

This approximation improves as h is made small. By taking the limit as $h \to 0$, we convert our approximations into the exact slope for the tangent line, and that slope is $f'(a)$, the derivative of f at a.

3.1 SUMMARY

- The *difference quotient* is the slope of the secant line through the points P and Q on the graph of f and equals

$$\frac{f(a+h) - f(a)}{h} \quad \text{with} \quad P = (a, f(a)) \quad \text{and} \quad Q = (a + h, f(a + h))$$

$$\frac{f(x) - f(a)}{x - a} \quad \text{with} \quad P = (a, f(a)) \quad \text{and} \quad Q = (x, f(x))$$

- The *derivative* $f'(a)$ is defined by the following equivalent limits:

$$f'(a) = \lim_{h \to 0} \frac{f(a+h) - f(a)}{h} = \lim_{x \to a} \frac{f(x) - f(a)}{x - a}$$

If the limit exists, we say that f is *differentiable* at $x = a$.

- By definition, the tangent line at $P = (a, f(a))$ is the line through P with slope $f'(a)$ [assuming that $f'(a)$ exists].
- Equation of the tangent line in point-slope form:

$$y - f(a) = f'(a)(x - a)$$

- To calculate $f'(a)$ using the definition $f'(a) = \lim\limits_{h \to 0} \frac{f(a+h) - f(a)}{h}$:

  Step 1. Write out the numerator of the difference quotient.

  Step 2. Divide by h and simplify.

  Step 3. Compute the derivative by taking the limit.

- For small h, we have the *difference quotient approximation*: $f'(x) \approx \frac{f(x+h) - f(x)}{h}$.

3.1 EXERCISES

Preliminary Questions

1. Which of the lines in Figure 12 are tangent to the curve?

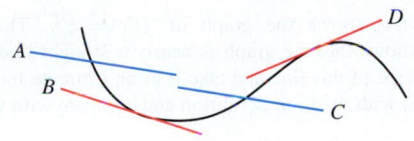

FIGURE 12

2. What are the two ways of writing the difference quotient?

3. Find a and h such that $\dfrac{f(a + h) - f(a)}{h}$ is equal to the slope of the secant line between $(3, f(3))$ and $(5, f(5))$.

4. Which derivative is approximated by $\dfrac{\tan\left(\frac{\pi}{4} + 0.0001\right) - 1}{0.0001}$?

5. What do the following quantities represent in terms of the graph of $f(x) = \sin x$?

(a) $\sin 1.3 - \sin 0.9$ (b) $\dfrac{\sin 1.3 - \sin 0.9}{0.4}$ (c) $f'(0.9)$

6. Choose (a) or (b). The derivative at a point is zero if the tangent line at that point is (a) horizontal (b) vertical.

7. Choose (a) or (b). The derivative at a point does not exist if the tangent line at that point is (a) horizontal (b) vertical.

Exercises

1. Let $f(x) = 5x^2$. Show that $f(3 + h) = 5h^2 + 30h + 45$. Then show that

$$\frac{f(3 + h) - f(3)}{h} = 5h + 30$$

and compute $f'(3)$ by taking the limit as $h \to 0$.

2. Let $f(x) = 2x^2 - 3x - 5$. Show that the secant line through $(2, f(2))$ and $(2 + h, f(2 + h))$ has slope $2h + 5$. Then use this formula to compute:
(a) The slope of the secant line through $(2, f(2))$ and $(3, f(3))$
(b) The slope of the tangent line at $x = 2$ (by taking a limit)

In Exercises 3–8, compute $f'(a)$ in two ways, using Eq. (1) and Eq. (2).

3. $f(x) = x^2 + 9x$, $a = 0$

4. $f(x) = x^2 + 9x$, $a = 2$

5. $f(x) = 3x^2 + 4x + 2$, $a = -1$

6. $f(x) = x^3$, $a = 2$

7. $f(x) = x^3 + 2x$, $a = 1$

8. $f(x) = \frac{1}{x}$, $a = 2$

In Exercises 9–12, refer to Figure 13.

9. Find the slope of the secant line through $(2, f(2))$ and $(2.5, f(2.5))$. Is it greater than or less than $f'(2)$? Explain.

10. Estimate $\dfrac{f(2 + h) - f(2)}{h}$ for $h = -0.5$. What does this quantity represent? Is it greater than or less than $f'(2)$? Explain.

11. Estimate $f'(1)$ and $f'(2)$.

12. Find a value of h for which $\dfrac{f(2 + h) - f(2)}{h} = 0$.

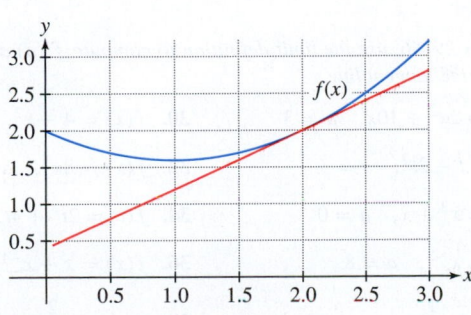

FIGURE 13

In Exercises 13–16, refer to Figure 14.

13. Determine $f'(a)$ for $a = 1, 2, 4, 7$.

14. For which values of x is $f'(x) < 0$?

15. Which is larger, $f'(5.5)$ or $f'(6.5)$?

16. Show that $f'(3)$ does not exist.

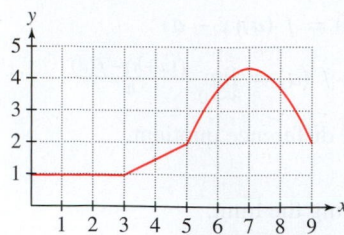

FIGURE 14 Graph of f.

In Exercises 17–20, use the limit definition to calculate the derivative of the linear function.

17. $f(x) = 7x - 9$

18. $f(x) = 12$

19. $g(t) = 8 - 3t$

20. $k(z) = 14z + 12$

21. Find an equation of the tangent line at $x = 3$, assuming that $f(3) = 5$ and $f'(3) = 2$.

22. Find $f(3)$ and $f'(3)$, assuming that the tangent line to $y = f(x)$ at $a = 3$ has equation $y = 5x + 2$.

23. Describe the tangent line at an arbitrary point on the graph of $y = 2x + 8$.

24. Suppose that $f(2 + h) - f(2) = 3h^2 + 5h$. Calculate:
(a) The slope of the secant line through $(2, f(2))$ and $(6, f(6))$
(b) $f'(2)$

25. Let $f(x) = \dfrac{1}{x}$. Does $f(-2 + h)$ equal $\dfrac{1}{-2 + h}$ or $\dfrac{1}{-2} + \dfrac{1}{h}$? Compute the difference quotient at $a = -2$ with $h = 0.5$.

26. Let $f(x) = \sqrt{x}$. Does $f(5 + h)$ equal $\sqrt{5 + h}$ or $\sqrt{5} + \sqrt{h}$? Compute the difference quotient at $a = 5$ with $h = 1$.

27. Let $f(x) = 1/\sqrt{x}$. Compute $f'(5)$ by showing that

$$\frac{f(5 + h) - f(5)}{h} = -\frac{1}{\sqrt{5}\sqrt{5 + h}(\sqrt{5 + h} + \sqrt{5})}$$

28. Find an equation of the tangent line to the graph of $f(x) = 1/\sqrt{x}$ at $x = 9$.

In Exercises 29–46, use the limit definition to compute $f'(a)$ and find an equation of the tangent line.

29. $f(x) = 2x^2 + 10x, \quad a = 3$

30. $f(x) = 4 - x^2, \quad a = -1$

31. $f(t) = t - 2t^2, \quad a = 3$

32. $f(x) = 8x^3, \quad a = 1$

33. $f(x) = x^3 + x, \quad a = 0$

34. $f(t) = 2t^3 + 4t, \quad a = 4$

35. $f(x) = x^{-1}, \quad a = 8$

36. $f(x) = x + x^{-1}, \quad a = 4$

37. $f(x) = \dfrac{1}{x + 3}, \quad a = -2$

38. $f(t) = \dfrac{2}{1 - t}, \quad a = -1$

39. $f(x) = \sqrt{x + 4}, \quad a = 1$

40. $f(t) = \sqrt{3t + 5}, \quad a = -1$

41. $f(x) = \dfrac{1}{\sqrt{x}}, \quad a = 4$

42. $f(x) = \dfrac{1}{\sqrt{2x + 1}}, \quad a = 4$

43. $f(t) = \sqrt{t^2 + 1}, \quad a = 3$

44. $f(x) = x^{-2}, \quad a = -1$

45. $f(x) = \dfrac{1}{x^2 + 1}, \quad a = 0$

46. $f(t) = t^{-3}, \quad a = 1$

47. Show that f is not differentiable at $x = 1$ and has a corner in its graph there.

$$f(x) = \begin{cases} 1 & x \le 1 \\ x^2 & x > 1 \end{cases}$$

48. Show that f is not differentiable at $x = 0$ and has a corner in its graph there.

$$f(x) = \begin{cases} x^3 & x \le 0 \\ x & x > 0 \end{cases}$$

In Exercises 49–51, sketch a graph of f and identify the points c such that $f'(c)$ does not exist. In which cases is there a corner at c?

49. $f(x) = |x + 3|$

50. $f(x) = x^{2/5}$

51. $f(x) = |x^2 - 4|$

52. Figure 15(A) shows the graph of $f(x) = \sqrt{x}$. The close-up in Figure 15(B) shows that the graph is nearly a straight line near $x = 16$. Estimate the slope of this line and take it as an estimate for $f'(16)$. Then compute $f'(16)$ with the limit definition and compare with your estimate.

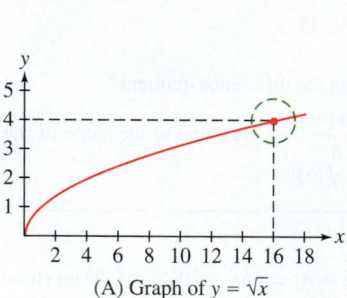
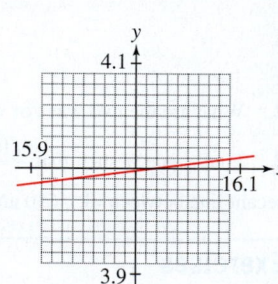

(A) Graph of $y = \sqrt{x}$

(B) Zoom view near $(16, 4)$

FIGURE 15

53. [GU] Let $f(x) = \dfrac{4}{1 + 2^x}$. Plot f over $[-2, 2]$. Then zoom in near $x = 0$ until the graph appears straight, and estimate the slope $f'(0)$.

54. [GU] Let $f(x) = \cot x$. Estimate $f'\left(\frac{\pi}{2}\right)$ graphically by zooming in on a plot of f near $x = \frac{\pi}{2}$.

55. Determine the intervals along the x-axis on which the derivative in Figure 16 is positive.

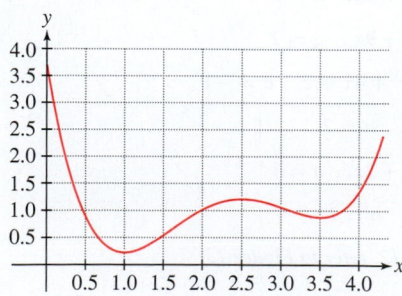

FIGURE 16

56. Sketch the graph of $f(x) = \sin x$ on $[0, \pi]$ and guess the value of $f'\left(\frac{\pi}{2}\right)$. Then calculate the difference quotient at $x = \frac{\pi}{2}$ for two small positive and negative values of h. Are these calculations consistent with your guess?

In Exercises 57–62, each limit represents a derivative $f'(a)$. Find $f(x)$ and a.

57. $\lim\limits_{h \to 0} \dfrac{(5+h)^3 - 125}{h}$

58. $\lim\limits_{x \to 5} \dfrac{x^3 - 125}{x - 5}$

59. $\lim\limits_{h \to 0} \dfrac{\sin\left(\frac{\pi}{6} + h\right) - 0.5}{h}$

60. $\lim\limits_{x \to \frac{1}{4}} \dfrac{x^{-1} - 4}{x - \frac{1}{4}}$

61. $\lim\limits_{h \to 0} \dfrac{5^{2+h} - 25}{h}$

62. $\lim\limits_{h \to 0} \dfrac{5^h - 1}{h}$

63. Apply the method of Example 6 to $f(x) = \sin x$ to determine $f'\left(\frac{\pi}{4}\right)$ accurately to four decimal places.

64. Apply the method of Example 6 to $f(x) = \cos x$ to determine $f'\left(\frac{\pi}{5}\right)$ accurately to four decimal places. Use a graph of f to explain how the method works in this case.

65. For each graph in Figure 17, determine whether $f'(1)$ is larger or smaller than the slope of the secant line between $x = 1$ and $x = 1 + h$ for $h > 0$. Explain.

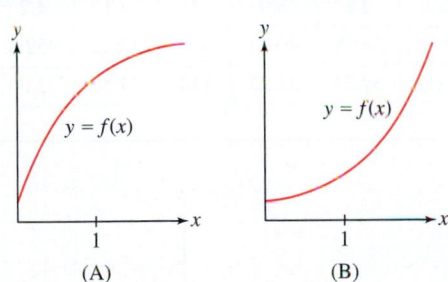

(A)   (B)

FIGURE 17

66. Refer to the graph of $f(x) = 2^x$ in Figure 18.
(a) Explain graphically why, for $h > 0$,

$$\frac{f(-h) - f(0)}{-h} \leq f'(0) \leq \frac{f(h) - f(0)}{h}$$

(b) Use (a) to show that $0.69314 \leq f'(0) \leq 0.69315$.
(c) Similarly, compute $f'(x)$ to four decimal places for $x = 1, 2, 3, 4$.
(d) Now compute the ratios $f'(x)/f'(0)$ for $x = 1, 2, 3, 4$. Can you guess an approximate formula for $f'(x)$?

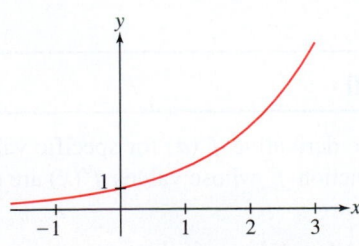

FIGURE 18 Graph of $f(x) = 2^x$.

67. GU Sketch the graph of $f(x) = x^{5/2}$ on $[0, 6]$.

(a) Use the sketch to justify the inequalities for $h > 0$:

$$\frac{f(4) - f(4-h)}{h} \leq f'(4) \leq \frac{f(4+h) - f(4)}{h}$$

(b) Use (a) to compute $f'(4)$ to four decimal places.
(c) Use a graphing utility to plot $y = f(x)$ and the tangent line at $x = 4$, utilizing your estimate for $f'(4)$.

68. GU Verify that $P = \left(1, \frac{1}{2}\right)$ lies on the graphs of both

$$f(x) = 1/(1 + x^2) \quad \text{and} \quad L(x) = \frac{1}{2} + m(x - 1)$$

for every slope m. Plot $y = f(x)$ and $y = L(x)$ on the same axes for several values of m until you find a value of m for which $y = L(x)$ appears tangent to the graph of f. What is your estimate for $f'(1)$?

69. GU Use a plot of $f(x) = x^x$ to estimate the value c such that $f'(c) = 0$. Find c to sufficient accuracy so that

$$\left| \frac{f(c+h) - f(c)}{h} \right| \leq 0.006 \quad \text{for} \quad h = \pm 0.001$$

70. GU Plot $f(x) = x^x$ and $y = 2x + a$ on the same set of axes for several values of a until the line becomes tangent to the graph. Then estimate the value c such that $f'(c) = 2$.

The vapor pressure of water at temperature T (in kelvins) is the atmospheric pressure P at which no net evaporation takes place. In Exercises 71–72, use the following table to estimate the indicated derivatives using the difference quotient approximation.

T (K)	293	303	313	323	333	343	353
P (atm)	0.0278	0.0482	0.0808	0.1311	0.2067	0.3173	0.4754

71. Estimate $P'(T)$ for $T = 293, 313, 333$. (Include proper units on the derivative.)

72. Estimate $P'(T)$ for $T = 303, 323, 343$. (Include proper units on the derivative.)

Let $P(t)$ represent the U.S. ethanol production as shown in Figure 19. In Exercises 73–74, estimate the indicated derivatives using the difference quotient approximation.

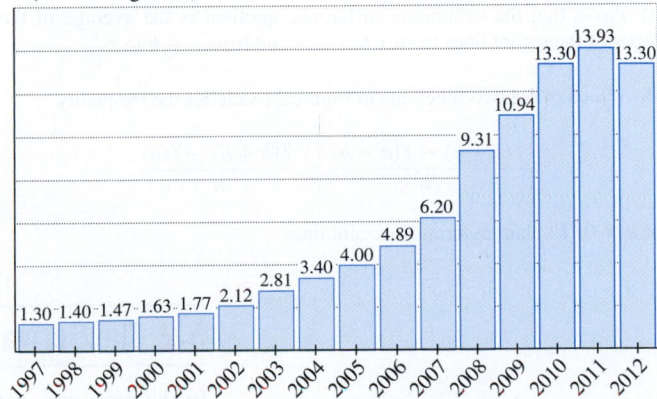

FIGURE 19 U.S. ethanol production.

73. Estimate $P'(t)$ for $t = 1997, 2001, 2005, 2009$. (Include proper units on the derivative.)

74. Estimate $P'(t)$ for $t = 1999, 2003, 2007, 2011$. (Include proper units on the derivative.)

In the remaining exercises, SDQ refers to the symmetric difference quotient derivative approximation that is based on the slope of the secant line between $(x - h, f(x - h))$ and $(x + h, f(x + h))$:

$$f'(x) \approx \frac{f(x + h) - f(x - h)}{2h}$$

75. With $P(T)$ as in Exercises 71 and 72, estimate $P'(T)$ for $T = 303$, 313, 333, 343, now using the SDQ.

76. With $P(t)$ as in Exercises 73 and 74, estimate $P'(t)$ for $t = 1999$, 2001, 2005, 2011, now using the SDQ.

In Exercises 77–78, traffic speed S along Katman Road (in kilometers per hour) varies as a function of traffic density q (number of cars per kilometer of road) according to the data:

q (density)	60	70	80	90	100
S (speed)	72.5	67.5	63.5	60	56

77. Estimate $S'(80)$ using the SDQ. (Include proper units on the derivative.)

78. 📝 Explain why $V = qS$, called *traffic volume*, is equal to the number of cars passing a point per hour. Use the data and the SDQ to estimate $V'(80)$. (Include proper units on the derivative.)

Exercises 79–81: The current (in amperes) at time t (in seconds) flowing in the circuit in Figure 20 is given by Kirchhoff's Law:

$$i(t) = Cv'(t) + R^{-1}v(t)$$

where $v(t)$ is the voltage (in volts, V), C the capacitance (in farads, F), and R the resistance (in ohms, Ω).

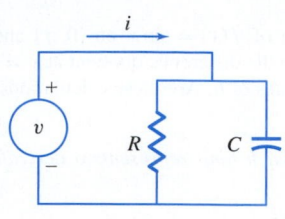

FIGURE 20

79. Calculate the current at $t = 3$ if

$$v(t) = 0.5t + 4 \text{ V}$$

where $C = 0.01$ F and $R = 100 \ \Omega$.

80. Use the following data and the SDQ to estimate $v'(10)$. Then estimate $i(10)$, assuming $C = 0.03$ and $R = 1000$.

t	9.8	9.9	10	10.1	10.2
$v(t)$	256.52	257.32	258.11	258.9	259.69

81. Assume that $R = 200 \ \Omega$ but C is unknown. Use the following data and the SDQ to estimate $v'(4)$, and deduce an approximate value for the capacitance C.

t	3.8	3.9	4	4.1	4.2
$v(t)$	388.8	404.2	420	436.2	452.8
$i(t)$	32.34	33.22	34.1	34.98	35.86

Further Insights and Challenges

82. The SDQ usually approximates the derivative much more closely than does the ordinary difference quotient. Let $f(x) = 2^x$ and $a = 0$. Compute the SDQ with $h = 0.001$ and the ordinary difference quotients with $h = \pm 0.001$. Compare with the actual value, which is $f'(0) = \ln 2$.

83. (a) Show that the symmetric difference quotient $\frac{f(x+h) - f(x-h)}{2h}$ is the slope of the secant line to the graph of f from $x - h$ to $x + h$. (Include an illustration.)

(b) Prove that the symmetric difference quotient is the average of the slopes of the secant lines from x to $x + h$ and from $x - h$ to x.

84. Which of the two functions in Figure 21 satisfies the inequality

$$\frac{f(a+h) - f(a-h)}{2h} \le \frac{f(a+h) - f(a)}{h}$$

for $h > 0$? Explain in terms of secant lines.

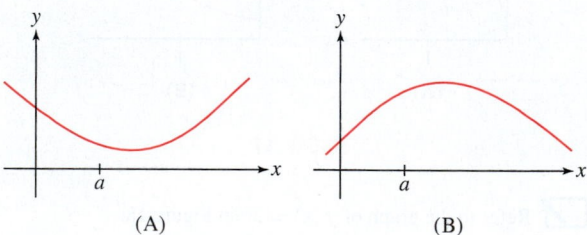

FIGURE 21

85. 📝 Show that if f is a quadratic polynomial, then the SDQ at $x = a$ (for any $h \ne 0$) is *equal* to $f'(a)$. Explain the graphical meaning of this result.

86. Let $f(x) = x^{-2}$. Compute $f'(1)$ by taking the limit of the SDQs (with $a = 1$) as $h \to 0$.

3.2 The Derivative as a Function

In the previous section, we computed the derivative $f'(a)$ for specific values of a. It is also useful to view the derivative as a function f' whose values $f'(x)$ are defined by the limit definition of the derivative:

$$f'(x) = \lim_{h \to 0} \frac{f(x + h) - f(x)}{h} \qquad \boxed{1}$$

If $y = f(x)$, we also write y' or $y'(x)$ for $f'(x)$.

Often, the domain of f' is clear from the context. If so, we usually do not mention the domain explicitly.

The domain of f' consists of all values of x in the domain of f for which the limit in Eq. (1) exists. We say that f is **differentiable** on (a, b) if $f'(x)$ exists for all x in (a, b). When $f'(x)$ exists for all x in the interval or intervals on which $f(x)$ is defined, we say simply that f is differentiable.

EXAMPLE 1 Prove that $f(x) = x^3 - 12x$ is differentiable. Compute $f'(x)$ and find $f'(-3)$, $f'(0)$, $f'(2)$, and $f'(3)$.

Solution We compute $f'(x)$ in three steps as in the previous section.

Step 1. **Write out the numerator of the difference quotient.**

$$f(x + h) - f(x) = \left((x + h)^3 - 12(x + h)\right) - \left(x^3 - 12x\right)$$

$$= (x^3 + 3x^2h + 3xh^2 + h^3 - 12x - 12h) - (x^3 - 12x)$$

$$= 3x^2h + 3xh^2 + h^3 - 12h$$

$$= h(3x^2 + 3xh + h^2 - 12) \qquad \text{(factor out } h\text{)}$$

Step 2. **Divide by h and simplify.**

$$\frac{f(x + h) - f(x)}{h} = \frac{h(3x^2 + 3xh + h^2 - 12)}{h} = 3x^2 + 3xh + h^2 - 12 \quad (h \neq 0)$$

Step 3. **Compute the limit.**

$$f'(x) = \lim_{h \to 0} \frac{f(x + h) - f(x)}{h} = \lim_{h \to 0} (3x^2 + 3xh + h^2 - 12) = 3x^2 - 12$$

In this limit, x is treated as a constant because it does not change as $h \to 0$. We see that the limit exists for all x, so f is differentiable and $f'(x) = 3x^2 - 12$.

Now evaluate:

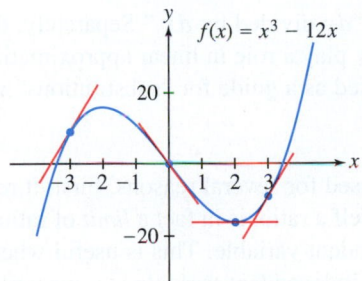

y
$f(x) = x^3 - 12x$

FIGURE 1 The derivative at each point is the slope of the tangent line.

- $f'(-3) = 3(-3)^2 - 12 = 15$
- $f'(0) = 3(0)^2 - 12 = -12$
- $f'(2) = 3(2)^2 - 12 = 0$
- $f'(3) = 3(3)^2 - 12 = 15$

These derivatives indicate the slope of the graph of f (and the tangent line to the graph) at the corresponding points, as shown in Figure 1. ■

In the previous example, we used the definition of $f'(x)$ to find an equation for the derivative as a function of x. Using the formula for $f'(x)$ we computed the derivative at specific values of x rather than computing each derivative separately using the limit definition. As we proceed with the development of the derivative, one goal will be to develop formulas and rules for the derivatives of functions so that we do not have to return to the limit definition for every derivative computation.

EXAMPLE 2 Prove that $y = x^{-2}$ is differentiable and calculate y'.

Solution The domain of $f(x) = x^{-2}$ is $\{x : x \neq 0\}$, so assume that $x \neq 0$. We compute $f'(x)$ directly, without the separate steps of the previous example:

$$f'(x) = \lim_{h \to 0} \frac{f(x + h) - f(x)}{h} = \lim_{h \to 0} \frac{\dfrac{1}{(x + h)^2} - \dfrac{1}{x^2}}{h}$$

$$= \lim_{h \to 0} \frac{\dfrac{x^2 - (x + h)^2}{x^2(x + h)^2}}{h} = \lim_{h \to 0} \frac{1}{h}\left(\frac{x^2 - (x + h)^2}{x^2(x + h)^2}\right)$$

$$= \lim_{h \to 0} \frac{1}{h}\left(\frac{-h(2x+h)}{x^2(x+h)^2}\right) = \lim_{h \to 0} -\frac{2x+h}{x^2(x+h)^2} \quad \text{(cancel } h)$$

$$= -\frac{2x+0}{x^2(x+0)^2} = -\frac{2x}{x^4} = -2x^{-3}$$

The limit exists for all $x \neq 0$, so $y = x^{-2}$ is differentiable and $y' = -2x^{-3}$. ■

Leibniz Notation

The "prime" notation y' and $f'(x)$ was introduced by the French mathematician Joseph Louis Lagrange (1736–1813). There is another standard notation for the derivative that we owe to Gottfried Wilhelm Leibniz:

$$\frac{df}{dx} \quad \text{or} \quad \frac{dy}{dx}$$

In Example 2, we showed that the derivative of $y = x^{-2}$ is $y' = -2x^{-3}$. In Leibniz notation, we would write

$$\frac{dy}{dx} = -2x^{-3} \quad \text{or} \quad \frac{d}{dx}x^{-2} = -2x^{-3}$$

To specify the value of the derivative for a fixed value of x, say, $x = 4$, we write

$$\left.\frac{df}{dx}\right|_{x=4} \quad \text{or} \quad \left.\frac{dy}{dx}\right|_{x=4}$$

You should not think of dy/dx as the fraction "dy divided by dx." Separately, the expressions dy and dx are called **differentials**. They play a role in linear approximation (Section 4.1), and relationships between them are used as a guide for "substitutions" we do later when working with integrals.

> **CONCEPTUAL INSIGHT**   Leibniz notation is widely used for several reasons. First, it reminds us that the derivative df/dx, although not itself a ratio, is in fact a *limit* of ratios $\Delta f/\Delta x$. Second, the notation specifies the independent variable. This is useful when variables other than x are used. For example, if the independent variable is t, we write df/dt. Third, we often think of d/dx as an "operator" that performs differentiation on functions. In other words, we apply the operator d/dx to f to obtain the derivative df/dx. We will see other advantages of Leibniz notation when we discuss the Chain Rule in Section 3.7.

We read $\frac{dy}{dx}$ and $\frac{d}{dx}y$ as "the derivative of y with respect to x."

Now we are ready to start assembling a collection of derivative rules and formulas that will enable us to compute derivatives of the most common functions in mathematics, the sciences, and engineering. We begin with two simple formulas that are consequences of Theorem 1 in the previous section:

$$\frac{d}{dx}x = 1 \quad \text{and} \quad \frac{d}{dx}c = 0 \quad \text{for any constant } c$$

The first indicates that the derivative with respect to x of x is 1, reflecting that the slope of the line $y = x$ is 1. The second, known as **the Constant Rule**, indicates that the derivative of a constant is 0. This makes sense, of course, since a constant does not change and therefore has a rate of change of zero. As simple as the latter is, we will find it quite useful as we work with derivatives throughout the book.

The next theorem will prove to be very valuable for differentiating polynomial functions and many other types of functions involving constant powers.

> **THEOREM 1 The Power Rule**   For all exponents n:
>
> $$\frac{d}{dx}x^n = nx^{n-1}$$

The Power Rule is valid for all exponents. We prove it here for positive integers n. See Exercise 93 for a proof for negative integers n and Exercise 87 in Section 3.9 for arbitrary n.

Proof We have

$$f'(x) = \lim_{h\to0}\frac{(x+h)^n - x^n}{h}$$

To simplify the difference quotient, we need to expand the $(x+h)^n$ term. However, we do not need to write all the terms to work through the limit. The binomial expansion formula helps here (see Section 1.1); it indicates that expanding $(x+h)^n$ results in a sum of terms $\frac{n!}{p!(n-p)!}x^{n-p}h^p$, one for each p from 0 to n. For $p=0$ we have x^n, for $p=1$ we have $nx^{n-1}h$, and for the rest of the terms, it is enough to observe that they are terms containing $x^q h^p$ for $p \geq 2$. Thus, for our purposes, we can express the expansion of $(x+h)^n$ as

$$(x+h)^n = x^n + nx^{n-1}h + [\text{terms with } x^q h^p, p \geq 2]$$

Now, we have

$$f'(x) = \lim_{h\to0}\frac{(x+h)^n - x^n}{h}$$
$$= \lim_{h\to0}\frac{(x^n + nx^{n-1}h + [\text{terms with } x^q h^p, p \geq 2]) - x^n}{h}$$
$$= \lim_{h\to0}\frac{nx^{n-1}h + [\text{terms with } x^q h^p, p \geq 2]}{h}$$
$$= \lim_{h\to0}(nx^{n-1} + [\text{terms with } x^q h^p, p \geq 1])$$
$$= nx^{n-1}$$

At the last stage, the limit of the terms containing $x^q h^p$ equal 0 as $h \to 0$ because those terms include a factor of h to at least the first power.

This proves that $f'(x) = nx^{n-1}$ for the case where n is a positive integer. ■

We make a few remarks before proceeding:

- The Power Rule in words: To differentiate x to a power, multiply by the power and reduce the power by one.

$$\frac{d}{dx}x^{\text{power}} = (\text{power})\, x^{\text{power}-1}$$

- The Power Rule is valid for all exponents, whether positive, negative, fractional, or irrational:

$$\frac{d}{dx}x^{-3/5} = -\frac{3}{5}x^{-8/5}, \qquad \frac{d}{dx}x^{\sqrt2} = \sqrt2\,x^{\sqrt2-1}$$

- The Power Rule can be applied with any variable, not just x. For example,

$$\frac{d}{dz}z^2 = 2z, \qquad \frac{d}{dt}t^{20} = 20t^{19}, \qquad \frac{d}{dr}r^{1/2} = \frac{1}{2}r^{-1/2}$$

CAUTION The Power Rule applies only to the power functions $y = x^n$. It does not apply to exponential functions such as $y = 2^x$. The derivative of $y = 2^x$ is not $x2^{x-1}$. A key difference between these two types of functions is that in x^n the base is the variable and the exponent is constant, while in 2^x those roles are reversed. We will introduce the derivative of $f(x) = e^x$ later in this section and the derivative of the general exponential $f(x) = b^x$ in Section 3.9.

Next, we state the Linearity Rules for derivatives, which are analogous to the Linearity Laws for limits:

THEOREM 2 Linearity Rules   Assume that f and g are differentiable. Then

Sum and Difference Rules: $f + g$ and $f - g$ are differentiable, and

$$(f + g)' = f' + g', \qquad (f - g)' = f' - g'$$

Constant Multiple Rule: For any constant c, cf is differentiable, and

$$(cf)' = cf'$$

Proof   To prove the Sum Rule, we use the definition

$$(f + g)'(x) = \lim_{h \to 0} \frac{(f(x + h) + g(x + h)) - (f(x) + g(x))}{h}$$

This difference quotient is equal to a sum ($h \neq 0$):

$$\frac{(f(x + h) + g(x + h)) - (f(x) + g(x))}{h} = \frac{f(x + h) - f(x)}{h} + \frac{g(x + h) - g(x)}{h}$$

Therefore, by the Sum Law for limits,

$$(f + g)'(x) = \lim_{h \to 0} \frac{f(x + h) - f(x)}{h} + \lim_{h \to 0} \frac{g(x + h) - g(x)}{h}$$

$$= f'(x) + g'(x)$$

as claimed. The Difference and Constant Multiple Rules are proved similarly.   ∎

In words, these rules state:

- The derivative of a sum is the sum of the derivatives.
- The derivative of a difference is the difference of the derivatives.
- The derivative of a constant times a function is the constant times the derivative of the function.

EXAMPLE 3   Find the points on the graph of $f(t) = t^3 - 12t + 4$ where the tangent line is horizontal.

Solution   We calculate the derivative:

$$\frac{df}{dt} = \frac{d}{dt}\left(t^3 - 12t + 4\right)$$

$$= \frac{d}{dt}t^3 - \frac{d}{dt}(12t) + \frac{d}{dt}4 \qquad \text{(Sum and Difference Rules)}$$

$$= \frac{d}{dt}t^3 - 12\frac{d}{dt}t + 0 \qquad \text{(Constant Multiple Rule and Constant Rule)}$$

$$= 3t^2 - 12 \qquad \text{(Power Rule)}$$

The tangent line is horizontal at points where the slope $f'(t)$ is zero, so we solve

$$3t^2 - 12 = 0 \quad \Rightarrow \quad t = \pm 2$$

Now $f(2) = -12$ and $f(-2) = 20$. Hence, the tangent lines are horizontal at $(2, -12)$ and $(-2, 20)$, as shown in Figure 2.   ∎

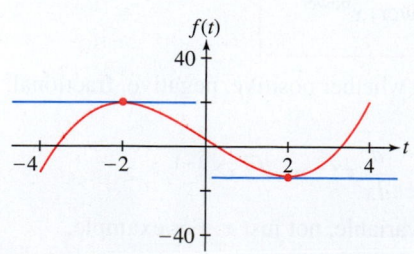

FIGURE 2 Graph of $f(t) = t^3 - 12t + 4$. Tangent lines at $t = \pm 2$ are horizontal.

EXAMPLE 4 Calculate $\dfrac{dg}{dt}\Big|_{t=1}$, where $g(t) = t^{-3} + 2\sqrt{t} - t^{-4/5}$.

Solution We differentiate term-by-term using the Power Rule and the Linearity Rules. Writing $\sqrt{t}$ as $t^{1/2}$, we have

$$\frac{dg}{dt} = \frac{d}{dt}\left(t^{-3} + 2t^{1/2} - t^{-4/5}\right) = -3t^{-4} + 2\left(\frac{1}{2}\right)t^{-1/2} - \left(-\frac{4}{5}\right)t^{-9/5}$$

$$= -3t^{-4} + t^{-1/2} + \frac{4}{5}t^{-9/5}$$

So $\dfrac{dg}{dt}\Big|_{t=1} = -3 + 1 + \dfrac{4}{5} = -\dfrac{6}{5}$ ∎

EXAMPLE 5 A power-law model relating the pulse rate P (in beats per minute) in mammals to body mass m (in kilograms) is given by $P = 200m^{-1/4}$ (see Figure 3). It is clear from the graph that, from species to species, as the mass increases, the pulse rate drops off. Furthermore the pulse rate drops off quickly for small mammals but relatively slowly for large mammals. Determine $P'(m)$ at the mass of a guinea pig (1 kg) and at the mass of cattle (500 kg).

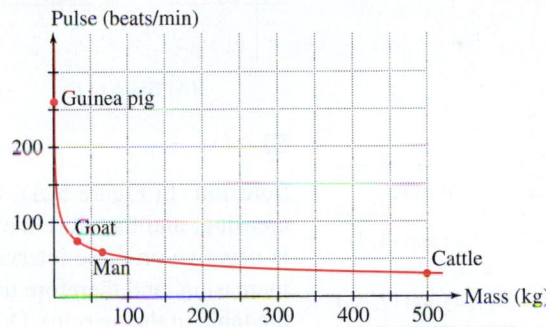

FIGURE 3

Solution Applying the Power Rule and the Constant Multiple Rule to $P(m) = 200m^{-1/4}$, we obtain $P'(m) = -50m^{-5/4}$. So,

- $P'(1) = -50$ beats per minute per kilogram
- $P'(500) = -50(500^{-5/4}) \approx -0.02$ beats per minute per kilogram

These values confirm our observation that P decreases rapidly for small m and decreases slowly for large m. ∎

The Derivative and Behavior of the Graph

The derivative f' gives us important information about the graph of f. For example, the sign of $f'(x)$ tells us whether the tangent line has positive or negative slope. When the tangent line has positive slope, it slopes upward and the graph must be increasing. When the tangent line has negative slope, it slopes downward and the graph must be decreasing. The magnitude of $f'(x)$ reveals how steep the slope is.

EXAMPLE 6 **f' and the Graph of f** How is the graph of $f(x) = x^3 - 12x^2 + 36x - 16$ related to the derivative $f'(x) = 3x^2 - 24x + 36$?

Solution The derivative $f'(x) = 3x^2 - 24x + 36 = 3(x-6)(x-2)$ is negative for $2 < x < 6$ and positive for $x < 2$ and $x > 6$ (Figure 4). The following table summarizes this sign information:

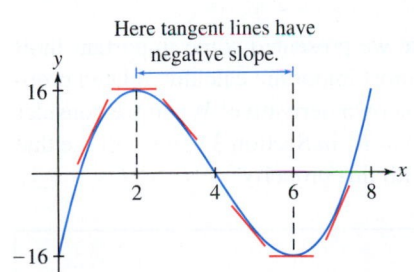

(A) Graph of $f(x) = x^3 - 12x^2 + 36x - 16$

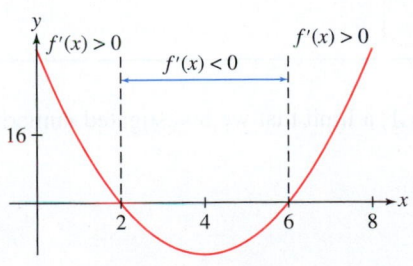

(B) Graph of the derivative $f'(x) = 3x^2 - 24x + 36$

FIGURE 4

Property of $f'(x)$	Property of the Graph of f
$f'(x) < 0$ for $2 < x < 6$	Tangent has negative slope for $2 < x < 6$ (graph is decreasing).
$f'(2) = f'(6) = 0$	Tangent is horizontal at $x = 2$ and $x = 6$.
$f'(x) > 0$ for $x < 2$ and $x > 6$	Tangent has positive slope for $x < 2$ and $x > 6$ (graph is increasing).

Note also that $f'(x) \to \infty$ as $|x|$ becomes large. This corresponds to the fact that the tangent lines to the graph of f get steeper as $|x|$ grows large.   ■

EXAMPLE 7   Identifying the Derivative   The graph of f (with some tangent lines included) is shown in Figure 5(A). Which graph, (B) or (C), is the graph of f'?

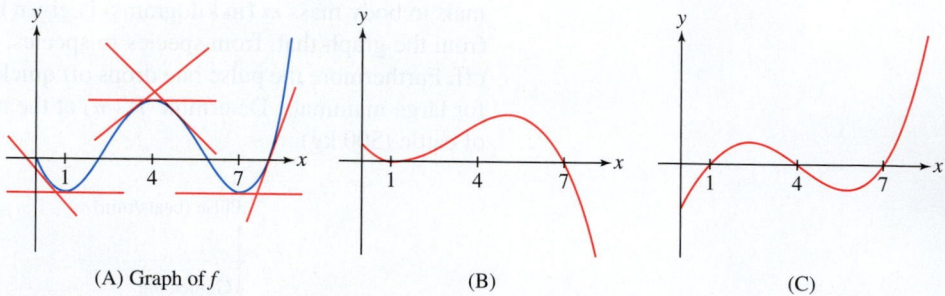

(A) Graph of f (B)   (C)

DF **FIGURE 5**

Slope of Tangent Line	Where
Negative	$(0, 1)$ and $(4, 7)$
Zero	$x = 1, 4, 7$
Positive	$(1, 4)$ and $(7, \infty)$

Solution   In Figure 5(A), we see that on the intervals $(0, 1)$ and $(4, 7)$, the graph is decreasing, and therefore the tangent lines to the graph have negative slope . Thus, $f'(x)$ is negative on these intervals. Similarly, on the intervals $(1, 4)$ and $(7, \infty)$, the graph is increasing, and therefore the tangent lines have positive slope and $f'(x)$ is positive (see the table in the margin). Only (C) has these properties, so (C) is the graph of f'.   ■

The Derivative of $f(x) = e^x$

In many books, e^x is denoted $\exp(x)$. Whenever we refer to the exponential function without specifying the base, the reference is to $f(x) = e^x$.

The number e was introduced in Section 1.6 where we presented some important limit properties related to it. We will see here one of the most important calculus-related properties of e: The exponential function $f(x) = e^x$ is its own derivative! When we consider the derivative of general exponential functions $f(x) = b^x$ in Section 3.9, we will see that $f(x) = e^x$ stands out as *the* exponential function with this property.

THEOREM 3 The Exponential Function Rule

$$\frac{d}{dx}e^x = e^x$$

$\boxed{2}$

In the proof we use the fact that $\displaystyle\lim_{h \to 0} \frac{e^h - 1}{h} = 1$, a limit that we investigated numerically and graphically in Section 2.5.

Proof First, for $f(x) = e^x$ we have

$$\frac{f(x + h) - f(x)}{h} = \frac{e^{x+h} - e^x}{h} = \frac{e^x e^h - e^x}{h} = \frac{e^x(e^h - 1)}{h}$$

It follows that,

$$f'(x) = \lim_{h \to 0} \frac{f(x+h) - f(x)}{h} = \lim_{h \to 0} \frac{e^x(e^h - 1)}{h}$$

$$= e^x \lim_{h \to 0} \left(\frac{e^h - 1}{h} \right)$$

$$= e^x$$

Therefore, for $f(x) = e^x$ we have $f'(x) = e^x$. ∎

Notice that in the proof, we could factor e^x outside the limit because e^x does not depend on h, the variable in the limit.

EXAMPLE 8 Find the tangent line to the graph of $f(x) = 3e^x - 5x^2$ at $x = 2$.

Solution We compute both $f'(2)$ and $f(2)$:

$$f'(x) = \frac{d}{dx}(3e^x - 5x^2) = 3\frac{d}{dx}e^x - 5\frac{d}{dx}x^2 = 3e^x - 10x$$

$$f'(2) = 3e^2 - 10(2) \approx 2.17$$

$$f(2) = 3e^2 - 5(2^2) \approx 2.17$$

Since $f'(2)$ is the slope and the tangent line passes through $(2, f(2))$, an equation of the tangent line is $y - f(2) = f'(2)(x - 2)$. Using these approximate values, we write the equation as (Figure 6)

$$y - 2.17 = 2.17(x - 2) \qquad \text{or} \qquad y = 2.17x - 2.17$$ ∎

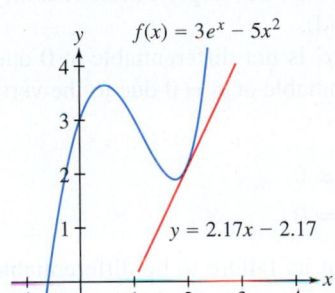

FIGURE 6

Differentiability, Continuity, and Local Linearity

In the rest of this section, we examine the concept of **differentiability** more closely. We begin by proving that a differentiable function is necessarily continuous. In particular, a function with a jump discontinuity cannot be differentiable. Figure 7 shows why: Although the secant lines from the right approach the line L (which is tangent to the right half of the graph), the secant lines from the left approach the vertical (and their slopes tend to ∞).

So the limit of the slopes of the secant lines does not exist and therefore the function is not differentiable at the point where the jump discontinuity occurs.

> **THEOREM 4** **Differentiability Implies Continuity** If f is differentiable at $x = c$, then f is continuous at $x = c$.

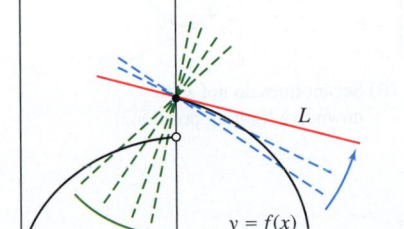

DF **FIGURE 7** Secant lines at a jump discontinuity.

Proof By definition, if f is differentiable at $x = c$, then the following limit exists:

$$f'(c) = \lim_{x \to c} \frac{f(x) - f(c)}{x - c}$$

We must prove that $\lim_{x \to c} f(x) = f(c)$, because this is the definition of continuity at $x = c$. To relate the two limits, consider the equation (valid for $x \neq c$)

$$f(x) - f(c) = (x - c)\frac{f(x) - f(c)}{x - c}$$

Both factors on the right approach a limit as $x \to c$, so

$$\lim_{x \to c} \big(f(x) - f(c)\big) = \lim_{x \to c} \left((x - c)\,\frac{f(x) - f(c)}{x - c}\right)$$

$$= \left(\lim_{x \to c}(x - c)\right)\left(\lim_{x \to c}\frac{f(x) - f(c)}{x - c}\right)$$

$$= 0 \cdot f'(c) = 0$$

by the Product Law for limits. The Sum Law now yields the desired conclusion:

$$\lim_{x \to c} f(x) = \lim_{x \to c}(f(x) - f(c)) + \lim_{x \to c} f(c) = 0 + f(c) = f(c) \qquad\blacksquare$$

Most of the functions encountered in this text are differentiable, but exceptions exist. As we saw in Example 5 in Section 3.1, the functions $f(x) = |x|$ and $g(x) = x^{1/3}$ are not differentiable at $x = 0$. Note that both of these functions are continuous at $x = 0$, and therefore they demonstrate that continuity at a point does not imply differentiability at the point (i.e., the converse of Theorem 4 does not hold).

Example 5 in Section 3.1 showed that $f(x) = |x|$ is not differentiable at 0 due to the corner in its graph, and $g(x) = x^{1/3}$ is not differentiable at $x = 0$ due to the vertical tangent. The function

$$h(x) = \begin{cases} x \sin \frac{1}{x} & \text{if } x \neq 0 \\ 0 & \text{if } x = 0 \end{cases}$$

is also continuous and not differentiable at $x = 0$, but its failure to be differentiable is more complicated than the situation with f and g (see Figure 8). In this case the secant lines from $(0, 0)$ to nearby points Q on the curve oscillate and do not settle down to a limiting tangent as Q approaches the origin (see Exercise 95).

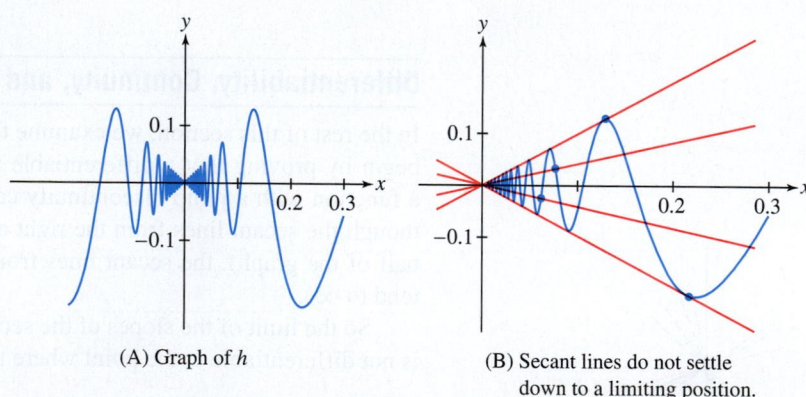

(A) Graph of h

(B) Secant lines do not settle down to a limiting position.

FIGURE 8

As we have seen with $f(x) = |x|$, $g(x) = x^{1/3}$, and h above, a function may not be differentiable at a point because of unusual behavior of the function near the point. On the other hand, as we see in the following Graphical Insight, differentiability at a point implies that a function behaves quite nicely near the point.

GRAPHICAL INSIGHT   Differentiability has an important graphical interpretation in terms of local linearity. We say that f is **locally linear** at $x = a$ if the graph looks more and more like a straight line as we zoom in on the point $(a, f(a))$. In this context, the adjective *linear* means "resembling a line," and *local* indicates that we are concerned only with the behavior of the graph near $(a, f(a))$. The graph of a locally linear function may be very wavy or *nonlinear*, as in Figure 9. But as soon as we zoom in on a sufficiently small piece of the graph, it begins to appear straight.

Not only does the graph look like a line as we zoom in on a point, but as Figure 9 suggests, the "zoom line" is the tangent line. Thus, the relation between differentiability and local linearity can be expressed as follows:

> If $f'(a)$ exists, then f is locally linear at $x = a$. That is, as we zoom in on the point $(a, f(a))$, the graph becomes nearly indistinguishable from its tangent line.

Local linearity gives us a graphical way to understand why $f(x) = |x|$ is not differentiable at $x = 0$. Figure 10 shows that the graph of $f(x) = |x|$ has a corner at $x = 0$, and this corner *does not disappear*, no matter how closely we zoom in on the origin. Since the graph does not straighten out under zooming, f is not locally linear at $x = 0$, reflecting that f is not differentiable at $x = 0$.

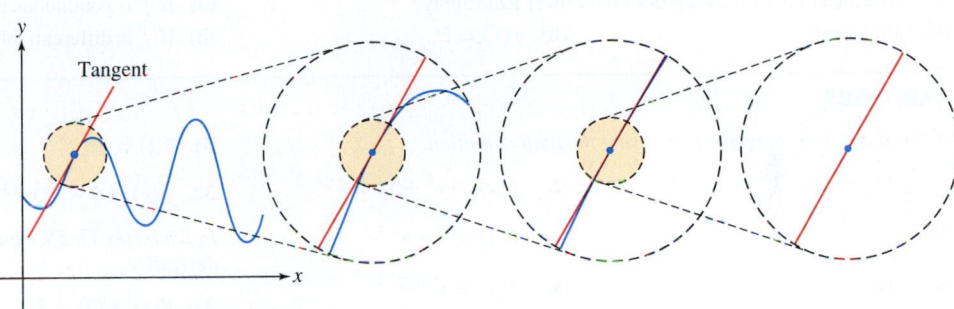

FIGURE 9 Local linearity: The graph looks more and more like the tangent line as we zoom in on a point.

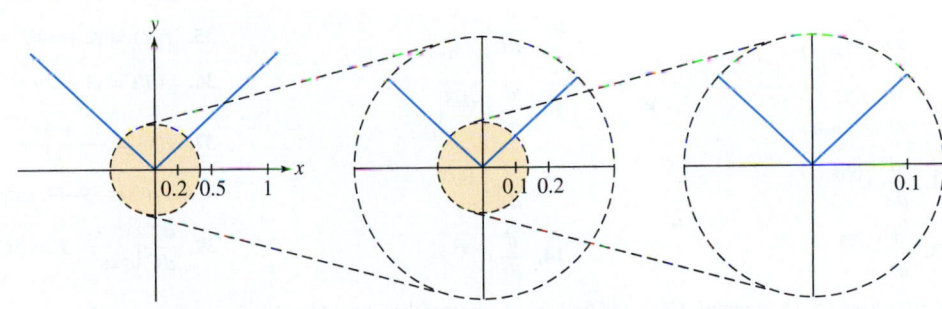

DF **FIGURE 10** The graph of $f(x) = |x|$ is not locally linear at $x = 0$. The corner does not disappear when we zoom in on the origin.

3.2 SUMMARY

- The derivative f' is the function whose value at x is the derivative $f'(x)$.
- We have several different notations for the derivative of $y = f(x)$:

$$y', \quad y'(x), \quad f'(x), \quad \frac{dy}{dx}, \quad \frac{df}{dx}$$

The value of the derivative at a particular point $x = a$ is written

$$y'(a), \quad f'(a), \quad \frac{dy}{dx}\bigg|_{x=a}, \quad \frac{df}{dx}\bigg|_{x=a}$$

- Derivative Rules

$$\text{The Constant Rule: } \frac{d}{dx}c = 0 \quad \text{The Power Rule: } \frac{d}{dx}x^n = nx^{n-1}$$

$$\text{The Exponential Function Rule: } \frac{d}{dx}e^x = e^x$$

$$\text{The Linearity Rules: } (f+g)' = f' + g' \quad \text{and} \quad (cf)' = cf'$$

- Differentiability implies continuity: If f is differentiable at $x = a$, then f is continuous at $x = a$. However, there exist continuous functions that are not differentiable.
- If $f'(a)$ exists, then f is locally linear in the following sense: As we zoom in on the point $(a, f(a))$, the graph becomes nearly indistinguishable from its tangent line.

3.2 EXERCISES

Preliminary Questions

1. What is the slope of the tangent line through the point $(2, f(2))$ if $f'(x) = x^3$?

2. Evaluate $(f - g)'(1)$ and $(3f + 2g)'(1)$, assuming that $f'(1) = 3$ and $g'(1) = 5$.

3. To which of the following does the Power Rule apply?

(a) $f(x) = x^2$ **(b)** $f(x) = 2^e$

(c) $f(x) = x^e$ **(d)** $f(x) = e^x$

(e) $f(x) = x^x$ **(f)** $f(x) = x^{-4/5}$

4. State whether each claim is true or false. If false, give an example demonstrating that it is false.

(a) If f is continuous at a, then f is differentiable at a.

(b) If f is differentiable at a, then f is continuous at a.

Exercises

In Exercises 1–6, compute $f'(x)$ using the limit definition.

1. $f(x) = 3x - 7$

2. $f(x) = x^2 + 3x$

3. $f(x) = x^3$

4. $f(x) = 1 - x^{-1}$

5. $f(x) = x - \sqrt{x}$

6. $f(x) = x^{-1/2}$

In Exercises 7–14, use the Power Rule to compute the derivative.

7. $\dfrac{d}{dx} x^4 \Big|_{x=-2}$

8. $\dfrac{d}{dt} t^{-3} \Big|_{t=4}$

9. $\dfrac{d}{dt} t^{2/3} \Big|_{t=8}$

10. $\dfrac{d}{dt} t^{-2/5} \Big|_{t=1}$

11. $\dfrac{d}{dx} x^{0.35}$

12. $\dfrac{d}{dx} x^{14/3}$

13. $\dfrac{d}{dt} t^{\sqrt{17}}$

14. $\dfrac{d}{dt} t^{-\pi^2}$

In Exercises 15–18, compute $f'(x)$ and find an equation of the tangent line to the graph at $x = a$.

15. $f(x) = x^4$, $a = 2$

16. $f(x) = x^{-2}$, $a = 5$

17. $f(x) = 5x - 32\sqrt{x}$, $a = 4$

18. $f(x) = \sqrt[3]{x}$, $a = 8$

19. Calculate:

(a) $\dfrac{d}{dx} 12e^x$ **(b)** $\dfrac{d}{dt}(25t - 8e^t)$ **(c)** $\dfrac{d}{dt} e^{t-3}$

Hint for (c): Write e^{t-3} as $e^{-3}e^t$.

20. Find an equation of the tangent line to $y = 24e^x$ at $x = 2$.

In Exercises 21–32, calculate the derivative.

21. $f(x) = 2x^3 - 3x^2 + 5$

22. $f(x) = 2x^3 - 3x^2 + 2x$

23. $f(x) = 4x^{5/3} - 3x^{-2} - 12$

24. $f(x) = x^{5/4} + 4x^{-3/2} + 11x$

25. $g(z) = 7z^{-5/14} + z^{-5} + 9$

26. $h(t) = 6\sqrt{t} + \dfrac{1}{\sqrt{t}}$

27. $f(s) = \sqrt[4]{s} + \sqrt[3]{s}$

28. $W(y) = 6y^4 + 7y^{2/3}$

29. $g(x) = e^2$

30. $f(x) = 3e^x - x^3$

31. $h(t) = 5e^{t-3}$

32. $f(x) = 9 - 12x^{1/3} + 8e^x$

In Exercises 33–38, expand or simplify the function, and then calculate the derivative.

33. $P(s) = (4s - 3)^2$

34. $Q(r) = (1 - 2r)(3r + 5)$

35. $f(x) = (2 - x)(2 + x)$

36. $g(w) = (1 + 2w)^3$

37. $g(x) = \dfrac{x^2 + 4x^{1/2}}{x^2}$ **38.** $s(t) = \dfrac{1 - 2t}{t^{1/2}}$

In Exercises 39–44, calculate the derivative indicated.

39. $\dfrac{dT}{dC}\Big|_{C=8}$, $T = 3C^{2/3}$

40. $\dfrac{dP}{dV}\Big|_{V=-2}$, $P = \dfrac{7}{V}$

41. $\dfrac{ds}{dz}\Big|_{z=2}$, $s = 4z - 16z^2$

42. $\dfrac{dR}{dW}\Big|_{W=1}$, $R = W^\pi$

43. $\dfrac{dr}{dt}\Big|_{t=4}$, $r = t - e^t$

44. $\dfrac{dp}{dh}\Big|_{h=4}$, $p = 7e^{h-2}$

45. Match the functions in graphs (A)–(D) with their derivatives (I)–(III) in Figure 11. Note that two of the functions have the same derivative. Explain why.

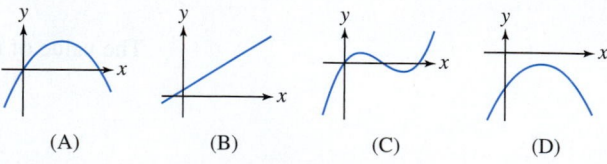

(A)   (B)   (C)   (D)

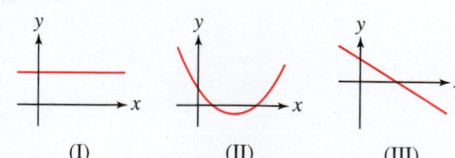

(I)   (II)   (III)

FIGURE 11

46. ✎ Of the two functions f and g in Figure 12, which is the derivative of the other? Justify your answer.

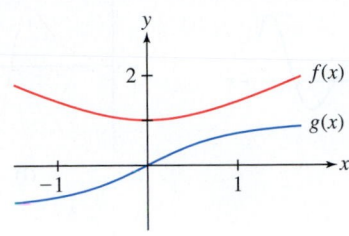

FIGURE 12

47. Assign the labels $y = f(x)$, $y = g(x)$, and $y = h(x)$ to the graphs in Figure 13 in such a way that $f'(x) = g(x)$ and $g'(x) = h(x)$.

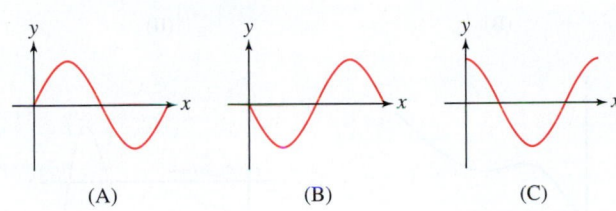

(A)   (B)   (C)

FIGURE 13

48. Prove each of the following using the definition of the derivative.

(a) The First-Power Rule: $\frac{d}{dx}x = 1$

(b) The Constant Rule: $\frac{d}{dx}c = 0$

49. Use the rules in Exercise 48 and the Linearity Rules to prove the first part of Theorem 1 in Section 3.1.

50. According to the *Peak Oil Theory*, first proposed in 1956 by geophysicist M. Hubbert, the total amount of crude oil $Q(t)$ produced worldwide up to time t has a graph like that in Figure 14.

(a) Sketch the derivative $Q'(t)$ for $1900 \leq t \leq 2150$. What does $Q'(t)$ represent?

(b) In which year (approximately) does $Q'(t)$ take on its maximum value?

(c) What is $L = \lim_{t \to \infty} Q(t)$? And what is its interpretation?

(d) What is the value of $\lim_{t \to \infty} Q'(t)$?

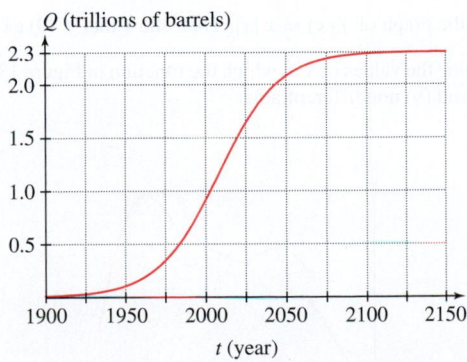

FIGURE 14 Total oil production up to time t.

51. ✎ Use the table of values of f to determine which of (A) or (B) in Figure 15 is the graph of f'. Explain.

x	0	0.5	1	1.5	2	2.5	3	3.5	4
$f(x)$	10	55	98	139	177	210	237	257	268

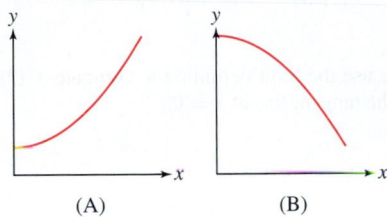

(A)   (B)

FIGURE 15 Which is the graph of f'?

52. Let R be a variable and r a constant. Compute the derivatives:

(a) $\dfrac{d}{dR} R$

(b) $\dfrac{d}{dR} r$

(c) $\dfrac{d}{dR} r^2 R^3$

53. Compute the derivatives, where c is a constant.

(a) $\dfrac{d}{dt} ct^3$

(b) $\dfrac{d}{dz} (5z + 4cz^2)$

(c) $\dfrac{d}{dy} (9c^2 y^3 - 24c)$

54. Find the points on the graph of $f(x) = 12x - x^3$ where the tangent line is horizontal.

55. Find the points on the graph of $y = x^2 + 3x - 7$ at which the slope of the tangent line is equal to 4.

56. Find the values of x where $y = x^3$ and $y = x^2 + 5x$ have parallel tangent lines.

57. Determine a and b such that

$$p(x) = x^2 + ax + b$$

satisfies $p(1) = 0$ and $p'(1) = 4$.

58. Find all values of x such that the tangent line to

$$y = 4x^2 + 11x + 2$$

is steeper than the tangent line to $y = x^3$.

59. Let $f(x) = x^3 - 3x + 1$. Show that $f'(x) \geq -3$ for all x and that, for every $m > -3$, there are precisely two points where $f'(x) = m$. Indicate the position of these points and the corresponding tangent lines for one value of m in a sketch of the graph of f.

60. Show that the tangent lines to $y = \frac{1}{3}x^3 - x^2$ at $x = a$ and at $x = b$ are parallel if $a = b$ or $a + b = 2$.

61. Compute the derivative of $f(x) = x^{3/2}$ using the limit definition. *Hint:* Show that

$$\frac{f(x+h) - f(x)}{h} = \frac{(x+h)^3 - x^3}{h} \left(\frac{1}{\sqrt{(x+h)^3} + \sqrt{x^3}} \right)$$

62. Compute the derivative of $f(x) = x^{1/3}$ using the limit definition. *Hint:* Multiply the numerator and denominator in the difference quotient $\frac{f(x+h)-f(x)}{h}$ by

$$(x + h)^{2/3} + (x + h)^{1/3}x^{1/3} + x^{2/3}$$

63. In each case use the limit definition to compute $f'(0)$, and then find the equation of the tangent line at $x = 0$.

(a) $f(x) = xe^x$

(b) $f(x) = x^2 e^x$

64. The average speed (in meters per second) of a gas molecule is

$$v_{\text{avg}} = \sqrt{\frac{8RT}{\pi M}}$$

where T is the temperature (in kelvins), M is the molar mass (in kilograms per mole), and $R = 8.31$. Calculate dv_{avg}/dT at $T = 300$ K for oxygen, which has a molar mass of 0.032 kg/mol.

65. The brightness b of the sun (in watts per square meter) at a distance of d meters from the sun is expressed as an inverse-square law in the form $b = \frac{L}{4\pi d^2}$ where L is the luminosity of the sun and equals 3.9×10^{26} watts. What is the derivative of b with respect to d at the earth's distance from the sun (1.5×10^{11} m)?

66. A power law model relating the kidney mass K in mammals (in kilograms) to the body mass m (in kilograms) is given by $K = 0.007m^{0.85}$. Calculate dK/dm at $m = 68$. Then calculate the derivative with respect to m of the relative kidney-to-mass ratio K/m at $m = 68$.

67. The Clausius–Clapeyron Law relates the *vapor pressure* of water P (in atmospheres) to the temperature T (in kelvins):

$$\frac{dP}{dT} = k\frac{P}{T^2}$$

where k is a constant. Estimate dP/dT for $T = 303, 313, 323, 333, 343$ using the data and the symmetric difference approximation

$$\frac{dP}{dT} \approx \frac{P(T + 10) - P(T - 10)}{20}$$

T (K)	293	303	313	323	333	343	353
P (atm)	0.0278	0.0482	0.0808	0.1311	0.2067	0.3173	0.4754

Do your estimates seem to confirm the Clausius–Clapeyron Law? What is the approximate value of k?

68. Let L be the tangent line to the hyperbola $xy = 1$ at $x = a$, where $a > 0$. Show that the area of the triangle bounded by L and the coordinate axes does not depend on a.

69. In the setting of Exercise 68, show that the point of tangency is the midpoint of the segment of L lying in the first quadrant.

70. Match functions (A)–(C) with their derivatives (I)–(III) in Figure 16.

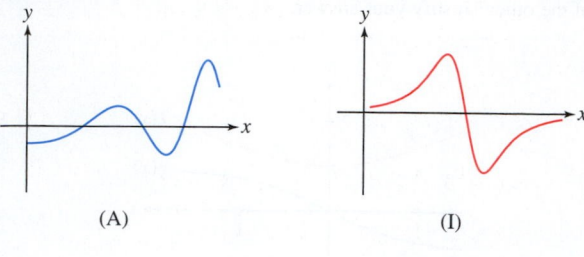

(A)   (I)

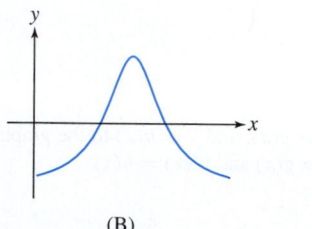
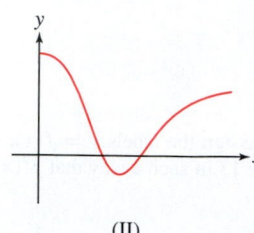

(B)   (II)

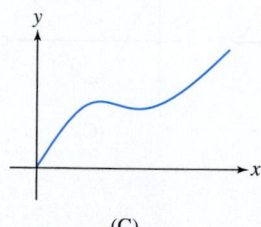
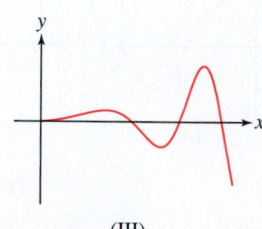

(C)   (III)

FIGURE 16

71. Make a rough sketch of the graph of the derivative of the function in Figure 17(A).

72. Graph the derivative of the function in Figure 17(B), omitting points where the derivative is not defined.

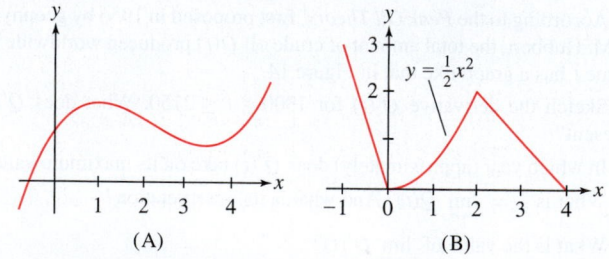

(A)   (B)

FIGURE 17

73. Sketch the graph of $f(x) = x|x|$. Then show that $f'(0)$ exists.

74. Determine the values of x at which the function in Figure 18 is: (a) discontinuous and (b) nondifferentiable.

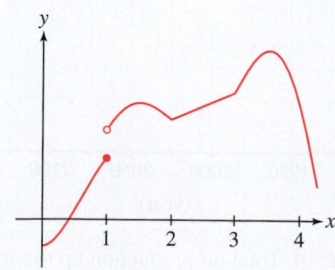

FIGURE 18

In Exercises 75–80, zoom in on a plot of f *at the point* $(a, f(a))$ *and state whether or not* f *appears to be differentiable at* $x = a$. *If it is nondifferentiable, state whether the tangent line appears to be vertical or does not exist.*

75. $f(x) = (x - 1)|x|$, $a = 0$

76. $f(x) = (x - 3)^{5/3}$, $a = 3$

77. $f(x) = (x - 3)^{1/3}$, $a = 3$

78. $f(x) = \sin(x^{1/3})$, $a = 0$

79. $f(x) = |\sin x|$, $a = 0$

80. $f(x) = |x - \sin x|$, $a = 0$

81. Find the coordinates of the point P in Figure 19 at which the tangent line passes through $(5, 0)$.

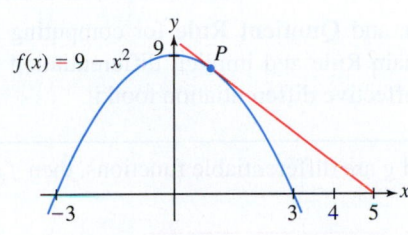

$f(x) = 9 - x^2$

FIGURE 19

82. GU Plot the derivative f' of $f(x) = 2x^3 - 10x^{-1}$ for $x > 0$ and observe that $f'(x) > 0$. What does the positivity of $f'(x)$ tell us about the graph of f itself? Plot f and confirm this conclusion.

Exercises 83–86 refer to Figure 20. Length QR *is called the* subtangent *at* P, *and length* RT *is called the* subnormal.

83. Calculate the subtangent of
$$f(x) = x^2 + 3x \quad \text{at } x = 2$$

84. Show that the subtangent of $f(x) = e^x$ is everywhere equal to 1.

85. Prove in general that the subnormal at P is $|f'(x)f(x)|$.

86. Show that $\overline{PQ}$ has length $|f(x)|\sqrt{1 + f'(x)^{-2}}$.

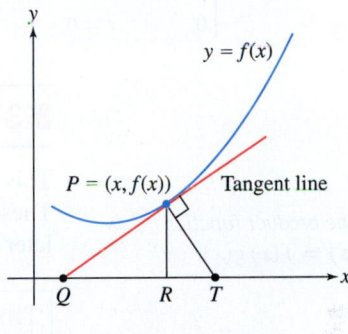

$y = f(x)$

$P = (x, f(x))$ Tangent line

FIGURE 20

87. Prove the following theorem of Apollonius of Perga (the Greek mathematician born in 262 BCE who gave the parabola, ellipse, and hyperbola their names): The subtangent of the parabola $y = x^2$ at $x = a$ is equal to $a/2$.

88. Show that the subtangent to $y = x^3$ at $x = a$ is equal to $a/3$.

89. ✎ Formulate and prove a generalization of Exercises 87 and 88 for $y = x^n$.

Further Insights and Challenges

90. Two small arches have the shape of parabolas. The first is the graph of $f(x) = 1 - x^2$ for $-1 \le x \le 1$ and the second is the graph of $g(x) = 4 - (x - 4)^2$ for $2 \le x \le 6$. A board is placed on top of these arches so it rests on both (Figure 21). What is the slope of the board? *Hint:* Find the tangent line to $y = f(x)$ that intersects $y = g(x)$ in exactly one point.

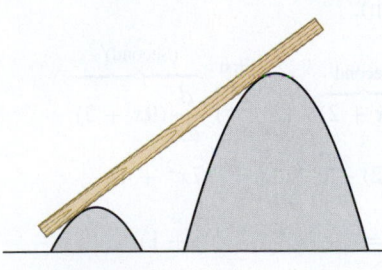

FIGURE 21

91. A vase is formed by rotating $y = x^2$ around the y-axis. If we drop in a marble, it will either touch the bottom point of the vase or be suspended above the bottom by touching the sides (Figure 22). How small must the marble be to touch the bottom?

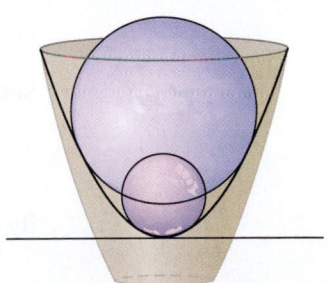

FIGURE 22

92. ✎ Let f be a differentiable function, and set the function $g(x) = f(x + c)$, where c is a constant. Use the limit definition to show that $g'(x) = f'(x + c)$. Explain this result graphically, recalling that the graph of g is obtained by shifting the graph of f c units to the left (if $c > 0$) or right (if $c < 0$).

93. Negative Exponents Let n be a whole number. Calculate the derivative of $f(x) = x^{-n}$ by showing that
$$\frac{f(x + h) - f(x)}{h} = \frac{-1}{x^n(x + h)^n} \cdot \frac{(x + h)^n - x^n}{h}$$

94. Verify the Power Rule for the exponent $1/n$, where n is a positive integer, using the following trick: Rewrite the difference quotient for $y = x^{1/n}$ at $x = b$ in terms of

$$u = (b + h)^{1/n} \quad \text{and} \quad a = b^{1/n}$$

95. Infinitely Rapid Oscillations   Define

$$h(x) = \begin{cases} x \sin \dfrac{1}{x} & x \neq 0 \\ 0 & x = 0 \end{cases}$$

Show that h is continuous at $x = 0$ but $h'(0)$ does not exist (see Figure 8).

96. For which value of λ does the equation $e^x = \lambda x$ have a unique solution? For which values of λ does it have at least one solution? For intuition, plot $y = e^x$ and the line $y = \lambda x$.

97. If $\lim\limits_{h \to 0^-} \dfrac{f(c+h) - f(c)}{h}$ and $\lim\limits_{h \to 0^+} \dfrac{f(c+h) - f(c)}{h}$ exist but are not equal, then f is not differentiable at c, and the graph of f has a corner at c. Prove that f is continuous at c.

3.3 Product and Quotient Rules

This section covers the **Product Rule** and **Quotient Rule** for computing derivatives. These two rules, together with the Chain Rule and implicit differentiation (covered in later sections), make up an extremely effective differentiation toolkit.

← **REMINDER** *The product function fg is defined by $(fg)(x) = f(x)\,g(x)$.*

THEOREM 1   Product Rule   If f and g are differentiable functions, then fg is differentiable and

$$(fg)'(x) = f'(x)\,g(x) + f(x)\,g'(x)$$

It may be helpful to remember the Product Rule in words: The derivative of a product of terms is equal to *the derivative of the first term times the second plus the first term times the derivative of the second*:

$$(\text{first})' \cdot \text{second} + \text{first} \cdot (\text{second})'$$

Be careful when taking the derivative of products. The product rule is *not* $(fg)' = f'g'$; that is, it does *not* say that the derivative of a product is the product of the derivatives.

We prove the Product Rule after presenting some examples.

EXAMPLE 1   Find the derivative of $h(x) = x^2(9x + 2)$.

Solution   This function is a product:

$$h(x) = \overbrace{x^2}^{\text{First}} \overbrace{(9x + 2)}^{\text{Second}}$$

By the Product Rule (in Leibniz notation),

$$h'(x) = \overbrace{\frac{d}{dx}(x^2)}^{(\text{First})'} \overbrace{(9x + 2)}^{\text{Second}} + \overbrace{(x^2)}^{\text{First}} \overbrace{\frac{d}{dx}(9x + 2)}^{(\text{Second})'}$$

$$= (2x)(9x + 2) + (x^2)(9) = 27x^2 + 4x \qquad \blacksquare$$

EXAMPLE 2   Find the derivative of $y = (2 + x^{-1})(x^{3/2} + 1)$.

Solution   Use the Product Rule:

Note how the prime notation is used in the solution to Example 2. We write $(x^{3/2} + 1)'$ to denote the derivative of $x^{3/2} + 1$.

$$y' = \overbrace{(2 + x^{-1})'(x^{3/2} + 1) + (2 + x^{-1})(x^{3/2} + 1)'}^{(\text{First})' \cdot \text{Second} + \text{First} \cdot (\text{Second})'}$$

$$= (-x^{-2})(x^{3/2} + 1) + (2 + x^{-1})\left(\frac{3}{2}x^{1/2}\right) \quad \text{(compute the derivatives)}$$

$$= -x^{-1/2} - x^{-2} + 3x^{1/2} + \frac{3}{2}x^{-1/2} = \frac{1}{2}x^{-1/2} - x^{-2} + 3x^{1/2} \quad \text{(simplify)} \qquad \blacksquare$$

In the previous two examples, we could have avoided the Product Rule by expanding the function. Thus, the result of Example 2 can be obtained as follows:

$$y = \left(2 + x^{-1}\right)\left(x^{3/2} + 1\right) = 2x^{3/2} + 2 + x^{1/2} + x^{-1}$$

$$y' = \frac{d}{dx}\left(2x^{3/2} + 2 + x^{1/2} + x^{-1}\right) = 3x^{1/2} + \tfrac{1}{2}x^{-1/2} - x^{-2}$$

In the next example, the function cannot be expanded, so we must use the Product Rule (or go back to the limit definition of the derivative).

EXAMPLE 3 Calculate $\dfrac{d}{dt}\left(t^2 e^t\right)$.

Solution Use the Product Rule and the formula $\dfrac{d}{dt} e^t = e^t$:

$$\frac{d}{dt}\left(t^2 e^t\right) = \overbrace{\left(\frac{d}{dt}\, t^2\right) e^t + t^2 \left(\frac{d}{dt}\, e^t\right)}^{\text{(First)}' \cdot \text{Second} + \text{First} \cdot \text{(Second)}'} = 2te^t + t^2(e^t) = (2t + t^2)e^t \qquad \blacksquare$$

EXAMPLE 4 Figure 1 depicts a rectangle whose length $L(t)$ and width $W(t)$ (measured in inches) are varying in time (t, in minutes). At $t = 5$, the length is 8, the width is 5, and they are changing according to $L'(5) = -4$ and $W'(5) = 3$. Compute $A'(5)$.

Solution Since the area is given by $A(t) = L(t)W(t)$, we can use the product rule to compute $A'(t)$. We have $A'(t) = L'(t)W(t) + L(t)W'(t)$. Therefore,

$$A'(5) = (-4)(5) + (8)(3) = 4 \qquad \blacksquare$$

It follows that the area of the rectangle in the example is increasing at a rate of 4 in.^{2}/min at $t = 5$. This may appear counterintuitive, given that the length is decreasing at a faster rate than the width is increasing. What is important, as the Product Rule demonstrates, is that the decreasing length acts over a short width of 5, contributing -20 to the rate of change of area, while the increasing width acts over a long length of 8, contributing 24 to the rate of change of area, resulting in an increasing area.

Proof of the Product Rule According to the limit definition of the derivative,

$$(fg)'(x) = \lim_{h \to 0} \frac{f(x + h)g(x + h) - f(x)g(x)}{h}$$

We can interpret the numerator as the area of the shaded region in Figure 2: the area of the larger rectangle $f(x + h)g(x + h)$ minus the area of the smaller rectangle $f(x)g(x)$. This shaded region is the union of two rectangular strips, so we obtain the following identity [which we can also obtain algebraically by adding and subtracting the term $f(x + h)g(x)$ from the left-hand side and then manipulating the result algebraically]:

$$f(x + h)g(x + h) - f(x)g(x) = \big(f(x + h) - f(x)\big)g(x) + f(x + h)\big(g(x + h) - g(x)\big)$$

Now use this identity to write $(fg)'(x)$ as a sum of two limits:

$$(fg)'(x) = \underbrace{\lim_{h \to 0} \frac{f(x + h) - f(x)}{h} g(x)}_{\text{We show that this equals } f'(x)g(x).} + \underbrace{\lim_{h \to 0} f(x + h) \frac{g(x + h) - g(x)}{h}}_{\text{We show that this equals } f(x)g'(x).} \qquad \boxed{1}$$

The use of the Sum Law is valid, provided that each limit on the right exists. To check that the first limit exists and to evaluate it, we note that f is differentiable. Thus,

$$\lim_{h \to 0} \frac{f(x + h) - f(x)}{h} g(x) = \lim_{h \to 0} \frac{f(x + h) - f(x)}{h} \lim_{h \to 0} g(x)$$

$$= f'(x)\, g(x) \qquad \boxed{2}$$

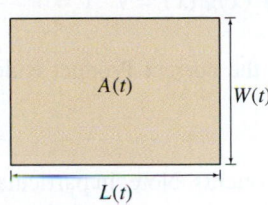

FIGURE 1 The length and width of the rectangle change in time.

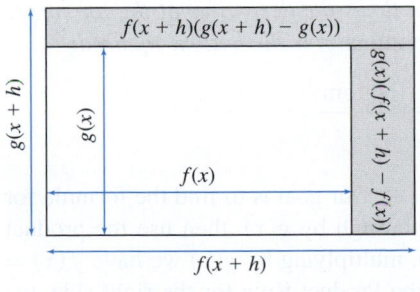

FIGURE 2

The second limit is similar, but using the facts that f is continuous (because it is differentiable) and g is differentiable:

$$\lim_{h\to 0} f(x+h) \frac{g(x+h)-g(x)}{h} = \lim_{h\to 0} f(x+h) \lim_{h\to 0} \frac{g(x+h)-g(x)}{h} = f(x)g'(x)$$

3

Using Eq. (2) and Eq. (3) in Eq. (1), we conclude that fg is differentiable and that $(fg)'(x) = f'(x)g(x) + f(x)g'(x)$ as claimed.   ∎

CONCEPTUAL INSIGHT   The Product Rule was first stated by the 29-year-old Leibniz in 1675, the year he developed some of his major ideas on calculus. To document his process of discovery for posterity, he recorded his thoughts and struggles, the moments of inspiration as well as the mistakes. In a manuscript dated November 11, 1675, Leibniz suggests *incorrectly* that $(fg)'$ equals $f'g'$. He then catches his error by taking $f(x) = g(x) = x$ and noticing that

$$(fg)'(x) = \left(x^2\right)' = 2x \qquad \text{is } not \text{ equal to} \qquad f'(x)g'(x) = 1 \cdot 1 = 1$$

Ten days later, on November 21, Leibniz writes down the correct Product Rule and comments, "*Now this is a really noteworthy theorem.*"

← **REMINDER** *The quotient function f/g is defined by*

$$\left(\frac{f}{g}\right)(x) = \frac{f(x)}{g(x)}$$

The next theorem states the rule for differentiating quotients. Note, in particular, that $(f/g)'$ is *not* equal to the quotient f'/g'; the derivative of the quotient is *not* the quotient of the derivatives.

THEOREM 2 Quotient Rule   If f and g are differentiable functions, then f/g is differentiable for all x such that $g(x) \neq 0$, and

$$\left(\frac{f}{g}\right)'(x) = \frac{g(x)f'(x) - f(x)g'(x)}{g(x)^2}$$

The numerator in the Quotient Rule is *the bottom times the derivative of the top minus the top times the derivative of the bottom.* The denominator is *the bottom squared*:

$$\frac{\text{bottom} \cdot (\text{top})' - \text{top} \cdot (\text{bottom})'}{\text{bottom}^2}$$

Proof of the Quotient Rule   Let $Q(x) = f(x)/g(x)$. Our goal is to find the formula for $Q'(x)$. First, we multiply the equation for $Q(x)$ through by $g(x)$, then use the product rule on the result, and finally solve for $Q'(x)$. So, multiplying by $g(x)$ we have $f(x) = Q(x) \cdot g(x)$. Differentiating both sides, utilizing the Product Rule for the right side, we obtain $f'(x) = Q'(x) \cdot g(x) + Q(x) \cdot g'(x)$. Solving for $Q'(x)$, we obtain

$$Q'(x) = \frac{f'(x) - Q(x) \cdot g'(x)}{g(x)} = \frac{f'(x) - \frac{f(x)}{g(x)} \cdot g'(x)}{g(x)} = \frac{f'(x)g(x) - f(x)g'(x)}{g(x)^2}$$

as we wanted to show.   ∎

An alternative proof appears in Exercises 66–68.

EXAMPLE 5 Compute the derivative of $f(x) = \dfrac{x}{1+x^2}$.

Solution Apply the Quotient Rule:

$$f'(x) = \frac{\overbrace{(1+x^2)}^{\text{Bottom}}\,\overbrace{(x)'}^{\text{(Top)}'} - \overbrace{(x)}^{\text{Top}}\,\overbrace{(1+x^2)'}^{\text{(Bottom)}'}}{(1+x^2)^2} = \frac{(1+x^2)(1) - (x)(2x)}{(1+x^2)^2}$$

$$= \frac{1 + x^2 - 2x^2}{(1+x^2)^2} = \frac{1 - x^2}{(1+x^2)^2} \qquad \blacksquare$$

EXAMPLE 6 Calculate $\dfrac{d}{dt}\left(\dfrac{e^t}{e^t + t}\right)$.

Note that it is not always the simplest method to apply the Quotient Rule. If we want to differentiate the function

$w(x) = \dfrac{\sqrt{x} - x^3}{x}$, it is easier to rewrite it

as $w(x) = x^{-1/2} - x^2$ and then differentiate it directly.

Solution Use the Quotient Rule and the formula $(e^t)' = e^t$:

$$\frac{d}{dt}\left(\frac{e^t}{e^t + t}\right) = \frac{(e^t + t)(e^t)' - e^t(e^t + t)'}{(e^t + t)^2} = \frac{(e^t + t)e^t - e^t(e^t + 1)}{(e^t + t)^2} = \frac{(t-1)e^t}{(e^t + t)^2} \qquad \blacksquare$$

EXAMPLE 7 Find the tangent line to the graph of $f(x) = \dfrac{3x^2 + x - 2}{4x^3 + 1}$ at $x = 1$.

Solution

$$f'(x) = \frac{d}{dx}\left(\frac{3x^2 + x - 2}{4x^3 + 1}\right) = \frac{\overbrace{(4x^3 + 1)}^{\text{Bottom}}\,\overbrace{(3x^2 + x - 2)'}^{\text{(Top)}'} - \overbrace{(3x^2 + x - 2)}^{\text{Top}}\,\overbrace{(4x^3 + 1)'}^{\text{(Bottom)}'}}{(4x^3 + 1)^2}$$

$$= \frac{(4x^3 + 1)(6x + 1) - (3x^2 + x - 2)(12x^2)}{(4x^3 + 1)^2}$$

$$= \frac{(24x^4 + 4x^3 + 6x + 1) - (36x^4 + 12x^3 - 24x^2)}{(4x^3 + 1)^2}$$

$$= \frac{-12x^4 - 8x^3 + 24x^2 + 6x + 1}{(4x^3 + 1)^2}$$

At $x = 1$,

$$f(1) = \frac{3 + 1 - 2}{4 + 1} = \frac{2}{5}$$

$$f'(1) = \frac{-12 - 8 + 24 + 6 + 1}{5^2} = \frac{11}{25}$$

An equation of the tangent line at $\left(1, \frac{2}{5}\right)$ is

$$y - \frac{2}{5} = \frac{11}{25}(x - 1) \qquad \text{or} \qquad y = \frac{11}{25}x - \frac{1}{25} \qquad \blacksquare$$

EXAMPLE 8 **Power Delivered by a Battery** The power that a battery supplies to an apparatus such as a laptop depends on the *internal resistance* of the battery. For a battery of voltage V and internal resistance r, the total power delivered to an apparatus of resistance R (Figure 3) is

$$P = \frac{V^2 R}{(R + r)^2}$$

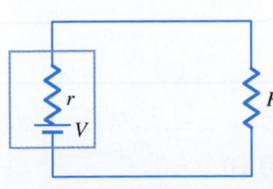

FIGURE 3 Apparatus of resistance R attached to a battery of voltage V.

(a) Calculate dP/dR, assuming that V and r are constants.

(b) Where, in the graph of P versus R, is the tangent line horizontal?

Solution

(a) Using the Constant Multiple Rule (V is a constant) and the Quotient Rule, we obtain

$$\frac{dP}{dR} = V^2 \frac{d}{dR}\left(\frac{R}{(R+r)^2}\right) = V^2 \frac{(R+r)^2 \frac{d}{dR}R - R\frac{d}{dR}(R+r)^2}{(R+r)^4} \qquad \boxed{4}$$

We have $\frac{d}{dR}R = 1$, and $\frac{d}{dR}r = 0$ because r is a constant. Therefore,

$$\frac{d}{dR}(R+r)^2 = \frac{d}{dR}(R^2 + 2rR + r^2)$$

$$= \frac{d}{dR}R^2 + 2r\frac{d}{dR}R + \frac{d}{dR}r^2$$

$$= 2R + 2r + 0 = 2(R+r) \qquad \boxed{5}$$

Using Eq. (5) in Eq. (4), we obtain

$$\frac{dP}{dR} = V^2 \frac{(R+r)^2 - 2R(R+r)}{(R+r)^4} = V^2 \frac{(R+r) - 2R}{(R+r)^3} = V^2 \frac{r - R}{(R+r)^3} \qquad \boxed{6}$$

(b) The tangent line is horizontal when the derivative is zero. We see from Eq. (6) that the derivative is zero when $r - R = 0$; that is, when $R = r$. ■

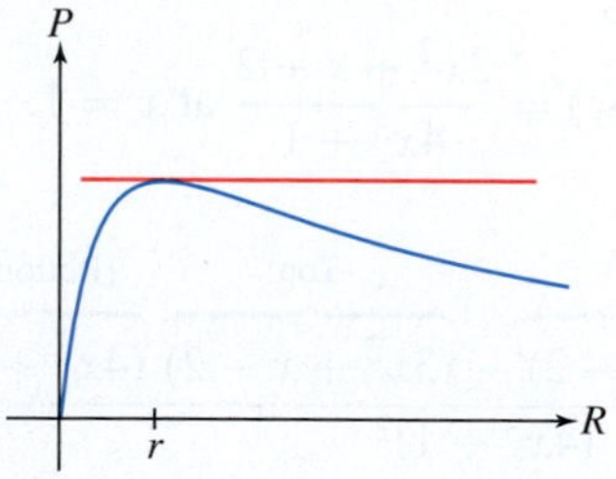

GRAPHICAL INSIGHT   Figure 4 shows that the point where the tangent line is horizontal is the *maximum point* on the graph. This proves an important result in circuit design: Maximum power is delivered when the resistance of the load (apparatus) is equal to the internal resistance of the battery.

FIGURE 4 Graph of power versus resistance:

$$P = \frac{V^2 R}{(R+r)^2}$$

3.3 SUMMARY

- Two basic rules of differentiation:

  Product Rule: $(fg)' = f'g + fg'$

  Quotient Rule: $\left(\dfrac{f}{g}\right)' = \dfrac{gf' - fg'}{g^2}$

- Remember: The derivative of fg is *not* equal to $f'g'$. Similarly, the derivative of f/g is *not* equal to f'/g'.

3.3 EXERCISES

Preliminary Questions

1. Are the following statements true or false? If false, state the correct version.

(a) fg denotes the function whose value at x is $f(g(x))$.

(b) f/g denotes the function whose value at x is $f(x)/g(x)$.

(c) The derivative of the product is the product of the derivatives.

(d) $\left.\dfrac{d}{dx}(fg)\right|_{x=4} = f(4)g'(4) - g(4)f'(4)$

(e) $\left.\dfrac{d}{dx}(fg)\right|_{x=0} = f'(0)g(0) + f(0)g'(0)$

2. Find $(f/g)'(1)$ if $f(1) = f'(1) = g(1) = 2$ and $g'(1) = 4$.

3. Find $g(1)$ if $f(1) = 0$, $f'(1) = 2$, and $(fg)'(1) = 10$.

Exercises

In Exercises 1–6, use the Product Rule to calculate the derivative.

1. $f(x) = x^3(2x^2 + 1)$

2. $f(x) = (3x - 5)(2x^2 - 3)$

3. $f(x) = x^2 e^x$

4. $f(x) = (2x - 9)(4e^x + 1)$

5. $\left.\dfrac{dh}{ds}\right|_{s=4}, \quad h(s) = (s^{-1/2} + 2s)(7 - s^{-1})$

6. $\left.\dfrac{dy}{dt}\right|_{t=2}, \quad y = (t - 8t^{-1})(e^t + t^2)$

In Exercises 7–12, use the Quotient Rule to calculate the derivative.

7. $f(x) = \dfrac{x}{x - 2}$

8. $f(x) = \dfrac{x + 4}{x^2 + x + 1}$

9. $\left.\dfrac{dg}{dt}\right|_{t=-2}, \quad g(t) = \dfrac{t^2 + 1}{t^2 - 1}$

10. $\left.\dfrac{dw}{dz}\right|_{z=9}, \quad w = \dfrac{z^2}{\sqrt{z} + z}$

11. $g(x) = \dfrac{1}{1 + e^x}$

12. $f(x) = \dfrac{e^x}{x^2 + 1}$

In Exercises 13–18, calculate the derivative in two ways. First use the Product or Quotient Rule; then rewrite the function algebraically and directly calculate the derivative.

13. $f(x) = x^3 x^{-3}$

14. $h(x) = \dfrac{x^5}{x^4}$

15. $f(t) = (2t + 1)(t^2 - 2)$

16. $f(x) = x^2(3 + x^{-1})$

17. $h(t) = \dfrac{t^2 - 1}{t - 1}$

18. $g(x) = \dfrac{x^3 + 2x^2 + 3x^{-1}}{x}$

In Exercises 19–40, calculate the derivative.

19. $f(x) = (x^3 + 5)(x^3 + x + 1)$

20. $f(x) = (4e^x - x^2)(x^3 + 1)$

21. $\left.\dfrac{dy}{dx}\right|_{x=3}, \quad y = \dfrac{1}{x + 10}$

22. $\left.\dfrac{dz}{dx}\right|_{x=-2}, \quad z = \dfrac{x}{3x^2 + 1}$

23. $f(x) = (\sqrt{x} + 1)(\sqrt{x} - 1)$

24. $f(x) = \dfrac{9x^{5/2} - 2}{x}$

25. $\left.\dfrac{dy}{dx}\right|_{x=2}, \quad y = \dfrac{x^4 - 4}{x^2 - 5}$

26. $f(x) = \dfrac{x^4 + e^x}{x + 1}$

27. $\left.\dfrac{dz}{dx}\right|_{x=1}, \quad z = \dfrac{1}{x^3 + 1}$

28. $f(x) = \dfrac{3x^3 - x^2 + 2}{\sqrt{x}}$

29. $h(t) = \dfrac{t}{(t + 1)(t^2 + 1)}$

30. $f(x) = x^{3/2}(2x^4 - 3x + x^{-1/2})$

31. $f(x) = x^2 e^2$

32. $h(x) = \pi^2(x - 1)$

33. $f(x) = (x + 3)(x - 1)(x - 5)$

34. $f(x) = e^x(x^2 + 1)(x + 4)$

35. $f(x) = \dfrac{e^x}{x + 1}$

36. $f(x) = \dfrac{x + 1}{e^x}$

37. $g(z) = \left(\dfrac{z^2 - 4}{z - 1}\right)\left(\dfrac{z^2 - 1}{z + 2}\right)$ *Hint:* Simplify first.

38. $\dfrac{d}{dx}\left((ax + b)(abx^2 + 1)\right)$ $(a, b$ constants)

39. $\dfrac{d}{dt}\left(\dfrac{xt - 4}{t^2 - x}\right)$ $(x$ constant)

40. $\dfrac{d}{dx}\left(\dfrac{ax + b}{cx + d}\right)$ $(a, b, c, d$ constants)

In Exercises 41–44, calculate $f'(x)$ in terms of $P(x)$, $Q(x)$, and $R(x)$, assuming that $P'(x) = Q(x)$, $Q'(x) = -R(x)$, and $R'(x) = P(x)$.

41. $f(x) = x R(x) + Q(x)$

42. $f(x) = Q(x)P(x)$

43. $f(x) = \dfrac{P(x)}{Q(x)} - x$

44. $f(x) = \dfrac{Q(x)R(x)}{P(x)}$

In Exercises 45–48, calculate the derivative using the values:

$f(4)$	$f'(4)$	$g(4)$	$g'(4)$
10	-2	5	-1

45. $(fg)'(4)$ and $(f/g)'(4)$

46. $F'(4)$, where $F(x) = x^2 f(x)$

47. $G'(4)$, where $G(x) = (g(x))^2$

48. $H'(4)$, where $H(x) = \dfrac{x}{g(x)f(x)}$

In Exercises 49 and 50, a rectangle's length $L(t)$ and width $W(t)$ (measured in inches) are varying in time (t, in minutes). Determine $A'(t)$ in each case. Is the area increasing or decreasing at that time?

49. At $t = 3$, we have $L(3) = 4$, $W(3) = 6$, $L'(3) = -4$, and $W'(3) = 5$.

50. At $t = 6$, we have $L(6) = 6$, $W(6) = 3$, $L'(6) = 5$, and $W'(6) = -2$.

51. Calculate $F'(0)$, where

$$F(x) = \dfrac{x^9 + x^8 + 4x^5 - 7x}{x^4 - 3x^2 + 2x + 1}$$

Hint: Do not calculate $F'(x)$. Instead, write $F(x) = f(x)/g(x)$ and express $F'(0)$ directly in terms of $f(0)$, $f'(0)$, $g(0)$, $g'(0)$.

52. Proceed as in Exercise 51 to calculate $F'(0)$, where

$$F(x) = \left(1 + x + x^{4/3} + x^{5/3}\right)\dfrac{3x^5 + 5x^4 + 5x + 1}{8x^9 - 7x^4 + 1}$$

53. Use the Product Rule to calculate $\dfrac{d}{dx} e^{2x}$.

54. **GU** Plot the derivative of $f(x) = x/(x^2 + 1)$ over $[-4, 4]$. Use the graph to determine the intervals on which $f'(x) > 0$ and $f'(x) < 0$. Then plot f and describe how the sign of $f'(x)$ is reflected in the graph of f.

55. **GU** Plot $f(x) = x/(x^2 - 1)$. Use the plot to determine whether $f'(x)$ is positive or negative on its domain $\{x : x \neq \pm 1\}$. Then compute $f'(x)$ and confirm your conclusion algebraically.

56. Let $P = V^2 R/(R + r)^2$ as in Example 8. Calculate dP/dr, assuming that r is variable and R is constant.

57. Find $a > 0$ such that the tangent line to the graph of

$$f(x) = x^2 e^{-x} \quad \text{at } x = a$$

passes through the origin (Figure 5).

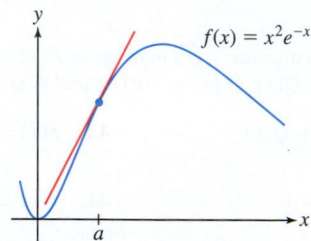

FIGURE 5

58. Current I (amperes), voltage V (volts), and resistance R (ohms) in a circuit are related by Ohm's Law, $I = V/R$.

(a) Calculate $\left.\dfrac{dI}{dR}\right|_{R=6}$ if V is constant with value $V = 24$.

(b) Calculate $\left.\dfrac{dV}{dR}\right|_{R=6}$ if I is constant with value $I = 4$.

59. The revenue per month earned by the Couture clothing chain at time t is $R(t) = N(t)S(t)$, where $N(t)$ is the number of stores and $S(t)$ is average revenue per store per month. Couture embarks on a two-part campaign: (A) to build new stores at a rate of five stores per month, and (B) to use advertising to increase average revenue per store at a rate of \$10,000 per month. Assume that $N(0) = 50$ and $S(0) = \$150,000$.

(a) Show that total revenue will increase at the rate

$$\frac{dR}{dt} = 5S(t) + 10{,}000N(t)$$

Note that the two terms in the Product Rule correspond to the separate effects of increasing the number of stores and the average revenue per store.

(b) Calculate $\left.\dfrac{dR}{dt}\right|_{t=0}$.

(c) If Couture can implement only one leg (A or B) of its expansion at $t = 0$, which choice will grow revenue most rapidly?

60. The **tip speed ratio** of a turbine is the ratio $R = T/W$, where T is the speed of the tip of a blade and W is the speed of the wind. (Engineers have found empirically that a turbine with n blades extracts maximum power from the wind when $R = 2\pi/n$.) Calculate dR/dt (t in minutes) if $W = 35$ km/h and W decreases at a rate of 4 km/h per minute, and the tip speed has constant value $T = 150$ km/h.

61. The curve $y = 1/(x^2 + 1)$ is called the *witch of Agnesi* (Figure 6) after the Italian mathematician Maria Agnesi (1718–1799). This strange name is the result of a mistranslation of the Italian word *la versiera*, meaning "that which turns." Find equations of the tangent lines at $x = \pm 1$.

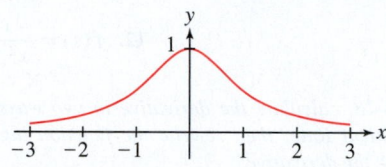

FIGURE 6 The witch of Agnesi.

62. Let $f(x) = g(x) = x$. Show that $(f/g)' \neq f'/g'$.

63. Use the Product Rule to show that $(f^2)' = 2ff'$.

64. Show that $(f^3)' = 3f^2 f'$.

Further Insights and Challenges

65. Let f, g, h be differentiable functions. Show that $(fgh)'(x)$ is equal to

$$f'(x)g(x)h(x) + f(x)g'(x)h(x) + f(x)g(x)h'(x)$$

Hint: Write fgh as $f(gh)$.

66. Prove the Quotient Rule using the limit definition of the derivative.

67. Derivative of the Reciprocal Use the limit definition to prove

$$\frac{d}{dx}\left(\frac{1}{f(x)}\right) = -\frac{f'(x)}{f^2(x)} \qquad \boxed{7}$$

Hint: Show that the difference quotient for $1/f(x)$ is equal to

$$\frac{f(x) - f(x + h)}{hf(x)f(x + h)}$$

68. Prove the Quotient Rule using Eq. (7) and the Product Rule.

69. Use the limit definition of the derivative to prove the following special case of the Product Rule:

$$\frac{d}{dx}(xf(x)) = f(x) + xf'(x)$$

70. Use the limit definition of the derivative to prove the following special case of the Quotient Rule:

$$\frac{d}{dx}\left(\frac{f(x)}{x}\right) = \frac{xf'(x) - f(x)}{x^2}$$

71. The Power Rule Revisited If you are familiar with *proof by induction*, use induction to prove the Power Rule for all whole numbers n. Show that the Power Rule holds for $n = 1$; then write x^n as $x \cdot x^{n-1}$ and use the Product Rule.

*Exercises 72 and 73: A basic fact of algebra states that c is a root of a polynomial f if and only if $f(x) = (x - c)g(x)$ for some polynomial g. We say that c is a **multiple root** if $f(x) = (x - c)^2 h(x)$, where h is a polynomial.*

72. Show that c is a multiple root of f if and only if c is a root of both f and f'.

73. Use Exercise 72 to determine whether $c = -1$ is a multiple root.
(a) $x^5 + 2x^4 - 4x^3 - 8x^2 - x + 2$
(b) $x^4 + x^3 - 5x^2 - 3x + 2$

74. 📝 Figure 7 is the graph of a polynomial with roots at A, B, and C. Which of these is a multiple root? Explain your reasoning using Exercise 72.

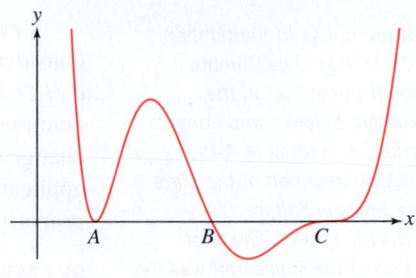

FIGURE 7

3.4 Rates of Change

In this section, we pause from building tools for computing the derivative and instead focus on the derivative as a rate of change, particularly in applied settings.

Recall the notation for the average rate of change of a function $y = f(x)$ over an interval $[x_0, x_1]$:

$$\Delta y = \text{change in } y = f(x_1) - f(x_0)$$

$$\Delta x = \text{change in } x = x_1 - x_0$$

$$\text{average rate of change} = \frac{\Delta y}{\Delta x} = \frac{f(x_1) - f(x_0)}{x_1 - x_0}$$

We usually omit the word "instantaneous" and refer to the derivative simply as the rate of change. This is shorter and also more accurate when applied to general rates, because the term "instantaneous" would seem to refer only to rates with respect to time.

In our prior discussion in Section 2.1, limits and derivatives had not yet been introduced. Now that we have them at our disposal, we can define the **instantaneous** rate of change of y with respect to x at $x = x_0$:

$$\text{instantaneous rate of change} = f'(x_0) = \lim_{\Delta x \to 0} \frac{\Delta y}{\Delta x} = \lim_{x_1 \to x_0} \frac{f(x_1) - f(x_0)}{x_1 - x_0}$$

Keep in mind the geometric interpretations: The average rate of change is the slope of the secant line (Figure 1), and the instantaneous rate of change is the slope of the tangent line (Figure 2).

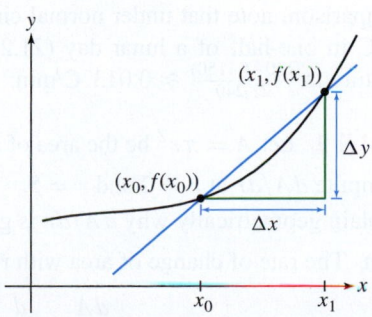

FIGURE 1 The average rate of change over $[x_0, x_1]$ is the slope of the secant line.

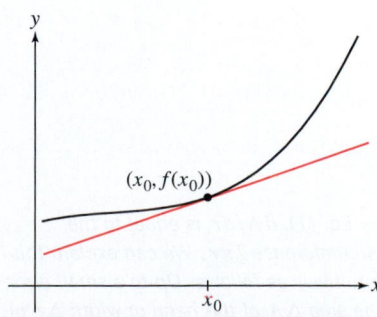

FIGURE 2 The instantaneous rate of change at x_0 is the slope of the tangent line.

FIGURE 3 A time lapse of a lunar eclipse.

Leibniz notation dy/dx is particularly convenient because it specifies that we are considering the rate of change of y with respect to the independent variable x. The rate dy/dx is measured in units of y per unit of x. For example, the rate of change of temperature with respect to time has units such as degrees per minute, whereas the rate of change of temperature with respect to altitude has units such as degrees per kilometer. In applications, it is important to be mindful of the units on rates of change and to interpret properly what the rate of change is communicating about the variables.

EXAMPLE 1   Surface Temperatures During an Eclipse   Since the moon has no atmosphere to help moderate the temperature, its surface experiences large extremes in temperature (typically between $-150°C$ and $120°C$). Furthermore, the primary source of heat is direct radiation from the sun, so a location experiences its highest temperatures when in direct sunlight and lowest when in darkness. With a "day" on the moon being 29.5 Earth days, a location on the moon will cycle between hottest and coldest temperatures over such a period of time. The only situation where this temperature-change cycle is disrupted is when the earth blocks the sun (that is, when a lunar eclipse is seen from Earth, as in Figure 3). During an eclipse, the temperature on the moon drops quickly, but then rebounds once the earth passes by the sun.

Table 1 contains data on the temperature (T, in °C) at a location on the moon t minutes into an eclipse.

TABLE 1

t	0	20	40	60	80	100	120	140	160	180	200	220	240	260	280	300
T	45	37	7	-33	-61	-78	-86	-91	-95	-97	-99	-101	-84	-33	12	37

(a) Calculate the average rate of change of temperature T from the start of the eclipse to the time t^* when the temperature was coldest and the average rate of change from t^* to the end of the eclipse.

(b) Use the difference quotient approximation to estimate the rate of change of the temperature 60, 160, and 260 min into the eclipse.

Solution

(a) From the table, we use $t^* = 220$. The average rate of change from eclipse start to t^* is $\frac{-101-45}{220-0} \approx -0.66°C/min$. The average rate of change from t^* to eclipse end is $\frac{37-(-101)}{300-220} \approx 1.73°C/min$.

(b) • The rate of change at $t = 60$ is approximately $\frac{-61-(-33)}{20} = -1.40°C/min$.

  • The rate of change at $t = 160$ is approximately $\frac{-97-(-95)}{20} = -0.10°C/min$.

  • The rate of change at $t = 260$ is approximately $\frac{12-(-33)}{20} = 2.25°C/min$.

For comparison, note that under normal circumstances if the moon heats from $-150°C$ to $120°C$ in one-half of a lunar day (21,240 min), then the average rate of change of temperature is $\frac{120-(-150)}{21,240} \approx 0.013°C/min$. ■

EXAMPLE 2   Let $A = \pi r^2$ be the area of a circle of radius r.

(a) Compute dA/dr at $r = 2$ and $r = 5$.

(b) Explain geometrically why dA/dr is greater at $r = 5$ than at $r = 2$.

Solution   The rate of change of area with respect to radius is the derivative

$$\frac{dA}{dr} = \frac{d}{dr}(\pi r^2) = 2\pi r \qquad \boxed{1}$$

(a) We have

$$\left.\frac{dA}{dr}\right|_{r=2} = 2\pi(2) \approx 12.57 \quad \text{and} \quad \left.\frac{dA}{dr}\right|_{r=5} = 2\pi(5) \approx 31.42$$

By Eq. (1), dA/dr is equal to the circumference $2\pi r$. We can explain this intuitively as follows: Up to a small error, the area ΔA of the band of width Δr in Figure 4 is equal to the circumference $2\pi r$ times the width Δr. Therefore, $\Delta A \approx 2\pi r \Delta r$ and

$$\frac{dA}{dr} = \lim_{\Delta r \to 0} \frac{\Delta A}{\Delta r} = 2\pi r$$

(b) The derivative dA/dr measures how the area of the circle changes when r increases. Figure 4 shows that when the radius increases by Δr, the area increases by a band of thickness Δr. The area of the band is greater at $r = 5$ than at $r = 2$. Therefore, the derivative is larger (and the tangent line is steeper) at $r = 5$. In general, for a fixed Δr, the change in area ΔA is greater when r is larger. ∎

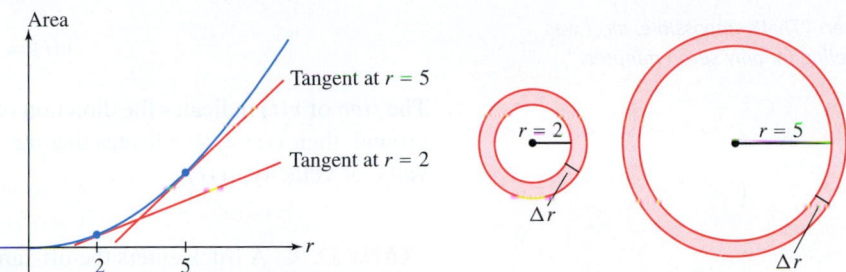

DF **FIGURE 4** The pink bands represent the change in area when r is increased by Δr.

Marginal Cost in Economics

Although $C(x)$ is meaningful only when x is a whole number, economists often treat $C(x)$ as a differentiable function of x so that the techniques of calculus can be applied. This is reasonable when the domain of C is large.

Let $C(x)$ denote the dollar cost (including labor and parts) of producing x units of a particular product. The number x of units manufactured is called the **production level**. To study the relation between costs and production, economists define the **marginal cost** at production level x_0 as the cost of producing one additional unit:

$$\text{marginal cost} = C(x_0 + 1) - C(x_0)$$

Note that if we use a difference quotient approximation for the derivative $C'(x_0)$ with $h = 1$, we obtain

$$C'(x_0) \approx \frac{C(x_0 + 1) - C(x_0)}{1} = C(x_0 + 1) - C(x_0)$$

and therefore we can use the derivative at x_0 as an approximation to the marginal cost.

EXAMPLE 3 **Cost of an Air Flight** Company data suggest that when there are 50 or more passengers, the total dollar cost of a certain flight is approximately $C(x) = 0.0005x^3 - 0.38x^2 + 120x$, where x is the number of passengers (Figure 5).

(a) Estimate the marginal cost of an additional passenger if the flight already has 150 passengers.

(b) Compare your estimate with the actual cost of an additional passenger.

(c) Is it more expensive to add a passenger when $x = 150$ or when $x = 200$?

Solution The derivative is $C'(x) = 0.0015x^2 - 0.76x + 120$.

(a) We estimate the marginal cost at $x = 150$ by the derivative

$$C'(150) = 0.0015(150)^2 - 0.76(150) + 120 = 39.75$$

Thus, it costs approximately \$39.75 to add one additional passenger.

(b) The actual cost of adding one additional passenger is

$$C(151) - C(150) \approx 11{,}177.10 - 11{,}137.50 = 39.60$$

Our estimate of \$39.75 is close enough for practical purposes.

(c) The marginal cost at $x = 200$ is approximately

$$C'(200) = 0.0015(200)^2 - 0.76(200) + 120 = 28$$

Since $39.75 > 28$, it is more expensive to add a passenger when $x = 150$ than when $x = 200$. ∎

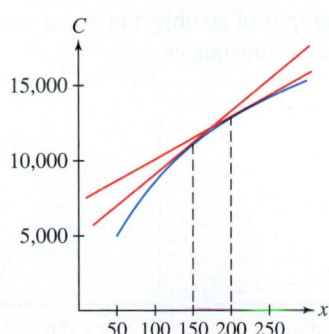

FIGURE 5 Cost of an air flight. The slopes of the tangent lines decrease as x increases, so the marginal cost decreases as well.

In his famous textbook Lectures on Physics, *Nobel laureate Richard Feynman (1918–1988) uses a dialogue to make a point about instantaneous velocity:*

Policeman: "My friend, you were going 75 miles an hour."

Driver: "That's impossible, sir, I was traveling for only seven minutes."

Linear Motion

Recall that *linear motion* is motion along a straight line. This includes horizontal motion along a straight highway and vertical motion of a falling object. Let $s(t)$ denote the position on a line, relative to the origin, at time t. **Velocity** is the rate of change of position with respect to time:

$$v(t) = \text{velocity} = \frac{ds}{dt}$$

The *sign* of $v(t)$ indicates the direction of motion. For example, if $s(t)$ is the height above ground, then $v(t) > 0$ indicates that the object is rising. **Speed** is defined as the absolute value of velocity, $|v(t)|$.

EXAMPLE 4   A truck enters the off-ramp of a highway at $t = 0$. Its position on the off-ramp after t seconds is $s(t) = 25t - 0.3t^3$ m for $0 \le t \le 5$.

(a) How fast is the truck going at the moment it enters the off-ramp?

(b) Is the truck speeding up or slowing down?

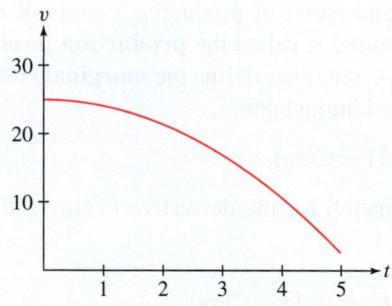

FIGURE 6 Graph of velocity $v(t) = 25 - 0.9t^2$.

Solution   The truck's velocity at time t is $v(t) = \dfrac{d}{dt}(25t - 0.3t^3) = 25 - 0.9t^2$.

(a) The truck enters the off-ramp with velocity $v(0) = 25$ m/s.

(b) Since $v(t) = 25 - 0.9t^2$ is decreasing and positive (Figure 6), the speed is decreasing and the truck is slowing down. ■

When we say "speeding up" or "slowing down" we typically are referring to the *speed* of an object, not its velocity. The relationship between speed and velocity is simple: Speed is the absolute value of velocity. Use care to apply velocity and speed properly when describing an object's motion. For instance, as the next example demonstrates, an object's velocity can increase while its speed decreases.

EXAMPLE 5   **Velocity and Speed**   Figure 7 shows graphs of an object in linear motion whose position s is changing in time t in four different circumstances.

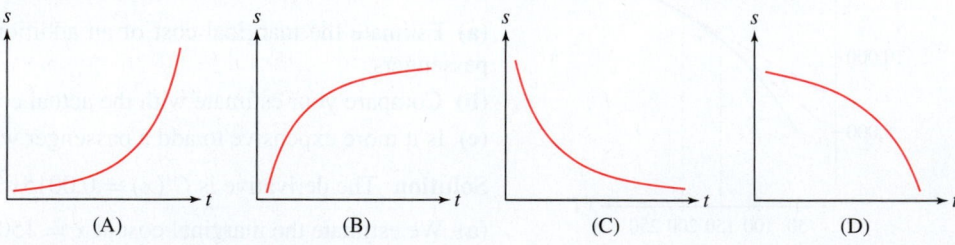

FIGURE 7

(a) In which cases is the velocity increasing? decreasing?

(b) In which cases is the speed increasing (so the object is speeding up)? decreasing (so the object is slowing down)?

◀ REMINDER *"Larger" means farther from 0, while "smaller" means closer to 0.*

Solution

(a)
- In Figure 7(A), the slope is positive and getting larger, so the velocity (or rate change) is increasing.
- In Figure 7(B), the slope is positive and getting smaller, so the velocity is decreasing.

- In Figure 7(C), the slope is negative and getting smaller; that is, getting closer to zero. Since the slope values are negative and approaching zero, the slope is increasing, and therefore the velocity is increasing.
- In Figure 7(D), the slope is negative and is getting larger in the negative direction, so the velocity is decreasing.

(b) Now we are considering the absolute value of the velocity; that is, the absolute value of the slope of the graph. It (and therefore speed) increases when the slope gets steeper, and that occurs in both Figures 7(A) and 7(D). Thus, in both of those cases the object is speeding up. On the other hand, in Figures 7(B) and 7(C), the slopes are getting less steep and therefore are getting smaller. Thus, in those cases the speed is decreasing and the object is slowing down.

Notice that Figure 7(C) depicts a situation where the velocity is increasing but the object is slowing down.   ■

Suppose s is the distance between a car and a wall during a crash test, and assume that during the test the car continued to speed up until it hit the wall. Which of the four graphs above best represents $s(t)$? This question, and others like it, are addressed in Exercises 15–16.

EXAMPLE 6 Describe the motion and velocities of a shuttle train that runs on a straight track at the airport, ferrying passengers from Terminal 1 to Terminal 2 according to the graph given in Figure 8. Assume that s represents the distance from Terminal 1 in meters, t represents time in minutes, and the terminals are 800 m apart.

Solution Note that the graph has portions resembling each of the four graphs in Example 5. Analyzing the motion:

- The train starts at rest, but then speeds up with increasing positive velocity for the first 2 min.
- Over the interval $[2,4]$, the velocity remains positive, but begins decreasing as the graph becomes less steep. The train is slowing down as it approaches Terminal 2.
- In the interval $[4,6]$, the graph is flat with slope 0. In this interval the train is stopped at Terminal 2.
- The train speeds up again at $t = 6$, now with negative velocity since the distance to Terminal 1 is decreasing. Furthermore, since the graph has a negative slope and is getting steeper, the velocity is decreasing and getting larger, indicating that the train is speeding up.
- Over the interval $[8,10]$, the velocity remains negative, but gets smaller as the graph become less steep. The train is slowing down as it approaches and arrives back at Terminal 1.   ■

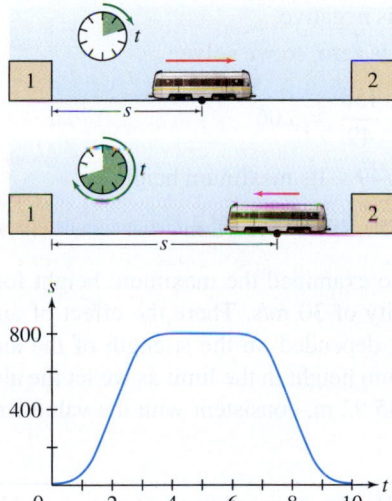

FIGURE 8

Motion Under the Influence of Gravity

Galileo discovered that the height $s(t)$ and velocity $v(t)$ at time t (seconds) of an object tossed vertically in the air near the earth's surface are accurately represented by the formulas

Galileo's formulas are valid only when air resistance is negligible. We assume this to be the case in all examples.

$$s(t) = s_0 + v_0 t - \frac{1}{2}gt^2, \qquad v(t) = \frac{ds}{dt} = v_0 - gt \qquad \boxed{2}$$

The constants s_0 and v_0 are the *initial values*:

- $s_0 = s(0)$, the position at time $t = 0$.
- $v_0 = v(0)$, the velocity at $t = 0$.
- $-g$ is the acceleration due to gravity on the surface of the earth (negative because the up direction is positive), where

$$g \approx 9.8 \text{ m/s}^2 \qquad \text{or} \qquad g \approx 32 \text{ ft/s}^2$$

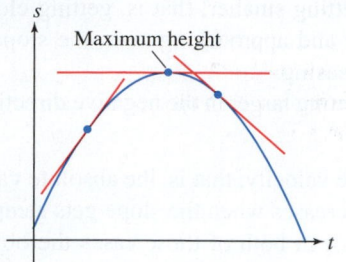

FIGURE 9 Maximum height occurs when $s'(t) = v(t) = 0$, where the tangent line is horizontal.

Galileo's formulas:

$$s(t) = s_0 + v_0 t - \frac{1}{2}gt^2$$

$$v(t) = \frac{ds}{dt} = v_0 - gt$$

A simple observation enables us to find the object's maximum height. Since velocity is positive as the object rises and negative as it falls back to Earth, the object reaches its maximum height at the moment of transition, when it is no longer rising and has not yet begun to fall. At that moment, its velocity is zero. In other words, *the maximum height is attained when $v(t) = 0$.* At this moment, the tangent line to the graph of s is horizontal (Figure 9).

EXAMPLE 7 **Finding the Maximum Height**   A projectile is launched upward from ground level with an initial velocity of 30 m/s.

(a) Find the velocity at $t = 2$ and at $t = 4$. Explain the change in sign.

(b) What is the projectile's maximum height and when does it reach that height?

Solution Apply Eq. (2) with $s_0 = 0$, $v_0 = 30$, and $g = 9.8$:

$$s(t) = 30t - 4.9t^2, \qquad v(t) = 30 - 9.8t$$

(a) Therefore,

$$v(2) = 30 - 9.8(2) = 10.4 \text{ m/s}, \qquad v(4) = 30 - 9.8(4) = -9.2 \text{ m/s}$$

At $t = 2$, the projectile is rising and its velocity $v(2)$ is positive (Figure 9). At $t = 4$, the projectile is on the way down and its velocity $v(4)$ is negative.

(b) Maximum height is attained when the velocity is zero, so we solve

$$30 - 9.8t = 0 \quad \Rightarrow \quad t = \frac{150}{49} \approx 3.06$$

The projectile reaches maximum height at $t = 150/49$ s. Its maximum height is

$$s(150/49) = 30(150/49) - 4.9(150/49)^2 \approx 45.92 \text{ m}$$

■

Note that in Example 7 in Section 2.5, we also examined the maximum height for a projectile launched upward with an initial velocity of 30 m/s. There the effect of air resistance was included, and the maximum height depended on the strength of the air resistance. We numerically investigated the maximum height in the limit as we let the air resistance go to zero. We obtained an estimate of 45.92 m, consistent with the value we obtained here, neglecting air resistance altogether.

HISTORICAL PERSPECTIVE

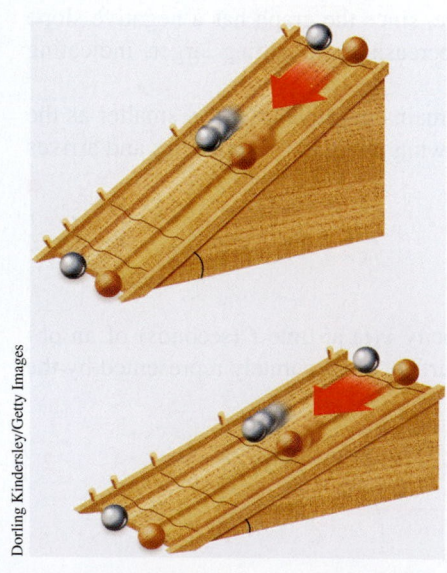

Galileo Galilei (1564–1642) discovered the laws of motion for falling objects on the earth's surface around 1600. This paved the way for Newton's general laws of motion. How did Galileo arrive at his formulas? The motion of a falling object is too rapid to measure directly, without modern photographic or electronic apparatus. To get around this difficulty, Galileo experimented with balls rolling down an incline (Figure 10). For a sufficiently flat incline, he was able to measure the motion with a water clock and found that the velocity of the rolling ball is proportional to

time. He then reasoned that motion in free-fall is just a faster version of motion down an incline and deduced the formula $v(t) = -gt$ for falling objects (assuming zero initial velocity).

Prior to Galileo, it had been assumed incorrectly that heavy objects fall more rapidly than lighter ones. Galileo realized that this was not true (as long as air resistance is negligible), and indeed, the formula $v(t) = -gt$ shows that the velocity depends on time but not on the weight of the object. Interestingly, 300 years later, another great physicist, Albert Einstein, was deeply puzzled by Galileo's discovery that all objects fall at the same rate regardless of their weight. He called this the Principle of Equivalence and sought to understand why it was true. In 1916, after a decade of intensive work, Einstein developed the General Theory of Relativity, which finally gave a full explanation of the Principle of Equivalence in terms of the geometry of space and time.

FIGURE 10 To explain the motion of falling objects, Galileo studied the motion of balls on an inclined plane.

Dorling Kindersley/Getty Images

Bettmann/Getty Images

3.4 SUMMARY

- The (instantaneous) rate of change of $y = f(x)$ with respect to x at $x = x_0$ is defined as the derivative

$$f'(x_0) = \lim_{\Delta x \to 0} \frac{\Delta y}{\Delta x} = \lim_{x_1 \to x_0} \frac{f(x_1) - f(x_0)}{x_1 - x_0}$$

- The rate dy/dx is measured in *units of y per unit of x*.
- Marginal cost is the cost of producing one additional unit. If $C(x)$ is the cost of producing x units, then the marginal cost at production level x_0 is $C(x_0 + 1) - C(x_0)$. The derivative $C'(x_0)$ is often a good estimate for marginal cost.
- For linear motion, velocity $v(t)$ is the rate of change of position $s(t)$ with respect to time—that is, $v(t) = s'(t)$.
- Galileo's formulas for an object rising or falling under the influence of gravity near Earth's surface ignoring air resistance (s_0 = initial position, v_0 = initial velocity):

$$s(t) = s_0 + v_0 t - \frac{1}{2} g t^2, \qquad v(t) = v_0 - g t$$

where $g \approx 9.8$ m/s², or $g \approx 32$ ft/s². Maximum height is attained when $v(t) = 0$.

3.4 EXERCISES

Preliminary Questions

1. Which units might be used for each rate of change?
(a) Pressure (in atmospheres) in a water tank with respect to depth
(b) The rate of a chemical reaction (change in concentration with respect to time with concentration in moles per liter)

2. Two trains travel from New Orleans to Memphis in 4 h. The first train travels at a constant velocity of 90 mph, but the velocity of the second train varies. What was the second train's average velocity during the trip?

3. Discuss how it is possible to be speeding up with a velocity that is decreasing.

4. Sketch the graph of a function that has an average rate of change equal to zero over the interval $[0, 1]$ but has instantaneous rates of change at 0 and 1 that are positive.

Exercises

In Exercises 1–8, find the rate of change.

1. Area of a square with respect to its side s when $s = 5$

2. Volume of a cube with respect to its side s when $s = 5$

3. Cube root $\sqrt[3]{x}$ with respect to x when $x = 1, 8, 27$

4. The reciprocal $1/x$ with respect to x when $x = 1, 2, 3$

5. The diameter of a circle with respect to radius

6. Surface area A of a sphere with respect to radius r ($A = 4\pi r^2$)

7. Volume V of a cylinder with respect to radius if the height is equal to the radius

8. Speed of sound v (in m/s) with respect to air temperature T (in kelvins), where $v = 20\sqrt{T}$

In Exercises 9–11, refer to Figure 11, the graph of distance s from the origin as a function of time for a car trip.

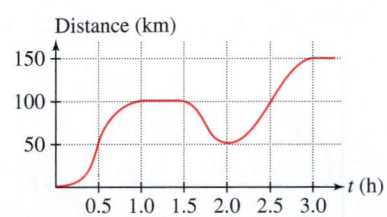

FIGURE 11 Distance from the origin versus time for a car trip.

9. Find the average velocity over each interval.
(a) $[0, 0.5]$ **(b)** $[0.5, 1]$ **(c)** $[1, 1.5]$ **(d)** $[1, 2]$

10. At what time is velocity at a maximum?

11. Match the descriptions (i)–(iii) with the intervals (a)–(c) in Figure 11.
(i) Velocity increasing
(ii) Velocity decreasing
(iii) Velocity negative
(a) $[0, 0.5]$ **(b)** $[2.5, 3]$ **(c)** $[1.5, 2]$

Exercises 12 and 13 refer to the data in Example 1. Approximate the derivative with the symmetric difference quotient (SDQ) approximation:
$$T'(t) \approx \frac{T(t+20) - T(t-20)}{40}.$$

12. (a) At what t does the SDQ approximation give the fastest rate of increase of temperature? What is the rate of change?

(b) At what t does the SDQ approximation give the fastest rate of decrease of temperature? What is the rate of change?

13. At what t does the SDQ approximation give the smallest (i.e., closest to 0) rate of change of temperature? What is the rate of change?

Exercises 14–16 refer to the four graphs of s as a function of t in Figure 7.

14. Sketch s' for each of the four graphs of s.

15. Match each situation with the graph that best represents it.

(a) Rocky slowed down his car as it approached the moose in the road. The distance from the car to the moose is s and the time since he spotted the moose is t.

(b) The rocket's speed increased after liftoff until the fuel was used up. The distance from the rocket to the launchpad is s and the time since liftoff is t.

(c) The increase in college costs slowed for the fourth year in a row. The cost of college is s and the time since the start of the 4-year period is t.

16. Match each situation with the graph that best represents it.

(a) Dusty's batting average increased over the first 10 games of the season but from game to game the amount of increase went down. Dusty's batting average is s and the time since the beginning of the season is t.

(b) In performing a crash test, the car continued to speed up until it hit the wall. The distance between the car and the wall is s and the time since the car started moving is t.

(c) The hurricane strengthened at an increasing rate over the first day of its development. The strength of the hurricane is s, and the time since it started developing is t.

17. Sketch a graph of velocity as a function of time for the shuttle train in Example 6.

18. Figure 12 shows the height y of a mass oscillating at the end of a spring, through one cycle of the oscillation. Sketch the graph of velocity as a function of time.

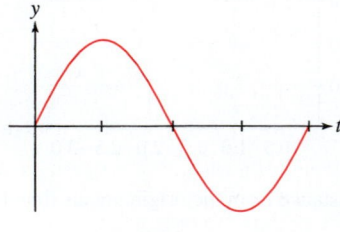

FIGURE 12

19. Fred X has to make a book delivery from his warehouse, 15 mi north of the city, to the Amazing Book Store 10 mi south of the city. Traffic is usually congested within 5 mi of the city. He leaves at noon, traveling due south through the city, and arrives at the store at 12:50. After 15 min at the store, he makes the return trip north to his warehouse, arriving at 2:00. Let s represent the distance from the warehouse in miles and t represent time in minutes since noon. Make sketches of the graphs of s and s' as functions of t for Fred's trip.

20. At the start of the 27th century, the population of Zosania was approximately 40 million. Early-century prosperity saw the population nearly double in the first three decades, but the growth slowed in the 30s and 40s and then leveled off completely during the war years in the 50s. A postwar boom saw another rapid population increase, but that turned around in a major decline resulting from the great famine of the 70s. A slow end-of-century rebound resulted in an increase of the population to approximately 90 million at century's end. Let P represent the population in millions and t represent time in years since the start of the century. Make sketches of the graphs of P and P' as functions of t for Zosania's population during the century.

21. The velocity (in centimeters per second) of blood molecules flowing through a capillary of radius 0.008 cm is $v = 6.4 \times 10^{-8} - 0.001r^2$, where r is the distance from the molecule to the center of the capillary. Find the rate of change of velocity with respect to r when $r = 0.004$ cm.

22. Figure 13 displays the voltage V across a capacitor as a function of time while the capacitor is being charged. Estimate the rate of change of voltage at $t = 20$ s. Indicate the values in your calculation and include proper units. Does voltage change more quickly or more slowly as time goes on? Explain in terms of tangent lines.

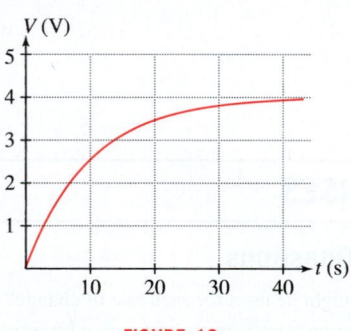

FIGURE 13

23. Use Figure 14 to estimate dT/dh at $h = 30$ and 70, where T is atmospheric temperature (in degrees Celsius) and h is altitude (in kilometers). Where is dT/dh equal to zero?

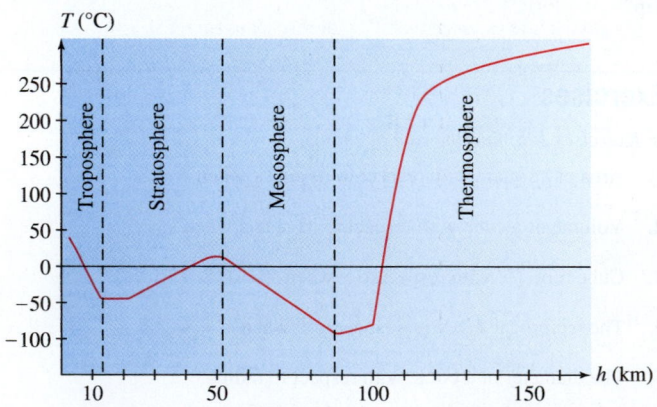

FIGURE 14 Atmospheric temperature versus altitude.

24. The earth exerts a gravitational force of $F(r) = (2.99 \times 10^{16})/r^2$ newtons on an object with a mass of 75 kg located r meters from the center of the earth. Find the rate of change of force with respect to distance r at the surface of the earth.

25. For the escape velocity relationship, $v_{esc} = (2.82 \times 10^7)r^{-1/2}$ m/s, calculate the rate of change of the escape velocity with respect to distance r from the center of the earth.

26. The power delivered by a battery to an apparatus of resistance R (in ohms) is $P = 2.25R/(R + 0.5)^2$ watts (W). Find the rate of change of power with respect to resistance for $R = 3\ \Omega$ and $R = 5\ \Omega$.

27. A particle moving along a line has position $s(t) = t^4 - 18t^2$ m at time t seconds. At which times does the particle pass through the origin? At which times is the particle instantaneously motionless (i.e., it has zero velocity)?

28. GU Plot the position of the particle in Exercise 27. What is the farthest distance to the left of the origin attained by the particle?

29. A projectile is launched in the air from the ground with an initial velocity $v_0 = 60$ m/s. What is the maximum height that the projectile reaches? (Compare your result with Exercise 37 in Section 2.5, where we considered maximum height when air resistance is included and we investigated the result of letting the air resistance go to 0.)

30. Find the velocity of an air conditioner accidentally dropped from a height of 300 m at the moment it hits the ground.

31. A ball tossed in the air vertically from ground level returns to Earth 4 s later. Find the initial velocity and maximum height of the ball.

32. Olivia is gazing out a window from the 10th floor of a building when a bucket (dropped by a window washer) passes by. She notes that it hits the ground 1.5 s later. Determine the floor from which the bucket was dropped if each floor is 5 m high and the window is in the middle of the 10th floor. Neglect air friction.

33. Show that for an object falling according to Galileo's formula, the average velocity over any time interval $[t_1, t_2]$ is equal to the average of the instantaneous velocities at t_1 and t_2.

34. An object falls under the influence of gravity near the earth's surface. Which of the following statements is true? Explain.
(a) Distance traveled increases by equal amounts in equal time intervals.
(b) Velocity increases by equal amounts in equal time intervals.
(c) The derivative of velocity increases with time.

35. By Faraday's Law, if a conducting wire of length ℓ meters moves at velocity v m/s perpendicular to a magnetic field of strength B (in teslas), a voltage of size $V = -B\ell v$ is induced in the wire. Assume that $B = 2$ and $\ell = 0.5$.
(a) Calculate dV/dv.
(b) Find the rate of change of V with respect to time t if $v(t) = 4t + 9$.

36. The voltage V, current I, and resistance R in a circuit are related by Ohm's Law: $V = IR$, where the units are volts, amperes, and ohms. Assume that voltage is constant with $V = 12$ volts (V). Calculate (specifying units):
(a) The average rate of change of I with respect to R for the interval from $R = 8$ to $R = 8.1$
(b) The rate of change of I with respect to R when $R = 8$
(c) The rate of change of R with respect to I when $I = 1.5$

37. Ethan finds that with h hours of tutoring, he is able to answer correctly $S(h)$ percent of the problems on a math exam. Which would you expect to be larger: $S'(3)$ or $S'(30)$? Explain.

38. Suppose $\theta(t)$ measures the angle between a clock's minute and hour hands. What is $\theta'(t)$ at 3 o'clock?

39. To determine drug dosages, doctors estimate a person's body surface area (BSA) (in meters squared) using the formula BSA $= \sqrt{hm}/60$, where h is the height in centimeters and m the mass in kilograms. Calculate the rate of change of BSA with respect to mass for a person of constant height $h = 180$. What is this rate at $m = 70$ and $m = 80$? Express your result in the correct units. Does BSA increase more rapidly with respect to mass at lower or higher body mass?

40. The atmospheric CO_2 level $A(t)$ at Mauna Loa, Hawaii, at time t (in parts per million by volume) is recorded by the Scripps Institution of Oceanography. Reading across, the annual values for the 4-year intervals are

1960	1964	1968	1972	1976	1980	1984
316.91	319.20	323.05	327.45	332.15	338.69	344.42

1988	1992	1996	2000	2004	2008	2012
351.48	356.37	362.64	369.48	377.38	385.34	393.87

(a) Estimate $A'(t)$ in 1962, 1970, 1978, 1986, 1994, 2002, and 2010.
(b) In which of the years from (a) did the approximation to $A'(t)$ take on its largest and smallest values?
(c) In which of these years does the approximation suggest that the CO_2 level was the most constant?

41. The tangent lines to the graph of $f(x) = x^2$ grow steeper as x increases. At what rate do the slopes of the tangent lines increase?

42. According to Kleiber's Law, the metabolic rate P (in kilocalories per day) and body mass m (in kilograms) of an animal are related by a *three-quarter-power law* $P = 73.3m^{3/4}$. Estimate the increase in metabolic rate when body mass increases from 60 to 61 kg.

43. The dollar cost of producing x bagels is given by the function $C(x) = 300 + 0.25x - 0.5(x/1000)^3$. Determine the cost of producing 2000 bagels and estimate the cost of the 2001st bagel. Compare your estimate with the actual cost of the 2001st bagel.

44. Suppose that for $x \geq 1000$, the dollar cost of producing x video cameras is $C(x) = 500x - 0.003x^2 + 10^{-8}x^3$.
(a) Estimate the marginal cost at production level $x = 5000$ and compare it with the actual cost $C(5001) - C(5000)$.
(b) Compare the marginal cost at $x = 5000$ with the average cost per camera, defined as $C(x)/x$.

45. According to Stevens's Law in psychology, the perceived magnitude of a stimulus is proportional (approximately) to a power of the actual intensity I of the stimulus. Experiments show that the *perceived brightness* B of a light satisfies $B = kI^{2/3}$, where I is the light intensity, whereas the *perceived heaviness* H of a weight W satisfies $H = kW^{3/2}$ (k is a constant that is different in the two cases). Compute dB/dI and dH/dW and state whether they are increasing or decreasing functions. Then explain the following statements:
(a) An increase in light intensity is felt more strongly when I is small than when I is large.
(b) An increase in load W is felt more strongly when W is large than when W is small.

46. Let $M(t)$ be the mass (in kilograms) of a plant as a function of time (in years). Recent studies by Niklas and Enquist have suggested that a remarkably wide range of plants (from algae and grass to palm trees) obey a *three-quarter-power growth law*—that is,

$$\frac{dM}{dt} = CM^{3/4} \quad \text{for some constant } C$$

(a) If a tree has a growth rate of 6 kg/yr when $M = 100$ kg, what is its growth rate when $M = 125$ kg?
(b) If $M = 0.5$ kg, how much more mass must the plant acquire to double its growth rate?

Further Insights and Challenges

*Exercises 47–49: The **Lorenz curve** $y = F(r)$ is used by economists to study income distribution in a given country (see Figure 15). By definition, $F(r)$ is the fraction of the total income that goes to the bottom rth part of the population, where $0 \le r \le 1$. For example, if $F(0.4) = 0.245$, then the bottom 40% of households receive 24.5% of the total income. Note that $F(0) = 0$ and $F(1) = 1$.*

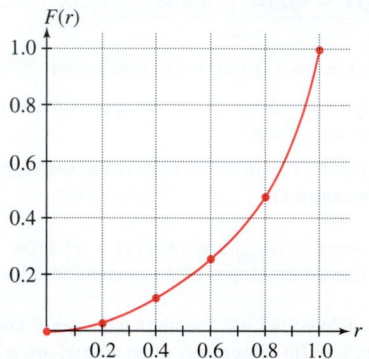

(A) Lorenz curve for the United States in 2010

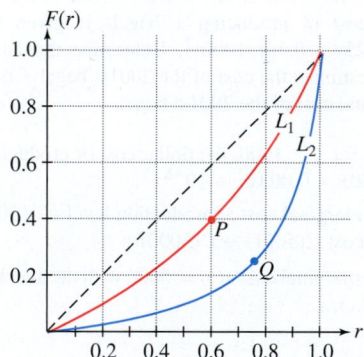

(B) Two Lorenz curves: The tangent lines at P and Q have slope 1.

FIGURE 15

47. [✎] Our goal is to find an interpretation for $F'(r)$. The average income for a group of households is the total income going to the group divided by the number of households in the group. The national average income is $A = T/N$, where N is the total number of households and T is the total income earned by the entire population.

(a) Show that the average income among households in the bottom rth part is equal to $(F(r)/r)A$.

(b) Show more generally that the average income of households belonging to an interval $[r, r + \Delta r]$ is equal to

$$\left(\frac{F(r + \Delta r) - F(r)}{\Delta r} \right) A$$

(c) Let $0 \le r \le 1$. A household belongs to the 100rth percentile if its income is greater than or equal to the income of $100r \%$ of all households. Pass to the limit as $\Delta r \to 0$ in (b) to derive the following interpretation: A household in the 100rth percentile has income $F'(r)A$. In particular, a household in the 100rth percentile receives more than the national average if $F'(r) > 1$ and less if $F'(r) < 1$.

(d) For the Lorenz curves L_1 and L_2 in Figure 15(B), what percentage of households have above-average income?

48. The following table provides values of $F(r)$ for the United States in 2010. Assume that the national average income was $A = \$66,000$.

r	0	0.2	0.4	0.6	0.8	1
$F(r)$	0	0.033	0.118	0.264	0.480	1

(a) What was the average income in the lowest 40% of households?

(b) Show that the average income of the households belonging to the interval $[0.4, 0.6]$ was $\$48,180$.

(c) Estimate $F'(0.5)$. Estimate the income of households in the 50th percentile. Was it greater or less than the national average?

49. Use Exercise 47(c) to prove:

(a) $F'(r)$ is an increasing function of r.

(b) Income is distributed equally (all households have the same income) if and only if $F(r) = r$ for $0 \le r \le 1$.

In Exercises 50 and 51, the average cost per unit at production level x is defined as

$$C_{avg}(x) = \frac{C(x)}{x},$$

where $C(x)$ is the cost of producing x units. Average cost is a measure of the efficiency of the production process.

50. The cost in dollars of producing alarm clocks is given by

$$C(x) = 50x^3 - 750x^2 + 3740x + 3750,$$

where x is in units of 1000.

(a) Calculate the average cost at $x = 4, 6, 8$, and 10.

(b) Use the graphical interpretation of average cost to find the production level x_0 at which average cost is lowest. What is the relation between average cost and marginal cost at x_0 (see Figure 16)?

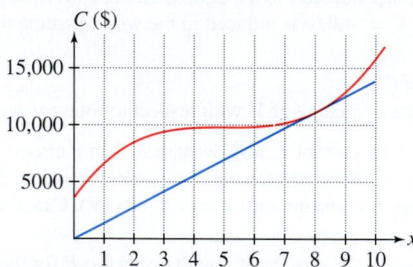

FIGURE 16 Cost function $C(x) = 50x^3 - 750x^2 + 3740x + 3750$.

51. Show that $C_{avg}(x)$ is equal to the slope of the line through the origin and the point $(x, C(x))$ on the graph of $y = C(x)$. Using this interpretation, determine whether average cost or marginal cost is greater at points A, B, C, D in Figure 17.

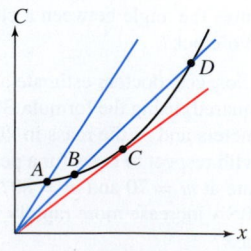

Production level

FIGURE 17 Graph of $y = C(x)$.

3.5 Higher Derivatives

Higher derivatives are obtained by repeatedly differentiating a function $y = f(x)$. If f' is differentiable, then the **second derivative**, denoted f'' or y'', is the derivative

$$f''(x) = \frac{d}{dx}\left(f'(x)\right)$$

For example, for $f(x) = x^2 + \frac{1}{x} + \sqrt{x}$, we have

$$f'(x) = 2x - x^{-2} + \frac{1}{2}x^{-1/2}$$

$$f''(x) = 2 + 2x^{-3} - \frac{1}{4}x^{-3/2}$$

The second derivative is the rate of change of $f'(x)$, so it is the rate of change of the rate of change of f.

The next example highlights the difference between the first and second derivatives.

EXAMPLE 1 Figure 1 and Table 1 show the number of cell phone subscribers $C(t)$ in the United States in year t. Discuss $C'(t)$ and $C''(t)$.

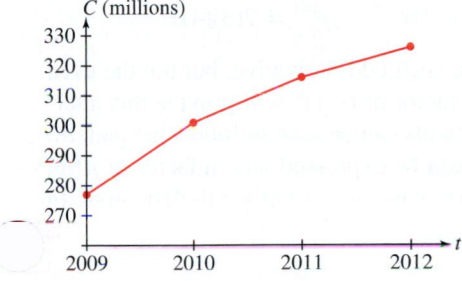

FIGURE 1 Number of cell phone subscribers C in the United States in millions.

TABLE 1 Number of Cell Phone Subscribers in the United States

Year	2009	2010	2011	2012
Number in millions	277	301	316	326
Change in C		24	15	10
Change in the change in C			−9	−5

Solution We will show that $C'(t)$ is positive but $C''(t)$ is negative. According to Table 1, the number of cell phone subscribers each year was greater than the previous year, so the rate of change $C'(t)$ is certainly positive. However, the amount of increase declined from 24 million in 2010 to 15 million in 2011 to 10 million in 2012. Thus, $C'(t)$ is positive, but $C'(t)$ decreases from one year to the next, and therefore its rate of change $C''(t)$ is negative. Figure 1 supports this conclusion: The slopes of the segments in the graph are positive [$C'(t)$ is positive], but the slopes decrease going from one segment to the next [$C''(t)$ is negative]. ■

The process of differentiation can be continued, provided that the derivatives exist. The third derivative, denoted $f'''(x)$ or $f^{(3)}(x)$, is the derivative of $f''(x)$. More generally, the nth derivative $f^{(n)}(x)$ is the derivative of the $(n-1)$st derivative. We use parentheses on the superscript for the derivative to distinguish $f^{(n)}$, the nth derivative of f, from f^n, the nth power of f. We call $f(x)$ the zeroth derivative and $f'(x)$ the first derivative. In Leibniz notation, we write

$$\frac{df}{dx}, \quad \frac{d^2 f}{dx^2}, \quad \frac{d^3 f}{dx^3}, \quad \frac{d^4 f}{dx^4}, \dots$$

- dy/dx has units of y per unit of x.
- d^2y/dx^2 has units of dy/dx per unit of x or units of y per unit of x squared.

EXAMPLE 2 Calculate $f'''(-1)$ for $f(x) = 3x^5 - 2x^2 + 7x^{-2}$.

Solution We must calculate the first three derivatives:

$$f'(x) = \frac{d}{dx}\left(3x^5 - 2x^2 + 7x^{-2}\right) = 15x^4 - 4x - 14x^{-3}$$

$$f''(x) = \frac{d}{dx}\left(15x^4 - 4x - 14x^{-3}\right) = 60x^3 - 4 + 42x^{-4}$$

$$f'''(x) = \frac{d}{dx}\left(60x^3 - 4 + 42x^{-4}\right) = 180x^2 - 168x^{-5}$$

At $x = -1$, $f'''(-1) = 180 + 168 = 348$. ■

Polynomials have a special property: Once n is large enough, the nth derivative is the zero function, and therefore so are all higher derivatives. More precisely, if f is a polynomial of degree k, then $f^{(n)}(x)$ is zero for $n > k$. Table 2 illustrates this property for $f(x) = x^5$. By contrast, the higher derivatives of a nonpolynomial function are never the zero function (see Exercise 85, Section 5.3).

TABLE 2   Derivatives of x^5

$f(x)$	$f'(x)$	$f''(x)$	$f'''(x)$	$f^{(4)}(x)$	$f^{(5)}(x)$	$f^{(6)}(x)$
x^5	$5x^4$	$20x^3$	$60x^2$	$120x$	120	0

EXAMPLE 3 Calculate the first four derivatives of $y = x^{-1}$. Then find the pattern and determine a general formula for $y^{(n)}$.

Solution By the Power Rule,

$$y'(x) = -x^{-2}, \quad y'' = 2x^{-3}, \quad y''' = -2(3)x^{-4}, \quad y^{(4)} = 2(3)(4)x^{-5}$$

First note that we have a leading negative sign on each odd derivative, but not the even derivatives. In a formula for the nth derivative, a factor of $(-1)^n$ will provide this alternating sign. Ignoring the negative sign, the coefficients can be seen to follow the pattern: $1, (1)(2), (1)(2)(3), (1)(2)(3)(4)$, and so on. This can be expressed with a factor of $n!$ in the nth derivative. Finally, we see that the power on x is $-n - 1$ in the nth derivative. In general, therefore, $y^{(n)}(x) = (-1)^n n! \, x^{-n-1}$. ∎

EXAMPLE 4 Calculate the first three derivatives of $f(x) = xe^x$. Then determine a general formula for $f^{(n)}(x)$.

Solution Use the Product Rule:

$$f'(x) = \frac{d}{dx}(xe^x) = e^x + xe^x = (1 + x)e^x$$

$$f''(x) = \frac{d}{dx}\big((1 + x)e^x\big) = e^x + (1 + x)e^x = (2 + x)e^x$$

$$f'''(x) = \frac{d}{dx}\big((2 + x)e^x\big) = e^x + (2 + x)e^x = (3 + x)e^x$$

We see that $f^n(x) = e^x + f^{n-1}(x)$, which leads to the general formula

$$f^{(n)}(x) = (n + x)e^x$$ ∎

A second derivative that you might be familiar with is acceleration. An object that is in linear motion with position $s(t)$ at time t has velocity $v(t) = s'(t)$ and acceleration $a(t) = v'(t) = s''(t)$. Thus, acceleration is the rate at which velocity changes and is measured in units of velocity per unit of time or "distance per time squared," such as m/s^2.

EXAMPLE 5   Acceleration Due to Gravity Find the acceleration $a(t)$ of a ball tossed vertically in the air from ground level with an initial velocity of 12 m/s. How does $a(t)$ describe the change in the ball's velocity as it rises and falls?

Solution The ball's height at time t is $s(t) = s_0 + v_0 t - 4.9t^2$ m by Galileo's formula. In our case, $s_0 = 0$ and $v_0 = 12$, so $s(t) = 12t - 4.9t^2$ m [Figure 2(A)]. Therefore, $v(t) = s'(t) = 12 - 9.8t$ m/s and the ball's acceleration is

$$a(t) = s''(t) = \frac{d}{dt}(12 - 9.8t) = -9.8 \text{ m/s}^2$$

← **REMINDER** n-factorial is the number

$$n! = n(n - 1)(n - 2)\ldots(2)(1)$$

Thus,

$$1! = 1, \quad 2! = (2)(1) = 2$$

$$3! = (3)(2)(1) = 6$$

By convention, we set $0! = 1$.

It is not always possible to find a simple formula for the higher derivatives of a function. In most cases, they become increasingly complicated.

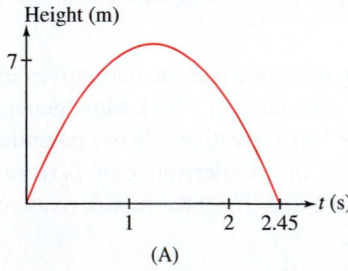

(A)

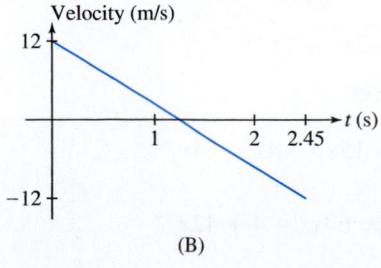

(B)

FIGURE 2 Height and velocity of a ball tossed vertically with initial velocity 12 m/s.

The acceleration is constant with value $-9.8 = -g$, where g (in m/s^2) is the acceleration due to gravity, as introduced in Section 3.4. As the ball rises and falls, its velocity decreases from 12 to -12 m/s at the constant rate $-g$ [Figure 2(B)]. ■

GRAPHICAL INSIGHT Can we visualize the rate represented by $f''(x)$? The second derivative is the rate at which $f'(x)$ is changing, so $f''(x)$ is large if the slopes of the tangent lines change rapidly, as in Figure 3(A). Similarly, $f''(x)$ is small if the slopes of the tangent lines change slowly—in this case, the curve is relatively flat, as in Figure 3(B). If f is a linear function [Figure 3(C)], then the tangent line does not change at all and $f''(x) = 0$. Thus, $f''(x)$ measures the "bending" or concavity of the graph.

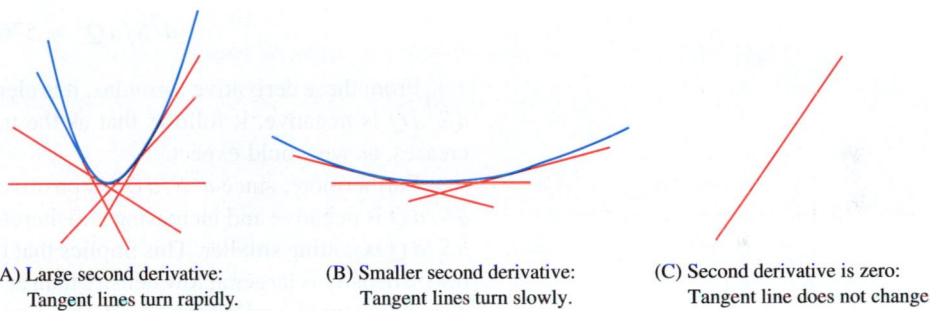

(A) Large second derivative:
Tangent lines turn rapidly.

(B) Smaller second derivative:
Tangent lines turn slowly.

(C) Second derivative is zero:
Tangent line does not change.

DF FIGURE 3

EXAMPLE 6 Identify curves I and II in Figure 4(B) as the graphs of f' or f'' for the function f in Figure 4(A).

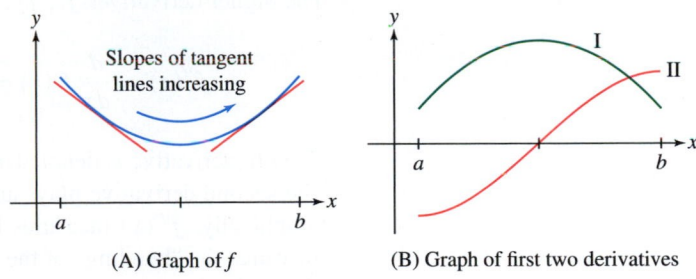

Slopes of tangent
lines increasing

(A) Graph of f

(B) Graph of first two derivatives

DF FIGURE 4

Solution The slopes of the tangent lines to the graph of f are *increasing* on the interval $[a, b]$. Therefore, f' is an increasing function and its graph must be II. Since $f''(x)$ is the rate of change of $f'(x)$, and $f'(x)$ is increasing, $f''(x)$ is positive and its graph must be I. ■

EXAMPLE 7 In a 1997 study, Boardman and Lave related the traffic speed S on a two-lane road to traffic density Q (number of cars per mile of road) by the formula

$$S = 2882Q^{-1} - 0.052Q + 31.73$$

for $60 \le Q \le 400$ (Figure 5).

Show that $dS/dQ < 0$ and $d^2S/dQ^2 > 0$ and interpret each in relation to the situation being modeled.

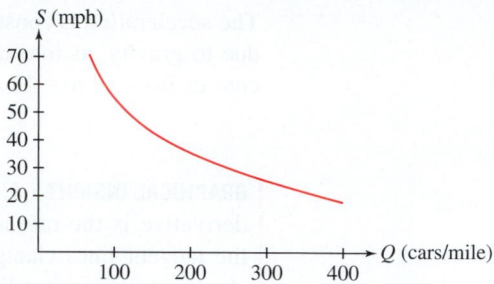

FIGURE 5 Speed as a function of traffic density.

Solution   Taking the derivatives:

$$dS/dQ = -2882Q^{-2} - 0.052$$

$$d^2S/dQ^2 = 5764Q^{-3}$$

From these derivative formulas, it is clear that $dS/dQ < 0$ and $d^2S/dQ^2 > 0$. Since dS/dQ is negative, it follows that as the traffic density increases, the traffic speed decreases, as we would expect.

Furthermore, since d^2S/dQ^2 is positive, it follows that dS/dQ is increasing. Given dS/dQ is negative and increasing and therefore is getting closer to zero, we conclude that dS/dQ is getting smaller. This implies that the speed decrease associated with increasing traffic density is larger at low density than at high density. This is to be expected as well— with low density and a high speed, the speed is going to drop off quickly as the density increases, but with higher density and an already low speed the speed is not going to drop off very much more as the density increases further. ■

3.5 SUMMARY

- The higher derivatives f', f'', f''',... are defined by successive differentiation:

$$f''(x) = \frac{d}{dx} f'(x) = \frac{d^2 f}{dx^2}, \qquad f'''(x) = \frac{d}{dx} f''(x) = \frac{d^3 f}{dx^3}, \ldots$$

The nth derivative is denoted $f^{(n)}(x) = \frac{d^n f}{dx^n}$.
- The second derivative plays an important role: It is the rate at which $f'(x)$ changes. Graphically, $f''(x)$ measures how fast the tangent lines change direction and thus measures the "bending" of the graph.
- If $s(t)$ is the position of an object at time t, then $s'(t)$ is velocity and $s''(t)$ is acceleration.

3.5 EXERCISES

Preliminary Questions

1. For each headline, rephrase as a statement about first and second derivatives and sketch a possible graph.

- "Stocks Go Higher, Though the Pace of Their Gains Slows"
- "Recent Rains Slow Roland Reservoir Water Level Drop"
- "Asteroid Approaching Earth at Rapidly Increasing Rate!!"

2. Sketch a graph of position as a function of time for an object that is slowing down and has positive acceleration.

3. Sketch a graph of position as a function of time for an object that is speeding up and has negative acceleration.

4. True or false? The third derivative of position with respect to time is zero for an object falling to Earth under the influence of gravity. Explain.

5. Which type of polynomial satisfies $f'''(x) = 0$ for all x?

6. What is the millionth derivative of $f(x) = e^x$?

7. What are the seventh and eighth derivatives of $f(x) = x^7$?

Exercises

In Exercises 1–16, calculate y'' and y'''.

1. $y = 14x^2$

2. $y = 7 - 2x$

3. $y = x^4 - 25x^2 + 2x$

4. $y = 4t^3 - 9t^2 + 7$

5. $y = \frac{4}{3}\pi r^3$

6. $y = \sqrt{x}$

7. $y = 20t^{4/5} - 6t^{2/3}$

8. $y = x^{-9/5}$

9. $y = z - \frac{4}{z}$

10. $y = 5t^{-3} + 7t^{-8/3}$

11. $y = \theta^2(2\theta + 7)$

12. $y = (x^2 + x)(x^3 + 1)$

13. $y = \frac{x - 4}{x}$

14. $y = \frac{1}{1 - x}$

15. $y = x^5 e^x$

16. $y = \frac{e^x}{x}$

In Exercises 17–26, calculate the derivative indicated.

17. $f^{(4)}(1), \quad f(x) = x^4$

18. $g'''(-1), \quad g(t) = -4t^{-5}$

19. $\left.\dfrac{d^2 y}{dt^2}\right|_{t=1}, \quad y = 4t^{-3} + 3t^2$

20. $\left.\dfrac{d^4 f}{dt^4}\right|_{t=1}, \quad f(t) = 6t^9 - 2t^5$

21. $\left.\dfrac{d^4 x}{dt^4}\right|_{t=16}, \quad x = t^{-3/4}$

22. $f'''(4), \quad f(t) = 2t^2 - t$

23. $f'''(-3), \quad f(x) = 4e^x - x^3$

24. $f''(1), \quad f(t) = \dfrac{t}{t + 1}$

25. $h''(1), \quad h(w) = \sqrt{w}e^w$

26. $g''(0), \quad g(s) = \dfrac{e^s}{s + 1}$

27. Calculate $y^{(k)}(0)$ for $0 \le k \le 5$, where $y = x^4 + ax^3 + bx^2 + cx + d$ (with a, b, c, d the constants).

28. Which of the following satisfy $f^{(k)}(x) = 0$ for all $k \ge 6$?

(a) $f(x) = 7x^4 + 4 + x^{-1}$

(b) $f(x) = x^3 - 2$

(c) $f(x) = \sqrt{x}$

(d) $f(x) = 1 - x^6$

(e) $f(x) = x^{9/5}$

(f) $f(x) = 2x^2 + 3x^5$

29. Use the result in Example 3 to find $\dfrac{d^6}{dx^6} x^{-1}$.

30. (a) Calculate the first five derivatives of $f(x) = \sqrt{x}$.

(b) Show that $f^{(n)}(x)$ is a multiple of $x^{-n+1/2}$.

(c) Show that $f^{(n)}(x)$ alternates in sign as $(-1)^{n-1}$ for $n \ge 1$.

(d) Find a formula for $f^{(n)}(x)$ for $n \ge 2$. *Hint:* Verify that the coefficient is $\pm 1 \cdot 3 \cdot 5 \cdots \dfrac{2n - 3}{2^n}$.

In Exercises 31–36, find a general formula for $f^{(n)}(x)$.

31. $f(x) = x^{-2}$

32. $f(x) = (x + 2)^{-1}$

33. $f(x) = x^{-1/2}$

34. $f(x) = x^{-3/2}$

35. $f(x) = xe^{-x}$

36. $f(x) = x^2 e^x$

37. (a) Find the acceleration at time $t = 5$ min of a helicopter whose height is $s(t) = 300t - 4t^3$ m.

(b) Plot the acceleration s'' for $0 \le t \le 6$. Is the helicopter speeding up or slowing down during this time interval? Explain.

38. Find an equation of the tangent line to the graph of $y = f'(x)$ at $x = 3$, where $f(x) = x^4$.

39. Figure 6 shows f, f', and f''. Determine which is which.

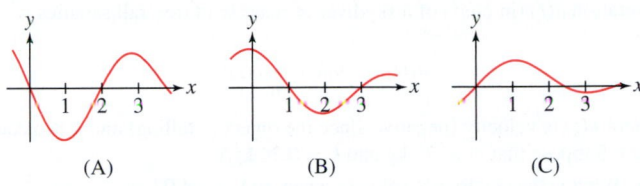

(A)   (B)   (C)

FIGURE 6

40. The second derivative f'' is shown in Figure 7. Which of (A) or (B) is the graph of f and which is f'?

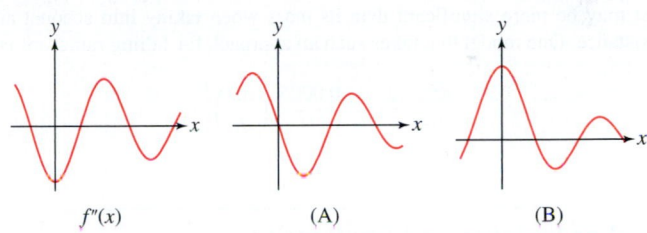

$f''(x)$ (A)   (B)

FIGURE 7

41. Figure 8 shows the graph of the position s of an object as a function of time t. Determine the intervals on which the acceleration is positive.

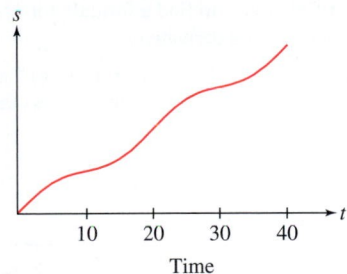

Time

FIGURE 8

42. Figure 9 shows the graph of the position s of an object as a function of time t. For each interval $[0, 10]$, $[10, 20]$, and so on, indicate whether the acceleration is negative, zero, or positive.

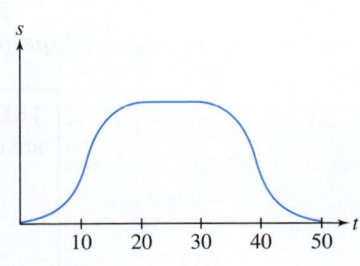

FIGURE 9

43. Find all values of n such that $y = x^n$ satisfies

$$x^2 y'' - 2xy' = 4y$$

44. Find all values of n such that $y = x^n$ satisfies

$$x^2 y'' - 12y = 0$$

45. According to one model that takes into account air resistance, the acceleration $a(t)$ (in m/s^2) of a skydiver of mass m in free-fall satisfies

$$a(t) = -9.8 + \frac{k}{m} v(t)^2$$

where $v(t)$ is velocity (negative since the object is falling) and k is a constant. Suppose that $m = 75$ kg and $k = 0.24$ kg/m.

(a) What is the skydiver's velocity when $a(t) = -4.9$?

(b) What is the skydiver's velocity when $a(t) = 0$? (This velocity is the terminal velocity, the velocity attained when air resistance balances gravity and the skydiver falls at a constant speed.)

46. ✎ In contrast to Exercise 45, the size of a falling lightweight object may be more significant than its mass when taking into account air resistance. One model that takes such an approach for falling raindrops is

$$\frac{d^2 s}{dt^2} = g - \frac{0.0005}{D} \left(\frac{ds}{dt} \right)^2$$

where $s(t)$ is the distance a raindrop has fallen (in meters), D is the raindrop diameter, and $g = 9.8$ m/s^2. Terminal velocity v_{term} is defined as the velocity at which the drop has zero acceleration (one can show that velocity approaches v_{term} as time proceeds).

(a) Show that $v_{\text{term}} = \sqrt{2000 g D}$.

(b) Find v_{term} for drops of diameter 10^{-3} and 10^{-4} m.

(c) In this model, do raindrops accelerate more rapidly at higher or lower velocities?

47. In a manufacturing process, a drill press automatically drills a hole into a sheet metal part on a conveyor. In the drilling operation the drill bit starts at rest directly above the part, descends quickly, drills a hole, and quickly returns to the start position. The maximum vertical speed of the drill bit is 4 in./s, and while drilling the hole, it must move no more than 2.6 in./s to avoid warping the metal. Let $s(t)$ be the drill bit's height (in inches) above the part as a function of time t in seconds. Sketch possible graphs of the drill bit's velocity [$s'(t)$] and acceleration [$s''(t)$].

48. CAS Use a computer algebra system to compute $f^{(k)}(x)$ for $k = 1$, 2, 3 for the following functions:

(a) $f(x) = (1 + x^3)^{5/3}$

(b) $f(x) = \dfrac{1 - x^4}{1 - 5x - 6x^2}$

49. CAS Let $f(x) = \dfrac{x + 2}{x - 1}$. Use a computer algebra system to compute the $f^{(k)}(x)$ for $1 \le k \le 4$. Can you find a general formula for $f^{(k)}(x)$?

Further Insights and Challenges

50. Find the 100th derivative of

$$p(x) = (x + x^5 + x^7)^{10} (1 + x^2)^{11} (x^3 + x^5 + x^7)$$

51. What is $p^{(99)}(x)$ for $p(x)$ in Exercise 50?

52. Use the Product Rule twice to find a formula for $(fg)''$ in terms of f and g and their first and second derivatives.

53. Use the Product Rule to find a formula for $(fg)'''$ and compare your result with the expansion of $(a + b)^3$. Then try to guess the general formula for $(fg)^{(n)}$.

54. Compute

$$\lim_{h \to 0} \frac{f(x + h) + f(x - h) - 2f(x)}{h^2}$$

for the following functions:

(a) $f(x) = x$

(b) $f(x) = x^2$

(c) $f(x) = x^3$

Based on these examples, what do you think the limit represents?

3.6 Trigonometric Functions

We can use the rules developed so far to differentiate functions involving powers of x, but we cannot yet handle the trigonometric functions. What is missing are the formulas for the derivatives of $\sin x$ and $\cos x$. Fortunately, their derivatives are simple—each is the derivative of the other up to a sign.

Recall our convention: *Angles are measured in radians, unless otherwise specified.*

THEOREM 1 Derivative of Sine and Cosine The functions $y = \sin x$ and $y = \cos x$ are differentiable and

$$\frac{d}{dx} \sin x = \cos x \qquad \text{and} \qquad \frac{d}{dx} \cos x = -\sin x$$

Proof We must go back to the definition of the derivative:

$$\frac{d}{dx}\sin x = \lim_{h\to 0}\frac{\sin(x+h)-\sin x}{h} \qquad \boxed{1}$$

We cannot cancel the h by rewriting the difference quotient, but we can use the addition formula (see marginal note) to write the numerator as a sum of two terms:

$$\sin(x+h)-\sin x = \sin x\cos h + \cos x\sin h - \sin x \qquad \text{(addition formula)}$$

$$= (\sin x\cos h - \sin x) + \cos x\sin h$$

$$= \sin x(\cos h - 1) + \cos x\sin h$$

This gives us

$$\frac{\sin(x+h)-\sin x}{h} = \frac{\sin x\,(\cos h - 1)}{h} + \frac{\cos x\,\sin h}{h}$$

$$\frac{d}{dx}\sin x = \lim_{h\to 0}\frac{\sin(x+h)-\sin x}{h} = \lim_{h\to 0}\frac{\sin x\,(\cos h - 1)}{h} + \lim_{h\to 0}\frac{\cos x\,\sin h}{h}$$

$$= (\sin x)\lim_{h\to 0}\frac{\cos h - 1}{h} + (\cos x)\lim_{h\to 0}\frac{\sin h}{h} \qquad \boxed{2}$$

REMINDER *Addition formula for* $\sin x$:

$$\sin(x+h) = \sin x\cos h + \cos x\sin h$$

We can take $\sin x$ and $\cos x$ outside the limits in Eq. (2) because they do not depend on the limiting variable h. The two limits are determined by Theorem 2 in Section 2.6, which indicates that

$$\lim_{h\to 0}\frac{1-\cos h}{h} = 0 \qquad \text{and} \qquad \lim_{h\to 0}\frac{\sin h}{h} = 1$$

It follows that $\lim\limits_{h\to 0}\frac{\cos h - 1}{h} = 0$, and therefore, Eq. (2) reduces to $\frac{d}{dx}\sin x = \cos x$, as desired. The formula $\frac{d}{dx}\cos x = -\sin x$ is proved similarly (see Exercise 57). ∎

CONCEPTUAL INSIGHT One property of $f(x) = \sin x$ that makes its derivative formula so simple is that $\lim\limits_{h\to 0}\frac{\sin h}{h} = 1$. The value of this limit depends on measuring angles in radians. The simplicity of this limit explains why we measure angles in radians when working with trigonometric functions in calculus. If instead we measure angles in degrees, then, as we pointed out in Section 2.2, $\lim\limits_{h\to 0}\frac{\sin h}{h} = \frac{\pi}{180}$. Nothing else in the previous proof would change, and we would end up with the unwieldy derivative formula $\frac{d}{dx}\sin x = \frac{\pi}{180}\cos x$.

EXAMPLE 1 For $f(x) = \sin x$, compute f' at $x = 0, \frac{\pi}{6}, \frac{\pi}{2},$ and $\frac{5\pi}{6}$.

Solution We have $f'(x) = \cos x$. Thus, $f'(0) = \cos(0) = 1$, $f'\left(\frac{\pi}{6}\right) = \cos\left(\frac{\pi}{6}\right) = \frac{\sqrt{3}}{2}$, $f'\left(\frac{\pi}{2}\right) = \cos\left(\frac{\pi}{2}\right) = 0$, and $f'\left(\frac{5\pi}{6}\right) = \cos\left(\frac{5\pi}{6}\right) = \frac{-\sqrt{3}}{2}$. ∎

Note, in Example 6 in Section 3.1, for $f(x) = \sin x$ we estimated $f'\left(\frac{\pi}{6}\right) \approx 0.8660$. Now we have the exact value $f'\left(\frac{\pi}{6}\right) = \frac{\sqrt{3}}{2}$.

GRAPHICAL INSIGHT The formula $(\sin x)' = \cos x$ seems reasonable when we compare the graphs in Figure 1. The tangent lines to the graph of $y = \sin x$ have positive slope on the interval $\left(-\frac{\pi}{2}, \frac{\pi}{2}\right)$, and on this interval, the derivative $y' = \cos x$ is positive. The tangent lines have negative slope on the interval $\left(\frac{\pi}{2}, \frac{3\pi}{2}\right)$, where $y' = \cos x$ is negative. The tangent lines are horizontal at $x = -\frac{\pi}{2}, \frac{\pi}{2}, \frac{3\pi}{2}$, where $\cos x = 0$.

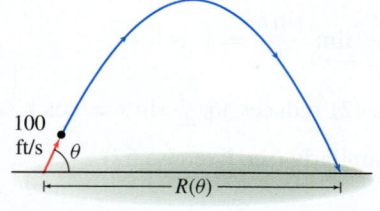

$y = \sin x$

$-\dfrac{\pi}{2}$ $\dfrac{\pi}{2}$ $\dfrac{3\pi}{2}$

$y' = \cos x$

DF FIGURE 1 The graphs of $y = \sin x$ and its derivative $y' = \cos x$.

EXAMPLE 2 Calculate $f''(x)$, where $f(x) = x \cos x$.

Solution By the Product Rule,

$$f'(x) = x' \cos x + x(\cos x)' = \cos x - x \sin x$$

$$f''(x) = (\cos x - x \sin x)' = -\sin x - \big(x'(\sin x) + x(\sin x)'\big)$$

$$= -2\sin x - x\cos x \qquad \blacksquare$$

EXAMPLE 3 A projectile is shot from ground level at 100 ft/s at a launch angle of θ that is between 0 and $\pi/2$ (Figure 2). Assume that the projectile is acted on by gravity, but not air resistance. Then, in a simple projectile-motion model (see Section 13.5) it can be shown that the projectile lands at a distance $R(\theta) = 625 \sin\theta \cos\theta$ ft from the launch point. What is the rate of change of the range with respect to the launch angle? For what angles does the range increase/decrease with increasing launch angle? What angle provides the maximum range, and what is that maximum range?

FIGURE 2 The range of the projectile is $R(\theta)$.

100 ft/s θ $R(\theta)$

Solution Using the Product Rule and the derivative rules for $\sin\theta$ and $\cos\theta$, we have

$$R'(\theta) = 625(\sin\theta)(-\sin\theta) + (\cos\theta)(\cos\theta)$$

$$= 625(\cos^2\theta - \sin^2\theta) = 625\cos 2\theta$$

where the last equality results from the double angle formula for $\cos 2\theta$.

Since $\cos 2\theta$ is positive for $0 \le \theta < \pi/4$ and is negative for $\pi/4 < \theta \le \pi/2$, it follows that the range increases with increasing launch angle for angles between 0 and $\pi/4$ and decreases with increasing launch angle for angles between $\pi/4$ and $\pi/2$. The maximum range occurs at $\theta = \pi/4$, and that maximum range is $R(\pi/4) = 312.5$ ft. $\qquad \blacksquare$

← **REMINDER** The standard trigonometric functions are defined in Section 1.4.

The derivatives of the other standard trigonometric functions can be computed using the Quotient Rule. We derive the formula for $(\tan x)'$ in Example 4 and the remaining formulas in Exercises 35–37.

THEOREM 2 Derivatives of Standard Trigonometric Functions

$$\frac{d}{dx}\tan x = \sec^2 x, \qquad \frac{d}{dx}\sec x = \sec x \tan x$$

$$\frac{d}{dx}\cot x = -\csc^2 x, \qquad \frac{d}{dx}\csc x = -\csc x \cot x$$

EXAMPLE 4 Derive the formula $\dfrac{d}{dx} \tan x = \sec^2 x$ (Figure 3).

Solution Use the Quotient Rule and the identity $\cos^2 x + \sin^2 x = 1$:

$$\frac{d}{dx} \tan x = \left(\frac{\sin x}{\cos x} \right)' = \frac{\cos x \cdot (\sin x)' - \sin x \cdot (\cos x)'}{\cos^2 x}$$

$$= \frac{\cos x \cos x - \sin x(-\sin x)}{\cos^2 x}$$

$$= \frac{\cos^2 x + \sin^2 x}{\cos^2 x} = \frac{1}{\cos^2 x} = \sec^2 x \qquad \blacksquare$$

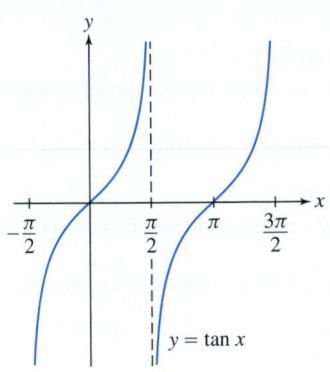

EXAMPLE 5 Determine y' for $y = \tan\theta \sec\theta$, and find an equation of the tangent line to the graph at $\theta = \frac{\pi}{4}$.

Solution By the Product Rule,

$$y' = (\tan\theta)' \sec\theta + \tan\theta (\sec\theta)' = \sec^2\theta \sec\theta + \tan\theta (\sec\theta \tan\theta)$$

$$= \sec^3\theta + \tan^2\theta \sec\theta$$

Now use the values $\sec\frac{\pi}{4} = \sqrt{2}$ and $\tan\frac{\pi}{4} = 1$ to compute

$$y\left(\frac{\pi}{4}\right) = \tan\left(\frac{\pi}{4}\right) \sec\left(\frac{\pi}{4}\right) = \sqrt{2}$$

$$y'\left(\frac{\pi}{4}\right) = \sec^3\left(\frac{\pi}{4}\right) + \tan^2\left(\frac{\pi}{4}\right) \sec\left(\frac{\pi}{4}\right) = 2\sqrt{2} + \sqrt{2} = 3\sqrt{2}$$

An equation of the tangent line (Figure 4) is $y - \sqrt{2} = 3\sqrt{2}\left(\theta - \frac{\pi}{4}\right)$. $\qquad \blacksquare$

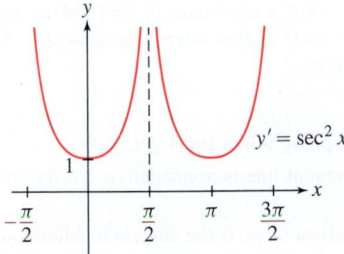

DF FIGURE 3 Graphs of $y = \tan x$ and its derivative $y' = \sec^2 x$.

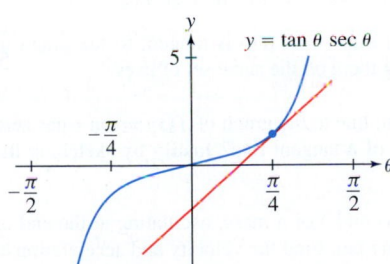

FIGURE 4 Tangent line to $y = \tan\theta \sec\theta$ at $\theta = \frac{\pi}{4}$.

3.6 SUMMARY

- The derivatives of the trigonometric functions:

$$\frac{d}{dx} \sin x = \cos x \qquad\qquad \frac{d}{dx} \cos x = -\sin x$$

$$\frac{d}{dx} \tan x = \sec^2 x \qquad\qquad \frac{d}{dx} \sec x = \sec x \tan x$$

$$\frac{d}{dx} \cot x = -\csc^2 x \qquad\qquad \frac{d}{dx} \csc x = -\csc x \cot x$$

3.6 EXERCISES

Preliminary Questions

1. Determine the sign ($+$ or $-$) that yields the correct formula for the following:

(a) $\dfrac{d}{dx}(\sin x + \cos x) = \pm \sin x \pm \cos x$

(b) $\dfrac{d}{dx} \sec x = \pm \sec x \tan x$

(c) $\dfrac{d}{dx} \cot x = \pm \csc^2 x$

2. Which of the following functions can be differentiated using the rules we have covered so far?

(a) $y = 3\cos x \cot x$ **(b)** $y = \cos(x^2)$ **(c)** $y = e^x \sin x$

3. For each, give an equation of the tangent line to the graph at $x = 0$.

(a) $y = \sin x$

(b) $y = \cos x$

4. How is the addition formula for sine used in deriving the formula $(\sin x)' = \cos x$?

Exercises

In Exercises 1–4, find an equation of the tangent line at the point indicated.

1. $y = \sin x$, $x = \frac{\pi}{4}$

2. $y = \cos x$, $x = \frac{\pi}{3}$

3. $y = \tan x$, $x = \frac{\pi}{4}$

4. $y = \sec x$, $x = \frac{\pi}{6}$

In Exercises 5–24, compute the derivative.

5. $f(x) = \sin x \cos x$

6. $f(x) = x^2 \cos x$

7. $f(x) = x \sin x$

8. $f(x) = 9 \sec x + 12 \cot x$

9. $H(t) = \sin t \sec^2 t$

10. $h(t) = 9 \csc t + t \cot t$

11. $f(\theta) = \tan \theta \sec \theta$

12. $k(\theta) = \theta^2 \sin^2 \theta$

13. $f(x) = (2x^4 - 4x^{-1}) \sec x$

14. $f(z) = z \tan z$

15. $y = \dfrac{\sec \theta}{\theta}$

16. $G(z) = \dfrac{1}{\tan z - \cot z}$

17. $R(y) = \dfrac{3 \cos y - 4}{\sin y}$

18. $f(x) = \dfrac{x}{\sin x + 2}$

19. $f(x) = \dfrac{1 + \tan x}{1 - \tan x}$

20. $f(\theta) = \theta \tan \theta \sec \theta$

21. $f(x) = e^x \sin x$

22. $h(t) = e^t \csc t$

23. $f(\theta) = e^\theta (5 \sin \theta - 4 \tan \theta)$

24. $f(x) = x e^x \cos x$

In Exercises 25–34, find an equation of the tangent line at the point specified.

25. $y = x^3 + \cos x$, $x = 0$

26. $y = \tan \theta$, $\theta = \frac{\pi}{6}$

27. $y = \dfrac{\sin t}{1 + \cos t}$, $t = \frac{\pi}{3}$

28. $y = \sin x + 3 \cos x$, $x = 0$

29. $y = 2(\sin \theta + \cos \theta)$, $\theta = \frac{\pi}{3}$

30. $y = \csc x - \cot x$, $x = \frac{\pi}{4}$

31. $y = e^x \cos x$, $x = 0$

32. $y = e^x \cos^2 x$, $x = \frac{\pi}{4}$

33. $y = e^t (1 - \cos t)$, $t = \frac{\pi}{2}$

34. $y = e^\theta \sec \theta$, $\theta = \frac{\pi}{4}$

In Exercises 35–37, use Theorem 1 to derive the formula.

35. $\dfrac{d}{dx} \cot x = -\csc^2 x$

36. $\dfrac{d}{dx} \sec x = \sec x \tan x$

37. $\dfrac{d}{dx} \csc x = -\csc x \cot x$

38. Show that both $y = \sin x$ and $y = \cos x$ satisfy $y'' = -y$.

In Exercises 39–42, calculate the higher derivative.

39. $f''(\theta)$, $f(\theta) = \theta \sin \theta$

40. $\dfrac{d^2}{dt^2} \cos^2 t$

41. y'', y''', $y = \tan x$

42. y'', y''', $y = e^t \sin t$

43. Calculate the first five derivatives of $f(x) = \cos x$. Then determine $f^{(8)}(x)$ and $f^{(37)}(x)$.

44. Calculate the first five derivatives of $f(x) = \sin x$. Then determine $f^{(9)}(x)$ and $f^{(102)}(x)$.

45. Let $f(x) = \sin x$. We can compute $f^{(n)}(x)$ as follows: First, express $n = 4m + r$ where m is a whole number and $r = 0, 1, 2,$ or 3. Then determine $f^{(n)}(x)$ from r. Explain how to do the latter step.

46. Let $f(x) = \cos x$. We can compute $f^{(n)}(x)$ as follows: First, express $n = 4m + r$ where m is a whole number and $r = 0, 1, 2,$ or 3. Then determine $f^{(n)}(x)$ from r. Explain how to do the latter step.

47. Let $f(x) = \sin^2 x$ and $g(x) = \cos^2 x$.

(a) Use an identity and prove $f'(x) = -g'(x)$ without directly computing $f'(x)$ and $g'(x)$.

(b) Now verify the result in (a) by directly computing $f'(x)$ and $g'(x)$.

48. Let $f(x) = \tan^2 x$ and $g(x) = \sec^2 x$.

(a) Use an identity and prove $f'(x) = g'(x)$ without directly computing $f'(x)$ and $g'(x)$.

(b) Now verify the result in (a) by directly computing $f'(x)$ and $g'(x)$.

49. Find the values of x between 0 and 2π where the tangent line to the graph of $y = \sin x \cos x$ is horizontal.

50. [GU] Plot the graph $f(\theta) = \sec \theta + \csc \theta$ over $[0, 2\pi]$ and determine the number of solutions to $f'(\theta) = 0$ in this interval graphically. Then compute $f'(\theta)$ and find the solutions.

51. [GU] Let $g(t) = t - \sin t$.

(a) Plot the graph of g with a graphing utility for $0 \leq t \leq 4\pi$.

(b) Show that the slope of the tangent line is nonnegative. Verify this on your graph.

(c) For which values of t in the given range is the tangent line horizontal?

52. [CAS] Let $f(x) = (\sin x)/x$ for $x \neq 0$ and $f(0) = 1$.

(a) Plot f on $[-3\pi, 3\pi]$.

(b) Show that $f'(c) = 0$ if $c = \tan c$. Approximate the smallest *positive* value c_0 such that $f'(c_0) = 0$.

(c) Verify that the horizontal line $y = f(c_0)$ is tangent to the graph of $y = f(x)$ at $x = c_0$ by plotting them on the same set of axes.

53. [✎] Show that no tangent line to the graph of $f(x) = \tan x$ has zero slope. What is the least slope of a tangent line? Justify by sketching the graph of $f'(x) = \sec^2 x$.

54. The height at time t (in seconds) of a mass, oscillating at the end of a spring, is $s(t) = 300 + 40 \sin t$ cm. Find the velocity and acceleration at $t = \frac{\pi}{3}$ s.

55. A projectile is launched from ground level with an initial velocity v_0 at an angle θ, where $0 \leq \theta \leq \pi/2$. Its horizontal range is

$$R = \left(\frac{2v_0^2}{g} \right) \sin \theta \cos \theta,$$

where g is the acceleration due to gravity. Calculate $dR/d\theta$. The maximum range occurs where $dR/d\theta = 0$. Show that that occurs at $\theta = \pi/4$ and that the maximum range is v_0^2/g.

56. The graph of $y = \sin x$ is shown in Figure 5, along with a tangent line at $x = 0$. Show that if $\frac{\pi}{2} < \theta < \pi$, then the distance along the x-axis between θ and the point where the tangent line intersects the x-axis is equal to $|\tan \theta|$.

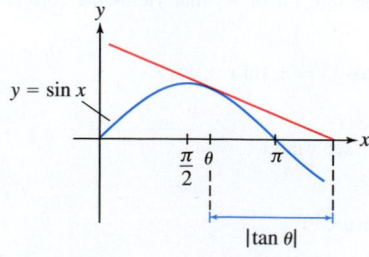

FIGURE 5

Further Insights and Challenges

57. Use the limit definition of the derivative and the addition law for the cosine function to prove that $(\cos x)' = -\sin x$.

58. Use the addition formula for the tangent

$$\tan(x+h) = \frac{\tan x + \tan h}{1 + \tan x \tan h}$$

to compute $(\tan x)'$ directly as a limit of the difference quotients. You will also need to show that $\lim\limits_{h\to 0} \dfrac{\tan h}{h} = 1$.

59. Verify the following identity and use it to give another proof of the formula $(\sin x)' = \cos x$:

$$\sin(x+h) - \sin x = 2\cos\left(x + \tfrac{1}{2}h\right)\sin\left(\tfrac{1}{2}h\right)$$

Hint: Use the addition formula for sine to prove that

$$\sin(a+b) - \sin(a-b) = 2\cos a \sin b$$

60. ✏️ Show that a nonzero polynomial function $y = f(x)$ *cannot* satisfy the equation $y'' = -y$. Use this to prove that neither $f(x) = \sin x$ nor $f(x) = \cos x$ is a polynomial. Can you think of another way to reach this conclusion by considering limits as $x \to \infty$?

61. Let $f(x) = x\sin x$ and $g(x) = x\cos x$.
(a) Show that $f'(x) = g(x) + \sin x$ and $g'(x) = -f(x) + \cos x$.
(b) Verify that $f''(x) = -f(x) + 2\cos x$ and $g''(x) = -g(x) - 2\sin x$.

(c) By further experimentation, try to find formulas for all higher derivatives of f and g. *Hint:* The kth derivative depends on whether $k = 4n$, $4n+1$, $4n+2$, or $4n+3$.

62. ✏️ Figure 6 shows the geometry behind the derivative formula $(\sin\theta)' = \cos\theta$. Segments $\overline{BA}$ and $\overline{BD}$ are parallel to the x- and y-axes. Let $\Delta\sin\theta = \sin(\theta+h) - \sin\theta$. Verify the following statements:
(a) $\Delta\sin\theta = BC$
(b) $\angle BDA = \theta$ *Hint:* $\overline{OA} \perp \overline{AD}$.
(c) $BD = (\cos\theta)AD$
Now explain the following intuitive argument: If h is small, then $BC \approx BD$ and $AD \approx h$, so $\Delta\sin\theta \approx (\cos\theta)h$ and $(\sin\theta)' = \cos\theta$.

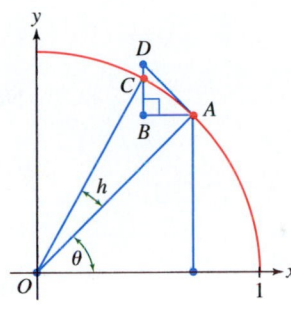

FIGURE 6

3.7 The Chain Rule

The **Chain Rule** is used to differentiate composite functions such as $y = \cos(x^3)$ and $y = \sqrt{x^4 + 1}$.

Recall that a *composite function* is obtained by evaluating one function at the output of another. The composite of f with g, denoted $f \circ g$, is defined by

$$(f \circ g)(x) = f\big(g(x)\big)$$

For convenience, we call f the *outside* function and g the *inside* function. Often, we write the composite function as $f(u)$, where $u = g(x)$. For example, $y = \cos(x^3)$ is the function $y = \cos u$, where $u = x^3$.

> The Chain Rule says
>
> $$\big(f(g(x))\big)' = \text{outside}'(\text{inside}) \cdot \text{inside}'$$
>
> Verbally, it is " the derivative of the outside function at the inside function times the derivative of the inside function."

THEOREM 1 Chain Rule   If f and g are differentiable, then the composite function $(f \circ g)(x) = f(g(x))$ is differentiable and

$$\boxed{\big(f(g(x))\big)' = f'\big(g(x)\big)\, g'(x)}$$

We will prove the Chain Rule at the end of the section.

EXAMPLE 1 Calculate the derivative of $y = \cos(x^3)$.

Solution As noted above, $y = \cos(x^3)$ is a composite $f(g(x))$, where

$$f(u) = \cos u, \qquad u = g(x) = x^3$$
$$f'(u) = -\sin u, \qquad g'(x) = 3x^2$$

Since $u = x^3$, $f'(g(x)) = f'(u) = f'(x^3) = -\sin(x^3)$. So, by the Chain Rule,

$$\frac{d}{dx}\cos(x^3) = \underbrace{-\sin(x^3)}_{f'(g(x))}\ \underbrace{(3x^2)}_{g'(x)} = -3x^2\sin(x^3).$$

Alternatively, we can build this derivative using the verbal description. Here, the outside function is $\cos \square$ and the inside function is x^3. We have

- The derivative of the outside: $-\sin \square$
- The derivative of the outside at the inside: $-\sin(x^3)$
- The derivative of the outside at the inside times the derivative of the inside: $(-\sin(x^3))(3x^2)$

So $\frac{d}{dx}\cos(x^3) = -3x^2\sin(x^3)$. ∎

EXAMPLE 2 Calculate the derivative of $y = \sqrt{x^4+1}$.

Solution The function $y = \sqrt{x^4+1}$ is a composite $f(g(x))$, where

$$f(u) = u^{1/2}, \qquad u = g(x) = x^4 + 1$$

$$f'(u) = \frac{1}{2}u^{-1/2}, \qquad g'(x) = 4x^3$$

Note that $f'(g(x)) = \frac{1}{2}(x^4+1)^{-1/2}$, so by the Chain Rule,

$$\frac{d}{dx}\sqrt{x^4+1} = \underbrace{\frac{1}{2}(x^4+1)^{-1/2}}_{f'(g(x))} \underbrace{(4x^3)}_{g'(x)} = \frac{4x^3}{2\sqrt{x^4+1}}$$

Alternatively, the outside is $\sqrt{\square} = \square^{1/2}$ and the inside is $x^4 + 1$. Therefore,

- The derivative of the outside: $\frac{1}{2}\square^{-1/2}$
- The derivative of the outside at the inside: $\frac{1}{2}(x^4+1)^{-1/2}$
- The derivative of the outside at the inside times the derivative of the inside: $\frac{1}{2}(x^4+1)^{-1/2}4x^3$

So $\frac{d}{dx}\sqrt{x^4+1} = \frac{4x^3}{2\sqrt{x^4+1}}$. ∎

EXAMPLE 3 Calculate the derivative of $y = e^{1+\sin x}$.

Solution The function $y = e^{1+\sin x}$ is a composite $f(g(x))$, where

$$f(u) = e^u, \qquad u = g(x) = 1 + \sin x$$

$$f'(u) = e^u, \qquad g'(x) = \cos x$$

Note that $f'(g(x)) = e^{1+\sin x}$, so by the Chain Rule,

$$\frac{d}{dx}e^{1+\sin x} = (\cos x)e^{1+\sin x}$$

Alternatively, the outside is $e^{\square}$ and the inside is $1 + \sin x$. Therefore,

- The derivative of the outside: $e^{\square}$
- The derivative of the outside at the inside: $e^{1+\sin x}$
- The derivative of the outside at the inside times the derivative of the inside: $(\cos x)e^{1+\sin x}$

So $\frac{d}{dx}e^{1+\sin x} = (\cos x)e^{1+\sin x}$. ∎

EXAMPLE 4 In Example 5 in Section 1.4 we introduced $L(t) = 12 + 3.1\sin(\frac{2\pi}{365}t)$ as a model for the length of a day in hours from sunrise to sunset in Orange City, Iowa, where t is the day in the year after the spring equinox on March 21. Determine $L'(t)$ and use it to calculate the rate that the length of the days are changing on July 1, August 9, September 15, and October 1.

Solution By the Chain Rule,

$$L'(t) = 3.1 \cos\left(\frac{2\pi}{365}t\right) \cdot \frac{2\pi}{365} = \frac{6.2\pi}{365} \cos\left(\frac{2\pi}{365}t\right)$$

We use this formula to compute L' for the given dates as follows:

July 1 corresponds to $t = 102$, and $L'(102) \approx -0.01$ h/day ≈ -0.6 min/day.

August 9 corresponds to $t = 142$, and $L'(142) \approx -0.04$ h/day ≈ -2.4 min/day.

September 15 corresponds to $t = 179$, and $L'(179) \approx -0.05$ h/day ≈ -3.0 min/day.

October 1 corresponds to $t = 195$, and $L'(195) \approx -0.05$ h/day ≈ -3.0 min/day.

These results may confirm your experience with the changing length of days in the summer. Once summer begins, the lengths of the days start to diminish ($L'(t) < 0$). As we see here, this rate of decrease gets larger throughout the summer, and as summer is ending and fall arrives the days are shortening fastest. ■

It is instructive to write the Chain Rule in Leibniz notation. Let

$$y = f(u) = f(g(x))$$

Then, by the Chain Rule, $\dfrac{dy}{dx} = f'(u)g'(x) = \dfrac{df}{du}\dfrac{du}{dx}$, or

$$\boxed{\frac{dy}{dx} = \frac{dy}{du}\frac{du}{dx}}$$

CONCEPTUAL INSIGHT In Leibniz notation, it appears as if we are multiplying fractions and the Chain Rule is simply a matter of "canceling the du." Since the expressions dy/du and du/dx are not fractions, this does not make sense literally, but it does suggest that derivatives behave *as if they were fractions* (this is reasonable because a derivative is a *limit* of fractions, namely of the difference quotients). Leibniz's form also emphasizes a key aspect of the Chain Rule: *Rates of change multiply*. To illustrate, suppose that (thanks to your knowledge of calculus) your salary increases twice as fast as your friend's. If your friend's salary increases $4000 per year, your salary will increase at the rate of 2×4000 or $8000/yr. In terms of derivatives,

$$\frac{d(\text{your salary})}{dt} = \frac{d(\text{your salary})}{d(\text{friend's salary})} \times \frac{d(\text{friend's salary})}{dt}$$

$$\$8000/\text{yr} = \qquad 2 \qquad \times \qquad \$4000/\text{yr}$$

Christiaan Huygens (1629–1695), one of the greatest scientists of his age, was Leibniz's teacher in mathematics and physics. He admired Isaac Newton greatly but did not accept Newton's theory of gravitation. He referred to it as the "improbable principle of attraction," because it did not explain how two masses separated by a distance could influence each other.

EXAMPLE 5 A spherical balloon has a radius r that is increasing at a rate of 3 cm/s. At what rate is the volume V of the balloon increasing when $r = 10$ cm?

Solution Because we are asked to determine the rate at which V is increasing, we must find dV/dt. We are given that $dr/dt = 3$ cm/s. The Chain Rule allows us to express dV/dt in terms of dV/dr and dr/dt:

$$\underbrace{\frac{dV}{dt}}_{\substack{\text{Rate of change of volume}\\\text{with respect to time}}} = \underbrace{\frac{dV}{dr}}_{\substack{\text{Rate of change of volume}\\\text{with respect to radius}}} \times \underbrace{\frac{dr}{dt}}_{\substack{\text{Rate of change of radius}\\\text{with respect to time}}}$$

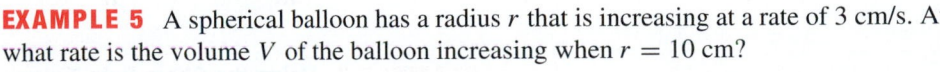

To compute dV/dr, we use the formula for the volume of a sphere, $V = \frac{4}{3}\pi r^3$:

$$\frac{dV}{dr} = \frac{d}{dr}\left(\frac{4}{3}\pi r^3\right) = 4\pi r^2$$

Because $dr/dt = 3$, we obtain

$$\frac{dV}{dt} = \frac{dV}{dr}\frac{dr}{dt} = 4\pi r^2(3) = 12\pi r^2$$

For $r = 10$,

$$\left.\frac{dV}{dt}\right|_{r=10} = (12\pi)10^2 = 1200\pi \approx 3770$$

The volume of the balloon is increasing at a rate of approximately 3770 cm^3/s. ∎

We now discuss some important special cases of the Chain Rule.

THEOREM 2 General Power and Exponential Rules   If g is differentiable, then

- $\dfrac{d}{dx}(g(x))^n = n(g(x))^{n-1}g'(x)$ (for any number n)

- $\dfrac{d}{dx}e^{g(x)} = e^{g(x)}g'(x)$

Proof Let $f(u) = u^n$. Then $(g(x))^n = f(g(x))$, and the Chain Rule yields

$$\frac{d}{dx}(g(x))^n = f'(g(x))g'(x) = n(g(x))^{n-1}g'(x)$$

On the other hand, $e^{g(x)} = h(g(x))$, where $h(u) = e^u$. We obtain

$$\frac{d}{dx}e^{g(x)} = h'(g(x))g'(x) = e^{g(x)}g'(x)$$ ∎

EXAMPLE 6 General Power and Exponential Rules   Find the derivatives of
(a) $y = (x^2 + 7x + 2)^{-1/3}$ and   **(b)** $y = e^{\cos t}$.

Solution Apply $\dfrac{d}{dx}g(x)^n = ng(x)^{n-1}g'(x)$ in (a) and $\dfrac{d}{dx}e^{g(x)} = e^{g(x)}g'(x)$ in (b).

(a)
$$\frac{d}{dx}(x^2 + 7x + 2)^{-1/3} = -\frac{1}{3}(x^2 + 7x + 2)^{-4/3}\frac{d}{dx}(x^2 + 7x + 2)$$
$$= -\frac{1}{3}(x^2 + 7x + 2)^{-4/3}(2x + 7)$$

(b)
$$\frac{d}{dt}e^{\cos t} = e^{\cos t}\frac{d}{dt}(\cos t) = -(\sin t)e^{\cos t}$$ ∎

EXAMPLE 7 Using the Chain Rule Twice   Calculate $\dfrac{d}{dx}\sqrt{1 + \sqrt{x^2 + 1}}$.

Solution In the computation that follows, we apply the Chain Rule, first to the square root of the inside function $u = 1 + \sqrt{x^2 + 1}$ and then to the derivative of the inside function:

$$\frac{d}{dx}\left(1 + (x^2 + 1)^{1/2}\right)^{1/2} = \frac{1}{2}\left(1 + (x^2 + 1)^{1/2}\right)^{-1/2}\frac{d}{dx}\left(1 + (x^2 + 1)^{1/2}\right)$$
$$= \frac{1}{2}(1 + (x^2 + 1)^{1/2})^{-1/2}\left(\frac{1}{2}(x^2 + 1)^{-1/2}(2x)\right)$$
$$= \frac{1}{2}x(x^2 + 1)^{-1/2}\left(1 + (x^2 + 1)^{1/2}\right)^{-1/2}$$ ∎

EXAMPLE 8 **The Surge Function** Surge functions have the form $S(t) = Ate^{-kt}$ for $A > 0$ and $k > 0$ (Figure 1). They model phenomena where a variable first increases to a maximum and then decreases slowly and levels off. One example is the intake of a drug into the bloodstream after ingestion and the subsequent elimination of the drug from the body.

From the form of $S(t)$ we can see that (because $e^{-kt} \approx 1$ for small t) the function initially grows like the linear function At, but subsequently the behavior is dominated by the e^{-kt} term, resulting in a decay toward 0.

There are two important features on the graph of S: The maximum of S (labeled M on the graph) and the point D where S is decreasing fastest. The maximum occurs where $S'(t) = 0$, and point D occurs where $S'(t)$ is minimum (the largest negative value of $S'(t)$). That minimum of $S'(t)$ is found where $S''(t) = 0$. Show that M occurs at $t = 1/k$, and D occurs at $t = 2/k$.

Solution Computing the derivatives, using the Product and Chain Rules, we have

$$S'(t) = Ae^{-kt} + Ate^{-kt}(-k) = (1 - kt)Ae^{-kt}$$

$$S''(t) = -kAe^{-kt} + (1 - kt)Ae^{-kt}(-k) = (-2k + k^2t)Ae^{-kt}$$

Now, since Ae^{-kt} is always positive, $S' = 0$ only when $1 - kt = 0$, that is, when $t = 1/k$. Similarly, $S'' = 0$ only when $-2k + k^2t = 0$, that is, when $t = 2/k$. ■

Proof of the Chain Rule The difference quotient for the composite $f \circ g$ is

$$\frac{f(g(x + h)) - f(g(x))}{h} \qquad (h \neq 0)$$

We express the difference quotient in a more complicated form that, we will see, results in the appearance of the appropriate terms in the Chain Rule formula when we take the limit as $h \to 0$.

$$\frac{f(g(x + h)) - f(g(x))}{h} = \frac{f(g(x + h)) - f(g(x))}{g(x + h) - g(x)} \times \frac{g(x + h) - g(x)}{h} \qquad \boxed{1}$$

This is legitimate only if the denominator $g(x + h) - g(x)$ is nonzero. Therefore, to continue our proof, we make the extra assumption that $g(x + h) - g(x) \neq 0$ for all h close to but not equal to 0. This assumption is not necessary, but without it, the argument is more technical (see Exercise 111).

Under our assumption, we may use Eq. (1) to write $(f \circ g)'(x)$ as a product of two limits:

$$(f \circ g)'(x) = \underbrace{\lim_{h \to 0} \frac{f(g(x + h)) - f(g(x))}{g(x + h) - g(x)}}_{\text{We show that this equals } f'(g(x)).} \times \underbrace{\lim_{h \to 0} \frac{g(x + h) - g(x)}{h}}_{\text{This is } g'(x).}$$

The second limit on the right is $g'(x)$. The Chain Rule will follow if we show that the first limit equals $f'(g(x))$. To verify this, set

$$k = g(x + h) - g(x)$$

Then $g(x + h) = g(x) + k$ and

$$\frac{f(g(x + h)) - f(g(x))}{g(x + h) - g(x)} = \frac{f(g(x) + k) - f(g(x))}{k}$$

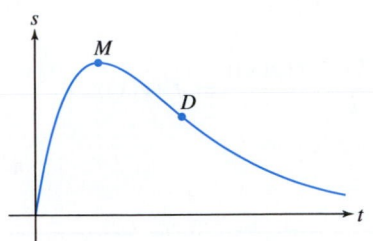

FIGURE 1 A Surge function, $S(t) = Ate^{-kt}$.

Notice that $S'(0) = A$, confirming that, at least at the outset, $S(t)$ grows like the linear term At.

The function g is continuous because it is differentiable. Therefore, $g(x + h)$ tends to $g(x)$ and $k = g(x + h) - g(x)$ tends to zero as $h \to 0$. Thus, we may rewrite the limit in terms of k to obtain the desired result:

$$\lim_{h \to 0} \frac{f(g(x + h)) - f(g(x))}{g(x + h) - g(x)} = \lim_{k \to 0} \frac{f(g(x) + k) - f(g(x))}{k} = f'(g(x))$$

It now follows that $(f \circ g)'(x) = f'(g(x))g'(x)$.   ■

3.7 SUMMARY

- The Chain Rule expresses $(f \circ g)'$ in terms of f' and g':

$$(f(g(x)))' = f'(g(x))\,g'(x)$$

- In Leibniz notation: $\dfrac{dy}{dx} = \dfrac{dy}{du}\dfrac{du}{dx}$, where $y = f(u)$ and $u = g(x)$

- General Power Rule: $\dfrac{d}{dx}(g(x))^n = n(g(x))^{n-1}g'(x)$

- General Exponential Rule: $\dfrac{d}{dx}e^{g(x)} = e^{g(x)}g'(x)$

3.7 EXERCISES

Preliminary Questions

1. Identify the outside and inside functions for each of these composite functions.
(a) $y = \sqrt{4x + 9x^2}$
(b) $y = \tan(x^2 + 1)$
(c) $y = \sec^5 x$
(d) $y = (1 + e^x)^4$

2. Which of the following can be differentiated *without* using the Chain Rule?
(a) $y = \tan(7x^2 + 2)$
(b) $y = \dfrac{x}{x + 1}$

(c) $y = \sqrt{x} \cdot \sec x$
(d) $y = x\sqrt{\sec x}$
(e) $y = xe^x$
(f) $y = \sin(e^x)$

3. Which is the derivative of $f(5x)$?
(a) $5f'(x)$
(b) $5f'(5x)$
(c) $f'(5x)$

4. Suppose that $f'(4) = g(4) = g'(4) = 1$. Do we have enough information to compute $F'(4)$, where $F(x) = f(g(x))$? If not, what is missing?

Exercises

In Exercises 1–4, fill in a table of the following type:

$f(g(x))$	$f'(u)$	$f'(g(x))$	$g'(x)$	$(f \circ g)'(x)$

1. $f(u) = u^{3/2}$, $g(x) = x^4 + 1$

2. $f(u) = u^3$, $g(x) = 3x + 5$

3. $f(u) = \tan u$, $g(x) = x^4$

4. $f(u) = u^4 + u$, $g(x) = \cos x$

In Exercises 5 and 6, write the function as a composite $f(g(x))$ and compute the derivative using the Chain Rule.

5. $y = (x + \sin x)^4$
6. $y = \cos(x^3)$

7. Calculate $\dfrac{d}{dx}\cos u$ for the following choices of $u(x)$:
(a) $u(x) = 9 - x^2$
(b) $u(x) = x^{-1}$
(c) $u(x) = \tan x$

8. Calculate $\dfrac{d}{dx}f(x^2 + 1)$ for the following choices of $f(u)$:
(a) $f(u) = \sin u$
(b) $f(u) = 3u^{3/2}$
(c) $f(u) = u^2 - u$

9. Compute $\dfrac{df}{dx}$ if $\dfrac{df}{du} = 2$ and $\dfrac{du}{dx} = 6$.

10. Compute $\dfrac{df}{dx}\Big|_{x=2}$ if $f(u) = u^2$, $u(2) = -5$, and $u'(2) = -5$.

11. Let $f(x) = (2x^2 - 5)^2$. Compute $f'(x)$ three different ways: 1) Multiplying out and then differentiating, 2) using the Product Rule, and 3) using the Chain Rule. Show that the results coincide.

12. Let $f(x) = (x + \sin x)^{-1}$. Compute $f'(x)$ separately using the Quotient Rule and the Chain Rule. Show that the results coincide.

In Exercises 13–24, compute the derivative using derivative rules that have been introduced so far.

13. $y = (x^4 + 5)^3$
14. $y = (8x^4 + 5)^3$
15. $y = \sqrt{7x - 3}$
16. $y = (4 - 2x - 3x^2)^5$

17. $y = (x^2 + 9x)^{-2}$

18. $y = (x^3 + 3x + 9)^{-4/3}$

19. $y = \cos^4 \theta$

20. $y = \cos(9\theta + 41)$

21. $y = (2\cos\theta + 5\sin\theta)^9$

22. $y = \sqrt{9 + x + \sin x}$

23. $y = e^{x-12}$

24. $y = e^{8x+9}$

In Exercises 25–28, compute the derivative of $f \circ g$.

25. $f(u) = \sin u, \quad g(x) = 2x + 1$

26. $f(u) = 2u + 1, \quad g(x) = \sin x$

27. $f(u) = e^u, \quad g(x) = x + x^{-1}$

28. $f(u) = \dfrac{u}{u-1}, \quad g(x) = \csc x$

In Exercises 29 and 30, find the derivatives of $f(g(x))$ and $g(f(u))$.

29. $f(u) = \cos u, \quad g(x) = x^2 + 1$

30. $f(u) = u^3, \quad g(x) = \dfrac{1}{x+1}$

In Exercises 31–44, use the Chain Rule to find the derivative.

31. $y = \sin(x^2)$

32. $y = \sin^2 x$

33. $y = \sqrt{t^2 + 9}$

34. $y = (t^2 + 3t + 1)^{-5/2}$

35. $y = (x^4 - x^3 - 1)^{2/3}$

36. $y = (\sqrt{x+1} - 1)^{3/2}$

37. $y = \left(\dfrac{x+1}{x-1}\right)^4$

38. $y = \cos^3(12\theta)$

39. $y = \sec \dfrac{1}{x}$

40. $y = \tan(\theta^2 - 4\theta)$

41. $y = \tan(\theta + \cos\theta)$

42. $y = e^{2x^2}$

43. $y = e^{2 - 9t^2}$

44. $y = \cos^3(e^{4\theta})$

In Exercises 45–74, compute the derivative using derivative rules that have been introduced so far.

45. $y = \tan(x^2 + 4x)$

46. $y = \sin(x^2 + 4x)$

47. $y = x\cos(1 - 3x)$

48. $y = \sin(x^2)\cos(x^2)$

49. $y = (4t + 9)^{1/2}$

50. $y = (z + 1)^4(2z - 1)^3$

51. $y = (x^3 + \cos x)^{-4}$

52. $y = \sin(\cos(\sin x))$

53. $y = \sqrt{\sin x \cos x}$

54. $y = (9 - (5 - 2x^4)^7)^3$

55. $y = (\cos 6x + \sin x^2)^{1/2}$

56. $y = \dfrac{(x+1)^{1/2}}{x+2}$

57. $y = \tan^3 x + \tan(x^3)$

58. $y = \sqrt{4 - 3\cos x}$

59. $y = \sqrt{\dfrac{z+1}{z-1}}$

60. $y = (\cos^3 x + 3\cos x + 7)^9$

61. $y = \dfrac{\cos(1 + x)}{1 + \cos x}$

62. $y = \sec(\sqrt{t^2 - 9})$

63. $y = \cot^7(x^5)$

64. $y = \dfrac{\cos(1/x)}{1 + x^2}$

65. $y = \left(1 + \cot^5(x^4 + 1)\right)^9$

66. $y = 4e^{-x} + 7e^{-2x}$

67. $y = (2e^{3x} + 3e^{-2x})^4$

68. $y = \cos(te^{-2t})$

69. $y = e^{(x^2 + 2x + 3)^2}$

70. $y = e^{e^x}$

71. $y = \sqrt{1 + \sqrt{1 + \sqrt{x}}}$

72. $y = \sqrt{\sqrt{x+1} + 1}$

73. $y = (kx + b)^{-1/3}; \quad k$ and b constants

74. $y = \dfrac{1}{\sqrt{kt^4 + b}}; \quad k$ and b constants

In Exercises 75–78, compute the higher derivative.

75. $\dfrac{d^2}{dx^2}\sin(x^2)$

76. $\dfrac{d^2}{dx^2}(x^2 + 9)^5$

77. $\dfrac{d^3}{dx^3}(9 - x)^8$

78. $\dfrac{d^3}{dx^3}\sin(2x)$

79. Assume that the average molecular velocity v of a gas in a particular container is given by $v(T) = 29\sqrt{T}$ m/s, where T is the temperature in kelvins. The temperature is related to the pressure (in atmospheres) by $T = 200P$. Find $\left.\dfrac{dv}{dP}\right|_{P=1.5}$.

80. The power P in a circuit is $P = Ri^2$, where R is the resistance and i is the current. Find dP/dt at $t = \frac{1}{3}$ if $R = 1000\ \Omega$ and i varies according to $i = \sin(4\pi t)$ (time in seconds).

81. An expanding sphere has radius $r = 0.4t$ cm at time t (in seconds). Let V be the sphere's volume. Find dV/dt when (a) $r = 3$ and (b) $t = 3$.

82. The function $L(t) = 12 + 5.5\sin(\frac{2\pi}{365}t)$ models the length of a day from sunrise to sunset in Moscow, Russia, where t is the day in the year after the spring equinox on March 21. Determine $L'(t)$, and use it to calculate the rate that the length of the days are changing on March 25, April 30, and June 10. Discuss what the results say about the changing day length in the spring in Moscow.

83. The function $L(t) = 12 + 3.4\sin(\frac{2\pi}{365}t)$ models the length of a day from sunrise to sunset in Sapporo, Japan, where t is the day in the year after the spring equinox on March 21. Determine $L'(t)$, and use it to calculate the rate that the length of the days are changing on December 1, January 1, and February 1. Discuss what the results say about the changing day length in the late fall and winter in Sapporo.

84. The general shape of the graph of $S(t) = At^2 e^{-kt}$ for $A, k > 0$ is shown in Figure 2(A). There are two points on the graph where the tangent line is horizontal and therefore where $S'(t) = 0$. Determine t in terms of k for these points by solving $S'(t) = 0$.

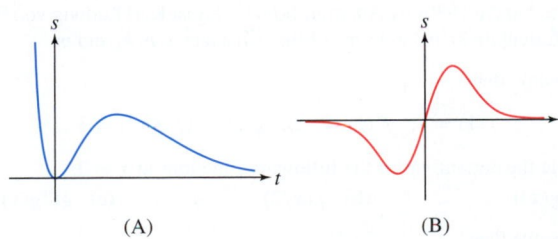

(A)   (B)

FIGURE 2

85. The general shape of the graph of $S(t) = Ate^{-kt^2}$ for $A, k > 0$ is shown in Figure 2(B). There are two points on the graph where the tangent line is horizontal and therefore where $S'(t) = 0$. Determine t in terms of k for these points by solving $S'(t) = 0$.

86. A 2005 study by the Fisheries Research Services in Aberdeen, Scotland, suggests that the average length of the species *Clupea harengus* (Atlantic herring) as a function of age t (in years) can be modeled by $L(t) = 32(1 - e^{-0.37t})$ cm for $0 \le t \le 13$. See Figure 3.

(a) How fast is the average length changing at age $t = 6$ yr?

(b) At what age is the average length changing at a rate of 5 cm/yr?

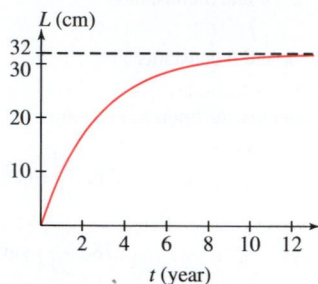

FIGURE 3 Average length of the species *Clupea harengus*.

87. A 1999 study by Starkey and Scarnecchia developed the following model for the average weight (in kilograms) at age t (in years) of channel catfish in the Lower Yellowstone River (Figure 4):

$$W(t) = \left(3.46293 - 3.32173e^{-0.03456t}\right)^{3.4026}$$

Find the rate at which average weight is changing at age $t = 10$.

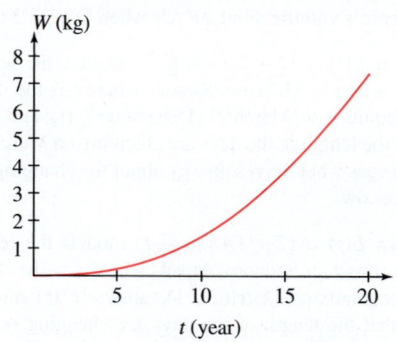

FIGURE 4 Average weight of channel catfish at age t.

88. The functions in Exercises 86 and 87 are examples of the **von Bertalanffy growth function**

$$M(t) = \left(a + (b - a)e^{kmt}\right)^{1/m} \qquad (m \ne 0)$$

introduced in the 1930s by Austrian-born biologist Karl Ludwig von Bertalanffy. Calculate $M'(0)$ in terms of the constants a, b, k, and m.

89. Assume that

$$f(1) = 4, \quad f'(1) = -3, \quad g(2) = 1, \quad g'(2) = 3$$

Calculate the derivatives of the following functions at $x = 2$:

(a) $f(g(x))$ **(b)** $f(x/2)$ **(c)** $g(2g(x))$

90. Assume that

$$f(0) = 2, \quad f'(0) = 3, \quad h(0) = -1, \quad h'(0) = 7$$

Calculate the derivatives of the following functions at $x = 0$:

(a) $(f(x))^3$ **(b)** $f(7x)$ **(c)** $f(4x)h(5x)$

91. Compute the derivative of $h(\sin x)$ at $x = \frac{\pi}{6}$, assuming that $h'(0.5) = 10$.

92. Let $F(x) = f(g(x))$, where the graphs of f and g are shown in Figure 5. Estimate $g'(2)$ and $f'(g(2))$ and compute $F'(2)$.

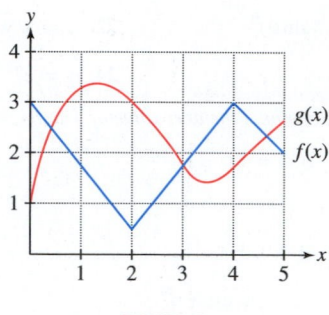

FIGURE 5

In Exercises 93–96, use the table of values to calculate the derivative of the function at the given point.

x	1	4	6
$f(x)$	4	0	6
$f'(x)$	5	7	4
$g(x)$	4	1	6
$g'(x)$	5	$\frac{1}{2}$	3

93. $f(g(x))$, $x = 6$ **94.** $e^{f(x)}$, $x = 4$

95. $g(\sqrt{x})$, $x = 16$ **96.** $f(2x + g(x))$, $x = 1$

97. The price (in dollars) of a computer component is $P = 2C - 18C^{-1}$, where C is the manufacturer's cost to produce it. Assume that cost at time t (in years) is $C = 9 + 3t^{-1}$. Determine the rate of change of price with respect to time at $t = 3$.

98. **GU** Plot the "astroid" $y = (4 - x^{2/3})^{3/2}$ for $0 \le x \le 8$. Show that the first-quadrant segment of each tangent line has length 8.

99. According to the U.S. standard atmospheric model, developed by the National Oceanic and Atmospheric Administration for use in aircraft and rocket design, atmospheric temperature T (in degrees Celsius), pressure P (kPa = 1000 pascals), and altitude h (in meters) are related by these formulas (valid in the troposphere $h \le 11,000$):

$$T = 15.04 - 0.00649h, \qquad P = 101.29\left(\frac{T + 273.1}{288.08}\right)^{5.256}$$

Use the Chain Rule to calculate dP/dh. Express the result in terms of h.

100. Climate scientists use the **Stefan–Boltzmann Law** $R = \sigma T^4$ to estimate the change in the earth's average temperature T (in kelvins) that results from a change in the radiation R (in joules per square meter per second) that the earth receives from the sun. Here, $\sigma = 5.67 \times 10^{-8}$ Js^{-1}m^{-2}K^{-4}. Calculate dR/dt, assuming that $T = 283$ and $\frac{dT}{dt} = 0.05$ K/year. What are the units of dR/dt?

101. In the setting of Exercise 100, calculate the rate of change of T (in K/yr) if $T = 283$ K and R increases at a rate of 0.5 Js^{-1}m^{-2} per yr.

102. **CAS** Use a computer algebra system to compute $f^{(k)}(x)$ for $k = 1, 2, 3$ for the following functions:

(a) $f(x) = \cot(x^2)$ **(b)** $f(x) = \sqrt{x^3 + 1}$

103. Use the Chain Rule to express the second derivative of $f \circ g$ in terms of the first and second derivatives of f and g.

104. Compute the second derivative of $\sin(g(x))$ at $x = 2$, assuming that $g(2) = \frac{\pi}{4}$, $g'(2) = 5$, and $g''(2) = 3$.

Further Insights and Challenges

105. Show that if f, g, and h are differentiable, then

$$[f(g(h(x)))]' = f'(g(h(x)))g'(h(x))h'(x)$$

106. [✎] Show that differentiation reverses parity: If f is even, then f' is odd, and if f is odd, then f' is even. *Hint:* Differentiate $f(-x)$.

107. (a) [✎] Sketch a graph of any even function f and explain graphically why f' is odd.

(b) Suppose that f' is even. Is f necessarily odd? *Hint:* Check whether this is true for linear functions.

108. Power Rule for Fractional Exponents Consider the functions $f(u) = u^q$ and $g(x) = x^{p/q}$. Assume that g is differentiable.

(a) Show that $f(g(x)) = x^p$ (recall the Laws of Exponents).

(b) Apply the Chain Rule and the Power Rule for whole-number exponents to show that $f'(g(x))\,g'(x) = px^{p-1}$.

(c) Use the result of (b) to derive the Power Rule for $x^{p/q}$.

109. Prove that for all whole numbers $n \geq 1$,

$$\frac{d^n}{dx^n}\sin x = \sin\left(x + \frac{n\pi}{2}\right)$$

Hint: Use the identity $\cos x = \sin\left(x + \frac{\pi}{2}\right)$.

110. A Discontinuous Derivative Use the limit definition to show that $g'(0)$ exists but $g'(0) \neq \lim_{x\to 0} g'(x)$, where

$$g(x) = \begin{cases} x^2 \sin\dfrac{1}{x} & x \neq 0 \\ 0 & x = 0 \end{cases}$$

111. Chain Rule This exercise proves the Chain Rule without the special assumption made in the text. For any number b, define a new function

$$F(u) = \frac{f(u) - f(b)}{u - b} \quad \text{for all } u \neq b$$

(a) Show that if we define $F(b) = f'(b)$, then F is continuous at $u = b$.

(b) Take $b = g(a)$. Show that if $x \neq a$, then for all u,

$$\frac{f(u) - f(g(a))}{x - a} = F(u)\frac{u - g(a)}{x - a} \qquad \boxed{2}$$

Note that both sides are zero if $u = g(a)$.

(c) Substitute $u = g(x)$ in Eq. (2) to obtain

$$\frac{f(g(x)) - f(g(a))}{x - a} = F(g(x))\frac{g(x) - g(a)}{x - a}$$

Derive the Chain Rule by computing the limit of both sides as $x \to a$.

3.8 Implicit Differentiation

We have developed techniques for calculating a derivative dy/dx when y is given in terms of x by a formula—such as $y = x^3 + 1$; that is, when y is expressed *explicitly* as a function of x. But suppose that y is instead related to x by an equation such as

$$y^4 + xy = x^3 - x + 2 \qquad \boxed{1}$$

In this case, we say that y is defined *implicitly* as a function of x. How can we find the slope of the tangent line at a point, such as $(1, 1)$, on the graph (Figure 1)? Although it may be difficult or even impossible to solve for y explicitly as a function of x, we can find dy/dx using the method of **implicit differentiation** (see Example 2).

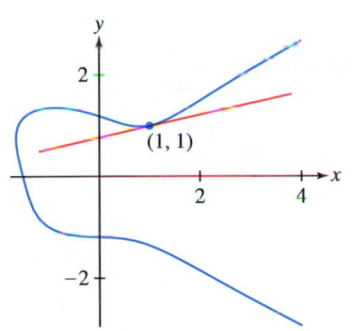

FIGURE 1 Graph of $y^4 + xy = x^3 - x + 2$.

To illustrate, we first compute the slope of the tangent line to the unit circle at $\left(\frac{3}{5}, \frac{4}{5}\right)$ (Figure 2). The equation of the unit circle is

$$x^2 + y^2 = 1$$

Compute dy/dx by taking the derivative of both sides of the equation:

$$\frac{d}{dx}(x^2 + y^2) = \frac{d}{dx}(1)$$

$$\frac{d}{dx}(x^2) + \frac{d}{dx}(y^2) = 0$$

$$2x + \frac{d}{dx}(y^2) = 0 \qquad \boxed{2}$$

DF **FIGURE 2** The tangent line to the unit circle $x^2 + y^2 = 1$ at P has slope $-\frac{3}{4}$.

How do we handle the term $\frac{d}{dx}(y^2)$? We use the Chain Rule. Think of y as a function $y = y(x)$. Then $y^2 = (y(x))^2$ and by the Chain Rule,

$$\frac{d}{dx}y^2 = \frac{d}{dx}(y(x))^2 = 2y(x)\frac{dy}{dx} = 2y\frac{dy}{dx}$$

Equation (2) becomes $2x + 2y\frac{dy}{dx} = 0$, and we can solve for $\frac{dy}{dx}$ if $y \neq 0$:

$$\boxed{\frac{dy}{dx} = -\frac{x}{y}}$$ **3**

EXAMPLE 1 Use Eq. (3) to find the slope of the tangent line at the point $P = \left(\frac{3}{5}, \frac{4}{5}\right)$ on the unit circle.

Solution Set $x = \frac{3}{5}$ and $y = \frac{4}{5}$ in Eq. (3):

$$\frac{dy}{dx}\bigg|_P = -\frac{x}{y} = -\frac{\frac{3}{5}}{\frac{4}{5}} = -\frac{3}{4}$$ ∎

In this particular example, we could have computed dy/dx directly, without implicit differentiation. The upper semicircle is the graph of $y = \sqrt{1 - x^2}$ and

$$\frac{dy}{dx} = \frac{d}{dx}\sqrt{1 - x^2} = \frac{1}{2}\left(1 - x^2\right)^{-1/2}\frac{d}{dx}\left(1 - x^2\right) = -\frac{x}{\sqrt{1 - x^2}}$$

This formula expresses dy/dx in terms of x alone, whereas Eq. (3) expresses dy/dx in terms of both x and y, as is typical when we use implicit differentiation. The two formulas agree because $y = \sqrt{1 - x^2}$.

Before presenting additional examples, let's examine again how the factor dy/dx arises when we differentiate an expression involving y with respect to x. It would not appear if we were differentiating with respect to y. Thus,

Notice what happens if we apply the Chain Rule to $\frac{d}{dy}\sin y$. The extra derivative factor appears, but it is equal to 1:

$$\frac{d}{dy}\sin y = (\cos y)\frac{dy}{dy} = \cos y$$

$$\frac{d}{dy}\sin y = \cos y \qquad \text{but} \qquad \frac{d}{dx}\sin y = (\cos y)\frac{dy}{dx}$$

$$\frac{d}{dy}y^4 = 4y^3 \qquad \text{but} \qquad \frac{d}{dx}y^4 = 4y^3\frac{dy}{dx}$$

Similarly, the Product Rule applied to xy yields

$$\frac{d}{dx}(xy) = \frac{dx}{dx}y + x\frac{dy}{dx} = y + x\frac{dy}{dx}$$

The Quotient Rule applied to t^2/y yields

$$\frac{d}{dt}\left(\frac{t^2}{y}\right) = \frac{y\frac{d}{dt}t^2 - t^2\frac{dy}{dt}}{y^2} = \frac{2ty - t^2\frac{dy}{dt}}{y^2}$$

EXAMPLE 2 Find an equation of the tangent line at the point $P = (1, 1)$ on the curve in Figure 1 with equation

$$y^4 + xy = x^3 - x + 2$$

Solution We break up the calculation into two steps.

Step 1. Differentiate both sides of the equation with respect to x.
Note that each occurrence of y in the original equation generates an additional $\frac{dy}{dx}$ upon differentiation.

$$\frac{d}{dx}y^4 + \frac{d}{dx}(xy) = \frac{d}{dx}\left(x^3 - x + 2\right)$$

$$4y^3\frac{dy}{dx} + \left(y + x\frac{dy}{dx}\right) = 3x^2 - 1$$ **4**

Step 2. **Solve for** $\dfrac{dy}{dx}$.

Move the terms involving dy/dx in Eq. (4) to the left and place the remaining terms on the right:

$$4y^3\frac{dy}{dx} + x\frac{dy}{dx} = 3x^2 - 1 - y$$

Then factor out dy/dx and divide:

$$\left(4y^3 + x\right)\frac{dy}{dx} = 3x^2 - 1 - y$$

$$\frac{dy}{dx} = \frac{3x^2 - 1 - y}{4y^3 + x} \qquad \boxed{5}$$

To find the derivative at $P = (1, 1)$, apply Eq. (5) with $x = 1$ and $y = 1$:

$$\left.\frac{dy}{dx}\right|_{(1,1)} = \frac{3\cdot 1^2 - 1 - 1}{4\cdot 1^3 + 1} = \frac{1}{5}$$

An equation of the tangent line is $y - 1 = \frac{1}{5}(x - 1)$ or $y = \frac{1}{5}x + \frac{4}{5}$. ∎

CONCEPTUAL INSIGHT The graph of an equation does not always define a function of x because there may be more than one y-value for a given value of x. Implicit differentiation works because the graph is generally made up of several pieces called **branches**, each of which does define a function (a proof of this fact relies on the Implicit Function Theorem from advanced calculus). For example, the branches of the unit circle $x^2 + y^2 = 1$ are the graphs of the functions $y = \sqrt{1 - x^2}$ and $y = -\sqrt{1 - x^2}$. Similarly, the graph in Figure 3 has an upper and a lower branch. In most examples, the branches are differentiable except at certain exceptional points where the tangent line may be vertical.

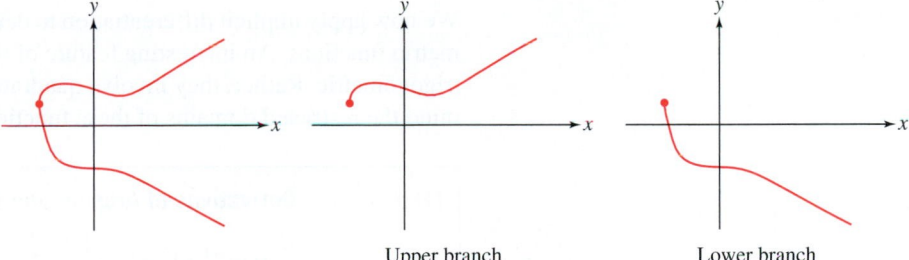

FIGURE 3 Each branch of the graph of $y^4 + xy = x^3 - x + 2$ defines a function of x.

Upper branch   Lower branch

EXAMPLE 3 Find the slope of the tangent line at the point $P = (1, 1)$ on the graph of $e^{x-y} = 2x^2 - y^2$.

Solution We follow the steps of the previous example, this time writing y' for dy/dx:

$$\frac{d}{dx}e^{x-y} = \frac{d}{dx}(2x^2 - y^2)$$

$$e^{x-y}(1 - y') = 4x - 2yy' \qquad \text{(Chain Rule applied to } e^{x-y} \text{ and } y^2\text{)}$$

$$e^{x-y} - e^{x-y}y' = 4x - 2yy'$$

$$(2y - e^{x-y})y' = 4x - e^{x-y} \qquad \text{(place all } y'\text{-terms on left)}$$

$$y' = \frac{4x - e^{x-y}}{2y - e^{x-y}}$$

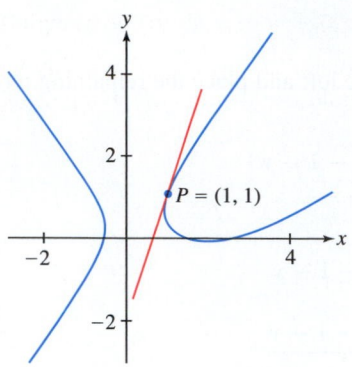

FIGURE 4 Graph of $e^{x-y} = 2x^2 - y^2$.

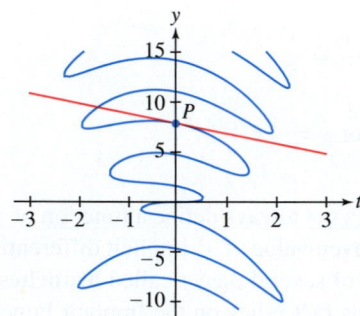

DF FIGURE 5 Graph of

$$y \cos(y + t + t^2) = t^3$$

The tangent line at $P = \left(0, \frac{5\pi}{2}\right)$ has slope -1.

The slope of the tangent line at $P = (1, 1)$ is (Figure 4)

$$\left.\frac{dy}{dx}\right|_{(1,1)} = \frac{4(1) - e^{1-1}}{2(1) - e^{1-1}} = \frac{4 - 1}{2 - 1} = 3$$ ∎

EXAMPLE 4   **Shortcut to Derivative at a Specific Point**   Calculate $\left.\dfrac{dy}{dt}\right|_P$ at the point $P = \left(0, \frac{5\pi}{2}\right)$ on the curve (Figure 5):

$$y \cos(y + t + t^2) = t^3$$

Solution   As before, differentiate both sides of the equation (we write y' for dy/dt):

$$\frac{d}{dt} y \cos(y + t + t^2) = \frac{d}{dt} t^3$$

$$y' \cos(y + t + t^2) - y \sin(y + t + t^2)(y' + 1 + 2t) = 3t^2 \qquad \boxed{6}$$

We could continue to solve for y' in terms of t and y, but that is not necessary since we are only interested in dy/dt at the point P. Instead, we can substitute $t = 0$, $y = \frac{5\pi}{2}$ directly in Eq. (6) and then solve for y':

$$y' \cos\left(\frac{5\pi}{2} + 0 + 0^2\right) - \left(\frac{5\pi}{2}\right) \sin\left(\frac{5\pi}{2} + 0 + 0^2\right)(y' + 1 + 0) = 0$$

$$0 - \left(\frac{5\pi}{2}\right)(1)(y' + 1) = 0$$

This gives us $y' + 1 = 0$ or $y' = -1$.   ∎

Derivatives of Inverse Trigonometric Functions

We now apply implicit differentiation to determine the derivatives of the inverse trigonometric functions. An interesting feature of these functions is that their derivatives are not trigonometric. Rather, they involve quadratic expressions and their square roots. Keep in mind the restricted domains of these functions.

THEOREM 1   Derivatives of Arcsine and Arccosine

$$\frac{d}{dx}(\sin^{-1} x) = \frac{1}{\sqrt{1 - x^2}}, \qquad \frac{d}{dx}(\cos^{-1} x) = -\frac{1}{\sqrt{1 - x^2}} \qquad \boxed{7}$$

← **REMINDER** In Example 7 of Section 1.5, we used the right triangle in Figure 6 in the computation

$$\cos(\sin^{-1} x) = \cos y = \frac{\text{adjacent}}{\text{hypotenuse}}$$

$$= \sqrt{1 - x^2}$$

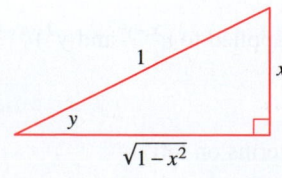

FIGURE 6 Right triangle constructed so that $\sin y = x$.

Proof   For $y = \sin^{-1} x$, our goal is to find $\frac{dy}{dx}$. By applying sine to both sides, we have

$$\sin y = x$$

Differentiating both sides of the equation with respect to x, and treating y as a function of x, we obtain

$$\cos y \frac{dy}{dx} = 1$$

$$\frac{dy}{dx} = \frac{1}{\cos y}$$

In order to determine an algebraic expression in x for $\cos y$, we construct a right triangle as in Figure 6 such that $\sin y = x$. We choose y to be its angle, and take its hypotenuse to be of length 1 and its opposite edge to have length x. Then, by the Pythagorean Theorem, its adjacent side must have length $\sqrt{1-x^2}$. We can therefore read off the triangle that $\cos y = \frac{\sqrt{1-x^2}}{1} = \sqrt{1-x^2}$. While the triangle demonstrates that this relationship holds for positive x and y, it is straightforward that it also holds for the corresponding negative values. Thus, over the domain of $f(x) = \sin^{-1} x$,

$$\frac{dy}{dx} = \frac{1}{\cos y} = \frac{1}{\sqrt{1-x^2}}$$

The derivation of $\frac{d}{dx}(\cos^{-1} x)$ is similar (see Exercise 47). ∎

EXAMPLE 5 Calculate $f'(\frac{1}{2})$, where $f(x) = \arcsin(x^2)$.

Solution Recall that $\arcsin x$ is another notation for $\sin^{-1} x$. By the Chain Rule,

$$\frac{d}{dx}\arcsin(x^2) = \frac{d}{dx}(\sin^{-1}(x^2)) = \frac{1}{\sqrt{1-(x^2)^2}}\frac{d}{dx}(x^2) = \frac{2x}{\sqrt{1-x^4}}$$

$$f'\left(\frac{1}{2}\right) = \frac{2(\frac{1}{2})}{\sqrt{1-(\frac{1}{2})^4}} = \frac{1}{\sqrt{\frac{15}{16}}} = \frac{4}{\sqrt{15}}$$ ∎

THEOREM 2 Derivatives of Inverse Trigonometric Functions

$$\frac{d}{dx}\tan^{-1} x = \frac{1}{x^2+1}, \qquad \frac{d}{dx}\cot^{-1} x = -\frac{1}{x^2+1}$$

$$\frac{d}{dx}\sec^{-1} x = \frac{1}{|x|\sqrt{x^2-1}}, \qquad \frac{d}{dx}\csc^{-1} x = -\frac{1}{|x|\sqrt{x^2-1}}$$

The proofs of the formulas in Theorem 2 are similar to the proof of Theorem 1. See Exercises 48 and 50.

EXAMPLE 6 Calculate $\left.\frac{d}{dx}(\csc^{-1}(e^x + 1))\right|_{x=0}$.

Solution Apply the Chain Rule using the formula $\frac{d}{du}\csc^{-1} u = -\frac{1}{|u|\sqrt{u^2-1}}$:

$$\frac{d}{dx}\csc^{-1}(e^x + 1) = -\frac{1}{|e^x + 1|\sqrt{(e^x + 1)^2 - 1}}\frac{d}{dx}(e^x + 1)$$

$$= -\frac{e^x}{(e^x + 1)\sqrt{e^{2x} + 2e^x}}$$

We have replaced $|e^x + 1|$ by $e^x + 1$ because this quantity is positive. Now we have

$$\left.\frac{d}{dx}\csc^{-1}(e^x + 1)\right|_{x=0} = -\frac{e^0}{(e^0 + 1)\sqrt{e^0 + 2e^0}} = -\frac{1}{2\sqrt{3}}$$ ∎

Finding Higher Order Derivatives Implicitly

By using the implicit derivative process repeatedly, we can find higher order derivatives of a function that is defined implicitly. We do so in the next example.

EXAMPLE 7 Find $\frac{d^2y}{dx^2}$ for $x^2 + 4y^2 = 7$.

Solution We differentiate with respect to x, writing y' for $\frac{dy}{dx}$:

$$2x + 8yy' = 0$$

Solving for y', we obtain

$$y' = \frac{-x}{4y}$$

Differentiating again with respect to x, we obtain

$$y'' = \frac{4y(-1) - (-x)(4y')}{16y^2} = \frac{-y + xy'}{4y^2}$$

Substituting in the fact that $y' = \frac{-x}{4y}$ yields

$$y'' = \frac{-y + x(-x/(4y))}{4y^2} = \frac{-4y^2 - x^2}{16y^3} = \frac{-7}{16y^3}$$

The last equality holds since $x^2 + 4y^2 = 7$ ∎

3.8 SUMMARY

- Implicit differentiation is used to compute dy/dx when x and y are related by an equation.

  Step 1. Take the derivative of both sides of the equation with respect to x, treating y as a function of x.

  Step 2. Solve for dy/dx by collecting the terms involving dy/dx on one side and the remaining terms on the other side of the equation.

- Remember to include the factor dy/dx when differentiating expressions involving y with respect to x. For instance,

$$\frac{d}{dx} \sin y = (\cos y) \frac{dy}{dx}$$

- Derivative formulas:

$$\frac{d}{dx} \sin^{-1} x = \frac{1}{\sqrt{1 - x^2}}, \qquad \frac{d}{dx} \cos^{-1} x = -\frac{1}{\sqrt{1 - x^2}}$$

$$\frac{d}{dx} \tan^{-1} x = \frac{1}{x^2 + 1}, \qquad \frac{d}{dx} \cot^{-1} x = -\frac{1}{x^2 + 1}$$

$$\frac{d}{dx} \sec^{-1} x = \frac{1}{|x|\sqrt{x^2 - 1}}, \qquad \frac{d}{dx} \csc^{-1} x = -\frac{1}{|x|\sqrt{x^2 - 1}}$$

3.8 EXERCISES

Preliminary Questions

1. Which differentiation rule is used to show $\dfrac{d}{dx} \sin y = \cos y \dfrac{dy}{dx}$?

2. One of (a)–(c) is incorrect. Find and correct the mistake.

(a) $\dfrac{d}{dy} \sin(y^2) = 2y \cos(y^2)$

(b) $\dfrac{d}{dx} \sin(x^2) = 2x \cos(x^2)$

(c) $\dfrac{d}{dx} \sin(y^2) = 2y \cos(y^2)$

3. On an exam, Jason was asked to differentiate the equation

$$x^2 + 2xy + y^3 = 7$$

Find the errors in Jason's answer: $2x + 2xy' + 3y^2 = 0$.

4. Which of (a) or (b) is equal to $\dfrac{d}{dx}(x \sin t)$?

(a) $(x \cos t)\dfrac{dt}{dx}$

(b) $(x \cos t)\dfrac{dt}{dx} + \sin t$

5. Determine which inverse trigonometric function g has the derivative

$$g'(x) = \frac{1}{x^2 + 1}$$

6. What does the following identity tell us about the derivatives of $\sin^{-1} x$ and $\cos^{-1} x$?

$$\sin^{-1} x + \cos^{-1} x = \frac{\pi}{2}$$

7. Assume that a is a constant and that y is implicitly a function of x. Compute the derivative with respect to x of each of a^2, x^2, and y^2.

Exercises

1. Show that if you differentiate both sides of $x^2 + 2y^3 = 6$, the result is $2x + 6y^2 \frac{dy}{dx} = 0$. Then solve for dy/dx and evaluate it at the point $(2, 1)$.

2. Show that if you differentiate both sides of $xy + 4x + 2y = 1$, the result is $(x + 2)\frac{dy}{dx} + y + 4 = 0$. Then solve for dy/dx and evaluate it at the point $(1, -1)$.

In Exercises 3–10, differentiate the expression with respect to x, assuming that y is implicitly a function of x.

3. $x^2 y^3$

4. $\dfrac{x^3}{y^2}$

5. $(x^2 + y^2)^{3/2}$

6. $\sqrt{x + y}$

7. $x\sqrt[3]{y}$

8. $\tan(xy)$

9. $\dfrac{y}{y + 1}$

10. $e^{y/x}$

In Exercises 11–28, calculate the derivative with respect to x of the other variable appearing in the equation.

11. $3y^3 + x^2 = 5$

12. $y^4 - 2y = 4x^3 + x$

13. $x^2 y + 2x^3 y = x + y$

14. $xy^2 + x^2 y^5 - x^3 = 3$

15. $x^3 R^5 = 1$

16. $x^4 + z^4 = 1$

17. $\dfrac{y}{x} + \dfrac{x}{y} = 2y$

18. $\sqrt{x + s} = \dfrac{1}{x} + \dfrac{1}{s}$

19. $y^{-2/3} + x^{3/2} = 1$

20. $x^{1/2} + y^{2/3} = -4y$

21. $y + \dfrac{1}{y} = x^2 + x$

22. $\sin(xt) = t$

23. $\sin(x + y) = x + \cos y$

24. $\tan(x^2 y) = (x + y)^3$

25. $xe^y = 2xy + y^3$

26. $e^{xy} = \sin(y^2)$

27. $e^x + e^y = x - y$

28. $e^{x^2 + y^2} = x + 4$

In Exercises 29–32, compute the derivative at the point indicated without using a calculator.

29. $y = \sin^{-1} x, \quad x = \frac{3}{5}$

30. $y = \tan^{-1} x, \quad x = \frac{1}{2}$

31. $y = \sec^{-1} x, \quad x = 4$

32. $y = \arccos(4x), \quad x = \frac{1}{5}$

In Exercises 33–46, find the derivative.

33. $y = \sin^{-1}(7x)$

34. $y = \arctan\left(\dfrac{x}{3}\right)$

35. $y = \cos^{-1}(x^2)$

36. $y = \sec^{-1}(t + 1)$

37. $y = x\tan^{-1} x$

38. $y = e^{\cos^{-1} x}$

39. $y = \arcsin(e^x)$

40. $y = \csc^{-1}(x^{-1})$

41. $y = \sqrt{1 - t^2} + \sin^{-1} t$

42. $y = \tan^{-1}\left(\dfrac{1 + t}{1 - t}\right)$

43. $y = (\tan^{-1} x)^3$

44. $y = \dfrac{\cos^{-1} x}{\sin^{-1} x}$

45. $y = \cos^{-1} t^{-1} - \sec^{-1} t$

46. $y = \cos^{-1}(x + \sin^{-1} x)$

47. Use Figure 7 to prove that $(\cos^{-1} x)' = -\dfrac{1}{\sqrt{1 - x^2}}$.

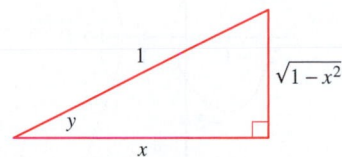

FIGURE 7 Right triangle with $y = \cos^{-1} x$.

48. Show that $(\tan^{-1} x)' = \cos^2(\tan^{-1} x)$ and then use Figure 8 to prove that $(\tan^{-1} x)' = (x^2 + 1)^{-1}$.

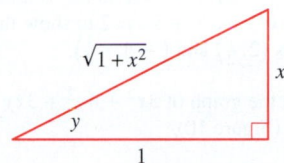

FIGURE 8 Right triangle with $y = \tan^{-1} x$.

49. Let $y = \sec^{-1} x$. Show that if $x \geq 1$, then $\tan y = \sqrt{x^2 - 1}$, and if $x \leq 1$, then $\tan y = -\sqrt{x^2 - 1}$. *Hint:* $\tan y \geq 0$ on $\left(0, \frac{\pi}{2}\right)$ and $\tan y \leq 0$ on $\left(\frac{\pi}{2}, \pi\right)$.

50. Use Exercise 49 to verify the formula

$$(\sec^{-1} x)' = \frac{1}{|x|\sqrt{x^2 - 1}}$$

51. Show that $x + yx^{-1} = 1$ and $y = x - x^2$ define the same curve [except that $(0, 0)$ is not a solution of the first equation] and that implicit differentiation yields $y' = yx^{-1} - x$ and $y' = 1 - 2x$. Explain why these formulas produce the same values for the derivative.

52. Use the method of Example 4 to compute $\frac{dy}{dx}\big|_P$ at $P = (2, 1)$ on the curve $y^2 x^3 + y^3 x^4 - 10x + y = 5$.

In Exercises 53 and 54, find dy/dx at the given point.

53. $(x + 2)^2 - 6(2y + 3)^2 = 3, \quad (1, -1)$

54. $\sin^2(3y) = x + y, \quad \left(\dfrac{2 - \pi}{4}, \dfrac{\pi}{4}\right)$

In Exercises 55–62, find an equation of the tangent line at the given point.

55. $xy + x^2 y^2 = 6, \quad (2, 1)$ **56.** $x^{2/3} + y^{2/3} = 2, \quad (1, 1)$

57. $x^2 + \sin y = xy^2 + 1, \quad (1, 0)$

58. $\sin(x - y) = x\cos\left(y + \frac{\pi}{4}\right), \quad \left(\frac{\pi}{4}, \frac{\pi}{4}\right)$

59. $2x^{1/2} + 4y^{-1/2} = xy, \quad (1, 4)$ **60.** $x^2 e^y + ye^x = 4, \quad (2, 0)$

61. $e^{2x - y} = \dfrac{x^2}{y}, \quad (2, 4)$

62. $y^2 e^{x^2 - 16} - xy^{-1} = 2, \quad (4, 2)$

63. Find the points on the graph of $y^2 = x^3 - 3x + 1$ (Figure 9) where the tangent line is horizontal, as follows:
(a) First show that $2yy' = 3x^2 - 3$, where $y' = dy/dx$.
(b) Do not solve for y'. Rather, set $y' = 0$ and solve for x. This yields two values of x where the slope may be zero.
(c) Show that the positive value of x does not correspond to a point on the graph.
(d) The negative value corresponds to the two points on the graph where the tangent line is horizontal. Find their coordinates.

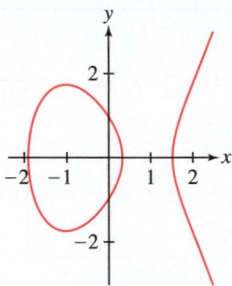

FIGURE 9 Graph of $y^2 = x^3 - 3x + 1$.

64. Show, by differentiating the equation, that if the tangent line at a point (x, y) on the curve $x^2 y - 2x + 8y = 2$ is horizontal, then $xy = 1$. Then substitute $y = x^{-1}$ in $x^2 y - 2x + 8y = 2$ to show that the tangent line is horizontal at the points $\left(2, \frac{1}{2}\right)$ and $\left(-4, -\frac{1}{4}\right)$.

65. Find all points on the graph of $3x^2 + 4y^2 + 3xy = 24$ where the tangent line is horizontal (Figure 10).

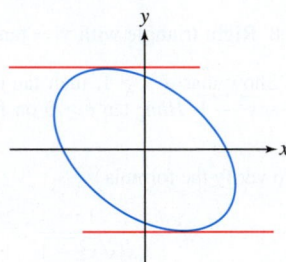

FIGURE 10 Graph of $3x^2 + 4y^2 + 3xy = 24$.

66. Show that no point on the graph of $x^2 - 3xy + y^2 = 1$ has a horizontal tangent line.

67. Figure 1 shows the graph of $y^4 + xy = x^3 - x + 2$. Find dy/dx at the two points on the graph with x-coordinate 0 and find an equation of the tangent line at each of those points.

68. Folium of Descartes The curve $x^3 + y^3 = 3xy$ (Figure 11) was first discussed in 1638 by the French philosopher-mathematician René Descartes, who called it the folium (meaning "leaf"). Descartes's scientific colleague Gilles de Roberval called it the jasmine flower. Both men believed incorrectly that the leaf shape in the first quadrant was repeated in each quadrant, giving the appearance of petals of a flower. Find an equation of the tangent line at the point $\left(\frac{2}{3}, \frac{4}{3}\right)$.

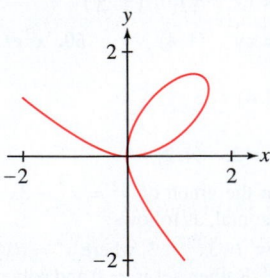

FIGURE 11 Folium of Descartes: $x^3 + y^3 = 3xy$.

69. Find a point on the folium $x^3 + y^3 = 3xy$ other than the origin at which the tangent line is horizontal.

70. 📝 (GU) Plot $x^3 + y^3 = 3xy + b$ for several values of b and describe how the graph changes as $b \to 0$. Then compute dy/dx at the point $(b^{1/3}, 0)$. How does this value change as $b \to \infty$? Do your plots confirm this conclusion?

71. Find the x-coordinates of the points where the tangent line is horizontal on the *trident curve* $xy = x^3 - 5x^2 + 2x - 1$, so named by Isaac Newton in his treatise on curves published in 1710 (Figure 12). *Hint:* $2x^3 - 5x^2 + 1 = (2x - 1)(x^2 - 2x - 1)$.

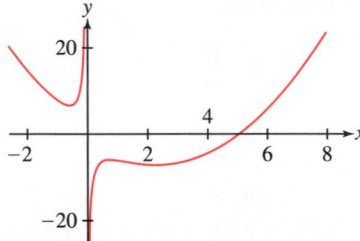

FIGURE 12 Trident curve: $xy = x^3 - 5x^2 + 2x - 1$.

72. Find an equation of the tangent line at each of the four points on the curve $(x^2 + y^2 - 4x)^2 = 2(x^2 + y^2)$ where $x = 1$. This curve (Figure 13) is an example of a *limaçon of Pascal*, named after the father of the French philosopher Blaise Pascal, who first described it in 1650.

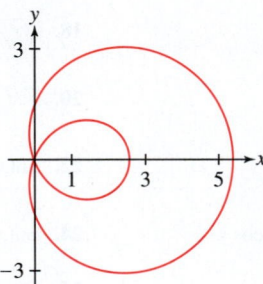

FIGURE 13 Limaçon: $(x^2 + y^2 - 4x)^2 = 2(x^2 + y^2)$.

73. Find the derivative, dy/dx, at the points where $x = 1$ on the folium $(x^2 + y^2)^2 = \frac{25}{4} xy^2$. See Figure 14.

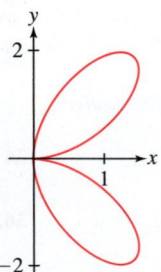

FIGURE 14 Folium curve: $(x^2 + y^2)^2 = \dfrac{25}{4} xy^2$.

74. (CAS) Plot $(x^2 + y^2)^2 = 12(x^2 - y^2) + 2$ for x and y between -4 and 4 using a computer algebra system. How many horizontal tangent lines does the curve appear to have? Find the points where these occur.

75. Calculate dx/dy for the equation $y^4 + 1 = y^2 + x^2$ and find the points on the graph where the tangent line is vertical.

76. Show that the tangent lines at $x = 1 \pm \sqrt{2}$ to the *conchoid* with equation $(x - 1)^2(x^2 + y^2) = 2x^2$ are vertical (Figure 15).

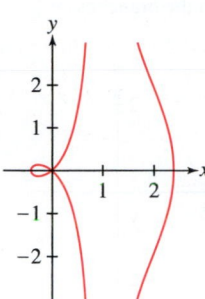

FIGURE 15 Conchoid: $(x - 1)^2(x^2 + y^2) = 2x^2$.

77. [CAS] Use a computer algebra system to plot $y^2 = x^3 - 4x$ for x and y between -4 and 4. Show that if $dx/dy = 0$, then $y = 0$. Conclude that the tangent line is vertical at the points where the curve intersects the x-axis. Does your plot confirm this conclusion?

78. Show that for all points P on the graph in Figure 16, the segments $\overline{OP}$ and $\overline{PR}$ have equal length.

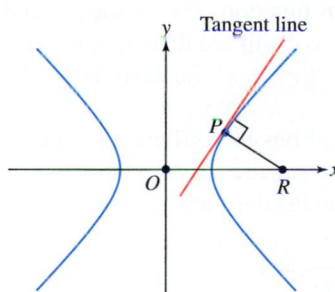

FIGURE 16 Graph of $x^2 - y^2 = a^2$.

In Exercises 79–80, first compute y' and y'' by implicit differentiation. Then solve the given equation for y, and compute y' and y'' by direct differentiation. Finally, show that the results obtained by each approach are the same.

79. $xy = y - 2$

80. $xy^3 = 8$

In Exercises 81–84, use implicit differentiation to calculate higher derivatives.

81. Consider the equation $y^3 - \frac{3}{2}x^2 = 1$.

(a) Show that $y' = x/y^2$ and differentiate again to show that

$$y'' = \frac{y^2 - 2xyy'}{y^4}$$

(b) Express y'' in terms of x and y using part (a).

82. Use the method of the previous exercise to show that $y'' = -y^{-3}$ on the circle $x^2 + y^2 = 1$.

83. Calculate y'' at the point $(1, 1)$ on the curve $xy^2 + y - 2 = 0$ by the following steps:

(a) Find y' by implicit differentiation and calculate y' at the point $(1, 1)$.

(b) Differentiate the expression for y' found in (a). Then compute y'' at $(1, 1)$ by substituting $x = 1$, $y = 1$, and the value of y' found in (a).

84. Use the method of the previous exercise to compute y'' at the point $(1, 1)$ on the curve

$$x^3 + y^3 = 3x + y - 2.$$

Exercises 85 and 86 explore the radius of curvature of curves. There can be many circles that are tangent to a curve at a particular point, but there is one that provides a "best fit" (Figure 17). This circle is called an **osculating circle** *of the curve. We define it formally in Section 13.4. The radius of the osculating circle is called the* **radius of curvature** *of the curve and can be computed using either of the formulas:*

$$r = \frac{(1 + (dy/dx)^2)^{3/2}}{|d^2y/dx^2|} \quad \text{or} \quad r = \frac{(1 + (dx/dy)^2)^{3/2}}{|d^2x/dy^2|}$$

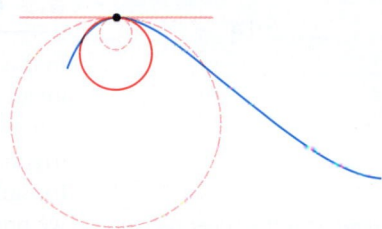

FIGURE 17 The osculating circle (solid red) at a point is the tangent circle that best fits the curve.

85. Consider the ellipse $x^2 + 4y^2 = 16$.

(a) Compute the radius of curvature in terms of x and y.

(b) Compute the radius of curvature at $(4, 0)$, $(2, \sqrt{3})$, and $(0, 2)$. Sketch the ellipse, plot these three points, and label them with the corresponding radius of curvature.

86. Consider the ellipse $9x^2 + y^2 = 36$.

(a) Compute the radius of curvature in terms of x and y.

(b) Compute the radius of curvature at $(-2, 0)$, $(1, -3\sqrt{3})$, and $(0, 6)$. Sketch the ellipse, plot these three points, and label them with the corresponding radius of curvature.

In Exercises 87–89, x and y are functions of a variable t. Use implicit differentiation to express dy/dt in terms of dx/dt, x, and y.

87. $x^2y = 3$

88. $x^3 - 6xy^2 = y$

89. $y^4 + 2x^2 = xy$

90. [✏️] The volume V and pressure P of gas in a piston (which vary in time t) satisfy $PV^{3/2} = C$, where C is a constant. Prove that

$$\frac{dP/dt}{dV/dt} = -\frac{3}{2}\frac{P}{V}$$

The ratio of the derivatives is negative. Could you have predicted this from the relation $PV^{3/2} = C$?

Further Insights and Challenges

91. Show that if P lies on the intersection of the two curves $x^2 - y^2 = c$ and $xy = d$ (c, d constants), then the tangents to the curves at P are perpendicular.

92. The *lemniscate curve* $(x^2 + y^2)^2 = 4(x^2 - y^2)$ was discovered by Jacob Bernoulli in 1694, who noted that it is "shaped like a figure 8, or a knot, or the bow of a ribbon." Find the coordinates of the four points at which the tangent line is horizontal (Figure 18).

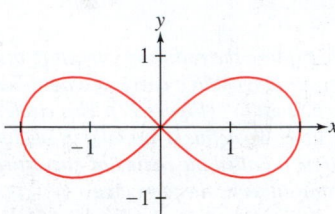

FIGURE 18 Lemniscate curve: $(x^2 + y^2)^2 = 4(x^2 - y^2)$.

93. Divide the curve in Figure 19 into five branches, each of which is the graph of a function. Sketch the branches.

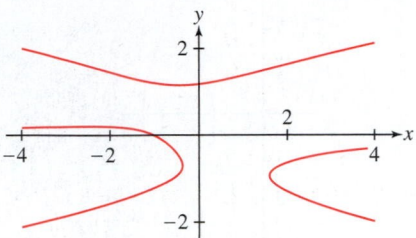

FIGURE 19 Graph of $y^5 - y = x^2y + x + 1$.

3.9 Derivatives of General Exponential and Logarithmic Functions

In this section, we develop derivative formulas for functions involving exponentials and logarithms, including the natural logarithm function $f(x) = \ln x$, the general exponential function $f(x) = b^x$, the general logarithm function $f(x) = \log_b x$, and the hyperbolic trigonometric functions and their inverses. We will see that all of the resulting derivative formulas arise from the derivative formula $\frac{d}{dx}e^x = e^x$ via the rules for differentiation that we presented previously.

To begin, implicit differentiation simplifies the differentiation of $f(x) = \ln x$. For $y = \ln x$, we want to determine dy/dx. We rewrite $y = \ln x$ as $e^y = x$, and then differentiate implicitly with respect to x to obtain the derivative:

$$e^y \frac{dy}{dx} = 1$$

$$\frac{dy}{dx} = \frac{1}{e^y} = \frac{1}{x}$$

> *It can also be shown that the power rule for $f(x) = x^n$ and the derivative rules for $f(x) = \sin x$ and $f(x) = \cos x$ (and therefore for all of the trigonometric functions) can be obtained via the derivative rule for e^x. (See Exercises 87 and 88.)*

THEOREM 1 Derivative of the Natural Logarithm

$$\frac{d}{dx} \ln x = \frac{1}{x} \qquad \text{for } x > 0 \qquad \boxed{1}$$

> *The two most important calculus facts about exponentials and logs are*
>
> $$\frac{d}{dx}e^x = e^x, \qquad \frac{d}{dx}\ln x = \frac{1}{x}$$

EXAMPLE 1 Differentiate **(a)** $y = x \ln x$ and **(b)** $y = (\ln x)^2$.

Solution

(a) Use the Product Rule:

$$\frac{d}{dx}(x \ln x) = x \cdot (\ln x)' + (x)' \cdot \ln x$$

$$= x \cdot \frac{1}{x} + \ln x = 1 + \ln x$$

(b) Use the General Power Rule:

$$\frac{d}{dx}(\ln x)^2 = 2 \ln x \cdot \frac{d}{dx} \ln x = \frac{2 \ln x}{x}$$ ∎

We obtain a useful formula for the derivative of $\ln(f(x))$ by applying the Chain Rule with $u = f(x)$:

$$\frac{d}{dx}\ln(f(x)) = \frac{d}{du}\ln(u)\frac{du}{dx} = \frac{1}{u}\cdot u' = \frac{1}{f(x)}f'(x)$$

$$\boxed{\frac{d}{dx}\ln(f(x)) = \frac{f'(x)}{f(x)}} \qquad \boxed{2}$$

EXAMPLE 2 Differentiate **(a)** $y = \ln(x^3 + 1)$ and **(b)** $y = \ln(\sqrt{\sin x})$.

Solution Use Eq. (2):

(a) $\dfrac{d}{dx}\ln(x^3 + 1) = \dfrac{(x^3 + 1)'}{x^3 + 1} = \dfrac{3x^2}{x^3 + 1}$

(b) The algebra is simpler if we write $\ln(\sqrt{\sin x}) = \ln\left((\sin x)^{1/2}\right) = \frac{1}{2}\ln(\sin x)$:

$$\frac{d}{dx}\ln(\sqrt{\sin x}) = \frac{1}{2}\frac{d}{dx}\ln(\sin x)$$

$$= \frac{1}{2}\frac{(\sin x)'}{\sin x} = \frac{1}{2}\frac{\cos x}{\sin x} = \frac{1}{2}\cot x \qquad ■$$

EXAMPLE 3 The **log wind profile** is a formula for determining wind speeds at different heights near the surface of the earth. Over open agricultural land with few buildings and obstructions, it is expressed as

$$v = v_0\frac{\ln(h/0.03)}{\ln(h_0/0.03)}$$

where v (in m/s) is the wind speed at height h (in m) and v_0 is the known wind speed at reference height h_0. Meteorologists believe this relationship is accurate for heights up to 200 m (Figure 1).

The value 0.03 is called a surface roughness factor. On different surfaces, this factor takes on different values. For example, 0.0002 is used over open water, and 0.4 is used over villages, forests, and rough uneven terrain.

Suppose $v_0 = 10$ m/s at $h_0 = 10$ m above a large corn field (thus with surface roughness factor 0.03). Determine v and dv/dh at $h = 60$ (a typical height of a wind-turbine tower).

Solution With $v_0 = 10$ at $h_0 = 10$, it follows that

$$10\frac{\ln(h/0.03)}{\ln(10/0.03)} \approx 1.72\ln\left(\frac{h}{0.03}\right) = 1.72(\ln h - \ln 0.03) \approx 1.72\ln h + 6.03$$

Therefore,

$$v(h) = 1.72\ln h + 6.03 \qquad \text{and} \qquad \frac{dv}{dh} = \frac{1.72}{h}$$

Thus, at $h = 60$ m, we have $v \approx 13.07$ m/s and $dv/dh \approx 0.03$ m/s per m. This rate of change indicates that at that height, the wind speed is increasing at three-hundredths of a meter per second for each meter increase in height. ■

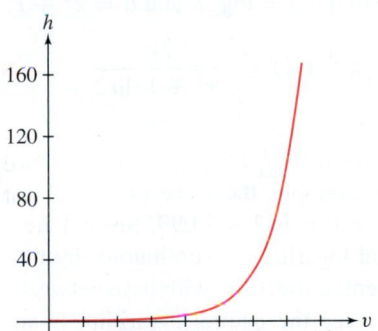

FIGURE 1 Graph of $v(h)$ with $v_0 = 10$ and $h_0 = 10$. The independent variable h is plotted on the vertical to depict how v is changing with height. The shape of the curve reflects the profile of the wind speed as the height increases.

The Derivative of $f(x) = b^x$ and $f(x) = \log_b x$

From the change-of-base formulas for exponential and logarithmic functions (see Section 1.6), we have $b^x = e^{x\ln b}$ and $\log_b x = \frac{\ln x}{\ln b}$. Differentiating these equations, using the Chain Rule on $e^{x\ln b}$ and the Constant Multiple Rule on $\frac{\ln x}{\ln b}$, we obtain

$$\frac{d}{dx}b^x = \frac{d}{dx}e^{x\ln b} = (\ln b)e^{x\ln b} = (\ln b)b^x$$

We have taken for granted that b^x is meaningful for all real numbers x, but we never specified how b^x is defined when x is irrational. For any rational number $x = m/n$,

$$b^x = b^{m/n} = \left(b^{1/n}\right)^m = \left(\sqrt[n]{b}\right)^m$$

When x is irrational, this definition does not apply and b^x cannot be defined directly in terms of roots and powers of b. Instead, limits are used. If x is irrational, we can find a sequence of rational numbers m/n whose limit is x, and we define

$$b^x = \lim_{m/n \to x} b^{m/n}$$

It can be shown (but we do not do so here) that this limit exists, that it does not depend on the choice of the sequence of rational numbers converging to x, and that the function $f(x) = b^x$ thus defined is continuous and differentiable.

$$\frac{d}{dx} \log_b x = \frac{d}{dx} \frac{\ln x}{\ln b} = \left(\frac{1}{\ln b}\right) \frac{1}{x} = \frac{1}{x \ln b}$$

Therefore, we have the following theorem:

> **THEOREM 2   Derivative of $f(x) = b^x$ and $f(x) = \log_b x$**
>
> $$\frac{d}{dx} b^x = (\ln b) b^x \qquad \frac{d}{dx} \log_b x = \frac{1}{x \ln b}$$
>
> 3

For example, $(10^x)' = (\ln 10) 10^x$ and $(\log_{10} x)' = \frac{1}{x \ln 10}$

EXAMPLE 4   Differentiate   **(a)** $f(x) = 4^{3x}$ and   **(b)** $f(x) = \log_2(x^2 + 1)$.

Solution

(a) The function $f(x) = 4^{3x}$ is a composite of $f(u) = 4^u$ and $u = 3x$:

$$\frac{d}{dx} 4^{3x} = \left(\frac{d}{du} 4^u\right) \frac{du}{dx} = (\ln 4) 4^u (3x)' = (\ln 4) 4^{3x} (3) = (3 \ln 4) 4^{3x}$$

(b) The function $f(x) = \log_2(x^2 + 1)$ is a composite of $f(u) = \log_2 u$ and $u = x^2 + 1$:

$$\frac{d}{dx} \log_2(x^2 + 1) = \left(\frac{d}{du} \log_2 u\right) \frac{du}{dx} = \frac{1}{u \ln 2} (x^2 + 1)' = \frac{2x}{(x^2 + 1) \ln 2} \qquad ∎$$

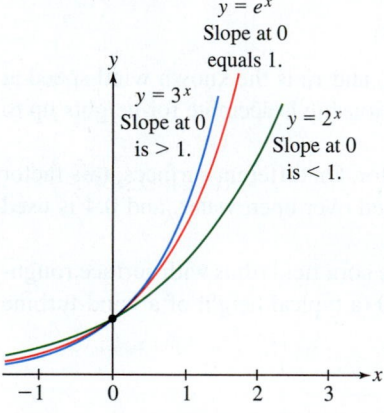

FIGURE 2 $f(x) = e^x$ is *the* exponential function whose graph has slope equal to 1 at $x = 0$.

GRAPHICAL INSIGHT   From $\frac{d}{dx} b^x = (\ln b) b^x$ it follows that $\frac{d}{dx} b^x \big|_{x=0} = \ln b$, so the slope of $y = b^x$ at $x = 0$ is $\ln b$ (see Figure 2). For example, the slope of $y = 2^x$ at $x = 0$ is $\ln 2 \approx 0.693$, and the slope of $y = 3^x$ at $x = 0$ is $\ln 3 \approx 1.099$. Since 1 lies between these two slope values, and since the natural logarithm is continuous, by the Intermediate Value Theorem there must be an exponential function, with base between 2 and 3, whose slope is 1 at $x = 0$. Furthermore, since the natural logarithm is an increasing function, there can be only one such exponential function. Of course, that function is $f(x) = e^x$. [Recall that in Section 1.6 we introduced this fact as a characterization of the number e; that is, we stated that $b = e$ is the unique base for which the slope of the tangent line to the graph of $y = b^x$ at $(0, 1)$ is equal to 1.]

CONCEPTUAL INSIGHT   We pointed out in Section 1.6 that the change-of-base formulas for exponential and logarithmic functions indicate that we can freely change between bases—an exponential with one base can be converted to an exponential with any other base; similarly for logarithms. From an algebraic perspective, no particular base is generally preferred over any other. Now we see that from a calculus perspective, there is a preferred base, e, for exponential and logarithmic functions. For $f(x) = e^x$ and $f(x) = \ln x$ we have the simple derivative formulas $\frac{d}{dx} e^x = e^x$ and $\frac{d}{dx} \ln x = \frac{1}{x}$, but for $f(x) = b^x$ and $f(x) = \log_b x$, we have to introduce a $\ln b$ term in their more awkward derivative formulas $\frac{d}{dx} b^x = (\ln b) b^x$ and $\frac{d}{dx} \log_b x = \frac{1}{x \ln b}$.

Logarithmic Differentiation

Logarithmic differentiation makes what would be a tedious derivative to take, involving multiple applications of the Product and Quotient Rules, into a relatively easy procedure.

The next example illustrates **logarithmic differentiation**. This technique saves work when the function is a product or quotient with several factors.

EXAMPLE 5   Find the derivative of

$$f(x) = \frac{(x + 1)^2 (2x^2 - 3)}{\sqrt{x^2 + 1}}$$

Solution In logarithmic differentiation, we differentiate $\ln(f(x))$ rather than $f(x)$ itself. First, we take the natural log of both sides of the equation:

$$\ln f(x) = \ln\left(\frac{(x+1)^2(2x^2-3)}{\sqrt{x^2+1}}\right)$$

Then we expand the right-hand side using the logarithm rules:

$$\ln(f(x)) = \ln\left((x+1)^2\right) + \ln\left(2x^2-3\right) - \ln\left(\sqrt{x^2+1}\right)$$

$$= 2\ln(x+1) + \ln\left(2x^2-3\right) - \frac{1}{2}\ln(x^2+1)$$

Next, use Eq. (2):

$$\frac{f'(x)}{f(x)} = \frac{d}{dx}\ln(f(x)) = 2\frac{d}{dx}\ln(x+1) + \frac{d}{dx}\ln\left(2x^2-3\right) - \frac{1}{2}\frac{d}{dx}\ln\left(x^2+1\right)$$

$$\frac{f'(x)}{f(x)} = \frac{2}{x+1} + \frac{4x}{2x^2-3} - \frac{1}{2}\frac{2x}{x^2+1}$$

Finally, multiply through by $f(x)$:

$$f'(x) = \left(\frac{2}{x+1} + \frac{4x}{2x^2-3} - \frac{x}{x^2+1}\right)\left(\frac{(x+1)^2(2x^2-3)}{\sqrt{x^2+1}}\right) \qquad\blacksquare$$

Logarithmic differentiation also allows us to take the derivative of functions of the form $y = f(x)^{g(x)}$, where both the base and the exponent depend on x.

EXAMPLE 6 Differentiate (for $x > 0$) **(a)** $f(x) = x^x$ and **(b)** $g(x) = x^{\sin x}$.

Solution The graphs of f and g are shown in Figure 3. We illustrate two different methods; either method could be used in either case.

(a) Method 1: Use the identity $x = e^{\ln x}$ to rewrite $f(x)$ as an exponential base e:

$$f(x) = x^x = (e^{\ln x})^x = e^{x\ln x}$$

$$f'(x) = (x\ln x)'e^{x\ln x} = (1 + \ln x)e^{x\ln x} = (1 + \ln x)x^x$$

(b) Method 2: Apply Eq. (2) to $\ln(g(x))$. Since $\ln(g(x)) = \ln(x^{\sin x}) = (\sin x)\ln x$,

$$\frac{g'(x)}{g(x)} = \frac{d}{dx}\ln(g(x)) = \frac{d}{dx}\left((\sin x)\ln x\right) = \frac{\sin x}{x} + (\cos x)\ln x$$

$$g'(x) = \left(\frac{\sin x}{x} + (\cos x)\ln x\right)g(x) = \left(\frac{\sin x}{x} + (\cos x)\ln x\right)x^{\sin x} \qquad\blacksquare$$

Note that since $f(x) = x^x$ has the variable x in both the base and the exponent, it is a rapidly growing function. It is evident from its graph that it decreases initially, reaches a minimum, and then increases from that point on. Where does that minimum occur and what is the minimum value? To answer this, we need to find the x where the slope of the graph of f is 0; that is, where $f'(x) = 0$. We solve:

$$(1 + \ln x)x^x = 0$$

$$1 + \ln x = 0 \quad (\text{since } x^x \text{ is never } 0)$$

$$\ln x = -1$$

$$x = e^{-1} = 1/e$$

So the minimum occurs at $x = 1/e$, and the minimum value is $(1/e)^{1/e} \approx 0.6922$.

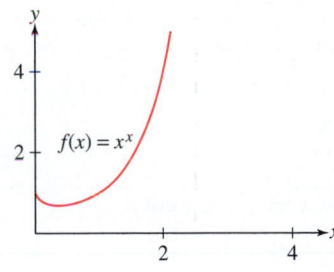

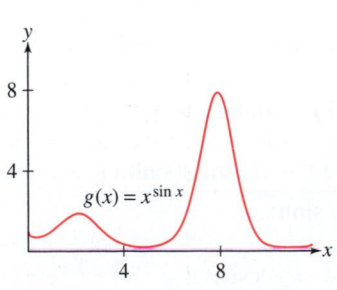

FIGURE 3 Graphs of $f(x) = x^x$ and $g(x) = x^{\sin x}$.

Derivatives of Hyperbolic Functions

Recall from Section 1.6 that the hyperbolic functions are special combinations of e^x and e^{-x}. The formulas for their derivatives are similar to those for the corresponding trigonometric functions, differing at most by a sign.

Consider the hyperbolic sine and cosine:

$$\sinh x = \frac{e^x - e^{-x}}{2}, \qquad \cosh x = \frac{e^x + e^{-x}}{2}$$

Their derivatives are

$$\frac{d}{dx}\sinh x = \cosh x, \qquad \frac{d}{dx}\cosh x = \sinh x$$

We can check this directly. For example,

$$\frac{d}{dx}\sinh x = \frac{d}{dx}\left(\frac{e^x - e^{-x}}{2}\right) = \frac{(e^x)' - (e^{-x})'}{2} = \frac{e^x + e^{-x}}{2} = \cosh x$$

Note the resemblance to the formulas $\frac{d}{dx}\sin x = \cos x$, $\frac{d}{dx}\cos x = -\sin x$. The derivatives of the other hyperbolic functions, which are computed in a similar fashion, also differ from their trigonometric counterparts by a sign at most.

← *REMINDER*

$$\tanh x = \frac{\sinh x}{\cosh x} = \frac{e^x - e^{-x}}{e^x + e^{-x}}$$

$$\operatorname{sech} x = \frac{1}{\cosh x} = \frac{2}{e^x + e^{-x}}$$

$$\coth x = \frac{\cosh x}{\sinh x} = \frac{e^x + e^{-x}}{e^x - e^{-x}}$$

$$\operatorname{csch} x = \frac{1}{\sinh x} = \frac{2}{e^x - e^{-x}}$$

Derivatives of Hyperbolic and Trigonometric Functions

$$\frac{d}{dx}\tanh x = \operatorname{sech}^2 x, \qquad \frac{d}{dx}\tan x = \sec^2 x$$

$$\frac{d}{dx}\coth x = -\operatorname{csch}^2 x, \qquad \frac{d}{dx}\cot x = -\csc^2 x$$

$$\frac{d}{dx}\operatorname{sech} x = -\operatorname{sech} x \tanh x, \qquad \frac{d}{dx}\sec x = \sec x \tan x$$

$$\frac{d}{dx}\operatorname{csch} x = -\operatorname{csch} x \coth x, \qquad \frac{d}{dx}\csc x = -\csc x \cot x$$

← *REMINDER* *Hyperbolic sine and cosine satisfy the basic identity (Section 1.6)*

$$\cosh^2 x - \sinh^2 x = 1$$

EXAMPLE 7   Verify $\dfrac{d}{dx}\coth x = -\operatorname{csch}^2 x$.

Solution   By the Quotient Rule and the identity $\cosh^2 x - \sinh^2 x = 1$,

$$\frac{d}{dx}\coth x = \left(\frac{\cosh x}{\sinh x}\right)' = \frac{(\sinh x)(\cosh x)' - (\cosh x)(\sinh x)'}{\sinh^2 x}$$

$$= \frac{\sinh^2 x - \cosh^2 x}{\sinh^2 x} = \frac{-1}{\sinh^2 x} = -\operatorname{csch}^2 x \qquad \blacksquare$$

EXAMPLE 8   Calculate   **(a)** $\dfrac{d}{dx}\cosh(3x^2 + 1)$ and   **(b)** $\dfrac{d}{dx}(\sinh x \tanh x)$.

Solution

(a) By the Chain Rule, $\frac{d}{dx}\cosh(3x^2 + 1) = 6x \sinh(3x^2 + 1)$.

(b) By the Product Rule,

$$\frac{d}{dx}(\sinh x \tanh x) = \sinh x \operatorname{sech}^2 x + \tanh x \cosh x = \operatorname{sech} x \tanh x + \sinh x \qquad \blacksquare$$

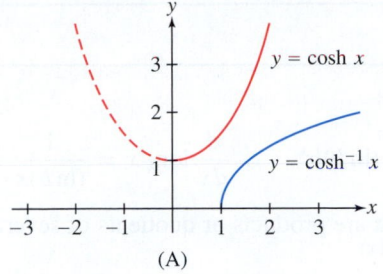

(A)

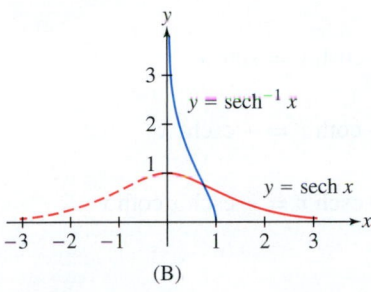

(B)

FIGURE 4

The graphs of $y = \cosh^{-1} x$ and $y = \mathrm{sech}^{-1} x$ have a vertical tangent at the endpoint $x = 1$ of their domains and therefore the derivative is undefined there.

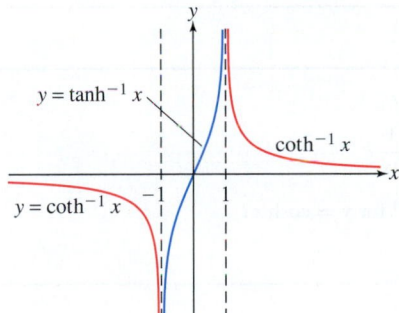

DF **FIGURE 5** The functions $y = \tanh^{-1} x$ and $y = \coth^{-1} x$ have disjoint domains.

Inverse Hyperbolic Functions

Recall that a function f with domain D has an inverse if it is one-to-one on D. Each of the hyperbolic functions except $f(x) = \cosh x$ and $f(x) = \mathrm{sech}\, x$ is one-to-one on its domain and therefore has a well-defined inverse. The functions $f(x) = \cosh x$ and $f(x) = \mathrm{sech}\, x$ are one-to-one on the restricted domain $\{x : x \geq 0\}$. We let $f(x) = \cosh^{-1} x$ and $f(x) = \mathrm{sech}^{-1} x$ denote the corresponding inverses (Figure 4). In reading the following table, keep in mind that the domain of the inverse is equal to the range of the function.

Inverse Hyperbolic Functions and Their Derivatives

Function	Domain	Derivative
$y = \sinh^{-1} x$	all x	$\dfrac{d}{dx} \sinh^{-1} x = \dfrac{1}{\sqrt{x^2 + 1}}$
$y = \cosh^{-1} x$	$x \geq 1$	$\dfrac{d}{dx} \cosh^{-1} x = \dfrac{1}{\sqrt{x^2 - 1}}$
$y = \tanh^{-1} x$	$|x| < 1$	$\dfrac{d}{dx} \tanh^{-1} x = \dfrac{1}{1 - x^2}$
$y = \coth^{-1} x$	$|x| > 1$	$\dfrac{d}{dx} \coth^{-1} x = \dfrac{1}{1 - x^2}$
$y = \mathrm{sech}^{-1} x$	$0 < x \leq 1$	$\dfrac{d}{dx} \mathrm{sech}^{-1} x = -\dfrac{1}{x\sqrt{1 - x^2}}$
$y = \mathrm{csch}^{-1} x$	$x \neq 0$	$\dfrac{d}{dx} \mathrm{csch}^{-1} x = -\dfrac{1}{|x|\sqrt{x^2 + 1}}$

The functions $y = \tanh^{-1} x$ and $y = \coth^{-1} x$ both have derivative $1/(1 - x^2)$. Note, however, that their domains are disjoint, $|x| < 1$ for $y = \tanh^{-1} x$ and $|x| > 1$ for $y = \coth^{-1} x$ (Figure 5).

EXAMPLE 9 Verify $\dfrac{d}{dx} \tanh^{-1} x = \dfrac{1}{1 - x^2}$.

Solution Let $y = \tanh^{-1} x$. Then $\tanh y = x$. Differentiating implicitly yields

$$\mathrm{sech}^2 y \frac{dy}{dx} = 1$$

$$\frac{dy}{dx} = \frac{1}{\mathrm{sech}^2 y}$$

To express the derivative in terms of x, we need to rewrite $\mathrm{sech}^2 y$ in terms of x. We have

$$\cosh^2 y - \sinh^2 y = 1 \qquad \text{(basic identity)}$$

$$1 - \tanh^2 y = \mathrm{sech}^2 y \quad \text{(divide by } \cosh^2 y\text{)}$$

$$1 - x^2 = \mathrm{sech}^2 y \quad \text{(because } x = \tanh y\text{)}$$

This gives the desired result:

$$\frac{d}{dx} \tanh^{-1} x = \frac{1}{\mathrm{sech}^2 y} = \frac{1}{1 - x^2}$$

■

3.9 SUMMARY

- Derivative formulas:

$$\frac{d}{dx}e^x = e^x, \qquad \frac{d}{dx}\ln x = \frac{1}{x}, \qquad \frac{d}{dx}b^x = (\ln b)b^x, \qquad \frac{d}{dx}\log_b x = \frac{1}{(\ln b)x}$$

- Use logarithmic differentiation on functions that are products or quotients of several factors, and on functions of the form $y = f(x)^{g(x)}$.
- Hyperbolic functions:

$$\frac{d}{dx}\sinh x = \cosh x, \qquad\qquad \frac{d}{dx}\cosh x = \sinh x$$

$$\frac{d}{dx}\tanh x = \operatorname{sech}^2 x, \qquad\qquad \frac{d}{dx}\coth x = -\operatorname{csch}^2 x$$

$$\frac{d}{dx}\operatorname{sech} x = -\operatorname{sech} x \tanh x, \qquad \frac{d}{dx}\operatorname{csch} x = -\operatorname{csch} x \coth x$$

- Inverse hyperbolic functions:

$$\frac{d}{dx}\sinh^{-1} x = \frac{1}{\sqrt{x^2+1}}, \qquad\qquad \frac{d}{dx}\cosh^{-1} x = \frac{1}{\sqrt{x^2-1}} \quad (x>1)$$

$$\frac{d}{dx}\tanh^{-1} x = \frac{1}{1-x^2} \quad (|x|<1), \qquad \frac{d}{dx}\coth^{-1} x = \frac{1}{1-x^2} \quad (|x|>1)$$

$$\frac{d}{dx}\operatorname{sech}^{-1} x = \frac{-1}{x\sqrt{1-x^2}} \quad (0<x<1), \qquad \frac{d}{dx}\operatorname{csch}^{-1} x = -\frac{1}{|x|\sqrt{x^2+1}} \quad (x \neq 0)$$

3.9 EXERCISES

Preliminary Questions

1. What is the slope of the tangent line to $y = 4^x$ at $x = 0$?

2. What is the rate of change of $y = \ln x$ at $x = 10$?

3. What is $b > 0$ if the tangent line to $y = b^x$ at $x = 0$ has slope 2?

4. What is b if $(\log_b x)' = \dfrac{1}{3x}$?

5. What are $y^{(100)}$ and $y^{(101)}$ for $y = \cosh x$?

Exercises

In Exercises 1–20, find the derivative.

1. $y = x \ln x$

2. $y = t \ln t - t$

3. $y = 2^{x^3}$

4. $y = \ln(x^5)$

5. $y = \ln(9x^2 - 8)$

6. $y = \ln(t 5^t)$

7. $y = (\ln x)^2$

8. $y = x^2 \ln x$

9. $y = e^{(\ln x)^2}$

10. $y = \dfrac{\ln x}{x}$

11. $y = \ln(\ln x)$

12. $y = \ln(\cot x)$

13. $y = \left(\ln(\ln x)\right)^3$

14. $y = \ln\left((\ln x)^3\right)$

15. $y = \ln\left((x+1)(2x+9)\right)$

16. $y = \ln\left(\dfrac{x+1}{x^3+1}\right)$

17. $y = 11^x$

18. $y = 7^{4x - x^2}$

19. $y = \dfrac{2^x - 3^{-x}}{x}$

20. $y = 16^{\sin x}$

In Exercises 21–24, compute the derivative.

21. $f'(x), \quad f(x) = \log_2 x$

22. $f'(3), \quad f(x) = \log_5 x$

23. $\dfrac{d}{dt}\log_3(\sin t)$

24. $\dfrac{d}{dt}\log_{10}(t + 2^t)$

In Exercises 25–36, find an equation of the tangent line at the point indicated.

25. $f(x) = 6^x, \quad x = 2$

26. $y = (\sqrt{2})^x, \quad x = 8$

27. $s(t) = 3^{9t}, \quad t = 2$

28. $y = \pi^{5x-2}, \quad x = 1$

29. $f(x) = 5^{x^2 - 2x}, \quad x = 1$

30. $s(t) = \ln t, \quad t = 5$

31. $s(t) = \ln(8 - 4t), \quad t = 1$

32. $f(x) = \ln(x^2), \quad x = 4$

33. $R(z) = \log_5(2z^2 + 7), \quad z = 3$

34. $y = \ln(\sin x), \quad x = \dfrac{\pi}{4}$

35. $f(w) = \log_2 w, \quad w = \dfrac{1}{8}$

36. $y = \log_2(1 + 4x^{-1}), \quad x = 4$

In Exercises 37–44, find the derivative using logarithmic differentiation as in Example 5.

37. $y = (x + 5)(x + 9)$

38. $y = (3x + 5)(4x + 9)$

39. $y = (x - 1)(x - 12)(x + 7)$

40. $y = \dfrac{x(x + 1)^3}{(3x - 1)^2}$

41. $y = \dfrac{x(x^2 + 1)}{\sqrt{x + 1}}$

42. $y = (2x + 1)(4x^2)\sqrt{x - 9}$

43. $y = \sqrt{\dfrac{x(x + 2)}{(2x + 1)(3x + 2)}}$

44. $y = (x^3 + 1)(x^4 + 2)(x^5 + 3)^2$

In Exercises 45–50, find the derivative using either method of Example 6.

45. $f(x) = x^{3x}$

46. $f(x) = x^{3^x}$

47. $f(x) = x^{e^x}$

48. $f(x) = x^{x^2}$

49. $f(x) = x^{\cos x}$

50. $f(x) = e^{x^x}$

In Exercises 51–74, calculate the derivative.

51. $y = \sinh(9x)$

52. $y = \sinh(x^2)$

53. $y = \cosh^2(9 - 3t)$

54. $y = \tanh(t^2 + 1)$

55. $y = \sqrt{\cosh x + 1}$

56. $y = \sinh x \tanh x$

57. $y = \dfrac{\coth t}{1 + \tanh t}$

58. $y = (\ln(\cosh x))^5$

59. $y = \sinh(\ln x)$

60. $y = e^{\coth x}$

61. $y = \tanh(e^x)$

62. $y = \sinh(\cosh^3 x)$

63. $y = \operatorname{sech}(\sqrt{x})$

64. $y = \ln(\coth x)$

65. $y = \operatorname{sech} x \coth x$

66. $y = x^{\sinh x}$

67. $y = \cosh^{-1}(3x)$

68. $y = \tanh^{-1}(e^x + x^2)$

69. $y = (\sinh^{-1}(x^2))^3$

70. $y = (\operatorname{csch}^{-1} 3x)^4$

71. $y = e^{\cosh^{-1} x}$

72. $y = \sinh^{-1}(\sqrt{x^2 + 1})$

73. $y = \tanh^{-1}(\ln t)$

74. $y = \ln(\tanh^{-1} x)$

In Exercises 75–77, prove the formula.

75. $\dfrac{d}{dx}(\operatorname{sech} x) = -\operatorname{sech} x \tanh x$

76. $\dfrac{d}{dt}\sinh^{-1} t = \dfrac{1}{\sqrt{t^2 + 1}}$

77. $\dfrac{d}{dt}\cosh^{-1} t = \dfrac{1}{\sqrt{t^2 - 1}}$ for $t > 1$

78. [✎] Use the formula $(\ln f(x))' = f'(x)/f(x)$ to show that $y = \ln x$ and $y = \ln(2x)$ have the same derivative. Is there a simpler explanation of this result?

79. Over rough uneven terrain, the log wind profile from Example 3 is expressed as

$$v = v_0 \frac{\ln(h/0.4)}{\ln(h_0/0.4)}$$

With $v_0 = 10$ m/s at $h_0 = 10$ m, determine v and dv/dh at $h = 60$.

80. Over open water, the log wind profile from Example 3 is expressed as

$$v = v_0 \frac{\ln(h/0.0002)}{\ln(h_0/0.0002)}$$

With $v_0 = 10$ m/s at $h_0 = 10$ m, determine v and dv/dh at $h = 60$.

81. In Exercises 46 in Section 2.3 and 41 and 42 in Section 2.5, similarities between $\lim\limits_{h \to 0} \frac{a^h - 1}{h}$ and $\ln a$ were investigated. Here, it is established that they are equal. Let $f(x) = a^x$. We have $f'(x) = (\ln a)a^x$. Set up and simplify the expression for $\lim\limits_{h \to 0} \frac{f(x+h) - f(x)}{h}$, and use the resulting expression to show that $\lim\limits_{h \to 0} \frac{a^h - 1}{h} = \ln a$.

82. The energy (in ergs) associated with an earthquake of moment magnitude M_w satisfies $\log_{10} E = 16.1 + 1.5M_w$. Calculate dE/dM for $M = 2, 5, 8$.

83. Show that for any constants M, k, and a, the function

$$y(t) = \frac{1}{2} M \left(1 + \tanh\left(\frac{k(t - a)}{2}\right)\right)$$

satisfies the **logistic equation**: $\dfrac{y'}{y} = k\left(1 - \dfrac{y}{M}\right)$.

84. Show that $V(x) = 2\ln(\tanh(x/2))$ satisfies the **Poisson–Boltzmann** equation $V''(x) = \sinh(V(x))$, which is used to describe electrostatic forces in certain molecules.

85. The Palermo Technical Impact Hazard Scale P is used to quantify the risk associated with the impact of an asteroid colliding with the earth:

$$P = \log_{10}\left(\frac{p_i E^{0.8}}{0.03T}\right)$$

where p_i is the probability of impact, T is the number of years until impact, and E is the energy of impact (in megatons of TNT). The risk is greater than a random event of similar magnitude if $P > 0$.

(a) Calculate dP/dT, assuming that $p_i = 2 \times 10^{-5}$ and $E = 2$ megatons.

(b) Use the derivative to estimate the change in P if T increases from 8 to 9 years.

Further Insights and Challenges

86. (a) Show that if f and g are differentiable, then

$$\frac{d}{dx}\ln(f(x)g(x)) = \frac{f'(x)}{f(x)} + \frac{g'(x)}{g(x)}$$ [4]

(b) Give a new proof of the Product Rule by observing that the left-hand side of Eq. (4) is equal to $\dfrac{(f(x)g(x))'}{f(x)g(x)}$.

87. For $x > 0$ and any real number n, we have $x^n = e^{n \ln x}$. Show that $\frac{d}{dx}e^{n \ln x} = nx^{n-1}$, thereby proving the Power Rule for all exponents (as long as $x > 0$).

88. The exponential function $f(x) = e^x$ can be defined for exponents that are complex numbers, $a + bi$, where $i = \sqrt{-1}$. Complex-valued functions may be differentiated just like real-valued functions, and previous derivative rules carry over in this setting. For example, $(e^{ax})' = ae^{ax}$ holds if a and x are complex numbers. Furthermore, in the complex-number domain there is an important formula (known as Euler's formula) relating exponential and trigonometric functions: $e^{ix} = \cos x + i \sin x$. (These topics are all addressed formally when we develop the theory of power series in Chapter 10.)

(a) Using Euler's formula, prove

$$\sin x = \frac{e^{ix} - e^{-ix}}{2i} \quad \text{and} \quad \cos x = \frac{e^{ix} + e^{-ix}}{2}$$

(b) Using the expressions from (a) and the derivative rule for e^x, prove the derivative rules for $\sin x$ and $\cos x$.

3.10 Related Rates

Suppose you are filling a bottle that has a cylindrical bottom and a top shaped like a cone. Water is flowing into the bottle from a faucet at a constant rate (Figure 1). Notice that the water level rises at a constant rate in the cylindrical part, but in the conical part, the level rises more rapidly. If you are not paying attention—expecting the water level to continue to rise at a constant rate—you might overfill the bottle.

To investigate this situation, we build a related rates model, a relationship between the rates of change of different variables that themselves are related. In this case, the rates that are related are the rate of increase of the volume of water in the bottle and the rate of increase of the height of the water. We investigate these rates of increase of volume and height for the cylindrical bottom in Example 1 and separately for the cone-shaped top in Example 2.

FIGURE 1

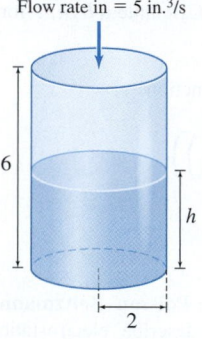

Flow rate in = 5 in.³/s

FIGURE 2

EXAMPLE 1   Filling a Cylindrical Container   Water flows into a cylindrical container at a rate of 5 in.^{3}/s. Assume that the container has a height of 6 in. and a base radius of 2 in. At what rate is the water level rising in the container?

Solution   Let V represent the volume of the water in the container in in.^{3}, and let h represent the height of the water in in. We draw a cylinder and label it with all the information we have (Figure 2). The goal is to find a relationship between the known rate, $\frac{dV}{dt}$, and the desired rate, $\frac{dh}{dt}$. That is,

$$\text{determine } \frac{dh}{dt} \text{ given that } \frac{dV}{dt} = 5 \text{ in.}^3/\text{s.}$$

First, we find a relationship between V and h and then, because we want a relationship between $\frac{dV}{dt}$ and $\frac{dh}{dt}$, we differentiate the relationship between V and h with respect to t.

The geometry is simple: The volume of the water in the container is the volume of a cylinder with height h and base radius 2. So $V = \pi 2^2 h = 4\pi h$. We differentiate with respect to t and obtain

$$\frac{dV}{dt} = 4\pi \frac{dh}{dt}$$

We are given that $\frac{dV}{dt} = 5$, so substituting that in the above equation and solving for $\frac{dh}{dt}$, we find

$$\frac{dh}{dt} = \frac{5}{4\pi} \approx 0.40$$

Thus, the water level is rising in the container at a rate of approximately 0.40 in./s. Note that, as expected, the water level is changing at a constant rate.   ∎

We summarize the steps that we typically follow in developing a related rates model and solving an associated problem.

Step 1. **Identify variables and the rates that are related.**

Step 2. **Find an equation relating the variables and differentiate it.**

Step 3. **Use given information to solve the problem.**

EXAMPLE 2   Filling a Conical Container   Water flows into a conical container at a rate of 5 in.^{3}/s. Assume that the container has a height of 4 in. and a base radius of 2 in. Show that the rate that the water level is rising depends on the level of the water in the container, rising faster the higher the water level.

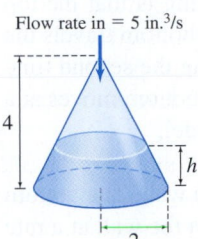

Flow rate in = 5 in.³/s

FIGURE 3

← **REMINDER** The volume of a cone with base radius r and height h is $\frac{1}{3}\pi r^2 h$.

Solution

Step 1. Identify variables and the rates that are related.

Let V represent the volume of the water in the container in in.³, and let h represent its height in in. Draw a cone with the given information (Figure 3). The goal:

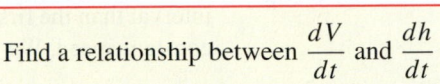

Find a relationship between $\dfrac{dV}{dt}$ and $\dfrac{dh}{dt}$

Step 2. Find an equation relating the variables and differentiate it.

We need to find a relationship between V and h that we can differentiate with respect to t. The geometry in this case is a little more involved than with the cylindrical container.

To express a relationship between the volume of the water and height, we regard the shaded volume in Figure 4 as the difference between the volume of the conical container and the volume of the conical space in the container above the water. The volume of the conical container is $\frac{1}{3}\pi(2^2)(4) = \frac{16\pi}{3}$ and the volume of the conical space is $\frac{1}{3}\pi r_s^2 h_s$, where r_s and h_s are the base radius and height, respectively, of the conical space. Note that $h_s = 4 - h$. Also note that there are similar triangles in Figure 4 from which we obtain $\frac{h_s}{r_s} = \frac{4}{2} = 2$. Thus, $r_s = \frac{1}{2}h_s = \frac{4-h}{2}$. Now, putting together the terms, we have

$$V = \frac{16\pi}{3} - \frac{1}{3}\pi\left(\frac{4-h}{2}\right)^2(4-h) = \frac{16\pi}{3} - \frac{1}{12}\pi(4-h)^3$$

Differentiating with respect to t, and using the Chain Rule on the $(4-h)^3$ term, we obtain

$$\frac{dV}{dt} = 0 - \frac{1}{12}\pi(3(4-h)^2)\left(-\frac{dh}{dt}\right) = \frac{1}{4}\pi(4-h)^2\frac{dh}{dt}$$

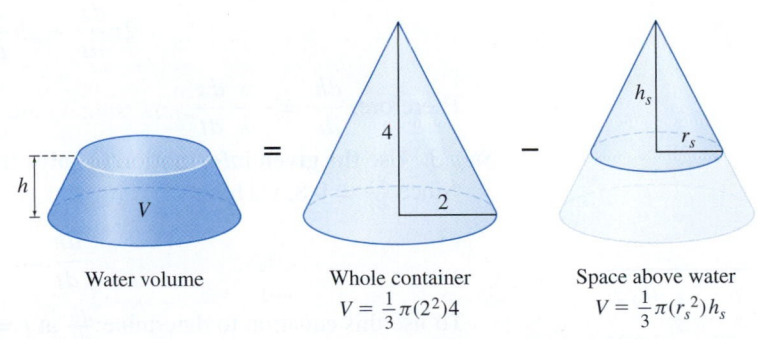

Water volume		Whole container		Space above water
V	$=$	$V = \frac{1}{3}\pi(2^2)4$	$-$	$V = \frac{1}{3}\pi(r_s^2)h_s$

FIGURE 4

Step 3. Use the given information to solve the problem.

Substituting in 5 for $\frac{dV}{dt}$ and solving for $\frac{dh}{dt}$, the result is

$$\frac{dh}{dt} = \frac{20}{\pi(4-h)^2}$$

As we expected, the rate of change dh/dt of the water level depends on the level h itself. In fact, it increases as the level increases. For example, at $h = 1, 2, 3$ in. we find $\frac{dh}{dt} \approx 0.71, 1.59, 6.37$ in./s, respectively. ∎

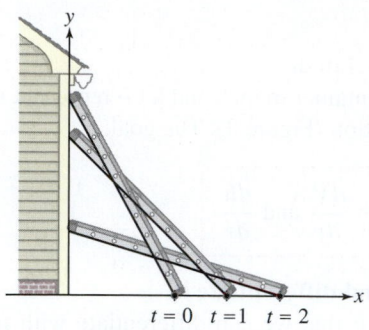

FIGURE 5 Positions of a ladder at times $t = 0, 1, 2$.

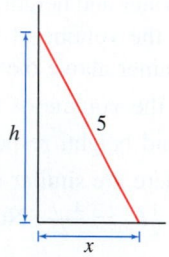

FIGURE 6 The variables x and h.

t	x	h	dh/dt
0	1.5	4.77	−0.25
1	2.3	4.44	−0.41
2	3.1	3.92	−0.63
3	3.9	3.13	−1.00

This table of values confirms that the top of the ladder is speeding up.

Next, we examine the "sliding ladder problem" where a ladder that is leaning against a wall has its bottom pulled away at constant velocity. The question is, "How fast does the top of the ladder move?" What is interesting and perhaps surprising is that the top and bottom travel at different speeds. Figure 5 shows this clearly: The bottom travels the same distance over each time interval, but the top travels farther during the second time interval than the first. In other words, the top is speeding up while the bottom moves at a constant speed. We will explore this further through a related rates model.

EXAMPLE 3 **Sliding Ladder Problem** A 5-m ladder leans against a wall. The bottom of the ladder is 1.5 m from the wall at time $t = 0$ and slides away from the wall at a rate of 0.8 m/s. Find the velocity of the top of the ladder at time $t = 1$.

Solution

Step 1. **Identify variables and the rates that are related.**
Since we are considering how the top and bottom of the ladder change position, we use the variables x for the distance from the bottom of the ladder to the wall and h for the distance from the ground to the top of the ladder (Figure 6). The velocity of the bottom is $dx/dt = 0.8$ m/s. The unknown velocity of the top is dh/dt. Our goal is to

$$\boxed{\text{compute } \frac{dh}{dt} \text{ at } t = 1, \text{ given that } \frac{dx}{dt} = 0.8 \text{ m/s and } x(0) = 1.5 \text{ m}}$$

Step 2. **Find an equation relating the variables and differentiate it.**
We wish to find a relationship between dx/dt and dh/dt, and therefore we need an equation relating x and h (Figure 6). This is provided by the Pythagorean Theorem:

$$x^2 + h^2 = 5^2$$

To calculate dh/dt, we differentiate both sides of this equation *with respect to t*:

$$\frac{d}{dt}x^2 + \frac{d}{dt}h^2 = \frac{d}{dt}25$$

$$2x\frac{dx}{dt} + 2h\frac{dh}{dt} = 0$$

Therefore, $\dfrac{dh}{dt} = -\dfrac{x}{h}\dfrac{dx}{dt}$.

Step 3. **Use the given information to solve the problem.**
Since $\frac{dx}{dt} = 0.8$, we have

$$\frac{dh}{dt} = -0.8\frac{x}{h} \qquad \boxed{1}$$

To use this equation to determine $\frac{dh}{dt}$ at $t = 1$, we need to find x and h at that time. Since the bottom slides away from the wall at 0.8 m/s and we are given $x(0) = 1.5$, we have $x(1) = 2.3$ and $h(1) = \sqrt{5^2 - 2.3^2} \approx 4.44$. We obtain

$$\left.\frac{dh}{dt}\right|_{t=1} = -0.8\frac{x(1)}{h(1)} \approx -0.8\frac{2.3}{4.44} \approx -0.41 \text{ m/s}$$

The negative value for $\frac{dh}{dt}$ reflects the fact that the top of the ladder is sliding *down* the wall. ∎

CONCEPTUAL INSIGHT A puzzling feature of Eq. (1) is that the velocity dh/dt, which is equal to $-0.8x/h$, becomes infinite as $h \to 0$ (as the top of the ladder gets close to the ground). Since this is impossible, our mathematical model must break down as $h \to 0$. In fact, using physics one can show that the ladder's top loses contact with the wall before reaching the bottom. From that moment on, the formula for dh/dt is no longer valid.

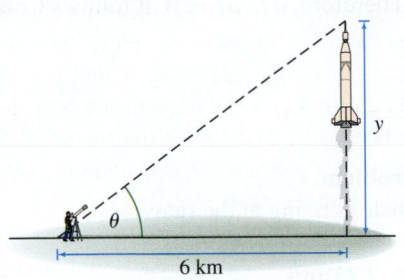

FIGURE 7 Tracking a rocket through a telescope.

EXAMPLE 4   Tracking a Rocket   A spy uses a telescope to track a rocket launched vertically from a launch pad 6 km away, as in Figure 7. At the moment when the angle θ between the telescope and the ground is $\frac{\pi}{3}$, the angle is changing at a rate of 0.9 radians per minute. What is the rocket's velocity at that moment?

Solution

Step 1. **Identify variables and the rates that are related.**

Let y be the height of the rocket at time t. We wish to determine the rocket's velocity dy/dt when $\theta = \frac{\pi}{3}$. We are given that $d\theta/dt = 0.9$. Thus, our goal is to

$$\text{compute } \frac{dy}{dt}\bigg|_{\theta=\frac{\pi}{3}}, \text{ given that } \frac{d\theta}{dt} = 0.9 \text{ rad/min when } \theta = \frac{\pi}{3}$$

Step 2. **Find an equation relating the variables and differentiate it.**

We want a relationship between dy/dt and $d\theta/dt$; therefore, we need to find a relation between θ and y. As we see in Figure 7,

$$\tan \theta = \frac{y}{6}$$

Now differentiate with respect to time:

$$\sec^2 \theta \frac{d\theta}{dt} = \frac{1}{6} \frac{dy}{dt}$$

$$\frac{dy}{dt} = \frac{6}{\cos^2 \theta} \frac{d\theta}{dt} \qquad \boxed{2}$$

Step 3. **Use the given information to solve the problem.**

At the given moment, $\theta = \frac{\pi}{3}$ and $d\theta/dt = 0.9$, so Eq. (2) yields

$$\frac{dy}{dt} = \frac{6}{\cos^2(\pi/3)}(0.9) = \frac{6}{(0.5)^2}(0.9) = 21.6 \text{ km/min}$$

The rocket's velocity at $\theta = \frac{\pi}{3}$ is 21.6 km/min, or approximately 1296 km/h.   ∎

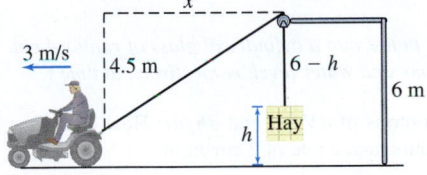

FIGURE 8

EXAMPLE 5   Farmer John's tractor, traveling at 3 m/s, pulls a rope attached to a bale of hay through a pulley. With dimensions as indicated in Figure 8, how fast is the bale rising when x, the horizontal distance from the tractor to the hay bale, is 5 meters?

Solution

Step 1. **Identify variables and the rates that are related.**

Let x be the horizontal distance from the tractor to the bale of hay, and let h be the height above ground of the top of the bale. We want to determine dh/dt when $x = 5$. We are given $dx/dt = 3$. Thus, our goal is to

$$\text{compute } \frac{dh}{dt}\bigg|_{x=5}, \text{ given that } \frac{dx}{dt} = 3 \text{ m/s}$$

Step 2. **Find an equation relating the variables and differentiate it.**

Let L be the total length of the rope. From Figure 8 (using the Pythagorean Theorem),

$$L = \sqrt{x^2 + 4.5^2} + (6 - h)$$

Differentiating with respect to t, we obtain

$$\frac{dL}{dt} = \frac{d}{dt}\left(\sqrt{x^2 + 4.5^2} + (6 - h)\right) = \frac{x\frac{dx}{dt}}{\sqrt{x^2 + 4.5^2}} - \frac{dh}{dt} \qquad \boxed{3}$$

CAUTION *A common mistake is to substitute particular variable values (such as $x = 5$ here) prior to differentiating. Doing so changes the variable to a constant and might prevent you from finding the proper relationship between the rates of change. Wait until Step 3, after differentiating, to substitute particular values to answer the posed problem.*

Now L, the length of the rope, is constant. Therefore, $dL/dt = 0$. It follows from Eq. (3) that

$$\frac{dh}{dt} = \frac{x\frac{dx}{dt}}{\sqrt{x^2 + 4.5^2}} \qquad \boxed{4}$$

Step 3. Use the given information to solve the problem.

Apply Eq. (4) with $x = 5$ and $dx/dt = 3$. The bale is rising at the rate

$$\frac{dh}{dt} = \frac{(5)(3)}{\sqrt{5^2 + 4.5^2}} \approx 2.23 \text{ m/s} \qquad \blacksquare$$

3.10 SUMMARY

- Related-rate models arise in situations where two or more variables are related and we wish to explore how the rate of change of one of the variables is related to the rates of change of the other variable(s).
- The following steps can be helpful in developing a related-rates model and solving an associated problem.

Step 1. Identify variables and the rates that are related.

Step 2. Find an equation relating the variables and differentiate it. (A diagram is often helpful in identifying variables and relationships between them.)

This gives us an equation relating the known and unknown derivatives. Remember not to substitute the specific values for the variables until after you have computed all derivatives.

Step 3. Use the given information to solve the problem.

3.10 EXERCISES

Preliminary Questions

1. If $\frac{dx}{dt} = 3$ and $y = x^2$, what is $\frac{dy}{dt}$ when $x = -3, 2, 5$?

2. If $\frac{dx}{dt} = 2$ and $y = x^3$, what is $\frac{dy}{dt}$ when $x = -4, 2, 6$?

3. Assign variables and restate the following problem in terms of known and unknown derivatives (but do not solve it): How fast is the volume of a cube increasing if its side increases at a rate of 0.5 cm/s?

4. What is the relation between dV/dt and dr/dt if $V = \left(\frac{4}{3}\right)\pi r^3$?

In Questions 5 and 6, water pours into a cylindrical glass of radius 4 cm. Let V and h denote the volume and water level, respectively, at time t.

5. Restate this question in terms of dV/dt and dh/dt: How fast is the water level rising if water pours in at a rate of 2 cm³/min?

6. Restate this question in terms of dV/dt and dh/dt: At what rate is water pouring in if the water level rises at a rate of 1 cm/min?

Exercises

In Exercises 1 and 2, consider a rectangular bathtub whose base is 18 ft².

1. How fast is the water level rising if water is filling the tub at a rate of 0.7 ft³/min?

2. At what rate is water pouring into the tub if the water level rises at a rate of 0.8 ft/min?

3. The radius of a circular oil slick expands at a rate of 2 m/min.

(a) How fast is the area of the oil slick increasing when the radius is 25 m?

(b) If the radius is 0 at time $t = 0$, how fast is the area increasing after 3 min?

4. At what rate is the diagonal of a cube increasing if its edges are increasing at a rate of 2 cm/s?

In Exercises 5–8, assume that the radius r of a sphere is expanding at a rate of 30 cm/min. The volume of a sphere is $V = \frac{4}{3}\pi r^3$ and its surface area is $4\pi r^2$. Determine the given rate.

5. Volume with respect to time when $r = 15$ cm

6. Volume with respect to time at $t = 2$ min, assuming that $r = 0$ at $t = 0$

7. Surface area with respect to time when $r = 40$ cm

8. Surface area with respect to time at $t = 2$ min, assuming that $r = 10$ at $t = 0$

9. A conical tank (as in Example 2) has height 3 m and radius 2 m at the base. Water flows in at a rate of 2 m³/min. How fast is the water level rising when the level is 1 m and when the level is 2 m?

10. (GU) A conical tank (as in Example 2) has height 8 m and radius 4 m at the base. Water flows in at a rate of 3 m³/min. Determine $\frac{dh}{dt}$ as a function of h, and provide a graph of this relationship.

In Exercises 11–14, refer to a 5-m ladder sliding down a wall, as in Figures 5 and 6. The variable h is the height of the ladder's top at time t, and x is the distance from the wall to the ladder's bottom.

11. Assume the bottom slides away from the wall at a rate of 0.8 m/s. Find the velocity of the top of the ladder at $t = 2$ s if the bottom is 1.5 m from the wall at $t = 0$ s.

12. Suppose that the top is sliding down the wall at a rate of 1.2 m/s. Calculate dx/dt when $h = 3$ m.

13. Suppose that $h(0) = 4$ and the top slides down the wall at a rate of 1.2 m/s. Calculate x and dx/dt at $t = 2$ s.

14. What is the relation between h and x at the moment when the top and bottom of the ladder move at the same speed?

15. The radius r and height h of a circular cone change at a rate of 2 cm/s. How fast is the volume of the cone increasing when $r = 10$ and $h = 20$?

16. A road perpendicular to a highway leads to a farmhouse located 2 km away (Figure 9). An automobile travels past the farmhouse at a speed of 80 km/h. How fast is the distance between the automobile and the farmhouse increasing when the automobile is 6 km past the intersection of the highway and the road?

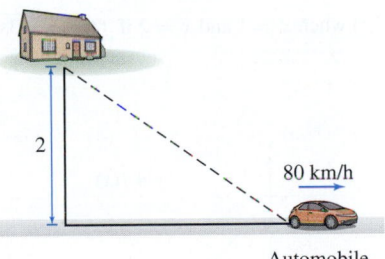

FIGURE 9

17. A man of height 1.8 m walks away from a 5-m lamppost at a speed of 1.2 m/s (Figure 10). Find the rate at which his shadow is increasing in length.

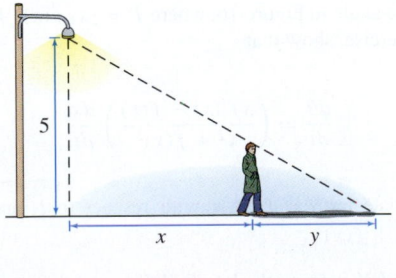

FIGURE 10

18. As Claudia walks away from a 264-cm lamppost, the tip of her shadow moves twice as fast as she does. What is Claudia's height?

19. At a given moment, a plane passes directly above a radar station at an altitude of 6 km.

(a) The plane's speed is 800 km/h. How fast is the distance between the plane and the station changing half a minute later?

(b) How fast is the distance between the plane and the station changing when the plane passes directly above the station?

20. In the setting of Exercise 19, let θ be the angle that the line through the radar station and the plane makes with the horizontal. How fast is θ changing 12 min after the plane passes over the radar station?

21. A hot air balloon rising vertically is tracked by an observer located 4 km from the lift-off point. At a certain moment, the angle between the observer's line of sight and the horizontal is $\frac{\pi}{5}$, and it is changing at a rate of 0.2 rad/min. How fast is the balloon rising at this moment?

22. A laser pointer is placed on a platform that rotates at a rate of 20 revolutions per minute. The beam hits a wall 8 m away, producing a dot of light that moves horizontally along the wall. Let θ be the angle between the beam and the line through the searchlight perpendicular to the wall (Figure 11). How fast is this dot moving when $\theta = \frac{\pi}{6}$?

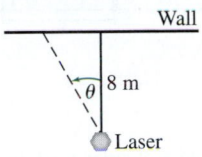

FIGURE 11

23. A rocket travels vertically at a speed of 1200 km/h. The rocket is tracked through a telescope by an observer located 16 km from the launching pad. Find the rate at which the angle between the telescope and the ground is increasing 3 min after lift-off.

24. Using a telescope, you track a rocket that was launched 4 km away, recording the angle θ between the telescope and the ground at half-second intervals. Estimate the velocity of the rocket if $\theta(10) = 0.205$ and $\theta(10.5) = 0.225$.

25. A police car traveling south toward Sioux Falls, Iowa, at 160 km/h pursues a truck traveling east away from Sioux Falls at 140 km/h (Figure 12). At time $t = 0$, the police car is 20 km north and the truck is 30 km east of Sioux Falls. Calculate the rate at which the distance between the vehicles is changing:

(a) At time $t = 0$ **(b)** 5 min later

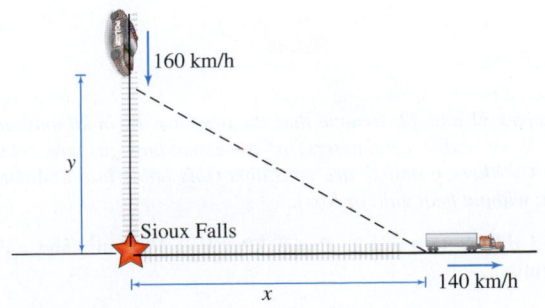

FIGURE 12

26. A car travels down a highway at 25 m/s. An observer stands 150 m from the highway.

(a) How fast is the distance from the observer to the car increasing when the car passes in front of the observer? Explain your answer without making any calculations.

(b) How fast is the distance increasing 20 s later?

27. In the setting of Example 5, at a certain moment, the tractor's speed is 3 m/s and the bale is rising at 2 m/s. How far is the tractor from the bale at this moment?

28. Placido pulls a rope attached to a wagon through a pulley at a rate of q m/s. With dimensions as in Figure 13:

(a) Find a formula for the speed of the wagon in terms of q and the variable x in the figure.

(b) Find the speed of the wagon when $x = 0.6$ if $q = 0.5$ m/s.

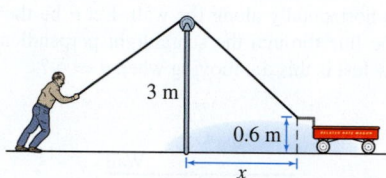

FIGURE 13

29. Julian is jogging around a circular track of radius 50 m. In a coordinate system with its origin at the center of the track, Julian's x-coordinate is changing at a rate of -1.25 m/s when his coordinates are $(40, 30)$. Find dy/dt at this moment.

30. A particle moves counterclockwise around the ellipse with equation $9x^2 + 16y^2 = 25$ (Figure 14).

(a) [icon] In which of the four quadrants is $dx/dt > 0$? Explain.

(b) Find a relation between dx/dt and dy/dt.

(c) At what rate is the x-coordinate changing when the particle passes the point $(1, 1)$ if its y-coordinate is increasing at a rate of 6 m/s?

(d) Find dy/dt when the particle is at the top and bottom of the ellipse.

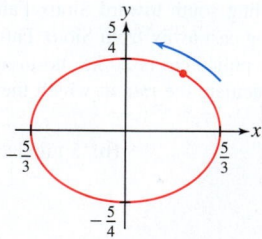

FIGURE 14

In Exercises 31 and 32, assume that the pressure P (in kilopascals) and volume V (in cubic centimeters) of an expanding gas are related by $PV^b = C$, where b and C are constants (this holds in an adiabatic expansion, without heat gain or loss).

31. Find dP/dt if $b = 1.2$, $P = 8$ kPa, $V = 100$ cm³, and $dV/dt = 20$ cm³/min.

32. Find b if $P = 25$ kPa, $dP/dt = 12$ kPa/min, $V = 100$ cm³, and $dV/dt = 20$ cm³/min.

33. The base x of the right triangle in Figure 15 increases at a rate of 5 cm/s, while the height remains constant at $h = 20$. How fast is the angle θ changing when $x = 20$?

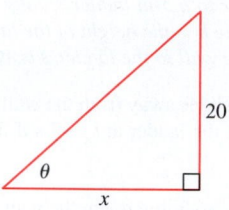

FIGURE 15

34. Two parallel paths 15 m apart run east–west through the woods. Brooke jogs east on one path at 10 km/h, while Jamail walks west on the other path at 6 km/h. If they pass each other at time $t = 0$, how far apart are they 3 s later, and how fast is the distance between them changing at that moment?

35. A particle travels along a curve $y = f(x)$ as in Figure 16. Let $L(t)$ be the particle's distance from the origin.

(a) Show that

$$\frac{dL}{dt} = \left(\frac{x + f(x)f'(x)}{\sqrt{x^2 + f(x)^2}} \right) \frac{dx}{dt}$$

if the particle's location at time t is $P = (x, f(x))$.

(b) Calculate $L'(t)$ when $x = 1$ and $x = 2$ if $f(x) = \sqrt{3x^2 - 8x + 9}$ and $dx/dt = 4$.

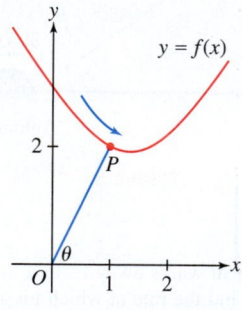

FIGURE 16

36. Let θ be the angle in Figure 16, where $P = (x, f(x))$. In the setting of the previous exercise, show that

$$\frac{d\theta}{dt} = \left(\frac{xf'(x) - f(x)}{x^2 + f(x)^2} \right) \frac{dx}{dt}$$

Hint: Differentiate $\tan\theta = f(x)/x$ with respect to t, and observe that $\cos\theta = x/\sqrt{x^2 + f(x)^2}$.

Exercises 37 and 38 refer to the baseball diamond (a square of side 90 ft) in Figure 17.

37. A baseball player runs from home plate toward first base at 20 ft/s. How fast is the player's distance from second base changing when the player is halfway to first base?

38. Player 1 runs to first base at a speed of 20 ft/s, while player 2 runs from second base to third base at a speed of 15 ft/s. Let s be the distance between the two players. How fast is s changing when player 1 is 30 ft from home plate and player 2 is 60 ft from second base?

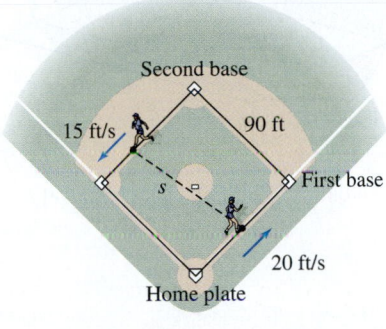

FIGURE 17

39. The conical watering pail in Figure 18 has a grid of holes. Water flows out through the holes at a rate of kA m³/min, where k is a constant and A is

the surface area of the part of the cone in contact with the water. This surface area is $A = \pi r \sqrt{h^2 + r^2}$ and the volume is $V = \frac{1}{3}\pi r^2 h$. Calculate the rate dh/dt at which the water level changes at $h = 0.3$ m, assuming that $k = 0.25$ m/min.

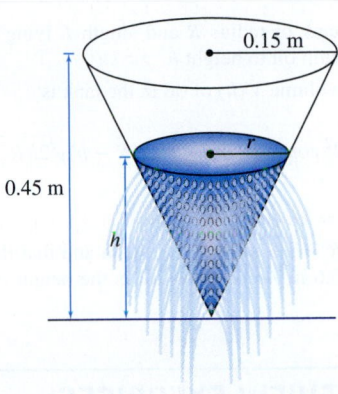

FIGURE 18

Further Insights and Challenges

40. A bowl contains water that evaporates at a rate proportional to the surface area of water exposed to the air (Figure 19). Let $A(h)$ be the cross-sectional area of the bowl at height h.

(a) Explain why $V(h + \Delta h) - V(h) \approx A(h)\Delta h$ if Δh is small.

(b) Use (a) to argue that $\dfrac{dV}{dh} = A(h)$.

(c) Show that the water level h decreases at a constant rate.

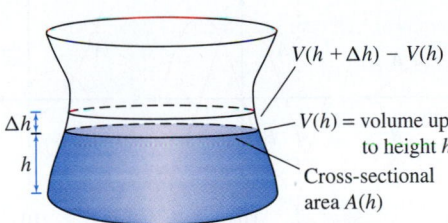

FIGURE 19

41. A roller coaster has the shape of the graph in Figure 20. Show that when the roller coaster passes the point $(x, f(x))$, the vertical velocity of the roller coaster is equal to $f'(x)$ times its horizontal velocity.

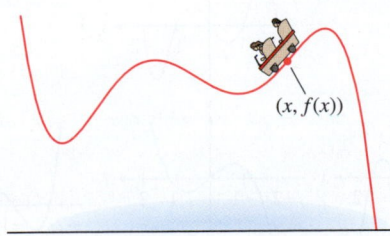

FIGURE 20 Graph of f as a roller coaster track.

42. Two trains leave a station at $t = 0$ and travel with constant velocity v along straight tracks that make an angle θ.

(a) Show that the trains are separating from each other at a rate $v\sqrt{2 - 2\cos\theta}$.

(b) What does this formula give for $\theta = \pi$?

43. As the wheel of radius r cm in Figure 21 rotates, the rod of length L attached at point P drives a piston back and forth in a straight line. Let x be the distance from the origin to point Q at the end of the rod, as shown in the figure.

(a) Use the Pythagorean Theorem to show that

$$L^2 = (x - r\cos\theta)^2 + r^2 \sin^2\theta \qquad \boxed{5}$$

(b) Differentiate Eq. (5) with respect to t to prove that

$$2(x - r\cos\theta)\left(\frac{dx}{dt} + r\sin\theta\frac{d\theta}{dt}\right) + 2r^2 \sin\theta\cos\theta\frac{d\theta}{dt} = 0$$

(c) Calculate the speed of the piston when $\theta = \frac{\pi}{2}$, assuming that $r = 10$ cm, $L = 30$ cm, and the wheel rotates at 4 revolutions per minute.

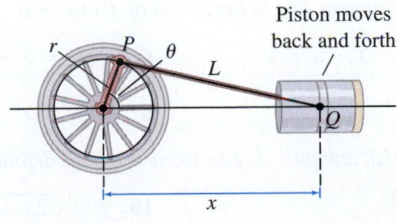

FIGURE 21

44. A spectator seated 300 m away from the center of a circular track of radius 100 m watches an athlete run laps at a speed of 5 m/s. How fast is the distance between the spectator and athlete changing when the runner is approaching the spectator and the distance between them is 250 m? *Hint:* The diagram for this problem is similar to Figure 21, with $r = 100$ and $x = 300$.

45. A cylindrical tank of radius R and length L lying horizontally as in Figure 22 is filled with oil to height h.

(a) Show that the volume $V(h)$ of oil in the tank is

$$V(h) = L \left(R^2 \cos^{-1}\left(1 - \frac{h}{R}\right) - (R - h)\sqrt{2hR - h^2} \right)$$

(b) Show that $\frac{dV}{dh} = 2L\sqrt{h(2R - h)}$.

(c) Suppose that $R = 1.5$ m and $L = 10$ m and that the tank is filled at a constant rate of 0.6 m³/min. How fast is the height h increasing when $h = 0.5$?

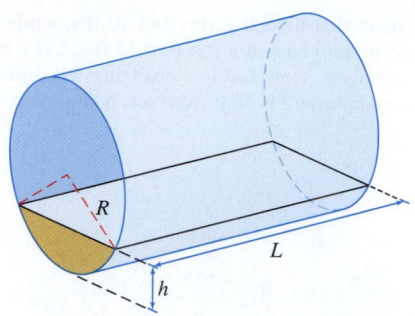

FIGURE 22 Oil in the tank has level h.

CHAPTER REVIEW EXERCISES

In Exercises 1–4, refer to the function f whose graph is shown in Figure 1.

1. Compute the average rate of change of $f(x)$ over $[0, 2]$. What is the graphical interpretation of this average rate?

2. For which value of h is $\dfrac{f(0.7 + h) - f(0.7)}{h}$ equal to the slope of the secant line between the points where $x = 0.7$ and $x = 1.1$?

3. Estimate $\dfrac{f(0.7 + h) - f(0.7)}{h}$ for $h = 0.3$. Is the value of this difference quotient greater than or less than $f'(0.7)$?

4. Estimate $f'(0.7)$ and $f'(1.1)$.

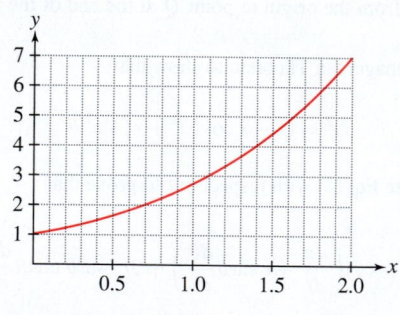

FIGURE 1

In Exercises 5–8, compute $f'(a)$ using the limit definition and find an equation of the tangent line to the graph of f at $x = a$.

5. $f(x) = x^2 - x$, $a = 1$

6. $f(x) = 5 - 3x$, $a = 2$

7. $f(x) = x^{-1}$, $a = 4$

8. $f(x) = x^3$, $a = -2$

In Exercises 9–12, compute dy/dx using the limit definition.

9. $y = 4 - x^2$

10. $y = \sqrt{2x + 1}$

11. $y = \dfrac{1}{2 - x}$

12. $y = \dfrac{1}{(x - 1)^2}$

In Exercises 13–16, express the limit as a derivative.

13. $\displaystyle\lim_{h \to 0} \frac{\sqrt{1 + h} - 1}{h}$

14. $\displaystyle\lim_{x \to -1} \frac{x^3 + 1}{x + 1}$

15. $\displaystyle\lim_{t \to \pi} \frac{\sin t \cos t}{t - \pi}$

16. $\displaystyle\lim_{\theta \to \pi} \frac{\cos\theta - \sin\theta + 1}{\theta - \pi}$

17. Find $f(4)$ and $f'(4)$ if the tangent line to the graph of f at $x = 4$ has equation $y = 3x - 14$.

18. Each graph in Figure 2 shows the graph of a function f and its derivative f'. Determine which is the function and which is the derivative.

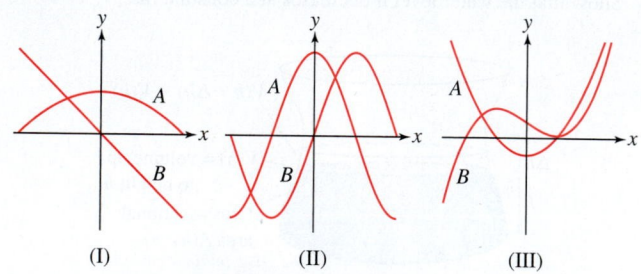

FIGURE 2 Graph of f.

19. Is (A), (B), or (C) the graph of the derivative of the function f shown in Figure 3?

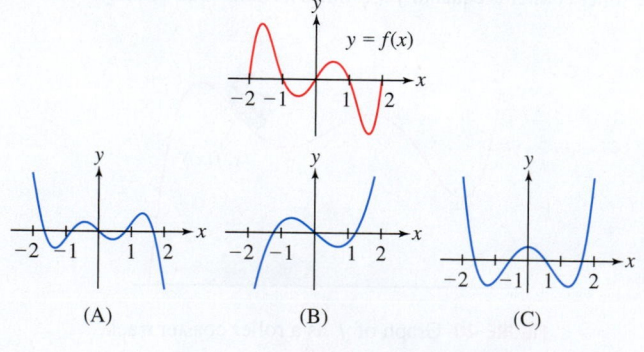

FIGURE 3

20. Sketch the graph of f' if the graph of f appears as in Figure 4.

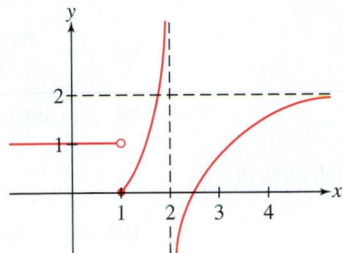

FIGURE 4

21. Sketch the graph of a continuous function f if the graph of f' appears as in Figure 5 and $f(0) = 0$.

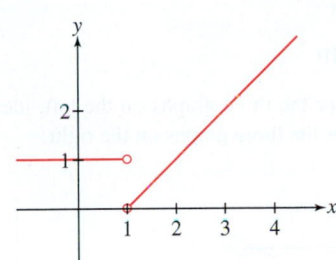

FIGURE 5

22. A patient is given 40 mg of a drug. After t hours, the amount of the drug (in mg) in the patient's bloodstream is given by $A(t) = 100t \left(0.3^t\right)$. At what rate was the amount of the drug in the bloodstream changing after 30 min? After 2 h?

23. A girl's height $h(t)$ (in centimeters) is measured at time t (in years) for $0 \le t \le 14$:

52, 75.1, 87.5, 96.7, 104.5, 111.8, 118.7, 125.2,
131.5, 137.5, 143.3, 149.2, 155.3, 160.8, 164.7

(a) What is the average growth rate over the 14-yr period?

(b) Is the average growth rate larger over the first half or the second half of this period?

(c) Estimate $h'(t)$ (in cm/yr) for $t = 3, 8$.

24. A planet's period P (number of days to complete one revolution around the sun) is approximately $0.199A^{3/2}$, where A is the average distance (in millions of kilometers) from the planet to the sun. Calculate P and dP/dA for Earth using the value $A = 150$.

In Exercises 25 and 26, use the following table of values for the number $A(t)$ of automobiles (in millions) manufactured in the United States in year t.

t	1970	1971	1972	1973	1974	1975	1976
$A(t)$	6.55	8.58	8.83	9.67	7.32	6.72	8.50

25. What is the interpretation of $A'(t)$? Estimate $A'(1971)$. Does $A'(1974)$ appear to be positive or negative?

26. Given the data, which of (A)–(C) in Figure 6 could be the graph of the derivative A'? Explain.

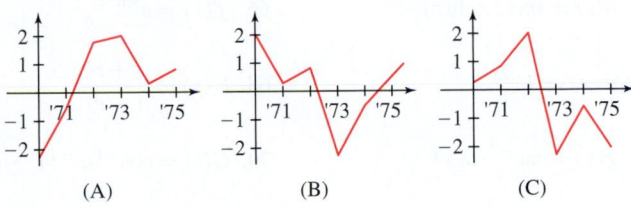

FIGURE 6

27. Which of the following is equal to $\dfrac{d}{dx} 2^x$?

(a) 2^x **(b)** $(\ln 2)2^x$ **(c)** $x2^{x-1}$ **(d)** $\dfrac{1}{\ln 2} 2^x$

28. Use the Chain Rule to show that if g is the inverse of f, then $g'(x) = 1/f'(g(x))$ for all x in the domain of g such that $f(g(x)) \neq 0$. Use this to obtain another method for finding the derivative of $\ln x$ using the derivative of e^x.

In Exercises 29–80, compute the derivative.

29. $y = 3x^5 - 7x^2 + 4$

30. $y = 4x^{-3/2}$

31. $y = t^{-7.3}$

32. $y = 4x^2 - x^{-2}$

33. $y = \dfrac{x+1}{x^2+1}$

34. $y = \dfrac{3t-2}{4t-9}$

35. $y = (x^4 - 9x)^6$

36. $y = (3t^2 + 20t^{-3})^6$

37. $y = (2 + 9x^2)^{3/2}$

38. $y = (x+1)^3(x+4)^4$

39. $y = \dfrac{z}{\sqrt{1-z}}$

40. $y = \left(1 + \dfrac{1}{x}\right)^3$

41. $y = \dfrac{x^4 + \sqrt{x}}{x^2}$

42. $y = \dfrac{1}{(1-x)\sqrt{2-x}}$

43. $y = \sqrt{x + \sqrt{x + \sqrt{x}}}$

44. $h(z) = \left(z + (z+1)^{1/2}\right)^{-3/2}$

45. $y = \tan(t^{-3})$

46. $y = 4\cos(2 - 3x)$

47. $y = \sin(2x)\cos^2 x$

48. $y = \sin\left(\dfrac{4}{\theta}\right)$

49. $y = \dfrac{t}{1 + \sec t}$

50. $y = z\csc(9z + 1)$

51. $y = \dfrac{8}{1 + \cot \theta}$

52. $y = \sin^{100} x$

53. $y = \cos(x^{100})$

54. $y = \cos(\cos(\cos(\theta)))$

55. $f(x) = 9e^{-4x}$

56. $f(x) = \dfrac{e^{-x}}{x}$

57. $g(t) = e^{4t - t^2}$

58. $g(t) = t^2 e^{1/t}$

59. $f(x) = \ln(4x^2 + 1)$

60. $f(x) = \ln(e^x - 4x)$

61. $G(s) = (\ln(s))^2$

62. $G(s) = \ln(s^2)$

63. $f(\theta) = \ln(\sin \theta)$

64. $f(\theta) = \sin(\ln \theta)$

65. $h(z) = \sec(z + \ln z)$

66. $f(x) = e^{\sin^2 x}$

67. $f(x) = 7^{-2x}$

68. $h(y) = \dfrac{1 + e^y}{1 - e^y}$

69. $g(x) = \tan^{-1}(\ln x)$

70. $G(s) = \cos^{-1}(s^{-1})$

71. $f(x) = \ln(\csc^{-1} x)$

72. $f(x) = e^{\sec^{-1} x}$

73. $R(s) = s^{\ln s}$

74. $f(x) = (\cos^2 x)^{\cos x}$

75. $G(t) = (\sin^2 t)^t$

76. $h(t) = t^{(t^t)}$

77. $g(t) = \sinh(t^2)$

78. $h(y) = y \tanh(4y)$

79. $g(x) = \tanh^{-1}(e^x)$

80. $g(t) = \sqrt{t^2 - 1}\,\sinh^{-1} t$

81. For which values of α is $f(x) = |x|^\alpha$ differentiable at $x = 0$?

82. Find $f'(2)$ if $f(g(x)) = e^{x^2}$, $g(1) = 2$, and $g'(1) = 4$.

In Exercises 83–88, use the following table of values to calculate the derivative of the given function at $x = 2$:

x	$f(x)$	$g(x)$	$f'(x)$	$g'(x)$
2	5	4	−3	9
4	3	2	−2	3

83. $S(x) = 3f(x) - 2g(x)$

84. $H(x) = f(x)g(x)$

85. $R(x) = \dfrac{f(x)}{g(x)}$

86. $G(x) = f(g(x))$

87. $F(x) = f(g(2x))$

88. $K(x) = f(x^2)$

89. Find the points on the graph of $f(x) = x^3 - 3x^2 + x + 4$ where the tangent line has slope 10.

90. Find the points on the graph of $x^{2/3} + y^{2/3} = 1$ where the tangent line has slope 1.

91. Find a such that the tangent lines to $y = x^3 - 2x^2 + x + 1$ at $x = a$ and $x = a + 1$ are parallel.

92. 📝 Use the table to compute the average rate of change of candidate A's percentage of votes over the intervals from day 20 to day 15, day 15 to day 10, and day 10 to day 5. If this trend continues over the last 5 days before the election, will candidate A win?

Days Before Election	20	15	10	5
Candidate A	44.8%	46.8%	48.3%	49.3%
Candidate B	55.2%	53.2%	51.7%	50.7%

In Exercises 93–98, calculate y''.

93. $y = 12x^3 - 5x^2 + 3x$

94. $y = x^{-2/5}$

95. $y = \sqrt{2x + 3}$

96. $y = \dfrac{4x}{x + 1}$

97. $y = \tan(x^2)$

98. $y = \sin^2(4x + 9)$

In Exercises 99–104, compute $\dfrac{dy}{dx}$.

99. $x^3 - y^3 = 4$

100. $4x^2 - 9y^2 = 36$

101. $y = xy^2 + 2x^2$

102. $\dfrac{y}{x} = x + y$

103. $y = \sin(x + y)$

104. $\tan(x + y) = xy$

In Exercises 105–106 compute $\dfrac{dy}{dx}$ and $\dfrac{d^2y}{dx^2}$.

105. $x^2 - 4y^2 = 8$

106. $6xy + y^2 = 10$

107. In Figure 7, for the three graphs on the left, identify f, f', and f''. Do the same for the three graphs on the right.

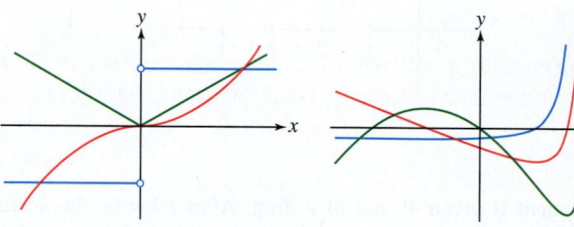

FIGURE 7

108. Let $f(x) = x^2 \sin(x^{-1})$ for $x \neq 0$ and $f(0) = 0$. Show that $f'(x)$ exists for all x (including $x = 0$) but that f' is not continuous at $x = 0$ (Figure 8).

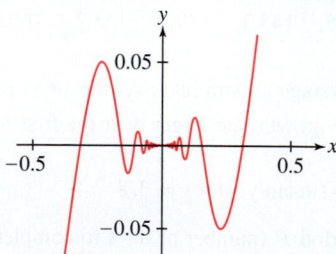

FIGURE 8 Graph of $f(x) = x^2 \sin(x^{-1})$.

In Exercises 109–114, use logarithmic differentiation to find the derivative.

109. $y = \dfrac{(x + 1)^3}{(4x - 2)^2}$

110. $y = \dfrac{(x + 1)(x + 2)^2}{(x + 3)(x + 4)}$

111. $y = e^{(x-1)^2} e^{(x-3)^2}$

112. $y = \dfrac{e^x \sin^{-1} x}{\ln x}$

113. $y = \dfrac{e^{3x}(x - 2)^2}{(x + 1)^2}$

114. $y = x^{\sqrt{x}}(x^{\ln x})$

115. How fast does the water level rise in the tank in Figure 9 when the water level is $h = 4$ m and water pours in at 20 m^3/min?

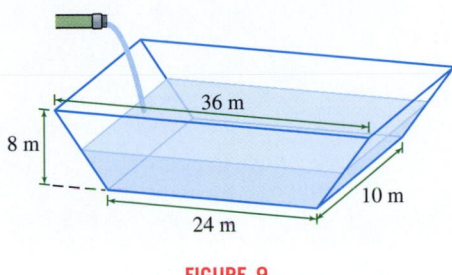

FIGURE 9

116. The minute hand of a clock is 8 cm long, and the hour hand is 5 cm long. How fast is the distance between the tips of the hands changing at 3 o'clock?

117. Chloe and Bao are in motorboats at the center of a lake. At time $t = 0$, Chloe begins traveling south at a speed of 50 km/h. One minute later, Bao takes off, heading east at a speed of 40 km/h. At what rate is the distance between them increasing at $t = 12$ min?

118. A bead slides down the curve $xy = 10$. Find the bead's horizontal velocity at time $t = 2$ s if the height of the bead at time t seconds is $y = 400 - 16t^2$ cm.

119. In Figure 10, x is increasing at 2 cm/s, y is increasing at 3 cm/s, and θ is decreasing such that the area of the triangle has the constant value 4 cm^2.

(a) How fast is θ decreasing when $x = 4$, $y = 4$?

(b) How fast is the distance between P and Q changing when $x = 4$, $y = 4$? *Hint:* Use the Law of Cosines.

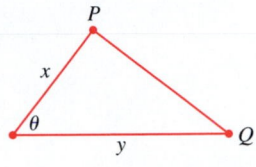

FIGURE 10

120. A light moving at 0.8 m/s approaches a man standing 4 m from a wall (Figure 11). The light is 1 m above the ground. How fast is the tip P of the man's shadow moving when the light is 7 m from the wall?

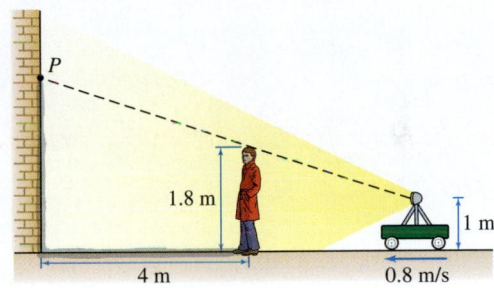

FIGURE 11

In northern forests, where the sun is relatively low in the sky, trees tend to be tall and narrow to maximize their exposure to the available light. In contrast, in equatorial regions trees tend to have broad flat tops because the sun is mostly overhead during the day. Optimization is an important characteristic of natural systems, as well as a key factor in decision-making processes in applied problem solving. In this chapter, we will use the derivative to analyze functions and find optimal solutions in a variety of applied settings.

4 APPLICATIONS OF THE DERIVATIVE

This chapter puts the derivative to work. The first and second derivatives are used to analyze functions and their graphs and to solve optimization problems (finding minimum and maximum values of a function). Newton's Method in Section 4.8 employs the derivative to approximate solutions of equations.

4.1 Linear Approximation and Applications

In this section, we introduce the process of linear approximation that uses the tangent line to the graph of a function f at $x = a$ to approximate $f(x)$ for x near a. These approximation methods are desirable because linear functions are usually easier to use and compute with than nonlinear ones. We introduce a few different formulas involving linear approximation. There are different settings and situations where each is useful. Keep in mind that they all come from the same basic idea that the tangent line approximates the function close to the point of tangency (Figure 1).

Linear Approximation

In some situations, we are interested in the effect of a small change. For example,

- How does a small change in angle affect the distance of a basketball shot? (Exercise 47)
- How are revenues at the box office affected by a small change in ticket prices? (Exercise 37)
- The cube root of 27 is 3. How much greater is the cube root of 27.2? (Exercise 7)

In each case, we have a function f and we're interested in the change

$$\Delta f = f(a + \Delta x) - f(a)$$

where Δx is small. The **Linear Approximation** uses the slope of the tangent line (i.e., the derivative) to estimate Δf without computing it exactly. By definition, the derivative is the limit

$$f'(a) = \lim_{\Delta x \to 0} \frac{f(a + \Delta x) - f(a)}{\Delta x} = \lim_{\Delta x \to 0} \frac{\Delta f}{\Delta x}$$

So when Δx is small, we have $\Delta f / \Delta x \approx f'(a)$, and thus,

$$\Delta f \approx f'(a)\Delta x$$

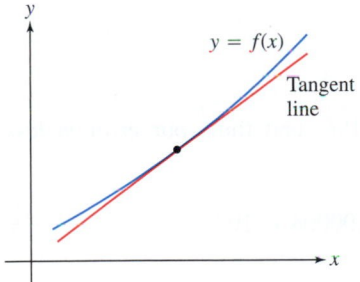

FIGURE 1 The tangent line approximates the graph of f near the point of tangency.

y = f(x)

Tangent line

REMINDER The notation $\approx$ means "approximately equal to." The accuracy of the Linear Approximation is discussed at the end of this section.

Linear Approximation of Δf If f is differentiable at $x = a$ and Δx is small, then

$$\Delta f \approx f'(a)\Delta x \qquad \boxed{1}$$

It is important to understand the different roles played by Δf and $f'(a)\Delta x$. The quantity of interest is the *actual change* Δf. We estimate it by $f'(a)\,\Delta x$, the change on the tangent line with slope $f'(a)$. The Linear Approximation tells us that up to a small error, Δf is approximately equal to $f'(a)\Delta x$ when Δx is small.

GRAPHICAL INSIGHT   As we indicated, the Linear Approximation is an approximation using a tangent line. In fact, it is sometimes called the **tangent line approximation**. Observe in Figure 2 that Δf is the vertical change in the graph from $x = a$ to $x = a + \Delta x$. For a line, the vertical change is equal to the slope times the horizontal change Δx, and since the tangent line has slope $f'(a)$, its vertical change is $f'(a)\Delta x$. So the Linear Approximation approximates Δf by the vertical change in the tangent line. When Δx is small, the two quantities are nearly equal.

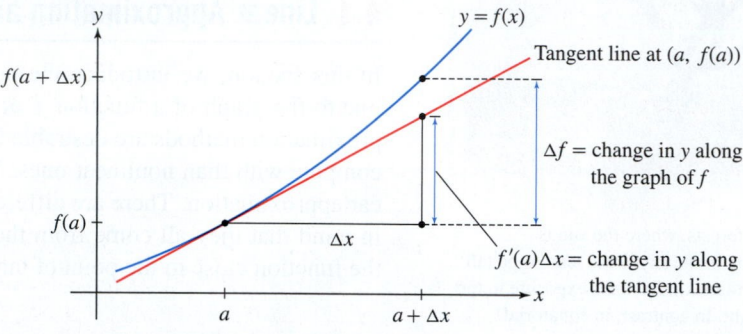

FIGURE 2 Graphical meaning of the Linear Approximation $\Delta f \approx f'(a)\Delta x$.

EXAMPLE 1   Use the Linear Approximation to estimate the change in $f(x) = 1/x$ as x goes from 10 to 10.2; that is, to estimate $\frac{1}{10.2} - \frac{1}{10}$. How accurate is the estimate?

Solution   We apply the Linear Approximation to $f(x) = \frac{1}{x}$ with $a = 10$ and $\Delta x = 0.2$: We have $f'(x) = -x^{-2}$ and $f'(10) = -0.01$, so Δf is approximated by

$$\Delta f \approx f'(10)\Delta x = (-0.01)(0.2) = -0.002$$

Since $\Delta f = \frac{1}{10.2} - \frac{1}{10}$ we have the approximation

$$\frac{1}{10.2} - \frac{1}{10} \approx -0.002$$

The error in the Linear Approximation is the quantity

$$\text{error} = \left| \Delta f - f'(a)\Delta x \right|$$

A calculator gives the value $\frac{1}{10.2} - \frac{1}{10} \approx -0.00196$, and thus, our error is less than 10^{-4}:

$$\text{error} \approx \left| -0.00196 - (-0.002) \right| = 0.00004 < 10^{-4}$$ ∎

EXAMPLE 2   Approximate how much greater $\sqrt[3]{8.1}$ is than $\sqrt[3]{8} = 2$, and then use the result to approximate $\sqrt[3]{8.1}$.

Solution   We are interested in $\sqrt[3]{8.1} - \sqrt[3]{8}$, so we apply the Linear Approximation to $f(x) = x^{1/3}$ with $a = 8$ and $\Delta x = 0.1$. We have

$$f'(x) = \frac{1}{3}x^{-2/3} \quad \text{and} \quad f'(8) = \left(\frac{1}{3}\right)8^{-2/3} = \left(\frac{1}{3}\right)\left(\frac{1}{4}\right) = \frac{1}{12}$$

Therefore, $\Delta f \approx f'(8)\Delta x = \frac{1}{12}(0.1) \approx 0.0083$, and since

$$\Delta f = f(a + \Delta x) - f(a) = \sqrt[3]{8 + 0.1} - \sqrt[3]{8} = \sqrt[3]{8.1} - 2$$

we have the approximation

$$\sqrt[3]{8.1} - 2 \approx 0.0083$$

Thus, $\sqrt[3]{8.1}$ is greater than $\sqrt[3]{8}$ by approximately 0.0083. It follows that

$$\sqrt[3]{8.1} \approx 2 + 0.0083 = 2.0083$$ ∎

Suppose that we measure the *diameter* D of a circle and use this result to compute the *area* of the circle. If our measurement of D is inexact, the area computation will also be inexact. What is the effect of the measurement error on the resulting area computation? This can be estimated using the Linear Approximation, as in the next example.

EXAMPLE 3   **Effect of an Inexact Measurement**   The Cheezy Pizza Parlor claims that its pizzas are circular with diameter 50 cm (Figure 3).

(a) What is the area of the pizza?

(b) Estimate the quantity of pizza lost or gained if the diameter is off by at most 1.2 cm.

Solution   First, we need a formula for the area A of a circle in terms of its diameter D. Since the radius is $r = D/2$, the area is

$$A(D) = \pi r^2 = \pi \left(\frac{D}{2}\right)^2 = \frac{\pi}{4} D^2$$

(a) If $D = 50$ cm, then the pizza has area $A(50) = \left(\frac{\pi}{4}\right)(50)^2 \approx 1963.5$ cm^2.

(b) If the actual diameter is equal to $50 + \Delta D$, then the loss or gain in pizza area is $A(50 + \Delta D) - A(50) = \Delta A$. We apply Linear Approximation to $A(D)$ with $D = 50$ and $\Delta D = \pm 1.2$. Observe that $A'(D) = \frac{\pi}{2} D$ and $A'(50) = 25\pi \approx 78.5$ cm, so the Linear Approximation yields

$$\Delta A \approx A'(D)\Delta D \approx (78.5)\,\Delta D$$

Because ΔD is at most ± 1.2 cm, the loss or gain in pizza is no more than around

$$\Delta A \approx \pm(78.5)(1.2) \approx \pm 94.2 \text{ cm}^2$$

This is a loss or gain of approximately 4.8% of the area of 1963.5 cm^2.   ■

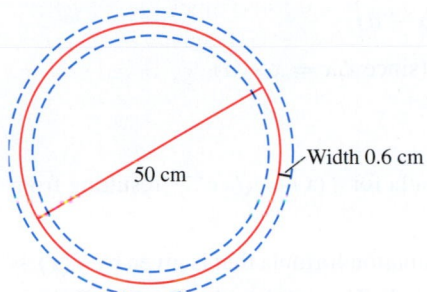

FIGURE 3 The border of the actual pizza lies between the dashed circles.

In this example, we interpret ΔA as the possible error in the computation of $A(D)$. This should not be confused with the error in the Linear Approximation. This latter error refers to the accuracy in using $A'(D)\,\Delta D$ to approximate ΔA.

Linearization

To approximate the function f itself rather than the change Δf, we use the **linearization** $L(x)$ **centered at** $x = a$, defined by

$$L(x) = f'(a)(x - a) + f(a)$$

Notice that $y = L(x)$ is the equation of the tangent line at $x = a$. For values of x close to a, $L(x)$ provides a good approximation to $f(x)$ (Figure 4).

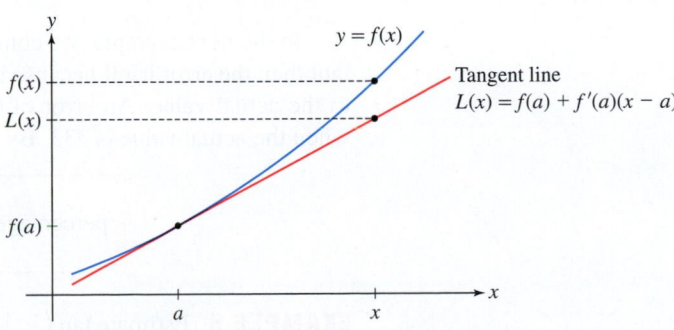

FIGURE 4 For x near a, the function value $f(x)$ is approximated by the tangent line value $L(x)$.

Approximating f by Its Linearization   If f is differentiable at a and x is close to a, then $f(x) \approx L(x)$, so

$$f(x) \approx f(a) + f'(a)(x - a)$$

Note that, by rearranging the terms in linearization formula, we obtain the Linear Approximation formula, $\Delta f \approx f'(a)\Delta x$, that we introduced previously. Specifically, with $\Delta x = x - a$ and $\Delta f = f(x) - f(a)$, we have

$$f(x) \approx f(a) + f'(a)(x - a)$$

$$f(x) - f(a) \approx f'(a)\,\Delta x \qquad \text{(since } \Delta x = x - a\text{)}$$

$$\Delta f \approx f'(a)\Delta x$$

EXAMPLE 4 Determine the approximation formula for $f(x) = \sqrt{x}e^{x-1}$ resulting from the linearization at $a = 1$.

Solution The linearization at $a = 1$ is the approximation formula that is given by $f(x) \approx f(1) + f'(1)(x - 1)$. Note that $f(1) = \sqrt{1}e^{1-1} = 1$. Then, using the Product Rule to compute the derivative, we obtain

$$f'(x) = \frac{1}{2}x^{-1/2}e^{x-1} + x^{1/2}e^{x-1} = \left(\frac{1}{2}x^{-1/2} + x^{1/2}\right)e^{x-1}$$

and therefore, $f'(1) = \left(\frac{1}{2} + 1\right)e^0 = \frac{3}{2}$. Thus,

$$f(1) + f'(1)(x - 1) = 1 + \frac{3}{2}(x - 1) = \frac{3}{2}x - \frac{1}{2}$$

This yields the approximation formula, valid for x close to 1:

$$\sqrt{x}e^{x-1} \approx \frac{3}{2}x - \frac{1}{2}$$

■

The following table compares values of the linearization to values obtained from a calculator for the function $f(x) = \sqrt{x}e^{x-1}$ in the previous example. Note that the error is large for $x = 2.5$, as expected, because 2.5 is not close to the center of the linearization $a = 1$ (Figure 5).

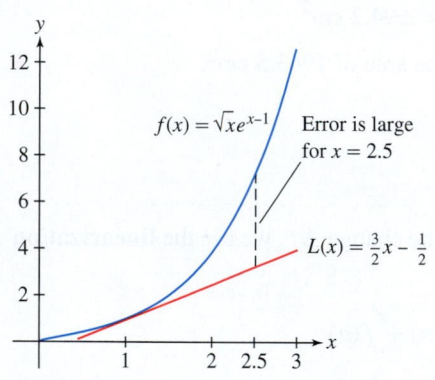

x	$\sqrt{x}e^{x-1}$	Linearization $\frac{3}{2}x - \frac{1}{2}$	Calculator	Error
1.1	$\sqrt{1.1}e^{0.1}$	$\frac{3}{2}(1.1) - \frac{1}{2} = 1.15$	1.15911	10^{-2}
0.999	$\sqrt{0.999}e^{-0.001}$	$\frac{3}{2}(0.999) - \frac{1}{2} = 0.9985$	0.998501	10^{-6}
2.5	$\sqrt{2.5}e^{1.5}$	$\frac{3}{2}(2.5) - \frac{1}{2} = 3.25$	7.086	3.84

FIGURE 5 Graph of $f(x) = \sqrt{x}e^{x-1}$ and its linearization at $a = 1$.

In the next example, we compute the **percentage error**, which is often more important than the error itself because it gives us a measure of how large the error is in relation to the actual value. An error of 0.1 is more significant when the actual value is 3 than when the actual value is 333. By definition,

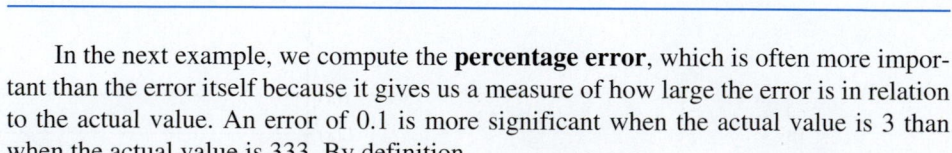

$$\text{percentage error} = \left|\frac{\text{error}}{\text{actual value}}\right| \times 100\%$$

EXAMPLE 5 Estimate $\tan\left(\frac{\pi}{4} + 0.02\right)$ and compute the percentage error.

Solution We use the linearization of $f(x) = \tan x$ at $a = \frac{\pi}{4}$ for our approximation. So we need to calculate the terms in $f(\pi/4) + f'(\pi/4)(x - \pi/4)$:

$$f\left(\frac{\pi}{4}\right) = \tan\left(\frac{\pi}{4}\right) = 1, \qquad f'\left(\frac{\pi}{4}\right) = \sec^2\left(\frac{\pi}{4}\right) = \left(\sqrt{2}\right)^2 = 2$$

$$f\left(\frac{\pi}{4}\right) + f'\left(\frac{\pi}{4}\right)\left(x - \frac{\pi}{4}\right) = 1 + 2\left(x - \frac{\pi}{4}\right)$$

So, for x near $\pi/4$, we have the approximation formula

$$\tan(x) \approx 1 + 2\left(x - \frac{\pi}{4}\right)$$

At $x = \frac{\pi}{4} + 0.02$, this approximation yields the estimate

$$\tan\left(\frac{\pi}{4} + 0.02\right) \approx 1 + 2\left(\frac{\pi}{4} + 0.02 - \frac{\pi}{4}\right) = 1.04$$

A calculator gives $\tan\left(\frac{\pi}{4} + 0.02\right) \approx 1.0408$, so

$$\text{percentage error} \approx \left|\frac{1.0408 - 1.04}{1.0408}\right| \times 100 \approx 0.08\%$$

■

Differential Form of Linear Approximation

Another way of expressing the Linear Approximation is via the **differentials** dx and dy that represent the change in x and y, respectively, on the tangent line to $f(x)$ at $x = a$. Since these differentials represent change on the tangent line, we have

$$dy = f'(a)dx \qquad \boxed{2}$$

As before, we let Δy represent the change in y on the graph of f. It follows—as in the previous approximations in this section—that with a small change in x, the change in y on the graph is approximately the change in y on the tangent line (Figure 6). Thus, $\Delta y \approx dy$, yielding the following:

> **Differential Form of Linear Approximation** If f is differentiable at a and dx is small, then
>
> $$\Delta y \approx dy = f'(a)dx \qquad \boxed{3}$$

As we mentioned before, in the Leibniz notation for the derivative, $\frac{dy}{dx}$ does not represent a fraction. It is via differentials, though, that the relationship $dy = \frac{dy}{dx}dx$ is made mathematically meaningful. We will find relationships like this to be very useful when simplifying computations involving integrals in subsequent chapters.

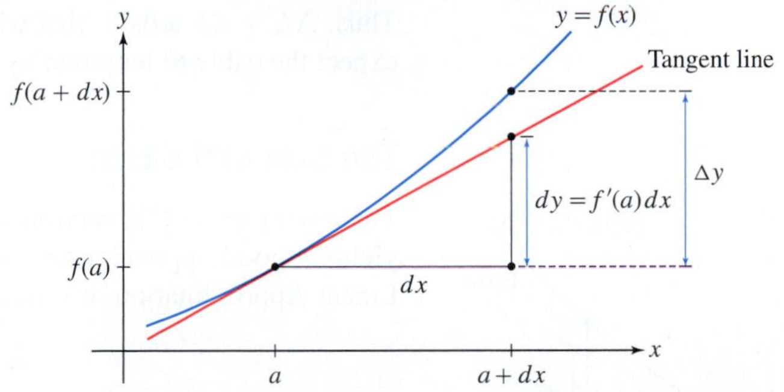

FIGURE 6 The approximation $\Delta y \approx dy = f'(a)dx$.

CONCEPTUAL INSIGHT At the start of the section, we observed that all of the approximation relationships presented here are based on the idea that the tangent line is a good approximation to the graph of the function near the point of tangency. The Linear Approximation, the linearization, and the Differential Form of Linear Approximation are illustrated in Figures 2, 4, and 6, respectively. Note that these figures all depict the graph of f and the tangent line at $x = a$. From figure to figure, various features are described or labeled differently in order to illustrate the important aspects of each approximation relationship.

You might wonder why we bother with the Differential Form of Linear Approximation. At this point, it just appears to be another way of expressing a relationship that we already had a perfectly good way of expressing. The intent here is to provide an initial glimpse into a tool, the differential, that is employed often by mathematicians, scientists, and engineers to express or approximate a small change involving related variables. A differential corresponds to the change on the tangent line (as we see here), on the tangent plane (see Section 14.4), or in the tangent space (in higher dimensions). Differentials provide a straightforward linear means for approximating and working with complicated relationships.

EXAMPLE 6   Thermal Expansion   Changes in temperature can have subtle effects on physical properties of objects that we might think normally are constant. A thin metal cable has length $L = 12$ cm when the temperature is $T = 21°C$. Estimate the change in length when T rises to $24°C$, assuming that

$$\frac{dL}{dT} = kL \qquad \boxed{4}$$

where $k = 1.7 \times 10^{-5}°C^{-1}$ (k is called the coefficient of thermal expansion).

Solution   How does the Linear Approximation apply here? We will use the differential dL to estimate the actual change in length ΔL when T increases from $21°C$ to $24°C$—that is, when $dT = \Delta T = 3°C$. By Eq. (2), the differential dL is

$$dL = \left(\frac{dL}{dT}\right) dT$$

By Eq. (4), since $L = 12$,

$$\left.\frac{dL}{dT}\right|_{L=12} = kL = (1.7 \times 10^{-5})(12) \approx 2 \times 10^{-4} \text{ cm/°C}$$

Therefore, with $dT = 3$, we have

$$dL = \left(\frac{dL}{dT}\right) dT \approx (2 \times 10^{-4})(3) = 6 \times 10^{-4} \text{ cm}$$

Thus, $\Delta L \approx dL$ tells us that when the temperature increases from $21°C$ to $24°C$, we can expect the cable to lengthen by approximately 0.0006 cm.   ■

The Size of the Error

The examples in this section may have convinced you that the Linear Approximation yields a good approximation to Δf when Δx is small, but if we want to rely on the Linear Approximation, we need to know more about the size of the error:

$$E = \text{error} = \left|\Delta f - f'(a)\Delta x\right|$$

Graphically the error E is the vertical gap between the graph of f and the tangent line (Figure 7). In Section 10.7, we will prove the following **Error Bound**:

$$\boxed{E \leq \frac{1}{2} K (\Delta x)^2} \qquad \boxed{5}$$

where K is the maximum value of $|f''(x)|$ on the interval from a to $a + \Delta x$.

The Error Bound tells us two important things. First, it says that the error is small when the second derivative (and hence K) is small. This makes sense, because $f''(x)$ measures how quickly the tangent lines change direction. When $|f''(x)|$ is smaller, the graph is flatter and the Linear Approximation is more accurate over a larger interval around $x = a$ (compare the graphs in Figure 8).

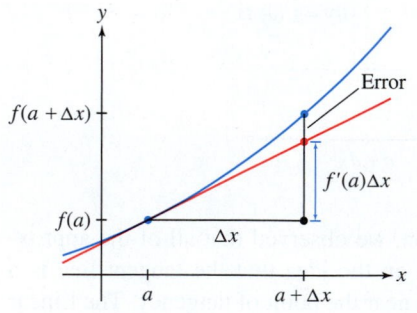

FIGURE 7 Graphical interpretation of the error in the Linear Approximation.

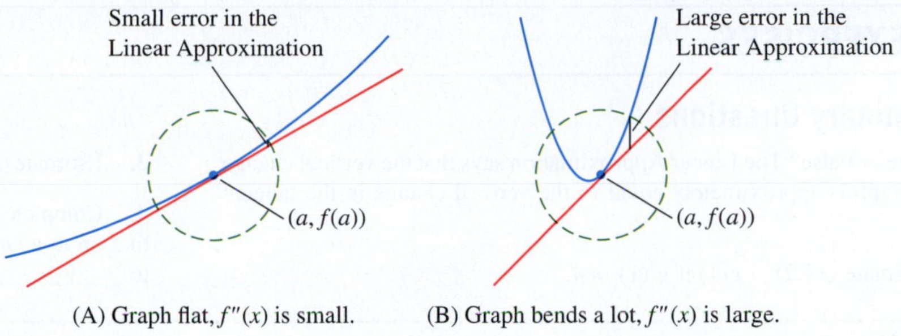

(A) Graph flat, $f''(x)$ is small.   (B) Graph bends a lot, $f''(x)$ is large.

FIGURE 8 The accuracy of the Linear Approximation depends on how much the curve bends.

Second, the Error Bound tells us that the error is of *order 2* in Δx, meaning that E is no larger than a constant times $(\Delta x)^2$. So if Δx is small, say, $\Delta x = 10^{-n}$, then E has a substantially smaller order of magnitude, since $(\Delta x)^2 = 10^{-2n}$. In particular, $E/\Delta x$ tends to zero (because $E/\Delta x < K\Delta x$), so the Error Bound tells us that the graph becomes nearly indistinguishable from its tangent line as we zoom in on the graph around $x = a$. This is a precise version of the "local linearity" property discussed in Section 3.2.

4.1 SUMMARY

The approximation formulas in this section are all based on the idea that the tangent line to the graph of a function f at $x = a$ can be used to approximate $f(x)$ for x near a.

- Let $\Delta f = f(a + \Delta x) - f(a)$. The *Linear Approximation* is the estimate

$$\Delta f \approx f'(a)\Delta x \quad \text{(for } \Delta x \text{ small)}$$

- The *linearization of $f(x)$ centered at $x = a$* is the function for the tangent line

$$L(x) = f(a) + f'(a)(x - a)$$

- The approximation based on linearization is

$$f(x) \approx f(a) + f'(a)(x - a) \quad \text{(for } x \text{ close to } a)$$

- Differential notation: $dx = \Delta x$ is the change in x, $dy = f'(a)dx$ is the change on the tangent line, and $\Delta y = f(a + \Delta x) - f(a)$ is the change in f. In this notation, the *Differential Form of Linear Approximation* is

$$\Delta y \approx dy = f'(a)dx \quad \text{(for } dx \text{ small)}$$

- The error in the Linear Approximation is the quantity

$$\text{error} = \left| \Delta f - f'(a)\Delta x \right|$$

The percentage error is often more significant because it is a measure of how large the error is in relation to the actual value:

$$\text{percentage error} = \left| \frac{\text{error}}{\text{actual value}} \right| \times 100\%$$

- The error E in the Linear Approximation is bounded as follows:

$$E \leq \frac{1}{2} K (\Delta x)^2$$

where K is the maximum value of $|f''(x)|$ on the interval from a to $a + \Delta x$.

4.1 EXERCISES

Preliminary Questions

1. True or False? The Linear Approximation says that the vertical change in the graph is approximately equal to the vertical change in the tangent line.

2. Estimate $g(1.2) - g(1)$ if $g'(1) = 4$.

3. Estimate $f(2.1)$ if $f(2) = 1$ and $f'(2) = 3$.

4. Complete the following sentence: The Linear Approximation shows that up to a small error, the change in output Δf is directly proportional to _____ .

Exercises

In Exercises 1–6, use Eq. (1) to estimate $\Delta f = f(3.02) - f(3)$.

1. $f(x) = x^2$

2. $f(x) = x^4$

3. $f(x) = x^{-1}$

4. $f(x) = \dfrac{1}{x+1}$

5. $f(x) = \sqrt{x+6}$

6. $f(x) = \tan \dfrac{\pi x}{3}$

7. The cube root of 27 is 3. How much larger is the cube root of 27.2? Estimate using the Linear Approximation.

8. The cube root of 64 is 4. How much smaller is the cube root of 63.6? Estimate using the Linear Approximation.

In Exercises 9–12, use Eq. (1) to estimate Δf. *Use a calculator to compute both the error and the percentage error.*

9. $f(x) = \sqrt{1+x}$, $a = 3$, $\Delta x = 0.2$

10. $f(x) = 2x^2 - x$, $a = 5$, $\Delta x = -0.4$

11. $f(x) = \dfrac{1}{1+x^2}$, $a = 3$, $\Delta x = 0.5$

12. $f(x) = \ln(x^2 + 1)$, $a = 1$, $\Delta x = 0.1$

In Exercises 13–20, using Linear Approximation, estimate Δf *for a change in x from $x = a$ to $x = b$. Use the estimate to approximate $f(b)$, and find the error using a calculator.*

13. $f(x) = \sqrt{x}$, $a = 25$, $b = 26$

14. $f(x) = x^{1/4}$, $a = 16$, $b = 16.5$

15. $f(x) = \dfrac{1}{\sqrt{x}}$, $a = 100$, $b = 101$

16. $f(x) = \dfrac{1}{\sqrt{x}}$, $a = 100$, $b = 98$

17. $f(x) = x^{1/3}$, $a = 8$, $b = 9$

18. $f(x) = \tan^{-1} x$, $a = 1$, $b = 1.05$

19. $f(x) = e^x$, $a = 0$, $b = -0.1$

20. $f(x) = \ln x$, $a = 1$, $b = 0.97$

In Exercises 21–28, find the linearization at $x = a$ and then use it to approximate $f(b)$.

21. $f(x) = x^4$, $a = 1$, $b = 0.96$

22. $f(x) = \dfrac{1}{x}$, $a = 2$, $b = 2.02$

23. $f(x) = \sin^2 x$, $a = \frac{\pi}{4}$, $b = \frac{1.1\pi}{4}$

24. $f(x) = \dfrac{x^2}{x-3}$, $a = 4$, $b = 4.1$

25. $f(x) = (1+x)^{-1/2}$, $a = 0$, $b = 0.08$

26. $f(x) = (1+x)^{-1/2}$, $a = 3$, $b = 2.88$

27. $f(x) = e^{\sqrt{x}}$, $a = 1$, $b = 0.85$

28. $f(x) = e^x \ln x$, $a = 1$, $b = 1.02$

In Exercises 29–32, estimate Δy *using differentials [Eq. (3)].*

29. $y = \cos x$, $a = \frac{\pi}{6}$, $dx = 0.014$

30. $y = \tan^2 x$, $a = \frac{\pi}{4}$, $dx = -0.02$

31. $y = \dfrac{10 - x^2}{2 + x^2}$, $a = 1$, $dx = 0.01$

32. $y = x^{1/3} e^{x-1}$, $a = 1$, $dx = 0.1$

33. Estimate $f(4.03)$ for $f(x)$ as in Figure 9.

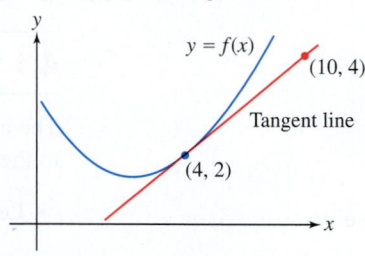

FIGURE 9

34. ✎ At a certain moment, an object in linear motion has velocity 100 m/s. Estimate the distance traveled over the next quarter-second, and explain how this is an application of the Linear Approximation.

35. Which is larger: $\sqrt{2.1} - \sqrt{2}$ or $\sqrt{9.1} - \sqrt{9}$? Explain using the Linear Approximation.

36. Estimate $\sin 61° - \sin 60°$ using the Linear Approximation. *Hint:* Express $\Delta\theta$ in radians.

37. Box office revenue at a cinema in Paris is $R(p) = 3600p - 10p^3$ euros per showing when the ticket price is p euros. Calculate $R(p)$ for $p = 9$ and use the Linear Approximation to estimate ΔR if p is raised or lowered by 0.5 euro.

38. The *stopping distance* for an automobile is $F(s) = 1.1s + 0.054s^2$ ft, where s is the speed in mph. Use the Linear Approximation to estimate the change in stopping distance per additional mph when $s = 35$ and when $s = 55$.

39. A thin silver wire has length $L = 18$ cm when the temperature is $T = 30°C$. Estimate ΔL when T decreases to $25°C$ if the coefficient of thermal expansion is $k = 1.9 \times 10^{-5}°C^{-1}$ (see Example 6).

40. At a certain moment, the temperature in a snake cage satisfies $dT/dt = 0.008°C/s$. Estimate the rise in temperature over the next 10 s.

41. The atmospheric pressure at altitude h (kilometers) for $11 \le h \le 25$ is approximately

$$P(h) = 128e^{-0.157h} \text{ kilopascals}$$

(a) Estimate ΔP at $h = 20$ when $\Delta h = 0.5$.

(b) Compute the actual change, and compute the percentage error in the Linear Approximation.

42. The resistance R of a copper wire at temperature $T = 20°C$ is $R = 15$ Ω. Estimate the resistance at $T = 22°C$, assuming that $dR/dT|_{T=20} = 0.06$ Ω/°C.

43. Newton's Law of Gravitation shows that if a person weighs w pounds on the surface of the earth, then his or her weight at distance x from the center of the earth is

$$W(x) = \frac{wR^2}{x^2} \quad \text{(for } x \geq R\text{)}$$

where $R = 3960$ miles is the radius of the earth (Figure 10).

(a) Show that the weight lost at altitude h miles above the earth's surface is approximately $\Delta W \approx -(0.0005w)h$. *Hint:* Use the Linear Approximation with $dx = h$.

(b) Estimate the weight lost by a 200-lb football player flying in a jet at an altitude of 7 miles.

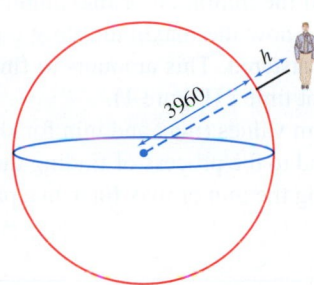

FIGURE 10 The distance to the center of the earth is $3960 + h$ miles.

44. Using Exercise 43(a), estimate the altitude at which a 130-lb pilot would weigh 129.5 lb.

45. A stone tossed vertically into the air with initial velocity v cm/s reaches a maximum height of $h = v^2/1960$ cm.

(a) Estimate Δh if $v = 700$ cm/s and $\Delta v = 1$ cm/s.

(b) Estimate Δh if $v = 1000$ cm/s and $\Delta v = 1$ cm/s.

(c) In general, does a 1-cm/s increase in v lead to a greater change in h at low or high initial velocities? Explain.

46. The side s of a square carpet is measured at 6 m. Estimate the maximum error in the area A of the carpet if s is accurate to within 2 cm.

In Exercises 47 and 48, use the following fact derived from Newton's Laws: An object released at an angle θ with initial velocity v ft/s travels a horizontal distance

$$s = \frac{1}{32}v^2 \sin 2\theta \text{ ft (Figure 11)}$$

47. A player located 18.1 ft from the basket launches a successful jump shot from a height of 10 ft (level with the rim of the basket), at an angle $\theta = 34°$ and initial velocity $v = 25$ ft/s.

(a) Show that $\Delta s \approx 0.255\Delta\theta$ ft for a small change of $\Delta\theta$.

(b) Is it likely that the shot would have been successful if the angle had been off by 2°?

(c) Estimate Δs if $\theta = 34°$, $v = 25$ ft/s, and $\Delta v = 2$.

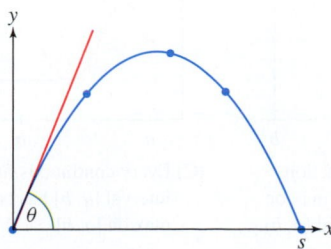

FIGURE 11 Trajectory of an object released at an angle θ.

48. A golfer hits a golf ball at an angle of $\theta = 23°$ with initial velocity $v = 120$ ft/s.

(a) Estimate Δs if the ball is hit at the same velocity but the angle is increased by 3°.

(b) Estimate Δs if the ball is hit at the same angle but the velocity is increased by 3 ft/s.

49. The radius of a spherical ball is measured at $r = 25$ cm. Estimate the maximum error in the volume and surface area if r is accurate to within 0.5 cm.

50. The dosage D of diphenhydramine for a dog of body mass w kg is $D = 4.7w^{2/3}$ mg. Estimate the maximum allowable error in w for a cocker spaniel of mass $w = 10$ kg if the percentage error in D must be less than 3%.

51. The volume (in liters) and pressure P (in atmospheres) of a certain gas satisfy $PV = 24$. A measurement yields $V = 4$ with a possible error of ± 0.3 L. Compute P and estimate the maximum error in this computation.

52. In the notation of Exercise 51, assume that a measurement yields $V = 4$. Estimate the maximum allowable error in V if P must have an error of less than 0.2 atm.

53. Approximate $f(2)$ if the linearization of $f(x)$ at $a = 2$ is $L(x) = 2x + 4$.

54. Compute the linearization of $f(x) = 3x - 4$ at $a = 0$ and $a = 2$. Prove more generally that a linear function coincides with its linearization at $x = a$ for all a.

55. Estimate $\sqrt{16.2}$ using the linearization $L(x)$ of $f(x) = \sqrt{x}$ at $a = 16$. Plot f and L on the same set of axes and from the plot indicate whether the estimate is greater than or less than the actual value.

56. [GU] Estimate $1/\sqrt{15}$ using a suitable linearization of $f(x) = 1/\sqrt{x}$. Plot f and L on the same set of axes and from the plot indicate whether the estimate is greater than or less than the actual value. Use a calculator to compute the percentage error.

In Exercises 57–65, approximate using linearization and use a calculator to compute the percentage error.

57. $\dfrac{1}{\sqrt{17}}$ **58.** $\dfrac{1}{101}$

59. $\dfrac{1}{(10.03)^2}$ **60.** $(17)^{1/4}$

61. $(64.1)^{1/3}$ **62.** $(1.2)^{5/3}$

63. $\cos^{-1}(0.52)$ **64.** $\ln 1.07$

65. $e^{-0.012}$

66. [GU] Compute the linearization $L(x)$ of $f(x) = x^2 - x^{3/2}$ at $a = 4$. Then plot $f - L$ and identify an interval I around $a = 4$ such that $|f(x) - L(x)| \leq 0.1$ for $x \in I$.

67. Show that the Linear Approximation to $f(x) = \sqrt{x}$ at $x = 9$ yields the estimate $\sqrt{9+h} - 3 \approx \frac{1}{6}h$. Set $K = 0.01$ and show that $|f''(x)| \leq K$ for $x \geq 9$. Then verify numerically that the error E satisfies Eq. (5) for $h = 10^{-n}$, for $1 \leq n \leq 4$.

68. [GU] The Linear Approximation to $f(x) = \tan x$ at $x = \frac{\pi}{4}$ yields the estimate $\tan\left(\frac{\pi}{4} + h\right) - 1 \approx 2h$. Set $K = 6.2$ and show, using a plot, that $|f''(x)| \leq K$ for $x \in [\frac{\pi}{4}, \frac{\pi}{4} + 0.1]$. Then verify numerically that the error E satisfies Eq. (5) for $h = 10^{-n}$, for $1 \leq n \leq 4$.

Further Insights and Challenges

69. Compute dy/dx at the point $P = (2, 1)$ on the curve $y^3 + 3xy = 7$ and show that the linearization at P is $L(x) = -\frac{1}{3}x + \frac{5}{3}$. Use $L(x)$ to estimate the y-coordinate of the point on the curve where $x = 2.1$.

70. Apply the method of Exercise 69 to $P = (0.5, 1)$ on $y^5 + y - 2x = 1$ to estimate the y-coordinate of the point on the curve where $x = 0.55$.

71. Apply the method of Exercise 69 to $P = (-1, 2)$ on $y^4 + 7xy = 2$ to estimate the solution of $y^4 - 7.7y = 2$ near $y = 2$.

72. Show that for any real number k, $(1 + \Delta x)^k \approx 1 + k\Delta x$ for small Δx. Estimate $(1.02)^{0.7}$ and $(1.02)^{-0.3}$.

73. Let $\Delta f = f(5 + h) - f(5)$, where $f(x) = x^2$. Verify directly that $E = |\Delta f - f'(5)h|$ satisfies (5) with $K = 2$.

74. Let $\Delta f = f(1 + h) - f(1)$, where $f(x) = x^{-1}$. Show directly that $E = |\Delta f - f'(1)h|$ is equal to $h^2/(1 + h)$. Then prove that $E \leq 2h^2$ if $-\frac{1}{2} \leq h \leq \frac{1}{2}$. *Hint:* In this case, $\frac{1}{2} \leq 1 + h \leq \frac{3}{2}$.

4.2 Extreme Values

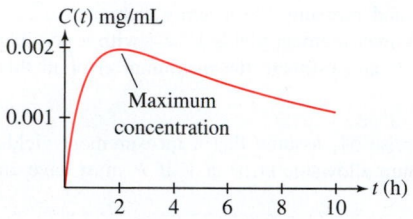

FIGURE 1 Drug concentration in bloodstream (see Exercise 78).

In many applications, it is important to find the minimum or maximum value of a function f. For example, a physician needs to know the maximum drug concentration in a patient's bloodstream when a drug is administered. This amounts to finding the highest point on the graph of C, the concentration at time t (Figure 1).

We refer to the maximum and minimum values (max and min for short) as **extreme values** or **extrema** (singular: extremum) and to the process of finding them as **optimization**. Sometimes, we are interested in finding the min or max for x in a particular interval I, rather than on the entire domain of f.

> *Often, we drop the word "absolute" and speak simply of the min or max on an interval I. When no interval is mentioned, it is understood that we refer to the extreme values on the entire domain of the function.*

DEFINITION Extreme Values on an Interval Let f be a function on an interval I and let $a \in I$. We say that $f(a)$ is the

- **Absolute minimum** of f on I if $f(a) \leq f(x)$ for all $x \in I$.
- **Absolute maximum** of f on I if $f(a) \geq f(x)$ for all $x \in I$.

Does every function have a minimum or maximum value? Clearly not, as we see by taking $f(x) = x$. Indeed, $f(x) = x$ increases without bound as $x \to \infty$ and decreases without bound as $x \to -\infty$. In fact, extreme values do not always exist even if we restrict ourselves to an interval I. Figure 2 illustrates what can go wrong if I is open or f has a discontinuity.

- **Discontinuity:** (A) shows a discontinuous function with no maximum value. The values of $f(x)$ get arbitrarily close to 3 from below, but 3 is not the maximum value because $f(x)$ never actually takes on the value 3.
- **Open interval:** In (B), $g(x)$ is defined on the *open* interval (a, b). It has no max because it tends to ∞ on the right, and it has no min because it tends to 10 on the left without ever reaching this value.

Fortunately, our next theorem guarantees that extreme values exist when the function is continuous and I is closed [Figure 2(C)].

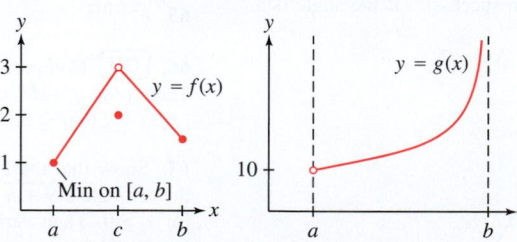
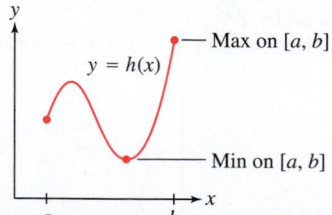

(A) Discontinuous function with no max on $[a, b]$, and a min at $x = a$.

(B) Continuous function with no min or max on the open interval (a, b).

(C) Every continuous function on a closed interval $[a, b]$ has both a min and a max on $[a, b]$.

FIGURE 2

← *REMINDER A closed, bounded interval is an interval I = [a, b] (endpoints included), where a and b are both finite. Often, we drop the word "bounded" and refer to I more simply as a closed interval. An open interval (a, b) (endpoints not included) may have one or two infinite endpoints.*

THEOREM 1 **Existence of Extrema on a Closed Interval** A continuous function f on a closed (bounded) interval $I = [a, b]$ takes on both a minimum and a maximum value on I.

CONCEPTUAL INSIGHT Why does Theorem 1 require a closed interval? Think of the graph of a continuous function as a string. If the interval is closed, the string is pinned down at the two endpoints and cannot fly off to infinity or approach a min/max without reaching it as in Figure 2(B). Intuitively, therefore, it must have a highest and lowest point. As with the Intermediate Value Theorem in Section 2.8, a rigorous proof of Theorem 1 relies on the *completeness property* of the real numbers (see Appendix B).

Local Extrema and Critical Points

We focus now on the problem of finding extreme values. A key concept is that of a local minimum or maximum.

DEFINITION **Local Extrema** We say that $f(c)$ is a

- **Local minimum** occurring at $x = c$ if $f(c)$ is the minimum value of f on some open interval (in the domain of f) containing c.
- **Local maximum** occurring at $x = c$ if $f(c)$ is the maximum value of f on some open interval (in the domain of f) containing c.

FIGURE 3 In the region surrounding Denali in Alaska, there are many local maxima, but there is one global maximum, the peak of Denali.

A local max occurs at $x = c$ if $(c, f(c))$ is the highest point on the graph within some small box [Figure 4(A)]. Thus, $f(c)$ is greater than or equal to all other *nearby* values, but it does not have to be the absolute maximum value of f (Figure 3). Local minima are similar. On the other hand, as Figure 4(B) illustrates, an absolute maximum of f on an interval $[a, b]$ need not be a local maximum of f in open intervals containing the point. In the figure, $f(a)$ is the absolute max on $[a, b]$ but is not a local max on open intervals containing a because $f(x)$ takes on greater values to the left of $x = a$.

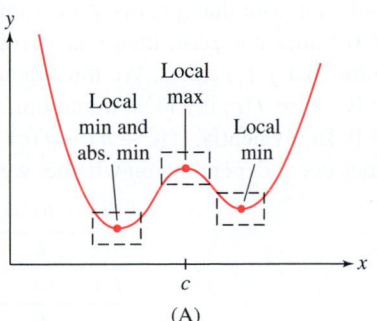
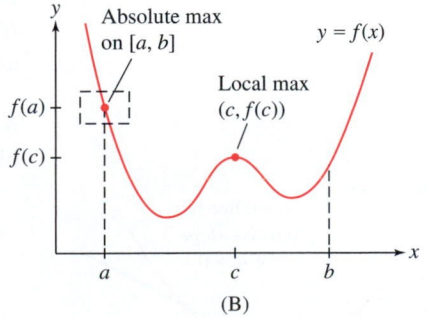

(A)   (B)

FIGURE 4

How do we find the local extrema? The crucial observation is that *the tangent line at a local min or max is horizontal* [Figure 5(A)]. In other words, if $f(c)$ is a local min or max, then $f'(c) = 0$. However, this assumes that f is differentiable. Otherwise, the tangent line may not exist, as in Figure 5(B). To take both possibilities into account, we define the notion of a critical point.

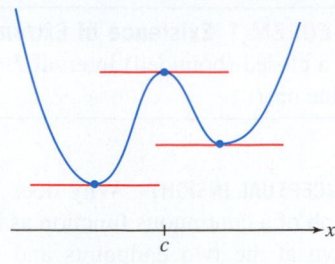
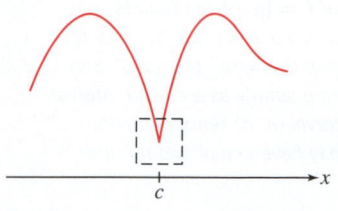

(A) Tangent line is horizontal at the local extrema.

(B) This local minimum occurs at a point where the function is not differentiable.

FIGURE 5

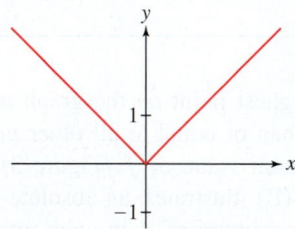

FIGURE 6 Graph of $f(x) = x^3 - 9x^2 + 24x - 10$.

> **DEFINITION Critical Points**   A number c in the domain of f is called a **critical point** if *either* $f'(c) = 0$ or $f'(c)$ does not exist.

EXAMPLE 1   Find the critical points of $f(x) = x^3 - 9x^2 + 24x - 10$.

Solution   The function f is differentiable everywhere (Figure 6). Therefore, the critical points are the solutions of $f'(x) = 0$:

$$f'(x) = 3x^2 - 18x + 24 = 3(x^2 - 6x + 8) = 3(x - 2)(x - 4)$$

To find the critical points, we solve $3(x - 2)(x - 4) = 0$. Thus, they are $x = 2$ and $x = 4$.   ∎

EXAMPLE 2   Nondifferentiable Function   Find the critical points of $f(x) = |x|$.

Solution   As we see in Figure 7, $f'(x) = -1$ for $x < 0$ and $f'(x) = 1$ for $x > 0$. Therefore, $f'(x) = 0$ has no solutions with $x \neq 0$. However, $f'(0)$ does not exist. Thus, $c = 0$ is a critical point.   ∎

The next theorem tells us that we can find local extrema by solving for the critical points. It is one of the most important results in calculus.

FIGURE 7 Graph of $f(x) = |x|$.

> **THEOREM 2   Fermat's Theorem on Local Extrema**   If $f(c)$ is a local min or max, then c is a critical point of f.

Proof   Suppose that $f(c)$ is a local minimum (the case of a local maximum is similar). If $f'(c)$ does not exist, then c is a critical point and there is nothing more to prove. So, assume that $f'(c)$ exists. We must then prove that $f'(c) = 0$.

Because $f(c)$ is a local minimum, we have $f(c + h) \geq f(c)$ for all sufficiently small $h \neq 0$. Equivalently, $f(c + h) - f(c) \geq 0$. Now divide this inequality by h. Two possibilities occur depending on whether we are dividing by a positive value or a negative one:

$$\frac{f(c + h) - f(c)}{h} \geq 0 \qquad \text{if } h > 0 \qquad \boxed{1}$$

$$\frac{f(c + h) - f(c)}{h} \leq 0 \qquad \text{if } h < 0 \qquad \boxed{2}$$

Figure 8 shows the graphical interpretation of these inequalities. Taking the one-sided limits of both sides of (1) and (2), we obtain

$$f'(c) = \lim_{h \to 0^+} \frac{f(c + h) - f(c)}{h} \geq \lim_{h \to 0^+} 0 = 0$$

$$f'(c) = \lim_{h \to 0^-} \frac{f(c + h) - f(c)}{h} \leq \lim_{h \to 0^-} 0 = 0$$

Thus, $f'(c) \geq 0$ and $f'(c) \leq 0$. The only possibility is $f'(c) = 0$ as claimed.   ∎

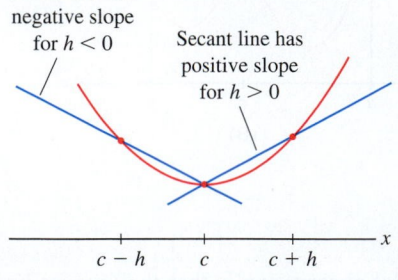

Secant line has negative slope for $h < 0$

Secant line has positive slope for $h > 0$

$c - h$ c $c + h$

FIGURE 8

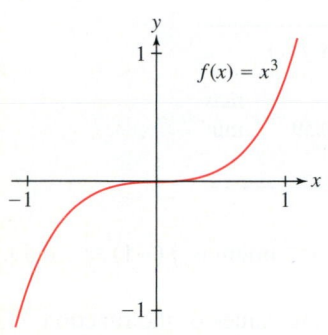

FIGURE 9 The tangent line at $(0,0)$ is horizontal, but $f(0)$ is not a local min or max.

CONCEPTUAL INSIGHT Theorem 2 indicates that a local max or min must be a critical point . However, the "converse" need not be true. That is, having a critical point c does not guarantee a local min or max occurs at c. For example, $f(x) = x^3$ has derivative $f'(x) = 3x^2$ and $f'(0) = 0$. So, 0 is a critical point, but $f(0)$ is neither a local min nor max (Figure 9). The origin is a point of inflection (studied in Section 4.4), where the tangent line crosses the graph.

Optimizing on a Closed Interval

Finally, we have all the tools needed for optimizing a continuous function on a closed interval. Theorem 1 guarantees that the extreme values exist, and the next theorem tells us where to find them, namely among the critical points or endpoints of the interval.

THEOREM 3 Extreme Values on a Closed Interval Assume that f is continuous on $[a, b]$ and let $f(c)$ be the minimum or maximum value on $[a, b]$. Then c is either a critical point or one of the endpoints a or b.

Proof If c is one of the endpoints a or b, there is nothing to prove. If not, then c belongs to the open interval (a, b). In this case, $f(c)$ is also a local min or max because it is the min or max on (a, b). By Fermat's Theorem, c is a critical point. ∎

EXAMPLE 3 Find the extrema of $f(x) = 2x^3 - 15x^2 + 24x + 7$ on $[0, 6]$.

Solution The extreme values occur at critical points or endpoints by Theorem 3, so we can break up the problem neatly into two steps.

Step 1. **Find the critical points.**
The function f is differentiable, so the critical points are solutions to $f'(x) = 0$.

$$f'(x) = 6x^2 - 30x + 24 = 6(x - 1)(x - 4)$$

The critical points satisfy $6(x - 1)(x - 4) = 0$, and therefore are $x = 1$ and 4.

Step 2. **Compare values of $f(x)$ at the critical points and endpoints.**

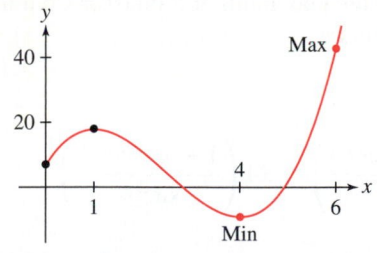

FIGURE 10 Extreme values of $f(x) = 2x^3 - 15x^2 + 24x + 7$ on $[0, 6]$.

x-value	Value of $f(x)$	
1 (critical point)	$f(1) = 18$	
4 (critical point)	$f(4) = -9$	min
0 (endpoint)	$f(0) = 7$	
6 (endpoint)	$f(6) = 43$	max

The maximum value of $f(x)$ on $[0, 6]$ is the greatest of the values in this table, namely $f(6) = 43$. Similarly, the minimum is $f(4) = -9$. See Figure 10. ∎

EXAMPLE 4 Function with a Cusp Find the extrema of $f(x) = 1 - (x - 1)^{2/3}$ on $[-1, 2]$.

Solution First, find the critical points:

$$f'(x) = -\frac{2}{3}(x - 1)^{-1/3} = -\frac{2}{3(x - 1)^{1/3}}$$

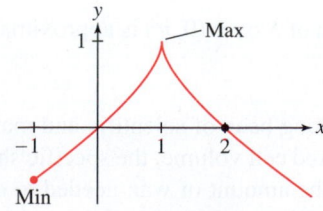

DF FIGURE 11 Extreme values of $f(x) = 1 - (x - 1)^{2/3}$ on $[-1, 2]$.

The equation $f'(x) = 0$ has no solutions because $f'(x)$ is never zero. However, $f'(x)$ does not exist at $x = 1$, so there is a critical point there (Figure 11).

Next, compare values of $f(x)$ at the critical points and endpoints:

x-value	Value of $f(x)$	
1 (critical point)	$f(1) = 1$	max
−1 (endpoint)	$f(-1) \approx -0.59$	min
2 (endpoint)	$f(2) = 0$	

So on $[-1, 2]$, the maximum of f is $f(1) = 1$ and the minimum is $f(-1) \approx -0.59$. ∎

EXAMPLE 5 **Logarithmic Example** Find the extreme values of the function $f(x) = x^2 - 8 \ln x$ on $[1, 4]$.

Solution First, we solve for the critical points. We have $f'(x) = 2x - 8/x$, so we solve

$$2x - \frac{8}{x} = 0 \quad \Rightarrow \quad 2x = \frac{8}{x} \quad \Rightarrow \quad x = \pm 2$$

The only critical point in the interval $[1, 4]$ is $x = 2$. Next, compare the values of $f(x)$ at the critical points and endpoints (Figure 12):

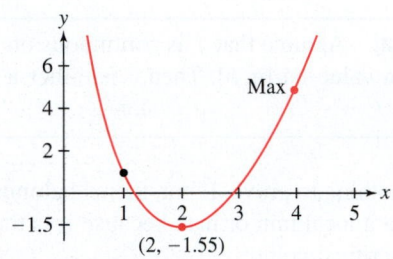

FIGURE 12 Extreme values of $f(x) = x^2 - 8 \ln x$ on $[1, 4]$.

x-value	Value of $f(x)$	
2 (critical point)	$f(2) \approx -1.55$	min
1 (endpoint)	$f(1) = 1$	
4 (endpoint)	$f(4) \approx 4.9$	max

We see that the minimum on $[1, 4]$ is $f(2) \approx -1.55$ and the maximum is $f(4) \approx 4.9$. ∎

EXAMPLE 6 **An Open-Interval Example** The function $S(\theta) = 240 + 24 \left(\frac{\sqrt{3} - \cos \theta}{\sin \theta} \right)$ arises in a model—that we describe after this example—of the geometry of a honeycomb cell. Figure 13 shows the graph of S for $0 < \theta < \pi$. As θ approaches 0 and π from inside the interval, $S(\theta) \to \infty$. Therefore, there is no absolute maximum of S on $(0, \pi)$, but the graph suggests that there is an absolute minimum. Find it.

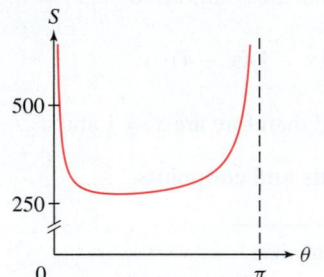

FIGURE 13 Graph of $S(\theta)$.

Solution Computing $S'(\theta)$, we have

$$S'(\theta) = 24 \left(\frac{(\sin \theta)(\sin \theta) - (\sqrt{3} - \cos \theta)(\cos \theta)}{\sin^2 \theta} \right) = 24 \left(\frac{1 - \sqrt{3} \cos \theta}{\sin^2 \theta} \right)$$

The derivative is defined for all θ in the interval and is zero where $1 - \sqrt{3} \cos \theta = 0$. Therefore the absolute minimum occurs at

$$\theta_m = \cos^{-1} \left(\frac{1}{\sqrt{3}} \right) \approx 0.96 \text{ radians} \approx 54.7°$$

Using $\theta_m = \cos^{-1} \left(\frac{1}{\sqrt{3}} \right)$, the minimum value can be shown to be exactly $24(10 + \sqrt{2})$.

Computing $S(0.96)$, we find that the absolute minimum of S over $(0, \pi)$ is approximately 273.94. ∎

Honeycomb Geometry The honeycomb of bees has long been of scientific and mathematical interest (Figure 14). Some believe that, for a fixed cell volume, the specific shape of the cell minimizes the cell's surface area and thus the amount of wax needed to construct it. Each cell has an open hexagonal top, six quadrilateral sides, and three rhombi on the bottom. Imagine that (as in Figure 15) the sides on the hexagonal top are 4 mm, three of the vertical sides are 10 mm, and the remaining dimensions can vary. Let θ be

FIGURE 14 What is the ideal honeycomb cell shape?

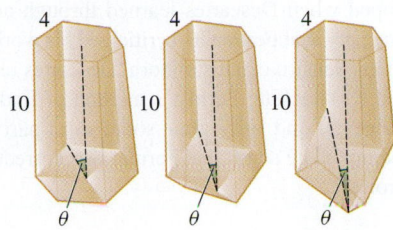

FIGURE 15 The honeycomb cell geometry.

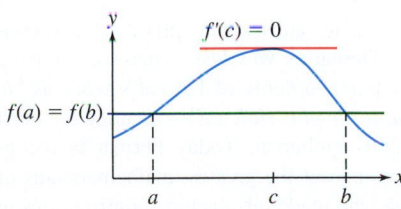

FIGURE 16 Rolle's Theorem: If $f(a) = f(b)$, then $f'(c) = 0$ for some c between a and b.

the angle between a vertical axis through the center of the cell and the bottom rhombi faces. Via geometry, two important facts about the shapes in the figure can be shown:

- The volume of the cell is independent of the angle θ.
- The surface area S of the cell (the total area of the six quadrilaterals and three rhombii) depends on θ according to

$$S(\theta) = 240 + 24 \left(\frac{\sqrt{3} - \cos\theta}{\sin\theta} \right)$$

The previous example indicates that for the cells we are considering, the minimum surface area occurs at $\theta \approx 54.7°$. How does this minimum compare with the actual cells that the bees construct?

In the early eighteenth century, astronomer Giacomo Maraldi made extensive measurements of bees' honeycomb and observed typical angle measurements consistent with the optimal angle we found. Subsequently, mathematicians Samuel Koenig and Colin Maclaurin performed a calculus-based analysis of the honeycomb geometry (as we have done here) supporting the idea that the bees are economical in their honeycomb construction. The question of why bees construct the honeycomb as they do is still unsettled, but calculus provides an interesting glimpse at the possibilities.

Rolle's Theorem

As an application of our optimization methods, we prove Rolle's Theorem: If f is differentiable and takes on the same value at two different points a and b, then somewhere between these two points the derivative is zero. Graphically, if the secant line between $x = a$ and $x = b$ is horizontal, then at least one tangent line between a and b is also horizontal (Figure 16).

THEOREM 4 Rolle's Theorem Assume that f is continuous on $[a, b]$ and differentiable on (a, b). If $f(a) = f(b)$, then there exists a number c between a and b such that $f'(c) = 0$.

Proof Since f is continuous and $[a, b]$ is closed, f has a min and a max in $[a, b]$. Where do they occur? If either the min or the max occurs at a point c in the *open* interval (a, b), then $f(c)$ is a local extreme value and $f'(c) = 0$ by Fermat's Theorem (Theorem 2). Otherwise, both the min and the max occur at the endpoints. However, $f(a) = f(b)$, so in this case, the min and max coincide and f is a constant function with zero derivative. Then, $f'(c) = 0$ for all c in (a, b). ∎

EXAMPLE 7 Illustrating Rolle's Theorem Verify Rolle's Theorem for

$$f(x) = x^4 - x^2 \quad \text{on} \quad [-2, 2]$$

Solution The hypotheses of Rolle's Theorem are satisfied because f is differentiable (and therefore continuous) everywhere, and $f(2) = f(-2)$:

$$f(2) = 2^4 - 2^2 = 12, \qquad f(-2) = (-2)^4 - (-2)^2 = 12$$

We must verify that $f'(c) = 0$ has a solution in $(-2, 2)$. Since

$$f'(x) = 4x^3 - 2x = 2x(2x^2 - 1)$$

we need to solve $2x(2x^2 - 1) = 0$. The solutions are $c = 0$ and $c = \pm 1/\sqrt{2} \approx \pm 0.707$. They all lie in $(-2, 2)$, so Rolle's Theorem is satisfied with three values of c. ∎

EXAMPLE 8 Using Rolle's Theorem Show that $f(x) = x^3 + 9x - 4$ has precisely one real root.

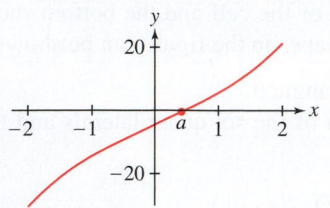

FIGURE 17 Graph of $f(x) = x^3 + 9x - 4$. This function has one real root.

Solution First, we note that $f(0) = -4$ is negative and $f(1) = 6$ is positive. By the Intermediate Value Theorem (Section 2.8), f has *at least* one root a in $[0, 1]$. If f had a second root b, then we would have $f(a) = f(b) = 0$. Rolle's Theorem would then imply that $f'(c) = 0$ for some $c \in (a, b)$. This is not possible because $f'(x) = 3x^2 + 9 \geq 9 > 0$, so $f'(c) = 0$ has no solutions. We conclude that a is *the only* real root of f (Figure 17). ∎

HISTORICAL PERSPECTIVE

© Bettmann/Getty Images

© Historical Picture Archive/Getty Images

Pierre de Fermat
(1601–1665)

René Descartes
(1596–1650)

Sometime in the 1630s, in the decade before Isaac Newton was born, the French mathematician Pierre de Fermat invented a general method for finding extreme values. Fermat said, in essence, that if you want to find extrema, you must set the derivative equal to zero and solve for the critical points, just as we have done in this section. He also described a general method for finding tangent lines that is not essentially different from our method of derivatives. For this reason, Fermat is often regarded as an inventor of calculus, together with Newton and Leibniz.

At around the same time, René Descartes (1596–1650) developed a different but less effective approach to finding tangent lines. Descartes, after whom Cartesian coordinates are named, was a profound thinker—the leading philosopher and scientist of his time in Europe. He is regarded today as the father of modern philosophy and the founder (along with Fermat) of analytic geometry. A dispute developed when Descartes learned through an intermediary that Fermat had criticized his work on optics. Sensitive and stubborn, Descartes retaliated by attacking Fermat's method of finding tangents, and only after some third-party refereeing did he admit that Fermat was correct. He wrote:

...Seeing the last method that you use for finding tangents to curved lines, I can reply to it in no other way than to say that it is very good and that, if you had explained it in this manner at the outset, I would have not contradicted it at all.

However, in subsequent private correspondence, Descartes was less generous, referring at one point to some of Fermat's work as "*le galimatias le plus ridicule*"—meaning the most ridiculous gibberish. Today Fermat is recognized as one of the greatest mathematicians of his age who made far-reaching contributions in several areas of mathematics.

4.2 SUMMARY

- The *extreme values* of f on an interval I are the minimum and maximum values of f for $x \in I$ (also called *absolute extrema* on I).
- Basic Theorem: If f is continuous on a closed interval $[a, b]$, then f has both a min and a max on $[a, b]$.
- $f(c)$ is a *local minimum* if $f(x) \geq f(c)$ for all x in some open interval around c. Local maxima are defined similarly.
- $x = c$ is a *critical point* of f if either $f'(c) = 0$ or $f'(c)$ does not exist.
- Fermat's Theorem on Local Extrema: If $f(c)$ is a local min or max, then c is a critical point.
- To find the extreme values of a continuous function f on a closed interval $[a, b]$:

  Step 1. Find the critical points of f in $[a, b]$.

  Step 2. Calculate $f(x)$ at the critical points in $[a, b]$ and at the endpoints. The min and max on $[a, b]$ are the least and greatest among the values computed in Step 2.

- Rolle's Theorem: If f is continuous on $[a, b]$ and differentiable on (a, b), and if $f(a) = f(b)$, then there exists c between a and b such that $f'(c) = 0$.

4.2 EXERCISES

Preliminary Questions

1. What is the definition of a critical point?

In Questions 2 and 3, which is the correct conclusion, (a) or (b)?

2. If f is not continuous on $[0, 1]$, then
(a) f has no extreme values on $[0, 1]$.
(b) f might not have any extreme values on $[0, 1]$.

3. If f is continuous but has no critical points in $[0, 1]$, then
(a) f has no min or max on $[0, 1]$.
(b) Either $f(0)$ or $f(1)$ is the minimum value on $[0, 1]$.

4. For each statement, indicate whether it is true or false. If false, correct the statement or explain why it is false.

(a) If $f'(c) = 0$, then $f(c)$ is either a local minimum or a local maximum.

(b) If $f(c)$ is the absolute maximum of f on an interval I, then $f'(c) = 0$.

(c) If f is differentiable and $f(c)$ is a local minimum of f, then $f'(c) = 0$.

(d) If there is one local minimum of f on an interval I, then it is the absolute minimum on I.

Exercises

1. The following refer to Figure 18.
(a) What are the critical points of f on $[0, 8]$?
(b) What are the maximum and minimum values of f on $[0, 8]$?
(c) What are the local maximum and minimum values of f, and where do they occur?
(d) Find a closed interval on which both the minimum and maximum values of f occur at critical points.
(e) Find an open interval on which f has neither a minimum nor a maximum value.
(f) Find an open interval on which f has a maximum value but no minimum value.

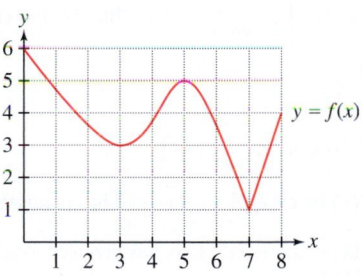

FIGURE 18

2. State whether $f(x) = x^{-1}$ (Figure 19) has a minimum or maximum value on the following intervals:
(a) $(0, 2)$
(b) $(1, 2)$
(c) $[1, 2]$

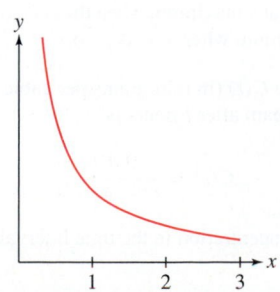

FIGURE 19 Graph of $f(x) = x^{-1}$.

In Exercises 3–20, find all critical points of the function.

3. $f(x) = x^2 - 2x + 4$

4. $f(x) = 7x - 2$

5. $f(x) = x^3 - \frac{9}{2}x^2 - 54x + 2$

6. $f(t) = 8t^3 - t^2$

7. $f(x) = x^{-1} - x^{-2}$

8. $g(z) = \dfrac{1}{z-1} - \dfrac{1}{z}$

9. $f(x) = \dfrac{x}{x^2 + 1}$

10. $f(x) = \dfrac{x^2}{x^2 - 4x + 8}$

11. $f(t) = t - 4\sqrt{t+1}$

12. $f(t) = 4t - \sqrt{t^2 + 1}$

13. $f(x) = xe^{2x}$

14. $f(x) = x + |2x + 1|$

15. $g(\theta) = \sin^2 \theta$

16. $R(\theta) = \cos \theta + \sin^2 \theta$

17. $f(x) = x \ln x$

18. $f(x) = x^2 \sqrt{1 - x^2}$

19. $f(x) = \sin^{-1} x - 2x$

20. $f(x) = \sec^{-1} x - \ln x$

21. Let $f(x) = 2x^2 - 8x + 7$.
(a) Find the critical point c of f and compute $f(c)$.
(b) Find the extreme values of f on $[0, 5]$.
(c) Find the extreme values of f on $[-4, 1]$.

22. Find the extreme values of $f(x) = 2x^3 - 9x^2 + 12x$ on $[0, 3]$ and $[0, 2]$.

23. Find the critical points of $f(x) = \sin x + \cos x$ and determine the extreme values on $\left[0, \frac{\pi}{2}\right]$.

24. Compute the critical points of $h(t) = (t^2 - 1)^{1/3}$. Check that your answer is consistent with Figure 20. Then find the extreme values of h on $[0, 1]$ and on $[0, 2]$.

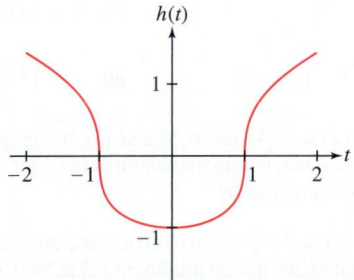

FIGURE 20 Graph of $h(t) = (t^2 - 1)^{1/3}$.

25. [GU] Plot $f(x) = 4\sqrt{x} - 2x + 3$ on $[0, 3]$ and indicate where it appears that the minimum and maximum occur. Then determine the minimum and maximum using calculus.

26. [CAS] Approximate the critical points of $g(x) = 5e^x - \tan x$ in $\left(-\frac{\pi}{2}, \frac{\pi}{2}\right)$.

In Exercises 27–60, find the minimum and maximum values of the function on the given interval by comparing values at the critical points and endpoints.

27. $y = 2x^2 + 4x + 5$, $[-2, 2]$

28. $y = 2x^2 + 4x + 5$, $[0, 2]$

29. $y = 6t - t^2$, $[0, 5]$

30. $y = 6t - t^2$, $[4, 6]$

31. $y = x^3 - 6x^2 + 8$, $[1, 6]$

32. $y = x^3 - 6x^2 + 8$, $[-1, 6]$

33. $y = x^3 - 6x^2 + 8$, $[1, 3]$

34. $y = x^3 - 6x^2 + 8$, $[-1, 3]$

35. $y = 2t^3 + 3t^2$, $[1, 2]$

36. $y = x^3 - 12x^2 + 21x$, $[0, 2]$

37. $y = z^5 - 80z$, $[-3, 3]$

38. $y = 2x^5 + 5x^2$, $[-2, 2]$

39. $y = \dfrac{x^2 + 1}{x - 4}$, $[5, 6]$

40. $y = \dfrac{1 - x}{x^2 + 3x}$, $[1, 4]$

41. $y = x - \dfrac{4x}{x + 1}$, $[0, 3]$

42. $y = 2\sqrt{x^2 + 1} - x$, $[0, 2]$

43. $y = (2 + x)\sqrt{2 + (2 - x)^2}$, $[0, 2]$

44. $y = \sqrt{1 + x^2} - 2x$, $[0, 1]$

45. $y = \sqrt{x + x^2} - 2\sqrt{x}$, $[0, 4]$

46. $y = (t - t^2)^{1/3}$, $[-1, 2]$

47. $y = \sin x \cos x$, $\left[0, \frac{\pi}{2}\right]$

48. $y = x + \sin x$, $[0, 2\pi]$

49. $y = \sqrt{2}\,\theta - \sec\theta$, $\left[0, \frac{\pi}{3}\right]$

50. $x^4 - 2x^2 + 1$, $[-3, 3]$

51. $y = x^3 + x^2 - x$, $[-2, 2]$

52. $y = \cos\theta + \sin\theta$, $[0, 2\pi]$

53. $y = \theta - 2\sin\theta$, $[0, 2\pi]$

54. $y = 4\sin^3\theta - 3\cos^2\theta$, $[0, 2\pi]$

55. $y = \tan x - 2x$, $[0, 1]$

56. $y = xe^{-x}$, $[0, 2]$

57. $y = \dfrac{\ln x}{x}$, $[1, 3]$

58. $y = 5\tan^{-1}x - x$, $[1, 5]$

59. $y = 3e^x - e^{2x}$, $\left[-\frac{1}{2}, 1\right]$

60. $y = x^3 - 24\ln x$, $\left[\frac{1}{2}, 3\right]$

61. **GU** Plot $f(x) = \dfrac{2 + x^2}{x}$ on $(0, 5)$ and use the graph to explain why there is a minimum value, but no maximum value, of f on $(0, 5)$. Use calculus to find the minimum value.

62. **GU** Plot $f(x) = \dfrac{4x - 1 - x^2}{x}$ on $(0, 3)$ and use the graph to explain why there is a maximum value, but no minimum value, of f on $(0, 3)$. Use calculus to find the maximum value.

63. Let $f(\theta) = 2\sin 2\theta + \sin 4\theta$.

(a) Show that θ is a critical point if $\cos 4\theta = -\cos 2\theta$.

(b) Show, using a unit circle, that $\cos\theta_1 = -\cos\theta_2$ if and only if $\theta_1 = \pi \pm \theta_2 + 2\pi k$ for an integer k.

(c) Show that $\cos 4\theta = -\cos 2\theta$ if and only if $\theta = \frac{\pi}{2} + \pi k$ or $\theta = \frac{\pi}{6} + \left(\frac{\pi}{3}\right)k$.

(d) Find the six critical points of f on $[0, 2\pi]$ and find the extreme values of f on this interval.

(e) **GU** Check your results against a graph of f.

64. **GU** Find the critical points of $f(x) = 2\cos 3x + 3\cos 2x$ in $[0, 2\pi]$. Check your answer against a graph of f.

In Exercises 65–68, find the critical points and the extreme values on $[0, 4]$. In Exercises 67 and 68, refer to Figure 21.

65. $y = |x - 2|$

66. $y = |3x - 9|$

67. $y = |x^2 + 4x - 12|$

68. $y = |\cos x|$

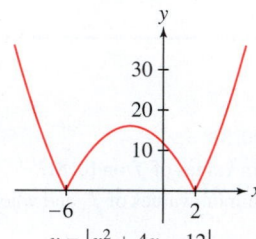
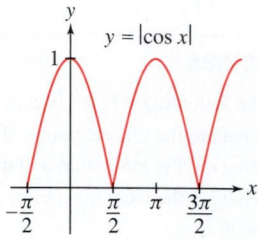

$y = |x^2 + 4x - 12|$

$y = |\cos x|$

FIGURE 21

In Exercises 69–72, verify Rolle's Theorem for the given interval by checking $f(a) = f(b)$ and then finding a value c in (a, b) such that $f'(c) = 0$.

69. $f(x) = x + x^{-1}$, $\left[\frac{1}{2}, 2\right]$

70. $f(x) = \sin x$, $\left[\frac{\pi}{4}, \frac{3\pi}{4}\right]$

71. $f(x) = \dfrac{x^2}{8x - 15}$, $[3, 5]$

72. $f(x) = \sin^2 x - \cos^2 x$, $\left[\frac{\pi}{4}, \frac{3\pi}{4}\right]$

73. Prove that $f(x) = x^5 + 2x^3 + 4x - 12$ has precisely one real root.

74. Prove that $f(x) = x^3 + 3x^2 + 6x$ has precisely one real root.

75. Prove that $f(x) = x^4 + 5x^3 + 4x$ has no root c satisfying $c > 0$. *Hint:* Note that $x = 0$ is a root and apply Rolle's Theorem.

76. Prove that $x = 4$ is the greatest root of $f(x) = x^4 - 8x^2 - 128$.

77. The position of a mass oscillating at the end of a spring is $s(t) = A\sin\omega t$, where A is the amplitude and ω is the angular frequency. Show that the speed $|v(t)|$ is at a maximum when the acceleration $a(t)$ is zero and that $|a(t)|$ is at a maximum when $v(t)$ is zero.

78. The concentration $C(t)$ (in milligrams per cubic centimeter) of a drug in a patient's bloodstream after t hours is

$$C(t) = \dfrac{0.016t}{t^2 + 4t + 4}$$

Find the maximum concentration in the time interval $[0, 8]$ and the time at which it occurs.

79. **CAS** **Antibiotic Levels** A study shows that the concentration $C(t)$ (in micrograms per milliliter) of antibiotic in a patient's blood serum after t hours is $C(t) = 120(e^{-0.2t} - e^{-bt})$, where $b \geq 1$ is a constant that depends on the particular combination of antibiotic agents used. Solve numerically for the value of b (to two decimal places) for which maximum concentration occurs at $t = 1$ h. You may assume that the maximum occurs at a critical point, as suggested by Figure 22.

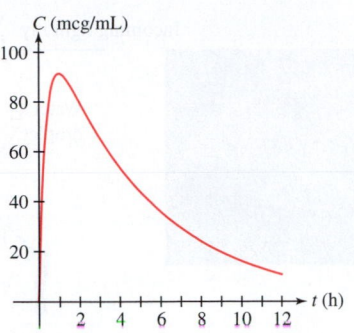

FIGURE 22 Graph of $C(t) = 120(e^{-0.2t} - e^{-bt})$ with b chosen so that the maximum occurs at $t = 1$ h.

80. [CAS] In the notation of Exercise 79, find the value of b (to two decimal places) for which the maximum value of C is equal to 100 mcg/ml.

81. In 1919, physicist Alfred Betz argued that the maximum efficiency of a wind turbine is around 59%. If wind enters a turbine with speed v_1 and exits with speed v_2, then the power extracted is the difference in kinetic energy per unit time:

$$P = \frac{1}{2}mv_1^2 - \frac{1}{2}mv_2^2 \quad \text{watts}$$

where m is the mass of wind flowing through the rotor per unit time (Figure 23). Betz assumed that $m = \rho A(v_1 + v_2)/2$, where ρ is the density of air and A is the area swept out by the rotor. Wind flowing undisturbed through the same area A would have mass per unit time $\rho A v_1$ and power $P_0 = \frac{1}{2}\rho A v_1^3$. The fraction of power extracted by the turbine is $F = P/P_0$.

(a) Show that F depends only on the ratio $r = v_2/v_1$ and is equal to $F(r) = \frac{1}{2}(1 - r^2)(1 + r)$, where $0 \le r \le 1$.

(b) Show that the maximum value of F, called the **Betz Limit**, is $16/27 \approx 0.59$.

(c) [✏️] Explain why Betz's formula for F is not meaningful for r close to zero. *Hint:* How much wind would pass through the turbine if v_2 were zero? Is this realistic?

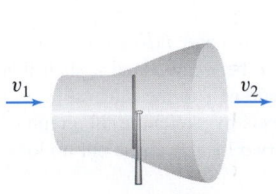

v_1 v_2

(A) Wind flowing through a turbine.

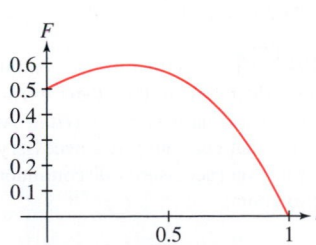

F

(B) F is the fraction of energy extracted by the turbine as a function of $r = v_2/v_1$.

FIGURE 23

82. [GU] The **Bohr radius** a_0 of the hydrogen atom is the value of r that minimizes the energy

$$E(r) = \frac{\hbar^2}{2mr^2} - \frac{e^2}{4\pi\epsilon_0 r}$$

where $\hbar, m, e,$ and ϵ_0 are physical constants. Show that $a_0 = 4\pi\epsilon_0\hbar^2/(me^2)$. Assume that the minimum occurs at a critical point, as suggested by Figure 24.

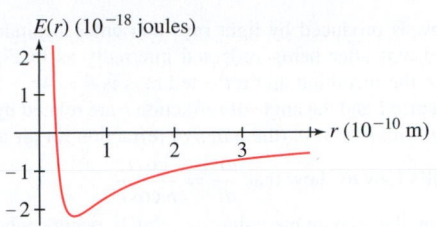

$E(r)$ (10^{-18} joules)

r (10^{-10} m)

FIGURE 24

83. The response of a circuit or other oscillatory system to an input of frequency ω ("omega") is described by the function

$$\phi(\omega) = \frac{1}{\sqrt{(\omega_0^2 - \omega^2)^2 + 4D^2\omega^2}}$$

Both ω_0 (the natural frequency of the system) and D (the damping factor) are positive constants. The graph of ϕ is called a **resonance curve**, and the positive frequency $\omega_r > 0$, where ϕ takes its maximum value, if it exists, is called the **resonant frequency**. Show that $\omega_r = \sqrt{\omega_0^2 - 2D^2}$ if $0 < D < \omega_0/\sqrt{2}$ and that no resonant frequency exists otherwise (Figure 25).

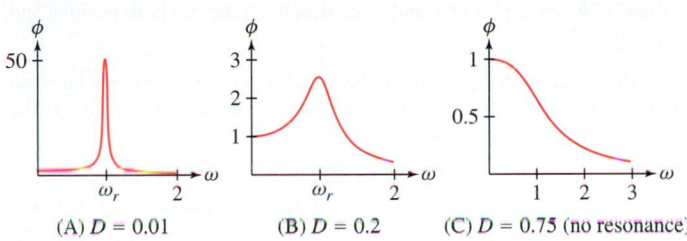

(A) $D = 0.01$ (B) $D = 0.2$ (C) $D = 0.75$ (no resonance)

FIGURE 25 Resonance curves with $\omega_0 = 1$.

84. Find the maximum and minimum of $y = (1 - x)e^{ax}$ on $[0, 1]$, where $0 < a$.

85. Find the maximum of $y = x^a - x^b$ on $[0, 1]$, where $0 < a < b$. In particular, find the maximum of $y = x^5 - x^{10}$ on $[0, 1]$.

In Exercises 86–88, plot the function using a graphing utility and find its critical points and extreme values on $[-5, 5]$.

86. [GU] $y = \dfrac{1}{1 + |x - 1|}$

87. [GU] $y = \dfrac{1}{1 + |x - 1|} + \dfrac{1}{1 + |x - 4|}$

88. [GU] $y = \dfrac{x}{|x^2 - 1| + |x^2 - 4|}$

89. (a) Use implicit differentiation to find the critical points on the curve $27x^2 = (x^2 + y^2)^3$.

(b) [GU] Plot the curve and the horizontal tangent lines on the same set of axes.

90. Sketch the graph of a continuous function on $(0, 4)$ with a minimum value but no maximum value.

91. Sketch the graph of a continuous function on $(0, 4)$ having a local minimum but no absolute minimum.

92. Sketch the graph of a function on $[0, 4]$ having

(a) Two local maxima and one local minimum

(b) An absolute minimum that occurs at an endpoint, and an absolute maximum that occurs at a critical point

93. Sketch the graph of a function f on $[0, 4]$ with a discontinuity such that f has an absolute minimum but no absolute maximum.

94. A rainbow is produced by light rays that enter a raindrop (assumed spherical) and exit after being reflected internally as in Figure 26. The angle between the incoming and reflected rays is $\theta = 4r - 2i$, where the angle of incidence i and the angle of refraction r are related by Snell's Law $\sin i = n \sin r$ with $n \approx 1.33$ (the index of refraction for air and water).

(a) Use Snell's Law to show that $\dfrac{dr}{di} = \dfrac{\cos i}{n \cos r}$.

(b) Show that the maximum value $\theta_{\max}$ of θ occurs when i satisfies $\cos i = \sqrt{\dfrac{n^2 - 1}{3}}$. *Hint:* Show that $\dfrac{d\theta}{di} = 0$ if $\cos i = \dfrac{n}{2} \cos r$. Then use Snell's Law to eliminate r.

(c) Show that $\theta_{\max} \approx 42.53°$.

FIGURE 26

Further Insights and Challenges

95. Show that the extreme values of $f(x) = a \sin x + b \cos x$ are $\pm\sqrt{a^2 + b^2}$.

96. Show, by considering its minimum, that $f(x) = x^2 - 2x + 3$ takes on only positive values. More generally, find the conditions on r and s under which the quadratic function $f(x) = x^2 + rx + s$ takes on only positive values. Give examples of r and s for which f takes on both positive and negative values.

97. Show that if the quadratic polynomial $f(x) = x^2 + rx + s$ takes on both positive and negative values, then its minimum value occurs at the midpoint between the two roots.

98. Generalize Exercise 97: Show that if the horizontal line $y = c$ intersects the graph of $f(x) = x^2 + rx + s$ at two points $(x_1, f(x_1))$ and $(x_2, f(x_2))$, then f takes its minimum value at the midpoint $M = \dfrac{x_1 + x_2}{2}$ (Figure 27).

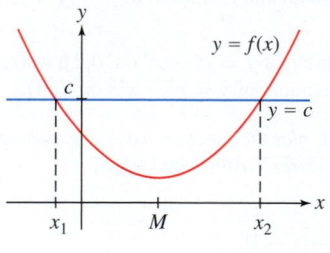

FIGURE 27

99. A cubic polynomial may have a local min and max, or it may have neither (Figure 28). Find conditions on the coefficients a and b of

$$f(x) = \frac{1}{3}x^3 + \frac{1}{2}ax^2 + bx + c$$

that ensure f has neither a local min nor a local max. *Hint:* Apply Exercise 96 to $f'(x)$.

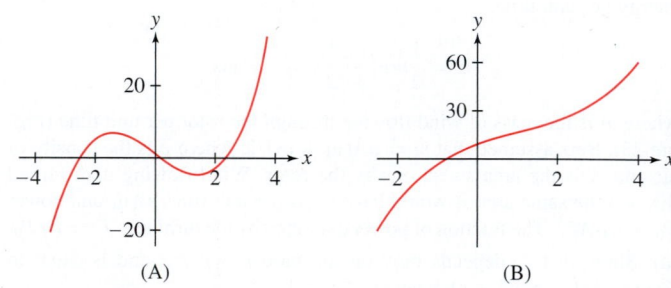

FIGURE 28 Cubic polynomials.

100. Find the min and max of

$$f(x) = x^p(1 - x)^q \quad \text{on } [0, 1]$$

where $p, q > 0$.

101. Prove that if f is continuous and $f(a)$ and $f(b)$ are local minima where $a < b$, then there exists a value c between a and b such that $f(c)$ is a local maximum. (*Hint:* Apply Theorem 1 to the interval $[a, b]$.) Show that continuity is a necessary hypothesis by sketching the graph of a function (necessarily discontinuous) with two local minima but no local maximum.

4.3 The Mean Value Theorem and Monotonicity

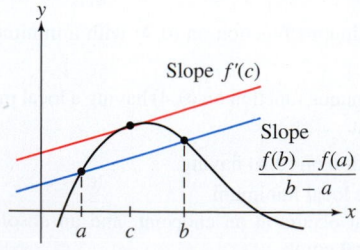

FIGURE 1 By the MVT, there exists at least one tangent line parallel to the secant line.

We have taken for granted that if $f'(x)$ is positive, the function f is increasing, and if $f'(x)$ is negative, f is decreasing. In this section, we prove this rigorously using an important result called the Mean Value Theorem (MVT). Then we develop a method for "testing" critical points—that is, for determining whether they correspond to local maxima, local minima, or neither.

The MVT says that a secant line between two points $(a, f(a))$ and $(b, f(b))$ on a graph is parallel to at least one tangent line in the interval (a, b) (Figure 1). Since the secant line between $(a, f(a))$ and $(b, f(b))$ has slope $\dfrac{f(b) - f(a)}{b - a}$ and since two lines are parallel if they have the same slope, the MVT is claiming that there exists a point c between a and b such that

$$\underbrace{f'(c)}_{\text{Slope of tangent line}} = \underbrace{\frac{f(b) - f(a)}{b - a}}_{\text{Slope of secant line}}$$

> **THEOREM 1   The Mean Value Theorem**   Assume that f is continuous on the closed interval $[a, b]$ and differentiable on (a, b). Then there exists at least one value c in (a, b) such that
>
> $$f'(c) = \frac{f(b) - f(a)}{b - a}$$

Rolle's Theorem (Section 4.2) is the special case of the MVT in which $f(a) = f(b)$. In this case, the conclusion is that $f'(c) = 0$.

GRAPHICAL INSIGHT   Imagine what happens when a secant line is moved parallel to itself. Eventually, it becomes a tangent line, as shown in Figure 2. This is the idea behind the MVT. We present a formal proof at the end of this section.

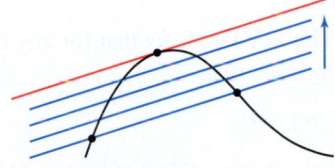

FIGURE 2 Move the secant line in a parallel fashion until it becomes tangent to the curve.

CONCEPTUAL INSIGHT   The conclusion of the MVT can be rewritten as

$$f(b) - f(a) = f'(c)(b - a)$$

We can think of this as a variation on the Linear Approximation, which says

$$f(b) - f(a) \approx f'(a)(b - a)$$

The MVT turns this approximation into an equality by replacing $f'(a)$ with $f'(c)$ for a suitable choice of c in (a, b).

EXAMPLE 1   Verify the MVT with $f(x) = \sqrt{x}$, $a = 1$, and $b = 9$.

Solution   First, compute the slope of the secant line (Figure 3):

$$\frac{f(b) - f(a)}{b - a} = \frac{\sqrt{9} - \sqrt{1}}{9 - 1} = \frac{3 - 1}{9 - 1} = \frac{1}{4}$$

We must find c such that $f'(c) = 1/4$. The derivative is $f'(x) = \frac{1}{2}x^{-1/2}$, and

$$f'(c) = \frac{1}{2\sqrt{c}} = \frac{1}{4} \quad \Rightarrow \quad 2\sqrt{c} = 4 \quad \Rightarrow \quad c = 4$$

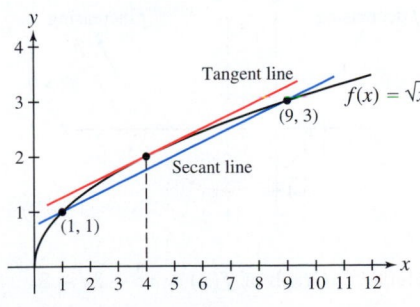

DF FIGURE 3 The tangent line at $c = 4$ is parallel to the secant line.

The value $c = 4$ lies in $(1, 9)$ and satisfies $f'(4) = \frac{1}{4}$. This verifies the MVT. ∎

As a first application, we prove that a function with zero derivative is constant.

> **COROLLARY**   If f is differentiable and $f'(x) = 0$ for all $x \in (a, b)$, then f is constant on (a, b). In other words, $f(x) = C$ for some constant C.

Proof   If a_1 and b_1 are any two distinct points in (a, b), then, by the MVT, there exists c between a_1 and b_1 such that

$$f(b_1) - f(a_1) = f'(c)(b_1 - a_1) = 0 \qquad \text{(since } f'(c) = 0\text{)}$$

Thus, $f(b_1) = f(a_1)$. This says that $f(x)$ is constant on (a, b). ∎

We say that f is "nondecreasing" if

$$f(x_1) \leq f(x_2) \quad \text{for} \quad x_1 \leq x_2$$

"Nonincreasing" is defined similarly. In Theorem 2, if we assume that $f'(x) \geq 0$ (instead of > 0), then f is nondecreasing on (a, b). If $f'(x) \leq 0$, then f is nonincreasing on (a, b).

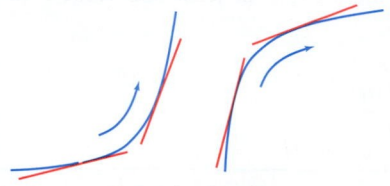

Increasing function:
Tangent lines have positive slope.

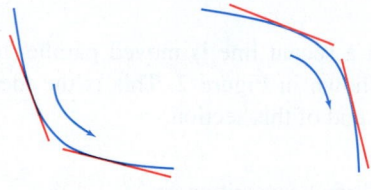

Decreasing function:
Tangent lines have negative slope.

FIGURE 4

Increasing/Decreasing Behavior of Functions

We prove now that the sign of the derivative determines whether a function f is increasing or decreasing. Recall that f is

- **Increasing on** (a, b) if $f(x_1) < f(x_2)$ for all $x_1, x_2 \in (a, b)$ such that $x_1 < x_2$.
- **Decreasing on** (a, b) if $f(x_1) > f(x_2)$ for all $x_1, x_2 \in (a, b)$ such that $x_1 < x_2$.

We say that f is **monotonic** on (a, b) if it is either increasing or decreasing on (a, b).

> **THEOREM 2 The Sign of the Derivative** Let f be a differentiable function on an open interval (a, b).
>
> - If $f'(x) > 0$ for $x \in (a, b)$, then f is increasing on (a, b).
> - If $f'(x) < 0$ for $x \in (a, b)$, then f is decreasing on (a, b).

Proof Suppose first that $f'(x) > 0$ for all $x \in (a, b)$. The MVT tells us that for any two points $x_1 < x_2$ in (a, b), there exists c between x_1 and x_2 such that

$$f(x_2) - f(x_1) = f'(c)(x_2 - x_1) > 0$$

The inequality holds because $f'(c)$ and $(x_2 - x_1)$ are both positive. Thus, $f(x_2) > f(x_1)$, as required. The case $f'(x) < 0$ is similar. ∎

GRAPHICAL INSIGHT Theorem 2 confirms our graphical intuition (Figure 4):

- $f'(x) > 0$ ⟹ tangent lines have positive slope ⟹ f increasing
- $f'(x) < 0$ ⟹ tangent lines have negative slope ⟹ f decreasing

EXAMPLE 2 Show that $f(x) = \ln x$ is increasing.

Solution The derivative $f'(x) = x^{-1}$ is positive on the domain $\{x : x > 0\}$, so $f(x) = \ln x$ is increasing (Figure 5). ∎

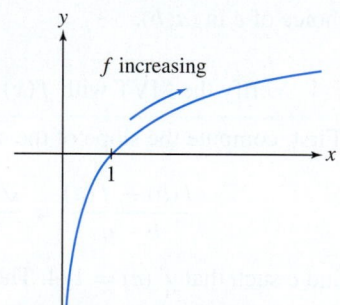

FIGURE 5 Graph of $f(x) = \ln x$.

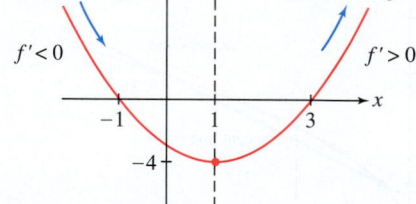

FIGURE 6 Graph of $f(x) = x^2 - 2x - 3$.

EXAMPLE 3 Find the intervals on which $f(x) = x^2 - 2x - 3$ is monotonic.

Solution The derivative $f'(x) = 2x - 2 = 2(x - 1)$ is positive for $x > 1$ and negative for $x < 1$. By Theorem 2, f is decreasing on the interval $(-\infty, 1)$ and increasing on the interval $(1, \infty)$, as confirmed in Figure 6. ∎

Testing Critical Points

There is a useful test for determining whether a critical point yields a min or max (or neither) based on the *sign change* of the derivative $f'(x)$.

To explain the term "sign change," suppose that a function g satisfies $g(c) = 0$. We say that $g(x)$ *changes from positive to negative* at $x = c$ if $g(x) > 0$ to the left of

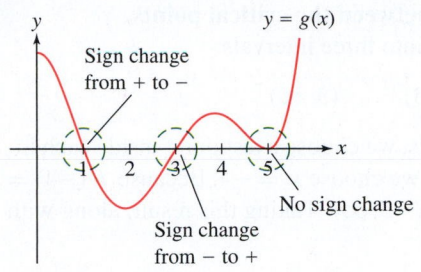

FIGURE 7

c and $g(x) < 0$ to the right of c for x within a small open interval around c (Figure 7). A sign change from negative to positive is defined similarly. Observe in Figure 7 that $g(5) = 0$ but $g(x)$ does not change sign at $x = 5$.

Now suppose that $f'(c) = 0$ and that $f'(x)$ changes sign at $x = c$, say, from $+$ to $-$. Then f is increasing to the left of c and decreasing to the right, so $f(c)$ is a local maximum. Similarly, if $f'(x)$ changes sign from $-$ to $+$, then $f(c)$ is a local minimum. See Figure 8(A). Figure 8(B) illustrates a case where $f'(c) = 0$ but $f'(x)$ does not change sign. In this case, $f'(x) > 0$ for all x near but not equal to c, so f is increasing and has neither a local min nor a local max at c.

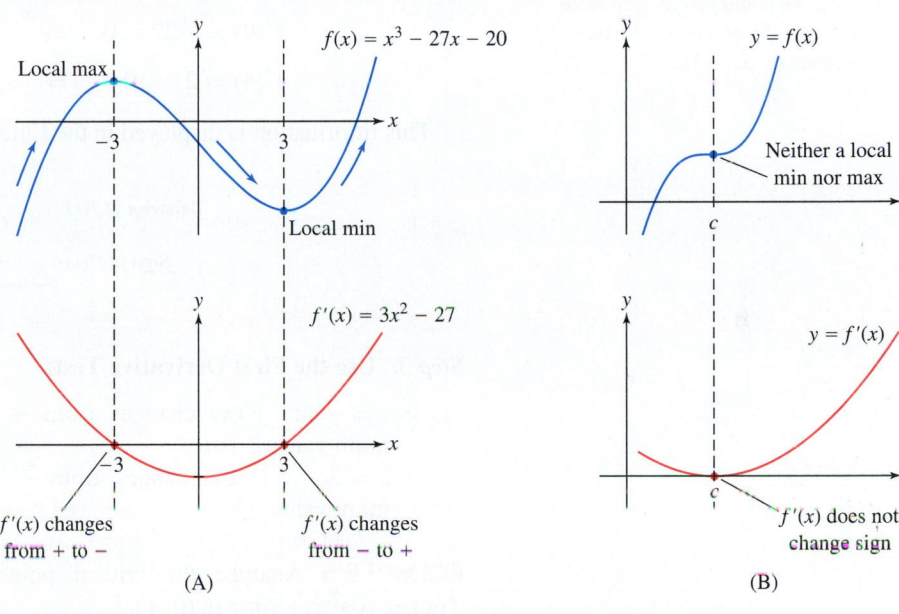

(A)   (B)

DF **FIGURE 8**

A similar analysis holds when $f'(c)$ does not exist and the possibilities for the sign of f' on either side of c are considered. As a result, we have the following theorem:

THEOREM 3 **First Derivative Test for Critical Points** Let c be a critical point of f. Then

- $f'(x)$ changes from $+$ to $-$ at c $\Rightarrow$ $f(c)$ is a local maximum.
- $f'(x)$ changes from $-$ to $+$ at c $\Rightarrow$ $f(c)$ is a local minimum.

To carry out the First Derivative Test, we make a useful observation: $f'(x)$ can change sign at a critical point, but it cannot change sign on the interval between two consecutive critical points as long as the function is defined over the whole interval. In such a case, we can determine the sign of $f'(x)$ on an interval between consecutive critical points by evaluating $f'(x)$ at any *test point* x_0 inside the interval. The sign of $f'(x_0)$ is the sign of $f'(x)$ on the entire interval. In a case where a function's domain is made up of separate intervals, this analysis of the sign of f' needs to be carried out individually on each of the intervals.

EXAMPLE 4 Analyze the critical points of $f(x) = x^3 - 27x - 20$.

Solution Our analysis will confirm the picture in Figure 8(A).

Step 1. **Find the critical points.**
    We have $f'(x) = 3x^2 - 27 = 3(x^2 - 9)$. The critical points satisfy $f'(c) = 0$ and therefore are $c = \pm 3$.

Step 2. **Find the sign of $f'(x)$ on the intervals between the critical points.**

The critical points $c = \pm 3$ divide the real line into three intervals:

$$(-\infty, -3), \qquad (-3, 3), \qquad (3, \infty)$$

To determine the sign of $f'(x)$ on these intervals, we choose a test point inside each interval and evaluate. For example, in $(-\infty, -3)$ we choose $x = -4$. Because $f'(-4) = 21 > 0$, $f'(x)$ is positive on the entire interval $(-3, \infty)$. Taking this result, along with the results from test points at 0 and 4, we have

$$f'(-4) = 21 > 0 \quad \Rightarrow \quad f'(x) > 0 \quad \text{for all } x \in (-\infty, -3)$$

$$f'(0) = -27 < 0 \quad \Rightarrow \quad f'(x) < 0 \quad \text{for all } x \in (-3, 3)$$

$$f'(4) = 21 > 0 \quad \Rightarrow \quad f'(x) > 0 \quad \text{for all } x \in (3, \infty)$$

This information is displayed in the following sign diagram:

> We chose the test points $-4, 0,$ and 4 arbitrarily. To find the sign of $f'(x)$ on $(-\infty, -3)$, we could just as well have computed $f'(-5)$ or any other value of f' in the interval $(-\infty, -3)$.

Behavior of $f(x)$ ↗ ↘ ↗

Sign of $f'(x)$ $+$ $-$ $+$

-3 0 3

Step 3. **Use the First Derivative Test.**

- $c = -3$: $f'(x)$ changes from $+$ to $-$ $\Rightarrow$ $f(-3) = 34$ is a local maximum value.
- $c = 3$: $f'(x)$ changes from $-$ to $+$ $\Rightarrow$ $f(3) = -74$ is a local minimum value. ∎

EXAMPLE 5 Analyze the critical points and the increase/decrease behavior of $f(x) = \cos^2 x + \sin x$ in $(0, \pi)$.

Solution First, find the critical points:

$$f'(x) = -2\cos x \sin x + \cos x = (\cos x)(1 - 2\sin x)$$

Therefore, the critical points are solutions to $\cos x = 0$ or $\sin x = \frac{1}{2}$. Since we are just examining the interval $(0, \pi)$, the critical points of interest are $\frac{\pi}{6}$, $\frac{\pi}{2}$, and $\frac{5\pi}{6}$. They divide $(0, \pi)$ into four intervals:

$$\left(0, \frac{\pi}{6}\right), \qquad \left(\frac{\pi}{6}, \frac{\pi}{2}\right), \qquad \left(\frac{\pi}{2}, \frac{5\pi}{6}\right), \qquad \left(\frac{5\pi}{6}, \pi\right)$$

We determine the sign of $f'(x)$ by evaluating $f'(x)$ at a test point inside each interval. Since $\frac{\pi}{6} \approx 0.52$, $\frac{\pi}{2} \approx 1.57$, $\frac{5\pi}{6} \approx 2.62$, and $\pi \approx 3.14$, we can use the following test points:

Interval	Test value	Sign of $f'(x)$	Behavior of $f(x)$
$\left(0, \frac{\pi}{6}\right)$	$f'(0.5) \approx 0.04$	$+$	↗
$\left(\frac{\pi}{6}, \frac{\pi}{2}\right)$	$f'(1) \approx -0.37$	$-$	↘
$\left(\frac{\pi}{2}, \frac{5\pi}{6}\right)$	$f'(2) \approx 0.34$	$+$	↗
$\left(\frac{5\pi}{6}, \pi\right)$	$f'(3) \approx -0.71$	$-$	↘

Now apply the First Derivative Test:

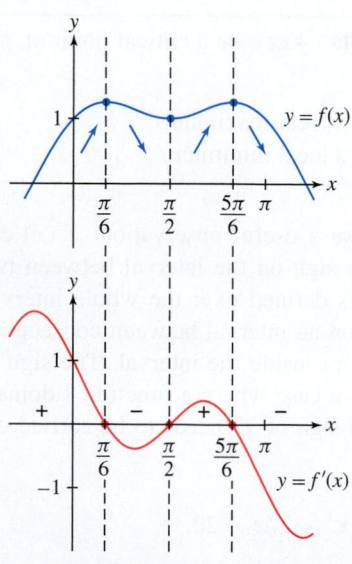

FIGURE 9 Graph of $f(x) = \cos^2 x + \sin x$ and its derivative.

- Local max at $c = \frac{\pi}{6}$ and $c = \frac{5\pi}{6}$ because $f'(x)$ changes from $+$ to $-$.

- Local min at $c = \frac{\pi}{2}$ because $f'(x)$ changes from $-$ to $+$.

The behavior of $f(x)$ and $f'(x)$ is reflected in the graphs in Figure 9. ∎

EXAMPLE 6 Analyze the critical points and the increase/decrease behavior of $f(x) = x^2 + \frac{1}{x^2}$.

Solution Note that f is undefined at $x = 0$, so we need to analyze f separately on $(-\infty, 0)$ and $(0, \infty)$. We have

$$f'(x) = 2x - \frac{2}{x^3}$$

The critical points are solutions to $x - \frac{1}{x^3} = 0$; that is, to $x^4 - 1 = 0$. They are $c = \pm 1$. Since we need to consider f separately on $(-\infty, 0)$ and $(0, \infty)$, there are four intervals on which we need to examine the sign of $f'(x)$: $(-\infty, -1)$, $(-1, 0)$, $(0, 1)$, and $(1, \infty)$. We determine the sign of $f'(x)$ by evaluating $f'(x)$ at a test point inside each interval.

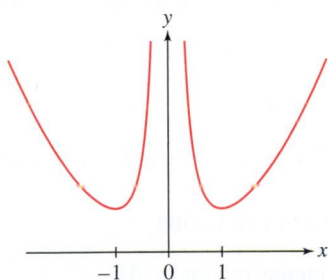

FIGURE 10 Graph of $f(x) = x^2 + \frac{1}{x^2}$.

Interval	Test value	Sign of $f'(x)$	Behavior of $f(x)$
$(-\infty, -1)$	$f'(-2) = -3.75$	$-$	↘
$(-1, 0)$	$f'(-0.5) = 15$	$+$	↗
$(0, 1)$	$f'(0.5) = -15$	$-$	↘
$(1, \infty)$	$f'(2) = 3.75$	$+$	↗

Applying the First Derivative Test, we see that both critical points are local minima. This is verified in the graph in Figure 10. ∎

EXAMPLE 7 **A Critical Point Where $f'(x)$ Is Undefined** Analyze the critical points of $f(x) = (1 - x)^{2/3}$.

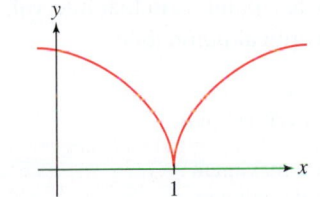

FIGURE 11 Graph of $f(x) = (1 - x)^{2/3}$.

Solution The derivative is $f'(x) = -\frac{2}{3}(1 - x)^{-1/3} = \frac{-2}{3(1-x)^{1/3}}$. The only critical point occurs at $c = 1$, when $f'(x)$ is undefined. For $x < 1$, $f'(x)$ is negative. For $x > 1$, $f'(x)$ is positive. So $f'(x)$ changes sign as we pass through $c = 1$, and by the First Derivative Test, $f(c)$ is a local minimum. See Figure 11. ∎

EXAMPLE 8 **Infinitely Many Critical Points, No Local Extrema** Analyze the critical points of $f(x) = x - \sin x$.

Solution We have $f'(x) = 1 - \cos x$, and therefore critical points occur at solutions to $\cos x = 1$; that is, at $n\pi$ for all even integers n. At none of the critical points does the sign of f' change since $f'(x) \geq 0$ for all x. Therefore, none of the critical points are local extrema (Figure 12). ∎

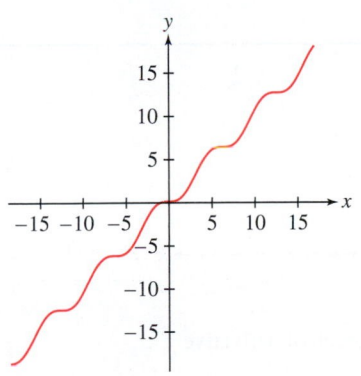

FIGURE 12 Graph of $f(x) = x - \sin x$.

Proof of the MVT Let $m = \dfrac{f(b) - f(a)}{b - a}$ be the slope of the secant line joining $(a, f(a))$ and $(b, f(b))$. The secant line has equation $y = mx + r$ for some r (Figure 13). Now consider the function

$$G(x) = f(x) - (mx + r)$$

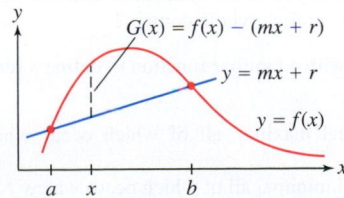

FIGURE 13 $G(x)$ is the vertical distance between the graph and the secant line.

As indicated in Figure 13, $G(x)$ is the vertical distance between the graph and the secant line at x (it is negative at points where the graph of f lies below the secant line). This distance is zero at the endpoints, and therefore, $G(a) = G(b) = 0$. By Rolle's Theorem (Section 4.2), there exists a point c in (a, b) such that $G'(c) = 0$. But $G'(x) = f'(x) - m$, so $G'(c) = f'(c) - m = 0$, and $f'(c) = m$ as desired. ∎

4.3 SUMMARY

- The Mean Value Theorem (MVT): If f is continuous on $[a, b]$ and differentiable on (a, b), then there exists at least one value c in (a, b) such that

$$f'(c) = \frac{f(b) - f(a)}{b - a}$$

This conclusion can also be written

$$f(b) - f(a) = f'(c)(b - a)$$

- Important corollary of the MVT: If $f'(x) = 0$ for all $x \in (a, b)$, then f is constant on (a, b).
- The *sign* of $f'(x)$ determines whether f is increasing or decreasing:

$$f'(x) > 0 \text{ for } x \in (a, b) \quad \Rightarrow \quad f \text{ is increasing on } (a, b)$$

$$f'(x) < 0 \text{ for } x \in (a, b) \quad \Rightarrow \quad f \text{ is decreasing on } (a, b)$$

- On an interval over which f is defined, the sign of $f'(x)$ can change only at the critical points, so f is *monotonic* (increasing or decreasing) on the intervals between the critical points.
- On an interval over which f is defined, to find the sign of $f'(x)$ on an interval between two critical points, calculate the sign of $f'(x_0)$ at any test point x_0 in that interval.
- First Derivative Test: If f is differentiable and c is a critical point, then

Sign change of $f'(x)$ at c	Type of critical point
From $+$ to $-$	Local maximum
From $-$ to $+$	Local minimum

4.3 EXERCISES

Preliminary Questions

1. For which value of m is the following statement correct? If $f(2) = 3$ and $f(4) = 9$, and f is differentiable, then f has a tangent line of slope m.

2. Assume f is differentiable. Which of the following statements does *not* follow from the MVT?

(a) If f has a secant line of slope 0, then f has a tangent line of slope 0.

(b) If $f(5) < f(9)$, then $f'(c) > 0$ for some $c \in (5, 9)$.

(c) If f has a tangent line of slope 0, then f has a secant line of slope 0.

(d) If $f'(x) > 0$ for all x, then every secant line has positive slope.

3. Can a function with the real numbers as its domain that takes on only negative values have a positive derivative? If so, sketch an example.

4. For f with derivative as in Figure 14:

(a) Is $f(c)$ a local minimum or maximum?

(b) Is f a decreasing function?

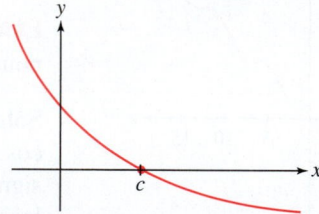

FIGURE 14 Graph of derivative f'.

5. Which of the six standard trigonometric functions have infinitely many local minima and infinitely many local maxima but no absolute maximum and no absolute minimum over their whole domain?

6. Compose the absolute value with a familiar function to define a function f that

- has infinitely many local maxima, all of which occur where $f' = 0$, and
- has infinitely many local minima, all of which occur where f' is undefined.

Exercises

In Exercises 1–8, find a point c satisfying the conclusion of the MVT for the given function and interval.

1. $y = x^{-1}$, $[2, 8]$

2. $y = \sqrt{x}$, $[9, 25]$

3. $y = \cos x - \sin x$, $[0, 2\pi]$

4. $y = \dfrac{x}{x+2}$, $[1, 4]$

5. $y = x^3$, $[-4, 5]$

6. $y = x \ln x$, $[1, 2]$

7. $y = e^{-2x}$, $[0, 3]$

8. $y = e^x - x$, $[-1, 1]$

In Exercises 9–12, find a point c satisfying the conclusion of the MVT for the given function and interval. Then draw the graph of the function, the secant line between the endpoints of the graph and the tangent line at $(c, f(c))$, to see that the secant and tangent lines are, in fact, parallel.

9. $y = x^2$, $[0, 1]$

10. $y = x^{2/3}$, $[0, 8]$

11. $y = e^x$, $[0, 1]$

12. $y = \sqrt{x}$, $[0, 3]$

13. **GU** Let $f(x) = x^5 + x^2$. The secant line between $(0, 0)$ and $(1, 2)$ has slope 2 (check this), so by the MVT, $f'(c) = 2$ for some $c \in (0, 1)$. Plot f and the secant line on the same axes. Then plot $y = 2x + b$ for different values of b until the line becomes tangent to the graph of f. Zoom in on the point of tangency to estimate the x-coordinate c of the point of tangency.

14. **GU** Plot the derivative of $f(x) = 3x^5 - 5x^3$. Describe its sign changes and use this to determine the local extreme values of f. Then graph f to confirm your conclusions.

15. Determine the intervals on which $f'(x)$ is positive and negative, assuming that Figure 15 is the graph of f.

16. Determine the intervals on which f is increasing or decreasing, assuming that Figure 15 is the graph of f'.

17. State whether $f(2)$ and $f(4)$ are local minima or local maxima, assuming that Figure 15 is the graph of f'.

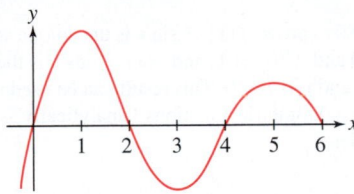

FIGURE 15

18. Figure 16 shows the graph of the derivative f' of a function f. Find the critical points of f and determine whether they are local minima, local maxima, or neither.

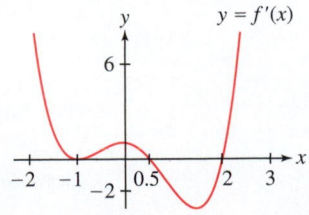

FIGURE 16

In Exercises 19–22, sketch the graph of a function f whose derivative f' has the given description.

19. $f'(x) > 0$ for $x > 3$ and $f'(x) < 0$ for $x < 3$

20. $f'(x) > 0$ for $x < 1$ and $f'(x) < 0$ for $x > 1$

21. $f'(x)$ is negative on $(1, 3)$ and positive everywhere else.

22. $f'(x)$ makes the sign transitions $+, -, +, -$.

In Exercises 23–26, find all critical points of f and use the First Derivative Test to determine whether they are local minima or local maxima.

23. $f(x) = 4 + 6x - x^2$

24. $f(x) = x^3 - 12x - 4$

25. $f(x) = \dfrac{x^2}{x+1}$

26. $f(x) = x^3 + x^{-3}$

In Exercises 27–58, find the critical points and the intervals on which the function is increasing or decreasing. Use the First Derivative Test to determine whether the critical point yields a local min or max (or neither).

27. $y = -x^2 + 7x - 17$

28. $y = 5x^2 + 6x - 4$

29. $y = x^3 - 12x^2$

30. $y = x(x - 2)^3$

31. $y = 3x^4 + 8x^3 - 6x^2 - 24x$

32. $y = x^2 + (10 - x)^2$

33. $y = \frac{1}{3}x^3 + \frac{3}{2}x^2 + 2x + 4$

34. $y = x^4 + x^3$

35. $y = x^5 + x^3 + 1$

36. $y = x^5 + x^3 + x$

37. $y = x^4 - 4x^{3/2}$ $(x > 0)$

38. $y = x^{5/2} - x^2$ $(x > 0)$

39. $y = x + x^{-1}$

40. $y = x^{-2} - 4x^{-1}$

41. $y = \dfrac{1}{x^2 + 1}$

42. $y = \dfrac{2x + 1}{x^2 + 1}$

43. $y = \dfrac{x^3}{x^2 + 1}$

44. $y = \dfrac{x^3}{x^2 - 3}$

45. $y = \theta + \sin\theta + \cos\theta$, $[0, 2\pi]$

46. $y = \sin\theta + \sqrt{3}\cos\theta$, $[0, 2\pi]$

47. $y = \sin^2\theta + \sin\theta$, $[0, 2\pi]$

48. $y = \theta - 2\cos\theta$, $[0, 2\pi]$

49. $y = x + e^{-x}$

50. $y = \dfrac{e^x}{x}$

51. $y = e^{-x}\cos x$, $\left[-\frac{\pi}{2}, \frac{\pi}{2}\right]$

52. $y = x^2 e^x$

53. $y = \tan^{-1}x - \frac{1}{2}x$

54. $y = (x^2 - 2x)e^x$

55. $y = x - \ln x^2$

56. $y = \dfrac{\ln x^2}{x}$

57. $y = x^{1/3}$

58. $y = x^{2/3} - x^2$

59. Find the maximum value of $f(x) = x^{-x}$ for $x > 0$.

60. Show that $f(x) = x^2 + bx + c$ is decreasing on $\left(-\infty, -\frac{b}{2}\right)$ and increasing on $\left(-\frac{b}{2}, \infty\right)$.

61. Show that $f(x) = x^3 - 2x^2 + 2x$ is an increasing function. *Hint:* Find the minimum value of f'.

62. Find conditions on a and b that ensure $f(x) = x^3 + ax + b$ is increasing on $(-\infty, \infty)$.

63. Ron's toll pass recorded him entering the tollway at mile 0 at 12:17 PM. He exited at mile 115 at 1:52 PM, and soon thereafter he was pulled over by the state police. "The speed limit on the tollway is 65 miles per hour," the trooper told Ron. "You exceeded that by more than five miles per hour this afternoon." "No way!" responded Ron. "I glance at the

speedometer frequently, and not once did it read over 65!" How did the trooper use the Mean Value Theorem to support her claim that Ron must have gone more than 70 miles per hour at some point?

64. Two days after he bought a speedometer for his bicycle, Lance brought it back to the Yellow Jersey Bike Shop. "There is a problem with this speedometer," Lance complained to the clerk. "Yesterday I cycled the 22-mile Rogadzo Road Trail in 78 minutes, and not once did the speedometer read above 15 miles per hour!" "Yeah?" responded the clerk. "What's the problem?" How did Lance use the Mean Value Theorem to explain his complaint?

65. Determine where $f(x) = (1,000 - x)^2 + x^2$ is decreasing. Use this to decide which is larger: $800^2 + 200^2$ or $600^2 + 400^2$.

66. Show that $f(x) = 1 - |x|$ satisfies the conclusion of the MVT on $[a, b]$ if both a and b are positive or negative, but not if $a < 0$ and $b > 0$.

67. Which values of c satisfy the conclusion of the MVT on the interval $[a, b]$ if f is a linear function?

68. Show that if f is any quadratic polynomial, then the midpoint $c = \dfrac{a + b}{2}$ satisfies the conclusion of the MVT on $[a, b]$ for any a and b.

69. Suppose that $f(0) = 2$ and $f'(x) \le 3$ for $x > 0$. Apply the MVT to the interval $[0, 4]$ to prove that $f(4) \le 14$. Prove more generally that $f(x) \le 2 + 3x$ for all $x > 0$.

70. Show that if $f(2) = -2$ and $f'(x) \ge 5$ for $x > 2$, then $f(4) \ge 8$.

71. Show that if $f(2) = 5$ and $f'(x) \ge 10$ for $x > 2$, then $f(x) \ge 10x - 15$ for all $x > 2$.

Further Insights and Challenges

72. Show that a cubic function $f(x) = x^3 + ax^2 + bx + c$ is increasing on $(-\infty, \infty)$ if $b > a^2/3$.

73. Prove that if $f(0) = g(0)$ and $f'(x) \le g'(x)$ for $x \ge 0$, then $f(x) \le g(x)$ for all $x \ge 0$. *Hint:* Show that the function given by $y = f(x) - g(x)$ is nonincreasing.

74. Use Exercise 73 to prove that $x \le \tan x$ for $0 \le x < \frac{\pi}{2}$ and $\sin x \le x$ for $x \ge 0$.

75. Use Exercises 73 and 74 to prove the following assertions for all $x \ge 0$ (each assertion follows from the previous one):

(a) $\cos x \ge 1 - \frac{1}{2}x^2$

(b) $\sin x \ge x - \frac{1}{6}x^3$

(c) $\cos x \le 1 - \frac{1}{2}x^2 + \frac{1}{24}x^4$

Can you guess the next inequality in the series?

76. Let $f(x) = e^{-x}$. Use the method of Exercise 75 to prove the following inequalities for $x \ge 0$:

(a) $e^{-x} \ge 1 - x$

(b) $e^{-x} \le 1 - x + \frac{1}{2}x^2$

(c) $e^{-x} \ge 1 - x + \frac{1}{2}x^2 - \frac{1}{6}x^3$

Can you guess the next inequality in the series?

77. Assume that f'' exists and $f''(x) = 0$ for all x. Prove that $f(x) = mx + b$, where $m = f'(0)$ and $b = f(0)$.

78. Define $f(x) = x^3 \sin\left(\frac{1}{x}\right)$ for $x \ne 0$ and $f(0) = 0$.

(a) Show that f' is continuous at $x = 0$ and that $x = 0$ is a critical point of f.

(b) [GU] Examine the graphs of f and f'. Can the First Derivative Test be applied?

(c) Show that $f(0)$ is neither a local min nor a local max.

79. Suppose that $f(x)$ satisfies the following equation (an example of a **differential equation**):

$$f''(x) = -f(x) \qquad \boxed{1}$$

(a) Show that $f(x)^2 + f'(x)^2 = f(0)^2 + f'(0)^2$ for all x. *Hint:* Show that the function on the left has zero derivative.

(b) Verify that $f(x) = \sin x$ and $f(x) = \cos x$ satisfy Eq. (1), and deduce that $\sin^2 x + \cos^2 x = 1$.

80. Suppose that functions f and g satisfy Eq. (1) and have the same initial values—that is, $f(0) = g(0)$ and $f'(0) = g'(0)$. Prove that $f(x) = g(x)$ for all x. *Hint:* Apply Exercise 79(a) to $f - g$.

81. Use Exercise 80 to prove $f(x) = \sin x$ is the unique solution of Eq. (1) such that $f(0) = 0$ and $f'(0) = 1$; and $g(x) = \cos x$ is the unique solution such that $g(0) = 1$ and $g'(0) = 0$. This result can be used to develop all the properties of the trigonometric functions "analytically"—that is, without reference to triangles.

4.4 The Second Derivative and Concavity

In the previous section, we studied the increasing/decreasing behavior of a function, as determined by the sign of the derivative. Another important property is concavity, which refers to the way the graph bends. Informally, a curve is *concave up* if it bends up and *concave down* if it bends down (Figure 1).

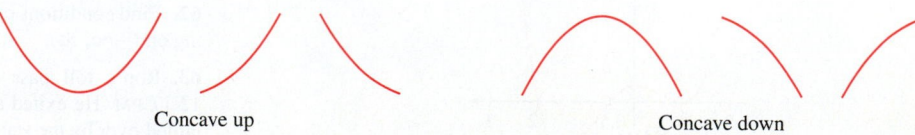

Concave up   Concave down

FIGURE 1

To analyze concavity in a precise fashion, let's examine how concavity is related to tangent lines and derivatives. Observe in Figure 2 that when f is concave up, f' is increasing (the slopes of the tangent lines increase as we move to the right). Similarly, when f is concave down, f' is decreasing. This suggests the following definition.

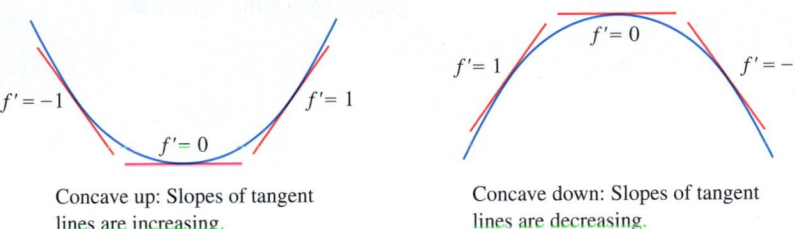

Concave up: Slopes of tangent lines are increasing.

Concave down: Slopes of tangent lines are decreasing.

FIGURE 2

DEFINITION **Concavity** Let f be a differentiable function on an open interval (a, b). Then

- f is **concave up** on (a, b) if f' is increasing on (a, b).
- f is **concave down** on (a, b) if f' is decreasing on (a, b).

EXAMPLE 1 **Concavity and Stock Prices** The stocks of two companies, Arenot Industries (AI) and Blurbenthal Business Associates (BBA), went up in value, and both currently sell for $75 (Figure 3). However, one is clearly a better investment than the other, assuming these trends continue in the same manner. Explain in terms of concavity.

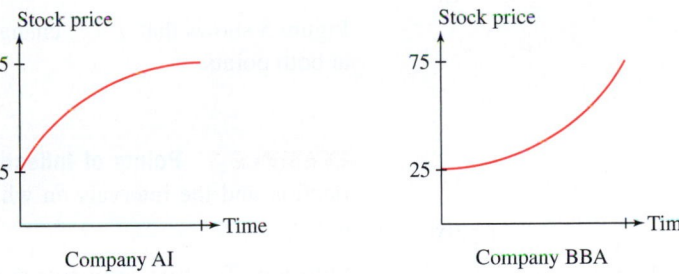

FIGURE 3

Solution The graph of Stock AI is concave down, so its growth rate (first derivative) is declining as time goes on. The graph of Stock BBA is concave up, so its growth rate is increasing. If these trends continue, Stock BBA is the better investment. ∎

The concavity of a function is determined by the *sign* of its second derivative. Indeed, if $f''(x) > 0$, then f' is increasing and hence f is concave up. Similarly, if $f''(x) < 0$, then f' is decreasing and f is concave down.

THEOREM 1 **Test for Concavity** Assume that $f''(x)$ exists for all $x \in (a, b)$.

- If $f''(x) > 0$ for all $x \in (a, b)$, then f is concave up on (a, b).
- If $f''(x) < 0$ for all $x \in (a, b)$, then f is concave down on (a, b).

CAUTION *A critical point c is just a single number, whereas a point of inflection $(c, f(c))$ is a point in the xy-plane.*

Of special interest are the points on the graph where the concavity changes. We say that $P = (c, f(c))$ is a **point of inflection** of f if the concavity changes from up to down or from down to up at $x = c$. Figure 4 shows a curve made up of two arcs—one is concave down and one is concave up (the word "arc" refers to a piece of a curve). The point P where the arcs are joined is a point of inflection. We will denote points of inflection in graphs by a solid square ■.

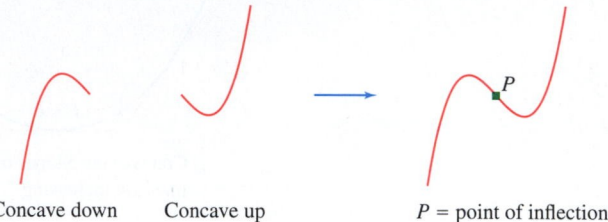

Concave down   Concave up P = point of inflection

FIGURE 4

According to Theorem 1, the concavity of f is determined by the sign of $f''(x)$. Therefore, a point of inflection is a point where $f''(x)$ changes sign.

> **THEOREM 2 Test for Inflection Points** If $f''(c) = 0$ or $f''(c)$ does not exist and $f''(x)$ changes sign at $x = c$, then f has a point of inflection at $x = c$.

EXAMPLE 2 Find the points of inflection of $f(x) = \cos x$ on $[0, 2\pi]$.

Solution We have

$$f''(x) = -\cos x, \quad \text{and} \quad f''(x) = 0 \quad \text{for } x = \frac{\pi}{2}, \frac{3\pi}{2}.$$

Figure 5 shows that $f''(x)$ changes sign at $x = \frac{\pi}{2}$ and $\frac{3\pi}{2}$, so f has a point of inflection at both points. ■

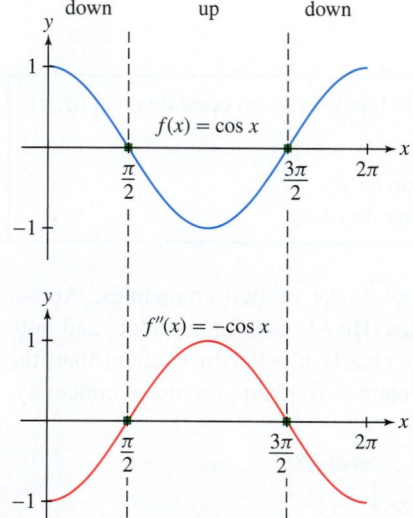

FIGURE 5

EXAMPLE 3 Points of Inflection and Intervals of Concavity Find the points of inflection and the intervals on which $f(x) = 3x^5 - 5x^4 + 1$ is concave up and concave down.

Solution The first derivative is $f'(x) = 15x^4 - 20x^3$ and

$$f''(x) = 60x^3 - 60x^2 = 60x^2(x - 1)$$

The zeros of $f''(x) = 60x^2(x - 1)$ are $x = 0$ and $x = 1$. They divide the x-axis into three intervals: $(-\infty, 0)$, $(0, 1)$, and $(1, \infty)$. We determine the sign of $f''(x)$ and the concavity of f by computing test values within each interval (Figure 6):

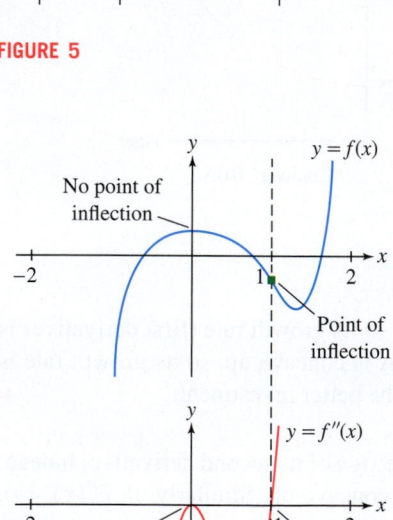

FIGURE 6 Graph of $f(x) = 3x^5 - 5x^4 + 1$ and its second derivative.

Interval	Test value	Sign of $f''(x)$	Behavior of $f(x)$
$(-\infty, 0)$	$f''(-1) = -120$	$-$	Concave down
$(0, 1)$	$f''\left(\frac{1}{2}\right) = -\frac{15}{2}$	$-$	Concave down
$(1, \infty)$	$f''(2) = 240$	$+$	Concave up

Since the concavity changes at $x = 1$ there is an inflection point there. The inflection point is $(1, -1)$. Note that, even though $f''(0) = 0$, there is not an inflection point at $x = 0$ because the concavity does not change at $x = 0$. ■

Usually, we find the inflection points by solving $f''(x) = 0$. However, an inflection point can also occur at a point $(c, f(c))$, where $f''(c)$ does not exist.

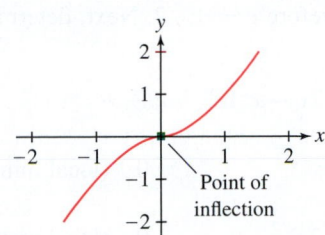

FIGURE 7 The concavity of $f(x) = x^{5/3}$ changes at $x = 0$ even though $f''(0)$ does not exist.

EXAMPLE 4 **A Case Where the Second Derivative Does Not Exist** Find the points of inflection of $f(x) = x^{5/3}$.

Solution In this case, $f'(x) = \frac{5}{3}x^{2/3}$ and $f''(x) = \frac{10}{9}x^{-1/3}$. Although $f''(0)$ does not exist, $f''(x)$ does change sign at $x = 0$:

$$f''(x) = \frac{10}{9x^{1/3}} \begin{cases} > 0 & \text{for } x > 0 \\ < 0 & \text{for } x < 0 \end{cases}$$

Therefore, the concavity of f changes at $x = 0$, and $(0,0)$ is a point of inflection (Figure 7). ■

GRAPHICAL INSIGHT Points of inflection are easy to spot on the graph of the first derivative f'. If $f''(c) = 0$ and $f''(x)$ changes sign at $x = c$, then the increasing/decreasing behavior of f' changes at $x = c$:

- If $f''(x)$ goes from positive to negative at $x = c$, then f' has a local max at $x = c$.
- If $f''(x)$ goes from negative to positive at $x = c$, then f' has a local min at $x = c$.

Thus, *inflection points of f occur where f' has a local min or max* (Figure 8).

Second Derivative Test for Critical Points

There is a simple test for critical points based on concavity. Suppose that $f'(c) = 0$. As we see in Figure 9, $f(c)$ is a local max if f is concave down, and it is a local min if f is concave up. Concavity is determined by the sign of $f''(x)$, so we obtain the Second Derivative Test in Theorem 3. (See Exercise 73 for a detailed proof.)

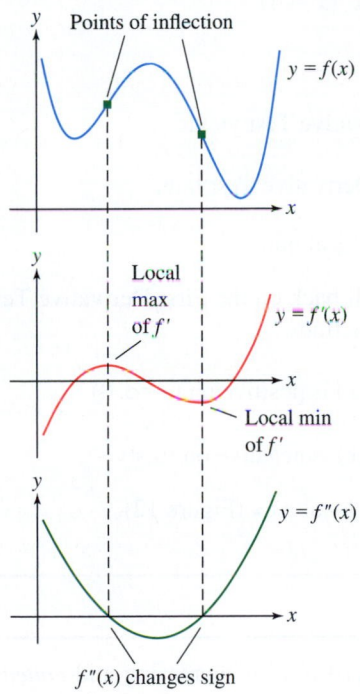

FIGURE 8

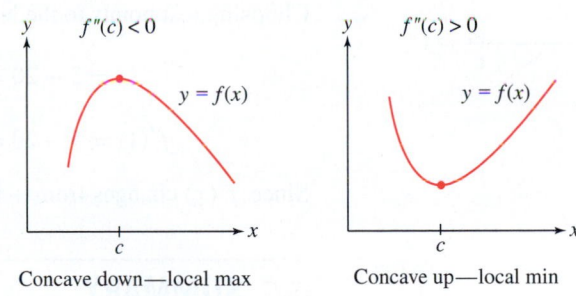

FIGURE 9 Concavity determines the type of the critical point.

THEOREM 3 **Second Derivative Test** Let c be a critical point of $f(x)$. If $f''(c)$ exists, then

- $f''(c) > 0$ ⇒ $f(c)$ is a local minimum.
- $f''(c) < 0$ ⇒ $f(c)$ is a local maximum.
- $f''(c) = 0$ ⇒ inconclusive: $f(c)$ may be a local min, a local max, or neither.

Mnemonic Device:

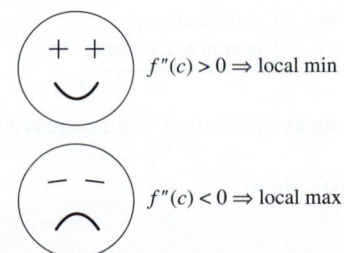

FIGURE 10

The mnemonic device appearing in Figure 10 provides an easy way to remember the test.

EXAMPLE 5 Analyze the critical points of $f(x) = (2x - x^2)e^x$.

Solution First, we have

$$f'(x) = e^x(2 - 2x) + (2x - x^2)e^x = (2 - x^2)e^x$$

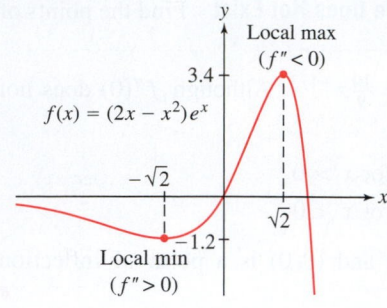

$f(x) = (2x - x^2)e^x$

Local max ($f'' < 0$)

3.4

$-\sqrt{2}$

$\sqrt{2}$

-1.2

Local min ($f'' > 0$)

FIGURE 11

To find the critical points, solve $(2 - x^2)e^x = 0$. Therefore $c = \pm\sqrt{2}$. Next, determine the sign of the second derivative at the critical points:

$$f''(x) = (-2x)e^x + (2 - x^2)e^x = (2 - 2x - x^2)e^x$$

$$f''(-\sqrt{2}) = \left(2 - 2(-\sqrt{2}) - (-\sqrt{2})^2\right)e^{-\sqrt{2}} = 2\sqrt{2}e^{-\sqrt{2}} \qquad > 0 \quad \text{(local min)}$$

$$f''(\sqrt{2}) = \left(2 - 2\sqrt{2} - (\sqrt{2})^2\right)e^{\sqrt{2}} = -2\sqrt{2}e^{\sqrt{2}} \qquad < 0 \quad \text{(local max)}$$

By the Second Derivative Test, f has a local min at $x = -\sqrt{2}$ and a local max at $x = \sqrt{2}$ (Figure 11). ∎

EXAMPLE 6   Second Derivative Test Inconclusive   Analyze the critical points of $f(x) = x^5 - 5x^4$.

Solution   The first two derivatives are

$$f'(x) = 5x^4 - 20x^3 = 5x^3(x - 4)$$

$$f''(x) = 20x^3 - 60x^2$$

The critical points are $c = 0, 4$, and the Second Derivative Test yields

$$f''(0) = 0 \quad \Rightarrow \quad \text{Second Derivative Test fails}$$

$$f''(4) = 320 > 0 \quad \Rightarrow \quad f(4) \text{ is a local min}$$

The Second Derivative Test fails at $x = 0$, so we fall back on the First Derivative Test. Choosing test points to the left and right of $x = 0$, we find

$$f'(-1) = 5 + 20 = 25 > 0 \quad \Rightarrow \quad f'(x) \text{ is positive on } (-\infty, 0)$$

$$f'(1) = 5 - 20 = -15 < 0 \quad \Rightarrow \quad f'(x) \text{ is negative on } (0, 4)$$

Since $f'(x)$ changes from $+$ to $-$ at $x = 0$, $f(0)$ is a local max (Figure 12). ∎

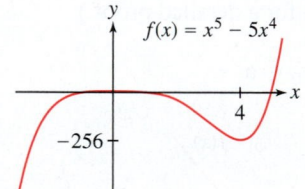

$f(x) = x^5 - 5x^4$

4

-256

FIGURE 12

4.4 SUMMARY

- A differentiable function f is *concave up* on (a, b) if f' is increasing and *concave down* if f' is decreasing on (a, b).
- The signs of the first two derivatives provide the following information:

First derivative	Second derivative
$f' > 0 \quad \Rightarrow \quad f$ is increasing	$f'' > 0 \quad \Rightarrow \quad f$ is concave up
$f' < 0 \quad \Rightarrow \quad f$ is decreasing	$f'' < 0 \quad \Rightarrow \quad f$ is concave down

- A *point of inflection* is a point $(c, f(c))$ where the concavity changes from concave up to concave down, or vice versa.
- Second Derivative Test: If $f'(c) = 0$ and $f''(c)$ exists, then
  - $f(c)$ is a local maximum value if $f''(c) < 0$
  - $f(c)$ is a local minimum value if $f''(c) > 0$
  - The test fails if $f''(c) = 0$

If this test fails, use the First Derivative Test.

4.4 EXERCISES

Preliminary Questions

1. If f is concave up, then f' is (choose one)

(a) increasing   **(b)** decreasing

2. What conclusion can you draw if $f'(c) = 0$ and $f''(c) < 0$?

3. True or false? If $f(c)$ is a local min, then $f''(c)$ must be positive.

4. True or false? If $f''(c) = 0$, then f has an inflection point at $x = c$.

5. The function $f(x) = \frac{x^2+1}{x}$ is concave down for $x < 0$ and concave up for $x > 0$. Is there an inflection point at $x = 0$? Explain.

6. Can a function have an inflection point at a critical point? Explain.

Exercises

1. Match the graphs in Figure 13 with the description:

(a) $f''(x) < 0$ for all x.   **(b)** $f''(x)$ goes from $+$ to $-$.

(c) $f''(x) > 0$ for all x.   **(d)** $f''(x)$ goes from $-$ to $+$.

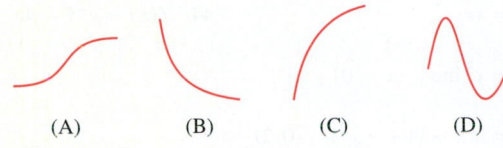

(A)   (B)   (C)   (D)

FIGURE 13

2. Match each statement with a graph in Figure 14 that represents company profits as a function of time.

(a) The outlook is great: The growth rate keeps increasing.

(b) We're losing money, but not as quickly as before.

(c) We're losing money, and it's getting worse as time goes on.

(d) We're doing well, but our growth rate is leveling off.

(e) Business had been cooling off, but now it's picking up.

(f) Business had been picking up, but now it's cooling off.

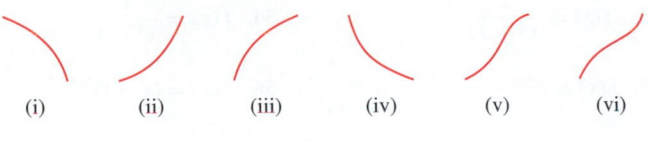

(i)   (ii)   (iii)   (iv)   (v)   (vi)

FIGURE 14

3. **GU** Plot $f(x) = (2x - x^2)e^x$ and indicate on the graph where it appears that inflection points occur. Then find the inflection points using calculus.

4. **GU** Plot $f(x) = x(x - 4)^3$ and indicate on the graph where it appears that inflection points occur. Then find the inflection points using calculus.

In Exercises 5–24, determine the intervals on which the function is concave up or down and find the points of inflection.

5. $y = x^2 - 4x + 3$

6. $y = t^3 - 6t^2 + 4$

7. $y = 10x^3 - x^5$

8. $y = 5x^2 + x^4$

9. $y = \theta - 2\sin\theta$, $[0, 2\pi]$

10. $y = \theta + \sin^2\theta$, $[0, \pi]$

11. $y = x(x - 8\sqrt{x})$ $(x \geq 0)$

12. $y = x^{7/2} - 35x^2$

13. $y = (x - 2)(1 - x^3)$

14. $y = x^{7/5}$

15. $y = \frac{1}{x^2 + 3}$

16. $y = \frac{x}{x^2 + 9}$

17. $f(x) = \frac{x^3}{1 + x}$

18. $w(t) = \frac{t^4 - 1}{t}$

19. $y = xe^{-3x}$

20. $y = (x^2 - 7)e^x$

21. $y = 2x^2 + \ln x$ $(x > 0)$

22. $y = x - \ln x$ $(x > 0)$

23. $f(t) = te^{-t^2}$

24. The Surge Function $S(t) = Ate^{-kt}$, with $A, k > 0$

25. The position of an ambulance in kilometers on a straight road over a period of 4 hours is given by the graph in Figure 15.

(a) Describe the motion of the ambulance.

(b) Explain what the fact that this graph is concave up tells us about the speed of the ambulance.

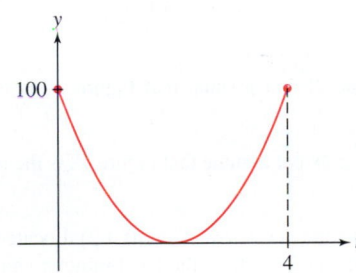

FIGURE 15

26. The position of a bicyclist on a straight road in kilometers over a period of 4 h is given by the graph in Figure 16, where inflection points occur when $t = 0.5$ and $t = 2$.

(a) Describe the motion of the bicyclist.

(b) Explain what the concavity of the graph over various intervals tells us about the speed of the bicyclist.

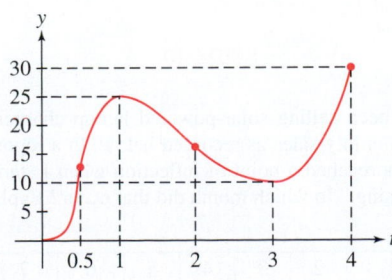

FIGURE 16

27. 📝 The growth of a sunflower during the first 100 days after sprouting is modeled well by the *logistic curve* $y = h(t)$ shown in Figure 17. Estimate the growth rate at the point of inflection and explain its significance. Then make a rough sketch of the first and second derivatives of h.

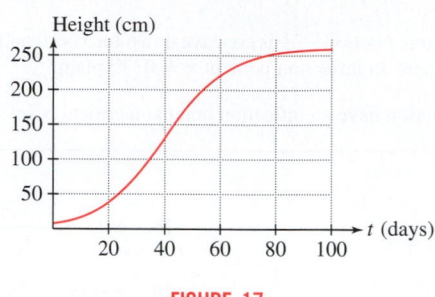

FIGURE 17

28. Assume that Figure 18 is the graph of f. Where do the points of inflection of f occur, and on which interval is f concave down?

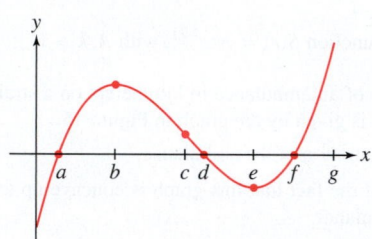

FIGURE 18

29. Repeat Exercise 28 but assume that Figure 18 is the graph of the *derivative* f'.

30. Repeat Exercise 28 but assume that Figure 18 is the graph of the *second derivative* f''.

31. Figure 19 shows the *derivative* f' on $[0, 1.2]$. Locate the points of inflection of f and the points where the local minima and maxima occur. Determine the intervals on which f has the following properties:

(a) Increasing **(b)** Decreasing
(c) Concave up **(d)** Concave down

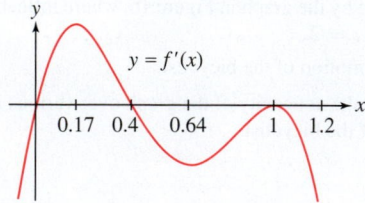

FIGURE 19

32. Leticia has been selling solar-powered laptop chargers through her Web site, with monthly sales as recorded below. In a report to investors, she states, "Sales reached a point of inflection when I started using pay-per-click advertising." In which month did that occur? Explain.

Month	1	2	3	4	5	6	7	8
Sales	2	30	50	60	90	150	230	340

In Exercises 33–46, find the critical points and apply the Second Derivative Test (or state that it fails).

33. $f(x) = x^3 - 12x^2 + 45x$

34. $f(x) = x^4 - 8x^2 + 1$

35. $f(x) = 3x^4 - 8x^3 + 6x^2$

36. $f(x) = x^5 - x^3$

37. $f(x) = \dfrac{x^2 - 8x}{x + 1}$

38. $f(x) = \dfrac{1}{x^2 - x + 2}$

39. $y = 6x^{3/2} - 4x^{1/2}$

40. $y = 9x^{7/3} - 21x^{1/2}$

41. $f(x) = \sin^2 x + \cos x$, $\quad [0, \pi]$

42. $y = \dfrac{1}{\sin x + 4}$, $\quad [0, 2\pi]$

43. $f(x) = xe^{-x^2}$

44. $f(x) = e^{-x} - 4e^{-2x}$

45. $f(x) = x^3 \ln x \quad (x > 0)$

46. $f(x) = \ln x + \ln(4 - x^2)$, $\quad (0, 2)$

In Exercises 47–62, find the intervals on which f is concave up or down, the points of inflection, the critical points, and the local minima and maxima.

47. $f(x) = x^3 - 2x^2 + x$

48. $f(x) = x^2(x - 4)$

49. $f(t) = t^2 - t^3$

50. $f(x) = 2x^4 - 3x^2 + 2$

51. $f(x) = x^2 - 8x^{1/2} \quad (x \geq 0)$

52. $f(x) = x^{3/2} - 4x^{-1/2} \quad (x > 0)$

53. $f(x) = \dfrac{x}{x^2 + 27}$

54. $f(x) = \dfrac{1}{x^4 + 1}$

55. $f(x) = x^{5/3} - x$

56. $f(x) = (x - 1)^{3/5}$

57. $f(\theta) = \theta + \sin \theta$, $\quad [0, 2\pi]$

58. $f(x) = \cos^2 x$, $\quad [0, \pi]$

59. $f(x) = \tan x$, $\quad \left(-\frac{\pi}{2}, \frac{\pi}{2}\right)$

60. $f(x) = e^{-x} \cos x$, $\quad \left[-\frac{\pi}{2}, \frac{3\pi}{2}\right]$

61. $y = (x^2 - 2)e^{-x} \quad (x > 0)$

62. $y = \ln(x^2 + 2x + 5)$

63. Sketch the graph of an increasing function such that $f''(x)$ changes from $+$ to $-$ at $x = 2$ and from $-$ to $+$ at $x = 4$. Do the same for a decreasing function.

In Exercises 64–66, sketch the graph of a function f satisfying all of the given conditions.

64. $f'(x) > 0$ and $f''(x) < 0$ for all x

65. **(i)** $f'(x) > 0$ for all x, and
(ii) $f''(x) < 0$ for $x < 0$ and $f''(x) > 0$ for $x > 0$

66. **(i)** $f'(x) < 0$ for $x < 0$ and $f'(x) > 0$ for $x > 0$, and
(ii) $f''(x) < 0$ for $|x| > 2$, and $f''(x) > 0$ for $|x| < 2$

67. An infectious flu spreads slowly at the beginning of an epidemic. The infection process accelerates until a majority of the susceptible individuals are infected, at which point the process slows down.

(a) If $R(t)$ is the number of individuals infected at time t, describe the concavity of the graph of R near the beginning and end of the epidemic.

(b) Describe the status of the epidemic on the day that R has a point of inflection.

68. Water is pumped into a sphere at a constant rate (Figure 20). Let $h(t)$ be the water level at time t. Sketch the graph of h (approximately, but with the correct concavity). Where does the point of inflection occur?

69. Water is pumped into a sphere of radius R at a variable rate in such a way that the water level rises at a constant rate (Figure 20). Let $V(t)$ be the volume of water in the tank at time t. Sketch the graph V (approximately, but with the correct concavity). Where does the point of inflection occur?

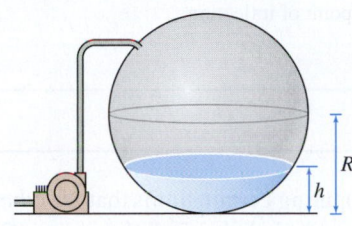

FIGURE 20

70. (Continuation of Exercise 69) If the sphere has radius R, the volume of water is

$$V = \pi\left(Rh^2 - \frac{1}{3}h^3\right),$$

where h is the water level. Assume the level rises at a constant rate of 1 (i.e., $h = t$).

(a) Find the inflection point of V. Does this agree with your conclusion in Exercise 69?

(b) [GU] Plot V for $R = 1$.

71. Image Processing The intensity of a pixel in a digital image is measured by a number u between 0 and 1. Often, images can be enhanced by rescaling intensities, as in the images of Amelia Earhart in Figure 21. When rescaling, pixels of intensity u are displayed with intensity $g(u)$ for a suitable function g. One common choice is the **sigmoidal correction**, defined for constants a, b by

$$g(u) = \frac{f(u) - f(0)}{f(1) - f(0)}, \qquad \text{where} \quad f(u) = \left(1 + e^{b(a-u)}\right)^{-1}$$

Original   Sigmoidal correction

FIGURE 21

Figure 22 shows that $g(u)$ reduces the intensity of low-intensity pixels [where $g(u) < u$] and increases the intensity of high-intensity pixels.

(a) Verify that $f'(u) > 0$ and use this to show that $g(u)$ increases from 0 to 1 for $0 \le u \le 1$.

(b) Where does $g(u)$ have a point of inflection?

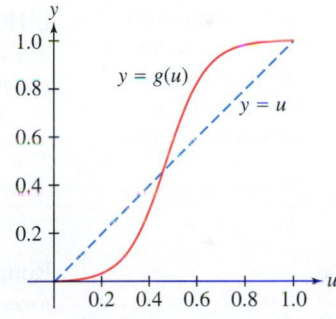

FIGURE 22 Sigmoidal correction with $a = 0.47$, $b = 12$.

72. Use graphical reasoning to determine whether the following statements are true or false. If false, modify the statement to make it correct.

(a) If f is increasing, then f^{-1} is decreasing.

(b) If f is decreasing, then f^{-1} is decreasing.

(c) If f is concave up, then f^{-1} is concave up.

(d) If f is concave down, then f^{-1} is concave up.

Further Insights and Challenges

In Exercises 73–75, assume that f is differentiable.

73. Proof of the Second Derivative Test Let c be a critical point such that $f''(c) > 0$ [the case $f''(c) < 0$ is similar].

(a) Show that $f''(c) = \lim\limits_{h \to 0} \dfrac{f'(c+h)}{h}$.

(b) Use (a) to show that there exists an open interval (a, b) containing c such that $f'(x) < 0$ if $a < x < c$ and $f'(x) > 0$ if $c < x < b$. Conclude that $f(c)$ is a local minimum.

74. Prove that if f'' exists and $f''(x) > 0$ for all x, then the graph of f "sits above" its tangent lines.

(a) For any c, set $G(x) = f(x) - f'(c)(x - c) - f(c)$. It is sufficient to prove that $G(x) \ge 0$ for all c. Explain why with a sketch.

(b) Show that $G(c) = G'(c) = 0$ and $G''(x) > 0$ for all x. Conclude that $G'(x) < 0$ for $x < c$ and $G'(x) > 0$ for $x > c$. Then deduce, using the MVT, that $G(x) > G(c)$ for $x \ne c$.

75. Assume that f'' exists and let c be a point of inflection of f.

(a) Use the method of Exercise 74 to prove that the tangent line at $x = c$ crosses the graph (Figure 23). *Hint:* Show that $G(x)$ changes sign at $x = c$.

(b) [GU] Verify this conclusion for $f(x) = \dfrac{x}{3x^2 + 1}$ by graphing f and the tangent line at each inflection point on the same set of axes.

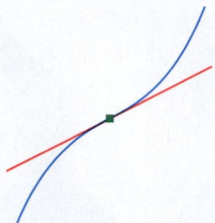

FIGURE 23 Tangent line crosses graph at point of inflection.

76. Let $C(x)$ be the cost of producing x units of a certain good. Assume that the graph of C is concave up.

(a) Show that the average cost $A(x) = C(x)/x$ is minimized at the production level x_0 such that average cost equals marginal cost—that is, $A(x_0) = C'(x_0)$.

(b) Show that the line through $(0, 0)$ and $(x_0, C(x_0))$ is tangent to the graph of C.

77. Let f be a polynomial of degree $n \geq 2$. Show that f has at least one point of inflection if n is odd. Then give an example to show that f need not have a point of inflection if n is even.

78. Critical and Inflection Points   If $f'(c) = 0$ and $f(c)$ is neither a local min nor a local max, must $x = c$ be a point of inflection? This is true for "reasonable" functions (including the functions studied in this text), but it is not true in general. Let

$$f(x) = \begin{cases} x^2 \sin \frac{1}{x} & \text{for } x \neq 0 \\ 0 & \text{for } x = 0 \end{cases}$$

(a) Use the limit definition of the derivative to show that $f'(0)$ exists and $f'(0) = 0$.

(b) Show that $f(0)$ is neither a local min nor a local max.

(c) Show that $f'(x)$ changes sign infinitely often near $x = 0$. Conclude that $x = 0$ is not a point of inflection.

4.5 L'Hôpital's Rule

L'Hôpital's Rule is named for the French mathematician Guillaume François Antoine Marquis de L'Hôpital (1661–1704), who wrote the first textbook on calculus in 1696. The name L'Hôpital is pronounced "Lo-pee-tal."

L'Hôpital's Rule is a valuable tool for computing certain limits that are otherwise difficult to evaluate, and also for determining "asymptotic behavior" (limits at infinity). We will use it for graph sketching in the next section.

Consider the limit of a quotient:

$$\lim_{x \to a} \frac{f(x)}{g(x)}$$

Roughly speaking, L'Hôpital's Rule states that *when $f(x)/g(x)$ has an indeterminate form of type $0/0$ or ∞/∞ at $x = a$, then we can replace $f(x)/g(x)$ by the quotient of the derivatives $f'(x)/g'(x)$.*

THEOREM 1 L'Hôpital's Rule   Assume that f and g are differentiable on an open interval containing a and that

$$f(a) = g(a) = 0$$

Also assume that $g'(x) \neq 0$ (except possibly at a). Then

$$\lim_{x \to a} \frac{f(x)}{g(x)} = \lim_{x \to a} \frac{f'(x)}{g'(x)}$$

if the limit on the right exists or is infinite (∞ or $-\infty$). This conclusion also holds if f and g are differentiable for x near (but not equal to) a and

$$\lim_{x \to a} f(x) = \pm\infty \qquad \text{and} \qquad \lim_{x \to a} g(x) = \pm\infty$$

Furthermore, this rule is valid for one-sided limits.

CAUTION When using L'Hôpital's Rule, be sure to take the derivative of the numerator and denominator separately:

$$\lim_{x \to a} \frac{f(x)}{g(x)} = \lim_{x \to a} \frac{f'(x)}{g'(x)}$$

Do not take the derivative of the function $y = f(x)/g(x)$ as a quotient, for example using the Quotient Rule.

EXAMPLE 1   Use L'Hôpital's Rule to evaluate $\lim_{x \to 2} \dfrac{x^3 - 8}{x^4 + 2x - 20}$.

Solution   Let $f(x) = x^3 - 8$ and $g(x) = x^4 + 2x - 20$. Both f and g are differentiable and $f(x)/g(x)$ is indeterminate of type $0/0$ at $a = 2$ because $f(2) = g(2) = 0$.

Furthermore, $g'(x) = 4x^3 + 2$ is nonzero near $x = 2$, so L'Hôpital's Rule applies. We may replace the numerator and denominator by their derivatives to obtain

$$\underbrace{\lim_{x \to 2} \frac{x^3 - 8}{x^4 + 2x - 2} = \lim_{x \to 2} \frac{(x^3 - 8)'}{(x^4 + 2x - 2)'}}_{\text{L'Hôpital's Rule}} = \lim_{x \to 2} \frac{3x^2}{4x^3 + 2} = \frac{3(2^2)}{4(2^3) + 2} = \frac{12}{34} = \frac{6}{17} \quad \blacksquare$$

EXAMPLE 2 Evaluate $\lim\limits_{x \to \pi/2} \dfrac{\cos^2 x}{1 - \sin x}$.

Solution Again, the quotient is indeterminate of type 0/0 at $x = \frac{\pi}{2}$ since

$$\cos^2\left(\frac{\pi}{2}\right) = 0, \qquad 1 - \sin\frac{\pi}{2} = 1 - 1 = 0$$

The other hypotheses are satisfied, so we may apply L'Hôpital's Rule:

$$\underbrace{\lim_{x \to \pi/2} \frac{\cos^2 x}{1 - \sin x} = \lim_{x \to \pi/2} \frac{(\cos^2 x)'}{(1 - \sin x)'} = \lim_{x \to \pi/2} \frac{-2\cos x \, \sin x}{-\cos x}}_{\text{L'Hôpital's Rule}} = \underbrace{\lim_{x \to \pi/2} (2\sin x) = 2}_{\text{Simplified}}$$

Note that the quotient $\dfrac{-2\cos x \, \sin x}{-\cos x}$ is also indeterminate at $x = \pi/2$. We removed this indeterminacy by cancelling the factor $-\cos x$. $\quad \blacksquare$

EXAMPLE 3 **The Form $0 \cdot \infty$** Evaluate $\lim\limits_{x \to 0^+} x \ln x$.

Solution This limit is one-sided because $f(x) = x \ln x$ is not defined for $x \le 0$. Furthermore, as $x \to 0^+$,

- x approaches 0.
- $\ln x$ approaches $-\infty$.

So $f(x)$ presents an indeterminate form of type $0 \cdot \infty$. To apply L'Hôpital's Rule, we rewrite our function as $f(x) = (\ln x)/x^{-1}$ so that $f(x)$ presents an indeterminate form of type $-\infty/\infty$. Then L'Hôpital's Rule applies:

$$\lim_{x \to 0^+} x \ln x = \underbrace{\lim_{x \to 0^+} \frac{\ln x}{x^{-1}} = \lim_{x \to 0^+} \frac{(\ln x)'}{(x^{-1})'} = \lim_{x \to 0^+} \left(\frac{x^{-1}}{-x^{-2}}\right)}_{\text{L'Hôpital's Rule}} = \underbrace{\lim_{x \to 0^+} (-x) = 0}_{\text{Simplified}} \quad \blacksquare$$

EXAMPLE 4 **Using L'Hôpital's Rule Twice** Evaluate $\lim\limits_{x \to 0} \dfrac{e^x - x - 1}{\cos x - 1}$.

Solution The limit is in the indeterminate form 0/0 since at $x = 0$, we have

$$e^x - x - 1 = e^0 - 0 - 1 = 0, \qquad \cos x - 1 = \cos 0 - 1 = 0$$

A first application of L'Hôpital's Rule gives

$$\lim_{x \to 0} \frac{e^x - x - 1}{\cos x - 1} = \lim_{x \to 0} \frac{(e^x - x - 1)'}{(\cos x - 1)'} = \lim_{x \to 0} \left(\frac{e^x - 1}{-\sin x}\right) = \lim_{x \to 0} \frac{1 - e^x}{\sin x}$$

This limit is again indeterminate of type 0/0, so we apply L'Hôpital's Rule a second time:

$$\lim_{x \to 0} \frac{1 - e^x}{\sin x} = \lim_{x \to 0} \frac{-e^x}{\cos x} = \frac{-e^0}{\cos 0} = -1$$

It follows that

$$\lim_{x \to 0} \frac{e^x - x - 1}{\cos x - 1} = -1 \quad \blacksquare$$

EXAMPLE 5   **Maximum Height Under Air Resistance**   In Example 7 in Section 2.5 we introduced a function

$$H(k) = \frac{30k - 9.8 \ln\left(\frac{150k}{49} + 1\right)}{k^2}$$

that gives the maximum height attained by a one kilogram ball launched upward at 30 m/s with gravity and air resistance acting on it. (The function is derived in Section 9.2.) The variable k reflects the strength of the air resistance. We investigated what happens to the maximum height as the air resistance approaches zero; that is, we investigated $\lim\limits_{k \to 0} H(k)$ numerically. Show this limit can be evaluated using L'Hôpital's Rule and find the limit.

Solution   The quotient $\dfrac{30k - 9.8 \ln\left(\frac{150k}{49} + 1\right)}{k^2}$ has the indeterminate form 0/0. To evaluate the limit, we need to use L'Hôpital's Rule twice:

$$\lim_{k \to 0} \frac{30k - 9.8 \ln\left(\frac{150k}{49} + 1\right)}{k^2} = \lim_{k \to 0} \frac{30 - \frac{30}{\frac{150k}{49} + 1}}{2k} = \lim_{k \to 0} \frac{\frac{4500/49}{\left(\frac{150k}{49} + 1\right)^2}}{2} = \frac{2250}{49} \approx 45.92 \qquad \blacksquare$$

This value of 45.92 m matches our previous numerical estimate and the result we obtained separately in Example 7 in Section 3.4 where we considered the launched projectile's height, ignoring air resistance altogether.

EXAMPLE 6   **Assumptions Matter**   Can L'Hôpital's Rule be applied to $\lim\limits_{x \to 1} \dfrac{x^2 + 1}{2x + 1}$?

Solution   The answer is no. The function does *not* have an indeterminate form because

$$\left.\frac{x^2 + 1}{2x + 1}\right|_{x=1} = \frac{1^2 + 1}{2 \cdot 1 + 1} = \frac{2}{3}$$

This limit can be evaluated directly by substitution: $\lim\limits_{x \to 1} \dfrac{x^2 + 1}{2x + 1} = \dfrac{2}{3}$. An incorrect application of L'Hôpital's Rule gives the wrong answer:

$$\lim_{x \to 1} \frac{(x^2 + 1)'}{(2x + 1)'} = \lim_{x \to 1} \frac{2x}{2} = 1 \quad \text{(not equal to original limit)} \qquad \blacksquare$$

EXAMPLE 7   **The Form $\infty - \infty$**   Evaluate $\lim\limits_{x \to 0} \left(\dfrac{1}{\sin x} - \dfrac{1}{x} \right)$.

Solution   Both $1/\sin x$ and $1/x$ become infinite at $x = 0$, so we have an indeterminate form of type $\infty - \infty$. We rewrite the function as

$$\frac{1}{\sin x} - \frac{1}{x} = \frac{x - \sin x}{x \sin x}$$

to obtain an indeterminate form of type 0/0. Applying L'Hôpital's Rule twice yields

$$\lim_{x \to 0} \left(\frac{1}{\sin x} - \frac{1}{x} \right) = \lim_{x \to 0} \frac{x - \sin x}{x \sin x} = \underbrace{\lim_{x \to 0} \frac{1 - \cos x}{x \cos x + \sin x}}_{\text{L'Hôpital's Rule}}$$

$$= \underbrace{\lim_{x \to 0} \frac{\sin x}{-x \sin x + 2 \cos x}}_{\text{L'Hôpital's Rule again}} = \frac{0}{2} = 0$$

This value of the limit is confirmed graphically in Figure 1. $\qquad \blacksquare$

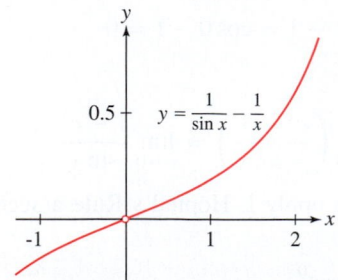

FIGURE 1 The graph confirms that $y = \dfrac{1}{\sin x} - \dfrac{1}{x}$ approaches 0 as $x \to 0$.

Limits of functions of the form $f(x)^{g(x)}$ can lead to the indeterminate forms 0^0, 1^∞, or ∞^0. These are indeterminate since the limit can take on a variety of values, depending on the relative rates at which the base and exponent approach their limits. In evaluating these limits, we use the change-of-base formula to write $f(x)^{g(x)} = e^{g(x) \ln f(x)}$ and then we obtain

$$\lim_{x \to a} f(x)^{g(x)} = \lim_{x \to a} e^{g(x) \ln f(x)} = e^{\lim_{x \to a} g(x) \ln f(x)}$$

The last equality is justified by the continuity of the exponential function.

◄— **REMINDER** *The change-of-base formula, changing an exponential base a to base e, is* $a^r = e^{r \ln a}$.

EXAMPLE 8 **The Form 0^0** Evaluate $\lim\limits_{x \to 0^+} x^x$.

Solution With $x^x = e^{x \ln x}$ by the change-of-base formula, it will be enough to consider the limit of $x \ln x$. Example 3 showed $\lim\limits_{x \to 0+} x \ln x = 0$. Therefore,

$$\lim_{x \to 0^+} x^x = \lim_{x \to 0^+} e^{x \ln x} = e^{\lim_{x \to 0^+} x \ln x} = e^0 = 1$$

This value for the limit is confirmed graphically in Figure 2. ■

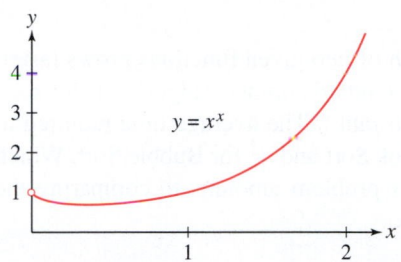

FIGURE 2 The function $y = x^x$ approaches 1 as $x \to 0^+$.

In Section 1.6, we pointed out that e is the value that $(1 + x)^{1/x}$ approaches as x approaches 0. This can be verified now by evaluating $\lim\limits_{x \to 0}(1 + x)^{1/x}$ using L'Hôpital's Rule.

EXAMPLE 9 **The Form 1^∞** Find $\lim\limits_{x \to 0}(1 + x)^{1/x}$.

Solution This has the indeterminate form 1^∞. We take the approach used in Example 8. Thus, we write $(1 + x)^{1/x} = e^{\frac{1}{x} \ln(1+x)}$ and consider $\lim\limits_{x \to 0} \frac{1}{x} \ln(1 + x)$. We obtain (using L'Hôpital's Rule for the first equality)

$$\lim_{x \to 0} \frac{\ln(1 + x)}{x} = \lim_{x \to 0} \frac{\frac{1}{1+x}}{1} = 1$$

Therefore,

$$\lim_{x \to 0}(1 + x)^{1/x} = \lim_{x \to 0} e^{\frac{1}{x} \ln(1+x)} = e^{\lim_{x \to 0} \frac{\ln(1+x)}{x}} = e^1 = e$$ ■

Note that if we substitute $x = \frac{1}{t}$ into $\lim\limits_{t \to \infty}\left(1 + \frac{1}{t}\right)^t$ we obtain the limit in the previous example. Therefore $\lim\limits_{t \to \infty}\left(1 + \frac{1}{t}\right)^t = e$. It is important to be familiar with these limits whose values are e:

$$e = \lim_{x \to 0}(1 + x)^{1/x} \quad \text{and} \quad e = \lim_{t \to \infty}\left(1 + \frac{1}{t}\right)^t$$

They arise in limit evaluations that we will see subsequently in the text.

CONCEPTUAL INSIGHT Exponential Limit Forms Knowing that $0 \cdot \infty$ is an indeterminate form, and using the exponential identity $a^x = e^{x \ln a}$, we can see why 0^0, 1^∞, and ∞^0 are indeterminate forms. A similar approach also shows why 0^∞ is not indeterminate and corresponds to a limit that equals 0.

The Form 0^0: If $\lim\limits_{x \to a} f(x)^{g(x)}$ is in the form 0^0, then $f(x) \to 0$ and $g(x) \to 0$. Therefore, in the limit, the equivalent exponential expression $e^{g(x) \ln f(x)}$ has an exponent in the indeterminate form $0(-\infty)$ since $g(x) \to 0$ and $\ln f(x) \to -\infty$ (because $f(x) \to 0$). Therefore, 0^0 is an indeterminate form.

Similar arguments can be made to demonstrate that 1^∞ and ∞^0 are indeterminate forms (see Exercise 61).

The Form 0^∞: If $\lim\limits_{x \to a} f(x)^{g(x)}$ is in the form 0^∞, then $f(x) \to 0$ and $g(x) \to \infty$. Therefore, in the limit, the equivalent exponential expression $e^{g(x)\ln f(x)}$ has an exponent in the form $(\infty)(-\infty)$. Since the limit of the exponent is $-\infty$, it follows that the limit of $e^{g(x)\ln f(x)}$ is 0, and therefore the limit of $f(x)^{g(x)}$ is as well. Thus, the form 0^∞ is not indeterminate but instead corresponds to a limit that is equal to 0.

Comparing Growth of Functions

Sometimes, we are interested in determining which of two given functions grows faster. For example, Quick Sort and Bubble Sort are two standard computer algorithms for sorting data (e.g., alphabetizing, ordering according to rank). The average time required to sort a list of size n is approxiamtely $n \ln n$ for Quick Sort and n^2 for Bubble Sort. Which algorithm is faster when the size n is large? This problem amounts to comparing the growth of $Q(x) = x \ln x$ and $B(x) = x^2$ as $x \to \infty$.

We say that $f(x)$ grows *faster* than $g(x)$ if

$$\lim_{x \to \infty} \frac{f(x)}{g(x)} = \infty \qquad \text{or, equivalently,} \qquad \lim_{x \to \infty} \frac{g(x)}{f(x)} = 0$$

To indicate that $f(x)$ grows faster than $g(x)$, we use the notation $g(x) \ll f(x)$. For example, $x \ll x^2$ because

$$\lim_{x \to \infty} \frac{x^2}{x} = \lim_{x \to \infty} x = \infty$$

To compare the growth of functions, we need a version of L'Hôpital's Rule that applies to limits at infinity.

THEOREM 2 L'Hôpital's Rule for Limits at Infinity Assume that f and g are differentiable in an interval (b, ∞) and that $g'(x) \neq 0$ for $x > b$. If $\lim\limits_{x \to \infty} f(x)$ and $\lim\limits_{x \to \infty} g(x)$ exist and either both are zero or both are infinite, then

$$\lim_{x \to \infty} \frac{f(x)}{g(x)} = \lim_{x \to \infty} \frac{f'(x)}{g'(x)}$$

provided that the limit on the right exists. A similar result holds for limits as $x \to -\infty$.

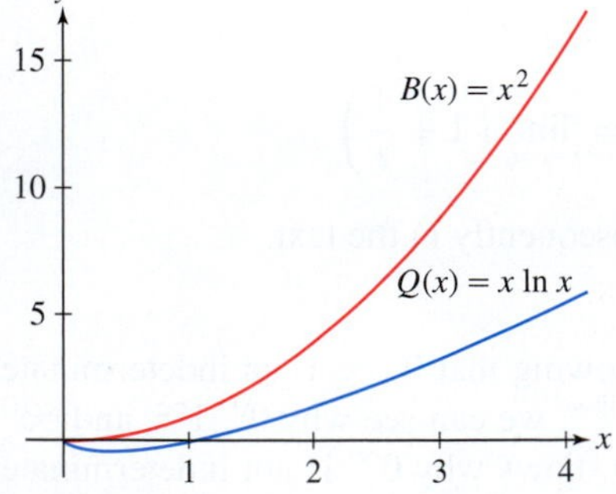

FIGURE 3

EXAMPLE 10 The Form $\dfrac{\infty}{\infty}$ Which of $B(x) = x^2$ or $Q(x) = x \ln x$ grows faster as $x \to \infty$?

Solution Both $B(x)$ and $Q(x)$ approach infinity as $x \to \infty$, so L'Hôpital's Rule applies to the quotient:

$$\lim_{x \to \infty} \frac{B(x)}{Q(x)} = \lim_{x \to \infty} \frac{x^2}{x \ln x} = \underbrace{\lim_{x \to \infty} \frac{x}{\ln x} = \lim_{x \to \infty} \frac{1}{x^{-1}}}_{\text{L'Hôpital's Rule}} = \lim_{x \to \infty} x = \infty$$

We conclude that $x \ln x \ll x^2$ (Figure 3). ∎

Note that this example implies that Quick Sort is a much faster sorting algorithm than Bubble Sort for large n.

In Section 1.6, we asserted that exponential functions increase more rapidly than the power functions. We now prove this by showing that $x^n \ll e^x$ for every exponent n (Figure 4).

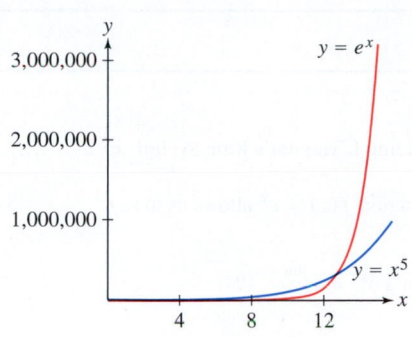

FIGURE 4 Graph illustrating that $x^5 \ll e^x$.

A full proof of L'Hôpital's Rule, without simplifying assumptions, is presented in a supplement on the text's Web site.

THEOREM 3 Growth of $f(x) = e^x$

$$x^n \ll e^x \qquad \text{for every exponent } n$$

In other words, $\lim\limits_{x\to\infty} \dfrac{e^x}{x^n} = \infty$ for all n.

Proof We first prove the theorem for positive integers n.

$$\lim_{x\to\infty}\frac{e^x}{x^n} = \lim_{x\to\infty}\frac{e^x}{nx^{n-1}} = \lim_{x\to\infty}\frac{e^x}{n(n-1)x^{n-2}} = \cdots = \lim_{x\to\infty}\frac{e^x}{n}$$

We applied L'Hôpital's Rule n times, each time obtaining an indeterminate form $\frac{\infty}{\infty}$, until the last stage shown. In $\lim\limits_{x\to\infty}\frac{e^x}{n}$ the numerator goes to ∞ and the denominator is constant (relative to x). Therefore, that limit is infinite, implying that $\lim\limits_{x\to\infty}\frac{e^x}{x^n} = \infty$ if n is a positive integer.

If n is any exponent, we can choose a natural number k such that $k > n$. It is easy to see that $x^n \ll x^k$, and because we also have $x^k \ll e^n$, it follows that $x^n \ll e^x$ for all exponents n. ∎

Proof of L'Hôpital's Rule

We prove L'Hôpital's Rule here only in the first case of Theorem 1—namely, in the case that $f(a) = g(a) = 0$. We also assume that f' and g' are continuous at $x = a$ and that $g'(a) \neq 0$. Then $g(x) \neq g(a)$ for x near a, but not equal to a, and

$$\frac{f(x)}{g(x)} = \frac{f(x) - f(a)}{g(x) - g(a)} = \frac{\dfrac{f(x) - f(a)}{x - a}}{\dfrac{g(x) - g(a)}{x - a}}$$

By the Quotient Law for Limits and the definition of the derivative,

$$\lim_{x\to a}\frac{f(x)}{g(x)} = \frac{\lim\limits_{x\to a}\dfrac{f(x) - f(a)}{x - a}}{\lim\limits_{x\to a}\dfrac{g(x) - g(a)}{x - a}} = \frac{f'(a)}{g'(a)} = \lim_{x\to a}\frac{f'(x)}{g'(x)} \qquad ∎$$

4.5 SUMMARY

- L'Hôpital's Rule: Assume that f and g are differentiable near a and that

$$f(a) = g(a) = 0$$

Assume also that $g'(x) \neq 0$ (except possibly at a). Then

$$\lim_{x\to a}\frac{f(x)}{g(x)} = \lim_{x\to a}\frac{f'(x)}{g'(x)}$$

provided that the limit on the right exists or is infinite (∞ or $-\infty$).
- L'Hôpital's Rule applies to indeterminate forms $0/0$ and $\pm\infty/\infty$. It can also apply to limits in any of the forms $0\cdot\infty$, $\infty - \infty$, 0^0, 1^∞, and ∞^0 by converting the expression to one in either the form $0/0$ or the form $\pm\infty/\infty$.
- L'Hôpital's Rule also applies to limits as $x\to\infty$ or $x\to-\infty$.
- In comparing the growth rates of functions, we say that $f(x)$ grows faster than $g(x)$, and we write $g \ll f$, if

$$\lim_{x\to\infty}\frac{f(x)}{g(x)} = \infty$$

4.5 EXERCISES

Preliminary Questions

1. What is wrong with applying L'Hôpital's Rule to $\lim\limits_{x\to 0} \dfrac{x^2 - 2x}{3x - 2}$?

2. Does L'Hôpital's Rule apply to $\lim\limits_{x\to a} f(x)g(x)$ if $f(x)$ and $g(x)$ both approach ∞ as $x \to a$?

3. What is wrong with saying, "To apply L'Hôpital's Rule to the limit $\lim\limits_{x\to 0} \dfrac{\ln(1-x)}{x}$, use the Quotient Rule to differentiate $\dfrac{\ln(1-x)}{x}$ and then take the limit."

4. What is wrong with applying L'Hôpital's Rule to $\lim\limits_{x\to 0^+} x^{\frac{1}{x}}$?

5. What property of the function $f(x) = e^x$ allows us to say

$$\lim_{x\to a} e^{g(x)} = e^{\lim\limits_{x\to a} g(x)}?$$

Exercises

In Exercises 1–10, evaluate the limit, using L'Hôpital's Rule where it applies.

1. $\lim\limits_{x\to 3} \dfrac{2x^2 - 5x - 3}{x - 4}$

2. $\lim\limits_{x\to -5} \dfrac{x^2 - 25}{5 - 4x - x^2}$

3. $\lim\limits_{x\to 4} \dfrac{x^3 - 64}{x^2 + 16}$

4. $\lim\limits_{x\to -1} \dfrac{x^4 + 2x + 1}{x^5 - 2x - 1}$

5. $\lim\limits_{x\to 9} \dfrac{x^{1/2} + x - 6}{x^{3/2} - 27}$

6. $\lim\limits_{x\to 3} \dfrac{\sqrt{x+1} - 2}{x^3 - 7x - 6}$

7. $\lim\limits_{x\to 0} \dfrac{\sin 4x}{x^2 + 3x + 1}$

8. $\lim\limits_{x\to 0} \dfrac{x^3}{\sin x - x}$

9. $\lim\limits_{x\to 0} \dfrac{\cos 2x - 1}{\sin 5x}$

10. $\lim\limits_{x\to 0} \dfrac{\cos x - \sin^2 x}{\sin x}$

In Exercises 11–16, use L'Hôpital's Rule to evaluate the limit.

11. $\lim\limits_{x\to \infty} \dfrac{9x + 4}{3 - 2x}$

12. $\lim\limits_{x\to -\infty} x \sin \dfrac{1}{x}$

13. $\lim\limits_{x\to \infty} \dfrac{\ln x}{x^{1/2}}$

14. $\lim\limits_{x\to \infty} \dfrac{x}{e^x}$

15. $\lim\limits_{x\to -\infty} \dfrac{\ln(x^4 + 1)}{x}$

16. $\lim\limits_{x\to \infty} \dfrac{x^2}{e^x}$

In Exercises 17–54, evaluate the limit.

17. $\lim\limits_{x\to 1} \dfrac{\sqrt{8+x} - 3x^{1/3}}{x^2 - 3x + 2}$

18. $\lim\limits_{x\to 4} \left[\dfrac{1}{\sqrt{x} - 2} - \dfrac{4}{x - 4} \right]$

19. $\lim\limits_{x\to -\infty} \dfrac{3x - 2}{1 - 5x}$

20. $\lim\limits_{x\to \infty} \dfrac{x^{2/3} + 3x}{x^{5/3} - x}$

21. $\lim\limits_{x\to -\infty} \dfrac{7x^2 + 4x}{9 - 3x^2}$

22. $\lim\limits_{x\to \infty} \dfrac{3x^3 + 4x^2}{4x^3 - 7}$

23. $\lim\limits_{x\to 1} \dfrac{(1 + 3x)^{1/2} - 2}{(1 + 7x)^{1/3} - 2}$

24. $\lim\limits_{x\to 8} \dfrac{x^{5/3} - 2x - 16}{x^{1/3} - 2}$

25. $\lim\limits_{x\to 0} \dfrac{\sin 2x}{\sin 7x}$

26. $\lim\limits_{x\to 0} \dfrac{\tan 4x}{\tan 5x}$

27. $\lim\limits_{x\to 0} \dfrac{\tan x}{x}$

28. $\lim\limits_{x\to 0} \left(\cot x - \dfrac{1}{x} \right)$

29. $\lim\limits_{x\to 0} \dfrac{\sin x - x \cos x}{x - \sin x}$

30. $\lim\limits_{x\to \pi/2} \left(x - \dfrac{\pi}{2} \right) \tan x$

31. $\lim\limits_{x\to 0} \dfrac{\cos(x + \frac{\pi}{2})}{\sin x}$

32. $\lim\limits_{x\to 0} \dfrac{x^2}{1 - \cos x}$

33. $\lim\limits_{x\to \pi/2} \dfrac{\cos x}{\sin(2x)}$

34. $\lim\limits_{x\to 0} \left(\dfrac{1}{x^2} - \csc^2 x \right)$

35. $\lim\limits_{x\to \pi/2} (\sec x - \tan x)$

36. $\lim\limits_{x\to 2} \dfrac{e^{x^2} - e^4}{x - 2}$

37. $\lim\limits_{x\to 1} \tan\left(\dfrac{\pi x}{2} \right) \ln x$

38. $\lim\limits_{x\to 1} \dfrac{x(\ln x - 1) + 1}{(x - 1) \ln x}$

39. $\lim\limits_{x\to 0} \dfrac{e^x - 1}{\sin x}$

40. $\lim\limits_{x\to 1} \dfrac{e^x - e}{\ln x}$

41. $\lim\limits_{x\to 0} \dfrac{e^{2x} - 1 - x}{x^2}$

42. $\lim\limits_{x\to \infty} \dfrac{e^{2x} - 1 - x}{x^2}$

43. $\lim\limits_{t\to 0^+} (\sin t)(\ln t)$

44. $\lim\limits_{x\to \infty} e^{-x}(x^3 - x^2 + 9)$

45. $\lim\limits_{x\to 0} \dfrac{a^x - 1}{x}$ $(a > 0)$

46. $\lim\limits_{x\to \infty} x^{1/x^2}$

47. $\lim\limits_{x\to 1} (1 + \ln x)^{1/(x-1)}$

48. $\lim\limits_{x\to 0^+} x^{\sin x}$

49. $\lim\limits_{x\to 0} (\cos x)^{3/x^2}$

50. $\lim\limits_{x\to \infty} \left(\dfrac{x}{x + 1} \right)^x$

51. $\lim\limits_{x\to 0} \dfrac{\sin^{-1} x}{x}$

52. $\lim\limits_{x\to 0} \dfrac{\tan^{-1} x}{\sin^{-1} x}$

53. $\lim\limits_{x\to 1} \dfrac{\tan^{-1} x - \frac{\pi}{4}}{\tan \frac{\pi}{4} x - 1}$

54. $\lim\limits_{x\to 0^+} \ln x \tan^{-1} x$

55. Evaluate $\lim\limits_{x\to \pi/2} \dfrac{\cos mx}{\cos nx}$, where $m, n \neq 0$ are integers.

56. Evaluate $\lim\limits_{x\to 1} \dfrac{x^m - 1}{x^n - 1}$ for any numbers $m, n \neq 0$.

57. Evaluate each of the following limits.

   (a) $\lim\limits_{x\to \infty} \left(1 + \dfrac{1}{x} \right)^{x^2}$

   (b) $\lim\limits_{x\to \infty} \left(1 + \dfrac{1}{x^2} \right)^x$

58. Show that $\lim\limits_{x\to \infty} \left(1 + \dfrac{r}{x} \right)^x = e^r$.

In Exercises 59–60, a ball is launched straight up in the air and is acted on by air resistance and gravity as in Example 5. The function M gives the maximum height that the projectile attains as a function of the air resistance parameter k. In each case, determine the maximum height as we let the air resistance term go to zero; that is, determine $\lim_{k \to 0} M(k)$.

59. A ball with a mass of 1 kilogram is launched upward with an initial velocity of 60 m/s, and

$$M(k) = \frac{60k - 9.8\ln(\frac{300k}{49} + 1)}{k^2}$$

(Compare with Exercises 37 in Section 2.5 and 29 in Section 3.4.)

60. A ball with a mass of 500 grams is launched upward with an initial velocity of 30 m/s, and

$$M(k) = \frac{15k - 2.45\ln(\frac{300k}{49} + 1)}{k^2}$$

(Compare with Exercise 38 in Section 2.5.)

61. In each case, show that the form is indeterminate by showing that if $\lim_{x \to c} f(x)^{g(x)}$ has the form, then the limit in the exponent in $e^{\lim_{x \to c} g(x)\ln f(x)}$ has a known indeterminate form.

(a) 1^∞

(b) ∞^0

62. GU Can L'Hôpital's Rule be applied to $\lim_{x \to 0+} x^{\sin(1/x)}$? Does a graphical or numerical investigation suggest that the limit exists?

63. Let $f(x) = x^{1/x}$ for $x > 0$.

(a) Calculate $\lim_{x \to 0+} f(x)$ and $\lim_{x \to \infty} f(x)$.

(b) Find the maximum value of f and determine the intervals on which f is increasing or decreasing.

64. (a) Use the results of Exercise 63 to prove that $x^{1/x} = c$ has a unique solution if $0 < c \leq 1$ or $c = e^{1/e}$, and has two solutions if $1 < c < e^{1/e}$, and no solutions if $c > e^{1/e}$.

(b) GU Plot the graph of $f(x) = x^{1/x}$ and verify that it confirms the conclusions of (a).

65. Determine whether $f \ll g$ or $g \ll f$ (or neither) for the functions $f(x) = \log_{10} x$ and $g(x) = \ln x$.

66. Show that $(\ln x)^3 \ll x^{1/3}$ and $(\ln x)^4 \ll x^{1/10}$.

67. Just as exponential functions are distinguished by their rapid rate of increase, the logarithm functions grow particularly slowly. Show that $\ln x \ll x^a$ for all $a > 0$.

68. Show that $(\ln x)^N \ll x^a$ for all N and all $a > 0$.

69. Determine whether $\sqrt{x} \ll e^{\sqrt{\ln x}}$ or $e^{\sqrt{\ln x}} \ll \sqrt{x}$. *Hint:* Use the substitution $u = \ln x$ instead of L'Hôpital's Rule.

70. Show that $\lim_{x \to \infty} x^n e^{-x} = 0$ for all whole numbers $n > 0$.

71. Assumptions Matter Suppose $f(x) = x(2 + \sin x)$ and let $g(x) = x^2 + 1$.

(a) Show directly that $\lim_{x \to \infty} f(x)/g(x) = 0$.

(b) Show that $\lim_{x \to \infty} f(x) = \lim_{x \to \infty} g(x) = \infty$, but $\lim_{x \to \infty} f'(x)/g'(x)$ does not exist.

Do (a) and (b) contradict L'Hôpital's Rule? Explain.

72. Let $H(b) = \lim_{x \to \infty} \frac{\ln(1 + b^x)}{x}$ for $b > 0$.

(a) Show that $H(b) = \ln b$ if $b \geq 1$.

(b) Determine $H(b)$ for $0 < b \leq 1$.

73. Let $G(b) = \lim_{x \to \infty} (1 + b^x)^{1/x}$.

(a) Use the result of Exercise 72 to evaluate $G(b)$ for all $b > 0$.

(b) GU Verify your result graphically by plotting $y = (1 + b^x)^{1/x}$ together with the horizontal line $y = G(b)$ for the values $b = 0.25, 0.5, 2, 3$.

74. Show that $\lim_{t \to \infty} t^k e^{-t^2} = 0$ for all k. *Hint:* Compare with $\lim_{t \to \infty} t^k e^{-t} = 0$.

In Exercises 75–77, let

$$f(x) = \begin{cases} e^{-1/x^2} & \text{for } x \neq 0 \\ 0 & \text{for } x = 0 \end{cases}$$

These exercises show that f has an unusual property: All of its derivatives at $x = 0$ exist and are equal to zero.

75. Show that $\lim_{x \to 0} \frac{f(x)}{x^k} = 0$ for all k. *Hint:* Let $t = x^{-1}$ and apply the result of Exercise 74.

76. Show that $f'(0)$ exists and is equal to zero. Also, verify that $f''(0)$ exists and is equal to zero.

77. Show that for $k \geq 1$ and $x \neq 0$,

$$f^{(k)}(x) = \frac{P(x)e^{-1/x^2}}{x^r}$$

for some polynomial $P(x)$ and some exponent $r \geq 1$. Use the result of Exercise 75 to show that $f^{(k)}(0)$ exists and is equal to zero for all $k \geq 1$.

Further Insights and Challenges

78. Show that L'Hôpital's Rule applies to $\lim_{x \to \infty} \frac{x}{\sqrt{x^2 + 1}}$ but that it does not help. Then evaluate the limit directly.

79. The Second Derivative Test for critical points fails if $f''(c) = 0$. This exercise develops a **Higher Derivative Test** based on the sign of the first nonzero derivative. Suppose that

$$f'(c) = f''(c) = \cdots = f^{(n-1)}(c) = 0, \quad \text{but} \quad f^{(n)}(c) \neq 0$$

(a) Show, by applying L'Hôpital's Rule n times, that

$$\lim_{x \to c} \frac{f(x) - f(c)}{(x - c)^n} = \frac{1}{n!} f^{(n)}(c)$$

where $n! = n(n-1)(n-2)\cdots(2)(1)$.

(b) Use (a) to show that if n is even, then $f(c)$ is a local minimum if $f^{(n)}(c) > 0$ and is a local maximum if $f^{(n)}(c) < 0$. *Hint:* If n is even, then $(x - c)^n > 0$ for $x \neq a$, so $f(x) - f(c)$ must be positive for x near c if $f^{(n)}(c) > 0$.

(c) Use (a) to show that if n is odd, then $f(c)$ is neither a local minimum nor a local maximum.

80. When a spring with natural frequency $\lambda/2\pi$ is driven with a sinusoidal force $\sin(\omega t)$ with $\omega \neq \lambda$, it oscillates according to

$$y(t) = \frac{1}{\lambda^2 - \omega^2}\left(\lambda \sin(\omega t) - \omega \sin(\lambda t)\right)$$

Let $y_0(t) = \lim_{\omega \to \lambda} y(t)$.

(a) Use L'Hôpital's Rule to determine $y_0(t)$.

(b) Show that $y_0(t)$ ceases to be periodic and that its amplitude $|y_0(t)|$ tends to ∞ as $t \to \infty$ (the system is said to be in **resonance**; eventually, the spring is stretched beyond its structural tolerance).

(c) CAS Plot y for $\lambda = 1$ and $\omega = 0.8, 0.9, 0.99$, and 0.999. Do the graphs confirm your conclusion in (b)?

81. We expended a lot of effort to evaluate $\lim\limits_{x \to 0} \dfrac{\sin x}{x}$ in Chapter 2. Show that we could have evaluated it easily using L'Hôpital's Rule. Then explain why this method would involve *circular reasoning*.

82. By a fact from algebra, if f, g are polynomials such that $f(a) = g(a) = 0$, then there are polynomials f_1, g_1 such that

$$f(x) = (x-a)f_1(x), \qquad g(x) = (x-a)g_1(x)$$

Use this to verify L'Hôpital's Rule directly for $\lim\limits_{x \to a} f(x)/g(x)$.

83. Patience Required   Use L'Hôpital's Rule to evaluate and check your answers numerically:

(a) $\lim\limits_{x \to 0+} \left(\dfrac{\sin x}{x} \right)^{1/x^2}$ **(b)** $\lim\limits_{x \to 0} \left(\dfrac{1}{\sin^2 x} - \dfrac{1}{x^2} \right)$

84. In the following cases, check that $x = c$ is a critical point and use Exercise 79 to determine whether $f(c)$ is a local minimum or a local maximum.

(a) $f(x) = x^5 - 6x^4 + 14x^3 - 16x^2 + 9x + 12 \quad (c = 1)$

(b) $f(x) = x^6 - x^3 \quad (c = 0)$

4.6 Analyzing and Sketching Graphs of Functions

In this section, our goal is to study graphs of functions f using the information provided by the first two derivatives f' and f''. You will see that you can acquire a good understanding of the properties of a graph without plotting a large number of points. Even though almost all graphs you may see are produced by computer (including, of course, the graphs in this textbook), the tools of calculus provide information beyond the image displayed on a computer. This information includes the exact locations of critical points and inflection points, the rates of increase and decrease over the function's domain, and the concavity of the function.

Most graphs are made up of smaller *arcs* that have one of the four basic shapes corresponding to the four possible sign combinations of f' and f'' (Figure 1). Since f' and f'' can each have sign $+$ or $-$, the sign combinations are

$$++ \qquad +- \qquad -+ \qquad --$$

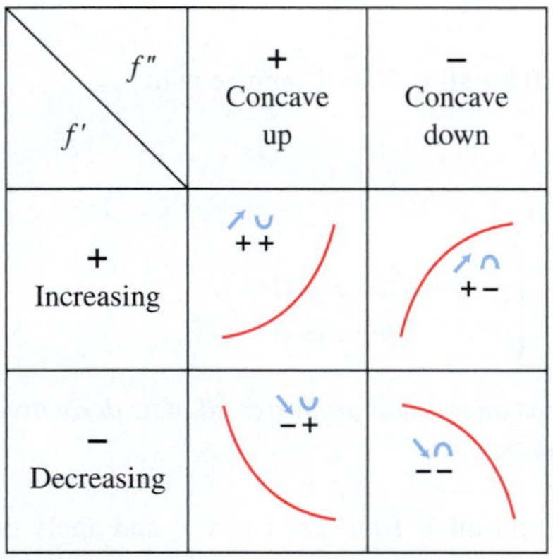

In this notation, the first sign refers to f' and the second sign to f''. For instance, $-+$ indicates that $f'(x) < 0$ and $f''(x) > 0$. We use a slanted arrow over the first sign to indicate whether the function is increasing or decreasing, and an upturned or downturned ∪ over the second sign to indicate the concavity.

In analyzing a graph, we focus on the **transition points**, where the basic shape changes due to a sign change in either f' (local min or max) or f'' (point of inflection). In this section, local extrema are indicated by solid dots, and points of inflection are indicated by green solid squares (Figure 2).

In examining the properties of a function, it is often useful to investigate the **asymptotic behavior**—that is, the behavior of $f(x)$ as x approaches either $\pm\infty$ or a vertical asymptote.

In the examples that follow, we use calculus to investigate the behavior of specific functions, and then we use the information we gather to construct a picture of the function's graph—that is, to "sketch the graph." The first three examples treat polynomials. Recall from Section 2.7 that the limits at infinity of a polynomial

$$f(x) = a_n x^n + a_{n-1} x^{n-1} + \cdots + a_1 x + a_0$$

(assuming that $a_n \neq 0$) are determined by

$$\lim_{x \to \infty} f(x) = a_n \lim_{x \to \infty} x^n$$

In general, the graph of a polynomial oscillates up and down a finite number of times and tends to positive or negative infinity as x tends to positive or negative infinity. Typical examples appear in Figure 3.

f' \\ f''	$+$ Concave up	$-$ Concave down
$+$ Increasing	⌣ $++$	⌢ $+-$
$-$ Decreasing	⌣ $-+$	⌢ $--$

FIGURE 1 The four basic shapes.

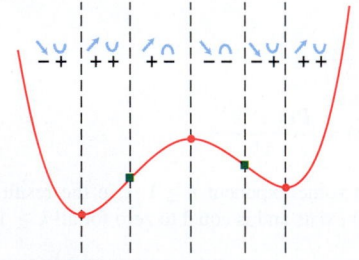

FIGURE 2 The graph of f with transition points and sign combinations of f' and f''.

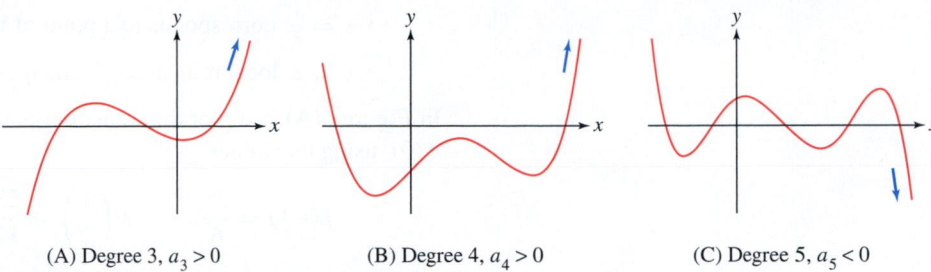

(A) Degree 3, $a_3 > 0$ (B) Degree 4, $a_4 > 0$ (C) Degree 5, $a_5 < 0$

FIGURE 3 Graphs of polynomials.

EXAMPLE 1 **Quadratic Polynomial** Investigate the behavior of $f(x) = x^2 - 4x + 3$ and sketch its graph.

Solution Note that $f(x) = (x - 1)(x - 3)$ so the graph intersects the x-axis at $x = 1$ and $x = 3$. We have $f'(x) = 2x - 4 = 2(x - 2)$. We can see directly that $f'(x)$ is negative for $x < 2$ and positive for $x > 2$, but let's confirm this using test values, as in previous sections:

Interval	Test value	Sign of f'
$(-\infty, 2)$	$f'(1) = -2$	$-$
$(2, \infty)$	$f'(3) = 2$	$+$

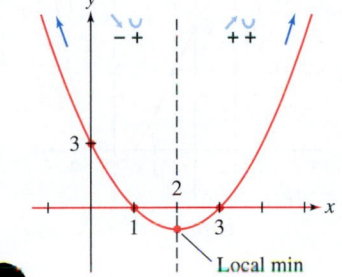

FIGURE 4 Graph of $f(x) = x^2 - 4x + 3$.

Furthermore, $f''(x) = 2$ is positive, so the graph is everywhere concave up. To sketch the graph, plot the local minimum $(2, -1)$, the y-intercept, and the roots $x = 1, 3$. Since the leading term of f is x^2, $f(x)$ tends to ∞ as $x \to \pm\infty$. This asymptotic behavior is noted by the arrows in Figure 4. ∎

EXAMPLE 2 **Cubic Polynomial** Investigate the behavior of the cubic function $f(x) = \frac{1}{3}x^3 - \frac{1}{2}x^2 - 2x + 3$ and sketch the graph.

Solution

Step 1. **Determine the signs of f' and f''.**
First, solve for the critical points:

$$f'(x) = x^2 - x - 2 = (x + 1)(x - 2)$$

The critical points are $c = -1, 2$, and they divide the x-axis into three intervals $(-\infty, -1)$, $(-1, 2)$, and $(2, \infty)$, on which we determine the sign of f' by computing test values:

Interval	Test value	Sign of f'
$(-\infty, -1)$	$f'(-2) = 4$	$+$
$(-1, 2)$	$f'(0) = -2$	$-$
$(2, \infty)$	$f'(3) = 4$	$+$

Next, $f''(x) = 2x - 1$, and therefore $x = \frac{1}{2}$ is the only solution to $f''(x) = 0$. We have

Interval	Test value	Sign of f''
$\left(-\infty, \frac{1}{2}\right)$	$f''(0) = -1$	$-$
$\left(\frac{1}{2}, \infty\right)$	$f''(1) = 1$	$+$

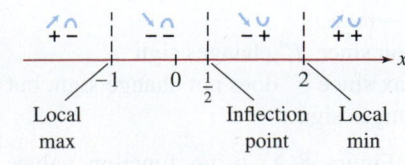

| Local max | | Inflection point | Local min |

FIGURE 5 Sign combinations of f' and f''.

Step 2. **Note transition points and sign combinations.**
This step merges the information about f' and f'' in a sign diagram (Figure 5). There are three transition points:

- $c = -1$: local max since f' changes from $+$ to $-$.

- $c = \frac{1}{2}$: corresponds to a point of inflection since f'' changes sign.
- $c = 2$: local min since f' changes from $-$ to $+$.

In Figure 6(A), we plot the transition points and, for added accuracy, the y-intercept $f(0)$, using the values

$$f(-1) = \frac{25}{6}, \qquad f\left(\frac{1}{2}\right) = \frac{23}{12}, \qquad f(0) = 3, \qquad f(2) = -\frac{1}{3}$$

Step 3. Draw arcs of appropriate shape and asymptotic behavior.
The leading term of $f(x)$ is $\frac{1}{3}x^3$. Therefore, $\lim\limits_{x \to \infty} f(x) = \infty$ and $\lim\limits_{x \to -\infty} f(x) = -\infty$.

To create the sketch, it remains only to connect the transition points by arcs of the appropriate concavity and asymptotic behavior, as in Figure 6(B) and (C). ■

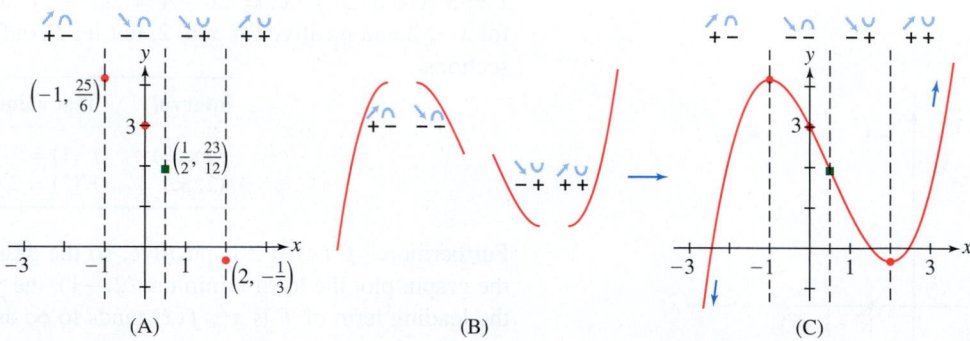

FIGURE 6 Graph of
$f(x) = \frac{1}{3}x^3 - \frac{1}{2}x^2 - 2x + 3$.

EXAMPLE 3 Investigate the behavior of $f(x) = 3x^4 - 8x^3 + 6x^2 + 1$ and sketch its graph.

Solution

Step 1. Determine the signs of f' and f''.
First, solve for the transition points:

$$f'(x) = 12x^3 - 24x^2 + 12x = 12x(x-1)^2, \quad \text{so } f' = 0 \;\Rightarrow\; x = 0, 1$$

$$f''(x) = 36x^2 - 48x + 12 = 12(x-1)(3x-1), \quad \text{so } f'' = 0 \;\Rightarrow\; x = \frac{1}{3}, 1$$

The signs of f' and f'' are recorded in the following tables:

Interval	Test value	Sign of f'	Interval	Test value	Sign of f''
$(-\infty, 0)$	$f'(-1) = -48$	$-$	$\left(-\infty, \frac{1}{3}\right)$	$f''(0) = 12$	$+$
$(0, 1)$	$f'\left(\frac{1}{2}\right) = \frac{3}{2}$	$+$	$\left(\frac{1}{3}, 1\right)$	$f''\left(\frac{1}{2}\right) = -3$	$-$
$(1, \infty)$	$f'(2) = 24$	$+$	$(1, \infty)$	$f''(2) = 60$	$+$

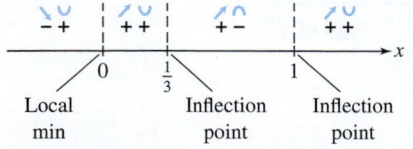

FIGURE 7

Step 2. Note transition points and sign combinations.
The transition points $c = 0, \frac{1}{3}, 1$ divide the x-axis into four intervals (Figure 7). The type of sign change determines the nature of the transition point:

- $c = 0$: local min since f' changes from $-$ to $+$.
- $c = \frac{1}{3}$: corresponds to a point of inflection since f'' changes sign.
- $c = 1$: neither a local min nor a local max since f' does not change sign, but it is a point of inflection since $f''(x)$ changes sign.

We plot the transition points $c = 0, \frac{1}{3}, 1$ in Figure 8(A) using function values $f(0) = 1$, $f\left(\frac{1}{3}\right) = \frac{38}{27}$, and $f(1) = 2$.

Step 3. **Draw arcs of appropriate shape and asymptotic behavior.**

Before drawing the arcs, we note that $f(x)$ has leading term $3x^4$, so $f(x)$ tends to ∞ as $x \to \infty$ and as $x \to -\infty$. We obtain Figure 8(B). ∎

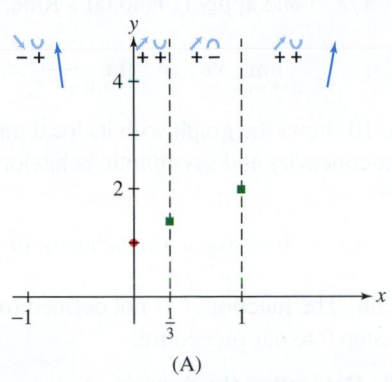

(A)

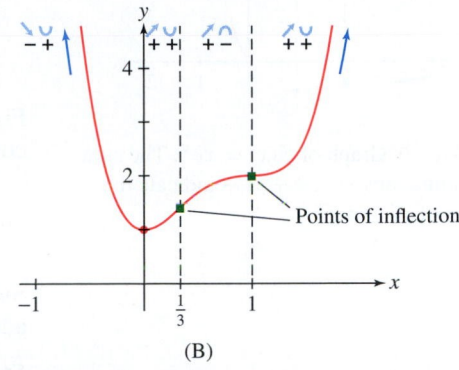

(B)

Points of inflection

FIGURE 8 $f(x) = 3x^4 - 8x^3 + 6x^2 + 1$.

EXAMPLE 4 Investigate the behavior of $f(x) = \cos x + \frac{1}{2}x$ over $[0, \pi]$, and sketch its graph.

Solution First, we find the transition points for x in $[0, \pi]$:

$$f'(x) = -\sin x + \frac{1}{2}, \quad \text{so } f'(x) = 0 \quad \Rightarrow \quad x = \frac{\pi}{6}, \frac{5\pi}{6}$$

$$f''(x) = -\cos x, \quad \text{so } f''(x) = 0 \quad \Rightarrow \quad x = \frac{\pi}{2}$$

The sign combinations are shown in the following tables:

Interval	Test value	Sign of f'
$(0, \frac{\pi}{6})$	$f'(\frac{\pi}{12}) \approx 0.24$	$+$
$(\frac{\pi}{6}, \frac{5\pi}{6})$	$f'(\frac{\pi}{2}) = -\frac{1}{2}$	$-$
$(\frac{5\pi}{6}, \pi)$	$f'(\frac{11\pi}{12}) \approx 0.24$	$+$

Interval	Test value	Sign of f''
$(0, \frac{\pi}{2})$	$f''(\frac{\pi}{4}) = -\frac{\sqrt{2}}{2}$	$-$
$(\frac{\pi}{2}, \pi)$	$f''(\frac{3\pi}{4}) = \frac{\sqrt{2}}{2}$	$+$

We record the sign changes and transition points in Figure 9 and sketch the graph using the values

$$f(0) = 1, \quad f\left(\frac{\pi}{6}\right) \approx 1.13, \quad f\left(\frac{\pi}{2}\right) \approx 0.79, \quad f\left(\frac{5\pi}{6}\right) \approx 0.44, \quad f(\pi) \approx 0.57 \quad ∎$$

FIGURE 9 $f(x) = \cos x + \frac{1}{2}x$.

EXAMPLE 5 Investigate the behavior of $f(x) = xe^x$ and sketch its graph.

Solution As usual, we solve for the transition points and determine the signs:

$$f'(x) = xe^x + e^x = (x + 1)e^x, \quad \text{so } f'(x) = 0 \quad \Rightarrow \quad x = -1$$

$$f''(x) = (x + 1)e^x + e^x = (x + 2)e^x, \quad \text{so } f''(x) = 0 \quad \Rightarrow \quad x = -2$$

Interval	Test value	Sign of f'
$(-\infty, -1)$	$f'(-2) = -e^{-2}$	$-$
$(-1, \infty)$	$f'(0) = e^0$	$+$

Interval	Test value	Sign of f''
$(-\infty, -2)$	$f''(-3) = -e^{-3}$	$-$
$(-2, \infty)$	$f''(0) = 2e^0$	$+$

The sign change of f' shows that $f(-1)$ is a local min. The sign change of f'' shows that f has a point of inflection at $x = -2$, where the graph changes from concave down to concave up.

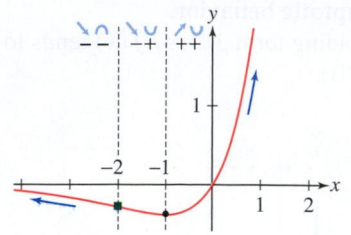

FIGURE 10 Graph of $f(x) = xe^x$. The sign combinations $--$, $-+$, $++$ indicate the signs of f' and f''.

The last pieces of information we need are the limits at infinity. Both x and e^x tend to ∞ as $x \to \infty$, so $\lim\limits_{x \to \infty} xe^x = \infty$. On the other hand, the limit as $x \to -\infty$ is indeterminate of type $\infty \cdot 0$ because x tends to $-\infty$ and e^x tends to zero. Therefore, we write $xe^x = x/e^{-x}$ and apply L'Hôpital's Rule:

$$\lim_{x \to -\infty} xe^x = \lim_{x \to -\infty} \frac{x}{e^{-x}} = \lim_{x \to -\infty} \frac{1}{-e^{-x}} = -\lim_{x \to -\infty} e^x = 0$$

Figure 10 shows the graph with its local minimum and point of inflection, drawn with the correct concavity and asymptotic behavior. ∎

EXAMPLE 6 Investigate the behavior of $f(x) = \dfrac{3x + 2}{2x - 4}$ and sketch its graph.

Solution The function f is not defined for all x. This plays a role in our analysis so we add a Step 0 to our procedure.

Step 0. **Determine the domain of f.**
Since $f(x)$ is not defined for $x = 2$, the domain of f consists of the two intervals $(-\infty, 2)$ and $(2, \infty)$. We must analyze f on these intervals separately.

Step 1. **Determine the signs of f' and f''.**
Calculation shows that

$$f'(x) = -\frac{4}{(x-2)^2}, \qquad f''(x) = \frac{8}{(x-2)^3}$$

Although $f'(x)$ is not defined at $x = 2$, it is not a critical point because $x = 2$ is not in the domain of f. In fact, $f'(x)$ is negative for $x \neq 2$, so f is decreasing and has no critical points.

On the other hand, $f''(x) > 0$ for $x > 2$ and $f''(x) < 0$ for $x < 2$, so the concavity of f changes at $x = 2$. However, there is not an inflection point at $x = 2$ because—as was the case above—$x = 2$ is not in the domain of f.

Step 2. **Note transition points and sign combinations.**
There are no transition points in the domain of f.

$(-\infty, 2)$ $f'(x) < 0$ and $f''(x) < 0$
$(2, \infty)$ $f'(x) < 0$ and $f''(x) > 0$

Step 3. **Draw arcs of appropriate shape and asymptotic behavior.**
The following limits as $x \to \pm\infty$, evaluated using L'Hôpital's Rule, show that $y = \frac{3}{2}$ is a horizontal asymptote:

$$\lim_{x \to \pm\infty} \frac{3x + 2}{2x - 4} = \lim_{x \to \pm\infty} \frac{3}{2} = \frac{3}{2}$$

The line $x = 2$ is a vertical asymptote because $f(x)$ has infinite one-sided limits

$$\lim_{x \to 2^-} \frac{3x + 2}{2x - 4} = -\infty, \qquad \lim_{x \to 2^+} \frac{3x + 2}{2x - 4} = \infty$$

To verify this, note that for x near 2, the numerator $3x + 2$ is positive while the denominator $2x - 4$ is small and negative for $x < 2$ and is small and positive for $x > 2$. Figure 11(A) summarizes the asymptotic behavior.

Now, to the left of $x = 2$, the graph is decreasing [$f'(x) < 0$], is concave down [$f''(x) < 0$], and approaches the asymptotes. The x-intercept is $x = -\frac{2}{3}$ because $f\left(-\frac{2}{3}\right) = 0$, and the y-intercept is $y = f(0) = -\frac{1}{2}$. We obtain the left part of the graph as shown in Figure 11(B). To the right of $x = 2$, the graph is decreasing [$f'(x) < 0$], is concave up [$f''(x) > 0$], and approaches the asymptotes as shown. ∎

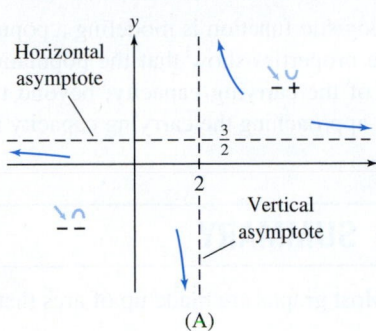

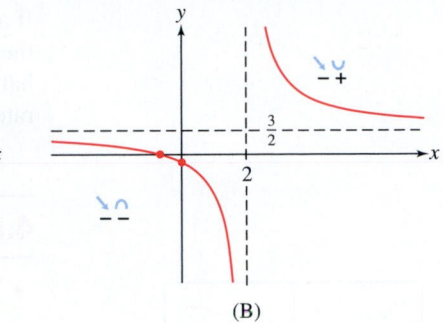

FIGURE 11 Graph of $y = \dfrac{3x+2}{2x-4}$.

(A)   (B)

EXAMPLE 7   A Logistic Function   Analyze the behavior of $P(x) = \dfrac{50}{1+2e^{-0.1x}}$ and sketch the graph.

Solution The function P is defined for all x. With some careful calculation and simplification, we find that

$$P'(x) = \frac{10e^{-0.1x}}{(1+2e^{-0.1x})^2}, \qquad P''(x) = \frac{e^{-0.1x}(2e^{-0.1x}-1)}{(1+2e^{-0.1x})^3}$$

First, note that $P'(x)$ is defined and positive for all x; therefore P is increasing for all x.

The sign of $P''(x)$ is equal to the sign of $2e^{-0.1x} - 1$ because the denominator and the other factor in the numerator are positive for all x. It follows that $P''(x) = 0$ when $2e^{-0.1x} - 1 = 0$. Solving for x:

$$2e^{-0.1x} = 1$$

$$e^{-0.1x} = \frac{1}{2}$$

$$-0.1x = \ln\frac{1}{2}$$

$$x = -10\ln\frac{1}{2} = 10\ln 2$$

Thus, $P''(x) = 0$ at $x = 10\ln 2 \approx 6.93$. Furthermore, $P''(x)$ is positive to the left of $10\ln 2$ and is negative to the right. Therefore there is an inflection point at $x = 10\ln 2$. Figure 12 summarizes the sign information.

The lines $P = 0$ and $P = 50$ are horizontal asymptotes because

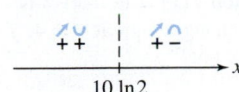

FIGURE 12 Signs of P' and P''.

$$\lim_{x\to\infty} \frac{50}{1+2e^{-0.1x}} = 50 \qquad \text{and} \qquad \lim_{x\to-\infty} \frac{50}{1+2e^{-0.1x}} = 0$$

The graph of P increases away from the asymptote $P = 0$ and is concave up until reaching $x = 10\ln 2 \approx 6.93$. From that point on, it continues to increase but is concave down and approaches the asymptote $P = 50$. Note that $P(10\ln 2) = 25$, so that at the inflection point, we are at half of the limiting value 50. The graph is sketched in Figure 13. ■

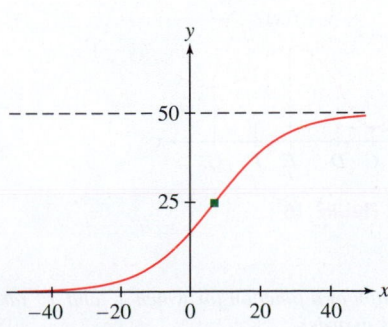

Properties that we observed for P in the previous example hold for general logistic functions $P(x) = \dfrac{M}{1+Ae^{-kx}}$ for M, A, and k all positive. In particular (see Exercise 73):

- $\displaystyle\lim_{x\to-\infty} P(x) = 0$ and $\displaystyle\lim_{x\to\infty} P(x) = M$, so P has horizontal asymptotes at $P = 0$ and $P = M$.
- P is increasing for all x.

FIGURE 13 The graph of the logistic function $P(x)$.

- There is an inflection point at $\left(\dfrac{\ln A}{k}, \dfrac{M}{2}\right)$, and P is concave up to the left of the inflection point, concave down to the right.

If a logistic function is modeling a population, such as in Example 2 in Section 2.7, then these properties show that the population increases at an increasing rate until it equals half of the carrying capacity; beyond that, it continues to increase but at a decreasing rate, approaching the carrying capacity in the long run.

4.6 SUMMARY

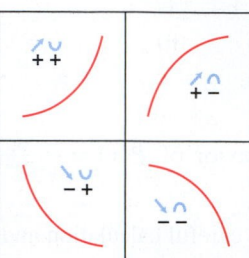

FIGURE 14 The four basic shapes.

- Most graphs are made up of arcs that have one of the four basic shapes (Figure 14):

Sign combination		Curve type
++	$f' > 0$, $f'' > 0$	Increasing and concave up
+−	$f' > 0$, $f'' < 0$	Increasing and concave down
−+	$f' < 0$, $f'' > 0$	Decreasing and concave up
−−	$f' < 0$, $f'' < 0$	Decreasing and concave down

- A *transition point* is a point in the domain of f at which either f' changes sign (local min or max) or f'' changes sign (point of inflection).
- It is convenient to break up the curve-sketching process into steps:

Step 0. Determine the domain of f.

Step 1. Determine the signs of f' and f''.

Step 2. Note transition points and sign combinations.

Step 3. Determine the asymptotic behavior of $f(x)$.

Step 4. Draw arcs of appropriate shape and asymptotic behavior.

4.6 EXERCISES

Preliminary Questions

1. Sketch an arc where f' and f'' have the sign combination ++. Do the same for −+.

2. If the sign combination of f' and f'' changes from ++ to +− at $x = c$, then (choose the correct answer)

(a) $f(c)$ is a local min.

(b) $f(c)$ is a local max.

(c) $(c, f(c))$ is a point of inflection.

3. The second derivative of the function $f(x) = (x - 4)^{-1}$ is $f''(x) = 2(x - 4)^{-3}$. Although $f''(x)$ changes sign at $x = 4$, f does not have a point of inflection at $x = 4$. Why not?

Exercises

1. Determine the sign combinations of f' and f'' for each interval A–G in Figure 15.

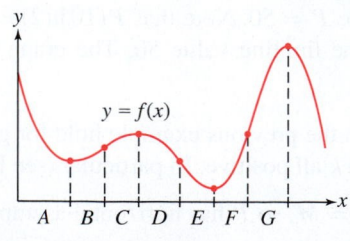

FIGURE 15

2. State the sign change at each transition point A–G in Figure 16. Example: $f'(x)$ goes from + to − at A.

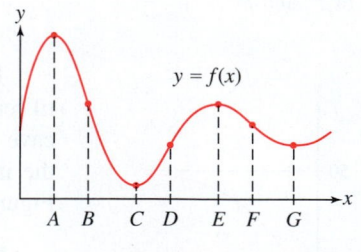

FIGURE 16

In Exercises 3–6, draw the graph of a function for which f' and f'' take on the given sign combinations in order.

3. ++, +−, −−

4. +−, −−, −+

5. −+, −−, −+

6. −+, ++, +−

7. Sketch the graph of a function that could have the graphs of f' and f'' appearing in Figure 17.

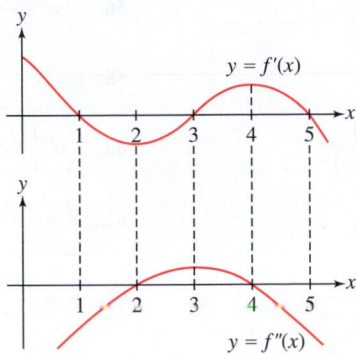

FIGURE 17

8. Sketch the graph of a function that could have the graphs of f' and f'' appearing in Figure 18.

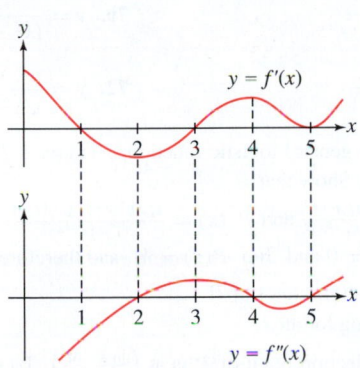

FIGURE 18

9. Investigate the behavior and sketch the graph of $y = x^2 - 5x + 4$.

10. Investigate the behavior and sketch the graph of $y = 12 - 5x - 2x^2$.

11. Investigate the behavior and sketch the graph of $f(x) = x^3 - 3x^2 + 2$. Include the zeros of f, which are $x = 1$ and $1 \pm \sqrt{3}$ (approximately $-0.73, 2.73$).

12. Show that $f(x) = x^3 - 3x^2 + 6x$ has a point of inflection but no local extreme values. Sketch the graph.

13. Extend the sketch of the graph of $f(x) = \cos x + \frac{1}{2}x$ in Example 4 to the interval $[0, 5\pi]$.

14. Investigate the behavior and sketch the graphs of $y = x^{2/3}$ and $y = x^{4/3}$.

In Exercises 15–36, find the transition points, intervals of increase/decrease, concavity, and asymptotic behavior. Then sketch the graph, with this information indicated.

15. $y = x^3 + 24x^2$

16. $y = x^3 - 3x + 5$

17. $y = x^2 - 4x^3$

18. $y = \frac{1}{3}x^3 + x^2 + 3x$

19. $y = 4 - 2x^2 + \frac{1}{6}x^4$

20. $y = 7x^4 - 6x^2 + 1$

21. $y = x^5 + 5x$

22. $y = x^5 - 15x^3$

23. $y = x^4 - 3x^3 + 4x$

24. $y = x^2(x - 4)^2$

25. $y = x^7 - 14x^6$

26. $y = x^6 - 9x^4$

27. $y = x - 4\sqrt{x}$

28. $y = \sqrt{x} + \sqrt{16 - x}$

29. $y = x(8 - x)^{1/3}$

30. $y = (x^2 - 4x)^{1/3}$

31. $y = xe^{-x^2}$

32. $y = (2x^2 - 1)e^{-x^2}$

33. $y = x - 2\ln x$

34. $y = x(4 - x) - 3\ln x$

35. $y = x^2 - 2\ln x$

36. $y = x - 2\ln(x^2 + 1)$

37. Investigate the behavior and sketch the graph of the function $f(x) = 18(x - 3)(x - 1)^{2/3}$ using the formulas

$$f'(x) = \frac{30\left(x - \frac{9}{5}\right)}{(x - 1)^{1/3}}, \qquad f''(x) = \frac{20\left(x - \frac{3}{5}\right)}{(x - 1)^{4/3}}$$

38. Investigate the behavior and sketch the graph of $f(x) = \dfrac{x}{x^2 + 1}$ using the formulas

$$f'(x) = \frac{1 - x^2}{(1 + x^2)^2}, \qquad f''(x) = \frac{2x(x^2 - 3)}{(x^2 + 1)^3}$$

CAS *In Exercises 39–42, sketch the graph of the function, indicating all transition points. If necessary, use a graphing utility or computer algebra system to locate the transition points numerically.*

39. $y = x^2 - 10\ln(x^2 + 1)$

40. $y = e^{-x/2}\ln x$

41. $y = x^4 - 4x^2 + x + 1$

42. $y = 2\sqrt{x} - \sin x, \quad 0 \le x \le 2\pi$

In Exercises 43–48, sketch the graph over the given interval, with all transition points indicated.

43. $y = x + \sin x, \quad [0, 2\pi]$

44. $y = \sin x + \cos x, \quad [0, 2\pi]$

45. $y = 2\sin x - \cos^2 x, \quad [0, 2\pi]$

46. $y = \sin x + \frac{1}{2}x, \quad [0, 2\pi]$

47. $y = \sin x + \sqrt{3}\cos x, \quad [0, \pi]$

48. $y = \sin x - \frac{1}{2}\sin 2x, \quad [0, \pi]$

49. ✏ Are all sign transitions possible? Explain with a sketch why the transitions $++ \to -+$ and $-- \to +-$ do not occur if the function is differentiable. (See Exercise 80 for a proof.)

50. Suppose that f is twice differentiable satisfying (i) $f(0) = 1$, (ii) $f'(x) > 0$ for all $x \ne 0$, and (iii) $f''(x) < 0$ for $x < 0$ and $f''(x) > 0$ for $x > 0$. Let $g(x) = f(x^2)$.

(a) Sketch a possible graph of f.

(b) Prove that g has no points of inflection and a unique local extreme value at $x = 0$. Sketch a possible graph of g.

In Exercises 51–52, draw the graph of a function f having the given limits at $\pm\infty$ and for which f' and f'' take on the given sign combinations in order.

51. $\displaystyle\lim_{x \to -\infty} f(x) = -\infty, \ \lim_{x \to \infty} f(x) = 0; \quad +-, \quad --, \quad -+, \quad ++, \quad +-$

52. $\displaystyle\lim_{x \to -\infty} f(x) = -1, \ \lim_{x \to \infty} f(x) = 1; \quad ++, \quad +-, \quad --, \quad -+$

53. Match the graphs in Figure 19 with the two functions $y = \dfrac{3x}{x^2 - 1}$ and $y = \dfrac{3x^2}{x^2 - 1}$. Explain.

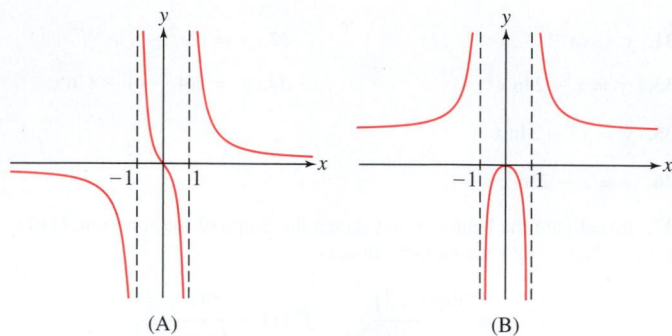

(A) (B)

FIGURE 19

54. Match the functions below with their graphs in Figure 20.

(a) $y = \dfrac{1}{x^2 - 1}$ **(b)** $y = \dfrac{x^2}{x^2 + 1}$

(c) $y = \dfrac{1}{x^2 + 1}$ **(d)** $y = \dfrac{x}{x^2 - 1}$

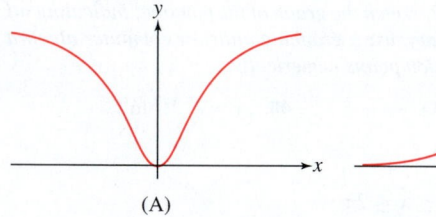 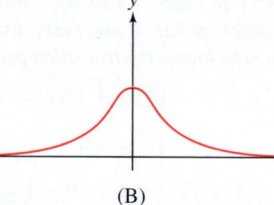

(A) (B)

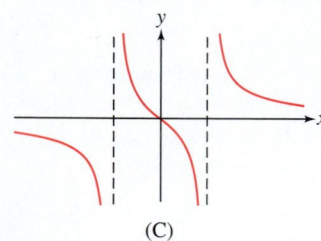 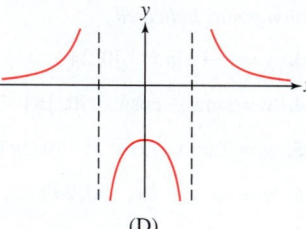

(C) (D)

FIGURE 20

In Exercises 55–72, sketch the graph of the function. Indicate the transition points and asymptotes.

55. $y = \dfrac{1}{3x - 1}$ **56.** $y = \dfrac{x - 2}{x - 3}$

57. $y = \dfrac{x + 3}{x - 2}$ **58.** $y = x + \dfrac{1}{x}$

59. $y = \dfrac{1}{x} + \dfrac{1}{x - 1}$ **60.** $y = \dfrac{1}{x} - \dfrac{1}{x - 1}$

61. $y = \dfrac{1}{x(x - 2)}$ **62.** $y = \dfrac{x}{x^2 - 9}$

63. $y = \dfrac{1}{x^2 - 6x + 8}$ **64.** $y = \dfrac{x^3 + 1}{x}$

65. $y = 1 - \dfrac{3}{x} + \dfrac{4}{x^3}$ **66.** $y = \dfrac{1}{x^2} + \dfrac{1}{(x - 2)^2}$

67. $y = \dfrac{1}{x^2} - \dfrac{1}{(x - 2)^2}$ **68.** $y = \dfrac{4}{x^2 - 9}$

69. $y = \dfrac{1}{(x^2 + 1)^2}$ **70.** $y = \dfrac{x^2}{(x^2 - 1)(x^2 + 1)}$

71. $y = \dfrac{1}{\sqrt{x^2 + 1}}$ **72.** $y = \dfrac{x}{\sqrt{x^2 + 1}}$

73. Consider the general logistic function, $P(x) = \dfrac{M}{1 + Ae^{-kx}}$, with A, M, and k all positive. Show that

(a) $P'(x) = \dfrac{MAke^{-kx}}{(1 + Ae^{-kx})^2}$ and $P''(x) = \dfrac{MAk^2e^{-kx}(Ae^{-kx} - 1)}{(1 + Ae^{-kx})^3}$

(b) $\lim\limits_{x \to -\infty} P(x) = 0$ and $\lim\limits_{x \to \infty} P(x) = M$, and therefore $P = 0$ and $P = M$ are horizontal asymptotes of P.

(c) P is increasing for all x.

(d) The only inflection point of P is at $\left(\dfrac{\ln A}{k}, \dfrac{M}{2} \right)$. To the left of it P is concave up, and to the right of it P is concave down.

74. Show that the function $R(x) = \dfrac{M}{\pi}\left(\dfrac{\pi}{2} + \tan^{-1} x \right)$, with $M > 0$, has the following properties (similar to the general logistic function):

(a) $\lim\limits_{x \to -\infty} R(x) = 0$ and $\lim\limits_{x \to \infty} R(x) = M$, and therefore $R = 0$ and $R = M$ are horizontal asymptotes of R.

(b) R is increasing for all x.

(c) R has a single inflection point. The value of R at the inflection point is $M/2$. To the left of the inflection point R is concave up, to the right R is concave down.

Further Insights and Challenges

*In Exercises 75–79, we explore functions whose graphs approach a non-horizontal line as $x \to \infty$. A line $y = ax + b$ is called a **slant asymptote** if*

$$\lim_{x \to \infty} (f(x) - (ax + b)) = 0$$

or

$$\lim_{x \to -\infty} (f(x) - (ax + b)) = 0$$

75. Let $f(x) = \dfrac{x^2}{x - 1}$ (Figure 21). Verify the following:

(a) $f(0)$ is a local max and $f(2)$ a local min.

(b) f is concave down on $(-\infty, 1)$ and concave up on $(1, \infty)$.

(c) $\lim\limits_{x \to 1-} f(x) = -\infty$ and $\lim\limits_{x \to 1+} f(x) = \infty$.

(d) $y = x + 1$ is a slant asymptote of f as $x \to \pm\infty$.

(e) The slant asymptote lies above the graph of f for $x < 1$ and below the graph for $x > 1$.

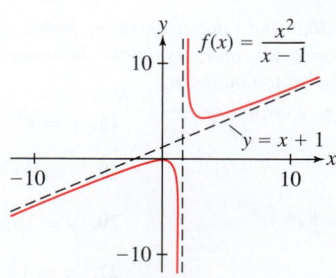

FIGURE 21

76. ☑ If $f(x) = P(x)/Q(x)$, where P and Q are polynomials of degrees $m + 1$ and m, then by long division, we can write

$$f(x) = (ax + b) + P_1(x)/Q(x)$$

where P_1 is a polynomial of degree $< m$. Show that $y = ax + b$ is the slant asymptote of $f(x)$. Use this procedure to find the slant asymptotes of the following functions:

(a) $y = \dfrac{x^2}{x + 2}$ **(b)** $y = \dfrac{x^3 + x}{x^2 + x + 1}$

77. Sketch the graph of

$$f(x) = \frac{x^2}{x + 1}$$

Proceed as in the previous exercise to find the slant asymptote.

78. Show that $y = 3x$ is a slant asymptote for $f(x) = 3x + x^{-2}$. Determine whether $f(x)$ approaches the slant asymptote from above or below, and make a sketch of the graph.

79. Sketch the graph of $f(x) = \dfrac{1 - x^2}{2 - x}$.

80. Assume that f' and f'' exist for all x and let c be a critical point of f. Show that $f(x)$ cannot make a transition from $++$ to $-+$ at $x = c$. *Hint:* Apply the MVT to $f'(x)$.

81. ☑ Assume that f'' exists and $f''(x) > 0$ for all x. Show that $f(x)$ cannot be negative for all x. *Hint:* Show that $f'(b) \neq 0$ for some b and use the result of Exercise 74 in Section 4.4.

4.7 Applied Optimization

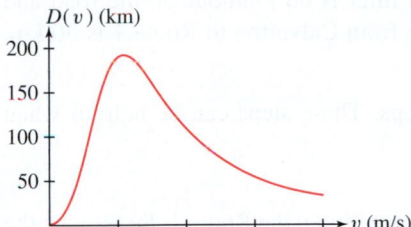

FIGURE 1 Physiology and aerodynamics are applied to obtain a plausible formula for bird migration distance D as a function of velocity v. The optimal velocity corresponds to the maximum point on the graph (see Exercise 69).

Optimization plays a role in a wide range of disciplines, including the physical sciences, economics, and biology. For example, scientists have studied how migrating birds choose an optimal velocity v that maximizes the distance D they can travel without stopping, given the energy that can be stored as body fat (Figure 1).

In many optimization problems, the first step is to write down the **objective function**. This is the function whose minimum or maximum we seek. Once we find the objective function, we can apply the techniques developed in this chapter. Our first examples require optimization on a closed interval $[a, b]$. Let's recall the steps for finding extrema developed in Section 4.2:

(i) Find the critical points of f in $[a, b]$.

(ii) Evaluate $f(x)$ at the critical points and the endpoints a and b.

(iii) The least and greatest values are the extreme values of f on $[a, b]$.

EXAMPLE 1 A piece of wire of length L is bent into the shape of a rectangle (Figure 2). Which dimensions produce the rectangle of maximum area?

FIGURE 2

An equation relating two or more independent variables in an optimization problem is called a constraint equation. The idea is that we cannot assume the variables take on any values we want; instead they are constrained to satisfy a specific equation. In Example 1, the constraint equation is

$$2x + 2y = L$$

Solution The rectangle has area $A = xy$, where x and y are the lengths of the sides. Since A depends on two variables x and y, we cannot find the maximum until we eliminate one of the variables. We can do this because the variables are related: The rectangle has perimeter $L = 2x + 2y$, so $y = \frac{1}{2}L - x$. This allows us to rewrite the area in terms of x alone to obtain the objective function

$$A(x) = x\left(\frac{1}{2}L - x\right) = \frac{1}{2}Lx - x^2$$

On which interval does the optimization take place? The sides of the rectangle are non-negative, so we require both $x \geq 0$ and $\frac{1}{2}L - x \geq 0$. Thus, $0 \leq x \leq \frac{1}{2}L$. Our problem is to maximize $A(x)$ on the closed interval $\left[0, \frac{1}{2}L\right]$.

We have $A'(x) = \frac{1}{2}L - 2x$. Solving $A'(x) = 0$, we obtain just a single critical point, $x = \frac{1}{4}L$. Comparing values of A, we find:

Endpoints: $\qquad A(0) = 0$

$$A\left(\frac{1}{2}L\right) = \frac{1}{2}L\left(\frac{1}{2}L - \frac{1}{2}L\right) = 0$$

Critical point: $\quad A\left(\frac{1}{4}L\right) = \left(\frac{1}{4}L\right)\left(\frac{1}{2}L - \frac{1}{4}L\right) = \frac{1}{16}L^2$

The greatest value occurs for $x = \frac{1}{4}L$, and in this case, $y = \frac{1}{2}L - \frac{1}{4}L = \frac{1}{4}L$. The rectangle of maximum area is the square of sides $x = y = \frac{1}{4}L$. ∎

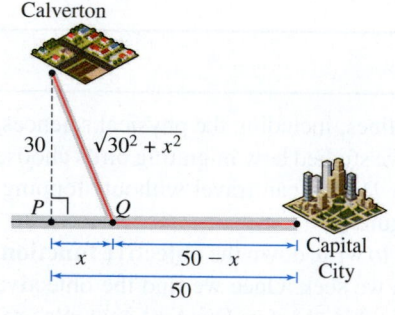

Calverton

30 $\sqrt{30^2 + x^2}$

P Q

x $50 - x$ Capital City

50

FIGURE 3

EXAMPLE 2 **Minimizing Travel Time** Your task is to build a road joining the small town of Calverton to Route 1 to enable drivers to reach Capital City in the shortest time (Figure 3). How should this be done if the speed limit is 60 km/hour on the road and 110 km/h on Route 1? The perpendicular distance from Calverton to Route 1 is 30 km, and Capital City is 50 km down Route 1.

Solution We will solve this problem in three steps. These steps can be helpful when solving other optimization problems.

Step 1. **Choose variables.**
We need to determine the point Q where the road will join the Route 1. So let x be the distance from Q to the point P where the perpendicular joins Route 1.

Step 2. **Find the objective function and the interval.**
Our objective function is the time $T(x)$ of the trip as a function of x. To find a formula for $T(x)$, recall that distance traveled at constant velocity v is $d = vt$, and the *time* required to travel a distance d is $t = d/v$. The road has length $\sqrt{30^2 + x^2}$ by the Pythagorean Theorem, so at velocity $v = 60$ km/h, it takes

$$\frac{\sqrt{30^2 + x^2}}{60} \text{ hours to travel from Calverton to } Q$$

The segment of Route 1 from Q to Capital City has length $50 - x$. At velocity $v = 110$ km/h, it takes

$$\frac{50 - x}{110} \text{ h to travel from } Q \text{ to the city}$$

The total number of hours for the trip is

$$T(x) = \frac{\sqrt{30^2 + x^2}}{60} + \frac{50 - x}{110}$$

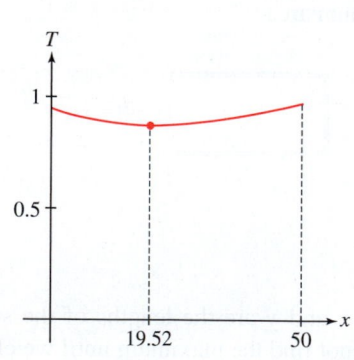

FIGURE 4 Graph of time of trip as function of x.

Our interval is $0 \le x \le 50$ because the road joins Route 1 somewhere between P and Capital City. So our task is to minimize T on $[0, 50]$ (Figure 4).

Step 3. **Optimize.**
Solve for the critical points:

$$T'(x) = \frac{x}{60\sqrt{30^2 + x^2}} - \frac{1}{110} = 0$$

$$110x = 60\sqrt{30^2 + x^2} \quad \Rightarrow \quad 11x = 6\sqrt{30^2 + x^2} \quad \Rightarrow$$

$$121x^2 = 36(30^2 + x^2) \quad \Rightarrow \quad 85x^2 = 32{,}400 \quad \Rightarrow \quad x = \sqrt{32{,}400/85} \approx 19.52$$

To find the minimum value of T, we compare the values of $T(x)$ at the critical point and the endpoints of $[0, 50]$:

$$T(0) \approx 0.95 \text{ h}, \qquad T(19.52) \approx 0.87 \text{ h}, \qquad T(50) \approx 0.97 \text{ h}$$

We conclude that the travel time is minimized if the road joins Route 1 at a distance $x \approx 19.52$ km along the highway from P. ■

EXAMPLE 3   Old Route 1 and Minimizing Travel Time   We revisit the situation in Example 2, considering a different pair of speeds along the road and Route 1. Suppose that Route 1 is old and in disrepair and we cannot expect to travel faster than 70 km/hour on it. Furthermore, assume that the new road will be designed for travel at 80 km/h. Now, how should the road be laid out in relation to Route 1?

Solution   Intuitively it seems clear that we should have the road go straight from Calverton to Capital City, completely avoiding Route 1. We will work out the solution and see how this case compares with the previous example. Taking the same approach used in the previous example, we find that the task is to determine the minimum of

$$T(x) = \frac{\sqrt{900 + x^2}}{80} + \frac{50 - x}{70}$$

over the interval $[0, 50]$.

In this case, T has no critical points (see Exercise 19). Thus, the minimum of T must occur at one of the endpoints (Figure 5). We have $T(0) \approx 1.09$ h, and $T(50) \approx 0.73$ h. So the minimum of T over $[0, 50]$ occurs at $x = 50$. Therefore, to minimize the time of the trip, the road should go directly from Calverton to Capital City, confirming our initial intuitive analysis. ■

In Example 2, the minimum occurred at an x between 0 and 50, and in Example 3, it occurred at $x = 50$. It is natural to ask whether there is a combination of speeds along the road and Route 1 so that the minimum occurs at $x = 0$? The answer is no (see Exercise 19). By choosing the Route 1 speed large enough in relation to the road speed, it is possible to have the minimum of T occur at a critical point as close as you like to $x = 0$, but there is no combination of speeds that results in a minimum at exactly $x = 0$.

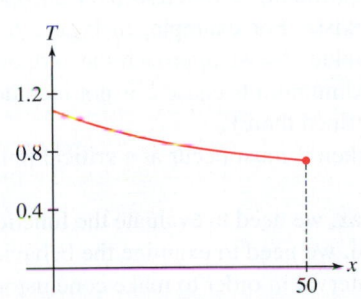

FIGURE 5 In this case $T(x)$ has no critical points in $[0, 50]$.

EXAMPLE 4   Optimal Price   All units in a 30-unit apartment building are rented out when the rent is set at $r = \$2000$ per month. A survey reveals that for each \$100 increase in rent, demand for apartments will decrease, such that one additional apartment becomes vacant. Suppose that each occupied unit costs \$200 per month in maintenance. Which rent r maximizes monthly profit?

Solution

Step 1.   Choose variables.
Our goal is to maximize the total monthly profit P. Let r be the monthly rent and let $N(r)$ be the number of occupied units when the rent is set at r.

Step 2.   Find the objective function and the interval.
Since one unit becomes vacant with each \$100 increase in rent above \$2000, we find that $(r - 2000)/100$ units are vacant when $r > 2000$. Therefore,

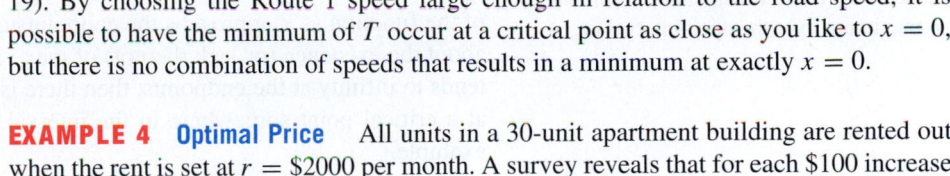

$$N(r) = 30 - \frac{1}{100}(r - 2000) = 50 - \frac{1}{100}r$$

Total monthly profit is equal to the number of occupied units times the profit per unit, which is $r - 200$ (because each unit costs \$200 in maintenance), so

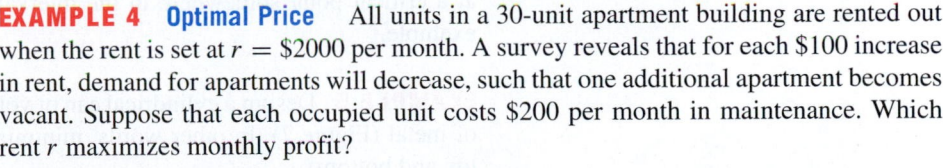

$$P(r) = N(r)(r - 200) = \left(50 - \frac{1}{100}r\right)(r - 200) = -10{,}000 + 52r - \frac{1}{100}r^2$$

Which interval of r-values should we consider? There is no reason to lower the rent below $r = 2000$ because all units are already occupied when $r = 2000$. On the other hand, for the upper limit of r we take the rent at which no units are occupied; that is, the r for which $N(r) = 0$. That occurs at $r = 100 \cdot 50 = 5000$. Therefore, we consider $P(r)$ over the interval $2000 \leq r \leq 5000$.

Step 3. Optimize.

Solve for the critical points:

$$P'(r) = 52 - \frac{1}{50}r \quad \text{so } P'(r) = 0 \quad \Rightarrow \quad r = 2600$$

and compare values at the critical point and the endpoints:

$$P(2000) = 54{,}000, \qquad P(2600) = 62{,}400, \qquad P(5000) = 0$$

We conclude that the profit is maximized when the rent is set at $r = \$2600$. In this case, 24 units are occupied. Note that if the maximum profit had occurred at a price that gave us a fractional number of units occupied, we could not have achieved that maximum. Instead, we would have taken the price corresponding to rounding the fractional number up or down to the integer number of units that maximized our profit.

■

Open Versus Closed Intervals

In contrast to the case of a closed interval, when optimizing a function over an open interval, there is no guarantee that a min or max exists. For example, in Figure 6, a minimum exists at $x = c$ but there is no maximum value. As we approach the endpoint at b, the function values increase, but there is no maximumum because b is not included in the interval (and furthermore the function is not defined there).

If a min or max does exist on an open interval, then it must occur at a critical point (because it is also a local min or max).

With a closed interval, to search for a min and max, we need to evaluate the function at the endpoints of the interval. With an open interval, we need to examine the behavior of the function as x approachs the endpoints of the interval in order to make conclusions about the existence (or lack thereof) of max values and min values. For example, if $f(x)$ tends to infinity at the endpoints, then there is no maximum, and a minimum must occur at a critical point somewhere in the interval. We consider such a situation in the next example.

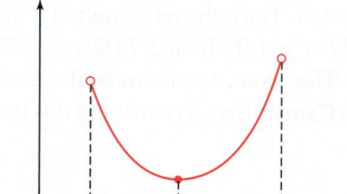

FIGURE 6 A function with a minimum but no maximum over the open interval (a, b).

EXAMPLE 5 Design a cylindrical can of volume 900 cm^3 so that it uses the least amount of metal (Figure 7). In other words, minimize the surface area of the can (including its top and bottom).

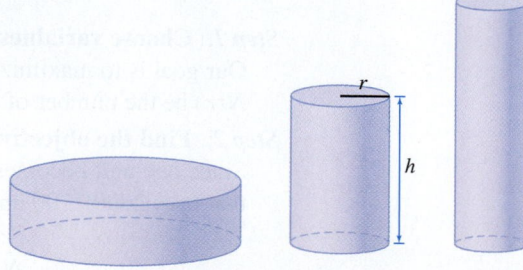

DF FIGURE 7 Cylinders with the same volume but different surface areas.

Solution

Step 1. Choose variables.

We want to find the radius and the height of the can with minimum surface area. Therefore, we let r be the radius and h the height. Furthermore, we denote the surface area of the can by A.

Step 2. **Find the objective function and the interval.**

We express A as a function of r and h:

$$A = \underbrace{\pi r^2}_{\text{Top}} + \underbrace{\pi r^2}_{\text{Bottom}} + \underbrace{2\pi rh}_{\text{Side}} = 2\pi r^2 + 2\pi rh$$

The can's volume is $V = \pi r^2 h$. Since we require that $V = 900$ cm³, we have the constraint equation $\pi r^2 h = 900$. Thus, $h = (900/\pi)r^{-2}$ and

$$A(r) = 2\pi r^2 + 2\pi r \left(\frac{900}{\pi r^2}\right) = 2\pi r^2 + \frac{1800}{r}$$

The radius r can take on any positive value, so we minimize $A(r)$ on $(0, \infty)$.

Step 3. **Optimize the function.**

Observe that $A(r)$ tends to infinity as r approaches the endpoints of $(0, \infty)$:

- $A(r) \to \infty$ as $r \to \infty$ (because of the r^2 term).
- $A(r) \to \infty$ as $r \to 0$ (because of the $1/r$ term).

Therefore, $A(r)$ must take on a minimum value at a critical point in $(0, \infty)$ (Figure 8). We solve in the usual way:

$$\frac{dA}{dr} = 4\pi r - \frac{1800}{r^2} = 0 \quad \Rightarrow \quad r^3 = \frac{450}{\pi} \quad \Rightarrow \quad r = \left(\frac{450}{\pi}\right)^{1/3} \approx 5.23 \text{ cm}$$

We also need to calculate the height:

$$h = \frac{900}{\pi r^2} = 2\left(\frac{450}{\pi}\right)r^{-2} = 2\left(\frac{450}{\pi}\right)\left(\frac{450}{\pi}\right)^{-2/3} = 2\left(\frac{450}{\pi}\right)^{1/3} \approx 10.46 \text{ cm}$$

Since we have a single critical point in our interval, it follows that we obtain the minimum of A there. Thus, the minimum surface area occurs when a can has radius approximately 5.23 cm and height approximately 10.46 cm. Notice that the optimal dimensions satisfy $h = 2r$. In other words, the optimal can is as tall as it is wide. ∎

EXAMPLE 6 **Optimization Problem with No Solution** Is it possible to design a cylinder of volume 900 cm³ with the largest possible surface area?

Solution The answer is no. In the previous example, we showed that a cylinder of volume 900 cm³ and radius r has surface area

$$A(r) = 2\pi r^2 + \frac{1800}{r}$$

This function has no maximum value because it tends to infinity as $r \to 0$ or $r \to \infty$ (Figure 8). This means that a cylinder of fixed volume has a large surface area if it is either very fat and short (r large) or very tall and skinny (r small). ∎

The **Principle of Least Distance** states that a light beam reflected in a mirror travels along the shortest path. More precisely, a beam traveling from A to B, as in Figure 9, is reflected at the point P for which the path APB has minimum length. In the next example, we show that this minimum occurs when *the angle of incidence is equal to the angle of reflection*, that is, $\theta_1 = \theta_2$.

EXAMPLE 7 Show that if P is the point for which the path APB in Figure 9 has minimal length, then $\theta_1 = \theta_2$.

Solution By the Pythagorean Theorem, the path APB has length

$$f(x) = AP + PB = \sqrt{x^2 + h_1^2} + \sqrt{(L-x)^2 + h_2^2}$$

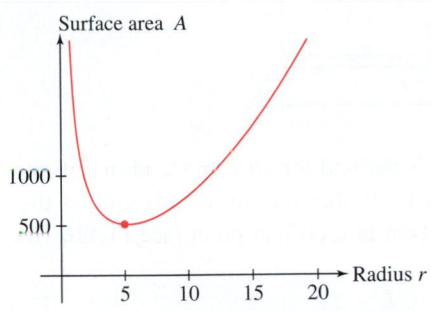

Surface area A

Radius r

FIGURE 8 Surface area increases as r tends to 0 or ∞. The minimum value exists.

In the case of a single critical point, as we have here, a second method for proving that the point corresponds to a minimum is to apply the First Derivative Test. Since $A'(r) < 0$ for $r < \left(\frac{450}{\pi}\right)^{1/3}$ and $A'(r) > 0$ for $r > \left(\frac{450}{\pi}\right)^{1/3}$, the critical point must be a local minimum and as the only critical point, the global minimum. A third method would be to apply the Second Derivative Test to show this is a local minimum and therefore, as the only extreme point, the global minimum.

---

*The Principle of Least Distance is also called **Heron's Principle** after the mathematician Heron of Alexandria (c. 100 CE). See Exercise 81 for an elementary proof that does not use calculus and would have been known to Heron. Exercise 56 develops Snell's Law, a more general optical law based on the Principle of Least Time.*

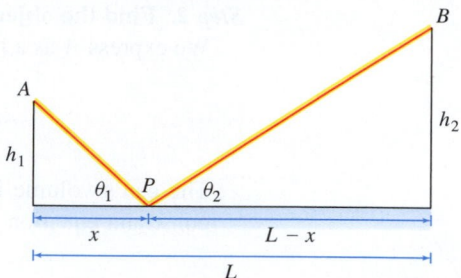

DF FIGURE 9 Reflection of a light beam in a mirror.

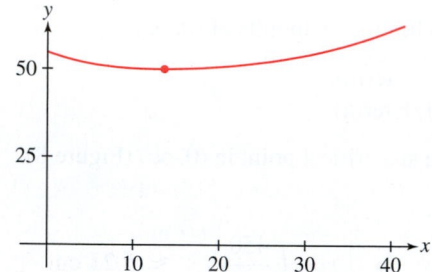

FIGURE 10 Graph of path length for $h_1 = 10$, $h_2 = 20$, $L = 40$.

with x, h_1, and h_2 as in the figure. The function f is defined for *all x* and tends to infinity as x approaches $\pm\infty$ (i.e., as P moves arbitrarily far to the right or left). It follows that f has an absolute minimum value, and it must occur at a critical point (see Figure 10). Taking the derivative:

$$f'(x) = \frac{x}{\sqrt{x^2 + h_1^2}} - \frac{L - x}{\sqrt{(L-x)^2 + h_2^2}} \qquad \boxed{1}$$

Since $f'(x)$ is defined for all x, critical points occur where $f'(x) = 0$. It is not necessary to solve for x because our goal is not to find critical points, but rather to show that $\theta_1 = \theta_2$ at the minimum. To do this, we set the derivative equal to 0 in Eq. (1) and rewrite as

$$\frac{x}{\sqrt{x^2 + h_1^2}} = \frac{L - x}{\sqrt{(L-x)^2 + h_2^2}} \qquad \boxed{2}$$

Note that the critical point x that satisfies Eq. (2) must lie between 0 and L because no $x < 0$ can satisfy this equation (otherwise, we would have a negative value on the left and a positive on the right) and no $x > L$ can satisfy this equation (for similar reasons). Since the critical point x lies in $[0, L]$ we can associate angles θ_1 and θ_2 with x as in Figure 9. We claim that $\theta_1 = \theta_2$. To see this, observe that with θ_1 and θ_2 as pictured, we have

$$\cos\theta_1 = \frac{x}{\sqrt{x^2 + h_1^2}} \quad \text{and} \quad \cos\theta_2 = \frac{L - x}{\sqrt{(L-x)^2 + h_2^2}}$$

Therefore, Eq. (2) implies that $\cos\theta_1 = \cos\theta_2$, and since θ_1 and θ_2 lie between 0 and $\frac{\pi}{2}$, we conclude that $\theta_1 = \theta_2$ as claimed. ∎

CONCEPTUAL INSIGHT Often, a maximum or minimum at a critical point represents the best compromise between "competing factors." In Example 4, we maximized profit by finding the best compromise between raising the rent and keeping the apartment units occupied. In Example 5, our solution minimizes surface area by finding the best compromise between height and radius. In Example 2, the solution represents a compromise between the slower speed on the road that leads to Route 1 and the faster speed along Route 1. On the other hand, in Example 3, since there is no compromise, a solution occurs at an endpoint of the interval rather than at a critical point. The faster speed along the road yields a road straight to the city, avoiding Route 1 altogether.

4.7 SUMMARY

• There are usually three main steps in solving an applied optimization problem:

Step 1. Choose variables.

Determine which quantities are relevant, often by drawing a diagram, and assign appropriate variables.

Step 2. Find the objective function and the interval.

Restate as an optimization problem for a function f over an interval. If f depends on more than one variable, use a *constraint equation* to write f as a function of just one variable.

Step 3. Optimize the objective function.

- If the interval is open, f does not necessarily take on a minimum or maximum value. But if it does, these must occur at critical points within the interval. To determine if a min or max exists, analyze the behavior of f as x approaches the endpoints of the interval.

4.7 EXERCISES

Preliminary Questions

1. The problem is to find the right triangle of perimeter 10 whose area is as large as possible. What is the constraint equation relating the base b and height h of the triangle?

2. Describe a way of showing that a continuous function on an open interval (a, b) has a minimum value.

3. Is there a rectangle of area 100 of largest perimeter? Explain.

Exercises

1. Find the dimensions x and y of the rectangle of maximum area that can be formed using 3 m of wire.

(a) What is the constraint equation relating x and y?

(b) Find a formula for the area in terms of x alone.

(c) What is the interval of optimization? Is it open or closed?

(d) Solve the optimization problem.

2. Wire of length 12 m is divided into two pieces and each piece is bent into a square. How should this be done in order to minimize the sum of the areas of the two squares?

(a) Express the sum of the areas of the squares in terms of the lengths x and y of the two pieces.

(b) What is the constraint equation relating x and y?

(c) What is the interval of optimization? Is it open or closed?

(d) Solve the optimization problem.

3. A rectangular bird sanctuary is being created with one side along a straight riverbank. The remaining three sides are to be enclosed with a protective fence. If there are 12 km of fence available, find the dimension of the rectangle to maximize the area of the sanctuary.

4. The rectangular bird sanctuary with one side along a straight river is to be constructed so that it contains 8 km^2 of area. Find the dimensions of the rectangle to minimize the amount of fence necessary to enclose the remaining three sides.

5. Find two positive real numbers such that the sum of the first number squared and the second number is 48 and their product is a maximum.

6. Find two positive real numbers such that they sum to 108 and the product of the first times the square of the second is a maximum.

7. A wire of length 12 m is divided into two pieces and the pieces are bent into a square and a circle. How should this be done in order to minimize the sum of their areas?

8. Find the positive number x such that the sum of x and its reciprocal is as small as possible. Does this problem require optimization over an open interval or a closed interval?

9. Find two positive real numbers such that they add to 40 and their product is as large as possible.

10. Find two positive real numbers x and y such that they add to 120 and $x^2 y$ is as large as possible.

11. Find two positive real numbers x and y such that their product is 800 and $x + 2y$ is as small as possible.

12. A flexible tube of length 4 m is bent into an L-shape. Where should the bend be made to minimize the distance between the two ends?

13. Find the dimensions of the box with square base with

(a) Volume 12 and the minimal surface area.

(b) Surface area 20 and maximal volume.

14. A jewelry box with a square base is to be built with copper-plated sides, nickel-plated bottom and top, and a volume of 40 cm^3. If nickel plating costs \$2 per cm^2 and copper plating costs \$1 per cm^2, find the dimensions of the box to minimize the cost of the materials.

15. A rancher will use 600 m of fencing to build a corral in the shape of a semicircle on top of a rectangle (Figure 11). Find the dimensions that maximize the area of the corral.

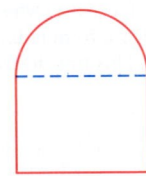

FIGURE 11

16. What is the maximum area of a rectangle inscribed in a right triangle with legs of length 3 and 4 as in Figure 12? The sides of the rectangle are parallel to the legs of the triangle.

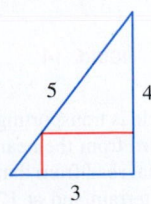

FIGURE 12

17. Find the dimensions of the rectangle of maximum area that can be inscribed in a circle of radius $r = 4$ (Figure 13).

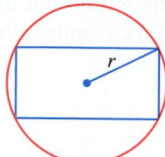

FIGURE 13

18. Find the dimensions x and y of the rectangle inscribed in a circle of radius r that maximizes the quantity xy^2.

19. In the setting of Examples 2 and 3, let r denote the speed along the road, and h denote the speed along the highway.
(a) Show that the travel-time function $T(x)$ has a critical point at

$$x = \frac{30}{\sqrt{(h/r)^2 - 1}}$$

and explain why this indicates that if $r \geq h$ there is no critical point.
(b) Explain why there cannot be a critical point at $x = 0$, but depending on the speeds, the critical point can be arbitrarily close to 0.

20. In the setting of Examples 2 and 3, replace 30 and 50 with general distances D and L, respectively. Also, let r denote the speed along the road, and h denote the speed along the highway. Show that the travel-time function $T(x)$ has a critical point at

$$x = \frac{D}{\sqrt{(h/r)^2 - 1}}.$$

21. In the article "Do Dogs Know Calculus?" the author Timothy Pennings explained how he noticed that when he threw a ball diagonally into Lake Michigan along a straight shoreline, his dog Elvis seemed to pick the optimal point in which to enter the water so as to minimize his time to reach the ball, as in Figure 14. He timed the dog and found Elvis could run at 6.4 m/s on the sand and swim at 0.91 m/s. If Tim stood at point A and threw the ball to a point B in the water, which was a perpendicular distance 10 m from point C on the shore, where C is a distance 15 m from where he stood, at what distance x from point C did Elvis enter the water if the dog effectively minimized his time to reach the ball?

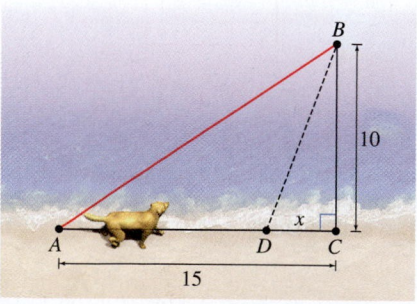

FIGURE 14

22. A four-wheel-drive vehicle is transporting an injured hiker to the hospital from a point that is 30 km from the nearest point on a straight road. The hospital is 50 km down that road from that nearest point. If the vehicle can drive at 30 kph over the terrain and at 120 kph on the road, how far down the road should the vehicle aim to reach the road to minimize the time it takes to reach the hospital?

23. Find the point on the line $y = x$ closest to the point $(1, 0)$. *Hint:* It is equivalent and easier to minimize the *square* of the distance.

24. Find the point P on the parabola $y = x^2$ closest to the point $(3, 0)$ (Figure 15).

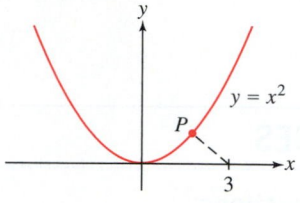

FIGURE 15

25. (CAS) Find a good numerical approximation to the coordinates of the point on the graph of $y = \ln x - x$ closest to the origin (Figure 16).

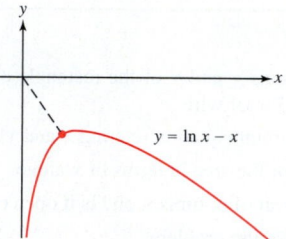

FIGURE 16

26. Problem of Tartaglia (1500–1557) Among all positive numbers a, b whose sum is 8, find those for which the product of the two numbers and their difference is largest.

27. Find the angle θ that maximizes the area of the isosceles triangle whose legs have length ℓ (Figure 17), using the fact the area is given by $A = \frac{1}{2}\ell^2 \sin\theta$.

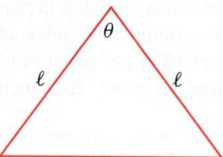

FIGURE 17

28. A right circular cone (Figure 18) has volume

$$V = \frac{\pi}{3}r^2 h$$

and surface area $S = \pi r \sqrt{r^2 + h^2}$. Find the dimensions of the cone with surface area 1 and maximal volume.

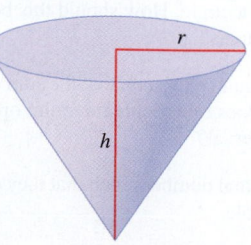

FIGURE 18

29. Find the area of the largest isosceles triangle that can be inscribed in a circle of radius 1 (Figure 19).

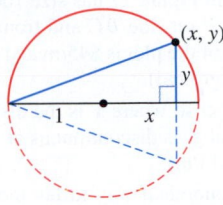

FIGURE 19

30. Find the radius and height of a cylindrical can of total surface area A whose volume is as large as possible. Does there exist a cylinder of surface area A and minimal total volume?

31. A poster of area 6000 cm² has blank margins of width 10 cm on the top and bottom and 6 cm on the sides. Find the dimensions that maximize the printed area.

32. According to postal regulations, a carton is classified as "oversized" if the sum of its height and girth (perimeter of its base) exceeds 108 in. Find the dimensions of a carton with a square base that is not oversized and has maximum volume.

33. Kepler's Wine Barrel Problem In his work *Nova stereometria doliorum vinariorum* (New Solid Geometry of a Wine Barrel), published in 1615, astronomer Johannes Kepler stated and solved the following problem: Find the dimensions of the cylinder of largest volume that can be inscribed in a sphere of radius R. *Hint:* Show that an inscribed cylinder has volume $2\pi x(R^2 - x^2)$, where x is one-half the height of the cylinder.

34. Find the angle θ that maximizes the area of the trapezoid with a base of length 4 and sides of length 2, as in Figure 20.

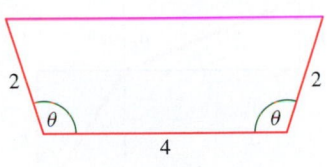

FIGURE 20

35. A landscape architect wishes to enclose a rectangular garden of area 1000 m² on one side by a brick wall costing \$90/m and on the other three sides by a metal fence costing \$30/m. Which dimensions minimize the total cost?

36. The amount of light reaching a point at a distance r from a light source A of intensity I_A is I_A/r^2. Suppose that a second light source B of intensity $I_B = 4I_A$ is located 10 m from A. Find the point on the segment joining A and B where the total amount of light is at a minimum.

37. Find the maximum area of a rectangle inscribed in the region bounded by the graph of $y = \dfrac{4 - x}{2 + x}$ and the axes (Figure 21).

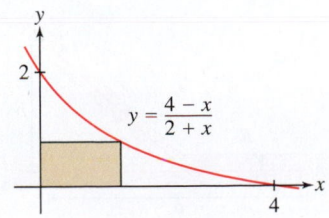

FIGURE 21

38. Find the maximum area of a triangle formed by the axes and a tangent line to the graph of $y = (x + 1)^{-2}$ with $x > 0$.

39. Find the maximum area of a rectangle circumscribed around a rectangle of sides L and H. *Hint:* Express the area in terms of the angle θ (Figure 22).

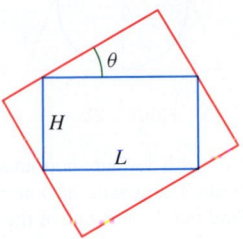

FIGURE 22

40. A contractor is engaged to build steps up the slope of a hill that has the shape of the graph of $y = \dfrac{x^2(120 - x)}{6400}$ for $0 \le x \le 80$ with x in meters (Figure 23). What is the maximum vertical rise of a stair if each stair has a horizontal length of $\frac{1}{3}$ m?

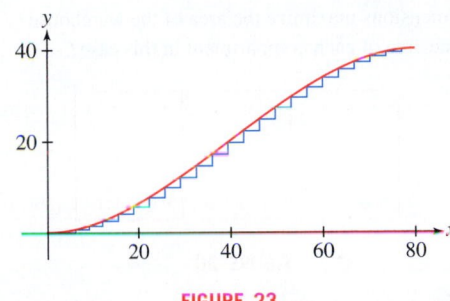

FIGURE 23

41. Find the equation of the line through $P = (4, 12)$ such that the triangle bounded by this line and the axes in the first quadrant has minimal area.

42. Let $P = (a, b)$ lie in the first quadrant. Find the slope of the line through P such that the triangle bounded by this line and the axes in the first quadrant has minimal area. Then show that P is the midpoint of the hypotenuse of this triangle.

43. Archimedes's Problem A spherical cap (Figure 24) of radius r and height h has volume $V = \pi h^2\left(r - \frac{1}{3}h\right)$ and surface area $S = 2\pi rh$. Prove that the hemisphere encloses the largest volume among all spherical caps of fixed surface area S.

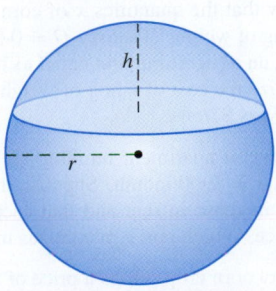

FIGURE 24

44. Find the isosceles triangle of smallest area (Figure 25) that circumscribes a circle of radius 1 (from Thomas Simpson's *The Doctrine and Application of Fluxions*, a calculus text that appeared in 1750).

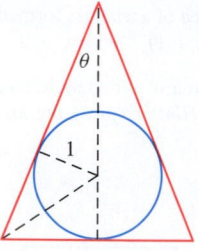

FIGURE 25

45. A box of volume 72 m³ with a square bottom and no top is constructed out of two different materials. The cost of the bottom is $40/m² and the cost of the sides is $30/m². Find the dimensions of the box that minimize total cost.

46. Find the dimensions of a cylinder of volume 1 m³ of minimal cost if the top and bottom are made of material that costs twice as much as the material for the side.

47. Your task is to design a rectangular industrial warehouse consisting of three separate spaces of equal size as in Figure 26. The wall materials cost $500 per linear meter and your company allocates $2,400,000 for that part of the project involving the walls.

(a) Which dimensions maximize the area of the warehouse?

(b) What is the area of each compartment in this case?

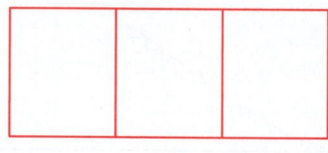

FIGURE 26

48. Suppose, in the previous exercise, that the warehouse consists of n separate spaces of equal size. Find a formula in terms of n for the maximum possible area of the warehouse.

49. According to a model developed by economists E. Heady and J. Pesek, if fertilizer made from N pounds of nitrogen and P lb of phosphate is used on an acre of farmland, then the yield of corn (in bushels per acre) is

$$Y = 7.5 + 0.6N + 0.7P - 0.001N^2 - 0.002P^2 + 0.001NP$$

A farmer intends to spend $30/acre on fertilizer. If nitrogen costs 25 cents/lb and phosphate costs 20 cents/lb, which combination of N and P produces the highest yield of corn?

50. Experiments show that the quantities x of corn and y of soybean required to produce a hog of weight Q satisfy $Q = 0.5x^{1/2}y^{1/4}$. The unit of x, y, and Q is the cwt, an agricultural unit equal to 100 lb. Find the values of x and y that minimize the cost of a hog of weight $Q = 2.5$ cwt if corn costs $3/cwt and soy costs $7/cwt.

51. All units in a 100-unit apartment building are rented out when the monthly rent is set at $r = \$900$/month. Suppose that one unit becomes vacant with each $10 increase in rent and that each occupied unit costs $80/mon in maintenance. Which rent r maximizes monthly profit?

52. An 8-billion-bushel corn crop brings a price of $2.40/bushel. A commodity broker uses the rule of thumb: If the crop is reduced by x percent, then the price increases by $10x$ cents. Which crop size results in maximum revenue and what is the price per bushel? *Hint:* Revenue is equal to price times crop size.

53. The monthly output of a Spanish light bulb factory is $P = 2LK^2$ (in millions), where L is the cost of labor and K is the cost of equipment (in millions of euros). The company needs to produce 1.7 million units per month. Which values of L and K would minimize the total cost $L + K$?

54. The rectangular plot in Figure 27 has size 100 m × 200 m. Pipe is to be laid from A to a point P on side BC and from there to C. The cost of laying pipe along the side of the plot is $45/m and the cost through the plot is $80/m (since it is underground).

(a) Let $f(x)$ be the total cost, where x is the distance from P to B. Determine $f(x)$, but note that f is discontinuous at $x = 0$ (when $x = 0$, the cost of the entire pipe is $45/m).

(b) What is the most economical way to lay the pipe? What if the cost along the sides is $65/m?

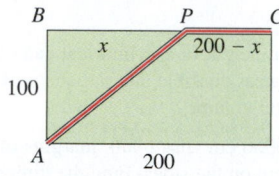

FIGURE 27

55. Brandon is on one side of a river that is 50 m wide and wants to reach a point 200 m downstream on the opposite side as quickly as possible by swimming diagonally across the river and then running the rest of the way. Find the best route if Brandon can swim at 1.5 m/s and run at 4 m/s.

56. Snell's Law When a light beam travels from a point A above a swimming pool to a point B below the water (Figure 28), it chooses the path that takes the *least time*. Let v_1 be the velocity of light in air and v_2 the velocity in water (it is known that $v_1 > v_2$). Prove Snell's Law of Refraction:

$$\frac{\sin\theta_1}{v_1} = \frac{\sin\theta_2}{v_2}$$

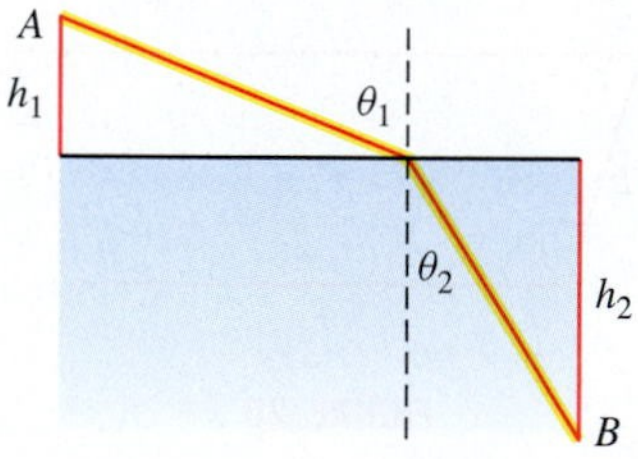

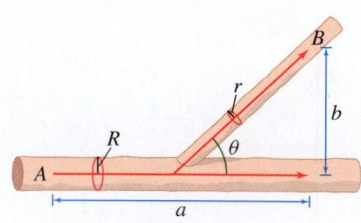

FIGURE 28

57. Vascular Branching A small blood vessel of radius r branches off at an angle θ from a larger vessel of radius R to supply blood along a path from A to B. According to Poiseuille's Law, the total resistance to blood flow is proportional to

$$T = \left(\frac{a - b\cot\theta}{R^4} + \frac{b\csc\theta}{r^4}\right)$$

where a and b are as in Figure 29. Show that the total resistance is minimized when $\cos\theta = (r/R)^4$.

FIGURE 29

In Exercises 58–59, a box (with no top) is to be constructed from a piece of cardboard with sides of length A and B by cutting out squares of length h from the corners and folding up the sides (Figure 30).

58. Find the value of h that maximizes the volume of the box if $A = 15$ and $B = 24$. What are the dimensions of this box?

59. Which values of A and B maximize the volume of the box if $h = 10$ cm and $AB = 900$ cm^2?

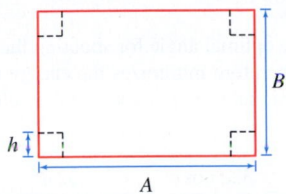

FIGURE 30

60. Which value of h maximizes the volume of the box if $A = B$?

61. Given n numbers $x_1, \ldots, x_n$, find the value of x minimizing the sum of the squares:

$$(x - x_1)^2 + (x - x_2)^2 + \cdots + (x - x_n)^2$$

First, solve for $n = 2, 3$ and then try it for arbitrary n.

62. A billboard of height b is mounted on the side of a building with its bottom edge at a distance h from the street as in Figure 31. At what distance x should an observer stand from the wall to maximize the angle of observation θ?

63. Solve Exercise 62 again using geometry rather than calculus. There is a unique circle passing through points B and C that is tangent to the street. Let R be the point of tangency. Note that the two angles labeled ψ in Figure 31 are equal because they subtend equal arcs on the circle.

(a) Show that the maximum value of θ is $\theta = \psi$. *Hint:* Show that $\psi = \theta + \angle PBA$, where A is the intersection of the circle with PC.

(b) Prove that this agrees with the answer to Exercise 62.

(c) Show that $\angle QRB = \angle RCQ$ for the maximal angle ψ.

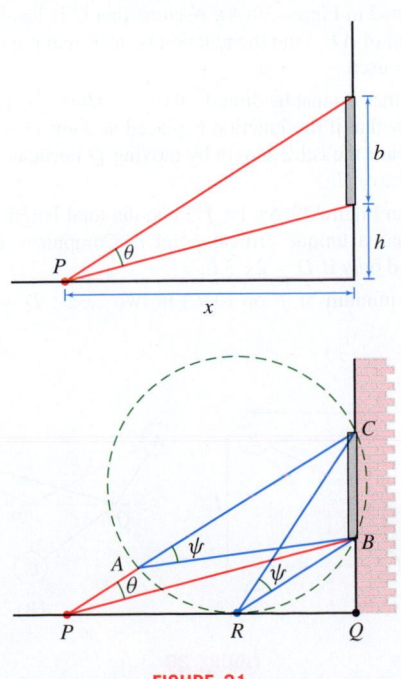

FIGURE 31

64. Optimal Delivery Schedule A gas station sells Q gallons of gasoline per year, which is delivered N times per year in equal shipments of Q/N gallons. The cost of each delivery is d dollars and the yearly storage costs are sQT, where T is the length of time (a fraction of a year) between shipments and s is a constant. Show that costs are minimized for $N = \sqrt{sQ/d}$. (*Hint:* $T = 1/N$.) Find the optimal number of deliveries if $Q = 2$ million gal, $d = \$8000$, and $s = 30$ cents/gal-year. Your answer should be a whole number, so compare costs for the two integer values of N nearest the optimal value.

65. Victor Klee's Endpoint Maximum Problem Given 40 m of straight fence, your goal is to build a rectangular enclosure using 80 additional meters of fence that encompasses the greatest area. Let $A(x)$ be the area of the enclosure, with x as in Figure 32.

(a) Find the maximum value of $A(x)$.

(b) Which interval of x values is relevant to our problem? Find the maximum value of $A(x)$ on this interval.

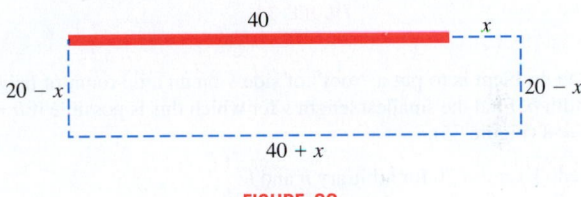

FIGURE 32

66. Let (a, b) be a fixed point in the first quadrant and let $S(d)$ be the sum of the distances from $(d, 0)$ to the points $(0, 0)$, (a, b), and $(a, -b)$.

(a) Find the value of d for which $S(d)$ is minimal. The answer depends on whether $b < \sqrt{3}a$ or $b \geq \sqrt{3}a$. *Hint:* Show that $d = 0$ when $b \geq \sqrt{3}a$.

(b) [GU] Let $a = 1$. Plot S for $b = 0.5$, $\sqrt{3}$, 3 and describe the position of the minimum.

67. The force F (in Newtons) required to move a box of mass m kg in motion by pulling on an attached rope (Figure 33) is

$$F(\theta) = \frac{fmg}{\cos\theta + f\sin\theta}$$

where θ is the angle between the rope and the horizontal, f is the coefficient of static friction, and $g = 9.8$ m/s^2. Find the angle θ that minimizes the required force F, assuming $f = 0.4$. *Hint:* Find the maximum value of $\cos\theta + f\sin\theta$.

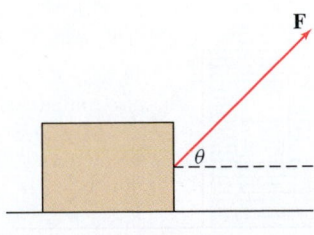

FIGURE 33

68. In the setting of Exercise 67, show that for any f the minimal force required is proportional to $1/\sqrt{1 + f^2}$.

69. Bird Migration Ornithologists have found that the power (in joules per second) consumed by a certain pigeon flying at velocity v m/s is described well by the function $P(v) = 17v^{-1} + 10^{-3}v^3$ joules/s. Assume that the pigeon can store 5×10^4 joules of usable energy as body fat.

(a) Show that at velocity v, a pigeon can fly a total distance of $D(v) = (5 \times 10^4)v/P(v)$ if it uses all of its stored energy.

(b) Find the velocity v_p that *minimizes* P.

(c) Migrating birds are smart enough to fly at the velocity that maximizes distance traveled rather than minimizes power consumption. Show that the velocity v_d which maximizes $D(v)$ satisfies $P'(v_d) = P(v_d)/v_d$. Show that v_d is obtained graphically as the velocity coordinate of the point where a line through the origin is tangent to the graph of P (Figure 34).

(d) Find v_d and the maximum distance $D(v_d)$.

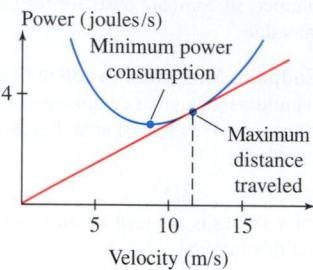

FIGURE 34

70. The problem is to put a "roof" of side s on an attic room of height h and width b. Find the smallest length s for which this is possible if $b = 27$ and $h = 8$ (Figure 35).

71. Redo Exercise 70 for arbitrary b and h.

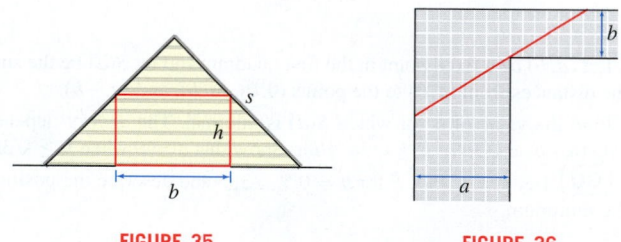

FIGURE 35   FIGURE 36

72. Find the maximum length of a pole that can be carried horizontally around a corner joining corridors of widths $a = 24$ and $b = 3$ (Figure 36).

73. Redo Exercise 72 for arbitrary widths a and b.

74. Find the minimum length ℓ of a beam that can clear a fence of height h and touch a wall located b ft behind the fence (Figure 37).

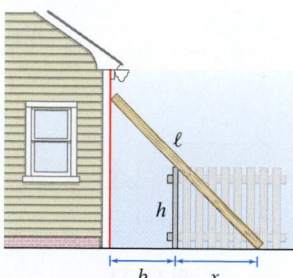

FIGURE 37

75. [✏️] A basketball player stands d feet from the basket. Let h and α be as in Figure 38. Using physics, one can show that if the player releases the ball at an angle θ, then the initial velocity required to make the ball go through the basket satisfies

$$v^2 = \frac{16d}{\cos^2 \theta (\tan \theta - \tan \alpha)}$$

(a) Explain why this formula is meaningful only for $\alpha < \theta < \frac{\pi}{2}$. Why does v approach infinity at the endpoints of this interval?

(b) [GU] Take $\alpha = \frac{\pi}{6}$ and plot v^2 as a function of θ for $\frac{\pi}{6} < \theta < \frac{\pi}{2}$. Verify that the minimum occurs at $\theta = \frac{\pi}{3}$.

(c) Set $F(\theta) = \cos^2 \theta (\tan \theta - \tan \alpha)$. Explain why v is minimized for θ such that $F(\theta)$ is maximized.

(d) Verify that $F'(\theta) = \cos(\alpha - 2\theta) \sec \alpha$ (you will need to use the addition formula for cosine) and show that the maximum value of F on $\left[\alpha, \frac{\pi}{2}\right]$ occurs at $\theta_0 = \frac{\alpha}{2} + \frac{\pi}{4}$.

(e) For a given α, the optimal angle for shooting the basket is θ_0 because it minimizes v^2 and therefore minimizes the energy required to make the shot (energy is proportional to v^2). Show that the velocity v_{opt} at the optimal angle θ_0 satisfies

$$v_{\text{opt}}^2 = \frac{32d \cos \alpha}{1 - \sin \alpha} = \frac{32 d^2}{-h + \sqrt{d^2 + h^2}}$$

(f) [GU] Show with a graph that for fixed d (say, $d = 15$ ft, the distance of a free throw), v_{opt}^2 is an increasing function of h. Use this to explain why taller players have an advantage and why it can help to jump while shooting.

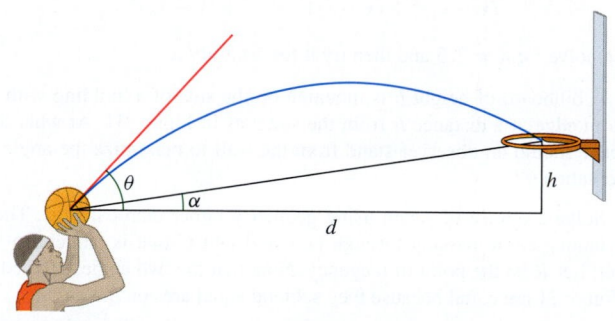

FIGURE 38

76. Three towns A, B, and C are to be joined by an underground fiber cable as illustrated in Figure 39(A). Assume that C is located directly below the midpoint of $\overline{AB}$. Find the junction point P that minimizes the total amount of cable used.

(a) First show that P must lie directly above C. *Hint:* Use the result of Example 7 to show that if the junction is placed at point Q in Figure 39(B), then we can reduce the cable length by moving Q horizontally over to the point P lying above C.

(b) With x as in Figure 39(A), let $f(x)$ be the total length of cable used. Show that f has a unique critical point c. Compute c and show that $0 \le c \le L$ if and only if $D \le 2\sqrt{3} L$.

(c) Find the minimum of f on $[0, L]$ in two cases: $D = 2$, $L = 4$ and $D = 8$, $L = 2$.

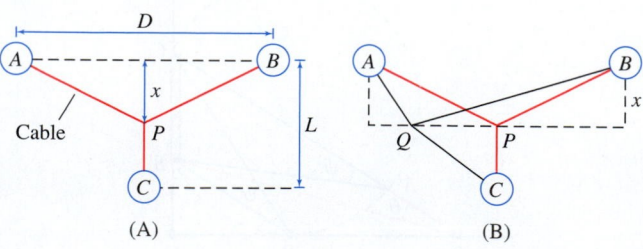

FIGURE 39

Further Insights and Challenges

77. Tom and Ali drive along a highway represented by the graph of f in Figure 40. During the trip, Ali views a billboard represented by the segment $\overline{BC}$ along the y-axis. Let Q be the y-intercept of the tangent line to $y = f(x)$. Show that θ is maximized at the value of x for which the angles $\angle QPB$ and $\angle QCP$ are equal. This generalizes Exercise 63 (c) [which corresponds to the case $f(x) = 0$]. *Hints:*

(a) Show that $d\theta/dx$ is equal to

$$(b - c) \cdot \frac{(x^2 + (xf'(x))^2) - (b - (f(x) - xf'(x)))(c - (f(x) - xf'(x)))}{(x^2 + (b - f(x))^2)(x^2 + (c - f(x))^2)}$$

(b) Show that the y-coordinate of Q is $f(x) - xf'(x)$.

(c) Show that the condition $d\theta/dx = 0$ is equivalent to

$$PQ^2 = BQ \cdot CQ$$

(d) Conclude that $\triangle QPB$ and $\triangle QCP$ are similar triangles.

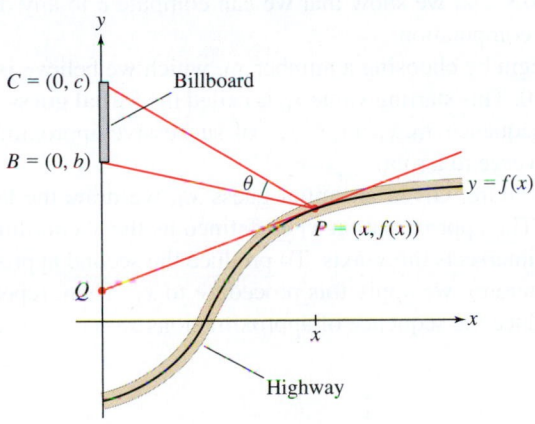

FIGURE 40

Seismic Prospecting *Exercises 78–80 are concerned with determining the thickness d of a layer of soil that lies on top of a rock formation. Geologists send two sound pulses from point A to point D separated by a distance s. The first pulse travels directly from A to D along the surface of the earth. The second pulse travels down to the rock formation, then along its surface, and then back up to D (path ABCD), as in Figure 41. The pulse travels with velocity v_1 in the soil and v_2 in the rock.*

78. (a) Show that the time required for the first pulse to travel from A to D is $t_1 = s/v_1$.

(b) Show that the time required for the second pulse is

$$t_2 = \frac{2d}{v_1} \sec \theta + \frac{s - 2d \tan \theta}{v_2}$$

provided that

$$\tan \theta \le \frac{s}{2d} \qquad \boxed{3}$$

(*Note:* If this inequality is not satisfied, then point B does not lie to the left of C.)

(c) Show that t_2 is minimized when $\sin \theta = v_1/v_2$.

79. In this exercise, assume that $v_2/v_1 \ge \sqrt{1 + 4(d/s)^2}$.

(a) Show that inequality (3) holds if $\sin \theta = v_1/v_2$.

(b) Show that the minimal time for the second pulse is

$$t_2 = \frac{2d}{v_1}(1 - k^2)^{1/2} + \frac{s}{v_2}$$

where $k = v_1/v_2$.

(c) Conclude that $\dfrac{t_2}{t_1} = \dfrac{2d(1 - k^2)^{1/2}}{s} + k.$

80. Continue with the assumption of the previous exercise.

(a) Find the thickness of the soil layer, assuming that $v_1 = 0.7v_2$, $t_2/t_1 = 1.3$, and $s = 400$ m.

(b) The times t_1 and t_2 are measured experimentally. The equation in Exercise 79(c) shows that t_2/t_1 is a linear function of $1/s$. What might you conclude if experiments were formed for several values of s and the points $(1/s, t_2/t_1)$ did *not* lie on a straight line?

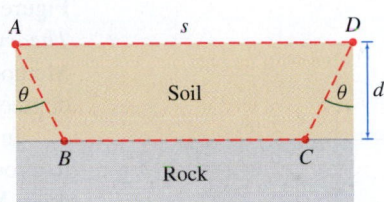

FIGURE 41

81. 🖊 In this exercise, we use Figure 42 to prove Heron's principle of Example 7 without calculus. By definition, C is the reflection of B across the line $\overline{MN}$ (so that $\overline{BC}$ is perpendicular to $\overline{MN}$ and $BN = CN$). Let P be the intersection of $\overline{AC}$ and $\overline{MN}$. Use geometry to justify the following:

(a) $\triangle PNB$ and $\triangle PNC$ are congruent and $\theta_1 = \theta_2$.

(b) The paths APB and APC have equal length.

(c) Similarly, AQB and AQC have equal length.

(d) The path APC is shorter than AQC for all $Q \ne P$.

Conclude that the shortest path AQB occurs for $Q = P$.

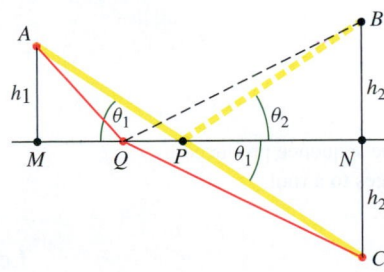

FIGURE 42

82. A jewelry designer plans to incorporate a component made of gold in the shape of a frustum of a cone of height 1 cm and fixed lower radius r (Figure 43). The upper radius x can take on any value between 0 and r. Note that $x = 0$ and $x = r$ correspond to a cone and cylinder, respectively. As a function of x, the surface area (not including the top and bottom) is $S(x) = \pi s(r + x)$, where s is the *slant height* as indicated in the figure. Which value of x yields the least expensive design [the minimum value of $S(x)$ for $0 \le x \le r$]?

(a) Show that $S(x) = \pi(r + x)\sqrt{1 + (r - x)^2}$.

(b) Show that if $r < \sqrt{2}$, then S is an increasing function. Conclude that the cone ($x = 0$) has minimal area in this case.

(c) Assume that $r > \sqrt{2}$. Show that S has two critical points $x_1 < x_2$ in $(0, r)$, and that $S(x_1)$ is a local maximum, and $S(x_2)$ is a local minimum.

(d) Conclude that the minimum occurs at $x = 0$ or x_2.

(e) Find the minimum in the cases $r = 1.5$ and $r = 2$.

(f) Challenge: Let $c = \sqrt{(5 + 3\sqrt{3})/4} \approx 1.597$. Prove that the minimum occurs at $x = 0$ (cone) if $\sqrt{2} < r < c$, but the minimum occurs at $x = x_2$ if $r > c$.

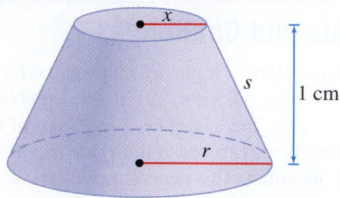

FIGURE 43 Frustum of height 1 cm.

4.8 Newton's Method

← **REMINDER** A "zero" or "root" of a function f is a solution of the equation $f(x) = 0$.

Newton's Method is a procedure for finding numerical approximations to zeros of functions. Numerical approximations are important because it is often impossible to find the zeros exactly. For example, the polynomial $f(x) = x^5 - x - 1$ has one real root c (see Figure 1), but we can prove, using an advanced branch of mathematics called *Galois Theory*, that there is no algebraic formula for this root. In this section, using Newton's Method, we show that $c \approx 1.1673$, and we show that we can compute c to any desired degree of accuracy with enough computation.

In Newton's Method, we begin by choosing a number x_0, which we believe is close to a root of the equation $f(x) = 0$. This starting value x_0 is called the **initial guess**. Newton's Method then produces a sequence $x_0, x_1, x_2, x_3 \ldots$ of successive approximations that, in favorable situations, converge to a root.

Figure 2 illustrates the procedure. Given an initial guess x_0, we draw the tangent line to the graph at $(x_0, f(x_0))$. The approximation x_1 is defined as the x-coordinate of the point where the tangent line intersects the x-axis. To produce the second approximation x_2 (also called the second iterate), we apply this procedure to x_1. Then, repeatedly applying this procedure, we produce the sequence of approximations $x_0, x_1, x_2, x_3 \ldots$.

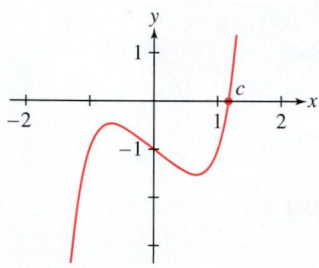

FIGURE 1 Graph of $y = x^5 - x - 1$. With Newton's Method, we can approximate the root c as accurately as we like.

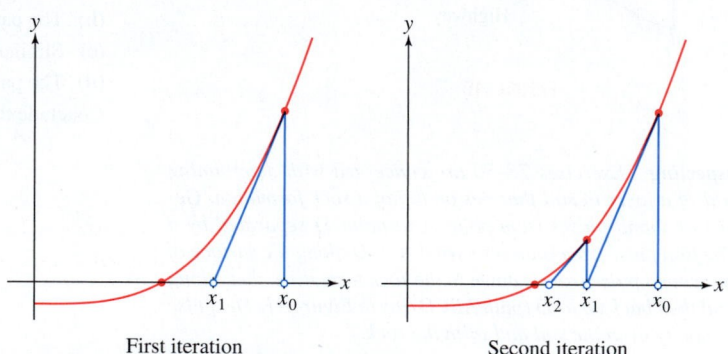

DF FIGURE 2 The sequence produced by iteration converges to a root.

First iteration   Second iteration

Let's derive a formula for x_1. The tangent line at $(x_0, f(x_0))$ has equation

$$y = f(x_0) + f'(x_0)(x - x_0)$$

The tangent line crosses the x-axis at x_1, where $y = 0$, that is, where

$$f(x_0) + f'(x_0)(x_1 - x_0) = 0$$

To solve for x_1, we first divide by $f'(x_0)$ (as long as it is not zero) to obtain $x_1 - x_0 = -f(x_0)/f'(x_0)$, and therefore,

$$x_1 = x_0 - \frac{f(x_0)}{f'(x_0)}$$

The second iterate x_2 is obtained by applying this formula to x_1 instead of x_0:

$$x_2 = x_1 - \frac{f(x_1)}{f'(x_1)}$$

and so on. Notice in Figure 2 that x_1 is closer to the root than x_0 is and that x_2 is closer still. This is typical: The successive approximations usually converge to the actual root. However, there are cases where Newton's Method fails (see Figure 4).

Newton's Method To approximate a root of $f(x) = 0$:

Step 1. **Choose an initial guess x_0 (close to the desired root if possible).**

Step 2. **Generate successive approximations $x_1, x_2, \ldots,$ where**

$$x_{n+1} = x_n - \frac{f(x_n)}{f'(x_n)} \qquad \boxed{1}$$

Newton's Method is an example of an iterative procedure. To "iterate" means to repeat, and in Newton's Method, we use Eq. (1) repeatedly to produce the sequence of approximations.

EXAMPLE 1 Calculate the first five approximations $x_1, \ldots, x_5$ to a root of $f(x) = x^5 - x - 1$ using the initial guess $x_0 = 1$.

Solution We have $f'(x) = 5x^4 - 1$. Therefore,

$$x_1 = x_0 - \frac{f(x_0)}{f'(x_0)} = x_0 - \frac{x_0^5 - x_0 - 1}{5x_0^4 - 1}$$

We compute the first two approximations as follows:

$$x_1 = x_0 - \frac{f(x_0)}{f'(x_0)} = 1 - \frac{1^5 - 1 - 1}{5(1)^4 - 1} = 1.25$$

$$x_2 = x_1 - \frac{f(x_1)}{f'(x_1)} = 1.25 - \frac{1.25^5 - 1.25 - 1}{5(1.25)^4 - 1} \approx 1.178459$$

Continuing, rounding to six decimal places at each stage, we obtain $x_3 \approx 1.167547$, $x_4 \approx 1.167304$, and $x_5 \approx 1.167304$. This suggests that, accurate to six decimal places, 1.167304 is a root of $f(x) = x^5 - x - 1$. ∎

We can check our approximation; evaluating $x^5 - x - 1$ at $x = 1.167304$, we obtain 0.00000018 (to eight decimal places), verifying that we have a good approximation to a root of $f(x) = x^5 - x - 1$.

How Many Iterations Are Required?

How many iterations of Newton's Method are required to approximate a root to within a given accuracy? There is no definitive answer, but in practice, it is usually safe to assume that if x_n and x_{n+1} agree to m decimal places, then the approximation x_n is correct to these m places.

EXAMPLE 2 [GU] Let c be the smallest positive solution of $\sin 3x = \cos x$.

(a) Use a computer-generated graph to choose an initial guess x_0 for c.

(b) Use Newton's Method to approximate c to within an error of at most 10^{-6}.

Solution

(a) A solution of $\sin 3x = \cos x$ is a zero of the function $f(x) = \sin 3x - \cos x$. Figure 3 shows that the smallest positive zero is approximately halfway between 0 and $\frac{\pi}{4}$. Because $\frac{\pi}{4} \approx 0.785$, a good initial guess is $x_0 = 0.4$.

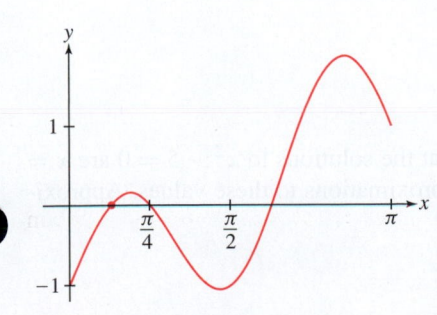

FIGURE 3 Graph of $f(x) = \sin 3x - \cos x$.

(b) Since $f'(x) = 3\cos 3x + \sin x$, Eq. (1) yields the formula

$$x_{n+1} = x_n - \frac{\sin 3x_n - \cos x_n}{3\cos 3x_n + \sin x_n}$$

With $x_0 = 0.4$ as the initial guess, the first four iterates are

$$x_1 \approx 0.3925647447$$

$$x_2 \approx 0.3926990382$$

$$x_3 \approx 0.3926990816987196$$

$$x_4 \approx 0.3926990816987241$$

Stopping here, we can be fairly confident that x_4 approximates the smallest positive root c to at least 12 places. In fact, $c = \frac{\pi}{8}$ and x_4 is accurate to 16 places.   ∎

There is no single "correct" initial guess. In Example 2, we chose $x_0 = 0.4$, but another possible choice is $x_0 = 0$, leading to the sequence

$$x_1 \approx 0.3333333333$$

$$x_2 \approx 0.3864547725$$

$$x_3 \approx 0.3926082513$$

$$x_4 \approx 0.3926990816$$

You can check, however, that $x_0 = 1$ yields a sequence converging to $\frac{\pi}{4}$, which is the second positive solution of $\sin 3x = \cos x$.

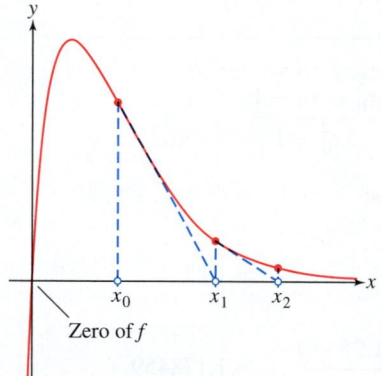

DF **FIGURE 4** Function has only one zero but the sequence of Newton iterates goes off to infinity.

Which Root Does Newton's Method Compute?

Sometimes, Newton's Method computes no root at all. In Figure 4, the iterates diverge to infinity. In practice, however, Newton's Method usually converges quickly, and if a particular choice of x_0 does not lead to a root, the best strategy is to try a different initial guess, consulting a graph if possible. If $f(x) = 0$ has more than one root, different initial guesses x_0 may lead to different roots.

EXAMPLE 3 Figure 5 shows that $f(x) = x^4 - 6x^2 + x + 5$ has four real roots.

(a) Show that with $x_0 = 0$, Newton's Method converges to the root near -2.

(b) Show that with $x_0 = -1$, Newton's Method converges to the root near -1.

Solution We have $f'(x) = 4x^3 - 12x + 1$ and

$$x_{n+1} = x_n - \frac{x_n^4 - 6x_n^2 + x_n + 5}{4x_n^3 - 12x_n + 1} = \frac{3x_n^4 - 6x_n^2 - 5}{4x_n^3 - 12x_n + 1}$$

(a) On the basis of Table 1, we can be confident that when $x_0 = 0$, Newton's Method converges to a root near -2.3. Notice in Figure 5 that this is not the closest root to x_0.

(b) Table 2 suggests that with $x_0 = -1$, Newton's Method converges to the root near -0.9.   ∎

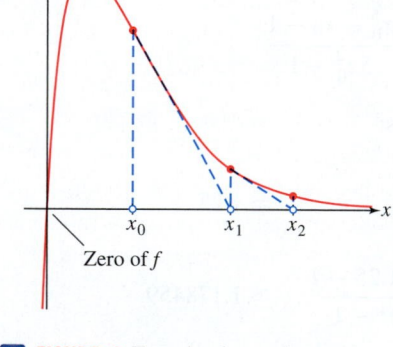

FIGURE 5 Graph of $f(x) = x^4 - 6x^2 + x + 5$.

TABLE 1			TABLE 2	
x_0	0		x_0	-1
x_1	-5		x_1	-0.8888888888
x_2	-3.9179954		x_2	-0.8882866140
x_3	-3.1669480		x_3	-0.88828656234358
x_4	-2.6871270		x_4	-0.888286562343575
x_5	-2.4363303			
x_6	-2.3572979			
x_7	-2.3495000			

EXAMPLE 4 **Approximating $\sqrt{5}$** We know that the solutions to $x^2 - 5 = 0$ are $x = \pm\sqrt{5}$. We can use Newton's method to obtain approximations to these values. Approximate $\sqrt{5}$ using an initial guess $x_0 = 2$.

Solution We have $f'(x) = 2x$. Therefore,

$$x_1 = x_0 - \frac{f(x_0)}{f'(x_0)} = x_0 - \frac{x_0^2 - 5}{2x_0}$$

We compute the successive approximations as follows:

$$x_1 = x_0 - \frac{f(x_0)}{f'(x_0)} = 2 - \frac{2^2 - 5}{2 \cdot 2} = 2.25$$

$$x_2 = x_1 - \frac{f(x_1)}{f'(x_1)} = 2.25 - \frac{2.25^2 - 5}{2 \cdot 2.25} \approx 2.23611$$

$$x_3 = x_2 - \frac{f(x_2)}{f'(x_2)} = 2.23611 - \frac{2.23611^2 - 5}{2 \cdot 2.23611} \approx 2.23606797789$$

Therefore, $\sqrt{5} \approx 2.23606797789$.

A calculator computation of $\sqrt{5}$ yields

$$\sqrt{5} = 2.23606797750\ldots$$

Observe that x_3 is accurate to within an error of less than 10^{-9}. This is impressive accuracy for just three iterations of Newton's Method. ∎

4.8 SUMMARY

- Newton's Method: To find a sequence of numerical approximations to a root of f, begin with an initial guess x_0. Then construct the sequence $x_0, x_1, x_2, \ldots$ using the formula

$$x_{n+1} = x_n - \frac{f(x_n)}{f'(x_n)}$$

You should choose the initial guess x_0 as close as possible to a root, possibly by referring to a graph. In favorable cases, the sequence converges rapidly to a root.
- If x_n and x_{n+1} agree to m decimal places, it is usually safe to assume that x_n agrees with a root to m decimal places.

4.8 EXERCISES

Preliminary Questions

1. How many iterations of Newton's Method are required to compute a root if f is a linear function?

2. What happens in Newton's Method if your initial guess happens to be a zero of f?

3. What happens in Newton's Method if your initial guess happens to be a local min or max of f?

4. Is the following a reasonable description of Newton's Method: "A root of the equation of the tangent line to the graph of f is used as an approximation to a root of f itself"? Explain.

Exercises

In this exercise set, all approximations should be carried out using Newton's Method.

In Exercises 1–6, apply Newton's Method to f and initial guess x_0 to calculate x_1, x_2, x_3.

1. $f(x) = x^2 - 6$, $x_0 = 2$

2. $f(x) = x^2 - 3x + 1$, $x_0 = 3$

3. $f(x) = x^3 - 10$, $x_0 = 2$

4. $f(x) = x^3 + x + 1$, $x_0 = -1$

5. $f(x) = \cos x - 4x$, $x_0 = 1$

6. $f(x) = 1 - x \sin x$, $x_0 = 7$

7. Use Figure 6 to choose an initial guess x_0 to the unique real root of $x^3 + 2x + 5 = 0$ and compute the first three Newton iterates.

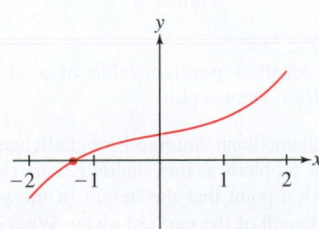

FIGURE 6 Graph of $y = x^3 + 2x + 5$.

8. Approximate a solution of $\sin x = \cos 2x$ in the interval $\left[0, \frac{\pi}{2}\right]$ to three decimal places. Then find the exact solution and compare with your approximation.

9. Approximate both solutions of $e^x = 5x$ to three decimal places (Figure 7).

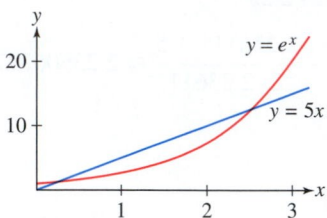

FIGURE 7 Graphs of $y = e^x$ and $y = 5x$.

10. The first positive solution of $\sin x = 0$ is $x = \pi$. Use Newton's Method to calculate π to four decimal places.

In Exercises 11–14, approximate to three decimal places using Newton's Method and compare with the value from a calculator.

11. $\sqrt{11}$ **12.** $5^{1/3}$ **13.** $2^{7/3}$ **14.** $3^{-1/4}$

15. Approximate the largest positive root of $f(x) = x^4 - 6x^2 + x + 5$ to within an error of at most 10^{-4}. Refer to Figure 5.

(GU) *In Exercises 16–21, approximate the value specified to three decimal places using Newton's Method. Use a plot to choose an initial guess.*

16. Largest positive root of $f(x) = x^3 - 5x + 1$

17. Negative root of $f(x) = x^5 - 20x + 10$

18. Positive solution of $\sin \theta = 0.8\theta$

19. Positive solution of $2 \tan^{-1} x = x$

20. The least positive solution of $x \cos x = 10$

21. Solution of $\ln(x + 4) = x$

22. Let x_1, x_2 be the estimates to a root obtained by applying Newton's Method with $x_0 = 1$ to the function graphed in Figure 8. Estimate the numerical values of x_1 and x_2, and draw the tangent lines used to obtain them.

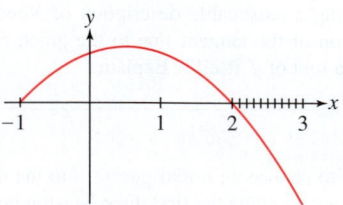

FIGURE 8

23. (GU) Find the smallest positive value of x at which $y = x$ and $y = \tan x$ intersect. *Hint:* Draw a plot.

24. In 1535, the mathematician Antonio Fior challenged his rival Niccolo Tartaglia to solve this problem: A tree stands 12 *braccia* high; it is broken into two parts at such a point that the height of the part left standing is the cube root of the length of the part cut away. What is the height of the part left standing? Show that this is equivalent to solving $x^3 + x = 12$ and finding the height to three decimal places. Tartaglia, who had discovered

the secret of solving the cubic equation, was able to determine the exact answer:

$$x = \frac{\left(\sqrt[3]{\sqrt{2919} + 54} - \sqrt[3]{\sqrt{2919} - 54}\right)}{\sqrt[3]{9}}$$

25. Find (to two decimal places) the coordinates of the point P in Figure 9 where the tangent line to $y = \cos x$ passes through the origin.

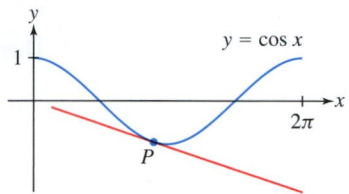

FIGURE 9

Newton's Method is often used to determine interest rates in financial calculations. In Exercises 26–28, r denotes a yearly interest rate expressed as a decimal (rather than as a percent).

26. If P dollars are deposited every month in an account earning interest at the yearly rate r, then the value S of the account after N years is

$$S = P \left(\frac{b^{12N+1} - b}{b - 1}\right), \qquad \text{where } b = 1 + \frac{r}{12}$$

You have decided to deposit $P = \$100$ per month.

(a) Determine S after 5 years if $r = 0.07$ (i.e., 7%).

(b) Show that to save \$10,000 after 5 years, you must earn interest at a rate r determined by the equation $b^{61} - 101b + 100 = 0$. Use Newton's Method to solve for b. Then find r. Note that $b = 1$ is a root, but you want the root satisfying $b > 1$.

27. If you borrow L dollars for N years at a yearly interest rate r, your monthly payment of P dollars is calculated using the equation

$$L = P \left(\frac{1 - b^{-12N}}{b - 1}\right), \qquad \text{where } b = 1 + \frac{r}{12}$$

(a) Find P if $L = \$5000$, $N = 3$, and $r = 0.08$ (8%).

(b) You are offered a loan of $L = \$5000$ to be paid back over 3 years with monthly payments of $P = \$200$. Use Newton's Method to compute b and find the implied interest rate r of this loan. *Hint:* Show that

$$(L/P)b^{12N+1} - (1 + L/P)b^{12N} + 1 = 0$$

28. If you deposit P dollars in a retirement fund every year for N years with the intention of then withdrawing Q dollars per year for M years, you must earn interest at a rate r satisfying

$$P(b^N - 1) = Q(1 - b^{-M}), \qquad \text{where } b = 1 + r$$

Assume that \$2000 is deposited each year for 30 years and the goal is to withdraw \$10,000 per year for 25 years. Use Newton's Method to compute b and then find r. Note that $b = 1$ is a root, but you want the root satisfying $b > 1$.

29. There is no simple formula for the position at time t of a planet P in its orbit (an ellipse) around the sun. Introduce the auxiliary circle and angle θ in Figure 10 (note that P determines θ because it is the central angle of point B on the circle). Let $a = OA$ and $e = OS/OA$ (the eccentricity of the orbit).

(a) Show that sector BSA has area $(a^2/2)(\theta - e \sin \theta)$.

(b) By Kepler's Second Law, the area of sector BSA is proportional to the time t elapsed since the planet passed point A, and because the circle has area πa^2, BSA has area $(\pi a^2)(t/T)$, where T is the period of the orbit. Deduce **Kepler's Equation**:

$$\frac{2\pi t}{T} = \theta - e\sin\theta$$

(c) The eccentricity of Mercury's orbit is approximately $e = 0.2$. Use Newton's Method to find θ after a quarter of Mercury's year has elapsed ($t = T/4$). Convert θ to degrees. Has Mercury covered more than a quarter of its orbit at $t = T/4$?

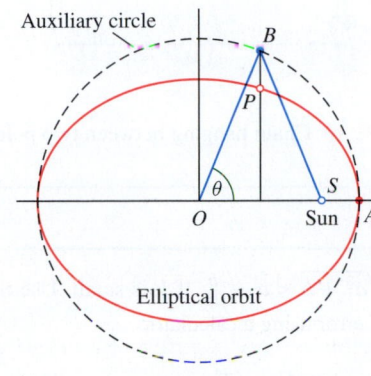

FIGURE 10

30. The roots of $f(x) = \frac{1}{3}x^3 - 4x + 1$ to three decimal places are -3.583, 0.251, and 3.332 (Figure 11). Determine the root to which Newton's Method converges for the initial choices $x_0 = 1.85$, 1.7, and 1.55. The answer shows that a small change in x_0 can have a significant effect on the outcome of Newton's Method.

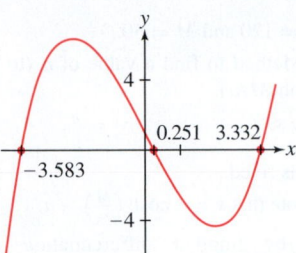

FIGURE 11 Graph of $f(x) = \frac{1}{3}x^3 - 4x + 1$.

31. Let $f(x) = xe^{-x}$.

(a) Show that a sequence obtained by applying Newton's Method to f satisfies $x_{n+1} = \frac{x_n^2}{x_n - 1}$.

(b) [CAS] Compute $x_1, x_2, \ldots, x_{10}$ separately with $x_0 = 0.8$ and $x_0 = 5$. Discuss what appears to be happening in each case. (*Note:* the only root of f is at $x = 0$.)

32. Let $f(x) = xe^{2x}$.

(a) Show that a sequence obtained by applying Newton's Method to f satisfies $x_{n+1} = \frac{2x_n^2}{2x_n + 1}$.

(b) [CAS] Compute $x_1, x_2, \ldots, x_{10}$ separately with $x_0 = -3$ and $x_0 = 7$. Discuss what appears to be happening in each case. (*Note:* the only root of f is at $x = 0$.)

33. What happens when you apply Newton's Method to find a zero of $f(x) = x^{1/3}$? Note that $x = 0$ is the only zero.

34. What happens when you apply Newton's Method to the equation $x^3 - 20x = 0$ with the unlucky initial guess $x_0 = 2$?

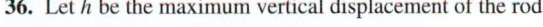

Further Insights and Challenges

35. Newton's Method can be used to compute reciprocals without performing division. Let $c > 0$ and set $f(x) = x^{-1} - c$.

(a) Show that $x - (f(x)/f'(x)) = 2x - cx^2$.

(b) Calculate the first three iterates of Newton's Method with $c = 10.3$ and the two initial guesses $x_0 = 0.1$ and $x_0 = 0.5$.

(c) Explain graphically why $x_0 = 0.5$ does not yield a sequence converging to $1/10.3$.

In Exercises 36 and 37, consider a metal rod of length L fastened at both ends. If you cut the rod and weld on an additional segment of length m, leaving the ends fixed, the rod will bow up into a circular arc of radius R (unknown), as indicated in Figure 12.

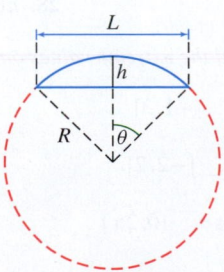

FIGURE 12 The bold circular arc has length $L + m$.

36. Let h be the maximum vertical displacement of the rod.

(a) Show that $L = 2R\sin\theta$ and conclude that

$$h = \frac{L(1 - \cos\theta)}{2\sin\theta}$$

(b) Show that $L + m = 2R\theta$ and then prove

$$\frac{\sin\theta}{\theta} = \frac{L}{L + m} \qquad \boxed{2}$$

37. Let $L = 3$ and $m = 1$. Apply Newton's Method to Eq. (2) to estimate θ, and use this to estimate h.

38. Quadratic Convergence to Square Roots Let $f(x) = x^2 - c$ and let $e_n = x_n - \sqrt{c}$ be the error in x_n.

(a) Show that $x_{n+1} = \frac{1}{2}(x_n + c/x_n)$ and $e_{n+1} = e_n^2/2x_n$.

(b) Show that if $x_0 > \sqrt{c}$, then $x_n > \sqrt{c}$ for all n. Explain graphically.

(c) Show that if $x_0 > \sqrt{c}$, then $e_{n+1} \le e_n^2/(2\sqrt{c})$.

*In Exercises 39–41, a flexible chain of length L is suspended between two poles of equal height separated by a distance $2M$ (Figure 13). By Newton's laws, the chain describes a **catenary** $y = a\cosh(\frac{x}{a})$ where a is the number such that $L = 2a\sinh(\frac{M}{a})$. The sag s is the vertical distance from the highest to the lowest point on the chain.*

39. Suppose that $L = 120$ and $M = 50$.

(a) Use Newton's Method to find a value of a (to two decimal places) satisfying $L = 2a \sinh(M/a)$.

(b) Compute the sag s.

40. Assume that M is fixed.

(a) Calculate $\frac{ds}{da}$. Note that $s = a \cosh\left(\frac{M}{a}\right) - a$.

(b) Calculate $\frac{da}{dL}$ by implicit differentiation using the relation $L = 2a \sinh\left(\frac{M}{a}\right)$.

(c) Use (a) and (b) and the Chain Rule to show that

$$\frac{ds}{dL} = \frac{ds}{da}\frac{da}{dL} = \frac{\cosh(M/a) - (M/a)\sinh(M/a) - 1}{2\sinh(M/a) - (2M/a)\cosh(M/a)}$$ **3**

41. Suppose that $L = 160$ and $M = 50$.

(a) Use Newton's Method to find a value of a (to two decimal places) satisfying $L = 2a \sinh(M/a)$.

(b) Use Eq. (3) and the Linear Approximation to estimate the increase in sag Δs for changes in length $\Delta L = 1$ and $\Delta L = 5$.

(c) [CAS] Compute $s(161) - s(160)$ and $s(165) - s(160)$ directly and compare with your estimates in (b).

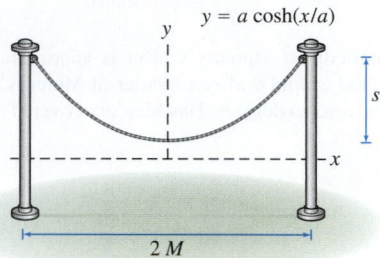

FIGURE 13 Chain hanging between two poles.

CHAPTER REVIEW EXERCISES

In Exercises 1–6, estimate using the Linear Approximation or linearization, and use a calculator to estimate the error.

1. $8.1^{1/3} - 2$

2. $\dfrac{1}{\sqrt{4.1}} - \dfrac{1}{2}$

3. $625^{1/4} - 624^{1/4}$

4. $\sqrt{101}$

5. $\dfrac{1}{1.02}$

6. $\sqrt[5]{33}$

In Exercises 7–12, find the linearization at the point indicated.

7. $y = \sqrt{x}, \quad a = 25$

8. $v(t) = 32t - 4t^2, \quad a = 2$

9. $A(r) = \frac{4}{3}\pi r^3, \quad a = 3$

10. $V(h) = 4h(2 - h)(4 - 2h), \quad a = 1$

11. $P(x) = e^{-x^2/2}, \quad a = 1$

12. $f(x) = \ln(x + e), \quad a = e$

In Exercises 13–16, use the Linear Approximation.

13. The position of an object in linear motion at time t is $s(t) = 0.4t^2 + (t + 1)^{-1}$. Estimate the distance traveled over the time interval $[4, 4.2]$.

14. A bond that pays $10,000 in 6 years is offered for sale at a price P. The percentage yield Y of the bond is

$$Y = 100\left(\left(\frac{10{,}000}{P}\right)^{1/6} - 1\right)$$

Verify that if $P = \$7500$, then $Y = 4.91\%$. Estimate the drop in yield if the price rises to $7700.

15. When a bus pass from Albuquerque to Los Alamos is priced at p dollars, a bus company takes in a monthly revenue of $R(p) = 1.5p - 0.01p^2$ (in thousands of dollars).

(a) Estimate ΔR if the price rises from $50 to $53.

(b) If $p = 80$, how will revenue be affected by a small increase in price? Explain using the Linear Approximation.

16. Show that $\sqrt{a^2 + b} \approx a + \frac{b}{2a}$ if b is small. Use this to estimate $\sqrt{26}$ and find the error using a calculator.

17. Use the Intermediate Value Theorem to show that $\sin x - \cos x = 3x$ has a solution, and use Rolle's Theorem to show that this solution is unique.

18. Show that $f(x) = 2x^3 + 2x + \sin x + 1$ has precisely one real root.

19. Verify the MVT for $f(x) = \ln x$ on $[1, 4]$.

20. Suppose that $f(1) = 5$ and $f'(x) \geq 2$ for $x \geq 1$. Use the MVT to show that $f(8) \geq 19$.

21. Use the MVT to prove that if $f'(x) \leq 2$ for $x > 0$ and $f(0) = 4$, then $f(x) \leq 2x + 4$ for all $x \geq 0$.

22. A function f has derivative $f'(x) = \dfrac{1}{x^4 + 1}$. Where on the interval $[1, 4]$ does f take on its maximum value?

In Exercises 23–28, find the critical points and determine whether they are minima, maxima, or neither.

23. $f(x) = x^3 - 4x^2 + 4x$

24. $s(t) = t^4 - 8t^2$

25. $f(x) = x^2(x + 2)^3$

26. $f(x) = x^{2/3}(1 - x)$

27. $g(\theta) = \sin^2\theta + \theta$

28. $h(\theta) = 2\cos 2\theta + \cos 4\theta$

In Exercises 29–36, find the extreme values on the interval.

29. $f(x) = x(10 - x), \quad [-1, 3]$

30. $f(x) = 6x^4 - 4x^6, \quad [-2, 2]$

31. $g(\theta) = \sin^2\theta - \cos\theta, \quad [0, 2\pi]$

32. $R(t) = \dfrac{t}{t^2 + t + 1}, \quad [0, 3]$

33. $f(x) = x^{2/3} - 2x^{1/3}$, $[-1, 3]$

34. $f(x) = 4x - \tan^2 x$, $\left[-\frac{\pi}{4}, \frac{\pi}{3}\right]$

35. $f(x) = x - 12 \ln x$, $[5, 40]$

36. $f(x) = e^x - 20x - 1$, $[0, 5]$

37. Find the critical points and extreme values of $f(x) = |x - 1| + |2x - 6|$ in $[0, 8]$.

38. Match the description of f with the graph of its *derivative* f' in Figure 1.

(a) f is increasing and concave up.

(b) f is decreasing and concave up.

(c) f is increasing and concave down.

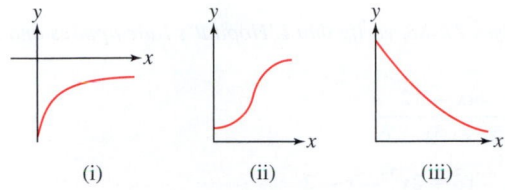

(i)   (ii)   (iii)

FIGURE 1 Graphs of the derivative.

In Exercises 39–44, find the points of inflection.

39. $y = x^3 - 4x^2 + 4x$

40. $y = x - 2 \cos x$

41. $y = \dfrac{x^2}{x^2 + 4}$

42. $y = \dfrac{x}{(x^2 - 4)^{1/3}}$

43. $f(x) = (x^2 - x)e^{-x}$

44. $f(x) = x(\ln x)^2$

In Exercises 45–54, sketch the graph, noting the transition points and asymptotic behavior.

45. $y = 12x - 3x^2$

46. $y = 8x^2 - x^4$

47. $y = x^3 - 2x^2 + 3$

48. $y = 4x - x^{3/2}$

49. $y = \dfrac{x}{x^3 + 1}$

50. $y = \dfrac{x}{(x^2 - 4)^{2/3}}$

51. $y = \dfrac{1}{|x + 2| + 1}$

52. $y = \sqrt{2 - x^3}$

53. $y = \sqrt{3} \sin x - \cos x$ on $[0, 2\pi]$

54. $y = 2x - \tan x$ on $[0, 2\pi]$

55. Draw a curve $y = f(x)$ for which f' and f'' have signs as indicated in Figure 2.

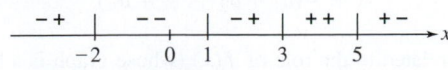

FIGURE 2

56. Find the dimensions of a cylindrical can with a bottom but no top of volume 4 m³ that uses the least amount of metal.

57. A rectangular open-topped box of height h with a square base of side b has volume $V = 4$ m³. Two of the side faces are made of material costing \$40/m². The remaining sides cost \$20/m². Which values of b and h minimize the cost of the box?

58. The corn yield on a certain farm is

$$Y = -0.118x^2 + 8.5x + 12.9 \quad \text{(bushels per acre)}$$

where x is the number of corn plants per acre (in thousands). Assume that corn seed costs \$1.25 (per thousand seeds) and that corn can be sold for \$1.50/bushel. Let $P(x)$ be the profit (revenue minus the cost of seeds) at planting level x.

(a) Compute $P(x_0)$ for the value x_0 that maximizes yield Y.

(b) Find the maximum value of $P(x)$. Does maximum yield lead to maximum profit?

59. Let $N(t)$ be the size of a tumor (in units of 10^6 cells) at time t (in days). According to the **Gompertz Model**, $dN/dt = N(a - b \ln N)$, where a, b are positive constants. Show that the maximum value of N is $e^{a/b}$ and that the tumor increases most rapidly when $N = e^{a/b-1}$.

60. A truck gets 10 miles per gallon (mpg) of diesel fuel traveling along an interstate highway at 50 mph. This mileage decreases by 0.15 mpg for each mile per hour increase above 50 mph.

(a) If the truck driver is paid \$30/h and diesel fuel costs $P = \$3$/gal, which speed v between 50 and 70 mph will minimize the cost of a trip along the highway? Notice that the actual cost depends on the length of the trip, but the optimal speed does not.

(b) ⟨GU⟩ Plot cost as a function of v (choose the length arbitrarily) and verify your answer to part (a).

(c) ⟨GU⟩ Do you expect the optimal speed v to increase or decrease if fuel costs go down to $P = \$2$/gal? Plot the graphs of cost as a function of v for $P = 2$ and $P = 3$ on the same axis and verify your conclusion.

61. Find the maximum volume of a right-circular cone placed upside-down in a right-circular cone of radius $R = 3$ and height $H = 4$ as in Figure 3. A cone of radius r and height h has volume $\frac{1}{3}\pi r^2 h$.

62. Redo Exercise 61 for arbitrary R and H.

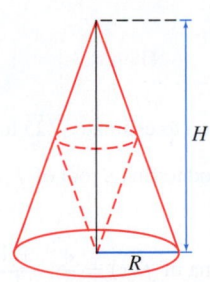

FIGURE 3

63. Show that the maximum area of a parallelogram $ADEF$ that is inscribed in a triangle ABC, as in Figure 4, is equal to one-half the area of $\triangle ABC$.

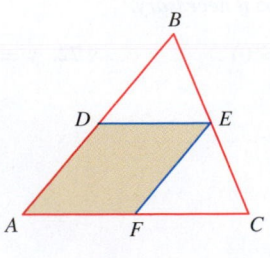

FIGURE 4

64. A box of volume 8 m³ with a square top and bottom is constructed out of two types of metal. The metal for the top and bottom costs $50/m² and the metal for the sides costs $30/m². Find the dimensions of the box that minimize total cost.

65. Let f be a function whose graph does not pass through the x-axis and let $Q = (a, 0)$. Let $P = (x_0, f(x_0))$ be the point on the graph closest to Q (Figure 5). Prove that $\overline{PQ}$ is perpendicular to the tangent line to the graph of x_0. *Hint:* Find the minimum value of the *square* of the distance from $(x, f(x))$ to $(a, 0)$.

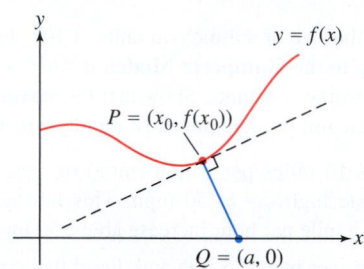

FIGURE 5

66. Take a circular piece of paper of radius R, remove a sector of angle θ (Figure 6), and fold the remaining piece into a cone-shaped cup. Which angle θ produces the cup of largest volume?

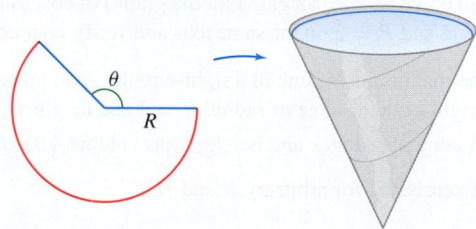

FIGURE 6

67. Use Newton's Method to estimate $\sqrt[3]{25}$ to four decimal places.

68. Use Newton's Method to find a root of $f(x) = x^2 - x - 1$ to four decimal places.

69. Find the local extrema of $f(x) = \dfrac{e^{2x} + 1}{e^{x+1}}$.

70. Find the points of inflection of $f(x) = \ln(x^2 + 1)$ and, at each point, determine whether the concavity changes from up to down or from down to up.

In Exercises 71–74, find the local extrema and points of inflection, and sketch the graph. Use L'Hôpital's Rule to determine the limits as $x \to 0+$ or $x \to \pm\infty$ if necessary.

71. $y = x \ln x$ $(x > 0)$

72. $y = e^{x-x^2}$

73. $y = x(\ln x)^2$ $(x > 0)$

74. $y = \tan^{-1}\left(\dfrac{x^2}{4}\right)$

75. [✎] Explain why L'Hôpital's Rule gives no information about $\displaystyle\lim_{x \to \infty} \dfrac{2x - \sin x}{3x + \cos 2x}$. Evaluate the limit by another method.

76. Let f be a differentiable function with inverse g which is also differentiable. Assume that $f(0) = 0$ and $f'(0) \neq 0$.

(a) Use the fact that $f(g(x)) = x$ and the Chain Rule to show that

$$g'(x) = \dfrac{1}{f'(g(x))}.$$

(b) Prove that

$$\lim_{x \to 0} \dfrac{f(x)}{g(x)} = f'(0)^2$$

In Exercises 77–88, verify that L'Hôpital's Rule applies and evaluate the limit.

77. $\displaystyle\lim_{x \to 3} \dfrac{4x - 12}{x^2 - 5x + 6}$

78. $\displaystyle\lim_{x \to -2} \dfrac{x^3 + 2x^2 - x - 2}{x^4 + 2x^3 - 4x - 8}$

79. $\displaystyle\lim_{x \to 0+} x^{1/2} \ln x$

80. $\displaystyle\lim_{t \to \infty} \dfrac{\ln(e^t + 1)}{t}$

81. $\displaystyle\lim_{\theta \to 0} \dfrac{2 \sin \theta - \sin 2\theta}{\sin \theta - \theta \cos \theta}$

82. $\displaystyle\lim_{x \to 0} \dfrac{\sqrt{4 + x} - 2\sqrt[8]{1 + x}}{x^2}$

83. $\displaystyle\lim_{t \to \infty} \dfrac{\ln(t + 2)}{\log_2 t}$

84. $\displaystyle\lim_{x \to 0} \left(\dfrac{e^x}{e^x - 1} - \dfrac{1}{x}\right)$

85. $\displaystyle\lim_{y \to 0} \dfrac{\sin^{-1} y - y}{y^3}$

86. $\displaystyle\lim_{x \to 1} \dfrac{\sqrt{1 - x^2}}{\cos^{-1} x}$

87. $\displaystyle\lim_{x \to 0} \dfrac{\sinh(x^2)}{\cosh x - 1}$

88. $\displaystyle\lim_{x \to 0} \dfrac{\tanh x - \sinh x}{\sin x - x}$

89. Let $f(x) = e^{-Ax^2/2}$, where $A > 0$ is a constant. Given any n numbers $a_1, a_2, \ldots, a_n$, set

$$\Phi(x) = f(x - a_1)f(x - a_2) \cdots f(x - a_n)$$

(a) Assume $n = 2$ and prove that Φ attains its maximum value at the average $x = \frac{1}{2}(a_1 + a_2)$. *Hint:* Calculate $\Phi'(x)$ using logarithmic differentiation.

(b) Show that for any n, Φ attains its maximum value at

$$x = \dfrac{1}{n}(a_1 + a_2 + \cdots + a_n).$$

This fact is related to the role of $f(x)$ (whose graph is a bell-shaped curve) in statistics.

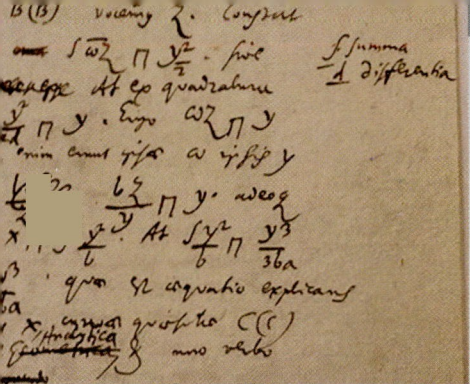

5 INTEGRATION

The last two chapters developed the derivative, one of the primary topics in calculus. The derivative was motivated by the basic problem of finding a line tangent to a curve at a given point. In addition, we saw that the derivative is much more significant than just a means for finding tangent lines.

In this chapter, we introduce the next primary topic, the definite integral. It too is motivated by a basic problem: finding the area under a curve. And in this case as well, once you become familiar with the definite integral you will realize that its importance goes well beyond computing area.

Computing tangent lines and areas may seem completely unrelated, and therefore the derivative and the definite integral may appear to be completely separate topics. They aren't. There is a deep connection between them that is revealed by the Fundamental Theorem of Calculus, discussed in Sections 5.4 and 5.5. This theorem expresses the "inverse" relationship between integration and differentiation. It plays a truly fundamental role in nearly all applications of calculus, both theoretical and practical.

5.1 Approximating and Computing Area

Why might we be interested in the area under a graph? Consider an object moving in a straight line with *constant positive velocity* v. The distance traveled over a time interval $[t_1, t_2]$ is equal to $v\Delta t$, where $\Delta t = (t_2 - t_1)$ is the time elapsed. This is the well-known formula

$$\text{distance traveled} = \underbrace{\text{velocity} \times \text{time elapsed}}_{v\Delta t} \qquad \boxed{1}$$

Because v is constant, the graph of velocity is a horizontal line (Figure 1) and $v\Delta t$ is equal to the area of the rectangular region under the graph of velocity over $[t_1, t_2]$. So we can write Eq. (1) as

$$\text{distance traveled} = \text{area under the graph of velocity over } [t_1, t_2] \qquad \boxed{2}$$

There is, however, an important difference between these two equations: Eq. (1) makes sense only if velocity v is constant, whereas Eq. (2) is correct *even if the velocity changes with time*. We examine the relationship in Eq. (2) further in Section 5.6. Thus, the advantage of expressing distance traveled as an area is that it enables us to deal with much more general types of motion.

To see why Eq. (2) might be true in general, let's consider the case where velocity changes over time but is constant on intervals. In other words, we assume that the object's velocity changes abruptly from one interval to the next as in Figure 2. The distance traveled over each time interval is equal to the area of the rectangle above that interval, so the total distance traveled is the sum of the areas of the rectangles. In Figure 2,

$$\text{distance traveled over } [0, 8] \text{ s} = \underbrace{10 + 15 + 30 + 10}_{\text{Sum of areas of rectangles}} = 65 \text{ m}$$

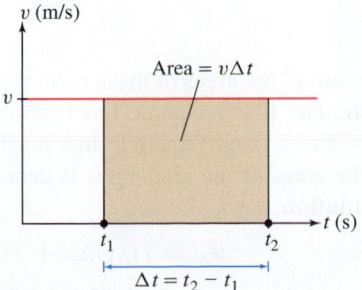

FIGURE 1 The rectangle has area $v\Delta t$, which is equal to the distance traveled.

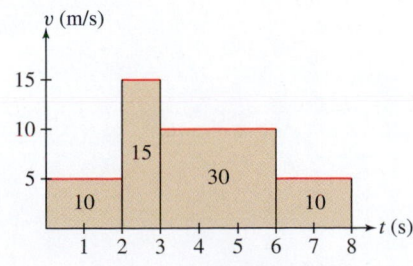

FIGURE 2 Distance traveled equals the sum of the areas of the rectangles.

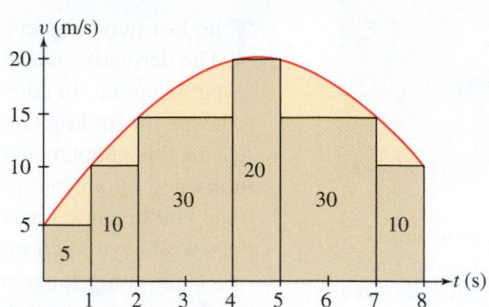

FIGURE 3 Distance traveled is equal to the area under the graph. It is *approximated* by the sum of the areas of the rectangles.

Our strategy when velocity changes continuously (Figure 3) is to *approximate* the area under the graph by a sum of areas of rectangles. We can continually improve the approximation by using thinner and thinner rectangles, and then take a limit to obtain an exact value of distance traveled. This idea leads to the concept of an integral.

Approximating Area by Rectangles

Our goal is to compute the area under the graph of a function f. In this section, we assume that f is continuous and *positive*, so that the graph of f lies above the x-axis (Figure 4). The first step is to approximate the area using rectangles.

Recall the two-step procedure for finding the slope of the tangent line (the derivative): First approximate the slope using secant lines and then compute the limit of these approximations. In integral calculus, there are also two steps:

- *First, approximate the area under the graph using rectangles.*
- *Then compute the exact area (the integral) as the limit of these approximations.*

To begin, choose a whole number N and divide $[a, b]$ into N subintervals of equal width, as in Figure 4(A). The full interval $[a, b]$ has width $b - a$, so each subinterval has width $\Delta x = (b - a)/N$. The right endpoints of the subintervals are

$$x_1 = a + \Delta x, \ x_2 = a + 2\Delta x, \ \ldots, \ x_{N-1} = a + (N-1)\Delta x, \ x_N = a + N\Delta x$$

Note that the last right endpoint is $x_N = b$ because $a + N\Delta x = a + N((b-a)/N) = b$. The general term x_j can be expressed as $x_j = a + j\Delta x$, indicating that it is obtained by adding j intervals of width Δx to a. Next, as in Figure 4(B), above each subinterval construct a rectangle whose height is the value of $f(x)$ at the *right endpoint* of the subinterval.

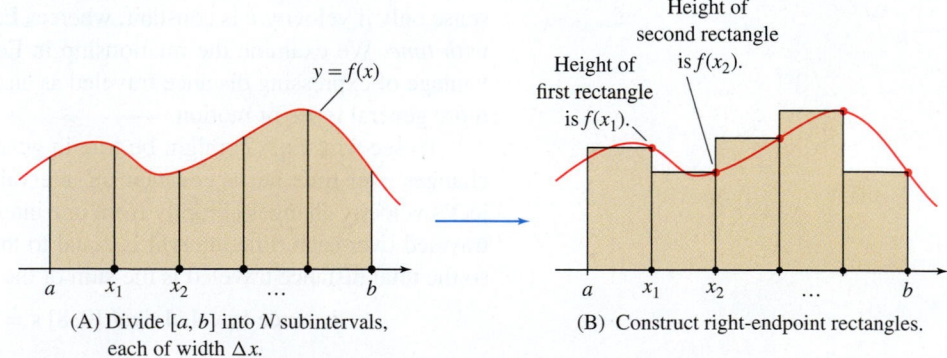

(A) Divide $[a, b]$ into N subintervals, each of width Δx.

(B) Construct right-endpoint rectangles.

FIGURE 4

The sum of the areas of these rectangles provides an *approximation* to the area under the graph. The first rectangle has base Δx and height $f(x_1)$, so its area is $f(x_1)\Delta x$. Similarly, the second rectangle has height $f(x_2)$ and area $f(x_2)\Delta x$, and so on. The sum of the areas of the rectangles is denoted R_N and is called the Nth **right-endpoint approximation**:

$$R_N = f(x_1)\Delta x + f(x_2)\Delta x + \cdots + f(x_N)\Delta x$$

Factoring out Δx, we obtain the formula

$$R_N = \Delta x \left(f(x_1) + f(x_2) + \cdots + f(x_N) \right)$$

To summarize:

a = left endpoint of interval $[a, b]$

b = right endpoint of interval $[a, b]$

N = number of subintervals in $[a, b]$

$$\Delta x = \frac{b - a}{N}$$

In words: R_N is equal to Δx times the sum of the function values at the right endpoints of the subintervals.

EXAMPLE 1 Calculate R_4 and R_6 for $f(x) = x^2$ on the interval $[1, 3]$.

Solution

Step 1. Determine Δx and the right endpoints.
To calculate R_4, divide $[1, 3]$ into four subintervals of width $\Delta x = \frac{3-1}{4} = \frac{1}{2}$. The right endpoints are the numbers $x_j = a + j\Delta x = 1 + j\left(\frac{1}{2}\right)$ for $j = 1, 2, 3, 4$. They are spaced at intervals of $\frac{1}{2}$ beginning at $\frac{3}{2}$, so, as we see in Figure 5(A), the right endpoints are $\frac{3}{2}, \frac{4}{2}, \frac{5}{2}, \frac{6}{2}$.

Step 2. Calculate Δx times the sum of function values.
R_4 is Δx times the sum of the function values at the right endpoints:

$$R_4 = \frac{1}{2}\left(f\left(\frac{3}{2}\right) + f\left(\frac{4}{2}\right) + f\left(\frac{5}{2}\right) + f\left(\frac{6}{2}\right)\right)$$

$$= \frac{1}{2}\left(\left(\frac{3}{2}\right)^2 + \left(\frac{4}{2}\right)^2 + \left(\frac{5}{2}\right)^2 + \left(\frac{6}{2}\right)^2\right) = \frac{43}{4} = 10.75$$

R_6 is similar: $\Delta x = \frac{3-1}{6} = \frac{1}{3}$, and the right endpoints are spaced at intervals of $\frac{1}{3}$ beginning at $\frac{4}{3}$ and ending at 3, as in Figure 5(B). Thus,

$$R_6 = \frac{1}{3}\left(f\left(\frac{4}{3}\right) + f\left(\frac{5}{3}\right) + f\left(\frac{6}{3}\right) + f\left(\frac{7}{3}\right) + f\left(\frac{8}{3}\right) + f\left(\frac{9}{3}\right)\right)$$

$$= \frac{1}{3}\left(\frac{16}{9} + \frac{25}{9} + \frac{36}{9} + \frac{49}{9} + \frac{64}{9} + \frac{81}{9}\right) = \frac{271}{27} \approx 10.037 \qquad \blacksquare$$

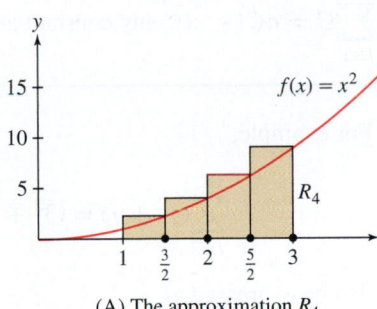

(A) The approximation R_4

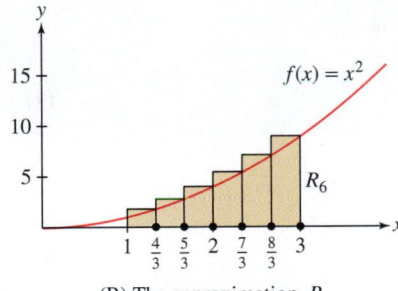

(B) The approximation R_6

DF **FIGURE 5**

The R_4 and R_6 values in the previous example are clearly overestimates of the area under $y = x^2$ between 1 and 3, although the approximation improves going from R_4 to R_6. Later in the section, we show how to obtain a general expression for R_N. With that, we can then take the limit as $N \to \infty$ to obtain an exact value of $8\frac{2}{3}$ as the area (see Exercise 55).

Summation Notation

Summation notation is a standard notation for writing sums in compact form. The sum of numbers $a_m, \ldots, a_n$ $(m \leq n)$ is denoted

$$\sum_{j=m}^{n} a_j = a_m + a_{m+1} + \cdots + a_n$$

The Greek letter $\sum$ (capital sigma) stands for "sum," and the notation $\sum\limits_{j=m}^{n}$ tells us to start the summation at $j = m$ and end it at $j = n$. For example,

$$\sum_{j=1}^{5} j^2 = 1^2 + 2^2 + 3^2 + 4^2 + 5^2 = 55$$

In this summation, the jth term is $a_j = j^2$. We refer to j^2 as the **general term**. The letter j is called the **summation index**. It is also referred to as a **dummy variable** because any other letter can be used instead. For example,

$$\sum_{k=4}^{6} \left(k^2 - 2k\right) = \overbrace{\left(4^2 - 2(4)\right)}^{k=4} + \overbrace{\left(5^2 - 2(5)\right)}^{k=5} + \overbrace{\left(6^2 - 2(6)\right)}^{k=6} = 47$$

$$\sum_{m=7}^{9} 1 = 1 + 1 + 1 = 3 \qquad (\text{because } a_7 = a_8 = a_9 = 1)$$

The usual commutative, associative, and distributive laws of addition give us the following rules for manipulating summations.

Linearity of Summations

- $\displaystyle\sum_{j=m}^{n}(a_j + b_j) = \sum_{j=m}^{n} a_j + \sum_{j=m}^{n} b_j$

- $\displaystyle\sum_{j=m}^{n} Ca_j = C \sum_{j=m}^{n} a_j \qquad (C \text{ any constant})$

- $\displaystyle\sum_{j=1}^{n} C = nC \qquad (C \text{ any constant and } n \geq 1)$

For example,

$$\sum_{j=3}^{5}(j^2 + j) = (3^2 + 3) + (4^2 + 4) + (5^2 + 5) = 62$$

can also be expressed as

$$\sum_{j=3}^{5} j^2 + \sum_{j=3}^{5} j = \left(3^2 + 4^2 + 5^2\right) + \left(3 + 4 + 5\right) = 50 + 12 = 62$$

The linearity properties can be used to write a single summation as a linear combination of several summations. For example,

$$\sum_{k=0}^{100}(7k^2 - 4k + 9) = \sum_{k=0}^{100} 7k^2 + \sum_{k=0}^{100}(-4k) + \sum_{k=0}^{100} 9$$

$$= 7\sum_{k=0}^{100} k^2 - 4\sum_{k=0}^{100} k + 9\sum_{k=0}^{100} 1$$

It is convenient to use summation notation when working with area approximations. For example, R_N is a sum with general term $f(x_j)$:

$$R_N = \Delta x\left[f(x_1) + f(x_2) + \cdots + f(x_N)\right]$$

The summation extends from $j = 1$ to $j = N$, so we can write R_N concisely as

$$R_N = \Delta x \sum_{j=1}^{N} f(x_j)$$

We shall make use of two other rectangular approximations to area: the left-endpoint and the midpoint approximations. Divide $[a, b]$ into N subintervals as before. In the **left-endpoint approximation** L_N, the heights of the rectangles are the values of $f(x)$ at the left endpoints [Figure 6(A)]. These left endpoints are

$$x_0 = a, \ x_1 = a + \Delta x, \ x_2 = a + 2\Delta x, \ \ldots, \ x_{N-1} = a + (N-1)\Delta x$$

and the sum of the areas of the left-endpoint rectangles is

$$L_N = \Delta x \Big(f(x_0) + f(x_1) + f(x_2) + \cdots + f(x_{N-1}) \Big)$$

Note that both R_N and L_N have general term $f(x_j)$, but the sum for L_N runs from $j = 0$ to $j = N - 1$ rather than from $j = 1$ to $j = N$:

$$L_N = \Delta x \sum_{j=0}^{N-1} f(x_j)$$

In the **midpoint approximation** M_N, the heights of the rectangles are the values of $f(x)$ at the midpoints of the subintervals rather than at the endpoints. As we see in Figure 6(B), the midpoints are

$$\frac{x_0 + x_1}{2}, \ \frac{x_1 + x_2}{2}, \ \ldots, \ \frac{x_{N-1} + x_N}{2}$$

The sum of the areas of the midpoint rectangles is

$$M_N = \Delta x \left(f\left(\frac{x_0 + x_1}{2}\right) + f\left(\frac{x_1 + x_2}{2}\right) + \cdots + f\left(\frac{x_{N-1} + x_N}{2}\right) \right)$$

In summation notation,

$$M_N = \Delta x \sum_{j=0}^{N-1} f\left(\frac{x_j + x_{j+1}}{2}\right)$$

← **REMINDER**

$$\Delta x = \frac{b - a}{N}$$

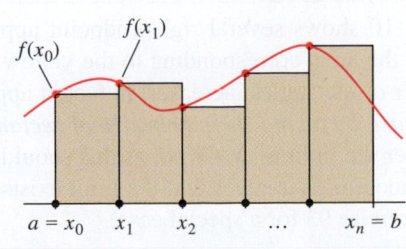

(A) Left-endpoint rectangles

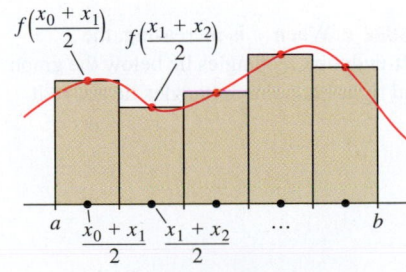

(B) Midpoint rectangles

FIGURE 6

EXAMPLE 2 Calculate R_6, L_6, and M_6 for $f(x) = x^{-1}$ on $[2, 4]$.

Solution In this case, $\Delta x = (b - a)/N = (4 - 2)/6 = \frac{1}{3}$. For the six intervals, the right endpoints are $\frac{7}{3}, \frac{8}{3}, 3, \frac{10}{3}, \frac{11}{3}$, and 4, and the left endpoints are $2, \frac{7}{3}, \frac{8}{3}, 3, \frac{10}{3}$, and $\frac{11}{3}$.

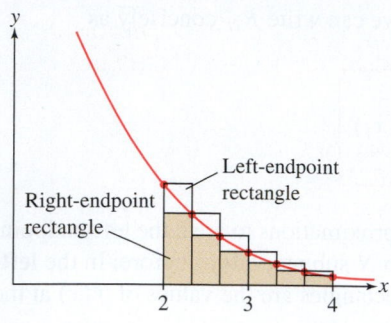

FIGURE 7 L_6 and R_6 for $f(x) = x^{-1}$ on $[2, 4]$.

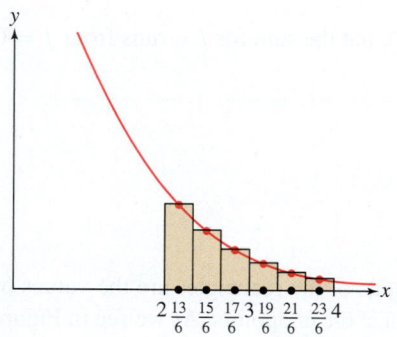

FIGURE 8 M_6 for $f(x) = x^{-1}$ on $[2, 4]$.

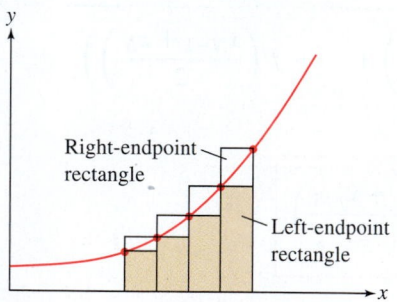

FIGURE 9 When f is increasing, the left-endpoint rectangles lie below the graph and right-endpoint rectangles lie above it.

Therefore (Figure 7),

$$R_6 = \frac{1}{3} \left(f\left(\frac{7}{3}\right) + f\left(\frac{8}{3}\right) + f(3) + f\left(\frac{10}{3}\right) + f\left(\frac{11}{3}\right) + f(4) \right)$$

$$= \frac{1}{3} \left(\frac{3}{7} + \frac{3}{8} + \frac{1}{3} + \frac{3}{10} + \frac{3}{11} + \frac{1}{4} \right) \approx 0.653$$

$$L_6 = \frac{1}{3} \left(f(2) + f\left(\frac{7}{3}\right) + f\left(\frac{8}{3}\right) + f(3) + f\left(\frac{10}{3}\right) + f\left(\frac{11}{3}\right) \right)$$

$$= \frac{1}{3} \left(\frac{1}{2} + \frac{3}{7} + \frac{3}{8} + \frac{1}{3} + \frac{3}{10} + \frac{3}{11} \right) \approx 0.737$$

The general term in M_6 is

$$f\left(\frac{x_j + x_{j+1}}{2}\right)$$

In this case, the midpoints are $\frac{13}{6}, \frac{15}{6}, \frac{17}{6}, \frac{19}{6}, \frac{21}{6}$ and $\frac{23}{6}$. Summing from $j = 0$ to 5, we obtain (Figure 8)

$$M_6 = \frac{1}{3} \left(f\left(\frac{13}{6}\right) + f\left(\frac{15}{6}\right) + f\left(\frac{17}{6}\right) + f\left(\frac{19}{6}\right) + f\left(\frac{21}{6}\right) + f\left(\frac{23}{6}\right) \right)$$

$$= \frac{1}{3} \left(\frac{6}{13} + \frac{6}{15} + \frac{6}{17} + \frac{6}{19} + \frac{6}{21} + \frac{6}{23} \right) \approx 0.692 \quad \blacksquare$$

GRAPHICAL INSIGHT **Monotonic Functions** Observe in Figure 7 that the left-endpoint rectangles for $f(x) = x^{-1}$ extend above the graph and the right-endpoint rectangles lie below it. The exact area A under the graph of f from 2 to 4 must lie between R_6 and L_6, and so, according to the previous example, $0.65 \le A \le 0.74$. More generally, *when f is monotonic (increasing or decreasing), the exact area lies between R_N and L_N* (Figure 9):

- f increasing $\implies$ $L_N \le$ area under graph $\le R_N$
- f decreasing $\implies$ $R_N \le$ area under graph $\le L_N$

Notice that M_6 lies between R_6 and L_6. This is always the case for a monotonic function (see Problem 93). In the upcoming sections, we will see how to determine the exact area in the example. It turns out to be $\ln 2 \approx 0.693$, and thus, M_6 provides the best estimate, which is understandable considering these observations.

Computing Area as the Limit of Approximations

Figure 10 shows several right-endpoint approximations. Notice that the *error* in computing the area, corresponding to the yellow region above the graph, gets smaller as the number of rectangles increases. In fact, it appears that *we can make the error as small as we please by taking the number N of rectangles large enough.* If so, it makes sense to consider the limit as $N \to \infty$, as this should give us the exact area under the curve. The next theorem guarantees that the limit exists (see Theorem 7 in Appendix D for a proof and Exercise 93 for a special case).

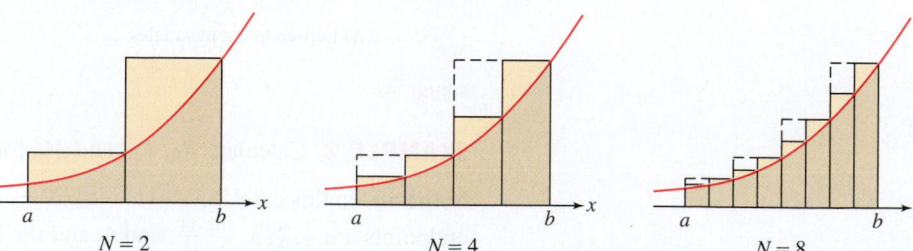

FIGURE 10 The error decreases as we use more rectangles.

In Theorem 1, it is not assumed that $f(x) \geq 0$. If $f(x)$ takes on negative values, the limit L no longer represents area under the graph, but we can interpret it as a "signed area," discussed in the next section.

THEOREM 1 If f is continuous on $[a,b]$, then the endpoint and midpoint approximations approach one and the same limit as $N \to \infty$. In other words, there is a value L such that

$$\lim_{N \to \infty} R_N = \lim_{N \to \infty} L_N = \lim_{N \to \infty} M_N = L$$

If $f(x) \geq 0$ on $[a,b]$, we define the area under the graph over $[a,b]$ to be L.

CONCEPTUAL INSIGHT In calculus, limits are used to define basic quantities that otherwise would not have a precise meaning. Theorem 1 allows us to *define* area as a limit L in much the same way that we define the slope of a tangent line as the limit of slopes of secant lines.

The next three examples illustrate Theorem 1 using formulas for **power sums**. The kth power sum is defined as the sum of the kth powers of the first N integers. We shall use the power sum formulas for $k = 1, 2, 3$.

A method for proving power sum formulas is developed in Exercises 47–51 of Section 1.3. Formulas (3)–(5) can also be verified using the method of induction.

Power Sums

$$\sum_{j=1}^{N} j = 1 + 2 + \cdots + N = \frac{N(N+1)}{2} = \frac{N^2}{2} + \frac{N}{2} \qquad \boxed{3}$$

$$\sum_{j=1}^{N} j^2 = 1^2 + 2^2 + \cdots + N^2 = \frac{N(N+1)(2N+1)}{6} = \frac{N^3}{3} + \frac{N^2}{2} + \frac{N}{6} \qquad \boxed{4}$$

$$\sum_{j=1}^{N} j^3 = 1^3 + 2^3 + \cdots + N^3 = \frac{N^2(N+1)^2}{4} = \frac{N^4}{4} + \frac{N^3}{2} + \frac{N^2}{4} \qquad \boxed{5}$$

For example, by Eq. (4),

$$\sum_{j=1}^{6} j^2 = 1^2 + 2^2 + 3^2 + 4^2 + 5^2 + 6^2 = \underbrace{\frac{6^3}{3} + \frac{6^2}{2} + \frac{6}{6}}_{\frac{N^3}{3} + \frac{N^2}{2} + \frac{N}{6} \text{ for } N = 6} = 91$$

As a first illustration, we verify this limit approach to area by computing the area of a right triangle, a figure whose area we can also compute geometrically.

EXAMPLE 3 Find the area A under the graph of $f(x) = x$ over $[0, 4]$ in three ways:

(a) Using geometry **(b)** $\lim_{N \to \infty} R_N$ **(c)** $\lim_{N \to \infty} L_N$

← *REMINDER*

$$R_N = \Delta x \sum_{j=1}^{N} f(x_j)$$

$$L_N = \Delta x \sum_{j=0}^{N-1} f(x_j)$$

$$\Delta x = \frac{b-a}{N}$$

$$x_j = a + j\Delta x$$

Solution The region under the graph is a right triangle with base $b = 4$ and height $h = 4$ (Figure 11).

(a) By geometry, $A = \frac{1}{2}bh = \left(\frac{1}{2}\right)(4)(4) = 8$.

(b) We compute this area again as a limit. Since $\Delta x = (b-a)/N = 4/N$ and $f(x) = x$,

$$f(x_j) = f(a + j\Delta x) = f\left(0 + j\left(\frac{4}{N}\right)\right) = \frac{4j}{N}$$

$$R_N = \Delta x \sum_{j=1}^{N} f(x_j) = \frac{4}{N} \sum_{j=1}^{N} \frac{4j}{N} = \frac{16}{N^2} \sum_{j=1}^{N} j$$

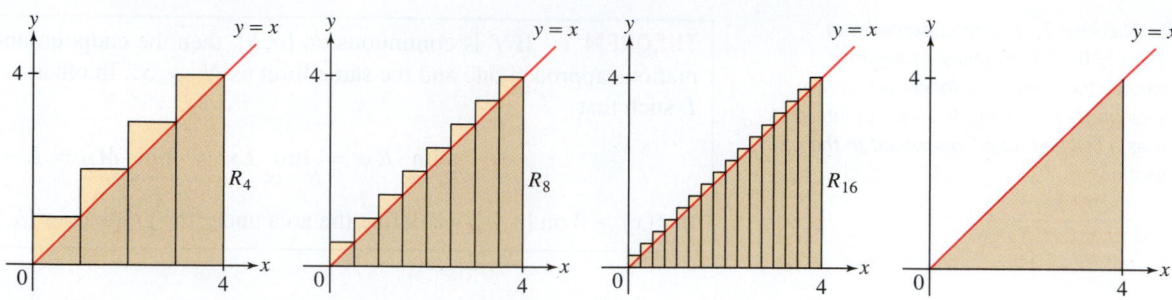

DF **FIGURE 11** The right-endpoint approximations approach the area of the triangle.

In the last equality, we factored out $4/N$ from the sum. This is valid because $4/N$ is a constant that does not depend on j. Now use formula (3):

$$R_N = \frac{16}{N^2} \sum_{j=1}^{N} j = \frac{16}{N^2} \underbrace{\left(\frac{N(N+1)}{2} \right)}_{\text{Formula for power sum}} = \frac{8}{N^2}\left(N^2 + N\right) = 8 + \frac{8}{N}$$

The second term $8/N$ tends to zero as N approaches ∞, so

$$A = \lim_{N \to \infty} R_N = \lim_{N \to \infty} \left(8 + \frac{8}{N}\right) = 8$$

As expected, this limit yields the same value as the formula $\frac{1}{2}bh$.

(c) The left-endpoint approximation is similar, but the sum begins at $j = 0$ and ends at $j = N - 1$:

In Eq. (6), we apply the formula

$$\sum_{j=1}^{N} j = \frac{N(N+1)}{2}$$

with $N - 1$ in place of N:

$$\sum_{j=1}^{N-1} j = \frac{(N-1)N}{2}$$

$$L_N = \frac{16}{N^2} \sum_{j=0}^{N-1} j = \frac{16}{N^2} \sum_{j=1}^{N-1} j = \frac{16}{N^2}\left(\frac{(N-1)N}{2} \right) = 8 - \frac{8}{N} \qquad \boxed{6}$$

Note in the second step that we replaced the sum beginning at $j = 0$ with a sum beginning at $j = 1$. This is valid because the term for $j = 0$ is zero and may be dropped. Again, we find that $A = \lim_{N \to \infty} L_N = \lim_{N \to \infty} (8 - 8/N) = 8$. ∎

In the next example, we compute the area under a curved graph. Unlike the previous example, it is not possible to compute this area directly using geometry.

EXAMPLE 4 Let A be the area under the graph of $f(x) = 2x^2 - x + 3$ over $[2, 4]$ (Figure 12). Compute A as the limit $\lim_{N \to \infty} R_N$.

Solution

Step 1. **Express R_N in terms of power sums.**
In this case, $\Delta x = (4 - 2)/N = 2/N$ and

$$R_N = \Delta x \sum_{j=1}^{N} f(x_j) = \Delta x \sum_{j=1}^{N} f(a + j \Delta x) = \frac{2}{N} \sum_{j=1}^{N} f\left(2 + \frac{2j}{N}\right)$$

Let's use algebra to simplify the general term. Since $f(x) = 2x^2 - x + 3$,

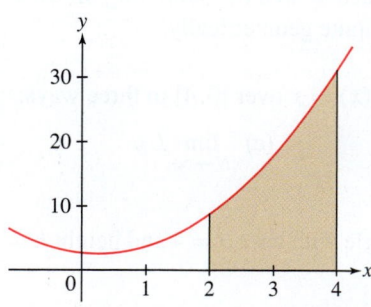

FIGURE 12 Area under the graph of $f(x) = 2x^2 - x + 3$ over $[2, 4]$.

$$f\left(2 + \frac{2j}{N}\right) = 2\left(2 + \frac{2j}{N}\right)^2 - \left(2 + \frac{2j}{N}\right) + 3$$

$$= 2\left(4 + \frac{8j}{N} + \frac{4j^2}{N^2}\right) - \left(2 + \frac{2j}{N}\right) + 3 = \frac{8}{N^2}j^2 + \frac{14}{N}j + 9$$

Now we can express R_N in terms of power sums:

$$R_N = \frac{2}{N}\sum_{j=1}^{N}\left(\frac{8}{N^2}j^2 + \frac{14}{N}j + 9\right) = \frac{2}{N}\sum_{j=1}^{N}\frac{8}{N^2}j^2 + \frac{2}{N}\sum_{j=1}^{N}\frac{14}{N}j + \frac{2}{N}\sum_{j=1}^{N}9$$

$$= \frac{16}{N^3}\sum_{j=1}^{N}j^2 + \frac{28}{N^2}\sum_{j=1}^{N}j + \frac{18}{N}\sum_{j=1}^{N}1 \qquad \boxed{7}$$

Step 2. Use the formulas for the power sums.

Using formulas (3) and (4) for the power sums in Eq. (7), we obtain

$$R_N = \frac{16}{N^3}\left(\frac{N^3}{3} + \frac{N^2}{2} + \frac{N}{6}\right) + \frac{28}{N^2}\left(\frac{N^2}{2} + \frac{N}{2}\right) + \frac{18}{N}(N)$$

$$= \left(\frac{16}{3} + \frac{8}{N} + \frac{8}{3N^2}\right) + \left(14 + \frac{14}{N}\right) + 18$$

$$= \frac{112}{3} + \frac{22}{N} + \frac{8}{3N^2}$$

Step 3. Calculate the limit.

$$A = \lim_{N\to\infty} R_N = \lim_{N\to\infty}\left(\frac{112}{3} + \frac{22}{N} + \frac{8}{3N^2}\right) = \frac{112}{3} \qquad ■$$

The area under the graph of any polynomial can be calculated using power sum formulas as in the examples above. For other functions, such as in the next example, the limit defining the area may be difficult or impossible to evaluate directly.

EXAMPLE 5 Let A be the area under the graph of $f(x) = \sin x$ on the interval $\left[\frac{\pi}{4}, \frac{3\pi}{4}\right]$ (Figure 13). Set up (but do not compute) the expression $A = \lim_{N\to\infty} R_N$ for determining the area.

Solution In this case, $\Delta x = (3\pi/4 - \pi/4)/N = \pi/(2N)$ and the area A is

$$A = \lim_{N\to\infty} R_N = \lim_{N\to\infty} \Delta x \sum_{j=1}^{N} f(a + j\Delta x) = \lim_{N\to\infty}\frac{\pi}{2N}\sum_{j=1}^{N}\sin\left(\frac{\pi}{4} + \frac{\pi j}{2N}\right) \qquad ■$$

The limit in the previous example is difficult to compute; however, it can be approximated by computing R_N for large N (see Exercise 83). In Section 5.4, we will see that this area is straightforward to compute (and is $\sqrt{2}$) using the definite integral and the Fundamental Theorem of Calculus.

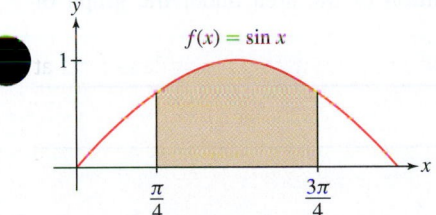

FIGURE 13 The area of this region is more difficult to compute as a limit of endpoint approximations.

HISTORICAL PERSPECTIVE

Jacob Bernoulli
(1654–1705)

We used the formulas for the kth power sums for $k = 1, 2, 3$. Do similar formulas exist for all powers k? This problem was studied in the seventeenth century and eventually solved around 1690 by the great Swiss mathematician Jacob Bernoulli. Of this discovery, he wrote

With the help of [these formulas] it took me less than half of a quarter of an hour to find that the 10th powers of the first 1000 numbers being added together will yield the sum

91409924241424243424241924242500

Bernoulli's formula has the general form

$$\sum_{j=1}^{n} j^k = \frac{1}{k+1}n^{k+1} + \frac{1}{2}n^k + \frac{k}{12}n^{k-1} + \cdots$$

The dots indicate terms involving smaller powers of n whose coefficients are expressed in terms of the so-called Bernoulli numbers. For example,

$$\sum_{j=1}^{n} j^4 = \frac{1}{5}n^5 + \frac{1}{2}n^4 + \frac{1}{3}n^3 - \frac{1}{30}n$$

These formulas are available on most computer algebra systems.

5.1 SUMMARY

Power Sums

$$\sum_{j=1}^{N} j = \frac{N(N+1)}{2} = \frac{N^2}{2} + \frac{N}{2}$$

$$\sum_{j=1}^{N} j^2 = \frac{N(N+1)(2N+1)}{6} = \frac{N^3}{3} + \frac{N^2}{2} + \frac{N}{6}$$

$$\sum_{j=1}^{N} j^3 = \frac{N^2(N+1)^2}{4} = \frac{N^4}{4} + \frac{N^3}{2} + \frac{N^2}{4}$$

- Approximations to the area under the graph of f over the interval $[a, b]$
$$\left(\Delta x = \frac{b-a}{N}, x_j = a + j\Delta x\right):$$

$$R_N = \Delta x \sum_{j=1}^{N} f(x_j) = \Delta x \big(f(x_1) + f(x_2) + \cdots + f(x_N)\big)$$

$$L_N = \Delta x \sum_{j=0}^{N-1} f(x_j) = \Delta x \big(f(x_0) + f(x_1) + \cdots + f(x_{N-1})\big)$$

$$M_N = \Delta x \sum_{j=0}^{N-1} f\left(\frac{x_j + x_{j+1}}{2}\right)$$

$$= \Delta x \left(f\left(\frac{x_0 + x_1}{2}\right) + \cdots + f\left(\frac{x_{N-1} + x_N}{2}\right)\right)$$

- If f is continuous on $[a, b]$, then the endpoint and midpoint approximations approach one and the same limit L:

$$\lim_{N \to \infty} R_N = \lim_{N \to \infty} L_N = \lim_{N \to \infty} M_N = L$$

- If $f(x) \geq 0$ on $[a, b]$, we take L as the definition of the area under the graph of $y = f(x)$ over $[a, b]$.

5.1 EXERCISES

Preliminary Questions

1. What are the right and left endpoints if $[2, 5]$ is divided into six subintervals?

2. The interval $[1, 5]$ is divided into eight subintervals.
(a) What is the left endpoint of the last subinterval?
(b) What are the right endpoints of the first two subintervals?

3. Which of the following pairs of sums are *not* equal?

(a) $\sum_{i=1}^{4} i, \quad \sum_{\ell=1}^{4} \ell$

(b) $\sum_{j=1}^{4} j^2, \quad \sum_{k=2}^{5} k^2$

(c) $\sum_{j=1}^{4} j, \quad \sum_{i=2}^{5} (i-1)$

(d) $\sum_{i=1}^{4} i(i+1), \quad \sum_{j=2}^{5} (j-1)j$

4. Explain: $\sum_{j=1}^{100} j = \sum_{j=0}^{100} j$ but $\sum_{j=1}^{100} 1$ is not equal to $\sum_{j=0}^{100} 1$.

5. Explain why $L_{100} \geq R_{100}$ for $f(x) = x^{-2}$ on $[3, 7]$.

Exercises

1. Figure 14 shows the velocity of an object over a 3-minute interval. Determine the distance traveled over the intervals $[0, 3]$ and $[1, 2.5]$ (remember to convert from kilometers per hour to kilometers per minute).

2. An ostrich (Figure 15) runs with velocity 20 km/hour for 2 minutes (min), 12 km/h for 3 min, and 40 km/h for another minute. Compute the total distance traveled and indicate with a graph how this quantity can be interpreted as an area.

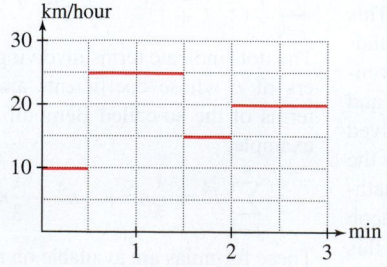

FIGURE 14

FIGURE 15 Ostriches can reach speeds as high as 70 km/h.

3. A rainstorm hit Portland, Maine, in October 1996, resulting in record rainfall. The rainfall rate $R(t)$ on October 21 is recorded, in centimeters per hour, in the following table, where t is the number of hours since midnight. Compute the total rainfall during this 24-hour period and indicate on a graph how this quantity can be interpreted as an area.

t (h)	0–2	2–4	4–9	9–12	12–20	20–24
$R(t)$ (cm/h)	0.5	0.3	1.0	2.5	1.5	0.6

4. The velocity of an object is $v(t) = 12t$ m/s. Use Eq. (2) and geometry to find the distance traveled over the time intervals $[0, 2]$ and $[2, 5]$.

5. Compute R_5 and L_5 over $[0, 1]$ using the following values:

x	0	0.2	0.4	0.6	0.8	1
$f(x)$	50	48	46	44	42	40

6. Compute R_6, L_6, and M_3 to estimate the distance traveled over $[0, 3]$ if the velocity at half-second intervals is as follows:

t (s)	0	0.5	1	1.5	2	2.5	3
v (m/s)	0	12	18	25	20	14	20

7. Let $f(x) = 2x + 3$.
(a) Compute R_6 and L_6 over $[0, 3]$.
(b) Use geometry to find the exact value of the area A, and compute the errors $|A - R_6|$ and $|A - L_6|$ in the approximations.

8. Repeat Exercise 7 for $f(x) = 20 - 3x$ over $[2, 4]$.

9. Calculate R_3 and L_3 for $f(x) = x^2 - x + 4$ over $[1, 4]$. Then sketch the graph of f and the rectangles that make up each approximation. Is the area under the graph larger or smaller than R_3? Is it larger or smaller than L_3?

10. Let $f(x) = \sqrt{x^2 + 1}$ and $\Delta x = \frac{1}{3}$. Sketch the graph of f and draw the right-endpoint rectangles whose area is represented by the sum

$$\sum_{i=1}^{6} f(1 + i\Delta x)\Delta x.$$

11. Estimate R_3, M_3, and L_6 over $[0, 1.5]$ for the function in Figure 16.

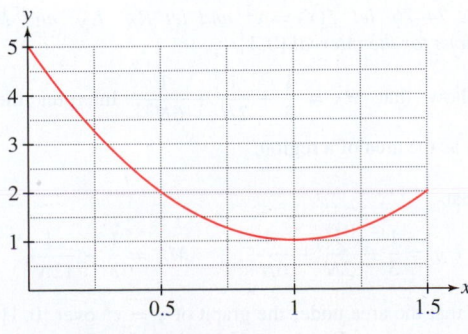

FIGURE 16

12. Calculate the area of the shaded rectangles in Figure 17. Which approximation do these rectangles represent?

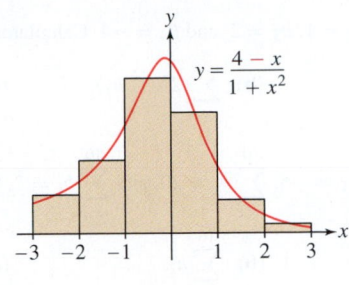

$$y = \frac{4 - x}{1 + x^2}$$

FIGURE 17

13. Let $f(x) = x^2$.
(a) Sketch the function over the interval $[0, 2]$ and the rectangles corresponding to L_4. Calculate the area contained within them.
(b) Sketch the function over the interval $[0, 2]$ again but with the rectangles corresponding to R_4. Calculate the area contained within them.
(c) Make a conclusion about the area under the curve $f(x) = x^2$ over the interval $[0, 2]$.

14. Let $f(x) = \sqrt{x}$.
(a) Sketch the function over the interval $[0, 4]$ and the rectangles corresponding to L_4. Calculate the area contained within them.
(b) Sketch the function over the interval $[0, 4]$ again but with the rectangles corresponding to R_4. Calculate the area contained within them.
(c) Make a conclusion about the area under the curve $f(x) = \sqrt{x}$ over the interval $[0, 4]$.

In Exercises 15–22, calculate the approximation for the given function and interval.

15. L_4, $f(x) = \sqrt{2 - x}$, $[0, 2]$

16. L_6, $f(x) = \sqrt{6x + 2}$, $[1, 3]$

17. R_6, $f(x) = 2x - x^2$, $[0, 2]$

18. R_5, $f(x) = x^2 + x$, $[-1, 1]$

19. M_5, $f(x) = \ln x$, $[1, 3]$

20. M_4, $f(x) = 2^{-x}$, $[1, 3]$

21. L_4, $f(x) = \cos^2 x$, $\left[\frac{\pi}{6}, \frac{\pi}{2}\right]$

22. L_6, $f(x) = x^2 + 3|x|$, $[-2, 1]$

In Exercises 23–28, write the sum in summation notation.

23. $4^7 + 5^7 + 6^7 + 7^7 + 8^7$

24. $(2^2 + 2) + (3^2 + 3) + (4^2 + 4) + (5^2 + 5)$

25. $(2^2 + 2) + (2^3 + 2) + (2^4 + 2) + (2^5 + 2)$

26. $1 + \frac{1}{2} + \frac{1}{4} + \frac{1}{8} + \frac{1}{16} + \frac{1}{32}$

27. $\dfrac{1}{2 \cdot 3} + \dfrac{2}{3 \cdot 4} + \cdots + \dfrac{n}{(n + 1)(n + 2)}$

28. $e^\pi + e^{\pi/2} + e^{\pi/3} + \cdots + e^{\pi/n}$

29. Calculate the sums:

(a) $\displaystyle\sum_{i=1}^{5} 9$ **(b)** $\displaystyle\sum_{i=0}^{5} 4$ **(c)** $\displaystyle\sum_{k=2}^{4} k^3$

30. Calculate the sums:

(a) $\displaystyle\sum_{j=3}^{4} \sin\left(j\frac{\pi}{2}\right)$ **(b)** $\displaystyle\sum_{k=3}^{5} \frac{1}{k - 1}$ **(c)** $\displaystyle\sum_{j=0}^{2} 3^{j-1}$

31. Let $b_1 = 4$, $b_2 = 1$, $b_3 = 2$, and $b_4 = -4$. Calculate:

(a) $\displaystyle\sum_{i=2}^{4} b_i$ **(b)** $\displaystyle\sum_{j=1}^{2} (2^{b_j} - b_j)$ **(c)** $\displaystyle\sum_{k=1}^{3} k b_k$

32. Assume that $a_1 = -5$, $\displaystyle\sum_{i=1}^{10} a_i = 20$, and $\displaystyle\sum_{i=1}^{10} b_i = 7$. Calculate:

(a) $\displaystyle\sum_{i=1}^{10} (4a_i + 3)$ **(b)** $\displaystyle\sum_{i=2}^{10} a_i$ **(c)** $\displaystyle\sum_{i=1}^{10} (2a_i - 3b_i)$

33. Calculate $\displaystyle\sum_{j=101}^{200} j$. *Hint:* Write as a difference of two sums and use formula (3).

34. Calculate $\displaystyle\sum_{j=1}^{30} (2j + 1)^2$. *Hint:* Expand and use formulas (3)–(4).

In Exercises 35–42, use linearity and formulas (3)–(5) to rewrite and evaluate the sums.

35. $\displaystyle\sum_{j=1}^{20} 8j^3$ **36.** $\displaystyle\sum_{k=1}^{30} (4k - 3)$

37. $\displaystyle\sum_{n=51}^{150} n^2$ **38.** $\displaystyle\sum_{k=101}^{200} k^3$

39. $\displaystyle\sum_{j=0}^{50} j(j - 1)$ **40.** $\displaystyle\sum_{j=2}^{30} \left(6j + \frac{4j^2}{3}\right)$

41. $\displaystyle\sum_{m=1}^{30} (4 - m)^3$ **42.** $\displaystyle\sum_{m=1}^{20} \left(5 + \frac{3m}{2}\right)^2$

In Exercises 43–46, use formulas (3)–(5) to evaluate the limit.

43. $\displaystyle\lim_{N \to \infty} \sum_{i=1}^{N} \frac{i}{N^2}$ **44.** $\displaystyle\lim_{N \to \infty} \sum_{j=1}^{N} \frac{j^3}{N^4}$

45. $\displaystyle\lim_{N \to \infty} \sum_{i=1}^{N} \frac{i^2 - i + 1}{N^3}$ **46.** $\displaystyle\lim_{N \to \infty} \sum_{i=1}^{N} \left(\frac{i^3}{N^4} - \frac{20}{N}\right)$

In Exercises 47–52, calculate the limit for the given function and interval. Verify your answer by using geometry.

47. $\displaystyle\lim_{N \to \infty} R_N$, $f(x) = 9x$, $[0, 2]$

48. $\displaystyle\lim_{N \to \infty} R_N$, $f(x) = 3x + 6$, $[1, 4]$

49. $\displaystyle\lim_{N \to \infty} L_N$, $f(x) = \frac{1}{2}x + 2$, $[0, 4]$

50. $\displaystyle\lim_{N \to \infty} L_N$, $f(x) = 4x - 2$, $[1, 3]$

51. $\displaystyle\lim_{N \to \infty} M_N$, $f(x) = x$, $[0, 2]$

52. $\displaystyle\lim_{N \to \infty} M_N$, $f(x) = 12 - 4x$, $[2, 6]$

53. Show, for $f(x) = 3x^2 + 4x$ over $[0, 2]$, that

$$R_N = \frac{2}{N} \sum_{j=1}^{N} \left(\frac{12j^2}{N^2} + \frac{8j}{N}\right)$$

Then evaluate $\displaystyle\lim_{N \to \infty} R_N$.

54. Show, for $f(x) = 3x^3 - x^2$ over $[1, 5]$, that

$$R_N = \frac{4}{N} \sum_{j=1}^{N} \left(\frac{192j^3}{N^3} + \frac{128j^2}{N^2} + \frac{28j}{N} + 2\right)$$

Then evaluate $\displaystyle\lim_{N \to \infty} R_N$.

In Exercises 55–62, find a formula for R_N and compute the area under the graph as a limit.

55. $f(x) = x^2$, $[1, 3]$ **56.** $f(x) = x^2$, $[-1, 5]$

57. $f(x) = 6x^2 - 4$, $[2, 5]$ **58.** $f(x) = x^2 + 7x$, $[6, 11]$

59. $f(x) = x^3 - x$, $[0, 2]$

60. $f(x) = 2x^3 + x^2$, $[-2, 2]$

61. $f(x) = 2x + 1$, $[a, b]$ (a, b constants with $a < b$)

62. $f(x) = x^2$, $[a, b]$ (a, b constants with $a < b$)

In Exercises 63–66, describe the area represented by the limits.

63. $\displaystyle\lim_{N \to \infty} \frac{1}{N} \sum_{j=1}^{N} \left(\frac{j}{N}\right)^4$ **64.** $\displaystyle\lim_{N \to \infty} \frac{3}{N} \sum_{j=1}^{N} \left(2 + \frac{3j}{N}\right)^4$

65. $\displaystyle\lim_{N \to \infty} \frac{5}{N} \sum_{j=0}^{N-1} e^{-2+5j/N}$

66. $\displaystyle\lim_{N \to \infty} \frac{\pi}{2N} \sum_{j=1}^{N} \sin\left(\frac{\pi}{3} - \frac{\pi}{4N} + \frac{j\pi}{2N}\right)$

In Exercises 67–72, express the area under the graph as a limit using the approximation indicated (in summation notation), but do not evaluate.

67. R_N, $f(x) = \sin x$ over $[0, \pi]$

68. R_N, $f(x) = x^{-1}$ over $[1, 7]$

69. L_N, $f(x) = \sqrt{2x + 1}$ over $[7, 11]$

70. L_N, $f(x) = \cos x$ over $\left[\frac{\pi}{8}, \frac{\pi}{4}\right]$

71. M_N, $f(x) = \tan x$ over $\left[\frac{1}{2}, 1\right]$

72. M_N, $f(x) = x^{-2}$ over $[3, 5]$

73. Evaluate $\displaystyle\lim_{N \to \infty} \frac{1}{N} \sum_{j=1}^{N} \sqrt{1 - \left(\frac{j}{N}\right)^2}$ by interpreting it as the area of part of a familiar geometric figure.

In Exercises 74–76, let $f(x) = x^2$ and let R_N, L_N, and M_N be the approximations for the interval $[0, 1]$.

74. ☑ Show that $R_N = \frac{1}{3} + \frac{1}{2N} + \frac{1}{6N^2}$. Interpret the quantity $\frac{1}{2N} + \frac{1}{6N^2}$ as the area of a region.

75. Show that

$$L_N = \frac{1}{3} - \frac{1}{2N} + \frac{1}{6N^2}, \qquad M_N = \frac{1}{3} - \frac{1}{12N^2}$$

Then, given that the area under the graph of $y = x^2$ over $[0, 1]$ is $\frac{1}{3}$, rank the three approximations R_N, L_N, and M_N in order of increasing accuracy (use Exercise 74).

76. For each of R_N, L_N, and M_N, find the smallest integer N for which the error is less than 0.001.

In Exercises 77–82, use the Graphical Insight on page 290 to obtain bounds on the area.

77. Let A be the area under $f(x) = \sqrt{x}$ over $[0, 1]$. By computing R_4 and L_4, prove that $0.51 \le A \le 0.77$. Explain your reasoning.

78. Use R_5 and L_5 to show that the area A under $y = x^{-2}$ over $[10, 13]$ satisfies $0.0218 \le A \le 0.0244$.

79. Use R_4 and L_4 to show that the area A under the graph of $y = \sin x$ over $\left[0, \frac{\pi}{2}\right]$ satisfies $0.79 \le A \le 1.19$.

80. Show that the area A under $f(x) = x^{-1}$ over $[1, 8]$ satisfies

$$\tfrac{1}{2} + \tfrac{1}{3} + \tfrac{1}{4} + \tfrac{1}{5} + \tfrac{1}{6} + \tfrac{1}{7} + \tfrac{1}{8} \le A \le 1 + \tfrac{1}{2} + \tfrac{1}{3} + \tfrac{1}{4} + \tfrac{1}{5} + \tfrac{1}{6} + \tfrac{1}{7}$$

81. **CAS** Show that the area A under $y = x^{1/4}$ over $[0, 1]$ satisfies $L_N \le A \le R_N$ for all N. Use a computer algebra system to calculate L_N and R_N for $N = 100$ and 200, and determine A to two decimal places.

82. **CAS** Show that the area A under $y = 4/(x^2 + 1)$ over $[0, 1]$ satisfies $R_N \le A \le L_N$ for all N. Determine A to at least three decimal places using a computer algebra system. Can you guess the exact value of A?

83. **CAS** Compute R_{100} from Example 5, approximating the area under the graph of $f(x) = \sin x$ between $\pi/4$ and $3\pi/4$.

84. **CAS** Compute R_{100} from Exercise 67, approximating the area under the graph of $f(x) = \sin x$ between 0 and π. Can you guess the exact value of the area?

85. **CAS** Compute R_{100}, approximating the area under the graph of $f(x) = \frac{1}{x}$ between 1 and e. Can you guess the exact value of the area?

86. **CAS** Separately for $n = 2, 3, 4,$ and 9, compute R_{100}, approximating the area under the graph of $f(x) = x^n$ between 0 and 1. Can you guess what the area is in general, expressed in terms of n?

87. In this exercise, we evaluate the area A under the graph of $y = e^x$ over $[0, 1]$ [Figure 18(A)] using the formula for a geometric sum (valid for $r \ne 1$):

$$1 + r + r^2 + \cdots + r^{N-1} = \sum_{j=0}^{N-1} r^j = \frac{r^N - 1}{r - 1} \qquad \boxed{8}$$

(a) Show that $L_N = \dfrac{1}{N} \displaystyle\sum_{j=0}^{N-1} e^{j/N}$.

(b) Apply Eq. (8) with $r = e^{1/N}$ to prove $L_N = \dfrac{e - 1}{N(e^{1/N} - 1)}$.

(c) Compute $A = \displaystyle\lim_{N \to \infty} L_N$ using L'Hôpital's Rule.

88. Use the result of Exercise 87 to show that the area B under the graph of $f(x) = \ln x$ over $[1, e]$ is equal to 1. *Hint:* Relate B in Figure 18(B) to the area A computed in Exercise 87.

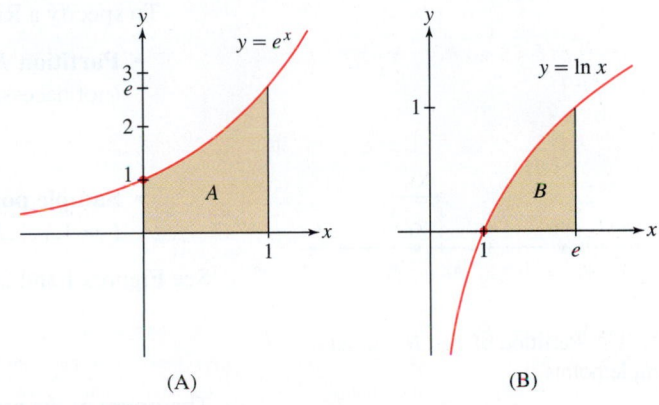

(A)   (B)

FIGURE 18

Further Insights and Challenges

89. Although the accuracy of R_N generally improves as N increases, this need not be true for small values of N. Draw the graph of a positive continuous function f on an interval such that R_1 is closer than R_2 to the exact area under the graph. Can such a function be monotonic?

90. Draw the graph of a positive continuous function on an interval such that R_2 and L_2 are both smaller than the exact area under the graph. Can such a function be monotonic?

91. Explain graphically: *The endpoint approximations are less accurate when $f'(x)$ is large.*

92. Prove that for any function f on $[a, b]$,

$$R_N - L_N = \frac{b - a}{N}(f(b) - f(a)) \qquad \boxed{9}$$

93. In this exercise, we prove that $\displaystyle\lim_{N \to \infty} R_N$ and $\displaystyle\lim_{N \to \infty} L_N$ exist and are equal if f is increasing (the case of f decreasing is similar). We use the concept of a least upper bound discussed in Appendix B.

(a) Explain with a graph why $L_N \le R_M$ for all $N, M \ge 1$.

(b) By (a), the sequence $\{L_N\}$ is bounded, so it has a least upper bound L. By definition, L is the smallest number such that $L_N \le L$ for all N. Show that $L \le R_M$ for all M.

(c) According to (b), $L_N \le L \le R_N$ for all N. Use Eq. (9) to show that $\displaystyle\lim_{N \to \infty} L_N = L$ and $\displaystyle\lim_{N \to \infty} R_N = L$.

94. Use Eq. (9) to show that if f is positive and monotonic, then the area A under its graph over $[a, b]$ satisfies

$$|R_N - A| \le \frac{b - a}{N} |f(b) - f(a)| \qquad \boxed{10}$$

In Exercises 95–96, use Eq. (10) to find N such that $|R_N - A| < 10^{-4}$ for the given function and interval.

95. $f(x) = \sqrt{x}, \quad [1, 4]$

96. $f(x) = \sqrt{9 - x^2}, \quad [0, 3]$

97. Prove that if f is positive and monotonic, then M_N lies between R_N and L_N and is closer to the actual area under the graph than both R_N and L_N. *Hint:* In the case that f is increasing, Figure 19 shows that the part of the error in R_N due to the ith rectangle is the sum of the areas $A + B + D$, and for M_N it is $|B - E|$. On the other hand, $A \ge E$.

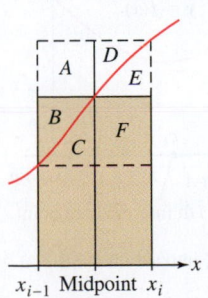

x_{i-1} Midpoint x_i

FIGURE 19

5.2 The Definite Integral

In the previous section, we saw that if f is continuous on an interval $[a, b]$, then the endpoint and midpoint approximations approach a common limit L as $N \to \infty$:

$$L = \lim_{N \to \infty} R_N = \lim_{N \to \infty} L_N = \lim_{N \to \infty} M_N \qquad \boxed{1}$$

In a moment, we will state formally that L is the *definite integral* of f over $[a, b]$. Before doing so, we introduce **Riemann sums**, sums that are structured similar to the approximating sums L_N, R_N, and M_N, and that generalize them.

Recall that R_N, L_N, and M_N use rectangles of equal width Δx, whose heights are the values of $f(x)$ at the endpoints or midpoints of the subintervals. In Riemann sums, we relax these requirements: The rectangles need not have equal width, $f(x)$ can have any real-number value, and the height (which could be negative) may be *any* value of $f(x)$ within the subinterval.

To specify a Riemann sum, we choose a partition and a set of sample points:

- **Partition** P of size N: a choice of points that divides $[a, b]$ into N subintervals (not necessarily of equal width):

$$P : a = x_0 < x_1 < x_2 < \cdots < x_N = b$$

- **Sample points** $C = \{c_1, \ldots, c_N\}$: c_i belongs to the subinterval $[x_{i-1}, x_i]$ for all $i = 1, \ldots, N$, (and could be any point in the subinterval).

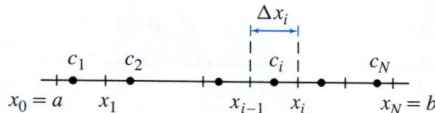

See Figures 1 and 2(A). The length of the ith subinterval $[x_{i-1}, x_i]$ is

$$\boxed{\Delta x_i = x_i - x_{i-1}}$$

FIGURE 1 Partition of size N and set of sample points.

The **norm** of the partition P, denoted $\|P\|$, is the maximum of the lengths Δx_i.

Given P and C, as Figure 2(B) illustrates, on each subinterval $[x_{i-1}, x_i]$ we have a rectangle whose height is $f(c_i)$. We allow the possibility that $f(c_i) < 0$. For each rectangle, we refer to $f(c_i)\Delta x_i$ as its **signed area**. Note that

- If $f(c_i) > 0$, the signed area is positive and is equal to the area of the rectangle
- If $f(c_i) < 0$ then the signed area is negative and is equal to the negative of the area of the rectangle

The Riemann sum is the sum of the $f(c_i)\Delta x_i$ terms that are determined by P and C, and it is expressed as

Keep in mind that R_N, L_N, and M_N are particular examples of Riemann sums in which $\Delta x_i = (b - a)/N$ for all i, and the sample points c_i are either endpoints or midpoints.

$$R(f, P, C) = \sum_{i=1}^{N} f(c_i)\Delta x_i = f(c_1)\Delta x_1 + f(c_2)\Delta x_2 + \cdots + f(c_N)\Delta x_N \qquad \boxed{2}$$

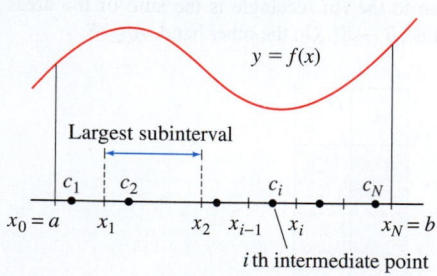

(A) Partition of $[a, b]$ into subintervals

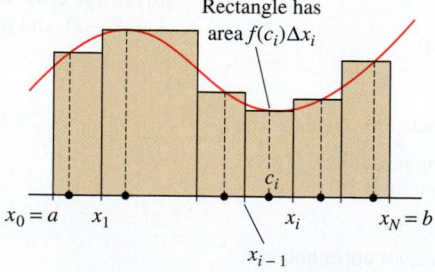

(B) Construct rectangle above each subinterval of height $f(c_i)$

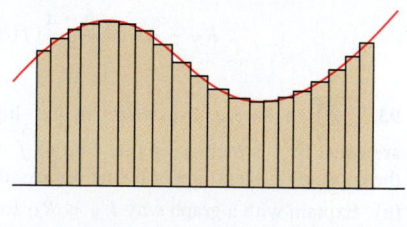

(C) Rectangles corresponding to a Riemann sum with $\|P\|$ all (a large number of rectangles).

FIGURE 2 Construction of $R(f, P, C)$.

EXAMPLE 1 Calculate the Riemann sum $R(f, P, C)$, where $f(x) = 8 + 12 \sin x - 4x$ on $[0, 4]$,

$$P : x_0 = 0, x_1 = 1, x_2 = 1.8, x_3 = 2.9, x_4 = 4$$

$$C : c_1 = 0.4, c_2 = 1.2, c_3 = 2, c_4 = 3.5$$

What is the norm $\|P\|$?

Solution The widths of the subintervals in the partition (Figure 3) are

$$\Delta x_1 = x_1 - x_0 = 1 - 0 = 1, \qquad \Delta x_2 = x_2 - x_1 = 1.8 - 1 = 0.8$$

$$\Delta x_3 = x_3 - x_2 = 2.9 - 1.8 = 1.1, \qquad \Delta x_4 = x_4 - x_3 = 4 - 2.9 = 1.1$$

The norm of the partition is $\|P\| = 1.1$ since the two longest subintervals have width 1.1. Using a calculator, we obtain

$$R(f, P, C) = f(0.4)\Delta x_1 + f(1.2)\Delta x_2 + f(2)\Delta x_3 + f(3.5)\Delta x_4$$

$$\approx 11.07(1) + 14.38(0.8) + 10.91(1.1) - 10.2(1.1) \approx 23.35 \qquad \blacksquare$$

Note in Figure 2(C) that as the norm $\|P\|$ tends to zero (meaning that the rectangles get thinner), the number of rectangles N tends to ∞ and the sums $R(f, P, C)$ approach a limiting value.

This leads to the following definition: f is **integrable** over $[a, b]$ if *all* of the Riemann sums (not just the endpoint and midpoint approximations) approach one and the same limit L as $\|P\|$ tends to zero. Formally, we write

$$L = \lim_{\|P\| \to 0} R(f, P, C) = \lim_{\|P\| \to 0} \sum_{i=1}^{N} f(c_i)\Delta x_i \qquad \boxed{3}$$

if $|R(f, P, C) - L|$ gets arbitrarily small as the norm $\|P\|$ tends to zero, no matter how we choose the partition and sample points. The limit L is called the **definite integral** of f over $[a, b]$.

DEFINITION **Definite Integral** The definite integral of f over $[a, b]$, denoted by the integral sign, is the limit of Riemann sums:

$$\int_a^b f(x)\, dx = \lim_{\|P\| \to 0} R(f, P, C) = \lim_{\|P\| \to 0} \sum_{i=1}^{N} f(c_i)\Delta x_i$$

When this limit exists, we say that f is integrable over $[a, b]$.

The definite integral is often called, more simply, the *integral* of f over $[a, b]$. The process of computing integrals is called **integration** and $f(x)$ is called the **integrand**. The endpoints a and b of $[a, b]$ are called the **limits of integration**. Finally, we remark that any variable may be used as a variable of integration. That is, that variable is a dummy variable. Thus, the following three integrals all denote the same quantity:

$$\int_a^b \sin x\, dx, \qquad \int_a^b \sin t\, dt, \qquad \int_a^b \sin u\, du$$

The next theorem assures us that continuous functions (and even functions with finitely many jump discontinuities) are integrable (see Appendix D for a proof). In practice, we rely on this theorem rather than attempting to prove directly that a given function is integrable.

THEOREM 1 If f is continuous on $[a, b]$, or if f is continuous except at finitely many jump discontinuities in $[a, b]$, then f is integrable over $[a, b]$.

FIGURE 3 Rectangles defined by a Riemann sum for $f(x) = 8 + 12 \sin x - 4x$.

One of the greatest mathematicians of the nineteenth century and perhaps second only to his teacher C. F. Gauss, Riemann transformed the fields of geometry, analysis, and number theory. Albert Einstein based his General Theory of Relativity on Riemann's geometry. The Riemann Hypothesis dealing with prime numbers is one of the great unsolved problems in present-day mathematics. The Clay Mathematics Institute has offered a $1 million prize for its solution.

Georg Friedrich Riemann (1826–1866)

A constant function $f(x) = K$ is integrable over every interval $[a, b]$. The following theorem provides the value of the integral and indicates that when K is positive, the integral is the area of the rectangle in Figure 4.

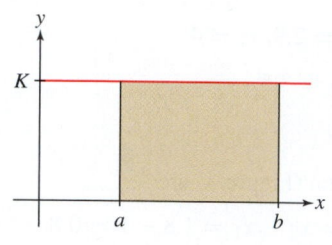

FIGURE 4 $\int_a^b K \, dx = K(b-a)$.

THEOREM 2 Integral of a Constant For any constant function $f(x) = K$,

$$\int_a^b f(x) \, dx = K(b-a)$$

4

Proof Let $R(f, P, C) = \sum_{i=1}^{N} f(c_i)\Delta x_i$ be a Riemann sum for f over $[a, b]$. Then, $f(c_i) = K$ for each c_i, and therefore,

$$R(f, P, C) = \sum_{i=1}^{N} f(c_i)\Delta x_i = \sum_{i=1}^{N} K \Delta x_i = K \sum_{i=1}^{N} \Delta x_i = K(b-a)$$

The latter equality holds because the sum of the Δx_i is the overall length $b - a$ of the interval $[a, b]$. Since every Riemann sum has the value $K(b-a)$, it follows that the integral does as well. Thus, $\int_a^b f(x) \, dx = K(b-a)$. ∎

The Definite Integral and Signed Area

We motivated the development of approximating sums, Riemann sums, and the definite integral with the problem of determining the area under the graph of a function. The definite integral provides an exact value for such areas. Specifically, if f is integrable and $f(x) \geq 0$ over $[a, b]$, then *the area* under the graph of f over $[a, b]$ is $\int_a^b f(x) \, dx$.

When the geometry of a region is simple (e.g., rectangles, triangles, circles) the definite integral corresponds to the area obtained by a geometric formula, as demonstrated for a rectangle in Theorem 2. In some instances, this correspondence enables us to compute a definite integral using geometric formulas (as in Example 2 below).

Allowing the possibility that $f(x)$ takes on both positive and negative values, we define the notion of signed area, where regions below the x-axis provide a negative contribution (Figure 5). Intuitively, the signed area of a region is the area above the x-axis minus the area below.

The $f(c_i)\Delta x_i$ terms in a Riemann sum are signed areas of rectangles. Therefore, a signed area such as in Figure 5 can be approximated by Riemann sums and is given by a definite integral. Thus, for all integrable f over $[a, b]$,

---

The **signed area** between the graph of f and the x-axis over $[a, b]$ is $\int_a^b f(x) \, dx$.

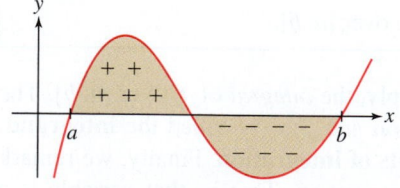

FIGURE 5 Signed area is the area above the x-axis minus the area below the x-axis.

EXAMPLE 2 Definite Intregrals Via Simple Geometry Calculate

$$\int_0^5 (3 - x) \, dx \quad \text{and} \quad \int_0^5 |3 - x| \, dx$$

Solution The region between $y = 3 - x$ and the x-axis consists of two triangles of areas $\frac{9}{2}$ and 2 [Figure 6(A)]. The triangle with area $\frac{9}{2}$ lies above the x-axis and therefore has signed area $\frac{9}{2}$. The second triangle lies below the x-axis, so it has signed area -2. In the graph of $y = |3 - x|$, both triangles lie above the x-axis [Figure 6(B)]. It follows that

$$\int_0^5 (3 - x) \, dx = \frac{9}{2} - 2 = \frac{5}{2} \qquad \int_0^5 |3 - x| \, dx = \frac{9}{2} + 2 = \frac{13}{2}$$ ∎

In the next example, the first integral is geometrically simple but the second is not. For the second, we need to take a limit of Riemann sums to arrive at a value. In Section 5.4, we will start to develop tools that make the computation of integrals such as the latter much simpler.

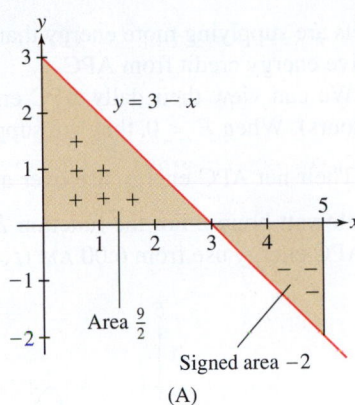

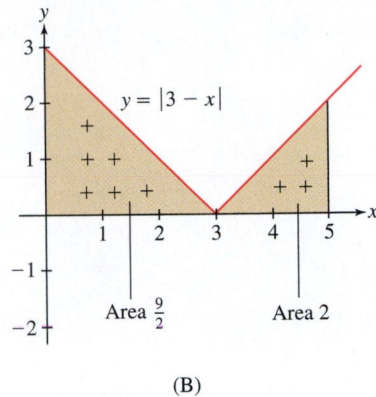

FIGURE 6

EXAMPLE 3 For $b > 0$, calculate

$$\text{(a)} \int_0^b x\, dx \quad \text{and} \quad \text{(b)} \int_0^b x^2\, dx$$

Solution The integrals represent the shaded areas in Figure 7.

(a) This integral represents the area of a triangle with base and height equal to b. Therefore,

$$\int_0^b x\, dx = \frac{1}{2}(b)(b) = \frac{b^2}{2}$$

(b) We evaluate by taking a limit of Riemann sums. Because $f(x) = x^2$ is continuous, it is also integrable by Theorem 1. It follows that the right-endpoint approximations R_N converge to the integral. Therefore, we compute $\int_0^b x^2\, dx$ by computing $\lim_{N \to \infty} R_N$.

We divide the interval $[0, b]$ into N subintervals of width $\Delta x = \frac{b-0}{N} = \frac{b}{N}$.

$$R_N = \Delta x \sum_{j=1}^{N} f(x_j) = \frac{b}{N} \sum_{j=1}^{N} f\left(\frac{jb}{N}\right) = \frac{b}{N} \sum_{j=1}^{N} \left(\frac{jb}{N}\right)^2$$

$$= \frac{b^3}{N^3} \sum_{j=1}^{N} j^2 = \frac{b^3}{N^3}\left(\frac{N^3}{3} + \frac{N^2}{2} + \frac{N}{6}\right) = \frac{b^3}{3} + \frac{b^3}{2N} + \frac{b^3}{6N^2}$$

Taking the limit of R_N as $N \to \infty$, we obtain $\int_0^b x^2\, dx = \frac{b^3}{3}$. ∎

The results of the previous example will be helpful in other definite integral computations in this section. Summarizing, if $b > 0$, then

$$\int_0^b x\, dx = \frac{b^2}{2} \quad \text{and} \quad \int_0^b x^2\, dx = \frac{b^3}{3} \qquad \boxed{5}$$

The integrals in Eq. (5) suggest a pattern. What do you think the value of $\int_0^b x^n\, dx$ is? The $n = 3$ case is considered in Exercise 50. We will compute this integral for general n in Section 5.4.

In cases where we do not have an exact representation of a function, such as in the following example, the best we can do to compute a definite integral is to approximate it using a Riemann sum.

EXAMPLE 4 A Grid-Connected Energy System At their home, Malina and Hors have solar panels and an energy system that is grid connected. They use energy from their panels and from the Apollo Power Company (APC) for their household needs. When the

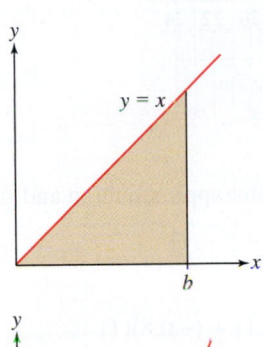

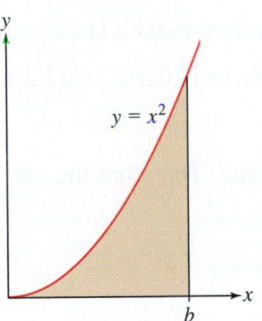

FIGURE 7

panels are supplying more energy than they need, the excess is fed into the grid and they receive energy credit from APC.

We can view their daily APC energy use E (in kilowatts) as a function of time t (in hours). When $E < 0$, they are supplying energy to APC.

Their net APC energy use over a period of time $a < t < b$ is given by $\int_a^b E(t)\,dt$ (in kilowatt-hours). For the function E shown in the graph in Figure 8, approximate the net APC energy use from 6:00 AM ($t = 6$) to 6:00 PM ($t = 18$).

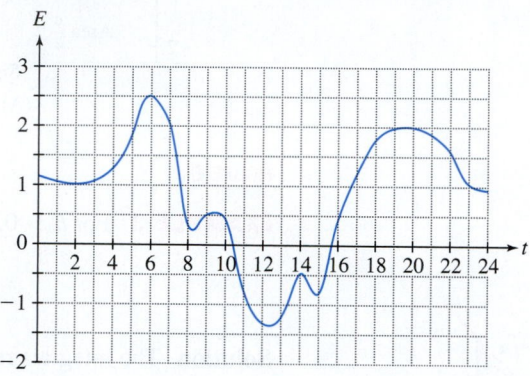

FIGURE 8 Energy use E in kilowatts as a function of time t in hours since midnight.

Solution   We approximate $\int_6^{18} E(t)\,dt$ with a left-endpoint approximation and $\Delta t = 1$. That sum is expressed as

$$L = E(6)\Delta t + E(7)\Delta t + \cdots + E(17)\Delta t$$

$$= (2.5)(1) + (2)(1) + (0.3)(1) + (0.5)(1) + (0.4)(1) + (-0.8)(1)$$

$$+ (-1.3)(1) + (-1.2)(1) + (-0.5)(1) + (-0.8)(1) + (0.3)(1) + (1.2)(1)$$

$$= 2.6$$

where the values of $E(t)$ are approximated from the graph. Therefore the net 12-hour APC energy use is

$$\int_6^{18} E(t)\,dt \approx 2.6 \text{ kilowatt-hours} \qquad \blacksquare$$

Properties of the Definite Integral

In the rest of this section, we discuss some basic properties of definite integrals, beginning with the linearity properties.

THEOREM 3   Linearity of the Definite Integral   If f and g are integrable over $[a, b]$, then $f + g$ and Cf are integrable (for any constant C), and

- $$\int_a^b \big(f(x) + g(x)\big)\,dx = \int_a^b f(x)\,dx + \int_a^b g(x)\,dx$$

- $$\int_a^b Cf(x)\,dx = C \int_a^b f(x)\,dx$$

Proof   These properties follow from the corresponding linearity properties of sums and limits. For example, Riemann sums are additive:

$$R(f + g, P, C) = \sum_{i=1}^N \big(f(c_i) + g(c_i)\big)\Delta x_i = \sum_{i=1}^N f(c_i)\Delta x_i + \sum_{i=1}^N g(c_i)\Delta x_i$$

$$= R(f, P, C) + R(g, P, C)$$

By the additivity of limits,

$$\int_a^b (f(x) + g(x))\, dx = \lim_{\|P\| \to 0} R(f + g, P, C)$$

$$= \lim_{\|P\| \to 0} R(f, P, C) + \lim_{\|P\| \to 0} R(g, P, C)$$

$$= \int_a^b f(x)\, dx + \int_a^b g(x)\, dx$$

The second property is proved similarly. ∎

EXAMPLE 5 Calculate $\int_0^3 (2x^2 - 3x - 7)\, dx$.

Solution

$$\int_0^3 (2x^2 - 3x - 7)\, dx = 2 \int_0^3 x^2\, dx - 3 \int_0^3 x\, dx + \int_0^3 (-7)\, dx \qquad \text{(linearity)}$$

$$= 2 \left(\frac{3^3}{3} \right) - 3 \left(\frac{3^2}{2} \right) - 7(3 - 0) = -\frac{33}{2} \qquad \text{[Eq. (5)]} \quad \blacksquare$$

So far we have used the notation $\int_a^b f(x)\, dx$ with the understanding that $a < b$.

Can we make sense of and compute integrals like $\int_5^1 f(x)\, dx$? The answer is yes. To do so rigorously involves expanding the idea of Riemann sum to allow Δx_i to be negative because we are essentially integrating in a negative direction. We do not pursue the details here. For our purposes, the following definition captures the idea properly.

> **DEFINITION Reversing the Limits of Integration** For $a < b$, we set
>
> $$\int_b^a f(x)\, dx = -\int_a^b f(x)\, dx$$
>
> 6

According to Eq. (6), the integral changes sign when the limits of integration are reversed.

For example, by Eqs. (5) and (6),

$$\int_5^0 x^2\, dx = -\int_0^5 x^2\, dx = -\frac{5^3}{3} = -\frac{125}{3}$$

When $a = b$, the interval $[a, b] = [a, a]$ has length zero and we define the definite integral to be zero:

$$\int_a^a f(x)\, dx = 0$$

EXAMPLE 6 Prove that, for all b (negative, zero, and positive),

$$\int_0^b x\, dx = \frac{1}{2}b^2 \quad \text{and} \quad \int_0^b x^2\, dx = \frac{1}{3}b^3 \qquad \boxed{7}$$

Solution These integral formulas hold for $b > 0$ by Eq. (5), and they hold for $b = 0$ by definition of a definite integral on an interval of length zero.

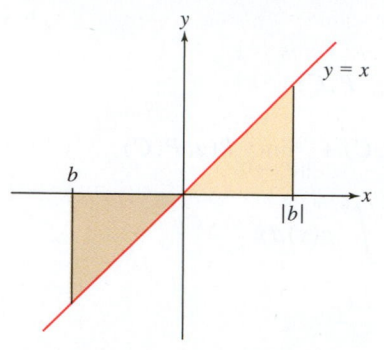

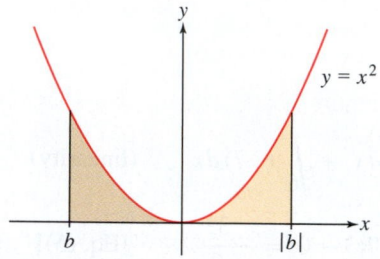

FIGURE 9

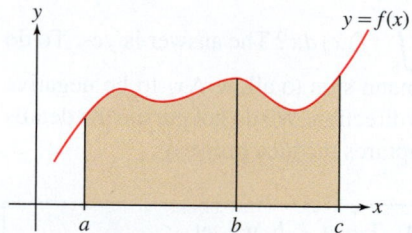

FIGURE 10 The area over $[a, c]$ is the *sum* of the areas over $[a, b]$ and $[b, c]$.

Now, we consider the situation where $b < 0$. In this case, the integrals can be determined from the signed area of the dark shaded regions in Figure 9. The signed area of the dark shaded triangle is $\int_b^0 x \, dx$. By symmetry and Eq. (5),

$$\int_b^0 x \, dx = -\int_0^{|b|} x \, dx = -\frac{1}{2}|b|^2$$

Therefore,

$$\int_0^b x \, dx = \frac{1}{2}|b|^2 = \frac{1}{2}b^2$$

In a similar way, by symmetry and Eq. (5), we have

$$\int_b^0 x^2 \, dx = \int_0^{|b|} x^2 \, dx = \frac{1}{3}|b|^3$$

Therefore,

$$\int_0^b x^2 \, dx = -\frac{1}{3}|b|^3 = \frac{1}{3}b^3 \qquad \blacksquare$$

Definite integrals satisfy an important additivity property: If f is an integrable function and $a \leq b \leq c$ as in Figure 10, then the integral from a to c is equal to the integral from a to b *plus* the integral from b to c. We state this in the next theorem (a formal proof can be given using Riemann sums).

> **THEOREM 4 Additivity for Adjacent Intervals** Let $a \leq b \leq c$, and assume that f is integrable. Then
> $$\int_a^c f(x) \, dx = \int_a^b f(x) \, dx + \int_b^c f(x) \, dx$$

This theorem remains true as stated even if the condition $a \leq b \leq c$ is not satisfied (Exercise 88).

EXAMPLE 7 Calculate $\int_4^7 x^2 \, dx$.

Solution Using Theorem 4, we can write

$$\int_0^4 x^2 \, dx + \int_4^7 x^2 \, dx = \int_0^7 x^2 \, dx$$

Solving this equation for the desired integral, and using Eq. (5), we have

$$\int_4^7 x^2 \, dx = \int_0^7 x^2 \, dx - \int_0^4 x^2 \, dx = \left(\frac{1}{3}\right)7^3 - \left(\frac{1}{3}\right)4^3 = 93 \qquad \blacksquare$$

Another basic property of the definite integral is that if $f(x) > g(x)$, then the integral of f is greater than the integral of g (Figure 11).

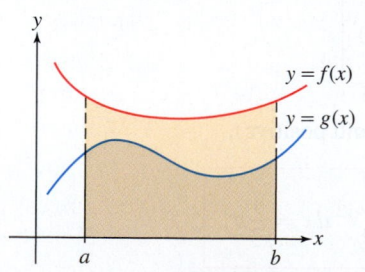

FIGURE 11 The integral of f is greater than the integral of g.

> **THEOREM 5 Comparison Theorem** If f and g are integrable and $g(x) \leq f(x)$ for x in $[a, b]$, then
> $$\int_a^b g(x) \, dx \leq \int_a^b f(x) \, dx$$

Proof If $g(x) \le f(x)$, then for any partition and choice of sample points, we have $g(c_i)\Delta x_i \le f(c_i)\Delta x_i$ for all i. Therefore, the Riemann sums satisfy

$$\sum_{i=1}^{N} g(c_i)\Delta x_i \le \sum_{i=1}^{N} f(c_i)\Delta x_i$$

Taking the limit as the norm $\|P\|$ tends to zero, we obtain

$$\int_a^b g(x)\,dx = \lim_{\|P\|\to 0}\sum_{i=1}^{N} g(c_i)\Delta x_i \le \lim_{\|P\|\to 0}\sum_{i=1}^{N} f(c_i)\Delta x_i = \int_a^b f(x)\,dx \qquad \blacksquare$$

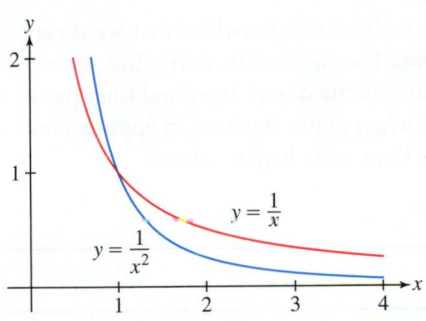

FIGURE 12

EXAMPLE 8 Prove the inequality $\displaystyle\int_1^4 \frac{1}{x^2}\,dx \le \int_1^4 \frac{1}{x}\,dx$.

Solution If $x \ge 1$, then $x^2 \ge x$, and $x^{-2} \le x^{-1}$ (Figure 12). Therefore, the inequality follows from the Comparison Theorem, applied with $g(x) = x^{-2}$ and $f(x) = x^{-1}$. $\blacksquare$

Suppose there are numbers m and M such that $m \le f(x) \le M$ for x in $[a, b]$. We call m and M **lower** and **upper bounds** for $f(x)$ on $[a, b]$. By the Comparison Theorem,

$$\int_a^b m\,dx \le \int_a^b f(x)\,dx \le \int_a^b M\,dx$$

By Theorem 2, it follows that

$$\boxed{m(b - a) \le \int_a^b f(x)\,dx \le M(b - a)} \qquad \boxed{8}$$

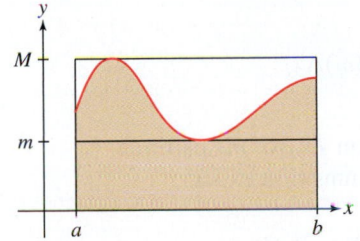

FIGURE 13 The integral $\displaystyle\int_a^b f(x)\,dx$ lies between the areas of the rectangles of heights m and M.

This says simply that if $f(x) \ge 0$ over $[a, b]$, then (as in Figure 13) the integral of f lies between the areas of two rectangles, one of height M enclosing the region associated with the integral, and one of height m enclosed in the region.

EXAMPLE 9 Prove the inequalities $\displaystyle\frac{3}{4} \le \int_{1/2}^2 \frac{1}{x}\,dx \le 3$.

Solution Because $f(x) = x^{-1}$ is decreasing (Figure 14), its minimum value on $\left[\frac{1}{2}, 2\right]$ is $m = f(2) = \frac{1}{2}$ and its maximum value is $M = f\left(\frac{1}{2}\right) = 2$. By Eq. (8),

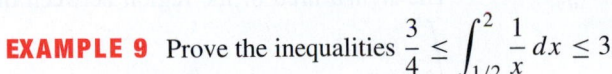

$$\frac{3}{4} = \underbrace{\frac{1}{2}\left(2 - \frac{1}{2}\right)}_{m(b-a)} \le \int_{1/2}^2 \frac{1}{x}\,dx \le \underbrace{2\left(2 - \frac{1}{2}\right)}_{M(b-a)} = 3 \qquad \blacksquare$$

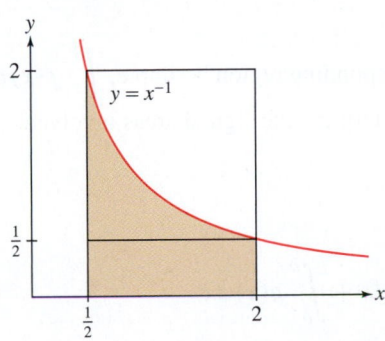

FIGURE 14

CONCEPTUAL INSIGHT Keep in mind that a definite integral $\displaystyle\int_a^b f(x)\,dx$ is defined as a limit of Riemann sums over finer and finer partitions of $[a, b]$, and each Riemann sum $R(f, P, C)$ is a sum of terms $f(c_i)\Delta x_i$ determined by a partition P. The situation that we saw with the derivative in Chapters 3 and 4 is being repeated in our development of the theory of the definite integral:

- **Definition:** We define the definite integral via a limit definition. Limits are needed in the definition in order to properly capture the concept.
- **Properties:** We establish properties of the definite integral by proving theorems based on the limit definition. Some properties provide insight into the workings of the definite integral, and some provide computational tools.

- **Computation:** Computing the definite integral directly from the limit definition is messy at best and generally very difficult. We establish rules that aid us significantly in carrying out definite integral computations. The most important is the first part of the Fundamental Theorem of Calculus that we introduce in Section 5.4.
- **Application:** While most of the definite integral computations that we do are simplified by computation rules, we cannot lose sight of its definition. Knowing that a definite integral is a limit of sums defined over finer and finer partitions of the domain will help us identify when to use this tool in applications. We will see plenty of instances in the sections and chapters ahead.

5.2 SUMMARY

- A Riemann sum $R(f, P, C)$ for the interval $[a, b]$ is defined by choosing a *partition*

$$P : a = x_0 < x_1 < x_2 < \cdots < x_N = b$$

and *sample points* $C = \{c_i\}$, where $c_i \in [x_{i-1}, x_i]$. Let $\Delta x_i = x_i - x_{i-1}$. Then

$$R(f, P, C) = \sum_{i=1}^{N} f(c_i) \Delta x_i$$

- The maximum of the widths Δx_i is called the norm $\|P\|$ of the partition.
- The *definite integral* is the limit of the Riemann sums (if it exists):

$$\int_a^b f(x) \, dx = \lim_{\|P\| \to 0} R(f, P, C)$$

We say that f is *integrable* over $[a, b]$ if the limit exists.
- Theorem: If f is continuous on $[a, b]$, then f is integrable over $[a, b]$.
- The *signed area* of the region between the graph of f and the x-axis over $[a, b]$ is

$$\int_a^b f(x) \, dx.$$

- Using geometry: When the geometry of the corresponding region is simple, $\int_a^b f(x) \, dx$ can be computed using geometric formulas to determine the signed areas involved.

- $\int_0^b x \, dx = \dfrac{1}{2} b^2$ and $\int_0^b x^2 \, dx = \dfrac{1}{3} b^3$

- Properties of definite integrals:

$$\int_a^b \big(f(x) + g(x) \big) \, dx = \int_a^b f(x) \, dx + \int_a^b g(x) \, dx$$

$$\int_a^b C f(x) \, dx = C \int_a^b f(x) \, dx \quad \text{for any constant } C$$

$$\int_b^a f(x) \, dx = - \int_a^b f(x) \, dx$$

$$\int_a^a f(x) \, dx = 0$$

$$\int_a^b f(x) \, dx + \int_b^c f(x) \, dx = \int_a^c f(x) \, dx \quad \text{for all } a, b, c$$

- Comparison Theorem: If $f(x) \le g(x)$ on $[a, b]$, then

$$\int_a^b f(x)\,dx \le \int_a^b g(x)\,dx$$

- If $m \le f(x) \le M$ on $[a, b]$, then

$$m(b - a) \le \int_a^b f(x)\,dx \le M(b - a)$$

5.2 EXERCISES

Preliminary Questions

1. What is $\int_3^5 dx$ [the function is $f(x) = 1$]?

2. Let $I = \int_2^7 f(x)\,dx$, where f is continuous. State whether the following are true or false:
(a) I is the area between the graph and the x-axis over $[2, 7]$.
(b) If $f(x) \ge 0$, then I is the area between the graph and the x-axis over $[2, 7]$.

(c) If $f(x) \le 0$, then $-I$ is the area between the graph of f and the x-axis over $[2, 7]$.

3. Explain graphically: $\int_0^\pi \cos x\,dx = 0$.

4. Which is negative, $\int_{-1}^{-5} 8\,dx$ or $\int_{-5}^{-1} 8\,dx$?

Exercises

In Exercises 1–10, draw a graph of the signed area represented by the integral and compute it using geometry.

1. $\displaystyle\int_{-3}^3 2x\,dx$

2. $\displaystyle\int_{-2}^3 (2x + 4)\,dx$

3. $\displaystyle\int_{-2}^1 (3x + 4)\,dx$

4. $\displaystyle\int_{-2}^1 4\,dx$

5. $\displaystyle\int_6^8 (7 - x)\,dx$

6. $\displaystyle\int_{\pi/2}^{3\pi/2} \sin x\,dx$

7. $\displaystyle\int_0^5 \sqrt{25 - x^2}\,dx$

8. $\displaystyle\int_{-2}^3 |x|\,dx$

9. $\displaystyle\int_{-2}^2 (2 - |x|)\,dx$

10. $\displaystyle\int_{-2}^5 (3 + x - 2|x|)\,dx$

11. Calculate $\displaystyle\int_0^{10} (8 - x)\,dx$ in two ways:

(a) As the limit $\lim\limits_{N \to \infty} R_N$

(b) By sketching the relevant signed area and using geometry

12. Calculate $\displaystyle\int_{-1}^4 (4x - 8)\,dx$ in two ways:

(a) As the limit $\lim\limits_{N \to \infty} R_N$

(b) By using geometry

In Exercises 13 and 14, refer to Figure 15.

13. Evaluate: **(a)** $\displaystyle\int_0^2 f(x)\,dx$ **(b)** $\displaystyle\int_0^6 f(x)\,dx$

14. Evaluate: **(a)** $\displaystyle\int_1^4 f(x)\,dx$ **(b)** $\displaystyle\int_1^6 |f(x)|\,dx$

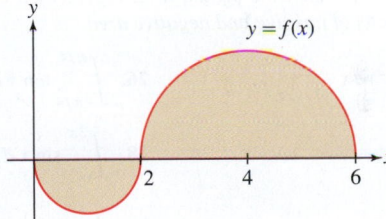

FIGURE 15 The two parts of the graph are semicircles.

In Exercises 15 and 16, refer to Figure 16.

15. Evaluate $\displaystyle\int_0^3 g(t)\,dt$ and $\displaystyle\int_3^5 g(t)\,dt$.

16. Find a, b, and c such that $\displaystyle\int_0^a g(t)\,dt$ and $\displaystyle\int_b^c g(t)\,dt$ are as large as possible.

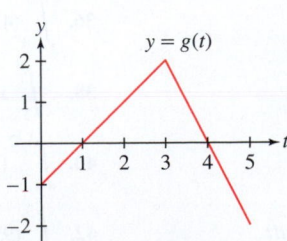

FIGURE 16

17. Describe the partition P and the set of sample points C for the Riemann sum shown in Figure 17. Compute the value of the Riemann sum.

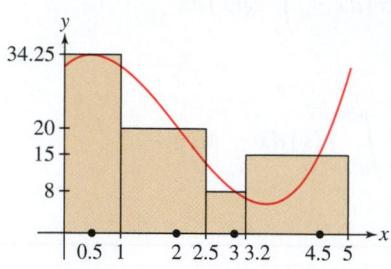

FIGURE 17

In Exercises 18–22, calculate the Riemann sum $R(f, P, C)$ for the given function, partition, and choice of sample points. Also, sketch the graph of f and the rectangles corresponding to $R(f, P, C)$.

18. $f(x) = x$, $\quad P = \{1, 1.2, 1.5, 2\}$, $\quad C = \{1.1, 1.4, 1.9\}$

19. $f(x) = 2x + 3$, $\quad P = \{-4, -1, 1, 4, 8\}$, $\quad C = \{-3, 0, 2, 5\}$

20. $f(x) = x^2 + x$, $\quad P = \{2, 3, 4.5, 5\}$, $\quad C = \{2, 3.5, 5\}$

21. $f(x) = \sin x$, $\quad P = \left\{0, \frac{\pi}{6}, \frac{\pi}{3}, \frac{\pi}{2}\right\}$, $\quad C = \{0.4, 0.7, 1.2\}$

22. $f(x) = x^2 + x$, $\quad P = \{0, 1, 2.5, 3.2, 5\}$, $\quad C = \{0.5, 2, 3, 4.5\}$

23. In Example 4, approximate the net APC energy use from midnight to noon.

24. In Example 4, approximate the net APC energy use from noon to midnight.

In Exercises 25–30, sketch the signed area represented by the integral. Indicate the regions of positive and negative area.

25. $\displaystyle\int_0^5 (4x - x^2)\, dx$

26. $\displaystyle\int_{-\pi/4}^{\pi/4} \tan x\, dx$

27. $\displaystyle\int_{\pi}^{2\pi} \sin x\, dx$

28. $\displaystyle\int_0^{3\pi} \sin x\, dx$

29. $\displaystyle\int_{1/2}^{2} \ln x\, dx$

30. $\displaystyle\int_{-1}^{1} \tan^{-1} x\, dx$

In Exercises 31–34, determine the sign of the integral without calculating it. Draw a graph if necessary.

31. $\displaystyle\int_{-2}^{1} x^4\, dx$

32. $\displaystyle\int_{-2}^{1} x^3\, dx$

33. $\boxed{\text{GU}}$ $\displaystyle\int_0^{2\pi} x \sin x\, dx$

34. $\boxed{\text{GU}}$ $\displaystyle\int_0^{2\pi} \frac{\sin x}{x}\, dx$

In Exercises 35–44, use properties of the integral and the formulas in the summary to calculate the integrals.

35. $\displaystyle\int_0^4 (6t - 3)\, dt$

36. $\displaystyle\int_{-3}^{2} (4x + 7)\, dx$

37. $\displaystyle\int_0^9 x^2\, dx$

38. $\displaystyle\int_2^5 x^2\, dx$

39. $\displaystyle\int_0^1 (u^2 - 2u)\, du$

40. $\displaystyle\int_0^{1/2} (12y^2 + 6y)\, dy$

41. $\displaystyle\int_{-3}^{1} (7t^2 + t + 1)\, dt$

42. $\displaystyle\int_{-3}^{3} (9x - 4x^2)\, dx$

43. $\displaystyle\int_{-a}^{1} (x^2 + x)\, dx$

44. $\displaystyle\int_a^{a^2} x^2\, dx$

In Exercises 45–48, calculate the integral, assuming that

$$\int_0^5 f(x)\, dx = 5, \qquad \int_0^5 g(x)\, dx = 12$$

45. $\displaystyle\int_0^5 (f(x) + g(x))\, dx$

46. $\displaystyle\int_0^5 \left(2f(x) - \frac{1}{3}g(x)\right) dx$

47. $\displaystyle\int_5^0 g(x)\, dx$

48. $\displaystyle\int_0^5 (f(x) - x)\, dx$

49. Assume $a < b$ and H is the Heaviside function given by

$$H(x) = \begin{cases} 0 & \text{when } x < 0 \\ 1 & \text{when } x \geq 0 \end{cases}$$

Find an expression for $\displaystyle\int_a^b H(x)\, dx$ in terms of a and b.

50. By computing the limit of right-endpoint approximations, prove that if $b > 0$, then $\displaystyle\int_0^b x^3\, dx = \frac{b^4}{4}$.

51. Using the result of Exercise 50, prove that for all b (negative, zero, and positive),

$$\boxed{\int_0^b x^3\, dx = \frac{b^4}{4}} \qquad \boxed{9}$$

In Exercises 52–56, evaluate the integral using the formulas in the summary and Eq. (9).

52. $\displaystyle\int_1^3 x^3\, dx$

53. $\displaystyle\int_0^3 (x - x^3)\, dx$

54. $\displaystyle\int_0^1 (2x^3 - x + 4)\, dx$

55. $\displaystyle\int_0^1 (12x^3 + 24x^2 - 8x)\, dx$

56. $\displaystyle\int_{-2}^{2} (2x^3 - 3x^2)\, dx$

In Exercises 57–60, calculate the integral, assuming that

$$\int_0^1 f(x)\, dx = 1, \qquad \int_0^2 f(x)\, dx = 4, \qquad \int_1^4 f(x)\, dx = 7$$

57. $\displaystyle\int_0^4 f(x)\, dx$

58. $\displaystyle\int_1^2 f(x)\, dx$

59. $\displaystyle\int_4^1 f(x)\, dx$

60. $\displaystyle\int_2^4 f(x)\, dx$

In Exercises 61–64, express each integral as a single integral.

61. $\displaystyle\int_0^3 f(x)\, dx + \int_3^7 f(x)\, dx$

62. $\displaystyle\int_2^9 f(x)\, dx - \int_4^9 f(x)\, dx$

63. $\displaystyle\int_2^9 f(x)\, dx - \int_2^5 f(x)\, dx$

64. $\displaystyle\int_7^3 f(x)\, dx + \int_3^9 f(x)\, dx$

In Exercises 65–66, prove the relationship for arbitrary a and b using the formulas in the summary.

65. $\int_a^b x \, dx = \dfrac{b^2 - a^2}{2}$ **66.** $\int_a^b x^2 \, dx = \dfrac{b^3 - a^3}{3}$

67. ✏️ Explain the difference in graphical interpretation between $\int_a^b f(x) \, dx$ and $\int_a^b |f(x)| \, dx$.

68. ✏️ Use the graphical interpretation of the definite integral to explain the inequality

$$\left| \int_a^b f(x) \, dx \right| \le \int_a^b |f(x)| \, dx$$

where f is continuous. Explain also why equality holds if and only if either $f(x) \ge 0$ for all x or $f(x) \le 0$ for all x.

69. Let $f(x) = x$. Find an interval $[a, b]$ such that

$$\left| \int_a^b f(x) \, dx \right| = \frac{1}{2} \quad \text{and} \quad \int_a^b |f(x)| \, dx = \frac{3}{2}$$

70. Evaluate $I = \int_0^{2\pi} \sin^2 x \, dx$ and $J = \int_0^{2\pi} \cos^2 x \, dx$ as follows. First, show with a graph that $I = J$. Then, prove that $I + J = 2\pi$.

In Exercises 71–74, calculate the integral.

71. $\int_0^6 |3 - x| \, dx$ **72.** $\int_1^3 |2x - 4| \, dx$

73. $\int_{-1}^1 |x^3| \, dx$ **74.** $\int_0^2 |x^2 - 1| \, dx$

75. Use the Comparison Theorem to show that

$$\int_0^1 x^5 \, dx \le \int_0^1 x^4 \, dx, \qquad \int_1^2 x^4 \, dx \le \int_1^2 x^5 \, dx$$

76. Prove that $\dfrac{1}{3} \le \int_4^6 \dfrac{1}{x} \, dx \le \dfrac{1}{2}$.

77. Prove that $0.0198 \le \int_{0.2}^{0.3} \sin x \, dx \le 0.0296$. *Hint:* Show that $0.198 \le \sin x \le 0.296$ for x in $[0.2, 0.3]$.

78. Prove that $0.277 \le \int_{\pi/8}^{\pi/4} \cos x \, dx \le 0.363$.

79. Prove that $0 \le \int_{\pi/4}^{\pi/2} \dfrac{\sin x}{x} \, dx \le \dfrac{\sqrt{2}}{2}$.

80. Find upper and lower bounds for $\int_0^1 \dfrac{dx}{\sqrt{5x^3 + 4}}$.

81. ✏️ Suppose that $f(x) \le g(x)$ on $[a, b]$. By the Comparison Theorem, $\int_a^b f(x) \, dx \le \int_a^b g(x) \, dx$. Is it also true that $f'(x) \le g'(x)$ for $x \in [a, b]$? If not, give a counterexample.

82. ✏️ State whether the following statement is true or false. If false, sketch the graph of a counterexample.

(a) If $f(x) > 0$, then $\int_a^b f(x) \, dx > 0$.

(b) If $\int_a^b f(x) \, dx > 0$, then $f(x) > 0$.

Further Insights and Challenges

83. Explain graphically: If f is an odd function, then $\int_{-a}^a f(x) \, dx = 0$.

84. Compute $\int_{-1}^1 (\sin x)(\sin^2 x + 1) \, dx$.

85. Let k and b be positive. Show, by comparing the right-endpoint approximations, that

$$\int_0^b x^k \, dx = b^{k+1} \int_0^1 x^k \, dx$$

86. Verify for $0 \le b \le 1$ by interpreting in terms of area:

$$\int_0^b \sqrt{1 - x^2} \, dx = \frac{1}{2} b\sqrt{1 - b^2} + \frac{1}{2} \sin^{-1} b$$

87. ✏️ Suppose that f and g are continuous functions such that, *for all a,*

$$\int_{-a}^a f(x) \, dx = \int_{-a}^a g(x) \, dx$$

Give an *intuitive* argument showing that $f(0) = g(0)$. Explain your idea with a graph.

88. Theorem 4 remains true without the assumption $a \le b \le c$. Verify this for the cases $b < a < c$ and $c < a < b$.

5.3 The Indefinite Integral

In earlier chapters, we have seen how useful it is to be able to find the derivative of a function. But what about the inverse problem? Given the derivative of an unknown function, can we find the function itself? For example, in physics we may know the velocity $v(t)$ (the derivative) and wish to compute the position $s(t)$ of an object. Since $s'(t) = v(t)$, this amounts to finding a function whose derivative is $v(t)$. A function F whose derivative is f is called an antiderivative of f. Antiderivatives will turn out to be the key to evaluating definite integrals.

> **DEFINITION Antiderivatives**   A function F is an antiderivative of f on an open interval (a, b) if $F'(x) = f(x)$ for all x in (a, b).

Examples:

- $F(x) = -\cos x$ is an antiderivative of $f(x) = \sin x$ because for all values of x,

$$F'(x) = \frac{d}{dx}(-\cos x) = \sin x = f(x)$$

- $F(x) = \frac{1}{3}x^3$ is an antiderivative of $f(x) = x^2$ because for all values of x,

$$F'(x) = \frac{d}{dx}\left(\frac{1}{3}x^3\right) = x^2 = f(x)$$

One critical observation is that antiderivatives are not unique. We are free to add a constant C because the derivative of a constant is zero, and so, if $F'(x) = f(x)$, then $(F(x) + C)' = f(x)$. For example, each of the following is an antiderivative of x^2:

$$\frac{1}{3}x^3, \qquad \frac{1}{3}x^3 + 5, \qquad \frac{1}{3}x^3 - 4$$

Are there any antiderivatives of f other than those obtained by adding a constant to a given antiderivative F? Our next theorem says that the answer is no if f is defined on an open interval (a, b).

> **THEOREM 1 The General Antiderivative**   Let $y = F(x)$ be an antiderivative of $y = f(x)$ on (a, b). Then every antiderivative on (a, b) is of the form $y = F(x) + C$ for some constant C.

Proof   Assume $y = G(x)$ is an antiderivative of $y = f(x)$, and set $H(x) = G(x) - F(x)$. Then $H'(x) = G'(x) - F'(x) = f(x) - f(x) = 0$. By the Corollary to the Mean Value Theorem in Section 4.3, $H(x)$ must be a constant—say, $H(x) = C$—and therefore $G(x) = F(x) + C$.   ∎

> **GRAPHICAL INSIGHT**   The graph of $y = F(x) + C$ is obtained by shifting the graph of $y = F(x)$ vertically by C units. Since vertical shifting moves the tangent lines without changing their slopes, it makes sense that the functions $y = F(x) + C$, for all possible C, have the same derivative (Figure 1). Theorem 1 asserts that these functions are all of the functions with the same derivative as F. That is, they all are in the form $y = F(x) + C$ and have a graph that is a vertical shift of the graph of $y = F(x)$.

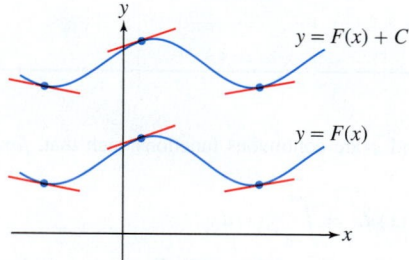

DF FIGURE 1 The tangent lines to the graphs of $y = F(x)$ and $y = F(x) + C$ are parallel.

We often describe the *general* antiderivative of a function in terms of an arbitrary constant C, as in the following example.

EXAMPLE 1   Find the general antiderivative of $f(x) = \cos x$.

Solution   The function $F(x) = \sin x$ is an antiderivative of $f(x) = \cos x$. The general antiderivative is $F(x) = \sin x + C$, where C is any constant.   ∎

The process of finding an antiderivative is called **antidifferentiation** or **integration**. In the next section we will see why the term "integration" is used when we discuss the connection between antiderivatives and areas under curves given by the Fundamental Theorem of Calculus. Anticipating this result, we begin using the following notation and additional terminology for the general antiderivative:

The terms "antiderivative" and "indefinite integral" are used interchangeably. In some textbooks, an antiderivative is called a primitive function.

NOTATION **Indefinite Integral** The notation

$$\int f(x)\,dx = F(x) + C \quad \text{means that} \quad F'(x) = f(x)$$

We say that $y = F(x) + C$ is the general antiderivative or **indefinite integral** of $y = f(x)$.

The expression $f(x)$ appearing in the integral sign is called the **integrand**. The symbol dx is a differential. It is part of the integral notation and serves to indicate the independent variable. The constant C is called the **constant of integration**.

Some indefinite integrals can be evaluated by reversing the familiar derivative formulas. For example, we obtain the indefinite integral of $y = x^n$ by reversing the Power Rule for derivatives.

THEOREM 2 **Power Rule for Integrals**

$$\int x^n\,dx = \frac{x^{n+1}}{n+1} + C \qquad \text{for } n \neq -1$$

Proof We just need to verify that $F(x) = \dfrac{x^{n+1}}{n+1}$ is an antiderivative of $f(x) = x^n$:

$$F'(x) = \frac{d}{dx}\left(\frac{x^{n+1}}{n+1} + C\right) = \frac{1}{n+1}(n+1)x^n = x^n \qquad \blacksquare$$

In words, the Power Rule for Integrals says that to integrate a power of x, "increase the power by one and divide by the new power." Here are some examples:

$$\int x^5\,dx = \frac{1}{6}x^6 + C, \qquad \int x^{-9}\,dx = -\frac{1}{8}x^{-8} + C, \qquad \int x^{3/5}\,dx = \frac{5}{8}x^{8/5} + C$$

The Power Rule is not valid for $n = -1$ because when $n = -1$, the expression $\dfrac{x^{n+1}}{n+1}$ is undefined.

Recall, however, that the derivative of the natural logarithm is $\dfrac{d}{dx}\ln x = \dfrac{1}{x}$. This shows that $F(x) = \ln x$ is an antiderivative of $y = \dfrac{1}{x}$. Thus, for $n = -1$, instead of the Power Rule we have

Notice that in integral notation, we treat dx as a movable variable, and thus, we write $\int \dfrac{1}{x}\,dx$ as $\int \dfrac{dx}{x}$.

$$\int \frac{dx}{x} = \ln x + C$$

This formula is valid for $x > 0$, where $\ln x$ is defined. We would like to have an antiderivative of $y = \dfrac{1}{x}$ on its full domain, namely on $\{x : x \neq 0\}$. To achieve this end, we extend $F(x)$ to an even function by setting $F(x) = \ln|x|$ (Figure 2). Then $F(x) = F(-x)$, and by the Chain Rule, $F'(x) = -F'(-x)$. For $x < 0$, we obtain

$$\frac{d}{dx}\ln|x| = F'(x) = -F'(-x) = -\frac{1}{-x} = \frac{1}{x}$$

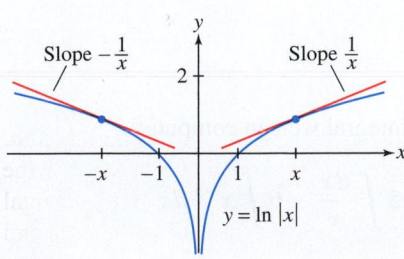

Slope $-\dfrac{1}{x}$ Slope $\dfrac{1}{x}$

$y = \ln|x|$

FIGURE 2

This proves that $\dfrac{d}{dx}\ln|x| = \dfrac{1}{x}$ for all $x \neq 0$.

> **THEOREM 3 Antiderivative of** $y = \dfrac{1}{x}$ The function $F(x) = \ln|x|$ is an antiderivative of $y = \dfrac{1}{x}$ in the domain $\{x : x \neq 0\}$; that is,
>
> $$\int \frac{dx}{x} = \ln|x| + C$$
>
> **1**

The indefinite integral obeys the usual linearity rules that allow us to integrate term by term. These rules follow from the linearity rules for the derivative (see Exercise 79).

> **THEOREM 4 Linearity of the Indefinite Integral**
>
> - **Sum Rule:** $\displaystyle \int (f(x) + g(x))\,dx = \int f(x)\,dx + \int g(x)\,dx$
>
> - **Multiples Rule:** $\displaystyle \int cf(x)\,dx = c \int f(x)\,dx$

EXAMPLE 2 Evaluate $\int (3x^4 - 5x^{2/3} + x^{-3})\,dx$.

Solution We integrate term by term and use the Power Rule:

$$\int (3x^4 - 5x^{2/3} + x^{-3})\,dx = \int 3x^4\,dx - \int 5x^{2/3}\,dx + \int x^{-3}\,dx \quad \text{(Sum Rule)}$$

$$= 3 \int x^4\,dx - 5 \int x^{2/3}\,dx + \int x^{-3}\,dx \quad \text{(Multiples Rule)}$$

$$= 3\left(\frac{x^5}{5}\right) - 5\left(\frac{x^{5/3}}{5/3}\right) + \frac{x^{-2}}{-2} + C \quad \text{(Power Rule)}$$

$$= \frac{3}{5}x^5 - 3x^{5/3} - \frac{1}{2}x^{-2} + C$$

When we break up an indefinite integral into a sum of several integrals as in Example 2, it is not necessary to include a separate constant of integration for each integral.

To check the answer, we verify that the derivative is equal to the integrand:

$$\frac{d}{dx}\left(\frac{3}{5}x^5 - 3x^{5/3} - \frac{1}{2}x^{-2} + C\right) = 3x^4 - 5x^{2/3} + x^{-3} \qquad \blacksquare$$

While we do not always do so in the text, it is often beneficial to check your work when computing antiderivatives. It is simply a matter of verifying that the derivative of your result is the function that you were antidifferentiating.

Although the linearity rules for the derivative carry over to linearity rules for indefinite integrals, there are no rules for directly computing indefinite integrals of products, quotients, and compositions of functions. At this point the best approach, if possible, is to convert the integrand algebraically so that the result is an integral that can be computed with the rules we have.

While there are no Product, Quotient, or Chain Rules for integrals, we will see that the Product Rule for derivatives leads to an important technique called Integration by Parts (Section 7.1) and the Chain Rule leads to the Substitution Method (Section 5.7).

EXAMPLE 3 Evaluate $\displaystyle \int \frac{5x^2 - 6}{x^3}\,dx$.

Solution We rewrite the integrand to produce an integral we can compute:

$$\int \frac{5x^2 - 6}{x^3}\,dx = \int \left(\frac{5}{x} - \frac{6}{x^3}\right)dx = 5 \int \frac{dx}{x} - 6 \int x^{-3}\,dx$$

$$= 5\ln|x| - 6\left(\frac{x^{-2}}{-2}\right) + C = 5\ln|x| + 3x^{-2} + C \qquad \blacksquare$$

The differentiation formulas for the trigonometric functions give us the following integration formulas. Each formula can be checked by differentiation.

Basic Trigonometric Integrals

$$\int \sin x \, dx = -\cos x + C, \qquad \int \cos x \, dx = \sin x + C$$

$$\int \sec^2 x \, dx = \tan x + C, \qquad \int \csc^2 x \, dx = -\cot x + C$$

$$\int \sec x \tan x \, dx = \sec x + C, \qquad \int \csc x \cot x \, dx = -\csc x + C$$

EXAMPLE 4 Evaluate $\int \left(\sin t + 20 \sec^2 t \right) dt$.

Solution

$$\int \left(\sin t + 20 \sec^2 t \right) dt = \int \sin t \, dt + 20 \int \sec^2 t \, dt$$

$$= -\cos t + 20 \tan t + C \qquad \blacksquare$$

Integrals Involving e^x

The formula $(e^x)' = e^x$ says that $f(x) = e^x$ is its own derivative. But this means that $f(x) = e^x$ is also *its own antiderivative*. In other words,

$$\boxed{\int e^x \, dx = e^x + C}$$

More generally, for any constant $k \neq 0$,

$$\boxed{\int e^{kx} \, dx = \frac{1}{k} e^{kx} + C}$$

EXAMPLE 5 Evaluate **(a)** $\int (3e^x - 4) \, dx$ and **(b)** $\int 12e^{7-3x} \, dx$.

Solution

(a) $\displaystyle \int (3e^x - 4) \, dx = 3 \int e^x \, dx - \int 4 \, dx = 3e^x - 4x + C$

(b) $\displaystyle \int 12e^{7-3x} \, dx = 12 \int e^7 e^{-3x} \, dx = 12e^7 \left(\frac{1}{-3} e^{-3x} \right) = -4e^{7-3x} + C \qquad \blacksquare$

CONCEPTUAL INSIGHT **Definite Versus Indefinite Integrals** While the definite integral and the indefinite integral have similar names and notation, it is important to realize that they are very different objects. They are about as similar as apples and bricks. The definite integral is a numerical value obtained as a limit of Riemann sums. The indefinite integral is a family of functions whose derivative is a given function. Make sure you understand these differences and use the names and notations properly.

Even though the definite integral and indefinite integral are very different, they are closely related to each other via the Fundamental Theorem of Calculus (FTC). We will learn more about the FTC relationships in the next two sections.

Differential Equations

We can think of an antiderivative as a solution to the **differential equation**

$$\frac{dy}{dx} = f(x) \qquad \boxed{2}$$

In general, a differential equation is an equation relating an unknown function and its derivatives. The unknown in Eq. (2) is a function $y = F(x)$ whose derivative is $f(x)$.

There are infinitely many solutions to Eq. (2)—all functions in the form $y = F(x) + C$, where $y = F(x)$ is an antiderivative of $y = f(x)$. However, we can specify a particular solution by imposing an **initial condition**—that is, by requiring that the solution satisfies $y(x_0) = y_0$ for some fixed values x_0 and y_0. A differential equation with an initial condition is called an **initial value problem**.

An initial condition is like the y-intercept of a line, which determines one particular line among all lines with the same slope. The graphs of the antiderivatives of $y = f(x)$ are all parallel (Figure 1), and the initial condition determines one of them.

EXAMPLE 6   Solve $\dfrac{dy}{dx} = 4x^7$ subject to the initial condition $y(0) = 4$.

Solution   First, find the general antiderivative:

$$y(x) = \int 4x^7 \, dx = \frac{1}{2}x^8 + C$$

Next, choose C so that the initial condition is satisfied: From $y(x) = \dfrac{1}{2}x^8 + C$ we have $y(0) = 0 + C$, and from the initial condition, we have $y(0) = 4$. This yields $C = 4$, and our solution is $y = \dfrac{1}{2}x^8 + 4$. ∎

As with antiderivative computations, it is always beneficial to check your solution to an initial value problem. There are two things to check: that the differential equation is satisfied, and that the initial condition is as well.

EXAMPLE 7   Solve the initial value problem $\dfrac{dy}{dt} = \sin t, \ y(0) = 2$.

Solution   First, find the general antiderivative:

$$y(t) = \int \sin t \, dt = -\cos t + C$$

Then solve for C: From $y(t) = -\cos t + C$ we have

$$y(0) = -\cos(0) + C = -1 + C$$

and from the initial condition we have $y(0) = 2$. So, $-1 + C = 2$, implying that $C = 3$. Therefore, the solution of the initial value problem is $y(t) = -\cos t + 3$. ∎

EXAMPLE 8   A car traveling with velocity 24 m/s begins to slow down at time $t = 0$ s with a constant acceleration of $a = -6$ m/s^2. Find **(a)** the velocity $v(t)$ at time t, and **(b)** the distance traveled before the car comes to a halt.

Negative acceleration is often referred to as deceleration.

Solution   **(a)** The derivative of velocity is acceleration, so *velocity is the antiderivative of acceleration*:

$$v(t) = \int a \, dt = \int (-6) \, dt = -6t + C$$

Relation between position, velocity, and acceleration:

$$s'(t) = v(t), \qquad s(t) = \int v(t) \, dt$$

$$v'(t) = a(t), \qquad v(t) = \int a(t) \, dt$$

The initial condition $v(0) = 24$ yields $C = 24$ and therefore $v(t) = -6t + 24$ m/s.
(b) Position is the antiderivative of velocity, so the car's position in meters is

$$s(t) = \int v(t) \, dt = \int (-6t + 24) \, dt = -3t^2 + 24t + C_1$$

where C_1 is a constant. We are not told where the car is at $t = 0$, so let us set $s(0) = 0$ for convenience, obtaining $C_1 = 0$. With this choice, $s(t) = -3t^2 + 24t$. This is the distance traveled from time $t = 0$.

The car comes to a halt when its velocity is zero, so we solve

$$0 = v(t) = -6t + 24 \quad \Rightarrow \quad t = 4 \text{ s}$$

The distance traveled before coming to a halt is $s(4) = -3(4^2) + 24(4) = 48$ m. ∎

Antidifferentiation is generally a more difficult process than differention. We will be developing indefinite integral formulas and techniques in the sections and chapters that follow. In this section, we derived a collection of integral formulas shown in the summary. At the end of this text there is a table of a few dozen integral formulas. Computer algebra systems such as WolframAlpha are excellent tools for computing integrals; they have made the process much easier than it used to be. Previously, extensive tables of integrals were relied on for computation of antiderivatives. Such tables are published by the CRC Press in their *Book of Standard Mathematical Tables*, containing over 50 pages of integral formulas.

5.3 SUMMARY

- F is called an *antiderivative* of f if $F'(x) = f(x)$.
- Any two antiderivatives of f on an interval (a, b) differ by a constant.
- The general antiderivative is denoted by the indefinite integral:

$$\int f(x)\, dx = F(x) + C$$

- Some integration formulas:

$$\int 0\, dx = C \qquad \int k\, dx = kx + C \qquad \int x^n\, dx = \frac{x^{n+1}}{n+1} + C \qquad (n \neq -1)$$

$$\int \sin x\, dx = -\cos x + C \qquad\qquad \int \cos x\, dx = \sin x + C$$

$$\int \sec^2 x\, dx = \tan x + C \qquad\qquad \int \csc^2 x\, dx = -\cot x + C$$

$$\int \sec x \tan x\, dx = \sec x + C \qquad\qquad \int \csc x \cot x\, dx = -\csc x + C$$

$$\int \frac{dx}{x} = \ln|x| + C \qquad\qquad \int e^{kx}\, dx = \frac{1}{k} e^{kx} + C \qquad (k \neq 0)$$

$$\int c f(x)\, dx = c \int f(x)\, dx \qquad\qquad \int (f(x) + g(x))\, dx = \int f(x)\, dx + \int g(x)\, dx$$

- To solve an initial value problem $\dfrac{dy}{dx} = f(x)$, $y(x_0) = y_0$, first, find the general antiderivative $y = F(x) + C$. Then, determine C using the initial condition $F(x_0) + C = y_0$.

5.3 EXERCISES

Preliminary Questions

1. Find an antiderivative of the function $f(x) = 0$.

2. Is there a difference between finding the general antiderivative of a function f and evaluating $\int f(x)\,dx$?

3. Jacques was told that f and g have the same derivative, and he wonders whether $f(x) = g(x)$. Does Jacques have sufficient information to answer his question?

4. Suppose that $F'(x) = f(x)$ and $G'(x) = g(x)$. Which of the following statements are true? Explain.
 (a) If $f = g$, then $F = G$.
 (b) If F and G differ by a constant, then $f = g$.
 (c) If f and g differ by a constant, then $F = G$.

5. Is $y = x$ a solution of the following initial value problem?
$$\frac{dy}{dx} = 1, \qquad y(0) = 1$$

Exercises

In Exercises 1–8, find the general antiderivative of f and check your answer by differentiating.

1. $f(x) = 18x^2$

2. $f(x) = x^{-3/5}$

3. $f(x) = 2x^4 - 24x^2 + 12x^{-1}$

4. $f(x) = 9x + 15x^{-2}$

5. $f(x) = 2\cos x - 9\sin x$

6. $f(x) = 4x^7 - 3\cos x$

7. $f(x) = 12e^x - 5x^{-2}$

8. $f(x) = e^x - 4\sin x$

9. Match functions (a)–(d) with their antiderivatives (i)–(iv).

 (a) $f(x) = \sin x$

 (i) $F(x) = \cos(1 - x)$

 (b) $f(x) = x\sin(x^2)$

 (ii) $F(x) = -\cos x$

 (c) $f(x) = \sin(1 - x)$

 (iii) $F(x) = -\frac{1}{2}\cos(x^2)$

 (d) $f(x) = x\sin x$

 (iv) $F(x) = \sin x - x\cos x$

In Exercises 10–37, evaluate the indefinite integral. Remember, there are no Product, Quotient, or Chain Rules for integration.

10. $\displaystyle\int (9x + 2)\,dx$

11. $\displaystyle\int (4 - 18x)\,dx$

12. $\displaystyle\int x^{-3}\,dx$

13. $\displaystyle\int t^{-6/11}\,dt$

14. $\displaystyle\int (5t^3 - t^{-3})\,dt$

15. $\displaystyle\int (18t^5 - 10t^4 - 28t)\,dt$

16. $\displaystyle\int 14s^{9/5}\,ds$

17. $\displaystyle\int (z^{-4/5} - z^{2/3} + z^{5/4})\,dz$

18. $\displaystyle\int \frac{3}{2}\,dx$

19. $\displaystyle\int \frac{1}{\sqrt[3]{x}}\,dx$

20. $\displaystyle\int \frac{dx}{x^{4/3}}$

21. $\displaystyle\int \frac{36\,dt}{t^3}$

22. $\displaystyle\int x(x^2 - 4)\,dx$

23. $\displaystyle\int (t^{1/2} + 1)(t + 1)\,dt$

24. $\displaystyle\int \frac{12 - z}{\sqrt{z}}\,dz$

25. $\displaystyle\int \frac{x^3 + 3x - 4}{x^2}\,dx$

26. $\displaystyle\int \left(\frac{1}{3}\sin x - \frac{1}{4}\cos x\right)dx$

27. $\displaystyle\int 12\sec x \tan x\,dx$

28. $\displaystyle\int (\theta + \sec^2 \theta)\,d\theta$

29. $\displaystyle\int \csc t \cot t\,dt$

30. $\displaystyle\int (t - \sin t)\,dt$

31. $\displaystyle\int (x^2 - \sec^2 x)\,dx$

32. $\displaystyle\int \tan\theta\cos\theta\,d\theta$

33. $\displaystyle\int \sec\theta(\sec\theta + \tan\theta)\,d\theta$

34. $\displaystyle\int \left(\frac{4}{x} - e^x\right)dx$

35. $\displaystyle\int (3e^{5x})\,dx$

36. $\displaystyle\int e^{3t-4}\,dt$ Hint: $e^{a+b} = e^a e^b$

37. $\displaystyle\int (8x - 4e^{5-2x})\,dx$ Hint: $e^{a+b} = e^a e^b$

38. In Figure 3, is graph (A) or graph (B) the graph of an antiderivative of $y = f(x)$?

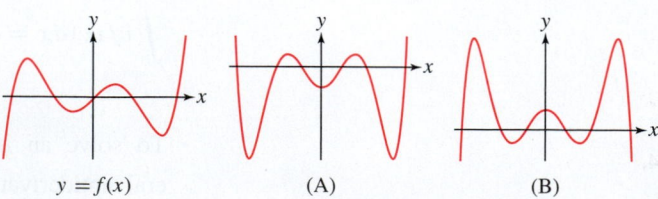

$y = f(x)$ (A) (B)

FIGURE 3

39. In Figure 4, which of graphs (A), (B), and (C) is *not* the graph of an antiderivative of $y = f(x)$? Explain.

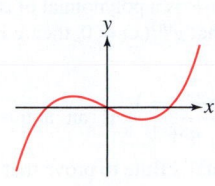

$y = f(x)$

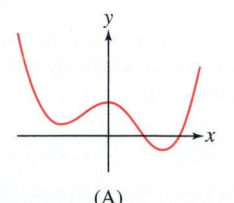 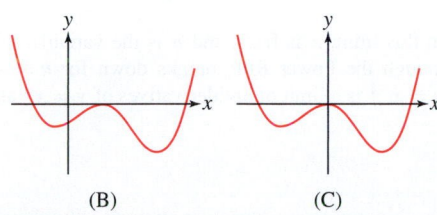

(A)   (B)   (C)

FIGURE 4

40. Verify that the function $F(x) = \frac{1}{3}(x+13)^3$ is an antiderivative of $f(x) = (x+13)^2$.

In Exercises 41–44, verify by differentiation.

41. $\int (x+13)^6 \, dx = \frac{1}{7}(x+13)^7 + C$

42. $\int (x+13)^{-5} \, dx = -\frac{1}{4}(x+13)^{-4} + C$

43. $\int (4x+13)^2 \, dx = \frac{1}{12}(4x+13)^3 + C$

44. $\int (ax+b)^n \, dx = \frac{1}{a(n+1)}(ax+b)^{n+1} + C$ (for $n \neq -1$)

In Exercises 45–46, we demonstrate that, in general, you cannot obtain an antiderivative of a product of functions by taking a product of antiderivatives of each.

45. Show that $G(x) = x^2 e^x$ is not an antiderivative of $f(x) = 2xe^x$ but $H(x) = 2xe^x - 2e^x$ is.

46. Show that $G(x) = 3x^2 \sin x$ is not an antiderivative of $f(x) = 6x \cos x$ but $H(x) = 6x \sin x + 6 \cos x$ is.

In Exercises 47–60, solve the initial value problem.

47. $\frac{dy}{dx} = x^3, \quad y(0) = 4$

48. $\frac{dy}{dt} = 3 - 2t, \quad y(0) = -5$

49. $\frac{dy}{dt} = 2t + 9t^2, \quad y(1) = 2$

50. $\frac{dy}{dx} = 8x^3 + 3x^2, \quad y(2) = 0$

51. $\frac{dy}{dt} = \sqrt{t}, \quad y(1) = 1$

52. $\frac{dz}{dt} = t^{-3/2}, \quad z(4) = -1$

53. $\frac{dy}{dx} = (3x+2)^3, \quad y(0) = 1$

54. $\frac{dy}{dt} = (4t+3)^{-2}, \quad y(1) = 0$

55. $\frac{dy}{dx} = \sin x, \quad y\left(\frac{\pi}{2}\right) = 1$

56. $\frac{dy}{dx} = \sec^2 x, \quad y\left(\frac{\pi}{4}\right) = 2$

57. $\frac{dy}{dx} = e^x, \quad y(2) = 0$

58. $\frac{dy}{dt} = e^{-t}, \quad y(0) = 0$

59. $\frac{dy}{dt} = 9e^{12-3t}, \quad y(4) = 7$

60. $\frac{dy}{dt} = t + 2e^{t-9}, \quad y(9) = 4$

In Exercises 61–67, first find f' and then find f.

61. $f''(x) = 12x, \quad f'(0) = 1, \quad f(0) = 2$

62. $f''(x) = x^3 - 2x, \quad f'(1) = 0, \quad f(1) = 2$

63. $f''(x) = x^3 - 2x + 1, \quad f'(0) = 1, \quad f(0) = 0$

64. $f''(x) = x^3 - 2x + 1, \quad f'(1) = 0, \quad f(1) = 4$

65. $f''(t) = t^{-3/2}, \quad f'(4) = 1, \quad f(4) = 4$

66. $f''(\theta) = \cos\theta, \quad f'\left(\frac{\pi}{2}\right) = 1, \quad f\left(\frac{\pi}{2}\right) = 6$

67. $f''(t) = t - \cos t, \quad f'(0) = 2, \quad f(0) = -2$

68. Show that $F(x) = \tan^2 x$ and $G(x) = \sec^2 x$ have the same derivative. What can you conclude about the relation between F and G? Verify this conclusion directly.

69. A particle located at the origin at $t = 1$ second moves along the x-axis with velocity $v(t) = (6t^2 - t)$ m/s. State the differential equation with its initial condition satisfied by the position $s(t)$ of the particle, and find $s(t)$.

70. A particle moves along the x-axis with velocity $v(t) = (6t^2 - t)$ m/s. Find the particle's position $s(t)$, assuming that $s(2) = 4$ m.

71. A water balloon is dropped from a high building. It falls for 5 seconds before hitting the ground. Determine the velocity it is traveling when it is about to hit the ground, assuming an acceleration due to gravity of -9.8 m/s^2 and no wind resistance.

72. A hammer is dropped and it falls for 2 seconds before hitting the ground. Determine how far it falls, assuming an acceleration due to gravity of -9.8 m/s^2 and no wind resistance.

73. A mass oscillates at the end of a spring. Let $s(t)$ be the displacement of the mass from the equilibrium position at time t. Assuming that the mass is located at the origin at $t = 0$ and has velocity $v(t) = \sin t$ m/s, state the differential equation with initial condition satisfied by $s(t)$, and find $s(t)$.

74. Beginning at $t = 0$ with initial velocity 4 m/s, a particle moves in a straight line with acceleration $a(t) = 3t^{1/2}$ m/s^2. Find the distance traveled after 25 s.

75. At time $t = 0$ a car traveling 25 m/s begins to accelerate at a constant rate of -4 m/s^2. After how many seconds does the car come to a stop and how far will the car have traveled between $t = 0$ and the time it stopped?

76. At time $t = 1$ second, a particle is traveling at 72 m/s and begins to accelerate at the rate $a(t) = -t^{-1/2}$ until it stops. How far does the particle travel from $t = 1$ until the time it stopped?

77. A 900-kg rocket is released from a space station. As it burns fuel, the rocket's mass decreases and its velocity increases. Let $v(m)$ be the velocity (in meters per second) as a function of mass m. Find the velocity when $m = 729$ kg if $dv/dm = -50m^{-1/2}$. Assume that $v(900) = 0$ m/s.

78. As water flows through a tube of radius $R = 10$ cm, the velocity v of an individual water particle depends only on its distance r from the center of the tube. The particles at the walls of the tube have zero velocity and $dv/dr = -0.06r$. Determine $v(r)$.

79. Verify the linearity properties of the indefinite integral stated in Theorem 4.

Further Insights and Challenges

80. Find constants c_1 and c_2 such that $F(x) = c_1 \sin 3x + c_2 x \cos 3x$ is an antiderivative of $f(x) = 2x \sin 3x$.

81. Find constants c_1 and c_2 such that $F(x) = c_1 x e^{-x} + c_2 e^{-x}$ is an antiderivative of $f(x) = 3x e^{-x}$.

82. Suppose that $F'(x) = f(x)$ and $G'(x) = g(x)$. Is it true that $y = F(x)G(x)$ is an antiderivative of $y = f(x)g(x)$? Confirm or provide a counterexample.

83. Suppose that $F'(x) = f(x)$.

(a) Show that $y = \frac{1}{2}F(2x)$ is an antiderivative of $y = f(2x)$.

(b) Find the general antiderivative of $y = f(kx)$ for $k \neq 0$.

84. Find an antiderivative for $f(x) = |x|$.

85. Using Theorem 1, prove that if $F'(x) = f(x)$, where f is a polynomial of degree $n - 1$, then F is a polynomial of degree n. Then prove that if g is any function such that $g^{(n)}(x) = 0$, then g is a polynomial of degree at most n.

86. Show that $F(x) = \dfrac{x^{n+1} - 1}{n + 1}$ is an antiderivative of $y = x^n$ for $n \neq -1$. Then use L'Hôpital's Rule to prove that

$$\lim_{n \to -1} F(x) = \ln x$$

In this limit, x is fixed and n is the variable. This result shows that, although the Power Rule breaks down for $n = -1$, the antiderivative of $y = x^{-1}$ is a limit of antiderivatives of $y = x^n$ as $n \to -1$.

5.4 The Fundamental Theorem of Calculus, Part I

The FTC was first stated clearly by Isaac Newton in 1666, although other mathematicians, including Newton's teacher Isaac Barrow, had discovered versions of it earlier.

Since we have so far introduced both derivatives and integrals, a very reasonable question is why they appear together in this topic called calculus. The answer is the Fundamental Theorem of Calculus (FTC), which is one of the most important theorems in all of mathematics. This foundational result reveals an unexpected connection between the two main operations of calculus: differentiation and integration. The theorem has two parts. Although they are closely related, we discuss them in separate sections to emphasize the different ways they are used. The first part of the Fundamental Theorem of Calculus will allow us to compute definite integrals without having to take limits of Riemann sums.

To explain FTC I, recall a result from Example 7 of Section 5.2:

$$\int_4^7 x^2\, dx = \left(\frac{1}{3}\right)7^3 - \left(\frac{1}{3}\right)4^3 = 93$$

Now, observe that $F(x) = \frac{1}{3}x^3$ is an antiderivative of $f(x) = x^2$, so we can write

$$\int_4^7 x^2\, dx = F(7) - F(4)$$

According to FTC I, this is no coincidence; this relation between the definite integral and the antiderivative holds in general.

◄ **REMINDER**

F is called an **antiderivative** of f if $F'(x) = f(x)$. We say also that F is an **indefinite integral** of f, and we use the notation

$$\int f(x)\, dx = F(x) + C$$

THEOREM 1 The Fundamental Theorem of Calculus, Part I   Assume that $a < b$ and that f is continuous on $[a, b]$. If F is an antiderivative of f on $[a, b]$, then

$$\int_a^b f(x)\, dx = F(b) - F(a) \qquad \boxed{1}$$

Proof   The quantity $F(b) - F(a)$ is the total change in F (also called the net change) over the interval $[a, b]$. Our task is to relate it to the integral of $F'(x) = f(x)$. There are two main steps.

Write total change as a sum of small changes: Given any partition P of $[a, b]$:

$$P : x_0 = a < x_1 < x_2 < \cdots < x_N = b$$

we can break up $F(b) - F(a)$ as a sum of changes over the intervals $[x_{i-1}, x_i]$:

$$F(b) - F(a) = (F(b) - F(x_{N-1})) + (F(x_{N-1}) - F(x_{N-2}))$$
$$+ \cdots + (F(x_2) - F(x_1)) + (F(x_1) - F(a))$$

On the right-hand side, $-F(x_{N-1})$ is canceled by $F(x_{N-1})$ in the second term, $-F(x_{N-2})$ is canceled by $F(x_{N-2})$ in the third term, and so on (Figure 1). In summation notation,

$$F(b) - F(a) = \sum_{i=1}^{N} \left(F(x_i) - F(x_{i-1}) \right) \qquad \boxed{2}$$

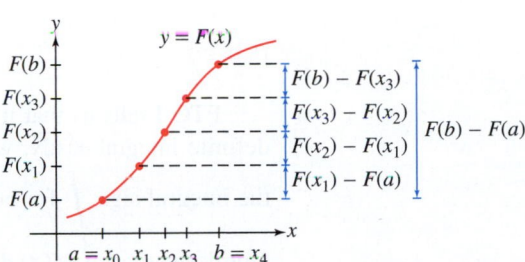

FIGURE 1 Note the cancellation when we write $F(b) - F(a)$ as a sum of small changes $F(x_i) - F(x_{i-1})$.

Interpret Eq. (2) as a Riemann sum: The Mean Value Theorem tells us that there is a point c_i^* in $[x_{i-1}, x_i]$ such that

$$F(x_i) - F(x_{i-1}) = F'(c_i^*)(x_i - x_{i-1}) = f(c_i^*)(x_i - x_{i-1}) = f(c_i^*)\,\Delta x_i$$

Therefore, Eq. (2) can be written

$$F(b) - F(a) = \sum_{i=1}^{N} f(c_i^*)\,\Delta x_i$$

This sum is the Riemann sum $R(f, P, C^*)$ with sample points $C^* = \{c_i^*\}$.

Now, f is integrable (Theorem 1, Section 5.2), so $R(f, P, C^*)$ approaches $\displaystyle\int_a^b f(x)\,dx$

as the norm $\|P\|$ tends to zero. On the other hand, $R(f, P, C^*)$ is *equal* to $F(b) - F(a)$ with our particular choice C^* of sample points. This proves the desired result:

$$F(b) - F(a) = \lim_{\|P\| \to 0} R(f, P, C^*) = \int_a^b f(x)\,dx \qquad \blacksquare$$

CONCEPTUAL INSIGHT **A Tale of Two Graphs** In the proof of FTC I, we used the MVT to write a small change in y in the graph of $y = F(x)$ in terms of the derivative $F'(x) = f(x)$:

$$F(x_i) - F(x_{i-1}) = f(c_i^*)\Delta x_i$$

But $f(c_i^*)\Delta x_i$ is the signed area of a thin rectangle that approximates a sliver of signed area under the graph of f (Figure 2). *This is the essence of the Fundamental Theorem*:

- The total change, $F(b) - F(a)$, is
- The sum of small changes, $F(x_i) - F(x_{i-1})$, which is
- The sum of the signed areas of rectangles from the graph of f, and that is
- A Riemann sum for f.

The Fundamental Theorem itself is then obtained by taking the limit as the widths of the rectangles tend to zero.

This change is equal to the area $f(c_i^*)\Delta x_i$ of this rectangle.

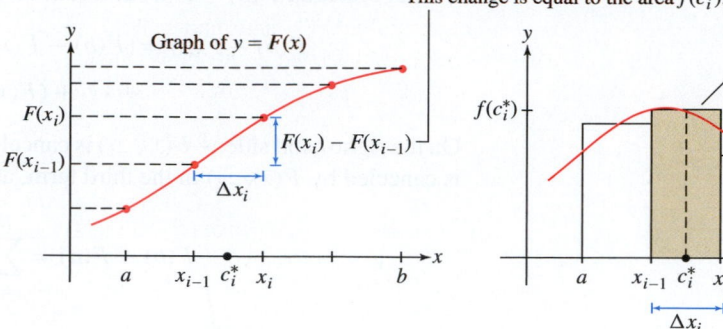

FIGURE 2

FTC I tells us that if we can find an antiderivative of f, then we can compute the definite integral easily, without calculating any limits. It is for this reason that we use the integral sign $\int$ for both the definite integral $\int_a^b f(x)\,dx$ and the indefinite integral (antiderivative) $\int f(x)\,dx$.

While Theorem 1 is stated with the assumption that $a < b$, it holds for general a and b as well. See Exercise 63.

NOTATION $F(b) - F(a)$ is denoted $F(x)\Big|_a^b$. In this notation, the FTC reads

$$\int_a^b f(x)\,dx = F(x)\Big|_a^b \quad \text{where} \quad \int f(x)\,dx = F(x) + C$$

This form of FTC I suggests a two-step approach for evaluating the definite integral $\int_a^b f(x)\,dx$:

- Compute $\int f(x)\,dx = F(x) + C$

- Evaluate $\int_a^b f(x)\,dx = F(x)\Big|_a^b = F(b) - F(a)$

In examples that follow, we will use this two-step process. As we become more familiar with antidifferentiation, we will usually carry out both steps at once.

← **REMINDER** *The Power Rule for Integrals (valid for $n \neq -1$) states*

$$\int x^n\,dx = \frac{x^{n+1}}{n+1} + C$$

EXAMPLE 1   Calculate the area under the graph of $f(x) = x^3$ over $[2, 4]$.

Solution   Since $\int x^3\,dx = \frac{1}{4}x^4 + C$, we have

$$\int_2^4 x^3\,dx = \frac{1}{4}x^4\Big|_2^4 = \frac{1}{4}4^4 - \frac{1}{4}2^4 = 60$$ ■

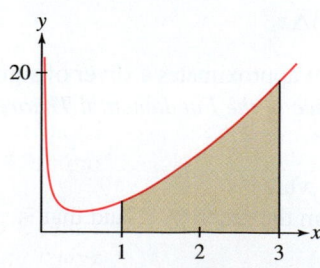

FIGURE 3 Region under the graph of $g(x) = x^{-3/4} + 3x^{5/3}$ over $[1, 3]$.

EXAMPLE 2   Find the area under $g(x) = x^{-3/4} + 3x^{5/3}$ over $[1, 3]$ (Figure 3).

Solution   To begin,

$$\int (x^{-3/4} + 3x^{5/3})\,dx = 4x^{1/4} + \tfrac{9}{8}x^{8/3} + C$$

Therefore, the area is equal to

$$\int_1^3 (x^{-3/4} + 3x^{5/3})\,dx = \left(4x^{1/4} + \frac{9}{8}x^{8/3}\right)\Big|_1^3$$

$$= \left(4 \cdot 3^{1/4} + \frac{9}{8} \cdot 3^{8/3}\right) - \left(4 \cdot 1^{1/4} + \frac{9}{8} \cdot 1^{8/3}\right)$$

$$\approx 26.325 - 5.125 = 21.2$$ ■

CONCEPTUAL INSIGHT   **Which Antiderivative?** As we have seen, antiderivatives are not unique. Does it matter then, which antiderivative is used in the FTC? The answer is no. If F and G are both antiderivatives of a continuous function f on $[a, b]$, then $F(x) = G(x) + C$ for some constant C, and

$$F(b) - F(a) = \underbrace{(G(b) + C) - (G(a) + C)}_{\text{The constant cancels.}} = G(b) - G(a)$$

The two antiderivatives yield the same value for the definite integral:

$$\int_a^b f(x)\, dx = F(b) - F(a) = G(b) - G(a)$$

In Section 5.2, we showed that $\int_0^b x\, dx = \frac{1}{2}b^2$ and $\int_0^b x^2\, dx = \frac{1}{3}b^3$. We asked you to conjecture what the result is for general $f(x) = x^n$. We are now in a position to consider the general case.

EXAMPLE 3   Calculate $\int_0^b x^n\, dx$, assuming that $n \neq -1$.

Solution   First, $\displaystyle\int x^n\, dx = \frac{1}{n+1}x^{n+1} + C$. Therefore,

$$\int_0^b x^n\, dx = \frac{1}{n+1}x^{n+1}\bigg|_0^b = \frac{1}{n+1}b^{n+1} \qquad\blacksquare$$

We know that the definite integral is equal to the *signed* area between the graph and the x-axis. Needless to say, the FTC "knows" this also: When you evaluate an integral using the FTC, you obtain the signed area.

EXAMPLE 4   Evaluate   **(a)** $\displaystyle\int_0^\pi \sin x\, dx$ **(b)** $\displaystyle\int_0^{2\pi} \sin x\, dx$ **(c)** $\displaystyle\int_{\pi/4}^{3\pi/4} \sin x\, dx$.

Solution

(a) $\displaystyle\int_0^\pi \sin x\, dx = -\cos x\bigg|_0^\pi = -\cos\pi - (-\cos 0) = -(-1) - (-1) = 2$

(b) $\displaystyle\int_0^{2\pi} \sin x\, dx = -\cos x\bigg|_0^{2\pi} = (-\cos(2\pi) - (-\cos 0)) = -1 - (-1) = 0$

(c) $\displaystyle\int_{\pi/4}^{3\pi/4} \sin x\, dx = -\cos x\bigg|_{\pi/4}^{3\pi/4} = -\cos(3\pi/4) - (-\cos(\pi/4))$

$$= -\left(-\frac{\sqrt{2}}{2}\right) - \left(-\frac{\sqrt{2}}{2}\right) = \sqrt{2} \qquad\blacksquare$$

It is an interesting fact that the area under one hump of the sine graph (Figure 4) is a nice whole-number value, 2. That is certainly worth remembering. Also, it is no surprise that the result of the second integral in this example is zero since it corresponds to the signed area from 0 to 2π, and that value is zero since the contribution from the hump below the axis exactly cancels the contribution from the hump above.

Finally, in Example 5 in Section 5.1, we introduced a general right-hand sum R_N for approximating the area associated with the integral in (c). We indicated that directly taking the limit of R_N to get an exact value is a difficult task, but now with FTC I, the process of obtaining the area couldn't be easier: Antidifferentiate $\sin x$, evaluate the result at the endpoints, and take the difference.

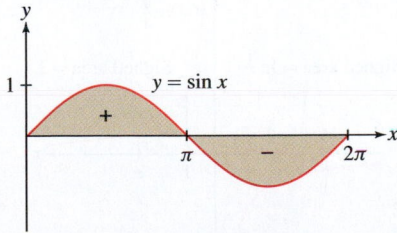

FIGURE 4 The area of one hump is 2. The signed area over $[0, 2\pi]$ is zero.

EXAMPLE 5 Probability density functions are used to calculate, via integration, the likelihood that an event has a value that lies within a specific range of values. The Eddington Bulb Company uses a probability density function $p(t) = 0.001e^{-0.001t}$ to determine the likelihood of failure of their Cool-Lite refrigerator bulb over any interval of hours of use. What is the probability that a Cool-Lite bulb lasts for more than 500 hours of use?

Solution The integral $\int_0^{500} 0.001e^{-0.001t}\, dt$ gives the probability that a Cool-Lite bulb fails before 500 h of use. Subtracting this value from 1 provides the desired probability. To begin,

$$\int 0.001e^{-0.001t}\, dt = 0.001\frac{1}{-0.001}e^{-0.001t} + C = -e^{-0.001t} + C$$

Therefore, for the definite integral,

$$\int_0^{500} 0.001e^{-0.001t}\, dt = -e^{-0.001t}\Big|_0^{500} = -e^{-0.5} - (-1) \approx 0.39$$

It follows that the probability that a Cool-Lite bulb lasts for more than 500 h of use is ≈ 0.61. ∎

Recall (Section 5.3) that $F(x) = \ln|x|$ is an antiderivative of $f(x) = x^{-1}$ in the domain $\{x : x \neq 0\}$. Therefore, FTC I yields the following formula that is valid if a and b are both positive or both negative [see Figure 5(A)]. Note that when $a < 0 < b$, FTC I does not apply since f is not continuous over $[a, b]$.

$$\int_a^b \frac{dx}{x} = \ln|b| - \ln|a| = \ln\frac{b}{a} \qquad \boxed{3}$$

EXAMPLE 6 Evaluate **(a)** $\int_2^8 \frac{dx}{x}$ **(b)** $\int_{-4}^{-2} \frac{dx}{x}$ **(c)** $\int_1^e \frac{dx}{x}$.

Solution By Eq. (3),

(a) $\displaystyle\int_2^8 \frac{dx}{x} = \ln\frac{8}{2} = \ln 4 \approx 1.39$

(b) $\displaystyle\int_{-4}^{-2} \frac{dx}{x} = \ln\left(\frac{-2}{-4}\right) = \ln\frac{1}{2} \approx -0.69$

(c) $\displaystyle\int_1^e \frac{dx}{x} = \ln\left(\frac{e}{1}\right) = \ln e = 1$ ∎

The signed areas represented by (a)–(c) are shown in (B) and (C) in Figure 5. When we introduced the number e in Section 1.6, we observed that e is *the* number for which the area under $y = 1/x$ from 1 to e is equal to 1. We have now verified this fact via a definite integral.

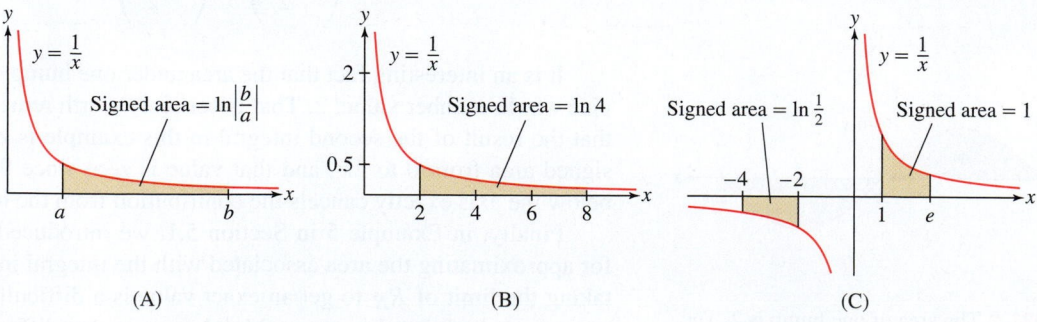

(A)   (B)   (C)

FIGURE 5

CONCEPTUAL INSIGHT FTC I reveals a valuable relationship for computing definite integrals, but it does not always help. For example, to use FTC I for $\int_0^4 \sqrt[3]{1 + \cos x}\, dx$, we need an antiderivative formula for $f(x) = \sqrt[3]{1 + \cos x}$. Unfortunately, even though f has an antiderivative, there is no formula for the antiderivative that helps determine the definite integral. (We elaborate on these points further in the next section.)

This does not mean that we cannot obtain a value for this definite integral. Remember, a definite integral is a limit of Riemann sums, so we can obtain an approximate value to the definite integral using Riemann sums directly. In this case, partitioning $[0, 4]$ into 1000 subintervals and computing (using technology!) the left-endpoint approximation, we obtain $\int_0^4 \sqrt[3]{1 + \cos x}\, dx \approx 3.190$. In Section 7.8, we introduce other techniques for approximating definite integrals.

5.4 SUMMARY

- The Fundamental Theorem of Calculus, Part I, states that

$$\int_a^b f(x)\, dx = F(b) - F(a)$$

  where F is an antiderivative of f.

- Two-step approach for using FTC I to evaluate the definite integral $\int_a^b f(x)\, dx$:

  – Compute $\int f(x)\, dx = F(x) + C$.

  – Evaluate $\int_a^b f(x)\, dx = F(x)\Big|_a^b = F(b) - F(a)$.

- Antiderivative formulas are helpful for evaluating definite integrals. See the summary in Section 5.3 and the table of integrals in this text's endleaf.

5.4 EXERCISES

Preliminary Questions

1. Suppose that $F'(x) = f(x)$ and $F(0) = 3$, $F(2) = 7$.

(a) What is the area under $y = f(x)$ over $[0, 2]$ if $f(x) \geq 0$?

(b) What is the graphical interpretation of $F(2) - F(0)$ if $f(x)$ takes on both positive and negative values?

2. Suppose that f is a *negative* function with antiderivative F such that $F(1) = 7$ and $F(3) = 4$. What is the area (a positive number) between the x-axis and the graph of f over $[1, 3]$?

3. Are the following statements true or false? Explain.

(a) FTC I is valid only for positive functions.

(b) To use FTC I, you have to choose the right antiderivative.

(c) If you cannot find an antiderivative of f, then the definite integral does not exist.

4. Evaluate $\int_2^9 f'(x)\, dx$, where f is differentiable and $f(2) = f(9) = 4$.

Exercises

In Exercises 1–4, sketch the region under the graph of the function and find its area using FTC I.

1. $f(x) = x^2$, $[0, 1]$

2. $f(x) = 2x - x^2$, $[0, 2]$

3. $f(x) = x^{-2}$, $[1, 2]$

4. $f(x) = \cos x$, $\left[0, \frac{\pi}{2}\right]$

In Exercises 5–40, evaluate the integral using FTC I.

5. $\int_3^6 x\, dx$

6. $\int_0^9 2\, dx$

7. $\int_0^1 (4x - 9x^2)\, dx$

8. $\int_{-3}^2 u^2\, du$

9. $\int_0^2 (12x^5 + 3x^2 - 4x)\, dx$

10. $\int_{-2}^2 (10x^9 + 3x^5)\, dx$

11. $\int_3^0 (2t^3 - 6t^2)\, dt$

12. $\int_{-1}^1 (5u^4 + u^2 - u)\, du$

13. $\int_0^4 \sqrt{y}\, dy$

14. $\int_1^8 x^{4/3}\, dx$

15. $\int_{1/16}^{1} t^{1/4}\, dt$

16. $\int_{4}^{1} t^{5/2}\, dt$

17. $\int_{1}^{3} \frac{dt}{t^2}$

18. $\int_{1}^{4} x^{-4}\, dx$

19. $\int_{1/2}^{1} \frac{8}{x^3}\, dx$

20. $\int_{-2}^{-1} \frac{1}{x^3}\, dx$

21. $\int_{1}^{2} (x^2 - x^{-2})\, dx$

22. $\int_{1}^{9} t^{-1/2}\, dt$

23. $\int_{1}^{27} \frac{t+1}{\sqrt{t}}\, dt$

24. $\int_{8/27}^{1} \frac{10t^{4/3} - 8t^{1/3}}{t^2}\, dt$

25. $\int_{-\pi/4}^{\pi} \sin\theta\, d\theta$

26. $\int_{0}^{13\pi} \sin x\, dx$

27. $\int_{0}^{\pi/3} \cos t\, dt$

28. $\int_{0}^{\pi/6} \sec\theta \tan\theta\, d\theta$

29. $\int_{\pi/4}^{3\pi/4} (2 - \csc^2 x)\, dx$

30. $\int_{0}^{1.57079} \sec^2 t\, dt$

31. $\int_{0}^{1} e^x\, dx$

32. $\int_{3}^{5} e^{-4x}\, dx$

33. $\int_{0}^{3} e^{1-6t}\, dt$

34. $\int_{2}^{3} e^{4t-3}\, dt$

35. $\int_{2}^{10} \frac{dx}{x}$

36. $\int_{-12}^{-4} \frac{dx}{x}$

37. $\int_{1}^{e} \frac{t+1}{t}\, dt$

38. $\int_{1}^{4} \frac{5t^2+4}{3t}\, dt$

39. $\int_{-2}^{0} (3x - 9e^{3x})\, dx$

40. $\int_{2}^{6} \left(x + \frac{1}{x}\right) dx$

41. In Example 5, what is the probability that the Cool-Lite bulb lasts for more than 100 hours of use? more than 1000 h of use?

42. In Example 5, after how many hours of use can we expect that 90% of the Cool-Lite bulbs have burned out?

In Exercises 43–48, write the integral as a sum of integrals without absolute values and evaluate.

43. $\int_{-2}^{1} |x|\, dx$

44. $\int_{0}^{5} |3 - x|\, dx$

45. $\int_{-2}^{3} |x^3|\, dx$

46. $\int_{0}^{3} |x^2 - 1|\, dx$

47. $\int_{0}^{\pi} |\cos x|\, dx$

48. $\int_{0}^{5} |x^2 - 4x + 3|\, dx$

In Exercises 49–54, evaluate the integral in terms of the constants.

49. $\int_{1}^{b} x^3\, dx$

50. $\int_{b}^{a} x^4\, dx$

51. $\int_{1}^{b} x^5\, dx$

52. $\int_{-x}^{x} (t^3 + t)\, dt$

53. $\int_{a}^{5a} \frac{dx}{x}$

54. $\int_{b}^{b^2} \frac{dx}{x}$

55. Calculate $\int_{-2}^{3} f(x)\, dx$, where

$$f(x) = \begin{cases} 12 - x^2 & \text{for } x \le 2 \\ x^3 & \text{for } x > 2 \end{cases}$$

56. Calculate $\int_{0}^{2\pi} f(x)\, dx$, where

$$f(x) = \begin{cases} \sin x & \text{for } x \le \pi \\ -2\sin x & \text{for } x > \pi \end{cases}$$

57. Use FTC I to show that $\int_{-1}^{1} x^n\, dx = 0$ if n is an odd whole number. Explain graphically.

58. **CAS** Plot the function $f(x) = 3\sin x - x$. Find the positive root of f to three decimal places and use it to find the area under the graph of f in the first quadrant.

59. Calculate $F(4)$ given that $F(1) = 3$ and $F'(x) = x^2$. *Hint:* Express $F(4) - F(1)$ as a definite integral.

60. Calculate $G(16)$, where $dG/dt = t^{-1/2}$ and $G(9) = -5$.

61. With $n > 0$, does $\int_{0}^{1} x^n\, dx$ get larger or smaller as n increases? Explain graphically.

62. With $k > 0$, does $\int_{0}^{1} e^{-kx}\, dx$ get larger or smaller as k increases? Explain graphically.

63. Theorem 1 is stated with the assumption that $a < b$. Prove that the FTC I relationship

$$\int_{a}^{b} f(x)\, dx = F(b) - F(a)$$

also holds for $a = b$ and for $b < a$ assuming that F is an antiderivative of f on $[b, a]$.

64. Show that the area of the shaded parabolic arch in Figure 6 is equal to four-thirds the area of the triangle shown.

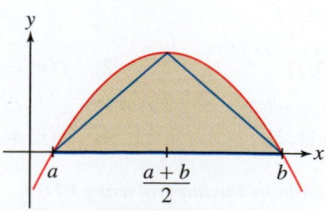

FIGURE 6 Graph of $y = (x - a)(b - x)$.

Further Insights and Challenges

65. Prove a famous result of Archimedes (generalizing Exercise 64): For $r < s$, the area of the shaded region in Figure 7 is equal to four-thirds the area of triangle $\triangle ACE$, where C is the point on the parabola at which the tangent line is parallel to secant line $\overline{AE}$.

(a) Show that C has x-coordinate $(r + s)/2$.

(b) Show that $ABDE$ has area $(s − r)^3/4$ by viewing it as a parallelogram of height $s − r$ and base of length $\overline{CF}$.

(c) Show that $\triangle ACE$ has area $(s − r)^3/8$ by observing that it has the same base and height as the parallelogram.

(d) Compute the shaded area as the area under the graph minus the area of a trapezoid, and prove Archimedes's result.

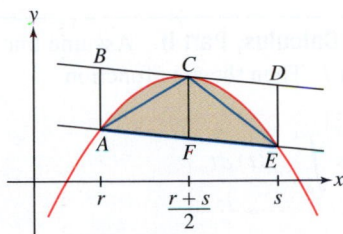

FIGURE 7 Graph of $f(x) = (x − a)(b − x)$.

66. (a) Apply the Comparison Theorem (Theorem 5 in Section 5.2) to the inequality $\sin x \le x$ (valid for $x \ge 0$) to prove that

$$1 - \frac{x^2}{2} \le \cos x \le 1$$

(b) Apply it again to prove that

$$x - \frac{x^3}{6} \le \sin x \le x \quad \text{(for } x \ge 0)$$

(c) Verify these inequalities for $x = 0.3$.

67. Use the method of Exercise 66 to prove that

$$1 - \frac{x^2}{2} \le \cos x \le 1 - \frac{x^2}{2} + \frac{x^4}{24}$$

$$x - \frac{x^3}{6} \le \sin x \le x - \frac{x^3}{6} + \frac{x^5}{120} \quad \text{(for } x \ge 0)$$

Verify these inequalities for $x = 0.1$. Why have we specified $x \ge 0$ for $\sin x$ but not for $\cos x$?

68. Calculate the next pair of inequalities for $\sin x$ and $\cos x$ by integrating the results of Exercise 67. Can you guess the general pattern?

69. Use Part I of the Fundamental Theorem of Calculus to prove that if $|f'(x)| \le K$ for $x \in [a, b]$, then $|f(x) - f(a)| \le K|x - a|$ for $x \in [a, b]$.

70. (a) Use Exercise 69 to prove that $|\sin a - \sin b| \le |a - b|$ for all a, b.

(b) Let $f(x) = \sin(x + a) - \sin x$. Use part (a) to show that the graph of f lies between the horizontal lines $y = \pm a$.

(c) **GU** Plot $y = f(x)$ and the lines $y = \pm a$ to verify (b) for $a = 0.5$ and $a = 0.2$.

5.5 The Fundamental Theorem of Calculus, Part II

Part I of the Fundamental Theorem says that we can compute definite integrals using indefinite integrals (antiderivatives). Part II does the opposite, providing a way to compute antiderivatives using definite integrals. Another interpretation of Part II of the FTC demonstrates how definite integration and differentiation are inverse processes. We will focus on that aspect of FTC II first, considering a motivating example before stating the theorem.

The idea behind this inverse relationship is as follows:

- Start with a function f.
- Create a new function A via definite integration of f. The function A is called an area function of f.
- Differentiate A to return to the original function f.

To begin, we introduce the **area function** of f with lower limit a:

$$A(x) = \int_a^x f(t)\,dt = \text{signed area from } a \text{ to } x$$

It would be more appropriate to call A a signed area function, but for simplicity we drop "signed." Keep in mind, though, that $A(x)$ is a signed area determined by a definite integral.

In the definition of $A(x)$, we use t as the variable of integration to avoid confusion with x, which is the upper limit of integration. In fact, t is a dummy variable and may be replaced by any other variable.

For $A(x)$ to be defined, f must be integrable over $[a, x]$ when $x > a$ or over $[x, a]$ when $x < a$. The definite integral defining $A(x)$ yields a function because we regard the upper limit x as a variable.

Let us consider an example with $f(x) = 4 - x^2$ and $a = 1$. The resulting area function is

$$A(x) = \int_1^x (4 - t^2)\,dt$$

We can obtain an expression for $A(x)$ by using FTC I to evaluate the definite integral:

$$A(x) = \int_1^x (4 - t^2)\, dt = \left(4t - \frac{1}{3}t^3\right)\Big|_1^x = \left(4x - \frac{1}{3}x^3\right) - \left(4(1) - \frac{1}{3}(1)^3\right) = 4x - \frac{1}{3}x^3 - \frac{11}{3}$$

Thus, $A(x) = 4x - \frac{1}{3}x^3 - \frac{11}{3}$. Now, notice that if we differentiate this area function, the result is f, the function we started with:

$$A'(x) = 4 - x^2 = f(x)$$

Thus, if we start with f, create an area function A, and then differentiate A, we return to f. Part II of the Fundamental Theorem of Calculus asserts that if f is continuous, then this relationship always holds:

THEOREM 1 Fundamental Theorem of Calculus, Part II   Assume that f is continuous on an open interval I and let a be in I. Then the area function

$$A(x) = \int_a^x f(t)\, dt$$

is an antiderivative of f on I; that is, $A'(x) = f(x)$. Equivalently,

$$\frac{d}{dx} \int_a^x f(t)\, dt = f(x)$$

Proof For simplicity we assume that f is nonnegative and increasing. (For the general case, see Exercise 54.) First, we use the additivity property of the definite integral to write the change in A over $[x, x + h]$ as an integral:

$$A(x+h) - A(x) = \int_a^{x+h} f(t)\, dt - \int_a^x f(t)\, dt = \int_x^{x+h} f(t)\, dt$$

In other words, $A(x + h) - A(x)$ is equal to the area of the thin sliver between the graph and the x-axis from x to $x + h$ in Figure 1.

In this proof,

$$A(x) = \int_a^x f(t)\, dt$$

$$A(x+h) - A(x) = \int_x^{x+h} f(t)\, dt$$

$$A'(x) = \lim_{h \to 0} \frac{A(x+h) - A(x)}{h}$$

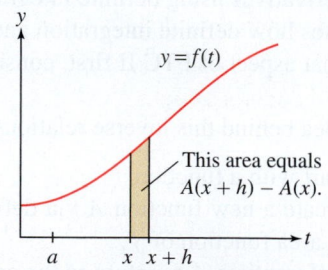

FIGURE 1 The area of the thin sliver equals $A(x + h) - A(x)$.

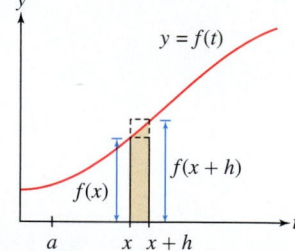

DF **FIGURE 2** The shaded sliver lies between the rectangles of heights $f(x)$ and $f(x + h)$.

Since f is nondecreasing, when $h > 0$, this thin sliver lies between the two rectangles of heights $f(x)$ and $f(x + h)$ in Figure 2, and we have

$$\underbrace{hf(x)}_{\text{Area of smaller rectangle}} \le \underbrace{A(x+h) - A(x)}_{\text{Area of sliver}} \le \underbrace{hf(x+h)}_{\text{Area of larger rectangle}}$$

Now divide by h to squeeze the difference quotient between $f(x)$ and $f(x + h)$:

$$f(x) \le \frac{A(x+h) - A(x)}{h} \le f(x+h)$$

We have $\displaystyle\lim_{h \to 0^+} f(x + h) = f(x)$ because f is continuous, and $\displaystyle\lim_{h \to 0^+} f(x) = f(x)$, so the Squeeze Theorem gives us

$$\lim_{h \to 0^+} \frac{A(x+h) - A(x)}{h} = f(x) \qquad \boxed{1}$$

A similar argument shows that for $h < 0$,

$$f(x+h) \le \frac{A(x+h) - A(x)}{h} \le f(x)$$

Again, the Squeeze Theorem gives us

$$\lim_{h \to 0^-} \frac{A(x+h) - A(x)}{h} = f(x) \qquad \boxed{2}$$

Equations (1) and (2) show that $A'(x)$ exists and that $A'(x) = f(x)$. ∎

EXAMPLE 1 Let $f(x) = 1 - 6x - \cos x$ and $a = -\pi$. Compute the area function $A(x) = \int_a^x f(t)\,dt$, and then verify the FTC II inverse relationship by showing that $A'(x) = f(x)$.

Solution

$$A(x) = \int_{-\pi}^x (1 - 6t - \cos t)\,dt = (t - 3t^2 - \sin t)\Big|_{-\pi}^x = x - 3x^2 - \sin x + \pi + 3\pi^2$$

Taking the derivative,

$$A'(x) = \frac{d}{dx}(x - 3x^2 - \sin x + \pi + 3\pi^2) = 1 - 6x - \cos x = f(x)$$

Thus, we see that when we start with f, compute an area function of it, and differentiate the area function, we obtain the function f back. ∎

EXAMPLE 2   A Numerically Approximate FTC II Verification   Consider the function f that is presented graphically in Figure 3. Here we will approximate an area function and its derivative to demonstrate (at least approximately) the FTC II inverse relationship. Let A be the area function defined by $A(x) = \int_0^x f(t)\,dt$.

(a) By estimating the corresponding signed areas, approximate $A(x)$ for $x = 0, 1, 2, \dots, 10$.

(b) Use the symmetric difference quotient approximation, $A'(x) \approx \frac{A(x+\Delta x) - A(x - \Delta x)}{2\Delta x}$, with $\Delta x = 1$ to approximate $A'(x)$ for $x = 1, 2, \dots, 9$. Plot these values of $A'(x)$ on the graph of f to demonstrate $A' \approx f$.

Solution
(a) First note that $A(0) = \int_0^0 f(t)\,dt = 0$. For other x, we can estimate $A(x)$ by counting the 1×1 squares (and partial squares) between the graph and the x-axis from 0 to x. For example, for $A(1)$ we have 4.4 squares between the graph and the x-axis from $x = 0$ to $x = 1$. Therefore, $A(1) \approx 4.4$. Continuing, we obtain the approximate values of $A(x)$ shown in the table below. Note that in determining the values for $A(8)$ through $A(10)$, each square below the x-axis contributes a value of -1 to the signed area.

(b) As a couple of examples, for $x = 1$ and $x = 9$, we obtain the following difference quotient approximations:

$$A'(1) \approx \frac{A(2) - A(0)}{2} = \frac{8.9 - 0}{2} = 4.45, \quad A'(9) \approx \frac{A(10) - A(8)}{2} = \frac{14.7 - 23.3}{2} = -4.3$$

The remaining values of $A'(x)$ are obtained similarly and are plotted with f in Figure 4.

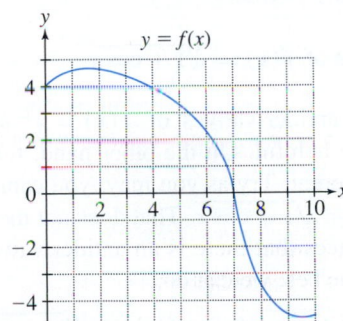

FIGURE 3

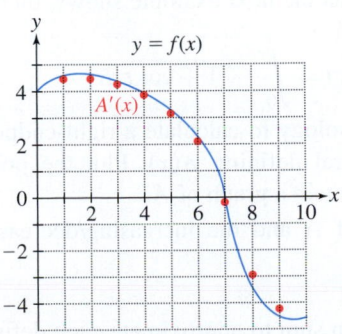

FIGURE 4

x	0	1	2	3	4	5	6	7	8	9	10
$A(x)$	0	4.4	8.9	13.3	17.4	21.0	23.7	25.2	23.3	19.3	14.7
$A'(x)$		4.45	4.45	4.25	3.85	3.15	2.1	−0.2	−2.95	−4.3	

We can see in the graph, at least approximately, the FTC II inverse relationship: If we start with a function f, compute an area function of f, and then take the derivative of the area function, we return to f. ∎

CONCEPTUAL INSIGHT The FTC shows that integration and differentiation are *inverse operations*. By FTC II, if you start with a continuous function f and form the integral $\int_a^x f(t)\, dt$, then you get back the original function by differentiating:

$$f(x) \xrightarrow{\text{Integrate}} \int_a^x f(t)\, dt \xrightarrow{\text{Differentiate}} \frac{d}{dx}\int_a^x f(t)\, dt = f(x)$$

On the other hand, by FTC I, if you differentiate first and then integrate, you also recover $f(x)$ [but only up to a constant $f(a)$]:

$$f(x) \xrightarrow{\text{Differentiate}} f'(x) \xrightarrow{\text{Integrate}} \int_a^x f'(t)\, dt = f(x) - f(a)$$

In addition to showing the inverse relationship between integration and differentiation, FTC II also reveals how area functions may be used to compute antiderivatives. For example,

- $A(x) = \displaystyle\int_0^x t^2\, dt$ is an antiderivative of $f(x) = x^2$,

- $A(x) = \displaystyle\int_0^x \cos t\, dt$ is an antiderivative of $f(x) = \cos x$,

- $A(x) = \displaystyle\int_0^x \sqrt[3]{1 + \cos t}\, dt$ is an antiderivative of $f(x) = \sqrt[3]{1 + \cos x}$.

Since we already have simple expressions for antiderivatives of $f(x) = x^2$ and $f(x) = \cos x$, the area-function version might not be helpful. On the other hand, for a function like $f(x) = \sqrt[3]{1 + \cos x}$, the situation is different. Try as you may, you cannot find an elementary function whose derivative is $f(x) = \sqrt[3]{1 + \cos x}$. That does not mean that there is no antiderivative. In fact, FTC II guarantees that there is an antiderivative, but expressing it in the area-function form might be the best we can do.

Unfortunately, the antiderivative $A(x) = \displaystyle\int_0^x \sqrt[3]{1 + \cos t}\, dt$ for $f(x) = \sqrt[3]{1 + \cos x}$ does not yield a formula that we can use to compute definite integrals involving f via FTC I. Instead, we need to consider alternative approaches such as numerical approximation.

While it might seem unfortunate that $A(x) = \displaystyle\int_0^x \sqrt[3]{1 + \cos t}\, dt$ is the best that we can do for an antiderivative of $f(x) = \sqrt[3]{1 + \cos x}$, as the next example shows, there is plenty that we can do to understand the behavior of A.

EXAMPLE 3 **Graphing an Area Function** Let $A(x) = \displaystyle\int_0^x \sqrt[3]{1 + \cos t}\, dt$.

(a) For $x = 1, 2, \ldots, 20$, with $\Delta x = 0.01$, use technology to calculate a right-endpoint Riemann sum that approximates the definite integral defining $A(x)$. Plot the points $(x, A(x))$ and connect them with a smooth curve to obtain a graph of A.

(b) Examine A' to determine the critical points of A and the increasing/decreasing behavior of the graph of A.

Solution

(a) We use $\Delta x = 0.01$ and a right-endpoint Riemann sum to approximate each definite integral. For example, to approximate $A(4)$, we compute R_{400} for the definite integral over $[0, 4]$. We obtain the following values for $A(x)$:

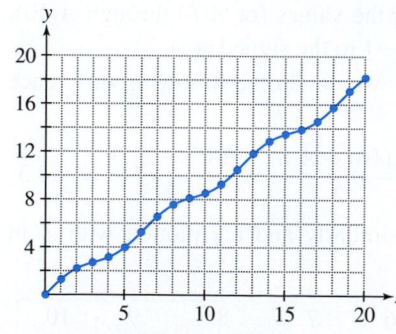

FIGURE 5 Graph of
$A(x) = \displaystyle\int_0^x \sqrt[3]{1 + \cos t}\, dt.$

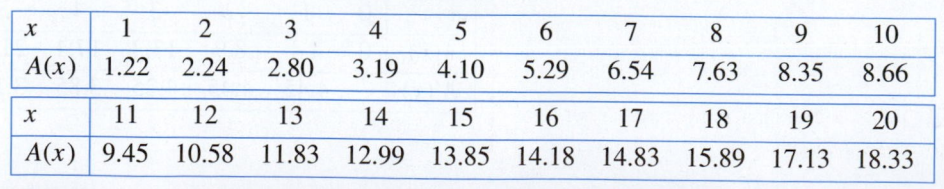

x	1	2	3	4	5	6	7	8	9	10
$A(x)$	1.22	2.24	2.80	3.19	4.10	5.29	6.54	7.63	8.35	8.66

x	11	12	13	14	15	16	17	18	19	20
$A(x)$	9.45	10.58	11.83	12.99	13.85	14.18	14.83	15.89	17.13	18.33

These points are plotted and joined with a smooth curve to obtain a graph of A in Figure 5.

(b) By FTC II, $A'(x) = \sqrt[3]{1 + \cos x}$. Thus, A has critical points when $\cos x = -1$; that is, at $n\pi$ for all odd n. Furthermore, except at the critical points, $A'(x)$ is positive, and therefore, A is an increasing function. ∎

CONCEPTUAL INSIGHT   Every continuous function on an open interval I is guaranteed by FTC II to have an antiderivative. Furthermore, we can obtain an antiderivative via an area function.

A number of important functions studied and employed by mathematicians, scientists, and engineers are defined by area functions. When no simple formula for such functions is available, we can examine their behavior by approximating the function via definite-integral estimates and by using the first and second derivatives to determine increasing and decreasing behavior and concavity.

The Gauss error function
$$\operatorname{erf}(x) = \frac{2}{\sqrt{\pi}} \int_0^x e^{-t^2}\, dt \text{ is defined via an}$$
area function. It is important in many areas of mathematics. We explore it in Exercise 53.

EXAMPLE 4   Given
$$A(x) = \int_2^x \sqrt{1 + t^3}\, dt$$
calculate or approximate $A(2)$, $A(3)$, $A'(2)$, and $A'(3)$.

Solution   First, $A(2) = \int_2^2 \sqrt{1 + t^3}\, dt = 0$. For $A(3) = \int_2^3 \sqrt{1 + t^3}\, dt$, we need to approximate the definite integral because we do not have a simple antiderivative that enables us to compute the integral using FTC I. We compute an approximating Riemann sum, R_{100}, and find $A(3) \approx 4.11$.

By FTC II, $A'(x) = \sqrt{1 + x^3}$. In particular,
$$A'(2) = \sqrt{1 + 2^3} = 3 \qquad \text{and} \qquad A'(3) = \sqrt{1 + 3^3} = \sqrt{28} \qquad ∎$$

When the upper limit of the integral is a *function* of x rather than x itself, we use FTC II together with the Chain Rule to differentiate a function defined via an integral.

EXAMPLE 5   **The FTC and the Chain Rule**   Find the derivative of
$$G(x) = \int_{-2}^{x^2} \sin t\, dt$$

Solution   FTC II does not apply directly because the upper limit is x^2 rather than x. It is necessary to recognize that G is a *composite function* with outer function $A(x) = \int_{-2}^x \sin t\, dt$:
$$G(x) = A(x^2) = \int_{-2}^{x^2} \sin t\, dt$$

FTC II tells us that $A'(x) = \sin x$, so by the Chain Rule,
$$G'(x) = A'(x^2) \cdot (x^2)' = \sin(x^2) \cdot (2x) = 2x \sin(x^2)$$

Alternatively, we may set $u = x^2$ and use the Chain Rule as follows:
$$\frac{dG}{dx} = \frac{d}{dx} \int_{-2}^{x^2} \sin t\, dt = \left(\frac{d}{du} \int_{-2}^{u} \sin t\, dt \right) \frac{du}{dx} = (\sin u)2x = 2x \sin(x^2) \qquad ∎$$

GRAPHICAL INSIGHT Another Tale of Two Graphs   FTC II tells us that $A'(x) = f(x)$, or in other words, $f(x)$ is the rate of change of $A(x)$. If we did not know this result, we might come to suspect it by comparing the graphs of A and f. Consider the following:

- Figure 6 shows that the increase in area ΔA for a given Δx is larger at x_2 than at x_1 because $f(x_2) > f(x_1)$. So the size of $f(x)$ determines how quickly $A(x)$ changes, as we would expect if $A'(x) = f(x)$.

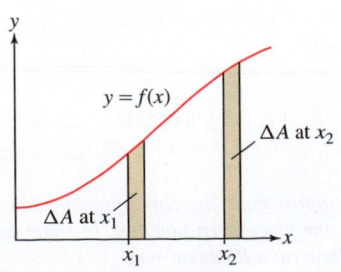

DF FIGURE 6 The change in area ΔA for a given Δx is larger when $f(x)$ is larger.

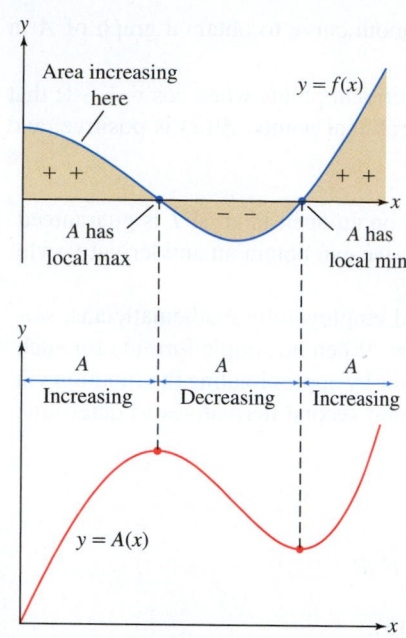

- Figure 7 shows that the sign of $f(x)$ determines whether A is increasing or decreasing. If $f(x) > 0$, then A is increasing because positive area is added as we move to the right. When $f(x)$ turns negative, A begins to decrease because we start adding negative area.
- A has a local max at points where $f(x)$ changes sign from $+$ to $-$ (the points where the area turns negative), and has a local min when $f(x)$ changes from $-$ to $+$. This agrees with the First Derivative Test.

These observations show that f behaves like A', as claimed by FTC II.

FIGURE 7 The sign of $f(x)$ determines the increasing/decreasing behavior of A.

5.5 SUMMARY

- The *area function* with lower limit a: $A(x) = \displaystyle\int_a^x f(t)\,dt$. It satisfies $A(a) = 0$.

- FTC II: $A'(x) = f(x)$, or equivalently, $\dfrac{d}{dx}\displaystyle\int_a^x f(t)\,dt = f(x)$.

- FTC II shows that every continuous function on an open interval I has an antiderivative on I—namely, its area function (with any lower limit in I).

- To differentiate the function $G(x) = \displaystyle\int_a^{g(x)} f(t)\,dt$, write $G(x) = A(g(x))$, where

  $A(x) = \displaystyle\int_a^x f(t)\,dt$. Then use the Chain Rule:

$$G'(x) = A'(g(x))g'(x) = f(g(x))g'(x)$$

5.5 EXERCISES

Preliminary Questions

1. Let $G(x) = \displaystyle\int_4^x \sqrt{t^3 + 1}\,dt$.

(a) Is the FTC II needed to calculate $G(4)$?

(b) Is the FTC II needed to calculate $G'(4)$?

2. Which of the following is an antiderivative F of $f(x) = x^2$ satisfying $F(2) = 0$?

(a) $\displaystyle\int_2^x 2t\,dt$ **(b)** $\displaystyle\int_0^2 t^2\,dt$ **(c)** $\displaystyle\int_2^x t^2\,dt$

3. Does every continuous function have an antiderivative? Explain.

4. Let $G(x) = \displaystyle\int_4^{x^3} \sin t\,dt$. Which of the following statements are correct?

(a) G is the composite function $\sin(x^3)$.

(b) G is the composite function $A(x^3)$, where

$$A(x) = \int_4^x \sin t\,dt$$

(c) $G(x)$ is too complicated to differentiate.

(d) The Product Rule is used to differentiate G.

(e) The Chain Rule is used to differentiate G.

(f) $G'(x) = 3x^2 \sin(x^3)$.

Exercises

In Exercises 1–8 compute an area function $A(x)$ of $f(x)$ with lower limit a. Then, to verify the FTC II inverse relationship, compute $A'(x)$ and show that it equals $f(x)$.

1. $f(x) = 4 - 2x$, $a = 0$

2. $f(x) = 4 - 2x$, $a = 5$

3. $f(x) = 4x + 6x^2$, $a = -1$

4. $f(x) = x^2 - 8$, $a = 3$

5. $f(x) = x^2 - \sin x$, $a = 0$

6. $f(x) = 1 - x + \cos x$, $a = 0$

7. $f(x) = e^{2x}$, $a = 0$

8. $f(x) = e^{-x}$, $a = -1$

In Exercises 9–12, compute or approximate the corresponding function values and derivative values for the given area function. In some cases, approximations will need to be done via a Riemann sum.

9. $F(x) = \displaystyle\int_0^x \sqrt{t^2 + t}\,dt$. Find $F(0)$, $F(3)$, $F'(0)$, and $F'(3)$.

10. $G(x) = \int_0^x \sqrt{4 - t^2}\, dt$. Find $G(0)$, $G(2)$, $G'(0)$, and $G'(1)$.

11. $F(x) = \int_{-2}^x \dfrac{du}{u^2 + 1}$. Find $F(-2)$, $F(2)$, $F'(0)$, and $F'(2)$.

12. $T(x) = \int_0^x \tan\theta\, d\theta$. Find $T(0)$, $T(\pi/3)$, $T'(0)$, and $T'(\pi/3)$.

In Exercises 13–22, find formulas for the functions represented by the integrals.

13. $\displaystyle\int_2^x u^4\, du$

14. $\displaystyle\int_2^x (12t^2 - 8t)\, dt$

15. $\displaystyle\int_0^x \sin u\, du$

16. $\displaystyle\int_{-\pi/4}^x \sec^2\theta\, d\theta$

17. $\displaystyle\int_4^x e^{3u}\, du$

18. $\displaystyle\int_x^0 e^{-t}\, dt$

19. $\displaystyle\int_1^{x^2} t\, dt$

20. $\displaystyle\int_{x/2}^{x/4} \sec^2 u\, du$

21. $\displaystyle\int_{3x}^{9x+2} e^{-u}\, du$

22. $\displaystyle\int_2^{\sqrt{x}} \dfrac{dt}{t}$

23. Verify $\displaystyle\int_0^x |t|\, dt = \dfrac{1}{2}x|x|$. *Hint:* Consider $x \geq 0$ and $x \leq 0$ separately.

24. Verify $\displaystyle\int_0^x |t|^3\, dt = \dfrac{1}{4}x|x|^3$. *Hint:* Consider $x \geq 0$ and $x \leq 0$ separately.

In Exercises 25–28, calculate the derivative.

25. $\dfrac{d}{dx} \displaystyle\int_0^x (t^5 - 9t^3)\, dt$

26. $\dfrac{d}{d\theta} \displaystyle\int_1^\theta \cot u\, du$

27. $\dfrac{d}{dt} \displaystyle\int_{100}^t \sec(5x - 9)\, dx$

28. $\dfrac{d}{ds} \displaystyle\int_{-2}^s \tan\left(\dfrac{1}{1 + u^2}\right) du$

29. Let $A(x) = \displaystyle\int_0^x f(t)\, dt$ for $f(x)$ in Figure 8.

(a) Calculate $A(2)$, $A(3)$, $A'(2)$, and $A'(3)$.

(b) Find formulas for $A(x)$ on $[0, 2]$ and $[2, 4]$, and sketch the graph of A.

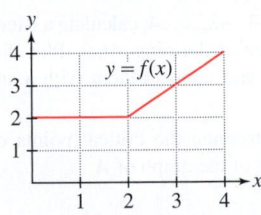

FIGURE 8

30. Make a rough sketch of the graph of $A(x) = \displaystyle\int_0^x g(t)\, dt$ for the function $g(x)$ in Figure 9.

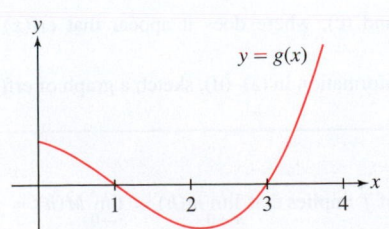

FIGURE 9

In Exercises 31–32, do the following:

- *For $x = 0, 1, 2, \ldots, 10$, approximate $A(x) = \displaystyle\int_0^x f(t)\, dt$.*

- *For $x = 1, 2, 3, \ldots, 9$, approximate $A'(x)$ using $\Delta x = 1$ and the symmetric difference quotient approximation,*

$$A'(x) \approx \frac{A(x + \Delta x) - A(x - \Delta x)}{2\Delta x}$$

- *Plot the values of $A'(x)$ on a graph of f to demonstrate $A' \approx f$.*

31. Use $f(x)$ from Figure 10(A).

32. Use $f(x)$ from Figure 10(B).

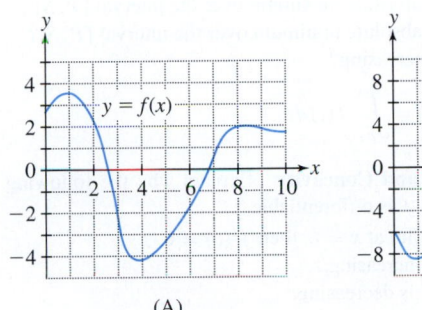 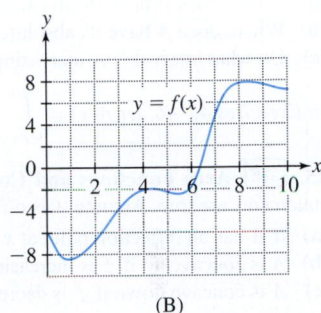

(A)   (B)

FIGURE 10

In Exercises 33–38, calculate the derivative.

33. $\dfrac{d}{dx} \displaystyle\int_0^{x^2} \dfrac{t\, dt}{t + 1}$

34. $\dfrac{d}{dx} \displaystyle\int_1^{1/x} \cos^3 t\, dt$

35. $\dfrac{d}{ds} \displaystyle\int_{-6}^{\cos s} u^4\, du$

36. $\dfrac{d}{dx} \displaystyle\int_{x^2}^{x^4} \sqrt{t}\, dt$

Hint for Exercise 36: $F(x) = A(x^4) - A(x^2)$.

37. $\dfrac{d}{dx} \displaystyle\int_{\sqrt{x}}^{x^2} \tan t\, dt$

38. $\dfrac{d}{du} \displaystyle\int_{-u}^{3u} \sqrt{x^2 + 1}\, dx$

In Exercises 39–42, with $f(x)$ as in Figure 11, let

$$A(x) = \int_0^x f(t)\, dt \quad \text{and} \quad B(x) = \int_2^x f(t)\, dt$$

39. Find the min and max of A on $[0, 6]$.

40. Find the min and max of B on $[0, 6]$.

41. Find formulas for $A(x)$ and $B(x)$ valid on $[2, 4]$.

42. Find formulas for $A(x)$ and $B(x)$ valid on $[4, 5]$.

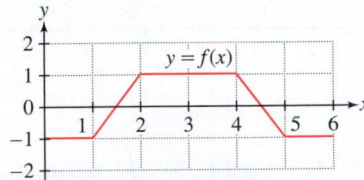

FIGURE 11

43. Let $A(x) = \displaystyle\int_0^x f(t)\, dt$, with $f(x)$ as in Figure 12.

(a) Does A have a local maximum at P?

(b) Where does A have a local minimum?

(c) Where does A have a local maximum?

(d) True or false? $A(x) < 0$ for all x in the interval shown.

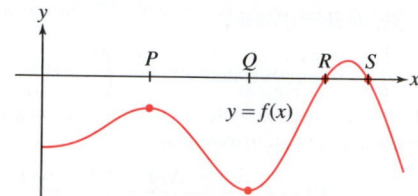

FIGURE 12 Graph of $y = f(x)$.

44. Let $A(x) = \int_0^x f(t)\,dt$, with $f(x)$ as in Figure 12.

(a) Where does A have its absolute maximum over the interval $[P, S]$?
(b) Where does A have its absolute minimum over the interval $[P, S]$?
(c) On what interval is A increasing?

In Exercises 45–46, let $A(x) = \int_a^x f(t)\,dt$.

45. 🖊 **Area Functions and Concavity** Explain why the following statements are true. Assume f is differentiable.

(a) If A has an inflection point at $x = c$, then $f'(c) = 0$.
(b) A is concave up if f is increasing.
(c) A is concave down if f is decreasing.

46. Match the property of A with the corresponding property of the graph of f. Assume f is differentiable.

Area function A

(a) A is decreasing.
(b) A has a local maximum.
(c) A is concave up.
(d) A goes from concave up to concave down.

Graph of f

(i) Lies below the x-axis.
(ii) Crosses the x-axis from positive to negative.
(iii) Has a local maximum.
(iv) f is increasing.

47. Let $A(x) = \int_0^x f(t)\,dt$, with $f(x)$ as in Figure 13. Determine:

(a) The intervals on which A is increasing and decreasing
(b) The values x where A has a local min or max
(c) The values of x where there are inflection points of A
(d) The intervals where A is concave up or concave down

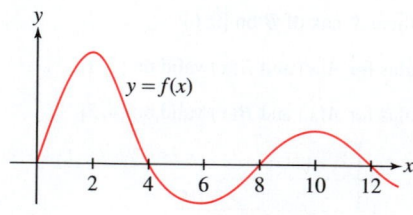

FIGURE 13

48. Let $f(x) = x^2 - 5x - 6$ and $F(x) = \int_0^x f(t)\,dt$.

(a) Find the critical points of F and determine whether they are local minima or local maxima.

(b) Find the points of inflection of F and determine whether the concavity changes from up to down or from down to up.

(c) 🔲 **GU** Plot $y = f(x)$ and $y = F(x)$ on the same set of axes and confirm your answers to (a) and (b).

49. Sketch the graph of an increasing function f such that both $f'(x)$ and $A(x) = \int_0^x f(t)\,dt$ are decreasing.

50. 🖊 Figure 14 shows the graph of $f(x) = x \sin x$. Let $F(x) = \int_0^x t \sin t\,dt$.

(a) Locate the local max and absolute max of F on $[0, 3\pi]$.
(b) Justify graphically: F has precisely one zero in $[\pi, 2\pi]$.
(c) How many zeros does F have in $[0, 3\pi]$?
(d) Find the inflection points of F on $[0, 3\pi]$. For each one, state whether the concavity changes from up to down or from down to up.

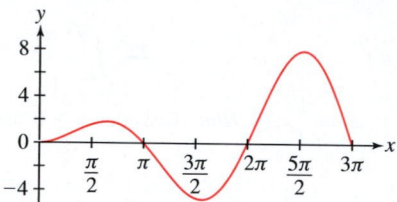

FIGURE 14 Graph of $f(x) = x \sin x$.

51. 🔲 **GU** Find the smallest positive critical point of

$$F(x) = \int_0^x \cos(t^{3/2})\,dt$$

and determine whether it is a local min or max. Then find the smallest positive inflection point of $F(x)$ and use a graph of $y = \cos(x^{3/2})$ to determine whether the concavity changes from up to down or from down to up.

52. Let $A(x) = \int_{-4}^x \sqrt[3]{4 - t^2}\,dt$.

(a) 🔲 **CAS** For $x = -3, -2, \ldots, 4$, calculate a Riemann sum that approximates the definite integral defining $A(x)$. Plot the points $(x, A(x))$ for $x = -4, -3, -2, \ldots, 4$ and connect them with a smooth curve to obtain a graph of A.

(b) Examine A' to determine the critical points of A and the increasing/decreasing behavior of the graph of A.

53. The Gauss error function is defined by $\text{erf}(x) = \dfrac{2}{\sqrt{\pi}} \int_0^x e^{-t^2}\,dt$.

(a) Explain why $\text{erf}(x)$ is an increasing function.
(b) Explain why $\text{erf}(x)$ is an odd function.
(c) 🔲 **CAS** Use Riemann sums to approximate $\text{erf}(x)$ for $x = 1/2, 1, 3/2, 2,$ and $5/2$.
(d) From (b) and (c), where does it appear that $\text{erf}(x)$ has horizontal asymptotes?
(e) From the information in (a)–(d), sketch a graph of $\text{erf}(x)$.

Further Insights and Challenges

54. Proof of FTC II The proof in the text assumes that f is nonnegative and increasing. To prove it for all continuous functions, let $m(h)$ and $M(h)$ denote the *minimum* and *maximum* of f on $[x, x + h]$ (Figure 15).

The continuity of f implies that $\displaystyle\lim_{h \to 0} m(h) = \lim_{h \to 0} M(h) = f(x)$. Show that for $h > 0$,

$$hm(h) \le A(x + h) - A(x) \le hM(h)$$

For $h < 0$, the inequalities are reversed. Prove that $A'(x) = f(x)$.

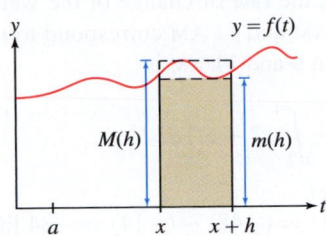

FIGURE 15 Graphical interpretation of $A(x + h) - A(x)$.

55. Proof of FTC I FTC I asserts that $\int_a^b f(t)\,dt = F(b) - F(a)$ if $F'(x) = f(x)$. Use FTC II to give a new proof of FTC I as follows. Set

$$A(x) = \int_a^x f(t)\,dt.$$

(a) Show that $F(x) = A(x) + C$ for some constant.

(b) Show that $F(b) - F(a) = A(b) - A(a) = \int_a^b f(t)\,dt$.

56. Can Every Antiderivative Be Expressed as an Integral? The area function $A(x) = \int_a^x f(t)\,dt$ is an antiderivative of f for every value of a. However, not all antiderivatives are obtained in this way. The general antiderivative of $f(x) = x$ is $F(x) = \frac{1}{2}x^2 + C$. Show that F is an area function if $C \le 0$ but not if $C > 0$.

57. Prove the formula

$$\frac{d}{dx} \int_{u(x)}^{v(x)} f(t)\,dt = f(v(x))v'(x) - f(u(x))u'(x)$$

58. Use the result of Exercise 57 to calculate

$$\frac{d}{dx} \int_{\ln x}^{e^x} \sin t\,dt$$

5.6 Net Change as the Integral of a Rate of Change

So far, we have seen how the definite integral can be used to compute area. That application barely scratches the surface of ways that this important tool from calculus can be applied. In this section, we use the integral to compute net change, a concept that arises in a broad range of applications.

Consider the following problem: Water flows into an empty bucket at a rate of $r(t)$ liters per second. How much water is in the bucket after 4 seconds? If the rate of water flow were *constant*—say, 1.5 L/s—we would have

$$\text{quantity of water} = \text{flow rate} \times \text{time elapsed} = (1.5)4 = 6 \text{ L}$$

Suppose, however, that the flow rate $r(t)$ varies as in Figure 1. Then *the quantity of water is equal to the area under the graph of* $y = r(t)$. To prove this, let $s(t)$ be the amount of water in the bucket at time t. Then $s'(t) = r(t)$ because $s'(t)$ is the rate at which the quantity of water is changing, that is, the rate that water is entering the bucket, $r(t)$. Furthermore, $s(0) = 0$ because the bucket is initially empty. By FTC I,

$$\underbrace{\int_0^4 s'(t)\,dt}_{\substack{\text{Area under the graph} \\ \text{of the flow rate}}} = s(4) - s(0) = \underbrace{s(4)}_{\substack{\text{Water in bucket} \\ \text{at } t = 4}}$$

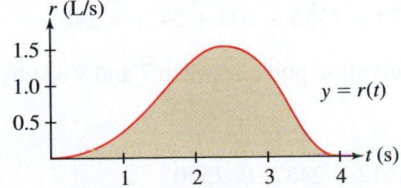

FIGURE 1 The quantity of water in the bucket is equal to the area under the graph of the flow rate $r(t)$.

In Theorem 1, the variable t does not have to be a time variable.

More generally, $s(t_2) - s(t_1)$ is the **net change** in $s(t)$ over the interval $[t_1, t_2]$. FTC I yields the following result:

> **THEOREM 1 Net Change as the Integral of a Rate of Change** The net change in $s(t)$ over an interval $[t_1, t_2]$ is given by the integral
>
> $$\underbrace{\int_{t_1}^{t_2} s'(t)\,dt}_{\text{Integral of the rate of change}} = \underbrace{s(t_2) - s(t_1)}_{\text{Net change over } [t_1, t_2]}$$

EXAMPLE 1 Water leaks from a tank at a rate of $2 + 5t$ L/hour, where t is the number of hours after 7 AM. How much water is lost between 9 and 11 AM?

Solution Let $s(t)$ be the quantity of water in the tank at time t. Since $2 + 5t$ represents the rate that the water is *leaving* that tank, the rate of change of the water *in* the tank is $-(2 + 5t)$. So $s'(t) = -(2 + 5t)$. Since 9 AM and 11 AM correspond to $t = 2$ and $t = 4$, respectively, the net change in $s(t)$ between 9 and 11 AM is

$$s(4) - s(2) = \int_2^4 s'(t)\,dt = -\int_2^4 (2 + 5t)\,dt$$

$$= -\left(2t + \frac{5}{2}t^2\right)\Big|_2^4 = (-48) - (-14) = -34 \text{ liters}$$

The tank lost 34 L between 9 and 11 AM. ∎

In the next example, we estimate an integral using numerical data. We shall compute the average of the left- and right-endpoint approximations, because this is usually more accurate than either endpoint approximation alone. This method of approximation is called the Trapezoidal Approximation; we investigate it further in Section 7.8.

EXAMPLE 2 **Traffic Flow** The number of cars per hour passing an observation point along a highway is called the traffic flow rate $q(t)$ (in cars per hour).

(a) Which quantity is represented by the integral $\int_{t_1}^{t_2} q(t)\,dt$?

(b) The flow rate on the Jocoro Highway is recorded at 15-minute intervals on Monday morning between 7:00 and 9:00 AM. Estimate the number of cars using the highway during this 2-hour period.

t	7:00	7:15	7:30	7:45	8:00	8:15	8:30	8:45	9:00
$q(t)$	1044	1297	1478	1844	1451	1378	1155	802	542

Solution

(a) The integral $\int_{t_1}^{t_2} q(t)\,dt$ represents the total number of cars that passed the observation point during the time interval $[t_1, t_2]$.

(b) The data values are spaced at intervals of $\Delta t = 0.25$ h. Thus,

$$L_N = 0.25\,(1044 + 1297 + 1478 + 1844 + 1451 + 1378 + 1155 + 802) \approx 2612$$

$$R_N = 0.25\,(1297 + 1478 + 1844 + 1451 + 1378 + 1155 + 802 + 542) \approx 2487$$

We estimate the number of cars that passed the observation point between 7 and 9 AM by taking the average of R_N and L_N:

$$\int_7^9 q(t)\,dt \approx \frac{1}{2}(R_N + L_N) = \frac{1}{2}(2612 + 2487) \approx 2550$$

Approximately 2550 cars used the Jocoro Highway during the time period. ∎

In Example 2, L_N is the product of Δt and the sum of the values of $q(t)$ at the left endpoints

$$7{:}00, \ 7{:}15, \ \dots, \ 8{:}45$$

and R_N is the product of Δt and the sum of the values of $q(t)$ at the right endpoints

$$7{:}15, \ \dots, \ 8{:}45, \ 9{:}00$$

The Integral of Velocity

Let $s(t)$ be the position at time t of an object in linear motion. Then the object's velocity is $v(t) = s'(t)$, and the integral of v is equal to the *net change in position* or *displacement* over a time interval $[t_1, t_2]$:

$$\int_{t_1}^{t_2} v(t)\,dt = \int_{t_1}^{t_2} s'(t)\,dt = \underbrace{s(t_2) - s(t_1)}_{\substack{\text{Displacement or net} \\ \text{change in position}}}$$

We must distinguish between displacement and *distance traveled*. If you travel 10 km and then return to your starting point, your displacement is zero but your distance traveled is 20 km. To compute distance traveled rather than displacement, we integrate the *speed* $|v(t)|$.

THEOREM 2 **The Integral of Velocity** For an object in linear motion with velocity $v(t)$, then

$$\text{displacement during } [t_1, t_2] = \int_{t_1}^{t_2} v(t)\, dt$$

$$\text{distance traveled during } [t_1, t_2] = \int_{t_1}^{t_2} |v(t)|\, dt$$

EXAMPLE 3 A particle has velocity $v(t) = t^3 - 10t^2 + 24t$ m/s. Compute:

(a) Displacement over $[0, 6]$ **(b)** Total distance traveled over $[0, 6]$

Indicate the particle's trajectory with a motion diagram.

Solution First, we compute the indefinite integral:

$$\int v(t)\, dt = \int (t^3 - 10t^2 + 24t)\, dt = \frac{1}{4}t^4 - \frac{10}{3}t^3 + 12t^2 + C$$

(a) The displacement over the time interval $[0, 6]$ is

$$\int_0^6 v(t)\, dt = \left(\frac{1}{4}t^4 - \frac{10}{3}t^3 + 12t^2 \right)\Big|_0^6 = 36 \text{ m}$$

(b) The factorization $v(t) = t(t - 4)(t - 6)$ shows that $v(t)$ changes sign at $t = 4$. It is positive on $[0, 4]$ and negative on $[4, 6]$, as we see in Figure 2. Therefore, the total distance traveled is

$$\int_0^6 |v(t)|\, dt = \int_0^4 v(t)\, dt - \int_4^6 v(t)\, dt$$

We evaluate these two integrals separately:

$$[0, 4]: \quad \int_0^4 v(t)\, dt = \left(\frac{1}{4}t^4 - \frac{10}{3}t^3 + 12t^2 \right)\Big|_0^4 = \frac{128}{3} \text{ m}$$

$$[4, 6]: \quad \int_4^6 v(t)\, dt = \left(\frac{1}{4}t^4 - \frac{10}{3}t^3 + 12t^2 \right)\Big|_4^6 = -\frac{20}{3} \text{ m}$$

The total distance traveled is $\frac{128}{3} - \left(-\frac{20}{3} \right) = \frac{148}{3} = 49\frac{1}{3}$ m.

Figure 3 is a motion diagram indicating the particle's trajectory. The particle travels $\frac{128}{3}$ m during the first 4 s and then backtracks $\frac{20}{3}$ m over the next 2 s. ∎

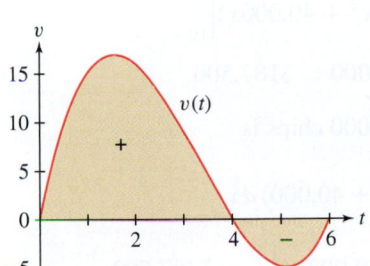

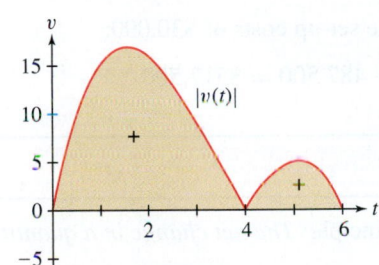

FIGURE 2 Top: graph of $v(t) = t^3 - 10t^2 + 24t$. Bottom: graph of $|v(t)|$.

FIGURE 3 Path of the particle along a straight line.

In Section 3.4, we defined the marginal cost at production level x_0 as the cost

$$C(x_0 + 1) - C(x_0)$$

of producing one additional unit. Since this marginal cost is approximated well by the derivative $C'(x_0)$ for large values of x_0 compared to 1, economists often refer to $C'(x)$ itself as the marginal cost.

Total Versus Marginal Cost

Let $C(x)$ represent a manufacturer's cost to produce x units of a particular product or commodity. The derivative $C'(x)$ is called the **marginal cost**. The cost of increasing production from a units to b units is the net change $C(b) - C(a)$, which is equal to the integral of the marginal cost:

$$\text{cost of increasing production from } a \text{ units to } b \text{ units} = \int_a^b C'(x)\, dx$$

EXAMPLE 4 The marginal cost of producing x computer chips (in units of 1000) is $C'(x) = 300x^2 - 4000x + 40,000$ (dollars per thousand chips).

(a) Find the cost of increasing production from 10,000 to 15,000 chips.

(b) Determine the total cost of producing 15,000 chips, assuming that it costs $30,000 to set up the manufacturing run [i.e., $C(0) = 30,000$].

Solution

(a) The cost of increasing production from 10,000 ($x = 10$) to 15,000 ($x = 15$) is

$$C(15) - C(10) = \int_{10}^{15} (300x^2 - 4000x + 40,000)\, dx$$

$$= (100x^3 - 2000x^2 + 40,000x)\Big|_{10}^{15}$$

$$= 487,500 - 300,000 = \$187,500$$

(b) The cost of increasing production from 0 to 15,000 chips is

$$C(15) - C(0) = \int_{0}^{15} (300x^2 - 4000x + 40,000)\, dx$$

$$= (100x^3 - 2000x^2 + 40,000x)\Big|_{0}^{15} = \$487,500$$

The total cost of producing 15,000 chips includes the set-up costs of $30,000:

$$C(15) = C(0) + 487,500 = 30,000 + 487,500 = \$517,500 \qquad \blacksquare$$

5.6 SUMMARY

- Many applications are based on the following principle: *The net change in a quantity $s(t)$ is equal to the integral of its rate of change:*

$$\underbrace{s(t_2) - s(t_1)}_{\text{Net change over } [t_1, t_2]} = \int_{t_1}^{t_2} s'(t)\, dt$$

- For an object traveling in a straight line at velocity $v(t)$,

$$\text{displacement during } [t_1, t_2] = \int_{t_1}^{t_2} v(t)\, dt$$

$$\text{total distance traveled during } [t_1, t_2] = \int_{t_1}^{t_2} |v(t)|\, dt$$

- If $C(x)$ is the cost of producing x units of a commodity, then $C'(x)$ is the marginal cost and the cost of increasing production from a units to b units is

$$C(b) - C(a) = \int_{a}^{b} C'(x)\, dx$$

5.6 EXERCISES

Preliminary Questions

1. A hot metal object is submerged in cold water. The rate at which the object cools (in degrees per minute) is a function $f(t)$ of time. Which quantity is represented by the integral $\int_0^T f(t)\, dt$?

2. A plane travels 560 km from Los Angeles to San Francisco in 1 hour (h). If the plane's velocity at time t is $v(t)$ km/h, what is the value of $\int_0^1 v(t)\, dt$?

3. Which of the following quantities would be naturally represented as derivatives and which as integrals?

(a) Velocity of a train

(b) Rainfall during a 6-month period

(c) Mileage per gallon of an automobile

(d) Increase in the U.S. population from 1990 to 2010

Exercises

1. Water flows into an empty reservoir at a rate of $3000 + 20t$ L per hour (t is in hours). What is the quantity of water in the reservoir after 5 h?

2. A population of insects increases at a rate of $200 + 10t + 0.25t^2$ insects per day (t in days). Find the insect population after 3 days, assuming that there are 35 insects at $t = 0$.

3. A survey shows that a mayoral candidate is gaining votes at a rate of $2000t + 1000$ votes per day, where t is the number of days since she announced her candidacy. How many supporters will the candidate have after 60 days, assuming that she had no supporters at $t = 0$?

4. A factory produces bicycles at a rate of $95 + 3t^2 - t$ bicycles per week (t in weeks). How many bicycles were produced from the beginning of week 2 to the end of week 3?

5. Find the displacement of a particle moving in a straight line with velocity $v(t) = 4t - 3$ m/s over the time interval $[2, 5]$.

6. Find the displacement over the time interval $[1, 6]$ of a helicopter whose (vertical) velocity at time t is $v(t) = 0.02t^2 + t$ m/s.

7. A cat falls from a tree (with zero initial velocity) at time $t = 0$. How far does the cat fall between $t = 0.5$ second and $t = 1$ s? Use Galileo's formula $v(t) = -9.8t$ m/s.

8. A projectile is released with an initial (vertical) velocity of 100 m/s. Use the formula $v(t) = 100 - 9.8t$ for velocity to determine the distance traveled during the first 15 seconds.

In Exercises 9–12, a particle moves in a straight line with the given velocity (in meters per second). Find the displacement and distance traveled over the time interval, and draw a motion diagram like Figure 3 (with distance and time labels).

9. $v(t) = 12 - 4t$, $[0, 5]$

10. $v(t) = 36 - 24t + 3t^2$, $[0, 10]$

11. $v(t) = t^{-2} - 1$, $[0.5, 2]$

12. $v(t) = \cos t$, $[0, 3\pi]$

13. Find the net change in velocity over $[1, 4]$ of an object with $a(t) = 8t - t^2$ m/s^2.

14. Show that if acceleration is constant, then the change in velocity is proportional to the length of the time interval.

15. The traffic flow rate past a certain point on a highway is $q(t) = 3000 + 2000t - 300t^2$ (t in hours), where $t = 0$ is 8 AM. How many cars pass by in the time interval from 8 to 10 AM?

16. The marginal cost of producing x tablet computers is

$$C'(x) = 120 - 0.06x + 0.00001x^2$$

What is the additional cost of producing 3000 units if the set-up cost is $90,000? If production is set at 3000 units, what is the cost of producing 200 additional units?

17. A small boutique produces wool sweaters at a marginal cost of $40 - 5\lfloor x/5 \rfloor$ for $0 \le x \le 20$, where $\lfloor x \rfloor$ is the greatest integer function. Find the cost of producing 20 sweaters. Then compute the average cost of the first 10 sweaters and the last 10 sweaters.

18. The rate (in liters per minute) at which water drains from a tank is recorded at half-minute intervals. Compute the average of the left- and right-endpoint approximations to estimate the total amount of water drained during the first 3 min.

t (min)	0	0.5	1	1.5	2	2.5	3
r (L/min)	50	48	46	44	42	40	38

19. The velocity of a car is recorded at half-second intervals (in feet per second). Use the average of the left- and right-endpoint approximations to estimate the total distance traveled during the first 4 s.

t	0	0.5	1	1.5	2	2.5	3	3.5	4
$v(t)$	0	12	20	29	38	44	32	35	30

20. To model the effects of a **carbon tax** on CO_2 emissions, policymakers study the *marginal cost of abatement $B(x)$*, defined as the cost of increasing CO_2 reduction from x to $x + 1$ tons (in units of ten thousand tons—Figure 4). Which quantity is represented by the area under the curve over $[0, 3]$ in Figure 4?

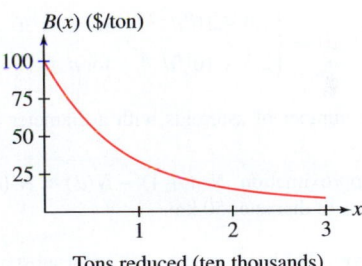

FIGURE 4 Marginal cost of abatement $B(x)$.

21. The snowfall rate R (in inches per hour) was tracked during a major 24-hour lake effect snowstorm in Buffalo, New York. The graph in Figure 5 shows R as a function of t (hours) during the storm. What quantity does $\int_0^{24} R(t)\,dt$ represent? Approximate the integral.

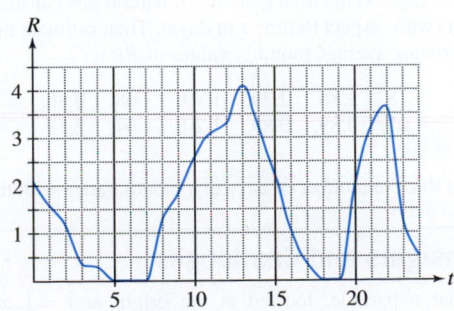

FIGURE 5 The snowfall rate during a major 24-h storm.

22. Figure 6 shows the migration rate $M(t)$ of Ireland in the period 1988–1998. This is the rate at which people (in thousands per year) moved into or out of the country.

(a) Is the following integral positive or negative? What does this quantity represent?

$$\int_{1988}^{1998} M(t)\,dt$$

(b) Did migration in the period 1988–1998 result in a net influx of people into Ireland or a net outflow of people from Ireland?

(c) During which 2 years could the Irish prime minister announce, "We've hit an inflection point. We are still losing population, but the trend is now improving."

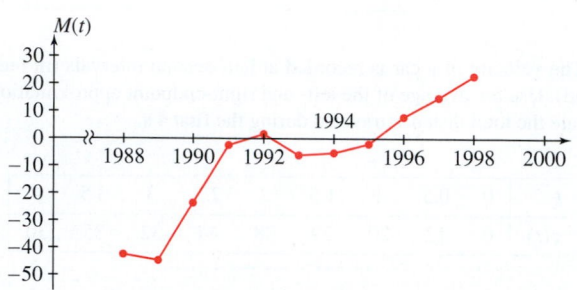

FIGURE 6 Irish migration rate (thousands per year).

23. Let $N(d)$ be the number of asteroids of diameter $\leq d$ kilometers. Data suggest that the diameters are distributed according to a piecewise power law:

$$N'(d) = \begin{cases} 1.9 \times 10^9 d^{-2.3} & \text{for } d < 70 \\ 2.6 \times 10^{12} d^{-4} & \text{for } d \geq 70 \end{cases}$$

(a) Compute the number of asteroids with a diameter between 0.1 and 100 km.

(b) Using the approximation $N(d+1) - N(d) \approx N'(d)$, estimate the number of asteroids of diameter 50 km.

24. Heat Capacity The heat capacity $C(T)$ of a substance is the amount of energy (in joules) required to raise the temperature of 1 g by 1°C at temperature T.

(a) Explain why the energy required to raise the temperature from T_1 to T_2 is the area under the graph of $C(T)$ over $[T_1, T_2]$.

(b) How much energy is required to raise the temperature from 50°C to 100°C if $C(T) = 6 + 0.2\sqrt{T}$?

25. Figure 7 shows the rate $R(t)$ of natural gas consumption (billions of cubic feet per day) in the mid-Atlantic states (New York, New Jersey, Pennsylvania). Express the total quantity of natural gas consumed in 2009 as an integral (with respect to time t in days). Then estimate this quantity, given the following average monthly values of $R(t)$:

$$3.18, \quad 2.86, \quad 2.39, \quad 1.49, \quad 1.08, \quad 0.80,$$
$$1.01, \quad 0.89, \quad 0.89, \quad 1.20, \quad 1.64, \quad 2.52$$

Keep in mind that the number of days in a month varies with the month.

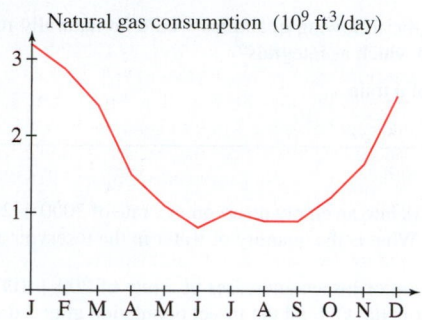

Natural gas consumption (10^9 ft^3/day)

FIGURE 7 Natural gas consumption in 2009 in the mid-Atlantic states.

26. Cardiac output is the rate R of volume of blood pumped by the heart per unit time (in liters per minute). Doctors measure R by injecting A mg of dye into a vein leading into the heart at $t = 0$ and recording the concentration $c(t)$ of dye (in milligrams per liter) pumped out at short regular time intervals (Figure 8).

(a) Explain: The quantity of dye pumped out in a small time interval $[t, t + \Delta t]$ is approximately $Rc(t)\Delta t$.

(b) Show that $A = R \displaystyle\int_0^T c(t)\,dt$, where T is large enough that all of the dye is pumped through the heart but not so large that the dye returns by recirculation.

(c) Assume $A = 5$ mg. Estimate R using the following values of $c(t)$ recorded at 1-second intervals

t	0	1	2	3	4	5	6	7	8	9	10
$c(t)$	0	0.4	2.8	6.5	9.8	8.9	6.1	4	2.3	1.1	0

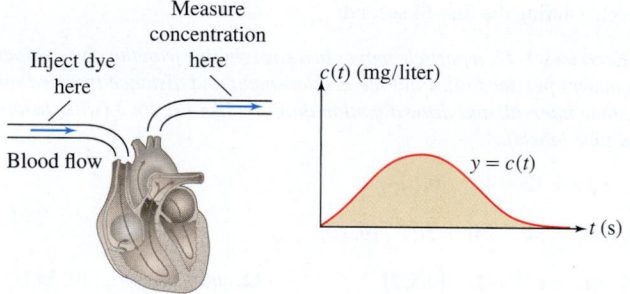

FIGURE 8

Exercises 27 and 28: A study suggests that the extinction rate $r(t)$ of marine animal families during the Phanerozoic Eon can be modeled by the function $r(t) = 3130/(t + 262)$ for $0 \leq t \leq 544$, where t is time elapsed (millions of years) since the beginning of the eon 544 million years ago. Thus, $t = 544$ refers to the present time, $t = 540$ is 4 million years ago, and so on.

27. Compute the average of R_N and L_N with $N = 5$ to estimate the total number of families that became extinct in the periods $100 \leq t \leq 150$ and $350 \leq t \leq 400$.

28. [CAS] Estimate the total number of extinct families from $t = 0$ to the present, using M_N with $N = 544$.

Further Insights and Challenges

29. Show that a particle, located at the origin at $t = 1$ and moving along the x-axis with velocity $v(t) = t^{-2}$, will never pass the point $x = 2$.

30. Show that a particle, located at the origin at $t = 1$ and moving along the x-axis with velocity $v(t) = t^{-1/2}$, moves arbitrarily far from the origin after sufficient time has elapsed.

31. In a free market economy, the demand curve is the graph of the function D that represents the demand for a specific product by the consumers in the economy at price q. It is not surprising that the curve is decreasing, as the demand drops as the price goes up. The supply curve is the graph of the function S that represents the supply of the product that the producers are willing to produce as a function of the price q. As the price goes up, the producers are willing to produce more of the item, and therefore, this curve is increasing. The point (p^*, q^*) at which the two curves cross is called the equilibrium point, where the supply and demand balance. Tradition in economics is to make the horizontal axis the quantity q of the item and the vertical axis the price p. We define $p = S(q)$ to correspond to the supply curve and $p = D(q)$ to correspond to the demand curve. In other words, we have inverted the formula for these two functions from giving quantity in terms of price to giving price in terms of quantity. The areas depicted in Figure 9 represent the excess supply and excess demand.

(a) The consumer surplus represents the savings on the part of consumers if they pay price p^* rather than the price greater than p^* that many were willing to pay. Write a formula for this consumer surplus. The formula will include a definite integral and it will depend on p^* and q^*.

(b) The producer surplus represents the savings on the part of producers if they sell at price p^* rather than the price less than p^* that some producers were willing to accept. Write a formula for this producer surplus.

(c) A variety of coffee shops in a town sell mocha latte supreme coffees. If the supply curve is given by $p = \dfrac{q}{100} + 1$ and the demand curve is given by $p = \dfrac{10}{q/100 + 1}$, determine the equilibrium point (p^*, q^*) and the consumer surplus and producer surplus when the mocha latte supreme coffees are sold at price p^*.

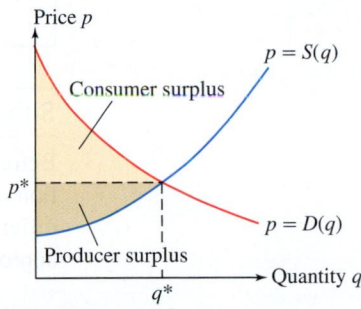

FIGURE 9 The supply and demand curves.

5.7 The Substitution Method

Integration (antidifferentiation) is generally more difficult than differentiation. There are no sure-fire methods, and many antiderivatives cannot be expressed in terms of elementary functions. However, there are a few important general techniques. One such technique is the **Substitution Method**, which is essentially an inverse of the Chain Rule for Differentiation.

Consider the integral $\displaystyle\int 2x \cos(x^2)\,dx$. We can evaluate it if we remember the Chain Rule calculation:

$$\frac{d}{dx}\sin(x^2) = 2x\cos(x^2)$$

This tells us that $\sin(x^2)$ is an antiderivative of $2x\cos(x^2)$, and therefore,

$$\int \underbrace{2x}_{\substack{\text{Derivative of}\\\text{inside function}}} \underbrace{\cos(x^2)}_{\substack{\text{Inside}\\\text{function}}}\,dx = \sin(x^2) + C$$

A similar Chain Rule calculation shows that

$$\int \underbrace{(1+3x^2)}_{\substack{\text{Derivative of}\\\text{inside function}}} \underbrace{\cos(x+x^3)}_{\substack{\text{Inside}\\\text{function}}}\,dx = \sin(x+x^3) + C$$

In both cases, the integrand is the product of a composite function and the derivative of the inside function. The Chain Rule does not help if the derivative of the inside function is missing. For instance, we cannot use the Chain Rule to compute $\displaystyle\int \cos(x+x^3)\,dx$ because the factor $(1+3x^2)$ does not appear.

In general, if $F'(u) = f(u)$, then by the Chain Rule,

$$\frac{d}{dx}F(u(x)) = F'(u(x))u'(x) = f(u(x))u'(x)$$

This translates into the following integration formula:

> **THEOREM 1 The Substitution Method**   If $F'(x) = f(x)$, and u is a differentiable function whose range includes the domain of f, then
>
> $$\int f(u(x))u'(x)\,dx = F(u(x)) + C$$

Substitution Using Differentials

Before proceeding to the examples, we discuss the procedure for carrying out substitution using differentials. We can treat the symbols du and dx that occur in integration as differentials, and in that way can use the relationship (introduced in Section 4.1 on linear approximation)

$$du = \frac{du}{dx}\,dx$$

to symbolically substitute into integrals in order to simplify them.
Equivalently, du and dx are related by

$$du = u'(x)\,dx \qquad \boxed{1}$$

The symbolic calculus of substitution using differentials was invented by Leibniz and is considered one of his most important achievements. It reduces the otherwise complicated process of transforming integrals to a convenient substitution procedure.

For example,

$$\text{If } u = x^2, \qquad \text{then} \qquad du = 2x\,dx$$
$$\text{If } u = \cos(x^3), \qquad \text{then} \qquad du = -3x^2\sin(x^3)\,dx$$

Now when the integrand has the form $f(u(x))\,u'(x)$, we can use Eq. (1) to rewrite the entire integral (including the dx term) in terms of u and its differential du:

$$\int \underbrace{f(u(x))}_{f(u)}\ \underbrace{u'(x)\,dx}_{du} = \int f(u)\,du$$

This equation is called the **Change of Variables Formula**. It transforms an integral in the variable x into a (hopefully simpler) integral in the new variable u.

In substitution, the key step is to choose the appropriate expression to assign to u. Often u can be chosen to be an inside function in a composition.

EXAMPLE 1   Evaluate $\displaystyle\int 3x^2 \sin(x^3)\,dx$.

Solution   The integrand contains the composite function $\sin(x^3)$, so we set $u = x^3$. The differential $du = 3x^2\,dx$ also appears, so we can carry out the substitution:

$$\int 3x^2 \sin(x^3)\,dx = \int \underbrace{\sin(x^3)}_{\sin u}\ \underbrace{3x^2\,dx}_{du} = \int \sin u\,du$$

Now evaluate the integral in the u-variable and replace u by x^3 in the answer:

$$\int 3x^2 \sin(x^3)\,dx = \int \sin u\,du = -\cos u + C = -\cos(x^3) + C$$

Let's check our answer by differentiating:

$$\frac{d}{dx}(-\cos(x^3)) = \sin(x^3)\frac{d}{dx}x^3 = 3x^2\sin(x^3)$$

∎

EXAMPLE 2 **Multiplying** du **by a Constant** Evaluate $\int x(x^2 + 9)^5 \, dx$.

Solution We let $u = x^2 + 9$ because the composite $u^5 = (x^2 + 9)^5$ appears in the integrand. The differential $du = 2x \, dx$ does not appear as is, but we can multiply by $\frac{1}{2}$ to obtain $\frac{1}{2} du = x \, dx$.

Now we can apply substitution:

$$\int x(x^2 + 9)^5 \, dx = \int \underbrace{(x^2 + 9)^5}_{u^5} \underbrace{x \, dx}_{\frac{1}{2} du} = \frac{1}{2} \int u^5 \, du = \frac{1}{12} u^6 + C$$

Finally, we express the answer in terms of x by substituting $u = x^2 + 9$:

$$\int x(x^2 + 9)^5 \, dx = \frac{1}{12} u^6 + C = \frac{1}{12}(x^2 + 9)^6 + C \qquad \blacksquare$$

Substitution Method:

(1) Choose u and compute du.

(2) Rewrite the integral in terms of u and du, and evaluate.

(3) Express the final answer in terms of x.

EXAMPLE 3 Evaluate $\int \dfrac{(x^2 + 2x) \, dx}{(x^3 + 3x^2 + 12)^6}$.

Solution The appearance of $(x^3 + 3x^2 + 12)^{-6}$ in the integrand suggests that we try $u = x^3 + 3x^2 + 12$. With this choice,

$$du = (3x^2 + 6x) \, dx = 3(x^2 + 2x) \, dx \quad \Rightarrow \quad \frac{1}{3} du = (x^2 + 2x) \, dx$$

$$\int \frac{(x^2 + 2x) \, dx}{(x^3 + 3x^2 + 12)^6} = \int \underbrace{(x^3 + 3x^2 + 12)^{-6}}_{u^{-6}} \underbrace{(x^2 + 2x) \, dx}_{\frac{1}{3} du}$$

$$= \frac{1}{3} \int u^{-6} \, du = \left(\frac{1}{3} \right) \left(\frac{u^{-5}}{-5} \right) + C$$

$$= -\frac{1}{15}(x^3 + 3x^2 + 12)^{-5} + C \qquad \blacksquare$$

> **CONCEPTUAL INSIGHT** An integration method that works for a given function may fail if we change the function even slightly. In the previous example, if we replace 2 by 2.1 and consider instead $\int \dfrac{(x^2 + 2.1x) \, dx}{(x^3 + 3x^2 + 12)^6}$, the Substitution Method does not work. The problem is that $(x^2 + 2.1x) \, dx$ is *not* a multiple of $du = (3x^2 + 6x) \, dx$.

EXAMPLE 4 Evaluate $\int \sin(7\theta + 5) \, d\theta$.

Solution Let $u = 7\theta + 5$. Then $du = 7 \, d\theta$ and $\frac{1}{7} du = d\theta$. We obtain

$$\int \underbrace{\sin(7\theta + 5)}_{\sin u} \underbrace{d\theta}_{\frac{1}{7} du} = \frac{1}{7} \int \sin u \, du = -\frac{1}{7} \cos u + C = -\frac{1}{7} \cos(7\theta + 5) + C \qquad \blacksquare$$

EXAMPLE 5 Evaluate $\int e^{-9t} \, dt$.

Solution Use the substitution $u = -9t$, $du = -9 \, dt$. Then $dt = -\frac{1}{9} du$:

$$\int e^{-9t} \, dt = \int e^u \left(-\frac{1}{9} \, du \right) = -\frac{1}{9} \int e^u \, du = -\frac{1}{9} e^u + C = -\frac{1}{9} e^{-9t} + C \qquad \blacksquare$$

Antiderivatives of $\sin\theta$ and $\cos\theta$ came naturally to us from the corresponding derivative formulas. However, we have not yet found an antiderivative for $\tan\theta$. We can obtain one now using the Substitution Method.

EXAMPLE 6   Integral of $\tan\theta$ Evaluate $\int \tan\theta \, d\theta$.

Solution In this case, the idea is to write $\tan\theta \, d\theta = \dfrac{\sin\theta \, d\theta}{\cos\theta}$ and to note that if $u = \cos\theta$, then $du = -\sin\theta \, d\theta$ so $-du = \sin\theta d\theta$:

$$\int \tan\theta \, d\theta = \int \frac{\sin\theta \, d\theta}{\cos\theta} = -\int \frac{du}{u} = -\ln|u| + C = -\ln|\cos\theta| + C$$

Alternate forms of this antiderivative are obtained using the identity $-\ln u = \ln\frac{1}{u}$. Thus, $-\ln|\cos\theta| = \ln\frac{1}{|\cos\theta|}$, and we have

$$\int \tan\theta \, d\theta = \ln\left|\frac{1}{\cos\theta}\right| + C = \ln|\sec\theta| + C \qquad \blacksquare$$

A similar approach shows that

$$\int \cot\theta \, d\theta = \ln|\sin\theta| + C$$

EXAMPLE 7   Additional Step Necessary   Evaluate $\int x\sqrt{5x+1}\,dx$.

Solution Since $\sqrt{5x+1}$ appears, we are tempted to set $u = 5x + 1$. Then

$$du = 5dx \quad \Rightarrow \quad \sqrt{5x+1}\,dx = \frac{1}{5}\sqrt{u}\,du$$

> The Substitution Method does not always work, even when the integral looks relatively simple. For example,
> $$\int \sin(x^2)\,dx \text{ cannot be evaluated}$$
> explicitly by substitution, or any other method. With experience, you will learn to recognize when substitution is likely to be successful.

Unfortunately, the integrand is not $\sqrt{5x+1}$ but $x\sqrt{5x+1}$. To take care of the extra factor of x, we solve $u = 5x + 1$ to obtain $x = \frac{1}{5}(u-1)$. Then

$$x\sqrt{5x+1}\,dx = \left(\frac{1}{5}(u-1)\right)\frac{1}{5}\sqrt{u}\,du = \frac{1}{25}(u^{3/2} - u^{1/2})\,du$$

$$\int x\sqrt{5x+1}\,dx = \frac{1}{25}\int (u^{3/2} - u^{1/2})\,du$$

$$= \frac{1}{25}\left(\frac{2}{5}u^{5/2} - \frac{2}{3}u^{3/2}\right) + C$$

$$= \frac{2}{125}(5x+1)^{5/2} - \frac{2}{75}(5x+1)^{3/2} + C \qquad \blacksquare$$

Change of Variables Formula for Definite Integrals

The Change of Variables Formula can be applied to definite integrals provided that the limits of integration are changed, as indicated in the next theorem.

> The new limits of integration with respect to the u-variable are $u(a)$ and $u(b)$. Think of it this way: As x varies from a to b, the variable $u = u(x)$ varies from $u(a)$ to $u(b)$.

THEOREM 2   Change of Variables Theorem for Definite Integrals   If u' is continuous on $[a, b]$, and f is continuous on the range of u, then

$$\int_a^b f(u(x))u'(x)\,dx = \int_{u(a)}^{u(b)} f(u)\,du$$

2

Proof If $F(x)$ is an antiderivative of $f(x)$, then $F(u(x))$ is an antiderivative of $f(u(x))u'(x)$. FTC I shows that the two integrals are equal:

$$\int_a^b f(u(x))u'(x)\,dx = F(u(b)) - F(u(a))$$

$$\int_{u(a)}^{u(b)} f(u)\,du = F(u(b)) - F(u(a)) \qquad \blacksquare$$

EXAMPLE 8 Evaluate $\displaystyle\int_0^2 x^2\sqrt{x^3+1}\,dx$.

Solution We use the substitution $u = x^3 + 1$. Thus, $du = 3x^2\,dx$, and therefore $x^2\,dx = \frac{1}{3}du$. By Eq. (2), the new limits of integration are

$$u(0) = 0^3 + 1 = 1 \qquad \text{and} \qquad u(2) = 2^3 + 1 = 9$$

Thus,

$$\int_0^2 x^2\sqrt{x^3+1}\,dx = \frac{1}{3}\int_1^9 \sqrt{u}\,du = \frac{2}{9}u^{3/2}\Big|_1^9 = \frac{52}{9}$$

This substitution shows that the area in (A) in Figure 1 is equal to one-third of the area in (B). $\qquad \blacksquare$

In the previous example, we can avoid changing the limits of integration by evaluating the indefinite integral in terms of x:

$$\int x^2\sqrt{x^3+1}\,dx = \frac{1}{3}\int \sqrt{u}\,du = \frac{2}{9}u^{3/2} + C = \frac{2}{9}(x^3+1)^{3/2} + C$$

This leads to the same result: $\displaystyle\int_0^2 x^2\sqrt{x^3+1}\,dx = \frac{2}{9}(x^3+1)^{3/2}\Big|_0^2 = \frac{52}{9}$.

EXAMPLE 9 Evaluate $\displaystyle\int_0^{\pi/4} \tan^3\theta \sec^2\theta\,d\theta$.

Solution The substitution $u = \tan\theta$ makes sense because $du = \sec^2\theta\,d\theta$, and therefore, $u^3\,du = \tan^3\theta \sec^2\theta\,d\theta$. The new limits of integration are

$$u(0) = \tan 0 = 0 \qquad \text{and} \qquad u\left(\frac{\pi}{4}\right) = \tan\left(\frac{\pi}{4}\right) = 1$$

Thus,

$$\int_0^{\pi/4} \tan^3\theta \sec^2\theta\,d\theta = \int_0^1 u^3\,du = \frac{u^4}{4}\Big|_0^1 = \frac{1}{4} \qquad \blacksquare$$

EXAMPLE 10 Calculate the area under the graph of $y = \dfrac{x}{x^2+1}$ over $[1,3]$.

Solution The area (Figure 2) is equal to $\displaystyle\int_1^3 \frac{x}{x^2+1}\,dx$. We use the substitution

$$u = x^2 + 1, \qquad du = 2x\,dx, \qquad \frac{1}{2}\frac{du}{u} = \frac{x\,dx}{x^2+1}$$

The new limits of integration are $u(1) = 1^2 + 1 = 2$ and $u(3) = 3^2 + 1 = 10$, so

$$\int_1^3 \frac{x}{x^2+1}\,dx = \frac{1}{2}\int_2^{10} \frac{du}{u} = \frac{1}{2}\ln|u|\Big|_2^{10} = \frac{1}{2}\ln 10 - \frac{1}{2}\ln 2 \approx 0.805 \qquad \blacksquare$$

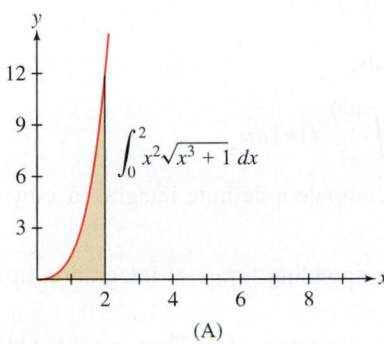

$$\int_0^2 x^2\sqrt{x^3+1}\,dx$$

(A)

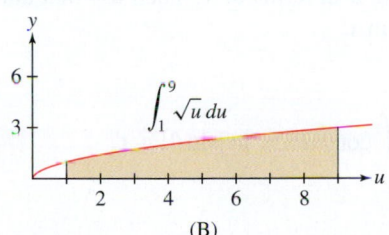

$$\int_1^9 \sqrt{u}\,du$$

(B)

FIGURE 1

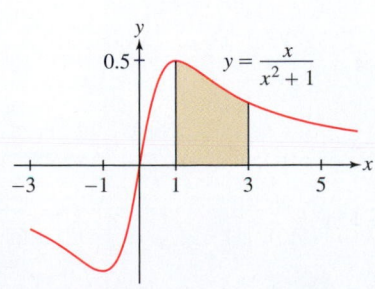

$y = \dfrac{x}{x^2+1}$

DF **FIGURE 2** Area under the graph of $y = \dfrac{x}{x^2+1}$ over $[1,3]$.

5.7 SUMMARY

- Try the Substitution Method when the integrand has the form $f(u(x))\, u'(x)$. If F is an antiderivative of f, then

$$\int f(u(x))\, u'(x)\, dx = F(u(x)) + C$$

- The differential of $u(x)$ is related to dx by $du = u'(x)\, dx$.
- The Substitution Method is expressed by the Change of Variables Formula:

$$\int f(u(x))\, u'(x)\, dx = \int f(u)\, du$$

- Change of Variables Formula for definite integrals:

$$\int_a^b f(u(x))\, u'(x)\, dx = \int_{u(a)}^{u(b)} f(u)\, du$$

- There are two approaches that can be taken to compute a definite integral in x by a change of variables to u.

  - Change the limits of integration to the corresponding limits of integration in u and compute the definite integral in u.
  - Compute the antiderivative first, expressing it in terms of x. Then use that antiderivative to compute the definite integral in x.

- Two new trigonometric integral formulas:

$$\int \tan\theta\, d\theta = \ln|\sec\theta| + C \quad \text{and} \quad \int \cot\theta\, d\theta = \ln|\sin\theta| + C$$

5.7 EXERCISES

Preliminary Questions

1. Which of the following integrals is a candidate for the Substitution Method?

(a) $\displaystyle\int 5x^4 \sin(x^5)\, dx$

(b) $\displaystyle\int \sin^5 x \, \cos x\, dx$

(c) $\displaystyle\int x^5 \sin x\, dx$

2. Find an appropriate choice of u for evaluating the following integrals by substitution:

(a) $\displaystyle\int x(x^2 + 9)^4\, dx$

(b) $\displaystyle\int x^2 \sin(x^3)\, dx$

(c) $\displaystyle\int \sin x \, \cos^2 x\, dx$

3. Which of the following is equal to $\displaystyle\int_0^2 x^2(x^3 + 1)\, dx$ for a suitable substitution?

(a) $\dfrac{1}{3}\displaystyle\int_0^2 u\, du$

(b) $\displaystyle\int_0^9 u\, du$

(c) $\dfrac{1}{3}\displaystyle\int_1^9 u\, du$

Exercises

In Exercises 1–6, calculate du.

1. $u = x^3 - x^2$

2. $u = 2x^4 + 8x^{-1}$

3. $u = \cos(x^2)$

4. $u = \tan x$

5. $u = e^{4x+1}$

6. $u = \ln(x^4 + 1)$

In Exercises 7–24, write the integral in terms of u and du. Then evaluate.

7. $\displaystyle\int (x + 8)^4\, dx, \quad u = x + 8$

8. $\displaystyle\int (x + 25)^{-2}\, dx, \quad u = x + 25$

9. $\displaystyle\int (3t - 4)^5\, dt, \quad u = 3t - 4$

10. $\displaystyle\int (8 - x)^{2/3}\, dx, \quad u = 8 - x$

11. $\displaystyle\int t\sqrt{t^2 + 1}\, dt, \quad u = t^2 + 1$

12. $\displaystyle\int (x^3 + 1) \cos(x^4 + 4x)\, dx, \quad u = x^4 + 4x$

13. $\displaystyle\int \dfrac{t^3}{(4 - 2t^4)^{11}}\, dt, \quad u = 4 - 2t^4$

14. $\int \sqrt{4x-1}\,dx, \quad u = 4x-1$

15. $\int x(x+1)^9\,dx, \quad u = x+1$

16. $\int x\sqrt{4x-1}\,dx, \quad u = 4x-1$

17. $\int x^2\sqrt{4-x}\,dx, \quad u = 4-x$

18. $\int \sin(4\theta-7)\,d\theta, \quad u = 4\theta-7$

19. $\int \sin\theta \cos^3\theta\,d\theta, \quad u = \cos\theta$

20. $\int \sec^2 x \tan x\,dx, \quad u = \tan x$

21. $\int x e^{-x^2}\,dx, \quad u = -x^2$

22. $\int (\sec^2 t)e^{\tan t}\,dt, \quad u = \tan t$

23. $\int \dfrac{(\ln x)^2\,dx}{x}, \quad u = \ln x$

24. $\int \dfrac{(\tan^{-1} x)^2\,dx}{x^2+1}, \quad u = \tan^{-1} x$

In Exercises 25–28, evaluate the integral in the form $a\sin(u(x)) + C$ for an appropriate choice of $u(x)$ and constant a.

25. $\int x^3 \cos(x^4)\,dx$

26. $\int x^2 \cos(x^3+1)\,dx$

27. $\int x^{1/2}\cos(x^{3/2})\,dx$

28. $\int \cos x \cos(\sin x)\,dx$

In Exercises 29–76, evaluate the indefinite integral.

29. $\int (4x+5)^9\,dx$

30. $\int \dfrac{dx}{(x-9)^5}$

31. $\int \dfrac{dt}{\sqrt{t+12}}$

32. $\int (9t+2)^{2/3}\,dt$

33. $\int \dfrac{x+1}{(x^2+2x)^3}\,dx$

34. $\int (x+1)(x^2+2x)^{3/4}\,dx$

35. $\int \dfrac{x}{\sqrt{x^2+9}}\,dx$

36. $\int \dfrac{2x^2+x}{(4x^3+3x^2)^2}\,dx$

37. $\int (2x^3-7)^2\,dx$

38. $\int (4-x^3)^2\,dx$

39. $\int x^2(2x^3-7)^3\,dx$

40. $\int 6x^2(4-x^3)^4\,dx$

41. $\int (3x+8)^{11}\,dx$

42. $\int x(3x+8)^{11}\,dx$

43. $\int x^2\sqrt{x^3+1}\,dx$

44. $\int x^5\sqrt{x^3+1}\,dx$

45. $\int \dfrac{dx}{(x+5)^3}$

46. $\int \dfrac{x^2\,dx}{(x+5)^3}$

47. $\int z^2(z^3+1)^{12}\,dz$

48. $\int (z^5+4z^2)(z^3+1)^{12}\,dz$

49. $\int (x+2)(x+1)^{1/4}\,dx$

50. $\int x^3(x^2-1)^{3/2}\,dx$

51. $\int \sin(8-3\theta)\,d\theta$

52. $\int \theta \sin(\theta^2)\,d\theta$

53. $\int \dfrac{\cos\sqrt{t}}{\sqrt{t}}\,dt$

54. $\int x^2 \sin(x^3+1)\,dx$

55. $\int \tan(4\theta+9)\,d\theta$

56. $\int \sin^8\theta \cos\theta\,d\theta$

57. $\int \cot x\,dx$

58. $\int x^{-1/5}\tan(x^{4/5})\,dx$

59. $\int \sec^2(4x+9)\,dx$

60. $\int \sec^2 x \tan^4 x\,dx$

61. $\int \dfrac{\sec^2(\sqrt{x})\,dx}{\sqrt{x}}$

62. $\int \dfrac{\cos 2x}{(1+\sin 2x)^2}\,dx$

63. $\int \sin 4x\sqrt{\cos 4x+1}\,dx$

64. $\int \cos x(3\sin x-1)\,dx$

65. $\int \sec\theta \tan\theta(\sec\theta-1)\,d\theta$

66. $\int (\tan t + \sin t)\cos t\,dt$

67. $\int e^{14x-7}\,dx$

68. $\int (x+1)e^{x^2+2x}\,dx$

69. $\int \dfrac{e^x\,dx}{(e^x+1)^4}$

70. $\int (\sec^2\theta)e^{\tan\theta}\,d\theta$

71. $\int \dfrac{e^t\,dt}{e^{2t}+2e^t+1}$

72. $\int \dfrac{dx}{x(\ln x)^2}$

73. $\int \dfrac{(\ln x)^4\,dx}{x}$

74. $\int \dfrac{x+\ln x}{x}\,dx$

75. $\int \dfrac{\tan(\ln x)}{x}\,dx$

76. $\int (\cot x)\ln(\sin x)\,dx$

77. Evaluate $\int \dfrac{dx}{(1+\sqrt{x})^3}$ using $u = 1+\sqrt{x}$. *Hint:* Show that $dx = 2(u-1)\,du$.

78. Can They Both Be Right? Hannah uses the substitution $u = \tan x$ and Akiva uses $u = \sec x$ to evaluate $\int \tan x \sec^2 x\,dx$. Show that they obtain different answers, and explain the apparent contradiction.

79. Evaluate $\int \sin x \cos x\,dx$ using substitution in two different ways: first using $u = \sin x$ and then using $u = \cos x$. Reconcile the two different answers.

80. Some Choices Are Better Than Others Evaluate

$$\int \sin x \cos^2 x\,dx$$

in two ways. First use $u = \sin x$ to show that

$$\int \sin x \cos^2 x\,dx = \int u\sqrt{1-u^2}\,du$$

and evaluate the integral on the right by a further substitution. Then show that $u = \cos x$ is a better choice than $u = \sin x$ to begin with.

81. What are the new limits of integration if we apply the substitution $u = 3x+\pi$ to the integral $\displaystyle\int_0^\pi \sin(3x+\pi)\,dx$?

82. Which of the following is the result of applying the substitution $u = 4x - 9$ to the integral $\int_2^8 (4x - 9)^{20}\, dx$?

(a) $\int_2^8 u^{20}\, du$

(b) $\frac{1}{4} \int_2^8 u^{20}\, du$

(c) $4 \int_{-1}^{23} u^{20}\, du$

(d) $\frac{1}{4} \int_{-1}^{23} u^{20}\, du$

In Exercises 83–84, compute the definite integral two ways:

- *Multiply out the integrand and then integrate directly without a substitution.*
- *Use the Change of Variables Formula with the substitution provided.*

83. $\int_1^4 (2x - 3)^3\, dx, \quad u = 2x - 3$

84. $\int_0^3 x(x^2 + 4)^2\, dx, \quad u = x^2 + 4$

In Exercises 85–98, use the Change of Variables Formula to evaluate the definite integral.

85. $\int_0^2 \dfrac{dx}{\sqrt{2x + 5}}$

86. $\int_1^6 \sqrt{x + 3}\, dx$

87. $\int_0^1 \dfrac{x}{(x^2 + 1)^3}\, dx$

88. $\int_{-1}^2 \sqrt{5x + 6}\, dx$

89. $\int_0^4 x\sqrt{x^2 + 9}\, dx$

90. $\int_1^2 \dfrac{4x + 12}{(x^2 + 6x + 1)^2}\, dx$

91. $\int_0^1 (x + 1)(x^2 + 2x)^5\, dx$

92. $\int_{10}^{17} (x - 9)^{-2/3}\, dx$

93. $\int_1^e \dfrac{\ln x}{x}\, dx$

94. $\int_0^{\sqrt{e-1}} \dfrac{x^3}{x^2 + 1}\, dx$

95. $\int_0^1 \theta \tan(\theta^2)\, d\theta$

96. $\int_0^{\pi/6} \sec^2\left(2x - \dfrac{\pi}{6}\right) dx$

97. $\int_0^{\pi/2} \cos^3 x \sin x\, dx$

98. $\int_{\pi/3}^{\pi/2} \cot^2 \dfrac{x}{2} \csc^2 \dfrac{x}{2}\, dx$

99. Evaluate $\int_0^2 r\sqrt{5 - \sqrt{4 - r^2}}\, dr$.

100. Find numbers a and b such that

$$\int_a^b (u^2 + 1)\, du = \int_{-\pi/4}^{\pi/4} \sec^4 \theta\, d\theta$$

and evaluate. *Hint:* Use the identity $\sec^2 \theta = \tan^2 \theta + 1$.

101. Wind engineers have found that wind speed v (in meters per second) at a given location follows a **Rayleigh distribution** of the type

$$W(v) = \dfrac{1}{32} v e^{-v^2/64}$$

This means that at a given moment in time, the probability that v lies between a and b is equal to the shaded area in Figure 3.

(a) Show that the probability that $v \in [0, b]$ is $1 - e^{-b^2/64}$.

(b) Calculate the probability that $v \in [2, 5]$.

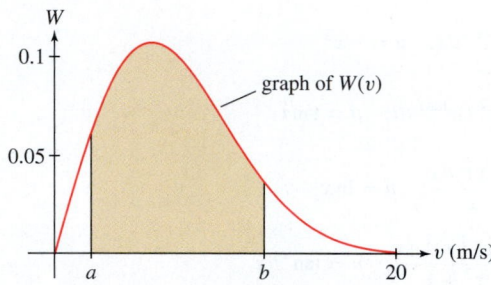

FIGURE 3 The shaded area is the probability that v lies between a and b.

102. Evaluate $\int_0^{\pi/2} \sin^n x \cos x\, dx$ for $n \geq 0$.

In Exercises 103–106, use substitution to evaluate the integral in terms of $f(x)$.

103. $\int f(x)^3 f'(x)\, dx$

104. $\int \dfrac{f'(x)}{f(x)^2}\, dx$

105. $\int \dfrac{f'(x)}{f(x)}\, dx$

106. $\int f'(-x + 7)\, dx$

107. Show that $\int_0^{\pi/6} f(\sin \theta)\, d\theta = \int_0^{1/2} f(u) \dfrac{1}{\sqrt{1 - u^2}}\, du$.

Further Insights and Challenges

108. Use the substitution $u = 1 + x^{1/n}$ to show that

$$\int \sqrt{1 + x^{1/n}}\, dx = n \int u^{1/2}(u - 1)^{n-1}\, du$$

Evaluate for $n = 2, 3$.

109. Evaluate $I = \int_0^{\pi/2} \dfrac{d\theta}{1 + \tan^{6000} \theta}$. *Hint:* Use substitution to show

that I is equal to $J = \int_0^{\pi/2} \dfrac{d\theta}{1 + \cot^{6000} \theta}$ and then check that $I + J = \int_0^{\pi/2} d\theta$.

110. Use substitution to prove that $\int_{-a}^a f(x)\, dx = 0$ if f is an odd function.

111. Prove that $\int_a^b \dfrac{1}{x}\, dx = \int_1^{b/a} \dfrac{1}{x}\, dx$ for $a, b > 0$. Then show that the regions under the hyperbola over the intervals $[1, 2], [2, 4], [4, 8], \ldots$ all have the same area (Figure 4).

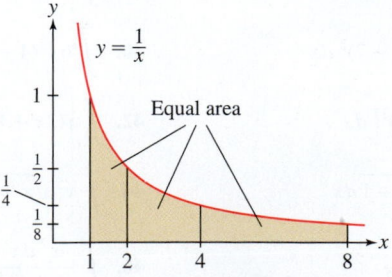

FIGURE 4 The area under $y = \dfrac{1}{x}$ over $[2^n, 2^{n+1}]$ is the same for all $n = 0, 1, 2, \ldots$.

112. Show that the two regions in Figure 5 have the same area. Then use the identity $\cos^2 u = \frac{1}{2}(1 + \cos 2u)$ to compute the second area.

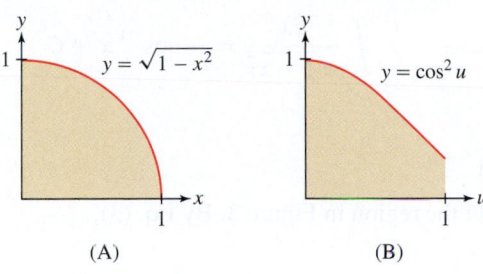

(A)   (B)

FIGURE 5

113. Area of an Ellipse Prove the formula $A = \pi ab$ for the area of the ellipse with equation (Figure 6)

$$\frac{x^2}{a^2} + \frac{y^2}{b^2} = 1$$

Hint: Use a change of variables to show that A is equal to ab times the area of the unit circle.

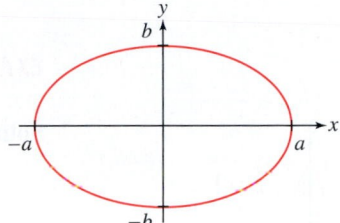

FIGURE 6 Graph of $\dfrac{x^2}{a^2} + \dfrac{y^2}{b^2} = 1$.

5.8 Further Integral Formulas

In Section 5.4, we used FTC I to show that when a and b have the same sign,

$$\int_a^b \frac{dx}{x} = \ln \frac{b}{a}$$

We obtain a formula for $\ln x$ as a definite integral by setting $a = 1$ and $b = x$ and keeping in mind that $\ln 1 = 0$:

$$\boxed{\ln x = \int_1^x \frac{dt}{t} \qquad \text{for } x > 0} \qquad \boxed{1}$$

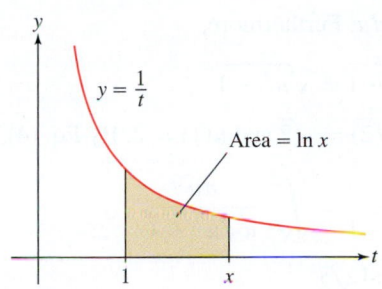

DF FIGURE 1

*It is possible to take Eq. (1) as the **definition** of $\ln x$ and to define e^x as the corresponding inverse function (see Exercises 82–83).*

Thus, $\ln x$ is equal to an area under the hyperbola $y = 1/t$ from 1 to x (Figure 1).

In a similar fashion, we can express $\sin^{-1} x$ as a definite integral using the derivative formula from Section 3.8 (Figure 2):

$$\frac{d}{dx}\sin^{-1} x = \frac{1}{\sqrt{1 - x^2}} \quad \Rightarrow \quad \int \frac{dx}{\sqrt{1 - x^2}} = \sin^{-1} x + C$$

Since $\sin^{-1} 0 = 0$, we have

$$\boxed{\sin^{-1} x = \int_0^x \frac{dt}{\sqrt{1 - t^2}} \qquad \text{for } -1 < x < 1}$$

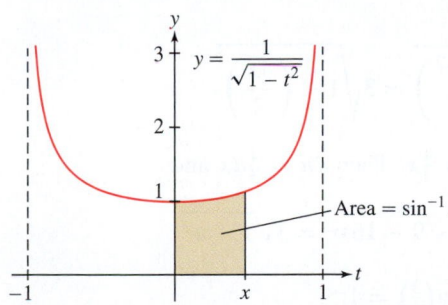

DF FIGURE 2

On the other hand, the derivative formulas from Section 3.8 yield integration formulas that are useful for evaluating new types of integrals.

Inverse Trigonometric Functions

$\dfrac{d}{dx}\sin^{-1} x = \dfrac{1}{\sqrt{1 - x^2}},$	$\displaystyle\int \dfrac{dx}{\sqrt{1 - x^2}} = \sin^{-1} x + C$ $\quad\boxed{2}$
$\dfrac{d}{dx}\tan^{-1} x = \dfrac{1}{x^2 + 1},$	$\displaystyle\int \dfrac{dx}{x^2 + 1} = \tan^{-1} x + C$ $\quad\boxed{3}$
$\dfrac{d}{dx}\sec^{-1} x = \dfrac{1}{|x|\sqrt{x^2 - 1}},$	$\displaystyle\int \dfrac{dx}{|x|\sqrt{x^2 - 1}} = \sec^{-1} x + C$ $\quad\boxed{4}$

In this list, we omit the integral formulas corresponding to the derivatives of $y = \cos^{-1} x$, $y = \cot^{-1} x$, and $y = \csc^{-1} x$ because their derivatives are the negative of the derivatives of $y = \sin^{-1} x$, $y = \tan^{-1} x$, and $y = \sec^{-1} x$, respectively. The omitted derivatives do not result in antiderivative formulas for new functions, just alternate formulas for functions for which we have a formula above. For example,

$$\frac{d}{dx}\cos^{-1}x = -\frac{1}{\sqrt{1-x^2}}, \qquad \int \frac{dx}{\sqrt{1-x^2}} = -\cos^{-1}x + C$$

EXAMPLE 1 Evaluate $\displaystyle\int_0^1 \frac{dx}{x^2+1}$.

Solution This integral is the area of the region in Figure 3. By Eq. (3),

$$\int_0^1 \frac{dx}{x^2+1} = \tan^{-1}x\Big|_0^1 = \tan^{-1}1 - \tan^{-1}0 = \frac{\pi}{4} - 0 = \frac{\pi}{4} \qquad \blacksquare$$

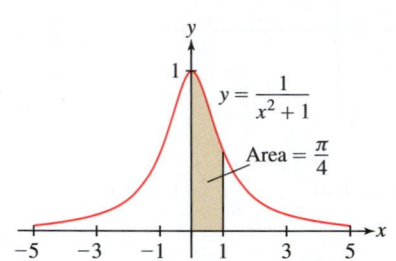

FIGURE 3 The shaded region has an area equal to $\tan^{-1}1 = \frac{\pi}{4}$.

EXAMPLE 2 Using Substitution Evaluate $\displaystyle\int_{1/\sqrt{2}}^1 \frac{dx}{x\sqrt{4x^2-1}}$.

Solution Notice that $\sqrt{4x^2-1}$ can be written as $\sqrt{(2x)^2-1}$, so it makes sense to try the substitution $u = 2x$. Thus, $x = \frac{1}{2}u$ and $dx = \frac{1}{2}du$. Furthermore,

$$u^2 = 4x^2 \qquad \text{and} \qquad \sqrt{4x^2-1} = \sqrt{u^2-1}$$

The new limits of integration are $u(1/\sqrt{2}) = 2(1/\sqrt{2}) = \sqrt{2}$ and $u(1) = 2$. By Eq. (4),

$$\int_{1/\sqrt{2}}^1 \frac{dx}{x\sqrt{4x^2-1}} = \int_{\sqrt{2}}^2 \frac{\frac{1}{2}du}{\frac{1}{2}u\sqrt{u^2-1}} = \int_{\sqrt{2}}^2 \frac{du}{u\sqrt{u^2-1}}$$

$$= \sec^{-1}2 - \sec^{-1}\sqrt{2}$$

$$= \frac{\pi}{3} - \frac{\pi}{4} = \frac{\pi}{12} \qquad \blacksquare$$

EXAMPLE 3 Using Substitution Evaluate $\displaystyle\int_0^{3/4} \frac{dx}{\sqrt{9-16x^2}}$.

Solution Let us first rewrite the integrand:

$$\sqrt{9-16x^2} = \sqrt{9\left(1 - \frac{16x^2}{9}\right)} = 3\sqrt{1 - \left(\frac{4x}{3}\right)^2}$$

Thus, it makes sense to use the substitution $u = \frac{4}{3}x$. Then $du = \frac{4}{3}dx$ and

$$x = \frac{3}{4}u, \qquad dx = \frac{3}{4}du, \qquad \sqrt{9-16x^2} = 3\sqrt{1-u^2}$$

The new limits of integration are $u(0) = 0$ and $u\left(\frac{3}{4}\right) = 1$:

$$\int_0^{3/4} \frac{dx}{\sqrt{9-16x^2}} = \int_0^1 \frac{\frac{3}{4}du}{3\sqrt{1-u^2}} = \frac{1}{4}\sin^{-1}x\Big|_0^1 = \frac{1}{4}(\sin^{-1}1 - \sin^{-1}0)$$

$$= \frac{1}{4}\left(\frac{\pi}{2}\right) = \frac{\pi}{8} \qquad \blacksquare$$

Integrals Involving $f(x) = b^x$

The exponential function $f(x) = e^x$ is particularly convenient because e^x is both its own derivative and its own antiderivative. For other bases b, we have

$$\frac{d}{dx}b^x = (\ln b)b^x \qquad \Rightarrow \qquad \frac{d}{dx}\left(\frac{b^x}{\ln b}\right) = b^x$$

This translates into the integral formula

$$\int b^x \, dx = \frac{b^x}{\ln b} + C \qquad \boxed{5}$$

EXAMPLE 4 Evaluate $\int_3^5 7^x \, dx$.

Solution Apply Eq. (5) with $b = 7$:

$$\int_3^5 7^x \, dx = \frac{7^x}{\ln 7}\Big|_3^5 = \frac{7^5 - 7^3}{\ln 7} \approx 8460.8 \qquad ■$$

EXAMPLE 5 Evaluate $\int_0^{\pi/2} (\cos\theta)10^{\sin\theta} \, d\theta$.

Solution Use the substitution $u = \sin\theta$, $du = \cos\theta \, d\theta$. The new limits of integration become $u(0) = 0$ and $u(\pi/2) = 1$:

$$\int_0^{\pi/2} (\cos\theta)10^{\sin\theta} \, d\theta = \int_0^1 10^u \, du = \frac{10^u}{\ln 10}\Big|_0^1 = \frac{10^1 - 10^0}{\ln 10} \approx 3.91 \qquad ■$$

5.8 SUMMARY

- Integral formula for the natural logarithm:

$$\ln x = \int_1^x \frac{dt}{t}$$

- Integral formulas involving inverse trigonometric functions:

$$\int \frac{dx}{\sqrt{1 - x^2}} = \sin^{-1} x + C$$

$$\int \frac{dx}{x^2 + 1} = \tan^{-1} x + C$$

$$\int \frac{dx}{|x|\sqrt{x^2 - 1}} = \sec^{-1} x + C$$

- Integrals of exponential functions ($b > 0$, $b \neq 1$):

$$\int e^x \, dx = e^x + C, \qquad \int b^x \, dx = \frac{b^x}{\ln b} + C$$

5.8 EXERCISES

Preliminary Questions

1. Find values for b such that $\int_1^b \frac{dx}{x}$ is equal to

(a) $\ln 3$ (b) 3

2. Find b such that $\int_0^b \frac{dx}{1 + x^2} = \frac{\pi}{3}$.

3. Which integral should be evaluated using substitution?

(a) $\int \frac{9 \, dx}{1 + x^2}$ (b) $\int \frac{dx}{1 + 9x^2}$

4. Which relation between x and u yields $\sqrt{16 + x^2} = 4\sqrt{1 + u^2}$?

Exercises

In Exercises 1–12, evaluate the definite integral.

1. $\displaystyle\int_1^9 \frac{dx}{x}$

2. $\displaystyle\int_4^{20} \frac{dx}{x}$

3. $\displaystyle\int_1^{e^3} \frac{1}{t}\,dt$

4. $\displaystyle\int_{-e^2}^{-e} \frac{1}{t}\,dt$

5. $\displaystyle\int_2^{12} \frac{dt}{3t+4}$

6. $\displaystyle\int_e^{e^3} \frac{dt}{t\ln t}$

7. $\displaystyle\int_1^{\sqrt{3}} \frac{dx}{x^2+1}$

8. $\displaystyle\int_2^7 \frac{x\,dx}{x^2+1}$

9. $\displaystyle\int_0^{1/2} \frac{dx}{\sqrt{1-x^2}}$

10. $\displaystyle\int_{-2}^{-2/\sqrt{3}} \frac{dx}{|x|\sqrt{x^2-1}}$

11. $\displaystyle\int_0^3 2^x\,dx$

12. $\displaystyle\int_0^3 2^{-x}\,dx$

13. Use the substitution $u = x/3$ to prove

$$\int \frac{dx}{9+x^2} = \frac{1}{3}\tan^{-1}\frac{x}{3} + C$$

14. Use the substitution $u = 2x$ to evaluate $\displaystyle\int \frac{dx}{4x^2+1}$.

In Exercises 15–34, calculate the integral.

15. $\displaystyle\int \frac{4}{4x^2+1}\,dx$

16. $\displaystyle\int \frac{4x}{4x^2+1}\,dx$

17. $\displaystyle\int_0^3 \frac{dx}{x^2+3}$

18. $\displaystyle\int_0^4 \frac{dt}{4t^2+9}$

19. $\displaystyle\int \frac{dt}{\sqrt{1-16t^2}}$

20. $\displaystyle\int_{-1/5}^{1/5} \frac{dx}{\sqrt{4-25x^2}}$

21. $\displaystyle\int \frac{dt}{\sqrt{5-3t^2}}$

22. $\displaystyle\int_{1/(2\sqrt{2})}^{1/2} \frac{dx}{x\sqrt{16x^2-1}}$

23. $\displaystyle\int \frac{dx}{x\sqrt{12x^2-3}}$

24. $\displaystyle\int \frac{x\,dx}{x^4+1}$

25. $\displaystyle\int \frac{dx}{x\sqrt{x^4-1}}$

26. $\displaystyle\int_{-1/2}^0 \frac{(x+1)\,dx}{\sqrt{1-x^2}}$

27. $\displaystyle\int_{-\ln 2}^0 \frac{e^x\,dx}{1+e^{2x}}$

28. $\displaystyle\int \frac{\ln(\cos^{-1}x)\,dx}{(\cos^{-1}x)\sqrt{1-x^2}}$

29. $\displaystyle\int \frac{\tan^{-1}x\,dx}{1+x^2}$

30. $\displaystyle\int_1^{\sqrt{3}} \frac{dx}{(\tan^{-1}x)(1+x^2)}$

31. $\displaystyle\int_0^{\log_4(3)} 4^x\,dx$

32. $\displaystyle\int_0^1 t\,5^{t^2}\,dt$

33. $\displaystyle\int 9^x \sin(9^x)\,dx$

34. $\displaystyle\int \frac{dx}{\sqrt{5^{2x}-1}}$

In Exercises 35–74, evaluate the integral using the methods covered in the text so far.

35. $\displaystyle\int y e^{y^2}\,dy$

36. $\displaystyle\int \frac{dx}{3x+5}$

37. $\displaystyle\int \frac{x\,dx}{\sqrt{4x^2+9}}$

38. $\displaystyle\int (x-x^{-2})^2\,dx$

39. $\displaystyle\int 7^{-x}\,dx$

40. $\displaystyle\int e^{9-12t}\,dt$

41. $\displaystyle\int \sec^2\theta\,\tan^7\theta\,d\theta$

42. $\displaystyle\int \sec^7\theta\,\tan\theta\,d\theta$

43. $\displaystyle\int \frac{\sqrt{1-\ln w}}{w}\,dw$

44. $\displaystyle\int \frac{\cos(\ln t)\,dt}{t}$

45. $\displaystyle\int \frac{t\,dt}{\sqrt{7-t^2}}$

46. $\displaystyle\int 2^x e^{4x}\,dx$

47. $\displaystyle\int \frac{(3x+2)\,dx}{x^2+4}$

48. $\displaystyle\int \tan(4x+1)\,dx$

49. $\displaystyle\int \frac{dx}{\sqrt{9-4x^2}}$

50. $\displaystyle\int e^t\sqrt{e^t+1}\,dt$

51. $\displaystyle\int (e^{-x}-4x)\,dx$

52. $\displaystyle\int (7-e^{10x})\,dx$

53. $\displaystyle\int \frac{e^{2x}-e^{4x}}{e^x}\,dx$

54. $\displaystyle\int \frac{dx}{x\sqrt{25x^2-1}}$

55. $\displaystyle\int \frac{(x+5)\,dx}{\sqrt{4-x^2}}$

56. $\displaystyle\int (t+1)\sqrt{t+1}\,dt$

57. $\displaystyle\int e^x\cos(e^x)\,dx$

58. $\displaystyle\int \frac{e^x}{\sqrt{e^x+1}}\,dx$

59. $\displaystyle\int \frac{dx}{\sqrt{9-16x^2}}$

60. $\displaystyle\int \frac{dx}{(4x-1)\ln(8x-2)}$

61. $\displaystyle\int e^x(e^{2x}+1)^3\,dx$

62. $\displaystyle\int \frac{dx}{x(\ln x)^5}$

63. $\displaystyle\int \frac{x^2\,dx}{x^3+2}$

64. $\displaystyle\int \frac{(3x-1)\,dx}{9-2x+3x^2}$

65. $\displaystyle\int \cot x\,dx$

66. $\displaystyle\int \frac{\cos x}{2\sin x+3}\,dx$

67. $\displaystyle\int \frac{4\ln x+5}{x}\,dx$

68. $\displaystyle\int (\sec\theta\,\tan\theta)5^{\sec\theta}\,d\theta$

69. $\displaystyle\int x3^{x^2}\,dx$

70. $\displaystyle\int \frac{\ln(\ln x)}{x\ln x}\,dx$

71. $\displaystyle\int \cot x\,\ln(\sin x)\,dx$

72. $\displaystyle\int \frac{t\,dt}{\sqrt{1-t^4}}$

73. $\displaystyle\int t^2\sqrt{t-3}\,dt$

74. $\displaystyle\int \cos x\,5^{-2\sin x}\,dx$

75. Use Figure 4 to prove

$$\int_0^x \sqrt{1-t^2}\,dt = \frac{1}{2}x\sqrt{1-x^2} + \frac{1}{2}\sin^{-1}x$$

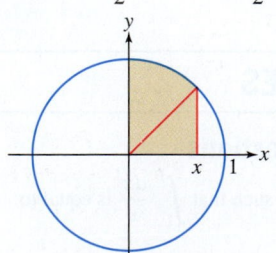

FIGURE 4

76. Use the substitution $u = \tan x$ to evaluate

$$\int \frac{dx}{1 + \sin^2 x}$$

Hint: Show that

$$\frac{dx}{1 + \sin^2 x} = \frac{du}{1 + 2u^2}$$

77. Prove

$$\int \sin^{-1} t \, dt = \sqrt{1 - t^2} + t \sin^{-1} t + C$$

78. (a) Verify for $r \neq 0$:

$$\int_0^T t e^{rt} \, dt = \frac{e^{rT}(rT - 1) + 1}{r^2} \qquad \boxed{6}$$

Hint: For fixed r, let $F(T)$ be the value of the integral on the left. By FTC II, $F'(t) = t e^{rt}$ and $F(0) = 0$. Show that the same is true of the function on the right.

(b) Use L'Hôpital's Rule to show that for fixed T, the limit as $r \to 0$ of the right-hand side of Eq. (6) is equal to the value of the integral for $r = 0$.

Further Insights and Challenges

79. Recall that if $f(t) \geq g(t)$ for $t \geq 0$, then for all $x \geq 0$,

$$\int_0^x f(t) \, dt \geq \int_0^x g(t) \, dt \qquad \boxed{7}$$

The inequality $e^t \geq 1$ holds for $t \geq 0$ because $e > 1$. Use (7) to prove that $e^x \geq 1 + x$ for $x \geq 0$. Then prove, by successive integration, the following inequalities (for $x \geq 0$):

$$e^x \geq 1 + x + \frac{1}{2}x^2, \qquad e^x \geq 1 + x + \frac{1}{2}x^2 + \frac{1}{6}x^3$$

80. Generalize Exercise 79; that is, use induction (if you are familiar with this method of proof) to prove that for all $n \geq 0$,

$$e^x \geq 1 + x + \frac{1}{2}x^2 + \frac{1}{6}x^3 + \cdots + \frac{1}{n}x^n \quad (x \geq 0)$$

81. Use Exercise 79 to show that $e^x/x^2 \geq x/6$ and conclude that $\lim_{x \to \infty} e^x/x^2 = \infty$. Then use Exercise 80 to prove more generally that $\lim_{x \to \infty} e^x/x^n = \infty$ for all n.

Exercises 82–84 develop an elegant approach to the exponential and logarithmic functions. Define a function $G(x)$ for $x > 0$:

$$G(x) = \int_1^x \frac{1}{t} \, dt$$

82. Defining $\ln x$ as an Integral This exercise proceeds as if we didn't know that $G(x) = \ln x$ and shows directly that G has all the basic properties of the logarithm. Prove the following statements:

(a) $\displaystyle\int_a^{ab} \frac{1}{t} \, dt = \int_1^b \frac{1}{t} \, dt$ for all $a, b > 0$. *Hint:* Use the substitution $u = t/a$.

(b) $G(ab) = G(a) + G(b)$. *Hint:* Break up the integral from 1 to ab into two integrals and use (a).

(c) $G(1) = 0$ and $G(a^{-1}) = -G(a)$ for $a > 0$.

(d) $G(a^n) = nG(a)$ for all $a > 0$ and integers n.

(e) $G(a^{1/n}) = \dfrac{1}{n}G(a)$ for all $a > 0$ and integers $n \neq 0$.

(f) $G(a^r) = rG(a)$ for all $a > 0$ and rational numbers r.

(g) G is increasing. *Hint:* Use FTC II.

(h) There exists a number a such that $G(a) > 1$. *Hint:* Show that $G(2) > 0$ and take $a = 2^m$ for $m > 1/G(2)$.

(i) $\displaystyle\lim_{x \to \infty} G(x) = \infty$ and $\displaystyle\lim_{x \to 0+} G(x) = -\infty$.

(j) There exists a unique number E such that $G(E) = 1$.

(k) $G(E^r) = r$ for every rational number r.

83. Defining e^x Use Exercise 82 to prove the following statements:

(a) G has an inverse with domain $\mathbf{R}$ and range $\{x : x > 0\}$. Denote the inverse by F.

(b) $F(x + y) = F(x)F(y)$ for all x, y. *Hint:* It suffices to show that $G(F(x)F(y)) = G(F(x + y))$.

(c) $F(r) = E^r$ for all numbers r. In particular, $F(0) = 1$.

(d) $F'(x) = F(x)$. *Hint:* Use the formula for the derivative of an inverse function that appears in Exercise 28 of the Chapter 3 Review Exercises. This shows that $E = e$ and $F(x) = e^x$ as defined in the text.

84. Defining b^x Let $b > 0$ and let $f(x) = F(xG(b))$ with F as in Exercise 83. Use Exercise 82 (f) to prove that $f(r) = b^r$ for every rational number r. This gives us a way of defining b^x for irrational x, namely $b^x = f(x)$. With this definition, $y = b^x$ is a differentiable function of x (because F is differentiable).

85. The formula $\displaystyle\int x^n \, dx = \frac{x^{n+1}}{n+1} + C$ is valid for $n \neq -1$. Show that the exceptional case $n = -1$ is a limit of the general case by applying L'Hôpital's Rule to the limit on the left:

$$\lim_{n \to -1} \int_1^x t^n \, dt = \int_1^x t^{-1} \, dt \qquad \text{(for fixed } x > 0\text{)}$$

Note that the integral on the left is equal to $\dfrac{x^{n+1} - 1}{n + 1}$.

86. [CAS] The integral inside the limit on the left in Exercise 85 is equal to $f_n(x) = \dfrac{x^{n+1} - 1}{n + 1}$ for $x \neq -1$. Investigate the limit graphically by plotting $y = f_n(x)$ for $n = 0, -0.3, -0.6$, and -0.9 together with $y = \ln x$ on a single plot.

87. ✎ **(a)** Explain why the shaded region in Figure 5 has area $\displaystyle\int_0^{\ln a} e^y \, dy$.

(b) Prove the formula $\displaystyle\int_1^a \ln x \, dx = a \ln a - \int_0^{\ln a} e^y \, dy$.

(c) Conclude that $\displaystyle\int_1^a \ln x \, dx = a \ln a - a + 1$.

(d) Use the result of (a) to find an antiderivative of $y = \ln x$.

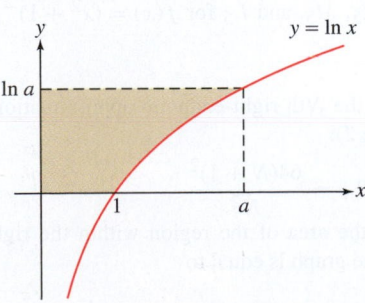

FIGURE 5

CHAPTER REVIEW EXERCISES

In Exercises 1–4, refer to the function f whose graph is shown in Figure 1.

1. Estimate L_4 and M_4 on $[0, 4]$.

2. Estimate R_4, L_4, and M_4 on $[1, 3]$.

3. Find an interval $[a, b]$ on which R_4 is larger than $\int_a^b f(x)\,dx$. Do the same for L_4.

4. Justify $\dfrac{3}{2} \le \int_1^2 f(x)\,dx \le \dfrac{9}{4}$.

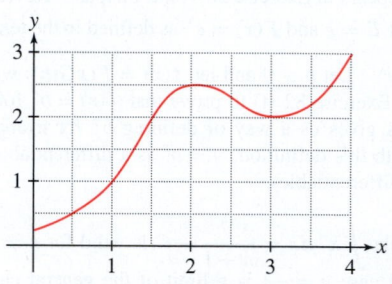

FIGURE 1

In Exercises 5–8, let $f(x) = x^2 + 3x$.

5. Calculate R_6, M_6, and L_6 for f on the interval $[2, 5]$. Sketch the graph of f and the corresponding rectangles for each approximation.

6. Use FTC I to evaluate $A(x) = \int_{-2}^x f(t)\,dt$.

7. Find a formula for R_N for f on $[2, 5]$ and compute $\int_2^5 f(x)\,dx$ by taking the limit.

8. Find a formula for L_N for f on $[0, 2]$ and compute $\int_0^2 f(x)\,dx$ by taking the limit.

9. Calculate R_5, M_5, and L_5 for $f(x) = (x^2 + 1)^{-1}$ on the interval $[0, 1]$.

10. Let R_N be the Nth right-endpoint approximation for $f(x) = x^3$ on $[0, 4]$ (Figure 2).

(a) Prove that $R_N = \dfrac{64(N+1)^2}{N^2}$.

(b) Prove that the area of the region within the right-endpoint rectangles above the graph is equal to

$$\frac{64(2N+1)}{N^2}$$

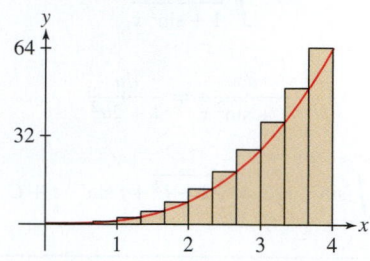

FIGURE 2 Approximation R_N for $f(x) = x^3$ on $[0, 4]$.

11. Which approximation to the area is represented by the shaded rectangles in Figure 3? Compute R_5 and L_5.

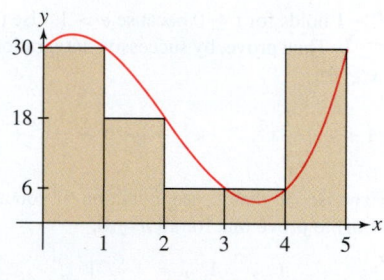

FIGURE 3

12. Calculate any two Riemann sums for $f(x) = x^2$ on the interval $[2, 5]$, but choose partitions with at least five subintervals of unequal widths and intermediate points that are neither endpoints nor midpoints.

In Exercises 13–16, express the limit as an integral (or multiple of an integral) and evaluate.

13. $\displaystyle \lim_{N \to \infty} \frac{\pi}{6N} \sum_{j=1}^N \sin\left(\frac{\pi}{3} + \frac{\pi j}{6N}\right)$

14. $\displaystyle \lim_{N \to \infty} \frac{3}{N} \sum_{k=0}^{N-1} \left(10 + \frac{3k}{N}\right)$

15. $\displaystyle \lim_{N \to \infty} \frac{5}{N} \sum_{j=1}^N \sqrt{4 + 5j/N}$

16. $\displaystyle \lim_{N \to \infty} \frac{1^k + 2^k + \cdots + N^k}{N^{k+1}} \quad (k > 0)$

In Exercises 17–30, calculate the indefinite integral.

17. $\displaystyle \int (4x^3 - 2x^2)\,dx$

18. $\displaystyle \int x^{9/4}\,dx$

19. $\displaystyle \int \sin(\theta - 8)\,d\theta$

20. $\displaystyle \int \cos(5 - 7\theta)\,d\theta$

21. $\displaystyle \int (4t^{-3} - 12t^{-4})\,dt$

22. $\displaystyle \int (9t^{-2/3} + 4t^{7/3})\,dt$

23. $\displaystyle \int \sec^2 x\,dx$

24. $\displaystyle \int \tan 3\theta \sec 3\theta\,d\theta$

25. $\displaystyle\int (y+2)^4\,dy$

26. $\displaystyle\int \frac{3x^3-9}{x^2}\,dx$

27. $\displaystyle\int (e^x - x)\,dx$

28. $\displaystyle\int e^{-4x}\,dx$

29. $\displaystyle\int 4x^{-1}\,dx$

30. $\displaystyle\int \sin(4x-9)\,dx$

In Exercises 31–36, solve the differential equation with the given initial condition.

31. $\dfrac{dy}{dx} = 4x^3,\quad y(1)=4$

32. $\dfrac{dy}{dt} = 3t^2 + \cos t,\quad y(0)=12$

33. $\dfrac{dy}{dx} = x^{-1/2},\quad y(1)=1$

34. $\dfrac{dy}{dx} = \sec^2 x,\quad y\!\left(\frac{\pi}{4}\right)=2$

35. $\dfrac{dy}{dx} = e^{-x},\quad y(0)=3$

36. $\dfrac{dy}{dx} = e^{4x},\quad y(1)=1$

37. Find $f(t)$ if $f''(t) = 1 - 2t$, $f(0)=2$, and $f'(0)=-1$.

38. At time $t = 0$, a driver begins decelerating at a constant rate of -10 m/s^2 and comes to a halt after traveling 500 m. Find the velocity at $t=0$.

In Exercises 39–42, use the given substitution to evaluate the integral.

39. $\displaystyle\int_0^2 \frac{dt}{4t+12},\quad u = 4t+12$

40. $\displaystyle\int \frac{(x^2+1)\,dx}{(x^3+3x)^4},\quad u = x^3+3x$

41. $\displaystyle\int_0^{\pi/6} \sin x \cos^4 x\,dx,\quad u = \cos x$

42. $\displaystyle\int \sec^2(2\theta)\tan(2\theta)\,d\theta,\quad u = \tan(2\theta)$

In Exercises 43–92, evaluate the integral.

43. $\displaystyle\int (20x^4 - 9x^3 - 2x)\,dx$

44. $\displaystyle\int_0^2 (12x^3 - 3x^2)\,dx$

45. $\displaystyle\int (2x^2 - 3x)^2\,dx$

46. $\displaystyle\int_0^1 (x^{7/3} - 2x^{1/4})\,dx$

47. $\displaystyle\int \frac{x^5 + 3x^4}{x^2}\,dx$

48. $\displaystyle\int_1^3 r^{-4}\,dr$

49. $\displaystyle\int_{-3}^3 |x^2 - 4|\,dx$

50. $\displaystyle\int_{-2}^4 |(x-1)(x-3)|\,dx$

51. $\displaystyle\int_1^4 \lfloor 2t \rfloor\,dt$

52. $\displaystyle\int_0^2 t - \lfloor t \rfloor\,dt$

53. $\displaystyle\int (10t - 7)^{14}\,dt$

54. $\displaystyle\int_2^3 \sqrt{7y - 5}\,dy$

55. $\displaystyle\int \frac{(2x^3 + 3x)\,dx}{(3x^4 + 9x^2)^5}$

56. $\displaystyle\int_{-3}^{-1} \frac{x\,dx}{(x^2 + 5)^2}$

57. $\displaystyle\int_0^5 15x\sqrt{x+4}\,dx$

58. $\displaystyle\int t^2\sqrt{t+8}\,dt$

59. $\displaystyle\int_0^1 \cos\!\left(\frac{\pi}{3}(t+2)\right)dt$

60. $\displaystyle\int_{\pi/2}^{\pi} \sin\!\left(\frac{5\theta - \pi}{6}\right)d\theta$

61. $\displaystyle\int t^2 \sec^2(9t^3 + 1)\,dt$

62. $\displaystyle\int \sin^2(3\theta)\cos(3\theta)\,d\theta$

63. $\displaystyle\int \csc^2(9 - 2\theta)\,d\theta$

64. $\displaystyle\int \sin\theta\sqrt{4 - \cos\theta}\,d\theta$

65. $\displaystyle\int_0^{\pi/3} \frac{\sin\theta}{\cos^{2/3}\theta}\,d\theta$

66. $\displaystyle\int \frac{\sec^2 t\,dt}{(\tan t - 1)^2}$

67. $\displaystyle\int e^{9-2x}\,dx$

68. $\displaystyle\int_1^3 e^{4x-3}\,dx$

69. $\displaystyle\int x^2 e^{x^3}\,dx$

70. $\displaystyle\int_0^{\ln 3} e^{x-e^x}\,dx$

71. $\displaystyle\int e^x 10^x\,dx$

72. $\displaystyle\int e^{-2x}\sin(e^{-2x})\,dx$

73. $\displaystyle\int \frac{e^{-x}\,dx}{(e^{-x} + 2)^3}$

74. $\displaystyle\int \sin\theta\cos\theta\, e^{\cos^2\theta + 1}\,d\theta$

75. $\displaystyle\int_0^{\pi/6} \tan 2\theta\,d\theta$

76. $\displaystyle\int_{\pi/3}^{2\pi/3} \cot\!\left(\frac{1}{2}\theta\right)d\theta$

77. $\displaystyle\int \frac{dt}{t(1 + (\ln t)^2)}$

78. $\displaystyle\int \frac{\cos(\ln x)\,dx}{x}$

79. $\displaystyle\int_1^e \frac{\ln x\,dx}{x}$

80. $\displaystyle\int \frac{dx}{x\sqrt{\ln x}}$

81. $\displaystyle\int \frac{dx}{4x^2 + 9}$

82. $\displaystyle\int_0^{0.8} \frac{dx}{\sqrt{1 - x^2}}$

83. $\displaystyle\int_4^{12} \frac{dx}{x\sqrt{x^2 - 1}}$

84. $\displaystyle\int_0^3 \frac{x\,dx}{x^2 + 9}$

85. $\displaystyle\int_0^3 \frac{dx}{x^2 + 9}$

86. $\displaystyle\int \frac{dx}{\sqrt{e^{2x} - 1}}$

87. $\displaystyle\int \frac{x\,dx}{\sqrt{1 - x^4}}$

88. $\displaystyle\int_{-5/\sqrt{2}}^{5/\sqrt{2}} \frac{dx}{\sqrt{25 - x^2}}$

89. $\displaystyle\int_0^4 \frac{dx}{2x^2 + 1}$

90. $\displaystyle\int_5^8 \frac{dx}{x\sqrt{x^2 - 16}}$

91. $\displaystyle\int_0^1 \frac{(\tan^{-1} x)^3\,dx}{1 + x^2}$

92. $\displaystyle\int \frac{\cos^{-1} t\,dt}{\sqrt{1 - t^2}}$

93. Combine to write as a single integral:

$$\int_0^8 f(x)\,dx + \int_{-2}^0 f(x)\,dx + \int_8^6 f(x)\,dx$$

94. Let $A(x) = \displaystyle\int_0^x f(x)\,dx$, where f is the function shown in Figure 4. Identify the location of the local minima, the local maxima, and points of inflection of A on the interval $[0, E]$, as well as the intervals where A is increasing, decreasing, concave up, or concave down. Where does the absolute maximum of A occur?

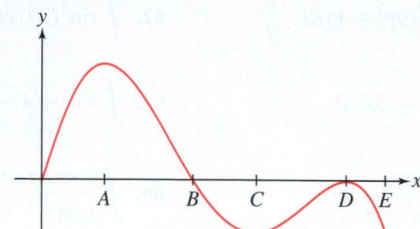

FIGURE 4

95. Find the local minima, the local maxima, and the inflection points of $A(x) = \displaystyle\int_3^x \dfrac{t\,dt}{t^2 + 1}$.

96. A particle starts at the origin at time $t = 0$ and moves with velocity $v(t)$ as shown in Figure 5.

(a) How many times does the particle return to the origin in the first 12 seconds?

(b) What is the particle's maximum distance from the origin?

(c) What is the particle's maximum distance to the left of the origin?

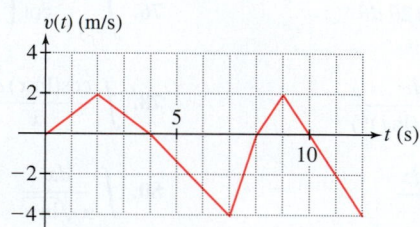

FIGURE 5

97. For the function f illustrated in Figure 6 do the following:

(a) For $x = 0, 1, 2, \ldots, 10$, approximate $A(x) = \displaystyle\int_0^x f(t)\,dt$.

(b) For $x = 1, 2, 3, \ldots, 9$, approximate $A'(x)$ using $\Delta x = 1$ and the symmetric difference quotient approximation,

$$A'(x) \approx \frac{A(x + \Delta x) - A(x - \Delta x)}{2\Delta x}$$

(c) Plot the values of $A'(x)$ on a graph of f to demonstrate $A' \approx f$.

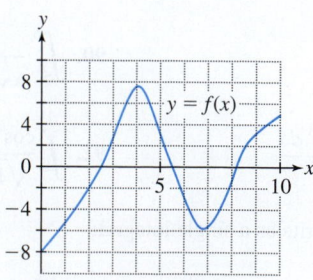

FIGURE 6

98. The sine integral function Si is an area function defined by $\mathrm{Si}(x) = \displaystyle\int_0^x \dfrac{\sin t}{t}\,dt$.

(a) Explain why Si has critical points at $n\pi$ for all nonzero integers n.

(b) [CAS] Use Riemann sums to approximate $\mathrm{Si}(x)$ for $x = \pi, 2\pi, \ldots, 8\pi$ and sketch a graph of $\mathrm{Si}(x)$ for $0 \le x \le 8\pi$.

99. On a typical day, a city consumes water at the rate of $r(t) = 100 + 72t - 3t^2$ (in thousands of gallons per hour), where t is the number of hours past midnight. What is the daily water consumption? How much water is consumed between 6 PM and midnight?

100. The learning curve in a certain bicycle factory is $L(x) = 12x^{-1/5}$ (in hours per bicycle), which means that it takes a bike mechanic $L(n)$ hours to assemble the nth bicycle. If a mechanic has produced 24 bicycles, how long does it take her or him to produce a subsequent batch of 12?

101. Cost engineers at NASA have the task of projecting the cost P of major space projects. It has been found that the cost C of developing a projection increases with P at the rate $dC/dP \approx 21P^{-0.65}$, where C is in thousands of dollars and P in millions of dollars. What is the cost of developing a projection for a project whose cost turns out to be $P = \$35$ million?

102. An astronomer estimates that in a certain constellation, the number of stars of magnitude m, per degree-squared of sky, is equal to $A(m) = 2.4 \times 10^{-6} m^{7.4}$ (fainter stars have higher magnitudes). Estimate the total number of stars of magnitude between 6 and 15 in a 1-degree-squared region of sky.

103. Evaluate $\displaystyle\int_{-8}^8 \dfrac{x^{15}\,dx}{3 + \cos^2 x}$, using the properties of odd functions.

104. Evaluate $\displaystyle\int_0^1 f(x)\,dx$, assuming that f is an even continuous function such that

$$\int_1^2 f(x)\,dx = 5, \qquad \int_{-2}^1 f(x)\,dx = 8$$

105. [GU] Plot the graph of $f(x) = \sin mx \sin nx$ on $[0, \pi]$ for the pairs $(m, n) = (2, 4)$, $(3, 5)$ and in each case guess the value of $I = \displaystyle\int_0^\pi f(x)\,dx$. Experiment with a few more values (including two cases with $m = n$) and formulate a conjecture for when I is zero.

106. Show that

$$\int x f(x)\,dx = x F(x) - G(x)$$

where $F'(x) = f(x)$ and $G'(x) = F(x)$. Use this to evaluate $\displaystyle\int x \cos x\,dx$.

107. Prove

$$2 \le \int_1^2 2^x\,dx \le 4 \qquad \text{and} \qquad \frac{1}{9} \le \int_1^2 3^{-x}\,dx \le \frac{1}{3}$$

108. [GU] Plot the graph of $f(x) = x^{-2} \sin x$, and show that

$$0.2 \le \int_1^2 f(x)\,dx \le 0.9.$$

109. Find upper and lower bounds for $\int_0^1 f(x)\,dx$, for $y = f(x)$ in Figure 7.

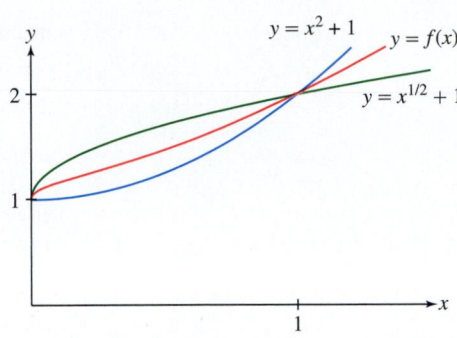

FIGURE 7

In Exercises 110–115, find the derivative.

110. $A'(x)$, where $A(x) = \int_3^x \sin(t^3)\,dt$

111. $A'(\pi)$, where $A(x) = \int_2^x \dfrac{\cos t}{1+t}\,dt$

112. $\dfrac{d}{dy} \int_{-2}^y 3^x\,dx$

113. $G'(x)$, where $G(x) = \int_{-2}^{\sin x} t^3\,dt$

114. $G'(2)$, where $G(x) = \int_0^{x^3} \sqrt{t+1}\,dt$

115. $H'(1)$, where $H(x) = \int_{4x^2}^9 \dfrac{1}{t}\,dt$

116. ✎ Explain with a graph: If f is increasing and concave up on $[a, b]$, then L_N is more accurate than R_N. Which is more accurate if f is increasing and concave down?

117. ✎ Explain with a graph: If f is linear on $[a, b]$, then the $\int_a^b f(x)\,dx = \frac{1}{2}(R_N + L_N)$ for all N.

118. In this exercise, we prove

$$x - \frac{x^2}{2} \le \ln(1 + x) \le x \qquad \text{(for } x > 0\text{)} \qquad \boxed{1}$$

(a) Show that $\ln(1 + x) = \int_0^x \dfrac{dt}{1+t}$ for $x > 0$.

(b) Verify that $1 - t \le \dfrac{1}{1+t} \le 1$ for all $t > 0$.

(c) Use (b) to prove Eq. (1).

(d) Verify Eq. (1) for $x = 0.5, 0.1,$ and 0.01.

119. Let

$$F(x) = x\sqrt{x^2 - 1} - 2\int_1^x \sqrt{t^2 - 1}\,dt$$

Prove that $F(x)$ and $y = \cosh^{-1} x$ differ by a constant by showing that they have the same derivative. Then prove they are equal by evaluating both at $x = 1$.

120. ✎ Let f be a positive increasing continuous function on $[a, b]$, where $0 \le a < b$ as in Figure 8. Show that the shaded region has area

$$I = bf(b) - af(a) - \int_a^b f(x)\,dx \qquad \boxed{2}$$

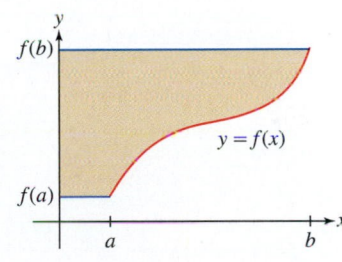

FIGURE 8

121. ✎ How can we interpret the quantity I in Eq. (2) if $a < b \le 0$? Explain with a graph.

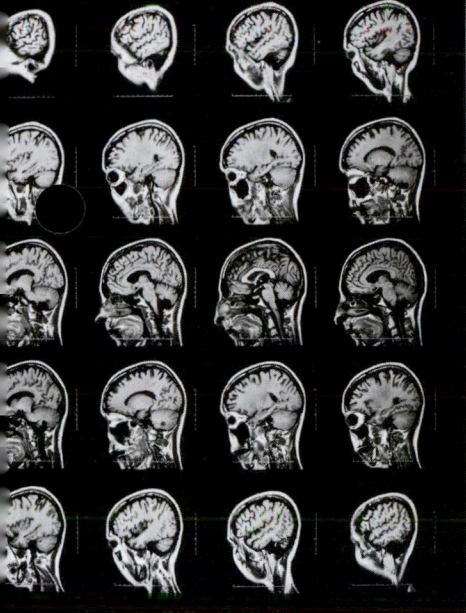

6 APPLICATIONS OF THE INTEGRAL

In the previous chapter, we used the integral to compute areas under curves and net change. In this chapter, we discuss some of the other quantities that are represented by integrals, including volume, average value, work, total mass, population, and fluid flow.

6.1 Area Between Two Curves

Sometimes we are interested in the area between two curves. Figure 1 shows projected electric power generation in the United States through renewable resources (wind, solar, biofuels, etc.) under two scenarios: with and without government stimulus spending. The area of the shaded region between the two graphs represents the additional energy projected to result from stimulus spending. How can we compute such an area?

The scans of the cranium show slices through the brain. Each slice reveals information about the structure and health of the brain. The images are combined to show the functioning of the entire brain. In much the same way, a definite integral takes slice-by-slice information about a function and sums it to analyze the function over the entire domain. In this chapter, we will see how this accumulating property of the definite integral may be applied in many different settings.

U. S. Renewable Generating Capacity Forecast Through 2030

Gigawatts (GW)

160 gigawatts with stimulus spending

133 gigawatts without stimulus spending

FIGURE 1 The area of the shaded region (which has units of *power × time*, or *energy*) represents the additional energy from renewable generating capacity projected to result from government stimulus spending in 2009–2010. *Source:* Energy Information Agency.

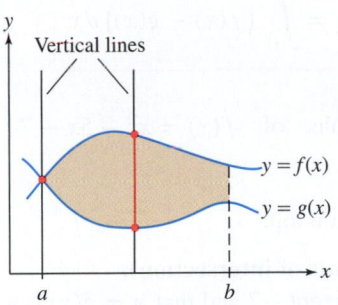

FIGURE 2 A vertically simple region.

Now suppose that we are given two functions $y = f(x)$ and $y = g(x)$ such that $f(x) \geq g(x)$ for all x in an interval $[a, b]$ (Figure 2). We call such a region **vertically simple** since any vertical line that intersects the region does so in a single point or a single vertical line segment with its lower endpoint on the graph of $y = g(x)$ and upper endpoint on the graph of $y = f(x)$. Then the graph of $y = f(x)$ lies above the graph of $y = g(x)$, and the area between the graphs is equal to the area under the top function minus the area under the bottom function (Figure 3):

$$\text{area between the graphs} = \int_a^b f(x)\,dx - \int_a^b g(x)\,dx$$

$$= \int_a^b \big(f(x) - g(x)\big)\,dx \qquad \boxed{1}$$

Figure 3 illustrates this formula in the case that both graphs lie above the x-axis. We see that the region between the graphs is obtained by removing the region under $y = g(x)$ from the region under $y = f(x)$.

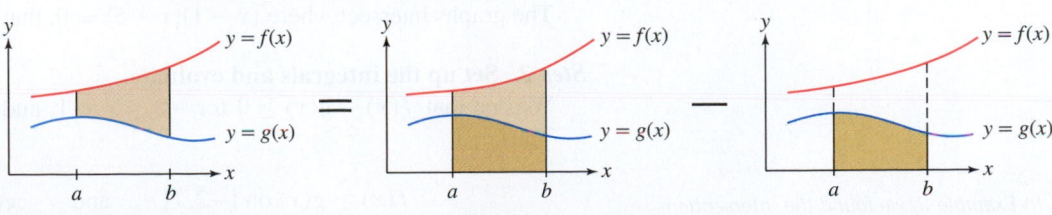

Region between the graphs

FIGURE 3 The area between the graphs is a difference of two areas.

EXAMPLE 1 Find the area of the region between the graphs of the functions as shown in Figure 4:

$$f(x) = x^2 - 4x + 10, \qquad g(x) = 4x - x^2, \qquad 1 \le x \le 3$$

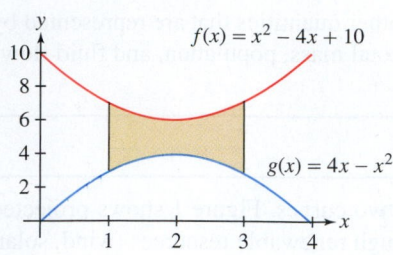

DF **FIGURE 4**

Solution Figure 4 shows that $f(x) \ge g(x)$ over $[1, 3]$, and therefore the region is vertically simple. By Eq. (1), the area between the graphs is

$$\int_1^3 \left(f(x) - g(x)\right) dx = \int_1^3 \left((x^2 - 4x + 10) - (4x - x^2)\right) dx$$

$$= \int_1^3 (2x^2 - 8x + 10) \, dx = \left(\frac{2}{3}x^3 - 4x^2 + 10x\right)\Big|_1^3$$

$$= 12 - \frac{20}{3} = \frac{16}{3} \qquad\blacksquare$$

Before continuing with more examples, we note that Eq. (1) remains valid whenever $f(x) \ge g(x)$, even if $f(x)$ and $g(x)$ are not assumed to be positive. As in Figure 5, we can simply shift the two functions up by adding to each a constant C big enough so that both functions are positive over the interval $[a, b]$. This does not change the area between them. Then by our previous result, we have

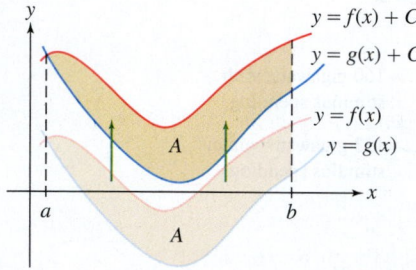

FIGURE 5 Shifting functions up to become positive-valued.

$$\text{area between the graphs} = \int_a^b \left((f(x) + C) - (g(x) + C)\right) dx$$

$$= \int_a^b \left(f(x) - g(x)\right) dx$$

Writing $y_{\text{top}} = f(x)$ for the top curve and $y_{\text{bot}} = g(x)$ for the bottom curve, we obtain

$$\boxed{\text{area between the graphs} = \int_a^b \left(y_{\text{top}} - y_{\text{bot}}\right) dx = \int_a^b \left(f(x) - g(x)\right) dx} \qquad \boxed{2}$$

EXAMPLE 2 Find the area between the graphs of $f(x) = x^2 - 5x - 7$ and $g(x) = x - 12$ over $[-2, 5]$.

Solution First, we must determine which graph lies on top.

Step 1. Sketch the region (especially, find any points of intersection).
We know that $y = f(x)$ is a parabola with y-intercept -7 and that $y = g(x)$ is a line with y-intercept -12 (Figure 6). To determine where the graphs intersect, we look for values of x where $f(x) = g(x)$, or equivalently, where $f(x) - g(x) = 0$:

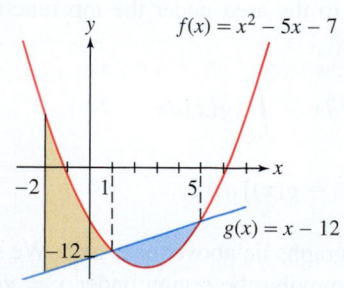

FIGURE 6

$$f(x) - g(x) = (x^2 - 5x - 7) - (x - 12) = x^2 - 6x + 5 = (x - 1)(x - 5)$$

The graphs intersect where $(x - 1)(x - 5) = 0$, that is, at $x = 1$ and $x = 5$.

Step 2. Set up the integrals and evaluate.
We see that $f(x) - g(x) \ge 0$ for $-2 \le x \le 1$, and $f(x) - g(x) \le 0$ for $1 \le x \le 5$. Thus,

$$f(x) \ge g(x) \text{ on } [-2, 1] \qquad \text{and} \qquad g(x) \ge f(x) \text{ on } [1, 5]$$

This tells us to subdivide our region into separate vertically simple regions, one over $[-2, 1]$ and the other over $[1, 5]$. Therefore, we write the area as a sum of integrals over the two intervals:

In Example 2, we found the intersection points of $y = f(x)$ and $y = g(x)$ algebraically. For more complicated functions, it may be necessary to use a computer algebra system.

$$\int_{-2}^{5} (y_{\text{top}} - y_{\text{bot}}) \, dx = \int_{-2}^{1} \left(f(x) - g(x) \right) dx + \int_{1}^{5} \left(g(x) - f(x) \right) dx$$

$$= \int_{-2}^{1} \left((x^2 - 5x - 7) - (x - 12) \right) dx$$

$$+ \int_{1}^{5} \left((x - 12) - (x^2 - 5x - 7) \right) dx$$

$$= \int_{-2}^{1} (x^2 - 6x + 5) \, dx + \int_{1}^{5} (-x^2 + 6x - 5) \, dx$$

$$= \left(\frac{1}{3}x^3 - 3x^2 + 5x \right) \Big|_{-2}^{1} + \left(-\frac{1}{3}x^3 + 3x^2 - 5x \right) \Big|_{1}^{5}$$

$$= \left(\frac{7}{3} - \frac{(-74)}{3} \right) + \left(\frac{25}{3} - \frac{(-7)}{3} \right) = \frac{113}{3} \qquad \blacksquare$$

EXAMPLE 3 **Calculating Area by Dividing the Region** Find the area of the region bounded by the graphs of $y = 8/x^2$, $y = 8x$, and $y = x$.

Solution

Step 1. **Sketch the region (especially, find any points of intersection).**
The curve $y = 8/x^2$ cuts off a region in the sector between the two lines $y = 8x$ and $y = x$ (Figure 7). We find the intersection of $y = 8/x^2$ and $y = 8x$ by solving

$$\frac{8}{x^2} = 8x \quad \Rightarrow \quad x^3 = 1 \quad \Rightarrow \quad x = 1$$

and the intersection of $y = 8/x^2$ and $y = x$ by solving

$$\frac{8}{x^2} = x \quad \Rightarrow \quad x^3 = 8 \quad \Rightarrow \quad x = 2$$

Step 2. **Set up the integrals and evaluate.**
Figure 7 shows that $y_{\text{bot}} = x$, but y_{top} changes at $x = 1$ from $y_{\text{top}} = 8x$ to $y_{\text{top}} = 8/x^2$. So, the region is not vertically simple. Therefore, we break up the regions into two parts, A and B, each vertically simple, and compute their areas separately.

$$\text{area of } A = \int_{0}^{1} (y_{\text{top}} - y_{\text{bot}}) \, dx = \int_{0}^{1} (8x - x) \, dx = \int_{0}^{1} 7x \, dx = \frac{7}{2}x^2 \Big|_{0}^{1} = \frac{7}{2}$$

$$\text{area of } B = \int_{1}^{2} (y_{\text{top}} - y_{\text{bot}}) \, dx = \int_{1}^{2} \left(\frac{8}{x^2} - x \right) dx = \left(-\frac{8}{x} - \frac{1}{2}x^2 \right) \Big|_{1}^{2} = \frac{5}{2}$$

The total area bounded by the curves is the sum $\frac{7}{2} + \frac{5}{2} = 6$. $\qquad \blacksquare$

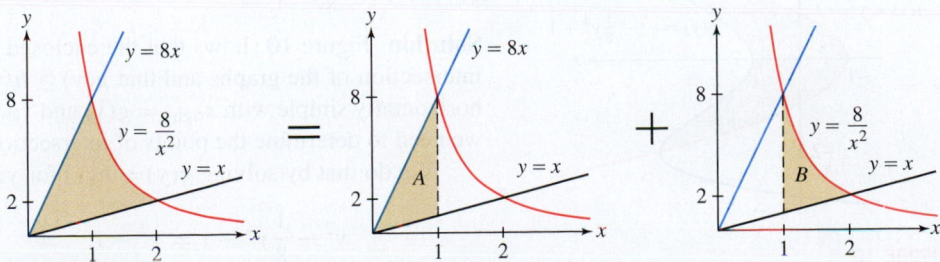

FIGURE 7 Area bounded by $y = 8/x^2$, $y = 8x$, and $y = x$ as a sum of two areas.

Integration Along the y-Axis

Suppose we are given x as a function of y, say, $x = g(y)$. What is the meaning of the integral $\int_c^d g(y)\,dy$? This integral can be interpreted as *signed area*, where regions to the *right* of the y-axis have positive area and regions to the *left* have negative area:

$$\int_c^d g(y)\,dy = \text{signed area between graph and } y\text{-axis for } c \le y \le d$$

In Figure 8(A), the part of the shaded region to the left of the y-axis has a negative signed area. The signed area of the entire region is

$$\underbrace{\int_{-6}^{6} (y^2 - 9)\,dy}_{\substack{\text{Area to the right of } y\text{-axis minus} \\ \text{area to the left of } y\text{-axis}}} = \left(\frac{1}{3}y^3 - 9y\right)\Bigg|_{-6}^{6} = 36$$

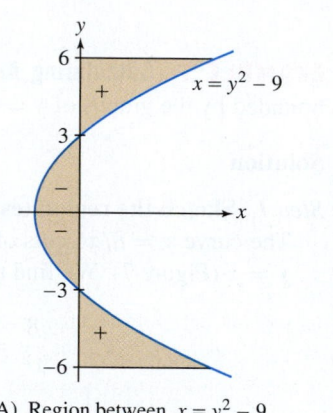

(A) Region between $x = y^2 - 9$ and the y-axis

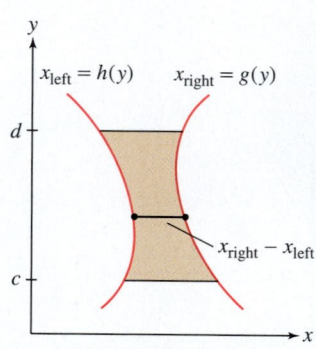

(B) Region between $x = h(y)$ and $x = g(y)$

FIGURE 8

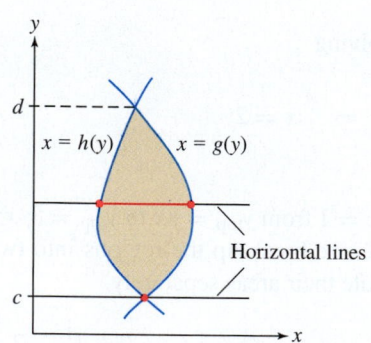

FIGURE 9 A horizontally simple region.

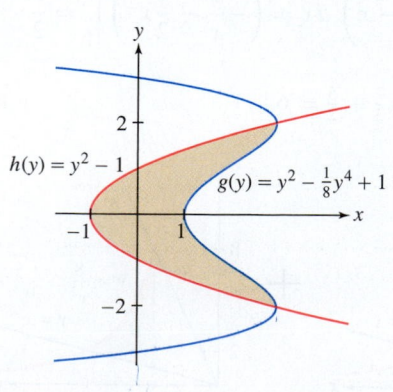

FIGURE 10

More generally, if $g(y) \ge h(y)$ as in Figure 8(B), then the graph of $x = g(y)$ lies to the right of the graph of $x = h(y)$. In this case, we write $x_{\text{right}} = g(y)$ and $x_{\text{left}} = h(y)$. We call the region **horizontally simple**, since every horizontal line that intersects the region in more than a single point does so in a single line segment such that the left endpoint is on the curve $x = h(y)$ and the right endpoint is on the curve $x = g(y)$, as in Figure 9. The formula for the area corresponding to Eq. (2) is

$$\text{area between the graphs} = \int_c^d (x_{\text{right}} - x_{\text{left}})\,dy = \int_c^d \big(g(y) - h(y)\big)\,dy \qquad \boxed{3}$$

EXAMPLE 4 Calculate the area enclosed by the graphs of $h(y) = y^2 - 1$ and $g(y) = y^2 - \frac{1}{8}y^4 + 1$.

Solution Figure 10 shows that the enclosed region stretches between the two points of intersection of the graphs and that $g(y) \ge h(y)$ over the region. Therefore, the region is horizontally simple with $x_{\text{right}} = g(y)$ and $x_{\text{left}} = h(y)$. To set up the integral for the area, we need to determine the points of intersection.

We do that by solving $g(y) = h(y)$ for y:

$$y^2 - \frac{1}{8}y^4 + 1 = y^2 - 1 \quad \Rightarrow \quad \frac{1}{8}y^4 - 2 = 0 \quad \Rightarrow \quad y = \pm 2$$

Now, we have

$$x_{\text{right}} - x_{\text{left}} = \left(y^2 - \frac{1}{8}y^4 + 1\right) - (y^2 - 1) = 2 - \frac{1}{8}y^4$$

The enclosed area is

$$\int_{-2}^{2} (x_{\text{right}} - x_{\text{left}}) \, dy = \int_{-2}^{2} \left(2 - \frac{1}{8}y^4\right) dy = \left(2y - \frac{1}{40}y^5\right)\Big|_{-2}^{2}$$

$$= \frac{16}{5} - \left(-\frac{16}{5}\right) = \frac{32}{5} \qquad \blacksquare$$

It would be more difficult to calculate the area of the region in Figure 10 as an integral with respect to x because the curves are not graphs of functions of x.

For many regions, we have a choice of whether to find the area by integrating with respect to x or with respect to y. The decision is usually based on two factors:

- How easy it is to obtain the curves as functions of one variable in terms of the other
- How easy it is to subdivide the region into simple regions and to integrate the functions involved

In the next example, we carry out the area computation, both integrating with respect to x and integrating with respect to y, demonstrating how these factors are involved.

EXAMPLE 5 Find the area of the region that is bounded by the three curves $y = x^2$, $y = (x - 2)^2$, and $y = 0$.

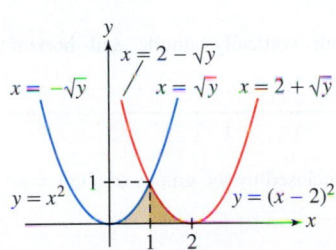

FIGURE 11 This region is horizontally simple but it is easier to cut it into two vertically simple regions.

Solution The area appears in Figure 11. Notice immediately that it is not vertically simple, since the top function changes over the interval $[0, 2]$. It is horizontally simple, but to calculate the area using the fact it is a horizontally simple region takes some work to set up the expressions for x_{right} and x_{left}.

We first compute the area of the region by splitting it into two vertically simple regions. In this case, the area is given by

$$\int_0^1 x^2 \, dx + \int_1^2 (x-2)^2 \, dx = \frac{x^3}{3}\Big|_0^1 + \frac{(x-2)^3}{3}\Big|_1^2 = \frac{1}{3} + 0 - \left(-\frac{1}{3}\right) = \frac{2}{3}$$

Now, let's redo the problem using the fact the region is horizontally simple. To determine x_{right} and x_{left}, we must invert the formulas for each of the parabolas. The left boundary of the region is the right side of the parabola given by $y = x^2$. Solving for x, we have $x = \pm\sqrt{y}$. The right side of the parabola corresponds to the positive choice; therefore, $x_{\text{left}} = \sqrt{y}$. The right boundary of the region is the left side of the parabola $y = (x - 2)^2$. We solve for x:

$$x - 2 = \pm\sqrt{y}$$
$$x = 2 \pm \sqrt{y}$$

The left side of the parabola is obtained by choosing the minus sign, and therefore $x_{\text{right}} = 2 - \sqrt{y}$.

Then the area is given by

$$\int_0^1 \left((2 - \sqrt{y}) - \sqrt{y}\right) dy = \int_0^1 \left(2 - 2y^{1/2}\right) dy = \left(2y - \frac{4}{3}y^{3/2}\right)\Big|_0^1 = \frac{2}{3} \qquad \blacksquare$$

6.1 SUMMARY

- If $f(x) \geq g(x)$ on $[a, b]$, then the region between the graphs is vertically simple and we have

$$\text{area between the graphs} = \int_a^b \left(y_{\text{top}} - y_{\text{bot}}\right) dx = \int_a^b \left(f(x) - g(x)\right) dx$$

- To calculate the area between $y = f(x)$ and $y = g(x)$, sketch the region to find y_{top}. If necessary, find points of intersection by solving $f(x) = g(x)$.

- Integral along the y-axis: $\displaystyle\int_c^d g(y)\,dy$ is equal to the signed area between the graph and the y-axis for $c \le y \le d$. The signed area to the right of the y-axis is positive and the signed area to the left is negative.

- If $g(y) \ge h(y)$ on $[c, d]$, then $x = g(y)$ lies to the right of $x = h(y)$ and the region is horizontally simple:

$$\text{area between the graphs} = \int_c^d \left(x_{\text{right}} - x_{\text{left}}\right) dy = \int_c^d \left(g(y) - h(y)\right) dy$$

6.1 EXERCISES

Preliminary Questions

1. What is the area interpretation of $\displaystyle\int_a^b \left(f(x) - g(x)\right) dx$ if $f(x) \ge g(x)$?

2. Is $\displaystyle\int_a^b \left(f(x) - g(x)\right) dx$ equal to the area between the graphs of f and g if $f(x) \ge 0$ but $g(x) \le 0$?

3. Suppose that $f(x) \ge g(x)$ on $[0, 3]$ and $g(x) \ge f(x)$ on $[3, 5]$. Express the area between the graphs over $[0, 5]$ as a sum of integrals.

4. Suppose that the graph of $x = f(y)$ lies to the left of the y-axis. Is $\displaystyle\int_a^b f(y)\,dy$ positive or negative?

5. Explain what $\displaystyle\int_a^b |f(x) - g(x)|\,dx$ represents.

6. Draw a region that is both vertically simple and horizontally simple.

Exercises

1. Find the area of the region between $y = 3x^2 + 12$ and $y = 4x + 4$ over $[-3, 3]$ (Figure 12).

3. Find the area of the region enclosed by the graphs of $f(x) = x^2 + 2$ and $g(x) = 2x + 5$ (Figure 14).

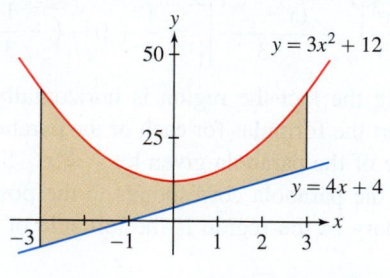

FIGURE 12

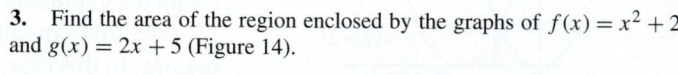

FIGURE 14

2. Find the area of the region between the graphs of $f(x) = 3x + 8$ and $g(x) = x^2 + 2x + 2$ over $[0, 2]$ (Figure 13).

4. Find the area of the region enclosed by the graphs of $f(x) = x^3 - 10x$ and $g(x) = 6x$ (Figure 15).

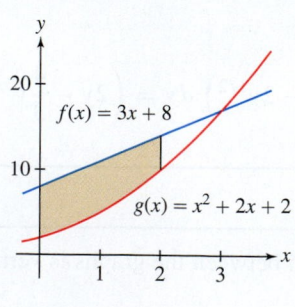

FIGURE 13

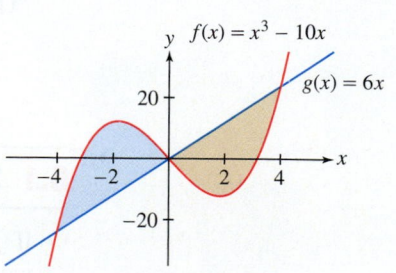

FIGURE 15

In Exercises 5 and 6, sketch the region between $y = \sin x$ and $y = \cos x$ over the interval and find its area.

5. $\left[\dfrac{\pi}{4}, \dfrac{\pi}{2}\right]$ **6.** $[0, \pi]$

In Exercises 7 and 8, let $f(x) = 20 + x - x^2$ and $g(x) = x^2 - 5x$.

7. Sketch the region enclosed by the graphs of f and g, and compute its area.

8. Sketch the region between the graphs of f and g over $[4, 8]$, and compute its area as a sum of two integrals.

9. Find the area between $y = e^x$ and $y = e^{2x}$ over $[0, 1]$.

10. Find the area of the region bounded by $y = e^x$ and $y = 12 - e^x$ and the y-axis.

11. Sketch the region bounded by the line $y = 2$ and the graph of $y = \sec^2 x$ for $-\frac{\pi}{4} < x < \frac{\pi}{4}$ and find its area.

12. Sketch the region bounded by

$$y = \sqrt{4 - x^2} \quad \text{and} \quad y = -\sqrt{4 - x^2}$$

for $-2 \le x \le 2$. Give a definite integral for the area of the region, but do not compute the integral. Instead, find the area using geometry.

In Exercises 13–16, determine whether or not the region bounded by the curves is vertically simple and/or horizontally simple.

13. $x = y^2, x = 2 - y^2$

14. $y = x^2, x = y^2$

15. $y = x, y = 2x, y = \frac{1}{x}$

16. In the first quadrant, $y = x$, $y = \sin\left(\frac{\pi}{2}x\right)$

In Exercises 17–20, find the area of the shaded region in Figures 16–19.

17.

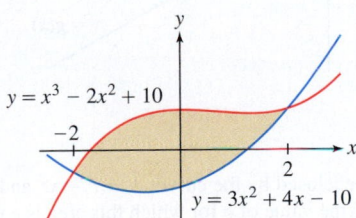

$y = x^3 - 2x^2 + 10$

$y = 3x^2 + 4x - 10$

FIGURE 16

18.

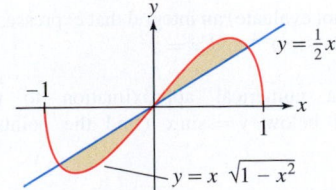

$y = \frac{1}{2}x$

$y = x\sqrt{1 - x^2}$

FIGURE 17

19.

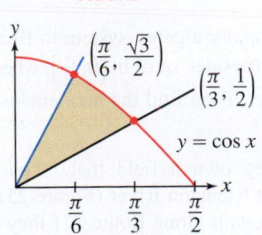

$\left(\dfrac{\pi}{6}, \dfrac{\sqrt{3}}{2}\right)$

$\left(\dfrac{\pi}{3}, \dfrac{1}{2}\right)$

$y = \cos x$

FIGURE 18

20.

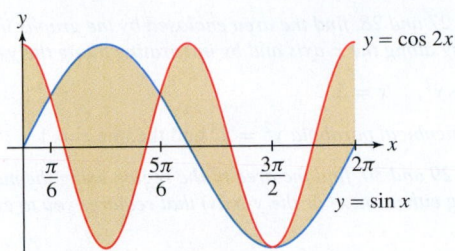

$y = \cos 2x$

$y = \sin x$

FIGURE 19

In Exercises 21 and 22, find the area between the graphs of $x = \sin y$ and $x = 1 - \cos y$ over the given interval (Figure 20).

21. $0 \le y \le \dfrac{\pi}{2}$ **22.** $-\dfrac{\pi}{2} \le y \le \dfrac{\pi}{2}$

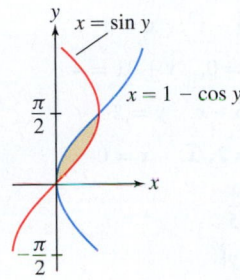

$x = \sin y$

$x = 1 - \cos y$

FIGURE 20

23. Find the area of the region lying to the right of $x = y^2 + 4y - 22$ and to the left of $x = 3y + 8$.

24. Find the area of the region lying to the right of $x = y^2 - 5$ and to the left of $x = 3 - y^2$.

25. Figure 21 shows the region enclosed by $x = y^3 - 26y + 10$ and $x = 40 - 6y^2 - y^3$. Match the equations with the curves and compute the area of the region.

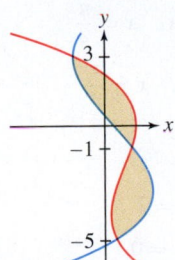

FIGURE 21

26. Figure 22 shows the region enclosed by $y = x^3 - 6x$ and $y = 8 - 3x^2$. Match the equations with the curves and compute the area of the region.

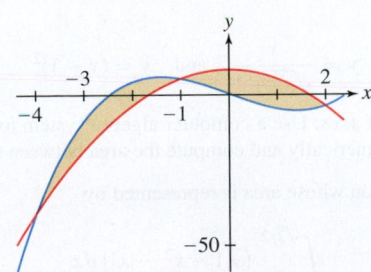

FIGURE 22 Region between $y = x^3 - 6x$ and $y = 8 - 3x^2$.

In Exercises 27 and 28, find the area enclosed by the graphs in two ways: by integrating along the x-axis and by integrating along the y-axis.

27. $x = 9 - y^2$, $x = 5$

28. The *semicubical parabola* $y^2 = x^3$ and the line $x = 1$

In Exercises 29 and 30, find the area of the region using the method (integration along either the x- or the y-axis) that requires you to evaluate just one integral.

29. Region between $y^2 = x + 5$ and $y^2 = 3 - x$

30. Region between $y = x$ and $x + y = 8$ over $[2, 3]$

In Exercises 31–50, sketch the region enclosed by the curves and compute its area as an integral along the x- or y-axis.

31. $y = 25 - x^2$, $y = x^2 - 25$

32. $y = x^2 - 6$, $y = 6 - x^3$, $x = 0$

33. $x + y = 4$, $x - y = 0$, $y + 3x = 4$

34. $y = 8 - 3x$, $y = 6 - x$, $y = 2$

35. $y = 15 - \sqrt{x}$, $y = 2\sqrt{x}$, $x = 0$

36. $y = \dfrac{x}{x^2 + 1}$, $y = \dfrac{x}{5}$

37. $x = |y|$, $x = 1 - |y|$

38. $y = |x|$, $y = \dfrac{x}{2} + 3$

39. $x = y^3 - 18y$, $y + 2x = 0$

40. $y = x\sqrt{x - 2}$, $y = -x\sqrt{x - 2}$, $x = 4$

41. $x = 2y$, $x + 1 = (y - 1)^2$

42. $x + y = 1$, $x^{1/2} + y^{1/2} = 1$

43. $y = \cos x$, $y = \cos 2x$, $x = 0$, $x = \dfrac{2\pi}{3}$

44. $y = \tan x$, $y = -\tan x$, $x = \dfrac{\pi}{4}$

45. $y = \sin x$, $y = \csc^2 x$, $x = \dfrac{\pi}{4}$

46. $x = \sin y$, $x = \dfrac{2}{\pi} y$

47. $y = e^x$, $y = e^{-x}$, $y = 2$

48. $y = 5e^x$, $y = e^{2x}$, $x = 0$

49. $y = \dfrac{\ln x}{x}$, $y = 0$, $x = e$

50. $y = \dfrac{\ln x}{x}$, $y = \dfrac{(\ln x)^2}{x}$

51. $\boxed{\text{CAS}}$ Plot

$$y = \frac{x}{\sqrt{x^2 + 1}} \quad \text{and} \quad y = (x - 1)^2$$

on the same set of axes. Use a computer algebra system to find the points of intersection numerically and compute the area between the curves.

52. Sketch a region whose area is represented by

$$\int_{-\sqrt{2}/2}^{\sqrt{2}/2} \left(\sqrt{1 - x^2} - |x| \right) dx$$

and determine the area using geometry.

53. $\boxed{}$ Beginning at the same time and location, Athletes 1 and 2 run for 30 seconds along a straight track with velocities $v_1(t)$ and $v_2(t)$ (in meters per second) as shown in Figure 23.

(a) What is represented by the area of the shaded region over $[0, 10]$?

(b) Which of the following is represented by the area of the shaded region over $[10, 30]$?

i. How far Athlete 2 is ahead of Athlete 1 at $t = 30$.

ii. How much further Athlete 2 ran than Athlete 1 did over the last 20 seconds.

(c) Who is ahead at the end of each 5-second interval, $t = 5, 10, \ldots, 30$?

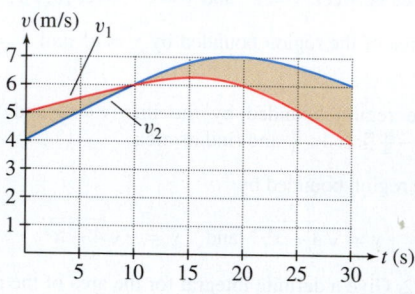

FIGURE 23

54. Express the area of the shaded region in Figure 24 as a sum of three integrals involving $f(x)$ and $g(x)$.

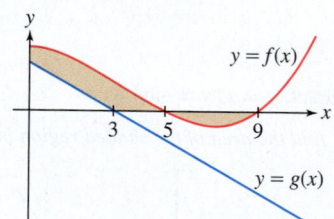

FIGURE 24

55. Find the area enclosed by the curves $y = c - x^2$ and $y = x^2 - c$ as a function of c. Find the value of c for which this area is equal to 1.

56. Set up (but do not evaluate) an integral that expresses the area between the circles $x^2 + y^2 = 2$ and $x^2 + (y - 1)^2 = 1$.

57. Set up (but do not evaluate) an integral that expresses the area between the graphs of $y = (1 + x^2)^{-1}$ and $y = x^2$.

58. $\boxed{\text{CAS}}$ Find a numerical approximation to the area above $y = 1 - (x/\pi)$ and below $y = \sin x$ (find the points of intersection numerically).

59. $\boxed{\text{CAS}}$ Find a numerical approximation to the area above $y = |x|$ and below $y = \cos x$.

60. $\boxed{\text{CAS}}$ Use a computer algebra system to find a numerical approximation to the number c (besides zero) in $\left[0, \frac{\pi}{2}\right]$, where the curves $y = \sin x$ and $y = \tan^2 x$ intersect. Then find the area enclosed by the graphs over $[0, c]$.

61. Lauren and Harvey own a field that is bordered by Route 271, Rogadzo Road, and the Riemann River (Figure 25). To estimate the area of the field, at 50-ft intervals along Route 271 they measured the distance from Route 271 to the river, parallel to Rogadzo Road. Their measurements (in feet) are shown in the figure and in the following table where x

represents the measurement location along Route 271 and y represents the distance from Route 271 to the Riemann River (both in feet).

x	0	50	100	150	200	250	300	350	400
y	260	265	215	205	250	305	295	240	150

Compute right-endpoint and left-endpoint Riemann sums to obtain approximations of the area of the field.

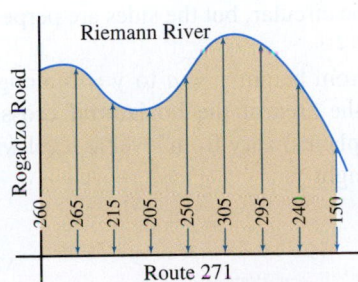

FIGURE 25

62. Referring to Figure 1 at the beginning of this section, estimate the projected number of additional joules produced in the years 2009–2030 as a result of government stimulus spending in 2009–2010. *Note:* One watt (W) is equal to 1 joule/second (J/s), and 1 gigawatt (GW) is 10^9 watts.

Exercises 63 and 64 use the notation and results of Exercises 47–49 of Section 3.4. For a given country, $F(r)$ is the fraction of total income that goes to the bottom rth fraction of households. The graph of $y = F(r)$ is called the Lorenz curve.

63. Let A be the area between $y = r$ and $y = F(r)$ over the interval $[0, 1]$ (Figure 26). The **Gini index** is the ratio $G = A/B$, where B is the area under $y = r$ over $[0, 1]$.

(a) Show that
$$G = 2 \int_0^1 (r - F(r))\, dr$$

(b) Calculate G if
$$F(r) = \begin{cases} \frac{1}{3}r & \text{for } 0 \le r \le \frac{1}{2} \\ \frac{5}{3}r - \frac{2}{3} & \text{for } \frac{1}{2} \le r \le 1 \end{cases}$$

(c) The Gini index is a measure of income distribution, with a lower value indicating a more equal distribution. Calculate G if $F(r) = r$ (in this case, all households have the same income by Exercise 49(b) of Section 3.4).

(d) What is G if all of the income goes to one household? *Hint:* In this extreme case, $F(r) = 0$ for $0 \le r < 1$.

64. Calculate the Gini index of the United States in the year 2010 from the Lorenz curve in Figure 26, which consists of segments joining the data points in the following table:

r	0	0.2	0.4	0.6	0.8	1
$F(r)$	0	0.033	0.118	0.264	0.480	1

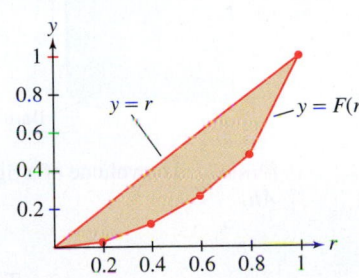

FIGURE 26 Lorenz curve for the United States in 2010.

Further Insights and Challenges

65. Find the line $y = mx$ that divides the area under the curve $y = x(1 - x)$ over $[0, 1]$ into two regions of equal area.

66. CAS Let c be the number such that the area under $y = \sin x$ over $[0, \pi]$ is divided in half by the line $y = cx$ (Figure 27). Find an equation for c and solve this equation *numerically* using a computer algebra system.

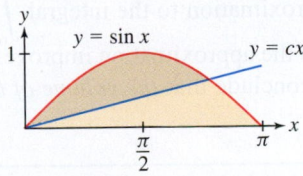

FIGURE 27

67. Explain geometrically (without calculation):
$$\int_0^1 x^n\, dx + \int_0^1 x^{1/n}\, dx = 1 \quad \text{(for } n > 0\text{)}$$

68. Let f be an increasing function with inverse g. Explain geometrically:
$$\int_0^a f(x)\, dx + \int_{f(0)}^{f(a)} g(x)\, dx = af(a)$$

6.2 Setting Up Integrals: Volume, Density, Average Value

Which quantities are represented by integrals? Roughly speaking, integrals represent quantities that are the "total amount" of something such as area, volume, or mass. We can approximate each quantity as a sum obtained by dividing an object into small pieces over which the quantity is easy to compute. We obtain an exact value by taking a limit; that is, by using an integral. There is a two-step procedure for computing such quantities: (1) approximate the quantity by a sum of N terms, and (2) pass to the limit as $N \to \infty$ to obtain an integral. We'll use this procedure often in this and other sections.

Volume

The terms "solid" and "solid body" refer to a solid three-dimensional object.

Our first example is the **volume** of a solid body. Before proceeding, let's recall that the volume of a *right cylinder* (Figure 1) is Ah, where A is the area of the base and h is the height, measured perpendicular to the base. Here, we use the "right cylinder" in the general sense; the base does not have to be circular, but the sides are perpendicular to the base.

Suppose that a solid body extends from height $y = a$ to $y = b$ along the y-axis as in Figure 2. Furthermore, assume that the area of the **horizontal cross sections** (the intersections of the solid with horizontal planes) vary from level to level within the solid. Let $A(y)$ be the cross-sectional area at height y.

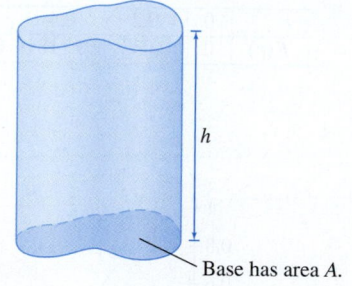

FIGURE 1 The volume of a right cylinder is Ah.

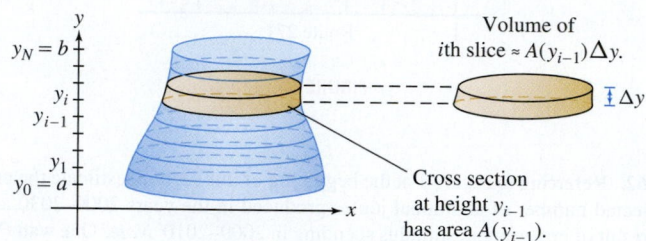

DF **FIGURE 2** Divide the solid into thin horizontal slices. Each slice is nearly a right cylinder whose volume can be approximated as area times height.

To compute the volume V of the body, divide the body into N horizontal slices of thickness $\Delta y = (b - a)/N$. The ith slice extends from y_{i-1} to y_i, where $y_i = a + i\Delta y$. Let V_i be the volume of the slice.

If N is very large, then Δy is very small and the slices are very thin. In this case, the ith slice is nearly a right cylinder of base $A(y_{i-1})$ and height Δy, and therefore, $V_i \approx A(y_{i-1})\Delta y$. The whole volume is obtained by summing the volumes of the slices, and therefore,

$$V = \sum_{i=1}^{N} V_i \approx \sum_{i=1}^{N} A(y_{i-1})\Delta y$$

The sum on the right is a left-endpoint approximation to the integral $\displaystyle\int_a^b A(y)\, dy$. If we assume that A is a continuous function, then the approximation improves in accuracy and converges to the integral as $N \to \infty$. We conclude that *the volume of the solid is equal to the integral of its cross-sectional area.*

Volume as the Integral of Cross-Sectional Area   Let $A(y)$ be the area of the horizontal cross section at height y of a solid body extending from $y = a$ to $y = b$. Then

$$\text{Volume of the solid body} = \int_a^b A(y)\, dy \qquad \boxed{1}$$

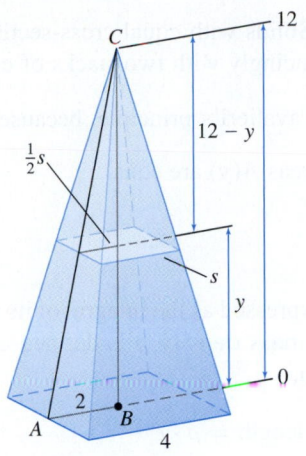

DF FIGURE 3 A horizontal cross section of the pyramid is a square.

EXAMPLE 1 **Volume of a Pyramid** Calculate the volume V of a pyramid of height 12 m whose base is a square of side 4 m.

Solution To use Eq. (1), we need a formula for the horizontal cross section $A(y)$.

Step 1. Find a formula for $A(y)$.

Figure 3 shows that the horizontal cross section at height y is a square. To find the side s of this square, apply the law of similar triangles to $\triangle ABC$ and to the triangle of height $12 - y$ whose base of length $\frac{1}{2}s$ lies on the cross section:

$$\frac{\text{base}}{\text{height}} = \frac{2}{12} = \frac{\frac{1}{2}s}{12 - y} \quad \Rightarrow \quad 2(12 - y) = 6s$$

We find that $s = \frac{1}{3}(12 - y)$, and therefore, $A(y) = s^2 = \frac{1}{9}(12 - y)^2$.

Step 2. Compute V as the integral of $A(y)$.

$$V = \int_0^{12} A(y)\,dy = \int_0^{12} \frac{1}{9}(12 - y)^2\,dy = -\frac{1}{27}(12 - y)^3 \Big|_0^{12} = 64 \text{ m}^3$$

We found the antiderivative $-\frac{1}{27}(12 - y)^3$ using a substitution $u = 12 - y$. The resulting relation $du = -dx$ introduces the negative sign appearing in the result.

This volume of 64 m³ agrees with the result obtained using the formula $V = \frac{1}{3}Ah$ for the volume of a pyramid of base A and height h, since $\frac{1}{3}Ah = \frac{1}{3}(4^2)(12) = 64$. ∎

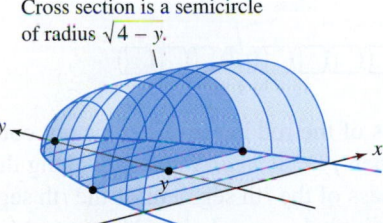

Cross section is a semicircle of radius $\sqrt{4 - y}$.

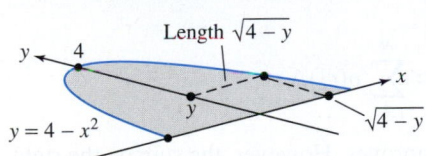

Length $\sqrt{4 - y}$

$y = 4 - x^2$

FIGURE 4

EXAMPLE 2 Compute the volume V of the solid in Figure 4, whose base (shown at the bottom in the figure) is the region between the parabola $y = 4 - x^2$ and the x-axis, and whose vertical cross sections perpendicular to the y-axis are semicircles.

Solution To find a formula for the area $A(y)$ of the cross section, observe that $y = 4 - x^2$ can be written $x = \pm\sqrt{4 - y}$. We see in Figure 4 that the cross section at y is a semicircle of radius $r = \sqrt{4 - y}$. This semicircle has area $A(y) = \frac{1}{2}\pi r^2 = \frac{\pi}{2}(4 - y)$. Therefore,

$$V = \int_0^4 A(y)\,dy = \frac{\pi}{2}\int_0^4 (4 - y)\,dy = \frac{\pi}{2}\left(4y - \frac{1}{2}y^2\right)\Big|_0^4 = 4\pi \quad ∎$$

In the next example, we compute volume using vertical rather than horizontal cross sections. This leads to an integral with respect to x rather than y.

EXAMPLE 3 **Volume of a Sphere: Vertical Cross Sections** Compute the volume of a sphere of radius R.

Solution As we see in Figure 5, the vertical cross section of the sphere at x is a circle whose radius r satisfies $x^2 + r^2 = R^2$ or $r = \sqrt{R^2 - x^2}$. Note that R is the fixed radius of the sphere, while r varies as the cross sections vary. The area of the cross section is $A(x) = \pi r^2 = \pi(R^2 - x^2)$. Therefore, the sphere has volume

$$\int_{-R}^{R} \pi(R^2 - x^2)\,dx = \pi\left(R^2 x - \frac{x^3}{3}\right)\Big|_{-R}^{R} = 2\left(\pi R^3 - \pi\frac{R^3}{3}\right) = \frac{4}{3}\pi R^3 \quad ∎$$

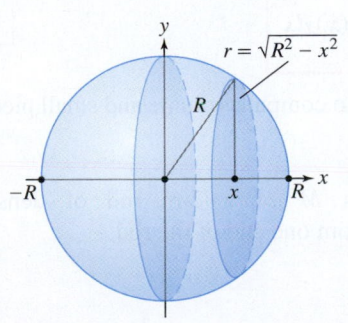

FIGURE 5

FIGURE 6 The two stacks of coins have equal cross sections, hence equal volumes by Cavalieri's principle.

Density

Next, we show that the total mass of an object can be expressed as the integral of its mass density. Consider a rod of length ℓ. The rod's **linear mass density** ρ is defined as the mass per unit length. If ρ is constant, then by definition,

The symbol ρ (lowercase Greek letter rho) is used often to denote density.

$$\text{total mass} = \text{linear mass density} \times \text{length} = \rho \cdot \ell \qquad \boxed{2}$$

For example, if $\ell = 10$ cm and $\rho = 9$ g/cm, then the total mass is $\rho\ell = 9 \cdot 10 = 90$ g.

Now, consider a rod extending along the x-axis from $x = a$ to $x = b$ whose density $y = \rho(x)$ is a continuous function of x, as in Figure 7. To compute the total mass M, we break up the rod into N small segments of length $\Delta x = (b-a)/N$. Then $M = \sum_{i=1}^{N} M_i$, where M_i is the mass of the ith segment.

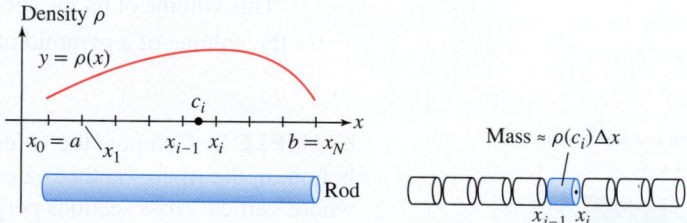

FIGURE 7 The total mass of the rod is equal to the integral of the mass density ρ along the rod.

We cannot use Eq. (2) directly to find the mass of the rod because $\rho(x)$ is not constant. However, we can argue that if Δx is small, then $\rho(x)$ is *nearly constant* along the ith segment, and we can use Eq. (2) to obtain the mass of the ith segment. If the ith segment extends from x_{i-1} to x_i and c_i is any sample point in $[x_{i-1}, x_i]$, then $M_i \approx \rho(c_i)\Delta x$ and

$$\text{total mass } M = \sum_{i=1}^{N} M_i \approx \sum_{i=1}^{N} \rho(c_i)\Delta x$$

As $N \to \infty$, the accuracy of the approximation improves. However, the sum on the right is a Riemann sum whose value approaches $\int_a^b \rho(x)\,dx$, and thus, it makes sense to define *the total mass of a rod as the integral of its linear mass density*:

$$\boxed{\text{Total mass } M = \int_a^b \rho(x)\,dx} \qquad \boxed{3}$$

Note the similarity in the way we use thin slices to compute volume and small pieces to compute total mass.

EXAMPLE 4 **Total Mass** Find the total mass M of a 2-m rod of density $\rho(x) = 1 + 2x - x^2$ kg/m, where x is the distance from one end of the rod.

Solution

$$M = \int_0^2 \rho(x)\,dx = \int_0^2 \left(1 + 2x - x^2\right)dx = \left(x + x^2 - \frac{1}{3}x^3\right)\Bigg|_0^2 = \frac{10}{3} \text{ kg} \qquad \blacksquare$$

In general, population density is a function $\rho(x, y)$ that depends not just on the distance to the origin (city center) but also on the coordinates (x, y) (the specific location). Total population is then computed using double integration, a topic in multivariable calculus.

In some situations, density is a function of distance to the origin. For example, in the study of urban populations, it might be assumed that the population density $\rho(r)$ (in people per square kilometer) depends only on the distance r from the center of a city. Such a density function is called a **radial density function**.

We now derive a formula for the total population P within a radius R of the city center, assuming a radial density $\rho(r)$. First, divide the circle of radius R into N thin rings of equal width $\Delta r = R/N$ as in Figure 8.

Let P_i be the population within the ith ring, so that the total population is given by $P = \sum_{i=1}^{N} P_i$. If the outer radius of the ith ring is r_i, then the circumference is $2\pi r_i$, and if Δr is small, the area of this ring is *approximately* $2\pi r_i \Delta r$ (outer circumference times width). Furthermore, the population density within the thin ring is nearly constant with value $\rho(r_i)$. With these approximations,

$$P_i \approx \underbrace{2\pi r_i \Delta r}_{\text{Area of ring}} \times \underbrace{\rho(r_i)}_{\substack{\text{Population} \\ \text{density}}} = 2\pi r_i \rho(r_i)\Delta r$$

Adding up the P_i, we obtain

$$P = \sum_{i=1}^{N} P_i \approx 2\pi \sum_{i=1}^{N} r_i \rho(r_i)\Delta r$$

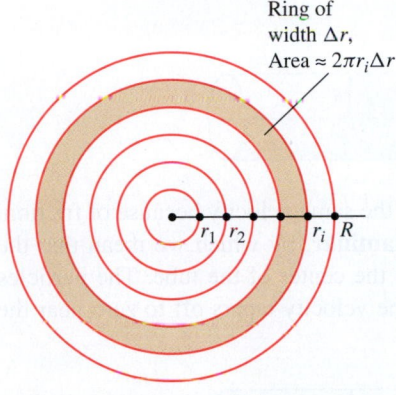

Ring of width Δr, Area $\approx 2\pi r_i \Delta r$

FIGURE 8 Dividing the circle of radius R into N thin rings of width $\Delta r = R/N$.

This last sum is a right-endpoint approximation to the integral $2\pi \int_0^R r\rho(r)\,dr$. As N tends to ∞, the approximation improves in accuracy and the sum converges to the integral. Thus, for a population with a radial density function ρ,

Remember that for a radial density function, the total population is obtained by integrating $2\pi r\rho(r)$ rather than $\rho(r)$.

$$\boxed{\text{Population } P \text{ within a radius } R = 2\pi \int_0^R r\rho(r)\,dr} \qquad \boxed{4}$$

EXAMPLE 5 **Computing Total Population** The population in the city of Isaactonia and its surrounding suburbs has radial density function $\rho(r) = 15(1 + r^2)^{-1/2}$, where r is the distance from the city center in kilometers and ρ has units of thousands per square kilometer. How many people live in the ring between 10 and 30 km from the city center?

Solution The population P (in thousands) within the ring is

$$P = 2\pi \int_{10}^{30} r\left(15(1 + r^2)^{-1/2}\right) dr = 2\pi(15) \int_{10}^{30} \frac{r}{(1 + r^2)^{1/2}}\,dr$$

Now use the substitution $u = 1 + r^2$, $du = 2r\,dr$. The limits of integration become $u(10) = 101$ and $u(30) = 901$:

$$P = 30\pi \int_{101}^{901} u^{-1/2}\left(\frac{1}{2}\right) du = 30\pi u^{1/2}\Big|_{101}^{901} \approx 1882 \text{ thousand}$$

In other words, the population in the ring is approximately 1.9 million people. ∎

Flow Rate

When fluid flows through a tube, the **flow rate** Q is the *volume per unit time* of fluid passing through the tube (Figure 9). The flow rate depends on the velocity of the fluid particles. If all particles of the fluid travel with the same velocity v (say, in units of cubic meters per minute), and the tube has radius R, then

$$\underbrace{\text{flow rate } Q}_{\text{Volume per unit time}} = \text{cross-sectional area} \times \text{velocity} = \pi R^2 v \text{ cm}^3/\text{min}$$

Why is this formula true? Let's fix an observation point P in the tube and ask: Which fluid particles flow past P in a 1-min interval? A particle travels v centimeters each minute, so it flows past P during this minute if it is located not more than v centimeters to the left of P (assuming the fluid flows from left to right). Therefore, the column of fluid flowing past P in a 1-min interval is a cylinder of radius R, length v, and volume $\pi R^2 v$ (Figure 9).

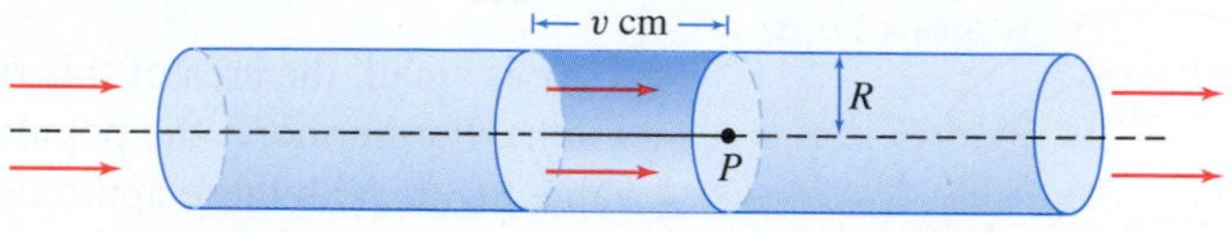

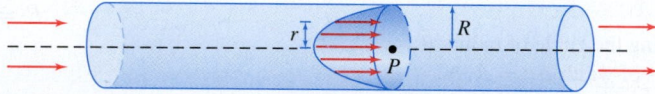

FIGURE 9 The column of fluid flowing past P in 1 unit of time is a cylinder of volume $\pi R^2 v$.

In reality, the fluid particles do not all travel at the same velocity because of friction. However, for a slowly moving fluid, the flow is **laminar**, by which we mean that the velocity $v(r)$ depends only on the distance r from the center of the tube. The particles at the center of the tube travel most quickly, and the velocity tapers off to zero near the walls of the tube (Figure 10).

FIGURE 10 Laminar flow: Velocity of fluid increases toward the center of the tube.

If the flow is laminar, we can express the flow rate Q as an integral. We divide the circular cross section of the tube into N thin concentric rings of width $\Delta r = R/N$ (Figure 11). The area of the ith ring is approximately $2\pi r_i \Delta r$ and the fluid particles flowing past this ring have a velocity that is nearly constant with value $v(r_i)$. Therefore, we can approximate the flow rate Q_i through the ith ring by

$$Q_i \approx \text{cross-sectional area} \times \text{velocity} \approx (2\pi r_i \Delta r)v(r_i)$$

We obtain

$$Q = \sum_{i=1}^{N} Q_i \approx 2\pi \sum_{i=1}^{N} r_i v(r_i)\Delta r$$

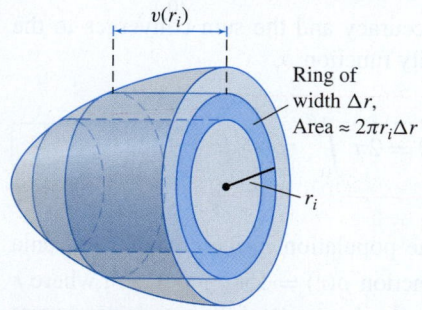

FIGURE 11 In a laminar flow, the fluid particles passing through a thin ring at distance r_i from the center all travel at nearly the same velocity $v(r_i)$.

The sum on the right is a right-endpoint approximation to the integral $2\pi \int_0^R r v(r)\,dr$. Once again, we let N tend to ∞ to obtain the formula for laminar flow with velocity $v(r)$,

$$\boxed{\text{Flow rate } Q = 2\pi \int_0^R r v(r)\,dr} \qquad \boxed{5}$$

Note the similarity of this formula and its derivation to that of population with a radial density function.

The French physician Jean Poiseuille (1799–1869) discovered the law of laminar flow that cardiologists use to study blood flow in humans. Poiseuille's Law highlights the danger of cholesterol buildup in blood vessels: The flow through a blood vessel of radius R is proportional to R^4, so if R is reduced by one-half, the flow is reduced by a factor of 16.

EXAMPLE 6 **Laminar Flow** According to **Poiseuille's Law**, the velocity of blood flowing in a blood vessel of radius R centimeters is $v(r) = k(R^2 - r^2)$, where r is the distance from the center of the vessel (in centimeters) and k is a constant. Calculate the flow Q as function of R, assuming that $k = 0.5$ (cm-s)$^{-1}$.

Solution By Eq. (5),

$$Q = 2\pi \int_0^R (0.5)r(R^2 - r^2)\,dr = \pi \left(R^2 \frac{r^2}{2} - \frac{r^4}{4} \right)\Bigg|_0^R = \frac{\pi}{4} R^4 \text{ cm}^3/\text{s}$$

Note that Q is proportional to R^4 (this is true for any value of k). ∎

CONCEPTUAL INSIGHT In this section, we saw a number of examples of Riemann sums and the definite integral at work. In each case there was a quantity that we were interested in computing over a whole domain. We cut the domain into small pieces over which the quantity was straightforward to compute. Adding those results yielded a Riemann sum approximation of the desired whole-domain quantity, and then passing to the limit resulted in a definite integral formula for determining the exact value.

In the remainder of this text we will see many more examples where this approach is carried out and Riemann sums and the definite integral are employed to compute a desired quantity.

Average Value

Next, we discuss the *average value* of a function. Recall that the average of N numbers $a_1, a_2, \ldots, a_N$ is the sum divided by N:

$$\frac{a_1 + a_2 + \cdots + a_N}{N} = \frac{1}{N} \sum_{j=1}^{N} a_j$$

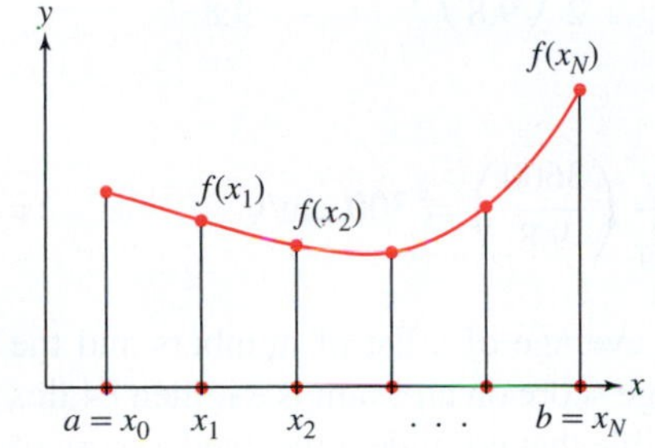

We cannot define the average value of a function f on an interval $[a, b]$ as a sum because there are infinitely many values of x to consider. But recall the formula for the right-endpoint approximation R_N (Figure 12):

$$R_N = \frac{b-a}{N} \left(f(x_1) + f(x_2) + \cdots + f(x_N) \right)$$

where $x_1, \ldots, x_N$ are the right endpoints of the subintervals. We see that R_N divided by $(b-a)$ is equal to the average of the equally spaced function values $f(x_i)$:

$$\frac{1}{b-a} R_N = \underbrace{\frac{f(x_1) + f(x_2) + \cdots + f(x_N)}{N}}_{\text{Average of the function values}}$$

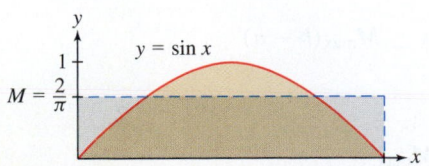

FIGURE 12 The average of the values of $f(x)$ at the points $x_1, x_2, \ldots, x_N$ is equal to $\dfrac{R_N}{b-a}$.

If N is large, it is reasonable to think of this quantity as an *approximation* to the average of $f(x)$ on $[a, b]$. Therefore, we define the average value itself as the limit:

$$\text{average value} = \lim_{N \to \infty} \frac{1}{b-a} R_N(f) = \frac{1}{b-a} \int_a^b f(x)\, dx$$

DEFINITION **Average Value** The **average value** of an integrable function f on $[a, b]$ is the quantity

$$\boxed{\text{Average value} = \frac{1}{b-a} \int_a^b f(x)\, dx}$$

6

The average value of a function is also called the **mean value**.

GRAPHICAL INSIGHT The average value M of a nonnegative function is the average height of its graph (Figure 13). The region under the graph has the same area as the rectangle of height M, because $\displaystyle\int_a^b f(x)\, dx = M(b-a)$.

FIGURE 13 The area under the graph is equal to the area of the rectangle whose height is the average value M.

EXAMPLE 7 Find the average value of $f(x) = \sin x$ on $[0, \pi]$.

Solution The average value of $f(x) = \sin x$ on $[0, \pi]$ is

$$\frac{1}{\pi} \int_0^\pi \sin x\, dx = -\frac{1}{\pi} \cos x \Big|_0^\pi = \frac{1}{\pi} \left(-(-1) - (-1) \right) = \frac{2}{\pi} \approx 0.637$$

This answer is reasonable because $\sin x$ varies from 0 to 1 on the interval $[0, \pi]$ and the average 0.637 lies somewhere between the two extremes (Figure 13). ■

EXAMPLE 8   Vertical Jump of a Bushbaby   The bushbaby (*Galago senegalensis*) is a small primate with remarkable jumping ability. Find the average speed during a jump if the initial vertical velocity is $v_0 = 600$ cm/s. Use Galileo's formula for the height $h(t) = v_0 t - \frac{1}{2}gt^2$ (in centimeters, where $g = 980$ cm/s^2).

Solution   The bushbaby's height is $h(t) = v_0 t - \frac{1}{2}gt^2 = t\left(v_0 - \frac{1}{2}gt\right)$. The height is zero at $t = 0$ and at $t = 2v_0/g = \frac{1200}{980} = \frac{6}{4.9}$ seconds, when the jump ends.

The bushbaby's velocity is $h'(t) = v_0 - gt = 600 - 980t$. The velocity is negative for $t > v_0/g = \frac{6}{9.8}$, so as we see in Figure 14, the integral of speed $|h'(t)|$ is equal to the sum of the areas of two triangles of base $\frac{6}{9.8}$ and height 600:

$$\int_0^{6/4.9} |600 - 980t|\, dt = \frac{1}{2}\left(\frac{6}{9.8}\right)(600) + \frac{1}{2}\left(\frac{6}{9.8}\right)(600) = \frac{3600}{9.8}$$

The average speed $\bar{s}$ is

$$\bar{s} = \frac{1}{\frac{6}{4.9}} \int_0^{6/4.9} |600 - 980t|\, dt = \frac{1}{\frac{6}{4.9}}\left(\frac{3600}{9.8}\right) = 300 \text{ cm/s}$$ ■

There is an important difference between the average of a list of numbers and the average value of a continuous function. If the average score on an exam is 84, then 84 lies between the highest and lowest scores, but it is possible that no student received a score of 84. By contrast, the Mean Value Theorem (MVT) for Integrals asserts that a continuous function always takes on its average value somewhere in the interval (Figure 15).

> **THEOREM 1   Mean Value Theorem for Integrals**   If f is continuous on $[a, b]$, then there exists a value $c \in [a, b]$ such that
>
> $$f(c) = \frac{1}{b-a}\int_a^b f(x)\, dx$$

For example, the average of $f(x) = \sin x$ on $[0, \pi]$ is $2/\pi$ by Example 7. We have $f(c) = 2/\pi$ for $c = \sin^{-1}(2/\pi) \approx 0.69$. Since 0.69 lies in $[0, \pi]$, $f(x) = \sin x$ indeed takes on its average value at a point in the interval.

Proof   Let $M = \dfrac{1}{b-a}\displaystyle\int_a^b f(x)\, dx$ be the average value. Because f is continuous, we can apply Theorem 1 of Section 4.2 to conclude that f takes on a minimum value m_{min} and a maximum value M_{max} on the closed interval $[a, b]$. Furthermore, by Eq. (8) of Section 5.2,

$$m_{min}(b - a) \le \int_a^b f(x)\, dx \le M_{max}(b - a)$$

Dividing by $(b - a)$, we find

$$m_{min} \le M \le M_{max}$$

In other words, the average value M lies between m_{min} and M_{max}. The Intermediate Value Theorem guarantees that $f(x)$ takes on every value between its min and max, so $f(c) = M$ for some c in $[a, b]$.   ■

A bushbaby can jump as high as 2 m (its center of mass rises more than five body lengths). By contrast, an Olympic high jumper rises a little more then one body length in a jump.

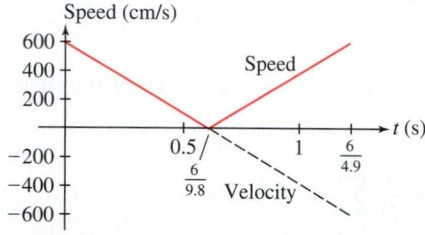

FIGURE 14 Graph of speed $|h'(t)| = |600 - 980t|$.

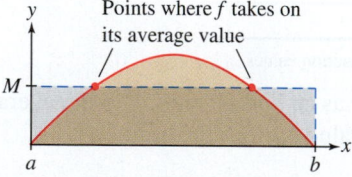

FIGURE 15 The function f takes on its average value M at the points where the upper edge of the rectangle intersects the graph.

6.2 SUMMARY

- Formulas:

Volume
$$V = \int_a^b A(y)\,dy, \qquad A(y) = \text{cross-sectional area}$$

Total Mass
$$M = \int_a^b \rho(x)\,dx, \qquad \rho(x) = \text{linear mass density}$$

Total Population
$$P = 2\pi \int_0^R r\rho(r)\,dr, \qquad \rho(r) = \text{radial density}$$

Laminar Flow Rate
$$Q = 2\pi \int_0^R rv(r)\,dr, \qquad v(r) = \text{velocity at radius } r$$

Average value
$$M = \frac{1}{b-a} \int_a^b f(x)\,dx, \qquad f = \text{any continuous function}$$

- The MVT for Integrals: If f is continuous on $[a,b]$ with average (or mean) value M, then $f(c) = M$ for some $c \in [a,b]$.

6.2 EXERCISES

Preliminary Questions

1. What is the average value of f on $[0,4]$ if the area between the graph of f and the x-axis is equal to 12?

2. Find the volume of a solid extending from $y = 2$ to $y = 5$ if every cross section has area $A(y) = 5$.

3. What is the definition of flow rate?

4. Which assumption about fluid velocity did we use to compute the flow rate as an integral?

5. The average value of f on $[1,4]$ is 5. Find $\int_1^4 f(x)\,dx$.

Exercises

1. Let V be the volume of a pyramid of height 20 whose base is a square of side 8.

(a) Use similar triangles as in Example 1 to find the area of the horizontal cross section at a height y.

(b) Calculate V by integrating the cross-sectional area.

2. Let V be the volume of a right circular cone of height 10 whose base is a circle of radius 4 [Figure 16(A)].

(a) Use similar triangles to find the area of a horizontal cross section at a height y.

(b) Calculate V by integrating the cross-sectional area.

3. Use the method of Exercise 2 to find the formula for the volume of a right circular cone of height h whose base is a circle of radius R [Figure 16(B)].

4. Calculate the volume of the ramp in Figure 17 in three ways by integrating the area of the cross sections:

(a) Perpendicular to the x-axis (rectangles)

(b) Perpendicular to the y-axis (triangles)

(c) Perpendicular to the z-axis (rectangles)

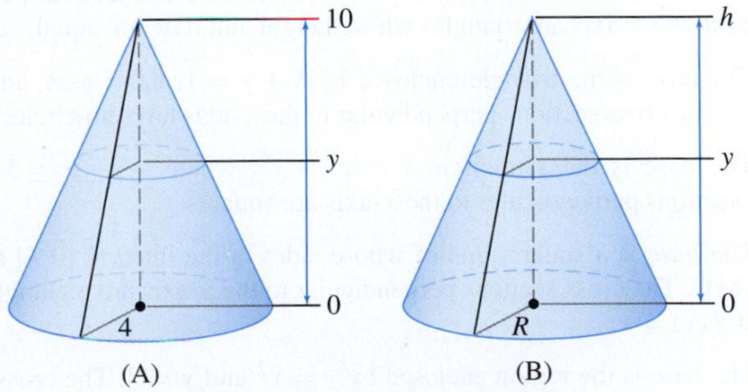

(A)   (B)

FIGURE 16 Right circular cones.

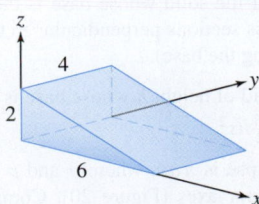

FIGURE 17 Ramp of length 6, width 4, and height 2.

5. Find the volume of liquid needed to fill a sphere of radius R to height h (Figure 18).

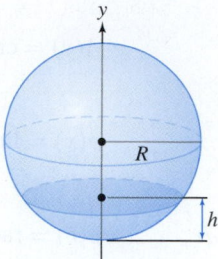

FIGURE 18 Sphere filled with liquid to height h.

6. Find the volume of the wedge in Figure 19(A) by integrating the area of vertical cross sections.

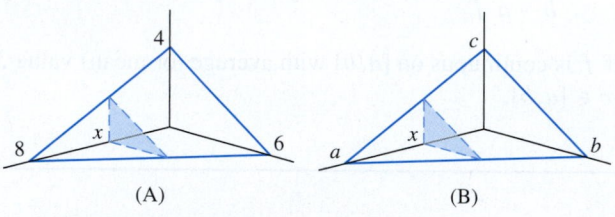

(A)   (B)

FIGURE 19

7. Derive a formula for the volume of the wedge in Figure 19(B) in terms of the constants a, b, and c.

8. Let B be the solid whose base is the unit circle $x^2 + y^2 = 1$ and whose vertical cross sections perpendicular to the x-axis are equilateral triangles. Show that the vertical cross sections have area $A(x) = \sqrt{3}(1 - x^2)$ and compute the volume of B.

In Exercises 9–14, find the volume of the solid with the given base and cross sections.

9. The base is the unit circle $x^2 + y^2 = 1$, and the cross sections perpendicular to the x-axis are triangles whose height and base are equal.

10. The base is the triangle enclosed by $x + y = 1$, the x-axis, and the y-axis. The cross sections perpendicular to the y-axis are semicircles.

11. The base is the semicircle $y = \sqrt{9 - x^2}$, where $-3 \le x \le 3$. The cross sections perpendicular to the x-axis are squares.

12. The base is a square, one of whose sides is the interval $[0, \ell]$ along the x-axis. The cross sections perpendicular to the x-axis are rectangles of height $f(x) = x^2$.

13. The base is the region enclosed by $y = x^2$ and $y = 3$. The cross sections perpendicular to the y-axis are squares.

14. The base is the region enclosed by $y = x^2$ and $y = 3$. The cross sections perpendicular to the y-axis are rectangles of height y^3.

15. Find the volume of the solid whose base is the region $|x| + |y| \le 1$ and whose vertical cross sections perpendicular to the y-axis are semicircles (with diameter along the base).

16. Show that a pyramid of height h whose base is an equilateral triangle of side s has volume $\frac{\sqrt{3}}{12}hs^2$.

17. The area of an ellipse is πab, where a and b are the lengths of the semimajor and semiminor axes (Figure 20). Compute the volume of a cone of height 12 whose base is an ellipse with semimajor axis $a = 6$ and semiminor axis $b = 4$.

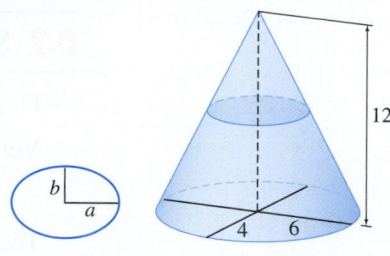

FIGURE 20

18. Find the volume V of a *regular* tetrahedron (Figure 21) whose face is an equilateral triangle of side s. The tetrahedron has height $h = \sqrt{2/3}s$.

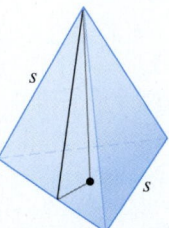

FIGURE 21   Regular tetrahedron.

19. A frustum of a pyramid is a pyramid with its top cut off [Figure 22(A)]. Let V be the volume of a frustum of height h whose base is a square of side a and whose top is a square of side b with $a > b \ge 0$.

(a) Show that if the frustum were continued to a full pyramid, it would have height $ha/(a - b)$ [Figure 22(B)].

(b) Show that the cross section at height x is a square whose side length is given by $s(x) = (1/h)(a(h - x) + bx)$.

(c) Show that $V = \frac{1}{3}h(a^2 + ab + b^2)$. A papyrus dating to the year 1850 BCE indicates that Egyptian mathematicians had discovered this formula almost 4000 years ago.

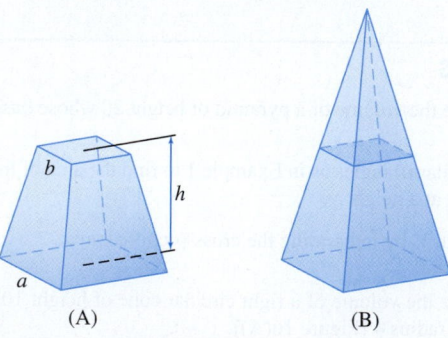

(A)   (B)

FIGURE 22

20. A plane inclined at an angle of $45°$ passes through a diameter of the base of a cylinder of radius r. Find the volume of the region within the cylinder and below the plane (Figure 23).

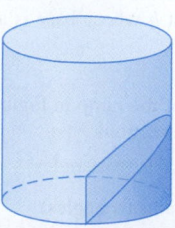

FIGURE 23

21. The solid S in Figure 24 is the intersection of two cylinders of radius r whose axes are perpendicular.

(a) The horizontal cross section of each cylinder at distance y from the central axis is a rectangular strip. Find the strip's width.

(b) Find the area of the horizontal cross section of S at distance y.

(c) Find the volume of S as a function of r.

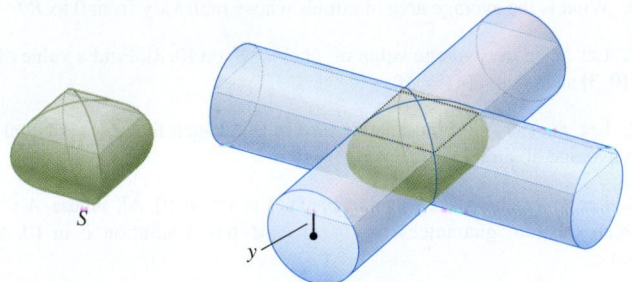

FIGURE 24 Two cylinders intersecting at right angles.

22. Let S be the intersection of two cylinders of radius r whose axes intersect at an angle θ. Find the volume of S as a function of r and θ.

23. Calculate the volume of a cylinder inclined at an angle $\theta = 30°$ with height 10 and base of radius 4 (Figure 25).

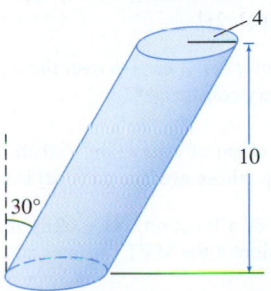

FIGURE 25 Cylinder inclined at an angle $\theta = 30°$.

24. The areas of cross sections of Lake Nogebow at 5-m intervals are given in the table below. Figure 26 shows a contour map of the lake. Estimate the volume V of the lake by taking the average of the right- and left-endpoint approximations to the integral of cross-sectional area.

Depth (m)	0	5	10	15	20
Area (million m²)	2.1	1.5	1.1	0.835	0.217

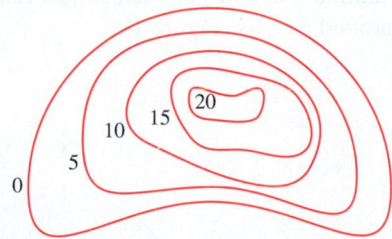

FIGURE 26 Depth contour map of Lake Nogebow.

25. Find the total mass of a 2-m rod whose linear density function is $\rho(x) = 12(x + 4)^{-1}$ kg/m for $0 \le x \le 2$.

26. Find the total mass of a 3-m rod whose linear density function is $\rho(x) = 3 + \cos(\pi x)$ kg/m for $0 \le x \le 3$.

27. A mineral deposit along a strip of length 6 cm has density $s(x) = 0.01x(6 - x)$ g/cm for $0 \le x \le 6$. Calculate the total mass of the deposit.

28. Charge is distributed along a glass tube of length 10 cm with linear charge density $\rho(x) = x(x^2 + 1)^{-2} \times 10^{-4}$ coulombs per centimeter (C/cm) for $0 \le x \le 10$. Calculate the total charge.

29. Calculate the population within a 10-mile radius of the city center if the radial population density is $\rho(r) = 4(1 + r^2)^{1/3}$ (in thousands per square mile).

30. Odzala National Park in the Republic of the Congo has a high density of gorillas. Suppose that the population density is given by the radial density function $\rho(r) = 52(1 + r^2)^{-2}$ gorillas/km², where r is the distance from a grassy clearing with a source of water. Calculate the number of gorillas within a 5-km radius of the clearing.

31. Table 1 lists the population density (in people per square kilometer) as a function of distance r (in kilometers) from the center of a rural town. Estimate the total population within a 1.2-km radius of the center by taking the average of the left- and right-endpoint approximations.

	TABLE 1	**Population Density**	
r	$\rho(r)$	r	$\rho(r)$
0.0	125.0	0.8	56.2
0.2	102.3	1.0	46.0
0.4	83.8	1.2	37.6
0.6	68.6		

32. Find the total mass of a circular plate of radius 20 cm whose mass density is the radial function $\rho(r) = 0.03 + 0.01 \cos(\pi r^2)$ g/cm².

33. Assume that the density of deer in a forest is given by the radial function $\rho(r) = 150(r^2 + 2)^{-2}$ deer per square kilometer, where r is the distance (in kilometers) to a small meadow. Calculate the number of deer in the region $2 \le r \le 5$ km.

34. Show that a circular plate of radius 2 cm with radial mass density $\rho(r) = \frac{4}{r}$ g/cm² has finite total mass, even though the density becomes infinite at the origin.

35. Find the flow rate through a tube of radius 4 cm, assuming that the velocity of fluid particles at a distance r centimeters from the center is $v(r) = (16 - r^2)$ cm/s.

36. The velocity of fluid particles flowing through a tube of radius 5 cm is $v(r) = (10 - 0.3r - 0.34r^2)$ cm/s, where r centimeters is the distance from the center. What quantity per second of fluid flows through the portion of the tube where $0 \le r \le 2$?

37. A solid rod of radius 1 cm is placed in a pipe of radius 3 cm so that their axes are aligned. Water flows through the pipe and around the rod. Find the flow rate if the velocity of the water is given by the radial function $v(r) = 0.5(r - 1)(3 - r)$ cm/s.

38. Let $v(r)$ be the velocity of blood in an arterial capillary of radius $R = 4 \times 10^{-5}$ m. Use Poiseuille's Law (Example 6) with $k = 10^6$ (m-s)$^{-1}$ to determine the velocity at the center of the capillary and the flow rate (use correct units).

In Exercises 39–50, calculate the average over the given interval.

39. $f(x) = x^3, \quad [0, 4]$

40. $f(x) = x^3, \quad [-1, 1]$

41. $f(x) = \cos x, \quad \left[0, \frac{\pi}{6}\right]$

42. $f(x) = \sec^2 x, \quad \left[\frac{\pi}{6}, \frac{\pi}{3}\right]$

43. $f(s) = s^{-2}, \quad [2, 5]$

44. $f(x) = \dfrac{\sin(\pi/x)}{x^2}, \quad [1, 2]$

45. $f(x) = 2x^3 - 6x^2, \quad [-1, 3]$

46. $f(x) = \dfrac{1}{x^2 + 1}, \quad [-1, 1]$

47. $f(x) = n \sin nx, \quad \left[0, \frac{\pi}{n}\right]$

48. $f(x) = \cosh nx$ for $n \neq 0$, $[-1, 1]$

49. $f(x) = x^n$ for $n \neq -1$, $[0, a]$

50. $f(x) = e^{nx}$ for $n \neq 0$, $[0, a]$

51. The temperature (in degrees Celsius) at time t (in hours) in an art museum varies according to $T(t) = 20 + 5\cos\left(\frac{\pi}{12}t\right)$. Find the average over the time periods $[0, 24]$ and $[2, 6]$.

52. A steel bar of length 3 m experiences extreme heat at its center, so that the temperature at coordinate x on the bar is given by $T(x) = 40\sin\left(\frac{\pi x}{3}\right) + 50°\text{C}$ where the bar sits along the interval $[0, 3]$ on the x-axis. Determine the average temperature of the bar.

53. The temperature in the town of Walla Walla during the month of July follows a pattern given by $T(t) = 10\sin\left(\frac{t\pi}{31}\right) + 14\sin\left(\frac{t\pi}{2}\right) + 73°\text{F}$. Here, t is measured in days, and there are 31 days in July. Explain why you might see a pattern like this and compute the average temperature during the month of July.

54. The door to the garage is left open, and over the next 4 hours the temperature in a house in degrees Celsius is given by $T(t) = 20e^{-t/4}$. Determine the average temperature over those 4 h.

55. A 10-cm copper wire with one end in an ice bath is heated at the other end, so that the temperature at each point x along the wire (in degrees Celsius) is given by $T(x) = 50\cos\frac{\pi x}{20}$. Find the average temperature over the wire.

56. A ball thrown in the air vertically from ground level with initial velocity 18 m/s has height $h(t) = 18t - 9.8t^2$ at time t (in seconds). Find the average height and the average speed over the time interval extending from the ball's release to its return to ground level.

57. Find the average speed over the time interval $[1, 5]$ (time in seconds) of a particle whose position at time t is $s(t) = t^3 - 6t^2$ m.

58. An object with zero initial velocity accelerates at a constant rate of 10 m/s². Find its average velocity during the first 15 seconds.

59. The acceleration of a particle is $a(t) = 60t - 4t^3$ m/s². Compute the average acceleration and the average speed over the time interval $[2, 6]$, assuming that the particle's initial velocity is zero.

60. What is the average area of circles whose radii vary from 0 to R?

61. Let M be the average value of $f(x) = x^4$ on $[0, 3]$. Find a value of c in $[0, 3]$ such that $f(c) = M$.

62. Let $f(x) = \sqrt{x}$. Find a value of c in $[4, 9]$ such that $f(c)$ is equal to the average of f on $[4, 9]$.

63. Let M be the average value of $f(x) = x^3$ on $[0, A]$, where $A > 0$. Which theorem guarantees that $f(c) = M$ has a solution c in $[0, A]$? Find c.

64. [CAS] Let $f(x) = 2\sin x - x$. Use a computer algebra system to plot f and estimate:

(a) The positive root α of f

(b) The average value M of f on $[0, \alpha]$

(c) A value $c \in [0, \alpha]$ such that $f(c) = M$

65. Which of $f(x) = x\sin^2 x$ and $g(x) = x^2\sin^2 x$ has a larger average value over $[0, 1]$? Over $[1, 2]$?

66. Find the average of $f(x) = ax + b$ over the interval $[-M, M]$, where a, b, and M are arbitrary constants.

67. Sketch the graph of a function f such that $f(x) \geq 0$ on $[0, 1]$ and $f(x) \leq 0$ on $[1, 2]$, whose average on $[0, 2]$ is negative.

68. Give an example of a function (necessarily discontinuous) that does not satisfy the conclusion of the MVT for Integrals.

Further Insights and Challenges

69. An object is tossed into the air vertically from ground level with initial velocity v_0 ft/s at time $t = 0$. Find the average speed of the object over the time interval $[0, T]$, where T is the time the object returns to Earth.

70. Review the MVT stated in Section 4.3 (Theorem 1) and show how it can be used, together with the Fundamental Theorem of Calculus, to prove the MVT for Integrals.

6.3 Volumes of Revolution: Disks and Washers

We use the terms "revolve" and "rotate" interchangeably.

A **solid of revolution** is a solid obtained by rotating a region in the plane about an axis. The sphere and right circular cone are familiar examples of such solids. Each of these is "swept out" as a plane region revolves around an axis (Figure 1).

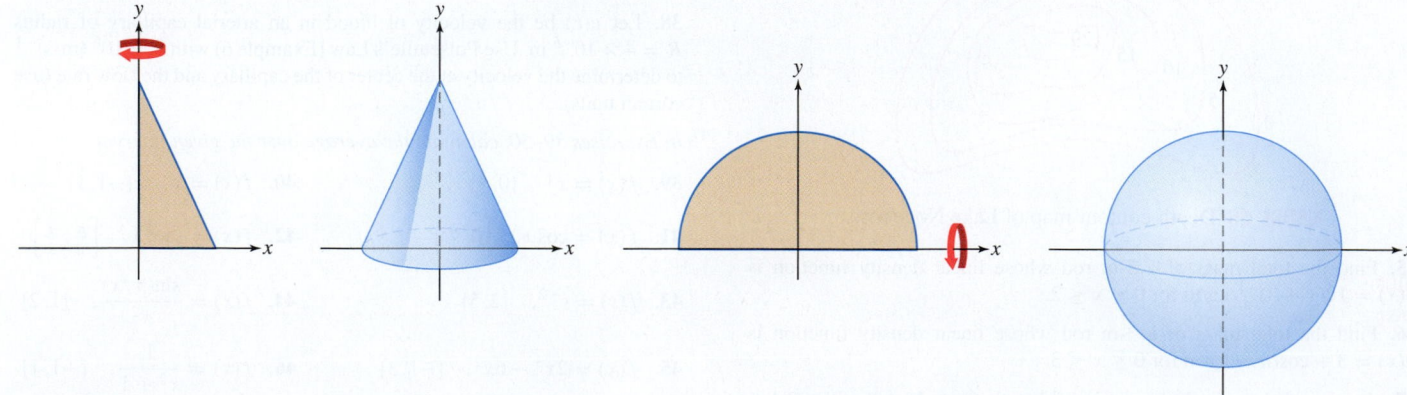

DF **FIGURE 1** The right circular cone and the sphere are solids of revolution.

This method for computing the volume is referred to as the disk method because the vertical slices of the solid are circular disks.

Suppose that $f(x) \geq 0$ for $a \leq x \leq b$. The solid obtained by rotating the region under the graph about the x-axis has a special feature: All vertical cross sections are circles (Figure 2). In fact, the vertical cross section at location x is a circle of radius $R = f(x)$ and thus,

$$\text{area of the vertical cross section} = \pi R^2 = \pi f(x)^2$$

We know from Section 6.2 that the total volume V is equal to the integral of cross-sectional area. Therefore, $V = \pi \int_a^b R^2 \, dx = \int_a^b \pi f(x)^2 \, dx$.

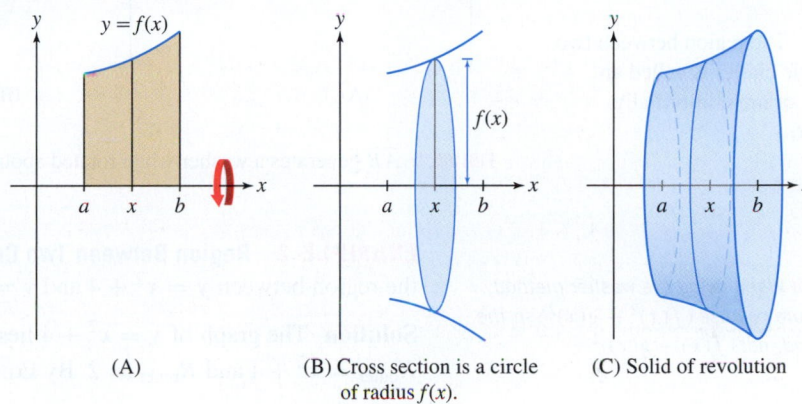

(A)   (B) Cross section is a circle   (C) Solid of revolution
                  of radius $f(x)$.

FIGURE 2

> **Volume of Revolution: Disk Method**   If f is continuous and $f(x) \geq 0$ on $[a, b]$, then the solid obtained by rotating the region under the graph about the x-axis has volume
>
> $$V = \pi \int_a^b R^2 \, dx = \pi \int_a^b f(x)^2 \, dx \qquad \boxed{1}$$

EXAMPLE 1   Calculate the volume V of the solid obtained by rotating the region under $y = x^2$ about the x-axis for $0 \leq x \leq 2$.

Solution   The solid is shown in Figure 3. By Eq. (1) with $f(x) = x^2$, its volume is

$$V = \pi \int_0^2 R^2 \, dx = \pi \int_0^2 (x^2)^2 \, dx = \pi \int_0^2 x^4 \, dx = \pi \frac{x^5}{5}\Big|_0^2 = \pi \frac{2^5}{5} = \frac{32}{5}\pi \quad \blacksquare$$

There are some useful variations on the formula for a volume of revolution. First, consider the region *between* two curves $y = f(x)$ and $y = g(x)$, where $f(x) \geq g(x) \geq 0$ as in Figure 5(A). When this region is rotated about the x-axis, segment $\overline{AB}$ sweeps out the **washer** shown in Figure 5(B). The inner and outer radii of this washer (also called an annulus; see Figure 4) are

$$R_{\text{outer}} = f(x), \qquad R_{\text{inner}} = g(x)$$

The washer has area $\pi R_{\text{outer}}^2 - \pi R_{\text{inner}}^2$ or $\pi(f(x)^2 - g(x)^2)$, and the volume of the solid of revolution [Figure 5(C)] is the integral of this cross-sectional area:

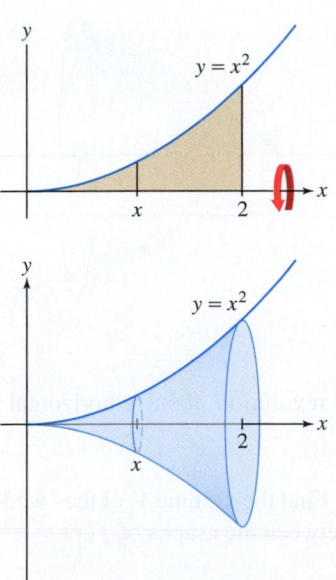

FIGURE 3 Region under $y = x^2$ rotated about the x-axis.

$$V = \pi \int_a^b \left(R_{\text{outer}}^2 - R_{\text{inner}}^2 \right) dx = \pi \int_a^b \left(f(x)^2 - g(x)^2 \right) dx \qquad \boxed{2}$$

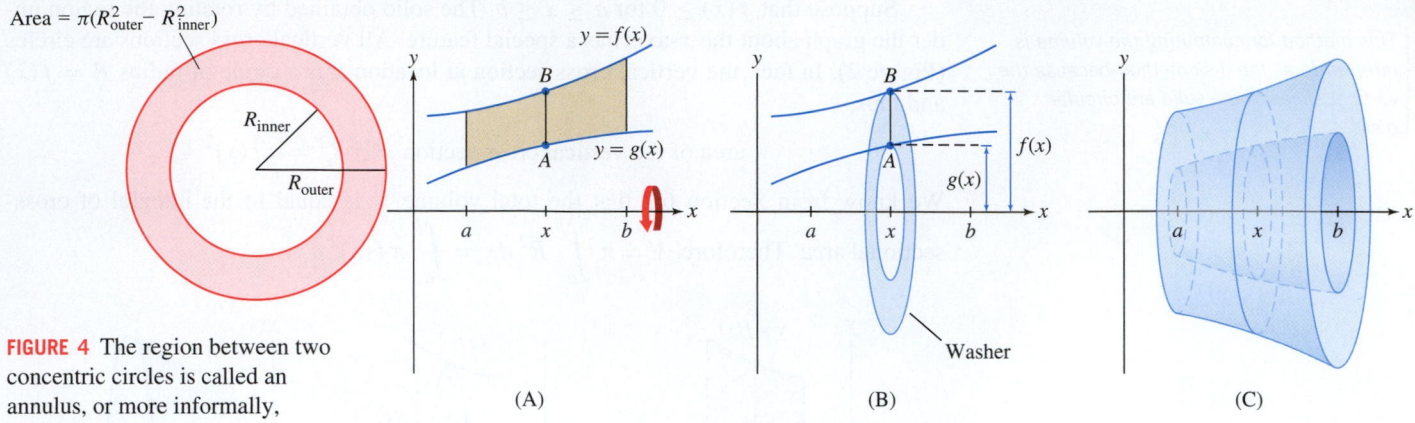

Area $= \pi(R_{outer}^2 - R_{inner}^2)$

FIGURE 4 The region between two concentric circles is called an annulus, or more informally, a washer.

R_{inner}

R_{outer}

$y = f(x)$

B

A

$y = g(x)$

a x b

(A)

B

A

$f(x)$

$g(x)$

a x b

Washer

(B)

a x b

(C)

FIGURE 5 $\overline{AB}$ generates a washer when rotated about the x-axis.

CAUTION *When using the washer method, make sure you use $(f(x)^2 - g(x)^2)$ in the integrand, not $(f(x) - g(x))^2$.*

EXAMPLE 2 **Region Between Two Curves** Find the volume V obtained by revolving the region between $y = x^2 + 4$ and $y = 2$ about the x-axis for $1 \le x \le 3$.

Solution The graph of $y = x^2 + 4$ lies above the graph of $y = 2$ (Figure 6). Therefore, $R_{outer} = x^2 + 4$ and $R_{inner} = 2$. By Eq. (2),

$$V = \pi \int_1^3 \left(R_{outer}^2 - R_{inner}^2\right) dx = \pi \int_1^3 \left((x^2 + 4)^2 - 2^2\right) dx$$

$$= \pi \int_1^3 \left(x^4 + 8x^2 + 12\right) dx = \pi \left(\frac{1}{5}x^5 + \frac{8}{3}x^3 + 12x\right)\Bigg|_1^3 = \frac{2126}{15}\pi \quad \blacksquare$$

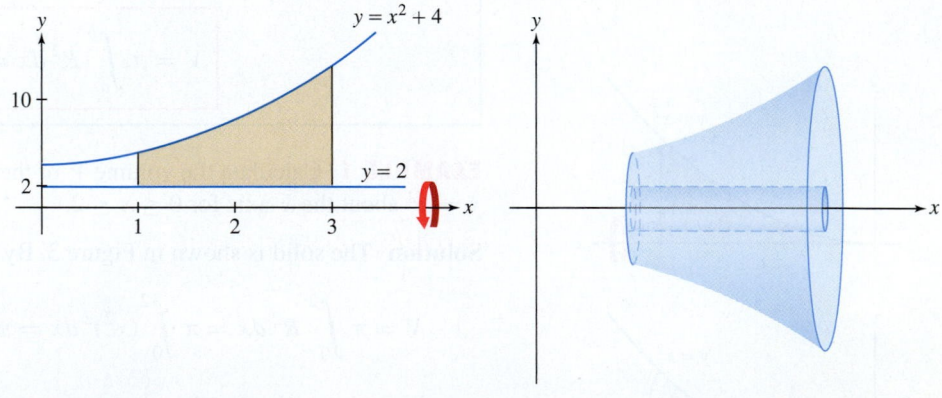

y

10

2

$y = x^2 + 4$

$y = 2$

1 2 3 x

y

x

FIGURE 6 The area between $y = x^2 + 4$ and $y = 2$ over $[1, 3]$ rotated about the x-axis.

In the next example, we calculate a volume of revolution about a horizontal axis parallel to the x-axis.

EXAMPLE 3 **Revolving About a Horizontal Axis** Find the volume V of the "wedding band" [Figure 7(C)] obtained by rotating the region between the graphs of $f(x) = x^2 + 2$ and $g(x) = 4 - x^2$ about the horizontal line $y = -3$.

When you set up the integral for a volume of revolution, visualize the cross sections. These cross sections are washers (or disks) whose inner and outer radii depend on the axis of rotation.

Solution First, let's find the points of intersection of the two graphs by solving

$$f(x) = g(x) \quad \Rightarrow \quad x^2 + 2 = 4 - x^2 \quad \Rightarrow \quad x^2 = 1 \quad \Rightarrow \quad x = \pm 1$$

Figure 7(A) shows that $g(x) \ge f(x)$ for $-1 \le x \le 1$.

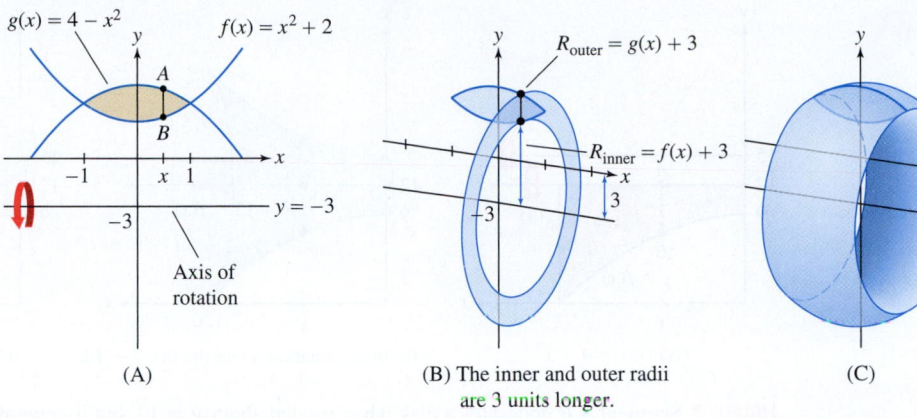

FIGURE 7

If we wanted to revolve about the x-axis, we would use Eq. (2). Since we want to revolve around $y = -3$, we must determine how the radii are affected. Figure 7(B) shows that when we rotate about $y = -3$, $\overline{AB}$ generates a washer whose outer and inner radii are both 3 units longer, and therefore we have

- $R_{\text{outer}} = g(x) - (-3) = (4 - x^2) + 3 = 7 - x^2$
- $R_{\text{inner}} = f(x) - (-3) = (x^2 + 2) + 3 = x^2 + 5$

The volume of revolution (about $y = -3$) is equal to the integral of the area of this washer:

> We get R_{outer} by subtracting $y = -3$ from $y = g(x)$ because vertical distance is the difference of the y-coordinates. Similarly, we subtract -3 from $f(x)$ to get R_{inner}.

$$V = \pi \int_{-1}^{1} \left(R_{\text{outer}}^2 - R_{\text{inner}}^2 \right) dx$$

$$= \pi \int_{-1}^{1} \left((7 - x^2)^2 - (x^2 + 5)^2 \right) dx$$

$$= \pi \int_{-1}^{1} \left((49 - 14x^2 + x^4) - (x^4 + 10x^2 + 25) \right) dx$$

$$= \pi \int_{-1}^{1} (24 - 24x^2) \, dx = \pi (24x - 8x^3) \Big|_{-1}^{1} = 32\pi$$ ■

EXAMPLE 4 Find the volume obtained by rotating the graphs of $f(x) = 9 - x^2$ and $y = 12$ for $0 \le x \le 3$ about

(a) The line $y = 12$ **(b)** The line $y = 15$

The resulting solids are illustrated in Figure 8.

Solution To set up the integrals, let's visualize the cross section. Is it a disk or a washer?

(a) Figure 8(B) shows that $\overline{AB}$ rotated about $y = 12$ generates a *disk* of radius

> In Figure 8, the length of $\overline{AB}$ is $12 - f(x)$ rather than $f(x) - 12$ because the line $y = 12$ lies above the graph of f.

$$R = \text{length of } \overline{AB} = 12 - f(x) = 12 - (9 - x^2) = 3 + x^2$$

The volume when we rotate about $y = 12$ is

$$V = \pi \int_{0}^{3} R^2 \, dx = \pi \int_{0}^{3} (3 + x^2)^2 \, dx = \pi \int_{0}^{3} (9 + 6x^2 + x^4) \, dx$$

$$= \pi \left(9x + 2x^3 + \frac{1}{5}x^5 \right) \Big|_{0}^{3} = \frac{648}{5}\pi$$

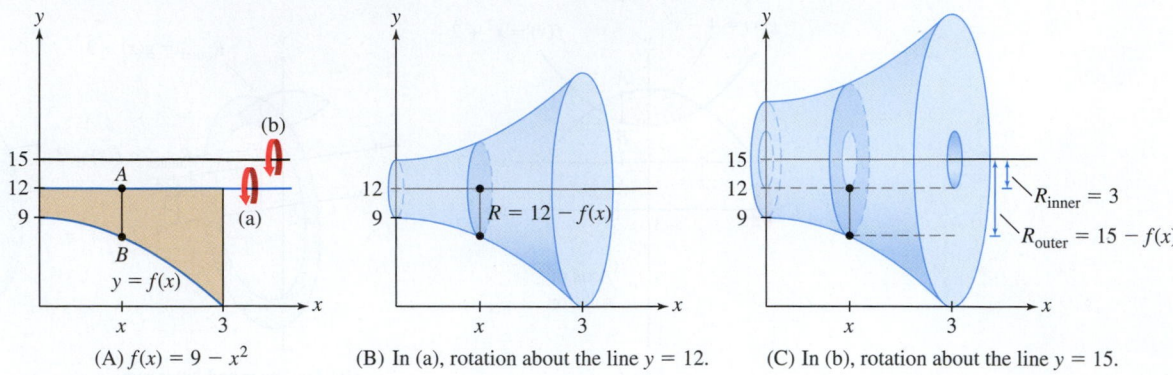

(A) $f(x) = 9 - x^2$

(B) In (a), rotation about the line $y = 12$.

(C) In (b), rotation about the line $y = 15$.

FIGURE 8 Segment $\overline{AB}$ generates a disk when rotated about $y = 12$, but it generates a washer when rotated about $y = 15$.

(b) Figure 8(C) shows that $\overline{AB}$ rotated about $y = 15$ generates a *washer*. The outer radius of this washer is the distance from B to the line $y = 15$:

$$R_{\text{outer}} = 15 - f(x) = 15 - (9 - x^2) = 6 + x^2$$

The inner radius is $R_{\text{inner}} = 3$, so the volume of revolution about $y = 15$ is

$$V = \pi \int_0^3 \left(R_{\text{outer}}^2 - R_{\text{inner}}^2 \right) dx = \pi \int_0^3 \left((6 + x^2)^2 - 3^2 \right) dx$$

$$= \pi \int_0^3 (27 + 12x^2 + x^4) \, dx$$

$$= \pi \left(27x + 4x^3 + \frac{1}{5}x^5 \right) \Big|_0^3 = \frac{1188}{5}\pi \qquad \blacksquare$$

We can use the disk and washer methods for solids of revolution about vertical axes, but it is necessary to describe the graph as a function of y—that is, $x = g(y)$.

EXAMPLE 5 **Revolving About a Vertical Axis** Find the volume of the solid obtained by rotating the region under the graph of $f(x) = 9 - x^2$ for $0 \le x \le 3$ about the vertical axis $x = -2$.

Solution Figure 9 shows that $\overline{AB}$ sweeps out a horizontal washer when rotated about the vertical line $x = -2$. We are going to integrate with respect to y, so we need the inner and outer radii of this washer as functions of y. Solving for x in $y = 9 - x^2$, we obtain $x^2 = 9 - y$, or $x = \pm\sqrt{9 - y}$. Since we are rotating the right half of the parabola, we choose the positive square root. Therefore,

$$R_{\text{outer}} = \sqrt{9 - y} + 2, \qquad R_{\text{inner}} = 2$$

$$R_{\text{outer}}^2 - R_{\text{inner}}^2 = \left(\sqrt{9 - y} + 2 \right)^2 - 2^2 = (9 - y) + 4\sqrt{9 - y} + 4 - 4$$

$$= 9 - y + 4\sqrt{9 - y}$$

The region extends from $y = 0$ to $y = 9$ along the y-axis, so

$$V = \pi \int_0^9 \left(R_{\text{outer}}^2 - R_{\text{inner}}^2 \right) dy = \pi \int_0^9 \left(9 - y + 4\sqrt{9 - y} \right) dy$$

$$= \pi \left(9y - \frac{1}{2}y^2 - \frac{8}{3}(9 - y)^{3/2} \right) \Big|_0^9 = \frac{225}{2}\pi \qquad \blacksquare$$

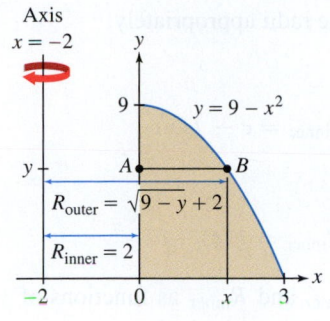

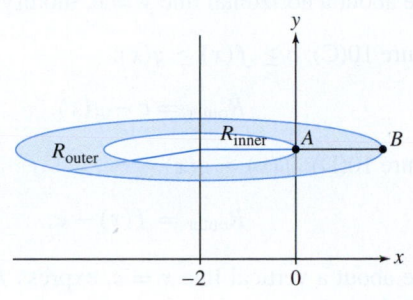

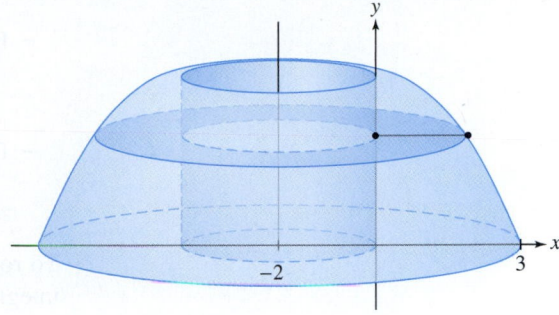

FIGURE 9

CONCEPTUAL INSIGHT A few different volume formulas were introduced in this section. Note that they all arise from the formula in the previous section for the volume as an integral of cross-sectional area $A(y)$:

$$V = \int_a^b A(y)\,dy$$

To compute the volumes in this section, we developed the cross-sectional area formulas for the different situations: disks or washers, and horizontal or vertical rotation axes. While they might appear to be distinct cases, we do not need to consider disks and washers separately. A disk is simply a washer with $R_{\text{inner}} = 0$.

6.3 SUMMARY

- Disk method: When you rotate the region between two graphs about an axis, the segments *perpendicular* to the axis generate disks or washers. The volume V of the solid of revolution is the integral of the areas of these disks or washers.
- Sketch the graphs to visualize the disks or washers.
- Figure 10(A): region between $y = f(x)$ and the x-axis, rotated about the x-axis.

    – Vertical cross section: a circle of radius $R = f(x)$ and area $\pi R^2 = \pi f(x)^2$:

$$V = \pi \int_a^b R^2\,dx = \pi \int_a^b f(x)^2\,dx$$

- Figure 10(B): region between $y = f(x)$ and $y = g(x)$, with $f(x) \geq g(x)$, rotated about the x-axis.

    – Vertical cross section: a washer of outer radius $R_{\text{outer}} = f(x)$ and inner radius $R_{\text{inner}} = g(x)$:

$$V = \pi \int_a^b \left(R_{\text{outer}}^2 - R_{\text{inner}}^2\right)dx = \pi \int_a^b \left(f(x)^2 - g(x)^2\right)dx$$

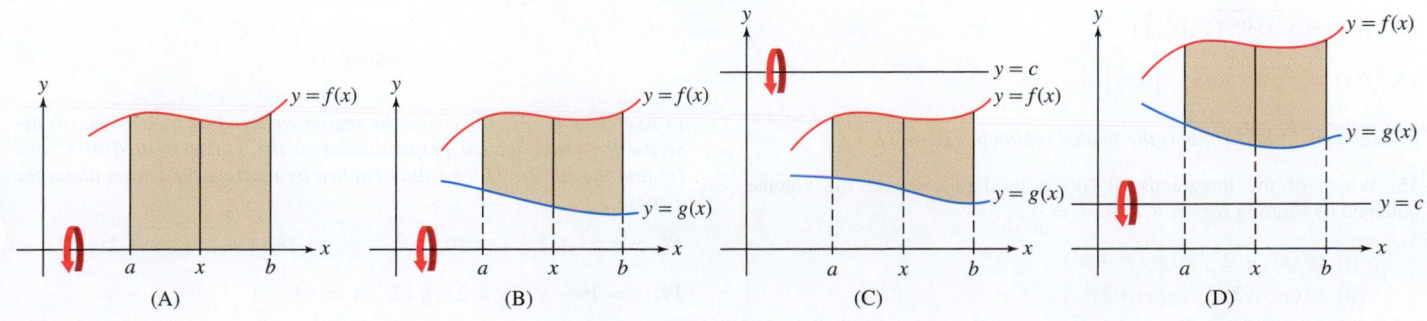

FIGURE 10

• To rotate about a horizontal line $y = c$, modify the radii appropriately.

– Figure 10(C): $c \geq f(x) \geq g(x)$:

$$R_{\text{outer}} = c - g(x), \qquad R_{\text{inner}} = c - f(x)$$

– Figure 10(D): $f(x) \geq g(x) \geq c$:

$$R_{\text{outer}} = f(x) - c, \qquad R_{\text{inner}} = g(x) - c$$

• To rotate about a vertical line $x = c$, express R_{outer} and R_{inner} as functions of y and integrate along the y-axis.

6.3 EXERCISES

Preliminary Questions

1. Which of the following is a solid of revolution?

(a) sphere   **(b)** pyramid   **(c)** cylinder   **(d)** cube

2. True or false? When the region under a single graph is rotated about the x-axis, the cross sections of the solid perpendicular to the x-axis are circular disks.

3. True or false? When the region between two graphs is rotated about the x-axis, the cross sections of the solid perpendicular to the x-axis are circular disks.

4. Which of the following integrals expresses the volume obtained by rotating the area between $y = f(x)$ and $y = g(x)$ over $[a, b]$ around the x-axis? [Assume $f(x) \geq g(x) \geq 0$.]

(a) $\pi \displaystyle\int_a^b \left(f(x) - g(x) \right)^2 dx$

(b) $\pi \displaystyle\int_a^b \left(f(x)^2 - g(x)^2 \right) dx$

Exercises

In Exercises 1–4, (a) sketch the solid obtained by revolving the region under the graph of f about the x-axis over the given interval, (b) describe the cross section perpendicular to the x-axis located at x, and (c) calculate the volume of the solid.

1. $f(x) = x + 1, \quad [0, 3]$

2. $f(x) = x^2, \quad [1, 3]$

3. $f(x) = \sqrt{x + 1}, \quad [1, 4]$

4. $f(x) = x^{-1}, \quad [1, 4]$

In Exercises 5–14, find the volume of revolution about the x-axis for the given function and interval.

5. $f(x) = 3x - x^2, \quad [0, 3]$

6. $f(x) = \dfrac{1}{x^2}, \quad [1, 4]$

7. $f(x) = x^{5/3}, \quad [1, 8]$

8. $f(x) = 4 - x^2, \quad [0, 2]$

9. $f(x) = \dfrac{2}{x + 1}, \quad [1, 3]$

10. $f(x) = \sqrt{x^4 + 1}, \quad [1, 3]$

11. $f(x) = e^x, \quad [0, 1]$

12. $f(x) = e^{-x}, \quad [0, 10]$

13. $f(x) = \sqrt{3 \cos x}, \quad [0, \frac{\pi}{4}]$

14. $f(x) = \sqrt{\cos x \sin x}, \quad [0, \frac{\pi}{2}]$

In Exercises 15 and 16, R is the shaded region in Figure 11.

15. Which of the integrands (i)–(iv) is used to compute the volume obtained by rotating region R about $y = -2$?

(i) $(f(x)^2 + 2^2) - (g(x)^2 + 2^2)$

(ii) $(f(x) + 2)^2 - (g(x) + 2)^2$

(iii) $(f(x)^2 - 2^2) - (g(x)^2 - 2^2)$

(iv) $(f(x) - 2)^2 - (g(x) - 2)^2$

16. Which of the integrands (i)–(iv) is used to compute the volume obtained by rotating R about $y = 9$ in Figure 11?

(i) $(9 + f(x))^2 - (9 + g(x))^2$

(ii) $(9 + g(x))^2 - (9 + f(x))^2$

(iii) $(9 - f(x))^2 - (9 - g(x))^2$

(iv) $(9 - g(x))^2 - (9 - f(x))^2$

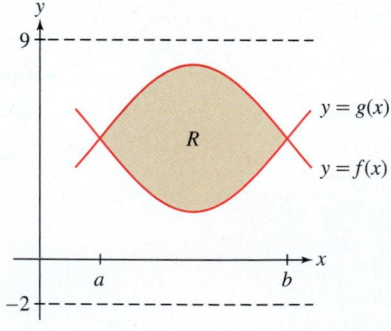

FIGURE 11

In Exercises 17–22, (a) sketch the region enclosed by the curves, (b) describe the cross section perpendicular to the x-axis located at x, and (c) find the volume of the solid obtained by rotating the region about the x-axis.

17. $y = x^2 + 2, \quad y = 10 - x^2$

18. $y = x^2, \quad y = 2x + 3$

19. $y = 16 - x, \quad y = 3x + 12, \quad x = -1$

20. $y = \dfrac{1}{x}, \quad y = \dfrac{5}{2} - x$

21. $y = \sec x$, $\quad y = 0$, $\quad x = -\dfrac{\pi}{4}$, $\quad x = \dfrac{\pi}{4}$

22. $y = \sec x$, $\quad y = 0$, $\quad x = 0$, $\quad x = \dfrac{\pi}{4}$

In Exercises 23–26, find the volume of the solid obtained by rotating the region enclosed by the graphs about the y-axis over the given interval.

23. $x = \sqrt{y}$, $\quad x = 0$; $\quad 1 \le y \le 4$

24. $x = \sqrt{\sin y}$, $\quad x = 0$; $\quad 0 \le y \le \pi$

25. $x = y^2$, $\quad x = \sqrt{y}$

26. $x = 4 - y$, $\quad x = 16 - y^2$

27. Rotation of the region in Figure 12 about the y-axis produces a solid with two types of different cross sections. Compute the volume as a sum of two integrals, one for $-12 \le y \le 4$ and one for $4 \le y \le 12$.

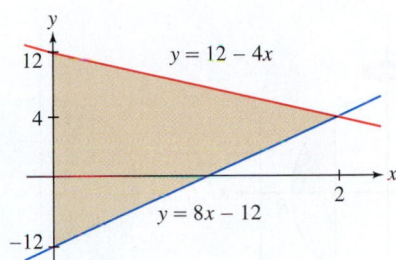

FIGURE 12

28. Let R be the region enclosed by $y = x^2 + 2$, $y = (x - 2)^2$ and the axes $x = 0$ and $y = 0$. Compute the volume V obtained by rotating R about the x-axis. *Hint:* Express V as a sum of two integrals.

In Exercises 29–34, find the volume of the solid obtained by rotating region A in Figure 13 about the given axis.

29. x-axis   **30.** $y = -2$ **31.** $y = 2$

32. y-axis   **33.** $x = -3$ **34.** $x = 2$

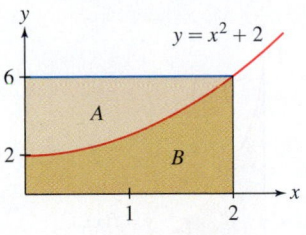

FIGURE 13

In Exercises 35–40, find the volume of the solid obtained by rotating region B in Figure 13 about the given axis.

35. x-axis   **36.** $y = -2$

37. $y = 6$ **38.** y-axis

Hint for Exercise 38: Express the volume as a sum of two integrals along the y-axis or use Exercise 32.

39. $x = 2$ **40.** $x = -3$

In Exercises 41–54, find the volume of the solid obtained by rotating the region enclosed by the graphs about the given axis.

41. $y = x^2$, $\quad y = 12 - x$, $\quad x = 0$, $\quad$ about $y = -2$ $\quad x \ge 0$

42. $y = x^2$, $\quad y = 12 - x$, $\quad x = 0$, $\quad$ about $y = 15$

43. $y = 16 - 2x$, $\quad y = 6$, $\quad x = 0$, $\quad$ about x-axis

44. $y = 32 - 2x$, $\quad y = 2 + 4x$, $\quad x = 0$, $\quad$ about y-axis

45. $y = \sec x$, $\quad y = 1 + \dfrac{3}{\pi} x$, $\quad$ about x-axis

46. $x = 2$, $\quad x = 3$, $\quad y = 16 - x^4$, $\quad y = 0$, $\quad$ about y-axis

47. $y = 2\sqrt{x}$, $\quad y = x$, $\quad$ about $x = -2$

48. $y = 2\sqrt{x}$, $\quad y = x$, $\quad$ about $y = 4$

49. $y = x^3$, $\quad y = x^{1/3}$, $\quad$ for $x \ge 0$, $\quad$ about y-axis

50. $y = x^2$, $\quad y = x^{1/2}$, $\quad$ about $x = -2$

51. $y = \dfrac{9}{x^2}$, $\quad y = 10 - x^2$, $\quad x \ge 0$, $\quad$ about $y = 12$

52. $y = \dfrac{9}{x^2}$, $\quad y = 10 - x^2$, $\quad x \ge 0$, $\quad$ about $x = -1$

53. $y = e^{-x}$, $\quad y = 1 - e^{-x}$, $\quad x = 0$, $\quad$ about $y = 4$

54. $y = \cosh x$, $\quad x = \pm 2$, $\quad$ about x-axis

55. The bowl in Figure 14(A) is 21 cm high, obtained by rotating the curve in Figure 14(B) as indicated. Estimate the volume capacity of the bowl shown by taking the average of right- and left-endpoint approximations to the integral with $N = 7$. The inner radii (in centimeters) starting from the top are 0, 4, 7, 8, 10, 13, 14, 20.

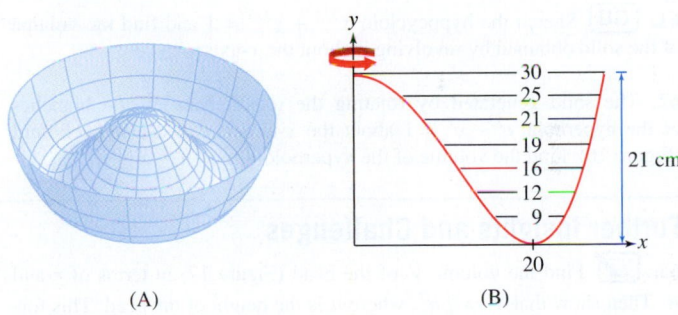

(A)   (B)

FIGURE 14

56. The region between the graphs of f and g over $[0, 1]$ is revolved about the line $y = -3$. Use the midpoint approximation with values from the following table to estimate the volume V of the resulting solid:

x	0.1	0.3	0.5	0.7	0.9
$f(x)$	8	7	6	7	8
$g(x)$	2	3.5	4	3.5	2

In Exercises 57–58, you assist your grandfather Umberto who wants to know the volume of his wine barrels. Knowing that you are taking a calculus course, he thought that you might be able to help. So he measured the circumference around each barrel at regular intervals from the bottom to the top and provided the measurements to you. "Can you figure out from this how many gallons each holds?" he asked.

57. With the following barrel circumference measurements, estimate the volume of the barrel in gallons.

Dist from Bottom (in.)	0	3	6	9	12	15	18	21	24
Circumference (in.)	30	36	38	40	41	39	38	35	28

58. With the following barrel circumference measurements, estimate the volume of the barrel in gallons.

Dist from Bottom (in.)	0	4	8	12	16	20	24	28	32	36
Circumference (in.)	62	70	75	79	83	83	77	74	68	62

59. Find the volume of the cone obtained by rotating the region under the segment joining $(0, h)$ and $(r, 0)$ about the y-axis.

60. The **torus** (doughnut-shaped solid) in Figure 15 is obtained by rotating the circle $(x - a)^2 + y^2 = b^2$ around the y-axis (assume that $a > b$). Show that it has volume $2\pi^2 ab^2$. *Hint:* After simplifying it, evaluate the integral by interpreting it as the area of a circle.

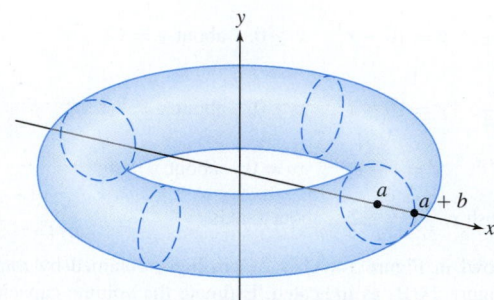

FIGURE 15 Torus obtained by rotating a circle about the y-axis.

61. $\boxed{\text{GU}}$ Sketch the hypocycloid $x^{2/3} + y^{2/3} = 1$ and find the volume of the solid obtained by revolving it about the x-axis.

62. The solid generated by rotating the region between the branches of the hyperbola $y^2 - x^2 = 1$ about the x-axis is called a **hyperboloid** (Figure 16). Find the volume of the hyperboloid for $-a \le x \le a$.

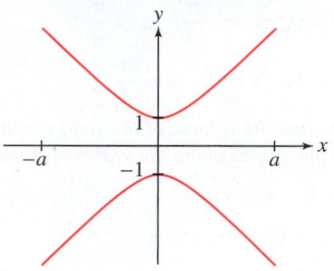

FIGURE 16 The hyperbola with equation $y^2 - x^2 = 1$.

63. A "bead" is formed by removing a cylinder of radius r from the center of a sphere of radius R (Figure 17). Find the volume of the bead with $r = 1$ and $R = 2$.

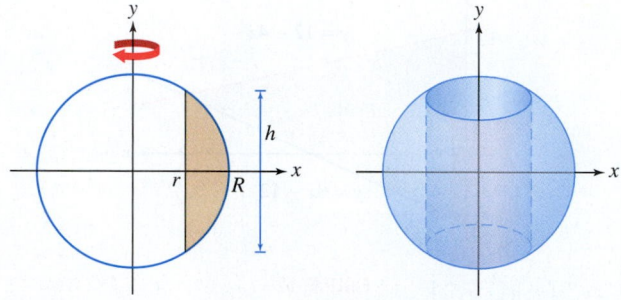

FIGURE 17 A bead is a sphere with a cylinder removed.

Further Insights and Challenges

64. $\boxed{\text{✎}}$ Find the volume V of the bead (Figure 17) in terms of r and R. Then show that $V = \frac{\pi}{6}h^3$, where h is the height of the bead. This formula has a surprising consequence: Since V can be expressed in terms of h alone, it follows that two beads of height 1 cm, one formed from a sphere the size of an orange and the other from a sphere the size of the earth, would have the same volume! Can you explain intuitively how this is possible?

65. The solid generated by rotating the region inside the ellipse with equation $\left(\frac{x}{a}\right)^2 + \left(\frac{y}{b}\right)^2 = 1$ around the x-axis is called an **ellipsoid**. Show that the ellipsoid has volume $\frac{4}{3}\pi ab^2$. What is the volume if the ellipse is rotated around the y-axis?

66. The curve $y = f(x)$ in Figure 18, called a **tractrix**, has the following property: The tangent line at each point (x, y) on the curve has slope

$$\frac{dy}{dx} = \frac{-y}{\sqrt{1 - y^2}}$$

Let R be the shaded region under the graph of $y = f(x)$ for $0 \le x \le a$ in Figure 18. Compute the volume V of the solid obtained by revolving R around the x-axis in terms of the constant $c = f(a)$. *Hint:* Use the substitution $u = f(x)$ to show that

$$V = \pi \int_c^1 u\sqrt{1 - u^2}\, du$$

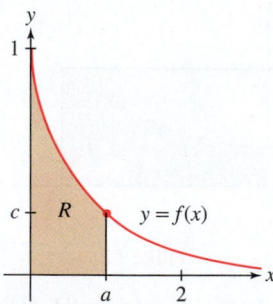

FIGURE 18 The tractrix.

67. Verify the formula

$$\int_{x_1}^{x_2} (x - x_1)(x - x_2)\, dx = \frac{1}{6}(x_1 - x_2)^3 \qquad \boxed{3}$$

Then prove that the solid obtained by rotating the shaded region in Figure 19 about the x-axis has volume $V = \frac{\pi}{6}BH^2$, with B and H as in the figure. *Hint:* Let x_1 and x_2 be the roots of $f(x) = ax + b - (mx + c)^2$, where $x_1 < x_2$. Show that

$$V = \pi \int_{x_1}^{x_2} f(x)\, dx$$

and use Eq. (3).

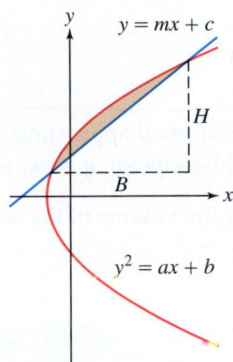

FIGURE 19 The line $y = mx + c$ intersects the parabola $y^2 = ax + b$ at two points above the x-axis.

68. Let R be the region in the unit circle lying above the cut with the line $y = mx + b$ (Figure 20). Assume that the points where the line intersects the circle lie above the x-axis. Use the method of Exercise 67 to show that the solid obtained by rotating R about the x-axis has volume $V = \frac{\pi}{6}hd^2$, with h and d as in the figure.

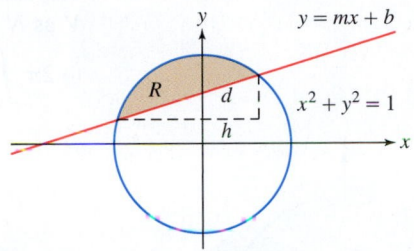

FIGURE 20

6.4 Volumes of Revolution: Cylindrical Shells

In the previous two sections, we computed the volume of solids by slicing them into parallel planar cross sections and integrating the cross-sectional area. The **Shell Method**, based on dividing a solid into concentric cylindrical shells, is more convenient in some cases, depending on the geometry of the solid under consideration.

Consider a cylindrical shell (Figure 1) of height h, with outer radius R and inner radius r. Because the shell is obtained by removing a cylinder of radius r from the wider cylinder of radius R, it has volume

$$\pi R^2 h - \pi r^2 h = \pi h (R^2 - r^2) = \pi h (R + r)(R - r) = \pi h (R + r)\Delta r$$

where $\Delta r = R - r$ is the width of the shell. If the shell is very thin, then R and r are nearly equal and we may approximate $(R + r)$ with $2R$ to obtain

volume of shell $\approx 2\pi R h \Delta r = 2\pi (\text{radius}) \times (\text{height of shell}) \times (\text{thickness})$ $\boxed{1}$

This is the product of surface area of the outer cylinder with the thickness Δr.

Now, let us rotate the region under $y = f(x)$ from $x = a$ to $x = b$ about the y-axis as in Figure 2. The resulting solid can be divided into thin concentric shells. More precisely, we divide $[a, b]$ into N subintervals of length $\Delta x = (b - a)/N$ with endpoints $x_0, x_1, \ldots, x_N$. When we rotate the thin strip of area above $[x_{i-1}, x_i]$ about the y-axis, we obtain a thin shell whose volume we denote by V_i. The volume of the solid is equal to the sum $V = \sum_{i=1}^{N} V_i$.

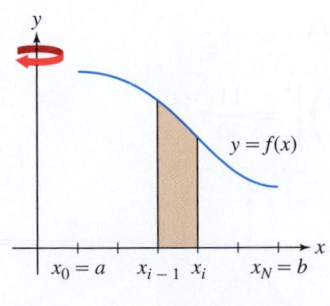 Width Δr

DF **FIGURE 1** The volume of the cylindrical shell is approximately $2\pi R h \Delta r$, where $\Delta r = R - r$.

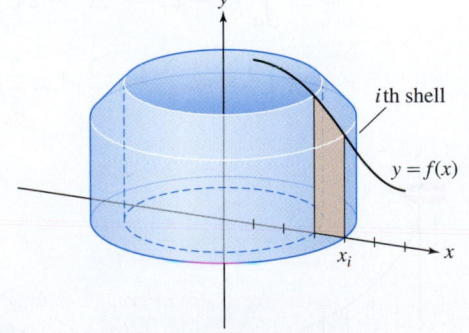
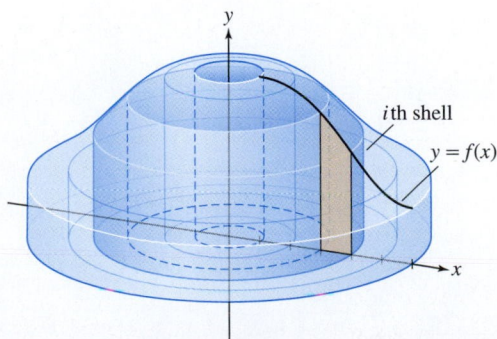

FIGURE 2 The shaded strip, when rotated about the y-axis, generates a "thin shell."

The top rim of the ith thin shell in Figure 2 is curved. However, when Δx is small, we can approximate this thin shell by the cylindrical shell (with flat rim) of height $f(x_i)$ and radius x_i. Then, using Eq. (1), we obtain

$$V_i \approx 2\pi \,(\text{radius})\,(\text{height of shell})\,(\text{thickness}) = 2\pi x_i \, f(x_i)\Delta x$$

$$V = \sum_{i=1}^{N} V_i \approx 2\pi \sum_{i=1}^{N} x_i \, f(x_i)\Delta x$$

The sum on the right is the volume of a cylindrical shell approximation that converges to V as $N \to \infty$ (Figure 3). This sum is also a right-endpoint approximation that converges to $2\pi \int_a^b x f(x)\,dx$. Thus, we obtain Eq. (2) for the volume of the solid.

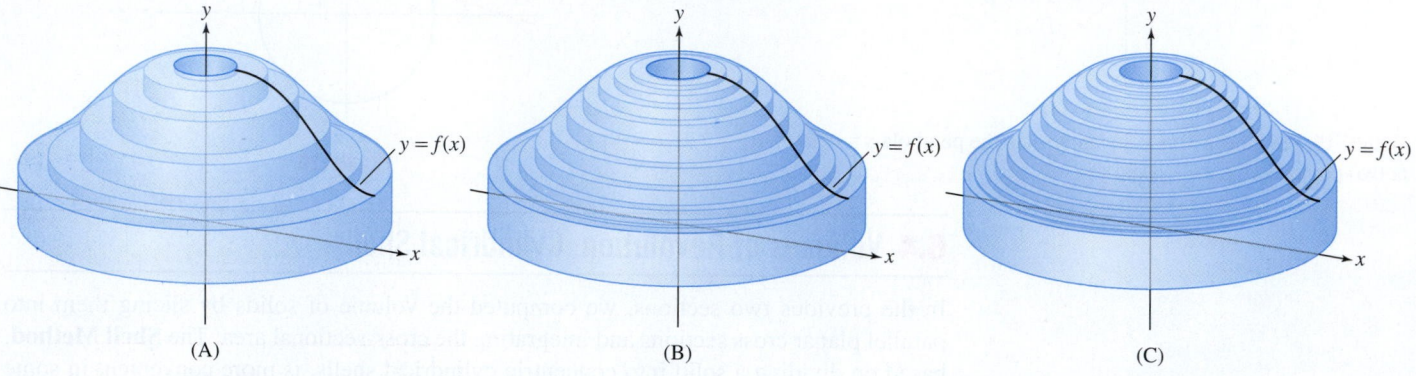

(A)   (B)   (C)

DF **FIGURE 3** Cylindrical shell approximations as $N \to \infty$.

Note: In the Shell Method, we integrate with respect to x when the region is rotated about the y-axis.

Volume of Revolution: The Shell Method   The solid obtained by rotating the region under $y = f(x)$ over the interval $[a, b]$ *about the y-axis* has volume

$$V = 2\pi \int_a^b (\text{radius})(\text{height of shell})\,dx = 2\pi \int_a^b x f(x)\,dx \qquad \boxed{2}$$

EXAMPLE 1   Find the volume V of the solid obtained by rotating the region under the graph of $f(x) = 1 - 2x + 3x^2 - 2x^3$ over $[0, 1]$ about the y-axis.

Solution   The solid is shown in Figure 4. By Eq. (2),

$$V = 2\pi \int_0^1 x f(x)\,dx = 2\pi \int_0^1 x(1 - 2x + 3x^2 - 2x^3)\,dx$$

$$= 2\pi \int_0^1 (x - 2x^2 + 3x^3 - 2x^4)\,dx$$

$$= 2\pi \left(\frac{1}{2}x^2 - \frac{2}{3}x^3 + \frac{3}{4}x^4 - \frac{2}{5}x^5 \right) \Bigg|_0^1 = \frac{11}{30}\pi \qquad \blacksquare$$

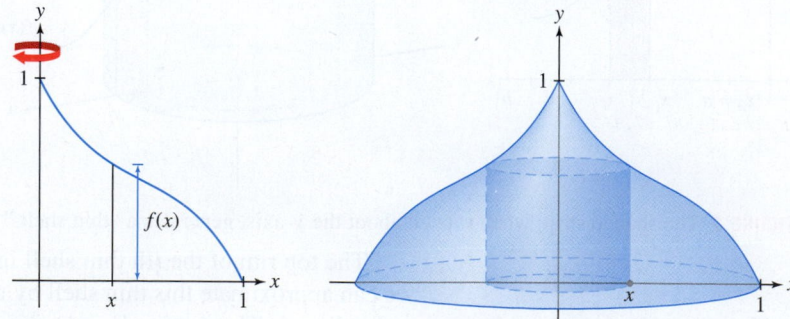

DF **FIGURE 4** The graph of $f(x) = 1 - 2x + 3x^2 - 2x^3$ rotated about the y-axis.

CONCEPTUAL INSIGHT **Shells Versus Disks and Washers**

- **Shell Method**: To calculate a volume, you must find the shell height, which is always *parallel* to the axis of rotation (Figure 5).
- **Disk and Washer Method**: To calculate a volume, you must find the disk radius or washer radii, which are always *perpendicular* to the axis of rotation.

Some volumes can be computed equally well using either the Shell Method or the Disk and Washer Method. In Example 1, however, the Shell Method is much easier because the shell height is $f(x)$. Using the Disk Method would have been more challenging because we would need to find an expression for the radius of the disk perpendicular to the y-axis (Figure 5). This would require finding the inverse $g(y) = f^{-1}(y)$, and that could be difficult or impossible.

In general: Use the Shell Method if finding the shell height is easier than finding the disk radius or washer radii. Use the Disk and Washer Method when finding the disk radius or washer radii is easier.

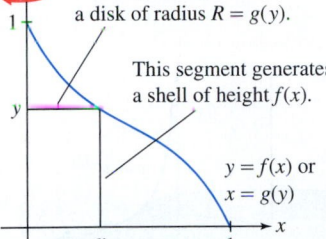

FIGURE 5 For rotation about the y-axis, the Shell Method uses $y = f(x)$ but the Disk Method requires the inverse function $x = g(y)$.

When we rotate the region between the graphs of two functions f and g satisfying $f(x) \geq g(x)$, the vertical segment at location x generates a cylindrical shell of radius x and height $f(x) - g(x)$ (Figure 6). Therefore, the volume is

$$V = 2\pi \int_a^b (\text{radius})(\text{height of shell})\,dx = 2\pi \int_a^b x\big(f(x) - g(x)\big)\,dx \qquad \boxed{3}$$

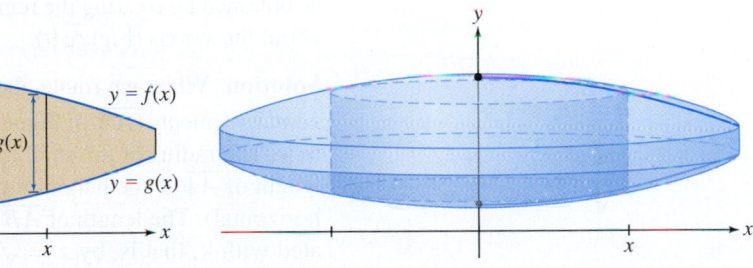

FIGURE 6 The vertical segment at location x generates a shell of radius x and height $f(x) - g(x)$.

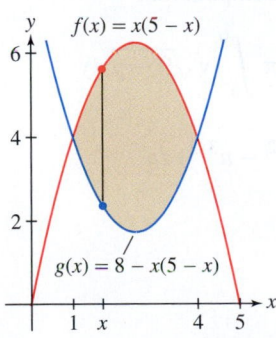

FIGURE 7

EXAMPLE 2 **Region Between Two Curves** Find the volume V obtained by rotating the region enclosed by the graphs of $f(x) = x(5 - x)$ and $g(x) = 8 - x(5 - x)$ about the y-axis.

Solution First, find the points of intersection by solving $x(5 - x) = 8 - x(5 - x)$. We obtain $0 = x^2 - 5x + 4 = (x - 1)(x - 4)$, so the curves intersect at $x = 1, 4$. Sketching the graphs (Figure 7), we see that $f(x) \geq g(x)$ on the interval $[1, 4]$ and

$$\text{height of shell} = f(x) - g(x) = x(5 - x) - \big(8 - x(5 - x)\big) = 10x - 2x^2 - 8$$

$$V = 2\pi \int_1^4 (\text{radius})(\text{height of shell})\,dx = 2\pi \int_1^4 x(10x - 2x^2 - 8)\,dx$$

$$= 2\pi \left(\frac{10}{3}x^3 - \frac{1}{2}x^4 - 4x^2 \right)\Bigg|_1^4 = 2\pi \left(\frac{64}{3} - \left(-\frac{7}{6}\right) \right) = 45\pi \qquad \blacksquare$$

The reasoning in Example 3 shows that if we rotate the region under $y = f(x)$ over $[a, b]$ about the vertical line $x = c$, then the volume is

$$V = 2\pi \int_a^b (x - c)f(x)\,dx \quad \text{if } c \leq a$$

$$V = 2\pi \int_a^b (c - x)f(x)\,dx \quad \text{if } c \geq b$$

EXAMPLE 3 **Rotating About a Vertical Axis** Use the Shell Method to calculate the volume V obtained by rotating the region under the graph of $f(x) = x^{-1/2}$ over $[1, 4]$ about the axis $x = -3$.

Solution If we were rotating this region about the y-axis (i.e., $x = 0$), we would use Eq. (3). To rotate it around the line $x = -3$, we must take into account that the radius of revolution is now 3 units longer.

Figure 8 shows that the radius of the shell at x is now $x - (-3) = x + 3$. The height of the shell is still $f(x) = x^{-1/2}$, so

$$V = 2\pi \int_1^4 (\text{radius})(\text{height of shell})\, dx$$

$$= 2\pi \int_1^4 (x + 3)x^{-1/2}\, dx = 2\pi \left(\frac{2}{3}x^{3/2} + 6x^{1/2} \right)\Bigg|_1^4 = \frac{64\pi}{3} \quad \blacksquare$$

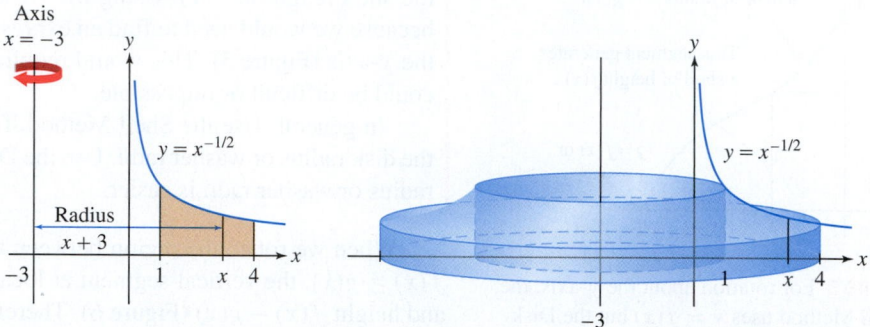

DF **FIGURE 8** Rotation about the axis $x = -3$.

The method of cylindrical shells can be applied to rotations about horizontal axes, but in this case, the graph must be described in the form $x = g(y)$.

EXAMPLE 4 **Rotating About the x-Axis** Use the Shell Method to compute the volume V obtained by rotating the region in the first quadrant between $y = 9 - x^2$ and the x-axis about the x-axis (Figure 9).

Solution When we rotate about the x-axis, the cylindrical shells are generated by horizontal segments ($\overline{AB}$ in Figure 9) and the Shell Method gives us an integral with respect to y. The radius of the shell is y, the distance from the rotation axis to the segment. The length of $\overline{AB}$ is the height of the shell (we use the term "height" even though the shell is horizontal). The length of $\overline{AB}$ is given by the positive value of x on the parabola associated with y, that is, by $x = \sqrt{9 - y}$. The volume is then obtained as follows, where we use a substitution $u = 9 - y$, $du = -dy$, in the integral computation:

REMINDER After making the substitution $u = 9 - y$, the limits of integration must be changed. Since $u(0) = 9$ and $u(9) = 0$, we change $\int_0^9$ to $\int_9^0$.

$$V = 2\pi \int_0^9 (\text{radius})(\text{height of shell})\, dy = 2\pi \int_0^9 y\sqrt{9 - y}\, dy$$

$$= -2\pi \int_9^0 (9 - u)\sqrt{u}\, du = 2\pi \int_0^9 (9u^{1/2} - u^{3/2})\, du$$

$$= 2\pi \left(6u^{3/2} - \frac{2}{5}u^{5/2} \right)\Bigg|_0^9 = \frac{648}{5}\pi \quad \blacksquare$$

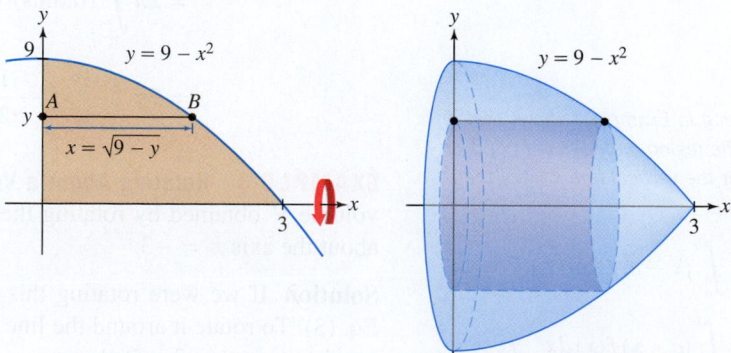

DF **FIGURE 9** Shell generated by a horizontal segment in the region under the graph of $y = 9 - x^2$.

6.4 SUMMARY

- Shell Method: When you rotate the region between two graphs about an axis, the segments *parallel* to the axis generate cylindrical shells [Figure 10(A)]. The volume V of the solid of revolution is the integral of the surface areas of these shells:

$$\text{surface area of shell} = 2\pi \, (\text{radius}) \, (\text{height of shell})$$

- Sketch the graphs to visualize the shells.
- Figure 10(B): Region between $y = f(x)$ (with $f(x) \geq 0$) and the y-axis, rotated about the y-axis:

$$V = 2\pi \int_a^b (\text{radius}) \, (\text{height of shell}) \, dx = 2\pi \int_a^b x f(x) \, dx$$

- Figure 10(C): Region between $y = f(x)$ and $y = g(x)$ (with $f(x) \geq g(x) \geq 0$), rotated about the y-axis:

$$V = 2\pi \int_a^b (\text{radius}) \, (\text{height of shell}) \, dx = 2\pi \int_a^b x(f(x) - g(x)) \, dx$$

- Rotation about a vertical axis $x = c$.

    – Figure 10(D): $c \leq a$, radius of shell is $(x - c)$:

$$V = 2\pi \int_a^b (x - c) f(x) \, dx$$

    – Figure 10(E): $c \geq a$, radius of shell is $(c - x)$:

$$V = 2\pi \int_a^b (c - x) f(x) \, dx$$

- Rotation about the x-axis using the Shell Method: Write the graph as $x = g(y)$:

$$V = 2\pi \int_c^d (\text{radius})(\text{height of shell}) \, dy = 2\pi \int_c^d y g(y) \, dy$$

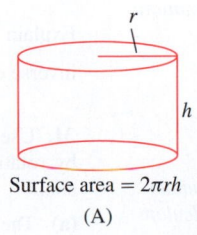

Surface area $= 2\pi r h$

(A)

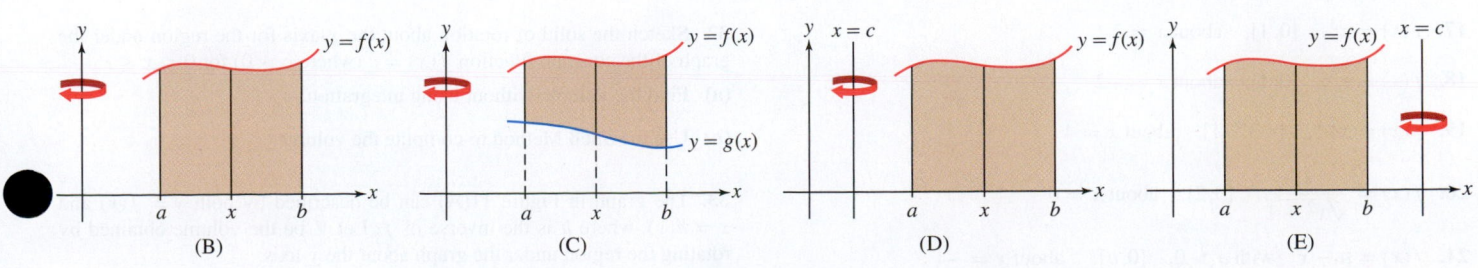

(B)   (C)   (D)   (E)

FIGURE 10

6.4 EXERCISES

Preliminary Questions

1. Consider the region under the graph of the constant function $f(x) = h$ over the interval $[0, r]$. Give the height and the radius of the cylinder generated when the region is rotated about

(a) The x-axis   **(b)** The y-axis

2. Let V be the volume of a solid of revolution about the y-axis.

(a) Does the Shell Method for computing V lead to an integral with respect to x or y?

(b) Does the Disk or Washer Method for computing V lead to an integral with respect to x or y?

3. If we rotate the region under the curve $y = 8$ between $x = 2$ and $x = 3$ about the x-axis, what answer should the Shell Method give us?

Exercises

In Exercises 1–6, sketch the solid obtained by rotating the region underneath the graph of the function over the given interval about the y-axis, and find its volume.

1. $f(x) = x^3$, $[0, 1]$

2. $f(x) = \sqrt{x}$, $[0, 4]$

3. $f(x) = x^{-1}$, $[1, 3]$

4. $f(x) = 4 - x^2$, $[0, 2]$

5. $f(x) = \sqrt{x^2 + 9}$, $[0, 3]$

6. $f(x) = \dfrac{x}{\sqrt{1 + x^3}}$, $[1, 4]$

In Exercises 7–14, use the Shell Method to compute the volume obtained by rotating the region enclosed by the graphs as indicated, about the y-axis.

7. $y = 3x - 2$, $y = 6 - x$, $x = 0$

8. $y = \sqrt{x}$, $y = x^2$

9. $y = x^2$, $y = 8 - x^2$, $x = 0$,   for $x \geq 0$

10. $y = 8 - x^3$, $y = 8 - 4x$,   for $x \geq 0$

11. $y = (x^2 + 1)^{-2}$, $y = 2 - (x^2 + 1)^{-2}$, $x = 2$

12. $y = 1 - |x - 1|$, $y = 0$

13. $y = e^{x^2}$, $y = 0$, $x = 1$, $x = 2$

14. $y = \sqrt{x^2 + 9}$, $y = 0$, $x = 0$, $x = 4$

In Exercises 15 and 16, use a graphing utility to find the points of intersection of the curves numerically and then compute the volume of rotation of the enclosed region about the y-axis.

15. $\boxed{\text{GU}}$ $y = \tfrac{1}{2}x^2$, $y = \sin(x^2)$, $x \geq 0$

16. $\boxed{\text{GU}}$ $y = e^{-x^2/2}$, $y = x$, $x = 0$

In Exercises 17–22, sketch the solid obtained by rotating the region underneath the graph of f over the interval about the given axis, and calculate its volume using the Shell Method.

17. $f(x) = x^3$, $[0, 1]$,   about $x = 2$

18. $f(x) = x^3$, $[0, 1]$,   about $x = -2$

19. $f(x) = x^{-4}$, $[-3, -1]$,   about $x = 4$

20. $f(x) = \dfrac{1}{\sqrt{x^2 + 1}}$, $[0, 2]$,   about $x = 0$

21. $f(x) = a - x$ with $a > 0$, $[0, a]$,   about $x = -1$

22. $f(x) = 1 - x^2$, $[-1, 1]$, $x = c$ with $c > 1$

In Exercises 23–28, sketch the enclosed region and use the Shell Method to calculate the volume of rotation about the x-axis.

23. $x = y$, $y = 0$, $x = 1$

24. $x = \tfrac{1}{4}y + 1$, $x = 3 - \tfrac{1}{4}y$, $y = 0$

25. $x = y(4 - y)$, $x = 0$

26. $x = y(4 - y)$, $x = (y - 2)^2$

27. $y = 4 - x^2$, $x = 0$, $y = 0$

28. $y = x^{1/3} - 2$, $y = 0$, $x = 27$

29. Determine which of the following is the appropriate integrand needed to determine the volume of the solid obtained by rotating around the vertical axis given by $x = -1$ the area that is between the curves $y = f(x)$ and $y = g(x)$ over the interval $[a, b]$, where $a \geq 0$ and $f(x) \geq g(x)$ over that interval.

(a) $x(f(x) - g(x))$ **(b)** $(x + 1)(f(x) - g(x))$

(c) $x((f(x) - 1) - (g(x) - 1))$ **(d)** $(x - 1)(f(x) - g(x))$

(e) $x(f(x + 1) - g(x + 1))$

30. Let $y = f(x)$ be a decreasing function on $[0, b]$, such that $f(b) = 0$. Explain why $2\pi \displaystyle\int_0^b x f(x)\, dx = \pi \int_0^{f(0)} (h(x))^2\, dx$, where h denotes the inverse of f.

31. Use both the Shell and Disk Methods to calculate the volume obtained by rotating the region under the graph of $f(x) = 8 - x^3$ for $0 \leq x \leq 2$ about

(a) The x-axis   **(b)** The y-axis

32. Sketch the solid of rotation about the y-axis for the region under the graph of the constant function $f(x) = c$ (where $c > 0$) for $0 \leq x \leq r$.

(a) Find the volume without using integration.

(b) Use the Shell Method to compute the volume.

33. The graph in Figure 11(A) can be described by both $y = f(x)$ and $x = h(y)$, where h is the inverse of f. Let V be the volume obtained by rotating the region under the graph about the y-axis.

(a) Describe the figures generated by rotating segments $\overline{AB}$ and $\overline{CB}$ about the y-axis.

(b) Set up integrals that compute V by the Shell and Disk Methods.

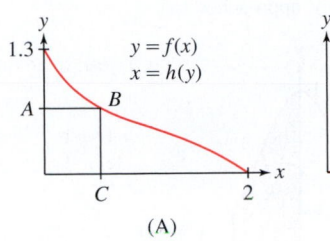 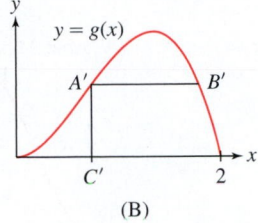

FIGURE 11

34. ✏️ Let W be the volume of the solid obtained by rotating the region under the graph in Figure 11(B) about the y-axis.

(a) Describe the figures generated by rotating segments $\overline{A'B'}$ and $\overline{A'C'}$ about the y-axis.

(b) Set up an integral that computes W by the Shell Method.

(c) Explain the difficulty in computing W by the Washer Method.

35. Let R be the region under the graph of $y = 9 - x^2$ for $0 \le x \le 2$. Use the Shell Method to compute the volume of rotation of R about the x-axis as a sum of two integrals along the y-axis. *Hint:* The shells generated depend on whether $y \in [0, 5]$ or $y \in [5, 9]$.

36. Let R be the region under the graph of $y = 4x^{-1}$ for $1 \le y \le 4$. Use the Shell Method to compute the volume of rotation of R about the y-axis is a sum of two integrals along the x-axis.

In Exercises 37–42, use the Shell Method to find the volume obtained by rotating region A in Figure 12 about the given axis.

37. y-axis

38. $x = -3$

39. $x = 2$

40. x-axis

41. $y = -2$

42. $y = 6$

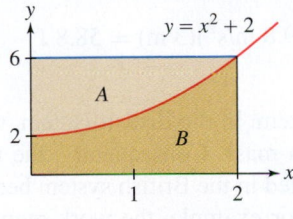

FIGURE 12

In Exercises 43–48, use the most convenient method (Disk or Shell Method) to find the volume obtained by rotating region B in Figure 12 about the given axis.

43. y-axis

44. $x = -3$

45. $x = 2$

46. x-axis

47. $y = -2$

48. $y = 8$

In Exercises 49–56, use the most convenient method (Disk or Shell Method) to find the given volume of rotation.

49. Region between $x = y(5 - y)$ and $x = 0$, rotated about the y-axis

50. Region between $x = y(5 - y)$ and $x = 0$, rotated about the x-axis

51. Region bounded by $y = x^2$ and $x = y^2$, rotated about the y-axis

52. Region bounded by $y = x^2$ and $x = y^2$, rotated about $x = 3$

53. Region in Figure 13, rotated about the x-axis

54. Region in Figure 13, rotated about the y-axis

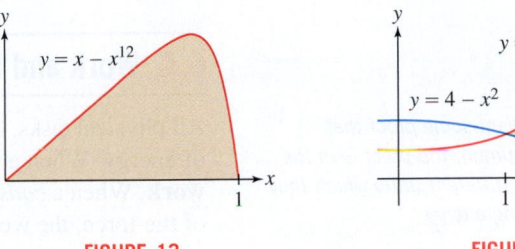

FIGURE 13   **FIGURE 14**

55. Region in Figure 14, rotated about $x = 4$

56. Region in Figure 14, rotated about $y = -2$

In Exercises 57–62, use the Shell Method to find the given volume of rotation.

57. A sphere of radius r

58. The "bead" formed by removing a cylinder of radius r from the center of a sphere of radius R (compare with Exercise 63 in Section 6.3)

59. The torus obtained by rotating the circle $(x - a)^2 + y^2 = b^2$ about the y-axis, where $a > b$ (compare with Exercise 60 in Section 6.3). *Hint:* Evaluate the integral by interpreting part of it as the area of a circle.

60. The "paraboloid" obtained by rotating the region between $y = x^2$ and $y = c$ $(c > 0)$ about the y-axis

61. The solid obtained by rotating the region bounded by $y = \sqrt{\ln x}$, the x-axis, and $x = e^4$ about the x-axis

62. The solid obtained by rotating the region bounded by $y = \sqrt{\sin^{-1} x}$, the x-axis, and $x = 1$ about the x-axis

63. Given a and b, $0 \le a \le b$, find a function f such that the volume obtained by rotating about the x-axis the region R under the graph of $y = f(x)$ over the interval $[a, b]$ equals the volume obtained by rotating that same region R about the y-axis.

Further Insights and Challenges

64. ✏️ The surface area of a sphere of radius r is $4\pi r^2$. Use this to derive the formula for the volume V of a sphere of radius R in a new way.

(a) Show that the volume of a thin spherical shell of inner radius r and thickness Δr is approximately $4\pi r^2 \Delta r$.

(b) Approximate V by decomposing the sphere of radius R into N thin spherical shells of thickness $\Delta r = R/N$.

(c) Show that the approximation is a Riemann sum that converges to an integral. Evaluate the integral.

65. Show that the solid (an **ellipsoid**) obtained by rotating the region R in Figure 15 about the y-axis has volume $\frac{4}{3}\pi a^2 b$.

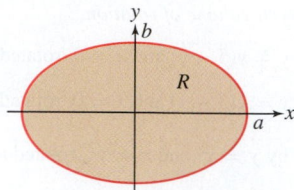

FIGURE 15 The ellipse $\left(\frac{x}{a}\right)^2 + \left(\frac{y}{b}\right)^2 = 1$.

66. The bell-shaped curve $y = f(x)$ in Figure 16 satisfies $dy/dx = -xy$. Use the Shell Method and the substitution $u = f(x)$ to show that the

solid obtained by rotating the region R about the y-axis has volume $V = 2\pi(1 - c)$, where $c = f(a)$. Observe that as $c \to 0$, the region R becomes infinite but the volume V approaches 2π.

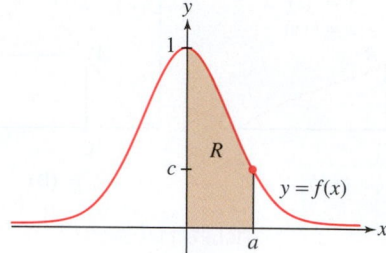

FIGURE 16 The bell-shaped curve.

6.5 Work and Energy

> For those who want some proof that physicists are human, the proof is in the idiocy of all the different units which they use for measuring energy.
>
> —Richard Feynman,
> *The Character of Physical Law*

Force F

A Distance d B

FIGURE 1 The work expended to move the object from A to B is $W = F \cdot d$.

All physical tasks, from running up a hill to turning on a computer, require an expenditure of energy. When a force is applied to an object to move it, the energy expended is called **work**. When a *constant* force F is applied to move the object a distance d in the direction of the force, the work W is defined as "force times distance" (Figure 1):

$$\boxed{W = F \cdot d} \qquad \boxed{1}$$

The International System (SI) unit of force is the *newton* (abbreviated N), defined as 1 kg-m/s^2. Energy and work are both measured in units of the *joule* (J), equal to 1 N-m. In the British system, the unit of force is the pound, and both energy and work are measured in foot-pounds. Another unit of energy is the *calorie*. One ft-lb is approximately 1.356 J or 0.324 calories.

To become familiar with the units, let's calculate the work W required to lift a 2-kg stone 3 m above the ground. Gravity acts on the stone of mass m with a force equal to $-mg$, where $g = 9.8$ m/s^2. Therefore, lifting the stone requires an upward vertical force $F = mg$, and the work expended is

$$W = \underbrace{(mg)h}_{F \cdot d} = (2 \text{ kg})(9.8 \text{ m/s}^2)(3 \text{ m}) = 58.8 \text{ J}$$

The kilogram is a unit of mass in the SI system. In the British system, we typically work with weight, which is a force rather than a mass. Consequently, the factor g does not appear when work against gravity is computed in the British system because, essentially, this factor is incorporated in the weight. For example, the work required to lift a 2-lb stone 3 ft is

$$W = \underbrace{(2 \text{ lb})(3 \text{ ft})}_{F \cdot d} = 6 \text{ ft-lb}$$

We are interested in the case where the force $F(x)$ varies as the object moves from a to b along the x-axis. Eq. (1) does not apply directly, but we can break up the task into a large number of smaller tasks for which Eq. (1) gives a good approximation. Divide $[a, b]$ into N subintervals of length $\Delta x = (b - a)/N$ as in Figure 2 and let W_i be the work required to move the object from x_{i-1} to x_i. If Δx is small, then the force $F(x)$ is nearly constant on the interval $[x_{i-1}, x_i]$ with value $F(x_i)$, so $W_i \approx F(x_i)\Delta x$. Summing the contributions, we obtain

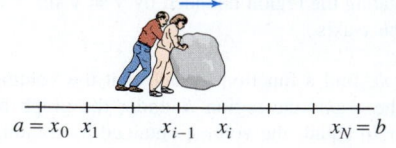

$a = x_0 \; x_1 \qquad x_{i-1} \; x_i \qquad x_N = b$

FIGURE 2 The work to move an object from x_{i-1} to x_i is approximately $F(x_i)\Delta x$.

$$W = \sum_{i=1}^{N} W_i \approx \underbrace{\sum_{i=1}^{N} F(x_i)\Delta x}_{\text{Right-endpoint approximation}}$$

The sum on the right is a right-endpoint approximation that converges to $\int_a^b F(x)\,dx$. This leads to the following definition.

> **DEFINITION Work**   The work performed in moving an object along the x-axis from a to b by applying a force $F(x)$ is
>
> $$W = \int_a^b F(x)\,dx \qquad \boxed{2}$$

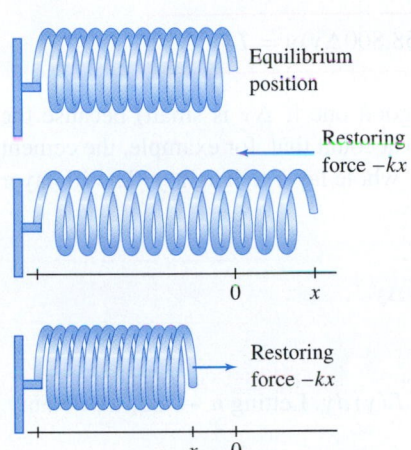

Equilibrium
position

Restoring
force $-kx$

0   x

Restoring
force $-kx$

x 0

FIGURE 3 Hooke's Law.

One typical calculation involves finding the work required to stretch or compress a spring. Assume that the free end of the spring has position $x = 0$ at equilibrium, when no force is acting (Figure 3). According to **Hooke's Law**, when the spring is stretched or compressed to position x, it exerts a restoring spring force $F(x) = -kx$, where $k > 0$ is the **spring constant**.

If we want to stretch the spring from $x = a$ to $x = b$, with $0 < a < b$, we must apply a force $F(x) = kx$ to counteract the force exerted by the spring. The work required to stretch the spring is $\int_a^b kx\,dx$. Similarly, if we wish to compress the spring from $x = a$ to $x = b$ with $b < a < 0$, the work required is also $\int_a^b kx\,dx$. In this latter case the integration is in the negative direction, but the applied force is also in the negative direction, so the result is a positive value of work.

EXAMPLE 1   Hooke's Law   Assuming a spring constant of $k = 400$ N/m, find the work required to

(a) Stretch the spring 10 cm beyond equilibrium

(b) Compress the spring 2 cm more when it is already compressed 3 cm

Solution   A force $F(x) = 400x$ N is required to stretch the spring (with x in meters). Note that centimeters must be converted to meters.

(a) The work required to stretch the spring 10 cm (0.1 m) beyond equilibrium is

$$W = \int_0^{0.1} 400x\,dx = 200x^2 \Big|_0^{0.1} = 2 \text{ J}$$

(b) If the spring is at position $x = -3$ cm, then the work W required to compress it further to $x = -5$ cm is

$$W = \int_{-0.03}^{-0.05} 400x\,dx = 200x^2 \Big|_{-0.03}^{-0.05} = 0.5 - 0.18 = 0.32 \text{ J} \qquad \blacksquare$$

In the next two examples, we are not moving a single object through a fixed distance, so we cannot apply Eq. (2). Rather, each thin layer of the object is moved through a different distance. The work performed is computed by "summing" (i.e., *integrating*) the work performed on the thin layers.

EXAMPLE 2   Building a Concrete Column   Compute the work (against gravity) required to build a concrete column of height 5 m and square base of side 2 m. Assume that concrete has density 1500 kg/m^3.

Solution   Think of the column as a stack of n thin layers of width $\Delta y = 5/n$. The work consists of lifting up these layers and placing them on the stack (Figure 4), but the work performed on a given layer depends on how high we lift it. First, let us compute the gravitational force on a thin layer of width Δy:

> On the earth's surface, work against gravity is equal to the force mg times the vertical distance through which the object is lifted. No work against gravity is done when an object is moved sideways.

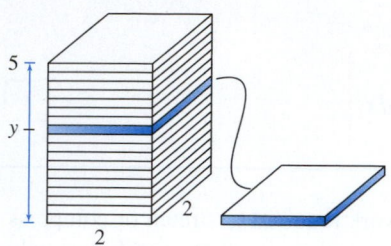

FIGURE 4 Total work is the sum of the work performed on each layer of the column.

> In Examples 2 and 3, the work performed on a thin layer is written
>
> $$L(y)\Delta y$$
>
> When we take the sum and let Δy approach zero, we obtain the integral of $L(y)$.
>
> Symbolically, the Δy "becomes" the dy of the integral. Note that
>
> $$L(y) = g \times density \times A(y)$$
> $$\times (vertical\ distance\ lifted)$$
>
> where $A(y)$ is the area of the cross section.

$$volume\ of\ layer = area \times width = 4\Delta y\ \text{m}^3$$

$$mass\ of\ layer = density \times volume = 1500 \cdot 4\Delta y\ \text{kg}$$

$$force\ on\ layer = g \times mass = 9.8 \cdot 1500 \cdot 4\Delta y = 58{,}800\,\Delta y\ \text{N}$$

The work performed in lifting this layer to height y is equal to the force times the distance y, which is $(58{,}800\Delta y)y$. Setting $L(y) = 58{,}800y$, we have

$$\boxed{\text{Work lifting layer to height } y \approx (58{,}800\Delta y)y = L(y)\Delta y}$$

This is only an approximation (although a very good one if Δy is small) because the layer has nonzero width and we are not taking into account that, for example, the cement particles at the top are lifted Δy prior to lifting the whole layer to height y. The ith layer is lifted to height y_i, so the total work performed is

$$W \approx \sum_{i=1}^{n} L(y_i)\,\Delta y$$

This sum is a right-endpoint approximation to $\int_0^5 L(y)\,dy$. Letting $n \to \infty$, we obtain

$$W = \int_0^5 L(y)\,dy = \int_0^5 58{,}800y\,dy = 58{,}800\frac{y^2}{2}\Big|_0^5 = 735{,}000\ \text{J} \quad\blacksquare$$

EXAMPLE 3 **Pumping Water out of a Tank** A spherical tank of radius 5 m is filled with water. Calculate the work W performed (against gravity) in pumping out the water through a spout of height 1 m at the top. The density of water is 1000 kg/m^3.

Solution The first step, as in the previous example, is to compute the work against gravity performed on a thin layer of water of width Δy. We place the origin of our coordinate system at the center of the sphere because this leads to a simple formula for the radius r of the cross section at height y (Figure 5).

Step 1. **Compute work performed on a layer.**
Figure 5 shows that the cross section at height y is a circle of radius $r = \sqrt{25 - y^2}$ and area $A(y) = \pi r^2 = \pi(25 - y^2)$. A thin layer has volume $A(y)\Delta y$ and mass obtained by multiplying this volume by the density 1000 kg/m^3. To lift this layer, we must exert a force against gravity equal to

$$force\ on\ layer = g \times \underbrace{density \times A(y)\Delta y}_{\text{Mass}} \approx (9.8)1000\pi(25 - y^2)\Delta y$$

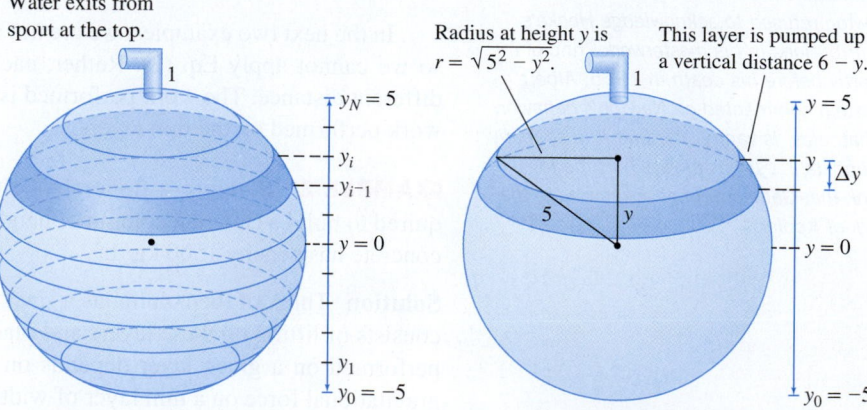

FIGURE 5 The sphere is divided into N thin layers.

The layer has to be lifted a vertical distance $6 - y$, so

$$\text{Work on layer} \approx \underbrace{9800\pi(25 - y^2)\Delta y}_{\text{Force against gravity}} \times \underbrace{(6 - y)}_{\text{Vertical distance lifted}} = L(y)\Delta y$$

where $L(y) = 9800\pi(25 - y^2)(6 - y) = 9800\pi(150 - 25y - 6y^2 + y^3)$.

Step 2. Compute total work.

Now, divide the sphere into N layers and let y_i be the height of the ith layer. The work performed on ith layer is approximately $L(y_i)\Delta y$, and therefore,

$$W \approx \sum_{i=1}^{N} L(y_i)\Delta y$$

This sum approaches the integral of $L(y)$ as $N \to \infty$ (i.e., $\Delta y \to 0$), so

$$W = \int_{-5}^{5} L(y)\,dy = 9800\pi \int_{-5}^{5} (150 - 25y - 6y^2 + y^3)\,dy$$

$$= 9800\pi \left(150y - \frac{25}{2}y^2 - 2y^3 + \frac{1}{4}y^4\right)\Big|_{-5}^{5} = 9{,}800{,}000\pi \approx 3.1 \times 10^7 \text{ J}$$

Note that the integral extends from -5 to 5 because the y-coordinate along the sphere varies from -5 to 5. ∎

How much energy is 3.1×10^7 joules? A liter of gasoline has an energy content of approximately 3.4×10^7 joules. Hence, the work required to pump the water out of the spout is equal to the energy content of roughly 0.9 L of gasoline.

6.5 SUMMARY

- Work performed to move an object:

$$\text{Constant force:} \quad W = F \cdot d, \qquad \text{Variable force:} \quad W = \int_{a}^{b} F(x)\,dx$$

- Hooke's Law: A spring stretched or compressed to position x from equilibrium exerts a restoring force $-kx$. An applied force $F(x) = kx$ is required to stretch or compress the spring further.

  To stretch a spring from a to b with $0 < a < b$ or to compress a spring from a to b with $b < a < 0$, the work performed is $W = \int_{a}^{b} kx\,dx$.

- To compute work against gravity by decomposing an object into N thin layers of thickness Δy, express the work performed on a thin layer as $L(y)\Delta y$, where

$$L(y) = g \times \text{density} \times A(y) \times (\text{vertical distance lifted})$$

  The total work performed is $W = \int_{a}^{b} L(y)\,dy$.

6.5 EXERCISES

Preliminary Questions

1. Why is integration needed to compute the work performed in stretching a spring?

2. Why is integration needed to compute the work performed in pumping water out of a tank but not to compute the work performed in lifting up the tank?

3. Which of the following represents the work required to stretch a spring (with spring constant k) a distance x beyond its equilibrium position: kx, $-kx$, $\frac{1}{2}mk^2$, $\frac{1}{2}kx^2$, or $\frac{1}{2}mx^2$?

4. What does it mean when the integral used to calculate work gives a negative answer?

Exercises

1. How much work is done raising a 4-kg mass to a height of 16 m above ground?

2. How much work is done raising a 4-lb mass to a height of 16 ft above ground?

In Exercises 3–6, compute the work (in joules) required to stretch or compress a spring as indicated, assuming a spring constant of $k = 800$ N/m.

3. Stretching from equilibrium to 12 cm past equilibrium

4. Compressing from equilibrium to 4 cm past equilibrium

5. Stretching from 5 to 15 cm past equilibrium

6. Compressing 4 cm more when it is already compressed 5 cm

In Exercises 7–10 we investigate nonlinear springs. A spring is linear if it obeys Hooke's Law, which indicates that the applied force to stretch the spring is $F(x) = kx$. For a linear spring, F' is constant. If, instead, F' is not constant, then the spring is called nonlinear. Furthermore, if $F'(x)$ increases as x increases, then the spring is said to be progressive, and if $F'(x)$ decreases as x increases, then the spring is said to be degressive.

7. Of the two statements (a) and (b), which describes a progressive spring, and which describes a degressive spring?

(a) To stretch the spring a fixed additional distance, a greater change in force is needed farther from equilibrium than closer to it.

(b) To stretch the spring a fixed additional distance, a greater change in force is needed closer to equilibrium than farther from it.

8. **(a)** Of the two applied force graphs in Figure 6, which describes a progressive spring, and which describes a degressive spring?

(b) For each, approximate the work required to stretch the spring from 4 to 9 cm.

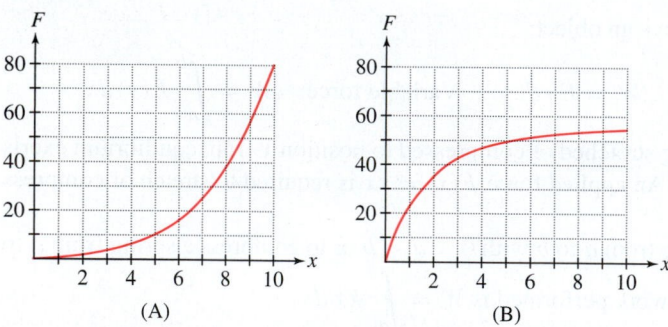

FIGURE 6 Applied force functions for nonlinear springs (F in N, x in cm).

9. Let $F(x) = 20\sqrt{3x}$ be the applied force function for a spring (with $F(x)$ in N and x in cm). Indicate whether the spring is progressive or degressive. Compute the work required to stretch the spring from 6 to 12 cm.

10. Let $F(x) = 10(e^{0.2x} - 1)$ be the applied force function for a spring (with $F(x)$ in N and x in cm). Indicate whether the spring is progressive or degressive. Compute the work required to stretch the spring from 6 to 12 cm.

In Exercises 11–14, use the method of Examples 2 and 3 to calculate the work against gravity required to build the structure out of a lightweight material of density 600 kg/m^3.

11. Solid box of height 3 m and square base of side 2 m

12. Cylindrical column of height 4 m and radius 0.8 m

13. Right circular cone of height 4 m and base of radius 1.2 m

14. Hemisphere of radius 0.8 m

15. Built around 2600 BCE, the Great Pyramid of Giza in Egypt (Figure 7) is 146 m high and has a square base of side 230 m. Find the work (against gravity) required to build the pyramid if the density of the stone is estimated at 2000 kg/m^3.

FIGURE 7 The Great Pyramid in Giza, Egypt.

16. Calculate the work (against gravity) required to build a box of height 3 m and square base of side 2 m out of material of variable density, assuming that the density at height y is $f(y) = 1000 - 100y$ kg/m^3.

In Exercises 17–22, calculate the work (in joules) required to pump all of the water out of a full tank. Distances are in meters, and the density of water is 1000 kg/m^3.

17. Rectangular tank in Figure 8; water exits from a small hole at the top

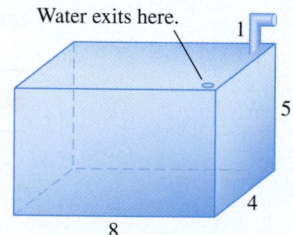

FIGURE 8

18. Rectangular tank in Figure 8; water exits through the spout

19. Hemisphere in Figure 9; water exits through the spout

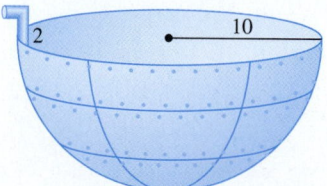

FIGURE 9

20. Conical tank in Figure 10; water exits through the spout

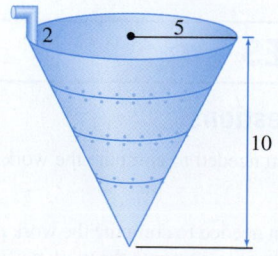

FIGURE 10

21. Horizontal cylinder in Figure 11; water exits from a small hole at the top. *Hint:* Evaluate the integral by interpreting part of it as the area of a circle.

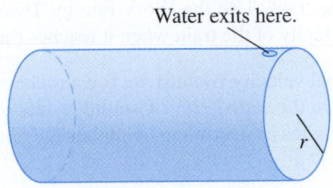

Water exits here.

FIGURE 11

22. Trough in Figure 12; water exits by pouring over the sides

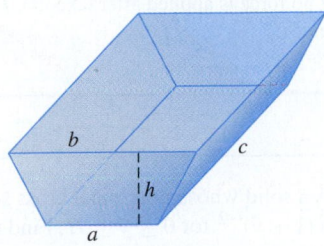

FIGURE 12

23. Find the work W required to empty the tank in Figure 8 through the hole at the top if the tank is half full of water.

24. Assume the tank in Figure 8 is full of water and let W be the work required to pump out half of the water through the hole at the top. Do you expect W to equal the work computed in Exercise 23? Explain and then compute W.

25. Assume the tank in Figure 10 is full. Find the work required to pump out half of the water. *Hint:* First, determine the level H at which the water remaining in the tank is equal to one-half the total capacity of the tank.

26. Assume that the tank in Figure 10 is full.
(a) Calculate the work $F(y)$ required to pump out water until the water level has reached level y.
(b) [CAS] Plot F.
(c) What is the significance of $F'(y)$ as a rate of change?
(d) [CAS] If your goal is to pump out all of the water, at which water level y_0 will half of the work be done?

27. Calculate the work required to lift a 10-m chain over the side of a building (Figure 13). Assume that the chain has a density of 8 kg/m. *Hint:* Break up the chain into N segments, estimate the work performed on a segment, and compute the limit as $N \to \infty$ as an integral.

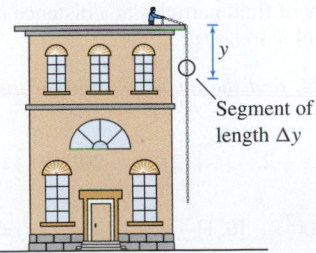

FIGURE 13 The small segment of the chain of length Δy located y meters from the top is lifted through a vertical distance y.

Segment of length Δy

28. How much work is done lifting a 3-m chain over the side of a building if the chain has mass density 4 kg/m?

29. A 6-m chain has mass 18 kg. Find the work required to lift the chain over the side of a building.

30. A 10-m chain with mass density 4 kg/m is initially coiled on the ground. How much work is performed in lifting the chain so that it is fully extended (and one end touches the ground)?

31. How much work is done lifting a 12-m chain that has mass density 3 kg/m (initially coiled on the ground) so that its top end is 10 m above the ground?

32. A 500-kg wrecking ball hangs from a 12-m cable of density 15 kg/m attached to a crane. Calculate the work done if the crane lifts the ball from ground level to 12 m in the air by drawing in the cable.

33. Calculate the work required to lift a 3-m chain over the side of a building if the chain has a variable density of $\rho(x) = x^2 - 3x + 10$ kg/m for $0 \le x \le 3$.

34. A 3-m chain with linear mass density $\rho(x) = 2x(4 - x)$ kg/m lies on the ground. Calculate the work required to lift the chain from its front end so that its bottom is 2 m above ground.

Exercises 35–37: The gravitational force between two objects of mass m and M, separated by a distance r, has magnitude GMm/r^2, where $G = 6.67 \times 10^{-11}$ m^{3}kg^{-1}s^{-1}.

35. Show that if two objects of mass M and m are separated by a distance r_1, then the work required to increase the separation to a distance r_2 is equal to $W = GMm(r_1^{-1} - r_2^{-1})$.

36. Use the result of Exercise 35 to calculate the work required to place a 2000-kg satellite in an orbit 1200 km above the surface of the earth. Assume that the earth is a sphere of radius $R_e = 6.37 \times 10^6$ m and mass $M_e = 5.98 \times 10^{24}$ kg. Treat the satellite as a point mass.

37. Use the result of Exercise 35 to compute the work required to move a 1500-kg satellite from an orbit 1000 to an orbit 1500 km above the surface of the earth.

38. The pressure P and volume V of the gas in a cylinder of length 0.8 m and radius 0.2 m, with a movable piston, are related by $PV^{1.4} = k$, where k is a constant (Figure 14). When the piston is fully extended, the gas pressure is 2000 kilopascals (kPa; 1 kilopascal is 10^3 newtons per square meter).
(a) Calculate k.
(b) The force on the piston is PA, where A is the piston's area. Calculate the force as a function of the length x of the column of gas.
(c) Calculate the work required to compress the gas column from 0.8 to 0.5 m.

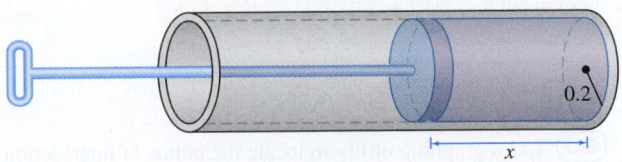

0.2

x

FIGURE 14 Gas in a cylinder with a piston.

Further Insights and Challenges

39. Work-Energy Theorem An object of mass m moves from x_1 to x_2 during the time interval $[t_1, t_2]$ due to a force $F(x)$ acting in the direction of motion. Let $x(t)$, $v(t)$, and $a(t)$ be the position, velocity, and acceleration at time t. The object's kinetic energy is $\text{KE} = \frac{1}{2}mv^2$.

(a) Use the Change of Variables Formula to show that the work performed is equal to

$$W = \int_{x_1}^{x_2} F(x)\,dx = \int_{t_1}^{t_2} F(x(t))v(t)\,dt$$

(b) Use Newton's Second Law, $F(x(t)) = ma(t)$, to show that

$$\frac{d}{dt}\left(\frac{1}{2}mv(t)^2\right) = F(x(t))v(t)$$

(c) Use the FTC to prove the Work-Energy Theorem: The change in kinetic energy during the time interval $[t_1, t_2]$ is equal to the work performed.

40. A model train of mass 0.5 kg is placed at one end of a straight 3-m electric track. Assume that a force $F(x) = (3x - x^2)$ N acts on the train at distance x along the track. Use the Work-Energy Theorem (Exercise 39) to determine the velocity of the train when it reaches the end of the track.

41. With what initial velocity v_0 must we fire a rocket so it attains a maximum height r above the earth? *Hint:* Use the results of Exercises 35 and 39. As the rocket reaches its maximum height, its KE decreases from $\frac{1}{2}mv_0^2$ to zero.

42. With what initial velocity must we fire a rocket so it attains a maximum height of $r = 20$ km above the surface of the earth?

43. Calculate **escape velocity,** the minimum initial velocity of an object to ensure that it will continue traveling into space and never fall back to earth (assuming that no force is applied after takeoff). *Hint:* Take the limit as $r \to \infty$ in Exercise 41.

CHAPTER REVIEW EXERCISES

1. Compute the area of the region in Figure 1(A) enclosed by $y = 2 - x^2$ and $y = -2$.

2. Compute the area of the region in Figure 1(B) enclosed by $y = 2 - x^2$ and $y = x$.

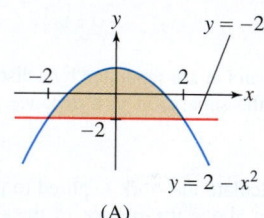
(A)

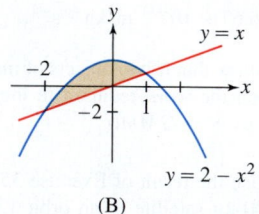
(B)

FIGURE 1

In Exercises 3–12, find the area of the region enclosed by the graphs of the functions.

3. $y = x^3 - 2x^2 + x$, $\quad y = x^2 - x$

4. $y = x^2 + 2x$, $\quad y = x^2 - 1$, $\quad h(x) = x^2 + x - 2$

5. $x = 4y$, $\quad x = 24 - 8y$, $\quad y = 0$

6. $x = y^2 - 9$, $\quad x = 15 - 2y$

7. $y = 4 - x^2$, $\quad y = 3x$, $\quad y = 4$

8. [GU] $x = \frac{1}{2}y$, $\quad x = y\sqrt{1 - y^2}$, $\quad 0 \le y \le 1$

9. $y = \sin x$, $\quad y = \cos x$, $\quad 0 \le x \le \frac{5\pi}{4}$

10. $f(x) = \sin x$, $\quad g(x) = \sin 2x$, $\quad \frac{\pi}{3} \le x \le \pi$

11. $y = e^x$, $\quad y = 1 - x$, $\quad x = 1$

12. $y = \cosh 1 - \cosh x$, $\quad y = \cosh x - \cosh 1$

13. [GU] Use a graphing utility to locate the points of intersection of $y = e^{-x}$ and $y = 1 - x^2$ and find the area between the two curves (approximately).

14. Figure 2 shows a solid whose horizontal cross section at height y is a circle of radius $(1 + y)^{-2}$ for $0 \le y \le H$. Find the volume of the solid.

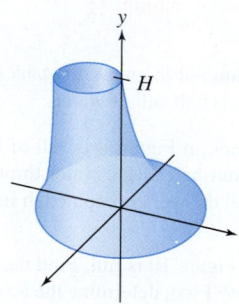

FIGURE 2

15. The base of a solid is the unit circle $x^2 + y^2 = 1$, and its cross sections perpendicular to the x-axis are rectangles of height 4. Find its volume.

16. The base of a solid is the triangle bounded by the axes and the line $2x + 3y = 12$, and its cross sections perpendicular to the y-axis have area $A(y) = (y + 2)$. Find its volume.

17. Find the total mass of a rod of length 1.2 m with linear density $\rho(x) = (1 + 2x + \frac{2}{9}x^3)$ kg/m.

18. Find the flow rate (in the correct units) through a pipe of diameter 6 cm if the velocity of fluid particles at a distance r from the center of the pipe is $v(r) = (3 - r)$ cm/s.

In Exercises 19–24, find the average value of the function over the interval.

19. $f(x) = x^3 - 2x + 2$, $\quad [-1, 2]$ **20.** $f(x) = |x|$, $\quad [-4, 4]$

21. $f(x) = x \cosh(x^2)$, $\quad [0, 1]$ **22.** $f(x) = \dfrac{e^x}{1 + e^{2x}}$, $\quad \left[0, \frac{1}{2}\right]$

23. $f(x) = \sqrt{9 - x^2}$, $\quad [0, 3]$ *Hint:* Use geometry to evaluate the integral.

24. $f(x) = x\lfloor x \rfloor$, [0, 3], where $\lfloor x \rfloor$ is the greatest integer function

25. Find $\int_2^5 g(t)\, dt$ if the average value of g on [2, 5] is 9.

26. The average value of R over $[0, x]$ is equal to x for all x. Use the FTC to determine $R(x)$.

27. Use the Washer Method to find the volume obtained by rotating the region in Figure 3 about the x-axis.

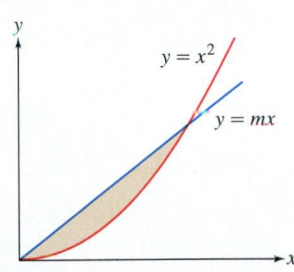

FIGURE 3

28. Use the Shell Method to find the volume obtained by rotating the region in Figure 3 about the x-axis.

In Exercises 29–40, use any method to find the volume of the solid obtained by rotating the region enclosed by the curves about the given axis.

29. $y = x^2 + 2$, $y = x + 4$, x-axis

30. $y = x^2 + 6$, $y = 8x - 1$, y-axis

31. $x = y^2 - 3$, $x = 2y$, axis $y = 4$

32. $y = 2x$, $y = 0$, $x = 8$, axis $x = -3$

33. $y = x^2 - 1$, $y = 2x - 1$, axis $x = -2$

34. $y = x^2 - 1$, $y = 2x - 1$, axis $y = 4$

35. $y = -x^2 + 4x - 3$, $y = 0$, axis $y = -1$

36. $y = -x^2 + 4x - 3$, $y = 0$, axis $x = 4$

37. $x = 4y - y^3$, $x = 0$, $y \geq 0$, x-axis

38. $y^2 = x^{-1}$, $x = 1$, $x = 3$, axis $y = -3$

39. $y = e^{-x^2/2}$, $y = -e^{-x^2/2}$, $x = 0$, $x = 1$, y-axis

40. $y = \sec x$, $y = \csc x$, $y = 0$, $x = 0$, $x = \frac{\pi}{2}$, x-axis

In Exercises 41–44, find the volume obtained by rotating the region about the given axis. The regions refer to the graph of the hyperbola $y^2 - x^2 = 1$ in Figure 4.

41. The shaded region between the upper branch of the hyperbola and the x-axis for $-c \leq x \leq c$, about the x-axis

42. The region between the upper branch of the hyperbola and the x-axis for $0 \leq x \leq c$, about the y-axis

43. The region between the upper branch of the hyperbola and the line $y = x$ for $0 \leq x \leq c$, about the x-axis

44. The region between the upper branch of the hyperbola and $y = 2$, about the y-axis

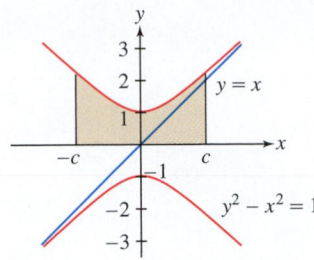

FIGURE 4

45. Let R be the intersection of the circles of radius 1 centered at $(1, 0)$ and $(0, 1)$. Express as an integral (but do not evaluate): **(a)** the area of R and **(b)** the volume of revolution of R about the x-axis.

46. Let R be the intersection of the circles of radius 1 centered at $(0, 0)$ and $(0, 1)$. Express an integral that gives the volume of revolution of R about the x-axis. (Do not evaluate the integral.)

47. Let $a > 0$. Show that the volume obtained when the region between $y = a\sqrt{x - ax^2}$ and the x-axis is rotated about the x-axis is independent of the constant a.

48. If 12 J of work are needed to stretch a spring 20 cm beyond equilibrium, how much work is required to compress it 6 cm beyond equilibrium?

49. A spring whose equilibrium length is 15 cm exerts a force of 50 N when it is stretched to 20 cm. Find the work required to stretch the spring from 22 to 24 cm.

50. If 18 ft-lb of work are needed to stretch a spring 1.5 ft beyond equilibrium, how far will the spring stretch if a 12-lb weight is attached to its end?

51. Let W be the work (against the Sun's gravitational force) required to transport an 80-kg person from Earth to Mars when the two planets are aligned with the Sun at their minimal distance of 55.7×10^6 km. Use Newton's Universal Law of Gravity (see Exercises 35–37 in Section 6.5) to express W as an integral and evaluate it. The Sun has mass $M_s = 1.99 \times 10^{30}$ kg, and the distance from the Sun to Earth is 149.6×10^6 km.

In Exercises 52 and 53, water is pumped into a spherical tank of radius 2 m from a source located 1 m below a hole at the bottom (Figure 5). The density of water is 1000 kg/m³.

52. Calculate the work required to fill the tank.

53. Calculate the work $F(h)$ required to fill the tank to level h meters in the sphere.

54. A tank of mass 20 kg containing 100 kg of water (density 1000 kg/m³) is raised vertically at a constant speed of 100 m/minute for 1 min, during which time it leaks water at a rate of 40 kg/min. Calculate the total work performed in raising the container.

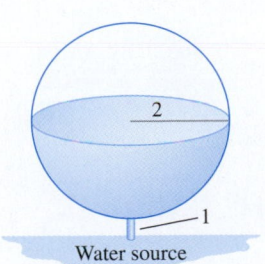

Water source

FIGURE 5

Xtremest/Shutterstock

7 TECHNIQUES OF INTEGRATION

The golden horn resembles an intriguing mathematical surface that we introduce in this chapter. The surface, known as Gabriel's Horn (after the archangel) or Torricelli's trumpet (after the seventeenth-century mathematician who first introduced it), encloses a finite volume yet has an infinite surface area.

In Section 5.7, we introduced substitution, one of the most important techniques of integration. In this chapter, we develop a second fundamental technique, Integration by Parts, as well as several techniques for treating particular classes of functions such as trigonometric and rational functions. However, there is no surefire method, and in fact, many important antiderivatives cannot be expressed in elementary terms. Therefore, we discuss numerical integration in the last section. Every definite integral can be approximated numerically to any desired degree of accuracy.

7.1 Integration by Parts

In this section, we derive a formula that often allows us to convert an integral that we cannot immediately evaluate into one that we can. The Integration by Parts formula is derived from the Product Rule.

Let u and v be functions of x:

$$\frac{d}{dx}(uv) = u\frac{dv}{dx} + \frac{du}{dx}v$$

According to this formula, uv is an antiderivative of the right-hand side, so

$$uv = \int u\frac{dv}{dx}\,dx + \int v\frac{du}{dx}\,dx$$

Moving the second integral on the right to the other side, we obtain

$$\int u\frac{dv}{dx}\,dx = uv - \int v\frac{du}{dx}\,dx$$

By letting $du = \frac{du}{dx}\,dx$ and $dv = \frac{dv}{dx}\,dx$, we find the following

We can keep track of our choices using the pattern

$$u = \square \qquad dv = \square$$
$$du = \square \qquad v = \square$$

The original integral is the product of the terms on the top line. The resulting expression is the product of the terms on the main diagonal minus the integral of the product of the terms on the bottom line. If we shade the terms that are multiplied together for each term, we have

$$\int \blacksquare\blacksquare_{\square\square} = \blacksquare_\square^\blacksquare \square - \int_{\blacksquare\blacksquare}^{\square\square}$$

In applying Eq. (1), any antiderivative v may be used.

> **Integration by Parts**
>
> $$\int u\,dv = uv - \int v\,du \qquad \boxed{1}$$

Because Integration by Parts applies to a product, we should consider using it when the integrand is a product of two functions. It is not, however, a product rule for antidifferentiation, because we cannot always use Integration by Parts to find an antiderivative of a product. Sometimes it works, as with $f(x) = x\cos x$ in Example 1; other times it does not, as with $f(x) = x\tan x$ (see Exercises 37 and 38). Furthermore, sometimes when the integrand is not expressed directly as a product, a clever application of Integration by Parts enables us to find an antiderivative (such as with $f(x) = \ln x$ in Example 3).

EXAMPLE 1 Evaluate $\int x\cos x\,dx$.

Solution The integrand is a product, so we try writing $x\cos x\,dx = u\,dv$ with

$$u = x \qquad dv = \cos x\,dx$$

Differentiating u and antidifferentiating dv, we find

$$du = \frac{du}{dx}\,dx = 1\,dx \qquad v = \sin x$$

By the Integration by Parts formula,

$$\int \underbrace{x}_{u}\underbrace{\cos x\,dx}_{dv} = \underbrace{x\sin x}_{uv} - \int \underbrace{\sin x}_{v}\underbrace{dx}_{du} = x\sin x + \cos x + C$$

Let's check the answer by taking the derivative:

$$\frac{d}{dx}(x \sin x + \cos x + C) = x \cos x + \sin x - \sin x = x \cos x$$

The key step in Integration by Parts is deciding how to write the integral as a product $u\,dv$. Keep in mind that Integration by Parts expresses $\int u\,dv$ in terms of uv and $\int v\,du$. *This is useful only if $v\,du$ is easier to integrate than $u\,dv$.* Here are two guidelines:

- Choose dv so that $v = \int dv$ can be evaluated.
- Choose u so that $\frac{du}{dx}$ is simpler than u itself.

The choices for u and dv that we made in Example 1 were good choices because they enabled us to transform the integral to a simpler one. A bad choice would be $u = \cos x$ and $dv = x\,dx$. Then, with $du = -\sin x\,dx$ and $v = \frac{1}{2}x^2$, the Integration by Parts formula yields

$$\int x \cos x\,dx = \underbrace{\frac{1}{2}x^2 \cos x}_{uv} - \int \underbrace{\left(\frac{1}{2}x^2\right)}_{v}\underbrace{(-\sin x)\,dx}_{du}$$

In this case, the resulting integral (essentially $\int x^2 \sin x\,dx$) is more complicated than the one we had initially.

EXAMPLE 2 **Integrating by Parts More Than Once** Evaluate $\int x^2 \cos x\,dx$.

In Example 2, it makes sense to take $u = x^2$ because Integration by Parts reduces the integration of $x^2 \cos x$ to the integration of $2x \sin x$, which is easier.

Solution Apply Integration by Parts a first time with $u = x^2$ and $dv = \cos x\,dx$:

$$\int \underbrace{x^2 \cos x\,dx}_{u\,dv} = \underbrace{x^2 \sin x}_{uv} - \int \underbrace{\sin x}_{v}\underbrace{2x\,dx}_{du} = x^2 \sin x - 2\int x \sin x\,dx \qquad \boxed{2}$$

Now apply it again to the integral on the right, this time with $u = x$ and $dv = \sin x\,dx$:

$$\int \underbrace{x \sin x\,dx}_{u\,dv} = \underbrace{-x \cos x}_{uv} - \int \underbrace{(-\cos x)}_{v}\underbrace{dx}_{du} = -x \cos x + \sin x + C$$

Using this result in Eq. (2), we obtain

$$\int x^2 \cos x\,dx = x^2 \sin x - 2\int x \sin x\,dx = x^2 \sin x - 2(-x \cos x + \sin x) + C$$

$$= x^2 \sin x + 2x \cos x - 2 \sin x + C$$

The function $f(x) = \ln x$ is one of the basic functions for which we have not yet seen an antiderivative formula. Now, with Integration by Parts, we can obtain one.

EXAMPLE 3 **Taking $dv = dx$** Evaluate $\int \ln x\,dx$.

Surprisingly, the choice $dv = dx$ is effective in some cases. Along with the integral of $\ln x$ in Example 3, it also works for the inverse trigonometric functions (see Exercise 6).

Solution The integrand is not a product, so at first glance, this integral does not look like a candidate for Integration by Parts. However, we can treat $\ln x\,dx$ as a product of $\ln x$ and dx. Then

$$u = \ln x \qquad dv = dx$$

$$du = \frac{1}{x}\,dx \qquad v = x$$

$$\int \underbrace{\ln x\,dx}_{u\,dv} = \underbrace{x \ln x}_{uv} - \int \underbrace{x\,\frac{1}{x}\,dx}_{v\,du} = x \ln x - \int dx$$

$$= x \ln x - x + C$$

Let's check the result of the previous example by differentiation:

$$\frac{d}{dx}(x \ln x - x) = x\left(\frac{1}{x}\right) + (1)\ln x - 1 = 1 + \ln x - 1 = \ln x$$

Now, to our table of integral formulas we can add the indefinite integral of $f(x) = \ln x$:

$$\int \ln x \, dx = x \ln x - x + C$$

There is a convenient definite-integral version of the Integration by Parts formula:

$$\int_a^b u \, dv = uv \Big|_a^b - \int_a^b v \, du$$

We employ this in the next example.

EXAMPLE 4   Red Fox Dispersal   You are investigating how the red fox disperses from its place of birth and have developed a model for determining the probability that breeding occurs within a particular distance from the birthplace. Specifically, for any x between 0 and 10, your model gives the likelihood that a fox breeds its first offspring within x km of its birthplace as $\int_0^x 0.2e^{-0.16t} \, dt$. With such a model, the average birth-to-breeding distance can be determined via the integral $\int_0^{10} x\left(0.2e^{-0.16x}\right) dx$. Compute this average.

The function $f(x) = 0.2e^{-0.16x}$ is a probability density function as introduced in Example 5 in Section 5.4. The average is also referred to as the "expected value" of the model. We investigate these concepts in more depth in Section 8.1.

Solution   To compute the average distance $\int_0^{10} x\left(0.2e^{-0.16x}\right) dx$, we use Integration by Parts with

$$u = x \qquad dv = 0.2e^{-0.16x} dx$$

$$du = dx \qquad v = \frac{0.2}{-0.16}e^{-0.16x} = -1.25e^{-0.16x}$$

Using the definite integral version of Integration by Parts, we have

$$\int_0^{10} x\left(0.2e^{-0.16x}\right) dx = -1.25xe^{-0.16x}\Big|_0^{10} + 1.25\int_0^{10} e^{-0.16x} \, dx$$

$$= -12.5e^{-1.6} + 0 + \frac{1.25}{-0.16}e^{-0.16x}\Big|_0^{10}$$

$$\approx -12.5e^{-1.6} - 7.81e^{-1.6} + 7.81 \approx 3.71$$

Thus, in the model, the average birth-to-breeding dispersal distance for the red fox is approximately 3.71 km.   ∎

EXAMPLE 5   Going in a Circle?   Evaluate $\int e^x \cos x \, dx$.

Solution   There are two reasonable ways of writing $e^x \cos x \, dx$ as $u \, dv$. Let's try setting $u = \cos x$. Then we have

$$u = \cos x \qquad dv = e^x \, dx$$

$$du = -\sin x \, dx \qquad v = e^x$$

In Example 5, the choice $u = e^x$, $dv = \cos x \, dx$ works equally well.

Thus,

$$\int \underbrace{e^x \cos x \, dx}_{u\,dv} = \underbrace{e^x \cos x}_{uv} - \int \underbrace{e^x(-\sin x) \, dx}_{v\,du}$$

Now use Integration by Parts on the integral on the right with $u = \sin x$:

$$u = \sin x \qquad dv = e^x\, dx$$

$$du = \cos x\, dx \qquad v = e^x$$

$$\int e^x \sin x\, dx = e^x \sin x - \int e^x \cos x\, dx \qquad \boxed{4}$$

Eq. (4) brings us back to our original integral of $e^x \cos x$, so it looks as if we're going in a circle. But we can substitute Eq. (4) in Eq. (3) and solve for the integral of $e^x \cos x$:

$$\int e^x \cos x\, dx = e^x \cos x + \int e^x \sin x\, dx = e^x \cos x + \left(e^x \sin x - \int e^x \cos x\, dx \right)$$

Now we can add $\int e^x \cos x\, dx$ to both sides. Note that we add a "$+\,C$" to the right side since we no longer have an integral on that side of the equation that will generate the necessary arbitrary constant:

$$2 \int e^x \cos x\, dx = e^x \cos x + e^x \sin x + C$$

Dividing an arbitrary constant by 2 still leaves an arbitrary constant, so we continue to denote it by C, absorbing the $\frac{1}{2}$ into the C.

$$\int e^x \cos x\, dx = \frac{1}{2} e^x (\cos x + \sin x) + C \qquad \blacksquare$$

A **reduction formula** expresses an integral for a given value of n in terms of a similar integral for a smaller value of n. It is sometimes necessary to apply a reduction formula repeatedly to reduce an integral to a simple one that can be computed.

EXAMPLE 6   A Reduction Formula   Derive the reduction formula

$$\boxed{\int \sin^n x\, dx = -\frac{1}{n} \sin^{n-1} x \cos x + \frac{n-1}{n} \int \sin^{n-2} x\, dx} \qquad \boxed{5}$$

Then use the reduction formula to evaluate $\int \sin^3 x\, dx$.

Solution   Although we do not know how to integrate $\sin^n x$, we do know how to integrate $\sin x$. So we apply Integration by Parts as follows:

$$u = \sin^{n-1} x \qquad\qquad dv = \sin x\, dx$$

$$du = (n-1) \sin^{n-2} x \cos x\, dx \qquad v = -\cos x$$

Then we have

$$\int \sin^n x\, dx = \underbrace{-\sin^{n-1} x\, \cos x}_{uv} - \int \underbrace{(-\cos x)(n-1)\sin^{n-2} x \cos x\, dx}_{v\,du}$$

$$= -\sin^{n-1} x \cos x + (n-1) \int \sin^{n-2} x\, \cos^2 x\, dx$$

Using the fact that $\cos^2 x = 1 - \sin^2 x$, we obtain

$$\int \sin^n x\, dx = -\sin^{n-1} x \cos x + (n-1) \int \sin^{n-2} x\, dx - (n-1) \int \sin^n x\, dx$$

Adding $(n-1) \int \sin^n x\, dx$ to both sides, and then dividing by n, results in the desired reduction formula:

$$n \int \sin^n x\, dx = -\sin^{n-1} x \cos x + (n-1) \int \sin^{n-2} x\, dx$$

$$\int \sin^n x\, dx = -\frac{1}{n} \sin^{n-1} x \cos x + \frac{n-1}{n} \int \sin^{n-2} x\, dx$$

Now applying the reduction formula in the case $n = 3$, we have

$$\int \sin^3 x \, dx = -\frac{1}{3} \sin^2 x \cos x + \frac{2}{3} \int \sin x \, dx = -\frac{1}{3} \sin^2 x \cos x - \frac{2}{3} \cos x + C \quad \blacksquare$$

Reduction formulas for $\int \cos^n x \, dx$, $\int x^n e^x \, dx$, and $\int (\ln x)^n \, dx$ are introduced and investigated in the exercises.

7.1 SUMMARY

- Integration by Parts formula: $\int u \, dv = uv - \int v \, du$.
- The key step is deciding how to write the integrand as a product $u \, dv$. Keep in mind that Integration by Parts is useful when $v \, du$ is easier (or, at least, not more difficult) to integrate than $u \, dv$. Here are some guidelines:
  - Choose u so that $\frac{du}{dx}$ is simpler than u itself.
  - Choose dv so that $v = \int dv$ can be evaluated.
  - Sometimes, $dv = dx$ is a good choice.
  - Good choices for u include x^n, $\ln x$, and inverse trigonometric functions.

7.1 EXERCISES

Preliminary Questions

1. Which derivative rule is used to derive the Integration by Parts formula?

2. For each of the following integrals, state whether substitution or Integration by Parts should be used:

$$\int x \cos(x^2) \, dx, \quad \int x \cos x \, dx, \quad \int x^2 e^x \, dx, \quad \int xe^{x^2} \, dx$$

3. Why is $u = \cos x$, $dv = x \, dx$ a poor choice for evaluating $\int x \cos x \, dx$?

Exercises

In Exercises 1–6, evaluate the integral using the Integration by Parts formula with the given choice of u and dv.

1. $\int x \sin x \, dx; \quad u = x, dv = \sin x \, dx$

2. $\int xe^{2x} \, dx; \quad u = x, dv = e^{2x} \, dx$

3. $\int (2x + 9)e^x \, dx; \quad u = 2x + 9, dv = e^x \, dx$

4. $\int x \cos 4x \, dx; \quad u = x, dv = \cos 4x \, dx$

5. $\int x^3 \ln x \, dx; \quad u = \ln x, dv = x^3 \, dx$

6. $\int \tan^{-1} x \, dx; \quad u = \tan^{-1} x, dv = dx$

In Exercises 7–34, evaluate using Integration by Parts.

7. $\int (4x - 3)e^{-x} \, dx$

8. $\int (2x + 1)e^x \, dx$

9. $\int x \, e^{5x+2} \, dx$

10. $\int x^2 e^x \, dx$

11. $\int x \cos 2x \, dx$

12. $\int x \sin(3 - x) \, dx$

13. $\int x^2 \sin x \, dx$

14. $\int x^2 \cos 3x \, dx$

15. $\int e^{-x} \sin x \, dx$

16. $\int e^x \sin 2x \, dx$

17. $\int e^{-5x} \sin x \, dx$

18. $\int e^{3x} \cos 4x \, dx$

19. $\int x \ln x \, dx$

20. $\int \frac{\ln x}{x^2} \, dx$

21. $\int x^2 \ln x \, dx$

22. $\int x^{-5} \ln x \, dx$

23. $\int (\ln x)^2 \, dx$

24. $\int x(\ln x)^2 \, dx$

25. $\int \cos^{-1} x \, dx$

26. $\int \sin^{-1} x \, dx$

27. $\int \sec^{-1} x \, dx$

28. $\int x5^x \, dx$

29. $\int 3^x \cos x \, dx$

30. $\int x \sinh x \, dx$

31. $\int x^2 \cosh x \, dx$

32. $\int \cos x \cosh x \, dx$

33. $\int \tanh^{-1} 4x \, dx$

34. $\int \sinh^{-1} x \, dx$

In Exercises 35–36, evaluate using substitution and then Integration by Parts.

35. $\int e^{\sqrt{x}} \, dx$ *Hint: Let $u = x^{1/2}$.*

36. $\int x^3 e^{x^2} \, dx$

37. 🖉 For $\int x \tan x \, dx$, try Integration by Parts with $u = x$, $dv = \tan x \, dx$ and with $u = \tan x$, $dv = x \, dx$, and describe the difficulty that you encounter in each case, keeping you from finding an antiderivative. (*Note: there is no antiderivative formula for $x \tan x$ involving elementary functions.*)

38. 🖉 For $\int x \sec x \, dx$, try Integration by Parts with $u = x$, $dv = \sec x \, dx$ and with $u = \sec x$, $dv = x \, dx$, and describe the difficulty that you encounter in each case, keeping you from finding an antiderivative. (*Note: there is no antiderivative formula for $x \sec x$ involving elementary functions.*)

In Exercises 39–48, evaluate using Integration by Parts, substitution, or both if necessary.

39. $\int x \cos 4x \, dx$

40. $\int \dfrac{\ln(\ln x) \, dx}{x}$

41. $\int \dfrac{x \, dx}{\sqrt{x+1}}$

42. $\int x^2 (x^3 + 9)^{15} \, dx$

43. $\int \cos x \ln(\sin x) \, dx$

44. $\int \sin \sqrt{x} \, dx$

45. $\int \sqrt{x} e^{\sqrt{x}} \, dx$

46. $\int \dfrac{\tan \sqrt{x} \, dx}{\sqrt{x}}$

47. $\int \dfrac{\ln(\ln x) \ln x \, dx}{x}$

48. $\int \sin(\ln x) \, dx$

In Exercises 49–58, compute the definite integral.

49. $\int_0^3 x e^{4x} \, dx$

50. $\int_0^{\pi/4} x \sin 2x \, dx$

51. $\int_1^2 x \ln x \, dx$

52. $\int_1^e \dfrac{\ln x \, dx}{x^2}$

53. $\int_0^1 x e^{-x} \, dx$

54. $\int_0^1 \dfrac{x^3}{\sqrt{9+x^2}} \, dx$

55. $\int_0^1 x 3^x \, dx$

56. $\int_0^1 x \cos(\pi x) \, dx$

57. $\int_0^{\pi} e^x \sin x \, dx$

58. $\int_0^1 \tan^{-1} x \, dx$

59. Robin has been tracking her archery accuracy. For $0 \le x \le 60$ the probability that an arrow that she shoots hits the target within x centimeters of the center is given by $\int_0^x 0.071 e^{-0.07t} \, dt$. The average distance of her shots from the center is given by $\int_0^{60} t \left(0.071 e^{-0.07t} \right) dt$. Compute the average distance.

60. When Darius is shooting for a bull's eye in darts, the probability that a throw lands within x millimeters of the center of the dart board is given by $\int_0^x 0.066 e^{-0.06t} \, dt$ for $0 \le x \le 40$.

(a) The bull's eye has a diameter of 12.5 mm. What is the probability that a throw is a bull's eye?

(b) The average distance of his throws from the center of the dart board is given by $\int_0^{40} t \left(0.066 e^{-0.06t} \right) dt$. Compute the average distance.

61. Use Eq. (5) to find $\int \sin^5 x \, dx$.

62. Derive the reduction formula
$$\int \cos^n x \, dx = \frac{1}{n} \cos^{n-1} x \sin x + \frac{n-1}{n} \int \cos^{n-2} x \, dx$$

63. Use the reduction formula from Exercise 62 to find $\int \cos^3 x \, dx$.

64. Derive the reduction formula
$$\int x^n e^x \, dx = x^n e^x - n \int x^{n-1} e^x \, dx$$

65. Use the reduction formula from Exercise 64 to find $\int x^3 e^x \, dx$.

66. Use substitution and the reduction formula from Exercise 64 to evaluate $\int x^4 e^{7x} \, dx$.

67. Find a reduction formula for $\int x^n e^{-x} \, dx$ similar to the formula appearing in Exercise 64.

68. Evaluate $\int x^n \ln x \, dx$ for $n \ne -1$. Which method should be used to evaluate $\int x^{-1} \ln x \, dx$?

69. Find the volume of the solid of revolution that results when the region under the graph of $f(x) = x \sqrt{\sin x}$ for $0 \le x \le \pi$ is revolved around the x-axis.

70. Find the volume of the solid of revolution that results when the region under the graph of $f(x) = \ln x$ for $1 \le x \le e$ is revolved around the x-axis.

71. Find the volume of the solid of revolution that results when the region under the graph of $f(x) = 3 \sin x$ for $0 \le x \le \pi$ is revolved around the y-axis.

72. Find the volume of the solid of revolution that results when the region under the graph of $f(x) = e^{-x}$ for $0 \le x \le 1$ is revolved around:

(a) the y-axis.   **(b)** the line $x = 1$.

In Exercises 73–80, indicate a good method for evaluating the integral (but do not evaluate). Your choices are algebraic manipulation, substitution (specify u and du), and Integration by Parts (specify u and dv). If it appears that the techniques you have learned thus far are not sufficient, state this.

73. $\int \sqrt{x} \ln x \, dx$

74. $\int \dfrac{x^2 - \sqrt{x}}{2x} \, dx$

75. $\int \dfrac{x^3 \, dx}{\sqrt{4 - x^2}}$

76. $\int \dfrac{dx}{\sqrt{4 - x^2}}$

77. $\int \dfrac{x + 2}{x^2 + 4x + 3} \, dx$

78. $\int \dfrac{dx}{(x+2)(x^2 + 4x + 3)}$

79. $\int x \sin(3x + 4) \, dx$

80. $\int x \cos(9x^2) \, dx$

81. Evaluate $\int (\sin^{-1} x)^2 \, dx$. *Hint: Use Integration by Parts first and then substitution.*

82. Evaluate $\int \dfrac{(\ln x)^2 \, dx}{x^2}$. *Hint: Use substitution first and then Integration by Parts.*

83. Evaluate $\int x^7 \cos(x^4)\, dx$.

84. Evaluate $\int_0^1 x^3 e^{-x^2}\, dx$.

85. Find the area of the region that lies under the graph of $y = (5 - x) \ln x$ and above the x-axis.

86. Find the area enclosed by $y = \ln x$ and $y = (\ln x)^2$.

87. The **present value** (PV) of an investment that provides income continuously at a rate $R(t)$ \$/year for T years, and earns interest at rate r, is $\int_0^T R(t) e^{-rt}\, dt$. We think of present value as the payment that we would need to receive at $t = 0$, instead of the investment income, so that at time T the payment's value (with accumulated interest) would be the same as the amount accumulated from the income stream (also accumulating interest). Find the PV if $R(t) = 5000 + 100t$ \$/year, $r = 0.05$, and $T = 10$ years.

88. Derive the reduction formula

$$\int (\ln x)^k\, dx = x(\ln x)^k - k \int (\ln x)^{k-1}\, dx \qquad \boxed{6}$$

89. Use Eq. (6) to calculate $\int (\ln x)^k\, dx$ for $k = 2, 3$.

90. Derive the reduction formulas

$$\int x^n \cos x\, dx = x^n \sin x - n \int x^{n-1} \sin x\, dx$$

$$\int x^n \sin x\, dx = -x^n \cos x + n \int x^{n-1} \cos x\, dx$$

91. Prove that $\int x b^x\, dx = b^x \left(\dfrac{x}{\ln b} - \dfrac{1}{(\ln b)^2} \right) + C$.

92. Define $P_n(x)$ by

$$\int x^n e^x\, dx = P_n(x)\, e^x + C$$

Use the reduction formula in Exercise 64 to prove that $P_n(x) = x^n - n P_{n-1}(x)$. Use this recursion relation to find $P_n(x)$ for $n = 1, 2, 3, 4$. Note that $P_0(x) = 1$.

Further Insights and Challenges

93. The Integration by Parts formula can be written

$$\int uv\, dx = uV - \int V\, du \qquad \boxed{7}$$

where $V(x)$ satisfies $V'(x) = v(x)$.

(a) Show directly that the right-hand side of Eq. (7) does not change if $V(x)$ is replaced by $V(x) + C$, where C is a constant.

(b) Use $u = \tan^{-1} x$ and $v = x$ in Eq. (7) to calculate $\int x \tan^{-1} x\, dx$,

but carry out the calculation twice: first with $V(x) = \frac{1}{2}x^2$ and then with $V(x) = \frac{1}{2}x^2 + \frac{1}{2}$. Which choice of $V(x)$ results in a simpler calculation?

94. Prove in two ways that

$$\int_0^a f(x)\, dx = af(a) - \int_0^a x f'(x)\, dx \qquad \boxed{8}$$

First use Integration by Parts. Then assume f is increasing. Use the substitution $u = f(x)$ to prove that $\int_0^a x f'(x)\, dx$ is equal to the area of the shaded region in Figure 1 and derive Eq. (8) a second time.

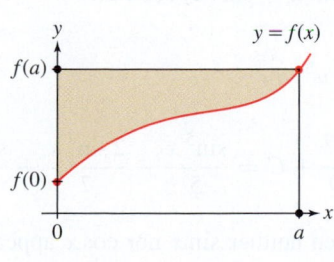

$y = f(x)$

$f(a)$

$f(0)$

0 a

FIGURE 1

95. Assume that $f(0) = f(1) = 0$ and that f'' exists. Prove

$$\int_0^1 f''(x) f(x)\, dx = -\int_0^1 f'(x)^2\, dx \qquad \boxed{9}$$

Use this to prove that if $f(0) = f(1) = 0$ and $f''(x) = \lambda f(x)$ for some constant λ, then $\lambda < 0$. Can you think of a function satisfying these conditions for some λ?

96. Set $I(a, b) = \int_0^1 x^a (1 - x)^b\, dx$, where a, b are whole numbers.

(a) Use substitution to show that $I(a, b) = I(b, a)$.

(b) Show that $I(a, 0) = I(0, a) = \dfrac{1}{a + 1}$.

(c) Prove that for $a \geq 1$ and $b \geq 0$,

$$I(a, b) = \frac{a}{b + 1} I(a - 1, b + 1)$$

(d) Use (b) and (c) to calculate $I(1, 1)$ and $I(3, 2)$.

(e) Show that $I(a, b) = \dfrac{a!\, b!}{(a + b + 1)!}$.

97. Let $I_n = \int x^n \cos(x^2)\, dx$ and $J_n = \int x^n \sin(x^2)\, dx$.

(a) Find a reduction formula that expresses I_n in terms of J_{n-2}. *Hint:* Write $x^n \cos(x^2)$ as $x^{n-1}(x \cos(x^2))$.

(b) ✏️ Use the result of (a) to show that I_n can be evaluated explicitly if n is odd.

(c) Evaluate I_3.

7.2 Trigonometric Integrals

In this section, we investigate integrals of various products of powers of trigonometric functions. We can often compute these integrals by combining substitution and Integration by Parts with trigonometric identities. In the section summary we expand on the table of integrals we have built so far, adding integral formulas employed or derived in this section and some other formulas similar to them. We begin with integrals of the form

$$\int \sin^m x \cos^n x \, dx$$

where m, n are whole numbers. The easier case is when at least one of m, n is *odd*.

EXAMPLE 1 **Odd Power of $\sin x$** Evaluate $\int \sin^3 x \, dx$.

Solution We did this integral in Example 6 of the last section. However, we will use a different method that is more broadly applicable. Because $\sin^3 x$ is an odd power, we split off one power of $\sin x$ and use the identity $\sin^2 x = 1 - \cos^2 x$ to convert the rest of the integrand into an expression in $\cos x$:

$$\sin^3 x = (\sin^2 x)(\sin x) = (1 - \cos^2 x) \sin x$$

We then use the substitution $u = \cos x$, $du = -\sin x \, dx$:

$$\int \sin^3 x \, dx = \int (1 - \cos^2 x) \sin x \, dx = -\int (1 - u^2) \, du$$

$$= \frac{u^3}{3} - u + C = \frac{\cos^3 x}{3} - \cos x + C \quad \blacksquare$$

The strategy of the previous example works when $\sin^m x$ appears with m odd, no matter what power of $\cos x$ is present. Similarly, if n is odd, we write $\cos^n x$ as a power of $(1 - \sin^2 x)$ times $\cos x$.

EXAMPLE 2 **Odd Power of $\cos x$** Evaluate $\int \sin^4 x \cos^5 x \, dx$.

Solution We take advantage of the fact that $\cos^5 x$ is an odd power to write

$$\sin^4 x \cos^5 x = \sin^4 x \, \cos^4 x (\cos x) = \sin^4 x (1 - \sin^2 x)^2 (\cos x)$$

$$= (\sin^4 x - 2 \sin^6 x + \sin^8 x) \cos x$$

This allows us to use the substitution $u = \sin x$, $du = \cos x \, dx$:

$$\int \sin^4 x \cos^5 x \, dx = \int (\sin^4 x - 2 \sin^6 x + \sin^8 x) \cos x \, dx$$

$$= \int (u^4 - 2u^6 + u^8) \, du$$

$$= \frac{u^5}{5} - \frac{2u^7}{7} + \frac{u^9}{9} + C = \frac{\sin^5 x}{5} - \frac{2 \sin^7 x}{7} + \frac{\sin^9 x}{9} + C \quad \blacksquare$$

We will need a different strategy when neither $\sin x$ nor $\cos x$ appears with an odd power.

EXAMPLE 3 Evaluate $\int \sin^2 x \, dx$.

Solution We utilize the trigonometric identity called the double angle formula $\sin^2 x = \frac{1}{2}(1 - \cos 2x)$. Then

$$\int \sin^2 x \, dx = \int \frac{1}{2}(1 - \cos 2x) \, dx = \frac{x}{2} - \frac{\sin 2x}{4} + C \quad \blacksquare$$

Using the trigonometric identities in the margin, we can also integrate $\cos^2 x$, obtaining the following:

$$\sin^2 x = \frac{1}{2}(1 - \cos 2x)$$

$$\cos^2 x = \frac{1}{2}(1 + \cos 2x)$$

$$\sin 2x = 2 \sin x \cos x$$

$$\cos 2x = \cos^2 x - \sin^2 x$$

$$\int \sin^2 x \, dx = \frac{x}{2} - \frac{\sin 2x}{4} + C = \frac{x}{2} - \frac{1}{2} \sin x \cos x + C \qquad \boxed{1}$$

$$\int \cos^2 x \, dx = \frac{x}{2} + \frac{\sin 2x}{4} + C = \frac{x}{2} + \frac{1}{2} \sin x \cos x + C \qquad \boxed{2}$$

EXAMPLE 4 Evaluate $\int \sin^4 x \, dx$.

Solution Double angle formulas will be of assistance to us here as well. Using $\sin^2 x = \frac{1}{2}(1 - \cos 2x)$, we obtain

$$\int \sin^4 x \, dx = \int (\sin^2 x)^2 \, dx = \int \left(\frac{1}{2}(1 - \cos 2x) \right)^2 dx$$

$$= \frac{1}{4} \int (1 - 2\cos 2x + \cos^2 2x) \, dx$$

As an alternative, the reduction formula for $\int \sin^n x \, dx$, Eq.(5) in the previous section, can be used to compute $\int \sin^2 x \, dx$ and $\int \sin^4 x \, dx$.

Applying the double angle formula to $\cos^2 2x$, we get

$$\int \sin^4 x \, dx = \frac{1}{4} \int (1 - 2\cos 2x + \cos^2 2x) \, dx$$

$$= \frac{1}{4} \int \left(1 - 2\cos 2x + \frac{1 + \cos 4x}{2} \right) dx$$

$$= \frac{1}{4} \int \left(\frac{3}{2} - 2\cos 2x + \frac{\cos 4x}{2} \right) dx$$

Integrating $\sin^m x \cos^n x$

Case 1: m **odd**

Split off one power of $\sin x$, and use $\sin^2 x = 1 - \cos^2 x$ to express the remaining powers of $\sin x$ in terms of $\cos x$. Then substitute $u = \cos x, du = -\sin x \, dx$.

By simple u-substitutions, this yields

$$\int \sin^4 x \, dx = \frac{1}{4} \left(\frac{3x}{2} - \sin 2x + \frac{\sin 4x}{8} \right) + C = \frac{3x}{8} - \frac{\sin 2x}{4} + \frac{\sin 4x}{32} + C \quad ■$$

Case 2: n **odd**

Split off one power of $\cos x$, and use $\cos^2 x = 1 - \sin^2 x$ to express the remaining powers of $\cos x$ in terms of $\sin x$. Then substitute $u = \sin x, du = \cos x \, dx$.

As indicated in the above margin comment, the reduction formula for $\int \sin^n x \, dx$ can be used for the previous two examples. There is also a corresponding reduction formula for $\int \cos^n x \, dx$. These two reduction formulas are (5) and (6) in the integral table at the end of this section. We employ the reduction formula for $\int \cos^n x \, dx$ in the next example.

Case 3: m, n **both even**

Either use the double angle formulas repeatedly or convert the integrand into an expression entirely in terms of $\sin x$ or $\cos x$ and then apply a reduction formula, either Eq. (5) or Eq. (6), in the integral table at the end of this section.

EXAMPLE 5   **Even Powers of $\sin x$ and $\cos x$**   Evaluate $\int \sin^2 x \cos^4 x \, dx$.

Solution Here, $m = 2$ and $n = 4$. Since $m < n$, we replace $\sin^2 x$ by $1 - \cos^2 x$:

$$\int \sin^2 x \cos^4 x \, dx = \int (1 - \cos^2 x)\cos^4 x \, dx = \int \cos^4 x \, dx - \int \cos^6 x \, dx$$

The reduction formula for $n = 6$ gives

$$\int \cos^6 x \, dx = \frac{1}{6} \cos^5 x \sin x + \frac{5}{6} \int \cos^4 x \, dx$$

Using this result in the right-hand side of the first equation in this solution, we obtain

$$\int \sin^2 x \cos^4 x \, dx = \int \cos^4 x \, dx - \left(\frac{1}{6} \cos^5 x \sin x + \frac{5}{6} \int \cos^4 x \, dx \right)$$

$$= -\frac{1}{6} \cos^5 x \sin x + \frac{1}{6} \int \cos^4 x \, dx$$

Next, we evaluate $\int \cos^4 x \, dx$ using the reduction formulas for $n = 4$ and $n = 2$:

$$\int \cos^4 x \, dx = \frac{1}{4} \cos^3 x \sin x + \frac{3}{4} \int \cos^2 x \, dx$$

$$= \frac{1}{4} \cos^3 x \sin x + \frac{3}{4} \left(\frac{1}{2} \cos x \sin x + \frac{1}{2} x \right) + C$$

$$= \frac{1}{4} \cos^3 x \sin x + \frac{3}{8} \cos x \sin x + \frac{3}{8} x + C$$

Altogether,

$$\int \sin^2 x \cos^4 x \, dx = -\frac{1}{6} \cos^5 x \sin x + \frac{1}{6} \left(\frac{1}{4} \cos^3 x \sin x + \frac{3}{8} \cos x \sin x + \frac{3}{8} x \right) + C$$

$$= -\frac{1}{6} \cos^5 x \sin x + \frac{1}{24} \cos^3 x \sin x + \frac{1}{16} \cos x \sin x + \frac{1}{16} x + C \quad \blacksquare$$

The results of trigonometric integrals can be expressed in more than one way. A computer algebra system provided the following:

$$\int \sin^2 x \cos^4 x \, dx = \frac{1}{16} x + \frac{1}{64} \sin 2x$$

$$-\frac{1}{64} \sin 4x - \frac{1}{192} \sin 6x + C$$

Trigonometric identities can be used to show that this result agrees with the one in Example 5.

So far, we have found antiderivative formulas for all of the standard trigonometric functions except $\sec x$ and $\csc x$. For $\sin x$ and $\cos x$, they were obtained directly from the corresponding derivative formulas. For $\tan x$, we used a substitution in Example 6 in Section 5.7 (and $\cot x$ can be handled similarly). In the next example, we obtain an antiderivative formula for $\sec x$. With a similar method we can find an antiderivative of $\csc x$ (see Exercise 68).

EXAMPLE 6 **Integral of Secant** Derive the formula

$$\int \sec x \, dx = \ln |\sec x + \tan x| + C$$

Solution To integrate $\sec x$, we multiply by an unusual fraction that equals one. This approach enables us to use a substitution that results in an antiderivative that we can compute.

Multiply the integrand by

$$1 = \frac{\sec x + \tan x}{\sec x + \tan x}$$

Then

$$\int \sec x \, dx = \int \sec x \left(\frac{\sec x + \tan x}{\sec x + \tan x} \right) dx = \int \frac{\sec^2 x + \sec x \tan x}{\sec x + \tan x} dx$$

Noting that the numerator is the derivative of the denominator, we let the denominator be u:

$$u = \sec x + \tan x \qquad \text{and} \qquad du = (\sec x \tan x + \sec^2 x) \, dx$$

Then our integral becomes

$$\int \frac{du}{u} = \ln |u| + C = \ln |\sec x + \tan x| + C \qquad \blacksquare$$

The integral $\int_0^x \sec t \, dt$ was first computed numerically in the 1590s by the English mathematician Edward Wright, decades before the invention of calculus. Although he did not invent the concept of an integral, Wright realized that the sums that approximate the integral hold the key to understanding the Mercator map projection, of great importance in sea navigation because it enabled sailors to reach their destinations along lines of fixed compass direction. The formula for the integral was first proved by James Gregory in 1668.

EXAMPLE 7 **Using a Table of Integrals** Evaluate $\int_0^{\pi/4} \tan^3 x \, dx$.

Solution We use Eq. (8) in the table at the end of the section. With $k = 3$,

$$\int_0^{\pi/4} \tan^3 x \, dx = \frac{\tan^2 x}{2} \Big|_0^{\pi/4} - \int_0^{\pi/4} \tan x \, dx = \left(\frac{1}{2} \tan^2 x - \ln |\sec x| \right) \Big|_0^{\pi/4}$$

$$= \left(\frac{1}{2} \tan^2 \frac{\pi}{4} - \ln \left| \sec \frac{\pi}{4} \right| \right) - \left(\frac{1}{2} \tan^2 0 - \ln |\sec 0| \right)$$

$$= \left(\frac{1}{2} (1)^2 - \ln \sqrt{2} \right) - \left(\frac{1}{2} 0^2 - \ln |1| \right) = \frac{1}{2} - \ln \sqrt{2} \qquad \blacksquare$$

In the margin at the top of the next page, we describe a method for integrating $\tan^m x \sec^n x$.

Integrating $\tan^m x \sec^n x$

Case 1: m odd and $n \geq 1$

Separate out the factor $\sec x \tan x$ and use the identity $\tan^2 x = \sec^2 x - 1$ to express the rest of the integrand in terms of $\sec x$. Then use the substitution $u = \sec x$, $du = \sec x \tan x \, dx$ to obtain an integral involving only powers of u.

Case 2: n even

Separate out a factor of $\sec^2 x$ and use the identity $\sec^2 x = 1 + \tan^2 x$ to express the rest of the integrand in terms of $\tan x$. Then substitute $u = \tan x$, $du = \sec^2 x \, dx$ to obtain an integral involving only powers of u.

Case 3: m even and n odd

Use the identity $\tan^2 x = \sec^2 x - 1$ to obtain an integral involving only powers of $\sec x$ and use the reduction formula given by Eq. (12).

EXAMPLE 8 Evaluate $\displaystyle\int \tan^3 x \sec^5 x \, dx$.

Solution We note that we have a copy of $\sec x \tan x$ in the integrand, and the remaining powers of $\tan x$ are even. So we separate out one copy of $\sec x \tan x$ and convert the rest of the integrand into powers of $\sec x$. Then since the derivative of $\sec x$ is $\sec x \tan x$, we are set up for a u-substitution.

The first step is to use the identity $\tan^2 x = \sec^2 x - 1$:

$$\int \tan^3 x \sec^5 x \, dx = \int (\sec^2 x - 1)(\sec^4 x)(\sec x \tan x) \, dx$$

$$= \int (\sec^6 x - \sec^4 x) \sec x \tan x \, dx$$

Letting $u = \sec x$, so $du = \sec x \tan x \, dx$, we have

$$\int \tan^3 x \sec^5 x \, dx = \int (u^6 - u^4) \, du = \frac{u^7}{7} - \frac{u^5}{5} + C$$

$$= \frac{\sec^7 x}{7} - \frac{\sec^5 x}{5} + C \quad\blacksquare$$

Note that the above method works whenever we integrate $\tan^m x \sec^n x$, where m is odd and $n > 0$.

EXAMPLE 9 Evaluate $\displaystyle\int \tan^2 x \sec^4 x \, dx$.

Solution In this case, since the derivative of $\tan x$ is $\sec^2 x$, we separate out $\sec^2 x$ and convert the rest of the integrand into powers of $\tan x$. Using the fact $\sec^2 x = \tan^2 x + 1$ yields

$$\int \tan^2 x \sec^4 x \, dx = \int (\tan^2 x)(\tan^2 x + 1)(\sec^2 x) \, dx = \int (\tan^4 x + \tan^2 x) \sec^2 x \, dx$$

Setting $u = \tan x$ and $du = \sec^2 x \, dx$, we have

$$\int \tan^2 x \sec^4 x \, dx = \int (u^4 + u^2) \, du = \frac{u^5}{5} + \frac{u^3}{3} + C = \frac{\tan^5 x}{5} + \frac{\tan^3 x}{3} + C \quad\blacksquare$$

The above method works to integrate $\tan^m x \sec^n x$ whenever $n > 0$ is even. The last case to consider is when m is even and n is odd. In that case we convert the integrand to be entirely in terms of powers of $\sec x$ and then apply reduction formulas. Examples of this case can be found in the exercises.

CONCEPTUAL INSIGHT We previously mentioned that different methods for evaluating an indefinite integral may yield different expressions. Using identities and simplification, we can show that the results are equivalent. Keep in mind that

$$\int f(x) \, dx = F(x) + C \quad \text{and} \quad \int f(x) \, dx = G(x) + C$$

do not imply that $F(x) = G(x)$. Instead, they indicate that the functions F and G differ by a constant. To verify that two such antiderivative results are equivalent, you need to show that $F(x) = G(x) + C$ for some constant C.

For example, separately using the substitutions $u = \sin x$ and $u = \cos x$, we obtain the results

$$\int 2 \sin x \cos x \, dx = \sin^2 x + C \quad \text{and} \quad \int 2 \sin x \cos x \, dx = -\cos^2 x + C$$

Of course, $\sin^2 x \neq -\cos^2 x$. However, since $\sin^2 x = -\cos^2 x + 1$ it follows that the results are equivalent. Specifically,

$$\int 2 \sin x \cos x \, dx = \sin^2 x + C = -\cos^2 x + 1 + C = -\cos^2 x + C$$

where, in the last equality, we express $1 + C$ as an arbitrary constant C.

The formulas in Eqs. (15)–(17) in the integral table at the end of the section give the integrals of the products $\sin mx \sin nx$, $\cos mx \cos nx$, and $\sin mx \cos nx$. These integrals appear in the theory of Fourier Series, which is a fundamental technique used extensively in engineering and physics.

EXAMPLE 10 **Integral of $\sin mx \cos nx$** Evaluate $\displaystyle\int_0^{\pi} \sin 4x \cos 3x \, dx$.

Solution Apply Eq. (16), with $m = 4$ and $n = 3$:

$$\int_0^{\pi} \sin 4x \cos 3x \, dx = \left(-\frac{\cos(4-3)x}{2(4-3)} - \frac{\cos(4+3)x}{2(4+3)} \right) \Big|_0^{\pi}$$

$$= \left(-\frac{\cos x}{2} - \frac{\cos 7x}{14} \right) \Big|_0^{\pi}$$

$$= \left(\frac{1}{2} + \frac{1}{14} \right) - \left(-\frac{1}{2} - \frac{1}{14} \right) = \frac{8}{7} \qquad \blacksquare$$

The following table of trigonometric integrals summarizes some of the integral formulas we have seen in this chapter and includes some other related formulas.

TABLE OF TRIGONOMETRIC INTEGRALS

$$\int \sin^2 x \, dx = \frac{x}{2} - \frac{\sin 2x}{4} + C = \frac{x}{2} - \frac{1}{2} \sin x \cos x + C \qquad \boxed{3}$$

$$\int \cos^2 x \, dx = \frac{x}{2} + \frac{\sin 2x}{4} + C = \frac{x}{2} + \frac{1}{2} \sin x \cos x + C \qquad \boxed{4}$$

$$\int \sin^n x \, dx = -\frac{\sin^{n-1} x \cos x}{n} + \frac{n-1}{n} \int \sin^{n-2} x \, dx \qquad \boxed{5}$$

$$\int \cos^n x \, dx = \frac{\cos^{n-1} x \sin x}{n} + \frac{n-1}{n} \int \cos^{n-2} x \, dx \qquad \boxed{6}$$

$$\int \tan x \, dx = \ln|\sec x| + C = -\ln|\cos x| + C \qquad \boxed{7}$$

$$\int \tan^m x \, dx = \frac{\tan^{m-1} x}{m-1} - \int \tan^{m-2} x \, dx \qquad \boxed{8}$$

$$\int \cot x \, dx = -\ln|\csc x| + C = \ln|\sin x| + C \qquad \boxed{9}$$

$$\int \cot^m x \, dx = -\frac{\cot^{m-1} x}{m-1} - \int \cot^{m-2} x \, dx \qquad \boxed{10}$$

$$\int \sec x \, dx = \ln|\sec x + \tan x| + C \qquad \boxed{11}$$

$$\int \sec^m x \, dx = \frac{\tan x \sec^{m-2} x}{m-1} + \frac{m-2}{m-1} \int \sec^{m-2} x \, dx \qquad \boxed{12}$$

$$\int \csc x \, dx = \ln|\csc x - \cot x| + C \qquad \boxed{13}$$

$$\int \csc^m x \, dx = -\frac{\cot x \csc^{m-2} x}{m-1} + \frac{m-2}{m-1} \int \csc^{m-2} x \, dx \qquad \boxed{14}$$

$$\int \sin mx \sin nx \, dx = \frac{\sin(m-n)x}{2(m-n)} - \frac{\sin(m+n)x}{2(m+n)} + C \quad (m \neq \pm n) \qquad \boxed{15}$$

$$\int \sin mx \cos nx \, dx = -\frac{\cos(m-n)x}{2(m-n)} - \frac{\cos(m+n)x}{2(m+n)} + C \quad (m \neq \pm n) \qquad \boxed{16}$$

$$\int \cos mx \cos nx \, dx = \frac{\sin(m-n)x}{2(m-n)} + \frac{\sin(m+n)x}{2(m+n)} + C \quad (m \neq \pm n) \qquad \boxed{17}$$

7.2 SUMMARY

- The integral $\int \sin^m x \cos^n x \, dx$ can be evaluated as in the marginal note on page 409.
- The integral $\int \tan^m x \sec^n x \, dx$ can be evaluated as in the marginal note on page 411.

7.2 EXERCISES

Preliminary Questions

1. Describe the technique used to evaluate $\int \sin^5 x \, dx$.

2. Describe a way of evaluating $\int \sin^6 x \, dx$.

3. Are reduction formulas needed to evaluate $\int \sin^7 x \cos^2 x \, dx$? Why or why not?

4. Describe a way of evaluating $\int \sin^6 x \cos^2 x \, dx$.

5. Which integral requires more work to evaluate?

$$\int \sin^{798} x \cos x \, dx \qquad \text{or} \qquad \int \sin^4 x \cos^4 x \, dx$$

Explain your answer.

Exercises

In Exercises 1–6, evaluate the integral.

1. $\int \cos^3 x \, dx$

2. $\int \sin^5 x \, dx$

3. $\int \sin^3 \theta \cos^2 \theta \, d\theta$

4. $\int \sin^5 x \cos x \, dx$

5. $\int \sin^3 t \cos^3 t \, dt$

6. $\int \sin^2 x \cos^5 x \, dx$

7. Compute the area under the graph of $y = \cos^3 x$ from $x = 0$ to $x = \frac{\pi}{2}$.

8. Compute the area under the graph of $y = \sin^5 x$ from $x = 0$ to $x = \pi$.

In Exercises 9–12, evaluate the integrals.

9. $\int \tan^3 x \sec x \, dx$

10. $\int \tan x \sec^3 x \, dx$

11. $\int \tan^2 x \sec^4 x \, dx$

12. $\int \tan^8 x \sec^2 x \, dx$

In Exercises 13–16, evaluate using methods similar to those that apply to the integrals of $\tan^m x \sec^n x$.

13. $\int \cot^3 x \, dx$

14. $\int \csc^4 x \, dx$

15. $\int \cot^5 x \csc^2 x \, dx$

16. $\int \cot^5 x \csc x \, dx$

17. Compute the area under the graph of $y = \tan^2 x$ from $x = 0$ to $x = \frac{\pi}{4}$.

18. Compute the area under the graph of $y = \sec^4 x$ from $x = 0$ to $x = \frac{\pi}{3}$.

In Exercises 19–46, evaluate the integral.

19. $\int \sin^6 x \, dx$

20. $\int \sin^2 x \cos^2 x \, dx$

21. $\int \cos^5 x \sin x \, dx$

22. $\int \cos^3(2 - x) \sin(2 - x) \, dx$

23. $\int \cos^4(3x + 2) \, dx$

24. $\int \cos^7 3x \, dx$

25. $\int \cos^3(\pi\theta) \sin^4(\pi\theta) \, d\theta$

26. $\int \cos^{498} y \sin^3 y \, dy$

27. $\int \frac{\cos^5 x}{\sin^3 x} \, dx$

28. $\int \frac{\sin^7 x}{\cos^4 x} \, dx$

29. $\int \csc^2(3 - 2x) \, dx$

30. $\int \csc^3 x \, dx$

31. $\int \tan x \sec^2 x \, dx$

32. $\int \tan^3 \theta \sec^3 \theta \, d\theta$

33. $\int \tan^5 x \sec^4 x \, dx$

34. $\int \tan^4 x \sec x \, dx$

35. $\int \tan^6 x \sec^4 x \, dx$

36. $\int \tan^2 x \sec^3 x \, dx$

37. $\int \cot^5 x \csc^5 x \, dx$

38. $\int \cot^2 x \csc^4 x \, dx$

39. $\int \sin 2x \cos 2x \, dx$

40. $\int \cos 4x \cos 6x \, dx$

41. $\int \sin 2x \cos^3 x \, dx$

42. $\int \sin^2 x \sec^4 x \, dx$

43. $\int t \cos^3(t^2) \, dt$

44. $\int \frac{\tan^3(\ln t)}{t} \, dt$

45. $\int \cos^2(\sin t) \cos t \, dt$

46. $\int e^x \tan^2(e^x) \, dx$

In Exercises 47–60, evaluate the definite integral.

47. $\int_0^{2\pi} \sin^2 x \, dx$

48. $\int_0^{\pi/2} \cos^3 x \, dx$

49. $\int_0^{\pi/2} \sin^5 x \, dx$

50. $\int_0^{\pi/2} \sin^2 x \cos^3 x \, dx$

51. $\int_0^{\pi/4} \frac{dx}{\cos x}$

52. $\int_{\pi/4}^{\pi/2} \frac{dx}{\sin x}$

53. $\int_0^{\pi/3} \tan x \, dx$

54. $\int_0^{\pi/4} \tan^5 x \, dx$

55. $\int_{-\pi/4}^{\pi/4} \sec^4 x \, dx$

56. $\int_{\pi/4}^{3\pi/4} \cot^4 x \csc^2 x \, dx$

57. $\int_0^{\pi} \sin 3x \cos 4x \, dx$

58. $\int_0^{\pi} \sin x \sin 3x \, dx$

59. $\int_0^{\pi/6} \sin 2x \cos 4x \, dx$

60. $\int_0^{\pi/4} \sin 7x \cos 2x \, dx$

61. For n a positive integer, compute the area under the graph of $y = \sin^n x \cos^3 x$ for $0 \le x \le \frac{\pi}{2}$.

62. For n a positive integer, compute the area under the graph of $y = \tan^n x \sec^4 x$ for $0 \le x \le \frac{\pi}{4}$.

63. Use the identities for $\sin 2x$ and $\cos 2x$ on page 409 to verify that the following formulas are equivalent:

$$\int \sin^4 x \, dx = \frac{1}{32} (12x - 8 \sin 2x + \sin 4x) + C$$

$$\int \sin^4 x \, dx = -\frac{1}{4} \sin^3 x \cos x - \frac{3}{8} \sin x \cos x + \frac{3}{8}x + C$$

64. Evaluate $\int \sin^2 x \cos^3 x \, dx$ using the method described in the text and verify that your result is equivalent to the following result produced by a computer algebra system:

$$\int \sin^2 x \cos^3 x \, dx = \frac{1}{30}(7 + 3 \cos 2x) \sin^3 x + C$$

65. Find the volume of the solid obtained by revolving $y = \sin x$ for $0 \le x \le \pi$ about the x-axis.

66. Find the volume of the solid obtained by revolving $y = \tan x$ for $0 \le x \le \frac{\pi}{6}$ about the x-axis.

67. Prove the reduction formula

$$\int \tan^k x \, dx = \frac{\tan^{k-1} x}{k-1} - \int \tan^{k-2} x \, dx$$

Hint: $\tan^k x = (\sec^2 x - 1) \tan^{k-2} x$.

68. Use the substitution $u = \csc x - \cot x$ to evaluate $\int \csc x \, dx$ (see Example 6).

69. Let $I_m = \int_0^{\pi/2} \sin^m x \, dx$.

(a) Show that $I_0 = \frac{\pi}{2}$ and $I_1 = 1$.

(b) Prove that, for $m \ge 2$,

$$I_m = \frac{m-1}{m} I_{m-2}$$

(c) Use (a) and (b) to compute I_m for $m = 2, 3, 4, 5$.

70. Evaluate $\int_0^{\pi} \sin^2 mx \, dx$ for m an arbitrary integer.

71. Evaluate $\int \sin x \ln(\sin x) \, dx$. *Hint:* Use Integration by Parts as a first step.

72. Total Energy A 100-watt (W) light bulb has resistance $R = 144$ ohms (Ω) when attached to household current, where the voltage varies as $V = V_0 \sin(2\pi f t)$ ($V_0 = 110$ V, $f = 60$ Hz). The energy (in joules) expended by the bulb over a period of T seconds is

$$U = \int_0^T P(t) \, dt$$

where $P = V^2/R$ (J/s) is the power. Compute U if the bulb remains on for 5 hours.

73. Let m, n be integers with $m \ne \pm n$. Use Eqs. (15)–(17) to prove the so-called **orthogonality relations** that play a basic role in the theory of Fourier Series (Figure 1):

$$\int_0^{\pi} \sin mx \sin nx \, dx = 0$$

$$\int_0^{\pi} \cos mx \cos nx \, dx = 0$$

$$\int_0^{2\pi} \sin mx \cos nx \, dx = 0$$

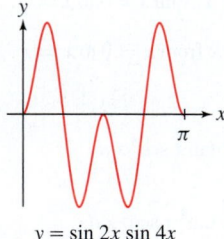
$y = \sin 2x \sin 4x$

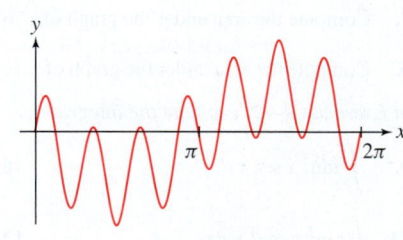
$y = \sin 3x \cos 4x$

FIGURE 1 The integrals are zero by the orthogonality relations.

Further Insights and Challenges

74. Use the trigonometric identity

$$\sin mx \cos nx = \frac{1}{2}\big(\sin(m-n)x + \sin(m+n)x\big)$$

to prove Eq. (16) in the table of integrals on page 412.

75. Use Integration by Parts to prove that (for $m \ne 1$)

$$\int \sec^m x \, dx = \frac{\tan x \sec^{m-2} x}{m-1} + \frac{m-2}{m-1} \int \sec^{m-2} x \, dx$$

76. Set $I_m = \int_0^{\pi/2} \sin^m x \, dx$. Use Exercise 69 to prove that

$$I_{2m} = \frac{2m-1}{2m} \frac{2m-3}{2m-2} \cdots \frac{1}{2} \cdot \frac{\pi}{2}$$

$$I_{2m+1} = \frac{2m}{2m+1} \frac{2m-2}{2m-1} \cdots \frac{2}{3}$$

Conclude that

$$\frac{\pi}{2} = \frac{2 \cdot 2}{1 \cdot 3} \cdot \frac{4 \cdot 4}{3 \cdot 5} \cdots \frac{2m \cdot 2m}{(2m-1)(2m+1)} \frac{I_{2m}}{I_{2m+1}}$$

77. This is a continuation of Exercise 76.

(a) Prove that $I_{2m+1} \le I_{2m} \le I_{2m-1}$. *Hint:*

$$\sin^{2m+1} x \le \sin^{2m} x \le \sin^{2m-1} x \quad \text{for} \quad 0 \le x \le \frac{\pi}{2}$$

(b) Show that $\dfrac{I_{2m-1}}{I_{2m+1}} = 1 + \dfrac{1}{2m}$.

(c) Show that $1 \le \dfrac{I_{2m}}{I_{2m+1}} \le 1 + \dfrac{1}{2m}$.

(d) Prove that $\displaystyle\lim_{m \to \infty} \frac{I_{2m}}{I_{2m+1}} = 1$.

(e) Finally, deduce the infinite product for $\frac{\pi}{2}$ discovered by English mathematician John Wallis (1616–1703):

$$\frac{\pi}{2} = \lim_{m \to \infty} \frac{2}{1} \cdot \frac{2}{3} \cdot \frac{4}{3} \cdot \frac{4}{5} \cdots \frac{2m \cdot 2m}{(2m-1)(2m+1)}$$

7.3 Trigonometric Substitution

Our next goal is to integrate functions involving one of the square root expressions:

$$\sqrt{a^2 - x^2}, \qquad \sqrt{x^2 + a^2}, \qquad \sqrt{x^2 - a^2}$$

More generally, we will see that we can also integrate functions involving $\sqrt{ax^2 + bx + c}$. In each case, a substitution transforms the integral into a trigonometric integral. For example, if we let $x = a \sin \theta$ in the first case (and assume $a > 0$ and $-\frac{\pi}{2} \le \theta \le \frac{\pi}{2}$) we can use the fact $\sin^2 \theta + \cos^2 \theta = 1$ to obtain

$$\sqrt{a^2 - x^2} = \sqrt{a^2 - a^2 \sin^2 \theta} = \sqrt{a^2(1 - \sin^2 \theta)} = \sqrt{a^2 \cos^2 \theta}$$

$$= |a \cos \theta| = a \cos \theta$$

◄— REMINDER $\sqrt{u^2} = |u|$

where the last equality holds since $a > 0$ and since $\cos \theta \ge 0$ (because $-\frac{\pi}{2} \le \theta \le \frac{\pi}{2}$). Therefore, using this substitution the term $\sqrt{a^2 - x^2}$ is changed to $a \cos \theta$. Integrals involving the other root functions listed above can also be simplified by applying a substitution and appropriate trigonometric identities as shown in the following examples.

EXAMPLE 1 Evaluate $\displaystyle\int \frac{1}{\sqrt{1 - x^2}} \, dx$.

Solution

Step 1. **Substitute to eliminate the square root.**

The integrand is defined for $-1 < x < 1$, so we may set $x = \sin \theta$, where $-\frac{\pi}{2} < \theta < \frac{\pi}{2}$. Then, with this substitution and the identity $\sin^2 x + \cos^2 x = 1$, we obtain

$$\sqrt{1 - x^2} = \sqrt{1 - \sin^2 \theta} = \sqrt{\cos^2 \theta} = |\cos \theta| = \cos \theta \qquad \boxed{1}$$

Step 2. **Evaluate the trigonometric integral.**

Since $x = \sin \theta$, we have $dx = \cos \theta \, d\theta$, and $\dfrac{1}{\sqrt{1 - x^2}} \, dx = \dfrac{1}{\cos \theta}(\cos \theta \, d\theta) = d\theta$.

Thus,

$$\int \frac{1}{\sqrt{1 - x^2}} \, dx = \int d\theta = \theta + C$$

Step 3. **Convert back to the original variable.**

Since $x = \sin \theta$ for $-\frac{\pi}{2} < \theta < \frac{\pi}{2}$, it follows that $\theta = \sin^{-1} x$, and

$$\int \frac{1}{\sqrt{1 - x^2}} \, dx = \sin^{-1} x + C$$

■

EXAMPLE 2 Evaluate $\displaystyle\int \sqrt{1 - x^2} \, dx$.

Solution

Step 1. **Substitute to eliminate the square root.**

As in the previous example, we set $x = \sin \theta$, where $-\frac{\pi}{2} \le \theta \le \frac{\pi}{2}$, and we have

$$\sqrt{1 - x^2} = \sqrt{1 - \sin^2 \theta} = \sqrt{\cos^2 \theta} = \cos \theta$$

◄— REMINDER

$$\int \cos^2 \theta \, d\theta = \frac{1}{2}\theta + \frac{1}{2}\sin \theta \cos \theta + C$$

Step 2. **Evaluate the trigonometric integral.**

Since $x = \sin \theta$, we have $dx = \cos \theta \, d\theta$, and $\sqrt{1 - x^2} \, dx = \cos \theta(\cos \theta \, d\theta)$. Thus,

$$\int \sqrt{1 - x^2} \, dx = \int \cos^2 \theta \, d\theta = \frac{1}{2}\theta + \frac{1}{2}\sin \theta \cos \theta + C$$

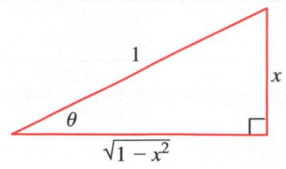

FIGURE 1 Right triangle with $x = \sin\theta$, from which $\cos\theta = \sqrt{1 - x^2}$.

Step 3. **Convert back to the original variable.**

It remains to express the answer in terms of x. We have $x = \sin\theta$ and $\theta = \sin^{-1} x$, but how do we convert the $\cos\theta$ term? We could express it as $\cos\left(\sin^{-1} x\right)$, but there is a better, clearer way to express that term using right-triangle trigonometry. In Figure 1, we have a right triangle for which our substitution relationship $x = \sin\theta$ holds. In that triangle, $\cos\theta = \sqrt{1 - x^2}$, and thus,

$$\int \sqrt{1 - x^2}\, dx = \frac{1}{2}\theta + \frac{1}{2}\sin\theta\cos\theta + C = \frac{1}{2}\sin^{-1} x + \frac{1}{2}x\sqrt{1 - x^2} + C \quad \blacksquare$$

The following summarizes the substitution approach we employed in the first two examples:

> **Integrals Involving $\sqrt{a^2 - x^2}$**   If $\sqrt{a^2 - x^2}$ occurs in an integral where $a > 0$, try the substitution (assuming $-\frac{\pi}{2} \le \theta \le \frac{\pi}{2}$)
>
> $$x = a\sin\theta, \qquad dx = a\cos\theta\, d\theta, \qquad \sqrt{a^2 - x^2} = a\cos\theta$$

This substitution approach can be used with integrands involving $(a^2 - x^2)^{n/2}$ for any integer n. The following example demonstrates a case with $n = 3$.

EXAMPLE 3   Integrand Involving $(a^2 - x^2)^{3/2}$ The graph of $f(x) = \dfrac{1}{(4 - x^2)^{3/2}}$ for $0 \le x < 2$ is shown in Figure 2. It has a vertical asymptote at $x = 2$. What is the area under the graph over $[0, 1.999]$?

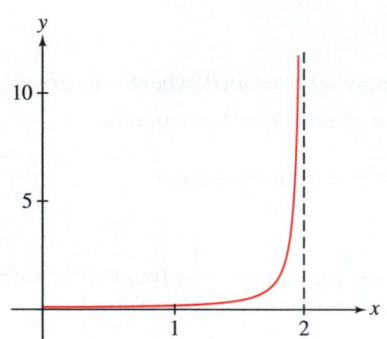

FIGURE 2 $f(x) = \frac{1}{(4-x^2)^{3/2}}$ for $0 \le x < 2$.

Solution We need to compute $\displaystyle\int_0^{1.999} \frac{1}{(4 - x^2)^{3/2}}\, dx$. We will compute the corresponding indefinite integral first, and then evaluate the definite integral.

Step 1. **Substitute to eliminate the square root.**

In this case, $a = 2$ since $\sqrt{4 - x^2} = \sqrt{2^2 - x^2}$. Therefore, we use

$$x = 2\sin\theta, \qquad dx = 2\cos\theta\, d\theta, \qquad \sqrt{4 - x^2} = 2\cos\theta$$

$$\int \frac{1}{(4 - x^2)^{3/2}}\, dx = \int \frac{1}{2^3\cos^3\theta}\, 2\cos\theta\, d\theta = \int \frac{1}{4\cos^2\theta}\, d\theta = \frac{1}{4}\int \sec^2\theta\, d\theta$$

Step 2. **Evaluate the trigonometric integral.**

$$\frac{1}{4}\int \sec^2\theta\, d\theta = \frac{1}{4}\tan\theta + C$$

Step 3. **Convert back to the original variable.**

We must write $\tan\theta$ in terms of x. Since $x = 2\sin\theta$, we have $\sin\theta = x/2$. Using the triangle in Figure 3 to determine an expression for $\tan\theta$, we obtain

$$\tan\theta = \frac{\text{opposite}}{\text{adjacent}} = \frac{x}{\sqrt{4 - x^2}}$$

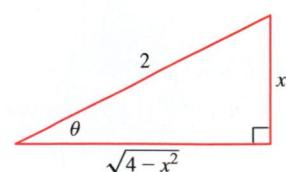

FIGURE 3 Right triangle with $\sin\theta = \frac{x}{2}$.

Thus, we have

$$\int \frac{1}{(4 - x^2)^{3/2}}\, dx = \frac{1}{4}\tan\theta + C = \frac{x}{4\sqrt{4 - x^2}} + C$$

Now, it follows that

$$\int_0^{1.999} \frac{1}{(4 - x^2)^{3/2}}\, dx = \left.\frac{x}{4\sqrt{4 - x^2}}\right|_0^{1.999} \approx 7.903$$

So the area under $f(x) = \dfrac{1}{(4 - x^2)^{3/2}}$ from $x = 0$ to $x = 1.999$ is approximately 7.903.

$\blacksquare$

What happens to the area in the previous example if we change the upper limit of the integration, letting the right side of the region approach the asymptote at $x = 2$? It turns out that as the upper limit approaches 2, the area gets larger and larger without bound (see Exercises 33 and 34). Interestingly, in contrast to the function in Example 3, there are functions that become infinite over an interval but the areas under the graphs of these functions are finite. We explore these ideas further in Section 7.7, investigating what are known as *improper* integrals.

When the integrand involves $\sqrt{x^2 + a^2}$, try the substitution $x = a \tan \theta$, assuming $a > 0$ and $-\frac{\pi}{2} \leq \theta \leq \frac{\pi}{2}$. Then

$$\sqrt{x^2 + a^2} = \sqrt{a^2 \tan^2 \theta + a^2} = \sqrt{a^2(\tan^2 \theta + 1)} = \sqrt{a^2 \sec^2 \theta}$$

$$= |a \sec \theta| = a \sec \theta$$

where the last equality holds since $a > 0$ and since $\sec \theta \geq 0$ for the values of θ involved.

Integrals Involving $\sqrt{x^2 + a^2}$ If $\sqrt{x^2 + a^2}$ occurs in an integral where $a > 0$, try the substitution (assuming $-\frac{\pi}{2} \leq \theta \leq \frac{\pi}{2}$)

$$x = a \tan \theta, \qquad dx = a \sec^2 \theta \, d\theta, \qquad \sqrt{x^2 + a^2} = a \sec \theta$$

EXAMPLE 4 Evaluate $\displaystyle\int \frac{1}{\sqrt{x^2 + 9}} \, dx$.

Solution We have the form $\sqrt{x^2 + a^2}$ with $a = 3$.

Step 1. **Substitute to eliminate the square root.**

$$x = 3 \tan \theta, \qquad dx = 3 \sec^2 \theta \, d\theta, \qquad \sqrt{x^2 + 9} = 3 \sec \theta$$

$$\int \frac{1}{\sqrt{x^2 + 9}} \, dx = \int \left(\frac{1}{3 \sec \theta} \right) 3 \sec^2 \theta \, d\theta = \int \sec \theta \, d\theta$$

Step 2. **Evaluate the trigonometric integral.**

$$\int \frac{1}{\sqrt{x^2 + 9}} \, dx = \int \sec \theta \, d\theta = \ln | \sec \theta + \tan \theta | + C$$

Step 3. **Convert back to the original variable.**

Since $x = 3 \tan \theta$, we have $\tan \theta = x/3$. The triangle in Figure 4 can be used to determine an expression for $\sec \theta$. It follows that

$$\sec \theta = \frac{\text{hypotenuse}}{\text{adjacent}} = \frac{\sqrt{x^2 + 9}}{3}$$

$$\int \frac{1}{\sqrt{x^2 + 9}} \, dx = \ln \left| \frac{\sqrt{x^2 + 9}}{3} + \frac{x}{3} \right| + C$$

Notice that we can rewrite this answer as follows, absorbing the constant term $(-\ln 3)$ into the arbitrary constant:

$$\ln \left| \frac{\sqrt{x^2 + 9}}{3} + \frac{x}{3} \right| + C = \ln |\sqrt{x^2 + 9} + x| - \ln 3 + C = \ln |\sqrt{x^2 + 9} + x| + C \quad \blacksquare$$

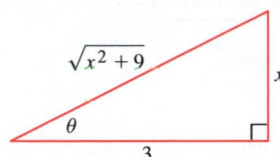

FIGURE 4 Right triangle with $\tan \theta = \frac{x}{3}$.

Our last substitution addresses the case where the integrand involves $\sqrt{x^2 - a^2}$. In this case, try the substitution $x = a \sec \theta$. We assume that $a > 0$. Furthermore, to ensure the square root is defined, we must have either $x \geq a$ or $x \leq -a$. In the first case, we assume $0 \leq \theta < \frac{\pi}{2}$ in the substitution, and in the second case, we assume $\pi \leq \theta < \frac{3\pi}{2}$. These choices for θ guarantee that $\tan \theta$ is positive in either case, and therefore,

Be aware that with this substitution, $\theta = \sec^{-1}(x/a)$ when $x > a$, but $\theta = \sec^{-1}(x/a) + \pi/2$ when $x < -a$.

$$\sqrt{x^2 - a^2} = \sqrt{a^2 \sec^2 \theta - a^2} = \sqrt{a^2(\sec^2 \theta - 1)} = \sqrt{a^2 \tan^2 \theta}$$

$$= |a \tan \theta| = a \tan \theta$$

Integrals Involving $\sqrt{x^2 - a^2}$ If $\sqrt{x^2 - a^2}$ occurs in an integral where $a > 0$, try the substitution (assuming $0 \le \theta < \frac{\pi}{2}$ with $x \ge a$, and $\pi \le \theta < \frac{3\pi}{2}$ with $x \le -a$)

$$x = a \sec \theta, \qquad dx = a \sec \theta \tan \theta \, d\theta, \qquad \sqrt{x^2 - a^2} = a \tan \theta$$

EXAMPLE 5 Evaluate $\displaystyle\int \frac{dx}{x^2\sqrt{x^2 - 9}}$.

Solution In this case, make the substitution

$$x = 3\sec\theta, \qquad dx = 3\sec\theta\tan\theta\,d\theta, \qquad \sqrt{x^2 - 9} = 3\tan\theta$$

$$\int \frac{dx}{x^2\sqrt{x^2 - 9}} = \int \frac{3\sec\theta\tan\theta\,d\theta}{(9\sec^2\theta)(3\tan\theta)} = \frac{1}{9}\int \cos\theta\,d\theta = \frac{1}{9}\sin\theta + C$$

Since $x = 3\sec\theta$, we have $\sec\theta = x/3$, and the triangle in Figure 5 shows that

$$\sin\theta = \frac{\text{opposite}}{\text{hypotenuse}} = \frac{\sqrt{x^2 - 9}}{x}$$

Therefore,

$$\int \frac{dx}{x^2\sqrt{x^2 - 9}} = \frac{1}{9}\sin\theta + C = \frac{\sqrt{x^2 - 9}}{9x} + C \qquad\blacksquare$$

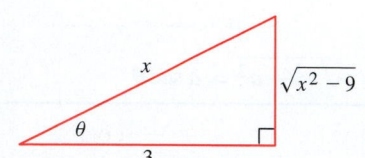

FIGURE 5 Right triangle with $\sec\theta = \frac{x}{3}$.

Square Roots of General Quadratic Functions

So far, we have dealt with the expressions $\sqrt{x^2 \pm a^2}$ and $\sqrt{a^2 - x^2}$. By completing the square (Section 1.2), we can treat the more general form $\sqrt{ax^2 + bx + c}$.

EXAMPLE 6 **Completing the Square** Evaluate $\displaystyle\int \frac{dx}{(x^2 + 2x + 3)^{3/2}}$.

Solution

Step 1. **Complete the square.**

$$x^2 + 2x + 3 = (x^2 + 2x + 1) + 2 = \underbrace{(x + 1)^2}_{u^2} + 2$$

Step 2. **Use substitution.**
Let $u = x + 1$, $du = dx$:

$$\int \frac{dx}{(x^2 + 2x + 3)^{3/2}} = \int \frac{du}{(u^2 + 2)^{3/2}} \qquad \boxed{2}$$

Step 3. **Trigonometric substitution.**
Evaluate the u-integral using trigonometric substitution:

$$u = \sqrt{2}\tan\theta, \qquad u^2 + 2 = 2\sec^2\theta, \qquad du = \sqrt{2}\sec^2\theta\,d\theta$$

$$\int \frac{du}{(u^2 + 2)^{3/2}} = \int \frac{\sqrt{2}\sec^2\theta\,d\theta}{2^{3/2}\sec^3\theta} = \frac{1}{2}\int \cos\theta\,d\theta$$

$$= \frac{1}{2}\sin\theta + C \qquad \boxed{3}$$

Since $\tan\theta = \dfrac{u}{\sqrt{2}}$, we use the right triangle in Figure 6 to obtain

$$\sin\theta = \left(\frac{\text{opposite}}{\text{hypotenuse}}\right) = \frac{u}{\sqrt{u^2 + 2}}$$

Thus, Eq. (3) becomes

$$\int \frac{du}{(u^2 + 2)^{3/2}} = \frac{1}{2}\frac{u}{\sqrt{u^2 + 2}} + C$$

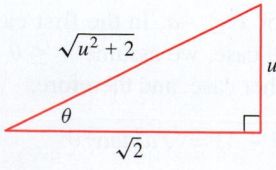

FIGURE 6 Right triangle with $\tan\theta = \frac{u}{\sqrt{2}}$.

Step 4. **Convert to the original variable.**

Since $u = x + 1$ and $u^2 + 2 = x^2 + 2x + 3$, Eq. (3) becomes

$$\int \frac{du}{(u^2 + 2)^{3/2}} = \frac{x + 1}{2\sqrt{x^2 + 2x + 3}} + C$$

Therefore, by Eq. (2):

$$\int \frac{dx}{(x^2 + 2x + 3)^{3/2}} = \frac{x + 1}{2\sqrt{x^2 + 2x + 3}} + C \qquad \blacksquare$$

7.3 SUMMARY

- Trigonometric substitution:

Square root form in integrand	Trigonometric substitution
$\sqrt{a^2 - x^2}$	$x = a \sin\theta, \quad dx = a\cos\theta\, d\theta, \quad \sqrt{a^2 - x^2} = a\cos\theta$
$\sqrt{x^2 + a^2}$	$x = a \tan\theta, \quad dx = a\sec^2\theta\, d\theta, \quad \sqrt{x^2 + a^2} = a\sec\theta$
$\sqrt{x^2 - a^2}$	$x = a \sec\theta, \quad dx = a\sec\theta\tan\theta\, d\theta, \quad \sqrt{x^2 - a^2} = a\tan\theta$

Step 1. Substitute to eliminate the square root.

Step 2. Evaluate the trigonometric integral.

Step 3. Convert back to the original variable.

- The three trigonometric substitutions correspond to three right triangles (Figure 7) that we use to express the trigonometric functions of θ in terms of x.

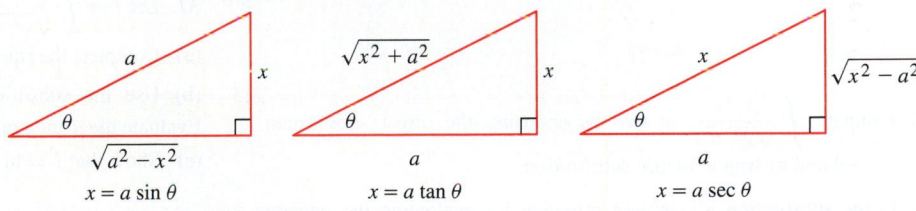

FIGURE 7 Right triangles used in trigonometric substitution.

- Integrands involving $\sqrt{x^2 + bx + c}$ are treated by completing the square (see Example 6).

7.3 EXERCISES

Preliminary Questions

1. State the trigonometric substitution appropriate to the given integral:

(a) $\int \sqrt{9 - x^2}\, dx$

(b) $\int x^2(x^2 - 16)^{3/2}\, dx$

(c) $\int x^2(x^2 + 16)^{3/2}\, dx$

(d) $\int (x^2 - 5)^{-2}\, dx$

2. Is trigonometric substitution needed to evaluate $\int x\sqrt{9 - x^2}\, dx$?

3. Express $\sin 2\theta$ in terms of $x = \sin\theta$.

4. Draw a triangle that would be used together with the substitution $x = 3\sec\theta$.

Exercises

In Exercises 1–4, evaluate the integral by following the steps given.

1. $I = \int \dfrac{dx}{\sqrt{9 - x^2}}$

(a) Show that the substitution $x = 3\sin\theta$ transforms I into $\int d\theta$, and evaluate I in terms of θ.

(b) Evaluate I in terms of x.

2. $I = \int \dfrac{dx}{x^2\sqrt{x^2 - 2}}$

(a) Show that the substitution $x = \sqrt{2}\sec\theta$ transforms the integral I into $\dfrac{1}{2}\int \cos\theta\, d\theta$, and evaluate I in terms of θ.

(b) Use a right triangle to show that with the above substitution, $\sin\theta = \sqrt{x^2 - 2}/x$.

(c) Evaluate I in terms of x.

3. $I = \displaystyle\int \frac{dx}{\sqrt{4x^2 + 9}}$

(a) Show that the substitution $x = \frac{3}{2}\tan\theta$ transforms I into

$$\frac{1}{2}\int \sec\theta\, d\theta.$$

(b) Evaluate I in terms of θ (refer to the table of integrals on page 412 in Section 7.2 if necessary).

(c) Express I in terms of x.

4. $I = \displaystyle\int \frac{dx}{(x^2 + 4)^2}$

(a) Show that the substitution $x = 2\tan\theta$ transforms the integral I into

$$\frac{1}{8}\int \cos^2\theta\, d\theta.$$

(b) Use the formula $\displaystyle\int \cos^2\theta\, d\theta = \frac{1}{2}\theta + \frac{1}{2}\sin\theta\cos\theta$ to evaluate I in terms of θ.

(c) Show that $\sin\theta = \dfrac{x}{\sqrt{x^2 + 4}}$ and $\cos\theta = \dfrac{2}{\sqrt{x^2 + 4}}$.

(d) Express I in terms of x.

In Exercises 5–10, use the indicated substitution to evaluate the integral.

5. $\displaystyle\int \sqrt{16 - 5x^2}\, dx, \quad x = \frac{4}{\sqrt{5}}\sin\theta$

6. $\displaystyle\int_0^{1/2} \frac{x^2}{\sqrt{1 - x^2}}\, dx, \quad x = \sin\theta$

7. $\displaystyle\int \frac{dx}{x\sqrt{x^2 - 9}}, \quad x = 3\sec\theta$

8. $\displaystyle\int_{1/2}^1 \frac{dx}{x^2\sqrt{x^2 + 4}}, \quad x = 2\tan\theta$

9. $\displaystyle\int \frac{dx}{(x^2 - 4)^{3/2}}, \quad x = 2\sec\theta$

10. $\displaystyle\int_0^1 \frac{dx}{(4 + 4x^2)^2}, \quad x = \tan\theta$

11. Evaluate $\displaystyle\int \frac{x\, dx}{\sqrt{x^2 - 4}}$ in two ways: using the direct substitution $u = x^2 - 4$ and by trigonometric substitution.

12. Is the substitution $u = x^2 - 4$ effective for evaluating the integral $\displaystyle\int \frac{x^2\, dx}{\sqrt{x^2 - 4}}$? If not, evaluate using trigonometric substitution.

13. Evaluate using the substitution $u = 1 - x^2$ or trigonometric substitution.

(a) $\displaystyle\int \frac{x}{\sqrt{1 - x^2}}\, dx$

(b) $\displaystyle\int x^2\sqrt{1 - x^2}\, dx$

(c) $\displaystyle\int x^3\sqrt{1 - x^2}\, dx$

(d) $\displaystyle\int \frac{x^4}{\sqrt{1 - x^2}}\, dx$

14. Evaluate:

(a) $\displaystyle\int_0^1 \frac{dt}{(t^2 + 1)^{3/2}}$

(b) $\displaystyle\int_0^1 \frac{t\, dt}{(t^2 + 1)^{3/2}}$

In Exercises 15–32, evaluate using trigonometric substitution. Refer to the table of trigonometric integrals as necessary.

15. $\displaystyle\int \frac{x^2\, dx}{\sqrt{9 - x^2}}$

16. $\displaystyle\int \frac{dt}{(16 - t^2)^{3/2}}$

17. $\displaystyle\int \frac{dx}{x\sqrt{x^2 + 16}}$

18. $\displaystyle\int \sqrt{12 + 4t^2}\, dt$

19. $\displaystyle\int \frac{dx}{\sqrt{x^2 - 9}}$

20. $\displaystyle\int \frac{dt}{t^2\sqrt{t^2 - 25}}$

21. $\displaystyle\int \frac{dy}{y^2\sqrt{5 - y^2}}$

22. $\displaystyle\int x^3\sqrt{9 - x^2}\, dx$

23. $\displaystyle\int \frac{dx}{\sqrt{25x^2 + 2}}$

24. $\displaystyle\int \frac{dt}{(9t^2 + 4)^2}$

25. $\displaystyle\int \frac{dz}{z^3\sqrt{z^2 - 4}}$

26. $\displaystyle\int \frac{dy}{\sqrt{y^2 - 9}}$

27. $\displaystyle\int \frac{x^2\, dx}{(6x^2 - 49)^{1/2}}$

28. $\displaystyle\int \frac{dx}{(x^2 - 4)^2}$

29. $\displaystyle\int_0^1 \frac{dt}{(t^2 + 9)^2}$

30. $\displaystyle\int_0^1 \frac{dx}{(x^2 + 1)^3}$

31. $\displaystyle\int \frac{x^2\, dx}{(x^2 - 1)^{3/2}}$

32. $\displaystyle\int \frac{x^2\, dx}{(x^2 + 1)^{3/2}}$

33. Compute $\displaystyle\int_0^{1.99999} \frac{1}{(4 - x^2)^{3/2}}\, dx$.

34. Compute $\displaystyle\int_0^r \frac{1}{(4 - x^2)^{3/2}}\, dx$, expressing the result in terms of r. Then, discuss what happens to the value of the definite integral as r approaches 2 from the left.

35. (a) Using a trigonometric substitution, compute the integral and show that for $a > 0$

$$\int \frac{dx}{x^2 + a^2} = \frac{1}{a}\tan^{-1}\frac{x}{a} + C$$

(b) Verify the formula via differentiation.

36. (a) Using a trigonometric substitution, compute the integral and show that for $a > 0$

$$\int \frac{dx}{(x^2 + a^2)^2} = \frac{1}{2a^2}\left(\frac{x}{x^2 + a^2} + \frac{1}{a}\tan^{-1}\frac{x}{a}\right) + C$$

(b) Verify the formula via differentiation.

37. Let $I = \displaystyle\int \frac{dx}{\sqrt{x^2 - 4x + 8}}$.

(a) Complete the square to show that $x^2 - 4x + 8 = (x - 2)^2 + 4$.

(b) Use the substitution $u = x - 2$ to show that $I = \displaystyle\int \frac{du}{\sqrt{u^2 + 2^2}}$. Evaluate the u-integral.

(c) Show that $I = \ln\left|\sqrt{(x - 2)^2 + 4} + x - 2\right| + C$.

38. Evaluate $\displaystyle\int \frac{dx}{\sqrt{12x - x^2}}$. First, complete the square to write $12x - x^2 = 36 - (x - 6)^2$.

In Exercises 39–44, evaluate the integral by completing the square and using trigonometric substitution.

39. $\displaystyle\int \frac{dx}{\sqrt{x^2 + 4x + 13}}$

40. $\displaystyle\int \frac{dx}{\sqrt{2 + x - x^2}}$

41. $\displaystyle\int \frac{dx}{\sqrt{x + 6x^2}}$

42. $\displaystyle\int \sqrt{x^2 - 4x + 7}\, dx$

43. $\displaystyle\int \sqrt{x^2 - 4x + 3}\, dx$

44. $\displaystyle\int \frac{dx}{(x^2 + 6x + 6)^2}$

In Exercises 45–48, evaluate using Integration by Parts as a first step.

45. $\displaystyle\int \sec^{-1}x\, dx$

46. $\displaystyle\int \frac{\sin^{-1}x}{x^2}\, dx$

47. $\displaystyle\int \ln(x^2 + 1)\, dx$

48. $\displaystyle\int x^2\ln(x^2 + 1)\, dx$

49. Find the average height of a point on the semicircle $y = \sqrt{1 - x^2}$ for $-1 \le x \le 1$.

50. Find the volume of the solid obtained by revolving the graph of $y = x\sqrt{1 - x^2}$ over $[0, 1]$ about the y-axis.

51. Find the volume of the solid obtained by revolving the region between the graph of $y^2 - x^2 = 1$ and the line $y = 2$ about the line $y = 2$.

52. Find the volume of revolution for the region in Exercise 51, but revolve around $y = 3$.

53. Compute $\int \dfrac{dx}{x^2 - 1}$ in two ways and verify that the answers agree: first via trigonometric substitution and then using the identity

$$\frac{1}{x^2 - 1} = \frac{1}{2}\left(\frac{1}{x-1} - \frac{1}{x+1}\right)$$

54. [CAS] You want to divide an 18-in. pizza equally among three friends using vertical slices at $\pm x$ as in Figure 8. Find an equation satisfied by x and find the approximate value of x using a computer algebra system.

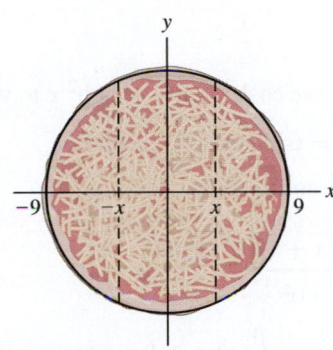

FIGURE 8 Dividing a pizza into three equal parts.

55. A charged wire creates an electric field at a point P located at a distance D from the wire (Figure 9). The component $E_\perp$ of the field perpendicular to the wire (in newtons per coulomb) is

$$E_\perp = \int_{x_1}^{x_2} \frac{k\lambda D}{(x^2 + D^2)^{3/2}}\, dx$$

where λ is the charge density (coulombs per meter), $k = 8.99 \times 10^9$ N·m^2/C^2 (Coulomb constant), and x_1, x_2 are as in the figure. Suppose that $\lambda = 6 \times 10^{-4}$ C/m, and $D = 3$ m. Find $E_\perp$ if (a) $x_1 = 0$ and $x_2 = 30$ m, and (b) $x_1 = -15$ m and $x_2 = 15$ m.

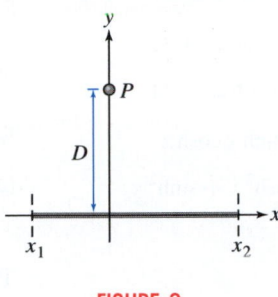

FIGURE 9

Further Insights and Challenges

56. Let $J_n = \displaystyle\int \frac{dx}{(x^2 + 1)^n}$.

(a) Compute J_1.

(b) Use Integration by Parts to prove

$$J_{n+1} = \left(1 - \frac{1}{2n}\right) J_n + \left(\frac{1}{2n}\right)\frac{x}{(x^2+1)^n}$$

(c) Use this recursion relation to calculate J_2 and J_3.

57. The area function $F(x) = \displaystyle\int_0^x \sqrt{1 - t^2}\, dt$ is an antiderivative of $f(x) = \sqrt{1 - x^2}$. Prove the formula

$$\int_0^x \sqrt{1 - t^2}\, dt = \frac{1}{2}\sin^{-1} x + \frac{1}{2}x\sqrt{1 - x^2} + C$$

using geometry by interpreting the integral as the area of part of the unit circle.

7.4 Integrals Involving Hyperbolic and Inverse Hyperbolic Functions

← **REMINDER**

$$\sinh x = \frac{e^x - e^{-x}}{2} \qquad \cosh x = \frac{e^x + e^{-x}}{2}$$

$$\frac{d}{dx}\sinh x = \cosh x \qquad \frac{d}{dx}\cosh x = \sinh x$$

$$\frac{d}{dx}\tanh x = \text{sech}^2 x$$

$$\frac{d}{dx}\coth x = -\text{csch}^2 x$$

$$\frac{d}{dx}\text{sech}\,x = -\text{sech}\,x \tanh x$$

$$\frac{d}{dx}\text{csch}\,x = -\text{csch}\,x \coth x$$

In Section 1.6, we noted the similarities between hyperbolic and trigonometric functions. We also saw in Section 3.9 that the formulas for their derivatives resemble each other, differing in at most a sign. The derivative formulas for the hyperbolic functions are equivalent to the following integral formulas.

Hyperbolic Integral Formulas

$$\int \sinh x\, dx = \cosh x + C, \qquad \int \cosh x\, dx = \sinh x + C$$

$$\int \text{sech}^2 x\, dx = \tanh x + C, \qquad \int \text{csch}^2 x\, dx = -\coth x + C$$

$$\int \text{sech}\,x \tanh x\, dx = -\text{sech}\,x + C, \qquad \int \text{csch}\,x \coth x\, dx = -\text{csch}\,x + C$$

EXAMPLE 1 Calculate $\int x \cosh(x^2)\, dx$.

Solution The substitution $u = x^2, du = 2x\, dx$ yields

$$\int x \cosh(x^2)\, dx = \frac{1}{2}\int \cosh u\, du = \frac{1}{2}\sinh u + C = \frac{1}{2}\sinh(x^2) + C \qquad \blacksquare$$

The techniques for computing trigonometric integrals discussed in Section 7.2 apply with little change to hyperbolic integrals. In place of trigonometric identities, we use the corresponding hyperbolic identities (see margin).

Hyperbolic identities:

$$\cosh^2 x - \sinh^2 x = 1$$

$$\cosh^2 x = 1 + \sinh^2 x$$

$$\cosh^2 x = \tfrac{1}{2}(\cosh 2x + 1)$$

$$\sinh^2 x = \tfrac{1}{2}(\cosh 2x - 1)$$

$$\sinh 2x = 2\sinh x \cosh x$$

$$\cosh 2x = \cosh^2 x + \sinh^2 x$$

EXAMPLE 2 **Powers of $\sinh x$ and $\cosh x$** Calculate: **(a)** $\int \sinh^4 x \cosh^5 x\, dx$

and **(b)** $\int \cosh^2 x\, dx$.

Solution

(a) Since $\cosh x$ appears to an odd power, use $\cosh^2 x = 1 + \sinh^2 x$ to write

$$\cosh^5 x = \cosh^4 x \cdot \cosh x = (\sinh^2 x + 1)^2 \cosh x$$

Then use the substitution $u = \sinh x,\, du = \cosh x\, dx$:

$$\int \sinh^4 x \cosh^5 x\, dx = \int \underbrace{\sinh^4 x}_{u^4}\, \underbrace{(\sinh^2 x + 1)^2}_{(u^2+1)^2}\, \underbrace{\cosh x\, dx}_{du}$$

$$= \int u^4(u^2 + 1)^2\, du = \int (u^8 + 2u^6 + u^4)\, du$$

$$= \frac{u^9}{9} + \frac{2u^7}{7} + \frac{u^5}{5} + C = \frac{\sinh^9 x}{9} + \frac{2\sinh^7 x}{7} + \frac{\sinh^5 x}{5} + C$$

(b) For $\int \cosh^2 x\, dx$, we use the identity $\cosh^2 x = \frac{1}{2}(\cosh 2x + 1)$:

$$\int \cosh^2 x\, dx = \frac{1}{2}\int (\cosh 2x + 1)\, dx = \frac{1}{2}\left(\frac{\sinh 2x}{2} + x\right) + C$$

$$= \frac{1}{4}\sinh 2x + \frac{1}{2}x + C \qquad \blacksquare$$

Hyperbolic substitution may be used as an alternative to trigonometric substitution to integrate functions involving the following square root expressions:

In trigonometric substitution, we treat $\sqrt{x^2 + a^2}$ using the substitution $x = a\tan\theta$ and $\sqrt{x^2 - a^2}$ using $x = a\sec\theta$. Identities can be used to show that the results coincide with those obtained from hyperbolic substitution (see Exercises 31–35).

Square root form	Hyperbolic substitution
$\sqrt{x^2 + a^2}$	$x = a\sinh u,\, dx = a\cosh u,\, \sqrt{x^2 + a^2} = a\cosh u$
$\sqrt{x^2 - a^2}$	$x = a\cosh u,\, dx = a\sinh u,\, \sqrt{x^2 - a^2} = a\sinh u$

EXAMPLE 3 **Hyperbolic Substitution** Calculate $\int \sqrt{x^2 + 16}\, dx$.

Solution

Step 1. **Substitute to eliminate the square root.**

Use the hyperbolic substitution $x = 4\sinh u,\, dx = 4\cosh u\, du$. Then

$$x^2 + 16 = 16(\sinh^2 u + 1) = (4\cosh u)^2$$

Furthermore, $4\cosh u > 0$, so $\sqrt{x^2 + 16} = 4\cosh u$, and thus,

$$\int \sqrt{x^2 + 16}\, dx = \int (4\cosh u)\, 4\cosh u\, du = 16\int \cosh^2 u\, du$$

Step 2. **Evaluate the hyperbolic integral.**

We evaluated the integral of $\cosh^2 u$ in Example 2(b):

$$\int \sqrt{x^2 + 16}\, dx = 16 \int \cosh^2 u\, du = 16 \left(\frac{1}{4} \sinh 2u + \frac{1}{2}u + C \right)$$

$$= 4 \sinh 2u + 8u + C \qquad \boxed{1}$$

Step 3. **Convert back to the original variable.**

To write the answer in terms of the original variable x, we note that

$$\sinh u = \frac{x}{4}, \qquad u = \sinh^{-1} \frac{x}{4}$$

Use the identities recalled in the margin to write

$$4 \sinh 2u = 4(2 \sinh u \cosh u) = 8 \sinh u \sqrt{\sinh^2 u + 1}$$

← *REMINDER*

$$\sinh 2u = 2 \sinh u \cosh u$$

$$\cosh u = \sqrt{\sinh^2 u + 1}$$

$$= 8 \left(\frac{x}{4}\right)\sqrt{\left(\frac{x}{4}\right)^2 + 1} = 2x\sqrt{\frac{x^2}{16} + 1} = \frac{1}{2}x\sqrt{x^2 + 16}$$

Now via Eq. (1), we have

$$\int \sqrt{x^2 + 16}\, dx = \frac{1}{2}x\sqrt{x^2 + 16} + 8 \sinh^{-1} \frac{x}{4} + C \qquad ■$$

In a similar manner, we could derive each of the following integral formulas. Alternatively, we can realize them from the corresponding derivative formulas for the inverse hyperbolic functions given in Section 3.9. Each formula is valid on the domain where the integrand and inverse hyperbolic function are defined.

THEOREM 1 **Integrals Involving Inverse Hyperbolic Functions**

$$\int \frac{dx}{\sqrt{x^2 + 1}} = \sinh^{-1} x + C$$

$$\int \frac{dx}{\sqrt{x^2 - 1}} = \cosh^{-1} x + C \qquad (\text{for } x > 1)$$

$$\int \frac{dx}{1 - x^2} = \tanh^{-1} x + C \qquad (\text{for } |x| < 1)$$

$$\int \frac{dx}{1 - x^2} = \coth^{-1} x + C \qquad (\text{for } |x| > 1)$$

$$\int \frac{dx}{x\sqrt{1 - x^2}} = -\operatorname{sech}^{-1} x + C \qquad (\text{for } 0 < x < 1)$$

$$\int \frac{dx}{|x|\sqrt{1 + x^2}} = -\operatorname{csch}^{-1} x + C \qquad (\text{for } x \neq 0)$$

EXAMPLE 4 Evaluate: (a) $\displaystyle\int_2^4 \frac{dx}{\sqrt{x^2 - 1}}$ and (b) $\displaystyle\int_{0.2}^{0.6} \frac{x\, dx}{1 - x^4}$.

Solution

(a) By Theorem 1,

$$\int_2^4 \frac{dx}{\sqrt{x^2 - 1}} = \cosh^{-1} x \Big|_2^4 = \cosh^{-1} 4 - \cosh^{-1} 2 \approx 0.75$$

(b) First, use the substitution $u = x^2$, $du = 2x\,du$. The new limits of integration become $u = (0.2)^2 = 0.04$ and $u = (0.6)^2 = 0.36$, so

$$\int_{0.2}^{0.6} \frac{x\,dx}{1 - x^4} = \int_{0.04}^{0.36} \frac{\frac{1}{2}du}{1 - u^2} = \frac{1}{2}\int_{0.04}^{0.36} \frac{du}{1 - u^2}$$

By Theorem 1, both $\tanh^{-1} u$ and $\coth^{-1} u$ are antiderivatives of $f(u) = (1 - u^2)^{-1}$. We use $\tanh^{-1} u$ because the interval of integration $[0.04, 0.36]$ is contained in the domain $(-1, 1)$ of $f(u) = \tanh^{-1} u$. If the limits of integration were contained in $(1, \infty)$ or $(-\infty, -1)$, we would use $\coth^{-1} u$. The result is

$$\frac{1}{2}\int_{0.04}^{0.36} \frac{du}{1 - u^2} = \frac{1}{2}\left(\tanh^{-1}(0.36) - \tanh^{-1}(0.04)\right) \approx 0.1684 \qquad \blacksquare$$

7.4 SUMMARY

• Integrals of hyperbolic functions:

$$\int \sinh x\,dx = \cosh x + C, \qquad \int \cosh x\,dx = \sinh x + C$$

$$\int \text{sech}^2 x\,dx = \tanh x + C, \qquad \int \text{csch}^2 x\,dx = -\coth x + C$$

$$\int \text{sech}\,x \tanh x\,dx = -\text{sech}\,x + C, \qquad \int \text{csch}\,x \coth x\,dx = -\text{csch}\,x + C$$

• Integrals involving inverse hyperbolic functions:

$$\int \frac{dx}{\sqrt{x^2 + 1}} = \sinh^{-1} x + C$$

$$\int \frac{dx}{\sqrt{x^2 - 1}} = \cosh^{-1} x + C \qquad (\text{for } x > 1)$$

$$\int \frac{dx}{1 - x^2} = \tanh^{-1} x + C \qquad (\text{for } |x| < 1)$$

$$\int \frac{dx}{1 - x^2} = \coth^{-1} x + C \qquad (\text{for } |x| > 1)$$

$$\int \frac{dx}{x\sqrt{1 - x^2}} = -\text{sech}^{-1} x + C \qquad (\text{for } 0 < x < 1)$$

$$\int \frac{dx}{|x|\sqrt{1 + x^2}} = -\text{csch}^{-1} x + C \qquad (\text{for } x \neq 0)$$

7.4 EXERCISES

Preliminary Questions

1. Which hyperbolic substitution can be used to evaluate the following integrals?

(a) $\displaystyle\int \frac{dx}{\sqrt{x^2 + 1}}$ 　　**(b)** $\displaystyle\int \frac{dx}{\sqrt{x^2 + 9}}$ 　　**(c)** $\displaystyle\int \frac{dx}{\sqrt{9x^2 + 1}}$

2. Which two of the hyperbolic integration formulas differ from their trigonometric counterparts by a minus sign?

3. Which antiderivative of $y = (1 - x^2)^{-1}$ should we use to evaluate the integral $\displaystyle\int_3^5 (1 - x^2)^{-1}\,dx$?

Exercises

In Exercises 1–16, calculate the integral.

1. $\displaystyle\int \cosh(3x)\, dx$

2. $\displaystyle\int x \sinh x^2\, dx$

3. $\displaystyle\int x \sinh x\, dx$

4. $\displaystyle\int \sinh^2 x \cosh x\, dx$

5. $\displaystyle\int \operatorname{sech}^2(1 - 2x)\, dx$

6. $\displaystyle\int \tanh(3x) \operatorname{sech}(3x)\, dx$

7. $\displaystyle\int \tanh x \operatorname{sech}^2 x\, dx$

8. $\displaystyle\int \frac{\cosh x}{3 \sinh x + 4}\, dx$

9. $\displaystyle\int \tanh x\, dx$

10. $\displaystyle\int x \operatorname{csch}(x^2) \coth(x^2)\, dx$

11. $\displaystyle\int \frac{\cosh x}{\sinh x}\, dx$

12. $\displaystyle\int \frac{\cosh x}{\sinh^2 x}\, dx$

13. $\displaystyle\int \sinh^2(4x - 9)\, dx$

14. $\displaystyle\int \sinh^3 x \cosh^6 x\, dx$

15. $\displaystyle\int \sinh^2 x \cosh^2 x\, dx$

16. $\displaystyle\int \tanh^3 x\, dx$

In Exercises 17–30, calculate the integral in terms of the inverse hyperbolic functions.

17. $\displaystyle\int \frac{dx}{\sqrt{x^2 - 1}}$

18. $\displaystyle\int \frac{dx}{\sqrt{9x^2 - 4}}$

19. $\displaystyle\int \frac{dx}{\sqrt{4 + x^2}}$

20. $\displaystyle\int \frac{dx}{\sqrt{1 + 3x^2}}$

21. $\displaystyle\int \sqrt{x^2 - 1}\, dx$

22. $\displaystyle\int \frac{x^2\, dx}{\sqrt{x^2 + 1}}$

23. $\displaystyle\int_{-1/2}^{1/2} \frac{dx}{1 - x^2}$

24. $\displaystyle\int_{4}^{5} \frac{dx}{1 - x^2}$

25. $\displaystyle\int_{0}^{1} \frac{dx}{\sqrt{1 + x^2}}$

26. $\displaystyle\int_{2}^{10} \frac{dx}{4x^2 - 1}$

27. $\displaystyle\int_{-3}^{-1} \frac{dx}{x\sqrt{x^2 + 16}}$

28. $\displaystyle\int_{0.2}^{0.8} \frac{dx}{x\sqrt{1 - x^2}}$

29. $\displaystyle\int \frac{\sqrt{x^2 - 1}\, dx}{x^2}$

30. $\displaystyle\int_{1}^{9} \frac{dx}{x\sqrt{x^4 + 1}}$

31. Verify the formulas

$$\sinh^{-1} x = \ln|x + \sqrt{x^2 + 1}|$$

$$\cosh^{-1} x = \ln|x + \sqrt{x^2 - 1}| \qquad \text{(for } x \geq 1\text{)}$$

32. Verify that $\tanh^{-1} x = \dfrac{1}{2} \ln\left|\dfrac{1 + x}{1 - x}\right|$ for $|x| < 1$.

33. Evaluate $\displaystyle\int \sqrt{x^2 + 16}\, dx$ using trigonometric substitution. Then use Exercise 31 to verify that your answer agrees with the answer in Example 3.

34. Evaluate $\displaystyle\int \sqrt{x^2 - 9}\, dx$ in two ways: using trigonometric substitution and using hyperbolic substitution. Then use Exercise 31 to verify that the two answers agree.

35. Prove the reduction formula for $n \geq 2$:

$$\int \cosh^n x\, dx = \frac{1}{n} \cosh^{n-1} x \sinh x + \frac{n - 1}{n} \int \cosh^{n-2} x\, dx \qquad \boxed{2}$$

36. Use Eq. (2) to evaluate $\displaystyle\int \cosh^4 x\, dx$.

In Exercises 37–40, evaluate the integral.

37. $\displaystyle\int \frac{\tanh^{-1} x\, dx}{x^2 - 1}$

38. $\displaystyle\int \sinh^{-1} x\, dx$

39. $\displaystyle\int \tanh^{-1} x\, dx$

40. $\displaystyle\int x \tanh^{-1} x\, dx$

41. (a) Compute the area under the graph of $y = \sinh x$ for $0 \leq x \leq 5$.

(b) Compute the area under the graph of $y = \sinh^{-1} x$ for $0 \leq x \leq \sinh 5$.

(c) Show that the sum of the areas in (a) and (b) is equal to $5 \sinh 5$.

(d) Refer to Figure 1 and explain why the sum of the areas in (a) and (b) is equal to $5 \sinh 5$.

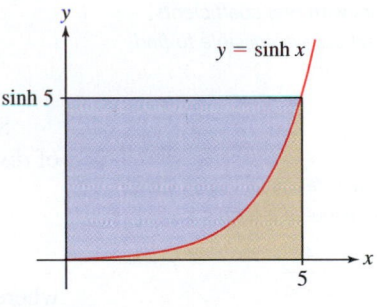

FIGURE 1

42. (a) Compute the area under the graph of $y = \tanh x$ for $0 \leq x \leq 4$.

(b) Compute the area under the graph of $y = \tanh^{-1} x$ for $0 \leq x \leq \tanh 4$.

(c) Show that the sum of the areas in (a) and (b) is equal to $4 \tanh 4$.

(d) Similar to Exercise 41(d), explain graphically why the sum of the areas in (a) and (b) is equal to $4 \tanh 4$.

Further Insights and Challenges

43. Show that if $u = \tanh(x/2)$, then

$$\cosh x = \frac{1 + u^2}{1 - u^2}, \qquad \sinh x = \frac{2u}{1 - u^2}, \qquad dx = \frac{2\, du}{1 - u^2}$$

Hint: For the first relation, use the identities

$$\sinh^2\left(\frac{x}{2}\right) = \frac{1}{2}(\cosh x - 1), \qquad \cosh^2\left(\frac{x}{2}\right) = \frac{1}{2}(\cosh x + 1)$$

In Exercises 44–46, evaluate using the substitution of Exercise 43.

44. $\displaystyle\int \operatorname{sech} x\, dx$

45. $\displaystyle\int \frac{dx}{1 + \cosh x}$

46. $\displaystyle\int \frac{dx}{1 - \cosh x}$

Exercises 47–50 refer to the function $gd(y) = \tan^{-1}(\sinh y)$, *called the* **Gudermannian**. *In a map of the earth constructed by Mercator projection, points located y radial units from the equator correspond to points on the globe of latitude* $gd(y)$.

47. Prove that $\dfrac{d}{dy} gd(y) = \operatorname{sech} y$.

48. Let $f(y) = 2\tan^{-1}(e^y) - \pi/2$. Prove that $gd(y) = f(y)$. *Hint:* Show that $gd'(y) = f'(y)$ and $f(0) = g(0)$.

49. Let $t(y) = \sinh^{-1}(\tan y)$. Show that $t(y)$ is the inverse of $gd(y)$ for $0 \le y < \pi/2$.

50. Verify that $t(y)$ in Exercise 49 satisfies $t'(y) = \sec y$, and find a value of a such that

$$t(y) = \int_a^y \frac{dt}{\cos t}$$

7.5 The Method of Partial Fractions

The Method of Partial Fractions is used to integrate rational functions:

$$f(x) = \frac{P(x)}{Q(x)}$$

where P and Q are polynomials. The idea is to write f as a sum of simpler rational functions that can be integrated directly. For example, expressing $\frac{1}{x^2-1}$ as a sum

$$\frac{1}{x^2 - 1} = \frac{\frac{1}{2}}{x - 1} - \frac{\frac{1}{2}}{x + 1}$$

enables us to evaluate the integral

$$\int \frac{dx}{x^2 - 1} = \frac{1}{2} \int \frac{dx}{x - 1} - \frac{1}{2} \int \frac{dx}{x + 1} = \frac{1}{2} \ln|x - 1| - \frac{1}{2} \ln|x + 1|$$

> It is a fact from algebra (known as the Fundamental Theorem of Algebra) that every polynomial Q with real coefficients can be written as a product of linear and quadratic factors with real coefficients. However, it is not always possible to find these factors explicitly.

A rational function P/Q is called **proper** if the degree of P [denoted $\deg(P)$] is *less than* the degree of Q. For example,

$$\underbrace{\frac{x^2 - 3x + 7}{x^4 - 16},\quad \frac{2x^2 + 7}{x - 5},\quad \frac{x - 2}{x - 5}}_{}$$

$$\text{Proper} \qquad\qquad \text{Not proper}$$

Suppose first that P/Q is proper and that the denominator $Q(x)$ factors as a product of *distinct linear factors*. In other words,

> Each distinct linear factor $(x - a)$ in the denominator contributes a term
>
> $$\frac{A}{x - a}$$
>
> to the partial fraction decomposition.

$$\frac{P(x)}{Q(x)} = \frac{P(x)}{(x - a_1)(x - a_2)\cdots(x - a_n)}$$

where the values $a_1, a_2, \ldots, a_n$ are all distinct and $\deg(P) < n$. Then there is a **partial fraction decomposition**:

$$\frac{P(x)}{Q(x)} = \frac{A_1}{(x - a_1)} + \frac{A_2}{(x - a_2)} + \cdots + \frac{A_n}{(x - a_n)}$$

for suitable constants $A_1, \ldots, A_n$. For example,

$$\frac{5x^2 + x - 28}{(x + 1)(x - 2)(x - 3)} = -\frac{2}{x + 1} + \frac{2}{x - 2} + \frac{5}{x - 3}$$

To obtain a partial fraction decomposition, we must find the constants $A_1, \ldots, A_n$. There are two different methods that we will employ: Undetermined Coefficients and Value Substitution. Ultimately, the process involves solving a system of linear equations in the constants.

In the examples that follow, given a fraction $\frac{P(x)}{Q(x)}$, we will see how these methods can be used to obtain a partial fraction decomposition. Once we have found the partial fraction decomposition, we can integrate the individual terms using techniques that we have previously seen.

EXAMPLE 1 Finding the Constants Evaluate $\displaystyle\int \frac{dx}{x^2 - 7x + 10}$.

Solution

Step 1. **Determine the form of the partial fraction decomposition.**

The denominator factors as $x^2 - 7x + 10 = (x-2)(x-5)$, so we look for a partial fraction decomposition:

$$\frac{1}{(x-2)(x-5)} = \frac{A}{x-2} + \frac{B}{x-5}$$

Step 2. **Determine the constants.**

To find A and B, first multiply by $(x-2)(x-5)$ to clear denominators:

$$1 = (x-2)(x-5)\left(\frac{A}{x-2} + \frac{B}{x-5}\right)$$

$$1 = A(x-5) + B(x-2) \qquad \boxed{1}$$

$$1 = (A+B)x + (-5A - 2B) \qquad \boxed{2}$$

> In the method of **Undetermined Coefficients**, we set up an equality between two polynomials and then equate coefficients of like powers of x to set up a system of equations that can be solved for the constants.

We will use the method of Undetermined Coefficients to determine the constants. Note that on the left side of Eq. (2), there are no x terms, and on the right side the coefficient of x is $A + B$. Furthermore, the constant terms are 1 on the left and $-5A - 2B$ on the right. Thus, equating coefficients of like powers of x we obtain

$$0 = A + B \qquad \text{(coefficient of } x\text{)}$$

$$1 = -5A - 2B \qquad \text{(constant terms)}$$

The first of these equations implies $B = -A$. Substituting $-A$ for B in the second equation and solving for A, we find $A = -\frac{1}{3}$. It follows that $B = \frac{1}{3}$. The resulting partial fraction decomposition is

$$\frac{1}{(x-2)(x-5)} = \frac{-\frac{1}{3}}{x-2} + \frac{\frac{1}{3}}{x-5}$$

Step 3. **Carry out the integration.**

$$\int \frac{dx}{(x-2)(x-5)} = -\frac{1}{3}\int \frac{dx}{x-2} + \frac{1}{3}\int \frac{dx}{x-5}$$

$$= -\frac{1}{3}\ln|x-2| + \frac{1}{3}\ln|x-5| + C \qquad \blacksquare$$

EXAMPLE 2 Evaluate $\displaystyle\int \frac{x^2+2}{(x-1)(2x-8)(x+2)}\,dx$.

Solution

Step 1. **Determine the form of the partial fraction decomposition.**

The decomposition has the form

> In Eq. (3), the linear factor $2x - 8$ does not have the form $(x - a)$ used previously, but the partial fraction decomposition can be carried out in the same way.

$$\frac{x^2+2}{(x-1)(2x-8)(x+2)} = \frac{A}{x-1} + \frac{B}{2x-8} + \frac{C}{x+2} \qquad \boxed{3}$$

Step 2. **Determine the constants.**

As before, multiply by $(x-1)(2x-8)(x+2)$ to clear denominators:

$$x^2 + 2 = A(2x-8)(x+2) + B(x-1)(x+2) + C(x-1)(2x-8) \qquad \boxed{4}$$

*In the method of **Value Substitution**, we strategically substitute values for x that provide simple equations to solve—either individually or as a system—for the constants.*

In this case, we will use Value Substitution to determine the constants. Note that in Eq. (4) the factor $x - 1$ appears in two terms on the right-hand side. Thus, if we substitute 1 for x in the equation, those terms drop out, leaving us with a simple equation to solve for A. Specifically, substituting $x = 1$, we obtain

$$1^2 + 2 = A(-6)(3) + 0 + 0$$

Therefore, $3 = -18A$, yielding $A = -\frac{1}{6}$. Similarly, substituting 4 for x, the two $2x - 8$ terms become 0, providing us with

$$4^2 + 2 = 0 + B(3)(6) + 0$$

From this, we obtain $B = 1$. Finally, C is determined by setting $x = -2$ in Eq. (4):

$$(-2)^2 + 2 = 0 + 0 + C(-3)(-12)$$

You can check your partial fraction decomposition by adding together the resulting fractions and verifying that the result is the fraction that you had to start.

Thus, $C = \frac{1}{6}$, and the resulting partial fraction decomposition is

$$\frac{x^2 + 2}{(x - 1)(2x - 8)(x + 2)} = -\frac{\frac{1}{6}}{x - 1} + \frac{1}{2x - 8} + \frac{\frac{1}{6}}{x + 2}$$

Step 3. Carry out the integration.

$$\int \frac{x^2 + 2}{(x - 1)(2x - 8)(x + 2)}\,dx = -\frac{1}{6}\int \frac{dx}{x - 1} + \int \frac{dx}{2x - 8} + \frac{1}{6}\int \frac{dx}{x + 2}$$

$$= -\frac{1}{6}\ln|x - 1| + \frac{1}{2}\ln|2x - 8| + \frac{1}{6}\ln|x + 2| + C \quad \blacksquare$$

When using Value Substitution to determine the constants, look for values of x that result in equations that are as simple as possible when the values are substituted for x. In the previous example, each constant was obtained directly by substituting an appropriate value for x. This is not always possible, but we can always obtain a system of equations in the constants that can be solved.

Now, what do we do if the denominator has a repeated linear factor? For instance, in the next example, $(x + 2)^2$ is a factor of the denominator. In the decomposition, we need to include terms $\frac{A}{x+2}$ and $\frac{B}{(x+2)^2}$.

In general, each factor $(x - a)^n$ contributes the following sum of terms to the partial fraction decomposition:

$$\frac{A_1}{(x - a)} + \frac{A_2}{(x - a)^2} + \cdots + \frac{A_n}{(x - a)^n}$$

EXAMPLE 3 **Repeated Linear Factors** Evaluate $\int \frac{3x - 9}{(x + 2)^2(x - 1)}\,dx$.

Solution

Step 1. Determine the form of the partial fraction expansion.
We are looking for a partial fraction decomposition of the form

$$\frac{3x - 9}{(x + 2)^2(x - 1)} = \frac{A}{x + 2} + \frac{B}{(x + 2)^2} + \frac{C}{x - 1}$$

Step 2. Determine the constants.
Let's clear denominators to obtain

$$3x - 9 = A(x - 1)(x + 2) + B(x - 1) + C(x + 2)^2 \qquad \boxed{5}$$

We use Value Substitution in this case. As in the previous example, by substituting appropriately, we can obtain single equations that yield values for B and C. We cannot obtain A directly in this manner, but a further substitution provides an equation that can be used to determine A, given the values of B and C.

We substitute into Eq. (5) as follows:

- Set $x = 1$. This gives $-6 = 9C$.
- Set $x = -2$. This gives $-15 = -3B$.
- Set $x = 0$. This gives $-9 = -2A - B + 4C$

Any substitution other than $x = 1$ or $x = -2$ can be used to obtain a third equation that enables us to determine A, given B and C.

We now have three equations that we can easily solve for A, B, and C. The first two equations yield $C = -\frac{2}{3}$ and $B = 5$, respectively. With those solutions, we then obtain $A = \frac{2}{3}$ from the third. Therefore, we have

$$\frac{3x - 9}{(x - 1)(x + 2)^2} = \frac{\frac{2}{3}}{x + 2} + \frac{5}{(x + 2)^2} - \frac{\frac{2}{3}}{x - 1}$$

Step 3. **Carry out the integration.**

$$\int \frac{3x - 9}{(x - 1)(x + 2)^2}\, dx = -\frac{2}{3}\int \frac{dx}{x - 1} + \frac{2}{3}\int \frac{dx}{x + 2} + 5\int \frac{dx}{(x + 2)^2}$$

$$= -\frac{2}{3}\ln|x - 1| + \frac{2}{3}\ln|x + 2| - \frac{5}{x + 2} + C \qquad \blacksquare$$

If P/Q is not proper—that is, if $\deg(P) \geq \deg(Q)$—we use long division to write

$$\frac{P(x)}{Q(x)} = g(x) + \frac{R(x)}{Q(x)}$$

where g is a polynomial and R/Q is proper. Then, integrating $P(x)/Q(x)$ involves evaluating an integral of the polynomial $g(x)$ (which is straightforward) and integrating $R(x)/Q(x)$ via a partial fraction decomposition, if possible.

Long division:

$$
\begin{array}{r}
x \\
x^2 - 4\,\big)\, \overline{x^3 + 1} \\
\underline{x^3 - 4x } \\
4x + 1
\end{array}
$$

The quotient $\dfrac{x^3 + 1}{x^2 - 4}$ is equal to x with remainder $4x + 1$.

EXAMPLE 4 **Long Division Necessary** Evaluate $\displaystyle\int \frac{x^3 + 1}{x^2 - 4}\, dx$.

Solution Using long division, we write

$$\frac{x^3 + 1}{x^2 - 4} = x + \frac{4x + 1}{x^2 - 4} = x + \frac{4x + 1}{(x - 2)(x + 2)}$$

It is not difficult to show that the second term has a partial fraction decomposition:

$$\frac{4x + 1}{(x - 2)(x + 2)} = \frac{\frac{9}{4}}{x - 2} + \frac{\frac{7}{4}}{x + 2}$$

Therefore,

$$\int \frac{(x^3 + 1)\, dx}{x^2 - 4} = \int x\, dx + \frac{9}{4}\int \frac{dx}{x - 2} + \frac{7}{4}\int \frac{dx}{x + 2}$$

$$= \frac{1}{2}x^2 + \frac{9}{4}\ln|x - 2| + \frac{7}{4}\ln|x + 2| + C \qquad \blacksquare$$

Quadratic Factors

A quadratic polynomial $x^2 + ax + b$ is called **irreducible** if it cannot be written as a product of two linear factors (without using complex numbers). If the denominator of a proper rational function has an irreducible quadratic factor $(x^2 + ax + b)^M$, then it contributes a sum of the following type to a partial fraction decomposition:

$$\frac{A_1 x + B_1}{x^2 + ax + b} + \frac{A_2 x + B_2}{(x^2 + ax + b)^2} + \cdots + \frac{A_M x + B_M}{(x^2 + ax + b)^M}$$

For example,

$$\frac{4 - 12x}{(x + 1)(x^2 + x + 4)^2} = \frac{1}{x + 1} - \frac{x}{x^2 + x + 4} - \frac{4x + 12}{(x^2 + x + 4)^2}$$

← REMINDER *If $b > 0$, then $x^2 + b$ is irreducible, but $x^2 - b$ is reducible because*

$$x^2 - b = \left(x + \sqrt{b}\right)\left(x - \sqrt{b}\right)$$

EXAMPLE 5 **Irreducible Versus Reducible Quadratic Factors** Evaluate:

(a) $\displaystyle\int \frac{18}{(x + 3)(x^2 + 9)} \, dx$ **(b)** $\displaystyle\int \frac{18}{(x + 3)(x^2 - 9)} \, dx$

Solution

(a) The quadratic factor $x^2 + 9$ is irreducible, so the partial fraction decomposition has the form

$$\frac{18}{(x + 3)(x^2 + 9)} = \frac{A}{x + 3} + \frac{Bx + C}{x^2 + 9}$$

Clear the denominators to obtain

$$18 = A(x^2 + 9) + (Bx + C)(x + 3)$$

We will use Undetermined Coefficients, so we multiply out the right side and then combine terms in like powers of x:

$$18 = Ax^2 + 9A + Bx^2 + 3Bx + Cx + 3C$$

$$18 = (A + B)x^2 + (3B + C)x + 9A + 3C$$

Now, equating coefficients of like powers of x:

$$0 = A + B \qquad \text{(coefficient of } x^2\text{)}$$

$$0 = 3B + C \qquad \text{(coefficient of } x\text{)}$$

$$18 = 9A + 3C \qquad \text{(constant terms)}$$

Into the third equation, we can substitute $A = -B$ from the first equation and $C = -3B$ from the second equation to obtain

$$18 = -9B - 9B$$

This yields $B = -1$. Now, knowing $B = -1$, we obtain $A = -B = 1$ and $C = -3B = 3$. We can then compute the integral:

In the second equality, the first integral is done with the substitution $u = x^2 + 9$. The second integral can be done with a trigonometric substitution $x = 3 \tan\theta$ or using the integral formula
$$\int \frac{dx}{x^2 + a^2} = \frac{1}{a}\tan^{-1}\frac{x}{a} + C. \text{ For the two integrals, we obtain}$$

$$\int \frac{x \, dx}{x^2 + 9} = \frac{1}{2}\int \frac{du}{u} = \frac{1}{2}\ln(x^2 + 9) + C$$

$$\int \frac{dx}{x^2 + 9} = \frac{1}{3}\tan^{-1}\frac{x}{3} + C$$

$$\int \frac{18 \, dx}{(x + 3)(x^2 + 9)} = \int \frac{dx}{x + 3} + \int \frac{(-x + 3) \, dx}{x^2 + 9}$$

$$= \int \frac{dx}{x + 3} - \int \frac{x \, dx}{x^2 + 9} + \int \frac{3 \, dx}{x^2 + 9}$$

$$= \ln|x + 3| - \frac{1}{2}\ln(x^2 + 9) + \tan^{-1}\frac{x}{3} + C$$

The last line comes from applying the formulas in the margin.

(b) The polynomial $x^2 - 9$ is reducible because $x^2 - 9 = (x - 3)(x + 3)$. Therefore, the partial fraction decomposition has the form

$$\frac{18}{(x + 3)(x^2 - 9)} = \frac{18}{(x + 3)^2(x - 3)} = \frac{A}{x - 3} + \frac{B}{x + 3} + \frac{C}{(x + 3)^2}$$

Clear the denominators:

$$18 = A(x + 3)^2 + B(x + 3)(x - 3) + C(x - 3)$$

We use Value Substitution. The substitutions $x = 3$ and $x = -3$ provide equations that yield values for A and C, respectively. An additional substitution (we will use $x = 0$) results in a third equation that will be used to determine B:

- Set $x = 3$. This gives $18 = 36A$.
- Set $x = -3$. This gives $18 = -6C$.
- Set $x = 0$. This gives $18 = 9A - 9B - 3C$.

The solutiona are $A = \frac{1}{2}$, $C = -3$, and $B = -\frac{1}{2}$. Determining the integral, we find

$$\int \frac{18}{(x+3)(x^2-9)}\,dx = \frac{1}{2}\int \frac{dx}{x-3} - \frac{1}{2}\int \frac{dx}{x+3} - 3\int \frac{dx}{(x+3)^2}$$

$$= \frac{1}{2}\ln|x-3| - \frac{1}{2}\ln|x+3| + 3(x+3)^{-1} + C \qquad \blacksquare$$

In the next example, we use both Value Substitution and Undetermined Coefficients to find the partial fraction decomposition. Due to the complexity of the polynomials involved, after Value Substitution determines one constant, Undetermined Coefficients is the best option for determining the remaining ones.

EXAMPLE 6 **Repeated Quadratic Factor** Evaluate $\int \frac{4-x}{x(x^2+2)^2}\,dx$.

Solution The partial fraction decomposition has the form

$$\frac{4-x}{x(x^2+2)^2} = \frac{A}{x} + \frac{Bx+C}{x^2+2} + \frac{Dx+E}{(x^2+2)^2}$$

Clear the denominators by multiplying through by $x(x^2+2)^2$:

$$4 - x = A(x^2+2)^2 + (Bx+C)\big(x(x^2+2)\big) + (Dx+E)x \qquad \boxed{6}$$

We compute A directly by setting $x = 0$. Then Eq. (6) reduces to $4 = 4A$, or $A = 1$.

We cannot find the remaining constants as simply as we determined A. The best option is to use Undetermined Coefficients to set up a system of four equations in the four unknowns B, C, D, and E. To begin, we substitute the known value $A = 1$ in Eq. (6) and expand:

$$4 - x = (x^4 + 4x^2 + 4) + (Bx^4 + 2Bx^2 + Cx^3 + 2Cx) + (Dx^2 + Ex)$$

$$-x = (1+B)x^4 + Cx^3 + (4+2B+D)x^2 + (2C+E)x$$

Now equate the coefficients on the two sides of the equation:

$$0 = 1 + B \qquad \text{(coefficient of } x^4)$$

$$0 = C \qquad \text{(coefficient of } x^3)$$

$$0 = 4 + 2B + D \qquad \text{(coefficient of } x^2)$$

$$-1 = 2C + E \qquad \text{(coefficient of } x)$$

These equations yield $B = -1$, $C = 0$, $D = -2$, and $E = -1$. Thus,

$$\int \frac{(4-x)\,dx}{x(x^2+2)^2} = \int \frac{dx}{x} - \int \frac{x\,dx}{x^2+2} - \int \frac{(2x+1)\,dx}{(x^2+2)^2}$$

$$= \ln|x| - \frac{1}{2}\ln(x^2+2) - \int \frac{(2x+1)\,dx}{(x^2+2)^2}$$

The middle integral was evaluated using the substitution $u = x^2 + 2$, $du = 2x\,dx$. The third integral separates as a sum:

$$\int \frac{(2x+1)\,dx}{(x^2+2)^2} = \underbrace{\int \frac{2x\,dx}{(x^2+2)^2}}_{\text{Use substitution } u = x^2+2.} + \int \frac{dx}{(x^2+2)^2}$$

$$= -(x^2+2)^{-1} + \int \frac{dx}{(x^2+2)^2} \qquad \boxed{7}$$

To evaluate the integral in Eq. (7), we use the trigonometric substitution

$$x = \sqrt{2}\tan\theta, \qquad dx = \sqrt{2}\sec^2\theta\,d\theta, \qquad x^2 + 2 = 2\tan^2\theta + 2 = 2\sec^2\theta$$

Referring to Figure 1, we obtain

$$\int \frac{dx}{(x^2+2)^2} = \int \frac{\sqrt{2}\sec^2\theta\,d\theta}{(2\tan^2\theta + 2)^2} = \int \frac{\sqrt{2}\sec^2\theta\,d\theta}{4\sec^4\theta}$$

$$= \frac{\sqrt{2}}{4} \int \cos^2\theta\,d\theta = \frac{\sqrt{2}}{4}\left(\frac{1}{2}\theta + \frac{1}{2}\sin\theta\cos\theta\right) + C$$

$$= \frac{\sqrt{2}}{8}\tan^{-1}\frac{x}{\sqrt{2}} + \frac{\sqrt{2}}{8}\frac{x}{\sqrt{x^2+2}}\frac{\sqrt{2}}{\sqrt{x^2+2}} + C$$

$$= \frac{1}{4\sqrt{2}}\tan^{-1}\frac{x}{\sqrt{2}} + \frac{1}{4}\frac{x}{x^2+2} + C$$

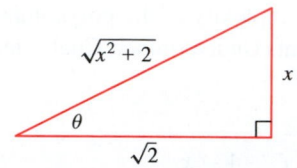

FIGURE 1 Right triangle with $\tan\theta = \frac{x}{\sqrt{2}}$.

Collecting all the terms, we have

$$\int \frac{4-x}{x(x^2+2)^2}\,dx = \ln|x| - \frac{1}{2}\ln(x^2+2) + \frac{1-\frac{1}{4}x}{x^2+2} - \frac{1}{4\sqrt{2}}\tan^{-1}\frac{x}{\sqrt{2}} + C \qquad \blacksquare$$

CONCEPTUAL INSIGHT   The examples in this section illustrate a general fact: The integral of a proper rational function can be expressed as a sum of rational functions, arctangents of linear or quadratic polynomials, and logarithms of linear or quadratic polynomials. Other types of functions, such as exponential and trigonometric functions, do not appear.

At the beginning of this section, we explained in a marginal comment that the Fundamental Theorem of Algebra guarantees that every polynomial can be factored into linear and quadratic terms. Thus, every proper rational function can be written as a sum of terms of the form $\frac{A}{(x-a)^n}$ and $\frac{Ax+B}{(x^2+bx+c)^n}$ for real numbers $a, b, c, A, B,$ and n. This property is simple in theory, but in practice, finding a sum can be difficult. Our challenge is to identify the factors of the denominator. Indeed, sometimes it is not possible to determine the exact factors and, at best, we can only approximate them. To find the numerators in the partial fraction decomposition once the denominator is factored, we can set up and solve a system of linear equations with the constants as unknowns.

Using a Computer Algebra System

Finding a partial fraction decomposition of a rational function P/Q often requires laborious computation. Fortunately, most computer algebra systems are able to compute partial fraction decompositions whenever the factors of Q can be determined exactly. Even though, in theory, all polynomials can be factored into a product of linear and quadratic terms, it is not always possible to determine the factors exactly. The rational function $f(x) = \frac{1}{x^5 + 2x + 2}$ is an example where an exact partial fraction decomposition cannot be found. Try it on a computer algebra system to see what results.

7.5 SUMMARY

The Method of Partial Fractions enables us to separate a complicated rational function into a sum of simpler rational terms that are easier to integrate than the original function.

Assume first that P/Q is a *proper* rational function [i.e., $\deg(P) < \deg(Q)$] and that $Q(x)$ can be factored explicitly as a product of linear and irreducible quadratic terms.

- If $Q(x)$ is equal to a product of powers of linear factors $(x-a)^M$ and irreducible quadratic factors $(x^2 + ax + b)^N$, then the partial fraction decomposition of the rational function $P(x)/Q(x)$ is a sum of terms of the following type:

$$(x-a)^M \quad \text{contributes} \quad \frac{A_1}{x-a} + \frac{A_2}{(x-a)^2} + \cdots + \frac{A_M}{(x-a)^M}$$

$$(x^2 + ax + b)^N \quad \text{contributes} \quad \frac{A_1 x + B_1}{x^2 + ax + b} + \frac{A_2 x + B_2}{(x^2 + ax + b)^2}$$

$$+ \cdots + \frac{A_N x + B_N}{(x^2 + ax + b)^N}$$

- The methods of Value Substitution and Undetermined Coefficients can be used to determine the constants A_i and B_i in the partial fraction decomposition.
- The individual terms in the decomposition then can be integrated. Substitution, completing the square, and trigonometric substitution may be needed to integrate the terms corresponding to $(x^2 + ax + b)^N$ (see Example 6).
- If P/Q is improper, use long division (see Example 4) to express P/Q as a sum of a polynomial and a proper rational function. The Method of Partial Fractions can then be applied to the proper rational function.

7.5 EXERCISES

Preliminary Questions

1. Suppose that $\int f(x)\,dx = \ln x + \sqrt{x+1} + C$. Can f be a rational function? Explain.

2. Which of the following are *proper* rational functions?

(a) $\dfrac{x}{x-3}$

(b) $\dfrac{4}{9-x}$

(c) $\dfrac{x^2 + 12}{(x+2)(x+1)(x-3)}$

(d) $\dfrac{4x^3 - 7x}{(x-3)(2x+5)(9-x)}$

3. Which of the following quadratic polynomials are irreducible? To check, complete the square if necessary.

(a) $x^2 + 5$

(b) $x^2 - 5$

(c) $x^2 + 4x + 6$

(d) $x^2 + 4x + 2$

4. Let P/Q be a proper rational function where $Q(x)$ factors as a product of distinct linear factors $(x - a_i)$. Then

$$\int \frac{P(x)\,dx}{Q(x)}$$

(choose the correct answer):

(a) is a sum of logarithmic terms $A_i \ln(x - a_i)$ for some constants A_i.

(b) may contain a term involving the arctangent.

Exercises

1. Match the rational functions (a)–(d) with the corresponding partial fraction decompositions (i)–(iv).

(a) $\dfrac{x^2 + 4x + 12}{(x+2)(x^2 + 4)}$

(b) $\dfrac{2x^2 + 8x + 24}{(x+2)^2(x^2 + 4)}$

(c) $\dfrac{x^2 - 4x + 8}{(x-1)^2(x-2)^2}$

(d) $\dfrac{x^4 - 4x + 8}{(x+2)(x^2 + 4)}$

(i) $x - 2 + \dfrac{4}{x+2} - \dfrac{4x - 4}{x^2 + 4}$

(ii) $\dfrac{-8}{x-2} + \dfrac{4}{(x-2)^2} + \dfrac{8}{x-1} + \dfrac{5}{(x-1)^2}$

(iii) $\dfrac{1}{x+2} + \dfrac{2}{(x+2)^2} + \dfrac{-x+2}{x^2 + 4}$

(iv) $\dfrac{1}{x+2} + \dfrac{4}{x^2 + 4}$

2. Determine the constants A, B:

$$\frac{2x - 3}{(x-3)(x-4)} = \frac{A}{x-3} + \frac{B}{x-4}$$

3. Clear the denominators in the following partial fraction decomposition and determine the constant B (substitute a value of x or use the method of undetermined coefficients).

$$\frac{3x^2 + 11x + 12}{(x+1)(x+3)^2} = \frac{1}{x+1} - \frac{B}{x+3} - \frac{3}{(x+3)^2}$$

4. Find the constants in the partial fraction decomposition

$$\frac{2x + 4}{(x-2)(x^2 + 4)} = \frac{A}{x-2} + \frac{Bx + C}{x^2 + 4}$$

In Exercises 5–8, evaluate using long division first to write $f(x)$ as the sum of a polynomial and a proper rational function.

5. $\displaystyle\int \frac{x\,dx}{3x-4}$

6. $\displaystyle\int \frac{(x^2+2)\,dx}{x+3}$

7. $\displaystyle\int \frac{(x^3+2x^2+1)\,dx}{x+2}$

8. $\displaystyle\int \frac{(x^3+1)\,dx}{x^2+1}$

In Exercises 9–46, evaluate the integral.

9. $\displaystyle\int \frac{dx}{(x-2)(x-4)}$

10. $\displaystyle\int \frac{(2-x)\,dx}{x+4}$

11. $\displaystyle\int \frac{dx}{x(3x+1)}$

12. $\displaystyle\int \frac{(2x-1)\,dx}{x^2-5x+6}$

13. $\displaystyle\int \frac{x^2\,dx}{x^2+9}$

14. $\displaystyle\int \frac{dx}{(x-2)(x-3)(x+2)}$

15. $\displaystyle\int \frac{(x^2+3x-44)\,dx}{(x+3)(x+5)(3x-2)}$

16. $\displaystyle\int \frac{3\,dx}{(x+1)(x^2+x)}$

17. $\displaystyle\int \frac{(x^2+11x)\,dx}{(x-1)(x+1)^2}$

18. $\displaystyle\int \frac{(4x^2-21x)\,dx}{(x-3)^2(2x+3)}$

19. $\displaystyle\int \frac{dx}{(x-1)^2(x-2)^2}$

20. $\displaystyle\int \frac{(x^2-8x)\,dx}{(x+1)(x+4)^3}$

21. $\displaystyle\int \frac{8\,dx}{x(x+2)^3}$

22. $\displaystyle\int \frac{(x^2-1)\,dx}{x^2+1}$

23. $\displaystyle\int \frac{dx}{2x^2-3}$

24. $\displaystyle\int \frac{dx}{(x-4)^2(x-1)}$

25. $\displaystyle\int \frac{dx}{x^3+x^2-x-1}$

26. $\displaystyle\int \frac{dx}{x^3-3x^2+4}$

27. $\displaystyle\int \frac{4x^2-20}{(2x+5)^3}\,dx$

28. $\displaystyle\int \frac{3x+6}{x^2(x-1)(x-3)}\,dx$

29. $\displaystyle\int \frac{dx}{x(x-1)^3}$

30. $\displaystyle\int \frac{(3x^2-2)\,dx}{x-4}$

31. $\displaystyle\int \frac{(x^2-x+1)\,dx}{x^2+x}$

32. $\displaystyle\int \frac{dx}{x(x^2-1)}$

33. $\displaystyle\int \frac{(3x^2-4x+5)\,dx}{(x-1)(x^2+1)}$

34. $\displaystyle\int \frac{x^2}{(x+1)(x^2+1)}\,dx$

35. $\displaystyle\int \frac{dx}{x(x^2+25)}$

36. $\displaystyle\int \frac{dx}{x^2(x^2+25)}$

37. $\displaystyle\int \frac{(6x^2+2)\,dx}{x^2+2x-3}$

38. $\displaystyle\int \frac{6x^2+7x-6}{(x^2-4)(x+2)}\,dx$

39. $\displaystyle\int \frac{10\,dx}{(x-1)^2(x^2+9)}$

40. $\displaystyle\int \frac{10\,dx}{(x+1)(x^2+9)^2}$

41. $\displaystyle\int \frac{dx}{x(x^2+8)^2}$

42. $\displaystyle\int \frac{100x\,dx}{(x-3)(x^2+1)^2}$

43. $\displaystyle\int \frac{dx}{(x+2)(x^2+4x+10)}$

44. $\displaystyle\int \frac{9\,dx}{(x+1)(x^2-2x+6)}$

45. $\displaystyle\int \frac{25\,dx}{x(x^2+2x+5)^2}$

46. $\displaystyle\int \frac{(x^2+3)\,dx}{(x^2+2x+3)^2}$

In Exercises 47–50, evaluate by using first substitution and then partial fractions if necessary.

47. $\displaystyle\int \frac{x\,dx}{x^4+1}$

48. $\displaystyle\int \frac{x\,dx}{(x+2)^4}$

49. $\displaystyle\int \frac{e^x\,dx}{e^{2x}-1}$

50. $\displaystyle\int \frac{\sec^2\theta\,d\theta}{\tan^2\theta-1}$

51. Evaluate $\displaystyle\int \frac{\sqrt{x}\,dx}{x-1}$. *Hint:* Use the substitution $u=\sqrt{x}$ (sometimes called a **rationalizing substitution**).

52. Evaluate $\displaystyle\int \frac{dx}{x^{1/2}-x^{1/3}}$. *Hint:* Use the substitution $u=x^{1/6}$.

53. Evaluate $\displaystyle\int \frac{dx}{x^{5/4}-4x^{3/4}}$.

54. Evaluate $\displaystyle\int \frac{dx}{x^{4/3}+x-2x^{2/3}}$.

55. Evaluate $\displaystyle\int \frac{dx}{x^2-1}$ in two ways: using partial fractions and using trigonometric substitution. Verify that the two answers agree.

56. [GU] Graph the equation $(x-40)y^2=10x(x-30)$ and find the volume of the solid obtained by revolving the region between the graph and the x-axis for $0\le x\le 30$ around the x-axis.

57. Show that the substitution $\theta=2\tan^{-1}t$ (Figure 2) yields the formulas

$$\cos\theta=\frac{1-t^2}{1+t^2}, \qquad \sin\theta=\frac{2t}{1+t^2}, \qquad d\theta=\frac{2\,dt}{1+t^2} \qquad \boxed{8}$$

This substitution transforms the integral of any rational function of $\cos\theta$ and $\sin\theta$ into an integral of a rational function of t (which can then be evaluated using partial fractions). Use it to evaluate $\displaystyle\int \frac{d\theta}{\cos\theta+\frac{3}{4}\sin\theta}$.

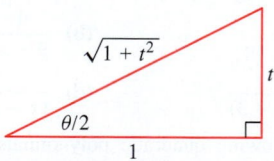

FIGURE 2

58. Use the substitution of Exercise 57 to evaluate $\displaystyle\int \frac{d\theta}{\cos\theta+\sin\theta}$.

Further Insights and Challenges

59. Prove the general formula

$$\int \frac{dx}{(x-a)(x-b)}=\frac{1}{a-b}\ln\left|\frac{x-a}{x-b}\right|+C$$

where a,b are constants such that $a\ne b$.

60. The method of partial fractions shows that

$$\int \frac{dx}{x^2-1}=\frac{1}{2}\ln|x-1|-\frac{1}{2}\ln|x+1|+C$$

A computer algebra system evaluates this integral as $-\tanh^{-1}x$, where $\tanh^{-1}x$ is the inverse hyperbolic tangent function. Can you reconcile the two answers?

61. Suppose that $Q(x)=(x-a)(x-b)$, where $a\ne b$, and let P/Q be a proper rational function so that

$$\frac{P(x)}{Q(x)}=\frac{A}{(x-a)}+\frac{B}{(x-b)}$$

(a) Show that $A = \dfrac{P(a)}{Q'(a)}$ and $B = \dfrac{P(b)}{Q'(b)}$.

(b) Use this result to find the partial fraction decomposition for $P(x) = 3x - 2$ and $Q(x) = x^2 - 4x - 12$.

62. Suppose that $Q(x) = (x - a_1)(x - a_2)\cdots(x - a_n)$, where the roots a_j are all distinct. Let P/Q be a proper rational function so that

$$\frac{P(x)}{Q(x)} = \frac{A_1}{(x - a_1)} + \frac{A_2}{(x - a_2)} + \cdots + \frac{A_n}{(x - a_n)}$$

(a) Show that $A_j = \dfrac{P(a_j)}{Q'(a_j)}$ for $j = 1, \ldots, n$.

(b) Use this result to find the partial fraction decomposition for $P(x) = 2x^2 - 1$, $Q(x) = x^3 - 4x^2 + x + 6 = (x + 1)(x - 2)(x - 3)$.

7.6 Strategies for Integration

In Chapter 5, and in the preceding sections of this chapter, we have seen a variety of techniques for evaluating various integrals. In a general setting, when confronted with a given integral, it will not appear in a particular section devoted to a particular technique of integration. Hence, it is important to be able to recognize which technique of integration is likely to apply. This section is devoted to that topic. In addition to considering how to recognize what technique to apply, we also discuss how tables of integrals and how computer algebra systems can be utilized to find an integral.

In general, there are no hard and fast rules for evaluating a given indefinite integral. But there are various heuristics that help us to determine which techniques are likely to apply. Here is an overview of the different approaches that we can employ to evaluate an integral.

Simplification: Do any algebraic simplification possible. Cancel terms in fractions when possible. For instance,

1. $\displaystyle\int \frac{x^3 - 1}{x - 1}\,dx = \int \frac{(x - 1)(x^2 + x + 1)}{x - 1}\,dx = \int (x^2 + x + 1)\,dx$

$$= \frac{x^3}{3} + \frac{x^2}{2} + x + C$$

2. $\displaystyle\int \frac{x - x^3}{\sqrt{x}}\,dx = \int (x^{1/2} - x^{5/2})\,dx = \frac{2}{3}x^{3/2} - \frac{2}{7}x^{7/2} + C$

3. $\displaystyle\int \frac{1}{\sin^2 x}\,dx = \int \csc^2 x\,dx = -\cot x + C$

4. $\displaystyle\int e^x \sinh x\,dx$. It is tempting to try to apply Integration by Parts to this integral, but this method does not work. Instead, we replace $\sinh x$ with its expression in terms of exponential functions to obtain

$$\int e^x \sinh x\,dx = \int e^x \left(\frac{e^x - e^{-x}}{2} \right) dx = \frac{1}{2}\int \left(e^{2x} - 1 \right) dx = \frac{e^{2x}}{4} - \frac{x}{2} + C$$

Substitution: Any time we recognize that our integral is of the form $\displaystyle\int f(g(x))g'(x)\,dx$, we can use substitution. The key to successfully substituting is that if we want to replace $g(x)$ by u, we need a $g'(x)$ to appear in the integrand in such a way that we can pair it with dx to obtain the necessary du.

So, for example, $\displaystyle\int x^2 \sin(x^3)\,dx$ is set up for substitution since if we let $u = x^3$, then $du = 3x^2\,dx$, and we have the requisite $x^2\,dx$ to make this work. We simply substitute $(1/3)du$ for $x^2\,dx$, and we will obtain an integral that we then can compute.

In the case of $\displaystyle\int e^{\sin x} \cos x\,dx$, we can use $u = \sin x$ and $du = \cos x\,dx$. In the case of $\displaystyle\int e^x \sqrt{e^x + 1}\,dx$, we can let $u = e^x + 1$ and then $du = e^x\,dx$, which we have.

Sometimes, some extra algebraic work is needed to carry out a substitution so that all of the x terms are changed to terms in the new variable u.

EXAMPLE 1 Find $\int x^3\sqrt{1+x^2}\,dx$.

Solution Our first inclination is to try the substitution $u = 1 + x^2$ since that expression is inside the square root. But then $du = 2x\,dx$, and we have an extra factor of x^2 left over. Instead of turning to another method of integration, though, we can note that since $u = 1 + x^2$, $x^2 = u - 1$. Thus, we have

$$\int x^3\sqrt{1+x^2}\,dx = \int x^2\sqrt{1+x^2}\,x\,dx = \int (u-1)\sqrt{u}\,\frac{du}{2}$$

$$= \frac{1}{2}\int \left(u^{3/2} - u^{1/2}\right)\,du = \frac{u^{5/2}}{5} - \frac{u^{3/2}}{3} + C$$

$$= \frac{(1+x^2)^{5/2}}{5} - \frac{(1+x^2)^{3/2}}{3} + C \qquad \blacksquare$$

This integral could also be evaluated using Integration by Parts, but that would be a more involved process.

Integration by Parts: As we have seen, if there is a product in the integrand, we can split the integral into two pieces u and dv and then apply Integration by Parts to obtain

$$\int u\,dv = uv - \int v\,du$$

For example, look at $\int xe^x\,dx$. In this case, it is obvious that Integration by Parts applies. If the x were not present, we could easily integrate e^x. So choosing $u = x$ and $dv = e^x\,dx$ will yield $\int xe^x\,dx = xe^x - \int e^x\,dx$. The remaining integral is now one we can compute.

In particular, we are looking for integrands that are products such that differentiating one term in the product and integrating the other yields an integral that is easier to evaluate. Here are some cases to look for:

1. $\int x^n f(x)\,dx$. Assuming $f(x)$ can be repeatedly integrated, we may use repeated applications of Integration by Parts, always choosing u equal to the remaining power of x, until we eliminate the powers of x, leaving something we can integrate. Candidates for $f(x)$ include $\sin x$, $\cos x$, e^x, a^x, $\sinh x$, $\cosh x$, $\sec^2 x$, and $\csc^2 x$, among others. Note that the same technique applies when we replace x^n in the integrand with a more complicated polynomial. Repeated application of Integration by Parts can be used to eliminate the polynomial.

2. $\int e^x \sin x\,dx$. Two applications of Integration by Parts, choosing $u = e^x$ both times, yields

$$\int e^x \sin x\,dx = -e^x \cos x + e^x \sin x - \int e^x \sin x\,dx$$

Instead of obtaining a simpler integral on the right, we obtained the integral we started with. Adding it to both sides, we get

$$2\int e^x \sin x\,dx = -e^x \cos x + e^x \sin x + C$$

$$\int e^x \sin x\,dx = \frac{1}{2}(-e^x \cos x + e^x \sin x) + C$$

3. In some special cases, we can use Integration by Parts for $\int f(x)\,dx$ by setting $u = f(x)$ and $dv = dx$. For example, this works with the integrals $\int \ln x\,dx$, $\int \tan^{-1} x\,dx$, and $\int \sin^{-1} x\,dx$.

4. Sometimes Integration by Parts is easier if we integrate, rather than differentiate, the x^n factor, as in the next example.

EXAMPLE 2 Compute $\int x^2 \ln x\,dx$.

Solution We proceed as follows:

$$u = \ln x \qquad dv = x^2 dx$$

$$du = \frac{1}{x}\,dx \qquad v = \frac{1}{3}x^3$$

$$\int x^2 \ln x\,dx = \frac{1}{3}x^3 \ln x - \int \left(\frac{1}{3}x^3\right)\frac{1}{x}\,dx = \frac{1}{3}x^3 \ln x - \frac{1}{3}\int x^2 dx$$

$$= \frac{1}{3}x^3 \ln x - \frac{1}{9}x^3 + C \qquad\blacksquare$$

For the previous example, using $u = x^2$ and $dv = \ln x\,dx$ also works, but this alternative is more complicated than the choice we used (see Exercise 60).

Special Techniques: Consider the form of the integral. Does it fall into one of the categories for which we have a special technique?

1. Is it a trigonometric integral? If it is of the form

$$\int \sin^n x \cos^m x\,dx, \quad \text{or} \quad \int \tan^n x \sec^m x\,dx$$

then we have specific rules that can be applied (as outlined in Section 7.2) to simplify the integral into one that can be computed.

2. Does the integrand contain

$$\sqrt{a^2 - x^2}, \quad \sqrt{x^2 + a^2}, \quad \text{or} \quad \sqrt{x^2 - a^2}$$

or more generally,

$$(a^2 - x^2)^{n/2}, \quad (x^2 + a^2)^{n/2}, \quad \text{or} \quad (x^2 - a^2)^{n/2}$$

for an integer n? If so, then try a trigonometric substitution, $x = a\sin\theta$, $x = a\tan\theta$, or $x = a\sec\theta$, respectively, as outlined in Section 7.3.

EXAMPLE 3 Evaluate $\int (x^2 + 16)^{3/2}\,dx$.

Solution Let $x = 4\tan\theta$. Then $(x^2 + 16)^{3/2} = (16\tan^2\theta + 16)^{3/2} = (16\sec^2\theta)^{3/2} = 64\sec^3\theta$ and $dx = 4\sec^2\theta\,d\theta$. Hence, we have

$$\int (x^2 + 16)^{3/2}\,dx = \int 64\sec^3\theta(4\sec^2\theta)\,d\theta = 256\int \sec^5\theta\,d\theta$$

Now, we can use the reduction formula in Eq. (12) from Section 7.2 to obtain

$$256 \int \sec^5 \theta \, d\theta = 256 \left(\frac{\tan \theta \sec^3 \theta}{4} + \frac{3}{4} \int \sec^3 \theta \, d\theta \right)$$

$$= 256 \left(\frac{\tan \theta \sec^3 \theta}{4} + \frac{3}{4} \left(\frac{\tan \theta \sec \theta}{2} + \frac{1}{2} \int \sec \theta \, d\theta \right) \right)$$

$$= 256 \left(\frac{\tan \theta \sec^3 \theta}{4} + \frac{3}{4} \left(\frac{\tan \theta \sec \theta}{2} + \frac{1}{2} \ln |\sec \theta + \tan \theta| \right) \right) + C$$

$$= 64 \tan \theta \sec^3 \theta + 96 \tan \theta \sec \theta + 96 \ln |\sec \theta + \tan \theta| + C$$

Since $x = 4 \tan \theta$, we know $\tan \theta = \frac{x}{4}$, and by the triangle in Figure 1, $\sec \theta = \frac{\sqrt{x^2+16}}{4}$. Thus, we have

$$\int (x^2 + 16)^{3/2} \, dx$$

$$= 64 \frac{x}{4} (x^2 + 16)^{3/2} + 96 \frac{x}{4} (x^2 + 16)^{1/2} + 96 \ln \left| (x^2 + 16)^{1/2} + \frac{x}{4} \right| + C$$

$$= 16x(x^2 + 16)^{3/2} + 24x(x^2 + 16)^{1/2} + 96 \ln \left| (x^2 + 16)^{1/2} + \frac{x}{4} \right| + C \qquad \blacksquare$$

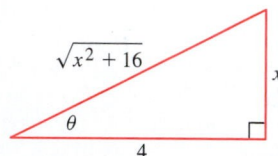

FIGURE 1 Since $x/4 = \tan \theta$, the hypotenuse is $\sqrt{x^2 + 16}$, and therefore $\sec \theta = \frac{\sqrt{x^2 + 16}}{4}$.

If the integral contains an expression of the form $\sqrt{ax^2 + bx + c}$, we can complete the square inside the square root in order to obtain one of the cases discussed above.

3. Is it a rational function $\dfrac{P}{Q}$ for polynomials P and Q? If so, we can often apply the Method of Partial Fractions. First, if P has a degree that is at least as large as the degree of Q, we divide $P(x)$ by $Q(x)$ to obtain a polynomial (which is easily integrated) together with a remainder term $\dfrac{R(x)}{Q(x)}$, to which the Method of Partial Fractions can be applied.

EXAMPLE 4 Find $\displaystyle\int \frac{x^2 + 2x + 10}{x^2 + x - 6} \, dx$.

Solution Noting that both the numerator and the denominator have degree 2, we divide to obtain

$$\frac{x^2 + 2x + 10}{x^2 + x - 6} = 1 + \frac{x + 16}{x^2 + x - 6}$$

Then we can write

$$\frac{x + 16}{x^2 + x - 6} = \frac{x + 16}{(x + 3)(x - 2)} = \frac{A}{x + 3} + \frac{B}{x - 2}$$

Clearing the denominator, this yields

$$x + 16 = A(x - 2) + B(x + 3)$$

We use Value Substitution to determine A and B. Substituting $x = 2$ yields $18 = 5B$, and therefore, $B = \frac{18}{5}$. Furthermore, substituting $x = -3$ results in $13 = -5A$, implying $A = -\frac{13}{5}$. Thus,

$$\int \frac{x^2 + 2x + 10}{x^2 + x - 6} \, dx = \int \left(1 + \frac{-13}{5(x + 3)} + \frac{18}{5(x - 2)} \right) dx$$

$$= x - \frac{13}{5} \ln |x + 3| + \frac{18}{5} \ln |x - 2| + C \qquad \blacksquare$$

Integral Table: Integral tables (such as the one on the final three pages of this text) contain a list of forms of many common integrals. Given a particular integral that we may want to evaluate, if we can get it into the form of one of the integrals in a table, then we can apply the given formula to evaluate the integral. Let's consider a couple of examples.

EXAMPLE 5 Evaluate $\displaystyle\int \frac{\sqrt{9 - 4x^2}}{x}\, dx$.

Solution Looking down the list of integrals given at the end of the text, we see the formula that appears to apply here is #69:

$$\int \frac{\sqrt{a^2 - u^2}}{u}\, du = \sqrt{a^2 - u^2} - a \ln \left| \frac{a + \sqrt{a^2 - u^2}}{u} \right| + C$$

We can put our integral in the correct form by factoring the 4 out of the square root:

$$\sqrt{9 - 4x^2} = 2\sqrt{9/4 - x^2}$$

Now we use formula #69 with $a = 3/2$:

$$\int \frac{\sqrt{9 - 4x^2}}{x}\, dx = 2\int \frac{\sqrt{9/4 - x^2}}{x}\, dx = 2\sqrt{9/4 - x^2} - 3 \ln \left| \frac{3/2 + \sqrt{9/4 - x^2}}{x} \right| + C$$

■

Notice that the integral in the previous example could also be evaluated by applying a trigonometric substitution. In fact, many of the formulas in the integral table result from integral techniques and formulas that we have developed.

EXAMPLE 6 Evaluate $\displaystyle\int 6x \tan^2(x^2)\, dx$.

Solution The integral table formula that seems applicable is #36:

$$\int \tan^2 u\, du = \tan u - u + C$$

Although the integrand does not have the form in this formula, observe that a substitution $u = x^2$ will put us in a position to use formula #36. With this substitution, we have $du = 2x\, dx$, so $3\, du = 6x\, dx$. Therefore,

$$\int 6x \tan^2(x^2)\, dx = 3\int \tan^2 u\, du = 3\tan u - 3u + C$$

$$= 3\tan(x^2) - 3x^2 + C$$

■

Computer Algebra System: In order to find these last integrals using an integral table, we needed to do some algebra or apply some integration techniques to get them to match the pattern for one of the formulas at the back of the book. Computers are particularly adept at attempting various rearrangements and integration techniques and then matching the resulting integrals. For this reason, computer algebra systems are very good at evaluating integrals. However, keep in mind that the output that they generate may not be the form that is the most convenient to use. For example, if we want to find $\displaystyle\int x(x - 3)^{15}\, dx$, we do a substitution for $u = x - 3$ (and therefore $x = u + 3$) and find

$$\int x(x - 3)^{15}\, dx = \int (u + 3)u^{15}\, du = \int u^{16} + 3u^{15}\, du = \frac{1}{17}u^{17} + \frac{3}{16}u^{16} + C$$

$$= \frac{1}{17}(x - 3)^{17} + \frac{3}{16}(x - 3)^{16} + C$$

On the other hand, the following answer was obtained from a computer algebra system:

$$\int x(x-3)^{15}\,dx = \frac{x^{17}}{17} - \frac{45x^{16}}{16} + 63x^{15} - \frac{1755x^{14}}{2} + 8505x^{13} - \frac{243{,}243x^{12}}{4}$$

$$+\,331{,}695x^{11} - \frac{2{,}814{,}669x^{10}}{2} + 4{,}691{,}115x^{9} - \frac{98{,}513{,}415x^{8}}{8}$$

$$+\,25{,}332{,}021x^{7} - \frac{80{,}601{,}885x^{6}}{2} + 48{,}361{,}131x^{5} - \frac{167{,}403{,}915x^{4}}{4}$$

$$+\,23{,}914{,}845x^{3} - \frac{14{,}348{,}907x^{2}}{2} + C$$

As you can see, the answer that we obtained by substitution is simpler to state and would be more convenient to use in computations.

Numerical Apprpoximation: If the integral is a definite integral, and finding an anti-derivative is difficult or impossible, then the integral can be approximated numerically. One approach would be to use an approximating Riemann sum. Other numerical methods are introduced in Section 7.8.

EXAMPLE 7 The Fresnel function $S(x) = \int_0^x \sin(t^2)\,dt$ (Figure 2) is important in the field of optics. It has a maximum at $x = \sqrt{\pi}$ [the least positive x at which $S'(x) = 0$]. Approximate the maximum value of S.

Solution We need to approximate $\int_0^{\sqrt{\pi}} \sin(t^2)\,dt$. Note, we cannot evaluate this definite integral via antidifferentiation since there is no antiderivative of $f(t) = \sin(t^2)$ that can be expressed in terms of elementary functions. Using technology to compute an approximating Riemann sum with 1000 subintervals, we obtain

$$S(\sqrt{\pi}) = \int_0^{\sqrt{\pi}} \sin(t^2)\,dt \approx 0.895 \qquad \blacksquare$$

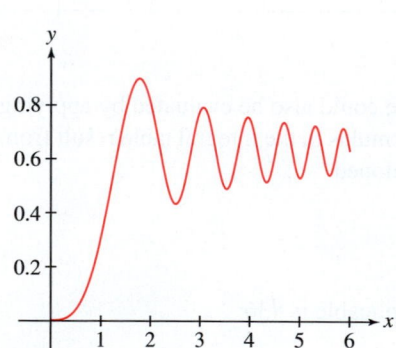

FIGURE 2 The Fresnel function $S(x) = \int_0^x \sin(t^2)\,dt$.

7.6 SUMMARY

Strategy for integration:

- **Simplification:** Do any algebraic simplification possible.
- **Substitution:** Consider possible substitutions, $u = g(x)$, keeping in mind that we will need the presence of $g'(x)$ since $du = g'(x)\,dx$.
- **Integration by Parts:** Consider Integration by Parts, choosing u to be a function that can be differentiated, dv to be something that can be integrated, and such that $\int v\,du$ is an easier integral than the original. Keep in mind that Integration by Parts can work when the integrand is not an obvious product, as long as u and dv are chosen properly.
- **Trigonometric Integral:** If the integral is a trigonometric integral of the form $\int \sin^n x \cos^m x\,dx$ or $\int \tan^m x \sec^m x\,dx$, then trigonometric identities and reduction formulas can be applied.
- **Trigonometric Substitution:** Does the integral contain $\sqrt{a^2 - x^2}$, $\sqrt{x^2 - a^2}$ or $\sqrt{x^2 + a^2}$? If so, then try a trigonometric substitution.
- **Partial Fractions:** Is the integrand a rational function $\dfrac{P}{Q}$ for polynomials P and Q? If so, the Method of Partial Fractions might enable you to separate the function into a sum of simpler rational terms that can be integrated.
- **Integral Table:** Determine whether the integral, after suitable manipulation, matches an integral in an integral table.

- **Computer Algebra System:** Consider using a computer algebra system to determine the integral.
- **Numerical Approximation:** If the integral is a definite integral, and none of the above methods work, approximate the integral numerically.

7.6 EXERCISES

Preliminary Questions

For each of the following, state what method applies and how one applies it, but do not evaluate the integral.

1. $\displaystyle\int x \sin x \, dx$

2. $\displaystyle\int \sqrt{1 + x^2} \, dx$

3. $\displaystyle\int \frac{1 + x^2}{1 - x^2} \, dx$

4. $\displaystyle\int \cos^2 x \sin x \, dx$

5. $\displaystyle\int x \ln x \, dx$

6. $\displaystyle\int \sqrt{1 - x^2} \, dx$

7. $\displaystyle\int \sin^3 x \cos^2 x \, dx$

For each of the following, find the formula in the integral table at the back of the text that can be applied to find the integral.

8. $\displaystyle\int \frac{3x^2 \, dx}{5x + 2}$

9. $\displaystyle\int \frac{\sqrt{25 + 16x^2}}{x^2} \, dx$

10. $\displaystyle\int \sec^3(4x) \, dx$

11. $\displaystyle\int \frac{x^2}{\sqrt{x^2 + 2x + 5}} \, dx$

Exercises

In Exercises 1–10, indicate a good method for evaluating the integral (but do not evaluate). Your choices are substitution (specify u and du), Integration by Parts (specify u and dv), a trigonometric method, or trigonometric substitution (specify). If it appears that these techniques are not sufficient, state this.

1. $\displaystyle\int \frac{x \, dx}{\sqrt{12 - 6x - x^2}}$

2. $\displaystyle\int \sqrt{4x^2 - 1} \, dx$

3. $\displaystyle\int \sin^3 x \cos^3 x \, dx$

4. $\displaystyle\int x \sec^2 x \, dx$

5. $\displaystyle\int \frac{dx}{\sqrt{9 - x^2}}$

6. $\displaystyle\int \sqrt{1 - x^3} \, dx$

7. $\displaystyle\int \sin^{3/2} x \, dx$

8. $\displaystyle\int x^2 \sqrt{x + 1} \, dx$

9. $\displaystyle\int \frac{dx}{(x + 1)(x + 2)^3}$

10. $\displaystyle\int \frac{dx}{(x + 12)^4}$

In Exercises 11–59, evaluate the integral using the appropriate method or combination of methods covered thus far in the text. You may use the integral tables at the end of the text, but do not use a computer algebra system.

11. $\displaystyle\int \frac{dx}{x^2\sqrt{4 - x^2}}$

12. $\displaystyle\int \frac{dx}{x(x - 1)^2}$

13. $\displaystyle\int \cos^2 4x \, dx$

14. $\displaystyle\int x \csc x \cot x \, dx$

15. $\displaystyle\int x \sin x \cot x \, dx$

16. $\displaystyle\int \frac{x \, dx}{(x^2 + 9)^2}$

17. $\displaystyle\int \frac{dx}{(x^2 + 9)^2}$

18. $\displaystyle\int \theta \sec^{-1} \theta \, d\theta$

19. $\displaystyle\int \tan^5 x \sec x \, dx$

20. $\displaystyle\int \frac{(3x^2 - 1) \, dx}{x(x^2 - 1)}$

21. $\displaystyle\int \ln(x^4 - 1) \, dx$

22. $\displaystyle\int \frac{x \, dx}{(x^2 - 1)^{3/2}}$

23. $\displaystyle\int \frac{x^2 \, dx}{(x^2 - 1)^{3/2}}$

24. $\displaystyle\int \frac{x^3 \, dx}{(x^2 - 1)^{3/2}}$

25. $\displaystyle\int \frac{(x + 1) \, dx}{(x^2 + 4x + 8)^2}$

26. $\displaystyle\int \frac{\sqrt{x} \, dx}{x^3 + 1}$

27. $\displaystyle\int \frac{x^{1/2} \, dx}{x^{1/3} + 1}$

28. $\displaystyle\int \frac{dx}{\sqrt{16 + x^2}}$

29. $\displaystyle\int \frac{e^x \, dx}{1 + e^x}$

30. $\displaystyle\int \frac{dt}{(1 + 4t^2)^{3/2}}$

31. $\displaystyle\int x^2 \ln x \, dx$

32. $\displaystyle\int \sec^5 y \tan y \, dy$

33. $\displaystyle\int \frac{dx}{x^2 + 2x + 5}$

34. $\displaystyle\int \frac{x^4 + 1}{x^2 + 1} \, dx$

35. $\displaystyle\int \sqrt{x^4 + x^7} \, dx$

36. $\displaystyle\int \sqrt{x^2 + 6x} \, dx$

37. $\displaystyle\int \frac{dx}{1 + e^x}$

38. $\displaystyle\int \frac{x^5}{x^3 - 1} \, dx$

39. $\displaystyle\int \frac{x}{\sqrt{x - 1}} \, dx$

40. $\displaystyle\int \frac{x}{\sqrt{x + 2}} \, dx$

41. $\displaystyle\int \frac{x^2}{\sqrt{x + 1}} \, dx$

42. $\displaystyle\int \sqrt{x^2 - 16} \, dx$

43. $\displaystyle\int (\sin x + \cos 2x)^2 \, dx$

44. $\displaystyle\int \sqrt{1 + \sqrt{x}} \, dx$

45. $\displaystyle\int \sin^2 x \tan x \, dx$

46. $\displaystyle\int \ln(x^2 - 9) \, dx$

47. $\displaystyle\int \ln(x^2 + 9) \, dx$

48. $\displaystyle\int \frac{dx}{x(x^2 - 6x - 7)}$

49. $\displaystyle\int \sin^5 x \cos^2 x \, dx$

50. $\displaystyle\int e^x \sqrt{e^{2x} - 1} \, dx$

51. $\displaystyle\int \cos^7 x \, dx$

52. $\displaystyle\int \frac{x^{11}}{x^4 - 1} \, dx$

53. $\displaystyle\int \frac{x^5}{x^4 - 1} \, dx$

54. $\displaystyle\int \tan x \sec^{5/4} x \, dx$

55. $\displaystyle\int (3 \sec x - \cos x)^2 \, dx$

56. $\displaystyle\int x^3 \ln x \, dx$

57. $\displaystyle\int \frac{(1 + \ln x)^2}{x} \, dx$

58. $\displaystyle\int \frac{e^x}{e^{2x} - 1} \, dx$

59. $\displaystyle\int \frac{dx}{\sqrt{x^2 - 36}}$

60. Use Integration by Parts to compute $\displaystyle\int x^2 \ln x \, dx$, setting $u = x^2$ and $dv = \ln x \, dx$.

7.7 Improper Integrals

The integrals we have studied so far represent signed areas of bounded regions. However, we also wish to consider unbounded regions, such as the region under the graph of $y = \frac{1}{1+x^2}$, for $-\infty < x < \infty$, shown in Figure 1. Integrals over such regions arise in applications and are represented by what are known as **improper integrals**.

There are two ways an integral can be improper: (1) The interval of integration is infinite, or (2) the integrand tends to infinity on a finite interval and therefore the graph of the function has a vertical asymptote. We deal first with improper integrals over infinite intervals. One or both endpoints may be infinite:

$$\int_{-\infty}^{a} f(x) \, dx, \qquad \int_{a}^{\infty} f(x) \, dx, \qquad \int_{-\infty}^{\infty} f(x) \, dx$$

How can an unbounded region have finite area? To answer this question, we must specify what we mean by the area of an unbounded region. Consider the area [Figure 2(A)] under the graph of $f(x) = e^{-x}$ over the finite interval $[0, R]$:

$$\int_{0}^{R} e^{-x} \, dx = -e^{-x} \Big|_{0}^{R} = -e^{-R} + e^0 = 1 - e^{-R}$$

As $R \to \infty$, this area approaches a finite value [Figure 2(B)]:

$$\int_{0}^{\infty} e^{-x} \, dx = \lim_{R \to \infty} \int_{0}^{R} e^{-x} \, dx = \lim_{R \to \infty} \left(1 - e^{-R}\right) = 1 \qquad \boxed{1}$$

It seems reasonable to take this limit as the *definition* of the area under the graph over the infinite interval $[0, \infty)$. Thus, the unbounded region in Figure 2(C) has area 1.

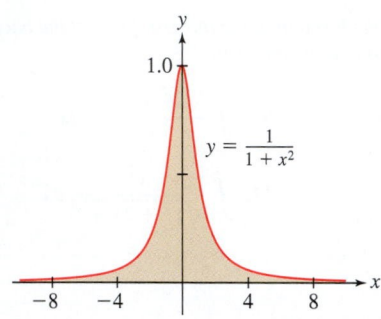

FIGURE 1 The region extends infinitely far in both directions, but the total area is finite (see Example 4).

*The great British mathematician G. H. Hardy (1877–1947) observed that in calculus, we learn to ask not "What is it?" but rather "How shall we **define** it?" We saw that tangent lines and areas under curves have no clear meaning until we define them precisely using limits. Here again, the key question is "How shall we define the area of an unbounded region?"*

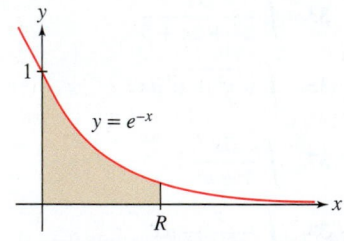
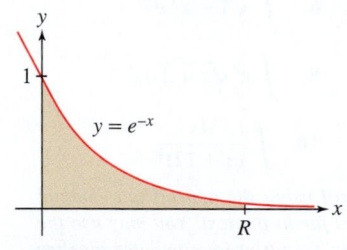
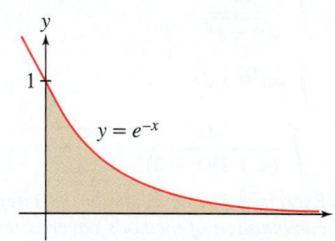

(A) Bounded region has area $1 - e^{-R}$ (B) Area approaches 1 as $R \to \infty$ (C) Unbounded region has area 1

DF FIGURE 2

←**REMINDER** When we say that $\displaystyle\lim_{R \to \infty} \int_{a}^{R} f(x) \, dx$ does not exist, we mean that the integral values do not approach any single real number as $R \to \infty$. This includes the possibilities that
$$\lim_{R \to \infty} \int_{a}^{R} f(x) \, dx = \infty \quad \text{or}$$
$$\lim_{R \to \infty} \int_{a}^{R} f(x) \, dx = -\infty.$$

DEFINITION   Improper Integral   Fix a number a and assume that f is integrable over $[a, b]$ for all $b > a$. The *improper integral* of f over $[a, \infty)$ is defined as the following limit (if it exists):

$$\int_{a}^{\infty} f(x) \, dx = \lim_{R \to \infty} \int_{a}^{R} f(x) \, dx$$

We say that the improper integral *converges* if the limit exists and that it *diverges* if the limit does not exist.

Similarly, we define

$$\int_{-\infty}^{a} f(x) \, dx = \lim_{R \to -\infty} \int_{R}^{a} f(x) \, dx$$

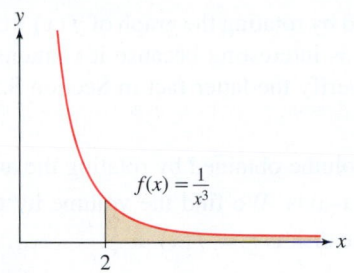

DF FIGURE 3 The area over $[2, \infty)$ is equal to $\frac{1}{8}$.

EXAMPLE 1 Show that $\displaystyle\int_2^\infty \frac{dx}{x^3}$ converges and compute its value.

Solution

$$\int_2^\infty \frac{dx}{x^3} = \lim_{R\to\infty} \int_2^R \frac{dx}{x^3} = \lim_{R\to\infty} -\frac{1}{2}x^{-2}\Big|_2^R = \lim_{R\to\infty} -\frac{1}{2}\left(R^{-2}\right) + \frac{1}{2}\left(2^{-2}\right)$$

$$= \lim_{R\to\infty} \left(\frac{1}{8} - \frac{1}{2R^2}\right) = \frac{1}{8}$$

We conclude that the unbounded shaded region in Figure 3 has area $\frac{1}{8}$. ∎

EXAMPLE 2 Determine whether $\displaystyle\int_{1000}^\infty \frac{dx}{x}$ converges.

Solution Since the integration is from 1000 to ∞, we take $R > 1000$ and compute the limit as $R \to \infty$:

$$\int_{1000}^\infty \frac{dx}{x} = \lim_{R\to\infty} \int_{1000}^R \frac{dx}{x} = \lim_{R\to\infty} \ln|x|\Big|_{1000}^R$$

$$= \lim_{R\to\infty} \ln R - \ln 1000 = \infty$$

The limit is infinite, so the improper integral diverges. We conclude that the area of the unbounded region in Figure 4 is infinite. ∎

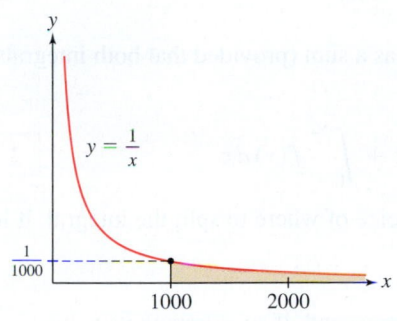

FIGURE 4 The integral of $f(x) = x^{-1}$ over $[1000, \infty)$ is infinite.

Note that in the previous example, even though $f(x) = 1/x$ is less than 0.001 in $(1000, \infty)$ and is decreasing to 0, the area under the graph is infinite. Such is the somewhat perplexing nature of improper integrals.

> **CONCEPTUAL INSIGHT** If you compare the unbounded shaded regions in Figures 3 and 4, you may wonder why one has finite area and the other has infinite area. Convergence of an improper integral depends on how rapidly $f(x)$ tends to zero as $x \to \infty$ (or $x \to -\infty$). The previous two examples show that $f(x) = x^{-3}$ tends to zero quickly enough that the integral converges, whereas $f(x) = x^{-1}$ does not.
>
> An improper integral of a power function $f(x) = x^{-p}$ is called a **p-integral**. Note that $f(x) = x^{-p}$ decreases more rapidly as p gets larger. Interestingly, our next theorem shows that the exponent $p = -1$ is the dividing line between convergence and divergence.

> **THEOREM 1** **The p-Integral over** $[a, \infty)$ For $a > 0$,
>
> $$\int_a^\infty \frac{dx}{x^p} = \begin{cases} \dfrac{a^{1-p}}{p-1} & \text{if } p > 1 \\[2mm] \text{diverges} & \text{if } p \le 1 \end{cases}$$

Proof Denote the p-integral by J. Then

$$J = \lim_{R\to\infty} \int_a^R x^{-p}\, dx = \lim_{R\to\infty} \frac{x^{1-p}}{1-p}\Big|_a^R = \lim_{R\to\infty} \left(\frac{R^{1-p}}{1-p} - \frac{a^{1-p}}{1-p}\right)$$

p-integrals are particularly important because they are often used to determine the convergence or divergence of more complicated improper integrals by means of the Comparison Test (see Example 10).

If $p > 1$, then $1 - p < 0$ and R^{1-p} tends to zero as $R \to \infty$. In this case, $J = \dfrac{a^{1-p}}{p-1}$.

If $p < 1$, then $1 - p > 0$ and R^{1-p} tends to ∞. In this case, J diverges. If $p = 1$, then

J diverges because $\displaystyle\lim_{R\to\infty} \int_a^R x^{-1}\, dx = \lim_{R\to\infty} (\ln R - \ln a) = \infty$. ∎

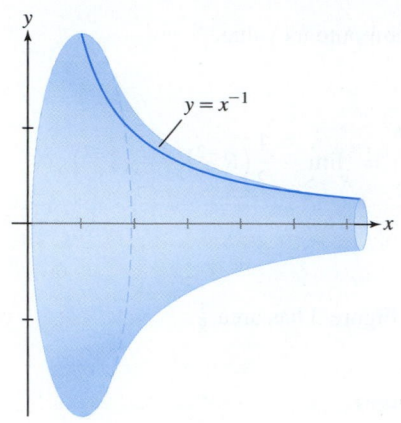

EXAMPLE 3 **Gabriel's Horn** is the surface obtained by rotating the graph of $f(x) = \frac{1}{x}$, for $x \geq 1$, about the x-axis (Figure 5). This surface is interesting because it contains a finite volume but has an infinite surface area. (We verify the latter fact in Section 8.2.) Compute the volume contained in Gabriel's Horn.

Solution The volume contained in the horn is the volume obtained by rotating the area under the graph of $f(x) = \frac{1}{x}$, for $x > 1$, about the x-axis. We find the volume by the Disk Method from Section 6.3 where the radius of the disk is $r = f(x) = \frac{1}{x}$:

$$\text{volume} = \pi \int_1^\infty r^2 \, dx = \pi \int_1^\infty \left(\frac{1}{x}\right)^2 \, dx = \lim_{R \to \infty} \pi \int_1^R x^{-2} \, dx$$

$$= \lim_{R \to \infty} -\pi x^{-1} \Big|_1^R = -\pi \lim_{R \to \infty} \left(\frac{1}{R} - 1\right) = \pi$$

Therefore, the volume contained in Gabriel's Horn is π. ∎

FIGURE 5 Gabriel's Horn contains a finite volume but has an infinite surface area. Therefore—paradoxically—it can be filled with a finite volume of paint but requires an infinite amount of paint to cover its surface. (A resolution to this paradox is presented in Exercise 63 in Section 10.2.)

A doubly infinite improper integral is defined as a sum (provided that both integrals on the right converge):

$$\int_{-\infty}^\infty f(x) \, dx = \int_{-\infty}^0 f(x) \, dx + \int_0^\infty f(x) \, dx \qquad \boxed{2}$$

We can use some number other than 0 as a choice of where to split the integral, if it is more convenient to do so.

EXAMPLE 4 Determine if $\displaystyle\int_{-\infty}^\infty \frac{1}{1+x^2} \, dx$ converges and, if so, compute its value.

Solution

$$\int_{-\infty}^\infty \frac{1}{1+x^2} \, dx = \int_{-\infty}^0 \frac{1}{1+x^2} \, dx + \int_0^\infty \frac{1}{1+x^2} \, dx$$

assuming both of these integrals converge. For the second of these,

$$\int_0^\infty \frac{1}{1+x^2} \, dx = \lim_{R \to \infty} \int_0^R \frac{1}{1+x^2} \, dx = \lim_{R \to \infty} \tan^{-1} x \Big|_0^R = \lim_{R \to \infty} \tan^{-1} R - 0 = \frac{\pi}{2}$$

Similarly,

$$\int_{-\infty}^0 \frac{1}{1+x^2} \, dx = \lim_{R \to -\infty} \int_R^0 \frac{1}{1+x^2} \, dx = \lim_{R \to -\infty} \tan^{-1} x \Big|_R^0 = 0 - \lim_{R \to -\infty} \tan^{-1} R = \frac{\pi}{2}$$

Thus, since both integrals converge,

$$\int_{-\infty}^\infty \frac{1}{1+x^2} \, dx = \int_{-\infty}^0 \frac{1}{1+x^2} \, dx + \int_0^\infty \frac{1}{1+x^2} \, dx = \frac{\pi}{2} + \frac{\pi}{2} = \pi \qquad ∎$$

Sometimes it is necessary to use L'Hôpital's Rule to determine the limits that arise in improper integrals.

EXAMPLE 5 **Using L'Hôpital's Rule** Calculate $\displaystyle\int_0^\infty xe^{-x} \, dx$.

Solution The integral corresponds to the area in Figure 6. First, we compute the associated indefinite integral using Integration by Parts with $u = x$ and $dv = e^{-x} \, dx$:

$$\int xe^{-x} \, dx = -xe^{-x} + \int e^{-x} \, dx = -xe^{-x} - e^{-x} + C = -(x+1)e^{-x} + C$$

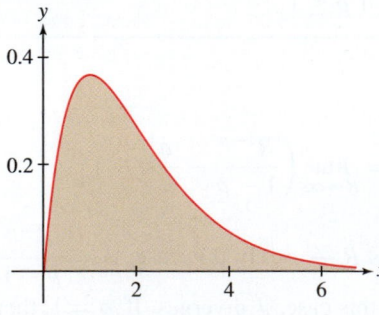

FIGURE 6 The integral of $f(x) = xe^{-x}$ over $[0, \infty)$ is the shaded area.

$$\int_0^R xe^{-x} \, dx = -(x+1)e^{-x} \Big|_0^R = -(R+1)e^{-R} + 1 = 1 - \frac{R+1}{e^R}$$

Now, compute the improper integral as a limit using L'Hôpital's Rule:

$$\int_0^\infty xe^{-x}\,dx = 1 - \lim_{R\to\infty}\frac{R+1}{e^R} = 1 - \underbrace{\lim_{R\to\infty}\frac{1}{e^R}}_{\text{L'Hôpital's Rule}} = 1 - 0 = 1 \qquad \blacksquare$$

Improper integrals arise in applications when it makes sense to treat certain large quantities as if they were infinite. In the next example, we determine the escape velocity of an object launched from Earth by assuming that the velocity is sufficient to take it "infinitely far" into space.

In physics, we speak of moving an object "infinitely far away." In practice, this means "very far away," but it is more convenient to work with an improper integral.

◀ **REMINDER** *The mass of the earth is*

$$M_e \approx 5.98 \cdot 10^{24}\text{ kg}$$

The radius of the earth is

$$r_e \approx 6.37 \cdot 10^6\text{ m}$$

The universal gravitational constant is

$$G \approx 6.67 \cdot 10^{-11}\text{ N-m}^2/\text{kg}^2$$

A newton is 1 kg-m/s^2 and a joule is 1 N-m.

EXAMPLE 6 **Escape Velocity** The earth exerts a gravitational force of magnitude $F(r) = GM_em/r^2$ on an object of mass m at distance r from the center of the earth.

(a) Find the work required to move the object infinitely far from the earth.

(b) Calculate the escape velocity v_{esc} on the earth's surface.

Solution This amounts to computing a p-integral with $p = 2$. Recall that work is the integral of force as a function of distance (Section 6.5).

(a) The work required to move an object from the earth's surface ($r = r_e$) to a distance R from the center is

$$\int_{r_e}^{R}\frac{GM_em}{r^2}\,dr = -\frac{GM_em}{r}\Big|_{r_e}^{R} = GM_em\left(\frac{1}{r_e} - \frac{1}{R}\right)\text{ joules}$$

The work moving the object "infinitely far away" is the improper integral

$$GM_em\int_{r_e}^\infty\frac{dr}{r^2} = \lim_{R\to\infty}GM_em\left(\frac{1}{r_e} - \frac{1}{R}\right) = \frac{GM_em}{r_e}\text{ joules}$$

(b) By the principle of Conservation of Energy, an object launched with velocity v_0 will escape the earth's gravitational field if its kinetic energy $\frac{1}{2}mv_0^2$ is at least as large as the work required to move the object to infinity—that is, if

$$\frac{1}{2}mv_0^2 \geq \frac{GM_em}{r_e} \quad\Rightarrow\quad v_0 \geq \left(\frac{2GM_e}{r_e}\right)^{1/2}$$

Escape velocity in miles per hour is approximately 25,000 mph.

Using the values recalled in the marginal note, we find that $v_0 \geq 11{,}200$ m/s. The minimal velocity is the escape velocity $v_{\text{esc}} = 11{,}200$ m/s. $\qquad \blacksquare$

In practice, the word "forever" means "a long but unspecified length of time." For example, if the investment pays out dividends for 100 years, then its present value is

$$\int_0^{100}6000e^{-0.04t}\,dt \approx \$147{,}253$$

The improper integral (\$150,000) gives an approximation to this value.

EXAMPLE 7 **Present Value of Future Income** If an investment pays a dividend continuously at a rate of $R(t)$ \$/year and earns interest at rate r (in decimal form), then the present value of the dividend income, for the period from $t = 0$ to $t = T$, is given by

$$PV = \int_0^T R(t)e^{-rt}\,dt.$$

We think of present value as the payment that we would need to receive at $t = 0$, instead of the dividend income, so that at time T the payment's value (with accumulated interest) would be the same as the amount accumulated from the dividend income (with its accumulated interest). It is essentially the present worth of the income we are about to receive up to time T.

Assuming that the dividend rate is \$6000/year, and the interest rate is 4%, compute the present value if the dividends continue forever.

Solution Over an infinite time interval,

$$PV = \int_0^\infty 6000e^{-0.04t}\,dt = \lim_{T\to\infty}\frac{6000e^{-0.04t}}{-0.04}\Big|_0^T = \frac{6000}{0.04} = \$150{,}000$$

Although an infinite amount of money is paid out during the infinite time interval, the total present value is finite. $\qquad \blacksquare$

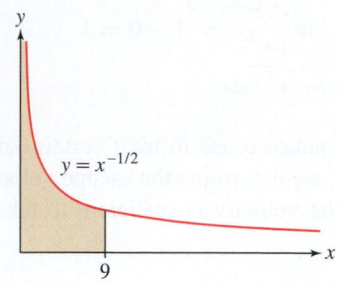

FIGURE 7 By Example 8(a), the shaded region has area 6.

Unbounded Functions

An integral over a finite interval $[a, b]$ is improper if the integrand is unbounded. In this case, the region in question is unbounded in the vertical direction. For example, $\int_0^9 \frac{dx}{\sqrt{x}}$ is improper because the integrand $f(x) = x^{-1/2}$ tends to ∞ as $x \to 0^+$ (Figure 7). Improper integrals of this type are defined as one-sided limits.

DEFINITION Unbounded Integrands If f is continuous on $[a, b)$ and $\lim\limits_{x \to b^-} f(x) = \pm\infty$, we define

$$\int_a^b f(x)\,dx = \lim_{R \to b^-} \int_a^R f(x)\,dx$$

Similarly, if f is continuous on $(a, b]$ and $\lim\limits_{x \to a^+} f(x) = \pm\infty$,

$$\int_a^b f(x)\,dx = \lim_{R \to a^+} \int_R^b f(x)\,dx$$

In both cases, we say that the improper integral *converges* if the limit exists and that it *diverges* otherwise.

Note that if there is a single point c in the interval $[a, b]$ such that $\lim\limits_{x \to c^-} f(x) = \pm\infty$, or $\lim\limits_{x \to c^+} f(x) = \pm\infty$, and if $\int_a^c f(x)\,dx$ and $\int_c^b f(x)\,dx$ both converge, then we define

$$\int_a^b f(x)\,dx = \int_a^c f(x)\,dx + \int_c^b f(x)\,dx.$$

EXAMPLE 8 Calculate: **(a)** $\int_0^9 \frac{dx}{\sqrt{x}}$ and   **(b)** $\int_0^{1/2} \frac{dx}{x}$.

Solution Both integrals are improper because the integrands have infinite discontinuities at $x = 0$. The first integral converges:

$$\int_0^9 \frac{dx}{\sqrt{x}} = \lim_{R \to 0^+} \int_R^9 x^{-1/2}dx = \lim_{R \to 0^+} 2x^{1/2} \Big|_R^9$$

$$= \lim_{R \to 0^+} (6 - 2R^{1/2}) = 6$$

The second integral diverges:

$$\int_0^{1/2} \frac{dx}{x} = \lim_{R \to 0^+} \int_R^{1/2} \frac{dx}{x} = \lim_{R \to 0^+} \left(\ln \frac{1}{2} - \ln R \right)$$

$$= \ln \frac{1}{2} - \lim_{R \to 0^+} \ln R = \infty \qquad\blacksquare$$

EXAMPLE 9 Calculate $\int_0^2 \frac{dx}{(x - 1)^{\frac{2}{3}}}$.

Solution This integral is improper with an infinite discontinuity at $x = 1$ (Figure 8). Therefore, we write

$$\int_0^2 \frac{dx}{(x - 1)^{\frac{2}{3}}} = \int_0^1 \frac{dx}{(x - 1)^{\frac{2}{3}}} + \int_1^2 \frac{dx}{(x - 1)^{\frac{2}{3}}}$$

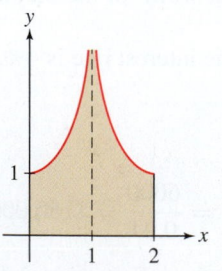

FIGURE 8 The unbounded shaded region has area 6.

We consider each integral individually:

$$\int_0^1 \frac{dx}{(x-1)^{\frac{2}{3}}} = \lim_{R \to 1^-} \int_0^R \frac{dx}{(x-1)^{\frac{2}{3}}} = \lim_{R \to 1^-} 3(x-1)^{\frac{1}{3}} \Big|_0^R$$

$$= \lim_{R \to 1^-} 3(R-1)^{\frac{1}{3}} - 3(-1)^{\frac{1}{3}} = 3$$

$$\int_1^2 \frac{dx}{(x-1)^{\frac{2}{3}}} = \lim_{R \to 1^+} \int_R^2 \frac{dx}{(x-1)^{\frac{2}{3}}} = \lim_{R \to 1^+} 3(x-1)^{\frac{1}{3}} \Big|_R^2$$

$$= 3(1)^{\frac{1}{3}} - \lim_{R \to 1^+} 3(R-1)^{\frac{1}{3}} = 3$$

Therefore, we obtain

Theorem 2 is valid for all exponents p. However, the integral is not improper if $p < 0$.

$$\int_0^2 \frac{dx}{(x-1)^{\frac{2}{3}}} = \int_0^1 \frac{dx}{(x-1)^{\frac{2}{3}}} + \int_1^2 \frac{dx}{(x-1)^{\frac{2}{3}}} = 3 + 3 = 6 \qquad \blacksquare$$

The proof of the next theorem is similar to the proof of Theorem 1 (see Exercise 52).

> **THEOREM 2** **The p-Integral over $[0, a]$**   For $a > 0$,
>
> $$\int_0^a \frac{dx}{x^p} = \begin{cases} \dfrac{a^{1-p}}{1-p} & \text{if } p < 1 \\ \text{diverges} & \text{if } p \geq 1 \end{cases}$$

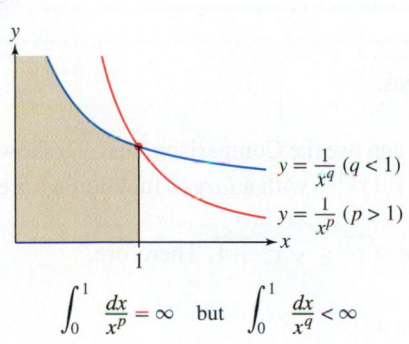

$$\int_0^1 \frac{dx}{x^p} = \infty \quad \text{but} \quad \int_0^1 \frac{dx}{x^q} < \infty$$

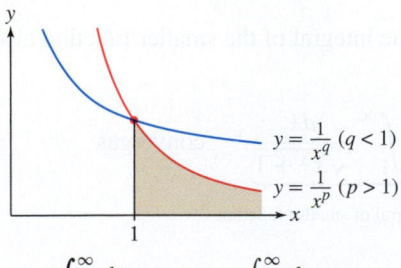

$$\int_1^\infty \frac{dx}{x^p} < \infty \quad \text{but} \quad \int_1^\infty \frac{dx}{x^q} = \infty$$

FIGURE 9

GRAPHICAL INSIGHT   The p-integrals $\int_1^\infty x^{-p} \, dx$ and $\int_0^1 x^{-p} \, dx$ have opposite behavior for $p \neq 1$. The first converges only for $p > 1$, and the second converges only for $p < 1$ (both diverge for $p = 1$). This is reflected in the graphs of $y = x^{-p}$ (with $p > 1$) and $y = x^{-q}$ (with $0 < q < 1$) in Figure 9.

Since $0 < q < 1$, the values of $f(x) = x^{-q}$ are arbitrarily large near $x = 0$ and decrease rapidly as x increases to 1, thereby resulting in the convergence of $\int_0^1 x^{-q} \, dx$. However, $f(x) = x^{-q}$ decreases slowly to 0 as $x \to \infty$, resulting in the divergence of $\int_1^\infty x^{-q} \, dx$.

The graphs for $q < 1$ and $p > 1$ switch relative positions and behaviors at the point of intersection $(1, 1)$.

Since $p > 1$, the values of $f(x) = x^{-p}$ are arbitrarily large near $x = 0$, but decrease slowly as $x \to 1$, resulting in the divergence of $\int_0^1 x^{-p} \, dx$. On the other hand, with $p > 1$, $f(x) = x^{-p}$ decreases rapidly to 0 as $x \to \infty$, resulting in the convergence of $\int_1^\infty x^{-p} \, dx$.

Comparing Integrals

Sometimes we are interested in determining whether an improper integral converges, even if we cannot find its exact value. For instance, the integral

$$\int_1^\infty \frac{e^{-x}}{x} \, dx$$

cannot be evaluated explicitly. However, if $x \geq 1$, then

$$0 \leq \frac{1}{x} \leq 1 \quad \Rightarrow \quad 0 \leq \frac{e^{-x}}{x} \leq e^{-x}$$

In other words, the graph of $y = e^{-x}/x$ lies *underneath* the graph of $y = e^{-x}$ for $x \geq 1$ (Figure 10). Therefore,

$$0 \quad \leq \quad \int_1^\infty \frac{e^{-x}}{x}\, dx \quad \leq \quad \underbrace{\int_1^\infty e^{-x}\, dx = e^{-1}}_{\text{Converges by direct computation}}$$

Since the larger integral converges, we can expect that the smaller integral also converges (and that its value is some positive number less than e^{-1}). This type of conclusion is stated in the next theorem. A proof is provided in a supplement on the text's Web site.

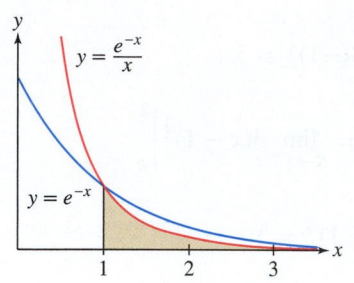

FIGURE 10 There is less area under $y = e^{-x}/x$ than $y = e^{-x}$ over the interval $[1, \infty)$.

THEOREM 3 Comparison Test for Improper Integrals
Assume that f and g are continuous functions such that $f(x) \geq g(x) \geq 0$ for $x \geq a$:

- If $\displaystyle\int_a^\infty f(x)\, dx$ converges, then $\displaystyle\int_a^\infty g(x)\, dx$ also converges.

- If $\displaystyle\int_a^\infty g(x)\, dx$ diverges, then $\displaystyle\int_a^\infty f(x)\, dx$ also diverges.

The Comparison Test is also valid for improper integrals of unbounded functions on a finite interval.

EXAMPLE 10 Show that $\displaystyle\int_1^\infty \frac{dx}{\sqrt{x^3 + 1}}$ converges.

Solution We cannot evaluate this integral, but we can use the Comparison Test. To show convergence, we must compare the integrand $(x^3 + 1)^{-1/2}$ with a *larger* function whose integral we can compute.

It makes sense to compare with $x^{-3/2}$ because $\sqrt{x^3} \leq \sqrt{x^3 + 1}$. Therefore,

$$\frac{1}{\sqrt{x^3 + 1}} \leq \frac{1}{\sqrt{x^3}} = x^{-3/2}$$

The integral of the larger function converges, so the integral of the smaller function also converges:

$$\underbrace{\int_1^\infty \frac{dx}{x^{3/2}}}_{p\text{-integral with } p > 1} \text{converges} \quad \Rightarrow \quad \underbrace{\int_1^\infty \frac{dx}{\sqrt{x^3 + 1}}}_{\text{Integral of smaller function}} \text{converges} \quad \blacksquare$$

What the Comparison Test says (for nonnegative functions):

- *If the integral of the larger function converges, then the integral of the smaller function also converges.*
- *If the integral of the smaller function diverges, then the integral of the larger function also diverges.*

EXAMPLE 11 Choosing the Right Comparison Does $\displaystyle\int_1^\infty \frac{dx}{\sqrt{x} + e^{3x}}$ converge?

Solution Since $\sqrt{x} \geq 0$, we have $\sqrt{x} + e^{3x} \geq e^{3x}$, and therefore,

$$\frac{1}{\sqrt{x} + e^{3x}} \leq \frac{1}{e^{3x}}$$

Furthermore,

$$\int_1^\infty \frac{dx}{e^{3x}} = \lim_{R \to \infty} -\frac{1}{3} e^{-3x} \Big|_1^R = \lim_{R \to \infty} \frac{1}{3}\left(e^{-3} - e^{-3R}\right) = \frac{1}{3} e^{-3} \quad \text{(converges)}$$

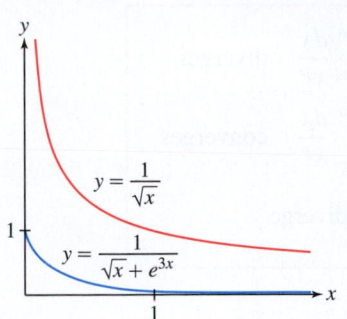

FIGURE 11 The divergence of the integral of the larger function says nothing about the integral of the smaller function.

Our integral converges by the Comparison Test:

$$\underbrace{\int_1^\infty \frac{dx}{e^{3x}}}_{\text{Integral of larger function}} \quad \text{converges} \quad \Rightarrow \quad \underbrace{\int_1^\infty \frac{dx}{\sqrt{x} + e^{3x}}}_{\text{Integral of smaller function}} \quad \text{also converges}$$

Had we not been thinking, we might have tried to use the inequality

$$\frac{1}{\sqrt{x} + e^{3x}} \leq \frac{1}{\sqrt{x}}$$

However, $\int_1^\infty \frac{dx}{\sqrt{x}}$ diverges (p-integral with $p < 1$), and this says nothing about the integral of the smaller function (Figure 11). ∎

EXAMPLE 12 **Endpoint Discontinuity** Does $\int_0^{0.5} \frac{dx}{x^8 + x^2}$ converge?

Solution This integrand has a discontinuity at $x = 0$, since $\lim\limits_{x \to 0^+} \frac{1}{x^8 + x^2} = +\infty$. We might try the comparison

$$x^8 + x^2 > x^2 \quad \Rightarrow \quad \frac{1}{x^8 + x^2} < \frac{1}{x^2}$$

However, the p-integral $\int_0^{0.5} \frac{dx}{x^2}$ diverges, so this says nothing about the integral involving the smaller function. But notice that if $0 < x < 0.5$, then $x^8 < x^2$, and therefore,

$$x^8 + x^2 < 2x^2 \quad \Rightarrow \quad \frac{1}{x^8 + x^2} > \frac{1}{2x^2}$$

Since $\int_0^{0.5} \frac{dx}{2x^2}$ diverges, $\int_0^{0.5} \frac{dx}{x^8 + x^2}$ also diverges. ∎

7.7 SUMMARY

- An *improper integral* is defined as the limit of definite integrals:

$$\int_a^\infty f(x)\,dx = \lim_{R \to \infty} \int_a^R f(x)\,dx$$

The improper integral *converges* if this limit exists, and it *diverges* otherwise.
- If f is continuous on $[a, b]$ and $\lim\limits_{x \to b^-} f(x) = \pm\infty$, then

$$\int_a^b f(x)\,dx = \lim_{R \to b^-} \int_a^R f(x)\,dx$$

- If f is continuous on $[a, b]$ and $\lim\limits_{x \to c^-} f(x) = \pm\infty$ or $\lim\limits_{x \to c^+} f(x) = \pm\infty$, where $a < c < b$ and if $\int_a^c f(x)\,dx$ and $\int_c^b f(x)\,dx$ converge, then

$$\int_a^b f(x)\,dx = \int_a^c f(x)\,dx + \int_c^b f(x)\,dx$$

- An improper integral of x^{-p} is called a p-integral. For $a > 0$,

$$p > 1: \quad \int_a^\infty \frac{dx}{x^p} \quad \text{converges} \quad \text{and} \quad \int_0^a \frac{dx}{x^p} \quad \text{diverges}$$

$$p < 1: \quad \int_a^\infty \frac{dx}{x^p} \quad \text{diverges} \quad \text{and} \quad \int_0^a \frac{dx}{x^p} \quad \text{converges}$$

$$p = 1: \quad \int_a^\infty \frac{dx}{x} \quad \text{and} \quad \int_0^a \frac{dx}{x} \quad \text{both diverge}$$

- The Comparison Test: Assume that f and g are continuous functions such that $f(x) \geq g(x) \geq 0$ for $x \geq a$. Then

$$\text{If} \quad \int_a^\infty f(x)\,dx \text{ converges,} \quad \text{then} \quad \int_a^\infty g(x)\,dx \text{ converges.}$$

$$\text{If} \quad \int_a^\infty g(x)\,dx \text{ diverges,} \quad \text{then} \quad \int_a^\infty f(x)\,dx \text{ diverges.}$$

- Remember that the Comparison Test provides no information if the integral of the larger function diverges or if the integral of the smaller function converges.
- The Comparison Test is also valid for improper integrals of functions with infinite discontinuities at an endpoint of the integral.

7.7 EXERCISES

Preliminary Questions

1. State whether each of the following integrals converges or diverges:

(a) $\int_1^\infty x^{-3}\,dx$

(b) $\int_0^1 x^{-3}\,dx$

(c) $\int_1^\infty x^{-2/3}\,dx$

(d) $\int_0^1 x^{-2/3}\,dx$

2. Is $\int_0^{\pi/2} \cot x\,dx$ an improper integral? Explain.

3. Find a value of $b > 0$ that makes $\int_0^b \frac{1}{x^2 - 4}\,dx$ an improper integral.

4. Which comparison would show that $\int_0^\infty \frac{dx}{x + e^x}$ converges?

5. Explain why it is not possible to draw any conclusions about the convergence of $\int_1^\infty \frac{e^{-x}}{x}\,dx$ by comparing with the integral $\int_1^\infty \frac{dx}{x}$.

Exercises

1. Which of the following integrals is improper? Explain your answer, but do not evaluate the integral.

(a) $\int_0^2 \frac{dx}{x^{1/3}}$

(b) $\int_1^\infty \frac{dx}{x^{0.2}}$

(c) $\int_{-1}^\infty e^{-x}\,dx$

(d) $\int_0^1 e^{-x}\,dx$

(e) $\int_0^\pi \sec x\,dx$

(f) $\int_0^\infty \sin x\,dx$

(g) $\int_0^1 \sin x\,dx$

(h) $\int_0^1 \frac{dx}{\sqrt{3 - x^2}}$

(i) $\int_1^\infty \ln x\,dx$

(j) $\int_0^3 \ln x\,dx$

2. Let $f(x) = x^{-4/3}$.

(a) Evaluate $\int_1^R f(x)\,dx$.

(b) Evaluate $\int_1^\infty f(x)\,dx$ by computing the limit

$$\lim_{R \to \infty} \int_1^R f(x)\,dx$$

3. Prove that $\int_1^\infty x^{-2/3}\,dx$ diverges by showing that

$$\lim_{R \to \infty} \int_1^R x^{-2/3}\,dx = \infty$$

4. Determine whether $\int_0^3 \frac{dx}{(3 - x)^{3/2}}$ converges by computing

$$\lim_{R \to 3-} \int_0^R \frac{dx}{(3 - x)^{3/2}}$$

In Exercises 5–40, determine whether the improper integral converges and, if so, evaluate it.

5. $\int_1^\infty \frac{dx}{x^{19/20}}$

6. $\int_1^\infty \frac{dx}{x^{20/19}}$

7. $\int_{-\infty}^4 e^{0.0001t}\,dt$

8. $\int_{20}^\infty \frac{dt}{t}$

9. $\displaystyle\int_0^5 \frac{dx}{x^{20/19}}$

10. $\displaystyle\int_0^5 \frac{dx}{x^{19/20}}$

11. $\displaystyle\int_0^4 \frac{dx}{\sqrt{4-x}}$

12. $\displaystyle\int_5^6 \frac{dx}{(x-5)^{3/2}}$

13. $\displaystyle\int_2^\infty x^{-3}\,dx$

14. $\displaystyle\int_0^\infty \frac{dx}{(x+1)^3}$

15. $\displaystyle\int_{-3}^\infty \frac{dx}{(x+4)^{3/2}}$

16. $\displaystyle\int_2^\infty e^{-2x}\,dx$

17. $\displaystyle\int_{-1}^1 \frac{dx}{x^{0.2}}$

18. $\displaystyle\int_2^\infty x^{-1/3}\,dx$

19. $\displaystyle\int_4^\infty e^{-3x}\,dx$

20. $\displaystyle\int_4^\infty e^{3x}\,dx$

21. $\displaystyle\int_{-\infty}^0 e^{3x}\,dx$

22. $\displaystyle\int_1^2 \frac{dx}{(x-1)^2}$

23. $\displaystyle\int_1^3 \frac{dx}{\sqrt{3-x}}$

24. $\displaystyle\int_{-4}^0 \frac{dx}{(x+2)^{1/3}}$

25. $\displaystyle\int_0^\infty \frac{dx}{1+x}$

26. $\displaystyle\int_{-\infty}^0 xe^{-x^2}\,dx$

27. $\displaystyle\int_0^\infty \frac{x\,dx}{(1+x^2)^2}$

28. $\displaystyle\int_3^6 \frac{x\,dx}{\sqrt{x-3}}$

29. $\displaystyle\int_0^\infty xe^{-3x}\,dx$

30. $\displaystyle\int_{-\infty}^0 x^2 e^x\,dx$

31. $\displaystyle\int_0^3 \frac{dx}{\sqrt{9-x^2}}$

32. $\displaystyle\int_0^1 \frac{e^{\sqrt{x}}\,dx}{\sqrt{x}}$

33. $\displaystyle\int_1^\infty \frac{e^{\sqrt{x}}\,dx}{\sqrt{x}}$

34. $\displaystyle\int_0^\pi \sec\theta\,d\theta$

35. $\displaystyle\int_0^\infty \sin x\,dx$

36. $\displaystyle\int_0^{\pi/2} \tan x\,dx$

37. $\displaystyle\int_0^1 \ln x\,dx$

38. $\displaystyle\int_1^2 \frac{dx}{x\ln x}$

39. $\displaystyle\int_0^1 \frac{\ln x}{x^2}\,dx$

40. $\displaystyle\int_1^\infty \frac{\ln x}{x^2}\,dx$

41. Let $I = \displaystyle\int_4^\infty \frac{dx}{(x-2)(x-3)}$.

(a) Show that for $R > 4$,

$$\int_4^R \frac{dx}{(x-2)(x-3)} = \ln\left|\frac{R-3}{R-2}\right| - \ln\frac{1}{2}$$

(b) Then show that $I = \ln 2$.

42. Evaluate the integral $I = \displaystyle\int_1^\infty \frac{dx}{x(2x+5)}$.

43. Evaluate $\displaystyle\int_0^1 \frac{dx}{x(2x+5)}$ or state that it diverges.

44. Evaluate $\displaystyle\int_2^\infty \frac{dx}{(x+3)(x+1)^2}$ or state that it diverges.

In Exercises 45–48, determine whether the doubly infinite improper integral converges and, if so, evaluate it. Use definition (2).

45. $\displaystyle\int_{-\infty}^\infty \frac{x\,dx}{1+x^2}$

46. $\displaystyle\int_{-\infty}^\infty e^{-|x|}\,dx$

47. $\displaystyle\int_{-\infty}^\infty xe^{-x^2}\,dx$

48. $\displaystyle\int_{-\infty}^\infty \frac{dx}{(x^2+1)^{3/2}}$

49. Determine whether $\displaystyle\int_{-1}^1 \frac{dx}{x^{1/3}}$ converges and, if so, to what.

50. Consider the integral $\displaystyle\int_{-\infty}^\infty x\,dx$.

(a) Show that it diverges.

(b) Show that $\displaystyle\lim_{R\to\infty} \int_{-R}^R x\,dx$ converges, thereby demonstrating that the definition of $\displaystyle\int_{-\infty}^\infty f(x)\,dx$ needs to be adhered to carefully.

51. For which values of a does $\displaystyle\int_0^\infty e^{ax}\,dx$ converge?

52. Show that $\displaystyle\int_0^1 \frac{dx}{x^p}$ converges if $p < 1$ and diverges if $p \ge 1$.

53. Sketch the region under the graph of the function $f(x) = \dfrac{1}{1+x^2}$ for $-\infty < x < \infty$, and show that its area is π.

54. Show that $\dfrac{1}{\sqrt{x^4+1}} \le \dfrac{1}{x^2}$ for all x, and use this to prove that $\displaystyle\int_1^\infty \frac{dx}{\sqrt{x^4+1}}$ converges.

55. Show that $\displaystyle\int_1^\infty \frac{dx}{x^3+4}$ converges by comparing with $\displaystyle\int_1^\infty x^{-3}\,dx$.

56. Show that $\displaystyle\int_2^\infty \frac{dx}{x^3-4}$ converges by comparing with $\displaystyle\int_2^\infty 2x^{-3}\,dx$.

57. ✏ Show that $0 \le e^{-x^2} \le e^{-x}$ for $x \ge 1$ (Figure 12). Use the Comparison Test to show that $\displaystyle\int_0^\infty e^{-x^2}\,dx$ converges. *Hint:* It suffices (why?) to make the comparison for $x \ge 1$ because

$$\int_0^\infty e^{-x^2}\,dx = \int_0^1 e^{-x^2}\,dx + \int_1^\infty e^{-x^2}\,dx$$

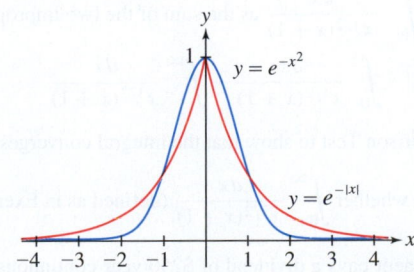

FIGURE 12 Comparison of $y = e^{-|x|}$ and $y = e^{-x^2}$.

58. Prove that $\displaystyle\int_{-\infty}^\infty e^{-x^2}\,dx$ converges by comparing with $\displaystyle\int_{-\infty}^\infty e^{-|x|}\,dx$ (Figure 12).

59. Show that $\displaystyle\int_1^\infty \frac{1-\sin x}{x^2}\,dx$ converges.

60. Let $a > 0$. Recall that $\displaystyle\lim_{x \to \infty} \frac{x^a}{\ln x} = \infty$ (by Exercise 67 in Section 4.5).

(a) Show that $x^a > 2 \ln x$ for all x sufficiently large.

(b) Show that $e^{-x^a} < x^{-2}$ for all x sufficiently large.

(c) Show that $\displaystyle\int_1^\infty e^{-x^a}\, dx$ converges.

In Exercises 61–75, use the Comparison Test to determine whether or not the integral converges.

61. $\displaystyle\int_1^\infty \frac{1}{\sqrt{x^5 + 2}}\, dx$

62. $\displaystyle\int_1^\infty \frac{dx}{(x^3 + 2x + 4)^{1/2}}$

63. $\displaystyle\int_3^\infty \frac{dx}{\sqrt{x} - 1}$

64. $\displaystyle\int_0^5 \frac{dx}{x^{1/3} + x^3}$

65. $\displaystyle\int_1^\infty e^{-(x + x^{-1})}\, dx$

66. $\displaystyle\int_0^1 \frac{|\sin x|}{\sqrt{x}}\, dx$

67. $\displaystyle\int_0^1 \frac{e^x}{x^2}\, dx$

68. $\displaystyle\int_1^\infty \frac{1}{x^4 + e^x}\, dx$

69. $\displaystyle\int_0^1 \frac{1}{x^4 + \sqrt{x}}\, dx$

70. $\displaystyle\int_1^\infty \frac{\ln x}{\sinh x}\, dx$

71. $\displaystyle\int_5^\infty \frac{1}{x^2 \ln x}\, dx$

72. $\displaystyle\int_1^\infty \frac{dx}{\sqrt{x^{1/3} + x^3}}$

73. $\displaystyle\int_0^1 \frac{dx}{(8x^2 + x^4)^{1/3}}$

74. $\displaystyle\int_1^\infty \frac{dx}{(x + x^2)^{1/3}}$

75. $\displaystyle\int_0^1 \frac{dx}{xe^x + x^2}$

Hint for Exercise 74: Show that for $x \geq 1$,

$$\frac{1}{(x + x^2)^{1/3}} \geq \frac{1}{2^{1/3} x^{2/3}}$$

Hint for Exercise 75: Show that for $0 \leq x \leq 1$,

$$\frac{1}{xe^x + x^2} \geq \frac{1}{(e + 1)x}$$

76. Use the Comparison Test to determine for what values of p this integral converges: $\displaystyle\int_5^\infty \frac{1}{x^p \ln x}\, dx$.

77. Consider $\displaystyle\int_0^\infty \frac{dx}{x^{1/2}(x + 1)}$ as the sum of the two improper integrals

$$\int_0^1 \frac{dx}{x^{1/2}(x + 1)} + \int_1^\infty \frac{dx}{x^{1/2}(x + 1)}$$

Use the Comparison Test to show that the integral converges.

78. Determine whether $\displaystyle\int_0^\infty \frac{dx}{x^{3/2}(x + 1)}$ (defined as in Exercise 77) converges.

79. An investment pays a dividend of \$250/year continuously forever. If the interest rate is 7%, what is the present value of the entire income stream generated by the investment?

80. An investment is expected to earn profits at a rate of $10{,}000e^{0.01t}$ dollars per year forever. Find the present value of the income stream if the interest rate is 4%.

81. Compute the present value of an investment that generates income at a rate of $5000te^{0.01t}$ dollars per year forever, assuming an interest rate of 6%.

82. Find the volume of the solid obtained by rotating the region below the graph of $y = e^{-x}$ about the x-axis for $0 \leq x < \infty$.

83. When a capacitor of capacitance C is charged by a source of voltage V, the power expended at time t is

$$P(t) = \frac{V^2}{R}(e^{-t/RC} - e^{-2t/RC})$$

where R is the resistance in the circuit. The total energy stored in the capacitor is

$$W = \int_0^\infty P(t)\, dt$$

Show that $W = \frac{1}{2}CV^2$.

84. Let $f(x) = e^{-0.05x}(1 + \sin x)$.

(a) [GU] Obtain a plot of $f(x)$ for $0 \leq x \leq 20$, and discuss the behavior of the function for positive and increasing x.

(b) $\displaystyle\int_0^\infty f(x)\, dx$ is the area above the positive x-axis and under the infinitely many "humps" of the graph of f. Compute this area.

85. Compute the volume of the solid obtained by rotating the region below the graph of $y = e^{-|x|/2}$ about the x-axis for $-\infty < x < \infty$.

86. For which integers p does $\displaystyle\int_0^{1/2} \frac{dx}{x(\ln x)^p}$ converge?

87. Conservation of Energy can be used to show that when a mass m oscillates at the end of a spring with spring constant k, the period of oscillation is

$$T = 4\sqrt{m} \int_0^{\sqrt{2E/k}} \frac{dx}{\sqrt{2E - kx^2}}$$

where E is the total energy of the mass. Show that this is an improper integral with value $T = 2\pi\sqrt{m/k}$.

*In Exercises 88–91, the **Laplace transform** of a function f is the function $\mathcal{L}f(s)$ of the variable s defined by the improper integral (if it converges):*

$$\mathcal{L}f(s) = \int_0^\infty f(x)e^{-sx}\, dx$$

Laplace transforms are widely used in physics and engineering.

88. Show that if $f(x) = C$, where C is a constant, then $\mathcal{L}f(s) = C/s$ for $s > 0$.

89. Show that if $f(x) = \sin \alpha x$, then $\mathcal{L}f(s) = \dfrac{\alpha}{s^2 + \alpha^2}$.

90. Compute $\mathcal{L}f(s)$, where $f(x) = e^{\alpha x}$ and $s > \alpha$.

91. Compute $\mathcal{L}f(s)$, where $f(x) = \cos \alpha x$ and $s > 0$.

92. [✎] When a radioactive substance decays, the fraction of atoms present at time t is $f(t) = e^{-kt}$, where $k > 0$ is the decay constant. It can be shown that the *average* life of an atom (until it decays) is $A = -\displaystyle\int_0^\infty tf'(t)\, dt$. Use Integration by Parts to show that $A = \displaystyle\int_0^\infty f(t)\, dt$ and compute A. What is the average decay time of radon-222, whose half-life is 3.825 days?

93. [✎] Let $J_n = \displaystyle\int_0^\infty x^n e^{-\alpha x}\, dx$, where $n \geq 1$ is an integer and $\alpha > 0$. Prove that

$$J_n = \frac{n}{\alpha} J_{n-1}$$

and $J_0 = 1/\alpha$. Use this to compute J_4. Show that $J_n = n/\alpha^{n+1}$.

94. Let $a > 0$ and $n > 1$. Define $f(x) = \dfrac{x^n}{e^{ax} - 1}$ for $x \neq 0$ and $f(0) = 0$.

(a) Use L'Hôpital's Rule to show that f is continuous at $x = 0$.

(b) Show that $\displaystyle\int_0^\infty f(x)\,dx$ converges. *Hint:* Show that $f(x) \leq 2x^n e^{-ax}$ if x is large enough. Then use the Comparison Test and Exercise 93.

95. ✏️ According to **Planck's Radiation Law**, the amount of electromagnetic energy with frequency between ν and $\nu + \Delta\nu$ that is radiated by a so-called black body at temperature T is proportional to $F(\nu)\,\Delta\nu$, where

$$F(\nu) = \left(\frac{8\pi h}{c^3}\right)\frac{\nu^3}{e^{h\nu/kT} - 1}$$

where c, h, k are physical constants. Use Exercise 94 to show that the total radiated energy

$$E = \int_0^\infty F(\nu)\,d\nu$$

is finite. To derive his law, Planck introduced the quantum hypothesis in 1900, which marked the birth of quantum mechanics.

Further Insights and Challenges

96. Consider $\displaystyle\int_0^1 x^p \ln x\,dx$.

(a) Show that the integral diverges for $p = -1$.

(b) Show that if $p \neq -1$, then

$$\int x^p \ln x\,dx = \frac{x^{p+1}}{p+1}\left(\ln x - \frac{1}{p+1}\right) + C$$

(c) Use L'Hôpital's Rule to show that the integral converges if $p > -1$ and diverges if $p < -1$.

97. Let

$$F(x) = \int_2^x \frac{dt}{\ln t} \quad \text{and} \quad G(x) = \frac{x}{\ln x}$$

Verify that L'Hôpital's Rule applies to the limit $L = \displaystyle\lim_{x \to \infty} \frac{F(x)}{G(x)}$ and evaluate L.

In Exercises 98–100, an improper integral $\displaystyle\int_a^\infty f(x)\,dx$ is called

absolutely convergent *if $\displaystyle\int_a^\infty |f(x)|\,dx$ converges. It can be shown that if an integral is absolutely convergent, then it is convergent.*

98. Show that $\displaystyle\int_1^\infty \frac{\sin x}{x^2}\,dx$ is absolutely convergent.

99. Show that $\displaystyle\int_1^\infty e^{-x^2}\cos x\,dx$ is absolutely convergent.

100. Let $f(x) = \sin x / x$ and consider $\displaystyle\int_0^\infty f(x)\,dx$. We define $f(0) = 1$. Then f is continuous and the integral is not improper at $x = 0$.

(a) Show that

$$\int_1^R \frac{\sin x}{x}\,dx = -\frac{\cos x}{x}\bigg|_1^R - \int_1^R \frac{\cos x}{x^2}\,dx$$

(b) Show that $\displaystyle\int_1^\infty (\cos x / x^2)\,dx$ converges. Conclude that the limit as $R \to \infty$ of the integral in (a) exists and is finite.

(c) Show that $\displaystyle\int_0^\infty f(x)\,dx$ converges.

It is known that $I = \frac{\pi}{2}$. However, the integral is *not* absolutely convergent. The convergence depends on cancellation, as shown in Figure 13.

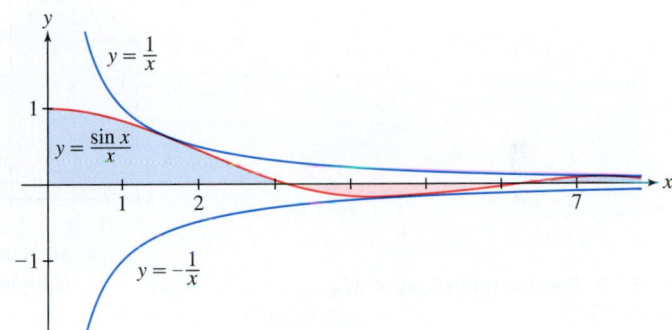

FIGURE 13 Convergence of $\int_1^\infty (\sin x/x)\,dx$ is due to the cancellation arising from the periodic change of sign.

101. The **gamma function**, which plays an important role in advanced applications, is defined for $n \geq 1$ by

$$\Gamma(n) = \int_0^\infty t^{n-1}e^{-t}\,dt$$

(a) Show that the integral defining $\Gamma(n)$ converges for $n \geq 1$ (it actually converges for all $n > 0$). *Hint:* Show that $t^{n-1}e^{-t} < t^{-2}$ for t sufficiently large.

(b) Show that $\Gamma(n+1) = n\Gamma(n)$ using Integration by Parts.

(c) Show that $\Gamma(n+1) = n!$ if $n \geq 1$ is an integer. *Hint:* Use (b) repeatedly. Thus, $\Gamma(n)$ provides a way of defining n-factorial when n is not an integer.

102. Use the results of Exercise 101 to show that the Laplace transform (see Exercises 88–91 above) of x^n is $\dfrac{n}{s^{n+1}}$.

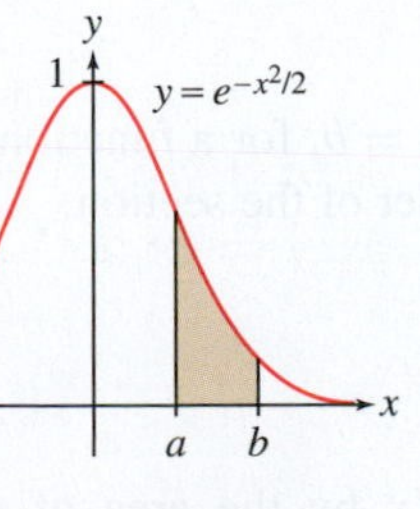

7.8 Numerical Integration

Numerical integration is the process of approximating a definite integral using well-chosen sums of function values. It is needed when we cannot find an antiderivative explicitly, as in the case of the Gaussian function $f(x) = e^{-x^2/2}$ (Figure 1).

In Section 5.1, we saw that we can approximate a definite integral by splitting the interval of integration $[a, b]$ into N subintervals, each of size Δx. Then we take the value of the function at each left-hand endpoint, multiply that by the width of the interval Δx, and sum over the intervals. This approximation is known as the left-endpoint approximation. Similarly, we saw a right-hand approximation. The third method that we introduced

FIGURE 1 Areas under the graph of $y = e^{-x^2/2}$ are approximated using numerical integration.

in that section, which we reconsider here, uses the midpoints of the intervals, and usually gives a better approximation.

The Midpoint Rule

To approximate the definite integral $\int_a^b f(x)\,dx$, we fix a whole number N and divide $[a, b]$ into N subintervals of length $\Delta x = (b - a)/N$. The endpoints of the subintervals are

$$x_0 = a, \qquad x_1 = a + \Delta x, \qquad x_2 = a + 2\Delta x, \qquad \dots, \qquad x_N = b$$

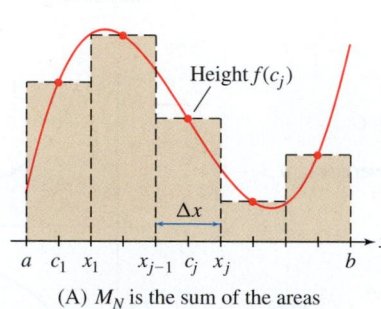
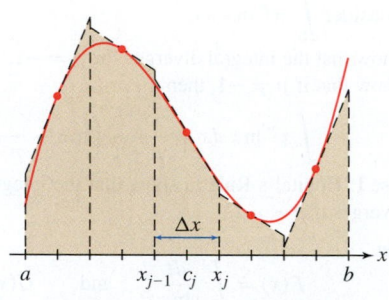

(A) M_N is the sum of the areas of the midpoint rectangles.

(B) M_N is also the sum of the areas of the tangential trapezoids.

FIGURE 2 Two interpretations of M_N.

The methods in this section apply for general integrable functions f. For simplicity in the development of the methods, we assume that the values of $f(x)$ are positive so that we can describe constructions and computations in terms of areas of rectangles, trapezoids, and other regions.

The midpoint approximation M_N is the sum of the areas of the rectangles of height $f(c_j)$ and base Δx, where c_j is the midpoint of the interval $[x_{j-1}, x_j]$ [Figure 2(A)]. Note that the midpoint of $[x_{j-1}, x_j]$ is $\frac{x_{j-1} + x_j}{2} = \frac{a + (j-1)\Delta x + a + j\Delta x}{2} = a + (j - \frac{1}{2})\Delta x$, and therefore $c_j = a + (j - \frac{1}{2})\Delta x$.

Midpoint Rule The Nth midpoint approximation to $\int_a^b f(x)\,dx$ is

$$M_N = \Delta x \big(f(c_1) + f(c_2) + \cdots + f(c_N) \big)$$

where $\Delta x = \dfrac{b - a}{N}$ and $c_j = a + \left(j - \frac{1}{2}\right) \Delta x$ is the midpoint of $[x_{j-1}, x_j]$.

GRAPHICAL INSIGHT M_N has a second interpretation as the sum of the areas of tangential trapezoids—that is, trapezoids whose top edges are tangent to the graph of f at the midpoints c_j [Figure 2(B)]. The trapezoids have the same area as the rectangles because the top edge of the trapezoid passes through the midpoint of the top edge of the rectangle, as shown in Figure 3.

With subinterval endpoints $a = x_0, x_1, x_2, \dots, x_N = b$, for a function f, we denote the function value $f(x_j)$ by y_j throughout the remainder of the section.

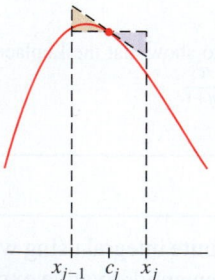

FIGURE 3 The rectangle and the trapezoid have the same area.

The Trapezoidal Rule

The **Trapezoidal Rule** T_N approximates $\int_a^b f(x)\,dx$ by the area of the trapezoids obtained by joining the points (x_0, y_0), (x_1, y_1), ..., (x_N, y_N) with line segments, as in Figure 4. The area of the jth trapezoid is $\frac{1}{2}\Delta x(y_{j-1} + y_j)$, and therefore,

$$T_N = \frac{1}{2}\Delta x(y_0 + y_1) + \frac{1}{2}\Delta x(y_1 + y_2) + \cdots + \frac{1}{2}\Delta x(y_{N-1} + y_N)$$

$$= \frac{1}{2}\Delta x\Big((y_0 + y_1) + (y_1 + y_2) + \cdots + (y_{N-1} + y_N)\Big)$$

Note that each value y_j occurs twice except for y_0 and y_N, so we obtain

$$T_N = \frac{1}{2}\Delta x\Big(y_0 + 2y_1 + 2y_2 + \cdots + 2y_{N-1} + y_N\Big)$$

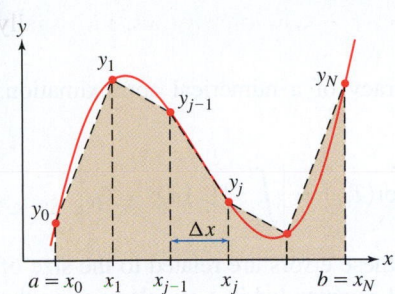

DF **FIGURE 4** T_N approximates the area under the graph by trapezoids.

T_N approximates the integral for any f that is integrable over $[a, b]$, so we have the following integral-approximation rule:

Trapezoidal Rule The Nth trapezoidal approximation to $\displaystyle\int_a^b f(x)\,dx$ is

$$T_N = \frac{1}{2}\,\Delta x\big(y_0 + 2y_1 + \cdots + 2y_{N-1} + y_N\big)$$

where $\Delta x = \dfrac{b-a}{N}$ and $y_j = f(x_j)$.

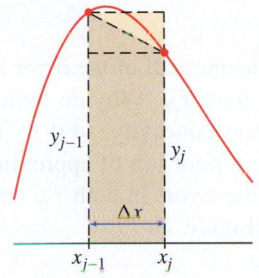

FIGURE 5 The shaded trapezoid has area $\frac{1}{2}\Delta x(y_{j-1} + y_j)$. This is the average of the areas of the left- and right-endpoint rectangles.

CONCEPTUAL INSIGHT We see in Figure 5 that the area of the jth trapezoid is equal to the average of the areas of the endpoint rectangles with heights y_{j-1} and y_j. It follows that T_N is equal to the average of the right- and left-endpoint approximations R_N and L_N introduced in Section 5.1:

$$T_N = \frac{1}{2}(R_N + L_N)$$

In general, this average is a better approximation than either R_N alone or L_N alone.

EXAMPLE 1 [CAS] Calculate T_8 for the integral $\displaystyle\int_1^3 \sin(x^2)\,dx$. Then use a computer algebra system to calculate T_N for $N = 50, 100, 500, 1000,$ and $10{,}000$.

Solution Divide $[1, 3]$ into $N = 8$ subintervals of length $\Delta x = \frac{3-1}{8} = \frac{1}{4}$. Then sum the function values at the endpoints (Figure 6) with the appropriate coefficients:

$$T_8 = \frac{1}{2}\left(\frac{1}{4}\right)\Big[\sin(1^2) + 2\sin(1.25^2) + 2\sin(1.5^2) + 2\sin(1.75^2)$$

$$+ 2\sin(2^2) + 2\sin(2.25^2) + 2\sin(2.5^2) + 2\sin(2.75^2) + \sin(3^2)\Big]$$

$$\approx 0.4281$$

In general, $\Delta x = (3 - 1)/N = 2/N$ and $x_j = 1 + 2j/N$. In summation notation,

$$T_N = \frac{1}{2}\left(\frac{2}{N}\right)\Big[\sin(1^2) + 2\underbrace{\sum_{j=1}^{N-1}\sin\left(\left(1 + \frac{2j}{N}\right)^2\right)}_{\text{Sum of terms with coefficient 2}} + \sin(3^2)\Big]$$

We evaluate the inner sum on a CAS. The results in Table 1 suggest that $\displaystyle\int_1^3 \sin(x^2)\,dx$ is approximately 0.4633. ∎

FIGURE 6 Division of $[1, 3]$ into $N = 8$ subintervals.

DF	**TABLE 1**
N	T_N
50	0.4624205
100	0.4630759
500	0.4632855
1000	0.4632920
10,000	0.4632942

Error Bounds

In applications, it is important to know the accuracy of a numerical approximation. We define the error in M_N and T_N by

$$\text{error}(M_N) = \left| \int_a^b f(x)\,dx - M_N \right|, \qquad \text{error}(T_N) = \left| \int_a^b f(x)\,dx - T_N \right|$$

According to the next theorem, the magnitudes of these errors are related to the size of the *second* derivative $f''(x)$. A proof of Theorem 1 is provided in a supplement on the text's Web site.

In the Error Bound, you can let K_2 be the maximum of $|f''(x)|$ on $[a,b]$, but if it is inconvenient to find this maximum exactly, take K_2 to be any number that is definitely larger than the maximum.

THEOREM 1 Error Bound for M_N and T_N Assume f'' exists and is continuous. Let K_2 be a number such that $|f''(x)| \le K_2$ for all x in $[a,b]$. Then

$$\text{error}(M_N) \le \frac{K_2(b-a)^3}{24N^2}, \qquad \text{error}(T_N) \le \frac{K_2(b-a)^3}{12N^2}$$

GRAPHICAL INSIGHT Note that the Error Bound for M_N is one-half of the Error Bound for T_N, suggesting that M_N is generally more accurate than T_N. Why do both Error Bounds depend on $f''(x)$? The second derivative measures concavity, so if $|f''(x)|$ is large, then the graph of f bends a lot and trapezoids do a poor job of approximating the region under the graph. Under such a circumstance, the errors in both T_N and M_N (which uses tangential trapezoids) are likely to be large (Figure 7).

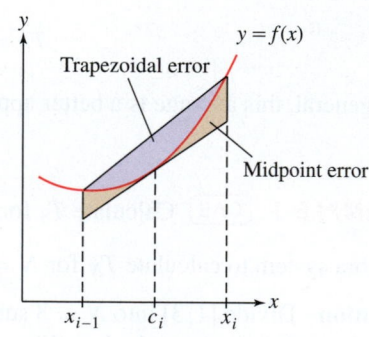

(A) $f''(x)$ is larger and the errors are larger.

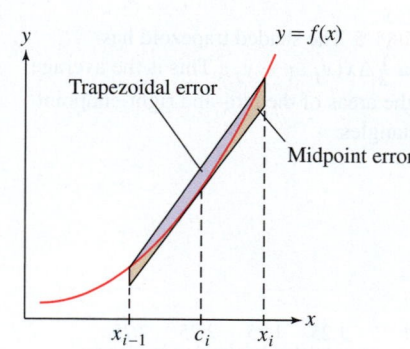

(B) $f''(x)$ is smaller and the errors are smaller.

FIGURE 7 M_N and T_N are more accurate when $|f''(x)|$ is small.

EXAMPLE 2 Checking the Error Bound Consider the integral $\int_1^4 \sqrt{x}\,dx$.

(a) Calculate M_6 and T_6 for the integral.

(b) Calculate the Error Bounds.

(c) Calculate the integral exactly and verify that the Error Bounds are satisfied.

Solution

(a) Divide $[1,4]$ into six subintervals of width $\Delta x = \frac{4-1}{6} = \frac{1}{2}$. Using the endpoints and midpoints shown in Figure 8, we obtain

$$M_6 = \frac{1}{2}\left(\sqrt{1.25} + \sqrt{1.75} + \sqrt{2.25} + \sqrt{2.75} + \sqrt{3.25} + \sqrt{3.75} \right) \approx 4.669245$$

$$T_6 = \frac{1}{2}\left(\frac{1}{2}\right)\left(\sqrt{1} + 2\sqrt{1.5} + 2\sqrt{2} + 2\sqrt{2.5} + 2\sqrt{3} + 2\sqrt{3.5} + \sqrt{4} \right) \approx 4.661488$$

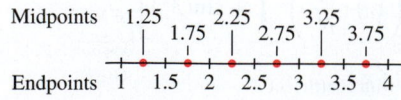

FIGURE 8 Interval $[1,4]$ divided into $N=6$ subintervals.

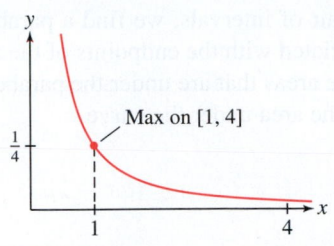

FIGURE 9 Graph of $y = |f''(x)| = \frac{1}{4}x^{-3/2}$ for $f(x) = \sqrt{x}$.

In Example 2, the actual error in T_6 is approximately twice as large as the error in M_6. In practice, this is often the case.

(b) Let $f(x) = \sqrt{x}$. We must find a number K_2 such that $|f''(x)| \le K_2$ for $1 \le x \le 4$. We have $f''(x) = -\frac{1}{4}x^{-3/2}$. The absolute value $|f''(x)| = \frac{1}{4}x^{-3/2}$ is decreasing on $[1, 4]$, so its maximum occurs at $x = 1$ (Figure 9). Thus, we may take $K_2 = |f''(1)| = \frac{1}{4}$. By Theorem 1,

$$\text{error}(M_6) \le \frac{K_2(b-a)^3}{24N^2} = \frac{\frac{1}{4}(4-1)^3}{24(6)^2} = \frac{1}{128} \approx 0.0078$$

$$\text{error}(T_6) \le \frac{K_2(b-a)^3}{12N^2} = \frac{\frac{1}{4}(4-1)^3}{12(6)^2} = \frac{1}{64} \approx 0.0156$$

(c) The exact value is $\displaystyle\int_1^4 \sqrt{x}\,dx = \frac{2}{3}x^{3/2}\Big|_1^4 = \frac{14}{3}$, so the actual errors are

$$\text{error}(M_6) \approx \left|\frac{14}{3} - 4.669245\right| \approx 0.00258 \quad \text{(less than Error Bound 0.0078)}$$

$$\text{error}(T_6) \approx \left|\frac{14}{3} - 4.661488\right| \approx 0.00518 \quad \text{(less than Error Bound 0.0156)}$$

The actual errors are less than the Error Bound, so Theorem 1 is verified. ∎

The Error Bound can be used to determine values of N so that M_N or T_N approximates an integral to a given accuracy.

EXAMPLE 3 **Obtaining the Desired Accuracy** Find N such that T_N approximates $\displaystyle\int_0^3 e^{-x^2}\,dx$ with an error of at most 10^{-4}. Then, for this value of N, use technology to determine T_N and discuss the resulting integral approximation.

A quick way to find a value for K_2 is to plot f'' using a graphing utility and find a bound for $|f''(x)|$ visually, as we do in Example 3.

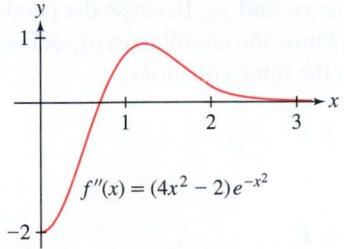

$f''(x) = (4x^2 - 2)e^{-x^2}$

FIGURE 10 Graph of the second derivative of $f(x) = e^{-x^2}$.

Solution Let $f(x) = e^{-x^2}$. To apply the Error Bound, we must find a number K_2 such that $|f''(x)| \le K_2$ for all $x \in [0, 3]$. We have $f'(x) = -2xe^{-x^2}$ and $f''(x) = (4x^2 - 2)e^{-x^2}$. A graphing utility was used to plot f'' (Figure 10). The graph shows that the maximum value of $|f''(x)|$ on $[0, 3]$ is $|f''(0)| = |-2| = 2$, so we take $K_2 = 2$ in the Error Bound:

$$\text{error}(T_N) \le \frac{K_2(b-a)^3}{12N^2} = \frac{2(3-0)^3}{12N^2} = \frac{9}{2N^2}$$

The error is at most 10^{-4} if

$$\frac{9}{2N^2} \le 10^{-4} \quad \Rightarrow \quad N^2 \ge \frac{9 \times 10^4}{2} \quad \Rightarrow \quad N \ge \frac{300}{\sqrt{2}} \approx 212.1$$

We conclude that T_{213} has an error of at most 10^{-4}.

A CAS shows that $T_{213} \approx 0.8862$. Since the error is at most 10^{-4}, we can infer that the actual value lies between 0.8861 and 0.8863, and therefore, $\displaystyle\int_0^3 e^{-x^2}\,dx \approx 0.886$, accurate to three decimal places. ∎

Simpson's Rule

As we have seen, the Midpoint Rule uses trapezoids that are tangent to the curve to approximate the area under the curve. The Trapezoidal Rule uses trapezoids with vertices on the curve to approximate the area. In both cases, the top edge of each trapezoid is a line segment. One might wonder whether we could do better using some other curve at the top of each region. In **Simpson's Rule**, we replace the line segments with parabolas, allowing us to obtain an approximation that is usually substantially more accurate.

To begin, we again subdivide $[a, b]$ into N subintervals, each of length $\Delta x = \frac{b-a}{N}$. However, we require N to be even. Then we pair up the resulting intervals, $[x_0, x_1]$ with

$[x_1, x_2]$, $[x_2, x_3]$ with $[x_3, x_4]$, and so on. For each pair of intervals, we find a parabola that passes through the three points on the curve associated with the endpoints of the two intervals, as in Figure 11. Then we take the sum of the areas that are under the parabolas and over the corresponding intervals to approximate the area under the curve.

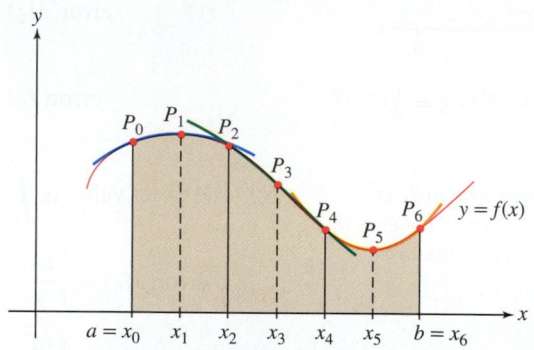

DF FIGURE 11 Approximating the area under the curve using parabolas (blue, through P_0, P_1, P_2; green, through P_2, P_3, P_4; orange, through P_4, P_5, P_6).

Our goal is to develop an approximation formula that, like the Trapezoidal Rule, expresses the approximation in terms of the function values, $y_i = f(x_i)$, at the subinterval endpoints. We begin with the case of a pair of intervals $[-\Delta x, 0]$ and $[0, \Delta x]$ centered around the origin. We assume that the corresponding three points on the curve are $P_0(-\Delta x, y_0)$, $P_1(0, y_1)$, and $P_2(\Delta x, y_2)$. See Figure 12.

Recall that the general equation for a parabola is $y = Cx^2 + Dx + E$ for constants C, D, and E. The area that is under the parabola and above the two intervals $[-\Delta x, 0]$ and $[0, \Delta x]$ is obtained by integrating:

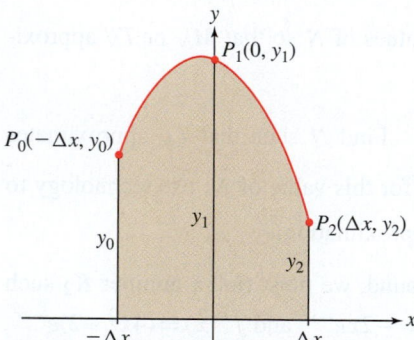

FIGURE 12 Finding the area under the parabola over the interval $[-\Delta x, \Delta x]$.

$$\text{area} = \int_{-\Delta x}^{\Delta x} Cx^2 + Dx + E \, dx = \frac{Cx^3}{3} + \frac{Dx^2}{2} + Ex \bigg|_{-\Delta x}^{\Delta x} \qquad \boxed{1}$$

$$= 2\left(\frac{C(\Delta x)^3}{3} + E\Delta x\right) = \frac{\Delta x}{3}(2C(\Delta x)^2 + 6E) \qquad \boxed{2}$$

We want to express this area in terms of the values y_0, y_1, and y_2. Because the parabola must pass through the three points P_0, P_1, and P_2, we know the coordinates of each must satisfy the equation of the parabola. Hence, we obtain the three equations:

$$y_0 = C(\Delta x)^2 - D\Delta x + E$$

$$y_1 = E$$

$$y_2 = C(\Delta x)^2 + D\Delta x + E$$

Multiplying the middle equation through by 4 and then adding the three equations yields

$$y_0 + 4y_1 + y_2 = 2C(\Delta x)^2 + 6E$$

Thus, from Eq. (2) it follows that the area under the parabola is $\frac{\Delta x}{3}(y_0 + 4y_1 + y_2)$.

This area depends only on the y-coordinates of the three points, so we obtain a similar expression if we site the parabola over any of the subsequent pairs of adjacent subintervals. Therefore, we can approximate the area under the curve by

$$\frac{\Delta x}{3}(y_0 + 4y_1 + y_2) + \frac{\Delta x}{3}(y_2 + 4y_3 + y_4) + \cdots + \frac{\Delta x}{3}(y_{N-2} + 4y_{N-1} + y_N)$$

Simplifying, we obtain the following approximation formula that is valid for any function f that is integrable over $[a, b]$:

> **Simpson's Rule** For N even, the Nth approximation to $\int_a^b f(x)\,dx$ by Simpson's Rule is
>
> $$S_N = \frac{1}{3}\Delta x\left[y_0 + 4y_1 + 2y_2 + \cdots + 4y_{N-3} + 2y_{N-2} + 4y_{N-1} + y_N\right] \qquad \boxed{3}$$
>
> where $\Delta x = \dfrac{b-a}{N}$ and $y_j = f(x_j)$.

As mentioned previously, we derived the approximation rules in this section in the case $f(x) \geq 0$ in order to make it easier to picture the computations in terms of area. The Midpoint Rule, the Trapezoidal Rule, and Simpson's Rule apply to any integrable function.

CONCEPTUAL INSIGHT Comparing Simpson's Rule to the Midpoint Rule and the Trapezoidal Rule, we see that Simpson's Rule is a linear combination of the other two rules. That is to say, Simpson's Rule is given by $S_N = \frac{2}{3}M_{N/2} + \frac{1}{3}T_{N/2}$. When a function is always concave up or always concave down, the value of the actual integral is sandwiched between $M_{N/2}$ and $T_{N/2}$. So a linear combination of the two should do better in this and many other cases. That $M_{N/2}$ is more heavily weighted in the linear combination is advantageous, as we have seen its Error Bound is half that of $T_{N/2}$.

EXAMPLE 4 Use Simpson's Rule with $N = 8$ to approximate $\int_2^4 \sqrt{1+x^3}\,dx$.

Solution We have $\Delta x = \frac{4-2}{8} = \frac{1}{4}$. Figure 13 shows the endpoints and coefficients needed to compute S_8 using Eq. (3):

$$S_8 = \frac{1}{3}\left(\frac{1}{4}\right)\Big[\sqrt{1+2^3} + 4\sqrt{1+2.25^3} + 2\sqrt{1+2.5^3} + 4\sqrt{1+2.75^3} + 2\sqrt{1+3^3}$$

$$+ 4\sqrt{1+3.25^3} + 2\sqrt{1+3.5^3} + 4\sqrt{1+3.75^3} + \sqrt{1+4^3}\Big]$$

$$\approx \frac{1}{12}\Big[3 + 4(3.52003) + 2(4.07738) + 4(4.66871) + 2(5.2915)$$

$$+ 4(5.94375) + 2(6.62382) + 4(7.33037) + 8.06226\Big] \approx 10.74159 \qquad \blacksquare$$

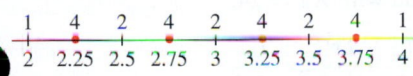

1 4 2 4 2 4 2 4 1

2 2.25 2.5 2.75 3 3.25 3.5 3.75 4

FIGURE 13 Coefficients for S_8 on $[2,4]$ shown above the corresponding endpoint.

The accuracy of Simpson's Rule is impressive. Using a computer algebra system, we find that the approximation in Example 4 has an error of less than 3×10^{-6}.

EXAMPLE 5 **Estimating Integrals from Numerical Data** The velocity (in kilometers per hour) of a Piper Cub aircraft traveling due west is recorded every minute during the first 10 minutes after takeoff. Use Simpson's Rule to estimate the distance traveled.

t (min)	0	1	2	3	4	5	6	7	8	9	10
$v(t)$ (km/h)	0	80	100	128	144	160	152	136	128	120	136

Solution The distance traveled is the integral of velocity. We convert from minutes to hours because velocity is given in kilometers per hour, and thus we apply Simpson's Rule, where the number of intervals is $N = 10$ and each interval has length $\Delta t = \frac{1}{60}$ h:

$$S_{10} = \left(\frac{1}{3}\right)\left(\frac{1}{60}\right)\Big(0 + 4(80) + 2(100) + 4(128) + 2(144) + 4(160)$$

$$+ 2(152) + 4(136) + 2(128) + 4(120) + 136\Big) \approx 20.4 \text{ km}$$

The distance traveled is approximately 20.4 km (Figure 14). $\qquad \blacksquare$

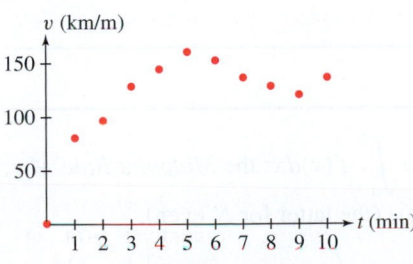

v (km/m)

FIGURE 14 Velocity of a Piper Cub.

We now state (without proof) the Error Bound for Simpson's Rule. Set

$$\text{error}(S_N) = \left|\int_a^b f(x) - S_N(f)\,dx\right|$$

The error involves the fourth derivative, which we assume exists and is continuous.

Although Simpson's Rule provides good approximations, more sophisticated techniques are implemented in computer algebra systems. These techniques are studied in the area of mathematics called numerical analysis.

THEOREM 2 Error Bound for S_N Let K_4 be a number such that $|f^{(4)}(x)| \leq K_4$ for all $x \in [a, b]$. Then

$$\text{error}(S_N) \leq \frac{K_4(b-a)^5}{180N^4}$$

EXAMPLE 6 Calculate S_8 for $\int_1^3 \frac{1}{x}\,dx$.

(a) Find a bound for the error in S_8.

(b) Find N such that S_N has an error of at most 10^{-6}. Then, for this value of N, use technology to determine S_N and discuss the resulting integral approximation.

Solution The width is $\Delta x = \frac{3-1}{8} = \frac{1}{4}$ and the endpoints in the partition of $[1, 3]$ are $1, 1.25, 1.5, \ldots, 2.75, 3$. Using Eq. (3) with $f(x) = x^{-1}$, we obtain

$$S_8 = \frac{1}{3}\left(\frac{1}{4}\right)\left[\frac{1}{1} + \frac{4}{1.25} + \frac{2}{1.5} + \frac{4}{1.75} + \frac{2}{2} + \frac{4}{2.25} + \frac{2}{2.5} + \frac{4}{2.75} + \frac{1}{3}\right]$$

$$\approx 1.09873$$

(a) The fourth derivative $f^{(4)}(x) = 24x^{-5}$ is decreasing, so the max of $|f^{(4)}(x)|$ on $[1, 3]$ is $|f^{(4)}(1)| = 24$. Therefore, we use the Error Bound with $K_4 = 24$:

$$\text{error}(S_N) \leq \frac{K_4(b-a)^5}{180N^4} = \frac{24(3-1)^5}{180N^4} = \frac{64}{15N^4}$$

$$\text{error}(S_8) \leq \frac{K_4(b-a)^5}{180(8)^4} = \frac{24(3-1)^5}{180(8^4)} \approx 0.001$$

(b) The error will be at most 10^{-6} if N satisfies

$$\text{error}(S_N) = \frac{64}{15N^4} \leq 10^{-6}$$

In other words,

$$N^4 \geq 10^6\left(\frac{64}{15}\right) \qquad \text{or} \qquad N \geq \left(\frac{10^6 \cdot 64}{15}\right)^{1/4} \approx 45.45$$

Thus, we may take $N = 46$. Using technology, we find that $S_{46} \approx 1.098612$, and therefore, we can conclude that $\int_1^3 \frac{1}{x}\,dx \approx 1.098612$ with an error less than 10^{-6}.

The value of the integral is $\ln 3$, and therefore, we have determined $\ln 3 \approx 1.098612$ with an error less than 10^{-6}. ■

7.8 SUMMARY

- We consider three numerical approximations to $\int_a^b f(x)\,dx$: the *Midpoint Rule* M_N, the *Trapezoidal Rule* T_N, and *Simpson's Rule* S_N (the latter for N even).

$$M_N = \Delta x\big(f(c_1) + f(c_2) + \cdots + f(c_N)\big) \qquad \left(c_j = a + \left(j - \frac{1}{2}\right)\Delta x\right)$$

$$T_N = \frac{1}{2}\Delta x\big(y_0 + 2y_1 + 2y_2 + \cdots + 2y_{N-1} + y_N\big)$$

$$S_N = \frac{1}{3}\Delta x\big[y_0 + 4y_1 + 2y_2 + \cdots + 4y_{N-3} + 2y_{N-2} + 4y_{N-1} + y_N\big]$$

where $\Delta x = (b-a)/N$ and $y_j = f(a + j\,\Delta x)$.

- M_N has two geometric interpretations; it may be interpreted either as the sum of the areas of the midpoint rectangles or as the sum of the areas of the tangential trapezoids.
- T_N is equal to the sum of the areas of the trapezoids obtained by connecting the points $(x_0, y_0), (x_1, y_1), \ldots, (x_N, y_N)$ with line segments.
- S_N is equal to $\frac{1}{3}T_{N/2} + \frac{2}{3}M_{N/2}$.
- Error Bounds:

$$\text{error}(M_N) \le \frac{K_2(b-a)^3}{24N^2}, \quad \text{error}(T_N) \le \frac{K_2(b-a)^3}{12N^2}, \quad \text{error}(S_N) \le \frac{K_4(b-a)^5}{180N^4}$$

where K_2 is any number such that $|f''(x)| \le K_2$ for all $x \in [a, b]$ and K_4 is any number such that $|f^{(4)}(x)| \le K_4$ for all $x \in [a, b]$.

7.8 EXERCISES

Preliminary Questions

1. What are T_1 and T_2 for a function on $[0, 2]$ such that $f(0) = 3$, $f(1) = 4$, and $f(2) = 3$?

2. For which graph in Figure 15 will T_N overestimate the integral? What about M_N?

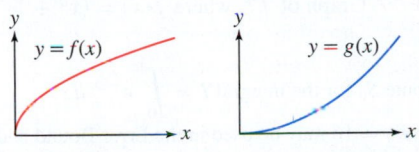

FIGURE 15

3. How large is the error when the Trapezoidal Rule is applied to a linear function? Explain graphically.

4. What is the maximum possible error if T_4 is used to approximate

$$\int_0^3 f(x)\,dx$$

where $|f''(x)| \le 2$ for all x.

5. What are the two graphical interpretations of the Midpoint Rule?

Exercises

In Exercises 1–4, calculate M_N and T_N for the value of N indicated and compare with the actual value of the integral.

1. $\displaystyle\int_1^3 x^2\,dx$, $N = 4$

2. $\displaystyle\int_0^4 \sqrt{x}\,dx$, $N = 4$

3. $\displaystyle\int_0^3 x^3\,dx$, $N = 6$

4. $\displaystyle\int_0^2 e^x\,dx$, $N = 6$

In Exercises 5–12, calculate M_N and T_N for the value of N indicated.

5. $\displaystyle\int_1^4 \frac{dx}{x}$, $N = 6$

6. $\displaystyle\int_1^2 \sqrt{x^4+1}\,dx$, $N = 5$

7. $\displaystyle\int_0^{\pi/2} \sqrt{\sin x}\,dx$, $N = 6$

8. $\displaystyle\int_0^{\pi/4} \sec x\,dx$, $N = 6$

9. $\displaystyle\int_1^2 \ln x\,dx$, $N = 5$

10. $\displaystyle\int_2^3 \frac{dx}{\ln x}$, $N = 5$

11. $\displaystyle\int_0^1 e^{-x^2}\,dx$, $N = 5$

12. $\displaystyle\int_{-2}^1 e^{x^2}\,dx$, $N = 6$

In Exercises 13–16, calculate S_N given by Simpson's Rule for the value of N indicated and compare with the actual value of the integral.

13. $\displaystyle\int_1^3 x^2\,dx$, $N = 4$

14. $\displaystyle\int_0^4 \sqrt{x}\,dx$, $N = 4$

15. $\displaystyle\int_0^3 e^{-x}\,dx$, $N = 6$

16. $\displaystyle\int_0^\pi \sin x\,dx$, $N = 6$

In Exercises 17–24, calculate S_N given by Simpson's Rule for the value of N indicated.

17. $\displaystyle\int_0^3 \frac{dx}{x^4+1}$, $N = 6$

18. $\displaystyle\int_0^1 \cos(x^2)\,dx$, $N = 6$

19. $\displaystyle\int_0^1 e^{-x^2}\,dx$, $N = 4$

20. $\displaystyle\int_1^2 e^{-x}\,dx$, $N = 6$

21. $\displaystyle\int_1^4 \ln x\,dx$, $N = 8$

22. $\displaystyle\int_2^4 \sqrt{x^4+1}\,dx$, $N = 8$

23. $\displaystyle\int_0^{\pi/4} \tan\theta\,d\theta$, $N = 10$

24. $\displaystyle\int_0^2 (x^2+1)^{-1/3}\,dx$, $N = 10$

In Exercises 25–28, calculate the approximation to the volume of the solid obtained by rotating the graph around the given axis.

25. $y = \cos x$; $\left[0, \frac{\pi}{2}\right]$; x-axis; M_8

26. $y = \cos x$; $\left[0, \frac{\pi}{2}\right]$; y-axis; S_8

27. $y = e^{-x^2}$; $[0, 1]$; x-axis; T_8

28. $y = e^{-x^2}$; $[0, 1]$; y-axis; S_8

29. The back of Jon's guitar (Figure 16) is 19 in. long. Jon measured the width at 1-in. intervals, beginning and ending $\frac{1}{2}$ in. from the ends, obtaining the results

6, 9, 10.25, 10.75, 10.75, 10.25, 9.75, 9.5, 10, 11.25,

12.75, 13.75, 14.25, 14.5, 14.5, 14, 13.25, 11.25, 9

Use the Midpoint Rule to estimate the area of the back.

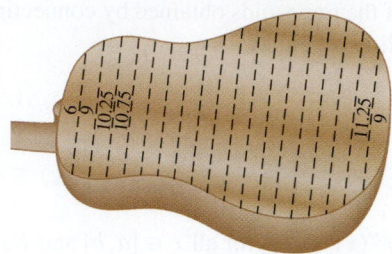

FIGURE 16 Back of guitar.

30. Use Simpson's Rule to determine the average temperature in a museum over a 3-hour period if the temperatures (in degrees Celsius), recorded at 15-minute intervals, are

$$21, \quad 21.3, \quad 21.5, \quad 21.8, \quad 21.6, \quad 21.2, \quad 20.8,$$
$$20.6, \quad 20.9, \quad 21.2, \quad 21.1, \quad 21.3, \quad 21.2$$

31. 📝 **Tsunami Arrival Times** Scientists estimate the arrival times of tsunamis (seismic ocean waves) based on the point of origin P and ocean depths. The speed s of a tsunami in miles per hour is approximately $s = \sqrt{15d}$, where d is the ocean depth in feet.

(a) Let $f(x)$ be the ocean depth x miles from P (in the direction of the coast). Argue using Riemann sums that the time T required for the tsunami to travel M miles toward the coast is

$$T = \int_0^M \frac{dx}{\sqrt{15 f(x)}}$$

(b) Use Simpson's Rule to estimate T if $M = 1000$ and the ocean depths (in feet), measured at 100-mile intervals starting from P, are

$$13{,}000, \quad 11{,}500, \quad 10{,}500, \quad 9000, \quad 8500,$$
$$7000, \quad 6000, \quad 4400, \quad 3800, \quad 3200, \quad 2000$$

32. Use S_8 to estimate $\int_0^{\pi/2} \frac{\sin x}{x}\, dx$, taking the value of $\frac{\sin x}{x}$ at $x = 0$ to be 1.

33. Calculate T_6 for the integral $I = \int_0^2 x^3\, dx$.

(a) Is T_6 too large or too small? Explain graphically.

(b) Show that $K_2 = |f''(2)|$ may be used in the Error Bound and find a bound for the error.

(c) Evaluate I and check that the actual error is less than the bound computed in (b).

34. Calculate M_4 for the integral $I = \int_0^1 x \sin(x^2)\, dx$.

(a) GU Use a plot of f'' to show that $K_2 = 3.2$ may be used in the Error Bound and find a bound for the error.

(b) CAS Evaluate I numerically and check that the actual error is less than the bound computed in (a).

In Exercises 35–38, state whether T_N or M_N underestimates or overestimates the integral and find a bound for the error (but do not calculate T_N or M_N).

35. $\int_1^4 \frac{1}{x}\, dx, \quad T_{10}$

36. $\int_0^2 e^{-x/4}\, dx, \quad T_{20}$

37. $\int_1^4 \ln x\, dx, \quad M_{10}$

38. $\int_0^{\pi/4} \cos x, \quad M_{20}$

CAS *In Exercises 39–42, use the Error Bound to find a value of N for which error$(T_N) \le 10^{-6}$. If you have a computer algebra system, calculate the corresponding approximation and confirm that the error satisfies the required bound.*

39. $\int_0^1 x^4\, dx$

40. $\int_0^3 (5x^4 - x^5)\, dx$

41. $\int_2^5 \frac{1}{x}\, dx$

42. $\int_0^3 e^{-x}\, dx$

43. Compute the Error Bound for the approximations T_{10} and M_{10} to $\int_0^3 (x^3 + 1)^{-1/2}\, dx$, using Figure 17 to determine a value of K_2. Then find a value of N such that the error in M_N is at most 10^{-6}.

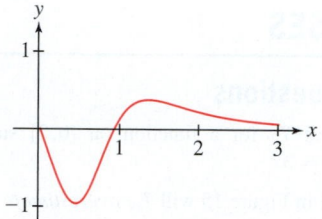

FIGURE 17 Graph of f'', where $f(x) = (x^3 + 1)^{-1/2}$.

44. **(a)** Compute S_6 for the integral $I = \int_0^1 e^{-2x}\, dx$.

(b) Show that $K_4 = 16$ may be used in the Error Bound and compute the Error Bound.

(c) Evaluate I and check that the actual error is less than the bound for the error computed in (b).

45. Calculate S_8 for $\int_1^5 \ln x\, dx$ and calculate the Error Bound. Then find a value of N such that S_N has an error of at most 10^{-6}.

46. Find a bound for the error in the approximation S_{10} to $\int_0^3 e^{-x^2}\, dx$ (use Figure 18 to determine a value of K_4). Then find a value of N such that S_N has an error of at most 10^{-6}.

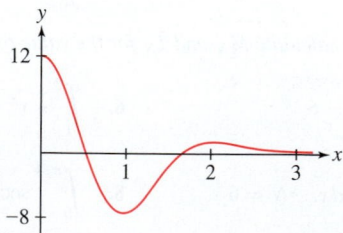

FIGURE 18 Graph of $f^{(4)}$, where $f(x) = e^{-x^2}$.

47. CAS Use a computer algebra system to compute and graph $f^{(4)}$ for $f(x) = \sqrt{1 + x^4}$, and find a bound for the error in the approximation S_{40} to $\int_0^5 f(x)\, dx$.

48. CAS Use a computer algebra system to compute and graph $f^{(4)}$ for $f(x) = \tan x - \sec x$, and find a bound for the error in the approximation S_{40} to $\int_0^{\pi/4} f(x)\, dx$.

In Exercises 49–52, use the Error Bound to find a value of N for which $Error(S_N) \le 10^{-9}$.

49. $\displaystyle\int_1^6 x^{4/3}\,dx$

50. $\displaystyle\int_0^4 xe^x\,dx$

51. $\displaystyle\int_0^1 e^{x^2}\,dx$

52. $\displaystyle\int_1^4 \sin(\ln x)\,dx$

53. CAS Show that $\displaystyle\int_0^1 \frac{dx}{1+x^2} = \frac{\pi}{4}$ [use Eq. (3) in Section 5.8].

(a) Use a computer algebra system to graph the function $f^{(4)}$ for $f(x) = (1+x^2)^{-1}$ and find its maximum on $[0, 1]$.

(b) Find a value of N such that S_N approximates the integral with an error of at most 10^{-6}. Calculate the corresponding approximation and confirm that you have computed $\frac{\pi}{4}$ to at least four places.

54. Let $J = \displaystyle\int_0^\infty e^{-x^2}\,dx$ and $J_N = \displaystyle\int_0^N e^{-x^2}\,dx$. Although e^{-x^2} has no elementary antiderivative, it is known that $J = \sqrt{\pi}/2$. Let T_N be the Nth trapezoidal approximation to J_N. Calculate T_4 and show that T_4 approximates J to three decimal places.

55. Let $f(x) = \sin(x^2)$ and $I = \displaystyle\int_0^1 f(x)\,dx$.

(a) Check that $f''(x) = 2\cos(x^2) - 4x^2\sin(x^2)$. Then show that $|f''(x)| \le 6$ for $x \in [0, 1]$. *Hint:* Note that $|2\cos(x^2)| \le 2$ and $|4x^2\sin(x^2)| \le 4$ for $x \in [0, 1]$.

(b) Show that Error(M_N) is at most $\dfrac{1}{4N^2}$.

(c) Find an N such that $|I - M_N| \le 10^{-3}$.

56. CAS The Error Bound for M_N is proportional to $1/N^2$, so the Error Bound decreases by $\frac{1}{4}$ if N is increased to $2N$. Compute the actual error in M_N for $\displaystyle\int_0^\pi \sin x\,dx$ for $N = 4, 8, 16, 32$, and 64. Does the actual error seem to decrease by $\frac{1}{4}$ as N is doubled?

57. CAS Observe that the Error Bound for T_N (which has 12 in the denominator) is twice as large as the Error Bound for M_N (which has 24 in the denominator). Compute the actual error in T_N for $\displaystyle\int_0^\pi \sin x\,dx$ for $N = 4, 8, 16, 32$, and 64 and compare it with the calculations of Exercise 56. Does the actual error in T_N seem to be roughly twice as large as the error in M_N in this case?

58. CAS Explain why the Error Bound for S_N decreases by $\frac{1}{16}$ if N is increased to $2N$. Compute the actual error in S_N for $\displaystyle\int_0^\pi \sin x\,dx$ for $N = 4, 8, 16, 32$, and 64. Does the actual error seem to decrease by $\frac{1}{16}$ as N is doubled?

59. Verify that S_2 yields the exact value of $\displaystyle\int_0^1 (x - x^3)\,dx$.

60. Verify that S_2 yields the exact value of $\displaystyle\int_a^b (x - x^3)\,dx$ for all $a < b$.

Further Insights and Challenges

61. Show that if $f(x) = rx + s$ is a linear function (r, s constants), then
$$T_N = \int_a^b f(x)\,dx$$
for all N and all endpoints a, b.

62. Show that if $f(x) = px^2 + qx + r$ is a quadratic polynomial, then
$$S_2 = \int_a^b f(x)\,dx.$$
In other words, show that
$$\int_a^b f(x)\,dx = \frac{b-a}{6}\left(y_0 + 4y_1 + y_2\right)$$
where $y_0 = f(a)$, $y_1 = f\left(\dfrac{a+b}{2}\right)$, and $y_2 = f(b)$. *Hint:* Show this first for $f(x) = 1, x, x^2$ and use linearity.

63. For N even, divide $[a, b]$ into N subintervals of width $\Delta x = \dfrac{b-a}{N}$. Set $x_j = a + j\,\Delta x$, $y_j = f(x_j)$, and
$$S_2^{2j} = \frac{b-a}{3N}\left(y_{2j} + 4y_{2j+1} + y_{2j+2}\right)$$

(a) Show that S_N is the sum of the approximations on the intervals $[x_{2j}, x_{2j+2}]$—that is, $S_N = S_2^0 + S_2^2 + \cdots + S_2^{N-2}$.

(b) By Exercise 62, $S_2^{2j} = \displaystyle\int_{x_{2j}}^{x_{2j+2}} f(x)\,dx$ if f is a quadratic polynomial. Use (a) to show that S_N is exact *for all N* if f is a quadratic polynomial.

64. Show that S_2 also gives the exact value for $\displaystyle\int_a^b x^3\,dx$ and conclude, as in Exercise 63, that S_N is exact for all cubic polynomials. Show by counterexample that S_2 is not exact for integrals of x^4.

65. Use the Error Bound for S_N to obtain another proof that Simpson's Rule is exact for all cubic polynomials.

66. **Sometimes Simpson's Rule Performs Poorly** Calculate M_{10} and S_{10} for the integral $\displaystyle\int_0^1 \sqrt{1 - x^2}\,dx$, whose value we know to be $\frac{\pi}{4}$ (one-quarter of the area of the unit circle).

(a) We usually expect S_N to be more accurate than M_N. Which of M_{10} and S_{10} is more accurate in this case?

(b) How do you explain the result of part (a)? *Hint:* The Error Bounds are not valid because $|f''(x)|$ and $|f^{(4)}(x)|$ tend to ∞ as $x \to 1$, but $|f^{(4)}(x)|$ goes to infinity more quickly.

CHAPTER REVIEW EXERCISES

1. Match the integrals (a)–(e) with their antiderivatives (i)–(v) on the basis of the general form (do not evaluate the integrals).

(a) $\displaystyle\int \frac{x\,dx}{\sqrt{x^2 - 4}}$

(b) $\displaystyle\int \frac{(2x + 9)\,dx}{x^2 + 4}$

(c) $\displaystyle\int \sin^3 x \cos^2 x\,dx$

(d) $\displaystyle\int \frac{dx}{x\sqrt{16x^2 - 1}}$

(e) $\displaystyle\int \frac{16\,dx}{x(x - 4)^2}$

(i) $\sec^{-1} 4x + C$

(ii) $\log|x| - \log|x-4| - \dfrac{4}{x-4} + C$

(iii) $\dfrac{1}{30}(3\cos^5 x - 3\cos^3 x \sin^2 x - 7\cos^3 x) + C$

(iv) $\dfrac{9}{2}\tan^{-1}\dfrac{x}{2} + \ln(x^2+4) + C$ **(v)** $\sqrt{x^2-4} + C$

2. Evaluate $\displaystyle\int \dfrac{x\,dx}{x+2}$ in two ways: using substitution and using the Method of Partial Fractions.

In Exercises 3–12, evaluate using the suggested method.

3. $\displaystyle\int \cos^3\theta \sin^8\theta\,d\theta$ [Write $\cos^3\theta$ as $\cos\theta(1-\sin^2\theta)$.]

4. $\displaystyle\int xe^{-12x}\,dx$ (Integration by Parts)

5. $\displaystyle\int \sec^3\theta \tan^4\theta\,d\theta$ (trigonometric identity, reduction formula)

6. $\displaystyle\int \dfrac{4x+4}{(x-5)(x+3)}\,dx$ (partial fractions)

7. $\displaystyle\int \dfrac{1}{x(x^2-1)^{3/2}}\,dx$ (trigonometric substitution)

8. $\displaystyle\int (1+x^2)^{-3/2}dx$ (trigonometric substitution)

9. $\displaystyle\int \dfrac{dx}{x^{3/2}+x^{1/2}}$ (substitution)

10. $\displaystyle\int \dfrac{dx}{x+x^{-1}}$ (rewrite integrand)

11. $\displaystyle\int x^{-2}\tan^{-1}x\,dx$ (Integration by Parts)

12. $\displaystyle\int \dfrac{dx}{x^2+4x-5}$ (complete the square, substitution, partial fractions)

In Exercises 13–64, evaluate using the appropriate method or combination of methods.

13. $\displaystyle\int_0^1 x^2 e^{4x}\,dx$

14. $\displaystyle\int \dfrac{x^2}{\sqrt{9-x^2}}\,dx$

15. $\displaystyle\int \cos^9 6\theta \sin^3 6\theta\,d\theta$

16. $\displaystyle\int \sec^2\theta \tan^4\theta\,d\theta$

17. $\displaystyle\int \dfrac{(6x+4)\,dx}{x^2-1}$

18. $\displaystyle\int_4^9 \dfrac{dt}{(t^2-1)^2}$

19. $\displaystyle\int \dfrac{d\theta}{\cos^4\theta}$

20. $\displaystyle\int \sin 2\theta \sin^2\theta\,d\theta$

21. $\displaystyle\int_0^1 \ln(4-2x)\,dx$

22. $\displaystyle\int (\ln(x+1))^2\,dx$

23. $\displaystyle\int \sin^5\theta\,d\theta$

24. $\displaystyle\int \cos^4(9x-2)\,dx$

25. $\displaystyle\int_0^{\pi/4} \sin 3x \cos 5x\,dx$

26. $\displaystyle\int \sin 2x \sec^2 x\,dx$

27. $\displaystyle\int \sqrt{\tan x}\,\sec^2 x\,dx$

28. $\displaystyle\int (\sec x + \tan x)^2\,dx$

29. $\displaystyle\int \sin^5\theta \cos^3\theta\,d\theta$

30. $\displaystyle\int \cot^3 x \csc x\,dx$

31. $\displaystyle\int \cot^2 x \csc^2 x\,dx$

32. $\displaystyle\int_{\pi/2}^{\pi} \cot^2\dfrac{\theta}{2}\,d\theta$

33. $\displaystyle\int_{\pi/4}^{\pi/2} \cot^2 x \csc^3 x\,dx$

34. $\displaystyle\int_4^6 \dfrac{dt}{(t-3)(t+4)}$

35. $\displaystyle\int \dfrac{dt}{(t-3)^2(t+4)}$

36. $\displaystyle\int \sqrt{x^2+9}\,dx$

37. $\displaystyle\int \dfrac{dx}{x\sqrt{x^2-4}}$

38. $\displaystyle\int_8^{27} \dfrac{dx}{x+x^{2/3}}$

39. $\displaystyle\int \dfrac{dx}{x^{3/2}+ax^{1/2}}$

40. $\displaystyle\int \dfrac{dx}{(x-b)^2+4}$

41. $\displaystyle\int \dfrac{(x^2-x)\,dx}{(x+2)^3}$

42. $\displaystyle\int \dfrac{(7x^2+x)\,dx}{(x-2)(2x+1)(x+1)}$

43. $\displaystyle\int \dfrac{16\,dx}{(x-2)^2(x^2+4)}$

44. $\displaystyle\int \dfrac{dx}{(x^2+25)^2}$

45. $\displaystyle\int \dfrac{dx}{x^2+8x+25}$

46. $\displaystyle\int \dfrac{dx}{x^2+8x+4}$

47. $\displaystyle\int \dfrac{x+4}{x^3-2x^2-x+2}\,dx$

48. $\displaystyle\int_0^1 t^2\sqrt{1-t^2}\,dt$

49. $\displaystyle\int \dfrac{dx}{x^4\sqrt{x^2+4}}$

50. $\displaystyle\int \dfrac{dx}{(x^2+5)^{3/2}}$

51. $\displaystyle\int (x+1)e^{4-3x}\,dx$

52. $\displaystyle\int x^{-2}\tan^{-1}x\,dx$

53. $\displaystyle\int x^3\cos(x^2)\,dx$

54. $\displaystyle\int x^2(\ln x)^2\,dx$

55. $\displaystyle\int x\tanh^{-1}x\,dx$

56. $\displaystyle\int \dfrac{\tan^{-1}t\,dt}{1+t^2}$

57. $\displaystyle\int \ln(x^2+9)\,dx$

58. $\displaystyle\int (\sin x)(\cosh x)\,dx$

59. $\displaystyle\int_0^1 \cosh 2t\,dt$

60. $\displaystyle\int \sinh^3 x \cosh x\,dx$

61. $\displaystyle\int \coth^2(1-4t)\,dt$

62. $\displaystyle\int_{-0.3}^{0.3} \dfrac{dx}{1-x^2}$

63. $\displaystyle\int_0^{3\sqrt{3}/2} \dfrac{dx}{\sqrt{9-x^2}}$

64. $\displaystyle\int \dfrac{\sqrt{x^2+1}\,dx}{x^2}$

65. Use the substitution $u = \tanh t$ to evaluate $\displaystyle\int \dfrac{dt}{\cosh^2 t + \sinh^2 t}$.

66. Find the volume obtained by rotating the region enclosed by $y = \ln x$ and $y = (\ln x)^2$ about the y-axis.

67. Let $I_n = \displaystyle\int \frac{x^n \, dx}{x^2 + 1}$.

(a) Prove that $I_n = \dfrac{x^{n-1}}{n-1} - I_{n-2}$.

(b) Use (a) to calculate I_n for $0 \le n \le 5$.

(c) Show that, in general,

$$I_{2n+1} = \frac{x^{2n}}{2n} - \frac{x^{2n-2}}{2n-2} + \cdots$$

$$+ (-1)^{n-1}\frac{x^2}{2} + (-1)^n \frac{1}{2}\ln(x^2 + 1) + C$$

$$I_{2n} = \frac{x^{2n-1}}{2n-1} - \frac{x^{2n-3}}{2n-3} + \cdots$$

$$+ (-1)^{n-1}x + (-1)^n \tan^{-1} x + C$$

68. Let $J_n = \displaystyle\int x^n e^{-x^2/2} \, dx$.

(a) Show that $J_1 = -e^{-x^2/2}$.

(b) Prove that $J_n = -x^{n-1}e^{-x^2/2} + (n-1)J_{n-2}$.

(c) Use (a) and (b) to compute J_3 and J_5.

In Exercises 69–78, determine whether the improper integral converges and, if so, evaluate it.

69. $\displaystyle\int_0^\infty \frac{dx}{(x+2)^2}$

70. $\displaystyle\int_4^\infty \frac{dx}{x^{2/3}}$

71. $\displaystyle\int_0^4 \frac{dx}{x^{2/3}}$

72. $\displaystyle\int_9^\infty \frac{dx}{x^{12/5}}$

73. $\displaystyle\int_{-\infty}^0 \frac{dx}{x^2 + 1}$

74. $\displaystyle\int_{-\infty}^9 e^{4x} \, dx$

75. $\displaystyle\int_0^{\pi/2} \cot\theta \, d\theta$

76. $\displaystyle\int_1^\infty \frac{dx}{(x+2)(2x+3)}$

77. $\displaystyle\int_0^\infty (5+x)^{-1/3} \, dx$

78. $\displaystyle\int_2^5 (5-x)^{-1/3} \, dx$

In Exercises 79–84, use the Comparison Test to determine whether the improper integral converges or diverges.

79. $\displaystyle\int_8^\infty \frac{dx}{x^2 - 4}$

80. $\displaystyle\int_8^\infty (\sin^2 x)e^{-x} \, dx$

81. $\displaystyle\int_3^\infty \frac{dx}{x^4 + \cos^2 x}$

82. $\displaystyle\int_1^\infty \frac{dx}{x^{1/3} + x^{2/3}}$

83. $\displaystyle\int_0^1 \frac{dx}{x^{1/3} + x^{2/3}}$

84. $\displaystyle\int_0^\infty e^{-x^3} \, dx$

85. Calculate the volume of the infinite solid obtained by rotating the region under $y = (x^2 + 1)^{-2}$ for $0 \le x < \infty$ about the y-axis.

86. Let R be the region under the graph of $y = (x+1)^{-1}$ for $0 \le x < \infty$. Which of the following quantities is finite?

(a) The area of R

(b) The volume of the solid obtained by rotating R about the x-axis

(c) The volume of the solid obtained by rotating R about the y-axis

87. Show that $\displaystyle\int_0^\infty x^n e^{-x^2} \, dx$ converges for all $n > 0$. *Hint:* First observe that $x^n e^{-x^2} < x^n e^{-x}$ for $x > 1$. Then show that $x^n e^{-x} < x^{-2}$ for x sufficiently large.

88. Compute the Laplace transform $\mathcal{L}f(s)$ of the function $f(x) = x$ for $s > 0$. See Exercises 88–91 in Section 7.7 for the definition of $\mathcal{L}f(s)$.

89. Compute the Laplace transform $\mathcal{L}f(s)$ of the function $f(x) = x^2 e^{\alpha x}$ for $s > \alpha$.

90. Estimate $\displaystyle\int_2^5 f(x) \, dx$ by computing T_2, M_3, T_6, and S_6 for a function f taking on the values in the following table:

x	2	2.5	3	3.5	4	4.5	5
$f(x)$	$\frac{1}{2}$	2	1	0	$-\frac{3}{2}$	-4	-2

91. State whether the approximation M_N or T_N is larger or smaller than the integral.

(a) $\displaystyle\int_0^\pi \sin x \, dx$

(b) $\displaystyle\int_\pi^{2\pi} \sin x \, dx$

(c) $\displaystyle\int_1^8 \frac{dx}{x^2}$

(d) $\displaystyle\int_2^5 \ln x \, dx$

92. The rainfall rate (in inches per hour) was measured hourly during a 10-hour thunderstorm with the following results:

$$0, \quad 0.41, \quad 0.49, \quad 0.32, \quad 0.3, \quad 0.23,$$
$$0.09, \quad 0.08, \quad 0.05, \quad 0.11, \quad 0.12$$

Use Simpson's Rule to estimate the total rainfall during the 10-h period.

In Exercises 93–98, compute the given approximation to the integral.

93. $\displaystyle\int_0^1 e^{-x^2} \, dx$, M_5

94. $\displaystyle\int_2^4 \sqrt{6t^3 + 1} \, dt$, T_3

95. $\displaystyle\int_{\pi/4}^{\pi/2} \sqrt{\sin\theta} \, d\theta$, M_4

96. $\displaystyle\int_1^4 \frac{dx}{x^3 + 1}$, T_6

97. $\displaystyle\int_0^1 e^{-x^2} \, dx$, S_4

98. $\displaystyle\int_5^9 \cos(x^2) \, dx$, S_8

99. The following table gives the area $A(h)$ of a horizontal cross section of a pond at depth h. Use the Trapezoidal Rule to estimate the volume V of the pond (Figure 1).

h (ft)	$A(h)$ (acres)	h (ft)	$A(h)$ (acres)
0	2.8	10	0.8
2	2.4	12	0.6
4	1.8	14	0.2
6	1.5	16	0.1
8	1.2	18	0

Area of horizontal
cross section is $A(h)$

FIGURE 1

100. Suppose that the second derivative of the function A in Exercise 99 satisfies $|A''(h)| \leq 1.5$. Use the Error Bound to find the maximum possible error in your estimate of the volume V of the pond.

101. Find a bound for the error $\left| M_{16} - \int_1^3 x^3 \, dx \right|$.

102. [GU] Let $f(x) = \sin(x^3)$. Find a bound for the error

$$\left| T_{24} - \int_0^{\pi/2} f(x) \, dx \right|$$

Hint: Find a bound K_2 for $|f''(x)|$ by plotting f'' with a graphing utility.

103. Find a value of N such that

$$\left| M_N - \int_0^{\pi/4} \tan x \, dx \right| \leq 10^{-4}$$

104. Find a value of N such that S_N approximates $\int_2^5 x^{-1/4} \, dx$ with an error of at most 10^{-2} (but do not calculate S_N).

8 FURTHER APPLICATIONS OF THE INTEGRAL

In this chapter, we examine a few more applications of the integral. In the first section we take a brief look at the importance of integration in probability theory. Following that we use the integral to define and compute the length of curves and the area of a surfaces of revolution. The last two sections address important physical applications involving pressure, force, and center of mass.

The Nurek Dam in Tajikistan, one of the tallest dams in the world, has a height of approximately 300 m. It has the typical dam thickness profile, where it is thicker at the bottom than at the top because the water pressure on the dam is greater at greater depths. Using a definite integral, we can sum the effect of the changing pressure at varying depths to calculate the overall force of the water on the dam.

8.1 Probability and Integration

What is the probability that a customer will arrive at the Rogadzo Pizza Parlor in the next 45 seconds? What is the probability of scoring above 90% on the NICE (National Integral Competency Exam)? Probabilities such as these are given by a number between 0 and 1, where 0 means there is no probability the event will occur and 1 means that the event is sure to happen. These probabilities are best described as areas under the graph of a function $y = p(x)$ called a **probability density function** (Figure 1). The methods of integration developed in this chapter are used extensively in the study of such functions.

In probability theory, the variable X that represents the phenomenon we are analyzing (time to arrival, exam score, etc.) is called a **random variable**. The probability that X lies in a given range $[a, b]$ is denoted

$$P(a \le X \le b)$$

For example, the probability of a customer arriving within the next 30 to 45 seconds is denoted $P(30 \le X \le 45)$.

We say that p is a **probability density function** for X if it is a continuous function such that

$$P(a \le X \le b) = \int_a^b p(x)\,dx$$

We assume that a probability density function p has a domain J that is an interval, and that J contains all of the possible values of the random variable X. We allow the possibility that J is an infinite interval. Furthermore, p must satisfy two important conditions. First, $p(x) \ge 0$ for all x in the domain, because a probability cannot be negative. Second,

$$\boxed{\int_J p(x)\,dx = 1} \qquad \boxed{1}$$

Notice that the integral in Eq. (1) is evaluated over the whole domain and represents the probability that the value of X is in the domain. This integral must equal 1 because it is certain (the probability is 1) that the value of X lies in J by assumption. That the integral of p over J is 1 ensures that $P(a \le X \le b)$ is a number in the interval $[0, 1]$ for any a and b in the domain.

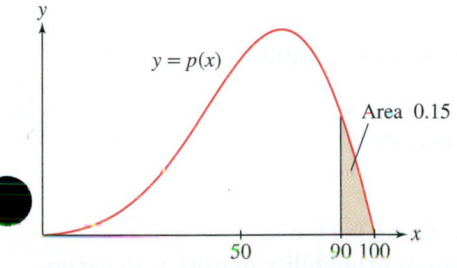

DF FIGURE 1 Probability density function for scores on an exam. The shaded region has area 0.15, so there is a 15% chance that a randomly chosen exam has a score at or above 90.

We write $P(X \le b)$ for the probability that X is at most b, and $P(X \ge b)$ for the probability that X is at least b.

← **REMINDER**

$$\int_{-\infty}^0 \frac{dx}{x^2+1} = \lim_{R \to -\infty} \tan^{-1} x \Big|_R^0$$
$$= \lim_{R \to -\infty} (\tan^{-1} 0 - \tan^{-1} R)$$
$$= 0 - \left(-\frac{\pi}{2}\right) = \frac{\pi}{2}$$

Similarly, $\int_0^\infty \frac{dx}{x^2+1} = \frac{\pi}{2}$.

EXAMPLE 1 Find a constant C for which $p(x) = \dfrac{C}{x^2+1}$ is a probability density function with domain $(-\infty, \infty)$. Then compute $P(1 \le X \le 4)$.

Solution We must choose C so that Eq. (1) is satisfied. The improper integral is a sum of two integrals (see Example 4 of Section 7.7):

$$\int_{-\infty}^\infty p(x)\,dx = C\int_{-\infty}^0 \frac{dx}{x^2+1} + C\int_0^\infty \frac{dx}{x^2+1} = C\frac{\pi}{2} + C\frac{\pi}{2} = C\pi$$

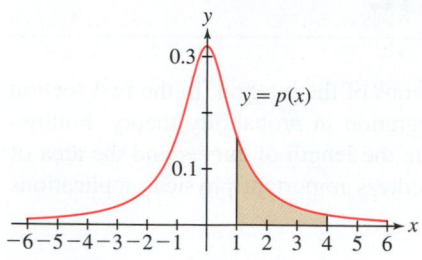

FIGURE 2 The probability density function
$p(x) = \dfrac{1}{\pi(x^2+1)}$.

Therefore, Eq. (1) is satisfied if $C\pi = 1$ or $C = \pi^{-1}$. We have

$$P(1 < X < 4) = \int_1^4 p(x)\,dx = \int_1^4 \frac{\pi^{-1}\,dx}{x^2+1} = \pi^{-1}(\tan^{-1} 4 - \tan^{-1} 1) \approx 0.17$$

Thus, X lies between 1 and 4 with probability approximately 0.17, or a 17% chance (Figure 2). ∎

CONCEPTUAL INSIGHT If X is a random variable with probability density function p, then the probability of X taking on any specific value a is *zero* because $\int_a^a p(x)\,dx = 0$. So what is the meaning of $p(a)$? We must think of it this way: The probability that X lies in a *small interval* $[a, a + \Delta x]$ is approximately $p(a)\Delta x$:

$$P(a \le X \le a + \Delta x) = \int_a^{a+\Delta x} p(x)\,dx \approx p(a)\Delta x$$

A probability density is similar to a linear mass density $\rho(x)$. The mass of a small segment $[a, a + \Delta x]$ is approximately $\rho(a)\Delta x$, but the mass of any particular point $x = a$ is zero.

The *mean* or *average value* of a random variable is the quantity

$$\boxed{\mu = \mu(X) = \int_J x p(x)\,dx}$$ **2**

if this integral exists. The symbol μ is a lowercase Greek letter mu.

In the next example, we consider the **exponential probability density** with parameter $r > 0$, defined on $[0, \infty)$ by

$$\boxed{p(t) = \frac{1}{r}e^{-t/r}}$$

This density function is often used to model "waiting times" between events that occur randomly. Exercise 12 asks you to verify that $p(t)$ satisfies Eq. (1).

EXAMPLE 2 **Mean of a Random Variable with Exponential Density** Let $r > 0$. Calculate the mean of a random variable with the exponential probability density $p(t) = \frac{1}{r}e^{-t/r}$ on $[0, \infty)$.

Solution The mean is the integral of $tp(t)$ over $[0, \infty)$. Using Integration by Parts with $u = t/r$ and $dv = e^{-t/r}\,dt$, we have $du = dt/r, v = -re^{-t/r}$, and

$$\int tp(t)\,dt = \int \left(\frac{t}{r}e^{-t/r}\right)dt = -te^{-t/r} + \int e^{-t/r}\,dt = -(r+t)e^{-t/r}$$

Thus the mean μ is given by

$$\mu = \int_0^\infty tp(t)\,dt = \int_0^\infty t\left(\frac{1}{r}e^{-t/r}\right)dt = \lim_{R \to \infty} -(r+t)e^{-t/r}\Big|_0^R$$

$$= \lim_{R \to \infty}\left(r - (r + R)e^{-R/r}\right) = r$$

The last equality holds since $(r + R)e^{-R/r} \to 0$ as $R \to \infty$. It follows that the mean of a random variable with the probability density function $p(t) = \frac{1}{r}e^{-t/r}$, over $[0, \infty)$, is r. ∎

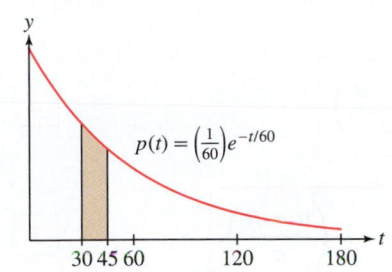

DF **FIGURE 3** Customer arrivals have an exponential distribution.

EXAMPLE 3   Waiting Time   The waiting time T between customer arrivals in a drive-through fast-food restaurant is a random variable with exponential probability density. If the average waiting time is 60 seconds, what is the probability that a customer will arrive within 30 to 45 s after another customer?

Solution   If the average waiting time is 60 s, then $r = 60$ and $p(t) = \frac{1}{60}e^{-t/60}$ because the mean of $\frac{1}{r}e^{-t/r}$ is r by the previous example. Therefore, the probability of waiting between 30 and 45 s for the next customer is

$$P(30 \leq T \leq 45) = \int_{30}^{45} \frac{1}{60}e^{-t/60} = -e^{-t/60}\Big|_{30}^{45} = -e^{-3/4} + e^{-1/2} \approx 0.134$$

This probability is the area of the shaded region in Figure 3.   ∎

The **normal density functions**, whose graphs are the familiar bell-shaped curves, appear in a surprisingly wide range of applications. The **standard normal** density is defined over $(-\infty, \infty)$ by

$$\boxed{p(x) = \frac{1}{\sqrt{2\pi}}e^{-x^2/2}} \qquad \boxed{3}$$

That $p(x)$ satisfies Eq. (1) is not easy to show. One problem is that p does not have an elementary antiderivative. In Exercise 57 in Section 15.4 we will see an approach using multivariable calculus.

More generally, we define the normal density function with mean μ and standard deviation σ:

$$\boxed{p(x) = \frac{1}{\sigma\sqrt{2\pi}}e^{-(x-\mu)^2/(2\sigma^2)}}$$

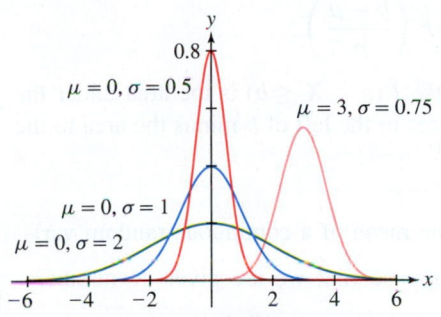

FIGURE 4 Normal density functions.

The standard deviation σ measures the spread; for larger values of σ, the graph is more spread out about the mean μ (Figure 4). The standard normal density in Eq. (3) has $\mu = 0$ and $\sigma = 1$. A random variable with a normal density function is said to have a **normal** or **Gaussian distribution**. Examples of data whose distribution is modeled well by a normal distribution include current sale prices for houses in Denver, heights of female children of age 11 in Egypt, and systolic blood pressure readings for adults in Frigento, Italy. The normal distribution is ubiquitous in everyday life. For example, Figure 5 shows a bar graph of data from a survey on the time of day that workers in the United States leave for work. Note the approximate bell-shaped curve generated by the data.

One difficulty with normal density functions is that they do not have elementary antiderivatives. As a result, we cannot evaluate the probabilities

$$P(a \leq X \leq b) = \frac{1}{\sigma\sqrt{2\pi}}\int_a^b e^{-(x-\mu)^2/(2\sigma^2)}\,dx$$

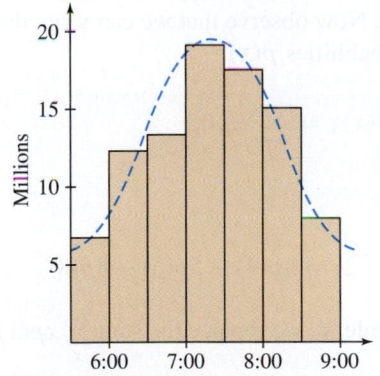

FIGURE 5 Time that people leave for work.

explicitly. However, the next theorem shows that these probabilities can all be expressed in terms of a single function called the **standard normal cumulative distribution function**:

$$\boxed{F(z) = \frac{1}{\sqrt{2\pi}}\int_{-\infty}^z e^{-x^2/2}\,dx}$$

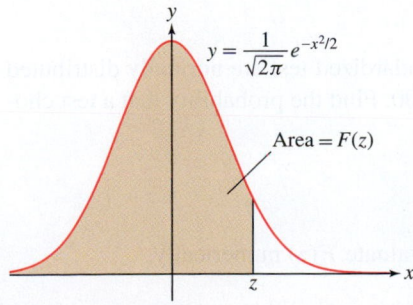

DF **FIGURE 6** $F(z)$ is the area of the shaded region.

Observe that $F(z)$ is equal to the shaded area under the graph in Figure 6. The function F is not an elementary function. Numerical values of $F(z)$, obtained by integral approximation, are typically available on scientific calculators and computer algebra systems.

THEOREM 1 If X has a normal distribution with mean μ and standard deviation σ, then for all $a \leq b$,

$$P(X \leq b) = F\left(\frac{b-\mu}{\sigma}\right)$$
4

$$P(a \leq X \leq b) = F\left(\frac{b-\mu}{\sigma}\right) - F\left(\frac{a-\mu}{\sigma}\right)$$
5

Proof We use two changes of variables, first $u = x - \mu$ and then $t = u/\sigma$:

$$P(X \leq b) = \frac{1}{\sigma\sqrt{2\pi}} \int_{-\infty}^{b} e^{-(x-\mu)^2/(2\sigma^2)} \, dx = \frac{1}{\sigma\sqrt{2\pi}} \int_{-\infty}^{b-\mu} e^{-u^2/(2\sigma^2)} \, du$$

$$= \frac{1}{\sqrt{2\pi}} \int_{-\infty}^{(b-\mu)/\sigma} e^{-t^2/2} \, dt = F\left(\frac{b-\mu}{\sigma}\right)$$

This proves Eq. (4). Equation (5) follows because $P(a \leq X \leq b)$ is the area under the graph between a and b, and this is equal to the area to the left of b minus the area to the left of a (Figure 7). ∎

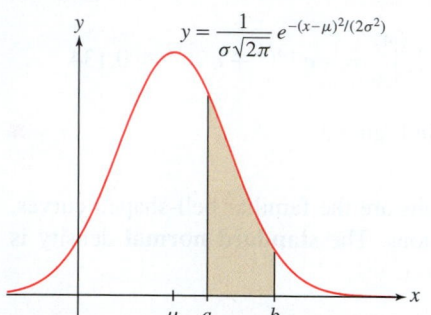

$$y = \frac{1}{\sigma\sqrt{2\pi}} e^{-(x-\mu)^2/(2\sigma^2)}$$

DF **FIGURE 7** The shaded region has area $F\left(\frac{b-\mu}{\sigma}\right) - F\left(\frac{a-\mu}{\sigma}\right)$.

CONCEPTUAL INSIGHT Why have we defined the mean of a continuous random variable X as the integral $\mu = \int_{-\infty}^{\infty} x p(x) \, dx$? Suppose first we are given N numbers $a_1, a_2, \ldots, a_N$, and for each value x, let $N(x)$ be the number of times x occurs among the a_j. Then a randomly chosen a_j has value x with probability $p(x) = N(x)/N$. For example, given the numbers $4, 4, 5, 5, 5, 8$, we have $N = 6$ and $N(5) = 3$. The probability of choosing a 5 is $p(5) = N(5)/N = \frac{3}{6} = \frac{1}{2}$. Now observe that we can write the mean (average value) of the a_j in terms of the probabilities $p(x)$:

$$\frac{a_1 + a_2 + \cdots + a_N}{N} = \frac{1}{N} \sum_x N(x)x = \sum_x x p(x)$$

For example,

$$\frac{4+4+5+5+5+8}{6} = \frac{1}{6}(2 \cdot 4 + 3 \cdot 5 + 1 \cdot 8) = 4p(4) + 5p(5) + 8p(8)$$

In defining the mean of a continuous random variable X, we replace the sum $\sum_x x p(x)$ with the integral $\mu = \int_{-\infty}^{\infty} x p(x) \, dx$. This makes sense because the integral is the limit of sums $\sum x_i p(x_i) \Delta x$, and as we have seen, $p(x_i) \Delta x$ is the approximate probability that X lies in $[x_i, x_i + \Delta x]$.

EXAMPLE 4 Assume that the scores X on a standardized test are normally distributed with mean $\mu = 500$ and standard deviation $\sigma = 100$. Find the probability that a test chosen at random has score

(a) at most 600.

(b) between 450 and 650.

Solution We use a computer algebra system to evaluate $F(z)$ numerically.

(a) Apply Eq. (4) with $\mu = 500$ and $\sigma = 100$:

$$P(x \leq 600) = F\left(\frac{600 - 500}{100}\right) = F(1) \approx 0.84$$

Thus, a randomly chosen score is 600 or less with a probability of 0.84, or 84%.

(b) Applying Eq. (5), we find that a randomly chosen score lies between 450 and 650 with a probability of 62.5%:

$$P(450 \le x \le 650) = F(1.5) - F(-0.5) \approx 0.933 - 0.308 = 0.625 \qquad \blacksquare$$

8.1 SUMMARY

- If X is a continuous random variable with probability density function p, then

$$P(a \le X \le b) = \int_a^b p(x)\,dx$$

- Probability densities with domain J satisfy two conditions: $p(x) \ge 0$ for x in J, and $\int_J p(x)\,dx = 1$.

- Mean (or average) value of X: $\mu = \int_J x p(x)\,dx$

- Exponential density function of mean r: $p(x) = \frac{1}{r} e^{-x/r}$

- Normal density of mean μ and standard deviation σ: $p(x) = \frac{1}{\sigma\sqrt{2\pi}} e^{-(x-\mu)^2/(2\sigma^2)}$

- Standard cumulative normal distribution function: $F(z) = \frac{1}{\sqrt{2\pi}} \int_{-\infty}^{z} e^{-t^2/2}\,dt$

- If X has a normal distribution of mean μ and standard deviation σ, then

$$P(X \le b) = F\left(\frac{b-\mu}{\sigma}\right)$$

$$P(a \le X \le b) = F\left(\frac{b-\mu}{\sigma}\right) - F\left(\frac{a-\mu}{\sigma}\right)$$

8.1 EXERCISES

Preliminary Questions

1. The function $p(x) = \cos x$ satisfies $\int_{-\pi/2}^{\pi} p(x)\,dx = 1$. Is p a probability density function on $[-\pi/2, \pi]$?

2. Estimate $P(2 \le X \le 2.1)$, assuming that the probability density function of X satisfies $p(2) = 0.2$.

3. Which exponential probability density has mean $\mu = \frac{1}{4}$?

Exercises

In Exercises 1–8, find a constant C such that p is a probability density function on the given interval, and compute the probability indicated.

1. $p(x) = \dfrac{C}{(x+1)^3}$ on $[0, \infty)$; $P(0 \le X \le 1)$

2. $p(x) = Cx(4-x)$ on $[0, 4]$; $P(3 \le X \le 4)$

3. $p(x) = \dfrac{C}{\sqrt{1-x^2}}$ on $(-1, 1)$; $P\left(-\frac{1}{2} \le X \le \frac{1}{2}\right)$

4. $p(x) = \dfrac{Ce^{-x}}{1+e^{-2x}}$ on $(-\infty, \infty)$; $P(X \le -4)$

5. $p(x) = C \sin x$ on $[0, \pi]$; $P\left(\frac{\pi}{4} \le X \le \frac{3\pi}{4}\right)$

6. $p(x) = C \ln x$ on $[1, e]$; $P(1 \le X \le 2)$

7. $p(x) = C\sqrt{1-x^2}$ on $(-1, 1)$; $P\left(-\frac{1}{2} \le X \le 1\right)$

8. $p(x) = Ce^{-x}e^{-e^{-x}}$ on $(-\infty, \infty)$; $P(-4 \le X \le 4)$

This function, called the **Gumbel density**, is used to model extreme events such as floods and earthquakes.

9. Verify that $p(x) = 3x^{-4}$ is a probability density function on $[1, \infty)$ and calculate its mean value.

10. Show that the density function $p(x) = \dfrac{2}{\pi(x^2+1)}$ on $[0, \infty)$ has infinite mean.

11. Verify that $p(t) = \frac{1}{50} e^{-t/50}$ satisfies the condition

$$\int_0^\infty p(t)\,dt = 1$$

12. Verify that for all $r > 0$, the exponential density function $p(t) = \frac{1}{r} e^{-t/r}$ satisfies the condition

$$\int_0^\infty p(t)\,dt = 1$$

13. The life X (in hours) of a battery in constant use is a random variable with exponential density. What is the probability that the battery will last more than 12 h if the average life is 8 h?

14. The time between incoming phone calls at a call center is a random variable with exponential density. There is a 50% probability of waiting 20 seconds or more between calls. What is the average time between calls?

15. The distance r between the electron and the nucleus in a hydrogen atom (in its lowest energy state) is a random variable with probability density $p(r) = 4a_0^{-3}r^2e^{-2r/a_0}$ for $r \geq 0$, where a_0 is the Bohr radius (Figure 8). Calculate the probability P that the electron is within one Bohr radius of the nucleus. The value of a_0 is approximately 5.29×10^{-11} m, but this value is not needed to compute P.

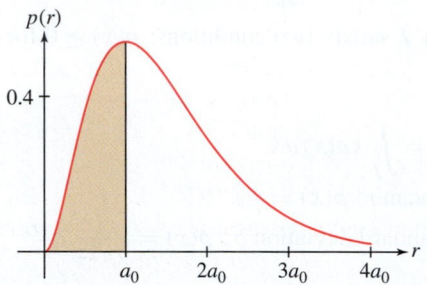

FIGURE 8 Probability density function $p(r) = 4a_0^{-3}r^2e^{-2r/a_0}$.

16. Show that the distance r between the electron and the nucleus in Exercise 15 has mean $\mu = 3a_0/2$.

In Exercises 17–22, $F(z)$ denotes the cumulative normal distribution function. Refer to a calculator or computer algebra system to obtain values of $F(z)$.

17. Express the area of region A in Figure 9 in terms of $F(z)$ and compute its value.

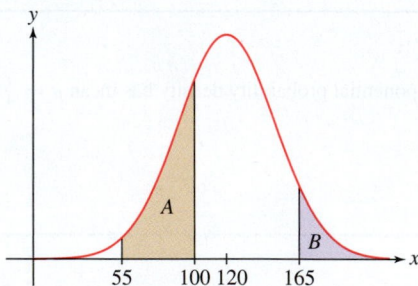

FIGURE 9 Normal density function with $\mu = 120$ and $\sigma = 30$.

18. Show that the area of region B in Figure 9 is equal to $1 - F(1.5)$ and compute its value. Verify numerically that this area is also equal to $F(-1.5)$ and explain why graphically.

19. Assume X has a standard normal distribution ($\mu = 0, \sigma = 1$). Express each of the following probabilities in terms of $F(z)$ and determine the value of each.

(a) $P(X \leq 1.2)$ **(b)** $P(X \geq -0.4)$

20. Assume X has a normal distribution with $\mu = 0$ and $\sigma = 5$. Express each of the following probabilities in terms of $F(z)$ and determine the value of each.

(a) $P(X \leq 1.2)$ **(b)** $P(X \geq -0.4)$

21. 🖊 Use a graph to show that $F(-z) = 1 - F(z)$ for all z. Then show that if $p(x)$ is a normal density function with mean μ and standard deviation σ, then for all $r \geq 0$,

$$P(\mu - r\sigma \leq X \leq \mu + r\sigma) = 2F(r) - 1$$

22. The average September rainfall in Erie, Pennsylvania, is a random variable X with mean $\mu = 102$ mm. Assume that the amount of rainfall is normally distributed with standard deviation $\sigma = 48$.

(a) Express $P(128 \leq X \leq 150)$ in terms of $F(z)$ and compute its value numerically.

(b) Let P be the probability that September rainfall will be at least 120 mm. Express P as an integral of an appropriate density function and compute its value numerically.

23. A bottling company produces bottles of fruit juice that are filled, on average, with 32 ounces of juice. Due to random fluctuations in the machinery, the actual volume of juice is normally distributed with a standard deviation of 0.4 oz. Let P be the probability of a bottle having less than 31 oz. Express P as an integral of an appropriate density function and compute its value numerically.

24. According to **Maxwell's Distribution Law**, in a gas of molecular mass m, the speed v of a molecule in a gas at temperature T (kelvins) is a random variable with density

$$p(v) = 4\pi \left(\frac{m}{2\pi kT}\right)^{3/2} v^2 e^{-mv^2/(2kT)} \quad (v \geq 0)$$

where k is Boltzmann's constant. Show that the average molecular speed is equal to $(8kT/\pi m)^{1/2}$. The average speed of oxygen molecules at room temperature is around 450 m/s.

25. Define the median of a probability distribution to be that value a such that $\int_a^\infty p(x)\,dx = \int_{-\infty}^a p(x)\,dx = \frac{1}{2}$. Show that if a probability function is symmetric about the line $x = m$, then m is both the mean and the median.

26. Define the quartiles of a probability function to be those values $a_1, a_2,$ and a_3 such that $P(-\infty < x \leq a_1) = P(a_1 \leq x \leq a_2) = P(a_2 \leq x \leq a_3) = P(a_3 \leq x < \infty) = \frac{1}{4}$. Find the quartile values for the probability function $p(x) = \frac{1}{1+x^2}$.

*In Exercises 27–30, calculate μ and σ, where σ is the **standard deviation**, defined by*

$$\sigma^2 = \int_{-\infty}^\infty (x - \mu)^2 p(x)\,dx$$

The smaller the value of σ, the more tightly clustered are the values of the random variable X about the mean μ. (The limits of integration need not be $\pm\infty$ if p is defined over a smaller domain.)

27. $p(x) = \dfrac{5}{2x^{7/2}}$ on $[1, \infty)$

28. $p(x) = \dfrac{1}{\pi\sqrt{1-x^2}}$ on $(-1, 1)$

29. $p(x) = \dfrac{1}{3}e^{-x/3}$ on $[0, \infty)$

30. $p(x) = \dfrac{1}{r}e^{-x/r}$ on $[0, \infty)$, where $r > 0$

Further Insights and Challenges

31. ✎ The time to decay of an atom in a radioactive substance is a random variable X. The law of radioactive decay states that if N atoms are present at time $t = 0$, then $Nf(t)$ atoms will be present at time t, where $f(t) = e^{-kt}$ ($k > 0$ is the decay constant). Explain the following statements:

(a) The fraction of atoms that decay in a small time interval $[t, t + \Delta t]$ is approximately $-f'(t)\Delta t$.

(b) The probability density function of X is $y = -f'(t)$.

(c) The average time to decay is $1/k$.

32. The half-life of radon-222 is 3.825 days. Use Exercise 31 to compute:

(a) the average time to decay of a radon-222 atom.

(b) the probability that a given atom will decay in the next 24 hours.

8.2 Arc Length and Surface Area

We have seen that integrals are used to compute total amounts (such as distance traveled, total mass, total cost). Another such quantity is the length of a curve (also called **arc length**). We derive a formula for arc length using our standard procedure: approximation followed by passage to a limit. In this case, we approximate the curve by a path made up of line segments connecting points on the curve. It is easy to find the length of a collection of line segments. We improve the approximation by using more, but smaller, segments. Then we take the limit of the sum of their lengths as the number of line segments grows.

To make this precise, consider the graph of $y = f(x)$ over an interval $[a, b]$. Choose a partition P of $[a, b]$ into N subintervals with endpoints

$$P : a = x_0 < x_1 < \cdots < x_N = b$$

Recall that the norm of the partition, $\|P\|$, is the length of the largest subinterval in the partition; that is, the largest of the distances $x_i - x_{i-1}$. Let $P_i = (x_i, f(x_i))$ be the point on the graph corresponding to x_i, and join the points P_{i-1} and P_i by a line segment L_i. The curve L, made up of the segments L_i, is called a **polygonal approximation** (Figure 1). The length of L, which we denote $|L|$, is the sum of the lengths $|L_i|$ of the segments:

FIGURE 1 A polygonal approximation L to $y = f(x)$.

$$|L| = |L_1| + |L_2| + \cdots + |L_N| = \sum_{i=1}^{N} |L_i|$$

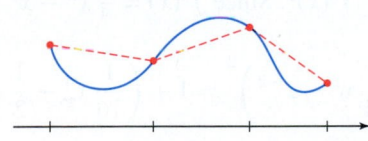 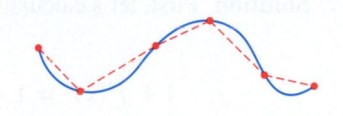 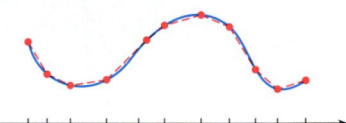

FIGURE 2 The polygonal approximations improve as the norm of the partition decreases.

The letter s is commonly used to denote arc length.

As may be expected, the polygonal approximations L approximate the curve more and more closely as the norm of the partition P decreases, as illustrated in Figure 2. Based on this idea, we define the arc length s of the graph to be the limit of the polygonal approximation lengths $|L|$ as $\|P\| \to 0$:

$$\text{arc length } s = \lim_{\|P\| \to 0} \sum_{i=1}^{N} |L_i|$$

To compute the arc length s, we express the limit of the polygonal approximations as an integral. Figure 3 shows that the segment L_i is the hypotenuse of a right triangle of base $\Delta x_i = x_i - x_{i-1}$ and height $|f(x_i) - f(x_{i-1})|$. By the Pythagorean Theorem,

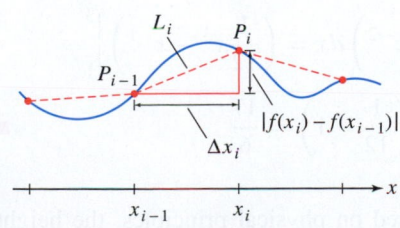

DF **FIGURE 3**

$$|L_i| = \sqrt{(\Delta x_i)^2 + (f(x_i) - f(x_{i-1}))^2}$$

We shall assume that f' exists and is continuous. Then, by the Mean Value Theorem, there is a value c_i in $[x_{i-1}, x_i]$ such that

$$f(x_i) - f(x_{i-1}) = f'(c_i)(x_i - x_{i-1}) = f'(c_i)\Delta x_i$$

and therefore,

$$|L_i| = \sqrt{(\Delta x_i)^2 + (f'(c_i)\Delta x_i)^2} = \sqrt{(\Delta x_i)^2(1 + f'(c_i)^2)} = \sqrt{1 + f'(c_i)^2}\,\Delta x_i$$

We find that the length $|L|$ is a Riemann sum for $\sqrt{1 + f'(x)^2}$:

➥ **REMINDER** A Riemann sum *for the* integral $\int_a^b g(x)\,dx$ *is a sum*

$$\sum_{i=1}^{N} g(c_i)\Delta x_i$$

where $x_0, x_1, \ldots, x_N$ *is a partition of* $[a, b]$, $\Delta x_i = x_i - x_{i-1}$, *and* c_i *is any number in* $[x_{i-1}, x_i]$.

$$|L| = |L_1| + |L_2| + \cdots + |L_N| = \sum_{i=1}^{N} \sqrt{1 + f'(c_i)^2}\,\Delta x_i$$

This function is continuous, and hence integrable, so the Riemann sums approach

$$\int_a^b \sqrt{1 + f'(x)^2}\,dx$$

as N becomes infinite.

Formula for Arc Length Assume that f' exists and is continuous on the interval $[a, b]$. Then the arc length s of $y = f(x)$ over $[a, b]$ is equal to

$$s = \int_a^b \sqrt{1 + f'(x)^2}\,dx \qquad \boxed{1}$$

In Exercises 24–26, we verify that Eq. (1) correctly gives the lengths of line segments and circles.

EXAMPLE 1 Find the arc length s of the graph of $f(x) = \frac{1}{12}x^3 + x^{-1}$ over the interval $[1, 3]$ (Figure 4).

Solution First, let's calculate $1 + f'(x)^2$. Since $f'(x) = \frac{1}{4}x^2 - x^{-2}$,

$$1 + f'(x)^2 = 1 + \left(\frac{1}{4}x^2 - x^{-2}\right)^2 = 1 + \left(\frac{1}{16}x^4 - \frac{1}{2} + x^{-4}\right)$$

$$= \frac{1}{16}x^4 + \frac{1}{2} + x^{-4} = \left(\frac{1}{4}x^2 + x^{-2}\right)^2$$

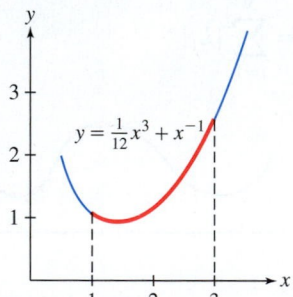

FIGURE 4 What is the arc length over $[1, 3]$?

Fortunately, since this expression for $1 + f'(x)^2$ is a square, the arc-length integral simplifies nicely and is easily computed:

$$s = \int_1^3 \sqrt{1 + f'(x)^2}\,dx = \int_1^3 \left(\frac{1}{4}x^2 + x^{-2}\right)dx = \left(\frac{1}{12}x^3 - x^{-1}\right)\Big|_1^3$$

$$= \left(\frac{9}{4} - \frac{1}{3}\right) - \left(\frac{1}{12} - 1\right) = \frac{17}{6} \qquad ■$$

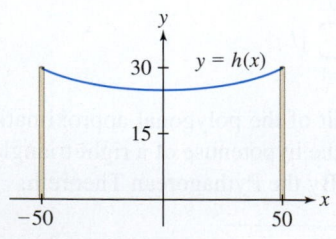

FIGURE 5 A hanging cable whose height is $h(x) = 250\cosh(0.004x) - 225$.

EXAMPLE 2 **Length of a Hanging Cable** Based on physical principles, the height of a cable hanging under its own weight is modeled well using the hyperbolic cosine function. Suppose we have a cable hanging from two poles that are 100 ft apart (located at $x = -50$ and $x = 50$) and such that its height is given by $h(x) = 250\cosh(0.004x) - 225$ (Figure 5). Note that in the middle, the height of the cable is $h(0) = 25$ ft, and at the poles, the height is $h(-50) = h(50) = 250 \cosh(0.2) - 225 \approx 30.0$ ft. So the cable hangs 5 ft lower in the middle than the height at the poles. What is the length of the cable?

◀ REMINDER $\cosh^2 x - \sinh^2 x = 1$ *and* $(\cosh x)' = \sinh x.$

Solution We have $h'(x) = (0.004)(250 \sinh(0.004x)) = \sinh(0.004x)$. Therefore,

$$1 + h'(x)^2 = 1 + \sinh^2(0.004x) = \cosh^2(0.004x)$$

Because hyperbolic cosine is a positive function, it follows that

$$\sqrt{1 + h'(x)^2} = \cosh(0.004x)$$

So, the length of the curve is

$$\int_{-50}^{50} \sqrt{1 + h'(x)^2}\, dx = \int_{-50}^{50} \cosh(0.004x)\, dx = \frac{1}{0.004} \sinh(0.004x)\Big|_{-50}^{50}$$

$$= 250(\sinh(0.2) - \sinh(-0.2)) \approx 100.67 \text{ ft}$$

It is interesting, and perhaps surprising, that with just 8 in. of length beyond the 100-ft direct distance from pole to pole, the cable hangs 5 ft lower at the middle than at the ends. ∎

In Examples 1 and 2, we were able to compute the arc length exactly because $1 + f'(x)^2$ could be expressed as a square that enabled us to simplify the integrand. Usually, $\sqrt{1 + f'(x)^2}$ does not have an elementary antiderivative and there is no explicit formula for the arc length. However, we can always approximate arc length by using numerical integration.

EXAMPLE 3 **No Exact Formula for Arc Length** Approximate the length s of $y = \sin x$ over $[0, \pi]$ using Simpson's Rule, computing S_N for $N = 6$.

Solution We have $y' = \cos x$ and $\sqrt{1 + (y')^2} = \sqrt{1 + \cos^2 x}$. The arc length is

$$s = \int_0^\pi \sqrt{1 + \cos^2 x}\, dx$$

This integral cannot be evaluated explicitly, so we approximate it by applying Simpson's Rule (Section 7.8) with $N = 6$. Divide $[0, \pi]$ into subintervals of width $\Delta x = \pi/6$. Then

$$S_6 = \frac{\Delta x}{3}\left(g(0) + 4g\left(\frac{\pi}{6}\right) + 2g\left(\frac{2\pi}{6}\right) + 4g\left(\frac{3\pi}{6}\right) + 2g\left(\frac{4\pi}{6}\right) + 4g\left(\frac{5\pi}{6}\right) + g(\pi)\right)$$

$$\approx \frac{\pi}{18}(1.4142 + 5.2915 + 2.2361 + 4 + 2.2361 + 5.2915 + 1.4142) \approx 3.82$$

Thus, $s \approx 3.82$ (Figure 6). ∎

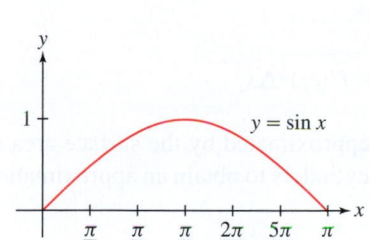

FIGURE 6 The arc length from 0 to π is approximately 3.82.

Surface Area

The surface area S of a surface of revolution (Figure 7) can be computed by an integral that is similar to the arc length integral. Suppose that $f(x) \geq 0$, so the graph lies above the x-axis. We revolve the graph around the x-axis to obtain a **surface of revolution** R. Our goal is to determine the surface area S of R. To do so, we start by creating another surface of revolution R^* by rotating a polygonal approximation to $y = f(x)$ about the x-axis. The result is a surface R^* that lies very close to R and whose surface area approximates S (Figure 8).

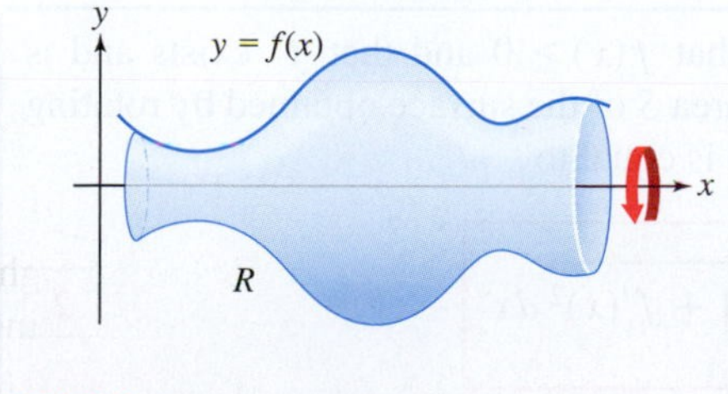

FIGURE 7 Surface obtained by revolving $y = f(x)$ about the x-axis.

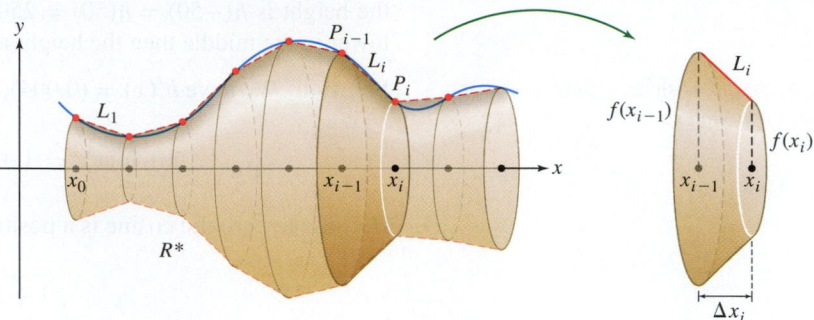

FIGURE 8 Rotating a polygonal approximation produces an approximation by slanted bands.

We will set up a Riemann sum that approximates the surface area of R^* and therefore that also approximates the surface area of R. Taking a limit results in a definite-integral formula for S, the surface area of R.

The surface R^* is made up of slanted bands, as shown in Figure 8. Consider the slanted band corresponding to the interval $[x_{i-1}, x_i]$. The segment L_i along the slanted band is a segment in the polygonal approximation for $y = f(x)$. As in the derivation of the arc length formula, the length of L_i can be expressed as

$$|L_i| = \sqrt{1 + f'(c_i)^2}\, \Delta x_i$$

for some c_i in $[x_{i-1}, x_i]$ and $\Delta x_i = x_i - x_{i-1}$.

Now, as Figure 9 illustrates, we can approximate the surface area of the single slanted band with the surface area of a cylinder of width $|L_i|$ and radius $f(b_i)$ *for any* b_i in $[x_{i-1}, x_i]$. Since we can use any b_i in $[x_{i-1}, x_i]$ for this approximation, we will use $b_i = c_i$ to match the value used in the expression for $|L_i|$.

The surface area of a cylinder of radius r and width w is $2\pi r w$; therefore, if we let $b_i = c_i$, the surface area of the cylinder in Figure 9 is

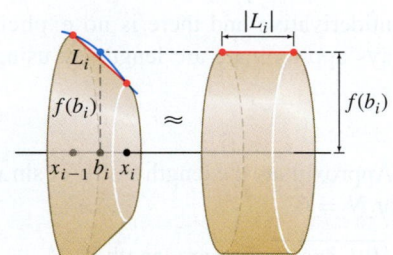

FIGURE 9 The surface area of the slanted band is approximated by the surface area of the cylinder.

$$2\pi f(c_i)|L_i| = 2\pi f(c_i)\sqrt{1 + f'(c_i)^2}\, \Delta x_i$$

The surface area of each slanted band in R^* is approximated by the surface area of such a cylinder. We add up the surface areas of these cylinders to obtain an approximation to the surface area of R^*:

$$\text{surface area of } R^* \approx 2\pi \sum_{i=1}^{N} f(c_i)\sqrt{1 + f'(c_i)^2}\, \Delta x_i$$

As the norm of the partition goes to zero, the error in this approximation of the surface area of R^* also goes to zero. Furthermore, the surface area of R^* approaches the surface area of R. Therefore, the sum on the right-hand side of the approximation approaches S in the limit. That sum is a Riemann sum that converges to the integral in the following definition:

> **Area of a Surface of Revolution** Assume that $f(x) \geq 0$ and that f' exists and is continuous on the interval $[a, b]$. The surface area S of the surface obtained by rotating the graph of f about the x-axis for $a \leq x \leq b$ is equal to
>
> $$S = 2\pi \int_a^b f(x)\sqrt{1 + f'(x)^2}\, dx \qquad \boxed{2}$$

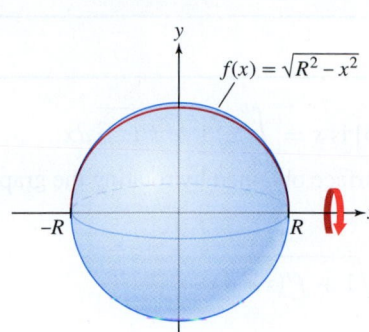

FIGURE 10 A sphere is obtained by revolving the red semicircle about the x-axis.

EXAMPLE 4 Calculate the surface area of a sphere of radius R.

Solution The graph of $f(x) = \sqrt{R^2 - x^2}$ is a semicircle of radius R (Figure 10). We obtain a sphere by rotating it about the x-axis. We have

$$f'(x) = -\frac{x}{\sqrt{R^2 - x^2}}, \qquad 1 + f'(x)^2 = 1 + \frac{x^2}{R^2 - x^2} = \frac{R^2}{R^2 - x^2}$$

The surface area integral gives us the usual formula for the surface area of a sphere:

$$S = 2\pi \int_{-R}^{R} f(x)\sqrt{1 + f'(x)^2}\,dx = 2\pi \int_{-R}^{R} \sqrt{R^2 - x^2}\,\frac{R}{\sqrt{R^2 - x^2}}\,dx$$

$$= 2\pi R \int_{-R}^{R} dx = 2\pi R(2R) = 4\pi R^2 \qquad \blacksquare$$

EXAMPLE 5 Find the surface area of the surface, called a paraboloid, that is obtained by rotating the graph of $f(x) = \sqrt{x}$ about the x-axis for $0 \le x \le 1$.

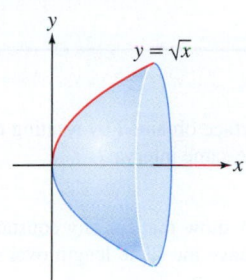

FIGURE 11 A paraboloid results when the top half of the parabola is revolved about the x-axis.

Solution The graph of $f(x) = \sqrt{x}$ is the top half of a parabola opening along the x-axis, which becomes a paraboloid when rotated about the x-axis (Figure 11). Then $f'(x) = \frac{1}{2\sqrt{x}}$ and hence, we obtain

$$S = 2\pi \int_0^1 f(x)\sqrt{1 + f'(x)^2}\,dx = 2\pi \int_0^1 \sqrt{x}\sqrt{1 + \left(\frac{1}{2\sqrt{x}}\right)^2}\,dx$$

$$= 2\pi \int_0^1 \frac{\sqrt{x}}{2\sqrt{x}}\sqrt{4x + 1}\,dx = \pi \int_0^1 \sqrt{4x + 1}\,dx$$

$$= \frac{\pi}{6}(4x + 1)^{3/2}\Big|_0^1 = \frac{\pi}{6}(5^{3/2} - 1) \approx 5.3304 \qquad \blacksquare$$

EXAMPLE 6 **A Corrugated Pipe** A corrugated pipe is obtained by rotating the graph of $f(x) = 1 + 0.1\sin(10x)$, for $0 \le x \le 10$, around the x-axis (Figure 12). What is the surface area of the pipe [assuming that x and $f(x)$ are in meters]?

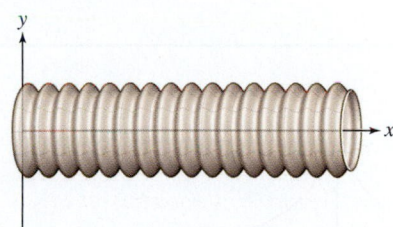

FIGURE 12 A corrugated pipe obtained by rotating the graph of $f(x) = 1 + 0.1 \sin(10x)$ about the x-axis.

Solution Note that $f'(x) = 10(0.1\cos(10x)) = \cos(10x)$. Substituting into the surface area formula, we have

$$S = 2\pi \int_0^{10} (1 + 0.1\sin(10x))\sqrt{1 + \cos^2(10x)}\,dx$$

We use numerical approximation to obtain a result here. Using Simpson's Rule with $N = 100$, a computer algebra system yields $S \approx 76.4\ \text{m}^2$. $\qquad \blacksquare$

EXAMPLE 7 **Gabriel's Horn** In Example 3 in Section 7.7, we introduced Gabriel's Horn (Figure 13), the surface obtained by rotating the graph of $f(x) = \frac{1}{x}$, for $x \ge 1$, about the x-axis. There we saw that the volume enclosed in the horn is π, and therefore is finite. Prove that the surface area of Gabriel's Horn is infinite.

Solution With $f(x) = \frac{1}{x} = x^{-1}$, we have $f'(x) = -x^{-2}$ and $f'(x)^2 = x^{-4}$. The surface area of Gabriel's Horn is

$$S = 2\pi \int_1^{\infty} x^{-1}\sqrt{1 + x^{-4}}\,dx$$

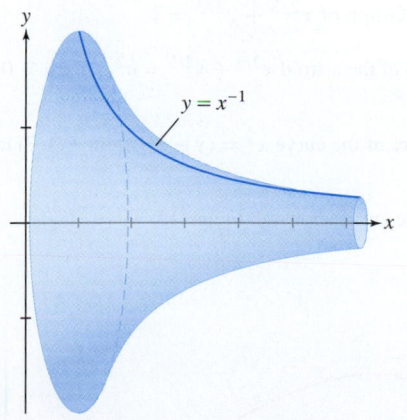

FIGURE 13 Gabriel's Horn is obtained by rotating the graph of $f(x) = \frac{1}{x}$ about the x-axis.

Now, $x^{-1}\sqrt{1 + x^{-4}} > x^{-1}$ over $[1, \infty)$, and $\int_1^{\infty} x^{-1}\,dx$ diverges (to infinity) by Theorem 1 in Section 7.7. By the Comparison Test for Improper Integrals (Theorem 3 in Section 7.7), it follows that S, the surface area of Gabriel's Horn, is infinite. $\qquad \blacksquare$

8.2 SUMMARY

- The *arc length* of $y = f(x)$ over the interval $[a, b]$ is $s = \displaystyle\int_a^b \sqrt{1 + f'(x)^2}\,dx$.
- Assume that $f(x) \geq 0$. The *surface area* of the surface obtained by rotating the graph of f about the x-axis for $a \leq x \leq b$ is

$$\text{surface area} = 2\pi \int_a^b f(x)\sqrt{1 + f'(x)^2}\,dx$$

- Use numerical integration to approximate arc length or surface area when the integral cannot be evaluated explicitly.

8.2 EXERCISES

Preliminary Questions

1. Which integral represents the length of the curve $y = \cos x$ between 0 and $\frac{\pi}{4}$?

$$\int_0^{\frac{\pi}{4}} \sqrt{1 + \cos^2 x}\,dx, \qquad \int_0^{\frac{\pi}{4}} \sqrt{1 + \sin^2 x}\,dx$$

2. By rotating the line $y = r$ about the x-axis, for x in the interval $[0, h]$, and applying the surface area formula, obtain the well-known fact that the surface area of a cylinder of radius r and length h is given by $2\pi rh$.

3. If $0 \leq f(x) \leq g(x)$ for x in the interval $[a, b]$, can the surface obtained by rotating the graph of $y = g(x)$ around the x-axis over the inter-val have less surface area than the surface obtained by rotating the graph of $y = f(x)$ around the x-axis over the same interval?

4. Use the formula for arc length to show that for any constant C, the graphs $y = f(x)$ and $y = f(x) + C$ have the same length over every in-terval $[a, b]$. Explain geometrically.

5. Use the formula for arc length to show that the length of a graph over $[1, 4]$ cannot be less than 3.

Exercises

1. Express the arc length of the curve $y = x^4$ between $x = 2$ and $x = 6$ as an integral (but do not evaluate).

2. Express the arc length of the curve $y = \tan x$ for $0 \leq x \leq \frac{\pi}{4}$ as an integral (but do not evaluate).

3. Find the arc length of $y = \frac{1}{12}x^3 + x^{-1}$ for $1 \leq x \leq 2$. *Hint:* Show that

$$1 + (y')^2 = \left(\tfrac{1}{4}x^2 + x^{-2}\right)^2.$$

4. Find the arc length of $y = \left(\dfrac{x}{2}\right)^4 + \dfrac{1}{2x^2}$ over $[1, 4]$. *Hint:* Show that $1 + (y')^2$ is a perfect square.

In Exercises 5–10, calculate the arc length over the given interval.

5. $y = 3x + 1$, $[0, 3]$
6. $y = 9 - 3x$, $[1, 3]$

7. $y = x^{3/2}$, $[1, 2]$
8. $y = \frac{1}{3}x^{3/2} - x^{1/2}$, $[2, 8]$

9. $y = \frac{1}{4}x^2 - \frac{1}{2}\ln x$, $[1, 2e]$
10. $y = \ln(\cos x)$, $\left[0, \frac{\pi}{4}\right]$

In Exercises 11–18, approximate the arc length of the curve over the inter-val using the Trapezoidal Rule T_N, the Midpoint Rule M_N, or Simpson's Rule S_N as indicated.

11. $y = \frac{1}{4}x^4$, $[1, 2]$, T_5
12. $y = \sin x$, $\left[0, \frac{\pi}{2}\right]$, M_8

13. $y = x^{-1}$, $[1, 2]$, S_8
14. $y = e^{-x^2}$, $[0, 2]$, S_8

15. $y = \ln x$, $[1, 3]$, M_6
16. $y = \cos x$, $[0, 2]$, T_8

17. (CAS) $y = x \sin x$, $[0, 10\pi]$, T_{100}

18. (CAS) $y = \frac{1}{1-x}$, $[0, 0.99]$, S_{100}

19. Calculate the length of the astroid $x^{2/3} + y^{2/3} = 1$ (Figure 14).

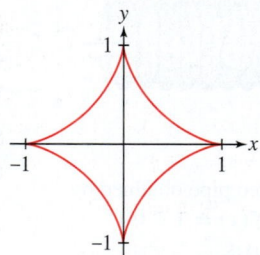

FIGURE 14 Graph of $x^{2/3} + y^{2/3} = 1$.

20. Show that the arc length of the astroid $x^{2/3} + y^{2/3} = a^{2/3}$ (for $a > 0$) is proportional to a.

21. Find the length of the arc of the curve $x^2 = (y - 2)^3$ from $P(1, 3)$ to $Q(8, 6)$.

22. Find the arc length of the curve shown in Figure 15.

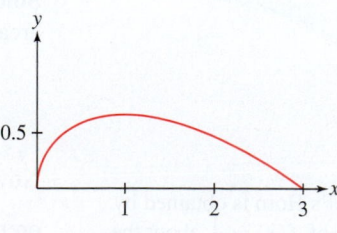

FIGURE 15 Graph of $9y^2 = x(x - 3)^2$.

23. Find the value of a such that the arc length of the *catenary* $y = \cosh x$ for $-a \le x \le a$ equals 10.

24. Calculate the arc length of the graph of $f(x) = mx + r$ over $[a, b]$ in two ways: using the Pythagorean theorem (Figure 16) and using the arc length integral.

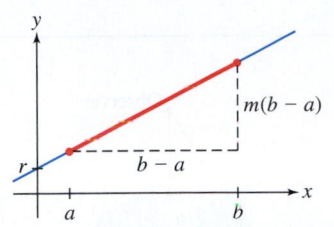

FIGURE 16

25. Show that the circumference of the unit circle is equal to

$$2 \int_{-1}^{1} \frac{dx}{\sqrt{1 - x^2}} \quad \text{(an improper integral)}$$

Evaluate, thus verifying that the circumference is 2π.

26. Generalize the result of Exercise 25 to show that the circumference of the circle of radius r is $2\pi r$.

27. Calculate the arc length of $y = x^2$ over $[0, a]$. *Hint:* Use trigonometric substitution. Evaluate for $a = 1$.

28. Express the arc length of $g(x) = \sqrt{x}$ over $[0, 1]$ as a definite integral. Then use the substitution $u = \sqrt{x}$ to show that this arc length is equal to the arc length of $y = x^2$ over $[0, 1]$ (but do not evaluate the integrals). Explain this result graphically.

29. Find the arc length of $y = e^x$ over $[0, a]$. *Hint:* Try the substitution $u = \sqrt{1 + e^{2x}}$ followed by partial fractions.

30. Show that the arc length of $y = \ln(f(x))$ for $a \le x \le b$ is

$$\int_a^b \frac{\sqrt{f(x)^2 + f'(x)^2}}{f(x)} \, dx \qquad \boxed{3}$$

31. Use Eq. (3) to compute the arc length of $y = \ln(\sin x)$ for $\frac{\pi}{4} \le x \le \frac{\pi}{2}$.

32. Use Eq. (3) to compute the arc length of $y = \ln\left(\dfrac{e^x + 1}{e^x - 1}\right)$ over $[1, 3]$.

33. Show that if $0 \le f'(x) \le 1$ for all x, then the arc length of $y = f(x)$ over $[a, b]$ is at most $\sqrt{2}(b - a)$. Show that for $f(x) = x$, the arc length equals $\sqrt{2}(b - a)$.

34. Use the Comparison Theorem (Section 5.2) to prove that the arc length of $y = x^{4/3}$ over $[1, 2]$ is not less than $\frac{5}{3}$.

35. Approximate the arc length of one-quarter of the unit circle (which we know is $\frac{\pi}{2}$) by computing the length of the polygonal approximation with $N = 4$ segments (Figure 17).

36. [CAS] A merchant intends to produce specialty carpets in the shape of the region in Figure 18, bounded by the axes and graph of $y = 1 - x^n$ (units in yards). Assume that material costs $\$50/\text{yd}^2$ and that it costs $50L$ dollars to cut the carpet, where L is the length of the curved side of the carpet. The carpet can be sold for $150A$ dollars, where A is the carpet's area. Using numerical integration with a computer algebra system, find the whole number n for which the merchant's profits are maximal.

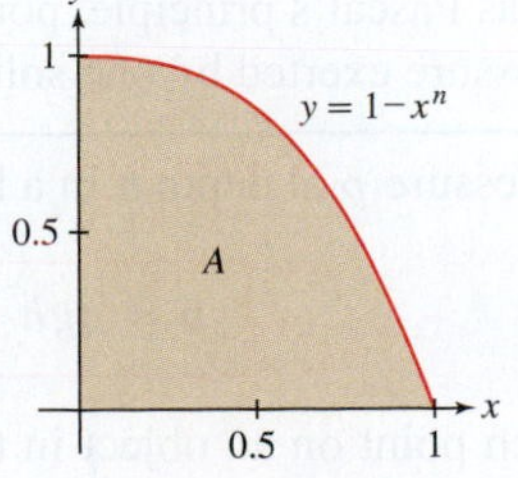

In Exercises 37–46, compute the surface area of revolution about the x-axis over the interval.

37. $y = x$, $[0, 4]$

38. $y = 4x + 3$, $[0, 1]$

39. $y = x^3$, $[0, 2]$

40. $y = x^3$, $[0, 10]$

41. $y = x^2$, $[0, 2]$

42. $y = x^2$, $[0, 10]$

43. $y = (4 - x^{2/3})^{3/2}$, $[0, 8]$

44. $y = e^{-x}$, $[0, 1]$

45. $y = \frac{1}{4}x^2 - \frac{1}{2}\ln x$, $[1, e]$

46. $y = \sin x$, $[0, \pi]$

[CAS] *In Exercises 47–50, use a computer algebra system to find the approximate surface area of the solid generated by rotating the curve about the x-axis.*

47. $y = x^{-1}$, $[1, 3]$

48. $y = x^4$, $[0, 1]$

49. $y = e^{-x^2/2}$, $[0, 2]$

50. $y = \tan x$, $\left[0, \frac{\pi}{4}\right]$

51. Find the area of the surface obtained by rotating $y = \cosh x$ over $[-\ln 2, \ln 2]$ around the x-axis.

52. Show that a spherical cap of height h and radius R (Figure 19) has surface area $2\pi R h$.

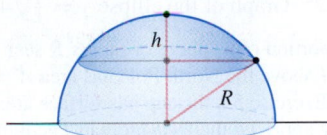

FIGURE 19

53. Find the surface area of the torus obtained by rotating the circle $x^2 + (y - b)^2 = r^2$ about the x-axis (Figure 20).

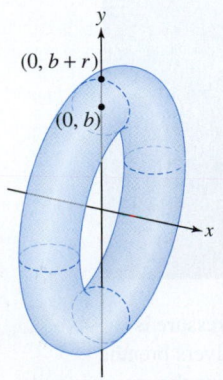

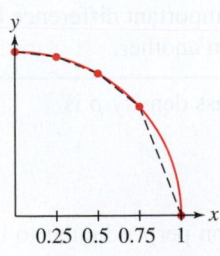

FIGURE 17 One-quarter of the unit circle.

FIGURE 18

FIGURE 20 Torus obtained by rotating a circle about the x-axis.

In Exercises 54–58, the graph of $y = f(x)$, for $a \leq x \leq b$, is rotated about the y-axis. In this situation, the surface area of the resulting surface is

$$S = 2\pi \int_a^b x\sqrt{1 + f'(x)^2}\, dx$$

Determine the surface area for each surface of revolution. If the surface area cannot be computed exactly, find an approximate value.

54. $f(x) = x^2$, $[1, 5]$

55. $f(x) = x^3$, $[0, 2]$

56. $f(x) = \sqrt{x}$, $[0, 4]$

57. $f(x) = e^x$, $[0, 3]$

58. $f(x) = \ln x$, $[1, 4]$

Further Insights and Challenges

59. Find the surface area of the ellipsoid obtained by rotating the ellipse $\left(\dfrac{x}{a}\right)^2 + \left(\dfrac{y}{b}\right)^2 = 1$ about the x-axis.

60. Show that if the arc length of $y = f(x)$ over $[0, a]$ is proportional to a, then $y = f(x)$ must be a linear function.

61. CAS Let L be the arc length of the upper half of the ellipse with equation

$$y = \frac{b}{a}\sqrt{a^2 - x^2}$$

(Figure 21) and let $\eta = \sqrt{1 - (b^2/a^2)}$. Use substitution to show that

$$L = a \int_{-\pi/2}^{\pi/2} \sqrt{1 - \eta^2 \sin^2\theta}\, d\theta$$

Use a computer algebra system to approximate L for $a = 2$, $b = 1$.

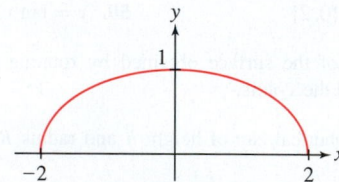

FIGURE 21 Graph of the ellipse $y = \frac{1}{2}\sqrt{4 - x^2}$.

62. Prove that the portion of a sphere of radius R seen by an observer located at a distance d above the North Pole has area $A = 2\pi d R^2/(d + R)$. *Hint:* According to Exercise 52, the cap has surface area $2\pi Rh$. Show that $h = dR/(d + R)$ by applying the Pythagorean Theorem to the three right triangles in Figure 22.

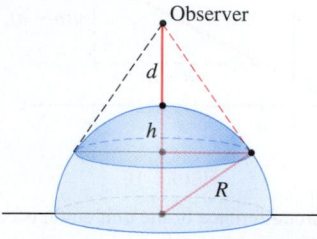

FIGURE 22 Spherical cap observed from a distance d above the North Pole.

63. ✏️ Suppose that the observer in Exercise 62 moves off to infinity— that is, $d \to \infty$. What do you expect the limiting value of the observed area to be? Check your guess by using the formula for the area in the previous exercise to calculate the limit.

64. ✏️ Let M be the total mass of a metal rod in the shape of the curve $y = f(x)$ over $[a, b]$ whose mass density $\rho(x)$ varies as a function of x. Use Riemann sums to justify the formula

$$M = \int_a^b \rho(x)\sqrt{1 + f'(x)^2}\, dx$$

65. ✏️ Let f be an increasing function on $[a, b]$ and let g be its inverse. Argue on the basis of arc length that the following equality holds:

$$\int_a^b \sqrt{1 + f'(x)^2}\, dx = \int_{f(a)}^{f(b)} \sqrt{1 + g'(y)^2}\, dy \qquad \boxed{4}$$

Then use the substitution $u = f(x)$ to prove Eq. (4).

8.3 Fluid Pressure and Force

Fluid force is the force on an object submerged in a fluid. Divers feel this force as they descend below the water surface (Figure 1). Our calculation of fluid force is based on two laws that determine the pressure exerted by a fluid:

- Fluid pressure p is proportional to depth.
- Fluid pressure does not act in a specific direction. Rather, a fluid exerts pressure on each side of an object in the perpendicular direction (Figure 2).

This second fact, known as Pascal's principle, points to an important difference between fluid pressure and the pressure exerted by one solid object on another.

> **Fluid Pressure** The pressure p at depth h in a fluid of mass density ρ is
>
> $$\boxed{p = \rho g h} \qquad \boxed{1}$$
>
> The pressure acts at each point on an object in the direction perpendicular to the object's surface at that point.

FIGURE 1 Since water pressure is proportional to depth, divers breathe compressed air to equalize the pressure and avoid lung injury.

Pressure, by definition, is force per unit area.

- The SI unit of pressure is the pascal (Pa) (1 Pa = 1 N/m² = 1 kg/ms²).
- Mass density (mass per unit volume) is denoted ρ (Greek rho).
- The factor ρg is the density by weight, where $g = 9.8$ m/s² is the acceleration due to gravity.

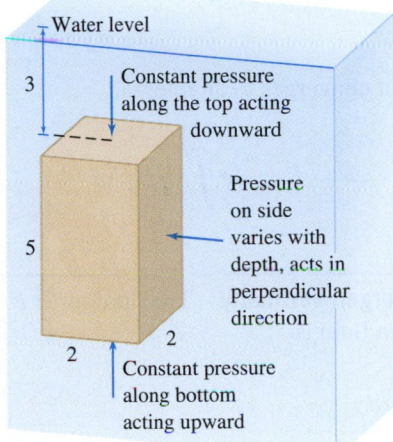

FIGURE 2 Fluid pressure acts on each side in the perpendicular direction.

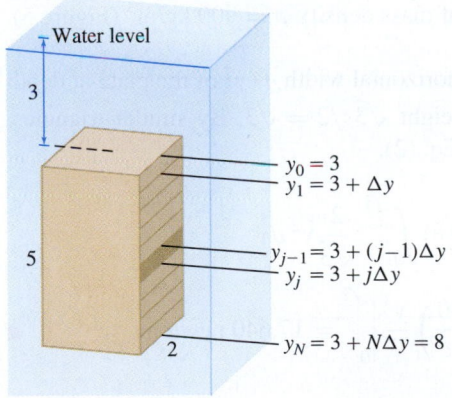

DF **FIGURE 3** The fluid pressure on each horizontal strip depends on the depth of the strip.

Our first example does not require integration because the pressure p is constant. In this case, the total force acting on a surface of area A is

$$\boxed{\text{force} = \text{pressure} \times \text{area} = pA}$$

EXAMPLE 1 Calculate the fluid force on the top and bottom of a box of dimensions $2 \times 2 \times 5$ m, submerged in a pool of water with its top 3 m below the water surface (Figure 2). The density of water is $\rho = 10^3$ kg/m³.

Solution The top of the box is located at depth $h = 3$ m, so by Eq. (1) with $g = 9.8$ m/s²,

$$\text{pressure on top } p = \rho g h = 10^3(9.8)(3) = 29{,}400 \text{ pascals}$$

The top has area $A = 4$ m² and the pressure is constant, so

$$\text{downward force on top} = pA = 29{,}400 \times 4 = 117{,}600 \text{ newtons}$$

The bottom of the box is at depth $h = 8$ m, so the total force on the bottom is

$$\text{upward force on bottom} = pA = \rho g A = 10^3(9.8)(8) \times 4 = 313{,}600 \text{ newtons} \quad \blacksquare$$

In the next example, the pressure varies with depth, and it is necessary to calculate the force as an integral.

EXAMPLE 2 **Calculating Force Using Integration** Calculate the fluid force F on the side of the box in Example 1.

Solution Since the pressure varies with depth, we divide the side of the box into N thin horizontal strips (Figure 3). Let F_j be the force on the jth strip. The total force F is equal to the sum of the forces on the strips:

$$F = F_1 + F_2 + \cdots + F_N$$

Step 1. Approximate the force on a strip.
We'll use the variable y to denote depth, where $y = 0$ at the water level and y is positive in the downward direction. Thus, a larger value of y denotes greater depth. Each strip is a rectangle of height $\Delta y = 5/N$ and length 2, so the area of a strip is $2\Delta y$. The bottom edge of the jth strip has depth $y_j = 3 + j\Delta y$.

If Δy is small, the pressure on the jth strip is nearly constant with value $\rho g y_j$ (because all points on the strip lie at nearly the same depth y_j), so we can approximate the force on the jth strip:

$$F_j \approx \underbrace{\rho g y_j}_{\text{Pressure}} \times \underbrace{(2\Delta y)}_{\text{Area}} = (\rho g) 2 y_j \Delta y$$

Step 2. Approximate total force as a Riemann sum.

$$F = F_1 + F_2 + \cdots + F_N \approx \rho g \sum_{j=1}^{N} 2 y_j \Delta y$$

The sum on the right is a Riemann sum that converges to the integral $\rho g \int_3^8 2y \, dy$.

The interval of integration is $[3, 8]$ because the box extends from $y = 3$ to $y = 8$ (the Riemann sum has been set up with $y_0 = 3$ and $y_N = 8$).

Step 3. Evaluate total force as an integral.
We obtain

$$F = \rho g \int_3^8 2y \, dy = (\rho g) y^2 \Big|_3^8 = (10^3)(9.8)(8^2 - 3^2) = 539{,}000 \text{ newtons} \quad \blacksquare$$

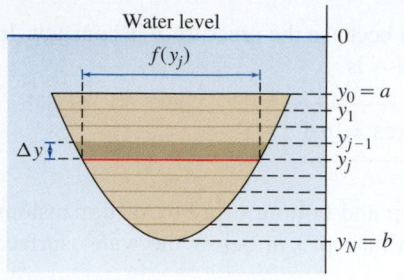

DF FIGURE 4 The area of the shaded strip depends on y and is approximately $f(y_j)\Delta y$.

Now, we'll add another complication: allowing the widths of the horizontal strips to vary with depth (Figure 4). Denote the width at depth y by $f(y)$:

$$f(y) = \text{width of the side at depth } y$$

As before, assume that the object extends from $y = a$ to $y = b$. Divide the flat side of the object into N horizontal strips of thickness $\Delta y = (b-a)/N$. If Δy is small, the jth strip is nearly rectangular of area $f(y_j)\Delta y$. Since the strip lies at depth $y_j = a + j\Delta y$, the force F_j on the jth strip can be approximated:

$$F_j \approx \underbrace{\rho g y_j}_{\text{Pressure}} \times \underbrace{f(y_j)\Delta y}_{\text{Area}} = (\rho g)y_j f(y_j)\Delta y$$

The force F is approximated by a Riemann sum that converges to an integral:

$$F = F_1 + \cdots + F_N \approx \rho g \sum_{j=1}^{N} y_j f(y_j)\Delta y \quad \Rightarrow \quad F = \rho g \int_a^b yf(y)\,dy$$

THEOREM 1 Fluid Force on a Flat Surface Submerged Vertically   The fluid force F on a flat side of an object submerged vertically in a fluid is

$$F = \rho g \int_a^b yf(y)\,dy \qquad \boxed{2}$$

where $f(y)$ is the horizontal width of the side at depth y, and the object extends from depth $y = a$ to depth $y = b$.

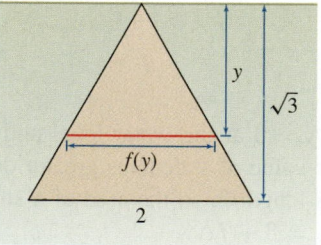

FIGURE 5 Triangular plate submerged in a tank of oil.

EXAMPLE 3   Calculate the fluid force F on one side of an equilateral triangular plate of side 2 m submerged vertically in a tank of oil of mass density $\rho = 900$ kg/m^3 (Figure 5).

Solution   To use Eq. (2), we need to find the horizontal width $f(y)$ of the plate at depth y. An equilateral triangle of side $s = 2$ has height $\sqrt{3}s/2 = \sqrt{3}$. By similar triangles, $y/f(y) = \sqrt{3}/2$ and thus $f(y) = 2y/\sqrt{3}$. By Eq. (2),

$$F = \rho g \int_0^{\sqrt{3}} y\, f(y)\,dy = (900)(9.8) \int_0^{\sqrt{3}} \frac{2}{\sqrt{3}} y^2\,dy$$

$$= \left(\frac{17{,}640}{\sqrt{3}}\right) \frac{y^3}{3}\bigg|_0^{\sqrt{3}} = 17{,}640 \text{ newtons} \qquad \blacksquare$$

The next example shows how to modify the force calculation when the side of the submerged object is inclined at an angle.

EXAMPLE 4   Force on an Inclined Surface   The side of a dam is inclined at an angle of $45°$. The dam has height 700 ft and width 1500 ft as in Figure 6. Calculate the force F on the dam if the reservoir is filled to the top of the dam. Water has weight density $w = 62.4$ pounds per cubic foot.

Solution   The vertical height of the dam is 700 ft, so we divide the vertical axis from 0 to 700 into N subintervals of length $\Delta y = 700/N$. This divides the face of the dam into N strips as in Figure 6. By trigonometry, each strip has a width equal to $\Delta y / \sin(45°) = \sqrt{2}\Delta y$. Therefore,

$$\text{area of each strip} = \text{length} \times \text{width} = 1500(\sqrt{2}\,\Delta y)$$

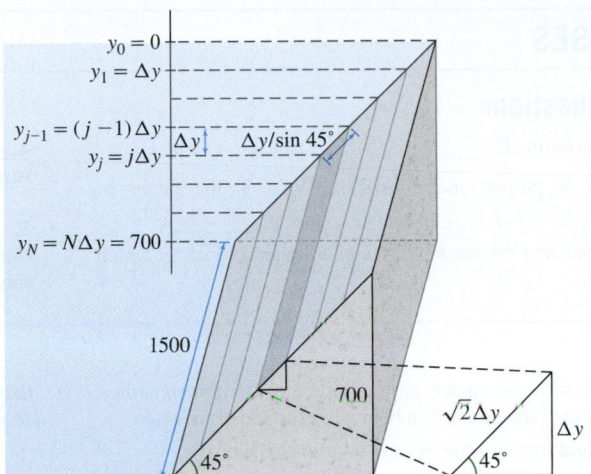

FIGURE 6

As usual, we approximate the force F_j on the jth strip. The term ρg is equal to weight per unit volume, so we use $w = 62.4$ lb/ft^3 in place of ρg:

$$F_j \approx \overbrace{wy_j}^{\text{Pressure}} \times \overbrace{1500\sqrt{2}\Delta y}^{\text{Area of strip}} = wy_j \times 1500\sqrt{2}\,\Delta y \text{ lb}$$

$$F = \sum_{j=1}^{N} F_j \approx \sum_{j=1}^{N} wy_j\left(1500\sqrt{2}\,\Delta y\right) = 1500\sqrt{2}\,w\sum_{j=1}^{N} y_j\Delta y$$

This is a Riemann sum for the integral $1500\sqrt{2}w\displaystyle\int_{0}^{700} y\,dy$. Therefore,

$$F = 1500\sqrt{2}w\int_{0}^{700} y\,dy = 1500\sqrt{2}(62.4)\frac{700^2}{2} \approx 3.24 \times 10^{10} \text{ lb} \qquad \blacksquare$$

8.3 SUMMARY

- If pressure is constant, then force = pressure × area.
- The fluid pressure at depth h is equal to $\rho g h$, where ρ is the fluid density (mass per unit volume) and $g = 9.8$ m/s^2 is the acceleration due to gravity. Fluid pressure acts on a surface in the direction perpendicular to the surface. Water has mass density 1000 kg/m^3.
- If an object is submerged vertically in a fluid and extends from depth $y = a$ to $y = b$, then the total fluid force on a side of the object is

$$F = \rho g \int_{a}^{b} yf(y)\,dy$$

where $f(y)$ is the horizontal width of the side at depth y.
- If fluid density is given as *weight* per unit volume, we use w in place of ρg. Water has weight density 62.4 lb/ft^3.

8.3 EXERCISES

Preliminary Questions

1. How is pressure defined?

2. Fluid pressure is proportional to depth. What is the factor of proportionality?

3. When fluid force acts on the side of a submerged object, in which direction does it act?

4. Why is fluid pressure on a surface calculated using thin horizontal strips rather than thin vertical strips?

5. If a thin plate is submerged horizontally, then the fluid force on one side of the plate is equal to pressure times area. Is this true if the plate is submerged vertically?

Exercises

1. A box of height 6 m and square base of side 3 m is submerged in a pool of water. The top of the box is 2 m below the surface of the water.

(a) Calculate the fluid force on the top and bottom of the box.

(b) Write a Riemann sum that approximates the fluid force on a side of the box by dividing the side into N horizontal strips of thickness $\Delta y = 6/N$.

(c) To which integral does the Riemann sum converge?

(d) Compute the fluid force on a side of the box.

2. A square plate that is 2 by 2 m is submerged in water so that its top edge is level with the surface of the water. Calculate the fuid force on one side of it.

3. If a rectangular plate that is 1 by 2 m is dipped into a pool of water so that initially its top edge of length 1 is even with the surface of the water, and then it is lowered so that its top edge is at a depth of 1 m, calculate the increase in fluid force on one side of it.

4. A plate in the shape of an isosceles triangle with base 1 m and height 2 m is submerged vertically in a tank of water so that its vertex touches the surface of the water (Figure 7).

(a) Show that the width of the triangle at depth y is $f(y) = \frac{1}{2}y$.

(b) Consider a thin strip of thickness Δy at depth y. Explain why the fluid force on a side of this strip is approximately equal to $\rho g \frac{1}{2} y^2 \Delta y$.

(c) Write an approximation for the total fluid force F on a side of the plate as a Riemann sum and indicate the integral to which it converges.

(d) Calculate F.

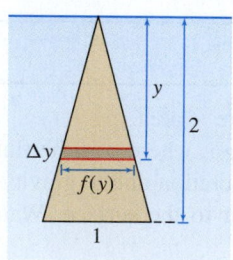

FIGURE 7

5. Repeat Exercise 4, but assume that the top of the triangle is located 3 m below the surface of the water.

6. The plate R in Figure 8, bounded by the parabola $y = x^2$ and $y = 1$, is submerged vertically in water (distance in meters).

(a) Show that the width of R at height y is $f(y) = 2\sqrt{y}$ and the fluid force on a side of a horizontal strip of thickness Δy at height y is approximately $(\rho g)2y^{1/2}(1 - y)\Delta y$.

(b) Write a Riemann sum that approximates the fluid force F on a side of R and use it to explain why

$$F = \rho g \int_0^1 2y^{1/2}(1 - y)\,dy$$

(c) Calculate F.

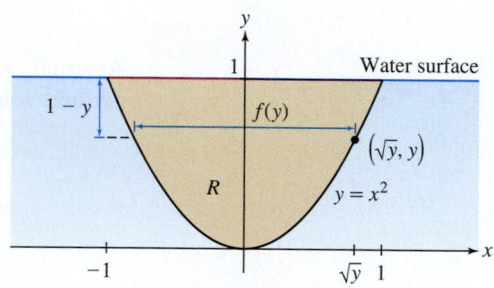

FIGURE 8

7. Let F be the fluid force on a side of a semicircular plate of radius r meters, submerged vertically in water so that its diameter is level with the water's surface (Figure 9).

(a) Show that the width of the plate at depth y is $2\sqrt{r^2 - y^2}$.

(b) Calculate F as a function of r using Eq. (2).

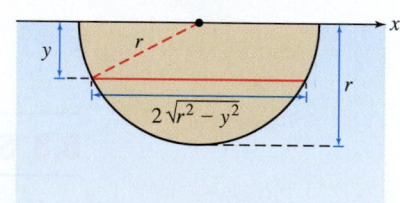

FIGURE 9

8. Calculate the force on one side of a circular plate with radius 2 m, submerged vertically in a tank of water so that the top of the circle is tangent to the water surface.

9. A semicircular plate of radius r meters, oriented as in Figure 9, is submerged in water so that its diameter is located at a depth of m meters. Calculate the fluid force on one side of the plate in terms of m and r.

10. A plate extending from depth $y = 2$ m to $y = 5$ m is submerged in a fluid of density $\rho = 850$ kg/m³. The horizontal width of the plate at depth y is $f(y) = 2(1 + y^2)^{-1}$. Calculate the fluid force on one side of the plate.

11. Figure 10 shows the wall of a dam on a water reservoir. Use the Trapezoidal Rule and the width and depth measurements in the figure to estimate the fluid force on the wall.

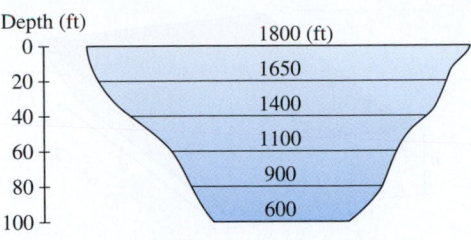

FIGURE 10

12. Assume in Figure 10 that the depth of water in the reservoir dropped 20 ft in a drought. Use the Trapezoidal Rule and the measurements in the figure to estimate the fluid force on the wall.

13. Calculate the fluid force on a side of the plate in Figure 11(A), submerged in water, assuming that the top of the plate is at a depth of $D = 2$ m.

14. Calculate the fluid force on a side of the plate in Figure 11(A), submerged in water, assuming that the top of the plate is at a depth of $D = 4$ m.

15. Calculate the fluid force on a side of the plate in Figure 11(B), submerged in a fluid of mass density $\rho = 800$ kg/m^3.

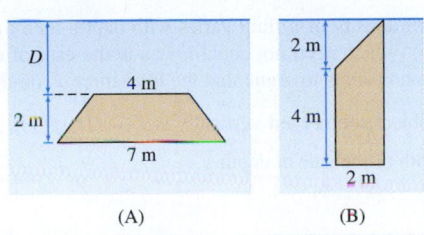

FIGURE 11

16. Find the fluid force on the side of the plate in Figure 12, submerged in a fluid of density $\rho = 1200$ kg/m^3. The top of the plate is level with the fluid surface. The edges of the plate are the curves $y = x^{1/3}$ and $y = -x^{1/3}$.

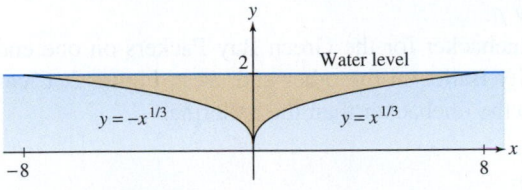

FIGURE 12

17. Let R be the plate in the shape of the region under $y = \sin x$ for $0 \le x \le \frac{\pi}{2}$ in Figure 13(A). If R is rotated counterclockwise by 90° and then submerged in a fluid of density 1100 kg/m^3 with its top edge level with the surface of the fluid as in Figure 13(B), find the fluid force on a side of R.

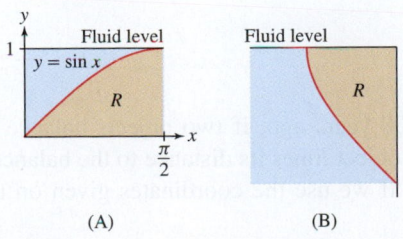

FIGURE 13

18. In the notation of Exercise 17, calculate the fluid force on a side of the plate R if it is oriented as in Figure 13(A).

19. Calculate the fluid force on one side of a plate in the shape of region A shown in Figure 14. The water surface is at $y = 1$, and the fluid has density $\rho = 900$ kg/m^3.

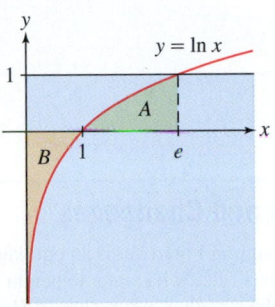

FIGURE 14

20. Calculate the fluid force on one side of the "infinite" plate B in Figure 14, assuming the fluid has density $\rho = 900$ kg/m^3.

21. Figure 15(A) shows a ramp inclined at 30° leading into a swimming pool. Calculate the fluid force on the ramp.

22. Calculate the fluid force on one side of the plate (an isosceles triangle) shown in Figure 15(B).

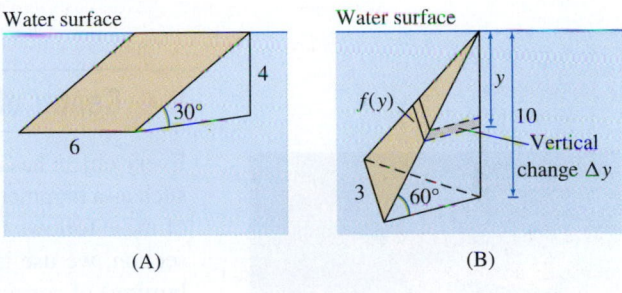

FIGURE 15

23. The massive Three Gorges Dam on China's Yangtze River has height 185 m (Figure 16). Calculate the force on the dam, assuming that the dam is a trapezoid of base 2000 m and upper edge 3000 m, inclined at an angle of 55° to the horizontal (Figure 17).

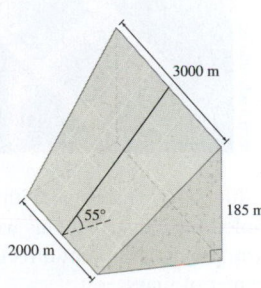

FIGURE 16 Three Gorges Dam on the Yangtze River.

FIGURE 17

24. A square plate of side 3 m is submerged in water at an incline of 30° with the horizontal. Calculate the fluid force on one side of the plate if the top edge of the plate lies at a depth of 6 m.

25. The trough in Figure 18 is filled with corn syrup, whose weight density is 90 lb/ft³. Calculate the force on the front side of the trough.

26. Calculate the fluid pressure on one of the slanted sides of the trough in Figure 18 when it is filled with corn syrup as in Exercise 25.

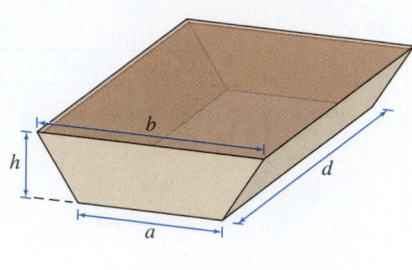

FIGURE 18

Further Insights and Challenges

27. The end of the trough in Figure 19 is an equilateral triangle of side 3. Assume that the trough is filled with water to height H. Calculate the fluid force on each side of the trough as a function of H and the length l of the trough.

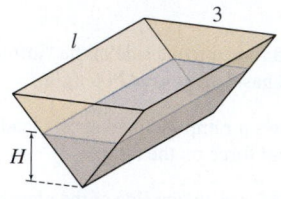

FIGURE 19

28. A rectangular plate of side ℓ is submerged vertically in a fluid of density w, with its top edge at depth h. Show that if the depth is increased by an amount Δh, then the force on a side of the plate increases by $wA\Delta h$, where A is the area of the plate.

29. Prove that the force on the side of a rectangular plate of area A submerged vertically in a fluid is equal to $p_0 A$, where p_0 is the fluid pressure at the center point of the rectangle.

30. ✒️ If the density of a fluid varies with depth, then the pressure at depth y is $p(y)$ (which need not equal wy as in the case of constant density). Use Riemann sums to argue that the total force F on the flat side of a submerged object submerged vertically is $F = \int_a^b f(y)p(y)\,dy$, where $f(y)$ is the width of the side at depth y.

8.4 Center of Mass

Every object has a balance point called the *center of mass* (Figure 1). When a rigid object such as a hammer is tossed in the air, it may rotate in a complicated fashion, but its center of mass follows the same simple parabolic trajectory as a stone tossed in the air. In this section, we use integration to compute the center of mass of a thin plate (also called a **lamina**) of constant mass density ρ.

Consider a seesaw with a linebacker for the Green Bay Packers on one end and a ballerina from the New York City Ballet on the other end, as in Figure 2. Clearly, the balance point $\bar{x}$ must be closer to the linebacker than the ballerina.

FIGURE 1 This acrobat with Cirque du Soleil must distribute his weight so that his arm provides support directly below his center of mass.

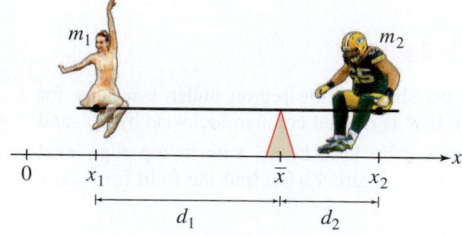

FIGURE 2

As Archimedes realized over 2000 years ago, if two objects balance on opposite sides of a lever, then the mass of each object times its distance to the balance point must be equal. In this case, $m_1 d_1 = m_2 d_2$. If we use the coordinates given on the line, this becomes

$$m_1(\bar{x} - x_1) = m_2(x_2 - \bar{x})$$

We solve this for $\bar{x}$:

$$m_1\bar{x} - m_1x_1 = m_2x_2 - m_2\bar{x}$$

$$m_1\bar{x} + m_2\bar{x} = m_1x_1 + m_2x_2$$

$$(m_1 + m_2)\bar{x} = m_1x_1 + m_2x_2$$

$$\bar{x} = \frac{m_1x_1 + m_2x_2}{m_1 + m_2}$$

In other words, the balance point, called the **center of mass (COM)**, occurs at the coordinate given by taking the sum of the product of each mass times its position on the line and then dividing this sum by the total mass. These quantities m_1x_1 and m_2x_2 are called **moments (with respect to the origin)**.

This idea generalizes to finitely many particles on a line. If we have particles of mass $m_1, m_2, \ldots, m_n$ at positions $x_1, x_2, \ldots, x_n$, respectively, then the COM is located at

$$\bar{x} = \frac{m_1x_1 + m_2x_2 + \ldots + m_nx_n}{m_1 + m_2 + \ldots + m_n}$$

EXAMPLE 1 On the line, we have a mass of 3 at -2, a mass of 2 at 1, and a mass of 15 at 6 (Figure 3). What is the location of the center of mass?

Solution We have

$$\bar{x} = \frac{(3)(-2) + (2)(1) + (15)(6)}{3 + 2 + 15} = 4.35$$

It makes sense that the COM is closest to the mass of 15 at 6 because most of the mass in the system is concentrated at $x = 6$. ∎

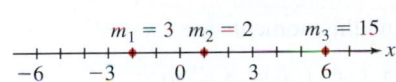

FIGURE 3

Now, how do we extend this idea to particles in the xy-plane (Figure 4)? Suppose we have a collection of particles with masses $m_1, m_2, \ldots, m_n$ at positions (x_1, y_1), $(x_2, y_2), \ldots, (x_n, y_n)$, respectively. The COM is a point in the plane with coordinates $(\bar{x}, \bar{y})$. These coordinates are determined in the same manner as the COM of the particles on the line. That is,

$$\bar{x} = \frac{m_1x_1 + m_2x_2 + \cdots + m_nx_n}{m_1 + m_2 + \cdots + m_n}, \quad \text{and} \quad \bar{y} = \frac{m_1y_1 + m_2y_2 + \cdots + m_ny_n}{m_1 + m_2 + \cdots + m_n}$$

Like the particles on the line, the terms in the numerators in these COM equations are referred to as moments. For a particle of mass m located at the point (x, y), the **moment with respect to the x-axis**, M_x, and the **moment with respect to the y-axis**, M_y, are given by

$$M_x = my \qquad \text{(mass times directed distance to x-axis)}$$

$$M_y = mx \qquad \text{(mass times directed distance to y-axis)}$$

Moments are additive: The moment of a system of n particles with coordinates (x_i, y_i) and mass m_i (Figure 5) is the sum

$$M_x = m_1y_1 + m_2y_2 + \cdots + m_ny_n = \sum_{i=1}^{n} m_iy_i$$

$$M_y = m_1x_1 + m_2x_2 + \cdots + m_nx_n = \sum_{i=1}^{n} m_ix_i$$

With this notation, the center of mass $(\bar{x}, \bar{y})$ is given by

$$\bar{x} = \frac{M_y}{M}, \qquad \bar{y} = \frac{M_x}{M}$$

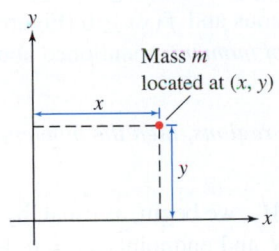

FIGURE 4

Mass m located at (x, y)

The directed distance from a point to an axis is the actual distance if the point is on the positive side of the axis, and is the negative of the distance if the point is on the negative side.

CAUTION The notation is potentially confusing: M_x is defined in terms of the directed distance to the x-axis (given by y-coordinates), and M_y is defined in terms of the directed distance to the y-axis (given by x-coordinates).

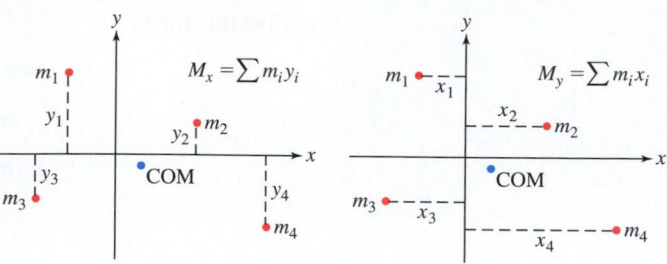

FIGURE 5

where $M = m_1 + m_2 + \cdots + m_n$ is the total mass of the system. That is, the x-coordinate of the COM of the system is the moment with respect to the y-axis divided by the total mass, and the y-coordinate is the moment with respect to the x-axis divided by the total mass.

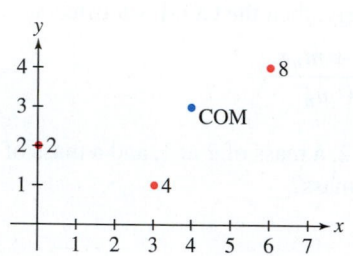

FIGURE 6

EXAMPLE 2 Find the COM of the system of three particles in Figure 6, having masses 2, 4, and 8 at locations $(0, 2)$, $(3, 1)$, and $(6, 4)$, respectively.

Solution The total mass is $M = 2 + 4 + 8 = 14$ and the moments are

$$M_x = m_1 y_1 + m_2 y_2 + m_3 y_3 = 2 \cdot 2 + 4 \cdot 1 + 8 \cdot 4 = 40$$

$$M_y = m_1 x_1 + m_2 x_2 + m_3 x_3 = 2 \cdot 0 + 4 \cdot 3 + 8 \cdot 6 = 60$$

Therefore, $\bar{x} = \frac{60}{14} = \frac{30}{7}$ and $\bar{y} = \frac{40}{14} = \frac{20}{7}$. The COM is $\left(\frac{30}{7}, \frac{20}{7}\right)$. ∎

Laminas (Thin Plates)

In this section, we restrict our attention to thin plates of constant mass density (also called "uniform density"). We use multiple integration in Section 15.5 to compute the COM when density is not constant.

Now consider a lamina (thin plate) of constant mass density ρ occupying the region under the graph of f over an interval $[a, b]$, where f is continuous and $f(x) \geq 0$ (Figure 7). In our calculations, we will use the principle of *additivity of moments* mentioned above for point masses:

If a region is decomposed into smaller, nonoverlapping regions, then the moment of the region is the sum of the moments of the smaller regions.

To compute the moment with respect to the y-axis, M_y, we begin, as usual, by dividing $[a, b]$ into N subintervals of width $\Delta x = (b - a)/N$ and endpoints $x_j = a + j\Delta x$. This divides the lamina into N vertical strips (Figure 8). If Δx is small, the jth strip is nearly rectangular of area $f(x_j)\Delta x$ and mass $\rho f(x_j)\Delta x$. Since all points in the strip lie at approximately the same distance x_j from the y-axis, the moment $M_{y,j}$ of the jth strip is approximately

$$M_{y,j} \approx (\text{mass}) \times (\text{directed distance to } y\text{-axis}) = (\rho f(x_j)\Delta x)x_j$$

By additivity of moments,

$$M_y = \sum_{j=1}^{N} M_{y,j} \approx \rho \sum_{j=1}^{N} x_j f(x_j)\Delta x$$

This is a Riemann sum whose value approaches $\rho \displaystyle\int_a^b x f(x)\, dx$ as $N \to \infty$, and thus,

$$M_y = \rho \int_a^b x f(x)\, dx$$

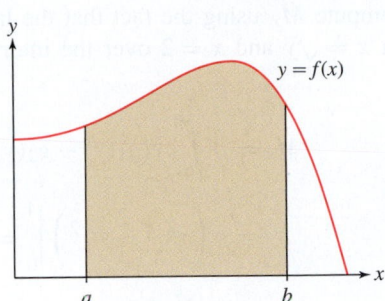

FIGURE 7 Lamina occupying the region under the graph of f over $[a, b]$.

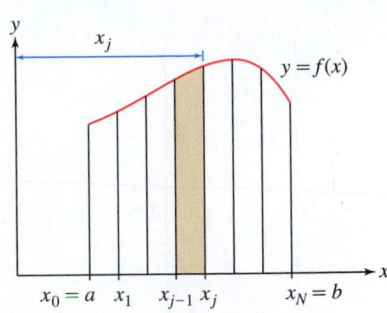

FIGURE 8 The shaded strip is nearly rectangular with area approximately $f(x_j)\Delta x$.

More generally, if the lamina occupies a vertically simple region between the graphs of two functions f_1 and f_2 over $[a, b]$, where $f_1(x) \geq f_2(x)$, then

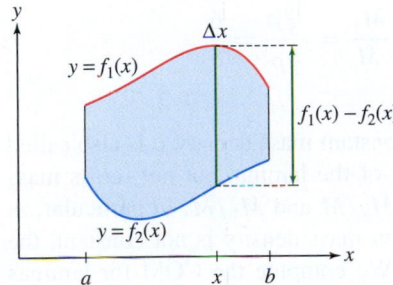

FIGURE 9

$$M_y = \rho \int_a^b x(\text{length of vertical cut})\, dx = \rho \int_a^b x\big(f_1(x) - f_2(x)\big)\, dx \qquad \boxed{1}$$

Think of the lamina as made up of vertical strips of length $f_1(x) - f_2(x)$ at distance x from the y-axis (Figure 9).

In a similar manner, we can compute the x-moment by dividing the lamina into horizontal strips. This computation is straightforward to set up when the lamina occupies a horizontally simple region between two curves $x = g_1(y)$ and $x = g_2(y)$ with $g_1(y) \geq g_2(y)$ over an interval $[c, d]$ along the y-axis (Figure 10):

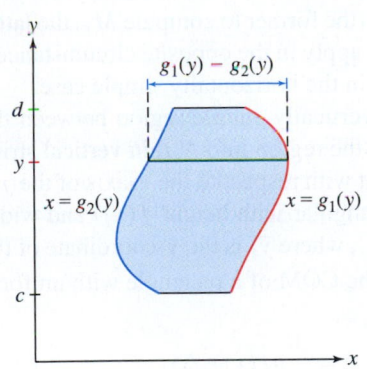

FIGURE 10

$$M_x = \rho \int_c^d y(\text{length of horizontal cut})\, dy = \rho \int_c^d y\big(g_1(y) - g_2(y)\big)\, dy \qquad \boxed{2}$$

The total mass of the lamina is $M = \rho A$, where A is the area of the lamina:

$$\text{For a vertically simple region:} \quad M = \rho A = \rho \int_a^b \big(f_1(x) - f_2(x)\big)\, dx$$

$$\text{For a horizontally simple region:} \quad M = \rho A = \rho \int_c^d \big(g_1(y) - g_2(y)\big)\, dy$$

The center-of-mass coordinates are the moments divided by the total mass:

$$\bar{x} = \frac{M_y}{M}, \qquad \bar{y} = \frac{M_x}{M}$$

The lamina will balance at the point $(\bar{x}, \bar{y})$ as in Figure 11.

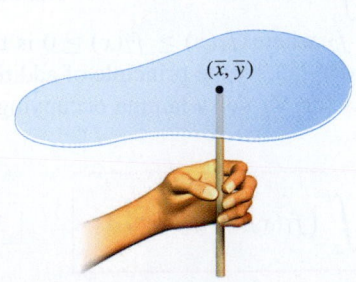

DF **FIGURE 11** A lamina balances at its center of mass $(\bar{x}, \bar{y})$.

EXAMPLE 3 Find the moments and COM of the lamina of uniform density ρ occupying the region underneath the graph of $f(x) = x^2$ and above the x-axis for $0 \leq x \leq 2$.

Solution The region is both vertically simple and horizontally simple. First, compute M_y using the fact that the region is vertically simple. By Eq. (1):

$$M_y = \rho \int_0^2 x f(x)\, dx = \rho \int_0^2 x(x^2)\, dx = \rho \left.\frac{x^4}{4}\right|_0^2 = 4\rho$$

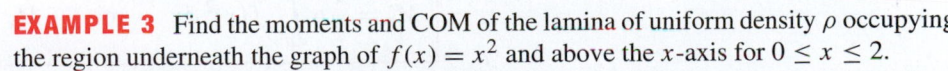

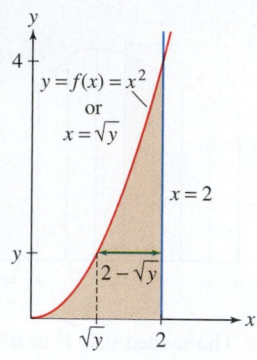

FIGURE 12 Lamina occupying the region under the graph of $f(x) = x^2$ over $[0, 2]$.

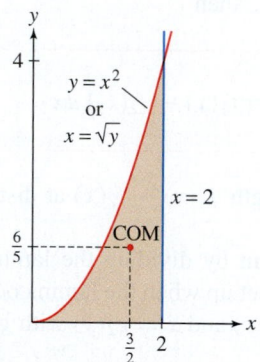

FIGURE 13 Center of mass of the lamina.

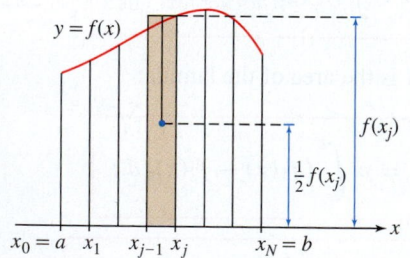

DF **FIGURE 14** Because the shaded strip is nearly rectangular, its COM has an approximate height of $\frac{1}{2}f(x_j)$.

Then compute M_x using the fact that the lamina occupies the vertically simple region between $x = \sqrt{y}$ and $x = 2$ over the interval $[0, 4]$ along the y-axis (Figure 12). By Eq. (2),

$$M_x = \rho \int_0^4 y\big(g_1(y) - g_2(y)\big)\, dy = \rho \int_0^4 y(2 - \sqrt{y})\, dy$$

$$= \rho \left(y^2 - \frac{2}{5}y^{5/2} \right) \Big|_0^4 = \rho \left(16 - \frac{2}{5} \cdot 32 \right) = \frac{16}{5}\rho$$

The plate has area $A = \int_0^2 x^2\, dx = \frac{8}{3}$ and total mass $M = \frac{8}{3}\rho$. Therefore as shown in Figure 13,

$$\bar{x} = \frac{M_y}{M} = \frac{4\rho}{\frac{8}{3}\rho} = \frac{3}{2}, \qquad \bar{y} = \frac{M_x}{M} = \frac{\frac{16}{5}\rho}{\frac{8}{3}\rho} = \frac{6}{5}$$

■

CONCEPTUAL INSIGHT   The COM of a lamina of constant mass density ρ is also called the **centroid**. The centroid depends on the shape of the lamina, but not on its mass density because the factor ρ cancels in the ratios M_x/M and M_y/M. In particular, *in calculating the centroid, we can take $\rho = 1$.* When mass density is not constant, the COM depends on both shape and mass density. We compute the COM for laminas with nonconstant density in Section 15.5.

So far, our process for computing the COM requires the lamina to occupy a region that is both vertically simple and horizontally simple (the former to compute M_y, the latter to compute M_x). Fortunately, there are formulas that apply in the opposite circumstances: for computing M_x in the vertically simple case, M_y in the horizontally simple case.

Let us consider M_x for a lamina occupying a vertically simple region between the x-axis and the graph of $y = f(x)$. As before, divide the region into N thin vertical strips of width Δx (see Figure 14). Let $M_{x,j}$ be the moment with respect to the x-axis of the jth strip and let m_j be its mass. The strip is nearly rectangular with height $f(x_j)$ and width Δx, so $m_j \approx \rho f(x_j)\,\Delta x$. Furthermore, $M_{x,j} = \overline{y_j} m_j$, where $\overline{y_j}$ is the y-coordinate of the COM of the strip. However, $\overline{y_j} \approx \frac{1}{2}f(x_j)$ because the COM of a rectangle with uniform mass density is located at its center. Thus,

$$M_{x,j} = m_j \overline{y_j} \approx \rho f(x_j)\Delta x \cdot \frac{1}{2}f(x_j) = \frac{1}{2}\rho f(x_j)^2 \Delta x$$

$$M_x = \sum_{j=1}^N M_{x,j} \approx \frac{1}{2}\rho \sum_{j=1}^N f(x_j)^2 \Delta x$$

This is a Riemann sum whose value approaches $\frac{1}{2}\rho \int_a^b f(x)^2\, dx$ as $N \to \infty$. The case of a region *between* the graphs of functions f_1 and f_2 where $f_1(x) \geq f_2(x) \geq 0$ is the difference of the moments corresponding to $f_1(x)$ and $f_2(x)$, by the principle of additivity of moments, so we obtain the following formulas for M_x for a lamina occupying a vertically simple region:

$$M_x = \frac{1}{2}\rho \int_a^b f(x)^2\, dx \qquad \text{or} \qquad M_x = \frac{1}{2}\rho \int_a^b \big(f_1(x)^2 - f_2(x)^2\big)\, dx \qquad \boxed{3}$$

The same idea holds for determining M_y in the horizontally simple case:

$$M_y = \frac{1}{2}\rho \int_c^d g(y)^2\, dy \qquad \text{or} \qquad M_y = \frac{1}{2}\rho \int_c^d \big(g_1(y)^2 - g_2(y)^2\big)\, dy \qquad \boxed{4}$$

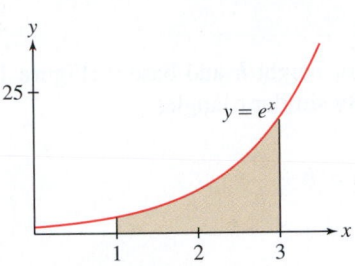

FIGURE 15 Region under the curve $y = e^x$ between $x = 1$ and $x = 3$.

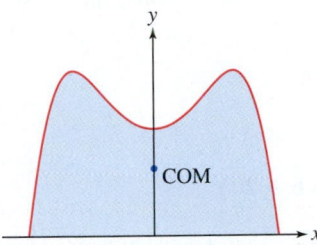

FIGURE 16 The COM of a symmetric plate lies on the axis of symmetry.

◀ **REMINDER** A region is symmetric with respect to a line if reflection across the line sends each point of the region to another point of the region.

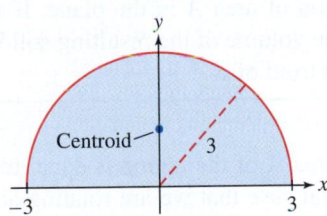

FIGURE 17 The half-disk under the graph of $f(x) = \sqrt{9 - x^2}$.

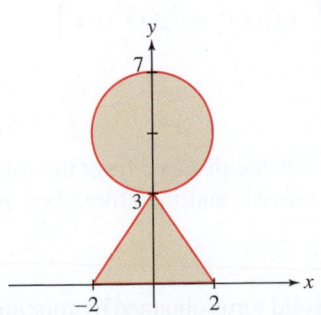

FIGURE 18 The moment of region R is the sum of the moments of the triangle and circle.

EXAMPLE 4 Find the centroid of the shaded region in Figure 15.

Solution The centroid does not depend on ρ, so we may set $\rho = 1$ and apply Eqs. (1) and (3) with $f(x) = e^x$:

$$M_x = \frac{1}{2} \int_1^3 f(x)^2 \, dx = \frac{1}{2} \int_1^3 e^{2x} \, dx = \frac{1}{4} e^{2x} \Big|_1^3 = \frac{e^6 - e^2}{4}$$

Using Integration by Parts, we get

$$M_y = \int_1^3 x f(x) \, dx = \int_1^3 x e^x \, dx = (x - 1)e^x \Big|_1^3 = 2e^3$$

The total mass is $M = \int_1^3 e^x \, dx = (e^3 - e)$. The centroid has coordinates

$$\bar{x} = \frac{M_y}{M} = \frac{2e^3}{e^3 - e} \approx 2.313, \qquad \bar{y} = \frac{M_x}{M} = \frac{e^6 - e^2}{4(e^3 - e)} \approx 5.701 \qquad ■$$

The symmetry properties of an object give information about its centroid (Figure 16). For instance, the centroid of a square or circular plate is located at its center. Here is a precise formulation (see Exercise 49):

> **THEOREM 1 Symmetry Principle** If a lamina is symmetric with respect to a line, then its centroid lies on that line.

EXAMPLE 5 Using Symmetry Find the centroid of the half-disk of radius 3, between the x-axis and the graph of $f(x) = \sqrt{9 - x^2}$, as shown in Figure 17.

Solution Symmetry cuts our work in half. The half-disk is symmetric with respect to the y-axis, so the centroid lies on the y-axis, and hence $\bar{x} = 0$. It remains to calculate M_x and $\bar{y}$. By Eq. (3) with $\rho = 1$,

$$M_x = \frac{1}{2} \int_{-3}^3 f(x)^2 \, dx = \frac{1}{2} \int_{-3}^3 (9 - x^2) \, dx = \frac{1}{2} \left(9x - \frac{1}{3} x^3 \right) \Big|_{-3}^3 = 9 - (-9) = 18$$

The half-disk has area (and mass) equal to $A = \frac{1}{2}\pi(3^2) = 9\pi/2$, so

$$\bar{y} = \frac{M_x}{M} = \frac{18}{9\pi/2} = \frac{4}{\pi} \approx 1.27 \qquad ■$$

EXAMPLE 6 Using Additivity and Symmetry Find the centroid of the region R in Figure 18.

Solution We set $\rho = 1$ because we are computing a centroid. The region R is symmetric with respect to the y-axis, and therefore, $\bar{x} = 0$. To find $\bar{y}$, we compute the moment M_x.

Step 1. Use additivity of moments.
Let M_x^{triangle} and M_x^{circle} be the x-moments of the triangle and the circle. Then

$$M_x = M_x^{\text{triangle}} + M_x^{\text{circle}}$$

Step 2. Moment of the circle.
To save work, we use the fact that the centroid of the circle is located at the center $(0, 5)$ by symmetry. Thus, $\bar{y}^{\text{circle}} = 5$ and we can solve for the moment:

$$\bar{y}^{\text{circle}} = \frac{M_x^{\text{circle}}}{M^{\text{circle}}} = \frac{M_x^{\text{circle}}}{4\pi} = 5 \quad \Rightarrow \quad M_x^{\text{circle}} = 20\pi$$

Here, the mass of the circle is its area $M^{\text{circle}} = \pi(2^2) = 4\pi$ (since $\rho = 1$).

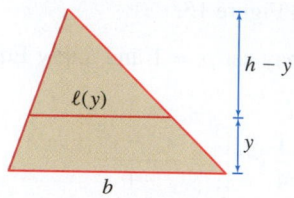

FIGURE 19 By similar triangles,
$\dfrac{\ell(y)}{h-y} = \dfrac{b}{h}$.

Step 3. **Moment of a triangle.**

Let's compute M_x^{triangle} for an arbitrary triangle of height h and base b (Figure 19). Let $\ell(y)$ be the width of the triangle at height y. By similar triangles,

$$\frac{\ell(y)}{h-y} = \frac{b}{h} \quad \Rightarrow \quad \ell(y) = b - \frac{b}{h}y$$

By Eq. (2),

$$M_x^{\text{triangle}} = \int_0^h y\,\ell(y)\,dy = \int_0^h y\left(b - \frac{b}{h}y\right) dy = \left(\frac{by^2}{2} - \frac{by^3}{3h}\right)\bigg|_0^h = \frac{bh^2}{6}$$

In our case, $b = 4$, $h = 3$, and $M_x^{\text{triangle}} = \dfrac{4 \cdot 3^2}{6} = 6$.

Step 4. **Computation of $\bar{y}$.**

$$M_x = M_x^{\text{triangle}} + M_x^{\text{circle}} = 6 + 20\pi$$

The triangle has mass $\frac{1}{2} \cdot 4 \cdot 3 = 6$, and the circle has mass 4π, so R has mass $M = 6 + 4\pi$ and

$$\bar{y} = \frac{M_x}{M} = \frac{6 + 20\pi}{6 + 4\pi} \approx 3.71$$

We end this section with the Theorem of Pappus, attributed to Pappus of Alexandria, a mathematician of the fourth century BCE:

THEOREM 2 Theorem of Pappus Let R be a region of area A in the plane. If we rotate R about an axis that is disjoint from R, then the volume of the resulting solid is the product of A with the distance traveled by the centroid of R.

Proof Since we assume a uniform density of 1, the area A of the region is equal to the mass M. We will prove the theorem only in the special case that we are rotating about the x-axis and that we have a region bounded by $y = f_1(x)$ and $y = f_2(x)$ for $a \le x \le b$ and $f_1(x) \ge f_2(x) > 0$. In this case, we know that the volume is given by

$$V = \pi \int_a^b (f_1(x)^2 - f_2(x)^2)\,dx = 2\pi \left(\frac{1}{2}\int_a^b (f_1(x)^2 - f_2(x)^2)\,dx\right)$$

$$= 2\pi M_x = A \cdot 2\pi \frac{M_x}{A} = A \cdot 2\pi \bar{y}$$

Thus $V = A \cdot 2\pi\bar{y}$. This is the desired result because $\bar{y}$ is the distance from the rotation axis to the centroid (since we are rotating about the x-axis), and therefore, $2\pi\bar{y}$ is the distance traveled by the centroid. ∎

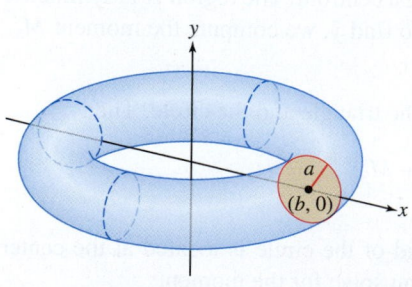

FIGURE 20 Rotating a disk about the y-axis to obtain a solid torus.

EXAMPLE 7 Find the formula for the volume of the solid torus obtained by rotating the disk of radius a centered at $(b, 0)$ about the y-axis, where $a < b$, as in Figure 20.

Solution The centroid of the disk occurs at its center. So, $(\bar{x}, \bar{y}) = (b, 0)$. The Theorem of Pappus then says that

$$V = A \cdot 2\pi\bar{x} = \pi a^2 2\pi b = 2\pi^2 a^2 b$$

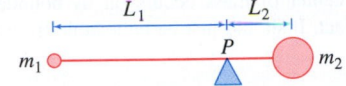

FIGURE 21 Archimedes's Law of the Lever:
$$m_1 L_1 = m_2 L_2$$

HISTORICAL PERSPECTIVE

Archimedes
(287–212 BCE)

We take it for granted that physical laws are best expressed as mathematical relationships. Think of $F = ma$ or the universal law of gravitation. However, the fundamental insight that mathematics could be used to formulate laws of nature (and not just for counting or measuring) developed gradually, beginning with the philosophers of ancient Greece and culminating some 2000 years later in the discoveries of Galileo and Newton. Archimedes was one of the first scientists (perhaps *the* first) to formulate a precise physical law. Concerning the principle of the lever, Archimedes wrote, "Commensurable magnitudes balance at distances reciprocally proportional to their weight." In other words, if weights of mass m_1 and m_2 are placed on a weightless lever at distances L_1 and L_2 from the fulcrum P (Figure 21), then the lever will balance if $m_1/m_2 = L_2/L_1$, or

$$m_1 L_1 = m_2 L_2$$

In our terminology, what Archimedes had discovered was the center of mass P of the system of weights (see Exercises 47 and 48).

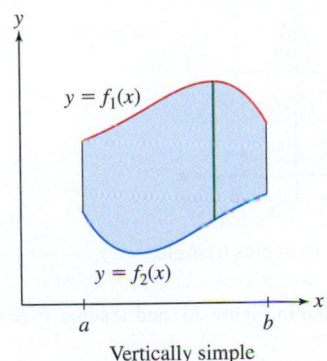

Vertically simple

Horizontally simple

FIGURE 22

8.4 SUMMARY

- The *moments* of a system of particles of mass m_j located at (x_j, y_j) are
$$M_x = m_1 y_1 + \cdots + m_n y_n, \qquad M_y = m_1 x_1 + \cdots + m_n x_n$$
The *center of mass* (COM) has coordinates
$$\bar{x} = \frac{M_y}{M} \qquad \text{and} \qquad \bar{y} = \frac{M_x}{M}$$
where $M = m_1 + \cdots + m_n$.
- Lamina (thin plate) of constant mass density ρ (Figure 22):

    – vertically simple region.   Mass: $M = \rho \int_a^b \big(f_1(x) - f_2(x)\big)\,dx$

    Moments: $M_x = \dfrac{1}{2}\rho \int_a^b \big(f_1(x)^2 - f_2(x)^2\big)\,dx, \quad M_y = \rho \int_a^b x\big(f_1(x) - f_2(x)\big)\,dx$

    – horizontally simple region.   Mass: $M = \rho \int_c^d \big(g_1(y) - g_2(y)\big)\,dy$

    Moments: $M_x = \rho \int_c^d y\big(g_1(y) - g_2(y)\big)\,dy, \quad M_y = \dfrac{1}{2}\rho \int_c^d \big(g_1(y)^2 - g_2(y)^2\big)\,dy$

    – The coordinates of the center of mass (also called the *centroid*) are
$$\bar{x} = \frac{M_y}{M}, \qquad \bar{y} = \frac{M_x}{M}$$

- Additivity: If a region is decomposed into smaller nonoverlapping regions, then the moment of the region is the sum of the moments of the smaller regions.
- Symmetry Principle: If a lamina of constant mass density is symmetric with respect to a given line, then the center of mass (centroid) lies on that line.
- The Theorem of Pappus: If a lamina is rotated about a disjoint axis, then the volume of the resulting solid of revolution is the area of the lamina times the distance traveled by the centroid.

8.4 EXERCISES

Preliminary Questions

1. What are the x- and y-moments of a lamina whose center of mass is located at the origin?

2. A thin plate has mass 3. What is the x-moment of the plate if its center of mass has coordinates $(2, 7)$?

3. The center of mass of a lamina of total mass 5 has coordinates $(2, 1)$. What are the lamina's x- and y-moments?

4. Explain how the Symmetry Principle is used to conclude that the centroid of a rectangle is the center of the rectangle.

5. Give an example of a plate such that its center of mass does not occur at any point on the plate.

6. Draw a plate such that its center of mass occurs on its boundary. (You do not need to verify this fact. It should just be believable from the drawing.)

Exercises

1. On a line, there are particles located at $-3, -1, 1, 2$, and 5. Their masses are 8, 2, 3, 2, and 1, respectively.

(a) What is the center of mass of the system?

(b) Keeping the other four masses the same, what would the mass at 5 need to be in order to have the center of mass be 0?

2. On a line, there are particles located at 1, 2, 3, 4, and 5. Their masses are 1, 2, 3, 4, and 5, respectively.

(a) What is the center of mass of the system?

(b) If we add a particle of mass 6 at 6, what is the center of mass?

(c) If we add particles of mass j at j for $j = 6$ to n, what is the center of mass? *Hint:* Power sums that were introduced in Section 5.1 will be helpful here.

3. Four particles are located at points $(1, 1), (1, 2), (4, 0)$, and $(3, 1)$.

(a) Find the moments M_x and M_y and the center of mass of the system, assuming that the particles have equal mass m.

(b) Find the center of mass of the system, assuming the particles have masses 3, 2, 5, and 7, respectively.

4. Find the center of mass for the system of particles of masses 4, 2, 5, and 1 located at $(1, 2), (-3, 2), (2, -1)$, and $(4, 0)$.

5. Point masses of equal size are placed at the vertices of the triangle with coordinates $(a, 0), (b, 0)$, and $(0, c)$. Show that the center of mass of the system of masses has coordinates $\left(\frac{1}{3}(a + b), \frac{1}{3}c\right)$.

6. Point masses of mass m_1, m_2, and m_3 are placed at the points $(-1, 0), (3, 0)$, and $(0, 4)$.

(a) Suppose that $m_1 = 6$. Find m_2 such that the center of mass lies on the y-axis.

(b) Suppose that $m_1 = 6$ and $m_2 = 4$. Find the value of m_3 such that $\overline{y} = 2$.

7. Sketch the lamina S of constant density $\rho = 3$ g/cm^2 occupying the region beneath the graph of $y = x^2$ for $0 \leq x \leq 3$.

(a) Use Eqs. (1) and (2) to compute M_x and M_y.

(b) Find the area and the center of mass of S.

8. Use Eqs. (1) and (3) to find the moments and center of mass of the lamina S of constant density $\rho = 2$ g/cm^2 occupying the region between $y = x^2$ and $y = 9x$ over $[0, 3]$. Sketch S, indicating the location of the center of mass.

9. Find the moments and center of mass of the lamina of uniform density ρ occupying the region underneath $y = x^3$ for $0 \leq x \leq 2$.

10. Calculate M_x (assuming $\rho = 1$) for the region underneath the graph of $y = 1 - x^2$ for $0 \leq x \leq 1$ in two ways, first using Eq. (2) and then using Eq. (3).

11. Let T be the triangular lamina in Figure 23 and assume $P = 6$.

(a) Show that the horizontal cut at height y has length $4 - \frac{2}{3}y$ and use Eq. (2) to compute M_x (with $\rho = 1$).

(b) Use the Symmetry Principle to show that $M_y = 0$ and find the center of mass.

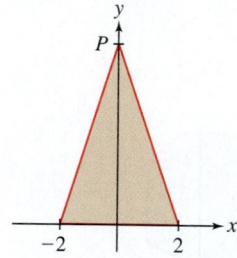

FIGURE 23 Isosceles triangle.

12. Let T be the triangular lamina in Figure 23, and assume $P = 8$ and $\rho = 1$. Find the center of mass.

In Exercises 13–21, find the centroid of the region lying underneath the graph of the function over the given interval.

13. $f(x) = 4x,$ $[0, 1]$

14. $f(x) = 6 - 2x,$ $[0, 3]$

15. $f(x) = \sqrt{x},$ $[1, 4]$

16. $f(x) = x^3,$ $[0, 1]$

17. $f(x) = 9 - x^2,$ $[0, 3]$

18. $f(x) = (1 + x^2)^{-1/2},$ $[0, 3]$

19. $f(x) = e^{-x},$ $[0, 4]$

20. $f(x) = \ln x,$ $[1, 2]$

21. $f(x) = \sin x,$ $[0, \pi]$

22. Calculate the moments and center of mass of the lamina occupying the region between the curves $y = x$ and $y = x^2$ for $0 \leq x \leq 1$.

23. Sketch the region between $y = x + 4$ and $y = 2 - x$ for $0 \leq x \leq 2$. Using symmetry, explain why the centroid of the region lies on the line $y = 3$. Verify this by computing the moments and the centroid.

In Exercises 24–29, find the centroid of the region lying between the graphs of the functions over the given interval.

24. $y = x,$ $y = \sqrt{x},$ $[0, 1]$

25. $y = x^2,$ $y = \sqrt{x},$ $[0, 1]$

26. $y = x^{-1},$ $y = 2 - x,$ $[1, 2]$

27. $y = e^x,$ $y = 1,$ $[0, 1]$

28. $y = \ln x,$ $y = x - 1,$ $[1, 3]$

29. $y = \sin x,$ $y = \cos x,$ $[0, \pi/4]$

30. Sketch the region enclosed by $y = x + 1$ and $y = (x - 1)^2$ and find its centroid.

31. Sketch the region enclosed by $y = 0, y = (x + 1)^3$, and $y = (1 - x)^3$, and find its centroid.

In Exercises 32–36, find the centroid of the region.

32. Top half of the ellipse $\left(\frac{x}{2}\right)^2 + \left(\frac{y}{4}\right)^2 = 1$

33. Top half of the ellipse $\left(\frac{x}{a}\right)^2 + \left(\frac{y}{b}\right)^2 = 1$ for arbitrary $a, b > 0$

34. Semicircle of radius r with center at the origin

35. Quarter of the unit circle lying in the first quadrant

36. Region between $y = x(a - x)$ and the x-axis for $a > 0$

37. Find the centroid of the shaded region of the semicircle of radius r in Figure 24. What is the centroid when $r = 1$ and $h = \frac{1}{2}$? *Hint:* Use geometry rather than integration to show that the *area* of the region is $r^2 \sin^{-1}(\sqrt{1 - h^2/r^2}) - h\sqrt{r^2 - h^2}$.

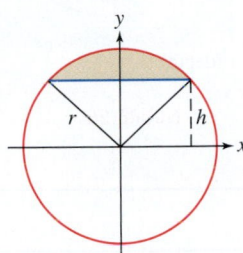

FIGURE 24

38. Sketch the region between $y = x^n$ and $y = x^m$ for $0 \le x \le 1$, where $m > n \ge 0$, and find the COM of the region. Find a pair (n, m) such that the COM lies outside the region.

39. Find the formula for the volume of a right circular cone of height H and radius R using the Theorem of Pappus as applied to the triangle bounded by the x-axis, the y-axis, and the line $y = \frac{-H}{R}x + H$, rotated about the y-axis.

40. Use the Theorem of Pappus to find the centroid of the half-disk bounded by $y = \sqrt{R^2 - x^2}$ and the x-axis.

In Exercises 41–43, use the additivity of moments to find the COM of the region.

41. Isosceles triangle of height 2 on top of a rectangle of base 4 and height 3 (Figure 25)

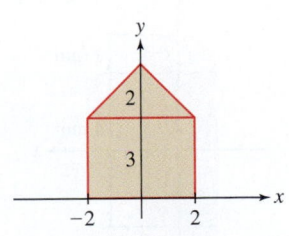

FIGURE 25

42. An ice cream cone consisting of a semicircle on top of an equilateral triangle of side 6 (Figure 26)

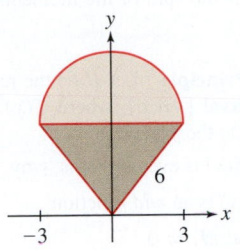

FIGURE 26

43. Three-quarters of the unit circle (remove the part in the fourth quadrant)

44. Let S be the lamina of mass density $\rho = 1$ obtained by removing a circle of radius r from the circle of radius $2r$ shown in Figure 27. Let M_x^S and M_y^S denote the moments of S. Similarly, let M_y^{big} and M_y^{small} be the y-moments of the larger and smaller circles.

(a) Use the Symmetry Principle to show that $M_x^S = 0$.

(b) Show that $M_y^S = M_y^{\text{big}} - M_y^{\text{small}}$ using the additivity of moments.

(c) Find M_y^{big} and M_y^{small} using the fact that the COM of a circle is its center. Then compute M_y^S using (b).

(d) Determine the COM of S.

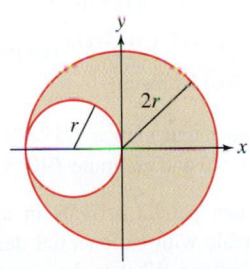

FIGURE 27

45. Find the COM of the laminas in Figure 28 obtained by removing squares of side 2 from a square of side 8.

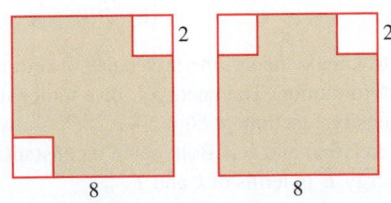

FIGURE 28

Further Insights and Challenges

46. A **median** of a triangle is a segment joining a vertex to the midpoint of the opposite side. Show that the centroid of a triangle lies on each of its medians, at a distance two-thirds down from the vertex. Then use this fact to prove that the three medians intersect at a single point. *Hint:* Simplify the calculation by assuming that one vertex lies at the origin and another on the x-axis.

47. Let P be the COM of a system of two weights with masses m_1 and m_2 separated by a distance d. Prove Archimedes's Law of the (weightless) Lever: P is the point on a line between the two weights such that $m_1 L_1 = m_2 L_2$, where L_j is the distance from mass j to P.

48. Find the COM of a system of two weights of masses m_1 and m_2 connected by a lever of length d whose mass density ρ is uniform. *Hint:* The moment of the system is the sum of the moments of the weights and the lever.

49. 📝 **Symmetry Principle**   Let R be the region under the graph of $y = f(x)$ over the interval $[-a, a]$, where $f(x) \geq 0$. Assume that R is symmetric with respect to the y-axis.

(a) Explain why $y = f(x)$ is even—that is, why $f(x) = f(-x)$.

(b) Show that $y = xf(x)$ is an *odd* function.

(c) Use (b) to prove that $M_y = 0$.

(d) Prove that the COM of R lies on the y-axis (a similar argument applies to symmetry with respect to the x-axis).

50. Prove directly that Eqs. (2) and (3) are equivalent in the following situation. Let f be a positive decreasing function on $[0, b]$ such that $f(b) = 0$. Set $d = f(0)$ and $g(y) = f^{-1}(y)$. Show that

$$\frac{1}{2} \int_0^b f(x)^2 \, dx = \int_0^d yg(y) \, dy$$

Hint: First apply the substitution $y = f(x)$ to the integral on the left and observe that $dx = g'(y) \, dy$. Then apply Integration by Parts.

51. Let R be a lamina of uniform density submerged in a fluid of density w (Figure 29). Prove the following law: The fluid force on one side of R is equal to the area of R times the fluid pressure on the centroid. *Hint:* Let $g(y)$ be the horizontal width of R at depth y. Express both the fluid pressure [Eq. (2) in Section 8.3] and y-coordinate of the centroid in terms of $g(y)$.

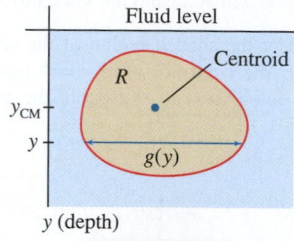

FIGURE 29

CHAPTER REVIEW EXERCISES

1. Compute $p(X \leq 1)$, where X is a continuous random variable with probability density $p(x) = \dfrac{1}{\pi(x^2 + 1)}$.

2. Show that $p(x) = \frac{1}{4}e^{-x/2} + \frac{1}{6}e^{-x/3}$ is a probability density over the domain $[0, \infty)$ and find its mean.

3. Find a constant C such that $p(x) = Cx^3 e^{-x^2}$ is a probability density over the domain $[0, \infty)$ and compute $P(0 \leq X \leq 1)$.

4. The interval between patient arrivals in an emergency department is a random variable with exponential density function $p(t) = 0.125e^{-0.125t}$ (t in minutes). What is the average time between patient arrivals? What is the probability of two patients arriving within 3 min of each other?

5. Calculate the following probabilities, assuming that X is normally distributed with mean $\mu = 40$ and $\sigma = 5$.

(a) $P(X \geq 45)$ **(b)** $P(0 \leq X \leq 40)$

6. According to kinetic theory, the molecules of ordinary matter are in constant random motion. The energy E of a molecule is a random variable with density function $p(E) = \frac{1}{kT}e^{-E/(kT)}$, where T is the temperature (in kelvins) and k is Boltzmann's constant. Compute the *mean* kinetic energy $\overline{E}$ in terms of k and T.

In Exercises 7–10, calculate the arc length over the given interval.

7. $y = \dfrac{x^5}{10} + \dfrac{x^{-3}}{6}$, $[1, 2]$

8. $y = e^{x/2} + e^{-x/2}$, $[0, 2]$

9. $y = 4x - 2$, $[-2, 2]$

10. $y = x^{2/3}$, $[1, 8]$

11. Show that the arc length of $y = 2\sqrt{x}$ over $[0, a]$ is equal to $\sqrt{a(a+1)} + \ln(\sqrt{a} + \sqrt{a+1})$. *Hint:* Apply the substitution $x = \tan^2 \theta$ to the arc length integral.

12. (CAS) Compute the trapezoidal approximation T_5 to the arc length s of $y = \tan x$ over $\left[0, \frac{\pi}{4}\right]$.

In Exercises 13–16, calculate the surface area of the solid obtained by rotating the curve over the given interval about the x-axis.

13. $y = x + 1$, $[0, 4]$

14. $y = \dfrac{2}{3}x^{3/4} - \dfrac{2}{5}x^{5/4}$, $[0, 1]$

15. $y = \dfrac{2}{3}x^{3/2} - \dfrac{1}{2}x^{1/2}$, $[1, 2]$ **16.** $y = \dfrac{1}{2}x^2$, $[0, 2]$

17. Compute the total surface area of the coin obtained by rotating the region in Figure 1 about the x-axis. The top and bottom parts of the region are semicircles with a radius of 1 mm.

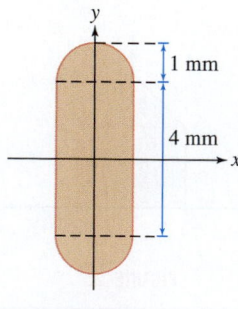

FIGURE 1

18. Calculate the fluid force on the side of a right triangle of height 3 m and base 2 m submerged in water vertically, with its upper vertex at the surface of the water.

19. Calculate the fluid force on the side of a right triangle of height 3 m and base 2 m submerged in water vertically, with its upper vertex located at a depth of 4 m.

20. A plate in the shape of the shaded region in Figure 2 is submerged in water. Calculate the fluid force on a side of the plate if the water surface is $y = 1$.

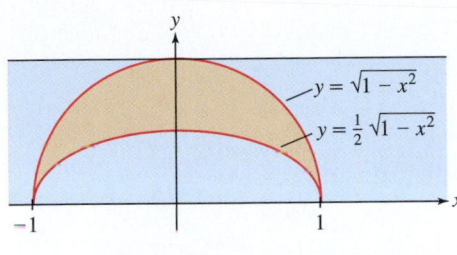

FIGURE 2

21. Figure 3 shows an object whose face is an equilateral triangle with 5-m sides. The object is 2 m thick and is submerged in water with its vertex 3 m below the water surface. Calculate the fluid force on both a triangular face and a slanted rectangular edge of the object.

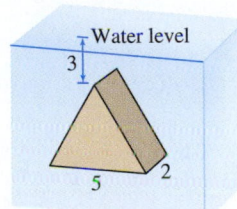

FIGURE 3

22. The end of a horizontal oil tank is an ellipse (Figure 4) with equation $(x/4)^2 + (y/3)^2 = 1$ (length in meters). Assume that the tank is filled with oil of density 900 kg/m^3.

(a) Calculate the total force F on the end of the tank when the tank is full.

(b) 🖊 Would you expect the total force on the lower half of the tank to be greater than, less than, or equal to $\frac{1}{2} F$? Explain. Then compute the force on the lower half exactly and confirm (or refute) your expectation.

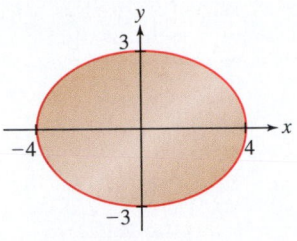

FIGURE 4

23. Calculate the moments and COM of the lamina occupying the region under $y = x(4 - x)$ for $0 \leq x \leq 4$, assuming a density of $\rho = 1200$ kg/m^3.

24. Sketch the region between $y = 4(x + 1)^{-1}$ and $y = 1$ for $0 \leq x \leq 3$, and find its centroid.

25. Find the centroid of the region between the semicircle $y = \sqrt{1 - x^2}$ and the top half of the ellipse $y = \frac{1}{2}\sqrt{1 - x^2}$ (Figure 2).

26. Find the centroid of the shaded region in Figure 5 bounded on the left by $x = 2y^2 - 2$ and on the right by a semicircle of radius 1. *Hint:* Use symmetry and additivity of moments.

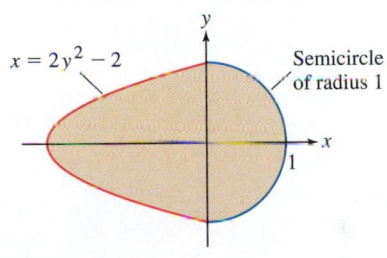

FIGURE 5

27. Use the Theorem of Pappus to find the volume of the solid of revolution obtained by rotating the region in the first quadrant bounded by $y = x^2$ and $y = \sqrt{x}$ about the y-axis.

28. Use the Theorem of Pappus to find a formula for the volume of the solid obtained by rotating the triangle with vertices $(1, 0)$, $(3, 0)$, and $(2, 2)$ about the y-axis.

9 INTRODUCTION TO DIFFERENTIAL EQUATIONS

Differential equations are among the most powerful tools we have for analyzing the world with mathematics. They are used to formulate the fundamental laws of nature (from Newton's laws to Maxwell's equations and the laws of quantum mechanics) and to model the most diverse physical phenomena. This chapter provides a brief introduction to some elementary techniques and applications of this important subject.

When a hot object, such as a glass blower's molten glass, is brought into a cooler environment, Newton's Law of Cooling says that the rate of change of the object's temperature is proportional to the temperature difference between the object and its surroundings. This relationship can be written as a differential equation that models the changing temperature of the object.

9.1 Solving Differential Equations

A **differential equation** is an equation that involves an unknown function and its first or higher derivatives. The following differential equations are among the ones that we will consider at various points in this chapter:

$$\frac{dy}{dx} = 1 - 6e^{2x}, \qquad \frac{dP}{dt} = rP - N, \qquad \frac{dy}{dt} + \frac{1}{t+30}y = 4$$

The **order** of a differential equation is the highest order derivative that appears in the equation. In this chapter, we restrict our attention to first-order differential equations. A solution to a differential equation is a function that satisfies the equation, and a primary goal when working with a differential equation is to find the solutions. The next example considers a simple differential equation like the ones we considered in Section 5.3.

EXAMPLE 1 Consider the differential equation $\frac{dy}{dx} = 1 - 6e^{2x}$.

(a) Find the solutions.

(b) Find the solution satisfying $y(0) = 5$.

Solution

A differential equation in the form $\frac{dy}{dx} = f(x)$ is said to be **directly integrable**. The solutions can be found by integration,

$$y(x) = \int f(x)\,dx.$$

(a) The form of the differential equation enables us to solve it directly by antidifferentiation:

$$y(x) = \int \left(1 - 6e^{2x}\right) dx = x - 3e^{2x} + C$$

(b) To satisfy $y(0) = 5$, we must have $5 = 0 - 3e^0 + C$. Therefore, $C = 8$. So the function

$$y(x) = x - 3e^{2x} + 8$$

satisfies the differential equation and the condition $y(0) = 5$. ∎

Generally, a differential equation has a family of solutions. For example, the solutions to the differential equation $\frac{dy}{dx} = 1 - 6e^{2x}$ in Example 1 are the functions $y(x) = x - 3e^{2x} + C$. The graphs of these solutions form a collection of curves in the xy-plane (Figure 1). An expression for the family of solutions to a differential equation, such as $y(x) = x - 3e^{2x} + C$ in Example 1, is called a **general solution**. For each value of C in the general solution, we obtain a **particular solution**. The function $y(x) = x - 3e^{2x} + 8$ is the particular solution to $\frac{dy}{dx} = 1 - 6e^{2x}$ satisfying $y(0) = 5$. A condition $y(x_0) = y_0$ for identifying a particular solution is called an **initial condition**. A problem consisting of a differential equation and an initial condition is called an **Initial Value Problem**.

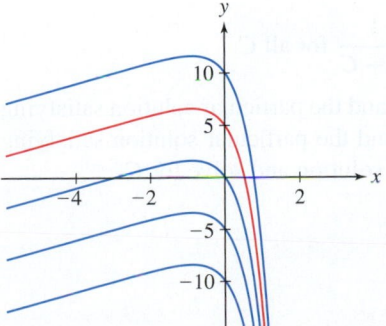

FIGURE 1 The family of solutions to $\frac{dy}{dx} = 1 - 6e^{2x}$.

Separation of Variables

A differential equation is called **separable** if it can be written in the form

$$\frac{dy}{dx} = f(x)g(y)$$

1

For example,

- $\dfrac{dy}{dx} = y \sin x$ and $\dfrac{dy}{dx} = x + xy$ are separable.

- $\dfrac{dy}{dx} = x + y$ is not separable because $x + y$ is not a *product* $f(x)g(y)$.

The manipulation of the dy and dx terms is symbolic. Like previous cases that we have seen of this type of manipulation, it can be justified mathematically via differentials.

Separable equations are solved using the method of **Separation of Variables**: Move the terms involving y to the left and those involving x to the right. Then set up integrals and evaluate.

$$\frac{dy}{dx} = f(x)g(y)$$

$$\frac{1}{g(y)}\, dy = f(x)\, dx$$

$$\int \frac{1}{g(y)}\, dy = \int f(x)\, dx$$

EXAMPLE 2 Find the general solution to $\dfrac{dy}{dx} = 6xy^2$ and the particular solution satisfying $y(0) = -1$.

Solution First, we observe that $y(x) = 0$ is a solution to the differential equation. Then, assuming $y \neq 0$, we use Separation of Variables as follows:

$$y^{-2}\, dy = 6x\, dx \qquad \text{(separate the variables)}$$

$$\int y^{-2}\, dy = \int 6x\, dx \qquad \text{(integrate)}$$

$$-y^{-1} = 3x^2 + C$$

$$y = \frac{-1}{3x^2 + C} \qquad \text{(solve for } y\text{)}$$

Thus, the general solution to $\dfrac{dy}{dx} = 6xy^2$ consists of the functions

$$y(x) = 0 \quad \text{and} \quad y(x) = \frac{-1}{3x^2 + C} \text{ for all } C$$

This family of solutions is sketched in Figure 2, and the particular solution satisfying the initial condition $y(0) = -1$ is highlighted. To find the particular solution satisfying $y(0) = -1$, we set $x = 0$ and $y = -1$ in the general solution and solve for C:

$$-1 = \frac{-1}{3(0)^2 + C}$$

$$-1 = \frac{-1}{C}$$

$$C = 1$$

Thus, the particular solution satisfying the given initial condition is

$$y = \frac{-1}{3x^2 + 1}$$

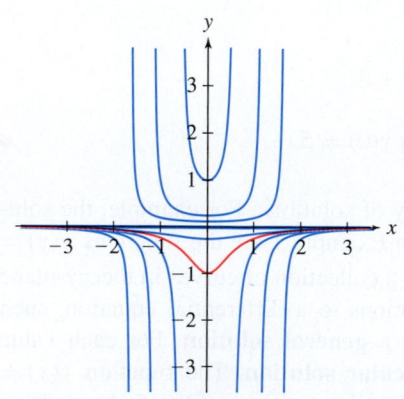

FIGURE 2 Solutions to $\dfrac{dy}{dx} = 6xy^2$.

■

It is beneficial to check your Initial Value Problem solution: First, $y(0) = \frac{-1}{0+1} = -1$, as required. Also,

$$\frac{dy}{dx} = \frac{d}{dx}\left(\frac{-1}{3x^2+1}\right) = \frac{6x}{(3x^2+1)^2} = 6xy^2$$

Therefore, $y = \frac{-1}{3x^2+1}$ satisfies the Initial Value Problem.

In the field of glaciology, a simple separable differential equation is used to model a glacier's thickness, the vertical distance from its surface to its base (Figure 3). A glacier is a river of ice, and like a river of water, it flows (but does so relatively slowly). Mass is added to a glacier by precipitation and is lost by sublimation, evaporation, and meltwater runoff. We assume that the amount of mass lost equals the amount that is gained, and therefore, the glacier's shape does not change. We model the glacier's thickness in meters by $T(x)$ where x is the distance in meters from the front of the glacier (the location where the glacier ends on land or water, also known as the **terminus**).

In a simple force-balance model, where fluid pressure forces in a glacier are balanced with a friction force at its base, the following differential equation for $T(x)$ results:

$$T\frac{dT}{dx} = \frac{\tau}{\rho g} \qquad \boxed{2}$$

In the equation, τ is the friction at the base of the glacier, ρ is the ice density, and g is acceleration due to gravity.

EXAMPLE 3 A Glacial Thickness Model Let $\rho = 917$ kg/m^3, $g = 9.8$ m/s^2, and $\tau = 75,000$ N/m^2 in Eq. (2). Use $T(0) = 0$ for an initial condition, and solve for $T(x)$. Then use $T(x)$ to determine the thickness of the glacier 1 km from its terminus.

Solution The differential equation that we need to solve is

$$T\frac{dT}{dx} = \frac{75,000}{(917)(9.8)}$$

It is a separable differential equation. We use the approximate value of 8.35 for the right-hand side, and proceed as follows:

$$\int T\,dT = \int 8.35\,dx$$

$$\frac{1}{2}T^2 = 8.35x + C$$

$$T(x) = \sqrt{16.7x + C}$$

Since $T(0) = 0$, we obtain $T(x) = \sqrt{16.7x}$ (Figure 4).

At a distance of 1 km from the terminus, the thickness is $T(1000) = \sqrt{16,700} \approx 129$ m. ■

Another model that results in a separable differential equation involves the water level in a container that has a leak through a hole in the bottom (Figure 5). We let $y(t)$ be the height of water in the container as a function of time, $A(y)$ be the cross-sectional area of the container at height y, and B be the area of the hole in the bottom of the container.

In a model where we assume the volume lost from the container equals the volume that flows out, we apply an important fluid-flow law known as Torricelli's Law to obtain the following differential equation:

$$\frac{dy}{dt} = -\frac{B\sqrt{2gy}}{A(y)} \qquad \boxed{3}$$

In the equation, g is acceleration due to gravity.

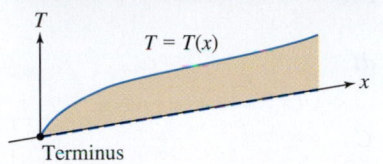

FIGURE 3 The glacier's thickness T is modeled as a function of distance x from the terminus.

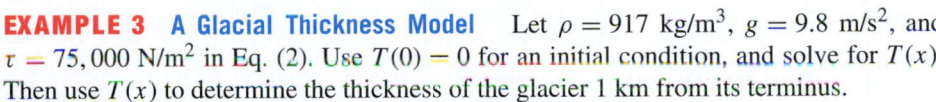

FIGURE 4 $T(x) = \sqrt{16.7x}$.

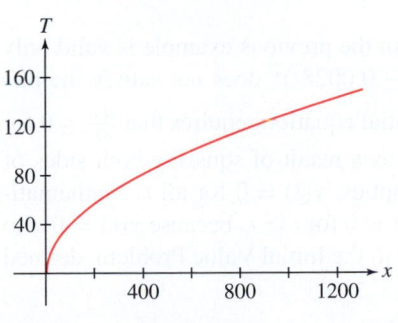

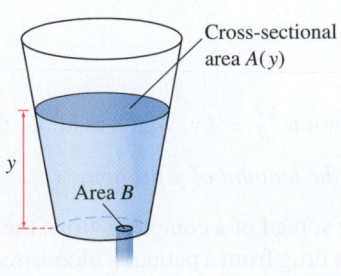

Cross-sectional area $A(y)$

Area B

FIGURE 5

NASA/Michael Studinger

EXAMPLE 4 A cylindrical container of height 400 cm and radius 100 cm is filled with water. In the bottom of the container is a square hole of side length 2 cm through which water leaks. Determine the water level $y(t)$, in cm, as a function of time t, in seconds. How long does it take for the tank to go from full to empty?

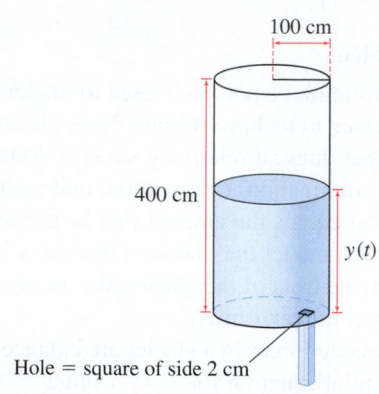

100 cm

400 cm

$y(t)$

Hole = square of side 2 cm

FIGURE 6

Solution The horizontal cross section of the cylinder is a circle of radius $r = 100$ cm and area $A(y) = \pi r^2 = 10{,}000\pi$ cm^2 (Figure 6). The hole has area $B = 4$ cm^2. With the units we are employing, we set $g = 980$ cm/s^2. Now, Eq. (3) becomes

$$\frac{dy}{dt} = -\frac{4(\sqrt{1960y})}{10{,}000\pi} \approx -0.0056\sqrt{y} \qquad \boxed{4}$$

Since the tank is full at $t = 0$, we have the initial condition $y(0) = 400$.

Now, $y(t) = 0$ is a solution to the differential equation, but it does not satisfy the initial condition. We assume $y \neq 0$ and solve using Separation of Variables:

$$\int \frac{dy}{\sqrt{y}} = -0.0056 \int dt$$

$$2y^{1/2} = -0.0056t + C \qquad \boxed{5}$$

At this point, rather than solving for general $y(t)$, it is convenient to determine the value of C for which the initial condition is satisfied. Substituting $y = 400$ and $t = 0$ into Eq. (5), we obtain $C = 40$. Thus, we have

$$2y^{1/2} = -0.0056t + 40 \qquad \boxed{6}$$

$$y(t) = (20 - 0.0028t)^2 \qquad \boxed{7}$$

To determine the time t_e that it takes to empty the tank, we solve

$$y(t_e) = (20 - 0.0028t_e)^2 = 0 \quad \Rightarrow \quad t_e \approx 7142 \text{ s}$$

So, the tank is empty after 7142 s, or nearly 2 hours (Figure 7). ∎

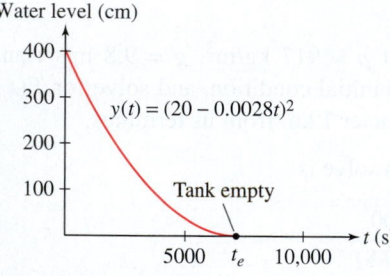

Water level (cm)

400
300
$y(t) = (20 - 0.0028t)^2$
200
100
Tank empty
5000 t_e 10,000
t (s)

FIGURE 7

Note that the solution $y(t) = (20 - 0.0028t)^2$ in the previous example is valid only for $0 \le t \le t_e$. For $t > t_e$ the function $y(t) = (20 - 0.0028t)^2$ does not satisfy the differential equation; it is increasing while the differential equation requires that $\frac{dy}{dt} \le 0$ for all t. This "extraneous solution" (for $t > t_e$) arose as a result of squaring both sides of Eq. (6). It is clear physically that after the tank empties, $y(t) = 0$ for all t. Mathematically, it also works to extend the solution so that $y(t) = 0$ for $t \ge t_e$ because $y(t) = 0$ also satisfies the differential equation. Thus, the solution to the Initial Value Problem, defined for all $t \ge 0$, is

$$y(t) = \begin{cases} (20 - 0.0028t)^2 & 0 \le t \le t_e \\ 0 & t \ge t_e \end{cases}$$

Exponential Growth and Decay

Many phenomena are modeled by the differential equation $\frac{dy}{dt} = ky$, which assumes that:

The rate of change of y is proportional to the amount of y present.

Examples include the growth of a population, the spread of a computer virus, the decay of a radioactive substance, and the elimination of a drug from a patient's bloodstream. For "growth" and "spread," the proportionality constant k in the differential equation is positive (y would be increasing); for "decay" and "elimination," it is negative.

EXAMPLE 5 Find the general solution to $\frac{dy}{dt} = ky$.

Solution First, note that $y(t) = 0$ is a solution to the differential equation. Then, assuming $y \neq 0$, we employ Separation of Variables:

$$\frac{1}{y} dy = k \, dt$$

$$\int \frac{1}{y} dy = \int k \, dt$$

$$\ln |y| = kt + C$$

$$|y| = e^{kt+C}$$

$$y = De^{kt}$$

Two simplifications occurred at the last step. First, we write $e^{kt+C} = e^{kt} e^{C} = De^{kt}$, where e to the power of an arbitrary constant is replaced by a positive constant term D. Second, if $|y| = De^{kt}$, then $y = \pm De^{kt}$, which can be expressed as $y = De^{kt}$, allowing D to be positive or negative. Furthermore, since $y = 0$ solves the differential equation, the possibility that $D = 0$ is also allowed. Therefore, the functions $y(t) = De^{kt}$, for all D, satisfy the differential equation. Summarizing,

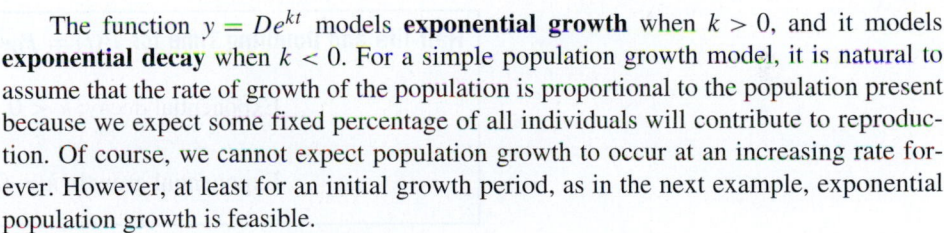

The general solution to $\frac{dy}{dt} = ky$ is $y(t) = De^{kt}$. **8**

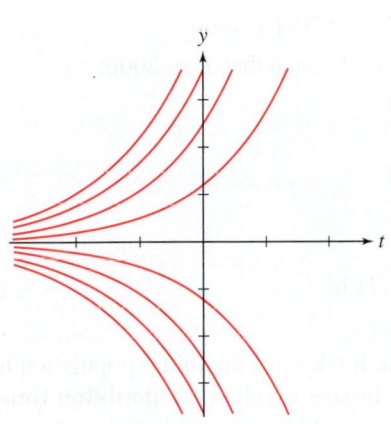

FIGURE 8 The general solution to $\frac{dy}{dt} = ky$ for $k > 0$.

The general solution with $k > 0$ is illustrated in Figure 8. ■

The function $y = De^{kt}$ models **exponential growth** when $k > 0$, and it models **exponential decay** when $k < 0$. For a simple population growth model, it is natural to assume that the rate of growth of the population is proportional to the population present because we expect some fixed percentage of all individuals will contribute to reproduction. Of course, we cannot expect population growth to occur at an increasing rate forever. However, at least for an initial growth period, as in the next example, exponential population growth is feasible.

EXAMPLE 6 In the laboratory, the *Escherichia coli* bacteria grows such that the rate of change of the population is proportional to the population present. Assume that 1000 bacteria are initially present, and 1500 are present after 1 hour.

(a) Determine $P(t)$, the population after t hours.

(b) How large is the population after 5 h?

(c) How long does it take for the population to double in size?

Solution

(a) Since the rate of change $P'(t)$ is proportional to $P(t)$, we have the differential equation $P'(t) = kP(t)$. Furthermore, we are given the initial conditions $P(0) = 1000$ and $P(1) = 1500$. Together, the differential equation and initial conditions make up the Initial Value Problem that we must solve.

Note that we have two initial conditions. Both are necessary because we need to determine the constant term D in the general solution to the differential equation and the term k in the differential equation itself.

The general solution to the differential equation is $P(t) = De^{kt}$, and the initial condition $P(0) = 1000$ implies that $D = 1000$. Thus, the solution is in the form

$$P(t) = 1000e^{kt}$$

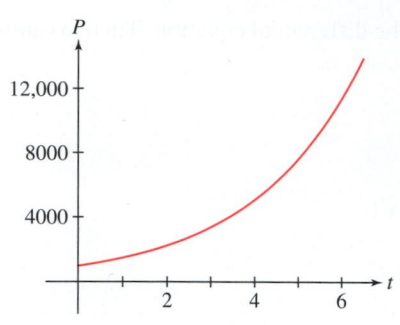

FIGURE 9 $P(t) = 1000e^{0.405t}$.

Exponential growth cannot continue over long periods of time. In this example, the population would grow to over 10^{94} bacteria cells after 3 weeks—much more than the estimated number of atoms in the observable universe. In typical population growth, an initial rapid growth phase is followed by a period in which growth slows. In Section 9.4, we adjust the population growth model to take into account a limited resource base for the population. This results in solutions where the population initially grows rapidly but then levels off.

To determine k, we substitute $P = 1500$ and $t = 1$ into $P(t) = 1000e^{kt}$ and solve for k:

$$1500 = 1000e^k$$
$$1.5 = e^k$$
$$\ln 1.5 = k$$
$$k \approx 0.405$$

Therefore, with an approximate value for k, we have that $P(t) = 1000e^{0.405t}$ (Figure 9).

(b) After 5 h the population is $P(5) = 1000e^{0.405(5)} \approx 7576$ bacteria.

(c) To determine when the population doubles, we find t such that $P = 2000$:

$$2000 = 1000e^{0.405t}$$
$$2 = e^{0.405t}$$
$$\ln 2 = 0.405t$$
$$t = \frac{\ln 2}{0.405} \approx 1.71 \text{ h}$$

■

In the previous example, we computed the time it takes for the initial population to double. In exponential growth, the time to double in size is called the **doubling time**. Similarly, in exponential decay, the **half-life** is the time it takes for an initial amount to decrease to half its size. Doubling time and half-life do not depend on the initial amount, but depend only on the value of the term k, as follows:

Half-life and Doubling Time for $P(t) = P_0e^{kt}$

Exponential decay: $k < 0$ half-life $= \dfrac{\ln 0.5}{k}$ | **9** |

Exponential growth: $k > 0$ doubling time $= \dfrac{\ln 2}{k}$ | **10** |

It is straightforward to verify these formulas. For example, suppose that $t = \frac{\ln 2}{k}$; then

$$P\left(\frac{\ln 2}{k}\right) = P_0e^{k\left(\frac{\ln 2}{k}\right)} = P_0e^{\ln 2} = 2P_0$$

Thus, $\frac{\ln 2}{k}$ is the value of t at which $P(t)$ is equal to twice the initial amount P_0.

EXAMPLE 7 A patient is administered 400 mg of penicillin, and after 50 minutes 300 mg remains in her bloodstream. Let $A(t)$ represent the amount of penicillin (in mg) in her bloodstream t minutes after the drug was administered. Assume that the drug is leaving her bloodstream at a rate proportional to the amount in her bloodstream. Determine $A(t)$ and the half-life of the resulting exponential decay.

Solution
Since the rate of change is proportional to the amount present, $A'(t) = kA(t)$. Furthermore, we are given the initial conditions $A(0) = 400$ and $A(50) = 300$, and therefore, we have an Initial Value Problem that we must solve.

The general solution to the differential equation is $A(t) = De^{kt}$, and the initial condition $A(0) = 400$ implies that $D = 400$. The solution is therefore in the form $A(t) = 400e^{kt}$.

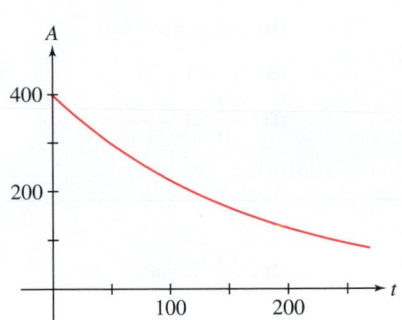

FIGURE 10 $A(t) = 400e^{-0.0058t}$.

Sofia Kovalevskaya (1850–1891) was a Russian mathematician who made important contributions to the field of differential equations. One of her major results is known as the Cauchy–Kovalevskaya Theorem. It asserts the existence and uniqueness of a class of solutions to differential equations for functions of multiple variables. A special case of the theorem was first proved by Augustin-Louis Cauchy in 1842. Kovalevskaya proved the general version of the theorem in 1875.

To determine k, we substitute $A = 300$ and $t = 50$ into $A(t) = 400e^{kt}$ and solve for k:

$$300 = 400e^{50k}$$

$$0.75 = e^{50k}$$

$$\ln 0.75 = 50k$$

$$k = \frac{\ln 0.75}{50} \approx -0.0058$$

Thus, with this approximate value for k, we have that $A(t) = 400e^{-0.0058t}$ (Figure 10). The half-life is $\frac{\ln 0.5}{k} \approx \frac{\ln 0.5}{-0.0058} \approx 120$ min. ■

If an Initial Value Problem models a physical relationship, we would want to know whether or not a solution exists and whether or not the solution is unique. Having no solution suggests that the model is not correct. Perhaps we would need to reevaluate the assumptions used to build the model. Having a unique solution would be desirable if we expect the phenomenon to change in only one way and we wish to make predictions.

Note that the Initial Value Problem, $\frac{dy}{dt} = ky$ and $y(0) = y_0$, has the solution $y = y_0 e^{kt}$ [obtained from Eq. (8) using the initial condition $y(0) = y_0$]. It is the only solution that is provided by the general solution formula. We can prove that this solution is unique by showing that all solutions to the differential equation are provided by the general solution (see Exercise 61).

9.1 SUMMARY

- A *directly integrable* differential equation is in the form $\frac{dy}{dx} = f(x)$. It is solved by direct antidifferentiation.
- A *separable first-order* differential equation is in the form

$$\frac{dy}{dx} = f(x)g(y)$$

Differential equations of this type are solved by *Separation of Variables*: Move all terms involving y to the left and all terms involving x to the right and integrate

$$\int \frac{1}{g(y)}\, dy = \int f(x)\, dx$$

- A simple differential equation model for exponential growth and decay is

$$\frac{dy}{dt} = ky$$

The general solution is $y(t) = De^{kt}$. Exponential growth occurs when $k > 0$, and the doubling time is $\frac{\ln 2}{k}$. Exponential decay occurs when $k < 0$, and the half-life is $\frac{\ln 0.5}{k}$.

9.1 EXERCISES

Preliminary Questions

1. Determine the order of the following differential equations:

(a) $x^5 y' = 1$

(b) $(y')^3 + x = 1$

(c) $y''' + x^4 y' = 2$

(d) $\sin(y'') + x = y$

2. Which of the following differential equations are directly integrable?

(a) $y' = x + y$

(b) $x\dfrac{dy}{dx} = 3$

(c) $\dfrac{dP}{dt} = 4P + 1$

(d) $\dfrac{dw}{dt} = \dfrac{2t}{1 + 4t}$

(e) $\dfrac{dx}{dt} = t^2 e^{-3t}$

(f) $t^2 \dfrac{dx}{dt} = x - 1$

3. Which of the following differential equations are first-order?

(a) $y' = x^2$

(b) $y'' = y^2$

(c) $(y')^3 + yy' = \sin x$

(d) $x^2 y' - e^x y = \sin y$

(e) $y'' + 3y' = \dfrac{y}{x}$

(f) $yy' + x + y = 0$

4. Which of the following differential equations are separable?

(a) $\dfrac{dy}{dx} = x - 2y$

(b) $xy' + 8ye^x = 0$

(c) $y' = x^2 y^2$

(d) $y' = 1 - y^2$

(e) $t\dfrac{dy}{dt} = 3\sqrt{1 + y}$

(f) $\dfrac{dP}{dt} = \dfrac{P + t}{t}$

Exercises

In Exercises 1–6, verify that the given function is a solution of the differential equation.

1. $y' - 8x = 0$, $\quad y = 4x^2$

2. $yy' + 4x = 0$, $\quad y = \sqrt{12 - 4x^2}$

3. $y' + 4xy = 0$, $\quad y = 25e^{-2x^2}$

4. $(x^2 - 1)y' + xy = 0$, $\quad y = 4(x^2 - 1)^{-1/2}$

5. $y'' - 2xy' + 8y = 0$, $\quad y = 4x^4 - 12x^2 + 3$

6. $y'' - 2y' + 5y = 0$, $\quad y = e^x \sin 2x$

7. The following differential equations appear similar but have very different solutions:

$$\frac{dy}{dx} = 0, \qquad \frac{dy}{dx} = 0.001$$

Solve both subject to the initial condition $y(1) = -1$.

8. The following differential equations appear similar but have very different solutions:

$$\frac{dy}{dx} = x, \qquad \frac{dy}{dx} = y$$

Solve both subject to the initial condition $y(1) = 2$.

9. Verify that $x^2 y' + e^{-y} = 0$ is separable.

(a) Write it as $e^y \, dy = -x^{-2} \, dx$.

(b) Integrate both sides to obtain $e^y = x^{-1} + C$.

(c) Verify that $y = \ln(x^{-1} + C)$ is the general solution.

(d) Find the particular solution satisfying $y(2) = 4$.

10. Consider the differential equation $y^3 y' - 9x^2 = 0$.

(a) Write it as $y^3 \, dy = 9x^2 \, dx$.

(b) Integrate both sides to obtain $\frac{1}{4} y^4 = 3x^3 + C$.

(c) Verify that $y = (12x^3 + C)^{1/4}$ is the general solution.

(d) Find the particular solution satisfying $y(1) = 2$.

In Exercises 11–28, use Separation of Variables to find the general solution.

11. $y' - \dfrac{6x^2}{y^3} = 0$

12. $y' + 4xy^2 = 0$

13. $y' + x^2 y = 0$

14. $y' - e^{x+y} = 0$

15. $\dfrac{dy}{dt} - 20t^4 e^{-y} = 0$

16. $t^3 y' + 4y^2 = 0$

17. $2y' + 5y = 4$

18. $\dfrac{dy}{dt} = 8\sqrt{y}$

19. $\sqrt{1 - x^2}\, y' = xy$

20. $y' = y^2(1 - x^2)$

21. $yy' = x$

22. $(\ln y)y' - ty = 0$

23. $\dfrac{dx}{dt} = (t + 1)(x^2 + 1)$

24. $(1 + x^2)y' = x^3 y$

25. $y' = x \sec y$

26. $\dfrac{dy}{d\theta} = \tan y$

27. $\dfrac{dy}{dt} = y \tan t$

28. $\dfrac{dx}{dt} = t \tan x$

In Exercises 29–42, solve the Initial Value Problem.

29. $y' + 2y = 0$, $\quad y(\ln 5) = 3$

30. $y' - 3y + 12 = 0$, $\quad y(2) = 1$

31. $yy' = xe^{-y^2}$, $\quad y(0) = -2$

32. $y^2 \dfrac{dy}{dx} = x^{-3}$, $\quad y(1) = 0$

33. $y' = (x - 1)(y - 2)$, $\quad y(2) = 4$

34. $y' = (x - 1)(y - 2)$, $\quad y(2) = 2$

35. $y' = x(y^2 + 1)$, $\quad y(0) = 0$

36. $(1 - t)\dfrac{dy}{dt} - y = 0$, $\quad y(2) = -4$

37. $\dfrac{dy}{dt} = ye^{-t}$, $\quad y(0) = 1$

38. $\dfrac{dy}{dt} = te^{-y}$, $\quad y(1) = 0$

39. $t^2 \dfrac{dy}{dt} - t = 1 + y + ty$, $\quad y(1) = 0$

40. $\sqrt{1 - x^2}\, y' = y^2 + 1$, $\quad y(0) = 0$

41. $y' = \tan y$, $\quad y(\ln 2) = \dfrac{\pi}{2}$

42. $y' = y^2 \sin x$, $\quad y(\pi) = 2$

In Example 3, we calculated the thickness of a glacier assuming the friction at the base of the glacier is constant. In Exercises 43–44, we consider cases where the friction varies along the length of the glacier.

43. (a) Solve the glacier thickness differential equation [Eq. (2)] for $T(x)$ with $\tau(x) = 75x$ N/m^2 and initial condition $T(0) = 0$.

(b) Sketch the graph of T for $0 \le x \le 1000$ m.

44. (a) Solve the glacier thickness differential equation [Eq. (2)] for $T(x)$ with $\tau(x) = 0.3x(1000 - x)$ N/m^2 and initial condition $T(0) = 0$.

(b) Sketch the graph of T for $0 \le x \le 1000$ m.

In Exercises 45–50, use the differential equation for a leaking container, Eq. (3).

45. Water leaks through a hole of area $B = 0.002$ m^2 at the bottom of a cylindrical tank that is filled with water and has height 3 m and a base of area 10 m^2. How long does it take (a) for half of the water to leak out and (b) for the tank to empty?

46. At $t = 0$, a conical tank of height 300 cm and top radius 100 cm [Figure 11(A)] is filled with water. Water leaks through a hole in the bottom of area $B = 3$ cm^2. Let $y(t)$ be the water level at time t.

(a) Show that the tank's cross-sectional area at height y is $A(y) = \dfrac{\pi}{9}y^2$.

(b) Solve the differential equation satisfied by $y(t)$

(c) How long does it take for the tank to empty?

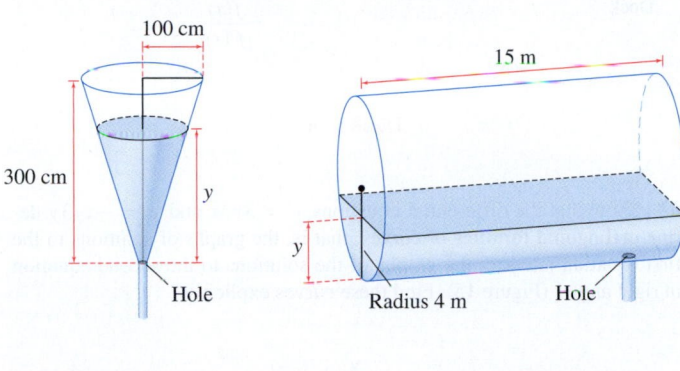

(A) Conical tank   (B) Horizontal tank

FIGURE 11

47. The tank in Figure 11(B) is a cylinder of radius 4 m and height 15 m. Assume that the tank is half-filled with water and that water leaks through a hole in the bottom of area $B = 0.001$ m^2. Determine the water level $y(t)$ and the time t_e when the tank is empty.

48. A tank has the shape of the parabola $y = x^2$, revolved around the y-axis. Water leaks from a hole of area $B = 0.0005$ m^2 at the bottom of the tank. Let $y(t)$ be the water level at time t. How long does it take for the tank to empty if it is initially filled to height $y_0 = 1$ m?

49. A tank has the shape of the parabola $y = ax^2$ (where a is a constant) revolved around the y-axis. Water drains from a hole of area B m^2 at the bottom of the tank.

(a) Show that the water level at time t is

$$y(t) = \left(y_0^{3/2} - \frac{3aB\sqrt{2g}}{2\pi}t \right)^{2/3}$$

where y_0 is the water level at time $t = 0$.

(b) Show that if the total volume of water in the tank has volume V at time $t = 0$, then $y_0 = \sqrt{2aV/\pi}$. *Hint:* Compute the volume of the tank as a volume of rotation.

(c) Show that the tank is empty at time

$$t_e = \left(\frac{2}{3B\sqrt{g}} \right) \left(\frac{2\pi V^3}{a} \right)^{1/4}$$

We see that for fixed initial water volume V, the time t_e is proportional to $a^{-1/4}$. A large value of a corresponds to a tall thin tank. Such a tank drains more quickly than a short wide tank of the same initial volume.

50. 📝 A cylindrical tank filled with water has height h and a base of area A. Water leaks through a hole in the bottom of area B.

(a) Show that the time required for the tank to empty is proportional to $A\sqrt{h}/B$.

(b) Show that the emptying time is proportional to $Vh^{-1/2}$, where V is the volume of the tank.

(c) Two tanks have the same volume and a hole of the same size, but they have different heights and bases. Which tank empties first: the taller or the shorter tank?

51. When cane sugar is dissolved in water, it converts to invert sugar over a period of several hours. The amount $A(t)$ of unconverted cane sugar at time t (in hours) satisfies $A' = -0.2A$. If there are initially 500 g of unconverted cane sugar, how much unconverted cane sugar remains after 5 h? After 10 h?

52. A certain RNA molecule replicates every 3 minutes. Find the differential equation for the number $N(t)$ of molecules present at time t (in minutes). How many molecules will be present after 1 h if there is one molecule at $t = 0$?

53. Assume that during periods of job growth, the rate of increase of the number employed is proportional to the number who are employed. At the outset of a job growth period, the number employed in a certain country grew from 11.3 million to 11.7 million in 10 weeks. Let $N(t)$ represent the number employed in millions as a function of t in weeks since the start of the growth period.

(a) Set up and solve an Initial Value Problem to determine $N(t)$.

(b) The growth period lasted for a year (52 weeks). What was the increase in the number of jobs?

54. Bismuth-210 decays at a rate proportional to the amount present. A sample of Bismuth-210 that initially had a mass of 1000 mg decayed 500 mg in 5 days. Let $M(t)$ be the mass of Bismuth-210 in milligrams t days after the initial sample of 1000 mg began to decay. Set up and solve an Initial Value Problem for determining $M(t)$.

55. (a) With $y(t) = y_0e^{kt}$, at what value of t (in terms of p and k) is $y(t) = py_0$?

(b) If $y(t) = y_0e^{1.5t}$, with t in hours, how long does it take for y to double? To triple? To increase 10-fold?

56. Drug Dosing Interval Let $y(t)$ be the drug concentration (in micrograms per milliliter) in a patient's body at time t. The initial concentration is $y(0) = L$. Additional doses that increase the concentration by an amount d are administered at regular time intervals of length T. In between doses, $y(t)$ decays exponentially—that is, $y' = -ky$. Find the value of T (in terms of k and d) for which the concentration varies between L and $L - d$ as in Figure 12.

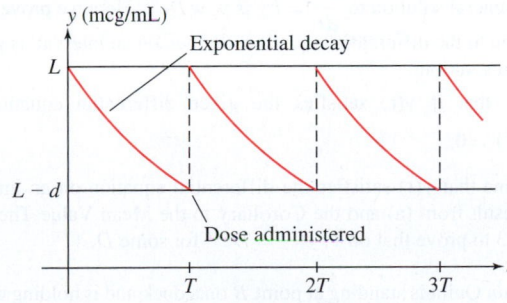

FIGURE 12 Drug concentration with periodic doses.

57. Figure 13 shows a circuit consisting of a resistor of R ohms, a capacitor of C farads, and a battery of voltage V. When the circuit is completed, the amount of charge $q(t)$ (in coulombs) on the plates of the capacitor varies according to the differential equation (t in seconds)

$$R\frac{dq}{dt} + \frac{1}{C}q = V$$

where R, C, and V are constants.

(a) Solve for $q(t)$, assuming that $q(0) = 0$.

(b) Sketch the graph of q.

(c) Show that $\lim_{t \to \infty} q(t) = CV$.

(d) Show that the capacitor charges to approximately 63% of its final value CV after a time period of length $\tau = RC$ (τ is called the time constant of the capacitor).

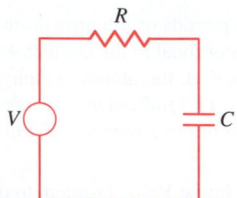

FIGURE 13 An RC circuit.

58. Assume in the circuit of Figure 13 that $R = 200$ ohms, $C = 0.02$ farad, and $V = 12$ volts. How many seconds does it take for the charge on the capacitor plates to reach half of its limiting value?

59. According to one hypothesis, the growth rate dV/dt of a cell's volume V is proportional to its surface area A. Since V has cubic units such as cm^3 and A has square units such as cm^2, we may assume roughly that $A \propto V^{2/3}$, and hence $dV/dt = kV^{2/3}$ for some constant k. If this hypothesis is correct, which dependence of volume on time would we expect to see (again, roughly speaking) in the laboratory?

(a) linear   **(b)** quadratic   **(c)** cubic

60. We might also guess that the volume V of a melting snowball decreases at a rate proportional to its surface area. Argue as in Exercise 59 to find a differential equation satisfied by V. Suppose the snowball has volume 1000 cm^3 and that it loses half of its volume after 5 minutes. According to this model, when will the snowball disappear?

61. The general solution to $\dfrac{dy}{dt} = ky$ is $y = De^{kt}$. Here we prove that every solution to the differential equation, defined on an interval, is given by the general solution.

(a) Show that if $y(t)$ satisfies the given differential equation, then $\dfrac{d}{dt}\left(ye^{-kt}\right) = 0$.

(b) Assume that $y(t)$ satisfies the differential equation on an interval I. Use the result from (a) and the Corollary to the Mean Value Theorem in Section 4.3 to prove that on I, $y(t) = De^{kt}$ for some D.

62. Captain Quint is standing at point B on a dock and is holding a rope of length ℓ attached to a boat at point A [Figure 14(A)]. As the captain walks along the dock, holding the rope taut, the boat moves along a curve called a **tractrix** (from the Latin *tractus*, meaning "pulled"). The segment from a point P on the curve to the x-axis along the tangent line has constant length ℓ [Figure 14(B)]. Let $y = f(x)$ be the equation of the tractrix.

(a) Show that $y^2 + (y/y')^2 = \ell^2$ and conclude $y' = -\dfrac{y}{\sqrt{\ell^2 - y^2}}$. Why must we choose the negative square root?

(b) Prove that the tractrix is the graph of

$$x = \ell \ln\left(\frac{\ell + \sqrt{\ell^2 - y^2}}{y}\right) - \sqrt{\ell^2 - y^2}$$

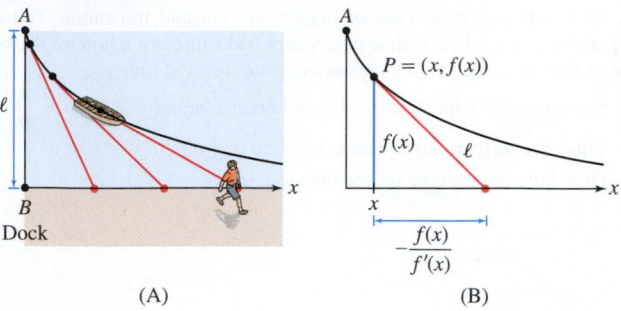

(A)   (B)

FIGURE 14

63. Show that the differential equations $y' = 3y/x$ and $y' = -x/3y$ define **orthogonal families** of curves; that is, the graphs of solutions to the first equation intersect the graphs of the solutions to the second equation in right angles (Figure 15). Find these curves explicitly.

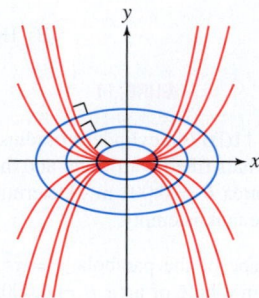

FIGURE 15 Two orthogonal families of curves.

64. Find the family of curves satisfying $y' = x/y$ and sketch several members of the family. Then find the differential equation for the orthogonal family (see Exercise 63), find its general solution, and add some members of this orthogonal family to your plot.

65. A 50-kg model rocket lifts off by expelling fuel downward at a rate of $k = 4.75$ kg/s for 10 s. The fuel leaves the end of the rocket with an exhaust velocity of $b = -100$ m/s. Let $m(t)$ be the mass of the rocket at time t. From the law of conservation of momentum, we find the following differential equation for the rocket's velocity $v(t)$ (in meters per second):

$$m(t)v'(t) = -9.8m(t) + b\frac{dm}{dt}$$

(a) Show that $m(t) = 50 - 4.75t$ kg.

(b) Solve for $v(t)$ and compute the rocket's velocity at rocket burnout (after 10 s).

66. Let $v(t)$ be the velocity of an object of mass m in free-fall near the earth's surface. If we assume that air resistance is proportional to v^2, then v satisfies the differential equation $m\dfrac{dv}{dt} = -g + kv^2$ for some constant $k > 0$.

(a) Set $\alpha = (g/k)^{1/2}$ and rewrite the differential equation as

$$\frac{dv}{dt} = -\frac{k}{m}(\alpha^2 - v^2)$$

Then solve using Separation of Variables with initial condition $v(0) = 0$.

(b) Show that the terminal velocity $\lim_{t \to \infty} v(t)$ is equal to $-\alpha$.

67. If a bucket of water spins about a vertical axis with constant angular velocity ω (in radians per second), the water climbs up the side of the bucket until it reaches an equilibrium position (Figure 16). Two forces act on a particle located at a distance x from the vertical axis: the gravitational force $-mg$ acting downward and the force of the bucket on the particle (transmitted indirectly through the liquid) in the direction perpendicular to the surface of the water. These two forces must combine to supply a centripetal force $m\omega^2 x$, and this occurs if the diagonal of the rectangle in Figure 16 is normal to the water's surface (i.e., perpendicular to the tangent line). Prove that if $y = f(x)$ is the equation of the curve obtained by taking a vertical cross section through the axis, then $-1/y' = -g/(\omega^2 x)$. Show that $y = f(x)$ is a parabola.

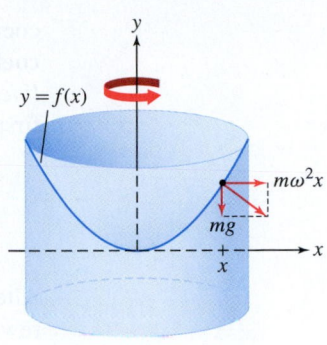

FIGURE 16

Further Insights and Challenges

68. In Section 6.2, we computed the volume V of a solid as the integral of cross-sectional area. Explain this formula in terms of differential equations. Let $V(y)$ be the volume of the solid up to height y, and let $A(y)$ be the cross-sectional area at height y as in Figure 17.

(a) Explain the following approximation for small Δy:

$$V(y + \Delta y) - V(y) \approx A(y)\,\Delta y \qquad \boxed{11}$$

(b) Use Eq. (11) to justify the differential equation $dV/dy = A(y)$. Then derive the formula

$$V = \int_a^b A(y)\,dy$$

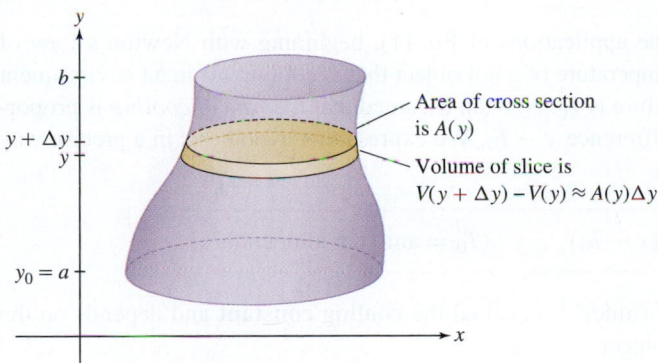

FIGURE 17

69. A basic theorem states that a *linear* differential equation of order n has a general solution that depends on n arbitrary constants. The following examples show that, in general, the theorem does not hold for nonlinear differential equations. In each case the differential equation is not linear because of the $(y')^2$ term.

(a) Show that $(y')^2 + y^2 = 0$ is a first-order equation with only one solution $y = 0$.

(b) Show that $(y')^2 + y^2 + 1 = 0$ is a first-order equation with no solutions.

70. Show that $y = Ce^{rx}$ is a solution of $y'' + ay' + by = 0$ if and only if r is a root of $P(r) = r^2 + ar + b$. Verify directly that $y = C_1 e^{3x} + C_2 e^{-x}$ is a solution of $y'' - 2y' - 3y = 0$ for any constants C_1, C_2.

71. A spherical tank of radius R is half-filled with water. Suppose that water leaks through a hole in the bottom of area B. Let $y(t)$ be the water level at time t (seconds).

(a) Show that $\dfrac{dy}{dt} = \dfrac{\sqrt{2g}\,B\sqrt{y}}{\pi(2Ry - y^2)}$.

(b) Show that for some constant C,

$$\frac{2\pi}{15B\sqrt{2g}}\left(10Ry^{3/2} - 3y^{5/2}\right) = C - t$$

(c) Use the initial condition $y(0) = R$ to compute C, and show that $C = t_e$, the time at which the tank is empty.

(d) Show that t_e is proportional to $R^{5/2}$ and inversely proportional to B.

9.2 Models Involving $y' = k(y - b)$

In this section we examine the differential equation

$$\boxed{\dfrac{dy}{dt} = k(y - b)} \qquad \boxed{1}$$

where k and b are constants. This differential equation describes a quantity y whose *rate of change is proportional to the difference* $y - b$. It arises in many different modeling situations. We will use it to model a cooling object in a fixed-temperature environment, vertical motion under the influence of gravity and air resistance, and the changing value of an annuity.

This differential equation can be written in the form $\frac{dy}{dt} = ky + c$ and is called **linear** because the relationship between $\frac{dy}{dt}$ and y is linear. Furthermore, because the coefficient k and the term c are constant, the differential equation is said to have **constant coefficients**. Thus, the differential equations we examine in this section are *first-order linear constant coefficient* differential equations. In Section 9.5, we examine the general first-order linear differential equation where the terms k and c could be functions of t.

We can use Separation of Variables to show that the general solution to Eq. (1) is

$$y(t) = b + Ce^{kt} \qquad \boxed{2}$$

Alternatively, we may observe that $(y - b)' = y'$ since b is a constant, so Eq. (1) may be rewritten

$$\frac{d}{dt}(y - b) = k(y - b)$$

In other words, $y - b$ satisfies the differential equation of an exponential function and thus $y - b = Ce^{kt}$, or $y = b + Ce^{kt}$, as claimed.

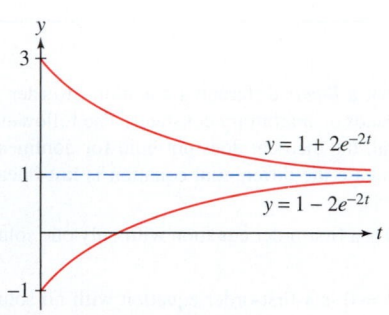

FIGURE 1 Two solutions to $y' = -2(y - 1)$ corresponding to $C = 2$ and $C = -2$.

> **GRAPHICAL INSIGHT** The behavior of the solution $y(t)$ as $t \to \infty$ depends on whether C and k are positive or negative:
>
> - When $k > 0$, e^{kt} tends to ∞ and, therefore, $y(t)$ tends to ∞ if $C > 0$ and $y(t)$ tends to $-\infty$ if $C < 0$.
> - When $k < 0$, we usually rewrite the differential equation as $y' = -k(y - b)$ with $k > 0$. In this case, $y(t) = b + Ce^{-kt}$ and $y(t)$ approaches the horizontal asymptote $y = b$ since Ce^{-kt} tends to zero as $t \to \infty$ (Figure 1). Note that $y(t)$ approaches the asymptote from above or below, depending on whether $C > 0$ or $C < 0$.

Newton's Law of Cooling implies that the object cools quickly when it is much hotter than its surroundings (when $y - T_0$ is large). The rate of cooling slows as y approaches T_0. When the object's initial temperature is less than T_0, y' is positive and Newton's Law models warming.

We now consider some applications of Eq. (1), beginning with Newton's Law of Cooling. Let $y(t)$ be the temperature of a hot object that is cooling off in an environment where the ambient temperature is T_0. Newton assumed that the *rate of cooling* is proportional to the temperature difference $y - T_0$. We express this hypothesis in a precise way by the differential equation

$$\boxed{y' = -k(y - T_0) \qquad (T_0 = \text{ambient temperature})}$$

The constant k, in units of $(\text{time})^{-1}$, is called the **cooling constant** and depends on the physical properties of the object.

EXAMPLE 1 **Newton's Law of Cooling** A hot metal bar with cooling constant $k = 2.1 \text{ min}^{-1}$ is submerged in a large tank of water held at temperature $T_0 = 10°\text{C}$. Let $y(t)$ be the bar's temperature at time t (in minutes).

(a) Find the differential equation satisfied by $y(t)$ and find its general solution.

(b) What is the bar's temperature after 1 min if its initial temperature was 180°C?

(c) What was the bar's initial temperature if it cooled to 80°C in 30 seconds?

Solution

(a) Since $k = 2.1 \text{ min}^{-1}$, $y(t)$ (with t in minutes) satisfies

$$y' = -2.1(y - 10)$$

By Eq. (2), the general solution is $y(t) = 10 + Ce^{-2.1t}$ for some constant C.

(b) If the initial temperature was 180°C, then $y(0) = 10 + C = 180$. Thus, $C = 170$ and $y(t) = 10 + 170e^{-2.1t}$ (Figure 2). After 1 min,

$$y(1) = 10 + 170e^{-2.1(1)} \approx 30.8°\text{C}$$

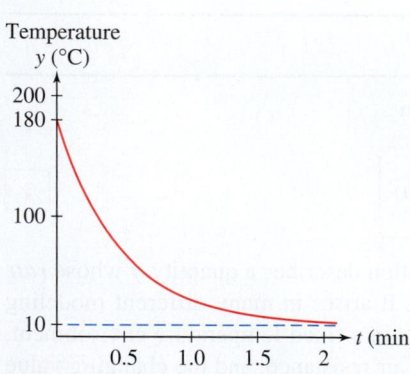

FIGURE 2 The temperature $y(t)$ of a metal bar as it cools when the ambient temperature is 10°C.

(c) If the temperature after 30 s is 80°C, then $y(0.5) = 80$, and we have

$$10 + Ce^{-2.1(0.5)} = 80 \quad \Rightarrow \quad Ce^{-1.05} = 70 \quad \Rightarrow \quad C = 70e^{1.05} \approx 200$$

It follows that $y(t) = 10 + 200e^{-2.1t}$ and the initial temperature was

$$y(0) = 10 + 200e^{-2.1(0)} = 10 + 200 = 210°C \qquad \blacksquare$$

The effect of air resistance depends on the physical situation. A high-speed bullet is affected differently than a skydiver. Our model is fairly realistic for a large object such as a skydiver falling from high altitudes.

The differential equation $y' = k(y - b)$ is also used to model vertical motion near the surface of the earth when air resistance is taken into account. Assume that the force due to air resistance is proportional to the velocity v and acts opposite to the direction of motion. We write this force as $-kv$, where $k > 0$. We take the upward direction to be positive, so $v < 0$ for a falling object and $-kv$ is an upward acting force, while $v > 0$ for a rising object and $-kv$ is a downward acting force.

The force due to gravity on an object of mass m is $-mg$, where g is the acceleration due to gravity, so the total force is $F = -mg - kv$. By Newton's Second Law of Motion,

$$F = ma = mv' \qquad (a = v' \text{ is the acceleration})$$

Thus, $mv' = -mg - kv$, which can be written as

In this model, k has units of mass per time, such as kilograms per second.

$$v' = -\frac{k}{m}\left(v + \frac{mg}{k}\right) \qquad \boxed{3}$$

This equation has the form $v' = -k(v - b)$ with k replaced by k/m and $b = -mg/k$. By Eq. (2), the general solution is

$$v(t) = -\frac{mg}{k} + Ce^{-(k/m)t} \qquad \boxed{4}$$

Since $Ce^{-(k/m)t}$ tends to zero as $t \to \infty$, $v(t)$ tends to a limiting terminal velocity:

$$\text{terminal velocity} = \lim_{t \to \infty} v(t) = -\frac{mg}{k} \qquad \boxed{5}$$

Without air resistance, the speed (absolute value of velocity) of a falling object would increase without bound until a sudden collision with the ground occurs. On the other hand, with air resistance the speed also increases, but levels off and approaches a limiting value of mg/k.

EXAMPLE 2 An 80-kg skydiver steps out of an airplane.

(a) What is her terminal velocity if $k = 8$ kg/s?

(b) What is her velocity after 30 s?

Solution

(a) By Eq. (5), with $k = 8$ kg/s and $g = 9.8$ m/s^2, the terminal velocity is

$$-\frac{mg}{k} = -\frac{(80)9.8}{8} = -98 \text{ m/s}$$

(b) With t in seconds, we have, by Eq. (4),

$$v(t) = -98 + Ce^{-(k/m)t} = -98 + Ce^{-(8/80)t} = -98 + Ce^{-0.1t}$$

We assume that the skydiver leaves the airplane with no initial vertical velocity, so $v(0) = -98 + C = 0$, and $C = 98$. Thus, we have $v(t) = -98(1 - e^{-0.1t})$ (Figure 3). The skydiver's velocity after 30 s is

$$v(30) = -98(1 - e^{-0.1(30)}) \approx -93.1 \text{ m/s} \qquad \blacksquare$$

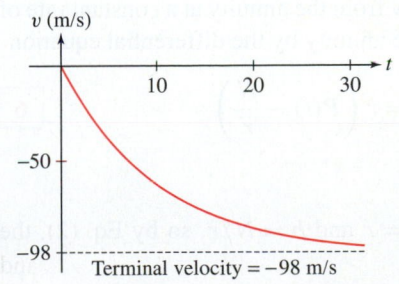

FIGURE 3 Velocity of 80-kg skydiver in free-fall with air resistance ($k = 8$ kg/s).

In Example 7 in Section 2.5 and Example 5 in Section 4.5, we investigated limits involving the maximum height attained by a 1-kg ball that is launched upward at 30 m/s and acted on by gravity and air resistance. We observed that the maximum height (in meters) depends on the strength of the air resistance (i.e., on the proportionality constant k), a relationship given by

$$H(k) = \frac{30k - 9.8 \ln\left(\frac{150k}{49} + 1\right)}{k^2}$$

Using Eq. (4), we can now derive this equation. In the next example, we find equations for the projectile's velocity $v(t)$ and height $y(t)$. The maximum height $H(k)$ is then found by determining the height when the velocity is zero (see Exercise 17).

EXAMPLE 3 A 1-kg ball is launched from the ground at 30 m/s and is acted on by air resistance that is expressed in the form $-kv$ and by gravity. Determine its velocity $v(t)$ and height $y(t)$.

Solution By Eq. (4), the projectile's velocity is

$$v(t) = -\frac{9.8}{k} + Ce^{-kt}$$

where C must be chosen so that $v(0) = 30$. That is, C must satisfy

$$30 = -\frac{9.8}{k} + C$$

Thus, $C = 30 + \frac{9.8}{k}$, and therefore,

$$v(t) = -\frac{9.8}{k} + \left(30 + \frac{9.8}{k}\right)e^{-kt}$$

Next, we take the antiderivative of $v(t)$ to find $y(t)$. We have

$$y(t) = -\frac{9.8}{k}t - \frac{1}{k}\left(30 + \frac{9.8}{k}\right)e^{-kt} + C$$

To satisfy $y(0) = 0$, we must have $C = \frac{1}{k}\left(30 + \frac{9.8}{k}\right)$. Using this equation to substitute for C in $y(t)$, and simplifying, we find

$$y(t) = -\frac{9.8}{k}t + \frac{30k + 9.8}{k^2}\left(1 - e^{-kt}\right)$$ ∎

An **annuity** is an investment in which an amount of money P_0, called the principal, is placed in an account that earns interest. Let $P(t)$ be the balance in the annuity (in dollars) after t years. When interest on the balance is **compounded continuously** at rate r, the rate of growth of the balance is proportional to the balance, and the proportionality constant is r. That is, $P'(t) = rP(t)$. If we withdraw from the annuity at a constant rate of N dollars per year, we can model the balance in the annuity by the differential equation

$$\underbrace{P'(t)}_{\substack{\text{Rate of}\\\text{change}}} = \underbrace{rP(t)}_{\substack{\text{Growth due}\\\text{to interest}}} - \underbrace{N}_{\substack{\text{Withdrawal}\\\text{rate}}} = r\left(P(t) - \frac{N}{r}\right) \qquad \boxed{6}$$

This equation has the form $y' = k(y - b)$ with $k = r$ and $b = N/r$, so by Eq. (2), the general solution is

$$P(t) = \frac{N}{r} + Ce^{rt} \qquad \boxed{7}$$

Since $P(0) = P_0$, we know $P_0 = \dfrac{N}{r} + C$ and, therefore, C is given by $C = P_0 - \dfrac{N}{r}$. Because e^{rt} tends to infinity as $t \to \infty$, the balance $P(t)$ tends to ∞ if $C > 0$. If $C < 0$, then $P(t)$ tends to $-\infty$ (i.e., the annuity eventually runs out of money). If $C = 0$, then $P(t)$ remains constant with value N/r.

EXAMPLE 4 Does an Annuity Pay Out Forever? An annuity earns continuously compounded interest at the rate $r = 0.07$, and withdrawals are made continuously at a rate of $N = \$500$/year.

(a) When will the annuity run out of money if the initial deposit is $P(0) = \$5000$?

(b) Show that the balance increases indefinitely if $P(0) = \$9000$.

Solution We have $N/r = \frac{500}{0.07} \approx 7143$, so $P(t) = 7143 + Ce^{0.07t}$ by Eq. (7).

(a) If $P(0) = 5000 = 7143 + Ce^0$, then $C = -2143$ and

$$P(t) = 7143 - 2143e^{0.07t}$$

The account runs out of money when $P(t) = 7143 - 2143e^{0.07t} = 0$, or

$$e^{0.07t} = \frac{7143}{2143} \quad \Rightarrow \quad 0.07t = \ln\left(\frac{7143}{2143}\right) \approx 1.2$$

The annuity money runs out at time $t = \frac{1.2}{0.07} \approx 17$ years.

(b) If $P(0) = 9000 = 7143 + Ce^0$, then $C = 1857$ and

$$P(t) = 7143 + 1857e^{0.07t}$$

Since the coefficient $C = 1857$ is positive, the account never runs out of money. In fact, $P(t)$ increases indefinitely as $t \to \infty$. Figure 4 illustrates the two cases. ■

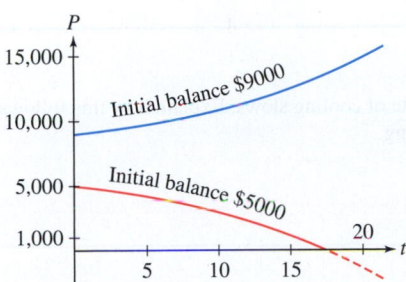

FIGURE 4 The balance in an annuity may increase indefinitely or decrease to zero (eventually becoming negative), depending on the size of initial deposit P_0, the interest rate, and the rate of withdrawal.

9.2 SUMMARY

- The general solution of $y' = k(y - b)$ is $y = b + Ce^{kt}$, where C is a constant.
- The following table describes the solutions to $y' = k(y - b)$ (see Figure 5):

Equation ($k > 0$)	Solution	Behavior as $t \to \infty$
$y' = k(y - b)$	$y(t) = b + Ce^{kt}$	$\lim\limits_{t\to\infty} y(t) = \begin{cases} \infty & \text{if } C > 0 \\ -\infty & \text{if } C < 0 \end{cases}$
$y' = -k(y - b)$	$y(t) = b + Ce^{-kt}$	$\lim\limits_{t\to\infty} y(t) = b$

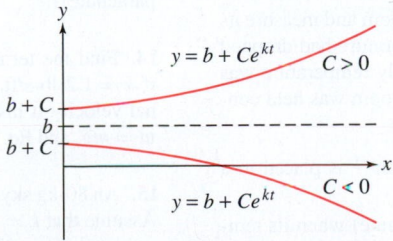

Solutions to $y' = k(y - b)$ with $k > 0$

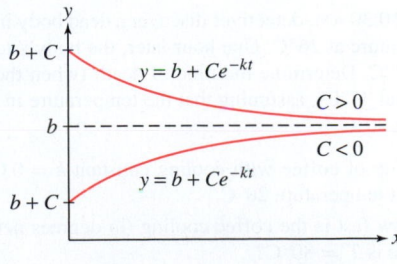

Solutions to $y' = -k(y - b)$ with $k > 0$

FIGURE 5

- Three applications:

  - Newton's Law of Cooling: $y' = -k(y - T_0)$, $y(t)$ = temperature of the object, T_0 = ambient temperature, k = cooling constant

  - Free-fall with air resistance: $v' = -\dfrac{k}{m}\left(v + \dfrac{mg}{k}\right)$, $v(t)$ = velocity, m = mass, k = air resistance constant, g = acceleration due to gravity

  - Continuous annuity: $P' = r\left(P - \dfrac{N}{r}\right)$, $P(t)$ = balance in the annuity, r = interest rate, N = withdrawal rate

9.2 EXERCISES

Preliminary Questions

1. Write a solution to $y' = 4(y - 5)$ that tends to $-\infty$ as $t \to \infty$.

2. Does $y' = -4(y - 5)$ have a solution that tends to ∞ as $t \to \infty$?

3. True or false? If $k > 0$, then all solutions of $y' = -k(y - b)$ approach the same limit as $t \to \infty$.

4. As an object cools, its rate of cooling slows. Explain how this follows from Newton's Law of Cooling.

Exercises

1. Find the general solution of $y' = 2(y - 10)$. Then find the two solutions satisfying $y(0) = 25$ and $y(0) = 5$, and sketch their graphs.

2. Verify directly that $y = 12 + Ce^{-3t}$ satisfies $y' = -3(y - 12)$ for all C. Then find the two solutions satisfying $y(0) = 20$ and $y(0) = 0$, and sketch their graphs.

3. Solve $y' = 4y + 24$ subject to $y(0) = 5$.

4. Solve $y' + 6y = 12$ subject to $y(2) = 10$.

In Exercises 5–12, use Newton's Law of Cooling.

5. A hot anvil with cooling constant $k = 0.02$ s^{-1} is submerged in a large pool of water whose temperature is $10°C$. Let $y(t)$ be the anvil's temperature t seconds later.

(a) What is the differential equation satisfied by $y(t)$?

(b) Find a formula for $y(t)$, assuming the object's initial temperature is $100°C$.

(c) How long does it take the object to cool down to $20°$?

6. Frank's automobile engine runs at $100°C$. On a day when the outside temperature is $21°C$, he turns off the ignition and notes that 5 minutes later, the engine has cooled to $70°C$.

(a) Determine the engine's cooling constant k.

(b) What is the formula for $y(t)$?

(c) When will the engine cool to $40°C$?

7. At 10:30 AM, detectives discover a dead body in a room and measure its temperature at $26°C$. One hour later, the body's temperature had dropped to $24.8°C$. Determine the time of death (when the body temperature was a normal $37°C$), assuming that the temperature in the room was held constant at $20°C$.

8. A cup of coffee with cooling constant $k = 0.09$ min^{-1} is placed in a room at temperature $20°C$.

(a) How fast is the coffee cooling (in degrees per minute) when its temperature is $T = 80°C$?

(b) Use the Linear Approximation to estimate the change in temperature over the next 6 s when $T = 80°C$.

(c) If the coffee is served at $90°C$, how long will it take to reach an optimal drinking temperature of $65°C$?

9. A cold metal bar at $-30°C$ is submerged in a pool maintained at a temperature of $40°C$. Half a minute later, the temperature of the bar is $20°C$. How long will it take for the bar to attain a temperature of $30°C$?

10. When a hot object is placed in a water bath whose temperature is $25°C$, it cools from $100°C$ to $50°C$ in 150 seconds. In another bath, the same cooling occurs in 120 s. Find the temperature of the second bath.

11. **GU** Objects A and B are placed in a warm bath at temperature $T_0 = 40°C$. Object A has initial temperature $-20°C$ and cooling constant $k = 0.004$ s^{-1}. Object B has initial temperature $0°C$ and cooling constant $k = 0.002$ s^{-1}. Plot the temperatures of A and B for $0 \le t \le 1000$. After how many seconds will the objects have the same temperature?

12. In Newton's Law of Cooling, the constant $\tau = 1/k$ is called the characteristic time. Show that τ is the time required for the temperature difference $(y - T_0)$ to decrease by the factor $e^{-1} \approx 0.37$. For example, if $y(0) = 100°C$ and $T_0 = 0°C$, then the object cools to $100/e \approx 37°C$ in time τ, to $100/e^2 \approx 13.5°C$ in time 2τ, and so on.

In Exercises 13–16, use Eq. (3) as a model for free-fall with air resistance.

13. A 60-kg skydiver jumps out of an airplane. What is her terminal velocity, in meters per second, assuming that $k = 10$ kg/s for free-fall (no parachute)?

14. Find the terminal velocity of a skydiver of weight $w = 192$ pounds if $k = 1.2$ lb-s/ft. How long does it take him to reach half of his terminal velocity if his initial velocity is zero? Mass and weight are related by $w = mg$, and Eq. (3) becomes $v' = -(kg/w)(v + w/k)$ with $g = 32$ ft/s^2.

15. An 80-kg skydiver jumps out of an airplane (with zero initial velocity). Assume that $k = 12$ kg/s with a closed parachute and $k = 70$ kg/s with an open parachute. What is the skydiver's velocity at $t = 25$ s if the parachute opens after 20 s of free-fall?

16. Does a heavier or a lighter skydiver reach terminal velocity more quickly?

17. As in Example 3, a 1-kg ball is launched upward at 30 m/s and is acted on by gravity and air resistance.

(a) Show that the ball's velocity is zero at time $t^* = \dfrac{1}{k} \ln\left(\dfrac{30k}{9.8} + 1\right)$.

(b) Show that $y(t^*) = \dfrac{30k - 9.8 \ln\left(\dfrac{150k}{49} + 1\right)}{k^2}$ (thereby establishing the formula for $H(k)$ given prior to the example).

18. A 500 g ball is launched upward at 60 m/s and is acted on by gravity and air resistance that can be expressed in the form $-kv$, where v is the ball's velocity.

(a) Determine the ball's velocity $v(t)$, expressed in terms of k.

(b) Determine the ball's height $y(t)$, expressed in terms of k.

(c) Determine $H(k)$, the maximum height reached by the ball, as a function of k.

In Exercises 19(a)–(f), use Eqs. (6) and (7) that describe the balance in a continuous annuity.

19. (a) A continuous annuity with withdrawal rate $N = \$5000$/year and interest rate $r = 5\%$ is funded by an initial deposit of $P_0 = \$50,000$.

i. What is the balance in the annuity after 10 years?

ii. When will the annuity run out of funds?

(b) Show that a continuous annuity with withdrawal rate $N = \$5000$/year and interest rate $r = 8\%$, funded by an initial deposit of $P_0 = \$75,000$, never runs out of money.

(c) Find the minimum initial deposit P_0 that will allow an annuity to pay out $\$6000$/year indefinitely if it earns interest at a rate of 5%.

(d) Find the minimum initial deposit P_0 necessary to fund an annuity for 20 years if withdrawals are made at a rate of $\$10,000$/year and interest is earned at a rate of 7%.

(e) An initial deposit of 100,000 euros is placed in an annuity with a French bank. What is the minimum interest rate the annuity must earn to allow withdrawals at a rate of 8000 euros/year to continue indefinitely?

(f) Show that a continuous annuity never runs out of money if the initial balance is greater than or equal to N/r, where N is the withdrawal rate and r the interest rate.

20. Sam borrows $\$10,000$ from a bank at an interest rate of 9% and pays back the loan continuously at a rate of N dollars per year. Let $P(t)$ denote the amount still owed at time t.

(a) Explain why $P(t)$ satisfies the differential equation

$$y' = 0.09y - N$$

(b) How long will it take Sam to pay back the loan if $N = \$1200$?

(c) Will the loan ever be paid back if $N = \$800$?

21. April borrows $\$18,000$ at an interest rate of 5% to purchase a new automobile. At what rate (in dollars per year) must she pay back the loan, if the loan must be paid off in 5 years? *Hint:* Set up the differential equation as in Exercise 20.

22. Let $N(t)$ be the fraction of the population who have heard a given piece of news t hours after its initial release. According to one model, the rate $N'(t)$ at which the news spreads is equal to k times the fraction of the population that has not yet heard the news, for some constant $k > 0$.

(a) Determine the differential equation satisfied by $N(t)$.

(b) Find the solution of this differential equation with the initial condition $N(0) = 0$ in terms of k.

(c) Suppose that half of the population is aware of an earthquake 8 hours after it occurs. Use the model to calculate k and estimate the percentage that will know about the earthquake 12 h after it occurs.

23. Current in a Circuit When the circuit in Figure 6 (which consists of a battery of V volts, a resistor of R ohms, and an inductor of L henries) is connected, the current $I(t)$ flowing in the circuit satisfies

$$L\frac{dI}{dt} + RI = V$$

with the initial condition $I(0) = 0$.

(a) Find a formula for $I(t)$ in terms of L, V, and R.

(b) Show that $\lim_{t \to \infty} I(t) = V/R$.

(c) Show that $I(t)$ reaches approximately 63% of its maximum value at the characteristic time $\tau = L/R$.

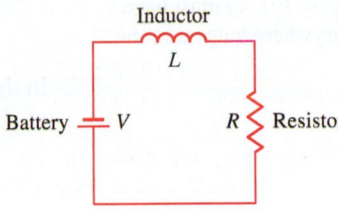

FIGURE 6 Current flow approaches the level $I_{\max} = V/R$.

Further Insights and Challenges

24. Show that the cooling constant of an object can be determined from two temperature readings $y(t_1)$ and $y(t_2)$ at times $t_1 \neq t_2$ by the formula

$$k = \frac{1}{t_1 - t_2} \ln\left(\frac{y(t_2) - T_0}{y(t_1) - T_0}\right)$$

25. Show that by Newton's Law of Cooling, the time required to cool an object from temperature A to temperature B is

$$t = \frac{1}{k} \ln\left(\frac{A - T_0}{B - T_0}\right)$$

where T_0 is the ambient temperature.

26. A projectile of mass m kg travels straight up from ground level with initial velocity v_0 m/s. Suppose that the velocity v satisfies $v'(t) = -9.8 - \dfrac{k}{m}v(t)$.

(a) Determine the velocity $v(t)$.

(b) Show that the projectile's velocity is zero at

$$t^* = \frac{m}{k} \ln\left(\frac{v_0 k}{9.8m} + 1\right)$$

(c) Determine the height $y(t)$.

(d) The maximum height is $y(t^*)$. Show that

$$y(t^*) = \frac{mv_0 k - 9.8m^2 \ln\left(\dfrac{v_0 k}{9.8m} + 1\right)}{k^2}$$

(e) If air resistance is negligible, then the maximum height reached by the projectile is $\lim_{k \to 0} y(t^*)$. Determine the limit.

9.3 Graphical and Numerical Methods

In the previous two sections, we focused on finding solutions to differential equations. Differential equations cannot always be solved explicitly. Fortunately, there are techniques for analyzing the solutions that do not rely on explicit formulas. Even for differential equations that have exact solutions, these techniques are valuable because they provide us with extra insight into the behavior of the solutions. In this section, we discuss the method of slope fields, which provides us with a good visual understanding of first-order equations. We also discuss Euler's Method for finding numerical approximations to solutions.

We use t as the independent variable. A first-order differential equation can then be written in the form

$$\frac{dy}{dt} = F(t, y)$$ **1**

where $F(t, y)$ is a function of t and y. The differential equation indicates that $\frac{dy}{dt}$, the slope of the graph of a solution $y = y(t)$, at a point (t, y), is given by $F(t, y)$.

It is useful to think of Eq. (1) as a set of instructions that "tells a solution" which direction to go in. Thus, a solution passing through a point (t, y) is "instructed" to do so in the direction of the slope $F(t, y)$ because the differential equation requires $F(t, y)$ to be the slope of the graph of the solution at that point. To visualize this set of instructions, we draw a **slope field**, which is an array of small segments of slope $F(t, y)$ at points (t, y) lying on a rectangular grid in the plane as in Figure 1. Solutions to the differential equation $\frac{dy}{dt} = F(t, y)$ must have graphs that are everywhere tangent to the slope field, and therefore, without knowing the solutions, we can picture how their graphs must appear within the slope field.

To illustrate, let's consider the differential equation

$$\frac{dy}{dt} = -ty$$

In this case, $F(t, y) = -ty$. To sketch a slope field, we first compute $F(t, y)$ for an array of points in the ty-plane:

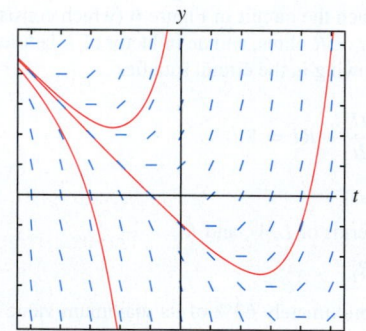

FIGURE 1 Each segment has slope $F(t, y)$, and solutions to $\frac{dy}{dt} = F(t, y)$ must have graphs that are everywhere tangent to the slope field.

	Values of **F(t, y)**		
$F(-1, 1) = 1$	$F(0, 1) = 0$	$F(1, 1) = -1$	$F(2, 1) = -2$
$F(-1, 0) = 0$	$F(0, 0) = 0$	$F(1, 0) = 0$	$F(2, 0) = 0$
$F(-1, -1) = -1$	$F(0, -1) = 0$	$F(1, -1) = 1$	$F(2, -1) = 2$

Then, at each point (t, y) plot a segment of slope $F(t, y)$ as in Figure 2.

A slope field is much easier to generate using technology than by hand. Many graphing calculators and computer algebra systems can produce slope fields. There are online programs available as well. We used technology to generate the slope field for $\frac{dy}{dt} = -ty$ in Figure 3(A).

We can use the slope field to see how solutions appear. The slope field enables us to visualize many solutions at a glance. Starting at any point, and going both to the left and to the right from the point, we can sketch the graph of a solution by drawing a curve that runs tangent to the slope segments at each point [Figure 3(B)]. The graph of a solution is also called an **integral curve**.

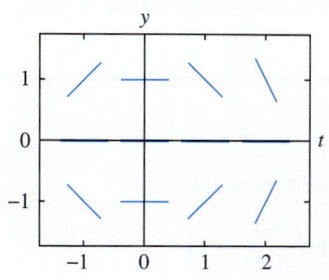

FIGURE 2 Slope field for $F(t, y) = -ty$.

EXAMPLE 1 **Using Isoclines** Draw the slope field for

$$\frac{dy}{dt} = y - t$$

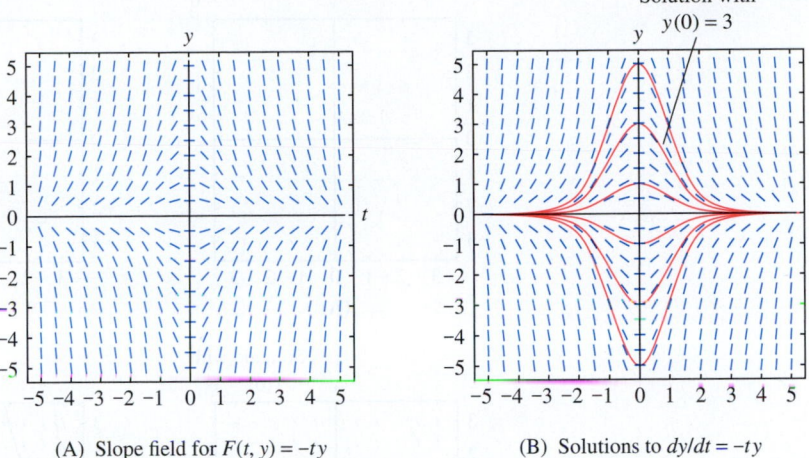

(A) Slope field for $F(t, y) = -ty$ (B) Solutions to $dy/dt = -ty$

FIGURE 3

and sketch the integral curves satisfying the initial conditions

$$\textbf{(a)} \ \ y(0) = 1 \quad \text{and} \quad \textbf{(b)} \ \ y(1) = -2.$$

Solution A good way to sketch the slope field of $\dfrac{dy}{dt} = F(t, y)$ is to choose several values c and identify the curve $F(t, y) = c$, called the **isocline** of slope c. The isocline is the curve consisting of all points where the slope field has slope c.

In our case, $F(t, y) = y - t$, so the isocline of fixed slope c has equation $y - t = c$, or $y = t + c$, which is a line. Consider the following values:

- $c = 0$: This isocline is $y - t = 0$, or $y = t$. We draw segments of slope $c = 0$ at points along the line $y = t$, as in Figure 4(A).
- $c = 1$: This isocline is $y - t = 1$, or $y = t + 1$. We draw segments of slope 1 at points along $y = t + 1$, as in Figure 4(B).
- $c = 2$: This isocline is $y - t = 2$, or $y = t + 2$. We draw segments of slope 2 at points along $y = t + 2$, as in Figure 4(C).
- $c = -1$: This isocline is $y - t = -1$, or $y = t - 1$ [Figure 4(C)].

A more detailed slope field is shown in Figure 4(D). To sketch the solution satisfying $y(0) = 1$, begin at the point $(t_0, y_0) = (0, 1)$ and draw the integral curve passing through $(0, 1)$ that is everywhere tangent to the slope field. The solution satisfying $y(1) = -2$ is sketched similarly. Figure 4(E) shows several other solutions (integral curves). ∎

GRAPHICAL INSIGHT Slope fields often let us see the *asymptotic* behavior of solutions (as $t \to \infty$) at a glance. Figure 4(E) suggests that the asymptotic behavior depends on the initial value. For example, if $y(0) > 1$, then $y(t)$ tends to ∞, and if $y(0) < 1$, then $y(t)$ tends to $-\infty$. We can check this using the general solution $y(t) = 1 + t + Ce^t$, where $y(0) = 1 + C$. If $y(0) > 1$, then $C > 0$ and $y(t)$ tends to ∞, but if $y(0) < 1$, then $C < 0$ and $y(t)$ tends to $-\infty$. The solution $y = 1 + t$ with initial condition $y(0) = 1$ is the straight line shown in Figure 4(D).

EXAMPLE 2 **Newton's Law of Cooling Revisited** The temperature $y(t)$ (in degrees Celsius) of an object placed in a refrigerator that is kept at $4°C$ satisfies $\dfrac{dy}{dt} = -0.5(y - 4)$ (t in minutes). Draw the slope field and describe the behavior of the solutions.

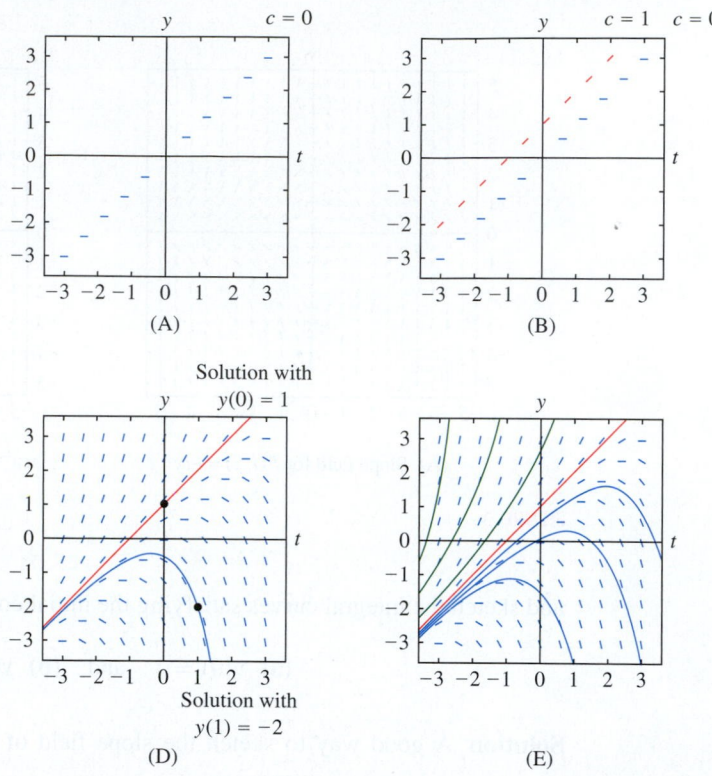

FIGURE 4 Drawing the slope field for $\dfrac{dy}{dt} = y - t$ using isoclines.

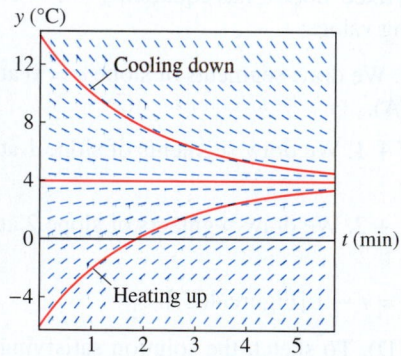

FIGURE 5 Slope field for $y' = -0.5(y - 4)$.

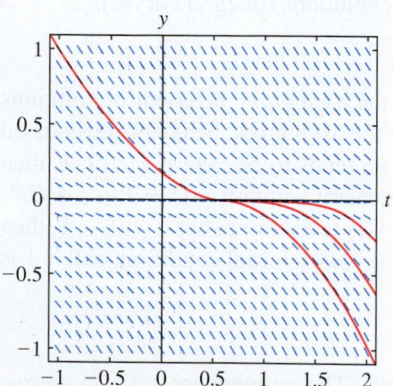

FIGURE 6 Overlapping integral curves for $\dfrac{dy}{dt} = -\sqrt{|y|}$ (uniqueness fails for this differential equation).

Solution The function $F(t, y) = -0.5(y - 4)$ depends only on y, so slopes of the segments in the slope field do not vary in the t-direction. The slope $F(t, y)$ is positive for $y < 4$ and negative for $y > 4$. More precisely, the slope at height y is $-0.5(y - 4) = -0.5y + 2$, so the segments grow steeper with positive slope as $y \to -\infty$, and they grow steeper with negative slope as $y \to \infty$ (Figure 5).

The slope field shows that if the initial temperature satisfies $y_0 > 4$, then $y(t)$ decreases to $y = 4$ as $t \to \infty$. In other words, the object cools down to 4°C when placed in the refrigerator. If $y_0 < 4$, then $y(t)$ increases to $y = 4$ as $t \to \infty$. The object warms up when placed in the refrigerator. If $y_0 = 4$, then y remains at 4°C for all time t. ∎

CONCEPTUAL INSIGHT Most first-order equations arising in applications have a uniqueness property: There is precisely one solution $y(t)$ satisfying a given initial condition $y(t_0) = y_0$. Graphically, this means that precisely one integral curve (solution) passes through the point (t_0, y_0). Thus, when uniqueness holds, distinct integral curves never cross or overlap. Figure 6 shows the slope field of $\dfrac{dy}{dt} = -\sqrt{|y|}$, where uniqueness fails. We can prove that once an integral curve touches the t-axis, it either remains on the t-axis or continues along the t-axis for a period of time before moving below the t-axis. Therefore, infinitely many integral curves pass through each point on the t-axis. However, the slope field does not show this clearly. Thus, when possible, it is important to obtain and analyze solutions rather than just rely on visual impressions alone.

Euler's Method

Euler's Method produces numerical approximations to the solution $y(t)$ of a first-order Initial Value Problem:

$$\frac{dy}{dt} = F(t, y), \qquad y(t_0) = y_0 \qquad \boxed{2}$$

Euler's Method is the simplest method for solving Initial Value Problems numerically, but it is not very efficient. Computer systems use more sophisticated schemes, making it possible to plot and analyze solutions to the complex systems of differential equations arising in areas such as weather prediction, aerodynamic modeling, and economic forecasting.

We begin by choosing a small number h, called the **time step**, and consider the sequence of times starting at the initial value t_0 and spaced at intervals of size h:

$$t_0, \quad t_1 = t_0 + h, \quad t_2 = t_0 + 2h, \quad t_3 = t_0 + 3h, \quad \cdots$$

In general, $t_k = t_0 + kh$ for $k = 0, 1, 2, \dots$. Euler's Method consists of computing a sequence of values $y_1, y_2, y_3, \dots, y_n$ successively using the formula

$$\boxed{y_k = y_{k-1} + hF(t_{k-1}, y_{k-1})} \qquad \boxed{3}$$

Each y_k is an approximation to the value of the solution to the Initial Value Problem at t_k; that is, $y_k \approx y(t_k)$. Starting with the initial value $y_0 = y(t_0)$, we compute

$$y_1 = y_0 + hF(t_0, y_0), \quad y_2 = y_1 + hF(t_1, y_1), \quad \text{etc.}$$

We connect the points (t_k, y_k) by segments to obtain an approximation to the graph of $y(t)$ (Figure 7).

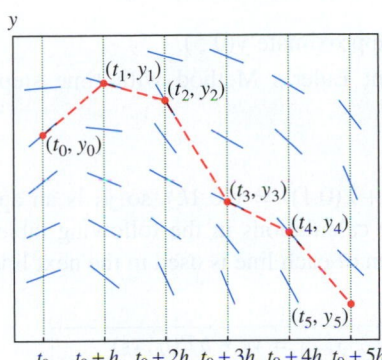

GRAPHICAL INSIGHT The values y_k are defined so that the line segment joining (t_{k-1}, y_{k-1}) to (t_k, y_k) has slope

$$\frac{y_k - y_{k-1}}{t_k - t_{k-1}} = \frac{(y_{k-1} + hF(t_{k-1}, y_{k-1})) - y_{k-1}}{h} = F(t_{k-1}, y_{k-1})$$

Thus, in Euler's Method, we move from (t_{k-1}, y_{k-1}) to (t_k, y_k) by traveling on a line segment in the direction specified by the slope field at (t_{k-1}, y_{k-1}) for a time interval of length h (Figure 7).

FIGURE 7 In Euler's Method, we move from one point to the next by traveling along the line indicated by the slope field. The line is not necessarily an integral curve, but provides an approximation to one.

EXAMPLE 3 Use Euler's Method with time step $h = 0.2$ and $n = 4$ steps to approximate the solution of $\dfrac{dy}{dt} = y - t^2$, $y(0) = 3$ at $t = 0.2, 0.4, 0.6,$ and 0.8.

Solution Our initial value at $t_0 = 0$ is $y_0 = 3$. The time values are $t_1 = 0.2$, $t_2 = 0.4$, $t_3 = 0.6$, and $t_4 = 0.8$. We use Eq. (3) with $F(t, y) = y - t^2$ to calculate

$$y(0.2) \approx y_1 = y_0 + hF(t_0, y_0) = 3 + 0.2(3 - (0)^2) = 3.6$$

$$y(0.4) \approx y_2 = y_1 + hF(t_1, y_1) = 3.6 + 0.2(3.6 - (0.2)^2) \approx 4.3$$

$$y(0.6) \approx y_3 = y_2 + hF(t_2, y_2) = 4.3 + 0.2(4.3 - (0.4)^2) \approx 5.14$$

$$y(0.8) \approx y_4 = y_3 + hF(t_3, y_3) = 5.14 + 0.2(5.14 - (0.6)^2) \approx 6.1 \qquad \blacksquare$$

For comparison, Figure 8(A) shows the exact solution, $y(t) = 2 + 2t + t^2 + e^t$, to the Initial Value Problem in the previous example, together with a plot of the points (t_k, y_k) for $k = 0, 1, 2, 3, 4$ connected by line segments.

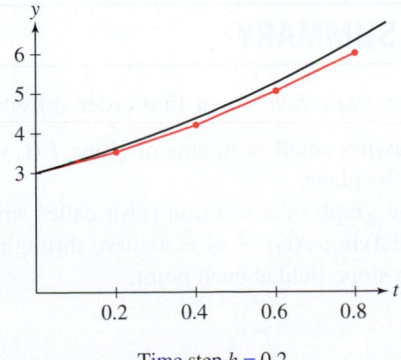

Time step $h = 0.2$

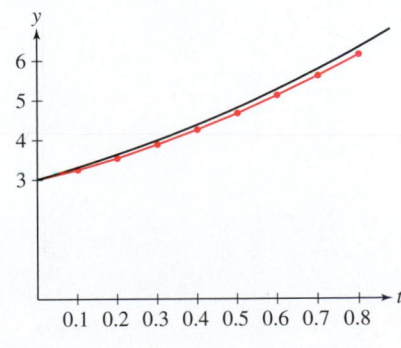

Time step $h = 0.1$

DF **FIGURE 8** Euler's Method applied to $\dfrac{dy}{dt} = y - t^2$, $y(0) = 3$.

CONCEPTUAL INSIGHT Figure 8(B) shows that the time step $h = 0.1$ gives a better approximation than $h = 0.2$. In general, the smaller the time step, the better the approximation. In fact, if we start at a point $(a, y(a))$ and use Euler's Method to approximate $(b, y(b))$ using N steps with $h = (b - a)/N$, then the error is roughly proportional to $1/N$ [provided that $F(t, y)$ is a well-behaved function]. This is similar to the error size in the Nth left- and right-endpoint approximations to an integral. What this means, however, is that Euler's Method is quite inefficient; to cut the error in half, it is necessary to double the number of steps, and to achieve n-digit accuracy requires roughly 10^n steps. Fortunately, there are several methods that improve on Euler's Method in much the same way as the Midpoint Rule and Simpson's Rule improve on the endpoint integral approximations (see Exercises 24–29).

EXAMPLE 4 (CAS) Let $y(t)$ be the solution of $\dfrac{dy}{dt} = \sin t \cos y$, $y(0) = 0$.

(a) Use Euler's Method with time step $h = 0.1$ to approximate $y(0.5)$.

(b) Use a computer algebra system to implement Euler's Method with time steps $h = 0.01, 0.001$, and 0.0001 to approximate $y(0.5)$.

Solution

(a) When $h = 0.1$, y_k is an approximation to $y(0 + k(0.1)) = y(0.1k)$, so y_5 is an approximation to $y(0.5)$. It is convenient to organize calculations in the following table. Note that the value y_{k+1} computed in the last column of each line is used in the next line to continue the process.

Euler's Method:

$$y_k = y_{k-1} + hF(t_{k-1}, y_{k-1})$$

t_k	y_k	$F(t_k, y_k) = \sin t_k \cos y_k$	$y_{k+1} = y_k + hF(t_k, y_k)$
$t_0 = 0$	$y_0 = 0$	$(\sin 0)\cos 0 = 0$	$y_1 = 0 + 0.1(0) = 0$
$t_1 = 0.1$	$y_1 = 0$	$(\sin 0.1)\cos 0 \approx 0.1$	$y_2 \approx 0 + 0.1(0.1) = 0.01$
$t_2 = 0.2$	$y_2 \approx 0.01$	$(\sin 0.2)\cos(0.01) \approx 0.2$	$y_3 \approx 0.01 + 0.1(0.2) = 0.03$
$t_3 = 0.3$	$y_3 \approx 0.03$	$(\sin 0.3)\cos(0.03) \approx 0.3$	$y_4 \approx 0.03 + 0.1(0.3) = 0.06$
$t_4 = 0.4$	$y_4 \approx 0.06$	$(\sin 0.4)\cos(0.06) \approx 0.4$	$y_5 \approx 0.06 + 0.1(0.4) = 0.10$

Thus, Euler's Method yields the approximation $y(0.5) \approx y_5 \approx 0.1$.

(b) When the number of steps is large, the calculations are too lengthy to do by hand, but they are easily carried out using a CAS. Note that for $h = 0.01$, the kth value y_k is an approximation to $y(0 + k(0.01)) = y(0.01k)$, and y_{50} gives an approximation to $y(0.5)$. Similarly, when $h = 0.001$, y_{500} is an approximation to $y(0.5)$, and when $h = 0.0001$, y_{5000} is an approximation to $y(0.5)$. Here are the results obtained using a CAS:

Time step $h = 0.01$	$y(0.5) \approx y_{50} \approx 0.1197$
Time step $h = 0.001$	$y(0.5) \approx y_{500} \approx 0.1219$
Time step $h = 0.0001$	$y(0.5) \approx y_{5000} \approx 0.1221$

The values appear to converge and we may assume that $y(0.5) \approx 0.12$. However, we see here that Euler's Method converges quite slowly. ∎

9.3 SUMMARY

- The *slope field* for a first-order differential equation $\dfrac{dy}{dt} = F(t, y)$ is obtained by drawing small segments of slope $F(t, y)$ at points (t, y) lying on a rectangular grid in the plane.
- The graph of a solution (also called an *integral curve* of the differential equation) satisfying $y(t_0) = y_0$ is a curve through (t_0, y_0) that runs tangent to the segments of the slope field at each point.

- Euler's Method: To approximate a solution to $\dfrac{dy}{dt} = F(t, y)$ with initial condition $y(t_0) = y_0$, fix a time step h and set $t_k = t_0 + kh$. Define $y_1, y_2, \ldots$ successively by the formula

$$\boxed{y_k = y_{k-1} + hF(t_{k-1}, y_{k-1})} \qquad \textbf{4}$$

The values $y_0, y_1, y_2, \ldots$ are approximations to the values $y(t_0), y(t_1), y(t_2), \ldots$.

9.3 EXERCISES

Preliminary Questions

1. What is the slope of the segment in the slope field for $\dfrac{dy}{dt} = ty + 1$ at the point $(2, 3)$?

2. What is the equation of the isocline of slope $c = 1$ for $\dfrac{dy}{dt} = y^2 - t$?

3. For which of the following differential equations are the slopes at points on a vertical line $t = C$ all equal?

(a) $\dfrac{dy}{dt} = \ln y$ **(b)** $\dfrac{dy}{dt} = \ln t$

4. Let $y(t)$ be the solution to $\dfrac{dy}{dt} = F(t, y)$ with $y(1) = 3$. How many iterations of Euler's Method are required to approximate $y(3)$ if the time step is $h = 0.1$?

Exercises

1. Figure 9 shows the slope field for $\dfrac{dy}{dt} = \sin y \sin t$. Sketch the graphs of the solutions with initial conditions $y(0) = 1$ and $y(0) = -1$. Show that $y(t) = 0$ is a solution and add its graph to the plot.

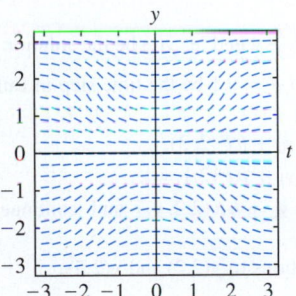

FIGURE 9 Slope field for $\dfrac{dy}{dt} = \sin y \sin t$.

2. Figure 10 shows the slope field for $\dfrac{dy}{dt} = y^2 - t^2$. Sketch the integral curve passing through the point $(0, -1)$, the curve through $(0, 0)$, and the curve through $(0, 2)$. Is $y(t) = 0$ a solution?

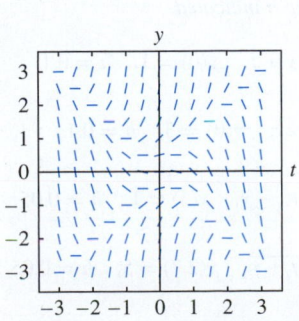

FIGURE 10 Slope field for $\dfrac{dy}{dt} = y^2 - t^2$.

3. Show that $f(t) = \frac{1}{2}\left(t - \frac{1}{2}\right)$ is a solution to $\dfrac{dy}{dt} = t - 2y$. Sketch the four solutions with $y(0) = \pm 0.5, \pm 1$ on the slope field in Figure 11. The slope field suggests that every solution approaches $f(t)$ as $t \to \infty$. Confirm this by showing that $y = f(t) + Ce^{-2t}$ is the general solution.

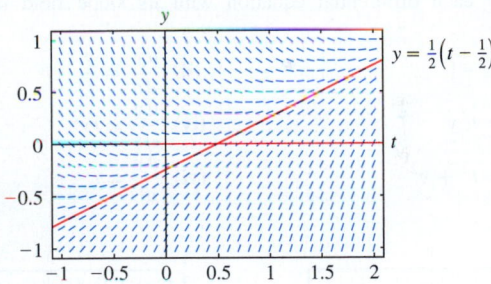

FIGURE 11 Slope field for $\dfrac{dy}{dt} = t - 2y$.

4. One of the slope fields in Figures 12(A) and (B) is the slope field for $\dfrac{dy}{dt} = t^2$. The other is for $\dfrac{dy}{dt} = y^2$. Identify which is which. In each case, sketch the solutions with initial conditions $y(0) = 1$, $y(0) = 0$, and $y(0) = -1$.

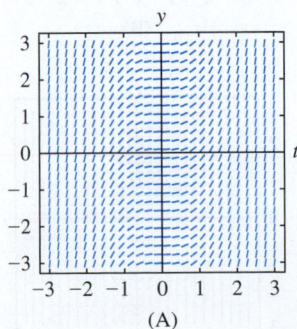

(A)

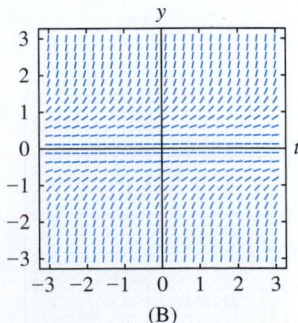

(B)

FIGURE 12

5. Consider the differential equation $\dfrac{dy}{dt} = t - y$.

(a) Sketch the slope field of the differential equation $\dfrac{dy}{dt} = t - y$ in the range $-1 \le t \le 3$, $-1 \le y \le 3$. As an aid, observe that the isocline of slope c is the line $t - y = c$, so the segments have slope c at points on the line $y = t - c$.

(b) Show that $y = t - 1 + Ce^{-t}$ is a solution for all C. Since $\lim\limits_{t\to\infty} e^{-t} = 0$, these solutions approach the particular solution $y = t - 1$ as $t \to \infty$. Explain how this behavior is reflected in your slope field.

6. Show that the isoclines of $\dfrac{dy}{dt} = 1/y$ are horizontal lines. Sketch the slope field for $-2 \le t \le 2$, $-2 \le y \le 2$ and plot the solutions with initial conditions $y(0) = 0$ and $y(0) = 1$.

7. Sketch the slope field for $\dfrac{dy}{dt} = y + t$ for $-2 \le t \le 2$, $-2 \le y \le 2$.

8. Sketch the slope field for $\dfrac{dy}{dt} = \dfrac{t}{y}$ for $-2 \le t \le 2$, $-2 \le y \le 2$.

9. Show that the isoclines of $\dfrac{dy}{dt} = t$ are vertical lines. Sketch the slope field for $-2 \le t \le 2$, $-2 \le y \le 2$ and plot the integral curves passing through $(0, -1)$ and $(0, 1)$.

10. Sketch the slope field of $\dfrac{dy}{dt} = ty$ for $-2 \le t \le 2$, $-2 \le y \le 2$. Based on the sketch, determine $\lim\limits_{t\to\infty} y(t)$, where $y(t)$ is a solution with $y(0) > 0$. What is $\lim\limits_{t\to\infty} y(t)$ if $y(0) < 0$?

11. Match each differential equation with its slope field in Figures 13(A)–(F).

(i) $\dfrac{dy}{dt} = -1$ **(ii)** $\dfrac{dy}{dt} = \dfrac{y}{t}$

(iii) $\dfrac{dy}{dt} = t^2 y$ **(iv)** $\dfrac{dy}{dt} = ty^2$

(v) $\dfrac{dy}{dt} = t^2 + y^2$ **(vi)** $\dfrac{dy}{dt} = t$

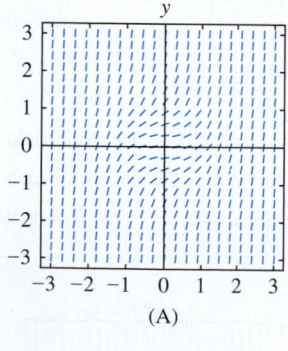

(A)

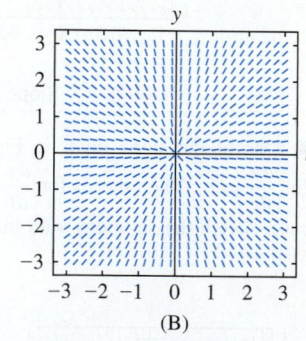

(B)

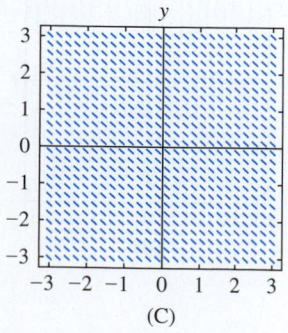

(C)

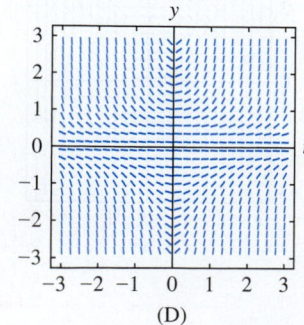

(D)

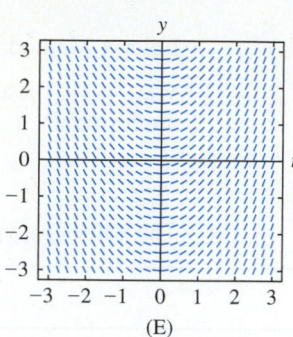

(E)

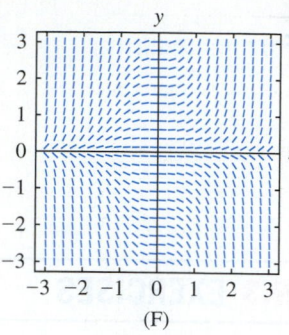

(F)

FIGURE 13

12. Sketch the solution of $\dfrac{dy}{dt} = ty^2$ satisfying $y(0) = 1$ in the appropriate slope field of Figure 13(A)–(F). Then show, using Separation of Variables, that if $y(t)$ is a solution such that $y(0) > 0$, then $y(t)$ tends to infinity as $t \to \sqrt{2/y(0)}$.

13. (a) Sketch the slope field of $\dfrac{dy}{dt} = t/y$ in the region $-2 \le t \le 2$, $-2 \le y \le 2$.

(b) Check that $y = \pm\sqrt{t^2 + C}$ is the general solution.

(c) Sketch the solutions on the slope field with initial conditions $y(0) = 1$ and $y(0) = -1$.

14. Sketch the slope field of $\dfrac{dy}{dt} = t^2 - y$ in the region $-3 \le t \le 3$, $-3 \le y \le 3$ and sketch the solutions satisfying $y(1) = 0$, $y(1) = 1$, and $y(1) = -1$.

15. Let $F(t, y) = t^2 - y$ and let $y(t)$ be the solution of $\dfrac{dy}{dt} = F(t, y)$ satisfying $y(2) = 3$. Let $h = 0.1$ be the time step in Euler's Method, and set $y_0 = y(2) = 3$.

(a) Calculate $y_1 = y_0 + hF(2, 3)$.

(b) Calculate $y_2 = y_1 + hF(2.1, y_1)$.

(c) Calculate $y_3 = y_2 + hF(2.2, y_2)$ and continue computing y_4, y_5, and y_6.

(d) Find approximations to $y(2.2)$ and $y(2.5)$.

16. Let $y(t)$ be the solution to $\dfrac{dy}{dt} = te^{-y}$ satisfying $y(0) = 0$.

(a) Use Euler's Method with time step $h = 0.1$ to approximate $y(0.1), y(0.2), \ldots, y(0.5)$.

(b) Use Separation of Variables to find $y(t)$ exactly.

(c) Compute the errors in the approximations to $y(0.1)$ and $y(0.5)$.

In Exercises 17–22, use Euler's Method to approximate the given value of $y(t)$ with the time step h indicated.

17. $y(0.5)$; $\dfrac{dy}{dt} = y + t$, $y(0) = 1$, $h = 0.1$

18. $y(0.7)$; $\dfrac{dy}{dt} = 2y$, $y(0) = 3$, $h = 0.1$

19. $y(3.3)$; $\dfrac{dy}{dt} = t^2 - y$, $y(3) = 1$, $h = 0.05$

20. $y(3)$; $\dfrac{dy}{dt} = \sqrt{t + y}$, $y(2.7) = 5$, $h = 0.05$

21. $y(2)$; $\dfrac{dy}{dt} = t \sin y$, $y(1) = 2$, $h = 0.2$

22. $y(5.2)$; $\dfrac{dy}{dt} = t - \sec y$, $y(4) = -2$, $h = 0.2$

Further Insights and Challenges

23. If f is continuous on $[a, b]$, then the solution to $\dfrac{dy}{dt} = f(t)$ with initial condition $y(a) = 0$ is $y(t) = \displaystyle\int_a^t f(u)\, du$. Show that Euler's Method with time step $h = (b - a)/N$ for N steps yields the Nth left-endpoint approximation to $y(b) = \displaystyle\int_a^b f(u)\, du$.

*Exercises 24–29: **Euler's Midpoint Method** is a variation on Euler's Method that is significantly more accurate in general. For time step h and initial value $y_0 = y(t_0)$, the values y_k are defined successively by*

$$y_k = y_{k-1} + h m_{k-1}$$

where $m_{k-1} = F\left(t_{k-1} + \dfrac{h}{2}, y_{k-1} + \dfrac{h}{2} F(t_{k-1}, y_{k-1})\right)$.

24. Apply both Euler's Method and the Euler Midpoint Method with $h = 0.1$ to estimate $y(1.5)$, where $y(t)$ satisfies $\dfrac{dy}{dt} = y$ with $y(0) = 1$. Find $y(t)$ exactly and compute the errors in these two approximations.

In Exercises 25–28, use Euler's Midpoint Method with the time step indicated to approximate the given value of $y(t)$.

25. $y(0.5)$; $\dfrac{dy}{dt} = y + t$, $y(0) = 1$, $h = 0.1$

26. $y(2)$; $\dfrac{dy}{dt} = t^2 - y$, $y(1) = 3$, $h = 0.2$

27. $y(0.25)$; $\dfrac{dy}{dt} = \cos(y + t)$, $y(0) = 1$, $h = 0.05$

28. $y(2.3)$; $\dfrac{dy}{dt} = y + t^2$, $y(2) = 1$, $h = 0.05$

29. Assume that f is continuous on $[a, b]$. Show that Euler's Midpoint Method applied to $\dfrac{dy}{dt} = f(t)$ with initial condition $y(a) = 0$ and time step $h = (b - a)/N$ for N steps yields the Nth midpoint approximation to

$$y(b) = \int_a^b f(u)\, du$$

9.4 The Logistic Equation

Mathematician Pierre-François Verhulst (1804–1849) first introduced the logistic equation in 1838 in a study of the population of his native country of Belgium.

In Section 9.1, we introduced the population growth model $dy/dt = ky$. The solutions to this differential equation (with $k > 0$) imply that the population grows exponentially. Although such a growth model might be valid over an initial short time period, no population can increase without limit because needed resources such as food or land are finite. To model a population subject to finite resources, we adjust the assumptions underlying the differential equation. If $y(t)$ represents the population at time t and A denotes the maximum population that the environment can support, we let $A - y(t)$ represent the room available for growth. Now we assume:

The rate of change of y is proportional to the amount y(t) present and the amount A − y(t) of room for growth.

Translating this relationship into a differential equation, we obtain what is known as the **logistic differential equation**:

$$\frac{dy}{dt} = Ky(A - y)$$

where K is a proportionality constant. For convenience, we rewrite the right side slightly:

$$Ky(A - y) = KyA\left(1 - \frac{y}{A}\right) = ky\left(1 - \frac{y}{A}\right)$$

In the last equality, we replaced the product of constant terms KA with constant k. Therefore, we have

$$\boxed{\frac{dy}{dt} = ky\left(1 - \frac{y}{A}\right)}$$

Here $k > 0$ and is called the **growth constant**, while $A > 0$ and is called the **carrying capacity**. Figure 1 shows a typical S-shaped solution of Eq. (1).

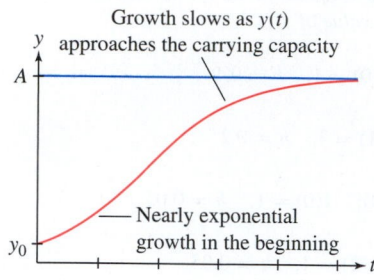

FIGURE 1 Solution of the logistic equation.

Growth slows as $y(t)$ approaches the carrying capacity

Nearly exponential growth in the beginning

CONCEPTUAL INSIGHT The logistic equation $\dfrac{dy}{dt} = ky(1 - y/A)$ differs from the exponential differential equation $\dfrac{dy}{dt} = ky$ only by the additional factor $(1 - y/A)$. As long as y is small relative to A, this factor is close to 1 and can be ignored, yielding $\dfrac{dy}{dt} \approx ky$. Thus, $y(t)$ grows nearly exponentially when the population is small (Figure 1). As $y(t)$ approaches A, the factor $(1 - y/A)$ tends to zero. This causes $\dfrac{dy}{dt}$ to decrease to zero and prevents $y(t)$ from exceeding the carrying capacity A.

The slope field in Figure 2 shows clearly that there are three families of solutions, depending on the initial value $y_0 = y(0)$.

- If $y_0 > A$, then $y(t)$ is decreasing and approaches A as $t \to \infty$.
- If $0 < y_0 < A$, then $y(t)$ is increasing and approaches A as $t \to \infty$.
- If $y_0 < 0$, then $y(t)$ is decreasing and there is a time t_b such that $\displaystyle\lim_{t \to t_b^-} y(t) = -\infty$.

Solutions of the logistic equation with $y_0 < 0$ are not relevant to populations because a population cannot be negative (see Exercise 18).

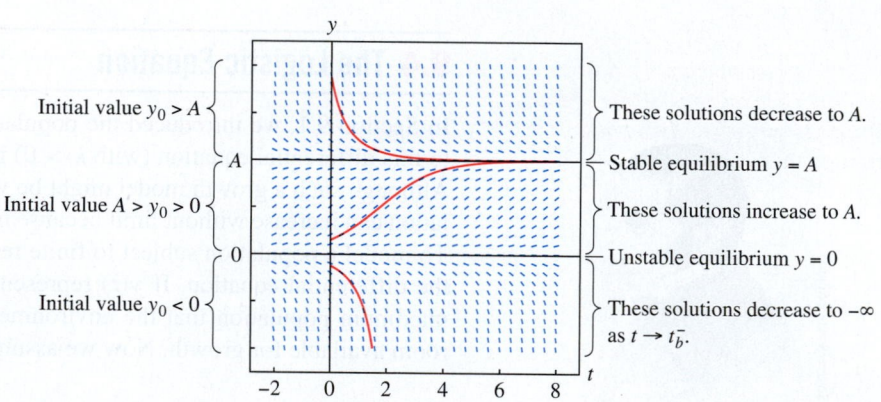

DF FIGURE 2 Slope field for $\dfrac{dy}{dt} = ky\left(1 - \dfrac{y}{A}\right)$.

Initial value $y_0 > A$ · These solutions decrease to A.
Stable equilibrium $y = A$
Initial value $A > y_0 > 0$ · These solutions increase to A.
Unstable equilibrium $y = 0$
Initial value $y_0 < 0$ · These solutions decrease to $-\infty$ as $t \to t_b^-$.

Equation (1) also has two constant solutions: $y = 0$ and $y = A$. They correspond to the roots of $ky(1 - y/A) = 0$, and they satisfy Eq. (1) because $\dfrac{dy}{dt} = 0$ when y is a constant. Constant solutions are called **equilibrium** or **steady-state** solutions. The equilibrium solution $y = A$ is a **stable equilibrium** because every solution with initial value y_0 close to A approaches the equilibrium $y = A$ as $t \to \infty$. By contrast, $y = 0$ is an **unstable equilibrium** because every nonequilibrium solution with initial value y_0 near $y = 0$ either increases to A or decreases to $-\infty$. These nonequilibrium solutions deviate away from the unstable equilibrium solution as time moves forward.

Having described the solutions qualitatively, let us now find the nonequilibrium solutions explicitly using Separation of Variables. Assuming that $y \neq 0$ and $y \neq A$, we have

$$\frac{dy}{dt} = ky\left(1 - \frac{y}{A}\right)$$

$$\frac{dy}{y\,(1 - y/A)} = k\,dt$$

$$\int \left(\frac{1}{y} - \frac{1}{y - A}\right) dy = \int k\,dt \qquad \boxed{2}$$

In Eq. (2), we use the partial fraction decomposition

$$\frac{1}{y\,(1 - y/A)} = \frac{1}{y} - \frac{1}{y - A}$$

$$\ln |y| - \ln |y - A| = kt + C$$

$$\left| \frac{y}{y-A} \right| = e^{kt+C} \quad \Rightarrow \quad \frac{y}{y-A} = \pm e^C e^{kt}$$

Since $\pm e^C$ takes on arbitrary nonzero values, we replace $\pm e^C$ with B (nonzero):

$$\frac{y}{y-A} = Be^{kt} \qquad \boxed{3}$$

For $t = 0$, this gives a useful relation between B and the initial value $y_0 = y(0)$:

$$\frac{y_0}{y_0 - A} = B \qquad \boxed{4}$$

To solve for y, multiply each side of Eq. (3) by $(y - A)$:

$$y = (y - A)Be^{kt}$$

$$y(1 - Be^{kt}) = -ABe^{kt}$$

$$y = \frac{ABe^{kt}}{Be^{kt} - 1}$$

As $B \neq 0$, we may divide by Be^{kt} to obtain the general nonequilibrium solution:

$$\boxed{\frac{dy}{dt} = ky\left(1 - \frac{y}{A}\right), \qquad y = \frac{A}{1 - e^{-kt}/B}} \qquad \boxed{5}$$

> Note that the solutions of the logistic equation are in the form of the logistic functions $y = \frac{M}{1 + Ae^{-kt}}$ we introduced and investigated in Sections 2.7 and 4.6.

We use this formula in the next two examples. In each case, the differential equation may instead be solved by Separation of Variables, the method used to derive this solution.

EXAMPLE 1 Solve $\dfrac{dy}{dt} = 0.3y(4 - y)$ with initial condition $y(0) = 1$.

Solution To apply Eq. (5), we must rewrite the equation in the form

$$\frac{dy}{dt} = 1.2y\left(1 - \frac{y}{4}\right)$$

Thus, $k = 1.2$ and $A = 4$, and the general solution is

$$y = \frac{4}{1 - e^{-1.2t}/B}$$

There are two ways to find B. One way is to solve $y(0) = 1$ for B directly. An easier way is to use Eq. (4):

$$B = \frac{y_0}{y_0 - A} = \frac{1}{1 - 4} = -\frac{1}{3}$$

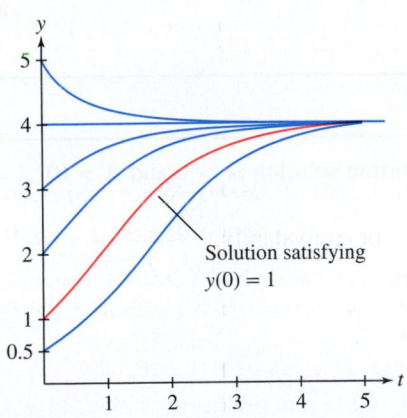

DF FIGURE 3 Several solutions of $\dfrac{dy}{dt} = 0.3y(4 - y)$.

We find that the particular solution is $y = \dfrac{4}{1 + 3e^{-1.2t}}$ (Figure 3). ∎

> The logistic equation may be too simple to describe a real deer population accurately, but it serves as a starting point for more sophisticated models used by ecologists, population biologists, and forestry professionals.

EXAMPLE 2 Deer Population A deer population grows logistically with growth constant $k = 0.4$ year^{-1} in a forest with a carrying capacity of 1000 deer.

(a) Find the deer population $P(t)$ if the initial population is $P_0 = 100$.

(b) How long does it take for the deer population to reach 500?

DF **FIGURE 4** Deer population as a function of t (in years).

Solution The time unit is the year because the unit of k is year^{-1}.

(a) Since $k = 0.4$ and $A = 1000$, $P(t)$ satisfies the differential equation

$$\frac{dP}{dt} = 0.4P\left(1 - \frac{P}{1000}\right)$$

The general solution is given by Eq. (5):

$$P(t) = \frac{1000}{1 - e^{-0.4t}/B}$$

6

Using Eq. (4) to compute B, we find (Figure 4)

$$B = \frac{P_0}{P_0 - A} = \frac{100}{100 - 1000} = -\frac{1}{9} \quad \Rightarrow \quad P(t) = \frac{1000}{1 + 9e^{-0.4t}}$$

(b) To find the time t when $P(t) = 500$, we could solve the equation

$$P(t) = \frac{1000}{1 + 9e^{-0.4t}} = 500$$

But it is easier to use Eq. (3):

$$\frac{P}{P - A} = Be^{kt}$$

$$\frac{P}{P - 1000} = -\frac{1}{9}e^{0.4t}$$

Set $P = 500$ and solve for t:

$$-\frac{1}{9}e^{0.4t} = \frac{500}{500 - 1000} = -1 \quad \Rightarrow \quad e^{0.4t} = 9 \quad \Rightarrow \quad 0.4t = \ln 9$$

This gives $t = (\ln 9)/0.4 \approx 5.5$ years. ∎

9.4 SUMMARY

- The *logistic equation* and its general nonequilibrium solution ($k > 0$ and $A > 0$):

$$\frac{dy}{dt} = ky\left(1 - \frac{y}{A}\right), \qquad y = \frac{A}{1 - e^{-kt}/B}, \qquad \text{or equivalently,} \qquad \frac{y}{y - A} = Be^{kt}$$

- Two equilibrium (constant) solutions:
  - $y = 0$ is an unstable equilibrium.
  - $y = A$ is a stable equilibrium.

- If the initial value $y_0 = y(0)$ satisfies $y_0 > 0$, then $y(t)$ approaches the stable equilibrium $y = A$; that is, $\lim_{t \to \infty} y(t) = A$.

9.4 EXERCISES

Preliminary Questions

1. Which of the following differential equations is a logistic differential equation?

(a) $\dfrac{dy}{dt} = 2y(1 - y^2)$

(b) $\dfrac{dy}{dt} = 2y\left(1 - \dfrac{y}{3}\right)$

(c) $\dfrac{dy}{dt} = 2y\left(1 - \dfrac{t}{4}\right)$

(d) $\dfrac{dy}{dt} = 2y(1 - 3y)$

2. What are the constant solutions to $\dfrac{dy}{dt} = ky\left(1 - \dfrac{y}{A}\right)$?

3. Is the logistic equation separable?

Exercises

1. Find the general solution of the logistic equation

$$\frac{dy}{dt} = 3y\left(1 - \frac{y}{5}\right)$$

Then find the particular solution satisfying $y(0) = 2$.

2. Find the solution of $\dfrac{dy}{dt} = 2y(3 - y)$, $y(0) = 10$.

In Exercises 3–4, for each of (a)–(c), give the solution $y(t)$ satisfying the initial condition. (Note: the general solution formula for the logistic equation, Eq. (5), applies when a solution is not constant.)

3. $\dfrac{dy}{dt} = 3y(6 - y)$

(a) $y(0) = 6$ (b) $y(0) = 4$ (c) $y(4) = 0$

4. $\dfrac{dy}{dt} = y\left(2 - \dfrac{y}{4}\right)$

(a) $y(0) = 6$ (b) $y(0) = 8$ (c) $y(0) = -2$

5. A population of squirrels lives in a forest with a carrying capacity of 2000. Assume logistic growth with growth constant $k = 0.6 \text{ yr}^{-1}$.

(a) Find a formula for the squirrel population $P(t)$, assuming an initial population of 500 squirrels.

(b) How long will it take for the squirrel population to double?

6. The population $P(t)$ of mosquito larvae growing in a tree hole increases according to the logistic equation with growth constant $k = 0.3 \text{ day}^{-1}$ and carrying capacity $A = 500$.

(a) Find a formula for the larvae population $P(t)$, assuming an initial population of $P_0 = 50$ larvae.

(b) After how many days will the larvae population reach 200?

7. Sunset Lake is stocked with 2000 rainbow trout, and after 1 year the population has grown to 4500. Assuming logistic growth with a carrying capacity of 20,000, find the growth constant k (specify the units) and determine when the population will increase to 10,000.

8. Spread of a Rumor A rumor spreads through a small town. Let $y(t)$ be the fraction of the population that has heard the rumor at time t and assume that the rate at which the rumor spreads is proportional to the product of the fraction y of the population that has heard the rumor and the fraction $1 - y$ that has not yet heard the rumor.

(a) Write the differential equation satisfied by y in terms of a proportionality factor k.

(b) Find k (in units of day^{-1}), assuming that 10% of the population knows the rumor at $t = 0$ and 40% knows it at $t = 2$ days.

(c) Using the assumptions of part (b), determine when 75% of the population will know the rumor.

9. A rumor spreads through a school with 1000 students. At 8 AM, 80 students have heard the rumor, and by noon, half the school has heard it. Using the logistic model of Exercise 8, determine when 90% of the students will have heard the rumor.

10. [GU] A simpler model for the spread of a rumor assumes that the rate at which the rumor spreads is proportional (with factor k) to the fraction of the population that has not yet heard the rumor.

(a) Compute the solutions to this model and the model of Exercise 8 with the values $k = 0.9$ and $y_0 = 0.1$.

(b) Graph the two solutions on the same axis.

(c) Which model seems more realistic? Why?

11. Let $k = 1$ and $A = 1$ in the logistic equation.

(a) Find the solutions satisfying $y_1(0) = 10$ and $y_2(0) = -1$.

(b) Find the time t when $y_1(t) = 5$.

(c) When does $y_2(t)$ become infinite?

12. A tissue culture grows until it has a maximum area of M square centimeters. The area $A(t)$ of the culture at time t may be modeled by the differential equation

$$\frac{dA}{dt} = k\sqrt{A}\left(1 - \frac{A}{M}\right) \qquad \boxed{7}$$

where k is a growth constant.

(a) Show that if we set $A = u^2$, then

$$\frac{du}{dt} = \frac{1}{2}k\left(1 - \frac{u^2}{M}\right)$$

Then find the general solution using Separation of Variables.

(b) Show that the general solution to Eq. (7) is

$$A(t) = M\left(\frac{Ce^{(k/\sqrt{M})t} - 1}{Ce^{(k/\sqrt{M})t} + 1}\right)^2$$

13. [GU] In the model of Exercise 12, let $A(t)$ be the area at time t (hours) of a growing tissue culture with initial size $A(0) = 1 \text{ cm}^2$, assuming that the maximum area is $M = 16 \text{ cm}^2$ and the growth constant is $k = 0.1$.

(a) Find a formula for $A(t)$. *Note:* The initial condition is satisfied for two values of the constant C. Choose the value of C for which $A(t)$ is increasing.

(b) Determine the area of the culture at $t = 10$ hours.

(c) [GU] Graph the solution using a graphing utility.

14. Show that if a tissue culture grows according to Eq. (7), then the growth rate reaches a maximum when $A = M/3$.

15. In 1751, Benjamin Franklin predicted that the U.S. population $P(t)$ would increase with growth constant $k = 0.028 \text{ year}^{-1}$. According to the census, the U.S. population was 5 million in 1800 and 76 million in 1900. Assuming logistic growth with $k = 0.028$, find the predicted carrying capacity for the U.S. population. *Hint:* Use Eqs. (3) and (4) to show that

$$\frac{P(t)}{P(t) - A} = \frac{P_0}{P_0 - A}e^{kt}$$

16. [✎] **Reverse Logistic Equation** Consider the following logistic equation (with $k, B > 0$):

$$\frac{dP}{dt} = -kP\left(1 - \frac{P}{B}\right) \qquad \boxed{8}$$

(a) Sketch the slope field of this equation.

(b) The general solution is $P(t) = B/(1 - e^{kt}/C)$, where C is a nonzero constant. Show that $P(0) > B$ if $C > 1$ and $0 < P(0) < B$ if $C < 0$.

(c) Show that Eq. (8) models an extinction–explosion population. That is, $P(t)$ tends to zero if the initial population satisfies $0 < P(0) < B$, and it tends to ∞ after a finite amount of time if $P(0) > B$.

(d) Show that $P = 0$ is a stable equilibrium and $P = B$ is an unstable equilibrium.

Further Insights and Challenges

In Exercises 17 and 18, let $y(t)$ be a solution of the logistic equation

$$\frac{dy}{dt} = ky\left(1 - \frac{y}{A}\right) \qquad \boxed{9}$$

where $A > 0$ and $k > 0$.

17. (a) Differentiate Eq. (9) with respect to t and use the Chain Rule to show that

$$\frac{d^2y}{dt^2} = k^2 y\left(1 - \frac{y}{A}\right)\left(1 - \frac{2y}{A}\right)$$

(b) Show that the graph of the function y is concave up if $0 < y < A/2$ and concave down if $A/2 < y < A$.

(c) Show that if $0 < y(0) < A/2$, then y has a point of inflection at $y = A/2$ (Figure 5).

(d) Assume that $0 < y(0) < A/2$. Find the time t when $y(t)$ reaches the inflection point.

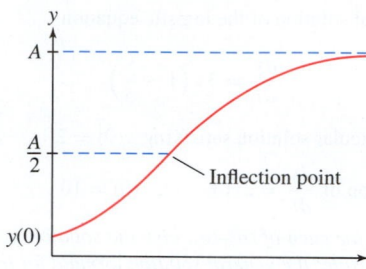

FIGURE 5 An inflection point occurs at $y = A/2$ in the logistic curve.

18. Let $y = \dfrac{A}{1 - e^{-kt}/B}$ be the general nonequilibrium solution to Eq. (9). If $y(t)$ has a vertical asymptote at $t = t_b$, that is, if $\lim\limits_{t \to t_b-} y(t) = \pm\infty$, we say that the solution "blows up" at $t = t_b$.

(a) Show that if $0 < y(0) < A$, then y does not blow up at any time t_b.

(b) Show that if $y(0) > A$, then y blows up at a time t_b, which is negative (and hence does not correspond to a real time).

(c) Show that y blows up at some positive time t_b if and only if $y(0) < 0$ (and hence does not correspond to a real population).

9.5 First-Order Linear Equations

In Section 9.2, we investigated the first-order linear constant coefficient differential equation $\frac{dy}{dt} = ky + c$. In this section, we work with a more general version of this differential equation, where the terms k and c are functions of the independent variable (which here is x). The method of "integrating factors" is used to solve these differential equations. Although some of these equations are separable and can be solved by Separation of Variables, the method that we present here applies to *all* first-order linear differential equations, whether separable or not (Figure 1).

A **first-order linear differential equation** is one that can be put in the following form:

$$\boxed{y' + P(x)y = Q(x)} \qquad \boxed{1}$$

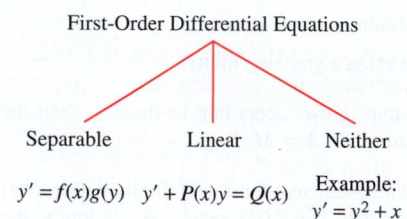

First-Order Differential Equations

Separable   Linear   Neither

$y' = f(x)g(y)$ $y' + P(x)y = Q(x)$ Example: $y' = y^2 + x$

FIGURE 1

To solve Eq. (1), we shall multiply through by a function $\alpha(x)$, called an **integrating factor**, that turns the left-hand side into the derivative of $\alpha(x)y$:

$$\alpha(x)\big(y' + P(x)y\big) = \big(\alpha(x)y\big)' \qquad \boxed{2}$$

Suppose we can find $\alpha(x)$ satisfying Eq. (2) that is nonzero. Then Eq. (1) yields

$$\alpha(x)\big(y' + P(x)y\big) = \alpha(x)Q(x)$$

$$\big(\alpha(x)y\big)' = \alpha(x)Q(x)$$

We can solve this equation by integration:

$$\alpha(x)y = \int \alpha(x)Q(x)\,dx \quad \text{or} \quad y = \frac{1}{\alpha(x)}\left(\int \alpha(x)Q(x)\,dx\right)$$

To find $\alpha(x)$, expand Eq. (2), using the Product Rule on the right-hand side:

$$\alpha(x)y' + \alpha(x)P(x)y = \alpha(x)y' + \alpha'(x)y \quad \Rightarrow \quad \alpha(x)P(x)y = \alpha'(x)y$$

Dividing by y, we obtain

$$\boxed{\frac{d\alpha}{dx} = \alpha(x)P(x)}$$

3

We solve this equation using Separation of Variables:

$$\frac{d\alpha}{\alpha} = P(x)\,dx \quad \Rightarrow \quad \int \frac{d\alpha}{\alpha} = \int P(x)\,dx$$

Therefore, $\ln|\alpha(x)| = \displaystyle\int P(x)\,dx$, and by exponentiation, $\alpha(x) = \pm e^{\int P(x)\,dx}$. Since we need just one solution of Eq. (3), we choose the positive sign in the expression for $\alpha(x)$.

> **THEOREM 1** The general solution of $y' + P(x)y = Q(x)$ is
>
> $$\boxed{y = \frac{1}{\alpha(x)}\left(\int \alpha(x)Q(x)\,dx\right)}$$
>
> **4**
>
> where $\alpha(x)$ is an integrating factor:
>
> $$\boxed{\alpha(x) = e^{\int P(x)\,dx}}$$
>
> **5**

In the formula for the integrating factor $\alpha(x)$, the integral $\int P(x)\,dx$ denotes any antiderivative of P.

To solve differential equations $y' + P(x)y = Q(x)$, we can either find an integrating factor and use Eq. (4) in Theorem 1, or we can use the method that yielded the solution in Eq. (4). That method is

- Find an integrating factor $\alpha(x) = e^{\int P(x)\,dx}$.
- Multiply both sides of the equation $y' + P(x)y = Q(x)$ by the integrating factor $\alpha(x)$. As a result, the left side of the equation is then the derivative of a product.
- Integrate both sides and solve for y to obtain the solution.

We follow this method to obtain the solution in the next example.

EXAMPLE 1 Solve $xy' - 3y = x^2$, $y(1) = 2$.

Solution First, divide by x to put the equation in the form $y' + P(x)y = Q(x)$:

$$y' - \frac{3}{x}y = x$$

Thus, $P(x) = -3x^{-1}$ and $Q(x) = x$.

Step 1. Find an integrating factor.
 In our case, $P(x) = -3x^{-1}$, and by Eq. (5),

$$\alpha(x) = e^{\int P(x)\,dx} = e^{\int (-3/x)\,dx} = e^{-3\ln x} = e^{\ln(x^{-3})} = x^{-3}$$

Step 2. Multiply the equation by the integrating factor.

$$x^{-3}\left(y' - \frac{3}{x}\,y\right) = x^{-3}(x)$$

$$(x^{-3}y)' = x^{-2}$$

Step 3. Integrate both sides.

$$x^{-3}y = -x^{-1} + C$$

CAUTION *When we determine the integrating factor, we can choose **one** antiderivative of $P(x)$, but at Step 3, when integrating both sides, we need to include a constant of integration since it becomes the constant term in the general solution. As in this example, when solving for y in Step 4, the constant might then be multiplied by a function of x and therefore might appear that way, rather than by itself, in the general solution.*

Step 4. **Solve for y.**

$$y = x^3(-x^{-1} + C)$$
$$= -x^2 + Cx^3$$

Step 5. **Solve the Initial Value Problem.**

Now solve for C using the initial condition $y(1) = 2$:

$$y(1) = -1^2 + C \cdot 1^3 = 2 \qquad \text{or} \qquad C = 3$$

Therefore, the solution of the Initial Value Problem is $y = -x^2 + 3x^3$.

Finally, let's check that $y = -x^2 + 3x^3$ satisfies our equation $xy' - 3y = x^2$:

$$xy' - 3y = x(-2x + 9x^2) - 3(-x^2 + 3x^3)$$
$$= (-2x^2 + 9x^3) + (3x^2 - 9x^3) = x^2 \qquad \blacksquare$$

EXAMPLE 2 Solve the Initial Value Problem: $y' + (1 - x^{-1})y = x^2$, $y(1) = 2$.

Solution This equation has the form $y' + P(x)y = Q(x)$ with $P(x) = 1 - x^{-1}$. By Eq. (5), an integrating factor is

$$\alpha(x) = e^{\int(1-x^{-1})\,dx} = e^{x - \ln x} = e^x e^{\ln x^{-1}} = x^{-1}e^x$$

By either multiplying by the integration factor and then integrating both sides of the resulting equation or by applying Eq. (4) with $Q(x) = x^2$, we obtain the general solution:

$$y = \alpha(x)^{-1}\left(\int \alpha(x)Q(x)\,dx\right) = xe^{-x}\left(\int (x^{-1}e^x)x^2\,dx\right)$$
$$= xe^{-x}\left(\int xe^x\,dx\right)$$

Integration by Parts shows that $\int xe^x\,dx = (x-1)e^x + C$, so we obtain

$$y = xe^{-x}\big((x-1)e^x + C\big) = x(x-1) + Cxe^{-x}$$

The initial condition $y(1) = 2$ gives

$$y(1) = 1(1-1) + Ce^{-1} = Ce^{-1} = 2 \quad \Rightarrow \quad C = 2e$$

The desired particular solution is

$$y = x(x-1) + (2e)xe^{-x} = x(x-1) + 2xe^{1-x} \qquad \blacksquare$$

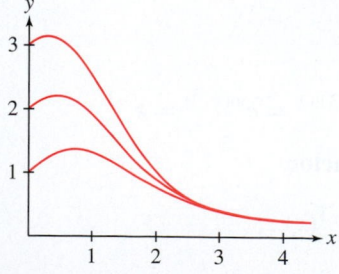

DF FIGURE 2 Solutions to $y' + xy = 1$ obtained numerically and plotted by computer.

CONCEPTUAL INSIGHT We have expressed the general solution of a first-order linear differential equation in terms of the integrals in Eqs. (4) and (5). Keep in mind, however, that it is not always possible to evaluate these integrals explicitly. For example, the general solution of $y' + xy = 1$ is

$$y = e^{-x^2/2}\left(\int e^{x^2/2}\,dx + C\right)$$

The integral $\int e^{x^2/2}\,dx$ cannot be evaluated in elementary terms. However, we can approximate the integral numerically and plot the solutions by computer (Figure 2).

In the next example, we use a differential equation to model a mixing problem, which has applications in biology, chemistry, and medicine.

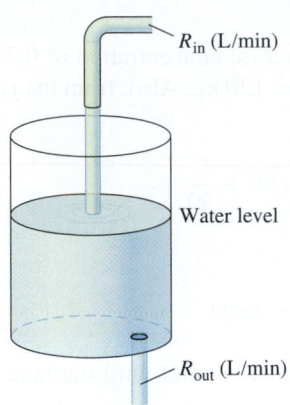

R_{in} (L/min)

Water level

R_{out} (L/min)

FIGURE 3

EXAMPLE 3 **A Mixing Problem** A tank contains 600 liters of water with a sucrose concentration of 0.2 kg/L. We begin adding water with a sucrose concentration of 0.1 kg/L at a rate of $R_{in} = 40$ L/min (Figure 3). The water mixes instantaneously and exits the bottom of the tank at a rate of $R_{out} = 20$ L/min. Let $y(t)$ be the quantity (in kilograms) of sucrose in the tank at time t (in minutes). Set up a differential equation for $y(t)$ and solve for $y(t)$.

Solution

Step 1. **Set up the differential equation.**

The derivative dy/dt is the difference of two rates of change, namely the rate at which sucrose enters the tank and the rate at which it leaves:

$$\frac{dy}{dt} = \text{sucrose rate in} - \text{sucrose rate out} \qquad \boxed{6}$$

The rate at which sucrose enters the tank is

$$\text{sucrose rate in} = \underbrace{(0.1 \text{ kg/L})(40 \text{ L/min})}_{\text{Concentration times water rate in}} = 4 \text{ kg/min}$$

Next, we compute the sucrose concentration in the tank at time t. Water flows in at 40 L/min and out at 20 L/min, so there is a net inflow of 20 L/min. The tank has 600 L at time $t = 0$, so it has $600 + 20t$ L at time t, and

$$\text{concentration at time } t = \frac{\text{kilograms of sucrose in tank}}{\text{liters of water in tank}} = \frac{y(t)}{600 + 20t} \text{ kg/L}$$

The rate at which sucrose leaves the tank is the product of the concentration and the rate at which water flows out:

$$\text{sucrose rate out} = \underbrace{\left(\frac{y}{600 + 20t} \, \frac{\text{kg}}{\text{L}} \right) \left(20 \, \frac{\text{L}}{\text{min}} \right)}_{\text{Concentration times water rate out}} = \frac{20y}{600 + 20t} = \frac{y}{t + 30} \text{ kg/min}$$

Now Eq. (6) gives us the differential equation

$$\frac{dy}{dt} = 4 - \frac{y}{t + 30} \qquad \boxed{7}$$

Step 2. **Find the general solution.**

We write Eq. (7) in standard form:

$$\frac{dy}{dt} + \underbrace{\frac{1}{t + 30}}_{P(t)} y = \underbrace{4}_{Q(t)} \qquad \boxed{8}$$

An integrating factor is

$$\alpha(t) = e^{\int P(t)\,dt} = e^{\int dt/(t+30)} = e^{\ln(t+30)} = t + 30$$

The general solution is

$$y(t) = \alpha(t)^{-1} \left(\int \alpha(t) Q(t)\, dt + C \right)$$

$$= \frac{1}{t + 30} \left(\int (t + 30)(4)\, dt + C \right)$$

$$= \frac{1}{t + 30} \left(2(t + 30)^2 + C \right) = 2t + 60 + \frac{C}{t + 30}$$

Summary:

sucrose rate in $= 4$ kg/min

sucrose rate out $= \dfrac{y}{t + 30}$ kg/min

$\dfrac{dy}{dt} = 4 - \dfrac{y}{t + 30}$

$\alpha(t) = t + 30$

$y(t) = 2t + 60 + \dfrac{C}{t + 30}$

Step 3. Solve the Initial Value Problem.

At $t = 0$, the tank contains 600 L of water with a sucrose concentration of 0.2 kg/L. Thus, the total sucrose at $t = 0$ is $y(0) = (600)(0.2) = 120$ kg. Also, from the general solution formula, we have

$$y(0) = 2(0) + 60 + \frac{C}{0+30} = 60 + \frac{C}{30}$$

Therefore,

$$60 + \frac{C}{30} = 120 \quad \Rightarrow \quad C = 1800$$

We obtain the following formula (t in minutes), which is valid until the tank overflows:

$$y(t) = 2t + 60 + \frac{1800}{t+30} \text{ kg sucrose}$$

9.5 SUMMARY

- A *first-order linear differential equation* is a differential equation in the form

$$y' + P(x)y = Q(x)$$

- The general solution is

$$y = \alpha(x)^{-1}\left(\int \alpha(x)Q(x)\,dx + C\right)$$

where $\alpha(x)$ is an *integrating factor*: $\alpha(x) = e^{\int P(x)\,dx}$.

9.5 EXERCISES

Preliminary Questions

1. Which of the following are first-order linear equations?
(a) $y' + x^2y = 1$ (b) $y' + xy^2 = 1$
(c) $x^5y' + y = e^x$ (d) $x^5y' + y = e^y$

2. If $\alpha(x)$ is an integrating factor for $y' + A(x)y = B(x)$, then $\alpha'(x)$ is equal to (choose the correct answer):

(a) $B(x)$ (b) $\alpha(x)A(x)$
(c) $\alpha(x)A'(x)$ (d) $\alpha(x)B(x)$

3. For what function P is the integrating factor $\alpha(x)$ equal to x?

4. For what function P is the integrating factor $\alpha(x)$ equal to e^x?

Exercises

1. Consider $y' + x^{-1}y = x^3$.
(a) Verify that $\alpha(x) = x$ is an integrating factor.
(b) Show that when multiplied by $\alpha(x)$, the differential equation can be written $(xy)' = x^4$.
(c) Conclude that xy is an antiderivative of x^4 and use this information to find the general solution.
(d) Find the particular solution satisfying $y(1) = 0$.

2. Consider $\frac{dy}{dt} + 2y = e^{-3t}$.
(a) Verify that $\alpha(t) = e^{2t}$ is an integrating factor.
(b) Use Eq. (4) to find the general solution.
(c) Find the particular solution with initial condition $y(0) = 1$.

3. Let $\alpha(x) = e^{x^2}$. Verify the identity
$$(\alpha(x)y)' = \alpha(x)(y' + 2xy)$$

and explain how it is used to find the general solution of
$$y' + 2xy = x$$

4. Find the solution of $y' - y = e^{2x}$, $y(0) = 1$.

In Exercises 5–20, find the general solution of the first-order linear differential equation.

5. $xy' + y = x$ 6. $xy' - y = x^2 - x$
7. $3xy' - y = x^{-1}$ 8. $y' + xy = x$
9. $y' + 3x^{-1}y = x + x^{-1}$ 10. $y' + x^{-1}y = \cos(x^2)$
11. $xy' = y - x$ 12. $xy' = x^{-2} - \frac{3y}{x}$

13. $y' + y = e^x$

14. $y' - y = e^x$

15. $y' + (\tan x)y = \cos x$

16. $y' + (\sec x)y = \cos x$

17. $e^x y' = 1 + 2e^x y$

18. $e^{2x} y' = 1 - e^x y$

19. $y' - (\ln x)y = x^x$

20. $y' + y = \cos x$

In Exercises 21–28, solve the Initial Value Problem.

21. $y' + 3y = e^{2x}$, $y(0) = -1$

22. $xy' + y = e^x$, $y(1) = 3$

23. $y' + \dfrac{1}{x+1}y = x^{-2}$, $y(1) = 2$

24. $y' + y = \sin x$, $y(0) = 1$

25. $(\sin x)y' = (\cos x)y + 1$, $y\left(\dfrac{\pi}{4}\right) = 0$

26. $y' + (\sec t)y = \sec t$, $y\left(\dfrac{\pi}{4}\right) = 1$

27. $y' + (\tanh x)y = 1$, $y(0) = 3$

28. $y' + \dfrac{x}{1+x^2}y = \dfrac{1}{(1+x^2)^{3/2}}$, $y(1) = 0$

29. The differential equation $\dfrac{dy}{dx} = x$ is directly integrable and also first-order linear. Show that solving the differential equation using Theorem 1 leads to solving it by direct integration.

30. The differential equation $\dfrac{dy}{dx} = 1 - y$ can be solved using Eq. (2) in Section 9.2 and using Theorem 1 in this section. Show that both approaches lead to the same general solution.

31. Find the general solution of $y' + ny = e^{mx}$ for all m, n. *Note:* The case $m = -n$ must be treated separately.

32. Find the general solution of $y' + ny = \cos x$ for all n.

In Exercises 33–36, a 1000-liter tank contains 500 L of water with a salt concentration of 10 g/L. Water with a salt concentration of 50 g/L flows into the tank at a rate of $R_{in} = 80$ L/min. The fluid mixes instantaneously and is pumped out at a specified rate R_{out}. Let $y(t)$ denote the quantity of salt in the tank at time t.

33. Assume that $R_{out} = 40$ L/min.

(a) Set up and solve the differential equation for $y(t)$.

(b) What is the salt concentration when the tank overflows?

34. Find the salt concentration when the tank overflows, assuming that $R_{out} = 60$ L/min.

35. Find the limiting salt concentration as $t \to \infty$, assuming that $R_{out} = 80$ L/min.

36. Assuming that $R_{out} = 120$ L/min, find $y(t)$. Then calculate the tank volume and the salt concentration at $t = 10$ min.

37. Water flows into a tank at the variable rate of $R_{in} = 20/(1+t)$ gal/min and out at the constant rate $R_{out} = 5$ gal/min. Let $V(t)$ be the volume of water in the tank at time t.

(a) Set up a differential equation for $V(t)$ and solve it with the initial condition $V(0) = 100$.

(b) Find the maximum value of V.

(c) [CAS] Plot $V(t)$ and estimate the time t when the tank is empty.

38. A stream feeds into a lake at a rate of 1000 m³/day. The stream is polluted with a toxin whose concentration is 5 g/m³. Assume that the lake has volume 10^6 m³ and that water flows out of the lake at the same rate of 1000 m³/day.

(a) Set up a differential equation for the concentration $c(t)$ of toxin in the lake and solve for $c(t)$, assuming that $c(0) = 0$. *Hint:* Find the differential equation for the quantity of toxin $y(t)$, and observe that $c(t) = y(t)/10^6$.

(b) What is the limiting concentration for large t?

In Exercises 39–42, consider a series circuit (Figure 4) consisting of a resistor of R ohms, an inductor of L henries, and a variable voltage source of $V(t)$ volts (time t in seconds). The current through the circuit $I(t)$ (in amperes) satisfies the differential equation

$$\frac{dI}{dt} + \frac{R}{L}I = \frac{1}{L}V(t)$$

9

39. Solve Eq. (9) with initial condition $I(0) = 0$, assuming that $R = 100$ ohms (Ω), $L = 5$ henries (H), and $V(t)$ is constant with $V(t) = 10$ volts (V).

40. Assume that $R = 110$ ohms, $L = 10$ henries, and $V(t) = e^{-t}$ volts.

(a) Solve Eq. (9) with initial condition $I(0) = 0$.

(b) Calculate t_m and $I(t_m)$, where t_m is the time at which $I(t)$ has a maximum value.

(c) [GU] Use a computer algebra system to sketch the graph of the solution for $0 \le t \le 3$.

41. Assume that $V(t) = V$ is constant and $I(0) = 0$.

(a) Solve for $I(t)$.

(b) Show that $\lim_{t\to\infty} I(t) = V/R$ and that $I(t)$ reaches approximately 63% of its limiting value after L/R seconds.

(c) How long does it take for $I(t)$ to reach 90% of its limiting value if $R = 500$ ohms, $L = 4$ henries, and $V = 20$ volts?

42. Solve for $I(t)$, assuming that $R = 500$ ohms, $L = 4$ henries, and $V = 20\cos(80)$ volts.

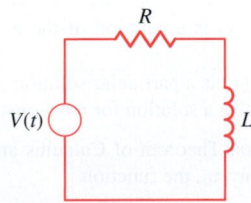

FIGURE 4 *RL* circuit.

43. [✎] Tank 1 in Figure 5 is filled with V_1 liters of water containing blue dye at an initial concentration of c_0 g/L. Water flows into the tank at a rate of R L/min, is mixed instantaneously with the dye solution, and flows out through the bottom at the same rate R. Let $c_1(t)$ be the dye concentration in the tank at time t.

(a) Explain why c_1 satisfies the differential equation

$$\frac{dc_1}{dt} = -\frac{R}{V_1}c_1$$

(b) Solve for $c_1(t)$ with $V_1 = 300$ L, $R = 50$, and $c_0 = 10$ g/L.

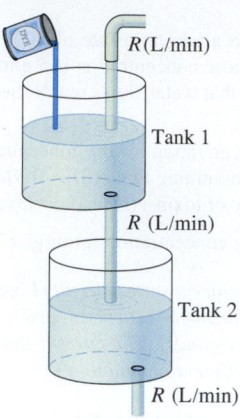

FIGURE 5

44. 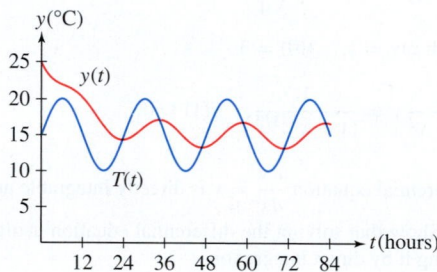 Continuing with the previous exercise, let tank 2 be another tank filled with V_2 gallons of water. Assume that the dye solution from tank 1 empties into tank 2 as in Figure 5, mixes instantaneously, and leaves tank 2 at the same rate R. Let $c_2(t)$ be the dye concentration in tank 2 at time t.

(a) Explain why c_2 satisfies the differential equation

$$\frac{dc_2}{dt} = \frac{R}{V_2}(c_1 - c_2)$$

(b) Use the solution to Exercise 43 to solve for $c_2(t)$ if $V_1 = 300$, $V_2 = 200$, $R = 50$, and $c_0 = 10$.

(c) Find the maximum concentration in tank 2.

(d) [GU] Plot the solution.

45. Let a, b, r be constants. Show that

$$y = Ce^{-kt} + a + bk\left(\frac{k \sin rt - r \cos rt}{k^2 + r^2}\right)$$

is a general solution of

$$\frac{dy}{dt} = -k\left(y - a - b \sin rt\right)$$

46. Assume that the outside temperature varies as

$$T(t) = 15 + 5 \sin(\pi t/12)$$

where $t = 0$ is 12 noon. A house is heated to $25°C$ at $t = 0$ and after that, its temperature $y(t)$ varies according to Newton's Law of Cooling (Figure 6):

$$\frac{dy}{dt} = -0.1\left(y(t) - T(t)\right)$$

Use Exercise 45 to solve for $y(t)$.

FIGURE 6 House temperature $y(t)$.

Further Insights and Challenges

47. Let $\alpha(x)$ be an integrating factor for $y' + P(x)y = Q(x)$. The differential equation $y' + P(x)y = 0$ is called the associated **homogeneous equation**.

(a) Show that $y = 1/\alpha(x)$ is a solution of the associated homogeneous equation.

(b) Show that if $y = f(x)$ is a particular solution of $y' + P(x)y = Q(x)$, then $f(x) + C/\alpha(x)$ is also a solution for any constant C.

48. Use the Fundamental Theorem of Calculus and the Product Rule to verify directly that for any x_0, the function

$$f(x) = \alpha(x)^{-1} \int_{x_0}^{x} \alpha(t)Q(t)\,dt$$

is a solution of the Initial Value Problem

$$y' + P(x)y = Q(x), \qquad y(x_0) = 0$$

where $\alpha(x)$ is an integrating factor [a solution to Eq. (3)].

49. Transient Currents Suppose the circuit described by Eq. (9) is driven by a sinusoidal voltage source $V(t) = V \sin \omega t$ (where V and ω are constant).

(a) Show that

$$I(t) = \frac{V}{R^2 + L^2\omega^2}(R \sin \omega t - L\omega \cos \omega t) + Ce^{-(R/L)t}$$

(b) Let $Z = \sqrt{R^2 + L^2\omega^2}$. Choose θ so that $Z \cos \theta = R$ and $Z \sin \theta = L\omega$. Use the addition formula for the sine function to show that

$$I(t) = \frac{V}{Z} \sin(\omega t - \theta) + Ce^{-(R/L)t}$$

This shows that the current in the circuit varies sinusoidally apart from a DC term (called the **transient current** in electronics) that decreases exponentially.

CHAPTER REVIEW EXERCISES

1. Which of the following differential equations are first-order linear?

(a) $y' = y^5 - 3x^4 y$

(b) $y' = x^5 - 3x^4 y$

(c) $y = y' - 3x\sqrt{y}$

(d) $\sin x \cdot y' = y - 1$

2. Find a value of c such that $y = x - 2 + e^{cx}$ is a solution of $2y' + y = x$.

In Exercises 3–6, solve using Separation of Variables.

3. $\dfrac{dy}{dt} = t^2 y^{-3}$

4. $xyy' = 1 - x^2$

5. $x\dfrac{dy}{dx} - 2y = 3$

6. $y' = \dfrac{xy^2}{x^2+1}$

In Exercises 7–10, solve the Initial Value Problem using Separation of Variables.

7. $y' = \cos^2 x$, $\quad y(0) = \dfrac{\pi}{4}$

8. $y' = \cos^2 y$, $\quad y(0) = \dfrac{\pi}{4}$

9. $y' = 6xy^2$, $\quad y(1) = 4$

10. $xyy' = 1$, $\quad y(3) = 2$

11. Figure 1 shows the slope field for $\dfrac{dy}{dt} = \sin y + ty$. Sketch the graphs of the solutions with the initial conditions $y(0) = 1$, $y(0) = 0$, and $y(0) = -1$.

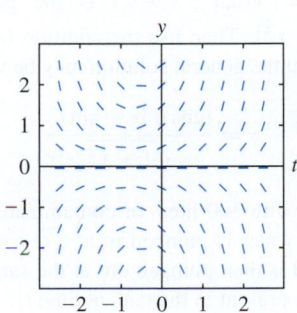

FIGURE 1

12. Sketch the slope field for $\dfrac{dy}{dt} = t^2 y$ for $-2 \le t \le 2$, $-2 \le y \le 2$.

13. Sketch the slope field for $\dfrac{dy}{dt} = y \sin t$ for $-2\pi \le t \le 2\pi$, $-2 \le y \le 2$.

14. Which of the equations (i)–(iii) corresponds to the slope field in Figure 2?

(i) $\dfrac{dy}{dt} = 1 - y^2$ $\qquad$ **(ii)** $\dfrac{dy}{dt} = 1 + y^2$ $\qquad$ **(iii)** $\dfrac{dy}{dt} = y^2$

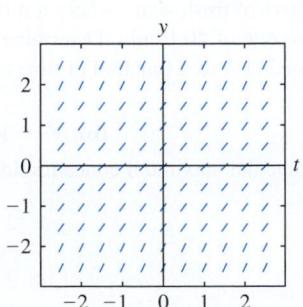

FIGURE 2

15. Let $y(t)$ be the solution to the differential equation with the slope field as shown in Figure 2, satisfying $y(0) = 0$. Sketch the graph of $y(t)$. Then use your answer to Exercise 14 to solve for $y(t)$.

16. Let $y(t)$ be the solution of $4\dfrac{dy}{dt} = y^2 + t$ satisfying $y(2) = 1$. Carry out Euler's Method with time step $h = 0.05$ for $n = 6$ steps.

17. Let $y(t)$ be the solution of $(x^3 + 1)\dfrac{dy}{dt} = y$ satisfying $y(0) = 1$. Compute approximations to $y(0.1)$, $y(0.2)$, and $y(0.3)$ using Euler's Method with time step $h = 0.1$.

In Exercises 18–21, solve using the method of integrating factors.

18. $\dfrac{dy}{dt} = y + t^2$, $\quad y(0) = 4$

19. $\dfrac{dy}{dx} = \dfrac{y}{2x} - x$, $\quad y(1) = 1$

20. $\dfrac{dy}{dt} = y - 3t$, $\quad y(-1) = 2$

21. $y' + 2y = 1 + e^{-x}$, $\quad y(0) = -4$

In Exercises 22–29, solve using an appropriate method.

22. $x^2 y' = x^2 + 1$, $\quad y(1) = 10$

23. $y' + (\tan x)y = \cos^2 x$, $\quad y(\pi) = 2$

24. $xy' = 2y + x - 1$, $\quad y\left(\dfrac{3}{2}\right) = 9$

25. $(y - 1)y' = t$, $\quad y(1) = -3$

26. $(\sqrt{y} + 1)y' = yte^{t^2}$, $\quad y(0) = 1$

27. $\dfrac{dw}{dx} = k\dfrac{1+w^2}{x}$, $\quad w(1) = 1$

28. $y' + \dfrac{3y-1}{t} = t + 2$ $\qquad$ **29.** $y' + \dfrac{y}{x} = \sin x$

30. Find the solutions to $y' = 4(y - 12)$ satisfying $y(0) = 20$ and $y(0) = 0$, and sketch their graphs.

31. Find the solutions to $y' = -2y + 8$ satisfying $y(0) = 3$ and $y(0) = 4$, and sketch their graphs.

32. Show that $y = \sin^{-1} x$ satisfies the differential equation $y' = \sec y$ with initial condition $y(0) = 0$.

33. What is the limit $\lim\limits_{t \to \infty} y(t)$ if $y(t)$ is a solution of each of the following?

(a) $\dfrac{dy}{dt} = -4(y - 12)$ $\qquad$ **(b)** $\dfrac{dy}{dt} = 4(y - 12)$

(c) $\dfrac{dy}{dt} = -4y - 12$

In Exercises 34–37, let $P(t)$ denote the balance at time t (years) of an annuity that earns 5% interest continuously compounded and pays out \$20,000/year continuously.

34. Find the differential equation satisfied by $P(t)$.

35. Determine $P(5)$ if $P(0) = \$200,000$.

36. When does the annuity run out of money if $P(0) = \$300,000$?

37. What is the minimum initial balance that will allow the annuity to make payments indefinitely?

38. State whether the differential equation can be solved using Separation of Variables, the method of integrating factors, both, or neither.

(a) $y' = y + x^2$ $\qquad\qquad$ **(b)** $xy' = y + 1$

(c) $y' = y^2 + x^2$ $\qquad\qquad$ **(d)** $xy' = y^2$

39. In the laboratory, the *Escherichia coli* bacteria grows such that the rate of change of the population is proportional to the population present. Assume that 500 bacteria are initially present, and 650 are present after 1 hour.

(a) Determine $P(t)$, the population after t hours.

(b) How long does it take for the population to double in size?

40. Uranium-238 is a radioactive material with a half-life of 4.468 billion years. With t in billions of years, let $M(t)$ be the mass in grams of a sample of uranium-238 that initially consisted of 100 g. Set up and solve an Initial Value Problem for determining $M(t)$.

41. Let A and B be constants. Prove that if $A > 0$, then all solutions of $\frac{dy}{dt} + Ay = B$ approach the same limit as $t \to \infty$.

42. At time $t = 0$, a tank of height 5 m in the shape of an inverted pyramid whose cross section at the top is a square of side 2 m is filled with water. Water flows through a hole at the bottom of area 0.002 m^2. Determine the time required for the tank to empty.

43. The trough in Figure 3 (dimensions in centimeters) is filled with water. At time $t = 0$ (in seconds), water begins leaking through a hole at the bottom of area 4 cm^2. Let $y(t)$ be the water height at time t. Find a differential equation for $y(t)$ and solve it to determine when the water level decreases to 60 cm.

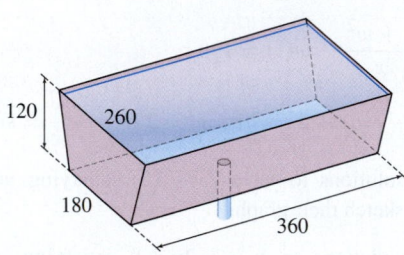

120   260
180   360

FIGURE 3

44. Find the solutions of the logistic equation $\frac{dy}{dt} = y(4 - y)$ satisfying the initial conditions:

(a) $y(0) = 1$ **(b)** $y(0) = 4$ **(c)** $y(0) = 6$

45. Let $y(t)$ be the solution of $\frac{dy}{dt} = 0.3y(2 - y)$ with $y(0) = 1$. Determine $\lim_{t \to \infty} y(t)$ without solving for y explicitly.

46. Suppose that $y' = ky(1 - y/8)$ has a solution satisfying $y(0) = 12$ and $y(10) = 24$. Find k.

47. A lake has a carrying capacity of 1000 fish. Assume that the fish population grows logistically with growth constant $k = 0.2$ day^{-1}. How many days will it take for the population to reach 900 fish if the initial population is 20 fish?

48. 📝 A rabbit population on an island increases exponentially with growth rate $k = 0.12$ months^{-1}. When the population reaches 300 rabbits (say, at time $t = 0$), wolves begin eating the rabbits at a rate of r rabbits per month.

(a) Find a differential equation satisfied by the rabbit population $P(t)$.

(b) How large can r be without the rabbit population becoming extinct?

49. Show that $y = \sin(\tan^{-1} x + C)$ is the general solution of $y' = \sqrt{1 - y^2}/(1 + x^2)$. Then use the addition formula for the sine function to show that the general solution may be written

$$y = \frac{(\cos C)x + \sin C}{\sqrt{1 + x^2}}$$

50. A tank is filled with 300 liters of contaminated water containing 3 kg of toxin. Pure water is pumped in at a rate of 40 L/min, mixes instantaneously, and is then pumped out at the same rate. Let $y(t)$ be the quantity of toxin present in the tank at time t.

(a) Find a differential equation satisfied by $y(t)$.

(b) Solve for $y(t)$.

(c) Find the time at which there is 0.01 kg of toxin present.

51. At $t = 0$, a tank of volume 300 liters is filled with 100 L of water containing salt at a concentration of 8 g/L. Fresh water flows in at a rate of 40 L/min, mixes instantaneously, and exits at the same rate. Let $c_1(t)$ be the salt concentration at time t.

(a) Find a differential equation satisfied by $c_1(t)$. *Hint:* Find the differential equation for the quantity of salt $y(t)$, and observe that $c_1(t) = y(t)/100$.

(b) Find the salt concentration $c_1(t)$ in the tank as a function of time.

52. The outflow of the tank in Exercise 51 is directed into a second tank containing V liters of fresh water where it mixes instantaneously and exits at the same rate of 40 L/min. Determine the salt concentration $c_2(t)$ in the second tank as a function of time in the following two cases:

(a) $V = 200$ **(b)** $V = 300$

In each case, determine the maximum concentration.

bilperry/Deposit Photos

Dr. Dan Russell

The figures below the photograph illustrate some of the infinitely many basic vibrational modes for a circular drumhead. A complex vibration can be understood as a combination of basic modes. Via the concept of infinite series, we can add contributions from infinitely many basic modes.

10 INFINITE SERIES

The theory of infinite series is a third branch of calculus, in addition to differential and integral calculus. Infinite series provide us with convenient and useful ways of expressing functions as infinite sums of simple functions. For example, we will see that we can express the exponential function as

$$e^x = 1 + x + \frac{x^2}{2!} + \frac{x^3}{3!} + \frac{x^4}{4!} + \cdots$$

The idea behind infinite series is that we add infinitely many numbers. We will see that although this is more complicated than adding finitely many numbers, sometimes adding infinitely many numbers yields a sum, but other times it does not. For example, we will learn that $\frac{1}{2} + \frac{1}{4} + \frac{1}{8} + \frac{1}{16} + \cdots$ adds up to 1, but $1 - 1 + 1 - 1 + 1 \cdots$ does not add up to any value (and is said to *diverge*). To make the idea of infinite series precise, we employ limits to determine what happens to the sum as we add more and more terms in a series.

We start the chapter with a section about sequences and their limits, important concepts behind infinite series. We then introduce infinite series in Section 10.2. After further developing infinite series in Sections 10.3 through 10.5, we close the chapter with three sections (Power Series, Taylor Polynomials, and Taylor Series) where we examine the idea of representing functions as infinite series.

10.1 Sequences

Limits and convergence played a fundamental role in the definitions of the derivative and definite integral. The limit concept will be significant throughout this chapter. We start by developing the basic ideas of a sequence of numbers and the limit of such a sequence.

A simple sequence of numbers arises if you eat half of a cake, and eat half of the remaining half, and continue eating half of what's left indefinitely (Figure 1). The fraction of the whole cake that remains after each step forms the sequence

$$\frac{1}{2}, \quad \frac{1}{4}, \quad \frac{1}{8}, \quad \cdots$$

This is the sequence of values of the function $f(n) = \dfrac{1}{2^n}$ for $n = 1, 2, \ldots$.

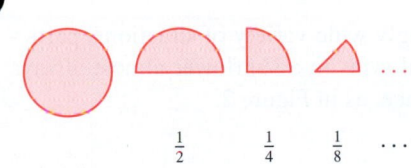

$\frac{1}{2}$ $\frac{1}{4}$ $\frac{1}{8}$ $\cdots$

FIGURE 1

> **DEFINITION Sequence** A **sequence** $\{a_n\}$ is an ordered collection of numbers defined by a function f on a set of sequential integers. The values $a_n = f(n)$ are called the **terms** of the sequence, and n is called the **index**. Informally, we think of a sequence $\{a_n\}$ as a list of terms:
>
> $$a_1, \quad a_2, \quad a_3, \quad a_4, \quad \cdots$$
>
> The sequence does not have to start at $n = 1$. It can start at $n = 0$, $n = 2$, or any other integer.

When a_n is given by a formula, we refer to a_n as the **general term**, and we refer to the set of the values n on which the sequence is defined as the **domain** of the sequence. For example, in the cake sequence, $a_n = \dfrac{1}{2^n}$ is the general term and the domain is $n \geq 1$.

Not all sequences are generated by a formula. For instance, the sequence of digits in the decimal expansion of π is

$$3, 1, 4, 1, 5, 9, 2, 6, \ldots$$

There is no specific formula for the nth digit of π and therefore, there is no formula for the general term in this sequence.

The following are examples of some sequences and their general terms.

General term	Domain	Sequence
$a_n = 1 - \dfrac{1}{n}$	$n \geq 1$	$0, \dfrac{1}{2}, \dfrac{2}{3}, \dfrac{3}{4}, \dfrac{4}{5}, \dots$
$b_n = \dfrac{364.5n^2}{n^2 - 4}$	$n \geq 3$	$656.1, 486, 433.9, 410.1, 396.9, \dots$
$c_n = \cos\left(\dfrac{n\pi}{2}\right)$	$n \geq 0$	$1, 0, -1, 0, 1, 0, \dots$
$d_n = (-1)^n n$	$n \geq 0$	$0, -1, 2, -3, 4, \dots$

The sequence b_n is the Balmer series of absorption wavelengths of the hydrogen atom in nanometers. It plays a key role in spectroscopy.

The sequence in the next example is defined *recursively*. For such a sequence, the first one or more terms may be given, and then the nth term is computed in terms of the preceding terms using some formula.

EXAMPLE 1 **The Fibonacci Sequence** We define the sequence by taking $F_1 = 1$, $F_2 = 1$, and $F_n = F_{n-1} + F_{n-2}$ for all integers $n > 2$. In other words, each subsequent term is obtained by adding together the two preceding terms. Determine the first 10 terms in the sequence.

Solution Given the first two terms, we can easily find each subsequent term by adding the previous two. The sequence is

$$1, 1, 2, 3, 5, 8, 13, 21, 34, 55, \dots$$ ∎

The Fibonacci sequence appears in a surprisingly wide variety of situations, particularly in nature. For instance, the number of spiral arms in a sunflower almost always turns out to be a number from the Fibonacci sequence, as in Figure 2.

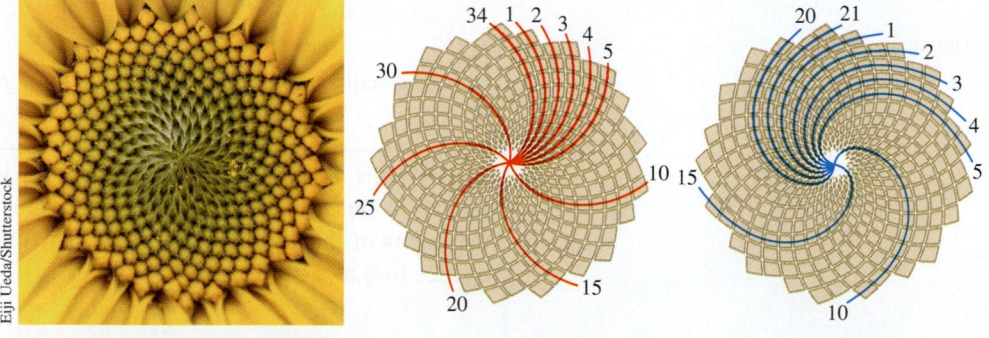

FIGURE 2 The number of spiral arms in a sunflower when counted at different angles are Fibonacci numbers. In the first case, we get 34, and in the second, we get 21.

Eiji Ueda/Shutterstock

EXAMPLE 2 **Recursive Sequence** Compute the three terms a_2, a_3, a_4 for the sequence defined recursively by

$$a_1 = 1, \qquad a_n = \frac{1}{2}\left(a_{n-1} + \frac{2}{a_{n-1}}\right)$$

Solution

You may recognize the sequence in Example 2 as the sequence of approximations to $\sqrt{2} \approx 1.4142136$ produced by Newton's Method with a starting value $a_1 = 1$ (see Section 4.8). As n tends to infinity, a_n approaches $\sqrt{2}$.

$$a_2 = \frac{1}{2}\left(a_1 + \frac{2}{a_1}\right) = \frac{1}{2}\left(1 + \frac{2}{1}\right) = \frac{3}{2} = 1.5$$

$$a_3 = \frac{1}{2}\left(a_2 + \frac{2}{a_2}\right) = \frac{1}{2}\left(\frac{3}{2} + \frac{2}{3/2}\right) = \frac{17}{12} \approx 1.4167$$

$$a_4 = \frac{1}{2}\left(a_3 + \frac{2}{a_3}\right) = \frac{1}{2}\left(\frac{17}{12} + \frac{2}{17/12}\right) = \frac{577}{408} \approx 1.414216$$ ∎

Our main goal is to study convergence of sequences. A sequence $\{a_n\}$ converges to a limit L if $|a_n - L|$ becomes arbitrarily small when n is sufficiently large. Here is the formal definition.

DEFINITION Limit of a Sequence We say $\{a_n\}$ **converges to a limit** L and write

$$\lim_{n \to \infty} a_n = L \qquad \text{or} \qquad a_n \to L$$

if, for every $\epsilon > 0$, there is a number M such that $|a_n - L| < \epsilon$ for all $n > M$.

- If no limit exists, we say that $\{a_n\}$ **diverges**.
- If the terms increase without bound, we say that $\{a_n\}$ **diverges to infinity**.

If $\{a_n\}$ converges, then its limit L is unique. To visualize the limit plot the points $(1, a_1), (2, a_2), (3, a_3), \ldots$, as in Figure 3. The sequence converges to L if, for every $\epsilon > 0$, the plotted points eventually remain within an ϵ-band around the horizontal line $y = L$. Figure 4 shows the plot of a sequence converging to $L = 1$. On the other hand, since it continually cycles through $1, 0, -1, 0$, $c_n = \cos\left(\frac{n\pi}{2}\right)$ in Figure 5 has no limit.

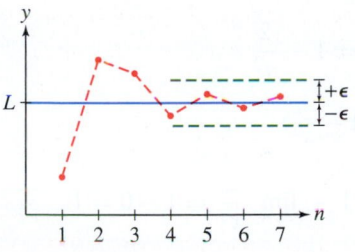

FIGURE 3 Plot of a sequence with limit L. For any ϵ, the dots eventually remain within an ϵ-band around L.

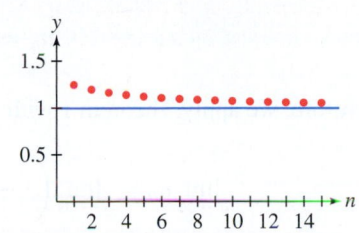

FIGURE 4 The sequence $a_n = \dfrac{n+4}{n+3}$.

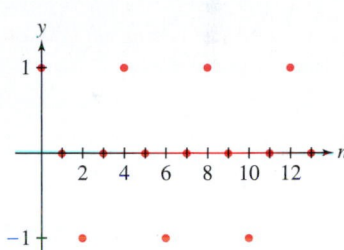

FIGURE 5 The sequence $c_n = \cos\left(\frac{n\pi}{2}\right)$ has no limit.

EXAMPLE 3 Proving Convergence Let $a_n = \dfrac{n+4}{n+3}$. Prove that $\lim_{n \to \infty} a_n = 1$.

Solution The definition requires us to find, for every $\epsilon > 0$, a number M such that

$$|a_n - 1| < \epsilon \qquad \text{for all } n > M$$

We have

$$|a_n - 1| = \left| \frac{n+4}{n+3} - 1 \right| = \frac{1}{n+3}$$

Therefore, $|a_n - 1| < \epsilon$ if

$$\frac{1}{n+3} < \epsilon \qquad \text{or} \qquad n > \frac{1}{\epsilon} - 3$$

In other words, $|a_n - 1| < \epsilon$ for all $n > \frac{1}{\epsilon} - 3$. This proves that $\lim_{n \to \infty} a_n = 1$. ∎

Note the following two facts about sequences:

- The limit does not change if we change or drop finitely many terms of the sequence.
- If C is a constant and $a_n = C$ for all n greater than some fixed value N, then $\lim_{n \to \infty} a_n = C$.

Many of the sequences we consider are defined by functions; that is, $a_n = f(n)$ for some function f. For example,

$$a_n = \frac{n-1}{n} \qquad \text{is defined by} \qquad f(x) = \frac{x-1}{x}$$

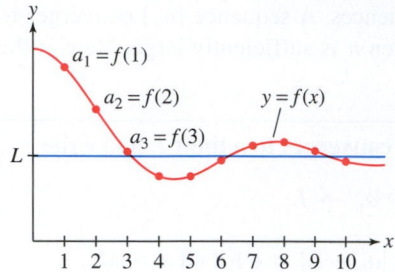

FIGURE 6 If $f(x)$ converges to L, then the sequence $a_n = f(n)$ also converges to L.

We will often use the fact that if $f(x)$ approaches a limit L as $x \to \infty$, then the sequence $a_n = f(n)$ approaches the same limit L (Figure 6). Indeed, if for all $\epsilon > 0$ we can find a positive real number M so that $|f(x) - L| < \epsilon$ for all $x > M$, then it follows automatically that $|f(n) - L| < \epsilon$ for all integers $n > M$.

THEOREM 1   Sequence Defined by a Function   If $\lim\limits_{x \to \infty} f(x)$ exists, then the sequence $a_n = f(n)$ converges to the same limit:

$$\lim_{n \to \infty} a_n = \lim_{x \to \infty} f(x)$$

EXAMPLE 4   Find the limit of the sequence

$$\frac{2^2 - 2}{2^2}, \quad \frac{3^2 - 2}{3^2}, \quad \frac{4^2 - 2}{4^2}, \quad \frac{5^2 - 2}{5^2}, \quad \cdots$$

Solution   This is the sequence with general term

$$a_n = \frac{n^2 - 2}{n^2} = 1 - \frac{2}{n^2}$$

Therefore, we apply Theorem 1 with $f(x) = 1 - \frac{2}{x^2}$:

$$\lim_{n \to \infty} a_n = \lim_{x \to \infty} \left(1 - \frac{2}{x^2}\right) = 1 - \lim_{x \to \infty} \frac{2}{x^2} = 1 - 0 = 1$$ ∎

EXAMPLE 5   Calculate $\lim\limits_{n \to \infty} \dfrac{n + \ln n}{n^2}$.

Solution   Apply Theorem 1, using L'Hôpital's Rule in the second step:

$$\lim_{n \to \infty} \frac{n + \ln n}{n^2} = \lim_{x \to \infty} \frac{x + \ln x}{x^2} = \lim_{x \to \infty} \frac{1 + (1/x)}{2x} = 0$$ ∎

The limit of the Balmer wavelengths b_n in the next example plays a role in physics and chemistry because it determines the ionization energy of the hydrogen atom. Figure 7 plots the sequence and the graph of a function f that defines the sequence. In Figure 8, the wavelengths are shown "crowding in" toward their limiting value.

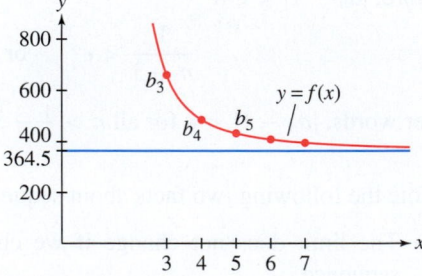

FIGURE 7   The sequence and the function approach the same limit.

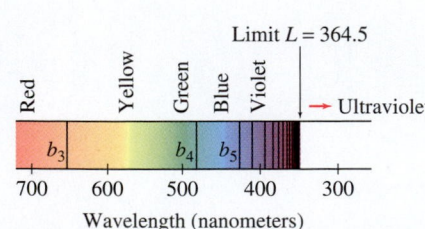

FIGURE 8

EXAMPLE 6   Balmer Wavelengths   Calculate the limit of the Balmer wavelengths $b_n = \dfrac{364.5n^2}{n^2 - 4}$ in nanometers, where $n \geq 3$.

Solution Apply Theorem 1 with $f(x) = \dfrac{364.5x^2}{x^2 - 4}$:

$$\lim_{n\to\infty} b_n = \lim_{x\to\infty} \frac{364.5x^2}{x^2 - 4} = \lim_{x\to\infty} \frac{364.5x^2 \frac{1}{x^2}}{(x^2 - 4)\frac{1}{x^2}}$$

$$= \lim_{x\to\infty} \frac{364.5}{1 - 4/x^2} = \frac{364.5}{\lim_{x\to\infty}(1 - 4/x^2)} = 364.5 \text{ nm} \qquad \blacksquare$$

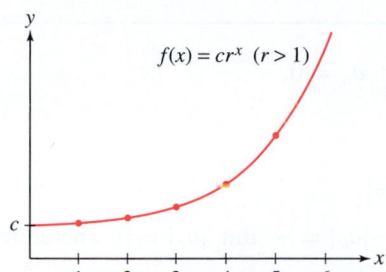

FIGURE 9 If $r > 1$, the geometric sequence $a_n = r^n$ diverges to ∞.

A **geometric sequence** is a sequence $a_n = cr^n$, where c and r are nonzero constants. Each term is r times the previous term; that is, $a_n/a_{n-1} = r$. The number r is called the **common ratio**. For instance, if $r = 3$ and $c = 2$, we obtain the sequence (starting at $n = 0$)

$$2, \quad 2 \cdot 3, \quad 2 \cdot 3^2, \quad 2 \cdot 3^3, \quad 2 \cdot 3^4, \quad 2 \cdot 3^5, \quad \ldots$$

In the next example, we determine when a geometric series converges. Recall that $\{a_n\}$ **diverges to** ∞ if the terms a_n increase beyond all bounds (Figure 9); that is,

$$\lim_{n\to\infty} a_n = \infty \quad \text{if, for every number } N, a_n > N \text{ for all sufficiently large } n$$

We define $\lim_{n\to\infty} a_n = -\infty$ similarly.

EXAMPLE 7 **Geometric Sequences with $r \geq 0$** Prove that for $r \geq 0$ and $c > 0$,

$$\lim_{n\to\infty} cr^n = \begin{cases} 0 & \text{if} & 0 \leq r < 1 \\ c & \text{if} & r = 1 \\ \infty & \text{if} & r > 1 \end{cases}$$

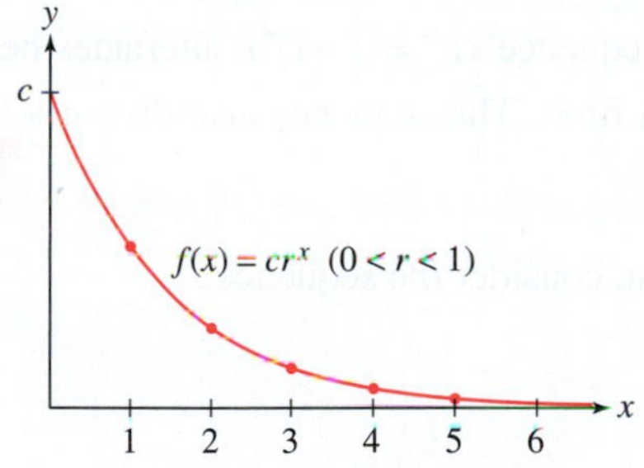

FIGURE 10 If $0 < r < 1$, then cr^x decreases to 0, and therefore the geometric sequence $a_n = r^n$ converges to 0.

Solution Set $f(r) = cr^x$. If $0 \leq r < 1$, then (Figure 10)

$$\lim_{n\to\infty} cr^n = \lim_{x\to\infty} f(x) = c \lim_{x\to\infty} r^x = 0$$

If $r > 1$, then since $c > 0$, both $f(x)$ and the sequence $\{cr^n\}$ diverge to ∞ (Figure 9). If $r = 1$, then $cr^n = c$ for all n, and the limit is c. $\qquad \blacksquare$

This last example will prove extremely useful when we consider geometric series in Section 10.2.

The limit laws we have used for functions also apply to sequences and are proved in a similar fashion.

THEOREM 2 **Limit Laws for Sequences** Assume that $\{a_n\}$ and $\{b_n\}$ are convergent sequences with

$$\lim_{n\to\infty} a_n = L, \qquad \lim_{n\to\infty} b_n = M$$

Then

(i) $\lim_{n\to\infty} (a_n \pm b_n) = \lim_{n\to\infty} a_n \pm \lim_{n\to\infty} b_n = L \pm M$

(ii) $\lim_{n\to\infty} a_n b_n = \left(\lim_{n\to\infty} a_n\right)\left(\lim_{n\to\infty} b_n\right) = LM$

(iii) $\lim_{n\to\infty} \dfrac{a_n}{b_n} = \dfrac{\lim_{n\to\infty} a_n}{\lim_{n\to\infty} b_n} = \dfrac{L}{M}$ if $M \neq 0$

(iv) $\lim_{n\to\infty} ca_n = c \lim_{n\to\infty} a_n = cL$ for any constant c

> **THEOREM 3 Squeeze Theorem for Sequences** Let $\{a_n\}$, $\{b_n\}$, $\{c_n\}$ be sequences such that for some number M,
>
> $$b_n \le a_n \le c_n \quad \text{for } n > M \qquad \text{and} \qquad \lim_{n\to\infty} b_n = \lim_{n\to\infty} c_n = L$$
>
> Then $\lim_{n\to\infty} a_n = L$.

EXAMPLE 8 Show that if $\lim_{n\to\infty} |a_n| = 0$, then $\lim_{n\to\infty} a_n = 0$.

Solution We have

$$-|a_n| \le a_n \le |a_n|$$

By hypothesis, $\lim_{n\to\infty} |a_n| = 0$, and thus also $\lim_{n\to\infty} -|a_n| = -\lim_{n\to\infty} |a_n| = 0$. Therefore, we can apply the Squeeze Theorem to conclude that $\lim_{n\to\infty} a_n = 0$. ∎

EXAMPLE 9 **Geometric Sequences with $r < 0$** Prove that for $c \ne 0$,

$$\lim_{n\to\infty} cr^n = \begin{cases} 0 & \text{if} \quad -1 < r < 0 \\ \text{diverges} & \text{if} \quad r \le -1 \end{cases}$$

Solution If $-1 < r < 0$, then $0 < |r| < 1$ and $\lim_{n\to\infty} |cr^n| = 0$ by Example 7. Thus, $\lim_{n\to\infty} cr^n = 0$ by Example 8. If $r = -1$, then the sequence $cr^n = (-1)^n c$ alternates between c and $-c$ and therefore does not approach a limit. The sequence also diverges if $r < -1$ because $|cr^n|$ grows arbitrarily large. ∎

← **REMINDER** $n!$ *(n-factorial) is the number*

$$n! = n(n-1)(n-2)\cdots 2 \cdot 1$$

For example, $4! = 4 \cdot 3 \cdot 2 \cdot 1 = 24$. *By definition,* $0! = 1$.

As another application of the Squeeze Theorem, consider the sequence

$$a_n = \frac{5^n}{n!}$$

Both the numerator and the denominator grow without bound, so it is not clear in advance whether $\{a_n\}$ converges. Figure 11 and Table 1 suggest that a_n increases initially and then tends to zero. In the next example, we verify that $a_n = R^n/n!$ converges to zero for all R. This fact is used in the discussion of Taylor series in Section 10.8.

EXAMPLE 10 Prove that $\lim_{n\to\infty} \dfrac{R^n}{n!} = 0$ for all R.

Solution Assume first that $R > 0$ and let M be the nonnegative integer such that

$$M \le R < M + 1$$

For $n > M$, we write $R^n/n!$ as a product of n factors:

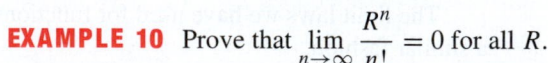

$$\frac{R^n}{n!} = \underbrace{\left(\frac{R}{1}\frac{R}{2}\cdots\frac{R}{M}\right)}_{\text{Call this constant } C.} \underbrace{\left(\frac{R}{M+1}\right)\left(\frac{R}{M+2}\right)\cdots\left(\frac{R}{n}\right)}_{\text{Each factor is less than 1.}} \le C\left(\frac{R}{n}\right) \qquad \boxed{1}$$

The first M factors are greater than or equal to 1 and the last $n - M$ factors are less than 1. If we lump together the first M factors and call the product C, and replace all the remaining factors except R/n with 1, we see that

$$0 \le \frac{R^n}{n!} \le \frac{CR}{n}$$

Since $CR/n \to 0$, the Squeeze Theorem gives us $\lim_{n\to\infty} R^n/n! = 0$ as claimed. If $R < 0$, the limit is also zero by Example 8 because $\left|R^n/n!\right|$ tends to zero. ∎

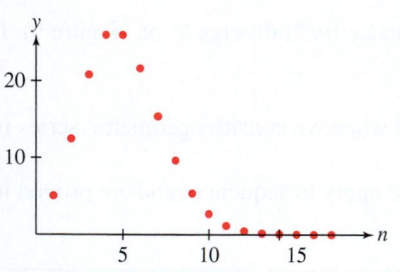

FIGURE 11 Graph of $a_n = \dfrac{5^n}{n!}$.

TABLE 1

n	$a_n = \dfrac{5^n}{n!}$
1	5
2	12.5
3	20.83
4	26.04
10	2.69
15	0.023
20	0.000039
50	2.92×10^{-30}

Given a sequence $\{a_n\}$ and a function f, we can form the new sequence $\{f(a_n)\}$. It is useful to know that if f is continuous and $a_n \to L$, then $f(a_n) \to f(L)$. A proof is given in Appendix D.

THEOREM 4 If f is continuous and $\lim_{n\to\infty} a_n = L$, then

$$\lim_{n\to\infty} f(a_n) = f\left(\lim_{n\to\infty} a_n\right) = f(L)$$

In other words, we may pass a limit of a sequence inside a continuous function.

EXAMPLE 11 Determine the limit of the sequence $a_n = \dfrac{3n}{n+1}$, and then apply Theorem 4 to determine the limits of the sequences $\{f(a_n)\}$ and $\{g(a_n)\}$, where $f(x) = e^x$ and $g(x) = x^2$.

Solution First,

$$L = \lim_{n\to\infty} a_n = \lim_{n\to\infty} \frac{3n}{n+1} = \lim_{n\to\infty} \frac{3}{1+n^{-1}} = 3$$

Now, with $f(x) = e^x$, we have $f(a_n) = e^{a_n} = e^{\frac{3n}{n+1}}$. According to Theorem 4,

$$\lim_{n\to\infty} f(a_n) = f\left(\lim_{n\to\infty} a_n\right) = e^{\lim_{n\to\infty} \frac{3n}{n+1}} = e^3$$

Finally, with $g(x) = x^2$, we have $g(a_n) = a_n^2$. According to Theorem 4,

$$\lim_{n\to\infty} g(a_n) = g\left(\lim_{n\to\infty} a_n\right) = \left(\lim_{n\to\infty} \frac{3n}{n+1}\right)^2 = 3^2 = 9 \qquad \blacksquare$$

Next, we define the concepts of a bounded sequence and a monotonic sequence, concepts of great importance for understanding convergence.

DEFINITION Bounded Sequences A sequence $\{a_n\}$ is

- **Bounded from above** if there is a number M such that $a_n \le M$ for all n. The number M is called an *upper bound.*
- **Bounded from below** if there is a number m such that $a_n \ge m$ for all n. The number m is called a *lower bound.*

The sequence $\{a_n\}$ is called **bounded** if it is bounded from above and below. A sequence that is not bounded is called an **unbounded sequence**.

Thus, for instance, the sequence given by $a_n = 3 - \frac{1}{n}$ is clearly bounded above by 3. It is also bounded below by 0, since all the terms are positive. Hence, this sequence is bounded.

Upper and lower bounds are not unique. If M is an upper bound, then any number greater than M is also an upper bound, and if m is a lower bound, then any number less than m is also a lower bound (Figure 12).

As we might expect, a convergent sequence $\{a_n\}$ is necessarily bounded because the terms a_n get closer and closer to the limit. This fact is stated in the next theorem.

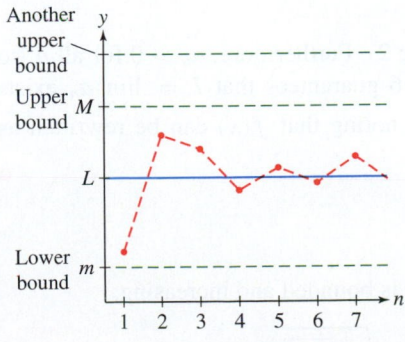

FIGURE 12 A convergent sequence is bounded.

THEOREM 5 Convergent Sequences Are Bounded If $\{a_n\}$ converges, then $\{a_n\}$ is bounded.

Proof Let $L = \lim\limits_{n\to\infty} a_n$. Then there exists $N > 0$ such that $|a_n - L| < 1$ for $n > N$. In other words,

$$L - 1 < a_n < L + 1 \qquad \text{for } n > N$$

If M is any number greater than $L + 1$ and also greater than the numbers $a_1, a_2, \ldots, a_N$, then $a_n < M$ for all n. Thus, M is an upper bound. Similarly, any number m less than $L - 1$ and also less than the numbers $a_1, a_2, \ldots, a_N$ is a lower bound. ∎

There are two ways that a sequence $\{a_n\}$ can diverge. One way is by being unbounded. For example, the unbounded sequence $a_n = n$ diverges:

$$1, \quad 2, \quad 3, \quad 4, \quad 5, \quad 6, \quad \ldots$$

However, a sequence can diverge even if it is bounded. This is the case with $a_n = (-1)^{n+1}$, whose terms a_n bounce back and forth but never settle down to approach a limit:

$$1, \quad -1, \quad 1, \quad -1, \quad 1, \quad -1, \quad \ldots$$

There is no surefire method for determining whether a sequence $\{a_n\}$ converges, unless the sequence happens to be both bounded and **monotonic**. By definition, $\{a_n\}$ is monotonic if it is either increasing or decreasing:

- $\{a_n\}$ is *increasing* if $a_n < a_{n+1}$ for all n.
- $\{a_n\}$ is *decreasing* if $a_n > a_{n+1}$ for all n.

Intuitively, if $\{a_n\}$ is increasing and bounded above by M, then the terms must bunch up near some limiting value L that is not greater than M (Figure 13). See Appendix B for a proof of the next theorem.

> The Fibonacci sequence $\{F_n\}$ diverges since it is unbounded ($F_n \geq n$ for all n), but the sequence defined by the ratios $a_n = \dfrac{F_{n+1}}{F_n}$ converges. The limit is an important number known as the **golden ratio** (see Exercises 33 and 34).

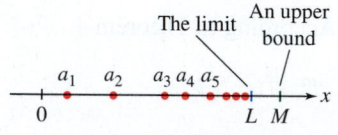

The limit   An upper bound

FIGURE 13 An increasing sequence with upper bound M approaches a limit L.

THEOREM 6 Bounded Monotonic Sequences Converge

- If $\{a_n\}$ is increasing and $a_n \leq M$, then $\{a_n\}$ converges and $\lim\limits_{n\to\infty} a_n \leq M$.
- If $\{a_n\}$ is decreasing and $a_n \geq m$, then $\{a_n\}$ converges and $\lim\limits_{n\to\infty} a_n \geq m$.

EXAMPLE 12 Verify that $a_n = \sqrt{n+1} - \sqrt{n}$ is decreasing and bounded below. Does $\lim\limits_{n\to\infty} a_n$ exist?

Solution The function $f(x) = \sqrt{x+1} - \sqrt{x}$ is decreasing because its derivative is negative:

$$f'(x) = \frac{1}{2\sqrt{x+1}} - \frac{1}{2\sqrt{x}} < 0 \qquad \text{for } x > 0$$

It follows that $a_n = f(n)$ is decreasing (see Table 2). Furthermore, $a_n > 0$ for all n, so the sequence has lower bound $m = 0$. Theorem 6 guarantees that $L = \lim\limits_{n\to\infty} a_n$ exists and $L \geq 0$. In fact, we can show that $L = 0$ by noting that $f(x)$ can be rewritten as $f(x) = \dfrac{1}{\sqrt{x+1} + \sqrt{x}}$. Hence, $\lim\limits_{x\to\infty} f(x) = 0$. ∎

DF **TABLE 2**

$a_n = \sqrt{n+1} - \sqrt{n}$
$a_1 \approx 0.4142$
$a_2 \approx 0.3178$
$a_3 \approx 0.2679$
$a_4 \approx 0.2361$
$a_5 \approx 0.2134$
$a_6 \approx 0.1963$
$a_7 \approx 0.1827$
$a_8 \approx 0.1716$

EXAMPLE 13 Show that the following sequence is bounded and increasing:

$$a_1 = \sqrt{2}, \qquad a_2 = \sqrt{2\sqrt{2}}, \qquad a_3 = \sqrt{2\sqrt{2\sqrt{2}}}, \qquad \ldots$$

Then prove that $L = \lim\limits_{n\to\infty} a_n$ exists and compute its value.

Solution

Step 1. Show that $\{a_n\}$ is bounded above.

We claim that $M = 2$ is an upper bound. We certainly have $a_1 < 2$ because $a_1 = \sqrt{2} \approx 1.414$. On the other hand,

$$\text{if} \quad a_n < 2, \qquad \text{then} \quad a_{n+1} < 2 \qquad \boxed{2}$$

is true because $a_{n+1} = \sqrt{2a_n} < \sqrt{2 \cdot 2} = 2$. Now, since $a_1 < 2$, we can apply (2) to conclude that $a_2 < 2$. Similarly, $a_2 < 2$ implies $a_3 < 2$, and so on. It follows that $a_n < 2$ for all n. (Formally speaking, this is a proof by induction.)

Step 2. Show that $\{a_n\}$ is increasing.

Since a_n is positive and $a_n < 2$, we have

$$a_{n+1} = \sqrt{2a_n} > \sqrt{a_n \cdot a_n} = a_n$$

This shows that $\{a_n\}$ is increasing. Since the sequence is bounded above and increasing, we conclude that the limit L exists.

Now that we know the limit L exists, we can find its value as follows. The idea is that L "contains a copy" of itself under the square root sign:

$$L = \sqrt{2\sqrt{2\sqrt{2\sqrt{2\cdots}}}} = \sqrt{2\left(\sqrt{2\sqrt{2\sqrt{2\cdots}}}\right)} = \sqrt{2L}$$

Thus, $L^2 = 2L$, which implies that $L = 2$ or $L = 0$. We eliminate $L = 0$ because the terms a_n are positive and increasing, so we must have $L = 2$ (see Table 3). ∎

In the previous example, the argument that $L = \sqrt{2L}$ is more formally expressed by noting that the sequence is defined recursively by

$$a_1 = \sqrt{2}, \qquad a_{n+1} = \sqrt{2a_n}$$

If a_n converges to L, then the sequence $b_n = a_{n+1}$ also converges to L (because it is the same sequence, with terms shifted one to the left). Then, applying Theorem 4 to $f(x) = \sqrt{x}$, we have

$$L = \lim_{n \to \infty} a_{n+1} = \lim_{n \to \infty} \sqrt{2a_n} = \sqrt{2 \lim_{n \to \infty} a_n} = \sqrt{2L}$$

DF TABLE 3
Recursive Sequence
$a_{n+1} = \sqrt{2a_n}$

a_1	1.4142
a_2	1.6818
a_3	1.8340
a_4	1.9152
a_5	1.9571
a_6	1.9785
a_7	1.9892
a_8	1.9946

10.1 SUMMARY

- A sequence $\{a_n\}$ *converges* to a limit L if, for every $\epsilon > 0$, there is a number M such that

$$|a_n - L| < \epsilon \qquad \text{for all } n > M$$

We write $\lim\limits_{n \to \infty} a_n = L$ or $a_n \to L$.
- If no limit exists, we say that $\{a_n\}$ *diverges*.
- In particular, if the terms increase without bound, we say that $\{a_n\}$ diverges to infinity.
- If $a_n = f(n)$ and $\lim\limits_{x \to \infty} f(x) = L$, then $\lim\limits_{n \to \infty} a_n = L$.
- A *geometric sequence* is a sequence $a_n = cr^n$, where c and r are nonzero. It converges to 0 for $-1 < r < 1$, converges to c for $r = 1$, and diverges otherwise.
- The Basic Limit Laws and the Squeeze Theorem apply to sequences.
- If f is continuous and $\lim\limits_{n \to \infty} a_n = L$, then $\lim\limits_{n \to \infty} f(a_n) = f(L)$.
- A sequence $\{a_n\}$ is

  - *bounded above* by M if $a_n \leq M$ for all n.
  - *bounded below* by m if $a_n \geq m$ for all n.

If $\{a_n\}$ is bounded above and below, $\{a_n\}$ is called *bounded*.
- A sequence $\{a_n\}$ is *monotonic* if it is increasing ($a_n < a_{n+1}$) or decreasing ($a_{n+1} < a_n$).
- Bounded monotonic sequences converge (Theorem 6).

10.1 EXERCISES

Preliminary Questions

1. What is a_4 for the sequence $a_n = n^2 - n$?

2. Which of the following sequences converge to zero?

(a) $\dfrac{n^2}{n^2 + 1}$
(b) 2^n
(c) $\left(\dfrac{-1}{2}\right)^n$

3. Let a_n be the nth decimal approximation to $\sqrt{2}$. That is, $a_1 = 1$, $a_2 = 1.4$, $a_3 = 1.41$, and so on. What is $\lim\limits_{n \to \infty} a_n$?

4. Which of the following sequences is defined recursively?

(a) $a_n = \sqrt{4 + n}$
(b) $b_n = \sqrt{4 + b_{n-1}}$

5. Theorem 5 says that every convergent sequence is bounded. Determine if the following statements are true or false, and if false, give a counterexample:

(a) If $\{a_n\}$ is bounded, then it converges.

(b) If $\{a_n\}$ is not bounded, then it diverges.

(c) If $\{a_n\}$ diverges, then it is not bounded.

Exercises

1. Match each sequence with its general term:

$a_1, a_2, a_3, a_4, \ldots$	General term
(a) $\frac{1}{2}, \frac{2}{3}, \frac{3}{4}, \frac{4}{5}, \ldots$	(i) $\cos \pi n$
(b) $-1, 1, -1, 1, \ldots$	(ii) $\dfrac{n!}{2^n}$
(c) $1, -1, 1, -1, \ldots$	(iii) $(-1)^{n+1}$
(d) $\frac{1}{2}, \frac{2}{4}, \frac{6}{8}, \frac{24}{16} \cdots$	(iv) $\dfrac{n}{n+1}$

2. Let $a_n = \dfrac{1}{2n - 1}$ for $n = 1, 2, 3, \ldots$. Write out the first three terms of the following sequences.

(a) $b_n = a_{n+1}$
(b) $c_n = a_{n+3}$
(c) $d_n = a_n^2$
(d) $e_n = 2a_n - a_{n+1}$

In Exercises 3–12, calculate the first four terms of the sequence, starting with $n = 1$.

3. $c_n = \dfrac{3^n}{n!}$

4. $b_n = \dfrac{(2n - 1)!}{n!}$

5. $a_1 = 2$, $a_{n+1} = 2a_n^2 - 3$

6. $b_1 = 1$, $b_n = b_{n-1} + \dfrac{1}{b_{n-1}}$

7. $b_n = 5 + \cos \pi n$

8. $c_n = (-1)^{2n+1}$

9. $c_n = 1 + \dfrac{1}{2} + \dfrac{1}{3} + \cdots + \dfrac{1}{n}$

10. $w_n = 1 + \dfrac{1}{2^2} + \dfrac{1}{3^2} + \cdots + \dfrac{1}{n^2}$

11. $b_1 = 2$, $b_2 = 3$, $b_n = 2b_{n-1} + b_{n-2}$

12. $a_n = \dfrac{F_{n+1}}{F_n}$ where F_n is the nth Fibonacci number.

13. Find a formula for the nth term of each sequence.

(a) $\dfrac{1}{1}, \dfrac{-1}{8}, \dfrac{1}{27}, \ldots$
(b) $\dfrac{2}{6}, \dfrac{3}{7}, \dfrac{4}{8}, \ldots$

14. Suppose that $\lim\limits_{n \to \infty} a_n = 4$ and $\lim\limits_{n \to \infty} b_n = 7$. Determine:

(a) $\lim\limits_{n \to \infty} (a_n + b_n)$
(b) $\lim\limits_{n \to \infty} a_n^3$
(c) $\lim\limits_{n \to \infty} \cos(\pi b_n)$
(d) $\lim\limits_{n \to \infty} (a_n^2 - 2a_n b_n)$

In Exercises 15–28, use Theorem 1 to determine the limit of the sequence or state that the sequence diverges.

15. $a_n = 5 - 2n$

16. $a_n = 20 - \dfrac{4}{n^2}$

17. $b_n = \dfrac{5n - 1}{12n + 9}$

18. $a_n = \dfrac{4 + n - 3n^2}{4n^2 + 1}$

19. $a_n = \left(\dfrac{1}{2}\right)^{-n}$

20. $z_n = \left(\dfrac{1}{3}\right)^n$

21. $c_n = 9^n$

22. $z_n = 10^{-1/n}$

23. $a_n = \dfrac{n}{\sqrt{n^2 + 1}}$

24. $a_n = \dfrac{n}{\sqrt{n^3 + 1}}$

25. $a_n = \ln\left(\dfrac{12n + 2}{-9 + 4n}\right)$

26. $r_n = \ln n - \ln(n^2 + 1)$

27. $z_n = \dfrac{n + 1}{e^n}$

28. $y_n = ne^{1/n}$

In Exercises 29–32, use Theorem 4 to determine the limit of the sequence.

29. $a_n = \sqrt{4 + \dfrac{1}{n}}$

30. $a_n = e^{4n/(3n+9)}$

31. $a_n = \cos^{-1}\left(\dfrac{n^3}{2n^3 + 1}\right)$

32. $a_n = \tan^{-1}(e^{-n})$

In Exercises 33–34 let $a_n = \dfrac{F_{n+1}}{F_n}$, where $\{F_n\}$ is the Fibonacci sequence. The sequence $\{a_n\}$ has a limit. We do not prove this fact, but investigate the value of the limit in these exercises.

33. **CAS** Estimate $\lim\limits_{n \to \infty} a_n$ to five decimal places by computing a_n for sufficiently large n.

34. Denote the limit of $\{a_n\}$ by L. Given that the limit exists, we can determine L as follows:

(a) Show that $a_{n+1} = 1 + \dfrac{1}{a_n}$.

(b) Given that $\{a_n\}$ converges to L, it follows that $\{a_{n+1}\}$ also converges to L (see Exercise 85). Show that $L^2 - L - 1 = 0$ and solve this equation to determine L. (The value of L is known as the **golden ratio**. It arises in many different situations in mathematics.)

35. Let $a_n = \dfrac{n}{n+1}$. Find a number M such that:

(a) $|a_n - 1| \le 0.001$ for $n \ge M$.

(b) $|a_n - 1| \le 0.00001$ for $n \ge M$.

Then use the limit definition to prove that $\lim_{n \to \infty} a_n = 1$.

36. Let $b_n = \left(\frac{1}{3}\right)^n$.

(a) Find a value of M such that $|b_n| \le 10^{-5}$ for $n \ge M$.

(b) Use the limit definition to prove that $\lim_{n \to \infty} b_n = 0$.

37. Use the limit definition to prove that $\lim_{n \to \infty} n^{-2} = 0$.

38. Use the limit definition to prove that $\lim_{n \to \infty} \dfrac{n}{n + n^{-1}} = 1$.

In Exercises 39–66, use the appropriate limit laws and theorems to determine the limit of the sequence or show that it diverges.

39. $a_n = 10 + \left(-\dfrac{1}{9}\right)^n$

40. $d_n = \sqrt{n+3} - \sqrt{n}$

41. $c_n = 1.01^n$

42. $b_n = e^{1-n^2}$

43. $a_n = 2^{1/n}$

44. $b_n = n^{1/n}$

45. $c_n = \dfrac{9^n}{n!}$

46. $a_n = \dfrac{8^{2n}}{n!}$

47. $a_n = \dfrac{3n^2 + n + 2}{2n^2 - 3}$

48. $a_n = \dfrac{\sqrt{n}}{\sqrt{n+4}}$

49. $a_n = \dfrac{\cos n}{n}$

50. $c_n = \dfrac{(-1)^n}{\sqrt{n}}$

51. $d_n = \ln 5^n - \ln n!$

52. $d_n = \ln(n^2 + 4) - \ln(n^2 - 1)$

53. $a_n = \left(2 + \dfrac{4}{n^2}\right)^{1/3}$

54. $b_n = \tan^{-1}\left(1 - \dfrac{2}{n}\right)$

55. $c_n = \ln\left(\dfrac{2n+1}{3n+4}\right)$

56. $c_n = \dfrac{n}{n + n^{1/n}}$

57. $y_n = \dfrac{e^n}{2^n}$

58. $a_n = \dfrac{n}{2^n}$

59. $y_n = \dfrac{e^n + (-3)^n}{5^n}$

60. $b_n = \dfrac{(-1)^n n^3 + 2^{-n}}{3n^3 + 4^{-n}}$

61. $a_n = n \sin \dfrac{\pi}{n}$

62. $b_n = \dfrac{n!}{\pi^n}$

63. $b_n = \dfrac{3 - 4^n}{2 + 7 \cdot 4^n}$

64. $a_n = \dfrac{3 - 4^n}{2 + 7 \cdot 3^n}$

65. $a_n = \left(1 + \dfrac{1}{n}\right)^n$

66. $a_n = \left(1 + \dfrac{1}{n^2}\right)^n$

In Exercises 67–70, find the limit of the sequence using L'Hôpital's Rule.

67. $a_n = \dfrac{(\ln n)^2}{n}$

68. $b_n = \sqrt{n} \ln\left(1 + \dfrac{1}{n}\right)$

69. $c_n = n\left(\sqrt{n^2 + 1} - n\right)$

70. $d_n = n^2\left(\sqrt[3]{n^3 + 1} - n\right)$

In Exercises 71–74, use the Squeeze Theorem to evaluate $\lim_{n \to \infty} a_n$ by verifying the given inequality.

71. $a_n = \dfrac{1}{\sqrt{n^4 + n^8}}$, $\dfrac{1}{\sqrt{2n^4}} \le a_n \le \dfrac{1}{\sqrt{2n^2}}$

72. $c_n = \dfrac{1}{\sqrt{n^2 + 1}} + \dfrac{1}{\sqrt{n^2 + 2}} + \cdots + \dfrac{1}{\sqrt{n^2 + n}}$,

$\dfrac{n}{\sqrt{n^2 + n}} \le c_n \le \dfrac{n}{\sqrt{n^2 + 1}}$

73. $a_n = (2^n + 3^n)^{1/n}$, $3 \le a_n \le (2 \cdot 3^n)^{1/n} = 2^{1/n} \cdot 3$

74. $a_n = (n + 10^n)^{1/n}$, $10 \le a_n \le (2 \cdot 10^n)^{1/n}$

75. ⬜ Which of the following statements is equivalent to the assertion $\lim_{n \to \infty} a_n = L$? Explain.

(a) For every $\epsilon > 0$, the interval $(L - \epsilon, L + \epsilon)$ contains at least one element of the sequence $\{a_n\}$.

(b) For every $\epsilon > 0$, the interval $(L - \epsilon, L + \epsilon)$ contains all but at most finitely many elements of the sequence $\{a_n\}$.

76. Show that $a_n = \dfrac{1}{2n+1}$ is decreasing.

77. Show that $a_n = \dfrac{3n^2}{n^2 + 2}$ is increasing. Find an upper bound.

78. Show that $a_n = \sqrt[3]{n+1} - n$ is decreasing.

79. Give an example of a divergent sequence $\{a_n\}$ such that $\lim_{n \to \infty} |a_n|$ converges.

80. Give an example of divergent sequences $\{a_n\}$ and $\{b_n\}$ such that $\{a_n + b_n\}$ converges.

81. Using the limit definition, prove that if $\{a_n\}$ converges and $\{b_n\}$ diverges, then $\{a_n + b_n\}$ diverges.

82. Use the limit definition to prove that if $\{a_n\}$ is a convergent sequence of integers with limit L, then there exists a number M such that $a_n = L$ for all $n \ge M$.

83. Theorem 1 states that if $\lim_{x \to \infty} f(x) = L$, then the sequence $a_n = f(n)$ converges and $\lim_{n \to \infty} a_n = L$. Show that the *converse* is false. In other words, find a function f such that $a_n = f(n)$ converges but $\lim_{x \to \infty} f(x)$ does not exist.

84. Use the limit definition to prove that the limit does not change if a finite number of terms are added or removed from a convergent sequence.

85. Let $b_n = a_{n+1}$. Use the limit definition to prove that if $\{a_n\}$ converges, then $\{b_n\}$ also converges and $\lim_{n \to \infty} a_n = \lim_{n \to \infty} b_n$.

86. Let $\{a_n\}$ be a sequence such that $\lim_{n \to \infty} |a_n|$ exists and is nonzero. Show that $\lim_{n \to \infty} a_n$ exists if and only if there exists an integer M such that the sign of a_n does not change for $n > M$.

87. Proceed as in Example 13 to show that the sequence $\sqrt{3}, \sqrt{3\sqrt{3}},$

$\sqrt{3\sqrt{3\sqrt{3}}}, \ldots$ is increasing and bounded above by $M = 3$. Then prove that the limit exists and find its value.

88. Let $\{a_n\}$ be a sequence defined recursively by

$$a_0 = 0, \qquad a_{n+1} = \sqrt{2 + a_n}$$

Thus, $a_1 = \sqrt{2}, \quad a_2 = \sqrt{2 + \sqrt{2}}, \quad a_3 = \sqrt{2 + \sqrt{2 + \sqrt{2}}}, \ldots$.

(a) Show that if $a_n < 2$, then $a_{n+1} < 2$. Conclude by induction that $a_n < 2$ for all n.

(b) Show that if $a_n < 2$, then $a_n \le a_{n+1}$. Conclude by induction that $\{a_n\}$ is increasing.

(c) Use (a) and (b) to conclude that $L = \lim_{n \to \infty} a_n$ exists. Then compute L by showing that $L = \sqrt{2 + L}$.

Further Insights and Challenges

89. Show that $\lim_{n \to \infty} \sqrt[n]{n!} = \infty$. *Hint:* Verify that $n! \geq (n/2)^{n/2}$ by observing that half of the factors of $n!$ are greater than or equal to $n/2$.

90. Let $b_n = \dfrac{\sqrt[n]{n!}}{n}$.

(a) Show that $\ln b_n = \dfrac{1}{n} \sum_{k=1}^{n} \ln \dfrac{k}{n}$.

(b) Show that $\ln b_n$ converges to $\displaystyle\int_0^1 \ln x \, dx$, and conclude that $b_n \to e^{-1}$.

91. Given positive numbers $a_1 < b_1$, define two sequences recursively by
$$a_{n+1} = \sqrt{a_n b_n}, \qquad b_{n+1} = \frac{a_n + b_n}{2}$$

(a) Show that $a_n \leq b_n$ for all n (Figure 14).
(b) Show that $\{a_n\}$ is increasing and $\{b_n\}$ is decreasing.

(c) Show that $b_{n+1} - a_{n+1} \leq \dfrac{b_n - a_n}{2}$.
(d) Prove that both $\{a_n\}$ and $\{b_n\}$ converge and have the same limit. This limit, denoted $\text{AGM}(a_1, b_1)$, is called the **arithmetic-geometric mean** of a_1 and b_1.
(e) Estimate $\text{AGM}(1, \sqrt{2})$ to three decimal places.

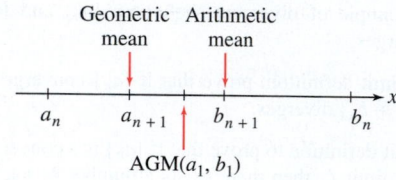

FIGURE 14

92. Let $c_n = \dfrac{1}{n} + \dfrac{1}{n+1} + \dfrac{1}{n+2} + \cdots + \dfrac{1}{2n}$.

(a) Calculate c_1, c_2, c_3, c_4.

(b) Use a comparison of rectangles with the area under $y = x^{-1}$ over the interval $[n, 2n]$ to prove that
$$\int_n^{2n} \frac{dx}{x} + \frac{1}{2n} \leq c_n \leq \int_n^{2n} \frac{dx}{x} + \frac{1}{n}$$

(c) Use the Squeeze Theorem to determine $\lim_{n \to \infty} c_n$.

93. Let $a_n = H_n - \ln n$, where H_n is the nth harmonic number:
$$H_n = 1 + \frac{1}{2} + \frac{1}{3} + \cdots + \frac{1}{n}$$

(a) Show that $a_n \geq 0$ for $n \geq 1$. *Hint:* Show that $H_n \geq \displaystyle\int_1^{n+1} \frac{dx}{x}$.
(b) Show that $\{a_n\}$ is decreasing by interpreting $a_n - a_{n+1}$ as an area.
(c) Prove that $\lim_{n \to \infty} a_n$ exists.

This limit, denoted γ, is known as *Euler's Constant*. It appears in many areas of mathematics, including analysis and number theory, and has been calculated to more than 100 million decimal places, but it is still not known whether γ is an irrational number. The first 10 digits are $\gamma \approx 0.5772156649$.

10.2 Summing an Infinite Series

Many quantities that arise in mathematics and its applications cannot be computed exactly. We cannot write down an exact decimal expression for the number π or for values of the sine function such as $\sin 1$. However, sometimes these quantities can be represented as infinite sums. For example, using Taylor series (Section 10.8), we can show that

$$\sin 1 = 1 - \frac{1}{3!} + \frac{1}{5!} - \frac{1}{7!} + \frac{1}{9!} - \frac{1}{11!} + \cdots \qquad \boxed{1}$$

Infinite sums of this type are called **infinite series**. We think of them as having been obtained by adding up all of the terms in a sequence of numbers.

But what precisely does Eq. (1) mean? How do we make sense of a sum of infinitely many terms? The idea is to examine finite sums of terms at the start of the series and see how they behave. We add progressively more terms and determine whether or not the sums approach a limiting value. More specifically, for the infinite series

$$a_1 + a_2 + a_3 + a_4 + \cdots + a_n + \cdots$$

define the **partial sums**:

$$S_1 = a_1$$

$$S_2 = a_1 + a_2$$

$$S_3 = a_1 + a_2 + a_3$$

$$\cdots$$

$$S_N = a_1 + a_2 + a_3 + \cdots + a_N$$

The idea then is to consider the *sequence* of values, $S_1, S_2, S_3, \ldots, S_N, \ldots$, and whether the limit of this sequence exists.

For example, here are the first five partial sums of the infinite series for $\sin 1$:

$$S_1 = 1$$

$$S_2 = 1 - \frac{1}{3!} = 1 - \frac{1}{6} \qquad\qquad \approx 0.833$$

$$S_3 = 1 - \frac{1}{3!} + \frac{1}{5!} = 1 - \frac{1}{6} + \frac{1}{120} \qquad\qquad \approx 0.841667$$

$$S_4 = 1 - \frac{1}{6} + \frac{1}{120} - \frac{1}{5040} \qquad\qquad \approx 0.841468$$

$$S_5 = 1 - \frac{1}{6} + \frac{1}{120} - \frac{1}{5040} + \frac{1}{362{,}880} \approx 0.8414709846$$

Compare these values with the value obtained from a calculator:

$$\sin 1 \approx 0.8414709848079$$

We see that S_5 differs from $\sin 1$ by less than 10^{-9}. This suggests that the partial sums converge to $\sin 1$, and in fact, in Section 10.8 we will prove that

$$\sin 1 = \lim_{N \to \infty} S_N$$

(see Example 2). It makes sense then to *define* the sum of an infinite series as a limit of partial sums.

In general, an infinite series is an expression of the form

$$\sum_{n=1}^{\infty} a_n = a_1 + a_2 + a_3 + a_4 + \cdots$$

where $\{a_n\}$ is any sequence. For example,

- *Infinite series may begin with any value for the index. For example,*

$$\sum_{n=3}^{\infty} \frac{1}{n} = \frac{1}{3} + \frac{1}{4} + \frac{1}{5} + \cdots$$

When it is not necessary to specify the starting point, we write simply $\sum a_n$.
- *Any letter may be used for the index. Thus, we may write a_m, a_k, a_i, and so on.*

Sequence	General term	Infinite series
$\dfrac{1}{3}, \dfrac{1}{9}, \dfrac{1}{27}, \ldots$	$a_n = \dfrac{1}{3^n}$	$\displaystyle\sum_{n=1}^{\infty} \dfrac{1}{3^n} = \dfrac{1}{3} + \dfrac{1}{9} + \dfrac{1}{27} + \dfrac{1}{81} + \cdots$
$\dfrac{1}{1}, \dfrac{1}{4}, \dfrac{1}{9}, \dfrac{1}{16}, \ldots$	$a_n = \dfrac{1}{n^2}$	$\displaystyle\sum_{n=1}^{\infty} \dfrac{1}{n^2} = \dfrac{1}{1} + \dfrac{1}{4} + \dfrac{1}{9} + \dfrac{1}{16} + \cdots$

The Nth partial sum S_N is the finite sum of the terms up to and including a_N:

$$S_N = \sum_{n=1}^{N} a_n = a_1 + a_2 + a_3 + \cdots + a_N$$

If the series begins at k, then $S_N = a_k + a_{k+1} + \cdots + a_N$.

DEFINITION **Convergence of an Infinite Series** An infinite series $\displaystyle\sum_{n=k}^{\infty} a_n$ converges to the sum S if the sequence of its partial sums $\{S_N\}$ converges to S:

$$\lim_{N \to \infty} S_N = S$$

In this case, we write $S = \displaystyle\sum_{n=k}^{\infty} a_n$.

- If the limit does not exist, we say that the infinite series diverges.
- If the limit is infinite, we say that the infinite series diverges to infinity.

We can investigate series numerically by computing several partial sums S_N. If the sequence of partial sums shows a trend of convergence to some number S, then we have evidence (but not proof) that the series converges to S. The next example treats a **telescoping series**, where the partial sums are particularly easy to evaluate.

EXAMPLE 1 **Telescoping Series** Investigate numerically:

$$\sum_{n=1}^{\infty} \frac{1}{n(n+1)} = \frac{1}{1(2)} + \frac{1}{2(3)} + \frac{1}{3(4)} + \frac{1}{4(5)} + \cdots$$

Then compute the sum of the series using the identity:

$$\frac{1}{n(n+1)} = \frac{1}{n} - \frac{1}{n+1}$$

DF TABLE 1 **Partial Sums for** $\displaystyle\sum_{n=1}^{\infty} \frac{1}{n(n+1)}$

N	S_N
10	0.90909
50	0.98039
100	0.990099
200	0.995025
300	0.996678

Solution The values of the partial sums listed in Table 1 suggest convergence to $S = 1$. To prove this, we observe that because of the identity, each partial sum collapses down to just two terms:

$$S_1 = \frac{1}{1(2)} = \frac{1}{1} - \frac{1}{2}$$

$$S_2 = \frac{1}{1(2)} + \frac{1}{2(3)} = \left(\frac{1}{1} - \frac{1}{2}\right) + \left(\frac{1}{2} - \frac{1}{3}\right) = 1 - \frac{1}{3}$$

$$S_3 = \frac{1}{1(2)} + \frac{1}{2(3)} + \frac{1}{3(4)} = \left(\frac{1}{1} - \frac{1}{2}\right) + \left(\frac{1}{2} - \frac{1}{3}\right) + \left(\frac{1}{3} - \frac{1}{4}\right) = 1 - \frac{1}{4}$$

In general,

$$S_N = \left(\frac{1}{1} - \frac{1}{2}\right) + \left(\frac{1}{2} - \frac{1}{3}\right) + \cdots + \left(\frac{1}{N-1} - \frac{1}{N}\right) + \left(\frac{1}{N} - \frac{1}{N+1}\right)$$

$$= 1 - \frac{1}{N+1}$$

2

In most cases (apart from telescoping series and the geometric series introduced later in this section), there is no simple formula like Eq. (2) for the partial sum S_N. Therefore, we shall develop techniques for evaluating infinite series that do not rely on formulas for S_N.

The sum S is the limit of the sequence of partial sums:

$$S = \lim_{N \to \infty} S_N = \lim_{N \to \infty}\left(1 - \frac{1}{N+1}\right) = 1 \qquad \blacksquare$$

It is important to keep in mind the difference between a sequence $\{a_n\}$ and an infinite series $\displaystyle\sum_{n=1}^{\infty} a_n$.

EXAMPLE 2 **Sequences Versus Series** Discuss the difference between $\{a_n\}$ and $\displaystyle\sum_{n=1}^{\infty} a_n$, where $a_n = \dfrac{1}{n(n+1)}$.

Make sure you understand the difference between sequences and series.

- With a sequence, we consider the limit of the individual terms a_n.
- With a series, we are interested in the sum of the terms

$$a_1 + a_2 + a_3 + \cdots$$

which is defined as the limit of the sequence of partial sums.

Solution The sequence is the list of numbers $\frac{1}{1(2)}, \frac{1}{2(3)}, \frac{1}{3(4)}, \ldots$. This sequence converges to zero:

$$\lim_{n \to \infty} a_n = \lim_{n \to \infty} \frac{1}{n(n+1)} = 0$$

The infinite series is the *sum* of the numbers a_n, defined as the limit of the sequence of partial sums. This sum is not zero. In fact, the sum is equal to 1 by Example 1:

$$\sum_{n=1}^{\infty} a_n = \sum_{n=1}^{\infty} \frac{1}{n(n+1)} = \frac{1}{1(2)} + \frac{1}{2(3)} + \frac{1}{3(4)} + \cdots = 1 \qquad \blacksquare$$

The next theorem shows that infinite series may be added or subtracted like ordinary sums, *provided that the series converge*.

THEOREM 1 **Linearity of Infinite Series** If $\sum a_n$ and $\sum b_n$ converge, then $\sum (a_n + b_n)$, $\sum (a_n - b_n)$, and $\sum c a_n$ also converge, the latter for any constant c. Furthermore,

$$\sum (a_n + b_n) = \sum a_n + \sum b_n$$

$$\sum (a_n - b_n) = \sum a_n - \sum b_n$$

$$\sum c a_n = c \sum a_n \qquad (c \text{ any constant})$$

Proof These rules follow from the corresponding linearity rules for limits. For example,

$$\sum_{n=1}^{\infty} (a_n + b_n) = \lim_{N \to \infty} \sum_{n=1}^{N} (a_n + b_n) = \lim_{N \to \infty} \left(\sum_{n=1}^{N} a_n + \sum_{n=1}^{N} b_n \right)$$

$$= \lim_{N \to \infty} \sum_{n=1}^{N} a_n + \lim_{N \to \infty} \sum_{n=1}^{\infty} b_n = \sum_{n=1}^{\infty} a_n + \sum_{n=1}^{\infty} b_n \qquad \blacksquare$$

A main goal in this chapter is to develop techniques for determining whether a series converges or diverges. It is easy to give examples of series that diverge:

- $\displaystyle\sum_{n=1}^{\infty} 1$ diverges to infinity (the partial sums increase without bound):

$$S_1 = 1, \quad S_2 = 1 + 1 = 2, \quad S_3 = 1 + 1 + 1 = 3, \quad S_4 = 1 + 1 + 1 + 1 = 4, \quad \dots$$

- $\displaystyle\sum_{n=1}^{\infty} (-1)^{n-1}$ diverges (the partial sums jump between 1 and 0):

$$S_1 = 1, \quad S_2 = 1 - 1 = 0, \quad S_3 = 1 - 1 + 1 = 1, \quad S_4 = 1 - 1 + 1 - 1 = 0, \quad \dots$$

Next, we study geometric series, which converge or diverge depending on the common ratio r.

A **geometric series** with common ratio $r \neq 0$ is a series defined by a geometric sequence $c r^n$, where $c \neq 0$. If the series begins at $n = 0$, then

$$\sum_{n=0}^{\infty} c r^n = c + cr + cr^2 + cr^3 + cr^4 + cr^5 + \cdots$$

For $r = \frac{1}{2}$ and $c = 1$, we have the following series:

$$\sum_{n=1}^{\infty} \frac{1}{2^n} = \frac{1}{2} + \frac{1}{4} + \frac{1}{8} + \frac{1}{16} + \cdots$$

Figure 1 demonstrates that adding successive terms in the series corresponds to moving stepwise from 0 to 1, where each step is a move to the right by half of the remaining distance. Thus it appears that the series converges to 1.

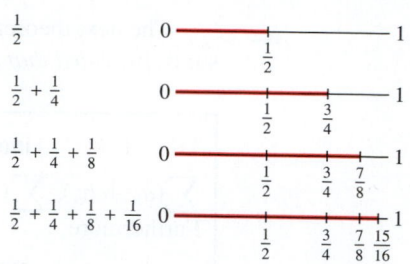

FIGURE 1 Partial sums of $\displaystyle\sum_{n=1}^{\infty} \frac{1}{2^n}$.

There is a simple formula for computing the partial sums of a geometric series:

THEOREM 2 **Partial Sums of a Geometric Series** For the geometric series $\sum_{n=0}^{\infty} cr^n$ with $r \neq 1$,

$$S_N = c + cr + cr^2 + cr^3 + \cdots + cr^N = \frac{c(1 - r^{N+1})}{1 - r}$$ **3**

Proof In the steps below, we start with the expression for S_N, multiply each side by r, take the difference between the first two lines, and then simplify:

$$S_N = c + cr + cr^2 + cr^3 + \cdots + cr^N$$

$$rS_N = cr + cr^2 + cr^3 + \cdots + cr^N + cr^{N+1}$$

$$S_N - rS_N = c - cr^{N+1}$$

$$S_N(1 - r) = c(1 - r^{N+1})$$

Since $r \neq 1$, we may divide by $(1 - r)$ to obtain

$$S_N = \frac{c(1 - r^{N+1})}{1 - r}$$ ∎

Now, the partial sum formula enables us to compute the sum of the geometric series when $|r| < 1$.

Geometric series are important because they

- *arise often in applications.*
- *can be evaluated explicitly.*
- *are used to study other, nongeometric series (by comparison).*

THEOREM 3 **Sum of a Geometric Series** Let $c \neq 0$. If $|r| < 1$, then

$$\sum_{n=0}^{\infty} cr^n = c + cr + cr^2 + cr^3 + \cdots = \frac{c}{1 - r}$$ **4**

If $|r| \geq 1$, then the geometric series diverges.

In words, the sum of a geometric series is the first term divided by 1 minus the common ratio.

Proof If $r = 1$, then the series certainly diverges because the partial sums $S_N = Nc$ grow arbitrarily large. If $r \neq 1$, then Eq. (3) yields

$$\lim_{N \to \infty} S_N = \lim_{N \to \infty} \frac{c(1 - r^{N+1})}{1 - r} = \frac{c}{1 - r} - \frac{c}{1 - r} \lim_{N \to \infty} r^{N+1}$$

If $|r| < 1$, then $\displaystyle\lim_{N \to \infty} r^{N+1} = 0$ and we obtain Eq. (4). If $|r| \geq 1$ and $r \neq 1$, then $\displaystyle\lim_{N \to \infty} r^{N+1}$ does not exist and the geometric series diverges. ∎

EXAMPLE 3 Evaluate $\sum_{n=0}^{\infty} 5^{-n}$.

Solution This is a geometric series with common ratio $r = 5^{-1}$ and first term $c = 1$. By Eq. (4),

$$\sum_{n=0}^{\infty} 5^{-n} = 1 + \frac{1}{5} + \frac{1}{5^2} + \frac{1}{5^3} + \cdots = \frac{1}{1 - 5^{-1}} = \frac{5}{4} \qquad \blacksquare$$

EXAMPLE 4 Evaluate $\sum_{n=3}^{\infty} 7\left(-\frac{3}{4}\right)^n = 7\left(-\frac{3}{4}\right)^3 + 7\left(-\frac{3}{4}\right)^4 + 7\left(-\frac{3}{4}\right)^5 + \cdots$.

Solution This is a geometric series with common ratio $r = -\frac{3}{4}$ and first term $c = 7\left(-\frac{3}{4}\right)^3$. Therefore, it converges to

$$\frac{c}{1-r} = \frac{7\left(-\frac{3}{4}\right)^3}{1 - \left(-\frac{3}{4}\right)} = -\frac{27}{16} \qquad \blacksquare$$

EXAMPLE 5 Find a fraction that has repeated decimal expansion $0.212121\ldots$.

You can check the result by dividing 7 by 33 on a calculator and seeing that the desired decimal expansion, $0.212121\ldots$, results.

Solution We can write this decimal as the series $\frac{21}{100} + \frac{21}{100^2} + \frac{21}{100^3} + \cdots$. This is a geometric series with $c = \frac{21}{100}$ and $r = \frac{1}{100}$. Thus, it converges to

$$\frac{c}{1-r} = \frac{\frac{21}{100}}{1 - \frac{1}{100}} = \frac{21}{99} = \frac{7}{33} \qquad \blacksquare$$

EXAMPLE 6 **A Probability Computation** Nina and Brook are participating in an archery competition where they take turns shooting at a target. The first one to hit the bullseye wins. Nina's success rate hitting the bullseye is 45%, while Brook's is 52%. Nina pointed out this difference, arguing that she should go first. Brook agreed to give the first turn to Nina. Should he have?

Solution We can answer this question by determining the probability that Nina wins the competition. It is done via a geometric series.

Nina wins in each of the following cases.

*Two events A and B are called **independent** if one of them occurring does not affect the probability of the other occurring. In such a case, the probability that A and B both occur is the product of the probabilities of each occurring individually. This idea applies to each case that leads to a win by Nina. For example, in the second case, the probability that Nina wins is the product of the probabilities of: Nina missing on Turn 1 (0.55), Brook missing on Turn 1 (0.48), and Nina hitting on Turn 2 (0.45).*

- By hitting the bullseye on her first turn (which happens with probability 0.45), or
- By having both players miss on their first turn and Nina hit on her second turn [which happens with probability (0.55)(0.48)(0.45)], or
- By having both players miss on their first two turns and Nina hit on her third [which happens with probability (0.55)(0.48)(0.55)(0.48)(0.45)], and so on...

There are infinitely many different cases that result in a win for Nina, and because they are distinct from each other (that is, no two of them can occur at the same time) the probability that some one of them occurs is the sum of each of the individual probabilities. That is, the probability that Nina hits the bullseye first is:

$$0.45 + (0.55)(0.48)(0.45) + (0.55)^2(0.48)^2(0.45) + \cdots$$

This is a geometric series with $c = 0.45$ and $r = (0.55)(0.48) = 0.264$. It follows that the probability that Nina wins is $\frac{0.45}{1 - 0.264} \approx 0.61$. Thus, Brook would have been wise not to let Nina go first. $\qquad \blacksquare$

EXAMPLE 7 Evaluate $\displaystyle\sum_{n=0}^{\infty}\frac{2+3^n}{5^n}$.

Solution Write the series as a sum of two geometric series. This is valid by Theorem 1 because both geometric series converge:

$$\sum_{n=0}^{\infty}\frac{2+3^n}{5^n}=\underbrace{\sum_{n=0}^{\infty}\frac{2}{5^n}+\sum_{n=0}^{\infty}\frac{3^n}{5^n}}_{\text{Both geometric series converge.}}=2\sum_{n=0}^{\infty}\frac{1}{5^n}+\sum_{n=0}^{\infty}\left(\frac{3}{5}\right)^n=2\cdot\frac{1}{1-\frac{1}{5}}+\frac{1}{1-\frac{3}{5}}=5$$

CONCEPTUAL INSIGHT Assumptions Matter   Knowing that a series converges, sometimes we can determine its sum through simple algebraic manipulation. For example, suppose we know that the geometric series with $r=1/2$ and $c=1/2$ converges. Let us say that the sum of the series is S, and we write

$$S=\frac{1}{2}+\frac{1}{4}+\frac{1}{8}+\cdots$$

$$2S=1+\frac{1}{2}+\frac{1}{4}+\frac{1}{8}+\cdots=1+S$$

Thus, $2S=1+S$, or $S=1$. Therefore, the sum of the series is 1.
Observe what happens when this approach is applied to a divergent series:

$$S=1+2+4+8+16+\cdots$$

$$2S=2+4+8+16+\cdots=S-1$$

This would yield $2S=S-1$, or $S=-1$, which is absurd because the series diverges. Thus, without the assumption that a series converges, we cannot employ such algebraic techniques to determine its sum.

The infinite series $\displaystyle\sum_{k=1}^{\infty}1$ diverges because the Nth partial sum $S_N=N$ diverges to infinity. It is less clear whether the following series converges or diverges:

$$\sum_{n=1}^{\infty}(-1)^{n+1}\frac{n}{n+1}=\frac{1}{2}-\frac{2}{3}+\frac{3}{4}-\frac{4}{5}+\frac{5}{6}-\cdots$$

We now introduce a useful test that allows us to conclude that this series diverges. The idea is that if the terms are not shrinking to 0 in size, then the series will not converge. This is typically the first test one applies when attempting to determine whether a series diverges.

The **nth Term Divergence Test** (also known as the **Divergence Test**) is often stated as follows:

If $\displaystyle\sum_{n=1}^{\infty}a_n$ converges, then $\displaystyle\lim_{n\to\infty}a_n=0$.

In practice, we use it to prove that a given series diverges. It is important to note that it does not say that if $\displaystyle\lim_{n\to\infty}a_n=0$, then $\displaystyle\sum_{n=1}^{\infty}a_n$ necessarily converges. We will see that even though $\displaystyle\lim_{n\to\infty}\frac{1}{n}=0$, the series $\displaystyle\sum_{n=1}^{\infty}\frac{1}{n}$ diverges.

THEOREM 4 nth Term Divergence Test   If $\displaystyle\lim_{n\to\infty}a_n\neq0$, then the series $\displaystyle\sum_{n=1}^{\infty}a_n$ diverges.

Proof First, note that $a_n=S_n-S_{n-1}$ because

$$S_n=\left(a_1+a_2+\cdots+a_{n-1}\right)+a_n=S_{n-1}+a_n$$

If $\displaystyle\sum_{n=1}^{\infty}a_n$ converges with sum S, then

$$\lim_{n\to\infty}a_n=\lim_{n\to\infty}(S_n-S_{n-1})=\lim_{n\to\infty}S_n-\lim_{n\to\infty}S_{n-1}=S-S=0$$

Therefore, if a_n does not converge to zero, $\displaystyle\sum_{n=1}^{\infty}a_n$ cannot converge. ∎

EXAMPLE 8 Prove the divergence of $\sum_{n=1}^{\infty} \dfrac{n}{4n+1}$.

Solution We have

$$\lim_{n\to\infty} a_n = \lim_{n\to\infty} \frac{n}{4n+1} = \lim_{n\to\infty} \frac{1}{4 + 1/n} = \frac{1}{4}$$

The nth term a_n does not converge to zero, so the series diverges by the nth Term Divergence Test (Theorem 4). ∎

EXAMPLE 9 Determine the convergence or divergence of

$$\sum_{n=1}^{\infty} (-1)^{n-1} \frac{n}{n+1} = \frac{1}{2} - \frac{2}{3} + \frac{3}{4} - \frac{4}{5} + \cdots$$

Solution The general term $a_n = (-1)^{n-1} \dfrac{n}{n+1}$ does not approach a limit. Indeed, $\dfrac{n}{n+1}$ tends to 1, so the odd terms a_{2n+1} tend to 1, and the even terms a_{2n} tend to -1. Because $\lim_{n\to\infty} a_n$ does not exist, the series diverges by the nth Term Divergence Test. ∎

The nth Term Divergence Test tells only part of the story. If a_n does not tend to zero, then $\sum a_n$ certainly diverges. But what if a_n does tend to zero? In this case, the series may converge or it may diverge. In other words, $\lim_{n\to\infty} a_n = 0$ is a *necessary* condition of convergence, but it is *not sufficient*. As we show in the next example, it is possible for a series to diverge even though its terms tend to zero.

EXAMPLE 10 **Sequence Tends to Zero, Yet the Series Diverges** Prove the divergence of

$$\sum_{n=1}^{\infty} \frac{1}{\sqrt{n}} = \frac{1}{\sqrt{1}} + \frac{1}{\sqrt{2}} + \frac{1}{\sqrt{3}} + \cdots$$

Solution The general term $1/\sqrt{n}$ tends to zero. However, because each term in the partial sum S_N is greater than or equal to $1/\sqrt{N}$, we have

$$S_N = \frac{1}{\sqrt{1}} + \frac{1}{\sqrt{2}} + \cdots + \frac{1}{\sqrt{N}}$$

$$\geq \underbrace{\frac{1}{\sqrt{N}} + \frac{1}{\sqrt{N}} + \cdots + \frac{1}{\sqrt{N}}}_{N \text{ terms}}$$

$$= N\left(\frac{1}{\sqrt{N}}\right) = \sqrt{N}$$

This shows that $S_N \geq \sqrt{N}$. But $\sqrt{N}$ increases without bound (Figure 2). Therefore, S_N also increases without bound. This proves that the series diverges. ∎

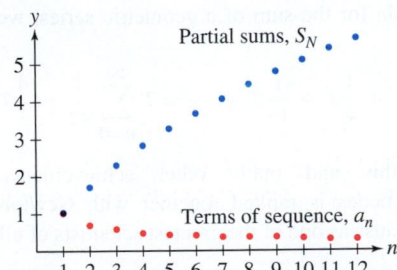

FIGURE 2 The partial sums of $\sum_{n=1}^{\infty} \frac{1}{\sqrt{n}}$ diverge even though the terms $a_n = 1/\sqrt{n}$ tend to zero.

10.2 SUMMARY

- An *infinite series* is an expression

$$\sum_{n=1}^{\infty} a_n = a_1 + a_2 + a_3 + a_4 + \cdots$$

We call a_n the *general term* of the series. An infinite series can begin at $n = k$ for any integer k.

- The Nth *partial sum* is the finite sum of the terms up to and including the Nth term:

$$S_N = \sum_{n=1}^{N} a_n = a_1 + a_2 + a_3 + \cdots + a_N$$

- By definition, the sum of an infinite series is the limit $S = \lim_{N \to \infty} S_N$. If the limit exists, we say that the infinite series is *convergent* or *converges* to the sum S. If the limit does not exist, we say that the infinite series *diverges*.
- If the sequence of partial sums of a series increases without bound, we say that the series diverges to infinity.
- nth Term Divergence Test: If $\lim_{n \to \infty} a_n \neq 0$, then $\sum_{n=1}^{\infty} a_n$ diverges. However, a series may diverge even if its general term a_n tends to zero.
- Partial sum of a geometric series:

$$c + cr + cr^2 + cr^3 + \cdots + cr^N = \frac{c(1 - r^{N+1})}{1 - r}$$

- Geometric series: Assume $c \neq 0$. If $|r| < 1$, then

$$\sum_{n=0}^{\infty} cr^n = c + cr + cr^2 + cr^3 + \cdots = \frac{c}{1 - r}$$

The geometric series diverges if $|r| \geq 1$.

Archimedes (287–212 BCE), who discovered the law of the lever, said, "Give me a place to stand on, and I can move the Earth" (quoted by Pappus of Alexandria c. 340 CE).

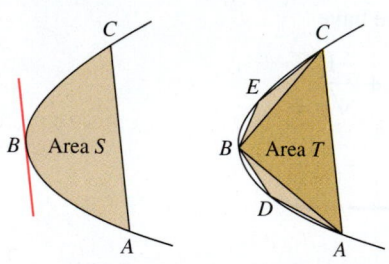

FIGURE 3 Archimedes showed that the area S of the parabolic segment is $\frac{4}{3}T$, where T is the area of $\triangle ABC$.

HISTORICAL PERSPECTIVE

Mechanics Magazine, London, 1824

Geometric series were used as early as the third century BCE by Archimedes in a brilliant argument for determining the area S of a "parabolic segment" (shaded region in Figure 3). Given two points A and C on a parabola, there is a point B between A and C where the tangent line is parallel to $\overline{AC}$ (apparently, Archimedes was aware of the Mean Value Theorem more than 2000 years before the invention of calculus). Let T be the area of triangle $\triangle ABC$. Archimedes proved that if D is chosen in a similar fashion relative to $\overline{AB}$ and E is chosen relative to $\overline{BC}$, then

$$\frac{1}{4}T = \text{area}(\triangle ADB) + \text{area}(\triangle BEC) \qquad \boxed{5}$$

This construction of triangles can be continued. The next step would be to construct the four

triangles on the segments $\overline{AD}$, $\overline{DB}$, $\overline{BE}$, $\overline{EC}$, of total area $\left(\frac{1}{4}\right)^2 T$. Then construct eight triangles of total area $\left(\frac{1}{4}\right)^3 T$, and so on. In this way, we obtain infinitely many triangles that completely fill up the parabolic segment. By the formula for the sum of a geometric series, we get

$$S = T + \frac{1}{4}T + \frac{1}{16}T + \cdots = T \sum_{n=0}^{\infty} \frac{1}{4^n} = \frac{4}{3}T$$

For this and many other achievements, Archimedes is ranked together with Newton and Gauss as one of the greatest scientists of all time.

The modern study of infinite series began in the seventeenth century with Newton, Leibniz, and their contemporaries. The divergence of $\sum_{n=1}^{\infty} 1/n$ (called the **harmonic series**) was known to the medieval scholar Nicole d'Oresme (1323–1382), but his proof was lost for centuries, and the result was rediscovered on more than one occasion. It was also known that the sum of the reciprocal squares $\sum_{n=1}^{\infty} 1/n^2$ converges, and in the 1640s, the Italian Pietro Mengoli put forward the challenge of finding its sum. Despite the efforts of the best mathematicians of the day, including Leibniz and the Bernoulli brothers Jakob and Johann, the problem resisted solution for nearly a century. In 1735, the great master Leonhard Euler (at the time, 28 years old) astonished his contemporaries by proving that

$$\frac{1}{1^2} + \frac{1}{2^2} + \frac{1}{3^2} + \frac{1}{4^2} + \frac{1}{5^2} + \frac{1}{6^2} + \cdots = \frac{\pi^2}{6}$$

We examine the convergence of this series in Exercises 85 and 91 in Section 10.3.

10.2 EXERCISES

Preliminary Questions

1. What role do partial sums play in defining the sum of an infinite series?

2. What is the sum of the following infinite series?

$$\frac{1}{4} + \frac{1}{8} + \frac{1}{16} + \frac{1}{32} + \frac{1}{64} + \cdots$$

3. What happens if you apply the formula for the sum of a geometric series to the following series? Is the formula valid?

$$1 + 3 + 3^2 + 3^3 + 3^4 + \cdots$$

4. Indicate whether or not the reasoning in the following statement is correct: $\sum_{n=1}^{\infty} \frac{1}{n^2} = 0$ because $\frac{1}{n^2}$ tends to zero.

5. Indicate whether or not the reasoning in the following statement is correct: $\sum_{n=1}^{\infty} \frac{1}{\sqrt{n}}$ converges because

$$\lim_{n \to \infty} \frac{1}{\sqrt{n}} = 0$$

6. Find an N such that $S_N > 25$ for the series $\sum_{n=1}^{\infty} 2$.

7. Does there exist an N such that $S_N > 25$ for the series $\sum_{n=1}^{\infty} 2^{-n}$? Explain.

8. Give an example of a divergent infinite series whose general term tends to zero.

Exercises

1. Find a formula for the general term a_n (not the partial sum) of the infinite series.

(a) $\dfrac{1}{3} + \dfrac{1}{9} + \dfrac{1}{27} + \dfrac{1}{81} + \cdots$

(b) $\dfrac{1}{1} + \dfrac{5}{2} + \dfrac{25}{4} + \dfrac{125}{8} + \cdots$

(c) $\dfrac{1}{1} - \dfrac{2^2}{2 \cdot 1} + \dfrac{3^3}{3 \cdot 2 \cdot 1} - \dfrac{4^4}{4 \cdot 3 \cdot 2 \cdot 1} + \cdots$

(d) $\dfrac{2}{1^2 + 1} + \dfrac{1}{2^2 + 1} + \dfrac{2}{3^2 + 1} + \dfrac{1}{4^2 + 1} + \cdots$

2. Write in summation notation:

(a) $1 + \dfrac{1}{4} + \dfrac{1}{9} + \dfrac{1}{16} + \cdots$

(b) $\dfrac{1}{9} + \dfrac{1}{16} + \dfrac{1}{25} + \dfrac{1}{36} + \cdots$

(c) $1 - \dfrac{1}{3} + \dfrac{1}{5} - \dfrac{1}{7} + \cdots$

(d) $\dfrac{125}{9} + \dfrac{625}{16} + \dfrac{3125}{25} + \dfrac{15{,}625}{36} + \cdots$

In Exercises 3–6, compute the partial sums S_2, S_4, and S_6.

3. $1 + \dfrac{1}{2^2} + \dfrac{1}{3^2} + \dfrac{1}{4^2} + \cdots$

4. $\sum_{k=1}^{\infty} (-1)^k k^{-1}$

5. $\dfrac{1}{1 \cdot 2} + \dfrac{1}{2 \cdot 3} + \dfrac{1}{3 \cdot 4} + \cdots$

6. $\sum_{j=1}^{\infty} \dfrac{1}{j!}$

7. The series $1 + \left(\frac{1}{5}\right) + \left(\frac{1}{5}\right)^2 + \left(\frac{1}{5}\right)^3 + \cdots$ converges to $\frac{5}{4}$. Calculate S_N for $N = 1, 2, \ldots$ until you find an S_N that approximates $\frac{5}{4}$ with an error less than 0.0001.

8. The series $\dfrac{1}{0!} - \dfrac{1}{1!} + \dfrac{1}{2!} - \dfrac{1}{3!} + \cdots$ is known to converge to e^{-1} (recall that $0! = 1$). Calculate S_N for $N = 1, 2, \ldots$ until you find an S_N that approximates e^{-1} with an error less than 0.001.

In Exercises 9 and 10, use a computer algebra system to compute S_{10}, S_{100}, S_{500}, and S_{1000} for the series. Do these values suggest convergence to the given value?

9. CAS

$$\frac{\pi - 3}{4} = \frac{1}{2 \cdot 3 \cdot 4} - \frac{1}{4 \cdot 5 \cdot 6} + \frac{1}{6 \cdot 7 \cdot 8} - \frac{1}{8 \cdot 9 \cdot 10} + \cdots$$

10. CAS

$$\frac{\pi^4}{90} = 1 + \frac{1}{2^4} + \frac{1}{3^4} + \frac{1}{4^4} + \cdots$$

11. Calculate S_3, S_4, and S_5 and then find the sum of the telescoping series

$$\sum_{n=1}^{\infty} \left(\frac{1}{n+1} - \frac{1}{n+2} \right)$$

12. Write $\sum_{n=3}^{\infty} \dfrac{1}{n(n-1)}$ as a telescoping series and find its sum.

13. Calculate S_3, S_4, and S_5 and then find the sum $\sum_{n=1}^{\infty} \dfrac{1}{4n^2 - 1}$ using the identity

$$\frac{1}{4n^2 - 1} = \frac{1}{2} \left(\frac{1}{2n - 1} - \frac{1}{2n + 1} \right)$$

14. Use partial fractions to rewrite $\sum_{n=1}^{\infty} \dfrac{1}{n(n+3)}$ as a telescoping series and find its sum.

15. Find the sum of $\dfrac{1}{1 \cdot 3} + \dfrac{1}{3 \cdot 5} + \dfrac{1}{5 \cdot 7} + \cdots$.

16. Find a formula for the partial sum S_N of $\sum_{n=1}^{\infty} (-1)^{n-1}$ and show that the series diverges.

In Exercises 17–22, use the nth Term Divergence Test (Theorem 4) to prove that the following series diverge.

17. $\displaystyle\sum_{n=1}^{\infty} \frac{n}{10n+12}$

18. $\displaystyle\sum_{n=1}^{\infty} \frac{n}{\sqrt{n^2+1}}$

19. $\displaystyle\frac{0}{1} - \frac{1}{2} + \frac{2}{3} - \frac{3}{4} + \cdots$

20. $\displaystyle\sum_{n=1}^{\infty} (-1)^n n^2$

21. $\displaystyle\cos\frac{1}{2} + \cos\frac{1}{3} + \cos\frac{1}{4} + \cdots$

22. $\displaystyle\sum_{n=0}^{\infty} (\sqrt{4n^2+1} - n)$

In Exercises 23–38, either use the formula for the sum of a geometric series to find the sum, or state that the series diverges.

23. $\displaystyle 1 + \frac{1}{6} + \frac{1}{36} + \frac{1}{216} + \cdots$

24. $\displaystyle \frac{4^3}{5^3} + \frac{4^4}{5^4} + \frac{4^5}{5^5} + \cdots$

25. $\displaystyle \frac{7}{3} + \frac{7}{3^2} + \frac{7}{3^3} + \frac{7}{3^4} + \cdots$

26. $\displaystyle \frac{7}{3} + \left(\frac{7}{3}\right)^2 + \left(\frac{7}{3}\right)^3 + \left(\frac{7}{3}\right)^4 + \cdots$

27. $\displaystyle\sum_{n=3}^{\infty} \left(\frac{3}{11}\right)^{-n}$

28. $\displaystyle\sum_{n=2}^{\infty} \frac{7 \cdot (-3)^n}{5^n}$

29. $\displaystyle\sum_{n=-4}^{\infty} \left(-\frac{4}{9}\right)^n$

30. $\displaystyle\sum_{n=0}^{\infty} \left(\frac{\pi}{e}\right)^n$

31. $\displaystyle\sum_{n=1}^{\infty} e^{-n}$

32. $\displaystyle\sum_{n=2}^{\infty} e^{3-2n}$

33. $\displaystyle\sum_{n=0}^{\infty} \frac{8+2^n}{5^n}$

34. $\displaystyle\sum_{n=0}^{\infty} \frac{3(-2)^n - 5^n}{8^n}$

35. $\displaystyle 5 - \frac{5}{4} + \frac{5}{4^2} - \frac{5}{4^3} + \cdots$

36. $\displaystyle \frac{2^3}{7} + \frac{2^4}{7^2} + \frac{2^5}{7^3} + \frac{2^6}{7^4} + \cdots$

37. $\displaystyle \frac{7}{8} - \frac{49}{64} + \frac{343}{512} - \frac{2401}{4096} + \cdots$

38. $\displaystyle \frac{25}{9} + \frac{5}{3} + 1 + \frac{3}{5} + \frac{9}{25} + \frac{27}{125} + \cdots$

In Exercises 39–44, determine a reduced fraction that has this decimal expansion.

39. $0.222\ldots$

40. $0.454545\ldots$

41. $0.313131\ldots$

42. $0.217217217\ldots$

43. $0.123333333\ldots$

44. $0.808888888\ldots$

45. Verify that $0.999999\ldots = 1$ by expressing the left side as a geometric series and determining the sum of the series.

46. The repeating decimal

$$0.012345678901234567890123456789\ldots$$

can be expressed as a fraction with denominator $1{,}111{,}111{,}111$. What is the numerator?

47. Which of the following are *not* geometric series?

(a) $\displaystyle\sum_{n=0}^{\infty} \frac{7^n}{29^n}$

(b) $\displaystyle\sum_{n=3}^{\infty} \frac{1}{n^4}$

(c) $\displaystyle\sum_{n=0}^{\infty} \frac{n^2}{2^n}$

(d) $\displaystyle\sum_{n=5}^{\infty} \pi^{-n}$

48. Use the method of Example 10 to show that $\displaystyle\sum_{k=1}^{\infty} \frac{1}{k^{1/3}}$ diverges.

49. Prove that if $\displaystyle\sum_{n=1}^{\infty} a_n$ converges and $\displaystyle\sum_{n=1}^{\infty} b_n$ diverges, then

$\displaystyle\sum_{n=1}^{\infty} (a_n + b_n)$ diverges. *Hint:* If not, derive a contradiction by writing

$$\sum_{n=1}^{\infty} b_n = \sum_{n=1}^{\infty} (a_n + b_n) - \sum_{n=1}^{\infty} a_n$$

50. Prove the divergence of $\displaystyle\sum_{n=0}^{\infty} \frac{9^n + 2^n}{5^n}$.

51. Give a counterexample to show that each of the following statements is false.

(a) If the general term a_n tends to zero, then $\displaystyle\sum_{n=1}^{\infty} a_n = 0$.

(b) The Nth partial sum of the infinite series defined by $\{a_n\}$ is a_N.

(c) If a_n tends to zero, then $\displaystyle\sum_{n=1}^{\infty} a_n$ converges.

(d) If a_n tends to L, then $\displaystyle\sum_{n=1}^{\infty} a_n = L$.

52. Suppose that $\displaystyle\sum_{n=1}^{\infty} a_n$ is an infinite series with partial sum $S_N = 5 - \dfrac{2}{N^2}$.

(a) What are the values of $\displaystyle\sum_{n=1}^{10} a_n$ and $\displaystyle\sum_{n=5}^{16} a_n$?

(b) What is the value of a_3?

(c) Find a general formula for a_n.

(d) Find the sum $\displaystyle\sum_{n=1}^{\infty} a_n$.

53. Consider the archery competition in Example 6.

(a) Assume that Nina goes first. Let p_n represent the probability that Brook wins on his nth turn. Give an expression for p_n.

(b) Use the result from (a) and a geometric series to determine the probability that Brook wins when Nina goes first.

(c) Now assume that Brook goes first. Use a geometric series to compute the probability that Brook wins the competition.

54. Consider the archery competition in Example 6. Assume that Nina's probability of hitting the bullseye on a turn is 0.45 and that Brook's probability is p. Assume that Nina goes first. For what value of p do both players have a probability of 1/2 of winning the competition?

55. Compute the total area of the (infinitely many) triangles in Figure 4.

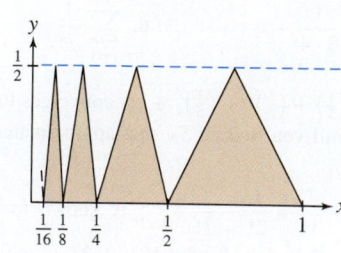

FIGURE 4

56. The winner of a lottery receives m dollars at the end of each year for N years. The present value (PV) of this prize in today's dollars is $PV = \sum_{i=1}^{N} m(1+r)^{-i}$, where r is the interest rate. Calculate PV if $m = \$50,000$, $r = 0.06$ (corresponding to 6%), and $N = 20$. What is PV if $N = \infty$?

57. If a patient takes a dose of D units of a particular drug, the amount of the dosage that remains in the patient's bloodstream after t days is De^{-kt}, where k is a positive constant depending on the particular drug.
(a) Show that if the patient takes a dose D every day for an extended period, the amount of drug in the bloodstream approaches $R = \dfrac{De^{-k}}{1 - e^{-k}}$.
(b) Show that if the patient takes a dose D once every t days for an extended period, the amount of drug in the bloodstream approaches $R = \dfrac{De^{-kt}}{1 - e^{-kt}}$.
(c) Suppose that it is considered dangerous to have more than S units of the drug in the bloodstream. What is the minimal time between doses that is safe? *Hint:* $D + R \leq S$.

58. In economics, the multiplier effect refers to the fact that when there is an injection of money to consumers, the consumers spend a certain percentage of it. That amount recirculates through the economy and adds additional income, which comes back to the consumers and of which they spend the same percentage. This process repeats indefinitely, circulating additional money through the economy. Suppose that in order to stimulate the economy, the government institutes a tax cut of $10 billion. If taxpayers are known to save 10% of any additional money they receive, and to spend 90%, how much total money will be circulated through the economy by that single $10 billion tax cut?

59. Find the total length of the infinite zigzag path in Figure 5 (each zag occurs at an angle of $\frac{\pi}{4}$).

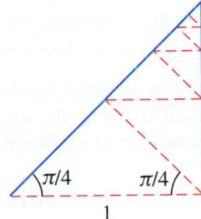

FIGURE 5

60. Evaluate $\sum_{n=1}^{\infty} \dfrac{1}{n(n+1)(n+2)}$. *Hint:* Find constants A, B, and C such that

$$\frac{1}{n(n+1)(n+2)} = \frac{A}{n} + \frac{B}{n+1} + \frac{C}{n+2}$$

and use the result to evaluate $\sum_{n=1}^{\infty} \dfrac{1}{n(n+1)(n+2)}$.

61. Show that if a is a positive integer, then

$$\sum_{n=1}^{\infty} \frac{1}{n(n+a)} = \frac{1}{a}\left(1 + \frac{1}{2} + \cdots + \frac{1}{a}\right)$$

62. A ball dropped from a height of 10 ft begins to bounce vertically. Each time it strikes the ground, it returns to two-thirds of its previous height. What is the total vertical distance traveled by the ball if it bounces infinitely many times?

63. [✎] In this exercise, we resolve the paradox of Gabriel's Horn (Example 3 in Section 7.7 and Example 7 in Section 8.2). Recall that the horn is the surface formed by rotating $y = \frac{1}{x}$ for $x \geq 1$ around the x-axis. The surface encloses a finite volume and has an infinite surface area. Thus, apparently we can fill the surface with a finite volume of paint, but an infinite volume of paint is required to paint the surface.
(a) Explain that if we can fill the horn with paint, then the paint must be Magic Paint that can be spread arbitrarily thin, thinner than the thickness of the molecules in normal paint.
(b) Explain that if we use Magic Paint, then we can paint the surface of the horn with a finite volume of paint, in fact with just a milliliter of it. *Hint:* A geometric series helps here. Use half of a milliliter to paint that part of the surface between $x = 1$ and $x = 2$.

64. A unit square is cut into nine equal regions as in Figure 6(A). The central subsquare is painted red. Each of the unpainted squares is then cut into nine equal subsquares and the central square of each is painted red as in Figure 6(B). This procedure is repeated for each of the resulting unpainted squares. After continuing this process an infinite number of times, what fraction of the total area of the original square is painted?

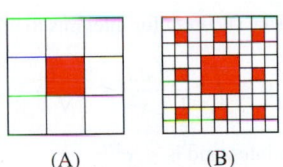

(A)   (B)

FIGURE 6

65. Let $\{b_n\}$ be a sequence and let $a_n = b_n - b_{n-1}$. Show that $\sum_{n=1}^{\infty} a_n$ converges if and only if $\lim_{n \to \infty} b_n$ exists.

66. Assumptions Matter Show, by giving counterexamples, that the assertions of Theorem 1 are not valid if the series $\sum_{n=0}^{\infty} a_n$ and $\sum_{n=0}^{\infty} b_n$ are not convergent.

Further Insights and Challenges

In Exercises 67–69, use the formula

$$1 + r + r^2 + \cdots + r^{N-1} = \frac{1 - r^N}{1 - r} \qquad \boxed{6}$$

67. Professor George Andrews of Pennsylvania State University observed that we can use Eq. (6) to calculate the derivative of $f(x) = x^N$ (for $N \geq 0$). Assume that $a \neq 0$ and let $x = ra$. Show that

$$f'(a) = \lim_{x \to a} \frac{x^N - a^N}{x - a} = a^{N-1} \lim_{r \to 1} \frac{r^N - 1}{r - 1}$$

and evaluate the limit.

68. Pierre de Fermat used geometric series to compute the area under the graph of $f(x) = x^N$ over $[0, A]$. For $0 < r < 1$, let $F(r)$ be the sum of the areas of the infinitely many right-endpoint rectangles with endpoints Ar^n, as in Figure 7. As r tends to 1, the rectangles become narrower and $F(r)$ tends to the area under the graph.

(a) Show that $F(r) = A^{N+1}\dfrac{1-r}{1-r^{N+1}}$.

(b) Use Eq. (6) to evaluate $\displaystyle\int_0^A x^N\,dx = \lim_{r\to 1} F(r)$.

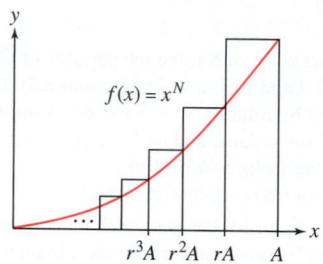

FIGURE 7

69. Verify the Gregory–Leibniz formula in part (d) as follows.

(a) Set $r = -x^2$ in Eq. (6) and rearrange to show that

$$\frac{1}{1+x^2} = 1 - x^2 + x^4 - \cdots + (-1)^{N-1}x^{2N-2} + \frac{(-1)^N x^{2N}}{1+x^2}$$

(b) Show, by integrating over $[0, 1]$, that

$$\frac{\pi}{4} = 1 - \frac{1}{3} + \frac{1}{5} - \frac{1}{7} + \cdots + \frac{(-1)^{N-1}}{2N-1} + (-1)^N \int_0^1 \frac{x^{2N}\,dx}{1+x^2}$$

(c) Use the Comparison Theorem for integrals to prove that

$$0 \le \int_0^1 \frac{x^{2N}\,dx}{1+x^2} \le \frac{1}{2N+1}$$

Hint: Observe that the integrand is $\le x^{2N}$.

(d) Prove that

$$\frac{\pi}{4} = 1 - \frac{1}{3} + \frac{1}{5} - \frac{1}{7} + \frac{1}{9} - \cdots$$

Hint: Use (b) and (c) to show that the partial sums S_N satisfy $\left|S_N - \frac{\pi}{4}\right| \le \frac{1}{2N+1}$, and thereby conclude that $\lim\limits_{N\to\infty} S_N = \frac{\pi}{4}$.

70. Cantor's Disappearing Table (following Larry Knop of Hamilton College) Take a table of length L (Figure 8). At Stage 1, remove the section of length $L/4$ centered at the midpoint. Two sections remain, each with length less than $L/2$. At Stage 2, remove sections of length $L/4^2$ from

each of these two sections (this stage removes $L/8$ of the table). Now four sections remain, each of length less than $L/4$. At Stage 3, remove the four central sections of length $L/4^3$, and so on.

(a) Show that at the Nth stage, each remaining section has length less than $L/2^N$ and that the total amount of table removed is

$$L\left(\frac{1}{4} + \frac{1}{8} + \frac{1}{16} + \cdots + \frac{1}{2^{N+1}}\right)$$

(b) Show that in the limit as $N \to \infty$, precisely one-half of the table remains.

This result is intriguing, because there are no nonzero intervals of table left (at each stage, the remaining sections have a length less than $L/2^N$). So, the table has "disappeared." However, we can place any object longer than $L/4$ on the table. The object will not fall through because it will not fit through any of the removed sections.

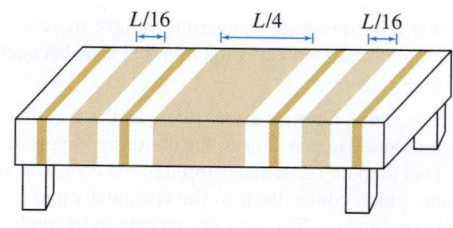

FIGURE 8

71. The **Koch snowflake** (described in 1904 by Swedish mathematician Helge von Koch) is an infinitely jagged "fractal" curve obtained as a limit of polygonal curves (it is continuous but has no tangent line at any point). Begin with an equilateral triangle (Stage 0) and produce Stage 1 by replacing each edge with four edges of one-third the length, arranged as in Figure 9. Continue the process: At the nth stage, replace each edge with four edges of one-third of the length of the edge from the $(n-1)$st stage.

(a) Show that the perimeter P_n of the polygon at the nth stage satisfies $P_n = \frac{4}{3}P_{n-1}$. Prove that $\lim\limits_{n\to\infty} P_n = \infty$. The snowflake has infinite length.

(b) Let A_0 be the area of the original equilateral triangle. Show that $(3)4^{n-1}$ new triangles are added at the nth stage, each with area $A_0/9^n$ (for $n \ge 1$). Show that the total area of the Koch snowflake is $\frac{8}{5}A_0$.

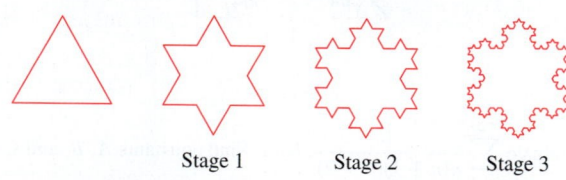

Stage 1   Stage 2   Stage 3

FIGURE 9

10.3 Convergence of Series with Positive Terms

The next three sections develop techniques for determining whether an infinite series converges or diverges. This is easier than finding the sum of an infinite series, which is possible only in special cases.

In this section, we consider **positive series** $\sum a_n$, where $a_n > 0$ for all n. We can visualize the terms of a positive series as rectangles of width 1 and height a_n (Figure 1). The partial sum

$$S_N = a_1 + a_2 + \cdots + a_N$$

is equal to the area of the first N rectangles.

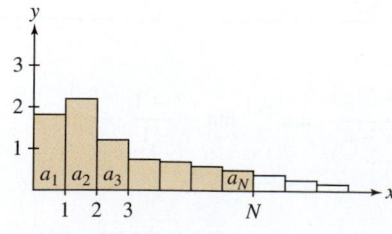

FIGURE 1 The partial sum S_N is the sum of the areas of the N shaded rectangles.

The key feature of positive series is that their partial sums form an increasing sequence

$$S_N < S_{N+1}$$

for all N. This is because S_{N+1} is obtained from S_N by adding a positive number:

$$S_{N+1} = (a_1 + a_2 + \cdots + a_N) + a_{N+1} = S_N + \underbrace{a_{N+1}}_{\text{Positive}}$$

Recall that an increasing sequence converges if it is bounded above. Otherwise, it diverges (Theorem 6, Section 10.1). It follows that a positive series behaves in one of two ways.

THEOREM 1 **Partial Sum Theorem for Positive Series** If $\displaystyle\sum_{n=1}^{\infty} a_n$ is a positive series, then either

(i) The partial sums S_N are bounded above. In this case, $\displaystyle\sum_{n=1}^{\infty} a_n$ converges. Or,

(ii) The partial sums S_N are not bounded above. In this case, $\displaystyle\sum_{n=1}^{\infty} a_n$ diverges.

- Theorem 1 remains true if $a_n \geq 0$. It is not necessary to assume that $a_n > 0$.
- It also remains true if $a_n > 0$ for all $n \geq M$ for some M, because the convergence or divergence of a series is not affected by the first M terms.

Assumptions Matter The theorem does not hold for nonpositive series. Consider

$$\sum_{n=1}^{\infty} (-1)^{n-1} = 1 - 1 + 1 - 1 + 1 - 1 + \cdots$$

The partial sums are bounded (because $S_N = 1$ or 0), but the series diverges.

Our first application of Theorem 1 is the following Integral Test. It is extremely useful because in many cases, integrals are easier to evaluate than series.

THEOREM 2 **Integral Test** Let $a_n = f(n)$, where f is a positive, decreasing, and continuous function of x for $x \geq 1$.

(i) If $\displaystyle\int_1^{\infty} f(x)\,dx$ converges, then $\displaystyle\sum_{n=1}^{\infty} a_n$ converges.

(ii) If $\displaystyle\int_1^{\infty} f(x)\,dx$ diverges, then $\displaystyle\sum_{n=1}^{\infty} a_n$ diverges.

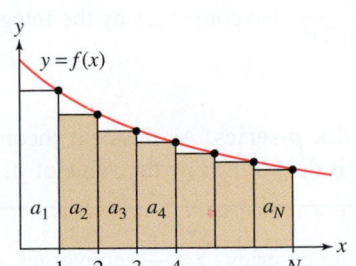

FIGURE 2

Proof Because f is decreasing, the shaded rectangles in Figure 2 lie below the graph of f, and therefore for all N,

$$\underbrace{a_2 + \cdots + a_N}_{\text{Area of shaded rectangles in Figure 2}} \leq \int_1^{N} f(x)\,dx \leq \int_1^{\infty} f(x)\,dx$$

If the improper integral on the right converges, then the sums $a_2 + \cdots + a_N$ are bounded above. That is, the partial sums S_N are bounded above, and therefore the infinite series converges by the Partial Sum Theorem for Positive Series (Theorem 1). This proves (i).

On the other hand, the rectangles in Figure 3 lie above the graph of f, so

$$\int_1^{N} f(x)\,dx \leq \underbrace{a_1 + a_2 + \cdots + a_{N-1}}_{\text{Area of shaded rectangles in Figure 3}}$$

1

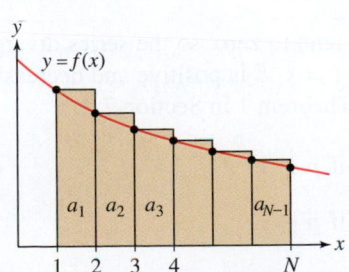

FIGURE 3

The Integral Test is valid for any series $\sum_{n=k}^{\infty} f(n)$, provided that for some $M > 0$, f is a positive, decreasing, and continuous function of x for $x \geq M$. The convergence of the series is determined by the convergence of

$$\int_M^{\infty} f(x)\,dx$$

The infinite series

$$\sum_{n=1}^{\infty} \frac{1}{n}$$

is called the harmonic series.

If $\int_1^{\infty} f(x)\,dx$ diverges, then $\int_1^{N} f(x)\,dx$ increases without bound as N increases. The inequality in (1) shows that S_N also increases without bound, and therefore, the series diverges. This proves (ii). ∎

EXAMPLE 1 **The Harmonic Series Diverges** Show that $\sum_{n=1}^{\infty} \frac{1}{n}$ diverges.

Solution Let $f(x) = \frac{1}{x}$. Then $f(n) = \frac{1}{n}$, and the Integral Test applies because f is positive, decreasing, and continuous for $x \geq 1$. The integral diverges:

$$\int_1^{\infty} \frac{dx}{x} = \lim_{R \to \infty} \int_1^{R} \frac{dx}{x} = \lim_{R \to \infty} \ln R = \infty$$

Therefore, the series $\sum_{n=1}^{\infty} \frac{1}{n}$ diverges. ∎

EXAMPLE 2 Does $\sum_{n=1}^{\infty} \frac{n}{(n^2+1)^2} = \frac{1}{2^2} + \frac{2}{5^2} + \frac{3}{10^2} + \cdots$ converge?

Solution The function $f(x) = \dfrac{x}{(x^2+1)^2}$ is positive and continuous for $x \geq 1$. It is decreasing because $f'(x)$ is negative:

$$f'(x) = \frac{1 - 3x^2}{(x^2+1)^3} < 0 \qquad \text{for } x \geq 1$$

Therefore, the Integral Test applies. Using the substitution $u = x^2 + 1$, $du = 2x\,dx$, we have

$$\int_1^{\infty} \frac{x}{(x^2+1)^2}\,dx = \lim_{R \to \infty} \int_1^{R} \frac{x}{(x^2+1)^2}\,dx = \lim_{R \to \infty} \frac{1}{2} \int_2^{R^2+1} \frac{du}{u^2}$$

$$= \lim_{R \to \infty} \frac{-1}{2u}\bigg|_2^{R^2+1} = \lim_{R \to \infty} \left(\frac{1}{4} - \frac{1}{2(R^2+1)} \right) = \frac{1}{4}$$

Thus, the integral converges, and therefore, $\sum_{n=1}^{\infty} \frac{n}{(n^2+1)^2}$ also converges by the Integral Test. ∎

The sum of the reciprocal powers n^{-p} is called a *p*-**series**. As the next theorem shows, the convergence or divergence of these series is determined by the value of p.

THEOREM 3 **Convergence of *p*-Series** The infinite series $\sum_{n=1}^{\infty} \frac{1}{n^p}$ converges if $p > 1$ and diverges otherwise.

Proof If $p \leq 0$, then the general term n^{-p} does not tend to zero, so the series diverges by the nth Term Divergence Test. If $p > 0$, then $f(x) = x^{-p}$ is positive and decreasing for $x \geq 1$, so the Integral Test applies. According to Theorem 1 in Section 7.7,

$$\int_1^{\infty} \frac{1}{x^p}\,dx = \begin{cases} \dfrac{1}{p-1} & \text{if } p > 1 \\[2mm] \infty & \text{if } p \leq 1 \end{cases}$$

Therefore, $\sum_{n=1}^{\infty} \frac{1}{n^p}$ converges for $p > 1$ and diverges for $p \leq 1$. ∎

Here are two examples of *p*-series:

$$p = \frac{1}{3}: \qquad \sum_{n=1}^{\infty} \frac{1}{\sqrt[3]{n}} = \frac{1}{\sqrt[3]{1}} + \frac{1}{\sqrt[3]{2}} + \frac{1}{\sqrt[3]{3}} + \frac{1}{\sqrt[3]{4}} + \cdots = \infty \quad \text{diverges}$$

$$p = 2: \qquad \sum_{n=1}^{\infty} \frac{1}{n^2} = \frac{1}{1} + \frac{1}{2^2} + \frac{1}{3^2} + \frac{1}{4^2} + \cdots \qquad \text{converges}$$

Another powerful method for determining convergence of positive series occurs via comparison with other series. Suppose that $0 \le a_n \le b_n$. Figure 4 suggests that if the larger sum $\sum b_n$ *converges*, then the smaller sum $\sum a_n$ also converges. Similarly, if the smaller sum *diverges*, then the larger sum also diverges.

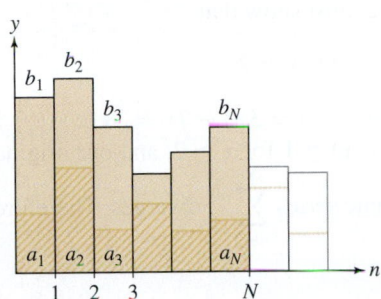

FIGURE 4 The convergence of $\sum b_n$ forces the convergence of $\sum a_n$.

> **THEOREM 4 Direct Comparison Test**
> Assume that there exists $M > 0$ such that $0 \le a_n \le b_n$ for $n \ge M$.
>
> **(i)** If $\displaystyle\sum_{n=1}^{\infty} b_n$ converges, then $\displaystyle\sum_{n=1}^{\infty} a_n$ also converges.
>
> **(ii)** If $\displaystyle\sum_{n=1}^{\infty} a_n$ diverges, then $\displaystyle\sum_{n=1}^{\infty} b_n$ also diverges.

Proof We can assume, without loss of generality, that $M = 1$. If $\displaystyle\sum_{n=1}^{\infty} b_n$ converges to S, then the partial sums of $\displaystyle\sum_{n=1}^{\infty} a_n$ are bounded above by S because

$$a_1 + a_2 + \cdots + a_N \le b_1 + b_2 + \cdots + b_N \le \sum_{n=1}^{\infty} b_n = S \qquad \boxed{2}$$

Note that the first inequality in (2) holds since $a_n \le b_n$ for all n, and the second holds since $b_n \ge 0$ for all n.

Under the assumption that $\displaystyle\sum_{n=1}^{\infty} b_n$ converges, it now follows that $\displaystyle\sum_{n=1}^{\infty} a_n$ converges by the Partial Sum Theorem for Positive Series (Theorem 1). This proves (i). On the other hand, if $\displaystyle\sum_{n=1}^{\infty} a_n$ diverges, then $\displaystyle\sum_{n=1}^{\infty} b_n$ must also diverge. Otherwise, we would have a contradiction to (i). ∎

A good analogy for the Direct Comparison Test, as in Figure 5, is one balloon containing the a_n terms, inside a balloon containing the b_n terms. As we add air in amounts corresponding to the subsequent terms in each series, the balloon with the a_n terms will always be smaller than the other since $a_n \le b_n$. If the bigger balloon does not contain enough air to pop, then the smaller balloon does not pop either. Thus, if the larger series converges, so does the smaller series. On the other hand, if the smaller balloon contains enough air to make it pop, then the bigger balloon must also pop, implying that if the smaller series diverges, the larger series diverges as well. But in the cases that the bigger balloon pops or the smaller balloon does not pop, nothing can be said about the other balloon.

EXAMPLE 3 Does $\displaystyle\sum_{n=1}^{\infty} \frac{1}{\sqrt{n}\,3^n}$ converge?

Solution For $n \ge 1$, we have

$$\frac{1}{\sqrt{n}\,3^n} \le \frac{1}{3^n}$$

The larger series $\displaystyle\sum_{n=1}^{\infty} \frac{1}{3^n}$ converges because it is a geometric series with $r = \frac{1}{3} < 1$. By the Direct Comparison Test, the smaller series $\displaystyle\sum_{n=1}^{\infty} \frac{1}{\sqrt{n}\,3^n}$ also converges. ∎

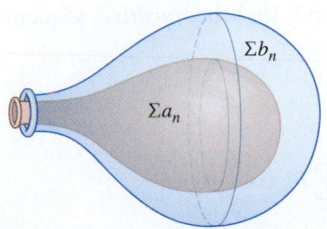

FIGURE 5 The smaller series is contained in the smaller balloon.

EXAMPLE 4 Does $\displaystyle\sum_{n=2}^{\infty} \frac{1}{(n^2+3)^{1/3}}$ converge?

Solution Let us show that

$$\frac{1}{n} \le \frac{1}{(n^2+3)^{1/3}} \qquad \text{for } n \ge 2$$

This inequality is equivalent to $(n^2 + 3) \le n^3$, so we must show that

$$f(x) = x^3 - (x^2 + 3) \ge 0 \qquad \text{for } x \ge 2$$

The function f is increasing because its derivative $f'(x) = 3x^2 - 2x = 3x\left(x - \frac{2}{3}\right)$ is positive for $x \ge 2$. Since $f(2) = 1$, it follows that $f(x) \ge 1$ for $x \ge 2$, and our original inequality follows. We know that the smaller harmonic series $\displaystyle\sum_{n=2}^{\infty} \frac{1}{n}$ diverges. Therefore, the larger series $\displaystyle\sum_{n=2}^{\infty} \frac{1}{(n^2+1)^{1/3}}$ also diverges. ■

EXAMPLE 5 Determine the convergence of

$$\sum_{n=2}^{\infty} \frac{1}{n(\ln n)^2}$$

Solution We might be tempted to compare $\displaystyle\sum_{n=2}^{\infty} \frac{1}{n(\ln n)^2}$ to the harmonic series $\displaystyle\sum_{n=2}^{\infty} \frac{1}{n}$ using the inequality (valid for $n \ge 3$ since $\ln 3 > 1$)

$$\frac{1}{n(\ln n)^2} \le \frac{1}{n}$$

However, $\displaystyle\sum_{n=2}^{\infty} \frac{1}{n}$ diverges, and this says nothing about the *smaller* series $\displaystyle\sum \frac{1}{n(\ln n)^2}$.

Fortunately, the Integral Test can be used. The substitution $u = \ln x$ yields

$$\int_2^{\infty} \frac{dx}{x(\ln x)^2} = \int_{\ln 2}^{\infty} \frac{du}{u^2} = \lim_{R \to \infty} \left(\frac{1}{\ln 2} - \frac{1}{R} \right) = \frac{1}{\ln 2} < \infty$$

The Integral Test shows that $\displaystyle\sum_{n=2}^{\infty} \frac{1}{n(\ln n)^2}$ converges. ■

The next test for convergence involves a comparison between two series $\sum a_n$ and $\sum b_n$ via a limit of the ratios, $\frac{a_n}{b_n}$, of the terms in the series.

CAUTION The Limit Comparison Test is not valid if the series are not positive. See Exercise 44 in Section 10.4.

> **THEOREM 5 Limit Comparison Test** Let $\{a_n\}$ and $\{b_n\}$ be *positive* sequences. Assume that the following limit exists:
>
> $$L = \lim_{n \to \infty} \frac{a_n}{b_n}$$
>
> - If $L > 0$, then $\sum a_n$ converges if and only if $\sum b_n$ converges.
> - If $L = \infty$ and $\sum a_n$ converges, then $\sum b_n$ converges.
> - If $L = 0$ and $\sum b_n$ converges, then $\sum a_n$ converges.

Proof Assume first that L is finite (possibly zero) and that $\sum b_n$ converges. Choose a positive number $R > L$. Then $0 \le a_n/b_n \le R$ for all n sufficiently large because a_n/b_n approaches L. Therefore, $a_n \le Rb_n$. The series $\sum Rb_n$ converges because it is a constant multiple of the convergent series $\sum b_n$. Thus, $\sum a_n$ converges by the Direct Comparison Test.

Next, suppose that L is nonzero (positive or infinite) and that $\sum a_n$ converges. Let $K = \lim_{n \to \infty} b_n/a_n$. Then either $K = L^{-1}$ (if L is finite) or $K = 0$ (if L is infinite). In either case, K is finite and we can apply the result of the previous paragraph with the roles of $\{a_n\}$ and $\{b_n\}$ reversed to conclude that $\sum b_n$ converges. ∎

CONCEPTUAL INSIGHT To remember the different cases of the Limit Comparison Test, you can think of it this way: If $L > 0$, then $a_n \approx Lb_n$ for large n. In other words, the series $\sum a_n$ and $\sum b_n$ are *roughly* multiples of each other, so one converges if and only if the other converges. If $L = \infty$, then a_n is much larger than b_n (for large n), so if $\sum a_n$ converges, $\sum b_n$ certainly converges. Finally, if $L = 0$, then b_n is much larger than a_n and the convergence of $\sum b_n$ yields the convergence of $\sum a_n$.

EXAMPLE 6 Show that $\displaystyle\sum_{n=2}^{\infty} \frac{n^2}{n^4 - n - 1}$ converges.

Solution If we divide the numerator and denominator by n, we can conclude that for large n,

$$\frac{n^2}{n^4 - n - 1} \approx \frac{1}{n^2}$$

To apply the Limit Comparison Test, we set

$$a_n = \frac{n^2}{n^4 - n - 1} \quad \text{and} \quad b_n = \frac{1}{n^2}$$

We observe that $\lim_{n \to \infty} \frac{a_n}{b_n}$ exists and is positive:

$$L = \lim_{n \to \infty} \frac{a_n}{b_n} = \lim_{n \to \infty} \frac{n^2}{n^4 - n - 1} \cdot \frac{n^2}{1} = \lim_{n \to \infty} \frac{1}{1 - n^{-3} - n^{-4}} = 1$$

Since $\displaystyle\sum_{n=2}^{\infty} \frac{1}{n^2}$ converges, our series $\displaystyle\sum_{n=2}^{\infty} \frac{n^2}{n^4 - n - 1}$ also converges by Theorem 5. ∎

EXAMPLE 7 Determine whether $\displaystyle\sum_{n=3}^{\infty} \frac{1}{\sqrt{n^2 + 4}}$ converges.

Solution Apply the Limit Comparison Test with $a_n = \dfrac{1}{\sqrt{n^2 + 4}}$ and $b_n = \dfrac{1}{n}$. Then

$$L = \lim_{n \to \infty} \frac{a_n}{b_n} = \lim_{n \to \infty} \frac{n}{\sqrt{n^2 + 4}} = \lim_{n \to \infty} \frac{1}{\sqrt{1 + 4/n^2}} = 1$$

Since $\displaystyle\sum_{n=3}^{\infty} \frac{1}{n}$ diverges and $L > 0$, the series $\displaystyle\sum_{n=3}^{\infty} \frac{1}{\sqrt{n^2 + 4}}$ also diverges. ∎

In the Limit Comparison Test, when attempting to find an appropriate b_n to compare with a_n, we typically keep only the largest power of n in the numerator and denominator of a_n, as we did in each of the previous examples.

10.3 SUMMARY

- The partial sums S_N of a positive series $\sum a_n$ form an increasing sequence.
- Partial Sum Theorem for Positive Series: A positive series converges if its partial sums S_N are bounded. Otherwise, it diverges.
- Integral Test: Assume that f is positive, decreasing, and continuous for $x > M$. Set $a_n = f(n)$. If $\int_M^\infty f(x)\,dx$ converges, then $\sum a_n$ converges, and if $\int_M^\infty f(x)\,dx$ diverges, then $\sum a_n$ diverges.
- p-Series: The series $\sum_{n=1}^\infty \dfrac{1}{n^p}$ converges if $p > 1$ and diverges if $p \le 1$.
- Direct Comparison Test: Assume there exists $M > 0$ such that $0 \le a_n \le b_n$ for all $n \ge M$. If $\sum b_n$ converges, then $\sum a_n$ converges, and if $\sum a_n$ diverges, then $\sum b_n$ diverges.
- Limit Comparison Test: Assume that $\{a_n\}$ and $\{b_n\}$ are positive and that the following limit exists:

$$L = \lim_{n\to\infty} \frac{a_n}{b_n}$$

  – If $L > 0$, then $\sum a_n$ converges if and only if $\sum b_n$ converges.
  – If $L = \infty$ and $\sum a_n$ converges, then $\sum b_n$ converges.
  – If $L = 0$ and $\sum b_n$ converges, then $\sum a_n$ converges.

10.3 EXERCISES

Preliminary Questions

1. For the series $\sum_{n=1}^\infty a_n$, if the partial sums S_N are increasing, then (choose the correct conclusion)

(a) $\{a_n\}$ is an increasing sequence.
(b) $\{a_n\}$ is a positive sequence.

2. What are the hypotheses of the Integral Test?

3. Which test would you use to determine whether $\sum_{n=1}^\infty n^{-3.2}$ converges?

4. Which test would you use to determine whether $\sum_{n=1}^\infty \dfrac{1}{2^n + \sqrt{n}}$ converges?

5. Ralph hopes to investigate the convergence of $\sum_{n=1}^\infty \dfrac{e^{-n}}{n}$ by comparing it with $\sum_{n=1}^\infty \dfrac{1}{n}$. Is Ralph on the right track?

Exercises

In Exercises 1–12, use the Integral Test to determine whether the infinite series is convergent.

1. $\displaystyle\sum_{n=1}^\infty \frac{1}{(n+1)^4}$

2. $\displaystyle\sum_{n=1}^\infty \frac{1}{n+3}$

3. $\displaystyle\sum_{n=1}^\infty n^{-1/3}$

4. $\displaystyle\sum_{n=5}^\infty \frac{1}{\sqrt{n-4}}$

5. $\displaystyle\sum_{n=25}^\infty \frac{n^2}{(n^3+9)^{5/2}}$

6. $\displaystyle\sum_{n=1}^\infty \frac{n}{(n^2+1)^{3/5}}$

7. $\displaystyle\sum_{n=1}^\infty \frac{1}{n^2+1}$

8. $\displaystyle\sum_{n=4}^\infty \frac{1}{n^2-1}$

9. $\displaystyle\sum_{n=1}^\infty \frac{1}{n(n+5)}$

10. $\displaystyle\sum_{n=1}^\infty n e^{-n^2}$

11. $\displaystyle\sum_{n=2}^\infty \frac{1}{n(\ln n)^{3/2}}$

12. $\displaystyle\sum_{n=1}^\infty \frac{\ln n}{n^2}$

13. Show that $\displaystyle\sum_{n=1}^\infty \frac{1}{n^3+8n}$ converges by using the Direct Comparison Test with $\displaystyle\sum_{n=1}^\infty n^{-3}$.

14. Show that $\displaystyle\sum_{n=2}^\infty \frac{1}{\sqrt{n^2-3}}$ diverges by comparing with $\displaystyle\sum_{n=2}^\infty n^{-1}$.

15. For $\displaystyle\sum_{n=1}^{\infty} \frac{1}{n + \sqrt{n}}$, verify that for $n \geq 1$,

$$\frac{1}{n + \sqrt{n}} \leq \frac{1}{n}, \qquad \frac{1}{n + \sqrt{n}} \leq \frac{1}{\sqrt{n}}$$

Can either inequality be used to show that the series diverges? Show that $\dfrac{1}{n + \sqrt{n}} \geq \dfrac{1}{2n}$ for $n \geq 1$ and conclude that the series diverges.

16. Which of the following inequalities can be used to study the convergence of $\displaystyle\sum_{n=2}^{\infty} \frac{1}{n^2 + \sqrt{n}}$? Explain.

$$\frac{1}{n^2 + \sqrt{n}} \leq \frac{1}{\sqrt{n}}, \qquad \frac{1}{n^2 + \sqrt{n}} \leq \frac{1}{n^2}$$

In Exercises 17–28, use the Direct Comparison Test to determine whether the infinite series is convergent.

17. $\displaystyle\sum_{n=1}^{\infty} \frac{1}{n2^n}$

18. $\displaystyle\sum_{n=1}^{\infty} \frac{n^3}{n^5 + 4n + 1}$

19. $\displaystyle\sum_{n=1}^{\infty} \frac{1}{n^{1/3} + 2^n}$

20. $\displaystyle\sum_{n=1}^{\infty} \frac{1}{\sqrt{n^3 + 2n - 1}}$

21. $\displaystyle\sum_{m=1}^{\infty} \frac{4}{m! + 4^m}$

22. $\displaystyle\sum_{n=4}^{\infty} \frac{\sqrt{n}}{n - 3}$

23. $\displaystyle\sum_{k=1}^{\infty} \frac{\sin^2 k}{k^2}$

24. $\displaystyle\sum_{k=2}^{\infty} \frac{k^{2/9}}{k^{10/9} - 1}$

25. $\displaystyle\sum_{n=1}^{\infty} \frac{2}{3^n + 3^{-n}}$

26. $\displaystyle\sum_{k=1}^{\infty} 2^{-k^2}$

27. $\displaystyle\sum_{n=1}^{\infty} \frac{1}{(n + 1)!}$

28. $\displaystyle\sum_{n=1}^{\infty} \frac{n!}{n^3}$

Exercise 29–34: For all $a > 0$ and $b > 1$, the inequalities $\ln n \leq n^a$, $n^a < b^n$ are true for n sufficiently large (this can be proved using L'Hôpital's Rule). Use this, together with the Direct Comparison Test, to determine whether the series converges or diverges.

29. $\displaystyle\sum_{n=1}^{\infty} \frac{\ln n}{n^3}$

30. $\displaystyle\sum_{m=2}^{\infty} \frac{1}{\ln m}$

31. $\displaystyle\sum_{n=1}^{\infty} \frac{(\ln n)^{100}}{n^{1.1}}$

32. $\displaystyle\sum_{n=1}^{\infty} \frac{1}{(\ln n)^{10}}$

33. $\displaystyle\sum_{n=1}^{\infty} \frac{n}{3^n}$

34. $\displaystyle\sum_{n=1}^{\infty} \frac{n^5}{2^n}$

35. Show that $\displaystyle\sum_{n=1}^{\infty} \sin \frac{1}{n^2}$ converges. *Hint:* Use $\sin x \leq x$ for $x \geq 0$.

36. Does $\displaystyle\sum_{n=2}^{\infty} \frac{\sin(1/n)}{\ln n}$ converge? *Hint:* By Theorem 3 in Section 2.6, $\sin(1/n) > (\cos(1/n))/n$. Thus, $\sin(1/n) > 1/(2n)$ for $n > 2$ [because $\cos(1/n) > \frac{1}{2}$].

In Exercises 37–46, use the Limit Comparison Test to prove convergence or divergence of the infinite series.

37. $\displaystyle\sum_{n=2}^{\infty} \frac{n^2}{n^4 - 1}$

38. $\displaystyle\sum_{n=2}^{\infty} \frac{1}{n(n - 1)}$

39. $\displaystyle\sum_{n=2}^{\infty} \frac{n}{\sqrt{n^3 + 1}}$

40. $\displaystyle\sum_{n=2}^{\infty} \frac{n^3}{\sqrt{n^7 + 2n^2 + 1}}$

41. $\displaystyle\sum_{n=3}^{\infty} \frac{3n + 5}{n(n - 1)(n - 2)}$

42. $\displaystyle\sum_{n=1}^{\infty} \frac{e^n + n}{e^{2n} - n^2}$

43. $\displaystyle\sum_{n=1}^{\infty} \frac{1}{\sqrt{n} + \ln n}$

44. $\displaystyle\sum_{n=1}^{\infty} \frac{\ln(n + 4)}{n^{5/2}}$

45. $\displaystyle\sum_{n=1}^{\infty} \left(1 - \cos \frac{1}{n}\right)$ *Hint:* Compare with $\displaystyle\sum_{n=1}^{\infty} n^{-2}$.

46. $\displaystyle\sum_{n=1}^{\infty} (1 - 2^{-1/n})$ *Hint:* Compare with the harmonic series.

In Exercises 47–76, determine convergence or divergence using any method covered so far.

47. $\displaystyle\sum_{n=4}^{\infty} \frac{1}{n^2 - 9}$

48. $\displaystyle\sum_{n=1}^{\infty} \frac{\cos^2 n}{n^2}$

49. $\displaystyle\sum_{n=1}^{\infty} \frac{\sqrt{n}}{4n + 9}$

50. $\displaystyle\sum_{n=1}^{\infty} \frac{n - \cos n}{n^3}$

51. $\displaystyle\sum_{n=1}^{\infty} \frac{n^2 - 1}{n^5 + 1}$

52. $\displaystyle\sum_{n=1}^{\infty} \frac{1}{n^2 + \sin n}$

53. $\displaystyle\sum_{n=5}^{\infty} (4/5)^{-n}$

54. $\displaystyle\sum_{n=1}^{\infty} \frac{1}{3n^2}$

55. $\displaystyle\sum_{n=2}^{\infty} \frac{1}{n^{3/2} \ln n}$

56. $\displaystyle\sum_{n=2}^{\infty} \frac{(\ln n)^{12}}{n^{9/8}}$

57. $\displaystyle\sum_{k=1}^{\infty} 4^{1/k}$

58. $\displaystyle\sum_{n=1}^{\infty} \frac{4^n}{5^n - 2n}$

59. $\displaystyle\sum_{n=2}^{\infty} \frac{1}{(\ln n)^4}$

60. $\displaystyle\sum_{n=1}^{\infty} \frac{2^n}{3^n - n}$

61. $\displaystyle\sum_{n=3}^{\infty} \frac{1}{n \ln n - n}$

62. $\displaystyle\sum_{n=3}^{\infty} \frac{1}{n(\ln n)^2 - n}$

63. $\displaystyle\sum_{n=1}^{\infty} \frac{1}{n^n}$

64. $\displaystyle\sum_{n=1}^{\infty} \frac{n^2 - 4n^{3/2}}{n^3}$

65. $\displaystyle\sum_{n=1}^{\infty} \frac{1 + (-1)^n}{n}$

66. $\displaystyle\sum_{n=1}^{\infty} \frac{2 + (-1)^n}{n^{3/2}}$

67. $\displaystyle\sum_{n=1}^{\infty} \sin \frac{1}{n}$

68. $\displaystyle\sum_{n=1}^{\infty} \frac{\sin(1/n)}{\sqrt{n}}$

69. $\displaystyle\sum_{n=1}^{\infty} \frac{2n + 1}{4^n}$

70. $\displaystyle\sum_{n=3}^{\infty} \frac{1}{e^{\sqrt{n}}}$

71. $\displaystyle\sum_{n=4}^{\infty} \frac{\ln n}{n^2 - 3n}$

72. $\displaystyle\sum_{n=1}^{\infty} \ln\left(1 + \frac{1}{n}\right)$

73. $\displaystyle\sum_{n=2}^{\infty} \frac{1}{n^{1/2} \ln n}$

74. $\displaystyle\sum_{n=1}^{\infty} \frac{1}{n^{3/2} - (\ln n)^4}$

75. $\displaystyle\sum_{n=2}^{\infty} \frac{4n^2 + 15n}{3n^4 - 5n^2 - 17}$

76. $\displaystyle\sum_{n=1}^{\infty} \frac{n}{4^{-n} + 5^{-n}}$

77. For which a does $\displaystyle\sum_{n=2}^{\infty} \frac{1}{n(\ln n)^a}$ converge?

78. For which a does $\displaystyle\sum_{n=2}^{\infty} \frac{1}{n^a \ln n}$ converge?

79. For which values of p does $\displaystyle\sum_{n=1}^{\infty} \frac{n^2}{(n^3 + 1)^p}$ converge?

80. For which values of p does $\displaystyle\sum_{n=1}^{\infty} \frac{e^x}{(1 + e^{2x})^p}$ converge?

Approximating Infinite Sums In Exercises 81–83, let $a_n = f(n)$, where f is a continuous, decreasing function such that $f(x) \geq 0$ and $\displaystyle\int_1^{\infty} f(x)\,dx$ converges.

81. Show that

$$\int_1^{\infty} f(x)\,dx \leq \sum_{n=1}^{\infty} a_n \leq a_1 + \int_1^{\infty} f(x)\,dx \qquad \boxed{3}$$

82. **CAS** Using the inequality in (3), show that

$$5 \leq \sum_{n=1}^{\infty} \frac{1}{n^{1.2}} \leq 6$$

This series converges slowly. Use a computer algebra system to verify that $S_N < 5$ for $N \leq 43{,}128$ and $S_{43,129} \approx 5.00000021$.

83. Assume $\displaystyle\sum_{n=1}^{\infty} a_n$ converges to S. Arguing as in Exercise 81, show that

$$\sum_{n=1}^{M} a_n + \int_{M+1}^{\infty} f(x)\,dx \leq S \leq \sum_{n=1}^{M+1} a_n + \int_{M+1}^{\infty} f(x)\,dx \qquad \boxed{4}$$

Conclude that

$$0 \leq S - \left(\sum_{n=1}^{M} a_n + \int_{M+1}^{\infty} f(x)\,dx\right) \leq a_{M+1} \qquad \boxed{5}$$

This provides a method for approximating S with an error of at most a_{M+1}.

84. **CAS** Use the inequalities in (4) from Exercise 83 with $M = 43{,}129$ to prove that

$$5.5915810 \leq \sum_{n=1}^{\infty} \frac{1}{n^{1.2}} \leq 5.5915839$$

85. **CAS** Use the inequalities in (4) from Exercise 83 with $M = 40{,}000$ to show that

$$1.644934066 \leq \sum_{n=1}^{\infty} \frac{1}{n^2} \leq 1.644934068$$

Is this consistent with Euler's result, according to which this infinite series has sum $\pi^2/6$?

86. **CAS** Use a CAS and the inequalities in (5) from Exercise 83 to determine the value of $\displaystyle\sum_{n=1}^{\infty} n^{-6}$ to within an error less than 10^{-4}. Check that your result is consistent with that of Euler, who proved that the sum is equal to $\pi^6/945$.

87. **CAS** Use a CAS and the inequalities in (5) from Exercise 85 to determine the value of $\displaystyle\sum_{n=1}^{\infty} n^{-5}$ to within an error less than 10^{-4}.

88. How far can a stack of identical books (of mass m and unit length) extend without tipping over? The stack will not tip over if the $(n + 1)$st book is placed at the bottom of the stack with its right edge located at or before the center of mass of the first n books (Figure 6). Let c_n be the center of mass of the first n books, measured along the x-axis, where we take the positive x-axis to the left of the origin as in Figure 7. Recall that if an object of mass m_1 has center of mass at x_1 and a second object of m_2 has center of mass x_2, then the center of mass of the system has x-coordinate

$$\frac{m_1 x_1 + m_2 x_2}{m_1 + m_2}$$

(a) Show that if the $(n + 1)$st book is placed with its right edge at c_n, then its center of mass is located at $c_n + \frac{1}{2}$.

(b) Consider the first n books as a single object of mass nm with center of mass at c_n and the $(n + 1)$st book as a second object of mass m. Show that if the $(n + 1)$st book is placed with its right edge at c_n, then

$$c_{n+1} = c_n + \frac{1}{2(n+1)}.$$

(c) Prove that $\displaystyle\lim_{n\to\infty} c_n = \infty$. Thus, by using enough books, the stack can be extended as far as desired without tipping over.

$$\frac{25}{24} \approx 1.04 \text{ book lengths}$$

$$\frac{1}{2} + \frac{1}{4} + \frac{1}{6} + \frac{1}{8} = \frac{25}{24}$$

FIGURE 6

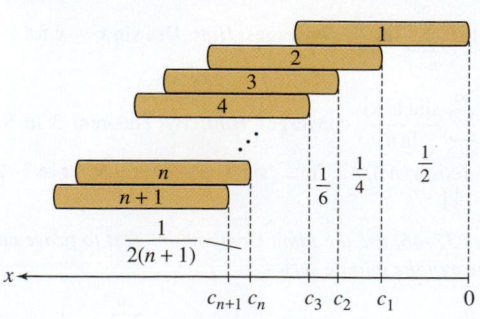

FIGURE 7

89. The following argument proves the divergence of the harmonic series $\sum_{n=1}^{\infty} 1/n$ without using the Integral Test. To begin, assume that the harmonic series converges to a value S.

(a) Prove that the following two series must also converge:

$$1 + \frac{1}{3} + \frac{1}{5} + \cdots \quad \text{and} \quad \frac{1}{2} + \frac{1}{4} + \frac{1}{6} + \cdots$$

(b) Prove that if S_1 and S_2 are the sums of the series on the left and right, respectively, then $S = S_1 + S_2$.

(c) Prove that $S_1 \geq S_2 + \frac{1}{2}$, and $S_2 = \frac{1}{2} S$. Explain how this leads to a contradiction and the conclusion that the harmonic series diverges.

Further Insights and Challenges

90. Consider the series $\sum_{n=2}^{\infty} a_n$, where $a_n = (\ln(\ln n))^{-\ln n}$.

(a) Show, by taking logarithms, that $a_n = n^{-\ln(\ln(\ln n))}$.

(b) Show that $\ln(\ln(\ln n)) \geq 2$ if $n > C$, where $C = e^{e^{e^2}}$.

(c) Show that the series converges.

91. Kummer's Acceleration Method Suppose we wish to approximate $S = \sum_{n=1}^{\infty} 1/n^2$. There is a similar telescoping series whose value can be computed exactly (Example 2 in Section 10.2):

$$\sum_{n=1}^{\infty} \frac{1}{n(n+1)} = 1$$

(a) Verify that

$$S = \sum_{n=1}^{\infty} \frac{1}{n(n+1)} + \sum_{n=1}^{\infty} \left(\frac{1}{n^2} - \frac{1}{n(n+1)} \right)$$

Thus for M large,

$$S \approx 1 + \sum_{n=1}^{M} \frac{1}{n^2(n+1)} \qquad \boxed{6}$$

(b) Explain what has been gained. Why is (6) a better approximation to S than $\sum_{n=1}^{M} 1/n^2$?

(c) [CAS] Compute

$$\sum_{n=1}^{1000} \frac{1}{n^2}, \qquad 1 + \sum_{n=1}^{100} \frac{1}{n^2(n+1)}$$

Which is a better approximation to S, whose exact value is $\pi^2/6$?

92. [CAS] The sum $S = \sum_{k=1}^{\infty} k^{-3}$ has been computed to more than 100 million digits. The first 30 digits are

$$S = 1.202056903159594285399738161511$$

Approximate S using Kummer's Acceleration Method of Exercise 91 with the similar series $\sum_{n=1}^{\infty} (n(n+1)(n+2))^{-1}$ and $M = 500$. According to Exercise 60 in Section 10.2, the similar series is a telescoping series with a sum of $\frac{1}{4}$.

10.4 Absolute and Conditional Convergence

In the previous section, we studied positive series, but we still lack the tools to analyze series with both positive and negative terms. One of the keys to understanding such series is the concept of absolute convergence.

> **DEFINITION Absolute Convergence** The series $\sum a_n$ **converges absolutely** if $\sum |a_n|$ converges.

EXAMPLE 1 Verify that the series

$$\sum_{n=1}^{\infty} \frac{(-1)^{n-1}}{n^2} = \frac{1}{1^2} - \frac{1}{2^2} + \frac{1}{3^2} - \frac{1}{4^2} + \cdots$$

converges absolutely.

Solution This series converges absolutely because taking the absolute value of each term, we obtain a *p*-series with $p = 2 > 1$:

$$\sum_{n=1}^{\infty} \left| \frac{(-1)^{n-1}}{n^2} \right| = \frac{1}{1^2} + \frac{1}{2^2} + \frac{1}{3^2} + \frac{1}{4^2} + \cdots \qquad \text{(convergent } p\text{-series)} \qquad \blacksquare$$

The next theorem tells us that if the series of absolute values converges, then the original series also converges.

> **THEOREM 1 Absolute Convergence Implies Convergence** If $\sum |a_n|$ converges, then $\sum a_n$ also converges.

Proof We have $-|a_n| \le a_n \le |a_n|$. By adding $|a_n|$ to all parts of the inequality, we get $0 \le |a_n| + a_n \le 2|a_n|$. If $\sum |a_n|$ converges, then $\sum 2|a_n|$ also converges, and therefore, $\sum (a_n + |a_n|)$ converges by the Direct Comparison Test. Our original series converges because it is the difference of two convergent series:

$$\sum a_n = \sum (a_n + |a_n|) - \sum |a_n| \qquad \blacksquare$$

EXAMPLE 2 Verify that $\sum_{n=1}^{\infty} \frac{(-1)^{n-1}}{n^2}$ converges.

Solution We showed that $\sum_{n=1}^{\infty} \frac{(-1)^{n-1}}{n^2}$ converges absolutely in Example 1. By Theorem 1, $\sum_{n=1}^{\infty} \frac{(-1)^{n-1}}{n^2}$ itself converges. $\qquad \blacksquare$

EXAMPLE 3 Does $\sum_{n=1}^{\infty} \frac{(-1)^{n-1}}{\sqrt{n}} = \frac{1}{\sqrt{1}} - \frac{1}{\sqrt{2}} + \frac{1}{\sqrt{3}} - \cdots$ converge absolutely?

Solution The series of absolute values is $\sum_{n=1}^{\infty} \frac{1}{\sqrt{n}}$, which is a *p*-series with $p = \frac{1}{2}$. It diverges because $p < 1$. Therefore, $\sum_{n=1}^{\infty} \frac{(-1)^{n-1}}{\sqrt{n}}$ does not converge absolutely. $\qquad \blacksquare$

The series in the previous example does not converge *absolutely*, but we still do not know whether or not it converges. A series $\sum a_n$ may converge without converging absolutely. In this case, we say that $\sum a_n$ is conditionally convergent.

> **DEFINITION Conditional Convergence** An infinite series $\sum a_n$ **converges conditionally** if $\sum a_n$ converges but $\sum |a_n|$ diverges.

If a series is not absolutely convergent, how can we determine whether it is conditionally convergent? This is often a difficult question, because we cannot use the Integral Test or the Direct Comparison Test since they apply only to positive series. However, convergence is guaranteed in the particular case of an **alternating series**

$$\sum_{n=1}^{\infty} (-1)^{n-1} b_n = b_1 - b_2 + b_3 - b_4 + \cdots$$

where the terms b_n are positive and decrease to zero (Figure 1).

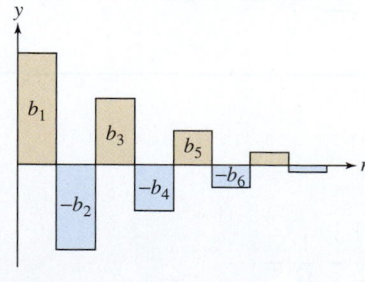

FIGURE 1 An alternating series with decreasing terms. The sum is the signed area, which is at most b_1.

Assumptions Matter The Alternating Series Test is not valid if we drop the assumption that b_n is decreasing (see Exercise 35).

THEOREM 2 Alternating Series Test Assume that $\{b_n\}$ is a positive sequence that is decreasing and converges to 0:

$$b_1 > b_2 > b_3 > b_4 > \cdots > 0, \qquad \lim_{n\to\infty} b_n = 0$$

Then the following alternating series converges:

$$\sum_{n=1}^{\infty} (-1)^{n-1} b_n = b_1 - b_2 + b_3 - b_4 + \cdots$$

Furthermore, if $S = \sum_{n=1}^{\infty} (-1)^{n-1} b_n$, then

$$0 < S < b_1 \quad \text{and} \quad S_p < S < S_q \quad \text{for } p \text{ even and } q \text{ odd}$$

The Alternating Series Test is the only test for conditional convergence developed in this text. Other tests, such as Abel's Criterion and the Dirichlet Test, are discussed in textbooks on analysis.

As we will see, this last fact allows the estimation of such a series to any level of accuracy needed.

Notice that under the same conditions, the series

$$\sum_{n=1}^{\infty} (-1)^n b_n = -b_1 + b_2 - b_3 + b_4 - \cdots$$

also converges since it is just -1 times the series appearing in the theorem.

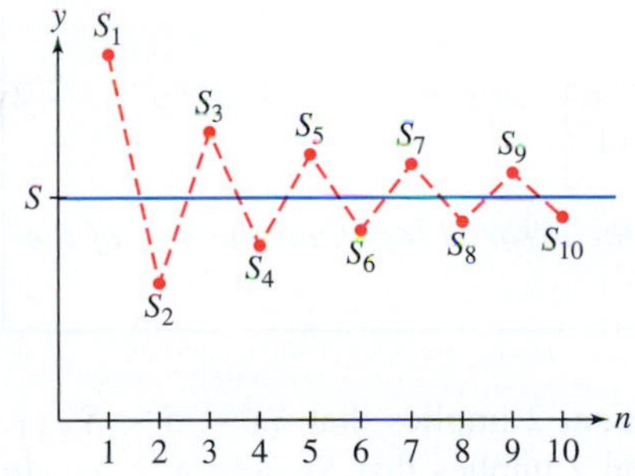

DF **FIGURE 2** The partial sums of an alternating series zigzag above and below the limit. The odd partial sums decrease and the even partial sums increase.

Proof We will prove that the partial sums zigzag above and below the sum S as in Figure 2. Note first that the even partial sums are increasing. Indeed, the odd-numbered terms occur with a plus sign and thus, for example,

$$S_4 + b_5 - b_6 = S_6$$

But $b_5 - b_6 > 0$ because b_n is decreasing, and therefore, $S_4 < S_6$. In general,

$$S_{2N} + (b_{2N+1} - b_{2N+2}) = S_{2N+2}$$

where $b_{2N+1} - b_{2N+2} > 0$. Thus, $S_{2N} < S_{2N+2}$ and

$$0 < S_2 < S_4 < S_6 < \cdots$$

Similarly,

$$S_{2N-1} - (b_{2N} - b_{2N+1}) = S_{2N+1}$$

Therefore, $S_{2N+1} < S_{2N-1}$, and the sequence of odd partial sums is decreasing:

$$\cdots < S_7 < S_5 < S_3 < S_1$$

Finally, $S_{2N} < S_{2N} + b_{2N+1} = S_{2N+1}$. The partial sums compare as follows:

$$0 < S_2 < S_4 < S_6 < \quad \cdots \quad < S_7 < S_5 < S_3 < S_1$$

Now, because bounded monotonic sequences converge (Theorem 6 of Section 10.1), the even and odd partial sums approach limits that are sandwiched in the middle:

$$0 < S_2 < S_4 < \cdots < \lim_{N\to\infty} S_{2N} \le \lim_{N\to\infty} S_{2N+1} < \cdots < S_5 < S_3 < S_1 \qquad \boxed{1}$$

These two limits must have a common value S because

$$\lim_{N\to\infty} S_{2N+1} - \lim_{N\to\infty} S_{2N} = \lim_{N\to\infty} (S_{2N+1} - S_{2N}) = \lim_{N\to\infty} b_{2N+1} = 0$$

Therefore, $\lim_{N \to \infty} S_N = S$ and the infinite series converges to S. From the inequalities in (1) we also see that $0 < S < S_1 = b_1$ and $S_p < S < S_q$ for all p even and q odd as claimed. ∎

EXAMPLE 4 Show that $\sum_{n=1}^{\infty} \dfrac{(-1)^{n-1}}{\sqrt{n}} = \dfrac{1}{\sqrt{1}} - \dfrac{1}{\sqrt{2}} + \dfrac{1}{\sqrt{3}} - \cdots$ converges conditionally. Furthermore show that if S is the sum of the series, then $0 \le S \le 1$.

Solution The terms $b_n = 1/\sqrt{n}$ are positive and decreasing, and $\lim_{n \to \infty} b_n = 0$. Therefore, the series converges by the Alternating Series Test. Furthermore, if S is the sum of the series, then $0 \le S \le 1$ because $b_1 = 1$. However, the positive series $\sum_{n=1}^{\infty} 1/\sqrt{n}$

diverges because it is a p-series with $p = \frac{1}{2} < 1$. Thus, $\sum_{n=1}^{\infty} \dfrac{(-1)^{n-1}}{\sqrt{n}}$ is conditionally convergent (Figure 3). ∎

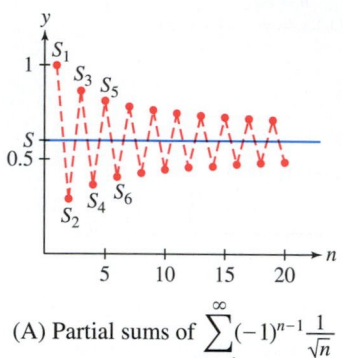

(A) Partial sums of $\displaystyle\sum_{n=1}^{\infty} (-1)^{n-1} \dfrac{1}{\sqrt{n}}$

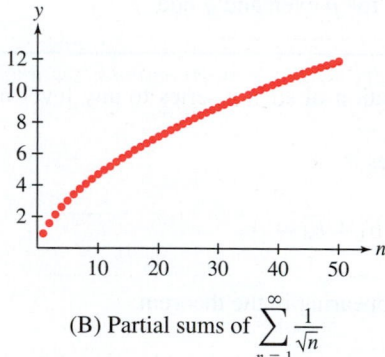

(B) Partial sums of $\displaystyle\sum_{n=1}^{\infty} \dfrac{1}{\sqrt{n}}$

FIGURE 3

The next corollary, which is based on the inequality $S_p < S < S_q$ in Theorem 2, gives us important information about the error involved in using a partial sum to approximate the sum of a convergent alternating series.

> **COROLLARY** Let $S = \sum_{n=1}^{\infty}(-1)^{n-1}b_n$, where $\{b_n\}$ is a positive decreasing sequence that converges to 0. Then
>
> $$\left| S - S_N \right| < b_{N+1} \qquad \boxed{2}$$
>
> In other words, *when we approximate S by S_N, the error is less than the size of the first omitted term b_{N+1}.*

Proof If N is even, then $N + 1$ is odd and Theorem 2 implies that $S_N < S < S_{N+1}$. Also, if N is odd, then $N + 1$ is even and Theorem 2 implies that $S_{N+1} < S < S_N$. In either case,

$$\left| S - S_N \right| < \left| S_{N+1} - S_N \right| = b_{N+1} \qquad ∎$$

EXAMPLE 5   Alternating Harmonic Series Show that $\sum_{n=1}^{\infty} \dfrac{(-1)^{n-1}}{n}$ converges conditionally. If S represents the sum, then

(a) Show that $|S - S_6| < \frac{1}{7}$.

(b) Find an N such that S_N approximates S with an error less than 10^{-3}.

Solution The terms $b_n = 1/n$ are positive and decreasing, and $\lim_{n \to \infty} b_n = 0$. Therefore, the series converges by the Alternating Series Test. The harmonic series $\sum_{n=1}^{\infty} 1/n$ diverges, so the series converges conditionally. Now, applying the inequality in (2), we have

$$\left| S - S_N \right| < b_{N+1} = \dfrac{1}{N + 1}$$

For $N = 6$, we obtain $|S - S_6| < b_7 = \frac{1}{7}$. We can make the error less than 10^{-3} by choosing N so that

$$\dfrac{1}{N + 1} \le 10^{-3} \quad \Rightarrow \quad N + 1 \ge 10^3 \quad \Rightarrow \quad N \ge 999$$

Therefore, with $N > 999$, S_N approximates S with error less than 10^{-3}. ∎

For the series in the previous example, a computer algebra system gives $S_{999} \approx 0.6937$, and therefore, S is within 10^{-3} of this value. In fact, it can be shown that $S = \ln 2$ (see Exercise 92 of Section 10.8). Thus, $S = \ln 2 \approx 0.6931$, which verifies the result in the example:

$$|S - S_{999}| \approx |\ln 2 - 0.6937| \approx 0.0006 < 10^{-3}$$

CONCEPTUAL INSIGHT The convergence of an infinite series $\sum a_n$ depends on two factors: (1) how quickly a_n tends to zero, and (2) how much cancellation takes place among the terms. Consider:

Harmonic series (diverges): $\qquad 1 + \dfrac{1}{2} + \dfrac{1}{3} + \dfrac{1}{4} + \dfrac{1}{5} + \cdots$

p-Series with $p = 2$ (converges): $\qquad 1 + \dfrac{1}{2^2} + \dfrac{1}{3^2} + \dfrac{1}{4^2} + \dfrac{1}{5^2} + \cdots$

Alternating harmonic series (converges): $\quad 1 - \dfrac{1}{2} + \dfrac{1}{3} - \dfrac{1}{4} + \dfrac{1}{5} - \cdots$

The harmonic series diverges because reciprocals $1/n$ do not tend to zero quickly enough. By contrast, the reciprocal squares $1/n^2$ tend to zero quickly enough for the p-series with $p = 2$ to converge. The alternating harmonic series converges, but only due to the cancellation among the terms.

10.4 SUMMARY

- $\sum a_n$ *converges absolutely* if the positive series $\sum |a_n|$ converges.
- Absolute convergence implies convergence: If $\sum |a_n|$ converges, then $\sum a_n$ also converges.
- $\sum a_n$ *converges conditionally* if $\sum a_n$ converges but $\sum |a_n|$ diverges.
- Alternating Series Test: If $\{b_n\}$ is positive and decreasing and $\lim\limits_{n \to \infty} b_n = 0$, then the alternating series

$$\sum_{n=1}^{\infty} (-1)^{n-1} b_n = b_1 - b_2 + b_3 - b_4 + b_5 - \cdots$$

converges. Furthermore, if S is the sum of the series, then $|S - S_N| < b_{N+1}$.
- We have developed two ways to handle nonpositive series: show absolute convergence if possible, or use the Alternating Series Test if applicable.

10.4 EXERCISES

Preliminary Questions

1. Give an example of a series such that $\sum a_n$ converges but $\sum |a_n|$ diverges.

2. Which of the following statements is equivalent to Theorem 1?

(a) If $\displaystyle\sum_{n=0}^{\infty} |a_n|$ diverges, then $\displaystyle\sum_{n=0}^{\infty} a_n$ also diverges.

(b) If $\displaystyle\sum_{n=0}^{\infty} a_n$ diverges, then $\displaystyle\sum_{n=0}^{\infty} |a_n|$ also diverges.

(c) If $\displaystyle\sum_{n=0}^{\infty} a_n$ converges, then $\displaystyle\sum_{n=0}^{\infty} |a_n|$ also converges.

3. Indicate whether or not the reasoning in the following statement is correct: Since $\displaystyle\sum_{n=1}^{\infty} (-1)^n \sqrt{n}$ is an alternating series, it must converge.

4. Suppose that b_n is positive, decreasing, and tends to 0, and let $S = \displaystyle\sum_{n=1}^{\infty} (-1)^{n-1} b_n$. What can we say about $|S - S_{100}|$ if $a_{101} = 10^{-3}$? Is S larger or smaller than S_{100}?

Exercises

1. Show that

$$\sum_{n=0}^{\infty} \frac{(-1)^n}{2^n}$$

converges absolutely.

2. Show that the following series converges conditionally:

$$\sum_{n=1}^{\infty} (-1)^{n-1} \frac{1}{n^{2/3}} = \frac{1}{1^{2/3}} - \frac{1}{2^{2/3}} + \frac{1}{3^{2/3}} - \frac{1}{4^{2/3}} + \cdots$$

In Exercises 3–10, determine whether the series converges absolutely, conditionally, or not at all.

3. $\displaystyle\sum_{n=1}^{\infty} \frac{(-1)^{n-1}}{n^{1/3}}$

4. $\displaystyle\sum_{n=1}^{\infty} \frac{(-1)^n \, n^4}{n^3 + 1}$

5. $\displaystyle\sum_{n=0}^{\infty} \frac{(-1)^n}{(1.001)^n}$

6. $\displaystyle\sum_{n=0}^{\infty} \frac{(-1)^n}{(0.999)^n}$

7. $\displaystyle\sum_{n=1}^{\infty} \frac{(-1)^n e^{-n}}{n^2}$

8. $\displaystyle\sum_{n=1}^{\infty} \frac{\sin(\frac{\pi n}{4})}{n^2}$

9. $\displaystyle\sum_{n=2}^{\infty} \frac{(-1)^n}{n \ln n}$

10. $\displaystyle\sum_{n=1}^{\infty} \frac{(-1)^n}{1 + \frac{1}{n}}$

11. Let $S = \displaystyle\sum_{n=1}^{\infty} (-1)^{n+1} \frac{1}{n^3}$.

(a) Calculate S_n for $1 \le n \le 10$.

(b) Use the inequality in (2) to show that $0.9 \le S \le 0.902$.

12. Use the inequality in (2) to approximate

$$\sum_{n=1}^{\infty} \frac{(-1)^{n+1}}{n!}$$

to four decimal places.

13. Approximate $\displaystyle\sum_{n=1}^{\infty} \frac{(-1)^{n+1}}{n^4}$ to three decimal places.

14. [CAS] Let

$$S = \sum_{n=1}^{\infty} (-1)^{n-1} \frac{n}{n^2 + 1}$$

Use a computer algebra system to calculate and plot the partial sums S_n for $1 \le n \le 100$. Observe that the partial sums zigzag above and below the limit.

In Exercises 15–16, find a value of N such that S_N approximates the series with an error of at most 10^{-5}. Using technology, compute this value of S_N.

15. $\displaystyle\sum_{n=1}^{\infty} \frac{(-1)^{n+1}}{n(n + 2)(n + 3)}$

16. $\displaystyle\sum_{n=1}^{\infty} \frac{(-1)^{n+1} \ln n}{n!}$

In Exercises 17–32, determine convergence or divergence by any method.

17. $\displaystyle\sum_{n=0}^{\infty} 7^{-n}$

18. $\displaystyle\sum_{n=1}^{\infty} \frac{1}{n^{7.5}}$

19. $\displaystyle\sum_{n=1}^{\infty} \frac{1}{5^n - 3^n}$

20. $\displaystyle\sum_{n=1}^{\infty} \frac{1}{n + \frac{1}{n}}$

21. $\displaystyle\sum_{n=1}^{\infty} \frac{1}{3n^4 + 12n}$

22. $\displaystyle\sum_{n=1}^{\infty} \frac{(-1)^n}{\sqrt{n^2 + 1}}$

23. $\displaystyle\sum_{n=1}^{\infty} \frac{1}{\sqrt{n^2 + 1}}$

24. $\displaystyle\sum_{n=0}^{\infty} \frac{(-1)^n n}{\sqrt{n^2 + 1}}$

25. $\displaystyle\sum_{n=1}^{\infty} \frac{3^n + (-2)^n}{5^n}$

26. $\displaystyle\sum_{n=1}^{\infty} \frac{(-1)^{n+1}}{(2n + 1)!}$

27. $\displaystyle\sum_{n=1}^{\infty} (-1)^n n^2 e^{-n^3/3}$

28. $\displaystyle\sum_{n=1}^{\infty} n e^{-n^3/3}$

29. $\displaystyle\sum_{n=2}^{\infty} \frac{(-1)^n}{n^{1/2}(\ln n)^2}$

30. $\displaystyle\sum_{n=2}^{\infty} \frac{1}{n(\ln n)^{1/4}}$

31. $\displaystyle\sum_{n=1}^{\infty} \frac{\ln n}{n^{1.05}}$

32. $\displaystyle\sum_{n=2}^{\infty} \frac{1}{(\ln n)^2}$

33. Show that

$$\frac{1}{2} - \frac{1}{2} + \frac{1}{3} - \frac{1}{3} + \frac{1}{4} - \frac{1}{4} + \cdots$$

converges by computing the partial sums. Does it converge absolutely?

34. The Alternating Series Test cannot be applied to

$$\frac{1}{2} - \frac{1}{3} + \frac{1}{2^2} - \frac{1}{3^2} + \frac{1}{2^3} - \frac{1}{3^3} + \cdots$$

Why not? Show that it converges by another method.

35. Assumptions Matter Show that the following series diverges:

$$\frac{1}{2} - \frac{1}{4} + \frac{1}{3} - \frac{1}{8} + \frac{1}{4} - \frac{1}{16} + \cdots + \left(\frac{1}{n} - \frac{1}{2^n}\right) + \cdots$$

(Note: This demonstrates that in the Alternating Series Test, we need the assumption that the sequence a_n is decreasing. It is not enough to assume only that a_n tends to zero.)

36. Determine whether the following series converges conditionally:

$$1 - \frac{1}{3} + \frac{1}{2} - \frac{1}{5} + \frac{1}{3} - \frac{1}{7} + \frac{1}{4} - \frac{1}{9} + \frac{1}{5} - \frac{1}{11} + \cdots$$

37. Prove that if $\sum a_n$ converges absolutely, then $\sum a_n^2$ also converges. Give an example where $\sum a_n$ is only conditionally convergent and $\sum a_n^2$ diverges.

Further Insights and Challenges

38. Prove the following variant of the Alternating Series Test: If $\{b_n\}$ is a positive, decreasing sequence with $\displaystyle\lim_{n\to\infty} b_n = 0$, then the series

$$b_1 + b_2 - 2b_3 + a_4 + b_5 - 2a_6 + \cdots$$

converges. *Hint:* Show that S_{3N} is increasing and bounded by $a_1 + a_2$, and continue as in the proof of the Alternating Series Test.

39. Use Exercise 38 to show that the following series converges:

$$\frac{1}{\ln 2} + \frac{1}{\ln 3} - \frac{2}{\ln 4} + \frac{1}{\ln 5} + \frac{1}{\ln 6} - \frac{2}{\ln 7} + \cdots$$

40. Prove the conditional convergence of

$$1 + \frac{1}{2} + \frac{1}{3} - \frac{3}{4} + \frac{1}{5} + \frac{1}{6} + \frac{1}{7} - \frac{3}{8} + \cdots$$

41. Show that the following series diverges:

$$1 + \frac{1}{2} + \frac{1}{3} - \frac{2}{4} + \frac{1}{5} + \frac{1}{6} + \frac{1}{7} - \frac{2}{8} + \cdots$$

Hint: Use the result of Exercise 40 to write the series as the sum of a convergent series and a divergent series.

42. Prove that

$$\sum_{n=1}^{\infty} (-1)^{n+1} \frac{(\ln n)^a}{n}$$

converges for all exponents a. *Hint:* Show that $f(x) = (\ln x)^a / x$ is decreasing for x sufficiently large.

43. We say that $\{b_n\}$ is a rearrangement of $\{a_n\}$ if $\{b_n\}$ has the same terms as $\{a_n\}$ but occurring in a different order. Show that if $\{b_n\}$ is a rearrangement of $\{a_n\}$ and $\sum_{n=1}^{\infty} a_n$ converges absolutely, then $\sum_{n=1}^{\infty} b_n$ also converges

absolutely. *Hint:* Prove that the partial sums $\sum_{n=1}^{N} |b_n|$ are bounded. (It can be shown further that the two series converge to the same value. This result does not hold if $\sum_{n=1}^{\infty} a_n$ is only conditionally convergent.)

44. Assumptions Matter In 1829, Lejeune Dirichlet pointed out that the great French mathematician Augustin Louis Cauchy made a mistake in a published paper by improperly assuming the Limit Comparison Test to be valid for nonpositive series. Here are Dirichlet's two series:

$$\sum_{n=1}^{\infty} \frac{(-1)^n}{\sqrt{n}}, \qquad \sum_{n=1}^{\infty} \frac{(-1)^n}{\sqrt{n}} \left(1 + \frac{(-1)^n}{\sqrt{n}} \right)$$

Explain how they provide a counterexample to the Limit Comparison Test when the series are not assumed to be positive.

10.5 The Ratio and Root Tests and Strategies for Choosing Tests

In the previous sections, we developed a number of theorems and tests that are used to investigate whether a series converges or diverges. In this section, we present two more tests, the Ratio Test and the Root Test. Then we outline a strategy for choosing which test to apply to determine if a specific series converges. We begin with the Ratio Test.

> **THEOREM 1 Ratio Test** Assume that the following limit exists:
>
> $$\rho = \lim_{n \to \infty} \left| \frac{a_{n+1}}{a_n} \right|$$
>
> **(i)** If $\rho < 1$, then $\sum a_n$ converges absolutely.
>
> **(ii)** If $\rho > 1$, then $\sum a_n$ diverges.
>
> **(iii)** If $\rho = 1$, the test is inconclusive.

The symbol ρ is a lowercase rho, the 17th letter of the Greek alphabet.

Proof The idea is to compare with a geometric series. If $\rho < 1$, we may choose a number r such that $\rho < r < 1$. Since $|a_{n+1}/a_n|$ converges to ρ, there exists a number M such that $|a_{n+1}/a_n| < r$ for all $n \geq M$. Therefore,

$$|a_{M+1}| < r|a_M|$$

$$|a_{M+2}| < r|a_{M+1}| < r(r|a_M|) = r^2|a_M|$$

$$|a_{M+3}| < r|a_{M+2}| < r^3|a_M|$$

In general, $|a_{M+n}| < r^n|a_M|$, and thus,

$$\sum_{n=M}^{\infty} |a_n| = \sum_{n=0}^{\infty} |a_{M+n}| \leq \sum_{n=0}^{\infty} |a_M| r^n = |a_M| \sum_{n=0}^{\infty} r^n$$

The geometric series on the right converges because $0 < r < 1$, so $\sum_{n=M}^{\infty} |a_n|$ converges by the Direct Comparison Test. Thus $\sum a_n$ converges absolutely.

If $\rho > 1$, choose r such that $1 < r < \rho$. Then there exists a number M such that $|a_{n+1}/a_n| > r$ for all $n \geq M$. Arguing as before with the inequalities reversed, we find that $|a_{M+n}| \geq r^n|a_M|$. Since r^n tends to ∞, the terms a_{M+n} do not tend to zero, and consequently, $\sum a_n$ diverges. Finally, Example 4 in this section shows that both convergence and divergence are possible when $\rho = 1$, so the test is inconclusive in this case. ∎

EXAMPLE 1 Prove that $\displaystyle\sum_{n=1}^{\infty} \frac{2^n}{n!}$ converges.

Solution Compute the ratio and its limit with $a_n = \dfrac{2^n}{n!}$. Note that $(n+1)! = (n+1)n!$. Thus,

$$\left| \frac{a_{n+1}}{a_n} \right| = \frac{2^{n+1}}{(n+1)!} \frac{n!}{2^n} = \frac{2^{n+1}}{2^n} \frac{n!}{(n+1)!} = \frac{2}{n+1}$$

We obtain

$$\rho = \lim_{n \to \infty} \left| \frac{a_{n+1}}{a_n} \right| = \lim_{n \to \infty} \frac{2}{n+1} = 0$$

Since $\rho < 1$, the series $\displaystyle\sum_{n=1}^{\infty} \frac{2^n}{n!}$ converges by the Ratio Test. ∎

EXAMPLE 2 Does $\displaystyle\sum_{n=1}^{\infty} \frac{n^2}{2^n}$ converge?

Solution Apply the Ratio Test with $a_n = \dfrac{n^2}{2^n}$:

$$\left| \frac{a_{n+1}}{a_n} \right| = \frac{(n+1)^2}{2^{n+1}} \frac{2^n}{n^2} = \frac{1}{2} \left(\frac{n^2 + 2n + 1}{n^2} \right) = \frac{1}{2} \left(1 + \frac{2}{n} + \frac{1}{n^2} \right)$$

We obtain

$$\rho = \lim_{n \to \infty} \left| \frac{a_{n+1}}{a_n} \right| = \frac{1}{2} \lim_{n \to \infty} \left(1 + \frac{2}{n} + \frac{1}{n^2} \right) = \frac{1}{2}$$

Since $\rho < 1$, the series converges by the Ratio Test. ∎

EXAMPLE 3 Does $\displaystyle\sum_{n=0}^{\infty} (-1)^n \frac{n!}{1000^n}$ converge?

Solution This series diverges by the Ratio Test because $\rho > 1$:

$$\rho = \lim_{n \to \infty} \left| \frac{a_{n+1}}{a_n} \right| = \lim_{n \to \infty} \frac{(n+1)!}{1000^{n+1}} \frac{1000^n}{n!} = \lim_{n \to \infty} \frac{n+1}{1000} = \infty$$ ∎

In the next example, we demonstrate why the Ratio Test is inconclusive in the case where $\rho = 1$.

EXAMPLE 4 **Ratio Test Inconclusive** Show that both convergence and divergence are possible when $\rho = 1$ by considering $\displaystyle\sum_{n=1}^{\infty} n^2$ and $\displaystyle\sum_{n=1}^{\infty} n^{-2}$.

Solution For $a_n = n^2$, we have

$$\rho = \lim_{n \to \infty} \left| \frac{a_{n+1}}{a_n} \right| = \lim_{n \to \infty} \frac{(n+1)^2}{n^2} = \lim_{n \to \infty} \frac{n^2 + 2n + 1}{n^2} = \lim_{n \to \infty} \left(1 + \frac{2}{n} + \frac{1}{n^2} \right) = 1$$

Furthermore, for $b_n = n^{-2}$,

$$\rho = \lim_{n \to \infty} \left| \frac{b_{n+1}}{b_n} \right| = \lim_{n \to \infty} \left| \frac{a_n}{a_{n+1}} \right| = \frac{1}{\displaystyle\lim_{n \to \infty} \left| \frac{a_{n+1}}{a_n} \right|} = 1$$

Thus, $\rho = 1$ in both cases, but, in fact, $\sum_{n=1}^{\infty} n^2$ diverges by the nth Term Divergence Test

since $\lim_{n \to \infty} n^2 = \infty$, and $\sum_{n=1}^{\infty} n^{-2}$ converges since it is a p-series with $p = 2 > 1$. This

shows that both convergence and divergence are possible when $\rho = 1$. ∎

Our next test is based on the limit of the nth roots $\sqrt[n]{|a_n|}$ rather than the ratios a_{n+1}/a_n. Its proof, like that of the Ratio Test, is based on a comparison with a geometric series (see Exercise 63).

THEOREM 2 Root Test Assume that the following limit exists:

$$L = \lim_{n \to \infty} \sqrt[n]{|a_n|}$$

  (i) If $L < 1$, then $\sum a_n$ converges absolutely.

  (ii) If $L > 1$, then $\sum a_n$ diverges.

  (iii) If $L = 1$, the test is inconclusive.

EXAMPLE 5 Does $\sum_{n=1}^{\infty} \left(\dfrac{n}{2n+3} \right)^n$ converge?

Solution We have $L = \lim_{n \to \infty} \sqrt[n]{a_n} = \lim_{n \to \infty} \dfrac{n}{2n+3} = \dfrac{1}{2}$. Since $L < 1$, the series converges by the Root Test. ∎

Determining Which Test to Apply

We end this section with a brief review of all of the tests we have introduced for determining convergence so far and how one decides which test to apply.

Let $\sum_{n=1}^{\infty} a_n$ be given. Keep in mind that the series for which convergence or divergence is known include the geometric series $\sum_{n=0}^{\infty} ar^n$, which converge for $|r| < 1$, and the

p-series $\sum_{n=0}^{\infty} \dfrac{1}{n^p}$, which converge for $p > 1$.

1. The nth Term Divergence Test Always check this test first. If $\lim_{n \to \infty} a_n \neq 0$, then the series diverges. But if $\lim_{n \to \infty} a_n = 0$, we do not know whether the series converges or diverges, and hence we move on to the next step.

2. Positive Series If all terms in the series are positive, try one of the following tests:

(a) The Direct Comparison Test Consider whether dropping terms in the numerator or denominator gives a series that we know either converges or diverges. If a larger series converges or a smaller series diverges, then the original series does the same. For example, $\sum_{n=1}^{\infty} \dfrac{1}{n^2 + \sqrt{n}}$ converges because $\dfrac{1}{n^2 + \sqrt{n}} < \dfrac{1}{n^2}$ and $\sum_{n=1}^{\infty} \dfrac{1}{n^2}$ converges (since it is a p-series with $p = 2 > 1$). On the other hand, this does not work for $\sum_{n=2}^{\infty} \dfrac{1}{n^2 - \sqrt{n}}$ since then the comparison series $\sum_{n=1}^{\infty} \dfrac{1}{n^2}$, while still converging, is smaller than the original series, so we cannot compare the series with $\sum \frac{1}{n^2}$ and apply the Direct Comparison Test. In this case, we can often apply the Limit Comparison Test as follows.

(b) **The Limit Comparison Test** Consider the dominant term in the numerator and denominator, and compare the original series to the ratio of those terms. For example, for $\sum_{n=2}^{\infty} \frac{1}{n^2 - \sqrt{n}}$, n^2 is dominant over $\sqrt{n}$ as it grows faster as n increases. So, we let $b_n = \frac{1}{n^2}$. Then

$$\lim_{n \to \infty} \frac{a_n}{b_n} = \lim_{n \to \infty} \frac{\frac{1}{n^2 - \sqrt{n}}}{\frac{1}{n^2}} = \lim_{n \to \infty} \frac{n^2}{n^2 - \sqrt{n}} = 1$$

The limit is a positive number, so the Limit Comparison Test applies. Since $\sum_{n=1}^{\infty} \frac{1}{n^2}$ converges, so does the original series.

(c) **The Ratio Test** The Ratio Test is often effective in the presence of a factorial such as $n!$ since in the ratio, the factorial disappears after cancellation. It is also effective when there are constants to the power n, such as 2^n, since in the ratio, the power n disappears after cancellation. For example, if the series is $\sum_{n=1}^{\infty} \frac{3^n}{n!}$, then applying the Ratio Test yields

$$\lim_{n \to \infty} \left| \frac{a_{n+1}}{a_n} \right| = \lim_{n \to \infty} \frac{\frac{3^{n+1}}{(n+1)!}}{\frac{3^n}{(n)!}} = \lim_{n \to \infty} \frac{3}{n+1} = 0 < 1$$

Therefore, the series converges.

(d) **The Root Test** The Root Test is often effective when there is a term of the form $f(n)^{g(n)}$. For example, $\sum_{n=1}^{\infty} \frac{2^n}{n^{2n}}$ is a good example since applying the Root Test yields

$$\lim_{n \to \infty} |a_n|^{1/n} = \lim_{n \to \infty} \left(\frac{2^n}{n^{2n}} \right)^{1/n} = \lim_{n \to \infty} \frac{2}{n^2} = 0 < 1$$

Thus, the series converges.

(e) **The Integral Test** When the other tests fail on a positive series, consider the Integral Test. If $a_n = f(n)$ is a decreasing function, then the series converges if and only if the improper integral $\int_1^{\infty} f(x)\,dx$ converges. For example, the other tests do not easily apply to $\sum_{n=2}^{\infty} \frac{1}{n \ln n}$. However, $f(x) = \frac{1}{n \ln n}$ is a decreasing function and $\int_2^{\infty} \frac{1}{x \ln x}\,dx = \ln(\ln x)\Big|_2^{\infty} = \infty$. Thus, the integral diverges, implying that the series does as well.

3. Series That Are Not Positive Series

(a) **Alternating Series Test** If the series is alternating of the form $\sum_{n=1}^{\infty} (-1)^{n-1} b_n$, show that $0 < b_{n+1} < b_n$ and $\lim_{n \to \infty} b_n = 0$. Then the Alternating Series Test shows the series converges.

(b) **Absolute Convergence** If the series $\sum a_n$ is not alternating, then see if $\sum |a_n|$, which is a positive series, converges using the tests for positive series. If so, the original series is absolutely convergent and therefore convergent.

10.5 SUMMARY

- Ratio Test: Assume that $\rho = \lim\limits_{n \to \infty} \left| \dfrac{a_{n+1}}{a_n} \right|$ exists.

  - Then $\sum a_n$ converges absolutely if $\rho < 1$.
  - Then $\sum a_n$ diverges if $\rho > 1$.
  - The test is inconclusive if $\rho = 1$.

- Root Test: Assume that $L = \lim\limits_{n \to \infty} \sqrt[n]{|a_n|}$ exists.

  - Then $\sum a_n$ converges absolutely if $L < 1$.
  - Then $\sum a_n$ diverges if $L > 1$.
  - The test is inconclusive if $L = 1$.

10.5 EXERCISES

Preliminary Questions

1. Consider the geometric series $\sum\limits_{n=0}^{\infty} cr^n$.

(a) In the Ratio Test, what do the terms $\left| \dfrac{a_{n+1}}{a_n} \right|$ equal?

(b) In the Root Test, what do the terms $\sqrt[n]{|a_n|}$ equal?

2. Consider the p-series $\sum\limits_{n=1}^{\infty} n^{-p}$.

(a) In the Ratio Test, what do the terms $\left| \dfrac{a_{n+1}}{a_n} \right|$ equal?

(b) What can be concluded from the Ratio Test?

3. Is the Ratio Test conclusive for $\sum\limits_{n=1}^{\infty} \dfrac{1}{n!}$? Is it conclusive for $\sum\limits_{n=1}^{\infty} \dfrac{1}{n+1}$?

4. Is the Root Test conclusive for $\sum\limits_{n=1}^{\infty} \dfrac{1}{2^n}$? Is it conclusive for

$$\sum\limits_{n=1}^{\infty} \left(1 + \frac{1}{n} \right)^{-n} ?$$

Exercises

In Exercises 1–20, apply the Ratio Test to determine convergence or divergence, or state that the Ratio Test is inconclusive.

1. $\sum\limits_{n=1}^{\infty} \dfrac{1}{5^n}$

2. $\sum\limits_{n=1}^{\infty} \dfrac{(-1)^{n-1} n}{5^n}$

3. $\sum\limits_{n=1}^{\infty} \dfrac{1}{n^n}$

4. $\sum\limits_{n=0}^{\infty} \dfrac{3n+2}{5n^3+1}$

5. $\sum\limits_{n=1}^{\infty} \dfrac{n}{n^2+1}$

6. $\sum\limits_{n=1}^{\infty} \dfrac{2^n}{n}$

7. $\sum\limits_{n=1}^{\infty} \dfrac{2^n}{n^{100}}$

8. $\sum\limits_{n=1}^{\infty} \dfrac{n^3}{3n^2}$

9. $\sum\limits_{n=1}^{\infty} \dfrac{10^n}{2^{n^2}}$

10. $\sum\limits_{n=1}^{\infty} \dfrac{e^n}{n!}$

11. $\sum\limits_{n=1}^{\infty} \dfrac{e^n}{n^n}$

12. $\sum\limits_{n=1}^{\infty} \dfrac{n^{40}}{n!}$

13. $\sum\limits_{n=0}^{\infty} \dfrac{n!}{6^n}$

14. $\sum\limits_{n=1}^{\infty} \dfrac{n!}{n^9}$

15. $\sum\limits_{n=2}^{\infty} \dfrac{1}{n^{3/2} \ln n}$

16. $\sum\limits_{n=1}^{\infty} \dfrac{1}{(2n)!}$

17. $\sum\limits_{n=1}^{\infty} \dfrac{n^2}{(2n+1)!}$

18. $\sum\limits_{n=1}^{\infty} \dfrac{(n!)^3}{(3n)!}$

19. $\sum\limits_{n=2}^{\infty} \dfrac{1}{2^n+1}$

20. $\sum\limits_{n=2}^{\infty} \dfrac{1}{\ln n}$

21. Show that $\sum\limits_{n=1}^{\infty} n^k \, 3^{-n}$ converges for all exponents k.

22. Show that $\sum\limits_{n=1}^{\infty} n^2 x^n$ converges if $|x| < 1$.

23. Show that $\sum\limits_{n=1}^{\infty} 2^n x^n$ converges if $|x| < \frac{1}{2}$.

24. Show that $\sum\limits_{n=1}^{\infty} \dfrac{r^n}{n!}$ converges for all r.

25. Show that $\sum\limits_{n=1}^{\infty} \dfrac{r^n}{n}$ converges if $|r| < 1$.

26. Is there any value of k such that $\sum\limits_{n=1}^{\infty} \dfrac{2^n}{n^k}$ converges?

In Exercises 27–28, the the following limit could be helpful:

$$\lim_{n\to\infty}\left(1+\frac{1}{n}\right)^n = e.$$

27. Does $\displaystyle\sum_{n=1}^{\infty}\frac{n!}{n^n}$ converge or diverge?

28. Does $\displaystyle\sum_{n=1}^{\infty}\frac{(2n)!}{n^n}$ converge or diverge?

In Exercises 29–33, assume that $|a_{n+1}/a_n|$ converges to $\rho=\frac{1}{3}$. What can you say about the convergence of the given series?

29. $\displaystyle\sum_{n=1}^{\infty} n^3 a_n$

30. $\displaystyle\sum_{n=1}^{\infty} 2^n a_n$

31. $\displaystyle\sum_{n=1}^{\infty} 3^n a_n$

32. $\displaystyle\sum_{n=1}^{\infty} 4^n a_n$

33. $\displaystyle\sum_{n=1}^{\infty} a_n^2$

34. Assume that $|a_{n+1}/a_n|$ converges to $\rho=4$. Does $\displaystyle\sum_{n=1}^{\infty} a_n^{-1}$ converge (assume that $a_n \neq 0$ for all n)?

35. Show that the Root Test is inconclusive for the *p*-series $\displaystyle\sum_{n=1}^{\infty}\frac{1}{n^p}$.

In Exercises 36–41, use the Root Test to determine convergence or divergence (or state that the test is inconclusive).

36. $\displaystyle\sum_{n=0}^{\infty}\frac{1}{10^n}$

37. $\displaystyle\sum_{n=1}^{\infty}\frac{1}{n^n}$

38. $\displaystyle\sum_{k=0}^{\infty}\left(\frac{k}{k+10}\right)^k$

39. $\displaystyle\sum_{k=0}^{\infty}\left(\frac{k}{3k+1}\right)^k$

40. $\displaystyle\sum_{n=1}^{\infty}\left(2+\frac{1}{n}\right)^{-n}$

41. $\displaystyle\sum_{n=4}^{\infty}\left(1+\frac{1}{n}\right)^{-n^2}$

42. Prove that $\displaystyle\sum_{n=1}^{\infty}\frac{2^{n^2}}{n!}$ diverges. *Hint:* Use $2^{n^2}=(2^n)^n$ and $n! \leq n^n$.

In Exercises 43–62, determine convergence or divergence using any method covered in the text so far.

43. $\displaystyle\sum_{n=1}^{\infty}\frac{2^n+4^n}{7^n}$

44. $\displaystyle\sum_{n=1}^{\infty}\frac{n^3}{n!}$

45. $\displaystyle\sum_{n=1}^{\infty}\frac{n}{2n+1}$

46. $\displaystyle\sum_{n=1}^{\infty} 2^{1/n}$

47. $\displaystyle\sum_{n=1}^{\infty}\frac{\sin n}{n^2}$

48. $\displaystyle\sum_{n=1}^{\infty}\frac{n!}{(2n)!}$

49. $\displaystyle\sum_{n=1}^{\infty}\frac{1}{n+\sqrt{n}}$

50. $\displaystyle\sum_{n=2}^{\infty}\frac{1}{n(\ln n)^3}$

51. $\displaystyle\sum_{n=1}^{\infty}\frac{n^3}{5^n}$

52. $\displaystyle\sum_{n=2}^{\infty}\frac{1}{n^2(\ln n)^3}$

53. $\displaystyle\sum_{n=2}^{\infty}\frac{1}{\sqrt{n^3-n^2}}$

54. $\displaystyle\sum_{n=1}^{\infty}\frac{n^2+4n}{3n^4+9}$

55. $\displaystyle\sum_{n=1}^{\infty} n^{-0.8}$

56. $\displaystyle\sum_{n=1}^{\infty}(0.8)^{-n}n^{-0.8}$

57. $\displaystyle\sum_{n=1}^{\infty} 4^{-2n+1}$

58. $\displaystyle\sum_{n=1}^{\infty}\frac{(-1)^{n-1}}{\sqrt{n}}$

59. $\displaystyle\sum_{n=1}^{\infty}\sin\frac{1}{n^2}$

60. $\displaystyle\sum_{n=1}^{\infty}(-1)^n\cos\frac{1}{n}$

61. $\displaystyle\sum_{n=1}^{\infty}\frac{(-2)^n}{\sqrt{n}}$

62. $\displaystyle\sum_{n=1}^{\infty}\left(\frac{n}{n+12}\right)^n$

Further Insights and Challenges

63. ✎ **Proof of the Root Test** Let $\displaystyle\sum_{n=0}^{\infty} a_n$ be a positive series, and assume that $L=\displaystyle\lim_{n\to\infty}\sqrt[n]{a_n}$ exists.

(a) Show that the series converges if $L<1$. *Hint:* Choose R with $L<R<1$ and show that $a_n \leq R^n$ for n sufficiently large. Then compare with the geometric series $\sum R^n$.

(b) Show that the series diverges if $L>1$.

64. Show that the Ratio Test does not apply, but verify convergence using the Direct Comparison Test for the series

$$\frac{1}{2}+\frac{1}{3^2}+\frac{1}{2^3}+\frac{1}{3^4}+\frac{1}{2^5}+\cdots$$

65. Let $\displaystyle\sum_{n=1}^{\infty}\frac{c^n n!}{n^n}$, where c is a constant.

(a) Prove that the series converges absolutely if $|c|<e$ and diverges if $|c|>e$.

(b) It is known that $\displaystyle\lim_{n\to\infty}\frac{e^n n!}{n^{n+1/2}}=\sqrt{2\pi}$. Verify this numerically.

(c) Use the Limit Comparison Test to prove that the series diverges for $c=e$.

10.6 Power Series

With series we can make sense of the idea of a polynomial of infinite degree:

$$F(x)=\sum_{n=0}^{\infty} a_n x^n = a_0 + a_1 x + a_2 x^2 + a_3 x^3 + \cdots$$

Specifically, a **power series** with center c is an infinite series

$$F(x)=\sum_{n=0}^{\infty} a_n (x-c)^n = a_0 + a_1(x-c) + a_2(x-c)^2 + a_3(x-c)^3 + \cdots$$

where x is a variable. For example,

$$F(x) = 1 - x + x^2 - x^3 + \cdots$$

$$G(x) = 1 + (x - 2) + 2(x - 2)^2 + 3(x - 2)^3 + \cdots \qquad \boxed{1}$$

are power series where $F(x)$ has center $c = 0$ and $G(x)$ has center $c = 2$.

A power series $F(x) = \displaystyle\sum_{n=0}^{\infty} a_n (x - c)^n$ converges for some values of x and may diverge for others. For example, if we set $x = \frac{9}{4}$ in the power series of Eq. (1), we obtain the infinite series

$$G\left(\frac{9}{4}\right) = 1 + \left(\frac{9}{4} - 2\right) + 2\left(\frac{9}{4} - 2\right)^2 + 3\left(\frac{9}{4} - 2\right)^3 + \cdots$$

$$= 1 + \left(\frac{1}{4}\right) + 2\left(\frac{1}{4}\right)^2 + 3\left(\frac{1}{4}\right)^3 + \cdots + n\left(\frac{1}{4}\right)^n + \cdots$$

This converges by the Ratio Test:

$$\lim_{n \to \infty} \left| \frac{a_{n+1}}{a_n} \right| = \lim_{n \to \infty} \left| \frac{\frac{n+1}{4^{n+1}}}{\frac{n}{4^n}} \right| = \lim_{n \to \infty} \frac{1}{4}\left(\frac{n+1}{n}\right)\frac{1/n}{1/n} = \lim_{n \to \infty} \frac{1}{4}\left(\frac{1 + 1/n}{1}\right) = \frac{1}{4}$$

On the other hand, the power series in Eq. (1) diverges for $x = 3$ by the nth Term Divergence Test:

$$G(3) = 1 + (3 - 2) + 2(3 - 2)^2 + 3(3 - 2)^3 + \cdots$$

$$= 1 + 1 + 2 + 3 + \cdots$$

There is a surprisingly simple way to describe the set of values x at which a power series $F(x)$ converges. According to our next theorem, either $F(x)$ converges absolutely for all values of x or there is a radius of convergence R such that

> $F(x)$ *converges absolutely when* $|x - c| < R$ *and diverges when* $|x - c| > R$.

This means that $F(x)$ converges for x in an **interval of convergence** consisting of the open interval $(c - R, c + R)$ and possibly one or both of the endpoints $c - R$ and $c + R$ (Figure 1). Note that $F(x)$ automatically converges at $x = c$ because

$$F(c) = a_0 + a_1(c - c) + a_2(c - c)^2 + a_3(c - c)^3 + \cdots = a_0$$

We set $R = 0$ if $F(x)$ converges only for $x = c$, and we set $R = \infty$ if $F(x)$ converges for all values of x.

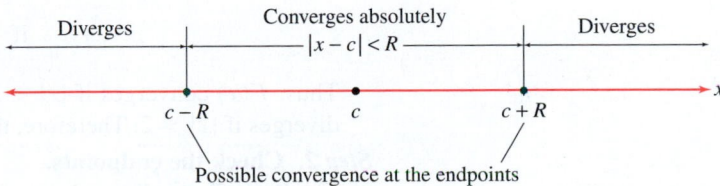

FIGURE 1 Interval of convergence of a power series.

THEOREM 1 Radius of Convergence Every power series

$$F(x) = \sum_{n=0}^{\infty} a_n (x - c)^n$$

has a radius of convergence R, which is either a nonnegative number $(R \geq 0)$ or infinity $(R = \infty)$. If R is finite, $F(x)$ converges absolutely when $|x - c| < R$ and diverges when $|x - c| > R$. If $R = \infty$, then $F(x)$ converges absolutely for all x.

Proof We assume that $c = 0$ to simplify the notation. If $F(x)$ converges only at $x = 0$, then $R = 0$. Otherwise, $F(x)$ converges for some nonzero value $x = B$. We claim that $F(x)$ must then converge absolutely for all $|x| < |B|$. To prove this, note that because

$$F(B) = \sum_{n=0}^{\infty} a_n B^n$$

converges, the general term $a_n B^n$ tends to zero. In particular, there exists $M > 0$ such that $|a_n B^n| < M$ for all n. Therefore,

$$\sum_{n=0}^{\infty} |a_n x^n| = \sum_{n=0}^{\infty} |a_n B^n| \, \left|\frac{x}{B}\right|^n < M \sum_{n=0}^{\infty} \left|\frac{x}{B}\right|^n$$

If $|x| < |B|$, then $|x/B| < 1$ and the series on the right is a convergent geometric series. By the Direct Comparison Test, the series on the left also converges. This proves that $F(x)$ converges absolutely if $|x| < |B|$.

Now let S be the set of numbers x such that $F(x)$ converges. Then S contains 0, and we have shown that if S contains a number $B \neq 0$, then S contains the open interval $(-|B|, |B|)$. If S is bounded, then S has a least upper bound $L > 0$ (see marginal note). In this case, there exist numbers $B \in S$ smaller than but arbitrarily close to L, and thus, S contains $(-B, B)$ for all $0 < B < L$. It follows that S contains the open interval $(-L, L)$. The set S cannot contain any number x with $|x| > L$, but S may contain one or both of the endpoints $x = \pm L$. So in this case, F has radius of convergence $R = L$. If S is not bounded, then S contains intervals $(-B, B)$ for B arbitrarily large. In this case, S is the entire real line $\mathbf{R}$, and the radius of convergence is $R = \infty$. ■

> Least Upper Bound Property: If S is a set of real numbers with an upper bound M (i.e., $x \leq M$ for all $x \in S$), then S has a least upper bound L. See Appendix B.

From Theorem 1, we see that there are two steps in determining the interval of convergence of F:

Step 1. Find the radius of convergence R (using the Ratio Test, in most cases).

Step 2. Check convergence at the endpoints (if $R \neq 0$ or ∞).

EXAMPLE 1   Using the Ratio Test   Where does $F(x) = \sum_{n=0}^{\infty} \dfrac{x^n}{2^n}$ converge?

Solution

Step 1. **Find the radius of convergence.**

Let $a_n = \dfrac{x^n}{2^n}$ and compute ρ from the Ratio Test:

$$\rho = \lim_{n\to\infty} \left|\frac{a_{n+1}}{a_n}\right| = \lim_{n\to\infty} \left|\frac{x^{n+1}}{2^{n+1}}\right| \cdot \left|\frac{2^n}{x^n}\right| = \lim_{n\to\infty} \frac{1}{2}|x| = \frac{1}{2}|x|$$

We find that

$$\rho < 1 \quad \text{if} \quad \frac{1}{2}|x| < 1, \quad \text{that is, if} \quad |x| < 2$$

Thus, $F(x)$ converges if $|x| < 2$. Similarly, $\rho > 1$ if $\frac{1}{2}|x| > 1$, or $|x| > 2$. So, $F(x)$ diverges if $|x| > 2$. Therefore, the radius of convergence is $R = 2$.

Step 2. **Check the endpoints.**

The Ratio Test is inconclusive for $x = \pm 2$, so we must check these cases directly:

$$F(2) = \sum_{n=0}^{\infty} \frac{2^n}{2^n} = 1 + 1 + 1 + 1 + 1 + 1 \cdots$$

$$F(-2) = \sum_{n=0}^{\infty} \frac{(-2)^n}{2^n} = 1 - 1 + 1 - 1 + 1 - 1 \cdots$$

Both series diverge. We conclude that $F(x)$ converges only for $|x| < 2$ (Figure 2). ■

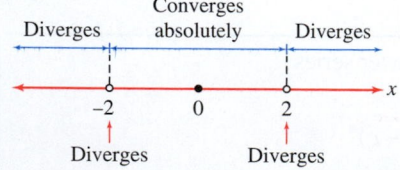

DF **FIGURE 2** The power series $\sum_{n=0}^{\infty} \dfrac{x^n}{2^n}$ has an interval of convergence $(-2, 2)$.

EXAMPLE 2 Where does $F(x) = \sum_{n=1}^{\infty} \frac{(-1)^n}{4^n n}(x-5)^n$ converge?

Solution We compute ρ with $a_n = \frac{(-1)^n}{4^n n}(x-5)^n$:

$$\rho = \lim_{n \to \infty} \left| \frac{a_{n+1}}{a_n} \right| = \lim_{n \to \infty} \left| \frac{(x-5)^{n+1}}{4^{n+1}(n+1)} \frac{4^n n}{(x-5)^n} \right|$$

$$= |x-5| \lim_{n \to \infty} \left| \frac{n}{4(n+1)} \right|$$

$$= \frac{1}{4}|x-5|$$

We find that

$$\rho < 1 \quad \text{if} \quad \frac{1}{4}|x-5| < 1, \quad \text{that is, if} \quad |x-5| < 4$$

Thus, $F(x)$ converges absolutely on the open interval $(1,9)$ of radius 4 with center $c = 5$. In other words, the radius of convergence is $R = 4$. Next, we check the endpoints:

$$x = 9: \quad \sum_{n=1}^{\infty} \frac{(-1)^n}{4^n n}(9-5)^n = \sum_{n=1}^{\infty} \frac{(-1)^n}{n} \quad \text{converges (Alternating Series Test)}$$

$$x = 1: \quad \sum_{n=1}^{\infty} \frac{(-1)^n}{4^n n}(-4)^n = \sum_{n=1}^{\infty} \frac{1}{n} \quad \text{diverges (harmonic series)}$$

We conclude that $F(x)$ converges for x in the half-open interval $(1,9]$ shown in Figure 3. ∎

Some power series contain only even powers or only odd powers of x. The Ratio Test can still be used to find the radius of convergence.

EXAMPLE 3 **An Even Power Series** Where does $\sum_{n=0}^{\infty} \frac{x^{2n}}{(2n)!}$ converge?

Solution Although this power series has only even powers of x, we can still apply the Ratio Test with $a_n = x^{2n}/(2n)!$. We have

$$a_{n+1} = \frac{x^{2(n+1)}}{(2(n+1))!} = \frac{x^{2n+2}}{(2n+2)!}$$

Furthermore, $(2n+2)! = (2n+2)(2n+1)(2n)!$, so

$$\rho = \lim_{n \to \infty} \left| \frac{a_{n+1}}{a_n} \right| = \lim_{n \to \infty} \frac{x^{2n+2}}{(2n+2)!} \frac{(2n)!}{x^{2n}} = |x|^2 \lim_{n \to \infty} \frac{1}{(2n+2)(2n+1)} = 0$$

Thus, $\rho = 0$ for all x, and $F(x)$ converges for all x. The radius of convergence is $R = \infty$. ∎

Geometric series are important examples of power series. Recall the formula $\sum_{n=0}^{\infty} r^n = 1/(1-r)$, valid for $|r| < 1$. Writing x in place of r, we obtain a power series expansion with radius of convergence $R = 1$:

$$\boxed{\frac{1}{1-x} = \sum_{n=0}^{\infty} x^n \qquad \text{for } |x| < 1}$$ 2

The next two examples show that we can modify this formula to find the power series expansions of other functions.

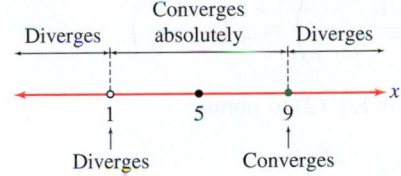

Diverges | Converges absolutely | Diverges

1   5   9

Diverges   Converges

DF **FIGURE 3** The power series $\sum_{n=1}^{\infty} \frac{(-1)^n}{4^n n}(x-5)^n$ has an interval of convergence $(1,9]$.

When a function f is represented by a power series on an interval I, we refer to it as the power series expansion of f on I.

EXAMPLE 4 **Geometric Series** Prove that

$$\frac{1}{1-2x} = \sum_{n=0}^{\infty} 2^n x^n \qquad \text{for } |x| < \frac{1}{2}$$

Solution Substitute $2x$ for x in Eq. (2):

$$\frac{1}{1-2x} = \sum_{n=0}^{\infty} (2x)^n = \sum_{n=0}^{\infty} 2^n x^n \qquad \boxed{3}$$

Expansion (2) is valid for $|x| < 1$, so Eq. (3) is valid for $|2x| < 1$, or $|x| < \frac{1}{2}$. ∎

EXAMPLE 5 Find a power series expansion with center $c = 0$ for

$$f(x) = \frac{1}{2+x^2}$$

and find the interval of convergence.

Solution We need to rewrite $f(x)$ so we can use Eq. (2). We have

$$\frac{1}{2+x^2} = \frac{1}{2}\left(\frac{1}{1+\frac{1}{2}x^2}\right) = \frac{1}{2}\left(\frac{1}{1-\left(-\frac{1}{2}x^2\right)}\right) = \frac{1}{2}\left(\frac{1}{1-u}\right)$$

where $u = -\frac{1}{2}x^2$. Now substitute $u = -\frac{1}{2}x^2$ for x in Eq. (2) to obtain

$$f(x) = \frac{1}{2+x^2} = \frac{1}{2}\sum_{n=0}^{\infty}\left(-\frac{x^2}{2}\right)^n$$

$$= \sum_{n=0}^{\infty} \frac{(-1)^n x^{2n}}{2^{n+1}}$$

This expansion is valid if $|-x^2/2| < 1$, or $|x| < \sqrt{2}$. The interval of convergence is $(-\sqrt{2}, \sqrt{2})$. ∎

Our next theorem tells us that within the interval of convergence, we can treat a power series as though it were a polynomial; that is, we can differentiate and integrate term by term.

Nina Karlovna Bari (1901–1961) was a Russian mathematician who studied trigonometric series, representations of functions using infinite sums of $\sin nx$ and $\cos nx$ terms (in contrast to the x^n terms used in power series). Series of this type are beneficial in fields such as signal processing where signals are broken down into sums of simple wave forms. In 1918, she enrolled as a student in the Department of Mathematics and Physics at Moscow State University and later she had a long career there as a professor.

The proof of Theorem 2 is somewhat technical and is omitted. See Exercise 70 for a proof that F is continuous.

THEOREM 2 **Term-by-Term Differentiation and Integration** Assume that

$$F(x) = \sum_{n=0}^{\infty} a_n(x-c)^n$$

has radius of convergence $R > 0$. Then F is differentiable on $(c - R, c + R)$. Furthermore, we can integrate and differentiate term by term. For $x \in (c - R, c + R)$,

$$F'(x) = \sum_{n=1}^{\infty} n a_n(x-c)^{n-1}$$

$$\int F(x)\, dx = A + \sum_{n=0}^{\infty} \frac{a_n}{n+1}(x-c)^{n+1} \qquad (A \text{ any constant})$$

For both the derivative series and the integral series the radius of convergence is also R.

Theorem 2 is a powerful tool for working with power series. The next two examples show how to use differentiation or antidifferentiation of power series representations of functions to obtain power series for other functions.

EXAMPLE 6 **Differentiating a Power Series** Prove that for $-1 < x < 1$,

$$\frac{1}{(1-x)^2} = 1 + 2x + 3x^2 + 4x^3 + 5x^4 + \cdots$$

Solution First, note that

$$\frac{1}{(1-x)^2} = \frac{d}{dx}\left(\frac{1}{1-x}\right)$$

For $\frac{1}{1-x}$, we have the following geometric series with radius of convergence $R = 1$:

$$\frac{1}{1-x} = 1 + x + x^2 + x^3 + x^4 + \cdots$$

By Theorem 2, we can differentiate term by term for $|x| < 1$ to obtain

$$\frac{d}{dx}\left(\frac{1}{1-x}\right) = \frac{d}{dx}(1 + x + x^2 + x^3 + x^4 + \cdots)$$

$$\frac{1}{(1-x)^2} = 1 + 2x + 3x^2 + 4x^3 + 5x^4 + \cdots \qquad \blacksquare$$

EXAMPLE 7 **Power Series for Arctangent** Prove that for $-1 < x < 1$,

$$\tan^{-1} x = \sum_{n=0}^{\infty} \frac{(-1)^n x^{2n+1}}{2n+1} = x - \frac{x^3}{3} + \frac{x^5}{5} - \frac{x^7}{7} + \cdots \qquad \boxed{4}$$

Solution Recall that $\tan^{-1} x$ is an antiderivative of $(1 + x^2)^{-1}$. We obtain a power series expansion of $(1 + x^2)^{-1}$ by substituting $-x^2$ for x in the geometric series of Eq. (2):

$$\frac{1}{1+x^2} = 1 - x^2 + x^4 - x^6 + \cdots$$

This expansion is valid for $|x^2| < 1$—that is, for $|x| < 1$. By Theorem 2, we can integrate this series term by term. The resulting expansion is also valid for $|x| < 1$:

$$\tan^{-1} x = \int \frac{dx}{1+x^2} = \int (1 - x^2 + x^4 - x^6 + \cdots)\, dx$$

$$= A + x - \frac{x^3}{3} + \frac{x^5}{5} - \frac{x^7}{7} + \cdots$$

Setting $x = 0$, we obtain $A = \tan^{-1} 0 = 0$. Thus, Eq. (4) is valid for $-1 < x < 1$. $\blacksquare$

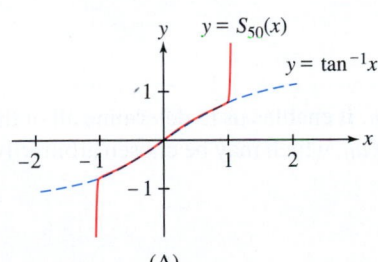

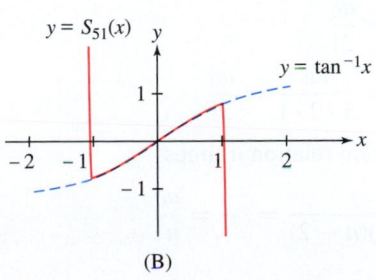

FIGURE 4 $y = S_{50}(x)$ and $y = S_{51}(x)$ are nearly indistinguishable from $y = \tan^{-1} x$ on $(-1, 1)$.

GRAPHICAL INSIGHT Let's examine the expansion of the previous example graphically. The partial sums of the power series for $f(x) = \tan^{-1} x$ are

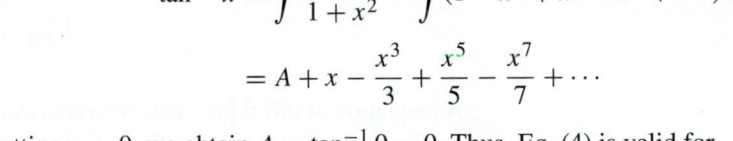

$$S_N(x) = x - \frac{x^3}{3} + \frac{x^5}{5} - \frac{x^7}{7} + \cdots + (-1)^N \frac{x^{2N-1}}{2N-1}$$

For large N, we can expect $S_N(x)$ to provide a good approximation to $f(x) = \tan^{-1} x$ on the interval $(-1, 1)$, where the power series expansion is valid. Figure 4 confirms this expectation: The graphs of $y = S_{50}(x)$ and $y = S_{51}(x)$ are nearly indistinguishable from the graph of $y = \tan^{-1} x$ on $(-1, 1)$. Thus, we may use the partial sums to approximate the arctangent. For example, using $S_5(x)$ to approximate $\tan^{-1} x$, we obtain $\tan^{-1}(0.3)$ is approximated by

$$S_5(0.3) = 0.3 - \frac{(0.3)^3}{3} + \frac{(0.3)^5}{5} - \frac{(0.3)^7}{7} + \frac{(0.3)^9}{9} \approx 0.2914569$$

Since the power series is an alternating series, the error in this approximation is less than the first omitted term, the term with $(0.3)^{11}$. Therefore,

$$\text{error} = |\tan^{-1}(0.3) - S_4(0.3)| < \frac{(0.3)^{11}}{11} \approx 1.61 \times 10^{-7}$$

Approximating $\tan^{-1} x$ with a partial sum $S_N(x)$ works well in the region $|x| < 1$. For $|x| > 1$, the situation changes drastically since the power series diverges and the partial sums deviate sharply from $\tan^{-1} x$.

Power Series Solutions of Differential Equations

Power series are a basic tool in the study of differential equations. To illustrate, consider the differential equation with initial condition

$$y' = y, \qquad y(0) = 1 \qquad \boxed{5}$$

From Example 5 in Section 9.1, it follows that $f(x) = e^x$ is a solution to this Initial Value Problem. Here, we take a different approach and find a solution in the form of a power series, $F(x) = \sum_{n=0}^{\infty} a_n x^n$. Ultimately, this approach will provide us with a power series representation of $f(x) = e^x$. We have

$$F(x) = \sum_{n=0}^{\infty} a_n x^n = a_0 + a_1 x + a_2 x^2 + a_3 x^3 + \cdots$$

$$F'(x) = \sum_{n=0}^{\infty} n a_n x^{n-1} = a_1 + 2a_2 x + 3a_3 x^2 + 4a_4 x^3 + \cdots$$

To satisfy the differential equation, we must have $F'(x) = F(x)$, and therefore,

$$a_0 = a_1, \quad a_1 = 2a_2, \quad a_2 = 3a_3, \quad a_3 = 4a_4, \quad \ldots$$

In other words, $F'(x) = F(x)$ if $a_{n-1} = n a_n$, or

$$\boxed{a_n = \frac{a_{n-1}}{n}}$$

An equation of this type is called a *recursion relation*. It enables us to determine all of the coefficients a_n successively from the first coefficient a_0, which may be chosen arbitrarily. For example,

$$n = 1: \qquad a_1 = \frac{a_0}{1}$$

$$n = 2: \qquad a_2 = \frac{a_1}{2} = \frac{a_0}{2 \cdot 1} = \frac{a_0}{2!}$$

$$n = 3: \qquad a_3 = \frac{a_2}{3} = \frac{a_1}{3 \cdot 2} = \frac{a_0}{3 \cdot 2 \cdot 1} = \frac{a_0}{3!}$$

To obtain a general formula for a_n, apply the recursion relation n times:

$$a_n = \frac{a_{n-1}}{n} = \frac{a_{n-2}}{n(n-1)} = \frac{a_{n-3}}{n(n-1)(n-2)} = \cdots = \frac{a_0}{n!}$$

We conclude that

$$F(x) = a_0 \sum_{n=0}^{\infty} \frac{x^n}{n!}$$

In Example 3, we showed that this power series has radius of convergence $R = \infty$, so $y = F(x)$ satisfies $y' = y$ for all x. Moreover, $F(0) = a_0$, so the initial condition $y(0) = 1$ is satisfied with $a_0 = 1$. Therefore,

$$F(x) = \sum_{n=0}^{\infty} \frac{x^n}{n!}$$

is the solution to the Initial Value Problem.

In Section 9.1, we showed that $f(x) = e^x$ is not just a solution to the Initial Value Problem in (5), but the *only* solution. The uniqueness of the solution implies that e^x and the power series solution we obtained must be equal. Thus, we have found a power series representation for e^x that is valid for all x:

$$e^x = \sum_{n=0}^{\infty} \frac{x^n}{n!} = 1 + x + \frac{x^2}{2!} + \frac{x^3}{3!} + \frac{x^4}{4!} + \cdots$$

In Section 10.8, we will see how to arrive at this power series representation of e^x via what are known as Taylor series.

In contrast to $y' = y$, the differential equation in the next example cannot be solved using any method that is simpler than the process of finding a power series solution. As with the solution of $y' = y$, the process involves solving a recursion relation that determines the coefficients a_n of a power series for the solution.

The solution in Example 8 is called the Bessel function of order 1. The Bessel function of order n is a solution of

$$x^2 y'' + xy' + (x^2 - n^2)y = 0$$

These functions have applications in many areas of physics and engineering.

EXAMPLE 8 Find a power series solution to the Initial Value Problem:

$$x^2 y'' + xy' + (x^2 - 1)y = 0, \qquad y'(0) = 1 \qquad \boxed{6}$$

Solution Assume that Eq. (6) has a power series solution $F(x) = \displaystyle\sum_{n=0}^{\infty} a_n x^n$. Then

$$y' = F'(x) = \sum_{n=0}^{\infty} n a_n x^{n-1} = a_1 + 2a_2 x + 3a_3 x^2 + \cdots$$

$$y'' = F''(x) = \sum_{n=0}^{\infty} n(n-1) a_n x^{n-2} = 2a_2 + 6a_3 x + 12a_4 x^2 + \cdots$$

Now, substitute the series for y, y', and y'' into the differential equation (6) to determine the recursion relation satisfied by the coefficients a_n:

$$x^2 y'' + xy' + (x^2 - 1)y$$

In Eq. (7), we combine the first three series into a single series using

$$n(n-1) + n - 1 = n^2 - 1$$

Also, we shift the fourth series by replacing n with $n - 2$. Consequently, the summation begins at $n - 2 = 0$; that is, at $n = 2$.

$$= x^2 \sum_{n=0}^{\infty} n(n-1) a_n x^{n-2} + x \sum_{n=0}^{\infty} n a_n x^{n-1} + (x^2 - 1) \sum_{n=0}^{\infty} a_n x^n$$

$$= \sum_{n=0}^{\infty} n(n-1) a_n x^n + \sum_{n=0}^{\infty} n a_n x^n - \sum_{n=0}^{\infty} a_n x^n + \sum_{n=0}^{\infty} a_n x^{n+2} \qquad \boxed{7}$$

$$= \sum_{n=0}^{\infty} (n^2 - 1) a_n x^n + \sum_{n=2}^{\infty} a_{n-2} x^n = 0$$

The differential equation is satisfied if

$$\sum_{n=0}^{\infty} (n^2 - 1) a_n x^n = - \sum_{n=2}^{\infty} a_{n-2} x^n$$

The first few terms on each side of this equation are

$$-a_0 + 0 \cdot x + 3a_2x^2 + 8a_3x^3 + 15a_4x^4 + \cdots = 0 + 0 \cdot x - a_0x^2 - a_1x^3 - a_2x^4 - \cdots$$

Matching up the coefficients of x^n, we find that

$$-a_0 = 0, \qquad 3a_2 = -a_0, \qquad 8a_3 = -a_1, \qquad 15a_4 = -a_2 \qquad \boxed{8}$$

In general, $(n^2 - 1)a_n = -a_{n-2}$, and this yields the recursion relation

$$\boxed{a_n = -\frac{a_{n-2}}{n^2 - 1} \qquad \text{for } n \geq 2} \qquad \boxed{9}$$

Note that $a_0 = 0$ by Eq. (8). The recursion relation forces all of the even coefficients a_2, $a_4, a_6, \ldots$ to be zero:

$$a_2 = \frac{a_0}{2^2 - 1} \quad \text{so } a_2 = 0, \qquad \text{and then} \qquad a_4 = \frac{a_2}{4^2 - 1} = 0 \text{ so } a_4 = 0, \qquad \text{and so on}$$

As for the odd coefficients, a_1 may be chosen arbitrarily. Because $F'(0) = a_1$, we set $a_1 = 1$ to obtain a solution $y = F(x)$ satisfying $F'(0) = 1$. Now, apply Eq. (9):

$$n = 3: \qquad a_3 = -\frac{a_1}{3^2 - 1} = -\frac{1}{3^2 - 1}$$

$$n = 5: \qquad a_5 = -\frac{a_3}{5^2 - 1} = \frac{1}{(5^2 - 1)(3^2 - 1)}$$

$$n = 7: \qquad a_7 = -\frac{a_5}{7^2 - 1} = -\frac{1}{(7^2 - 1)(3^2 - 1)(5^2 - 1)}$$

This shows the general pattern of coefficients. To express the coefficients in a compact form, let $n = 2k + 1$. Then the denominator in the recursion relation (9) can be written

$$n^2 - 1 = (2k + 1)^2 - 1 = 4k^2 + 4k = 4k(k + 1)$$

and

$$a_{2k+1} = -\frac{a_{2k-1}}{4k(k + 1)}$$

Applying this recursion relation k times, we obtain the closed formula

$$a_{2k+1} = (-1)^k \left(\frac{1}{4k(k + 1)}\right)\left(\frac{1}{4(k - 1)k}\right) \cdots \left(\frac{1}{4(1)(2)}\right) = \frac{(-1)^k}{4^k \, k! \, (k + 1)!}$$

Thus, we obtain a power series representation of our solution:

$$F(x) = \sum_{k=0}^{\infty} \frac{(-1)^k}{4^k k!(k + 1)!} x^{2k+1}$$

A straightforward application of the Ratio Test shows that F has an infinite radius of convergence. Therefore, $F(x)$ is a solution of the Initial Value Problem for all x. ∎

10.6 SUMMARY

- A *power series* is an infinite series of the form

$$F(x) = \sum_{n=0}^{\infty} a_n(x - c)^n$$

The constant c is called the *center* of $F(x)$.

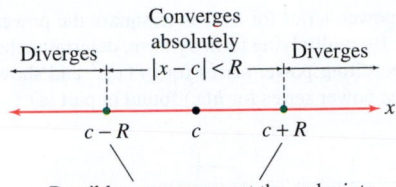

FIGURE 5 Interval of convergence of a power series.

- Every power series $F(x)$ has a *radius of convergence R* (Figure 5) such that
  - $F(x)$ converges absolutely for $|x - c| < R$ and diverges for $|x - c| > R$.
  - $F(x)$ may converge or diverge at the endpoints $c - R$ and $c + R$.

We set $R = 0$ if $F(x)$ converges only for $x = c$ and $R = \infty$ if $F(x)$ converges for all x.

- The *interval of convergence* of F consists of the open interval $(c - R, c + R)$ and possibly one or both endpoints $c - R$ and $c + R$.
- In many cases, the Ratio Test can be used to find the radius of convergence R. It is necessary to check convergence at the endpoints separately.
- If $R > 0$, then F is differentiable and has antiderivatives on $(c - R, c + R)$. The derivative and antiderivatives can be obtained by directly differentiating and anti-differentiating, respectively, the power series for F:

$$F'(x) = \sum_{n=1}^{\infty} na_n(x - c)^{n-1}, \qquad \int F(x)\,dx = A + \sum_{n=0}^{\infty} \frac{a_n}{n+1}(x - c)^{n+1}$$

(A is any constant.) These two power series have the same radius of convergence R.

- The expansion $\dfrac{1}{1 - x} = \displaystyle\sum_{n=0}^{\infty} x^n$ is valid for $|x| < 1$. It can be used to derive expansions of other related functions by substitution, integration, or differentiation.

10.6 EXERCISES

Preliminary Questions

1. Suppose that $\sum a_n x^n$ converges for $x = 5$. Must it also converge for $x = 4$? What about $x = -3$?

2. Suppose that $\sum a_n(x - 6)^n$ converges for $x = 10$. At which of the points (a)–(d) must it also converge?
   (a) $x = 8$ (b) $x = 11$ (c) $x = 3$ (d) $x = 0$

3. What is the radius of convergence of $F(3x)$ if $F(x)$ is a power series with radius of convergence $R = 12$?

4. The power series $F(x) = \displaystyle\sum_{n=1}^{\infty} nx^n$ has radius of convergence $R = 1$. What is the power series expansion of $F'(x)$ and what is its radius of convergence?

Exercises

1. Use the Ratio Test to determine the radius of convergence R of $\displaystyle\sum_{n=0}^{\infty} \frac{x^n}{2^n}$. Does it converge at the endpoints $x = \pm R$?

2. Use the Ratio Test to show that $\displaystyle\sum_{n=1}^{\infty} \frac{x^n}{\sqrt{n}2^n}$ has radius of convergence $R = 2$. Then determine whether it converges at the endpoints $R = \pm 2$.

3. Show that the power series (a)–(c) have the same radius of convergence. Then show that (a) diverges at both endpoints, (b) converges at one endpoint but diverges at the other, and (c) converges at both endpoints.
   (a) $\displaystyle\sum_{n=1}^{\infty} \frac{x^n}{3^n}$ (b) $\displaystyle\sum_{n=1}^{\infty} \frac{x^n}{n3^n}$ (c) $\displaystyle\sum_{n=1}^{\infty} \frac{x^n}{n^2 3^n}$

4. Repeat Exercise 3 for the following series:
   (a) $\displaystyle\sum_{n=1}^{\infty} \frac{(x-5)^n}{9^n}$ (b) $\displaystyle\sum_{n=1}^{\infty} \frac{(x-5)^n}{n9^n}$ (c) $\displaystyle\sum_{n=1}^{\infty} \frac{(x-5)^n}{n^2 9^n}$

5. Show that $\displaystyle\sum_{n=0}^{\infty} n^n x^n$ diverges for all $x \neq 0$.

6. For which values of x does $\displaystyle\sum_{n=0}^{\infty} n!\,x^n$ converge?

7. Use the Ratio Test to show that $\displaystyle\sum_{n=0}^{\infty} \frac{x^{2n}}{3^n}$ has radius of convergence $R = \sqrt{3}$.

8. Show that $\displaystyle\sum_{n=0}^{\infty} \frac{x^{3n+1}}{64^n}$ has radius of convergence $R = 4$.

In Exercises 9–34, find the interval of convergence.

9. $\displaystyle\sum_{n=0}^{\infty} nx^n$

10. $\displaystyle\sum_{n=1}^{\infty} \frac{2^n}{n}x^n$

11. $\displaystyle\sum_{n=1}^{\infty} (-1)^n \frac{x^{2n+1}}{2^n n}$

12. $\displaystyle\sum_{n=0}^{\infty} (-1)^n \frac{n}{4^n} x^{2n}$

13. $\displaystyle\sum_{n=4}^{\infty} \frac{x^n}{n^5}$

14. $\displaystyle\sum_{n=8}^{\infty} n^7 x^n$

15. $\displaystyle\sum_{n=0}^{\infty} \frac{x^n}{(n!)^2}$

16. $\displaystyle\sum_{n=0}^{\infty} \frac{8^n}{n!} x^n$

17. $\displaystyle\sum_{n=0}^{\infty} \frac{(2n)!}{(n!)^3} x^n$

18. $\displaystyle\sum_{n=0}^{\infty} \frac{4^n}{(2n+1)!} x^{2n-1}$

19. $\displaystyle\sum_{n=0}^{\infty} \frac{(-1)^n x^n}{\sqrt{n^2+1}}$

20. $\displaystyle\sum_{n=0}^{\infty} \frac{x^n}{n^4+2}$

21. $\displaystyle\sum_{n=15}^{\infty} \frac{x^{2n+1}}{3n+1}$

22. $\displaystyle\sum_{n=9}^{\infty} \frac{x^n}{n-4\ln n}$

23. $\displaystyle\sum_{n=2}^{\infty} \frac{x^n}{\ln n}$

24. $\displaystyle\sum_{n=2}^{\infty} \frac{x^{3n+2}}{\ln n}$

25. $\displaystyle\sum_{n=1}^{\infty} n(x-3)^n$

26. $\displaystyle\sum_{n=1}^{\infty} \frac{(-5)^n (x-3)^n}{n^2}$

27. $\displaystyle\sum_{n=1}^{\infty} (-1)^n n^5 (x-7)^n$

28. $\displaystyle\sum_{n=0}^{\infty} 27^n (x-1)^{3n+2}$

29. $\displaystyle\sum_{n=1}^{\infty} \frac{2^n}{3n}(x+3)^n$

30. $\displaystyle\sum_{n=0}^{\infty} \frac{(x-4)^n}{n!}$

31. $\displaystyle\sum_{n=0}^{\infty} \frac{(-5)^n}{n!}(x+10)^n$

32. $\displaystyle\sum_{n=10}^{\infty} n!\,(x+5)^n$

33. $\displaystyle\sum_{n=12}^{\infty} e^n (x-2)^n$

34. $\displaystyle\sum_{n=2}^{\infty} \frac{(x+4)^n}{(n\ln n)^2}$

In Exercises 35–40, use Eq. (2) to expand the function in a power series with center $c = 0$ and determine the interval of convergence.

35. $f(x) = \dfrac{1}{1-3x}$

36. $f(x) = \dfrac{1}{1+3x}$

37. $f(x) = \dfrac{1}{3-x}$

38. $f(x) = \dfrac{1}{4+3x}$

39. $f(x) = \dfrac{1}{1-x^3}$

40. $f(x) = \dfrac{1}{1-x^4}$

41. Differentiate the power series in Exercise 39 to obtain a power series for $g(x) = \dfrac{3x^2}{(1-x^3)^2}$.

42. Differentiate the power series in Exercise 40 to obtain a power series for $g(x) = \dfrac{4x^3}{(1-x^4)^2}$.

43. **(a)** Divide the power series in Exercise 41 by $3x^2$ to obtain a power series for $h(x) = \dfrac{1}{(1-x^3)^2}$ and use the Ratio Test to show that the radius of convergence is 1.

(b) Another way to obtain a power series for $h(x)$ is to square the power series for $f(x)$ in Exercise 39. By multiplying term by term, determine the terms up to degree 9 in the resulting power series for $(f(x))^2$ and show that they match the terms in the power series for $h(x)$ found in part (a).

44. **(a)** Divide the power series in Exercise 42 by $4x^3$ to obtain a power series for $h(x) = \dfrac{1}{(1-x^4)^2}$ and use the Ratio Test to show that the radius of convergence is 1.

(b) Another way to obtain a power series for $h(x)$ is to square the power series for $f(x)$ in Exercise 40. By multiplying term by term, determine the terms up to degree 12 in the resulting power series for $(f(x))^2$ and show that they match the terms in the power series for $h(x)$ found in part (a).

45. Use the equalities
$$\frac{1}{1-x} = \frac{1}{-3-(x-4)} = \frac{-\frac{1}{3}}{1+\left(\frac{x-4}{3}\right)}$$
to show that for $|x-4| < 3$,
$$\frac{1}{1-x} = \sum_{n=0}^{\infty} (-1)^{n+1} \frac{(x-4)^n}{3^{n+1}}$$

46. Use the method of Exercise 45 to expand $1/(1-x)$ in power series with centers $c = 2$ and $c = -2$. Determine the interval of convergence for each.

47. Use the method of Exercise 45 to expand $1/(4-x)$ in a power series with center $c = 5$. Determine the interval of convergence.

48. Find a power series that converges only for x in $[2, 6)$.

49. Apply integration to the expansion
$$\frac{1}{1+x} = \sum_{n=0}^{\infty} (-1)^n x^n = 1 - x + x^2 - x^3 + \cdots$$
to prove that for $-1 < x < 1$,
$$\ln(1+x) = \sum_{n=1}^{\infty} \frac{(-1)^{n-1} x^n}{n} = x - \frac{x^2}{2} + \frac{x^3}{3} - \frac{x^4}{4} + \cdots$$

50. Use the result of Exercise 49 to prove that
$$\ln\frac{3}{2} = \frac{1}{2} - \frac{1}{2\cdot 2^2} + \frac{1}{3\cdot 2^3} - \frac{1}{4\cdot 2^4} + \cdots$$
Use the fact that this is an alternating series to find an N such that the partial sum S_N approximates $\ln\frac{3}{2}$ to within an error of at most 10^{-3}. Confirm by using a calculator to compute both S_N and $\ln\frac{3}{2}$.

51. Let $F(x) = (x+1)\ln(1+x) - x$.
(a) Apply integration to the result of Exercise 49 to prove that the following power series holds for $F(x)$ for $-1 < x < 1$,
$$F(x) = \sum_{n=1}^{\infty} (-1)^{n+1} \frac{x^{n+1}}{n(n+1)}$$

(b) Evaluate at $x = \frac{1}{2}$ to prove
$$\frac{3}{2}\ln\frac{3}{2} - \frac{1}{2} = \frac{1}{1\cdot 2\cdot 2^2} - \frac{1}{2\cdot 3\cdot 2^3} + \frac{1}{3\cdot 4\cdot 2^4} - \frac{1}{4\cdot 5\cdot 2^5} + \cdots$$

(c) Use a calculator to verify that the partial sum S_4 approximates the left-hand side with an error no greater than the term a_5 of the series.

52. Prove that for $|x| < 1$,
$$\int \frac{dx}{x^4+1} = A + x - \frac{x^5}{5} + \frac{x^9}{9} - \cdots$$
Use the first two terms to approximate $\displaystyle\int_0^{1/2} dx/(x^4+1)$ numerically. Use the fact that you have an alternating series to show that the error in this approximation is at most 0.00022.

53. Use the result of Example 7 to show that
$$F(x) = \frac{x^2}{1\cdot 2} - \frac{x^4}{3\cdot 4} + \frac{x^6}{5\cdot 6} - \frac{x^8}{7\cdot 8} + \cdots$$
is an antiderivative of $f(x) = \tan^{-1} x$ satisfying $F(0) = 0$. What is the radius of convergence of this power series?

54. Verify that function $F(x) = x \tan^{-1} x - \frac{1}{2} \ln(x^2 + 1)$ is an antiderivative of $f(x) = \tan^{-1} x$ satisfying $F(0) = 0$. Then use the result of Exercise 53 with $x = \frac{1}{\sqrt{3}}$ to show that

$$\frac{\pi}{6\sqrt{3}} - \frac{1}{2}\ln\frac{4}{3} = \frac{1}{1\cdot 2(3)} - \frac{1}{3\cdot 4(3^2)} + \frac{1}{5\cdot 6(3^3)} - \frac{1}{7\cdot 8(3^4)} + \cdots$$

Use a calculator to compare the value of the left-hand side with the partial sum S_4 of the series on the right.

55. Evaluate $\displaystyle\sum_{n=1}^{\infty} \frac{n}{2^n}$. *Hint:* Use differentiation to show that

$$(1-x)^{-2} = \sum_{n=1}^{\infty} nx^{n-1} \quad (\text{for } |x| < 1)$$

56. Use the power series for $(1 + x^2)^{-1}$ and differentiation to prove that for $|x| < 1$,

$$\frac{2x}{(x^2+1)^2} = \sum_{n=1}^{\infty} (-1)^{n-1}(2n)x^{2n-1}$$

57. Show that the following series converges absolutely for $|x| < 1$ and compute its sum:

$$F(x) = 1 - x - x^2 + x^3 - x^4 - x^5 + x^6 - x^7 - x^8 + \cdots$$

Hint: Write $F(x)$ as a sum of three geometric series with common ratio x^3.

58. Show that for $|x| < 1$,

$$\frac{1+2x}{1+x+x^2} = 1 + x - 2x^2 + x^3 + x^4 - 2x^5 + x^6 + x^7 - 2x^8 + \cdots$$

Hint: Use the hint from Exercise 57.

59. Find all values of x such that $\displaystyle\sum_{n=1}^{\infty} \frac{x^{n^2}}{n!}$ converges.

60. Find all values of x such that the following series converges:

$$F(x) = 1 + 3x + x^2 + 27x^3 + x^4 + 243x^5 + \cdots$$

61. Find a power series $P(x) = \displaystyle\sum_{n=0}^{\infty} a_n x^n$ satisfying the differential equation $y' = -y$ with initial condition $y(0) = 1$. Then use Eq. (8) in Section 9.1 to conclude that $P(x) = e^{-x}$.

62. Let $C(x) = 1 - \dfrac{x^2}{2!} + \dfrac{x^4}{4!} - \dfrac{x^6}{6!} + \cdots$.

(a) Show that $C(x)$ has an infinite radius of convergence.

(b) Prove that $C(x)$ and $f(x) = \cos x$ are both solutions of $y'' = -y$ with initial conditions $y(0) = 1$, $y'(0) = 0$. [This Initial Value Problem has a unique solution, so it follows that $C(x) = \cos x$ for all x.]

63. Use the power series for $y = e^x$ to show that

$$\frac{1}{e} = \frac{1}{2!} - \frac{1}{3!} + \frac{1}{4!} - \cdots$$

Use the fact that this is an alternating series to find an N such that the partial sum S_N approximates e^{-1} to within an error of at most 10^{-3}. Confirm this using a calculator to compute both S_N and e^{-1}.

64. Let $P(x) = \displaystyle\sum_{n=0}^{\infty} a_n x^n$ be a power series solution to $y' = 2xy$ with initial condition $y(0) = 1$.

(a) Show that the odd coefficients a_{2k+1} are all zero.

(b) Prove that $a_{2k} = a_{2k-2}/k$ and use this result to determine the coefficients a_{2k}.

65. Find a power series $P(x)$ satisfying the differential equation

$$y'' - xy' + y = 0 \qquad \boxed{10}$$

with initial condition $y(0) = 1$, $y'(0) = 0$. What is the radius of convergence of the power series?

66. Find a power series satisfying Eq. (10) with initial condition $y(0) = 0$, $y'(0) = 1$.

67. Prove that

$$J_2(x) = \sum_{k=0}^{\infty} \frac{(-1)^k}{2^{2k+2}\, k!\, (k+3)!} x^{2k+2}$$

is a solution of the Bessel differential equation of order 2:

$$x^2 y'' + xy' + (x^2 - 4)y = 0$$

68. ✏ Why is it impossible to expand $f(x) = |x|$ as a power series that converges in an interval around $x = 0$? Explain using Theorem 2.

Further Insights and Challenges

69. Suppose that the coefficients of $F(x) = \displaystyle\sum_{n=0}^{\infty} a_n x^n$ are *periodic*; that is, for some whole number $M > 0$, we have $a_{M+n} = a_n$. Prove that $F(x)$ converges absolutely for $|x| < 1$ and that

$$F(x) = \frac{a_0 + a_1 x + \cdots + a_{M-1} x^{M-1}}{1 - x^M}$$

Hint: Use the hint for Exercise 57.

70. Continuity of Power Series Let $F(x) = \displaystyle\sum_{n=0}^{\infty} a_n x^n$ be a power series with radius of convergence $R > 0$.

(a) Prove the inequality

$$|x^n - y^n| \le n|x-y|(|x|^{n-1} + |y|^{n-1}) \qquad \boxed{11}$$

Hint: $x^n - y^n = (x-y)(x^{n-1} + x^{n-2}y + \cdots + y^{n-1})$.

(b) Choose R_1 with $0 < R_1 < R$. Show that the infinite series

$$M = \sum_{n=0}^{\infty} 2n|a_n| R_1^n \text{ converges.}$$ *Hint:* Show that $n|a_n| R_1^n < |a_n| x^n$ for all n sufficiently large if $R_1 < x < R$.

(c) Use the inequality in (11) to show that if $|x| < R_1$ and $|y| < R_1$, then $|F(x) - F(y)| \le M|x - y|$.

(d) Prove that if $|x| < R$, then F is continuous at x. *Hint:* Choose R_1 such that $|x| < R_1 < R$. Show that if $\epsilon > 0$ is given, then $|F(x) - F(y)| \le \epsilon$ for all y such that $|x - y| < \delta$, where δ is any positive number that is less than ϵ/M and $R_1 - |x|$ (see Figure 6).

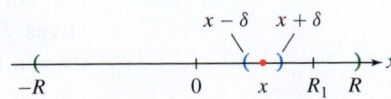

FIGURE 6 If $x > 0$, choose $\delta > 0$ less than ϵ/M and $R_1 - x$.

10.7 Taylor Polynomials

Using power series, we have seen how we can express some functions as polynomials of infinite degree. We saw that we can take power series for specific functions and manipulate them by substitution, differentiation, integration, and algebraic operations to obtain power series for other functions.

Next, we consider how we can obtain a power series for a specific given function. To do so, first we introduce Taylor polynomials, special polynomial functions that turn out to be partial sums of the power series of a function. The Taylor polynomials are important in their own right since they are useful tools for approximating functions. In the next section, we extend these Taylor polynomials to Taylor series representations of functions.

Many functions are difficult to work with. For instance, $f(x) = \sin(x^2)$ cannot be integrated using elementary functions. Nor can $f(x) = e^{-x^2}$. In fact, even simple functions like $f(x) = \sin x$, $f(x) = \cos x$, $f(x) = e^x$, and $f(x) = \ln x$ can only be evaluated exactly at relatively few values of x and otherwise they must be numerically approximated. On the other hand, polynomials such as $f(x) = 3x^4 - 7x^3 + 2x - 4$ can be easily differentiated and integrated. They can be evaluated at any value of x using just multiplication and addition. Thus, given a function, it is natural to ask if there is a way to accurately approximate the function using a polynomial function.

We have worked with a simple polynomial approximation of a function before. In Section 4.1, we used the linearization $L(x) = f(a) + f'(a)(x - a)$ to approximate $f(x)$ near a point $x = a$:

$$f(x) \approx f(a) + f'(a)(x - a)$$

We refer to $L(x)$ as a "first-order" approximation to $f(x)$ at $x = a$ because $f(x)$ and $L(x)$ have the same value and the same first derivative at $x = a$ (Figure 1):

$$L(a) = f(a), \qquad L'(a) = f'(a)$$

A first-order approximation is useful only in a small interval around $x = a$. In this section, we achieve greater accuracy over larger intervals using higher-order approximations (Figure 2). These higher-order approximations will simply be polynomials with higher powers, the Taylor polynomials. Along with using Taylor polynomials to approximate functions, we will develop tools for estimating the error in the approximation.

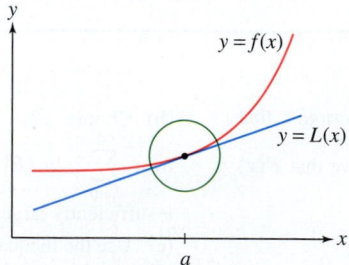

FIGURE 1 The linear approximation $L(x)$ is a first-order approximation to f.

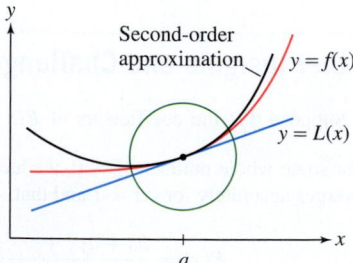

FIGURE 2 A second-order approximation is more accurate over a larger interval.

In what follows, assume that f is defined on an open interval I and that all derivatives $f^{(k)}$ exist on I. Let $a \in I$. We say that two functions f and g **agree to order n** at $x = a$ if their derivatives up to order n at $x = a$ are equal:

$$f(a) = g(a), \quad f'(a) = g'(a), \quad f''(a) = g''(a), \quad \ldots, \quad f^{(n)}(a) = g^{(n)}(a)$$

We also say that g **approximates f to order n** at $x = a$.

Define the nth **Taylor polynomial T_n of f centered at $x = a$** as follows:

$$T_n(x) = f(a) + \frac{f'(a)}{1!}(x-a) + \frac{f''(a)}{2!}(x-a)^2 + \cdots + \frac{f^{(n)}(a)}{n!}(x-a)^n$$

◄ REMINDER *k-factorial is the number*
$k! = k(k-1)(k-2)\cdots(2)(1)$. Thus,

$$1! = 1, \quad 2! = (2)1 = 2$$

$$3! = (3)(2)1 = 6$$

By convention, we define $0! = 1$.

The first few Taylor polynomials are

$$T_0(x) = f(a)$$

$$T_1(x) = f(a) + f'(a)(x-a)$$

$$T_2(x) = f(a) + f'(a)(x-a) + \frac{1}{2}f''(a)(x-a)^2$$

$$T_3(x) = f(a) + f'(a)(x-a) + \frac{1}{2}f''(a)(x-a)^2 + \frac{1}{6}f'''(a)(x-a)^3$$

Note that T_0 is a constant function equal to the value of f at a, and that T_1 is the linearization of f at a. Note also that T_n is obtained from T_{n-1} by adding on a term of degree n:

$$T_n(x) = T_{n-1}(x) + \frac{f^{(n)}(a)}{n!}(x-a)^n$$

The next theorem justifies our definition of T_n.

THEOREM 1 The polynomial T_n centered at a agrees with f to order n at $x = a$, and it is the only polynomial of degree at most n with this property.

The verification of Theorem 1 is left to the exercises (Exercises 76–77), but we'll illustrate the idea by checking that T_2 agrees with f to order $n = 2$:

$$T_2(x) = f(a) + f'(a)(x-a) + \frac{1}{2}f''(a)(x-a)^2, \quad T_2(a) = f(a)$$

$$T_2'(x) = f'(a) + f''(a)(x-a), \qquad\qquad\qquad T_2'(a) = f'(a)$$

$$T_2''(x) = f''(a), \qquad\qquad\qquad\qquad\qquad T_2''(a) = f''(a)$$

This shows that the value and the derivatives of order up to $n = 2$ at $x = a$ are equal. Before proceeding to the examples, we write T_n in summation notation:

$$T_n(x) = \sum_{j=0}^{n} \frac{f^{(j)}(a)}{j!}(x-a)^j$$

By convention, we regard f as the *zeroth* derivative, and thus $f^{(0)}$ is f itself. When $a = 0$, T_n is also called the nth **Maclaurin polynomial**.

EXAMPLE 1 **Maclaurin Polynomials for** $f(x) = e^x$ Plot the third and fourth Maclaurin polynomials for $f(x) = e^x$. Compare with the linear approximation.

Solution All higher derivatives coincide with f itself: $f^{(k)}(x) = e^x$. Therefore,

$$f(0) = f'(0) = f''(0) = f'''(0) = f^{(4)}(0) = e^0 = 1$$

The third Maclaurin polynomial (the case $a = 0$) is

$$T_3(x) = f(0) + f'(0)x + \frac{1}{2}f''(0)x^2 + \frac{1}{3!}f'''(0)x^3 = 1 + x + \frac{1}{2}x^2 + \frac{1}{6}x^3$$

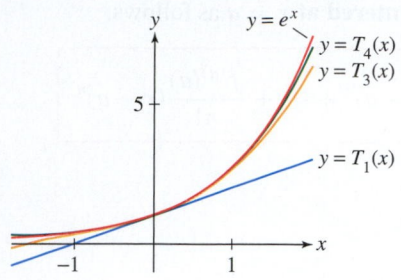

y

$y = e^x$

$y = T_4(x)$

$y = T_3(x)$

5

$y = T_1(x)$

x

−1   1

FIGURE 3 Maclaurin polynomials for $f(x) = e^x$.

We obtain $T_4(x)$ by adding the term of degree 4 to $T_3(x)$:

$$T_4(x) = T_3(x) + \frac{1}{4!}f^{(4)}(0)x^4 = 1 + x + \frac{1}{2}x^2 + \frac{1}{6}x^3 + \frac{1}{24}x^4$$

Figure 3 shows that T_3 and T_4 approximate $f(x) = e^x$ much more closely than the linear approximation T_1 on an interval around $a = 0$. Higher-degree Maclaurin polynomials would provide even better approximations on larger intervals.   ■

EXAMPLE 2   For objects near the surface of the earth, to two decimal places the acceleration due to gravity is $g = 9.81$ m/s^2. For objects at higher altitudes, Newton's Law of Gravitation says that the acceleration due to gravity is

$$G(h) = \frac{g}{(1 + \frac{h}{6370})^2}$$

where $G(h)$ is in m/s^2, h is the altitude above the surface of the earth in km, and 6370 is the radius of the earth in km.

(a) Find a power series representation of G as a function of h. For what values of h is the power series valid?

(b) Use the third Maclaurin polynomial to approximate the acceleration due to gravity on an object at an altitude of 1000 km, and estimate the error in the approximation.

Solution

(a) We use the power series representation for $\frac{1}{(1-x)^2}$ from Example 6 in the previous section. Substituting $\frac{-h}{6370}$ for x, we obtain

$$G(h) = g\sum_{n=0}^{\infty}\frac{(n+1)(-h)^n}{6370^n} = 9.81 - \frac{19.6h}{6370} + \frac{29.4h^2}{6370^2} - \frac{39.2h^3}{6370^3} + \cdots$$

The power series for $\frac{1}{(1-x)^2}$ is valid for $-1 < x < 1$, and therefore, the series for $G(h)$ holds for $-1 < -\frac{h}{6370} < 1$. Since altitude is nonnegative, it follows that the power series for $G(h)$ is valid for $0 \le h < 6370$.

(b) Using the third Maclaurin polynomial,

$$G(1000) \approx 9.81 - \frac{(19.6)(1000)}{6370} + \frac{29.4(1000^2)}{6370^2} - \frac{39.2(1000^3)}{6370^3} \approx 7.30 \text{ m/s}^2$$

Since we have an alternating series, we can apply the corollary to Theorem 2 in Section 10.4 and use the fourth power term in the series to estimate the error in our approximation. Thus,

$$\text{error} \le \frac{49.0(1000^4)}{6370^4} \approx 0.03$$ ■

EXAMPLE 3   Computing Taylor Polynomials   Compute the Taylor polynomial T_4 centered at $a = 3$ for $f(x) = \sqrt{x+1}$.

Solution   First, evaluate the derivatives up to degree 4 at $a = 3$:

$$f(x) = (x+1)^{1/2}, \qquad\qquad f(3) = 2$$

$$f'(x) = \frac{1}{2}(x+1)^{-1/2}, \qquad\qquad f'(3) = \frac{1}{4}$$

$$f''(x) = -\frac{1}{4}(x+1)^{-3/2}, \qquad\qquad f''(3) = -\frac{1}{32}$$

$$f'''(x) = \frac{3}{8}(x+1)^{-5/2}, \qquad\qquad f'''(3) = \frac{3}{256}$$

$$f^{(4)}(x) = -\frac{15}{16}(x+1)^{-7/2}, \qquad\qquad f^{(4)}(3) = -\frac{15}{2048}$$

Then compute the coefficients $\dfrac{f^{(j)}(3)}{j!}$:

$$\text{constant term} = f(3) = 2$$

$$\text{coefficient of } (x-3) = f'(3) = \frac{1}{4}$$

$$\text{coefficient of } (x-3)^2 = \frac{f''(3)}{2!} = -\frac{1}{32} \cdot \frac{1}{2!} = -\frac{1}{64}$$

$$\text{coefficient of } (x-3)^3 = \frac{f'''(3)}{3!} = \frac{3}{256} \cdot \frac{1}{3!} = \frac{1}{512}$$

$$\text{coefficient of } (x-3)^4 = \frac{f^{(4)}(3)}{4!} = -\frac{15}{2048} \cdot \frac{1}{4!} = -\frac{5}{16,384}$$

The Taylor polynomial T_4 centered at $a = 3$ is (see Figure 4)

$$T_4(x) = 2 + \frac{1}{4}(x-3) - \frac{1}{64}(x-3)^2 + \frac{1}{512}(x-3)^3 - \frac{5}{16,384}(x-3)^4 \qquad \blacksquare$$

The first term $f(a)$ in the Taylor polynomial T_n is called the constant term.

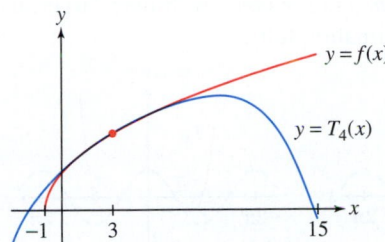

FIGURE 4 Graphs of $f(x) = \sqrt{x+1}$ and T_4 centered at $x = 3$.

EXAMPLE 4 **Finding a General Formula for** T_n Find the Taylor polynomials T_n of $f(x) = \ln x$ centered at $a = 1$.

Solution For $f(x) = \ln x$, the constant term of T_n at $a = 1$ is zero because $f(1) = \ln 1 = 0$. Next, we compute the derivatives:

$$f'(x) = x^{-1}, \qquad f''(x) = -x^{-2}, \qquad f'''(x) = 2x^{-3}, \qquad f^{(4)}(x) = -3 \cdot 2x^{-4}$$

After computing several derivatives of $f(x) = \ln x$, we begin to discern the pattern. For many functions of interest, however, the derivatives follow no simple pattern and there is no convenient formula for the general Taylor polynomial.

Similarly, $f^{(5)}(x) = 4 \cdot 3 \cdot 2x^{-5}$. The general pattern is that $f^{(k)}(x)$ is a multiple of x^{-k}, with a coefficient $\pm(k-1)!$ that alternates in sign:

$$f^{(k)}(x) = (-1)^{k-1}(k-1)! \, x^{-k} \qquad \boxed{1}$$

The coefficient of $(x-1)^k$ in T_n is

$$\frac{f^{(k)}(1)}{k!} = \frac{(-1)^{k-1}(k-1)!}{k!} = \frac{(-1)^{k-1}}{k} \qquad (\text{for } k \geq 1)$$

Taylor polynomials for $\ln x$ at $a = 1$:

$$T_1(x) = (x-1)$$

$$T_2(x) = (x-1) - \frac{1}{2}(x-1)^2$$

$$T_3(x) = (x-1) - \frac{1}{2}(x-1)^2 + \frac{1}{3}(x-1)^3$$

Thus, the coefficients for $k \geq 1$ form a sequence $1, -\frac{1}{2}, \frac{1}{3}, -\frac{1}{4}, \ldots,$ and

$$T_n(x) = (x-1) - \frac{1}{2}(x-1)^2 + \frac{1}{3}(x-1)^3 - \cdots + (-1)^{n-1}\frac{1}{n}(x-1)^n \qquad \blacksquare$$

EXAMPLE 5 **Cosine** Find the Maclaurin polynomials of $f(x) = \cos x$.

Solution The derivatives form a repeating pattern of period 4:

$$f(x) = \cos x, \qquad f'(x) = -\sin x, \qquad f''(x) = -\cos x, \qquad f'''(x) = \sin x,$$

$$f^{(4)}(x) = \cos x, \qquad f^{(5)}(x) = -\sin x, \qquad \cdots$$

In general, $f^{(j+4)}(x) = f^{(j)}(x)$. The derivatives at $x = 0$ also form a pattern:

$f(0)$	$f'(0)$	$f''(0)$	$f'''(0)$	$f^{(4)}(0)$	$f^{(5)}(0)$	$f^{(6)}(0)$	$f^{(7)}(0)$
1	0	−1	0	1	0	−1	0

Scottish mathematician Colin Maclaurin (1698–1746) was a professor in Edinburgh. Newton was so impressed by his work that he once offered to pay part of Maclaurin's salary.

Therefore, the coefficients of the odd powers x^{2k+1} are zero, and the coefficients of the even powers x^{2k} alternate in sign with value $(-1)^k/(2k)!$:

$$T_0(x) = T_1(x) = 1, \qquad T_2(x) = T_3(x) = 1 - \frac{1}{2!}x^2$$

$$T_4(x) = T_5(x) = 1 - \frac{x^2}{2} + \frac{x^4}{4!}$$

$$T_{2n}(x) = T_{2n+1}(x) = 1 - \frac{1}{2}x^2 + \frac{1}{4!}x^4 - \frac{1}{6!}x^6 + \cdots + (-1)^n \frac{1}{(2n)!}x^{2n}$$

Figure 5 shows that as n increases, T_n approximates $f(x) = \cos x$ well over larger and larger intervals, but outside this interval, the approximation fails.   ∎

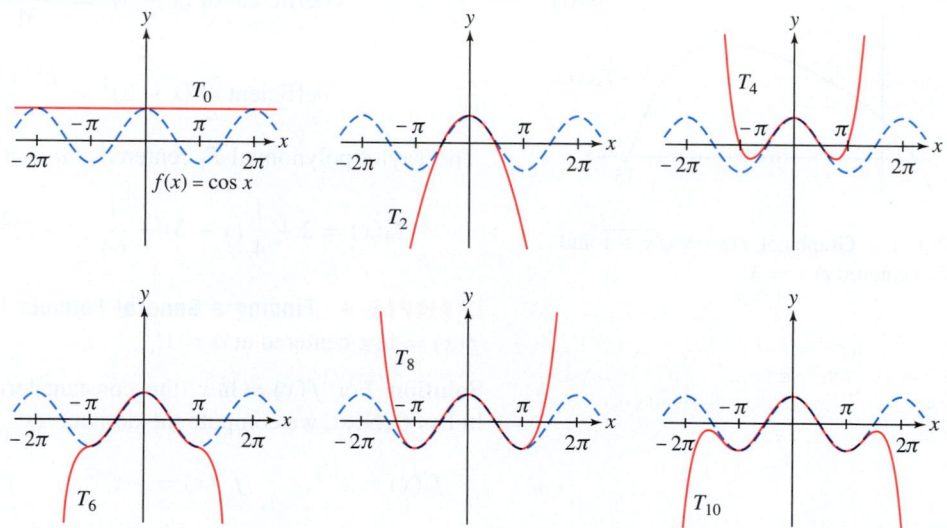

DF FIGURE 5 Maclaurin polynomials for $f(x) = \cos x$. The graph of f is shown as a dashed curve.

The Error Bound

To use Taylor polynomials effectively to approximate a function, we need a way to estimate the size of the error in the approximation. This is provided by the next theorem, which shows that when approximating f with T_n, the size of this error depends on the size of the $(n+1)$st derivative.

THEOREM 2 Error Bound   Assume that $f^{(n+1)}$ exists and is continuous. Let K be a number such that $|f^{(n+1)}(u)| \le K$ for all u between a and x. Then

$$\boxed{|f(x) - T_n(x)| \le K \frac{|x-a|^{n+1}}{(n+1)!}}$$

where T_n is the nth Taylor polynomial centered at $x = a$.

A proof of Theorem 2 is presented at the end of this section.

EXAMPLE 6   Using the Error Bound   Apply the Error Bound to

$$|\ln 1.2 - T_3(1.2)|$$

where T_3 is the third Taylor polynomial for $f(x) = \ln x$ at $a = 1$. Check your result with a calculator.

Solution

Step 1. Find a value of K.

To use the Error Bound with $n = 3$, we must find a value of K such that $|f^{(4)}(u)| \le K$ for all u between $a = 1$ and $x = 1.2$. As we computed in Example 4, $f^{(4)}(x) = -6x^{-4}$. The absolute value $|f^{(4)}(x)|$ is decreasing for $x > 0$, so its maximum value on $[1, 1.2]$ is $|f^{(4)}(1)| = 6$. Therefore, we may take $K = 6$.

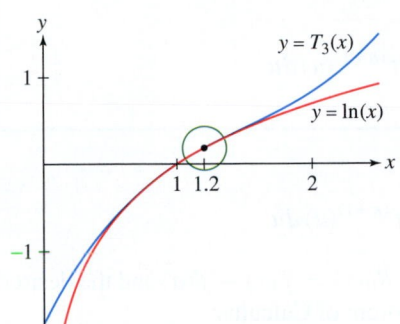

To use the Error Bound, it is not necessary to find the smallest possible value of K. In this example, we take $K = 1$. This works for all n, but for odd n we could have used the smaller value $K = \sin 0.2 \approx 0.2$.

Step 2. **Apply the Error Bound.**

$$| \ln 1.2 - T_3(1.2)| \le K \frac{|x - a|^{n+1}}{(n + 1)!} = 6 \frac{|1.2 - 1|^4}{4!} \approx 0.0004$$

Step 3. **Check the result.**

Recall from Example 4 that

$$T_3(x) = (x - 1) - \frac{1}{2}(x - 1)^2 + \frac{1}{3}(x - 1)^3$$

The following values from a calculator confirm that the error is at most 0.0004:

$$| \ln 1.2 - T_3(1.2)| \approx |0.182667 - 0.182322| \approx 0.00035 < 0.0004$$

Observe in Figure 6 that $y = \ln x$ and $y = T_3(x)$ are indistinguishable near $x = 1.2$. ∎

EXAMPLE 7 **Approximating with a Given Accuracy** Let T_n be the nth Maclaurin polynomial for $f(x) = \cos x$. Find a value of n such that

$$| \cos 0.2 - T_n(0.2)| < 10^{-5}$$

Solution

Step 1. **Find a value of K.**

Since $|f^{(n)}(x)|$ is $|\cos x|$ or $|\sin x|$, depending on whether n is even or odd, we have $|f^{(n)}(u)| \le 1$ for all u. Thus, we may apply the Error Bound with $K = 1$.

Step 2. **Find a value of n.**

The Error Bound gives us

$$| \cos 0.2 - T_n(0.2)| \le K \frac{|0.2 - 0|^{n+1}}{(n + 1)!} = \frac{|0.2|^{n+1}}{(n + 1)!}$$

To make the error less than 10^{-5}, we must choose n so that

$$\frac{|0.2|^{n+1}}{(n + 1)!} < 10^{-5}$$

It's not possible to solve this inequality for n, but we can find a suitable n by checking several values:

n	2	3	4
$\dfrac{\|0.2\|^{n+1}}{(n+1)!}$	$\dfrac{0.2^3}{3!} \approx 0.0013$	$\dfrac{0.2^4}{4!} \approx 6.67 \times 10^{-5}$	$\dfrac{0.2^5}{5!} \approx 2.67 \times 10^{-6} < 10^{-5}$

We see that the error is less than 10^{-5} for $n = 4$. ∎

CONCEPTUAL INSIGHT The term K in the Error Bound usually depends on n, the number of terms in the Taylor polynomial. However, in some instances, K can be chosen independent of n. For example, if $f(x) = \sin x$ or $f(x) = \cos x$, then we can let K equal 1 for all n (since the absolute value of all derivatives of these functions is no larger than 1). Because the $(n + 1)!$ term in the denominator of the Error Bound grows very rapidly and dominates the fraction, the error goes to 0 as n increases. Thus, for these functions, the more terms in the Taylor polynomial, the better the approximation. Therefore, if we include infinitely many terms, we can ask if the resulting series and f are equal. This naturally leads to the subject of the next section, Taylor Series.

The rest of this section is devoted to a proof of the Error Bound (Theorem 2). Define the *nth remainder*:

$$R_n(x) = f(x) - T_n(x)$$

The error in $T_n(x)$ is the absolute value $|R_n(x)|$. As a first step in proving the Error Bound, we show that $R_n(x)$ can be represented as an integral.

> **THEOREM 3 Taylor's Theorem** Assume that $f^{(n+1)}$ exists and is continuous. Then
>
> $$R_n(x) = \frac{1}{n!}\int_a^x (x-u)^n f^{(n+1)}(u)\,du \qquad \boxed{2}$$

Proof Set

$$I_n(x) = \frac{1}{n!}\int_a^x (x-u)^n f^{(n+1)}(u)\,du$$

Our goal is to show that $R_n(x) = I_n(x)$. For $n = 0$, $R_0(x) = f(x) - f(a)$ and the desired result is just a restatement of the Fundamental Theorem of Calculus:

$$I_0(x) = \int_a^x f'(u)\,du = f(x) - f(a) = R_0(x)$$

Exercise 70 reviews this proof for the special case $n = 2$.

To prove the formula for $n > 0$, we apply Integration by Parts to $I_n(x)$ with

$$h(u) = \frac{1}{n!}(x-u)^n, \qquad g(u) = f^{(n)}(u)$$

Then $g'(u) = f^{(n+1)}(u)$, and so

$$I_n(x) = \int_a^x h(u)\,g'(u)\,du = h(u)g(u)\Big|_a^x - \int_a^x h'(u)g(u)\,du$$

$$= \frac{1}{n!}(x-u)^n f^{(n)}(u)\Big|_a^x - \frac{1}{n!}\int_a^x (-n)(x-u)^{n-1} f^{(n)}(u)\,du$$

$$= -\frac{1}{n!}(x-a)^n f^{(n)}(a) + I_{n-1}(x)$$

This can be rewritten as

$$I_{n-1}(x) = \frac{f^{(n)}(a)}{n!}(x-a)^n + I_n(x)$$

Now, apply this relation n times, noting that $I_0(x) = f(x) - f(a)$:

$$f(x) = f(a) + I_0(x)$$

$$= f(a) + \frac{f'(a)}{1!}(x-a) + I_1(x)$$

$$= f(a) + \frac{f'(a)}{1!}(x-a) + \frac{f''(a)}{2!}(x-a)^2 + I_2(x)$$

$$\vdots$$

$$= f(a) + \frac{f'(a)}{1!}(x-a) + \cdots + \frac{f^{(n)}(a)}{n!}(x-a)^n + I_n(x)$$

This shows that $f(x) = T_n(x) + I_n(x)$ and hence $I_n(x) = R_n(x)$, as desired. ∎

Proof Now, we can prove Theorem 2. Assume first that $x \geq a$. Then

$$|f(x) - T_n(x)| = |R_n(x)| = \left|\frac{1}{n!}\int_a^x (x-u)^n f^{(n+1)}(u)\,du\right|$$

To establish the inequality in (3), we use the inequality

$$\left|\int_a^b f(x)\,dx\right| \leq \int_a^b |f(x)|\,dx$$

which is valid for all integrable functions.

$$\leq \frac{1}{n!}\int_a^x \left|(x-u)^n f^{(n+1)}(u)\right|\,du \qquad \boxed{3}$$

$$\leq \frac{K}{n!}\int_a^x |x-u|^n\,du \qquad \boxed{4}$$

$$= \frac{K}{n!}\frac{-(x-u)^{n+1}}{n+1}\Big|_{u=a}^x = K\frac{|x-a|^{n+1}}{(n+1)!}$$

Note that the absolute value is not needed in the inequality in (4) because $x - u \geq 0$ for $a \leq u \leq x$. If $x \leq a$, we must interchange the upper and lower limits of the integrals in (3) and (4). ∎

10.7 SUMMARY

- The nth *Taylor polynomial* centered at $x = a$ for the function f is

$$T_n(x) = f(a) + \frac{f'(a)}{1!}(x-a)^1 + \frac{f''(a)}{2!}(x-a)^2 + \cdots + \frac{f^{(n)}(a)}{n!}(x-a)^n$$

When $a = 0$, T_n is also called the nth *Maclaurin polynomial*.
- If $f^{(n+1)}$ exists and is continuous, then we have the *Error Bound*

$$|T_n(x) - f(x)| \leq K \frac{|x-a|^{n+1}}{(n+1)!}$$

where K is a number such that $|f^{(n+1)}(u)| \leq K$ for all u between a and x.
- For reference, we include a table of standard Maclaurin and Taylor polynomials.

$f(x)$	a	Maclaurin or Taylor Polynomial
e^x	0	$T_n(x) = 1 + x + \dfrac{x^2}{2!} + \dfrac{x^3}{3!} + \cdots + \dfrac{x^n}{n!}$
$\sin x$	0	$T_{2n+1}(x) = T_{2n+2}(x) = x - \dfrac{x^3}{3!} + \cdots + (-1)^n \dfrac{x^{2n+1}}{(2n+1)!}$
$\cos x$	0	$T_{2n}(x) = T_{2n+1}(x) = 1 - \dfrac{x^2}{2!} + \dfrac{x^4}{4!} - \cdots + (-1)^n \dfrac{x^{2n}}{(2n)!}$
$\ln x$	1	$T_n(x) = (x-1) - \dfrac{1}{2}(x-1)^2 + \cdots + \dfrac{(-1)^{n-1}}{n}(x-1)^n$
$\dfrac{1}{1-x}$	0	$T_n(x) = 1 + x + x^2 + \cdots + x^n$

10.7 EXERCISES

Preliminary Questions

1. What is T_3 centered at $a = 3$ for a function f such that $f(3) = 9$, $f'(3) = 8$, $f''(3) = 4$, and $f'''(3) = 12$?

2. The dashed graphs in Figure 7 are Taylor polynomials for a function f. Which of the two is a Maclaurin polynomial?

3. For which value of x does the Maclaurin polynomial T_n satisfy $T_n(x) = f(x)$, no matter what f is?

4. Let T_n be the Maclaurin polynomial of a function f satisfying $|f^{(4)}(x)| \leq 1$ for all x. Which of the following statements follow from the Error Bound?

(a) $|T_4(2) - f(2)| \leq \dfrac{2}{3}$

(b) $|T_3(2) - f(2)| \leq \dfrac{2}{3}$

(c) $|T_3(2) - f(2)| \leq \dfrac{1}{3}$

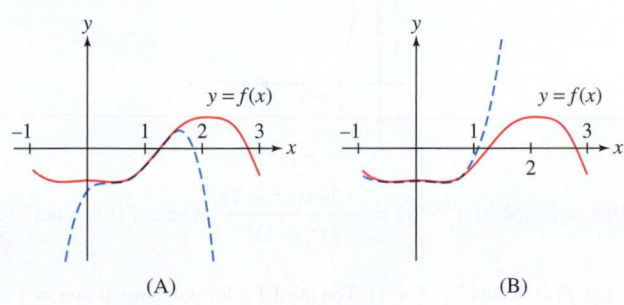

(A)   (B)

FIGURE 7

Exercises

In Exercises 1–16, calculate the Taylor polynomials T_2 and T_3 centered at $x = a$ for the given function and value of a.

1. $f(x) = \sin x$, $a = 0$

2. $f(x) = \sin x$, $a = \dfrac{\pi}{2}$

3. $f(x) = \dfrac{1}{1+x}$, $a = 2$

4. $f(x) = \dfrac{1}{1+x^2}$, $a = -1$

5. $f(x) = x^4 - 2x$, $a = 3$

6. $f(x) = \dfrac{x^2+1}{x+1}$, $a = -2$

7. $f(x) = \sqrt{x}$, $a = 1$

8. $f(x) = \sqrt{x}$, $a = 9$

9. $f(x) = \tan x$, $a = 0$

10. $f(x) = \tan x$, $a = \dfrac{\pi}{4}$

11. $f(x) = e^{-x} + e^{-2x}$, $a = 0$

12. $f(x) = e^{2x}$, $a = \ln 2$

13. $f(x) = x^2 e^{-x}$, $a = 1$

14. $f(x) = \cosh 2x$, $a = 0$

15. $f(x) = \dfrac{\ln x}{x}$, $a = 1$

16. $f(x) = \ln(x+1)$, $a = 0$

17. Show that the second Taylor polynomial for $f(x) = px^2 + qx + r$, centered at $a = 1$, is $f(x)$.

18. Show that the third Maclaurin polynomial for $f(x) = (x-3)^3$ is $f(x)$.

19. Show that the nth Maclaurin polynomial for $f(x) = e^x$ is

$$T_n(x) = 1 + \frac{x}{1!} + \frac{x^2}{2!} + \cdots + \frac{x^n}{n!}$$

20. Show that the nth Taylor polynomial for $f(x) = \dfrac{1}{x+1}$ at $a = 1$ is

$$T_n(x) = \frac{1}{2} - \frac{(x-1)}{4} + \frac{(x-1)^2}{8} + \cdots + (-1)^n \frac{(x-1)^n}{2^{n+1}}$$

21. Show that the Maclaurin polynomials for $f(x) = \sin x$ are

$$T_{2n+1}(x) = T_{2n+2}(x) = x - \frac{x^3}{3!} + \frac{x^5}{5!} - \cdots + (-1)^n \frac{x^{2n+1}}{(2n+1)!}$$

22. Show that the Maclaurin polynomials for $f(x) = \ln(1+x)$ are

$$T_n(x) = x - \frac{x^2}{2} + \frac{x^3}{3} + \cdots + (-1)^{n-1} \frac{x^n}{n}$$

In Exercises 23–30, find T_n centered at $x = a$ for all n.

23. $f(x) = \dfrac{1}{1+x}$, $a = 0$

24. $f(x) = \dfrac{1}{x-1}$, $a = 4$

25. $f(x) = e^x$, $a = 1$

26. $f(x) = e^x$, $a = -2$

27. $f(x) = x^{-2}$, $a = 1$

28. $f(x) = x^{-2}$, $a = 2$

29. $f(x) = \cos x$, $a = \dfrac{\pi}{4}$

30. $f(\theta) = \sin 3\theta$, $a = 0$

In Exercises 31–34, find T_2 and use a calculator to compute the error $|f(x) - T_2(x)|$ for the given values of a and x.

31. $y = e^x$, $a = 0$, $x = -0.5$

32. $y = \cos x$, $a = 0$, $x = \dfrac{\pi}{12}$

33. $y = x^{-2/3}$, $a = 1$, $x = 1.2$

34. $y = e^{\sin x}$, $a = \dfrac{\pi}{2}$, $x = 1.5$

35. **GU** Compute T_3 for $f(x) = \sqrt{x}$ centered at $a = 1$. Then use a plot of the error $|f(x) - T_3(x)|$ to find a value $c > 1$ such that the error on the interval $[1, c]$ is at most 0.25.

36. **CAS** Plot $f(x) = 1/(1+x)$ together with the Taylor polynomials T_n at $a = 1$ for $1 \le n \le 4$ on the interval $[-2, 8]$ (be sure to limit the upper plot range).

(a) Over which interval does T_4 appear to approximate f closely?

(b) What happens for $x < -1$?

(c) Use a computer algebra system to produce and plot T_{30} together with f on $[-2, 8]$. Over which interval does T_{30} appear to give a close approximation?

37. Let T_3 be the Maclaurin polynomial of $f(x) = e^x$. Use the Error Bound to find the maximum possible value of $|f(1.1) - T_3(1.1)|$. Show that we can take $K = e^{1.1}$.

38. Let T_2 be the Taylor polynomial of $f(x) = \sqrt{x}$ centered at $a = 4$. Apply the Error Bound to find the maximum possible value of the error $|f(3.9) - T_2(3.9)|$.

In Exercises 39–42, compute the Taylor polynomial indicated and use the Error Bound to find the maximum possible size of the error. Verify your result with a calculator.

39. $f(x) = \cos x$, $a = 0$; $|\cos 0.25 - T_5(0.25)|$

40. $f(x) = x^{11/2}$, $a = 1$; $|f(1.2) - T_4(1.2)|$

41. $f(x) = x^{-1/2}$, $a = 4$; $|f(4.3) - T_3(4.3)|$

42. $f(x) = \sqrt{1+x}$, $a = 8$; $|\sqrt{9.02} - T_3(8.02)|$

43. Calculate the Maclaurin polynomial T_3 for $f(x) = \tan^{-1} x$. Compute $T_3\left(\frac{1}{2}\right)$ and use the Error Bound to find a bound for $\left|\tan^{-1} \frac{1}{2} - T_3\left(\frac{1}{2}\right)\right|$. Refer to the graph in Figure 8 to find an acceptable value of K. Verify your result by computing $\left|\tan^{-1} \frac{1}{2} - T_3\left(\frac{1}{2}\right)\right|$ using a calculator.

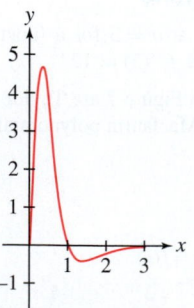

FIGURE 8 Graph of $f^{(4)}(x) = \dfrac{-24x(x^2 - 1)}{(x^2 + 1)^4}$, where $f(x) = \tan^{-1} x$.

44. Let $f(x) = \ln(x^3 - x + 1)$. The third Taylor polynomial at $a = 1$ is

$$T_3(x) = 2(x-1) + (x-1)^2 - \frac{7}{3}(x-1)^3$$

Find the maximum possible value of $|f(1.1) - T_3(1.1)|$, using the graph in Figure 9 to find an acceptable value of K. Verify your result by computing $|f(1.1) - T_3(1.1)|$ using a calculator.

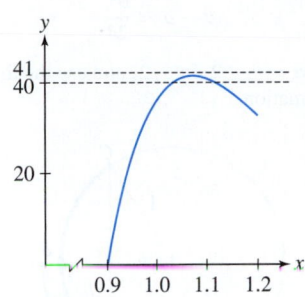

FIGURE 9 Graph of $f^{(4)}$, where $f(x) = \ln(x^3 - x + 1)$.

45. **GU** Let T_2 be the Taylor polynomial at $a = 0.5$ for $f(x) = \cos(x^2)$. Use the Error Bound to find the maximum possible value of $|f(0.6) - T_2(0.6)|$. Plot $f^{(3)}$ to find an acceptable value of K.

46. **GU** Calculate the Maclaurin polynomial T_2 for $f(x) = \operatorname{sech} x$ and use the Error Bound to estimate the error $\left| f\left(\frac{1}{2}\right) - T_2\left(\frac{1}{2}\right) \right|$. Plot f''' to find an acceptable value of K.

In Exercises 47–50, use the Error Bound to find a value of n for which the given inequality is satisfied. Then verify your result using a calculator.

47. $|\cos 0.1 - T_n(0.1)| \le 10^{-7}, \quad a = 0$

48. $|\ln 1.3 - T_n(1.3)| \le 10^{-4}, \quad a = 1$

49. $|\sqrt{1.3} - T_n(1.3)| \le 10^{-6}, \quad a = 1$

50. $|e^{-0.1} - T_n(-0.1)| \le 10^{-6}, \quad a = 0$

51. Let $f(x) = e^{-x}$ and $T_3(x) = 1 - x + \dfrac{x^2}{2} - \dfrac{x^3}{6}$.
(a) Use the Error Bound to show that for all $x \ge 0$,
$$|f(x) - T_3(x)| \le \frac{x^4}{24}$$

(b) **GU** Illustrate this inequality by plotting $y = f(x) - T_3(x)$ and $y = x^4/24$ together over $[0, 1]$.

52. Use the Error Bound with $n = 4$ to show that
$$\left| \sin x - \left(x - \frac{x^3}{6} \right) \right| \le \frac{|x|^5}{120} \quad \text{(for all } x\text{)}$$

53. Let T_n be the Taylor polynomial for $f(x) = \ln x$ at $a = 1$, and let $c > 1$. Show that
$$|\ln c - T_n(c)| \le \frac{|c - 1|^{n+1}}{n + 1}$$

Then find a value of n such that $|\ln 1.5 - T_n(1.5)| \le 10^{-2}$.

54. Let $n \ge 1$. Show that if $|x|$ is small, then
$$(x + 1)^{1/n} \approx 1 + \frac{x}{n} + \frac{1 - n}{2n^2} x^2$$

Use this approximation with $n = 6$ to estimate $1.5^{1/6}$.

55. Verify that the third Maclaurin polynomial for $f(x) = e^x \sin x$ is equal to the product of the third Maclaurin polynomials of $f(x) = e^x$ and $f(x) = \sin x$ (after discarding terms of degree greater than 3 in the product).

56. Find the fourth Maclaurin polynomial for $f(x) = \sin x \cos x$ by multiplying the fourth Maclaurin polynomials for $f(x) = \sin x$ and $f(x) = \cos x$.

57. Find the Maclaurin polynomials T_n for $f(x) = \cos(x^2)$. You may use the fact that $T_n(x)$ is equal to the sum of the terms up to degree n obtained by substituting x^2 for x in the nth Maclaurin polynomial of $\cos x$.

58. Find the Maclaurin polynomials of $1/(1 + x^2)$ by substituting $-x^2$ for x in the Maclaurin polynomials of $1/(1 - x)$.

59. Let $f(x) = 3x^3 + 2x^2 - x - 4$. Calculate T_j for $j = 1, 2, 3, 4, 5$ at both $a = 0$ and $a = 1$. Show that $T_3(x) = f(x)$ in both cases.

60. Let T_n be the nth Taylor polynomial at $x = a$ for a polynomial f of degree n. Based on the result of Exercise 59, guess the value of $|f(x) - T_n(x)|$. Prove that your guess is correct using the Error Bound.

61. Let $s(t)$ be the distance of a truck to an intersection. At time $t = 0$, the truck is 60 m from the intersection, travels away from it with a velocity of 24 m/s, and begins to slow down with an acceleration of $a = -3$ m/s^2. Determine the second Maclaurin polynomial of s, and use it to estimate the truck's distance from the intersection after 4 s.

62. A bank owns a portfolio of bonds whose value $P(r)$ depends on the interest rate r (measured in percent; e.g., $r = 5$ means a 5% interest rate). The bank's quantitative analyst determines that

$$P(5) = 100{,}000, \qquad \left.\frac{dP}{dr}\right|_{r=5} = -40{,}000, \qquad \left.\frac{d^2 P}{dr^2}\right|_{r=5} = 50{,}000$$

In finance, this second derivative is called **bond convexity**. Find the second Taylor polynomial of $P(r)$ centered at $r = 5$ and use it to estimate the value of the portfolio if the interest rate moves to $r = 5.5\%$.

63. A narrow, negatively charged ring of radius R exerts a force on a positively charged particle P located at distance x above the center of the ring of magnitude

$$F(x) = -\frac{kx}{(x^2 + R^2)^{3/2}}$$

where $k > 0$ is a constant (Figure 10).
(a) Compute the third-degree Maclaurin polynomial for F.
(b) Show that $F \approx -(k/R^3)x$ to second order. This shows that when x is small, $F(x)$ behaves like a restoring force similar to the force exerted by a spring.
(c) Show that $F(x) \approx -k/x^2$ when x is large by showing that
$$\lim_{x \to \infty} \frac{F(x)}{-k/x^2} = 1$$

Thus, $F(x)$ behaves like an inverse square law, and the charged ring looks like a point charge from far away.

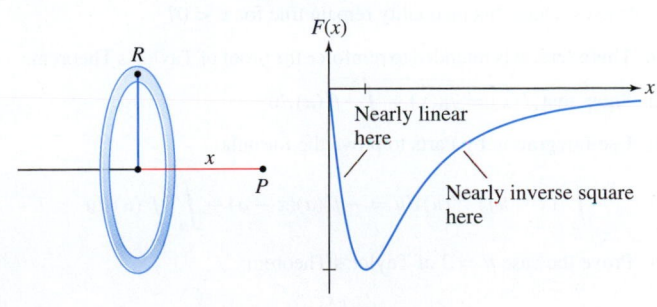

FIGURE 10

64. A light wave of wavelength λ travels from A to B by passing through an aperture (circular region) located in a plane that is perpendicular to $\overline{AB}$ (see Figure 11 for the notation). Let $f(r) = d' + h'$; that is, $f(r)$ is the distance $AC + CB$ as a function of r.

(a) Show that $f(r) = \sqrt{d^2 + r^2} + \sqrt{h^2 + r^2}$, and use the Maclaurin polynomial of order 2 to show that

$$f(r) \approx d + h + \frac{1}{2}\left(\frac{1}{d} + \frac{1}{h}\right)r^2$$

(b) The **Fresnel zones**, used to determine the optical disturbance at B, are the concentric bands bounded by the circles of radius R_n such that $f(R_n) = d + h + n\lambda/2$. Show that R_n can be approximated by $R_n \approx \sqrt{n\lambda L}$, where $L = (d^{-1} + h^{-1})^{-1}$.

(c) Estimate the radii R_1 and R_{100} for blue light ($\lambda = 475 \times 10^{-7}$ cm) if $d = h = 100$ cm.

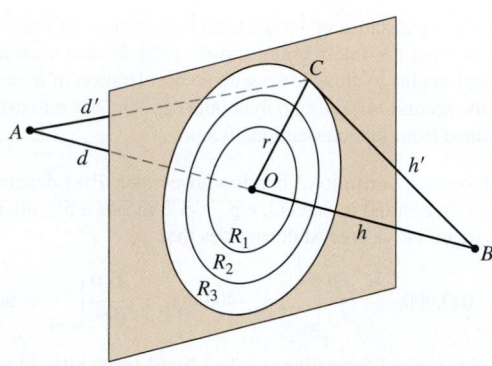

FIGURE 11 The Fresnel zones are the regions between the circles of radius R_n.

65. Referring to Figure 12, let a be the length of the chord $\overline{AC}$ of angle θ of the unit circle. Derive the following approximation for the excess of the arc over the chord:

$$\theta - a \approx \frac{\theta^3}{24}$$

Hint: Show that $\theta - a = \theta - 2\sin(\theta/2)$ and use the third Maclaurin polynomial as an approximation.

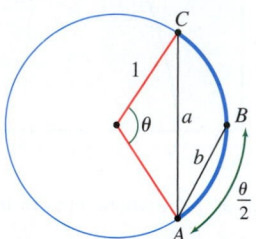

FIGURE 12 Unit circle.

66. To estimate the length θ of a circular arc of the unit circle, the seventeenth-century Dutch scientist Christian Huygens used the approximation $\theta \approx (8b - a)/3$, where a is the length of the chord $\overline{AC}$ of angle θ and b is the length of the chord $\overline{AB}$ of angle $\theta/2$ (Figure 12).

(a) Prove that $a = 2\sin(\theta/2)$ and $b = 2\sin(\theta/4)$, and show that the Huygens' approximation amounts to the approximation

$$\theta \approx \frac{16}{3}\sin\frac{\theta}{4} - \frac{2}{3}\sin\frac{\theta}{2}$$

(b) Compute the fifth Maclaurin polynomial of the function on the right.

(c) Use the Error Bound to show that the error in the Huygens' approximation is less than $0.00022|\theta|^5$.

Further Insights and Challenges

67. Show that the nth Maclaurin polynomial of $f(x) = \arcsin x$ for n odd is

$$T_n(x) = x + \frac{1}{2}\frac{x^3}{3} + \frac{1 \cdot 3}{2 \cdot 4}\frac{x^5}{5} + \cdots + \frac{1 \cdot 3 \cdot 5 \cdots (n-2)}{2 \cdot 4 \cdot 6 \cdots (n-1)}\frac{x^n}{n}$$

68. Let $x \geq 0$ and assume that $f^{(n+1)}(t) \geq 0$ for $0 \leq t \leq x$. Use Taylor's Theorem to show that the nth Maclaurin polynomial T_n satisfies

$$T_n(x) \leq f(x), \quad \text{for all } x \geq 0$$

69. Use Exercise 68 to show that for $x \geq 0$ and all n,

$$e^x \geq 1 + x + \frac{x^2}{2!} + \cdots + \frac{x^n}{n!}$$

Sketch the graphs of $y = e^x$, $y = T_1(x)$, and $y = T_2(x)$ on the same coordinate axes. Does this inequality remain true for $x < 0$?

70. This exercise is intended to reinforce the proof of Taylor's Theorem.

(a) Show that $f(x) = T_0(x) + \displaystyle\int_a^x f'(u)\,du$.

(b) Use Integration by Parts to prove the formula

$$\int_a^x (x - u)f^{(2)}(u)\,du = -f'(a)(x - a) + \int_a^x f'(u)\,du$$

(c) Prove the case $n = 2$ of Taylor's Theorem:

$$f(x) = T_1(x) + \int_a^x (x - u)f^{(2)}(u)\,du$$

In Exercises 71–75, we estimate integrals using Taylor polynomials. Exercise 72 is used to estimate the error.

71. Find the fourth Maclaurin polynomial T_4 for $f(x) = e^{-x^2}$, and calculate $I = \displaystyle\int_0^{1/2} T_4(x)\,dx$ as an estimate for $\displaystyle\int_0^{1/2} e^{-x^2}\,dx$. A CAS yields the value $I \approx 0.461281$. How large is the error in your approximation? *Hint:* T_4 is obtained by substituting $-x^2$ in the second Maclaurin polynomial for e^x.

72. Approximating Integrals Let $L > 0$. Show that if two functions f and g satisfy $|f(x) - g(x)| < L$ for all $x \in [a, b]$, then

$$\left| \int_a^b f(x)\,dx - \int_a^b g(x)\,dx \right| dx < L(b - a)$$

73. Let T_4 be the fourth Maclaurin polynomial for $f(x) = \cos x$.

(a) Show that

$$|\cos x - T_4(x)| \leq \frac{\left(\frac{1}{2}\right)^6}{6!} \quad \text{for all } \quad x \in \left[0, \frac{1}{2}\right]$$

Hint: $T_4(x) = T_5(x)$.

(b) Evaluate $\displaystyle\int_0^{1/2} T_4(x)\,dx$ as an approximation to $\displaystyle\int_0^{1/2} \cos x\,dx$. Use Exercise 72 to find a bound for the size of the error.

74. Let $Q(x) = 1 - x^2/6$. Use the Error Bound for $f(x) = \sin x$ to show that

$$\left| \frac{\sin x}{x} - Q(x) \right| \le \frac{|x|^4}{5!}$$

Then calculate $\int_0^1 Q(x)\,dx$ as an approximation to $\int_0^1 (\sin x/x)\,dx$ and find a bound for the error.

75. (a) Compute the sixth Maclaurin polynomial T_6 for $f(x) = \sin(x^2)$ by substituting x^2 in $P(x) = x - x^3/6$, the third Maclaurin polynomial for $f(x) = \sin x$.

(b) Show that $|\sin(x^2) - T_6(x)| \le \dfrac{|x|^{10}}{5!}$.

Hint: Substitute x^2 for x in the Error Bound for $|\sin x - P(x)|$, noting that P is also the fourth Maclaurin polynomial for $f(x) = \sin x$.

(c) Use T_6 to approximate $\int_0^{1/2} \sin(x^2)\,dx$ and find a bound for the error.

76. Prove by induction that for all k,

$$\frac{d^j}{dx^j}\left(\frac{(x-a)^k}{k!} \right) = \frac{k(k-1)\cdots(k-j+1)(x-a)^{k-j}}{k!}$$

$$\frac{d^j}{dx^j}\left(\frac{(x-a)^k}{k!} \right)\Bigg|_{x=a} = \begin{cases} 1 & \text{for } k = j \\ 0 & \text{for } k \neq j \end{cases}$$

Use this to prove that T_n agrees with f at $x = a$ to order n.

77. Let a be any number and let

$$P(x) = a_n x^n + a_{n-1} x^{n-1} + \cdots + a_1 x + a_0$$

be a polynomial of degree n or less.

(a) Show that if $P^{(j)}(a) = 0$ for $j = 0, 1, \ldots, n$, then $P(x) = 0$, that is, $a_j = 0$ for all j. *Hint:* Use induction, noting that if the statement is true for degree $n - 1$, then $P'(x) = 0$.

(b) Prove that T_n is the only polynomial of degree n or less that agrees with f at $x = a$ to order n. *Hint:* If Q is another such polynomial, apply (a) to $P(x) = T_n(x) - Q(x)$.

10.8 Taylor Series

In this section, we extend the Taylor polynomial to the Taylor series of a given function f, obtained by including terms of all orders in the Taylor polynomial.

> **DEFINITION   Taylor Series**   If f is infinitely differentiable at $x = c$, then the Taylor series for $f(x)$ centered at c is the power series
>
> $$T(x) = f(c) + f'(c)(x-c) + \frac{f''(c)}{2!}(x-c)^2 + \cdots = \sum_{n=0}^{\infty} \frac{f^{(n)}(c)}{n!}(x-c)^n$$

← **REMINDER** f is called infinitely differentiable if $f^{(n)}$ exists for all n.

While this definition enables us to construct a power series using information from the function f, we do not yet know whether this series defines a function that *equals f*. The next two theorems settle the matter.

> **THEOREM 1   Taylor Series Expansion**   If $f(x)$ is represented by a power series centered at c in an interval $|x - c| < R$ with $R > 0$, then that power series is the Taylor series
>
> $$T(x) = \sum_{n=0}^{\infty} \frac{f^{(n)}(c)}{n!}(x-c)^n$$

English mathematician Brook Taylor (1685–1731) made important contributions to calculus and physics, as well as to the theory of linear perspective used in drawing.

Proof   Suppose that $f(x)$ is represented by a power series centered at $x = c$ on an interval $(c - R, c + R)$ with $R > 0$:

$$f(x) = \sum_{n=0}^{\infty} a_n(x-c)^n = a_0 + a_1(x-c) + a_2(x-c)^2 + \cdots$$

According to Theorem 2 in Section 10.6, we can compute the derivatives of f by differentiating the series term by term:

$$f(x) = a_0 + a_1(x-c) + a_2(x-c)^2 + a_3(x-c)^3 + \cdots$$
$$f'(x) = a_1 + 2a_2(x-c) + 3a_3(x-c)^2 + 4a_4(x-c)^3 + \cdots$$
$$f''(x) = 2a_2 + 2\cdot 3a_3(x-c) + 3\cdot 4a_4(x-c)^2 + 4\cdot 5a_5(x-c)^3 + \cdots$$

In general,

$$f^{(k)}(x) = k!a_k + \left(2 \cdot 3 \cdots (k+1)\right)a_{k+1}(x-c) + \cdots$$

Setting $x = c$ in each of these series, we find that

$$f(c) = a_0, \quad f'(c) = a_1, \quad f''(c) = 2a_2, \quad \ldots, \quad f^{(k)}(c) = k!a_k, \quad \ldots$$

It follows that $a_k = \dfrac{f^{(k)}(c)}{k!}$. Therefore, $f(x) = T(x)$, where $T(x)$ is the Taylor series of $f(x)$ centered at $x = c$. ∎

In the special case $c = 0$, $T(x)$ is also called the **Maclaurin series**:

$$f(x) = \sum_{n=0}^{\infty} \frac{f^{(n)}(0)}{n!}x^n = f(0) + f'(0)x + \frac{f''(0)}{2!}x^2 + \frac{f'''(0)}{3!}x^3 + \frac{f^{(4)}(0)}{4!}x^4 + \cdots$$

EXAMPLE 1 Find the Taylor series for $f(x) = x^{-3}$ centered at $c = 1$.

Solution It often helps to create a table, as in Table 1, to see the pattern. The derivatives of $f(x)$ are $f'(x) = -3x^{-4}$, $f''(x) = (-3)(-4)x^{-5}$, and in general,

$$f^{(n)}(x) = (-1)^n (3)(4) \cdots (n+2)x^{-3-n}$$

Note that $(3)(4) \cdots (n+2) = \frac{1}{2}(n+2)!$. Therefore,

$$f^{(n)}(1) = (-1)^n \frac{1}{2}(n+2)!$$

Noting that $(n+2)! = (n+2)(n+1)n!$, we write the coefficients of the Taylor series as

$$a_n = \frac{f^{(n)}(1)}{n!} = \frac{(-1)^n \frac{1}{2}(n+2)!}{n!} = (-1)^n \frac{(n+2)(n+1)}{2}$$

The Taylor series for $f(x) = x^{-3}$ centered at $c = 1$ is

$$T(x) = 1 - 3(x-1) + 6(x-1)^2 - 10(x-1)^3 + \cdots$$

$$= \sum_{n=0}^{\infty} (-1)^n \frac{(n+2)(n+1)}{2}(x-1)^n$$

∎

TABLE 1

n	$f^{(n)}(x)$	$\dfrac{f^{(n)}(x)}{n!}$	$\dfrac{f^{(n)}(1)}{n!}$
0	x^{-3}	x^{-3}	1
1	$-3x^{-4}$	$-3x^{-4}$	-3
2	$12x^{-5}$	$6x^{-5}$	6
3	$-60x^{-6}$	$-10x^{-6}$	-10
4	$360x^{-7}$	$15x^{-7}$	15

Theorem 1 tells us that if we want to represent a function f by a power series centered at c, then the only candidate for the job is the Taylor series:

$$T(x) = \sum_{n=0}^{\infty} \frac{f^{(n)}(c)}{n!}(x-c)^n$$

See Exercise 100 for an example where a Taylor series $T(x)$ converges but does not converge to $f(x)$.

However, *there is no guarantee that $T(x)$ converges to $f(x)$*, even if $T(x)$ converges. To study convergence, we consider the kth partial sum, which is the Taylor polynomial of degree k:

$$T_k(x) = f(c) + f'(c)(x-c) + \frac{f''(c)}{2!}(x-c)^2 + \cdots + \frac{f^{(k)}(c)}{k!}(x-c)^k$$

In Section 10.7, we defined the remainder

$$R_k(x) = f(x) - T_k(x)$$

Since $T(x)$ is the limit of the partial sums $T_k(x)$, we see that

The Taylor series converges to $f(x)$ if and only if $\displaystyle\lim_{k \to \infty} R_k(x) = 0$.

There is no general method for determining whether $R_k(x)$ tends to zero, but the following theorem can be applied in some important cases.

THEOREM 2 Let $I = (c - R, c + R)$, where $R > 0$, and assume that f is infinitely differentiable on I. Suppose there exists $K > 0$ such that all derivatives of f are bounded by K on I:

$$|f^{(k)}(x)| \leq K \qquad \text{for all} \quad k \geq 0 \quad \text{and} \quad x \in I$$

Then f is represented by its Taylor series in I:

$$f(x) = \sum_{n=0}^{\infty} \frac{f^{(n)}(c)}{n!}(x - c)^n \qquad \text{for all} \quad x \in I$$

Proof According to the Error Bound for Taylor polynomials (Theorem 2 in Section 10.7),

$$|R_k(x)| = |f(x) - T_k(x)| \leq K\frac{|x - c|^{k+1}}{(k + 1)!}$$

If $x \in I$, then $|x - c| < R$ and

$$|R_k(x)| \leq K\frac{R^{k+1}}{(k + 1)!}$$

We showed in Example 10 of Section 10.1 that $R^k / k!$ tends to zero as $k \to \infty$. Therefore, $\lim_{k \to \infty} R_k(x) = 0$ for all $x \in (c - R, c + R)$, as required. ∎

Taylor expansions were studied throughout the seventeenth and eighteenth centuries by Gregory, Leibniz, Newton, Maclaurin, Taylor, Euler, and others. These developments were anticipated by the great Hindu mathematician Madhava (c. 1340–1425), who discovered the expansions of sine and cosine two centuries earlier.

EXAMPLE 2 **Expansions of Sine and Cosine** Show that the following Maclaurin expansions are valid for all x:

$$\sin x = \sum_{n=0}^{\infty} (-1)^n \frac{x^{2n+1}}{(2n + 1)!} = x - \frac{x^3}{3!} + \frac{x^5}{5!} - \frac{x^7}{7!} + \cdots$$

$$\cos x = \sum_{n=0}^{\infty} (-1)^n \frac{x^{2n}}{(2n)!} = 1 - \frac{x^2}{2!} + \frac{x^4}{4!} - \frac{x^6}{6!} + \cdots$$

Solution Recall that the derivatives of $f(x) = \sin x$ and their values at $x = 0$ form a repeating pattern of period 4:

$f(x)$	$f'(x)$	$f''(x)$	$f'''(x)$	$f^{(4)}(x)$	$\cdots$
$\sin x$	$\cos x$	$-\sin x$	$-\cos x$	$\sin x$	$\cdots$
0	1	0	-1	0	$\cdots$

In other words, the even derivatives are zero and the odd derivatives alternate in sign: $f^{(2n+1)}(0) = (-1)^n$. Therefore, the nonzero Taylor coefficients for $\sin x$ are

$$a_{2n+1} = \frac{(-1)^n}{(2n + 1)!}$$

For $f(x) = \cos x$, the situation is reversed. The odd derivatives are zero and the even derivatives alternate in sign: $f^{(2n)}(0) = (-1)^n \cos 0 = (-1)^n$. Therefore, the nonzero Taylor coefficients for $\cos x$ are $a_{2n} = (-1)^n/(2n)!$.

We can apply Theorem 2 with $K = 1$ and any value of R because both sine and cosine satisfy $|f^{(n)}(x)| \leq 1$ for all x and n. The conclusion is that the Taylor series converges to $f(x)$ for $|x| < R$. Since R is arbitrary, the Taylor expansions hold for all x. ∎

EXAMPLE 3 **Taylor Expansion of** $f(x) = e^x$ **at** $x = c$ Find the Taylor series $T(x)$ of $f(x) = e^x$ at $x = c$.

Solution We have $f^{(n)}(c) = e^c$ for all x. Thus,

$$T(x) = \sum_{n=0}^{\infty} \frac{e^c}{n!}(x - c)^n$$

Because $f(x) = e^x$ is increasing for all $R > 0$, $|f^{(k)}(x)| \le e^{c+R}$ for $x \in (c - R, c + R)$. Applying Theorem 2 with $K = e^{c+R}$, we conclude that $T(x)$ converges to $f(x)$ for all $x \in (c - R, c + R)$. Since R is arbitrary, the Taylor expansion holds for all x. For $c = 0$, we obtain the standard Maclaurin series

$$e^x = 1 + x + \frac{x^2}{2!} + \frac{x^3}{3!} + \cdots$$

Shortcuts to Finding Taylor Series

There are several methods for generating new Taylor series from known ones. First of all, we can differentiate and integrate Taylor series term by term within its interval of convergence, by Theorem 2 of Section 10.6. We can also multiply two Taylor series or substitute one Taylor series into another (we omit the proofs of these facts).

EXAMPLE 4 Find the Maclaurin series for $f(x) = x^2 e^x$.

In Example 4, we can also write the Maclaurin series as

$$\sum_{n=0}^{\infty} \frac{x^{n+2}}{n!}$$

Solution Multiply the known Maclaurin series for e^x by x^2:

$$x^2 e^x = x^2 \left(1 + x + \frac{x^2}{2!} + \frac{x^3}{3!} + \frac{x^4}{4!} + \frac{x^5}{5!} + \cdots \right)$$

$$= x^2 + x^3 + \frac{x^4}{2!} + \frac{x^5}{3!} + \frac{x^6}{4!} + \frac{x^7}{5!} + \cdots = \sum_{n=2}^{\infty} \frac{x^n}{(n-2)!}$$

EXAMPLE 5 **Substitution** Find the Maclaurin series for $f(x) = e^{-x^2}$.

Solution Substitute $-x^2$ for x in the Maclaurin series for e^x:

$$e^{-x^2} = \sum_{n=0}^{\infty} \frac{(-x^2)^n}{n!} = \sum_{n=0}^{\infty} \frac{(-1)^n x^{2n}}{n!} = 1 - x^2 + \frac{x^4}{2!} - \frac{x^6}{3!} + \frac{x^8}{4!} - \cdots \qquad \boxed{1}$$

The Taylor expansion of e^x is valid for all x, so this expansion is also valid for all x.

EXAMPLE 6 **Integration** Find the Maclaurin series for $f(x) = \ln(1 + x)$.

Solution We integrate the geometric series with common ratio $-x$ (valid for $|x| < 1$):

$$\frac{1}{1+x} = 1 - x + x^2 - x^3 + \cdots$$

$$\ln(1 + x) = \int \frac{dx}{1+x} = A + x - \frac{x^2}{2} + \frac{x^3}{3} - \frac{x^4}{4} + \cdots = A + \sum_{n=1}^{\infty} (-1)^{n-1} \frac{x^n}{n}$$

The constant of integration A on the right is zero because $\ln(1 + x) = 0$ for $x = 0$, so

$$\ln(1 + x) = \sum_{n=1}^{\infty} (-1)^{n-1} \frac{x^n}{n}$$

This expansion is valid for $|x| < 1$. It also holds for $x = 1$ (see Exercise 92).

In many cases, there is no convenient formula for the coefficients of a Taylor series for a given function, but we can still compute as many coefficients as desired, as the next example demonstrates.

EXAMPLE 7 Multiplying Taylor Series Write out the terms up to degree 5 in the Maclaurin series for $f(x) = e^x \cos x$.

Solution We multiply the fifth-order Maclaurin polynomials of e^x and $\cos x$ together, dropping the terms of degree greater than 5:

$$\left(1 + x + \frac{x^2}{2} + \frac{x^3}{6} + \frac{x^4}{24} + \frac{x^5}{120}\right)\left(1 - \frac{x^2}{2} + \frac{x^4}{24}\right)$$

Distributing the term on the left (and ignoring products that result in terms of degree greater than 5), we obtain

$$\left(1 + x + \frac{x^2}{2} + \frac{x^3}{6} + \frac{x^4}{24} + \frac{x^5}{120}\right) - \left(1 + x + \frac{x^2}{2} + \frac{x^3}{6}\right)\left(\frac{x^2}{2}\right) + (1 + x)\left(\frac{x^4}{24}\right)$$

$$= 1 + x - \frac{x^3}{3} - \frac{x^4}{6} - \frac{x^5}{30}$$

We conclude that the Maclaurin series for $f(x) = e^x \cos x$ (with the terms up to degree 5) appears as

$$e^x \cos x = 1 + x - \frac{x^3}{3} - \frac{x^4}{6} - \frac{x^5}{30} + \cdots$$ ■

In the next example, we express a definite integral of $\sin(x^2)$ as an infinite series. This is useful because the definite integral cannot be evaluated directly by finding an antiderivative of $\sin(x^2)$.

EXAMPLE 8 Let $J = \displaystyle\int_0^1 \sin(x^2)\,dx$.

(a) Express J as an infinite series.

(b) Determine J to within an error less than 10^{-4}.

Solution

(a) The Maclaurin expansion for $f(x) = \sin x$ is valid for all x, so we have

$$\sin x = \sum_{n=0}^{\infty} \frac{(-1)^n}{(2n+1)!} x^{2n+1} \quad \Rightarrow \quad \sin(x^2) = \sum_{n=0}^{\infty} \frac{(-1)^n}{(2n+1)!} x^{4n+2}$$

We obtain an infinite series for J by integration:

$$J = \int_0^1 \sin(x^2)\,dx = \sum_{n=0}^{\infty} \frac{(-1)^n}{(2n+1)!} \int_0^1 x^{4n+2}dx = \sum_{n=0}^{\infty} \frac{(-1)^n}{(2n+1)!}\left(\frac{1}{4n+3}\right)$$

$$= \frac{1}{3} - \frac{1}{42} + \frac{1}{1320} - \frac{1}{75,600} + \cdots$$ $\boxed{2}$

(b) The infinite series for J is an alternating series with decreasing terms, so the sum of the first N terms is accurate to within an error that is less than the $(N+1)$st term. The absolute value of the fourth term $1/75,600$ is smaller than 10^{-4}, so we obtain the desired accuracy using the first three terms of the series for J:

$$J \approx \frac{1}{3} - \frac{1}{42} + \frac{1}{1320} \approx 0.31028$$

The error satisfies

$$\left|J - \left(\frac{1}{3} - \frac{1}{42} + \frac{1}{1320}\right)\right| < \frac{1}{75,600} \approx 1.3 \times 10^{-5}$$

The percentage error is less than 0.005% with just three terms. ■

The next example demonstrates how power series can be used to assist in the evaluation of limits.

EXAMPLE 9 Determine $\lim\limits_{x\to 0}\dfrac{x-\sin x}{x^3\cos x}$.

Solution This limit is of indeterminate form $\dfrac{0}{0}$, so we could use L'Hôpital's Rule repeatedly. However, instead, we will work with the Maclaurin series. We have

$$\sin x = x - \frac{x^3}{3!} + \frac{x^5}{5!} - \cdots$$

$$\cos x = 1 - \frac{x^2}{2!} + \frac{x^4}{4!} - \cdots$$

Hence, the limit becomes

$$\lim_{x\to 0}\frac{x-\sin x}{x^3\cos x} = \lim_{x\to 0}\frac{x - \left(x - \frac{x^3}{3!} + \frac{x^5}{5!} - \cdots\right)}{x^3\left(1 - \frac{x^2}{2!} + \frac{x^4}{4!} - \cdots\right)}$$

$$= \lim_{x\to 0}\frac{\frac{x^3}{3!} - \frac{x^5}{5!} + \cdots}{x^3\left(1 - \frac{x^2}{2!} + \frac{x^4}{4!} - \cdots\right)}$$

$$= \lim_{x\to 0}\frac{x^3\left(\frac{1}{3!} - \frac{x^2}{5!} + \cdots\right)}{x^3\left(1 - \frac{x^2}{2!} + \frac{x^4}{4!} - \cdots\right)}$$

$$= \lim_{x\to 0}\frac{\frac{1}{3!} - \frac{x^2}{5!} + \cdots}{1 - \frac{x^2}{2!} + \frac{x^4}{4!} - \cdots}$$

$$= \frac{1}{3!} = \frac{1}{6}$$

Binomial Series

Isaac Newton discovered an important generalization of the Binomial Theorem around 1665. For any number a (integer or not) and integer $n \geq 0$, we define the **binomial coefficient**:

$$\binom{a}{n} = \frac{a(a-1)(a-2)\cdots(a-n+1)}{n!}, \qquad \binom{a}{0} = 1$$

For example,

$$\binom{6}{3} = \frac{6\cdot 5\cdot 4}{3\cdot 2\cdot 1} = 20, \qquad \binom{\frac{4}{3}}{3} = \frac{\frac{4}{3}\cdot\frac{1}{3}\cdot\left(-\frac{2}{3}\right)}{3\cdot 2\cdot 1} = -\frac{4}{81}$$

Let

$$f(x) = (1+x)^a$$

The **Binomial Theorem** of algebra (see Appendix C) states that for any whole number a,

$$(r+s)^a = r^a + \binom{a}{1}r^{a-1}s + \binom{a}{2}r^{a-2}s^2 + \cdots + \binom{a}{a-1}rs^{a-1} + s^a$$

Setting $r = 1$ and $s = x$, we obtain the expansion of $f(x)$:

$$(1+x)^a = 1 + \binom{a}{1}x + \binom{a}{2}x^2 + \cdots + \binom{a}{a-1}x^{a-1} + x^a$$

We derive Newton's generalization by computing the Maclaurin series of $f(x)$ without assuming that a is a whole number. Observe that the derivatives follow a pattern:

$$f(x) = (1+x)^a \qquad\qquad f(0) = 1$$

$$f'(x) = a(1+x)^{a-1} \qquad\qquad f'(0) = a$$

$$f''(x) = a(a-1)(1+x)^{a-2} \qquad\qquad f''(0) = a(a-1)$$

$$f'''(x) = a(a-1)(a-2)(1+x)^{a-3} \qquad\qquad f'''(0) = a(a-1)(a-2)$$

In general, $f^{(n)}(0) = a(a-1)(a-2)\cdots(a-n+1)$ and

$$\frac{f^{(n)}(0)}{n!} = \frac{a(a-1)(a-2)\cdots(a-n+1)}{n!} = \binom{a}{n}$$

When a is a positive whole number, $\binom{a}{n}$ is zero for $n > a$, and in this case, the binomial series breaks off at degree n. The binomial series is an infinite series when a is not a positive whole number.

Hence, the Maclaurin series for $f(x) = (1+x)^a$ is the binomial series

$$\sum_{n=0}^{\infty} \binom{a}{n} x^n = 1 + ax + \frac{a(a-1)}{2!}x^2 + \frac{a(a-1)(a-2)}{3!}x^3 + \cdots + \binom{a}{n}x^n + \cdots$$

The Ratio Test shows that this series has radius of convergence $R = 1$ (Exercise 94), and an additional argument (developed in Exercise 95) shows that it converges to $(1+x)^a$ for $|x| < 1$.

THEOREM 3   The Binomial Series   For any exponent a and for $|x| < 1$,

$$(1+x)^a = 1 + \frac{a}{1!}x + \frac{a(a-1)}{2!}x^2 + \frac{a(a-1)(a-2)}{3!}x^3 + \cdots + \binom{a}{n}x^n + \cdots$$

EXAMPLE 10   Find the terms through degree 4 in the Maclaurin expansion of

$$f(x) = (1+x)^{4/3}$$

Solution   The binomial coefficients $\binom{a}{n}$ for $a = \dfrac{4}{3}$ for $0 < n < 4$ are

$$1, \quad \frac{\frac{4}{3}}{1!} = \frac{4}{3}, \quad \frac{\frac{4}{3}\left(\frac{1}{3}\right)}{2!} = \frac{2}{9}, \quad \frac{\frac{4}{3}\left(\frac{1}{3}\right)\left(-\frac{2}{3}\right)}{3!} = -\frac{4}{81}, \quad \frac{\frac{4}{3}\left(\frac{1}{3}\right)\left(-\frac{2}{3}\right)\left(-\frac{5}{3}\right)}{4!} = \frac{5}{243}$$

Therefore, $(1+x)^{4/3} \approx 1 + \frac{4}{3}x + \frac{2}{9}x^2 - \frac{4}{81}x^3 + \frac{5}{243}x^4 + \cdots$. ∎

EXAMPLE 11   Find the Maclaurin series for

$$f(x) = \frac{1}{\sqrt{1-x^2}}$$

Solution   First, let's find the coefficients in the binomial series for $(1+x)^{-1/2}$:

$$1, \quad \frac{-\frac{1}{2}}{1!} = -\frac{1}{2}, \quad \frac{-\frac{1}{2}\left(-\frac{3}{2}\right)}{1\cdot2} = \frac{1\cdot3}{2\cdot4}, \quad \frac{-\frac{1}{2}\left(-\frac{3}{2}\right)\left(-\frac{5}{2}\right)}{1\cdot2\cdot3} = \frac{1\cdot3\cdot5}{2\cdot4\cdot6}$$

The general pattern is

$$\binom{-\frac{1}{2}}{n} = \frac{-\frac{1}{2}\left(-\frac{3}{2}\right)\left(-\frac{5}{2}\right)\cdots\left(-\frac{2n-1}{2}\right)}{1\cdot2\cdot3\cdots n} = (-1)^n\frac{1\cdot3\cdot5\cdots(2n-1)}{2\cdot4\cdot6\cdot2n}$$

Thus, the following binomial expansion is valid for $|x| < 1$:

$$\frac{1}{\sqrt{1+x}} = 1 + \sum_{n=1}^{\infty}(-1)^n\frac{1\cdot3\cdot5\cdots(2n-1)}{2\cdot4\cdot6\cdots(2n)}x^n = 1 - \frac{1}{2}x + \frac{1\cdot3}{2\cdot4}x^2 - \cdots$$

If $|x| < 1$, then $|x|^2 < 1$, and we can substitute $-x^2$ for x to obtain

$$\frac{1}{\sqrt{1-x^2}} = 1 + \sum_{n=1}^{\infty}\frac{1\cdot3\cdot5\cdots(2n-1)}{2\cdot4\cdot6\cdots2n}x^{2n} = 1 + \frac{1}{2}x^2 + \frac{1\cdot3}{2\cdot4}x^4 + \cdots \qquad \boxed{3}$$

∎

Taylor series are particularly useful for studying the so-called *special functions* (such as Bessel and hypergeometric functions) that appear in a wide range of physics and engineering applications. One example is the following **elliptic integral of the first kind**, defined for $|k| < 1$:

$$E(k) = \int_0^{\pi/2} \frac{dt}{\sqrt{1 - k^2\sin^2 t}}$$

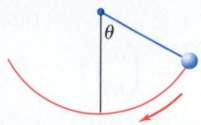

FIGURE 1 Pendulum released at an angle θ.

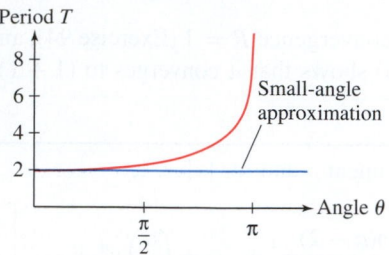

FIGURE 2 The period T of a 1-m pendulum as a function of the angle θ at which it is released.

For comparison, the value given by a computer algebra system to seven places is $E(\frac{1}{2}) \approx 1.6856325$.

Leonhard Euler (1707–1783). Euler (pronounced "oi-ler") ranks among the greatest mathematicians of all time. His work (printed in more than 70 volumes) contains fundamental contributions to almost every aspect of the mathematics and physics of his time. The French mathematician Pierre Simon de Laplace once declared: "Read Euler, he is our master in everything."

This function is used in physics to compute the period T of pendulum of length L (meters) released from an angle θ (Figure 1). When θ is small, we can use the small-angle approximation $T \approx 2\pi \sqrt{L/g}$ where g is the acceleration due to gravity, 9.8 m/s². This approximation breaks down for large angles (Figure 2). The exact value of the period is $T = 4\sqrt{L/g}\, E(k)$, where $k = \sin \frac{1}{2}\theta$.

EXAMPLE 12 **Elliptic Function**   Find the Maclaurin series for $E(k)$ and estimate $E(k)$ for $k = \sin \frac{\pi}{6}$.

Solution   Substitute $x = k \sin t$ in the Taylor expansion (3):

$$\frac{1}{\sqrt{1 - k^2 \sin^2 t}} = 1 + \frac{1}{2}k^2 \sin^2 t + \frac{1 \cdot 3}{2 \cdot 4}k^4 \sin^4 t + \frac{1 \cdot 3 \cdot 5}{2 \cdot 4 \cdot 6}k^6 \sin^6 t + \cdots$$

This expansion is valid because $|k| < 1$ and hence, $|x| = |k \sin t| < 1$. Thus, $E(k)$ is equal to

$$\int_0^{\pi/2} \frac{dt}{\sqrt{1 - k^2 \sin^2 t}} = \int_0^{\pi/2} dt + \sum_{n=1}^{\infty} \frac{1 \cdot 3 \cdots (2n-1)}{2 \cdot 4 \cdot (2n)} \left(\int_0^{\pi/2} \sin^{2n} t \, dt \right) k^{2n}$$

According to Exercise 76 in Section 7.2,

$$\int_0^{\pi/2} \sin^{2n} t \, dt = \left(\frac{1 \cdot 3 \cdots (2n-1)}{2 \cdot 4 \cdot (2n)} \right) \frac{\pi}{2}$$

This yields

$$E(k) = \frac{\pi}{2} + \frac{\pi}{2} \sum_{n=1}^{\infty} \left(\frac{1 \cdot 3 \cdots (2n-1)^2}{2 \cdot 4 \cdots (2n)} \right)^2 k^{2n}$$

We approximate $E(k)$ for $k = \sin\left(\frac{\pi}{6}\right) = \frac{1}{2}$ using the first five terms:

$$E\left(\frac{1}{2}\right) \approx \frac{\pi}{2} \left(1 + \left(\frac{1}{2}\right)^2 \left(\frac{1}{2}\right)^2 + \left(\frac{1 \cdot 3}{2 \cdot 4}\right)^2 \left(\frac{1}{2}\right)^4 \right.$$

$$\left. + \left(\frac{1 \cdot 3 \cdot 5}{2 \cdot 4 \cdot 6}\right)^2 \left(\frac{1}{2}\right)^6 + \left(\frac{1 \cdot 3 \cdot 5 \cdot 7}{2 \cdot 4 \cdot 6 \cdot 8}\right)^2 \left(\frac{1}{2}\right)^8 \right)$$

$$\approx 1.68517 \qquad \blacksquare$$

Euler's Formula

Euler's formula expresses a surprising relationship between the exponential function, $f(x) = e^x$, and the basic trigonometric functions, $g(x) = \sin x$ and $h(x) = \cos x$. This formula holds for all complex numbers. The complex numbers are numbers in the form $a + bi$, where a and b are real numbers and i is defined to be the square root of -1, that is, $i^2 = -1$. We can add and multiply complex numbers:

$$(a + bi) + (c + di) = (a + c) + (b + d)i$$

$$(a + bi)(c + di) = ac + adi + bci + bdi^2 = (ac - bd) + (ad + bc)i$$

As a result, we can compute polynomial functions of a complex variable $z = a + bi$:

$$f(z) = a_0 + a_1 z + a_2 z^2 + \cdots + a_n z^n$$

We can measure the distance d between complex numbers:

$$d(a + bi, c + di) = \sqrt{(c - a)^2 + (d - b)^2}$$

With such a measure of distance, we can address the convergence of a sequence or a series of complex numbers to a limit that is a complex number, just as we do with real numbers.

In particular, the power series associated with e^x, $\sin x$, and $\cos x$ can each be shown to converge for all complex numbers (the proofs are essentially the same as the proofs

In the field of complex variables, the letter z, rather than x, is often used as a general variable.

for the real-number case). In the field of complex variables, these series are often used to *define* the corresponding functions for all complex numbers z:

$$e^z = 1 + z + \frac{z^2}{2} + \frac{z^3}{3!} + \cdots, \quad \sin z = z - \frac{z^3}{3!} + \frac{z^5}{5!} + \cdots, \quad \cos z = 1 - \frac{z^2}{2} + \frac{z^4}{4!} + \cdots$$

We can combine these series to obtain Euler's Formula:

> **THEOREM 4** **Euler's Formula** For all complex numbers z,
>
> $$e^{iz} = \cos z + i \sin z$$

Proof The key to the proof lies in the pattern of the powers of i. First, note that

$$i^0 = 1, \quad i^1 = i, \quad i^2 = -1, \quad i^3 = i^2 i = (-1)i = -i$$

Furthermore, $i^4 = i^2 i^2 = (-1)(-1) = 1$ and $i^5 = i^4 i = (1)i = i$, and the cycle of values $1, i, -1, -i$ is now repeating. Therefore, starting with $n = 0$, the values of i^n repeatedly cycle through $1, i, -1, -i$. We have

$$e^{iz} = 1 + iz + \frac{(iz)^2}{2} + \frac{(iz)^3}{3!} + \frac{(iz)^4}{4!} + \frac{(iz)^5}{5!} + \cdots$$

$$= 1 + iz - \frac{z^2}{2} - i\frac{z^3}{3!} + \frac{z^4}{4!} + i\frac{z^5}{5!} + \cdots$$

$$= 1 - \frac{z^2}{2} + \frac{z^4}{4!} + \cdots + i\left(z - \frac{z^3}{3!} + \frac{z^5}{5!} + \cdots\right)$$

$$= \cos z + i \sin z \qquad \blacksquare$$

Euler's Formula is particularly useful in electrical engineering. Periodic signals are often expressed in terms of sine and cosine functions. Mathematical operations and computations involving combinations of signals are often more conveniently approached with the signals expressed in terms of complex exponential functions such as $f(z) = e^{iz}$.

If we substitute $z = \pi$ into Euler's Formula, we obtain

$$e^{i\pi} = \cos \pi + i \sin \pi = -1 + i(0) = -1$$

Rearranging, this relation is expressed as

$$\boxed{e^{i\pi} + 1 = 0}$$

This equation is known as **Euler's Identity** and is particularly pleasing because it relates five of the more important numbers used in mathematics and its applications.

In Table 2, we provide a list of useful Maclaurin series and the values of x for which they converge.

TABLE 2

$f(x)$	Maclaurin series	Converges to $f(x)$ for
e^x	$\displaystyle\sum_{n=0}^{\infty} \frac{x^n}{n!} = 1 + x + \frac{x^2}{2!} + \frac{x^3}{3!} + \frac{x^4}{4!} + \cdots$	All x
$\sin x$	$\displaystyle\sum_{n=0}^{\infty} \frac{(-1)^n x^{2n+1}}{(2n+1)!} = x - \frac{x^3}{3!} + \frac{x^5}{5!} - \frac{x^7}{7!} + \cdots$	All x
$\cos x$	$\displaystyle\sum_{n=0}^{\infty} \frac{(-1)^n x^{2n}}{(2n)!} = 1 - \frac{x^2}{2!} + \frac{x^4}{4!} - \frac{x^6}{6!} + \cdots$	All x
$\dfrac{1}{1-x}$	$\displaystyle\sum_{n=0}^{\infty} x^n = 1 + x + x^2 + x^3 + x^4 + \cdots$	$|x| < 1$

TABLE 2 (continued)

$f(x)$	Maclaurin series	Converges to $f(x)$ for
$\dfrac{1}{1+x}$	$\displaystyle\sum_{n=0}^{\infty}(-1)^n x^n = 1 - x + x^2 - x^3 + x^4 - \cdots$	$\|x\| < 1$
$\ln(1+x)$	$\displaystyle\sum_{n=1}^{\infty}\frac{(-1)^{n-1}x^n}{n} = x - \frac{x^2}{2} + \frac{x^3}{3} - \frac{x^4}{4} + \cdots$	$\|x\| < 1$ and $x = 1$
$\tan^{-1} x$	$\displaystyle\sum_{n=0}^{\infty}\frac{(-1)^n x^{2n+1}}{2n+1} = x - \frac{x^3}{3} + \frac{x^5}{5} - \frac{x^7}{7} + \cdots$	$\|x\| \leq 1$
$(1+x)^a$	$\displaystyle\sum_{n=0}^{\infty}\binom{a}{n}x^n = 1 + ax + \frac{a(a-1)}{2!}x^2 + \frac{a(a-1)(a-2)}{3!}x^3 + \cdots$	$\|x\| < 1$

10.8 SUMMARY

- *Taylor series* of $f(x)$ centered at $x = c$:

$$T(x) = \sum_{n=0}^{\infty}\frac{f^{(n)}(c)}{n!}(x-c)^n$$

The partial sum $T_k(x)$ is the kth Taylor polynomial.
- *Maclaurin series* ($c = 0$):

$$T(x) = \sum_{n=0}^{\infty}\frac{f^{(n)}(0)}{n!}x^n$$

- If $f(x)$ is represented by a power series $\displaystyle\sum_{n=0}^{\infty} a_n(x-c)^n$ for $|x - c| < R$ with $R > 0$, then this power series is necessarily the Taylor series centered at $x = c$.
- A function f is represented by its Taylor series $T(x)$ if and only if the remainder $R_k(x) = f(x) - T_k(x)$ tends to zero as $k \to \infty$.
- Let $I = (c - R, c + R)$ with $R > 0$. Suppose that there exists $K > 0$ such that $|f^{(k)}(x)| < K$ for all $x \in I$ and all k. Then f is represented by its Taylor series on I; that is, $f(x) = T(x)$ for $x \in I$.
- A good way to find the Taylor series of a function is to start with known Taylor series and apply one of the following operations: differentiation, integration, multiplication, or substitution.
- For any exponent a, the binomial expansion is valid for $|x| < 1$:

$$(1+x)^a = 1 + ax + \frac{a(a-1)}{2!}x^2 + \frac{a(a-1)(a-2)}{3!}x^3 + \cdots + \binom{a}{n}x^n + \cdots$$

10.8 EXERCISES

Preliminary Questions

1. Determine $f(0)$ and $f'''(0)$ for a function f with Maclaurin series

$$T(x) = 3 + 2x + 12x^2 + 5x^3 + \cdots$$

2. Determine $f(-2)$ and $f^{(4)}(-2)$ for a function with Taylor series

$$T(x) = 3(x + 2) + (x + 2)^2 - 4(x + 2)^3 + 2(x + 2)^4 + \cdots$$

3. What is the easiest way to find the Maclaurin series for the function $f(x) = \sin(x^2)$?

4. Find the Taylor series for f centered at $c = 3$ if $f(3) = 4$ and $f'(x)$ has a Taylor expansion

$$f'(x) = \sum_{n=1}^{\infty}\frac{(x-3)^n}{n}$$

5. Let $T(x)$ be the Maclaurin series of $f(x)$. Which of the following guarantees that $f(2) = T(2)$?

(a) $T(x)$ converges for $x = 2$.

(b) The remainder $R_k(2)$ approaches a limit as $k \to \infty$.

(c) The remainder $R_k(2)$ approaches zero as $k \to \infty$.

Exercises

1. Write out the first four terms of the Maclaurin series of $f(x)$ if

$$f(0) = 2, \quad f'(0) = 3, \quad f''(0) = 4, \quad f'''(0) = 12$$

2. Write out the first four terms of the Taylor series of $f(x)$ centered at $c = 3$ if

$$f(3) = 1, \quad f'(3) = 2, \quad f''(3) = 12, \quad f'''(3) = 3$$

In Exercises 3–18, find the Maclaurin series and find the interval on which the expansion is valid.

3. $f(x) = \dfrac{1}{1 + 10x}$

4. $f(x) = \dfrac{x^2}{1 - x^3}$

5. $f(x) = \cos 3x$

6. $f(x) = \sin(2x)$

7. $f(x) = \sin(x^2)$

8. $f(x) = e^{4x}$

9. $f(x) = \ln(1 - x^2)$

10. $f(x) = (1 - x)^{-1/2}$

11. $f(x) = \tan^{-1}(x^2)$

12. $f(x) = x^2 e^{x^2}$

13. $f(x) = e^{x-2}$

14. $f(x) = \dfrac{1 - \cos x}{x}$

15. $f(x) = \ln(1 - 5x)$

16. $f(x) = (x^2 + 2x)e^x$

17. $f(x) = \sinh x$

18. $f(x) = \cosh x$

In Exercises 19–30, find the terms through degree 4 of the Maclaurin series of $f(x)$. Use multiplication and substitution as necessary.

19. $f(x) = e^x \sin x$

20. $f(x) = e^x \ln(1 - x)$

21. $f(x) = \dfrac{\sin x}{1 - x}$

22. $f(x) = \dfrac{1}{1 + \sin x}$

23. $f(x) = (1 + x)^{1/4}$

24. $f(x) = (1 + x)^{-3/2}$

25. $f(x) = e^x \tan^{-1} x$

26. $f(x) = \sin(x^3 - x)$

27. $f(x) = e^{\sin x}$

28. $f(x) = e^{(e^x)}$

29. $f(x) = \cosh(x^2)$

30. $f(x) = \sinh(x)\cosh(x)$

In Exercises 31–40, find the Taylor series centered at c and the interval on which the expansion is valid.

31. $f(x) = \dfrac{1}{x}, \quad c = 1$

32. $f(x) = e^{3x}, \quad c = -1$

33. $f(x) = \dfrac{1}{1 - x}, \quad c = 5$

34. $f(x) = \sin x, \quad c = \dfrac{\pi}{2}$

35. $f(x) = x^4 + 3x - 1, \quad c = 2$

36. $f(x) = x^4 + 3x - 1, \quad c = 0$

37. $f(x) = \dfrac{1}{x^2}, \quad c = 4$

38. $f(x) = \sqrt{x}, \quad c = 4$

39. $f(x) = \dfrac{1}{1 - x^2}, \quad c = 3$

40. $f(x) = \dfrac{1}{3x - 2}, \quad c = -1$

41. Use the identity $\cos^2 x = \frac{1}{2}(1 + \cos 2x)$ to find the Maclaurin series for $f(x) = \cos^2 x$.

42. Show that for $|x| < 1$,

$$\tanh^{-1} x = x + \frac{x^3}{3} + \frac{x^5}{5} + \cdots$$

Hint: Recall that $\dfrac{d}{dx} \tanh^{-1} x = \dfrac{1}{1 - x^2}$.

43. Use the Maclaurin series for $\ln(1 + x)$ and $\ln(1 - x)$ to show that

$$\frac{1}{2} \ln\left(\frac{1 + x}{1 - x}\right) = x + \frac{x^3}{3} + \frac{x^5}{5} + \cdots$$

for $|x| < 1$. What can you conclude by comparing this result with that of Exercise 42?

44. Differentiate the Maclaurin series for $\dfrac{1}{1 - x}$ twice to find the Maclaurin series of $\dfrac{1}{(1 - x)^3}$.

45. Show, by integrating the Maclaurin series for $f(x) = \dfrac{1}{\sqrt{1 - x^2}}$, that for $|x| < 1$,

$$\sin^{-1} x = x + \sum_{n=1}^{\infty} \frac{1 \cdot 3 \cdot 5 \cdots (2n - 1)}{2 \cdot 4 \cdot 6 \cdots (2n)} \frac{x^{2n+1}}{2n + 1}$$

46. Use the first five terms of the Maclaurin series in Exercise 45 to approximate $\sin^{-1} \frac{1}{2}$. Compare the result with the calculator value.

47. How many terms of the Maclaurin series of $f(x) = \ln(1 + x)$ are needed to compute $\ln 1.2$ to within an error of at most 0.0001? Make the computation and compare the result with the calculator value.

48. Show that

$$\pi - \frac{\pi^3}{3!} + \frac{\pi^5}{5!} - \frac{\pi^7}{7!} + \cdots$$

converges to zero. How many terms must be computed to get within 0.01 of zero?

49. Use the Maclaurin expansion for e^{-t^2} to express the function $F(x) = \displaystyle\int_0^x e^{-t^2}\, dt$ as an alternating power series in x (Figure 3).

(a) How many terms of the Maclaurin series are needed to approximate the integral for $x = 1$ to within an error of at most 0.001?

(b) CAS Carry out the computation and check your answer using a computer algebra system.

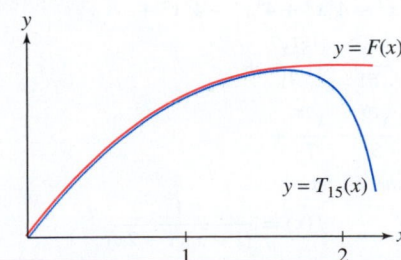

FIGURE 3 The Maclaurin polynomial $T_{15}(x)$ for $F(x) = \displaystyle\int_0^x e^{-t^2}\, dt$.

50. Let $F(x) = \displaystyle\int_0^x \frac{\sin t\, dt}{t}$. Show that

$$F(x) = x - \frac{x^3}{3 \cdot 3!} + \frac{x^5}{5 \cdot 5!} - \frac{x^7}{7 \cdot 7!} + \cdots$$

Evaluate $F(1)$ to three decimal places.

In Exercises 51–54, express the definite integral as an infinite series and find its value to within an error of at most 10^{-4}.

51. $\int_0^1 \cos(x^2)\,dx$

52. $\int_0^1 \tan^{-1}(x^2)\,dx$

53. $\int_0^1 e^{-x^3}\,dx$

54. $\int_0^1 \dfrac{dx}{\sqrt{x^4+1}}$

In Exercises 55–58, express the integral as an infinite series.

55. $\int_0^x \dfrac{1-\cos t}{t}\,dt$, for all x

56. $\int_0^x \dfrac{t-\sin t}{t}\,dt$, for all x

57. $\int_0^x \ln(1+t^2)\,dt$, for $|x| < 1$

58. $\int_0^x \dfrac{dt}{\sqrt{1-t^4}}$, for $|x| < 1$

59. Which function has Maclaurin series $\displaystyle\sum_{n=0}^{\infty}(-1)^n 2^n x^n$?

60. Which function has the following Maclaurin series?

$$\sum_{k=0}^{\infty} \frac{(-1)^k}{3^{k+1}}(x-3)^k$$

For which values of x is the expansion valid?

61. Using Maclaurin series, determine to exactly what value the following series converges:

$$\sum_{n=0}^{\infty}(-1)^n \frac{(\pi)^{2n}}{(2n)!}$$

62. Using Maclaurin series, determine to exactly what value the following series converges:

$$\sum_{n=0}^{\infty} \frac{(\ln 5)^n}{n!}$$

In Exercises 61–64, use Theorem 2 to prove that the $f(x)$ is represented by its Maclaurin series for all x.

63. $f(x) = \sin(x/2) + \cos(x/3)$

64. $f(x) = e^{-x}$

65. $f(x) = \sinh x$

66. $f(x) = (1+x)^{100}$

In Exercises 67–70, find the functions with the following Maclaurin series (refer to Table 2 prior to the section summary).

67. $1 + x^3 + \dfrac{x^6}{2!} + \dfrac{x^9}{3!} + \dfrac{x^{12}}{4!} + \cdots$

68. $1 - 4x + 4^2 x^2 - 4^3 x^3 + 4^4 x^4 - 4^5 x^5 + \cdots$

69. $1 - \dfrac{5^3 x^3}{3!} + \dfrac{5^5 x^5}{5!} - \dfrac{5^7 x^7}{7!} + \cdots$

70. $x^4 - \dfrac{x^{12}}{3} + \dfrac{x^{20}}{5} - \dfrac{x^{28}}{7} + \cdots$

In Exercises 71 and 72, let

$$f(x) = \frac{1}{(1-x)(1-2x)}$$

71. Find the Maclaurin series of $f(x)$ using the identity

$$f(x) = \frac{2}{1-2x} - \frac{1}{1-x}$$

72. Find the Taylor series for $f(x)$ at $c = 2$. *Hint:* Rewrite the identity of Exercise 71 as

$$f(x) = \frac{2}{-3 - 2(x-2)} - \frac{1}{-1-(x-2)}$$

73. When a voltage V is applied to a series circuit consisting of a resistor R and an inductor L, the current at time t is

$$I(t) = \left(\frac{V}{R}\right)\left(1 - e^{-Rt/L}\right)$$

Expand $I(t)$ in a Maclaurin series. Show that $I(t) \approx \dfrac{Vt}{L}$ for small t.

74. Use the result of Exercise 73 and your knowledge of alternating series to show that

$$\frac{Vt}{L}\left(1 - \frac{R}{2L}t\right) \le I(t) \le \frac{Vt}{L} \qquad \text{(for all } t\text{)}$$

75. Find the Maclaurin series for $f(x) = \cos(x^3)$ and use it to determine $f^{(6)}(0)$.

76. Find $f^{(7)}(0)$ and $f^{(8)}(0)$ for $f(x) = \tan^{-1}x$ using the Maclaurin series.

77. [✎] Use substitution to find the first three terms of the Maclaurin series for $f(x) = e^{x^{20}}$. How does the result show that $f^{(k)}(0) = 0$ for $1 \le k \le 19$?

78. Use the binomial series to find $f^{(8)}(0)$ for $f(x) = \sqrt{1-x^2}$.

79. Does the Maclaurin series for $f(x) = (1+x)^{3/4}$ converge to $f(x)$ at $x = 2$? Give numerical evidence to support your answer.

80. [✎] Explain the steps required to verify that the Maclaurin series for $f(x) = e^x$ converges to $f(x)$ for all x.

81. [GU] Let $f(x) = \sqrt{1+x}$.
(a) Use a graphing calculator to compare the graph of f with the graphs of the first five Taylor polynomials for f. What do they suggest about the interval of convergence of the Taylor series?
(b) Investigate numerically whether or not the Taylor expansion for f is valid for $x = 1$ and $x = -1$.

82. Use the first five terms of the Maclaurin series for the elliptic integral $E(k)$ to estimate the period T of a 1-m pendulum released at an angle $\theta = \frac{\pi}{4}$ (see Example 12).

83. Use Example 12 and the approximation $\sin x \approx x$ to show that the period T of a pendulum released at an angle θ has the following second-order approximation:

$$T \approx 2\pi\sqrt{\frac{L}{g}}\left(1 + \frac{\theta^2}{16}\right)$$

In Exercises 84–87, the limits can be done using multiple L'Hôpital's Rule steps. Power series provide an alternative approach. In each case substitute in the Maclaurin series for the trig function or the inverse trig function involved, simplify, and compute the limit.

84. $\displaystyle\lim_{x\to 0} \frac{\cos x - 1 + \frac{x^2}{2}}{x^4}$

85. $\displaystyle\lim_{x\to 0} \frac{\sin x - x + \frac{x^3}{6}}{x^5}$

86. $\displaystyle\lim_{x\to 0} \frac{\tan^{-1}x - x\cos x - \frac{1}{6}x^3}{x^5}$

87. $\displaystyle\lim_{x\to 0}\left(\frac{\sin(x^2)}{x^4} - \frac{\cos x}{x^2}\right)$

88. Use Euler's Formula to express each of the following in $a + bi$ form.
(a) $e^{\frac{\pi}{4}i}$ **(b)** $4e^{\frac{5\pi}{3}i}$ **(c)** $ie^{\frac{-\pi}{2}i}$

89. Use Euler's Formula to express each of the following in $a + bi$ form.
(a) $-e^{\frac{3\pi}{4}i}$ **(b)** $e^{2\pi i}$ **(c)** $3ie^{\frac{-\pi}{3}i}$

In Exercises 90–91, use Euler's Formula to prove that the identity holds. Note the similarity between these relationships and the definitions of the hyperbolic sine and cosine functions.

90. $\cos z = \dfrac{e^{iz} + e^{-iz}}{2}$

91. $\sin z = \dfrac{e^{iz} - e^{-iz}}{2i}$

Further Insights and Challenges

92. In this exercise, we show that the Maclaurin expansion of $f(x) = \ln(1 + x)$ is valid for $x = 1$.

(a) Show that for all $x \neq -1$,

$$\frac{1}{1+x} = \sum_{n=0}^{N}(-1)^n x^n + \frac{(-1)^{N+1}x^{N+1}}{1+x}$$

(b) Integrate from 0 to 1 to obtain

$$\ln 2 = \sum_{n=1}^{N}\frac{(-1)^{n-1}}{n} + (-1)^{N+1}\int_0^1 \frac{x^{N+1}\,dx}{1+x}$$

(c) Verify that the integral on the right tends to zero as $N \to \infty$ by showing that it is smaller than $\int_0^1 x^{N+1}dx$.

(d) Prove the formula

$$\ln 2 = 1 - \frac{1}{2} + \frac{1}{3} - \frac{1}{4} + \cdots$$

93. Let $g(t) = \dfrac{1}{1+t^2} - \dfrac{t}{1+t^2}$.

(a) Show that $\displaystyle\int_0^1 g(t)\,dt = \frac{\pi}{4} - \frac{1}{2}\ln 2$.

(b) Show that $g(t) = 1 - t - t^2 + t^3 + t^4 - t^5 - t^6 + \cdots$.

(c) Evaluate $S = 1 - \frac{1}{2} - \frac{1}{3} + \frac{1}{4} + \frac{1}{5} - \frac{1}{6} - \frac{1}{7} + \cdots$.

In Exercises 94 and 95, we investigate the convergence of the binomial series

$$T_a(x) = \sum_{n=0}^{\infty}\binom{a}{n}x^n$$

94. Prove that $T_a(x)$ has radius of convergence $R = 1$ if a is not a whole number. What is the radius of convergence if a is a whole number?

95. By Exercise 94, $T_a(x)$ converges for $|x| < 1$, but we do not yet know whether $T_a(x) = (1 + x)^a$.

(a) Verify the identity

$$a\binom{a}{n} = n\binom{a}{n} + (n+1)\binom{a}{n+1}$$

(b) Use (a) to show that $y = T_a(x)$ satisfies the differential equation $(1 + x)y' = ay$ with initial condition $y(0) = 1$.

(c) Prove $T_a(x) = (1 + x)^a$ for $|x| < 1$ by showing that the derivative of the ratio $\dfrac{T_a(x)}{(1+x)^a}$ is zero.

96. The function $G(k) = \displaystyle\int_0^{\pi/2}\sqrt{1 - k^2 \sin^2 t}\,dt$ is called an **elliptic integral of the second kind**. Prove that for $|k| < 1$,

$$G(k) = \frac{\pi}{2} - \frac{\pi}{2}\sum_{n=1}^{\infty}\left(\frac{1 \cdot 3 \cdots (2n-1)}{2 \cdots 4 \cdot (2n)}\right)^2 \frac{k^{2n}}{2n-1}$$

97. Assume that $a < b$ and let L be the arc length (circumference) of the ellipse $\left(\frac{x}{a}\right)^2 + \left(\frac{y}{b}\right)^2 = 1$ shown in Figure 4. There is no explicit formula for L, but it is known that $L = 4bG(k)$, with $G(k)$ as in Exercise 96 and $k = \sqrt{1 - a^2/b^2}$. Use the first three terms of the expansion of Exercise 96 to estimate L when $a = 4$ and $b = 5$.

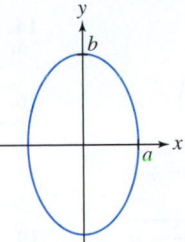

FIGURE 4 The ellipse $\left(\frac{x}{a}\right)^2 + \left(\frac{y}{b}\right)^2 = 1$.

98. Use Exercise 96 to prove that if $a < b$ and a/b is near 1 (a nearly circular ellipse), then

$$L \approx \frac{\pi}{2}\left(3b + \frac{a^2}{b}\right)$$

Hint: Use the first two terms of the series for $G(k)$.

99. Irrationality of e Prove that e is an irrational number using the following argument by contradiction. Suppose that $e = M/N$, where M, N are nonzero integers.

(a) Show that $M!\,e^{-1}$ is a whole number.

(b) Use the power series for $f(x) = e^x$ at $x = -1$ to show that there is an integer B such that $M!\,e^{-1}$ equals

$$B + (-1)^{M+1}\left(\frac{1}{M+1} - \frac{1}{(M+1)(M+2)} + \cdots\right)$$

(c) Use your knowledge of alternating series with decreasing terms to conclude that $0 < |M!\,e^{-1} - B| < 1$ and observe that this contradicts (a). Hence, e is not equal to M/N.

100. Use the result of Exercise 75 in Section 4.5 to show that the Maclaurin series of the function

$$f(x) = \begin{cases} e^{-1/x^2} & \text{for } x \neq 0 \\ 0 & \text{for } x = 0 \end{cases}$$

is $T(x) = 0$. This provides an example of a function f whose Maclaurin series converges but does not converge to $f(x)$ (except at $x = 0$).

CHAPTER REVIEW EXERCISES

1. Let $a_n = \dfrac{n-3}{n!}$ and $b_n = a_{n+3}$. Calculate the first three terms in each sequence.

(a) a_n^2

(b) b_n

(c) $a_n b_n$

(d) $2a_{n+1} - 3a_n$

2. Prove that $\displaystyle\lim_{n\to\infty}\frac{2n-1}{3n+2} = \frac{2}{3}$ using the limit definition.

In Exercises 3–8, compute the limit (or state that it does not exist) assuming that $\displaystyle\lim_{n\to\infty} a_n = 2$.

3. $\displaystyle\lim_{n\to\infty}(5a_n - 2a_n^2)$

4. $\displaystyle\lim_{n\to\infty}\frac{1}{a_n}$

5. $\displaystyle\lim_{n\to\infty}e^{a_n}$

6. $\displaystyle\lim_{n\to\infty}\cos(\pi a_n)$

7. $\displaystyle\lim_{n\to\infty}(-1)^n a_n$

8. $\displaystyle\lim_{n\to\infty}\frac{a_n + n}{a_n + n^2}$

In Exercises 9–22, determine the limit of the sequence or show that the sequence diverges.

9. $a_n = \sqrt{n+5} - \sqrt{n+2}$

10. $a_n = \dfrac{3n^3 - n}{1 - 2n^3}$

11. $a_n = 2^{1/n^2}$

12. $a_n = \dfrac{10^n}{n!}$

13. $b_m = 1 + (-1)^m$

14. $b_m = \dfrac{1 + (-1)^m}{m}$

15. $b_n = \tan^{-1}\left(\dfrac{n+2}{n+5}\right)$

16. $a_n = \dfrac{100^n}{n!} - \dfrac{3 + \pi^n}{5^n}$

17. $b_n = \sqrt{n^2 + n} - \sqrt{n^2 + 1}$

18. $c_n = \sqrt{n^2 + n} - \sqrt{n^2 - n}$

19. $b_m = \left(1 + \dfrac{1}{m}\right)^{3m}$

20. $c_n = \left(1 + \dfrac{3}{n}\right)^n$

21. $b_n = n\big(\ln(n+1) - \ln n\big)$

22. $c_n = \dfrac{\ln(n^2 + 1)}{\ln(n^3 + 1)}$

23. Use the Squeeze Theorem to show that $\displaystyle\lim_{n\to\infty} \dfrac{\arctan(n^2)}{\sqrt{n}} = 0$.

24. Give an example of a divergent sequence $\{a_n\}$ such that $\{\sin a_n\}$ is convergent.

25. Calculate $\displaystyle\lim_{n\to\infty} \dfrac{a_{n+1}}{a_n}$, where $a_n = \dfrac{1}{2}3^n - \dfrac{1}{3}2^n$.

26. Define $a_{n+1} = \sqrt{a_n + 6}$ with $a_1 = 2$.
(a) Compute a_n for $n = 2, 3, 4, 5$.
(b) Show that $\{a_n\}$ is increasing and is bounded by 3.
(c) Prove that $\displaystyle\lim_{n\to\infty} a_n$ exists and find its value.

27. Calculate the partial sums S_4 and S_7 of the series $\displaystyle\sum_{n=1}^{\infty} \dfrac{n-2}{n^2 + 2n}$.

28. Find the sum $1 - \dfrac{1}{4} + \dfrac{1}{4^2} - \dfrac{1}{4^3} + \cdots$.

29. Find the sum $\dfrac{4}{9} + \dfrac{8}{27} + \dfrac{16}{81} + \dfrac{32}{243} + \cdots$.

30. Use series to determine a reduced fraction that has decimal expansion $0.121212\cdots$.

31. Use series to determine a reduced fraction that has decimal expansion $0.108108108\cdots$.

32. Find the sum $\displaystyle\sum_{n=2}^{\infty} \left(\dfrac{2}{e}\right)^n$.

33. Find the sum $\displaystyle\sum_{n=-1}^{\infty} \dfrac{2^{n+3}}{3^n}$.

34. Show that $\displaystyle\sum_{n=1}^{\infty} \big(b - \tan^{-1} n^2\big)$ diverges if $b \neq \dfrac{\pi}{2}$.

35. Give an example of divergent series $\displaystyle\sum_{n=1}^{\infty} a_n$ and $\displaystyle\sum_{n=1}^{\infty} b_n$ such that $\displaystyle\sum_{n=1}^{\infty}(a_n + b_n) = 1$.

36. Let $S = \displaystyle\sum_{n=1}^{\infty} \left(\dfrac{1}{n} - \dfrac{1}{n+2}\right)$. Compute S_N for $N = 1, 2, 3, 4$. Find S by showing that

$$S_N = \dfrac{3}{2} - \dfrac{1}{N+1} - \dfrac{1}{N+2}$$

37. Evaluate $S = \displaystyle\sum_{n=3}^{\infty} \dfrac{1}{n(n+3)}$.

38. Find the total area of the infinitely many circles on the interval $[0, 1]$ in Figure 1.

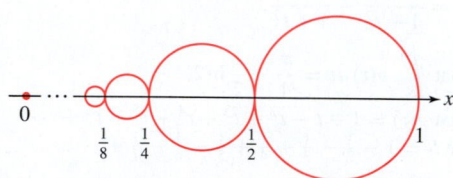

FIGURE 1

In Exercises 39–42, use the Integral Test to determine whether the infinite series converges.

39. $\displaystyle\sum_{n=1}^{\infty} \dfrac{n^2}{n^3 + 1}$

40. $\displaystyle\sum_{n=1}^{\infty} \dfrac{n^2}{(n^3 + 1)^{1.01}}$

41. $\displaystyle\sum_{n=1}^{\infty} \dfrac{1}{(n+2)(\ln(n+2))^3}$

42. $\displaystyle\sum_{n=1}^{\infty} \dfrac{n^3}{e^{n^4}}$

In Exercises 43–50, use the Direct Comparison or Limit Comparison Test to determine whether the infinite series converges.

43. $\displaystyle\sum_{n=1}^{\infty} \dfrac{1}{(n+1)^2}$

44. $\displaystyle\sum_{n=1}^{\infty} \dfrac{1}{\sqrt{n} + n}$

45. $\displaystyle\sum_{n=2}^{\infty} \dfrac{n^2 + 1}{n^{3.5} - 2}$

46. $\displaystyle\sum_{n=1}^{\infty} \dfrac{1}{n - \ln n}$

47. $\displaystyle\sum_{n=2}^{\infty} \dfrac{n}{\sqrt{n^5 + 5}}$

48. $\displaystyle\sum_{n=1}^{\infty} \dfrac{1}{3^n - 2^n}$

49. $\displaystyle\sum_{n=1}^{\infty} \dfrac{n^{10} + 10^n}{n^{11} + 11^n}$

50. $\displaystyle\sum_{n=1}^{\infty} \dfrac{n^{20} + 21^n}{n^{21} + 20^n}$

51. Determine the convergence of $\displaystyle\sum_{n=1}^{\infty} \dfrac{2^n + n}{3^n - 2}$ using the Limit Comparison Test with $b_n = \left(\dfrac{2}{3}\right)^n$.

52. Determine the convergence of $\displaystyle\sum_{n=1}^{\infty} \dfrac{\ln n}{1.5^n}$ using the Limit Comparison Test with $b_n = \dfrac{1}{1.4^n}$.

53. Let $a_n = 1 - \sqrt{1 - \frac{1}{n}}$. Show that $\lim_{n\to\infty} a_n = 0$ and that $\sum_{n=1}^{\infty} a_n$ diverges. *Hint:* Show that $a_n \geq \frac{1}{2n}$.

54. Determine whether $\sum_{n=2}^{\infty} \left(1 - \sqrt{1 - \frac{1}{n^2}}\right)$ converges.

55. Consider $\sum_{n=1}^{\infty} \frac{n}{(n^2+1)^2}$.

(a) Show that the series converges.

(b) [CAS] Use the inequality in (4) from Exercise 83 of Section 10.3 with $M = 99$ to approximate the sum of the series. What is the maximum size of the error?

In Exercises 56–59, determine whether the series converges absolutely. If it does not, determine whether it converges conditionally.

56. $\sum_{n=1}^{\infty} \frac{(-1)^n}{\sqrt[3]{n} + 2n}$

57. $\sum_{n=1}^{\infty} \frac{(-1)^n}{n^{1.1} \ln(n+1)}$

58. $\sum_{n=1}^{\infty} \frac{\cos\left(\frac{\pi}{4} + \pi n\right)}{\sqrt{n}}$

59. $\sum_{n=1}^{\infty} \frac{\cos\left(\frac{\pi}{4} + 2\pi n\right)}{\sqrt{n}}$

60. [CAS] Use a computer algebra system to approximate $\sum_{n=1}^{\infty} \frac{(-1)^n}{n^3 + \sqrt{n}}$ to within an error of at most 10^{-5}.

61. Catalan's constant is defined by $K = \sum_{k=0}^{\infty} \frac{(-1)^k}{(2k+1)^2}$.

(a) How many terms of the series are needed to calculate K with an error of less than 10^{-6}?

(b) [CAS] Carry out the calculation.

62. Give an example of conditionally convergent series $\sum_{n=1}^{\infty} a_n$ and $\sum_{n=1}^{\infty} b_n$ such that $\sum_{n=1}^{\infty}(a_n + b_n)$ converges absolutely.

63. Let $\sum_{n=1}^{\infty} a_n$ be an absolutely convergent series. Determine whether the following series are convergent or divergent:

(a) $\sum_{n=1}^{\infty} \left(a_n + \frac{1}{n^2}\right)$

(b) $\sum_{n=1}^{\infty} (-1)^n a_n$

(c) $\sum_{n=1}^{\infty} \frac{1}{1 + a_n^2}$

(d) $\sum_{n=1}^{\infty} \frac{|a_n|}{n}$

64. Let $\{a_n\}$ be a positive sequence such that $\lim_{n\to\infty} \sqrt[n]{a_n} = \frac{1}{2}$. Determine whether the following series converge or diverge:

(a) $\sum_{n=1}^{\infty} 2a_n$

(b) $\sum_{n=1}^{\infty} 3^n a_n$

(c) $\sum_{n=1}^{\infty} \sqrt{a_n}$

In Exercises 65–72, apply the Ratio Test to determine convergence or divergence, or state that the Ratio Test is inconclusive.

65. $\sum_{n=1}^{\infty} \frac{n^5}{5^n}$

66. $\sum_{n=1}^{\infty} \frac{\sqrt{n+1}}{n^8}$

67. $\sum_{n=1}^{\infty} \frac{1}{n 2^n + n^3}$

68. $\sum_{n=1}^{\infty} \frac{n^4}{n!}$

69. $\sum_{n=1}^{\infty} \frac{2^{n^2}}{n!}$

70. $\sum_{n=4}^{\infty} \frac{\ln n}{n^{3/2}}$

71. $\sum_{n=1}^{\infty} \left(\frac{n}{2}\right)^n \frac{1}{n!}$

72. $\sum_{n=1}^{\infty} \left(\frac{n}{4}\right)^n \frac{1}{n!}$

In Exercises 73–76, apply the Root Test to determine convergence or divergence, or state that the Root Test is inconclusive.

73. $\sum_{n=1}^{\infty} \frac{1}{4^n}$

74. $\sum_{n=1}^{\infty} \left(\frac{2}{n}\right)^n$

75. $\sum_{n=1}^{\infty} \left(\frac{3}{4n}\right)^n$

76. $\sum_{n=1}^{\infty} \left(\cos \frac{1}{n}\right)^{n^3}$

In Exercises 77–100, determine convergence or divergence using any method covered in the text.

77. $\sum_{n=1}^{\infty} \left(\frac{2}{3}\right)^n$

78. $\sum_{n=1}^{\infty} \frac{\pi^{7n}}{e^{8n}}$

79. $\sum_{n=1}^{\infty} e^{-0.02n}$

80. $\sum_{n=1}^{\infty} n e^{-0.02n}$

81. $\sum_{n=1}^{\infty} \frac{(-1)^{n-1}}{\sqrt{n} + \sqrt{n+1}}$

82. $\sum_{n=10}^{\infty} \frac{1}{n(\ln n)^{3/2}}$

83. $\sum_{n=2}^{\infty} \frac{(-1)^n}{\ln n}$

84. $\sum_{n=1}^{\infty} \frac{n!}{(2n)!}$

85. $\sum_{n=2}^{\infty} \frac{n}{1 + 100n}$

86. $\sum_{n=2}^{\infty} \frac{n^3 - 2n^2 + n - 4}{2n^4 + 3n^3 - 4n^2 - 1}$

87. $\sum_{n=1}^{\infty} \frac{\cos n}{n^{3/2}}$

88. $\sum_{n=1}^{\infty} \frac{n}{\sqrt{n^{3/2} + 1}}$

89. $\sum_{n=1}^{\infty} \left(\frac{n}{5n+2}\right)^n$

90. $\sum_{n=1}^{\infty} \frac{e^n}{n!}$

91. $\sum_{n=1}^{\infty} \frac{1}{n\sqrt{n} + \ln n}$

92. $\sum_{n=1}^{\infty} \frac{1}{\sqrt[3]{n}(1 + \sqrt{n})}$

93. $\sum_{n=1}^{\infty} \left(\frac{1}{\sqrt{n}} - \frac{1}{\sqrt{n+1}}\right)$

94. $\sum_{n=1}^{\infty} \left(\ln n - \ln(n+1)\right)$

95. $\sum_{n=1}^{\infty} \frac{1}{n + \sqrt{n}}$

96. $\sum_{n=2}^{\infty} \frac{\cos(\pi n)}{n^{2/3}}$

97. $\sum_{n=2}^{\infty} \frac{1}{n^{\ln n}}$

98. $\sum_{n=2}^{\infty} \frac{1}{\ln^3 n}$

99. $\sum_{n=1}^{\infty} \sin^2 \frac{\pi}{n}$

100. $\sum_{n=0}^{\infty} \frac{2^{2n}}{n!}$

In Exercises 101–106, find the interval of convergence of the power series.

101. $\displaystyle\sum_{n=0}^{\infty} \frac{2^n x^n}{n!}$

102. $\displaystyle\sum_{n=0}^{\infty} \frac{x^n}{n+1}$

103. $\displaystyle\sum_{n=0}^{\infty} \frac{n^6}{n^8+1}(x-3)^n$

104. $\displaystyle\sum_{n=0}^{\infty} nx^n$

105. $\displaystyle\sum_{n=0}^{\infty} (nx)^n$

106. $\displaystyle\sum_{n=2}^{\infty} \frac{(2x-3)^n}{n \ln n}$

107. Expand $f(x) = \dfrac{2}{4-3x}$ as a power series centered at $c = 0$. Determine the values of x for which the series converges.

108. Prove that

$$\sum_{n=0}^{\infty} ne^{-nx} = \frac{e^{-x}}{(1-e^{-x})^2}$$

Hint: Express the left-hand side as the derivative of a geometric series.

109. Let $F(x) = \displaystyle\sum_{k=0}^{\infty} \frac{x^{2k}}{2^k \cdot k!}$.

(a) Show that $F(x)$ has infinite radius of convergence.

(b) Show that $y = F(x)$ is a solution of

$$y'' = xy' + y, \qquad y(0) = 1, \qquad y'(0) = 0$$

(c) [CAS] Plot the partial sums S_N for $N = 1, 3, 5, 7$ on the same set of axes.

110. Find a power series $P(x) = \displaystyle\sum_{n=0}^{\infty} a_n x^n$ that satisfies the Laguerre differential equation

$$xy'' + (1-x)y' - y = 0$$

with initial condition satisfying $P(0) = 1$.

111. Use power series to evaluate $\displaystyle\lim_{x \to 0} \frac{x^2 e^x}{\cos x - 1}$.

112. Use power series to evaluate $\displaystyle\lim_{x \to 0} \frac{x^2(1 - \ln(x+1))}{\sin x - x}$.

In Exercises 113–118, find the Taylor polynomial at $x = a$ for the given function.

113. $f(x) = x^3$, $\quad T_3$, $\quad a = 1$

114. $f(x) = 3(x+2)^3 - 5(x+2)$, $\quad T_3$, $\quad a = -2$

115. $f(x) = x \ln(x)$, $\quad T_4$, $\quad a = 1$

116. $f(x) = (3x+2)^{1/3}$, $\quad T_3$, $\quad a = 2$

117. $f(x) = xe^{-x^2}$, $\quad T_4$, $\quad a = 0$

118. $f(x) = \ln(\cos x)$, $\quad T_3$, $\quad a = 0$

119. Find the nth Maclaurin polynomial for $f(x) = e^{3x}$.

120. Use the fifth Maclaurin polynomial of $f(x) = e^x$ to approximate $\sqrt{e}$. Use a calculator to determine the error.

121. Use the third Taylor polynomial of $f(x) = \tan^{-1} x$ at $a = 1$ to approximate $f(1.1)$. Use a calculator to determine the error.

122. Let T_4 be the Taylor polynomial for $f(x) = \sqrt{x}$ at $a = 16$. Use the Error Bound to find the maximum possible size of $|f(17) - T_4(17)|$.

123. Find n such that $|e - T_n(1)| < 10^{-8}$, where T_n is the nth Maclaurin polynomial for $f(x) = e^x$.

124. Let T_4 be the Taylor polynomial for $f(x) = x \ln x$ at $a = 1$ computed in Exercise 115. Use the Error Bound to find a bound for $|f(1.2) - T_4(1.2)|$.

125. Verify that $T_n(x) = 1 + x + x^2 + \cdots + x^n$ is the nth Maclaurin polynomial of $f(x) = 1/(1-x)$. Show using substitution that the nth Maclaurin polynomial for $f(x) = 1/(1 - x/4)$ is

$$T_n(x) = 1 + \frac{1}{4}x + \frac{1}{4^2}x^2 + \cdots + \frac{1}{4^n}x^n$$

What is the nth Maclaurin polynomial for $g(x) = \dfrac{1}{1+x}$?

126. Let $f(x) = \dfrac{5}{4 + 3x - x^2}$ and let a_k be the coefficient of x^k in the Maclaurin polynomial T_n for $k \le n$.

(a) Show that $f(x) = \left(\dfrac{1/4}{1 - x/4} + \dfrac{1}{1+x} \right)$.

(b) Use Exercise 125 to show that $a_k = \dfrac{1}{4^{k+1}} + (-1)^k$.

(c) Compute T_3.

In Exercises 127–136, find the Taylor series centered at c.

127. $f(x) = e^{4x}$, $\quad c = 0$

128. $f(x) = e^{2x}$, $\quad c = -1$

129. $f(x) = x^4$, $\quad c = 2$

130. $f(x) = x^3 - x$, $\quad c = -2$

131. $f(x) = \sin x$, $\quad c = \pi$

132. $f(x) = e^{x-1}$, $\quad c = -1$

133. $f(x) = \dfrac{1}{1-2x}$, $\quad c = -2$

134. $f(x) = \dfrac{1}{(1-2x)^2}$, $\quad c = -2$

135. $f(x) = \ln \dfrac{x}{2}$, $\quad c = 2$

136. $f(x) = x \ln \left(1 + \dfrac{x}{2}\right)$, $\quad c = 0$

In Exercises 137–140, find the first three terms of the Maclaurin series of $f(x)$ and use it to calculate $f^{(3)}(0)$.

137. $f(x) = (x^2 - x)e^{x^2}$

138. $f(x) = \tan^{-1}(x^2 - x)$

139. $f(x) = \dfrac{1}{1 + \tan x}$

140. $f(x) = (\sin x)\sqrt{1+x}$

141. Calculate $\dfrac{\pi}{2} - \dfrac{\pi^3}{2^3 3!} + \dfrac{\pi^5}{2^5 5!} - \dfrac{\pi^7}{2^7 7!} + \cdots$.

142. Find the Maclaurin series of the function $F(x) = \displaystyle\int_0^x \frac{e^t - 1}{t}\,dt$.

Tom Brakefield/Getty Images

11 PARAMETRIC EQUATIONS, POLAR COORDINATES, AND CONIC SECTIONS

We can study the interaction between two animal species with populations $q(t)$ and $p(t)$, where each population is a function of time, to investigate how the two populations change. Combining the functions in the form $(q(t), p(t))$ yields a parametric representation of a curve in the qp-plane. Tracing this curve as t changes creates a story about this interaction and its impact on population size.

This chapter introduces two important new tools. First, we consider parametric equations, which describe curves in a form that is particularly useful for analyzing motion and is indispensable in fields such as computer graphics and computer-aided design. We then study polar coordinates, an alternative to rectangular coordinates that simplifies computations in many applications. The chapter closes with a discussion of the conic sections (ellipses, hyperbolas, and parabolas).

11.1 Parametric Equations

We use the term "particle" when we treat an object as a moving point, ignoring its internal structure.

Imagine a particle moving along a curve $\mathcal{C}$ in the plane as in Figure 1. We would like to be able to describe the particle's motion along the curve. To express this motion mathematically, we consider how its coordinates x and y are changing in time, that is, how they depend on a time variable t. Thus, x and y are both functions of time, t, and the location of the particle at t is given by

$$c(t) = (x(t), y(t))$$

This representation of the curve $\mathcal{C}$ is called a **parametrization** with **parameter** t, and $\mathcal{C}$ is called a **parametric curve**.

In a parametrization, we often use t for the parameter, thinking of the dependent variables as changing in time, but we are free to use any other variable (such as s or θ). In plots of parametric curves, the direction of motion is often indicated by an arrow as in Figure 1.

Specific equations defining a parametrization, such as $x = 2t - 4$ and $y = 3 + t^2$ in the next example, are called **parametric equations**.

EXAMPLE 1 Sketch the curve with parametric equations

$$x = 2t - 4, \qquad y = 3 + t^2 \qquad \boxed{1}$$

Solution First compute the x- and y-coordinates for several values of t as in Table 1, and plot the corresponding points (x, y) as in Figure 2. Then join the points by a smooth curve, indicating the direction of motion (direction of increasing t) with an arrow. ∎

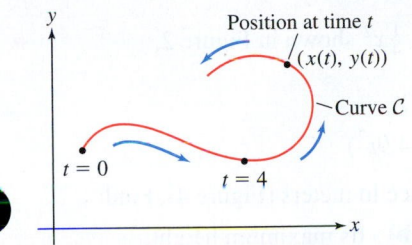

FIGURE 1 Particle moving along a curve $\mathcal{C}$ in the plane.

TABLE 1

t	$x = 2t - 4$	$y = 3 + t^2$
-2	-8	7
0	-4	3
2	0	7
4	4	19

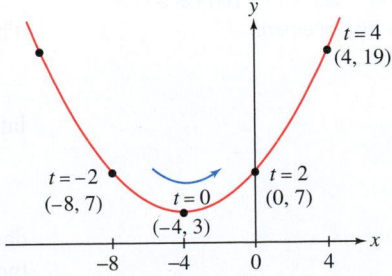

Graphing calculators or CAS software can be used to sketch and examine parametric curves.

DF **FIGURE 2** The parametric curve $x = 2t - 4$, $y = 3 + t^2$.

CONCEPTUAL INSIGHT The graph of $y = x^2$ can be parametrized in a simple way. We take $x = t$ and $y = t^2$. Then, since $y = t^2$ and $t = x$, it follows that $y = x^2$. Therefore, the parabola $y = x^2$ is parametrized by $c(t) = (t, t^2)$. More generally, we can

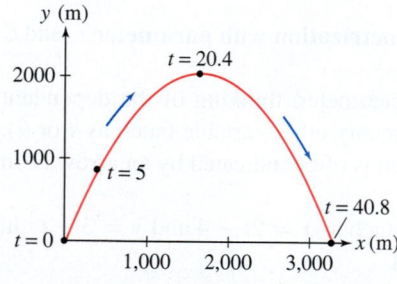

FIGURE 3 The parametric curve
$x = 5\cos(3t)\cos\left(\frac{2}{3}\sin(5t)\right)$,
$y = 4\sin(3t)\cos\left(\frac{2}{3}\sin(5t)\right)$.

parametrize the graph of $y = f(x)$ by taking $x = t$ and $y = f(t)$. Therefore, $c(t) = (t, f(t))$ parametrizes the graph of $y = f(x)$. For another example, the graph of $y = e^x$ is parametrized by $c(t) = (t, e^t)$. An advantage of parametric equations is that they enable us to describe curves that are not graphs of functions. For example, the curve in Figure 3 is not of the form $y = f(x)$ but it can be expressed parametrically.

As we have just noted, a parametric curve $c(t)$ need not be the graph of a function. If it is, however, it may be possible to find the function f by "eliminating the parameter" as in the next example.

EXAMPLE 2 Eliminating the Parameter Describe the parametric curve

$$c(t) = (2t - 4, 3 + t^2)$$

of the previous example in the form $y = f(x)$.

Solution We eliminate the parameter by solving for y as a function of x. First, express t in terms of x: Since $x = 2t - 4$, we have $t = \frac{1}{2}x + 2$. Then substitute

$$y = 3 + t^2 = 3 + \left(\frac{1}{2}x + 2\right)^2 = 7 + 2x + \frac{1}{4}x^2$$

Thus, $c(t)$ traces out the graph of $f(x) = 7 + 2x + \frac{1}{4}x^2$ shown in Figure 2. ∎

EXAMPLE 3 A model rocket follows the trajectory

$$c(t) = (80t, 200t - 4.9t^2)$$

until it hits the ground, with t in seconds and distance in meters (Figure 4). Find:

(a) The rocket's height at $t = 5$ s. **(b)** Its maximum height.

Solution The height of the rocket at time t is $y(t) = 200t - 4.9t^2$.

(a) The height at $t = 5$ s is

$$y(5) = 200(5) - 4.9(5^2) = 877.5 \text{ m}$$

(b) The maximum height occurs at the critical point of $y(t)$ found as follows:

$$y'(t) = \frac{d}{dt}(200t - 4.9t^2) = 200 - 9.8t$$

Thus, $y' = 0$ when $200 - 9.8t = 0$, that is, for

$$t = \frac{200}{9.8} \approx 20.4 \text{ s}$$

The rocket's maximum height is $y(20.4) = 200(20.4) - 4.9(20.4)^2 \approx 2041$ m. ∎

We now discuss parametrizations of lines and circles. They will appear frequently in later chapters.

To begin, note that the parametric equations

$$x = t, \qquad y = mt \qquad -\infty < t < \infty$$

describe a line that passes through the origin at $t = 0$ and has slope m (since $y = mx$ for these equations). Translating $x(t)$ by a and $y(t)$ by b, we instead have a parametrization of the line of slope m passing through (a, b) at $t = 0$:

DF FIGURE 4 Trajectory of rocket.

CAUTION *The graph of height versus time for an object tossed in the air is a parabola (by Galileo's formula). But keep in mind that Figure 4 is not a graph of height versus time. It shows the actual path of the rocket (which has both a vertical and a horizontal displacement).*

Parametrization of a Line The line through $P = (a, b)$ of slope m is parametrized by

$$\boxed{x = a + t, \qquad y = b + mt \qquad -\infty < t < \infty} \qquad \boxed{2}$$

EXAMPLE 4 Parametrization of a Line Find parametric equations for the line through $P = (3, -1)$ and $Q = (5, -8)$.

Solution The slope of the line is $m = \dfrac{-8 - (-1)}{5 - 3} = -\dfrac{7}{2}$. Using $(a, b) = (3, -1)$ in Eq. (3), we obtain the parametrization

$$x = 3 + t, \qquad y = -1 - \frac{7}{2}t$$

We get another parametrization using $(a, b) = (5, -8)$, in which case

$$x = 5 + t, \qquad y = -8 - \frac{7}{2}t$$

The parametrizations here are different parametrizations of the same line. We can think of the line as a road and a parametrization as a trip along the road. Thus, the different parametrizations correspond to different trips along the same road. ∎

If $p/q = m$, then the equations

$$x = a + qt, \qquad y = b + pt \qquad -\infty < t < \infty$$

also parametrize the line of slope m passing through (a, b) (see Exercise 46).

The circle of radius R centered at the origin has the parametrization

$$x = R \cos \theta, \qquad y = R \sin \theta$$

The parameter θ represents the angle corresponding to the point (x, y) on the circle (Figure 5). The circle is traversed once in the counterclockwise direction as θ varies over a half-open interval of length 2π such as $[0, 2\pi)$ or $[-\pi, \pi)$.

More generally, the circle of radius R with center (a, b) has parametrization (Figure 5)

$$\boxed{x = a + R \cos \theta, \qquad y = b + R \sin \theta} \qquad \boxed{3}$$

As a check, let's verify that a point (x, y) given by Eq. (5) satisfies the equation of the circle of radius R centered at (a, b):

$$(x - a)^2 + (y - b)^2 = (a + R \cos \theta - a)^2 + (b + R \sin \theta - b)^2$$

$$= R^2 \cos^2 \theta + R^2 \sin^2 \theta = R^2$$

In general, to **translate** (meaning "to move") a parametric curve horizontally a units and vertically b units, replace $c(t) = (x(t), y(t))$ by $c(t) = (a + x(t), b + y(t))$.

Suppose we have a parametrization $c(t) = (x(t), y(t))$, where $x(t)$ is an even function and $y(t)$ is an odd function, that is, $x(-t) = x(t)$ and $y(-t) = -y(t)$. In this case, $c(-t)$ is the *reflection* of $c(t)$ across the x-axis:

$$c(-t) = (x(-t), y(-t)) = (x(t), -y(t))$$

The curve, therefore, is *symmetric* with respect to the x-axis. We apply this remark in the next example.

EXAMPLE 5   **Parametrization of an Ellipse**   Verify that the ellipse with equation $\left(\frac{x}{a}\right)^2 + \left(\frac{y}{b}\right)^2 = 1$ is parametrized by

$$\boxed{c(t) = (a \cos t, \, b \sin t) \qquad (\text{for } -\pi \le t < \pi)}$$

Plot the case $a = 4, b = 2$.

Solution To verify that $c(t)$ parametrizes the ellipse, show that the equation of the ellipse is satisfied with $x = a \cos t$, $y = b \sin t$:

$$\left(\frac{x}{a}\right)^2 + \left(\frac{y}{b}\right)^2 = \left(\frac{a \cos t}{a}\right)^2 + \left(\frac{b \sin t}{b}\right)^2 = \cos^2 t + \sin^2 t = 1$$

To plot the case $a = 4$, $b = 2$, we connect the points for the t-values in Table 2 [see Figure 6(A)]. This gives us the top half of the ellipse for $0 \le t \le \pi$. Then we observe that $x(t) = 4 \cos t$ is even and $y(t) = 2 \sin t$ is odd. As noted earlier, this tells us that the bottom half of the ellipse is obtained by symmetry with respect to the x-axis, as in Figure 6(B). Alternatively, we could also evaluate $x(t)$ and $y(t)$ for negative values of t between $-\pi$ and 0 to determine the bottom portion of the ellipse. ∎

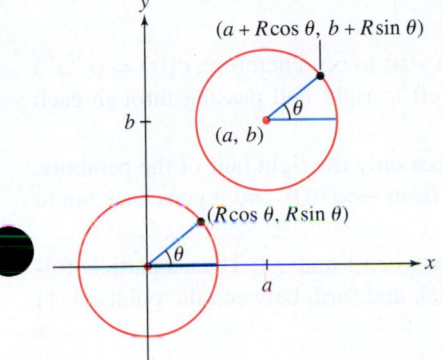

FIGURE 5 Parametrization of a circle of radius R with center (a, b).

TABLE 2

t	$x(t) = 4 \cos t$	$y(t) = 2 \sin t$
0	4	0
$\dfrac{\pi}{6}$	$2\sqrt{3}$	1
$\dfrac{\pi}{3}$	2	$\sqrt{3}$
$\dfrac{\pi}{2}$	0	2
$\dfrac{2\pi}{3}$	-2	$\sqrt{3}$
$\dfrac{5\pi}{6}$	$-2\sqrt{3}$	1
π	-4	0

A parametric curve $c(t)$ is also called a **path**. This term emphasizes that $c(t)$ describes not just an underlying curve $\mathcal{C}$, but a particular way of moving along the curve.

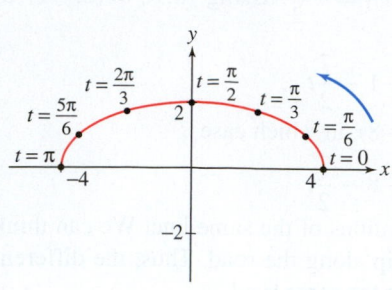

(A)

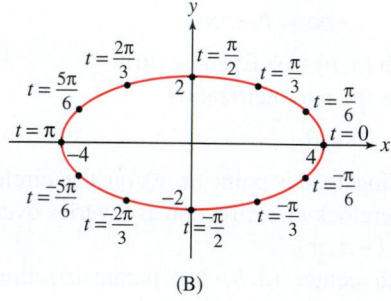

(B)

FIGURE 6 Ellipse with parametric equations $x = 4\cos t$, $y = 2\sin t$.

> **CONCEPTUAL INSIGHT** The parametric equations for the ellipse in Example 5 illustrate a key difference between the path $c(t)$ and its underlying curve $\mathcal{C}$. The curve $\mathcal{C}$ is an ellipse in the plane, whereas $c(t)$ describes a particular, counterclockwise motion of a particle along the ellipse. If we let t vary from 0 to 4π, then the particle goes around the ellipse twice.
>
> A key feature of parametrizations of curves is that they are not unique. In fact, every curve can be parametrized in infinitely many different ways. For instance, the parabola $y = x^2$ is parametrized not only by (t, t^2) but also by (t^3, t^6) or (t^5, t^{10}), and so on.

EXAMPLE 6 **Different Paths on the Parabola** $y = x^2$ Describe the motion of a particle along each of the following paths:

(a) $c_1(t) = (t^3, t^6)$ **(b)** $c_2(t) = (t^2, t^4)$ **(c)** $c_3(t) = (\cos t, \cos^2 t)$

Solution Each of these parametrizations satisfies $y = x^2$, so all three parametrize portions of the parabola $y = x^2$.

(a) As t varies from $-\infty$ to ∞, t^3 also varies from $-\infty$ to ∞. Therefore, $c_1(t) = (t^3, t^6)$ traces the entire parabola $y = x^2$, moving from left to right and passing through each point once [Figure 7(A)].

(b) Since $x = t^2 \geq 0$, the path $c_2(t) = (t^2, t^4)$ traces only the right half of the parabola. The particle comes in toward the origin as t varies from $-\infty$ to 0, and it goes back out to the right as t varies from 0 to ∞ [Figure 7(B)].

(c) As t varies from $-\infty$ and ∞, $\cos t$ oscillates between 1 and -1. Thus, a particle following the path $c_3(t) = (\cos t, \cos^2 t)$ oscillates back and forth between the points $(1, 1)$ and $(-1, 1)$ on the parabola [Figure 7(C)]. ∎

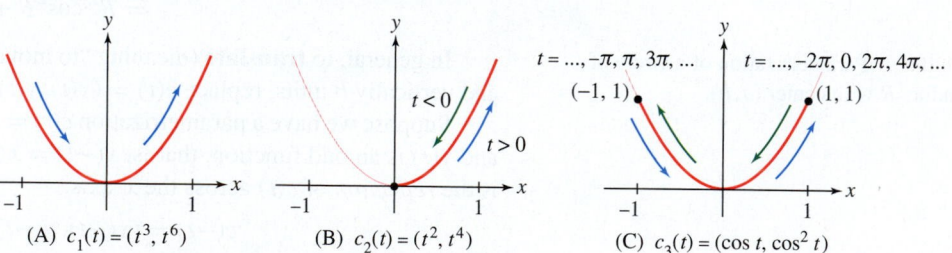

FIGURE 7 Three parametrizations of portions of the parabola.

(A) $c_1(t) = (t^3, t^6)$ (B) $c_2(t) = (t^2, t^4)$ (C) $c_3(t) = (\cos t, \cos^2 t)$

A **cycloid** is a curve traced by a point on the circumference of a rolling wheel as in Figure 8. Cycloids are particularly interesting because they satisfy the "brachistochrone property" (see the marginal note).

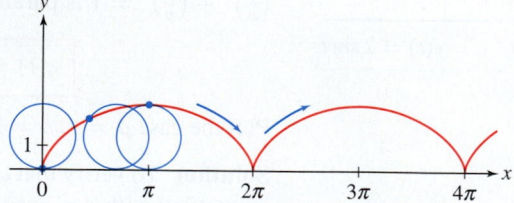

FIGURE 8 A cycloid.

A stellar cast of mathematicians (including Galileo, Pascal, Newton, Leibniz, Huygens, and Bernoulli) studied the cycloid and discovered many of its remarkable properties. A slide designed so that an object sliding down (without friction) reaches the bottom in the least time must have the shape of an inverted cycloid. This is the brachistochrone property, a term derived from the Greek brachistos, "shortest," and chronos, "time."

EXAMPLE 7 **Parametrizing the Cycloid** Find parametric equations for the cycloid generated by a point P on the unit circle.

Solution The point P is located at the origin at $t = 0$. We parametrize the path with the parameter t representing the angle, in radians, through which the wheel has rotated [Figure 9(A)]. At time t, the circle has rolled t units along the x-axis and the center C

of the circle then has coordinates $(t, 1)$ as in the figure. Figure 9(B) shows that P is $\sin t$ units to the left of C and $\cos t$ units down from C, giving us the parametric equations

$$x(t) = t - \sin t, \qquad y(t) = 1 - \cos t$$

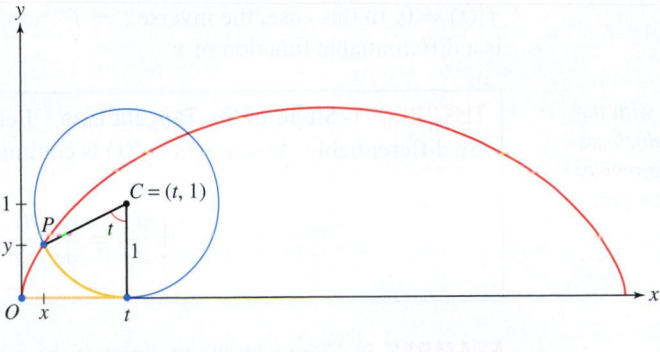

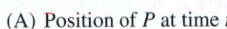

(A) Position of P at time t

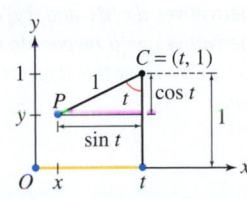

(B) P has coordinates
$x = t - \sin t,\ y = 1 - \cos t$

FIGURE 9

The argument in Example 7 shows in a similar fashion that the cycloid generated by a circle of radius R has parametric equations

$$x = Rt - R\sin t, \qquad y = R - R\cos t$$

We usually think of parametric equations as representing a particle or object in motion, but we can use parametric equations in any situation when two (or even more) variables depend on a particular independent variable. In the next example, population sizes $p(t)$ and $q(t)$ of two interacting animal species vary in time t.

EXAMPLE 8 **A Predator–Prey Model** Let $p(t)$ and $q(t)$ represent the changing (in time t) population sizes of a predator and its prey, respectively. Put together as $(q(t), p(t))$, these functions give a parametric representation of a curve in the qp-plane. The curve in Figure 10 represents a simple model of how the populations might change over time. Discuss how the populations change as time increases at points A and B on the curve. Determine whether the curve is traced clockwise or counterclockwise with increasing t.

Solution Notice that at point A on the curve, the predator population p is close to its maximum value. We expect significant consumption of the prey species is occurring then and therefore q is decreasing. Thus, as t increases, $(q(t), p(t))$ moves to the left through A. Furthermore, at point B, the prey species is near a minimum. In this situation, the resources for the predator are low, and therefore, we expect the predator population, p, is decreasing. It follows that as t increases, $(q(t), p(t))$ moves down through B. Both of these situations support the conclusion that the curve is traced in the counterclockwise direction.

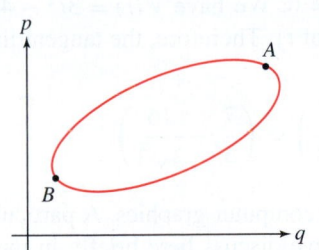

FIGURE 10 A parametric curve $(q(t), p(t))$ providing a simple model of a predator–prey interaction.

Tangent Lines to Parametric Curves

Just as we use tangent lines to the graph of $y = f(x)$ to determine the rate of change of the function f, we would like to be able to determine how y changes with x when the curve is described by parametric equations. The slope of the tangent line is the derivative dy/dx. We have to use the Chain Rule to compute dy/dx because y is not given explicitly as a function of x. Write $x = f(t)$, $y = g(t)$. Then, by the Chain Rule,

$$g'(t) = \frac{dy}{dt} = \frac{dy}{dx}\frac{dx}{dt} = \frac{dy}{dx}f'(t)$$

If $f'(t) \neq 0$, we can divide by $f'(t)$ to obtain

$$\frac{dy}{dx} = \frac{g'(t)}{f'(t)}$$

This calculation is valid if $f(t)$ and $g(t)$ are differentiable, $f'(t)$ is continuous, and $f'(t) \neq 0$. In this case, the inverse $t = f^{-1}(x)$ exists, and the composite $y = g(f^{-1}(x))$ is a differentiable function of x.

THEOREM 1 Slope of the Tangent Line Let $c(t) = (x(t), y(t))$, where $x(t)$ and $y(t)$ are differentiable. Assume that $x'(t)$ is continuous and $x'(t) \neq 0$. Then

$$\frac{dy}{dx} = \frac{dy/dt}{dx/dt} = \frac{y'(t)}{x'(t)} \qquad \boxed{6}$$

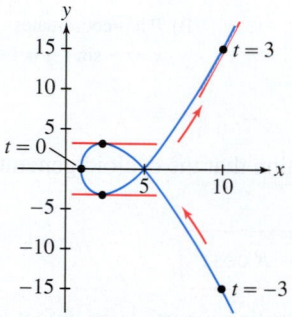

FIGURE 11 The paramteric curve $c(t) = (t^2 + 1, t^3 - 4t)$.

EXAMPLE 9 Figure 11 shows a plot of the parametric curve $c(t) = (t^2 + 1, t^3 - 4t)$.

(a) Find an equation of the tangent line at $t = 3$.

(b) There appear to be two points where the tangent is horizontal. Find them.

Solution We have

$$\frac{dy}{dx} = \frac{y'(t)}{x'(t)} = \frac{(t^3 - 4t)'}{(t^2 + 1)'} = \frac{3t^2 - 4}{2t}$$

(a) The slope at $t = 3$ is

$$\frac{dy}{dx} = \frac{3t^2 - 4}{2t}\bigg|_{t=3} = \frac{3(3)^2 - 4}{2(3)} = \frac{23}{6}$$

Since $c(3) = (10, 15)$, the equation of the tangent line in point-slope form is

$$y - 15 = \frac{23}{6}(x - 10)$$

(b) The slope dy/dx is zero if $y'(t) = 0$ and $x'(t) \neq 0$. We have $y'(t) = 3t^2 - 4 = 0$ for $t = \pm 2/\sqrt{3}$ [and $x'(t) = 2t \neq 0$ for these values of t]. Therefore, the tangent line is horizontal at the points

$$c\left(-\frac{2}{\sqrt{3}}\right) = \left(\frac{7}{3}, \frac{16}{3\sqrt{3}}\right), \qquad c\left(\frac{2}{\sqrt{3}}\right) = \left(\frac{7}{3}, -\frac{16}{3\sqrt{3}}\right) \qquad ■$$

Parametric curves are widely used in the field of computer graphics. A particularly important class of curves are **Bézier curves**, which we discuss here briefly in the cubic case. Figure 12 shows two examples of Bézier curves determined by four "control points":

$$P_0 = (a_0, b_0), \qquad P_1 = (a_1, b_1), \qquad P_2 = (a_2, b_2), \qquad P_3 = (a_3, b_3)$$

The Bézier curve $c(t) = (x(t), y(t))$ is defined for $0 \leq t \leq 1$ by

$$x(t) = a_0(1-t)^3 + 3a_1 t(1-t)^2 + 3a_2 t^2(1-t) + a_3 t^3 \qquad \boxed{7}$$

$$y(t) = b_0(1-t)^3 + 3b_1 t(1-t)^2 + 3b_2 t^2(1-t) + b_3 t^3 \qquad \boxed{8}$$

Note that $c(0) = (a_0, b_0)$ and $c(1) = (a_3, b_3)$, so the Bézier curve begins at P_0 and ends at P_3 (Figure 12). It can also be shown that the Bézier curve is contained within the quadrilateral (shown in blue) with vertices P_0, P_1, P_2, P_3. However, $c(t)$ does not pass through P_1 and P_2. Instead, these intermediate control points determine the slopes of the tangent lines at P_0 and P_3, as we show in the next example (also, see Exercises 73 and 76).

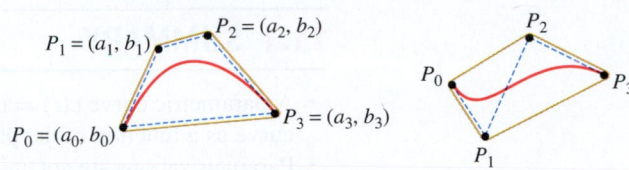

FIGURE 12 Cubic Bézier curves specified by four control points.

EXAMPLE 10 Show that the Bézier curve is tangent to segment $\overline{P_0 P_1}$ at P_0.

Solution The Bézier curve passes through P_0 at $t = 0$, so we must show that the slope of the tangent line at $t = 0$ is equal to the slope of $\overline{P_0 P_1}$. To find the slope, we compute the derivatives:

$$x'(t) = -3a_0(1-t)^2 + 3a_1(1 - 4t + 3t^2) + 3a_2(2t - 3t^2) + 3a_3t^2$$

$$y'(t) = -3b_0(1-t)^2 + 3b_1(1 - 4t + 3t^2) + 3b_2(2t - 3t^2) + 3b_3t^2$$

Evaluating at $t = 0$, we obtain $x'(0) = 3(a_1 - a_0)$, $y'(0) = 3(b_1 - b_0)$, and

$$\left.\frac{dy}{dx}\right|_{t=0} = \frac{y'(0)}{x'(0)} = \frac{3(b_1 - b_0)}{3(a_1 - a_0)} = \frac{b_1 - b_0}{a_1 - a_0}$$

This is equal to the slope of the line through $P_0 = (a_0, b_0)$ and $P_1 = (a_1, b_1)$ as claimed (provided that $a_1 \neq a_0$). ∎

Area Under a Parametric Curve

As we know, the area under a curve given by $y = h(x)$ when $h(x) \geq 0$ for $a \leq x \leq b$ is given by

$$A = \int_a^b h(x)\,dx$$

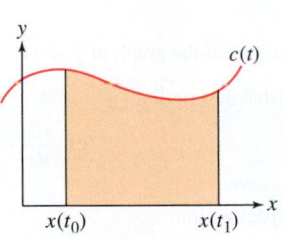

FIGURE 13 Finding area under a parametric curve $c(t)$.

When the curve $y = h(x)$ is traced once by a parametric curve $c(t) = (x(t), y(t))$ as in Figure 13, where $x(t_0) = a$ and $x(t_1) = b$, then we can substitute, replacing $y = h(x)$ by $y(t)$ and dx by $x'(t)dt$, yielding a formula for the area A under the curve:

$$A = \int_{t_0}^{t_1} y(t)x'(t)\,dt \qquad \boxed{9}$$

EXAMPLE 11 The parametric curve $c(t) = (t^2, 4t - t^3)$ is shown in Figure 14. Determine the area enclosed within the loop.

Solution The curve is symmetric about the x-axis. Therefore, the desired area is twice the area A between the top half of the loop and the x-axis. The area A can be computed as the area under a parametric curve.

Note that the curve crosses the x-axis when $y = 0$, and therefore, at $t = -2, 0, 2$. Furthermore, $c(0) = (0,0)$, $c(2) = (4,0)$, and $y \geq 0$ for $0 \leq t \leq 2$. Thus, the path traces the top half of the loop (which can be considered as the graph of a function) one time as t goes from 0 to 2. It follows that A, the area between the top half of the loop and the x-axis, is given by

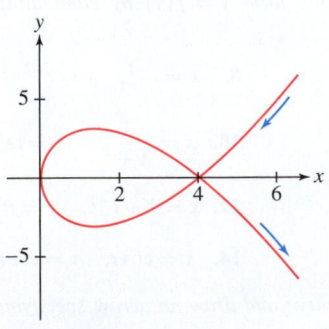

FIGURE 14 The parametric curve $c(t) = (t^2, 4t - t^3)$.

$$A = \int_0^2 \underbrace{(4t - t^3)}_{y(t)} \underbrace{(2t)}_{x'(t)}\,dt = \int_0^2 (8t^2 - 2t^4)\,dt = \left(\frac{8}{3}t^3 - \frac{2}{5}t^5\right)\Bigg|_0^2 = \frac{128}{15}$$

Therefore, the area enclosed in the loop is $2A = 256/15 \approx 17.07$. ∎

11.1 SUMMARY

- A parametric curve $c(t) = (x(t), y(t))$ describes the path of a particle moving along a curve as a function of the parameter t.
- Parametrizations are not unique: Every curve $\mathcal{C}$ can be parametrized in infinitely many ways. Furthermore, the path $c(t)$ may traverse all or part of $\mathcal{C}$ more than once.
- Slope of the tangent line at $c(t)$:

$$\frac{dy}{dx} = \frac{dy/dt}{dx/dt} = \frac{y'(t)}{x'(t)} \qquad \text{[valid if } x'(t) \neq 0]$$

- Do not confuse the slope of the tangent line dy/dx with the derivatives dy/dt and dx/dt, with respect to t.
- Standard parametrizations:

  - Line of slope $m = s/r$ through $P = (a, b)$: $c(t) = (a + rt, b + st)$
  - Circle of radius R centered at $P = (a, b)$: $c(t) = (a + R\cos t, b + R\sin t)$
  - Cycloid generated by a circle of radius R: $c(t) = (R(t - \sin t), R(1 - \cos t))$
  - Graph of $y = f(x)$: $c(t) = (t, f(t))$

- Area under a parametric curve $c(t) = (x(t), y(t))$ that does not dip below the x-axis and that traces once the graph of a function is given by $A = \displaystyle\int_{t_0}^{t_1} y(t) x'(t)\, dt$.

11.1 EXERCISES

Preliminary Questions

1. Describe the shape of the curve $x = 3\cos t$, $y = 3\sin t$.

2. How does $x = 4 + 3\cos t$, $y = 5 + 3\sin t$ differ from the curve in the previous question?

3. What is the maximum height of a particle whose path has parametric equations $x = t^9$, $y = 4 - t^2$?

4. Can the parametric curve $(t, \sin t)$ be represented as a graph $y = f(x)$? What about $(\sin t, t)$?

5. (a) Describe the path of an ant that is crawling along the plane according to $c_1(t) = (f(t), f(t))$, where $f(t)$ is an increasing function.

(b) Compare that path to the path of a second ant crawling according to $c_2(t) = (f(2t), f(2t))$.

6. Find three different parametrizations of the graph of $y = x^3$.

7. Match the derivatives with a verbal description:

(a) $\dfrac{dx}{dt}$ (b) $\dfrac{dy}{dt}$ (c) $\dfrac{dy}{dx}$

(i) Slope of the tangent line to the curve

(ii) Vertical rate of change with respect to time

(iii) Horizontal rate of change with respect to time

Exercises

1. Find the coordinates at times $t = 0, 2, 4$ of a particle following the path $x = 1 + t^3$, $y = 9 - 3t^2$.

2. Find the coordinates at $t = 0, \frac{\pi}{4}, \pi$ of a particle moving along the path $c(t) = (\cos 2t, \sin^2 t)$.

3. Show that the path traced by the model rocket in Example 3 is a parabola by eliminating the parameter.

4. Use the table of values to sketch the parametric curve $(x(t), y(t))$, indicating the direction of motion.

t	-3	-2	-1	0	1	2	3
x	-15	0	3	0	-3	0	15
y	5	0	-3	-4	-3	0	5

5. Graph the parametric curves. Include arrows indicating the direction of motion.

(a) (t, t), $-\infty < t < \infty$

(b) $(\sin t, \sin t)$, $0 \leq t \leq 2\pi$

(c) (e^t, e^t), $-\infty < t < \infty$

(d) (t^3, t^3), $-1 \leq t \leq 1$

6. Give two different parametrizations of the line through $(4, 1)$ with slope 2.

In Exercises 7–14, express in the form $y = f(x)$ by eliminating the parameter.

7. $x = t + 3$, $y = 4t$

8. $x = t^{-1}$, $y = t^{-2}$

9. $x = t^3 - 1$, $y = t^2 + 1$

10. $x = \dfrac{1}{1+t}$, $y = te^t$

11. $x = e^{-2t}$, $y = 6e^{4t}$

12. $x = 1 + t^{-1}$, $y = t^2$

13. $x = \ln t$, $y = 2 - t$

14. $x = \cos t$, $y = \csc t \cot t$

In Exercises 15–18, graph the curve and draw an arrow specifying the direction corresponding to motion.

15. $x = \frac{1}{2}t$, $y = 2t^2$

16. $x = 2 + 4t$, $y = 3 + 2t$

17. $x = \pi t$, $y = \sin t$

18. $x = t^2$, $y = t^3$

19. Match the parametrizations (a)–(d) with their plots in Figure 15, and draw an arrow indicating the direction of motion.

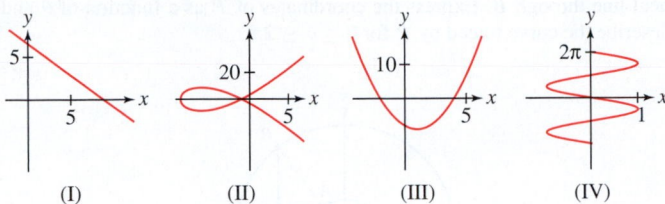

FIGURE 15

(a) $c(t) = (\sin t, -t)$

(b) $c(t) = (t^2 - 9, 8t - t^3)$

(c) $c(t) = (1 - t, t^2 - 9)$

(d) $c(t) = (4t + 2, 5 - 3t)$

20. Find an interval of t-values such that $c(t) = (\cos t, \sin t)$ traces the lower half of the unit circle.

21. A particle follows the trajectory

$$x(t) = \frac{1}{4}t^3 + 2t, \qquad y(t) = 20t - t^2$$

with t in seconds and distance in centimeters.

(a) What is the particle's maximum height?

(b) When does the particle hit the ground and how far from the origin does it land?

22. Find an interval of t-values such that $c(t) = (2t + 1, 4t - 5)$ parametrizes the segment from $(0, -7)$ to $(7, 7)$.

In Exercises 23–38, find parametric equations for the given curve.

23. $y = 9 - 4x$

24. $y = 8x^2 - 3x$

25. $4x - y^2 = 5$

26. $x^2 + y^2 = 49$

27. $(x + 9)^2 + (y - 4)^2 = 49$

28. $\left(\frac{x}{5}\right)^2 + \left(\frac{y}{12}\right)^2 = 1$

29. Line of slope 8 through $(-4, 9)$

30. Line through $(2, 5)$ perpendicular to $y = 3x$

31. Line through $(3, 1)$ and $(-5, 4)$

32. Line through $\left(\frac{1}{3}, \frac{1}{6}\right)$ and $\left(-\frac{7}{6}, \frac{5}{3}\right)$

33. Segment joining $(1, 1)$ and $(2, 3)$

34. Segment joining $(-3, 0)$ and $(0, 4)$

35. Circle of radius 4 with center $(3, 9)$

36. Ellipse of Exercise 28, with its center translated to $(7, 4)$

37. $y = x^2$, translated so that the minimum occurs at $(-4, -8)$

38. $y = \cos x$, translated so that a maximum occurs at $(3, 5)$

In Exercises 39–42, find a parametrization $c(t)$ of the curve satisfying the given condition.

39. $y = 3x - 4$, $c(0) = (2, 2)$

40. $y = 3x - 4$, $c(3) = (2, 2)$

41. $y = x^2$, $c(0) = (3, 9)$

42. $x^2 + y^2 = 4$, $c(0) = (1, \sqrt{3})$

43. Find a parametrization of the top half of the ellipse $4x^2 + 5y^2 = 100$, starting at $(-5, 0)$ and ending at $(5, 0)$.

44. Find a parametrization of the right branch $(x > 0)$ of the hyperbola

$$\left(\frac{x}{a}\right)^2 - \left(\frac{y}{b}\right)^2 = 1$$

using $\cosh t$ and $\sinh t$. How can you parametrize the branch $x < 0$?

45. Describe $c(t) = (\sec t, \tan t)$ for $0 \le t < \frac{\pi}{2}$ in the form $y = f(x)$. Specify the domain of x.

46. Show that $x = a + qt$, $y = b + pt$, with $q \ne 0$, parametrizes a line with slope $m = p/q$. What are the x- and y-intercepts of the line?

47. The graphs of $x(t)$ and $y(t)$ as functions of t are shown in Figure 16(A). Which of (I)–(III) is the plot of $c(t) = (x(t), y(t))$? Explain.

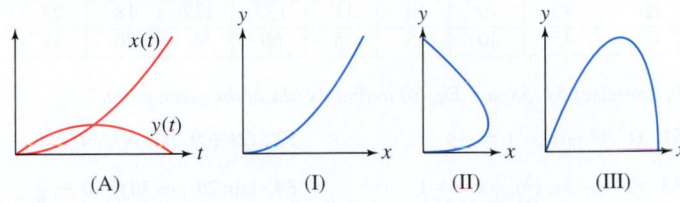

FIGURE 16

48. Which graph, (I) or (II), is the graph of $x(t)$ and which is the graph of $y(t)$ for the parametric curve in Figure 17(A)?

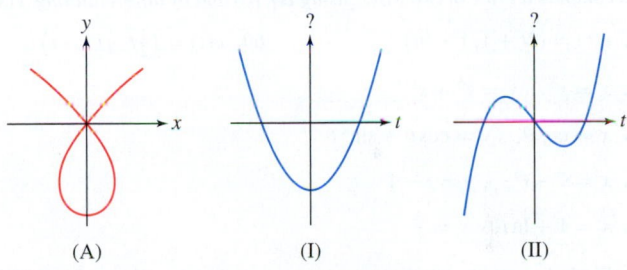

FIGURE 17

49. Figure 18 shows a parametric curve $c(t) = (q(t), p(t))$ that models the changing population sizes of a predator (p) and its prey (q).

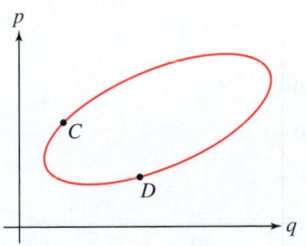

FIGURE 18

(a) Discuss how you expect the predator and prey populations to change as time increases at points C and D on the parametric curve.

(b) As functions of t, sketch graphs of $p(t)$ and $q(t)$ for three cycles around the parametric curve, beginning at point C.

(c) Both graphs in (b) should show oscillations between minimum and maximum values. Indicate which (predator or prey) has its peaks shortly after the other has its peaks, and explain why that makes sense in terms of an interaction between a predator and its prey.

50. For many years, the Hudson's Bay Company in Canada kept records of the number of snowshoe hare and lynx pelts traded each year. It is natural to expect that these values are roughly proportional to the sizes of the populations. Data for odd years between 1861 and 1891 appear in the table, where the number of pelts for lynx, L, and snowshoe hares, H, are shown (both in thousands). Plot the data on an LH-coordinate system, connecting consecutive data points by a segment to create a parametric curve traced out by the data.

Year	1861	1863	1865	1867	1869	1871	1873	1875
H	36	150	110	60	7	10	70	100
L	6	6	65	70	40	9	20	34

Year	1877	1879	1881	1883	1885	1887	1889	1891
H	92	70	10	11	137	137	18	22
L	45	40	15	15	60	80	26	18

In Exercises 51–58, use Eq. (6) to find dy/dx at the given point.

51. $(t^3, t^2 - 1)$, $t = -4$

52. $(2t + 9, 7t - 9)$, $t = 1$

53. $(s^{-1} - 3s, s^3)$, $s = -1$

54. $(\sin 2\theta, \cos 3\theta)$, $\theta = \frac{\pi}{6}$

55. $(\sin^3 \theta, \cos \theta)$, $\theta = \frac{\pi}{4}$

56. $(\sec \theta, \tan \theta)$, $t = \frac{\pi}{4}$

57. $(\ln t, \frac{1}{t})$, $t = 4$

58. (e^t, t^2), $t = 1$

In Exercises 59–64, find an equation $y = f(x)$ for the parametric curve and compute dy/dx in two ways: using Eq. (6) and by differentiating $f(x)$.

59. $c(t) = (2t + 1, 1 - 9t)$

60. $c(t) = \left(\frac{1}{2}t, \frac{1}{4}t^2 - t\right)$

61. $x = s^3$, $y = s^6 + s^{-3}$

62. $x = \cos\theta$, $y = \cos\theta + \sin^2\theta$

63. $x = 1 - e^t$, $y = t - 1$

64. $x = 1 + \ln t$, $y = \frac{1}{t}$

65. Find the points on the parametric curve $c(t) = (3t^2 - 2t, t^3 - 6t)$ where the tangent line has slope 3.

66. Find the equation of the tangent line to the cycloid generated by a circle of radius 4 at $t = \frac{\pi}{2}$.

In Exercises 67–70, let $c(t) = (t^2 - 9, t^2 - 8t)$ (see Figure 19).

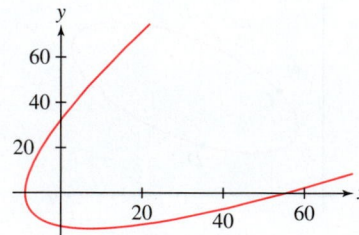

FIGURE 19 Plot of $c(t) = (t^2 - 9, t^2 - 8t)$.

67. Draw an arrow indicating the direction of motion, and determine the interval(s) of t-values corresponding to the portion(s) of the curve in each of the four quadrants.

68. Find the equation of the tangent line at $t = 4$.

69. Find the points where the tangent has slope $\frac{1}{2}$.

70. Find the points where the tangent is horizontal or vertical.

71. Let A and B be the points where the ray of angle θ intersects the two concentric circles of radii $r < R$ centered at the origin (Figure 20). Let P be the point of intersection of the horizontal line through A and the vertical line through B. Express the coordinates of P as a function of θ and describe the curve traced by P for $0 \le \theta \le 2\pi$.

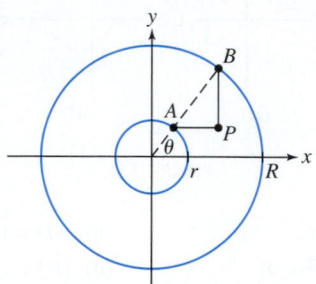

FIGURE 20

72. A 10-ft ladder slides down a wall as its bottom B is pulled away from the wall (Figure 21). Using the angle θ as a parameter, find the parametric equations for the path followed by (a) the top of the ladder A, (b) the bottom of the ladder B, and (c) the point P on the ladder, located 4 ft from the top. Show that P describes an ellipse.

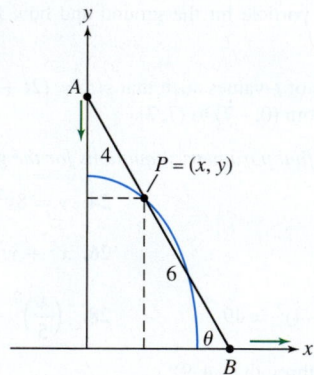

FIGURE 21

In Exercises 73–76, refer to the Bézier curve defined by Eqs. (7) and (8).

73. Show that the Bézier curve with control points

$$P_0 = (1, 4), \quad P_1 = (3, 12), \quad P_2 = (6, 15), \quad P_3 = (7, 4)$$

has parametrization

$$c(t) = (1 + 6t + 3t^2 - 3t^3, 4 + 24t - 15t^2 - 9t^3)$$

Verify that the slope at $t = 0$ is equal to the slope of the segment $\overline{P_0 P_1}$.

74. Find an equation of the tangent line to the Bézier curve in Exercise 73 at $t = \frac{1}{3}$.

75. [CAS] Find and plot the Bézier curve $c(t)$ with control points

$$P_0 = (3, 2), \quad P_1 = (0, 2), \quad P_2 = (5, 4), \quad P_3 = (2, 4)$$

76. Show that a cubic Bézier curve is tangent to the segment $\overline{P_2 P_3}$ at P_3.

77. A launched projectile follows the trajectory

$$x = at, \quad y = bt - 16t^2 \quad (a, b > 0)$$

Show that the projectile is launched at an angle $\theta = \tan^{-1}\left(\frac{b}{a}\right)$ and lands at a distance $\frac{ab}{16}$ from the origin.

78. [CAS] Plot $c(t) = (t^3 - 4t, t^4 - 12t^2 + 48)$ for $-3 \le t \le 3$. Find the points where the tangent line is horizontal or vertical.

79. [CAS] Plot the astroid $x = \cos^3 \theta$, $y = \sin^3 \theta$ and find the equation of the tangent line at $\theta = \frac{\pi}{3}$.

80. Find the equation of the tangent line at $t = \frac{\pi}{4}$ to the cycloid generated by the unit circle with parametric equation (4).

81. Find the points with a horizontal tangent line on the cycloid with parametric equation (4).

82. Property of the Cycloid Prove that the tangent line at a point P on the cycloid always passes through the top point on the rolling circle as indicated in Figure 22. Assume the generating circle of the cycloid has radius 1.

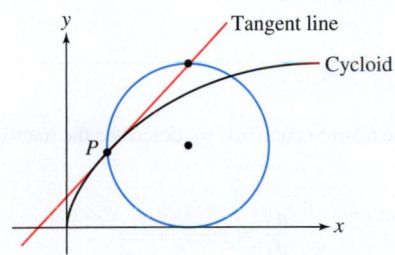

FIGURE 22

83. A *curtate cycloid* (Figure 23) is the curve traced by a point at a distance h from the center of a circle of radius R rolling along the x-axis where $h < R$. Show that this curve has parametric equations $x = Rt - h \sin t$, $y = R - h \cos t$.

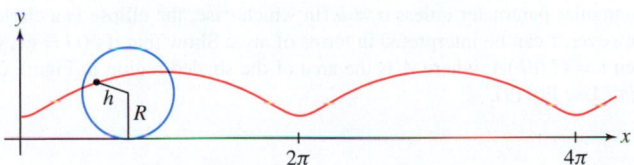

FIGURE 23 Curtate cycloid.

84. [CAS] Use a computer algebra system to explore what happens when $h > R$ in the parametric equations of Exercise 83. Describe the result.

85. [✎] Show that the line of slope t through $(-1, 0)$ intersects the unit circle in the point with coordinates

$$x = \frac{1 - t^2}{t^2 + 1}, \qquad y = \frac{2t}{t^2 + 1} \qquad \boxed{10}$$

Conclude that these equations parametrize the unit circle with the point $(-1, 0)$ excluded (Figure 24). Show further that $t = y/(x + 1)$.

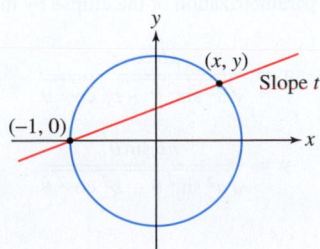

FIGURE 24 Unit circle.

86. The **folium of Descartes** is the curve with equation $x^3 + y^3 = 3axy$, where $a \ne 0$ is a constant (Figure 25).

(a) Show that the line $y = tx$ intersects the folium at the origin and at one other point P for all $t \ne -1, 0$. Express the coordinates of P in terms of t to obtain a parametrization of the folium. Indicate the direction of the parametrization on the graph.

(b) Describe the interval of t-values parametrizing the parts of the curve in quadrants I, II, and IV. Note that $t = -1$ is a point of discontinuity of the parametrization.

(c) Calculate dy/dx as a function of t and find the points with horizontal or vertical tangent.

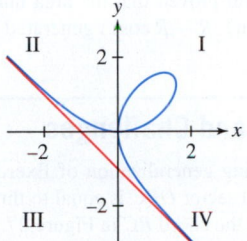

FIGURE 25 Folium $x^3 + y^3 = 3axy$.

87. Use the results of Exercise 86 to show that the asymptote of the folium is the line $x + y = -a$. *Hint:* Show that $\lim_{t \to -1} (x + y) = -a$.

88. Find a parametrization of $x^{2n+1} + y^{2n+1} = ax^n y^n$, where a and n are constants.

89. Second Derivative for a Parametrized Curve Given a parametrized curve $c(t) = (x(t), y(t))$, show that

$$\frac{d}{dt}\left(\frac{dy}{dx}\right) = \frac{x'(t)y''(t) - y'(t)x''(t)}{x'(t)^2}$$

Use this to prove the formula

$$\frac{d^2 y}{dx^2} = \frac{x'(t)y''(t) - y'(t)x''(t)}{x'(t)^3} \qquad \boxed{11}$$

90. The second derivative of $y = x^2$ is $dy^2/d^2x = 2$. Verify that Eq. (11) applied to $c(t) = (t, t^2)$ yields $dy^2/d^2x = 2$. In fact, any parametrization may be used. Check that $c(t) = (t^3, t^6)$ and $c(t) = (\tan t, \tan^2 t)$ also yield $dy^2/d^2x = 2$.

In Exercises 91–94, use Eq. (11) to find $d^2 y/dx^2$.

91. $x = t^3 + t^2$, $y = 7t^2 - 4$, $t = 2$

92. $x = s^{-1} + s$, $y = 4 - s^{-2}$, $s = 1$

93. $x = 8t + 9$, $y = 1 - 4t$, $t = -3$

94. $x = \cos \theta$, $y = \sin \theta$, $\theta = \frac{\pi}{4}$

95. Use Eq. (11) to find the t-intervals on which $c(t) = (t^2, t^3 - 4t)$ is concave up.

96. Use Eq. (11) to find the t-intervals on which $c(t) = (t^2, t^4 - 4t)$ is concave up.

97. Calculate the area under $y = x^2$ over $[0, 1]$ using Eq. (9) with the parametrizations (t^3, t^6) and (t^2, t^4).

98. What does Eq. (9) say if $c(t) = (t, f(t))$?

99. Consider the curve $c(t) = (t^2, t^3)$ for $0 \le t \le 1$.
(a) Find the area under the curve using Eq. (9).
(b) Find the area under the curve by expressing y as a function of x and finding the area using the standard method.

100. Compute the area under the parametrized curve $c(t) = (e^t, t)$ for $0 \le t \le 1$ using Eq. (9).

101. Compute the area under the parametrized curve given by $c(t) = (\sin t, \cos^2 t)$ for $0 \le t \le \pi/2$ using Eq. (9).

102. Sketch the graph of $c(t) = (\ln t, 2 - t)$ for $1 \le t \le 2$ and compute the area under the graph using Eq. (9).

103. Galileo tried unsuccessfully to find the area under a cycloid. Around 1630, Gilles de Roberval proved that the area under one arch of the cycloid $c(t) = (Rt - R \sin t, R - R \cos t)$ generated by a circle of radius R

is equal to three times the area of the circle (Figure 26). Verify Roberval's result using Eq. (9).

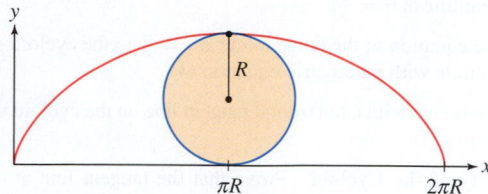

FIGURE 26 The area of one arch of the cycloid equals three times the area of the generating circle.

Further Insights and Challenges

104. Prove the following generalization of Exercise 103: For all $t > 0$, the area of the cycloidal sector OPC is equal to three times the area of the circular segment cut by the chord PC in Figure 27.

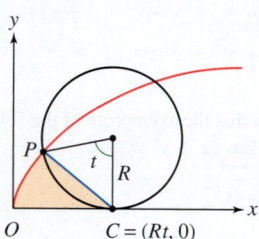

(A) Cycloidal sector OPC (B) Circular segment cut by the chord PC

FIGURE 27

105. ✍ Derive the formula for the slope of the tangent line to a parametric curve $c(t) = (x(t), y(t))$ using a method different from that presented in the text. Assume that $x'(t_0)$ and $y'(t_0)$ exist and $x'(t_0) \ne 0$. Show that

$$\lim_{h \to 0} \frac{y(t_0 + h) - y(t_0)}{x(t_0 + h) - x(t_0)} = \frac{y'(t_0)}{x'(t_0)}$$

Then explain why this limit is equal to the slope dy/dx. Draw a diagram showing that the ratio in the limit is the slope of a secant line.

106. Verify that the **tractrix** curve ($\ell > 0$)

$$c(t) = \left(t - \ell \tanh \frac{t}{\ell}, \ell \operatorname{sech} \frac{t}{\ell}\right)$$

has the following property: For all t, the segment from $c(t)$ to $(t, 0)$ is tangent to the curve and has length ℓ (Figure 28).

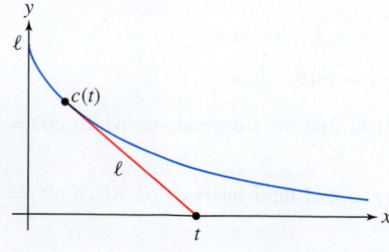

FIGURE 28 The tractrix $c(t) = \left(t - \ell \tanh \dfrac{t}{\ell}, \ell \operatorname{sech} \dfrac{t}{\ell}\right)$.

107. In Exercise 62 of Section 9.1, we described the tractrix by the differential equation

$$\frac{dy}{dx} = -\frac{y}{\sqrt{\ell^2 - y^2}}$$

Show that the parametric curve $c(t)$ identified as the tractrix in Exercise 106 satisfies this differential equation. Note that the derivative on the left is taken with respect to x, not t.

In Exercises 108 and 109, refer to Figure 29.

108. In the parametrization $c(t) = (a \cos t, b \sin t)$ of an ellipse, t is *not* an angular parameter unless $a = b$ (in which case, the ellipse is a circle). However, t can be interpreted in terms of area: Show that if $c(t) = (x, y)$, then $t = (2/ab)A$, where A is the area of the shaded region in Figure 29. *Hint:* Use Eq. (9).

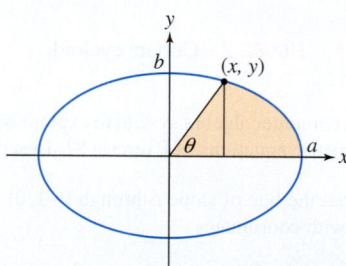

FIGURE 29 The parameter θ on the ellipse $\left(\dfrac{x}{a}\right)^2 + \left(\dfrac{y}{b}\right)^2 = 1$.

109. Show that the parametrization of the ellipse by the angle θ is

$$x = \frac{ab \cos \theta}{\sqrt{a^2 \sin^2 \theta + b^2 \cos^2 \theta}}$$

$$y = \frac{ab \sin \theta}{\sqrt{a^2 \sin^2 \theta + b^2 \cos^2 \theta}}$$

11.2 Arc Length and Speed

We now derive a formula for the arc length s of a curve in parametric form. Recall that in Section 8.2, arc length of a curve $\mathcal{C}$ was defined as the limit of the lengths of polygonal approximations of $\mathcal{C}$ (Figure 1).

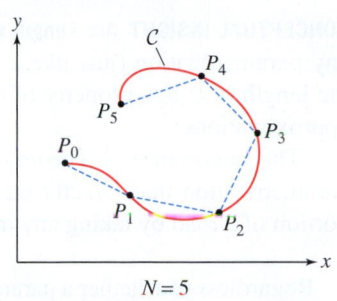 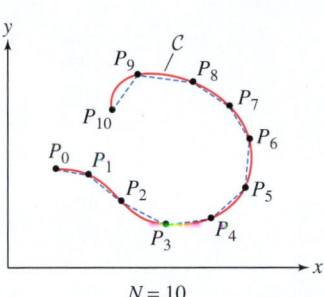

$N = 5$ $N = 10$

FIGURE 1 Polygonal approximations to a curve $\mathcal{C}$ for $N = 5$ and $N = 10$.

To compute the length of $\mathcal{C}$ via a parametrization, we need to assume that the parametrization **directly traverses** $\mathcal{C}$, that is, the path traces $\mathcal{C}$ from one end to the other without changing direction along the way. Thus, assume that $c(t) = (x(t), y(t))$ is a parametrization that directly traverses $\mathcal{C}$ for $a \leq t \leq b$. We construct a polygonal approximation L consisting of the N segments obtained by joining points

$$P_0 = c(t_0), \quad P_1 = c(t_1), \quad \ldots, \quad P_N = c(t_N)$$

corresponding to a choice of values $t_0 = a < t_1 < t_2 < \cdots < t_N = b$. By the distance formula, if L_i is the segment joining P_{i-1} and P_i, then

$$|L_i| = \sqrt{\left(x(t_i) - x(t_{i-1})\right)^2 + \left(y(t_i) - y(t_{i-1})\right)^2} \qquad \boxed{1}$$

Now assume that $x(t)$ and $y(t)$ are differentiable. According to the Mean Value Theorem, there are values t_i^* and t_i^{**} in the interval $[t_{i-1}, t_i]$ such that

$$x(t_i) - x(t_{i-1}) = x'(t_i^*)\Delta t_i, \qquad y(t_i) - y(t_{i-1}) = y'(t_i^{**})\Delta t_i$$

where $\Delta t_i = t_i - t_{i-1}$, and therefore,

$$|L_i| = \sqrt{x'(t_i^*)^2 \Delta t_i^2 + y'(t_i^{**})^2 \Delta t_i^2} = \sqrt{x'(t_i^*)^2 + y'(t_i^{**})^2}\, \Delta t_i$$

The length of the polygonal approximation L is equal to the sum

$$\sum_{i=1}^{N} |L_i| = \sum_{i=1}^{N} \sqrt{x'(t_i^*)^2 + y'(t_i^{**})^2}\, \Delta t_i \qquad \boxed{2}$$

This is *nearly* a Riemann sum for the function $\sqrt{x'(t)^2 + y'(t)^2}$. It would be a true Riemann sum if the intermediate values t_i^* and t_i^{**} were equal. Although they are not necessarily equal, it can be shown (and we will take for granted) that if $x'(t)$ and $y'(t)$ are continuous, then the sum in Eq. (2) still approaches the integral as the widths Δt_i tend to 0. Thus,

$$s = \lim_{\Delta t_i \to 0} \sum_{i=1}^{N} |L_i| = \int_a^b \sqrt{x'(t)^2 + y'(t)^2}\, dt$$

Because of the square root, the arc length integral cannot be evaluated explicitly except in special cases, but we can always approximate it numerically.

THEOREM 1 Arc Length Let $c(t) = (x(t), y(t))$ be a parametrization that directly traverses $\mathcal{C}$ for $a \leq t \leq b$. Assume that $x'(t)$ and $y'(t)$ exist and are continuous. Then the arc length s of $\mathcal{C}$ is equal to

$$s = \int_a^b \sqrt{x'(t)^2 + y'(t)^2}\, dt \qquad \boxed{3}$$

The graph of a function $y = f(x)$ has parametrization $c(t) = (t, f(t))$. In this case,

$$\sqrt{x'(t)^2 + y'(t)^2} = \sqrt{1 + f'(t)^2}$$

and Eq. (3) reduces to the arc length formula derived in Section 8.2.

> **CONCEPTUAL INSIGHT   Arc Length via Parametrizations**   A curve $\mathcal{C}$ exists independent of any parametrization (just like a road exists independent of any trip taken on it), and the length of $\mathcal{C}$ is a property of $\mathcal{C}$, defined as the limit of the lengths of its polygonal approximations.
>
> The importance of Theorem 1 is that we can compute the length of $\mathcal{C}$ from any parametrization that directly traverses $\mathcal{C}$, just like we can determine the length of a portion of a road by taking any trip along it without doubling back.

Regardless of whether a parametrization $c(t) = (x(t), y(t))$ directly traverses a curve, if $x'(t)$ and $y'(t)$ exist and are continuous, then the integral $\displaystyle\int_a^b \sqrt{x'(t)^2 + y'(t)^2}\, dt$ exists and can be interpreted as the **distance traveled** along the path from $t = a$ to $t = b$. Of course, this distance might not equal the length of the underlying curve. For example, over $0 \le t \le 10$, the path $c(t) = (\cos(2\pi t), \cos(2\pi t))$ cycles 10 times from $(1, 1)$ to $(-1, -1)$ and back along the line $y = x$. The length of the underlying curve is $\sqrt{2}$, but the distance traveled is $20\sqrt{2}$.

As mentioned above, the arc length integral can be evaluated explicitly only in special cases. The circle (Example 1) and the cycloid (Example 3) are two such cases.

EXAMPLE 1   Use Eq. (3) to calculate the arc length of a circle of radius R.

Solution   With the parametrization $x = R\cos\theta$, $y = R\sin\theta$,

$$x'(\theta)^2 + y'(\theta)^2 = (-R\sin\theta)^2 + (R\cos\theta)^2 = R^2(\sin^2\theta + \cos^2\theta) = R^2$$

We obtain the expected result:

$$s = \int_0^{2\pi} \sqrt{x'(\theta)^2 + y'(\theta)^2}\, d\theta = \int_0^{2\pi} R\, d\theta = 2\pi R$$ ∎

EXAMPLE 2   Find the arc length of the curve given in parametric form by $c(t) = (t^2, t^3)$ for $0 \le t \le 1$.

Solution   The arc length of this curve is given by

$$s = \int_0^1 \sqrt{x'(t)^2 + y'(t)^2}\, dt = \int_0^1 \sqrt{(2t)^2 + (3t^2)^2}\, dt$$

$$= \int_0^1 t\sqrt{4 + 9t^2}\, dt$$

Letting $u = 4 + 9t^2$, and therefore $du = 18t\, dt$, we obtain

$$s = \frac{1}{18}\int_4^{13} \sqrt{u}\, du = \frac{2}{3}\frac{u^{3/2}}{18}\Big|_4^{13} = \frac{1}{27}(13^{3/2} - 4^{3/2}) \approx 1.4397$$ ∎

EXAMPLE 3   Length of the Cycloid   Calculate the length s of one arch of the cycloid generated by a circle of radius $R = 2$ (Figure 2).

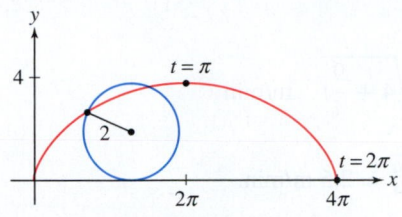

FIGURE 2 One arch of the cycloid generated by a circle of radius 2.

← **REMINDER**

$$\frac{1 - \cos t}{2} = \sin^2 \frac{t}{2}$$

*In Chapter 13, we will discuss not just the speed but also the velocity of a particle moving along a curved path. Velocity indicates speed **and** direction and is represented by a vector.*

Solution We use the parametrization of the cycloid in Eq. (5) of Section 11.1:

$$x(t) = 2(t - \sin t), \qquad y(t) = 2(1 - \cos t)$$

$$x'(t) = 2(1 - \cos t), \qquad y'(t) = 2 \sin t$$

Thus,

$$x'(t)^2 + y'(t)^2 = 2^2(1 - \cos t)^2 + 2^2 \sin^2 t$$

$$= 4 - 8 \cos t + 4 \cos^2 t + 4 \sin^2 t$$

$$= 8 - 8 \cos t$$

$$= 16 \sin^2 \frac{t}{2} \qquad \text{(Use the identity recalled in the margin.)}$$

One arch of the cycloid is traced as t varies from 0 to 2π, so

$$s = \int_0^{2\pi} \sqrt{x'(t)^2 + y'(t)^2} \, dt = \int_0^{2\pi} 4 \sin \frac{t}{2} \, dt = -8 \cos \frac{t}{2} \Big|_0^{2\pi} = -8(-1) + 8 = 16$$

Note that because $\sin \frac{t}{2} \geq 0$ for $0 \leq t \leq 2\pi$, we did not need an absolute value when taking the square root of $16 \sin^2 \frac{t}{2}$. ∎

Now consider a particle moving along a path $c(t)$. The distance traveled by the particle over the time interval $[t_0, t]$ is given by the arc length integral:

$$s(t) = \int_{t_0}^t \sqrt{x'(u)^2 + y'(u)^2} \, du$$

The speed of the particle is the rate of change of distance traveled with respect to time. Therefore, speed equals $s'(t)$, and using the Fundamental Theorem of Calculus, we can express it as

$$\text{speed} = \frac{ds}{dt} = \frac{d}{dt} \int_{t_0}^t \sqrt{x'(u)^2 + y'(u)^2} \, du = \sqrt{x'(t)^2 + y'(t)^2}$$

THEOREM 2 **Speed Along a Parametrized Path** The speed of $c(t) = (x(t), y(t))$ is

$$\text{speed} = \frac{ds}{dt} = \sqrt{x'(t)^2 + y'(t)^2}$$

The next example illustrates the difference between distance traveled along a path and **displacement** (also called the net change in position). The displacement along a path is the distance between the initial point $c(t_0)$ and the endpoint $c(t_1)$. The distance traveled is greater than or equal to the displacement (Figure 3). When the particle moves in one direction on a line, distance traveled equals displacement.

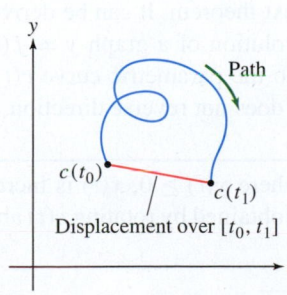

FIGURE 3 The distance along the path is greater than or equal to the displacement.

EXAMPLE 4 A particle travels along the path $c(t) = (2t, 1 + t^{3/2})$. Find:

(a) The particle's speed at $t = 1$ (assume units of meters and minutes).

(b) The distance traveled s and displacement d during the interval $0 \leq t \leq 4$.

Solution We have

$$x'(t) = 2, \qquad y'(t) = \frac{3}{2}t^{1/2}$$

The speed at time t is

$$s'(t) = \sqrt{x'(t)^2 + y'(t)^2} = \sqrt{4 + \frac{9}{4}t} \quad \text{m/min}$$

(a) The particle's speed at $t = 1$ is $s'(1) = \sqrt{4 + \frac{9}{4}} = 2.5$ m/min.

(b) The distance traveled in the first 4 min is

$$s = \int_0^4 \sqrt{4 + \frac{9}{4}t}\, dt = \frac{8}{27}\left(4 + \frac{9}{4}t\right)^{3/2}\Bigg|_0^4 = \frac{8}{27}(13^{3/2} - 8) \approx 11.52 \text{ m}$$

The displacement d is the distance from the initial point $c(0) = (0, 1)$ to the endpoint $c(4) = (8, 1 + 4^{3/2}) = (8, 9)$ (see Figure 4):

$$d = \sqrt{(8 - 0)^2 + (9 - 1)^2} = 8\sqrt{2} \approx 11.31 \text{ m} \quad \blacksquare$$

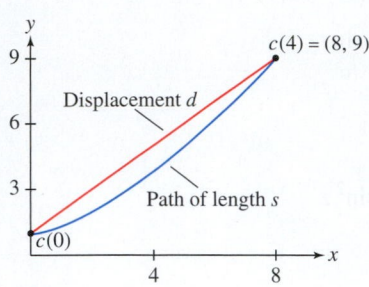

FIGURE 4 The path $c(t) = (2t, 1 + t^{3/2})$.

In physics, we often describe the path of a particle moving with constant speed along a circle of radius R in terms of a constant ω (lowercase Greek omega) as follows:

$$c(t) = (R\cos\omega t,\ R\sin\omega t)$$

The constant ω, called the *angular velocity*, is the rate of change with respect to time of the particle's angle θ (Figure 5).

EXAMPLE 5 **Angular Velocity** Calculate the speed of the circular path of radius R and angular velocity ω. What is the speed if $R = 3$ m and $\omega = 4$ radians per second (rad/s)?

Solution We have $x = R\cos\omega t$ and $y = R\sin\omega t$, and

$$x'(t) = -\omega R\sin\omega t, \qquad y'(t) = \omega R\cos\omega t$$

The particle's speed is

$$\frac{ds}{dt} = \sqrt{x'(t)^2 + y'(t)^2} = \sqrt{(-\omega R\sin\omega t)^2 + (\omega R\cos\omega t)^2}$$

$$= \sqrt{\omega^2 R^2(\sin^2\omega t + \cos^2\omega t)} = |\omega|R$$

Thus, the speed is constant with value $|\omega|R$. If $R = 3$ m and $\omega = 4$ rad/s, then the speed is $|\omega|R = 3(4) = 12$ m/s. $\quad \blacksquare$

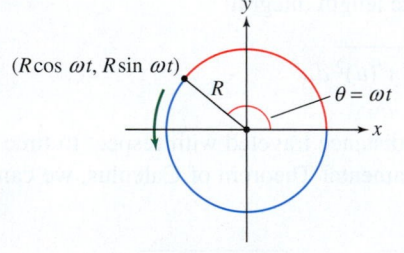

FIGURE 5 A particle moving on a circle of radius R with angular velocity ω has speed $|\omega|R$.

Consider the surface obtained by rotating a parametric curve $c(t) = (x(t), y(t))$ about the x-axis. The surface area is given by Eq. (4) in the next theorem. It can be derived in much the same way as the formula for a surface of revolution of a graph $y = f(x)$ in Section 8.2. In this theorem, we assume that $y(t) \geq 0$ so the parametric curve $c(t)$ lies above the x-axis, and that $x(t)$ is increasing so the curve does not reverse direction.

> **THEOREM 3** **Surface Area** Let $c(t) = (x(t), y(t))$, where $y(t) \geq 0$, $x(t)$ is increasing, and $x'(t)$ and $y'(t)$ are continuous. Then the surface obtained by rotating $c(t)$ about the x-axis for $a \leq t \leq b$ has surface area
>
> $$S = 2\pi \int_a^b y(t)\sqrt{x'(t)^2 + y'(t)^2}\, dt \qquad \boxed{4}$$

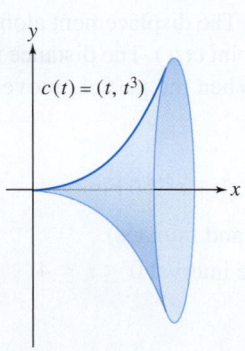

FIGURE 6 Surface generated by revolving the curve about the x-axis.

EXAMPLE 6 Calculate the surface area of the surface obtained by rotating the parametric curve $c(t) = (t, t^3)$ about the x-axis for $0 \leq t \leq 1$. The surface appears as in Figure 6.

Solution We have $x'(t) = 1$ and $y'(t) = 3t^2$.
Therefore,

$$S = 2\pi \int_0^1 t^3 \sqrt{1 + (3t^2)^2} \, dt = 2\pi \int_0^1 t^3 \sqrt{1 + 9t^4} \, dt$$

With the substitution $u = 1 + 9t^4$ and $du = 36t^3 \, dt$, we obtain

$$S = 2\pi \frac{1}{36} \int_1^{10} \sqrt{u} \, du = \frac{\pi}{18} \left(\frac{2}{3}\right) u^{3/2} \Big|_1^{10} = \frac{\pi}{27}(10^{3/2} - 1) \approx 3.5631 \qquad \blacksquare$$

11.2 SUMMARY

- Arc length of $\mathcal{C}$: If $c(t) = (x(t), y(t))$ directly traverses $\mathcal{C}$ for $a \le t \le b$, then

$$s = \text{arc length of } \mathcal{C} = \int_a^b \sqrt{x'(t)^2 + y'(t)^2} \, dt$$

- The distance traveled along the path $c(t)$, for $a \le t \le b$, is

$$\int_a^b \sqrt{x'(t)^2 + y'(t)^2} \, dt$$

- The displacement of $c(t)$ over $a \le t \le b$ is the distance from the starting point $c(a)$ to the endpoint $c(b)$. Displacement is less than or equal to distance traveled.
- Distance traveled as a function of t, starting at t_0:

$$s(t) = \int_{t_0}^t \sqrt{x'(u)^2 + y'(u)^2} \, du$$

- Speed at time t:

$$\frac{ds}{dt} = \sqrt{x'(t)^2 + y'(t)^2}$$

- Surface area of the surface obtained by rotating $c(t) = (x(t), y(t))$ about the x-axis for $a \le t \le b$ [assuming $y(t) \ge 0$, $x(t)$ is increasing, and $x'(t)$ and $y'(t)$ are continuous]:

$$S = 2\pi \int_a^b y(t) \sqrt{x'(t)^2 + y'(t)^2} \, dt$$

11.2 EXERCISES

Preliminary Questions

1. What is the definition of arc length?

2. Can the distance traveled by a particle ever be less than its displacement? When are they equal?

3. What is the interpretation of $\sqrt{x'(t)^2 + y'(t)^2}$ for a particle following the trajectory $(x(t), y(t))$?

4. A particle travels along a path from $(0, 0)$ to $(3, 4)$. What is the displacement? Can the distance traveled be determined from the information given?

5. A particle traverses the parabola $y = x^2$ with constant speed 3 cm/s. What is the distance traveled during the first minute? *Hint:* Only simple computation is necessary.

6. If the straight line segment given by $c(t) = (t, 3)$ for $0 \le t \le 2$ is rotated around the x-axis, what surface area results? *Hint:* Only simple computation is necessary.

Exercises

In Exercises 1–2, use Eq. (3) to find the length of the path over the given interval, and verify your answer using geometry.

1. $(3t - 1, 2 - 2t)$, $0 \le t \le 5$

2. $(1 + 5t, t - 5)$, $-3 \le t \le 3$

In Exercises 3–8, use Eq. (3) to find the length of the path over the given interval.

3. $(2t^2, 3t^2 - 1)$, $0 \le t \le 4$

4. $(3t, 4t^{3/2})$, $0 \le t \le 1$

5. $(3t^2, 4t^3)$, $1 \le t \le 4$

6. $(t^3 + 1, t^2 - 3)$, $0 \le t \le 1$

7. $(\sin 3t, \cos 3t)$, $0 \le t \le \pi$

8. $(\sin\theta - \theta\cos\theta, \cos\theta + \theta\sin\theta)$, $0 \le \theta \le 2$

In Exercises 9 and 10, find the length of the path. The following identity should be helpful:

$$\frac{1 - \cos t}{2} = \sin^2 \frac{t}{2}$$

9. $(2\cos t - \cos 2t, 2\sin t - \sin 2t)$, $0 \le t \le \frac{\pi}{2}$

10. $(5(\theta - \sin\theta), 5(1 - \cos\theta))$, $0 \le \theta \le 2\pi$

11. Show that one arch of a cycloid generated by a circle of radius R has length $8R$.

12. Find the length of the spiral $c(t) = (t\cos t, t\sin t)$ for $0 \le t \le 2\pi$ to three decimal places (Figure 7). *Hint:* Use the formula

$$\int \sqrt{1 + t^2}\, dt = \frac{1}{2} t\sqrt{1 + t^2} + \frac{1}{2}\ln\left(t + \sqrt{1 + t^2}\right)$$

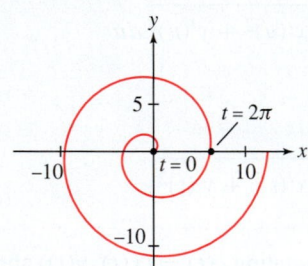

FIGURE 7 The spiral $c(t) = (t\cos t, t\sin t)$.

13. Find the length of the parabola given by $c(t) = (t, t^2)$ for $0 \le t \le 1$. See the hint for Exercise 12.

14. **CAS** Find a numerical approximation to the length of $c(t) = (\cos 5t, \sin 3t)$ for $0 \le t \le 2\pi$ (Figure 8).

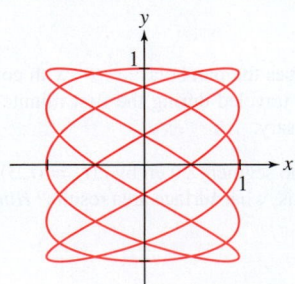

FIGURE 8

In Exercises 15–20, determine the speed $\frac{ds}{dt}$ at time t (assume units of meters and seconds).

15. (t^3, t^2), $t = 2$

16. $(3\sin 5t, 8\cos 5t)$, $t = \frac{\pi}{4}$

17. $(5t + 1, 4t - 3)$, $t = 9$

18. $(\ln(t^2 + 1), t^3)$, $t = 1$

19. (t^2, e^t), $t = 0$

20. $(\sin^{-1} t, \tan^{-1} t)$, $t = 0$

21. Find the minimum speed of a particle with parametric trajectory $c(t) = (t^3 - 4t, t^2 + 1)$ for $t \ge 0$. *Hint:* It is easier to find the minimum of the square of the speed.

22. Find the minimum speed of a particle with trajectory $c(t) = (t^3, t^{-2})$ for $t \ge 0.5$.

23. Find the speed of the cycloid $c(t) = (4t - 4\sin t, 4 - 4\cos t)$ at points where the tangent line is horizontal.

24. Calculate the arc length integral $s(t)$ for the *logarithmic spiral* $c(t) = (e^t \cos t, e^t \sin t)$.

CAS *In Exercises 25–28, plot the curve and use the Midpoint Rule with $N = 10, 20, 30,$ and 50 to approximate its length.*

25. $c(t) = (\cos t, e^{\sin t})$ for $0 \le t \le 2\pi$

26. $c(t) = (t - \sin 2t, 1 - \cos 2t)$ for $0 \le t \le 2\pi$

27. The ellipse $\left(\dfrac{x}{5}\right)^2 + \left(\dfrac{y}{3}\right)^2 = 1$

28. $x = \sin 2t$, $y = \sin 3t$ for $0 \le t \le 2\pi$

29. If you unwind thread from a stationary circular spool, keeping the thread taut at all times, then the endpoint traces a curve C called the **involute** of the circle (Figure 9). Observe that $\overline{PQ}$ has length $R\theta$. Show that C is parametrized by

$$c(\theta) = \big(R(\cos\theta + \theta\sin\theta),\ R(\sin\theta - \theta\cos\theta)\big)$$

Then find the length of the involute for $0 \le \theta \le 2\pi$.

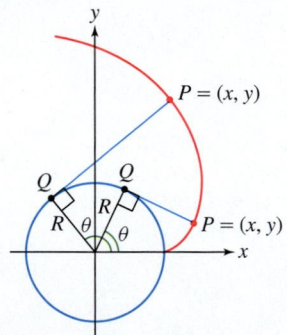

FIGURE 9 Involute of a circle.

30. Let $a > b$ and set

$$k = \sqrt{1 - \frac{b^2}{a^2}}$$

Use a parametric representation to show that the ellipse $\left(\dfrac{x}{a}\right)^2 + \left(\dfrac{y}{b}\right)^2 = 1$ has length $L = 4aG\left(\frac{\pi}{2}, k\right)$, where

$$G(\theta, k) = \int_0^\theta \sqrt{1 - k^2\sin^2 t}\, dt$$

is the *elliptic integral of the second kind.*

In Exercises 31–38, use Eq. (4) to compute the surface area of the given surface.

31. The cone generated by revolving $c(t) = (t, mt)$ about the x-axis for $0 \leq t \leq A$

32. A sphere of radius R

33. The surface generated by revolving the curve $c(t) = (t^2, t)$ about the x-axis for $0 \leq t \leq 1$

34. The surface generated by revolving the curve $c(t) = (t, e^t)$ about the x-axis for $0 \leq t \leq 1$

35. The surface generated by revolving the curve $c(t) = (\sin^2 t, \cos^2 t)$ about the x-axis for $0 \leq t \leq \frac{\pi}{2}$

36. The surface generated by revolving the curve $c(t) = (t, \sin t)$ about the x-axis for $0 \leq t \leq \pi$. *Hint:* After a substitution, use #84 in the table in the front or back of the text for the integral of $\sqrt{1 + u^2}$.

37. The surface generated by revolving one arch of the cycloid $c(t) = (t - \sin t, 1 - \cos t)$ about the x-axis

38. The surface generated by revolving the astroid $c(t) = (\cos^3 t, \sin^3 t)$ about the x-axis for $0 \leq t \leq \frac{\pi}{2}$

39. CAS Use Simpson's Rule and $N = 30$ to approximate the surface area of the surface generated by revolving $c(t) = (t^2, e^{-t}), 0 \leq t \leq 2$ about the x-axis.

40. CAS Use Simpson's Rule and $N = 50$ to approximate the surface area of the surface generated by revolving $c(t) = ((t+1)^3, \ln t), 1 \leq t \leq 5$ about the x-axis.

Further Insights and Challenges

41. CAS Let $b(t)$ be the *Butterfly Curve*:

$$x(t) = \sin t \left(e^{\cos t} - 2\cos 4t - \sin\left(\frac{t}{12}\right)^5 \right)$$

$$y(t) = \cos t \left(e^{\cos t} - 2\cos 4t - \sin\left(\frac{t}{12}\right)^5 \right)$$

(a) Use a computer algebra system to plot $b(t)$ and the speed $s'(t)$ for $0 \leq t \leq 12\pi$.

(b) Approximate the length $b(t)$ for $0 \leq t \leq 10\pi$.

42. CAS Let $a \geq b > 0$ and set $k = \dfrac{2\sqrt{ab}}{a-b}$. Show that the **trochoid**

$$x = at - b\sin t, \qquad y = a - b\cos t, \quad 0 \leq t \leq T$$

has length $2(a-b)G\left(\frac{T}{2}, k\right)$, with $G(\theta, k)$ as in Exercise 30.

43. A satellite orbiting at a distance R from the center of the earth follows the circular path $x(t) = R\cos \omega t, y(t) = R\sin \omega t$.

(a) Show that the period T (the time of one revolution) is $T = 2\pi/\omega$.

(b) According to Newton's Laws of Motion and Gravity,

$$x''(t) = -Gm_e \frac{x}{R^3}, \qquad y''(t) = -Gm_e \frac{y}{R^3}$$

where G is the universal gravitational constant and m_e is the mass of the earth. Prove that $R^3/T^2 = Gm_e/4\pi^2$. Thus, R^3/T^2 has the same value for all orbits (a special case of Kepler's Third Law).

44. The acceleration due to gravity on the surface of the earth is

$$g = \frac{Gm_e}{R_e^2} = 9.8 \text{ m/s}^2, \quad \text{where } R_e = 6378 \text{ km}$$

Use Exercise 43(b) to show that a satellite orbiting at the earth's surface would have period $T_e = 2\pi\sqrt{R_e/g} \approx 84.5$ minutes. Then estimate the distance R_m from the moon to the center of the earth. Assume that the period of the moon (sidereal month) is $T_m \approx 27.43$ days.

11.3 Polar Coordinates

Polar coordinates are appropriate when distance from the origin or angle plays a role. For example, the gravitational force exerted on a planet by the sun depends only on the distance r from the sun and is conveniently described in polar coordinates.

The rectangular coordinates that we have utilized up to now provide a useful way to represent points in the plane. However, there are a variety of situations where a different coordinate system is more natural. In polar coordinates, we label a point P by coordinates (r, θ), where r is the distance to the origin O and θ is the angle between $\overline{OP}$ and the positive x-axis (Figure 1). By convention, an angle is positive if the corresponding rotation is counterclockwise. We call r the **radial coordinate** and θ the **angular coordinate**.

The point P in Figure 2 has polar coordinates $(r, \theta) = \left(4, \frac{2\pi}{3}\right)$. It is located at distance $r = 4$ from the origin (so it lies on the circle of radius 4), and it lies on the ray of angle $\theta = \frac{2\pi}{3}$. Notice that it can also be described by $(r, \theta) = \left(4, \frac{-4\pi}{3}\right)$. Unlike Cartesian coordinates, polar coordinates are not unique, as we will discuss in more detail shortly.

Figure 3 shows the two families of **grid lines** in polar coordinates:

Circle centered at O $\longleftrightarrow$ $r = $ constant

Ray starting at O $\longleftrightarrow$ $\theta = $ constant

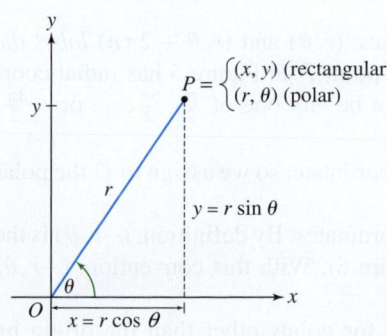

DF **FIGURE 1**

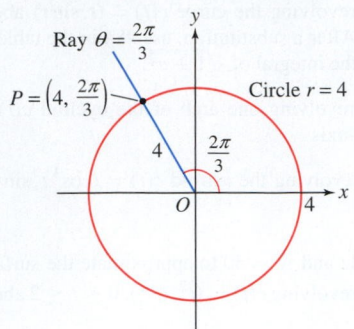

FIGURE 2

Every point in the plane other than the origin lies at the intersection of the two grid lines and these two grid lines determine its polar coordinates. For example, point Q in Figure 3 lies on the circle $r = 3$ and the ray $\theta = \frac{5\pi}{6}$, so $Q = \left(3, \frac{5\pi}{6}\right)$ in polar coordinates.

Figure 1 shows that polar and rectangular coordinates are related by the equations $x = r \cos \theta$ and $y = r \sin \theta$. On the other hand, $r^2 = x^2 + y^2$ by the distance formula, and $\tan \theta = y/x$ if $x \neq 0$. This yields the conversion formulas:

Polar to Rectangular	Rectangular to Polar
$x = r \cos \theta$	$r = \sqrt{x^2 + y^2}$
$y = r \sin \theta$	$\tan \theta = \dfrac{y}{x} \quad (x \neq 0)$

Note, we do not write $\theta = \tan^{-1} \frac{y}{x}$ since that relationship holds only for $-\frac{\pi}{2} < \theta < \frac{\pi}{2}$.

EXAMPLE 1 From Polar to Rectangular Coordinates Find the rectangular coordinates of point Q in Figure 3.

Solution The point $Q = (r, \theta) = \left(3, \frac{5\pi}{6}\right)$ has rectangular coordinates

$$x = r \cos \theta = 3 \cos \left(\frac{5\pi}{6}\right) = 3 \left(-\frac{\sqrt{3}}{2}\right) = -\frac{3\sqrt{3}}{2}$$

$$y = r \sin \theta = 3 \sin \left(\frac{5\pi}{6}\right) = 3 \left(\frac{1}{2}\right) = \frac{3}{2}$$

∎

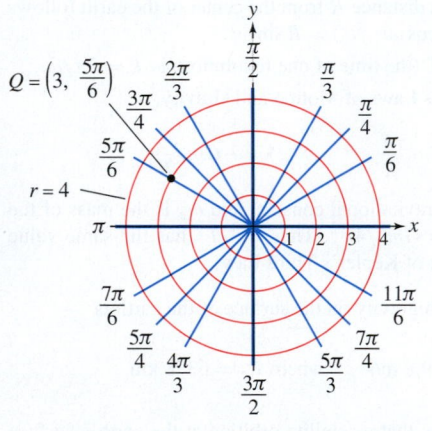

FIGURE 3 Grid lines in polar coordinates.

EXAMPLE 2 From Rectangular to Polar Coordinates Find polar coordinates for the point P in Figure 4.

Solution Since $P = (x, y) = (3, 2)$,

$$r = \sqrt{x^2 + y^2} = \sqrt{3^2 + 2^2} = \sqrt{13} \approx 3.6$$

$$\tan \theta = \frac{y}{x} = \frac{2}{3}$$

and (see the margin comment below Figure 4) because P lies in the first quadrant,

$$\theta = \tan^{-1} \frac{y}{x} = \tan^{-1} \frac{2}{3} \approx 0.588$$

Thus, P has polar coordinates $(r, \theta) \approx (3.6, 0.588)$.

∎

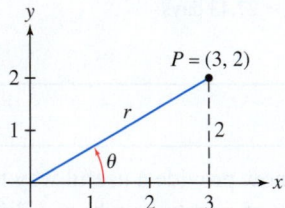

FIGURE 4 The polar coordinates of P satisfy $r = \sqrt{3^2 + 2^2}$ and $\tan \theta = \frac{2}{3}$.

If $r > 0$, a θ-coordinate of $P = (x, y)$ is

$$\theta = \begin{cases} \tan^{-1} \dfrac{y}{x} & \text{if } x > 0 \\[2mm] \tan^{-1} \dfrac{y}{x} + \pi & \text{if } x < 0 \\[2mm] \pm \dfrac{\pi}{2} & \text{if } x = 0 \end{cases}$$

A few remarks are in order before proceeding:

- The angular coordinate is not unique because (r, θ) and $(r, \theta + 2\pi n)$ *label the same point* for any integer n. For instance, point P in Figure 5 has radial coordinate $r = 2$, but its angular coordinate can be any one of $\frac{\pi}{2}, \frac{5\pi}{2}, \ldots$ or $-\frac{3\pi}{2}, -\frac{7\pi}{2}, \ldots$.
- The origin O has no well-defined angular coordinate, so we assign to O the polar coordinates $(0, \theta)$ for any angle θ.
- By convention, we allow *negative* radial coordinates. By definition, $(-r, \theta)$ is the reflection of (r, θ) through the origin (Figure 6). With this convention, $(-r, \theta)$ and $(r, \theta + \pi)$ represent the same point.
- We may specify unique polar coordinates for points other than the origin by placing restrictions on r and θ. We commonly choose $r > 0$ and $0 \leq \theta < 2\pi$, but other choices are sometimes made.

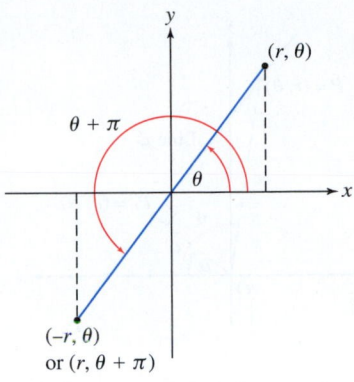

FIGURE 5 The angular coordinate of $P = (0, 2)$ is $\frac{\pi}{2}$ or any angle $\frac{\pi}{2} + 2\pi n$, where n is an integer.

FIGURE 6 Relation between (r, θ) and $(-r, \theta)$.

When determining the angular coordinate of a point $P = (x, y)$, remember that there are two angles between 0 and 2π satisfying $\tan \theta = y/x$. One of these angles is paired with $r = \sqrt{x^2 + y^2}$ to provide polar coordinates for P, the other is paired with $-r$.

EXAMPLE 3 **Choosing θ Correctly** Find two polar representations of $P = (-1, 1)$, one with $r > 0$ and one with $r < 0$.

Solution The point $P = (x, y) = (-1, 1)$ has polar coordinates (r, θ), where

$$r = \sqrt{(-1)^2 + 1^2} = \sqrt{2}, \qquad \tan \theta = \frac{y}{x} = -1$$

Now, $\tan^{-1}(-1) = -\frac{\pi}{4}$, an angle that places us in the fourth quadrant if we use $r = \sqrt{2}$ (Figure 7). But P is in the second quadrant, and therefore the correct angle is

$$\theta = \tan^{-1} \frac{y}{x} + \pi = -\frac{\pi}{4} + \pi = \frac{3\pi}{4}$$

If we wish to use the negative radial coordinate $r = -\sqrt{2}$, then the angle becomes $\theta = -\frac{\pi}{4}$ or $\frac{7\pi}{4}$. Thus,

$$P = \left(\sqrt{2}, \frac{3\pi}{4}\right) \qquad \text{or} \qquad \left(-\sqrt{2}, \frac{7\pi}{4}\right) \qquad \blacksquare$$

A curve is described in polar coordinates by an equation involving r and θ, which we call a **polar equation**. By convention, we allow solutions with $r < 0$.

A line through the origin O has the simple equation $\theta = \theta_0$, where θ_0 is the angle between the line and the x-axis (Figure 8). Indeed, the points with $\theta = \theta_0$ are (r, θ_0), where r is arbitrary (positive, negative, or zero).

EXAMPLE 4 **Line Through the Origin** Find a polar equation of the line through the origin of slope $\frac{3}{2}$ (Figure 9).

Solution A line of slope m makes an angle θ_0 with the x-axis, where $m = \tan \theta_0$. In our case, $\theta_0 = \tan^{-1} \frac{3}{2} \approx 0.98$. An equation of the line is $\theta = \tan^{-1} \frac{3}{2}$ or $\theta \approx 0.98$. $\qquad \blacksquare$

To describe lines that do not pass through the origin, we note that any such line has a unique point P_0 that is *closest* to the origin. The next example shows how to write the polar equation of the line in terms of P_0 (Figure 10).

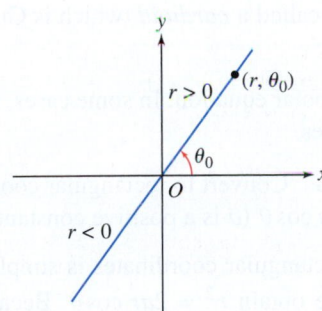

$P = (-1, 1)$

$(1, -1)$

FIGURE 7

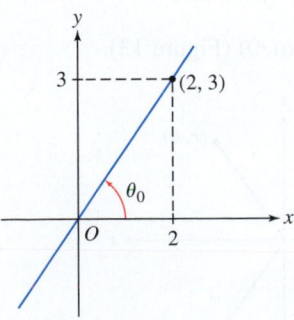

$r > 0 \quad (r, \theta_0)$

$r < 0$

FIGURE 8 Lines through O with polar equation $\theta = \theta_0$.

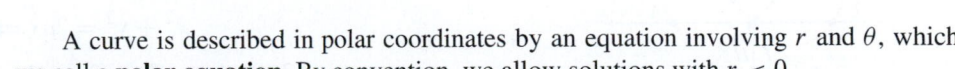

$3 \quad (2, 3)$

θ_0

2

FIGURE 9 Line of slope $\frac{3}{2}$ through the origin.

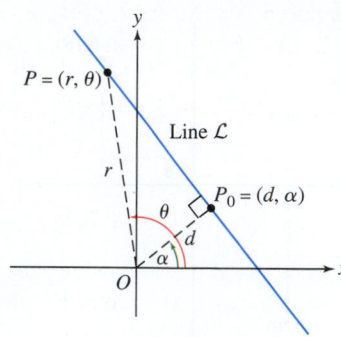

DF **FIGURE 10** P_0 is the point on $\mathcal{L}$ closest to the origin.

EXAMPLE 5 **Line Not Passing Through the Origin** Show that

$$\boxed{r = d \sec(\theta - \alpha)} \qquad \boxed{1}$$

is the polar equation of the line $\mathcal{L}$ whose point closest to the origin is $P_0 = (d, \alpha)$.

Solution The segment OP_0 in Figure 10 is perpendicular to $\mathcal{L}$. If $P = (r, \theta)$ is any point on $\mathcal{L}$ other than P_0, then $\triangle OPP_0$ is a right triangle. Therefore, $d/r = \cos(\theta - \alpha)$, or $r = d \sec(\theta - \alpha)$, as claimed. ∎

EXAMPLE 6 Find the polar equation of the line $\mathcal{L}$ tangent to the circle $r = 4$ at the point with polar coordinates $P_0 = \left(4, \frac{\pi}{3}\right)$.

Solution The point on $\mathcal{L}$ closest to the origin is P_0 itself (Figure 11). Therefore, we take $(d, \alpha) = \left(4, \frac{\pi}{3}\right)$ in Eq. (1) to obtain the equation $r = 4 \sec\left(\theta - \frac{\pi}{3}\right)$. ∎

EXAMPLE 7 Sketch the curve corresponding to $r = 1 + \sin\theta$.

Solution If we let θ vary from 0 to 2π, we see all possible values of the function, and then it will repeat. So, we consider values between 0 and 2π.

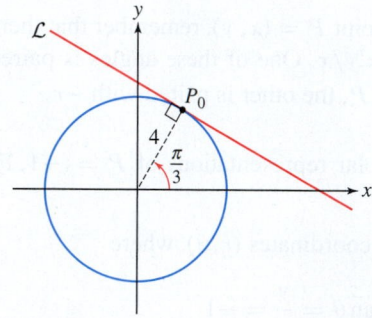

FIGURE 11 The tangent line has equation $r = 4 \sec\left(\theta - \frac{\pi}{3}\right)$.

	A	B	C	D	E	F	G	H
θ	0	$\frac{\pi}{4}$	$\frac{\pi}{2}$	$\frac{3\pi}{4}$	π	$\frac{5\pi}{4}$	$\frac{3\pi}{2}$	$\frac{7\pi}{4}$
$r = 1 + \sin\theta$	1	1.707	2	1.707	1	0.293	0	0.293

For each of the given angles, we plot the point as in Figure 12, and then we connect the points with a smooth curve. The resulting curve is called a *cardioid*, which is Greek for the "heart" that it resembles. ∎

Often, it is hard to guess the shape of a graph of a polar equation. In some cases, it is helpful to rewrite the equation in rectangular coordinates.

EXAMPLE 8 **Converting to Rectangular Coordinates** Convert to rectangular coordinates and identify the curve with polar equation $r = 2a \cos\theta$ (a is a positive constant).

Solution In this case, the process of converting to rectangular coordinates is simple if we first multiply both sides of the equation by r. We obtain $r^2 = 2ar \cos\theta$. Because $r^2 = x^2 + y^2$ and $x = r \cos\theta$, this equation becomes

$$x^2 + y^2 = 2ax \Rightarrow x^2 - 2ax + y^2 = 0 \Rightarrow$$

$$x^2 - 2ax + a^2 + y^2 = a^2 \Rightarrow (x - a)^2 + y^2 = a^2$$

This is the equation of the circle of radius a and center $(a, 0)$ (Figure 13). ∎

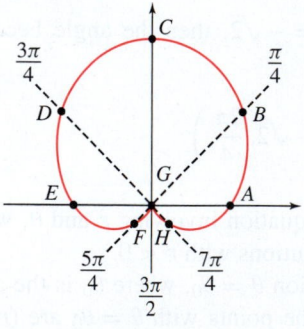

FIGURE 12 The cardioid given by $r = 1 + \sin\theta$.

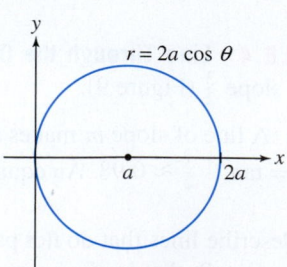

FIGURE 13

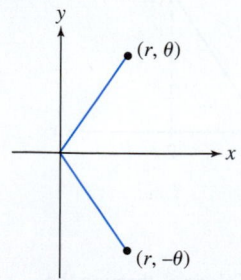

FIGURE 14 The points (r, θ) and $(r, -\theta)$ are symmetric with respect to the x-axis.

A similar conversion shows that the polar equation $r = 2a \sin\theta$ corresponds to the circle $x^2 + (y-a)^2 = a^2$ whose radius is a and center is on the y-axis at $(0, a)$.

In the next example, we make use of symmetry. Note that the points (r, θ) and $(r, -\theta)$ are symmetric with respect to the x-axis (Figure 14).

EXAMPLE 9 **Symmetry About the x-Axis** Sketch the *limaçon* curve $r = 2\cos\theta - 1$.

Solution Since $f(\theta) = \cos\theta$ is periodic with period 2π, it suffices to consider angles $-\pi \le \theta \le \pi$. We will sketch the curve for $0 \le \theta \le \pi$ and then use symmetry to obtain the complete graph.

We take two different approaches to sketching the curve for $0 \le \theta \le \pi$. The first involves plotting points and connecting with a curve; the second involves analyzing r versus θ on a rectangular system and sketching the curve from that information.

Step 1. Plot points and connect.

To get started, we plot points A–G on a grid and join them by a smooth curve (Figure 15).

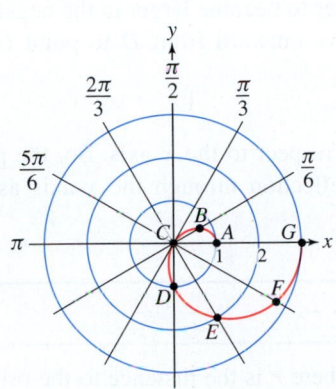

DF **FIGURE 15** Plotting $r = 2\cos\theta - 1$ using a grid.

		A	B	C	D	E	F	G
θ		0	$\dfrac{\pi}{6}$	$\dfrac{\pi}{3}$	$\dfrac{\pi}{2}$	$\dfrac{2\pi}{3}$	$\dfrac{5\pi}{6}$	π
$r = 2\cos\theta - 1$		1	0.73	0	-1	-2	-2.73	-3

Step 1. Alternate–Analyze r versus θ on a rectangular system.

Figure 16(A) shows the graph of r in terms of θ on a rectangular system. From it we see that

As θ increases from 0 to $\frac{\pi}{3}$, r decreases from 1 to 0.

As θ varies from $\frac{\pi}{3}$ to π, r is *negative* and varies from 0 to -3.

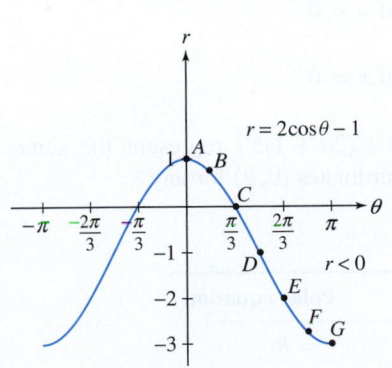

(A) Variation of r as a function of θ

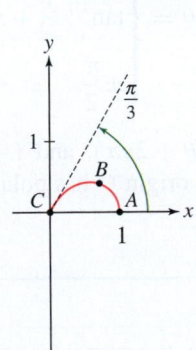

(B) θ varies from 0 to $\pi/3$; r varies from 1 to 0.

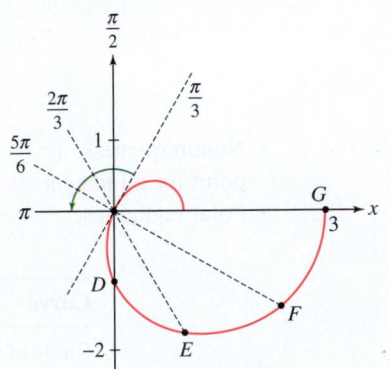

(C) θ varies from $\pi/3$ to π, but r is negative and varies from 0 to -3.

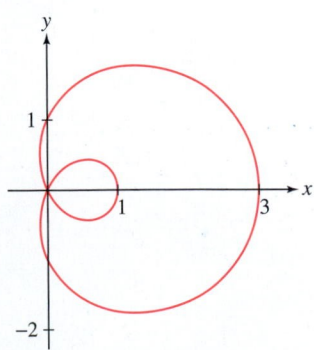

(D) The entire limaçon.

FIGURE 16 The curve $r = 2\cos\theta - 1$ is called a *limaçon*, from the Latin word for "snail." It was first described in 1525 by the German artist Albrecht Dürer.

This information guides us as we sketch the graph in Figure 16 (B–C) in the following manner:

- The graph begins at point A in Figure 16(B) and moves in toward point C, at the origin, as θ increases from 0 to $\frac{\pi}{3}$.
- Since r is negative for $\frac{\pi}{3} \le \theta \le \pi$, the curve continues into the third and fourth quadrants (rather than into the first and second quadrants).

  As θ increases from $\frac{\pi}{3}$ to $\frac{\pi}{2}$, r becomes larger in the negative direction, and the graph moves outward from the origin to point $D = \left(-1, \frac{\pi}{2}\right)$ in Figure 16(C).

  As θ increases from $\frac{\pi}{2}$ to π, r continues to become larger in the negative direction, and the graph continues to move outward from D to point $G = (-3, \pi)$.

Step 2. Use symmetry to complete the graph.

Since $r(\theta) = r(-\theta)$, the curve is symmetric with respect to the x-axis. So, the part of the curve with $-\pi \le \theta \le 0$ is obtained by reflection through the x-axis as in Figure 16(D). ∎

11.3 SUMMARY

- A point $P = (x, y)$ has polar coordinates (r, θ), where r is the distance to the origin and θ is the angle between the positive x-axis and the segment $\overline{OP}$, measured in the counterclockwise direction.
- Conversions between polar and rectangular coordinates:

$$x = r \cos \theta, \qquad r = \sqrt{x^2 + y^2}$$

$$y = r \sin \theta, \qquad \tan \theta = \frac{y}{x} \quad (x \ne 0)$$

- The angular coordinate θ must be chosen so that (r, θ) lies in the proper quadrant. If $r > 0$, then

$$\theta = \begin{cases} \tan^{-1} \dfrac{y}{x} & \text{if } x > 0 \\[2mm] \tan^{-1} \dfrac{y}{x} + \pi & \text{if } x < 0 \\[2mm] \pm \dfrac{\pi}{2} & \text{if } x = 0 \end{cases}$$

- Nonuniqueness: (r, θ), $(r, \theta + 2n\pi)$, and $(-r, \theta + (2n + 1)\pi)$ represent the same point for all integers n. The origin O has polar coordinates $(0, \theta)$ for any θ.
- Polar equations:

Curve	Polar equation
Circle of radius R, center at the origin	$r = R$
Line through origin of slope $m = \tan \theta_0$	$\theta = \theta_0$
Line on which $P_0 = (d, \alpha)$ is the point closest to the origin	$r = d \sec(\theta - \alpha)$
Circle of radius a, center at $(a, 0)$ $(x - a)^2 + y^2 = a^2$	$r = 2a \cos \theta$
Circle of radius a, center at $(0, a)$ $x^2 + (y - a)^2 = a^2$	$r = 2a \sin \theta$

11.3 EXERCISES

Preliminary Questions

1. Points P and Q with the same radial coordinate (choose the correct answer):

(a) lie on the same circle with the center at the origin.

(b) lie on the same ray based at the origin.

2. Give two polar representations for the point $(x, y) = (0, 1)$, one with negative r and one with positive r.

3. Describe each of the following curves:

(a) $r = 2$ **(b)** $r^2 = 2$ **(c)** $r\cos\theta = 2$

4. If $f(-\theta) = f(\theta)$, then the curve $r = f(\theta)$ is symmetric with respect to the (choose the correct answer):

(a) x-axis.   **(b)** y-axis.   **(c)** origin.

Exercises

1. Find polar coordinates for each of the seven points plotted in Figure 17. [Choose $r \ge 0$ and θ in $[0, 2\pi)$.]

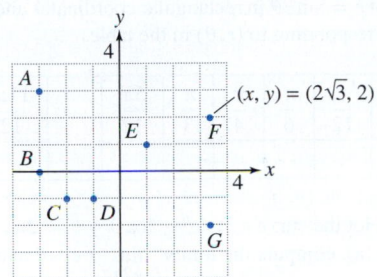

FIGURE 17

2. Plot the points with polar coordinates:

(a) $\left(2, \frac{\pi}{6}\right)$ **(b)** $\left(4, \frac{3\pi}{4}\right)$ **(c)** $\left(3, -\frac{\pi}{2}\right)$ **(d)** $\left(0, \frac{\pi}{6}\right)$

3. Convert from rectangular to polar coordinates:

(a) $(1, 0)$ **(b)** $(3, \sqrt{3})$ **(c)** $(-2, 2)$ **(d)** $(-1, \sqrt{3})$

4. Convert from rectangular to polar coordinates using a calculator (make sure your choice of θ gives the correct quadrant):

(a) $(2, 3)$ **(b)** $(4, -7)$ **(c)** $(-3, -8)$ **(d)** $(-5, 2)$

5. Convert from polar to rectangular coordinates:

(a) $\left(3, \frac{\pi}{6}\right)$ **(b)** $\left(6, \frac{3\pi}{4}\right)$ **(c)** $\left(0, \frac{\pi}{5}\right)$ **(d)** $\left(5, -\frac{\pi}{2}\right)$

6. Which of the following are possible polar coordinates for the point P with rectangular coordinates $(0, -2)$?

(a) $\left(2, \frac{\pi}{2}\right)$ **(b)** $\left(2, \frac{7\pi}{2}\right)$

(c) $\left(-2, -\frac{3\pi}{2}\right)$ **(d)** $\left(-2, \frac{7\pi}{2}\right)$

(e) $\left(-2, -\frac{\pi}{2}\right)$ **(f)** $\left(2, -\frac{7\pi}{2}\right)$

7. Describe each tan-shaded sector in Figure 18 by inequalities in r and θ.

8. Describe each green-shaded sector in Figure 18 by inequalities in r and θ.

9. Find an equation in polar coordinates of the line through the origin with slope $\frac{1}{\sqrt{3}}$.

10. Find an equation in polar coordinates of the line through the origin with slope $1 - \sqrt{2}$.

11. What is the slope of the line $\theta = \frac{3\pi}{5}$?

12. One of $r = 2\sec\theta$ and $r = 2\csc\theta$ is a horizontal line, and the other is a vertical line. Convert each to rectangular coordinates to show which is which.

In Exercises 13–18, convert to an equation in rectangular coordinates.

13. $r = 7$ **14.** $r = \sin\theta$

15. $r = 2\sin\theta$ **16.** $r = 2\csc\theta - \sec\theta$

17. $r = \dfrac{1}{\cos\theta - \sin\theta}$ **18.** $r = \dfrac{1}{2 - \cos\theta}$

In Exercises 19–24, convert to an equation in polar coordinates of the form $r = f(\theta)$.

19. $x^2 + y^2 = 5$ **20.** $x = 5$

21. $y = x^2$ **22.** $xy = 1$

23. $e^{\sqrt{x^2+y^2}} = 1$ **24.** $\ln x = 1$

25. Match each equation with its description:

(a) $r = 2$ **(i)** Vertical line
(b) $\theta = 2$ **(ii)** Horizontal line
(c) $r = 2\sec\theta$ **(iii)** Circle
(d) $r = 2\csc\theta$ **(iv)** Line through origin

26. Suppose that $P = (x, y)$ has polar coordinates (r, θ). Find the polar coordinates for the points:

(a) $(x, -y)$ **(b)** $(-x, -y)$ **(c)** $(-x, y)$ **(d)** (y, x)

27. Find the values of θ in the plot of $r = 4\cos\theta$ corresponding to points A, B, C, D in Figure 19. Then indicate the portion of the graph traced out as θ varies in the following intervals:

(a) $0 \le \theta \le \frac{\pi}{2}$ **(b)** $\frac{\pi}{2} \le \theta \le \pi$ **(c)** $\pi \le \theta \le \frac{3\pi}{2}$

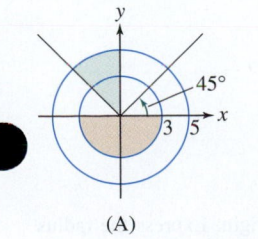

(A)

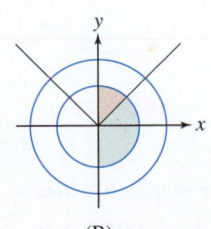

(B)

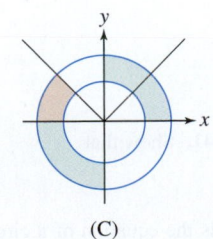

(C)

FIGURE 18

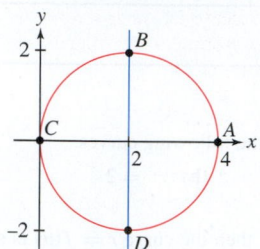

FIGURE 19 Plot of $r = 4\cos\theta$.

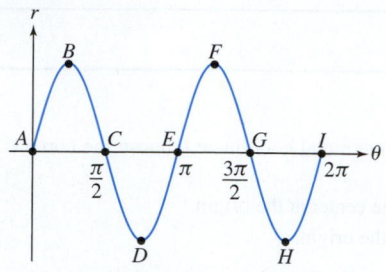

(A) Graph of r as a function of θ, where $r = \sin 2\theta$

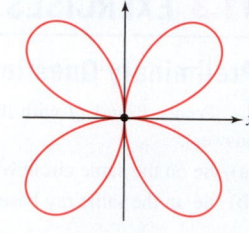

(B) Graph of $r = \sin 2\theta$ in polar coordinates

FIGURE 21

28. Match each equation in rectangular coordinates with its equation in polar coordinates:

(a) $x^2 + y^2 = 4$ **(i)** $r^2(1 - 2\sin^2\theta) = 4$

(b) $x^2 + (y - 1)^2 = 1$ **(ii)** $r(\cos\theta + \sin\theta) = 4$

(c) $x^2 - y^2 = 4$ **(iii)** $r = 2\sin\theta$

(d) $x + y = 4$ **(iv)** $r = 2$

29. What are the polar equations of the lines parallel to the line with equation $r\cos\left(\theta - \frac{\pi}{3}\right) = 1$?

30. Show that the circle with its center at $\left(\frac{1}{2}, \frac{1}{2}\right)$ in Figure 20 has polar equation $r = \sin\theta + \cos\theta$ and find the values of θ between 0 and π corresponding to points A, B, C, and D.

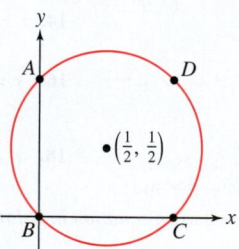

FIGURE 20 Plot of $r = \sin\theta + \cos\theta$.

31. Sketch the curve $r = \frac{1}{2}\theta$ (the spiral of Archimedes) for θ between 0 and 2π by plotting the points for $\theta = 0, \frac{\pi}{4}, \frac{\pi}{2}, \ldots, 2\pi$.

32. Sketch $r = 3\cos\theta - 1$ (see Example 9).

33. Sketch the cardioid curve $r = 1 + \cos\theta$.

34. Show that the cardioid of Exercise 33 has equation

$$(x^2 + y^2 - x)^2 = x^2 + y^2$$

in rectangular coordinates.

35. Figure 21 displays the graphs of $r = \sin 2\theta$ in r versus θ rectangular coordinates and in polar coordinates, where it is a "rose with four petals." Identify:

(a) The points in (B) corresponding to points A–I in (A).

(b) The parts of the curve in (B) corresponding to the angle intervals $\left[0, \frac{\pi}{2}\right], \left[\frac{\pi}{2}, \pi\right], \left[\pi, \frac{3\pi}{2}\right]$, and $\left[\frac{3\pi}{2}, 2\pi\right]$.

36. Sketch the curve $r = \sin 3\theta$ by filling in the table of r-values below and plotting the corresponding points of the curve. Notice that the three petals of the curve correspond to the angle intervals $\left[0, \frac{\pi}{3}\right], \left[\frac{\pi}{3}, \frac{2\pi}{3}\right]$, and $\left[\frac{\pi}{3}, \pi\right]$. Then plot $r = \sin 3\theta$ in rectangular coordinates and label the points on this graph corresponding to (r, θ) in the table.

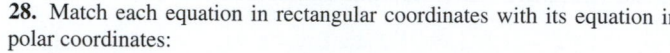

θ	0	$\frac{\pi}{12}$	$\frac{\pi}{6}$	$\frac{\pi}{4}$	$\frac{\pi}{3}$	$\frac{5\pi}{12}$	$\cdots$	$\frac{11\pi}{12}$	π
r									

37. (a) [GU] Plot the curve $r = \frac{\theta}{2\pi - \theta}$ for $0 \le \theta < 2\pi$.

(b) With r as in (a), compute the limits $\lim\limits_{\theta \to 2\pi^-} r\cos\theta$ and $\lim\limits_{\theta \to 2\pi^-} r\sin\theta$.

(c) [✎] Explain how the limits in (b) show that the curve approaches a horizontal asymptote as θ approaches 2π from the left. What is the asymptote?

38. (a) [GU] Plot the curve $r = \frac{1}{\pi - \theta}$ for $0 \le \theta \le 2\pi$.

(b) With r as in (a), compute the limits $\lim\limits_{\theta \to \pi^-} r\cos\theta$, $\lim\limits_{\theta \to \pi^+} r\cos\theta$, $\lim\limits_{\theta \to \pi^-} r\sin\theta$, and $\lim\limits_{\theta \to \pi^+} r\sin\theta$.

(c) [✎] Explain how the limits in (b) show that the curve approaches a horizontal asymptote as θ approaches π, both from the left and from the right. What is the asymptote?

39. [GU] Plot the **cissoid** $r = 2\sin\theta\tan\theta$ and show that its equation in rectangular coordinates is

$$y^2 = \frac{x^3}{2 - x}$$

40. Prove that $r = 2a\cos\theta$ is the equation of the circle in Figure 22 using only the fact that a triangle inscribed in a circle with one side a diameter is a right triangle.

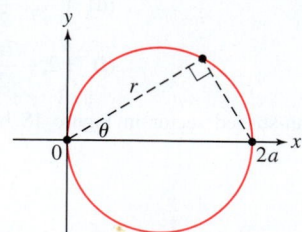

FIGURE 22

41. Show that

$$r = a\cos\theta + b\sin\theta$$

is the equation of a circle passing through the origin. Express the radius and center (in rectangular coordinates) in terms of a and b and write the equation in rectangular coordinates.

42. Use the previous exercise to write the equation of the circle of radius 5 and center $(3, 4)$ in the form $r = a \cos \theta + b \sin \theta$.

43. Use the identity $\cos 2\theta = \cos^2 \theta - \sin^2 \theta$ to find a polar equation of the hyperbola $x^2 - y^2 = 1$.

44. Find an equation in rectangular coordinates for the curve $r^2 = \cos 2\theta$.

45. Show that $\cos 3\theta = \cos^3 \theta - 3 \cos \theta \sin^2 \theta$ and use this identity to find an equation in rectangular coordinates for the curve $r = \cos 3\theta$.

46. Use the addition formula for the cosine to show that the line $\mathcal{L}$ with polar equation $r \cos(\theta - \alpha) = d$ has the equation in rectangular coordinates $(\cos \alpha)x + (\sin \alpha)y = d$. Show that $\mathcal{L}$ has slope $m = -\cot \alpha$ and y-intercept $d / \sin \alpha$.

In Exercises 47–50, find an equation in polar coordinates of the line $\mathcal{L}$ with the given description.

47. The point on $\mathcal{L}$ closest to the origin has polar coordinates $\left(2, \frac{\pi}{9}\right)$.

48. The point on $\mathcal{L}$ closest to the origin has rectangular coordinates $(-2, 2)$.

49. $\mathcal{L}$ is tangent to the circle $r = 2\sqrt{10}$ at the point with rectangular coordinates $(-2, -6)$.

50. $\mathcal{L}$ has slope 3 and is tangent to the unit circle in the fourth quadrant.

51. Show that every line that does not pass through the origin has a polar equation of the form

$$r = \frac{b}{\sin \theta - a \cos \theta}$$

where $b \neq 0$.

52. By the Law of Cosines, the distance d between two points (Figure 23) with polar coordinates (r, θ) and (r_0, θ_0) is

$$d^2 = r^2 + r_0^2 - 2rr_0 \cos(\theta - \theta_0)$$

Use this distance formula to show that

$$r^2 - 10r \cos\left(\theta - \frac{\pi}{4}\right) = 56$$

is the equation of the circle of radius 9 whose center has polar coordinates $\left(5, \frac{\pi}{4}\right)$.

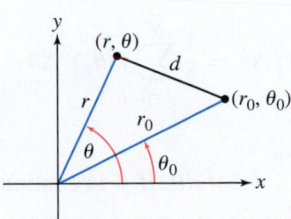

FIGURE 23

53. For $a > 0$, a **lemniscate curve** is the set of points P such that the product of the distances from P to $(a, 0)$ and $(-a, 0)$ is a^2. Show that the equation of the lemniscate is

$$(x^2 + y^2)^2 = 2a^2(x^2 - y^2)$$

Then find the equation in polar coordinates. To obtain the simplest form of the equation, use the identity $\cos 2\theta = \cos^2 \theta - \sin^2 \theta$. Use graph plotting software and plot the lemniscate for $a = 2$.

54. 🖉 Let c be a fixed constant. Explain the relationship between the graphs of:

(a) $y = f(x + c)$ and $y = f(x)$ (rectangular).

(b) $r = f(\theta + c)$ and $r = f(\theta)$ (polar).

(c) $y = f(x) + c$ and $y = f(x)$ (rectangular).

(d) $r = f(\theta) + c$ and $r = f(\theta)$ (polar).

55. The Slope of the Tangent Line in Polar Coordinates   Show that a polar curve $r = f(\theta)$ has parametric equations

$$x = f(\theta) \cos \theta, \qquad y = f(\theta) \sin \theta$$

Then apply Theorem 1 of Section 11.1 to prove

$$\frac{dy}{dx} = \frac{f(\theta) \cos \theta + f'(\theta) \sin \theta}{-f(\theta) \sin \theta + f'(\theta) \cos \theta} \qquad \boxed{2}$$

where $f'(\theta) = df/d\theta$.

56. Use Eq. (2) to find the slope of the tangent line to $r = \sin \theta$ at $\theta = \frac{\pi}{3}$.

57. Use Eq. (2) to find the slope of the tangent line to $r = \theta$ at $\theta = \frac{\pi}{2}$ and $\theta = \pi$.

58. Find the equation in rectangular coordinates of the tangent line to $r = 4 \cos 3\theta$ at $\theta = \frac{\pi}{6}$.

59. Find the polar coordinates of the points on the lemniscate $r^2 = \cos 2\theta$ in Figure 24 where the tangent line is horizontal.

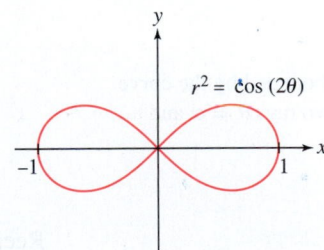

FIGURE 24

60. Find the polar coordinates of the points on the cardioid $r = 1 + \cos \theta$ where the tangent line is horizontal (see Figure 25(A)).

61. Use Eq. (2) to show that for $r = \sin \theta + \cos \theta$,

$$\frac{dy}{dx} = \frac{\cos 2\theta + \sin 2\theta}{\cos 2\theta - \sin 2\theta}$$

Then calculate the slopes of the tangent lines at points A, B, C in Figure 20.

Further Insights and Challenges

62. [✏️] Let $y = f(x)$ be a periodic function of period 2π—that is, $f(x) = f(x + 2\pi)$. Explain how this periodicity is reflected in the graph of:

(a) $y = f(x)$ in rectangular coordinates.

(b) $r = f(\theta)$ in polar coordinates.

63. [GU] Use a graphing utility to convince yourself that the polar equations $r = f_1(\theta) = 2\cos\theta - 1$ and $r = f_2(\theta) = 2\cos\theta + 1$ have the same graph. Then explain why. *Hint:* Show that the points $(f_1(\theta + \pi), \theta + \pi)$ and $(f_2(\theta), \theta)$ coincide.

64. [GU] We investigate how the shape of the limaçon curve given by $r = b + \cos\theta$ depends on the constant b (see Figure 25).

(a) Argue as in Exercise 63 to show that the constants b and $-b$ yield the same curve.

(b) Plot the limaçon for $b = 0, 0.2, 0.5, 0.8, 1$ and describe how the curve changes.

(c) Plot the limaçon for $b = 1.2, 1.5, 1.8, 2, 2.4$ and describe how the curve changes.

(d) Use Eq. (2) to show that

$$\frac{dy}{dx} = -\left(\frac{b\cos\theta + \cos 2\theta}{b + 2\cos\theta}\right)\csc\theta$$

(e) Find the points where the tangent line is vertical. Note that there are three cases: $0 \le b < 2$, $b = 2$, and $b > 2$. Do the plots constructed in (b) and (c) reflect your results?

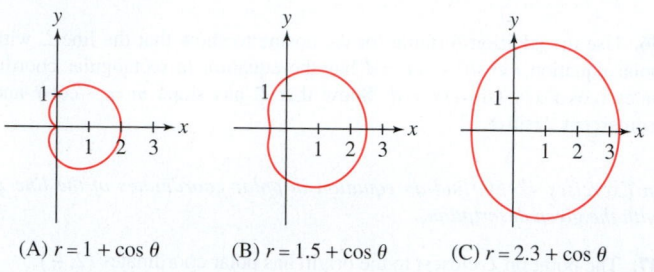

(A) $r = 1 + \cos\theta$ (B) $r = 1.5 + \cos\theta$ (C) $r = 2.3 + \cos\theta$

FIGURE 25

11.4 Area and Arc Length in Polar Coordinates

For a function f, if $f(\theta) > 0$ for $\alpha < \theta < \beta$, then the polar-coordinates region bounded by the curve $r = f(\theta)$ and the two rays $\theta = \alpha$ and $\theta = \beta$ is a sector as in Figure 1(A). We can compute this area via a polar-coordinates integral; we will see how in the first part of this section.

To derive a formula for the area, divide the region into N narrow sectors of angle $\Delta\theta = (\beta - \alpha)/N$ corresponding to a partition of the interval $[\alpha, \beta]$ as in Figure 1(B):

$$\theta_0 = \alpha < \theta_1 < \theta_2 < \cdots < \theta_N = \beta$$

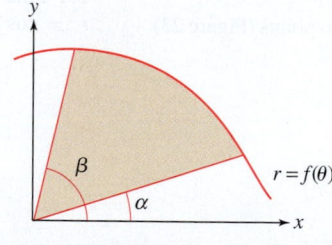

(A) Region $\alpha \le \theta \le \beta$

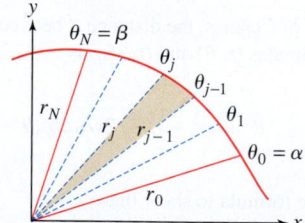

(B) Region divided into narrow sectors

[DF] **FIGURE 1** Area bounded by the curve $r = f(\theta)$ and the two rays $\theta = \alpha$ and $\theta = \beta$.

Recall that a circular sector of angle $\Delta\theta$ and radius r has area $\frac{1}{2}r^2\Delta\theta$ (Figure 2). If $\Delta\theta$ is small, the jth narrow sector (Figure 3) is nearly a circular sector of radius $r_j = f(\theta_j)$, so its area is *approximately* $\frac{1}{2}r_j^2\Delta\theta$. The total area is approximated by the sum

$$\text{area of region} \approx \sum_{j=1}^{N}\frac{1}{2}r_j^2\Delta\theta = \frac{1}{2}\sum_{j=1}^{N}f(\theta_j)^2\Delta\theta \qquad \boxed{1}$$

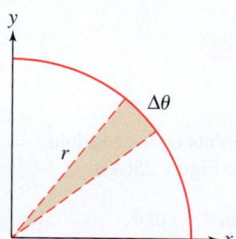

FIGURE 2 The area of a circular sector is $\frac{1}{2}r^2\Delta\theta$.

This is a Riemann sum for the integral $\frac{1}{2}\int_{\alpha}^{\beta}f(\theta)^2\,d\theta$. If f is continuous, then the sum approaches the integral as $N \to \infty$, and we obtain the following formula.

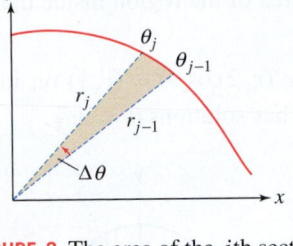

DF **FIGURE 3** The area of the jth sector is approximately $\frac{1}{2}r_j^2 \Delta\theta$.

THEOREM 1 **Area in Polar Coordinates** If f is a continuous function, then the area bounded by a curve in polar form $r = f(\theta)$ and the rays $\theta = \alpha$ and $\theta = \beta$ (with $\alpha < \beta$) is equal to

$$\frac{1}{2} \int_\alpha^\beta r^2 \, d\theta = \frac{1}{2} \int_\alpha^\beta f(\theta)^2 \, d\theta \qquad \boxed{2}$$

We know that $r = R$ defines a circle of radius R. By Eq. (2), the area is equal to

$\frac{1}{2} \int_0^{2\pi} R^2 \, d\theta = \frac{1}{2}R^2(2\pi) = \pi R^2$, as expected.

EXAMPLE 1 Use Theorem 1 to compute the area of the right semicircle with equation $r = 4\sin\theta$.

Solution The equation $r = 4\sin\theta$ defines a circle of radius 2 tangent to the x-axis at the origin. The right semicircle is swept out as θ varies from 0 to $\frac{\pi}{2}$ as in Figure 4(A). By Eq. (2), the area of the right semicircle is

◄ REMINDER In Eq. (3), we use the identity

$$\sin^2\theta = \frac{1}{2}(1 - \cos 2\theta)$$

$$\frac{1}{2} \int_0^{\pi/2} r^2 \, d\theta = \frac{1}{2} \int_0^{\pi/2} (4\sin\theta)^2 \, d\theta = 8 \int_0^{\pi/2} \sin^2\theta \, d\theta \qquad \boxed{3}$$

$$= 8 \int_0^{\pi/2} \frac{1}{2}(1 - \cos 2\theta) \, d\theta$$

$$= (4\theta - 2\sin 2\theta)\Big|_0^{\pi/2} = 4\left(\frac{\pi}{2}\right) - 0 = 2\pi \qquad ■$$

CAUTION *Keep in mind that the integral* $\frac{1}{2} \int_\alpha^\beta r^2 \, d\theta$ *does* **not** *compute the area* **under** *a curve as in Figure 4(B), but rather computes the area swept out by a radial segment as θ varies from α to β, as in Figure 4(A).*

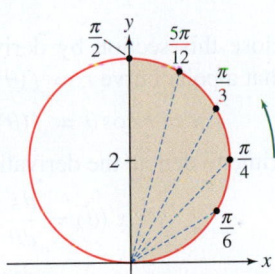

(A) The polar integral computes the area swept out by a radial segment.

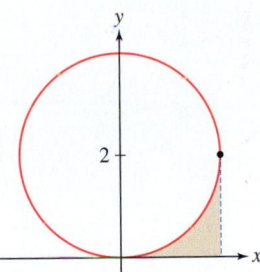

(B) The ordinary integral in rectangular coordinates computes the area underneath a curve.

FIGURE 4

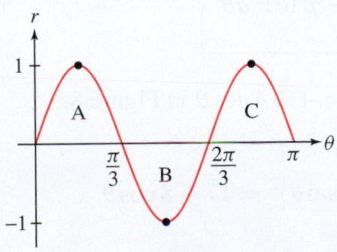

FIGURE 5 Graph of $r = \sin 3\theta$ in r versus θ rectangular coordinates.

EXAMPLE 2 Sketch $r = \sin 3\theta$ and compute the area of one "petal."

Solution To sketch the curve, we first graph $r = \sin 3\theta$ in r versus θ rectangular coordinates. Figure 5 shows that the radius r varies from 0 to 1 and back to 0 as θ varies from 0 to $\frac{\pi}{3}$. This gives petal A in Figure 6. Petal B is traced as θ varies from $\frac{\pi}{3}$ to $\frac{2\pi}{3}$ (with $r \le 0$), and petal C is traced for $\frac{2\pi}{3} \le \theta \le \pi$. We find that the area of petal A [using Eq. (2) in the margin to evaluate the integral] is equal to

$$\frac{1}{2} \int_0^{\pi/3} (\sin 3\theta)^2 \, d\theta = \frac{1}{2} \int_0^{\pi/3} \left(\frac{1 - \cos 6\theta}{2}\right) d\theta = \left(\frac{1}{4}\theta - \frac{1}{24}\sin 6\theta\right)\Big|_0^{\pi/3} = \frac{\pi}{12} \qquad ■$$

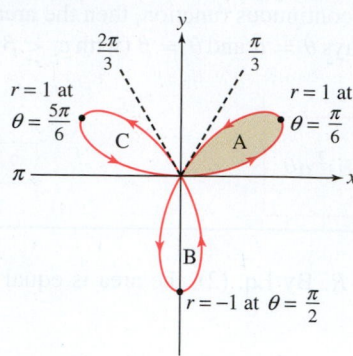

FIGURE 6 Graph of polar curve $r = \sin 3\theta$, a "rose with three petals."

EXAMPLE 3   Area Between Two Curves   Find the area of the region inside the circle $r = 2 \cos \theta$ but outside the circle $r = 1$ [Figure 7(A)].

Solution The two circles intersect at the points where $(r, 2 \cos \theta) = (r, 1)$ or, in other words, when $2 \cos \theta = 1$. This yields $\cos \theta = \frac{1}{2}$, which has solutions $\theta = \pm \frac{\pi}{3}$.

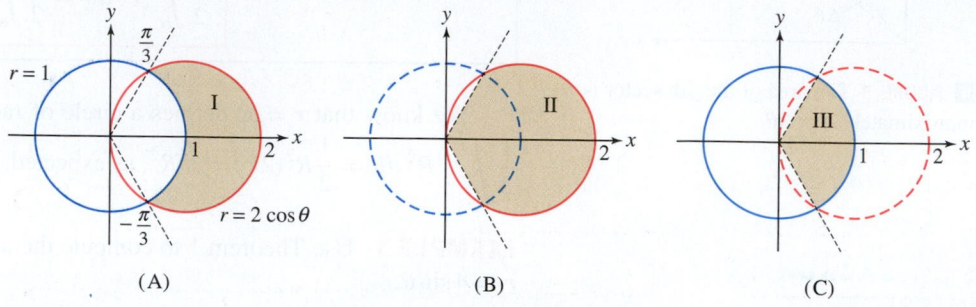

(A)   (B)   (C)

FIGURE 7 Region I is the difference of regions II and III.

We see in Figure 7 that region I is the difference of regions II and III in Figures 7(B) and (C). Therefore,

$$\text{area of I} = \text{area of II} - \text{area of III}$$

← **REMINDER** In Eq. (4), we use the identity

$$\cos^2 \theta = \frac{1}{2}(1 + \cos 2\theta)$$

$$= \frac{1}{2} \int_{-\pi/3}^{\pi/3} (2 \cos \theta)^2 \, d\theta - \frac{1}{2} \int_{-\pi/3}^{\pi/3} (1)^2 \, d\theta$$

$$= \frac{1}{2} \int_{-\pi/3}^{\pi/3} (4 \cos^2 \theta - 1) \, d\theta = \frac{1}{2} \int_{-\pi/3}^{\pi/3} (2 \cos 2\theta + 1) \, d\theta \qquad \boxed{4}$$

$$= \frac{1}{2}(\sin 2\theta + \theta) \Big|_{-\pi/3}^{\pi/3} = \frac{\sqrt{3}}{2} + \frac{\pi}{3} \approx 1.91 \qquad \blacksquare$$

We close this section by deriving a formula for arc length in polar coordinates. Observe that a polar curve $r = f(\theta)$ has a parametrization with θ as a parameter:

$$x = r \cos \theta = f(\theta) \cos \theta, \qquad y = r \sin \theta = f(\theta) \sin \theta$$

Using a prime to denote the derivative with respect to θ, we have

$$x'(\theta) = \frac{dx}{d\theta} = -f(\theta) \sin \theta + f'(\theta) \cos \theta$$

$$y'(\theta) = \frac{dy}{d\theta} = f(\theta) \cos \theta + f'(\theta) \sin \theta$$

Recall from Section 11.2 that arc length is obtained by integrating $\sqrt{x'(\theta)^2 + y'(\theta)^2}$. Straightforward algebra shows that $x'(\theta)^2 + y'(\theta)^2 = f(\theta)^2 + f'(\theta)^2$; thus,

$$\boxed{\text{arc length } s = \int_{\alpha}^{\beta} \sqrt{f(\theta)^2 + f'(\theta)^2} \, d\theta} \qquad \boxed{5}$$

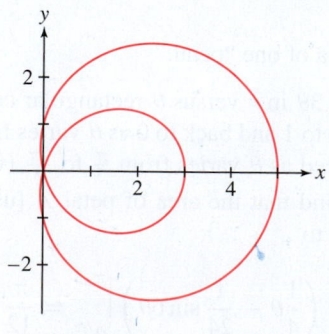

FIGURE 8 Graph of $r = 1 + 4 \cos \theta$.

EXAMPLE 4 Find the total length of the limaçon $r = 1 + 4 \cos \theta$ in Figure 8.

Solution In this case, $f(\theta) = 1 + 4 \cos \theta$ and

$$f(\theta)^2 + f'(\theta)^2 = (1 + 4 \cos \theta)^2 + (-4 \sin \theta)^2 = 17 + 8 \cos \theta$$

The total length of this limaçon is

$$\int_0^{2\pi} \sqrt{f(\theta)^2 + f'(\theta)^2} \, d\theta = \int_0^{2\pi} \sqrt{17 + 8 \cos \theta} \, d\theta$$

Using numerical approximation, we find that the length is approximately 25.53. $\blacksquare$

11.4 SUMMARY

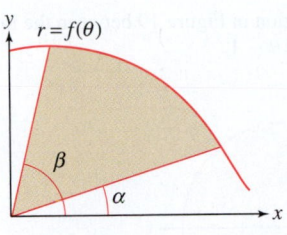

FIGURE 9 Region bounded by the polar curve $r = f(\theta)$ and the rays $\theta = \alpha$, $\theta = \beta$.

- Area of the sector bounded by a polar curve $r = f(\theta)$ and two rays $\theta = \alpha$ and $\theta = \beta$ (Figure 9):

$$\text{area} = \frac{1}{2} \int_{\alpha}^{\beta} f(\theta)^2 \, d\theta$$

- Arc length of the polar curve $r = f(\theta)$ for $\alpha \le \theta \le \beta$:

$$\text{arc length} = \int_{\alpha}^{\beta} \sqrt{f(\theta)^2 + f'(\theta)^2} \, d\theta$$

11.4 EXERCISES

Preliminary Questions

1. Polar coordinates are suited to finding the area (choose one):
(a) under a curve between $x = a$ and $x = b$.
(b) bounded by a curve and two rays through the origin.

2. Is the formula for area in polar coordinates valid if $f(\theta)$ takes negative values?

3. The horizontal line $y = 1$ has polar equation $r = \csc \theta$. Which area is represented by the integral $\frac{1}{2} \int_{\pi/6}^{\pi/2} \csc^2 \theta \, d\theta$ (Figure 10)?

(a) $\square ABCD$ (b) $\triangle ABC$ (c) $\triangle ACD$

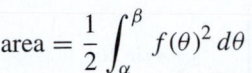

FIGURE 10

Exercises

1. Sketch the region bounded by the circle $r = 5$ and the rays $\theta = \frac{\pi}{2}$ and $\theta = \pi$, and compute its area as an integral in polar coordinates.

2. Sketch the region bounded by the line $r = \sec \theta$ and the rays $\theta = 0$ and $\theta = \frac{\pi}{3}$. Compute its area in two ways: as an integral in polar coordinates and using geometry.

3. Calculate the area of the circle $r = 4 \sin \theta$ as an integral in polar coordinates (see Figure 4). Be careful to choose the correct limits of integration.

4. Find the area of the shaded triangle in Figure 11 as an integral in polar coordinates. Then find the rectangular coordinates of P and Q, and compute the area via geometry.

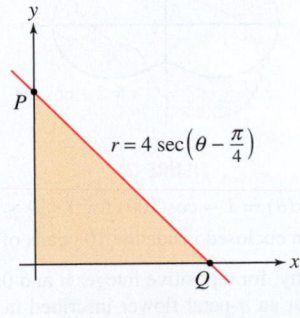

FIGURE 11

5. Find the area of the shaded region in Figure 12. Note that θ varies from 0 to $\frac{\pi}{2}$.

6. Which interval of θ-values corresponds to the shaded region in Figure 13? Find the area of the region.

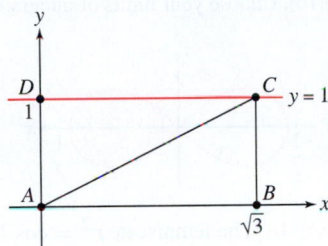

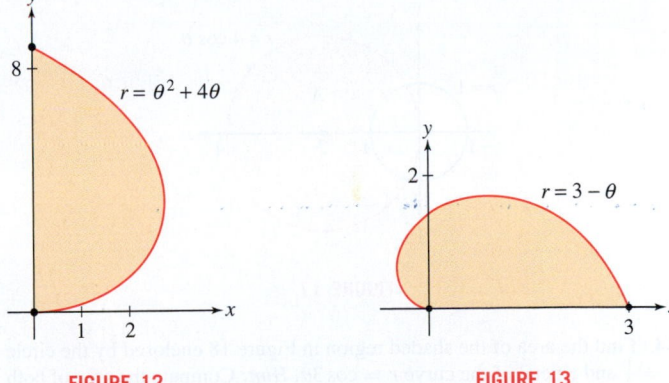

FIGURE 12   **FIGURE 13**

7. Find the total area enclosed by the cardioid in Figure 14.

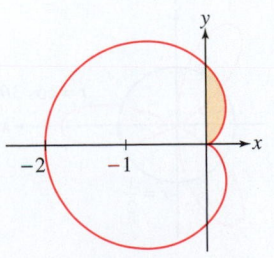

FIGURE 14 The cardioid $r = 1 - \cos \theta$.

8. Find the area of the shaded region in Figure 14.

9. Find the area of one leaf of the four-petaled rose $r = \sin 2\theta$ (Figure 15). Then prove that the total area of the rose is equal to one-half the area of the circumscribed circle.

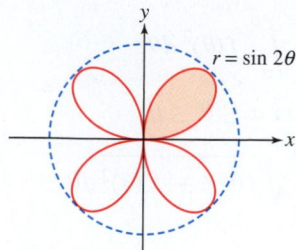

FIGURE 15 Four-petaled rose $r = \sin 2\theta$.

10. Find the area enclosed by one loop of the lemniscate with equation $r^2 = \cos 2\theta$ (Figure 16). Choose your limits of integration carefully.

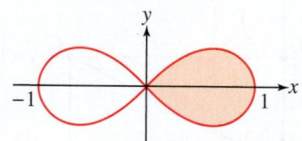

FIGURE 16 The lemniscate $r^2 = \cos 2\theta$.

11. Find the area of the intersection of the circles $r = 2\sin\theta$ and $r = 2\cos\theta$.

12. Find the area of the intersection of the circles $r = \sin\theta$ and $r = \sqrt{3}\cos\theta$.

13. Find the area of region A in Figure 17.

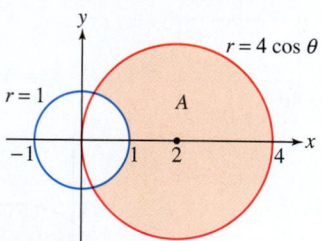

FIGURE 17

14. Find the area of the shaded region in Figure 18 enclosed by the circle $r = \frac{1}{2}$ and a petal of the curve $r = \cos 3\theta$. *Hint:* Compute the area of both the petal and the region inside the petal and outside the circle.

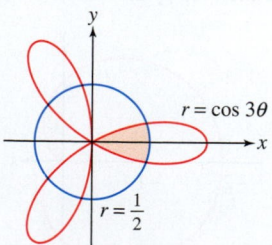

FIGURE 18

15. Find the area of the inner loop of the limaçon with polar equation $r = 2\cos\theta - 1$ (Figure 19).

16. Find the area of the shaded region in Figure 19 between the inner and outer loop of the limaçon $r = 2\cos\theta - 1$.

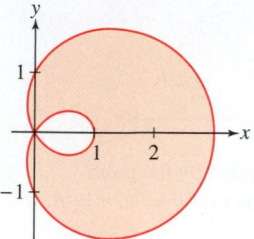

FIGURE 19 The limaçon $r = 2\cos\theta - 1$.

17. Find the area of the part of the circle $r = \sin\theta + \cos\theta$ in the fourth quadrant (see Exercise 30 in Section 11.3).

18. Find the area of the region inside the circle $r = 2\sin\left(\theta + \frac{\pi}{4}\right)$ and above the line $r = \sec\left(\theta - \frac{\pi}{4}\right)$.

19. Find the area between the two curves in Figure 20(A).

20. Find the area between the two curves in Figure 20(B).

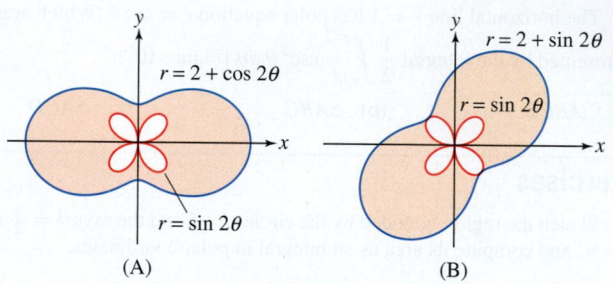

(A)   (B)

FIGURE 20

21. Find the area inside both curves in Figure 21.

22. Find the area of the region that lies inside one but not both of the curves in Figure 21.

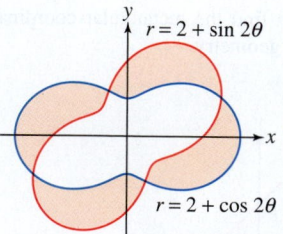

FIGURE 21

23. (a) $\boxed{\text{GU}}$ Plot $r(\theta) = 1 - \cos(10\theta)$ for $0 \le \theta \le 2\pi$.

(b) Compute the area enclosed inside the 10 petals of the graph of $r(\theta)$.

(c) Explain why, for a positive integer n and $0 \le \theta \le 2\pi$, $r_n(\theta) = 1 - \cos(n\theta)$ traces out an n-petal flower inscribed in a circle of radius 2 centered at the origin.

(d) Show that the area enclosed inside the n petals of the graph of $r_n(\theta)$ is independent of n and equals $\frac{3}{8}A$, where A is the area of the circle of radius 2.

24. (a) [GU] Plot the spiral $r(\theta) = \theta$ for $0 \le \theta \le 8\pi$.

(b) On your plot, shade in the region that represents the increase in area enclosed by the curve as θ goes from 6π to 8π. Compute the shaded area.

(c) Show that the increase in area enclosed by the graph of $r(\theta)$ as θ goes from $2n\pi$ to $2(n + 1)\pi$ is $8n\pi^3$.

25. Calculate the total length of the circle $r = 4 \sin \theta$ as an integral in polar coordinates.

26. Sketch the segment $r = \sec \theta$ for $0 \le \theta \le A$. Then compute its length in two ways: as an integral in polar coordinates and using trigonometry.

In Exercises 27–34, compute the length of the polar curve.

27. The length of $r = \theta^2$ for $0 \le \theta \le \pi$

28. The spiral $r = \theta$ for $0 \le \theta \le A$

29. The curve $r = \sin \theta$ for $0 \le \theta \le \pi$

30. The equiangular spiral $r = e^{\theta}$ for $0 \le \theta \le 2\pi$

31. $r = \sqrt{1 + \sin 2\theta}$ for $0 \le \theta \le \pi/4$

32. The cardioid $r = 1 - \cos \theta$ in Figure 14

33. $r = \cos^2 \theta$

34. $r = 1 + \theta$ for $0 \le \theta \le \pi/2$

In Exercises 35–38, express the length of the curve as an integral but do not evaluate it.

35. $r = e^{\theta} + 1$, $0 \le \theta \le \pi/2$.

36. $r = (2 - \cos \theta)^{-1}$, $0 \le \theta \le 2\pi$

37. $r = \sin^3 \theta$, $0 \le \theta \le 2\pi$

38. $r = \sin \theta \cos \theta$, $0 \le \theta \le \pi$

In Exercises 39–42, use a computer algebra system to calculate the total length to two decimal places.

39. [CAS] The three-petal rose $r = \cos 3\theta$ in Figure 18

40. [CAS] The curve $r = 2 + \sin 2\theta$ in Figure 21

41. [CAS] The curve $r = \theta \sin \theta$ in Figure 22 for $0 \le \theta \le 4\pi$

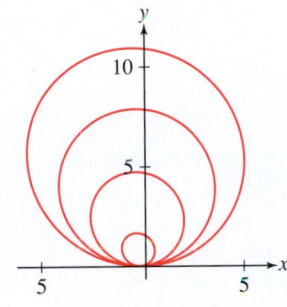

FIGURE 22 $r = \theta \sin \theta$ for $0 \le \theta \le 4\pi$.

42. [CAS] $r = \sqrt{\theta}$, $0 \le \theta \le 4\pi$

Further Insights and Challenges

43. Suppose that the polar coordinates of a moving particle at time t are $(r(t), \theta(t))$. Prove that the particle's speed is equal to $\sqrt{(dr/dt)^2 + r^2(d\theta/dt)^2}$.

44. [✎] Compute the speed at time $t = 1$ of a particle whose polar coordinates at time t are $r = t$, $\theta = t$ (use Exercise 43). What would the speed be if the particle's rectangular coordinates were $x = t$, $y = t$? Why is the speed increasing in one case and constant in the other?

11.5 Conic Sections

The conics were first studied by the ancient Greek mathematicians, beginning possibly with Menaechmus (c. 380–320 BCE) and including Archimedes (287–212 BCE) and Apollonius (c. 262–190 BCE).

Three familiar families of curves—ellipses, hyperbolas, and parabolas—appear throughout mathematics and its applications. They are called **conic sections** because they are obtained as the intersection of a cone with a suitable plane (Figure 1). Our goal in this section is to derive equations for the conic sections from their geometric definitions as curves in the plane.

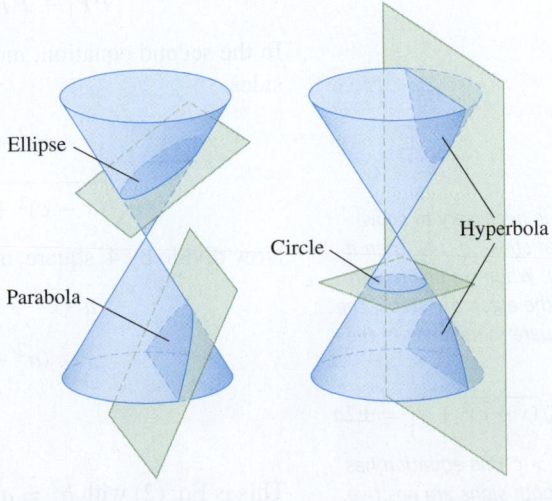

[DF] **FIGURE 1** The conic sections are curves that result from the intersection of a plane and a cone.

An **ellipse** is an oval-shaped curve [Figure 2(A)] consisting of all points P such that the sum of the distances to two fixed points F_1 and F_2 is a constant $K > 0$:

$$PF_1 + PF_2 = K \qquad \boxed{1}$$

We assume always that K is greater than the distance $F_1 F_2$ between the foci, because the ellipse reduces to the line segment $\overline{F_1 F_2}$ if $K = F_1 F_2$, and it has no points at all if $K < F_1 F_2$.

The points F_1 and F_2 are called the **foci** (plural of "focus") of the ellipse. Note that if the foci coincide, then Eq. (1) reduces to $2PF_1 = K$ and we obtain a circle of radius $\frac{1}{2} K$ centered at F_1.

We use the following terminology:

- The midpoint of $\overline{F_1 F_2}$ is the **center** of the ellipse.
- The line through the foci is the **focal axis**.
- The line through the center perpendicular to the focal axis is the **conjugate axis**.

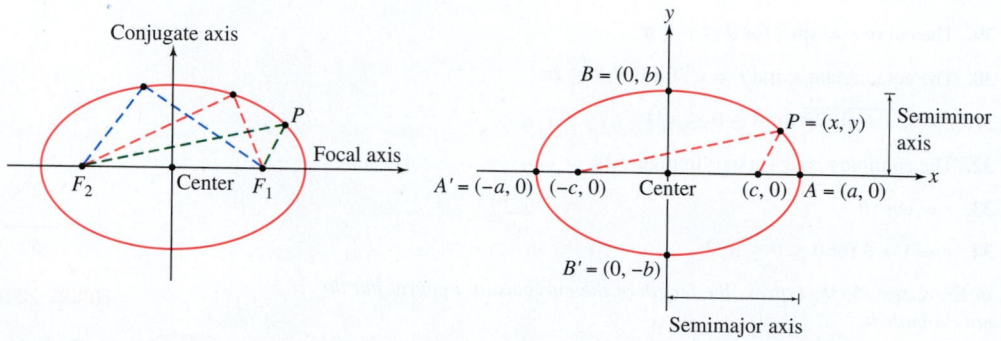

(A) The ellipse consists of all points P such that $PF_1 + PF_2 = K$

(B) Ellipse in standard position:

$$\left(\frac{x}{a}\right)^2 + \left(\frac{y}{b}\right)^2 = 1$$

FIGURE 2

The ellipse is said to be in **standard position** (and is called a **standard ellipse**) if the focal and conjugate axes are the x- and y-axes, as shown in Figure 2(B). In this case, the foci have coordinates $F_1 = (c, 0)$ and $F_2 = (-c, 0)$ for some $c > 0$. Let us prove that the equation of this ellipse has the particularly simple form

$$\left(\frac{x}{a}\right)^2 + \left(\frac{y}{b}\right)^2 = 1 \qquad \boxed{2}$$

where $a = K/2$ and $b = \sqrt{a^2 - c^2}$.

By the distance formula, $P = (x, y)$ lies on the ellipse in Figure 2(B) if

$$PF_1 + PF_2 = \sqrt{(x - c)^2 + y^2} + \sqrt{(x + c)^2 + y^2} = 2a \qquad \boxed{3}$$

In the second equation, move the first term on the left over to the right and square both sides:

$$(x + c)^2 + y^2 = 4a^2 - 4a\sqrt{(x - c)^2 + y^2} + (x - c)^2 + y^2$$

$$4a\sqrt{(x - c)^2 + y^2} = 4a^2 + (x - c)^2 - (x + c)^2 = 4a^2 - 4cx$$

Strictly speaking, it is necessary to show that if $P = (x, y)$ satisfies Eq. (4), then it also satisfies Eq. (3). When we begin with Eq. (4) and reverse the algebraic steps, the process of taking square roots leads to the relation

$$\sqrt{(x - c)^2 + y^2} \pm \sqrt{(x + c)^2 + y^2} = \pm 2a$$

However, because $a > c$ this equation has no solutions unless both signs are positive.

Now divide by 4, square, and simplify:

$$a^2(x^2 - 2cx + c^2 + y^2) = a^4 - 2a^2cx + c^2x^2$$

$$(a^2 - c^2)x^2 + a^2y^2 = a^4 - a^2c^2 = a^2(a^2 - c^2)$$

$$\frac{x^2}{a^2} + \frac{y^2}{a^2 - c^2} = 1 \qquad \boxed{4}$$

This is Eq. (2) with $b^2 = a^2 - c^2$ as claimed.

The ellipse intersects the axes in four points A, A', B, B' called **vertices**. Vertices A and A' along the focal axis are called the **focal vertices**. Following common usage, the numbers a and b are referred to as the **semimajor axis** and the **semiminor axis** (even though they are numbers rather than axes).

THEOREM 1 **Ellipse in Standard Position**   Let $a > b > 0$, and set $c = \sqrt{a^2 - b^2}$. The ellipse $PF_1 + PF_2 = 2a$ with foci $F_1 = (c, 0)$ and $F_2 = (-c, 0)$ has equation

$$\left(\frac{x}{a}\right)^2 + \left(\frac{y}{b}\right)^2 = 1$$
$$\boxed{5}$$

Furthermore, the ellipse has

- semimajor axis a, semiminor axis b.
- focal vertices $(\pm a, 0)$, minor vertices $(0, \pm b)$.

If $b > a > 0$, then Eq. (5) defines an ellipse with foci $(0, \pm c)$, where $c = \sqrt{b^2 - a^2}$.

EXAMPLE 1   Find the equation of the ellipse with foci $(\pm\sqrt{11}, 0)$ and semimajor axis $a = 6$. Then find the semiminor axis and sketch the graph.

Solution   The foci are $(\pm c, 0)$ with $c = \sqrt{11}$, and the semimajor axis is $a = 6$, so we can use the relation $c = \sqrt{a^2 - b^2}$ to find b:

$$b^2 = a^2 - c^2 = 6^2 - (\sqrt{11})^2 = 25 \quad \Rightarrow \quad b = 5$$

Thus, the semiminor axis is $b = 5$ and the ellipse has equation $\left(\frac{x}{6}\right)^2 + \left(\frac{y}{5}\right)^2 = 1$. To sketch this ellipse, plot the vertices $(\pm 6, 0)$ and $(0, \pm 5)$ and connect them as in Figure 3. ∎

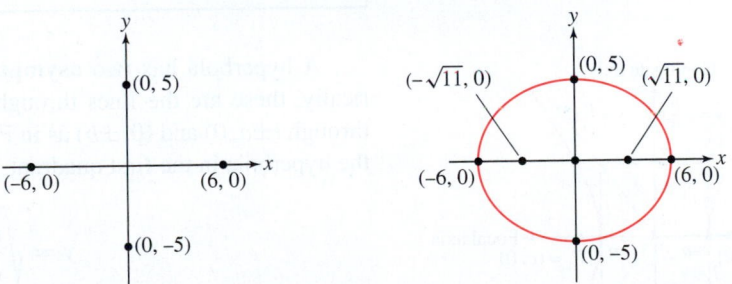

FIGURE 3

To write the equation of an ellipse with axes parallel to the x- and y-axes and center translated to the point $C = (h, k)$, replace x by $x - h$ and y by $y - k$ in the equation (Figure 4):

$$\left(\frac{x - h}{a}\right)^2 + \left(\frac{y - k}{b}\right)^2 = 1$$

EXAMPLE 2   **Translating an Ellipse**   Find an equation of the ellipse with center at $C = (6, 7)$, vertical focal axis, semimajor axis 5, and semiminor axis 3. Where are the foci located?

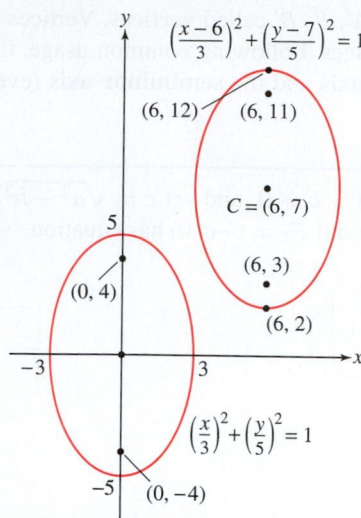

FIGURE 4 An ellipse with a vertical major axis and center at the origin, and a translation of it to an ellipse with center $C = (6, 7)$.

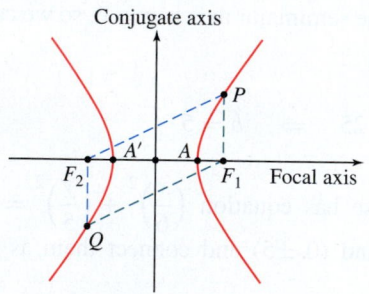

FIGURE 5 A hyperbola with center $(0, 0)$.

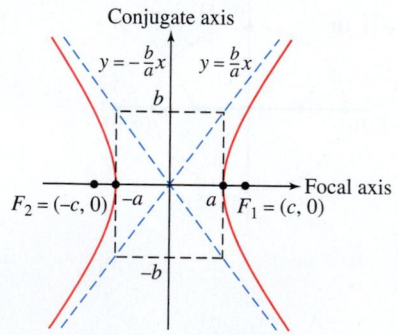

DF FIGURE 6 Hyperbola in standard position.

Solution Since the focal axis is vertical, we have $a = 3$ and $b = 5$, so that $a < b$ (Figure 4). The ellipse centered at the origin would have equation $\left(\frac{x}{3}\right)^2 + \left(\frac{y}{5}\right)^2 = 1$. When the center is translated to $(h, k) = (6, 7)$, the equation becomes

$$\left(\frac{x - 6}{3}\right)^2 + \left(\frac{y - 7}{5}\right)^2 = 1$$

Furthermore, $c = \sqrt{b^2 - a^2} = \sqrt{5^2 - 3^2} = 4$, so the foci are located ± 4 vertical units above and below the center—that is, $F_1 = (6, 11)$ and $F_2 = (6, 3)$. ■

A **hyperbola** is the set of all points P such that the difference of the distances from P to two foci F_1 and F_2 is $\pm K$:

$$PF_1 - PF_2 = \pm K \qquad \boxed{6}$$

We assume that K is less than the distance $F_1 F_2$ between the foci (the hyperbola has no points if $K > F_1 F_2$). Note that a hyperbola consists of two branches corresponding to the choices of sign $\pm$ (Figure 5).

As before, the midpoint of $\overline{F_1 F_2}$ is the **center** of the hyperbola, the line through F_1 and F_2 is called the **focal axis**, and the line through the center perpendicular to the focal axis is called the **conjugate axis**. The **vertices** are the points where the focal axis intersects the hyperbola; they are labeled A and A' in Figure 5. The hyperbola is said to be in **standard position** (and is called a **standard hyperbola**) when the focal and conjugate axes are the x- and y-axes, respectively, as in Figure 6. The next theorem can be proved in much the same way as Theorem 1.

THEOREM 2 Hyperbola in Standard Position Let $a > 0$ and $b > 0$, and set $c = \sqrt{a^2 + b^2}$. The hyperbola $PF_1 - PF_2 = \pm 2a$ with foci $F_1 = (c, 0)$ and $F_2 = (-c, 0)$ has equation

$$\boxed{\left(\frac{x}{a}\right)^2 - \left(\frac{y}{b}\right)^2 = 1} \qquad \boxed{7}$$

A hyperbola has two **asymptotes** that we claim are the lines $y = \pm \frac{b}{a} x$. Geometrically, these are the lines through opposite corners of the rectangle whose sides pass through $(\pm a, 0)$ and $(0, \pm b)$ as in Figure 6. To prove the claim, consider a point (x, y) on the hyperbola in the first quadrant. By Eq. (7),

$$y = \sqrt{\frac{b^2}{a^2} x^2 - b^2} = \frac{b}{a} \sqrt{x^2 - a^2}$$

The following limit shows that a point (x, y) on the hyperbola approaches the line $y = \frac{b}{a} x$ as $x \to \infty$:

$$\lim_{x \to \infty} \left(y - \frac{b}{a} x \right) = \frac{b}{a} \lim_{x \to \infty} \left(\sqrt{x^2 - a^2} - x \right)$$

$$= \frac{b}{a} \lim_{x \to \infty} \left(\sqrt{x^2 - a^2} - x \right) \left(\frac{\sqrt{x^2 - a^2} + x}{\sqrt{x^2 - a^2} + x} \right)$$

$$= \frac{b}{a} \lim_{x \to \infty} \left(\frac{-a^2}{\sqrt{x^2 - a^2} + x} \right) = 0$$

The asymptotic behavior in the remaining quadrants is similar.

EXAMPLE 3 Find the foci of the hyperbola $9x^2 - 4y^2 = 36$. Sketch its graph and asymptotes.

Solution First, divide by 36 to write the equation in standard form:

$$\frac{x^2}{4} - \frac{y^2}{9} = 1 \quad \text{or} \quad \left(\frac{x}{2}\right)^2 - \left(\frac{y}{3}\right)^2 = 1$$

Thus, $a = 2$, $b = 3$, and $c = \sqrt{a^2 + b^2} = \sqrt{4 + 9} = \sqrt{13}$. The foci are

$$F_1 = (\sqrt{13}, 0), \qquad F_2 = (-\sqrt{13}, 0)$$

To sketch the graph, we draw the rectangle through the points $(\pm 2, 0)$ and $(0, \pm 3)$ as in Figure 7. The diagonals of the rectangle are the asymptotes $y = \pm\frac{3}{2}x$. The hyperbola passes through the vertices $(\pm 2, 0)$ and approaches the asymptotes. ∎

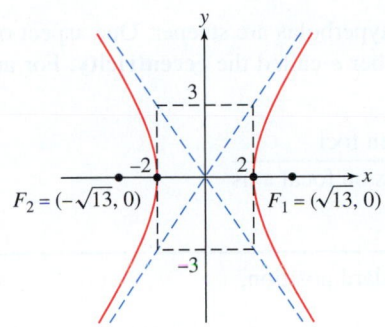

FIGURE 7 The hyperbola $9x^2 - 4y^2 = 36$.

Unlike the ellipse and hyperbola, which are defined in terms of two foci, a **parabola** is the set of points P equidistant from a focus F and a line $\mathcal{D}$ called the **directrix**:

$$PF = PD \qquad \boxed{8}$$

Here, when we speak of the *distance* from a point P to a line $\mathcal{D}$, we mean the distance from P to the point Q on $\mathcal{D}$ closest to P, obtained by dropping a perpendicular from P to $\mathcal{D}$ (Figure 8). We denote this distance by PD.

The line through the focus F perpendicular to $\mathcal{D}$ is called the **axis** of the parabola. The **vertex** is the point where the parabola intersects its axis. We say that the parabola is in **standard position** (and is a **standard parabola**) if, for some c, the focus is $F = (0, c)$ and the directrix is $y = -c$, as shown in Figure 8. We prove in Exercise 75 that the vertex is then located at the origin and the equation of the parabola is $y = x^2/4c$. If $c < 0$, then the parabola opens downward.

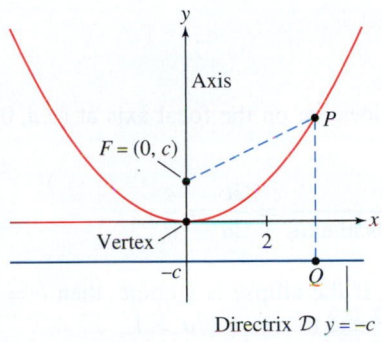

DF **FIGURE 8** Parabola with focus $(0, c)$ and directrix $y = -c$.

> **THEOREM 3 Parabola in Standard Position** Let $c \neq 0$. The parabola with focus $F = (0, c)$ and directrix $y = -c$ has equation
>
> $$y = \frac{1}{4c}x^2 \qquad \boxed{9}$$
>
> The vertex is located at the origin. The parabola opens upward if $c > 0$ and downward if $c < 0$.

EXAMPLE 4 The standard parabola with directrix $y = -2$ is translated so that its vertex is located at $(2, 8)$. Find its equation, directrix, and focus.

Solution By Eq. (9) with $c = 2$, the standard parabola with directrix $y = -2$ has equation $y = \frac{1}{8}x^2$ (Figure 9). The focus of this standard parabola is $(0, c) = (0, 2)$, which is 2 units above the vertex $(0, 0)$.

To obtain the equation when the parabola is translated with vertex at $(2, 8)$, we replace x by $x - 2$ and y by $y - 8$:

$$y - 8 = \frac{1}{8}(x - 2)^2 \quad \text{or} \quad y = \frac{1}{8}x^2 - \frac{1}{2}x + \frac{17}{2}$$

The vertex has moved up 8 units, so the directrix also moves up 8 units to become $y = 6$. The new focus is 2 units above the new vertex $(2, 8)$, so the new focus is $(2, 10)$. ∎

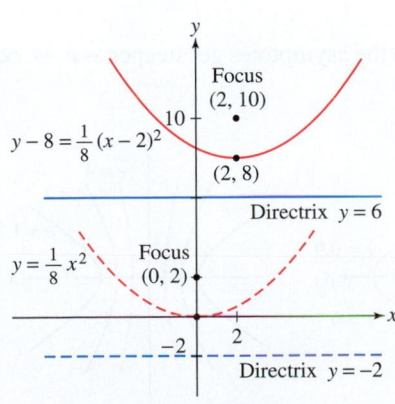

FIGURE 9 A parabola and its translate.

Eccentricity

Some ellipses are flatter than others, just as some hyperbolas are steeper. One aspect of the shape of a conic section is measured by a number e called the **eccentricity**. For an ellipse or hyperbola,

$$e = \frac{\text{distance betweeen foci}}{\text{distance between vertices on focal axis}}$$

⬅ *REMINDER*

Standard ellipse:

$$\left(\frac{x}{a}\right)^2 + \left(\frac{y}{b}\right)^2 = 1, \quad c = \sqrt{a^2 - b^2}$$

Standard hyperbola:

$$\left(\frac{x}{a}\right)^2 - \left(\frac{y}{b}\right)^2 = 1, \quad c = \sqrt{a^2 + b^2}$$

THEOREM 4 For ellipses and hyperbolas in standard position,

$$e = \frac{c}{a}$$

1. An ellipse has eccentricity $0 \le e < 1$ (with a circle having eccentricity 0).
2. A hyperbola has eccentricity $e > 1$.

A parabola is defined to have eccentricity $e = 1$.

Proof The foci are located at $(\pm c, 0)$ and the vertices are on the focal axis at $(\pm a, 0)$. Therefore,

$$e = \frac{\text{distance between foci}}{\text{distance between vertices on focal axis}} = \frac{2c}{2a} = \frac{c}{a}$$

For an ellipse, $c = \sqrt{a^2 - b^2}$ and so $e = c/a < 1$. If the ellipse is a circle, then $c = 0$ and therefore $e = 0$. For a hyperbola, $c = \sqrt{a^2 + b^2}$ and thus $e = c/a > 1$. ∎

How the eccentricity determines the shape of a conic is summarized in Figure 10. Consider the ratio b/a of the semiminor axis to the semimajor axis of an ellipse. The ellipse is nearly circular if b/a is close to 1, whereas it is elongated and flat if b/a is small. Now

$$\frac{b}{a} = \frac{\sqrt{a^2 - c^2}}{a} = \sqrt{1 - \frac{c^2}{a^2}} = \sqrt{1 - e^2}$$

This shows that b/a gets smaller (and the ellipse gets flatter) as $e \to 1$ [Figure 10(B)]. The most round ellipse is the circle, with $e = 0$.

Similarly, for a hyperbola,

$$\frac{b}{a} = \sqrt{1 + e^2}$$

The ratios $\pm b/a$ are the slopes of the asymptotes, so the asymptotes get steeper as $e \to \infty$ [Figure 10(C)].

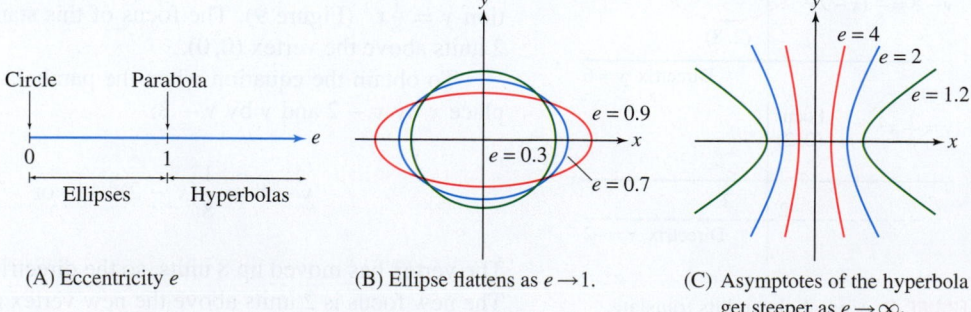

(A) Eccentricity e

(B) Ellipse flattens as $e \to 1$.

(C) Asymptotes of the hyperbola get steeper as $e \to \infty$.

DF FIGURE 10

CONCEPTUAL INSIGHT There is a more precise way to explain how eccentricity determines the shape of a conic. We can prove that if two conics C_1 and C_2 have the same eccentricity e, then there is a change of scale that makes C_1 *congruent* to C_2. Changing the scale means changing the units along the x- and y-axes by a common positive factor. A curve scaled by a factor of 10 has the same shape but is 10 times as large. By "congruent," we mean that after scaling, it is possible to move C_1 by a rigid motion (involving rotation and translation, but no stretching or bending) so that it lies directly on top of C_2.

All circles ($e = 0$) have the same shape because scaling by a factor $r > 0$ transforms a circle of radius R into a circle of radius rR. Similarly, any two parabolas ($e = 1$) become congruent after suitable scaling. However, an ellipse of eccentricity $e = 0.5$ cannot be made congruent to an ellipse of eccentricity $e = 0.8$ by scaling (see Exercise 76).

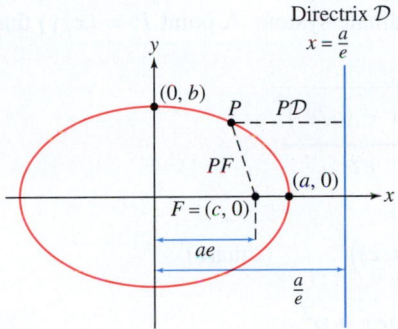

FIGURE 11 The ellipse consists of points P such that $PF = ePD$.

Eccentricity can be used to give a unified focus-directrix definition of the conic sections with $e > 0$. Given a point F (the focus), a line $\mathcal{D}$ (the directrix), and a number $e > 0$, we consider the set of all points P such that

$$PF = ePD \qquad \boxed{10}$$

For $e = 1$, this is our definition of a parabola. According to the next theorem, Eq. (10) defines a conic section of eccentricity e for all $e > 0$ (Figures 11 and 12).

THEOREM 5 **Focus-Directrix Relationship**

Ellipse

- If $0 < e < 1$, then the set of points satisfying Eq. (10) is an ellipse, and xy-coordinate axes can be chosen, and a, b defined, so that the ellipse has eccentricity e and is in standard position with equation

$$\left(\frac{x}{a}\right)^2 + \left(\frac{y}{b}\right)^2 = 1$$

- Conversely, if $a > b > 0$ and $c = \sqrt{a^2 - b^2}$, then the ellipse

$$\left(\frac{x}{a}\right)^2 + \left(\frac{y}{b}\right)^2 = 1$$

satisfies Eq. (10) with $F = (c, 0)$, $e = \frac{c}{a}$, and vertical directrix $x = \frac{a}{e}$.

Hyperbola

- If $e > 1$, then the set of points satisfying Eq. (10) is a hyperbola, and xy-coordinate axes can be chosen, and a, b defined, so that the hyperbola has eccentricity e and is in standard position with equation

$$\left(\frac{x}{a}\right)^2 - \left(\frac{y}{b}\right)^2 = 1$$

- Conversely, if $a, b > 0$ and $c = \sqrt{a^2 + b^2}$, the hyperbola

$$\left(\frac{x}{a}\right)^2 - \left(\frac{y}{b}\right)^2 = 1$$

satisfies Eq. (10) with $F = (c, 0)$, $e = \frac{c}{a}$, and vertical directrix $x = \frac{a}{e}$.

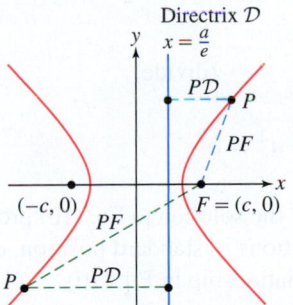

FIGURE 12 The hyperbola consists of points P such that $PF = ePD$.

Proof We prove the first part of the hyperbola relationship. The remaining parts of the theorem are proven in Exercises 65, 66, and 68. We are assuming we have a set of points satisfying Eq. (10) with $e > 1$. Let d denote the distance between the focus and directrix. We want to set up our x- and y-axes and define a, b, c so that the focus is located at $(c, 0)$

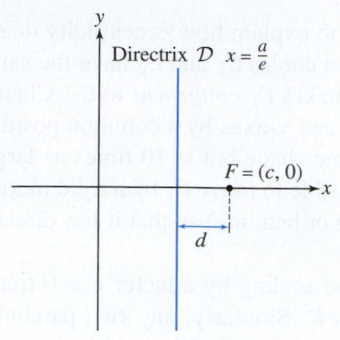

FIGURE 13

and the directrix is $x = \frac{a}{e}$ as in Figure 13. What works (and we will see why in the steps that follow) is to set

$$c = \frac{d}{1 - e^{-2}}, \qquad a = \frac{c}{e}, \qquad b = \sqrt{c^2 - a^2}$$

With the coordinate axes set up so that the focus is located at $(c, 0)$, the directrix is then the line

$$x = c - d = c - c(1 - e^{-2})$$
$$= c\,e^{-2} = \frac{a}{e}$$

which is what we wanted.

Now, we need to show that the equation $PF = e\,PD$ is in the form $\left(\frac{x}{a}\right)^2 - \left(\frac{y}{b}\right)^2 = 1$ in the coordinate system. This establishes that the resulting curve is a hyperbola with an equation in the standard-position form in the coordinate system. A point $P = (x, y)$ that satisfies $PF = e\,PD$ then satisfies

$$\underbrace{\sqrt{(x - c)^2 + y^2}}_{PF} = \underbrace{e\sqrt{\left(x - (a/e)\right)^2}}_{PD}$$

Algebraic manipulation yields

$$(x - c)^2 + y^2 = e^2\left(x - (a/e)\right)^2 \qquad \text{(square)}$$

$$x^2 - 2cx + c^2 + y^2 = e^2 x^2 - 2aex + a^2$$

$$x^2 - 2a\hspace{-2pt}\cancel{ex} + a^2 e^2 + y^2 = e^2 x^2 - 2a\hspace{-2pt}\cancel{ex} + a^2 \qquad \text{(use } c = ae\text{)}$$

$$(e^2 - 1)x^2 - y^2 = a^2(e^2 - 1) \qquad \text{(rearrange)}$$

$$\frac{x^2}{a^2} - \frac{y^2}{a^2(e^2 - 1)} = 1 \qquad \text{(divide)}$$

This is the desired equation because $a^2(e^2 - 1) = c^2 - a^2 = b^2$. ∎

Note that Theorem 5 indicates that for every $e > 0$, the solution to Eq. (10) produces a conic section in standard position. Also, all conic sections in standard position, except for the circle, can be obtained via the focus-directrix relationship in Eq. (10).

EXAMPLE 5 Find the equation, foci, and directrix of the standard ellipse with eccentricity $e = 0.8$ and focal vertices $(\pm 10, 0)$.

Solution The vertices are $(\pm a, 0)$, with $a = 10$ (Figure 14). By Theorem 5,

$$c = ae = 10(0.8) = 8, \qquad b = \sqrt{a^2 - c^2} = \sqrt{10^2 - 8^2} = 6$$

Thus, our ellipse has equation

$$\left(\frac{x}{10}\right)^2 + \left(\frac{y}{6}\right)^2 = 1$$

The foci are $(\pm c, 0) = (\pm 8, 0)$ and the directrix is $x = \frac{a}{e} = \frac{10}{0.8} = 12.5$. ∎

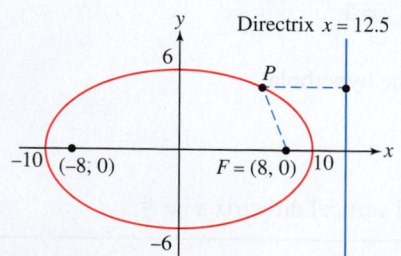

FIGURE 14 Ellipse of eccentricity $e = 0.8$ with focus at $(8, 0)$.

In Section 13.6, we discuss the famous law of Johannes Kepler stating that the orbit of a planet around the sun is an ellipse with one focus at the sun. In this discussion, we will need to write the equation of an ellipse in polar coordinates. To derive the polar equations of the conic sections, it is convenient to use the focus-directrix definition with

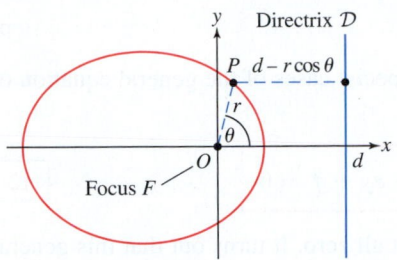

FIGURE 15 Focus-directrix definition of the ellipse in polar coordinates.

focus F at the origin O and vertical line $x = d$ as directrix $\mathcal{D}$ (Figure 15). Note from the figure that if $P = (r, \theta)$, then

$$PF = r, \qquad P\mathcal{D} = d - r\cos\theta$$

Thus, the focus-directrix equation of the ellipse $PF = e\,P\mathcal{D}$ becomes $r = e(d - r\cos\theta)$, or $r(1 + e\cos\theta) = ed$. This proves the following result, which is also valid for the parabola and hyperbola (see Exercise 69).

> **THEOREM 6** **Polar Equation of a Conic Section** The conic section of eccentricity $e > 0$ with focus at the origin and directrix $x = d$ has polar equation
>
> $$r = \frac{ed}{1 + e\cos\theta}$$ **11**

EXAMPLE 6 Find the eccentricity, directrix, and focus of the conic section

$$r = \frac{24}{4 + 3\cos\theta}$$

Solution First, we write the equation in the standard form:

$$r = \frac{24}{4 + 3\cos\theta} = \frac{6}{1 + \frac{3}{4}\cos\theta}$$

Comparing with Eq. (11), we see that $e = \frac{3}{4}$ and $ed = 6$. Therefore, $d = 8$. Since $e < 1$, the conic is an ellipse. By Theorem 6, the directrix is the line $x = 8$ and the focus is the origin. ∎

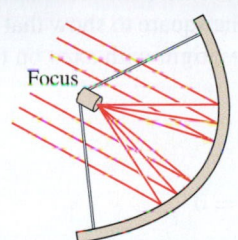

FIGURE 16 The parabolic shape of this radio telescope directs the incoming signal to the focus.

Reflective Properties of Conic Sections

The conic sections have numerous geometric properties. Especially important are the *reflective properties*, which are used in optics and communications (e.g., in antenna and telescope design; Figure 16). We describe these properties here briefly without proof (but see Exercises 70–72 for proofs of the reflective property of ellipses).

- **Ellipse:** The segments $\overline{F_1 P}$ and $\overline{F_2 P}$ make equal angles with the tangent line at a point P on the ellipse. Therefore, a beam of light originating at focus F_1 is reflected off the ellipse toward the second focus F_2 [Figure 17(A)]. See also Figure 18.
- **Hyperbola:** The tangent line at a point P on the hyperbola bisects the angle formed by the segments $\overline{F_1 P}$ and $\overline{F_2 P}$. Therefore, a beam of light directed toward F_2 is reflected off the hyperbola toward the second focus F_1 [Figure 17(B)].
- **Parabola:** The segment $\overline{FP}$ and the line through P parallel to the axis make equal angles with the tangent line at a point P on the parabola [Figure 17(C)]. Therefore, a beam of light approaching P from above in the axial direction is reflected off the parabola toward the focus F.

FIGURE 18 The ellipsoidal dome of the National Statuary Hall in the U.S. Capitol Building creates a "whisper chamber." Legend has it that John Quincy Adams would locate at one focus in order to eavesdrop on conversations taking place at the other focus.

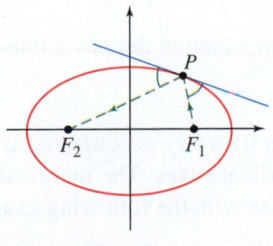

(A) Ellipse

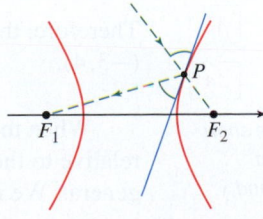

(B) Hyperbola

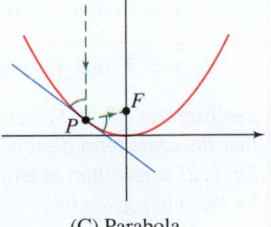

(C) Parabola

FIGURE 17

General Equations of Degree 2

The equations of the standard conic sections are special cases of the general equation of degree 2 in x and y:

$$ax^2 + bxy + cy^2 + dx + ey + f = 0 \qquad \boxed{12}$$

Here, a, b, c, d, e, f are constants with a, b, c not all zero. It turns out that this general equation of degree 2 does not give rise to any new types of curves. Apart from certain "degenerate cases," Eq. (12) defines a conic section that is not necessarily in standard position: It need not be centered at the origin, and its focal and conjugate axes may be rotated relative to the coordinate axes. For example, the equation

$$6x^2 - 8xy + 8y^2 - 12x - 24y + 38 = 0$$

defines an ellipse with its center at $(3, 3)$ whose axes are rotated (Figure 19).

We say that Eq. (12) is **degenerate** if the set of solutions is a pair of intersecting lines, a pair of parallel lines, a single line, a point, or the empty set. For example,

- $x^2 - y^2 = 0$ defines a pair of intersecting lines $y = x$ and $y = -x$.
- $x^2 - x = 0$ defines a pair of parallel lines $x = 0$ and $x = 1$.
- $x^2 = 0$ defines a single line (the y-axis).
- $x^2 + y^2 = 0$ has just one solution, $(0, 0)$.
- $x^2 + y^2 = -1$ has no solutions.

Now assume that Eq. (12) is nondegenerate. The term bxy is called the *cross term*. When the cross term is zero (i.e., when $b = 0$), we can complete the square to show that Eq. (12) defines a translate of a conic section that is centered at the origin with axes on the coordinate axes. This is illustrated in the next example.

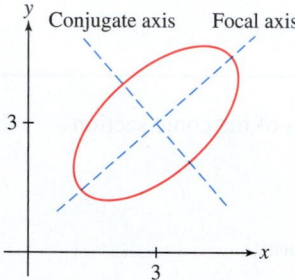

FIGURE 19 The ellipse with equation $6x^2 - 8xy + 8y^2 - 12x - 24y + 38 = 0$.

EXAMPLE 7 **Completing the Square** Show that

$$4x^2 + 9y^2 + 24x - 72y + 144 = 0$$

represents an ellipse that is a translation of an ellipse in standard position (Figure 20).

Solution Since there is no cross term, we may complete the square of the terms involving x and y separately:

$$4x^2 + 9y^2 + 24x - 72y + 144 = 0$$

$$4(x^2 + 6x + 9 - 9) + 9(y^2 - 8y + 16 - 16) + 144 = 0$$

$$4(x + 3)^2 - 4(9) + 9(y - 4)^2 - 9(16) + 144 = 0$$

$$4(x + 3)^2 + 9(y - 4)^2 = 36$$

This quadratic equation can be rewritten in the form

$$\left(\frac{x + 3}{3}\right)^2 + \left(\frac{y - 4}{2}\right)^2 = 1$$

Therefore, the given equation defines a translate of a standard ellipse to one centered at $(-3, 4)$. ∎

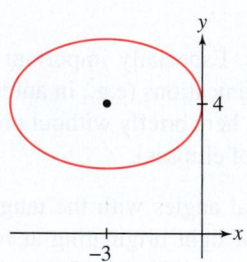

FIGURE 20 The ellipse with equation $4x^2 + 9y^2 + 24x - 72y + 144 = 0$.

If $(\tilde{x}, \tilde{y})$ are coordinates relative to axes rotated by an angle θ as in Figure 21, then

$$x = \tilde{x} \cos\theta - \tilde{y} \sin\theta \qquad \boxed{13}$$

$$y = \tilde{x} \sin\theta + \tilde{y} \cos\theta \qquad \boxed{14}$$

See Exercise 77. In Exercise 78, we show that the cross term disappears when Eq. (12) is rewritten in terms of $\tilde{x}$ and $\tilde{y}$ for the angle given by

$$\theta = \frac{1}{2} \cot^{-1} \frac{a - c}{b} \qquad \boxed{15}$$

When the cross term bxy is nonzero, Eq. (12) defines a conic whose axes are rotated relative to the coordinate axes. The marginal note describes how this may be verified in general. We illustrate with the following example.

EXAMPLE 8 Show that $2xy = 1$ defines a conic section whose focal and conjugate axes are rotated relative to the coordinate axes.

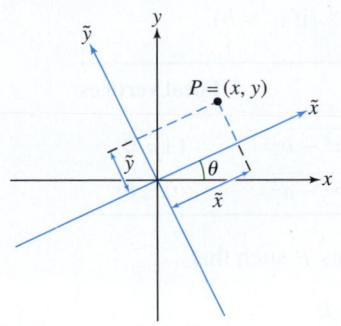

FIGURE 21

Solution Figure 22(A) shows axes labeled $\tilde{x}$ and $\tilde{y}$ that are rotated by 45° relative to the standard coordinate axes. A point P with coordinates (x, y) may also be described by coordinates $(\tilde{x}, \tilde{y})$ relative to these rotated axes. Applying Eqs. (13) and (14) with $\theta = \frac{\pi}{4}$, we find that (x, y) and $(\tilde{x}, \tilde{y})$ are related by the formulas

$$x = \frac{\tilde{x} - \tilde{y}}{\sqrt{2}}, \qquad y = \frac{\tilde{x} + \tilde{y}}{\sqrt{2}}$$

Therefore, if $P = (x, y)$ satisfies $2xy = 1$, then

$$2xy = 2\left(\frac{\tilde{x} - \tilde{y}}{\sqrt{2}}\right)\left(\frac{\tilde{x} + \tilde{y}}{\sqrt{2}}\right) = \tilde{x}^2 - \tilde{y}^2 = 1$$

Thus, the coordinates $(\tilde{x}, \tilde{y})$ satisfy the equation of the standard hyperbola $\tilde{x}^2 - \tilde{y}^2 = 1$ whose focal and conjugate axes are the $\tilde{x}$- and $\tilde{y}$-axes, respectively, as shown in Figure 22(B). ∎

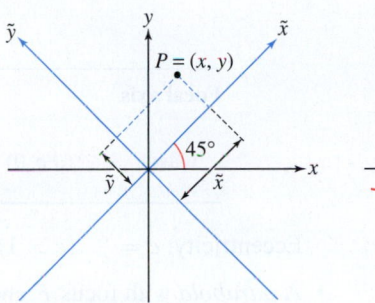

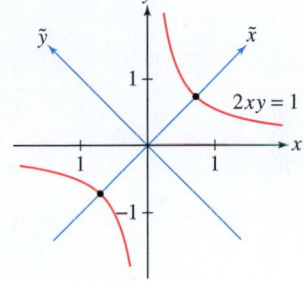

(A) The point $P = (x, y)$ may also be described by coordinates $(\tilde{x}, \tilde{y})$ relative to the rotated axis.

(B) The hyperbola $2xy = 1$ has the standard form $\tilde{x}^2 - \tilde{y}^2 = 1$ relative to the $\tilde{x}$-, $\tilde{y}$- axes.

FIGURE 22 The $\tilde{x}$- and $\tilde{y}$-axes are rotated at a 45° angle relative to the x- and y-axes.

We conclude our discussion of conics by stating the Discriminant Test. Suppose that the equation

$$ax^2 + bxy + cy^2 + dx + ey + f = 0$$

is nondegenerate and thus defines a conic section. According to the Discriminant Test, the type of conic is determined by the **discriminant** D:

$$\boxed{D = b^2 - 4ac}$$

We have the following cases:

- $D < 0$: ellipse or circle
- $D = 0$: parabola
- $D > 0$: hyperbola

For example, the discriminant of the equation $2xy = 1$ is

$$D = b^2 - 4ac = 2^2 - 0 = 4 > 0$$

According to the Discriminant Test, $2xy = 1$ defines a hyperbola. This agrees with our conclusion in Example 8.

11.5 SUMMARY

- An *ellipse* with foci F_1 and F_2 is the set of points P such that $PF_1 + PF_2 = K$, where K is a constant such that $K > F_1 F_2$. The equation in standard position is

$$\left(\frac{x}{a}\right)^2 + \left(\frac{y}{b}\right)^2 = 1$$

The vertices of the ellipse are $(\pm a, 0)$ and $(0, \pm b)$.
Eccentricity: $e = \frac{c}{a}$ $(0 \le e < 1)$. Directrix: $x = \frac{a}{e}$ (if $a > b$).

	Focal axis	Foci	Focal vertices
$a > b$	x-axis	$(\pm c, 0)$ with $c = \sqrt{a^2 - b^2}$	$(\pm a, 0)$
$a < b$	y-axis	$(0, \pm c)$ with $c = \sqrt{b^2 - a^2}$	$(0, \pm b)$

- A *hyperbola* with foci F_1 and F_2 is the set of points P such that

$$PF_1 - PF_2 = \pm K$$

where K is a constant such that $0 < K < F_1 F_2$. The equation in standard position is

$$\left(\frac{x}{a}\right)^2 - \left(\frac{y}{b}\right)^2 = 1$$

Focal axis	Foci	Vertices	Asymptotes
x-axis	$(\pm c, 0)$ with $c = \sqrt{a^2 + b^2}$	$(\pm a, 0)$	$y = \pm \frac{b}{a}x$

Eccentricity: $e = \frac{c}{a}$ $(e > 1)$. Directrix: $x = \frac{a}{e}$.

- A *parabola* with focus F and directrix $\mathcal{D}$ is the set of points P such that $PF = PD$. The equation in standard position is

$$y = \frac{1}{4c}x^2$$

Focus $F = (0, c)$, directrix $y = -c$, and vertex at the origin $(0, 0)$.

- *Focus-directrix definition* of conic with focus F and directrix $\mathcal{D}$: $PF = ePD$.
- To translate a conic section h units horizontally and k units vertically, replace x by $x - h$ and y by $y - k$ in the equation.
- Polar equation of conic of eccentricity $e > 0$, focus at the origin, directrix $x = d$:

$$r = \frac{ed}{1 + e\cos\theta}$$

- The equation $ax^2 + bxy + cy^2 + dx + ey + f = 0$ is *degenerate* if it has no solution or represents a point, a line, or a pair of lines. Otherwise, the equation is *nondegenerate* and represents a conic section.

  In the nondegenerate case, the *discriminant* $D = b^2 - 4ac$ determines the conic section via the Discriminant Test:

$$D < 0: \text{ellipse or circle}, \quad D = 0: \text{parabola}, \quad D > 0: \text{hyperbola}$$

11.5 EXERCISES

Preliminary Questions

1. Decide if the equation defines an ellipse, a hyperbola, a parabola, or no conic section at all.

(a) $4x^2 - 9y^2 = 12$

(b) $-4x + 9y^2 = 0$

(c) $4y^2 + 9x^2 = 12$

(d) $4x^3 + 9y^3 = 12$

2. For which conic sections do the vertices lie between the foci?

3. What are the foci of $\left(\frac{x}{a}\right)^2 + \left(\frac{y}{b}\right)^2 = 1$ if $a < b$?

4. What is the geometric interpretation of b/a in the equation of a hyperbola in standard position?

Exercises

In Exercises 1–6, find the vertices and foci of the conic section.

1. $\left(\dfrac{x}{9}\right)^2 + \left(\dfrac{y}{4}\right)^2 = 1$

2. $\dfrac{x^2}{9} + \dfrac{y^2}{4} = 1$

3. $\left(\dfrac{x}{4}\right)^2 - \left(\dfrac{y}{9}\right)^2 = 1$

4. $\dfrac{x^2}{4} - \dfrac{y^2}{9} = 36$

5. $\left(\dfrac{x-4}{6}\right)^2 - \left(\dfrac{y+3}{5}\right)^2 = 1$

6. $\left(\dfrac{x-4}{5}\right)^2 + \left(\dfrac{y+3}{6}\right)^2 = 1$

In Exercises 7–10, find the equation of the ellipse obtained by translating (as indicated) the ellipse

$$\left(\dfrac{x-8}{6}\right)^2 + \left(\dfrac{y+4}{3}\right)^2 = 1$$

7. Translated with center at the origin

8. Translated with center at $(-2, -12)$

9. Translated to the right 6 units

10. Translated down 4 units

In Exercises 11–14, find the equation of the given ellipse.

11. Vertices $(\pm 3, 0)$ and $(0, \pm 5)$

12. Foci $(+6, 0)$ and focal vertices $(\pm 10, 0)$

13. Foci $(0, \pm 10)$ and eccentricity $e = \frac{3}{5}$

14. Vertices $(4, 0)$, $(28, 0)$ and eccentricity $e = \frac{2}{3}$

In Exercises 15–20, find the equation of the given hyperbola.

15. Vertices $(\pm 3, 0)$ and foci $(\pm 5, 0)$

16. Vertices $(\pm 3, 0)$ and asymptotes $y = \pm \frac{1}{2}x$

17. Foci $(\pm 3, 0)$ and eccentricity $e = 3$

18. Foci $(0, \pm 5)$ and eccentricity $e = 1.5$

19. Vertices $(-3, 0)$, $(7, 0)$ and eccentricity $e = 3$

20. Vertices $(0, -6)$, $(0, 4)$ and foci $(0, -9)$, $(0, 7)$

In Exercises 21–28, find the equation of the parabola with the given properties.

21. Vertex $(0, 0)$, focus $\left(\frac{1}{12}, 0\right)$

22. Vertex $(0, 0)$, focus $(0, 2)$

23. Vertex $(0, 0)$, directrix $y = -5$

24. Vertex $(3, 4)$, directrix $y = -2$

25. Focus $(0, 4)$, directrix $y = -4$

26. Focus $(0, -4)$, directrix $y = 4$

27. Focus $(2, 0)$, directrix $x = -2$

28. Focus $(-2, 0)$, vertex $(2, 0)$

In Exercises 29–38, find the vertices, foci, center (if an ellipse or a hyperbola), and asymptotes (if a hyperbola).

29. $x^2 + 4y^2 = 16$

30. $4x^2 + y^2 = 16$

31. $\left(\dfrac{x-3}{4}\right)^2 - \left(\dfrac{y+5}{7}\right)^2 = 1$

32. $3x^2 - 27y^2 = 12$

33. $4x^2 - 3y^2 + 8x + 30y = 215$

34. $y = 4x^2$

35. $y = 4(x - 4)^2$

36. $8y^2 + 6x^2 - 36x - 64y + 134 = 0$

37. $4x^2 + 25y^2 - 8x - 10y = 20$

38. $16x^2 + 25y^2 - 64x - 200y + 64 = 0$

In Exercises 39–42, use the Discriminant Test to determine the type of the conic section (in each case, the equation is nondegenerate). Use a graphing utility or computer algebra system to plot the curve.

39. $4x^2 + 5xy + 7y^2 = 24$

40. $x^2 - 2xy + y^2 + 24x - 8 = 0$

41. $2x^2 - 8xy + 3y^2 - 4 = 0$

42. $2x^2 - 3xy + 5y^2 - 4 = 0$

43. Show that the "conic" $x^2 + 3y^2 - 6x + 12y + 23 = 0$ has no points.

44. For which values of a does the conic $3x^2 + 2y^2 - 16y + 12x = a$ have at least one point?

45. Show that $\dfrac{b}{a} = \sqrt{1 - e^2}$ for a standard ellipse of eccentricity e.

46. Show that the eccentricity of a hyperbola in standard position is $e = \sqrt{1 + m^2}$, where $\pm m$ are the slopes of the asymptotes.

47. Explain why the dots in Figure 23 lie on a parabola. Where are the focus and directrix located?

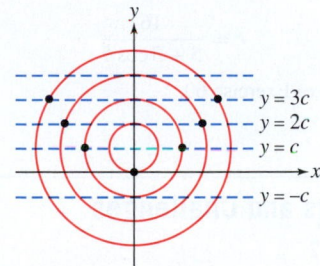

FIGURE 23

48. Find the equation of the ellipse consisting of points P such that $PF_1 + PF_2 = 12$, where $F_1 = (4, 0)$ and $F_2 = (-2, 0)$.

49. A **latus rectum** of a conic section is a chord through a focus parallel to the directrix. Find the area bounded by the parabola $y = x^2/(4c)$ and its latus rectum (refer to Figure 8).

50. Show that the tangent line at a point $P = (x_0, y_0)$ on the hyperbola $\left(\dfrac{x}{a}\right)^2 - \left(\dfrac{y}{b}\right)^2 = 1$ has equation

$$Ax - By = 1$$

where $A = \dfrac{x_0}{a^2}$ and $B = \dfrac{y_0}{b^2}$.

In Exercises 51–54, find the polar equation of the conic with the given eccentricity and directrix, and focus at the origin.

51. $e = \frac{1}{2}$, $x = 3$

52. $e = \frac{1}{2}$, $x = -3$

53. $e = 1$, $x = 4$

54. $e = \frac{3}{2}$, $x = -4$

In Exercises 55–58, identify the type of conic, the eccentricity, and the equation of the directrix.

55. $r = \dfrac{8}{1 + 4\cos\theta}$

56. $r = \dfrac{8}{4 + \cos\theta}$

57. $r = \dfrac{8}{4 + 3\cos\theta}$

58. $r = \dfrac{12}{4 + 3\cos\theta}$

59. Find a polar equation for the hyperbola with focus at the origin, directrix $x = -2$, and eccentricity $e = 1.2$.

60. Let $\mathcal{C}$ be the ellipse $r = de/(1 + e\cos\theta)$, where $e < 1$. Show that the x-coordinates of the points in Figure 24 are as follows:

Point	A	C	F_2	A'
x-coordinate	$\dfrac{de}{e+1}$	$-\dfrac{de^2}{1-e^2}$	$-\dfrac{2de^2}{1-e^2}$	$-\dfrac{de}{1-e}$

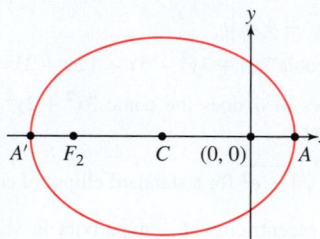

FIGURE 24

61. Find an equation in rectangular coordinates of the conic

$$r = \frac{16}{5 + 3\cos\theta}$$

Hint: Use the results of Exercise 60.

62. Let $e > 1$. Show that the vertices of the hyperbola

$$r = \frac{de}{1 + e\cos\theta}$$

have x-coordinates $\dfrac{ed}{e+1}$ and $\dfrac{ed}{e-1}$.

63. Kepler's First Law states that planetary orbits are ellipses with the sun at one focus. The orbit of Pluto has eccentricity $e \approx 0.25$. Its **perihelion** (closest distance to the sun) is approximately 2.7 billion miles. Find the **aphelion** (farthest distance from the sun).

64. Kepler's Third Law states that the ratio $T/a^{3/2}$ is equal to a constant C for all planetary orbits around the sun, where T is the period (time for a complete orbit) and a is the semimajor axis.

(a) Compute C in units of days and kilometers, given that the semimajor axis of the earth's orbit is 150×10^6 km.

(b) Compute the period of Saturn's orbit, given that its semimajor axis is approximately 1.43×10^9 km.

(c) Saturn's orbit has eccentricity $e = 0.056$. Find the perihelion and aphelion of Saturn (see Exercise 63).

65. Prove that if $a > b > 0$ and $c = \sqrt{a^2 - b^2}$, then a point $P = (x, y)$ on the ellipse

$$\left(\frac{x}{a}\right)^2 + \left(\frac{y}{b}\right)^2 = 1$$

satisfies $PF = ePD$ with $F = (c, 0)$, $e = \frac{c}{a}$, and vertical directrix $\mathcal{D}$ at $x = \frac{a}{e}$.

66. Prove that if $a, b > 0$ and $c = \sqrt{a^2 + b^2}$, then a point $P = (x, y)$ on the hyperbola

$$\left(\frac{x}{a}\right)^2 - \left(\frac{y}{b}\right)^2 = 1$$

satisfies $PF = ePD$ with $F = (c, 0)$, $e = \frac{c}{a}$, and vertical directrix $\mathcal{D}$ at $x = \frac{a}{e}$.

Further Insights and Challenges

67. Prove Theorem 2.

68. Prove Theorem 5 in the case $0 < e < 1$. *Hint:* Repeat the proof of Theorem 5, but set $c = d/(e^{-2} - 1)$.

69. Verify that if $e > 1$, then Eq. (11) defines a hyperbola of eccentricity e, with its focus at the origin and directrix at $x = d$.

Reflective Property of the Ellipse *In Exercises 70–72, we prove that the focal radii at a point on an ellipse make equal angles with the tangent line $\mathcal{L}$. Let $P = (x_0, y_0)$ be a point on the ellipse in Figure 25 with foci $F_1 = (-c, 0)$ and $F_2 = (c, 0)$, and eccentricity $e = c/a$.*

70. Show that the equation of the tangent line at P is $Ax + By = 1$, where

$$A = \frac{x_0}{a^2} \quad \text{and} \quad B = \frac{y_0}{b^2}.$$

71. Points R_1 and R_2 in Figure 25 are defined so that $\overline{F_1 R_1}$ and $\overline{F_2 R_2}$ are perpendicular to the tangent line.

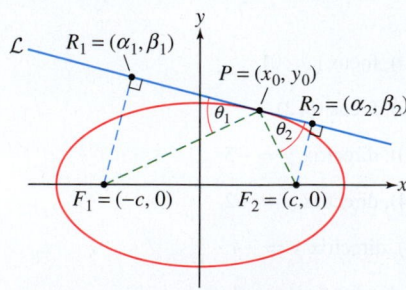

FIGURE 25 The ellipse $\left(\dfrac{x}{a}\right)^2 + \left(\dfrac{y}{b}\right)^2 = 1$.

(a) Show, with A and B as in Exercise 70, that

$$\frac{\alpha_1 + c}{\beta_1} = \frac{\alpha_2 - c}{\beta_2} = \frac{A}{B}$$

(b) Use (a) and the distance formula to show that

$$\frac{F_1 R_1}{F_2 R_2} = \frac{\beta_1}{\beta_2}$$

(c) Use (a) and the equation of the tangent line in Exercise 70 to show that

$$\beta_1 = \frac{B(1 + Ac)}{A^2 + B^2}, \qquad \beta_2 = \frac{B(1 - Ac)}{A^2 + B^2}$$

72. (a) Prove that $PF_1 = a + x_0 e$ and $PF_2 = a - x_0 e$. *Hint:* Show that $PF_1^2 - PF_2^2 = 4x_0 c$. Then use the defining property $PF_1 + PF_2 = 2a$ and the relation $e = c/a$.

(b) Verify that $\dfrac{F_1 R_1}{PF_1} = \dfrac{F_2 R_2}{PF_2}$.

(c) Show that $\sin\theta_1 = \sin\theta_2$. Conclude that $\theta_1 = \theta_2$.

73. ✎ Here is another proof of the Reflective Property.
(a) Figure 25 suggests that $\mathcal{L}$ is the unique line that intersects the ellipse only in the point P. Assuming this, prove that $QF_1 + QF_2 > PF_1 + PF_2$ for all points Q on the tangent line other than P.
(b) Use the Principle of Least Distance (Example 6 in Section 4.7) to prove that $\theta_1 = \theta_2$.

74. Show that the length QR in Figure 26 is independent of the point P.

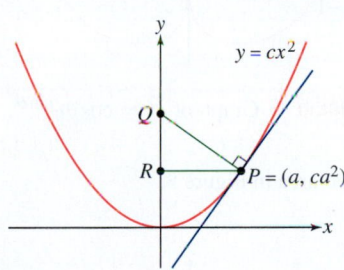

FIGURE 26

75. Show that $y = x^2/4c$ is the equation of a parabola with directrix $y = -c$, focus $(0, c)$, and the vertex at the origin, as stated in Theorem 3.

76. Consider two ellipses in standard position:

$$E_1 : \quad \left(\frac{x}{a_1}\right)^2 + \left(\frac{y}{b_1}\right)^2 = 1$$

$$E_2 : \quad \left(\frac{x}{a_2}\right)^2 + \left(\frac{y}{b_2}\right)^2 = 1$$

We say that E_1 is similar to E_2 under scaling if there exists a factor $r > 0$ such that for all (x, y) on E_1, the point (rx, ry) lies on E_2. Show that E_1 and E_2 are similar under scaling if and only if they have the same eccentricity. Show that any two circles are similar under scaling.

77. ✎ Derive Eqs. (13) and (14) in the text as follows. Write the coordinates of P with respect to the rotated axes in Figure 21 in polar form $\tilde{x} = r\cos\alpha$, $\tilde{y} = r\sin\alpha$. Explain why P has polar coordinates $(r, \alpha + \theta)$ with respect to the standard x- and y-axes, and derive Eqs. (13) and (14) using the addition formulas for cosine and sine.

78. If we rewrite the general equation of degree 2 (Eq. 12) in terms of variables $\tilde{x}$ and $\tilde{y}$ that are related to x and y by Eqs. (13) and (14), we obtain a new equation of degree 2 in $\tilde{x}$ and $\tilde{y}$ of the same form but with different coefficients:

$$a'\tilde{x}^2 + b'\tilde{x}\tilde{y} + c'\tilde{y}^2 + d'\tilde{x} + e'\tilde{y} + f' = 0$$

(a) Show that $b' = b\cos 2\theta + (c - a)\sin 2\theta$.

(b) Show that if $b \neq 0$, then we obtain $b' = 0$ for

$$\theta = \frac{1}{2}\cot^{-1}\frac{a - c}{b}$$

This proves that it is always possible to eliminate the cross term bxy by rotating the axes through a suitable angle.

CHAPTER REVIEW EXERCISES

1. Which of the following curves pass through the point $(1, 4)$?
(a) $c(t) = (t^2, t + 3)$ 　　　　 (b) $c(t) = (t^2, t - 3)$
(c) $c(t) = (t^2, 3 - t)$ 　　　　 (d) $c(t) = (t - 3, t^2)$

2. Find parametric equations for the line through $P = (2, 5)$ perpendicular to the line $y = 4x - 3$.

3. Find parametric equations for the circle of radius 2 with center $(1, 1)$. Use the equations to find the points of intersection of the circle with the x- and y-axes.

4. Find a parametrization $c(t)$ of the line $y = 5 - 2x$ such that $c(0) = (2, 1)$.

5. Find a parametrization $c(\theta)$ of the unit circle such that $c(0) = (-1, 0)$.

6. Find a path $c(t)$ that traces the parabolic arc $y = x^2$ from $(0, 0)$ to $(3, 9)$ for $0 \leq t \leq 1$.

7. Find a path $c(t)$ that traces the line $y = 2x + 1$ from $(1, 3)$ to $(3, 7)$ for $0 \leq t \leq 1$.

8. Sketch the graph $c(t) = (1 + \cos t, \sin 2t)$ for $0 \leq t \leq 2\pi$ and draw arrows specifying the direction of motion.

In Exercises 9–12, express the parametric curve in the form $y = f(x)$.

9. $c(t) = (4t - 3, 10 - t)$ 　　　　 **10.** $c(t) = (t^3 + 1, t^2 - 4)$

11. $c(t) = \left(3 - \dfrac{2}{t}, t^3 + \dfrac{1}{t}\right)$ 　　　　 **12.** $x = \tan t$, 　 $y = \sec t$

In Exercises 13–16, calculate dy/dx at the point indicated.

13. $c(t) = (t^3 + t, t^2 - 1), \quad t = 3$

14. $c(\theta) = (\tan^2 \theta, \cos \theta), \quad \theta = \frac{\pi}{4}$

15. $c(t) = (e^t - 1, \sin t), \quad t = 20$

16. $c(t) = (\ln t, 3t^2 - t), \quad P = (0, 2)$

17. **CAS** Find the point on the cycloid $c(t) = (t - \sin t, 1 - \cos t)$ where the tangent line has slope $\frac{1}{2}$.

18. Find the points on $(t + \sin t, t - 2 \sin t)$ where the tangent is vertical or horizontal.

19. Find the equation of the Bézier curve with control points

$$P_0 = (-1, -1), \quad P_1 = (-1, 1), \quad P_2 = (1, 1), \quad P_3(1, -1)$$

20. Find the speed at $t = \frac{\pi}{4}$ of a particle whose position at time t seconds is $c(t) = (\sin 4t, \cos 3t)$.

21. Find the speed (as a function of t) of a particle whose position at time t seconds is $c(t) = (\sin t + t, \cos t + t)$. What is the particle's maximal speed?

22. Find the length of $(3e^t - 3, 4e^t + 7)$ for $0 \le t \le 1$.

In Exercises 23 and 24, let $c(t) = (e^{-t} \cos t, e^{-t} \sin t)$.

23. Show that $c(t)$ for $0 \le t < \infty$ has finite length and calculate its value.

24. Find the first positive value of t_0 such that the tangent line to $c(t_0)$ is vertical, and calculate the speed at $t = t_0$.

25. **CAS** Plot $c(t) = (\sin 2t, 2 \cos t)$ for $0 \le t \le \pi$. Express the length of the curve as a definite integral, and approximate it using a computer algebra system.

26. Convert the points $(x, y) = (1, -3), (3, -1)$ from rectangular to polar coordinates.

27. Convert the points $(r, \theta) = \left(1, \frac{\pi}{6}\right), \left(3, \frac{5\pi}{4}\right)$ from polar to rectangular coordinates.

28. Write $(x + y)^2 = xy + 6$ as an equation in polar coordinates.

29. Write $r = \dfrac{2 \cos \theta}{\cos \theta - \sin \theta}$ as an equation in rectangular coordinates.

30. Show that $r = \dfrac{4}{7 \cos \theta - \sin \theta}$ is the polar equation of a line.

31. **GU** Convert the equation

$$9(x^2 + y^2) = (x^2 + y^2 - 2y)^2$$

to polar coordinates, and plot it with a graphing utility.

32. Calculate the area of the circle $r = 3 \sin \theta$ bounded by the rays $\theta = \frac{\pi}{3}$ and $\theta = \frac{2\pi}{3}$.

33. Calculate the area of one petal of $r = \sin 4\theta$ (see Figure 1).

34. The equation $r = \sin(n\theta)$, where $n \ge 2$ is even, is a "rose" of $2n$ petals (Figure 1). Compute the total area of the flower, and show that it does not depend on n.

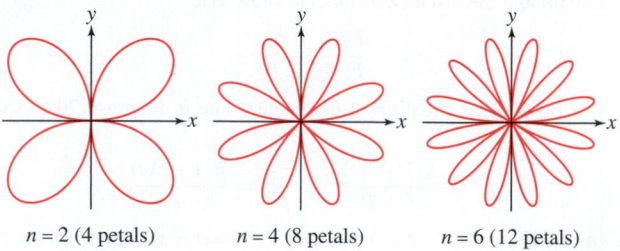

$n = 2$ (4 petals) $n = 4$ (8 petals) $n = 6$ (12 petals)

FIGURE 1 Plot of $r = \sin(n\theta)$.

35. Calculate the total area enclosed by the curve $r^2 = \cos \theta e^{\sin \theta}$ (Figure 2).

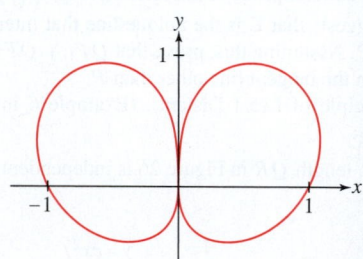

FIGURE 2 Graph of $r^2 = \cos \theta e^{\sin \theta}$.

36. Find the shaded area in Figure 3.

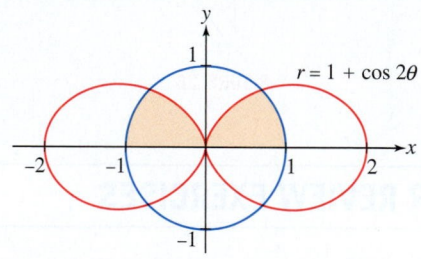

$r = 1 + \cos 2\theta$

FIGURE 3

37. Find the area enclosed by the cardioid $r = a(1 + \cos \theta)$, where $a > 0$.

38. Calculate the length of the curve with polar equation $r = \theta$ in Figure 4.

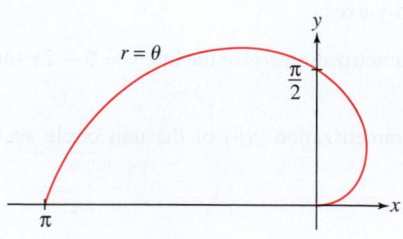

$r = \theta$

FIGURE 4

39. [CAS] The graph of $r = e^{0.5\theta} \sin\theta$ for $0 \le \theta \le 2\pi$ is shown in Figure 5. Use a computer algebra system to approximate the difference in length between the outer and inner loops.

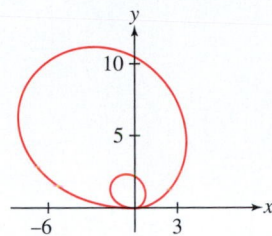

FIGURE 5

40. [✎] Show that $r = f_1(\theta)$ and $r = f_2(\theta)$ define the same curves in polar coordinates if $f_1(\theta) = -f_2(\theta + \pi)$. Use this to show that the following define the same conic section:

$$r = \frac{de}{1 - e\cos\theta}, \qquad r = \frac{-de}{1 + e\cos\theta}$$

In Exercises 41–44, identify the conic section. Find the vertices and foci.

41. $\left(\frac{x}{3}\right)^2 + \left(\frac{y}{2}\right)^2 = 1$

42. $x^2 - 2y^2 = 4$

43. $\left(2x + \frac{1}{2}y\right)^2 = 4 - (x - y)^2$

44. $(y - 3)^2 = 2x^2 - 1$

In Exercises 45–50, find the equation of the conic section indicated.

45. Ellipse with vertices $(\pm 8, 0)$, foci $(\pm\sqrt{3}, 0)$

46. Ellipse with foci $(\pm 8, 0)$, eccentricity $\frac{1}{8}$

47. Hyperbola with vertices $(\pm 8, 0)$, asymptotes $y = \pm\frac{3}{4}x$

48. Hyperbola with foci $(2, 0)$ and $(10, 0)$, eccentricity $e = 4$

49. Parabola with focus $(8, 0)$, directrix $x = -8$

50. Parabola with vertex $(4, -1)$, directrix $x = 15$

51. Find the asymptotes of the hyperbola $3x^2 + 6x - y^2 - 10y = 1$.

52. Show that the "conic section" with equation $x^2 - 4x + y^2 + 5 = 0$ has no points.

53. Show that the relation $\frac{dy}{dx} = (e^2 - 1)\frac{x}{y}$ holds on a standard ellipse or hyperbola of eccentricity e.

54. The orbit of Jupiter is an ellipse with the sun at a focus. Find the eccentricity of the orbit if the perihelion (closest distance to the sun) equals 740×10^6 km and the aphelion (farthest distance from the sun) equals 816×10^6 km.

55. Refer to Figure 25 in Section 11.5. Prove that the product of the perpendicular distances $F_1 R_1$ and $F_2 R_2$ from the foci to a tangent line of an ellipse is equal to the square b^2 of the semiminor axes.

12 VECTOR GEOMETRY

Vectors play a role in nearly all areas of mathematics and its applications. In physical settings, they are used to represent quantities that have both magnitude and direction, such as velocity and force. Newtonian mechanics, quantum physics, and special and general relativity all depend fundamentally on vectors. We could not understand electricity and magnetism without this basic mathematical concept. Computer graphics depend on vectors to help depict how light reflects off objects in a scene, and they provide a means for changing an observer's point of view. Fields such as economics and statistics use vectors to represent information in a manner that may be efficiently manipulated.

This chapter introduces the basic geometric and algebraic properties of vectors, setting the stage for the development of multivariable calculus in the chapters ahead.

Engineers use vectors to analyze the forces on the cables in a suspension bridge, such as the Penobscot Narrows Bridge in Maine. Tension forces in the horizontal direction must balance so that there is no net force on the bridge towers. In the vertical direction, the tension forces support the weight of the bridge deck. The cables must be designed so that each can support the combined vertical and horizontal tension forces.

12.1 Vectors in the Plane

Recall that the plane is the set of points $\{(x, y): x, y \in \mathbf{R}\}$. We occasionally denote the plane by $\mathbf{R}^2$. This notation represents the idea that the plane is a "product" of two copies of the real line $\mathbf{R}$, where one copy represents the points' x-coordinates and the other represents the y-coordinates. (We extend this notation and idea to three-dimensional space in the next section.)

A two-dimensional **vector v** is determined by two points in the plane: an initial point P (also called the tail or basepoint) and a terminal point Q (also called the head or tip). We write

$$\mathbf{v} = \overrightarrow{PQ}$$

and we draw $\mathbf{v}$ as an arrow pointing from P to Q. This vector is said to be based at P. Figure 1(A) shows the vector with initial point $P = (2, 2)$ and terminal point $Q = (7, 5)$. The **length** or **magnitude** of $\mathbf{v}$, denoted $\|\mathbf{v}\|$, is the distance from P to Q.

The vector $\mathbf{v} = \overrightarrow{OR}$ pointing from the origin to a point R is called the **position vector** of R. Figure 1(B) shows the position vector of the point $R = (3, 5)$.

NOTATION *In this text, vectors are represented by boldface letters such as* **v**, **w**, **a**, **b**, **F**.

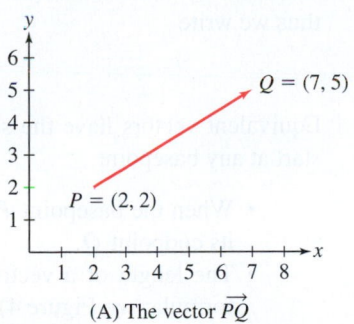
(A) The vector $\overrightarrow{PQ}$

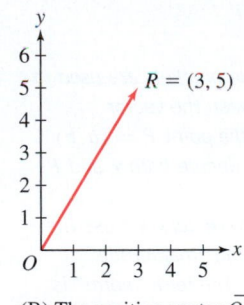
(B) The position vector $\overrightarrow{OR}$

DF FIGURE 1

We now introduce some vector terminology.

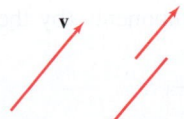

(A) Vectors parallel to **v**

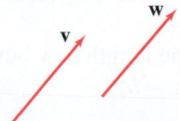

(B) **w** is a translation of **v**.

FIGURE 2

- Two vectors **v** and **w** of nonzero length are called **parallel** if the lines through **v** and **w** are parallel. Parallel vectors point either in the same or in opposite directions [Figure 2(A)].
- A vector **v** is said to undergo a **translation** when it is moved to begin at a new point without changing its length or direction. The resulting vector **w** is called a translation of **v** [Figure 2(B)]. A translation **w** of a vector **v** has the same length and direction as **v** but a different basepoint.

In almost all situations, it is convenient to treat vectors with the same length and direction as equivalent, even if they have different basepoints. With this in mind, we say that

- **v** and **w** are **equivalent** if **w** is a translation of **v** [Figure 3(A)].

Note that no pair of the vectors in Figure 3(B) are equivalent. Every vector can be translated so that its tail is at the origin [Figure 3(C)]. Therefore,

Every vector $\mathbf{v}$ is equivalent to a unique vector $\mathbf{v}_0$ based at the origin.

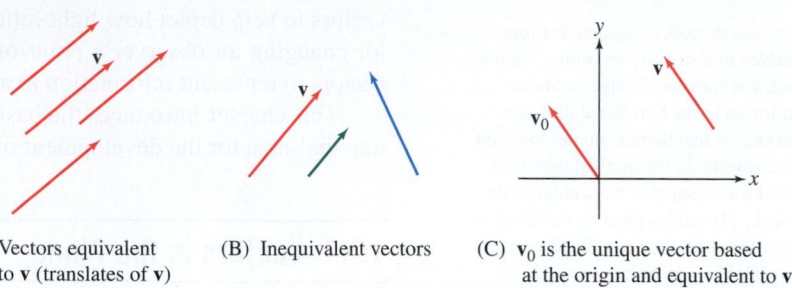

(A) Vectors equivalent to $\mathbf{v}$ (translates of $\mathbf{v}$)

(B) Inequivalent vectors

(C) $\mathbf{v}_0$ is the unique vector based at the origin and equivalent to $\mathbf{v}$.

FIGURE 3

To work algebraically, we define the components of a vector (Figure 4).

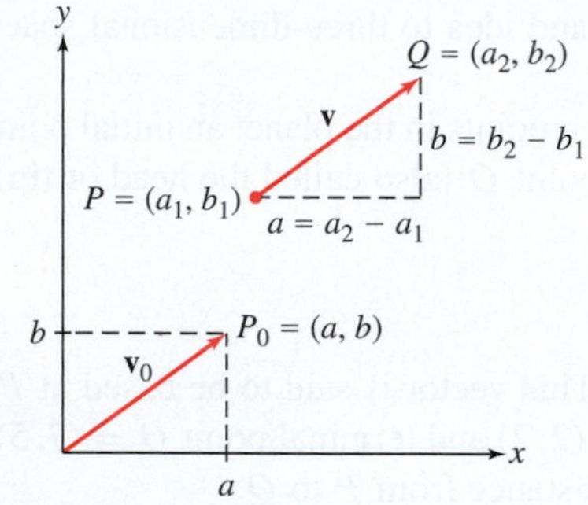

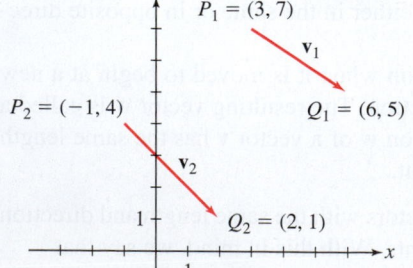

FIGURE 4 Both of the vectors $\mathbf{v}$ and $\mathbf{v}_0$ have components $\langle a, b \rangle$.

> • In this text, angle brackets are used to distinguish between the vector $\mathbf{v} = \langle a, b \rangle$ and the point $P = (a, b)$. Some textbooks denote both $\mathbf{v}$ and P by (a, b).
> • When referring to vectors, we use the terms "length" and "magnitude" interchangeably. The term "norm" is also commonly used.

DEFINITION Components of a Vector The components of $\mathbf{v} = \overrightarrow{PQ}$, where $P = (a_1, b_1)$ and $Q = (a_2, b_2)$, are the quantities

$$a = a_2 - a_1 \quad (x\text{-component}), \qquad b = b_2 - b_1 \quad (y\text{-component})$$

The pair of components is denoted $\langle a, b \rangle$.

The pair of components $\langle a, b \rangle$ determine the length and direction of $\mathbf{v}$, but not its basepoint. Therefore, *two vectors are equivalent if and only if they have the same components*. Nevertheless, the standard practice is to describe a vector by its components, and thus we write

$$\mathbf{v} = \langle a, b \rangle$$

Equivalent vectors have the same length and point in the same direction, but they can start at any basepoint.

• When the basepoint $P = (0, 0)$, the components of $\mathbf{v}$ are just the coordinates of its endpoint Q.
• The length of a vector $\mathbf{v} = \langle a, b \rangle$ in terms of its components (by the distance formula; see Figure 4) is

$$\|\mathbf{v}\| = \|\overrightarrow{PQ}\| = \sqrt{a^2 + b^2}$$

• The **zero vector** (whose head and tail coincide) is the vector $\mathbf{0} = \langle 0, 0 \rangle$ of length zero. It is the only vector that lacks a direction.
• For a vector $\mathbf{v}$, the vector $-\mathbf{v}$ is the vector with the same length as $\mathbf{v}$ but pointing in the opposite direction. If $\mathbf{v} = \langle a, b \rangle$, then $-\mathbf{v} = \langle -a, -b \rangle$.

EXAMPLE 1 Determine whether $\mathbf{v}_1 = \overrightarrow{P_1 Q_1}$ and $\mathbf{v}_2 = \overrightarrow{P_2 Q_2}$ are equivalent, where

$$P_1 = (3, 7), \quad Q_1 = (6, 5) \quad \text{and} \quad P_2 = (-1, 4), \quad Q_2 = (2, 1)$$

What is the magnitude of $\mathbf{v}_1$?

Solution We can test for equivalence by computing the components (Figure 5):

$$\mathbf{v}_1 = \langle 6 - 3, 5 - 7 \rangle = \langle 3, -2 \rangle, \qquad \mathbf{v}_2 = \langle 2 - (-1), 1 - 4 \rangle = \langle 3, -3 \rangle$$

FIGURE 5

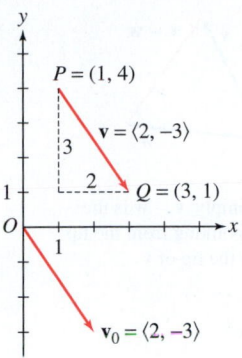

FIGURE 6 The vectors **v** and $\mathbf{v}_0$ have the same components but different basepoints.

The components of $\mathbf{v}_1$ and $\mathbf{v}_2$ are not the same, so $\mathbf{v}_1$ and $\mathbf{v}_2$ are not equivalent. Since $\mathbf{v}_1 = \langle 3, -2 \rangle$, its magnitude is

$$\|\mathbf{v}_1\| = \sqrt{3^2 + (-2)^2} = \sqrt{13}$$ ■

EXAMPLE 2 Sketch the vector $\mathbf{v} = \langle 2, -3 \rangle$ based at $P = (1, 4)$ and the vector $\mathbf{v}_0$ equivalent to **v** based at the origin.

Solution The vector $\mathbf{v} = \langle 2, -3 \rangle$, which is based at $P = (1, 4)$, has the terminal point $Q = (1 + 2, 4 - 3) = (3, 1)$, located 2 units to the right and 3 units down from P as shown in Figure 6. The vector $\mathbf{v}_0$ that is equivalent to **v** and based at O has terminal point $(2, -3)$. ■

Vector Algebra

We now define two basic vector operations: vector addition and scalar multiplication.

   The vector sum $\mathbf{v} + \mathbf{w}$ is defined when **v** and **w** have the same basepoint: Translate **w** to the equivalent vector $\mathbf{w}'$ whose tail coincides with the head of **v**. The sum $\mathbf{v} + \mathbf{w}$ is the vector pointing from the tail of **v** to the head of $\mathbf{w}'$ [Figure 7(A)]. Alternatively, we can use the **Parallelogram Law**: $\mathbf{v} + \mathbf{w}$ is the vector pointing from the basepoint to the opposite vertex of the parallelogram formed by **v** and **w** [Figure 7(B)].

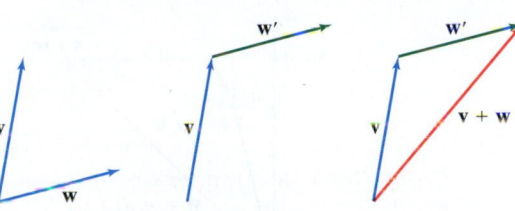

(A) The vector sum $\mathbf{v} + \mathbf{w}$

(B) Addition via the Parallelogram Law

FIGURE 7

   To add several vectors $\mathbf{v}_1, \mathbf{v}_2, \ldots, \mathbf{v}_n$, translate the vectors to $\mathbf{v}_1 = \mathbf{v}_1', \mathbf{v}_2', \ldots, \mathbf{v}_n'$ so that they lie head to tail as in Figure 8. The vector sum $\mathbf{v} = \mathbf{v}_1 + \mathbf{v}_2 + \cdots + \mathbf{v}_n$ is the vector whose initial point is the initial point of $\mathbf{v}_1$ and terminal point is the terminal point of $\mathbf{v}_n'$.

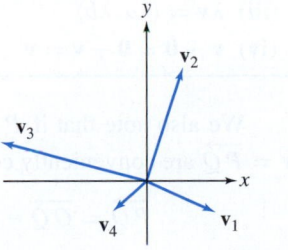
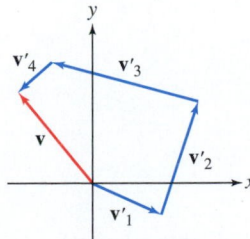

DF FIGURE 8 The sum $\mathbf{v} = \mathbf{v}_1 + \mathbf{v}_2 + \mathbf{v}_3 + \mathbf{v}_4$.

CAUTION Remember that the vector $\mathbf{v} - \mathbf{w}$ *points in the direction from the tip of* **w** *to the tip of* **v** *(not from the tip of* **v** *to the tip of* **w***).*

NOTATION λ *(pronounced "lambda") is the 11th letter in the Greek alphabet. We use the symbol* λ *often (but not exclusively) to denote a scalar.*

Vector subtraction $\mathbf{v} - \mathbf{w}$ is carried out by adding $-\mathbf{w}$ to **v** as in Figure 9(A). Or, more simply, draw the vector pointing from **w** to **v** as in Figure 9(B).

   The term "**scalar**" is another word for real number, and we often speak of scalar versus vector quantities. Thus, the number 8 is a scalar, while $\langle 8, 2 \rangle$ is a vector. If λ is a scalar and **v** is a nonzero vector, the **scalar multiple** $\lambda \mathbf{v}$ is defined as follows (Figure 10):

- It has length $|\lambda| \, \|\mathbf{v}\|$.
- It points in the same direction as **v** if $\lambda > 0$.
- It points in the opposite direction if $\lambda < 0$.

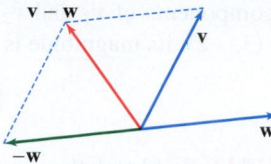

(A) **v** − **w** equals **v** plus (−**w**).

(B) More simply, **v** − **w** is the vector pointing from the tip of **w** to the tip of **v**.

FIGURE 9 Vector subtraction.

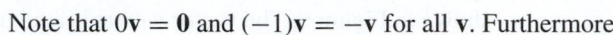

FIGURE 10 Vectors **v** and **2v** are based at *P* but **2v** is twice as long. Vectors **v** and −**v** have the same length but opposite directions.

Note that $0\mathbf{v} = \mathbf{0}$ and $(-1)\mathbf{v} = -\mathbf{v}$ for all **v**. Furthermore,

$$\|\lambda \mathbf{v}\| = |\lambda| \, \|\mathbf{v}\|$$

A vector **w** is parallel to **v** if and only if $\mathbf{w} = \lambda \mathbf{v}$ for some nonzero scalar λ.

Vector addition and scalar multiplication operations are easily performed using components. To add or subtract two vectors **v** and **w**, we add or subtract their components. This follows from the Parallelogram Law as indicated in Figure 11(A).

Similarly, to multiply **v** by a scalar λ, we multiply the components of **v** by λ [Figure 11(B)]. Indeed, if $\mathbf{v} = \langle a, b \rangle$ is nonzero, $\langle \lambda a, \lambda b \rangle$ has length $|\lambda| \, \|\mathbf{v}\|$. It points in the same direction as $\langle a, b \rangle$ if $\lambda > 0$, and in the opposite direction if $\lambda < 0$.

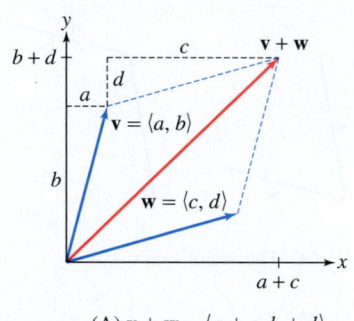

(A) $\mathbf{v} + \mathbf{w} = \langle a + c, b + d \rangle$

(B) $\lambda \mathbf{v} = \langle \lambda a, \lambda b \rangle$

FIGURE 11 Vector operations using components.

Vector Operations Using Components If $\mathbf{v} = \langle a, b \rangle$ and $\mathbf{w} = \langle c, d \rangle$, then:

(i) $\mathbf{v} + \mathbf{w} = \langle a + c, b + d \rangle$

(ii) $\mathbf{v} - \mathbf{w} = \langle a - c, b - d \rangle$

(iii) $\lambda \mathbf{v} = \langle \lambda a, \lambda b \rangle$

(iv) $\mathbf{v} + \mathbf{0} = \mathbf{0} + \mathbf{v} = \mathbf{v}$

We also note that if $P = (a_1, b_1)$ and $Q = (a_2, b_2)$, then components of the vector $\mathbf{v} = \overrightarrow{PQ}$ are conveniently computed as the difference

$$\overrightarrow{PQ} = \overrightarrow{OQ} - \overrightarrow{OP} = \langle a_2, b_2 \rangle - \langle a_1, b_1 \rangle = \langle a_2 - a_1, b_2 - b_1 \rangle$$

EXAMPLE 3 For $\mathbf{v} = \langle 1, 4 \rangle$, $\mathbf{w} = \langle 3, 2 \rangle$, calculate:

(a) $\mathbf{v} + \mathbf{w}$

(b) $5\mathbf{v}$

Solution

$$\mathbf{v} + \mathbf{w} = \langle 1, 4 \rangle + \langle 3, 2 \rangle = \langle 1 + 3, 4 + 2 \rangle = \langle 4, 6 \rangle$$

$$5\mathbf{v} = 5 \langle 1, 4 \rangle = \langle 5, 20 \rangle$$

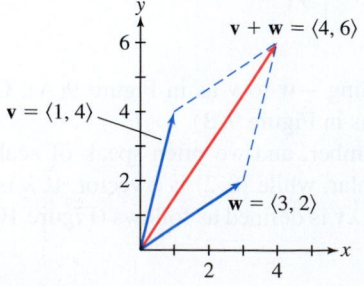

DF **FIGURE 12**

The vector sum is illustrated in Figure 12. ∎

Vector operations obey the usual laws of algebra.

> **THEOREM 1 Basic Properties of Vector Algebra**   For all vectors $\mathbf{u}, \mathbf{v}, \mathbf{w}$ and for all scalars λ,
>
> **Commutative Law:** $\mathbf{v} + \mathbf{w} = \mathbf{w} + \mathbf{v}$
> **Associative Law:** $\mathbf{u} + (\mathbf{v} + \mathbf{w}) = (\mathbf{u} + \mathbf{v}) + \mathbf{w}$
> **Distributive Law for Scalars:** $\lambda(\mathbf{v} + \mathbf{w}) = \lambda\mathbf{v} + \lambda\mathbf{w}$

These properties are verified easily using components. For example, we can check that vector addition is commutative:

$$\langle v_1, v_2 \rangle + \langle w_1, w_2 \rangle = \underbrace{\langle v_1 + w_1, v_2 + w_2 \rangle = \langle w_1 + v_1, w_2 + v_2 \rangle}_{\text{Commutativity of addition of real numbers}} = \langle w_1, w_2 \rangle + \langle v_1, v_2 \rangle$$

A **linear combination** of vectors $\mathbf{v}$ and $\mathbf{w}$ is a vector

$$r\mathbf{v} + s\mathbf{w}$$

where r and s are scalars. If $\mathbf{v}$ and $\mathbf{w}$ are not parallel, then every vector $\mathbf{u}$ in the plane can be expressed as a linear combination $\mathbf{u} = r\mathbf{v} + s\mathbf{w}$ [Figure 13(A)]. The parallelogram $\mathcal{P}$ whose vertices are the origin and the terminal points of $\mathbf{v}$, $\mathbf{w}$ and $\mathbf{v} + \mathbf{w}$ is called the **parallelogram spanned** by $\mathbf{v}$ and $\mathbf{w}$ [Figure 13(B)]. It consists of the linear combinations $r\mathbf{v} + s\mathbf{w}$ with $0 \le r \le 1$ and $0 \le s \le 1$.

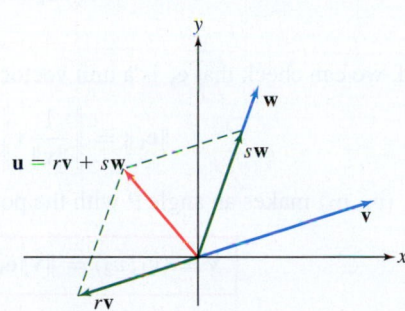

(A) The vector $\mathbf{u}$ can be expressed as a linear combination $\mathbf{u} = r\mathbf{v} + s\mathbf{w}$. In this example, $r < 0$.

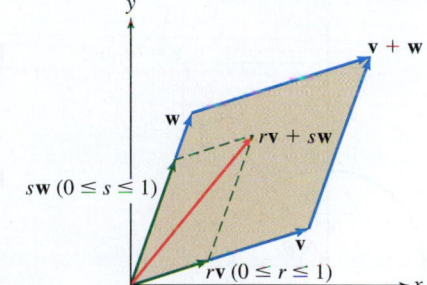

(B) The parallelogram $\mathcal{P}$ spanned by $\mathbf{v}$ and $\mathbf{w}$ consists of all linear combinations $r\mathbf{v} + s\mathbf{w}$ with $0 \le r, s \le 1$.

FIGURE 13

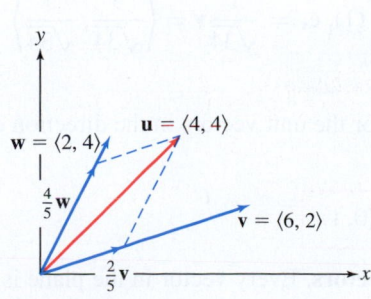

FIGURE 14

EXAMPLE 4   Linear Combinations   Express the vector $\mathbf{u} = \langle 4, 4 \rangle$ in Figure 14 as a linear combination of $\mathbf{v} = \langle 6, 2 \rangle$ and $\mathbf{w} = \langle 2, 4 \rangle$.

Solution   We must find r and s such that $r\mathbf{v} + s\mathbf{w} = \langle 4, 4 \rangle$, or

$$r \langle 6, 2 \rangle + s \langle 2, 4 \rangle = \langle 6r + 2s, 2r + 4s \rangle = \langle 4, 4 \rangle$$

The components must be equal, so we have a system of two linear equations:

$$6r + 2s = 4$$
$$2r + 4s = 4$$

Subtracting the equations, we obtain $4r - 2s = 0$ or $s = 2r$. Setting $s = 2r$ in the first equation yields $6r + 4r = 4$ or $r = \frac{2}{5}$, and then $s = 2r = \frac{4}{5}$. Therefore,

$$\mathbf{u} = \langle 4, 4 \rangle = \frac{2}{5} \langle 6, 2 \rangle + \frac{4}{5} \langle 2, 4 \rangle \qquad \blacksquare$$

CONCEPTUAL INSIGHT In general, to write a vector $\mathbf{u} = \langle u_1, u_2 \rangle$ as a linear combination of two other vectors $\mathbf{v} = \langle v_1, v_2 \rangle$ and $\mathbf{w} = \langle w_1, w_2 \rangle$, we have to solve a system of two linear equations in two unknowns r and s:

$$r\mathbf{v} + s\mathbf{w} = \mathbf{u} \quad \Leftrightarrow \quad r\langle v_1, v_2 \rangle + s\langle w_1, w_2 \rangle = \langle u_1, u_2 \rangle \quad \Leftrightarrow \quad \begin{cases} rv_1 + sw_1 = u_1 \\ rv_2 + sw_2 = u_2 \end{cases}$$

On the other hand, vectors give us a way of visualizing the system of equations geometrically. The solution is represented by a parallelogram as in Figure 14. This relation between vectors and systems of linear equations extends to any number of variables and is the starting point for the important subject of linear algebra.

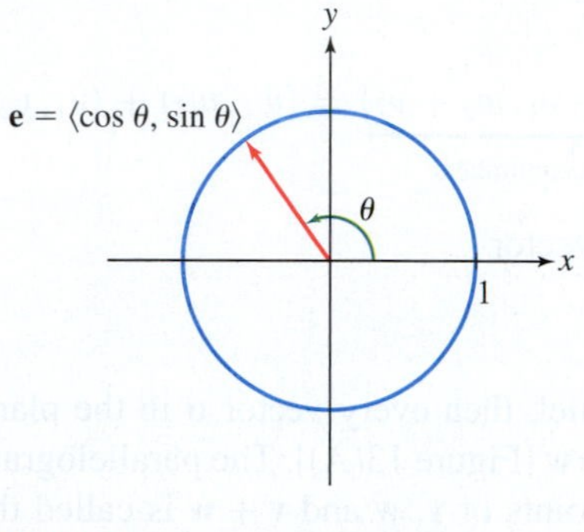

A vector of length 1 is called a **unit vector**. Unit vectors are often used to indicate direction, when it is not necessary to specify length. The head of a unit vector $\mathbf{e}$ based at the origin lies on the unit circle, and $\mathbf{e}$ can be given as

$$\mathbf{e} = \langle \cos\theta, \sin\theta \rangle$$

where θ is the angle between $\mathbf{e}$ and the positive x-axis (Figure 15). The fact that its length is 1 when represented in this way follows immediately from the trigonometric identity $\sin^2\theta + \cos^2\theta = 1$.

We can always scale a nonzero vector $\mathbf{v} = \langle v_1, v_2 \rangle$ to obtain a unit vector pointing in the same direction (Figure 16):

$$\boxed{\mathbf{e_v} = \frac{1}{\|\mathbf{v}\|}\mathbf{v}} \qquad \boxed{1}$$

Indeed, we can check that $\mathbf{e_v}$ is a unit vector as follows:

$$\|\mathbf{e_v}\| = \left\| \frac{1}{\|\mathbf{v}\|}\mathbf{v} \right\| = \frac{1}{\|\mathbf{v}\|}\|\mathbf{v}\| = 1$$

If $\mathbf{v} = \langle v_1, v_2 \rangle$ makes an angle θ with the positive x-axis, then

$$\boxed{\mathbf{v} = \langle v_1, v_2 \rangle = \|\mathbf{v}\|\mathbf{e_v} = \|\mathbf{v}\|\langle \cos\theta, \sin\theta \rangle} \qquad \boxed{2}$$

Note that the relation $\mathbf{v} = \|\mathbf{v}\|\langle \cos\theta, \sin\theta \rangle$ in Eq. (2) concisely reflects the direction and magnitude of a vector. The angle θ determines the direction of $\mathbf{v}$ and $\|\mathbf{v}\|$ gives the magnitude.

EXAMPLE 5 Find the unit vector in the direction of $\mathbf{v} = \langle 3, 5 \rangle$.

Solution $\|\mathbf{v}\| = \sqrt{3^2 + 5^2} = \sqrt{34}$, and thus by Eq. (1), $\mathbf{e_v} = \frac{1}{\sqrt{34}}\mathbf{v} = \left\langle \frac{3}{\sqrt{34}}, \frac{5}{\sqrt{34}} \right\rangle$. ∎

It is customary to introduce a special notation for the unit vectors in the direction of the positive x- and y-axes (Figure 17):

$$\boxed{\mathbf{i} = \langle 1, 0 \rangle, \qquad \mathbf{j} = \langle 0, 1 \rangle}$$

The vectors $\mathbf{i}$ and $\mathbf{j}$ are called the **standard basis vectors**. Every vector in the plane is a linear combination of $\mathbf{i}$ and $\mathbf{j}$ (Figure 17):

$$\mathbf{v} = \langle a, b \rangle = a\mathbf{i} + b\mathbf{j}$$

For example, $\langle 4, -2 \rangle = 4\mathbf{i} - 2\mathbf{j}$ and $\langle 5, 7 \rangle = 5\mathbf{i} + 7\mathbf{j}$.

For vectors represented in this form, vector addition is performed by adding the $\mathbf{i}$ and $\mathbf{j}$ coefficients. For example,

$$(4\mathbf{i} - 2\mathbf{j}) + (5\mathbf{i} + 7\mathbf{j}) = (4 + 5)\mathbf{i} + (-2 + 7)\mathbf{j} = 9\mathbf{i} + 5\mathbf{j}$$

FIGURE 15 The head of a unit vector lies on the unit circle.

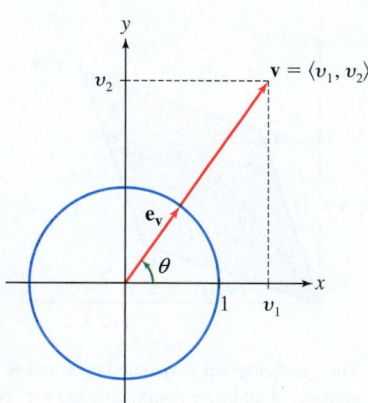

FIGURE 16 Unit vector in the direction of $\mathbf{v}$.

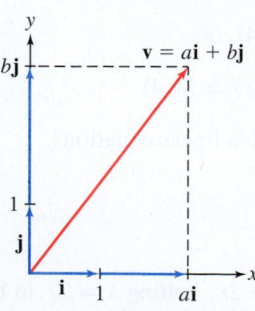

FIGURE 17

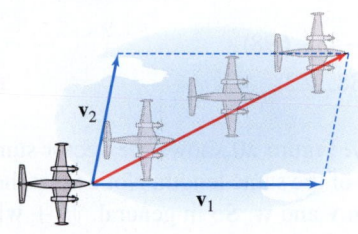

FIGURE 18 When an airplane traveling with velocity $\mathbf{v}_1$ encounters a wind of velocity $\mathbf{v}_2$, its resultant velocity is the vector sum $\mathbf{v}_1 + \mathbf{v}_2$.

CONCEPTUAL INSIGHT   It is often said that quantities such as force and velocity are vectors because they have both magnitude and direction, but there is more to this statement than meets the eye. A vector quantity must obey the Law of Vector Addition, so if we say that force is a vector, we are really claiming that forces add according to the Parallelogram Law. In other words, if forces $\mathbf{F}_1$ and $\mathbf{F}_2$ act on an object, then the resultant force is the vector sum $\mathbf{F}_1 + \mathbf{F}_2$. A similar situation holds for velocities, as illustrated in Figure 18. This aspect of force and velocity is a physical fact that must be verified experimentally. It was well known to scientists and engineers long before the vector concept was introduced formally in the 1800s.

EXAMPLE 6   Find the forces on cables 1 and 2 in Figure 19(A).

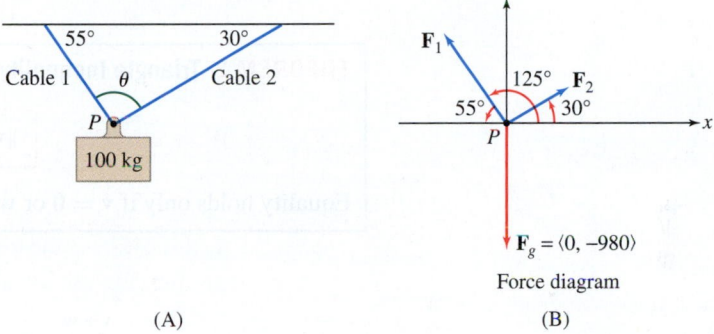

(A)   (B)

FIGURE 19

Newton's second law indicates $F = ma$, where F is the force on an object, m is its mass, and a is its acceleration. With acceleration due to gravity equal to 9.8 m/sec^2, the force due to gravity on a 100-kg mass is $(9.8)(100) = 980$ newtons.

Solution   Three forces act on the point P in Figure 19(A): the force $\mathbf{F}_g$ due to gravity that acts vertically downward with a magnitude of 980 newtons (on a 100-kg mass), and two unknown forces $\mathbf{F}_1$ and $\mathbf{F}_2$ acting through cables 1 and 2, as indicated in Figure 19(B).

Since the point P is not in motion, the net force on P is zero:

$$\mathbf{F}_1 + \mathbf{F}_2 + \mathbf{F}_g = \mathbf{0}$$

We use this fact to determine $\mathbf{F}_1$ and $\mathbf{F}_2$.

Let $f_1 = \|\mathbf{F}_1\|$ and $f_2 = \|\mathbf{F}_2\|$ be the magnitudes of the unknown forces. Because $\mathbf{F}_1$ makes an angle of $125°$ (the supplement of $55°$) with the positive x-axis, and $\mathbf{F}_2$ makes an angle of $30°$, we can use Eq. (2) and the table in the margin to write these vectors in component form:

$$\mathbf{F}_1 = f_1 \langle \cos 125°, \sin 125° \rangle \approx f_1 \langle -0.573, 0.819 \rangle$$

$$\mathbf{F}_2 = f_2 \langle \cos 30°, \sin 30° \rangle \approx f_2 \langle 0.866, 0.5 \rangle$$

$$\mathbf{F}_g = \langle 0, -980 \rangle$$

θ	$\cos\theta$	$\sin\theta$
$125°$	-0.573	0.819
$30°$	0.866	0.5

Since the sum of the forces is the zero vector, using the approximate values for the forces, we have:

$$f_1 \langle -0.573, 0.819 \rangle + f_2 \langle 0.866, 0.5 \rangle + \langle 0, -980 \rangle = \langle 0, 0 \rangle$$

We will solve for f_1 and f_2. Equating vector components gives us two equations in two unknowns:

$$-0.573 f_1 + 0.866 f_2 = 0, \qquad 0.819 f_1 + 0.5 f_2 - 980 = 0$$

By the first equation, $f_2 = \left(\frac{0.573}{0.866} \right) f_1$. Substitution in the second equation yields

$$0.819 f_1 + 0.5 \left(\frac{0.573}{0.866} \right) f_1 - 980 \approx 1.15 f_1 - 980 = 0$$

Therefore, the force magnitudes in newtons are

$$f_1 \approx \frac{980}{1.15} \approx 852 \text{ newtons} \qquad \text{and} \qquad f_2 \approx \left(\frac{0.573}{0.866} \right) 852 \approx 564 \text{ newtons}$$

Hence,

$$\mathbf{F}_1 \approx 852 \langle -0.573, 0.819 \rangle \approx \langle -488, 698 \rangle$$
$$\mathbf{F}_2 \approx 564 \langle 0.866, 0.5 \rangle \approx \langle 488, 282 \rangle$$

■

We close this section with the Triangle Inequality. Figure 20 shows the vector sum $\mathbf{v} + \mathbf{w}$ for a fixed vector $\mathbf{v}$ and three different vectors $\mathbf{w}$ of the same length. Notice that the length $\|\mathbf{v} + \mathbf{w}\|$ varies depending on the angle between $\mathbf{v}$ and $\mathbf{w}$. So in general, $\|\mathbf{v} + \mathbf{w}\|$ is not equal to the sum $\|\mathbf{v}\| + \|\mathbf{w}\|$. What we can say is that $\|\mathbf{v} + \mathbf{w}\|$ is *at most* equal to the sum $\|\mathbf{v}\| + \|\mathbf{w}\|$. This corresponds to the fact that the length of one side of a triangle is at most the sum of the lengths of the other two sides. A formal proof may be given using the dot product (see Exercise 98 in Section 12.3).

THEOREM 2 Triangle Inequality For any two vectors $\mathbf{v}$ and $\mathbf{w}$,

$$\|\mathbf{v} + \mathbf{w}\| \leq \|\mathbf{v}\| + \|\mathbf{w}\|$$

Equality holds only if $\mathbf{v} = \mathbf{0}$ or $\mathbf{w} = \mathbf{0}$, or if $\mathbf{w} = \lambda \mathbf{v}$, where $\lambda > 0$.

DF **FIGURE 20** The length of $\mathbf{v} + \mathbf{w}$ depends on the angle between $\mathbf{v}$ and $\mathbf{w}$.

12.1 SUMMARY

- A *vector* $\mathbf{v} = \overrightarrow{PQ}$ is determined by a basepoint P (the "tail") and a terminal point Q (the "head" or "tip").
- Components of $\mathbf{v} = \overrightarrow{PQ}$, where $P = (a_1, b_1)$ and $Q = (a_2, b_2)$: $\mathbf{v} = \langle a, b \rangle$ with $a = a_2 - a_1, b = b_2 - b_1$.
- Length or magnitude: $\|\mathbf{v}\| = \sqrt{a^2 + b^2}$.
- The *length* $\|\mathbf{v}\|$ is the distance from P to Q.
- The *position vector* of $P_0 = (a, b)$ is the vector $\mathbf{v} = \langle a, b \rangle$ pointing from the origin O to P_0.
- Vectors $\mathbf{v}$ and $\mathbf{w}$ are *equivalent* if they have the same magnitude and direction. Two vectors are equivalent if and only if they have the same components.
- The *zero vector* is the vector $\mathbf{0} = \langle 0, 0 \rangle$ of length 0.
- *Vector addition* is defined geometrically by the *Parallelogram Law*. In components,

$$\langle v_1, v_2 \rangle + \langle w_1, w_2 \rangle = \langle v_1 + w_1, v_2 + w_2 \rangle$$

- Scalar multiplication: $\lambda \mathbf{v}$ is the vector of length $|\lambda| \, \|\mathbf{v}\|$ in the same direction as $\mathbf{v}$ if $\lambda > 0$, and in the opposite direction if $\lambda < 0$. In components,

$$\lambda \langle v_1, v_2 \rangle = \langle \lambda v_1, \lambda v_2 \rangle$$

- Nonzero vectors $\mathbf{v}$ and $\mathbf{w}$ are *parallel* if $\mathbf{w} = \lambda \mathbf{v}$ for some scalar λ.
- Unit vector making an angle θ with the positive x-axis: $\mathbf{e} = \langle \cos \theta, \sin \theta \rangle$.
- Unit vector in the direction of $\mathbf{v} \neq \mathbf{0}$: $\mathbf{e_v} = \dfrac{1}{\|\mathbf{v}\|} \mathbf{v}$.

- If $\mathbf{v} = \langle v_1, v_2 \rangle$ makes an angle θ with the positive x-axis, then

$$v_1 = \|\mathbf{v}\| \cos \theta, \qquad v_2 = \|\mathbf{v}\| \sin \theta, \qquad \mathbf{e_v} = \langle \cos \theta, \sin \theta \rangle$$

- Standard basis vectors: $\mathbf{i} = \langle 1, 0 \rangle$ and $\mathbf{j} = \langle 0, 1 \rangle$.
- Every vector $\mathbf{v} = \langle a, b \rangle$ is a linear combination $\mathbf{v} = a\mathbf{i} + b\mathbf{j}$.
- Triangle Inequality: $\|\mathbf{v} + \mathbf{w}\| \le \|\mathbf{v}\| + \|\mathbf{w}\|$.

12.1 EXERCISES

Preliminary Questions

1. Answer true or false. Every nonzero vector is:

(a) equivalent to a vector based at the origin.

(b) equivalent to a unit vector based at the origin.

(c) parallel to a vector based at the origin.

(d) parallel to a unit vector based at the origin.

2. What is the length of $-3\mathbf{a}$ if $\|\mathbf{a}\| = 5$?

3. Suppose that $\mathbf{v}$ has components $\langle 3, 1 \rangle$. How, if at all, do the components change if you translate $\mathbf{v}$ horizontally 2 units to the left?

4. What are the components of the zero vector based at $P = (3, 5)$?

5. True or false?

(a) The vectors $\mathbf{v}$ and $-2\mathbf{v}$ are parallel.

(b) The vectors $\mathbf{v}$ and $-2\mathbf{v}$ point in the same direction.

6. Explain the commutativity of vector addition in terms of the Parallelogram Law.

Exercises

1. Sketch the vectors $\mathbf{v}_1, \mathbf{v}_2, \mathbf{v}_3, \mathbf{v}_4$ with tail P and head Q, and compute their lengths. Are any two of these vectors equivalent?

	$\mathbf{v}_1$	$\mathbf{v}_2$	$\mathbf{v}_3$	$\mathbf{v}_4$
P	$(2,4)$	$(-1,3)$	$(-1,3)$	$(4,1)$
Q	$(4,4)$	$(1,3)$	$(2,4)$	$(6,3)$

2. Sketch the vector $\mathbf{b} = \langle 3, 4 \rangle$ based at $P = (-2, -1)$.

3. What is the terminal point of the vector $\mathbf{a} = \langle 1, 3 \rangle$ based at $P = (2, 2)$? Sketch $\mathbf{a}$ and the vector $\mathbf{a}_0$ based at the origin and equivalent to $\mathbf{a}$.

4. Let $\mathbf{v} = \overrightarrow{PQ}$, where $P = (1, 1)$ and $Q = (2, 2)$. What is the head of the vector $\mathbf{v}'$ equivalent to $\mathbf{v}$ based at $(2, 4)$? What is the head of the vector $\mathbf{v}_0$ equivalent to $\mathbf{v}$ based at the origin? Sketch $\mathbf{v}$, $\mathbf{v}_0$, and $\mathbf{v}'$.

In Exercises 5–8, refer to the unit vectors in Figure 21.

5. Find the components of $\mathbf{u}$.

6. Find the components of $\mathbf{v}$.

7. Find the components of $\mathbf{w}$.

8. Find the components of $\mathbf{q}$.

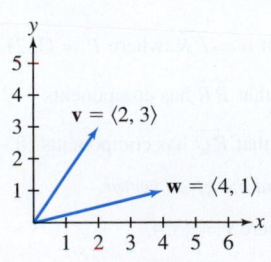

FIGURE 21

In Exercises 9–12, find the components of $\overrightarrow{PQ}$.

9. $P = (3, 2)$, $Q = (2, 7)$

10. $P = (-3, -5)$, $Q = (4, -6)$

11. $P = (1, -7)$, $Q = (0, 17)$

12. $P = (0, 2)$, $Q = (5, 0)$

In Exercises 13–20, calculate.

13. $\langle 2, 1 \rangle + \langle 3, 4 \rangle$

14. $\langle -4, 6 \rangle - \langle 3, -2 \rangle$

15. $5 \langle 6, 2 \rangle$

16. $4(\langle 1, 1 \rangle + \langle 3, 2 \rangle)$

17. $\left\langle -\frac{1}{2}, \frac{5}{3} \right\rangle + \left\langle 3, \frac{10}{3} \right\rangle$

18. $2.7 \langle -1.4, 0.8 \rangle - 3.3 \langle 3.1, -2.2 \rangle$

19. $\langle 2e, 1 - 2\pi \rangle - \langle 2e - \pi, 8 - 2\pi \rangle$

20. $\langle \ln 6, \sin^2 3 \rangle + \langle 1 - \ln 3, \cos^2 3 \rangle$

21. Which of the vectors (A)–(C) in Figure 22 is equivalent to $\mathbf{v} - \mathbf{w}$?

 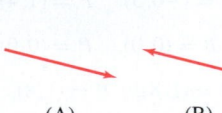 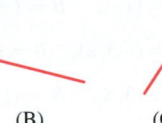

(A)   (B)   (C)

FIGURE 22

22. Sketch $\mathbf{v} + \mathbf{w}$ and $\mathbf{v} - \mathbf{w}$ for the vectors in Figure 23.

FIGURE 23

23. Sketch $2\mathbf{v}$, $-\mathbf{w}$, $\mathbf{v} + \mathbf{w}$, and $2\mathbf{v} - \mathbf{w}$ for the vectors in Figure 24.

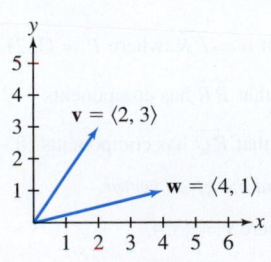

FIGURE 24

24. Sketch $\mathbf{v} = \langle 1, 3 \rangle$, $\mathbf{w} = \langle 2, -2 \rangle$, $\mathbf{v} + \mathbf{w}$, $\mathbf{v} - \mathbf{w}$.

25. Sketch $\mathbf{v} = \langle 0, 2 \rangle$, $\mathbf{w} = \langle -2, 4 \rangle$, $3\mathbf{v} + \mathbf{w}$, $2\mathbf{v} - 2\mathbf{w}$.

26. Sketch $\mathbf{v} = \langle -2, 1 \rangle$, $\mathbf{w} = \langle 2, 2 \rangle$, $\mathbf{v} + 2\mathbf{w}$, $\mathbf{v} - 2\mathbf{w}$.

27. Sketch the vector $\mathbf{v}$ such that $\mathbf{v} + \mathbf{v}_1 + \mathbf{v}_2 = \mathbf{0}$ for $\mathbf{v}_1$ and $\mathbf{v}_2$ in Figure 25(A).

28. Sketch the vector sum $\mathbf{v} = \mathbf{v}_1 + \mathbf{v}_2 + \mathbf{v}_3 + \mathbf{v}_4$ in Figure 25(B).

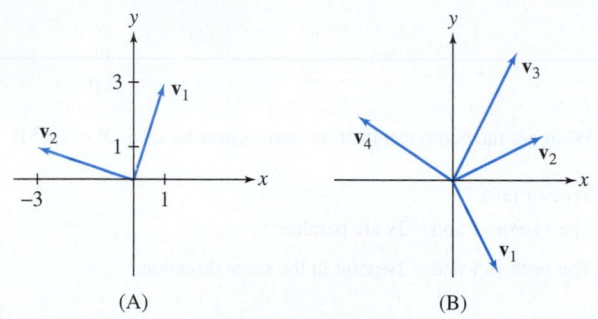

(A)   (B)

FIGURE 25

29. Let $\mathbf{v} = \overrightarrow{PQ}$, where $P = (-2, 5)$, $Q = (1, -2)$. Which of the following vectors with the given tails and heads are equivalent to $\mathbf{v}$?

(a) $(-3, 3)$, $(0, 4)$ (b) $(0, 0)$, $(3, -7)$
(c) $(-1, 2)$, $(2, -5)$ (d) $(4, -5)$, $(1, 4)$

30. Which of the following vectors are parallel to $\mathbf{v} = \langle 6, 9 \rangle$ and which point in the same direction?

(a) $\langle 12, 18 \rangle$ (b) $\langle 3, 2 \rangle$ (c) $\langle 2, 3 \rangle$
(d) $\langle -6, -9 \rangle$ (e) $\langle -24, -27 \rangle$ (f) $\langle -24, -36 \rangle$

In Exercises 31–34, sketch the vectors $\overrightarrow{AB}$ and $\overrightarrow{PQ}$, and determine whether they are equivalent.

31. $A = (1, 1)$, $B = (3, 7)$, $P = (4, -1)$, $Q = (6, 5)$

32. $A = (1, 4)$, $B = (-6, 3)$, $P = (1, 4)$, $Q = (6, 3)$

33. $A = (-3, 2)$, $B = (0, 0)$, $P = (0, 0)$, $Q = (3, -2)$

34. $A = (5, 8)$, $B = (1, 8)$, $P = (1, 8)$, $Q = (-3, 8)$

In Exercises 35–38, are $\overrightarrow{AB}$ and $\overrightarrow{PQ}$ parallel? And if so, do they point in the same direction?

35. $A = (1, 1)$, $B = (3, 4)$, $P = (1, 1)$, $Q = (7, 10)$

36. $A = (-3, 2)$, $B = (0, 0)$, $P = (0, 0)$, $Q = (3, 2)$

37. $A = (2, 2)$, $B = (-6, 3)$, $P = (9, 5)$, $Q = (17, 4)$

38. $A = (5, 8)$, $B = (2, 2)$, $P = (2, 2)$, $Q = (-3, 8)$

In Exercises 39–42, let $R = (-2, 7)$. Calculate the following:

39. The length of $\overrightarrow{OR}$

40. The components of $\mathbf{u} = \overrightarrow{PR}$, where $P = (1, 2)$

41. The point P such that $\overrightarrow{PR}$ has components $\langle -2, 7 \rangle$

42. The point Q such that $\overrightarrow{RQ}$ has components $\langle 8, -3 \rangle$

In Exercises 43–52, find the given vector.

43. Unit vector $\mathbf{e}_\mathbf{v}$, where $\mathbf{v} = \langle 3, 4 \rangle$

44. Unit vector $\mathbf{e}_\mathbf{w}$, where $\mathbf{w} = \langle 24, 7 \rangle$

45. Vector of length 4 in the direction of $\mathbf{u} = \langle -1, -1 \rangle$

46. Vector of length 3 in the direction of $\mathbf{v} = 4\mathbf{i} + 3\mathbf{j}$

47. Vector of length 2 in the direction opposite to $\mathbf{v} = \mathbf{i} - \mathbf{j}$

48. Unit vector in the direction opposite to $\mathbf{v} = \langle -2, 4 \rangle$

49. Unit vector $\mathbf{e}$ making an angle of $\frac{4\pi}{7}$ with the x-axis

50. Vector $\mathbf{v}$ of length 2 making an angle of $30°$ with the x-axis

51. A unit vector pointing in the direction from $(1, 1)$ to $(0, 3)$

52. A unit vector pointing in the direction from $(-3, 4)$ to the origin

53. Find all scalars λ such that $\lambda \langle 2, 3 \rangle$ has length 1.

54. Find a vector $\mathbf{v}$ satisfying $3\mathbf{v} + \langle 5, 20 \rangle = \langle 11, 17 \rangle$.

55. What are the coordinates of the point P in the parallelogram in Figure 26(A)?

56. What are the coordinates a and b in the parallelogram in Figure 26(B)?

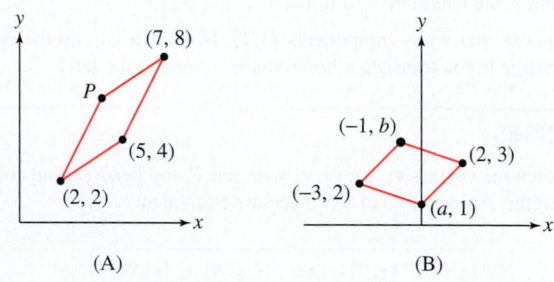

(A)   (B)

FIGURE 26

57. Let $\mathbf{v} = \overrightarrow{AB}$ and $\mathbf{w} = \overrightarrow{AC}$, where A, B, C are three distinct points in the plane. Match (a)–(d) with (i)–(iv). *Hint:* Draw a picture.

(a) $-\mathbf{w}$ (b) $-\mathbf{v}$ (c) $\mathbf{w} - \mathbf{v}$ (d) $\mathbf{v} - \mathbf{w}$
(i) $\overrightarrow{CB}$ (ii) $\overrightarrow{CA}$ (iii) $\overrightarrow{BC}$ (iv) $\overrightarrow{BA}$

58. Find the components and length of the following vectors:

(a) $4\mathbf{i} + 3\mathbf{j}$ (b) $2\mathbf{i} - 3\mathbf{j}$ (c) $\mathbf{i} + \mathbf{j}$ (d) $\mathbf{i} - 3\mathbf{j}$

In Exercises 59–62, calculate the linear combination.

59. $3\mathbf{j} + (9\mathbf{i} + 4\mathbf{j})$ 60. $-\frac{3}{2}\mathbf{i} + 5\left(\frac{1}{2}\mathbf{j} - \frac{1}{2}\mathbf{i}\right)$

61. $(3\mathbf{i} + \mathbf{j}) - 6\mathbf{j} + 2(\mathbf{j} - 4\mathbf{i})$ 62. $3(3\mathbf{i} - 4\mathbf{j}) + 5(\mathbf{i} + 4\mathbf{j})$

63. For each of the position vectors $\mathbf{u}$ with endpoints A, B, and C in Figure 27, indicate with a diagram the multiples $r\mathbf{v}$ and $s\mathbf{w}$ such that $\mathbf{u} = r\mathbf{v} + s\mathbf{w}$. A sample is shown for $\mathbf{u} = \overrightarrow{OQ}$.

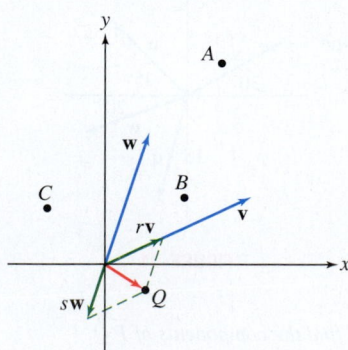

FIGURE 27

64. Sketch the parallelogram spanned by $\mathbf{v} = \langle 1, 4 \rangle$ and $\mathbf{w} = \langle 5, 2 \rangle$. Add the vector $\mathbf{u} = \langle 2, 3 \rangle$ to the sketch and express $\mathbf{u}$ as a linear combination of $\mathbf{v}$ and $\mathbf{w}$.

In Exercises 65 and 66, express $\mathbf{u}$ *as a linear combination* $\mathbf{u} = r\mathbf{v} + s\mathbf{w}$. *Then sketch* $\mathbf{u}, \mathbf{v}, \mathbf{w}$, *and the parallelogram formed by* $r\mathbf{v}$ *and* $s\mathbf{w}$.

65. $\mathbf{u} = \langle 3, -1 \rangle$; $\mathbf{v} = \langle 2, 1 \rangle$, $\mathbf{w} = \langle 1, 3 \rangle$

66. $\mathbf{u} = \langle 6, -2 \rangle$; $\mathbf{v} = \langle 1, 1 \rangle$, $\mathbf{w} = \langle 1, -1 \rangle$

67. Calculate the magnitude of the force on cables 1 and 2 shown in Figure 28.

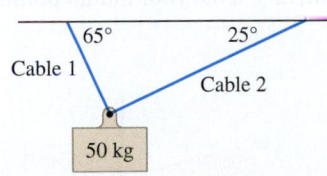

Cable 1   65°   25°   Cable 2

50 kg

FIGURE 28

68. Determine the magnitude of the forces $\mathbf{F}_1$ and $\mathbf{F}_2$ in Figure 29, assuming that there is no net force on the object.

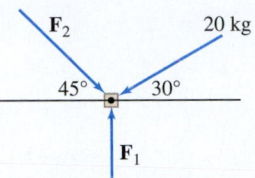

$\mathbf{F}_2$ 20 kg

45°   30°

$\mathbf{F}_1$

FIGURE 29

69. A plane flying due east at 200 km/h encounters a 40-km/h wind blowing in the northeast direction. The resultant velocity of the plane is the vector sum $\mathbf{v} = \mathbf{v}_1 + \mathbf{v}_2$, where $\mathbf{v}_1$ is the velocity vector of the plane and $\mathbf{v}_2$ is the velocity vector of the wind (Figure 30). The angle between $\mathbf{v}_1$ and $\mathbf{v}_2$ is $\frac{\pi}{4}$. Determine the resultant *speed* of the plane (the length of the vector $\mathbf{v}$).

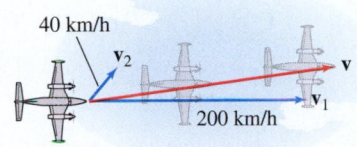

40 km/h

$\mathbf{v}_2$ $\mathbf{v}$

200 km/h $\mathbf{v}_1$

FIGURE 30

Further Insights and Challenges

In Exercises 70–72, refer to Figure 31, which shows a robotic arm consisting of two segments of lengths L_1 *and* L_2.

70. Find the components of the vector $\mathbf{r} = \overrightarrow{OP}$ in terms of θ_1 and θ_2.

71. Let $L_1 = 5$ and $L_2 = 3$. Find $\mathbf{r}$ for $\theta_1 = \frac{\pi}{3}, \theta_2 = \frac{\pi}{4}$.

72. Let $L_1 = 5$ and $L_2 = 3$. Show that the set of points reachable by the robotic arm with $\theta_1 = \theta_2$ is an ellipse.

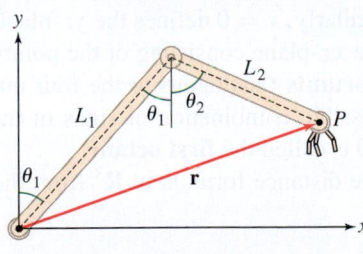

y

L_2

L_1 θ_1 θ_2 P

θ_1

$\mathbf{r}$

x

FIGURE 31

73. Use vectors to prove that the diagonals $\overline{AC}$ and $\overline{BD}$ of a parallelogram bisect each other (Figure 32). *Hint:* Observe that the midpoint of $\overline{BD}$ is the terminal point of $\mathbf{w} + \frac{1}{2}(\mathbf{v} - \mathbf{w})$.

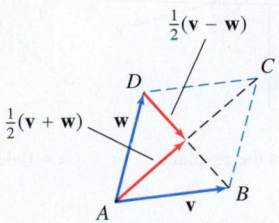

$\frac{1}{2}(\mathbf{v} - \mathbf{w})$

D C

$\frac{1}{2}(\mathbf{v} + \mathbf{w})$ $\mathbf{w}$

A $\mathbf{v}$ B

FIGURE 32

74. Use vectors to prove that the segments joining the midpoints of opposite sides of a quadrilateral bisect each other (Figure 33). *Hint:* Show that the midpoints of these segments are the terminal points of

$$\frac{1}{4}(2\mathbf{u} + \mathbf{v} + \mathbf{z}) \quad \text{and} \quad \frac{1}{4}(2\mathbf{v} + \mathbf{w} + \mathbf{u})$$

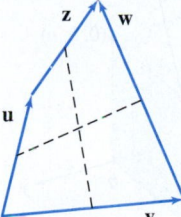

$\mathbf{z}$ $\mathbf{w}$

$\mathbf{u}$

$\mathbf{v}$

FIGURE 33

75. Prove that two nonzero vectors $\mathbf{v} = \langle a, b \rangle$ and $\mathbf{w} = \langle c, d \rangle$ are perpendicular if and only if

$$ac + bd = 0$$

12.2 Three-Dimensional Space: Surfaces, Vectors, and Curves

This section introduces three-dimensional space and extends the vector concepts introduced in the previous section to three dimensions. We begin with some introductory remarks about the three-dimensional coordinate system.

By convention, we label the axes as in Figure 1(A), where the positive sides of the axes are labeled x, y, and z. This labeling satisfies the **right-hand rule**, which means that when you position your right hand so that your fingers curl from the positive x-axis toward the positive y-axis, your thumb points in the positive z-direction. The axes in Figure 1(B) are not labeled according to the right-hand rule because when your fingers curl from the positive x-axis toward the positive y-axis, your thumb points in the negative z-direction.

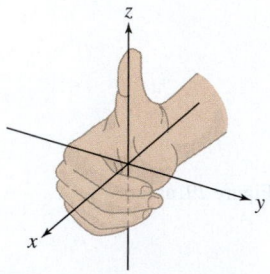 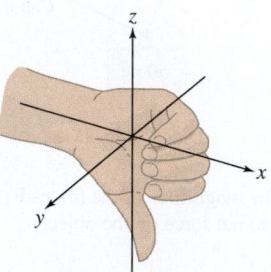

(A) This axis labeling satisfies the right-hand rule.

(B) This axis labeling does not satisfy the right-hand rule because the thumb points in the negative z-direction.

FIGURE 1

Each point in space has unique coordinates (a, b, c) relative to the axes (Figure 2). We denote the set of all triples (a, b, c) by $\mathbf{R}^3$, and we refer to this set as **3-space** or **three-dimensional space**. The **coordinate planes** in $\mathbf{R}^3$ are defined by setting one of the coordinates equal to zero (Figure 3). The xy-plane consists of the points $(a, b, 0)$ and is defined by the equation $z = 0$. Similarly, $x = 0$ defines the yz-plane consisting of the points $(0, b, c)$, and $y = 0$ defines the xz-plane consisting of the points $(a, 0, c)$. The coordinate planes divide $\mathbf{R}^3$ into eight **octants** (analogous to the four quadrants in the plane). Each octant corresponds to a possible combination of signs of the coordinates. The set of points (a, b, c) with $a, b, c > 0$ is called the **first octant**.

As in two dimensions, we derive the distance formula in $\mathbf{R}^3$ from the Pythagorean Theorem.

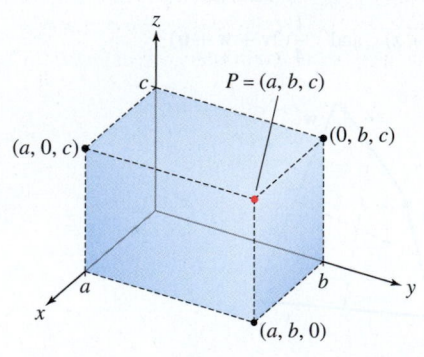

FIGURE 2

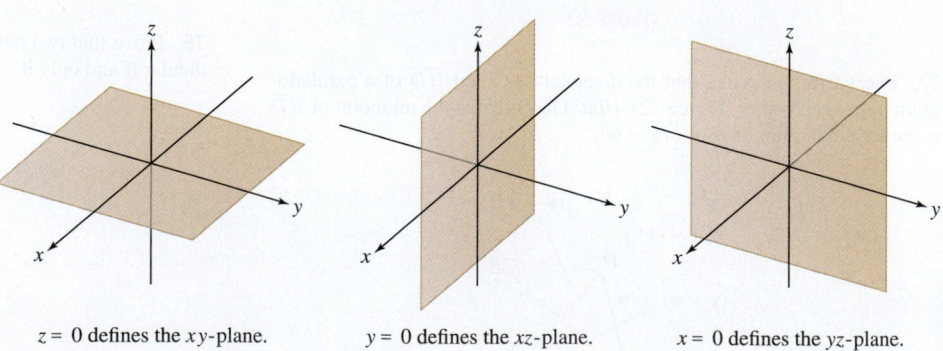

$z = 0$ defines the xy-plane.

$y = 0$ defines the xz-plane.

$x = 0$ defines the yz-plane.

FIGURE 3

THEOREM 1 Distance Formula in $\mathbf{R}^3$ The distance $|P - Q|$ between the points $P = (a_1, b_1, c_1)$ and $Q = (a_2, b_2, c_2)$ is

$$|P - Q| = \sqrt{(a_2 - a_1)^2 + (b_2 - b_1)^2 + (c_2 - c_1)^2}$$ **1**

Proof First, apply the distance formula in the plane to the points P and R [Figure 4(A)]:

$$|P - R|^2 = (a_2 - a_1)^2 + (b_2 - b_1)^2$$

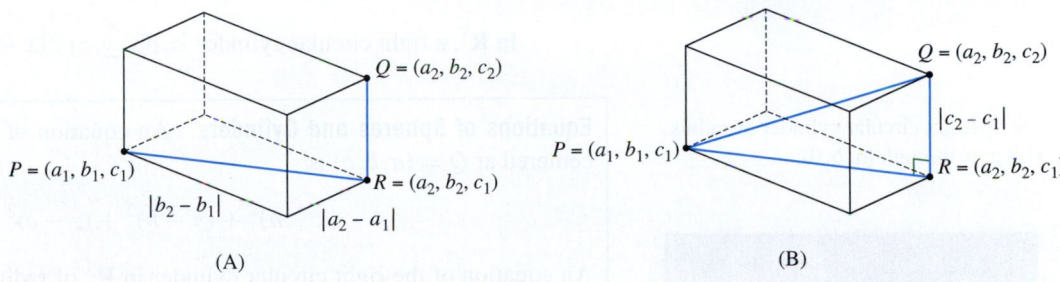

(A)   (B)

FIGURE 4

Then observe that $\triangle PRQ$ is a right triangle [Figure 4(B)] and use the Pythagorean Theorem:

$$|P - Q|^2 = |P - R|^2 + |R - Q|^2 = (a_2 - a_1)^2 + (b_2 - b_1)^2 + (c_2 - c_1)^2 \quad \blacksquare$$

Surfaces

Surfaces in $\mathbf{R}^3$ will play an important role in our development of multivariable calculus. Planes are the most basic surfaces. The equation $x = a$ defines a plane parallel to the yz-plane, while $y = b$ and $z = c$ define planes parallel to the xz-plane and xy-plane, respectively. For example, the planes $x = 3$, $y = -5$, and $z = 10$ are illustrated in Figure 5. We explore planes further in Section 5 in this chapter.

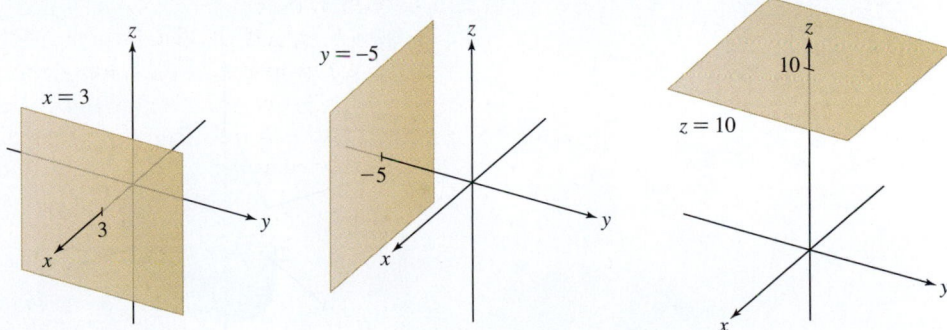

FIGURE 5 Planes in 3-space.

Spheres and cylinders are some other surfaces we will encounter. The sphere of radius R with center $Q = (a, b, c)$ consists of all points $P = (x, y, z)$ located a distance R from Q (Figure 6). By the distance formula, the coordinates of $P = (x, y, z)$ must satisfy

$$\sqrt{(x - a)^2 + (y - b)^2 + (z - c)^2} = R$$

On squaring both sides, we obtain the standard equation of the sphere:

$$(x - a)^2 + (y - b)^2 + (z - c)^2 = R^2$$

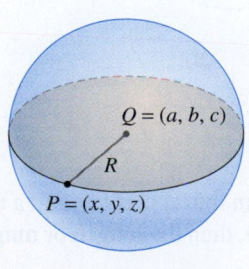

FIGURE 6 Sphere of radius R centered at (a, b, c).

Now consider the equation

$$(x - a)^2 + (y - b)^2 = R^2 \qquad \boxed{2}$$

In the xy-plane, Eq. (2) defines the circle of radius R with center (a, b). However, as an equation in $\mathbf{R}^3$, it defines the right circular cylinder of radius R whose central axis is the vertical line through $(a, b, 0)$ (Figure 7). Indeed, a point (x, y, z) satisfies Eq. (2) for any value of z if (x, y) lies on the circle. It is usually clear from the context which of the following is intended with Eq. (2):

In $\mathbf{R}^2$, a circle $= \{(x, y) : (x - a)^2 + (y - b)^2 = R^2\}$

In $\mathbf{R}^3$, a right circular cylinder $= \{(x, y, z) : (x - a)^2 + (y - b)^2 = R^2\}$

FIGURE 7 Right circular cylinder of radius R with axis through $(a, b, 0)$.

Equations of Spheres and Cylinders   An equation of the sphere in $\mathbf{R}^3$ of radius R centered at $Q = (a, b, c)$ is

$$(x - a)^2 + (y - b)^2 + (z - c)^2 = R^2 \qquad \boxed{3}$$

An equation of the right circular cylinder in $\mathbf{R}^3$ of radius R whose central axis is the vertical line through $(a, b, 0)$ is

$$(x - a)^2 + (y - b)^2 = R^2 \qquad \boxed{4}$$

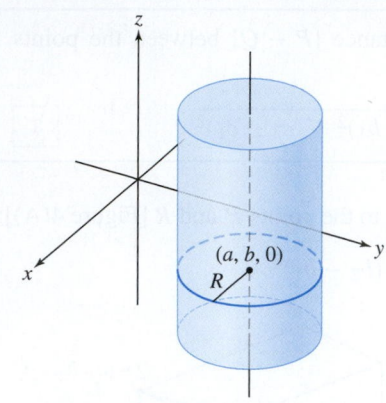

Maryam Mirzakhani (1977–2017) was an Iranian mathematician who studied a special class of surfaces known as hyperbolic. By examining different types of surface curves, she was able to prove important new theorems about the properties of hyperbolic surfaces. For her accomplishments, she was awarded the Fields Medal, the highest honor for a mathematician, in 2014.

EXAMPLE 1   Describe the sets of points defined by the following conditions:

(a) $x^2 + y^2 + z^2 = 4, \quad y \geq 0$ **(b)** $(x - 3)^2 + (y - 2)^2 = 1, \quad z \geq -1$

Solution

(a) The equation $x^2 + y^2 + z^2 = 4$ defines a sphere of radius 2 centered at the origin. The inequality $y \geq 0$ holds for points lying on the positive side of the xz-plane. We obtain the right hemisphere of radius 2 illustrated in Figure 8(A).

(b) The equation $(x - 3)^2 + (y - 2)^2 = 1$ defines a cylinder of radius 1 whose central axis is the vertical line through $(3, 2, 0)$. The part of the cylinder where $z \geq -1$ is the upper part of the cylinder, on and above the plane $z = -1$, as shown in Figure 8(B). ∎

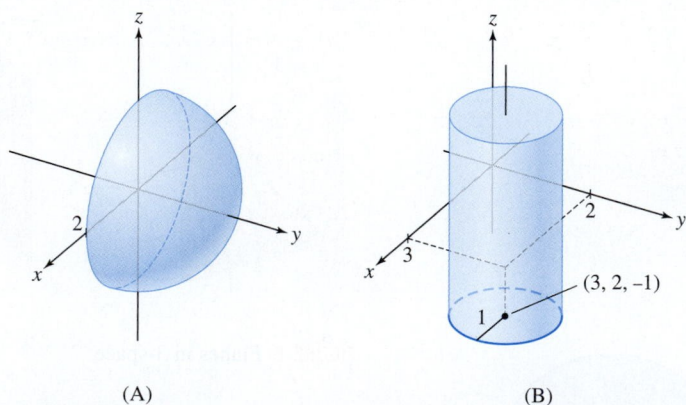

(A)   (B)

DF FIGURE 8 Hemisphere and upper cylinder.

Vectors in 3-Space

As in the plane, a vector $\mathbf{v} = \overrightarrow{PQ}$ in $\mathbf{R}^3$ is determined by an initial point P and a terminal point Q (Figure 9). If $P = (a_1, b_1, c_1)$ and $Q = (a_2, b_2, c_2)$, then the **length** or **magnitude** of $\mathbf{v} = \overrightarrow{PQ}$, denoted $\|\mathbf{v}\|$, is the distance from P to Q:

$$\|\mathbf{v}\| = \|\overrightarrow{PQ}\| = \sqrt{(a_2 - a_1)^2 + (b_2 - b_1)^2 + (c_2 - c_1)^2}$$

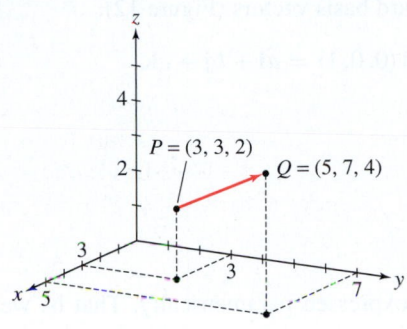

FIGURE 9 A vector $\overrightarrow{PQ}$ in 3-space.

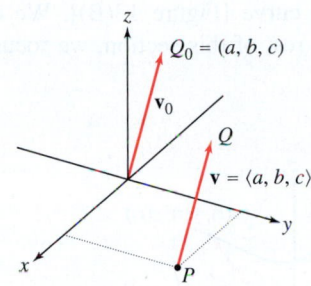

FIGURE 10 A vector $\mathbf{v}$ and its translation based at the origin.

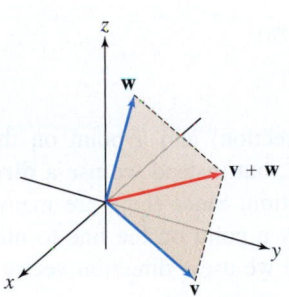

FIGURE 11 Vector addition is defined by the Parallelogram Law.

The terminology and basic properties discussed in the previous section carry over to $\mathbf{R}^3$ with little change.

- A vector $\mathbf{v}$ is said to undergo a **translation** if it is moved without changing direction or magnitude.
- Two vectors $\mathbf{v}$ and $\mathbf{w}$ are **equivalent** if $\mathbf{w}$ is a translation of $\mathbf{v}$; that is, $\mathbf{v}$ and $\mathbf{w}$ have the same length and direction.
- The **position vector** of a point Q_0 is the vector $\mathbf{v}_0 = \overrightarrow{OQ_0}$ based at the origin (Figure 10).
- A vector $\mathbf{v} = \overrightarrow{PQ}$ with components $\langle a, b, c \rangle$ is equivalent to the vector $\mathbf{v}_0 = \overrightarrow{OQ_0}$ based at the origin with $Q_0 = (a, b, c)$ (Figure 10).
- The **components** of $\mathbf{v} = \overrightarrow{PQ}$, where $P = (a_1, b_1, c_1)$ and $Q = (a_2, b_2, c_2)$, are the differences $a = a_2 - a_1, b = b_2 - b_1, c = c_2 - c_1$; that is,

$$\mathbf{v} = \overrightarrow{PQ} = \langle a, b, c \rangle = \langle a_2 - a_1, b_2 - b_1, c_2 - c_1 \rangle$$

For example, if $P = (3, -4, -4)$ and $Q = (2, 5, -1)$, then

$$\mathbf{v} = \overrightarrow{PQ} = \langle 2 - 3, 5 - (-4), -1 - (-4) \rangle = \langle -1, 9, 3 \rangle$$

- Two vectors are equivalent if and only if they have the same components.
- Vector addition and scalar multiplication are defined as in the two-dimensional case. Vector addition is defined by the Parallelogram Law (Figure 11).
- In terms of components, if $\mathbf{v} = \langle v_1, v_2, v_3 \rangle$ and $\mathbf{w} = \langle w_1, w_2, w_3 \rangle$, then

$$\lambda \mathbf{v} = \lambda \langle v_1, v_2, v_3 \rangle = \langle \lambda v_1, \lambda v_2, \lambda v_3 \rangle$$

$$\mathbf{v} + \mathbf{w} = \langle v_1, v_2, v_3 \rangle + \langle w_1, w_2, w_3 \rangle = \langle v_1 + w_1, v_2 + w_2, v_3 + w_3 \rangle$$

- Two nonzero vectors $\mathbf{v}$ and $\mathbf{w}$ are **parallel** if $\mathbf{v} = \lambda \mathbf{w}$ for some scalar λ.
- Vector addition is commutative, is associative, and satisfies the distributive property with respect to scalar multiplication (Theorem 1 in Section 12.1).

EXAMPLE 2 **Vector Calculations** Given $\mathbf{v} = \langle 3, -1, 2 \rangle$ and $\mathbf{w} = \langle 4, 6, -8 \rangle$, determine the following:

(a) $\|\mathbf{v}\|$,

(b) A unit vector in the direction of $\mathbf{v}$,

(c) $6\mathbf{v} - \frac{1}{2}\mathbf{w}$,

(d) Whether $\mathbf{v}$ and $\mathbf{w}$ are parallel or not.

Solution

(a) $\|\mathbf{v}\| = \sqrt{3^2 + (-1)^2 + 2^2} = \sqrt{14}$

(b) A unit vector in the direction of $\mathbf{v}$ is $\frac{1}{\sqrt{14}} \langle 3, -1, 2 \rangle = \left\langle \frac{3}{\sqrt{14}}, \frac{-1}{\sqrt{14}}, \frac{2}{\sqrt{14}} \right\rangle$.

(c) $6\mathbf{v} - \frac{1}{2}\mathbf{w} = 6 \langle 3, -1, 2 \rangle - \frac{1}{2} \langle 4, 6, -8 \rangle = \langle 18, -6, 12 \rangle - \langle 2, 3, -4 \rangle = \langle 16, -9, 16 \rangle$

(d) Is there a scalar λ such that $\mathbf{v} = \lambda \mathbf{w}$; that is, such that $\langle 3, -1, 2 \rangle = \lambda \langle 4, 6, -8 \rangle$? Considering components, this requires that

$$3 = \lambda 4, \quad -1 = \lambda 6, \quad 2 = \lambda(-8)$$

The first equation implies that $\lambda = \frac{3}{4}$, but this value does not satisfy either of the other two equations. So, there is no λ satisfying $\mathbf{v} = \lambda \mathbf{w}$, implying that the vectors are not parallel. ∎

The **standard basis vectors** in $\mathbf{R}^3$ are

$$\mathbf{i} = \langle 1, 0, 0 \rangle, \qquad \mathbf{j} = \langle 0, 1, 0 \rangle, \qquad \mathbf{k} = \langle 0, 0, 1 \rangle$$

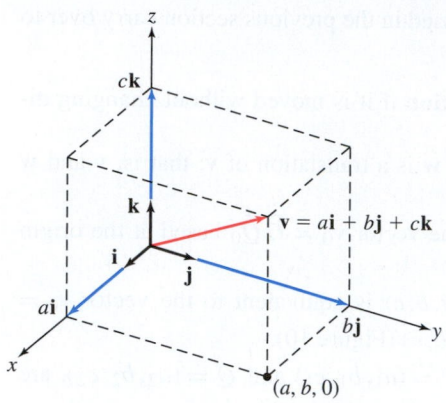

FIGURE 12 Writing $\mathbf{v} = \langle a, b, c \rangle$ as the sum $a\mathbf{i} + b\mathbf{j} + c\mathbf{k}$.

Every vector is a **linear combination** of the standard basis vectors (Figure 12):

$$\langle a, b, c \rangle = a \langle 1, 0, 0 \rangle + b \langle 0, 1, 0 \rangle + c \langle 0, 0, 1 \rangle = a\mathbf{i} + b\mathbf{j} + c\mathbf{k}$$

For example, $\langle -9, -4, 17 \rangle = -9\mathbf{i} - 4\mathbf{j} + 17\mathbf{k}$.

Curves and Lines

Most curves in 3-space that we consider will be expressed parametrically. That is, we will describe a curve using a set of three equations for $x(t)$, $y(t)$, $z(t)$ that we can think of as representing the coordinates of a particle traveling through space and tracing out a curve [Figure 13(A)]. Alternatively, taken together, the equations for $x(t)$, $y(t)$, $z(t)$ form the components of a vector $\mathbf{r}(t) = \langle x(t), y(t), z(t) \rangle$ with base at the origin. In this instance, the tips of the vectors trace out the curve [Figure 13(B)]. We consider general curves in 3-space in Chapter 13. For the rest of this section, we focus on lines.

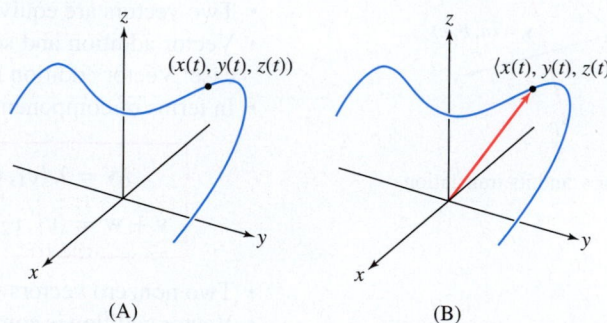

FIGURE 13 A curve is traced by a point (A) or by the tip of a vector (B).

(A)   (B)

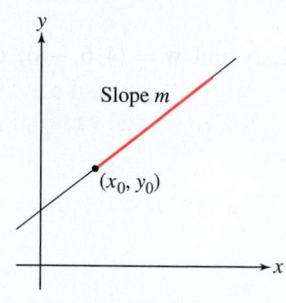

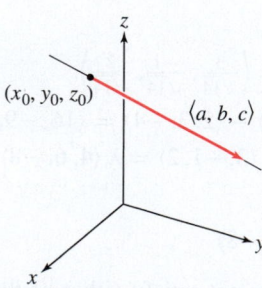

FIGURE 14 A line is determined by a point and slope in the plane, by a point and direction vector in 3-space.

In the plane, a line is identified by a slope (direction) and a point on the line (Figure 14). The same idea carries over to lines in $\mathbf{R}^3$, but instead we use a **direction vector** $\langle a, b, c \rangle$ parallel to the line to determine direction. Since there are many lines in any given direction (all parallel) we need to specify a point on the line to uniquely determine the line. Thus, to represent a line in 3-space we use a direction vector and a point on the line.

Given a direction vector $\mathbf{v} = \langle a, b, c \rangle$ for a line $\mathcal{L}$, and a point $P_0 = (x_0, y_0, z_0)$ on it, we first describe $\mathcal{L}$ geometrically, and then translate that to two algebraic representations, one as a vector equation and another as parametric equations in x, y, z.

Geometric Description of a Line The line $\mathcal{L}$ through $P_0 = (x_0, y_0, z_0)$ and parallel to $\mathbf{v} = \langle a, b, c \rangle$ consists of the tips of all vectors based at P_0 that are parallel to $\mathbf{v}$. For example, the tips of the vectors $\mathbf{v}$, $\mathbf{v}_1$, $\mathbf{v}_2$, and $\mathbf{v}_3$ in Figure 15 all lie on $\mathcal{L}$.

Let $P = (x, y, z)$ represent an arbitrary point on $\mathcal{L}$. By the Geometric Description of $\mathcal{L}$, the vector based at P_0 with tip at P is parallel to $\mathbf{v}$. That is, the vector

$$\overrightarrow{P_0P} = \langle x - x_0, y - y_0, z - z_0 \rangle$$

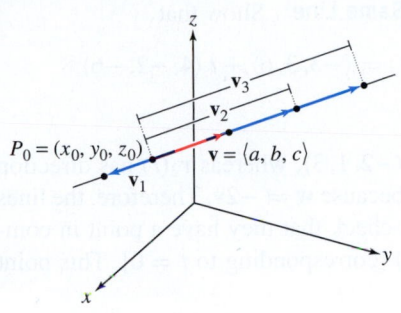

FIGURE 15

must be parallel to $\mathbf{v}$, a relationship that can be expressed as $\overrightarrow{P_0P} = t\mathbf{v}$ for some real number t. Thus,

$$\langle x - x_0, y - y_0, z - z_0 \rangle = t \langle a, b, c \rangle \quad \text{or}$$

$$\langle x, y, z \rangle = \langle x_0, y_0, z_0 \rangle + t \langle a, b, c \rangle$$

Now, letting $\mathbf{r}(t) = \langle x(t), y(t), z(t) \rangle$ and $\mathbf{r}_0 = \langle x_0, y_0, z_0 \rangle$, we have

---

Equations of a Line   The line $\mathcal{L}$ through point $P_0 = (x_0, y_0, z_0)$ in the direction of vector $\mathbf{v} = \langle a, b, c \rangle$ is described by

Vector parametrization $\mathbf{r}(t)$ of $\mathcal{L}$:

$$\mathbf{r}(t) = \mathbf{r}_0 + t\mathbf{v} = \langle x_0, y_0, z_0 \rangle + t \langle a, b, c \rangle \qquad \boxed{5}$$

where $\mathbf{r}_0 = \overrightarrow{OP_0} = \langle x_0, y_0, z_0 \rangle$.

Parametric equations of $\mathcal{L}$:

$$x(t) = x_0 + at, \quad y(t) = y_0 + bt, \quad z(t) = z_0 + ct \qquad \boxed{6}$$

The parameter t takes on values $-\infty < t < \infty$.

EXAMPLE 3 Find a vector parametrization and parametric equations for the line through $P_0 = (3, -1, 4)$ with direction vector $\mathbf{v} = \langle 2, 1, 7 \rangle$.

Solution By Eq. (5), the following is a vector parametrization:

$$\mathbf{r}(t) = \underbrace{\langle 3, -1, 4 \rangle}_{\text{The vector } \overrightarrow{OP_0}} + \underbrace{t \langle 2, 1, 7 \rangle}_{\text{Direction vector } \mathbf{v}} = \langle 3 + 2t, -1 + t, 4 + 7t \rangle$$

The corresponding parametric equations are $x = 3 + 2t, y = -1 + t, z = 4 + 7t$. ■

EXAMPLE 4   Parametric Equations of the Line Through Two Points   Find parametric equations for the line through $P = (1, 0, 2)$ and $Q = (2, 5, -1)$. Use them to find parametric equations for the line segment between P and Q.

Solution We can take our direction vector to be $\mathbf{v} = \overrightarrow{PQ} = \langle 2 - 1, 5 - 0, -1 - 2 \rangle = \langle 1, 5, -3 \rangle$.
Hence, we obtain

$$\mathbf{r}(t) = \langle 1, 0, 2 \rangle + t \langle 1, 5, -3 \rangle = \langle 1 + t, 5t, 2 - 3t \rangle$$

Thus, the parametric equations for the line are $x = 1 + t$, $y = 5t$, $z = 2 - 3t$, where $-\infty < t < \infty$.

To obtain parametric equations for the line segment between P and Q, we note that $\mathbf{r}(0) = \langle 1, 0, 2 \rangle = \overrightarrow{OP}$ and $\mathbf{r}(1) = \langle 2, 5, -1 \rangle = \overrightarrow{OQ}$. Therefore, if we use the same parametric equations, but restrict t so that $0 \leq t \leq 1$, we obtain parametric equations for the line segment. ■

We can think of a curve as a road. A parametrization describes a trip on the road. Different parametrizations can describe different trips while the road stays fixed.

The parametrization of a line $\mathcal{L}$ is not unique. We are free to choose any point P_0 on $\mathcal{L}$ and we may replace a direction vector $\mathbf{v}$ by any nonzero scalar multiple $\lambda \mathbf{v}$.

Two lines in $\mathbf{R}^3$ coincide if they are parallel and pass through a common point, so we can always check whether two parametrizations describe the same line, as in the following example.

EXAMPLE 5 **Different Parametrizations of the Same Line** Show that

$$\mathbf{r}_1(t) = \langle 1, 1, 0 \rangle + t \langle -2, 1, 3 \rangle \quad \text{and} \quad \mathbf{r}_2(t) = \langle -3, 3, 6 \rangle + t \langle 4, -2, -6 \rangle$$

parametrize the same line.

Solution The line $\mathbf{r}_1(t)$ has direction vector $\mathbf{v} = \langle -2, 1, 3 \rangle$, whereas $\mathbf{r}_2(t)$ has direction vector $\mathbf{w} = \langle 4, -2, -6 \rangle$. These vectors are parallel because $\mathbf{w} = -2\mathbf{v}$. Therefore, the lines described by $\mathbf{r}_1(t)$ and $\mathbf{r}_2(t)$ are parallel. We must check that they have a point in common. Choose any point on $\mathbf{r}_1(t)$, say, $P = (1, 1, 0)$ [corresponding to $t = 0$]. This point lies on $\mathbf{r}_2(t)$ if there is a value of t such that

$$\langle 1, 1, 0 \rangle = \langle -3, 3, 6 \rangle + t \langle 4, -2, -6 \rangle \qquad \boxed{7}$$

This yields three equations:

$$1 = -3 + 4t, \qquad 1 = 3 - 2t, \qquad 0 = 6 - 6t$$

All three are satisfied with $t = 1$. Therefore, P also lies on $\mathbf{r}_2(t)$. We conclude that $\mathbf{r}_1(t)$ and $\mathbf{r}_2(t)$ parametrize the same line. If Eq. (7) had no solution, we would conclude that $\mathbf{r}_1(t)$ and $\mathbf{r}_2(t)$ are parallel but do not coincide. ■

EXAMPLE 6 **Intersection of Two Lines** Determine whether the following two lines intersect:

$$\mathbf{r}_1(t) = \langle 1, 0, 1 \rangle + t \langle 3, 3, 5 \rangle$$

$$\mathbf{r}_2(t) = \langle 3, 6, 1 \rangle + t \langle 4, -2, 7 \rangle$$

Solution The two lines intersect if there exist parameter values t_1 and t_2 such that $\mathbf{r}_1(t_1) = \mathbf{r}_2(t_2)$—that is, if

$$\langle 1, 0, 1 \rangle + t_1 \langle 3, 3, 5 \rangle = \langle 3, 6, 1 \rangle + t_2 \langle 4, -2, 7 \rangle \qquad \boxed{8}$$

This is equivalent to three equations for the components:

$$x = 1 + 3t_1 = 3 + 4t_2, \qquad y = 3t_1 = 6 - 2t_2, \qquad z = 1 + 5t_1 = 1 + 7t_2 \qquad \boxed{9}$$

Let's solve the first two equations for t_1 and t_2. Subtracting the second equation from the first, we get $1 = 6t_2 - 3$ or $t_2 = \frac{2}{3}$. Using this value in the second equation, we get $t_1 = 2 - \frac{2}{3}t_2 = \frac{14}{9}$. The values $t_1 = \frac{14}{9}$ and $t_2 = \frac{2}{3}$ satisfy the first two equations, and thus $\mathbf{r}_1(t_1)$ and $\mathbf{r}_2(t_2)$ have the same x- and y-coordinates (Figure 16). However, they do not have the same z-coordinates because t_1 and t_2 do not satisfy the third equation in (9):

$$1 + 5\left(\frac{14}{9}\right) \neq 1 + 7\left(\frac{2}{3}\right)$$

Therefore, Eq. (8) has no solution and the lines do not intersect. ■

CAUTION We use different parameter values, t_1 and t_2, in Eq. 8 because an intersection need not occur at the same parameter value, only at the same point.

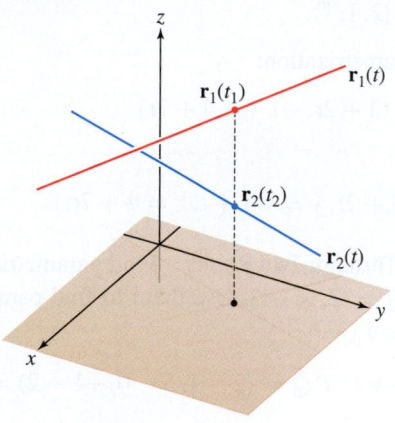

FIGURE 16 The lines $\mathbf{r}_1(t)$ and $\mathbf{r}_2(t)$ do not intersect, but the particular points $\mathbf{r}_1(t_1)$ and $\mathbf{r}_2(t_2)$ have the same x- and y-coordinates.

12.2 SUMMARY

- The axes in $\mathbf{R}^3$ are labeled so that they satisfy the *right-hand rule*: When the fingers of your right hand curl from the positive x-axis toward the positive y-axis, your thumb points in the positive z-direction (Figure 17).

- Sphere of radius R and center (a, b, c): $(x - a)^2 + (y - b)^2 + (z - c)^2 = R^2$
- Cylinder of radius R with central axis through $(a, b, 0)$: $(x - a)^2 + (y - b)^2 = R^2$
- The notation, terminology, and basic properties for vectors in the plane carry over to vectors in $\mathbf{R}^3$.
- The length (or magnitude) of $\mathbf{v} = \overrightarrow{PQ}$, where $P = (a_1, b_1, c_1)$ and $Q = (a_2, b_2, c_2)$, is

$$\|\mathbf{v}\| = \|\overrightarrow{PQ}\| = \sqrt{(a_2 - a_1)^2 + (b_2 - b_1)^2 + (c_2 - c_1)^2}$$

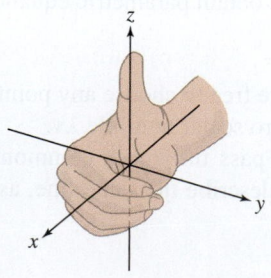

FIGURE 17

- The line through $P_0 = (x_0, y_0, z_0)$ with direction vector $\mathbf{v} = \langle a, b, c \rangle$:
  - Geometrically: The line formed by the tips of all vectors based at P_0 that are parallel to $\mathbf{v}$.
  - Algebraically:

$$\text{Vector parametrization:} \quad \mathbf{r}(t) = \overrightarrow{OP_0} + t\mathbf{v} = \langle x_0, y_0, z_0 \rangle + t\langle a, b, c \rangle$$

$$\text{Parametric equations:} \quad x = x_0 + at, \quad y = y_0 + bt, \quad z = z_0 + ct$$

- To obtain the line through $P = (a_1, b_1, c_1)$ and $Q = (a_2, b_2, c_2)$, take direction vector $\mathbf{v} = \overrightarrow{PQ} = \langle a_2 - a_1, b_2 - b_1, c_2 - c_1 \rangle$, and use the vector parametrization or the parametric equations for a line. The segment $\overline{PQ}$ is parametrized by $\mathbf{r}(t)$ for $0 \leq t \leq 1$.

12.2 EXERCISES

Preliminary Questions

1. What is the terminal point of the vector $\mathbf{v} = \langle 3, 2, 1 \rangle$ based at the point $P = (1, 1, 1)$?

(a) $\langle 3, 2, 1 \rangle$ **(b)** $\langle 1, 1, 1 \rangle$ **(c)** $\langle 2, 1, 0 \rangle$

5. How many different direction vectors does a line have?

2. What are the components of the vector $\mathbf{v} = \langle 3, 2, 1 \rangle$ based at the point $P = (1, 1, 1)$?

6. True or false? If $\mathbf{v}$ is a direction vector for a line $\mathcal{L}$, then $-\mathbf{v}$ is also a direction vector for $\mathcal{L}$.

3. If $\mathbf{v} = -3\mathbf{w}$, then (choose the correct answer):

(a) $\mathbf{v}$ and $\mathbf{w}$ are parallel.

(b) $\mathbf{v}$ and $\mathbf{w}$ point in the same direction.

7. What is the radius of the sphere $x^2 + y^2 + z^2 = 5$?

8. Which of the following points are on the cylinder $(x - 1)^2 + y^2 = 1$?

4. Which of the following is a direction vector for the line through $P = (3, 2, 1)$ and $Q = (1, 1, 1)$?

(a) $(1, 0, 0)$ **(b)** $(0, 0, 0)$ **(c)** $(0, 0, -1)$

(d) $(0, -1, 1)$ **(e)** $(1, -1, 1)$ **(f)** $(1, 1, 0)$

Exercises

1. Sketch the vector $\mathbf{v} = \langle 1, 3, 2 \rangle$ and compute its length.

2. Let $\mathbf{v} = \overrightarrow{P_0 Q_0}$, where $P_0 = (1, -2, 5)$ and $Q_0 = (0, 1, -4)$. Which of the following vectors (with tail P and head Q) are equivalent to $\mathbf{v}$?

	$\mathbf{v}_1$	$\mathbf{v}_2$	$\mathbf{v}_3$	$\mathbf{v}_4$
P	$(1, 2, 4)$	$(1, 5, 4)$	$(0, 0, 0)$	$(2, 4, 5)$
Q	$(0, 5, -5)$	$(0, -8, 13)$	$(-1, 3, -9)$	$(1, 7, 4)$

3. Sketch the vector $\mathbf{v} = \langle 1, 1, 0 \rangle$ based at $P = (0, 1, 1)$. Describe this vector in the form $\overrightarrow{PQ}$ for some point Q, and sketch the vector $\mathbf{v}_0$ based at the origin equivalent to $\mathbf{v}$.

4. Determine whether the coordinate systems (A)–(C) in Figure 18 satisfy the right-hand rule.

In Exercises 5–8, find the components of the vector $\overrightarrow{PQ}$.

5. $P = (1, 0, 1)$, $Q = (2, 1, 0)$

6. $P = (-3, -4, 2)$, $Q = (1, -4, 3)$

7. $P = (4, 6, 0)$, $Q = \left(-\frac{1}{2}, \frac{9}{2}, 1\right)$

8. $P = \left(-\frac{1}{2}, \frac{9}{2}, 1\right)$, $Q = (4, 6, 0)$

In Exercises 9–12, let $R = (1, 4, 3)$.

9. Calculate the length of $\overrightarrow{OR}$.

10. Find the point Q such that $\mathbf{v} = \overrightarrow{RQ}$ has components $\langle 4, 1, 1 \rangle$, and sketch $\mathbf{v}$.

11. Find the point P such that $\mathbf{w} = \overrightarrow{PR}$ has components $\langle 3, -2, 3 \rangle$, and sketch $\mathbf{w}$.

12. Find the components of $\mathbf{u} = \overrightarrow{PR}$, where $P = (1, 2, 2)$.

13. Let $\mathbf{v} = \langle 4, 8, 12 \rangle$. Which of the following vectors is parallel to $\mathbf{v}$? Which point in the same direction?

(a) $\langle 2, 4, 6 \rangle$ **(b)** $\langle -1, -2, 3 \rangle$

(c) $\langle -7, -14, -21 \rangle$ **(d)** $\langle 6, 10, 14 \rangle$

In Exercises 14–17, determine whether $\overrightarrow{AB}$ is equivalent to $\overrightarrow{PQ}$.

14. $A = (1, 1, 1)$ $B = (3, 3, 3)$
      $P = (1, 4, 5)$ $Q = (3, 6, 7)$

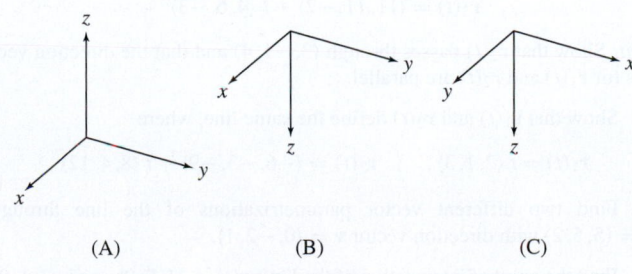

(A)   **(B)**   **(C)**

FIGURE 18

15. $A = (1, 4, 1)$ $B = (-2, 2, 0)$
$P = (2, 5, 7)$ $Q = (-3, 2, 1)$

16. $A = (0, 0, 0)$ $B = (-4, 2, 3)$
$P = (4, -2, -3)$ $Q = (0, 0, 0)$

17. $A = (1, 1, 0)$ $B = (3, 3, 5)$
$P = (2, -9, 7)$ $Q = (4, -7, 13)$

In Exercises 18–23, calculate the linear combinations.

18. $5 \langle 2, 2, -3 \rangle + 3 \langle 1, 7, 2 \rangle$

19. $-2 \langle 8, 11, 3 \rangle + 4 \langle 2, 1, 1 \rangle$

20. $6(4\mathbf{j} + 2\mathbf{k}) - 3(2\mathbf{i} + 7\mathbf{k})$

21. $\frac{1}{2} \langle 4, -2, 8 \rangle - \frac{1}{3} \langle 12, 3, 3 \rangle$

22. $5(\mathbf{i} + 2\mathbf{j}) - 3(2\mathbf{j} + \mathbf{k}) + 7(2\mathbf{k} - \mathbf{i})$

23. $4 \langle 6, -1, 1 \rangle - 2 \langle 1, 0, -1 \rangle + 3 \langle -2, 1, 1 \rangle$

In Exercises 24–27, determine whether or not the two vectors are parallel.

24. $\mathbf{u} = \langle 1, -2, 5 \rangle$, $\mathbf{v} = \langle -2, 4, -10 \rangle$

25. $\mathbf{u} = \langle 4, 2, -6 \rangle$, $\mathbf{v} = \langle 2, -1, 3 \rangle$

26. $\mathbf{u} = \langle 4, 2, -6 \rangle$, $\mathbf{v} = \langle 2, 1, 3 \rangle$

27. $\mathbf{u} = \langle -3, 1, 4 \rangle$, $\mathbf{v} = \langle 6, -2, 8 \rangle$

In Exercises 28–31, find the given vector.

28. $\mathbf{e_v}$, where $\mathbf{v} = \langle 1, 1, 2 \rangle$

29. $\mathbf{e_w}$, where $\mathbf{w} = \langle 4, -2, -1 \rangle$

30. Unit vector in the direction of $\mathbf{u} = \langle 1, 0, 7 \rangle$

31. Unit vector in the direction opposite to $\mathbf{v} = \langle -4, 4, 2 \rangle$

32. Sketch the following vectors, and find their components and lengths:

(a) $4\mathbf{i} + 3\mathbf{j} - 2\mathbf{k}$

(b) $\mathbf{i} + \mathbf{j} + \mathbf{k}$

(c) $4\mathbf{j} + 3\mathbf{k}$

(d) $12\mathbf{i} + 8\mathbf{j} - \mathbf{k}$

In Exercises 33–36, describe the surface.

33. $x^2 + y^2 + (z - 2)^2 = 4$, with $z \geq 2$

34. $x^2 + y^2 + z^2 = 9$, with $x, y, z \geq 0$

35. $x^2 + y^2 = 7$, with $|z| \leq 7$

36. $x^2 + y^2 = 4$, with $y, z \geq 0$

In Exercises 37–42, give an equation for the indicated surface.

37. The sphere of radius 3 centered at $(0, 0, -3)$

38. The sphere centered at the origin passing through $(1, 2, -3)$

39. The sphere centered at $(6, -3, 11)$ passing through $(0, 1, -4)$

40. The sphere with diameter $\overline{PQ}$ where $P = (1, 1, -3)$ and $Q = (1, 7, 1)$

41. The cylinder passing through the origin with the vertical line through $(1, -1, 0)$ as its central axis

42. The cylinder passing through $(0, 2, 1)$ with the vertical line through $(1, 0, 0)$ as its central axis

In Exercises 43–50, find a vector parametrization for the line with the given description.

43. Passes through $P = (1, 2, -8)$, direction vector $\mathbf{v} = \langle 2, 1, 3 \rangle$

44. Passes through $P = (4, 0, 8)$, direction vector $\mathbf{v} = \langle 1, 0, 1 \rangle$

45. Passes through $P = (4, 0, 8)$, direction vector $\mathbf{v} = 7\mathbf{i} + 4\mathbf{k}$

46. Passes through O, direction vector $\mathbf{v} = \langle 3, -1, -4 \rangle$

47. Passes through $(1, 1, 1)$ and $(3, -5, 2)$

48. Passes through $(-2, 0, -2)$ and $(4, 3, 7)$

49. Passes through O and $(4, 1, 1)$

50. Passes through $(1, 1, 1)$ parallel to the line through $(2, 0, -1)$ and $(4, 1, 3)$

In Exercises 51–54, find parametric equations for the lines with the given description.

51. Perpendicular to the xy-plane, passes through the origin

52. Perpendicular to the yz-plane, passes through $(0, 0, 2)$

53. Parallel to the line through $(1, 1, 0)$ and $(0, -1, -2)$, passes through $(0, 0, 4)$

54. Passes through $(1, -1, 0)$ and $(0, -1, 2)$

55. Which of the following is a parametrization of the line through $P = (4, 9, 8)$ perpendicular to the xz-plane (Figure 19)?

(a) $\mathbf{r}(t) = \langle 4, 9, 8 \rangle + t \langle 1, 0, 1 \rangle$

(b) $\mathbf{r}(t) = \langle 4, 9, 8 \rangle + t \langle 0, 0, 1 \rangle$

(c) $\mathbf{r}(t) = \langle 4, 9, 8 \rangle + t \langle 0, 1, 0 \rangle$

(d) $\mathbf{r}(t) = \langle 4, 9, 8 \rangle + t \langle 1, 1, 0 \rangle$

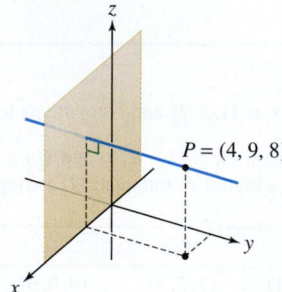

FIGURE 19

56. Find a parametrization of the line through $P = (4, 9, 8)$ perpendicular to the yz-plane.

57. Show that $\mathbf{r}_1(t)$ and $\mathbf{r}_2(t)$ define the same line, where

$$\mathbf{r}_1(t) = \langle 3, -1, 4 \rangle + t \langle 8, 12, -6 \rangle$$

$$\mathbf{r}_2(t) = \langle 11, 11, -2 \rangle + t \langle 4, 6, -3 \rangle$$

Hint: Show that $\mathbf{r}_2(t)$ passes through $(3, -1, 4)$ and that the direction vectors for $\mathbf{r}_1(t)$ and $\mathbf{r}_2(t)$ are parallel.

58. Show that $\mathbf{r}_1(t)$ and $\mathbf{r}_2(t)$ define the same line, where

$$\mathbf{r}_1(t) = t \langle 2, 1, 3 \rangle, \qquad \mathbf{r}_2(t) = \langle -6, -3, -9 \rangle + t \langle 8, 4, 12 \rangle$$

59. Find two different vector parametrizations of the line through $P = (5, 5, 2)$ with direction vector $\mathbf{v} = \langle 0, -2, 1 \rangle$.

60. Find the point of intersection of the lines $\mathbf{r}(t) = \langle 1, 0, 0 \rangle + t \langle -3, 1, 0 \rangle$ and $\mathbf{s}(t) = \langle 0, 1, 1 \rangle + t \langle 2, 0, 1 \rangle$.

61. Show that the lines $\mathbf{r}_1(t) = \langle -1, 2, 2 \rangle + t \langle 4, -2, 1 \rangle$ and $\mathbf{r}_2(t) = \langle 0, 1, 1 \rangle + t \langle 2, 0, 1 \rangle$ do not intersect.

62. Determine whether the lines $\mathbf{r}_1(t) = \langle 2, 1, 1 \rangle + t \langle -4, 0, 1 \rangle$ and $\mathbf{r}_2(s) = \langle -4, 1, 5 \rangle + s \langle 2, 1, -2 \rangle$ intersect, and if so, find the point of intersection.

63. Determine whether the lines $\mathbf{r}_1(t) = \langle 0, 1, 1 \rangle + t \langle 1, 1, 2 \rangle$ and $\mathbf{r}_2(s) = \langle 2, 0, 3 \rangle + s \langle 1, 4, 4 \rangle$ intersect, and if so, find the point of intersection.

64. Find the intersection of the lines $\mathbf{r}_1(t) = \langle -1, 1 \rangle + t \langle 2, 4 \rangle$ and $\mathbf{r}_2(s) = \langle 2, 1 \rangle + s \langle -1, 6 \rangle$ in the plane.

65. A meteor follows a trajectory $\mathbf{r}(t) = \langle 2, 1, 4 \rangle + t \langle 3, 2, -1 \rangle$ km with t in minutes, near the surface of the earth, which is represented by the xy-plane. Determine at what time the meteor hits the ground.

66. A laser's beam shines along the ray given by $\mathbf{r}_1(t) = \langle 1, 2, 4 \rangle + t \langle 2, 1, -1 \rangle$ for $t \geq 0$. A second laser's beam shines along the ray given by $\mathbf{r}_2(s) = \langle 6, 3, -1 \rangle + s \langle -5, 2, c \rangle$ for $s \geq 0$, where the value of c allows for the adjustment of the z-coordinate of its direction vector. Find the value of c that will make the two beams intersect.

67. The line with vector parametrization $\mathbf{r}(t) = \langle 3, 1, -4 \rangle + t \langle -2, -2, 3 \rangle$ intersects the sphere $(x - 1)^2 + (y + 3)^2 + z^2 = 8$ in two points. Find them. *Hint:* Determine t such that the point $(x(t), y(t), z(t))$ satisfies the equation of the sphere, and then find the corresponding points on the line.

68. Show that the line with vector parametrization $\mathbf{r}(t) = \langle 3, 5, 6 \rangle + t \langle 1, -2, -1 \rangle$ does not intersect the sphere of radius 5 centered at the origin.

69. Find the components of the vector $\mathbf{v}$ whose tail and head are the midpoints of segments $\overline{AC}$ and $\overline{BC}$ in Figure 20. [Note that the midpoint of (a_1, a_2, a_3) and (b_1, b_2, b_3) is $\left(\dfrac{a_1 + b_1}{2}, \dfrac{a_2 + b_2}{2}, \dfrac{a_3 + b_3}{2} \right)$.]

70. Find the components of the vector $\mathbf{w}$ whose tail is C and head is the midpoint of $\overline{AB}$ in Figure 20.

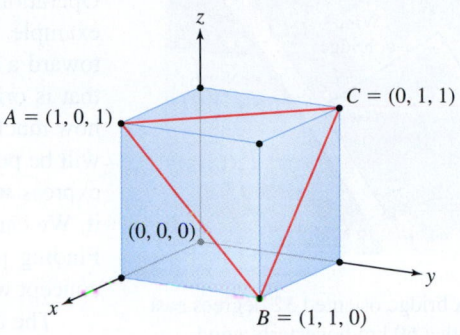

FIGURE 20

71. A box that weighs 1000 kg is hanging from a crane at the dock. The crane has a square 20 m by 20 m framework as in Figure 21, with four cables, each of the same length, supporting the box. The box hangs 10 m below the level of the framework. Find the magnitude of the force acting on each cable.

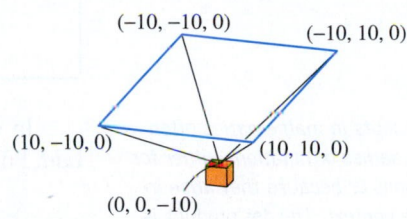

FIGURE 21

Further Insights and Challenges

*In Exercises 72–78, we consider the equations of a line in **symmetric form**, when $a \neq 0$, $b \neq 0$, $c \neq 0$.*

$$\frac{x - x_0}{a} = \frac{y - y_0}{b} = \frac{z - z_0}{c} \qquad \boxed{10}$$

72. Let $\mathcal{L}$ be the line through $P_0 = (x_0, y_0, z_0)$ with direction vector $\mathbf{v} = \langle a, b, c \rangle$. Show that $\mathcal{L}$ is defined by the symmetric equations (10). *Hint:* Use the vector parametrization to show that every point on $\mathcal{L}$ satisfies (10).

73. Find the symmetric equations of the line through $P_0 = (-2, 3, 3)$ with direction vector $\mathbf{v} = \langle 2, 4, 3 \rangle$.

74. Find the symmetric equations of the line through $P = (1, 1, 2)$ and $Q = (-2, 4, 0)$.

75. Find the symmetric equations of the line

$$x = 3 + 2t, \quad y = 4 - 9t, \quad z = 12t$$

76. Find a vector parametrization for the line

$$\frac{x - 5}{9} = \frac{y + 3}{7} = z - 10$$

77. Find a vector parametrization for the line $\dfrac{x}{2} = \dfrac{y}{7} = \dfrac{z}{8}$.

78. Show that the line in the plane through (x_0, y_0) of slope m has symmetric equations

$$x - x_0 = \frac{y - y_0}{m}$$

79. A median of a triangle is a segment joining a vertex to the midpoint of the opposite side. Referring to Figure 22(A), prove that three medians of triangle ABC intersect at the terminal point P of the vector $\frac{1}{3}(\mathbf{u} + \mathbf{v} + \mathbf{w})$. The point P is the *centroid* of the triangle. *Hint:* Show, by parametrizing the segment $\overline{AA'}$, that P lies two-thirds of the way from A to A'. It will follow similarly that P lies on the other two medians.

80. A median of a tetrahedron is a segment joining a vertex to the centroid of the opposite face. The tetrahedron in Figure 22(B) has vertices at the origin and at the terminal points of vectors $\mathbf{u}$, $\mathbf{v}$, and $\mathbf{w}$. Show that the medians intersect at the terminal point of $\frac{1}{4}(\mathbf{u} + \mathbf{v} + \mathbf{w})$.

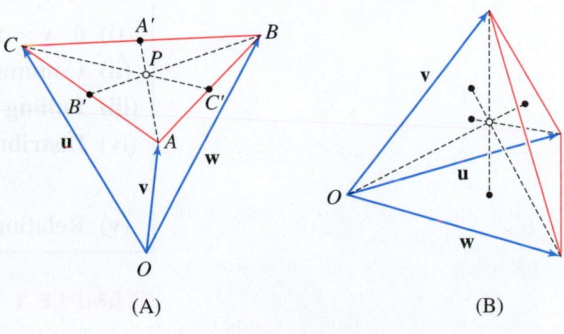

(A)   (B)

FIGURE 22

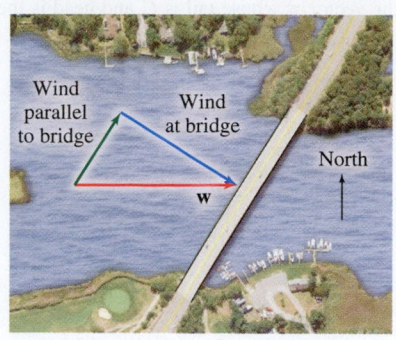

FIGURE 1 A bridge oriented 32 degrees east of north with a 60 km/h westerly wind.

Labels in figure: Wind parallel to bridge; Wind at bridge; North; **w**

12.3 Dot Product and the Angle Between Two Vectors

Operations on vectors are widely used in engineering and other scientific disciplines. For example, an operation called the dot product can be applied to analyze the wind blowing toward a bridge. Suppose a 60 km/h wind **w** is blowing from the west toward a bridge that is oriented 32 degrees east of north (Figure 1). A civil engineer needs to compute how much of the wind is blowing directly at the bridge to determine whether large trucks will be permitted on the bridge under such wind conditions. To analyze this problem we express **w** as a sum of vectors, one parallel to the bridge and a second perpendicular to it. We can compute the parallel part by projecting **w** onto a vector parallel to the bridge. Finding projections of one vector onto another is easily done with the dot product, a concept we introduce in this section.

The dot product is one of two important products that we define on pairs of vectors. The other, cross product, is introduced in the next section. Dot product is defined as follows:

> **DEFINITION** Dot Product The **dot product** $\mathbf{v} \cdot \mathbf{w}$ of two vectors
>
> $$\mathbf{v} = \langle v_1, v_2, v_3 \rangle, \qquad \mathbf{w} = \langle w_1, w_2, w_3 \rangle$$
>
> is the scalar defined by
>
> $$\mathbf{v} \cdot \mathbf{w} = v_1 w_1 + v_2 w_2 + v_3 w_3$$

Important concepts in mathematics often have multiple names or notations either for historical reasons or because they arise in more than one context. The dot product is also called the scalar product or inner product, and in many texts, $\mathbf{v} \cdot \mathbf{w}$ is denoted $(\mathbf{v}, \mathbf{w})$ or $\langle \mathbf{v}, \mathbf{w} \rangle$.

In words, to compute the dot product, *multiply the corresponding components and add.* For example,

$$\langle 2, 3, 1 \rangle \cdot \langle -4, 2, 5 \rangle = 2(-4) + 3(2) + 1(5) = -8 + 6 + 5 = 3$$

The dot product of vectors $\mathbf{v} = \langle v_1, v_2 \rangle$ and $\mathbf{w} = \langle w_1, w_2 \rangle$ in $\mathbf{R}^2$ is defined similarly:

$$\mathbf{v} \cdot \mathbf{w} = v_1 w_1 + v_2 w_2$$

We will see in a moment that the dot product is closely related to the angle between **v** and **w**. Before getting to this, we describe some elementary properties of dot products.

First, the dot product is *commutative*: $\mathbf{v} \cdot \mathbf{w} = \mathbf{w} \cdot \mathbf{v}$, because the components can be multiplied in either order. Second, the dot product of a vector with itself is the square of the length: If $\mathbf{v} = \langle v_1, v_2, v_3 \rangle$, then

$$\mathbf{v} \cdot \mathbf{v} = v_1 v_1 + v_2 v_2 + v_3 v_3 = v_1^2 + v_2^2 + v_3^2 = \|\mathbf{v}\|^2$$

The dot product appears in a very wide range of applications. For example, to determine how closely a Web document matches a search input, a dot product is used to develop a numerical score by which candidate documents can be ranked. (See The Anatomy of a Large-Scale Hypertextual Web Search Engine by Sergey Brin and Lawrence Page.)

The dot product also satisfies the Distributive Law and a scalar-multiplication property as summarized in the next theorem (see Exercises 94 and 95).

> **THEOREM 1** Properties of the Dot Product
>
> (i) $\mathbf{0} \cdot \mathbf{v} = \mathbf{v} \cdot \mathbf{0} = 0$
>
> (ii) **Commutativity:** $\mathbf{v} \cdot \mathbf{w} = \mathbf{w} \cdot \mathbf{v}$
>
> (iii) **Pulling out scalars:** $(\lambda \mathbf{v}) \cdot \mathbf{w} = \mathbf{v} \cdot (\lambda \mathbf{w}) = \lambda (\mathbf{v} \cdot \mathbf{w})$
>
> (iv) **Distributive Law:** $\mathbf{u} \cdot (\mathbf{v} + \mathbf{w}) = \mathbf{u} \cdot \mathbf{v} + \mathbf{u} \cdot \mathbf{w}$
>
> $$(\mathbf{v} + \mathbf{w}) \cdot \mathbf{u} = \mathbf{v} \cdot \mathbf{u} + \mathbf{w} \cdot \mathbf{u}$$
>
> (v) **Relation with length:** $\mathbf{v} \cdot \mathbf{v} = \|\mathbf{v}\|^2$

EXAMPLE 1 Verify the Distributive Law $\mathbf{u} \cdot (\mathbf{v} + \mathbf{w}) = \mathbf{u} \cdot \mathbf{v} + \mathbf{u} \cdot \mathbf{w}$ for

$$\mathbf{u} = \langle 4, 3, 3 \rangle, \qquad \mathbf{v} = \langle 1, 2, 2 \rangle, \qquad \mathbf{w} = \langle 3, -2, 5 \rangle$$

Solution We compute both sides and check that they are equal:

$$\mathbf{u} \cdot (\mathbf{v} + \mathbf{w}) = \langle 4,3,3 \rangle \cdot \big(\langle 1,2,2 \rangle + \langle 3,-2,5 \rangle \big)$$
$$= \langle 4,3,3 \rangle \cdot \langle 4,0,7 \rangle = 4(4) + 3(0) + 3(7) = 37$$
$$\mathbf{u} \cdot \mathbf{v} + \mathbf{u} \cdot \mathbf{w} = \langle 4,3,3 \rangle \cdot \langle 1,2,2 \rangle + \langle 4,3,3 \rangle \cdot \langle 3,-2,5 \rangle$$
$$= \big(4(1) + 3(2) + 3(2)\big) + \big(4(3) + 3(-2) + 3(5)\big)$$
$$= 16 + 21 = 37 \qquad\blacksquare$$

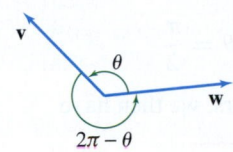

FIGURE 2 By convention, the angle θ between two vectors is chosen so that $0 \le \theta \le \pi$.

As mentioned earlier, the dot product $\mathbf{v} \cdot \mathbf{w}$ is related to the angle θ between $\mathbf{v}$ and $\mathbf{w}$. This angle θ is not uniquely defined because, as we see in Figure 2, both θ and $2\pi - \theta$ can serve as an angle between $\mathbf{v}$ and $\mathbf{w}$. Furthermore, any multiple of 2π may be added to θ. All of these angles have the same cosine, so it does not matter which angle we use in the next theorem. However, we shall adopt the following convention:

The angle between two vectors is chosen to satisfy $0 \le \theta \le \pi$.

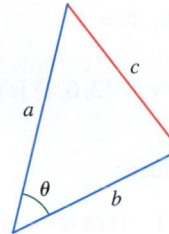

THEOREM 2 Dot Product and the Angle Let θ be the angle between two nonzero vectors $\mathbf{v}$ and $\mathbf{w}$. Then

$$\mathbf{v} \cdot \mathbf{w} = \|\mathbf{v}\| \, \|\mathbf{w}\| \cos \theta \quad \text{or} \quad \cos \theta = \frac{\mathbf{v} \cdot \mathbf{w}}{\|\mathbf{v}\| \, \|\mathbf{w}\|} \qquad \boxed{1}$$

Proof According to the Law of Cosines, the three sides of a triangle satisfy (Figure 3)

$$c^2 = a^2 + b^2 - 2ab \cos \theta$$

If two sides of the triangle are $\mathbf{v}$ and $\mathbf{w}$, then the third side can be expressed as $\mathbf{v} - \mathbf{w}$, as in the figure, and the Law of Cosines gives

$$\|\mathbf{v} - \mathbf{w}\|^2 = \|\mathbf{v}\|^2 + \|\mathbf{w}\|^2 - 2 \cos \theta \|\mathbf{v}\| \, \|\mathbf{w}\| \qquad \boxed{2}$$

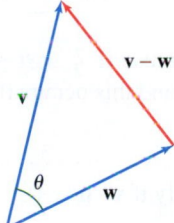

FIGURE 3

Now, by property (v) of Theorem 1 and the Distributive Law,

$$\|\mathbf{v} - \mathbf{w}\|^2 = (\mathbf{v} - \mathbf{w}) \cdot (\mathbf{v} - \mathbf{w}) = \mathbf{v} \cdot \mathbf{v} - 2\mathbf{v} \cdot \mathbf{w} + \mathbf{w} \cdot \mathbf{w}$$
$$= \|\mathbf{v}\|^2 + \|\mathbf{w}\|^2 - 2\mathbf{v} \cdot \mathbf{w} \qquad \boxed{3}$$

Comparing Eq. (2) and Eq. (3), we obtain $-2 \cos \theta \|\mathbf{v}\| \, \|\mathbf{w}\| = -2\mathbf{v} \cdot \mathbf{w}$, and Eq. (1) follows. $\qquad\blacksquare$

By definition of the arccosine, the angle $\theta = \cos^{-1} x$ is the angle in the interval $[0, \pi]$ satisfying $\cos \theta = x$. Thus, for nonzero vectors $\mathbf{v}$ and $\mathbf{w}$, we have

$$\cos \theta = \frac{\mathbf{v} \cdot \mathbf{w}}{\|\mathbf{v}\| \, \|\mathbf{w}\|} \quad \text{or} \quad \theta = \cos^{-1}\left(\frac{\mathbf{v} \cdot \mathbf{w}}{\|\mathbf{v}\| \, \|\mathbf{w}\|} \right)$$

EXAMPLE 2 Find the angle θ between the vectors $\mathbf{v} = \langle 3,6,2 \rangle$ and $\mathbf{w} = \langle 4,2,4 \rangle$ shown in Figure 4.

Solution Compute $\cos \theta$ using the dot product:

$$\|\mathbf{v}\| = \sqrt{3^2 + 6^2 + 2^2} = \sqrt{49} = 7, \qquad \|\mathbf{w}\| = \sqrt{4^2 + 2^2 + 4^2} = \sqrt{36} = 6$$
$$\cos \theta = \frac{\mathbf{v} \cdot \mathbf{w}}{\|\mathbf{v}\| \|\mathbf{w}\|} = \frac{\langle 3,6,2 \rangle \cdot \langle 4,2,4 \rangle}{7(6)} = \frac{3(4) + 6(2) + 2(4)}{42} = \frac{32}{42} = \frac{16}{21}$$

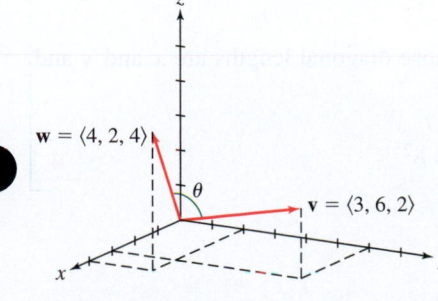

FIGURE 4

The angle itself is $\theta = \cos^{-1}\left(\frac{16}{21} \right) \approx 0.705$ radians (Figure 4). $\qquad\blacksquare$

The terms "orthogonal" and "perpendicular" are synonymous and are used interchangeably, although we usually use "orthogonal" when dealing with vectors.

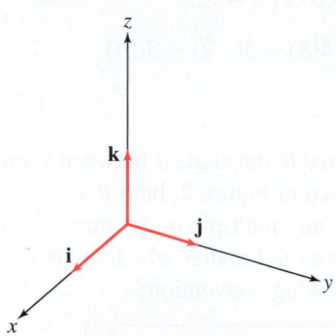

FIGURE 5 The standard basis vectors are mutually orthogonal and have length 1.

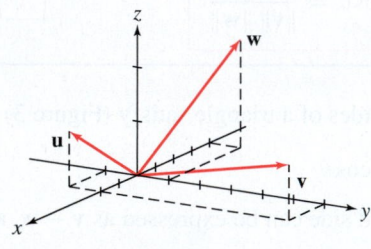

FIGURE 6 Vectors **v**, **w**, and **u** in Example 3.

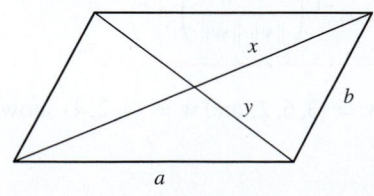

FIGURE 7

Two nonzero vectors **v** and **w** are called **perpendicular** or **orthogonal** if the angle between them is $\frac{\pi}{2}$. In this case, we write $\mathbf{v} \perp \mathbf{w}$.

We can use the dot product to test whether **v** and **w** are orthogonal. Because an angle θ between 0 and π satisfies $\cos\theta = 0$ if and only if $\theta = \frac{\pi}{2}$, we see that if **v** and **w** are nonzero, then

$$\mathbf{v} \cdot \mathbf{w} = \|\mathbf{v}\| \, \|\mathbf{w}\| \cos\theta = 0 \quad \Leftrightarrow \quad \theta = \frac{\pi}{2}$$

Defining the zero vector to be orthogonal to all other vectors, we then have

$$\boxed{\mathbf{v} \perp \mathbf{w} \quad \text{if and only if} \quad \mathbf{v} \cdot \mathbf{w} = 0}$$

The standard basis vectors are mutually orthogonal and have length 1 (Figure 5). In terms of dot products, because $\mathbf{i} = \langle 1,0,0 \rangle$, $\mathbf{j} = \langle 0,1,0 \rangle$, and $\mathbf{k} = \langle 0,0,1 \rangle$,

$$\mathbf{i} \cdot \mathbf{j} = \mathbf{i} \cdot \mathbf{k} = \mathbf{j} \cdot \mathbf{k} = 0, \qquad \mathbf{i} \cdot \mathbf{i} = \mathbf{j} \cdot \mathbf{j} = \mathbf{k} \cdot \mathbf{k} = 1$$

EXAMPLE 3 **Testing for Orthogonality** Determine whether $\mathbf{v} = \langle 2,6,1 \rangle$ is orthogonal to $\mathbf{u} = \langle 2,-1,1 \rangle$ or $\mathbf{w} = \langle -4,1,2 \rangle$ (Figure 6).

Solution We test for orthogonality by computing the dot products:

$$\mathbf{v} \cdot \mathbf{u} = \langle 2,6,1 \rangle \cdot \langle 2,-1,1 \rangle = 2(2) + 6(-1) + 1(1) = -1 \quad \text{(not orthogonal)}$$

$$\mathbf{v} \cdot \mathbf{w} = \langle 2,6,1 \rangle \cdot \langle -4,1,2 \rangle = 2(-4) + 6(1) + 1(2) = 0 \quad \text{(orthogonal)} \quad \blacksquare$$

By definition, the angle θ between vectors **v** and **u** is obtuse if $\frac{\pi}{2} < \theta \le \pi$, and in this interval $\cos\theta < 0$. Furthermore, θ is acute if $0 < \theta < \frac{\pi}{2}$, and this occurs if $\cos\theta > 0$. From Eq. (1), we then have the following:

> The angle θ between **v** and **u** is obtuse if and only if $\mathbf{v} \cdot \mathbf{u} < 0$.
>
> The angle θ between **v** and **u** is acute if and only if $\mathbf{v} \cdot \mathbf{u} > 0$.

EXAMPLE 4 **Testing for Obtuseness and Acuteness** Determine whether the angles between the vector $\mathbf{v} = \langle 3,1,-2 \rangle$ and the vectors $\mathbf{u} = \langle \frac{1}{2}, \frac{1}{2}, 5 \rangle$ and $\mathbf{w} = \langle 4,-3,0 \rangle$ are obtuse or acute.

Solution
We have

$$\mathbf{v} \cdot \mathbf{u} = \langle 3,1,-2 \rangle \cdot \left\langle \frac{1}{2}, \frac{1}{2}, 5 \right\rangle = \frac{3}{2} + \frac{1}{2} - 10 = -8 < 0 \quad \text{(angle is obtuse)}$$

$$\mathbf{v} \cdot \mathbf{w} = \langle 3,1,-2 \rangle \cdot \langle 4,-3,0 \rangle = 12 - 3 + 0 = 9 > 0 \quad \text{(angle is acute)} \quad \blacksquare$$

Dot product and its properties can be used to prove geometric relationships, as we demonstrate next.

EXAMPLE 5 Figure 7 shows a parallelogram whose diagonal lengths are x and y and whose side lengths are a and b. Prove that

$$\frac{x^2 + y^2}{2} = a^2 + b^2 \qquad \boxed{4}$$

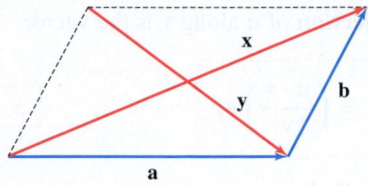

FIGURE 8

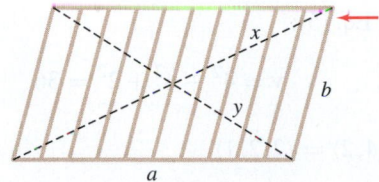

FIGURE 9 The carpenter adjusts the frame to obtain a rectangle.

Solution We represent the sides and diagonals of the parallelogram by vectors, as shown in Figure 8. Note that $\mathbf{x} = \mathbf{a} + \mathbf{b}$ and $\mathbf{y} = \mathbf{a} - \mathbf{b}$. We have

$$x^2 + y^2 = \|\mathbf{x}\|^2 + \|\mathbf{y}\|^2 = \mathbf{x} \cdot \mathbf{x} + \mathbf{y} \cdot \mathbf{y}$$

$$= (\mathbf{a} + \mathbf{b}) \cdot (\mathbf{a} + \mathbf{b}) + (\mathbf{a} - \mathbf{b}) \cdot (\mathbf{a} - \mathbf{b})$$

$$= \mathbf{a} \cdot \mathbf{a} + 2\mathbf{a} \cdot \mathbf{b} + \mathbf{b} \cdot \mathbf{b} + \mathbf{a} \cdot \mathbf{a} - 2\mathbf{a} \cdot \mathbf{b} + \mathbf{b} \cdot \mathbf{b}$$

$$= 2\|\mathbf{a}\|^2 + 2\|\mathbf{b}\|^2 = 2(a^2 + b^2)$$

The desired result now follows. ∎

EXAMPLE 6 **An Application to Carpentry** When a carpenter constructs a frame for an *a*-by-*b* rectangular structure, such as a backyard deck, she measures the diagonals (Figure 9). If they are equal, then the frame is rectangular. If not, she needs to adjust the frame. Since the sides have lengths a and b, the desired diagonal length is $\sqrt{a^2 + b^2}$, and that equals $\sqrt{\dfrac{x^2 + y^2}{2}}$ by Eq. (4).

Instead of computing a square root of a sum of squares, she "splits the difference" between the two diagonal measurements and adjusts the frame until that amount has been added to the shorter diagonal. Specifically, if $x > y$, she takes $\dfrac{x-y}{2}$ and adds that to y for the desired measurement. The resulting measurement is $y + \dfrac{x-y}{2} = \dfrac{x+y}{2}$, instead of $\sqrt{\dfrac{x^2 + y^2}{2}}$. These are not necessarily equal, but in Section 14.4, we extend the idea of linearization, introduced in Section 4.1, to multivariable calculus and show that $\dfrac{x+y}{2}$ is the linearization of $\sqrt{\dfrac{x^2 + y^2}{2}}$ for values of x near y.

If the carpenter's diagonal measurements are $198\frac{3}{4}$ and $187\frac{1}{4}$ inches, find the diagonal length that produces a rectangular frame, and compute the split-difference approximation for comparison.

Solution For $\sqrt{\dfrac{x^2 + y^2}{2}}$, we obtain

$$\sqrt{\frac{(198.75)^2 + (187.25)^2}{2}} \approx 193.09 \text{ in.}$$

With the split-difference method, we have a difference of $11\frac{1}{2}$ in., which, when split, gives $5\frac{3}{4}$ in. When that is added to $187\frac{1}{4}$ in., we obtain 193 in. ∎

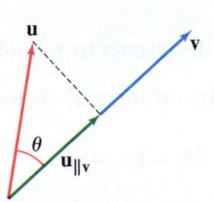

DF **FIGURE 10** The projection $\mathbf{u}_{\|\mathbf{v}}$ of $\mathbf{u}$ along $\mathbf{v}$ has length $\|\mathbf{u}\| \cos \theta$.

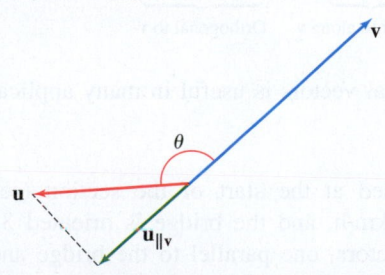

DF **FIGURE 11** When θ is obtuse, $\mathbf{u}_{\|\mathbf{v}}$ and $\mathbf{v}$ point in opposite directions.

Another important use of the dot product is in finding the projection $\mathbf{u}_{\|\mathbf{v}}$ of a vector $\mathbf{u}$ along a nonzero vector $\mathbf{v}$. The projection $\mathbf{u}_{\|\mathbf{v}}$ is the vector parallel to $\mathbf{v}$ obtained by dropping a perpendicular from $\mathbf{u}$ to the line through $\mathbf{v}$ as in Figures 10 and 11. We think of $\mathbf{u}_{\|\mathbf{v}}$ as that part of $\mathbf{u}$ that is parallel to $\mathbf{v}$. How do we determine $\mathbf{u}_{\|\mathbf{v}}$?

Referring to Figures 10 and 11, we see by trigonometry that $\mathbf{u}_{\|\mathbf{v}}$ has length $\|\mathbf{u}\| |\cos \theta|$. If θ is acute, then $\mathbf{u}_{\|\mathbf{v}}$ is a positive multiple of $\mathbf{v}$, and thus, since $\cos \theta > 0$, we have $\mathbf{u}_{\|\mathbf{v}} = (\|\mathbf{u}\| \cos \theta)\mathbf{e_v}$, where $\mathbf{e_v} = \dfrac{1}{\|\mathbf{v}\|}\mathbf{v}$ is the unit vector in the direction of $\mathbf{v}$. Similarly, if θ is obtuse, then $\mathbf{u}_{\|\mathbf{v}}$ is a negative multiple of $\mathbf{e_v}$ and again $\mathbf{u}_{\|\mathbf{v}} = (\|\mathbf{u}\| \cos \theta)\mathbf{e_v}$ since $\cos \theta < 0$. Therefore,

$$\mathbf{u}_{\|\mathbf{v}} = (\|\mathbf{u}\| \cos \theta)\mathbf{e_v} = \|\mathbf{u}\| \cos \theta \frac{\mathbf{v}}{\|\mathbf{v}\|} = \frac{\|\mathbf{u}\| \|\mathbf{v}\| \cos \theta}{\|\mathbf{v}\|^2}\mathbf{v} = \left(\frac{\mathbf{u} \cdot \mathbf{v}}{\mathbf{v} \cdot \mathbf{v}}\right)\mathbf{v}$$

This formula provides us with the desired expression for the projection. We present three equivalent expressions for $\mathbf{u}_{\|\mathbf{v}}$:

Projection of u along v Assume $\mathbf{v} \neq \mathbf{0}$. The **projection** of $\mathbf{u}$ along $\mathbf{v}$ is the vector

$$\mathbf{u}_{||\mathbf{v}} = \left(\frac{\mathbf{u} \cdot \mathbf{v}}{\mathbf{v} \cdot \mathbf{v}}\right) \mathbf{v} = \left(\frac{\mathbf{u} \cdot \mathbf{v}}{\|\mathbf{v}\|^2}\right) \mathbf{v} = \left(\frac{\mathbf{u} \cdot \mathbf{v}}{\|\mathbf{v}\|}\right) \mathbf{e_v}$$

$\boxed{5}$

This is sometimes denoted $\text{proj}_{\mathbf{v}}\mathbf{u}$. The scalar $\dfrac{\mathbf{u} \cdot \mathbf{v}}{\|\mathbf{v}\|}$ is called the **component** or the **scalar component** of $\mathbf{u}$ along $\mathbf{v}$ and is sometimes denoted $\text{comp}_{\mathbf{v}}\mathbf{u}$.

EXAMPLE 7 Find the projection of $\mathbf{u} = \langle 5, 1, -3 \rangle$ along $\mathbf{v} = \langle 4, 4, 2 \rangle$.

Solution It is convenient to use the first formula in Eq. (5):

$$\mathbf{u} \cdot \mathbf{v} = \langle 5, 1, -3 \rangle \cdot \langle 4, 4, 2 \rangle = 20 + 4 - 6 = 18, \qquad \mathbf{v} \cdot \mathbf{v} = 4^2 + 4^2 + 2^2 = 36$$

$$\mathbf{u}_{||\mathbf{v}} = \left(\frac{\mathbf{u} \cdot \mathbf{v}}{\mathbf{v} \cdot \mathbf{v}}\right) \mathbf{v} = \left(\frac{18}{36}\right) \langle 4, 4, 2 \rangle = \langle 2, 2, 1 \rangle$$ ■

We show now that if $\mathbf{v} \neq \mathbf{0}$, then every vector $\mathbf{u}$ can be written as the sum of the projection $\mathbf{u}_{||\mathbf{v}}$ and a vector $\mathbf{u}_{\perp\mathbf{v}}$ that is orthogonal to $\mathbf{v}$ (see Figure 12). In fact, if we set

$$\boxed{\mathbf{u}_{\perp\mathbf{v}} = \mathbf{u} - \mathbf{u}_{||\mathbf{v}}}$$

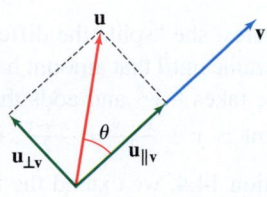

DF FIGURE 12 Decomposition of $\mathbf{u}$ as a sum $\mathbf{u} = \mathbf{u}_{||\mathbf{v}} + \mathbf{u}_{\perp\mathbf{v}}$ of vectors parallel and orthogonal to $\mathbf{v}$.

then we have the following **decomposition** of $\mathbf{u}$ with respect to $\mathbf{v}$:

$$\boxed{\mathbf{u} = \mathbf{u}_{||\mathbf{v}} + \mathbf{u}_{\perp\mathbf{v}}}$$

$\boxed{6}$

Equation (6) expresses $\mathbf{u}$ as a sum of vectors, one parallel to $\mathbf{v}$ and one perpendicular to $\mathbf{v}$. We must verify, however, that $\mathbf{u}_{\perp\mathbf{v}}$ is perpendicular to $\mathbf{v}$. We do this by showing that the dot product is zero:

$$\mathbf{u}_{\perp\mathbf{v}} \cdot \mathbf{v} = (\mathbf{u} - \mathbf{u}_{||\mathbf{v}}) \cdot \mathbf{v} = \left(\mathbf{u} - \left(\frac{\mathbf{u} \cdot \mathbf{v}}{\mathbf{v} \cdot \mathbf{v}}\right) \mathbf{v}\right) \cdot \mathbf{v} = \mathbf{u} \cdot \mathbf{v} - \left(\frac{\mathbf{u} \cdot \mathbf{v}}{\mathbf{v} \cdot \mathbf{v}}\right) (\mathbf{v} \cdot \mathbf{v}) = 0$$

EXAMPLE 8 Find the decomposition of $\mathbf{u} = \langle 5, 1, -3 \rangle$ with respect to $\mathbf{v} = \langle 4, 4, 2 \rangle$.

Solution In Example 7, we showed that $\mathbf{u}_{||\mathbf{v}} = \langle 2, 2, 1 \rangle$. The orthogonal vector is

$$\mathbf{u}_{\perp\mathbf{v}} = \mathbf{u} - \mathbf{u}_{||\mathbf{v}} = \langle 5, 1, -3 \rangle - \langle 2, 2, 1 \rangle = \langle 3, -1, -4 \rangle$$

The decomposition of $\mathbf{u}$ with respect to $\mathbf{v}$ is

$$\mathbf{u} = \langle 5, 1, -3 \rangle = \mathbf{u}_{||\mathbf{v}} + \mathbf{u}_{\perp\mathbf{v}} = \underbrace{\langle 2, 2, 1 \rangle}_{\text{Projection along } \mathbf{v}} + \underbrace{\langle 3, -1, -4 \rangle}_{\text{Orthogonal to } \mathbf{v}}$$ ■

The decomposition into parallel and orthogonal vectors is useful in many applications, as we see in the next two examples.

EXAMPLE 9 Let us return to the problem posed at the start of the section (see Figure 13). We have a wind vector $\mathbf{w} = \langle 60, 0 \rangle$ km/h, and the bridge is oriented 32 degrees east of north. Express $\mathbf{w}$ as a sum of vectors, one parallel to the bridge and one perpendicular to it. Also, compute the magnitude of the vector perpendicular to the bridge to determine the speed of the part of the wind blowing directly at the bridge.

Solution To begin, note that $\mathbf{u} = \langle \cos 58°, \sin 58° \rangle$ is a unit vector parallel to the bridge. It is shown, but not drawn to scale, in Figure 13. The goal is to decompose $\mathbf{w}$ as the sum of $\mathbf{w}_{||\mathbf{u}}$ and $\mathbf{w}_{\perp\mathbf{u}}$.

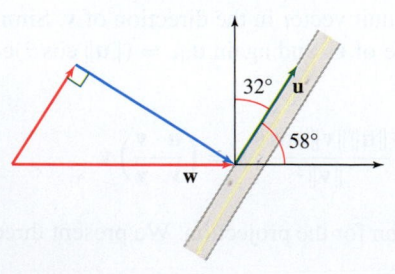

FIGURE 13

For $\mathbf{w}_{||\mathbf{u}} = \left(\frac{\mathbf{w} \cdot \mathbf{u}}{\mathbf{u} \cdot \mathbf{u}}\right)\mathbf{u}$, note that $\mathbf{u} \cdot \mathbf{u} = 1$ since $\mathbf{u}$ is a unit vector. Also, $\mathbf{w} \cdot \mathbf{u} = \langle 60, 0 \rangle \cdot \langle \cos 58°, \sin 58° \rangle = 60 \cos 58°$. Therefore,

$$\mathbf{w}_{||\mathbf{u}} = 60\cos 58° \langle \cos 58°, \sin 58° \rangle \approx \langle 16.85, 26.96 \rangle$$

and then

$$\mathbf{w}_{\perp \mathbf{u}} = \mathbf{w} - \mathbf{w}_{||\mathbf{u}} \approx \langle 43.15, -26.96 \rangle$$

So we have the decomposition:

$$\mathbf{w} = \langle 16.85, 26.96 \rangle + \langle 43.15, -26.96 \rangle$$

where, approximately, $\langle 16.85, 26.96 \rangle$ is along the bridge and $\langle 43.15, -26.96 \rangle$ is perpendicular to it. The magnitude of the perpendicular part of the wind, the part blowing directly at the bridge, is approximately $\sqrt{(43.15)^2 + (-26.96)^2} \approx 50.9$ km/h. ∎

EXAMPLE 10 What is the minimum force you must apply to pull a 20-kg wagon up a frictionless ramp inclined at an angle $\theta = 15°$?

Solution Let $\mathbf{v}$ be a vector in the direction of the ramp, and let $\mathbf{F}_g$ be the force on the wagon due to gravity. It has magnitude $20g$ newtons, or N, with $g = 9.8$. Referring to Figure 14, we decompose $\mathbf{F}_g$ as a sum

$$\mathbf{F}_g = \mathbf{F}_{||\mathbf{v}} + \mathbf{F}_{\perp \mathbf{v}}$$

where $\mathbf{F}_{||\mathbf{v}}$ is the projection along the ramp and $\mathbf{F}_{\perp \mathbf{v}}$, called the normal force, is the force perpendicular to the ramp. The normal force $\mathbf{F}_{\perp \mathbf{v}}$ is canceled by the ramp pushing back against the wagon in the opposite direction, and thus (because there is no friction) you need only pull against $\mathbf{F}_{||\mathbf{v}}$.

Notice that the angle between $\mathbf{F}_g$ and the ramp is the complementary angle $90° - \theta$. Since $\mathbf{F}_{||\mathbf{v}}$ is parallel to the ramp, the angle between $\mathbf{F}_g$ and $\mathbf{F}_{||\mathbf{v}}$ is also $90° - \theta$, or $75°$, and

$$\|\mathbf{F}_{||\mathbf{v}}\| = \|\mathbf{F}_g\| \cos(75°) \approx 20(9.8)(0.26) \approx 51 \text{ newtons}$$

Since gravity pulls the wagon down the ramp with a 51-newton force, it takes a minimum force of 51 newtons to pull the wagon up the ramp. ∎

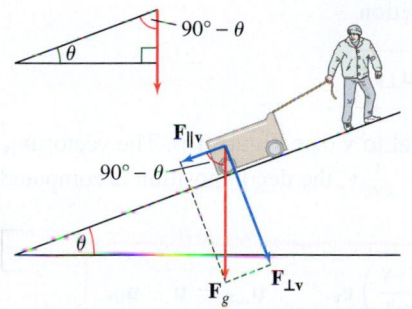

FIGURE 14 The angle between $\mathbf{F}_g$ and $\mathbf{F}_{||\mathbf{v}}$ is $90° - \theta$.

GRAPHICAL INSIGHT It seems that we are using the term "component" in two ways. We say that a vector $\mathbf{u} = \langle a, b \rangle$ has components a and b. On the other hand, $\mathbf{u} \cdot \mathbf{e}$ is called the component of $\mathbf{u}$ along the unit vector $\mathbf{e}$.

In fact, these two notions of component are the same. The components a and b are the dot products of $\mathbf{u}$ with the standard unit vectors:

$$\mathbf{u} \cdot \mathbf{i} = \langle a, b \rangle \cdot \langle 1, 0 \rangle = a$$
$$\mathbf{u} \cdot \mathbf{j} = \langle a, b \rangle \cdot \langle 0, 1 \rangle = b$$

and we have the decomposition [Figure 15(A)]

$$\mathbf{u} = a\mathbf{i} + b\mathbf{j}$$

But any two orthogonal unit vectors $\mathbf{e}$ and $\mathbf{f}$ give rise to a rotated coordinate system, and we see in Figure 15(B) that

$$\mathbf{u} = (\mathbf{u} \cdot \mathbf{e})\mathbf{e} + (\mathbf{u} \cdot \mathbf{f})\mathbf{f}$$

In other words, $\mathbf{u} \cdot \mathbf{e}$ and $\mathbf{u} \cdot \mathbf{f}$ really are the components when we express $\mathbf{u}$ relative to the rotated system.

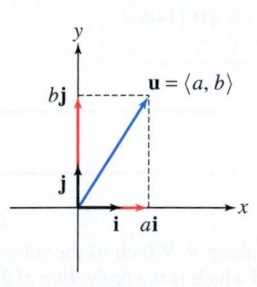

(A)

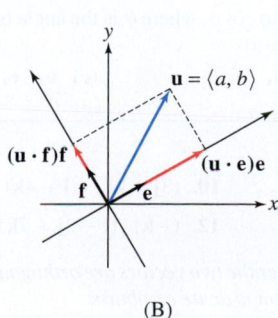

(B)

FIGURE 15

12.3 SUMMARY

- The *dot product* of $\mathbf{v} = \langle a_1, b_1, c_1 \rangle$ and $\mathbf{w} = \langle a_2, b_2, c_2 \rangle$ is

$$\mathbf{v} \cdot \mathbf{w} = a_1 a_2 + b_1 b_2 + c_1 c_2$$

- Basic Properties of the Dot Product:

    - Commutativity: $\mathbf{v} \cdot \mathbf{w} = \mathbf{w} \cdot \mathbf{v}$
    - Pulling out scalars: $(\lambda \mathbf{v}) \cdot \mathbf{w} = \mathbf{v} \cdot (\lambda \mathbf{w}) = \lambda(\mathbf{v} \cdot \mathbf{w})$
    - Distributive Law: $\mathbf{u} \cdot (\mathbf{v} + \mathbf{w}) = \mathbf{u} \cdot \mathbf{v} + \mathbf{u} \cdot \mathbf{w}$
      $$(\mathbf{v} + \mathbf{w}) \cdot \mathbf{u} = \mathbf{v} \cdot \mathbf{u} + \mathbf{w} \cdot \mathbf{u}$$

    - $\mathbf{v} \cdot \mathbf{v} = \|\mathbf{v}\|^2$
    - $\mathbf{v} \cdot \mathbf{w} = \|\mathbf{v}\| \, \|\mathbf{w}\| \cos\theta$, where θ is the angle between $\mathbf{v}$ and $\mathbf{w}$ that satisfies $0 \le \theta \le \pi$.

- Test for orthogonality: $\mathbf{v} \perp \mathbf{w}$ if and only if $\mathbf{v} \cdot \mathbf{w} = 0$.
- The angle between $\mathbf{v}$ and $\mathbf{w}$ is acute if $\mathbf{v} \cdot \mathbf{w} > 0$ and obtuse if $\mathbf{v} \cdot \mathbf{w} < 0$.
- Assume $\mathbf{v} \ne \mathbf{0}$. Every vector $\mathbf{u}$ has a decomposition

$$\mathbf{u} = \mathbf{u}_{\|\mathbf{v}} + \mathbf{u}_{\perp \mathbf{v}}$$

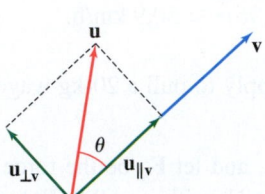

FIGURE 16

where $\mathbf{u}_{\|\mathbf{v}}$ is parallel to $\mathbf{v}$, and $\mathbf{u}_{\perp \mathbf{v}}$ is orthogonal to $\mathbf{v}$ (see Figure 16). The vector $\mathbf{u}_{\|\mathbf{v}}$ is called the *projection* of $\mathbf{u}$ along $\mathbf{v}$. With $\mathbf{e}_{\mathbf{v}} = \frac{1}{\|\mathbf{v}\|} \mathbf{v}$, the decomposition is computed as follows:

$$\mathbf{u}_{\|\mathbf{v}} = \left(\frac{\mathbf{u} \cdot \mathbf{v}}{\mathbf{v} \cdot \mathbf{v}} \right) \mathbf{v} = \left(\frac{\mathbf{u} \cdot \mathbf{v}}{\|\mathbf{v}\|^2} \right) \mathbf{v} = \left(\frac{\mathbf{u} \cdot \mathbf{v}}{\|\mathbf{v}\|} \right) \mathbf{e}_{\mathbf{v}}, \qquad \mathbf{u}_{\perp \mathbf{v}} = \mathbf{u} - \mathbf{u}_{\|\mathbf{v}}$$

The coefficient $\dfrac{\mathbf{u} \cdot \mathbf{v}}{\|\mathbf{v}\|}$ is called the *component* of $\mathbf{u}$ along $\mathbf{v}$:

$$\text{component of } \mathbf{u} \text{ along } \mathbf{v} = \frac{\mathbf{u} \cdot \mathbf{v}}{\|\mathbf{v}\|} = \|\mathbf{u}\| \cos\theta$$

12.3 EXERCISES

Preliminary Questions

1. Is the dot product of two vectors a scalar or a vector?

2. What can you say about the angle between $\mathbf{a}$ and $\mathbf{b}$ if $\mathbf{a} \cdot \mathbf{b} < 0$?

3. Which property of dot products allows us to conclude that if $\mathbf{v}$ is orthogonal to both $\mathbf{u}$ and $\mathbf{w}$, then $\mathbf{v}$ is orthogonal to $\mathbf{u} + \mathbf{w}$?

4. Which is the projection of $\mathbf{v}$ along $\mathbf{v}$: (a) $\mathbf{v}$ or (b) $\mathbf{e}_{\mathbf{v}}$?

5. Let $\mathbf{u}_{\|\mathbf{v}}$ be the projection of $\mathbf{u}$ along $\mathbf{v}$. Which of the following is the projection $\mathbf{u}$ along the vector $2\mathbf{v}$ and which is the projection of $2\mathbf{u}$ along $\mathbf{v}$?

(a) $\frac{1}{2} \mathbf{u}_{\|\mathbf{v}}$ (b) $\mathbf{u}_{\|\mathbf{v}}$ (c) $2\mathbf{u}_{\|\mathbf{v}}$

6. Which of the following is equal to $\cos\theta$, where θ is the angle between $\mathbf{u}$ and $\mathbf{v}$?

(a) $\mathbf{u} \cdot \mathbf{v}$ (b) $\mathbf{u} \cdot \mathbf{e}_{\mathbf{v}}$ (c) $\mathbf{e}_{\mathbf{u}} \cdot \mathbf{e}_{\mathbf{v}}$

Exercises

In Exercises 1–12, compute the dot product.

1. $\langle 1, 2, 1 \rangle \cdot \langle 4, 3, 5 \rangle$

2. $\langle 3, -2, 2 \rangle \cdot \langle 1, 0, 1 \rangle$

3. $\langle 0, 1, 1 \rangle \cdot \langle -7, 41, -39 \rangle$

4. $\langle 1, -1, 1 \rangle \cdot \langle -2, 4, -6 \rangle$

5. $\langle 3, 1 \rangle \cdot \langle 4, -7 \rangle$

6. $\left\langle \frac{1}{6}, \frac{1}{2} \right\rangle \cdot \left\langle 3, \frac{1}{2} \right\rangle$

7. $\mathbf{k} \cdot \mathbf{j}$

8. $\mathbf{k} \cdot \mathbf{k}$

9. $(\mathbf{i} + \mathbf{j}) \cdot (\mathbf{j} + \mathbf{k})$

10. $(3\mathbf{j} + 2\mathbf{k}) \cdot (\mathbf{i} - 4\mathbf{k})$

11. $(\mathbf{i} + \mathbf{j} + \mathbf{k}) \cdot (3\mathbf{i} + 2\mathbf{j} - 5\mathbf{k})$

12. $(-\mathbf{k}) \cdot (\mathbf{i} - 2\mathbf{j} + 7\mathbf{k})$

In Exercises 13–18, determine whether the two vectors are orthogonal and, if not, whether the angle between them is acute or obtuse.

13. $\langle 1, 1, 1 \rangle, \quad \langle 1, -2, -2 \rangle$

14. $\langle 0, 2, 4 \rangle, \quad \langle -5, 0, 0 \rangle$

15. $\langle 1, 2, 1 \rangle, \quad \langle 7, -3, -1 \rangle$

16. $\langle 0, 2, 4 \rangle, \quad \langle 3, 1, 0 \rangle$

17. $\left\langle \frac{12}{5}, -\frac{4}{5} \right\rangle$, $\left\langle \frac{1}{2}, -\frac{7}{4} \right\rangle$

18. $\langle 12, 6 \rangle$, $\langle 2, -4 \rangle$

In Exercises 19–22, find the cosine of the angle between the vectors.

19. $\langle 0, 3, 1 \rangle$, $\langle 4, 0, 0 \rangle$

20. $\langle 1, 1, 1 \rangle$, $\langle 2, -1, 2 \rangle$

21. $\mathbf{i} + \mathbf{j}$, $\mathbf{j} + 2\mathbf{k}$

22. $3\mathbf{i} + \mathbf{k}$, $\mathbf{i} + \mathbf{j} + \mathbf{k}$

In Exercises 23–30, find the angle between the vectors.

23. $\langle 2, \sqrt{2} \rangle$, $\langle 1 + \sqrt{2}, 1 - \sqrt{2} \rangle$

24. $\langle 5, \sqrt{3} \rangle$, $\langle \sqrt{3}, 2 \rangle$

25. $\langle 1, 1, 1 \rangle$, $\langle 1, 0, 1 \rangle$

26. $\langle 3, 1, 1 \rangle$, $\langle 2, -4, 2 \rangle$

27. $\langle 0, 1, 1 \rangle$, $\langle 1, -1, 0 \rangle$

28. $\langle 1, 1, -1 \rangle$, $\langle 1, -2, -1 \rangle$

29. $\mathbf{i}$, $3\mathbf{i} + 2\mathbf{j} + \mathbf{k}$

30. $\mathbf{i} + \mathbf{k}$, $\mathbf{j} - \mathbf{k}$

31. Find all values of b for which the vectors are orthogonal.

(a) $\langle b, 3, 2 \rangle$, $\langle 1, b, 1 \rangle$

(b) $\langle 4, -2, 7 \rangle$, $\langle b^2, b, 0 \rangle$

32. Find a vector that is orthogonal to $\langle -1, 2, 2 \rangle$.

33. Find two vectors that are not multiples of each other and are both orthogonal to $\langle 2, 0, -3 \rangle$.

34. Find a vector that is orthogonal to $\mathbf{v} = \langle 1, 2, 1 \rangle$ but not to $\mathbf{w} = \langle 1, 0, -1 \rangle$.

35. Find $\mathbf{v} \cdot \mathbf{e}$, where $\|\mathbf{v}\| = 3$, $\mathbf{e}$ is a unit vector, and the angle between $\mathbf{e}$ and $\mathbf{v}$ is $\frac{2\pi}{3}$.

36. Assume that $\mathbf{v}$ lies in the yz-plane. Which of the following dot products is equal to zero for all choices of $\mathbf{v}$?

(a) $\mathbf{v} \cdot \langle 0, 2, 1 \rangle$

(b) $\mathbf{v} \cdot \mathbf{k}$

(c) $\mathbf{v} \cdot \langle -3, 0, 0 \rangle$

(d) $\mathbf{v} \cdot \mathbf{j}$

In Exercises 37–40, simplify the expression.

37. $(\mathbf{v} - \mathbf{w}) \cdot \mathbf{v} + \mathbf{v} \cdot \mathbf{w}$

38. $(\mathbf{v} + \mathbf{w}) \cdot (\mathbf{v} + \mathbf{w}) - 2\mathbf{v} \cdot \mathbf{w}$

39. $(\mathbf{v} + \mathbf{w}) \cdot \mathbf{v} - (\mathbf{v} + \mathbf{w}) \cdot \mathbf{w}$

40. $(\mathbf{v} + \mathbf{w}) \cdot \mathbf{v} - (\mathbf{v} - \mathbf{w}) \cdot \mathbf{w}$

In Exercises 41–44, use the properties of the dot product to evaluate the expression, assuming that $\mathbf{u} \cdot \mathbf{v} = 2$, $\|\mathbf{u}\| = 1$, and $\|\mathbf{v}\| = 3$.

41. $\mathbf{u} \cdot (4\mathbf{v})$

42. $(\mathbf{u} + \mathbf{v}) \cdot \mathbf{v}$

43. $2\mathbf{u} \cdot (3\mathbf{u} - \mathbf{v})$

44. $(\mathbf{u} + \mathbf{v}) \cdot (\mathbf{u} - \mathbf{v})$

45. Find the angle between $\mathbf{v}$ and $\mathbf{w}$ if $\mathbf{v} \cdot \mathbf{w} = -\|\mathbf{v}\| \, \|\mathbf{w}\|$.

46. Find the angle between $\mathbf{v}$ and $\mathbf{w}$ if $\mathbf{v} \cdot \mathbf{w} = \frac{1}{2}\|\mathbf{v}\| \, \|\mathbf{w}\|$.

47. Assume that $\|\mathbf{v}\| = 3$, $\|\mathbf{w}\| = 5$, and the angle between $\mathbf{v}$ and $\mathbf{w}$ is $\theta = \frac{\pi}{3}$.

(a) Use the relation $\|\mathbf{v} + \mathbf{w}\|^2 = (\mathbf{v} + \mathbf{w}) \cdot (\mathbf{v} + \mathbf{w})$ to show that $\|\mathbf{v} + \mathbf{w}\|^2 = 3^2 + 5^2 + 2\mathbf{v} \cdot \mathbf{w}$.

(b) Find $\|\mathbf{v} + \mathbf{w}\|$.

48. Assume that $\|\mathbf{v}\| = 2$, $\|\mathbf{w}\| = 3$, and the angle between $\mathbf{v}$ and $\mathbf{w}$ is $120°$. Determine:

(a) $\mathbf{v} \cdot \mathbf{w}$

(b) $\|2\mathbf{v} + \mathbf{w}\|$

(c) $\|2\mathbf{v} - 3\mathbf{w}\|$

49. Show that if $\mathbf{e}$ and $\mathbf{f}$ are unit vectors such that $\|\mathbf{e} + \mathbf{f}\| = \frac{3}{2}$, then $\|\mathbf{e} - \mathbf{f}\| = \frac{\sqrt{7}}{2}$. *Hint:* Show that $\mathbf{e} \cdot \mathbf{f} = \frac{1}{8}$.

50. Find $\|2\mathbf{e} - 3\mathbf{f}\|$, assuming that $\mathbf{e}$ and $\mathbf{f}$ are unit vectors such that $\|\mathbf{e} + \mathbf{f}\| = \sqrt{3/2}$.

51. Find the angle θ in the triangle in Figure 17.

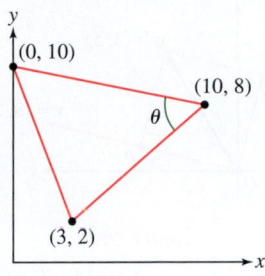

FIGURE 17

52. Find all three angles in the triangle in Figure 18.

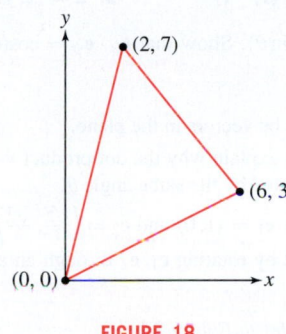

FIGURE 18

53. (a) Draw $\mathbf{u}_{\|\mathbf{v}}$ and $\mathbf{v}_{\|\mathbf{u}}$ for the vectors appearing as in Figure 19.

(b) Which of $\mathbf{u}_{\|\mathbf{v}}$ and $\mathbf{v}_{\|\mathbf{u}}$ has the greater magnitude?

FIGURE 19

54. Let $\mathbf{u}$ and $\mathbf{v}$ be two nonzero vectors.

(a) Is it possible for the component of $\mathbf{u}$ along $\mathbf{v}$ to have the opposite sign from the component of $\mathbf{v}$ along $\mathbf{u}$? Why or why not?

(b) What must be true of the vectors if either of these two components is 0?

In Exercises 55–62, find the projection of $\mathbf{u}$ along $\mathbf{v}$.

55. $\mathbf{u} = \langle 2, 5 \rangle$, $\mathbf{v} = \langle 1, 1 \rangle$

56. $\mathbf{u} = \langle 2, -3 \rangle$, $\mathbf{v} = \langle 1, 2 \rangle$

57. $\mathbf{u} = \langle -1, 2, 0 \rangle$, $\mathbf{v} = \langle 2, 0, 1 \rangle$

58. $\mathbf{u} = \langle 1, 1, 1 \rangle$, $\mathbf{v} = \langle 1, 1, 0 \rangle$

59. $\mathbf{u} = 5\mathbf{i} + 7\mathbf{j} - 4\mathbf{k}$, $\mathbf{v} = \mathbf{k}$

60. $\mathbf{u} = \mathbf{i} + 29\mathbf{k}$, $\mathbf{v} = \mathbf{j}$

61. $\mathbf{u} = \langle a, b, c \rangle$, $\mathbf{v} = \mathbf{i}$

62. $\mathbf{u} = \langle a, a, b \rangle$, $\mathbf{v} = \mathbf{i} - \mathbf{j}$

In Exercises 63 and 64, compute the component of $\mathbf{u}$ along $\mathbf{v}$.

63. $\mathbf{u} = \langle 3, 2, 1 \rangle$, $\mathbf{v} = \langle 1, 0, 1 \rangle$

64. $\mathbf{u} = \langle 3, 0, 9 \rangle$, $\mathbf{v} = \langle 1, 2, 2 \rangle$

65. Find the length of $\overline{OP}$ in Figure 20.

66. Find $\|\mathbf{u}_{\perp \mathbf{v}}\|$ in Figure 20.

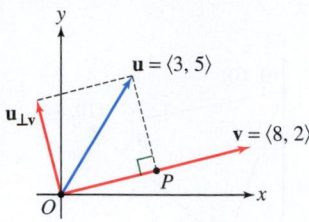

FIGURE 20

In Exercises 67–72, find the decomposition $\mathbf{a} = \mathbf{a}_{\|\mathbf{b}} + \mathbf{a}_{\perp \mathbf{b}}$ with respect to $\mathbf{b}$.

67. $\mathbf{a} = \langle 1, 0 \rangle$, $\mathbf{b} = \langle 1, 1 \rangle$ **68.** $\mathbf{a} = \langle 2, -3 \rangle$, $\mathbf{b} = \langle 5, 0 \rangle$

69. $\mathbf{a} = \langle 4, -1, 0 \rangle$, $\mathbf{b} = \langle 0, 1, 1 \rangle$ **70.** $\mathbf{a} = \langle 4, -1, 5 \rangle$, $\mathbf{b} = \langle 2, 1, 1 \rangle$

71. $\mathbf{a} = \langle x, y \rangle$, $\mathbf{b} = \langle 1, -1 \rangle$ **72.** $\mathbf{a} = \langle x, y, z \rangle$, $\mathbf{b} = \langle 1, 1, 1 \rangle$

73. Let $\mathbf{e}_\theta = \langle \cos\theta, \sin\theta \rangle$. Show that $\mathbf{e}_\theta \cdot \mathbf{e}_\psi = \cos(\theta - \psi)$ for any two angles θ and ψ.

74. 📝 Let $\mathbf{v}$ and $\mathbf{w}$ be vectors in the plane.
(a) Use Theorem 2 to explain why the dot product $\mathbf{v} \cdot \mathbf{w}$ does not change if both $\mathbf{v}$ and $\mathbf{w}$ are rotated by the same angle θ.
(b) Sketch the vectors $\mathbf{e}_1 = \langle 1, 0 \rangle$ and $\mathbf{e}_2 = \left(\frac{\sqrt{2}}{2}, \frac{\sqrt{2}}{2} \right)$, and determine the vectors $\mathbf{e}_1', \mathbf{e}_2'$ obtained by rotating $\mathbf{e}_1, \mathbf{e}_2$ through an angle $\frac{\pi}{4}$. Verify that $\mathbf{e}_1 \cdot \mathbf{e}_2 = \mathbf{e}_1' \cdot \mathbf{e}_2'$.

In Exercises 75–78, refer to Figure 21.

75. Find the angle between $\overline{AB}$ and $\overline{AC}$.

76. Find the angle between $\overline{AB}$ and $\overline{AD}$.

77. Calculate the projection of $\overrightarrow{AC}$ along $\overrightarrow{AD}$.

78. Calculate the projection of $\overrightarrow{AD}$ along $\overrightarrow{AB}$.

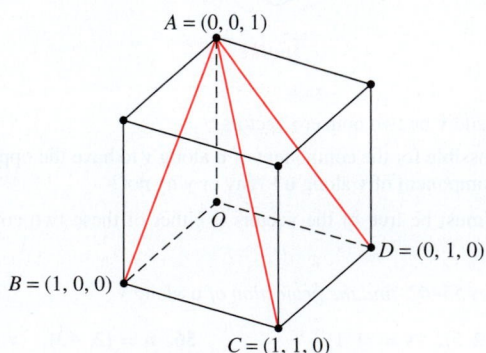

FIGURE 21 Unit cube in $\mathbf{R}^3$.

In Exercises 79–80, as in Example 6, assume that the carpenter's diagonal measurements are x and y, and compute the diagonal length that produces a rectangular frame. Compare the result with the corresponding split-difference approximation.

79. $x = 234\frac{1}{2}$ inches and $y = 223$ in.

80. $x = 87.2$ cm and $y = 82.7$ cm

81. The methane molecule CH_4 consists of a carbon molecule bonded to four hydrogen molecules that are spaced as far apart from each other as possible. The hydrogen atoms then sit at the vertices of a tetrahedron, with the carbon atom at its center, as in Figure 22. We can model this with the carbon atom at the point $(\frac{1}{2}, \frac{1}{2}, \frac{1}{2})$ and the hydrogen atoms at $(0, 0, 0), (1, 1, 0), (1, 0, 1),$ and $(0, 1, 1)$. Use the dot product to find the bond angle α formed between any two of the line segments from the carbon atom to the hydrogen atoms.

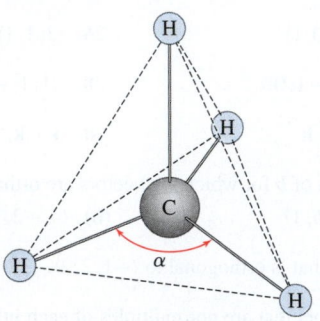

FIGURE 22 A methane molecule.

82. Iron forms a crystal lattice where each central atom appears at the center of a cube, the corners of which correspond to additional iron atoms, as in Figure 23. Use the dot product to find the angle β between the line segments from the central atom to two adjacent outer atoms. *Hint:* Take the central atom to be situated at the origin and the corner atoms to occur at $(\pm 1, \pm 1, \pm 1)$.

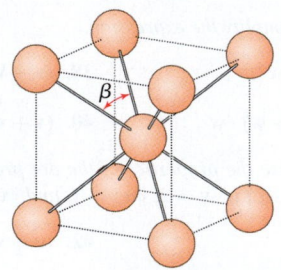

FIGURE 23 An iron crystal.

83. 📝 Let $\mathbf{v}$ and $\mathbf{w}$ be nonzero vectors and set $\mathbf{u} = \mathbf{e}_\mathbf{v} + \mathbf{e}_\mathbf{w}$. Use the dot product to show that the angle between $\mathbf{u}$ and $\mathbf{v}$ is equal to the angle between $\mathbf{u}$ and $\mathbf{w}$. Explain this result geometrically with a diagram.

84. 📝 Let $\mathbf{v}$, $\mathbf{w}$, and $\mathbf{a}$ be nonzero vectors such that $\mathbf{v} \cdot \mathbf{a} = \mathbf{w} \cdot \mathbf{a}$. Is it true that $\mathbf{v} = \mathbf{w}$? Either prove this or give a counterexample.

85. In Example 9, assume that the wind is out of the north at 45 km/h. Express the corresponding wind vector as a sum of vectors, one parallel to the bridge and one perpendicular to it. Also, compute the magnitude of the perpendicular term to determine the speed of the part of the wind blowing directly at the bridge.

86. A plane flies with velocity $\mathbf{v} = \langle 220, -90, 10 \rangle$ km/h. A wind is blowing out of the northeast with velocity $\mathbf{w} = \langle -30, -30, 0 \rangle$ km/h. Express the wind vector as a sum of vectors, one parallel to the plane's velocity, one perpendicular to it. Is the parallel part of the wind blowing with or against the plane?

87. Calculate the force (in newtons) required to push a 40-kg wagon up a 10° incline (Figure 24).

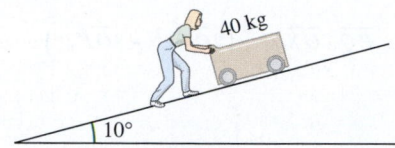

FIGURE 24

88. A force **F** is applied to each of two ropes (of negligible weight) attached to opposite ends of a 40-kg wagon and making an angle of 35° with the horizontal (Figure 25). What is the maximum magnitude of **F** (in newtons) that can be applied without lifting the wagon off the ground?

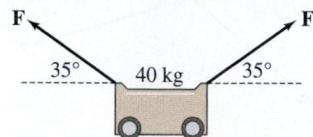

FIGURE 25

89. A light beam travels along the ray determined by a unit vector **L**, strikes a flat surface at point P, and is reflected along the ray determined by a unit vector **R**, where $\theta_1 = \theta_2$ (Figure 26). Show that if **N** is the unit vector orthogonal to the surface, then

$$\mathbf{R} = 2(\mathbf{L} \cdot \mathbf{N})\mathbf{N} - \mathbf{L}$$

Incoming light   Reflected light

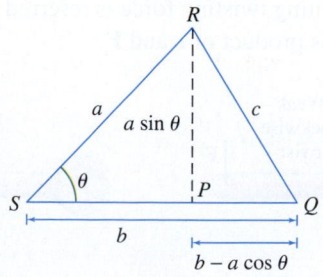

FIGURE 26

90. Let P and Q be antipodal (opposite) points on a sphere of radius r centered at the origin and let R be a third point on the sphere (Figure 27). Prove that $\overline{PR}$ and $\overline{QR}$ are orthogonal.

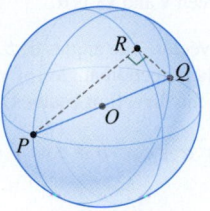

FIGURE 27

91. Prove that $\|\mathbf{v} + \mathbf{w}\|^2 - \|\mathbf{v} - \mathbf{w}\|^2 = 4\mathbf{v} \cdot \mathbf{w}$.

92. Use Exercise 91 to show that **v** and **w** are orthogonal if and only if $\|\mathbf{v} - \mathbf{w}\| = \|\mathbf{v} + \mathbf{w}\|$.

93. A rhombus is a parallelogram in which all four sides have equal length. Show that the diagonals of a parallelogram are perpendicular if and only if the parallelogram is a rhombus. *Hint:* Take an approach similar to the solution in Example 5, and consider $\mathbf{x} \cdot \mathbf{y}$.

94. Verify the Distributive Law:

$$\mathbf{u} \cdot (\mathbf{v} + \mathbf{w}) = \mathbf{u} \cdot \mathbf{v} + \mathbf{u} \cdot \mathbf{w}$$

95. Verify that $(\lambda \mathbf{v}) \cdot \mathbf{w} = \lambda(\mathbf{v} \cdot \mathbf{w})$ for any scalar λ.

Further Insights and Challenges

96. Prove the Law of Cosines, $c^2 = a^2 + b^2 - 2ab \cos \theta$, by referring to Figure 28. *Hint:* Consider the right triangle $\triangle PQR$.

FIGURE 28

97. In this exercise, we prove the Cauchy–Schwarz inequality: If **v** and **w** are any two vectors, then

$$|\mathbf{v} \cdot \mathbf{w}| \leq \|\mathbf{v}\| \, \|\mathbf{w}\| \qquad \boxed{7}$$

(a) Let $f(x) = \|x\mathbf{v} + \mathbf{w}\|^2$ where x is a scalar value. Show that $f(x)$ may be written $f(x) = ax^2 + bx + c$, where $a = \|\mathbf{v}\|^2$, $b = 2\mathbf{v} \cdot \mathbf{w}$, and $c = \|\mathbf{w}\|^2$.

(b) Conclude that $b^2 - 4ac \leq 0$. *Hint:* Observe that $f(x) \geq 0$ for all x.

98. Use (7) to prove the Triangle Inequality:

$$\|\mathbf{v} + \mathbf{w}\| \leq \|\mathbf{v}\| + \|\mathbf{w}\|$$

Hint: First use the Triangle Inequality for numbers to prove

$$|(\mathbf{v} + \mathbf{w}) \cdot (\mathbf{v} + \mathbf{w})| \leq |(\mathbf{v} + \mathbf{w}) \cdot \mathbf{v}| + |(\mathbf{v} + \mathbf{w}) \cdot \mathbf{w}|$$

99. This exercise gives another proof of the relation between the dot product and the angle θ between two vectors $\mathbf{v} = \langle a_1, b_1 \rangle$ and $\mathbf{w} = \langle a_2, b_2 \rangle$ in the plane. Observe that $\mathbf{v} = \|\mathbf{v}\| \langle \cos \theta_1, \sin \theta_1 \rangle$ and $\mathbf{w} = \|\mathbf{w}\| \langle \cos \theta_2, \sin \theta_2 \rangle$, with θ_1 and θ_2 as in Figure 29. Then use the addition formula for the cosine to show that

$$\mathbf{v} \cdot \mathbf{w} = \|\mathbf{v}\| \, \|\mathbf{w}\| \cos \theta$$

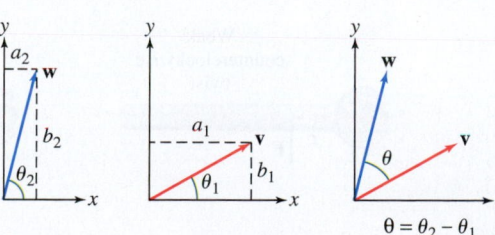

FIGURE 29

100. Let $\mathbf{v} = \langle x, y \rangle$ and

$$\mathbf{v}_\theta = \langle x \cos\theta + y \sin\theta, -x \sin\theta + y \cos\theta \rangle$$

Prove that the angle between $\mathbf{v}$ and $\mathbf{v}_\theta$ is θ.

101. Let $\mathbf{v}$ be a nonzero vector. The angles α, β, γ between $\mathbf{v}$ and the unit vectors $\mathbf{i}, \mathbf{j}, \mathbf{k}$ are called the direction angles of $\mathbf{v}$ (Figure 30). The cosines of these angles are called the **direction cosines** of $\mathbf{v}$. Prove that

$$\cos^2\alpha + \cos^2\beta + \cos^2\gamma = 1$$

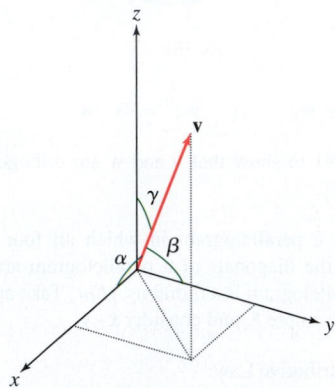

FIGURE 30 Direction angles of $\mathbf{v}$.

102. Find the direction cosines of $\mathbf{v} = \langle 3, 6, -2 \rangle$.

103. The set of all points $X = (x, y, z)$ equidistant from two points P, Q in $\mathbf{R}^3$ is a plane (Figure 31). Show that X lies on this plane if

$$\overrightarrow{PQ} \cdot \overrightarrow{OX} = \frac{1}{2}\left(\|\overrightarrow{OQ}\|^2 - \|\overrightarrow{OP}\|^2 \right) \qquad \boxed{8}$$

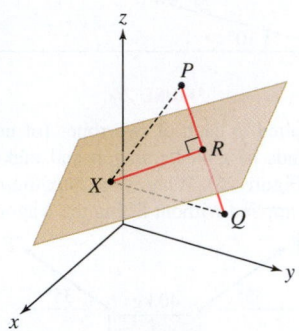

FIGURE 31

Hint: If R is the midpoint of $\overline{PQ}$, then X is equidistant from P and Q if and only if $\overrightarrow{XR}$ is orthogonal to $\overrightarrow{PQ}$.

104. Sketch the plane consisting of all points $X = (x, y, z)$ equidistant from the points $P = (0, 1, 0)$ and $Q = (0, 0, 1)$. Use Eq. (8) to show that X lies on this plane if and only if $y = z$.

105. Use Eq. (8) to find the equation of the plane consisting of all points $X = (x, y, z)$ equidistant from $P = (2, 1, 1)$ and $Q = (1, 0, 2)$.

12.4 The Cross Product

Some applications of vectors require another operation called the cross product. In physics and engineering, the cross product is used to compute torque, a twisting force that causes an object to rotate. Figure 1 displays diagrams of a force $\mathbf{F}$ of varying strength and direction applied to a wrench to turn a bolt. The vector $\mathbf{r}$ is referred to as a position vector and indicates the location of the force relative to the turning axis in the bolt. The resulting twist on the bolt varies from weak to strong, and the direction of the twist is either clockwise or counterclockwise. Since the twist has magnitude and direction, it is naturally represented by a vector. The resulting twisting force is referred to as the **torque** on the bolt and is calculated using the cross product of $\mathbf{r}$ and $\mathbf{F}$.

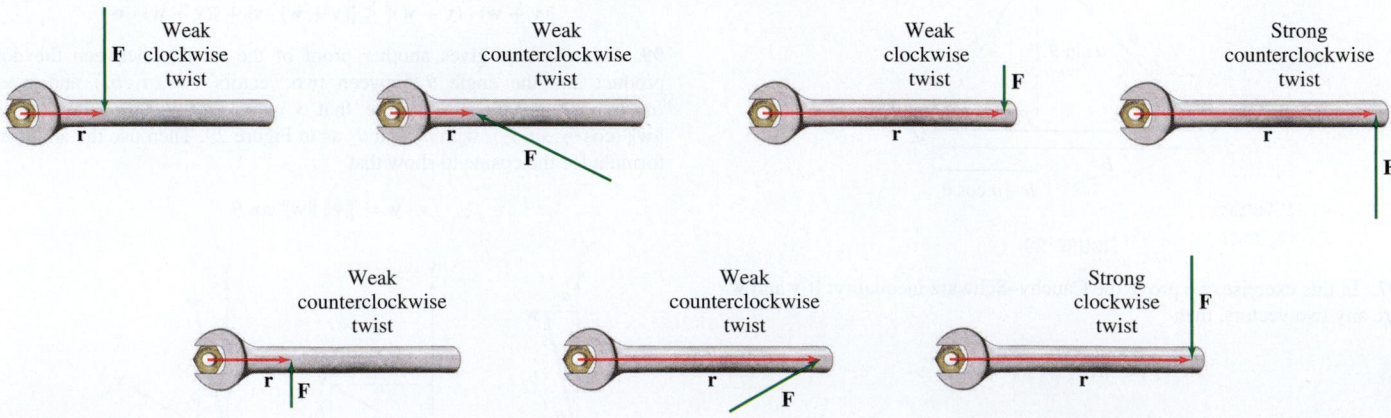

FIGURE 1

Unlike the dot product $\mathbf{v} \cdot \mathbf{w}$ (which is a scalar), the cross product $\mathbf{v} \times \mathbf{w}$ is a vector. It is defined algebraically using what is known as a 3×3 "determinant." We introduce

2×2 and 3×3 determinants next, and then show how they are used to define the cross product.

An $n \times n$ **matrix** is an array consisting of n rows and n columns of numbers (or vectors, as we will see in the definition of cross product). The determinant of a 2×2 matrix is denoted and defined as follows:

$$\begin{vmatrix} a & b \\ c & d \end{vmatrix} = ad - bc \qquad \boxed{1}$$

Note that the determinant is the difference of the diagonal products. For example,

$$\begin{vmatrix} 3 & 2 \\ \frac{1}{2} & 4 \end{vmatrix} = \begin{vmatrix} 3 & 2 \\ \frac{1}{2} & 4 \end{vmatrix} - \begin{vmatrix} 3 & 2 \\ \frac{1}{2} & 4 \end{vmatrix} = 3 \cdot 4 - 2 \cdot \frac{1}{2} = 11$$

The determinant of a 3×3 matrix is denoted and defined by

$$\begin{vmatrix} a_1 & b_1 & c_1 \\ a_2 & b_2 & c_2 \\ a_3 & b_3 & c_3 \end{vmatrix} = a_1 \underbrace{\begin{vmatrix} b_2 & c_2 \\ b_3 & c_3 \end{vmatrix}}_{(1,1)\text{-minor}} - b_1 \underbrace{\begin{vmatrix} a_2 & c_2 \\ a_3 & c_3 \end{vmatrix}}_{(1,2)\text{-minor}} + c_1 \underbrace{\begin{vmatrix} a_2 & b_2 \\ a_3 & b_3 \end{vmatrix}}_{(1,3)\text{-minor}} \qquad \boxed{2}$$

This formula expresses the 3×3 determinant in terms of 2×2 determinants called **minors**. The minors are obtained by crossing out the first row and one of the three columns of the 3×3 matrix. For example, the minor labeled $(1, 2)$ above is obtained as follows:

$$\begin{vmatrix} a_1 & b_1 & c_1 \\ a_2 & b_2 & c_2 \\ a_3 & b_3 & c_3 \end{vmatrix} \qquad \text{to obtain the } (1,2)\text{-minor} \qquad \underbrace{\begin{vmatrix} a_2 & c_2 \\ a_3 & c_3 \end{vmatrix}}_{(1,2)\text{-minor}}$$

Cross out row 1 and column 2

The theory of matrices and determinants is part of linear algebra, a subject of great importance throughout mathematics. In this section, we discuss just a few basic definitions and facts from linear algebra needed for our treatment of multivariable calculus.

EXAMPLE 1 **A 3×3 Determinant** Calculate $\begin{vmatrix} 2 & 4 & 3 \\ 0 & 1 & -7 \\ -1 & 5 & 3 \end{vmatrix}$.

Solution

$$\begin{vmatrix} 2 & 4 & 3 \\ 0 & 1 & -7 \\ -1 & 5 & 3 \end{vmatrix} = 2\begin{vmatrix} 1 & -7 \\ 5 & 3 \end{vmatrix} - 4\begin{vmatrix} 0 & -7 \\ -1 & 3 \end{vmatrix} + 3\begin{vmatrix} 0 & 1 \\ -1 & 5 \end{vmatrix}$$

$$= 2(38) - 4(-7) + 3(1) = 107 \qquad \blacksquare$$

Later in this section, we will see how determinants are related to area and volume. First, we introduce the cross product, which is defined as a determinant whose first row has the vector entries $\mathbf{i}, \mathbf{j}, \mathbf{k}$.

CAUTION Note in Eq. (3) that the middle term comes with a minus sign.

> **DEFINITION** **The Cross Product** The cross product of vectors $\mathbf{v} = \langle v_1, v_2, v_3 \rangle$ and $\mathbf{w} = \langle w_1, w_2, w_3 \rangle$ is the vector
>
> $$\mathbf{v} \times \mathbf{w} = \begin{vmatrix} \mathbf{i} & \mathbf{j} & \mathbf{k} \\ v_1 & v_2 & v_3 \\ w_1 & w_2 & w_3 \end{vmatrix} = \begin{vmatrix} v_2 & v_3 \\ w_2 & w_3 \end{vmatrix}\mathbf{i} - \begin{vmatrix} v_1 & v_3 \\ w_1 & w_3 \end{vmatrix}\mathbf{j} + \begin{vmatrix} v_1 & v_2 \\ w_1 & w_2 \end{vmatrix}\mathbf{k} \qquad \boxed{3}$$

The cross product differs fundamentally from the dot product in that $\mathbf{u} \times \mathbf{v}$ is a vector, whereas $\mathbf{u} \cdot \mathbf{v}$ is a number.

EXAMPLE 2 Calculate $\mathbf{v} \times \mathbf{w}$, where $\mathbf{v} = \langle -2, 1, 4 \rangle$ and $\mathbf{w} = \langle 3, 2, 5 \rangle$.

Solution

$$\mathbf{v} \times \mathbf{w} = \begin{vmatrix} \mathbf{i} & \mathbf{j} & \mathbf{k} \\ -2 & 1 & 4 \\ 3 & 2 & 5 \end{vmatrix} = \begin{vmatrix} 1 & 4 \\ 2 & 5 \end{vmatrix}\mathbf{i} - \begin{vmatrix} -2 & 4 \\ 3 & 5 \end{vmatrix}\mathbf{j} + \begin{vmatrix} -2 & 1 \\ 3 & 2 \end{vmatrix}\mathbf{k}$$

$$= (-3)\mathbf{i} - (-22)\mathbf{j} + (-7)\mathbf{k} = \langle -3, 22, -7 \rangle \qquad \blacksquare$$

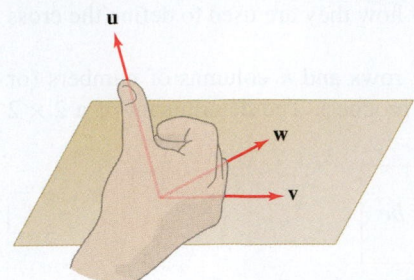

FIGURE 2 $\{\mathbf{v}, \mathbf{w}, \mathbf{u}\}$ forms a right-handed system.

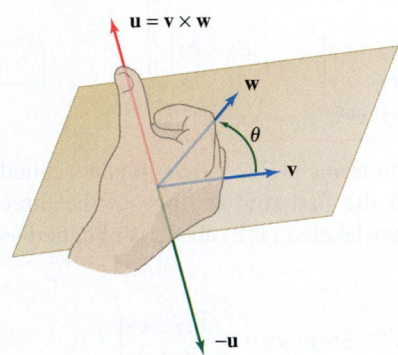

FIGURE 3 There are two vectors orthogonal to $\mathbf{v}$ and $\mathbf{w}$ with length $\|\mathbf{v}\| \, \|\mathbf{w}\| \sin \theta$. The right-hand rule determines which is $\mathbf{v} \times \mathbf{w}$.

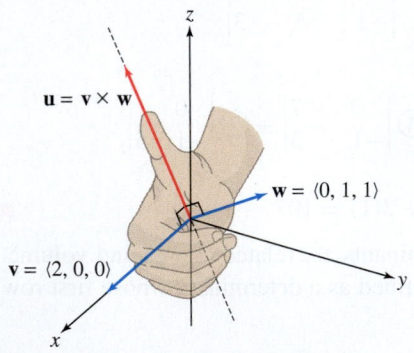

FIGURE 4 The direction of $\mathbf{u} = \mathbf{v} \times \mathbf{w}$ is determined by the right-hand rule. Thus, $\mathbf{u}$ has a positive z-component.

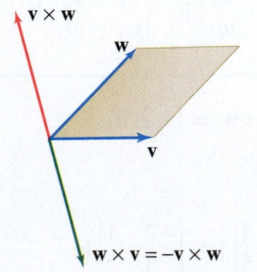

FIGURE 5

Formula (3) gives no hint of the geometric meaning of the cross product. However, there is a simple way to visualize the vector $\mathbf{v} \times \mathbf{w}$ using the **right-hand rule**. Suppose that $\mathbf{v}$, $\mathbf{w}$, and $\mathbf{u}$ are nonzero vectors that do not all lie in a plane. We say that $\{\mathbf{v}, \mathbf{w}, \mathbf{u}\}$ forms a **right-handed system** if the direction of $\mathbf{u}$ is determined by the right-hand rule: *When the fingers of your right hand curl from* $\mathbf{v}$ *to* $\mathbf{w}$, *your thumb points to the same side of the plane spanned by* $\mathbf{v}$ *and* $\mathbf{w}$ *as* $\mathbf{u}$ (Figure 2). The following theorem describes the cross product geometrically. The first two parts are proved at the end of the section.

> **THEOREM 1 Geometric Description of the Cross Product** Given two nonzero nonparallel vectors $\mathbf{v}$ and $\mathbf{w}$ with angle θ between them, the cross product $\mathbf{v} \times \mathbf{w}$ is the unique vector with the following three properties:
>
> **(i)** $\mathbf{v} \times \mathbf{w}$ is orthogonal to $\mathbf{v}$ and $\mathbf{w}$.
> **(ii)** $\mathbf{v} \times \mathbf{w}$ has length $\|\mathbf{v}\| \, \|\mathbf{w}\| \sin \theta$.
> **(iii)** $\{\mathbf{v}, \mathbf{w}, \mathbf{v} \times \mathbf{w}\}$ forms a right-handed system.

How do the three properties in Theorem 1 determine $\mathbf{v} \times \mathbf{w}$? By property (i), $\mathbf{v} \times \mathbf{w}$ lies on the line orthogonal to $\mathbf{v}$ and $\mathbf{w}$. By property (ii), $\mathbf{v} \times \mathbf{w}$ is one of the two vectors on this line of length $\|\mathbf{v}\| \, \|\mathbf{w}\| \sin \theta$. Finally, property (iii) tells us which of these two vectors is $\mathbf{v} \times \mathbf{w}$—namely, the vector for which $\{\mathbf{v}, \mathbf{w}, \mathbf{v} \times \mathbf{w}\}$ forms a right-handed system (Figure 3).

EXAMPLE 3 Let $\mathbf{v} = \langle 2, 0, 0 \rangle$ and $\mathbf{w} = \langle 0, 1, 1 \rangle$. Determine $\mathbf{u} = \mathbf{v} \times \mathbf{w}$ using the geometric properties of the cross product rather than Eq. (3).

Solution We use Theorem 1. First, by property (i), $\mathbf{u} = \mathbf{v} \times \mathbf{w}$ is orthogonal to $\mathbf{v}$ and $\mathbf{w}$. Since $\mathbf{v}$ lies along the x-axis, $\mathbf{u}$ must lie in the yz-plane (Figure 4). In other words, $\mathbf{u} = \langle 0, b, c \rangle$. But $\mathbf{u}$ is also orthogonal to $\mathbf{w} = \langle 0, 1, 1 \rangle$, so $\mathbf{u} \cdot \mathbf{w} = b + c = 0$ and thus $\mathbf{u} = \langle 0, b, -b \rangle$.

Next, direct computation shows that $\|\mathbf{v}\| = 2$ and $\|\mathbf{w}\| = \sqrt{2}$. Furthermore, the angle between $\mathbf{v}$ and $\mathbf{w}$ is $\theta = \frac{\pi}{2}$ since $\mathbf{v} \cdot \mathbf{w} = 0$. By property (ii),

$$\|\mathbf{u}\| = \sqrt{b^2 + (-b)^2} = |b|\sqrt{2} \quad \text{is equal to} \quad \|\mathbf{v}\| \, \|\mathbf{w}\| \sin \frac{\pi}{2} = 2\sqrt{2}$$

Therefore, $|b| = 2$ and $b = \pm 2$. Finally, property (iii) tells us that $\mathbf{u}$ points in the positive z-direction (Figure 4). Thus, $b = -2$ and $\mathbf{u} = \langle 0, -2, 2 \rangle$. You can verify that the formula for the cross product yields the same answer. ∎

One of the most striking properties of the cross product is that it is *anticommutative*. Reversing the order changes the sign:

$$\boxed{\mathbf{w} \times \mathbf{v} = -\mathbf{v} \times \mathbf{w}} \qquad \boxed{4}$$

We verify this using Eq. (3). When we interchange $\mathbf{v}$ and $\mathbf{w}$, each of the 2×2 determinants changes sign. For example,

$$\begin{vmatrix} v_1 & v_2 \\ w_1 & w_2 \end{vmatrix} = v_1 w_2 - v_2 w_1 = -(v_2 w_1 - v_1 w_2) = -\begin{vmatrix} w_1 & w_2 \\ v_1 & v_2 \end{vmatrix}$$

Anticommutativity also follows from the geometric description of the cross product. By properties (i) and (ii) in Theorem 1, $\mathbf{v} \times \mathbf{w}$ and $\mathbf{w} \times \mathbf{v}$ are both orthogonal to $\mathbf{v}$ and $\mathbf{w}$ and have the same length. However, $\mathbf{v} \times \mathbf{w}$ and $\mathbf{w} \times \mathbf{v}$ point in opposite directions by the right-hand rule, and thus $\mathbf{v} \times \mathbf{w} = -\mathbf{w} \times \mathbf{v}$ (Figure 5). In particular, $\mathbf{v} \times \mathbf{v} = -\mathbf{v} \times \mathbf{v}$ and hence $\mathbf{v} \times \mathbf{v} = \mathbf{0}$.

The next theorem lists these and some further properties of cross products (the proofs are given as Exercises 53–56).

Note an important distinction between the dot product and cross product of a vector with itself:

$$\mathbf{v} \times \mathbf{v} = \mathbf{0}$$

$$\mathbf{v} \cdot \mathbf{v} = \|\mathbf{v}\|^2$$

> **THEOREM 2 Basic Properties of the Cross Product**
>
> (i) $\mathbf{w} \times \mathbf{v} = -\mathbf{v} \times \mathbf{w}$
>
> (ii) $\mathbf{v} \times \mathbf{v} = \mathbf{0}$
>
> (iii) $\mathbf{v} \times \mathbf{w} = \mathbf{0}$ if and only if either $\mathbf{w} = \lambda\mathbf{v}$ for some scalar λ or $\mathbf{v} = \mathbf{0}$
>
> (iv) $(\lambda\mathbf{v}) \times \mathbf{w} = \mathbf{v} \times (\lambda\mathbf{w}) = \lambda(\mathbf{v} \times \mathbf{w})$
>
> (v) $(\mathbf{u} + \mathbf{v}) \times \mathbf{w} = \mathbf{u} \times \mathbf{w} + \mathbf{v} \times \mathbf{w}$
>
> $\mathbf{u} \times (\mathbf{v} + \mathbf{w}) = \mathbf{u} \times \mathbf{v} + \mathbf{u} \times \mathbf{w}$

The cross product of any two of the standard basis vectors $\mathbf{i}$, $\mathbf{j}$, and $\mathbf{k}$ is equal to the third, possibly with a minus sign. More precisely (see Exercise 57),

$$\mathbf{i} \times \mathbf{j} = \mathbf{k}, \qquad \mathbf{j} \times \mathbf{k} = \mathbf{i}, \qquad \mathbf{k} \times \mathbf{i} = \mathbf{j} \qquad \boxed{5}$$

$$\mathbf{j} \times \mathbf{i} = -\mathbf{k}, \qquad \mathbf{k} \times \mathbf{j} = -\mathbf{i}, \qquad \mathbf{i} \times \mathbf{k} = -\mathbf{j} \qquad \boxed{6}$$

Furthermore, we also have:

$$\mathbf{i} \times \mathbf{i} = \mathbf{j} \times \mathbf{j} = \mathbf{k} \times \mathbf{k} = \mathbf{0}$$

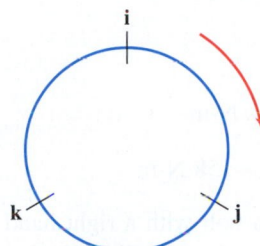

FIGURE 6 Circle for computing the cross products of the basis vectors.

An easy way to remember the relations (5) and (6) is to draw $\mathbf{i}$, $\mathbf{j}$, and $\mathbf{k}$ in a circle as in Figure 6. Starting at any vector, go around the circle in the clockwise direction and you obtain one of the relations (5). For example, starting at $\mathbf{i}$ and moving clockwise yields $\mathbf{i} \times \mathbf{j} = \mathbf{k}$. If you go around in the counterclockwise direction, you obtain the relations (6). Thus, starting at $\mathbf{k}$ and going counterclockwise gives the relation $\mathbf{k} \times \mathbf{j} = -\mathbf{i}$.

EXAMPLE 4   Using the ijk Relations   Compute $(2\mathbf{i} + \mathbf{k}) \times (3\mathbf{j} + 5\mathbf{k})$.

Solution We use the properties of the cross product:

$$(2\mathbf{i} + \mathbf{k}) \times (3\mathbf{j} + 5\mathbf{k}) = (2\mathbf{i}) \times (3\mathbf{j}) + (2\mathbf{i}) \times (5\mathbf{k}) + \mathbf{k} \times (3\mathbf{j}) + \mathbf{k} \times (5\mathbf{k})$$

$$= 6(\mathbf{i} \times \mathbf{j}) + 10(\mathbf{i} \times \mathbf{k}) + 3(\mathbf{k} \times \mathbf{j}) + 5(\mathbf{k} \times \mathbf{k})$$

$$= 6\mathbf{k} - 10\mathbf{j} - 3\mathbf{i} + 5(\mathbf{0}) = -3\mathbf{i} - 10\mathbf{j} + 6\mathbf{k} \qquad \blacksquare$$

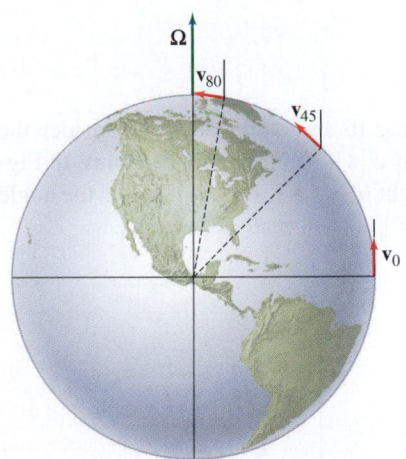

FIGURE 7

The cross product can be used to demonstrate a significant property of the large-scale motion in the earth's atmosphere. Meteorologists study the properties and motion of small volumes (called parcels) of air to help forecast the weather and understand our climate. The rotation of the earth impacts the movement of air parcels via what is known as the Coriolis force. For a parcel of mass m with velocity $\mathbf{v}$, the Coriolis force is $\mathbf{F}_c = -2m\boldsymbol{\Omega} \times \mathbf{v}$, where $\boldsymbol{\Omega}$ is the angular velocity of the rotating earth (Figure 7). The vector $\boldsymbol{\Omega}$ is parallel to the rotation axis of the earth and has a magnitude of approximately 7.3×10^{-5} s^{-1}, reflecting that the earth completes one rotation through 2π radians in a day. The next example demonstrates that in the northern hemisphere, the impact of the Coriolis force increases with increasing latitude.

EXAMPLE 5   The Coriolis Force in Meteorology   We consider three parcels of air with mass 2 kg, moving directly north at 20 m/s, one at the equator, at a latitude of 45° north, and at a latitude of 80° north. The parcel velocities have the same magnitude and the same direction relative to the surface of the earth, but not the same direction relative to the earth's axis (Figure 7). In Figure 8 we illustrate that the angle that each velocity vector makes with a line parallel to the earth's axis (and therefore with $\boldsymbol{\Omega}$) is equal to the latitude. Find the magnitude of the Coriolis force for each parcel.

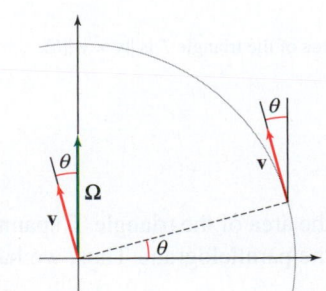

FIGURE 8

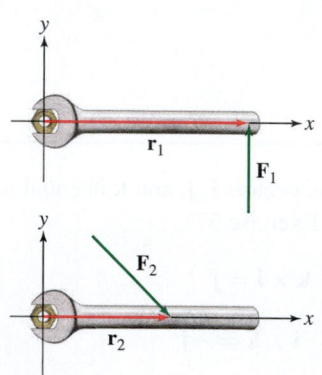

FIGURE 9

Solution Let $\mathbf{v}_0$, $\mathbf{v}_{45}$, and $\mathbf{v}_{80}$ represent the parcel velocities at the different latitudes.

- At the equator, $\mathbf{v}_0 \| \mathbf{\Omega}$, implying that $\mathbf{\Omega} \times \mathbf{v}_0 = \mathbf{0}$ and therefore $\|\mathbf{F}_c\| = 0$.
- At latitude $45°$, we have $\|\mathbf{F}_c\| = 2m\|\mathbf{\Omega}\|\|\mathbf{v}_{45}\|\sin(45°) \approx 0.0041$ N.
- At latitude $80°$, we have $\|\mathbf{F}_c\| = 2m\|\mathbf{\Omega}\|\|\mathbf{v}_{80}\|\sin(80°) \approx 0.0058$ N. ∎

In Section 14.5, we will show how the Coriolis force steers air currents so that they tend to circulate around low-pressure systems rather than flow directly into them.

EXAMPLE 6   Torque on a Bolt Figure 9 shows two forces applied to a wrench to turn a bolt. The **torque** on the bolt is the vector $\boldsymbol{\tau} = \mathbf{r} \times \mathbf{F}$, where $\mathbf{F}$ is the force applied to the wrench, and $\mathbf{r}$ is a position vector, directed from the axis of the bolt to the point where the force is applied. In the figure, assume the z-axis is pointing out of the page. Compute the torque in each case where $\mathbf{F}_1 = \langle 0, 60, 0 \rangle$, $\mathbf{F}_2 = \langle 50, -50, 0 \rangle$ (both in newtons), and $\mathbf{r}_1 = \langle 0.5, 0, 0 \rangle$, $\mathbf{r}_2 = \langle 0.3, 0, 0 \rangle$ (both in meters).

Solution We have

$$\boldsymbol{\tau}_1 = \mathbf{r}_1 \times \mathbf{F}_1 = 0.5\mathbf{i} \times 60\mathbf{j} = 30\mathbf{k} \text{ N-m}$$

$$\boldsymbol{\tau}_2 = \mathbf{r}_2 \times \mathbf{F}_2 = 0.3\mathbf{i} \times (50\mathbf{i} - 50\mathbf{j}) = -15\mathbf{k} \text{ N-m}$$ ∎

That $\boldsymbol{\tau}_1$ is in the positive z-direction indicates that a bolt with a right-hand thread turns upward out of the page with that combination of $\mathbf{F}_1$ and $\mathbf{r}_1$. Similarly in the second case, the bolt turns into the page. Also note that, even though the force is greater in magnitude in the second case, the resulting torque is smaller in magnitude. This is due to the force in the second case being applied closer to the bolt and more obliquely than the force in the first case.

Cross Products, Area, and Volume

Cross products and determinants are closely related to area and volume. Consider the parallelogram $\mathcal{P}$ spanned by nonzero vectors $\mathbf{v}$ and $\mathbf{w}$ with a common basepoint. In Figure 10(A), we see that $\mathcal{P}$ has base $b = \|\mathbf{v}\|$ and height $h = \|\mathbf{w}\|\sin\theta$, where θ is the angle between $\mathbf{v}$ and $\mathbf{w}$. Therefore, $\mathcal{P}$ has area $A = bh = \|\mathbf{v}\|\,\|\mathbf{w}\|\sin\theta = \|\mathbf{v} \times \mathbf{w}\|$.

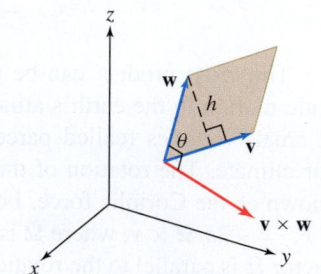

(A) The area of the parallelogram $\mathcal{P}$ is
$\|\mathbf{v} \times \mathbf{w}\| = \|\mathbf{v}\|\,\|\mathbf{w}\|\sin\theta$.

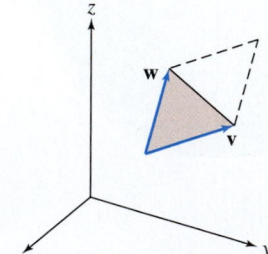

(B) The area of the triangle $\mathcal{T}$ is $\|\mathbf{v} \times \mathbf{w}\|/2$.

FIGURE 10

Notice also, as in Figure 10(B), that we also know the area of the triangle $\mathcal{T}$ spanned by nonzero vectors $\mathbf{v}$ and $\mathbf{w}$ is exactly half the area of the parallelogram. Thus, we have the following:

Areas If $\mathcal{P}$ is the parallelogram spanned by $\mathbf{v}$ and $\mathbf{w}$, and $\mathcal{T}$ is the triangle spanned by $\mathbf{v}$ and $\mathbf{w}$, then

$$\text{area}(\mathcal{P}) = \|\mathbf{v} \times \mathbf{w}\| \quad \text{and} \quad \text{area}(\mathcal{T}) = \frac{\|\mathbf{v} \times \mathbf{w}\|}{2}$$

7

A "parallelepiped" is the solid spanned by three vectors. Each face is a parallelogram.

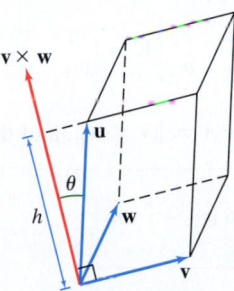

FIGURE 11 The volume of the parallelepiped is $|\mathbf{u} \cdot (\mathbf{v} \times \mathbf{w})|$.

We use the following notation for the determinant of the matrix whose rows are the vectors $\mathbf{u}, \mathbf{v}, \mathbf{w}$:

$$\det \begin{pmatrix} \mathbf{u} \\ \mathbf{v} \\ \mathbf{w} \end{pmatrix} = \begin{vmatrix} u_1 & u_2 & u_3 \\ v_1 & v_2 & v_3 \\ w_1 & w_2 & w_3 \end{vmatrix}$$

It is awkward to write the absolute value of a determinant in the notation on the right, but we may denote it as

$$\left| \det \begin{pmatrix} \mathbf{u} \\ \mathbf{v} \\ \mathbf{w} \end{pmatrix} \right|$$

Next, consider the **parallelepiped** $\mathcal{D}$ spanned by three nonzero vectors $\mathbf{u}, \mathbf{v}, \mathbf{w}$ in $\mathbf{R}^3$ (the three-dimensional prism in Figure 11). The base of $\mathcal{D}$ is the parallelogram spanned by $\mathbf{v}$ and $\mathbf{w}$, so the area of the base is $\|\mathbf{v} \times \mathbf{w}\|$. The height of $\mathcal{D}$ is $h = \|\mathbf{u}\| \cdot |\cos\theta|$, where θ is the angle between $\mathbf{u}$ and $\mathbf{v} \times \mathbf{w}$. Therefore,

$$\text{volume of } \mathcal{D} = (\text{area of base})(\text{height}) = \|\mathbf{v} \times \mathbf{w}\| \cdot \|\mathbf{u}\| \cdot |\cos\theta|$$

But, $\|\mathbf{v} \times \mathbf{w}\| \|\mathbf{u}\| \cos\theta$ is equal to the dot product of $\mathbf{v} \times \mathbf{w}$ and $\mathbf{u}$. This proves the formula

$$\text{volume of } \mathcal{D} = |\mathbf{u} \cdot (\mathbf{v} \times \mathbf{w})|$$

The quantity $\mathbf{u} \cdot (\mathbf{v} \times \mathbf{w})$, called the **scalar triple product**, can be expressed as a determinant. Let

$$\mathbf{u} = \langle u_1, u_2, u_3 \rangle, \qquad \mathbf{v} = \langle v_1, v_2, v_3 \rangle, \qquad \mathbf{w} = \langle w_1, w_2, w_3 \rangle$$

Then

$$\mathbf{u} \cdot (\mathbf{v} \times \mathbf{w}) = \mathbf{u} \cdot \left(\begin{vmatrix} v_2 & v_3 \\ w_2 & w_3 \end{vmatrix} \mathbf{i} - \begin{vmatrix} v_1 & v_3 \\ w_1 & w_3 \end{vmatrix} \mathbf{j} + \begin{vmatrix} v_1 & v_2 \\ w_1 & w_2 \end{vmatrix} \mathbf{k} \right)$$

$$= u_1 \begin{vmatrix} v_2 & v_3 \\ w_2 & w_3 \end{vmatrix} - u_2 \begin{vmatrix} v_1 & v_3 \\ w_1 & w_3 \end{vmatrix} + u_3 \begin{vmatrix} v_1 & v_2 \\ w_1 & w_2 \end{vmatrix}$$

$$= \begin{vmatrix} u_1 & u_2 & u_3 \\ v_1 & v_2 & v_3 \\ w_1 & w_2 & w_3 \end{vmatrix} = \det \begin{pmatrix} \mathbf{u} \\ \mathbf{v} \\ \mathbf{w} \end{pmatrix}$$

8

We obtain the following volume formula:

THEOREM 3 Volume via Scalar Triple Product and Determinants Let $\mathbf{u}, \mathbf{v}, \mathbf{w}$ be nonzero vectors in $\mathbf{R}^3$. Then the parallelepiped $\mathcal{D}$ spanned by $\mathbf{u}, \mathbf{v}$, and $\mathbf{w}$ has volume

$$V = |\mathbf{u} \cdot (\mathbf{v} \times \mathbf{w})| = \left| \det \begin{pmatrix} \mathbf{u} \\ \mathbf{v} \\ \mathbf{w} \end{pmatrix} \right|$$

9

EXAMPLE 7 Let $\mathbf{v} = \langle 1, 4, 5 \rangle$ and $\mathbf{w} = \langle -2, -1, 2 \rangle$. Calculate:

(a) The area A of the parallelogram spanned by $\mathbf{v}$ and $\mathbf{w}$

(b) The volume V of the parallelepiped in Figure 12

Solution We compute the cross product and apply Theorem 3:

$$\mathbf{v} \times \mathbf{w} = \begin{vmatrix} 4 & 5 \\ -1 & 2 \end{vmatrix} \mathbf{i} - \begin{vmatrix} 1 & 5 \\ -2 & 2 \end{vmatrix} \mathbf{j} + \begin{vmatrix} 1 & 4 \\ -2 & -1 \end{vmatrix} \mathbf{k} = \langle 13, -12, 7 \rangle$$

(a) The area of the parallelogram spanned by $\mathbf{v}$ and $\mathbf{w}$ is

$$A = \|\mathbf{v} \times \mathbf{w}\| = \sqrt{13^2 + (-12)^2 + 7^2} = \sqrt{362} \approx 19$$

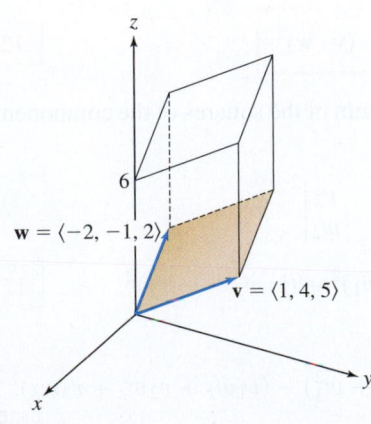

FIGURE 12

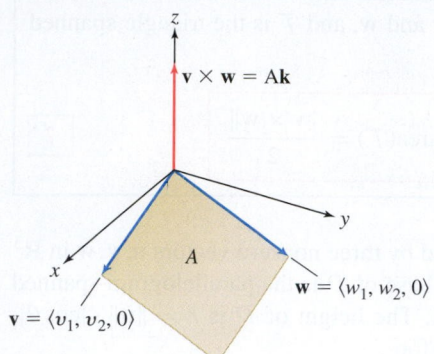

FIGURE 13 Parallelogram spanned by $\mathbf{v}$ and $\mathbf{w}$ in the xy-plane.

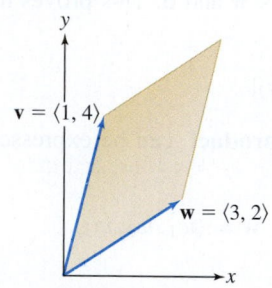

FIGURE 14

The third property in Theorem 1 is more subtle than the first two. It cannot be verified by algebra alone.

(b) The vertical leg of the parallelepiped is the vector $6\mathbf{k}$, so by Eq. (9),

$$V = |(6\mathbf{k}) \cdot (\mathbf{v} \times \mathbf{w})| = |\langle 0, 0, 6 \rangle \cdot \langle 13, -12, 7 \rangle| = 6(7) = 42 \quad \blacksquare$$

In $\mathbf{R}^2$, we can compute the area A of the parallelogram spanned by vectors $\mathbf{v} = \langle v_1, v_2 \rangle$ and $\mathbf{w} = \langle w_1, w_2 \rangle$ by regarding $\mathbf{v}$ and $\mathbf{w}$ as vectors in $\mathbf{R}^3$ with a zero component in the z-direction (Figure 13). Thus, we write $\mathbf{v} = \langle v_1, v_2, 0 \rangle$ and $\mathbf{w} = \langle w_1, w_2, 0 \rangle$. The cross product $\mathbf{v} \times \mathbf{w}$ is a vector pointing in the z-direction:

$$\mathbf{v} \times \mathbf{w} = \begin{vmatrix} \mathbf{i} & \mathbf{j} & \mathbf{k} \\ v_1 & v_2 & 0 \\ w_1 & w_2 & 0 \end{vmatrix} = \begin{vmatrix} v_2 & 0 \\ w_2 & 0 \end{vmatrix} \mathbf{i} - \begin{vmatrix} v_1 & 0 \\ w_1 & 0 \end{vmatrix} \mathbf{j} + \begin{vmatrix} v_1 & v_2 \\ w_1 & w_2 \end{vmatrix} \mathbf{k} = \begin{vmatrix} v_1 & v_2 \\ w_1 & w_2 \end{vmatrix} \mathbf{k}$$

By Eq. (7), the parallelogram spanned by $\mathbf{v}$ and $\mathbf{w}$ has area $A = \|\mathbf{v} \times \mathbf{w}\|$, and thus

$$A = \left| \det \begin{pmatrix} \mathbf{v} \\ \mathbf{w} \end{pmatrix} \right| = \left| \det \begin{pmatrix} v_1 & v_2 \\ w_1 & w_2 \end{pmatrix} \right| = |v_1 w_2 - v_2 w_1| \qquad \boxed{10}$$

EXAMPLE 8 Compute the area A of the parallelogram in Figure 14.

Solution We have $\begin{vmatrix} \mathbf{v} \\ \mathbf{w} \end{vmatrix} = \begin{vmatrix} 1 & 4 \\ 3 & 2 \end{vmatrix} = 1 \cdot 2 - 3 \cdot 4 = -10$. The area is the absolute value $A = |-10| = 10$. $\quad \blacksquare$

Proofs of Cross-Product Properties

We now derive the first two properties of the cross product listed in Theorem 1. Let

$$\mathbf{v} = \langle v_1, v_2, v_3 \rangle, \qquad \mathbf{w} = \langle w_1, w_2, w_3 \rangle$$

We prove that $\mathbf{v} \times \mathbf{w}$ is orthogonal to $\mathbf{v}$ by showing that $\mathbf{v} \cdot (\mathbf{v} \times \mathbf{w}) = 0$. By Eq. (8),

$$\mathbf{v} \cdot (\mathbf{v} \times \mathbf{w}) = \det \begin{pmatrix} \mathbf{v} \\ \mathbf{v} \\ \mathbf{w} \end{pmatrix} = v_1 \begin{vmatrix} v_2 & v_3 \\ w_2 & w_3 \end{vmatrix} - v_2 \begin{vmatrix} v_1 & v_3 \\ w_1 & w_3 \end{vmatrix} + v_3 \begin{vmatrix} v_1 & v_2 \\ w_1 & w_2 \end{vmatrix} \qquad \boxed{11}$$

Straightforward algebra (left to the reader) shows that the right-hand side of Eq. (11) is equal to zero. This shows that $\mathbf{v} \cdot (\mathbf{v} \times \mathbf{w}) = 0$ and thus $\mathbf{v} \times \mathbf{w}$ is orthogonal to $\mathbf{v}$ as claimed. Interchanging the roles of $\mathbf{v}$ and $\mathbf{w}$, we conclude also that $\mathbf{w} \times \mathbf{v}$ is orthogonal to $\mathbf{w}$, and since $\mathbf{v} \times \mathbf{w} = -\mathbf{w} \times \mathbf{v}$, it follows that $\mathbf{v} \times \mathbf{w}$ is orthogonal to $\mathbf{w}$. This proves part (i) of Theorem 1. To prove (ii), we shall use the following identity:

$$\|\mathbf{v} \times \mathbf{w}\|^2 = \|\mathbf{v}\|^2 \|\mathbf{w}\|^2 - (\mathbf{v} \cdot \mathbf{w})^2 \qquad \boxed{12}$$

To verify this identity, we compute $\|\mathbf{v} \times \mathbf{w}\|^2$ as the sum of the squares of the components of $\mathbf{v} \times \mathbf{w}$:

$$\|\mathbf{v} \times \mathbf{w}\|^2 = \begin{vmatrix} v_2 & v_3 \\ w_2 & w_3 \end{vmatrix}^2 + \begin{vmatrix} v_1 & v_3 \\ w_1 & w_3 \end{vmatrix}^2 + \begin{vmatrix} v_1 & v_2 \\ w_1 & w_2 \end{vmatrix}^2$$

$$= (v_2 w_3 - v_3 w_2)^2 + (v_1 w_3 - v_3 w_1)^2 + (v_1 w_2 - v_2 w_1)^2 \qquad \boxed{13}$$

On the other hand,

$$\|\mathbf{v}\|^2 \|\mathbf{w}\|^2 - (\mathbf{v} \cdot \mathbf{w})^2 = (v_1^2 + v_2^2 + v_3^2)(w_1^2 + w_2^2 + w_3^2) - (v_1 w_1 + v_2 w_2 + v_3 w_3)^2$$

$$\boxed{14}$$

Again, algebra (left to the reader) shows that the right side of Eq. (13) is equal to the right side of Eq. (14), proving Eq. (12).

Now let θ be the angle between $\mathbf{v}$ and $\mathbf{w}$. By Eq. (12),

$$\|\mathbf{v} \times \mathbf{w}\|^2 = \|\mathbf{v}\|^2\|\mathbf{w}\|^2 - (\mathbf{v} \cdot \mathbf{w})^2 = \|\mathbf{v}\|^2\|\mathbf{w}\|^2 - \|\mathbf{v}\|^2\|\mathbf{w}\|^2 \cos^2 \theta$$

$$= \|\mathbf{v}\|^2\|\mathbf{w}\|^2(1 - \cos^2 \theta) = \|\mathbf{v}\|^2\|\mathbf{w}\|^2 \sin^2 \theta$$

Therefore, $\|\mathbf{v} \times \mathbf{w}\| = \|\mathbf{v}\|\|\mathbf{w}\| \sin \theta$. Note that $\sin \theta \geq 0$ since, by convention, θ lies between 0 and π. This proves (ii).

12.4 SUMMARY

- Determinants of sizes 2×2 and 3×3:

$$\begin{vmatrix} a_{11} & a_{12} \\ a_{21} & a_{22} \end{vmatrix} = a_{11}a_{22} - a_{12}a_{21}$$

$$\begin{vmatrix} a_{11} & a_{12} & a_{13} \\ a_{21} & a_{22} & a_{23} \\ a_{31} & a_{32} & a_{33} \end{vmatrix} = a_{11} \begin{vmatrix} a_{22} & a_{23} \\ a_{32} & a_{33} \end{vmatrix} - a_{12} \begin{vmatrix} a_{21} & a_{23} \\ a_{31} & a_{33} \end{vmatrix} + a_{13} \begin{vmatrix} a_{21} & a_{22} \\ a_{31} & a_{32} \end{vmatrix}$$

- The *cross product* of $\mathbf{v} = \langle v_1, v_2, v_3 \rangle$ and $\mathbf{w} = \langle w_1, w_2, w_3 \rangle$ is the determinant

$$\mathbf{v} \times \mathbf{w} = \begin{vmatrix} \mathbf{i} & \mathbf{j} & \mathbf{k} \\ v_1 & v_2 & v_3 \\ w_1 & w_2 & w_3 \end{vmatrix} = \begin{vmatrix} v_2 & v_3 \\ w_2 & w_3 \end{vmatrix}\mathbf{i} - \begin{vmatrix} v_1 & v_3 \\ w_1 & w_3 \end{vmatrix}\mathbf{j} + \begin{vmatrix} v_1 & v_2 \\ w_1 & w_2 \end{vmatrix}\mathbf{k}$$

- The cross product $\mathbf{v} \times \mathbf{w}$ is the unique vector with the following three properties:

   (i) $\mathbf{v} \times \mathbf{w}$ is orthogonal to $\mathbf{v}$ and $\mathbf{w}$.
   (ii) $\mathbf{v} \times \mathbf{w}$ has length $\|\mathbf{v}\| \|\mathbf{w}\| \sin \theta$ (where θ is the angle between $\mathbf{v}$ and $\mathbf{w}$).
   (iii) $\{\mathbf{v}, \mathbf{w}, \mathbf{v} \times \mathbf{w}\}$ is a right-handed system.

- Properties of the cross product:

   (i) $\mathbf{w} \times \mathbf{v} = -\mathbf{v} \times \mathbf{w}$
   (ii) $\mathbf{v} \times \mathbf{w} = \mathbf{0}$ if and only if $\mathbf{w} = \lambda\mathbf{v}$ for some scalar or $\mathbf{v} = \mathbf{0}$
   (iii) $(\lambda\mathbf{v}) \times \mathbf{w} = \mathbf{v} \times (\lambda\mathbf{w}) = \lambda(\mathbf{v} \times \mathbf{w})$
   (iv) $(\mathbf{u} + \mathbf{v}) \times \mathbf{w} = \mathbf{u} \times \mathbf{w} + \mathbf{v} \times \mathbf{w}$ and $\mathbf{v} \times (\mathbf{u} + \mathbf{w}) = \mathbf{v} \times \mathbf{u} + \mathbf{v} \times \mathbf{w}$

- Cross products of standard basis vectors (Figure 15):

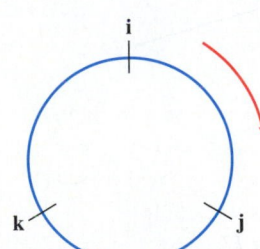

$$\mathbf{i} \times \mathbf{j} = \mathbf{k}, \qquad \mathbf{j} \times \mathbf{k} = \mathbf{i}, \qquad \mathbf{k} \times \mathbf{i} = \mathbf{j}$$

$$\mathbf{j} \times \mathbf{i} = -\mathbf{k}, \qquad \mathbf{k} \times \mathbf{j} = -\mathbf{i}, \qquad \mathbf{i} \times \mathbf{k} = -\mathbf{j}$$

$$\mathbf{i} \times \mathbf{i} = \mathbf{0}, \qquad \mathbf{j} \times \mathbf{j} = \mathbf{0}, \qquad \mathbf{k} \times \mathbf{k} = \mathbf{0}$$

FIGURE 15 Circle for computing the cross products of the basis vectors.

- The parallelogram spanned by $\mathbf{v}$ and $\mathbf{w}$ has area $\|\mathbf{v} \times \mathbf{w}\|$.
- The triangle spanned by $\mathbf{v}$ and $\mathbf{w}$ has area $\dfrac{\|\mathbf{v} \times \mathbf{w}\|}{2}$.
- Cross-product identity: $\|\mathbf{v} \times \mathbf{w}\|^2 = \|\mathbf{v}\|^2\|\mathbf{w}\|^2 - (\mathbf{v} \cdot \mathbf{w})^2$.
- The *scalar triple product* is defined by $\mathbf{u} \cdot (\mathbf{v} \times \mathbf{w})$. We have

$$\mathbf{u} \cdot (\mathbf{v} \times \mathbf{w}) = \det \begin{pmatrix} \mathbf{u} \\ \mathbf{v} \\ \mathbf{w} \end{pmatrix}$$

- The parallelepiped spanned by $\mathbf{u}$, $\mathbf{v}$, and $\mathbf{w}$ has volume $|\mathbf{u} \cdot (\mathbf{v} \times \mathbf{w})|$.

12.4 EXERCISES

Preliminary Questions

1. What is the $(1, 3)$ minor of the matrix $\begin{vmatrix} 3 & 4 & 2 \\ -5 & -1 & 1 \\ 4 & 0 & 3 \end{vmatrix}$?

2. The angle between two unit vectors $\mathbf{e}$ and $\mathbf{f}$ is $\frac{\pi}{6}$. What is the length of $\mathbf{e} \times \mathbf{f}$?

3. What is $\mathbf{u} \times \mathbf{w}$, assuming that $\mathbf{w} \times \mathbf{u} = \langle 2, 2, 1 \rangle$?

4. Find the cross product without using the formula:
(a) $\langle 4, 8, 2 \rangle \times \langle 4, 8, 2 \rangle$
(b) $\langle 4, 8, 2 \rangle \times \langle 2, 4, 1 \rangle$

5. What are $\mathbf{i} \times \mathbf{j}$ and $\mathbf{i} \times \mathbf{k}$?

6. When is the cross product $\mathbf{v} \times \mathbf{w}$ equal to zero?

7. Which of the following are meaningful and which are not? Explain.
(a) $(\mathbf{u} \cdot \mathbf{v}) \times \mathbf{w}$
(b) $(\mathbf{u} \times \mathbf{v}) \cdot \mathbf{w}$
(c) $\|\mathbf{w}\|(\mathbf{u} \cdot \mathbf{v})$
(d) $\|\mathbf{w}\|(\mathbf{u} \times \mathbf{v})$

8. Which of the following vectors is equal to $\mathbf{j} \times \mathbf{i}$?
(a) $\mathbf{i} \times \mathbf{k}$
(b) $-\mathbf{k}$
(c) $\mathbf{i} \times \mathbf{j}$

Exercises

In Exercises 1–4, calculate the 2×2 determinant.

1. $\begin{vmatrix} 1 & 2 \\ 4 & 3 \end{vmatrix}$

2. $\begin{vmatrix} \frac{2}{3} & \frac{1}{6} \\ -5 & 2 \end{vmatrix}$

3. $\begin{vmatrix} -6 & 9 \\ 1 & 1 \end{vmatrix}$

4. $\begin{vmatrix} 9 & 25 \\ 5 & 14 \end{vmatrix}$

In Exercises 5–8, calculate the 3×3 determinant.

5. $\begin{vmatrix} 1 & 2 & 1 \\ 4 & -3 & 0 \\ 1 & 0 & 1 \end{vmatrix}$

6. $\begin{vmatrix} 1 & 0 & 1 \\ -2 & 0 & 3 \\ 1 & 3 & -1 \end{vmatrix}$

7. $\begin{vmatrix} 1 & 2 & 3 \\ 2 & 4 & 6 \\ -3 & -4 & 2 \end{vmatrix}$

8. $\begin{vmatrix} 1 & 0 & 0 \\ 0 & 0 & -1 \\ 0 & 1 & 0 \end{vmatrix}$

In Exercises 9–14, calculate $\mathbf{v} \times \mathbf{w}$.

9. $\mathbf{v} = \langle 1, 2, 1 \rangle$, $\mathbf{w} = \langle 3, 1, 1 \rangle$

10. $\mathbf{v} = \langle 2, 0, 0 \rangle$, $\mathbf{w} = \langle -1, 0, 1 \rangle$

11. $\mathbf{v} = \langle \frac{2}{3}, 1, \frac{1}{2} \rangle$, $\mathbf{w} = \langle 4, -6, 3 \rangle$

12. $\mathbf{v} = \langle 1, 1, 0 \rangle$, $\mathbf{w} = \langle 0, 1, 1 \rangle$

13. $\mathbf{v} = \langle 1, 2, 3 \rangle$, $\mathbf{w} = \langle 1, 2, 3.01 \rangle$

14. $\mathbf{v} = \langle 2.4, -1.25, 3 \rangle$, $\mathbf{w} = \langle -7.68, 4, -9.6 \rangle$

In Exercises 15–18, use the relations in Eqs. (5) and (6) to calculate the cross product.

15. $(\mathbf{i} + \mathbf{j}) \times \mathbf{k}$

16. $(\mathbf{j} - \mathbf{k}) \times (\mathbf{j} + \mathbf{k})$

17. $(\mathbf{i} - 3\mathbf{j} + 2\mathbf{k}) \times (\mathbf{j} - \mathbf{k})$

18. $(2\mathbf{i} - 3\mathbf{j} + 4\mathbf{k}) \times (\mathbf{i} + \mathbf{j} - 7\mathbf{k})$

In Exercises 19–24, calculate the cross product assuming that

$$\mathbf{u} \times \mathbf{v} = \langle 1, 1, 0 \rangle, \quad \mathbf{u} \times \mathbf{w} = \langle 0, 3, 1 \rangle, \quad \mathbf{v} \times \mathbf{w} = \langle 2, -1, 1 \rangle$$

19. $\mathbf{v} \times \mathbf{u}$

20. $\mathbf{v} \times (\mathbf{u} + \mathbf{v})$

21. $\mathbf{w} \times (\mathbf{u} + \mathbf{v})$

22. $(3\mathbf{u} + 4\mathbf{w}) \times \mathbf{w}$

23. $(\mathbf{u} - 2\mathbf{v}) \times (\mathbf{u} + 2\mathbf{v})$

24. $(\mathbf{v} + \mathbf{w}) \times (3\mathbf{u} + 2\mathbf{v})$

25. Let $\mathbf{v} = \langle a, b, c \rangle$. Calculate $\mathbf{v} \times \mathbf{i}$, $\mathbf{v} \times \mathbf{j}$, and $\mathbf{v} \times \mathbf{k}$.

26. Find $\mathbf{v} \times \mathbf{w}$, where $\mathbf{v}$ and $\mathbf{w}$ are vectors of length 3 in the xz-plane, oriented as in Figure 16, and $\theta = \frac{\pi}{6}$.

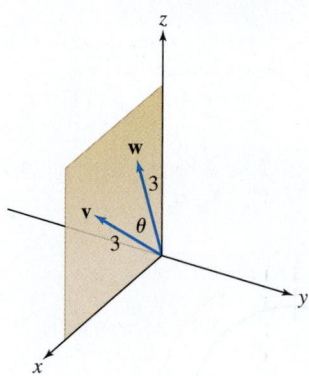

FIGURE 16

In Exercises 27 and 28, refer to Figure 17.

27. Which of $\mathbf{u}$ and $-\mathbf{u}$ is equal to $\mathbf{v} \times \mathbf{w}$?

28. Which of the following form a right-handed system?
(a) $\{\mathbf{v}, \mathbf{w}, \mathbf{u}\}$
(b) $\{\mathbf{w}, \mathbf{v}, \mathbf{u}\}$
(c) $\{\mathbf{v}, \mathbf{u}, \mathbf{w}\}$
(d) $\{\mathbf{u}, \mathbf{v}, \mathbf{w}\}$
(e) $\{\mathbf{w}, \mathbf{v}, -\mathbf{u}\}$
(f) $\{\mathbf{v}, -\mathbf{u}, \mathbf{w}\}$

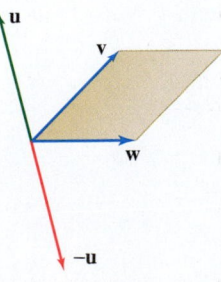

FIGURE 17

29. Let $\mathbf{v} = \langle 3, 0, 0 \rangle$ and $\mathbf{w} = \langle 0, 1, -1 \rangle$. Determine $\mathbf{u} = \mathbf{v} \times \mathbf{w}$ using the geometric properties of the cross product rather than the formula.

30. What are the possible angles θ between two unit vectors $\mathbf{e}$ and $\mathbf{f}$ if $\|\mathbf{e} \times \mathbf{f}\| = \frac{1}{2}$?

31. Show that if $\mathbf{v}$ and $\mathbf{w}$ lie in the yz-plane, then $\mathbf{v} \times \mathbf{w}$ is a multiple of $\mathbf{i}$.

32. Find the two unit vectors orthogonal to both $\mathbf{a} = \langle 3, 1, 1 \rangle$ and $\mathbf{b} = \langle -1, 2, 1 \rangle$.

33. Let $\mathbf{e}$ and $\mathbf{e}'$ be unit vectors in $\mathbf{R}^3$ such that $\mathbf{e} \perp \mathbf{e}'$. Use the geometric properties of the cross product to compute $\mathbf{e} \times (\mathbf{e}' \times \mathbf{e})$.

34. Determine the magnitude of each Coriolis force on a 1.5-kg parcel of air with wind $\mathbf{v}$.

(a) $\mathbf{v}$ is 25 m/s toward the east at the equator

(b) $\mathbf{v}$ is 25 m/s toward the east at 45° N

(c) $\mathbf{v}$ is 35 m/s toward the south at 30° N

(d) $\mathbf{v}$ is 35 m/s toward the south at 60° N

35. Determine the magnitude of each Coriolis force on a 2.3-kg parcel of air with wind $\mathbf{v}$.

(a) $\mathbf{v}$ is 20 m/s toward the west at the equator

(b) $\mathbf{v}$ is 20 m/s toward the west at 60° N

(c) $\mathbf{v}$ is 40 m/s toward the south at the equator

(d) $\mathbf{v}$ is 40 m/s toward the south at 45° S

In Exercises 36 and 37, a force $\mathbf{F}$ *(in newtons) on an electron moving at velocity* $\mathbf{v}$ *meters per second in a uniform magnetic field* $\mathbf{B}$ *(in teslas) is given by* $\mathbf{F} = q(\mathbf{v} \times \mathbf{B})$, *where* $q = -1.6 \times 10^{-19}$ *coulombs is the charge on the electron.*

36. Calculate the force $\mathbf{F}$ on an electron moving with velocity 10^5 m/s in the direction $\mathbf{i}$ in a uniform magnetic field $\mathbf{B}$, where $\mathbf{B} = 0.0004\mathbf{i} + 0.0001\mathbf{j}$ teslas.

37. Assume an electron moves with velocity $\mathbf{v}$ in the plane and $\mathbf{B}$ is a uniform magnetic field pointing directly out of the page. Which of the two vectors, $\mathbf{F}_1$ or $\mathbf{F}_2$, in Figure 18 represents the force on the electron? Remember that q is negative.

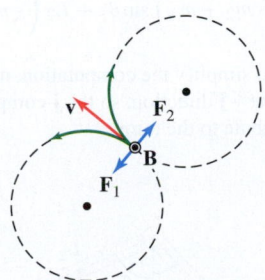

FIGURE 18 The magnetic field vector $\mathbf{B}$ points directly out of the page.

38. Calculate the scalar triple product $\mathbf{u} \cdot (\mathbf{v} \times \mathbf{w})$, where $\mathbf{u} = \langle 1, 1, 0 \rangle$, $\mathbf{v} = \langle 3, -2, 2 \rangle$, and $\mathbf{w} = \langle 4, -1, 2 \rangle$.

39. Verify identity (12) for vectors $\mathbf{v} = \langle 3, -2, 2 \rangle$ and $\mathbf{w} = \langle 4, -1, 2 \rangle$.

40. Find the volume of the parallelepiped spanned by $\mathbf{u}$, $\mathbf{v}$, and $\mathbf{w}$ in Figure 19.

41. Find the area of the parallelogram spanned by $\mathbf{v}$ and $\mathbf{w}$ in Figure 19.

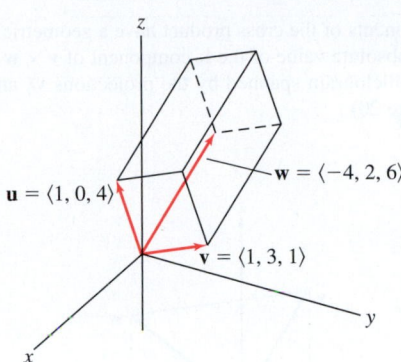

FIGURE 19

42. Calculate the volume of the parallelepiped spanned by

$$\mathbf{u} = \langle 2, 2, 1 \rangle, \qquad \mathbf{v} = \langle 1, 0, 3 \rangle, \qquad \mathbf{w} = \langle 0, -4, 0 \rangle$$

43. Sketch and compute the volume of the parallelepiped spanned by

$$\mathbf{u} = \langle 1, 0, 0 \rangle, \qquad \mathbf{v} = \langle 0, 2, 0 \rangle, \qquad \mathbf{w} = \langle 1, 1, 2 \rangle$$

44. Sketch the parallelogram spanned by $\mathbf{u} = \langle 1, 1, 1 \rangle$ and $\mathbf{v} = \langle 0, 0, 4 \rangle$, and compute its area.

45. Calculate the area of the parallelogram spanned by $\mathbf{u} = \langle 1, 0, 3 \rangle$ and $\mathbf{v} = \langle 2, 1, 1 \rangle$.

46. Find the area of the parallelogram determined by the vectors $\langle a, 0, 0 \rangle$ and $\langle 0, b, c \rangle$.

47. Sketch the triangle with vertices at the origin O, $P = (3, 3, 0)$, and $Q = (0, 3, 3)$, and compute its area using cross products.

48. Use the cross product to find the area of the triangle with vertices $P = (1, 1, 5)$, $Q = (3, 4, 3)$, and $R = (1, 5, 7)$.

49. Use the cross product to find the area of the triangle in the xy-plane defined by $(1, 2)$, $(3, 4)$, and $(-2, 2)$.

50. Use the cross product to find the area of the quadrilateral in the xy-plane defined by $(0, 0)$, $(1, -1)$, $(3, 1)$, and $(2, 4)$.

51. Check that the four points $P(2, 4, 4)$, $Q(3, 1, 6)$, $R(2, 8, 0)$, and $S(7, 2, 1)$ all lie in a plane. Then use vectors to find the area of the quadrilateral they define.

52. Use the cross product to find the area of the triangle with vertices $(a, 0, 0)$, $(0, b, 0)$, and $(0, 0, c)$.

In Exercises 53–55, prove each of the identities using the formula for the cross product.

53. $\mathbf{v} \times \mathbf{w} = -\mathbf{w} \times \mathbf{v}$

54. $(\lambda \mathbf{v}) \times \mathbf{w} = \lambda(\mathbf{v} \times \mathbf{w})$ (λ a scalar)

55. $(\mathbf{u} + \mathbf{v}) \times \mathbf{w} = \mathbf{u} \times \mathbf{w} + \mathbf{v} \times \mathbf{w}$

56. Use the geometric description in Theorem 1 to prove Theorem 2 (iii): $\mathbf{v} \times \mathbf{w} = \mathbf{0}$ if and only if $\mathbf{w} = \lambda \mathbf{v}$ for some scalar λ or $\mathbf{v} = \mathbf{0}$.

57. Prove $\mathbf{i} \times \mathbf{j} = \mathbf{k}$ and $\mathbf{k} \times \mathbf{j} = -\mathbf{i}$ by each of the following methods:

(a) Using the definition of cross product as a determinant

(b) Using the geometric description of the cross product in Theorem 1

58. Using standard basis vectors, find an example demonstrating that the cross product is not associative.

59. The components of the cross product have a geometric interpretation. Show that the absolute value of the **k**-component of $\mathbf{v} \times \mathbf{w}$ is equal to the area of the parallelogram spanned by the projections $\mathbf{v}_0$ and $\mathbf{w}_0$ onto the xy-plane (Figure 20).

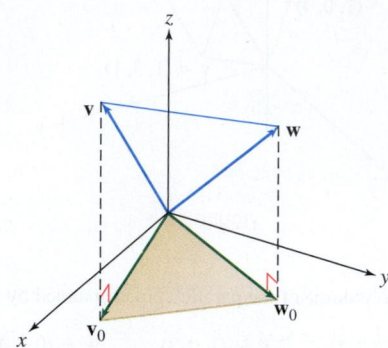

FIGURE 20

60. [icon] Formulate and prove analogs of the result in Exercise 59 for the **i**- and **j**-components of $\mathbf{v} \times \mathbf{w}$.

61. [icon] Show that three points P, Q, R are collinear (lie on a line) if and only if $\overrightarrow{PQ} \times \overrightarrow{PR} = \mathbf{0}$.

62. Use the result of Exercise 61 to determine whether the points P, Q, and R are collinear, and if not, find a vector perpendicular to the plane containing them.

(a) $P = (2, 1, 0)$, $Q = (1, 5, 2)$, $R = (-1, 13, 6)$

(b) $P = (2, 1, 0)$, $Q = (-3, 21, 10)$, $R = (5, -2, 9)$

(c) $P = (1, 1, 0)$, $Q = (1, -2, -1)$, $R = (3, 2, -4)$

63. Solve the equation $\langle 1, 1, 1 \rangle \times \mathbf{X} = \langle 1, -1, 0 \rangle$, where $\mathbf{X} = \langle x, y, z \rangle$. *Note:* There are infinitely many solutions.

64. [icon] Explain geometrically why $\langle 1, 1, 1 \rangle \times \mathbf{X} = \langle 1, 0, 0 \rangle$ has no solution, where $\mathbf{X} = \langle x, y, z \rangle$.

65. [icon] Let $\mathbf{X} = \langle x, y, z \rangle$. Show that $\mathbf{i} \times \mathbf{X} = \mathbf{v}$ has a solution if and only if $\mathbf{v}$ is contained in the yz-plane (the **i**-component is zero).

66. [icon] Suppose that vectors $\mathbf{u}$, $\mathbf{v}$, and $\mathbf{w}$ are mutually orthogonal—that is, $\mathbf{u} \perp \mathbf{v}$, $\mathbf{u} \perp \mathbf{w}$, and $\mathbf{v} \perp \mathbf{w}$. Prove that $(\mathbf{u} \times \mathbf{v}) \times \mathbf{w} = \mathbf{0}$ and $\mathbf{u} \times (\mathbf{v} \times \mathbf{w}) = \mathbf{0}$.

In Exercises 67–70, the torque about the origin O due to a force $\mathbf{F}$ acting on an object with position vector $\mathbf{r}$ is the vector quantity $\boldsymbol{\tau} = \mathbf{r} \times \mathbf{F}$. If several forces $\mathbf{F}_j$ act at positions $\mathbf{r}_j$, then the net torque (units: N-m or lb-ft) is the sum

$$\boldsymbol{\tau} = \sum \mathbf{r}_j \times \mathbf{F}_j$$

67. Calculate the torque $\boldsymbol{\tau}$ about O acting at the point P on the mechanical arm in Figure 21(A), assuming that a 25-newton force acts as indicated.

68. Calculate the net torque about O at P, assuming that a 30-kg mass is attached at P [Figure 21(B)]. The force $\mathbf{F}_g$ due to gravity on a mass m has magnitude $9.8m$ m/s^2 in the downward direction.

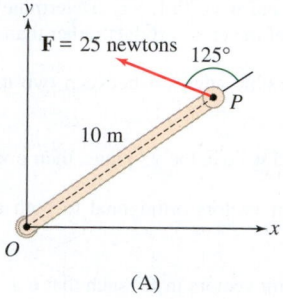

(A)

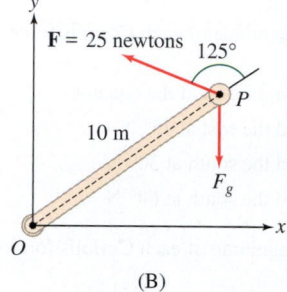

(B)

FIGURE 21

69. Let $\boldsymbol{\tau}$ be the net torque about O acting on the robotic arm of Figure 22, here taking into account the weight of the arms themselves. Assume that the arms have mass m_1 and m_2 (in kilograms) and that a weight of m_3 kg is located at the endpoint P. In calculating the torque, we may assume that the entire mass of each arm segment lies at the midpoint of the arm (its center of mass). Show that the position vectors of the masses m_1, m_2, and m_3 are

$$\mathbf{r}_1 = \frac{1}{2}L_1(\sin\theta_1\mathbf{i} + \cos\theta_1\mathbf{j})$$

$$\mathbf{r}_2 = L_1(\sin\theta_1\mathbf{i} + \cos\theta_1\mathbf{j}) + \frac{1}{2}L_2(\sin\theta_2\mathbf{i} - \cos\theta_2\mathbf{j})$$

$$\mathbf{r}_3 = L_1(\sin\theta_1\mathbf{i} + \cos\theta_1\mathbf{j}) + L_2(\sin\theta_2\mathbf{i} - \cos\theta_2\mathbf{j})$$

Then show that

$$\boldsymbol{\tau} = -g\left(L_1\left(\frac{1}{2}m_1 + m_2 + m_3\right)\sin\theta_1 + L_2\left(\frac{1}{2}m_2 + m_3\right)\sin\theta_2\right)\mathbf{k}$$

where $g = 9.8$ m/s^2. To simplify the computation, note that all three gravitational forces act in the $-\mathbf{j}$ direction, so the **j**-components of the position vectors $\mathbf{r}_i$ do not contribute to the torque.

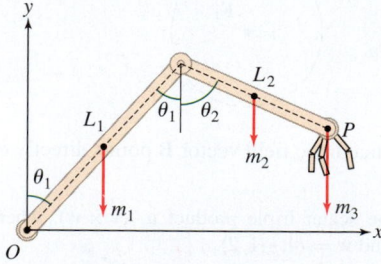

FIGURE 22

70. Continuing with Exercise 69, suppose that $L_1 = 3$ m, $L_2 = 2$ m, $m_1 = 15$ kg, $m_2 = 20$ kg, and $m_3 = 18$ kg. If the angles θ_1, θ_2 are equal (say, to θ), what is the maximum allowable value of θ if we assume that the robotic arm can sustain a maximum torque of 1200 N-m?

Further Insights and Challenges

71. Show that 3×3 determinants can be computed using the **diagonal rule**: Repeat the first two columns of the matrix and form the products of the numbers along the six diagonals indicated. Then add the products for the diagonals that slant from left to right and subtract the products for the diagonals that slant from right to left.

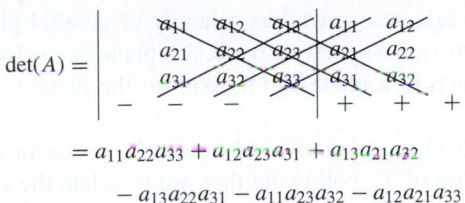

$$\det(A) = \begin{vmatrix} a_{11} & a_{12} & a_{13} \\ a_{21} & a_{22} & a_{23} \\ a_{31} & a_{32} & a_{33} \end{vmatrix} \begin{matrix} a_{11} & a_{12} \\ a_{21} & a_{22} \\ a_{31} & a_{32} \end{matrix}$$

$$= a_{11}a_{22}a_{33} + a_{12}a_{23}a_{31} + a_{13}a_{21}a_{32}$$
$$- a_{13}a_{22}a_{31} - a_{11}a_{23}a_{32} - a_{12}a_{21}a_{33}$$

72. Use the diagonal rule to calculate $\begin{vmatrix} 2 & 4 & 3 \\ 0 & 1 & -7 \\ -1 & 5 & 3 \end{vmatrix}$.

73. Prove that $\mathbf{v} \times \mathbf{w} = \mathbf{v} \times \mathbf{u}$ if and only if $\mathbf{u} = \mathbf{w} + \lambda \mathbf{v}$ for some scalar λ. Assume that $\mathbf{v} \neq \mathbf{0}$.

74. Use Eq. (12) to prove the Cauchy–Schwarz inequality:

$$|\mathbf{v} \cdot \mathbf{w}| \leq \|\mathbf{v}\| \, \|\mathbf{w}\|$$

Show that equality holds if and only if $\mathbf{w}$ is a multiple of $\mathbf{v}$ or at least one of $\mathbf{v}$ and $\mathbf{w}$ is zero.

75. Show that if $\mathbf{u}$, $\mathbf{v}$, and $\mathbf{w}$ are nonzero vectors and $(\mathbf{u} \times \mathbf{v}) \times \mathbf{w} = \mathbf{0}$, then either (i) $\mathbf{u}$ and $\mathbf{v}$ are parallel, or (ii) $\mathbf{w}$ is orthogonal to $\mathbf{u}$ and $\mathbf{v}$.

76. Suppose that $\mathbf{u}$, $\mathbf{v}$, $\mathbf{w}$ are nonzero and

$$(\mathbf{u} \times \mathbf{v}) \times \mathbf{w} = \mathbf{u} \times (\mathbf{v} \times \mathbf{w}) = \mathbf{0}$$

Show that $\mathbf{u}$, $\mathbf{v}$, and $\mathbf{w}$ are either mutually parallel or mutually perpendicular. *Hint:* Use Exercise 75.

77. ✎ Let $\mathbf{a}$, $\mathbf{b}$, $\mathbf{c}$ be nonzero vectors. Assume that $\mathbf{b}$ and $\mathbf{c}$ are not parallel, and set

$$\mathbf{v} = \mathbf{a} \times (\mathbf{b} \times \mathbf{c}), \qquad \mathbf{w} = (\mathbf{a} \cdot \mathbf{c})\mathbf{b} - (\mathbf{a} \cdot \mathbf{b})\mathbf{c}$$

(a) Prove that:

(i) $\mathbf{v}$ lies in the plane spanned by $\mathbf{b}$ and $\mathbf{c}$.

(ii) $\mathbf{v}$ is orthogonal to $\mathbf{a}$.

(b) Prove that $\mathbf{w}$ also satisfies (i) and (ii). Conclude that $\mathbf{v}$ and $\mathbf{w}$ are parallel.

(c) Show algebraically that $\mathbf{v} = \mathbf{w}$ (Figure 23).

78. Use Exercise 77 to prove the identity

$$(\mathbf{a} \times \mathbf{b}) \times \mathbf{c} - \mathbf{a} \times (\mathbf{b} \times \mathbf{c}) = (\mathbf{a} \cdot \mathbf{b})\mathbf{c} - (\mathbf{b} \cdot \mathbf{c})\mathbf{a}$$

79. Show that if $\mathbf{a}$, $\mathbf{b}$ are nonzero vectors such that $\mathbf{a} \perp \mathbf{b}$, then there exists a vector $\mathbf{X}$ such that

$$\mathbf{a} \times \mathbf{X} = \mathbf{b} \qquad \boxed{15}$$

Hint: Show that if $\mathbf{X}$ is orthogonal to $\mathbf{b}$ and is not a multiple of $\mathbf{a}$, then $\mathbf{a} \times \mathbf{X}$ is a multiple of $\mathbf{b}$.

80. Show that if $\mathbf{a}$, $\mathbf{b}$ are nonzero vectors such that $\mathbf{a} \perp \mathbf{b}$, then the set of all solutions of Eq. (15) is a line with $\mathbf{a}$ as direction vector. *Hint:* Let $\mathbf{X}_0$ be any solution (which exists by Exercise 79), and show that every other solution is of the form $\mathbf{X}_0 + \lambda \mathbf{a}$ for some scalar λ.

81. Assume that $\mathbf{v}$ and $\mathbf{w}$ lie in the first quadrant in $\mathbf{R}^2$ as in Figure 24. Use geometry to prove that the area of the parallelogram is equal to $\det \begin{pmatrix} \mathbf{v} \\ \mathbf{w} \end{pmatrix}$.

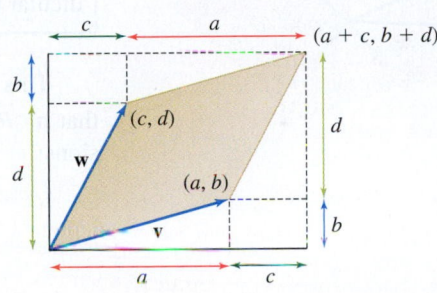

FIGURE 24

82. Consider the tetrahedron spanned by vectors $\mathbf{a}$, $\mathbf{b}$, and $\mathbf{c}$ as in Figure 25(A). Let A, B, C be the faces containing the origin O, and let D be the fourth face opposite O. For each face F, let $\mathbf{v}_F$ be the vector that is perpendicular to the face, pointing outside the tetrahedron, of magnitude equal to twice the area of F. Prove the relations

$$\mathbf{v}_A + \mathbf{v}_B + \mathbf{v}_C = \mathbf{a} \times \mathbf{b} + \mathbf{b} \times \mathbf{c} + \mathbf{c} \times \mathbf{a}$$

$$\mathbf{v}_A + \mathbf{v}_B + \mathbf{v}_C + \mathbf{v}_D = \mathbf{0}$$

Hint: Show that $\mathbf{v}_D = (\mathbf{c} - \mathbf{b}) \times (\mathbf{b} - \mathbf{a})$.

83. In the notation of Exercise 82, suppose that $\mathbf{a}, \mathbf{b}, \mathbf{c}$ are mutually perpendicular as in Figure 25(B). Let S_F be the area of face F. Prove the following three-dimensional version of the Pythagorean Theorem:

$$S_A^2 + S_B^2 + S_C^2 = S_D^2$$

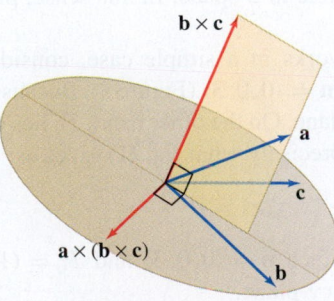

FIGURE 23

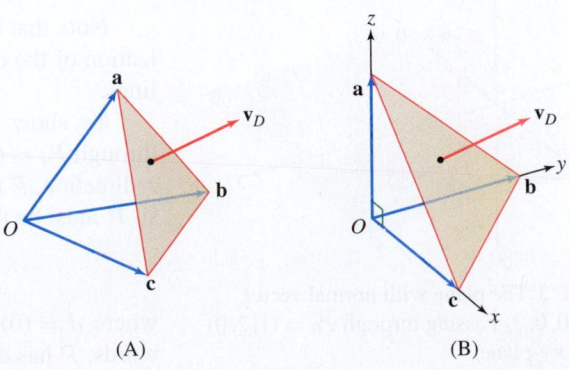

FIGURE 25 The vector $\mathbf{v}_D$ is perpendicular to the face.

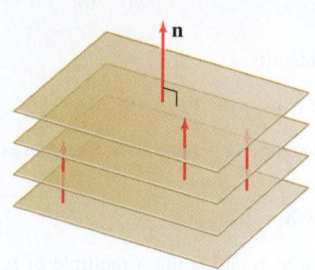

FIGURE 1 The vector **n** is perpendicular to a family of parallel planes.

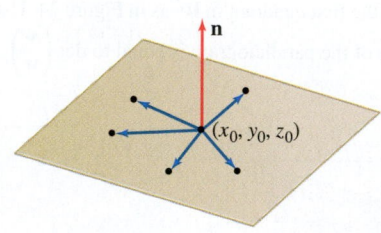

FIGURE 2

12.5 Planes in 3-Space

A linear equation $ax + by = c$ in two variables defines a line in $\mathbf{R}^2$. In this section, we show that a linear equation $ax + by + cz = d$ in three variables defines a plane in $\mathbf{R}^3$.

To identify a plane in $\mathbf{R}^3$, we need to identify how the plane is oriented in space. A vector **n** that is perpendicular to a plane is called a **normal vector** to the plane. This vector determines how the plane is oriented as well as a family of parallel planes that all have **n** as a normal vector (Figure 1). To specify a particular plane, we select a point on the plane. Thus, a normal vector **n** to a plane and a point on the plane completely determine the plane.

Given a normal vector $\mathbf{n} = \langle a, b, c \rangle$ to a plane $\mathcal{P}$, and a point $P_0 = (x_0, y_0, z_0)$ on $\mathcal{P}$, we begin with a geometric description of $\mathcal{P}$. Following that, we translate the geometric description into a variety of equations for $\mathcal{P}$ in x, y, z.

> **Geometric Description of a Plane**   The plane $\mathcal{P}$ through $P_0 = (x_0, y_0, z_0)$ with normal vector $\mathbf{n} = \langle a, b, c \rangle$ consists of the tips of all vectors based at P_0 that are perpendicular to **n** (Figure 2).

This geometric description indicates that the plane $\mathcal{P}$ is the set of $P = (x, y, z)$ such that $\mathbf{n} \cdot \overrightarrow{P_0 P} = 0$. This equation of the plane is equivalent to each of the following versions:

$$\langle a, b, c \rangle \cdot \langle x - x_0, y - y_0, z - z_0 \rangle = 0$$

$$a(x - x_0) + b(y - y_0) + c(z - z_0) = 0$$

$$ax + by + cz = ax_0 + by_0 + cz_0$$

$$\mathbf{n} \cdot \langle x, y, z \rangle = ax_0 + by_0 + cz_0$$

Now, for simplicity, set $d = ax_0 + by_0 + cz_0$. We have

> **Equations of a Plane**   The plane through the point $P_0 = (x_0, y_0, z_0)$ with normal vector $\mathbf{n} = \langle a, b, c \rangle$ is described by
>
> **Vector form:** $\qquad\qquad\qquad\qquad \mathbf{n} \cdot \langle x, y, z \rangle = d \qquad \boxed{1}$
>
> **Scalar forms:** $\qquad a(x - x_0) + b(y - y_0) + c(z - z_0) = 0 \qquad \boxed{2}$
>
> $$ax + by + cz = d \qquad \boxed{3}$$
>
> where $d = ax_0 + by_0 + cz_0$.

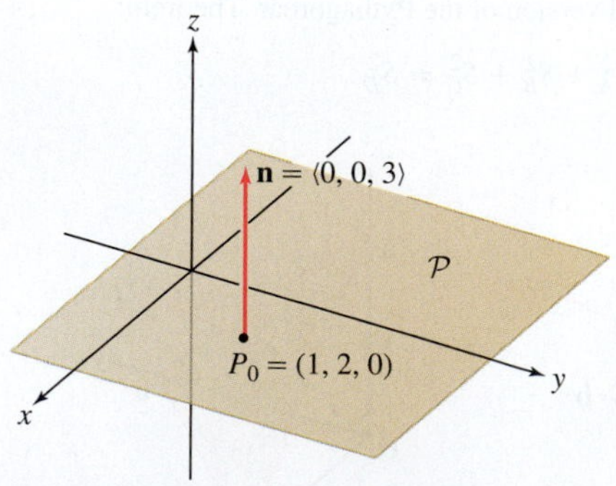

Note that the equation $ax + by + cz = d$ of a plane in 3-space is the direct generalization of the equation $ax + by = c$ of a line in 2-space. In this sense, planes generalize lines.

To show how the plane equation works in a simple case, consider the plane $\mathcal{P}$ through $P_0 = (1, 2, 0)$ with normal vector $\mathbf{n} = \langle 0, 0, 3 \rangle$ (Figure 3). Because **n** points in the z-direction, $\mathcal{P}$ must be parallel to the xy-plane. On the other hand, P_0 lies on the xy-plane, so $\mathcal{P}$ must be the xy-plane itself. This is precisely what Eq. (1) gives us:

$$\langle 0, 0, 3 \rangle \cdot \langle x, y, z \rangle = d$$

FIGURE 3 The plane with normal vector $\mathbf{n} = \langle 0, 0, 3 \rangle$ passing through $P_0 = (1, 2, 0)$ is the xy-plane.

where $d = (0)(1) + (0)(2) + (3)(0) = 0$ since $\mathbf{n} = \langle 0, 0, 3 \rangle$ and $P_0 = (1, 2, 0)$. In other words, $\mathcal{P}$ has equation $z = 0$, so $\mathcal{P}$ is the xy-plane.

EXAMPLE 1 Find an equation of the plane through $P_0 = (3, 1, 0)$ with normal vector $\mathbf{n} = \langle 3, 2, -5 \rangle$.

Solution Using Eq. (2), we obtain

$$3(x - 3) + 2(y - 1) - 5z = 0 \quad \text{or} \quad 3x + 2y - 5z = 11$$

Alternatively, we can compute

$$d = \mathbf{n} \cdot \overrightarrow{OP_0} = \langle 3, 2, -5 \rangle \cdot \langle 3, 1, 0 \rangle = 11$$

and write the equation as $\langle 3, 2, -5 \rangle \cdot \langle x, y, z \rangle = 11$, or $3x + 2y - 5z = 11$. ∎

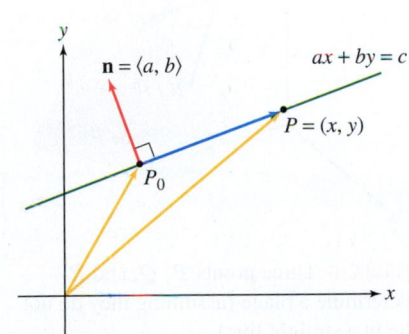

FIGURE 4 A line in $\mathbf{R}^2$ with normal vector $\mathbf{n}$.

CONCEPTUAL INSIGHT Keep in mind that the components of a normal vector are "lurking" inside the equation $ax + by + cz = d$, because $\mathbf{n} = \langle a, b, c \rangle$. The same is true for lines in $\mathbf{R}^2$. The line $ax + by = c$ in Figure 4 has normal vector $\mathbf{n} = \langle a, b \rangle$ because the line has slope $-a/b$ and the vector $\mathbf{n}$ has slope b/a (lines are orthogonal if the product of their slopes is -1).

Note that if $\mathbf{n}$ is normal to a plane $\mathcal{P}$, then so is every nonzero scalar multiple $\lambda \mathbf{n}$. When we use $\lambda \mathbf{n}$ instead of $\mathbf{n}$, the resulting equation for $\mathcal{P}$ changes by a factor of λ. For example, the following two equations define the same plane:

$$x + y + z = 1, \qquad 4x + 4y + 4z = 4$$

The first equation uses the normal $\langle 1, 1, 1 \rangle$, and the second uses the normal $\langle 4, 4, 4 \rangle$.

On the other hand, two planes $\mathcal{P}$ and $\mathcal{P}'$ are parallel if they have a common normal vector. The following planes are parallel because each is normal to $\mathbf{n} = \langle 1, 1, 1 \rangle$:

$$x + y + z = 1, \qquad x + y + z = 2, \qquad 4x + 4y + 4z = 7$$

In general, a family of parallel planes is obtained by choosing a normal vector $\mathbf{n} = \langle a, b, c \rangle$ and varying the constant d in the equation

$$ax + by + cz = d$$

The unique plane in this family through the origin has equation $ax + by + cz = 0$.

EXAMPLE 2 **Parallel Planes** Let $\mathcal{P}$ have equation $7x - 4y + 2z = -10$. Find an equation of the plane parallel to $\mathcal{P}$ passing through:

(a) The origin.   **(b)** $Q = (2, -1, 3)$.

Solution The planes parallel to $\mathcal{P}$ have an equation of the form (Figure 5)

$$7x - 4y + 2z = d \qquad \boxed{4}$$

(a) For $d = 0$, we get the plane through the origin: $7x - 4y + 2z = 0$.
(b) The point $Q = (2, -1, 3)$ satisfies Eq. (4) with

$$d = 7(2) - 4(-1) + 2(3) = 24$$

Therefore, the plane parallel to $\mathcal{P}$ through Q has equation $7x - 4y + 2z = 24$. ∎

Points that lie on a line are called **collinear**. If we are given three points P, Q, and R that are not collinear, then there is just one plane passing through P, Q, and R (Figure 6). The next example shows how to find an equation of this plane.

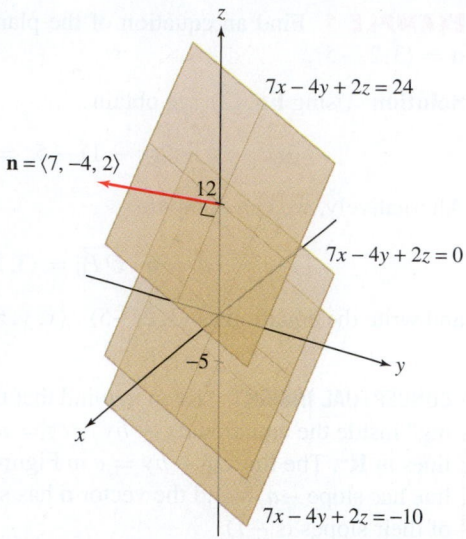

DF **FIGURE 5** Parallel planes with normal vector $\mathbf{n} = \langle 7, -4, 2 \rangle$.

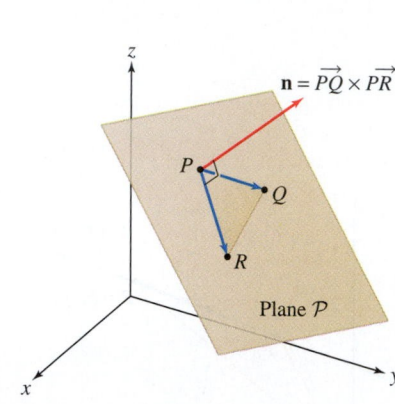

FIGURE 6 Three points P, Q, and R determine a plane (assuming they do not lie in a straight line).

EXAMPLE 3   **The Plane Determined by Three Points**   Find an equation of the plane $\mathcal{P}$ determined by the points

$$P = (1, 0, -1), \qquad Q = (2, 2, 1), \qquad R = (4, 1, 2)$$

Solution

Step 1. **Find a normal vector.**

The vectors $\overrightarrow{PQ}$ and $\overrightarrow{PR}$ lie in the plane $\mathcal{P}$, so their cross product is normal to $\mathcal{P}$:

$$\overrightarrow{PQ} = \langle 2, 2, 1 \rangle - \langle 1, 0, -1 \rangle = \langle 1, 2, 2 \rangle$$

$$\overrightarrow{PR} = \langle 4, 1, 2 \rangle - \langle 1, 0, -1 \rangle = \langle 3, 1, 3 \rangle$$

> In Example 3, we could have instead used the vectors $\overrightarrow{QP}$ and $\overrightarrow{QR}$ (or $\overrightarrow{RP}$ and $\overrightarrow{RQ}$) to find a normal vector to $\mathcal{P}$.

$$\mathbf{n} = \overrightarrow{PQ} \times \overrightarrow{PR} = \begin{vmatrix} \mathbf{i} & \mathbf{j} & \mathbf{k} \\ 1 & 2 & 2 \\ 3 & 1 & 3 \end{vmatrix} = 4\mathbf{i} + 3\mathbf{j} - 5\mathbf{k} = \langle 4, 3, -5 \rangle$$

By Eq. (3), $\mathcal{P}$ has equation $4x + 3y - 5z = d$ for some d.

Step 2. **Choose a point on the plane and compute d.**

Now, choose any one of the three points—say, $P = (1, 0, -1)$—and compute

$$d = \mathbf{n} \cdot \overrightarrow{OP} = \langle 4, 3, -5 \rangle \cdot \langle 1, 0, -1 \rangle = 9$$

> **CAUTION** When you find a normal vector to the plane containing points P, Q, R, be sure to compute a cross product such as $\overrightarrow{PQ} \times \overrightarrow{PR}$. A common mistake is to use a cross product such as $\overrightarrow{OP} \times \overrightarrow{OQ}$ or $\overrightarrow{OP} \times \overrightarrow{OR}$, which need not be normal to the plane.

We conclude that $\mathcal{P}$ has equation $4x + 3y - 5z = 9$.   ■

To check our result, simply verify that the three points we were given satisfy this equation of the plane.

Example 3 shows that three noncollinear points determine a plane. A plane $\mathcal{P}$ also can be determined by any of the following:

- Two lines that intersect in a single point;
- A line and a point not on the line;
- Two distinct parallel lines.

For each of these situations, think about how to determine a normal vector to the plane and a point on the plane. We examine these cases in the exercises.

EXAMPLE 4 **Intersection of a Plane and a Line** Find the point P where the plane $3x - 9y + 2z = 7$ and the line $\mathbf{r}(t) = \langle 1, 2, 1 \rangle + t \langle -2, 0, 1 \rangle$ intersect.

Solution The line has parametric equations

$$x = 1 - 2t, \qquad y = 2, \qquad z = 1 + t$$

Substitute in the equation of the plane and solve for t:

$$3x - 9y + 2z = 3(1 - 2t) - 9(2) + 2(1 + t) = 7$$

Simplification yields $-4t - 13 = 7$ or $t = -5$. Therefore, P has coordinates

$$x = 1 - 2(-5) = 11, \qquad y = 2, \qquad z = 1 + (-5) = -4$$

The plane and line intersect at the point $P = (11, 2, -4)$. ∎

If we think of t representing time on a path on the line, then $t = -5$ is the time that the path meets the plane. The point $P = (11, 2, -4)$ is the location on the path at that time.

The intersection of a plane $\mathcal{P}$ with a coordinate plane or a plane parallel to a coordinate plane is called a **trace**. The trace is a line unless $\mathcal{P}$ is parallel to the coordinate plane (in which case, the trace is empty or is $\mathcal{P}$ itself).

EXAMPLE 5 **Traces of the Plane** Graph the plane $-2x + 3y + z = 6$ and then find its traces in the coordinate planes.

Solution To draw the plane, we determine its intersections with the coordinate axes. To find where it intersects the x-axis, we set $y = z = 0$ and obtain

$$-2x = 6 \qquad \text{so} \qquad x = -3$$

It intersects the y-axis when $x = z = 0$, giving

$$3y = 6 \qquad \text{so} \qquad y = 2$$

It intersects the z-axis when $x = y = 0$, giving

$$z = 6$$

Thus, the plane appears as in Figure 7.

We obtain the trace in the xy-plane by setting $z = 0$ in the equation of the plane. Therefore, the trace is the line $-2x + 3y = 6$ in the xy-plane (Figure 7). Similarly, the trace in the xz-plane is obtained by setting $y = 0$, which gives the line $-2x + z = 6$ in the xz-plane. Finally, the trace in the yz-plane is $3y + z = 6$. ∎

In most cases, you can picture a plane by determining where it intersects the coordinate axes. However, if the plane is parallel to one of the axes and does not intersect it, or the plane passes through the origin, and therefore intersects all three axes there, we can use the normal vector to determine how the plane is oriented.

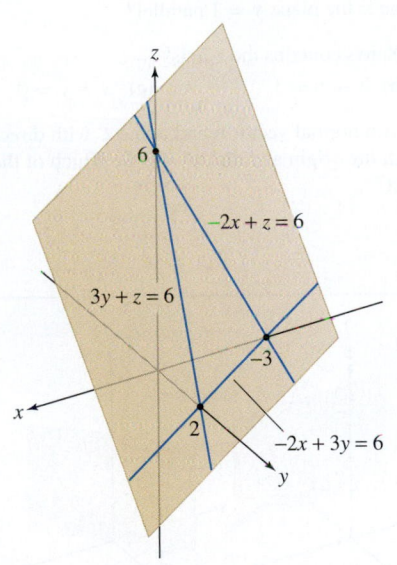

FIGURE 7 The three blue lines are the traces of the plane $-2x + 3y + z = 6$ in the coordinate planes.

12.5 SUMMARY

- Plane through $P_0 = (x_0, y_0, z_0)$ with normal vector $\mathbf{n} = \langle a, b, c \rangle$:

  – Geometrically: The tips of all vectors based at P_0 that are perpendicular to $\mathbf{n}$
  – Algebraically:

    Vector form: $\mathbf{n} \cdot \langle x, y, z \rangle = d$

    Scalar forms: $a(x - x_0) + b(y - y_0) + c(z - z_0) = 0$

    $$ax + by + cz = d$$

  where $d = \mathbf{n} \cdot \langle x_0, y_0, z_0 \rangle = ax_0 + by_0 + cz_0$.

- The family of parallel planes with given normal vector $\mathbf{n} = \langle a, b, c \rangle$ consists of all planes with equation $ax + by + cz = d$ for some d.
- A plane is determined in each of the following cases. In each case a point on the plane and a normal vector to the plane can be determined in order to find an equation for the plane.

     – Three noncollinear points
     – Two lines that intersect in a point
     – A line and a point not on it
     – Two distinct parallel lines

- The intersection of a plane $\mathcal{P}$ with a coordinate plane or a plane parallel to a coordinate plane is called a *trace*. The trace in the yz-plane is obtained by setting $x = 0$ in the equation of the plane (and similarly for the traces in the xz- and xy-planes).

12.5 EXERCISES

Preliminary Questions

1. What is the equation of the plane parallel to $3x + 4y - z = 5$ passing through the origin?

2. The vector $\mathbf{k}$ is normal to which of the following planes?
(a) $x = 1$ **(b)** $y = 1$ **(c)** $z = 1$

3. Which of the following planes is not parallel to the plane $x + y + z = 1$?
(a) $2x + 2y + 2z = 1$ **(b)** $x + y + z = 3$
(c) $x - y + z = 0$

4. To which coordinate plane is the plane $y = 1$ parallel?

5. Which of the following planes contains the z-axis?
(a) $z = 1$ **(b)** $x + y = 1$ **(c)** $x + y = 0$

6. Suppose that a plane $\mathcal{P}$ with normal vector $\mathbf{n}$ and a line $\mathcal{L}$ with direction vector $\mathbf{v}$ both pass through the origin and that $\mathbf{n} \cdot \mathbf{v} = 0$. Which of the following statements is correct?
(a) $\mathcal{L}$ is contained in $\mathcal{P}$.
(b) $\mathcal{L}$ is orthogonal to $\mathcal{P}$.

Exercises

In Exercises 1–8, write the equation of the plane with normal vector $\mathbf{n}$ passing through the given point in the scalar form $ax + by + cz = d$.

1. $\mathbf{n} = \langle 1, 3, 2 \rangle$, $(4, -1, 1)$

2. $\mathbf{n} = \langle -1, 2, 1 \rangle$, $(3, 1, 9)$

3. $\mathbf{n} = \langle -1, 2, 1 \rangle$, $(4, 1, 5)$

4. $\mathbf{n} = \langle 2, -4, 1 \rangle$, $\left(\frac{1}{3}, \frac{2}{3}, 1 \right)$

5. $\mathbf{n} = \mathbf{i}$, $(3, 1, -9)$

6. $\mathbf{n} = \mathbf{j}$, $\left(-5, \frac{1}{2}, \frac{1}{2} \right)$

7. $\mathbf{n} = \mathbf{k}$, $(6, 7, 2)$

8. $\mathbf{n} = \mathbf{i} - \mathbf{k}$, $(4, 2, -8)$

9. Write the equation of any plane through the origin.

10. Write the equations of any two distinct planes with normal vector $\mathbf{n} = \langle 3, 2, 1 \rangle$ that do not pass through the origin.

11. Which of the following statements are true of a plane that is parallel to the yz-plane?

(a) $\mathbf{n} = \langle 0, 0, 1 \rangle$ is a normal vector.

(b) $\mathbf{n} = \langle 1, 0, 0 \rangle$ is a normal vector.

(c) The equation has the form $ay + bz = d$.

(d) The equation has the form $x = d$.

12. Find a normal vector $\mathbf{n}$ and an equation for the planes in Figures 8 (A)–(C).

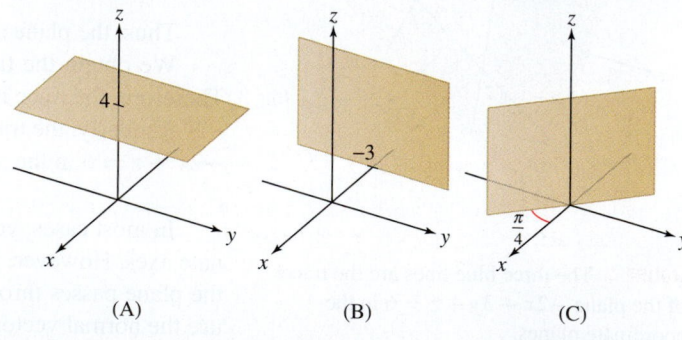

(A)   **(B)**   **(C)**

FIGURE 8

In Exercises 13–16, find a vector normal to the plane with the given equation.

13. $9x - 4y - 11z = 2$ **14.** $x - z = 0$

15. $3(x - 4) - 8(y - 1) + 11z = 0$

16. $x = 1$

In Exercises 17–20, find the equation of the plane with the given description.

17. Passes through O and is parallel to $4x - 9y + z = 3$

18. Passes through $(4, 1, 9)$ and is parallel to $x + y + z = 3$

19. Passes through $(4, 1, 9)$ and is parallel to $x = 3$

20. Passes through $P = (3, 5, -9)$ and is parallel to the xz-plane

In Exercises 21–24, find an equation of the plane passing through the three points given.

21. $P = (2, -1, 4)$, $Q = (1, 1, 1)$, $R = (3, 1, -2)$

22. $P = (5, 1, 1)$, $Q = (1, 1, 2)$, $R = (2, 1, 1)$

23. $P = (1, 0, 0)$, $Q = (0, 1, 1)$, $R = (2, 0, 1)$

24. $P = (2, 0, 0)$, $Q = (0, 4, 0)$, $R = (0, 0, 2)$

25. ✎ In each case, describe how to find a normal vector to the plane:
(a) Three noncollinear points are given. The plane contains all three points.
(b) Two lines are given that intersect in a point. The plane contains the lines.

26. ✎ In each case, describe how to find a normal vector to the plane:
(a) A line and a point that is not on the line are given. The plane contains the line and the point.
(b) Two lines are given that are parallel and distinct. The plane contains the lines.

27. In each case, determine whether or not the lines have a single point of intersection. If they do, give an equation of a plane containing them.
(a) $\mathbf{r}_1(t) = \langle t, 2t - 1, t - 3 \rangle$ and $\mathbf{r}_2(t) = \langle 4, 2t - 1, -1 \rangle$
(b) $\mathbf{r}_1(t) = \langle 3t, 2t + 1, t - 5 \rangle$ and $\mathbf{r}_2(t) = \langle 4t, 4t - 3, -1 \rangle$

28. In each case, determine whether or not the lines have a single point of intersection. If they do, give an equation of a plane containing them.
(a) $\mathbf{r}_1(t) = \langle 5t, 2t - 1, 2t - 2 \rangle$ and $\mathbf{r}_2(t) = \langle t - 5, -t + 4, t - 7 \rangle$
(b) $\mathbf{r}_1(t) = \langle 3t, -2t + 1, t - 3 \rangle$ and $\mathbf{r}_2(t) = \langle 2t - 1, -t, -t - 1 \rangle$

29. In each case, determine whether or not the point lies on the line. If it does not, give an equation of a plane containing the point and the line.
(a) $(2, 2, -1)$ and $\mathbf{r}(t) = \langle 4t, 6t - 1, -1 \rangle$
(b) $(3, -3, 2)$ and $\mathbf{r}(t) = \langle 4t + 3, 4t - 3, t + 1 \rangle$

30. In each case, determine whether or not the point lies on the line. If it does not, give an equation of a plane containing the point and the line.
(a) $(-7, 10, -3)$ and $\mathbf{r}(t) = \langle 1 - 4t, 6t - 5, t - 5 \rangle$
(b) $(-1, 5, 9)$ and $\mathbf{r}(t) = \langle 4t + 3, t + 6, 5 - 4t \rangle$

31. In each case, determine whether or not the lines are distinct parallel lines. If they are, give an equation of a plane containing them.
(a) $\mathbf{r}_1(t) = \langle t, 2t - 1, t - 3 \rangle$ and $\mathbf{r}_2(t) = \langle 3t - 3, 6t - 1, 3t - 1 \rangle$
(b) $\mathbf{r}_1(t) = \langle 3t, 2t + 1, t - 5 \rangle$ and $\mathbf{r}_2(t) = \langle -6t, 1 - 4t, 2t - 3 \rangle$

32. In each case, determine whether or not the lines are distinct parallel lines. If they are, give an equation of a plane containing them.
(a) $\mathbf{r}_1(t) = \langle 2t + 1, -2t - 1, 3t - 7 \rangle$ and $\mathbf{r}_2(t) = \langle 7 - 6t, 6t - 7, 2 - 9t \rangle$
(b) $\mathbf{r}_1(t) = \langle -4t, 2t + 1, 8t + 5 \rangle$ and $\mathbf{r}_2(t) = \langle 2t - 2, -t + 4, 5 - 4t \rangle$

In Exercises 33–37, draw the plane given by the equation.

33. $x + y + z = 4$

34. $3x + 2y - 6z = 12$

35. $12x - 6y + 4z = 6$

36. $x + 2y = 6$

37. $x + y + z = 0$

38. Let a, b, c be constants. Which two of the following equations define the plane passing through $(a, 0, 0)$, $(0, b, 0)$, $(0, 0, c)$?
(a) $ax + by + cz = 1$
(b) $bcx + acy + abz = abc$
(c) $bx + cy + az = 1$
(d) $\dfrac{x}{a} + \dfrac{y}{b} + \dfrac{z}{c} = 1$

39. Find an equation of the plane $\mathcal{P}$ in Figure 9.

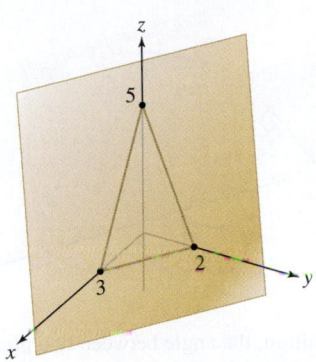

FIGURE 9

40. Verify that the plane $x - y + 5z = 10$ and the line $\mathbf{r}(t) = \langle 1, 0, 1 \rangle + t \langle -2, 1, 1 \rangle$ intersect at $P = (-3, 2, 3)$.

In Exercises 41–44, find the intersection of the line and the plane.

41. $x + y + z = 14$, $\mathbf{r}(t) = \langle 1, 1, 0 \rangle + t \langle 0, 2, 4 \rangle$

42. $2x + y = 3$, $\mathbf{r}(t) = \langle 2, -1, -1 \rangle + t \langle 1, 2, -4 \rangle$

43. $z = 12$, $\mathbf{r}(t) = t \langle -6, 9, 36 \rangle$

44. $x - z = 6$, $\mathbf{r}(t) = \langle 1, 0, -1 \rangle + t \langle 4, 9, 2 \rangle$

In Exercises 45–50, find the trace of the plane in the given coordinate plane.

45. $3x - 9y + 4z = 5$, yz

46. $3x - 9y + 4z = 5$, xz

47. $3x + 4z = -2$, xy

48. $3x + 4z = -2$, xz

49. $-x + y = 4$, xz

50. $-x + y = 4$, yz

51. Does the plane $x = 5$ have a trace in the yz-plane? Explain.

52. Give equations for two distinct planes whose trace in the xy-plane has equation $4x + 3y = 8$.

53. Give equations for two distinct planes whose trace in the yz-plane has equation $y = 4z$.

54. Find parametric equations for the line through $P_0 = (3, -1, 1)$ perpendicular to the plane $3x + 5y - 7z = 29$.

55. Find all planes in $\mathbf{R}^3$ whose intersection with the xz-plane is the line with equation $3x + 2z = 5$.

56. Find all planes in $\mathbf{R}^3$ whose intersection with the xy-plane is the line $\mathbf{r}(t) = t \langle 2, 1, 0 \rangle$.

In Exercises 57–62, compute the angle between the two planes, defined as the angle θ (between 0 and π) between their normal vectors (Figure 10).

57. Planes with normals $\mathbf{n}_1 = \langle 1, 0, 1 \rangle$, $\mathbf{n}_2 = \langle -1, 1, 1 \rangle$

58. Planes with normals $\mathbf{n}_1 = \langle 1, 2, 1 \rangle$, $\mathbf{n}_2 = \langle 4, 1, 3 \rangle$

59. $2x + 3y + 7z = 2$ and $4x - 2y + 2z = 4$

60. $x - 3y + z = 3$ and $2x - 3z = 4$

61. $3(x - 1) - 5y + 2(z - 12) = 0$ and the plane with normal $\mathbf{n} = \langle 1, 0, 1 \rangle$

62. The plane through $(1, 0, 0)$, $(0, 1, 0)$, and $(0, 0, 1)$ and the yz-plane

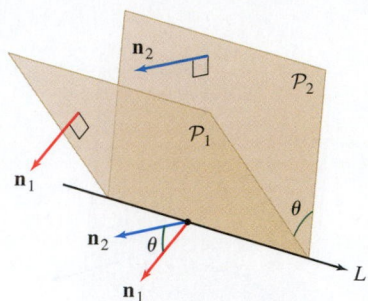

FIGURE 10 By definition, the angle between two planes is the angle between their normal vectors.

63. Find an equation of a plane making an angle of $\frac{\pi}{2}$ with the plane $3x + y - 4z = 2$.

64. Let $\mathcal{P}_1$ and $\mathcal{P}_2$ be planes with normal vectors $\mathbf{n}_1$ and $\mathbf{n}_2$. Assume that the planes are not parallel, and let $\mathcal{L}$ be their intersection (a line). Show that $\mathbf{n}_1 \times \mathbf{n}_2$ is a direction vector for $\mathcal{L}$.

65. Find a plane that is perpendicular to the two planes $x + y = 3$ and $x + 2y - z = 4$.

66. Let $\mathcal{L}$ be the line of intersection of the planes $x + y + z = 1$ and $x + 2y + 3z = 1$. Use Exercise 64 to find a direction vector for $\mathcal{L}$. Then find a point P on $\mathcal{L}$ by *inspection*, and write down the parametric equations for $\mathcal{L}$.

67. Let $\mathcal{L}$ denote the line of intersection of the planes $x - y - z = 1$ and $2x + 3y + z = 2$. Find parametric equations for the line $\mathcal{L}$. *Hint:* To find a point on $\mathcal{L}$, substitute an arbitrary value for z (say, $z = 2$) and then solve the resulting pair of equations for x and y.

68. Find parametric equations for the line of intersection of the planes $2x + y - 3z = 0$ and $x + y = 1$.

69. Vectors $\mathbf{v}$ and $\mathbf{w}$, each of length 12, lie in the plane $x + 2y - 2z = 0$. The angle between $\mathbf{v}$ and $\mathbf{w}$ is $\pi/6$. This information determines $\mathbf{v} \times \mathbf{w}$ up to a sign ± 1. What are the two possible values of $\mathbf{v} \times \mathbf{w}$?

70. The plane

$$\frac{x}{2} + \frac{y}{4} + \frac{z}{3} = 1$$

intersects the x-, y-, and z-axes in points P, Q, and R. Find the area of the triangle $\triangle PQR$.

71. In this exercise, we show that the orthogonal distance D from the plane $\mathcal{P}$ with equation $ax + by + cz = d$ to the origin O is equal to (Figure 11)

$$D = \frac{|d|}{\sqrt{a^2 + b^2 + c^2}}$$

Let $\mathbf{n} = \langle a, b, c \rangle$, and let P be the point where the line through the origin with direction vector $\mathbf{n}$ intersects $\mathcal{P}$. By definition, the orthogonal distance from P to O is the distance from P to O.

(a) Show that P is the terminal point of $\mathbf{v} = \left(\dfrac{d}{\mathbf{n} \cdot \mathbf{n}} \right) \mathbf{n}$.

(b) Show that the distance from P to O is D.

72. Use Exercise 71 to compute the orthogonal distance from the plane $x + 2y + 3z = 5$ to the origin.

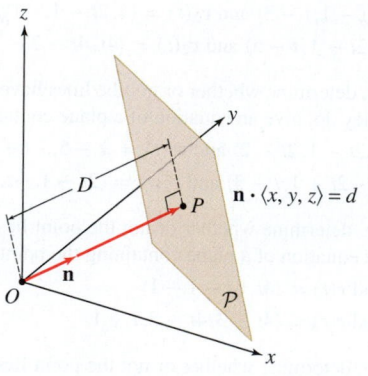

FIGURE 11

Further Insights and Challenges

In Exercises 73 and 74, let $\mathcal{P}$ be a plane with equation

$$ax + by + cz = d$$

and normal vector $\mathbf{n} = \langle a, b, c \rangle$. For any point Q, there is a unique point P on $\mathcal{P}$ that is closest to Q, and is such that $\overline{PQ}$ is orthogonal to $\mathcal{P}$ (Figure 12).

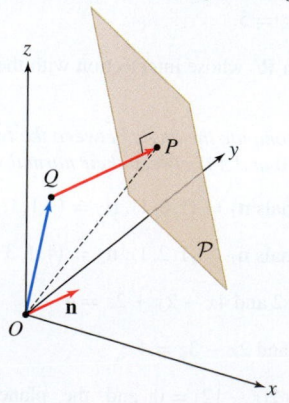

FIGURE 12

73. Show that the point P on $\mathcal{P}$ closest to Q is determined by the equation

$$\overrightarrow{OP} = \overrightarrow{OQ} + \left(\frac{d - \overrightarrow{OQ} \cdot \mathbf{n}}{\mathbf{n} \cdot \mathbf{n}} \right) \mathbf{n} \qquad \boxed{5}$$

74. By definition, the distance from $Q = (x_1, y_1, z_1)$ to the plane $\mathcal{P}$ is the distance to the point P on $\mathcal{P}$ closest to Q. Prove

$$\text{distance from } Q \text{ to } \mathcal{P} = \frac{|ax_1 + by_1 + cz_1 - d|}{\|\mathbf{n}\|} \qquad \boxed{6}$$

75. Use Eq. (5) to find the point P nearest to $Q = (2, 1, 2)$ on the plane $x + y + z = 1$.

76. Find the point P nearest to $Q = (-1, 3, -1)$ on the plane

$$x - 4z = 2$$

77. Use Eq. (6) to find the distance from $Q = (1, 1, 1)$ to the plane $2x + y + 5z = 2$.

78. Find the distance from $Q = (1, 2, 2)$ to the plane $\mathbf{n} \cdot \langle x, y, z \rangle = 3$, where $\mathbf{n} = \langle \frac{3}{5}, \frac{4}{5}, 0 \rangle$.

79. What is the distance from $Q = (a, b, c)$ to the plane $x = 0$? Visualize your answer geometrically and explain without computation. Then verify that Eq. (6) yields the same answer.

80. The equation of a plane $\mathbf{n} \cdot \langle x, y, z \rangle = d$ is said to be in **normal form** if $\mathbf{n}$ is a unit vector. Show that in this case, $|d|$ is the distance from the plane to the origin. Write the equation of the plane $4x - 2y + 4z = 24$ in normal form.

12.6 A Survey of Quadric Surfaces

Quadric surfaces are the surface analogs of conic sections. Recall that a conic section is a curve in $\mathbf{R}^2$ defined by a quadratic equation in two variables. A quadric surface is defined by a quadratic equation in *three* variables:

> To ensure that Eq. (1) is genuinely quadratic, we assume that the degree-2 coefficients A, B, C, D, E, F are not all zero.

$$Ax^2 + By^2 + Cz^2 + Dxy + Eyz + Fzx + ax + by + cz + d = 0 \qquad \boxed{1}$$

Like conic sections, quadric surfaces are classified into a small number of types. When the coordinate axes are chosen to coincide with the axes of the quadric, the equation of the quadric has a simple form. The quadric is then said to be in **standard position**. In standard position, the coefficients D, E, F are all zero. In this short survey of quadric surfaces, we restrict our attention to quadrics in standard position. The idea here is not to memorize the formulas for the various quadric surfaces, but rather to be able to recognize and graph them using cross sections obtained by slicing the surface with certain planes, as we describe in this section.

The surface analogs of ellipses are the egg-shaped **ellipsoids** (Figure 1). In standard form, an ellipsoid has the equation

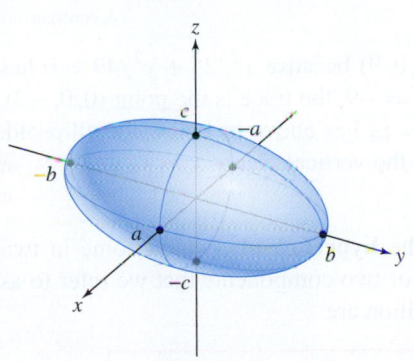

DF FIGURE 1 Ellipsoid with equation $\left(\dfrac{x}{a}\right)^2 + \left(\dfrac{y}{b}\right)^2 + \left(\dfrac{z}{c}\right)^2 = 1$.

Ellipsoid	$\left(\dfrac{x}{a}\right)^2 + \left(\dfrac{y}{b}\right)^2 + \left(\dfrac{z}{c}\right)^2 = 1$

For $a = b = c$, this equation is equivalent to $x^2 + y^2 + z^2 = a^2$ and the ellipsoid is a sphere of radius a.

Surfaces are often represented graphically by a mesh of curves called **traces**, obtained by intersecting the surface with planes parallel to one of the coordinate planes (Figure 2), yielding certain cross sections of the surface. Algebraically, this corresponds to holding one of the three variables constant. For example, the intersection of the horizontal plane $z = z_0$ with the surface is a horizontal trace curve.

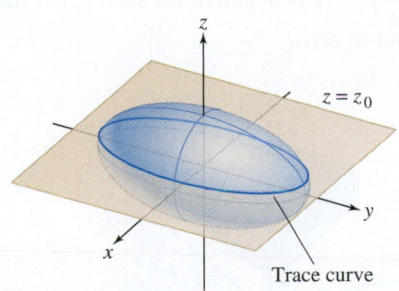

DF FIGURE 2 The intersection of the plane $z = z_0$ with an ellipsoid is an ellipse.

EXAMPLE 1 **The Traces of an Ellipsoid** Describe the traces of the ellipsoid

$$\left(\frac{x}{5}\right)^2 + \left(\frac{y}{7}\right)^2 + \left(\frac{z}{9}\right)^2 = 1$$

Solution First, we observe that the traces in the coordinate planes are ellipses [Figure 3(A)]:

xy-trace (set $z = 0$, blue in figure): $\qquad \left(\dfrac{x}{5}\right)^2 + \left(\dfrac{y}{7}\right)^2 = 1$

yz-trace (set $x = 0$, green in figure): $\qquad \left(\dfrac{y}{7}\right)^2 + \left(\dfrac{z}{9}\right)^2 = 1$

xz-trace (set $y = 0$, red in figure): $\qquad \left(\dfrac{x}{5}\right)^2 + \left(\dfrac{z}{9}\right)^2 = 1$

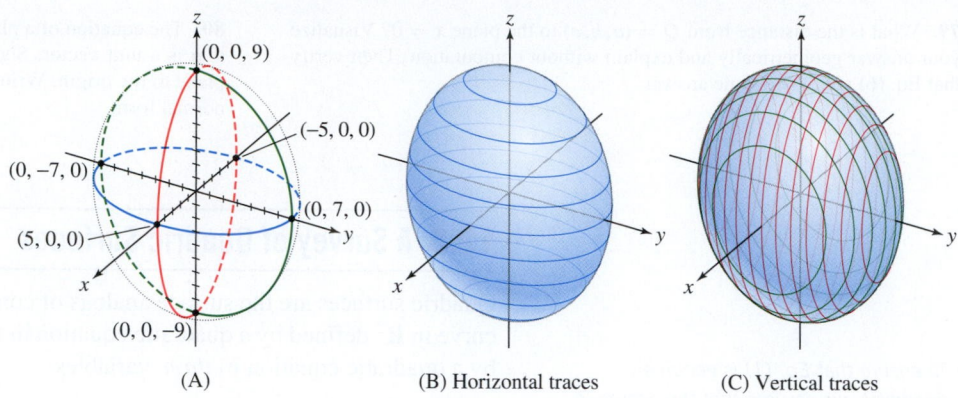

(A)   (B) Horizontal traces   (C) Vertical traces

FIGURE 3 The ellipsoid $\left(\dfrac{x}{5}\right)^2 + \left(\dfrac{y}{7}\right)^2 + \left(\dfrac{z}{9}\right)^2 = 1$.

In fact, all the traces of an ellipsoid are ellipses (or just single points). For example, the horizontal trace defined by setting $z = z_0$ is the ellipse [Figure 3(B)]

$$\text{trace at height } z_0: \quad \left(\frac{x}{5}\right)^2 + \left(\frac{y}{7}\right)^2 + \left(\frac{z_0}{9}\right)^2 = 1 \quad \text{or} \quad \frac{x^2}{25} + \frac{y^2}{49} = \underbrace{1 - \frac{z_0^2}{81}}_{\text{A constant}}$$

The trace at height $z_0 = 9$ is the single point $(0, 0, 9)$ because $x^2/25 + y^2/49 = 0$ has only one solution: $x = 0$, $y = 0$. Similarly, for $z_0 = -9$, the trace is the point $(0, 0, -9)$. If $|z_0| > 9$, then $1 - z_0^2/81 < 0$ and the plane $z = z_0$ lies above or below the ellipsoid. The trace has no points in this case. The traces in the vertical planes $x = x_0$ and $y = y_0$ have a similar description [Figure 3(C)]. ∎

The surface analogs of the hyperbolas are the **hyperboloids**, which come in two types, depending on whether the surface has one or two components that we refer to as **sheets** (Figure 4). Their equations in standard position are

> **Hyperboloids** One sheet: $\left(\dfrac{x}{a}\right)^2 + \left(\dfrac{y}{b}\right)^2 = \left(\dfrac{z}{c}\right)^2 + 1$
>
> Two sheets: $\left(\dfrac{x}{a}\right)^2 + \left(\dfrac{y}{b}\right)^2 = \left(\dfrac{z}{c}\right)^2 - 1$

2

Notice that a hyperboloid of two sheets does not contain any points whose z-coordinate satisfies $-c < z < c$ because the right-hand side $\left(\dfrac{z}{c}\right)^2 - 1$ is negative for such z, but the left-hand side of the equation is greater than or equal to zero.

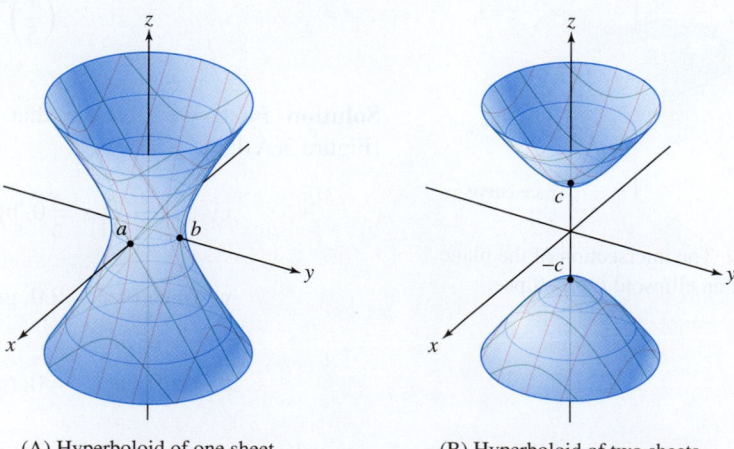

FIGURE 4 Hyperboloids of one and two sheets.

(A) Hyperboloid of one sheet   (B) Hyperboloid of two sheets

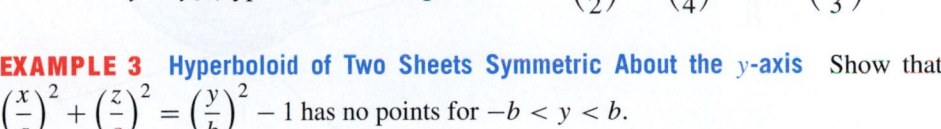

FIGURE 5 The hyperboloid $\left(\frac{x}{2}\right)^2 + \left(\frac{y}{3}\right)^2 = \left(\frac{z}{4}\right)^2 + 1$.

EXAMPLE 2 **The Traces of a Hyperboloid of One Sheet** Determine the traces of the hyperboloid $\left(\frac{x}{2}\right)^2 + \left(\frac{y}{3}\right)^2 = \left(\frac{z}{4}\right)^2 + 1$.

Solution The horizontal traces are ellipses and the vertical traces (parallel to either the yz-plane or the xz-plane) are hyperbolas or pairs of crossed lines (Figure 5):

Trace $z = z_0$ (ellipse, blue in figure): $\qquad \left(\frac{x}{2}\right)^2 + \left(\frac{y}{3}\right)^2 = \left(\frac{z_0}{4}\right)^2 + 1$

Trace $x = x_0$ (hyperbola, green in figure): $\qquad \left(\frac{y}{3}\right)^2 - \left(\frac{z}{4}\right)^2 = 1 - \left(\frac{x_0}{2}\right)^2$

Trace $y = y_0$ (hyperbola, red in figure): $\qquad \left(\frac{x}{2}\right)^2 - \left(\frac{z}{4}\right)^2 = 1 - \left(\frac{y_0}{3}\right)^2$ ■

EXAMPLE 3 **Hyperboloid of Two Sheets Symmetric About the y-axis** Show that $\left(\frac{x}{a}\right)^2 + \left(\frac{z}{c}\right)^2 = \left(\frac{y}{b}\right)^2 - 1$ has no points for $-b < y < b$.

Solution This equation does not have the same form as Eq. (2) because the variables y and z have been interchanged. This hyperboloid is symmetric about the y-axis rather than the z-axis (Figure 6). The left-hand side of the equation is always ≥ 0. Thus, there are no solutions with $|y| < b$ because the right-hand side is $\left(\frac{y}{b}\right)^2 - 1 < 0$. Therefore, the hyperboloid has two sheets, corresponding to $y \geq b$ and $y \leq -b$. ■

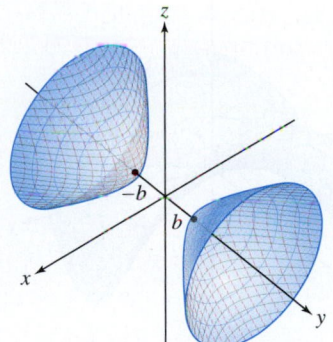

FIGURE 6 The hyperboloid of two sheets $\left(\frac{x}{a}\right)^2 + \left(\frac{z}{c}\right)^2 = \left(\frac{y}{b}\right)^2 - 1$.

The following equation defines an **elliptic cone** (Figure 7):

$$\boxed{\textbf{Elliptic cone:} \qquad \left(\frac{x}{a}\right)^2 + \left(\frac{y}{b}\right)^2 = \left(\frac{z}{c}\right)^2}$$

An elliptic cone is a transition case between a hyperboloid of one sheet and a hyperboloid of two sheets (Figure 8). The hyperboloid of one sheet is pinched at its narrowest part to form an elliptic cone, and then the two parts of the elliptic cone separate to form the two components in a hyperboloid of two sheets.

To visualize the elliptic cone, first consider its intersection with the xz-plane. When $y = 0$, we have $\left(\frac{x}{a}\right)^2 = \left(\frac{z}{c}\right)^2$, which is the pair of diagonal lines $z = \pm\left(\frac{c}{a}\right)x$. Similarly, intersecting with the yz-plane, where $x = 0$, we obtain the pair of diagonal lines $z = \pm\left(\frac{c}{b}\right)y$. Next, to see how these pairs of lines fit in the elliptic cone, we slice the surface parallel to the xy-plane. For instance, in the plane $z = 1$, the trace is given by

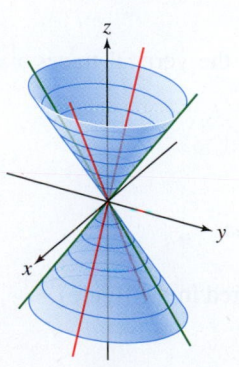

FIGURE 7 Elliptic cone $\left(\frac{x}{a}\right)^2 + \left(\frac{y}{b}\right)^2 = \left(\frac{z}{c}\right)^2$.

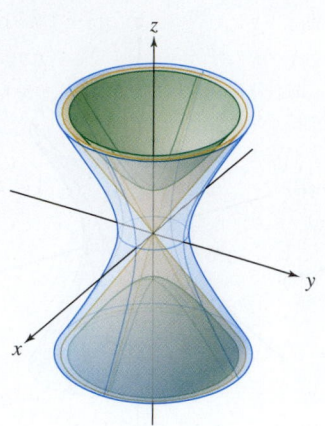

FIGURE 8 The elliptic cone is a transition case between the two types of hyperboloids.

$\left(\frac{x}{a}\right)^2 + \left(\frac{y}{b}\right)^2 = \left(\frac{1}{c}\right)^2$. This is the equation of an ellipse. In the plane $z = 2$, the trace is a larger ellipse. We obtain similar ellipses when we slice with the planes $z = -1$ and $z = -2$. Hence, the resulting surface is the elliptic cone appearing in Figure 7.

The third main family of quadric surfaces are the **paraboloids**. There are two types—elliptic and hyperbolic. In standard position, their equations are

> **Paraboloids**   Elliptic: $z = \left(\frac{x}{a}\right)^2 + \left(\frac{y}{b}\right)^2$
>
>   Hyperbolic: $z = \left(\frac{x}{a}\right)^2 - \left(\frac{y}{b}\right)^2$

3

Let's compare their traces (Figure 9):

	Elliptic paraboloid	Hyperbolic paraboloid
Horizontal Traces	Ellipses	Hyperbolas
Vertical traces	Upward parabolas	Upward and downward parabolas

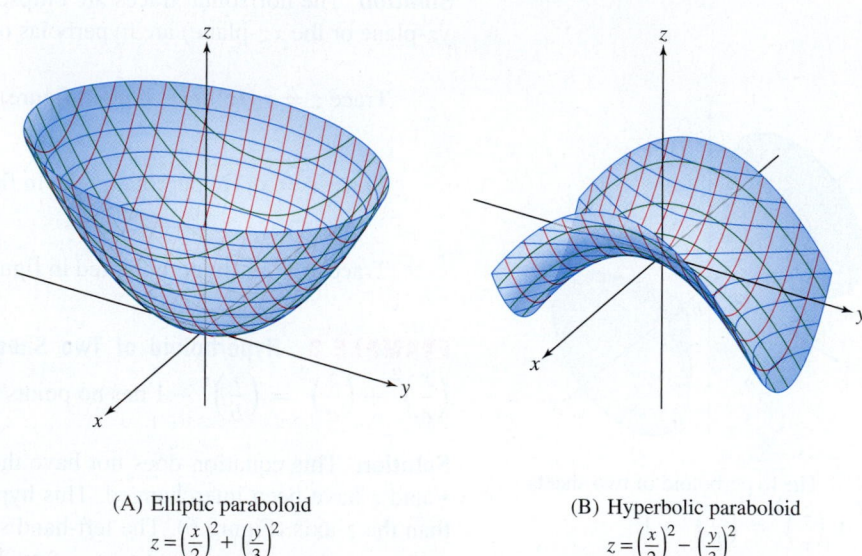

(A) Elliptic paraboloid
$z = \left(\frac{x}{2}\right)^2 + \left(\frac{y}{3}\right)^2$

(B) Hyperbolic paraboloid
$z = \left(\frac{x}{2}\right)^2 - \left(\frac{y}{3}\right)^2$

FIGURE 9

Paraboloids play an important role in the analysis of functions of two variables. The minimum of the elliptic paraboloid at the origin and the hyperbolic paraboloid's saddle shape are models for local behavior near critical points of functions of two variables. We explore this topic further in Section 14.7.

Notice, for example, that for the hyperbolic paraboloid, the vertical traces $x = x_0$ are downward parabolas (green in the figure)

$$\underbrace{z = -\left(\frac{y}{b}\right)^2 + \left(\frac{x_0}{a}\right)^2}_{\text{Trace } x = x_0 \text{ of hyperbolic paraboloid}}$$

whereas the vertical traces $y = y_0$ are upward parabolas (red in the figure)

$$\underbrace{z = \left(\frac{x}{a}\right)^2 - \left(\frac{y_0}{b}\right)^2}_{\text{Trace } y = y_0 \text{ of hyperbolic paraboloid}}$$

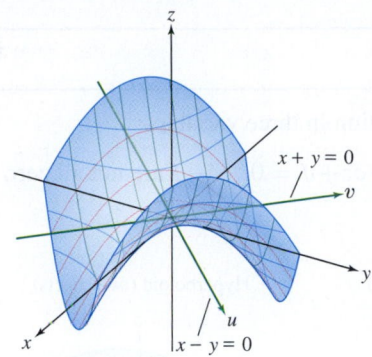

DF FIGURE 10 The hyperbolic paraboloid is defined by $z = 4xy$ or $z = u^2 - v^2$.

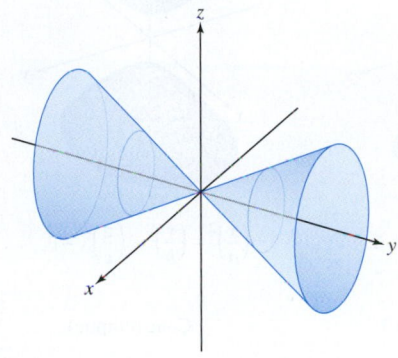

FIGURE 11 Elliptic cone on y-axis.

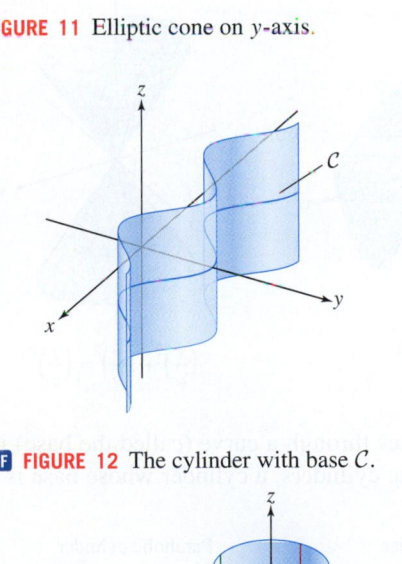

DF FIGURE 12 The cylinder with base $\mathcal{C}$.

It is somewhat surprising to realize that if we slice the hyperbolic paraboloid with the plane $z = 0$, we obtain $0 = \left(\frac{x}{a}\right)^2 - \left(\frac{y}{b}\right)^2$, which yields the pair of diagonal lines given by $y = \pm \left(\frac{b}{a}\right) x$ as the trace in the xy-plane. It is not obvious that this surface would contain these two diagonal lines, but in fact it does.

EXAMPLE 4 **Alternative Form of a Hyperbolic Paraboloid** Show that $z = 4xy$ is a hyperbolic paraboloid by writing the equation in terms of the variables $u = x + y$ and $v = x - y$.

Solution Note that $u + v = 2x$ and $u - v = 2y$. Therefore,

$$4xy = (u + v)(u - v) = u^2 - v^2$$

and thus the equation is $z = u^2 - v^2$ in the coordinates $\{u, v, z\}$. These coordinates are obtained by rotating the coordinates $\{x, y, z\}$ by $45°$ about the z-axis (Figure 10). ∎

EXAMPLE 5 Without referring back to the formulas, use traces to determine and graph the quadric surface given by $x^2 + 2z^2 - y^2 = 0$.

Solution We first slice with the coordinate planes. When $x = 0$, we obtain $2z^2 - y^2 = 0$, and therefore the trace in the yz-plane is the pair of diagonal lines $z = \pm \frac{y}{\sqrt{2}}$. When $y = 0$, we have $x^2 + 2z^2 = 0$, which has the solution $x = z = 0$. Hence, the xz-plane intersects the surface only at the origin. When $z = 0$, we obtain $x^2 - y^2 = 0$, which generates two diagonal lines in the xy-plane given by $y = \pm x$. At this point, the structure of the surface might not yet be clear. So, we will slice with planes parallel to the xz-plane. For instance, if we set $y = 1$, we obtain $x^2 + 2z^2 = 1$, which is an ellipse. If we set $y = 2$, we obtain $x^2 + 2z^2 = 4$, which is a larger ellipse. A similar pair of ellipses is obtained when we slice with the planes $y = -1$ and $y = -2$, respectively. Hence, we can now see that the surface is an elliptic cone opening along the y-axis, as in Figure 11. ∎

Further examples of quadric surfaces are the **quadratic cylinders**. We use the term "cylinder" in the following sense: Given a curve $\mathcal{C}$ in the xy-plane, the cylinder with base $\mathcal{C}$ is the surface consisting of all vertical lines passing through $\mathcal{C}$ (Figure 12). Equations of cylinders involve just two of the variables x, y, and z. A quadratic cylinder is a cylinder whose base curve is a conic section. The equation $x^2 + y^2 = r^2$ defines a circular cylinder of radius r with the z-axis as the central axis. Figure 13 shows a circular cylinder and three other types of quadratic cylinders.

The ellipsoids, hyperboloids, paraboloids, and quadratic cylinders are called **nondegenerate** quadric surfaces. There are also a certain number of "degenerate" quadric surfaces. For example, $x^2 + y^2 + z^2 = 0$ is a quadric that reduces to a single point $(0, 0, 0)$, and $(x + y + z)^2 = 1$ reduces to the union of the two planes $x + y + z = \pm 1$.

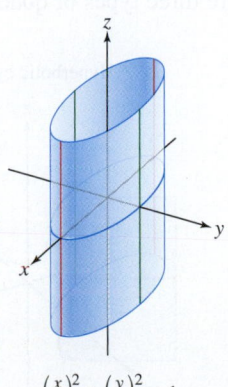

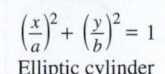

$x^2 + y^2 = r^2$

Right circular cylinder of radius r

$\left(\frac{x}{a}\right)^2 + \left(\frac{y}{b}\right)^2 = 1$

Elliptic cylinder

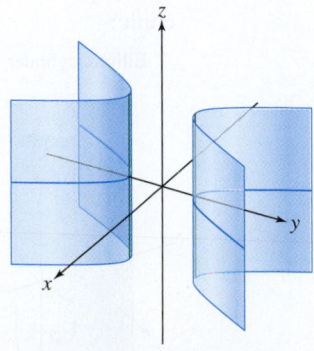

$\left(\frac{y}{b}\right)^2 - \left(\frac{x}{a}\right)^2 = 1$

Hyperbolic cylinder

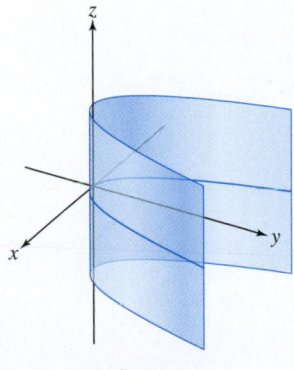

$y = ax^2$

Parabolic cylinder

DF FIGURE 13

12.6 SUMMARY

- A *quadric surface* is defined by a quadratic equation in three variables:

$$Ax^2 + By^2 + Cz^2 + Dxy + Eyz + Fzx + ax + by + cz + d = 0 \quad A\text{–}F \text{ are not all zero}$$

- Quadric surfaces in standard position:

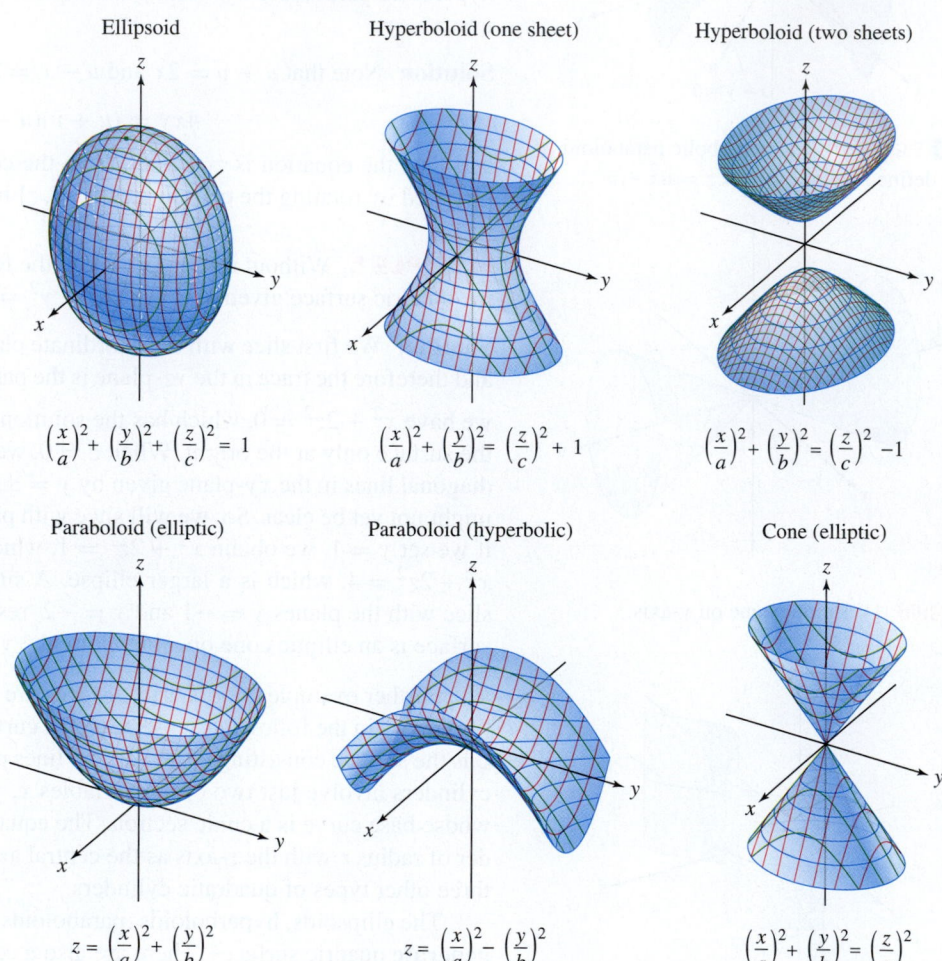

Ellipsoid

$$\left(\frac{x}{a}\right)^2 + \left(\frac{y}{b}\right)^2 + \left(\frac{z}{c}\right)^2 = 1$$

Hyperboloid (one sheet)

$$\left(\frac{x}{a}\right)^2 + \left(\frac{y}{b}\right)^2 = \left(\frac{z}{c}\right)^2 + 1$$

Hyperboloid (two sheets)

$$\left(\frac{x}{a}\right)^2 + \left(\frac{y}{b}\right)^2 = \left(\frac{z}{c}\right)^2 - 1$$

Paraboloid (elliptic)

$$z = \left(\frac{x}{a}\right)^2 + \left(\frac{y}{b}\right)^2$$

Paraboloid (hyperbolic)

$$z = \left(\frac{x}{a}\right)^2 - \left(\frac{y}{b}\right)^2$$

Cone (elliptic)

$$\left(\frac{x}{a}\right)^2 + \left(\frac{y}{b}\right)^2 = \left(\frac{z}{c}\right)^2$$

- A (vertical) cylinder consists of all vertical lines through a curve (called the base) in the xy-plane. There are three types of quadratic cylinders, a cylinder whose base is a conic:

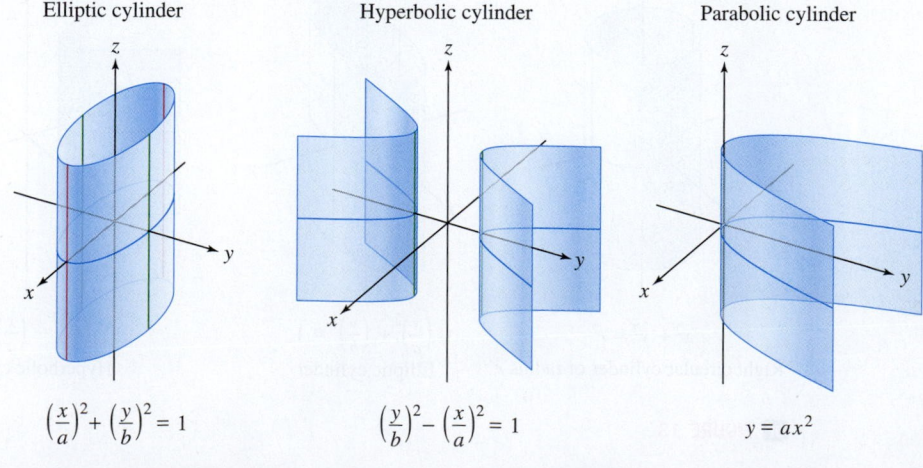

Elliptic cylinder

$$\left(\frac{x}{a}\right)^2 + \left(\frac{y}{b}\right)^2 = 1$$

Hyperbolic cylinder

$$\left(\frac{y}{b}\right)^2 - \left(\frac{x}{a}\right)^2 = 1$$

Parabolic cylinder

$$y = ax^2$$

12.6 EXERCISES

Preliminary Questions

1. True or false? All traces of an ellipsoid are ellipses.

2. True or false? All traces of a hyperboloid are hyperbolas.

3. Which quadric surfaces have both hyperbolas and parabolas as traces?

4. Is there any quadric surface whose traces are all parabolas?

5. A surface is called **bounded** if there exists $M > 0$ such that every point on the surface lies at a distance of at most M from the origin. Which of the quadric surfaces are bounded?

6. What is the definition of a parabolic cylinder?

Exercises

In Exercises 1–8, state whether the given equation defines an ellipsoid or hyperboloid, and if a hyperboloid, whether it is of one or two sheets.

1. $\left(\dfrac{x}{2}\right)^2 + \left(\dfrac{y}{3}\right)^2 + \left(\dfrac{z}{5}\right)^2 = 1$

2. $\left(\dfrac{x}{5}\right)^2 + \left(\dfrac{y}{5}\right)^2 - \left(\dfrac{z}{7}\right)^2 = 1$

3. $x^2 + 3y^2 + 9z^2 = 1$

4. $-\left(\dfrac{x}{2}\right)^2 - \left(\dfrac{y}{3}\right)^2 + \left(\dfrac{z}{5}\right)^2 = 1$

5. $x^2 - 3y^2 + 9z^2 = 1$

6. $x^2 - 3y^2 - 9z^2 = 1$

7. $x^2 + y^2 = 4 - 4z^2$

8. $x^2 + 3y^2 = 9 + z^2$

In Exercises 9–16, state whether the given equation defines an elliptic paraboloid, a hyperbolic paraboloid, or an elliptic cone.

9. $z = \left(\dfrac{x}{4}\right)^2 + \left(\dfrac{y}{3}\right)^2$

10. $z^2 = \left(\dfrac{x}{4}\right)^2 + \left(\dfrac{y}{3}\right)^2$

11. $z = \left(\dfrac{x}{9}\right)^2 - \left(\dfrac{y}{12}\right)^2$

12. $4z = 9x^2 + 5y^2$

13. $3x^2 - 7y^2 = z$

14. $3x^2 + 7y^2 = 14z^2$

15. $y^2 = 5x^2 - 4z^2$

16. $y = 3x^2 - 4z^2$

In Exercises 17–24, state the type of the quadric surface and describe the trace obtained by intersecting with the given plane.

17. $x^2 + \left(\dfrac{y}{4}\right)^2 + z^2 = 1, \quad y = 0$

18. $x^2 + \left(\dfrac{y}{4}\right)^2 + z^2 = 1, \quad y = 5$

19. $x^2 + \left(\dfrac{y}{4}\right)^2 + z^2 = 1, \quad z = \dfrac{1}{4}$

20. $\left(\dfrac{x}{2}\right)^2 + \left(\dfrac{y}{5}\right)^2 - 5z^2 = 1, \quad x = 0$

21. $\left(\dfrac{x}{3}\right)^2 + \left(\dfrac{y}{5}\right)^2 - 5z^2 = 1, \quad y = 1$

22. $4x^2 + \left(\dfrac{y}{3}\right)^2 - 2z^2 = -1, \quad z = 1$

23. $y = 3x^2, \quad z = 27$

24. $y = 3x^2, \quad y = 27$

25. Match each of the ellipsoids in Figure 14 with the correct equation:
(a) $x^2 + 4y^2 + 4z^2 = 16$
(b) $4x^2 + y^2 + 4z^2 = 16$
(c) $4x^2 + 4y^2 + z^2 = 16$

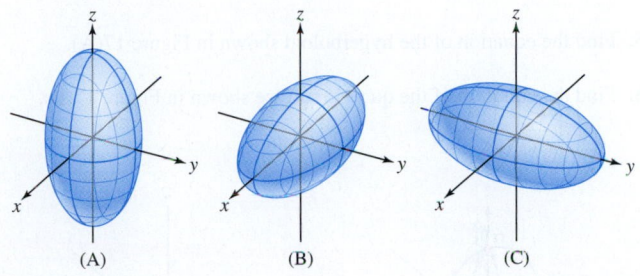

(A) (B) (C)

FIGURE 14

26. Describe the surface that is obtained when, in the equation $\pm 8x^2 \pm 3y^2 \pm z^2 = 1$, we choose (a) all plus signs, (b) one minus sign, and (c) two minus signs.

27. What is the equation of the surface obtained when the elliptic paraboloid $z = \left(\dfrac{x}{2}\right)^2 + \left(\dfrac{y}{4}\right)^2$ is rotated about the x-axis by $90°$? Refer to Figure 15.

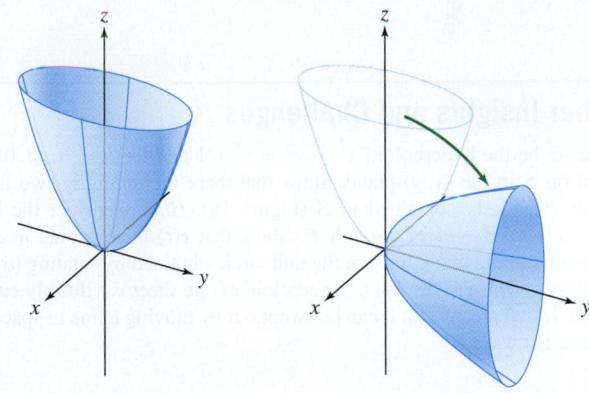

FIGURE 15

28. Describe the intersection of the horizontal plane $z = h$ and the hyperboloid $-x^2 - 4y^2 + 4z^2 = 1$. For which values of h is the intersection empty?

In Exercises 29–42, sketch the given surface.

29. $x^2 + y^2 - z^2 = 1$

30. $\left(\dfrac{x}{4}\right)^2 + \left(\dfrac{y}{8}\right)^2 + \left(\dfrac{z}{12}\right)^2 = 1$

31. $z = \left(\dfrac{x}{4}\right)^2 + \left(\dfrac{y}{8}\right)^2$

32. $z = \left(\dfrac{x}{4}\right)^2 - \left(\dfrac{y}{8}\right)^2$

33. $z^2 = \left(\dfrac{x}{4}\right)^2 + \left(\dfrac{y}{8}\right)^2$

34. $y = -x^2$

35. $-x^2 - y^2 + 9z^2 = 9$

36. $x^2 + 36y^2 = 1$

37. $xy = 1$

38. $x = 2y^2 - z^2$

39. $x = 1 + y^2 + z^2$

40. $x^2 - 4y^2 = z$

41. $x^2 + 9y^2 + 4z^2 = 36$

42. $y^2 - 4x^2 - z^2 = 4$

43. Find the equation of the ellipsoid passing through the points marked in Figure 16(A).

44. Find the equation of the elliptic cylinder passing through the points marked in Figure 16(B).

45. Find the equation of the hyperboloid shown in Figure 17(A).

46. Find the equation of the quadric surface shown in Figure 17(B).

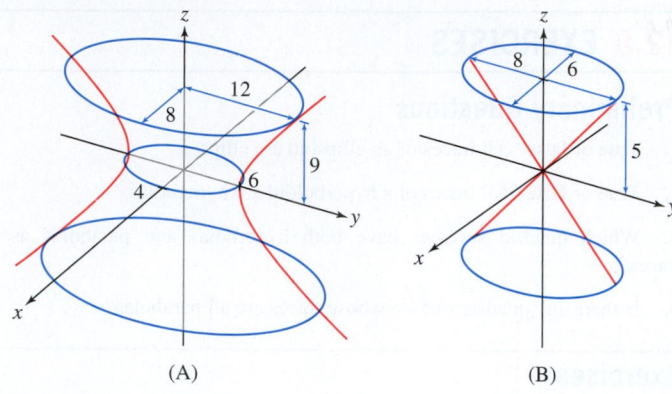

(A)   (B)

FIGURE 17

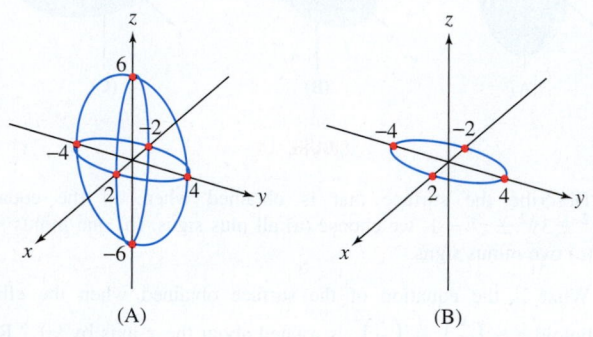

(A)   (B)

FIGURE 16

47. Determine the vertical traces of elliptic and parabolic cylinders in standard form.

48. What is the equation of a hyperboloid of one or two sheets in standard form if every horizontal trace is a circle?

49. Let $\mathcal{C}$ be an ellipse in a horizonal plane lying above the xy-plane. Which type of quadric surface is made up of all lines passing through the origin and a point on $\mathcal{C}$?

50. The eccentricity of a conic section is defined in Section 11.5. Show that the horizontal traces of the ellipsoid

$$\left(\frac{x}{a}\right)^2 + \left(\frac{y}{b}\right)^2 + \left(\frac{z}{c}\right)^2 = 1$$

are ellipses of the same eccentricity (apart from the traces at height $h = \pm c$, which reduce to a single point). Find the eccentricity.

Further Insights and Challenges

51. Let $\mathcal{S}$ be the hyperboloid $x^2 + y^2 = z^2 + 1$ and let $P = (\alpha, \beta, 0)$ be a point on $\mathcal{S}$ in the (x, y)-plane. Show that there are precisely two lines through P entirely contained in $\mathcal{S}$ (Figure 18). *Hint:* Consider the line $\mathbf{r}(t) = \langle \alpha + at, \beta + bt, t \rangle$ through P. Show that $\mathbf{r}(t)$ is contained in $\mathcal{S}$ if (a, b) is one of the two points on the unit circle obtained by rotating (α, β) through $\pm\frac{\pi}{2}$. This proves that a hyperboloid of one sheet is a **doubly ruled surface**, which means that it can be swept out by moving a line in space in two different ways.

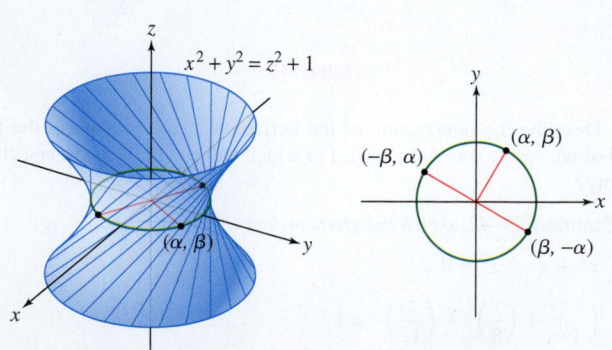

FIGURE 18

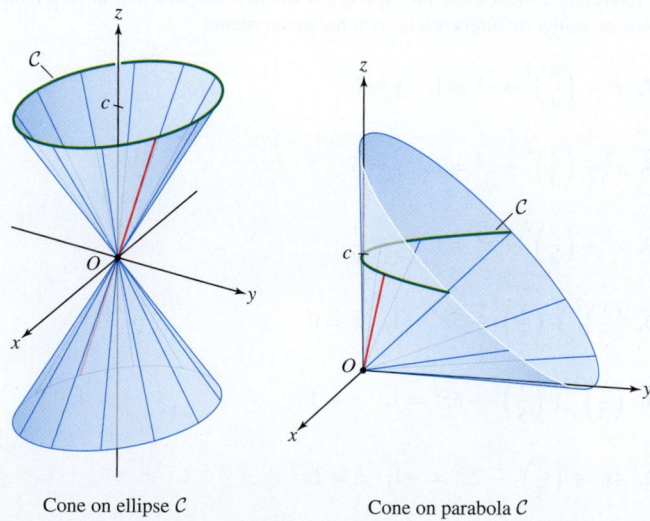

Cone on ellipse $\mathcal{C}$ Cone on parabola $\mathcal{C}$
                              (half of cone shown)

FIGURE 19

In Exercises 52 and 53, let C be a curve in $\mathbf{R}^3$ not passing through the origin. The cone on C is the surface consisting of all lines passing through the origin and a point on C [Figure 19(A)].

52. Show that the elliptic cone $\left(\frac{z}{c}\right)^2 = \left(\frac{x}{a}\right)^2 + \left(\frac{y}{b}\right)^2$ is, in fact, a cone on the ellipse C consisting of all points (x, y, c) such that $\left(\frac{x}{a}\right)^2 + \left(\frac{y}{b}\right)^2 = 1$.

53. Let a and c be nonzero constants and let C be the parabola at height c consisting of all points (x, ax^2, c) [Figure 19(B)]. Let S be the cone consisting of all lines passing through the origin and a point on C. This exercise shows that S is also an elliptic cone.

(a) Show that S has equation $yz = acx^2$.

(b) Show that under the change of variables $y = u + v$ and $z = u - v$, this equation becomes $acx^2 = u^2 - v^2$ or $u^2 = acx^2 + v^2$ (the equation of an elliptic cone in the variables x, v, u).

12.7 Cylindrical and Spherical Coordinates

This section introduces two generalizations of polar coordinates to $\mathbf{R}^3$: cylindrical and spherical coordinates. These coordinate systems are commonly used in problems having symmetry about an axis or rotational symmetry. For example, the magnetic field generated by a current flowing in a long, straight wire is conveniently expressed in cylindrical coordinates (Figure 1). We will also see the benefits of cylindrical and spherical coordinates when we study change of variables for multiple integrals.

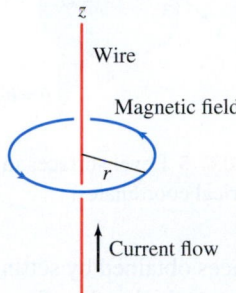

FIGURE 1 The magnetic field generated by a current flowing in a long, straight wire is conveniently expressed in cylindrical coordinates.

Cylindrical Coordinates

In cylindrical coordinates, we replace the x- and y-coordinates of a point $P = (x, y, z)$ by polar coordinates. Thus, the **cylindrical coordinates** of P are (r, θ, z), where (r, θ) are polar coordinates of the projection $Q = (x, y, 0)$ of P onto the xy-plane (Figure 2). Note that the points at fixed distance r from the z-axis make up a cylinder; hence, the name cylindrical coordinates.

We convert between rectangular and cylindrical coordinates using the rectangular-polar formulas of Section 11.3. In cylindrical coordinates, we usually assume $r \geq 0$.

Cylindrical to rectangular	Rectangular to cylindrical
$x = r \cos \theta$	$r = \sqrt{x^2 + y^2}$
$y = r \sin \theta$	$\tan \theta = \dfrac{y}{x}$
$z = z$	$z = z$

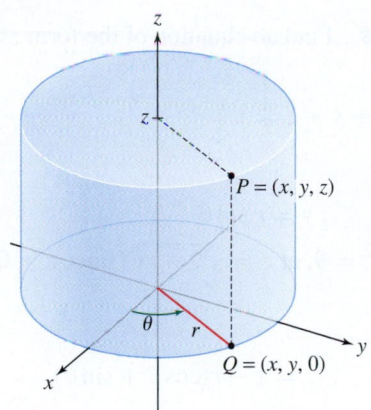

DF **FIGURE 2** P has cylindrical coordinates (r, θ, z).

EXAMPLE 1 **Converting from Cylindrical to Rectangular Coordinates** Find the rectangular coordinates of the point P with cylindrical coordinates $(r, \theta, z) = \left(2, \frac{3\pi}{4}, 5\right)$.

Solution Converting to rectangular coordinates is straightforward (Figure 3):

$$x = r \cos \theta = 2 \cos \frac{3\pi}{4} = 2\left(-\frac{\sqrt{2}}{2}\right) = -\sqrt{2}$$

$$y = r \sin \theta = 2 \sin \frac{3\pi}{4} = 2\left(\frac{\sqrt{2}}{2}\right) = \sqrt{2}$$

The z-coordinate is unchanged, so $(x, y, z) = (-\sqrt{2}, \sqrt{2}, 5)$. ∎

EXAMPLE 2 **Converting from Rectangular to Cylindrical Coordinates** Find cylindrical coordinates for the point with rectangular coordinates $(x, y, z) = (-3\sqrt{3}, -3, 5)$.

Solution We have $r = \sqrt{x^2 + y^2} = \sqrt{(-3\sqrt{3})^2 + (-3)^2} = 6$. The angle θ satisfies

$$\tan \theta = \frac{y}{x} = \frac{-3}{-3\sqrt{3}} = \frac{1}{\sqrt{3}} \quad \Rightarrow \quad \theta = \frac{\pi}{6} \quad \text{or} \quad \frac{7\pi}{6}$$

The correct choice is $\theta = \frac{7\pi}{6}$ because the projection $Q = (-3\sqrt{3}, -3, 0)$ lies in the third quadrant (Figure 4). The cylindrical coordinates are $(r, \theta, z) = \left(6, \frac{7\pi}{6}, 5\right)$. ∎

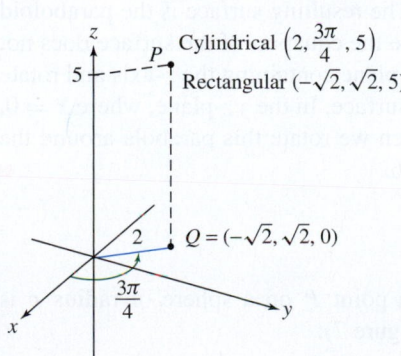

DF **FIGURE 3**

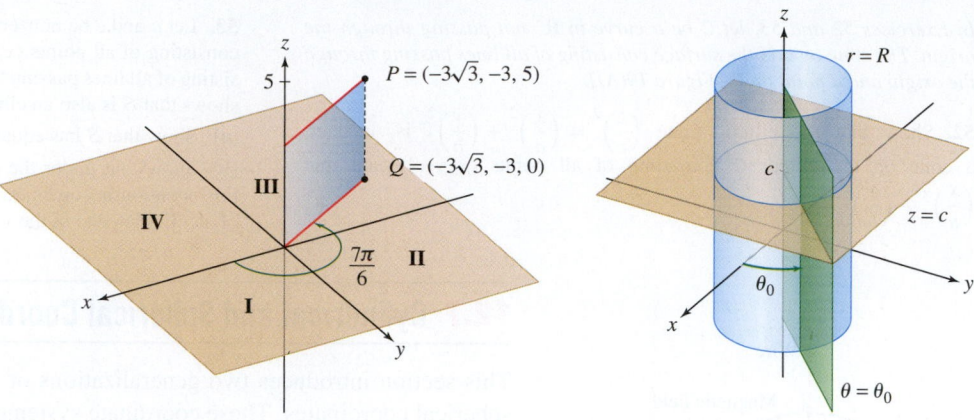

FIGURE 4 The projection Q lies in the third quadrant. Therefore, $\theta = \frac{7\pi}{6}$.

DF **FIGURE 5** Level surfaces in cylindrical coordinates.

Level Surfaces in Cylindrical Coordinates:

$r = R$ Cylinder of radius R with the z-axis as axis of symmetry

$\theta = \theta_0$ Half-plane through the z-axis making an angle θ_0 with the xz-plane

$z = c$ Horizontal plane at height c

The **level surfaces** of a coordinate system are the surfaces obtained by setting one of the coordinates equal to a constant. In rectangular coordinates, the level surfaces are the planes $x = x_0$, $y = y_0$, and $z = z_0$. In cylindrical coordinates, the level surfaces come in three types (Figure 5). The surface $r = R$ is the cylinder of radius R consisting of all points located a distance R from the z-axis. The equation $\theta = \theta_0$ defines the half-plane of all points that project onto the ray $\theta = \theta_0$ in the (x, y)-plane. Finally, $z = c$ is the horizontal plane at height c.

EXAMPLE 3 **Equations in Cylindrical Coordinates**   Find an equation of the form $z = f(r, \theta)$ for the surfaces:

(a) $x^2 + y^2 + z^2 = 9$, with $z \geq 0$ **(b)** $x + y + z = 1$

Solution We use the formulas

$$x^2 + y^2 = r^2, \qquad x = r\cos\theta, \qquad y = r\sin\theta$$

(a) The equation $x^2 + y^2 + z^2 = 9$ becomes $r^2 + z^2 = 9$, or $z = \sqrt{9 - r^2}$ (since $z \geq 0$). This is the upper half of a sphere of radius 3.

(b) The plane $x + y + z = 1$ becomes

$$z = 1 - x - y = 1 - r\cos\theta - r\sin\theta \qquad \text{or} \qquad z = 1 - r(\cos\theta + \sin\theta) \qquad ■$$

EXAMPLE 4 **Graphing Equations in Cylindrical Coordinates**   Graph the surface corresponding to the equation in cylindrical coordinates given by $z = r^2$.

Solution We consider two straightforward ways of picturing the surface. First, convert to rectangular coordinates to obtain $z = x^2 + y^2$. The resulting surface is the paraboloid illustrated in Figure 6. Alternatively, note that since the equation of the surface does not depend on θ we can graph its intersection with any plane containing the z-axis and rotate the resulting curve around the z-axis to obtain the surface. In the yz-plane, where $x = 0$, $z = r^2$ corresponds with the parabola $z = y^2$. When we rotate this parabola around the z-axis, we obtain the circular paraboloid in Figure 6. ■

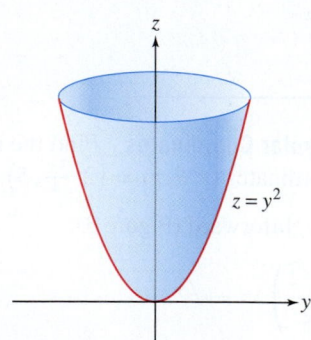

FIGURE 6 The surface $z = r^2$ is the paraboloid $z = x^2 + y^2$ and can also be seen to be the result of rotating the curve $z = y^2$ around the z-axis.

Spherical Coordinates

Spherical coordinates make use of the fact that a point P on a sphere of radius ρ is determined by two angular coordinates θ and ϕ (Figure 7):

- θ is the polar angle of the projection Q of P onto the xy-plane.
- ϕ is the **angle of declination**, which measures how much the ray through P declines from the vertical.

Spherical Coordinates:

ρ = distance from origin

θ = polar angle in the xy-plane

ϕ = angle of declination from the vertical

In some textbooks, θ is referred to as the azimuthal angle and ϕ as the polar angle.

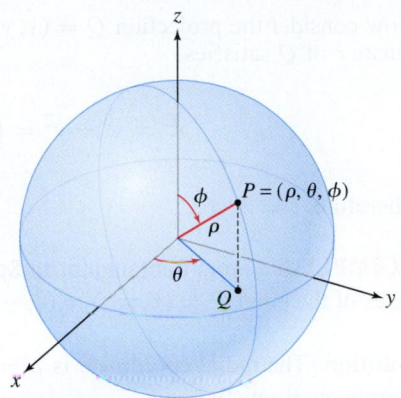

DF **FIGURE 7** Spherical coordinates (ρ, θ, ϕ).

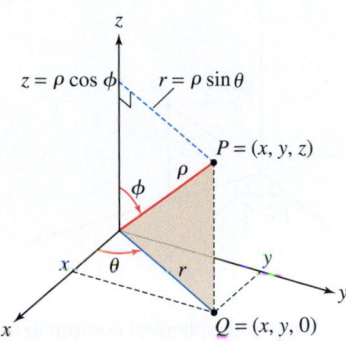

DF **FIGURE 8**

- The symbol ϕ (usually pronounced fee, but sometimes pronounced fie) is the 21st letter of the Greek alphabet.
- We use ρ (written out as "rho" and pronounced row) for the radial coordinate, although r is also used to denote distance from the origin in other contexts.

Thus, P is determined by the triple (ρ, θ, ϕ), which are called **spherical coordinates**. Typically, we restrict the coordinates so that $\rho \geq 0$ and $0 \leq \phi \leq \pi$.

Suppose that $P = (x, y, z)$ in rectangular coordinates. Since ρ is the distance from P to the origin,

$$\rho = \sqrt{x^2 + y^2 + z^2}$$

On the other hand, we see in Figure 8 that

$$\tan \theta = \frac{y}{x}, \qquad \cos \phi = \frac{z}{\rho}$$

The radial coordinate r of $Q = (x, y, 0)$ is $r = \rho \sin \phi$, and therefore

$$x = r \cos \theta = \rho \sin \phi \cos \theta, \qquad y = r \sin \theta = \rho \sin \phi \sin \theta, \qquad z = \rho \cos \phi$$

Spherical to rectangular	Rectangular to spherical
$x = \rho \sin \phi \cos \theta$	$\rho = \sqrt{x^2 + y^2 + z^2}$
$y = \rho \sin \phi \sin \theta$	$\tan \theta = \dfrac{y}{x}$
$z = \rho \cos \phi$	$\cos \phi = \dfrac{z}{\rho}$

EXAMPLE 5 **From Spherical to Rectangular Coordinates** Find the rectangular coordinates of $P = (\rho, \theta, \phi) = \left(3, \frac{\pi}{3}, \frac{\pi}{4}\right)$, and find the radial coordinate r of its projection Q onto the xy-plane.

Solution By the formulas discussed,

$$x = \rho \sin \phi \cos \theta = 3 \sin \frac{\pi}{4} \cos \frac{\pi}{3} = 3 \left(\frac{\sqrt{2}}{2}\right) \frac{1}{2} = \frac{3\sqrt{2}}{4}$$

$$y = \rho \sin \phi \sin \theta = 3 \sin \frac{\pi}{4} \sin \frac{\pi}{3} = 3 \left(\frac{\sqrt{2}}{2}\right) \frac{\sqrt{3}}{2} = \frac{3\sqrt{6}}{4}$$

$$z = \rho \cos \phi = 3 \cos \frac{\pi}{4} = 3 \frac{\sqrt{2}}{2} = \frac{3\sqrt{2}}{2}$$

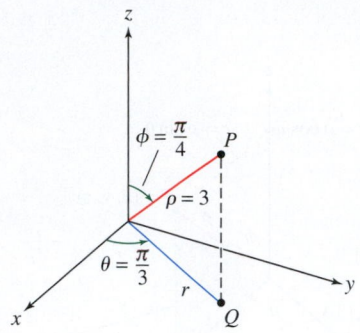

FIGURE 9 Point with spherical coordinates $\left(3, \frac{\pi}{3}, \frac{\pi}{4}\right)$.

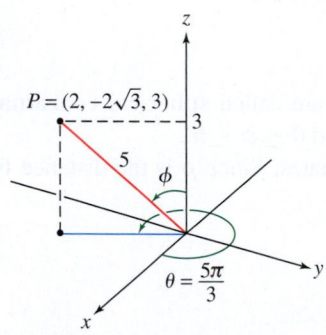

DF **FIGURE 10** Point with rectangular coordinates $(2, -2\sqrt{3}, 3)$.

Now consider the projection $Q = (x, y, 0) = \left(\frac{3\sqrt{2}}{4}, \frac{3\sqrt{6}}{4}, 0\right)$ (Figure 9). The radial coordinate r of Q satisfies

$$r^2 = x^2 + y^2 = \left(\frac{3\sqrt{2}}{4}\right)^2 + \left(\frac{3\sqrt{6}}{4}\right)^2 = \frac{9}{2}$$

Therefore, $r = 3/\sqrt{2}$. ∎

EXAMPLE 6 **From Rectangular to Spherical Coordinates** Find the spherical coordinates of the point $P = (x, y, z) = (2, -2\sqrt{3}, 3)$.

Solution The radial coordinate is $\rho = \sqrt{2^2 + (-2\sqrt{3})^2 + 3^2} = \sqrt{25} = 5$. The angular coordinate θ satisfies

$$\tan \theta = \frac{y}{x} = \frac{-2\sqrt{3}}{2} = -\sqrt{3} \quad \Rightarrow \quad \theta = \frac{2\pi}{3} \text{ or } \frac{5\pi}{3}$$

Since the point $(x, y) = (2, -2\sqrt{3})$ lies in the fourth quadrant, the correct choice is $\theta = \frac{5\pi}{3}$ (Figure 10). Finally, $\cos \phi = \frac{z}{\rho} = \frac{3}{5}$ and so $\phi = \cos^{-1} \frac{3}{5} \approx 0.93$. Therefore, P has spherical coordinates $\left(5, \frac{5\pi}{3}, 0.93\right)$. ∎

Figure 11 shows the three types of level surfaces in spherical coordinates. Notice that if $\phi \neq 0, \frac{\pi}{2}$ or π, then the level surface $\phi = \phi_0$ is the right circular cone consisting of points P such that $\overline{OP}$ makes an angle ϕ_0 with the z-axis. There are three exceptional cases: $\phi = \frac{\pi}{2}$ defines the xy-plane, $\phi = 0$ is the positive z-axis, and $\phi = \pi$ is the negative z-axis.

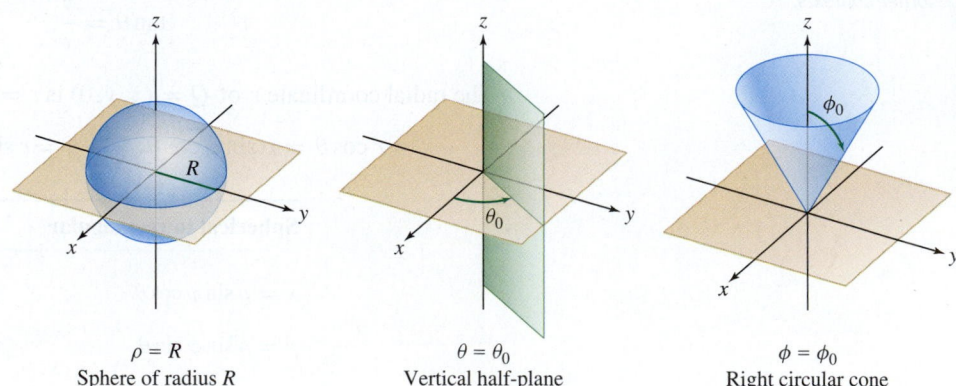

$\rho = R$	$\theta = \theta_0$	$\phi = \phi_0$
Sphere of radius R	Vertical half-plane	Right circular cone

FIGURE 11

EXAMPLE 7 **Finding an Equation in Spherical Coordinates** Find an equation of the form $\rho = f(\theta, \phi)$ for the following surfaces:

(a) $x^2 + y^2 + z^2 = 9$ **(b)** $z = x^2 - y^2$

Solution

(a) The equation $x^2 + y^2 + z^2 = 9$ defines the sphere of radius 3 centered at the origin. Since $\rho^2 = x^2 + y^2 + z^2$, the equation in spherical coordinates is $\rho = 3$.

(b) To convert $z = x^2 - y^2$ to spherical coordinates, we substitute the formulas for x, y, and z in terms of ρ, θ, and ϕ:

$$\overbrace{\rho \cos \phi}^{z} = \overbrace{(\rho \sin \phi \cos \theta)^2}^{x^2} - \overbrace{(\rho \sin \phi \sin \theta)^2}^{y^2}$$

$$\cos \phi = \rho \sin^2 \phi (\cos^2 \theta - \sin^2 \theta) \qquad \text{(divide by } \rho \text{ and factor)}$$

$$\cos \phi = \rho \sin^2 \phi \cos 2\theta \qquad \text{(since } \cos^2 \theta - \sin^2 \theta = \cos 2\theta\text{)}$$

Solving for ρ, we obtain $\rho = \dfrac{\cos\phi}{\sin^2\phi\,\cos 2\theta}$, which is valid for $\phi \neq 0, \pi$, and when $\theta \neq \pi/4, 3\pi/4, 5\pi/4, 7\pi/4$. ∎

The angular coordinates (θ, ϕ) on a sphere of fixed radius are closely related to the longitude-latitude system used to identify points on the surface of the earth (Figure 12). By convention, in this system, we use degrees rather than radians.

- A **longitude** is a half-circle stretching from the North to the South Pole (Figure 13). The axes are chosen so that $\theta = 0$ passes through Greenwich, England (this longitude is called the *prime meridian*). We designate the longitude by an angle between 0 and 180° together with a label E or W, according to whether it lies to the east or west of the prime meridian.
- The set of points on the sphere satisfying $\phi = \phi_0$ is a horizontal circle called a **latitude**. We measure latitudes from the equator and use the label N or S to specify the Northern or Southern Hemisphere. Thus, in the upper hemisphere $0 \le \phi_0 \le 90°$, and a spherical coordinate ϕ_0 corresponds to the latitude $(90° - \phi_0)$ N. In the lower hemisphere $90° \le \phi_0 \le 180°$, and ϕ_0 corresponds to the latitude $(\phi_0 - 90°)$ S.

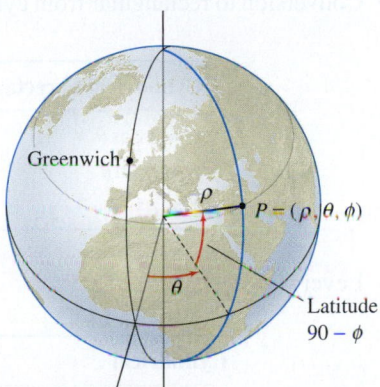

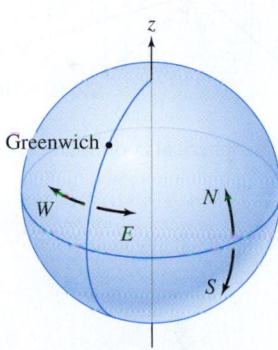

FIGURE 12 Longitude and latitude provide spherical coordinates on the surface of the earth.

FIGURE 13 Latitude is measured from the equator and is labeled N (north) in the upper hemisphere, and S (south) in the lower hemisphere.

EXAMPLE 8 **Spherical Coordinates via Longitude and Latitude** Find the angles (θ, ϕ) for Nairobi (1.17° S, 36.48° E) and Ottawa (45.27° N, 75.42° W).

Solution For Nairobi, $\theta = 36.48°$ since the longitude lies to the east of Greenwich. Nairobi's latitude is south of the equator, so $1.17 = \phi_0 - 90$ and $\phi_0 = 91.17°$.

For Ottawa, we have $\theta = 360 - 75.42 = 284.58°$ because 75.42° W refers to 75.42° in the negative θ direction. Since the latitude of Ottawa is north of the equator, $45.27 = 90 - \phi_0$ and $\phi_0 = 44.73°$. ∎

EXAMPLE 9 **Graphing Equations in Spherical Coordinates** Graph the surface corresponding to the equation in spherical coordinates given by $\rho = \sec\phi$.

Solution We could plug in values for ϕ, obtain the corresponding values for ρ, and then plot points, but it would be difficult to obtain an accurate representation of the surface in this way. Instead, notice that we can rewrite the equation:

$$\rho = \frac{1}{\cos\phi}$$

$$\rho\cos\phi = 1$$

From our conversion equations, we see that this is $z = 1$. Hence, our surface is simply the horizontal plane at height $z = 1$, as in Figure 14. ∎

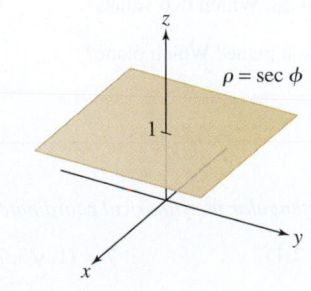

FIGURE 14 This plane is the graph of $\rho = \sec\phi$.

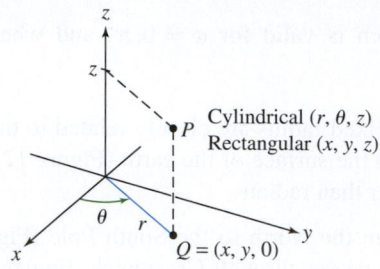

FIGURE 15 Cylindrical coordinates (r, θ, z).

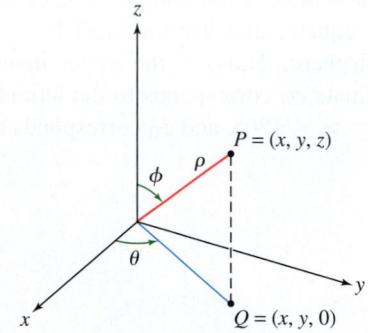

FIGURE 16 Spherical coordinates (ρ, θ, ϕ).

12.7 SUMMARY

- Conversion from rectangular to cylindrical (r, θ, z) and spherical (ρ, θ, ϕ) coordinates (Figures 15 and 16):

Rectangular to cylindrical	Rectangular to spherical
$r = \sqrt{x^2 + y^2}$	$\rho = \sqrt{x^2 + y^2 + z^2}$
$\tan \theta = \dfrac{y}{x}$	$\tan \theta = \dfrac{y}{x}$
$z = z$	$\cos \phi = \dfrac{z}{\rho}$

The angles are chosen so that

$$0 \leq \theta < 2\pi \quad \text{(cylindrical or spherical)}, \qquad 0 \leq \phi \leq \pi \quad \text{(spherical)}$$

- Conversion to rectangular from cylindrical (r, θ, z) and spherical (ρ, θ, ϕ) coordinates:

Cylindrical to rectangular	Spherical to rectangular
$x = r \cos \theta$	$x = \rho \sin \phi \cos \theta$
$y = r \sin \theta$	$y = \rho \sin \phi \sin \theta$
$z = z$	$z = \rho \cos \phi$

- Level surfaces:

Cylindrical		Spherical	
$r = R$:	cylinder of radius R	$\rho = R$:	sphere of radius R
$\theta = \theta_0$:	vertical half-plane	$\theta = \theta_0$:	vertical half-plane
$z = c$:	horizontal plane	$\phi = \phi_0$:	right circular cone

12.7 EXERCISES

Preliminary Questions

1. Describe the surfaces $r = R$ in cylindrical coordinates and $\rho = R$ in spherical coordinates.

2. Which statement about cylindrical coordinates is correct?
(a) If $\theta = 0$, then P lies on the z-axis.
(b) If $\theta = 0$, then P lies in the xz-plane.

3. Which statement about spherical coordinates is correct?

(a) If $\phi = 0$, then P lies on the z-axis.
(b) If $\phi = 0$, then P lies in the xy-plane.

4. The level surface $\phi = \phi_0$ in spherical coordinates, usually a cone, reduces to a half-line for two values of ϕ_0. Which two values?

5. For which value of ϕ_0 is $\phi = \phi_0$ a plane? Which plane?

Exercises

In Exercises 1–4, convert from cylindrical to rectangular coordinates.

1. $(4, \pi, 4)$

2. $\left(2, \dfrac{\pi}{3}, -8\right)$

3. $\left(0, \dfrac{\pi}{5}, \dfrac{1}{2}\right)$

4. $\left(1, \dfrac{\pi}{2}, -2\right)$

In Exercises 5–10, convert from rectangular to cylindrical coordinates.

5. $(1, -1, 1)$

6. $(2, 2, 1)$

7. $(1, \sqrt{3}, 7)$

8. $\left(\dfrac{3}{2}, \dfrac{3\sqrt{3}}{2}, 9\right)$

9. $\left(\dfrac{5}{\sqrt{2}}, \dfrac{5}{\sqrt{2}}, 2\right)$

10. $(3, 3\sqrt{3}, 2)$

In Exercises 11–16, describe the set in cylindrical coordinates.

11. $x^2 + y^2 \leq 3$

12. $x^2 + y^2 + z^2 \leq 10$

13. $y^2 + z^2 \leq 4, \quad x = 0$

14. $x^2 + y^2 + z^2 = 9, \quad y \geq 0, \quad z \geq 0$

15. $x^2 + y^2 \leq 9, \quad x \geq y$

16. $y^2 + z^2 \leq 9, \quad x \geq y$

In Exercises 17–26, sketch the set (described in cylindrical coordinates).

17. $r = 4$

18. $\theta = \dfrac{\pi}{3}$

19. $z = -2$

20. $r = 2, \quad z = 3$

21. $1 \leq r \leq 3, \quad 0 \leq z \leq 4$

22. $z = r$

23. $r = \sin\theta$ (*Hint:* Convert to rectangular.)

24. $1 \leq r \leq 3, \quad 0 \leq \theta \leq \dfrac{\pi}{2}, \quad 0 \leq z \leq 4$

25. $z^2 + r^2 \leq 4$

26. $r \leq 3, \quad \pi \leq \theta \leq \dfrac{3\pi}{2}, \quad z = 4$

In Exercises 27–32, find an equation of the form $r = f(\theta, z)$ in cylindrical coordinates for the following surfaces.

27. $z = x + y$

28. $x^2 + y^2 + z^2 = 2$

29. $\dfrac{x^2}{yz} = 1$

30. $x^2 - y^2 = 4$

31. $x^2 + y^2 = 4$

32. $z = 3xy$

In Exercises 33–38, convert from spherical to rectangular coordinates.

33. $\left(3, 0, \dfrac{\pi}{2}\right)$

34. $\left(2, \dfrac{\pi}{4}, \dfrac{\pi}{3}\right)$

35. $(3, \pi, 0)$

36. $\left(5, \dfrac{3\pi}{4}, \dfrac{\pi}{4}\right)$

37. $\left(6, \dfrac{\pi}{6}, \dfrac{5\pi}{6}\right)$

38. $(0.5, 3.7, 2)$

In Exercises 39–44, convert from rectangular to spherical coordinates.

39. $(\sqrt{3}, 0, 1)$

40. $\left(\dfrac{\sqrt{3}}{2}, \dfrac{3}{2}, 1\right)$

41. $(1, 1, 1)$

42. $(1, -1, 1)$

43. $\left(\dfrac{1}{2}, \dfrac{\sqrt{3}}{2}, \sqrt{3}\right)$

44. $\left(\dfrac{\sqrt{2}}{2}, \dfrac{\sqrt{2}}{2}, \sqrt{3}\right)$

In Exercises 45 and 46, convert from cylindrical to spherical coordinates.

45. $(2, 0, 2)$

46. $(3, \pi, \sqrt{3})$

In Exercises 47 and 48, convert from spherical to cylindrical coordinates.

47. $\left(4, 0, \dfrac{\pi}{4}\right)$

48. $\left(2, \dfrac{\pi}{3}, \dfrac{\pi}{6}\right)$

In Exercises 49–54, describe the given set in spherical coordinates.

49. $x^2 + y^2 + z^2 \leq 100$

50. $x^2 + y^2 + z^2 = 1, \quad z \geq 0$

51. $x^2 + y^2 + z^2 = 10, \quad x \geq 0, \quad y \geq 0, \quad z \geq 0$

52. $x^2 + y^2 + z^2 \leq 1, \quad x = y, \quad x \geq 0, \quad y \geq 0$

53. $y^2 + z^2 \leq 4, \quad x = 0$

54. $x^2 + y^2 = 3z^2$

In Exercises 55–64, sketch the set of points (described in spherical coordinates).

55. $\rho = 4$

56. $\phi = \dfrac{\pi}{4}$

57. $\rho = 2, \quad \theta = \dfrac{\pi}{4}$

58. $\rho = 2, \quad \phi = \dfrac{\pi}{4}$

59. $\rho = 2, \quad 0 \leq \phi \leq \dfrac{\pi}{2}$

60. $\theta = \dfrac{\pi}{2}, \quad \phi = \dfrac{\pi}{4}, \quad \rho \geq 1$

61. $\rho \leq 2, \quad 0 \leq \theta \leq \dfrac{\pi}{2}, \quad \dfrac{\pi}{2} \leq \phi \leq \pi$

62. $\rho = 1, \quad \dfrac{\pi}{3} \leq \phi \leq \dfrac{2\pi}{3}$

63. $\rho = \csc\phi$

64. $\rho = \csc\phi \cot\phi$

In Exercises 65–72, find an equation of the form $\rho = f(\theta, \phi)$ in spherical coordinates for the following surfaces.

65. $x^2 + y^2 = 9$

66. $x = 3$

67. $z = 2$

68. $z^2 = 3(x^2 + y^2)$

69. $x = z^2$

70. $z = x^2 + y^2$

71. $x^2 - y^2 = 4$

72. $xy = z$

73. [icon] Which of (a)–(c) is the equation of the cylinder of radius R in spherical coordinates? Refer to Figure 17.

(a) $R\rho = \sin\phi$

(b) $\rho \sin\phi = R$

(c) $\rho = R \sin\phi$

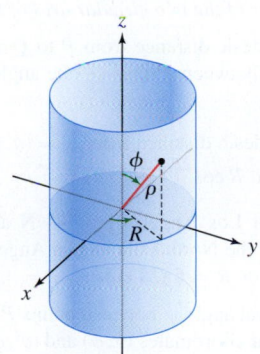

FIGURE 17

74. Let $P_1 = (1, -\sqrt{3}, 5)$ and $P_2 = (-1, \sqrt{3}, 5)$ in rectangular coordinates. In which quadrants do the projections of P_1 and P_2 onto the xy-plane lie? Find the polar angle θ of each point.

75. Determine the spherical angles (θ, ϕ) for the cities Helsinki, Finland (60.1° N, 25.0° E), and São Paulo, Brazil (23.52° S, 46.52° W).

76. Find the longitude and latitude for the points on the globe with angular coordinates $(\theta, \phi) = (\pi/8, 7\pi/12)$ and $(4, 2)$.

77. Consider a rectangular coordinate system with its origin at the center of the earth, z-axis through the North Pole, and x-axis through the prime meridian. Find the rectangular coordinates of Sydney, Australia (34° S, 151° E), and Bogotá, Colombia (4° 32′ N, 74° 15′ W). A minute is $1/60°$. Assume that the earth is a sphere of radius $R = 6370$ km.

78. Find the equation in rectangular coordinates of the quadric surface consisting of the two cones $\phi = \dfrac{\pi}{4}$ and $\phi = \dfrac{3\pi}{4}$.

79. Find an equation of the form $z = f(r, \theta)$ in cylindrical coordinates for $z^2 = x^2 - y^2$.

80. Show that $\rho = 2 \cos \phi$ is the equation of a sphere with its center on the z-axis. Find its radius and center.

81. An apple modeled by taking all the points in and on a sphere of radius 2 inches is cored with a vertical cylinder of radius 1 in. Use inequalities in cylindrical coordinates to describe the set of all points that remain in the apple once the core is removed.

82. Repeat Exercise 81 using inequalities in spherical coordinates.

83. ✏️ Explain the following statement: If the equation of a surface in cylindrical or spherical coordinates does not involve the coordinate θ, then the surface is rotationally symmetric with respect to the z-axis.

84. CAS Plot the surface $\rho = 1 - \cos \phi$. Then plot the trace of S in the xz-plane and explain why S is obtained by rotating this trace.

85. Find equations $r = g(\theta, z)$ (cylindrical) and $\rho = f(\theta, \phi)$ (spherical) for the hyperboloid $x^2 + y^2 = z^2 + 1$ (Figure 18). Do there exist points on the hyperboloid with $\phi = 0$ or π? Which values of ϕ occur for points on the hyperboloid?

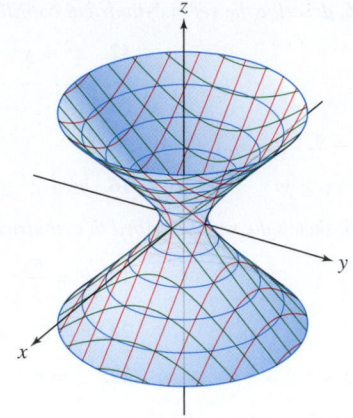

FIGURE 18 The hyperboloid $x^2 + y^2 = z^2 + 1$.

Further Insights and Challenges

*In Exercises 86–90, a **great circle** on a sphere S with center O and radius R is a circle obtained by intersecting S with a plane that passes through O (Figure 19). If P and Q are not antipodal (on opposite sides), there is a unique great circle through P and Q on S (intersect S with the plane through O, P, and Q). The geodesic distance from P to Q is defined as the length of the smaller of the two circular arcs of this great circle.*

86. Show that the geodesic distance from P to Q is equal to $R\psi$, where ψ is the *central angle* between P and Q (the angle between the vectors $\mathbf{v} = \overrightarrow{OP}$ and $\mathbf{u} = \overrightarrow{OQ}$).

87. Show that the geodesic distance from $Q = (a, b, c)$ to the North Pole $P = (0, 0, R)$ is equal to $R \cos^{-1}\left(\dfrac{c}{R}\right)$.

88. The coordinates of Los Angeles are 34° N and 118° W. Find the geodesic distance from the North Pole to Los Angeles, assuming that the earth is a sphere of radius $R = 6370$ km.

89. Show that the central angle ψ between points P and Q on a sphere (of any radius) with angular coordinates (θ, ϕ) and (θ', ϕ') is equal to

$$\psi = \cos^{-1}\left(\sin\phi \sin\phi' \cos(\theta - \theta') + \cos\phi \cos\phi'\right)$$

Hint: Compute the dot product of $\overrightarrow{OP}$ and $\overrightarrow{OQ}$. Check this formula by computing the geodesic distance between the North and South Poles.

90. Use Exercise 89 to find the geodesic distance between Los Angeles (34° N, 118° W) and Bombay (19° N, 72.8° E).

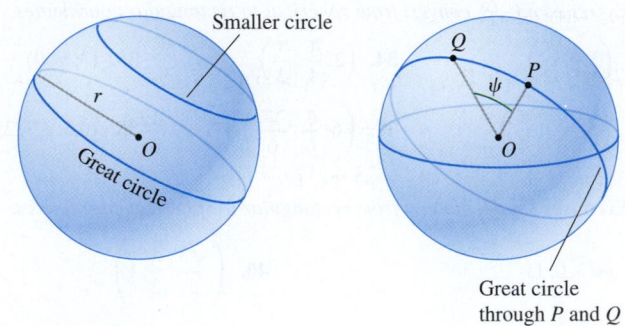

FIGURE 19

CHAPTER REVIEW EXERCISES

In Exercises 1–6, let $\mathbf{v} = \langle -2, 5 \rangle$ and $\mathbf{w} = \langle 3, -2 \rangle$.

1. Calculate $5\mathbf{w} - 3\mathbf{v}$ and $5\mathbf{v} - 3\mathbf{w}$.

2. Sketch $\mathbf{v}$, $\mathbf{w}$, and $2\mathbf{v} - 3\mathbf{w}$.

3. Find the unit vector in the direction of $\mathbf{v}$.

4. Find the length of $\mathbf{v} + \mathbf{w}$.

5. Express $\mathbf{i}$ as a linear combination $r\mathbf{v} + s\mathbf{w}$.

6. Find a scalar α such that $\|\mathbf{v} + \alpha\mathbf{w}\| = 6$.

7. If $P = (1, 4)$ and $Q = (-3, 5)$, what are the components of $\overrightarrow{PQ}$? What is the length of $\overrightarrow{PQ}$?

8. Let $A = (2, -1)$, $B = (1, 4)$, and $P = (2, 3)$. Find the point Q such that $\overrightarrow{PQ}$ is equivalent to $\overrightarrow{AB}$. Sketch $\overrightarrow{PQ}$ and $\overrightarrow{AB}$.

9. Find the vector with length 3 making an angle of $\frac{7\pi}{4}$ with the positive x-axis.

10. Calculate $3(\mathbf{i} - 2\mathbf{j}) - 6(\mathbf{i} + 6\mathbf{j})$.

11. Find the value of β for which $\mathbf{w} = \langle -2, \beta \rangle$ is parallel to $\mathbf{v} = \langle 4, -3 \rangle$.

12. Let $P = (1, 4, -3)$.

(a) Find the point Q such that $\overrightarrow{PQ}$ is equivalent to $\langle 3, -1, 5 \rangle$.

(b) Find a unit vector $\mathbf{e}$ equivalent to $\overrightarrow{PQ}$.

13. Let $\mathbf{w} = \langle 2, -2, 1 \rangle$ and $\mathbf{v} = \langle 4, 5, -4 \rangle$. Solve for $\mathbf{u}$ if $\mathbf{v} + 5\mathbf{u} = 3\mathbf{w} - \mathbf{u}$.

14. Let $\mathbf{v} = 3\mathbf{i} - \mathbf{j} + 4\mathbf{k}$. Find the length of $\mathbf{v}$ and the vector $2\mathbf{v} + 3(4\mathbf{i} - \mathbf{k})$.

15. Find a parametrization $\mathbf{r}_1(t)$ of the line passing through $(1, 4, 5)$ and $(-2, 3, -1)$. Then find a parametrization $\mathbf{r}_2(t)$ of the line parallel to $\mathbf{r}_1$ passing through $(1, 0, 0)$.

16. Let $\mathbf{r}_1(t) = \mathbf{v}_1 + t\mathbf{w}_1$ and $\mathbf{r}_2(t) = \mathbf{v}_2 + t\mathbf{w}_2$ be parametrizations of lines $\mathcal{L}_1$ and $\mathcal{L}_2$. For each statement (a)–(e), provide a proof if the statement is true and a counterexample if it is false.
(a) If $\mathcal{L}_1 = \mathcal{L}_2$, then $\mathbf{v}_1 = \mathbf{v}_2$ and $\mathbf{w}_1 = \mathbf{w}_2$.
(b) If $\mathcal{L}_1 = \mathcal{L}_2$ and $\mathbf{v}_1 = \mathbf{v}_2$, then $\mathbf{w}_1 = \mathbf{w}_2$.
(c) If $\mathcal{L}_1 = \mathcal{L}_2$ and $\mathbf{w}_1 = \mathbf{w}_2$, then $\mathbf{v}_1 = \mathbf{v}_2$.
(d) If $\mathcal{L}_1$ is parallel to $\mathcal{L}_2$, then $\mathbf{w}_1 = \mathbf{w}_2$.
(e) If $\mathcal{L}_1$ is parallel to $\mathcal{L}_2$, then $\mathbf{w}_1 = \lambda \mathbf{w}_2$ for some scalar λ.

17. Find a and b such that the lines $\mathbf{r}_1 = \langle 1, 2, 1 \rangle + t\langle 1, -1, 1 \rangle$ and $\mathbf{r}_2 = \langle 3, -1, 1 \rangle + t\langle a, b, -2 \rangle$ are parallel.

18. Find a such that the lines $\mathbf{r}_1 = \langle 1, 2, 1 \rangle + t\langle 1, -1, 1 \rangle$ and $\mathbf{r}_2 = \langle 3, -1, 1 \rangle + t\langle a, 4, -2 \rangle$ intersect.

19. Sketch the vector sum $\mathbf{v} = \mathbf{v}_1 - \mathbf{v}_2 + \mathbf{v}_3$ for the vectors in Figure 1(A).

20. Sketch the sums $\mathbf{v}_1 + \mathbf{v}_2 + \mathbf{v}_3$, $\mathbf{v}_1 + 2\mathbf{v}_2$, and $\mathbf{v}_2 - \mathbf{v}_3$ for the vectors in Figure 1(B).

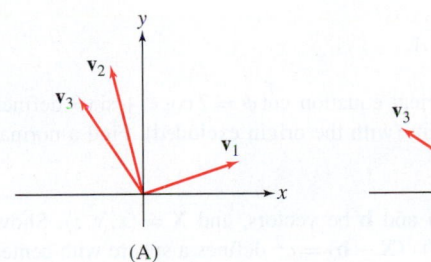 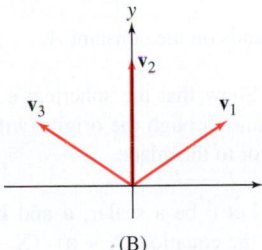

(A)   (B)

FIGURE 1

In Exercises 21–26, let $\mathbf{v} = \langle 1, 3, -2 \rangle$ *and* $\mathbf{w} = \langle 2, -1, 4 \rangle$.

21. Compute $\mathbf{v} \cdot \mathbf{w}$.

22. Compute the angle between $\mathbf{v}$ and $\mathbf{w}$.

23. Compute $\mathbf{v} \times \mathbf{w}$.

24. Find the area of the parallelogram spanned by $\mathbf{v}$ and $\mathbf{w}$.

25. Find the volume of the parallelepiped spanned by $\mathbf{v}$, $\mathbf{w}$, and $\mathbf{u} = \langle 1, 2, 6 \rangle$.

26. Find all the vectors orthogonal to both $\mathbf{v}$ and $\mathbf{w}$.

27. Use vectors to prove that the line connecting the midpoints of two sides of a triangle is parallel to the third side.

28. Let $\mathbf{v} = \langle 1, -1, 3 \rangle$ and $\mathbf{w} = \langle 4, -2, 1 \rangle$.
(a) Find the decomposition $\mathbf{v} = \mathbf{v}_{\parallel \mathbf{w}} + \mathbf{v}_{\perp \mathbf{w}}$ with respect to $\mathbf{w}$.
(b) Find the decomposition $\mathbf{w} = \mathbf{w}_{\parallel \mathbf{v}} + \mathbf{w}_{\perp \mathbf{v}}$ with respect to $\mathbf{v}$.

29. Calculate the component of $\mathbf{v} = \langle -2, \frac{1}{2}, 3 \rangle$ along $\mathbf{w} = \langle 1, 2, 2 \rangle$.

30. Calculate the magnitude of the forces on the two ropes in Figure 2.

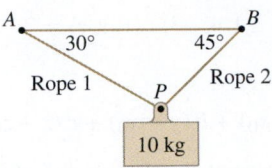

FIGURE 2

31. A 50-kg wagon is pulled to the right by a force $\mathbf{F}_1$ making an angle of $30°$ with the ground. At the same time, the wagon is pulled to the left by a horizontal force $\mathbf{F}_2$.
(a) Find the magnitude of $\mathbf{F}_1$ in terms of the magnitude of $\mathbf{F}_2$ if the wagon does not move.
(b) What is the maximal magnitude of $\mathbf{F}_1$ that can be applied to the wagon without lifting it?

32. Let $\mathbf{v}$, $\mathbf{w}$, and $\mathbf{u}$ be the vectors in $\mathbf{R}^3$. Which of the following is a scalar?
(a) $\mathbf{v} \times (\mathbf{u} + \mathbf{w})$
(b) $(\mathbf{u} + \mathbf{w}) \cdot (\mathbf{v} \times \mathbf{w})$
(c) $(\mathbf{u} \times \mathbf{w}) + (\mathbf{w} - \mathbf{v})$

In Exercises 33–36, let $\mathbf{v} = \langle 1, 2, 4 \rangle$, $\mathbf{u} = \langle 6, -1, 2 \rangle$, *and* $\mathbf{w} = \langle 1, 0, -3 \rangle$. *Calculate the given quantity.*

33. $\mathbf{v} \times \mathbf{w}$

34. $\mathbf{w} \times \mathbf{u}$

35. $\det \begin{pmatrix} \mathbf{u} \\ \mathbf{v} \\ \mathbf{w} \end{pmatrix}$

36. $\mathbf{v} \cdot (\mathbf{u} \times \mathbf{w})$

37. Use the cross product to find the area of the triangle whose vertices are $(1, 3, -1)$, $(2, -1, 3)$, and $(4, 1, 1)$.

38. Calculate $\|\mathbf{v} \times \mathbf{w}\|$ if $\|\mathbf{v}\| = 2$, $\mathbf{v} \cdot \mathbf{w} = 3$, and the angle between $\mathbf{v}$ and $\mathbf{w}$ is $\frac{\pi}{6}$.

39. Show that if the vectors $\mathbf{v}$, $\mathbf{w}$ are orthogonal, then $\|\mathbf{v} + \mathbf{w}\|^2 = \|\mathbf{v}\|^2 + \|\mathbf{w}\|^2$.

40. Find the angle between $\mathbf{v}$ and $\mathbf{w}$ if $\|\mathbf{v} + \mathbf{w}\| = \|\mathbf{v}\| = \|\mathbf{w}\|$.

41. Find $\|\mathbf{e} - 4\mathbf{f}\|$, assuming that $\mathbf{e}$ and $\mathbf{f}$ are unit vectors such that $\|\mathbf{e} + \mathbf{f}\| = \sqrt{3}$.

42. Find the area of the parallelogram spanned by vectors $\mathbf{v}$ and $\mathbf{w}$ such that $\|\mathbf{v}\| = \|\mathbf{w}\| = 2$ and $\mathbf{v} \cdot \mathbf{w} = 1$.

43. Show that the equation $\langle 1, 2, 3 \rangle \times \mathbf{v} = \langle -1, 2, a \rangle$ has no solution for $a \neq -1$.

44. Prove with a diagram the following: If $\mathbf{e}$ is a unit vector orthogonal to $\mathbf{v}$, then $\mathbf{e} \times (\mathbf{v} \times \mathbf{e}) = (\mathbf{e} \times \mathbf{v}) \times \mathbf{e} = \mathbf{v}$.

45. Use the identity

$$\mathbf{u} \times (\mathbf{v} \times \mathbf{w}) = (\mathbf{u} \cdot \mathbf{w})\mathbf{v} - (\mathbf{u} \cdot \mathbf{v})\mathbf{w}$$

to prove that

$$\mathbf{u} \times (\mathbf{v} \times \mathbf{w}) + \mathbf{v} \times (\mathbf{w} \times \mathbf{u}) + \mathbf{w} \times (\mathbf{u} \times \mathbf{v}) = \mathbf{0}$$

46. Find an equation of the plane through $(1, -3, 5)$ with normal vector $\mathbf{n} = \langle 2, 1, -4 \rangle$.

47. Write the equation of the plane $\mathcal{P}$ with vector equation

$$\langle 1, 4, -3 \rangle \cdot \langle x, y, z \rangle = 7$$

in the form

$$a\,(x - x_0) + b\,(y - y_0) + c\,(z - z_0) = 0$$

Hint: You must find a point $P = (x_0, y_0, z_0)$ on $\mathcal{P}$.

48. Find all the planes parallel to the plane passing through the points $(1, 2, 3)$, $(1, 2, 7)$, and $(1, 1, -3)$.

49. Find the plane through $P = (4, -1, 9)$ containing the line $\mathbf{r}(t) = \langle 1, 4, -3 \rangle + t \langle 2, 1, 1 \rangle$.

50. Find the intersection of the line $\mathbf{r}(t) = \langle 3t + 2, 1, -7t \rangle$ and the plane $2x - 3y + z = 5$.

51. Find the trace of the plane $3x - 2y + 5z = 4$ in the xy-plane.

52. Find the line of intersection of the plane $x + y + z = 1$ and the plane $3x - 2y + z = 5$.

In Exercises 53–58, determine the type of the quadric surface.

53. $\left(\dfrac{x}{3}\right)^2 + \left(\dfrac{y}{4}\right)^2 + 2z^2 = 1$

54. $\left(\dfrac{x}{3}\right)^2 - \left(\dfrac{y}{4}\right)^2 + 2z^2 = 1$

55. $\left(\dfrac{x}{3}\right)^2 + \left(\dfrac{y}{4}\right)^2 - 2z = 0$

56. $\left(\dfrac{x}{3}\right)^2 - \left(\dfrac{y}{4}\right)^2 - 2z = 0$

57. $\left(\dfrac{x}{3}\right)^2 - \left(\dfrac{y}{4}\right)^2 - 2z^2 = 0$

58. $\left(\dfrac{x}{3}\right)^2 - \left(\dfrac{y}{4}\right)^2 - 2z^2 = 1$

59. Determine the type of the quadric surface $ax^2 + by^2 - z^2 = 1$ if:
(a) $a < 0, \quad b < 0$
(b) $a > 0, \quad b > 0$
(c) $a > 0, \quad b < 0$

60. Describe the traces of the surface

$$\left(\dfrac{x}{2}\right)^2 - y^2 + \left(\dfrac{z}{2}\right)^2 = 1$$

in the three coordinate planes.

61. Convert $(x, y, z) = (3, 4, -1)$ from rectangular to cylindrical and spherical coordinates.

62. Convert $(r, \theta, z) = \left(3, \frac{\pi}{6}, 4\right)$ from cylindrical to spherical coordinates.

63. Convert the point $(\rho, \theta, \phi) = \left(3, \frac{\pi}{6}, \frac{\pi}{3}\right)$ from spherical to cylindrical coordinates.

64. Describe the set of all points $P = (x, y, z)$ satisfying $x^2 + y^2 \le 4$ in both cylindrical and spherical coordinates.

65. Sketch the graph of the cylindrical equation $z = 2r \cos \theta$ and write the equation in rectangular coordinates.

66. Write the surface $x^2 + y^2 - z^2 = 2(x + y)$ as an equation $r = f(\theta, z)$ in cylindrical coordinates.

67. Show that the cylindrical equation

$$r^2(1 - 2\sin^2 \theta) + z^2 = 1$$

is a hyperboloid of one sheet.

68. Sketch the graph of the spherical equation $\rho = 2 \cos \theta \sin \phi$ and write the equation in rectangular coordinates.

69. Describe how the surface with spherical equation

$$\rho^2(1 + A \cos^2 \phi) = 1$$

depends on the constant A.

70. Show that the spherical equation $\cot \phi = 2 \cos \theta + \sin \theta$ defines a plane through the origin (with the origin excluded). Find a normal vector to this plane.

71. Let c be a scalar, $\mathbf{a}$ and $\mathbf{b}$ be vectors, and $\mathbf{X} = \langle x, y, z \rangle$. Show that the equation $(\mathbf{X} - \mathbf{a}) \cdot (\mathbf{X} - \mathbf{b}) = c^2$ defines a sphere with center $\mathbf{m} = \frac{1}{2}(\mathbf{a} + \mathbf{b})$ and radius R, where $R^2 = c^2 + \left\| \frac{1}{2}(\mathbf{a} - \mathbf{b}) \right\|^2$.

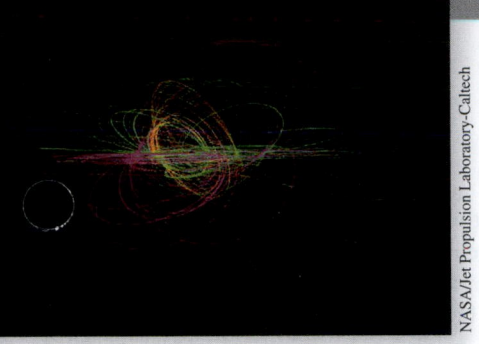

13 CALCULUS OF VECTOR-VALUED FUNCTIONS

The Cassini spacecraft was launched in 1997 and, after arriving at Saturn in 2004, spent nearly 13 years orbiting and investigating the planet. The path it followed in orbit around Saturn is illustrated in the figure. Vector-valued functions are used to design and control the trajectory of the spacecraft, and the concepts of calculus are invaluable in the process.

I n this chapter, we study vector-valued functions and their derivatives, and we use them to analyze curves and motion in 3-space. Although many techniques from single-variable calculus carry over to the vector setting, there are important new aspects to the derivative. For a real-valued function f, the derivative $f'(x)$ is a numerical value that indicates the rate of change of f at x. By contrast, the derivative of a vector-valued function is a vector. It identifies the magnitude and the direction of the rate of change of the function. To develop these new concepts, we begin with an introduction to vector-valued functions.

13.1 Vector-Valued Functions

Consider a particle moving in $\mathbf{R}^3$ whose coordinates at time t are $(x(t), y(t), z(t))$. It is convenient to represent the particle's path by the **vector-valued function**

$$\mathbf{r}(t) = \langle x(t), y(t), z(t) \rangle = x(t)\mathbf{i} + y(t)\mathbf{j} + z(t)\mathbf{k} \qquad \boxed{1}$$

Think of $\mathbf{r}(t)$ as a moving vector that points from the origin to the position of the particle at time t (Figure 1).

Functions with real number values are often called scalar-valued *to distinguish them from vector-valued functions.*

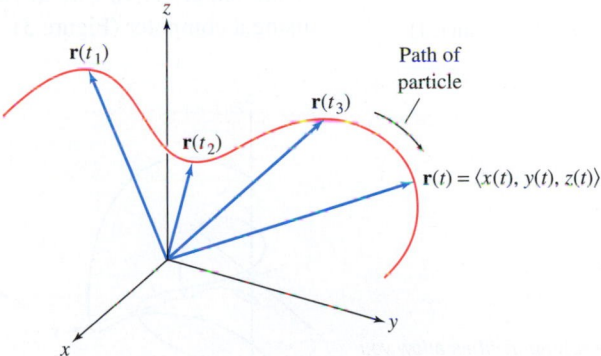

DF FIGURE 1

More generally, a vector-valued function is any function $\mathbf{r}(t)$ of the form in Eq. (1) whose domain $\mathcal{D}$ is a set of real numbers and whose range is a set of position vectors. The variable t is called a **parameter**, and the functions $x(t), y(t), z(t)$ are called the **components** or **coordinate functions**. We usually take as domain the set of all values of t for which $\mathbf{r}(t)$ is defined—that is, all values of t that belong to the domains of all three coordinate functions $x(t), y(t), z(t)$. For example,

We often use t for the parameter, thinking of it as representing time, but we are free to use any other variable such as s or θ. It is best to avoid writing $\mathbf{r}(x)$ or $\mathbf{r}(y)$ to prevent confusion with the x- and y-components of $\mathbf{r}$.

$$\mathbf{r}(t) = \langle t^2, e^t, 4 - 7t \rangle, \qquad \text{domain } \mathcal{D} = \mathbf{R}$$

$$\mathbf{r}(s) = \langle \sqrt{s}, e^s, s^{-1} \rangle, \qquad \text{domain } \mathcal{D} = \{s \in \mathbf{R} : s > 0\}$$

The terminal point of a vector-valued function $\mathbf{r}(t)$ traces a path in $\mathbf{R}^3$ as t varies. We refer to $\mathbf{r}(t)$ either as a path or as a **vector parametrization** of a path.

We have already studied special cases of vector parametrizations. In Chapter 12, we described lines in $\mathbf{R}^3$ using vector parametrizations. Recall that

$$\mathbf{r}(t) = \langle x_0, y_0, z_0 \rangle + t\mathbf{v} = \langle x_0 + ta, y_0 + tb, z_0 + tc \rangle$$

parametrizes the line through $P = (x_0, y_0, z_0)$ in the direction of the vector $\mathbf{v} = \langle a, b, c \rangle$. In Chapter 11, we studied parametrized curves in the plane $\mathbf{R}^2$ in the form

$$c(t) = (x(t), y(t))$$

Such a curve is described equally well by the vector-valued function $\mathbf{r}(t) = \langle x(t), y(t) \rangle$ The difference lies only in whether we visualize the path as traced by a moving point $c(t)$ or the tip of a moving vector $\mathbf{r}(t)$. The advantage of the vector form is that we can define a vector-valued derivative, a vector that specifies both the magnitude and direction of a rate of change in position of a point on the path.

It is important to distinguish between the path parametrized by $\mathbf{r}(t)$ and the underlying curve $\mathcal{C}$ traced by $\mathbf{r}(t)$. The curve $\mathcal{C}$ is the set of all points $(x(t), y(t), z(t))$ as t ranges over the domain of $\mathbf{r}(t)$. The path is a particular way of traversing the curve; it may traverse the curve several times, reverse direction, move back and forth, etc.

EXAMPLE 1 **The Path Versus the Curve** Describe the path

$$\mathbf{r}(t) = \langle \cos t, \sin t, 1 \rangle, \qquad -\infty < t < \infty$$

How are the path and the curve $\mathcal{C}$ traced by $\mathbf{r}(t)$ different?

Solution As t varies from $-\infty$ to ∞, the endpoint of the vector $\mathbf{r}(t)$ moves around a unit circle at height $z = 1$ infinitely many times in the counterclockwise direction when viewed from above (Figure 2). The underlying curve $\mathcal{C}$ traced by $\mathbf{r}(t)$ is the circle itself. ∎

> **REMINDER** As we indicated previously, with parametrizations, we can think of the curve as a road on which the parametrization travels, and the path as a particular trip on the road.

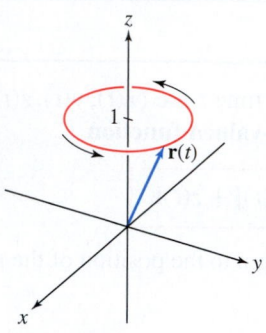

FIGURE 2 Plot of $\mathbf{r}(t) = \langle \cos t, \sin t, 1 \rangle$.

A curve in $\mathbf{R}^3$ is also referred to as a **space curve** (as opposed to a curve in $\mathbf{R}^2$, which is called a **plane curve**). Space curves can be quite complicated and difficult to sketch by hand. An effective way to visualize a space curve is to plot it from different viewpoints using a computer (Figure 3).

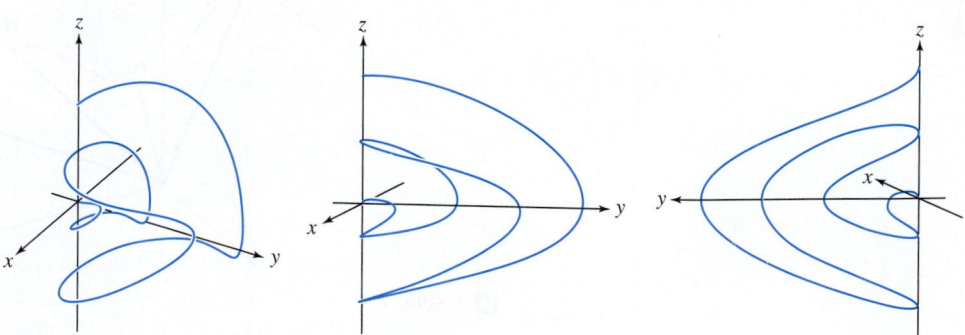

> Some computer graphing utilities allow you to plot a space curve and rotate in different directions so that you can examine it from any viewpoint.

FIGURE 3 The curve $\mathbf{r}(t) = \langle t \sin 2t \cos t, t \sin^2 t, t \cos t \rangle$ for $0 \le t \le 4\pi$, seen from three different viewpoints.

The projections onto the coordinate planes are another aid in visualizing space curves. The projection of a path $\mathbf{r}(t) = \langle x(t), y(t), z(t) \rangle$ onto the xy-plane is the path $\mathbf{p}(t) = \langle x(t), y(t), 0 \rangle$ (Figure 4). Similarly, the projections onto the yz- and xz-planes are the paths $\langle 0, y(t), z(t) \rangle$ and $\langle x(t), 0, z(t) \rangle$, respectively.

EXAMPLE 2 **Helix** Describe the curve traced by $\mathbf{r}(t) = \langle -\sin t, \cos t, t \rangle$ for $t \ge 0$ in terms of its projections onto the coordinate planes.

Solution The projections are as follows (Figure 4):

(A) xy-plane (set $z = 0$): the path traced by $\mathbf{p}(t) = \langle -\sin t, \cos t, 0 \rangle$, which goes counterclockwise around the unit circle starting at $\mathbf{p}(0) = (0, 1, 0)$
(B) xz-plane (set $y = 0$): the path $\langle -\sin t, 0, t \rangle$, which is a sine wave in the z-direction
(C) yz-plane (set $x = 0$): the path $\langle 0, \cos t, t \rangle$, which is a cosine wave in the z-direction

The function $\mathbf{r}(t)$ describes a point moving above the unit circle in the xy-plane, while its height $z = t$ increases linearly, resulting in the helix of Figure 4. ∎

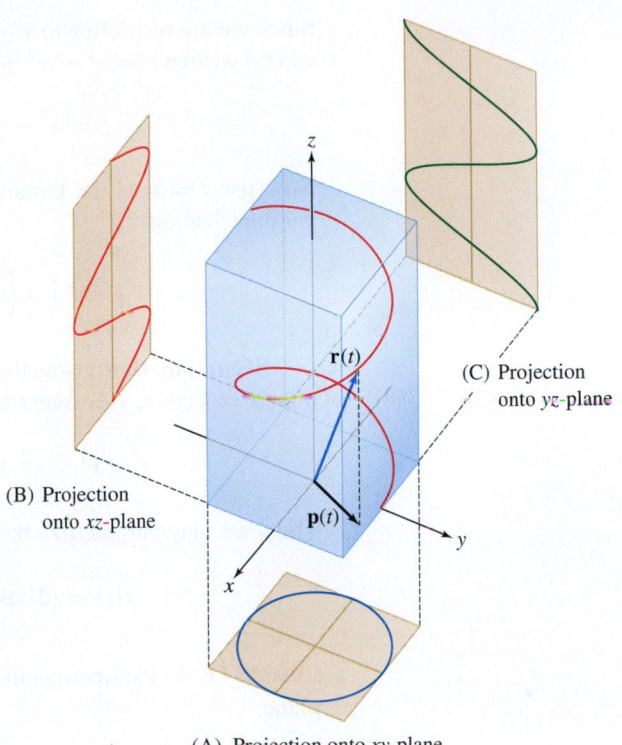

FIGURE 4 Projections of the helix $\mathbf{r}(t) = \langle -\sin t, \cos t, t \rangle$.

(B) Projection onto xz-plane

(C) Projection onto yz-plane

$\mathbf{r}(t)$

$\mathbf{p}(t)$

(A) Projection onto xy-plane

Every curve can be parametrized in infinitely many ways (because there are infinitely many ways that a particle can traverse a curve as a function of time). The next example describes two very different parametrizations of the same curve.

EXAMPLE 3 Parametrizing the Intersection of Surfaces Parametrize the curve $\mathcal{C}$ obtained as the part of the intersection of the surfaces $x^2 - y^2 = z - 1$ and $x^2 + y^2 = 4$ where $y \geq 0$ (Figure 5).

Solution We have to express the coordinates (x, y, z) of a point on the curve as functions of a parameter t. We will demonstrate two different methods for doing this.

First method: Solve the given equations for y and z in terms of x. First, solve for y:

$$x^2 + y^2 = 4 \quad \Rightarrow \quad y^2 = 4 - x^2 \quad \Rightarrow \quad y = \pm\sqrt{4 - x^2}$$

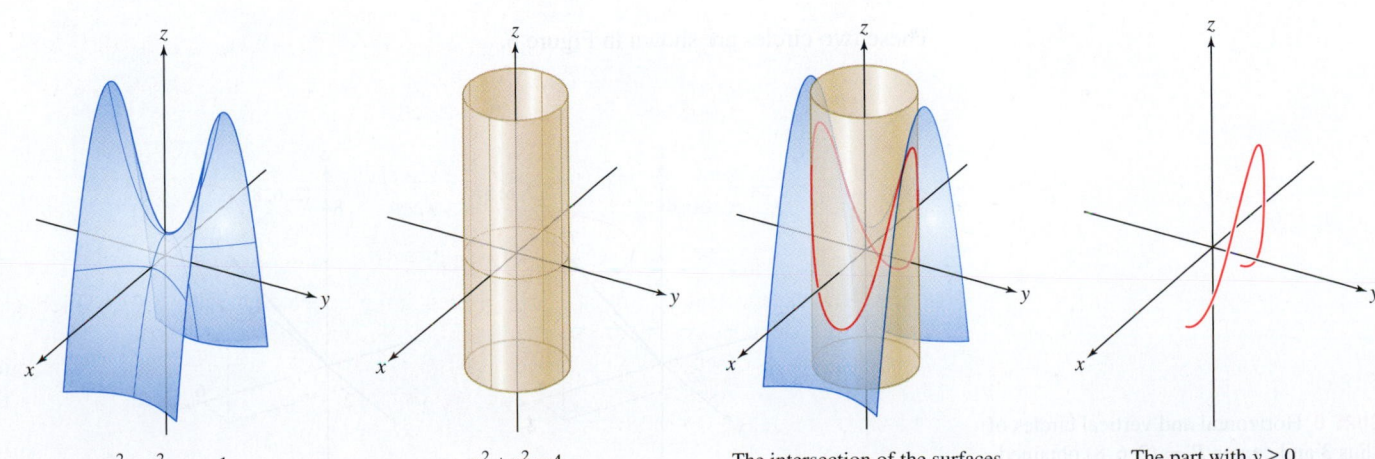

$x^2 - y^2 = z - 1$

$x^2 + y^2 = 4$

The intersection of the surfaces

The part with $y \geq 0$

FIGURE 5

Since we are restricting to $y \geq 0$, we take just $y = \sqrt{4 - x^2}$. The equation $x^2 - y^2 = z - 1$ can be written $z = x^2 - y^2 + 1$. Thus, we can substitute $y^2 = 4 - x^2$ to solve for z:

$$z = x^2 - y^2 + 1 = x^2 - (4 - x^2) + 1 = 2x^2 - 3$$

Now use $t = x$ as the parameter. Then $y = \sqrt{4 - t^2}$, $z = 2t^2 - 3$. Thus, we have the parametrization

$$\mathbf{r}(t) = \left\langle t, \sqrt{4 - t^2}, 2t^2 - 3 \right\rangle, \qquad -2 \leq t \leq 2$$

Second method: Note that $x^2 + y^2 = 4$, with $y \geq 0$, has a trigonometric parametrization: $x = 2\cos t$, $y = 2\sin t$ for $0 \leq t \leq \pi$. The equation $x^2 - y^2 = z - 1$ gives us

$$z = x^2 - y^2 + 1 = 4\cos^2 t - 4\sin^2 t + 1 = 4\cos 2t + 1$$

Thus, we may parametrize the curve by the vector-valued function:

$$\mathbf{r}(t) = \langle 2\cos t, 2\sin t, 4\cos 2t + 1 \rangle, \qquad 0 \leq t \leq \pi \qquad ∎$$

EXAMPLE 4 Parametrize the circle of radius 3 with its center $P = (2, 6, 8)$ located in a plane:

(a) parallel to the xy-plane.   **(b)** parallel to the xz-plane.

Solution **(a)** A circle of radius R in the xy-plane centered at the origin has parametrization $\langle R\cos t, R\sin t \rangle$. To place this circle of radius R in a three-dimensional coordinate system, we use the parametrization $\langle R\cos t, R\sin t, 0 \rangle$.

Thus, the circle of radius 3 in the xy-plane centered at $(0, 0, 0)$ has parametrization $\langle 3\cos t, 3\sin t, 0 \rangle$. To move this circle in a parallel fashion so that its center lies at $P = (2, 6, 8)$, we translate by the vector $\langle 2, 6, 8 \rangle$:

$$\mathbf{r}_1(t) = \langle 2, 6, 8 \rangle + \langle 3\cos t, 3\sin t, 0 \rangle = \langle 2 + 3\cos t, 6 + 3\sin t, 8 \rangle$$

(b) The parametrization $\langle 3\cos t, 0, 3\sin t \rangle$ gives us a circle of radius 3 centered at the origin in the xz-plane. To move the circle in a parallel fashion so that its center lies at $(2, 6, 8)$, we translate by the vector $\langle 2, 6, 8 \rangle$:

$$\mathbf{r}_2(t) = \langle 2, 6, 8 \rangle + \langle 3\cos t, 0, 3\sin t \rangle = \langle 2 + 3\cos t, 6, 8 + 3\sin t \rangle$$

These two circles are shown in Figure 6.   ∎

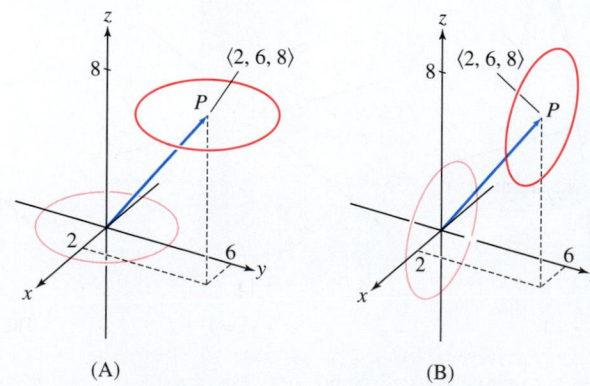

FIGURE 6 Horizontal and vertical circles of radius 3 and center $P = (2, 6, 8)$ obtained by translating two different circles centered at the origin.

(A)   (B)

13.1 SUMMARY

- A *vector-valued function* is a function of the form

$$\mathbf{r}(t) = \langle x(t), y(t), z(t) \rangle = x(t)\mathbf{i} + y(t)\mathbf{j} + z(t)\mathbf{k}$$

- We often think of t as time and $\mathbf{r}(t)$ as a moving vector whose terminal point traces out a path as a function of time. We refer to $\mathbf{r}(t)$ as a *vector parametrization* of the path, or simply as a path.
- The underlying curve $\mathcal{C}$ traced by $\mathbf{r}(t)$ is the set of all points $(x(t), y(t), z(t))$ in $\mathbf{R}^3$ for t in the domain of $\mathbf{r}(t)$. A curve in $\mathbf{R}^3$ is also called a *space curve*.
- Every curve $\mathcal{C}$ can be parametrized in infinitely many ways.
- The projection of $\mathbf{r}(t)$ onto the xy-plane is the curve traced by $\langle x(t), y(t), 0 \rangle$. The projection onto the xz-plane is $\langle x(t), 0, z(t) \rangle$, and the projection onto the yz-plane is $\langle 0, y(t), z(t) \rangle$.

13.1 EXERCISES

Preliminary Questions

1. Which one of the following does *not* parametrize a line?

(a) $\mathbf{r}_1(t) = \langle 8 - t, 2t, 3t \rangle$

(b) $\mathbf{r}_2(t) = t^3\mathbf{i} - 7t^3\mathbf{j} + t^3\mathbf{k}$

(c) $\mathbf{r}_3(t) = \langle 8 - 4t^3, 2 + 5t^2, 9t^3 \rangle$

2. What is the projection of $\mathbf{r}(t) = t\mathbf{i} + t^4\mathbf{j} + e^t\mathbf{k}$ onto the xz-plane?

3. Which projection of $\langle \cos t, \cos 2t, \sin t \rangle$ is a circle?

4. What is the center of the circle with the following parametrization?

$$\mathbf{r}(t) = (-2 + \cos t)\mathbf{i} + 2\mathbf{j} + (3 - \sin t)\mathbf{k}$$

5. How do the paths $\mathbf{r}_1(t) = \langle \cos t, \sin t \rangle$ and $\mathbf{r}_2(t) = \langle \sin t, \cos t \rangle$ around the unit circle differ?

6. Which three of the following vector-valued functions parametrize the same space curve?

(a) $(-2 + \cos t)\mathbf{i} + 9\mathbf{j} + (3 - \sin t)\mathbf{k}$

(b) $(2 + \cos t)\mathbf{i} - 9\mathbf{j} + (-3 - \sin t)\mathbf{k}$

(c) $(-2 + \cos 3t)\mathbf{i} + 9\mathbf{j} + (3 - \sin 3t)\mathbf{k}$

(d) $(-2 - \cos t)\mathbf{i} + 9\mathbf{j} + (3 + \sin t)\mathbf{k}$

(e) $(2 + \cos t)\mathbf{i} + 9\mathbf{j} + (3 + \sin t)\mathbf{k}$

Exercises

1. What is the domain of $\mathbf{r}(t) = e^t\mathbf{i} + \dfrac{1}{t}\mathbf{j} + (t + 1)^{-3}\mathbf{k}$?

2. What is the domain of $\mathbf{r}(s) = e^s\mathbf{i} + \sqrt{s}\mathbf{j} + \cos s\,\mathbf{k}$?

3. Evaluate $\mathbf{r}(2)$ and $\mathbf{r}(-1)$ for $\mathbf{r}(t) = \langle \sin \frac{\pi}{2}t, t^2, (t^2 + 1)^{-1} \rangle$.

4. Does either of $P = (4, 11, 20)$ or $Q = (-1, 6, 16)$ lie on the path $\mathbf{r}(t) = \langle 1 + t, 2 + t^2, t^4 \rangle$?

5. Find a vector parametrization of the line through $P = (3, -5, 7)$ in the direction $\mathbf{v} = \langle 3, 0, 1 \rangle$.

6. Find a direction vector for the line with parametrization $\mathbf{r}(t) = (4 - t)\mathbf{i} + (2 + 5t)\mathbf{j} + \frac{1}{2}t\mathbf{k}$.

7. Determine whether the space curve given by $\mathbf{r}(t) = \langle \sin t, \cos t/2, t \rangle$ intersects the z-axis, and if it does, determine where.

8. Determine whether the curve given by $\mathbf{r}(t) = \langle t^2, t^2 - 2t - 3, t - 3 \rangle$ intersects the x-axis, and if it does, determine where.

9. Determine whether the space curve given by $\mathbf{r}(t) = \langle t, t^3, t^2 + 1 \rangle$ intersects the xy-plane, and if it does, determine where.

10. Show that the path given by $\mathbf{r}(t) = \langle \cos t, \cos(2t), \sin t \rangle$ intersects the xy-plane infinitely many times, but the underlying space curve intersects the xy-plane only twice.

11. Show that the space curve given by $\mathbf{r}(t) = \langle 1 - \cos(2t), t + \sin t, t^2 \rangle$ intersects the yz-plane in infinitely many points but does not cross through it.

12. Show that the path given by $\mathbf{r}(t) = \langle e^{-t} \sin t, e^{-t} \cos t, e^{-t} \rangle$ intersects the sphere $x^2 + y^2 + z^2 = 4$ once, traveling from outside the sphere to inside as t goes from $-\infty$ to ∞.

13. Match the space curves in Figure 7 with their projections onto the xy-plane in Figure 8.

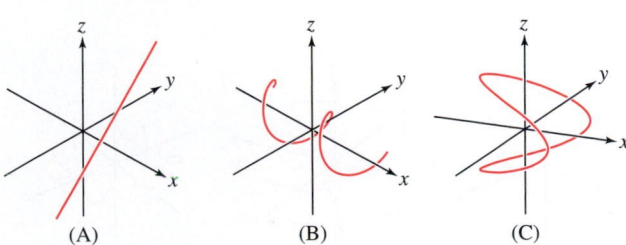

(A)   (B)   (C)

FIGURE 7

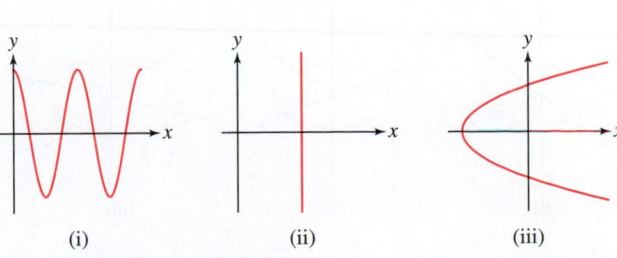

(i)   (ii)   (iii)

FIGURE 8

14. Match the space curves in Figure 7 with the following vector-valued functions:

(a) $\mathbf{r}_1(t) = \langle \cos 2t, \cos t, \sin t \rangle$ **(b)** $\mathbf{r}_2(t) = \langle t, \cos 2t, \sin 2t \rangle$

(c) $\mathbf{r}_3(t) = \langle 1, t, t \rangle$

15. Match the vector-valued functions (a)–(f) with the space curves (i)–(vi) in Figure 9.

(a) $\mathbf{r}(t) = \langle t + 15, e^{0.08t} \cos t, e^{0.08t} \sin t \rangle$

(b) $\mathbf{r}(t) = \langle \cos t, \sin t, \sin 12t \rangle$ **(c)** $\mathbf{r}(t) = \left\langle t, t, \dfrac{25t}{1 + t^2} \right\rangle$

(d) $\mathbf{r}(t) = \langle \cos^3 t, \sin^3 t, \sin 2t \rangle$ **(e)** $\mathbf{r}(t) = \langle t, t^2, 2t \rangle$

(f) $\mathbf{r}(t) = \langle \cos t, \sin t, \cos t \sin 12t \rangle$

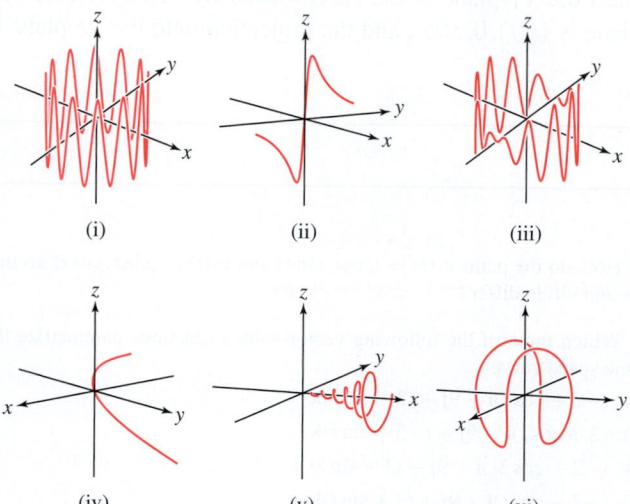

(i)   (ii)   (iii)

(iv)   (v)   (vi)

FIGURE 9

16. Which of the following curves have the same projection onto the xy-plane?

(a) $\mathbf{r}_1(t) = \langle t, t^2, e^t \rangle$ **(b)** $\mathbf{r}_2(t) = \langle e^t, t^2, t \rangle$

(c) $\mathbf{r}_3(t) = \langle t, t^2, \cos t \rangle$

17. Match the space curves (A)–(C) in Figure 10 with their projections (i)–(iii) onto the xy-plane.

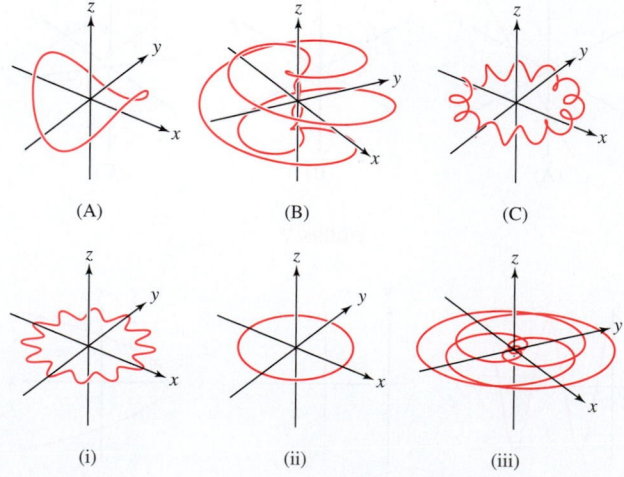

(A)   (B)   (C)

(i)   (ii)   (iii)

FIGURE 10

18. Describe the projections of the circle $\mathbf{r}(t) = \langle \sin t, 0, 4 + \cos t \rangle$ onto the coordinate planes.

In Exercises 19–22, the function $\mathbf{r}(t)$ traces a circle. Determine the radius, center, and plane containing the circle.

19. $\mathbf{r}(t) = (9 \cos t)\mathbf{i} + (9 \sin t)\mathbf{j}$

20. $\mathbf{r}(t) = 7\mathbf{i} + (12 \cos t)\mathbf{j} + (12 \sin t)\mathbf{k}$

21. $\mathbf{r}(t) = \langle \sin t, 0, 4 + \cos t \rangle$

22. $\mathbf{r}(t) = \langle 6 + 3 \sin t, 9, 4 + 3 \cos t \rangle$

23. Consider the curve $\mathcal{C}$ given by

$$\mathbf{r}(t) = \langle \cos(2t) \sin t, \sin(2t), \cos(2t) \cos t \rangle$$

(a) Show that $\mathcal{C}$ lies on the sphere of radius 1 centered at the origin.

(b) Show that $\mathcal{C}$ intersects the x-axis, the y-axis, and the z-axis.

24. Show that the curve $\mathcal{C}$ that is parametrized by

$$\mathbf{r}(t) = \langle t^2 - 1, t - 2t^2, 4 - 6t \rangle$$

lies on a plane as follows:

(a) Show that the points on the curve at $t = 0, 1$, and 2 do not lie on a line, and find an equation of the plane that they determine.

(b) Show that for all t, the points on $\mathcal{C}$ satisfy the equation of the plane in (a).

25. Let $\mathcal{C}$ be the curve given by $\mathbf{r}(t) = \langle t \cos t, t \sin t, t \rangle$.

(a) Show that $\mathcal{C}$ lies on the cone $x^2 + y^2 = z^2$.

(b) Sketch the cone and make a rough sketch of $\mathcal{C}$ on the cone.

26. $\boxed{\text{CAS}}$ Use a computer algebra system to plot the projections onto the xy- and xz-planes of the curve $\mathbf{r}(t) = \langle t \cos t, t \sin t, t \rangle$ in Exercise 25.

In Exercises 27 and 28, let

$$\mathbf{r}(t) = \langle \sin t, \cos t, \sin t \cos 2t \rangle$$

be a parametrization of the curve shown in Figure 11.

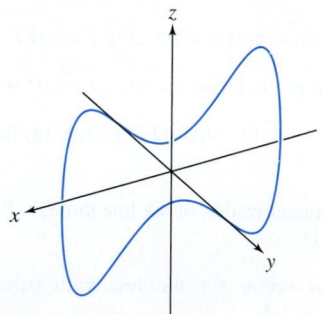

FIGURE 11

27. Find the points where $\mathbf{r}(t)$ intersects the xy-plane.

28. Show that the projection of $\mathbf{r}(t)$ onto the xz-plane is the curve

$$z = x - 2x^3 \quad \text{for} \quad -1 \le x \le 1$$

29. Parametrize the part of the intersection of the surfaces

$$y^2 - z^2 = x - 2, \qquad y^2 + z^2 = 9$$

where $z \ge 0$ using $t = y$ as the parameter.

30. Find a parametrization of the entire intersection of the surfaces in Exercise 29 using trigonometric functions.

31. Viviani's Curve C is the intersection of the surfaces (Figure 12)

$$x^2 + y^2 = z^2, \qquad y = z^2$$

(a) Separately parametrize each of the two parts of C corresponding to $x \geq 0$ and $x \leq 0$, taking $t = z$ as the parameter.

(b) Describe the projection of C onto the xy-plane.

(c) Show that C lies on the sphere of radius 1 with its center $(0, 1, 0)$. This curve looks like a figure eight lying on a sphere [Figure 12(B)].

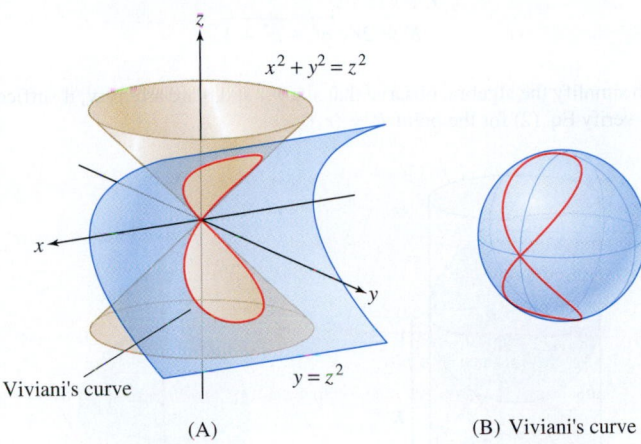

(A)

(B) Viviani's curve viewed from the negative y-axis

FIGURE 12 Viviani's curve is the intersection of the surfaces $x^2 + y^2 = z^2$ and $y = z^2$.

32. (a) Show that any point on $x^2 + y^2 = z^2$ can be written in the form $(z \cos \theta, z \sin \theta, z)$ for some θ.

(b) Use this to find a parametrization of Viviani's curve (Exercise 31) with θ as the parameter.

33. Use sine and cosine to parametrize the intersection of the cylinder $x^2 + y^2 = 1$ and the plane $x + y + z = 1$. Then describe the projections of this curve onto the three coordinate planes.

34. Use hyperbolic functions to parametrize the intersection of the surfaces $x^2 - y^2 = 4$ and $z = xy$.

35. Use sine and cosine to parametrize the intersection of the surfaces $x^2 + y^2 = 1$ and $z = 4x^2$ (Figure 13).

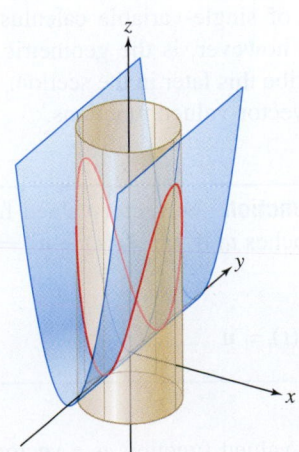

FIGURE 13 Intersection of the surfaces $x^2 + y^2 = 1$ and $z = 4x^2$.

In Exercises 36–38, two paths $\mathbf{r}_1(t)$ and $\mathbf{r}_2(t)$ intersect if there is a point P lying on both curves. We say that $\mathbf{r}_1(t)$ and $\mathbf{r}_2(t)$ collide if $\mathbf{r}_1(t_0) = \mathbf{r}_2(t_0)$ at some time t_0.

36. Which of the following statements are true?

(a) If $\mathbf{r}_1(t)$ and $\mathbf{r}_2(t)$ intersect, then they collide.

(b) If $\mathbf{r}_1(t)$ and $\mathbf{r}_2(t)$ collide, then they intersect.

(c) Intersection depends only on the underlying curves traced by $\mathbf{r}_1$ and $\mathbf{r}_2$, but collision depends on the actual parametrizations.

37. Determine whether $\mathbf{r}_1(t)$ and $\mathbf{r}_2(t)$ collide or intersect, giving the co-ordinates of the corresponding points if they exist:

$$\mathbf{r}_1(t) = \langle t^2 + 3, t + 1, 6t^{-1} \rangle, \qquad \mathbf{r}_2(t) = \langle 4t, 2t - 2, t^2 - 7 \rangle$$

38. Determine whether $\mathbf{r}_1(t)$ and $\mathbf{r}_2(t)$ collide or intersect, giving the co-ordinates of the corresponding points if they exist:

$$\mathbf{r}_1(t) = \langle t, t^2, t^3 \rangle, \qquad \mathbf{r}_2(t) = \langle 4t + 6, 4t^2, 7 - t \rangle$$

In Exercises 39–48, find a parametrization of the curve.

39. The vertical line passing through the point $(3, 2, 0)$

40. The line passing through $(1, 0, 4)$ and $(4, 1, 2)$

41. The line through the origin whose projection on the xy-plane is a line of slope 3 and whose projection on the yz-plane is a line of slope 5 (i.e., $\Delta z / \Delta y = 5$)

42. The circle of radius 1 with center $(2, -1, 4)$ in a plane parallel to the xy-plane

43. The circle of radius 2 with center $(1, 2, 5)$ in a plane parallel to the yz-plane

44. The ellipse $\left(\dfrac{x}{2}\right)^2 + \left(\dfrac{y}{3}\right)^2 = 1$ in the xy-plane, translated to have center $(9, -4, 0)$

45. The intersection of the plane $y = \frac{1}{2}$ with the sphere $x^2 + y^2 + z^2 = 1$

46. The intersection of the surfaces

$$z = x^2 - y^2 \qquad \text{and} \qquad z = x^2 + xy - 1$$

47. The ellipse $\left(\dfrac{x}{2}\right)^2 + \left(\dfrac{z}{3}\right)^2 = 1$ in the xz-plane, translated to have center $(3, 1, 5)$ [Figure 14(A)]

48. The ellipse $\left(\dfrac{y}{2}\right)^2 + \left(\dfrac{z}{3}\right)^2 = 1$, translated to have center $(3, 1, 5)$ [Figure 14(B)]

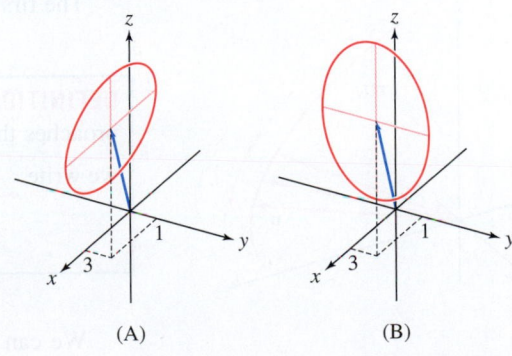

(A)   (B)

FIGURE 14 The ellipses described in Exercises 47 and 48.

Further Insights and Challenges

49. Sketch the curve parametrized by $\mathbf{r}(t) = \langle |t| + t, |t| - t \rangle$.

50. Find the maximum height above the xy-plane of a point on $\mathbf{r}(t) = \langle e^t, \sin t, t(4 - t) \rangle$.

51. ✎ Let $\mathcal{C}$ be the curve obtained by intersecting a cylinder of radius r and a plane. Insert two spheres of radius r into the cylinder above and below the plane, and let F_1 and F_2 be the points where the plane is tangent to the spheres [Figure 15(A)]. Let K be the vertical distance between the equators of the two spheres. Rediscover Archimedes's proof that $\mathcal{C}$ is an ellipse by showing that every point P on $\mathcal{C}$ satisfies

$$PF_1 + PF_2 = K \qquad \boxed{2}$$

Hint: If two lines through a point P are tangent to a sphere and intersect the sphere at Q_1 and Q_2 as in Figure 15(B), then the segments $\overline{PQ_1}$ and $\overline{PQ_2}$ have equal length. Use this to show that $PF_1 = PR_1$ and $PF_2 = PR_2$.

52. Assume that the cylinder in Figure 15 has equation $x^2 + y^2 = r^2$ and the plane has equation $z = ax + by$. Find a vector parametrization $\mathbf{r}(t)$ of the curve of intersection using the trigonometric functions $y = \cos t$ and $y = \sin t$.

53. CAS Now reprove the result of Exercise 51 using vector geometry. Assume that the cylinder has equation $x^2 + y^2 = r^2$ and the plane has equation $z = ax + by$.

(a) Show that the upper and lower spheres in Figure 15 have centers

$$C_1 = \left(0, 0, r\sqrt{a^2 + b^2 + 1}\right)$$

$$C_2 = \left(0, 0, -r\sqrt{a^2 + b^2 + 1}\right)$$

(b) Show that the points where the plane is tangent to the sphere are

$$F_1 = \frac{r}{\sqrt{a^2 + b^2 + 1}}\left(a, b, a^2 + b^2\right)$$

$$F_2 = \frac{-r}{\sqrt{a^2 + b^2 + 1}}\left(a, b, a^2 + b^2\right)$$

Hint: Show that $\overline{C_1 F_1}$ and $\overline{C_2 F_2}$ have length r and are orthogonal to the plane.

(c) Verify, with the aid of a computer algebra system, that Eq. (2) holds with

$$K = 2r\sqrt{a^2 + b^2 + 1}$$

To simplify the algebra, observe that since a and b are arbitrary, it suffices to verify Eq. (2) for the point $P = (r, 0, ar)$.

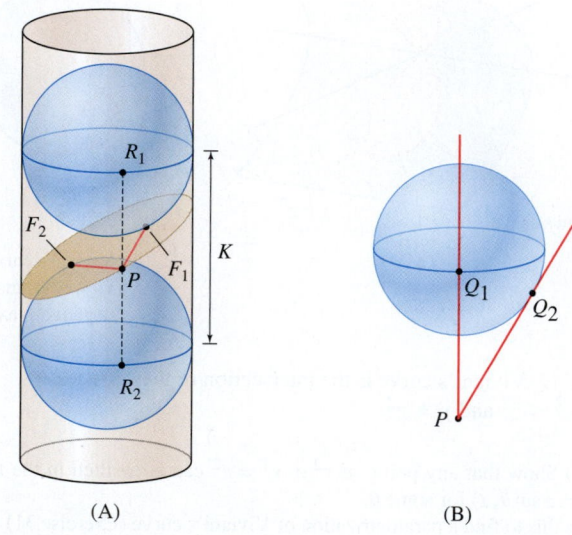

(A)   (B)

FIGURE 15

13.2 Calculus of Vector-Valued Functions

In this section, we extend differentiation and integration to vector-valued functions. This is straightforward because the techniques of single-variable calculus carry over with little change. What is new and important, however, is the geometric interpretation of the derivative as a tangent vector. We describe this later in the section.

The first step is to define the limits of vector-valued functions.

DEFINITION **Limit of a Vector-Valued Function** A vector-valued function $\mathbf{r}(t)$ approaches the limit $\mathbf{u}$ (a vector) as t approaches t_0 if $\lim_{t \to t_0} \|\mathbf{r}(t) - \mathbf{u}\| = 0$. In this case, we write

$$\lim_{t \to t_0} \mathbf{r}(t) = \mathbf{u}$$

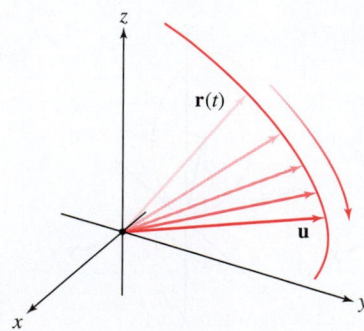

DF **FIGURE 1** The vector-valued function $\mathbf{r}(t)$ approaches the vector $\mathbf{u}$ as $t \to t_0$.

We can visualize the limit of a vector-valued function as a vector $\mathbf{r}(t)$ moving toward the limit vector $\mathbf{u}$ (Figure 1). According to the next theorem, vector limits may be computed componentwise.

> **THEOREM 1** **Vector-Valued Limits Are Computed Componentwise** A vector-valued function $\mathbf{r}(t) = \langle x(t), y(t), z(t) \rangle$ approaches a limit as $t \to t_0$ if and only if each component approaches a limit, and in this case,
>
> $$\lim_{t \to t_0} \mathbf{r}(t) = \left\langle \lim_{t \to t_0} x(t), \lim_{t \to t_0} y(t), \lim_{t \to t_0} z(t) \right\rangle \qquad \boxed{1}$$

Proof Let $\mathbf{u} = \langle a, b, c \rangle$ and consider the square of the length

$$\|\mathbf{r}(t) - \mathbf{u}\|^2 = (x(t) - a)^2 + (y(t) - b)^2 + (z(t) - c)^2 \qquad \boxed{2}$$

The term on the left approaches zero if and only if each term on the right approaches zero (because these terms are nonnegative). It follows that $\|\mathbf{r}(t) - \mathbf{u}\|$ approaches zero if and only if $|x(t) - a|$, $|y(t) - b|$, and $|z(t) - c|$ tend to zero. Therefore, $\mathbf{r}(t)$ approaches a limit $\mathbf{u}$ as $t \to t_0$ if and only if $x(t)$, $y(t)$, and $z(t)$ converge to the components a, b, and c, respectively. ∎

EXAMPLE 1 Calculate $\lim_{t \to 3} \mathbf{r}(t)$, where $\mathbf{r}(t) = \langle t^2, 1 - t, t^{-1} \rangle$.

The Limit Laws of scalar functions remain valid in the vector-valued case. They are verified by applying the Limit Laws to the components.

Solution By Theorem 1,

$$\lim_{t \to 3} \mathbf{r}(t) = \lim_{t \to 3} \langle t^2, 1 - t, t^{-1} \rangle = \left\langle \lim_{t \to 3} t^2, \lim_{t \to 3} (1 - t), \lim_{t \to 3} t^{-1} \right\rangle = \left\langle 9, -2, \frac{1}{3} \right\rangle \qquad ∎$$

Continuity of vector-valued functions is defined in the same way as in the scalar case. A vector-valued function $\mathbf{r}(t) = \langle x(t), y(t), z(t) \rangle$ is **continuous** at t_0 if

$$\lim_{t \to t_0} \mathbf{r}(t) = \mathbf{r}(t_0)$$

By Theorem 1, $\mathbf{r}(t)$ is continuous at t_0 if and only if the components $x(t)$, $y(t)$, $z(t)$ are continuous at t_0.

We define the derivative of $\mathbf{r}(t)$ as the limit of the difference quotient:

$$\mathbf{r}'(t) = \frac{d}{dt}\mathbf{r}(t) = \lim_{h \to 0} \frac{\mathbf{r}(t + h) - \mathbf{r}(t)}{h} \qquad \boxed{3}$$

In Leibniz notation, the derivative is written $d\mathbf{r}/dt$.

We say that $\mathbf{r}(t)$ is **differentiable at** t if the limit in Eq. (3) exists, and we say that $\mathbf{r}$ is **differentiable** if it is differentiable at all t in its domain. Notice that the components of the difference quotient are themselves difference quotients:

$$\lim_{h \to 0} \frac{\mathbf{r}(t + h) - \mathbf{r}(t)}{h} = \lim_{h \to 0} \left\langle \frac{x(t + h) - x(t)}{h}, \frac{y(t + h) - y(t)}{h}, \frac{z(t + h) - z(t)}{h} \right\rangle$$

and by Theorem 1, $\mathbf{r}(t)$ is differentiable if and only if the components are differentiable. In this case, $\mathbf{r}'(t)$ is equal to the vector of derivatives $\langle x'(t), y'(t), z'(t) \rangle$.

> **THEOREM 2** **Vector-Valued Derivatives Are Computed Componentwise** A vector-valued function $\mathbf{r}(t) = \langle x(t), y(t), z(t) \rangle$ is differentiable if and only if each component is differentiable. In this case,
>
> $$\mathbf{r}'(t) = \frac{d}{dt}\mathbf{r}(t) = \langle x'(t), y'(t), z'(t) \rangle$$

By Theorems 1 and 2, vector-valued limits and derivatives are computed componentwise, so they are no more difficult to compute than ordinary limits and derivatives.

Here are some vector-valued derivatives, computed componentwise:

$$\frac{d}{dt}\langle t^2, t^3, \sin t \rangle = \langle 2t, 3t^2, \cos t \rangle, \qquad \frac{d}{dt}\langle \cos t, -1, e^{2t} \rangle = \langle -\sin t, 0, 2e^{2t} \rangle$$

Higher order derivatives are defined by repeated differentiation:

$$\mathbf{r}''(t) = \frac{d}{dt}\,\mathbf{r}'(t), \quad \mathbf{r}'''(t) = \frac{d}{dt}\,\mathbf{r}''(t), \quad \dots$$

EXAMPLE 2 Calculate $\mathbf{r}''(3)$, where $\mathbf{r}(t) = \langle \ln t, t, t^2 \rangle$.

Solution We perform the differentiation componentwise:

$$\mathbf{r}'(t) = \frac{d}{dt}\,\langle \ln t, t, t^2 \rangle = \langle t^{-1}, 1, 2t \rangle$$

$$\mathbf{r}''(t) = \frac{d}{dt}\,\langle t^{-1}, 1, 2t \rangle = \langle -t^{-2}, 0, 2 \rangle$$

Therefore, $\mathbf{r}''(3) = \langle -\frac{1}{9}, 0, 2 \rangle$. ■

The differentiation rules of single-variable calculus carry over to the vector setting.

Differentiation Rules   Assume that $\mathbf{r}(t)$, $\mathbf{r}_1(t)$, and $\mathbf{r}_2(t)$ are differentiable. Then

- **Sum Rule:** $(\mathbf{r}_1(t) + \mathbf{r}_2(t))' = \mathbf{r}_1'(t) + \mathbf{r}_2'(t)$
- **Constant Multiple Rule:** For any constant c, $(c\,\mathbf{r}(t))' = c\,\mathbf{r}'(t)$.
- **Scalar Product Rule:** For any differentiable scalar-valued function f,

$$\frac{d}{dt}\big(f(t)\mathbf{r}(t)\big) = f'(t)\mathbf{r}(t) + f(t)\mathbf{r}'(t)$$

- **Chain Rule:** For any differentiable scalar-valued function g,

$$\frac{d}{dt}\,\mathbf{r}(g(t)) = \mathbf{r}'(g(t))g'(t)$$

Proof Each rule is proved by applying the single-variable differentiation rules to the components. For example, to prove the Scalar Product Rule (we consider vector-valued functions in the plane, to keep the notation simple), we write

$$f(t)\mathbf{r}(t) = f(t)\,\langle x(t), y(t)\rangle = \langle f(t)x(t), f(t)y(t)\rangle$$

Now apply the Product Rule to each component:

$$\frac{d}{dt}\,f(t)\mathbf{r}(t) = \left\langle \frac{d}{dt}\,f(t)x(t), \frac{d}{dt}\,f(t)y(t) \right\rangle$$

$$= \langle f'(t)x(t) + f(t)x'(t), f'(t)y(t) + f(t)y'(t)\rangle$$

$$= \langle f'(t)x(t), f'(t)y(t)\rangle + \langle f(t)x'(t), f(t)y'(t)\rangle$$

$$= f'(t)\,\langle x(t), y(t)\rangle + f(t)\langle x'(t), y'(t)\rangle = f'(t)\mathbf{r}(t) + f(t)\mathbf{r}'(t)$$

The remaining proofs are left as exercises (Exercises 73–74). ■

EXAMPLE 3 Let $\mathbf{r}(t) = \langle t^2, 5t, 1 \rangle$ and $f(t) = e^{3t}$. Calculate:

(a) $\dfrac{d}{dt}\,f(t)\mathbf{r}(t)$

(b) $\dfrac{d}{dt}\,\mathbf{r}(f(t))$

Solution We have $\mathbf{r}'(t) = \langle 2t, 5, 0 \rangle$ and $f'(t) = 3e^{3t}$.

(a) By the Scalar Product Rule,

$$\frac{d}{dt}\,f(t)\mathbf{r}(t) = f'(t)\mathbf{r}(t) + f(t)\mathbf{r}'(t) = 3e^{3t}\langle t^2, 5t, 1\rangle + e^{3t}\langle 2t, 5, 0\rangle$$

$$= \langle (3t^2 + 2t)e^{3t}, (15t + 5)e^{3t}, 3e^{3t}\rangle$$

Note that we could have first found $f(t)\mathbf{r}(t) = e^{3t}\langle t^2, 5t, 1\rangle = \langle e^{3t}t^2, e^{3t}5t, e^{3t}\rangle$, and then differentiated to obtain the same answer. In that case, we would have needed to use the single-variable product rule when differentiating the x- and y-components.

(b) By the Chain Rule,

$$\frac{d}{dt}\mathbf{r}(f(t)) = \mathbf{r}'(f(t))f'(t) = \mathbf{r}'(e^{3t})3e^{3t} = \langle 2e^{3t}, 5, 0\rangle 3e^{3t} = \langle 6e^{6t}, 15e^{3t}, 0\rangle \qquad \blacksquare$$

In addition to the derivative rule for the product of a scalar function f and a vector-valued function $\mathbf{r}$ stated earlier in this section, there are product rules for the dot and cross products. These rules are very important in applications, as we will see.

> **THEOREM 3   Product Rules for Dot and Cross Products**   Assume that $\mathbf{r}_1(t)$ and $\mathbf{r}_2(t)$ are differentiable. Then
>
> Dot Product Rule: $\dfrac{d}{dt}\big(\mathbf{r}_1(t) \cdot \mathbf{r}_2(t)\big) = \mathbf{r}_1'(t) \cdot \mathbf{r}_2(t) + \mathbf{r}_1(t) \cdot \mathbf{r}_2'(t)$ **4**
>
> Cross Product Rule: $\dfrac{d}{dt}\big(\mathbf{r}_1(t) \times \mathbf{r}_2(t)\big) = \big(\mathbf{r}_1'(t) \times \mathbf{r}_2(t)\big) + \big(\mathbf{r}_1(t) \times \mathbf{r}_2'(t)\big)$ **5**

CAUTION *Order is important in the Cross Product Rule. The first term in Eq. (5) must be written as*

$$\mathbf{r}_1'(t) \times \mathbf{r}_2(t)$$

not $\mathbf{r}_2(t) \times \mathbf{r}_1'(t)$. *Remember, cross product is not commutative. Similarly, the second term is* $\mathbf{r}_1(t) \times \mathbf{r}_2'(t)$. *Why is order not a concern for dot products?*

We have seen three product rules involving vector-valued functions: the Scalar Product Rule, the Dot Product Rule, and the Cross Product Rule. Each has the same form as the single-variable product rule: *The derivative of the first term "times" the second, plus the first term "times" the derivative of the second.* The type of product referred to by "times" in each case is different—scalar multiplication in the first case, dot product in the second, cross product in the third.

Proof We prove Eq. (4) for vector-valued functions in the plane. If $\mathbf{r}_1(t) = \langle x_1(t), y_1(t)\rangle$ and $\mathbf{r}_2(t) = \langle x_2(t), y_2(t)\rangle$, then

$$\frac{d}{dt}\big(\mathbf{r}_1(t) \cdot \mathbf{r}_2(t)\big) = \frac{d}{dt}\big(x_1(t)x_2(t) + y_1(t)y_2(t)\big)$$

$$= x_1'(t)x_2(t) + x_1(t)x_2'(t) + y_1'(t)y_2(t) + y_1(t)y_2'(t)$$

$$= \big(x_1'(t)x_2(t) + y_1'(t)y_2(t)\big) + \big(x_1(t)x_2'(t) + y_1(t)y_2'(t)\big)$$

$$= \mathbf{r}_1'(t) \cdot \mathbf{r}_2(t) + \mathbf{r}_1(t) \cdot \mathbf{r}_2'(t)$$

The proof of Eq. (5) is left as an exercise (Exercise 75). $\blacksquare$

In the next example and throughout this chapter, *all vector-valued functions are assumed differentiable, unless otherwise stated.*

EXAMPLE 4 Prove that $\dfrac{d}{dt}\big(\mathbf{r}(t) \times \mathbf{r}'(t)\big) = \mathbf{r}(t) \times \mathbf{r}''(t)$.

Solution By the Cross Product Rule,

$$\frac{d}{dt}\big(\mathbf{r}(t) \times \mathbf{r}'(t)\big) = \underbrace{\mathbf{r}'(t) \times \mathbf{r}'(t)}_{\text{Equals } \mathbf{0}} + \mathbf{r}(t) \times \mathbf{r}''(t) = \mathbf{r}(t) \times \mathbf{r}''(t)$$

Here, $\mathbf{r}'(t) \times \mathbf{r}'(t) = \mathbf{0}$ because the cross product of a vector with itself is zero. $\blacksquare$

The Derivative as a Tangent Vector

The derivative vector $\mathbf{r}'(t_0)$ has an important geometric property: It points in the direction tangent to the path traced by $\mathbf{r}(t)$ at $t = t_0$.

To understand why, consider the difference quotient, where $\Delta\mathbf{r} = \mathbf{r}(t_0 + h) - \mathbf{r}(t_0)$ and $\Delta t = h$ with $h > 0$:

$$\frac{\Delta\mathbf{r}}{\Delta t} = \frac{\mathbf{r}(t_0 + h) - \mathbf{r}(t_0)}{h} \qquad \boxed{6}$$

The vector $\Delta\mathbf{r}$ points from the head of $\mathbf{r}(t_0)$ to the head of $\mathbf{r}(t_0 + h)$ as in Figure 2(A). The difference quotient $\Delta\mathbf{r}/\Delta t$ is a positive scalar multiple of $\Delta\mathbf{r}$ and therefore points in the same direction [Figure 2(B)].

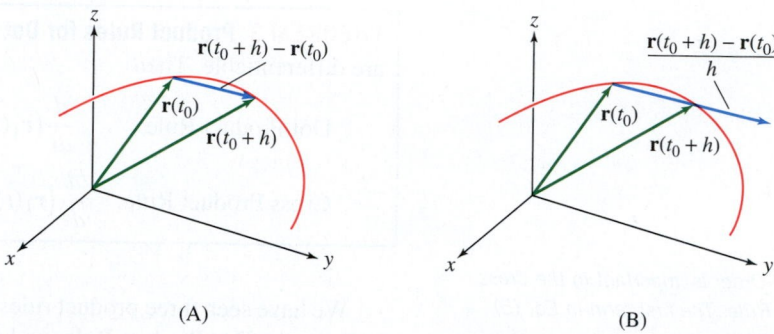

DF **FIGURE 2** The difference quotient points in the direction of $\Delta\mathbf{r} = \mathbf{r}(t_0 + h) - \mathbf{r}(t_0)$.

If we think of $\mathbf{r}(t)$ as indicating the position of a particle moving along a curve, then $\mathbf{r}'(t)$ gives the rate of change of position with respect to time, which is the velocity of the particle. Since the velocity vector is tangent to the curve, it indicates the (instantaneous) direction of motion of the particle. In the next section we will see that it also indicates the particle's speed.

As $h = \Delta t$ tends to zero, $\Delta\mathbf{r}$ also tends to zero but the quotient $\Delta\mathbf{r}/\Delta t$ approaches a vector $\mathbf{r}'(t_0)$ (assuming it exists), which, if nonzero, points in the direction tangent to the curve. Figure 3 illustrates the limiting process. We refer to $\mathbf{r}'(t_0)$ as the **tangent vector** or the **velocity vector** at $\mathbf{r}(t_0)$.

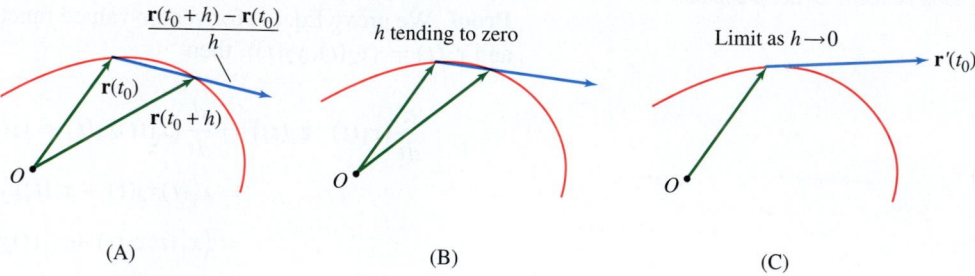

DF **FIGURE 3** The difference quotient converges to a vector $\mathbf{r}'(t_0)$, tangent to the curve.

The tangent vector $\mathbf{r}'(t_0)$ (if it exists and is nonzero) is a direction vector for the tangent line to the curve. The **tangent line** then is the line with vector parametrization:

$$\boxed{\text{Tangent line at } \mathbf{r}(t_0): \qquad \mathbf{L}(t) = \mathbf{r}(t_0) + t\,\mathbf{r}'(t_0)} \qquad \boxed{7}$$

EXAMPLE 5 **Plotting Tangent Vectors** [CAS] Plot $\mathbf{r}(t) = \langle\cos t, \sin t, 4\cos^2 t\rangle$ together with its tangent vectors at $t = \frac{\pi}{4}$ and $\frac{3\pi}{2}$. Find a parametrization of the tangent line at $t = \frac{\pi}{4}$.

Solution The derivative is $\mathbf{r}'(t) = \langle-\sin t, \cos t, -8\cos t \sin t\rangle$, and thus the tangent vectors at $t = \frac{\pi}{4}$ and $\frac{3\pi}{2}$ are

$$\mathbf{r}'\left(\frac{\pi}{4}\right) = \left\langle-\frac{\sqrt{2}}{2}, \frac{\sqrt{2}}{2}, -4\right\rangle, \qquad \mathbf{r}'\left(\frac{3\pi}{2}\right) = \langle 1, 0, 0\rangle$$

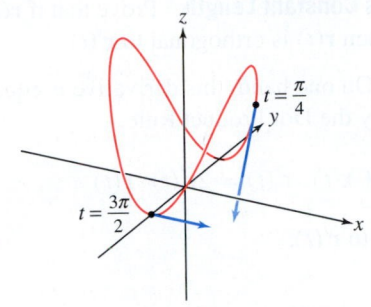

FIGURE 4 Tangent vectors to

$$\mathbf{r}(t) = \langle \cos t, \sin t, 4\cos^2 t \rangle$$

at $t = \frac{\pi}{4}$ and $\frac{3\pi}{2}$.

Figure 4 shows a plot of $\mathbf{r}(t)$ with $\mathbf{r}'\!\left(\frac{\pi}{4}\right)$ based at $\mathbf{r}\!\left(\frac{\pi}{4}\right)$ and $\mathbf{r}'\!\left(\frac{3\pi}{2}\right)$ based at $\mathbf{r}\!\left(\frac{3\pi}{2}\right)$.

At $t = \frac{\pi}{4}$, $\mathbf{r}\!\left(\frac{\pi}{4}\right) = \left\langle \frac{\sqrt{2}}{2}, \frac{\sqrt{2}}{2}, 2 \right\rangle$, and thus the tangent line is parametrized by

$$\mathbf{L}(t) = \mathbf{r}\left(\frac{\pi}{4}\right) + t\,\mathbf{r}'\left(\frac{\pi}{4}\right) = \left\langle \frac{\sqrt{2}}{2}, \frac{\sqrt{2}}{2}, 2 \right\rangle + t \left\langle -\frac{\sqrt{2}}{2}, \frac{\sqrt{2}}{2}, -4 \right\rangle \qquad ∎$$

There are some important differences between vector- and scalar-valued derivatives. The tangent line to a plane curve $y = f(x)$ is horizontal at x_0 exactly when $f'(x_0) = 0$. But in a vector parametrization of a plane curve, the tangent vector $\mathbf{r}'(t_0) = \langle x'(t_0), y'(t_0) \rangle$ is horizontal and nonzero if $y'(t_0) = 0$ but $x'(t_0) \neq 0$.

In the case where a vector-valued function $\mathbf{r}(t)$ describes a particle moving along a curve, if $\mathbf{r}'(t_0) = 0$, the particle has momentarily stopped at t_0. It could subsequently continue to move in the same direction, or move in any other direction from that point, including reversing itself and returning along the path upon which it arrived. We will see such instantaneous-stop behavior in the next example.

EXAMPLE 6 **Tangent Vectors on the Cycloid**   Recall from Example 7 in Section 11.1 that a cycloid is a curve traced out by a point on the rim of a rolling wheel as the center of the wheel moves horizontally. We assume the point begins on the ground and the center of the wheel is moving to the right at a speed of 1. If the radius of the wheel is 1, then the resulting cycloid is traced out by

$$\mathbf{r}(t) = \langle t - \sin t, 1 - \cos t \rangle, \qquad \text{for } t \geq 0$$

Find the points where:

(a) $\mathbf{r}'(t)$ is horizontal and nonzero.   **(b)** $\mathbf{r}'(t)$ is the zero vector.

Solution   The tangent vector is $\mathbf{r}'(t) = \langle 1 - \cos t, \sin t \rangle$. The y-component of $\mathbf{r}'(t)$ is zero if $\sin t = 0$—that is, if $t = 0, \pi, 2\pi, \dots$. Therefore (see Figure 5),

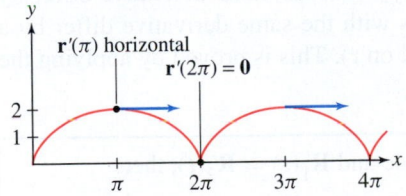

FIGURE 5 Points on the cycloid where $\mathbf{r}'$ is **0** or horizontal.

- At $\mathbf{r}(0) = \langle 0, 0 \rangle$, we have $\mathbf{r}'(0) = \langle 1 - \cos 0, \sin 0 \rangle = \langle 0, 0 \rangle$, so $\mathbf{r}'$ is the zero vector.
- At $\mathbf{r}(\pi) = \langle \pi, 2 \rangle$, we have $\mathbf{r}'(\pi) = \langle 1 - \cos \pi, \sin \pi \rangle = \langle 2, 0 \rangle$, so $\mathbf{r}'$ is horizontal and nonzero.

By periodicity, we conclude that $\mathbf{r}'(t)$ is nonzero and horizontal for $t = \pi, 3\pi, 5\pi, \dots$ [and therefore at $(\pi, 2), (3\pi, 2), (5\pi, 2), \dots$] and $\mathbf{r}'(t) = 0$ for $t = 0, 2\pi, 4\pi, \dots$ [and therefore at $(0, 0), (2\pi, 0), (4\pi, 0), \dots$]. Note that, while the center of the wheel moves with a constant speed of 1, the point on the rim that traces out the cycloid has speed 2 at $t = \pi, 3\pi, 5\pi, \dots$ and has speed 0 at $t = 0i, 2\pi, 4\pi, \dots$ ∎

> **CONCEPTUAL INSIGHT**   The cycloid in Figure 5 has sharp points called **cusps** at points where $x = 0, 2\pi, 4\pi, \dots$. If we represent the cycloid as the graph of a function $y = f(x)$, then $f'(x)$ does not exist at these points. By contrast, the vector derivative $\mathbf{r}'(t) = \langle 1 - \cos t, \sin t \rangle$ exists *for all t*, but $\mathbf{r}'(t) = 0$ at the cusps. In general, $\mathbf{r}'(t)$ is a direction vector for the tangent line whenever $\mathbf{r}'(t)$ exists and is nonzero. If $\mathbf{r}'(t) = 0$, then either the curve does not have a tangent line or the curve has a tangent line and $\mathbf{r}'(t)$ (being the zero vector) is not a direction vector for it.

The next example establishes an important property of vector-valued functions that will be used in later sections in this chapter.

EXAMPLE 7 **Orthogonality of r and r′ when r Has Constant Length** Prove that if $\mathbf{r}(t)$ and $\mathbf{r}'(t)$ are nonzero and $\mathbf{r}(t)$ has constant length, then $\mathbf{r}(t)$ is orthogonal to $\mathbf{r}'(t)$.

Solution We prove this by considering $\frac{d}{dt}\|\mathbf{r}(t)\|^2$. On one hand, this derivative is equal to 0 because $\|\mathbf{r}(t)\|$ is constant. On the other hand, by the Dot Product Rule,

$$\frac{d}{dt}\|\mathbf{r}(t)\|^2 = \frac{d}{dt}\big(\mathbf{r}(t)\cdot\mathbf{r}(t)\big) = \mathbf{r}'(t)\cdot\mathbf{r}(t) + \mathbf{r}(t)\cdot\mathbf{r}'(t) = 2\mathbf{r}'(t)\cdot\mathbf{r}(t)$$

It follows that $\mathbf{r}'(t)\cdot\mathbf{r}(t) = 0$, and $\mathbf{r}(t)$ is orthogonal to $\mathbf{r}'(t)$. ∎

GRAPHICAL INSIGHT The result of Example 7 has a geometric explanation. A vector parametrization $\mathbf{r}(t)$ consisting of vectors of constant length R traces a curve on the surface of a sphere of radius R with its center at the origin (Figure 6). Thus, $\mathbf{r}'(t)$ is tangent to this sphere. But any line that is tangent to a sphere at a point P is orthogonal to the radial vector through P, and thus $\mathbf{r}(t)$ is orthogonal to $\mathbf{r}'(t)$.

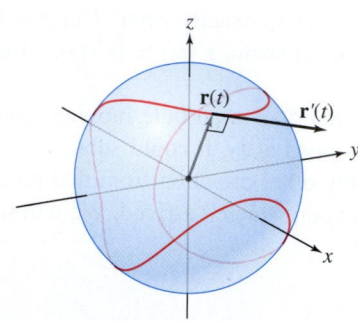

← **REMINDER** $\|\mathbf{r}(t)\|$ represents the length of the vector $\mathbf{r}(t)$.

FIGURE 6 $\mathbf{r}'(t)$ is orthogonal to $\mathbf{r}(t)$ if $\mathbf{r}(t)$ has fixed length.

Vector-Valued Integration

The integral of a vector-valued function can be defined in terms of Riemann sums as in Chapter 5. We will define it more simply via componentwise integration (the two definitions are equivalent). In other words,

$$\int_a^b \mathbf{r}(t)\,dt = \left\langle \int_a^b x(t)\,dt, \int_a^b y(t)\,dt, \int_a^b z(t)\,dt \right\rangle$$

The integral exists if each of the components $x(t)$, $y(t)$, $z(t)$ is integrable. For example,

$$\int_0^\pi \langle 1, t, \sin t\rangle\,dt = \left\langle \int_0^\pi 1\,dt, \int_0^\pi t\,dt, \int_0^\pi \sin t\,dt \right\rangle = \left\langle \pi, \frac{1}{2}\pi^2, 2 \right\rangle$$

Vector-valued integrals obey the same linearity rules as scalar-valued integrals (see Exercise 76).

An **antiderivative** of $\mathbf{r}(t)$ is a vector-valued function $\mathbf{R}(t)$ such that $\mathbf{R}'(t) = \mathbf{r}(t)$. In the single-variable case, two functions f_1 and f_2 with the same derivative differ by a constant. Similarly, two vector-valued functions with the same derivative differ by a *constant vector* (i.e., a vector that does not depend on t). This is proved by applying the scalar result to each component of $\mathbf{r}(t)$.

THEOREM 4 If $\mathbf{R}_1(t)$ and $\mathbf{R}_2(t)$ are differentiable and $\mathbf{R}_1'(t) = \mathbf{R}_2'(t)$, then

$$\mathbf{R}_1(t) = \mathbf{R}_2(t) + \mathbf{c}$$

for some constant vector $\mathbf{c}$.

The general antiderivative of $\mathbf{r}(t)$ is written

$$\int \mathbf{r}(t)\,dt = \mathbf{R}(t) + \mathbf{c}$$

where $\mathbf{c} = \langle c_1, c_2, c_3\rangle$ is an arbitrary constant vector. For example,

$$\int \langle 1, t, \sin t\rangle\,dt = \left\langle t, \frac{1}{2}t^2, -\cos t \right\rangle + \mathbf{c} = \left\langle t + c_1, \frac{1}{2}t^2 + c_2, -\cos t + c_3 \right\rangle$$

EXAMPLE 8 **Finding Position via Vector-Valued Differential Equations** The path of a particle satisfies

$$\frac{d\mathbf{r}}{dt} = \left\langle 1 - 6\sin 3t, \frac{1}{5}t \right\rangle$$

Find the particle's location at $t = 4$ if $\mathbf{r}(0) = \langle 4, 1\rangle$.

Solution The general solution is obtained by integration:

$$\mathbf{r}(t) = \int \left\langle 1 - 6\sin 3t, \frac{1}{5}t \right\rangle dt = \left\langle t + 2\cos 3t, \frac{1}{10}t^2 \right\rangle + \mathbf{c}$$

From the general solution and from the initial condition, we have two expressions for $\mathbf{r}(0)$:

$$\mathbf{r}(0) = \langle 2, 0 \rangle + \mathbf{c} \quad \text{and} \quad \mathbf{r}(0) = \langle 4, 1 \rangle$$

Therefore, we have $\mathbf{c} = \langle 2, 1 \rangle$, and it follows that

$$\mathbf{r}(t) = \left\langle t + 2\cos 3t, \frac{1}{10}t^2 \right\rangle + \langle 2, 1 \rangle = \left\langle t + 2\cos 3t + 2, \frac{1}{10}t^2 + 1 \right\rangle$$

The path is illustrated in Figure 7. The particle's position at $t = 4$ is

$$\mathbf{r}(4) = \left\langle 4 + 2\cos 12 + 2, \frac{1}{10}(4^2) + 1 \right\rangle \approx \langle 7.69, 2.6 \rangle \qquad \blacksquare$$

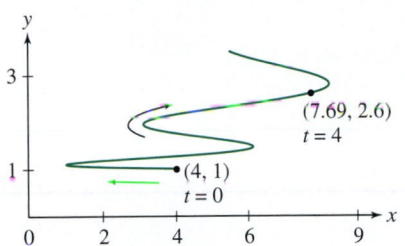

FIGURE 7 Particle path

$$\mathbf{r}(t) = \left\langle t + 2\cos 3t + 2, \frac{1}{10}t^2 + 1 \right\rangle$$

The Fundamental Theorem of Calculus from single-variable calculus naturally carries over to vector-valued functions:

Fundamental Theorem of Calculus for Vector-Valued Functions

Part I: If $\mathbf{r}(t)$ is continuous on $[a, b]$, and $\mathbf{R}(t)$ is an antiderivative of $\mathbf{r}(t)$, then

$$\int_a^b \mathbf{r}(t)\, dt = \mathbf{R}(b) - \mathbf{R}(a)$$

Part II: Assume that $\mathbf{r}(t)$ is continuous on an open interval I and let a be in I. Then

$$\frac{d}{dt} \int_a^t \mathbf{r}(s)\, ds = \mathbf{r}(t)$$

13.2 SUMMARY

- Limits, differentiation, and integration of vector-valued functions are performed componentwise.
- Differentiation rules:

  - Sum Rule: $(\mathbf{r}_1(t) + \mathbf{r}_2(t))' = \mathbf{r}_1'(t) + \mathbf{r}_2'(t)$
  - Constant Multiple Rule: $(c\,\mathbf{r}(t))' = c\,\mathbf{r}'(t)$
  - Chain Rule: $\dfrac{d}{dt}\mathbf{r}(g(t)) = g'(t)\mathbf{r}'(g(t))$

- Product Rules:

  - Scalar times vector: $\dfrac{d}{dt}\big(f(t)\mathbf{r}(t)\big) = f'(t)\mathbf{r}(t) + f(t)\mathbf{r}'(t)$

  - Dot product: $\dfrac{d}{dt}\big(\mathbf{r}_1(t) \cdot \mathbf{r}_2(t)\big) = \mathbf{r}_1'(t) \cdot \mathbf{r}_2(t) + \mathbf{r}_1(t) \cdot \mathbf{r}_2'(t)$

  - Cross product: $\dfrac{d}{dt}\big(\mathbf{r}_1(t) \times \mathbf{r}_2(t)\big) = \big(\mathbf{r}_1'(t) \times \mathbf{r}_2(t)\big) + \big(\mathbf{r}_1(t) \times \mathbf{r}_2'(t)\big)$

- The derivative $\mathbf{r}'(t_0)$ is called the *tangent vector* or *velocity vector*.
- If $\mathbf{r}'(t_0)$ is nonzero, then it points in the direction tangent to the curve at $\mathbf{r}(t_0)$. The tangent line at $\mathbf{r}(t_0)$ has vector parametrization $\mathbf{L}(t) = \mathbf{r}(t_0) + t\mathbf{r}'(t_0)$

- If $\mathbf{R}_1'(t) = \mathbf{R}_2'(t)$, then $\mathbf{R}_1(t) = \mathbf{R}_2(t) + \mathbf{c}$ for some constant vector $\mathbf{c}$.
- The Fundamental Theorem of Calculus for vector-valued functions: If $\mathbf{r}(t)$ is continuous, then

  - if $\mathbf{R}'(t) = \mathbf{r}(t)$, then $\displaystyle\int_a^b \mathbf{r}(t)\,dt = \mathbf{R}(b) - \mathbf{R}(a)$,

  - $\displaystyle\frac{d}{dt}\int_a^t \mathbf{r}(s)\,ds = \mathbf{r}(t)$

13.2 EXERCISES

Preliminary Questions

1. State the three forms of the Product Rule for vector-valued functions.

In Questions 2–6, indicate whether the statement is true or false, and if it is false, provide a correct statement.

2. The derivative of a vector-valued function is defined as the limit of a difference quotient, just as in the scalar-valued case.

3. The integral of a vector-valued function is obtained by integrating each component.

4. The terms "velocity vector" and "tangent vector" for a path $\mathbf{r}(t)$ mean the same thing.

5. The derivative of a vector-valued function is the slope of the tangent line, just as in the scalar case.

6. The derivative of the cross product is the cross product of the derivatives.

7. State whether the following derivatives of vector-valued functions $\mathbf{r}_1(t)$ and $\mathbf{r}_2(t)$ are scalars or vectors:

(a) $\dfrac{d}{dt}\mathbf{r}_1(t)$ (b) $\dfrac{d}{dt}\big(\mathbf{r}_1(t)\cdot\mathbf{r}_2(t)\big)$ (c) $\dfrac{d}{dt}\big(\mathbf{r}_1(t)\times\mathbf{r}_2(t)\big)$

Exercises

In Exercises 1–6, evaluate the limit.

1. $\displaystyle\lim_{t\to 3}\left\langle t^2, 4t, \frac{1}{t}\right\rangle$

2. $\displaystyle\lim_{t\to\pi}\sin 2t\,\mathbf{i} + \cos t\,\mathbf{j} + \tan 4t\,\mathbf{k}$

3. $\displaystyle\lim_{t\to 0} e^{2t}\mathbf{i} + \ln(t+1)\mathbf{j} + 4\mathbf{k}$

4. $\displaystyle\lim_{t\to 0}\left\langle \frac{1}{t+1}, \frac{e^t-1}{t}, 4t\right\rangle$

5. Evaluate $\displaystyle\lim_{h\to 0}\frac{\mathbf{r}(t+h)-\mathbf{r}(t)}{h}$ for $\mathbf{r}(t) = \langle t^{-1}, \sin t, 4\rangle$.

6. Evaluate $\displaystyle\lim_{t\to 0}\frac{\mathbf{r}(t)}{t}$ for $\mathbf{r}(t) = \langle \sin t, 1-\cos t, -2t\rangle$.

In Exercises 7–12, compute the derivative.

7. $\mathbf{r}(t) = \langle t, t^2, t^3\rangle$

8. $\mathbf{r}(t) = \langle 2t^2, \sqrt{2t}, 2-t^{-2}\rangle$

9. $\mathbf{r}(s) = \langle e^{1-s}, 1-s, \ln(1-s)\rangle$

10. $\mathbf{b}(t) = \langle e^{3t-4}, e^{6-t}, (t+1)^{-1}\rangle$ 11. $\mathbf{c}(t) = t^{-1}\mathbf{i} - e^{2t}\mathbf{k}$

12. $\mathbf{a}(\theta) = (\cos 3\theta)\mathbf{i} + (\sin^2\theta)\mathbf{j} + (\tan\theta)\mathbf{k}$

13. Calculate $\mathbf{r}'(t)$ and $\mathbf{r}''(t)$ for $\mathbf{r}(t) = \langle t, t^2, t^3\rangle$.

14. Sketch the curve parametrized by $\mathbf{r}(t) = \langle 1-t^2, t\rangle$ for $-1 \le t \le 1$. Compute the tangent vector at $t = 1$ and add it to the sketch.

15. Sketch the curve parametrized by $\mathbf{r}_1(t) = \langle t, t^2\rangle$ together with its tangent vector at $t = 1$. Then do the same for $\mathbf{r}_2(t) = \langle t^3, t^6\rangle$.

16. Sketch the cycloid $\mathbf{r}(t) = \langle t - \sin t, 1-\cos t\rangle$ together with its tangent vectors at $t = \frac{\pi}{3}$ and $\frac{3\pi}{4}$.

17. Determine the value of t between 0 and 2π such that the tangent vector to the cycloid $\mathbf{r}(t) = \langle t - \sin t, 1-\cos t\rangle$ is parallel to $\langle\sqrt{3}, 1\rangle$.

18. Determine the values of t between 0 and 2π such that the tangent vector to the cycloid $\mathbf{r}(t) = \langle t - \sin t, 1-\cos t\rangle$ is a unit vector.

In Exercises 19–22, evaluate the derivative by using the appropriate Product Rule, where

$$\mathbf{r}_1(t) = \langle t^2, t^3, t\rangle, \qquad \mathbf{r}_2(t) = \langle e^{3t}, e^{2t}, e^t\rangle$$

19. $\dfrac{d}{dt}\big(\mathbf{r}_1(t)\cdot\mathbf{r}_2(t)\big)$ 20. $\dfrac{d}{dt}\big(t^4\mathbf{r}_1(t)\big)$

21. $\dfrac{d}{dt}\big(\mathbf{r}_1(t)\times\mathbf{r}_2(t)\big)$

22. $\dfrac{d}{dt}\big(\mathbf{r}(t)\cdot\mathbf{r}_1(t)\big)\Big|_{t=2}$, assuming that

$$\mathbf{r}(2) = \langle 2,1,0\rangle, \qquad \mathbf{r}'(2) = \langle 1,4,3\rangle$$

In Exercises 23 and 24, let

$$\mathbf{r}_1(t) = \langle t^2, 1, 2t\rangle, \qquad \mathbf{r}_2(t) = \langle 1, 2, e^t\rangle$$

23. Compute $\dfrac{d}{dt}\mathbf{r}_1(t)\cdot\mathbf{r}_2(t)\Big|_{t=1}$ in two ways:

(a) Calculate $\mathbf{r}_1(t)\cdot\mathbf{r}_2(t)$ and differentiate.
(b) Use the Dot Product Rule.

24. Compute $\dfrac{d}{dt}\mathbf{r}_1(t)\times\mathbf{r}_2(t)\Big|_{t=1}$ in two ways:

(a) Calculate $\mathbf{r}_1(t)\times\mathbf{r}_2(t)$ and differentiate.
(b) Use the Cross Product Rule.

In Exercises 25–28, evaluate $\dfrac{d}{dt}\mathbf{r}(g(t))$ using the Chain Rule.

25. $\mathbf{r}(t) = \langle t^2, 1-t\rangle, \quad g(t) = e^t$

26. $\mathbf{r}(t) = \langle t^2, t^3\rangle, \quad g(t) = \sin t$

27. $\mathbf{r}(t) = \langle e^t, e^{2t}, 4 \rangle$, $g(t) = 4t + 9$

28. $\mathbf{r}(t) = \langle 4\sin 2t, 6\cos 2t \rangle$, $g(t) = t^2$

29. Let $\mathbf{r}(t) = \langle t^2, 1 - t, 4t \rangle$. Calculate the derivative of $\mathbf{r}(t) \cdot \mathbf{a}(t)$ at $t = 2$, assuming that $\mathbf{a}(2) = \langle 1, 3, 3 \rangle$ and $\mathbf{a}'(2) = \langle -1, 4, 1 \rangle$.

30. Let $\mathbf{v}(s) = s^2\mathbf{i} + 2s\mathbf{j} + 9s^{-2}\mathbf{k}$. Evaluate $\dfrac{d}{ds}\mathbf{v}(g(s))$ at $s = 4$, assuming that $g(4) = 3$ and $g'(4) = -9$.

In Exercises 31–36, find a parametrization of the tangent line at the point indicated.

31. $\mathbf{r}(t) = \langle t^2, t^4 \rangle$, $t = -2$

32. $\mathbf{r}(t) = \langle \cos t, \sin 2t \rangle$, $t = \frac{\pi}{3}$

33. $\mathbf{r}(t) = \langle 1 - t^2, 5t, 2t^3 \rangle$, $t = 2$

34. $\mathbf{r}(t) = \langle 6t, 4t^2, 2t^3 \rangle$, $t = -2$

35. $\mathbf{r}(s) = 4s^{-1}\mathbf{i} - \frac{8}{3}s^{-3}\mathbf{k}$, $s = 2$

36. $\mathbf{r}(s) = (\ln s)\mathbf{i} + s^{-1}\mathbf{j} + 9s\mathbf{k}$, $s = 1$

37. Use Example 4 to calculate $\dfrac{d}{dt}(\mathbf{r} \times \mathbf{r}')$, where $\mathbf{r}(t) = \langle t, t^2, e^t \rangle$.

38. Let $\mathbf{r}(t) = \langle 3\cos t, 5\sin t, 4\cos t \rangle$. Show that $\|\mathbf{r}(t)\|$ is constant and conclude, using Example 7, that $\mathbf{r}(t)$ and $\mathbf{r}'(t)$ are orthogonal. Then compute $\mathbf{r}'(t)$ and verify directly that $\mathbf{r}'(t)$ is orthogonal to $\mathbf{r}(t)$.

39. Show that the *derivative of the norm* is not equal to the *norm of the derivative* by verifying that $\|\mathbf{r}(t)\|' \neq \|\mathbf{r}(t)'\|$ for $\mathbf{r}(t) = \langle t, 1, 1 \rangle$.

40. Show that $\dfrac{d}{dt}(\mathbf{a} \times \mathbf{r}) = \mathbf{a} \times \mathbf{r}'$ for any constant vector $\mathbf{a}$.

In Exercises 41–48, evaluate the integrals.

41. $\displaystyle\int_{-2}^{2} \langle t^2 + 4t, 4t^3 - t \rangle \, dt$ **42.** $\displaystyle\int_{0}^{1} \left\langle \frac{1}{1 + s^2}, \frac{s}{1 + s^2} \right\rangle ds$

43. $\displaystyle\int_{-2}^{2} (u^3\mathbf{i} + u^5\mathbf{j}) \, du$ **44.** $\displaystyle\int_{0}^{1} \left(te^{-t^2}\mathbf{i} + t\ln(t^2 + 1)\mathbf{j} \right) dt$

45. $\displaystyle\int_{0}^{\pi} \langle -\sin t, 6t, 2t + \cos 2t \rangle \, dt$ **46.** $\displaystyle\int_{1/2}^{1} \left\langle \frac{1}{u^2}, \frac{1}{u^4}, \frac{1}{u^5} \right\rangle du$

47. $\displaystyle\int_{1}^{4} \left(t^{-1}\mathbf{i} + 4\sqrt{t}\,\mathbf{j} - 8t^{3/2}\mathbf{k} \right) dt$ **48.** $\displaystyle\int_{0}^{t} (3s\mathbf{i} + 6s^2\mathbf{j} + 9\mathbf{k}) \, ds$

In Exercises 49–56, find both the general solution of the differential equation and the solution with the given initial condition.

49. $\dfrac{d\mathbf{r}}{dt} = \langle 1 - 2t, 4t \rangle$, $\mathbf{r}(0) = \langle 3, 1 \rangle$

50. $\mathbf{r}'(t) = \mathbf{i} - \mathbf{j}$, $\mathbf{r}(0) = 2\mathbf{i} + 3\mathbf{k}$

51. $\mathbf{r}'(t) = t^2\mathbf{i} + 5t\mathbf{j} + \mathbf{k}$, $\mathbf{r}(1) = \mathbf{j} + 2\mathbf{k}$

52. $\mathbf{r}'(t) = \langle \sin 3t, \sin 3t, t \rangle$, $\mathbf{r}\!\left(\frac{\pi}{2}\right) = \left\langle 2, 4, \frac{\pi^2}{4} \right\rangle$

53. $\mathbf{r}''(t) = 16\mathbf{k}$, $\mathbf{r}(0) = \langle 1, 0, 0 \rangle$, $\mathbf{r}'(0) = \langle 0, 1, 0 \rangle$

54. $\mathbf{r}''(t) = \langle e^{2t-2}, t^2 - 1, 1 \rangle$, $\mathbf{r}(1) = \langle 0, 0, 1 \rangle$, $\mathbf{r}'(1) = \langle 2, 0, 0 \rangle$

55. $\mathbf{r}''(t) = \langle 0, 2, 0 \rangle$, $\mathbf{r}(3) = \langle 1, 1, 0 \rangle$, $\mathbf{r}'(3) = \langle 0, 0, 1 \rangle$

56. $\mathbf{r}''(t) = \langle e^t, \sin t, \cos t \rangle$, $\mathbf{r}(0) = \langle 1, 0, 1 \rangle$, $\mathbf{r}'(0) = \langle 0, 2, 2 \rangle$

57. Find the location at $t = 3$ of a particle whose path (Figure 8) satisfies

$$\frac{d\mathbf{r}}{dt} = \left\langle 2t - \frac{1}{(t + 1)^2}, 2t - 4 \right\rangle, \qquad \mathbf{r}(0) = \langle 3, 8 \rangle$$

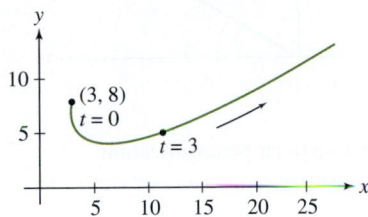

FIGURE 8 Particle path.

58. Find the location and velocity at $t = 4$ of a particle whose path satisfies

$$\frac{d\mathbf{r}}{dt} = \left\langle 2t^{-1/2}, 6, 8t \right\rangle, \qquad \mathbf{r}(1) = \langle 4, 9, 2 \rangle$$

59. A fighter plane, which can shoot a laser beam straight ahead, travels along the path $\mathbf{r}(t) = \langle 5 - t, 21 - t^2, 3 - t^3/27 \rangle$. Show that there is precisely one time t at which the pilot can hit a target located at the origin.

60. The plane of Exercise 59 travels along $\mathbf{r}(t) = \langle t - t^3, 12 - t^2, 3 - t \rangle$. Show that the pilot cannot hit any target on the x-axis.

61. Find all solutions to $\mathbf{r}'(t) = \mathbf{v}$ with initial condition $\mathbf{r}(1) = \mathbf{w}$, where $\mathbf{v}$ and $\mathbf{w}$ are constant vectors in $\mathbf{R}^3$.

62. Let $\mathbf{u}$ be a constant vector in $\mathbf{R}^3$. Find the solution of the equation $\mathbf{r}'(t) = (\sin t)\mathbf{u}$ satisfying $\mathbf{r}'(0) = \mathbf{0}$.

63. Find all solutions to $\mathbf{r}'(t) = 2\mathbf{r}(t)$, where $\mathbf{r}(t)$ is a vector-valued function in 3-space.

64. Show that $\mathbf{w}(t) = \langle \sin(3t + 4), \sin(3t - 2), \cos 3t \rangle$ satisfies the differential equation $\mathbf{w}''(t) = -9\mathbf{w}(t)$.

65. Prove that the **Bernoulli spiral** (Figure 9) with parametrization $\mathbf{r}(t) = \langle e^t \cos 4t, e^t \sin 4t \rangle$ has the property that the angle ψ between the position vector and the tangent vector is constant. Find the angle ψ in degrees.

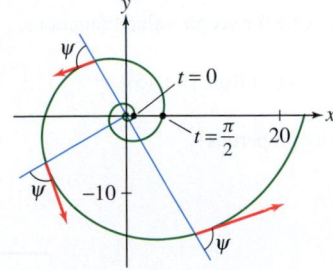

FIGURE 9 Bernoulli spiral.

66. A curve in polar form $r = f(\theta)$ has parametrization

$$\mathbf{r}(\theta) = f(\theta) \langle \cos \theta, \sin \theta \rangle$$

Let ψ be the angle between the radial and tangent vectors (Figure 10). Prove that

$$\tan \psi = \frac{r}{dr/d\theta} = \frac{f(\theta)}{f'(\theta)}$$

Hint: Compute $\mathbf{r}(\theta) \times \mathbf{r}'(\theta)$ and $\mathbf{r}(\theta) \cdot \mathbf{r}'(\theta)$.

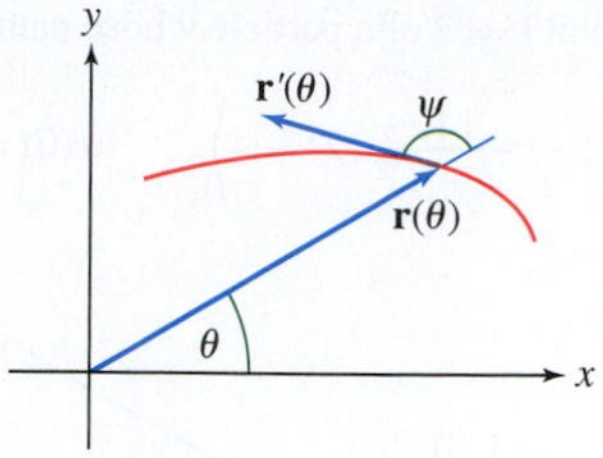

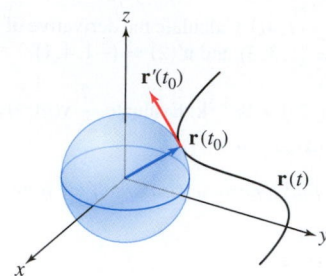

FIGURE 10 Curve with polar parametrization $\mathbf{r}(\theta) = f(\theta) \langle \cos\theta, \sin\theta \rangle$.

67. Prove that if $\|\mathbf{r}(t)\|$ takes on a local minimum or maximum value at t_0, then $\mathbf{r}(t_0)$ is orthogonal to $\mathbf{r}'(t_0)$. Explain how this result is related to Figure 11. *Hint:* Observe that if $\|\mathbf{r}(t_0)\|$ is a minimum, then $\mathbf{r}(t)$ is tangent at t_0 to the sphere of radius $\|\mathbf{r}(t_0)\|$ centered at the origin.

68. Newton's Second Law of Motion in vector form states that $\mathbf{F} = \dfrac{d\mathbf{p}}{dt}$, where $\mathbf{F}$ is the force acting on an object of mass m and $\mathbf{p} = m\mathbf{r}'(t)$ is the object's momentum. The analogs of force and momentum for rotational motion are the **torque** $\tau = \mathbf{r} \times \mathbf{F}$ and **angular momentum**

$$\mathbf{J} = \mathbf{r}(t) \times \mathbf{p}(t)$$

Use the Second Law to prove that $\tau = \dfrac{d\mathbf{J}}{dt}$.

FIGURE 11

69. Use FTC I from single-variable calculus to prove the first part of the Fundamental Theorem of Calculus for Vector-Valued Functions.

70. Use FTC II from single-variable calculus to prove the second part of the Fundamental Theorem of Calculus for Vector-Valued Functions.

Further Insights and Challenges

71. Let $\mathbf{r}(t) = \langle x(t), y(t) \rangle$ trace a plane curve $\mathcal{C}$. Assume that $x'(t_0) \neq 0$. Show that the slope of the tangent vector $\mathbf{r}'(t_0)$ is equal to the slope dy/dx of the curve at $\mathbf{r}(t_0)$.

72. Prove that $\dfrac{d}{dt}(\mathbf{r} \cdot (\mathbf{r}' \times \mathbf{r}'')) = \mathbf{r} \cdot (\mathbf{r}' \times \mathbf{r}''')$.

73. Prove the Sum and Constant Multiple for derivatives of vector-valued functions.

74. Prove the Chain Rule for vector-valued functions.

75. Prove the Cross Product Rule [Eq. (5)].

76. Prove the linearity properties

$$\int c\mathbf{r}(t)\, dt = c \int \mathbf{r}(t)\, dt \qquad (c \text{ any constant})$$

$$\int (\mathbf{r}_1(t) + \mathbf{r}_2(t))\, dt = \int \mathbf{r}_1(t)\, dt + \int \mathbf{r}_2(t)\, dt$$

77. Prove the Substitution Rule [where g is a differentiable scalar function with an inverse]:

$$\int_a^b \mathbf{r}(g(t))g'(t)\, dt = \int_{g^{-1}(a)}^{g^{-1}(b)} \mathbf{r}(u)\, du$$

78. Prove that if $\|\mathbf{r}(t)\| \leq K$ for $t \in [a, b]$, then

$$\left\| \int_a^b \mathbf{r}(t)\, dt \right\| \leq K(b - a)$$

13.3 Arc Length and Speed

In Section 11.2, we derived a formula for the arc length of a plane curve given in parametric form. This discussion applies to curves in 3-space with only minor changes.

Recall that arc length is defined as the limit of the lengths of polygonal approximations. Let $\mathcal{C}$ be the curve parametrized by

$$\mathbf{r}(t) = \langle x(t), y(t), z(t) \rangle, \qquad a \leq t \leq b$$

We assume that the parametrization directly traverses $\mathcal{C}$; that is, the path traces $\mathcal{C}$ from one end to the other without changing direction along the way. To produce a polygonal approximation of $\mathcal{C}$, we choose a partition $a = t_0 < t_1 < t_2 < \cdots < t_N = b$ and join the

◄ REMINDER *The length of a curve is referred to as the arc length.*

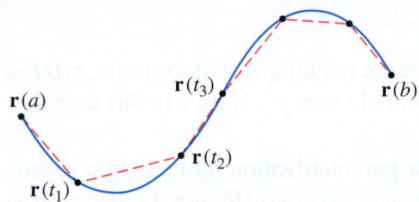

FIGURE 1 Polygonal approximation to the curve traced by $\mathbf{r}(t)$ for $a \leq t \leq b$.

terminal points of the vectors $\mathbf{r}(t_j)$ by segments, as in Figure 1. As in Section 11.2, we find that if $\mathbf{r}'(t)$ exists and is continuous on the interval $[a, b]$, then the lengths of the polygonal approximations approach a limit L as the maximum of the widths $|t_j - t_{j-1}|$ tends to zero. This limit is the length s of $\mathcal{C}$ and is computed by the integral in the next theorem.

> **THEOREM 1 Arc Length** Let $\mathbf{r}(t)$ directly traverse $\mathcal{C}$ for $a \leq t \leq b$. Assume that $\mathbf{r}'(t)$ exists and is continuous. Then the arc length s of $\mathcal{C}$ is equal to
>
> $$s = \int_a^b \|\mathbf{r}'(t)\| \, dt = \int_a^b \sqrt{x'(t)^2 + y'(t)^2 + z'(t)^2} \, dt \qquad \boxed{1}$$

Keep in mind that the arc length s in Eq. (1) is the length of $\mathcal{C}$ only if $\mathbf{r}(t)$ directly traverses $\mathcal{C}$. In general, the integral represents the distance traveled by a particle whose path is traced by $\mathbf{r}(t)$.

EXAMPLE 1 Find the arc length s of the helix given by $\mathbf{r}(t) = \langle \cos 3t, \sin 3t, 3t \rangle$ for $0 \leq t \leq 2\pi$.

Solution The derivative is $\mathbf{r}'(t) = \langle -3 \sin 3t, 3 \cos 3t, 3 \rangle$, and

$$\|\mathbf{r}'(t)\|^2 = 9 \sin^2 3t + 9 \cos^2 3t + 9 = 9(\sin^2 3t + \cos^2 3t) + 9 = 18$$

Therefore, $s = \displaystyle\int_0^{2\pi} \|\mathbf{r}'(t)\| \, dt = \int_0^{2\pi} \sqrt{18} \, dt = 6\sqrt{2}\pi$. ∎

Speed, by definition, is the rate of change of distance traveled with respect to time t. To calculate the speed, we define the **arc length function**:

$$s(t) = \int_a^t \|\mathbf{r}'(u)\| \, du$$

Thus, $s(t)$ is the distance traveled during the time interval $[a, t]$. By Part II of Fundamental Theorem of Calculus,

$$\text{Speed at time } t = \frac{ds}{dt} = \|\mathbf{r}'(t)\|$$

This relationship provides us with the second of the two important features of the velocity vector, $\mathbf{r}'(t)$: It points in the direction of motion, and its magnitude is the speed (Figure 2). We often denote the velocity vector by $\mathbf{v}(t)$ and the speed by $v(t)$:

$$\mathbf{v}(t) = \mathbf{r}'(t), \qquad v(t) = \|\mathbf{v}(t)\|$$

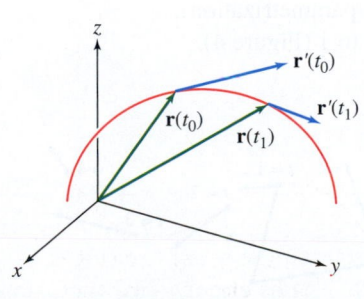

DF FIGURE 2 The velocity vector $\mathbf{r}'(t_0)$ is longer than $\mathbf{r}'(t_1)$, indicating that the particle is moving faster at t_0 than at t_1.

EXAMPLE 2 With distance measured in feet, find the speed at time $t = 2$ seconds for a particle whose position vector is

$$\mathbf{r}(t) = t^3 \mathbf{i} - e^t \mathbf{j} + 4t \mathbf{k}$$

Solution The velocity vector is $\mathbf{v}(t) = \mathbf{r}'(t) = 3t^2 \mathbf{i} - e^t \mathbf{j} + 4\mathbf{k}$, and at $t = 2$,

$$\mathbf{v}(2) = 12\mathbf{i} - e^2 \mathbf{j} + 4\mathbf{k}$$

The particle's speed is $v(2) = \|\mathbf{v}(2)\| = \sqrt{12^2 + (-e^2)^2 + 4^2} \approx 14.65$ ft/s. ∎

Keep in mind that a parametrization $\mathbf{r}(t)$ describes not just a curve, but also how a particle traverses the curve, possibly speeding up, slowing down, or reversing direction along the way. Changing the parametrization amounts to describing a different way of traversing the same underlying curve.

Arc Length Parametrization

We have seen that parametrizations of a given curve are not unique. For example, $\mathbf{r}_1(t) = \langle t, t^2 \rangle$ and $\mathbf{r}_2(u) = \langle u^3, u^6 \rangle$ both parametrize the parabola $y = x^2$. Notice in this case that $\mathbf{r}_2(u)$ is obtained by substituting $t = u^3$ in $\mathbf{r}_1(t)$.

Given a parametrization $\mathbf{r}(t)$ we obtain a new parametrization by making a substitution $t = g(u)$—that is, by replacing $\mathbf{r}(t)$ with $\mathbf{r}_1(u) = \mathbf{r}(g(u))$ (Figure 3). If $t = g(u)$ increases from a to b as u varies from c to d, then the path $\mathbf{r}(t)$ for $a \leq t \leq b$ is also parametrized by $\mathbf{r}_1(u)$ for $c \leq u \leq d$.

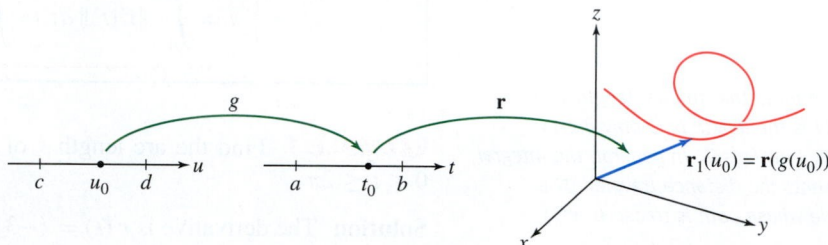

FIGURE 3 The curve is parametrized by $\mathbf{r}(t)$ and by $\mathbf{r}_1(u) = \mathbf{r}(g(u))$.

EXAMPLE 3 Let $\mathcal{C}$ be the curve parametrized by $\mathbf{r}(t) = (t^2, \sin t, t)$ for $3 \leq t \leq 9$. Give a different parametrization of $\mathcal{C}$ using the parameter u, where $t = g(u) = e^u$.

Solution Substituting $t = e^u$ in $\mathbf{r}(t)$, we obtain the parametrization

$$\mathbf{r}_1(u) = \mathbf{r}(g(u)) = \langle e^{2u}, \sin e^u, e^u \rangle$$

Because $u = \ln t$, the parameter t varies from 3 to 9 as u varies from $\ln 3$ to $\ln 9$. Therefore, $\mathcal{C}$ is parametrized by $\mathbf{r}_1(u)$ for $\ln 3 \leq u \leq \ln 9$. ∎

A way of parametrizing a curve that is useful in studying properties of the curve is to choose a starting point and travel along the curve at unit speed (say, 1 m/s). A parametrization of this type is called an **arc length parametrization** with the parameter often denoted by s. It is defined by the property that the speed has constant value 1:

$$\|\mathbf{r}'(s)\| = 1 \qquad \text{for all } s$$

*Arc length parametrizations are also called **unit speed parametrizations**. We will use arc length parametrizations to define curvature in Section 13.4.*

There are three important properties of an arc length parametrization:

- The parameter s corresponds to the arc length of the curve that is traced from the starting point (and thus the name *arc length* parametrization).
- Every velocity vector $\mathbf{r}'(s)$ has length equal to 1 (Figure 4).

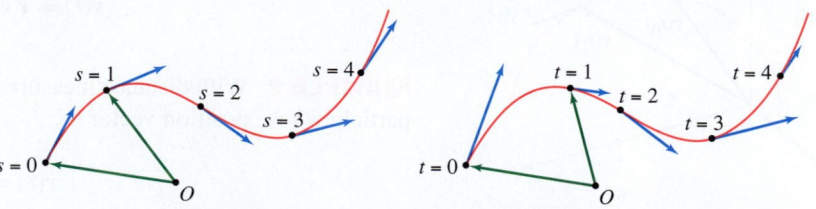

(A) An arc length parametrization:
All velocity vectors have length 1,
so speed is 1.

(B) Not an arc length parametrization:
Lengths of velocity vectors vary,
so the speed is changing.

FIGURE 4

- The arc length of the curve that is traced over any interval $[a, b]$ is equal to $b - a$, the length of the interval:

$$\text{distance traveled over } [a, b] = \int_a^b \|\mathbf{r}'(s)\| \, ds = \int_a^b 1 \, dt = b - a$$

Finding an arc length parametrization: Start with any parametrization $\mathbf{r}(t)$ such that $\mathbf{r}'(t) \neq \mathbf{0}$ for all t.

Step 1. **Form the arc length integral**

$$s = g(t) = \int_0^t \|\mathbf{r}'(u)\| \, du$$

Step 2. **Determine the inverse of** $g(t)$. Note that, because $\|\mathbf{r}'(t)\| \neq 0$, $s = g(t)$ is an increasing function, and g has an inverse $t = g^{-1}(s)$.

Step 3. **Take the new parametrization**

$$\mathbf{r}_1(s) = \mathbf{r}(g^{-1}(s))$$

This is our arc length parametrization.

In most cases, we cannot evaluate the arc length integral $s = g(t)$ explicitly, and we cannot find a formula for its inverse $g^{-1}(s)$ either. So although arc length parametrizations exist in general, we can find them explicitly only in special cases.

EXAMPLE 4 **Finding an Arc Length Parametrization** Find an arc length parametrization of the helix that is traced by the parametrization $\mathbf{r}(t) = \langle \cos 4t, \sin 4t, 3t \rangle$.

Solution *Step 1.* First, we evaluate the arc length function

$$\|\mathbf{r}'(t)\| = \|\langle -4 \sin 4t, 4 \cos 4t, 3 \rangle\| = \sqrt{16 \sin^2 4t + 16 \cos^2 4t + 3^2} = 5$$

$$s = g(t) = \int_0^t \|\mathbf{r}'(u)\| \, du = \int_0^t 5 \, du = 5t$$

Step 2. Then we observe that the inverse of $s = 5t$ is $t = s/5$; that is, $g^{-1}(s) = s/5$.

Step 3. Substituting $\frac{s}{5}$ for each appearance of t in the original parametrization, we obtain the arc length parametrization

$$\mathbf{r}_1(s) = \mathbf{r}(g^{-1}(s)) = \mathbf{r}\left(\frac{s}{5}\right) = \left\langle \cos \frac{4s}{5}, \sin \frac{4s}{5}, \frac{3s}{5} \right\rangle$$

As a check, let's verify that $\mathbf{r}_1(s)$ has unit speed:

$$\|\mathbf{r}_1'(s)\| = \left\| \left\langle -\frac{4}{5} \sin \frac{4s}{5}, \frac{4}{5} \cos \frac{4s}{5}, \frac{3}{5} \right\rangle \right\| = \sqrt{\frac{16}{25} \sin^2 \frac{4s}{5} + \frac{16}{25} \cos^2 \frac{4s}{5} + \frac{9}{25}} = 1 \quad \blacksquare$$

13.3 SUMMARY

- If $\mathbf{r}(t) = \langle x(t), y(t), z(t) \rangle$ directly traverses $\mathcal{C}$ for $a \leq t \leq b$, then the arc length s of $\mathcal{C}$ is

$$s = \int_a^b \|\mathbf{r}'(t)\| \, dt = \int_a^b \sqrt{x'(t)^2 + y'(t)^2 + z'(t)^2} \, dt$$

In general, whether $\mathbf{r}(t)$ directly traverses a curve or not, the integral represents the distance traveled on the path $\mathbf{r}(t)$ over $[a, b]$.

- Arc length function: $s(t) = \int_a^t \|\mathbf{r}'(u)\| \, du$

- Speed is the derivative of distance traveled with respect to time:

$$v(t) = \frac{ds}{dt} = \|\mathbf{r}'(t)\|$$

- The velocity vector $\mathbf{v}(t) = \mathbf{r}'(t)$ points in the direction of motion [provided that $\mathbf{r}'(t) \neq \mathbf{0}$] and its magnitude $v(t) = \|\mathbf{r}'(t)\|$ is the object's speed.
- We say that $\mathbf{r}(s)$ is an *arc length parametrization* if $\|\mathbf{r}'(s)\| = 1$ for all s. In this case, the arc length of the curve that is traced over $[a, b]$ is $b - a$.
- If $\mathbf{r}(t)$ is a parametrization of a curve $\mathcal{C}$ such that $\mathbf{r}'(t) \neq \mathbf{0}$ for all t, then

$$\mathbf{r}_1(s) = \mathbf{r}(g^{-1}(s))$$

is an arc length parametrization of $\mathcal{C}$, where $t = g^{-1}(s)$ is the inverse of the arc length function $s = g(t)$.

13.3 EXERCISES

Preliminary Questions

1. At a given instant, a car on a roller coaster has velocity vector $\mathbf{v} = \langle 25, -35, 10 \rangle$ (in miles per hour). What would the velocity vector be if the speed were doubled? What would it be if the car's direction were reversed but its speed remained unchanged?

2. Two cars travel in the same direction along the same roller coaster (at different times). Which of the following statements about their velocity vectors at a given point P on the roller coaster are true?

(a) The velocity vectors are identical.

(b) The velocity vectors point in the same direction but may have different lengths.

(c) The velocity vectors may point in opposite directions.

3. Starting at the origin, a mosquito flies along a parabola with speed $v(t) = t^2$. Let $L(t)$ be the total distance traveled at time t.

(a) How fast is $L(t)$ changing at $t = 2$?

(b) Is $L(t)$ equal to the mosquito's distance from the origin?

4. What is the length of the path traced by $\mathbf{r}(t)$ for $4 \leq t \leq 10$ if $\mathbf{r}(t)$ is an arc length parametrization?

Exercises

In Exercises 1–8, compute the length of the curve traced by $\mathbf{r}(t)$ over the given interval.

1. $\mathbf{r}(t) = \langle 3t, 4t - 3, 6t + 1 \rangle, \quad 0 \leq t \leq 3$

2. $\mathbf{r}(t) = \langle 2t, 2 - 4t, 5 \rangle, \quad 5 \leq t \leq 10$

3. $\mathbf{r}(t) = \langle 2t, \ln t, t^2 \rangle, \quad 1 \leq t \leq 4$

4. $\mathbf{r}(t) = \langle \cos t, \sin t, t^{3/2} \rangle, \quad 0 \leq t \leq 2\pi$

5. $\mathbf{r}(t) = \langle t, 4t^{3/2}, 2t^{3/2} \rangle, \quad 0 \leq t \leq 3$

6. $\mathbf{r}(t) = \langle 2t^2 + 1, 2t^2 - 1, t^3 \rangle, \quad 0 \leq t \leq 2$

7. $\mathbf{r}(t) = \langle t \cos t, t \sin t, 3t \rangle, \quad 0 \leq t \leq 2\pi$

8. $\mathbf{r}(t) = t\mathbf{i} + 2t\mathbf{j} + (t^2 - 3)\mathbf{k}, \quad 0 \leq t \leq 2$

9. ⬚CAS The curve shown in Figure 5(A) is parametrized by $\mathbf{r}(t) = \langle \cos(7t), \sin(7t), 2\cos t \rangle$ for $0 \leq t \leq 2\pi$. Approximate its length.

10. ⬚CAS The curve shown in Figure 5(B) is parametrized by $\mathbf{r}(t) = \langle 2\cos t, 2\sin t, \cos(19t) \rangle$ for $0 \leq t \leq 2\pi$. Approximate its length.

In Exercises 11 and 12, compute the arc length function $s(t) = \int_a^t \|\mathbf{r}'(u)\| \, du$ for the given value of a.

11. $\mathbf{r}(t) = \langle t^2, 2t^2, t^3 \rangle, \quad a = 0$

12. $\mathbf{r}(t) = \langle 4t^{1/2}, \ln t, 2t \rangle, \quad a = 1$

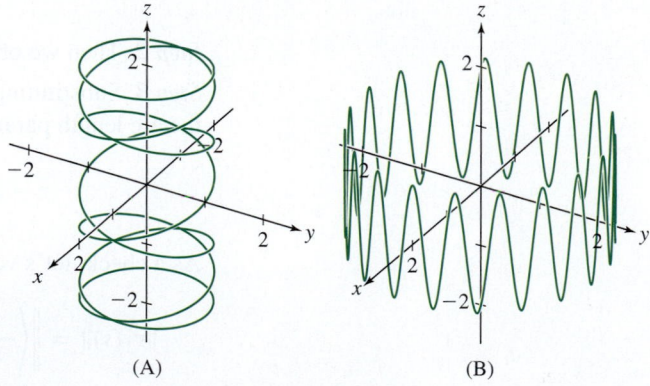

(A)   (B)

FIGURE 5

In Exercises 13–18, find the speed at the given value of t.

13. $\mathbf{r}(t) = \langle 2t + 3, 4t - 3, 5 - t \rangle, \quad t = 4$

14. $\mathbf{r}(t) = \langle -t, 2t^2, -3t^3 \rangle, \quad t = 2$

15. $\mathbf{r}(t) = \langle t, \ln t, (\ln t)^2 \rangle, \quad t = 1$

16. $\mathbf{r}(t) = \langle e^{t-3}, 12, 3t^{-1} \rangle, \quad t = 3$

17. $\mathbf{r}(t) = \langle \sin 3t, \cos 4t, \cos 5t \rangle, \quad t = \frac{\pi}{2}$

18. $\mathbf{r}(t) = \langle \cosh t, \sinh t, t \rangle, \quad t = 0$

19. At an air show, a jet has a trajectory following the curve $y = x^2$. If when the jet is at the point $(1, 1)$, it has a speed of 500 km/h, determine its velocity vector at this point.

20. What is the velocity vector of a particle traveling to the right along the hyperbola $y = x^{-1}$ with constant speed 5 cm/s when the particle's location is $\left(2, \frac{1}{2}\right)$?

21. A bee with velocity vector $\mathbf{r}'(t)$ starts out at the origin at $t = 0$ and flies around for T seconds. Where is the bee located at time T if $\int_0^T \mathbf{r}'(u)\,du = \mathbf{0}$? What does the quantity $\int_0^T \|\mathbf{r}'(u)\|\,du$ represent?

22. The DNA molecule comes in the form of a double helix, meaning two helices that wrap around one another. Suppose a single one of the helices has a radius of 10Å (1 angstrom Å $= 10^{-8}$ cm) and one full turn of the helix has a height of 34Å.

(a) Show that $\mathbf{r}(t) = \left(10\cos t, 10\sin t, \dfrac{34t}{2\pi}\right)$ is a parametrization of the helix.

(b) Find the arc length of one full turn of the helix.

23. Let

$$\mathbf{r}(t) = \left\langle R\cos\left(\frac{2\pi Nt}{h}\right), R\sin\left(\frac{2\pi Nt}{h}\right), t\right\rangle, \qquad 0 \le t \le h$$

(a) Show that $\mathbf{r}(t)$ parametrizes a helix of radius R and height h making N complete turns.

(b) Guess which of the two springs in Figure 6 uses more wire.

(c) Compute the lengths of the two springs and compare.

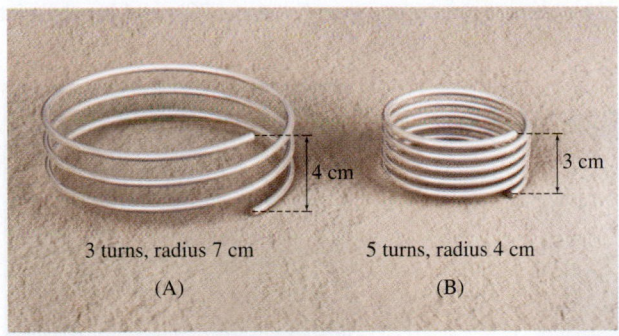

3 turns, radius 7 cm   5 turns, radius 4 cm

(A)   (B)

FIGURE 6 Which spring uses more wire?

24. Use Exercise 23 to find a general formula for the length of a helix of radius R and height h that makes N complete turns.

25. The cycloid generated by the unit circle has parametrization

$$\mathbf{r}(t) = \langle t - \sin t, 1 - \cos t\rangle$$

(a) Find the value of t in $[0, 2\pi]$ where the speed is at a maximum.

(b) Show that one arch of the cycloid has length 8. Recall the identity $\sin^2(t/2) = (1 - \cos t)/2$.

26. Which of the following is an arc length parametrization of a circle of radius 4 centered at the origin?

(a) $\mathbf{r}_1(t) = \langle 4\sin t, 4\cos t\rangle$

(b) $\mathbf{r}_2(t) = \langle 4\sin 4t, 4\cos 4t\rangle$

(c) $\mathbf{r}_3(t) = \left\langle 4\sin\frac{t}{4}, 4\cos\frac{t}{4}\right\rangle$

27. Let $\mathbf{r}(t) = \langle 3t + 1, 4t - 5, 2t\rangle$.

(a) Evaluate the arc length integral $s = g(t) = \displaystyle\int_0^t \|\mathbf{r}'(u)\|\,du$.

(b) Find the inverse $g^{-1}(s)$ of $g(t)$.

(c) Verify that $\mathbf{r}_1(s) = \mathbf{r}(g^{-1}(s))$ is an arc length parametrization.

28. Find an arc length parametrization of the line $y = 4x + 9$.

29. Let $\mathbf{r}(t) = \mathbf{w} + t\mathbf{v}$ be a parametrization of a line.

(a) Show that the arc length function $s = g(t) = \displaystyle\int_0^t \|\mathbf{r}'(u)\|\,du$ is given by $s = t\|\mathbf{v}\|$. This shows that $\mathbf{r}(t)$ is an arc length parametrizaton if and only if $\mathbf{v}$ is a unit vector.

(b) Find an arc length parametrization of the line with $\mathbf{w} = \langle 1, 2, 3\rangle$ and $\mathbf{v} = \langle 3, 4, 5\rangle$.

30. Find an arc length parametrization of the circle in the plane $z = 9$ with radius 4 and center $(1, 4, 9)$.

31. Find a path that traces the circle in the plane $y = 10$ with radius 4 and center $(2, 10, -3)$ with constant speed 8.

32. Find an arc length parametrization of the curve parametrized by $\mathbf{r}(t) = \left\langle t, \frac{2}{3}t^{3/2}, \frac{2}{\sqrt{3}}t^{3/2}\right\rangle$, with the parameter s measuring from $(0, 0, 0)$.

33. Find an arc length parametrization of the curve parametrized by $\mathbf{r}(t) = \left\langle \cos t, \sin t, \frac{2}{3}t^{3/2}\right\rangle$, with the parameter s measuring from $(1, 0, 0)$.

34. Find an arc length parametrization of the curve parametrized by $\mathbf{r}(t) = \langle e^t \sin t, e^t \cos t, e^t\rangle$.

35. Find an arc length parametrization of the curve parametrized by $\mathbf{r}(t) = \langle t^2, t^3\rangle$.

36. Find an arc length parametrization of the cycloid with parametrization $\mathbf{r}(t) = \langle t - \sin t, 1 - \cos t\rangle$.

37. Find an arc length parametrization of the line $y = mx$ for an arbitrary slope m.

38. Express the arc length L of $y = x^3$ for $0 \le x \le 8$ as an integral in two ways, using the parametrizations $\mathbf{r}_1(t) = \langle t, t^3\rangle$ and $\mathbf{r}_2(t) = \langle t^3, t^9\rangle$. Do not evaluate the integrals, but use substitution to show that they yield the same result.

39. The curve known as the **Bernoulli spiral** (Figure 7) has parametrization $\mathbf{r}(t) = \langle e^t \cos 4t, e^t \sin 4t\rangle$.

(a) Evaluate $s = g(t) = \displaystyle\int_{-\infty}^t \|\mathbf{r}'(u)\|\,du$. It is convenient to take lower limit $-\infty$ because $\displaystyle\lim_{t \to -\infty} \mathbf{r}(t) = \langle 0, 0\rangle$.

(b) Use (a) to find an arc length parametrization of the spiral.

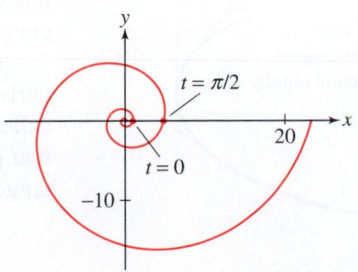

FIGURE 7 Bernoulli spiral.

Further Insights and Challenges

40. Prove that the length of a curve as computed using the arc length integral does not depend on its parametrization. More precisely, let $\mathcal{C}$ be the curve traced by $\mathbf{r}(t)$ for $a \leq t \leq b$. Let $f(s)$ be a differentiable function such that $f'(s) > 0$ and $f(c) = a$ and $f(d) = b$. Then $\mathbf{r}_1(s) = \mathbf{r}(f(s))$ parametrizes $\mathcal{C}$ for $c \leq s \leq d$. Verify that

$$\int_a^b \|\mathbf{r}'(t)\| \, dt = \int_c^d \|\mathbf{r}_1'(s)\| \, ds$$

41. The unit circle with the point $(-1, 0)$ removed has parametrization (see Exercise 85 in Section 11.1)

$$\mathbf{r}(t) = \left\langle \frac{1 - t^2}{1 + t^2}, \frac{2t}{1 + t^2} \right\rangle, \qquad -\infty < t < \infty$$

Use this parametrization to compute the length of the unit circle as an improper integral. *Hint:* The expression for $\|\mathbf{r}'(t)\|$ simplifies.

42. The involute of a circle (Figure 8), traced by a point at the end of a thread unwinding from a circular spool of radius R, has parametrization (see Exercise 29 in Section 11.2)

$$\mathbf{r}(\theta) = \langle R(\cos\theta + \theta\sin\theta), R(\sin\theta - \theta\cos\theta) \rangle$$

Find an arc length parametrization of the involute.

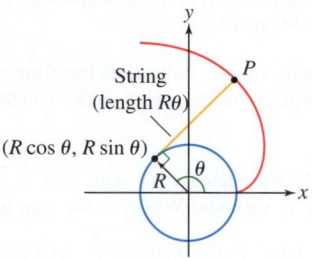

FIGURE 8 The involute of a circle.

43. The curve $\mathbf{r}(t) = \langle t - \tanh t, \operatorname{sech} t \rangle$ is called a **tractrix** (see Exercise 106 in Section 11.1).

(a) Show that $s(t) = \displaystyle\int_0^t \|\mathbf{r}'(u)\| \, du$ is equal to $s(t) = \ln(\cosh t)$.

(b) Show that $t = g(s) = \ln(e^s + \sqrt{e^{2s} - 1})$ is an inverse of $s(t)$ and verify that

$$\mathbf{r}_1(s) = \left\langle \tanh^{-1}\left(\sqrt{1 - e^{-2s}}\right) - \sqrt{1 - e^{-2s}}, e^{-s} \right\rangle$$

is an arc length parametrization of the tractrix.

13.4 Curvature

Curvature is a measure of how much a curve bends. It is used to study geometric properties of curves and motion along curves, and has applications in diverse areas such as roller coaster design (Figure 1), optics, eye surgery (see Exercise 70), and biochemistry (Figure 2).

FIGURE 1 Curvature is a key ingredient in roller coaster design.

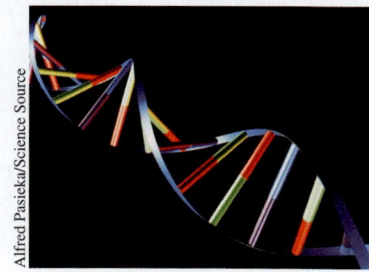

FIGURE 2 Biochemists study the effect of the curvature of DNA strands on biological processes.

A natural way to define curvature is to consider the rate at which the direction along the curve is changing (Figure 3). To indicate the direction along a curve, we use unit tangent vectors. We introduce them next, and then employ them to precisely define curvature.

Consider a path with parametrization $\mathbf{r}(t) = \langle x(t), y(t), z(t) \rangle$. We assume that the derivative $\mathbf{r}'(t) \neq \mathbf{0}$ for all t in the domain of $\mathbf{r}(t)$. A parametrization with this property is called **regular**. At every point P along the path, there is a **unit tangent vector** $\mathbf{T} = \mathbf{T}_P$ that points in the direction of motion of the parametrization. We write $\mathbf{T}(t)$ for the unit tangent vector at the terminal point of $\mathbf{r}(t)$:

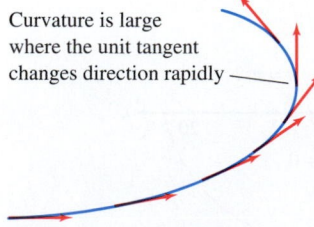

Curvature is large where the unit tangent changes direction rapidly

DF **FIGURE 3** The unit tangent vector varies in direction but not in length.

$$\text{Unit tangent vector} = \mathbf{T}(t) = \frac{\mathbf{r}'(t)}{\|\mathbf{r}'(t)\|}$$

For example, if $\mathbf{r}(t) = \langle t, t^2, t^3 \rangle$, then $\mathbf{r}'(t) = \langle 1, 2t, 3t^2 \rangle$, and the unit tangent vector at $P = (1, 1, 1)$, which is the terminal point of $\mathbf{r}(1) = \langle 1, 1, 1 \rangle$, is

$$\mathbf{T}_P = \frac{\langle 1, 2, 3 \rangle}{\|\langle 1, 2, 3 \rangle\|} = \frac{\langle 1, 2, 3 \rangle}{\sqrt{1^2 + 2^2 + 3^2}} = \left\langle \frac{1}{\sqrt{14}}, \frac{2}{\sqrt{14}}, \frac{3}{\sqrt{14}} \right\rangle$$

Now imagine following the path and observing how the unit tangent vector $\mathbf{T}$ changes direction as in Figure 3. A change in $\mathbf{T}$ indicates that the path is bending, and the more rapidly $\mathbf{T}$ changes, the more the path bends. Thus, $\left\|\dfrac{d\mathbf{T}}{dt}\right\|$ would seem to be a good measure of curvature. However, $\left\|\dfrac{d\mathbf{T}}{dt}\right\|$ depends on how fast you walk (when you walk faster, the unit tangent vector changes more quickly). Therefore, we assume that you walk at unit speed. In other words, curvature is the magnitude $\kappa(s) = \left\|\dfrac{d\mathbf{T}}{ds}\right\|$, where s is the parameter of an arc length parametrization. Recall that $\mathbf{r}(s)$ is an arc length parametrization if $\|\mathbf{r}'(s)\| = 1$ for all s.

> **DEFINITION Curvature** Let $\mathbf{r}(s)$ be an arc length parametrization and $\mathbf{T}$ the unit tangent vector. The **curvature** of the underlying curve at $\mathbf{r}(s)$ is the quantity (denoted by a lowercase Greek letter "kappa")
>
> $$\kappa(s) = \left\|\frac{d\mathbf{T}}{ds}\right\| \qquad \boxed{1}$$

By definition, curvature is the magnitude of the rate of change of the unit tangent vector $\mathbf{T}$ with respect to distance traveled s along the curve.

EXAMPLE 1   The Curvature of a Circle of Radius R Is $1/R$ Compute the curvature of a circle of radius R.

Solution   Assume the circle is centered at the origin, so that it has parametrization $\mathbf{r}(\theta) = \langle R\cos\theta, R\sin\theta \rangle$ (Figure 4). Since curvature is defined using an arc length parametrization, we need to find an arc length parametrization of the circle. First, we compute the arc length function:

$$s(\theta) = \int_0^\theta \|\mathbf{r}'(u)\|\, du = \int_0^\theta R\, du = R\theta$$

Thus, $s = R\theta$, and the inverse of the arc length function $s = g(\theta)$ is $\theta = g^{-1}(s) = s/R$. In Section 13.3, we showed that $\mathbf{r}_1(s) = \mathbf{r}(g^{-1}(s))$ is an arc length parametrization. In our case, we obtain

$$\mathbf{r}_1(s) = \mathbf{r}(g(s)) = \mathbf{r}\left(\frac{s}{R}\right) = \left\langle R\cos\frac{s}{R}, R\sin\frac{s}{R} \right\rangle$$

Now, with this parametrization, the unit tangent vector and its derivative are

$$\mathbf{T}(s) = \frac{d\mathbf{r}_1}{ds} = \frac{d}{ds}\left\langle R\cos\frac{s}{R}, R\sin\frac{s}{R} \right\rangle = \left\langle -\sin\frac{s}{R}, \cos\frac{s}{R} \right\rangle$$

$$\frac{d\mathbf{T}}{ds} = -\frac{1}{R}\left\langle \cos\frac{s}{R}, \sin\frac{s}{R} \right\rangle$$

By definition of curvature,

$$\kappa(s) = \left\|\frac{d\mathbf{T}}{ds}\right\| = \frac{1}{R}\left\|\left\langle \cos\frac{s}{R}, \sin\frac{s}{R} \right\rangle\right\| = \frac{1}{R}$$

This shows that the curvature is $1/R$ at all points on the circle. ∎

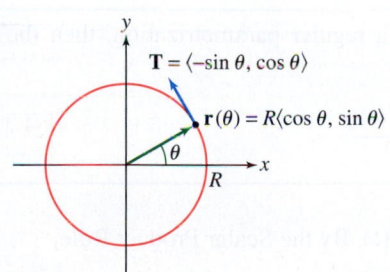

FIGURE 4 The unit tangent vector at a point on a circle of radius R.

Example 1 shows that a circle of large radius R has small curvature $1/R$. This makes sense because your direction of motion changes slowly when you walk at unit speed along a circle of large radius.

In practice, it is often difficult, if not impossible, to find an arc length parametrization explicitly. Fortunately, we can compute curvature using any regular parametrization $\mathbf{r}(t)$. We next develop some alternative formulas for curvature that do not rely on an arc length parametrization.

Since arc length s is a function of time t, the derivatives of $\mathbf{T}$ with respect to t and s are related by the Chain Rule. Denoting the derivative with respect to t by a prime, we have

$$\mathbf{T}'(t) = \frac{d\mathbf{T}}{dt} = \frac{d\mathbf{T}}{ds}\frac{ds}{dt} = v(t)\frac{d\mathbf{T}}{ds}$$

where $v(t) = \dfrac{ds}{dt} = \|\mathbf{r}'(t)\|$ is the speed of $\mathbf{r}(t)$. Since curvature is the magnitude $\left\|\dfrac{d\mathbf{T}}{ds}\right\|$, we obtain

$$\|\mathbf{T}'(t)\| = v(t)\kappa(t)$$

This yields an alternative formula for the curvature in the case where we do not have an arc length parametrization:

$$\boxed{\kappa(t) = \frac{1}{v(t)}\|\mathbf{T}'(t)\|} \qquad \boxed{2}$$

We can directly apply this formula to find curvature, but we can also use it to derive another option for calculations.

> **THEOREM 1 Formula for Curvature** If $\mathbf{r}(t)$ is a regular parametrization, then the curvature of the underlying curve at $\mathbf{r}(t)$ is
>
> $$\kappa(t) = \frac{\|\mathbf{r}'(t) \times \mathbf{r}''(t)\|}{\|\mathbf{r}'(t)\|^3} \qquad \boxed{3}$$

Proof Since $v(t) = \|\mathbf{r}'(t)\|$, we have $\mathbf{r}'(t) = v(t)\mathbf{T}(t)$. By the Scalar Product Rule,

$$\mathbf{r}''(t) = v'(t)\mathbf{T}(t) + v(t)\mathbf{T}'(t)$$

Now compute the following cross product, using the fact that $\mathbf{T}(t) \times \mathbf{T}(t) = \mathbf{0}$:

$$\mathbf{r}'(t) \times \mathbf{r}''(t) = v(t)\mathbf{T}(t) \times \big(v'(t)\mathbf{T}(t) + v(t)\mathbf{T}'(t)\big)$$

$$= v(t)^2\mathbf{T}(t) \times \mathbf{T}'(t) \qquad \boxed{4}$$

Since $\|\mathbf{T}(t)\|$ is constant, Example 7 in Section 13.2 implies that $\mathbf{T}(t)$ and $\mathbf{T}'(t)$ are orthogonal. Then, using the geometric interpretation of cross product and the fact that $\|\mathbf{T}(t)\| = 1$, it follows that

$$\|\mathbf{T}(t) \times \mathbf{T}'(t)\| = \|\mathbf{T}(t)\|\,\|\mathbf{T}'(t)\|\sin\frac{\pi}{2} = \|\mathbf{T}'(t)\|$$

Equation (4) yields $\|\mathbf{r}'(t) \times \mathbf{r}''(t)\| = v(t)^2\|\mathbf{T}'(t)\|$. Using Eq. (2), we obtain

$$\|\mathbf{r}'(t) \times \mathbf{r}''(t)\| = v(t)^2\|\mathbf{T}'(t)\| = v(t)^3\kappa(t) = \|\mathbf{r}'(t)\|^3\kappa(t)$$

This yields the desired formula. ∎

It is a quick consequence of Theorem 1 that—as we might expect—the curvature of a line is zero. A line in 3-space has a parametrization $\mathbf{r}(t) = \langle x_0 + at, y_0 + bt, z_0 + ct\rangle$. This is a regular parametrization since $\mathbf{r}'(t) = \langle a, b, c\rangle$ and at least one of a, b, or c must be nonzero. For this parametrization, $\mathbf{r}''(t) = \mathbf{0}$, and by Eq. (3) the curvature is zero.

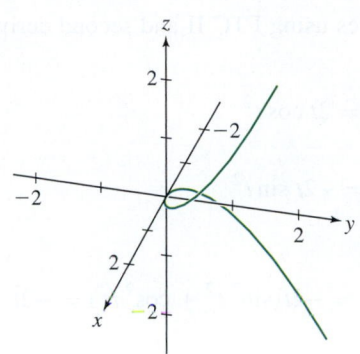

FIGURE 5 The twisted cubic $\mathbf{r}(t) = \langle t, t^2, t^3 \rangle$.

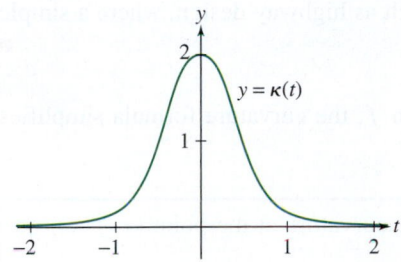

FIGURE 6 Graph of the curvature $\kappa(t)$ of the twisted cubic.

FIGURE 7

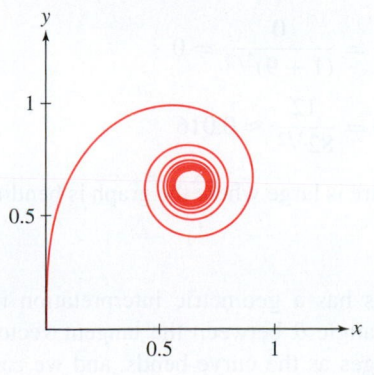

FIGURE 8 The Cornu spiral.

EXAMPLE 2 **Twisted Cubic Curve** [CAS] Calculate the curvature $\kappa(t)$ of the twisted cubic $\mathbf{r}(t) = \langle t, t^2, t^3 \rangle$ (Figure 5). Then plot the graph of $\kappa(t)$ and determine where the curvature is largest.

Solution The derivatives are

$$\mathbf{r}'(t) = \langle 1, 2t, 3t^2 \rangle, \qquad \mathbf{r}''(t) = \langle 0, 2, 6t \rangle$$

The parametrization is regular because $\mathbf{r}'(t) \neq \mathbf{0}$ for all t, so we may use Eq. (3):

$$\mathbf{r}'(t) \times \mathbf{r}''(t) = \begin{vmatrix} \mathbf{i} & \mathbf{j} & \mathbf{k} \\ 1 & 2t & 3t^2 \\ 0 & 2 & 6t \end{vmatrix} = 6t^2\mathbf{i} - 6t\mathbf{j} + 2\mathbf{k}$$

$$\kappa(t) = \frac{\|\mathbf{r}'(t) \times \mathbf{r}''(t)\|}{\|\mathbf{r}'(t)\|^3} = \frac{\sqrt{36t^4 + 36t^2 + 4}}{(1 + 4t^2 + 9t^4)^{3/2}}$$

The graph of $\kappa(t)$ in Figure 6 shows that the curvature is largest at $t = 0$, when the curve is passing through the origin. ■

Via Theorem 1, we can obtain a straightforward formula for computing the curvature of a plane curve given parametrically by $\mathbf{r}(t) = \langle x(t), y(t) \rangle$:

> **THEOREM 2** **Curvature of a Plane Curve** Assume $\mathbf{r}(t) = \langle x(t), y(t) \rangle$ is a regular parametrization of a plane curve. At the point $(x(t), y(t))$, the curvature is given by
>
> $$\kappa(t) = \frac{|x'(t)y''(t) - y'(t)x''(t)|}{\big(x'(t)^2 + y'(t)^2\big)^{3/2}} \qquad \boxed{5}$$

Proof We consider the curve as a curve in 3-space with a vector parametrization given by $\mathbf{r}(t) = \langle x(t), y(t), 0 \rangle$. Then $\mathbf{r}'(t) = \langle x'(t), y'(t), 0 \rangle$ and $\mathbf{r}''(t) = \langle x''(t), y''(t), 0 \rangle$. Therefore,

$$\mathbf{r}'(t) \times \mathbf{r}''(t) = \begin{vmatrix} \mathbf{i} & \mathbf{j} & \mathbf{k} \\ x'(t) & y'(t) & 0 \\ x''(t) & y''(t) & 0 \end{vmatrix} = (x'(t)y''(t) - y'(t)x''(t))\mathbf{k}$$

Since $\|\mathbf{r}'(t)\| = \big(x'(t)^2 + y'(t)^2\big)^{1/2}$, Eq. (3) yields

$$\kappa(t) = \frac{\|\mathbf{r}'(x) \times \mathbf{r}''(x)\|}{\|\mathbf{r}'(x)\|^3} = \frac{|x'(t)y''(t) - y'(t)x''(t)|}{\big(x'(t)^2 + y'(t)^2\big)^{3/2}} \qquad ■$$

EXAMPLE 3 **Highway Curvature Transitions** Highway engineers design roads to achieve continuous and simple transitions between road segments with different curvatures, such as between the straight highways in Figure 7 and the circular parts of the entrance and exit ramps. Curve segments commonly used in such transitions are taken from the **Cornu spiral** (Figure 8) that is defined parametrically by $\mathbf{r}(t) = \langle x(t), y(t) \rangle$, $t \geq 0$, where

$$x(t) = \int_0^t \sin u^2 \, du, \qquad y(t) = \int_0^t \cos u^2 \, du$$

Show that $\kappa(t) = 2t$, and therefore curvature changes linearly as a function of t along the Cornu spiral.

Solution We use Eq. (5). Computing first derivatives using FTC II and second derivatives using the Chain Rule, we have

$$x'(t) = \sin t^2 \quad \text{and} \quad x''(t) = 2t \cos t^2$$

$$y'(t) = \cos t^2 \quad \text{and} \quad y''(t) = -2t \sin t^2$$

Therefore,

$$x'(t)y''(t) - y'(t)x''(t) = -2t \sin^2 t^2 - 2t \cos^2 t^2 = -2t(\sin^2 t^2 + \cos^2 t^2) = -2t$$

$$x'(t)^2 + y'(t)^2 = \sin^2 t^2 + \cos^2 t^2 = 1$$

It follows that $\kappa(t) = \frac{|-2t|}{1} = 2t$ and curvature is a linear function of t. Note that by changing linearly, the curvature changes in a relatively simple manner. Therefore, the Cornu spiral is natural to employ in situations, such as highway design, where a simple transition in curvature is desired. ∎

When a plane curve is the graph of a function f, the curvature formula simplifies further:

THEOREM 3 Curvature of the Graph of f The curvature at the point $(x, f(x))$ on the graph of $y = f(x)$ is equal to

$$\kappa(x) = \frac{|f''(x)|}{\left(1 + f'(x)^2\right)^{3/2}} \qquad \boxed{6}$$

We can prove this theorem using Eq. (5) and a parametrization with $x = t$ and $y = f(t)$ (see Exercise 28).

EXAMPLE 4 Compute the curvature of $f(x) = x^3 - 3x^2 + 4$ at $x = 0, 1, 2, 3$.

Solution We apply Eq. (6):

$$f'(x) = 3x^2 - 6x = 3x(x - 2), \qquad f''(x) = 6x - 6$$

$$\kappa(x) = \frac{|f''(x)|}{\left(1 + f'(x)^2\right)^{3/2}} = \frac{|6x - 6|}{\left(1 + 9x^2(x-2)^2\right)^{3/2}}$$

We obtain the following values:

$$\kappa(0) = \frac{6}{(1+0)^{3/2}} = 6, \qquad \kappa(1) = \frac{0}{(1+9)^{3/2}} = 0$$

$$\kappa(2) = \frac{6}{(1+0)^{3/2}} = 6, \qquad \kappa(3) = \frac{12}{82^{3/2}} \approx 0.016$$

Figure 9 shows that, as we might expect, the curvature is large where the graph is bending more. ∎

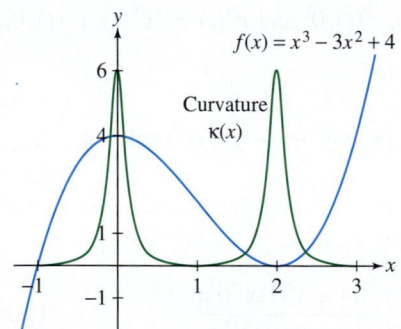

DF **FIGURE 9** Graph of $f(x) = x^3 - 3x^2 + 4$ and the curvature $\kappa(x)$.

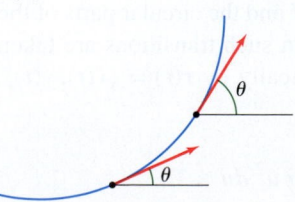

DF **FIGURE 10** The angle θ changes as the curve bends.

CONCEPTUAL INSIGHT Curvature for plane curves has a geometric interpretation in terms of the angle of inclination, defined as the angle θ between the tangent vector and the horizontal (Figure 10). The angle θ changes as the curve bends, and we can show that the curvature κ is the rate of change of θ with respect to distance traveled along the curve (see Exercise 71).

The Normal Vector

We noted earlier that $\mathbf{T}'(t)$ and $\mathbf{T}(t)$ are orthogonal. The unit vector in the direction of $\mathbf{T}'(t)$, assuming it is nonzero, is called the **normal vector** and denoted $\mathbf{N}(t)$ or simply $\mathbf{N}$:

$$\text{Normal vector:} \qquad \mathbf{N}(t) = \frac{\mathbf{T}'(t)}{\|\mathbf{T}'(t)\|} \qquad \boxed{7}$$

This pair $\mathbf{T}$ and $\mathbf{N}$ of orthogonal unit vectors play a critical role in understanding a given space curve. Since $\|\mathbf{T}'(t)\| = v(t)\kappa(t)$ by Eq. (2), we have

$$\mathbf{T}'(t) = v(t)\kappa(t)\mathbf{N}(t) \qquad \boxed{8}$$

Intuitively, $\mathbf{N}$ points in the direction in which the curve is turning (Figure 11). This is particularly clear for a plane curve. In this case, there are two unit vectors orthogonal to $\mathbf{T}$, and of these two shown in the figure, $\mathbf{N}$ is the vector that points to the inside of the curve.

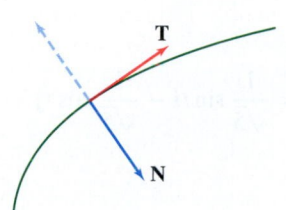

FIGURE 11 For a plane curve, the normal vector points in the direction of bending.

EXAMPLE 5 **Normal Vector to a Helix** Consider the helix $\mathbf{r}(t) = \langle 2\cos t, 2\sin t, t \rangle$. Show that for all t, the normal vector is parallel to the xy-plane and points toward the z-axis.

Solution Since $\mathbf{r}(t) = \langle 2\cos t, 2\sin t, t \rangle$, the tangent vector $\mathbf{r}'(t) = \langle -2\sin t, 2\cos t, 1 \rangle$ has length $\|\mathbf{r}'(t)\| = \sqrt{(-2\sin t)^2 + (2\cos t)^2 + 1} = \sqrt{5}$, so

$$\mathbf{T}(t) = \frac{\mathbf{r}'(t)}{\|\mathbf{r}'(t)\|} = \frac{1}{\sqrt{5}} \langle -2\sin t, 2\cos t, 1 \rangle$$

$$\mathbf{T}'(t) = \frac{1}{\sqrt{5}} \langle -2\cos t, -2\sin t, 0 \rangle$$

$$\|\mathbf{T}'(t)\| = \frac{1}{\sqrt{5}} \sqrt{(-2\cos t)^2 + (-2\sin t)^2 + 0} = \frac{2}{\sqrt{5}}$$

$$\mathbf{N}(t) = \frac{\mathbf{T}'(t)}{\|\mathbf{T}'(t)\|} = \langle -\cos t, -\sin t, 0 \rangle$$

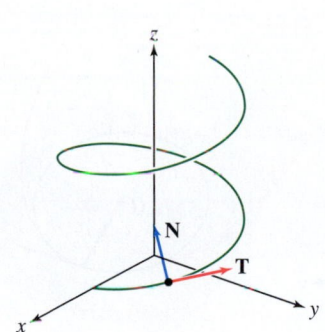

FIGURE 12 The normal vector is parallel to the xy-plane and points toward the z-axis.

Since the z-component is 0, it follows that $\mathbf{N}(t)$ is parallel to the xy-plane. Furthermore, at the point on the curve where $\mathbf{r}(t) = \langle 2\cos t, 2\sin t, t \rangle$, the x- and y-components of the normal $\mathbf{N}(t) = \langle -\cos t, -\sin t, 0 \rangle$ are the same negative scalar multiple of the corresponding components of $\mathbf{r}(t)$. Thus, $\mathbf{N}(t)$ points toward the z-axis from the point (Figure 12). ∎

The Frenet Frame

At a point P on a curve, the vectors $\mathbf{T}$ and $\mathbf{N}$ determine a plane. The normal vector $\mathbf{B}$ to this plane, which we call the **binormal vector**, is defined by

$$\text{Binormal vector:} \qquad \mathbf{B} = \mathbf{T} \times \mathbf{N} \qquad \boxed{9}$$

By the Geometric Description of the Cross Product (Theorem 1 in Section 12.4), we can conclude:

- $\mathbf{B}$ is orthogonal to both $\mathbf{T}$ and $\mathbf{N}$,
- $\mathbf{B}$ is a unit vector since $\|\mathbf{B}\| = \|\mathbf{T}\| \, \|\mathbf{N}\| \sin \pi/2 = (1)(1)(1) = 1$.
- $\{\mathbf{T}, \mathbf{N}, \mathbf{B}\}$ forms a right-handed system.

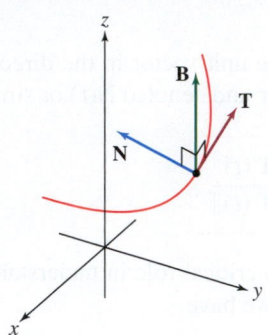

DF **FIGURE 13** The Frenet frame at a point on a curve.

Science History Images / Alamy

Katherine Johnson was a physicist and mathematician who conducted technical work at the National Aeronautics and Space Administration during the early decades of the manned space program. During this time, she calculated the trajectories, launch windows, and emergency back-up return paths for numerous historic flights including the 1969 *Apollo 11* flight that involved the first lunar landing. In 2015, Johnson received the Presidential Medal of Freedom.

This set of mutually perpendicular unit vectors is called the **Frenet frame**, after the French geometer Jean Frenet (1816–1900). It is an important tool in the field of differential geometry. As we move along a space curve, the Frenet frame moves and twists along with us, as in Figure 13. It provides a moving coordinate system, with origin at the current location, that is used to study trajectories of objects such as spacecraft, satellites, and asteroids.

EXAMPLE 6 In Example 5, for the helix $\mathbf{r}(t) = \langle 2\cos t, 2\sin t, t \rangle$, we determined $\mathbf{T}(t)$ and $\mathbf{N}(t)$. Compute $\mathbf{B}(t)$ to complete the Frenet frame for this curve.

Solution From Example 5, we have

$$\mathbf{T}(t) = \frac{1}{\sqrt{5}}\langle -2\sin t, 2\cos t, 1 \rangle \quad \text{and} \quad \mathbf{N}(t) = \langle -\cos t, -\sin t, 0 \rangle$$

Therefore,

$$\mathbf{B}(t) = \mathbf{T}(t) \times \mathbf{N}(t) = \begin{vmatrix} \mathbf{i} & \mathbf{j} & \mathbf{k} \\ -\frac{2}{\sqrt{5}}\sin t & \frac{2}{\sqrt{5}}\cos t & \frac{1}{\sqrt{5}} \\ -\cos t & -\sin t & 0 \end{vmatrix} = \frac{1}{\sqrt{5}}\sin t\,\mathbf{i} - \frac{1}{\sqrt{5}}\cos t\,\mathbf{j} + \frac{2}{\sqrt{5}}\mathbf{k}$$

∎

The Osculating Plane and Circle

At a point P on a curve parametrized by $\mathbf{r}(t)$, the unit tangent vector and the normal vector define a plane through P, called the **osculating plane** at P [Figure 14(A)]. Within the osculating plane, there is a best-fit circle that is tangent to the curve at P and has the same curvature at P as the curve [Figure 14(B)].

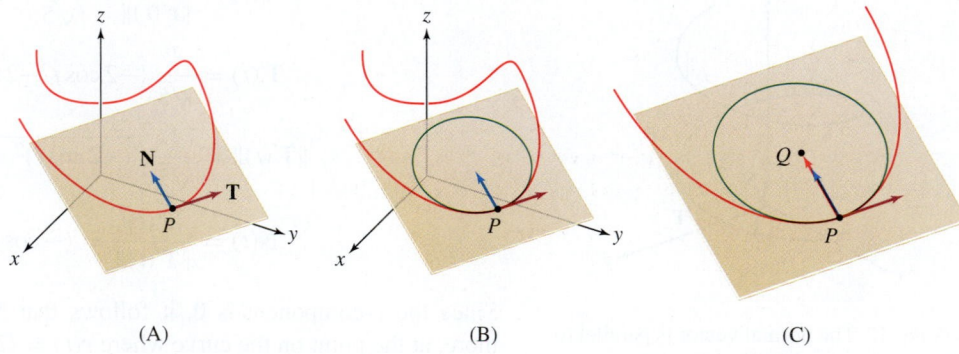

(A)   (B)   (C)

FIGURE 14

Recall that the curvature and radius of a circle are related by $R = 1/\kappa$. Since we want the osculating circle to have the same curvature as the curve at P, it has radius $R = 1/\kappa_P$.

Specifically, assume that κ_P, the curvature of the curve at P, is nonzero. The **osculating circle**, denoted Osc_P, lies in the osculating plane and is the circle of radius $R = 1/\kappa_P$ through P whose center Q lies in the direction of the unit normal $\mathbf{N}$ [Figure 14 (C)].

To find the center Q, first note that with O representing the origin, we have $\overrightarrow{OQ} = \overrightarrow{OP} + \overrightarrow{PQ}$ (Figure 15). Now, if P is the point on the curve corresponding to $\mathbf{r}(t_0)$, then $\overrightarrow{OP} = \mathbf{r}(t_0)$. Furthermore, $\overrightarrow{PQ}$ points in the direction of $\mathbf{N}$ and has length equal to $1/\kappa_P$, the radius of the circle. Thus, the center Q of the osculating circle at P is determined by

$$\overrightarrow{OQ} = \mathbf{r}(t_0) + \frac{1}{\kappa_P}\mathbf{N} \qquad \boxed{10}$$

Among all circles tangent to the curve at P, Osc_P is the best approximation to the curve (see Exercise 81). We refer to $R = 1/\kappa_P$ as the **radius of curvature** at P. The center Q of Osc_P is called the **center of curvature** at P.

For a curve in the plane $\mathbf{R}^2$, the osculating plane is $\mathbf{R}^2$ itself. In this setting, determining osculating circles is relatively straightforward as we see in the following example.

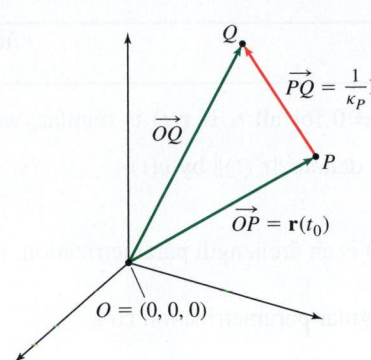

FIGURE 15

EXAMPLE 7 Determine the equations of the osculating circles to $y = x^2$ at $x = 0$ and at $x = 3/2$.

Solution Let $f(x) = x^2$.

Step 1. Find the radius.
Apply Eq. (6) to $f(x) = x^2$ to compute the curvature:

$$\kappa(x) = \frac{|f''(x)|}{\left(1 + f'(x)^2\right)^{3/2}} = \frac{2}{\left(1 + 4x^2\right)^{3/2}}$$

Therefore,

$$\kappa(0) = 2, \quad \kappa(3/2) = \frac{1}{5\sqrt{10}}$$

So the radius of the osculating circle at $x = 0$ is $1/2$, and at $x = 3/2$ is $5\sqrt{10}$.

Step 2. Find N.
Next, we parametrize the curve by $\mathbf{r}(x) = \langle x, f(x)\rangle = \langle x, x^2\rangle$. For a plane curve, we can find $\mathbf{N}$ without computing $\mathbf{T}'$. The vector $\mathbf{r}'(x) = \langle 1, 2x\rangle$ is tangent to the curve, and we know that $\langle 2x, -1\rangle$ is orthogonal to $\mathbf{r}'(x)$ (because their dot product is zero). Therefore, $\mathbf{N}(x)$ is the unit vector in one of the two directions $\pm\langle 2x, -1\rangle$. Since $\mathbf{N}(x)$ must point inside the curve $y = x^2$, it follows that for all x, $\mathbf{N}(x)$ is in the direction of $\langle -2x, 1\rangle$ (Figure 16). Thus,

$$\mathbf{N}(x) = \frac{\langle -2x, 1\rangle}{\|\langle -2x, 1\rangle\|} = \frac{\langle -2x, 1\rangle}{\sqrt{1 + 4x^2}}$$

Therefore,

$$\mathbf{N}(0) = \langle 0, 1\rangle \quad \text{and} \quad \mathbf{N}\left(\frac{3}{2}\right) = \frac{1}{\sqrt{10}}\langle -3, 1\rangle$$

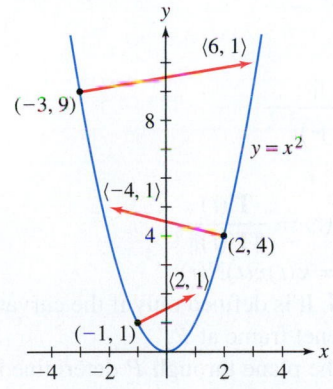

FIGURE 16 At (x, x^2) the vector $\langle -2x, 1\rangle$ is normal to the parabola and points inside it.

Step 3. Find the center Q.
Apply Eq. (10) for $x = 0$:

$$\overrightarrow{OQ} = \mathbf{r}(0) + \frac{1}{\kappa(0)}\mathbf{N}(0) = \langle 0, 0\rangle + \frac{1}{2}\langle 0, 1\rangle = \left\langle 0, \frac{1}{2}\right\rangle$$

Now, for $x = 3/2$:

$$\overrightarrow{OQ} = \mathbf{r}\left(\frac{3}{2}\right) + \frac{1}{\kappa(3/2)}\mathbf{N}\left(\frac{3}{2}\right) = \left\langle \frac{3}{2}, \frac{9}{4}\right\rangle + 5\sqrt{10}\left(\frac{\langle -3, 1\rangle}{\sqrt{10}}\right) = \left\langle -\frac{27}{2}, \frac{29}{4}\right\rangle$$

Step 4. Determine the equation of the osculating circle.
See Figure 17. At $x = 0$, the osculating circle has center $(0, 1/2)$ and radius $1/2$. The equation of the circle is

$$x^2 + \left(y - \frac{1}{2}\right)^2 = \frac{1}{4}$$

At $x = 3/2$, the osculating circle has center $(-27/2, 29/4)$ and radius $5\sqrt{10}$. Its equation is

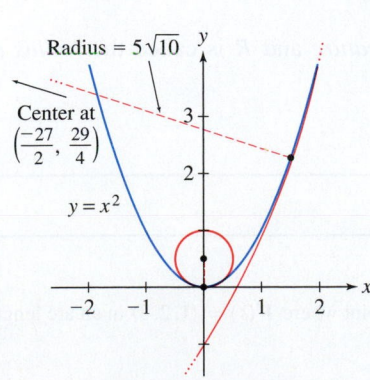

FIGURE 17 The osculating circles at $x = 0$ and $x = 3/2$.

$$\left(x + \frac{27}{2}\right)^2 + \left(y - \frac{29}{4}\right)^2 = 250$$

■

13.4 SUMMARY

- A parametrization $\mathbf{r}(t)$ is called *regular* if $\mathbf{r}'(t) \neq \mathbf{0}$ for all t. If $\mathbf{r}(t)$ is regular, we define the *unit tangent vector* $\mathbf{T}(t) = \dfrac{\mathbf{r}'(t)}{\|\mathbf{r}'(t)\|}$. We denote $\|\mathbf{r}'(t)\|$ by $v(t)$.

- *Curvature* is defined by $\kappa(s) = \left\|\dfrac{d\mathbf{T}}{ds}\right\|$, where $\mathbf{r}(s)$ is an arc length parametrization. It can be computed by $\kappa(t) = \dfrac{1}{v(t)}\left\|\dfrac{d\mathbf{T}}{dt}\right\|$ for any regular parametrization $\mathbf{r}(t)$.

- In practice, it is easier to compute curvature using the following formulas:

    - For a regular parametrization:

$$\kappa(t) = \frac{\|\mathbf{r}'(t) \times \mathbf{r}''(t)\|}{\|\mathbf{r}'(t)\|^3}$$

    - For a regular parametrization $\langle x(t), y(t) \rangle$ of a plane curve:

$$\kappa(t) = \frac{|x'(t)y''(t) - y'(t)x''(t)|}{\left(x'(t)^2 + y'(t)^2\right)^{3/2}}$$

    - At a point on a graph $y = f(x)$ in the plane:

$$\kappa(x) = \frac{|f''(x)|}{\left(1 + f'(x)^2\right)^{3/2}}$$

- If $\|\mathbf{T}'(t)\| \neq 0$, we define the *unit normal vector* $\mathbf{N}(t) = \dfrac{\mathbf{T}'(t)}{\|\mathbf{T}'(t)\|}$.

- $\mathbf{T}'(t)$ and the unit normal are also related by $\mathbf{T}'(t) = \kappa(t)v(t)\mathbf{N}(t)$

- The *binormal vector* at P is defined by $\mathbf{B} = \mathbf{T} \times \mathbf{N}$. It is defined only if the curvature at P is nonzero. Together $\mathbf{T}$, $\mathbf{N}$, and $\mathbf{B}$ form the Frenet frame at P.

- The *osculating plane* at a point P on a curve $\mathcal{C}$ is the plane through P determined by the vectors $\mathbf{T}$ and $\mathbf{N}$ at P.

- The *osculating circle* Osc_P is the circle in the osculating plane through P of radius $R = 1/\kappa_P$ whose center Q lies in the normal direction $\mathbf{N}$ at P. If P corresponds to $\mathbf{r}(t_0)$, then the center is found via:

$$\overrightarrow{OQ} = \mathbf{r}(t_0) + \frac{1}{\kappa_P}\mathbf{N}$$

The center of Osc_P is called the *center of curvature* and R is called the *radius of curvature*.

13.4 EXERCISES

Preliminary Questions

1. What is the unit tangent vector of $\mathbf{r}(t)$ if the underlying curve is a line with direction vector $\mathbf{w} = \langle 2, 1, -2 \rangle$ and x is decreasing along $\mathbf{r}(t)$?

2. What is the curvature of a circle of radius 4?

3. Which has larger curvature, a circle of radius 2 or a circle of radius 4?

4. What is the curvature of $\mathbf{r}(t) = \langle 2 + 3t, 7t, 5 - t \rangle$?

5. What is the curvature at a point where $\mathbf{T}'(s) = \langle 1, 2, 3 \rangle$ in an arc length parametrization $\mathbf{r}(s)$?

6. What is the radius of curvature of a circle of radius 4?

7. What is the radius of curvature at P if $\kappa_P = 9$?

Exercises

In Exercises 1–6, calculate $\mathbf{r}'(t)$ *and* $\mathbf{T}(t)$, *and evaluate* $\mathbf{T}(1)$.

1. $\mathbf{r}(t) = \langle 4t^2, 9t \rangle$ **2.** $\mathbf{r}(t) = \langle e^t, t^2 \rangle$

3. $\mathbf{r}(t) = \langle 3 + 4t, 3 - 5t, 9t \rangle$ **4.** $\mathbf{r}(t) = \langle 1 + 2t, t^2, 3 - t^2 \rangle$

5. $\mathbf{r}(t) = \langle \cos \pi t, \sin \pi t, t \rangle$ **6.** $\mathbf{r}(t) = \langle e^t, e^{-t}, t^2 \rangle$

In Exercises 7–10, use Eq. (3) to calculate the curvature function $\kappa(t)$.

7. $\mathbf{r}(t) = \langle 1, e^t, t \rangle$ **8.** $\mathbf{r}(t) = \langle 4 \cos t, t, 4 \sin t \rangle$

9. $\mathbf{r}(t) = \langle 4t + 1, 4t - 3, 2t \rangle$ **10.** $\mathbf{r}(t) = \langle t^{-1}, 1, t \rangle$

In Exercises 11–14, use Eq. (3) to evaluate the curvature at the given point.

11. $\mathbf{r}(t) = \langle 1/t, 1/t^2, t^2 \rangle$, $t = -1$

12. $\mathbf{r}(t) = \langle 3 - t, e^{t-4}, 8t - t^2 \rangle$, $t = 4$

13. $\mathbf{r}(t) = \langle \cos t, \sin t, t^2 \rangle$, $t = \frac{\pi}{2}$

14. $\mathbf{r}(t) = \langle \cosh t, \sinh t, t \rangle$, $t = 0$

In Exercises 15–18, find the curvature of the plane curve at the point indicated.

15. $y = e^t$, $t = 3$ **16.** $y = \cos x$, $x = 0$

17. $y = t^4$, $t = 2$ **18.** $y = t^n$, $t = 1$

19. Find the curvature of $\mathbf{r}(t) = \langle 2 \sin t, \cos 3t, t \rangle$ at $t = \frac{\pi}{3}$ and $t = \frac{\pi}{2}$.

20. [CAS] Find the curvature function $\kappa(x)$ for $y = \sin x$. Use a computer algebra system to plot $\kappa(x)$ for $0 \le x \le 2\pi$. Prove that the curvature takes its maximum at $x = \frac{\pi}{2}$ and $\frac{3\pi}{2}$. *Hint:* As a shortcut to finding the max, observe that the maximum of the numerator and the minimum of the denominator of $\kappa(x)$ occur at the same points.

21. Show that the tractrix $\mathbf{r}(t) = \langle t - \tanh t, \operatorname{sech} t \rangle$ has the curvature function $\kappa(t) = \operatorname{sech} t$.

22. Show that curvature at an inflection point of a plane curve $y = f(x)$ is zero.

23. Find the value(s) of α such that the curvature of $y = e^{\alpha x}$ at $x = 0$ is as large as possible.

24. Find the point of maximum curvature on $y = e^x$.

25. Show that the curvature function of the parametrization $\mathbf{r}(t) = \langle a \cos t, b \sin t \rangle$ of the ellipse $\left(\frac{x}{a} \right)^2 + \left(\frac{y}{b} \right)^2 = 1$ is

$$\kappa(t) = \frac{ab}{(b^2 \cos^2 t + a^2 \sin^2 t)^{3/2}} \qquad \boxed{11}$$

26. Use a sketch to predict where the points of minimal and maximal curvature occur on an ellipse. Then use Eq. (11) to confirm or refute your prediction.

27. In the notation of Exercise 25, assume that $a \ge b$. Show that $b/a^2 \le \kappa(t) \le a/b^2$ for all t.

28. Use Eq. (5) and a parametrization $x = t$ and $y = f(t)$ to prove that the curvature of the graph of $y = f(x)$ is given by

$$\kappa(x) = \frac{|f''(x)|}{\left(1 + f'(x)^2\right)^{3/2}}$$

In Exercises 29–32, use Eq. (5) to compute the curvature at the given point.

29. $\langle t^2, t^3 \rangle$, $t = 2$ **30.** $\langle \cosh s, s \rangle$, $s = 0$

31. $\langle t \cos t, \sin t \rangle$, $t = \pi$ **32.** $\langle \sin 3s, 2 \sin 4s \rangle$, $s = \frac{\pi}{2}$

33. Let $s(t) = \displaystyle\int_{-\infty}^{t} \|\mathbf{r}'(u)\| \, du$ where $\mathbf{r}(t) = \langle e^t \cos 4t, e^t \sin 4t \rangle$. Show that the radius of curvature is proportional to $s(t)$. (This curve is known as the Bernoulli spiral and was introduced in Exercise 39 in Section 13.3.)

34. The curve that is parametrized by $x(t) = \cos^3 t$ and $y(t) = \sin^3 t$, with $0 \le t \le 2\pi$, is called a hypocycloid (Figure 18).
(a) Show that the curvature is $\kappa(t) = \frac{1}{|3 \sin t \cos t|}$.
(b) What is the minimum curvature and where on the curve does it occur?
(c) For what t is the curvature undefined? At what points on the hypocycloid does that occur? Explain what is happening on the traced-out path at those points.

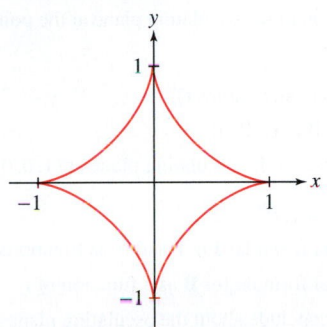

FIGURE 18 A hypocycloid.

35. [CAS] Plot the **clothoid** $\mathbf{r}(t) = \langle x(t), y(t) \rangle$, and compute its curvature $\kappa(t)$ where

$$x(t) = \int_0^t \sin \frac{u^3}{3} \, du, \qquad y(t) = \int_0^t \cos \frac{u^3}{3} \, du$$

36. Find the normal vector $\mathbf{N}(\theta)$ to $\mathbf{r}(\theta) = R \langle \cos \theta, \sin \theta \rangle$, the circle of radius R. Does $\mathbf{N}(\theta)$ point inside or outside the circle? Draw $\mathbf{N}(\theta)$ at $\theta = \frac{\pi}{4}$ with $R = 4$.

37. Find the normal vector $\mathbf{N}(t)$ to $\mathbf{r}(t) = \langle 4, \sin 2t, \cos 2t \rangle$.

38. Sketch the graph of $\mathbf{r}(t) = \langle t, t^3 \rangle$. Since $\mathbf{r}'(t) = \langle 1, 3t^2 \rangle$, the unit normal $\mathbf{N}(t)$ points in one of the two directions $\pm \langle -3t^2, 1 \rangle$. Which sign is correct at $t = 1$? Which is correct at $t = -1$?

39. Find the normal vectors to $\mathbf{r}(t) = \langle t, \cos t \rangle$ at $t = \frac{\pi}{4}$ and $t = \frac{3\pi}{4}$.

40. Find the normal vector to the Cornu spiral (Example 3) at $t = \sqrt{\pi}$.

In Exercises 41–44, find $\mathbf{T}, \mathbf{N},$ *and* $\mathbf{B}$ *for the curve at the indicated point. Hint: After finding* $\mathbf{T}'$, *plug in the specific value for t before computing* $\mathbf{N}$ *and* $\mathbf{B}$.

41. $\mathbf{r}(t) = \langle 0, t, t^2 \rangle$ at $(0, 1, 1)$. In this case, draw the curve and the three resultant vectors in 3-space.

42. $\mathbf{r}(t) = \langle \cos t, \sin t, 2 \rangle$ at $(1, 0, 2)$. In this case, draw the curve and the three resultant vectors in 3-space.

43. $\mathbf{r}(t) = \left\langle t, t^2, \frac{2}{3} t^3 \right\rangle$ at $(1, 1, \frac{2}{3})$ **44.** $\mathbf{r}(t) = \langle t, t, e^t \rangle$ at $(0, 0, 1)$

45. Find the normal vector to the clothoid (Exercise 35) at $t = \pi^{1/3}$.

46. Method for Computing N Let $v(t) = \|\mathbf{r}'(t)\|$. Show that

$$\mathbf{N}(t) = \frac{v(t)\mathbf{r}''(t) - v'(t)\mathbf{r}'(t)}{\|v(t)\mathbf{r}''(t) - v'(t)\mathbf{r}'(t)\|} \qquad \boxed{12}$$

Hint: **N** is the unit vector in the direction $\mathbf{T}'(t)$. Differentiate $\mathbf{T}(t) = \mathbf{r}'(t)/v(t)$ to show that $v(t)\mathbf{r}''(t) - v'(t)\mathbf{r}'(t)$ is a positive multiple of $\mathbf{T}'(t)$.

In Exercises 47–52, use Eq. (12) to find **N** *at the point indicated.*

47. $\langle t^2, t^3 \rangle$, $\quad t = 1$

48. $\langle t - \sin t, 1 - \cos t \rangle$, $\quad t = \pi$

49. $\langle t^2/2, t^3/3, t \rangle$, $\quad t = 1$ **50.** $\langle t^{-1}, t, t^2 \rangle$, $\quad t = -1$

51. $\langle t, e^t, t \rangle$, $\quad t = 0$ **52.** $\langle \cosh t, \sinh t, t^2 \rangle$, $\quad t = 0$

53. Let $\mathbf{r}(t) = \left\langle t, \frac{4}{3}t^{3/2}, t^2 \right\rangle$.

(a) Find $\mathbf{T}$, $\mathbf{N}$, and $\mathbf{B}$ at the point corresponding to $t = 1$.

(b) Find the equation of the osculating plane at the point corresponding to $t = 1$.

54. Let $\mathbf{r}(t) = \langle \cos t, \sin t, \ln(\cos t) \rangle$.

(a) Find $\mathbf{T}$, $\mathbf{N}$, and $\mathbf{B}$ at $(1, 0, 0)$.

(b) Find the equation of the osculating plane at $(1, 0, 0)$.

55. Let $\mathbf{r}(t) = \langle t, 1 - t, t^2 \rangle$.

(a) Find the general formulas for $\mathbf{T}$ and $\mathbf{N}$ as functions of t.

(b) Find the general formula for $\mathbf{B}$ as a function of t.

(c) What can you conclude about the osculating planes of the curve based on your answer to b?

56. (a) What does it mean for a space curve to have a constant unit tangent vector $\mathbf{T}$?

(b) What does it mean for a space curve to have a constant normal vector $\mathbf{N}$?

(c) What does it mean for a space curve to have a constant binormal vector $\mathbf{B}$?

57. Let $f(x) = x^2$. Show that the center of the osculating circle at (x_0, x_0^2) is given by $\left(-4x_0^3, \frac{1}{2} + 3x_0^2 \right)$.

58. Use Eq. (10) to find the center of curvature of $\mathbf{r}(t) = \langle t^2, t^3 \rangle$ at $t = 1$.

In Exercises 59–68, find an equation of the osculating circle at the point indicated or indicate that none exists.

59. $y = x^2$, $\quad x = 1$ **60.** $y = x^2$, $\quad x = 2$

61. $y = \sin x$, $\quad x = \frac{\pi}{2}$ **62.** $y = \sin x$, $\quad x = \pi$

63. $y = e^x$, $\quad x = 0$ **64.** $y = \ln x$, $\quad x = 1$

65. $\mathbf{r}(t) = \langle \cos t, \sin t \rangle$, $\quad t = \frac{\pi}{4}$ **66.** $\mathbf{r}(t) = \langle t^2, 1 - 2t^2 \rangle$, $\quad t = 2$

67. $\mathbf{r}(t) = \langle 1 - \sin t, 1 - 2\cos t \rangle$, $\quad t = \pi$

68. $\mathbf{r}(t) = \langle \cosh t, \sinh t \rangle$, $\quad t = 0$

69. Figure 19 shows the graph of the half-ellipse $y = \pm\sqrt{2rx - px^2}$, where r and p are positive constants. Show that the radius of curvature at the origin is equal to r. *Hint:* One way of proceeding is to write the ellipse in the form of Exercise 25 and apply Eq. (11).

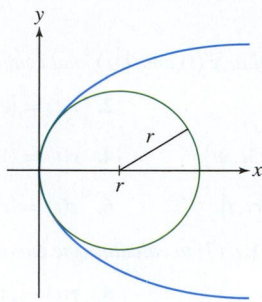

FIGURE 19 The curve $y = \pm\sqrt{2rx - px^2}$ and the osculating circle at the origin.

70. In a recent study of laser eye surgery by Gatinel, Hoang-Xuan, and Azar, a vertical cross section of the cornea is modeled by the half-ellipse of Exercise 69. Show that the half-ellipse can be written in the form $x = f(y)$, where $f(y) = p^{-1}\left(r - \sqrt{r^2 - py^2} \right)$. During surgery, tissue is removed to a depth $t(y)$ at height y for $-S \le y \le S$, where $t(y)$ is given by Munnerlyn's equation (for some $R > r$):

$$t(y) = \sqrt{R^2 - S^2} - \sqrt{R^2 - y^2} - \sqrt{r^2 - S^2} + \sqrt{r^2 - y^2}$$

After surgery, the cornea's cross section has the shape $x = f(y) + t(y)$ (Figure 20). Show that after surgery, the radius of curvature at the point P (where $y = 0$) is R.

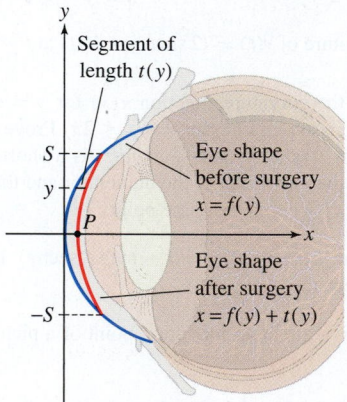

FIGURE 20 Contour of cornea before and after surgery.

71. The **angle of inclination** at a point P on a plane curve is the angle θ between the unit tangent vector $\mathbf{T}$ and the x-axis (Figure 21). Assume that $\mathbf{r}(s)$ is an arc length parametrization, and let $\theta = \theta(s)$ be the angle of inclination at $\mathbf{r}(s)$. Prove that

$$\kappa(s) = \left| \frac{d\theta}{ds} \right| \qquad \boxed{13}$$

Hint: Observe that $\mathbf{T}(s) = \langle \cos \theta(s), \sin \theta(s) \rangle$.

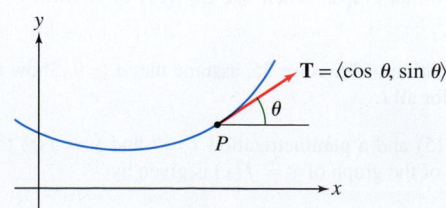

FIGURE 21 The curvature at P is the quantity $|d\theta/ds|$.

72. A particle moves along the path $y = x^3$ with unit speed. How fast is the tangent turning (i.e., how fast is the angle of inclination changing) when the particle passes through the point $(2, 8)$?

73. Let $\theta(x)$ be the angle of inclination at a point on the graph $y = f(x)$ (see Exercise 71).

(a) Use the relation $f'(x) = \tan\theta$ to prove that $\dfrac{d\theta}{dx} = \dfrac{f''(x)}{(1 + f'(x)^2)}$.

(b) Use the arc length integral to show that $\dfrac{ds}{dx} = \sqrt{1 + f'(x)^2}$.

(c) Now give a proof of Eq. (6) using Eq. (13).

74. Use the parametrization $\mathbf{r}(\theta) = \langle f(\theta)\cos\theta, f(\theta)\sin\theta \rangle$ to show that a curve $r = f(\theta)$ in polar coordinates has curvature

$$\kappa(\theta) = \frac{|f(\theta)^2 + 2f'(\theta)^2 - f(\theta)f''(\theta)|}{\left(f(\theta)^2 + f'(\theta)^2\right)^{3/2}} \qquad \boxed{14}$$

In Exercises 75–77, use Eq. (14) to find the curvature of the curve given in polar form.

75. $f(\theta) = 2\cos\theta$ **76.** $f(\theta) = \theta$ **77.** $f(\theta) = e^\theta$

78. Use Eq. (14) to find the curvature of the general Bernoulli spiral $r = ae^{b\theta}$ in polar form (a and b are constants).

Further Insights and Challenges

83. Viviani's curve is given by $\mathbf{r}(t) = \langle 1 + \cos t, \sin t, 2\sin(t/2) \rangle$. Show that its curvature is

$$\kappa(t) = \frac{\sqrt{13 + 3\cos t}}{(3 + \cos t)^{3/2}}$$

84. Let $\mathbf{r}(s)$ be an arc length parametrization of a closed curve $\mathcal{C}$ of length L. We call $\mathcal{C}$ an **oval** if $d\theta/ds > 0$ (see Exercise 71). Observe that $-\mathbf{N}$ points to the *outside* of $\mathcal{C}$. For $k > 0$, the curve $\mathcal{C}_1$ defined by $\mathbf{r}_1(s) = \mathbf{r}(s) - k\mathbf{N}$ is called the expansion of $c(s)$ in the normal direction.

(a) Show that $\|\mathbf{r}_1'(s)\| = \|\mathbf{r}'(s)\| + k\kappa(s)$.

(b) As P moves around the oval counterclockwise, θ increases by 2π [Figure 22(A)]. Use this and a change of variables to prove that $\displaystyle\int_0^L \kappa(s)\, ds = 2\pi$.

(c) Show that $\mathcal{C}_1$ has length $L + 2\pi k$.

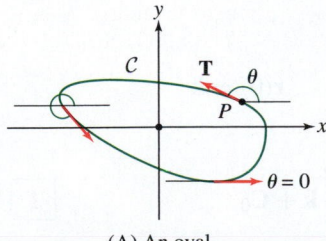

(A) An oval

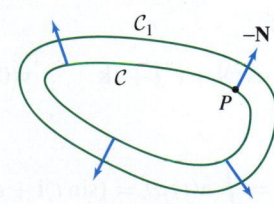

(B) $\mathcal{C}_1$ is the expansion of $\mathcal{C}$ in the normal direction.

FIGURE 22 As P moves around the oval, θ increases by 2π.

In Exercises 85–93, we investigate the binormal vector further.

85. Let $\mathbf{r}(t) = \langle x(t), y(t), 0 \rangle$. Assuming that $\mathbf{T}(t) \times \mathbf{N}(t)$ is nonzero, there are two possibilities for the vector $\mathbf{B}(t)$. What are they? Explain.

79. Show that both $\mathbf{r}'(t)$ and $\mathbf{r}''(t)$ lie in the osculating plane for a vector function $\mathbf{r}(t)$. *Hint:* Differentiate $\mathbf{r}'(t) = v(t)\mathbf{T}(t)$.

80. Show that

$$\gamma(s) = \mathbf{r}(t_0) + \frac{1}{\kappa}\mathbf{N} + \frac{1}{\kappa}\big((\sin\kappa s)\mathbf{T} - (\cos\kappa s)\mathbf{N}\big)$$

is an arc length parametrization of the osculating circle at $\mathbf{r}(t_0)$.

81. Two vector-valued functions $\mathbf{r}_1(s)$ and $\mathbf{r}_2(s)$ are said to *agree to order 2* at s_0 if

$$\mathbf{r}_1(s_0) = \mathbf{r}_2(s_0), \quad \mathbf{r}_1'(s_0) = \mathbf{r}_2'(s_0), \quad \mathbf{r}_1''(s_0) = \mathbf{r}_2''(s_0)$$

Let $\mathbf{r}(s)$ be an arc length parametrization of a curve $\mathcal{C}$, and let P be the terminal point of $\mathbf{r}(0)$. Let $\gamma(s)$ be the arc length parametrization of the osculating circle given in Exercise 80. Show that $\mathbf{r}(s)$ and $\gamma(s)$ agree to order 2 at $s = 0$ (in fact, the osculating circle is the unique circle that approximates $\mathcal{C}$ to order 2 at P).

82. Let $\mathbf{r}(t) = \langle x(t), y(t), z(t) \rangle$ be a path with curvature $\kappa(t)$ and define the scaled path $\mathbf{r}_1(t) = \langle \lambda x(t), \lambda y(t), \lambda z(t) \rangle$, where $\lambda \neq 0$ is a constant. Prove that curvature varies inversely with the scale factor. That is, prove that the curvature $\kappa_1(t)$ of $\mathbf{r}_1(t)$ is $\kappa_1(t) = \lambda^{-1}\kappa(t)$. This explains why the curvature of a circle of radius R is proportional to $1/R$ (in fact, it is equal to $1/R$). *Hint:* Use Eq. (3).

86. Follow steps (a)–(c) to prove that there is a number τ called the **torsion** such that

$$\frac{d\mathbf{B}}{ds} = -\tau\mathbf{N} \qquad \boxed{15}$$

(a) Show that $\dfrac{d\mathbf{B}}{ds} = \mathbf{T} \times \dfrac{d\mathbf{N}}{ds}$ and conclude that $d\mathbf{B}/ds$ is orthogonal to $\mathbf{T}$.

(b) Differentiate $\mathbf{B} \cdot \mathbf{B} = 1$ with respect to s to show that $d\mathbf{B}/ds$ is orthogonal to $\mathbf{B}$.

(c) Conclude that $d\mathbf{B}/ds$ is a multiple of $\mathbf{N}$.

87. Show that if $\mathcal{C}$ is contained in a plane $\mathcal{P}$, then $\mathbf{B}$ is a unit vector normal to $\mathcal{P}$. Conclude that $\tau = 0$ for a plane curve.

88. 🖊 Torsion means twisting. Is this an appropriate name for τ? Explain by interpreting τ geometrically.

89. Use the identity

$$\mathbf{a} \times (\mathbf{b} \times \mathbf{c}) = (\mathbf{a} \cdot \mathbf{c})\mathbf{b} - (\mathbf{a} \cdot \mathbf{b})\mathbf{c}$$

to prove

$$\mathbf{N} \times \mathbf{B} = \mathbf{T}, \qquad \mathbf{B} \times \mathbf{T} = \mathbf{N} \qquad \boxed{16}$$

90. Follow steps (a)–(b) to prove

$$\frac{d\mathbf{N}}{ds} = -\kappa\mathbf{T} + \tau\mathbf{B} \qquad \boxed{17}$$

(a) Show that $d\mathbf{N}/ds$ is orthogonal to $\mathbf{N}$. Conclude that $d\mathbf{N}/ds$ lies in the plane spanned by $\mathbf{T}$ and $\mathbf{B}$, and hence, $d\mathbf{N}/ds = a\mathbf{T} + b\mathbf{B}$ for some scalars a, b.

(b) Use $\mathbf{N} \cdot \mathbf{T} = 0$ to show that $\mathbf{T} \cdot \dfrac{d\mathbf{N}}{ds} = -\mathbf{N} \cdot \dfrac{d\mathbf{T}}{ds}$ and compute a. Compute b similarly. Equations (15) and (17) together with $d\mathbf{T}/dt = \kappa\mathbf{N}$ are called the **Frenet formulas**.

91. Show that $\mathbf{r}' \times \mathbf{r}''$ is a multiple of $\mathbf{B}$. Conclude that

$$\mathbf{B} = \frac{\mathbf{r}' \times \mathbf{r}''}{\|\mathbf{r}' \times \mathbf{r}''\|} \qquad \boxed{18}$$

92. Use the formula from the preceding problem to find **B** for the space curve given by $\mathbf{r}(t) = \langle \sin t, -\cos t, \sin t \rangle$. Conclude that the space curve lies in a plane.

93. The vector **N** can be computed using $\mathbf{N} = \mathbf{B} \times \mathbf{T}$ [Eq. (16)] with **B**, as in Eq. (18). Use this method to find **N** in the following cases:

(a) $\mathbf{r}(t) = \langle \cos t, t, t^2 \rangle$ at $t = 0$

(b) $\mathbf{r}(t) = \langle t^2, t^{-1}, t \rangle$ at $t = 1$

FIGURE 1 The trajectory of a comet is analyzed using vector calculus.

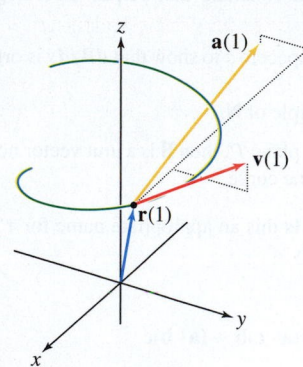

FIGURE 2

13.5 Motion in 3-Space

In this section, we study the motion of an object traveling along a path $\mathbf{r}(t)$. That object could be a variety of things, including a particle, a baseball, or comet Hale-Bopp (Figure 1). Recall that the velocity vector is the derivative

$$\mathbf{v}(t) = \mathbf{r}'(t) = \lim_{h \to 0} \frac{\mathbf{r}(t + h) - \mathbf{r}(t)}{h}$$

As we have seen, $\mathbf{v}(t)$ points in the direction of motion (if it is nonzero), and its magnitude $v(t) = \|\mathbf{v}(t)\|$ is the object's speed. The **acceleration vector** is the second derivative $\mathbf{r}''(t)$, which we shall denote as $\mathbf{a}(t)$. In summary, from $\mathbf{r}(t)$, we have

$$\boxed{\mathbf{v}(t) = \mathbf{r}'(t), \qquad v(t) = \|\mathbf{v}(t)\|, \qquad \mathbf{a}(t) = \mathbf{r}''(t)}$$

EXAMPLE 1 Calculate and plot the velocity and acceleration vectors at $t = 1$ of $\mathbf{r}(t) = \langle \sin 2t, -\cos 2t, \sqrt{t+1} \rangle$. Then find the speed at $t = 1$ (Figure 2).

Solution

$$\mathbf{v}(t) = \mathbf{r}'(t) = \left\langle 2\cos 2t, 2\sin 2t, \frac{1}{2}(t+1)^{-1/2} \right\rangle, \qquad \mathbf{v}(1) \approx \langle -0.83, 1.82, 0.35 \rangle$$

$$\mathbf{a}(t) = \mathbf{r}''(t) = \left\langle -4\sin 2t, 4\cos 2t, -\frac{1}{4}(t+1)^{-3/2} \right\rangle, \qquad \mathbf{a}(1) \approx \langle -3.64, -1.66, -0.089 \rangle$$

The speed at $t = 1$ is

$$\|\mathbf{v}(1)\| \approx \sqrt{(-0.83)^2 + (1.82)^2 + (0.35)^2} \approx 2.03 \qquad \blacksquare$$

If an object's acceleration is given, we can solve for $\mathbf{v}(t)$ and $\mathbf{r}(t)$ by integrating twice:

$$\mathbf{v}(t) = \int \mathbf{a}(t)\, dt \quad \text{and then} \quad \mathbf{r}(t) = \int \mathbf{v}(t)\, dt$$

Arbitrary constants arise in each of these antiderivatives. To determine specific functions $\mathbf{v}(t)$ and $\mathbf{r}(t)$, initial conditions need to be provided.

EXAMPLE 2 Find $\mathbf{r}(t)$ if

$$\mathbf{a}(t) = (\cos t)\mathbf{i} + e^t \mathbf{j} + t\mathbf{k}, \qquad \mathbf{v}(0) = \mathbf{i}, \qquad \mathbf{r}(0) = \mathbf{k}$$

Solution We obtain

$$\mathbf{v}(t) = \int \mathbf{a}(t)\, dt = (\sin t)\mathbf{i} + e^t \mathbf{j} + \frac{t^2}{2}\mathbf{k} + \mathbf{C}_0 \qquad \boxed{1}$$

With the initial condition for **v**, we can determine $\mathbf{C}_0$. Specifically, from the initial condition and from Eq. (1) we have, respectively, the following two expressions for $\mathbf{v}(0)$:

$$\mathbf{v}(0) = \mathbf{i} \quad \text{and} \quad \mathbf{v}(0) = \mathbf{j} + \mathbf{C}_0$$

It follows that $\mathbf{i} = \mathbf{j} + \mathbf{C}_0$, and therefore $\mathbf{C}_0 = \mathbf{i} - \mathbf{j}$. Thus,

$$\mathbf{v}(t) = (\sin t)\mathbf{i} + e^t \mathbf{j} + \frac{t^2}{2}\mathbf{k} + \mathbf{i} - \mathbf{j} = (\sin t + 1)\mathbf{i} + (e^t - 1)\mathbf{j} + \frac{t^2}{2}\mathbf{k}$$

CAUTION While the initial condition determines the constant $\mathbf{C}_0$ that arises in the antiderivative, the constant is not necessarily equal to the value in the initial condition. To find $\mathbf{C}_0$, we need to find two expressions for $\mathbf{v}(0)$ and use them to solve for $\mathbf{C}_0$.

Now, taking another antiderivative, we obtain

$$\mathbf{r}(t) = \int \mathbf{v}(t)\,dt = (-\cos t + t)\mathbf{i} + (e^t - t)\mathbf{j} + \frac{t^3}{6}\mathbf{k} + \mathbf{C}_1 \qquad \boxed{2}$$

The initial condition and Eq. (2) provide the following two expressions for $\mathbf{r}(0)$:

$$\mathbf{r}(0) = \mathbf{k} \quad \text{and} \quad \mathbf{r}(0) = -\mathbf{i} + \mathbf{j} + \mathbf{C}_1$$

This implies that $\mathbf{k} = -\mathbf{i} + \mathbf{j} + \mathbf{C}_1$, and therefore $\mathbf{C}_1 = \mathbf{i} - \mathbf{j} + \mathbf{k}$. We now have

$$\mathbf{r}(t) = (-\cos t + t)\mathbf{i} + (e^t - t)\mathbf{j} + \frac{t^3}{6}\mathbf{k} + \mathbf{i} - \mathbf{j} + \mathbf{k}$$

$$= (-\cos t + t + 1)\mathbf{i} + (e^t - t - 1)\mathbf{j} + \left(\frac{t^3}{6} + 1\right)\mathbf{k} \qquad \blacksquare$$

Near the surface of the earth, gravity imparts an acceleration of approximately $9.8 \text{ m/s}^2 \approx 32 \text{ ft/s}^2$ in the downward direction. This means that if we have a projectile moving near the surface of the earth that has no additional means of acquiring acceleration, we know that its acceleration vector $\mathbf{a}(t)$ is determined by gravity. When a projectile's motion occurs within a vertical plane, we can model the motion in the xy-plane, using x for the horizontal motion and y for the vertical. In that case, we use $\mathbf{a}(t) = -9.8\mathbf{j} \text{ m/s}^2$ or $\mathbf{a}(t) = -32\mathbf{j} \text{ ft/s}^2$.

On the moon, the acceleration due to gravity is 1.6 m/s^2. During his moonwalk, *Apollo 14* astronaut Alan Shepard used a makeshift club to hit a golf ball. It went "miles and miles and miles," he said. In Exercise 21, we explore how far it might have gone.

EXAMPLE 3 A projectile is launched from the ground at angle θ with a speed of 100 ft/s. Show that the projectile lands at a distance of $625 \sin\theta \cos\theta$ ft from the launch point. (This launch-to-landing distance is called the **range** of the projectile.)

Solution Assume that the launch point is at the origin, and let $\mathbf{r}(t)$ be the position vector. For an initial condition, we have $\mathbf{r}(0) = \mathbf{0}$. Furthermore, since the launch speed is 100 ft/s with an angle of θ, we have the initial condition $\mathbf{v}(0) = 100\cos\theta\,\mathbf{i} + 100\sin\theta\,\mathbf{j}$ (Figure 3).

We assume that we have an acceleration vector $\mathbf{a}(t) = -32\mathbf{j}$. We determine $\mathbf{r}(t)$ by integrating twice:

$$\mathbf{v}(t) = \int \mathbf{a}(t)\,dt = -32t\mathbf{j} + \mathbf{C}_0 \qquad \boxed{3}$$

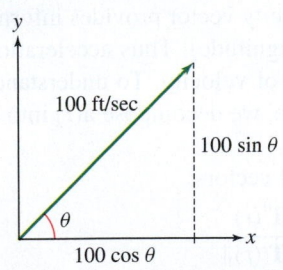

FIGURE 3 The launch velocity is $\mathbf{v}(0) = 100\cos\theta\,\mathbf{i} + 100\sin\theta\,\mathbf{j}$.

From the initial condition and from Eq. (3), we have

$$\mathbf{v}(0) = 100\cos\theta\,\mathbf{i} + 100\sin\theta\,\mathbf{j} \quad \text{and} \quad \mathbf{v}(0) = \mathbf{C}_0$$

Hence, $\mathbf{C}_0 = 100\cos\theta\,\mathbf{i} + 100\sin\theta\,\mathbf{j}$.

Therefore,

$$\mathbf{v}(t) = 100\cos\theta\,\mathbf{i} + (100\sin\theta - 32t)\mathbf{j}$$

Integrating again:

$$\mathbf{r}(t) = \int \mathbf{v}(t)\,dt = (100\cos\theta)t\mathbf{i} + ((100\sin\theta)t - 16t^2)\mathbf{j} + \mathbf{C}_1 \qquad \boxed{4}$$

Eq. (4) implies that $\mathbf{r}(0) = \mathbf{C}_1$. From the initial condition $\mathbf{r}(0) = \mathbf{0}$, we get $\mathbf{C}_1 = \mathbf{0}$. Therefore,

$$\mathbf{r}(t) = (100\cos\theta)t\mathbf{i} + ((100\sin\theta)t - 16t^2)\mathbf{j}$$

Now, the projectile is at ground level when the y-component of $\mathbf{r}(t)$ is zero. Therefore, we solve

$$(100\sin\theta)t - 16t^2 = 0$$

$$t(100\sin\theta - 16t) = 0$$

In Example 3 in Section 3.6, we showed that in this case, the maximum range occurs when $\theta = \pi/4$. In fact, given any initial speed, the maximum range occurs with $\theta = \pi/4$; see Exercise 23.

We have solutions at $t = 0$ (when the projectile was launched) and at $t = \frac{100 \sin \theta}{16} = \frac{25 \sin \theta}{4}$ (when the projectile returns to the ground). The distance from the launch point to the landing point is the x-component of $\mathbf{r}(t)$ evaluated at the time of landing. That is, the projectile lands at $(100 \cos \theta) \left(\frac{25 \sin \theta}{4} \right) = 625 \cos \theta \sin \theta$ ft from the launch point.   ■

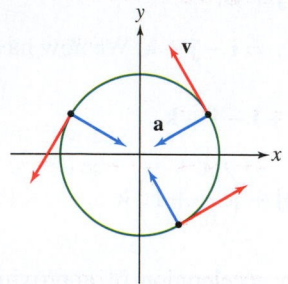

FIGURE 4 In uniform circular motion, $\mathbf{v}$ has constant length but turns continuously. The acceleration $\mathbf{a}$ is centripetal, pointing toward the center of the circle.

In general, acceleration is the rate of change of velocity with respect to time. In linear motion, acceleration is zero if the speed is constant. By contrast, in two or three dimensions, the acceleration can be nonzero even when the object's speed is constant. This happens when $v(t) = \|\mathbf{v}(t)\|$ is constant but the *direction* of $\mathbf{v}(t)$ is changing. The simplest example is **uniform circular motion**, in which an object travels in a circular path at constant speed (Figure 4).

EXAMPLE 4   Uniform Circular Motion   Find $\mathbf{a}(t)$ and $\|\mathbf{a}(t)\|$ for the motion of a particle around a circle of radius R with constant speed v.

Solution   Assume that the particle follows the circular path $\mathbf{r}(t) = R \langle \cos \omega t, \sin \omega t \rangle$ for some constant ω. Then the velocity and speed of the particle are

The constant ω (lowercase Greek omega) is called the angular speed because the particle's angle along the circle changes at a rate of ω radians per unit time.

$$\mathbf{v}(t) = R\omega \langle -\sin \omega t, \cos \omega t \rangle, \qquad v = \|\mathbf{v}(t)\| = R|\omega|$$

Thus, $|\omega| = v/R$; accordingly,

$$\mathbf{a}(t) = \mathbf{v}'(t) = -R\omega^2 \langle \cos \omega t, \sin \omega t \rangle, \qquad \|\mathbf{a}(t)\| = R\omega^2 = R \left(\frac{v}{R} \right)^2 = \frac{v^2}{R}$$

The vector $\mathbf{a}(t)$ is called the **centripetal acceleration**: It has length v^2/R and points toward the center of the circle, in this case the origin [because $\mathbf{a}(t)$ is a negative multiple of the position vector $\mathbf{r}(t)$], as in Figure 4.   ■

Understanding the Acceleration Vector

Acceleration is the rate of change of velocity, and a velocity vector provides information about the direction of motion and the speed (via its magnitude). Thus acceleration can involve change in either the direction or the magnitude of velocity. To understand how the acceleration vector $\mathbf{a}(t)$ encodes both types of change, we decompose $\mathbf{a}(t)$ into a sum of tangential and normal components.

Recall the definition of unit tangent and unit normal vectors:

$$\mathbf{T}(t) = \frac{\mathbf{v}(t)}{\|\mathbf{v}(t)\|}, \qquad \mathbf{N}(t) = \frac{\mathbf{T}'(t)}{\|\mathbf{T}'(t)\|}$$

Thus, $\mathbf{v}(t) = v(t)\mathbf{T}(t)$, where $v(t) = \|\mathbf{v}(t)\|$, so by the Scalar Product Rule,

When you make a left turn in an automobile at constant speed, your tangential acceleration is zero [because $v'(t) = 0$] and you will not be pushed back against your seat. But the car seat (via friction) pushes you to the left toward the car door, causing you to accelerate in the normal direction. Due to inertia, you feel as if you are being pushed to the right toward the passenger's seat. This force is proportional to κv^2, so a sharp turn (large κ) or high speed (large v) produces a strong normal force.

$$\mathbf{a}(t) = \frac{d\mathbf{v}}{dt} = \frac{d}{dt} v(t)\mathbf{T}(t) = v'(t)\mathbf{T}(t) + v(t)\mathbf{T}'(t)$$

Furthermore, $\mathbf{T}'(t) = v(t)\kappa(t)\mathbf{N}(t)$ by Eq. (8) of Section 13.4, where $\kappa(t)$ is the curvature. Thus, we can write

$$\boxed{\mathbf{a} = a_{\mathbf{T}}\mathbf{T} + a_{\mathbf{N}}\mathbf{N}, \qquad a_{\mathbf{T}} = v'(t), \qquad a_{\mathbf{N}} = \kappa(t)v(t)^2}$$ 5

The coefficient $a_{\mathbf{T}}(t)$ is called the **tangential component** and $a_{\mathbf{N}}(t)$ the **normal component** of acceleration (Figure 5).

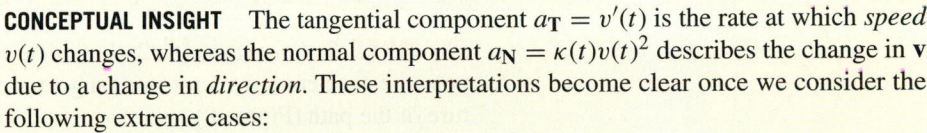

FIGURE 5 Decomposition of **a** into tangential and normal components.

*The normal component a_N is often called the **centripetal acceleration**. In the case of uniform circular motion it is directed toward the center of the circle.*

CONCEPTUAL INSIGHT The tangential component $a_T = v'(t)$ is the rate at which *speed* $v(t)$ changes, whereas the normal component $a_N = \kappa(t)v(t)^2$ describes the change in **v** due to a change in *direction*. These interpretations become clear once we consider the following extreme cases:

- A particle travels in a straight line. Then direction does not change [$\kappa(t) = 0$] and $\mathbf{a}(t) = v'(t)\mathbf{T}$ is parallel to the direction of motion.
- A particle travels with constant speed along a curved path. Then $v'(t) = 0$ and the acceleration vector $\mathbf{a}(t) = \kappa(t)v(t)^2\mathbf{N}$ is normal to the direction of motion.

General motion combines both tangential and normal acceleration.

EXAMPLE 5 The Giant Ferris Wheel in Vienna has radius $R = 30$ m (Figure 6). Assume that at time $t = t_0$, a person in a seat at the bottom of the wheel has a speed of 40 m/min that is slowing at a rate of 15 m/min^2. Find the acceleration vector **a** for the person.

Solution At the bottom of the wheel, $\mathbf{T} = \langle 1, 0 \rangle$ and $\mathbf{N} = \langle 0, 1 \rangle$. We are told that $a_T = v' = -15$ at time t_0. The curvature of the wheel is $\kappa = 1/R = 1/30$, so the normal component is $a_N = \kappa v^2 = v^2/R = (40)^2/30 \approx 53.3$. Therefore (Figure 7),

$$\mathbf{a} \approx -15\mathbf{T} + 53.3\mathbf{N} = \langle -15, 53.3 \rangle \ \text{m/min}^2 \qquad ■$$

FIGURE 6 The Giant Ferris Wheel in Vienna, Austria, erected in 1897 to celebrate the 50th anniversary of the coronation of Emperor Franz Joseph I.

The following theorem provides useful formulas for the tangential and normal components.

> **THEOREM 1 Tangential and Normal Components of Acceleration** In the decomposition $\mathbf{a} = a_T\mathbf{T} + a_N\mathbf{N}$, we have
>
> $$a_T = \mathbf{a} \cdot \mathbf{T} = \frac{\mathbf{a} \cdot \mathbf{v}}{\|\mathbf{v}\|}, \qquad a_N = \mathbf{a} \cdot \mathbf{N} = \sqrt{\|\mathbf{a}\|^2 - |a_T|^2} \qquad \boxed{6}$$
>
> and
>
> $$a_T\mathbf{T} = \left(\frac{\mathbf{a} \cdot \mathbf{v}}{\mathbf{v} \cdot \mathbf{v}} \right)\mathbf{v}, \qquad a_N\mathbf{N} = \mathbf{a} - a_T\mathbf{T} = \mathbf{a} - \left(\frac{\mathbf{a} \cdot \mathbf{v}}{\mathbf{v} \cdot \mathbf{v}} \right)\mathbf{v} \qquad \boxed{7}$$

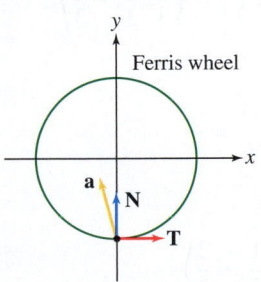

FIGURE 7

Proof To begin, note that $\mathbf{T} \cdot \mathbf{T} = 1$ and $\mathbf{N} \cdot \mathbf{T} = 0$. Thus,

$$\mathbf{a} \cdot \mathbf{T} = (a_T\mathbf{T} + a_N\mathbf{N}) \cdot \mathbf{T} = a_T$$

$$\mathbf{a} \cdot \mathbf{N} = (a_T\mathbf{T} + a_N\mathbf{N}) \cdot \mathbf{N} = a_N$$

and since $\mathbf{T} = \dfrac{\mathbf{v}}{\|\mathbf{v}\|}$, we have

$$a_T\mathbf{T} = (\mathbf{a} \cdot \mathbf{T})\mathbf{T} = \left(\frac{\mathbf{a} \cdot \mathbf{v}}{\|\mathbf{v}\|} \right)\frac{\mathbf{v}}{\|\mathbf{v}\|} = \left(\frac{\mathbf{a} \cdot \mathbf{v}}{\mathbf{v} \cdot \mathbf{v}} \right)\mathbf{v}$$

and

$$a_{\mathbf{N}} \mathbf{N} = \mathbf{a} - a_{\mathbf{T}} \mathbf{T} = \mathbf{a} - \left(\frac{\mathbf{a} \cdot \mathbf{v}}{\mathbf{v} \cdot \mathbf{v}} \right) \mathbf{v}$$

Finally, the vectors $a_{\mathbf{T}} \mathbf{T}$ and $a_{\mathbf{N}} \mathbf{N}$ are the sides of a right triangle with hypotenuse $\mathbf{a}$ as in Figure 5, so by the Pythagorean Theorem,

$$\|\mathbf{a}\|^2 = |a_{\mathbf{T}}|^2 + |a_{\mathbf{N}}|^2 \quad \Rightarrow \quad a_{\mathbf{N}} = \sqrt{\|\mathbf{a}\|^2 - |a_{\mathbf{T}}|^2} \qquad \blacksquare$$

Keep in mind that $a_{\mathbf{N}} \geq 0$ but $a_{\mathbf{T}}$ is positive or negative, depending on whether the object is speeding up or slowing down along the curve.

EXAMPLE 6 For $\mathbf{r}(t) = \langle t^2, 2t, \ln t \rangle$, determine the acceleration $\mathbf{a}(t)$. At $t = \frac{1}{2}$, decompose the acceleration vector into tangential and normal components, and find the curvature of the path (Figure 8).

Solution First, we compute the tangential components $\mathbf{T}$ and $a_{\mathbf{T}}$. We have

$$\mathbf{v}(t) = \mathbf{r}'(t) = \langle 2t, 2, t^{-1} \rangle, \qquad \mathbf{a}(t) = \mathbf{r}''(t) = \langle 2, 0, -t^{-2} \rangle$$

At $t = \frac{1}{2}$,

$$\mathbf{v} = \mathbf{r}'\left(\frac{1}{2}\right) = \left\langle 2\left(\frac{1}{2}\right), 2, \left(\frac{1}{2}\right)^{-1} \right\rangle = \langle 1, 2, 2 \rangle$$

$$\mathbf{a} = \mathbf{r}''\left(\frac{1}{2}\right) = \left\langle 2, 0, -\left(\frac{1}{2}\right)^{-2} \right\rangle = \langle 2, 0, -4 \rangle$$

Thus,

$$\mathbf{T} = \frac{\mathbf{v}}{\|\mathbf{v}\|} = \frac{\langle 1, 2, 2 \rangle}{\sqrt{1^2 + 2^2 + 2^2}} = \left\langle \frac{1}{3}, \frac{2}{3}, \frac{2}{3} \right\rangle$$

and by Eq. (6),

$$a_{\mathbf{T}} = \mathbf{a} \cdot \mathbf{T} = \langle 2, 0, -4 \rangle \cdot \left\langle \frac{1}{3}, \frac{2}{3}, \frac{2}{3} \right\rangle = -2$$

Next, we use Eq. (7):

$$a_{\mathbf{N}} \mathbf{N} = \mathbf{a} - a_{\mathbf{T}} \mathbf{T} = \langle 2, 0, -4 \rangle - (-2)\left\langle \frac{1}{3}, \frac{2}{3}, \frac{2}{3} \right\rangle = \left\langle \frac{8}{3}, \frac{4}{3}, -\frac{8}{3} \right\rangle$$

This vector has length

$$a_{\mathbf{N}} = \|a_{\mathbf{N}} \mathbf{N}\| = \sqrt{\frac{64}{9} + \frac{16}{9} + \frac{64}{9}} = 4$$

and thus,

$$\mathbf{N} = \frac{a_{\mathbf{N}} \mathbf{N}}{a_{\mathbf{N}}} = \frac{\left\langle \frac{8}{3}, \frac{4}{3}, -\frac{8}{3} \right\rangle}{4} = \left\langle \frac{2}{3}, \frac{1}{3}, -\frac{2}{3} \right\rangle$$

Finally, we obtain the decomposition

$$\mathbf{a} = \langle 2, 0, -4 \rangle = a_{\mathbf{T}} \mathbf{T} + a_{\mathbf{N}} \mathbf{N} = -2\mathbf{T} + 4\mathbf{N}$$

Now, since $a_{\mathbf{N}} = 4$ at $t = \frac{1}{2}$, and we know $a_{\mathbf{N}} = \kappa v^2$ from Eq. (5), to obtain the curvature at $t = \frac{1}{2}$ divide 4 by the square of the speed. With $\mathbf{v} = \langle 1, 2, 2 \rangle$ at $t = \frac{1}{2}$, we have $v^2 = 9$, and therefore $\kappa(1/2) = 4/9$. $\qquad \blacksquare$

EXAMPLE 7   Nonuniform Circular Motion Figure 9 shows the acceleration vectors of three particles moving *counterclockwise* around a circle. In each case, state whether the particle's speed v around the circle is increasing, decreasing, or momentarily constant.

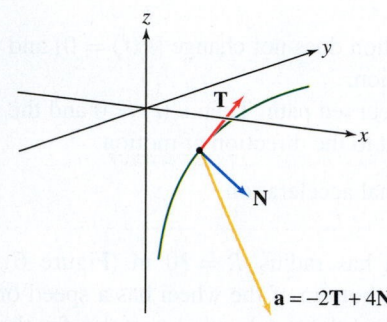

DF **FIGURE 8** The vectors $\mathbf{T}$, $\mathbf{N}$, and $\mathbf{a}$ at $t = \frac{1}{2}$ on the curve given by $\mathbf{r}(t) = \langle t^2, 2t, \ln t \rangle$.

Summary of steps in Example 6:

$$\mathbf{T} = \frac{\mathbf{v}}{\|\mathbf{v}\|}$$

$$a_{\mathbf{T}} = \mathbf{a} \cdot \mathbf{T}$$

$$a_{\mathbf{N}} \mathbf{N} = \mathbf{a} - a_{\mathbf{T}} \mathbf{T}$$

$$a_{\mathbf{N}} = \|a_{\mathbf{N}} \mathbf{N}\|$$

$$\mathbf{N} = \frac{a_{\mathbf{N}} \mathbf{N}}{a_{\mathbf{N}}}$$

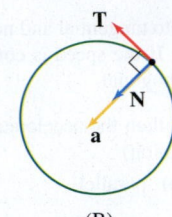
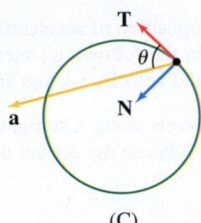

(A)   (B)   (C)

FIGURE 9 Acceleration vectors of particles moving counterclockwise (in the direction of **T**) around a circle.

Solution The rate of change of speed depends on the angle θ between **a** and **T**:

$$v' = a_{\mathbf{T}} = \mathbf{a} \cdot \mathbf{T} = \|\mathbf{a}\| \, \|\mathbf{T}\| \cos\theta = \|\mathbf{a}\| \cos\theta$$

Here, the first equality follows from Eq. (5), the second from Eq. (6), the third from the geometric interpretation of the dot product, and the last since **T** is a unit vector.

- In (A), θ is obtuse, so $\cos\theta < 0$ and $v' < 0$. The particle's speed is decreasing.
- In (B), $\theta = \frac{\pi}{2}$, so $\cos\theta = 0$ and $v' = 0$. The particle's speed is momentarily constant.
- In (C), θ is acute, so $\cos\theta > 0$ and $v' > 0$. The particle's speed is increasing. ∎

13.5 SUMMARY

- For an object whose path is described by a vector-valued function $\mathbf{r}(t)$,

$$\mathbf{v}(t) = \mathbf{r}'(t), \qquad v(t) = \|\mathbf{v}(t)\|, \qquad \mathbf{a}(t) = \mathbf{r}''(t)$$

- The *velocity vector* $\mathbf{v}(t)$ points in the direction of motion. Its length $v(t) = \|\mathbf{v}(t)\|$ is the object's speed.
- The *acceleration vector* **a** is the sum of a tangential component (reflecting change in speed along the path) and a normal component (reflecting change in direction):

$$\mathbf{a}(t) = a_{\mathbf{T}}(t)\mathbf{T}(t) + a_{\mathbf{N}}(t)\mathbf{N}(t)$$

Unit tangent vector	$\mathbf{T}(t) = \dfrac{\mathbf{v}(t)}{\|\mathbf{v}(t)\|}$
Unit normal vector	$\mathbf{N}(t) = \dfrac{\mathbf{T}'(t)}{\|\mathbf{T}'(t)\|}$
Tangential component	$a_{\mathbf{T}} = v'(t) = \mathbf{a} \cdot \mathbf{T} = \dfrac{\mathbf{a} \cdot \mathbf{v}}{\|\mathbf{v}\|}$
	$a_{\mathbf{T}}\mathbf{T} = \left(\dfrac{\mathbf{a} \cdot \mathbf{v}}{\mathbf{v} \cdot \mathbf{v}}\right)\mathbf{v}$
Normal component	$a_{\mathbf{N}} = \kappa(t)v(t)^2 = \sqrt{\|\mathbf{a}\|^2 - |a_{\mathbf{T}}|^2}$
	$a_{\mathbf{N}}\mathbf{N} = \mathbf{a} - a_{\mathbf{T}}\mathbf{T} = \mathbf{a} - \left(\dfrac{\mathbf{a} \cdot \mathbf{v}}{\mathbf{v} \cdot \mathbf{v}}\right)\mathbf{v}$

13.5 EXERCISES

Preliminary Questions

1. If a particle travels with constant speed, must its acceleration vector be zero? Explain.

2. For a particle in uniform circular motion around a circle, which of the vectors $\mathbf{v}(t)$ or $\mathbf{a}(t)$ always points toward the center of the circle?

3. Two objects travel to the right along the parabola $y = x^2$ with nonzero speed. Which of the following statements must be true?

(a) Their velocity vectors point in the same direction.

(b) Their velocity vectors have the same length.

(c) Their acceleration vectors point in the same direction.

4. Use the decomposition of acceleration into tangential and normal components to explain the following statement: If the speed is constant, then the acceleration and velocity vectors are orthogonal.

5. If a particle travels along a straight line, then the acceleration and velocity vectors are (choose the correct description):

(a) orthogonal.   **(b)** parallel.

6. What is the length of the acceleration vector of a particle traveling around a circle of radius 2 cm with constant speed 4 cm/s?

7. Two cars are racing around a circular track. If, at a certain moment, both of their speedometers read 110 mph, then the two cars have the same (choose one):

(a) $a_\mathbf{T}$ **(b)** $a_\mathbf{N}$

Exercises

1. Use the table to calculate the difference quotients $\dfrac{\mathbf{r}(1+h)-\mathbf{r}(1)}{h}$ for $h = -0.2, -0.1, 0.1, 0.2$. Then estimate the velocity and speed at $t = 1$.

$\mathbf{r}(0.8)$	$\langle 1.557, 2.459, -1.970 \rangle$
$\mathbf{r}(0.9)$	$\langle 1.559, 2.634, -1.740 \rangle$
$\mathbf{r}(1)$	$\langle 1.540, 2.841, -1.443 \rangle$
$\mathbf{r}(1.1)$	$\langle 1.499, 3.078, -1.035 \rangle$
$\mathbf{r}(1.2)$	$\langle 1.435, 3.342, -0.428 \rangle$

2. Draw the vectors $\mathbf{r}(2+h)-\mathbf{r}(2)$ and $\dfrac{\mathbf{r}(2+h)-\mathbf{r}(2)}{h}$ for $h = 0.5$ for the path in Figure 10. Draw $\mathbf{v}(2)$ (using a rough estimate for its length).

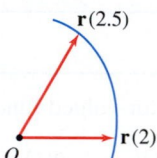

FIGURE 10

In Exercises 3–6, calculate the velocity and acceleration vectors and the speed at the time indicated.

3. $\mathbf{r}(t) = \langle t^3, 1-t, 4t^2 \rangle$, $t = 1$ **4.** $\mathbf{r}(t) = e^t\mathbf{j} - \cos(2t)\mathbf{k}$, $t = 0$

5. $\mathbf{r}(\theta) = \langle \sin\theta, \cos\theta, \cos 3\theta \rangle$, $\theta = \frac{\pi}{3}$

6. $\mathbf{r}(s) = \left\langle \dfrac{1}{1+s^2}, \dfrac{s}{1+s^2} \right\rangle$, $s = 2$

7. Find $\mathbf{a}(t)$ for a particle moving around a circle of radius 8 cm at a constant speed of $v = 4$ cm/s (see Example 4). Draw the path, and on it, draw the acceleration vector at $t = \frac{\pi}{4}$.

8. Sketch the path $\mathbf{r}(t) = \langle 1-t^2, 1-t \rangle$ for $-2 \le t \le 2$, indicating the direction of motion. Draw the velocity and acceleration vectors at $t = 0$ and $t = 1$.

9. Sketch the path $\mathbf{r}(t) = \langle t^2, t^3 \rangle$ together with the velocity and acceleration vectors at $t = 1$.

10. [✎] The paths $\mathbf{r}(t) = \langle t^2, t^3 \rangle$ and $\mathbf{r}_1(t) = \langle t^4, t^6 \rangle$ trace the same curve, and $\mathbf{r}_1(1) = \mathbf{r}(1)$. Do you expect either the velocity vectors or the acceleration vectors of these paths at $t = 1$ to point in the same direction? Explain. Compute these vectors and draw them on a single plot of the curve.

In Exercises 11–14, find $\mathbf{v}(t)$ given $\mathbf{a}(t)$ and the initial velocity.

11. $\mathbf{a}(t) = \langle t, 4 \rangle$, $\mathbf{v}(0) = \langle \frac{1}{3}, -2 \rangle$

12. $\mathbf{a}(t) = \langle e^t, 0, t+1 \rangle$, $\mathbf{v}(0) = \langle 1, -3, \sqrt{2} \rangle$

13. $\mathbf{a}(t) = \mathbf{k}$, $\mathbf{v}(0) = \mathbf{i}$ **14.** $\mathbf{a}(t) = t^2\mathbf{k}$, $\mathbf{v}(0) = \mathbf{i} - \mathbf{j}$

In Exercises 15–18, find $\mathbf{r}(t)$ and $\mathbf{v}(t)$ given $\mathbf{a}(t)$ and the initial velocity and position.

15. $\mathbf{a}(t) = \langle t, 4 \rangle$, $\mathbf{v}(0) = \langle 3, -2 \rangle$, $\mathbf{r}(0) = \langle 0, 0 \rangle$

16. $\mathbf{a}(t) = \langle e^t, 2t, t+1 \rangle$, $\mathbf{v}(0) = \langle 1, 0, 1 \rangle$, $\mathbf{r}(0) = \langle 2, 1, 1 \rangle$

17. $\mathbf{a}(t) = t\mathbf{k}$, $\mathbf{v}(0) = \mathbf{i}$, $\mathbf{r}(0) = \mathbf{j}$

18. $\mathbf{a}(t) = \cos t\mathbf{k}$, $\mathbf{v}(0) = \mathbf{i} - \mathbf{j}$, $\mathbf{r}(0) = \mathbf{i}$

19. A projectile is launched from the ground at an angle of $45°$. What initial speed must the projectile have in order to hit the top of a 120-m tower located 180 m away?

20. Find the initial velocity vector $\mathbf{v}_0$ of a projectile released with initial speed 100 m/s that reaches a maximum height of 300 m.

21. Assume that astronaut Alan Shepard hit his golf shot on the moon (acceleration due to gravity = 1.6 m/s^2) with a modest initial speed of 35 m/s at an angle of $30°$. How far did the ball travel?

22. Golfer Judy Robinson hit a golf ball on the planet Priplanus with an initial speed of 50 m/s at an angle of $40°$. It landed exactly 2 km away. What is the acceleration due to gravity on Priplanus?

23. Show that a projectile launched at an angle θ with initial speed v_0 travels a distance $(v_0^2/g)\sin 2\theta$ before hitting the ground. Conclude that the maximum distance (for a given v_0) is attained at $\theta = \frac{\pi}{4}$.

24. Show that a projectile launched at an angle θ will hit the top of an h-meter tower located d meters away if its initial speed is

$$v_0 = \frac{\sqrt{g/2}\,d\sec\theta}{\sqrt{d\tan\theta - h}}$$

25. A quarterback throws a football while standing at the very center of the field on the 50-yard line. The ball leaves his hand at a height of 5 ft and has initial velocity $\mathbf{v}_0 = 40\mathbf{i} + 35\mathbf{j} + 32\mathbf{k}$ ft/s. Assume an acceleration of 32 ft/s^2 due to gravity and that the $\mathbf{i}$ vector points down the field toward the endzone and the $\mathbf{j}$ vector points to the sideline. The field is 150 ft in width and 300 ft in length.

(a) Determine the position function that gives the position of the ball t seconds after it is thrown.

(b) The ball is caught by a player 5 ft above the ground. Is the player in bounds or out of bounds when he receives the ball? Assume the player is standing vertically with both toes on the ground at the time of reception.

26. A soccer ball is kicked from ground level with (x, y)-coordinates $(85, 20)$ on the soccer field shown in Figure 11 and with an initial velocity $\mathbf{v}_0 = 10\mathbf{i} - 5\mathbf{j} + 25\mathbf{k}$ ft/s. Assume an acceleration of 32 ft/s^2 due to gravity and that the goal net has a height of 8 ft and a total width of 24 ft.

(a) Determine the position function that gives the position of the ball t seconds after it is hit.

(b) Does the ball go in the goal before hitting the ground? Explain why or why not.

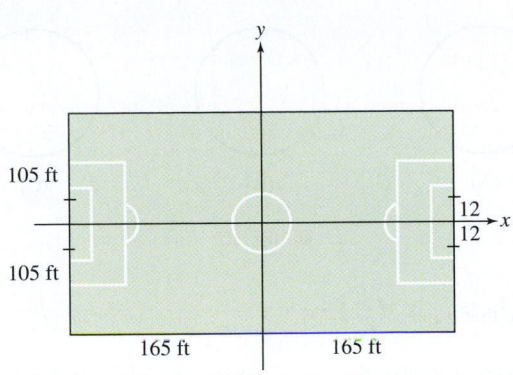

FIGURE 11

27. A constant force $\mathbf{F} = \langle 5, 2 \rangle$ (in newtons) acts on a 10-kg mass. Find the position of the mass at $t = 10$ seconds if it is located at the origin at $t = 0$ and has initial velocity $\mathbf{v}_0 = \langle 2, -3 \rangle$ (in meters per second).

28. A force $\mathbf{F} = \langle 24t, 16 - 8t \rangle$ (in newtons) acts on a 4-kg mass. Find the position of the mass at $t = 3$ s if it is located at $(10, 12)$ at $t = 0$ and has zero initial velocity.

29. A particle follows a path $\mathbf{r}(t)$ for $0 \le t \le T$, beginning at the origin O. The vector $\overline{\mathbf{v}} = \dfrac{1}{T} \displaystyle\int_0^T \mathbf{r}'(t)\, dt$ is called the **average velocity** vector. Suppose that $\overline{\mathbf{v}} = \mathbf{0}$. Answer and explain the following:

(a) Where is the particle located at time T if $\overline{\mathbf{v}} = \mathbf{0}$?

(b) Is the particle's average speed necessarily equal to zero?

30. At a certain moment, a moving particle has velocity $\mathbf{v} = \langle 2, 2, -1 \rangle$ and acceleration $\mathbf{a} = \langle 0, 4, 3 \rangle$. Find $\mathbf{T}$, $\mathbf{N}$, and the decomposition of $\mathbf{a}$ into tangential and normal components.

31. At a certain moment, a particle moving along a path has velocity $\mathbf{v} = \langle 12, 20, 20 \rangle$ and acceleration $\mathbf{a} = \langle 2, 1, -3 \rangle$. Is the particle speeding up or slowing down?

In Exercises 32–35, use Eq. (6) to find the coefficients a_T and a_N as a function of t (or at the specified value of t).

32. $\mathbf{r}(t) = \langle t^2, t^3 \rangle$

33. $\mathbf{r}(t) = \langle t, \cos t, \sin t \rangle$

34. $\mathbf{r}(t) = \langle t^{-1}, \ln t, t^2 \rangle, \quad t = 1$

35. $\mathbf{r}(t) = \langle e^{2t}, t, e^{-t} \rangle, \quad t = 0$

In Exercises 36–43, find the decomposition of $\mathbf{a}(t)$ into tangential and normal components at the point indicated, as in Example 6.

36. $\mathbf{r}(t) = \langle e^t, 1 - t \rangle, \quad t = 0$

37. $\mathbf{r}(t) = \left\langle \frac{1}{3}t^3, 1 - 3t \right\rangle, \quad t = -1$

38. $\mathbf{r}(t) = \left\langle t, \frac{1}{2}t^2, \frac{1}{6}t^3 \right\rangle, \quad t = 1$

39. $\mathbf{r}(t) = \left\langle t, \frac{1}{2}t^2, \frac{1}{6}t^3 \right\rangle, \quad t = 4$

40. $\mathbf{r}(t) = \langle 4 - t, t + 1, t^2 \rangle, \quad t = 2$

41. $\mathbf{r}(t) = \langle t, e^t, te^t \rangle, \quad t = 0$

42. $\mathbf{r}(\theta) = \langle \cos \theta, \sin \theta, \theta \rangle, \quad \theta = 0$

43. $\mathbf{r}(t) = \langle t, \cos t, t \sin t \rangle, \quad t = \frac{\pi}{2}$

44. Let $\mathbf{r}(t) = \langle t^2, 4t - 3 \rangle$. Find $\mathbf{T}(t)$ and $\mathbf{N}(t)$, and show that the decomposition of $\mathbf{a}(t)$ into tangential and normal components is

$$\mathbf{a}(t) = \left(\frac{2t}{\sqrt{t^2 + 4}} \right) \mathbf{T} + \left(\frac{4}{\sqrt{t^2 + 4}} \right) \mathbf{N}$$

45. Find the components a_T and a_N of the acceleration vector of a particle moving along a circular path of radius $R = 100$ cm with constant speed $v_0 = 5$ cm/s.

46. In the notation of Example 5, find the acceleration vector for a person seated in a car at (a) the highest point of the Ferris wheel and (b) the two points level with the center of the wheel.

47. Suppose that the Ferris wheel in Example 5 is rotating clockwise and that the point P at angle 45° has acceleration vector $\mathbf{a} = \langle 0, -50 \rangle$ m/min^2 pointing down, as in Figure 12. Determine the speed and tangential component of the acceleration of the Ferris wheel.

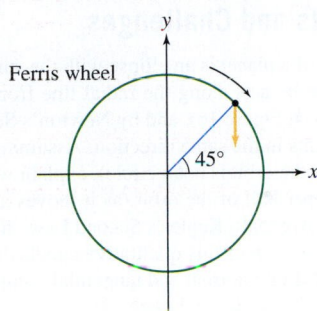

FIGURE 12

48. At time t_0, a moving particle has velocity vector $\mathbf{v} = 2\mathbf{i}$ and acceleration vector $\mathbf{a} = 3\mathbf{i} + 18\mathbf{k}$. Determine the curvature $\kappa(t_0)$ of the particle's path at time t_0.

49. A satellite orbits the earth at an altitude 400 km above the earth's surface, with constant speed $v = 28{,}000$ km/h. Find the magnitude of the satellite's acceleration (in kilometers per square hour), assuming that the radius of the earth is 6378 km (Figure 13).

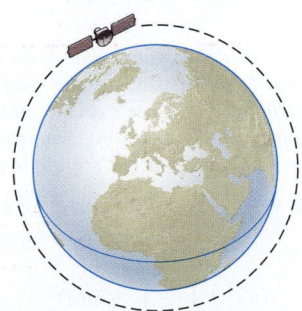

FIGURE 13 Satellite orbit.

50. A car proceeds along a circular path of radius $R = 300$ m centered at the origin. Starting at rest, its speed increases at a rate of t m/s^2. Find the acceleration vector $\mathbf{a}$ at time $t = 3$ s and determine its decomposition into normal and tangential components.

51. A particle follows a path $\mathbf{r}_1(t)$ on the helical curve with parametrization $\mathbf{r}(\theta) = \langle \cos \theta, \sin \theta, \theta \rangle$. When it is at position $\mathbf{r}\left(\frac{\pi}{2}\right)$, its speed is 3 m/s and its speed is increasing at a rate of $\frac{1}{2}$ m/s^2. Find its acceleration vector $\mathbf{a}$ at this moment. *Note:* The particle's acceleration vector does not coincide with $\mathbf{r}''(\theta)$.

52. 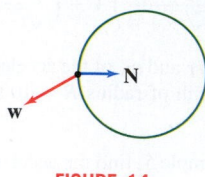 Explain why the vector **w** in Figure 14 cannot be the acceleration vector of a particle moving along the circle. *Hint:* Consider the sign of **w** · **N**.

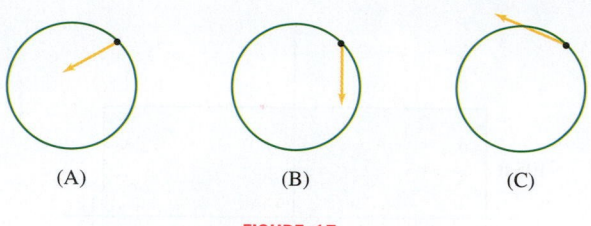

(A)   (B)   (C)

FIGURE 15

FIGURE 14

53. 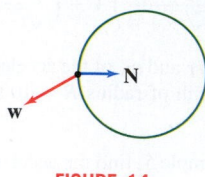 Figure 15 shows the acceleration vectors of a particle moving clockwise around a circle. In each case, state whether the particle is speeding up, slowing down, or momentarily at constant speed. Explain.

54. Prove that $a_{\mathbf{N}} = \dfrac{\|\mathbf{a} \times \mathbf{v}\|}{\|\mathbf{v}\|}$.

55. Suppose that $\mathbf{r}(t)$ lies on a sphere of radius R for all t. Let $\mathbf{J} = \mathbf{r} \times \mathbf{r}'$. Show that $\mathbf{r}' = (\mathbf{J} \times \mathbf{r})/\|\mathbf{r}\|^2$. *Hint:* Observe that $\mathbf{r}$ and $\mathbf{r}'$ are perpendicular.

Further Insights and Challenges

56. 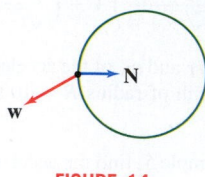 The orbit of a planet is an ellipse with the sun at one focus. The sun's gravitational force acts along the radial line from the planet to the sun (the dashed lines in Figure 16), and by Newton's Second Law, the acceleration vector points in the same direction. Assuming that the orbit has positive eccentricity (the orbit is not a circle), explain why the planet must slow down in the upper half of the orbit (as it moves away from the sun) and speed up in the lower half. Kepler's Second Law, discussed in the next section, is a precise version of this qualitative conclusion. *Hint:* Consider the decomposition of **a** into normal and tangential components.

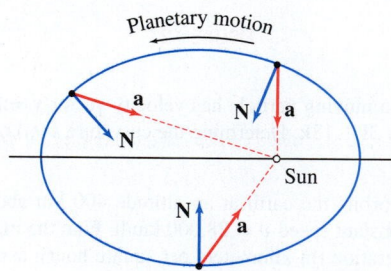

FIGURE 16 Elliptical orbit of a planet around the sun.

In Exercises 57–61, we consider an automobile of mass m traveling along a curved but level road. To avoid skidding, the road must supply a frictional force $\mathbf{F} = m\mathbf{a}$, where $\mathbf{a}$ is the car's acceleration vector. The maximum magnitude of the frictional force is μmg, where μ is the coefficient of friction and $g = 9.8$ m/s². Let v be the car's speed in meters per second.

57. Show that the car will not skid if the curvature κ of the road is such that (with $R = 1/\kappa$)

$$(v')^2 + \left(\frac{v^2}{R}\right)^2 \leq (\mu g)^2 \qquad \boxed{8}$$

Note that braking ($v' < 0$) and speeding up ($v' > 0$) contribute equally to skidding.

58. Suppose that the maximum radius of curvature along a curved highway is $R = 180$ m. How fast can an automobile travel (at constant speed) along the highway without skidding if the coefficient of friction is $\mu = 0.5$?

59. Beginning at rest, an automobile drives around a circular track of radius $R = 300$ m, accelerating at a rate of 0.3 m/s². After how many seconds will the car begin to skid if the coefficient of friction is $\mu = 0.6$?

60. You want to reverse your direction in the shortest possible time by driving around a semicircular bend (Figure 17). If you travel at the maximum possible *constant speed* v that will not cause skidding, is it faster to hug the inside curve (radius r) or the outside curb (radius R)? *Hint:* Use Eq. (8) to show that at maximum speed, the time required to drive around the semicircle is proportional to the square root of the radius.

61. What is the smallest radius R about which an automobile can turn without skidding at 100 km/h if $\mu = 0.75$ (a typical value)?

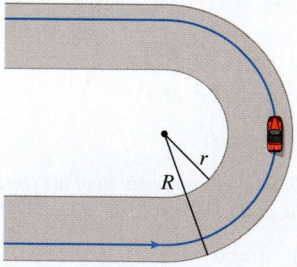

FIGURE 17 Car going around the bend.

13.6 Planetary Motion According to Kepler and Newton

In this section, we derive Kepler's laws of planetary motion, a feat first accomplished by Isaac Newton and published by him in 1687. No event was more emblematic of the scientific revolution. It demonstrated the power of mathematics to make the natural world comprehensible and it led succeeding generations of scientists to seek and discover mathematical laws governing other phenomena, such as electricity and magnetism, thermodynamics, and atomic processes.

According to Kepler, the planetary orbits are ellipses with the sun at one focus. Furthermore, if we imagine a radial vector $\mathbf{r}(t)$ pointing from the sun to the planet, as in Figure 1, then this radial vector sweeps out area at a constant rate or, as Kepler stated in his Second Law, the radial vector sweeps out equal areas in equal times (Figure 2). Kepler's Third Law determines the **period** T of the orbit, defined as the time required to complete one full revolution. These laws are valid not just for planets orbiting the sun, but for any body orbiting about another body according to the inverse-square law of gravitation.

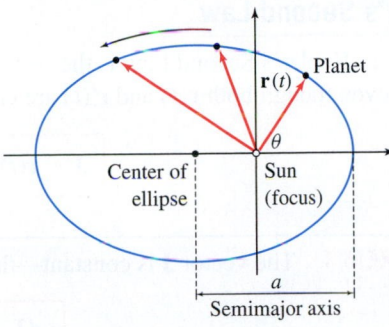

FIGURE 1 The planet travels along an ellipse with the sun at one focus.

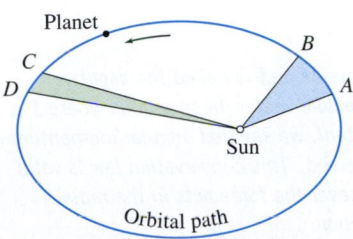

DF **FIGURE 2** The two shaded regions have equal areas, and by Kepler's Second Law, the planet sweeps them out in equal times. To do so, the planet must travel faster going from A to B than from C to D.

Kepler's Three Laws

(i) **Law of Ellipses:** The orbit of a planet is an ellipse with the sun at one focus.

(ii) **Law of Equal Area in Equal Time:** The position vector pointing from the sun to the planet sweeps out equal areas in equal times.

(iii) **Law of the Period of Motion:** $T^2 = \left(\dfrac{4\pi^2}{GM}\right) a^3$, where

- a is the semimajor axis of the ellipse (Figure 1) in meters.
- G is the universal gravitational constant: 6.673×10^{-11} m^3 kg^{-1} s^{-2}.
- M is the mass of the sun, approximately 1.989×10^{30} kg.
- T is the period of the orbit, in seconds.

Kepler's version of the Third Law stated only that T^2 is proportional to a^3. Newton discovered that the constant of proportionality is equal to $4\pi^2/(GM)$, and he observed that if you can measure T and a through observation, then you can use the Third Law to solve for the mass M. This method is used by astronomers to find the masses of the planets (by measuring T and a for moons revolving around the planet) as well as the masses of binary stars and galaxies. See Exercises 2–5.

Our derivation makes a few simplifying assumptions. We treat the sun and planet as point masses and ignore the gravitational attraction of the planets on each other. And although both the sun and the planet revolve around their mutual center of mass, we ignore the sun's motion and assume that the planet revolves around the center of the sun. This is justified because the sun is much more massive than the planet.

We place the sun at the origin of the coordinate system. Let $\mathbf{r} = \mathbf{r}(t)$ be the position vector of a planet of mass m, as in Figure 1, and let (Figure 3)

$$\mathbf{e}_r = \frac{\mathbf{r}(t)}{\|\mathbf{r}(t)\|}$$

be the unit radial vector at time t ($\mathbf{e}_r$ is the unit vector that points to the planet as it moves around the sun). By Newton's Universal Law of Gravitation (the inverse-square law), the sun attracts the planet with a gravitational force

$$\mathbf{F}(\mathbf{r}(t)) = -\left(\frac{km}{\|\mathbf{r}(t)\|^2}\right) \mathbf{e}_r$$

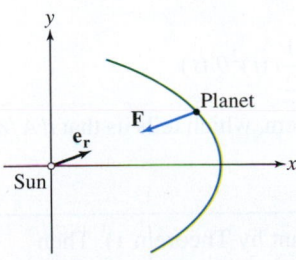

FIGURE 3 The gravitational force $\mathbf{F}$, directed from the planet to the sun, is a negative multiple of $\mathbf{e}_r$.

where $k = GM$. Combining the Law of Gravitation with Newton's Second Law of Motion $\mathbf{F}(\mathbf{r}(t)) = m\mathbf{r}''(t)$, we obtain

$$\mathbf{r}''(t) = -\frac{k}{\|\mathbf{r}(t)\|^2}\,\mathbf{e}_r \qquad \boxed{1}$$

Kepler's Laws are a consequence of this *differential equation*.

Kepler's Second Law

The key to Kepler's Second Law is the fact that the following cross product is a constant vector [even though both $\mathbf{r}(t)$ and $\mathbf{r}'(t)$ are changing in time]:

$$\mathbf{J} = \mathbf{r}(t) \times \mathbf{r}'(t)$$

In physics, $m\mathbf{J}$ is called the **angular momentum** vector. In situations where $\mathbf{J}$ is constant, we say that angular momentum is conserved. *This conservation law is valid whenever the force acts in the radial direction.*

> **THEOREM 1** The vector $\mathbf{J}$ is constant—that is,
>
> $$\frac{d\mathbf{J}}{dt} = 0 \qquad \boxed{2}$$

Proof By the Cross Product Rule (Theorem 3 in Section 13.2),

$$\frac{d\mathbf{J}}{dt} = \frac{d}{dt}\big(\mathbf{r}(t) \times \mathbf{r}'(t)\big) = \mathbf{r}'(t) \times \mathbf{r}'(t) + \mathbf{r}(t) \times \mathbf{r}''(t)$$

The cross product of parallel vectors is zero, so the first term is certainly zero. The second term is also zero because $\mathbf{r}''(t)$ is a multiple of $\mathbf{e}_r$ by Eq. (1), and hence also of $\mathbf{r}(t)$. ∎

← *REMINDER*

- $\mathbf{a} \times \mathbf{b}$ *is orthogonal to both* $\mathbf{a}$ *and* $\mathbf{b}$.
- $\mathbf{a} \times \mathbf{b} = 0$ *if* $\mathbf{a}$ *and* $\mathbf{b}$ *are parallel, that is, one is a multiple of the other.*

How can we use Eq. (2)? First of all, the cross product $\mathbf{J}$ is orthogonal to both $\mathbf{r}(t)$ and $\mathbf{r}'(t)$. Because $\mathbf{J}$ is constant, $\mathbf{r}(t)$ and $\mathbf{r}'(t)$ are confined to the fixed plane orthogonal to $\mathbf{J}$. This proves that the *motion of a planet around the sun takes place in a plane.*

We can choose coordinates so that the sun is at the origin and the planet moves in the counterclockwise direction (Figure 4). Let (r, θ) be the polar coordinates of the planet, where $r = r(t)$ and $\theta = \theta(t)$ are functions of time. Note that $r(t) = \|\mathbf{r}(t)\|$.

Recall from Section 11.4 (Theorem 1) that the area swept out by the planet's radial vector, from 0 to θ, is

$$A = \frac{1}{2}\int_0^\theta r^2\,d\theta$$

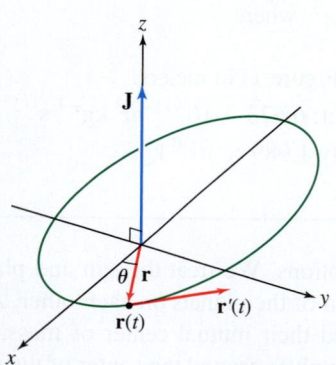

DF **FIGURE 4** The orbit is contained in the plane orthogonal to $\mathbf{J}$ (but we have not proven yet that the orbit is an ellipse).

Kepler's Second Law states that this area is swept out at a constant rate. But, this rate is simply dA/dt. We will prove that dA/dt is constant. By the Fundamental Theorem of Calculus, $\dfrac{dA}{d\theta} = \dfrac{1}{2}r^2$, and by the Chain Rule,

$$\frac{dA}{dt} = \frac{dA}{d\theta}\frac{d\theta}{dt} = \frac{1}{2}\theta'(t)r(t)^2 = \frac{1}{2}r(t)^2\theta'(t)$$

Thus, Kepler's Second Law follows from the next theorem, which tells us that dA/dt has the constant value $\frac{1}{2}\|\mathbf{J}\|$.

> **THEOREM 2** Let $J = \|\mathbf{J}\|$ ($\mathbf{J}$ and hence J are constant by Theorem 1). Then
>
> $$r(t)^2\theta'(t) = J \qquad \boxed{3}$$

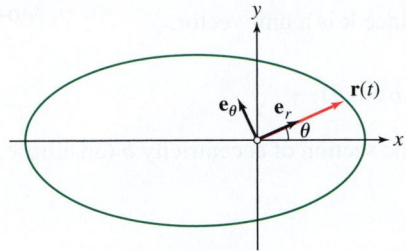

FIGURE 5 The unit vectors $\mathbf{e}_r$ and $\mathbf{e}_\theta$ are orthogonal, and rotate around the origin along with the planet.

Proof We note that in polar coordinates, $\mathbf{e}_r = \langle \cos\theta, \sin\theta \rangle$. We also define the unit vector $\mathbf{e}_\theta = \langle -\sin\theta, \cos\theta \rangle$ that is orthogonal to $\mathbf{e}_r$ (Figure 5). In summary,

$$\boxed{r(t) = \|\mathbf{r}(t)\|, \quad \mathbf{e}_r = \langle \cos\theta, \sin\theta \rangle, \quad \mathbf{e}_\theta = \langle -\sin\theta, \cos\theta \rangle, \quad \mathbf{e}_r \cdot \mathbf{e}_\theta = 0}$$

We see directly that the derivatives of $\mathbf{e}_r$ and $\mathbf{e}_\theta$ with respect to θ are

$$\boxed{\frac{d}{d\theta}\mathbf{e}_r = \mathbf{e}_\theta, \qquad \frac{d}{d\theta}\mathbf{e}_\theta = -\mathbf{e}_r} \qquad \boxed{4}$$

The time derivative of $\mathbf{e}_r$ is computed using the Chain Rule:

$$\mathbf{e}_r' = \left(\frac{d\theta}{dt}\right)\left(\frac{d}{d\theta}\mathbf{e}_r\right) = \theta'(t)\,\mathbf{e}_\theta \qquad \boxed{5}$$

Now apply the Product Rule to $\mathbf{r} = r\mathbf{e}_r$:

$$\mathbf{r}' = \frac{d}{dt}r\mathbf{e}_r = r'\mathbf{e}_r + r\mathbf{e}_r' = r'\mathbf{e}_r + r\theta'\mathbf{e}_\theta$$

Using $\mathbf{e}_r \times \mathbf{e}_r = \mathbf{0}$, we obtain

$$\mathbf{J} = \mathbf{r} \times \mathbf{r}' = r\mathbf{e}_r \times (r'\mathbf{e}_r + r\theta'\mathbf{e}_\theta) = r^2\theta'(\mathbf{e}_r \times \mathbf{e}_\theta)$$

To compute cross products of vectors in the plane, such as $\mathbf{r}$, $\mathbf{e}_r$, and $\mathbf{e}_\theta$, we treat them as vectors in 3-space with a z-component equal to zero. The cross product is then a multiple of $\mathbf{k}$.

It is straightforward to check that $\mathbf{e}_r \times \mathbf{e}_\theta = \mathbf{k}$, and since $\mathbf{k}$ is a unit vector, $J = \|\mathbf{J}\| = |r^2\theta'|$. However, $\theta' > 0$ because the planet moves in the counterclockwise direction, so $J = r^2\theta'$. This proves Theorem 2. ∎

Proof of the Law of Ellipses

We show that the orbit of a planet is indeed an ellipse with the sun as one of the foci.

Let $\mathbf{v} = \mathbf{r}'(t)$ be the velocity vector. Then $\mathbf{r}'' = \mathbf{v}'$ and Eq. (1) may be written

$$\frac{d\mathbf{v}}{dt} = -\frac{k}{r(t)^2}\mathbf{e}_r \qquad \boxed{6}$$

◀ *REMINDER Eq. (1) states*

$$\mathbf{r}''(t) = -\frac{k}{r(t)^2}\mathbf{e}_r$$

where $r(t) = \|\mathbf{r}(t)\|$.

On the other hand, by the Chain Rule and the relation $r(t)^2\theta'(t) = J$ of Eq. (3),

$$\frac{d\mathbf{v}}{dt} = \frac{d\theta}{dt}\frac{d\mathbf{v}}{d\theta} = \theta'(t)\frac{d\mathbf{v}}{d\theta} = \frac{J}{r(t)^2}\frac{d\mathbf{v}}{d\theta}$$

Together with Eq. (6), this yields $J\dfrac{d\mathbf{v}}{d\theta} = -k\mathbf{e}_r$, or

$$\frac{d\mathbf{v}}{d\theta} = -\frac{k}{J}\mathbf{e}_r = -\frac{k}{J}\langle\cos\theta, \sin\theta\rangle$$

This is a first-order differential equation that no longer involves time t. We can solve it by integration:

$$\mathbf{v} = -\frac{k}{J}\int \langle\cos\theta, \sin\theta\rangle\, d\theta = \frac{k}{J}\langle -\sin\theta, \cos\theta\rangle + \mathbf{c} = \frac{k}{J}\mathbf{e}_\theta + \mathbf{c} \qquad \boxed{7}$$

where $\mathbf{c}$ is an arbitrary constant vector.

We are still free to rotate our coordinate system in the plane of motion, so we may assume that $\mathbf{c}$ points along the y-axis. We can then write $\mathbf{c} = \langle 0, (k/J)b \rangle$ for some scalar constant b. We finish the proof by computing $\mathbf{J} = \mathbf{r} \times \mathbf{v}$:

$$\mathbf{J} = \mathbf{r} \times \mathbf{v} = r\mathbf{e}_r \times \left(\frac{k}{J}\mathbf{e}_\theta + \mathbf{c}\right) = \frac{k}{J}r\left(\mathbf{e}_r \times \mathbf{e}_\theta + \mathbf{e}_r \times \langle 0, b \rangle\right)$$

Direct calculation yields

$$\mathbf{e}_r \times \mathbf{e}_\theta = \mathbf{k}, \qquad \mathbf{e}_r \times \langle 0, b \rangle = (b\cos\theta)\mathbf{k}$$

◀ **REMINDER** *The equation of a conic section in polar coordinates is discussed in Section 11.5.*

so our equation becomes $\mathbf{J} = \dfrac{k}{J} r(1 + b\cos\theta)\mathbf{k}$. Since $\mathbf{k}$ is a unit vector,

$$J = \|\mathbf{J}\| = \frac{k}{J} r\left(1 + b\cos\theta\right)$$

Solving for r, we obtain the polar equation of a conic section of eccentricity b (an ellipse, parabola, or hyperbola):

$$r = \frac{J^2/k}{1 + b\cos\theta}$$

This result shows that if a planet travels around the sun in a bounded orbit, then the orbit must be an ellipse. There are also "open orbits" that are either parabolic or hyperbolic. They describe comets that pass by the sun and then continue into space, never to return. In our derivation, we assumed implicitly that $\mathbf{J} \neq \mathbf{0}$. If $\mathbf{J} = \mathbf{0}$, then $\theta'(t) = 0$. In this case, the orbit is a straight line, and the planet falls directly into the sun.

Kepler's Third Law is verified in Exercises 23 and 24.

CONCEPTUAL INSIGHT We exploited the fact that $\mathbf{J}$ is constant to prove the Law of Ellipses without ever finding a formula for the position vector $\mathbf{r}(t)$ of the planet as a function of time t. In fact, $\mathbf{r}(t)$ cannot be expressed in terms of elementary functions. This illustrates an important principle: Sometimes it is possible to describe solutions of a differential equation even if we cannot write them down explicitly.

HISTORICAL PERSPECTIVE

The Hubble Space Telescope produced this image of the Antenna galaxies, a pair of spiral galaxies that began to collide hundreds of millions of years ago.

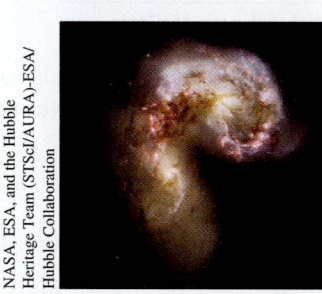

NASA, ESA, and the Hubble Heritage Team (STScI/AURA)-ESA/Hubble Collaboration

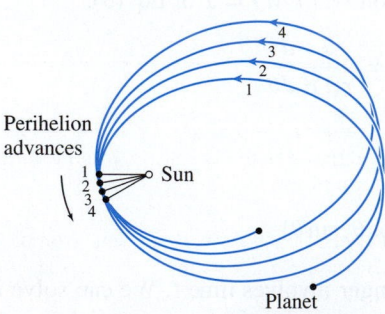

FIGURE 6 The perihelion of an orbit shifts slowly over time. For Mercury, the semimajor axis makes a full revolution approximately once every 24,000 years.

The astronomers of the ancient world (Babylon, Egypt, and Greece) mapped out the nighttime sky with impressive accuracy, but their models of planetary motion were based on the erroneous assumption that the planets revolve around the earth. Although the Greek astronomer Aristarchus (310–230 BCE) had suggested that the earth revolves around the sun, this idea was rejected and forgotten for nearly 18 centuries, until the Polish astronomer Nicolaus Copernicus (1473–1543) introduced a revolutionary set of ideas about the solar system, including the hypothesis that the planets revolve around the sun. Copernicus paved the way for the next generation, most notably Tycho Brahe (1546–1601), Galileo Galilei (1564–1642), and Johannes Kepler (1571–1630).

The German astronomer Johannes Kepler was the son of a mercenary soldier who apparently left his family when Johannes was 5 and may have died at war. Johannes was raised by his mother in his grandfather's inn. Kepler's mathematical brilliance earned him a scholarship at the University of Tübingen, and at the age of 29, he went to work for the Danish astronomer Tycho Brahe, who had compiled the most complete and accurate data on planetary motion then available. When Brahe died in 1601, Kepler succeeded him as Imperial Mathematician to the Holy Roman Emperor, and in 1609, he formulated the first two of his laws of planetary motion in a work entitled *Astronomia Nova (New Astronomy)*.

In the centuries since Kepler's death, as observational data improved, astronomers found that the planetary orbits are not exactly elliptical. Furthermore, the perihelion (the point on the orbit closest to the sun) shifts slowly over time (Figure 6). Most of these deviations can be explained by the mutual pull of the planets, but the perihelion shift of Mercury is larger than can be accounted for by Newton's Laws. On November 18, 1915, Albert Einstein made a discovery about which he later wrote to a friend, "I was beside myself with ecstasy for days." He had been working for a decade on his famous **General Theory of Relativity**, a theory that would replace Newton's Law of Gravitation with a new set of much more complicated equations called the Einstein Field Equations. On that 18th of November, Einstein showed that Mercury's perihelion shift was accurately explained by his new theory. At the time, this was the only substantial piece of evidence that the General Theory of Relativity was correct.

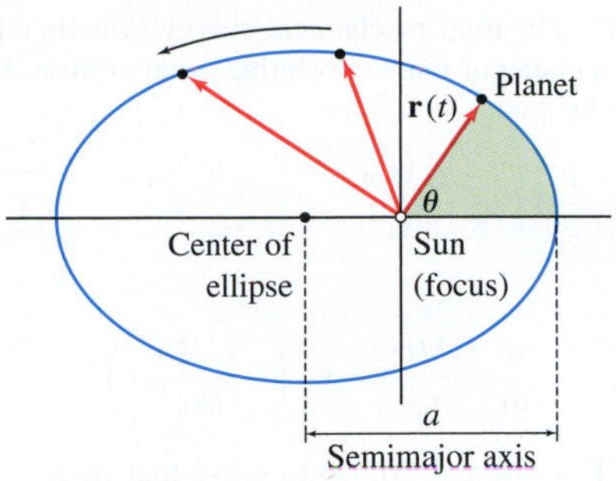

FIGURE 7 Planetary orbit.

Constants:

• Gravitational constant:

$G \approx 6.673 \times 10^{-11}$ m^3 kg^{-1} s^{-2}

• Mass of the sun:

$M \approx 1.989 \times 10^{30}$ kg

• $k = GM \approx 1.327 \times 10^{20}$

13.6 SUMMARY

• Kepler's three laws of planetary motion:

  – Law of Ellipses
  – Law of Equal Area in Equal Time
  – Law of the Period $T^2 = \left(\dfrac{4\pi^2}{GM}\right) a^3$, where T is the period (time to complete one full revolution) and a is the semimajor axis (Figure 7)

• According to Newton's Universal Law of Gravitation and Second Law of Motion, the position vector $\mathbf{r}(t)$ of a planet satisfies the differential equation

$$\mathbf{r}''(t) = -\frac{k}{r(t)^2}\mathbf{e}_r, \qquad \text{where } r(t) = \|\mathbf{r}(t)\|, \quad \mathbf{e}_r = \frac{\mathbf{r}(t)}{\|\mathbf{r}(t)\|}$$

• Properties of $\mathbf{J} = \mathbf{r}(t) \times \mathbf{r}'(t)$:

  – $\mathbf{J}$ is a constant of planetary motion.
  – Let $J = \|\mathbf{J}\|$. Then $J = r(t)^2\theta'(t)$.
  – The planet sweeps out area at the rate $\dfrac{dA}{dt} = \dfrac{1}{2}J$.

• A planetary orbit has polar equation $r = \dfrac{J^2/k}{1 + e\cos\theta}$, where e is the eccentricity of the orbit.

13.6 EXERCISES

Preliminary Questions

1. Describe the relation between the vector $\mathbf{J} = \mathbf{r} \times \mathbf{r}'$ and the rate at which the radial vector sweeps out area.

2. Equation (1) shows that $\mathbf{r}''$ is proportional to $\mathbf{r}$. Explain how this fact is used to prove Kepler's Second Law.

3. How is the period T affected if the semimajor axis a is increased four-fold?

Exercises

1. Kepler's Third Law states that T^2/a^3 has the same value for each planetary orbit. Do the data in the following table support this conclusion? Estimate the length of Jupiter's period, assuming that $a = 77.8 \times 10^{10}$ m.

Planet	Mercury	Venus	Earth	Mars
a (10^{10} m)	5.79	10.8	15.0	22.8
T (years)	0.241	0.615	1.00	1.88

2. Finding the Mass of a Star Using Kepler's Third Law, show that if a planet revolves around a star with period T and semimajor axis a, then the mass of the star is $M = \left(\dfrac{4\pi^2}{G}\right)\left(\dfrac{a^3}{T^2}\right)$.

3. Ganymede, one of Jupiter's moons discovered by Galileo, has an orbital period of 7.154 days and a semimajor axis of 1.07×10^9 m. Use Exercise 2 to estimate the mass of Jupiter.

4. An astronomer observes a planet orbiting a star with a period of 9.5 years and a semimajor axis of 3×10^8 km. Find the mass of the star using Exercise 2.

5. Mass of the Milky Way The sun revolves around the center of mass of the Milky Way galaxy in an orbit that is approximately circular, of radius $a \approx 2.8 \times 10^{17}$ km and velocity $v \approx 250$ km/s. Use the result of Exercise 2 to estimate the mass of the portion of the Milky Way inside the sun's orbit (place all of this mass at the center of the orbit).

6. A satellite orbiting above the equator of the earth is **geosynchronous** if the period is $T = 24$ hours (in this case, the satellite stays over a fixed point on the equator). Use Kepler's Third Law to show that in a circular geosynchronous orbit, the distance from the center of the earth is $R \approx 42{,}246$ km. Then compute the altitude h of the orbit above the earth's surface. The earth has mass $M \approx 5.974 \times 10^{24}$ kg and radius $R \approx 6371$ km.

7. Show that a planet in a circular orbit travels at constant speed. *Hint:* Use the facts that $\mathbf{J}$ is constant and that $\mathbf{r}(t)$ is orthogonal to $\mathbf{r}'(t)$ for a circular orbit.

8. Verify that the circular orbit

$$\mathbf{r}(t) = \langle R\cos\omega t, R\sin\omega t\rangle$$

satisfies the differential equation, Eq. (1), provided that $\omega^2 = kR^{-3}$. Then deduce Kepler's Third Law $T^2 = \left(\dfrac{4\pi^2}{k}\right) R^3$ for this orbit.

9. Prove that if a planetary orbit is circular of radius R, then $vT = 2\pi R$, where v is the planet's speed (constant by Exercise 7) and T is the period. Then use Kepler's Third Law to prove that $v = \sqrt{\dfrac{k}{R}}$.

10. Find the velocity of a satellite in geosynchronous orbit about the earth. *Hint:* Use Exercises 6 and 9.

11. A communications satellite orbiting the earth has initial position $\mathbf{r}(0) = \langle 29{,}000, 20{,}000, 0 \rangle$ (in kilometers) and initial velocity $\mathbf{r}'(0) = \langle 1, 1, 1 \rangle$ (in kilometers per second), where the origin is the earth's center. Find the equation of the plane containing the satellite's orbit. *Hint:* This plane is orthogonal to $\mathbf{J}$.

12. Assume that the earth's orbit is circular of radius $R = 150 \times 10^6$ km (it is nearly circular with eccentricity $e = 0.017$). Find the rate at which the earth's radial vector sweeps out area in units of square kilometers per second. What is the magnitude of the vector $\mathbf{J} = \mathbf{r} \times \mathbf{r}'$ for the earth (in units of square kilometers per second)?

In Exercises 13–19, the perihelion and aphelion are the points on the orbit closest to and farthest from the sun, respectively (Figure 8). The distance from the sun at the perihelion is denoted r_{per} and the speed at this point is denoted v_{per}. Similarly, we write r_{ap} and v_{ap} for the distance and speed at the aphelion. The semimajor axis is denoted a.

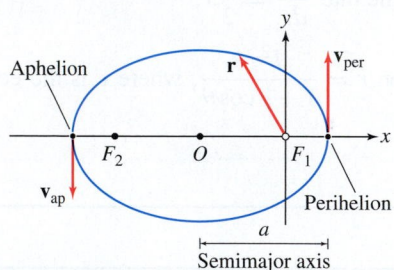

FIGURE 8 $\mathbf{r}$ and $\mathbf{v} = \mathbf{r}'$ are perpendicular at the perihelion and aphelion.

13. Use the polar equation of an ellipse

$$r = \frac{p}{1 + e \cos \theta}$$

to show that $r_{\text{per}} = a(1 - e)$ and $r_{\text{ap}} = a(1 + e)$. *Hint:* Use the fact that $r_{\text{per}} + r_{\text{ap}} = 2a$.

14. Use the result of Exercise 13 to prove the formulas

$$e = \frac{r_{\text{ap}} - r_{\text{per}}}{r_{\text{ap}} + r_{\text{per}}}, \qquad p = \frac{2 r_{\text{ap}} r_{\text{per}}}{r_{\text{ap}} + r_{\text{per}}}$$

15. Use the fact that $\mathbf{J} = \mathbf{r} \times \mathbf{r}'$ is constant to prove

$$v_{\text{per}}(1 - e) = v_{\text{ap}}(1 + e)$$

Hint: $\mathbf{r}$ is perpendicular to $\mathbf{r}'$ at the perihelion and aphelion.

16. Compute r_{per} and r_{ap} for the orbit of Mercury, which has eccentricity $e = 0.244$ (see the table in Exercise 1 for the semimajor axis).

17. Conservation of Energy The total mechanical energy (kinetic energy plus potential energy) of a planet of mass m orbiting a sun of mass M with position $\mathbf{r}$ and speed $v = \|\mathbf{r}'\|$ is

$$E = \frac{1}{2} m v^2 - \frac{GMm}{\|\mathbf{r}\|} \qquad \boxed{8}$$

(a) Prove the equations

$$\frac{d}{dt} \frac{1}{2} m v^2 = \mathbf{v} \cdot (m\mathbf{a}), \qquad \frac{d}{dt} \frac{GMm}{\|\mathbf{r}\|} = \mathbf{v} \cdot \left(-\frac{GMm}{\|\mathbf{r}\|^3} \mathbf{r} \right)$$

(b) Then use Newton's Law $\mathbf{F} = m\mathbf{a}$ and Eq. (1) to prove that energy is conserved—that is, $\dfrac{dE}{dt} = 0$.

18. Show that the total energy [Eq. (8)] of a planet in a circular orbit of radius R is $E = -\dfrac{GMm}{2R}$. *Hint:* Use Exercise 9.

19. Prove that $v_{\text{per}} = \sqrt{\left(\dfrac{GM}{a} \right) \dfrac{1 + e}{1 - e}}$ as follows:

(a) Use Conservation of Energy (Exercise 17) to show that

$$v_{\text{per}}^2 - v_{\text{ap}}^2 = 2GM \left(r_{\text{per}}^{-1} - r_{\text{ap}}^{-1} \right)$$

(b) Show that $r_{\text{per}}^{-1} - r_{\text{ap}}^{-1} = \dfrac{2e}{a(1 - e^2)}$ using Exercise 13.

(c) Show that $v_{\text{per}}^2 - v_{\text{ap}}^2 = 4 \dfrac{e}{(1 + e)^2} v_{\text{per}}^2$ using Exercise 15. Then solve for v_{per} using (a) and (b).

20. Show that a planet in an elliptical orbit has total mechanical energy $E = -\dfrac{GMm}{2a}$, where a is the semimajor axis. *Hint:* Use Exercise 19 to compute the total energy at the perihelion.

21. Prove that $v^2 = GM \left(\dfrac{2}{r} - \dfrac{1}{a} \right)$ at any point on an elliptical orbit, where $r = \|\mathbf{r}\|$, v is the velocity, and a is the semimajor axis of the orbit.

22. 📝 Two space shuttles A and B orbit the earth along the solid trajectory in Figure 9. Hoping to catch up to B, the pilot of A applies a forward thrust to increase her shuttle's kinetic energy. Use Exercise 20 to show that shuttle A will move off into a larger orbit as shown in the figure. Then use Kepler's Third Law to show that A's orbital period T will increase (and she will fall farther and farther behind B)!

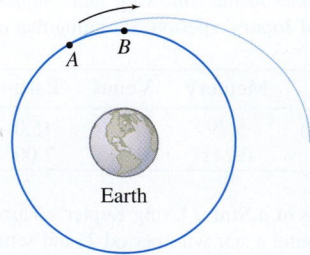

FIGURE 9

Further Insights and Challenges

Exercises 23 and 24 prove Kepler's Third Law. Figure 10 shows an elliptical orbit with polar equation

$$r = \frac{p}{1 + e \cos \theta}$$

where $p = J^2 / k$. The origin of the polar coordinates occurs at F_1. Let a and b be the semimajor and semiminor axes, respectively.

23. This exercise shows that $b = \sqrt{pa}$.

(a) Show that $CF_1 = ae$. *Hint:* $r_{\text{per}} = a(1 - e)$ by Exercise 13.

(b) Show that $a = \dfrac{p}{1 - e^2}$.

(c) Show that $F_1 A + F_2 A = 2a$. Conclude that $F_1 B + F_2 B = 2a$ and hence $F_1 B = F_2 B = a$.

(d) Use the Pythagorean Theorem to prove that $b = \sqrt{pa}$.

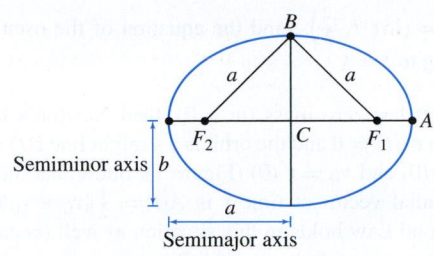

Seminimor axis b

Semimajor axis

FIGURE 10

24. The area A of the ellipse is $A = \pi ab$.

(a) Prove, using Kepler's First Law, that $A = \frac{1}{2}JT$, where T is the period of the orbit.

(b) Use Exercise 23 to show that $A = (\pi \sqrt{p})a^{3/2}$.

(c) Deduce Kepler's Third Law: $T^2 = \dfrac{4\pi^2}{GM}a^3$.

25. ✎ According to Eq. (7), the velocity vector of a planet as a function of the angle θ is

$$\mathbf{v}(\theta) = \frac{k}{J}\mathbf{e}_\theta + \mathbf{c}$$

Use this to explain the following statement: As a planet revolves around the sun, its velocity vector traces out a circle of radius k/J with its center at the terminal point of $\mathbf{c}$ (Figure 11). This beautiful but hidden property of orbits was discovered by William Rowan Hamilton in 1847.

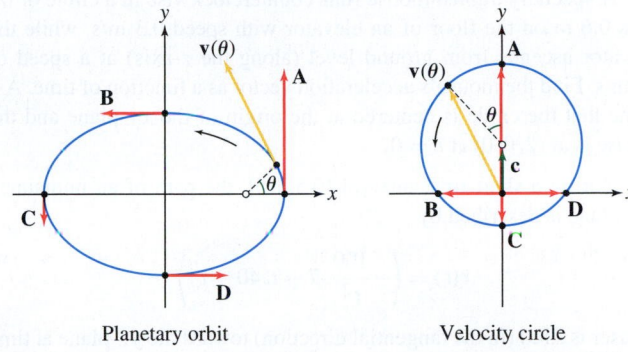

Planetary orbit   Velocity circle

FIGURE 11 The velocity vector traces out a circle as the planet travels along its orbit.

CHAPTER REVIEW EXERCISES

1. Determine the domains of the vector-valued functions.

(a) $\mathbf{r}_1(t) = \langle t^{-1}, (t+1)^{-1}, \sin^{-1} t \rangle$

(b) $\mathbf{r}_2(t) = \langle \sqrt{8 - t^3}, \ln t, e^{\sqrt{t}} \rangle$

2. Sketch the paths $\mathbf{r}_1(\theta) = \langle \theta, \cos \theta \rangle$ and $\mathbf{r}_2(\theta) = \langle \cos \theta, \theta \rangle$ in the xy-plane.

3. Find a vector parametrization of the intersection of the surfaces $x^2 + y^4 + 2z^3 = 6$ and $x = y^2$ in $\mathbf{R}^3$.

4. Find a vector parametrization using trigonometric functions of the intersection of the plane $x + y + z = 1$ and the elliptical cylinder $\left(\dfrac{y}{3}\right)^2 + \left(\dfrac{z}{8}\right)^2 = 1$ in $\mathbf{R}^3$.

In Exercises 5–10, calculate the derivative indicated.

5. $\mathbf{r}'(t)$, $\mathbf{r}(t) = \langle 1 - t, t^{-2}, \ln t \rangle$

6. $\mathbf{r}'''(t)$, $\mathbf{r}(t) = \langle t^3, 4t^2, 7t \rangle$

7. $\mathbf{r}'(0)$, $\mathbf{r}(t) = \langle e^{2t}, e^{-4t^2}, e^{6t} \rangle$

8. $\mathbf{r}''(-3)$, $\mathbf{r}(t) = \langle t^{-2}, (t+1)^{-1}, t^3 - t \rangle$

9. $\dfrac{d}{dt} e^t \langle 1, t, t^2 \rangle$

10. $\dfrac{d}{d\theta} \mathbf{r}(\cos \theta)$, $\mathbf{r}(s) = \langle s, 2s, s^2 \rangle$

In Exercises 11–14, calculate the derivative at $t = 3$, assuming that

$$\mathbf{r}_1(3) = \langle 1, 1, 0 \rangle, \qquad \mathbf{r}_2(3) = \langle 1, 1, 0 \rangle$$

$$\mathbf{r}_1'(3) = \langle 0, 0, 1 \rangle, \qquad \mathbf{r}_2'(3) = \langle 0, 2, 4 \rangle$$

11. $\dfrac{d}{dt}(6\mathbf{r}_1(t) - 4 \cdot \mathbf{r}_2(t))$

12. $\dfrac{d}{dt}(e^t \mathbf{r}_2(t))$

13. $\dfrac{d}{dt}(\mathbf{r}_1(t) \cdot \mathbf{r}_2(t))$

14. $\dfrac{d}{dt}(\mathbf{r}_1(t) \times \mathbf{r}_2(t))$

15. Calculate $\displaystyle\int_0^3 \langle 4t + 3, t^2, -4t^3 \rangle \, dt$.

16. Calculate $\displaystyle\int_0^\pi \langle \sin \theta, \theta, \cos 2\theta \rangle \, d\theta$.

17. A particle located at $(1, 1, 0)$ at time $t = 0$ follows a path whose velocity vector is $\mathbf{v}(t) = \langle 1, t, 2t^2 \rangle$. Find the particle's location at $t = 2$.

18. Find the vector-valued function $\mathbf{r}(t) = \langle x(t), y(t) \rangle$ in $\mathbf{R}^2$ satisfying $\mathbf{r}'(t) = -\mathbf{r}(t)$ with initial conditions $\mathbf{r}(0) = \langle 1, 2 \rangle$.

19. Calculate $\mathbf{r}(t)$, assuming that

$$\mathbf{r}''(t) = \langle 4 - 16t, 12t^2 - t \rangle, \qquad \mathbf{r}'(0) = \langle 1, 0 \rangle, \qquad \mathbf{r}(0) = \langle 0, 1 \rangle$$

20. Solve $\mathbf{r}''(t) = \langle t^2 - 1, t + 1, t^3 \rangle$ subject to the initial conditions $\mathbf{r}(0) = \langle 1, 0, 0 \rangle$ and $\mathbf{r}'(0) = \langle -1, 1, 0 \rangle$

21. Compute the length of the path

$$\mathbf{r}(t) = \langle \sin 2t, \cos 2t, 3t - 1 \rangle \quad \text{for } 1 \le t \le 3$$

22. [CAS] For the path $\mathbf{r}(t) = \langle \ln t, t, e^t \rangle$, with $1 \le t \le 2$, express the length as a definite integral, and use a computer algebra system to find its value to two decimal places.

23. Find an arc length parametrization of a helix of height 20 cm that makes four full rotations over a circle of radius 5 cm.

24. Find the minimum speed of a particle with trajectory $\mathbf{r}(t) = \langle t, e^{t-3}, e^{4-t} \rangle$.

25. A projectile fired at an angle of $60°$ lands 400 m away. What was its initial speed?

26. A specially trained mouse runs counterclockwise in a circle of radius 0.6 m on the floor of an elevator with speed 0.3 m/s, while the elevator ascends from ground level (along the z-axis) at a speed of 12 m/s. Find the mouse's acceleration vector as a function of time. Assume that the circle is centered at the origin of the xy-plane and the mouse is at $(2, 0, 0)$ at $t = 0$.

27. During a short time interval $[0.5, 1.5]$, the path of an unmanned spy plane is described by

$$\mathbf{r}(t) = \left\langle -\frac{100}{t^2}, 7 - t, 40 - t^2 \right\rangle$$

A laser is fired (in the tangential direction) toward the yz-plane at time $t = 1$. Which point in the yz-plane does the laser beam hit?

28. A force $\mathbf{F} = \langle 12t + 4, 8 - 24t \rangle$ (in newtons) acts on a 2-kg mass. Find the position of the mass at $t = 2$ s if it is located at $(4, 6)$ at $t = 0$ and has initial velocity $\langle 2, 3 \rangle$ in meters per second.

29. Find the unit tangent vector to $\mathbf{r}(t) = \langle \sin t, t, \cos t \rangle$ at $t = \pi$.

30. Find the unit tangent vector to $\mathbf{r}(t) = \langle t^2, \tan^{-1} t, t \rangle$ at $t = 1$.

31. Calculate $\kappa(1)$ for $\mathbf{r}(t) = \langle \ln t, t \rangle$.

32. Calculate $\kappa\left(\frac{\pi}{4}\right)$ for $\mathbf{r}(t) = \langle \tan t, \sec t, \cos t \rangle$.

In Exercises 33 and 34, write the acceleration vector $\mathbf{a}$ at the point indicated as a sum of tangential and normal components.

33. $\mathbf{r}(\theta) = \langle \cos \theta, \sin 2\theta \rangle, \quad \theta = \frac{\pi}{4}$

34. $\mathbf{r}(t) = \langle t^2, 2t - t^2, t \rangle, \quad t = 2$

35. At a certain time t_0, the path of a moving particle is tangent to the y-axis in the positive direction. The particle's speed at time t_0 is 4 m/s, and its acceleration vector is $\mathbf{a} = \langle 5, 4, 12 \rangle$. Determine the curvature of the path at t_0.

36. Give an equation for the osculating circle to $y = x^2 - x^3$ at $x = 1$.

37. Give an equation for the osculating circle to $y = \sqrt{x}$ at $x = 4$.

38. Let $\mathbf{r}(t) = \langle \cos t, \sin t, 2t \rangle$.
(a) Find $\mathbf{T}, \mathbf{N}$, and $\mathbf{B}$ at the point corresponding to $t = \frac{\pi}{2}$. *Hint:* Evaluate $\mathbf{T}$ at $t = \frac{\pi}{2}$ before finding $\mathbf{N}$ and $\mathbf{B}$.
(b) Find the equation of the osculating plane at the point corresponding to $t = \frac{\pi}{2}$.

39. Let $\mathbf{r}(t) = \left\langle \ln t, t, \frac{t^2}{2} \right\rangle$. Find the equation of the osculating plane corresponding to $t = 1$.

40. If a planet has zero mass ($m = 0$), then Newton's laws of motion reduce to $\mathbf{r}''(t) = \mathbf{0}$ and the orbit is a straight line $\mathbf{r}(t) = \mathbf{r}_0 + t\mathbf{v}_0$, where $\mathbf{r}_0 = \mathbf{r}(0)$ and $\mathbf{v}_0 = \mathbf{r}'(0)$ (Figure 1). Show that the area swept out by the radial vector at time t is $A(t) = \frac{1}{2}\|\mathbf{r}_0 \times \mathbf{v}_0\|t$, and thus Kepler's Second Law holds in this situation as well (because the rate of change of swept-out area is constant).

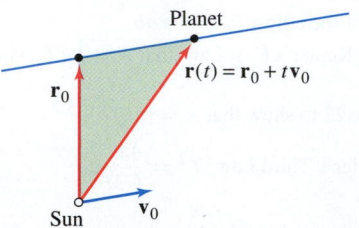

FIGURE 1

41. Suppose the orbit of a planet is an ellipse of eccentricity $e = c/a$ and period T (Figure 2). Use Kepler's Second Law to show that the time required to travel from A' to B' is equal to

$$\left(\frac{1}{4} + \frac{e}{2\pi} \right) T$$

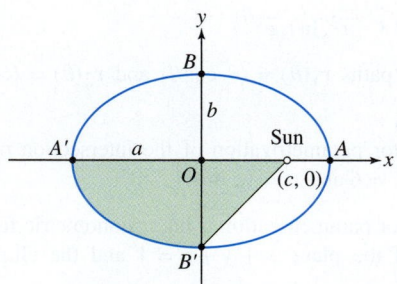

FIGURE 2

42. The period of Mercury is approximately 88 days, and its orbit has eccentricity 0.205. How much longer does it take Mercury to travel from A' to B' than from B' to A (Figure 2)?

GOES 12 Satellite, NASA, NOAA

14 DIFFERENTIATION IN SEVERAL VARIABLES

The circulation of weather systems around areas of low pressure can be understood using the gradient vector, an important tool arising in multivariable differentiation.

I n this chapter, we extend the concepts and techniques of differential calculus to functions of several variables. As we will see, a function f that depends on two or more variables has not just one derivative but rather a set of *partial derivatives*, one for each variable. The partial derivatives are the components of the gradient vector, which provides valuable insight into the function's behavior. In the last two sections, we apply the tools we have developed to optimization in several variables.

14.1 Functions of Two or More Variables

A familiar example of a function of two variables is the area A of a rectangle, equal to the product xy of the base x and height y. We write

$$A(x, y) = xy$$

or $A = f(x, y)$, where $f(x, y) = xy$. An example in three variables is the distance from a point $P = (x, y, z)$ to the origin:

$$g(x, y, z) = \sqrt{x^2 + y^2 + z^2}$$

An important but less familiar example is the density of seawater, denoted ρ, which is a function of salinity S and temperature T and is a key factor in the makeup of ocean current systems (Figure 1). Although there is no simple formula for $\rho(S, T)$, scientists determine values of the function experimentally. According to Table 1, if $S = 32$ (in parts per thousand or ppt) and $T = 10°C$, then

$$\rho(32, 10) = 1.0246 \text{ kg/m}^3$$

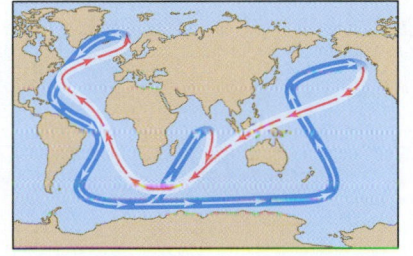

FIGURE 1 The global climate is influenced by the ocean "conveyer belt," a system of deep currents driven by variations in seawater density.

TABLE 1 Seawater Density ρ (kg/m^3) as a Function of Temperature and Salinity.

°C	Salinity (ppt)		
	32	32.5	33
5	1.0253	1.0257	1.0261
10	1.0246	1.0250	1.0254
15	1.0237	1.0240	1.0244
20	1.0224	1.0229	1.0232

A function of n variables is a function f that assigns a real number $f(x_1, \dots, x_n)$ to each n-tuple $(x_1, \dots, x_n)$ in a domain in $\mathbf{R}^n$. Sometimes we write $f(P)$ for the value of f at a point $P = (x_1, \dots, x_n)$. When f is defined by an algebraic expression involving $x_1, \dots, x_n$, we usually take as the domain the set of all n-tuples for which $f(x_1, \dots, x_n)$ is defined. The range of f is the set of all values $f(x_1, \dots, x_n)$ for $(x_1, \dots, x_n)$ in the domain. Since we focus on functions of two or three variables, we shall often use the variables x, y, and z (rather than x_1, x_2, x_3).

EXAMPLE 1 Sketch the domains of:

(a) $f(x, y) = \sqrt{9 - x^2 - y}$

(b) $g(x, y, z) = x\sqrt{y} + \ln(z - 1)$

What are the ranges of these functions?

Solution

(a) $f(x, y) = \sqrt{9 - x^2 - y}$ is defined only when $9 - x^2 - y \geq 0$, or $y \leq 9 - x^2$. Thus, the domain consists of all points (x, y) lying on or below the parabola $y = 9 - x^2$ [Figure 2(A)]:

$$\mathcal{D} = \{(x, y) : y \leq 9 - x^2\}$$

To determine the range, note that f is a nonnegative function and that $f(0, y) = \sqrt{9 - y}$. Since $9 - y$ can be any nonnegative number, $f(0, y)$ takes on all nonnegative values. Therefore, the range of f is the infinite interval $[0, \infty)$.

(b) $g(x, y, z) = x\sqrt{y} + \ln(z - 1)$ is defined only when both $\sqrt{y}$ and $\ln(z - 1)$ are defined. Therefore, both $y \geq 0$ and $z > 1$ are required, so the domain of the function is given by $\{(x, y, z) : y \geq 0, z > 1\}$ [Figure 2(B)]. The range of g is the entire real line $\mathbf{R}$. Indeed, for the particular choices $y = 1$ and $z = 2$, we have $g(x, 1, 2) = x\sqrt{1} + \ln 1 = x$, and since x is arbitrary, we see that g takes on all values. ∎

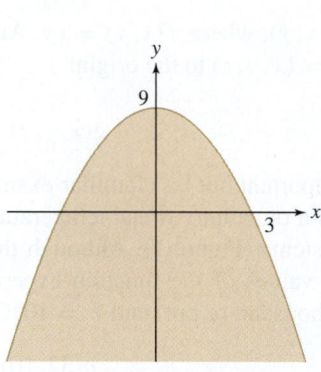

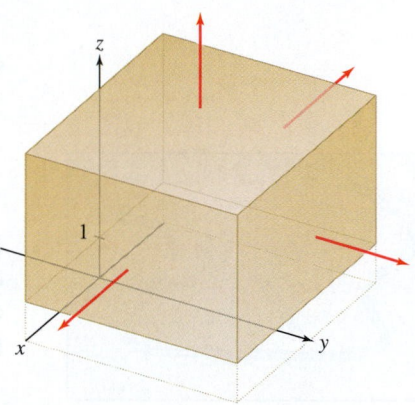

(A) The domain of $f(x, y) = \sqrt{9 - x^2 - y}$ is the set of all points lying below the parabola $y = 9 - x^2$.

(B) The domain of $g(x, y, z) = x\sqrt{y} + \ln(z - 1)$ is the set of points with $y \geq 0$ and $z > 1$. The domain continues out to infinity in the directions indicated by the arrows.

FIGURE 2

Graphing Functions of Two Variables

In single-variable calculus, we use graphs to visualize the important features of a function [Figure 3(A)]. Graphs play a similar role for functions of two variables. The graph of a function f of two variables consists of all points $(a, b, f(a, b))$ in $\mathbf{R}^3$ for (a, b) in the domain $\mathcal{D}$ of f. Assuming that f is continuous (as defined in the next section), the graph is a surface whose *height* above or below the xy-plane at (a, b) is the value of the function $f(a, b)$ [Figure 3(B)]. We often write $z = f(x, y)$ to stress that the z-coordinate of a point on the graph is a function of x and y.

EXAMPLE 2 Sketch the graph of $f(x, y) = 2x^2 + 5y^2$.

Solution The graph is a paraboloid (Figure 4), which we saw in Section 12.6. We sketch the graph using the fact that the horizontal cross section at height $z = c$ is the ellipse $2x^2 + 5y^2 = c$. ∎

Plotting more complicated graphs by hand can be difficult. Fortunately, graphing technology (e.g., graphing calculators, computer algebra systems) eliminates the labor and greatly enhances our ability to explore functions graphically. Graphs can be rotated and viewed from different perspectives (Figure 5).

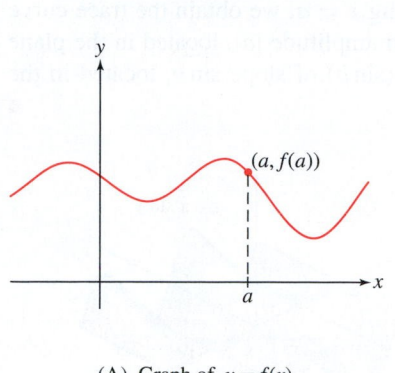

(A) Graph of $y = f(x)$

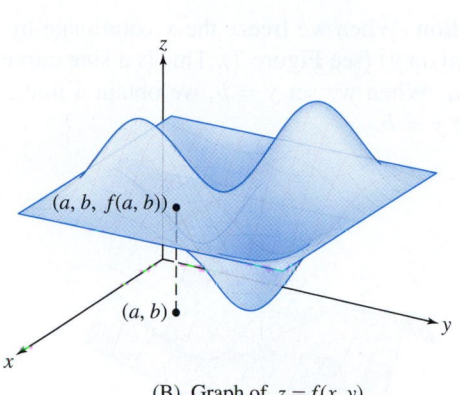

(B) Graph of $z = f(x, y)$

FIGURE 3

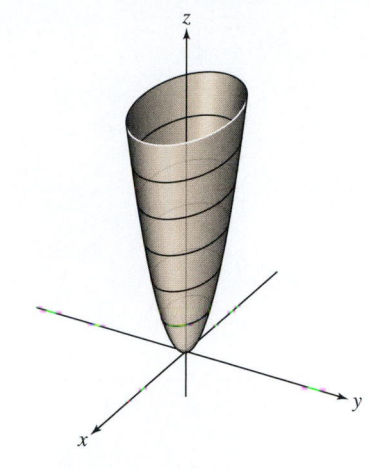

FIGURE 4 Graph of $f(x, y) = 2x^2 + 5y^2$.

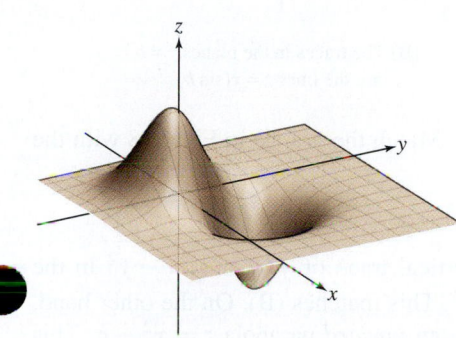

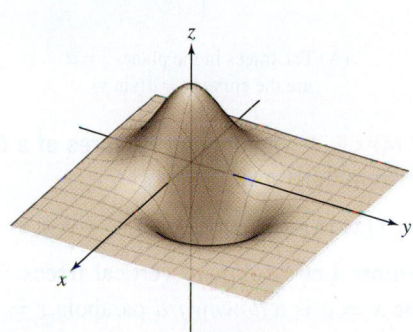

DF **FIGURE 5** Different views of the graph of $g(x, y) = e^{-x^2 - y^2} - e^{-(x-1)^2 - (y-1)^2}$.

Traces

One way of analyzing the graph of a function $f(x, y)$ is to freeze the x-coordinate by setting $x = a$ and examine the resulting curve given by $z = f(a, y)$. Similarly, we may set $y = b$ and consider the curve $z = f(x, b)$. Curves of this type are called **vertical traces**. They are obtained by intersecting the graph with planes parallel to a vertical coordinate plane (Figure 6):

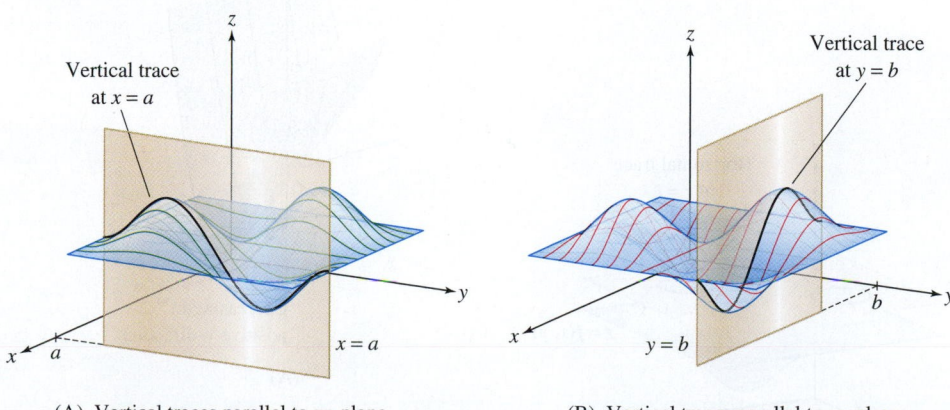

(A) Vertical traces parallel to yz-plane

(B) Vertical traces parallel to xz-plane

DF **FIGURE 6**

- **Vertical trace in the plane** $x = a$: Intersection of the graph with the vertical plane $x = a$, consisting of all points $(a, y, f(a, y))$
- **Vertical trace in the plane** $y = b$: Intersection of the graph with the vertical plane $y = b$, consisting of all points $(x, b, f(x, b))$

EXAMPLE 3 Describe the vertical traces of $f(x, y) = x(\sin y)$.

Solution When we freeze the x-coordinate by setting $x = a$, we obtain the trace curve $z = a(\sin y)$ (see Figure 7). This is a sine curve with amplitude $|a|$, located in the plane $x = a$. When we set $y = b$, we obtain a line $z = x(\sin b)$ of slope $\sin b$, located in the plane $y = b$. ∎

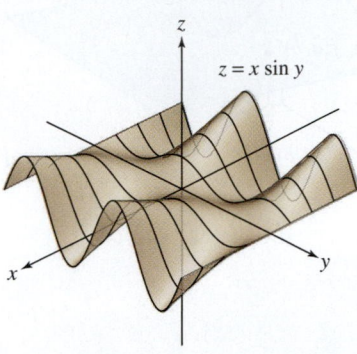

(A) The traces in the planes $x = a$ are the curves $z = a(\sin y)$.

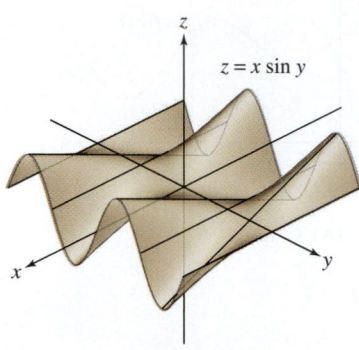

(B) The traces in the planes $y = b$ are the lines $z = x(\sin b)$.

DF FIGURE 7 Vertical traces of $f(x, y) = x(\sin y)$.

EXAMPLE 4   Identifying Features of a Graph   Match the graphs in Figure 8 with the following functions:

   (i) $f(x, y) = x - y^2$ (ii) $g(x, y) = x^2 - y$

Solution Let's compare vertical traces. The vertical trace of $f(x, y) = x - y^2$ in the plane $x = a$ is a *downward* parabola $z = a - y^2$. This matches (B). On the other hand, the vertical trace of $g(x, y)$ in the plane $y = b$ is an *upward* parabola $z = x^2 - b$. This matches (A).

   Notice also that $f(x, y) = x - y^2$ is an increasing function of x [i.e., $f(x, y)$ increases as x increases] as in (B), whereas $g(x, y) = x^2 - y$ is a decreasing function of y as in (A). ∎

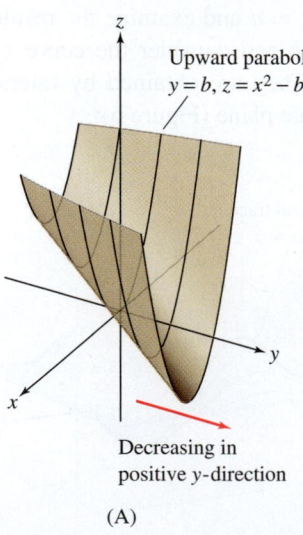

Upward parabolas
$y = b, z = x^2 - b$

Decreasing in positive y-direction

(A)

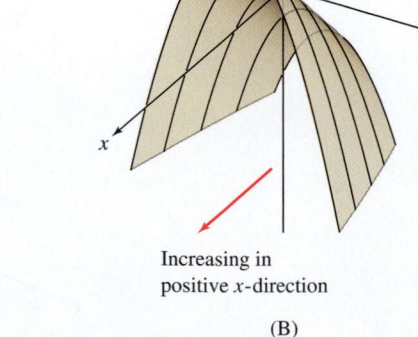

Downward parabolas
$x = a, z = a - y^2$

Increasing in positive x-direction

(B)

FIGURE 8

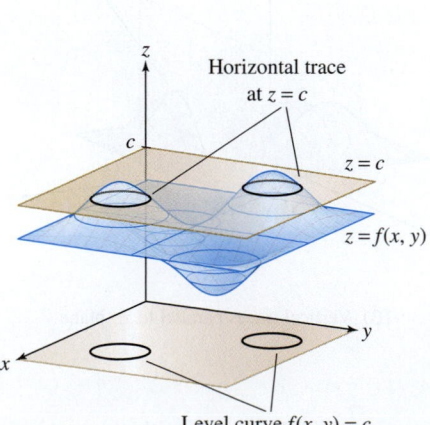

Horizontal trace at $z = c$

$z = c$

$z = f(x, y)$

Level curve $f(x, y) = c$

DF FIGURE 9 The level curve consists of all points (x, y) where the function takes on the value c.

Level Curves and Contour Maps

In addition to vertical traces, the graph of $f(x, y)$ has horizontal traces. These traces and their associated level curves are especially important in analyzing the behavior of the function (Figure 9):

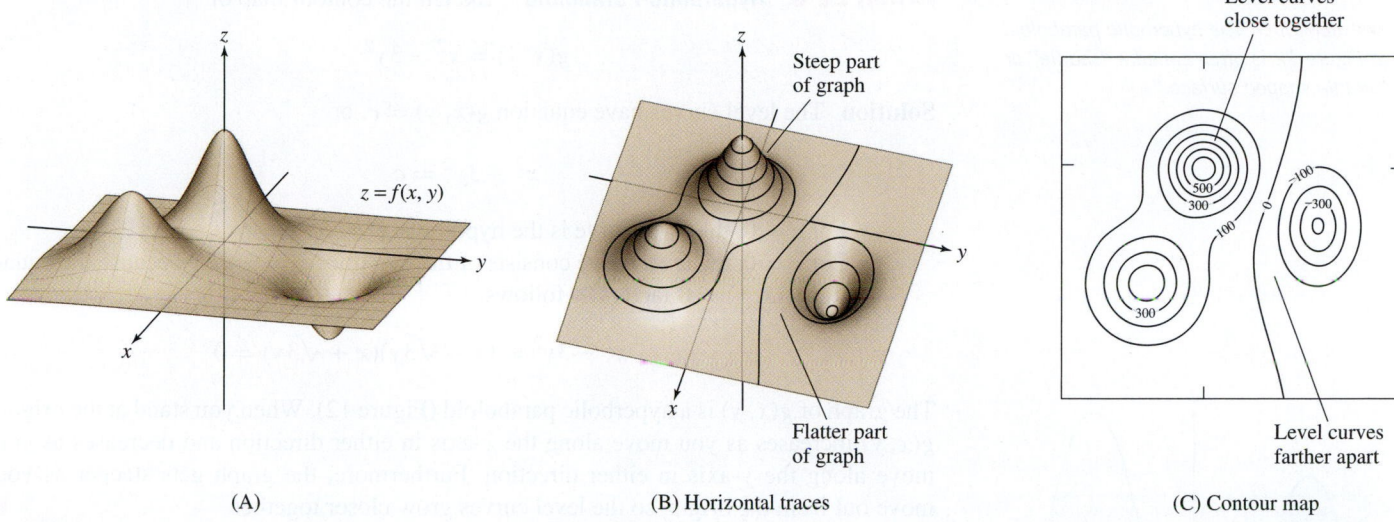

(A)

(B) Horizontal traces

(C) Contour map

Steep part of graph

Flatter part of graph

Level curves close together

Level curves farther apart

$z = f(x, y)$

FIGURE 10

*On contour maps, level curves are often referred to as **contour lines**. When we refer to level curves on a contour map, we mean the curves that are actually displayed. Keep in mind that between the displayed level curves there are additional curves associated with other values of f.*

- **Horizontal trace at height c:** Intersection of the graph with the horizontal plane $z = c$, consisting of the points $(x, y, f(x, y))$ such that $f(x, y) = c$
- **Level curve:** The curve $f(x, y) = c$ in the xy-plane

Thus, the level curve corresponding to c consists of all points (x, y) in the domain of f in the xy-plane where the function takes the value c. Each level curve is the projection onto the xy-plane of the horizontal trace on the graph that lies above it.

A **contour map** is a plot in the domain in the xy-plane that shows the level curves $f(x, y) = c$ for equally spaced values of c. The interval m between the values of c is called the **contour interval**. When you move from one level curve to the next, the value of $f(x, y)$ (and hence the height of the graph) changes by $\pm m$.

Figure 10 compares the graph of a function $f(x, y)$ in (A) and its horizontal traces in (B) with the contour map in (C). The contour map in (C) has contour interval $m = 100$.

It is important to understand how the contour map indicates the steepness of the graph. If the level curves are close together, then a small move from one level curve to the next in the xy-plane leads to a large change in height. In other words, *the level curves are close together if the graph is steep* (Figure 10). Similarly, the graph is flatter when the level curves are farther apart.

EXAMPLE 5 **Elliptic Paraboloid** Sketch the contour map of

$$f(x, y) = x^2 + 3y^2$$

and comment on the spacing of the contour curves.

Solution The level curves have equation $f(x, y) = c$, or

$$x^2 + 3y^2 = c$$

- For $c > 0$, the level curve is an ellipse.
- For $c = 0$, the level curve is just the point $(0, 0)$ because $x^2 + 3y^2 = 0$ only for $(x, y) = (0, 0)$.
- There is no level curve for $c < 0$ because $f(x, y)$ is never negative.

The graph of $f(x, y)$ is an elliptic paraboloid (Figure 11). As we move away from the origin, $f(x, y)$ increases more rapidly. The graph gets steeper, and the level curves become closer together.

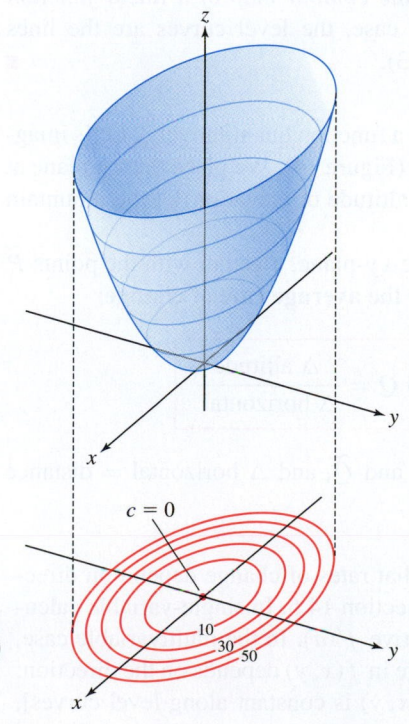

FIGURE 11 $f(x, y) = x^2 + 3y^2$. Contour interval $m = 10$.

← *REMINDER The hyperbolic paraboloid in Figure 12 is often called a "saddle" or "saddle-shaped surface."*

EXAMPLE 6   **Hyperbolic Paraboloid**   Sketch the contour map of

$$g(x, y) = x^2 - 3y^2$$

Solution   The level curves have equation $g(x, y) = c$, or

$$x^2 - 3y^2 = c$$

- For $c \neq 0$, the level curve is the hyperbola $x^2 - 3y^2 = c$.
- For $c = 0$, the level curve consists of the two lines $x = \pm\sqrt{3}y$ because the equation $g(x, y) = 0$ factors as follows:

$$x^2 - 3y^2 = (x - \sqrt{3}y)(x + \sqrt{3}y) = 0$$

The graph of $g(x, y)$ is a hyperbolic paraboloid (Figure 12). When you stand at the origin, $g(x, y)$ increases as you move along the x-axis in either direction and decreases as you move along the y-axis in either direction. Furthermore, the graph gets steeper as you move out from the origin, so the level curves grow closer together. ∎

EXAMPLE 7   **Contour Map of a Linear Function**   Sketch the graph of

$$f(x, y) = 12 - 2x - 3y$$

and the associated contour map with contour interval $m = 4$.

Solution   Note that if we set $z = f(x, y)$, we can write the equation as $2x + 3y + z = 12$. As we discussed in Section 12.5, this is the equation of a plane. To plot the graph, we find the intercepts of the plane with the axes (Figure 13). The graph intercepts the z-axis at $z = f(0, 0) = 12$. To find the x-intercept, we set $y = z = 0$ to obtain $12 - 2x - 3(0) = 0$, or $x = 6$. Similarly, solving $12 - 3y = 0$ gives the y-intercept $y = 4$. The graph is the plane determined by the three intercepts.

In general, the level curves of a linear function $f(x, y) = qx + ry + s$ are the lines with equation $qx + ry + s = c$. Therefore, *the contour map of a linear function consists of equally spaced parallel lines*. In our case, the level curves are the lines $12 - 2x - 3y = c$, or $2x + 3y = 12 - c$ (Figure 13). ∎

How can we measure steepness of the graph of a function quantitatively? Let's imagine the surface given by $z = f(x, y)$ as a mountain (Figure 14). We place the xy-plane at sea level, so that $f(a, b)$ is the height (also called altitude or elevation) of the mountain above sea level at the point (a, b) in the plane.

Figure 14(A) shows two points P and Q in the xy-plane, together with the points $\widetilde{P}$ and $\widetilde{Q}$ on the graph that lie above them. We define the **average rate of change**:

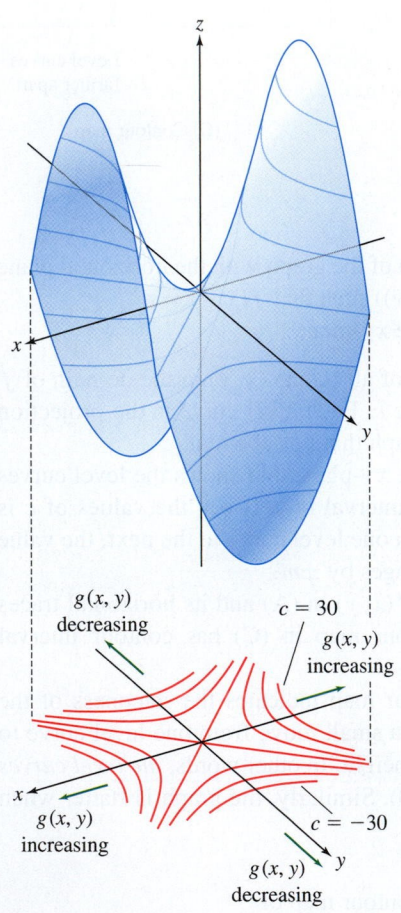

$$\text{average rate of change from } P \text{ to } Q = \frac{\Delta \text{ altitude}}{\Delta \text{ horizontal}}$$

where Δ altitude = change in the height from $\widetilde{P}$ and $\widetilde{Q}$, and Δ horizontal = distance from P to Q.

FIGURE 12 $g(x, y) = x^2 - 3y^2$. Contour interval $m = 10$.

CONCEPTUAL INSIGHT   We will discuss the idea that rates of change depend on direction when we come to directional derivatives in Section 14.5. In single-variable calculus, we measure the rate of change by the derivative $f'(a)$. In the multivariable case, there is no single rate of change because the change in $f(x, y)$ depends on the direction: The rate is zero along a level curve [because $f(x, y)$ is constant along level curves], and the rate is nonzero in directions pointing from one level curve to the next [Figure 14(B)].

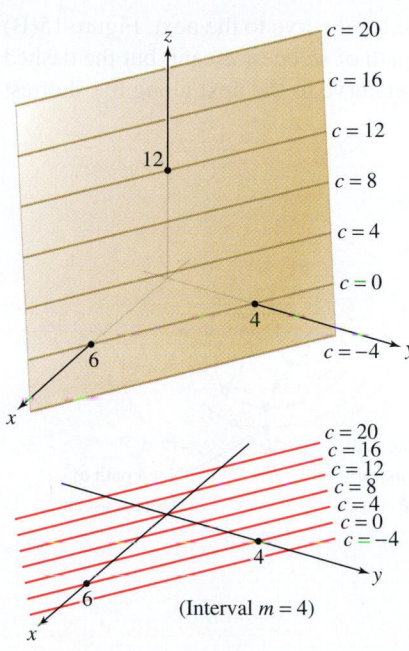

FIGURE 13 Graph and contour map of $f(x, y) = 12 - 2x - 3y$.

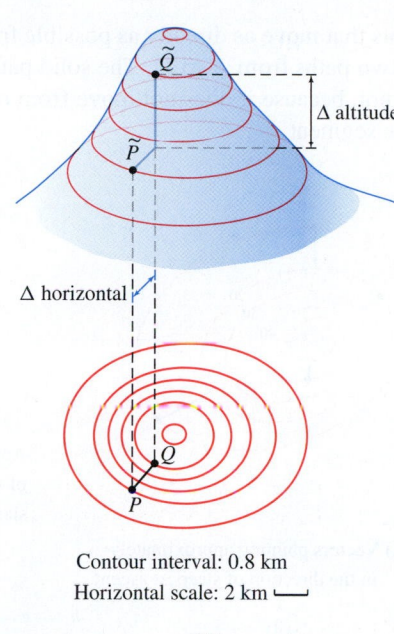

FIGURE 14

Contour interval: 0.8 km
Horizontal scale: 2 km

(A)

Contour interval: 100 m
Horizontal scale: 200 m

(B)

A contour map is like a topographic map that hikers would use to help understand the terrain that they encounter. They are both two-dimensional representations of the features of three-dimensional structures.

EXAMPLE 8 **Average Rate of Change Depends on Direction** Compute the average rate of change from A to the points B, C, and D in Figure 14(B).

Solution The contour interval in Figure 14(B) is $m = 100$ m. Segments $\overline{AB}$ and $\overline{AC}$ both span two level curves, so the change in altitude is 200 m in both cases. The horizontal scale shows that $\overline{AB}$ corresponds to a horizontal change of 200 m, and $\overline{AC}$ corresponds to a horizontal change of 400 m. On the other hand, there is no change in altitude from A to D. Therefore,

$$\text{average rate of change from } A \text{ to } B = \frac{\Delta \text{ altitude}}{\Delta \text{ horizontal}} = \frac{200}{200} = 1.0$$

$$\text{average rate of change from } A \text{ to } C = \frac{\Delta \text{ altitude}}{\Delta \text{ horizontal}} = \frac{200}{400} = 0.5$$

$$\text{average rate of change from } A \text{ to } D = \frac{\Delta \text{ altitude}}{\Delta \text{ horizontal}} = 0$$

We see here explicitly that the average rate varies according to the direction. ■

A path of steepest descent is the same as a path of steepest ascent traversed in the opposite direction. Water flowing down a mountain approximately follows a path of steepest descent.

When we walk up a mountain, the incline at each moment depends on the path we choose. If we walk around the mountain, our altitude does not change at all. On the other hand, at each point there is a *steepest* direction in which the altitude increases most rapidly. On a contour map, the steepest direction is approximately the direction that takes us to the closest point on the next highest level curve [Figure 15(A)]. We say "approximately" because the terrain may vary between level curves. A **path of steepest ascent** is a path that begins at a point P and, everywhere along the way, points in the steepest direction. We can approximate the path of steepest ascent by drawing a sequence of segments that move as directly as possible from one level curve to the next. Figure 15(B) shows two paths from P to Q. The solid path is a path of steepest ascent, but the dashed path is not, because it does not move from one level curve to the next along the shortest possible segment.

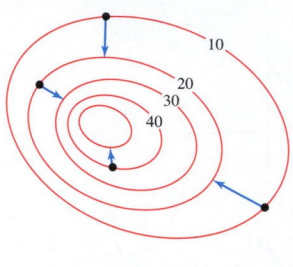

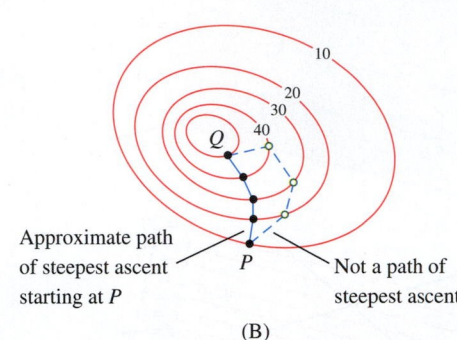

(A) Vectors pointing approximately in the direction of steepest ascent

Approximate path of steepest ascent starting at P

Not a path of steepest ascent

(B)

FIGURE 15

More Than Two Variables

There are many modeling situations where it is necessary to use a function of more than two variables. For instance, we might want to keep track of temperature at the various points in a room using a function $T(x, y, z)$ that depends on the three variables corresponding to the coordinates of each point. In making quantitative models of the economy, functions often depend on more than 100 variables.

Unfortunately, it is not possible to draw the graph of a function of more than two variables. The graph of a function $f(x, y, z)$ would consist of the set of points $(x, y, z, f(x, y, z))$ in four-dimensional space $\mathbf{R}^4$. However, just as we can use contour maps to visualize a three-dimensional mountain using curves on a two-dimensional plane, it is possible to draw the **level surfaces** of a function of three variables $f(x, y, z)$. These are the surfaces with equation $f(x, y, z) = c$ for different values of c. For example, the level surfaces of

$$f(x, y, z) = x^2 + y^2 + z^2$$

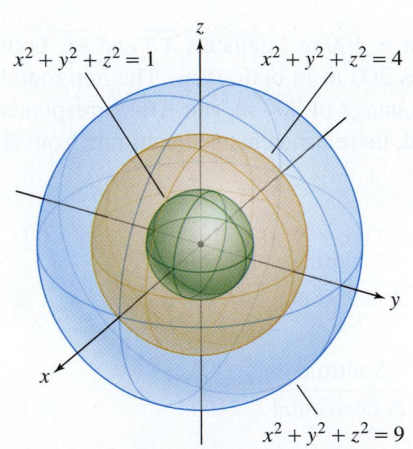

FIGURE 16 The level surfaces of $f(x, y, z) = x^2 + y^2 + z^2$ are spheres.

are the spheres with equation $x^2 + y^2 + z^2 = c$ (Figure 16). In the case of a function $T(x, y, z)$ that represents temperature of points in space, we call the level surfaces corresponding to $T(x, y, z) = k$ the **isotherms**. These are the collections of points, all of which have the same temperature k.

For functions of four or more variables, we can no longer visualize the graph or the level surfaces. We must rely on intuition developed through the study of functions of two and three variables.

EXAMPLE 9 Describe the level surfaces of $g(x, y, z) = x^2 + y^2 - z^2$.

Solution The level surface for $c = 0$ is the cone $x^2 + y^2 - z^2 = 0$. For $c \neq 0$, the level surfaces are the hyperboloids $x^2 + y^2 - z^2 = c$. The hyperboloid has one sheet if $c > 0$ and it lies outside the cone. The hyperboloid has two sheets if $c < 0$, one sheet lies inside the upper part of the cone, the other lies inside the lower part (Figure 17). ■

SECTION 14.1 | Functions of Two or More Variables 797

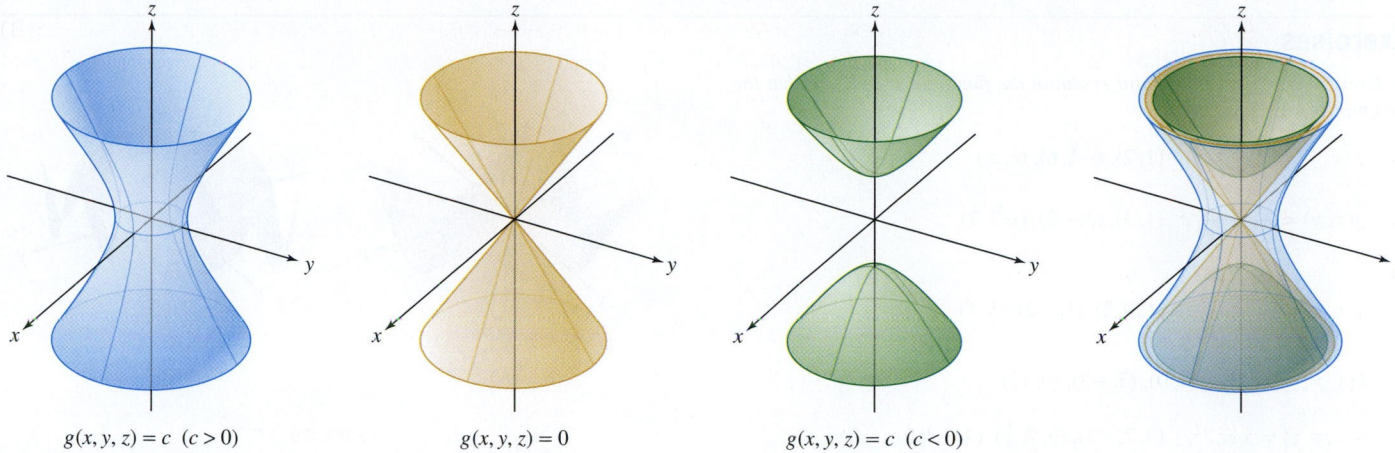

$$g(x, y, z) = c \ (c > 0) \qquad g(x, y, z) = 0 \qquad g(x, y, z) = c \ (c < 0)$$

FIGURE 17 Level surfaces of $g(x, y, z) = x^2 + y^2 - z^2$.

14.1 SUMMARY

- The domain $\mathcal{D}$ of a function $f(x_1, \ldots, x_n)$ of n variables is the set of n-tuples $(a_1, \ldots, a_n)$ in $\mathbf{R}^n$ for which $f(a_1, \ldots, a_n)$ is defined. The range of f is the set of values taken on by f.
- The graph of a continuous real-valued function $f(x, y)$ is the surface in $\mathbf{R}^3$ consisting of the points $(a, b, f(a, b))$ for (a, b) in the domain $\mathcal{D}$ of f.
- A *vertical trace* is a curve obtained by intersecting the graph with a vertical plane $x = a$ or $y = b$.
- A *level curve* is a curve in the xy-plane defined by an equation $f(x, y) = c$. The level curve $f(x, y) = c$ is the projection onto the xy-plane of the *horizontal trace* curve, obtained by intersecting the graph with the horizontal plane $z = c$.
- A *contour map* shows the level curves $f(x, y) = c$ for equally spaced values of c. The spacing m is called the *contour interval*.
- When reading a contour map, keep in mind:
  - Your altitude does not change when you hike along a level curve.
  - Your altitude increases or decreases by m (the contour interval) when you hike from one level curve to the next.
- The spacing of the level curves indicates steepness: They are closer together where the graph is steeper.
- The *average rate of change* from P to Q is the ratio $\dfrac{\Delta\text{altitude}}{\Delta\text{horizontal}}$.
- A direction of steepest ascent at a point P is a direction along which $f(x, y)$ increases most rapidly. The steepest direction is obtained (approximately) by drawing the segment from P to the nearest point on the next level curve.
- Level surfaces can be used to understand a function $f(x, y, z)$. In the case where the function represents temperature, we call the level surfaces isotherms.

14.1 EXERCISES

Preliminary Questions

1. What is the difference between a horizontal trace and a level curve? How are they related?

2. Describe the trace of $f(x, y) = x^2 - \sin(x^3 y)$ in the xz-plane.

3. Is it possible for two different level curves of a function to intersect? Explain.

4. Describe the contour map of $f(x, y) = x$ with contour interval 1.

5. How will the contour maps of

$$f(x, y) = x \quad \text{and} \quad g(x, y) = 2x$$

with contour interval 1 look different?

Exercises

In Exercises 1–6, at each point evaluate the function or indicate that the function is undefined there.

1. $f(x, y) = x + yx^3$, $(1, 2), (-1, 6), (e, \pi)$

2. $g(x, y) = \dfrac{y}{x^2 - y^2}$, $(1, 3), (3, -3), (\sqrt{2}, 2)$

3. $h(x, y) = \dfrac{\sqrt{x - y^2}}{x - y}$, $(20, 2), (1, -2), (1, 1)$

4. $k(x, y) = xe^{-y}$, $(1, 0), (3, -3), (0, 12)$

5. $h(x, y, z) = xyz^{-2}$, $(3, 7, -2), (3, 2, \frac{1}{4}), (4, -4, 0)$

6. $w(r, s, t) = \dfrac{r - s}{\sin t}$, $(2, 2, \frac{\pi}{2}), (\pi, \pi, \pi), (-2, 2, \frac{\pi}{6})$

In Exercises 7–14, sketch the domain of the function.

7. $f(x, y) = 12x - 5y$

8. $f(x, y) = \sqrt{81 - x^2}$

9. $f(x, y) = \ln(4x^2 - y)$

10. $h(x, t) = \dfrac{1}{x + t}$

11. $g(y, z) = \dfrac{1}{z + y^2}$

12. $f(x, y) = \sin \dfrac{y}{x}$

13. $F(I, R) = \sqrt{IR}$

14. $f(x, y) = \cos^{-1}(x + y)$

In Exercises 15–18, describe the domain and range of the function.

15. $f(x, y, z) = xz + e^y$

16. $f(x, y, z) = x\sqrt{y + z}e^{z/x}$

17. $P(r, s, t) = \sqrt{16 - r^2 s^2 t^2}$

18. $g(r, s) = \cos^{-1}(rs)$

19. Match graphs (A) and (B) in Figure 18 with the functions:

(i) $f(x, y) = -x + y^2$

(ii) $g(x, y) = x + y^2$

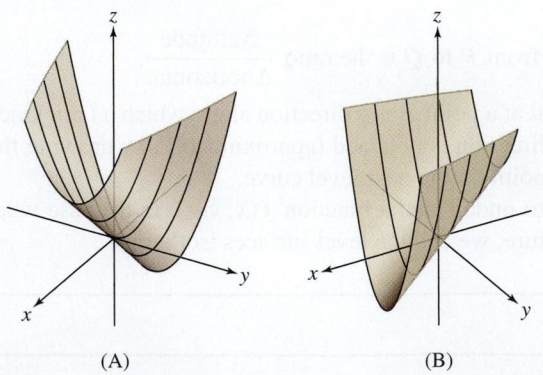

(A)　　　　(B)

FIGURE 18

20. Match each of graphs (A) and (B) in Figure 19 with one of the following functions:

(i) $f(x, y) = (\cos x)(\cos y)$

(ii) $g(x, y) = \cos(x^2 + y^2)$

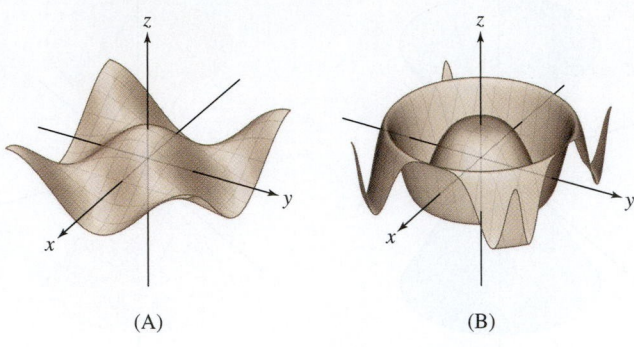

(A)　　　　(B)

FIGURE 19

21. Match the functions (a)–(f) with their graphs (A)–(F) in Figure 20.

(a) $f(x, y) = |x| + |y|$

(b) $f(x, y) = \cos(x - y)$

(c) $f(x, y) = \dfrac{-1}{1 + 9x^2 + y^2}$

(d) $f(x, y) = \cos(y^2)e^{-0.1(x^2 + y^2)}$

(e) $f(x, y) = \dfrac{-1}{1 + 9x^2 + 9y^2}$

(f) $f(x, y) = \cos(x^2 + y^2)e^{-0.1(x^2 + y^2)}$

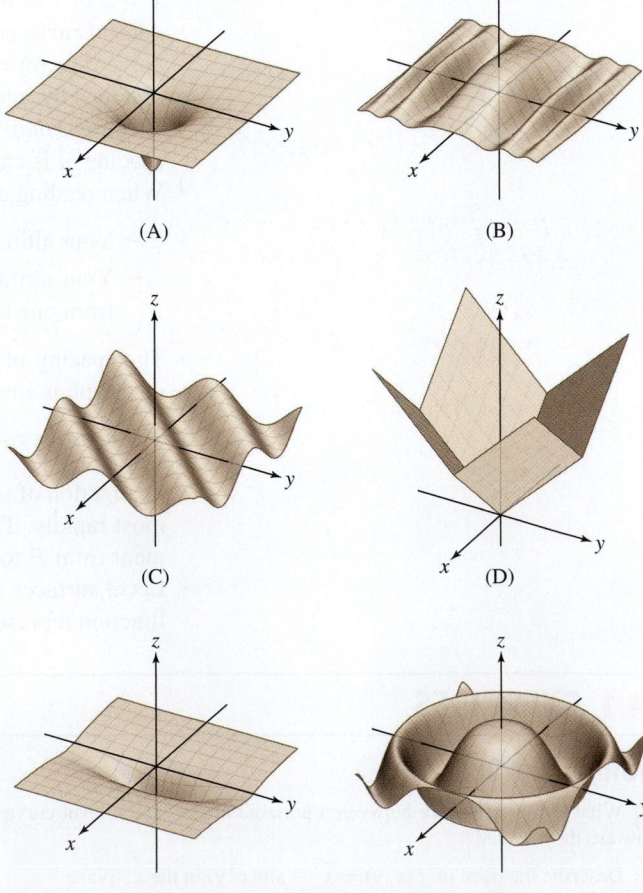

(A)　　　　(B)

(C)　　　　(D)

(E)　　　　(F)

FIGURE 20

22. Match the functions (a)–(d) with their contour maps (A)–(D) in Figure 21.

(a) $f(x, y) = 3x + 4y$

(b) $g(x, y) = x^3 - y$

(c) $h(x, y) = 4x - 3y$

(d) $k(x, y) = x^2 - y$

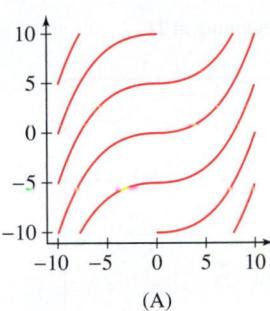

(A)

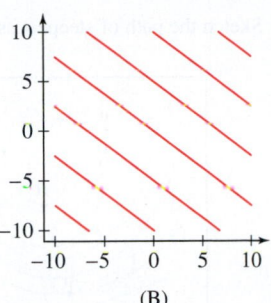

(B)

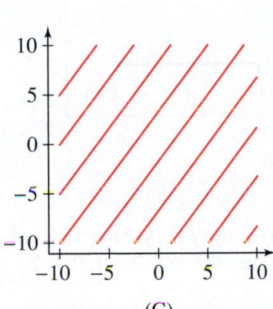

(C)

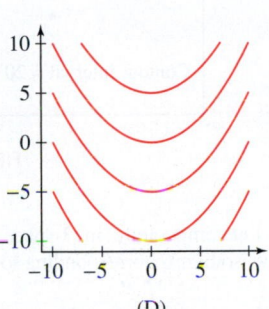

(D)

FIGURE 21

In Exercises 23–28, sketch the graph and draw several vertical and horizontal traces.

23. $f(x, y) = 12 - 3x - 4y$

24. $f(x, y) = \sqrt{4 - x^2 - y^2}$

25. $f(x, y) = x^2 + 4y^2$

26. $f(x, y) = y^2$

27. $f(x, y) = \sin(x - y)$

28. $f(x, y) = \dfrac{1}{x^2 + y^2 + 1}$

29. Sketch contour maps of $f(x, y) = x + y$ with contour intervals $m = 1$ and 2.

30. Sketch the contour map of $f(x, y) = x^2 + y^2$ with level curves $c = 0$, 4, 8, 12, 16.

In Exercises 31–38, draw a contour map of $f(x, y)$ with an appropriate contour interval, showing at least six level curves.

31. $f(x, y) = x^2 - y$

32. $f(x, y) = \dfrac{y}{x^2}$

33. $f(x, y) = \dfrac{y}{x}$

34. $f(x, y) = xy$

35. $f(x, y) = x^2 + 4y^2$

36. $f(x, y) = x + 2y - 1$

37. $f(x, y) = x^2$

38. $f(x, y) = 3x^2 - y^2$

39. [✎] Find the linear function whose contour map (with contour interval $m = 6$) is shown in Figure 22. What is the linear function if $m = 3$ (and the curve labeled $c = 6$ is relabeled $c = 3$)?

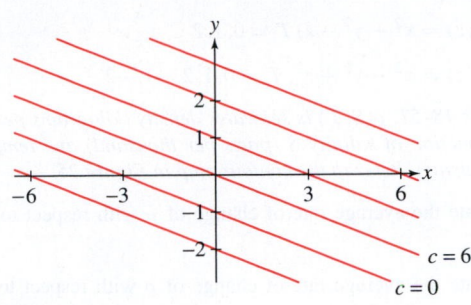

FIGURE 22 Contour map with contour interval $m = 6$.

40. Use the contour map in Figure 23 to calculate the average rate of change:

(a) from A to B.

(b) from A to C.

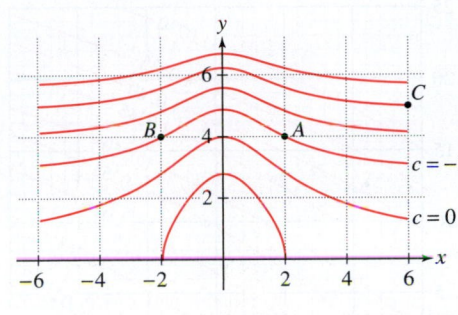

FIGURE 23

Exercises 41–43 refer to the map in Figure 24.

41. (a) At which of A–C is pressure increasing in the northern direction?

(b) At which of A–C is pressure increasing in the westerly direction?

42. For each of A–C indicate in which of the four cardinal directions, N, S, E, or W, pressure is increasing the greatest.

43. Rank the following states in order from greatest change in pressure across the state to least: Arkansas, Colorado, North Dakota, Wisconsin.

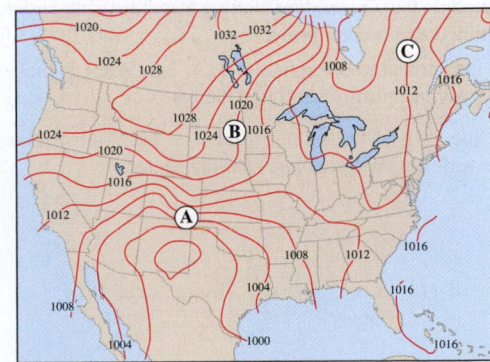

FIGURE 24 Atmospheric pressure (in millibars) over North America on March 26, 2009.

In Exercises 44–47, let $T(x, y, z)$ denote temperature at each point in space. Draw level surfaces (also called isotherms) corresponding to the fixed temperatures given.

44. $T(x, y, z) = 2x + 3y - z$, $T = 0, 1, 2$

45. $T(x, y, z) = x - y + 2z$, $T = 0, 1, 2$

46. $T(x, y, z) = x^2 + y^2 - z$, $T = 0, 1, 2$

47. $T(x, y, z) = x^2 - y^2 + z^2$, $T = 0, 1, 2, -1, -2$

In Exercises 48–51, $\rho(S, T)$ is seawater density (kilograms per cubic meter) as a function of salinity S (parts per thousand) and temperature T (degrees Celsius). Refer to the contour map in Figure 25.

48. Calculate the average rate of change of ρ with respect to T from B to A.

49. Calculate the average rate of change of ρ with respect to S from B to C.

50. At a fixed level of salinity, is seawater density an increasing or a decreasing function of temperature?

51. Does water density appear to be more sensitive to a change in temperature at point A or point B?

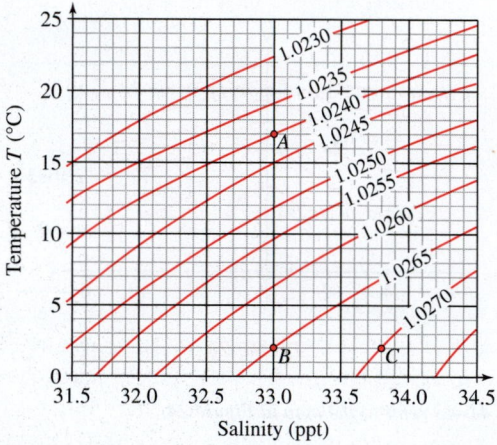

FIGURE 25 Contour map of seawater density $\rho(S, T)$ (kilograms per cubic meter).

In Exercises 52–55, refer to Figure 26.

52. Find the change in seawater density from A to B.

53. Estimate the average rate of change from A to B and from A to C.

54. Estimate the average rate of change from A to points i, ii, and iii.

55. Sketch the path of steepest ascent beginning at D.

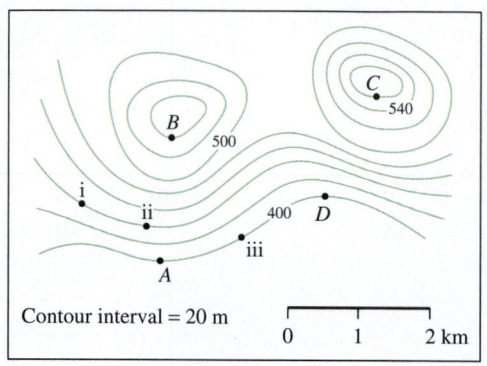

FIGURE 26

56. Let temperature in 3-space be given by $T(x, y, z) = x^2 + y^2 - z$. Draw isotherms corresponding to temperatures $T = -2, -1, 0, 1, 2$.

57. Let temperature in 3-space be given by $T(x, y, z) = \frac{x^2}{4} + \frac{y^2}{9} + z^2$. Draw isotherms corresponding to temperatures $T = 0, 1, 2$.

58. Let temperature in 3-space be given by $T(x, y, z) = x^2 - y^2 - z$. Draw isotherms corresponding to temperatures $T = -1, 0, 1$.

59. Let temperature in 3-space be given by $T(x, y, z) = x^2 - y^2 - z^2$. Draw isotherms corresponding to temperatures $T = -2, -1, 0, 1, 2$.

Further Insights and Challenges

60. 📝 The function $f(x, t) = t^{-1/2} e^{-x^2/t}$, whose graph is shown in Figure 27, models the temperature along a metal bar after an intense burst of heat is applied at its center point.

(a) Sketch the vertical traces at times $t = 1, 2, 3$. What do these traces tell us about the way heat diffuses through the bar?

(b) Sketch the vertical traces $x = c$ for $c = \pm 0.2, \pm 0.4$. Describe how temperature varies in time at points near the center.

61. Let

$$f(x, y) = \frac{x}{\sqrt{x^2 + y^2}} \quad \text{for } (x, y) \neq (0, 0)$$

Write f as a function $f(r, \theta)$ in polar coordinates, and use this to find the level curves of f.

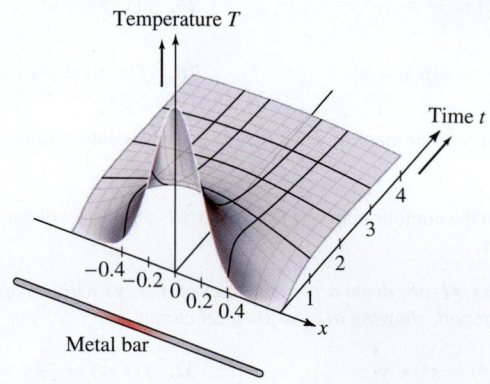

FIGURE 27 Graph of $f(x, t) = t^{-1/2} e^{-x^2/t}$ beginning shortly after $t = 0$.

14.2 Limits and Continuity in Several Variables

This section develops limits and continuity in the multivariable setting. We focus on functions of two variables, but similar definitions and results apply to functions of three or more variables.

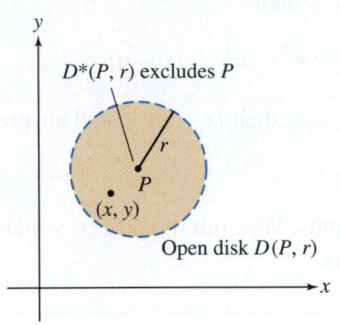

FIGURE 1 The open disk $D(P, r)$ consists of points (x, y) at distance $< r$ from P. It does not include the boundary circle.

Recall that on the real number line, a number x is close to a if the distance $|x - a|$ is small. In the plane, a point (x, y) is close to another point $P = (a, b)$ if the distance

$$d((x, y), (a, b)) = \sqrt{(x - a)^2 + (y - b)^2}$$ between them is small.

Note that if we take all the points that are a distance of less than r from $P = (a, b)$, as in Figure 1, this is a disk $D(P, r)$ centered at P that does not include its boundary. If we insist also that $d((x, y), (a, b)) \neq 0$, then we get a punctured disk that does not include P and that we denote $D^*(P, r)$.

Now assume that $f(x, y)$ is defined near P but not necessarily at P itself. In other words, $f(x, y)$ is defined for all (x, y) in some punctured disk $D^*(P, r)$ with $r > 0$. We say that $f(x, y)$ approaches the limit L as (x, y) approaches $P = (a, b)$ if $|f(x, y) - L|$ becomes arbitrarily small for (x, y) sufficiently close to $P = (a, b)$ [Figure 2(A)]. In this case, we write

$$\lim_{(x,y) \to P} f(x, y) = \lim_{(x,y) \to (a,b)} f(x, y) = L$$

Here is the formal definition.

DEFINITION **Limit**   Assume that $f(x, y)$ is defined near $P = (a, b)$. Then

$$\lim_{(x,y) \to P} f(x, y) = L$$

if, for any $\epsilon > 0$, there exists $\delta > 0$ such that if (x, y) satisfies

$$0 < d((x, y), (a, b)) < \delta, \quad \text{then} \quad |f(x, y) - L| < \epsilon$$

This is similar to the definition of the limit in one variable, but there is an important difference. In a one-variable limit, we require that $f(x)$ tend to L as x approaches a from two directions—the left and the right [Figure 2(B)]. In a multivariable limit, $f(x, y)$ must tend to L as (x, y) approaches P from infinitely many different directions [Figure 2(C)].

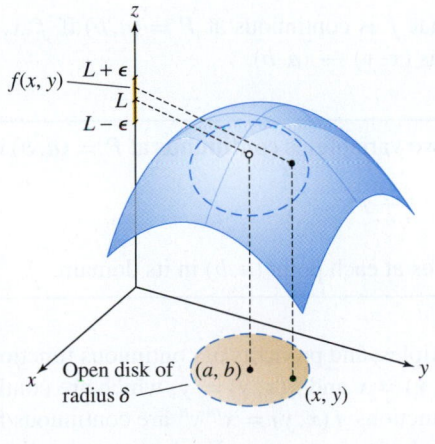

(A) $|f(x, y) - L| < \epsilon$ for all (x, y) inside the punctured disk

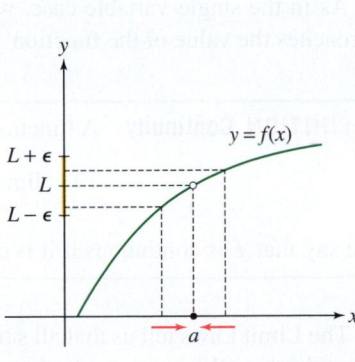

(B) In one variable, we can approach a from only two possible directions.

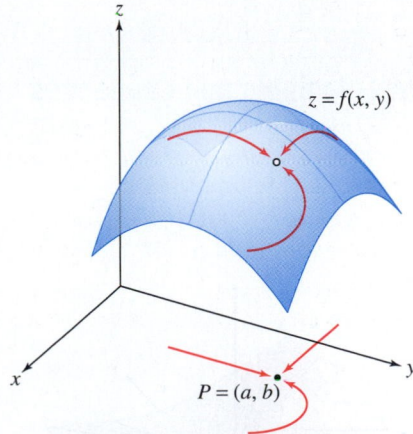

(C) In two variables, (x, y) can approach $P = (a, b)$ along any direction or path.

FIGURE 2

EXAMPLE 1 Show that **(a)** $\lim\limits_{(x,y) \to (a,b)} x = a$ and **(b)** $\lim\limits_{(x,y) \to (a,b)} y = b$.

Solution Let $P = (a, b)$. To verify (a), let $f(x, y) = x$ and $L = a$. We must show that for any $\epsilon > 0$, we can find $\delta > 0$ such that

$$\text{If} \quad 0 < d((x, y), (a, b)) < \delta, \quad \text{then} \quad |f(x, y) - L| = |x - a| < \epsilon \qquad \boxed{1}$$

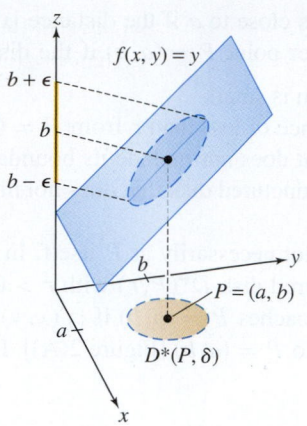

DF **FIGURE 3** If $|y - b| < \delta$ for $\delta = \epsilon$, then $|f(x, y) - b| < \epsilon$. Therefore,

$$\lim_{(x,y)\to(a,b)} y = b.$$

In fact, we can choose $\delta = \epsilon$, for if $d((x, y), (a, b)) < \epsilon$, then

$$(x - a)^2 + (y - b)^2 < \epsilon^2 \quad \Rightarrow \quad (x - a)^2 < \epsilon^2 \quad \Rightarrow \quad |x - a| < \epsilon$$

In other words, for any $\epsilon > 0$, if $0 < d((x, y), (a, b)) < \epsilon$, then $|x - a| < \epsilon$. This proves (a). The limit (b) is similar (see Figure 3). ∎

The following theorem lists the basic laws for limits. We omit the proofs, which are similar to the proofs of the single-variable Limit Laws.

THEOREM 1 Limit Laws   Assume that $\lim\limits_{(x,y)\to P} f(x, y)$ and $\lim\limits_{(x,y)\to P} g(x, y)$ exist.

(i) Sum Law:

$$\lim_{(x,y)\to P} (f(x, y) + g(x, y)) = \lim_{(x,y)\to P} f(x, y) + \lim_{(x,y)\to P} g(x, y)$$

(ii) Constant Multiple Law: For any number k,

$$\lim_{(x,y)\to P} kf(x, y) = k \lim_{(x,y)\to P} f(x, y)$$

(iii) Product Law:

$$\lim_{(x,y)\to P} f(x, y)\, g(x, y) = \left(\lim_{(x,y)\to P} f(x, y) \right) \left(\lim_{(x,y)\to P} g(x, y) \right)$$

(iv) Quotient Law: If $\lim\limits_{(x,y)\to P} g(x, y) \neq 0$, then

$$\lim_{(x,y)\to P} \frac{f(x, y)}{g(x, y)} = \frac{\lim\limits_{(x,y)\to P} f(x, y)}{\lim\limits_{(x,y)\to P} g(x, y)}$$

As in the single-variable case, we say that f is continuous at $P = (a, b)$ if $f(x, y)$ approaches the value of the function $f(a, b)$ as $(x, y) \to (a, b)$.

DEFINITION Continuity   A function f of two variables is **continuous** at $P = (a, b)$ if

$$\lim_{(x,y)\to(a,b)} f(x, y) = f(a, b)$$

We say that f is continuous if it is continuous at each point (a, b) in its domain.

The Limit Laws tell us that all sums, multiples, and products of continuous functions are continuous. When we apply them to $f(x, y) = x$ and $g(x, y) = y$, which are continuous by Example 1, we find that the power functions $f(x, y) = x^m y^n$ are continuous for all whole numbers m, n and that all polynomials are continuous. Furthermore, a rational function $h(x, y)/g(x, y)$, where h and g are polynomials, is continuous at all points (a, b) where $g(a, b) \neq 0$. As in the single-variable case, we can evaluate limits of continuous functions using substitution.

EXAMPLE 2 Evaluating Limits by Substitution   Show that

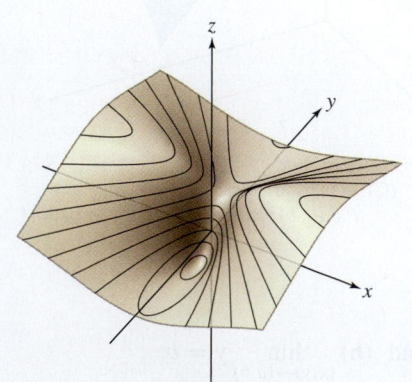

FIGURE 4 Top view of the graph $f(x, y) = \dfrac{3x + y}{x^2 + y^2 + 1}$.

$$f(x, y) = \frac{3x + y}{x^2 + y^2 + 1}$$

is continuous (Figure 4). Then evaluate $\lim\limits_{(x,y)\to(1,2)} f(x, y)$.

Solution The function f is continuous at all points (a, b) because it is a rational function whose denominator $Q(x, y) = x^2 + y^2 + 1$ is never zero. Therefore, we can evaluate the limit by substitution:

$$\lim_{(x,y)\to(1,2)} \frac{3x + y}{x^2 + y^2 + 1} = \frac{3(1) + 2}{1^2 + 2^2 + 1} = \frac{5}{6}$$ ∎

If f is a product $f(x, y) = h(x)g(y)$, where $h(x)$ and $g(y)$ are continuous, then the limit is a product of limits by the Product Law:

$$\lim_{(x,y)\to(a,b)} f(x, y) = \lim_{(x,y)\to(a,b)} h(x)g(y) = \left(\lim_{x\to a} h(x)\right)\left(\lim_{y\to b} g(y)\right)$$

EXAMPLE 3 **Product Functions** Evaluate $\lim_{(x,y)\to(3,0)} x^3 \frac{\sin y}{y}$.

Solution Since $\lim_{x\to 3} x^3$ and $\lim_{y\to 0} \frac{\sin y}{y}$ both exist, the desired limit can be expressed as a product of limits:

$$\lim_{(x,y)\to(3,0)} x^3 \frac{\sin y}{y} = \left(\lim_{x\to 3} x^3\right)\left(\lim_{y\to 0} \frac{\sin y}{y}\right) = (3^3)(1) = 27$$ ∎

Composition is another important way to build functions. If f is a function of two variables and $G(u)$ a function of one variable, then the composite function $G \circ f$ is the function of two variables given by $G(f(x, y))$. According to the next theorem, a composition of continuous functions is continuous.

> **THEOREM 2** **A Composition of Continuous Functions Is Continuous** If a function of two variables f is continuous at (a, b) and a function of one variable G is continuous at $c = f(a, b)$, then the composite function $G(f(x, y))$ is continuous at (a, b).

EXAMPLE 4 Write $H(x, y) = e^{-x^2+2y}$ as a composite function and evaluate

$$\lim_{(x,y)\to(1,2)} H(x, y)$$

Solution We have $H(x, y) = G(f(x, y))$, where $G(u) = e^u$ and $f(x, y) = -x^2 + 2y$. Both f and G are continuous, so H is also continuous and

$$\lim_{(x,y)\to(1,2)} H(x, y) = \lim_{(x,y)\to(1,2)} e^{-x^2+2y} = e^{-(1)^2+2(2)} = e^3$$ ∎

As we indicated previously, if a limit $\lim_{(x,y)\to(a,b)} f(x, y)$ exists and equals L, then $f(x, y)$ tends to L as (x, y) approaches (a, b) along any path. In the next example, we prove that a limit *does not exist* by showing that $f(x, y)$ approaches *different limits* when $(0, 0)$ is approached along different lines through the origin. We use three different methods on the problem to demonstrate a variety of approaches one can take.

EXAMPLE 5 **Showing a Limit Does Not Exist** Examine $\lim_{(x,y)\to(0,0)} \frac{x^2}{x^2 + y^2}$ numerically. Then prove that the limit does not exist.

Solution If the limit existed, we would expect the values of $f(x, y)$ in Table 1 to get closer to a limiting value L as (x, y) gets close to $(0, 0)$. However, the table suggests that:

- As (x, y) approaches $(0, 0)$ along the x-axis, $f(x, y)$ approaches 1.
- As (x, y) approaches $(0, 0)$ along the y-axis, $f(x, y)$ approaches 0.
- As (x, y) approaches $(0, 0)$ along the line $y = x$, $f(x, y)$ approaches 0.5.

Therefore, $f(x, y)$ does not seem to approach any fixed value L as $(x, y) \to (0, 0)$.

TABLE 1 **Values of** $f(x, y) = \dfrac{x^2}{x^2 + y^2}$

y \ x	−0.5	−0.4	−0.3	−0.2	−0.1	0	0.1	0.2	0.3	0.4	0.5
0.5	**0.5**	0.39	0.265	0.138	0.038	**0**	0.038	0.138	0.265	0.39	**0.5**
0.4	0.61	**0.5**	0.36	0.2	0.059	**0**	0.059	0.2	0.36	**0.5**	0.61
0.3	0.735	0.64	**0.5**	0.308	0.1	**0**	0.1	0.308	**0.5**	0.64	0.735
0.2	0.862	0.8	0.692	**0.5**	0.2	**0**	0.2	**0.5**	0.692	0.8	0.862
0.1	0.962	0.941	0.9	0.8	**0.5**	**0**	**0.5**	0.8	0.9	0.941	0.962
0	**1**	**1**	**1**	**1**	**1**		**1**	**1**	**1**	**1**	**1**
−0.1	0.962	0.941	0.9	0.8	**0.5**	**0**	**0.5**	0.8	0.9	0.941	0.962
−0.2	0.862	0.8	0.692	**0.5**	0.2	**0**	0.2	**0.5**	0.692	0.8	0.862
−0.3	0.735	0.640	**0.5**	0.308	0.1	**0**	0.1	0.308	**0.5**	0.640	0.735
−0.4	0.610	**0.5**	0.360	0.2	0.059	**0**	0.059	0.2	0.36	**0.5**	0.61
−0.5	**0.5**	0.39	0.265	0.138	0.038	**0**	0.038	0.138	0.265	0.390	**0.5**

Now, let's prove that the limit does not exist. We demonstrate three different methods.

First Method We show that $f(x, y)$ approaches different limits along the x- and y-axes (Figure 5):

Limit along x-axis: $\quad \displaystyle\lim_{x \to 0} f(x, 0) = \lim_{x \to 0} \frac{x^2}{x^2 + 0^2} = \lim_{x \to 0} 1 = 1$

Limit along y-axis: $\quad \displaystyle\lim_{y \to 0} f(0, y) = \lim_{y \to 0} \frac{0^2}{0^2 + y^2} = \lim_{y \to 0} 0 = 0$

These two limits are different, and hence, $\displaystyle\lim_{(x,y) \to (0,0)} f(x, y)$ does not exist.

Second Method If we set $y = mx$, we have restricted ourselves to the line through the origin with slope m. Then the limit becomes

$$\lim_{x \to 0} f(x, mx) = \lim_{x \to 0} \frac{x^2}{x^2 + (mx)^2} = \frac{1}{1 + m^2}$$

This clearly depends on the slope m, and therefore gives different values when the origin is approached along lines of differing slope. For instance, when $m = 0$, so that we are approaching along the x-axis, we have a limit of 1. But when $m = 1$, so that we are approaching along the line $y = x$, the limit is $\frac{1}{2}$. Hence, the overall limit does not exist. The contour map in Figure 5 shows the variety of limits that occur as we approach the origin along different lines.

Third Method We convert to polar coordinates, setting $x = r \cos\theta$ and $y = r \sin\theta$. Then for any path that approaches $(0, 0)$, it must be the case that r approaches 0. Different linear paths can be considered by fixing θ at various values and having r approach 0.

Hence, we can consider

$$\lim_{r \to 0} \frac{x^2}{x^2 + y^2} = \lim_{r \to 0} \frac{(r \cos\theta)^2}{(r \cos\theta)^2 + (r \sin\theta)^2} = \lim_{r \to 0} \cos^2\theta$$

This result depends on θ. For instance, fixing θ at 0 would mean we are approaching $(0, 0)$ along the positive x-axis, and that gives a limit of 1. Fixing θ at $\pi/2$ would mean we are approaching $(0, 0)$ along the positive y-axis, and that gives a limit of 0. Since different values of θ yield different results, the overall limit does not exist. ∎

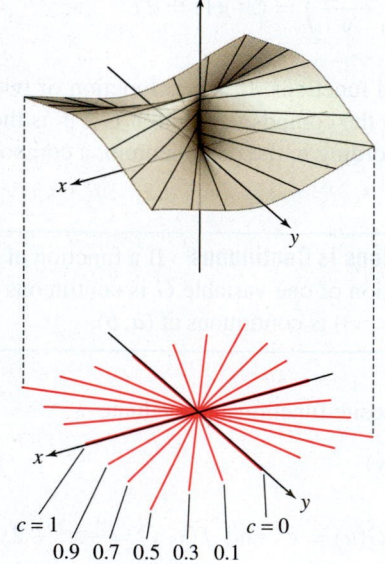

FIGURE 5 Graph and contour map of $f(x, y) = \dfrac{x^2}{x^2 + y^2}$.

$c = 1 \qquad c = 0$
$0.9 \quad 0.7 \quad 0.5 \quad 0.3 \quad 0.1$

EXAMPLE 6 **Verifying a Limit** Calculate $\lim\limits_{(x,y)\to(0,0)} f(x,y)$, where $f(x,y)$ is defined for $(x,y) \neq (0,0)$ by

$$f(x,y) = \frac{xy^2}{x^2+y^2}$$

as in Figure 6.

Solution Since substitution yields the indeterminate form of type $\frac{0}{0}$, we need to try an alternate method. We convert to polar coordinates:

$$x = r\cos\theta, \qquad y = r\sin\theta$$

Keep in mind that for any path approaching $(0,0)$, r approaches 0. Then $x^2+y^2 = r^2$ and for $r \neq 0$,

$$0 \le \left|\frac{xy^2}{x^2+y^2}\right| = \left|\frac{(r\cos\theta)(r\sin\theta)^2}{r^2}\right| = r|\cos\theta\sin^2\theta| \le r$$

As (x,y) approaches $(0,0)$, the variable r also approaches 0, so the desired conclusion follows from the Squeeze Theorem:

$$0 \le \lim_{(x,y)\to(0,0)}\left|\frac{xy^2}{x^2+y^2}\right| \le \lim_{r\to 0} r = 0$$

Therefore, $\lim\limits_{(x,y)\to(0,0)} \dfrac{xy^2}{x^2+y^2} = 0.$ ∎

Converting to polar coordinates enabled us to evaluate the previous two limits. In the next example, converting to polar coordinates does not help because it does not result in a useful simplification.

EXAMPLE 7 Determine whether or not the following limit exists:

$$\lim_{(x,y)\to(0,0)} \frac{x^2 y}{x^4+y^2}$$

Solution We first consider paths along lines through the origin where $y = mx$. Then the limit becomes

$$\lim_{x\to 0} \frac{x^2(mx)}{x^4+(mx)^2} = \lim_{x\to 0} \frac{xm}{x^2+m^2} = 0$$

Thus, all paths along lines through the origin yield the same limit. However, this does not mean that all paths through the origin yield the same limit. By examining the form of $\frac{x^2 y}{x^4+y^2}$ you might notice that this expression simplifies greatly if we consider curves $y = ax^2$. For example, if we consider $y = x^2$ (in the first case) and $y = 2x^2$ (in the second), then we obtain

$$\lim_{x\to 0} \frac{x^2(x^2)}{x^4+(x^2)^2} = \frac{1}{2} \quad \text{and} \quad \lim_{x\to 0} \frac{x^2(2x^2)}{x^4+(2x^2)^2} = \frac{2}{5}$$

Since these limits are not equal (and furthermore do not equal the limit obtained along lines), the overall limit does not exist. ∎

To prove a limit does not exist, we only need to find two paths that yield different limits. However, to prove that a limit *does* exist at a point, it is not enough just to consider the limit along a set of paths approaching the point. Instead, as we did in Examples 2, 3, 4, and 6, we employ limit laws and theorems to prove a limit exists.

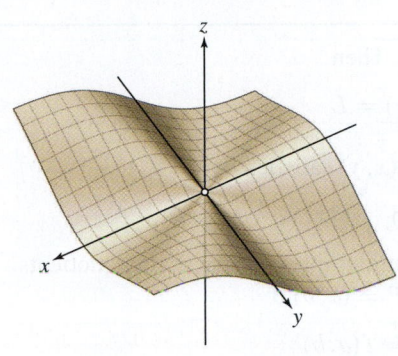

FIGURE 6 Graph of $f(x,y) = \dfrac{xy^2}{x^2+y^2}$.

14.2 SUMMARY

- Suppose that $f(x, y)$ is defined near $P = (a, b)$. Then

$$\lim_{(x,y) \to (a,b)} f(x, y) = L$$

if, for any $\epsilon > 0$, there exists $\delta > 0$ such that if (x, y) satisfies

$$0 < d((x, y), (a, b)) < \delta, \quad \text{then} \quad |f(x, y) - L| < \epsilon$$

- There are algebraic limit laws for sums, constant multiples, products, and quotients.
- A function f of two variables is *continuous* at $P = (a, b)$ if

$$\lim_{(x,y) \to (a,b)} f(x, y) = f(a, b)$$

- To prove that a limit does not exist, it is enough to show that the limits obtained along two different paths are not equal.

14.2 EXERCISES

Preliminary Questions

1. What is the difference between $D(P, r)$ and $D^*(P, r)$?

2. Suppose that $f(x, y)$ is continuous at $(2, 3)$ and that $f(2, y) = y^3$ for $y \neq 3$. What is the value $f(2, 3)$?

3. Suppose that $Q(x, y)$ is a function such that $1/Q(x, y)$ is continuous for all (x, y). Which of the following statements are true?

(a) $Q(x, y)$ is continuous for all (x, y).

(b) $Q(x, y)$ is continuous for $(x, y) \neq (0, 0)$.

(c) $Q(x, y) \neq 0$ for all (x, y).

4. Suppose that $f(x, 0) = 3$ for all $x \neq 0$ and $f(0, y) = 5$ for all $y \neq 0$. What can you conclude about $\lim_{(x,y) \to (0,0)} f(x, y)$?

Exercises

In Exercises 1–8, evaluate the limit using continuity.

1. $\displaystyle \lim_{(x,y) \to (1,2)} (x^2 + y)$

2. $\displaystyle \lim_{(x,y) \to (\frac{4}{9}, \frac{2}{9})} \frac{x}{y}$

3. $\displaystyle \lim_{(x,y) \to (-2,1)} (x^2 y - 3x^4 y^3)$

4. $\displaystyle \lim_{(x,y) \to (0,1)} \frac{e^x}{x - 4y}$

5. $\displaystyle \lim_{(x,y) \to (\frac{\pi}{4},0)} \tan x \cos y$

6. $\displaystyle \lim_{(x,y) \to (2,3)} \tan^{-1}(x^2 - y)$

7. $\displaystyle \lim_{(x,y) \to (1,1)} \frac{e^{x^2} - e^{-y^2}}{x + y}$

8. $\displaystyle \lim_{(x,y) \to (1,0)} \ln(x - y)$

In Exercises 9–12, assume that

$$\lim_{(x,y) \to (2,5)} f(x, y) = 3, \qquad \lim_{(x,y) \to (2,5)} g(x, y) = 7$$

to find the limit.

9. $\displaystyle \lim_{(x,y) \to (2,5)} \big(g(x, y) - 2f(x, y)\big)$

10. $\displaystyle \lim_{(x,y) \to (2,5)} f(x, y)^2 g(x, y)$

11. $\displaystyle \lim_{(x,y) \to (2,5)} e^{f(x,y)^2 - g(x,y)}$

12. $\displaystyle \lim_{(x,y) \to (2,5)} \frac{f(x, y)}{f(x, y) + g(x, y)}$

13. Does $\displaystyle \lim_{(x,y) \to (0,0)} \frac{y^2}{x^2 + y^2}$ exist? Explain.

14. Let $f(x, y) = xy/(x^2 + y^2)$. Show that $f(x, y)$ approaches zero along the x- and y-axes. Then prove that $\lim_{(x,y) \to (0,0)} f(x, y)$ does not exist by showing that the limit along the line $y = x$ is nonzero.

15. Let $f(x, y) = \dfrac{x^3 + y^3}{xy^2}$. Set $y = mx$ and show that the resulting limit depends on m, and therefore the limit $\lim_{(x,y) \to (0,0)} f(x, y)$ does not exist.

16. Let $f(x, y) = \dfrac{2x^2 + 3y^2}{xy}$. Set $y = mx$ and show that the resulting limit depends on m, and therefore the limit $\lim_{(x,y) \to (0,0)} f(x, y)$ does not exist.

17. Prove that

$$\lim_{(x,y) \to (0,0)} \frac{x}{x^2 + y^2}$$

does not exist by considering the limit along the x-axis.

18. Let $f(x, y) = x^3/(x^2 + y^2)$ and $g(x, y) = x^2/(x^2 + y^2)$. Using polar coordinates, prove that

$$\lim_{(x,y) \to (0,0)} f(x, y) = 0$$

and that $\lim_{(x,y) \to (0,0)} g(x, y)$ does not exist. *Hint:* Show that $g(x, y) = \cos^2 \theta$ and observe that $\cos \theta$ can take on any value between -1 and 1 as $(x, y) \to (0, 0)$.

In Exercises 19–22, use any method to evaluate the limit or show that it does not exist.

19. $\lim\limits_{(x,y)\to(0,0)} \dfrac{x^2 - y^2}{\sqrt{x^2 + y^2}}$

20. $\lim\limits_{(x,y)\to(0,0)} \dfrac{x^2 - y^2}{x^2 + y^2}$

21. $\lim\limits_{(x,y)\to(0,0)} \dfrac{xy}{3x^2 + 2y^2}$

22. $\lim\limits_{(x,y)\to(0,0)} \dfrac{x^4 - y^4}{x^4 + x^2y^2 + y^4}$

In Exercises 23–24, show that the limit does not exist by approaching the origin along one or more of the coordinate axes.

23. $\lim\limits_{(x,y,z)\to(0,0,0)} \dfrac{x + y + z}{x^2 + y^2 + z^2}$

24. $\lim\limits_{(x,y,z)\to(0,0,0)} \dfrac{x^2 - y^2 + z^2}{x^2 + y^2 + z^2}$

25. Use the Squeeze Theorem to evaluate

$$\lim\limits_{(x,y)\to(4,0)} (x^2 - 16) \cos\left(\dfrac{1}{(x-4)^2 + y^2}\right)$$

26. Evaluate $\lim\limits_{(x,y)\to(0,0)} \tan x \sin\left(\dfrac{1}{|x| + |y|}\right)$.

In Exercises 27–42, evaluate the limit or determine that it does not exist.

27. $\lim\limits_{(z,w)\to(-2,1)} \dfrac{z^4 \cos(\pi w)}{e^{z+w}}$

28. $\lim\limits_{(z,w)\to(-1,2)} (z^2 w - 9z)$

29. $\lim\limits_{(x,y)\to(4,2)} \dfrac{y - 2}{\sqrt{x^2 - 4}}$

30. $\lim\limits_{(x,y)\to(0,0)} \dfrac{x^2 + y^2}{1 + y^2}$

31. $\lim\limits_{(x,y)\to(3,4)} \dfrac{1}{\sqrt{x^2 + y^2}}$

32. $\lim\limits_{(x,y)\to(0,0)} \dfrac{xy}{\sqrt{x^2 + y^2}}$

33. $\lim\limits_{(x,y)\to(\pi,0)} \dfrac{\cos x}{\sin y}$

34. $\lim\limits_{(x,y)\to(0,0)} \dfrac{\cos x \sin y}{y}$

35. $\lim\limits_{(x,y)\to(1,-3)} e^{x-y} \ln(x - y)$

36. $\lim\limits_{(x,y)\to(0,0)} \dfrac{|x|}{|x| + |y|}$

37. $\lim\limits_{(x,y)\to(-3,-2)} (x^2 y^3 + 4xy)$

38. $\lim\limits_{(x,y)\to(2,1)} e^{x^2 - y^2}$

39. $\lim\limits_{(x,y)\to(0,0)} \tan(x^2 + y^2) \tan^{-1}\left(\dfrac{1}{x^2 + y^2}\right)$

40. $\lim\limits_{(x,y)\to(0,0)} (x + y + 2)e^{-1/(x^2+y^2)}$

41. $\lim\limits_{(x,y)\to(0,0)} \dfrac{x^2 + y^2}{\sqrt{x^2 + y^2 + 1} - 1}$

42. $\lim\limits_{(x,y)\to(1,1)} \dfrac{x^2 + y^2 - 2}{|x - 1| + |y - 1|}$

Hint: Rewrite the limit in terms of $u = x - 1$ and $v = y - 1$.

43. Let $f(x, y) = \dfrac{x^3 + y^3}{x^2 + y^2}$.

(a) Show that

$$|x^3| \le |x|(x^2 + y^2), \quad |y^3| \le |y|(x^2 + y^2)$$

(b) Show that $|f(x, y)| \le |x| + |y|$.

(c) Use the Squeeze Theorem to prove that $\lim\limits_{(x,y)\to(0,0)} f(x, y) = 0$.

44. Let $a, b \ge 0$. Show that $\lim\limits_{(x,y)\to(0,0)} \dfrac{x^a y^b}{x^2 + y^2} = 0$ if $a + b > 2$ and that the limit does not exist if $a + b \le 2$.

45. Figure 7 shows the contour maps of two functions. Explain why the limit $\lim\limits_{(x,y)\to P} f(x, y)$ in (A) does not exist. Does $\lim\limits_{(x,y)\to Q} g(x, y)$ appear to exist in (B)? If so, what is its limit?

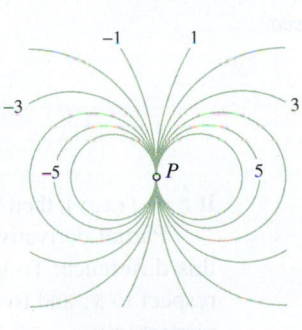
(A) Contour map of $f(x, y)$

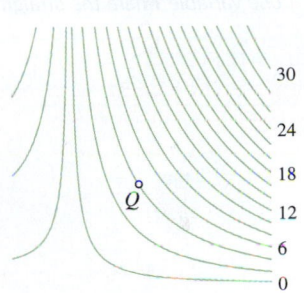
(B) Contour map of $g(x, y)$

FIGURE 7

Further Insights and Challenges

46. Evaluate $\lim\limits_{(x,y)\to(0,2)} (1 + x)^{y/x}$.

47. Is the following function continuous?

$$f(x, y) = \begin{cases} x^2 + y^2 & \text{if } x^2 + y^2 < 1 \\ 1 & \text{if } x^2 + y^2 \ge 1 \end{cases}$$

48. [CAS] The function $f(x, y) = \sin(xy)/xy$ is defined for $xy \ne 0$.

(a) Is it possible to extend the domain of f to all of $\mathbf{R}^2$ so that the result is a continuous function?

(b) Use a computer algebra system to plot f. Does the result support your conclusion in (a)?

49. Prove that the function

$$f(x, y) = \begin{cases} \dfrac{(2^x - 1)(\sin y)}{xy} & \text{if } xy \ne 0 \\ \ln 2 & \text{if } xy = 0 \end{cases}$$

is continuous at $(0, 0)$.

50. Prove that if $f(x)$ is continuous at $x = a$ and $g(y)$ is continuous at $y = b$, then $F(x, y) = f(x)g(y)$ is continuous at (a, b).

51. Consider the function $f(x, y) = \dfrac{x^3 y}{x^6 + 2y^2}$.

(a) Show that as $(x, y) \to (0, 0)$ along any line $y = mx$, the limit equals 0.

(b) Show that as $(x, y) \to (0, 0)$ along the curve $y = x^3$, the limit does not equal 0, and hence, $\lim\limits_{(x,y)\to(0,0)} f(x, y)$ does not exist.

14.3 Partial Derivatives

We have stressed that a function f of two or more variables does not have a unique rate of change because each variable may affect f in different ways. For example, the current I in a circuit is a function of both voltage V and resistance R given by Ohm's Law:

$$I(V, R) = \frac{V}{R}$$

The current I is *increasing* as a function of V (when R is fixed) but *decreasing* as a function of R (when V is fixed).

The **partial derivatives** are the rates of change with respect to each variable separately. A function $f(x, y)$ of two variables has two partial derivatives, denoted f_x and f_y, defined by the following limits (if they exist):

$$f_x(a, b) = \lim_{h \to 0} \frac{f(a + h, b) - f(a, b)}{h}, \qquad f_y(a, b) = \lim_{k \to 0} \frac{f(a, b + k) - f(a, b)}{k}$$

The partial derivative symbol ∂ is a rounded "d." It is used to distinguish derivatives of a function of multiple variables from derivatives of functions of one variable where the straight "d" is used.

Thus, f_x is the derivative of $f(x, b)$ as a function of x alone, and f_y is the derivative of $f(a, y)$ as a function of y alone. We refer to f_x as **the partial derivative of f with respect to x** or **the x-derivative of f**. We refer to f_y similarly. The Leibniz notation for partial derivatives is

$$\frac{\partial f}{\partial x} = f_x, \qquad\qquad \frac{\partial f}{\partial y} = f_y$$

$$\left.\frac{\partial f}{\partial x}\right|_{(a,b)} = f_x(a, b), \qquad \left.\frac{\partial f}{\partial y}\right|_{(a,b)} = f_y(a, b)$$

If $z = f(x, y)$, then we also write $\partial z / \partial x$ and $\partial z / \partial y$.

Partial derivatives are computed just like ordinary derivatives in one variable with this difference: To compute f_x, treat y as a constant and take the derivative of f with respect to x, and to compute f_y, treat x as a constant and take the derivative of f with respect to y.

EXAMPLE 1 Compute the partial derivatives of $f(x, y) = x^2 y^5$.

Solution

$$\frac{\partial f}{\partial x} = \underbrace{\frac{\partial}{\partial x}\left(x^2 y^5\right) = y^5 \frac{\partial}{\partial x}\left(x^2\right)}_{\text{Treat } y^5 \text{ as a constant.}} = y^5(2x) = 2xy^5$$

$$\frac{\partial f}{\partial y} = \underbrace{\frac{\partial}{\partial y}\left(x^2 y^5\right) = x^2 \frac{\partial}{\partial y}\left(y^5\right)}_{\text{Treat } x^2 \text{ as a constant.}} = x^2(5y^4) = 5x^2 y^4 \qquad \blacksquare$$

GRAPHICAL INSIGHT The partial derivatives at $P = (a, b)$ are the slopes of the tangent lines to the trace curves through the graph of $f(x, y)$ at the point $(a, b, f(a, b))$ in Figure 1(A). To compute $f_x(a, b)$, we set $y = b$ and differentiate in the x-direction. This gives us the slope of the tangent line to the trace curve in the plane $y = b$ [Figure 1(B)]. Similarly, $f_y(a, b)$ is the slope of the trace curve in the plane $x = a$ [Figure 1(C)].

The differentiation rules from calculus of one variable (the Product, Quotient, and Chain Rules) are valid for partial derivatives.

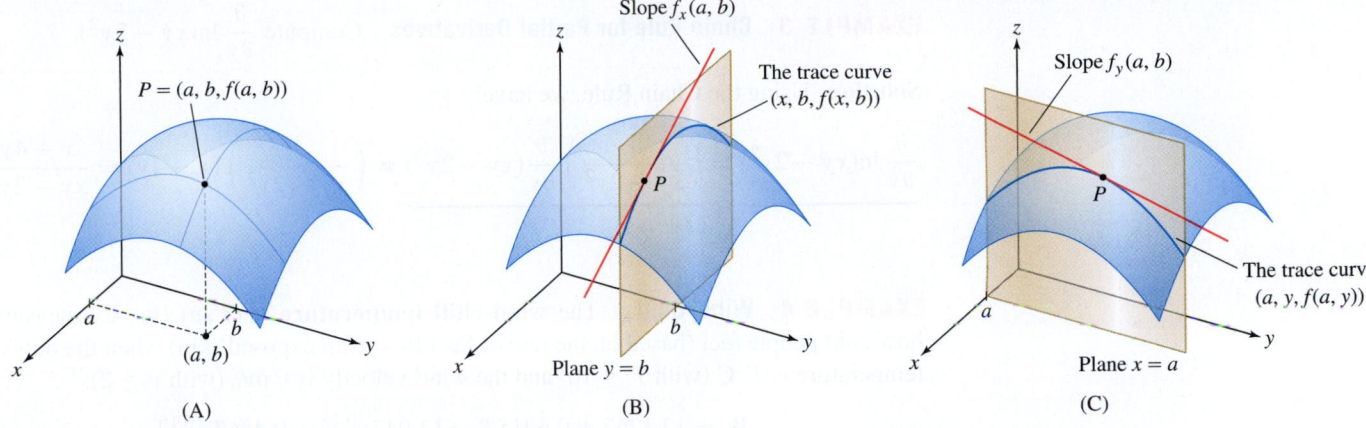

Slope $f_x(a, b)$

$P = (a, b, f(a, b))$

The trace curve
$(x, b, f(x, b))$

Slope $f_y(a, b)$

The trace curve
$(a, y, f(a, y))$

(a, b)

Plane $y = b$

Plane $x = a$

(A)   (B)   (C)

DF **FIGURE 1** The partial derivatives are the slopes of the trace curves.

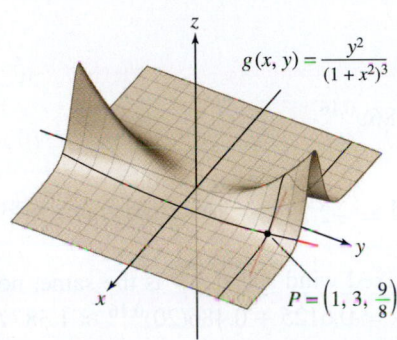

$g(x, y) = \dfrac{y^2}{(1 + x^2)^3}$

$P = \left(1, 3, \dfrac{9}{8}\right)$

FIGURE 2 The slopes of the tangent lines to the trace curves are $g_x(1, 3)$ and $g_y(1, 3)$.

CAUTION *It is not necessary to use the Quotient Rule to compute the partial derivative in Eq. (1). The denominator does not depend on y, so we treat it as a constant when differentiating with respect to y.*

EXAMPLE 2 Calculate $g_x(1, 3)$ and $g_y(1, 3)$, where $g(x, y) = \dfrac{y^2}{(1 + x^2)^3}$.

Solution To calculate g_x, treat y (and therefore y^2) as a constant and differentiate with respect to x:

$$g_x(x, y) = \frac{\partial}{\partial x}\left(\frac{y^2}{(1 + x^2)^3}\right) = y^2 \frac{\partial}{\partial x}(1 + x^2)^{-3} = \frac{-6xy^2}{(1 + x^2)^4}$$

$$g_x(1, 3) = \frac{-6(1)3^2}{(1 + 1^2)^4} = -\frac{27}{8}$$

To calculate g_y, treat x [and therefore $(1 + x^2)^3$] as a constant and differentiate with respect to y:

$$g_y(x, y) = \frac{\partial}{\partial y}\left(\frac{y^2}{(1 + x^2)^3}\right) = \frac{1}{(1 + x^2)^3}\frac{\partial}{\partial y}y^2 = \frac{2y}{(1 + x^2)^3} \qquad \boxed{1}$$

$$g_y(1, 3) = \frac{2(3)}{(1 + 1^2)^3} = \frac{3}{4}$$

These partial derivatives are the slopes of the trace curves through the point $P = \left(1, 3, \frac{9}{8}\right)$ shown in Figure 2. Note that in the figure, g is decreasing as x increases through P, consistent with our determination that $g_x(1, 3) < 0$. Similarly, g is increasing as y increases through P, reflecting that $g_y(1, 3) > 0$. ■

The Chain Rule was used in Example 2 to compute $g_x(x, y)$. We use the Chain Rule to compute partial derivatives of a composite function like $f(x, y) = \sin(3x^2 + 4y)$ in the same way the Chain Rule is applied in the single variable case. For example, to compute the partial derivative of f with respect to x, we take the derivative of the outside function at the inside function [yielding $\cos(3x^2 + 4y)$] and multiply by the derivative (with respect to x) of the inside function; that is, by $6x$. Therefore,

$$\frac{\partial}{\partial x}\sin(3x^2 + 4y) = \cos(3x^2 + 4y)\frac{\partial}{\partial x}(3x^2 + 4y) = 6x\cos(3x^2 + 4y)$$

In multivariable calculus, there are a number of different ways compositions of functions can arise. Consequently, there are multiple possibilities for chain-rule derivative computations. We examine other possibilities for multivariable chain rules in Sections 14.5 and 14.6.

EXAMPLE 3   **Chain Rule for Partial Derivatives**   Compute $\dfrac{\partial}{\partial y}\ln(xy - 2y^2)$.

Solution   Using the Chain Rule, we have:

$$\underbrace{\frac{\partial}{\partial y}\ln(xy - 2y^2) = \left(\frac{1}{xy - 2y^2}\right)\frac{\partial}{\partial y}(xy - 2y^2)}_{\text{Chain Rule}} = \left(\frac{1}{xy - 2y^2}\right)(x - 4y) = \frac{x - 4y}{xy - 2y^2}$$

EXAMPLE 4   **Wind Chill**   The **wind-chill temperature** $W(T, v)$ (in °C) measures how cold people feel (based on the rate of heat loss from exposed skin) when the outside temperature is $T°C$ (with $T \le 10$) and the wind velocity is v m/s (with $v \ge 2$):

$$W = 13.1267 + 0.6215T - 13.947v^{0.16} + 0.486Tv^{0.16}$$

Calculate $\dfrac{\partial W}{\partial T}$ and $\dfrac{\partial W}{\partial v}$. Show that at a fixed wind speed, the impact on wind chill of a changing temperature does not depend on the temperature, but the impact of a changing wind speed is larger the colder the temperature.

Solution   Computing the partial derivatives:

$$\frac{\partial W}{\partial T} = 0.6125 + 0.486v^{0.16}$$

$$\frac{\partial W}{\partial v} = -13.947(0.16)v^{-0.84} + 0.486T(0.16)v^{-0.84} = -2.2315v^{-0.84} + 0.0778Tv^{-0.84}$$

Note that $\dfrac{\partial W}{\partial T}$ does not depend on T, so that at a fixed wind speed $\dfrac{\partial W}{\partial T}$ is the same, no matter the temperature. For example, at 20 m/s, $\dfrac{\partial W}{\partial T} = 0.6125 + 0.486(20)^{0.16} \approx 1.3877$ (°C per °C) at all values of T.

On the other hand, at a fixed wind speed, $\dfrac{\partial W}{\partial v}$ decreases as T decreases. For example (in units of °C per m/s),

$$\frac{\partial W}{\partial v}\Big|_{(5,10)} \approx -0.2663 \qquad \frac{\partial W}{\partial v}\Big|_{(-5,10)} \approx -0.3788 \qquad \frac{\partial W}{\partial v}\Big|_{(-15,10)} \approx -0.4912$$

Therefore at a fixed wind speed, an increase in wind speed has a larger cooling effect at colder temperatures.

Partial derivatives are defined for functions of any number of variables. We compute the partial derivative with respect to any one of the variables by differentiating with respect to that variable while holding the remaining variables constant.

EXAMPLE 5   **More Than Two Variables**   Calculate $f_z(0, 0, 1, 1)$, where

$$f(x, y, z, w) = \frac{e^{xz+y}}{z^2 + w}$$

In Example 5, the calculation

$$\frac{\partial}{\partial z}e^{xz+y} = xe^{xz+y}$$

follows from the Chain Rule, just like

$$\frac{d}{dz}e^{4z+2} = 4e^{4z+2}$$

Solution   Use the Quotient Rule, treating x, y, and w as constants and differentiating with respect to z:

$$f_z(x, y, z, w) = \frac{\partial}{\partial z}\left(\frac{e^{xz+y}}{z^2 + w}\right) = \frac{(z^2 + w)\frac{\partial}{\partial z}e^{xz+y} - e^{xz+y}\frac{\partial}{\partial z}(z^2 + w)}{(z^2 + w)^2}$$

$$= \frac{(z^2 + w)xe^{xz+y} - 2ze^{xz+y}}{(z^2 + w)^2} = \frac{(z^2x + wx - 2z)e^{xz+y}}{(z^2 + w)^2}$$

$$f_z(0, 0, 1, 1) = \frac{-2e^0}{(1^2 + 1)^2} = -\frac{1}{2}$$

In the next example, we estimate a partial derivative numerically. Since f_x and f_y are limits of difference quotients, we have the following approximations when Δx and Δy are small:

$$f_x(a,b) \approx \frac{\Delta f}{\Delta x} = \frac{f(a+\Delta x, b) - f(a,b)}{\Delta x}$$

$$f_y(a,b) \approx \frac{\Delta f}{\Delta y} = \frac{f(a, b+\Delta y) - f(a,b)}{\Delta y}$$

These approximation formulas are multivariable versions of the difference quotient approximation introduced in Section 3.1.

Similar approximations are valid in any number of variables.

EXAMPLE 6   Estimating Partial Derivatives Using Contour Maps   Seawater density depends on salinity and temperature and can be expressed as a function $\rho(S,T)$, where ρ is in kilograms per cubic meter, salinity S is in parts per thousand, and temperature T is in degrees Celsius. Use the contour map of seawater density appearing in Figure 3 to estimate $\partial\rho/\partial T$ and $\partial\rho/\partial S$ at $A = (33, 15)$.

Solution   We estimate $\partial\rho/\partial T$ at A in two steps.

Step 1.   **Choose ΔT, and estimate or evaluate $\rho(33, 15 + \Delta T)$.**
With S held constant at 33, a change in T moves us vertically on the contour map from the point A. Any choice of small ΔT can be used to make our estimate. We choose $\Delta T = 2$ because the corresponding point (B on the contour map) lies on a level curve near A, and at B we can evaluate ρ, rather than estimate it. With $\Delta T = 2$, we have
$\rho(33, 15 + \Delta T) = \rho(33, 17) = 1.0240$.

Step 2.   **Compute the difference quotient and make the approximation.**

$$\left.\frac{\partial\rho}{\partial T}\right|_{(33,15)} \approx \frac{\rho(33,17) - \rho(33,15)}{2} = \frac{1.0240 - 1.0245}{2} = \frac{-0.0005}{2}$$

$$= -0.00025 \text{ kg-m}^{-3}/^\circ\text{C}$$

We estimate $\partial\rho/\partial S$ in a similar way, using $\Delta S \approx 0.7$ to put us at point C on a level curve of ρ on the contour map. We obtain

$$\left.\frac{\partial\rho}{\partial S}\right|_{(33,15)} \approx \frac{\rho(33.7, 15) - \rho(33,15)}{0.7} = \frac{1.0250 - 1.0245}{0.7} = \frac{0.0005}{0.7}$$

$$\approx 0.0007 \text{ kg-m}^{-3}/\text{ppt}$$ ∎

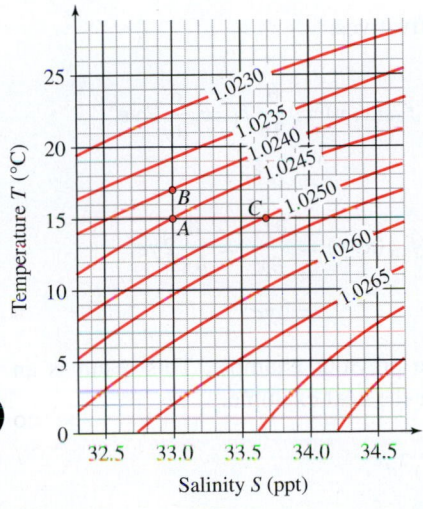

FIGURE 3 Contour map of seawater density as a function of temperature and salinity.

Higher Order Partial Derivatives

The higher order partial derivatives are the derivatives of derivatives. The *second-order* partial derivatives of f are the partial derivatives of f_x and f_y. We write f_{xx} for the x-derivative of f_x and f_{yy} for the y-derivative of f_y:

$$f_{xx} = \frac{\partial}{\partial x}\left(\frac{\partial f}{\partial x}\right), \qquad f_{yy} = \frac{\partial}{\partial y}\left(\frac{\partial f}{\partial y}\right)$$

We also have the *mixed partials*:

$$f_{xy} = \frac{\partial}{\partial y}\left(\frac{\partial f}{\partial x}\right), \qquad f_{yx} = \frac{\partial}{\partial x}\left(\frac{\partial f}{\partial y}\right)$$

The process can be continued. For example, f_{xyx} is the x-derivative of f_{xy}, and f_{xyy} is the y-derivative of f_{xy} (perform the differentiation in the order of the subscripts from left to right). The Leibniz notation for higher order partial derivatives is

$$f_{xx} = \frac{\partial^2 f}{\partial x^2}, \qquad f_{xy} = \frac{\partial^2 f}{\partial y \partial x}, \qquad f_{yx} = \frac{\partial^2 f}{\partial x \partial y}, \qquad f_{yy} = \frac{\partial^2 f}{\partial y^2}$$

Higher order partial derivatives are defined for functions of three or more variables in a similar manner.

EXAMPLE 7 Calculate the second-order partial derivatives of $f(x, y) = x^3 + y^2 e^x$.

Solution First, we compute the first-order partial derivatives:

$$f_x(x, y) = \frac{\partial}{\partial x}(x^3 + y^2 e^x) = 3x^2 + y^2 e^x, \qquad f_y(x, y) = \frac{\partial}{\partial y}(x^3 + y^2 e^x) = 2y e^x$$

Then we can compute the second-order partial derivatives:

$$f_{xx}(x, y) = \frac{\partial}{\partial x} f_x = \frac{\partial}{\partial x}(3x^2 + y^2 e^x) \qquad\qquad f_{yy}(x, y) = \frac{\partial}{\partial y} f_y = \frac{\partial}{\partial y} 2y e^x$$

$$= 6x + y^2 e^x, \qquad\qquad\qquad\qquad\qquad = 2 e^x$$

$$f_{xy}(x, y) = \frac{\partial f_x}{\partial y} = \frac{\partial}{\partial y}(3x^2 + y^2 e^x) \qquad\qquad f_{yx}(x, y) = \frac{\partial f_y}{\partial x} = \frac{\partial}{\partial x} 2y e^x$$

$$= 2y e^x, \qquad\qquad\qquad\qquad\qquad\qquad = 2y e^x \qquad \blacksquare$$

It is not a coincidence that $f_{xy} = f_{yx}$ in the previous example. This result is an example of a general theorem that we present after the next example.

> Remember how the subscripts are used in partial derivatives. The notation f_{xyy} indicates that we first differentiate with respect to x and then differentiate twice with respect to y.

EXAMPLE 8 Calculate f_{xyy} for $f(x, y) = x^3 + y^2 e^x$.

Solution By the previous example, $f_{xy} = 2y e^x$. Therefore,

$$f_{xyy} = \frac{\partial}{\partial y} f_{xy} = \frac{\partial}{\partial y} 2y e^x = 2 e^x \qquad \blacksquare$$

The next theorem, named for the French mathematician Alexis Clairaut (Figure 4), indicates that in a mixed partial derivative, the order in which the derivatives are taken does not matter, provided that the mixed partial derivatives are continuous. A proof of the theorem is provided in Appendix D.

> The hypothesis of Clairaut's Theorem, that f_{xy} and f_{yx} are continuous, is almost always satisfied in practice, but see Exercise 80 for an example where the mixed partial derivatives are not equal.

THEOREM 1 Clairaut's Theorem: Equality of Mixed Partials If f_{xy} and f_{yx} both exist and are continuous on a disk D, then $f_{xy}(a, b) = f_{yx}(a, b)$ for all $(a, b) \in D$. Therefore, on D,

$$\frac{\partial^2 f}{\partial x \, \partial y} = \frac{\partial^2 f}{\partial y \, \partial x}$$

EXAMPLE 9 Check that $\dfrac{\partial^2 W}{\partial U \partial T} = \dfrac{\partial^2 W}{\partial T \partial U}$ for $W = e^{U/T}$.

Solution We compute both mixed partial derivatives and observe that they are equal:

$$\frac{\partial W}{\partial T} = e^{U/T} \frac{\partial}{\partial T}\left(\frac{U}{T}\right) = -U T^{-2} e^{U/T}, \qquad\qquad \frac{\partial W}{\partial U} = e^{U/T} \frac{\partial}{\partial U}\left(\frac{U}{T}\right) = T^{-1} e^{U/T}$$

$$\frac{\partial}{\partial U} \frac{\partial W}{\partial T} = -T^{-2} e^{U/T} - U T^{-3} e^{U/T}, \qquad\qquad \frac{\partial}{\partial T} \frac{\partial W}{\partial U} = -T^{-2} e^{U/T} - U T^{-3} e^{U/T} \qquad \blacksquare$$

Although Clairaut's Theorem is stated for f_{xy} and f_{yx}, it implies more generally that partial differentiation may be carried out in any order, provided that the derivatives in question are continuous (see Exercise 71). For example, we can compute f_{xyxy} by differentiating f twice with respect to x and twice with respect to y, in any order. Thus,

$$f_{xyxy} = f_{xxyy} = f_{yyxx} = f_{yxyx} = f_{xyyx} = f_{yxxy}$$

EXAMPLE 10 **Choosing the Order Wisely** Calculate the partial derivative g_{zzwx}, where $g(x, y, z, w) = x^3 w^2 z^2 + \sin\left(\dfrac{xy}{z^2}\right)$.

Solution Let's take advantage of the fact that the derivatives may be calculated in any order. If we differentiate with respect to w first, the second term disappears because it does not depend on w:

$$g_w = \frac{\partial}{\partial w}\left(x^3 w^2 z^2 + \sin\left(\frac{xy}{z^2}\right)\right) = 2x^3 w z^2$$

Next, differentiate twice with respect to z and once with respect to x:

$$g_{wz} = \frac{\partial}{\partial z} 2x^3 w z^2 = 4x^3 w z$$

$$g_{wzz} = \frac{\partial}{\partial z} 4x^3 w z = 4x^3 w$$

$$g_{wzzx} = \frac{\partial}{\partial x} 4x^3 w = 12x^2 w$$

We conclude that $g_{zzwx} = g_{wzzx} = 12x^2 w$.

FIGURE 4 Alexis Clairaut (1713–1765) was a brilliant French mathematician who presented his first paper to the Paris Academy of Sciences at the age of 13. In 1752, Clairaut won a prize for an essay on lunar motion that Euler praised (surely an exaggeration) as "the most important and profound discovery that has ever been made in mathematics."

A **partial differential equation** (PDE) is a differential equation involving functions of several variables and their partial derivatives. A solution to a PDE is a function that satisfies the equation. The heat equation in the next example is a PDE that models temperature as heat spreads through an object. There are infinitely many solutions, depending on the initial temperature distribution in the object. The particular function in the example describes temperature at times $t > 0$ along a metal rod when the center point is given a burst of heat at $t = 0$ (Figure 5).

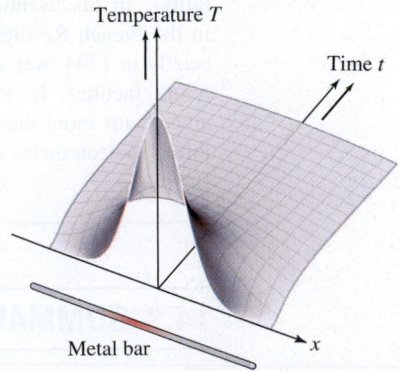

FIGURE 5 The plot of

$$u(x, t) = \frac{1}{2\sqrt{\pi t}} e^{-(x^2/4t)}$$

illustrates the diffusion of a burst of heat over time.

EXAMPLE 11 **The Heat Equation** Show that $u(x, t) = \dfrac{1}{2\sqrt{\pi t}} e^{-(x^2/4t)}$, defined for $t > 0$, satisfies the heat equation

$$\frac{\partial u}{\partial t} = \frac{\partial^2 u}{\partial x^2}$$

$\boxed{2}$

Solution We write $u(x,t) = \dfrac{1}{2\sqrt{\pi}}t^{-1/2}e^{-(x^2/4t)}$. We first compute $\dfrac{\partial^2 u}{\partial x^2}$:

$$\frac{\partial u}{\partial x} = \frac{\partial}{\partial x}\left(\frac{1}{2\sqrt{\pi}}t^{-1/2}e^{-(x^2/4t)}\right) = -\frac{1}{4\sqrt{\pi}}xt^{-3/2}e^{-(x^2/4t)}$$

$$\frac{\partial^2 u}{\partial x^2} = \frac{\partial}{\partial x}\left(-\frac{1}{4\sqrt{\pi}}xt^{-3/2}e^{-(x^2/4t)}\right) = -\frac{1}{4\sqrt{\pi}}t^{-3/2}e^{-(x^2/4t)} + \frac{1}{8\sqrt{\pi}}x^2t^{-5/2}e^{-(x^2/4t)}$$

Then compute $\partial u/\partial t$ and observe that it equals $\partial^2 u/\partial x^2$ as required:

$$\frac{\partial u}{\partial t} = \frac{\partial}{\partial t}\left(\frac{1}{2\sqrt{\pi}}t^{-1/2}e^{-(x^2/4t)}\right) = -\frac{1}{4\sqrt{\pi}}t^{-3/2}e^{-(x^2/4t)} + \frac{1}{8\sqrt{\pi}}x^2t^{-5/2}e^{-(x^2/4t)} \quad\blacksquare$$

HISTORICAL PERSPECTIVE

Hulton Archive/Getty Images

NLM/Science Source

Joseph Fourier
(1768–1830)

Adolf Fick
(1829–1901)

The general heat equation, of which Eq. (2) is a special case, was first introduced in 1807 by French mathematician Jean Baptiste Joseph Fourier. As a young man, Fourier was unsure whether to enter the priesthood or pursue mathematics, but he must have been very ambitious. He wrote in a letter, "Yesterday was my 21st birthday, at that age Newton and Pascal had already acquired many claims to immortality." In his twenties, Fourier got involved in the French Revolution and was imprisoned briefly in 1794 over an incident involving different factions. In 1798 he was summoned, along with more than 150 other scientists, to join Napoleon on his unsuccessful campaign in Egypt.

Fourier's true impact, however, lay in his mathematical contributions. The heat equation is applied throughout the physical sciences and engineering, from the study of heat flow through the earth's oceans and atmosphere to the use of heat probes to destroy tumors and treat heart disease.

Fourier also introduced a striking new technique—known as the **Fourier transform**—for solving his equation, based on the idea that a periodic function can be expressed as a (possibly infinite) sum of sines and cosines. Leading mathematicians of the day, including Lagrange and Laplace, initially raised objections because this technique was not easy to justify rigorously. Nevertheless, the Fourier transform turned out to be one of the most important mathematical discoveries of the nineteenth century. A Web search on the term "Fourier transform" reveals its vast range of modern applications.

In 1855, the German physiologist Adolf Fick showed that the heat equation describes not only heat conduction but also a wide range of diffusion processes, such as osmosis, ion transport at the cellular level, and the motion of pollutants through air or water. The heat equation thus became a basic tool in chemistry, molecular biology, and environmental science, where it is often called **Fick's Second Law**.

14.3 SUMMARY

- The partial derivatives of $f(x, y)$ are defined as the limits

$$f_x(a, b) = \frac{\partial f}{\partial x}\bigg|_{(a,b)} = \lim_{h\to 0}\frac{f(a+h, b) - f(a, b)}{h}$$

$$f_y(a, b) = \frac{\partial f}{\partial y}\bigg|_{(a,b)} = \lim_{k\to 0}\frac{f(a, b+k) - f(a, b)}{k}$$

- Compute f_x by holding y constant and differentiating with respect to x, and compute f_y by holding x constant and differentiating with respect to y.
- $f_x(a, b)$ is the slope at $x = a$ of the tangent line to the trace curve $z = f(x, b)$. Similarly, $f_y(a, b)$ is the slope at $y = b$ of the tangent line to the trace curve $z = f(a, y)$.
- Approximating partial derivatives: For small Δx and Δy,

$$f_x(a, b) \approx \frac{\Delta f}{\Delta x} = \frac{f(a + \Delta x, b) - f(a, b)}{\Delta x}$$

$$f_y(a, b) \approx \frac{\Delta f}{\Delta y} = \frac{f(a, b + \Delta y) - f(a, b)}{\Delta y}$$

Similar approximations are valid in any number of variables.
- The second-order partial derivatives are

$$\frac{\partial^2}{\partial x^2} f = f_{xx}, \qquad \frac{\partial^2}{\partial y\, \partial x} f = f_{xy}, \qquad \frac{\partial^2}{\partial x\, \partial y} f = f_{yx}, \qquad \frac{\partial^2}{\partial y^2} f = f_{yy}$$

- Clairaut's Theorem states that mixed partials are equal—that is, $f_{xy} = f_{yx}$ provided that f_{xy} and f_{yx} are continuous.
- More generally, higher order partial derivatives may be computed in any order. For example, $f_{xyyz} = f_{yxzy}$ if f is a function of x, y, z whose fourth-order partial derivatives are continuous.

14.3 EXERCISES

Preliminary Questions

1. Patricia derived the following *incorrect* formula by misapplying the Product Rule:

$$\frac{\partial}{\partial x}(x^2 y^2) = x^2(2y) + y^2(2x)$$

What was her mistake and what is the correct calculation?

2. Explain why it is not necessary to use the Quotient Rule to compute $\dfrac{\partial}{\partial x}\left(\dfrac{x + y}{y + 1}\right)$. Should the Quotient Rule be used to compute $\dfrac{\partial}{\partial y}\left(\dfrac{x + y}{y + 1}\right)$?

3. Which of the following partial derivatives should be evaluated without using the Quotient Rule?

(a) $\dfrac{\partial}{\partial x}\dfrac{xy}{y^2 + 1}$ (b) $\dfrac{\partial}{\partial y}\dfrac{xy}{y^2 + 1}$ (c) $\dfrac{\partial}{\partial x}\dfrac{y^2}{y^2 + 1}$

4. What is f_x, where $f(x, y, z) = (\sin yz)e^{z^3 - z^{-1}\sqrt{y}}$?

5. Assuming the hypotheses of Clairaut's Theorem are satisfied, which of the following partial derivatives are equal to f_{xxy}?

(a) f_{xyx} (b) f_{yyx} (c) f_{xyy} (d) f_{yxx}

Exercises

1. Use the limit definition of the partial derivative to verify the formulas

$$\frac{\partial}{\partial x} xy^2 = y^2, \qquad \frac{\partial}{\partial y} xy^2 = 2xy$$

2. Use the limit definition of the partial derivative to verify the formulas

$$\frac{\partial}{\partial x}\left(\frac{x}{y}\right) = \frac{1}{y}, \qquad \frac{\partial}{\partial y}\left(\frac{x}{y}\right) = \frac{-x}{y^2}$$

3. Use the Product Rule to compute f_x and f_y for $f(x, y) = (x^2 - y)(x - y^2)$.

4. Use the Product Rule to compute f_x and f_y for $f(x, y) = xye^x \sin y$.

5. Use the Quotient Rule to compute $\dfrac{\partial}{\partial y}\dfrac{y}{x + y}$.

6. Use the Chain Rule to compute $\dfrac{\partial}{\partial u}\ln(u^2 + uv)$.

7. Calculate $f_z(2, 3, 1)$, where $f(x, y, z) = xyz$.

8. ✏️ Explain the relation between the following two formulas (c is a constant):

$$\frac{d}{dx}\sin(cx) = c\cos(cx), \qquad \frac{\partial}{\partial x}\sin(xy) = y\cos(xy)$$

9. The plane $y = 1$ intersects the surface $z = x^4 + 6xy - y^4$ in a certain curve. Find the slope of the tangent line to this curve at the point $P = (1, 1, 6)$.

10. Determine whether the partial derivatives $\partial f/\partial x$ and $\partial f/\partial y$ are positive or negative at the point P on the graph in Figure 6.

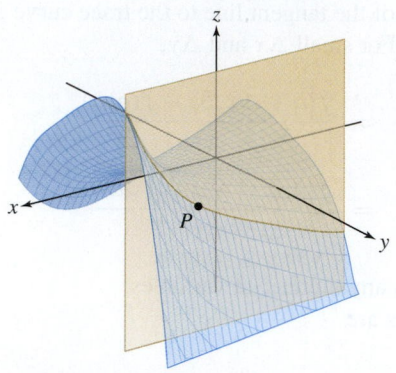

FIGURE 6

In Exercises 11–14, refer to Figure 7.

11. Estimate f_x and f_y at point A.

12. Is f_x positive or negative at B?

13. Starting at point B, in which compass direction (N, NE, SW, etc.) does f increase most rapidly?

14. At which of A, B, or C is f_y the least?

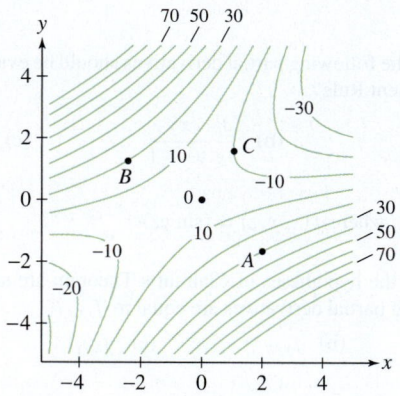

FIGURE 7 Contour map of $f(x, y)$.

In Exercises 15–42, compute the first-order partial derivatives.

15. $z = x^2 + y^2$

16. $z = x^4 y^3$

17. $z = x^4 y + xy^{-2}$

18. $V = \pi r^2 h$

19. $z = \dfrac{x}{y}$

20. $z = \dfrac{x}{x - y}$

21. $z = \sqrt{9 - x^2 - y^2}$

22. $z = \dfrac{x}{\sqrt{x^2 + y^2}}$

23. $z = (\sin x)(\cos y)$

24. $z = \tan(uv^3)$

25. $z = \cos \dfrac{1 - x}{y}$

26. $\theta = \tan^{-1}(xy^2)$

27. $w = \ln(x^2 - y^2)$

28. $P = \sin(2s - 3t)$

29. $W = e^{r+s}$

30. $Q = re^{\theta}$

31. $z = e^{xy}$

32. $R = e^{-v^2/k}$

33. $z = e^{-x^2 - y^2}$

34. $P = e^{\sqrt{y^2 + z^2}}$

35. $U = \dfrac{e^{-rt}}{r}$

36. $z = y^x$

37. $z = \sinh(x^2 y)$

38. $z = \cosh(t - \cos x)$

39. $w = xy^2 z^3$

40. $w = \dfrac{x}{y + z}$

41. $Q = \dfrac{L}{M} e^{-Lt/M}$

42. $w = \dfrac{x}{(x^2 + y^2 + z^2)^{3/2}}$

In Exercises 43–46, compute the given partial derivatives.

43. $f(x, y) = 3x^2 y + 4x^3 y^2 - 7xy^5$, $\quad f_x(1, 2)$

44. $f(x, y) = \sin(x^2 - y)$, $\quad f_y(0, \pi)$

45. $g(u, v) = u \ln(u + v)$, $\quad g_u(1, 2)$

46. $h(x, z) = e^{xz - x^2 z^3}$, $\quad h_z(3, 0)$

47. The **heat index** I is a measure of how hot it feels when the relative humidity is H (as a percentage) and the actual air temperature is T (in degrees Fahrenheit). An approximate formula for the heat index that is valid for (T, H) near $(90, 40)$ is

$$I(T, H) = 45.33 + 0.6845T + 5.758H - 0.00365T^2$$
$$- 0.1565HT + 0.001HT^2$$

(a) Calculate I at $(T, H) = (95, 50)$.

(b) Which partial derivative tells us the increase in I per degree increase in T when $(T, H) = (95, 50)$? Calculate this partial derivative.

48. Calculate $\partial P/\partial T$ and $\partial P/\partial V$, where pressure P, volume V, and temperature T are related by the Ideal Gas Law, $PV = nRT$ (R and n are constants).

49. [✐] Use the contour map of $f(x, y)$ in Figure 8 to explain the following statements:

(a) f_y is larger at P than at Q, and f_x is more negative at P than at Q.

(b) $f_x(x, y)$ is decreasing as a function of y; that is, for any fixed value $x = a$, $f_x(a, y)$ is decreasing in y.

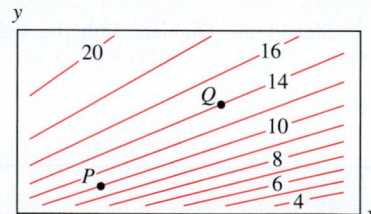

FIGURE 8

50. Estimate the partial derivatives at P of the function whose contour map is shown in Figure 9.

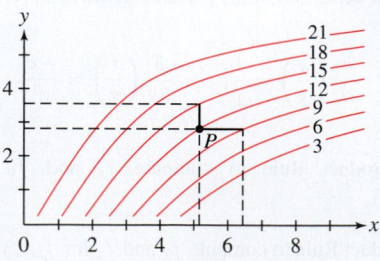

FIGURE 9

51. Over most of the earth, a magnetic compass does not point to true (geographic) north; instead, it points at some angle east or west of true north. The angle D between magnetic north and true north is called the **magnetic declination**. Use Figure 10 to determine which of the following statements is true:

(a) $\left.\dfrac{\partial D}{\partial y}\right|_A > \left.\dfrac{\partial D}{\partial y}\right|_B$ **(b)** $\left.\dfrac{\partial D}{\partial x}\right|_C > 0$ **(c)** $\left.\dfrac{\partial D}{\partial y}\right|_C > 0$

Note that the horizontal axis increases from right to left because of the way longitude is measured.

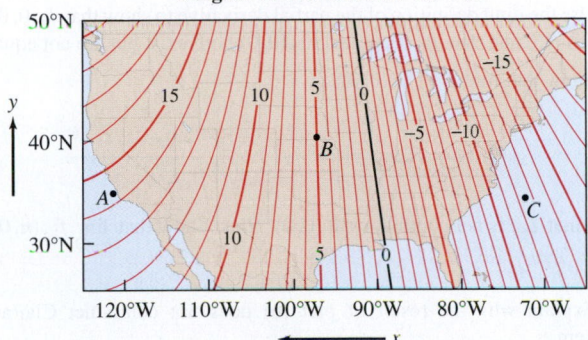

Magnetic Declination for the U.S.

FIGURE 10 Contour interval $1°$.

52. Refer to Table 1.

(a) Using difference quotients, approximate $\partial \rho / \partial T$ and $\partial \rho / \partial S$ at the points $(S, T) = (30, 2)$, $(32, 6)$, and $(35, 10)$.

(b) For fixed salinity $S = 32$, determine whether the quotients $\Delta \rho / \Delta T$ are increasing or decreasing as T increases. What can you conclude about the sign of $\partial^2 \rho / \partial T^2$ and the concavity of ρ as a function of T?

TABLE 1 Seawater Density ρ as a Function of Temperature T and Salinity S

T \ S	30	31	32	33	34	35	36
12	22.75	23.51	24.27	25.07	25.82	26.6	27.36
10	23.07	23.85	24.62	25.42	26.17	26.99	27.73
8	23.36	24.15	24.93	25.73	26.5	27.28	29.09
6	23.62	24.44	25.22	26	26.77	27.55	28.35
4	23.85	24.62	25.42	26.23	27	27.8	28.61
2	24	24.78	25.61	26.38	27.18	28.01	28.78
0	24.11	24.92	25.72	26.5	27.34	28.12	28.91

In Exercises 53–58, compute the derivatives indicated.

53. $f(x, y) = 3x^2 y - 6xy^4$, $\dfrac{\partial^2 f}{\partial x^2}$ and $\dfrac{\partial^2 f}{\partial y^2}$

54. $g(x, y) = \dfrac{xy}{x - y}$, $\dfrac{\partial^2 g}{\partial x \, \partial y}$

55. $h(u, v) = \dfrac{u}{u + 4v}$, $h_{vv}(u, v)$

56. $h(x, y) = \ln(x^3 + y^3)$, $h_{xy}(x, y)$

57. $f(x, y) = x \ln(y^2)$, $f_{yy}(2, 3)$

58. $g(x, y) = xe^{-xy}$, $g_{xy}(-3, 2)$

59. Compute f_{xyxzy} for

$$f(x, y, z) = y \sin(xz) \sin(x + z) + (x + z^2) \tan y + x \tan\left(\dfrac{z + z^{-1}}{y - y^{-1}}\right)$$

Hint: Use a well-chosen order of differentiation on each term.

60. Let

$$f(x, y, u, v) = \dfrac{x^2 + e^y v}{3y^2 + \ln(2 + u^2)}$$

What is the fastest way to show that $f_{uvxyvu}(x, y, u, v) = 0$ for all (x, y, u, v)?

In Exercises 61–68, compute the derivative indicated.

61. $f(u, v) = \cos(u + v^2)$, f_{uuv}

62. $g(x, y, z) = x^4 y^5 z^6$, g_{xxyz}

63. $F(r, s, t) = r(s^2 + t^2)$, F_{rst}

64. $u(x, t) = t^{-1/2} e^{-(x^2/4t)}$, u_{xx}

65. $F(\theta, u, v) = \sinh(uv + \theta^2)$, $F_{uu\theta}$

66. $R(u, v, w) = \dfrac{u}{v + w}$, R_{uvw}

67. $g(x, y, z) = \sqrt{x^2 + y^2 + z^2}$, g_{xyz}

68. $u(x, t) = \text{sech}^2(x - t)$, u_{xxx}

69. Find a function such that $\dfrac{\partial f}{\partial x} = 2xy$ and $\dfrac{\partial f}{\partial y} = x^2$.

70. [✎] Prove that there does not exist any function $f(x, y)$ such that $\dfrac{\partial f}{\partial x} = xy$ and $\dfrac{\partial f}{\partial y} = x^2$. *Hint:* Consider Clairaut's Theorem.

71. Assume that f_{xy} and f_{yx} are continuous and that f_{yxx} exists. Show that f_{xyx} also exists and that $f_{yxx} = f_{xyx}$.

72. Show that $u(x, t) = \sin(nx) e^{-n^2 t}$ satisfies the heat equation for any constant n:

$$\dfrac{\partial u}{\partial t} = \dfrac{\partial^2 u}{\partial x^2} \qquad \boxed{3}$$

73. Find all values of A and B such that $f(x, t) = e^{Ax + Bt}$ satisfies Eq. (3).

74. The function

$$f(x, t) = \dfrac{1}{2\sqrt{\pi t}} e^{-x^2/4t}$$

describes the temperature profile along a metal rod at time $t > 0$ when a burst of heat is applied at the origin (see Example 11). A small bug sitting on the rod at distance x from the origin feels the temperature rise and fall as heat diffuses through the bar. Show that the bug feels the maximum temperature at time $t = \frac{1}{2} x^2$.

*In Exercises 75–78, the **Laplace operator** Δ is defined by $\Delta f = f_{xx} + f_{yy}$. A function $u(x, y)$ satisfying the Laplace equation $\Delta u = 0$ is called **harmonic**.*

75. Show that the following functions are harmonic:

(a) $u(x, y) = x$

(b) $u(x, y) = e^x \cos y$

(c) $u(x, y) = \tan^{-1} \dfrac{y}{x}$

(d) $u(x, y) = \ln(x^2 + y^2)$

76. Find all harmonic polynomials $u(x, y)$ of degree 3, that is, $u(x, y) = ax^3 + bx^2 y + cxy^2 + dy^3$.

77. Show that if $u(x, y)$ is harmonic, then the partial derivatives $\partial u / \partial x$ and $\partial u / \partial y$ are harmonic.

78. Find all constants a, b such that $u(x, y) = \cos(ax)e^{by}$ is harmonic.

79. Show that $u(x, t) = \operatorname{sech}^2(x - t)$ satisfies the **Korteweg–deVries equation** (which arises in the study of water waves):

$$4u_t + u_{xxx} + 12uu_x = 0$$

Further Insights and Challenges

80. Assumptions Matter This exercise shows that the hypotheses of Clairaut's Theorem are needed. Let

$$f(x, y) = xy\frac{x^2 - y^2}{x^2 + y^2}$$

for $(x, y) \neq (0, 0)$ and $f(0, 0) = 0$.

(a) Verify for $(x, y) \neq (0, 0)$:

$$f_x(x, y) = \frac{y(x^4 + 4x^2 y^2 - y^4)}{(x^2 + y^2)^2}$$

$$f_y(x, y) = \frac{x(x^4 - 4x^2 y^2 - y^4)}{(x^2 + y^2)^2}$$

(b) Use the limit definition of the partial derivative to show that $f_x(0, 0) = f_y(0, 0) = 0$ and that $f_{yx}(0, 0)$ and $f_{xy}(0, 0)$ both exist but are not equal.

(c) Show that for $(x, y) \neq (0, 0)$:

$$f_{xy}(x, y) = f_{yx}(x, y) = \frac{x^6 + 9x^4 y^2 - 9x^2 y^4 - y^6}{(x^2 + y^2)^3}$$

Show that f_{xy} is not continuous at $(0, 0)$. *Hint:* Show that $\lim\limits_{h \to 0} f_{xy}(h, 0) \neq \lim\limits_{h \to 0} f_{xy}(0, h)$.

(d) Explain why the result of part (b) does not contradict Clairaut's Theorem.

14.4 Differentiability, Tangent Planes, and Linear Approximation

In this section, we explore the important concept of differentiability for functions of more than one variable, along with the related ideas of the tangent plane and linear approximation. In single-variable calculus, a function f is differentiable if the derivative f' exists. By extension, one might expect that a function $f(x, y)$ would be differentiable if the partial derivatives $f_x(x, y)$ and $f_y(x, y)$ exist. Unfortunately, the existence of partial derivatives is not a strong enough condition for differentiability.

First, we will show why the existence of the partial derivatives is not sufficient. Differentiability of $f(x, y)$ at (a, b) should ensure that there is a tangent plane to the graph of $f(x, y)$ at $P = (a, b, f(a, b))$ as illustrated in Figure 1.

If $f(x, y)$ has partial derivatives $f_x(a, b)$ and $f_y(a, b)$ at (a, b), then these derivatives determine lines that are tangent to the graph of $f(x, y)$ at P. Figure 2(A) shows that one

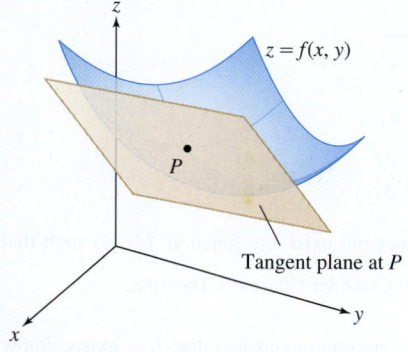

FIGURE 1 Tangent plane to the graph of $f(x, y)$.

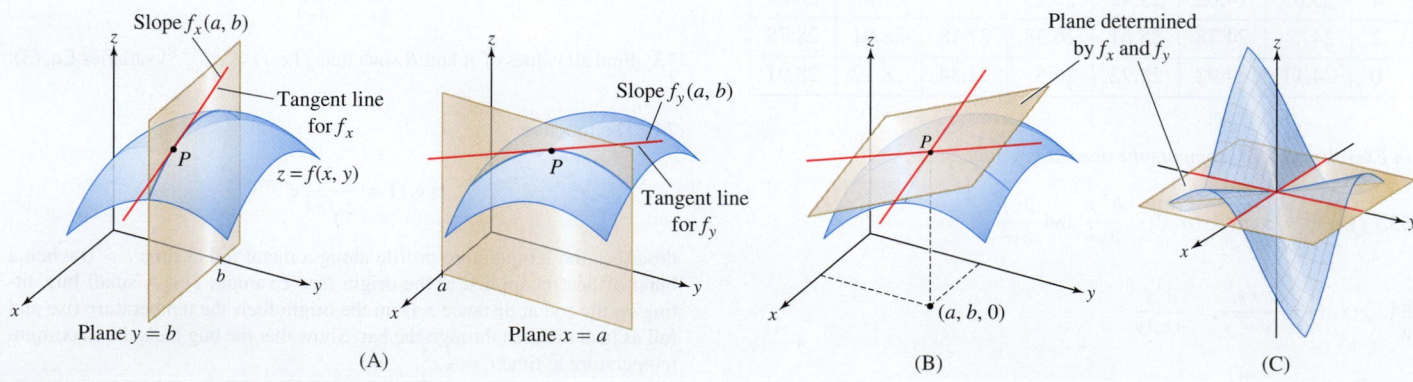

FIGURE 2 Is the plane determined by f_x and f_y tangent to the graph?

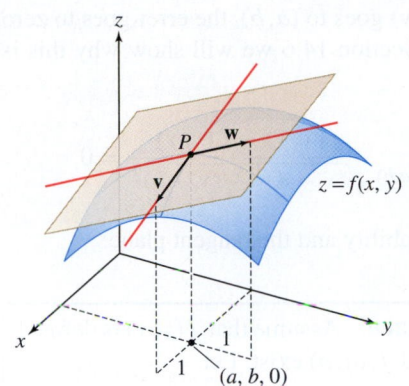

FIGURE 3 The vectors **v** and **w** are parallel to the plane determined by f_x and f_y.

of these tangent lines lies in the plane $y = b$, and the other lies in the plane $x = a$. We refer to these lines as the **tangent line for** f_x and the **tangent line for** f_y, respectively. These two tangent lines determine a plane that is certainly a good candidate for a tangent plane to the graph [Figure 2(B)]. We refer to this plane as the **plane determined by** f_x **and** f_y. Unfortunately, this plane might not be fully tangent to the graph at P because other lines through P in this plane might not be tangent to the graph as in Figure 2(C). We will give an example of just such a situation later in the section.

To identify a condition that guarantees that the plane determined by f_x and f_y is tangent to the graph, we first need an equation of this plane. (We will also use this condition to define differentiability.)

We begin by finding a normal vector to the plane determined by f_x and f_y. To do that, we find direction vectors for the tangent lines for f_x and f_y (which are parallel to the plane) and then take their cross product (which results in a vector normal to the plane).

Consider first the tangent line for f_x, which lies in the plane $y = b$. In that plane, the line has slope $f_x(a, b)$. Therefore, if we move on the line 1 unit in the positive x-direction from P, then we move $f_x(a, b)$ units in the z-direction. It follows that the vector $\mathbf{v} = \langle 1, 0, f_x(a, b) \rangle$ is a direction vector for this line, as in Figure 3.

Similarly, the vector $\mathbf{w} = \langle 0, 1, f_y(a, b) \rangle$ is a direction vector for the tangent line for f_y. As indicated previously, a normal vector to the plane determined by f_x and f_y is obtained by taking a cross product of the vectors **v** and **w**. It is convenient to compute this cross product as $\mathbf{w} \times \mathbf{v}$:

$$\mathbf{w} \times \mathbf{v} = \begin{vmatrix} \mathbf{i} & \mathbf{j} & \mathbf{k} \\ 0 & 1 & f_y(a,b) \\ 1 & 0 & f_x(a,b) \end{vmatrix} = \langle f_x(a,b), f_y(a,b), -1 \rangle$$

Now, $P = (a, b, f(a, b))$ lies on the plane determined by f_x and f_y, and the vector $\langle f_x(a,b), \ f_y(a,b), -1 \rangle$ is normal to it. Therefore, the plane has equation

◄ REMINDER A plane through the point $P = (x_0, y_0, z_0)$ with normal vector $\mathbf{n} = \langle A, B, C \rangle$ has equation $A(x - x_0) + B(y - y_0) + C(z - z_0) = 0$

$$f_x(a,b)(x - a) + f_y(a,b)(y - b) - (z - f(a,b)) = 0$$

or

$$z = f(a,b) + f_x(a,b)(x - a) + f_y(a,b)(y - b)$$

Next, we use this equation for the plane determined by f_x and f_y to identify a condition that ensures this plane is fully tangent to the graph of $f(x, y)$ [and thus that $f(x, y)$ is differentiable]. Let

$$L(x,y) = f(a,b) + f_x(a,b)(x - a) + f_y(a,b)(y - b)$$

We refer to $L(x, y)$ as the **linearization of** $f(x, y)$ **centered at** (a, b). The linearization can be used to approximate $f(x, y)$ near (a, b). (Later in this section, we will develop this idea further and present some examples.) The graph of $L(x, y)$ is the plane determined by f_x and f_y. Figure 4 demonstrates that the difference $f(x, y) - L(x, y)$ is the error $e(x, y)$ obtained when approximating $f(x, y)$ by $L(x, y)$. As (x, y) approaches (a, b), this error approaches zero because the two functions are continuous and agree at (a, b). If the error goes to zero fast enough that the graph of $f(x, y)$ flattens and becomes approximately a plane, then the plane determined by f_x and f_y is an actual tangent plane. (We will explain why when we introduce directional derivatives in the next section.)

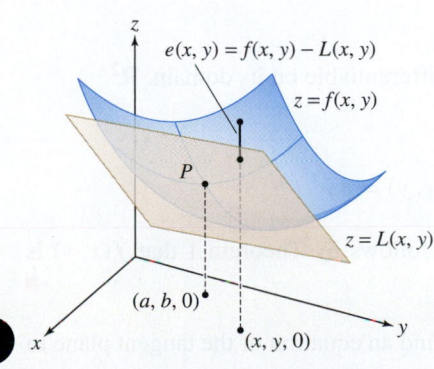

FIGURE 4 The error approximating $f(x, y)$ with $L(x, y)$.

What suffices for "fast enough" is that as (x, y) goes to (a, b), the error goes to zero faster than the distance from (x, y) to (a, b). (In Section 14.6 we will show why this is so.) That is,

$$\lim_{(x,y)\to(a,b)} \frac{e(x, y)}{\sqrt{(x - a)^2 + (y - b)^2}} = \lim_{(x,y)\to(a,b)} \frac{f(x, y) - L(x, y)}{\sqrt{(x - a)^2 + (y - b)^2}} = 0$$

This now brings us to definitions of differentiability and the tangent plane:

DEFINITION Differentiability and the Tangent Plane   Assume that $f(x, y)$ is defined in a disk D containing (a, b) and that $f_x(a, b)$ and $f_y(a, b)$ exist. Let

$$L(x, y) = f(a, b) + f_x(a, b)(x - a) + f_y(a, b)(y - b)$$

- $f(x, y)$ is **differentiable** at (a, b) if

$$\lim_{(x,y)\to(a,b)} \frac{f(x, y) - L(x, y)}{\sqrt{(x - a)^2 + (y - b)^2}} = 0$$

- If $f(x, y)$ is differentiable at (a, b), then the **tangent plane** to the graph at $(a, b, f(a, b))$ is the plane with equation $z = L(x, y)$. Explicitly, the equation of the tangent plane is

$$\boxed{z = f(a, b) + f_x(a, b)(x - a) + f_y(a, b)(y - b)} \qquad \boxed{1}$$

The definition of differentiability extends to functions of n-variables, and Theorem 1 holds in this setting: If all of the partial derivatives of $f(x_1, \ldots, x_n)$ exist and are continuous on an open domain $\mathcal{D}$, then $f(x_1, \ldots, x_n)$ is differentiable on $\mathcal{D}$.

If $f(x, y)$ is differentiable at all points in a domain $\mathcal{D}$, we say that $f(x, y)$ is **differentiable on $\mathcal{D}$**.

To prove that a particular function is differentiable, we need to prove that the limit in the definition holds. That can be tedious to verify (see Exercise 43), but fortunately, this is rarely necessary. The following theorem provides conditions that imply differentiability and are easy to verify. It assures us that most functions arising in practice are differentiable on their domains. See Appendix D for a proof.

THEOREM 1 Confirming Differentiability   If $f_x(x, y)$ and $f_y(x, y)$ exist and are continuous on an open disk D, then $f(x, y)$ is differentiable on D.

EXAMPLE 1   Show that $f(x, y) = 5x + 4y^2$ is differentiable on its domain, $\mathbf{R}^2$.

Solution   The partial derivatives are

$$f_x(x, y) = 5, \qquad f_y(x, y) = 8y$$

These are continuous functions over all of $\mathbf{R}^2$. It follows by Theorem 1 that $f(x, y)$ is differentiable for all (x, y). ∎

EXAMPLE 2   Let $f(x, y) = x^2 + 2y^2 - y - 4$. Find an equation of the tangent plane to the graph of f at $P = (1, 2, f(1, 2))$.

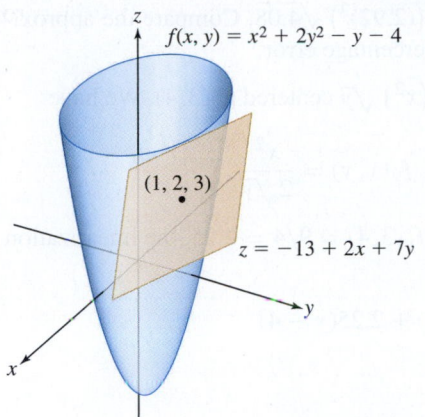

FIGURE 5 The graph of
$f(x, y) = x^2 + 2y^2 - y - 4$ and the
tangent plane at $P = (1, 2, 3)$.

Solution We have $f_x(x, y) = 2x$ and $f_y(x, y) = 4y - 1$. From that,

$$f_x(1, 2) = 2 \quad \text{and} \quad f_y(1, 2) = 7$$

These values, along with $f(1, 2) = 3$, enable us to determine the equation of the tangent plane:

$$z = \underbrace{3 + 2(x - 1) + 7(y - 2)}_{f(a, b) + f_x(a, b)(x - a) + f_y(a, b)(y - b)} = -13 + 2x + 7y$$

The tangent plane through $P = (1, 2, 3)$ has equation $z = -13 + 2x + 7y$ (Figure 5). ∎

EXAMPLE 3 Find an equation of the tangent plane to the graph of $f(x, y) = xy^3 + x^2$ at $(2, -2, f(2, -2))$.

Solution The partial derivatives of $f(x, y)$ are $f_x(x, y) = y^3 + 2x$ and $f_y(x, y) = 3xy^2$.
With $f(2, -2) = -12$, $f_x(2, -2) = -4$, and $f_y(2, -2) = 24$, the tangent plane through $(2, -2, -12)$ has equation

$$z = -12 - 4(x - 2) + 24(y + 2)$$

This can be rewritten as $z = 44 - 4x + 24y$. ∎

In single-variable calculus, a function f that is differentiable at a has a graph that, as you continuously zoom in at $(a, f(a))$, appears more and more like the tangent line to the graph at $(a, f(a))$.

A similar situation holds for differentiable functions of two variables. Specifically, if $f(x, y)$ is differentiable at (a, b), then as you continuously zoom in on the graph at $P = (a, b, f(a, b))$, it appears more and more like the tangent plane to the graph at P (Figure 6).

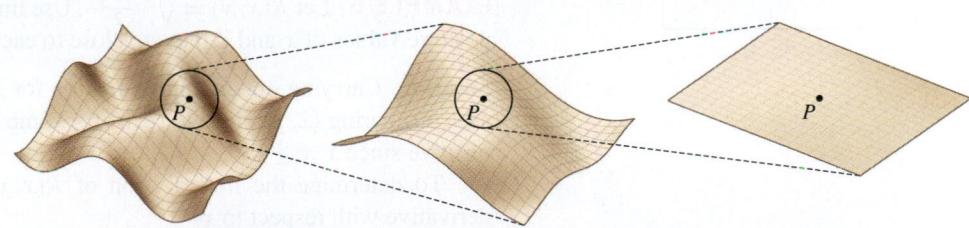

FIGURE 6 The graph looks more and more like the tangent plane at P as we zoom in.

As Figure 6 suggests, differentiability at (a, b) implies that in a small region around P, the graph of $f(x, y)$ is nearly indistinguishable from the tangent plane at P. That is to say, we can approximate $f(x, y)$ near (a, b) by the linearization $L(x, y)$. We have obtained the multivariable version of approximation by linearization:

Approximating $f(x, y)$ by Its Linearization If $f(x, y)$ is differentiable at (a, b), and (x, y) is close to (a, b), then $f(x, y) \approx L(x, y)$. Thus,

$$\boxed{f(x, y) \approx f(a, b) + f_x(a, b)(x - a) + f_y(a, b)(y - b)} \quad \boxed{2}$$

← **REMINDER** *The percentage error is equal to*

$$\left|\frac{\text{error}}{\text{actual value}}\right| \times 100\%$$

EXAMPLE 4 Use linearization to approximate $\left((2.92)^2\right)\sqrt{4.08}$. Compare the approximation with a calculator value and estimate the percentage error.

Solution We use the linearization of $f(x, y) = \left(x^2\right)\sqrt{y}$ centered at $(3, 4)$. We have

$$f_x(x, y) = (2x)\sqrt{y} \quad \text{and} \quad f_y(x, y) = \frac{x^2}{2\sqrt{y}}$$

Then with $f(3, 4) = 18$, $f_x(3, 4) = 12$, and $f_y(3, 4) = 9/4 = 2.25$, the linearization centered at $(3, 4)$ is

$$L(x, y) = 18 + 12(x - 3) + 2.25(y - 4)$$

Therefore,

$$\left((2.92)^2\right)\sqrt{4.08} \approx 18 + 12(2.92 - 3) + 2.25(4.08 - 4) = 17.24$$

A calculator yields $\left((2.92)^2\right)\sqrt{4.08} \approx 17.2225$ rounded to four decimal places.
The percentage error is

$$\frac{17.24 - 17.2225}{17.2225} \times 100\% \approx 0.10\% \qquad\blacksquare$$

Recall in Examples 5 and 6 in Section 12.3, we examined a problem where we had a parallelogram of side lengths a and b and diagonal lengths x and y (Figure 7). We straightened the parallelogram into a rectangle that also had side lengths a and b and asked the question "How are the diagonal lengths in the rectangle related to the original diagonal lengths x and y?" Using vector geometry, we were able to answer that the rectangle diagonal lengths are $\sqrt{\frac{x^2+y^2}{2}}$.

We went on to explore the situation where a carpenter, wanting to "square-up" a parallelogram with diagonal lengths x and y, uses the simpler expression $\frac{x+y}{2}$ for the target diagonal length. We are now in a position to show that the carpenter's formula is obtained via linearization of $R(x, y) = \sqrt{\frac{x^2+y^2}{2}}$ for positive values of x and y close to each other.

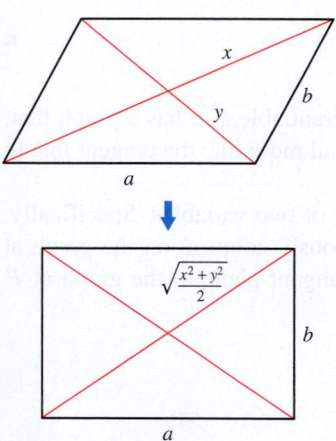

FIGURE 7

EXAMPLE 5 Let $R(x, y) = \sqrt{\frac{x^2+y^2}{2}}$. Use linearization to show that $R(x, y) \approx \frac{x+y}{2}$ for positive values of x and y that are close to each other.

Solution Carrying out an approximation for x and y close to each other means that we are considering (x, y) close to (a, a) for some a. Furthermore, we are assuming that a is positive since $x, y \geq 0$.

To determine the linearization of $R(x, y)$ centered at (a, a), compute the partial derivative with respect to x:

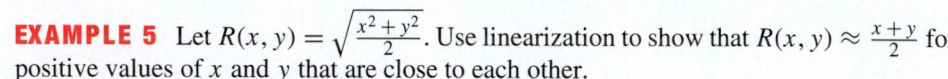

$$R_x(x, y) = \frac{1}{2}\left(\frac{x^2+y^2}{2}\right)^{-1/2}\left(\frac{2x}{2}\right) = \left(\frac{x}{2}\right)\sqrt{\frac{2}{x^2+y^2}} = \frac{x}{\sqrt{2(x^2+y^2)}}$$

The partial derivative with respect to y is obtained similarly:

$$R_y(x, y) = \frac{y}{\sqrt{2(x^2+y^2)}}$$

Since the linearization is centered at (a, a), we need $R(x, y)$, R_x, and R_y evaluated there:

$$R(a, a) = \sqrt{\frac{a^2+a^2}{2}} = \sqrt{a^2} = a \quad \text{since } a > 0$$

$$R_x(a, a) = \frac{a}{\sqrt{2(a^2+a^2)}} = \frac{1}{2} \quad \text{and similarly,} \quad R_y(a, a) = \frac{1}{2}$$

Therefore, we have by Eq. (2):

$$R(x, y) \approx a + \frac{1}{2}(x - a) + \frac{1}{2}(y - a) = \frac{x + y}{2} \qquad \blacksquare$$

Like functions of a single variable, the approximation by linearization has associated linear approximation formulas for the change in f. If we let Δx and Δy represent small changes in x and y, and set $x = a + \Delta x$ and $y = b + \Delta y$, then from the linearization formula, Eq. (2), we obtain the **Linear Approximation**:

$$\boxed{f(a + \Delta x, b + \Delta y) \approx f(a, b) + f_x(a, b)\Delta x + f_y(a, b)\Delta y} \qquad \boxed{3}$$

We can also write the Linear Approximation in terms of the change in f:

$$\Delta f = f(x, y) - f(a, b)$$

$$\boxed{\Delta f \approx f_x(a, b)\Delta x + f_y(a, b)\Delta y} \qquad \boxed{4}$$

BMI is one factor used to assess the risk of certain diseases such as diabetes and high blood pressure. The range $18.5 \leq I \leq 24.9$ is considered normal for adults over 20 years of age.

EXAMPLE 6 **Body Mass Index** A person's BMI is $I = W/H^2$, where W is the body weight (in kilograms) and H is the body height (in meters). Estimate the change in a child's BMI if (W, H) changes from $(40, 1.45)$ to $(41.5, 1.47)$.

Solution To begin, we compute the partial derivatives:

$$\frac{\partial I}{\partial W} = \frac{\partial}{\partial W}\left(\frac{W}{H^2}\right) = \frac{1}{H^2}, \qquad \frac{\partial I}{\partial H} = \frac{\partial}{\partial H}\left(\frac{W}{H^2}\right) = -\frac{2W}{H^3}$$

At $(W, H) = (40, 1.45)$, we have

$$\left.\frac{\partial I}{\partial W}\right|_{(40, 1.45)} = \frac{1}{1.45^2} \approx 0.48, \qquad \left.\frac{\partial I}{\partial H}\right|_{(40, 1.45)} = -\frac{2(40)}{1.45^3} \approx -26.24$$

If (W, H) changes from $(40, 1.45)$ to $(41.5, 1.47)$, then

$$\Delta W = 41.5 - 40 = 1.5, \qquad \Delta H = 1.47 - 1.45 = 0.02$$

Therefore, by Eq. (4),

$$\Delta I \approx \left.\frac{\partial I}{\partial W}\right|_{(40,1.45)} \Delta W + \left.\frac{\partial I}{\partial H}\right|_{(40,1.45)} \Delta H \approx 0.48(1.5) - 26.24(0.02) \approx 0.2$$

We find that BMI increases by approximately 0.2. $\qquad \blacksquare$

CONCEPTUAL INSIGHT Linear Approximation for estimating the change in $f(x, y)$ is similar to the corresponding Linear Approximation for a function f of a single variable:

$$\underbrace{\Delta f}_{\substack{\text{Change} \\ \text{in } f}} \approx \underbrace{f'(a)}_{\substack{\text{Rate of} \\ \text{change at } a}} \underbrace{(x - a)}_{\substack{\text{Change} \\ \text{in } x}}$$

$$\underbrace{\Delta f}_{\substack{\text{Change} \\ \text{in } f}} \approx \underbrace{f_x(a, b)}_{\substack{\text{Rate of change} \\ \text{with respect to} \\ x \text{ at } (a, b)}} \underbrace{(x - a)}_{\substack{\text{Change} \\ \text{in } x}} + \underbrace{f_y(a, b)}_{\substack{\text{Rate of change} \\ \text{with respect to} \\ y \text{ at } (a, b)}} \underbrace{(y - b)}_{\substack{\text{Change} \\ \text{in } y}}$$

For a function of two variables, we have contributions to the change in f from changes in each independent variable.

We define differentials and approximation via differentials like the Differential Form of Linear Approximation for functions of a single variable:

> **Differentials and Linear Approximation** Assume that f is differentiable at (a, b), and let $dx = \Delta x$, $dy = \Delta y$. Then the differential df is defined by
>
> $$df = f_x(x, y)\, dx + f_y(x, y)\, dy \qquad \boxed{5}$$
>
> Figure 8 shows that df represents the change in height of the tangent plane for given changes dx and dy in x and y.
>
> With Δf representing the actual change in $f(x, y)$, it follows that $\Delta f \approx df$, and we obtain the Differential Form of Linear Approximation:
>
> $$\Delta f \approx df = f_x(x, y)\, dx + f_y(x, y)\, dy \qquad \boxed{6}$$

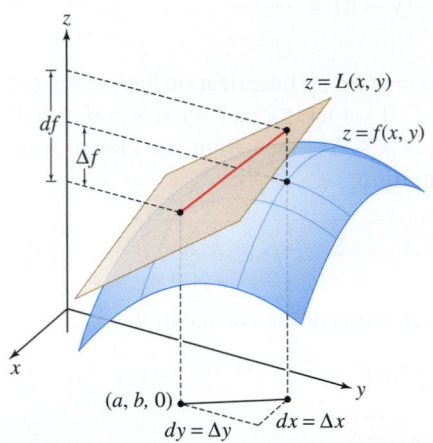

FIGURE 8 The differential df is the change in height of the tangent plane.

Each of the approximations we have presented extends to functions of any number of variables. For example, if f is a function of three variables x, y, and z, then

- Approximation via linearization: $f(x, y, z) \approx f(a, b, c) + f_x(a, b, c)(x - a) + f_y(a, b, c)(y - b) + f_z(a, b, c)(z - c)$
- Linear Approximation: $\Delta f \approx f_x(a, b, c)\Delta x + f_y(a, b, c)\Delta y + f_z(a, b, c)\Delta z$

These approximation formulas will be helpful in some of the exercises.

Assumptions Matter

The mere existence of the partial derivatives does not guarantee differentiability. The function $g(x, y)$ in Figure 9 shows what can go wrong. It is defined by

$$g(x, y) = \begin{cases} \dfrac{2xy(x + y)}{x^2 + y^2} & (x, y) \neq (0, 0) \\ 0 & (x, y) = (0, 0) \end{cases}$$

The graph contains the x- and y-axes—in other words, $g(x, 0) = 0$ and $g(0, y) = 0$—and therefore the partial derivatives $g_x(0, 0)$ and $g_y(0, 0)$ are both zero. This implies that at the origin $(0, 0)$, the plane determined by g_x and g_y is the xy-plane. However,

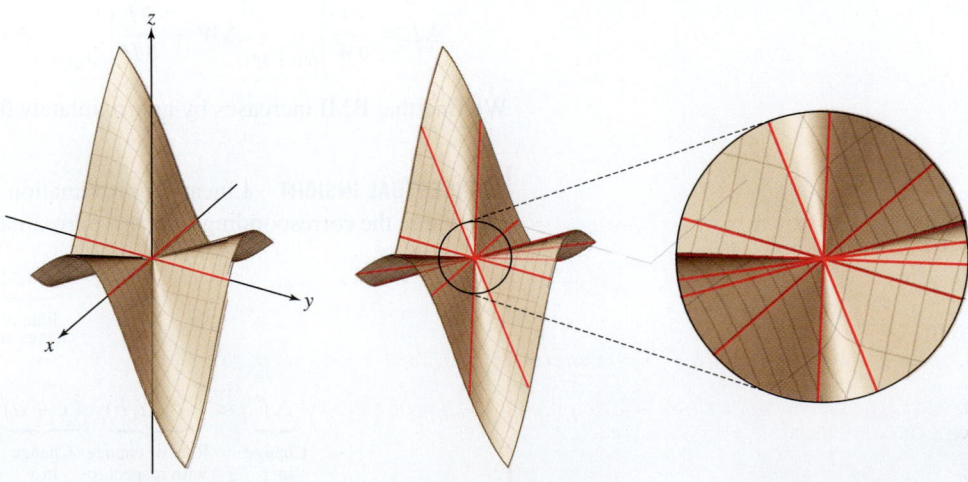

FIGURE 9 The function $g(x, y)$ is not differentiable at $(0, 0)$.

(A) The horizontal trace at $z = 0$ includes the x- and y-axes.

(B) But the graph also contains nonhorizontal lines through the origin.

(C) Thus, the graph does not appear any flatter as we zoom in on the origin.

Figure 9(B) shows that the graph also contains lines through the origin that do not lie in the xy-plane (in fact, the graph is composed entirely of lines through the origin). As we zoom in on the origin, these lines remain at an angle to the xy-plane, and the graph does not flatten out. Thus, $g(x, y)$ is not differentiable at $(0, 0)$, and there is no tangent plane there.

Furthermore, since $g(x, y)$ is not differentiable at $(0, 0)$, the assumptions of Theorem 1 must not be satisfied there. In particular, while the partial derivatives $g_x(x, y)$ and $g_y(x, y)$ exist, they must not be continuous at the origin (see Exercise 47 for details).

14.4 SUMMARY

- $f(x, y)$ is *differentiable* at (a, b) if $f_x(a, b)$ and $f_y(a, b)$ exist and

$$\lim_{(x,y)\to(a,b)} \frac{f(x, y) - L(x, y)}{\sqrt{(x - a)^2 + (y - b)^2}} = 0$$

where $L(x, y) = f(a, b) + f_x(a, b)(x - a) + f_y(a, b)(y - b)$.
- Result used in practice: *If $f_x(x, y)$ and $f_y(x, y)$ exist and are continuous in a disk D containing (a, b), then $f(x, y)$ is differentiable at (a, b).*
- If $f(x, y)$ is differentiable at (a, b), the equation of the tangent plane to $z = f(x, y)$ at (a, b) is

$$z = f(a, b) + f_x(a, b)(x - a) + f_y(a, b)(y - b)$$

- If $f(x, y)$ is differentiable at (a, b), the *linearization* of f centered at (a, b) is the function

$$L(x, y) = f(a, b) + f_x(a, b)(x - a) + f_y(a, b)(y - b)$$

- The approximation by linearization is $f(x, y) \approx L(x, y)$, or

$$f(x, y) \approx L(x, y) = f(a, b) + f_x(a, b)(x - a) + f_y(a, b)(y - b)$$

- *Linear Approximation*:

$$f(a + \Delta x, b + \Delta y) \approx f(a, b) + f_x(a, b)\Delta x + f_y(a, b)\Delta y$$

$$\Delta f \approx f_x(a, b)\, \Delta x + f_y(a, b)\, \Delta y$$

- *Differential Form of Linear Approximation*: $\Delta f \approx df$, where

$$df = f_x(x, y)\, dx + f_y(x, y)\, dy = \frac{\partial f}{\partial x} dx + \frac{\partial f}{\partial y} dy$$

14.4 EXERCISES

Preliminary Questions

1. How is the linearization of $f(x, y)$ centered at (a, b) defined?

2. If f is differentiable at (a, b) and $f_x(a, b) = f_y(a, b) = 0$, what can we conclude about the tangent plane at (a, b)?

In Exercises 3–5, assume that

$$f(2, 3) = 8, \qquad f_x(2, 3) = 5, \qquad f_y(2, 3) = 7$$

3. Which of (a)–(b) is the linearization of f centered at $(2, 3)$?

(a) $L(x, y) = 8 + 5x + 7y$

(b) $L(x, y) = 8 + 5(x - 2) + 7(y - 3)$

4. Estimate $f(2, 3.1)$.

5. Estimate Δf at $(2, 3)$ if $\Delta x = -0.3$ and $\Delta y = 0.2$.

6. In the derivation of the equation for the plane determined by f_x and f_y, we used $\mathbf{w} \times \mathbf{v}$ for a normal vector to the plane. How would the choice of $\mathbf{v} \times \mathbf{w}$ for a normal vector have affected the resultant equation?

Exercises

1. Find an equation of the tangent plane to the graph of $f(x, y) = 2x^2 - 4xy^2$ at $(-1, 2)$.

2. Find the equation of the plane in Figure 10, which is tangent to the graph at $(x, y) = (1, 0.8)$.

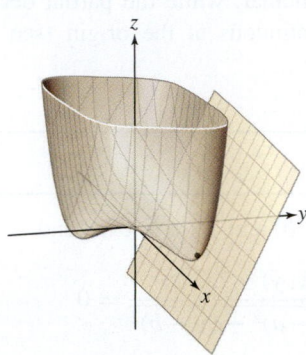

FIGURE 10 Graph of $f(x, y) = 0.2x^4 + y^6 - xy$.

In Exercises 3–10, find an equation of the tangent plane at the given point.

3. $f(x, y) = xy^2 + x^3y^2$, $(-1, 2)$

4. $f(x, y) = \dfrac{y}{\sqrt{x}}$, $(4, -3)$

5. $f(x, y) = x^2 + y^{-2}$, $(4, 1)$

6. $G(u, w) = \sin(uw)$, $\left(\frac{\pi}{6}, 1\right)$

7. $F(r, s) = r^2 s^{-1/2} + s^{-3}$, $(2, 1)$

8. $g(x, y) = e^{x/y}$, $(2, 1)$

9. $f(x, y) = \operatorname{sech}(x - y)$, $(\ln 4, \ln 2)$

10. $f(x, y) = \ln(4x^2 - y^2)$, $(1, 1)$

11. Find the points on the graph of $z = 3x^2 - 4y^2$ at which the vector $\mathbf{n} = \langle 3, 2, 2 \rangle$ is normal to the tangent plane.

12. Find the points on the graph of $z = xy^3 + 8y^{-1}$ where the tangent plane is parallel to $2x + 7y + 2z = 0$.

13. Find the points on the graph of $f(x, y) = 3x^2 - xy - y^2$ at which the tangent plane is horizontal.

14. Find the points on the graph of $f(x, y) = (x + 1)y^2$ at which the tangent plane is horizontal.

15. Find the linearization $L(x, y)$ of $f(x, y) = x^2 y^3$ at $(a, b) = (2, 1)$. Use it to estimate $f(2.01, 1.02)$ and $f(1.97, 1.01)$, and compare with values obtained using a calculator.

16. Write the Linear Approximation to $f(x, y) = x(1 + y)^{-1}$ at $(a, b) = (8, 1)$ in the form

$$f(a + h, b + k) \approx f(a, b) + f_x(a, b)h + f_y(a, b)k$$

Use it to estimate $\dfrac{7.98}{2.02}$ and compare with the value obtained using a calculator.

17. Let $f(x, y) = x^3 y^{-4}$. Use Eq. (4) to estimate the change

$$\Delta f = f(2.03, 0.9) - f(2, 1)$$

18. Use the Linear Approximation to $f(x, y) = \sqrt{x/y}$ at $(9, 4)$ to estimate $\sqrt{9.1/3.9}$.

19. Use the Linear Approximation of $f(x, y) = e^{x^2 + y}$ at $(0, 0)$ to estimate $f(0.01, -0.02)$. Compare with the value obtained using a calculator.

20. Let $f(x, y) = x^2/(y^2 + 1)$. Use the Linear Approximation at an appropriate point (a, b) to estimate $f(4.01, 0.98)$.

21. Find the linearization of $f(x, y, z) = z\sqrt{x + y}$ centered at $(8, 4, 5)$.

22. Find the linearization of $f(x, y, z) = xy/z$ centered at $(2, 1, 2)$. Use it to estimate $f(2.05, 0.9, 2.01)$ and compare with the value obtained from a calculator.

23. Estimate $f(2.1, 3.8)$ assuming that

$$f(2, 4) = 5, \qquad f_x(2, 4) = 0.3, \qquad f_y(2, 4) = -0.2$$

24. Estimate $f(1.02, 0.01, -0.03)$ assuming that

$$f(1, 0, 0) = -3, \qquad f_x(1, 0, 0) = -2$$
$$f_y(1, 0, 0) = 4, \qquad f_z(1, 0, 0) = 2$$

In Exercises 25–30, use the Linear Approximation to estimate the value. Compare with the value given by a calculator.

25. $(2.01)^3(1.02)^2$

26. $\dfrac{4.1}{7.9}$

27. $\sqrt{3.01^2 + 3.99^2}$

28. $\dfrac{0.98^2}{2.01^3 + 1}$

29. $\sqrt{(1.9)(2.02)(4.05)}$

30. $\dfrac{8.01}{\sqrt{(1.99)(2.01)}}$

31. Suppose that the plane tangent to $z = f(x, y)$ at $(-2, 3, 4)$ has equation $4x + 2y + z = 2$. Estimate $f(-2.1, 3.1)$.

32. The vector $\mathbf{n} = \langle 2, -3, 6 \rangle$ is normal to the tangent plane to $z = h(x, y)$ at $(1, -3, 5)$. Estimate $h(0.85, -3.08)$.

In Exercises 33–36, let $I = W/H^2$ denote the BMI described in Example 6.

33. A child has weight $W = 34$ kg and height $H = 1.3$ m. Use the Linear Approximation to estimate the change in I if (W, H) changes to $(36, 1.32)$.

34. Suppose that $(W, H) = (34, 1.3)$. Use the Linear Approximation to estimate the increase in H required to keep I constant if W increases to 35.

35. (a) Show that $\Delta I \approx 0$ if $\Delta H / \Delta W \approx H/2W$.

(b) Suppose that $(W, H) = (25, 1.1)$. What increase in H will leave I (approximately) constant if W is increased by 1 kg?

36. Estimate the change in height that will decrease I by 1 if $(W, H) = (25, 1.1)$, assuming that W remains constant.

37. A cylinder of radius r and height h has volume $V = \pi r^2 h$.

(a) Use the Linear Approximation to show that

$$\frac{\Delta V}{V} \approx \frac{2\Delta r}{r} + \frac{\Delta h}{h}$$

(b) Estimate the percentage increase in V if r and h are each increased by 2%.

(c) The volume of a certain cylinder V is determined by measuring r and h. Which will lead to a greater error in V: a 1% error in r or a 1% error in h?

38. Use the Linear Approximation to show that if $I = x^a y^b$, then

$$\frac{\Delta I}{I} \approx a\frac{\Delta x}{x} + b\frac{\Delta y}{y}$$

39. The monthly payment for a home loan is given by a function $f(P, r, N)$, where P is the principal (initial size of the loan), r the interest rate, and N the length of the loan in months. Interest rates are expressed as a decimal: A 6% interest rate is denoted by $r = 0.06$. If $P = \$100{,}000$, $r = 0.06$, and $N = 240$ (a 20-year loan), then the monthly payment is $f(100{,}000, 0.06, 240) = 716.43$. Furthermore, at these values, we have

$$\frac{\partial f}{\partial P} = 0.0071, \qquad \frac{\partial f}{\partial r} = 5769, \qquad \frac{\partial f}{\partial N} = -1.5467$$

Estimate:

(a) The change in monthly payment per $1000 increase in loan principal

(b) The change in monthly payment if the interest rate increases to $r = 6.5\%$ and $r = 7\%$

(c) The change in monthly payment if the length of the loan increases to 24 years

40. Automobile traffic passes a point P on a road of width w feet at an average rate of R vehicles per second. Although the arrival of automobiles is irregular, traffic engineers have found that the average waiting time T until there is a gap in traffic of at least t seconds is approximately $T = te^{Rt}$ seconds. A pedestrian walking at a speed of 3.5 ft/s (5.1 miles per hour) requires $t = w/3.5$ s to cross the road. Therefore, the average time the pedestrian will have to wait before crossing is $f(w, R) = (w/3.5)e^{wR/3.5}$ s.

(a) What is the pedestrian's average waiting time if $w = 25$ ft and $R = 0.2$ vehicle per second?

(b) Use the Linear Approximation to estimate the increase in waiting time if w is increased to 27 ft.

(c) Estimate the waiting time if the width is increased to 27 ft and R decreases to 0.18.

(d) What is the rate of increase in waiting time per 1-ft increase in width when $w = 30$ ft and $R = 0.3$ vehicle per second?

41. The volume V of a right-circular cylinder is computed using the values 3.5 m for diameter and 6.2 m for height. Use the Linear Approximation to estimate the maximum error in V if each of these values has a possible error of at most 5%. Recall that $V = \pi r^2 h$.

Further Insights and Challenges

42. Show that if $f(x, y)$ is differentiable at (a, b), then the function of one variable $f(x, b)$ is differentiable at $x = a$. Use this to prove that $f(x, y) = \sqrt{x^2 + y^2}$ is *not* differentiable at $(0, 0)$.

43. This exercise shows directly (without using Theorem 1) that the function $f(x, y) = 5x + 4y^2$ from Example 1 is differentiable at $(a, b) = (2, 1)$.

(a) Show that $f(x, y) = L(x, y) + e(x, y)$ with $e(x, y) = 4(y - 1)^2$.

(b) Show that

$$0 \le \frac{e(x, y)}{\sqrt{(x - 2)^2 + (y - 1)^2}} \le 4|y - 1|$$

(c) Verify that $f(x, y)$ is differentiable.

44. Show directly, as in Exercise 43, that $f(x, y) = xy^2$ is differentiable at $(0, 2)$.

45. Differentiability Implies Continuity Use the definition of differentiability to prove that if f is differentiable at (a, b), then f is continuous at (a, b).

46. Let $f(x)$ be a function of one variable defined near $x = a$. Given a number M, set

$$L(x) = f(a) + M(x - a), \qquad e(x) = f(x) - L(x)$$

Thus, $f(x) = L(x) + e(x)$. We say that f is locally linear at $x = a$ if M can be chosen so that $\lim\limits_{x \to a} \dfrac{e(x)}{|x - a|} = 0$.

(a) Show that if $f(x)$ is differentiable at $x = a$, then $f(x)$ is locally linear with $M = f'(a)$.

(b) Show conversely that if f is locally linear at $x = a$, then $f(x)$ is differentiable and $M = f'(a)$.

47. Assumptions Matter Define $g(x, y) = 2xy(x + y)/(x^2 + y^2)$ for $(x, y) \ne 0$ and $g(0, 0) = 0$. In this exercise, we show that $g(x, y)$ is continuous at $(0, 0)$ and that $g_x(0, 0)$ and $g_y(0, 0)$ exist, but $g(x, y)$ is not differentiable at $(0, 0)$.

(a) Show using polar coordinates that $g(x, y)$ is continuous at $(0, 0)$.

(b) Use the limit definitions to show that $g_x(0, 0)$ and $g_y(0, 0)$ exist and that both are equal to zero.

(c) Show that the linearization of $g(x, y)$ at $(0, 0)$ is $L(x, y) = 0$.

(d) Show that if $g(x, y)$ were differentiable at $(0, 0)$, we would have $\lim\limits_{h \to 0} \dfrac{g(h, h)}{h} = 0$. Then observe that this is not the case because $g(h, h) = 2h$. This shows that $g(x, y)$ is not differentiable at $(0, 0)$.

14.5 The Gradient and Directional Derivatives

For a function $f(x, y)$, the rate of change in the x direction is given by f_x, and the rate of change in the y direction is given by f_y. These partial derivatives give rates of change in the directions of the vectors $\mathbf{i}$ and $\mathbf{j}$, respectively. What if we want to know the rate of change of f in some other direction, say in the direction of the vector $\langle 2, -1 \rangle$?

To formally express a rate of change in any given direction, we will define the directional derivative. Before doing that, we introduce the gradient vector, an important vector that is used in a variety of situations, including computing directional derivatives. The components of the gradient of a function f are the partial derivatives of f.

The gradient of a function of n variables is the vector

$$\nabla f = \left\langle \frac{\partial f}{\partial x_1}, \frac{\partial f}{\partial x_2}, \ldots, \frac{\partial f}{\partial x_n} \right\rangle$$

Initially, we can think of it as a convenient method for keeping track of the collection of first partial derivatives. We will soon see, though, it is much more than just that.

DEFINITION **The Gradient** The gradient of a function $f(x, y)$ at a point $P = (a, b)$ is the vector

$$\boxed{\nabla f_P = \langle f_x(a, b), f_y(a, b) \rangle}$$

In three variables, for $f(x, y, z)$ and $P = (a, b, c)$,

$$\boxed{\nabla f_P = \langle f_x(a, b, c), f_y(a, b, c), f_z(a, b, c) \rangle}$$

The symbol ∇, called "del," is an upside-down Greek delta. It was popularized by the Scottish physicist P. G. Tait (1831–1901), who called the symbol "nabla," because of its resemblance to an ancient Assyrian harp. The great physicist James Clerk Maxwell was reluctant to adopt this term and would refer to the gradient simply as the "slope." He wrote jokingly to his friend Tait in 1871, "Still harping on that nabla?"

We also write $\nabla f_{(a,b)}$ or $\nabla f(a, b)$ for the gradient of f at $P = (a, b)$. Sometimes, we omit reference to the point P and write

$$\nabla f = \left\langle \frac{\partial f}{\partial x}, \frac{\partial f}{\partial y} \right\rangle \qquad \text{or} \qquad \nabla f = \left\langle \frac{\partial f}{\partial x}, \frac{\partial f}{\partial y}, \frac{\partial f}{\partial z} \right\rangle$$

The gradient ∇f assigns a vector ∇f_P to each point in the domain of f, as in Figure 1.

EXAMPLE 1 **Drawing Gradient Vectors** Let $f(x, y) = x^2 + y^2$. Calculate the gradient ∇f, draw several gradient vectors, and compute ∇f_P at $P = (1, 1)$.

Solution The partial derivatives are $f_x(x, y) = 2x$ and $f_y(x, y) = 2y$, so

$$\nabla f = \langle 2x, 2y \rangle$$

The gradient attaches the vector $\langle 2x, 2y \rangle$ to the point (x, y). As we see in Figure 1, these vectors point away from the origin. At the particular point $(1, 1)$,

$$\nabla f_P = \nabla f(1, 1) = \langle 2, 2 \rangle \qquad\blacksquare$$

EXAMPLE 2 **Gradient in Three Variables** Calculate $\nabla f_{(3, -2, 4)}$, where

$$f(x, y, z) = ze^{2x+3y}$$

Solution The partial derivatives and the gradient are

$$\frac{\partial f}{\partial x} = 2ze^{2x+3y}, \qquad \frac{\partial f}{\partial y} = 3ze^{2x+3y}, \qquad \frac{\partial f}{\partial z} = e^{2x+3y}$$

$$\nabla f = \langle 2ze^{2x+3y}, 3ze^{2x+3y}, e^{2x+3y} \rangle$$

Therefore, $\nabla f_{(3,-2,4)} = \langle 2 \cdot 4e^0, 3 \cdot 4e^0, e^0 \rangle = \langle 8, 12, 1 \rangle$. $\qquad\blacksquare$

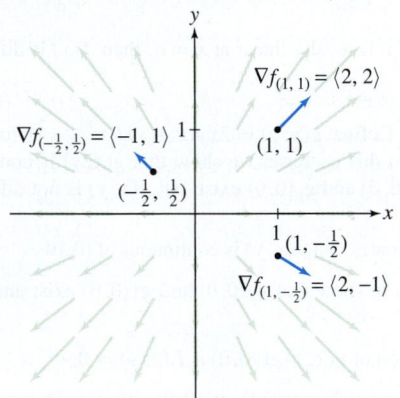

FIGURE 1 Gradient vectors of $f(x, y) = x^2 + y^2$ at several points (vectors not drawn to scale).

The following theorem lists some useful properties of the gradient. The proofs are left as exercises (see Exercises 66–68).

THEOREM 1 **Properties of the Gradient** If $f(x, y, z)$ and $g(x, y, z)$ are differentiable and c is a constant, then

  (i) $\nabla(f + g) = \nabla f + \nabla g$

  (ii) $\nabla(cf) = c\nabla f$

  (iii) **Product Rule for Gradients:** $\nabla(fg) = f\nabla g + g\nabla f$

  (iv) **Chain Rule for Gradients:** If $F(t)$ is a differentiable function of one variable, then

$$\nabla(F(f(x, y, z))) = F'(f(x, y, z))\nabla f \qquad \boxed{1}$$

The gradient is what is known as a differential operator. Note that the properties of the gradient in Theorem 1 resemble properties of derivatives we have previously seen.

EXAMPLE 3   **Using the Chain Rule for Gradients**   Find the gradient of

$$g(x, y, z) = (x^2 + y^2 + z^2)^8$$

Solution   The function g is a composite $g(x, y, z) = F(f(x, y, z))$ with $F(t) = t^8$ and $f(x, y, z) = x^2 + y^2 + z^2$. Apply Eq. (1):

$$\nabla g = \nabla\big((x^2 + y^2 + z^2)^8\big) = 8(x^2 + y^2 + z^2)^7 \nabla(x^2 + y^2 + z^2)$$

$$= 8(x^2 + y^2 + z^2)^7 \langle 2x, 2y, 2z \rangle$$

$$= 16(x^2 + y^2 + z^2)^7 \langle x, y, z \rangle \qquad ■$$

Introduction to the Chain Rule for Paths

Section 14.6 introduces general chain rules in multivariable calculus. There are a number of different chain rules because there are a number of different ways we can compose functions of multiple variables. We consider a particular one here because it is an important application of the gradient vector that we will need later in this section when working with directional derivatives.

We use the Chain Rule for Paths when we are given a function f along a parametric path given by $x(t)$ and $y(t)$ in the plane or by $x(t)$, $y(t)$, and $z(t)$ in 3-space. For notational simplicity, we let $\mathbf{r}(t)$ represent both the vector $\langle x(t), y(t) \rangle$ and the point $(x(t), y(t))$. In the former case, the path is traced out by the tips of the vectors, in the latter by the points. We follow a similar notational convention with $x(t)$, $y(t)$, and $z(t)$ in the three-dimensional case.

A function f that is defined along a path $\mathbf{r}(t)$ results in a composition $f(\mathbf{r}(t))$. The Chain Rule for Paths is used to find the derivative of these composite functions.

As an example, suppose that $T(x, y)$ is the temperature at location (x, y). Now imagine that Alexa is riding a bike along a path $\mathbf{r}(t)$ (Figure 2). We suppose that Alexa carries a thermometer with her and checks it as she rides. Her location at time t is $\mathbf{r}(t)$, so her temperature reading at time t is the composite function

$$T(\mathbf{r}(t)) = \text{Alexa's temperature at time } t$$

The temperature reading varies as Alexa's location changes, and the rate at which it changes is the derivative

$$\frac{d}{dt} T(\mathbf{r}(t))$$

The Chain Rule for Paths tells us that this derivative is simply the dot product of the temperature gradient ∇T, evaluated at $\mathbf{r}(t)$, and Alexa's velocity vector $\mathbf{r}'(t)$.

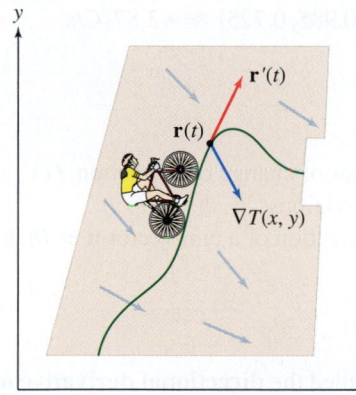

FIGURE 2 Alexa's temperature changes at the rate $\nabla T_{\mathbf{r}(t)} \cdot \mathbf{r}'(t)$.

> **THEOREM 2   Chain Rule for Paths**   If f and $\mathbf{r}(t)$ are differentiable, then
>
> $$\frac{d}{dt} f(\mathbf{r}(t)) = \nabla f_{\mathbf{r}(t)} \cdot \mathbf{r}'(t)$$

CAUTION Do not confuse the Chain Rule for Paths with the Chain Rule for Gradients stated in Theorem 1 above. They are different rules for different types of compositions.

For two variables, the Chain Rule for Paths is

$$\frac{d}{dt} f(\mathbf{r}(t)) = \left\langle \frac{\partial f}{\partial x}, \frac{\partial f}{\partial y} \right\rangle \cdot \langle x'(t), y'(t) \rangle = \frac{\partial f}{\partial x} \frac{dx}{dt} + \frac{\partial f}{\partial y} \frac{dy}{dt}$$

We prove the Chain Rule for Paths in the next section when we address multivariable calculus chain rules more generally.

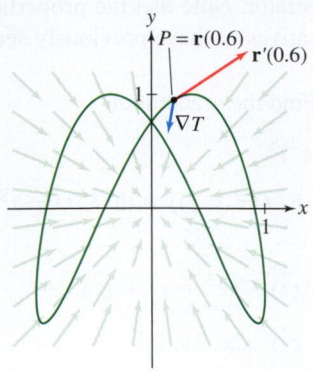

FIGURE 3 Gradient vectors ∇T and the path $\mathbf{r}(t) = \langle \cos(t - 2), \sin 2t \rangle$.

EXAMPLE 4 The temperature at location (x, y) is $T(x, y) = 20 + 10e^{-0.3(x^2 + y^2)} {}^\circ\text{C}$. A bug follows the path

$$\mathbf{r}(t) = \langle \cos(t - 2), \sin 2t \rangle$$

(t in seconds) as in Figure 3. What is the rate of change of temperature with respect to time that the bug experiences at $t = 0.6$ seconds?

Solution At $t = 0.6$ s, the bug is at location

$$\mathbf{r}(0.6) = \langle \cos(-1.4), \sin 1.2 \rangle \approx \langle 0.170, 0.932 \rangle$$

By the Chain Rule for Paths, the rate of change of temperature is the dot product

$$\left.\frac{dT}{dt}\right|_{t=0.6} = \nabla T_{\mathbf{r}(0.6)} \cdot \mathbf{r}'(0.6)$$

We compute the vectors

$$\nabla T = \left\langle -6xe^{-0.3(x^2+y^2)}, -6ye^{-0.3(x^2+y^2)} \right\rangle$$

$$\mathbf{r}'(t) = \langle -\sin(t - 2), 2\cos 2t \rangle$$

and evaluate at $\mathbf{r}(0.6) = \langle 0.170, 0.932 \rangle$:

$$\nabla T_{\mathbf{r}(0.6)} \approx \langle -0.779, -4.272 \rangle$$

$$\mathbf{r}'(0.6) \approx \langle 0.985, 0.725 \rangle$$

Therefore, the rate of change is

$$\left.\frac{dT}{dt}\right|_{t=0.6} \nabla T_{\mathbf{r}(0.6)} \cdot \mathbf{r}'(t) \approx \langle -0.779, -4.272 \rangle \cdot \langle 0.985, 0.725 \rangle \approx -3.87{}^\circ\text{C/s} \quad \blacksquare$$

Directional Derivatives

Now we are ready to introduce methods to compute rates of change of a function $f(x, y)$ in directions other than the positive x and positive y directions.

Consider a line through a point $P = (a, b)$ in the direction of a unit vector $\mathbf{u} = \langle h, k \rangle$ (see Figure 4):

$$\mathbf{r}(t) = \langle a + th, b + tk \rangle$$

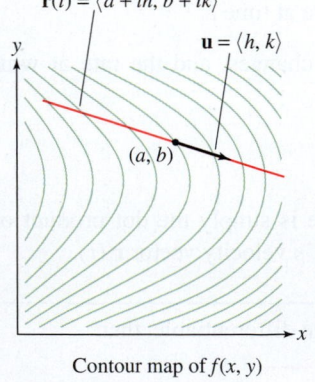

FIGURE 4 The directional derivative $D_{\mathbf{u}} f(a, b)$ is the rate of change of f along the linear path through P with direction vector $\mathbf{u}$.

The derivative with respect to t of $f(\mathbf{r}(t))$ at $t = 0$ is called the **directional derivative of** f **with respect to u at** P, and is denoted $D_{\mathbf{u}} f(P)$ or $D_{\mathbf{u}} f(a, b)$:

$$D_{\mathbf{u}} f(a, b) = \left.\frac{d}{dt} f(\mathbf{r}(t))\right|_{t=0} = \lim_{t \to 0} \frac{f(a + th, b + tk) - f(a, b)}{t}$$

Directional derivatives of functions of three or more variables are defined in a similar way.

DEFINITION Directional Derivative The directional derivative of f at $P = (a, b)$ in the direction of a unit vector $\mathbf{u} = \langle h, k \rangle$ is the limit (assuming it exists)

$$D_{\mathbf{u}} f(P) = D_{\mathbf{u}} f(a, b) = \lim_{t \to 0} \frac{f(a + th, b + tk) - f(a, b)}{t}$$

Note that the partial derivatives are the directional derivatives with respect to the standard unit vectors $\mathbf{i} = \langle 1, 0 \rangle$ and $\mathbf{j} = \langle 0, 1 \rangle$. For example,

$$D_{\mathbf{i}} f(a, b) = \lim_{t \to 0} \frac{f(a + t(1), b + t(0)) - f(a, b)}{t} = \lim_{t \to 0} \frac{f(a + t, b) - f(a, b)}{t}$$

$$= f_x(a, b)$$

Thus, we have

$$f_x(a, b) = D_{\mathbf{i}} f(a, b), \qquad f_y(a, b) = D_{\mathbf{j}} f(a, b)$$

CONCEPTUAL INSIGHT The directional derivative $D_{\mathbf{u}} f(P)$ is the rate of change of f per *unit change* in the horizontal direction of $\mathbf{u}$ at $P = (a, b)$ (Figure 5). This is the slope of the tangent line at $Q = (a, b, f(a, b))$ to the trace curve obtained when we intersect the graph with the vertical plane through P in the direction $\mathbf{u}$. With $\mathbf{u} = \langle h, k \rangle$, the vector $\mathbf{v} = \langle h, k, D_{\mathbf{u}} f(P) \rangle$ points along this line from Q.

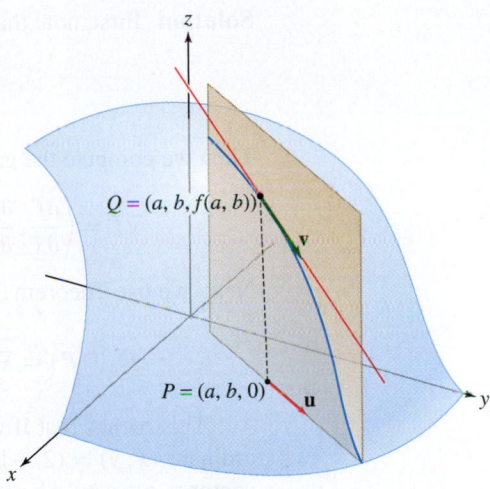

FIGURE 5 $D_{\mathbf{u}} f(a, b)$ is the slope of the tangent line to the trace curve through Q in the vertical plane through P in the direction $\mathbf{u}$. The vector $\mathbf{v} = \langle h, k, D_{\mathbf{u}} f(P) \rangle$ is parallel to this line.

Typically, we do not compute directional derivatives using the definition. For differentiable functions, the following theorem provides a more convenient approach using the gradient vector. The theorem is proved using the Chain Rule for Paths.

> **THEOREM 3 Computing the Directional Derivative** If f is differentiable at P and $\mathbf{u}$ is a unit vector, then the directional derivative in the direction of $\mathbf{u}$ is given by
>
> $$\boxed{D_{\mathbf{u}} f(P) = \nabla f_P \cdot \mathbf{u}} \qquad \boxed{2}$$

For a function $f(x, y)$ and unit vector $\mathbf{u} = \langle h, k \rangle$, computing the dot product in Eq. (2) yields

$$D_{\mathbf{u}} f(a, b) = \nabla f_{(a,b)} \cdot \mathbf{u} = f_x(a, b)h + f_y(a, b)k$$

Theorem 3 holds in all dimensions. In particular, for $f(x, y, z)$ and unit vector $\mathbf{u} = \langle h, k, m \rangle$, we have

$$D_{\mathbf{u}} f(a, b, c) = \nabla f_{(a,b,c)} \cdot \mathbf{u} = f_x(a, b, c)h + f_y(a, b, c)k + f_z(a, b, c)m$$

Proof We prove the theorem for functions of two variables, $f(x, y)$, and do so using a composition of functions and the Chain Rule for Paths. Let $P = (a, b)$ and $\mathbf{u} = \langle h, k \rangle$. Furthermore, let $\mathbf{r}(t) = \langle a + th, b + tk \rangle$ represent the line through P in the direction of $\mathbf{u}$, and consider the composite function $f(\mathbf{r}(t))$. By definition of the directional derivative,

$$D_{\mathbf{u}} f(a, b) = \frac{d}{dt} f(\mathbf{r}(t)) \bigg|_{t=0}$$

By the Chain Rule for Paths,

$$\frac{d}{dt} f(\mathbf{r}(t)) \bigg|_{t=0} = \nabla f_{\mathbf{r}(0)} \cdot \mathbf{r}'(0) = \nabla f_{(a,b)} \cdot \langle h, k \rangle = \nabla f_P \cdot \mathbf{u}$$

Therefore,

$$D_{\mathbf{u}} f(P) = \nabla f_P \cdot \mathbf{u}$$

∎

EXAMPLE 5 Let $f(x, y) = xe^y$, $P = (2, -1)$, and $\mathbf{v} = \langle 2, 3 \rangle$. Calculate the directional derivative in the direction of $\mathbf{v}$.

Solution First, note that $\mathbf{v}$ is *not* a unit vector. So, we first replace it with the vector

$$\mathbf{u} = \frac{\mathbf{v}}{\|\mathbf{v}\|} = \frac{\langle 2, 3 \rangle}{\sqrt{13}} = \left\langle \frac{2}{\sqrt{13}}, \frac{3}{\sqrt{13}} \right\rangle$$

Then we compute the gradient at $P = (2, -1)$:

$$\nabla f = \left\langle \frac{\partial f}{\partial x}, \frac{\partial f}{\partial y} \right\rangle = \langle e^y, xe^y \rangle \quad \Rightarrow \quad \nabla f_P = \nabla f_{(2,-1)} = \left\langle e^{-1}, 2e^{-1} \right\rangle$$

Next, we use Theorem 3:

$$D_{\mathbf{u}} f(P) = \nabla f_P \cdot \mathbf{u} = \left\langle e^{-1}, 2e^{-1} \right\rangle \cdot \left\langle \frac{2}{\sqrt{13}}, \frac{3}{\sqrt{13}} \right\rangle = \frac{8e^{-1}}{\sqrt{13}} \approx 0.82.$$

∎

This means that if we think of this function as representing a mountain, then at coordinate $(x, y) = (2, -1)$, we should expect that if we head 1 unit in the direction of vector $\mathbf{v}$, we would have to climb in the vertical direction by approximately 0.82 unit.

EXAMPLE 6 Find the rate of change of pressure at the point $Q = (1, 2, 1)$ in the direction of $\mathbf{v} = \langle 0, 1, 1 \rangle$, assuming that the pressure (in millibars) is given by

$$f(x, y, z) = 1000 + 0.01(yz^2 + x^2 z - xy^2) \qquad (x, y, z \text{ in kilometers})$$

Solution First, compute the gradient at $Q = (1, 2, 1)$:

$$\nabla f = 0.01 \left\langle 2xz - y^2, z^2 - 2xy, 2yz + x^2 \right\rangle$$

$$\nabla f_Q = \nabla f_{(1,2,1)} = \langle -0.02, -0.03, 0.05 \rangle$$

Then we compute a unit vector, $\mathbf{u}$, in the direction of $\mathbf{v}$:

$$\mathbf{u} = \frac{\mathbf{v}}{\|\mathbf{v}\|} = \left\langle 0, \frac{1}{\sqrt{2}}, \frac{1}{\sqrt{2}} \right\rangle$$

Next,

$$D_{\mathbf{u}} f(Q) = \nabla f_Q \cdot \mathbf{u} = \langle -0.02, -0.03, 0.05 \rangle \cdot \left\langle 0, \frac{1}{\sqrt{2}}, \frac{1}{\sqrt{2}} \right\rangle \approx 0.014 \text{ millibars/km}$$

Thus, we expect that as we move in the direction of $\mathbf{v}$ from Q, the pressure should increase by about 0.014 millibars/km.

∎

GRAPHICAL INSIGHT Given a function $f(x, y)$ that is differentiable at $P = (a, b)$, Theorem 3 guarantees the tangent plane to the graph of f at $Q = (a, b, f(a, b))$ is tangent to the graph in all directions, not just the directions determined by the partial derivatives.

Recall that the plane determined by f_x and f_y is defined as the plane through Q determined by the vectors $\mathbf{v}_1 = \langle 1, 0, f_x(a, b) \rangle$ and $\mathbf{v}_2 = \langle 0, 1, f_y(a, b) \rangle$. Given a unit vector $\mathbf{u} = \langle h, k \rangle$, the vector $\mathbf{v} = \langle h, k, D_{\mathbf{u}} f(P) \rangle$, based at Q, is tangent to the graph as illustrated in Figure 5 and explained in the corresponding Conceptual Insight.

In general, we cannot be sure that $\mathbf{v}$ is in the plane determined by f_x and f_y. However, if f is differentiable at (a, b), then we can show that it is. First, using Theorem 3 and assuming f is differentiable at (a, b), we show that $\mathbf{v} = h\mathbf{v}_1 + k\mathbf{v}_2$:

$$\mathbf{v} = \langle h, k, D_{\mathbf{u}} f(a, b) \rangle$$
$$= \langle h, k, hf_x(a, b) + kf_y(a, b) \rangle \quad \text{by Theorem 3}$$
$$= h \langle 1, 0, f_x(a, b) \rangle + k \langle 0, 1, f_y(a, b) \rangle$$
$$= h\mathbf{v}_1 + k\mathbf{v}_2$$

Since $\mathbf{v} = h\mathbf{v}_1 + k\mathbf{v}_2$, it follows that $\mathbf{v}$ is a linear combination of $\mathbf{v}_1$ and $\mathbf{v}_2$, implying that these vectors, based at Q, all lie in the same plane. Thus, when f is differentiable at (a, b), the tangent plane is tangent to the graph in all directions, justifying calling it a tangent plane.

Properties of the Gradient

Here we explore some properties of the gradient. We demonstrate how it provides important information about the behavior of functions and how it arises naturally in the development of mathematical models.

First, suppose that $\nabla f_P \neq \mathbf{0}$ and let $\mathbf{u}$ be a unit vector (Figure 6). By the properties of the dot product and the fact that $\mathbf{u}$ is a unit vector,

$$D_{\mathbf{u}} f(P) = \nabla f_P \cdot \mathbf{u} = \|\nabla f_P\| \|\mathbf{u}\| \cos\theta = \|\nabla f_P\| \cos\theta \qquad \boxed{3}$$

where θ is the angle between ∇f_P and $\mathbf{u}$. In other words, *the rate of change in a given direction varies with the cosine of the angle θ between the gradient and the direction.*

Because the cosine takes values between -1 and 1, we have

$$-\|\nabla f_P\| \leq D_{\mathbf{u}} f(P) \leq \|\nabla f_P\|$$

Since $\cos 0 = 1$, the maximum value of $D_{\mathbf{u}} f(P)$ occurs for $\theta = 0$—that is, when $\mathbf{u}$ points in the direction of ∇f_P. In other words, *the gradient vector points in the direction of the maximum rate of increase, and this maximum rate is $\|\nabla f_P\|$.* Similarly, f decreases most rapidly in the opposite direction, $-\nabla f_P$, because $\cos\theta = -1$ for $\theta = \pi$. The rate of fastest decrease is $-\|\nabla f_P\|$. The directional derivative is zero in directions orthogonal to the gradient because $\cos\frac{\pi}{2} = 0$.

Another key property is that gradient vectors are normal to level curves (Figure 7). To prove this, suppose that P lies on the level curve $f(x, y) = k$. We parametrize this level curve by a path $\mathbf{r}(t)$ such that $\mathbf{r}(0) = P$ and $\mathbf{r}'(0) \neq \mathbf{0}$ (this is possible whenever $\nabla f_P \neq \mathbf{0}$). Then $f(\mathbf{r}(t)) = k$ for all t, so by the Chain Rule,

$$\nabla f_P \cdot \mathbf{r}'(0) = \frac{d}{dt} f(\mathbf{r}(t)) \Big|_{t=0} = \frac{d}{dt} k = 0$$

This proves that ∇f_P is orthogonal to $\mathbf{r}'(0)$, and since $\mathbf{r}'(0)$ is tangent to the level curve, we conclude that ∇f_P is normal to the level curve (Figure 7). We encapsulate these remarks in the following theorem.

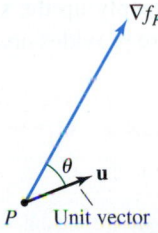

FIGURE 6 $D_{\mathbf{u}} f(P) = \|\nabla f_P\| \cos\theta$.

◄ REMINDER

- The terms "normal" and "orthogonal" both mean "perpendicular."
- We say that a vector is normal to a curve at a point P if it is normal to the tangent line to the curve at P.

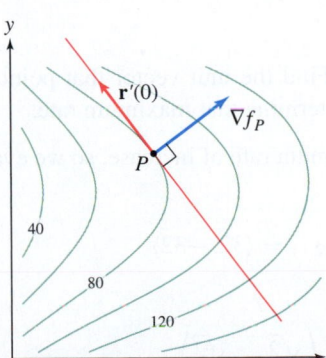

FIGURE 7 Contour map of $f(x, y)$. The gradient at P is orthogonal to the level curve through P and points in the direction of maximum increase of $f(x, y)$.

> **THEOREM 4 Interpretation of the Gradient**   Assume that $\nabla f_P \neq \mathbf{0}$. Let $\mathbf{u}$ be a unit vector making an angle θ with ∇f_P. Then
>
> $$D_{\mathbf{u}} f(P) = \|\nabla f_P\| \cos \theta \qquad \boxed{4}$$
>
> - ∇f_P points in the direction of fastest rate of increase of f at P, and that rate of increase is $\|\nabla f_P\|$.
> - $-\nabla f_P$ points in the direction of fastest rate of decrease at P, and that rate of decrease is $-\|\nabla f_P\|$.
> - ∇f_P is normal to the level curve (or surface) of f at P.

GRAPHICAL INSIGHT   At each point P, there is a unique direction in which $f(x, y)$ increases most rapidly (per unit distance). Theorem 4 tells us that this direction of fastest increase is perpendicular to the level curves and that it is specified by the gradient vector (Figure 7). For most functions, the direction of maximum rate of increase varies from point to point, as does the maximum rate of increase itself.

Figure 8 shows a contour map of $f(x, y) = y^2 - x^2$ along with gradient vectors at various points. The graph of f is a hyperbolic paraboloid (saddle). At each point on a level curve, the gradient vector must point in the direction in the domain of f that yields the steepest increase on the saddle. If we were actually on the saddle, each gradient vector tells us the horizontal direction we should take to go most steeply up the saddle. Note that at $(0, 0)$, the gradient vector is the zero vector, and therefore provides no information about directions of increase from $(0, 0)$.

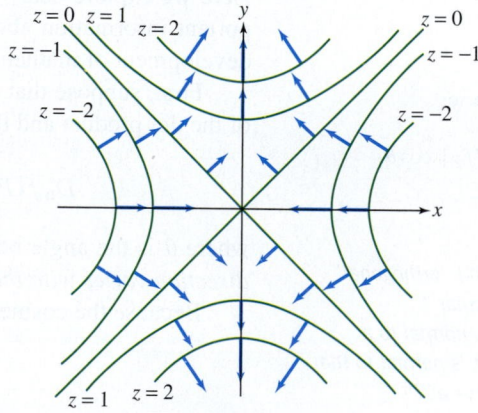

FIGURE 8 Contour map of $f(x, y) = y^2 - x^2$ and the corresponding gradient vectors at each point.

EXAMPLE 7   Let $f(x, y) = x^4 y^{-2}$ and $P = (2, 1)$. Find the unit vector that points in the direction of maximum rate of increase at P and determine that maximum rate.

Solution   The gradient points in the direction of maximum rate of increase, so we evaluate the gradient at P:

$$\nabla f = \left\langle 4x^3 y^{-2}, -2x^4 y^{-3} \right\rangle, \qquad \nabla f_{(2,1)} = \langle 32, -32 \rangle$$

The unit vector in this direction is

$$\mathbf{u} = \frac{\langle 32, -32 \rangle}{\|\langle 32, -32 \rangle\|} = \frac{\langle 32, -32 \rangle}{32\sqrt{2}} = \left\langle \frac{\sqrt{2}}{2}, -\frac{\sqrt{2}}{2} \right\rangle$$

The maximum rate, which is the rate in this direction, is given by

$$\|\nabla f_{(2,1)}\| = \sqrt{(32^2 + (-32)^2} = 32\sqrt{2}$$

EXAMPLE 8 The altitude of a mountain at (x, y) is

$$f(x, y) = 2500 + 100(x + y^2)e^{-0.3y^2}$$

where x, y are in units of 100 m.

(a) Find the directional derivative of f at $P = (-1, -1)$ in the direction of unit vector $\mathbf{u}$ making an angle of $\theta = \frac{\pi}{4}$ with the gradient (Figure 9).

(b) What is the interpretation of this derivative?

Solution First compute $\|\nabla f_P\|$:

$$f_x(x, y) = 100e^{-0.3y^2}, \qquad f_y(x, y) = 100y(2 - 0.6x - 0.6y^2)e^{-0.3y^2}$$

$$f_x(-1, -1) = 100e^{-0.3} \approx 74, \qquad f_y(-1, -1) = -200e^{-0.3} \approx -148$$

Hence, $\nabla f_P \approx \langle 74, -148 \rangle$ and

$$\|\nabla f_P\| \approx \sqrt{74^2 + (-148)^2} \approx 165.5$$

Apply Eq. (4) with $\theta = \pi/4$:

$$D_{\mathbf{u}} f(P) = \|\nabla f_P\| \cos \theta \approx 165.5 \left(\frac{\sqrt{2}}{2} \right) \approx 116.7$$

Recall that x and y are measured in units of 100 m. Therefore, the interpretation is the following: If you stand on the mountain at the point lying above $(-1, -1)$ and begin climbing so that your horizontal displacement is in the direction of $\mathbf{u}$, then your altitude increases at a rate of 116.7 m per 100 m of horizontal displacement, or 1.167 m per meter of horizontal displacement. ∎

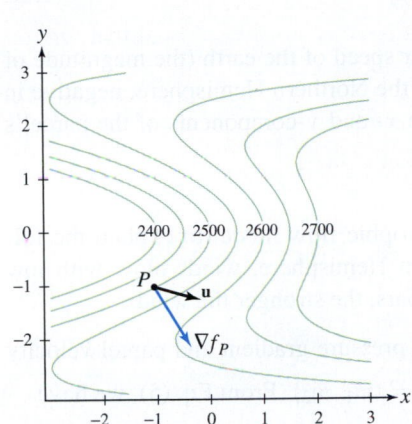

FIGURE 9 Contour map of the function $f(x, y)$ in Example 8.

Like level curves on a contour map, isobars on a weather map represent curves of constant air pressure p [Figure 10(A)]. On a small volume (parcel) of air, a force known as the pressure gradient force is determined by the gradient of p and the volume V of the parcel. The force equals $-V\nabla p$. Note that the force is directed from higher pressures to lower (because of the negative sign) and is stronger when the isobars are closer together. When you add wind vectors to the weather map [Figure 10(B)], or you look at an image of a large cyclone (such as the one pictured at the start of the chapter), the winds appear to circulate around a low pressure rather than flow directly toward it. This wind-steering effect is caused by the Coriolis force.

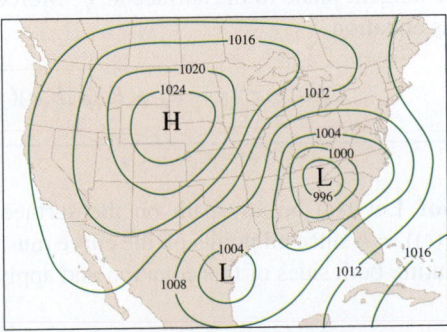
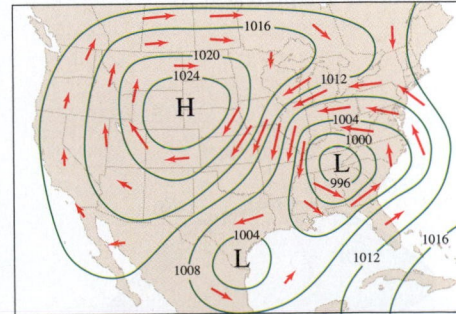

(A) Weather map with isobars   (B) Wind vectors included

FIGURE 10

Geostrophic flow is a simple approximation to large-scale flow in the atmosphere. In this model, we assume that the pressure gradient force is balanced by the Coriolis force, and that vertical motion is negligible. In a local coordinate system on the surface of the earth, with the positive x-, y-, and z-axes pointing east, north, and up, respectively, this force balance results in the equations

$$V\frac{\partial p}{\partial x} = 2m\omega(\sin L)w_2 \quad \text{and} \quad V\frac{\partial p}{\partial y} = -2m\omega(\sin L)w_1 \qquad \boxed{5}$$

where m is the mass of the parcel, ω is the angular speed of the earth (the magnitude of the angular velocity), L is the latitude (positive in the Northern Hemisphere, negative in the Southern Hemisphere), and w_1 and w_2 are the x- and y-components of the parcel's velocity, respectively.

EXAMPLE 9   Geostrophic Flow   Use the geostrophic flow model to explain the following atmospheric phenomenon: In the Northern Hemisphere, winds blow with low pressure to the left, and the closer together the isobars, the stronger the winds.

Solution   Ignoring vertical motion, we regard the pressure gradient and parcel velocity as two-dimensional vectors, $\nabla p = \left\langle \frac{\partial p}{\partial x}, \frac{\partial p}{\partial y} \right\rangle$ and $\mathbf{w} = \langle w_1, w_2 \rangle$. From Eq. (5), we have

$$\mathbf{w} = \langle w_1, w_2 \rangle = C\left\langle -\frac{\partial p}{\partial y}, \frac{\partial p}{\partial x} \right\rangle$$

for a constant C. Since $\sin L$ is positive (along with m, ω, and V) so is C. The vector $\left\langle -\frac{\partial p}{\partial y}, \frac{\partial p}{\partial x} \right\rangle$ is ∇p rotated counterclockwise by $90°$. So the parcel velocity vector (i.e., the wind vector) is proportional in magnitude to the pressure gradient and points in the direction $90°$ counterclockwise to it (Figure 11). It follows that in the Northern Hemisphere, the wind blows with lower pressure to the left and is strongest when the magnitude of the pressure gradient is largest, that is, when the isobars are closest together. In particular, in the Northern Hemisphere, winds circulate counterclockwise around low pressure systems and clockwise around high pressure ones.   ∎

Another use of the gradient is in finding normal vectors on a surface with equation $F(x, y, z) = k$, where k is a constant.

THEOREM 5   Gradient as a Normal Vector   Let $P = (a, b, c)$ be a point on the surface given by $F(x, y, z) = k$ and assume that $\nabla F_P \neq \mathbf{0}$. Then ∇F_P is a vector normal to the tangent plane to the surface at P. Moreover, the tangent plane to the surface at P has equation

$$F_x(a, b, c)(x - a) + F_y(a, b, c)(y - b) + F_z(a, b, c)(z - c) = 0$$

Proof   Let $\mathbf{r}(t)$ be any path on the surface such that $\mathbf{r}(0) = P$ and $\mathbf{r}'(0) \neq \mathbf{0}$. Then $F(\mathbf{r}(t)) = k$ since all points on the curve must satisfy the equation $F(x, y, z) = k$. Differentiating both sides of this equation and applying the Chain Rule for Paths, we have

$$\nabla F_P \cdot \mathbf{r}'(0) = 0$$

Hence, ∇F_P is perpendicular to $\mathbf{r}'(0)$, which we know to be tangent to the curve given by $\mathbf{r}(t)$ at P and thus tangent to the surface at P. However, we can take $\mathbf{r}(t)$ to pass through P from any direction, as in Figure 12, and hence ∇F_P must be perpendicular to tangent vectors pointing in any direction, and therefore perpendicular to the entire tangent plane at P.

The right-hand side Coriolis force terms arise from the force vector $\mathbf{F}_c = -2m\mathbf{\Omega} \times \mathbf{w}$, where $\mathbf{\Omega}$ is the angular velocity vector of the earth, and $\mathbf{w}$ is the velocity of the parcel (Exercise 44).

Given a vector $\mathbf{v} = \langle a, b \rangle$, the vector $\langle -b, a \rangle$ is obtained by rotating $\mathbf{v}$ counterclockwise by $90°$.

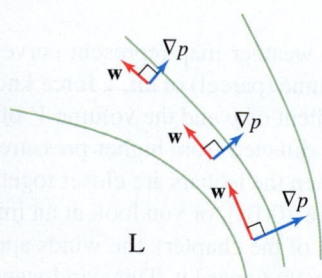

FIGURE 11   The wind vectors are proportional to ∇p, rotated $90°$.

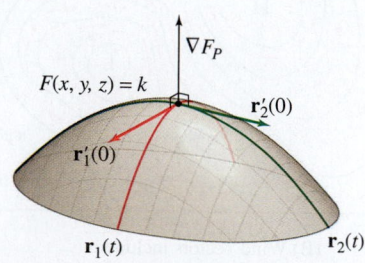

FIGURE 12 ∇F_P is normal to the surface $F(x, y, z) = k$ at P.

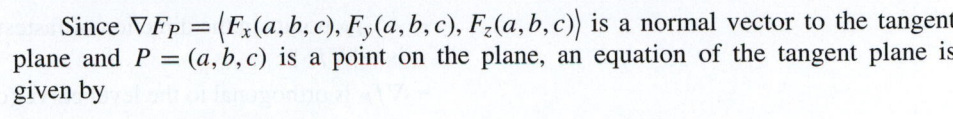

Since $\nabla F_P = \langle F_x(a,b,c), F_y(a,b,c), F_z(a,b,c) \rangle$ is a normal vector to the tangent plane and $P = (a,b,c)$ is a point on the plane, an equation of the tangent plane is given by

$$F_x(a,b,c)(x-a) + F_y(a,b,c)(y-b) + F_z(a,b,c)(z-c) = 0 \qquad ■$$

EXAMPLE 10 **Normal Vector and Tangent Plane** Find an equation of the tangent plane to the surface $4x^2 + 9y^2 - z^2 = 16$ at $P = (2,1,3)$.

Solution Let $F(x,y,z) = 4x^2 + 9y^2 - z^2$. Then

$$\nabla F = \langle 8x, 18y, -2z \rangle, \qquad \nabla F_P = \nabla F_{(2,1,3)} = \langle 16, 18, -6 \rangle$$

The vector $\langle 16, 18, -6 \rangle$ is normal to the surface $F(x,y,z) = 16$ at P (Figure 13), so the tangent plane at P has equation

$$16(x-2) + 18(y-1) - 6(z-3) = 0 \qquad \text{or} \qquad 16x + 18y - 6z = 32 \qquad ■$$

Notice how this equation for the tangent plane relates to Eq. (2) of Section 14.4, where we found that an equation for the tangent plane to a surface given by $z = f(x,y)$ at a point $(a, b, f(a,b))$ is given by

$$z = f(a,b) + f_x(a,b)(x-a) + f_y(a,b)(y-b)$$

To apply our new formula for the tangent plane to this situation, we take $F(x,y,z) = f(x,y) - z = 0$. Then note that $F_x = f_x$, $F_y = f_y$, and $F_z = -1$. Hence, our new formula yields

$$f_x(a,b)(x-a) + f_y(a,b)(y-b) + (-1)(z-c) = 0$$

where $c = f(a,b)$. This agrees exactly with Eq. (2) from Section 14.4.

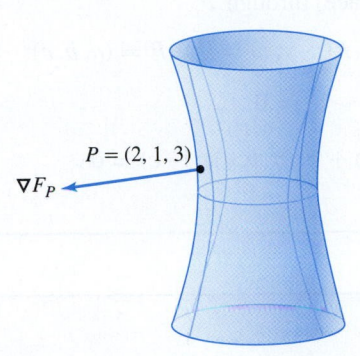

DF **FIGURE 13** The gradient vector ∇F_P is normal to the surface at P.

14.5 SUMMARY

- The *gradient* of a function f is the vector of partial derivatives:

$$\nabla f = \left\langle \frac{\partial f}{\partial x}, \frac{\partial f}{\partial y} \right\rangle \qquad \text{or} \qquad \nabla f = \left\langle \frac{\partial f}{\partial x}, \frac{\partial f}{\partial y}, \frac{\partial f}{\partial z} \right\rangle$$

- Chain Rule for Paths:

$$\frac{d}{dt} f(\mathbf{r}(t)) = \nabla f_{\mathbf{r}(t)} \cdot \mathbf{r}'(t)$$

- For $\mathbf{u} = \langle h, k \rangle$, a unit vector, $D_{\mathbf{u}} f$ is the *directional derivative with respect to* $\mathbf{u} = \langle h, k \rangle$:

$$D_{\mathbf{u}} f(a,b) = \lim_{t \to 0} \frac{f(a + th, b + tk) - f(a,b)}{t}$$

This definition extends to three or more variables.

- For differentiable f, the directional derivative can be computed using the gradient:

$$D_{\mathbf{u}} f(a,b) = \nabla f_{(a,b)} \cdot \mathbf{u}$$

- $D_{\mathbf{u}} f(a,b) = \|\nabla f_{(a,b)}\| \cos \theta$, where θ is the angle between $\nabla f_{(a,b)}$ and $\mathbf{u}$.
- Basic geometric properties of the gradient (assume $\nabla f_P \neq \mathbf{0}$):

  - ∇f_P points in the direction of fastest rate of increase, and that rate of increase is $\|\nabla f_P\|$.

- $-\nabla f_P$ points in the direction of fastest rate of decrease, and that rate of decrease is $-\|\nabla f_P\|$.
- ∇f_P is orthogonal to the level curve (or surface) through P.

• Equation of the tangent plane to the level surface $F(x, y, z) = k$ at $P = (a, b, c)$:

$$\nabla F_P \cdot \langle x - a, y - b, z - c \rangle = 0$$

$$F_x(a, b, c)(x - a) + F_y(a, b, c)(y - b) + F_z(a, b, c)(z - c) = 0$$

14.5 EXERCISES

Preliminary Questions

1. Which of the following is a possible value of the gradient ∇f of a function $f(x, y)$ of two variables?

(a) 5 **(b)** $\langle 3, 4 \rangle$ **(c)** $\langle 3, 4, 5 \rangle$

2. True or false? A differentiable function increases at the rate $\|\nabla f_P\|$ in the direction of ∇f_P.

3. Describe the two main geometric properties of the gradient ∇f.

4. You are standing at a point where the temperature gradient vector is pointing in the northeast (NE) direction. In which direction(s) should you walk to avoid a change in temperature?

(a) NE **(b)** NW **(c)** SE **(d)** SW

5. What is the rate of change of $f(x, y)$ at $(0, 0)$ in the direction making an angle of $45°$ with the x-axis if $\nabla f(0, 0) = \langle 2, 4 \rangle$?

Exercises

1. Let $f(x, y) = xy^2$ and $\mathbf{r}(t) = \left\langle \frac{1}{2}t^2, t^3 \right\rangle$.

(a) Calculate ∇f and $\mathbf{r}'(t)$.

(b) Use the Chain Rule for Paths to evaluate $\dfrac{d}{dt} f(\mathbf{r}(t))$ at $t = 1$ and $t = -1$.

2. Let $f(x, y) = e^{xy}$ and $\mathbf{r}(t) = \langle t^3, 1 + t \rangle$.

(a) Calculate ∇f and $\mathbf{r}'(t)$.

(b) Use the Chain Rule for Paths to calculate $\dfrac{d}{dt} f(\mathbf{r}(t))$.

(c) Write out the composite $f(\mathbf{r}(t))$ as a function of t and differentiate. Check that the result agrees with part (b).

3. Figure 14 shows the level curves of a function $f(x, y)$ and a path $\mathbf{r}(t)$, traversed in the direction indicated. State whether the derivative $\dfrac{d}{dt} f(\mathbf{r}(t))$ is positive, negative, or zero at points A–D.

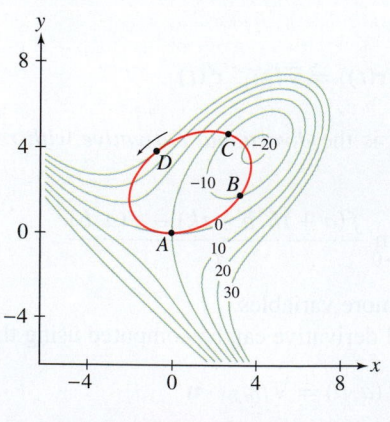

FIGURE 14

4. Let $f(x, y) = x^2 + y^2$ and $\mathbf{r}(t) = \langle \cos t, \sin t \rangle$.

(a) Find $\dfrac{d}{dt} f(\mathbf{r}(t))$ without making any calculations. Explain.

(b) Verify your answer to (a) using the Chain Rule.

In Exercises 5–8, calculate the gradient.

5. $f(x, y) = \cos(x^2 + y)$

6. $g(x, y) = \dfrac{x}{x^2 + y^2}$

7. $h(x, y, z) = xyz^{-3}$

8. $r(x, y, z, w) = xze^{yw}$

In Exercises 9–20, use the Chain Rule to calculate $\dfrac{d}{dt} f(\mathbf{r}(t))$ at the value of t given.

9. $f(x, y) = 3x - 7y$, $\mathbf{r}(t) = \langle \cos t, \sin t \rangle$, $t = 0$

10. $f(x, y) = 2x + 3y$, $\mathbf{r}(t) = \langle t^3, t^2 \rangle$, $t = -2$

11. $f(x, y) = x^2 - 3xy$, $\mathbf{r}(t) = \langle \cos t, \sin t \rangle$, $t = 0$

12. $f(x, y) = x^2 - 3xy$, $\mathbf{r}(t) = \langle \cos t, \sin t \rangle$, $t = \frac{\pi}{2}$

13. $f(x, y) = \sin(xy)$, $\mathbf{r}(t) = \langle e^{2t}, e^{3t} \rangle$, $t = 0$

14. $f(x, y) = \cos(y - x)$, $\mathbf{r}(t) = \langle e^t, e^{2t} \rangle$, $t = \ln 3$

15. $f(x, y) = x - xy$, $\mathbf{r}(t) = \langle t^2, t^2 - 4t \rangle$, $t = 4$

16. $f(x, y) = 3xe^{-y}$, $\mathbf{r}(t) = \langle 2t^2, t^2 - 2t \rangle$, $t = 0$

17. $f(x, y) = \ln x + \ln y$, $\mathbf{r}(t) = \langle \cos t, t^2 \rangle$, $t = \frac{\pi}{4}$

18. $g(x, y, z) = xye^z$, $\mathbf{r}(t) = \langle t^2, t^3, t - 1 \rangle$, $t = 1$

19. $g(x, y, z) = xyz^{-1}$, $\mathbf{r}(t) = \langle e^t, t, t^2 \rangle$, $t = 1$

20. $g(x, y, z, w) = x + 2y + 3z + 5w$, $\mathbf{r}(t) = \langle t^2, t^3, t, t - 2 \rangle$, $t = 1$

In Exercises 21–30, calculate the directional derivative in the direction of $\mathbf{v}$ at the given point. Remember to use a unit vector in your directional derivative computation.

21. $f(x, y) = x^2 + y^3$, $\mathbf{v} = \langle 4, 3 \rangle$, $P = (1, 2)$

22. $f(x, y) = xy^3 - x^2$, $\mathbf{v} = \mathbf{i} - \mathbf{j}$, $P = (2, -1)$

23. $f(x, y) = x^2 y^3$, $\mathbf{v} = \mathbf{i} + \mathbf{j}$, $P = \left(\frac{1}{6}, 3 \right)$

24. $f(x, y) = \sin(x - y)$, $\mathbf{v} = \langle 1, 1 \rangle$, $P = \left(\frac{\pi}{2}, \frac{\pi}{6} \right)$

25. $f(x, y) = \tan^{-1}(xy)$, $\mathbf{v} = \langle 1, 1 \rangle$, $P = (3, 4)$

26. $f(x, y) = e^{xy - y^2}$, $\mathbf{v} = \langle 12, -5 \rangle$, $P = (2, 2)$

27. $f(x, y) = \ln(x^2 + y^2)$, $\mathbf{v} = 3\mathbf{i} - 2\mathbf{j}$, $P = (1, 0)$

28. $g(x, y, z) = z^2 - xy + 2y^2$, $\mathbf{v} = \langle 1, -2, 2 \rangle$, $P = (2, 1, -3)$

29. $g(x, y, z) = xe^{-yz}$, $\mathbf{v} = \langle 1, 1, 1 \rangle$, $P = (1, 2, 0)$

30. $g(x, y, z) = x \ln(y + z)$, $\mathbf{v} = 2\mathbf{i} - \mathbf{j} + \mathbf{k}$, $P = (2, e, e)$

31. Find the directional derivative of $f(x, y) = x^2 + 4y^2$ at the point $P = (3, 2)$ in the direction pointing to the origin.

32. Find the directional derivative of $f(x, y, z) = xy + z^3$ at the point $P = (3, -2, -1)$ in the direction pointing to the origin.

In Exercises 33–36, determine the direction in which f has maximum rate of increase from P, and give the rate of change in that direction.

33. $f(x, y) = xe^{-y}$, $P = (2, 0)$

34. $f(x, y) = x^2 - xy + y^2$, $P = (-1, 4)$

35. $f(x, y, z) = \dfrac{xy}{z}$, $P = (1, -1, 3)$

36. $f(x, y, z) = x^2 y \sqrt{z}$, $P = (1, 5, 9)$

37. Suppose that $\nabla f_P = \langle 2, -4, 4 \rangle$. Is f increasing or decreasing at P in the direction $\mathbf{v} = \langle 2, 1, 3 \rangle$?

38. Let $f(x, y) = xe^{x^2 - y}$ and $P = (1, 1)$.

(a) Calculate $\| \nabla f_P \|$.

(b) Find the rate of change of f in the direction ∇f_P.

(c) Find the rate of change of f in the direction of a vector making an angle of $45°$ with ∇f_P.

39. Let $f(x, y, z) = \sin(xy + z)$ and $P = (0, -1, \pi)$. Calculate $D_{\mathbf{u}} f(P)$, where $\mathbf{u}$ is a unit vector making an angle $\theta = 30°$ with ∇f_P.

40. Let $T(x, y)$ be the temperature at location (x, y) on a thin sheet of metal. Assume that $\nabla T = \langle y - 4, x + 2y \rangle$. Let $\mathbf{r}(t) = \langle t^2, t \rangle$ be a path on the sheet. Find the values of t such that

$$\frac{d}{dt} T(\mathbf{r}(t)) = 0$$

41. Find a vector normal to the surface $x^2 + y^2 - z^2 = 6$ at $P = (3, 1, 2)$.

42. Find a vector normal to the surface $3z^3 + x^2 y - y^2 x = 1$ at $P = (1, -1, 1)$.

43. Find the two points on the ellipsoid

$$\frac{x^2}{4} + \frac{y^2}{9} + z^2 = 1$$

where the tangent plane is normal to $\mathbf{v} = \langle 1, 1, -2 \rangle$.

44. Assume we have a local coordinate system at latitude L on the earth's surface with east, north, and up as the x-, y-, and z- directions, respectively. In this coordinate system, the earth's angular velocity vector is $\mathbf{\Omega} = \langle 0, \omega \cos L, \omega \sin L \rangle$. Let $\mathbf{w} = \langle w_1, w_2, 0 \rangle$ be a wind vector.

(a) Determine the components of the Coriolis force vector $\mathbf{F}_c = -2m\mathbf{\Omega} \times \mathbf{w}$.

(b) The equation $-V\nabla p + \mathbf{F}_c = \mathbf{0}$ results from balancing the pressure gradient force, $-V\nabla p$, and the Coriolis force. Show that the x- and y- components of this equation result in Eq. 5.

45. Use the geostrophic flow model to explain the following: In the Southern Hemisphere, winds blow with low pressure to the right, and the closer together the isobars, the stronger the winds. In particular, winds blow clockwise around low pressure systems and counterclockwise around high pressure systems.

In Exercises 46–49, find an equation of the tangent plane to the surface at the given point.

46. $x^2 + 3y^2 + 4z^2 = 20$, $P = (2, 2, 1)$

47. $xz + 2x^2 y + y^2 z^3 = 11$, $P = (2, 1, 1)$

48. $x^2 + z^2 e^{y - x} = 13$, $P = \left(2, 3, \dfrac{3}{\sqrt{e}} \right)$

49. $\ln(1 + 4x^2 + 9y^4) - 0.1z^2 = 0$, $P = (3, 1, 6.1876)$

50. Verify what is clear from Figure 15: Every tangent plane to the cone $x^2 + y^2 - z^2 = 0$ passes through the origin.

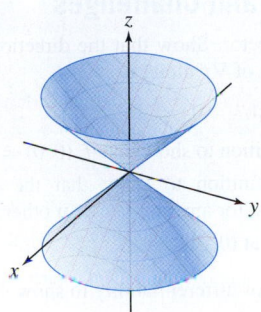

FIGURE 15 Graph of $x^2 + y^2 - z^2 = 0$.

51. (CAS) Use a computer algebra system to produce a contour plot of $f(x, y) = x^2 - 3xy + y - y^2$ together with its gradient vector field on the domain $[-4, 4] \times [-4, 4]$.

52. Find a function $f(x, y, z)$ such that ∇f is the constant vector $\langle 1, 3, 1 \rangle$.

53. Find a function $f(x, y, z)$ such that $\nabla f = \langle 2x, 1, 2 \rangle$.

54. Find a function $f(x, y, z)$ such that $\nabla f = \langle x, y^2, z^3 \rangle$.

55. Find a function $f(x, y, z)$ such that $\nabla f = \langle z, 2y, x \rangle$.

56. Find a function $f(x, y)$ such that $\nabla f = \langle y, x \rangle$.

57. Show that there does not exist a function $f(x, y)$ such that $\nabla f = \langle y^2, x \rangle$. Hint: Use Clairaut's Theorem $f_{xy} = f_{yx}$.

58. Let $\Delta f = f(a + h, b + k) - f(a, b)$ be the change in f at $P = (a, b)$. Set $\Delta \mathbf{v} = \langle h, k \rangle$. Show that the Linear Approximation can be written

$$\Delta f \approx \nabla f_P \cdot \Delta \mathbf{v} \qquad \boxed{6}$$

59. Use Eq. (6) to estimate

$$\Delta f = f(3.53, 8.98) - f(3.5, 9)$$

assuming that $\nabla f_{(3.5, 9)} = \langle 2, -1 \rangle$.

60. Find a unit vector $\mathbf{n}$ that is normal to the surface $z^2 - 2x^4 - y^4 = 16$ at $P = (2, 2, 8)$ that points in the direction of the xy-plane (in other words, if you travel in the direction of $\mathbf{n}$, you will eventually cross the xy-plane).

61. Suppose, in the previous exercise, that a particle located at the point $P = (2, 2, 8)$ travels toward the xy-plane in the direction normal to the surface.

(a) Through which point Q on the xy-plane will the particle pass?

(b) Suppose the axes are calibrated in centimeters. Determine the path $\mathbf{r}(t)$ of the particle if it travels at a constant speed of 8 cm/s. How long will it take the particle to reach Q?

62. Let $f(x, y) = \tan^{-1}\dfrac{x}{y}$ and $\mathbf{u} = \left\langle \dfrac{\sqrt{2}}{2}, \dfrac{\sqrt{2}}{2} \right\rangle$.

(a) Calculate the gradient of f.

(b) Calculate $D_{\mathbf{u}} f(1, 1)$ and $D_{\mathbf{u}} f(\sqrt{3}, 1)$.

(c) Show that the lines $y = mx$ for $m \neq 0$ are level curves for f.

(d) Verify that ∇f_P is orthogonal to the level curve through P for $P = (x, y) \neq (0, 0)$.

63. ✎ Suppose that the intersection of two surfaces $F(x, y, z) = 0$ and $G(x, y, z) = 0$ is a curve C, and let P be a point on C. Explain why the vector $\mathbf{v} = \nabla F_P \times \nabla G_P$ is a direction vector for the tangent line to C at P.

64. Let C be the curve of intersection of the spheres $x^2 + y^2 + z^2 = 3$ and $(x - 2)^2 + (y - 2)^2 + z^2 = 3$. Use the result of Exercise 63 to find parametric equations of the tangent line to C at $P = (1, 1, 1)$.

65. Let C be the curve obtained as the intersection of the two surfaces $x^3 + 2xy + yz = 7$ and $3x^2 - yz = 1$. Find the parametric equations of the tangent line to C at $P = (1, 2, 1)$.

66. Prove the linearity relations for gradients:

(a) $\nabla(f + g) = \nabla f + \nabla g$

(b) $\nabla(cf) = c\nabla f$

67. Prove the Chain Rule for Gradients in Theorem 1.

68. Prove the Product Rule for Gradients in Theorem 1.

Further Insights and Challenges

69. Let $\mathbf{u}$ be a unit vector. Show that the directional derivative $D_{\mathbf{u}} f$ is equal to the component of ∇f along $\mathbf{u}$.

70. Let $f(x, y) = (xy)^{1/3}$.

(a) Use the limit definition to show that $f_x(0, 0) = f_y(0, 0) = 0$.

(b) Use the limit definition to show that the directional derivative $D_{\mathbf{u}} f(0, 0)$ does not exist for any unit vector $\mathbf{u}$ other than $\mathbf{i}$ and $\mathbf{j}$.

(c) Is f differentiable at $(0, 0)$?

71. Use the definition of differentiability to show that if $f(x, y)$ is differentiable at $(0, 0)$ and

$$f(0, 0) = f_x(0, 0) = f_y(0, 0) = 0$$

then

$$\lim_{(x, y) \to (0, 0)} \frac{f(x, y)}{\sqrt{x^2 + y^2}} = 0 \qquad \boxed{7}$$

72. This exercise shows that there exists a function that is not differentiable at $(0, 0)$ even though all directional derivatives at $(0, 0)$ exist. Define $f(x, y) = x^2 y/(x^2 + y^2)$ for $(x, y) \neq 0$ and $f(0, 0) = 0$.

(a) Use the limit definition to show that $D_{\mathbf{v}} f(0, 0)$ exists for all vectors $\mathbf{v}$. Show that $f_x(0, 0) = f_y(0, 0) = 0$.

(b) Prove that f is *not* differentiable at $(0, 0)$ by showing that Eq. (7) does not hold.

73. Prove that if $f(x, y)$ is differentiable and $\nabla f_{(x, y)} = \mathbf{0}$ for all (x, y), then f is constant.

74. Prove the following Quotient Rule, where f, g are differentiable:

$$\nabla\left(\frac{f}{g}\right) = \frac{g\nabla f - f\nabla g}{g^2}$$

In Exercises 75–77, a path $\mathbf{r}(t) = \langle x(t), y(t) \rangle$ follows the gradient of a function $f(x, y)$ if the tangent vector $\mathbf{r}'(t)$ points in the direction of ∇f for all t. In other words, $\mathbf{r}'(t) = k(t)\nabla f_{\mathbf{r}(t)}$ for some positive function $k(t)$. Note that in this case, $\mathbf{r}(t)$ crosses each level curve of $f(x, y)$ at a right angle.

75. Show that if the path $\mathbf{r}(t) = \langle x(t), y(t) \rangle$ follows the gradient of $f(x, y)$, then

$$\frac{y'(t)}{x'(t)} = \frac{f_y}{f_x}$$

76. Find a path of the form $\mathbf{r}(t) = (t, g(t))$ passing through $(1, 2)$ that follows the gradient of $f(x, y) = 2x^2 + 8y^2$ (Figure 16). *Hint:* Use Separation of Variables.

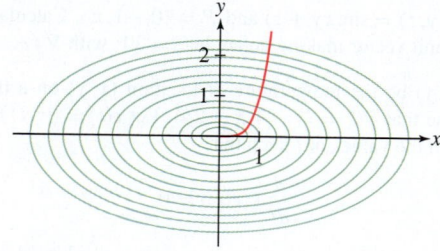

FIGURE 16 The path $\mathbf{r}(t)$ is orthogonal to the level curves of $f(x, y) = 2x^2 + 8y^2$.

77. (CAS) Find the curve $y = g(x)$ passing through $(0, 1)$ that crosses each level curve of $f(x, y) = y \sin x$ at a right angle. Using a computer algebra system, graph $y = g(x)$ together with the level curves of f.

14.6 Multivariable Calculus Chain Rules

We have seen a few different chain rule formulas for functions involving multiple variables. In this section, we show how they all fall under a general scheme for identifying the structure of a composite function and determining the type of chain rule formula from the structure.

To begin, we return to the Chain Rule for Paths and prove it. In the proof, we use the limit condition in the definition of differentiability, demonstrating why that limit is a necessary and important part of the concept of differentiability.

The Chain Rule for Paths applies to compositions $f(\mathbf{r}(t))$, where f and $\mathbf{r}$ are differentiable. We primarily consider the cases where f is a function of x and y, and $\mathbf{r}(t)$ is a path in the plane, or f is a function of x, y, and z, and $\mathbf{r}(t)$ is a path in 3-space.

← REMINDER *We regard $\mathbf{r}(t)$ as representing both a vector, $\langle x(t), y(t) \rangle$ in the plane or $\langle x(t), y(t), z(t) \rangle$ in 3-space, and a point $(x(t), y(t))$ or $(x(t), y(t), z(t))$.*

THEOREM 1 Chain Rule for Paths If f and $\mathbf{r}(t)$ are differentiable, then

$$\frac{d}{dt} f(\mathbf{r}(t)) = \nabla f_{\mathbf{r}(t)} \cdot \mathbf{r}'(t)$$

In the cases of two and three variables, this chain rule states:

$$\frac{d}{dt} f(\mathbf{r}(t)) = \left\langle \frac{\partial f}{\partial x}, \frac{\partial f}{\partial y} \right\rangle \cdot \langle x'(t), y'(t) \rangle = \frac{\partial f}{\partial x} \frac{dx}{dt} + \frac{\partial f}{\partial y} \frac{dy}{dt}$$

$$\frac{d}{dt} f(\mathbf{r}(t)) = \left\langle \frac{\partial f}{\partial x}, \frac{\partial f}{\partial y}, \frac{\partial f}{\partial z} \right\rangle \cdot \langle x'(t), y'(t), z'(t) \rangle = \frac{\partial f}{\partial x} \frac{dx}{dt} + \frac{\partial f}{\partial y} \frac{dy}{dt} + \frac{\partial f}{\partial z} \frac{dz}{dt}$$

We prove the theorem for the two-variable case.

Proof By definition,

$$\frac{d}{dt} f(\mathbf{r}(t)) = \lim_{h \to 0} \frac{f(x(t+h), y(t+h)) - f(x(t), y(t))}{h}$$

To calculate this derivative, set

$$\Delta f = f(x(t+h), y(t+h)) - f(x(t), y(t))$$

$$\Delta x = x(t+h) - x(t), \qquad \Delta y = y(t+h) - y(t)$$

$$e(x(t+h), y(t+h)) = f(x(t+h), y(t+h)) - (f(x(t), y(t)) + f_x(x(t), y(t))\Delta x$$
$$+ f_y(x(t), y(t))\Delta y)$$

The last term is the error, as in Section 14.4, in approximating f with its linearization centered at $(x(t), y(t))$. Putting these terms together, we have

$$\Delta f = f_x(x(t), y(t))\Delta x + f_y(x(t), y(t))\Delta y + e(x(t+h), y(t+h))$$

Now, set $h = \Delta t$ and divide by Δt:

$$\frac{\Delta f}{\Delta t} = f_x(x(t), y(t))\frac{\Delta x}{\Delta t} + f_y(x(t), y(t))\frac{\Delta y}{\Delta t} + \frac{e(x(t+\Delta t), y(t+\Delta t))}{\Delta t}$$

We show below that the last term tends to zero as $\Delta t \to 0$. Given that, we obtain the desired result:

$$\frac{d}{dt} f(\mathbf{r}(t)) = \lim_{\Delta t \to 0} \frac{\Delta f}{\Delta t}$$

$$= f_x(x(t), y(t)) \lim_{\Delta t \to 0} \frac{\Delta x}{\Delta t} + f_y(x(t), y(t)) \lim_{\Delta t \to 0} \frac{\Delta y}{\Delta t}$$

$$= f_x(x(t), y(t))\frac{dx}{dt} + f_y(x(t), y(t))\frac{dy}{dt}$$

$$= \nabla f_{\mathbf{r}(t)} \cdot \mathbf{r}'(t)$$

We verify that the last term tends to zero as follows:

$$\lim_{\Delta t \to 0} \frac{e(x(t+\Delta t), y(t+\Delta t))}{\Delta t} = \lim_{\Delta t \to 0} \frac{e(x(t+\Delta t), y(t+\Delta t))}{\sqrt{(\Delta x)^2 + (\Delta y)^2}} \left(\frac{\sqrt{(\Delta x)^2 + (\Delta y)^2}}{\Delta t} \right)$$

$$= \underbrace{\left(\lim_{\Delta t \to 0} \frac{e(x(t+\Delta t), y(t+\Delta t))}{\sqrt{(\Delta x)^2 + (\Delta y)^2}} \right)}_{\text{Zero}} \lim_{\Delta t \to 0} \left(\sqrt{\left(\frac{\Delta x}{\Delta t} \right)^2 + \left(\frac{\Delta y}{\Delta t} \right)^2} \right) = 0$$

The first limit is zero, as indicated, because f is differentiable. The second limit is equal to $\sqrt{x'(t)^2 + y'(t)^2}$, a finite value, and therefore the product is zero. ∎

EXAMPLE 1 Calculate $\dfrac{d}{dt} f(\mathbf{r}(t)) \Big|_{t=\pi/2}$, where

$$f(x, y, z) = xy + z^2 \qquad \text{and} \qquad \mathbf{r}(t) = \langle \cos t, \sin t, t \rangle$$

Solution We have $\mathbf{r}\left(\frac{\pi}{2}\right) = \langle \cos \frac{\pi}{2}, \sin \frac{\pi}{2}, \frac{\pi}{2} \rangle = \langle 0, 1, \frac{\pi}{2} \rangle$. Compute the gradient:

$$\nabla f = \left\langle \frac{\partial f}{\partial x}, \frac{\partial f}{\partial y}, \frac{\partial f}{\partial z} \right\rangle = \langle y, x, 2z \rangle, \qquad \nabla f_{\mathbf{r}(\pi/2)} = \nabla f\left(0, 1, \frac{\pi}{2}\right) = \langle 1, 0, \pi \rangle$$

Then compute the tangent vector:

$$\mathbf{r}'(t) = \langle -\sin t, \cos t, 1 \rangle, \qquad \mathbf{r}'\left(\frac{\pi}{2}\right) = \left\langle -\sin \frac{\pi}{2}, \cos \frac{\pi}{2}, 1 \right\rangle = \langle -1, 0, 1 \rangle$$

By the Chain Rule,

$$\frac{d}{dt} f(\mathbf{r}(t)) \Big|_{t=\pi/2} = \nabla f_{\mathbf{r}(\pi/2)} \cdot \mathbf{r}'\left(\frac{\pi}{2}\right) = \langle 1, 0, \pi \rangle \cdot \langle -1, 0, 1 \rangle = \pi - 1 \quad ∎$$

Next, let's consider the case of more general composite functions. Suppose, for example, that x, y, z are differentiable functions of s and t—say, $x = x(s,t)$, $y = y(s,t)$, and $z = z(s,t)$. The composition

$$f(x(s,t), y(s,t), z(s,t)) \qquad \boxed{1}$$

is then a function of s and t. We refer to s and t as the **independent variables**.

EXAMPLE 2 Given that $f(x, y, z) = xy + z$ and $x = s^2$, $y = st$, $z = t^2$, find the composite function.

Solution We can keep track of which variable depends on which other variable by using a chart as in Figure 1. The composite function is given by

$$f(x(s,t), y(s,t), z(s,t)) = xy + z = (s^2)(st) + t^2 = s^3 t + t^2 \quad ∎$$

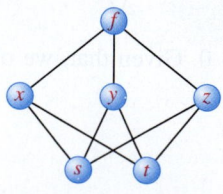

FIGURE 1 Keeping track of the relationships between the variables.

The Chain Rule expresses the derivatives of f with respect to the independent variables. For example, the partial derivatives of $f(x(s,t), y(s,t), z(s,t))$ are

$$\frac{\partial f}{\partial s} = \frac{\partial f}{\partial x} \frac{\partial x}{\partial s} + \frac{\partial f}{\partial y} \frac{\partial y}{\partial s} + \frac{\partial f}{\partial z} \frac{\partial z}{\partial s} \qquad \boxed{2}$$

$$\frac{\partial f}{\partial t} = \frac{\partial f}{\partial x} \frac{\partial x}{\partial t} + \frac{\partial f}{\partial y} \frac{\partial y}{\partial t} + \frac{\partial f}{\partial z} \frac{\partial z}{\partial t} \qquad \boxed{3}$$

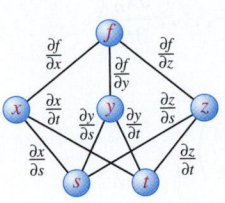

FIGURE 2 Keeping track of the relationships between the variables.

Note that we can obtain the formula for $\dfrac{\partial f}{\partial s}$ by labeling each edge in Figure 1 with the partial derivative of the top variable with respect to the bottom variable as in Figure 2. Then to obtain the formula for $\dfrac{\partial f}{\partial s}$, we consider each of the paths along the edges down from f to s: the first through x, the second through y, and the third through z. Each path contributes a term to the formula, and those terms are added together. The first term, through x, is the product of the partial derivatives labeling the path's edges, giving $\dfrac{\partial f}{\partial x}\dfrac{\partial x}{\partial s}$. Similarly, the second term is $\dfrac{\partial f}{\partial y}\dfrac{\partial y}{\partial s}$, and the third term is $\dfrac{\partial f}{\partial z}\dfrac{\partial z}{\partial s}$. Thus, we obtain the formula

$$\frac{\partial f}{\partial s} = \frac{\partial f}{\partial x}\frac{\partial x}{\partial s} + \frac{\partial f}{\partial y}\frac{\partial y}{\partial s} + \frac{\partial f}{\partial z}\frac{\partial z}{\partial s}$$

We obtain the formula for $\dfrac{\partial f}{\partial t}$ in a similar manner. To prove these formulas, we observe that $\dfrac{\partial f}{\partial s}$, when evaluated at a point (s_0, t_0), is equal to the derivative with respect to s on the path obtained by fixing $t = t_0$ and letting s vary. That path is

$$\mathbf{r}(s) = \langle x(s, t_0), y(s, t_0), z(s, t_0)\rangle$$

Fixing $t = t_0$ and taking the derivative with respect to s, we obtain

$$\frac{\partial f}{\partial s}(s_0, t_0) = \frac{d}{ds} f(\mathbf{r}(s))\Big|_{s=s_0}$$

The tangent vector is

$$\mathbf{r}'(s) = \left\langle \frac{\partial x}{\partial s}(s, t_0), \frac{\partial y}{\partial s}(s, t_0), \frac{\partial z}{\partial s}(s, t_0)\right\rangle$$

Therefore, by the Chain Rule for Paths,

$$\frac{\partial f}{\partial s}\Big|_{(s_0,t_0)} = \frac{d}{ds} f(\mathbf{r}(s))\Big|_{s=s_0} = \nabla f \cdot \mathbf{r}'(s_0) = \frac{\partial f}{\partial x}\frac{\partial x}{\partial s} + \frac{\partial f}{\partial y}\frac{\partial y}{\partial s} + \frac{\partial f}{\partial z}\frac{\partial z}{\partial s}$$

The derivatives on the right are evaluated at (s_0, t_0). This proves Eq. (2). A similar argument proves Eq. (3), as well as the general case of a function $f(x_1, \ldots, x_n)$, where the variables x_i depend on independent variables $t_1, \ldots, t_m$.

> **THEOREM 2** **General Version of the Chain Rule** Let $f(x_1, \ldots, x_n)$ be a differentiable function of n variables. Suppose that each of the variables $x_1, \ldots, x_n$ is a differentiable function of m independent variables $t_1, \ldots, t_m$. Then, for $k = 1, \ldots, m$,
>
> $$\frac{\partial f}{\partial t_k} = \frac{\partial f}{\partial x_1}\frac{\partial x_1}{\partial t_k} + \frac{\partial f}{\partial x_2}\frac{\partial x_2}{\partial t_k} + \cdots + \frac{\partial f}{\partial x_n}\frac{\partial x_n}{\partial t_k} \qquad \boxed{4}$$

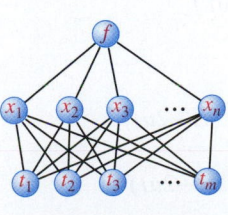

FIGURE 3 Keeping track of the dependencies between the variables.

The term "primary derivative" is not standard. We use it in this section only, to clarify the structure of the Chain Rule.

We keep track of the dependencies between the variables as in Figure 3. As an aid to remembering the Chain Rule, we will refer to

$$\frac{\partial f}{\partial x_1}, \quad \ldots \quad, \quad \frac{\partial f}{\partial x_n}$$

as the **primary derivatives**. They are the components of the gradient ∇f. By Eq. (4), the derivative of f with respect to the independent variable t_k is equal to a sum of n terms:

$$j\text{th term:} \quad \frac{\partial f}{\partial x_j}\frac{\partial x_j}{\partial t_k} \quad \text{for } j = 1, 2, \ldots, n$$

Note that we can write Eq. (4) as a dot product:

$$\frac{\partial f}{\partial t_k} = \left\langle \frac{\partial f}{\partial x_1}, \frac{\partial f}{\partial x_2}, \dots, \frac{\partial f}{\partial x_n} \right\rangle \cdot \left\langle \frac{\partial x_1}{\partial t_k}, \frac{\partial x_2}{\partial t_k}, \dots, \frac{\partial x_n}{\partial t_k} \right\rangle$$

$$\boxed{\frac{\partial f}{\partial t_k} = \nabla f \cdot \left\langle \frac{\partial x_1}{\partial t_k}, \frac{\partial x_2}{\partial t_k}, \dots, \frac{\partial x_n}{\partial t_k} \right\rangle} \qquad \boxed{5}$$

EXAMPLE 3   Using the Chain Rule   Let $f(x, y, z) = xy + z$. Calculate $\dfrac{\partial f}{\partial s}$, where

$$x = s^2, \quad y = st, \quad z = t^2$$

Solution   We keep track of the dependencies of the variables as in Figure 2.

Step 1. Compute the primary derivatives.

$$\frac{\partial f}{\partial x} = y, \qquad \frac{\partial f}{\partial y} = x, \qquad \frac{\partial f}{\partial z} = 1$$

Step 2. Apply the Chain Rule.

$$\frac{\partial f}{\partial s} = \frac{\partial f}{\partial x}\frac{\partial x}{\partial s} + \frac{\partial f}{\partial y}\frac{\partial y}{\partial s} + \frac{\partial f}{\partial z}\frac{\partial z}{\partial s} = y\frac{\partial}{\partial s}(s^2) + x\frac{\partial}{\partial s}(st) + \frac{\partial}{\partial s}(t^2)$$

$$= (y)(2s) + (x)(t) + 0$$

$$= 2sy + xt$$

This expresses the derivative in terms of both sets of variables. If desired, we can substitute $x = s^2$ and $y = st$ to write the derivative in terms of s and t:

$$\frac{\partial f}{\partial s} = 2ys + xt = 2(st)s + (s^2)t = 3s^2 t$$

To check this result, recall that in Example 2, we computed the composite function:

$$f(x(s,t), y(s,t), z(s,t)) = f(s^2, st, t^2) = s^3 t + t^2$$

From this, we see directly that $\dfrac{\partial f}{\partial s} = 3s^2 t$, confirming our result.   ■

EXAMPLE 4   Evaluating the Derivative   Let $f(x, y) = e^{xy}$. Evaluate $\dfrac{\partial f}{\partial t}$ at $(s, t, u) = (2, 3, -1)$, where $x = st$, $y = s - ut^2$.

Solution   We keep track of the dependencies of the variables as in Figure 4. We can use either Eq. (4) or Eq. (5). We'll use the dot product form in Eq. (5). We have

$$\nabla f = \left\langle \frac{\partial f}{\partial x}, \frac{\partial f}{\partial y} \right\rangle = \left\langle y e^{xy}, x e^{xy} \right\rangle, \qquad \left\langle \frac{\partial x}{\partial t}, \frac{\partial y}{\partial t} \right\rangle = \langle s, -2ut \rangle$$

and the Chain Rule gives us

$$\frac{\partial f}{\partial t} = \nabla f \cdot \left\langle \frac{\partial x}{\partial t}, \frac{\partial y}{\partial t} \right\rangle = \left\langle y e^{xy}, x e^{xy} \right\rangle \cdot \langle s, -2ut \rangle$$

$$= y e^{xy}(s) + x e^{xy}(-2ut)$$

$$= (ys - 2xut) e^{xy}$$

To finish the problem, we do not have to rewrite $\dfrac{\partial f}{\partial t}$ in terms of s, t, u. For $(s, t, u) = (2, 3, -1)$, we obtain

$$x = st = 2(3) = 6, \qquad y = s - ut^2 = 2 - (-1)(3^2) = 11$$

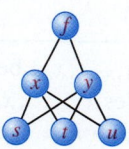

FIGURE 4 Keeping track of the dependencies between the variables.

With $(s, t, u) = (2, 3, -1)$ and $(x, y) = (6, 11)$, we have

$$\frac{\partial f}{\partial t}\bigg|_{(2,3,-1)} = (ys - 2xut)e^{xy}\bigg|_{(2,3,-1)} = \left((11)(2) - 2(6)(-1)(3)\right)e^{6(11)} = 58e^{66} \quad \blacksquare$$

EXAMPLE 5 **Polar Coordinates** Let $f(x, y)$ be a function of two variables, and let (r, θ) be polar coordinates.

(a) Express $\dfrac{\partial f}{\partial \theta}$ in terms of $\dfrac{\partial f}{\partial x}$ and $\dfrac{\partial f}{\partial y}$.

(b) Evaluate $\dfrac{\partial f}{\partial \theta}$ at $(x, y) = (1, 1)$ for $f(x, y) = x^2 y$.

Solution

(a) Since $x = r \cos \theta$ and $y = r \sin \theta$,

$$\frac{\partial x}{\partial \theta} = -r \sin \theta, \qquad \frac{\partial y}{\partial \theta} = r \cos \theta$$

By the Chain Rule,

$$\frac{\partial f}{\partial \theta} = \frac{\partial f}{\partial x}\frac{\partial x}{\partial \theta} + \frac{\partial f}{\partial y}\frac{\partial y}{\partial \theta} = -r \sin \theta \frac{\partial f}{\partial x} + r \cos \theta \frac{\partial f}{\partial y}$$

*If you have studied quantum mechanics, you may recognize the right-hand side of Eq. (6) as the **angular momentum** operator (with respect to the z-axis) applied to the function f.*

Since $x = r \cos \theta$ and $y = r \sin \theta$, we can write $\dfrac{\partial f}{\partial \theta}$ in terms of x and y alone:

$$\frac{\partial f}{\partial \theta} = x \frac{\partial f}{\partial y} - y \frac{\partial f}{\partial x} \qquad \boxed{6}$$

(b) Apply Eq. (6) to $f(x, y) = x^2 y$:

$$\frac{\partial f}{\partial \theta} = x \frac{\partial}{\partial y}(x^2 y) - y \frac{\partial}{\partial x}(x^2 y) = x^3 - 2xy^2$$

$$\frac{\partial f}{\partial \theta}\bigg|_{(x,y)=(1,1)} = 1^3 - 2(1)(1^2) = -1 \qquad \blacksquare$$

Notice that the General Version of the Chain Rule encompasses the Chain Rule for Paths. In the Chain Rule for Paths, there is just one independent variable, which is the parameter for the path.

Implicit Differentiation

In single-variable calculus, we used implicit differentiation to compute dy/dx when y is defined implicitly as a function of x through an equation $f(x, y) = 0$. This method also works for functions of several variables. Suppose that z is defined implicitly by an equation

$$F(x, y, z) = 0$$

Here we are treating z as a dependent variable with independent variables x and y. We could switch the roles of the variables and similarly work with $y(x, z)$, where y is dependent while x and z are independent. Likewise, we could work with $x(y, z)$.

Thus, $z = z(x, y)$ is a function of x and y. We may not be able to solve explicitly for $z(x, y)$, but we can treat $F(x, y, z)$ as a composite function with x and y as independent variables and use the Chain Rule to differentiate implicitly with respect to x:

$$\frac{\partial F}{\partial x}\frac{\partial x}{\partial x} + \frac{\partial F}{\partial y}\frac{\partial y}{\partial x} + \frac{\partial F}{\partial z}\frac{\partial z}{\partial x} = 0$$

We have $\partial x/\partial x = 1$ and also $\partial y/\partial x = 0$, since y does not depend on x. Thus,

$$\frac{\partial F}{\partial x} + \frac{\partial F}{\partial z}\frac{\partial z}{\partial x} = F_x + F_z \frac{\partial z}{\partial x} = 0$$

If $F_z \neq 0$, we may solve for $\partial z/\partial x$ (we compute $\partial z/\partial y$ similarly):

$$\boxed{\frac{\partial z}{\partial x} = -\frac{F_x}{F_z}, \qquad \frac{\partial z}{\partial y} = -\frac{F_y}{F_z}} \qquad \boxed{7}$$

EXAMPLE 6 Calculate $\partial z/\partial x$ and $\partial z/\partial y$ at $P = (1, 1, 1)$, where

$$F(x, y, z) = x^2 + y^2 - 2z^2 + 12x - 8z - 4 = 0$$

What is the graphical interpretation of these partial derivatives?

Solution We have

$$F_x = 2x + 12, \qquad F_y = 2y, \qquad F_z = -4z - 8$$

Hence,

$$\frac{\partial z}{\partial x} = -\frac{F_x}{F_z} = \frac{2x + 12}{4z + 8}, \qquad \frac{\partial z}{\partial y} = -\frac{F_y}{F_z} = \frac{2y}{4z + 8}$$

The derivatives at $P = (1, 1, 1)$ are

$$\frac{\partial z}{\partial x}\bigg|_{(1,1,1)} = \frac{2(1) + 12}{4(1) + 8} = \frac{14}{12} = \frac{7}{6}, \qquad \frac{\partial z}{\partial y}\bigg|_{(1,1,1)} = \frac{2(1)}{4(1) + 8} = \frac{2}{12} = \frac{1}{6}$$

Figure 5 shows the surface $F(x, y, z) = 0$. The surface as a whole is not the graph of a function $f(x, y)$ because it fails the Vertical Line Test [that is, for some (x, y) there is more than one point (x, y, z) on the surface]. However, a small patch near P may be represented as a graph of a function $z = f(x, y)$, and the partial derivatives $\dfrac{\partial z}{\partial x}$ and $\dfrac{\partial z}{\partial y}$ are equal to f_x and f_y. Implicit differentiation has enabled us to compute these partial derivatives without finding $f(x, y)$ explicitly. ∎

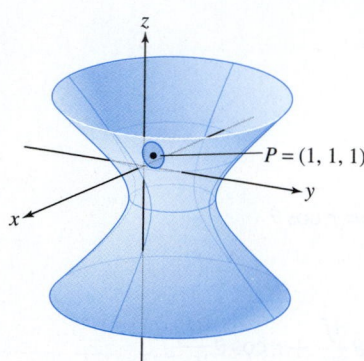

FIGURE 5 The surface
$x^2 + y^2 - 2z^2 + 12x - 8z - 4 = 0$.
A small patch of the surface around P can be represented as the graph of a function of x and y.

Assumptions Matter Implicit differentiation is based on the assumption that we can solve the equation $F(x, y, z) = 0$ for z in the form $z = f(x, y)$. Otherwise, the partial derivatives $\dfrac{\partial z}{\partial x}$ and $\dfrac{\partial z}{\partial y}$ would have no meaning. The Implicit Function Theorem of advanced calculus guarantees that this can be done (at least near a point P) if F has continuous partial derivatives and $F_z(P) \neq 0$. Why is this condition necessary? Recall that the gradient vector $\nabla F_P = \langle F_x(P), F_y(P), F_z(P) \rangle$ is normal to the surface at P, so $F_z(P) = 0$ means that the tangent plane at P is vertical. To see what can go wrong, consider the cylinder (shown in Figure 6):

$$F(x, y, z) = x^2 + y^2 - 1 = 0$$

In this particular case, F_z is 0 for all (x, y, z). The z-coordinate on the cylinder does not depend on x or y, so it is impossible to represent the cylinder as a graph $z = f(x, y)$ and the derivatives $\dfrac{\partial z}{\partial x}$ and $\dfrac{\partial z}{\partial y}$ do not exist.

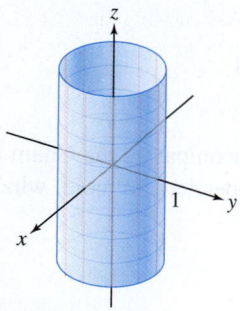

FIGURE 6 Graph of the cylinder
$x^2 + y^2 - 1 = 0$.

14.6 SUMMARY

- If $f(x, y, z)$ is a function of x, y, z, and if x, y, z depend on two other variables, say, s and t, then

$$f(x, y, z) = f(x(s, t), y(s, t), z(s, t))$$

is a composite function of s and t. We refer to s and t as the *independent variables*.

- The *Chain Rule* expresses the partial derivatives with respect to the independent variables s and t in terms of the *primary derivatives*:

$$\frac{\partial f}{\partial x}, \qquad \frac{\partial f}{\partial y}, \qquad \frac{\partial f}{\partial z}$$

Namely,

$$\frac{\partial f}{\partial s} = \frac{\partial f}{\partial x}\frac{\partial x}{\partial s} + \frac{\partial f}{\partial y}\frac{\partial y}{\partial s} + \frac{\partial f}{\partial z}\frac{\partial z}{\partial s}, \qquad \frac{\partial f}{\partial t} = \frac{\partial f}{\partial x}\frac{\partial x}{\partial t} + \frac{\partial f}{\partial y}\frac{\partial y}{\partial t} + \frac{\partial f}{\partial z}\frac{\partial z}{\partial t}$$

- In general, if $f(x_1, \ldots, x_n)$ is a function of n variables and if $x_1, \ldots, x_n$ depend on the independent variables $t_1, \ldots, t_m$, then

$$\frac{\partial f}{\partial t_k} = \frac{\partial f}{\partial x_1}\frac{\partial x_1}{\partial t_k} + \frac{\partial f}{\partial x_2}\frac{\partial x_2}{\partial t_k} + \cdots + \frac{\partial f}{\partial x_n}\frac{\partial x_n}{\partial t_k}$$

- The Chain Rule can be expressed as a dot product:

$$\frac{\partial f}{\partial t_k} = \underbrace{\left\langle \frac{\partial f}{\partial x_1}, \frac{\partial f}{\partial x_2}, \ldots, \frac{\partial f}{\partial x_n} \right\rangle}_{\nabla f} \cdot \left\langle \frac{\partial x_1}{\partial t_k}, \frac{\partial x_2}{\partial t_k}, \ldots, \frac{\partial x_n}{\partial t_k} \right\rangle$$

- Implicit differentiation is used to find the partial derivatives $\partial z/\partial x$ and $\partial z/\partial y$ when z is defined implicitly by an equation $F(x, y, z) = 0$:

$$\frac{\partial z}{\partial x} = -\frac{F_x}{F_z}, \qquad \frac{\partial z}{\partial y} = -\frac{F_y}{F_z}$$

14.6 EXERCISES

Preliminary Questions

1. Let $f(x, y) = xy$, where $x = uv$ and $y = u + v$.
(a) What are the primary derivatives of f?
(b) What are the independent variables?

In Exercises 2 and 3, suppose that $f(u, v) = ue^v$, where $u = rs$ and $v = r + s$.

2. The composite function $f(u, v)$ is equal to:
(a) rse^{r+s} (b) re^s (c) rse^{rs}

3. What is the value of $f(u, v)$ at $(r, s) = (1, 1)$?

4. According to the Chain Rule, $\partial f/\partial r$ is equal to (choose the correct answer):
(a) $\dfrac{\partial f}{\partial x}\dfrac{\partial x}{\partial r} + \dfrac{\partial f}{\partial x}\dfrac{\partial x}{\partial s}$

(b) $\dfrac{\partial f}{\partial x}\dfrac{\partial x}{\partial r} + \dfrac{\partial f}{\partial y}\dfrac{\partial y}{\partial r}$

(c) $\dfrac{\partial f}{\partial r}\dfrac{\partial r}{\partial x} + \dfrac{\partial f}{\partial s}\dfrac{\partial s}{\partial x}$

5. Suppose that x, y, z are functions of the independent variables u, v, w. Which of the following terms appear in the Chain Rule expression for $\partial f/\partial w$?
(a) $\dfrac{\partial f}{\partial v}\dfrac{\partial x}{\partial v}$ (b) $\dfrac{\partial f}{\partial w}\dfrac{\partial w}{\partial x}$ (c) $\dfrac{\partial f}{\partial z}\dfrac{\partial z}{\partial w}$

6. With notation as in the previous exercise, does $\partial x/\partial v$ appear in the Chain Rule expression for $\partial f/\partial u$?

Exercises

1. Let $f(x, y, z) = x^2 y^3 + z^4$ and $x = s^2$, $y = st^2$, and $z = s^2 t$.
(a) Calculate the primary derivatives $\dfrac{\partial f}{\partial x}, \dfrac{\partial f}{\partial y}, \dfrac{\partial f}{\partial z}$.

(b) Calculate $\dfrac{\partial x}{\partial s}, \dfrac{\partial y}{\partial s}, \dfrac{\partial z}{\partial s}$.

(c) Compute $\dfrac{\partial f}{\partial s}$ using the Chain Rule:

$$\frac{\partial f}{\partial s} = \frac{\partial f}{\partial x}\frac{\partial x}{\partial s} + \frac{\partial f}{\partial y}\frac{\partial y}{\partial s} + \frac{\partial f}{\partial z}\frac{\partial z}{\partial s}$$

Express the answer in terms of the independent variables s, t.

2. Let $f(x, y) = x\cos(y)$ and $x = u^2 + v^2$ and $y = u - v$.
(a) Calculate the primary derivatives $\dfrac{\partial f}{\partial x}, \dfrac{\partial f}{\partial y}$.

(b) Use the Chain Rule to calculate $\partial f/\partial v$. Leave the answer in terms of both the dependent and the independent variables.
(c) Determine (x, y) for $(u, v) = (2, 1)$ and evaluate $\partial f/\partial v$ at $(u, v) = (2, 1)$.

In Exercises 3–10, use the Chain Rule to calculate the partial derivatives. Express the answer in terms of the independent variables.

3. $\dfrac{\partial f}{\partial s}, \dfrac{\partial f}{\partial r}$; $f(x, y, z) = xy + z^2$, $x = s^2$, $y = 2rs$, $z = r^2$

4. $\dfrac{\partial f}{\partial r}, \dfrac{\partial f}{\partial t}$; $f(x, y, z) = xy + z^2$, $x = r + s - 2t$, $y = 3rt$, $z = s^2$

5. $\dfrac{\partial g}{\partial x}, \dfrac{\partial g}{\partial y}$; $g(\theta, \phi) = \tan(\theta + \phi)$, $\theta = xy$, $\phi = x + y$

6. $\dfrac{\partial R}{\partial v}, \dfrac{\partial R}{\partial w}$; $R(x, y) = (x - 2y)^3$, $x = w^2$, $y = v^w$

7. $\dfrac{\partial F}{\partial y}$; $F(u, v) = e^{u+v}$, $u = x^2$, $v = xy$

8. $\dfrac{\partial f}{\partial u}$; $f(x, y) = x^2 + y^2$, $x = e^{u+v}$, $y = u + v$

9. $\dfrac{\partial h}{\partial t_2}$; $h(x, y) = \dfrac{x}{y}$, $x = t_1 t_2$, $y = t_1^2 t_2$

10. $\dfrac{\partial f}{\partial \theta}$; $f(x, y, z) = xy - z^2$, $x = r\cos\theta$, $y = \cos^2\theta$, $z = r$

In Exercises 11–16, use the Chain Rule to evaluate the partial derivative at the point specified.

11. $\partial f/\partial u$ and $\partial f/\partial v$ at $(u, v) = (-1, -1)$, where $f(x, y, z) = x^3 + yz^2$, $x = u^2 + v$, $y = u + v^2$, $z = uv$

12. $\partial f/\partial s$ at $(r, s) = (1, 0)$, where $f(x, y) = \ln(xy)$, $x = 3r + 2s$, $y = 5r + 3s$

13. $\partial g/\partial\theta$ at $(r, \theta) = \left(2\sqrt{2}, \frac{\pi}{4}\right)$, where $g(x, y) = 1/(x + y^2)$, $x = r\cos\theta$, $y = r\sin\theta$

14. dg/ds at $s = 4$, where $g(x, y) = x^2 - y^2$, $x = s^2 + 1$, $y = 1 - 2s$

15. $\partial g/\partial u$ at $(u, v) = (0, 1)$, where $g(x, y) = x^2 - y^2$, $x = e^u\cos v$, $y = e^u\sin v$

16. $\dfrac{\partial h}{\partial q}$ at $(q, r) = (3, 2)$, where $h(u, v) = ue^v$, $u = q^3$, $v = qr^2$

17. Given $f(x, y)$ and $y = y(x)$, we can define a composite function $g(x) = f(x, y(x))$.

(a) Show that $g'(x) = \dfrac{\partial f}{\partial x} + \dfrac{\partial f}{\partial y} y'(x)$.

(b) Let $f(x, y) = x^3 - xy^2$ and $y(x) = 1 - x$. With $g(x) = f(x, y(x))$, use the formula in (a) to determine $g'(x)$, expressing the result in terms of x only.

(c) With $f(x, y)$ and $y(x)$ as in (b), give an expression for $g(x)$ in terms of x. Then compute $g'(x)$ from $g(x)$, and show that the result coincides with the one from (b).

18. Let $f(x, y) = 4 - x^2 y^2 + e^{2x}$ and $y(x) = \dfrac{e^x}{x}$. Define $g(x) = f(x, y(x))$.

(a) Use the derivative formula from Exercise 17(a) to prove that $g'(x) = 0$ and therefore that g is a constant function.

(b) Express $g(x)$ directly in terms of x, and simplify to show that g is indeed a constant function.

19. A baseball player hits the ball and then runs down the first base line at 20 ft/s. The first baseman fields the ball and then runs toward first base along the second base line at 18 ft/s as in Figure 7.

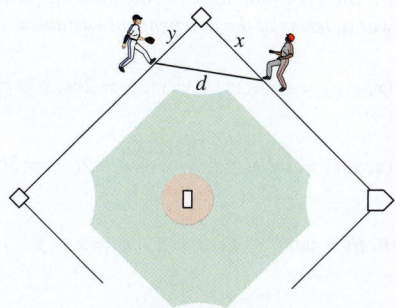

FIGURE 7

Determine how fast the distance between the two players is changing at a moment when the hitter is 8 ft from first base and the first baseman is 6 ft from first base.

20. Jessica and Matthew are running toward the point P along the straight paths that make a fixed angle of θ (Figure 8). Suppose that Matthew runs with velocity v_a meters per second and Jessica with velocity v_b meters per second. Let $f(x, y)$ be the distance from Matthew to Jessica when Matthew is x meters from P and Jessica is y meters from P.

(a) Show that $f(x, y) = \sqrt{x^2 + y^2 - 2xy\cos\theta}$.

(b) Assume that $\theta = \pi/3$. Use the Chain Rule to determine the rate at which the distance between Matthew and Jessica is changing when $x = 30$, $y = 20$, $v_a = 4$ m/s, and $v_b = 3$ m/s.

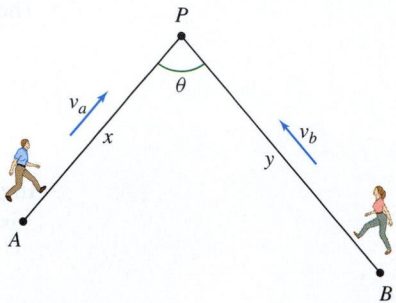

FIGURE 8

21. Two spacecraft are following paths in space given by $\mathbf{r}_1 = \langle \sin t, t, t^2\rangle$ and $\mathbf{r}_2 = \langle \cos t, 1 - t, t^3\rangle$. If the temperature for points in space is given by $T(x, y, z) = x^2 y(1 - z)$, use the Chain Rule to determine the rate of change of the difference D in the temperatures the two spacecraft experience at time $t = \pi$.

22. The Law of Cosines states that $c^2 = a^2 + b^2 - 2ab\cos\theta$, where a, b, c are the sides of a triangle and θ is the angle opposite the side of length c.

(a) Compute $\partial\theta/\partial a$, $\partial\theta/\partial b$, and $\partial\theta/\partial c$ using implicit differentiation.

(b) Suppose that $a = 10$, $b = 16$, $c = 22$. Estimate the change in θ if a and b are increased by 1 and c is increased by 2.

23. Let $u = u(x, y)$, and let (r, θ) be polar coordinates. Verify the relation

$$\|\nabla u\|^2 = u_r^2 + \frac{1}{r^2}u_\theta^2 \qquad \boxed{8}$$

Hint: Compute the right-hand side by expressing u_θ and u_r in terms of u_x and u_y.

24. Let $u(r, \theta) = r^2\cos^2\theta$. Use Eq. (8) to compute $\|\nabla u\|^2$. Then compute $\|\nabla u\|^2$ directly by observing that $u(x, y) = x^2$, and compare.

25. Let $x = s + t$ and $y = s - t$. Show that for any differentiable function $f(x, y)$,

$$\left(\frac{\partial f}{\partial x}\right)^2 - \left(\frac{\partial f}{\partial y}\right)^2 = \frac{\partial f}{\partial s}\frac{\partial f}{\partial t}$$

26. Express the derivatives

$$\frac{\partial f}{\partial\rho}, \frac{\partial f}{\partial\theta}, \frac{\partial f}{\partial\phi} \quad \text{in terms of} \quad \frac{\partial f}{\partial x}, \frac{\partial f}{\partial y}, \frac{\partial f}{\partial z}$$

where (ρ, θ, ϕ) are spherical coordinates.

27. Suppose that z is defined implicitly as a function of x and y by the equation $F(x, y, z) = xz^2 + y^2 z + xy - 1 = 0$.

(a) Calculate F_x, F_y, F_z.

(b) Use Eq. (7) to calculate $\dfrac{\partial z}{\partial x}$ and $\dfrac{\partial z}{\partial y}$.

28. Calculate $\partial z/\partial x$ and $\partial z/\partial y$ at the points $(3, 2, 1)$ and $(3, 2, -1)$, where z is defined implicitly by the equation $z^4 + z^2 x^2 - y - 8 = 0$.

In Exercises 29–34, calculate the partial derivative using implicit differentiation.

29. $\dfrac{\partial z}{\partial x}, \quad x^2 y + y^2 z + xz^2 = 10$

30. $\dfrac{\partial w}{\partial z}, \quad x^2 w + w^3 + wz^2 + 3yz = 0$

31. $\dfrac{\partial z}{\partial y}, \quad e^{xy} + \sin(xz) + y = 0$

32. $\dfrac{\partial r}{\partial t}$ and $\dfrac{\partial t}{\partial r}, \quad r^2 = te^{s/r}$

33. $\dfrac{\partial w}{\partial y}, \quad \dfrac{1}{w^2 + x^2} + \dfrac{1}{w^2 + y^2} = 1$ at $(x, y, w) = (1, 1, 1)$

34. $\partial U/\partial T$ and $\partial T/\partial U, \quad (TU - V)^2 \ln(W - UV) = \ln 2$ at $(T, U, V, W) = (1, 1, 2, 4)$

35. Let $\mathbf{r} = \langle x, y, z \rangle$ and $e_{\mathbf{r}} = \mathbf{r}/\|\mathbf{r}\|$. Show that if a function $f(x, y, z) = F(r)$ depends only on the distance from the origin $r = \|\mathbf{r}\| = \sqrt{x^2 + y^2 + z^2}$, then

$$\nabla f = F'(r) e_{\mathbf{r}} \qquad \boxed{9}$$

36. Let $f(x, y, z) = e^{-x^2 - y^2 - z^2} = e^{-r^2}$, with r as in Exercise 35. Compute ∇f directly and using Eq. (9).

37. Use Eq. (9) to compute $\nabla \left(\dfrac{1}{r} \right)$.

38. Use Eq. (9) to compute $\nabla(\ln r)$.

39. Figure 9 shows the graph of the equation

$$F(x, y, z) = x^2 + y^2 - z^2 - 12x - 8z - 4 = 0$$

(a) Use the quadratic formula to solve for z as a function of x and y. This gives two formulas, depending on the choice of sign.

(b) Which formula defines the portion of the surface satisfying $z \geq -4$? Which formula defines the portion satisfying $z \leq -4$?

(c) Calculate $\partial z/\partial x$ using the formula $z = f(x, y)$ (for both choices of sign) and again via implicit differentiation. Verify that the two answers agree.

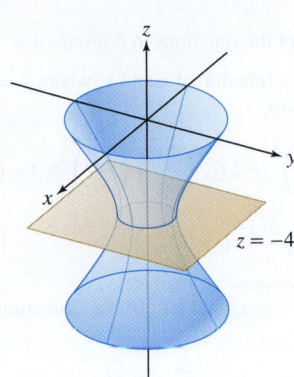

FIGURE 9 Graph of $x^2 + y^2 - z^2 - 12x - 8z - 4 = 0$.

40. For all $x > 0$, there is a unique value $y = r(x)$ that solves the equation $y^3 + 4xy = 16$.

(a) Show that $dy/dx = -4y/(3y^2 + 4x)$.

(b) Let $g(x) = f(x, r(x))$, where $f(x, y)$ is a function satisfying

$$f_x(1, 2) = 8, \quad f_y(1, 2) = 10$$

Use the Chain Rule to calculate $g'(1)$. Note that $r(1) = 2$ because $(x, y) = (1, 2)$ satisfies $y^3 + 4xy = 16$.

41. The pressure P, volume V, and temperature T of a van der Waals gas with n molecules (n constant) are related by the equation

$$\left(P + \frac{an^2}{V^2} \right)(V - nb) = nRT$$

where a, b, and R are constant. Calculate $\partial P/\partial T$ and $\partial V/\partial P$.

42. When x, y, and z are related by an equation $F(x, y, z) = 0$, we sometimes write $(\partial z/\partial x)_y$ in place of $\partial z/\partial x$ to indicate that in the differentiation, z is treated as a function of x with y held constant (and similarly for the other variables).

(a) Use Eq. (7) to prove the **cyclic relation**

$$\left(\frac{\partial z}{\partial x} \right)_y \left(\frac{\partial x}{\partial y} \right)_z \left(\frac{\partial y}{\partial z} \right)_x = -1 \qquad \boxed{10}$$

(b) Verify Eq. (10) for $F(x, y, z) = x + y + z = 0$.

(c) Verify the cyclic relation for the variables P, V, T in the Ideal Gas Law $PV - nRT = 0$ (n and R are constants).

43. Show that if $f(x)$ is differentiable and $c \neq 0$ is a constant, then $u(x, t) = f(x - ct)$ satisfies the so-called **advection equation**

$$\frac{\partial u}{\partial t} + c \frac{\partial u}{\partial x} = 0$$

Further Insights and Challenges

*In Exercises 44–47, a function $f(x, y, z)$ is called **homogeneous of degree** n if $f(\lambda x, \lambda y, \lambda z) = \lambda^n f(x, y, z)$ for all $\lambda \in \mathbf{R}$.*

44. Show that the following functions are homogeneous and determine their degree:

(a) $f(x, y, z) = x^2 y + xyz$

(b) $f(x, y, z) = 3x + 2y - 8z$

(c) $f(x, y, z) = \ln \left(\dfrac{xy}{z^2} \right)$

(d) $f(x, y, z) = z^4$

45. Prove that if $f(x, y, z)$ is homogeneous of degree n, then $f_x(x, y, z)$ is homogeneous of degree $n - 1$. *Hint:* Either use the limit definition or apply the Chain Rule to $f(\lambda x, \lambda y, \lambda z)$.

46. Prove that if $f(x, y, z)$ is homogeneous of degree n, then

$$x \frac{\partial f}{\partial x} + y \frac{\partial f}{\partial y} + z \frac{\partial f}{\partial z} = nf \qquad \boxed{11}$$

Hint: Let $F(t) = f(tx, ty, tz)$ and calculate $F'(1)$ using the Chain Rule.

47. Verify Eq. (11) for the functions in Exercise 44.

48. Suppose that f is a function of x and y, where $x = g(t, s)$, $y = h(t, s)$. Show that f_{tt} is equal to

$$f_{xx}\left(\frac{\partial x}{\partial t}\right)^2 + 2f_{xy}\left(\frac{\partial x}{\partial t}\right)\left(\frac{\partial y}{\partial t}\right) + f_{yy}\left(\frac{\partial y}{\partial t}\right)^2$$

$$+ f_x\frac{\partial^2 x}{\partial t^2} + f_y\frac{\partial^2 y}{\partial t^2} \qquad \boxed{12}$$

49. Let $r = \sqrt{x_1^2 + \cdots + x_n^2}$ and let $g(r)$ be a function of r. Prove the formulas

$$\frac{\partial g}{\partial x_i} = \frac{x_i}{r}g_r, \qquad \frac{\partial^2 g}{\partial x_i^2} = \frac{x_i^2}{r^2}g_{rr} + \frac{r^2 - x_i^2}{r^3}g_r$$

50. Prove that if $g(r)$ is a function of r as in Exercise 49, then

$$\frac{\partial^2 g}{\partial x_1^2} + \cdots + \frac{\partial^2 g}{\partial x_n^2} = g_{rr} + \frac{n-1}{r}g_r$$

In Exercises 51–55, the **Laplace operator** *is defined by* $\Delta f = f_{xx} + f_{yy}$. *A function* $f(x, y)$ *satisfying the Laplace equation* $\Delta f = 0$ *is called*

harmonic. A function $f(x, y)$ *is called* **radial** *if* $f(x, y) = g(r)$, *where* $r = \sqrt{x^2 + y^2}$.

51. Use Eq. (12) to prove that in polar coordinates (r, θ),

$$\Delta f = f_{rr} + \frac{1}{r^2}f_{\theta\theta} + \frac{1}{r}f_r \qquad \boxed{13}$$

52. Use Eq. (13) to show that $f(x, y) = \ln r$ is harmonic.

53. Verify that $f(x, y) = x$ and $f(x, y) = y$ are harmonic using both the rectangular and polar expressions for Δf.

54. Verify that $f(x, y) = \tan^{-1}\frac{y}{x}$ is harmonic using both the rectangular and polar expressions for Δf.

55. Use the Product Rule to show that

$$f_{rr} + \frac{1}{r}f_r = r^{-1}\frac{\partial}{\partial r}\left(r\frac{\partial f}{\partial r}\right)$$

Use this formula to show that if f is a radial harmonic function, then $rf_r = C$ for some constant C. Conclude that $f(x, y) = C\ln r + b$ for some constant b.

14.7 Optimization in Several Variables

Recall that optimization is the process of finding the extreme values of a function. This amounts to finding the highest and lowest points on the graph over a given domain. As we saw in the one-variable case, it is important to distinguish between *local* and *global* extreme values. A local extreme value is a value $f(a, b)$ that is a maximum or minimum in some small open disk around (a, b) (Figure 1).

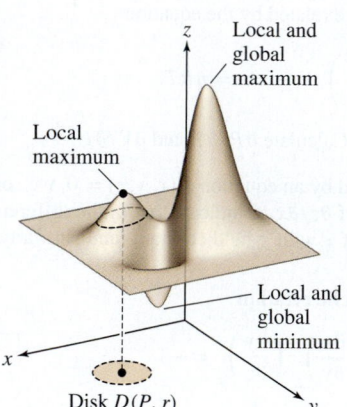

FIGURE 1 $f(x, y)$ has a local maximum at P.

Local and global maximum

Local maximum

Local and global minimum

Disk $D(P, r)$

DEFINITION Local Extreme Values A function $f(x, y)$ has a **local extremum** at $P = (a, b)$ if there exists an open disk $D(P, r)$ such that

- **Local maximum:** $f(x, y) \le f(a, b)$ for all $(x, y) \in D(P, r)$
- **Local minimum:** $f(x, y) \ge f(a, b)$ for all $(x, y) \in D(P, r)$

Fermat's Theorem for functions of one variable states that if $f(a)$ is a local extreme value, then a is a critical point and thus the tangent line (if it exists) is horizontal at $x = a$. A similar result holds for functions of two variables, but in this case, it is the *tangent plane* that must be horizontal (Figure 2). The tangent plane to $z = f(x, y)$ at $P = (a, b)$ has equation

$$z = f(a, b) + f_x(a, b)(x - a) + f_y(a, b)(y - b)$$

Thus, the tangent plane is horizontal if $f_x(a, b) = f_y(a, b) = 0$—that is, if the equation reduces to $z = f(a, b)$. This leads to the following definition of a critical point, where we take into account the possibility that one or both partial derivatives do not exist.

> **REMINDER** The term "extremum" (the plural is "extrema") means a minimum or maximum value.

> • More generally, $(a_1, \ldots, a_n)$ is a critical point of $f(x_1, \ldots, x_n)$ if each partial derivative satisfies
> $$f_{x_j}(a_1, \ldots, a_n) = 0$$
> or does not exist.

DEFINITION Critical Point A point $P = (a, b)$ in the domain of $f(x, y)$ is called a **critical point** if:

- $f_x(a, b) = 0$ or $f_x(a, b)$ does not exist, and
- $f_y(a, b) = 0$ or $f_y(a, b)$ does not exist.

FIGURE 2 The tangent line or plane is horizontal at a local extremum.

(A)

(B)

As in the single-variable case, we have the following:

• *Theorem 1 holds in any number of variables: Local extrema occur at critical points.*

> **THEOREM 1 Fermat's Theorem** If $f(x, y)$ has a local minimum or maximum at $P = (a, b)$, then (a, b) is a critical point of $f(x, y)$.

Proof If $f(x, y)$ has a local minimum at $P = (a, b)$, then $f(x, y) \geq f(a, b)$ for all (x, y) near (a, b). In particular, there exists $r > 0$ such that $f(x, b) \geq f(a, b)$ if $|x - a| < r$. In other words, $g(x) = f(x, b)$ has a local minimum at $x = a$. By Fermat's Theorem for functions of one variable, either $g'(a) = 0$ or $g'(a)$ does not exist. Since $g'(a) = f_x(a, b)$, we conclude that either $f_x(a, b) = 0$ or $f_x(a, b)$ does not exist. Similarly, $f_y(a, b) = 0$ or $f_y(a, b)$ does not exist. Therefore, $P = (a, b)$ is a critical point. The case of a local maximum is similar. ∎

In most cases, the partial derivatives exist for the functions $f(x, y)$ we encounter. In such cases, finding the critical points amounts to solving the simultaneous equations $f_x(x, y) = 0$ and $f_y(x, y) = 0$.

EXAMPLE 1 Show that $f(x, y) = 11x^2 - 2xy + 2y^2 + 3y$ has one critical point. Use Figure 3 to determine whether it corresponds to a local minimum or maximum.

Solution The partial derivatives are

$$f_x(x, y) = 22x - 2y, \qquad f_y(x, y) = -2x + 4y + 3$$

Set the partial derivatives equal to zero and solve:

$$22x - 2y = 0$$

$$-2x + 4y + 3 = 0$$

By the first equation, $y = 11x$. Substituting $y = 11x$ in the second equation gives

$$-2x + 4(11x) + 3 = 42x + 3 = 0$$

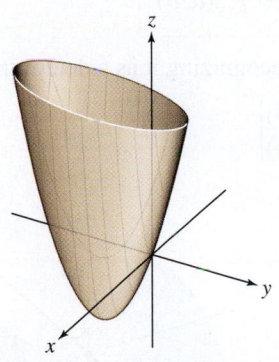

DF **FIGURE 3** Graph of $f(x, y) = 11x^2 - 2xy + 2y^2 + 3y$.

Thus, $x = -\frac{1}{14}$ and $y = -\frac{11}{14}$. There is just one critical point, $P = \left(-\frac{1}{14}, -\frac{11}{14}\right)$. Figure 3 shows that $f(x, y)$ has a local minimum at P (that is, in fact, a global minimum). ∎

As the next example demonstrates, computational software can be of assistance in finding critical points.

EXAMPLE 2 **CAS** Determine the critical points of

$$f(x, y) = \frac{x - y}{2x^2 + 8y^2 + 3}$$

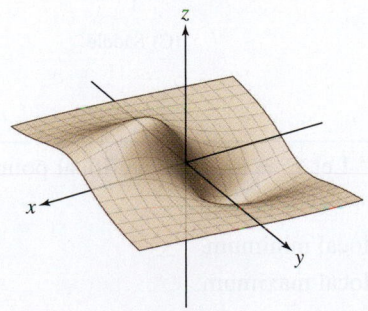

FIGURE 4 Graph of $f(x, y) = \dfrac{x - y}{2x^2 + 8y^2 + 3}$.

Are they local minima or maxima? Refer to Figure 4.

Solution We use a CAS to compute the partial derivatives, obtaining

$$f_x(x, y) = \frac{-2x^2 + 8y^2 + 4xy + 3}{(2x^2 + 8y^2 + 3)^2} \qquad f_y(x, y) = \frac{-2x^2 + 8y^2 - 16xy - 3}{(2x^2 + 8y^2 + 3)^2}$$

To determine where the partial derivatives are zero, we set the numerators equal to zero:

$$-2x^2 + 8y^2 + 4xy + 3 = 0$$

$$-2x^2 + 8y^2 - 16xy - 3 = 0$$

Figure 4 suggests that $f(x, y)$ has a local max with $x > 0$ and a local min with $x < 0$. Using a CAS to solve the resulting system of equations, we have solutions at $\left(\sqrt{\frac{6}{5}}, -\sqrt{\frac{3}{40}}\right)$ and $\left(-\sqrt{\frac{6}{5}}, \sqrt{\frac{3}{40}}\right)$. The former is the local maximum we see in the figure; the latter is the local minimum. ∎

We know that in one variable, a function f may have a point of inflection rather than a local extremum at a critical point. A similar phenomenon occurs in several variables. Each of the functions in Figure 5 has a critical point at $(0, 0)$. However, the function in Figure 5(C) has a **saddle point**, a critical point that is neither a local minimum nor a local maximum. If you stand at the saddle point and begin walking, some directions such as the $+\mathbf{j}$ or $-\mathbf{j}$ directions take you uphill and other directions such as the $+\mathbf{i}$ or $-\mathbf{i}$ directions take you downhill.

As in the one-variable case, there is a Second Derivative Test determining the type of a critical point (a, b) of a function $f(x, y)$ in two variables. This test relies on the sign of the **discriminant** $D = D(a, b)$, defined as follows:

The discriminant is also referred to as the "Hessian determinant."

$$D = D(a, b) = f_{xx}(a, b) f_{yy}(a, b) - f_{xy}^2(a, b)$$

We can remember the formula for the discriminant by recognizing it as a determinant:

$$D = \begin{vmatrix} f_{xx}(a, b) & f_{xy}(a, b) \\ f_{yx}(a, b) & f_{yy}(a, b) \end{vmatrix}$$

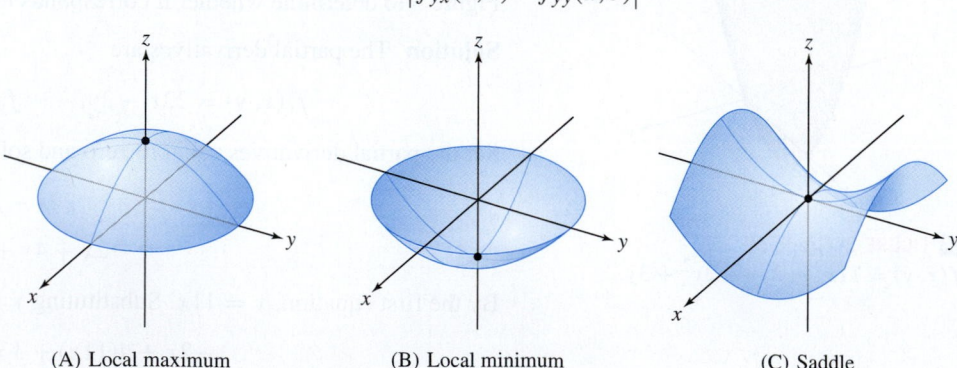

(A) Local maximum   (B) Local minimum   (C) Saddle

FIGURE 5

THEOREM 2 Second Derivative Test for $f(x, y)$ Let $P = (a, b)$ be a critical point of $f(x, y)$. Assume that f_{xx}, f_{yy}, f_{xy} are continuous near P. Then

(i) If $D > 0$ and $f_{xx}(a, b) > 0$, then $f(a, b)$ is a local minimum.

(ii) If $D > 0$ and $f_{xx}(a, b) < 0$, then $f(a, b)$ is a local maximum.

(iii) If $D < 0$, then f has a saddle point at (a, b).

(iv) If $D = 0$, the test is inconclusive.

If $D > 0$, then $f_{xx}(a, b)$ and $f_{yy}(a, b)$ must have the same sign, so the sign of $f_{yy}(a, b)$ also determines whether $f(a, b)$ is a local minimum or a local maximum in the $D > 0$ case.

A proof of this theorem is discussed at the end of this section.

EXAMPLE 3 **Applying the Second Derivative Test** Find the critical points of

$$f(x, y) = (x^2 + y^2)e^{-x}$$

and analyze them using the Second Derivative Test.

Solution

Step 1. **Find the critical points.**

Compute the partial derivatives:

$$f_x(x, y) = -(x^2 + y^2)e^{-x} + 2xe^{-x}$$

$$= (2x - x^2 - y^2)e^{-x}$$

$$f_y(x, y) = 2ye^{-x}$$

Set them equal to zero:

$$(2x - x^2 - y^2)e^{-x} = 0$$

$$2ye^{-x} = 0$$

The solution to the second equation is $y = 0$. Now, substitute $y = 0$ in the first equation to obtain

$$(2x - x^2)e^{-x} = 0$$

The solutions to this equation are $x = 0, 2$, and therefore the critical points are $(0, 0)$ and $(2, 0)$ (Figure 6).

Step 2. **Compute the second-order partials.**

$$f_{xx}(x, y) = \frac{\partial}{\partial x}\left((2x - x^2 - y^2)e^{-x}\right) = (2 - 4x + x^2 + y^2)e^{-x}$$

$$f_{yy}(x, y) = \frac{\partial}{\partial y}(2ye^{-x}) = 2e^{-x}$$

$$f_{xy}(x, y) = f_{yx}(x, y) = \frac{\partial}{\partial x}(2ye^{-x}) = -2ye^{-x}$$

Step 3. **Apply the Second Derivative Test.**

Critical point	f_{xx}	f_{yy}	f_{xy}	Discriminant $D = f_{xx}f_{yy} - f_{xy}^2$	Type
$(0, 0)$	2	2	0	$(2)(2) - 0^2 = 4$	Local minimum since $D > 0$ and $f_{xx} > 0$
$(2, 0)$	$-2e^{-2}$	$2e^{-2}$	0	$(-2e^{-2})(2e^{-2}) - 0^2 = -4e^{-4}$	Saddle since $D < 0$

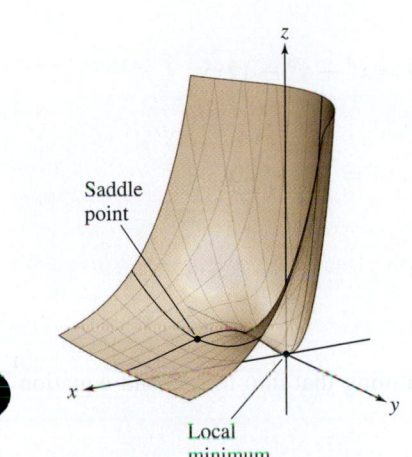

Saddle point

Local minimum

FIGURE 6 Graph of $f(x, y) = (x^2 + y^2)e^{-x}$.

GRAPHICAL INSIGHT We can also read off the type of critical point from the contour map. For example, consider the function depicted in Figure 7. Notice that the level curves encircle the local minimum at P, with f increasing in all directions emanating from P. By contrast, f has a saddle point at Q: The neighborhood near Q is divided into four regions in which $f(x, y)$ alternately increases and decreases.

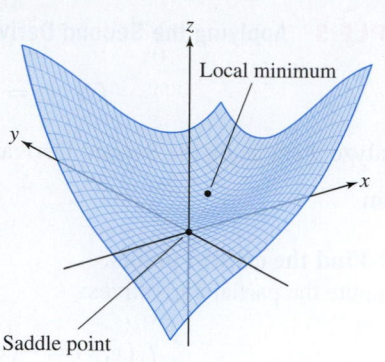

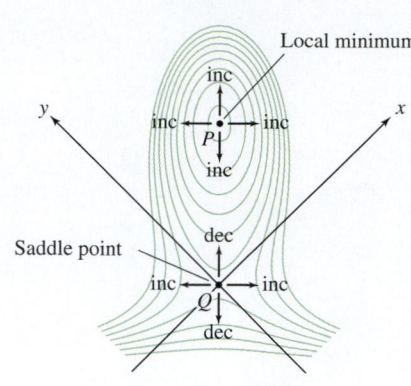

FIGURE 7 $f(x, y) = x^3 + y^3 - 12xy.$

In the next example, we confirm the observations from the Graphical Insight using the Second Derivative Test.

EXAMPLE 4 Analyze the critical points of $f(x, y) = x^3 + y^3 - 12xy$.

Solution We have the following partial derivatives:

$$f_x(x, y) = 3x^2 - 12y, \qquad f_y(x, y) = 3y^2 - 12x$$

Set the partial derivatives equal to zero:

$$3x^2 - 12y = 0$$
$$3y^2 - 12x = 0$$

From the first equation, we obtain $y = \frac{1}{4}x^2$. Substituting that into the second equation and simplifying yields

$$3\left(\frac{1}{4}x^2\right)^2 - 12x = \frac{3}{16}x(x^3 - 64) = 0$$

This equation has solutions $x = 0, 4$. Then, since $y = \frac{1}{4}x^2$, the critical points are $(0, 0)$ and $(4, 4)$.

Now, computing the second partial derivatives, we obtain

$$f_{xx}(x, y) = 6x, \qquad f_{yy}(x, y) = 6y, \qquad f_{xy}(x, y) = -12$$

The Second Derivative Test confirms what we see in Figure 7: f has a local min at $(4, 4)$ and a saddle at $(0, 0)$.

Critical point	f_{xx}	f_{yy}	f_{xy}	Discriminant $D = f_{xx}f_{yy} - f_{xy}^2$	Type
$(0, 0)$	0	0	-12	$(0)(0) - (-12)^2 = -144$	Saddle since $D < 0$
$(4, 4)$	24	24	-12	$(24)(24) - (-12)^2 = 432$	Local minimum since $D > 0$ and $f_{xx} > 0$

∎

EXAMPLE 5 **When the Second Derivative Test Fails** Analyze the critical points of $f(x, y) = 3xy^2 - x^3$.

Solution We have the following partial derivatives:

$$f_x(x, y) = 3y^2 - 3x^2, \qquad f_y(x, y) = 6xy$$

Setting them equal to zero:

$$3y^2 - 3x^2 = 0$$
$$6xy = 0$$

From the second equation, either $x = 0$ or $y = 0$. From the first equation, we find that the only critical point is $(0, 0)$.

Next, we compute the second partial derivatives:

$$f_{xx}(x, y) = -6x, \qquad f_{yy}(x, y) = 6x, \qquad f_{xy}(x, y) = 6y$$

Applying the Second Derivative Test, we obtain

Critical point	f_{xx}	f_{yy}	f_{xy}	Discriminant $D = f_{xx} f_{yy} - f_{xy}^2$	Type
$(0, 0)$	0	0	0	0	No information since $D = 0$

Thus, we need to analyze this critical point by examining the graph more carefully. Consider the vertical trace in the xz-plane obtained by setting $y = 0$. The resulting curve is $z = -x^3$ (Figure 8). When it passes through the origin, it has neither a local maximum nor local minimum, so the critical point $(0, 0)$ is a saddle point, not an extreme point.

Graphs can take on a variety of different shapes at a saddle point. The graph of $f(x, y)$ is called a "monkey saddle" because a monkey can sit on this saddle with room for each of its legs and its tail. ∎

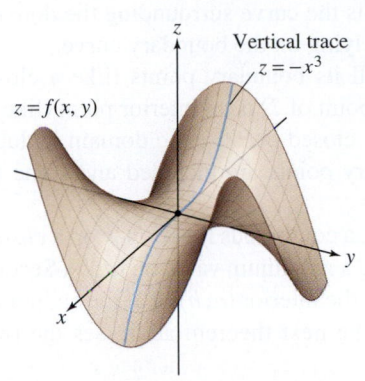

FIGURE 8 Graph of a "monkey saddle" with equation $f(x, y) = 3xy^2 - x^3$.

Global Extrema

Often we are interested in finding the minimum or maximum value of a function f on a given domain $\mathcal{D}$. These are called **global** or **absolute extreme values**. However, global extrema do not always exist. The function $f(x, y) = x + y$ has a maximum value on the unit square $\mathcal{D}_1$ in Figure 9 [the max is $f(1, 1) = 2$], but it has no maximum value on the entire plane $\mathbf{R}^2$.

To state conditions that guarantee the existence of global extrema, we need a few definitions. First, we say that a domain $\mathcal{D}$ is **bounded** if there is a number $M > 0$ such that $\mathcal{D}$ is contained in a disk of radius M centered at the origin. In other words, no point of $\mathcal{D}$ is more than a distance M from the origin [Figures 11(A) and 11(B)]. Next, a point P is called

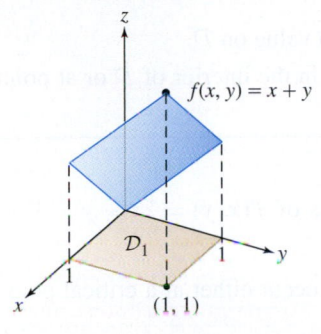

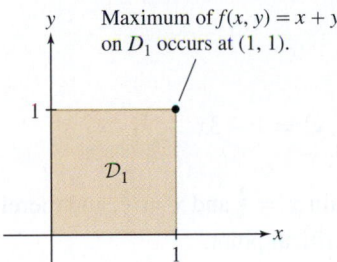

- An **interior point** of $\mathcal{D}$ if $\mathcal{D}$ contains some open disk $D(P, r)$ centered at P.
- A **boundary point** of $\mathcal{D}$ if every disk centered at P contains points in $\mathcal{D}$ and points not in $\mathcal{D}$.

DF FIGURE 9

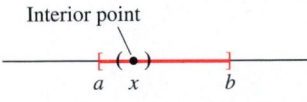

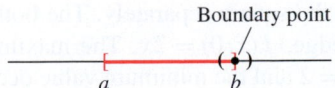

FIGURE 10 Interior and boundary points of an interval $[a, b]$.

> **CONCEPTUAL INSIGHT** To understand the concept of interior and boundary points, think of the familiar case of an interval $I = [a, b]$ in the real line $\mathbf{R}$ (Figure 10). Every point x in the open interval (a, b) is an *interior point* of I (because there exists a small open interval around x entirely contained in I). The two endpoints a and b are *boundary points* (because every open interval containing a or b also contains points not in I).

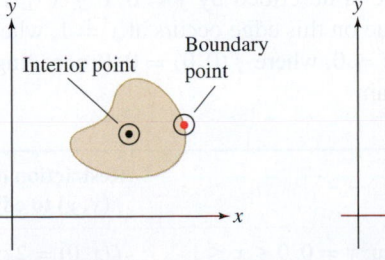

(A) This domain is bounded and closed (contains all boundary points).

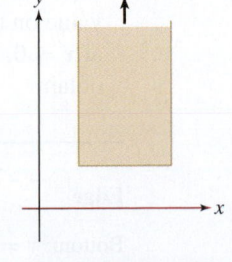

(B) An unbounded domain (contains points arbitrarily far from the origin).

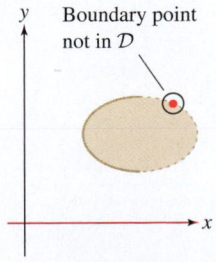

(C) A nonclosed domain (contains some but not all boundary points).

FIGURE 11 Domains in $\mathbf{R}^2$.

The **interior** of $\mathcal{D}$ is the set of all interior points, and the **boundary** of $\mathcal{D}$ is the set of all boundary points. In Figure 11(C), the boundary is the curve surrounding the domain. The interior consists of all points in the domain not lying on the boundary curve.

A domain $\mathcal{D}$ is called **closed** if $\mathcal{D}$ contains all its boundary points (like a closed interval in **R**). A domain $\mathcal{D}$ is called **open** if every point of $\mathcal{D}$ is an interior point (like an open interval in **R**). The domain in Figure 11(A) is closed because the domain includes its boundary curve. In Figure 11(C), some boundary points are included and some are excluded, so the domain is neither open nor closed.

In Section 4.2, we stated two basic results. First, a continuous function f on a *closed, bounded interval* $[a, b]$ takes on both a minimum and a maximum value on $[a, b]$. Second, these extreme values occur either at critical points in the interior (a, b) or at the endpoints. Analogous results are valid in several variables. The next theorem addresses the two-dimensional case.

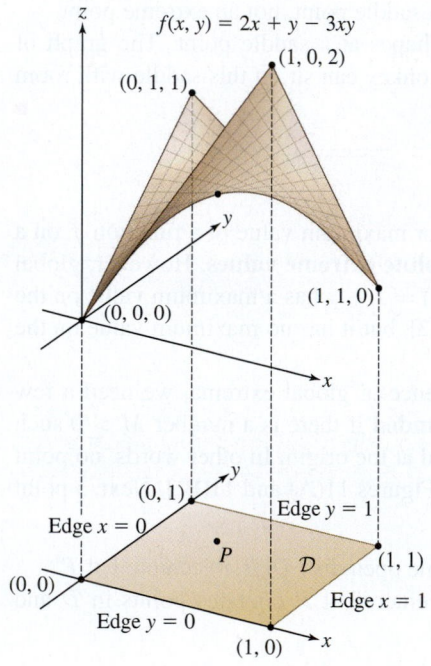

$$f(x, y) = 2x + y - 3xy$$

FIGURE 12

> **THEOREM 3 Existence and Location of Global Extrema**   Let $f(x, y)$ be a continuous function on a closed, bounded domain $\mathcal{D}$ in $\mathbf{R}^2$. Then
>
> **(i)** $f(x, y)$ takes on both a minimum and a maximum value on $\mathcal{D}$.
>
> **(ii)** The extreme values occur either at critical points in the interior of $\mathcal{D}$ or at points on the boundary of $\mathcal{D}$.

EXAMPLE 6   Find the maximum and minimum values of $f(x, y) = 2x + y - 3xy$ on the unit square $\mathcal{D} = \{(x, y) : 0 \le x, y \le 1\}$.

Solution   By Theorem 3, the maximum and minimum occur either at a critical point or on the boundary of the square (Figure 12).

Step 1. **Examine the critical points.**

$$f_x(x, y) = 2 - 3y, \qquad f_y(x, y) = 1 - 3x$$

Setting the partial derivatives equal to zero, we obtain $y = \frac{2}{3}$ and $x = \frac{1}{3}$, and therefore there is a unique critical point $P = \left(\frac{1}{3}, \frac{2}{3}\right)$. At the critical point,

$$f(P) = f\left(\frac{1}{3}, \frac{2}{3}\right) = 2\left(\frac{1}{3}\right) + \left(\frac{2}{3}\right) - 3\left(\frac{1}{3}\right)\left(\frac{2}{3}\right) = \frac{2}{3}$$

Step 2. **Check the boundary.**

We do this by checking each of the four edges of the square separately. The bottom edge is described by $y = 0$, $0 \le x \le 1$. On this edge, $f(x, 0) = 2x$. The maximum value on this edge occurs at $x = 1$, where $f(1, 0) = 2$ and the minimum value occurs at $x = 0$, where $f(0, 0) = 0$. Proceeding in a similar fashion with the other edges, we obtain

Edge	Restriction of $f(x, y)$ to edge	Maximum of $f(x, y)$ on edge	Minimum of $f(x, y)$ on edge
Bottom: $y = 0, 0 \le x \le 1$	$f(x, 0) = 2x$	$f(1, 0) = 2$	$f(0, 0) = 0$
Top: $y = 1, 0 \le x \le 1$	$f(x, 1) = 1 - x$	$f(0, 1) = 1$	$f(1, 1) = 0$
Left: $x = 0, 0 \le y \le 1$	$f(0, y) = y$	$f(0, 1) = 1$	$f(0, 0) = 0$
Right: $x = 1, 0 \le y \le 1$	$f(1, y) = 2 - 2y$	$f(1, 0) = 2$	$f(1, 1) = 0$

Step 3. Compare.

The maximum of f on the boundary is $f(1,0) = 2$. This is greater than the value $f(P) = \frac{2}{3}$ at the critical point, so the maximum of f on the unit square is 2. Similarly, the minimum of f is 0. ∎

EXAMPLE 7 Find the maximum and minimum values of the function $f(x, y) = xy$ on the disk $\mathcal{D} = \{(x, y) : x^2 + y^2 \le 1\}$.

Solution

Step 1. Examine the critical points.

$$f_x(x, y) = y, \qquad f_y(x, y) = x$$

There is a unique critical point $P = (0,0)$ in the interior of the disk, and $f(0,0) = 0$.

Step 2. Check the boundary.

As in Figure 13, we subdivide the boundary into two arcs labeled I and II. The first is given by $y = +\sqrt{1 - x^2}, -1 \le x \le 1$. Restricting f to this part of the boundary, we have $f(x, \sqrt{1 - x^2}) = x\sqrt{1 - x^2}$. Candidates for the maximum and minimum of f on Arc I are obtained by examining $g(x) = x\sqrt{1 - x^2}$ over $[-1, 1]$. To find critical points of g, we start with the derivative,

$$g'(x) = \sqrt{1 - x^2} - x \frac{x}{\sqrt{1 - x^2}}$$

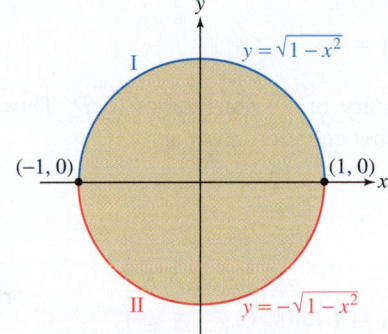

$y = \sqrt{1 - x^2}$

$(-1, 0)$ $(1, 0)$

II $y = -\sqrt{1 - x^2}$

FIGURE 13 Dividing the boundary of the domain into arcs.

Setting $g'(x)$ equal to zero and simplifying, we obtain $1 - 2x^2 = 0$, and therefore $x = \pm\frac{1}{\sqrt{2}}$. Since $y = \sqrt{1 - x^2}$, the corresponding points on Arc I are $\left(\frac{1}{\sqrt{2}}, \frac{1}{\sqrt{2}}\right)$ and $\left(-\frac{1}{\sqrt{2}}, \frac{1}{\sqrt{2}}\right)$. We also obtain candidates for the maximum and minimum from the endpoints of Arc I, $(-1, 0)$ and $(1, 0)$.

Restricting f to Arc II, which is given by $y = -\sqrt{1 - x^2}, -1 \le x \le 1$, our function becomes $f(x, -\sqrt{1 - x^2}) = -x\sqrt{1 - x^2}$. We need to examine $h(x) = -x\sqrt{1 - x^2}$ over $[-1, 1]$. As with Arc I, we obtain the following candidates for the maximum and minimum of f on Arc II: $\left(\frac{1}{\sqrt{2}}, -\frac{1}{\sqrt{2}}\right)$, $\left(-\frac{1}{\sqrt{2}}, -\frac{1}{\sqrt{2}}\right)$, $(-1, 0)$, and $(1, 0)$.

Step 3. Compare.

Evaluating f at the interior critical point and each of the candidate points from the two arcs, we find

$$f(0,0) = 0, \, f\left(\tfrac{1}{\sqrt{2}}, \tfrac{1}{\sqrt{2}}\right) = \tfrac{1}{2}, \, f\left(-\tfrac{1}{\sqrt{2}}, \tfrac{1}{\sqrt{2}}\right) = -\tfrac{1}{2}, \, f\left(\tfrac{1}{\sqrt{2}}, -\tfrac{1}{\sqrt{2}}\right) = -\tfrac{1}{2},$$

$$f\left(-\tfrac{1}{\sqrt{2}}, -\tfrac{1}{\sqrt{2}}\right) = \tfrac{1}{2}, \, f(1,0) = 0, \, f(-1,0) = 0$$

Comparing these values, we see that the maximum value of $\frac{1}{2}$ over the disk occurs at the two boundary points $\left(\frac{1}{\sqrt{2}}, \frac{1}{\sqrt{2}}\right)$ and $\left(-\frac{1}{\sqrt{2}}, -\frac{1}{\sqrt{2}}\right)$, and the minimum value of $-\frac{1}{2}$ occurs at the two boundary points $\left(-\frac{1}{\sqrt{2}}, \frac{1}{\sqrt{2}}\right)$ and $\left(\frac{1}{\sqrt{2}}, -\frac{1}{\sqrt{2}}\right)$. ∎

EXAMPLE 8   Box of Maximum Volume Find the maximum volume of a box inscribed in the tetrahedron bounded by the coordinate planes and the plane $\frac{1}{3}x + y + z = 1$ (Figure 14).

Solution

Step 1. Find a function to be maximized.

Let $P = (x, y, z)$ be the corner of the box lying on the front face of the tetrahedron as in the figure. Then the box has sides of lengths x, y, z and volume $V = xyz$. Using $\frac{1}{3}x + y + z = 1$, or $z = 1 - \frac{1}{3}x - y$, we express V in terms of x and y:

$$V(x, y) = xyz = xy\left(1 - \frac{1}{3}x - y\right) = xy - \frac{1}{3}x^2 y - xy^2$$

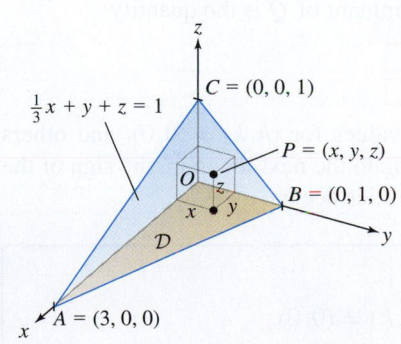

$\frac{1}{3}x + y + z = 1$

$C = (0, 0, 1)$

$P = (x, y, z)$

O z

$B = (0, 1, 0)$

x y

$\mathcal{D}$

$A = (3, 0, 0)$

FIGURE 14 Maximize the volume of the inscribed box.

Our problem is to maximize V, but which domain $\mathcal{D}$ should we choose? We let $\mathcal{D}$ be the shaded triangle $\triangle OAB$ in the xy-plane in Figure 14. Then the corner point $P = (x, y, z)$ of each possible box lies above a point (x, y) in $\mathcal{D}$. Because $\mathcal{D}$ is closed and bounded, the maximum exists and occurs at a critical point inside $\mathcal{D}$ or on the boundary of $\mathcal{D}$.

Step 2. Examine the critical points.

First, compute and simplify the partial derivatives:

$$\frac{\partial V}{\partial x} = y - \frac{2}{3}xy - y^2 = y\left(1 - \frac{2}{3}x - y\right)$$

$$\frac{\partial V}{\partial y} = x - \frac{1}{3}x^2 - 2xy = x\left(1 - \frac{1}{3}x - 2y\right)$$

To find the critical points, we need to solve:

$$y\left(1 - \frac{2}{3}x - y\right) = 0$$

$$x\left(1 - \frac{1}{3}x - 2y\right) = 0$$

If $x = 0$ or $y = 0$, then (x, y) lies on the boundary of $\mathcal{D}$, not interior to $\mathcal{D}$. Thus, assume that x and y are both nonzero. Then the first equation gives us

$$1 - \frac{2}{3}x - y = 0 \quad \Rightarrow \quad y = 1 - \frac{2}{3}x$$

Substituting into the second equation, we obtain

$$1 - \frac{1}{3}x - 2\left(1 - \frac{2}{3}x\right) = 0 \quad \Rightarrow \quad x - 1 = 0 \quad \Rightarrow \quad x = 1$$

For $x = 1$, we have $y = 1 - \frac{2}{3}x = \frac{1}{3}$. Therefore, $\left(1, \frac{1}{3}\right)$ is a critical point, and

$$V\left(1, \frac{1}{3}\right) = (1)\frac{1}{3} - \frac{1}{3}(1)^2\frac{1}{3} - (1)\left(\frac{1}{3}\right)^2 = \frac{1}{9}$$

Step 3. Check the boundary.

We have $V(x, y) = 0$ for all points on the boundary of $\mathcal{D}$ (because the three edges of the boundary are defined by $x = 0$, $y = 0$, and $1 - \frac{1}{3}x - y = 0$). Clearly, then, the maximum occurs at the critical point, and the maximum volume is $\frac{1}{9}$. ■

We close the section with a proof of the Second Derivative Test. The proof is based on completing the square for quadratic forms. A **quadratic form** is a function

$$\boxed{Q(h, k) = ah^2 + 2bhk + ck^2}$$

where a, b, c are constants (not all zero). The discriminant of Q is the quantity

$$\boxed{D = ac - b^2}$$

Some quadratic forms take on only positive values for $(h, k) \neq (0, 0)$, and others take on both positive and negative values. According to the next theorem, the sign of the discriminant determines which of these two possibilities occurs.

THEOREM 4 With $Q(h, k)$ and D as above:

(i) If $D > 0$ and $a > 0$, then $Q(h, k) > 0$ for $(h, k) \neq (0, 0)$.

(ii) If $D > 0$ and $a < 0$, then $Q(h, k) < 0$ for $(h, k) \neq (0, 0)$.

(iii) If $D < 0$, then $Q(h, k)$ takes on both positive and negative values.

To illustrate Theorem 4, consider

$$Q(h,k) = h^2 + 2hk + 2k^2$$

It has a positive discriminant

$$D = (1)(2) - 1 = 1$$

We can see directly that $Q(h,k)$ takes on only positive values for $(h,k) \neq (0,0)$ by writing $Q(h,k)$ as

$$Q(h,k) = (h+k)^2 + k^2$$

Proof Assume first that $a \neq 0$ and rewrite $Q(h,k)$ by "completing the square":

$$Q(h,k) = ah^2 + 2bhk + ck^2 = a\left(h + \frac{b}{a}k\right)^2 + \left(c - \frac{b^2}{a}\right)k^2$$

$$= a\left(h + \frac{b}{a}k\right)^2 + \frac{D}{a}k^2 \qquad \boxed{1}$$

If $D > 0$ and $a > 0$, then $D/a > 0$ and both terms in Eq. (1) are nonnegative. Furthermore, if $Q(h,k) = 0$, then each term in Eq. (1) must equal zero. Thus, $k = 0$ and $h + \frac{b}{a}k = 0$, and then, necessarily, $h = 0$. This shows that $Q(h,k) > 0$ if $(h,k) \neq (0,0)$, and (i) is proved. Part (ii) follows similarly. To prove (iii), note that if $a \neq 0$ and $D < 0$, then the coefficients of the squared terms in Eq. (1) have opposite signs and $Q(h,k)$ takes on both positive and negative values. Finally, if $a = 0$ and $D < 0$, then $Q(h,k) = 2bhk + ck^2$ with $b \neq 0$. In this case, $Q(h,k)$ again takes on both positive and negative values. ∎

Proof of the Second Derivative Test Assume that $f(x,y)$ has a critical point at $P = (a,b)$. We shall analyze $f(x,y)$ by considering the restriction of $f(x,y)$ to the line (Figure 15) through $P = (a,b)$ in the direction of a unit vector $\langle h,k \rangle$:

$$F(t) = f(a+th, b+tk)$$

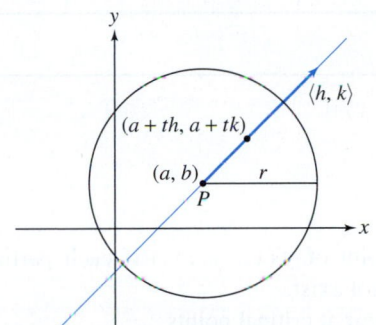

Then $F(0) = f(a,b)$. By the Chain Rule,

$$F'(t) = f_x(a+th, b+tk)h + f_y(a+th, b+tk)k$$

Because P is a critical point, we have $f_x(a,b) = f_y(a,b) = 0$, and therefore

$$F'(0) = f_x(a,b)h + f_y(a,b)k = 0$$

FIGURE 15 Line through P in the direction of $\langle h,k \rangle$.

Thus, $t = 0$ is a critical point of $F(t)$.

Now apply the Chain Rule again:

$$F''(t) = \frac{d}{dt}\left(f_x(a+th, b+tk)h + f_y(a+th, b+tk)k\right)$$

$$= \left(f_{xx}(a+th, b+tk)h^2 + f_{xy}(a+th, b+tk)hk\right)$$

$$+ \left(f_{yx}(a+th, b+tk)kh + f_{yy}(a+th, b+tk)k^2\right)$$

$$= f_{xx}(a+th, b+tk)h^2 + 2f_{xy}(a+th, b+tk)hk + f_{yy}(a+th, b+tk)k^2 \qquad \boxed{2}$$

We see that $F''(t)$ is the value at (h,k) of a quadratic form whose discriminant is equal to $D(a+th, b+tk)$. Here, we set

$$D(r,s) = f_{xx}(r,s)f_{yy}(r,s) - f_{xy}(r,s)^2$$

Note that the discriminant of $f(x,y)$ at the critical point $P = (a,b)$ is $D = D(a,b)$.

Case 1: $D(a,b) > 0$ and $f_{xx}(a,b) > 0$. We must prove that $f(a,b)$ is a local minimum. Consider a small disk of radius r around P (Figure 15). Because the second derivatives are continuous near P, we can choose $r > 0$ so that for every unit vector $\langle h,k \rangle$,

$$D(a+th, b+tk) > 0 \qquad \text{for } |t| < r$$

$$f_{xx}(a+th, b+tk) > 0 \qquad \text{for } |t| < r$$

Then $F''(t)$ is positive for $|t| < r$ by Theorem 4(i). This tells us that $F(t)$ is concave up, and hence $F(0) < F(t)$ if $0 < |t| < |r|$ (see Exercise 74 in Section 4.4). Because

$F(0) = f(a, b)$, we may conclude that $f(a, b)$ is the minimum value of f along each segment of radius r through (a, b). Therefore, $f(a, b)$ is a local minimum value of f as claimed. The case that $D(a, b) > 0$ and $f_{xx}(a, b) < 0$ is similar.

Case 2: $D(a, b) < 0$. For $t = 0$, Eq. (2) yields

$$F''(0) = f_{xx}(a, b)h^2 + 2f_{xy}(a, b)hk + f_{yy}(a, b)k^2$$

Since $D(a, b) < 0$, this quadratic form takes on both positive and negative values by Theorem 4(iii). Choose $\langle h, k \rangle$ for which $F''(0) > 0$. By the Second Derivative Test in one variable, $F(0)$ is a local minimum of $F(t)$, and hence there is a value $r > 0$ such that $F(0) < F(t)$ for all $0 < |t| < r$. However, we can also choose $\langle h, k \rangle$ so that $F''(0) < 0$, in which case, $F(0) > F(t)$ for $0 < |t| < r$ for some $r > 0$. Because $F(0) = f(a, b)$, we conclude that $f(a, b)$ is a local min in some directions and a local max in other directions. Therefore, f has a saddle point at $P = (a, b)$. ∎

14.7 SUMMARY

- We say that $P = (a, b)$ is a *critical point* of $f(x, y)$ if

    - $f_x(a, b) = 0$ or $f_x(a, b)$ does not exist, and
    - $f_y(a, b) = 0$ or $f_y(a, b)$ does not exist.

   In n-variables, $P = (a_1, \ldots, a_n)$ is a critical point of $f(x_1, \ldots, x_n)$ if each partial derivative $f_{x_j}(a_1, \ldots, a_n)$ either is zero or does not exist.
- The local minimum or maximum values of f occur at critical points.
- The *discriminant* of $f(x, y)$ at $P = (a, b)$ is the quantity

$$D(a, b) = f_{xx}(a, b)f_{yy}(a, b) - f_{xy}^2(a, b)$$

- Second Derivative Test: If $P = (a, b)$ is a critical point of $f(x, y)$, then

$$D(a, b) > 0, \quad f_{xx}(a, b) > 0 \quad \Rightarrow \quad f(a, b) \text{ is a local minimum}$$

$$D(a, b) > 0, \quad f_{xx}(a, b) < 0 \quad \Rightarrow \quad f(a, b) \text{ is a local maximum}$$

$$D(a, b) < 0 \quad \Rightarrow \quad \text{saddle point}$$

$$D(a, b) = 0 \quad \Rightarrow \quad \text{test inconclusive}$$

- A point P is an *interior* point of a domain $\mathcal{D}$ if $\mathcal{D}$ contains some open disk $D(P, r)$ centered at P. A point P is a *boundary point* of $\mathcal{D}$ if every open disk $D(P, r)$ contains points in $\mathcal{D}$ and points not in $\mathcal{D}$. The *interior* of $\mathcal{D}$ is the set of all interior points, and the *boundary* is the set of all boundary points. A domain is *closed* if it contains all of its boundary points and *open* if it is equal to its interior.
- Existence and location of global extrema: If f is continuous and $\mathcal{D}$ is closed and bounded, then

    - f takes on both a minimum and a maximum value on $\mathcal{D}$.
    - The extreme values occur either at critical points in the interior of $\mathcal{D}$ or at points on the boundary of $\mathcal{D}$.

   To determine the extreme values, first find the critical points in the interior of $\mathcal{D}$. Then compare the values of f at the critical points with the minimum and maximum values of f on the boundary.

14.7 EXERCISES

Preliminary Questions

1. The functions $f(x, y) = x^2 + y^2$ and $g(x, y) = x^2 - y^2$ both have a critical point at $(0, 0)$. How is the behavior of the two functions at the critical point different?

2. Identify the points indicated in the contour maps as local minima, local maxima, saddle points, or neither (Figure 16).

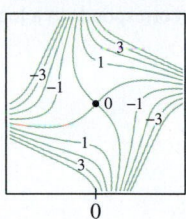
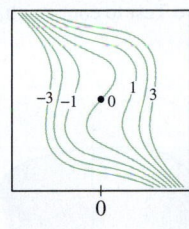
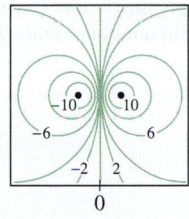

FIGURE 16

3. Let $f(x, y)$ be a continuous function on a domain $\mathcal{D}$ in $\mathbf{R}^2$. Determine which of the following statements are true:

(a) If $\mathcal{D}$ is closed and bounded, then f takes on a maximum value on $\mathcal{D}$.

(b) If $\mathcal{D}$ is neither closed nor bounded, then f does not take on a maximum value of $\mathcal{D}$.

(c) $f(x, y)$ need not have a maximum value on the domain $\mathcal{D}$ defined by $0 \leq x \leq 1, 0 \leq y \leq 1$.

(d) A continuous function takes on neither a minimum nor a maximum value on the open quadrant

$$\{(x, y) : x > 0, y > 0\}$$

Exercises

1. Let $P = (a, b)$ be a critical point of $f(x, y) = x^2 + y^4 - 4xy$.

(a) First use $f_x(x, y) = 0$ to show that $a = 2b$. Then use $f_y(x, y) = 0$ to show that $P = (0, 0)$, $(2\sqrt{2}, \sqrt{2})$, or $(-2\sqrt{2}, -\sqrt{2})$.

(b) Referring to Figure 17, determine the local minima and saddle points of $f(x, y)$ and find the absolute minimum value of $f(x, y)$.

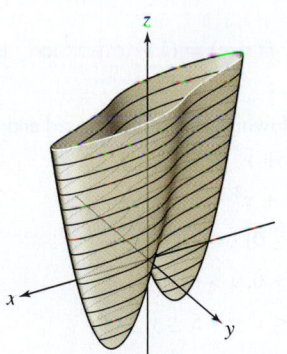

FIGURE 17

2. Find the critical points of the functions

$$f(x, y) = x^2 + 2y^2 - 4y + 6x, \qquad g(x, y) = x^2 - 12xy + y$$

Use the Second Derivative Test to determine the local minimum, local maximum, and saddle points. Match $f(x, y)$ and $g(x, y)$ with their graphs in Figure 18.

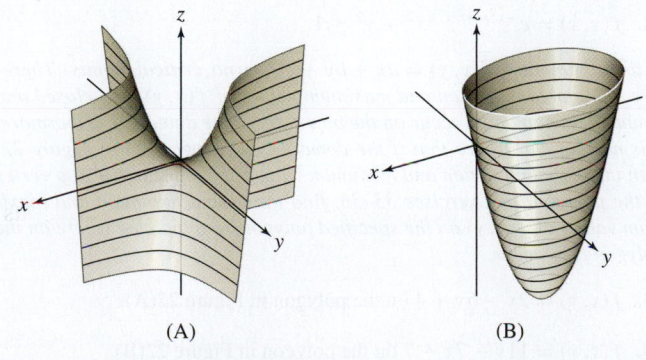

FIGURE 18

3. Find the critical points of

$$f(x, y) = 8y^4 + x^2 + xy - 3y^2 - y^3$$

Use the contour map in Figure 19 to determine their nature (local minimum, local maximum, or saddle point).

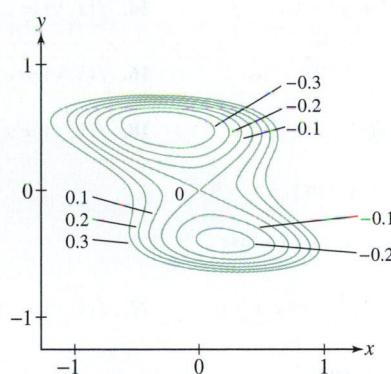

FIGURE 19 Contour map of $f(x, y) = 8y^4 + x^2 + xy - 3y^2 - y^3$.

4. Use the contour map in Figure 20 to determine whether the critical points A, B, C, D are local minima, local maxima, or saddle points.

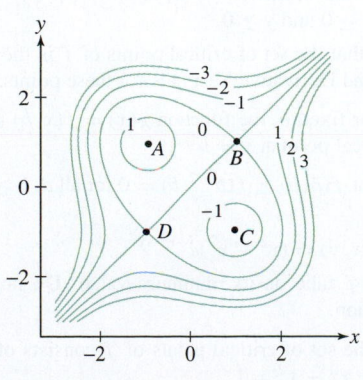

FIGURE 20

5. Let $f(x, y) = y^2 x - yx^2 + xy$.

(a) Show that the critical points (x, y) satisfy the equations

$$y(y - 2x + 1) = 0, \qquad x(2y - x + 1) = 0$$

(b) Show that f has three critical points where $x = 0$ or $y = 0$ (or both) and one critical point where x and y are nonzero.

(c) Use the Second Derivative Test to determine the nature of the critical points.

6. Show that $f(x, y) = \sqrt{x^2 + y^2}$ has one critical point P and that f is nondifferentiable at P. Does f have a minimum, maximum, or saddle point at P?

In Exercises 7–23, find the critical points of the function. Then use the Second Derivative Test to determine whether they are local minima, local maxima, or saddle points (or state that the test fails).

7. $f(x, y) = x^2 + y^2 - xy + x$

8. $f(x, y) = x^3 - xy + y^3$

9. $f(x, y) = x^3 + 2xy - 2y^2 - 10x$

10. $f(x, y) = x^3 y + 12x^2 - 8y$

11. $f(x, y) = 4x - 3x^3 - 2xy^2$

12. $f(x, y) = x^3 + y^4 - 6x - 2y^2$

13. $f(x, y) = x^4 + y^4 - 4xy$

14. $f(x, y) = e^{x^2 - y^2 + 4y}$

15. $f(x, y) = xye^{-x^2 - y^2}$

16. $f(x, y) = e^x - xe^y$

17. $f(x, y) = \sin(x + y) - \cos x$

18. $f(x, y) = x \ln(x + y)$

19. $f(x, y) = \ln x + 2 \ln y - x - 4y$

20. $f(x, y) = (x + y) \ln(x^2 + y^2)$

21. $f(x, y) = x - y^2 - \ln(x + y)$

22. $f(x, y) = (x - y)e^{x^2 - y^2}$

23. $f(x, y) = (x + 3y)e^{y - x^2}$

24. Show that $f(x, y) = x^2$ has infinitely many critical points (as a function of two variables) and that the Second Derivative Test fails for all of them. What is the minimum value of f? Does $f(x, y)$ have any local maxima?

25. Prove that the function $f(x, y) = \frac{1}{3}x^3 + \frac{2}{3}y^{3/2} - xy$ satisfies $f(x, y) \geq 0$ for $x \geq 0$ and $y \geq 0$.

(a) First, verify that the set of critical points of f is the parabola $y = x^2$ and that the Second Derivative Test fails for these points.

(b) Show that for fixed b, the function $g(x) = f(x, b)$ is concave up for $x > 0$ with a critical point at $x = b^{1/2}$.

(c) Conclude that $f(a, b) \geq f(b^{1/2}, b) = 0$ for all $a, b \geq 0$.

26. ✏️ Let $f(x, y) = (x^2 + y^2)e^{-x^2 - y^2}$.

(a) Where does f take on its minimum value? Do not use calculus to answer this question.

(b) Verify that the set of critical points of f consists of the origin $(0, 0)$ and the unit circle $x^2 + y^2 = 1$.

(c) The Second Derivative Test fails for points on the unit circle (this can be checked by some lengthy algebra). Prove, however, that f takes on its maximum value on the unit circle by analyzing the function $g(t) = te^{-t}$ for $t > 0$.

27. ⬛CAS Use a computer algebra system to find a numerical approximation to the critical point of

$$f(x, y) = (1 - x + x^2)e^{y^2} + (1 - y + y^2)e^{x^2}$$

Apply the Second Derivative Test to confirm that it corresponds to a local minimum as in Figure 21.

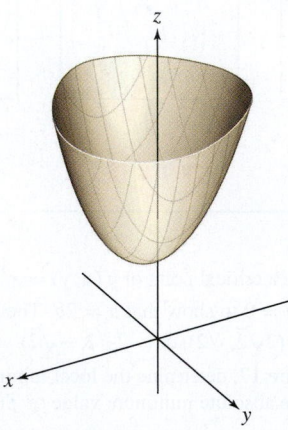

FIGURE 21 Plot of $f(x, y) = (1 - x + x^2)e^{y^2} + (1 - y + y^2)e^{x^2}$.

28. Which of the following domains are closed and which are bounded?

(a) $\{(x, y) \in \mathbf{R}^2 : x^2 + y^2 \leq 1\}$

(b) $\{(x, y) \in \mathbf{R}^2 : x^2 + y^2 < 1\}$

(c) $\{(x, y) \in \mathbf{R}^2 : x \geq 0\}$

(d) $\{(x, y) \in \mathbf{R}^2 : x > 0, y > 0\}$

(e) $\{(x, y) \in \mathbf{R}^2 : 1 \leq x \leq 4, 5 \leq y \leq 10\}$

(f) $\{(x, y) \in \mathbf{R}^2 : x > 0, x^2 + y^2 \leq 10\}$

📝 *In Exercises 29–32, determine the global extreme values of the function on the given set without using calculus.*

29. $f(x, y) = x + y, \quad 0 \leq x \leq 1, \quad 0 \leq y \leq 1$

30. $f(x, y) = 2x - y, \quad 0 \leq x \leq 1, \quad 0 \leq y \leq 3$

31. $f(x, y) = (x^2 + y^2 + 1)^{-1}, \quad 0 \leq x \leq 3, \quad 0 \leq y \leq 5$

32. $f(x, y) = e^{-x^2 - y^2}, \quad x^2 + y^2 \leq 1$

A linear function $f(x, y) = ax + by + c$ has no critical points. Therefore, the global minimum and maximum values of $f(x, y)$ on a closed and bounded domain must occur on the boundary of the domain. Furthermore, it is not difficult to see that if the domain is a polygon, as in Figure 22, then the global minimum and maximum values of f must occur at a vertex of the polygon. In Exercises 33–36, find the global minimum and maximum values of $f(x, y)$ on the specified polygon, and indicate where on the polygon they occur.

33. $f(x, y) = 2x - 6y + 4$ on the polygon in Figure 22(A).

34. $f(x, y) = 11y - 7x + 7$ on the polygon in Figure 22(B).

35. $f(x, y) = 12 + 5y - 20x$ on the polygon in Figure 22(A).

36. $f(x, y) = 3x - 6y - 8$ on the domain where $|x| + |y| \le 3$.

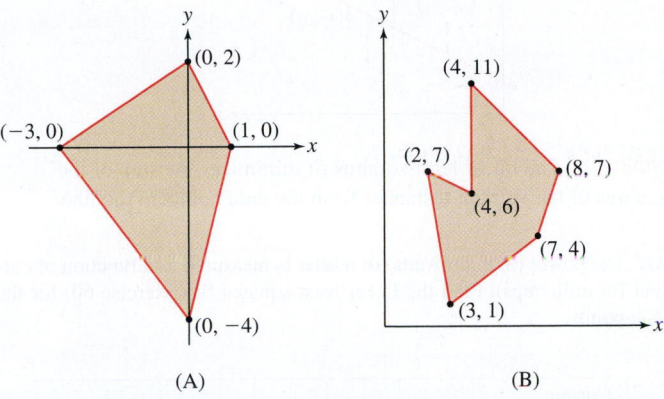

FIGURE 22

37. Assumptions Matter Show that $f(x, y) = xy$ does not have a global minimum or a global maximum on the domain

$$\mathcal{D} = \{(x, y) : 0 < x < 1, 0 < y < 1\}$$

Explain why this does not contradict Theorem 3.

38. Find a continuous function that does not have a global maximum on the domain $\mathcal{D} = \{(x, y) : x + y \ge 0, x + y \le 1\}$. Explain why this does not contradict Theorem 3.

39. Find the maximum of

$$f(x, y) = x + y - x^2 - y^2 - xy$$

on the square, $0 \le x \le 2, 0 \le y \le 2$ (Figure 23).

(a) First, locate the critical point of f in the square, and evaluate f at this point.

(b) On the bottom edge of the square, $y = 0$ and $f(x, 0) = x - x^2$. Find the extreme values of f on the bottom edge.

(c) Find the extreme values of f on the remaining edges.

(d) Find the greatest among the values computed in (a), (b), and (c).

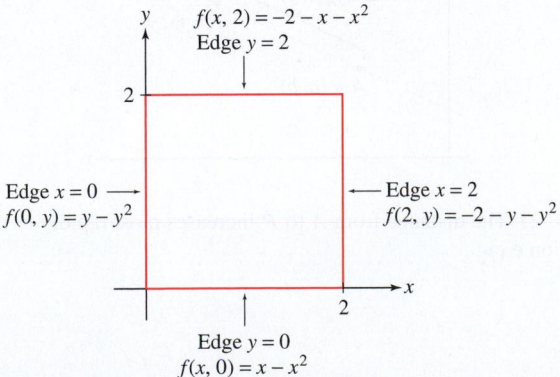

FIGURE 23 The function $f(x, y) = x + y - x^2 - y^2 - xy$ on the boundary segments of the square $0 \le x \le 2, 0 \le y \le 2$.

40. Find the maximum of $f(x, y) = y^2 + xy - x^2$ on the square domain $0 \le x \le 2, 0 \le y \le 2$.

In Exercises 41–49, determine the global extreme values of the function on the given domain.

41. $f(x, y) = x^3 - 2y, \quad 0 \le x \le 1, \quad 0 \le y \le 1$

42. $f(x, y) = 5x - 3y, \quad y \ge x - 2, \quad y \ge -x - 2, \quad y \le 3$

43. $f(x, y) = x^2 + 2y^2, \quad 0 \le x \le 1, \quad 0 \le y \le 1$

44. $f(x, y) = x^3 + x^2y + 2y^2, \quad x, y \ge 0, \quad x + y \le 1$

45. $f(x, y) = x^2 + xy^2 + y^2, \quad x, y \ge 0, \quad x + y \le 1$

46. $f(x, y) = x^3 + y^3 - 3xy, \quad 0 \le x \le 1, \quad 0 \le y \le 1$

47. $f(x, y) = x^2 + y^2 - 2x - 4y, \quad x \ge 0, \quad 0 \le y \le 3, \quad y \ge x$

48. $f(x, y) = (4y^2 - x^2)e^{-x^2-y^2}, \quad x^2 + y^2 \le 2$

49. $f(x, y) = x^2 + 2xy^2, \quad x^2 + y^2 \le 1$

50. Find the maximum volume of a box inscribed in the tetrahedron bounded by the coordinate planes and the plane

$$x + \frac{1}{2}y + \frac{1}{3}z = 1$$

51. Find the volume of the largest box of the type shown in Figure 24, with one corner at the origin and the opposite corner at a point $P = (x, y, z)$ on the paraboloid

$$z = 1 - \frac{x^2}{4} - \frac{y^2}{9} \quad \text{with } x, y, z \ge 0$$

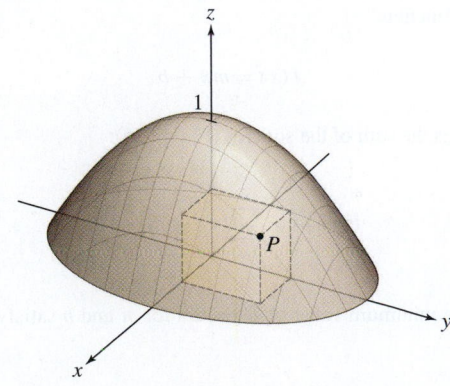

FIGURE 24

52. Find the point on the plane

$$z = x + y + 1$$

closest to the point $P = (1, 0, 0)$. *Hint:* Minimize the square of the distance.

53. Show that the sum of the squares of the distances from a point $P = (c, d)$ to n fixed points $(a_1, b_1), \ldots, (a_n, b_n)$ is minimized when c is the average of the x-coordinates a_i and d is the average of the y-coordinates b_i.

54. Show that the rectangular box (including the top and bottom) with fixed volume $V = 27$ m³ and smallest possible surface area is a cube (Figure 25).

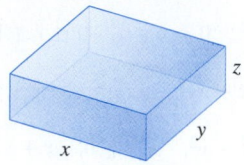

FIGURE 25 Rectangular box with dimensions x, y, z.

55. 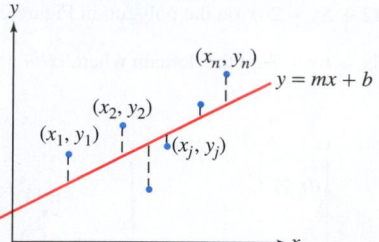 Consider a rectangular box B that has a bottom and sides but no top and has minimal surface area among all boxes with fixed volume $V = 64$ m³.

(a) Do you think B is a cube as in the solution to Exercise 54? If not, how would its shape differ from a cube?

(b) Find the dimensions of B and compare with your response to (a).

56. Find three positive numbers that sum to 150 with the greatest possible product of the three.

57. A 120-m long fence is to be cut into pieces to make three enclosures, each of which is square. How should the fence be cut up in order to minimize the total area enclosed by the fence?

58. A box with a volume of 8 m³ is to be constructed with a gold-plated top, silver-plated bottom, and copper-plated sides. If gold plate costs \$120 per square meter, silver plate costs \$40 per square meter, and copper plate costs \$10 per square meter, find the dimensions that will minimize the cost of the materials for the box.

59. Find the maximum volume of a cylindrical can such that the sum of its height and its circumference is 120 cm.

60. Given n data points $(x_1, y_1), \ldots, (x_n, y_n)$, the **linear least-squares fit** is the linear function

$$f(x) = mx + b$$

that minimizes the sum of the squares (Figure 26):

$$E(m, b) = \sum_{j=1}^{n} (y_j - f(x_j))^2$$

Show that the minimum value of E occurs for m and b satisfying the two equations

$$m \left(\sum_{j=1}^{n} x_j \right) + bn = \sum_{j=1}^{n} y_j$$

$$m \sum_{j=1}^{n} x_j^2 + b \sum_{j=1}^{n} x_j = \sum_{j=1}^{n} x_j y_j$$

FIGURE 26 The linear least-squares fit minimizes the sum of the squares of the vertical distances from the data points to the line.

61. The power (in microwatts) of a laser is measured as a function of current (in milliamps). Find the linear least-squares fit (Exercise 60) for the data points.

Current (milliamps)	1.0	1.1	1.2	1.3	1.4	1.5
Laser power (microwatts)	0.52	0.56	0.82	0.78	1.23	1.50

62. Let $A = (a, b)$ be a fixed point in the plane, and let $f_A(P)$ be the distance from A to the point $P = (x, y)$. For $P \neq A$, let $\mathbf{e}_{AP}$ be the unit vector pointing from A to P (Figure 27):

$$\mathbf{e}_{AP} = \frac{\overrightarrow{AP}}{\|\overrightarrow{AP}\|}$$

Show that

$$\nabla f_A(P) = \mathbf{e}_{AP}$$

Note that we can derive this result without calculation: Because $\nabla f_A(P)$ points in the direction of maximal increase, it must point directly away from A at P, and because the distance $f_A(x, y)$ increases at a rate of 1 as you move away from A along the line through A and P, $\nabla f_A(P)$ must be a unit vector.

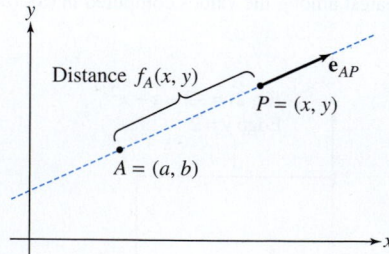

FIGURE 27 The distance from A to P increases most rapidly in the direction $\mathbf{e}_{AP}$.

Further Insights and Challenges

63. In this exercise, we prove that for all $x, y \geq 0$:

$$\frac{1}{\alpha}x^\alpha + \frac{1}{\beta}x^\beta \geq xy$$

where $\alpha \geq 1$ and $\beta \geq 1$ are numbers such that $\alpha^{-1} + \beta^{-1} = 1$. To do this, we prove that the function

$$f(x, y) = \alpha^{-1}x^\alpha + \beta^{-1}y^\beta - xy$$

satisfies $f(x, y) \geq 0$ for all $x, y \geq 0$.

(a) Show that the set of critical points of $f(x, y)$ is the curve $y = x^{\alpha-1}$ (Figure 28). Note that this curve can also be described as $x = y^{\beta-1}$. What is the value of $f(x, y)$ at points on this curve?

(b) Verify that the Second Derivative Test fails. Show, however, that for fixed $b > 0$, the function $g(x) = f(x, b)$ is concave up with a critical point at $x = b^{\beta-1}$.

(c) Conclude that for all $x > 0$, $f(x, b) \geq f(b^{\beta-1}, b) = 0$.

64. The following problem was posed by Pierre de Fermat: Given three points $A = (a_1, a_2)$, $B = (b_1, b_2)$, and $C = (c_1, c_2)$ in the plane, find the point $P = (x, y)$ that minimizes the sum of the distances

$$f(x, y) = AP + BP + CP$$

Let $\mathbf{e}, \mathbf{f}, \mathbf{g}$ be the unit vectors pointing from P to the points A, B, C as in Figure 29.

(a) Use Exercise 62 to show that the condition $\nabla f(P) = \mathbf{0}$ is equivalent to

$$\mathbf{e} + \mathbf{f} + \mathbf{g} = \mathbf{0} \qquad \boxed{3}$$

(b) Show that $f(x, y)$ is differentiable except at points A, B, C. Conclude that the minimum of $f(x, y)$ occurs either at a point P satisfying Eq. (3) or at one of the points A, B, or C.

(c) Prove that Eq. (3) holds if and only if P is the **Fermat point**, defined as the point P for which the angles between the segments $\overline{AP}, \overline{BP}, \overline{CP}$ are all 120° (Figure 29).

(d) Show that the Fermat point does not exist if one of the angles in $\triangle ABC$ is greater than 120°. Where does the minimum occur in this case?

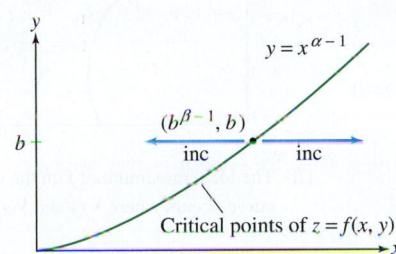

FIGURE 28 The critical points of $f(x, y) = \alpha^{-1}x^\alpha + \beta^{-1}y^\beta - xy$ form a curve $y = x^{\alpha-1}$.

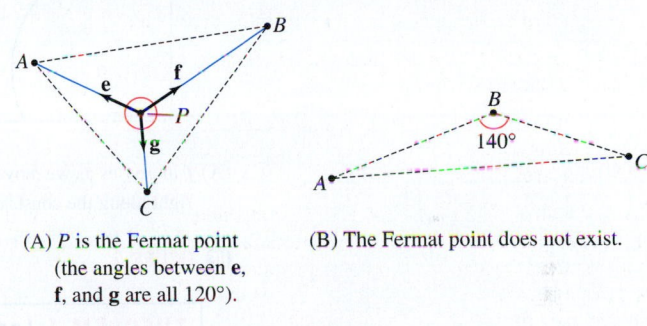

(A) P is the Fermat point (the angles between $\mathbf{e}$, $\mathbf{f}$, and $\mathbf{g}$ are all 120°).

(B) The Fermat point does not exist.

FIGURE 29

14.8 Lagrange Multipliers: Optimizing with a Constraint

Some optimization problems involve finding the extreme values of a function $f(x, y)$ subject to a constraint $g(x, y) = 0$. Suppose that we want to find the point on the line $2x + 3y = 6$ closest to the origin (Figure 1). The distance from (x, y) to the origin is $f(x, y) = \sqrt{x^2 + y^2}$, so our problem is

$$\text{Minimize } f(x, y) = \sqrt{x^2 + y^2} \quad \text{subject to} \quad g(x, y) = 2x + 3y - 6 = 0$$

We are not seeking the minimum value of $f(x, y)$ (which is 0), but rather the minimum among all points (x, y) that lie on the line.

The method of **Lagrange multipliers** is a general procedure for solving optimization problems with a constraint. Here is a description of the main idea.

DF FIGURE 1 Finding the minimum of

$$f(x, y) = \sqrt{x^2 + y^2}$$

on the line $2x + 3y = 6$.

GRAPHICAL INSIGHT   Imagine standing at point Q in Figure 2(A). We want to increase the value of f while remaining on the constraint curve $g(x, y) = 0$. The gradient vector ∇f_Q points in the direction of *maximum* increase, but we cannot move in the gradient direction because that would take us off the constraint curve. However, the gradient points to the right, so we can still increase f somewhat by moving to the right along the constraint curve.

We keep moving to the right until we arrive at the point P, where ∇f_P is orthogonal to the constraint curve [Figure 2(B)]. Once at P, we cannot increase f further by moving either to the right or to the left along the constraint curve. Thus, $f(P)$ is a local maximum subject to the constraint.

Now, the vector ∇g_P is also orthogonal to the constraint curve because it is the gradient of $g(x, y)$ at P and therefore is orthogonal to the level curve through P. Thus, ∇f_P and ∇g_P are parallel. In other words, $\nabla f_P = \lambda \nabla g_P$ for some scalar λ (called a **Lagrange multiplier**). Graphically, this means that a local max subject to the constraint occurs at points P where the level curves of f and g are tangent. The same holds for a local min subject to a constraint.

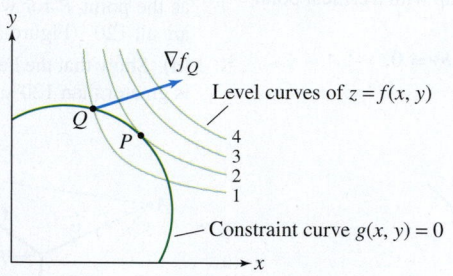

(A) f increases as we move to the right along the constraint curve.

(B) The local maximum of f on the constraint curve occurs where ∇f_P and ∇g_P are parallel.

DF FIGURE 2

THEOREM 1 Lagrange Multipliers   Assume that $f(x, y)$ and $g(x, y)$ are differentiable functions. If $f(x, y)$ has a local minimum or a local maximum on the constraint curve $g(x, y) = 0$ at $P = (a, b)$, and if $\nabla g_P \neq \mathbf{0}$, then there is a scalar λ such that

$$\boxed{\nabla f_P = \lambda \nabla g_P}   \qquad \boxed{1}$$

In Theorem 1, the assumption $\nabla g_P \neq \mathbf{0}$ guarantees (by the Implicit Function Theorem of advanced calculus) that we can parametrize the curve $g(x, y) = 0$ near P by a path $\mathbf{r}(t)$ such that $\mathbf{r}(0) = P$ and $\mathbf{r}'(0) \neq \mathbf{0}$.

Proof   Let $\mathbf{r}(t)$ be a parametrization of the constraint curve $g(x, y) = 0$ near P, chosen so that $\mathbf{r}(0) = P$ and $\mathbf{r}'(0) \neq \mathbf{0}$. Then $f(\mathbf{r}(0)) = f(P)$, and by assumption, $f(\mathbf{r}(t))$ has a local min or max at $t = 0$. Thus, $t = 0$ is a critical point of $f(\mathbf{r}(t))$ and

$$\underbrace{\left. \frac{d}{dt} f(\mathbf{r}(t)) \right|_{t=0} = \nabla f_P \cdot \mathbf{r}'(0)}_{\text{Chain Rule}} = 0$$

This shows that ∇f_P is orthogonal to the tangent vector $\mathbf{r}'(0)$ to the curve $g(x, y) = 0$. The gradient ∇g_P is also orthogonal to $\mathbf{r}'(0)$ [because ∇g_P is orthogonal to the level curve $g(x, y) = 0$ at P]. We conclude that ∇f_P and ∇g_P are parallel, and hence ∇f_P is a multiple of ∇g_P as claimed.   ∎

We refer to Eq. (1) as the **Lagrange condition**. When we write this condition in terms of components, we obtain the **Lagrange equations**:

$$f_x(a, b) = \lambda g_x(a, b)$$
$$f_y(a, b) = \lambda g_y(a, b)$$

A point $P = (a, b)$ satisfying these equations is called a **critical point** for the optimization problem with constraint and $f(a, b)$ is called a **critical value**.

EXAMPLE 1 Find the extreme values of $f(x, y) = 2x + 5y$ on the ellipse

$$\left(\frac{x}{4}\right)^2 + \left(\frac{y}{3}\right)^2 = 1$$

Solution

Step 1. Write out the Lagrange equations.
The constraint curve is $g(x, y) = 0$, where $g(x, y) = (x/4)^2 + (y/3)^2 - 1$. We have

$$\nabla f = \langle 2, 5 \rangle, \qquad \nabla g = \left\langle \frac{x}{8}, \frac{2y}{9} \right\rangle$$

The Lagrange equations $\nabla f_P = \lambda \nabla g_P$ are

$$\langle 2, 5 \rangle = \lambda \left\langle \frac{x}{8}, \frac{2y}{9} \right\rangle \quad \Rightarrow \quad 2 = \frac{\lambda x}{8}, \qquad 5 = \frac{\lambda(2y)}{9} \qquad \boxed{2}$$

Step 2. Solve for λ in terms of x and y.
Equation (2) gives us two equations for λ:

$$\lambda = \frac{16}{x}, \qquad \lambda = \frac{45}{2y} \qquad \boxed{3}$$

To justify dividing by x and y, note that x and y must be nonzero, because $x = 0$ or $y = 0$ would violate Eq. (2).

Step 3. Solve for x and y using the constraint.
The two expressions for λ must be equal, so we obtain $\dfrac{16}{x} = \dfrac{45}{2y}$ or $y = \dfrac{45}{32}x$. Now substitute this in the constraint equation and solve for x:

$$\left(\frac{x}{4}\right)^2 + \left(\frac{\frac{45}{32}x}{3}\right)^2 = 1$$

$$x^2 \left(\frac{1}{16} + \frac{225}{1024}\right) = x^2 \left(\frac{289}{1024}\right) = 1$$

Thus, $x = \pm\sqrt{\dfrac{1024}{289}} = \pm\dfrac{32}{17}$, and since $y = \dfrac{45x}{32}$, the critical points are $P = \left(\dfrac{32}{17}, \dfrac{45}{17}\right)$ and $Q = \left(-\dfrac{32}{17}, -\dfrac{45}{17}\right)$.

Step 4. Calculate the critical values.

$$f(P) = f\left(\frac{32}{17}, \frac{45}{17}\right) = 2\left(\frac{32}{17}\right) + 5\left(\frac{45}{17}\right) = 17$$

and $f(Q) = -17$. We conclude that the maximum of $f(x, y)$ on the ellipse is 17 and the minimum is -17 (Figure 3). ■

Level curve of $f(x, y) = 2x + 5y$

Constraint curve $g(x, y) = 0$

DF **FIGURE 3** The min and max occur where a level curve of f is tangent to the constraint curve $g(x, y) = 0$.

Assumptions Matter According to Theorem 3 in Section 14.7, a continuous function on a closed, bounded domain takes on extreme values. This tells us that if the constraint curve is closed and bounded (as in the previous example, where the constraint curve is an ellipse), then every continuous function $f(x, y)$ takes on both a minimum and a maximum value subject to the constraint. Be aware, however, that extreme values need not exist if the constraint curve is not bounded. For example, the constraint $x - y = 0$

is a line which is unbounded. The function $f(x, y) = xy^2$ has neither a minimum nor a maximum subject to $x - y = 0$ because every point (a, a) satisfies the constraint, yet $f(a, a) = a^3$ can be arbitrarily large positive (so there is no maximum) and arbitrarily large negative (so there is no minimum).

EXAMPLE 2 **Cobb–Douglas Production Function** By investing x units of labor and y units of capital, a watch manufacturer can produce $P(x, y) = 50x^{0.4}y^{0.6}$ watches. Find the maximum number of watches that can be produced on a budget of \$20,000 if labor costs \$100 per unit and capital costs \$200 per unit.

Solution The total cost of x units of labor and y units of capital is $100x + 200y$. Our task is to maximize the function $P(x, y) = 50x^{0.4}y^{0.6}$ subject to the following budget constraint (Figure 4):

$$g(x, y) = 100x + 200y - 20{,}000 = 0 \qquad \boxed{4}$$

Step 1. Write out the Lagrange equations.

$$P_x(x, y) = \lambda g_x(x, y): \quad 20x^{-0.6}y^{0.6} = 100\lambda$$

$$P_y(x, y) = \lambda g_y(x, y): \quad 30x^{0.4}y^{-0.4} = 200\lambda$$

Step 2. Solve for λ in terms of x and y.
These equations yield two expressions for λ that must be equal:

$$\lambda = \frac{1}{5}\left(\frac{y}{x}\right)^{0.6} = \frac{3}{20}\left(\frac{y}{x}\right)^{-0.4} \qquad \boxed{5}$$

Step 3. Solve for x and y using the constraint.
Multiply Eq. (5) by $5(y/x)^{0.4}$ to obtain $y/x = 15/20$, or $y = \frac{3}{4}x$. Then substitute in Eq. (4):

$$100x + 200y = 100x + 200\left(\frac{3}{4}x\right) = 20{,}000 \quad \Rightarrow \quad 250x = 20{,}000$$

We obtain $x = \dfrac{20{,}000}{250} = 80$ and $y = \frac{3}{4}x = 60$. The critical point is $A = (80, 60)$.

Step 4. Calculate the critical values.
Since $P(x, y)$ is increasing as a function of x and y, ∇P points to the northeast, and it is clear that $P(x, y)$ takes on a maximum value at A (Figure 4). The maximum is $P(80, 60) = 50(80)^{0.4}(60)^{0.6} = 3365.87$, or roughly 3365 watches, with a cost per watch of $\dfrac{20{,}000}{3365}$ or about \$5.94. ∎

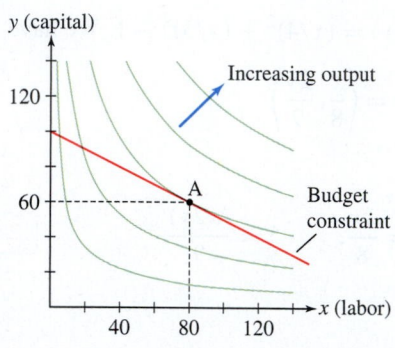

FIGURE 4 Contour plot of the Cobb–Douglas production function $P(x, y) = 50x^{0.4}y^{0.6}$. The level curves of a production function are called *isoquants*.

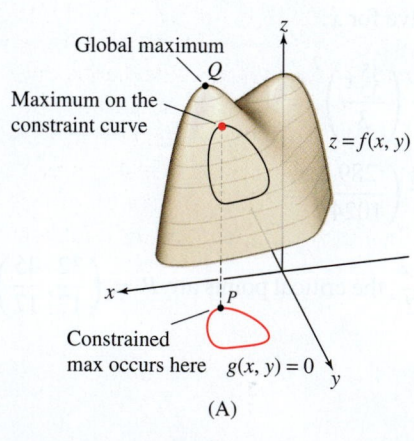

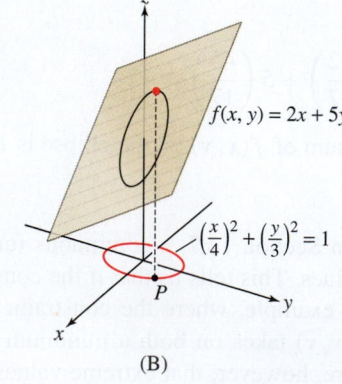

FIGURE 5

> **GRAPHICAL INSIGHT** In an ordinary optimization problem without constraint, the global maximum value is the height of the highest point on the surface $z = f(x, y)$ [point Q in Figure 5(A)]. When a constraint is given, we restrict our attention to the curve on the surface lying above the constraint curve $g(x, y) = 0$. The maximum value subject to the constraint is the height of the highest point on this curve. Figure 5(B) shows the optimization problem solved in Example 1.

The method of Lagrange multipliers is valid in any number of variables. Imagine, for instance, that we are trying to find the maximum temperature $f(x, y, z)$ for points on a surface S in 3-space given by $g(x, y, z) = 0$, as in Figure 6. This surface is a level surface for the function g, and therefore, ∇g_P is perpendicular to the tangent planes to this surface at every point P on the surface. Consider the level surfaces for temperature, which we have called the isotherms. They appear as surfaces in 3-space, and their intersections with S yield the level sets of temperature on S. If, as in the figure, the temperature increases as we move to the right on the surface, then it is apparent that the maximum temperature for the surface occurs when the last isotherm intersects the surface in just a single point and hence that isotherm is tangent to the surface. That is to say, the last isotherm and the surface share the same tangent plane at their single point of intersection.

However, as we know, ∇f_P is always perpendicular to the tangent plane to the level surfaces for f at each point P on a level surface. So at the hottest point on the surface, ∇g_P and ∇f_P are both perpendicular to the same tangent plane. Hence, they must be parallel, and one must be a multiple of the other. Thus, at that point, $\nabla f_P = \lambda \nabla g_P$. A similar argument holds for the minimum temperature on the surface.

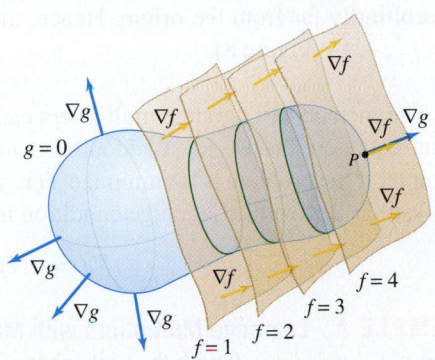

FIGURE 6 As we move to the right, temperature increases, attaining a maximum on the surface of $f = 4$ at P.

DF **FIGURE 7** As we move to the right, temperature increases and then decreases.

There is one other situation to consider. Imagine that as we move left to right across our surface, temperature first increases to $f = 4$ and then it decreases again, as in Figure 7. There is a collection of points with the maximal temperature of $f = 4$. In this case, ∇f must point to the right on isotherms that are to the left of $f = 4$ since this is the direction of increasing temperature, and ∇f must point to the left on isotherms that are to the right of $f = 4$ since this is the direction of increasing temperature. Hence, in order for the right-pointing gradient vectors to become left-pointing gradient vectors in a continuous manner, they must be equal to $\mathbf{0}$ on the $f = 4$ isotherm. This makes sense, since on that isotherm there is no direction of increasing temperature. So for all of the points on the surface with maximal temperature, of which there are many, the equation $\nabla f = \lambda \nabla g$ is satisfied, but by taking $\lambda = 0$.

In the next example, we consider a problem in three variables.

EXAMPLE 3 **Lagrange Multipliers in Three Variables** Find the point on the plane $\dfrac{x}{2} + \dfrac{y}{4} + \dfrac{z}{4} = 1$ closest to the origin in $\mathbf{R}^3$.

Solution Our task is to minimize the distance $d = \sqrt{x^2 + y^2 + z^2}$ subject to the constraint $\dfrac{x}{2} + \dfrac{y}{4} + \dfrac{z}{4} = 1$. But, finding the minimum distance d is the same as finding the minimum square of the distance d^2, so our problem can be stated:

$$\text{Minimize } f(x, y, z) = x^2 + y^2 + z^2 \quad \text{subject to} \quad g(x, y, z) = \frac{x}{2} + \frac{y}{4} + \frac{z}{4} - 1 = 0$$

The Lagrange condition is

$$\underbrace{\langle 2x, 2y, 2z \rangle}_{\nabla f} = \lambda \underbrace{\left\langle \frac{1}{2}, \frac{1}{4}, \frac{1}{4} \right\rangle}_{\nabla g}$$

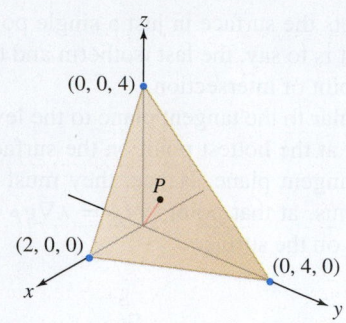

FIGURE 8 Point P closest to the origin on the plane.

This yields

$$\lambda = 4x = 8y = 8z \quad \Rightarrow \quad z = y = \frac{x}{2}$$

Substituting in the constraint equation, we obtain

$$\frac{x}{2} + \frac{y}{4} + \frac{z}{4} = \frac{2z}{2} + \frac{z}{4} + \frac{z}{4} = \frac{3z}{2} = 1 \quad \Rightarrow \quad z = \frac{2}{3}$$

Thus, $x = 2z = \frac{4}{3}$ and $y = z = \frac{2}{3}$. This critical point must correspond to the minimum of f. There is no maximum of f on the plane since there are points on the plane that are arbitrarily far from the origin. Hence, the point on the plane closest to the origin is $P = \left(\frac{4}{3}, \frac{2}{3}, \frac{2}{3}\right)$ (Figure 8). ■

The method of Lagrange multipliers can be used when there is more than one constraint equation, but we must add another multiplier for each additional constraint. For example, if the problem is to minimize $f(x, y, z)$ subject to constraints $g(x, y, z) = 0$ and $h(x, y, z) = 0$, then the Lagrange condition is

$$\nabla f = \lambda \nabla g + \mu \nabla h$$

EXAMPLE 4   Lagrange Multipliers with Multiple Constraints   The intersection of the plane $x + \frac{1}{2}y + \frac{1}{3}z = 0$ with the unit sphere $x^2 + y^2 + z^2 = 1$ is a great circle (Figure 9). Find the point on this great circle with the greatest x-coordinate.

The intersection of a sphere with a plane through its center is called a **great circle**.

Solution   Our task is to maximize the function $f(x, y, z) = x$ subject to the two constraint equations

$$g(x, y, z) = x + \frac{1}{2}y + \frac{1}{3}z = 0, \qquad h(x, y, z) = x^2 + y^2 + z^2 - 1 = 0$$

The Lagrange condition is

$$\nabla f = \lambda \nabla g + \mu \nabla h$$

$$\langle 1, 0, 0 \rangle = \lambda \left\langle 1, \frac{1}{2}, \frac{1}{3} \right\rangle + \mu \langle 2x, 2y, 2z \rangle$$

Note that μ cannot be zero, since, if it were, the Lagrange condition would become $\langle 1, 0, 0 \rangle = \lambda \langle 1, \frac{1}{2}, \frac{1}{3} \rangle$, and this equation is not satisfied for any value of λ. Now, the Lagrange condition gives us three equations:

$$\lambda + 2\mu x = 1, \qquad \frac{1}{2}\lambda + 2\mu y = 0, \qquad \frac{1}{3}\lambda + 2\mu z = 0$$

The last two equations yield $\lambda = -4\mu y$ and $\lambda = -6\mu z$. Because $\mu \neq 0$,

$$-4\mu y = -6\mu z \quad \Rightarrow \quad y = \frac{3}{2}z$$

Now use this relation in the first constraint equation:

$$x + \frac{1}{2}y + \frac{1}{3}z = x + \frac{1}{2}\left(\frac{3}{2}z\right) + \frac{1}{3}z = 0 \quad \Rightarrow \quad x = -\frac{13}{12}z$$

Finally, we can substitute in the second constraint equation:

$$x^2 + y^2 + z^2 - 1 = \left(-\frac{13}{12}z\right)^2 + \left(\frac{3}{2}z\right)^2 + z^2 - 1 = 0$$

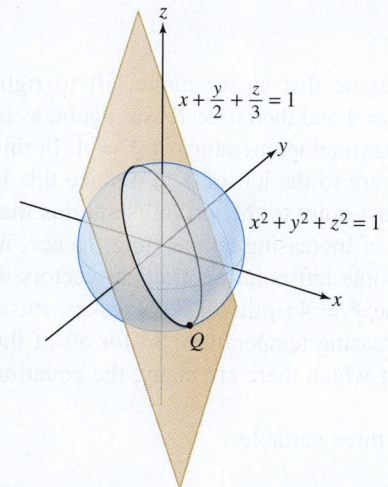

FIGURE 9 The plane intersects the sphere in a great circle. Q is the point on this great circle with the greatest x-coordinate.

to obtain $\frac{637}{144}z^2 = 1$ or $z = \pm\frac{12}{7\sqrt{13}}$. Since $x = -\frac{13}{12}z$ and $y = \frac{3}{2}z$, the critical points are

$$P = \left(-\frac{\sqrt{13}}{7}, \frac{18}{7\sqrt{13}}, \frac{12}{7\sqrt{13}}\right), \qquad Q = \left(\frac{\sqrt{13}}{7}, -\frac{18}{7\sqrt{13}}, -\frac{12}{7\sqrt{13}}\right)$$

The critical point with the greatest x-coordinate [the maximum value of $f(x, y, z)$] is Q with x-coordinate $\dfrac{\sqrt{13}}{7} \approx 0.515$. ∎

14.8 SUMMARY

- Method of Lagrange multipliers: The local extreme values of $f(x, y)$ subject to a constraint $g(x, y) = 0$ occur at points P (called critical points) satisfying the Lagrange condition $\nabla f_P = \lambda \nabla g_P$. This condition is equivalent to the *Lagrange equations*

$$f_x(x, y) = \lambda g_x(x, y), \qquad f_y(x, y) = \lambda g_y(x, y)$$

- If the constraint curve $g(x, y) = 0$ is bounded [e.g., if $g(x, y) = 0$ is a circle or ellipse], then global minimum and maximum values of f subject to the constraint exist.
- Lagrange condition for a function of three variables $f(x, y, z)$ subject to two constraints $g(x, y, z) = 0$ and $h(x, y, z) = 0$:

$$\nabla f = \lambda \nabla g + \mu \nabla h$$

14.8 EXERCISES

Preliminary Questions

1. Suppose that the maximum of $f(x, y)$ subject to the constraint $g(x, y) = 0$ occurs at a point $P = (a, b)$ such that $\nabla f_P \neq 0$. Which of the following statements is true?

(a) ∇f_P is tangent to $g(x, y) = 0$ at P.

(b) ∇f_P is orthogonal to $g(x, y) = 0$ at P.

2. Figure 10 shows a constraint $g(x, y) = 0$ and the level curves of a function f. In each case, determine whether f has a local minimum, a local maximum, or neither at the labeled point.

3. On the contour map in Figure 11:

(a) Identify the points where $\nabla f = \lambda \nabla g$ for some scalar λ.

(b) Identify the minimum and maximum values of $f(x, y)$ subject to $g(x, y) = 0$.

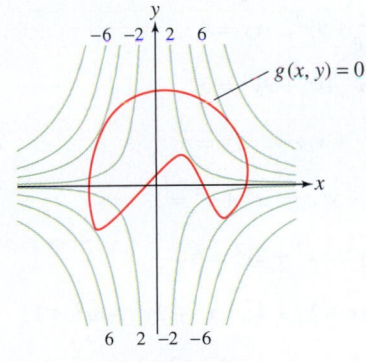

Contour plot of $f(x, y)$
(contour interval 2)

FIGURE 11

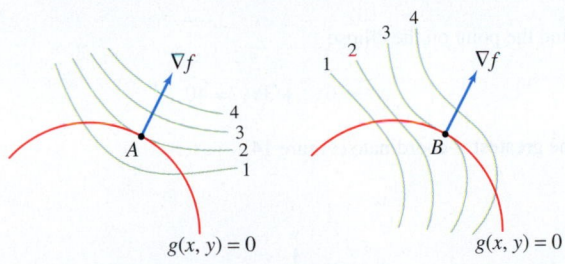

FIGURE 10

Exercises

In this exercise set, use the method of Lagrange multipliers unless otherwise stated.

1. Find the extreme values of the function $f(x, y) = 2x + 4y$ subject to the constraint $g(x, y) = x^2 + y^2 - 5 = 0$.

(a) Show that the Lagrange equation $\nabla f = \lambda \nabla g$ gives $\lambda x = 1$ and $\lambda y = 2$.

(b) Show that these equations imply $\lambda \neq 0$ and $y = 2x$.

(c) Use the constraint equation to determine the possible critical points (x, y).

(d) Evaluate $f(x, y)$ at the critical points and determine the minimum and maximum values.

2. Find the extreme values of $f(x, y) = x^2 + 2y^2$ subject to the constraint $g(x, y) = 4x - 6y = 25$.

(a) Show that the Lagrange equations yield $2x = 4\lambda$, $4y = -6\lambda$.

(b) Show that if $x = 0$ or $y = 0$, then the Lagrange equations give $x = y = 0$. Since $(0, 0)$ does not satisfy the constraint, you may assume that x and y are nonzero.

(c) Use the Lagrange equations to show that $y = -\frac{3}{4}x$.

(d) Substitute in the constraint equation to show that there is a unique critical point P.

(e) Does P correspond to a minimum or maximum value of f? Refer to Figure 12 to justify your answer. *Hint:* Do the values of $f(x, y)$ increase or decrease as (x, y) moves away from P along the line $g(x, y) = 0$?

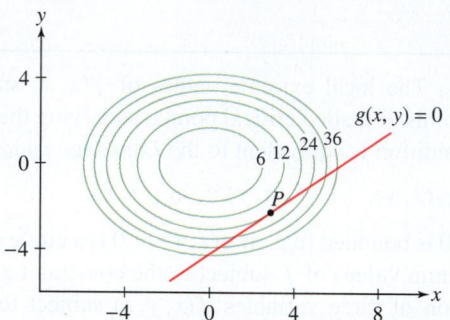

FIGURE 12 Level curves of $f(x, y) = x^2 + 2y^2$ and graph of the constraint $g(x, y) = 4x - 6y - 25 = 0$.

3. Apply the method of Lagrange multipliers to the function $f(x, y) = (x^2 + 1)y$ subject to the constraint $x^2 + y^2 = 5$. *Hint:* First show that $y \neq 0$; then treat the cases $x = 0$ and $x \neq 0$ separately.

In Exercises 4–15, find the minimum and maximum values of the function subject to the given constraint.

4. $f(x, y) = 2x + 3y$, $\quad x^2 + y^2 = 4$

5. $f(x, y) = x^2 + y^2$, $\quad 2x + 3y = 6$

6. $f(x, y) = 4x^2 + 9y^2$, $\quad xy = 4$

7. $f(x, y) = xy$, $\quad 4x^2 + 9y^2 = 32$

8. $f(x, y) = x^2 y + x + y$, $\quad xy = 4$

9. $f(x, y) = x^2 + y^2$, $\quad x^4 + y^4 = 1$

10. $f(x, y) = x^2 y^4$, $\quad x^2 + 2y^2 = 6$

11. $f(x, y, z) = 3x + 2y + 4z$, $\quad x^2 + 2y^2 + 6z^2 = 1$

12. $f(x, y, z) = x^2 - y - z$, $\quad x^2 - y^2 + z = 0$

13. $f(x, y, z) = xy + 2z$, $\quad x^2 + y^2 + z^2 = 36$

14. $f(x, y, z) = x^2 + y^2 + z^2$, $\quad x + 3y + 2z = 36$

15. $f(x, y, z) = xy + xz$, $\quad x^2 + y^2 + z^2 = 4$

16. ✏️ Let

$$f(x, y) = x^3 + xy + y^3, \qquad g(x, y) = x^3 - xy + y^3$$

(a) Show that there is a unique point $P = (a, b)$ on $g(x, y) = 1$ where $\nabla f_P = \lambda \nabla g_P$ for some scalar λ.

(b) Refer to Figure 13 to determine whether $f(P)$ is a local minimum or a local maximum of f subject to the constraint.

(c) Does Figure 13 suggest that $f(P)$ is a global extremum subject to the constraint?

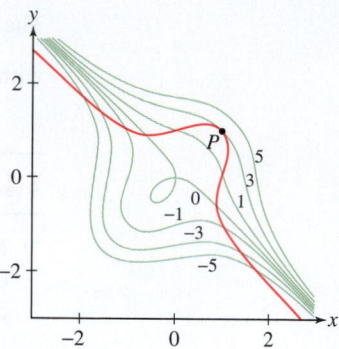

FIGURE 13 Contour map of $f(x, y) = x^3 + xy + y^3$ and graph of the constraint $g(x, y) = x^3 - xy + y^3 = 1$.

17. Find the point (a, b) on the graph of $y = e^x$ where the value ab is the least.

18. Find the rectangular box of maximum volume if the sum of the lengths of the edges is 300 cm.

19. The surface area of a right-circular cone of radius r and height h is $S = \pi r \sqrt{r^2 + h^2}$, and its volume is $V = \frac{1}{3} \pi r^2 h$.

(a) Determine the ratio h/r for the cone with given surface area S and maximum volume V.

(b) What is the ratio h/r for a cone with given volume V and minimum surface area S?

(c) Does a cone with given volume V and maximum surface area exist?

20. In Example 1, we found the maximum of $f(x, y) = 2x + 5y$ on the ellipse $(x/4)^2 + (y/3)^2 = 1$. Solve this problem again without using Lagrange multipliers. First, show that the ellipse is parametrized by $x = 4 \cos t$, $y = 3 \sin t$. Then find the maximum value of $f(4 \cos t, 3 \sin t)$ using single-variable calculus. Is one method easier than the other?

21. Find the point on the ellipse

$$x^2 + 6y^2 + 3xy = 40$$

with the greatest x-coordinate (Figure 14).

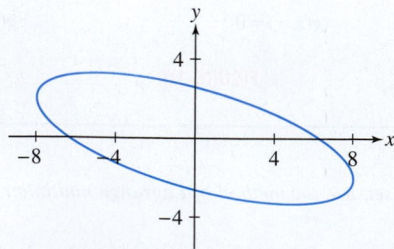

FIGURE 14 Graph of $x^2 + 6y^2 + 3xy = 40$.

22. Use Lagrange multipliers to find the maximum area of a rectangle inscribed in the ellipse (Figure 15):

$$\frac{x^2}{a^2} + \frac{y^2}{b^2} = 1$$

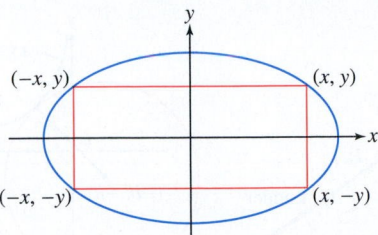

FIGURE 15 Rectangle inscribed in the ellipse $\dfrac{x^2}{a^2} + \dfrac{y^2}{b^2} = 1$.

23. Find the point (x_0, y_0) on the line $4x + 9y = 12$ that is closest to the origin.

24. Show that the point (x_0, y_0) closest to the origin on the line $ax + by = c$ has coordinates

$$x_0 = \frac{ac}{a^2 + b^2}, \qquad y_0 = \frac{bc}{a^2 + b^2}$$

25. Find the maximum value of $f(x, y) = x^a y^b$ for $x \ge 0, y \ge 0$ on the line $x + y = 1$, where $a, b > 0$ are constants.

26. Show that the maximum value of $f(x, y) = x^2 y^3$ on the unit circle is $\dfrac{6}{25}\sqrt{\dfrac{3}{5}}$.

27. Find the maximum value of $f(x, y) = x^a y^b$ for $x \ge 0, y \ge 0$ on the unit circle, where $a, b > 0$ are constants.

28. Find the maximum value of $f(x, y, z) = x^a y^b z^c$ for $x, y, z \ge 0$ on the unit sphere, where $a, b, c > 0$ are constants.

29. Show that the minimum distance from the origin to a point on the plane $ax + by + cz = d$ is

$$\frac{|d|}{\sqrt{a^2 + b^2 + c^2}}$$

30. Antonio has $5.00 to spend on a lunch consisting of hamburgers ($1.50 each) and french fries ($1.00 per order). Antonio's satisfaction from eating x_1 hamburgers and x_2 orders of french fries is measured by a function $U(x_1, x_2) = \sqrt{x_1 x_2}$. How much of each type of food should he purchase to maximize his satisfaction? (Assume that fractional amounts of each food can be purchased.)

31. [✏] Let Q be the point on an ellipse closest to a given point P outside the ellipse. It was known to the Greek mathematician Apollonius (third century BCE) that $\overline{PQ}$ is perpendicular to the tangent to the ellipse at Q (Figure 16). Explain in words why this conclusion is a consequence of the method of Lagrange multipliers. *Hint:* The circles centered at P are level curves of the function to be minimized.

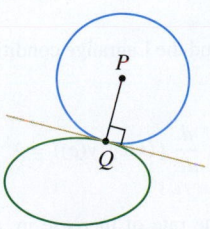

FIGURE 16

32. [✏] In a contest, a runner starting at A must touch a point P along a river and then run to B in the shortest time possible (Figure 17). The runner should choose the point P that minimizes the total length of the path.

(a) Define a function

$$f(x, y) = AP + PB, \quad \text{where } P = (x, y)$$

Rephrase the runner's problem as a constrained optimization problem, assuming that the river is given by an equation $g(x, y) = 0$.

(b) Explain why the level curves of $f(x, y)$ are ellipses.

(c) Use Lagrange multipliers to justify the following statement: The ellipse through the point P minimizing the length of the path is tangent to the river.

(d) Identify the point on the river in Figure 17 for which the length is minimal.

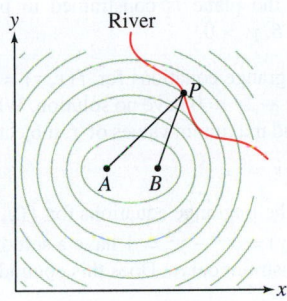

FIGURE 17

In Exercises 33 and 34, let V be the volume of a can of radius r and height h, and let S be its surface area (including the top and bottom).

33. Find r and h that minimize S subject to the constraint $V = 54\pi$.

34. [✏] Show that for both of the following two problems, $P = (r, h)$ is a Lagrange critical point if $h = 2r$:

- Minimize surface area S for fixed volume V.
- Maximize volume V for fixed surface area S.

Then use the contour plots in Figure 18 to explain why S has a minimum for fixed V but no maximum and, similarly, V has a maximum for fixed S but no minimum.

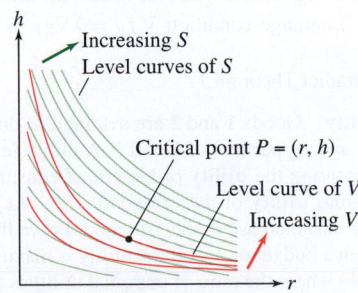

FIGURE 18

35. Figure 19 depicts a tetrahedron whose faces lie in the coordinate planes and in the plane with equation $\dfrac{x}{a} + \dfrac{y}{b} + \dfrac{z}{c} = 1$ $(a, b, c > 0)$. The volume of the tetrahedron is given by $V = \frac{1}{6}abc$. Find the minimum value of V among all planes passing through the point $P = (1, 1, 1)$.

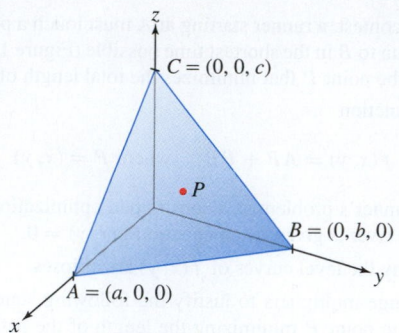

FIGURE 19

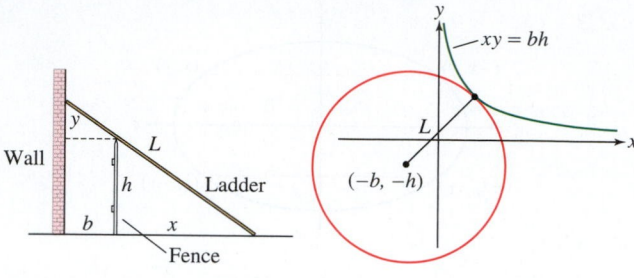

FIGURE 20

36. With the same set-up as in the previous problem, find the plane that minimizes V if the plane is constrained to pass through a point $P = (\alpha, \beta, \gamma)$ with $\alpha, \beta, \gamma > 0$.

37. Show that the Lagrange equations for $f(x, y) = x + y$ subject to the constraint $g(x, y) = x + 2y = 0$ have no solution. What can you conclude about the minimum and maximum values of f subject to $g = 0$? Show this directly.

38. Show that the Lagrange equations for $f(x, y) = 2x + y$ subject to the constraint $g(x, y) = x^2 - y^2 = 1$ have a solution but that f has no min or max on the constraint curve. Does this contradict Theorem 1?

39. Let L be the minimum length of a ladder that can reach over a fence of height h to a wall located a distance b behind the wall.

(a) Use Lagrange multipliers to show that $L = (h^{2/3} + b^{2/3})^{3/2}$ (Figure 20). *Hint:* Show that the problem amounts to minimizing $f(x, y) = (x + b)^2 + (y + h)^2$ subject to $y/b = h/x$ or $xy = bh$.

(b) Show that the value of L is also equal to the radius of the circle with center $(-b, -h)$ that is tangent to the graph of $xy = bh$.

40. Find the maximum value of $f(x, y, z) = xy + xz + yz - xyz$ subject to the constraint $x + y + z = 1$, for $x \geq 0, y \geq 0, z \geq 0$.

41. Find the minimum of $f(x, y, z) = x^2 + y^2 + z^2$ subject to the two constraints $x + y + z = 1$ and $x + 2y + 3z = 6$.

42. Find the maximum of $f(x, y, z) = z$ subject to the two constraints $x^2 + y^2 = 1$ and $x + y + z = 1$.

43. Find the point lying on the intersection of the plane $x + \frac{1}{2}y + \frac{1}{4}z = 0$ and the sphere $x^2 + y^2 + z^2 = 9$ with the greatest z-coordinate.

44. Find the maximum of $f(x, y, z) = x + y + z$ subject to the two constraints $x^2 + y^2 + z^2 = 9$ and $\frac{1}{4}x^2 + \frac{1}{4}y^2 + 4z^2 = 9$.

45. The cylinder $x^2 + y^2 = 1$ intersects the plane $x + z = 1$ in an ellipse. Find the point on such an ellipse that is farthest from the origin.

46. Find the minimum and maximum of $f(x, y, z) = y + 2z$ subject to two constraints, $2x + z = 4$ and $x^2 + y^2 = 1$.

47. Find the minimum value of $f(x, y, z) = x^2 + y^2 + z^2$ subject to two constraints, $x + 2y + z = 3$ and $x - y = 4$.

Further Insights and Challenges

48. Suppose that both $f(x, y)$ and the constraint function $g(x, y)$ are linear. Use contour maps to explain why $f(x, y)$ does not have a maximum subject to $g(x, y) = 0$ unless $g = af + b$ for some constants a, b.

49. Assumptions Matter Consider the problem of minimizing $f(x, y) = x$ subject to $g(x, y) = (x - 1)^3 - y^2 = 0$.

(a) Show, without using calculus, that the minimum occurs at $P = (1, 0)$.

(b) Show that the Lagrange condition $\nabla f_P = \lambda \nabla g_P$ is not satisfied for any value of λ.

(c) Does this contradict Theorem 1?

50. Marginal Utility Goods 1 and 2 are available at dollar prices of p_1 per unit of Good 1 and p_2 per unit of Good 2. A utility function $U(x_1, x_2)$ is a function representing the **utility** or benefit of consuming x_j units of good j. The **marginal utility** of the jth good is $\partial U/\partial x_j$, the rate of increase in utility per unit increase in the jth good. Prove the following law of economics: Given a budget of L dollars, utility is maximized at the consumption level (a, b) where the ratio of marginal utility is equal to the ratio of prices:

$$\frac{\text{marginal utility of Good 1}}{\text{marginal utility of Good 2}} = \frac{U_{x_1}(a, b)}{U_{x_2}(a, b)} = \frac{p_1}{p_2}$$

51. Consider the utility function $U(x_1, x_2) = x_1 x_2$ with budget constraint $p_1 x_1 + p_2 x_2 = c$.

(a) Show that the maximum of $U(x_1, x_2)$ subject to the budget constraint is equal to $c^2/(4p_1 p_2)$.

(b) Calculate the value of the Lagrange multiplier λ occurring in (a).

(c) Prove the following interpretation: λ is the rate of increase in utility per unit increase in total budget c.

52. This exercise shows that the multiplier λ may be interpreted as a rate of change in general. Assume that the maximum of $f(x, y)$ subject to $g(x, y) = c$ occurs at a point P. Then P depends on the value of c, so we may write $P = (x(c), y(c))$ and we have $g(x(c), y(c)) = c$.

(a) Show that

$$\nabla g(x(c), y(c)) \cdot \langle x'(c), y'(c) \rangle = 1$$

Hint: Differentiate the equation $g(x(c), y(c)) = c$ with respect to c using the Chain Rule.

(b) Use the Chain Rule and the Lagrange condition $\nabla f_P = \lambda \nabla g_P$ to show that

$$\frac{d}{dc} f(x(c), y(c)) = \lambda$$

(c) Conclude that λ is the rate of increase in f per unit increase in the "budget level" c.

53. Let $B > 0$. Show that the maximum of

$$f(x_1, \ldots, x_n) = x_1 x_2 \cdots x_n$$

subject to the constraints $x_1 + \cdots + x_n = B$ and $x_j \geq 0$ for $j = 1, \ldots, n$ occurs for $x_1 = \cdots = x_n = B/n$. Use this to conclude that

$$(a_1 a_2 \cdots a_n)^{1/n} \leq \frac{a_1 + \cdots + a_n}{n}$$

for all positive numbers $a_1, \ldots, a_n$.

54. Let $B > 0$. Show that the maximum of $f(x_1, \ldots, x_n) = x_1 + \cdots + x_n$ subject to $x_1^2 + \cdots + x_n^2 = B^2$ is $\sqrt{n} B$. Conclude that

$$|a_1| + \cdots + |a_n| \leq \sqrt{n}(a_1^2 + \cdots + a_n^2)^{1/2}$$

for all numbers $a_1, \ldots, a_n$.

55. Given constants E, E_1, E_2, E_3, consider the maximum of

$$S(x_1, x_2, x_3) = x_1 \ln x_1 + x_2 \ln x_2 + x_3 \ln x_3$$

subject to two constraints:

$$x_1 + x_2 + x_3 = N, \qquad E_1 x_1 + E_2 x_2 + E_3 x_3 = E$$

Show that there is a constant μ such that $x_i = A^{-1} e^{\mu E_i}$ for $i = 1, 2, 3$, where $A = N^{-1}(e^{\mu E_1} + e^{\mu E_2} + e^{\mu E_3})$.

56. Boltzmann Distribution Generalize Exercise 55 to n variables: Show that there is a constant μ such that the maximum of

$$S = x_1 \ln x_1 + \cdots + x_n \ln x_n$$

subject to the constraints

$$x_1 + \cdots + x_n = N, \qquad E_1 x_1 + \cdots + E_n x_n = E$$

occurs for $x_i = A^{-1} e^{\mu E_i}$, where

$$A = N^{-1}(e^{\mu E_1} + \cdots + e^{\mu E_n})$$

This result lies at the heart of statistical mechanics. It is used to determine the distribution of velocities of gas molecules at temperature T; x_i is the number of molecules with kinetic energy E_i; $\mu = -(kT)^{-1}$, where k is Boltzmann's constant. The quantity S is called the **entropy**.

CHAPTER REVIEW EXERCISES

1. Given $f(x, y) = \dfrac{\sqrt{x^2 - y^2}}{x + 3}$:

(a) Sketch the domain of f.

(b) Calculate $f(3, 1)$ and $f(-5, -3)$.

(c) Find a point satisfying $f(x, y) = 1$.

2. Find the domain and range of:

(a) $f(x, y, z) = \sqrt{x - y} + \sqrt{y - z}$

(b) $f(x, y) = \ln(4x^2 - y)$

3. Sketch the graph $f(x, y) = x^2 - y + 1$ and describe its vertical and horizontal traces.

4. ⬚CAS⬚ Use a graphing utility to draw the graph of the function $\cos(x^2 + y^2)e^{1-xy}$ in the domains $[-1, 1] \times [-1, 1]$, $[-2, 2] \times [-2, 2]$, and $[-3, 3] \times [-3, 3]$, and explain its behavior.

5. Match the functions (a)–(d) with their graphs in Figure 1.

(a) $f(x, y) = x^2 + y$

(b) $f(x, y) = x^2 + 4y^2$

(c) $f(x, y) = \sin(4xy)e^{-x^2 - y^2}$

(d) $f(x, y) = \sin(4x)e^{-x^2 - y^2}$

6. Referring to the contour map in Figure 2:

(a) Estimate the average rate of change of elevation from A to B and from A to D.

(b) Estimate the directional derivative at A in the direction of **v**.

(c) What are the signs of f_x and f_y at D?

(d) At which of the labeled points are both f_x and f_y negative?

7. Describe the level curves of:

(a) $f(x, y) = e^{4x - y}$

(b) $f(x, y) = \ln(4x - y)$

(c) $f(x, y) = 3x^2 - 4y^2$

(d) $f(x, y) = x + y^2$

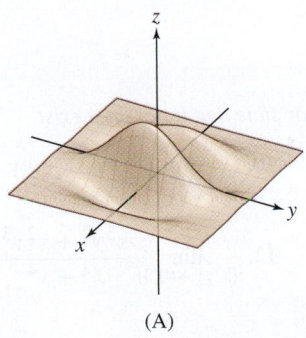

(A)

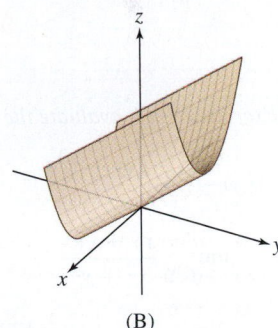

(B)

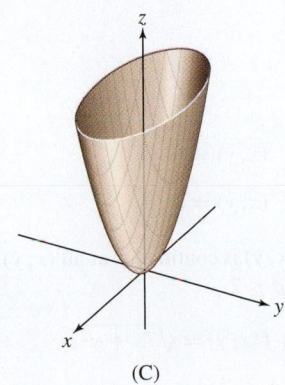

(C)

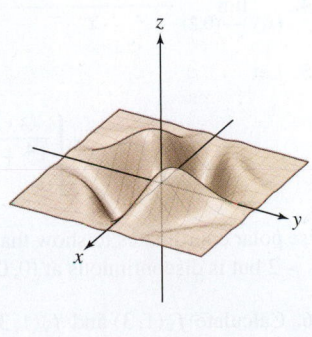

(D)

FIGURE 1

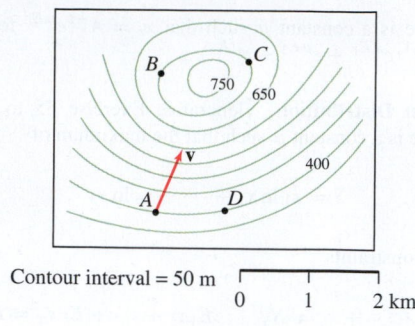

Contour interval = 50 m

0   1   2 km

FIGURE 2

8. Match each function (a)–(c) with its contour graph (i)–(iii) in Figure 3:

(a) $f(x, y) = xy$

(b) $f(x, y) = e^{xy}$

(c) $f(x, y) = \sin(xy)$

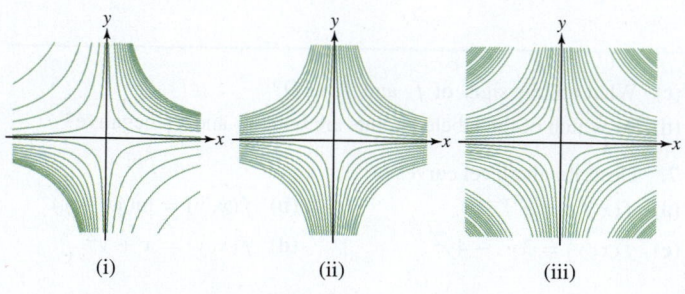

(i)   (ii)   (iii)

FIGURE 3

In Exercises 9–14, evaluate the limit or state that it does not exist.

9. $\displaystyle\lim_{(x,y)\to(1,-3)} (xy + y^2)$

10. $\displaystyle\lim_{(x,y)\to(1,-3)} \ln(3x + y)$

11. $\displaystyle\lim_{(x,y)\to(0,0)} \frac{xy + xy^2}{x^2 + y^2}$

12. $\displaystyle\lim_{(x,y)\to(0,0)} \frac{x^3y^2 + x^2y^3}{x^4 + y^4}$

13. $\displaystyle\lim_{(x,y)\to(1,-3)} (2x + y)e^{-x+y}$

14. $\displaystyle\lim_{(x,y)\to(0,2)} \frac{(e^x - 1)(e^y - 1)}{x}$

15. Let

$$f(x, y) = \begin{cases} \dfrac{(xy)^p}{x^4 + y^4} & (x, y) \neq (0,0) \\ 0 & (x, y) = (0,0) \end{cases}$$

Use polar coordinates to show that $f(x, y)$ is continuous at all (x, y) if $p > 2$ but is discontinuous at $(0,0)$ if $p \leq 2$.

16. Calculate $f_x(1, 3)$ and $f_y(1, 3)$ for $f(x, y) = \sqrt{7x + y^2}$.

In Exercises 17–20, compute f_x and f_y.

17. $f(x, y) = 2x + y^2$

18. $f(x, y) = 4xy^3$

19. $f(x, y) = \sin(xy)e^{-x-y}$

20. $f(x, y) = \ln(x^2 + xy^2)$

21. Calculate f_{xxyz} for $f(x, y, z) = y\sin(x + z)$.

22. Fix $c > 0$. Show that for any constants α, β, the function $u(t, x) = \sin(\alpha ct + \beta)\sin(\alpha x)$ satisfies the wave equation

$$\frac{\partial^2 u}{\partial t^2} = c^2 \frac{\partial^2 u}{\partial x^2}$$

23. Find an equation of the tangent plane to the graph of $f(x, y) = xy^2 - xy + 3x^3y$ at $P = (1, 3)$.

24. Suppose that $f(4, 4) = 3$ and $f_x(4, 4) = f_y(4, 4) = -1$. Use the Linear Approximation to estimate $f(4.1, 4)$ and $f(3.88, 4.03)$.

25. Use a Linear Approximation of $f(x, y, z) = \sqrt{x^2 + y^2 + z}$ to estimate $\sqrt{7.1^2 + 4.9^2 + 69.5}$. Compare with a calculator value.

26. The plane $z = 2x - y - 1$ is tangent to the graph of $z = f(x, y)$ at $P = (5, 3)$.

(a) Determine $f(5, 3)$, $f_x(5, 3)$, and $f_y(5, 3)$.

(b) Approximate $f(5.2, 2.9)$.

27. Figure 4 shows the contour map of a function $f(x, y)$ together with a path $\mathbf{r}(t)$ in the counterclockwise direction. The points $\mathbf{r}(1)$, $\mathbf{r}(2)$, and $\mathbf{r}(3)$ are indicated on the path. Let $g(t) = f(\mathbf{r}(t))$. Which of statements (i)–(iv) are true? Explain.

(i) $g'(1) > 0$.

(ii) $g(t)$ has a local minimum for some $1 \leq t \leq 2$.

(iii) $g'(2) = 0$.

(iv) $g'(3) = 0$.

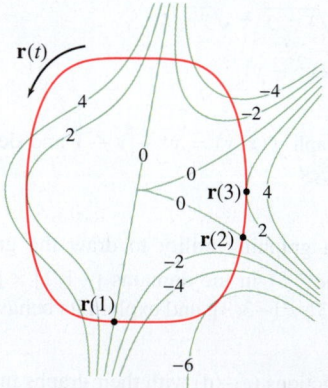

FIGURE 4

28. Jason earns $S(h, c) = 20h\left(1 + \frac{c}{100}\right)^{1.5}$ dollars per month at a used car lot, where h is the number of hours worked and c is the number of cars sold. He has already worked 160 hours and sold 69 cars. Right now Jason wants to go home but wonders how much more he might earn if he stays another 10 minutes with a customer who is considering buying a car. Use the Linear Approximation to estimate how much extra money Jason will earn if he sells his 70th car during these 10 min.

In Exercises 29–32, compute $\dfrac{d}{dt} f(\mathbf{r}(t))$ at the given value of t.

29. $f(x, y) = x + e^y$, $\mathbf{r}(t) = \left\langle 3t - 1, t^2 \right\rangle$ at $t = 2$

30. $f(x, y, z) = xz - y^2$, $\mathbf{r}(t) = \left\langle t, t^3, 1 - t \right\rangle$ at $t = -2$

31. $f(x, y) = xe^{3y} - ye^{3x}$, $\mathbf{r}(t) = \langle e^t, \ln t \rangle$ at $t = 1$

32. $f(x, y) = \tan^{-1} \frac{y}{x}$, $\mathbf{r}(t) = \langle \cos t, \sin t \rangle$, $t = \frac{\pi}{3}$

In Exercises 33–36, compute the directional derivative at P in the direction of $\mathbf{v}$.

33. $f(x, y) = x^3 y^4$, $P = (3, -1)$, $\mathbf{v} = 2\mathbf{i} + \mathbf{j}$

34. $f(x, y, z) = zx - xy^2$, $P = (1, 1, 1)$, $\mathbf{v} = \langle 2, -1, 2 \rangle$

35. $f(x, y) = e^{x^2 + y^2}$, $P = \left(\dfrac{\sqrt{2}}{2}, \dfrac{\sqrt{2}}{2} \right)$, $\mathbf{v} = \langle 3, -4 \rangle$

36. $f(x, y, z) = \sin(xy + z)$, $P = (0, 0, 0)$, $\mathbf{v} = \mathbf{j} + \mathbf{k}$

37. Find the unit vector $\mathbf{e}$ at $P = (0, 0, 1)$ pointing in the direction along which $f(x, y, z) = xz + e^{-x^2 + y}$ increases most rapidly.

38. Find an equation of the tangent plane at $P = (0, 3, -1)$ to the surface with equation

$$ze^x + e^{z+1} = xy + y - 3$$

39. Let $n \neq 0$ be an integer and r an arbitrary constant. Show that the tangent plane to the surface $x^n + y^n + z^n = r$ at $P = (a, b, c)$ has equation

$$a^{n-1}x + b^{n-1}y + c^{n-1}z = r$$

40. Let $f(x, y) = (x - y)e^x$. Use the Chain Rule to calculate $\partial f / \partial u$ and $\partial f / \partial v$ (in terms of u and v), where $x = u - v$ and $y = u + v$.

41. Let $f(x, y, z) = x^2 y + y^2 z$. Use the Chain Rule to calculate $\partial f / \partial s$ and $\partial f / \partial t$ (in terms of s and t), where

$$x = s + t, \quad y = st, \quad z = 2s - t$$

42. Let P have spherical coordinates $(\rho, \theta, \phi) = \left(2, \frac{\pi}{4}, \frac{\pi}{4} \right)$. Calculate $\left. \dfrac{\partial f}{\partial \phi} \right|_P$ assuming that

$$f_x(P) = 4, \quad f_y(P) = -3, \quad f_z(P) = 8$$

Recall that $x = \rho \cos \theta \sin \phi$, $y = \rho \sin \theta \sin \phi$, $z = \rho \cos \phi$.

43. Let $g(u, v) = f(u^3 - v^3, v^3 - u^3)$. Prove that

$$v^2 \frac{\partial g}{\partial u} + u^2 \frac{\partial g}{\partial v} = 0$$

44. Let $f(x, y) = g(u)$, where $u = x^2 + y^2$ and $g(u)$ is differentiable. Prove that

$$\left(\frac{\partial f}{\partial x} \right)^2 + \left(\frac{\partial f}{\partial y} \right)^2 = 4u \left(\frac{dg}{du} \right)^2$$

45. Calculate $\partial z / \partial x$, where $xe^z + ze^y = x + y$.

46. Let $f(x, y) = x^4 - 2x^2 + y^2 - 6y$.

(a) Find the critical points of f and use the Second Derivative Test to determine whether they are a local minima or a local maxima.

(b) Find the minimum value of f without calculus by completing the square.

In Exercises 47–50, find the critical points of the function and analyze them using the Second Derivative Test.

47. $f(x, y) = x^4 - 4xy + 2y^2$

48. $f(x, y) = x^3 + 2y^3 - xy$

49. $f(x, y) = e^{x+y} - xe^{2y}$

50. $f(x, y) = \sin(x + y) - \dfrac{1}{2}(x + y^2)$

51. Prove that $f(x, y) = (x + 2y)e^{xy}$ has no critical points.

52. Find the global extrema of $f(x, y) = x^3 - xy - y^2 + y$ on the square $[0, 1] \times [0, 1]$.

53. Find the global extrema of $f(x, y) = 2xy - x - y$ on the domain $\{y \leq 4, y \geq x^2\}$.

54. Find the maximum of $f(x, y, z) = xyz$ subject to the constraint $g(x, y, z) = 2x + y + 4z = 1$.

55. Use Lagrange multipliers to find the minimum and maximum values of $f(x, y) = 3x - 2y$ on the circle $x^2 + y^2 = 4$.

56. Find the minimum value of $f(x, y) = xy$ subject to the constraint $5x - y = 4$ in two ways: using Lagrange multipliers and setting $y = 5x - 4$ in $f(x, y)$.

57. Find the minimum and maximum values of $f(x, y) = x^2 y$ on the ellipse $4x^2 + 9y^2 = 36$.

58. Find the point in the first quadrant on the curve $y = x + x^{-1}$ closest to the origin.

59. Find the extreme values of $f(x, y, z) = x + 2y + 3z$ subject to the two constraints $x + y + z = 1$ and $x^2 + y^2 + z^2 = 1$.

60. Find the minimum and maximum values of $f(x, y, z) = x - z$ on the intersection of the cylinders $x^2 + y^2 = 1$ and $x^2 + z^2 = 1$ (Figure 5).

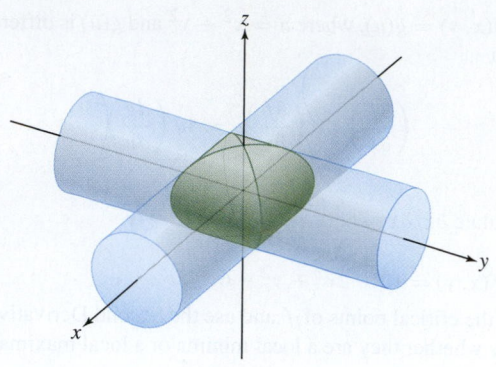

FIGURE 5

61. Use Lagrange multipliers to find the dimensions of a cylindrical can with a bottom but no top, of fixed volume V with minimum surface area.

62. Find the dimensions of the box of maximum volume with its sides parallel to the coordinate planes that can be inscribed in the ellipsoid (Figure 6)

$$\left(\frac{x}{a}\right)^2 + \left(\frac{y}{b}\right)^2 + \left(\frac{z}{c}\right)^2 = 1$$

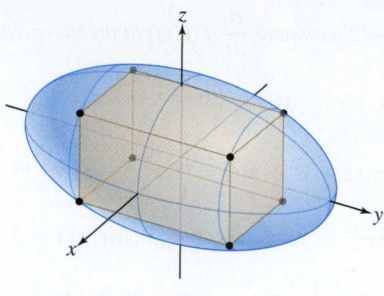

FIGURE 6

63. Given n nonzero numbers $\sigma_1, \ldots, \sigma_n$, show that the minimum value of

$$f(x_1, \ldots, x_n) = x_1^2 \sigma_1^2 + \cdots + x_n^2 \sigma_n^2$$

subject to $x_1 + \cdots + x_n = 1$ is c, where $c = \left(\displaystyle\sum_{j=1}^{n} \sigma_j^{-2}\right)^{-1}$.

15 MULTIPLE INTEGRATION

Integrals of functions of several variables, called **multiple integrals**, are a natural extension of the single-variable integrals studied in the first part of the text. They are used to compute many quantities that appear in applications, such as volumes, masses, heat flow, total charge, and net force.

The volcanic-rock columns making up Devil's Tower in Wyoming resemble the columns of volume in a Riemann sum representation of the volume under the graph of a function of two variables. As in the single-variable case, we define integrals in two and three variables as limits of Riemann sums.

15.1 Integration in Two Variables

The integral of a function of two variables $f(x, y)$, called a **double integral**, is denoted

$$\iint_{\mathcal{D}} f(x, y)\, dA$$

When $f(x, y) \geq 0$ on a domain $\mathcal{D}$ in the xy-plane, the integral represents the volume of the solid region between the graph of $f(x, y)$ and the xy-plane (Figure 1). More generally, the integral represents a signed volume, where positive contributions arise from regions above the xy-plane and negative contributions from regions below.

There are many similarities between double integrals and single integrals:

- Double integrals are defined as limits of Riemann sums.
- Double integrals are evaluated using the Fundamental Theorem of Calculus (but we have to use it twice—see the discussion of iterated integrals below).

An important difference, however, is that the domains of integration of double integrals are often more complicated. In one variable, the domain of integration is simply an interval $[a, b]$. In two variables, the domain $\mathcal{D}$ is a plane region whose boundary can be made up of a number of different curves and segments (e.g., $\mathcal{D}$ in Figure 1 and $\mathcal{R}$ in Figure 2).

In this section, we focus on the simplest case where the domain is a rectangle, leaving more general domains for Section 15.2. Let

$$\mathcal{R} = [a, b] \times [c, d]$$

denote the rectangle in the plane (Figure 2) consisting of all points (x, y) such that

$$\mathcal{R}: \quad a \leq x \leq b, \qquad c \leq y \leq d$$

Like integrals in one variable, double integrals are defined through a three-step process: subdivision, summation, and passage to the limit. Figure 3 illustrates the subdivision step which itself has three steps:

1. Subdivide $[a, b]$ and $[c, d]$ by choosing partitions:

$$a = x_0 < x_1 < \cdots < x_N = b, \qquad c = y_0 < y_1 < \cdots < y_M = d$$

where N and M are positive integers.

2. Create an $N \times M$ grid of subrectangles $\mathcal{R}_{ij}$.

3. Choose a sample point P_{ij} in each $\mathcal{R}_{ij}$.

Note that $\mathcal{R}_{ij} = [x_{i-1}, x_i] \times [y_{j-1}, y_j]$, so $\mathcal{R}_{ij}$ has area

$$\Delta A_{ij} = \Delta x_i\, \Delta y_j$$

where $\Delta x_i = x_i - x_{i-1}$ and $\Delta y_j = y_j - y_{j-1}$.

The next step in defining the integral is summation where we form a Riemann sum with the function values $f(P_{ij})$:

$$S_{N,M} = \sum_{i=1}^{N} \sum_{j=1}^{M} f(P_{ij})\, \Delta A_{ij} = \sum_{i=1}^{N} \sum_{j=1}^{M} f(P_{ij})\, \Delta x_i\, \Delta y_j$$

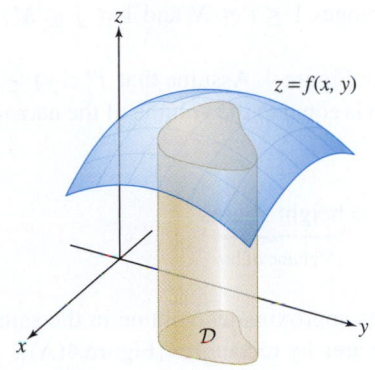

FIGURE 1 The double integral of $f(x, y)$ over the domain $\mathcal{D}$ yields the volume of the solid region between the graph of $f(x, y)$ and the xy-plane over $\mathcal{D}$.

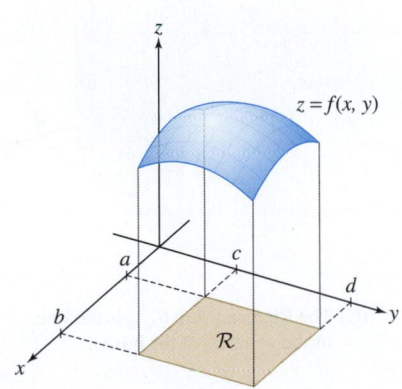

FIGURE 2

Keep in mind that a Riemann sum depends on the choice of partition and sample points. It would be more proper to write

$$S_{N,M}(\{P_{ij}\}, \{x_i\}, \{y_j\})$$

but we write $S_{N,M}$ to keep the notation simple.

879

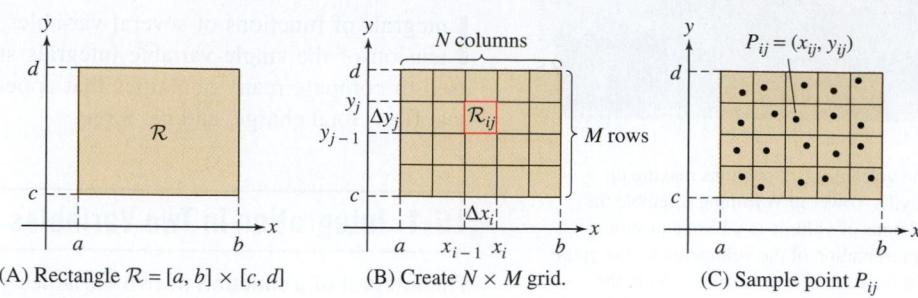

(A) Rectangle $\mathcal{R} = [a, b] \times [c, d]$ (B) Create $N \times M$ grid.   (C) Sample point P_{ij}

FIGURE 3

The double summation runs over all i and j in the ranges $1 \leq i \leq N$ and $1 \leq j \leq M$, a total of NM terms.

The geometric interpretation of $S_{N,M}$ is shown in Figure 4. Assume that $f(x, y) \geq 0$ over $\mathcal{R}$. Each individual term $f(P_{ij}) \, \Delta A_{ij}$ of the sum is equal to the volume of the narrow box of height $f(P_{ij})$ above $\mathcal{R}_{ij}$:

$$f(P_{ij}) \, \Delta A_{ij} = f(P_{ij}) \, \Delta x_i \, \Delta y_j = \underbrace{\text{height} \times \text{area}}_{\text{Volume of box}}$$

The sum $S_{N,M}$ of the volumes of these narrow boxes approximates volume in the same way that Riemann sums in one variable approximate area by rectangles [Figure 4(A)].

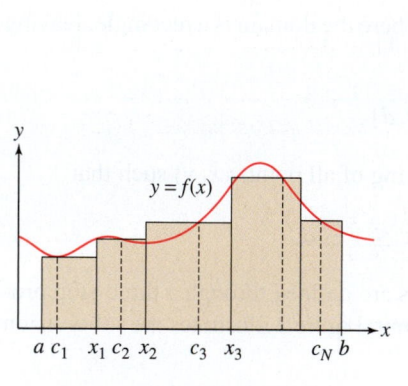

(A) In one variable, a Riemann sum approximates the area under the curve by a sum of areas of rectangles.

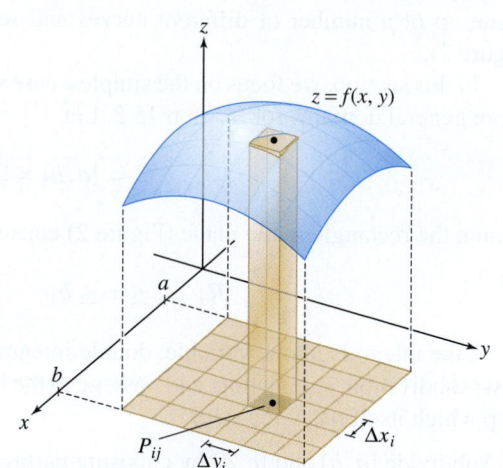

(B) The volume of the box is $f(P_{ij})\Delta A_{ij}$, where $\Delta A_{ij} = \Delta x_i \Delta y_j$.

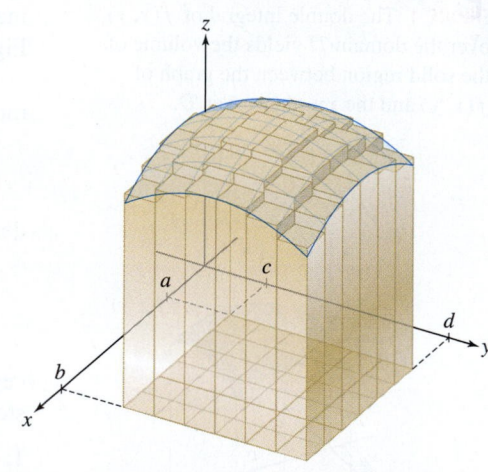

(C) The Riemann sum $S_{N,M}$ is the sum of the volumes of the boxes.

DF **FIGURE 4**

When $f(P_{ij}) < 0$, the term $f(P_{ij}) \, \Delta A_{ij}$ is the signed volume of a narrow box lying below the xy-plane. Generally, we can think of the Riemann sum $S_{N,M}$ as a sum of signed volumes of narrow boxes, some lying above the xy-plane, some below.

The final step in defining the double integral is passing to the limit. We write $\mathcal{P} = \{\{x_i\}, \{y_j\}\}$ for the partition and $\|\mathcal{P}\|$ for the maximum of the widths Δx_i, Δy_j. The following definition makes precise the idea of the Riemann sums converging to a limit as the subrectangles get smaller and smaller:

Limit of Riemann Sums   The Riemann sum $S_{N,M}$ approaches a limit L as $\|\mathcal{P}\| \to 0$ if, for all $\epsilon > 0$, there exists $\delta > 0$ such that

$$|L - S_{N,M}| < \epsilon$$

for all partitions satisfying $\|\mathcal{P}\| < \delta$ and all choices of sample points. We write

$$\lim_{\|\mathcal{P}\| \to 0} S_{N,M} = \lim_{\|\mathcal{P}\| \to 0} \sum_{i=1}^{N} \sum_{j=1}^{M} f(P_{ij}) \, \Delta A_{ij} = L$$

For example, Figure 5 shows that the Riemann sums converge to the volume under the graph of $z = 24 - 3x^2 - y^2$ over $\mathcal{R} = [0, 2] \times [0, 3]$ because the narrower the boxes, the better the collection of them fills out the solid region.

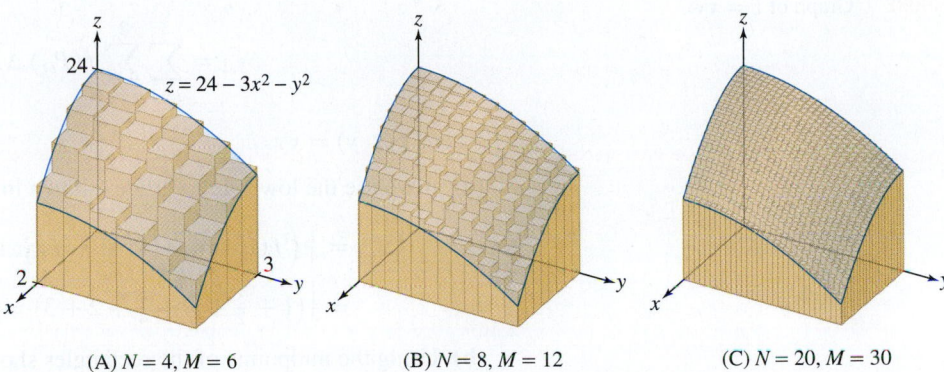

DF **FIGURE 5** Approximations to the volume under $z = 24 - 3x^2 - y^2$.

(A) $N = 4, M = 6$ (B) $N = 8, M = 12$ (C) $N = 20, M = 30$

If the limit of Riemann sums exists, then we obtain the double integral:

DEFINITION Double Integral over a Rectangle   The double integral of $f(x, y)$ over a rectangle $\mathcal{R}$ is defined as the limit

$$\iint_{\mathcal{R}} f(x, y) \, dA = \lim_{\|\mathcal{P}\| \to 0} \sum_{i=1}^{N} \sum_{j=1}^{M} f(P_{ij}) \Delta A_{ij}$$

If this limit exists, we say that $f(x, y)$ is **integrable** over $\mathcal{R}$.

The double integral enables us to define the volume V of the solid region between the graph of a positive function $f(x, y)$ and the rectangle $\mathcal{R}$ by

$$V = \iint_{\mathcal{R}} f(x, y) \, dA$$

If $f(x, y)$ takes on both positive and negative values, the double integral defines the signed volume. So in Figure 6, where each V_i represents the actual volume indicated,

$$\iint_{\mathcal{R}} f(x, y) \, dA = V_1 + V_2 - V_3 - V_4.$$

It is often convenient to work with partitions that are **regular**, that is, partitions whose intervals $[a, b]$ and $[c, d]$ are each divided into subintervals of equal length. In other words, the partition is regular if $\Delta x_i = \Delta x$ and $\Delta y_j = \Delta y$ for all i and j, where

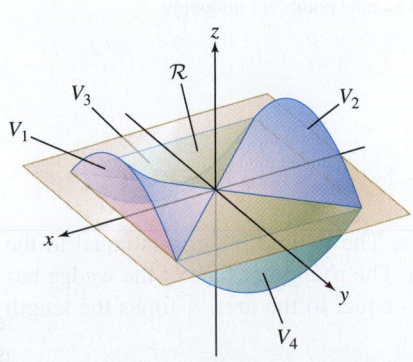

FIGURE 6 $\iint_{\mathcal{R}} f(x, y) \, dA$ is the signed volume of the region between the graph of $z = f(x, y)$ and the rectangle $\mathcal{R}$.

$$\Delta x = \frac{b - a}{N}, \qquad \Delta y = \frac{d - c}{M}$$

For a regular partition, $\|\mathcal{P}\|$ tends to zero as N and M tend to ∞.

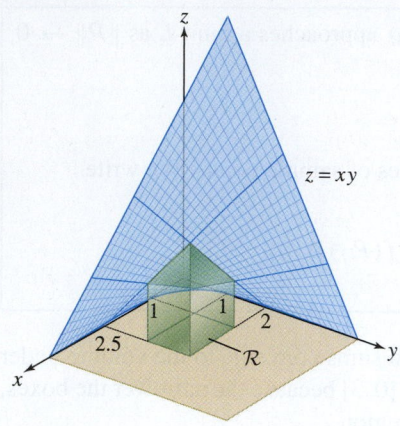

FIGURE 7 Graph of $z = xy$.

EXAMPLE 1 **Estimating a Double Integral**   Let $\mathcal{R} = [1, 2.5] \times [1, 2]$. Calculate $S_{3,2}$ for the integral (Figure 7)

$$\iint_{\mathcal{R}} xy \, dA$$

using the following two choices of sample points:

(a) Lower-left vertex   **(b)** Midpoint of rectangle

Solution Since we use the regular partition to compute $S_{3,2}$, each subrectangle has sides of length

$$\Delta x = \frac{2.5 - 1}{3} = \frac{1}{2}, \qquad \Delta y = \frac{2 - 1}{2} = \frac{1}{2}$$

and area $\Delta A = \Delta x \, \Delta y = \frac{1}{4}$. The corresponding Riemann sum is

$$S_{3,2} = \sum_{i=1}^{3} \sum_{j=1}^{2} f(P_{ij}) \, \Delta A = \frac{1}{4} \sum_{i=1}^{3} \sum_{j=1}^{2} f(P_{ij})$$

where $f(x, y) = xy$.

(a) If we use the lower-left vertices shown in Figure 8(A), the Riemann sum is

$$S_{3,2} = \tfrac{1}{4} \left(f(1,1) + f\left(1, \tfrac{3}{2}\right) + f\left(\tfrac{3}{2}, 1\right) + f\left(\tfrac{3}{2}, \tfrac{3}{2}\right) + f(2,1) + f\left(2, \tfrac{3}{2}\right) \right)$$

$$= \tfrac{1}{4}\left(1 + \tfrac{3}{2} + \tfrac{3}{2} + \tfrac{9}{4} + 2 + 3\right) = \tfrac{1}{4}\left(\tfrac{45}{4}\right) = 2.8125$$

(b) Using the midpoints of the rectangles shown in Figure 8(B), we obtain

$$S_{3,2} = \tfrac{1}{4}\left(f\left(\tfrac{5}{4}, \tfrac{5}{4}\right) + f\left(\tfrac{5}{4}, \tfrac{7}{4}\right) + f\left(\tfrac{7}{4}, \tfrac{5}{4}\right) + f\left(\tfrac{7}{4}, \tfrac{7}{4}\right) + f\left(\tfrac{9}{4}, \tfrac{5}{4}\right) + f\left(\tfrac{9}{4}, \tfrac{7}{4}\right) \right)$$

$$= \tfrac{1}{4}\left(\tfrac{25}{16} + \tfrac{35}{16} + \tfrac{35}{16} + \tfrac{49}{16} + \tfrac{45}{16} + \tfrac{63}{16}\right) = \tfrac{1}{4}\left(\tfrac{252}{16}\right) = 3.9375 \quad \blacksquare$$

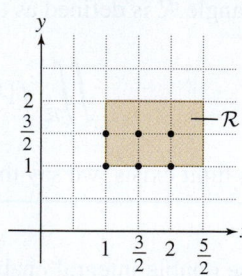

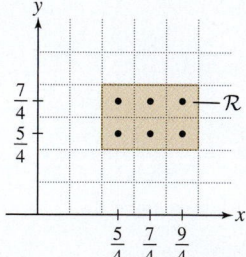

(A) Sample points are the lower-left vertices.   (B) Sample points are midpoints.

FIGURE 8

EXAMPLE 2 Use geometry to evaluate $\iint_{\mathcal{R}} (8 - 2y) \, dA$, where $\mathcal{R} = [0,3] \times [0,4]$.

Solution Figure 9 shows the graph of $z = 8 - 2y$. The double integral is equal to the volume V of the solid wedge underneath the graph. The triangular face of the wedge has area $A = \frac{1}{2}(8)4 = 16$. The volume of the wedge is equal to the area A times the length $\ell = 3$; that is, $V = \ell A = 3(16) = 48$. Therefore,

$$\iint_{\mathcal{R}} (8 - 2y) \, dA = 48 \quad \blacksquare$$

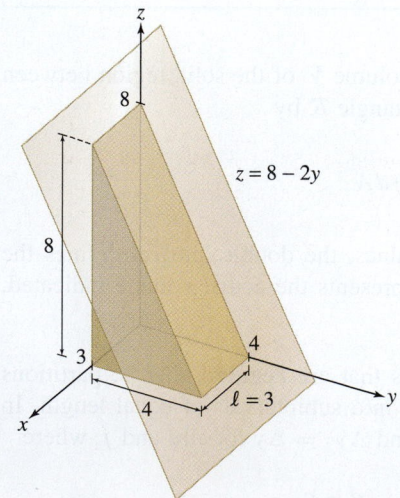

DF FIGURE 9 Solid wedge under the graph of $z = 8 - 2y$.

The next theorem assures us that continuous functions are integrable.

THEOREM 1 Continuous Functions Are Integrable If a function f of two variables is continuous on a rectangle $\mathcal{R}$, then $f(x, y)$ is integrable over $\mathcal{R}$.

As in the single-variable case, we often make use of the linearity properties of the double integral. They follow from the definition of the double integral as a limit of Riemann sums.

THEOREM 2 Linearity of the Double Integral Assume that $f(x, y)$ and $g(x, y)$ are integrable over a rectangle $\mathcal{R}$. Then

(i) $\displaystyle\iint_{\mathcal{R}} \left(f(x, y) + g(x, y)\right) dA = \iint_{\mathcal{R}} f(x, y)\, dA + \iint_{\mathcal{R}} g(x, y)\, dA$

(ii) For any constant C, $\displaystyle\iint_{\mathcal{R}} Cf(x, y)\, dA = C \iint_{\mathcal{R}} f(x, y)\, dA$

If $f(x, y) = C$ is a constant function, then

$$\iint_{\mathcal{R}} C\, dA = C \cdot \text{area}(\mathcal{R})$$

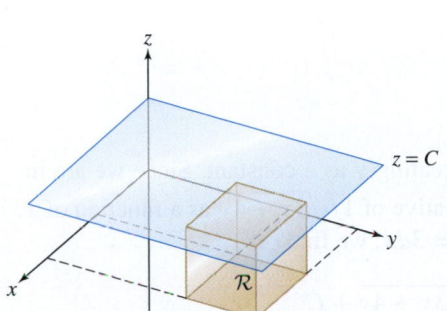

FIGURE 10 The double integral of $f(x, y) = C$ over a rectangle $\mathcal{R}$ is $C \cdot \text{area}(\mathcal{R})$.

The double integral is the signed volume of the box bounded by the rectangle $\mathcal{R}$ in the xy-plane and the plane $z = C$ (Figure 10). That signed volume is C times the area of the rectangle, and therefore the integral equals $C \cdot \text{area}(\mathcal{R})$.

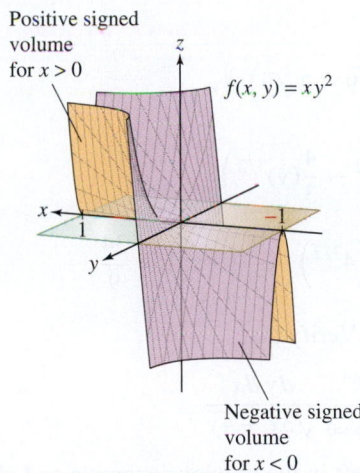

Positive signed volume for $x > 0$

$f(x, y) = xy^2$

Negative signed volume for $x < 0$

FIGURE 11 The signed volumes on either side of the xy-plane cancel.

EXAMPLE 3 Arguing by Symmetry Use symmetry to show that $\displaystyle\iint_{\mathcal{R}} xy^2\, dA = 0$, where $\mathcal{R} = [-1, 1] \times [-1, 1]$.

Solution The double integral is the signed volume of the region between the graph of $f(x, y) = xy^2$ and the xy-plane (Figure 11). Note that $f(x, y)$ takes opposite values at (x, y) and $(-x, y)$:

$$f(-x, y) = -xy^2 = -f(x, y)$$

Because of symmetry, the (negative) signed volume of the region below the xy-plane where $-1 \le x \le 0$ cancels with the (positive) signed volume of the region above the xy-plane where $0 \le x \le 1$. The net result is $\displaystyle\iint_{\mathcal{R}} xy^2\, dA = 0$. ∎

Iterated Integrals

Our main tool for evaluating double integrals is the Fundamental Theorem of Calculus, Part I (FTC I), as in the single-variable case. To use FTC I, we express the double integral as an **iterated integral**, which is an expression of the form

$$\int_a^b \left(\int_c^d f(x, y)\, dy \right) dx$$

Iterated integrals are evaluated in two steps.

Step 1. Hold x constant and evaluate the inner integral with respect to y. This gives us a function of x alone:

$$S(x) = \int_c^d f(x, y)\, dy$$

Step 2. Integrate the resulting function $S(x)$ with respect to x.

EXAMPLE 4 Evaluate $\displaystyle\int_2^4 \left(\int_1^9 ye^x \, dy \right) dx$.

We often omit the parentheses in the notation for an iterated integral:

$$\int_a^b \int_c^d f(x,y) \, dy \, dx$$

The order of the variables in $dy\,dx$ tells us to integrate first with respect to y between the limits $y = c$ and $y = d$.

Solution First, evaluate the inner integral, treating x as a constant:

$$S(x) = \int_1^9 ye^x \, dy = e^x \int_1^9 y \, dy = e^x \left(\frac{1}{2} y^2 \right) \Big|_{y=1}^9 = e^x \left(\frac{81 - 1}{2} \right) = 40e^x$$

Then integrate $S(x)$ with respect to x:

$$\int_2^4 \left(\int_1^9 ye^x \, dy \right) dx = \int_2^4 40e^x \, dx = 40e^x \Big|_2^4 = 40(e^4 - e^2) \qquad \blacksquare$$

EXAMPLE 5 Evaluate $\displaystyle\int_{y=0}^4 \int_{x=0}^3 \frac{dx \, dy}{\sqrt{3x + 4y}}$.

Here we integrate first with respect to x. Sometimes for clarity, as in this case, we include the variables in the limits of integration.

Solution We evaluate the inner integral first, treating y as a constant. Since we are integrating with respect to x, we need an antiderivative of $1/\sqrt{3x + 4y}$ as a function of x. Using the substitution $u = 3x + 4y$, so that $du = 3dx$, we find

$$\int \frac{dx}{\sqrt{3x + 4y}} = \frac{2}{3} \sqrt{3x + 4y} + C$$

Thus, we have

$$\int_{x=0}^3 \frac{dx}{\sqrt{3x + 4y}} = \frac{2}{3} \sqrt{3x + 4y} \Big|_{x=0}^3 = \frac{2}{3} \left(\sqrt{4y + 9} - \sqrt{4y} \right)$$

For the integral of $\sqrt{4y + 9}$, we use the substitution $u = 4y + 9$, $du = 4dy$.

Therefore, we obtain

$$\int_{y=0}^4 \int_{x=0}^3 \frac{dx \, dy}{\sqrt{3x + 4y}} = \frac{2}{3} \int_{y=0}^4 \left(\sqrt{4y + 9} - 2\sqrt{y} \right) dy$$

$$= \frac{2}{3} \left(\frac{1}{6}(4y + 9)^{3/2} - \frac{4}{3}(y)^{3/2} \right) \Big|_{y=0}^4$$

$$= \frac{1}{9} \left(25^{3/2} \right) - \frac{8}{9} \left(4^{3/2} \right) - \frac{1}{9} \left(9^{3/2} \right) = \frac{34}{9} \qquad \blacksquare$$

EXAMPLE 6 **Reversing the Order of Integration** Verify that

$$\int_{y=0}^4 \int_{x=0}^3 \frac{dx \, dy}{\sqrt{3x + 4y}} = \int_{x=0}^3 \int_{y=0}^4 \frac{dy \, dx}{\sqrt{3x + 4y}}$$

Solution We evaluated the iterated integral on the left in the previous example and obtained a value of $\dfrac{34}{9}$. We compute the integral on the right and verify that the result is also $\dfrac{34}{9}$:

$$\int_{y=0}^4 \frac{dy}{\sqrt{3x + 4y}} = \frac{1}{2} \sqrt{3x + 4y} \Big|_{y=0}^4 = \frac{1}{2} \left(\sqrt{3x + 16} - \sqrt{3x} \right)$$

$$\int_{x=0}^3 \int_{y=0}^4 \frac{dy \, dx}{\sqrt{3x + 4y}} = \frac{1}{2} \int_0^3 \left(\sqrt{3x + 16} - \sqrt{3x} \right) dx$$

$$= \frac{1}{2} \left(\frac{2}{9}(3x + 16)^{3/2} - \frac{2}{9}(3x)^{3/2} \right) \Big|_{x=0}^3$$

$$= \frac{1}{9} \left(25^{3/2} - 9^{3/2} - 16^{3/2} \right) = \frac{34}{9} \qquad \blacksquare$$

The previous example illustrates a general fact: The value of an iterated integral does not depend on the order in which the integration is performed. This is part of Fubini's Theorem. Even more important, Fubini's Theorem states that a double integral over a rectangle can be evaluated as an iterated integral.

> **THEOREM 3 Fubini's Theorem** The double integral of a continuous function $f(x, y)$ over a rectangle $\mathcal{R} = [a, b] \times [c, d]$ is equal to the iterated integral (in either order):
>
> $$\iint_{\mathcal{R}} f(x, y)\, dA = \int_{x=a}^{b} \int_{y=c}^{d} f(x, y)\, dy\, dx = \int_{y=c}^{d} \int_{x=a}^{b} f(x, y)\, dx\, dy$$

CAUTION *When you reverse the order of integration in an iterated integral over a rectangle, remember to interchange the limits of integration (the inner limits become the outer limits). However, in contrast, over nonrectangular regions, the process is more complicated, and reversing the order of integration involves more than simply interchanging the limits. We examine the nonrectangular case in the next section.*

3	$f(P_{13})$	$f(P_{23})$	$f(P_{33})$
2	$f(P_{12})$	$f(P_{22})$	$f(P_{32})$
1	$f(P_{11})$	$f(P_{21})$	$f(P_{31})$
$j\,\backslash\,i$	1	2	3

Proof We sketch the proof. We can compute the double integral as a limit of Riemann sums that use a regular partition of $\mathcal{R}$ and sample points $P_{ij} = (x_i, y_j)$, where $\{x_i\}$ are sample points for a regular partition of $[a, b]$, and $\{y_j\}$ are sample points for a regular partition of $[c, d]$:

$$\iint_{\mathcal{R}} f(x, y)\, dA = \lim_{N,M \to \infty} \sum_{i=1}^{N} \sum_{j=1}^{M} f(x_i, y_j)\Delta y\Delta x$$

Here, $\Delta x = (b - a)/N$ and $\Delta y = (d - c)/M$. Fubini's Theorem stems from the elementary fact that we can add up the values in the sum in any order. So if we list the values $f(P_{ij})$ in an $N \times M$ array as shown in the margin, we can add up the columns first and then add up the sums from the columns. This yields

First, sum the columns.

$$\iint_{\mathcal{R}} f(x, y)\, dA = \lim_{N,M \to \infty} \sum_{i=1}^{N} \underbrace{\left(\sum_{j=1}^{M} f(x_i, y_j)\Delta y \right) \Delta x}$$

Then add up the column sums.

For fixed i, $f(x_i, y)$ is a continuous function of y and the inner sum on the right is a Riemann sum that approaches the single integral $\int_{c}^{d} f(x_i, y)\, dy$. In other words, setting $S(x) = \int_{c}^{d} f(x, y)\, dy$, we have, for each fixed x_i,

$$\lim_{M \to \infty} \sum_{j=1}^{M} f(x_i, y_j)\Delta y = \int_{c}^{d} f(x_i, y)\, dy = S(x_i)$$

To complete the proof, we take two facts for granted. First, $S(x)$ is a continuous function for $a \le x \le b$. Second, the limit as $N, M \to \infty$ may be computed by taking the limit first with respect to M and then with respect to N. Granting this, we get

$$\iint_{\mathcal{R}} f(x, y)\, dA = \lim_{N \to \infty} \sum_{i=1}^{N} \left(\lim_{M \to \infty} \sum_{j=1}^{M} f(x_i, y_j)\Delta y \right) \Delta x = \lim_{N \to \infty} \sum_{i=1}^{N} S(x_i)\Delta x$$

$$= \int_{a}^{b} S(x)\, dx = \int_{a}^{b} \left(\int_{c}^{d} f(x, y)\, dy \right) dx$$

Note that the sums on the right in the first line are Riemann sums for $S(x)$ that converge to the integral of $S(x)$ in the second line. This proves Fubini's Theorem for the order $dy\, dx$. A similar argument applies to the order $dx\, dy$. ∎

The term dA in a double integral is often referred to as an **area element**. Fubini's Theorem indicates that we have two choices for how we can express dA when we compute a double integral as an iterated integral: either as $dA = dy\,dx$, where we integrate first with respect to y, or as $dA = dx\,dy$, where we integrate first with respect to x.

GRAPHICAL INSIGHT   Assume that $f(x, y) \geq 0$ on a rectangle $\mathcal{R}$, and therefore the double integral of f over $\mathcal{R}$ is the volume of a solid S bounded between $\mathcal{R}$ and the graph of f (Figure 12). When we write the integral as an iterated integral in the order $dy\,dx$, then for each fixed value $x = x_0$, the inner integral is the area of the cross section of S in the vertical plane $x = x_0$ perpendicular to the x-axis as in Figure 12(A). That is,

$$S(x_0) = \int_c^d f(x_0, y)\,dy = \begin{array}{l}\text{area of cross section in vertical plane}\\ x = x_0 \text{ perpendicular to the } x\text{-axis}\end{array}$$

What Fubini's Theorem says is that the volume V of S can be calculated as the integral of cross-sectional area $S(x)$:

$$V = \int_a^b \int_c^d f(x, y)\,dy\,dx = \int_a^b S(x)\,dx = \text{integral of cross-sectional area}$$

Similarly, the iterated integral in the order $dx\,dy$ calculates V as the integral of cross sections perpendicular to the y-axis as in Figure 12(B).

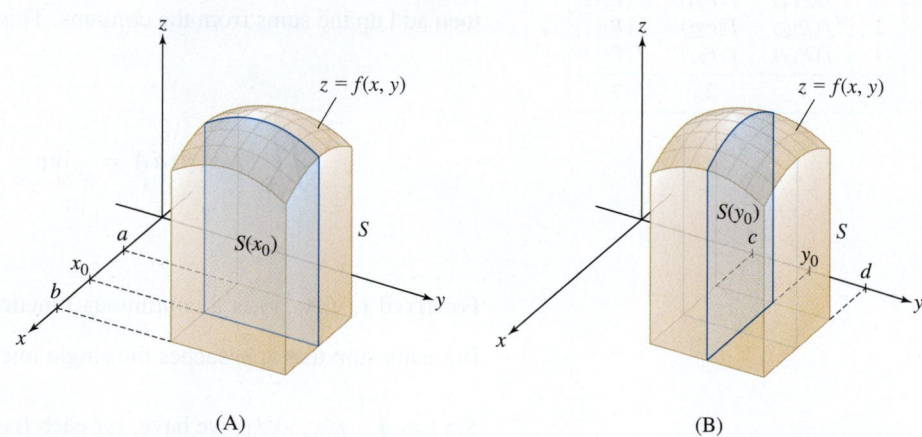

(A)   (B)

FIGURE 12

EXAMPLE 7   Find the volume V of the solid region enclosed between the graph of $f(x, y) = 16 - x^2 - 3y^2$ and the rectangle $\mathcal{R} = [0, 3] \times [0, 1]$ as shown in Figure 13.

Solution   The volume V is equal to the double integral of $f(x, y)$, which we write as an iterated integral:

$$V = \iint_{\mathcal{R}} (16 - x^2 - 3y^2)\,dA = \int_{x=0}^3 \int_{y=0}^1 (16 - x^2 - 3y^2)\,dy\,dx$$

We evaluate the inner integral first and then compute V:

$$\int_{y=0}^1 (16 - x^2 - 3y^2)\,dy = (16y - x^2 y - y^3)\Big|_{y=0}^1 = 15 - x^2$$

$$V = \int_{x=0}^3 (15 - x^2)\,dx = \left(15x - \frac{1}{3}x^3\right)\Big|_0^3 = 36$$

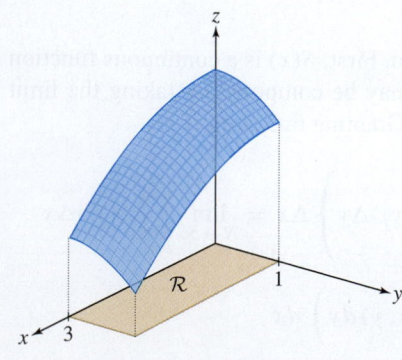

FIGURE 13 Graph of $f(x, y) = 16 - x^2 - 3y^2$ over $\mathcal{R} = [0, 3] \times [0, 1]$.

Multiple integration may be used to model the rate at which heat is transported by currents in the ocean. Imagine we have an xy-plane oriented vertically in the ocean (Figure 14) and a rectangular region $\mathcal{R}$ in it. Assume that the water directly crosses the region with speed $s(x, y)$, measured in meters per second (the corresponding velocity vectors are illustrated in the figure). Furthermore, assume that the temperature of the crossing water depends on x and y, and is given by $T(x, y)$ in degrees centigrade. Then the rate at which heat flows across $\mathcal{R}$ is given by the double integral

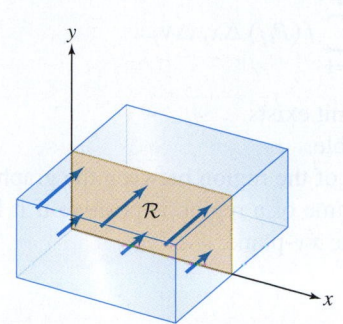

$$H = \iint_{\mathcal{R}} \rho c T(x, y) s(x, y) \, dA$$

FIGURE 14 The current carries heat across $\mathcal{R}$.

where, for the ocean water, ρ is the density (1025 kg/m^3) and c is the specific heat [3850 J/(kg°C)]. If the sides of $\mathcal{R}$ are measured in kilometers, then the resulting rate at which heat is transported is in units of megawatts.

The Gulf Stream is an important Atlantic Ocean current that flows northward along the coast of the United States and then eastward toward Europe. The following example provides a rough estimate of the heat transport rate of the Gulf Stream through a rectangular cross section. The speed and temperature vary widely along the current, as do the current's width and depth, but the values we use are representative.

EXAMPLE 8 Assume that the Gulf Stream is 100 km wide and the temperature varies from 15°C at the outer edges to 20°C in the middle. We model the temperature across it with $T(x, y) = 15 + 0.2x - 0.002x^2$. Further, we assume that it is 1 km deep and that the speed varies from bottom to top according to $s(x, y) = 0.5 + 1.5y$ m/s. Determine the rate of heat transport across the 100-km by 1-km rectangular section through the Gulf Stream.

Solution We compute H via

$$H = \iint_{\mathcal{R}} \rho c T(x, y) s(x, y) \, dA = \rho c \int_{x=0}^{100} \left(\int_{y=0}^{1} (15 + 0.2x - 0.002x^2)(0.5 + 1.5y) \, dy \right) dx$$

$$= \rho c \int_{x=0}^{100} (15 + 0.2x - 0.002x^2) \left(\int_{y=0}^{1} (0.5 + 1.5y) \, dy \right) dx$$

Now,

$$\int_{y=0}^{1} (0.5 + 1.5y) \, dy = \left(0.5y + 0.75y^2 \right) \Big|_0^1 = 1.25$$

Therefore

$$H = 1.25 \rho c \int_{x=0}^{100} (15 + 0.2x - 0.002x^2) \, dx$$

$$= 1.25(1025)(3850) \left(15x + 0.1x^2 - \frac{0.002}{3} x^3 \right) \Big|_0^{100} \approx 9.04 \times 10^9 \text{ megawatts} \quad \blacksquare$$

In Section 16.5, we generalize the flow-rate computation in the previous example to surface regions that are not necessarily flat and rectangular, and to flows that do not necessarily cross the surface directly.

15.1 SUMMARY

- A *Riemann sum* for $f(x, y)$ on a rectangle $\mathcal{R} = [a, b] \times [c, d]$ is a sum of the form

$$S_{N,M} = \sum_{i=1}^{N} \sum_{j=1}^{M} f(P_{ij}) \, \Delta x_i \, \Delta y_j$$

corresponding to partitions of $[a, b]$ and $[c, d]$, and choice of sample points P_{ij} in the subrectangle $\mathcal{R}_{ij}$.

- The double integral of $f(x, y)$ over $\mathcal{R}$ is defined as the limit (if it exists)

$$\iint_{\mathcal{R}} f(x, y)\, dA = \lim_{\|\mathcal{P}\| \to 0} \sum_{i=1}^{N} \sum_{j=1}^{M} f(P_{ij})\, \Delta x_i\, \Delta y_j$$

We say that $f(x, y)$ is *integrable* over $\mathcal{R}$ if this limit exists.
- A continuous function on a rectangle $\mathcal{R}$ is integrable.
- The double integral is equal to the *signed volume* of the region between the graph of $z = f(x, y)$ and the rectangle $\mathcal{R}$. The signed volume of a region is positive if it lies above the xy-plane and negative if it lies below the xy-plane.
- If $f(x, y) = C$ is a constant function, then

$$\iint_{\mathcal{R}} C\, dA = C \cdot \text{area}(\mathcal{R})$$

- Fubini's Theorem: The double integral of a continuous function $f(x, y)$ over a rectangle $\mathcal{R} = [a, b] \times [c, d]$ can be evaluated as an iterated integral (in either order):

$$\iint_{\mathcal{R}} f(x, y)\, dA = \int_{x=a}^{b} \int_{y=c}^{d} f(x, y)\, dy\, dx = \int_{y=c}^{d} \int_{x=a}^{b} f(x, y)\, dx\, dy$$

15.1 EXERCISES

Preliminary Questions

1. If $S_{8,4}$ is a Riemann sum for a double integral over a rectangle $\mathcal{R} = [1, 5] \times [2, 10]$ using a regular partition, what is the area of each subrectangle? How many subrectangles are there?

2. Estimate the double integral of a continuous function f over the small rectangle $\mathcal{R} = [0.9, 1.1] \times [1.9, 2.1]$ if $f(1, 2) = 4$.

3. What is the integral of the constant function $f(x, y) = 5$ over the rectangle $[-2, 3] \times [2, 4]$?

4. What is the interpretation of $\displaystyle\iint_{\mathcal{R}} f(x, y)\, dA$ if $f(x, y)$ takes on both positive and negative values on $\mathcal{R}$?

5. Which of (a) or (b) is equal to $\displaystyle\int_{1}^{2} \int_{4}^{5} f(x, y)\, dy\, dx$?

(a) $\displaystyle\int_{1}^{2} \int_{4}^{5} f(x, y)\, dx\, dy$ **(b)** $\displaystyle\int_{4}^{5} \int_{1}^{2} f(x, y)\, dx\, dy$

6. For which of the following functions is the double integral over the rectangle in Figure 15 equal to zero? Explain your reasoning.

(a) $f(x, y) = x^2 y$ **(b)** $f(x, y) = xy^2$

(c) $f(x, y) = \sin x$ **(d)** $f(x, y) = e^x$

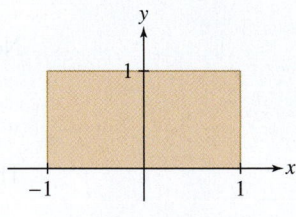

FIGURE 15

Exercises

1. Compute the Riemann sum $S_{4,3}$ to estimate the double integral of $f(x, y) = xy$ over $\mathcal{R} = [1, 3] \times [1, 2.5]$. Use the regular partition and upper-right vertices of the subrectangles as sample points.

2. Compute the Riemann sum with $N = M = 2$ to estimate the integral of $\sqrt{x + y}$ over $\mathcal{R} = [0, 1] \times [0, 1]$. Use the regular partition and midpoints of the subrectangles as sample points.

In Exercises 3–6, compute the Riemann sums for the double integral
$\displaystyle\iint_{\mathcal{R}} f(x, y)\, dA$, *where* $\mathcal{R} = [1, 4] \times [1, 3]$, *for the grid and two choices of sample points shown in Figure 16.*

3. $f(x, y) = 2x + y$ **4.** $f(x, y) = 7$

5. $f(x, y) = 4x$ **6.** $f(x, y) = x - 2y$

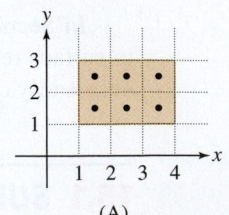
(A)

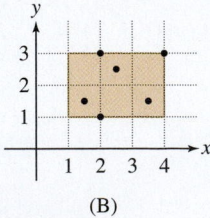
(B)

FIGURE 16

7. Let $\mathcal{R} = [0, 1] \times [0, 1]$. Estimate $\displaystyle\iint_{\mathcal{R}} (x + y)\, dA$ by computing two different Riemann sums, each with at least six rectangles.

8. Evaluate $\iint_{\mathcal{R}} 4 \, dA$, where $\mathcal{R} = [2,5] \times [4,7]$.

9. Evaluate $\iint_{\mathcal{R}} (15 - 3x) \, dA$, where $\mathcal{R} = [0,5] \times [0,3]$, and sketch the corresponding solid region (see Example 2).

10. Evaluate $\iint_{\mathcal{R}} (-5) \, dA$, where $\mathcal{R} = [2,5] \times [4,7]$.

11. The following table gives the approximate height at quarter-meter intervals of a mound of gravel. Estimate the volume of the mound by computing the average of the two Riemann sums $S_{4,3}$ with lower-left and upper-right vertices of the subrectangles as sample points.

0.75	0.1	0.2	0.2	0.15	0.1
0.5	0.2	0.3	0.5	0.4	0.2
0.25	0.15	0.2	0.4	0.3	0.2
0	0.1	0.15	0.2	0.15	0.1
$y \diagdown x$	0	0.25	0.5	0.75	1

12. Use the following table to compute a Riemann sum $S_{3,3}$ for $f(x, y)$ on the square $\mathcal{R} = [0, 1.5] \times [0.5, 2]$. Use the regular partition and sample points of your choosing.

Values of $f(x, y)$					
2	2.6	2.17	1.86	1.62	1.44
1.5	2.2	1.83	1.57	1.37	1.22
1	1.8	1.5	1.29	1.12	1
0.5	1.4	1.17	1	0.87	0.78
0	1	0.83	0.71	0.62	0.56
$y \diagdown x$	0	0.5	1	1.5	2

13. [CAS] Let $S_{N,N}$ be the Riemann sum for $\int_0^1 \int_0^1 e^{x^3 - y^3} \, dy \, dx$ using the regular partition and the lower-left vertex of each subrectangle as sample points. Use a computer algebra system to calculate $S_{N,N}$ for $N = 25$, 50, 100.

14. [CAS] Let $S_{N,M}$ be the Riemann sum for

$$\int_0^4 \int_0^2 \ln(1 + x^2 + y^2) \, dy \, dx$$

using the regular partition and the upper-right vertex of each subrectangle as sample points. Use a computer algebra system to calculate $S_{2N,N}$ for $N = 25, 50, 100$.

In Exercises 15–18, use symmetry to evaluate the double integral.

15. $\iint_{\mathcal{R}} x^3 \, dA$, $\mathcal{R} = [-4, 4] \times [0, 5]$

16. $\iint_{\mathcal{R}} (1 - x) \, dA$, $\mathcal{R} = [0, 2] \times [-7, 7]$

17. $\iint_{\mathcal{R}} \sin x \, dA$, $\mathcal{R} = [0, 2\pi] \times [0, 2\pi]$

18. $\iint_{\mathcal{R}} (2 + x^2 y) \, dA$, $\mathcal{R} = [0, 1] \times [-1, 1]$

In Exercises 19–36, evaluate the iterated integral.

19. $\int_1^3 \int_0^2 x^3 y \, dy \, dx$

20. $\int_0^2 \int_1^3 x^3 y \, dx \, dy$

21. $\int_4^9 \int_{-3}^8 1 \, dx \, dy$

22. $\int_{-4}^{-1} \int_4^8 (-5) \, dx \, dy$

23. $\int_{-1}^1 \int_0^{\pi} x^2 \sin y \, dy \, dx$

24. $\int_{-1}^1 \int_0^{\pi} x^2 \sin y \, dx \, dy$

25. $\int_2^6 \int_1^4 x^2 \, dx \, dy$

26. $\int_2^6 \int_1^4 y^2 \, dx \, dy$

27. $\int_0^1 \int_0^2 (x + 4y^3) \, dx \, dy$

28. $\int_0^2 \int_0^2 (x^2 - y^2) \, dy \, dx$

29. $\int_0^4 \int_0^9 \sqrt{x + 4y} \, dx \, dy$

30. $\int_0^{\pi/4} \int_{\pi/4}^{\pi/2} \cos(2x + y) \, dy \, dx$

31. $\int_1^2 \int_0^4 \frac{dy \, dx}{x + y}$

32. $\int_1^2 \int_2^4 e^{3x - y} \, dy \, dx$

33. $\int_0^4 \int_0^5 \frac{dy \, dx}{\sqrt{x + y}}$

34. $\int_0^8 \int_1^2 \frac{x \, dx \, dy}{\sqrt{x^2 + y}}$

35. $\int_1^2 \int_1^3 \frac{\ln(xy) \, dy \, dx}{y}$

36. $\int_0^1 \int_2^3 \frac{1}{(x + 4y)^3} \, dx \, dy$

In Exercises 37–44, evaluate the integral.

37. $\iint_{\mathcal{R}} \frac{x}{y} \, dA$, $\mathcal{R} = [-2, 4] \times [1, 3]$

38. $\iint_{\mathcal{R}} x^2 y \, dA$, $\mathcal{R} = [-1, 1] \times [0, 2]$

39. $\iint_{\mathcal{R}} \cos x \sin 2y \, dA$, $\mathcal{R} = \left[0, \frac{\pi}{2}\right] \times \left[0, \frac{\pi}{2}\right]$

40. $\iint_{\mathcal{R}} \frac{y}{x + 1} \, dA$, $\mathcal{R} = [0, 2] \times [0, 4]$

41. $\iint_{\mathcal{R}} e^x \sin y \, dA$, $\mathcal{R} = [0, 2] \times \left[0, \frac{\pi}{4}\right]$

42. $\iint_{\mathcal{R}} e^{3x + 4y} \, dA$, $\mathcal{R} = [0, 1] \times [1, 2]$

43. $\iint_{\mathcal{R}} x \ln y \, dA$, $\mathcal{R} = [0, 3] \times [1, e]$

44. $\iint_{\mathcal{R}} x^2 \tan y \, dA$, $\mathcal{R} = [0, 2] \times \left[0, \frac{\pi}{3}\right]$

45. Let $f(x, y) = mxy^2$, where m is a constant. Find a value of m such that $\iint_{\mathcal{R}} f(x, y) \, dA = 1$, where $\mathcal{R} = [0, 1] \times [0, 2]$.

46. Evaluate

$$I = \int_1^3 \int_0^1 ye^{xy} \, dy \, dx$$

You will need Integration by Parts and the formula

$$\int e^x (x^{-1} - x^{-2}) \, dx = x^{-1} e^x + C$$

Then evaluate I again using Fubini's Theorem to change the order of integration (i.e., integrate first with respect to x). Which method is easier?

47. (a) Which is easier, antidifferentiating $y\sqrt{1+xy}$ with respect to x or with respect to y? Explain.

(b) Evaluate $\iint_{\mathcal{R}} y\sqrt{1+xy} \, dA$, where $\mathcal{R} = [0,1] \times [0,1]$.

48. (a) Which is easier, antidifferentiating xe^{xy} with respect to x or with respect to y? Explain.

(b) Evaluate $\iint_{\mathcal{R}} xe^{xy} \, dA$, where $\mathcal{R} = [0,1] \times [0,1]$.

49. (a) Which is easier, antidifferentiating $\frac{y}{1+xy}$ with respect to x or with respect to y? Explain.

(b) Evaluate $\iint_{\mathcal{R}} \frac{y}{1+xy} \, dA$, where $\mathcal{R} = [0,1] \times [0,1]$.

50. Calculate a Riemann sum $S_{3,3}$ on the square $\mathcal{R} = [0,3] \times [0,3]$ for the function $f(x,y)$ whose contour plot is shown in Figure 17. Choose sample points and use the plot to find the values of $f(x,y)$ at these points.

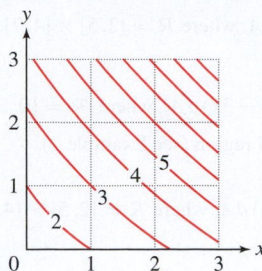

FIGURE 17 Contour plot of $f(x,y)$.

51. 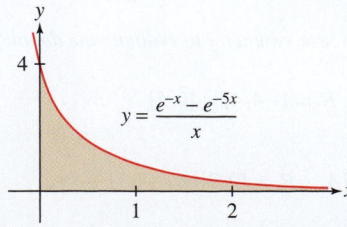 Using Fubini's Theorem, argue that the solid in Figure 18 has volume AL, where A is the area of the front face of the solid.

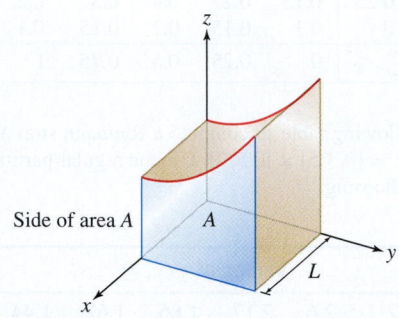

Side of area A

FIGURE 18

Further Insights and Challenges

52. Prove the following extension of the Fundamental Theorem of Calculus to two variables: If $\dfrac{\partial^2 F}{\partial x \, \partial y} = f(x,y)$, then

$$\iint_{\mathcal{R}} f(x,y) \, dA = F(b,d) - F(a,d) - F(b,c) + F(a,c)$$

where $\mathcal{R} = [a,b] \times [c,d]$.

53. Let $F(x,y) = x^{-1} e^{xy}$. Show that $\dfrac{\partial^2 F}{\partial x \, \partial y} = ye^{xy}$ and use the result of Exercise 52 to evaluate $\iint_{\mathcal{R}} ye^{xy} \, dA$ for $\mathcal{R} = [1,3] \times [0,1]$.

54. Find a function $F(x,y)$ satisfying $\dfrac{\partial^2 F}{\partial x \, \partial y} = 6x^2 y$ and use the result of Exercise 52 to evaluate $\iint_{\mathcal{R}} 6x^2 y \, dA$ for $\mathcal{R} = [0,1] \times [0,4]$.

55. In this exercise, we use double integration to evaluate the following improper integral for $a > 0$ a positive constant:

$$I(a) = \int_0^\infty \frac{e^{-x} - e^{-ax}}{x} \, dx$$

(a) Use L'Hôpital's Rule to show that $f(x) = \dfrac{e^{-x} - e^{-ax}}{x}$, though not defined at $x = 0$, can be defined and made continuous at $x = 0$ by assigning the value $f(0) = a - 1$.

(b) Prove that $|f(x)| \le e^{-x} + e^{-ax}$ for $x > 1$ (use the Triangle Inequality), and apply the Comparison Theorem to show that $I(a)$ converges.

(c) Show that $I(a) = \int_0^\infty \int_1^a e^{-xy} \, dy \, dx$.

(d) Prove, by interchanging the order of integration, that

$$I(a) = \ln a - \lim_{T \to \infty} \int_1^a \frac{e^{-Ty}}{y} \, dy \qquad \boxed{1}$$

(e) Use the Comparison Theorem to show that the limit in Eq. (1) is zero and therefore that $I(a) = \ln a$ (see Figure 19, for example). *Hint:* If $a \ge 1$, show that $e^{-Ty}/y \le e^{-T}$ for $y \ge 1$, and if $a < 1$, show that $e^{-Ty}/y \le e^{-aT}/a$ for $a \le y \le 1$.

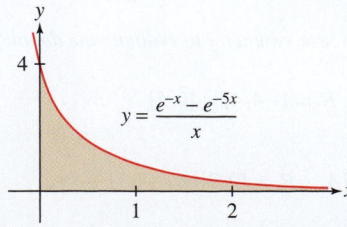

FIGURE 19 The shaded region has area $\ln 5$.

15.2 Double Integrals over More General Regions

In the previous section, we restricted our attention to integrals on rectangular domains. Now, we shall treat the more general case of domains $\mathcal{D}$ whose boundaries are simple closed curves (a curve is simple if it does not intersect itself and closed if it begins and ends at the same point). We assume that the boundary of $\mathcal{D}$ is smooth as in Figure 1(A) or consists of finitely many smooth curves, joined together with possible corners, as in Figure 1(B). A boundary curve of this type is called **piecewise smooth**. We also assume that $\mathcal{D}$ is a closed domain; that is, $\mathcal{D}$ contains its boundary.

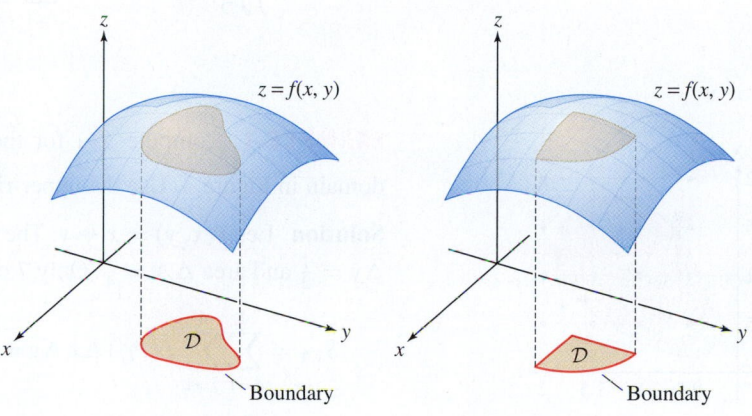

(A) $\mathcal{D}$ has a smooth boundary.

(B) $\mathcal{D}$ has a piecewise smooth boundary, consisting of three smooth curves joined at the corners.

FIGURE 1

Fortunately, we do not need to start from the beginning to define the double integral over a domain $\mathcal{D}$ of this type. Given a function $f(x, y)$ on $\mathcal{D}$, we choose a rectangle $\mathcal{R} = [a, b] \times [c, d]$ containing $\mathcal{D}$ and define a new function $\tilde{f}(x, y)$ that agrees with $f(x, y)$ on $\mathcal{D}$ and is zero outside of $\mathcal{D}$ (Figure 2):

$$\tilde{f}(x, y) = \begin{cases} f(x, y) & \text{if } (x, y) \in \mathcal{D} \\ 0 & \text{if } (x, y) \notin \mathcal{D} \end{cases}$$

The double integral of f over $\mathcal{D}$ is defined as the integral of $\tilde{f}$ over $\mathcal{R}$:

$$\iint_{\mathcal{D}} f(x, y)\, dA = \iint_{\mathcal{R}} \tilde{f}(x, y)\, dA \qquad \boxed{1}$$

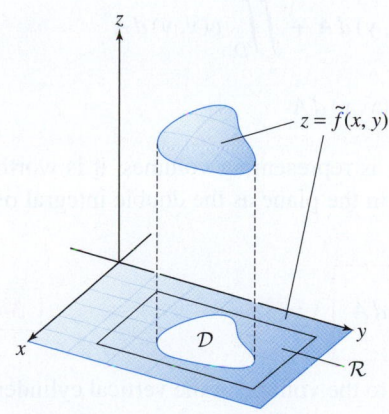

FIGURE 2 The function $\tilde{f}$ is zero outside of $\mathcal{D}$.

We say that f is **integrable** over $\mathcal{D}$ if the integral of $\tilde{f}$ over $\mathcal{R}$ exists. The value of the integral does not depend on the particular choice of $\mathcal{R}$ because $\tilde{f}$ is zero outside of $\mathcal{D}$.

This definition seems reasonable because the integral of $\tilde{f}$ involves only the values of f on $\mathcal{D}$. However, $\tilde{f}$ is likely to be discontinuous because its values could jump to zero beyond the boundary. Despite this possible discontinuity, the next theorem guarantees that the integral of $\tilde{f}$ over $\mathcal{R}$ exists if our original function f is continuous on $\mathcal{D}$.

THEOREM 1 If $f(x, y)$ is continuous on a closed domain $\mathcal{D}$ whose boundary is a simple closed piecewise smooth curve, then $\displaystyle\iint_{\mathcal{D}} f(x, y)\, dA$ exists.

In Theorem 1, we define continuity on $\mathcal{D}$ to mean that f is defined and continuous on some open set containing $\mathcal{D}$.

As in the previous section, the double integral defines the signed volume between the graph of $f(x, y)$ and the xy-plane.

We can approximate the double integral by Riemann sums for the function $\tilde{f}$ on a rectangle $\mathcal{R}$ containing $\mathcal{D}$. Because $\tilde{f}(P) = 0$ for points P in $\mathcal{R}$ that do not belong to $\mathcal{D}$, any such Riemann sum reduces to a sum over those sample points that lie in $\mathcal{D}$:

$$\iint_{\mathcal{D}} f(x, y)\, dA \approx \sum_{i=1}^{N} \sum_{j=1}^{M} \tilde{f}(P_{ij})\, \Delta x_i\, \Delta y_j = \underbrace{\sum f(P_{ij})\, \Delta x_i\, \Delta y_j}_{\substack{\text{Sum only over points} \\ P_{ij} \text{ that lie in } \mathcal{D}}} \qquad \boxed{2}$$

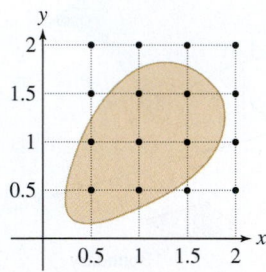

FIGURE 3 Domain $\mathcal{D}$.

EXAMPLE 1 Compute $S_{4,4}$ for the integral $\displaystyle\iint_{\mathcal{D}} (x + y)\, dA$, where $\mathcal{D}$ is the shaded domain in Figure 3. Use the upper-right corners of the squares as sample points.

Solution Let $f(x, y) = x + y$. The subrectangles in Figure 3 have sides of length $\Delta x = \Delta y = \frac{1}{2}$ and area $\Delta A = \frac{1}{4}$. Only 7 of the 16 sample points lie in $\mathcal{D}$, so

$$S_{4,4} = \sum_{i=1}^{4} \sum_{j=1}^{4} \tilde{f}(P_{ij})\, \Delta x\, \Delta y = \frac{1}{4}\big(f(0.5, 0.5) + f(1, 0.5) + f(0.5, 1) + f(1, 1)$$

$$+ f(1.5, 1) + f(1, 1.5) + f(1.5, 1.5)\big)$$

$$= \frac{1}{4}\big(1 + 1.5 + 1.5 + 2 + 2.5 + 2.5 + 3 \big) = \frac{7}{2} \qquad \blacksquare$$

The linearity properties of the double integral carry over to general domains: If $f(x, y)$ and $g(x, y)$ are integrable and C is a constant, then

$$\iint_{\mathcal{D}} (f(x, y) + g(x, y))\, dA = \iint_{\mathcal{D}} f(x, y)\, dA + \iint_{\mathcal{D}} g(x, y)\, dA$$

$$\iint_{\mathcal{D}} Cf(x, y)\, dA = C \iint_{\mathcal{D}} f(x, y)\, dA$$

Although we usually think of double integrals as representing volumes, it is worth noting that we can express the *area* of a domain $\mathcal{D}$ in the plane as the double integral of the constant function $f(x, y) = 1$:

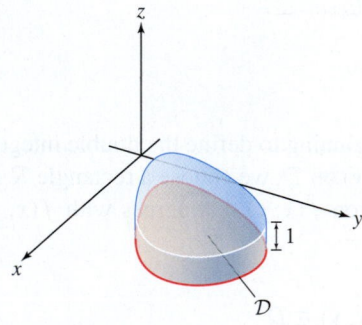

FIGURE 4 The volume of the cylinder of height 1 with $\mathcal{D}$ as its base is equal to the area of $\mathcal{D}$.

$$\boxed{\text{area}(\mathcal{D}) = \iint_{\mathcal{D}} 1\, dA} \qquad \boxed{3}$$

Indeed, as we see in Figure 4, the area of $\mathcal{D}$ is equal to the volume of the vertical cylinder of height 1 with $\mathcal{D}$ as base. More generally, for any constant C,

$$\iint_{\mathcal{D}} C\, dA = C \cdot \text{area}(\mathcal{D}) \qquad \boxed{4}$$

CONCEPTUAL INSIGHT Equation (3) tells us that we can approximate the area of a domain $\mathcal{D}$ by a Riemann sum for $\displaystyle\iint_{\mathcal{D}} 1\, dA$. In this case, $f(x, y) = 1$, and we obtain a Riemann sum by creating a grid and adding up the areas $\Delta x_i\, \Delta y_j$ of those rectangles in the grid that are contained in $\mathcal{D}$ or that intersect the boundary of $\mathcal{D}$ (Figure 5). The finer the grid, the better the approximation. The exact area is the limit as the sides of the rectangles tend to zero.

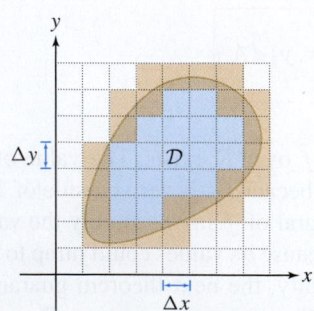

FIGURE 5 The area of $\mathcal{D}$ is approximated by the sum of the areas of the rectangles contained in $\mathcal{D}$.

Now, how do we compute double integrals over nonrectangular regions? We address this question next, in two special cases where the region lies between two graphs. In each case, the computation involves an iterated integral.

Regions Between Two Graphs

When $\mathcal{D}$ is a region between two graphs in the xy-plane, we can evaluate double integrals over $\mathcal{D}$ as iterated integrals. Recall from Section 6.1 that $\mathcal{D}$ is vertically simple if it is the region between the graphs of two continuous functions $y = g_1(x)$ and $y = g_2(x)$ over a fixed interval of x-values [Figure 6(A)]:

$$\mathcal{D} = \{(x, y) : a \le x \le b, \quad g_1(x) \le y \le g_2(x)\}$$

Similarly, $\mathcal{D}$ is **horizontally simple** [Figure 6(B)] if

$$\mathcal{D} = \{(x, y) : c \le y \le d, \quad h_1(y) \le x \le h_2(y)\}$$

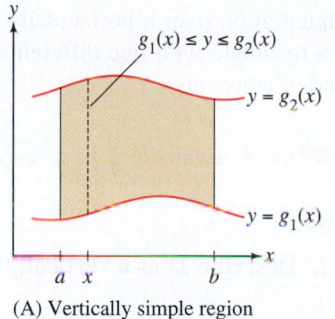

(A) Vertically simple region

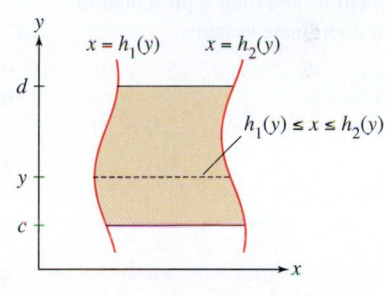

(B) Horizontally simple region

FIGURE 6

THEOREM 2 If $\mathcal{D}$ is vertically simple with description

$$a \le x \le b, \qquad g_1(x) \le y \le g_2(x)$$

then

> *When you write a double integral over a vertically simple region as an iterated integral, the inner integral is an integral over the dashed segment shown in Figure 6(A). For a horizontally simple region, the inner integral is an integral over the dashed segment shown in Figure 6(B).*

$$\iint_{\mathcal{D}} f(x, y)\, dA = \int_a^b \int_{g_1(x)}^{g_2(x)} f(x, y)\, dy\, dx$$

If $\mathcal{D}$ is a horizontally simple region with description

$$c \le y \le d, \qquad h_1(y) \le x \le h_2(y)$$

then

$$\iint_{\mathcal{D}} f(x, y)\, dA = \int_c^d \int_{h_1(y)}^{h_2(y)} f(x, y)\, dx\, dy$$

Proof We sketch the proof, assuming that $\mathcal{D}$ is vertically simple (the horizontally simple case is similar). Choose a rectangle $\mathcal{R} = [a, b] \times [c, d]$ containing $\mathcal{D}$, and let $\tilde{f}$ be the function shown in Figure 2 that equals f on $\mathcal{D}$ and otherwise equals zero. We wish to

Winifred Edgerton Merrill (1862–1951) was the first American woman to earn a Ph.D. in mathematics, awarded by Columbia University in 1886. Her thesis was a study of geometric interpretations of multiple integrals and their representation in different coordinate systems.

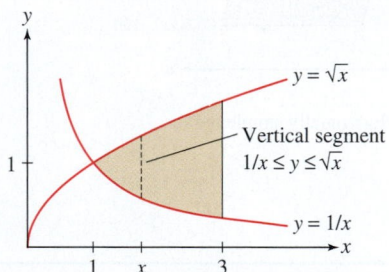

FIGURE 7 Domain between $y = \sqrt{x}$ and $y = 1/x$.

employ Fubini's Theorem next. However, the theorem states that the function must be continuous, but $\tilde{f}$ may not be continuous on $\mathcal{R}$. It is possible, however, to prove that Fubini's Theorem also holds for functions such as $\tilde{f}$ that are continuous on a domain in $\mathcal{R}$ and that are zero outside the domain. Therefore, using Fubini's Theorem (for the second equality) we have

$$\iint_{\mathcal{D}} f(x,y)\, dA = \iint_{\mathcal{R}} \tilde{f}(x,y)\, dA = \int_a^b \int_c^d \tilde{f}(x,y)\, dy\, dx \qquad \boxed{5}$$

By definition, $\tilde{f}(x,y)$ is zero outside $\mathcal{D}$, so for fixed x, $\tilde{f}(x,y)$ is zero unless y satisfies $g_1(x) \le y \le g_2(x)$. Therefore,

$$\int_c^d \tilde{f}(x,y)\, dy = \int_{g_1(x)}^{g_2(x)} f(x,y)\, dy$$

Substituting in Eq. (5), we obtain the desired equality:

$$\iint_{\mathcal{D}} f(x,y)\, dA = \int_a^b \int_{g_1(x)}^{g_2(x)} f(x,y)\, dy\, dx \qquad \blacksquare$$

Integration over a horizontally or vertically simple region is similar to integration over a rectangle with one difference: The limits of the inner integral may be functions instead of constants.

EXAMPLE 2 Evaluate $\displaystyle\iint_{\mathcal{D}} x^2 y\, dA$, where $\mathcal{D}$ is the region in Figure 7.

Solution

Step 1. **Describe $\mathcal{D}$ as a vertically simple region.**

$$\underbrace{1 \le x \le 3}_{\substack{\text{Limits of outer}\\\text{integral}}}, \qquad \underbrace{\frac{1}{x} \le y \le \sqrt{x}}_{\substack{\text{Limits of inner}\\\text{integral}}}$$

In this case, $g_1(x) = 1/x$ and $g_2(x) = \sqrt{x}$.

Step 2. **Set up the iterated integral.**

$$\iint_{\mathcal{D}} x^2 y\, dA = \int_1^3 \int_{y=1/x}^{\sqrt{x}} x^2 y\, dy\, dx$$

Notice that the inner integral is an integral over a vertical segment between the graphs of $y = 1/x$ and $y = \sqrt{x}$.

Step 3. **Compute the iterated integral.**

First, we evaluate the inner integral. We compute it using FTC I, finding an antiderivative with respect to y, holding x constant, and then taking the difference of the antiderivative evaluated at the limits. Note that the limits depend on x, and therefore the result of the inner integral will be expressed as a function of x:

$$\int_{y=1/x}^{\sqrt{x}} x^2 y\, dy = \frac{1}{2} x^2 y^2 \Big|_{y=1/x}^{\sqrt{x}} = \frac{1}{2} x^2 (\sqrt{x})^2 - \frac{1}{2} x^2 \left(\frac{1}{x}\right)^2 = \frac{1}{2} x^3 - \frac{1}{2}$$

We complete the calculation by computing the outer integral:

$$\iint_{\mathcal{D}} x^2 y\, dA = \int_1^3 \left(\frac{1}{2} x^3 - \frac{1}{2}\right) dx = \left(\frac{1}{8} x^4 - \frac{1}{2} x\right)\Big|_1^3$$

$$= \frac{69}{8} - \left(-\frac{3}{8}\right) = 9 \qquad \blacksquare$$

EXAMPLE 3 **A Volume Integral** Find the volume V of the region under the plane $z = 2x + 3y$ and above the triangle $\mathcal{D}$ in the xy-plane in Figure 8.

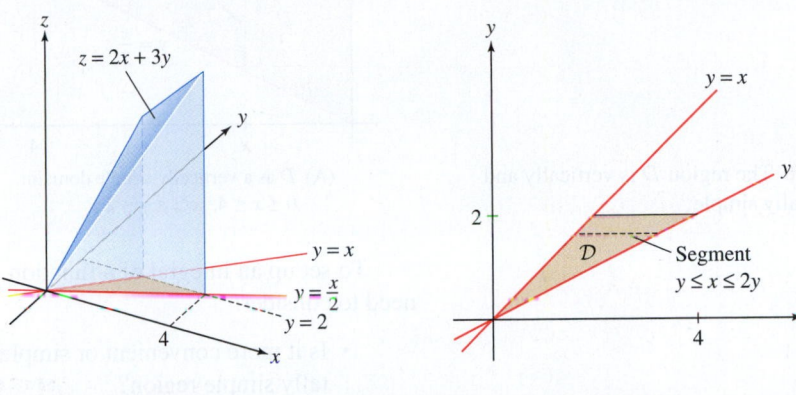

FIGURE 8   FIGURE 9

Solution The triangle $\mathcal{D}$ is bounded by the lines $y = x/2$, $y = x$, and $y = 2$. Figure 9 shows that $\mathcal{D}$ is a horizontally simple region described by

$$\mathcal{D}: 0 \le y \le 2, \quad y \le x \le 2y$$

The domain $\mathcal{D}$ is also a vertically simple region, but the upper curve is not given by a single formula, requiring $\mathcal{D}$ to be divided into two domains over which separate integrals would be set up. The formula switches from $y = x$ to $y = 2$. Therefore, it is more convenient to consider $\mathcal{D}$ as a horizontally simple region.

The volume is equal to the double integral of $f(x, y) = 2x + 3y$ over $\mathcal{D}$:

$$V = \iint_{\mathcal{D}} f(x, y)\, dA = \int_0^2 \int_{x=y}^{2y} (2x + 3y)\, dx\, dy$$

$$= \int_0^2 \left. (x^2 + 3yx) \right|_{x=y}^{2y} dy = \int_0^2 \left((4y^2 + 6y^2) - (y^2 + 3y^2) \right) dy$$

$$= \int_0^2 6y^2\, dy = 2y^3 \Big|_0^2 = 16 \qquad \blacksquare$$

The next example shows that in some cases, one iterated integral is easier to evaluate than the other.

EXAMPLE 4 **Choosing the Best Iterated Integral** Evaluate $\displaystyle\iint_{\mathcal{D}} e^{y^2}\, dA$ for $\mathcal{D}$ in Figure 10.

Solution First, let's try describing $\mathcal{D}$ as a vertically simple domain. Referring to Figure 10(A), we have

$$\mathcal{D}: 0 \le x \le 4, \quad \frac{1}{2}x \le y \le 2 \quad \Rightarrow \quad \iint_{\mathcal{D}} e^{y^2}\, dA = \int_{x=0}^4 \int_{y=x/2}^2 e^{y^2}\, dy\, dx$$

The inner integral cannot be evaluated because we have no explicit antiderivative for e^{y^2}. Therefore, we try describing $\mathcal{D}$ as horizontally simple [Figure 10(B)]:

$$\mathcal{D}: 0 \le y \le 2, \quad 0 \le x \le 2y$$

This leads to an iterated integral that can be evaluated:

$$\int_0^2 \int_{x=0}^{2y} e^{y^2}\, dx\, dy = \int_0^2 \left(x e^{y^2} \Big|_{x=0}^{2y} \right) dy = \int_0^2 2y e^{y^2}\, dy$$

$$= e^{y^2} \Big|_0^2 = e^4 - 1 \qquad \blacksquare$$

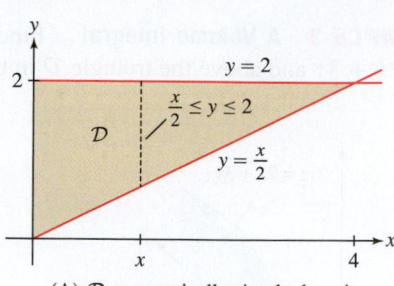

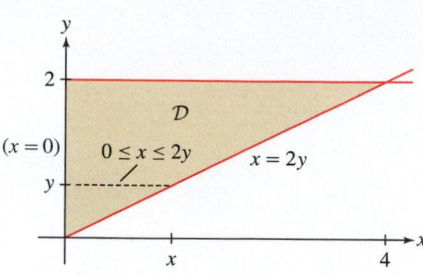

(A) $\mathcal{D}$ as a vertically simple domain:
$0 \le x \le 4,\ x/2 \le y \le 2$

(B) $\mathcal{D}$ as a horizontally simple domain:
$0 \le y \le 2,\ 0 \le x \le 2y$

FIGURE 10 The region $\mathcal{D}$ is vertically and horizontally simple.

To set up an integral of a function $f(x, y)$ over a domain $\mathcal{D}$ there are two issues you need to consider:

- Is it more convenient or simpler to express $\mathcal{D}$ as a vertically simple or a horizontally simple region?
- To compute the inner integral, is it easier to find an antiderivative of f with respect to y first or with respect to x first?

EXAMPLE 5   **Changing the Order of Integration**   Sketch the domain of integration $\mathcal{D}$ corresponding to

$$\int_1^9 \int_{\sqrt{y}}^3 x e^y \, dx \, dy$$

Then change the order of integration and evaluate.

Solution The limits of integration give us inequalities that describe the domain $\mathcal{D}$ as a horizontally simple region such that

$$1 \le y \le 9, \qquad \sqrt{y} \le x \le 3$$

We sketch the region in Figure 11. Now observe that $\mathcal{D}$ is also vertically simple:

$$1 \le x \le 3, \qquad 1 \le y \le x^2$$

so we can rewrite our integral and evaluate:

$$\int_1^9 \int_{x=\sqrt{y}}^3 x e^y \, dx \, dy = \int_1^3 \int_{y=1}^{x^2} x e^y \, dy \, dx = \int_1^3 \left(\int_{y=1}^{x^2} x e^y \, dy \right) dx$$

$$= \int_1^3 \left(x e^y \Big|_{y=1}^{x^2} \right) dx = \int_1^3 (x e^{x^2} - ex) \, dx = \frac{1}{2} (e^{x^2} - ex^2) \Big|_1^3$$

$$= \frac{1}{2}(e^9 - 9e) - 0 = \frac{1}{2}(e^9 - 9e) \qquad \blacksquare$$

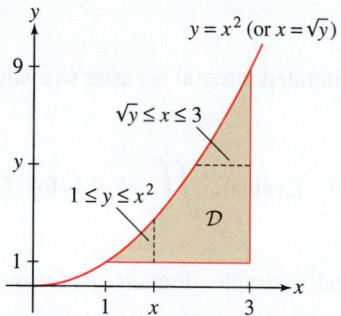

FIGURE 11 Describing $\mathcal{D}$ as horizontally simple and vertically simple.

Example 5 demonstrates that changing the order of integration is more complicated when the domain is not a rectangle. We cannot simply switch the dx and dy and exchange the limits on the inner and outer integrals. First, we need to determine what curves form the boundary of $\mathcal{D}$ and then (if possible) rewrite the integrals with appropriate limits of integration regarding $\mathcal{D}$ as vertically simple (if switching to a $dy\,dx$ integral) or horizontally simple (if switching to $dx\,dy$).

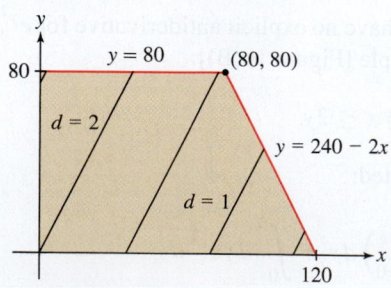

FIGURE 12 The snowpack in Jocoro Provincial Park on April 1.

EXAMPLE 6   **Estimating Snowpack**   Estimating the winter snowpack is important in situations where the subsequent spring melt could cause flooding or affect water-use management. Assume that Jocoro Provincial Park occupies an area modeled by the domain $\mathcal{D}$ in Figure 12. This domain is bordered by the coordinate axes, and the lines $y = 80$ and $y = 240 - 2x$ (where units are in kilometers). Furthermore, assume that snow depth measurements (in meters) were taken over the park on April 1 and that the depth measurements are modeled by the function $d(x, y) = -0.014x + 0.007y + 2.0$ that is displayed via a contour map in the figure. Estimate the volume of the snowpack in the park in cubic meters.

Solution The snowpack volume is represented by the volume under the graph of $d(x, y)$ over $\mathcal{D}$, so we compute the corresponding integral. The domain is best expressed as horizontally simple:

$$0 \le y \le 80, \qquad 0 \le x \le 120 - \frac{1}{2}y$$

Integrating, we have

$$\text{volume} = \int_0^{80} \int_{x=0}^{120 - \frac{1}{2}y} (-0.014x + 0.007y + 2.0)\, dx\, dy$$

$$= \int_0^{80} (-0.007x^2 + 0.007yx + 2.0x) \Big|_{x=0}^{120 - \frac{1}{2}y} dy$$

$$= \int_0^{80} \left(-0.007 \left(120 - \frac{1}{2}y\right)^2 + 0.007y \left(120 - \frac{1}{2}y\right) + 2.0 \left(120 - \frac{1}{2}y\right) \right) dy$$

$$= \int_0^{80} (-0.00525y^2 + 0.68y + 139.2)\, dy = 12{,}416$$

Since x and y are in kilometers and d is in meters, we need to multiply the result by 10^6 to convert to a quantity in cubic meters. It follows that our estimate for the snowpack volume in the park is 12.416 billion cubic meters. ■

EXAMPLE 7 **A Volume Enclosed Between Two Surfaces** Find the volume V of the solid bounded above by the paraboloid $z = 8 - x^2 - y^2$ and below by the paraboloid $z = x^2 + y^2$ over the domain $\mathcal{D} = \{(x, y) : -1 \le x \le 1, -1 \le y \le 1\}$.

Solution The solid region is shown in Figure 13. It is obtained from the solid region between the paraboloid $z = 8 - x^2 - y^2$ and $\mathcal{D}$ by removing the solid region between the paraboloid $z = x^2 + y^2$ and $\mathcal{D}$. The desired volume is therefore the difference in the corresponding volumes and is obtained as follows:

$$V = \int_{-1}^{1} \int_{-1}^{1} (8 - x^2 - y^2)\, dy\, dx - \int_{-1}^{1} \int_{-1}^{1} (x^2 + y^2)\, dy\, dx$$

$$= \int_{-1}^{1} \int_{-1}^{1} ((8 - x^2 - y^2) - (x^2 + y^2))\, dy\, dx$$

$$= \int_{-1}^{1} \int_{-1}^{1} (8 - 2x^2 - 2y^2)\, dy\, dx = \int_{-1}^{1} \left(8y - 2x^2y - \frac{2y^3}{3} \right) \Big|_{-1}^{1} dx$$

$$= \int_{-1}^{1} \left(16 - 4x^2 - \frac{4}{3} \right) dx = 26\frac{2}{3} \qquad ■$$

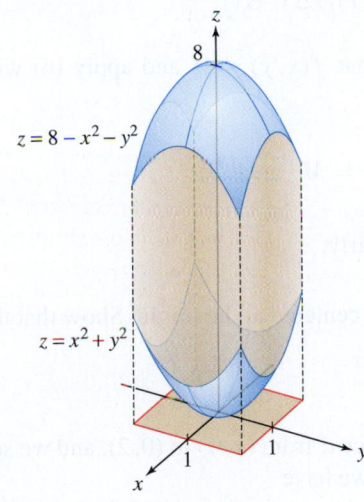

$z = 8 - x^2 - y^2$

$z = x^2 + y^2$

FIGURE 13 Finding the volume of a solid sandwiched between two paraboloids above a square.

We can generalize the idea in the previous example to compute the volume of a solid region Q in space that is sandwiched between surfaces and defined on a domain $\mathcal{D}$ in the xy-plane as in Figure 14. The surfaces are graphs of functions $z_1(x, y)$ and $z_2(x, y)$ with $z_1(x, y) \le z_2(x, y)$ on $\mathcal{D}$, and the volume is obtained by

$$V = \iint_{\mathcal{D}} z_2(x, y)\, dA - \iint_{\mathcal{D}} z_1(x, y)\, dA$$

$$= \iint_{\mathcal{D}} (z_2(x, y) - z_1(x, y))\, dA$$

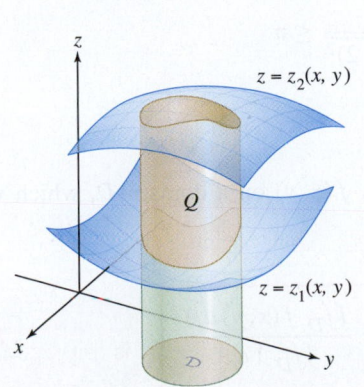

$z = z_2(x, y)$

Q

$z = z_1(x, y)$

$\mathcal{D}$

FIGURE 14 Finding the volume of a solid Q sandwiched between two surfaces above a domain $\mathcal{D}$.

The second equality holds by the linearity of the integral.

In the next theorem, part (a) is a formal statement of the fact that larger functions have larger integrals, a fact that we also noted in the single-variable case. Part (b) is useful for estimating integrals.

THEOREM 3 Let $f(x, y)$ and $g(x, y)$ be integrable functions on $\mathcal{D}$.

(a) If $f(x, y) \leq g(x, y)$ for all $(x, y) \in \mathcal{D}$, then

$$\iint_{\mathcal{D}} f(x, y)\, dA \leq \iint_{\mathcal{D}} g(x, y)\, dA \qquad \boxed{6}$$

(b) If $m \leq f(x, y) \leq M$ for all $(x, y) \in \mathcal{D}$, then

$$m \cdot \text{area}(\mathcal{D}) \leq \iint_{\mathcal{D}} f(x, y)\, dA \leq M \cdot \text{area}(\mathcal{D}) \qquad \boxed{7}$$

Proof If $f(x, y) \leq g(x, y)$, then every Riemann sum for $f(x, y)$ is less than or equal to the corresponding Riemann sum for g:

$$\sum f(P_{ij})\, \Delta x_i\, \Delta y_j \leq \sum g(P_{ij})\, \Delta x_i\, \Delta y_j$$

We obtain (6) by taking the limit. Now suppose that $f(x, y) \leq M$ and apply (6) with $g(x, y) = M$:

$$\iint_{\mathcal{D}} f(x, y)\, dA \leq \iint_{\mathcal{D}} M\, dA = M \cdot \text{area}(\mathcal{D})$$

This proves half of (7). The other half follows similarly. ∎

EXAMPLE 8 Assume that $\mathcal{D}$ is the disk of radius 1 centered at the origin. Show that the value of $\displaystyle\iint_{\mathcal{D}} \frac{dA}{\sqrt{x^2 + (y-2)^2}}$ lies between $\dfrac{\pi}{3}$ and π.

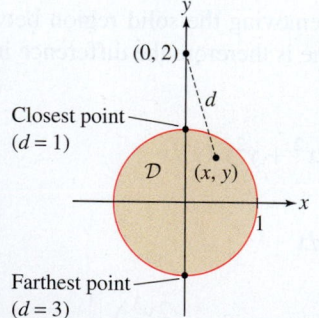

Closest point
$(d = 1)$

$\mathcal{D}$ (x, y)

d

$(0, 2)$

Farthest point
$(d = 3)$

DF **FIGURE 15** The distance d from (x, y) to $(0, 2)$ varies from 1 to 3 for (x, y) in the unit disk.

Solution The quantity $\sqrt{x^2 + (y-2)^2}$ is the distance d from (x, y) to $(0, 2)$, and we see from Figure 15 that $1 \leq d \leq 3$. Taking reciprocals, we have

$$\frac{1}{3} \leq \frac{1}{\sqrt{x^2 + (y-2)^2}} \leq 1$$

We apply (7) with $m = \frac{1}{3}$ and $M = 1$, using the fact that $\text{area}(\mathcal{D}) = \pi$, to obtain

$$\frac{\pi}{3} \leq \iint_{\mathcal{D}} \frac{dA}{\sqrt{x^2 + (y-2)^2}} \leq \pi$$ ∎

Average Value

The **average value** (or **mean value**) of a function $f(x, y)$ on a domain $\mathcal{D}$, which we denote by $\overline{f}$, is the quantity

← **REMINDER** Equation (8) is similar to the definition of an average value in one variable:

$$\overline{f} = \frac{1}{b-a} \int_a^b f(x)\, dx = \frac{\int_a^b f(x)\, dx}{\int_a^b 1\, dx}$$

$$\boxed{\overline{f} = \frac{1}{\text{area}(\mathcal{D})} \iint_{\mathcal{D}} f(x, y)\, dA = \frac{\iint_{\mathcal{D}} f(x, y)\, dA}{\iint_{\mathcal{D}} 1\, dA}} \qquad \boxed{8}$$

Equivalently, $\overline{f}$ is the value satisfying the relation

$$\iint_{\mathcal{D}} f(x, y)\, dA = \overline{f} \cdot \text{area}(\mathcal{D)}$$

9

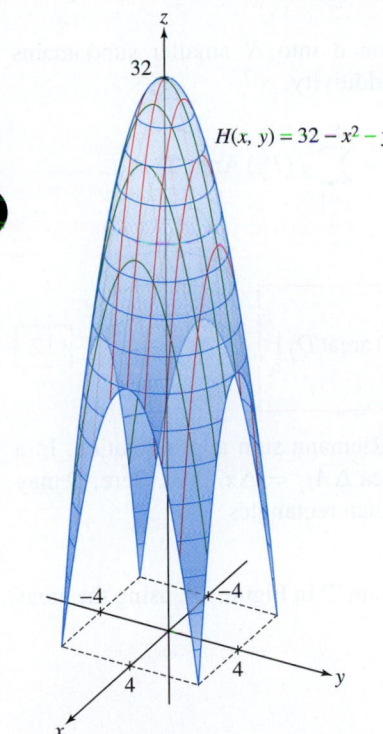

FIGURE 16

GRAPHICAL INSIGHT Equation 9 implies that if $f(x, y) \geq 0$ on a domain $\mathcal{D}$, then the solid region under the graph of f over $\mathcal{D}$ has the same volume as the cylinder with base $\mathcal{D}$ and height $\overline{f}$ (Figure 16).

EXAMPLE 9 An architect needs to know the average height $\overline{H}$ of the ceiling of a pagoda whose base $\mathcal{D}$ is the square $[-4, 4] \times [-4, 4]$ and roof is the graph of

$$H(x, y) = 32 - x^2 - y^2$$

where distances are in feet (Figure 17). Calculate $\overline{H}$.

Solution First, we compute the integral of $H(x, y)$ over $\mathcal{D}$:

$$\iint_{\mathcal{D}} (32 - x^2 - y^2)\, dA = \int_{-4}^{4} \int_{-4}^{4} (32 - x^2 - y^2)\, dy\, dx$$

$$= \int_{-4}^{4} \left(32y - x^2 y - \frac{1}{3} y^3 \right) \Bigg|_{y=-4}^{4} dx = \int_{-4}^{4} \left(\frac{640}{3} - 8x^2 \right) dx$$

$$= \left(\frac{640}{3} x - \frac{8}{3} x^3 \right) \Bigg|_{-4}^{4} = \frac{4096}{3}$$

The area of $\mathcal{D}$ is $8 \times 8 = 64$, so the average height of the pagoda's ceiling is

$$\overline{H} = \frac{1}{\text{Area}(\mathcal{D})} \iint_{\mathcal{D}} H(x, y)\, dA = \frac{1}{64} \left(\frac{4096}{3} \right) = \frac{64}{3} \approx 21.3 \text{ ft}$$ ∎

The following Mean Value Theorem for Double Integrals states that a continuous function on a domain $\mathcal{D}$ must take on its average value at some point P in $\mathcal{D}$, provided that $\mathcal{D}$ is closed, bounded, and also **connected** (see Exercise 69 for a proof). By definition, $\mathcal{D}$ is connected if any two points in $\mathcal{D}$ can be joined by a curve in $\mathcal{D}$ (Figure 18).

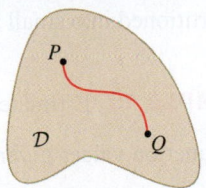

(A) Connected domain: Any two points can be joined by a curve lying entirely in $\mathcal{D}$.

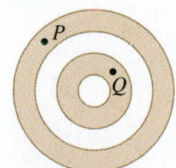

(B) Nonconnected domain: Points P and Q cannot be joined by a curve in $\mathcal{D}$.

FIGURE 17 Pagoda with ceiling $H(x, y) = 32 - x^2 - y^2$.

FIGURE 18

THEOREM 4 Mean Value Theorem for Double Integrals If $f(x, y)$ is continuous and $\mathcal{D}$ is closed, bounded, and connected, then there exists a point $P \in \mathcal{D}$ such that

$$\iint_{\mathcal{D}} f(x, y)\, dA = f(P)\, \text{area}(\mathcal{D})$$

10

Equivalently, $f(P) = \overline{f}$, where $\overline{f}$ is the average value of f on $\mathcal{D}$.

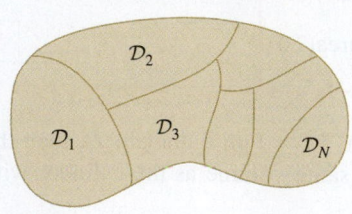

FIGURE 19 The region $\mathcal{D}$ is a union of smaller domains.

Decomposing the Domain into Smaller Domains

Double integrals are additive with respect to the domain: If $\mathcal{D}$ is the union of domains $\mathcal{D}_1, \mathcal{D}_2, \ldots, \mathcal{D}_N$ that do not overlap except possibly on boundary curves (Figure 19), then

$$\iint_{\mathcal{D}} f(x, y)\, dA = \iint_{\mathcal{D}_1} f(x, y)\, dA + \cdots + \iint_{\mathcal{D}_N} f(x, y)\, dA$$

Additivity may be used to evaluate double integrals over domains $\mathcal{D}$ that are not horizontally or vertically simple but can be decomposed into finitely many horizontally or vertically simple domains.

We close this section with a simple but useful remark. If $f(x, y)$ is a continuous function on a *small* domain $\mathcal{D}$, then

$$\iint_{\mathcal{D}} f(x, y)\, dA \approx \underbrace{f(P)\, \text{area}(\mathcal{D})}_{\text{Function value} \times \text{area}}$$

11

where P is any sample point in $\mathcal{D}$. In fact, we can choose P so that Eq. (11) is an equality by Theorem 4. But if $\mathcal{D}$ is small enough, then f is nearly constant on $\mathcal{D}$, and Eq. (11) holds as a good approximation for all $P \in \mathcal{D}$.

If the domain $\mathcal{D}$ is not small, we may partition it into N smaller subdomains $\mathcal{D}_1, \ldots, \mathcal{D}_N$ and choose sample points P_j in $\mathcal{D}_j$. By additivity,

$$\iint_{\mathcal{D}} f(x, y)\, dA = \sum_{j=1}^{N} \iint_{\mathcal{D}_j} f(x, y)\, dA \approx \sum_{j=1}^{N} f(P_j)\, \text{Area}(\mathcal{D}_j)$$

and thus we have the approximation

$$\iint_{\mathcal{D}} f(x, y)\, dA \approx \sum_{j=1}^{N} f(P_j)\, \text{area}(\mathcal{D}_j)$$

12

We can think of Eq. (12) as a generalization of the Riemann sum approximation. In a Riemann sum, $\mathcal{D}$ is partitioned by rectangles $\mathcal{R}_{ij}$ of area $\Delta A_{ij} = \Delta x_i\, \Delta y_j$. Here, $\mathcal{D}$ may be partitioned into small regions having shapes other than rectangles.

EXAMPLE 10 Estimate $\displaystyle\iint_{\mathcal{D}} f(x, y)\, dA$ for the domain $\mathcal{D}$ in Figure 20, using the areas and function values given in the accompanying table.

Solution

$$\iint_{\mathcal{D}} f(x, y)\, dA \approx \sum_{j=1}^{4} f(P_j)\, \text{Area}(\mathcal{D}_j)$$

$$= (1.8)(1) + (2.2)(1) + (2.1)(0.9) + (2.4)(1.2) \approx 8.8 \qquad \blacksquare$$

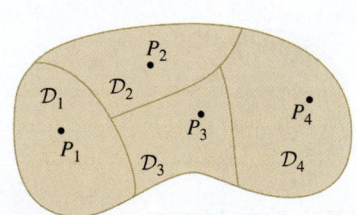

FIGURE 20

j	1	2	3	4
Area($\mathcal{D}_j$)	1	1	0.9	1.2
$f(P_j)$	1.8	2.2	2.1	2.4

15.2 SUMMARY

- We assume that $\mathcal{D}$ is a closed, bounded domain whose boundary is a simple closed curve that either is smooth or has a finite number of corners. The double integral is defined by

$$\iint_{\mathcal{D}} f(x, y)\, dA = \iint_{\mathcal{R}} \tilde{f}(x, y)\, dA$$

where $\mathcal{R}$ is a rectangle containing $\mathcal{D}$ and $\tilde{f}(x, y) = f(x, y)$ if $(x, y) \in \mathcal{D}$, and $\tilde{f}(x, y) = 0$ otherwise. The value of the integral does not depend on the choice of $\mathcal{R}$.

- The double integral defines the signed volume between the graph of $f(x, y)$ and the xy-plane, where the signed volumes of regions below the xy-plane are negative.

- For any constant C, $\displaystyle\iint_{\mathcal{D}} C \, dA = C \cdot \text{area}(\mathcal{D})$.

- If $\mathcal{D}$ is vertically or horizontally simple, $\displaystyle\iint_{\mathcal{D}} f(x, y) \, dA$ can be evaluated as an iterated integral:

Vertically simple domain $a \le x \le b, \qquad g_1(x) \le y \le g_2(x)$	$\displaystyle\int_a^b \int_{g_1(x)}^{g_2(x)} f(x, y) \, dy \, dx$
Horizontally simple domain $c \le y \le d, \quad h_1(y) \le x \le h_2(y)$	$\displaystyle\int_c^d \int_{h_1(y)}^{h_2(y)} f(x, y) \, dx \, dy$

- If $f(x, y) \le g(x, y)$ on $\mathcal{D}$, then $\displaystyle\iint_{\mathcal{D}} f(x, y) \, dA \le \iint_{\mathcal{D}} g(x, y) \, dA$.

- If m is the minimum value and M the maximum value of f on $\mathcal{D}$, then

$$m \cdot \text{area}(\mathcal{D}) \le \iint_{\mathcal{D}} f(x, y) \, dA \le M \cdot \text{area}(\mathcal{D})$$

- If $z_1(x, y) \le z_2(x, y)$ for all points in $\mathcal{D}$, then the volume V of the solid region between the surfaces given by $z = z_1(x, y)$ and $z = z_2(x, y)$ over $\mathcal{D}$ is given by

$$V = \iint_{\mathcal{D}} (z_2(x, y) - z_1(x, y)) \, dA$$

- The *average value* of f on $\mathcal{D}$ is

$$\overline{f} = \frac{1}{\text{area}(\mathcal{D})} \iint_{\mathcal{D}} f(x, y) \, dA = \frac{\iint_{\mathcal{D}} f(x, y) \, dA}{\iint_{\mathcal{D}} 1 \, dA}$$

- Mean Value Theorem for Double Integrals: If $f(x, y)$ is continuous and $\mathcal{D}$ is closed, bounded, and connected, then there exists a point $P \in \mathcal{D}$ such that

$$\iint_{\mathcal{D}} f(x, y) \, dA = f(P) \cdot \text{area}(\mathcal{D})$$

Equivalently, $f(P) = \overline{f}$, where $\overline{f}$ is the average value of f on $\mathcal{D}$.
- Additivity with respect to the domain: If $\mathcal{D}$ is a union of nonoverlapping (except possibly on their boundaries) domains $\mathcal{D}_1, \ldots, \mathcal{D}_N$, then

$$\iint_{\mathcal{D}} f(x, y) \, dA = \sum_{j=1}^{N} \iint_{\mathcal{D}_j} f(x, y) \, dA$$

- If the domains $\mathcal{D}_1, \ldots, \mathcal{D}_N$ are small and P_j is a sample point in $\mathcal{D}_j$, then

$$\iint_{\mathcal{D}} f(x, y) \, dA \approx \sum_{j=1}^{N} f(P_j) \text{area}(\mathcal{D}_j)$$

15.2 EXERCISES

Preliminary Questions

1. Which of the following expressions do not make sense?

(a) $\int_0^1 \int_1^x f(x,y)\,dy\,dx$

(b) $\int_0^1 \int_1^y f(x,y)\,dy\,dx$

(c) $\int_0^1 \int_x^y f(x,y)\,dy\,dx$

(d) $\int_0^1 \int_x^1 f(x,y)\,dy\,dx$

2. Draw a domain in the plane that is neither vertically nor horizontally simple.

3. Which of the four regions in Figure 21 is the domain of integration for

$$\int_{-\sqrt{2}/2}^{0} \int_{-x}^{\sqrt{1-x^2}} f(x,y)\,dy\,dx?$$

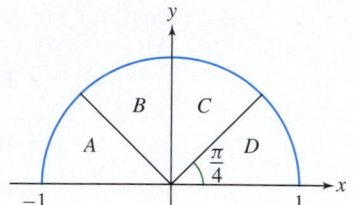

FIGURE 21

4. Let $\mathcal{D}$ be the unit disk. If the maximum value of $f(x,y)$ on $\mathcal{D}$ is 4, then the largest possible value of $\iint_{\mathcal{D}} f(x,y)\,dA$ is (choose the correct answer):

(a) 4   **(b)** 4π **(c)** $\dfrac{4}{\pi}$

Exercises

1. Calculate the Riemann sum for $f(x,y) = x - y$ and the shaded domain $\mathcal{D}$ in Figure 22 with two choices of sample points, • and ○. Which do you think is a better approximation to the integral of f over $\mathcal{D}$? Why?

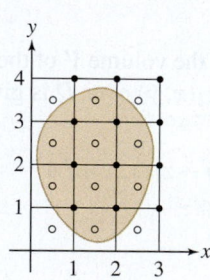

FIGURE 22

2. Approximate values of $f(x,y)$ at sample points on a grid are given in Figure 23. Estimate $\iint_{\mathcal{D}} f(x,y)\,dx\,dy$ for the shaded domain by computing the Riemann sum with the given sample points.

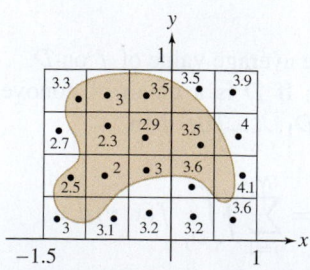

FIGURE 23

3. Express the domain $\mathcal{D}$ in Figure 24 as both a vertically simple region and a horizontally simple region, and evaluate the integral of $f(x,y) = xy$ over $\mathcal{D}$ as an iterated integral in two ways.

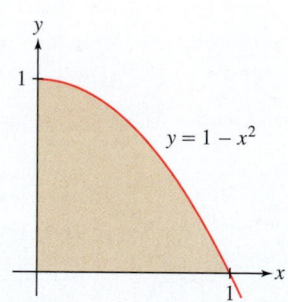

FIGURE 24

4. Sketch the domain

$$\mathcal{D} : 0 \le x \le 1, \quad x^2 \le y \le 4 - x^2$$

and evaluate $\iint_{\mathcal{D}} y\,dA$ as an iterated integral.

In Exercises 5–7, compute the double integral of $f(x,y) = x^2 y$ over the given shaded domain in Figure 25.

5. (A)   **6.** (B)   **7.** (C)

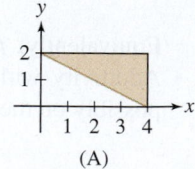 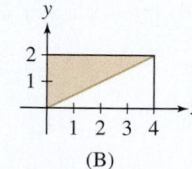 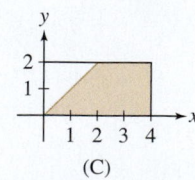

(A)   (B)   (C)

FIGURE 25

8. Sketch the domain $\mathcal{D}$ defined by $x + y \le 12$, $x \ge 4$, $y \ge 4$ and compute $\iint_{\mathcal{D}} e^{x+y}\,dA$.

9. Integrate $f(x,y) = x$ over the region bounded by $y = x^2$ and $y = x + 2$.

10. Sketch the region $\mathcal{D}$ between $y = x^2$ and $y = x(1-x)$. Express $\mathcal{D}$ as a simple region and calculate the integral of $f(x, y) = 2y$ over $\mathcal{D}$.

11. Evaluate $\iint_{\mathcal{D}} \dfrac{y}{x} \, dA$, where $\mathcal{D}$ is the shaded part inside the semicircle of radius 2 in Figure 26.

12. Calculate the double integral of $f(x, y) = y^2$ over the rhombus $\mathcal{R}$ in Figure 27.

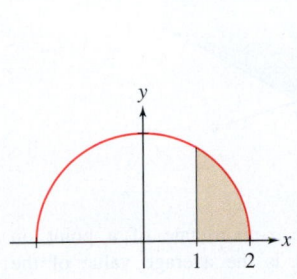

FIGURE 26 $y = \sqrt{4 - x^2}$.

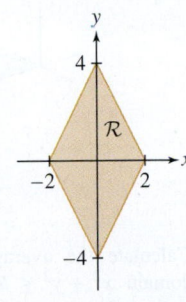

FIGURE 27 $|x| + \frac{1}{2}|y| \le 1$.

13. Calculate the double integral of $f(x, y) = x + y$ over the domain $\mathcal{D} = \{(x, y) : x^2 + y^2 \le 4, y \ge 0\}$.

14. Integrate $f(x, y) = (x + y + 1)^{-2}$ over the triangle with vertices $(0, 0)$, $(4, 0)$, and $(0, 8)$.

15. Calculate the integral of $f(x, y) = x$ over the region $\mathcal{D}$ bounded above by $y = x(2 - x)$ and below by $x = y(2 - y)$. *Hint:* Apply the quadratic formula to the lower boundary curve to solve for y as a function of x.

16. Integrate $f(x, y) = x$ over the region bounded by $y = x$, $y = 4x - x^2$, and $y = 0$ in two ways: as a vertically simple region and as a horizontally simple region.

In Exercises 17–24, compute the double integral of $f(x, y)$ over the domain $\mathcal{D}$ indicated.

17. $f(x, y) = x^3 y$; $0 \le x \le 5$, $x \le y \le 2x + 3$

18. $f(x, y) = -2$; $0 \le x \le 3$, $1 \le y \le e^x$

19. $f(x, y) = x$; $0 \le x \le 1$, $1 \le y \le e^{x^2}$

20. $f(x, y) = \cos(2x + y)$; $\frac{1}{2} \le x \le \frac{\pi}{2}$, $1 \le y \le 2x$

21. $f(x, y) = 6xy - x^2$; bounded below by $y = x^2$, above by $y = \sqrt{x}$

22. $f(x, y) = \sin x$; bounded by $x = 0$, $x = 1$, $y = 0$, $y = \cos x$

23. $f(x, y) = e^{x+y}$; bounded by $y = x - 1$, $y = 12 - x$ for $2 \le y \le 4$

24. $f(x, y) = (x + y)^{-1}$; bounded by $y = x$, $y = 1$, $y = e$, $x = 0$

In Exercises 25–28, sketch the domain of integration and express as an iterated integral in the opposite order.

25. $\displaystyle\int_0^4 \int_x^4 f(x, y) \, dy \, dx$

26. $\displaystyle\int_4^9 \int_{\sqrt{y}}^3 f(x, y) \, dx \, dy$

27. $\displaystyle\int_4^9 \int_2^{\sqrt{y}} f(x, y) \, dx \, dy$

28. $\displaystyle\int_0^1 \int_{e^x}^e f(x, y) \, dy \, dx$

29. Sketch the domain $\mathcal{D}$ corresponding to

$$\int_0^4 \int_{\sqrt{y}}^2 \sqrt{4x^2 + 5y} \, dx \, dy$$

Then change the order of integration and evaluate.

30. Change the order of integration and evaluate

$$\int_0^1 \int_0^{\pi/2} x \cos(xy) \, dx \, dy$$

Explain the simplification achieved by changing the order.

31. Compute the integral of $f(x, y) = (\ln y)^{-1}$ over the domain $\mathcal{D}$ bounded by $y = e^x$ and $y = e^{\sqrt{x}}$. *Hint:* Choose the order of integration that enables you to evaluate the integral.

32. Evaluate by changing the order of integration:

$$\int_0^4 \int_{\sqrt{x}}^2 \sin y^3 \, dy \, dx$$

In Exercises 33–36, sketch the domain of integration. Then change the order of integration and evaluate. Explain the simplification achieved by changing the order.

33. $\displaystyle\int_0^1 \int_y^1 \dfrac{\sin x}{x} \, dx \, dy$

34. $\displaystyle\int_0^4 \int_{\sqrt{y}}^2 \sqrt{x^3 + 1} \, dx \, dy$

35. $\displaystyle\int_0^1 \int_{y=x}^1 x e^{y^3} \, dy \, dx$

36. $\displaystyle\int_0^1 \int_{y=x^{2/3}}^1 x e^{y^4} \, dy \, dx$

37. Sketch the domain $\mathcal{D}$ where $0 \le x \le 2$, $0 \le y \le 2$, and x or y is greater than 1. Then compute $\iint_{\mathcal{D}} e^{x+y} \, dA$.

38. Calculate $\iint_{\mathcal{D}} e^x \, dA$, where $\mathcal{D}$ is bounded by the lines $y = x + 1$, $y = x$, $x = 0$, and $x = 1$.

In Exercises 39–42, calculate the double integral of $f(x, y)$ over the triangle indicated in Figure 28.

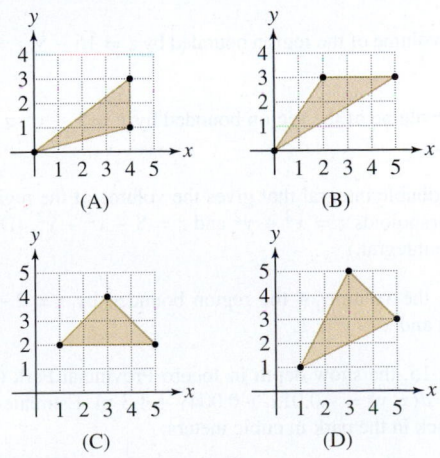

FIGURE 28

39. $f(x, y) = e^{x^2}$,   (A)

40. $f(x, y) = 1 - 2x$,   (B)

41. $f(x, y) = \dfrac{x}{y^2}$,   (C)

42. $f(x, y) = x + 1$,   (D)

43. Calculate the double integral of $f(x, y) = \dfrac{\sin y}{y}$ over the region $\mathcal{D}$ in Figure 29.

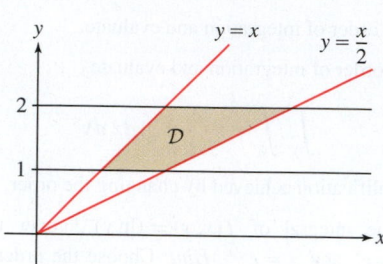

FIGURE 29

44. Evaluate $\displaystyle\iint_{\mathcal{D}} x \, dA$ for $\mathcal{D}$ in Figure 30.

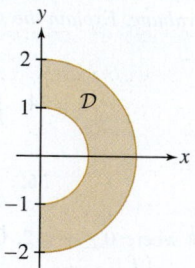

FIGURE 30

45. Find the volume of the region bounded by $z = 40 - 10y$, $z = 0$, $y = 0$, and $y = 4 - x^2$.

46. Find the volume of the region enclosed by $z = 1 - y^2$ and $z = y^2 - 1$ for $0 \le x \le 2$.

47. Find the volume of the region bounded by $z = 16 - y$, $z = y$, $y = x^2$, and $y = 8 - x^2$.

48. Find the volume of the region bounded by $y = 1 - x^2$, $z = 1$, $y = 0$, and $z + y = 2$.

49. Set up a double integral that gives the volume of the region bounded by the two paraboloids $z = x^2 + y^2$ and $z = 8 - x^2 - y^2$. (Do not evaluate the double integral.)

50. Compute the volume of the region bounded by $z = 2 - y^2$, $z = y$, $x = 0$, $y = 0$, and $x + y = 1$.

51. On April 15, the snow depth in Jocoro Provincial Park (Example 6) was given by $d(x, y) = -0.01x + 0.004y + 1.3$ m. Estimate the volume of the snowpack in the park in cubic meters.

52. On May 1, there was no snow in Jocoro Provincial Park (Example 6) east of Highway 55 (i.e., for $x \ge 100$). Otherwise, the snow depth was given by $d(x, y) = -0.008x + 0.002y + 0.8$ m. Estimate the volume of the snowpack in the park in cubic meters.

53. Calculate the average value of $f(x, y) = e^{x+y}$ on the square domain $[0, 1] \times [0, 1]$.

54. Calculate the average y-coordinate of the points in the region given by $0 \le x \le 1$, $0 \le y \le x^2$.

55. Find the average height of the "ceiling" in Figure 31 defined by $z = y^2 \sin x$ for $0 \le x \le \pi$, $0 \le y \le 1$.

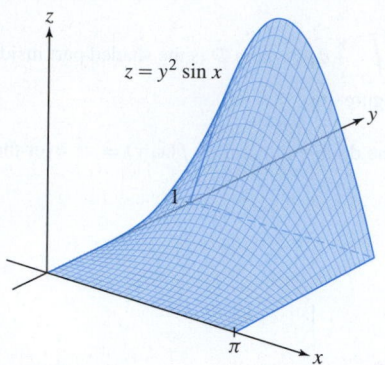

FIGURE 31

56. Calculate the average value of the x-coordinate of a point on the domain $x^2 + y^2 \le R^2$, $x \ge 0$. What is the average value of the y-coordinate?

57. What is the average value of the linear function

$$f(x, y) = mx + ny + p$$

on the ellipse $\left(\dfrac{x}{a}\right)^2 + \left(\dfrac{y}{b}\right)^2 \le 1$? Argue by symmetry rather than calculation.

58. Find the average of the square of the distance from the origin to a point in the domain $\mathcal{D}$ in Figure 32.

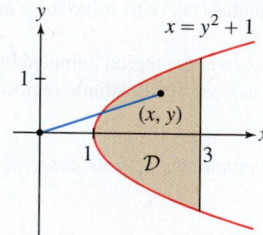

FIGURE 32

59. Let $\mathcal{D}$ be the rectangle $0 \le x \le 2$, $-\frac{1}{8} \le y \le \frac{1}{8}$, and let $f(x, y) = \sqrt{x^3 + 1}$. Prove that

$$\iint_{\mathcal{D}} f(x, y) \, dA \le \frac{3}{2}$$

60. (a) Use the inequality $\sin \theta \le \theta$ for $\theta \ge 0$ to show that

$$\int_0^1 \int_0^1 \sin(xy) \, dx \, dy \le \frac{1}{4}$$

(b) Use a computer algebra system to evaluate the double integral to three decimal places.

61. Prove the inequality $\displaystyle\iint_{\mathcal{D}} \dfrac{dA}{4 + x^2 + y^2} \le \pi$, where $\mathcal{D}$ is the disk $x^2 + y^2 \le 4$.

62. Let $\mathcal{D}$ be the domain bounded by $y = x^2 + 1$ and $y = 2$. Prove the inequality

$$\frac{4}{3} \le \iint_{\mathcal{D}} (x^2 + y^2) \, dA \le \frac{20}{3}$$

63. Let $\overline{f}$ be the average of $f(x, y) = xy^2$ on $\mathcal{D} = [0, 1] \times [0, 4]$. Find a point $P \in \mathcal{D}$ such that $f(P) = \overline{f}$ (the existence of such a point is guaranteed by the Mean Value Theorem for Double Integrals).

64. Verify the Mean Value Theorem for Double Integrals for $f(x, y) = e^{x-y}$ on the triangle bounded by $y = 0$, $x = 1$, and $y = x$.

In Exercises 65 and 66, use the approximation in (12) to estimate the double integral.

65. The following table lists the areas of the subdomains $\mathcal{D}_j$ of the domain $\mathcal{D}$ in Figure 33 and the values of a function $f(x, y)$ at sample points $P_j \in \mathcal{D}_j$. Estimate $\displaystyle\iint_{\mathcal{D}} f(x, y)\, dA$.

j	1	2	3	4	5	6
Area($\mathcal{D}_j$)	1.2	1.1	1.4	0.6	1.2	0.8
$f(P_j)$	9	9.1	9.3	9.1	8.9	8.8

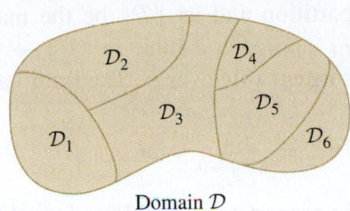

Domain $\mathcal{D}$

FIGURE 33

66. The domain $\mathcal{D}$ between the circles of radii 5 and 5.2 in the first quadrant in Figure 34 is divided into six subdomains of angular width $\Delta\theta = \frac{\pi}{12}$, and the values of a function $f(x, y)$ at sample points are given. Compute the area of the subdomains and estimate $\displaystyle\iint_{\mathcal{D}} f(x, y)\, dA$.

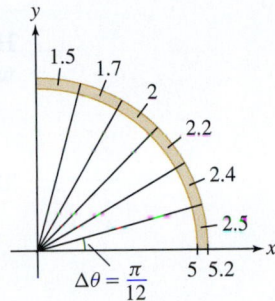

FIGURE 34

67. According to Eq. (3), the area of a domain $\mathcal{D}$ is equal to $\displaystyle\iint_{\mathcal{D}} 1\, dA$. Prove that if $\mathcal{D}$ is the region between two curves $y = g_1(x)$ and $y = g_2(x)$ with $g_2(x) \leq g_1(x)$ for $a \leq x \leq b$, then

$$\iint_{\mathcal{D}} 1\, dA = \int_a^b (g_1(x) - g_2(x))\, dx$$

Further Insights and Challenges

68. Let $\mathcal{D}$ be a closed connected domain and let $P, Q \in \mathcal{D}$. The Intermediate Value Theorem (IVT) states that if f is continuous on $\mathcal{D}$, then $f(x, y)$ takes on every value between $f(P)$ and $f(Q)$ at some point in $\mathcal{D}$.
(a) Show, by constructing a counterexample, that the IVT is false if $\mathcal{D}$ is not connected.
(b) Prove the IVT as follows: Let $\mathbf{r}(t)$ be a path such that $\mathbf{r}(0) = P$ and $\mathbf{r}(1) = Q$ (such a path exists because $\mathcal{D}$ is connected). Apply the IVT in one variable to the composite function $f(\mathbf{r}(t))$.

69. Use the fact that a continuous function on a closed bounded domain $\mathcal{D}$ attains both a minimum value m and a maximum value M, together with Theorem 3, to prove that the average value $\overline{f}$ lies between m and M. Then use the IVT in Exercise 68 to prove the Mean Value Theorem for Double Integrals.

70. Let $G(t) = \displaystyle\int_0^t \int_0^x f(y)\, dy\, dx$ where $f(y)$ is a function of y alone.
(a) Use the Fundamental Theorem of Calculus to prove that $G''(t) = f(t)$.
(b) Show, by changing the order in the double integral, that $G(t) = \displaystyle\int_0^t (t - y) f(y)\, dy$. This illustrates that the "second antiderivative" of $f(y)$ can be expressed as a single integral.

15.3 Triple Integrals

Triple integrals of functions $f(x, y, z)$ of three variables are a fairly straightforward generalization of double integrals. In the first simple case, instead of a rectangle in the plane, our domain is a box (Figure 1)

$$\mathcal{B} = [a, b] \times [c, d] \times [p, q]$$

consisting of all points (x, y, z) in $\mathbf{R}^3$ such that

$$a \leq x \leq b, \qquad c \leq y \leq d, \qquad p \leq z \leq q$$

To integrate over this box, we subdivide the box (as usual) into subboxes

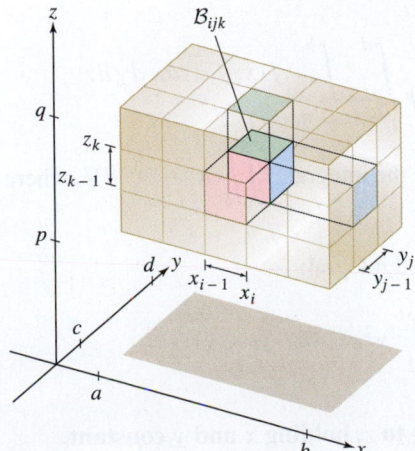

FIGURE 1 The box $\mathcal{B} = [a, b] \times [c, d] \times [p, q]$ decomposed into smaller boxes.

$$\mathcal{B}_{ijk} = [x_{i-1}, x_i] \times [y_{j-1}, y_j] \times [z_{k-1}, z_k]$$

by choosing partitions of the three intervals

$$a = x_0 < x_1 < \cdots < x_N = b$$

$$c = y_0 < y_1 < \cdots < y_M = d$$

$$p = z_0 < z_1 < \cdots < z_L = q$$

Here, N, M, and L are positive integers. The volume of $\mathcal{B}_{ijk}$ is $\Delta V_{ijk} = \Delta x_i \, \Delta y_j \, \Delta z_k$, where

$$\Delta x_i = x_i - x_{i-1}, \qquad \Delta y_j = y_j - y_{j-1}, \qquad \Delta z_k = z_k - z_{k-1}$$

Then we choose a sample point P_{ijk} in each subbox $\mathcal{B}_{ijk}$ and form the Riemann sum:

$$S_{N,M,L} = \sum_{i=1}^{N} \sum_{j=1}^{M} \sum_{k=1}^{L} f(P_{ijk}) \, \Delta V_{ijk}$$

We write $\mathcal{P} = \{\{x_i\}, \{y_j\}, \{z_k\}\}$ for the partition and let $\|\mathcal{P}\|$ be the maximum of the widths $\Delta x_i, \Delta y_j, \Delta z_k$. If the sums $S_{N,M,L}$ approach a limit as $\|\mathcal{P}\| \to 0$ for arbitrary choices of sample points, we say that f is **integrable** over $\mathcal{B}$. The limit value is denoted

$$\iiint_{\mathcal{B}} f(x, y, z) \, dV = \lim_{\|\mathcal{P}\| \to 0} S_{N,M,L}$$

> The term dA, used in double integrals and referred to as an area element, suggests small areas are involved in integrals over domains in the plane. Similarly, the dV used in triple integrals is called a **volume element** and suggests small volumes are involved when integrating over a domain in $\mathbf{R}^3$.

Triple integrals have many of the same properties as double and single integrals. The linearity properties are satisfied, and continuous functions are integrable over a box $\mathcal{B}$. Furthermore, triple integrals can be evaluated as iterated integrals.

THEOREM 1   Fubini's Theorem for Triple Integrals   The triple integral of a continuous function $f(x, y, z)$ over a box $\mathcal{B} = [a,b] \times [c,d] \times [p,q]$ is equal to the iterated integral:

$$\iiint_{\mathcal{B}} f(x, y, z) \, dV = \int_{x=a}^{b} \int_{y=c}^{d} \int_{z=p}^{q} f(x, y, z) \, dz \, dy \, dx$$

Furthermore, the iterated integral may be evaluated in any order.

As noted in the theorem, we are free to evaluate the iterated integral over a box in any order (there are six different orders). For instance,

$$\int_{x=a}^{b} \int_{y=c}^{d} \int_{z=p}^{q} f(x, y, z) \, dz \, dy \, dx = \int_{z=p}^{q} \int_{y=c}^{d} \int_{x=a}^{b} f(x, y, z) \, dx \, dy \, dz$$

EXAMPLE 1   Integration over a Box   Calculate the integral $\displaystyle\iiint_{\mathcal{B}} x^2 e^{y+3z} \, dV$, where $\mathcal{B} = [1,4] \times [0,3] \times [2,6]$.

Solution   We write this triple integral as an iterated integral:

$$\iiint_{\mathcal{B}} x^2 e^{y+3z} \, dV = \int_{1}^{4} \int_{0}^{3} \int_{2}^{6} x^2 e^{y+3z} \, dz \, dy \, dx$$

Step 1.   Evaluate the inner integral with respect to z, holding x and y constant.

$$\int_{z=2}^{6} x^2 e^{y+3z} \, dz = \frac{1}{3} x^2 e^{y+3z} \Big|_{z=2}^{6} = \frac{1}{3} x^2 e^{y+18} - \frac{1}{3} x^2 e^{y+6} = \frac{1}{3}(e^{18} - e^6) x^2 e^y$$

Step 2. **Evaluate the middle integral with respect to y, holding x constant.**

$$\int_{y=0}^{3} \frac{1}{3}(e^{18} - e^6)x^2 e^y \, dy = \frac{1}{3}(e^{18} - e^6)x^2 \int_{y=0}^{3} e^y \, dy = \frac{1}{3}(e^{18} - e^6)(e^3 - 1)x^2$$

Step 3. **Evaluate the outer integral with respect to x.**

$$\iiint_{\mathcal{B}} (x^2 e^{y+3z}) \, dV = \frac{1}{3}(e^{18} - e^6)(e^3 - 1) \int_{x=1}^{4} x^2 \, dx = 7(e^{18} - e^6)(e^3 - 1) \quad \blacksquare$$

Next, instead of a box, we integrate over a solid region $\mathcal{W}$ that is enclosed between two surfaces $z = z_1(x, y)$ and $z = z_2(x, y)$ over a domain $\mathcal{D}$ in the xy-plane (Figure 2):

$$\mathcal{W} = \{(x, y, z) : (x, y) \in \mathcal{D} \quad \text{and} \quad z_1(x, y) \leq z \leq z_2(x, y)\} \qquad \boxed{1}$$

In this case, the region $\mathcal{W}$ is called z-**simple**. Furthermore, the domain $\mathcal{D}$ is referred to as the **projection** of $\mathcal{W}$ onto the xy-plane. We can similarly define x-**simple** for regions enclosed between surfaces $x = x_1(y, z)$ and $x = x_2(y, z)$, as well as y-**simple** for regions between surfaces $y = y_1(x, z)$ and $y = y_2(x, z)$.

As in the case of double integrals, we define the triple integral of $f(x, y, z)$ over $\mathcal{W}$ by

$$\iiint_{\mathcal{W}} f(x, y, z) \, dV = \iiint_{\mathcal{B}} \tilde{f}(x, y, z) \, dV$$

where $\mathcal{B}$ is a box containing $\mathcal{W}$, and $\tilde{f}$ is the function that is equal to f on $\mathcal{W}$ and equal to zero outside of $\mathcal{W}$. The triple integral exists, assuming that $z_1(x, y)$, $z_2(x, y)$, and the integrand f are continuous. In practice, we evaluate triple integrals as iterated integrals. This is justified by the following theorem, whose proof is similar to that of Theorem 2 in Section 15.2.

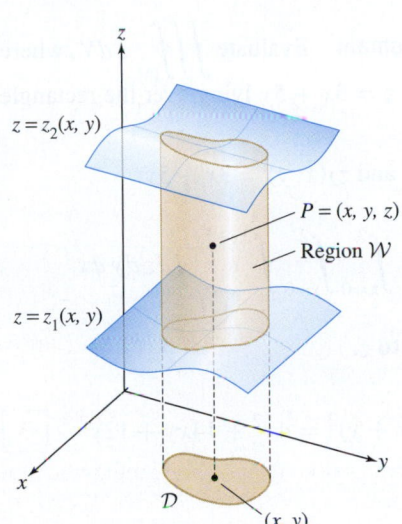

FIGURE 2 The point $P = (x, y, z)$ is in the z-simple region $\mathcal{W}$ if $(x, y) \in \mathcal{D}$ and $z_1(x, y) \leq z \leq z_2(x, y)$.

THEOREM 2 The triple integral of a continuous function f over the region

$$\mathcal{W} : (x, y) \in \mathcal{D}, \quad z_1(x, y) \leq z \leq z_2(x, y)$$

is equal to the iterated integral

$$\iiint_{\mathcal{W}} f(x, y, z) \, dV = \iint_{\mathcal{D}} \left(\int_{z=z_1(x,y)}^{z_2(x,y)} f(x, y, z) \, dz \right) dA$$

More generally, integrals of functions of n variables (for any n) arise naturally in many different contexts. For example, the average distance between two points in a ball in $\mathbf{R}^3$ is expressed as a six-fold integral because we integrate over all possible coordinates of the two points. Each point has three coordinates for a total of six variables.

Note that the inner integral on the right side in the theorem is a single-variable integral with respect to z, and the outer integral is a double integral over x and y. Typically, we would compute that double integral as a double iterated integral.

One thing missing from our discussion so far is a geometric interpretation of triple integrals. A double integral represents the signed volume of the three-dimensional region between a graph $z = f(x, y)$ and the xy-plane. The graph of a function $f(x, y, z)$ of three variables lives in *four-dimensional space*, and thus, a triple integral represents a signed "volume" of a four-dimensional region. Such a region might be hard or impossible to visualize. On the other hand, triple integrals can be used to compute many different types of quantities in a three-dimensional setting. Some examples are mass, center of mass, moments of inertia, heat content, and total charge (see Section 15.5).

Furthermore, the volume V of a region $\mathcal{W}$ is defined as the triple integral of the constant function $f(x, y, z) = 1$:

Equation 2 is analogous to the fact we have already seen that the area A of a region $\mathcal{R}$ in the xy-plane is given by taking the double integral of 1 over the region

$$A = \iint_{\mathcal{R}} 1 \, dA$$

$$V = \iiint_{\mathcal{W}} 1 \, dV \qquad \boxed{2}$$

In particular, if $\mathcal{W}$ is a z-simple region between $z = z_1(x, y)$ and $z = z_2(x, y)$, then

$$\iiint_{\mathcal{W}} 1 \, dV = \iint_{\mathcal{D}} \left(\int_{z=z_1(x,y)}^{z_2(x,y)} 1 \, dz \right) dA = \iint_{\mathcal{D}} (z_2(x, y) - z_1(x, y)) \, dA$$

Thus, the triple integral for V is equal to the double integral defining the volume of the region between the two surfaces, as we saw in the previous section.

EXAMPLE 2 **Solid Region over a Rectangular Domain** Evaluate $\iiint_{\mathcal{W}} z \, dV$, where $\mathcal{W}$ is the region between the planes $z = x + y$ and $z = 3x + 5y$ lying over the rectangle $\mathcal{D} = [0, 3] \times [0, 2]$ (Figure 3).

Solution Apply Theorem 2 with $z_1(x, y) = x + y$ and $z_2(x, y) = 3x + 5y$:

$$\iiint_{\mathcal{W}} z \, dV = \iint_{\mathcal{D}} \left(\int_{z=x+y}^{3x+5y} z \, dz \right) dA = \int_{x=0}^{3} \int_{y=0}^{2} \int_{z=x+y}^{3x+5y} z \, dz \, dy \, dx$$

Step 1. **Evaluate the inner integral with respect to z.**

$$\int_{z=x+y}^{3x+5y} z \, dz = \frac{1}{2} z^2 \Big|_{z=x+y}^{3x+5y} = \frac{1}{2}(3x + 5y)^2 - \frac{1}{2}(x + y)^2 = 4x^2 + 14xy + 12y^2 \quad \boxed{3}$$

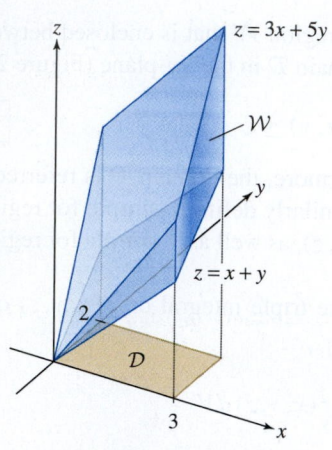

DF **FIGURE 3** Region $\mathcal{W}$ between the planes $z = x + y$ and $z = 3x + 5y$ lying over $\mathcal{D} = [0, 3] \times [0, 2]$.

Step 2. **Evaluate the integral with respect to y.**

$$\int_{y=0}^{2} (4x^2 + 14xy + 12y^2) \, dy = (4x^2 y + 7xy^2 + 4y^3) \Big|_{y=0}^{2} = 8x^2 + 28x + 32$$

Step 3. **Evaluate the integral with respect to x.**

$$\iiint_{\mathcal{W}} z \, dV = \int_{x=0}^{3} (8x^2 + 28x + 32) \, dx = \left(\frac{8}{3} x^3 + 14x^2 + 32x \right) \Big|_0^3$$

$$= 72 + 126 + 96 = 294 \qquad \blacksquare$$

EXAMPLE 3 **Solid Region over a Triangular Domain** Evaluate $\iiint_{\mathcal{W}} z \, dV$, where $\mathcal{W}$ is the region in Figure 4.

Solution This is similar to the previous example, but now $\mathcal{W}$ lies over the triangle $\mathcal{D}$ in the xy-plane defined by

$$0 \le x \le 1, \qquad 0 \le y \le 1 - x$$

Thus, the triple integral is equal to the iterated integral:

$$\iiint_{\mathcal{W}} z \, dV = \iint_{\mathcal{D}} \left(\int_{z=x+y}^{3x+5y} z \, dz \right) dA = \underbrace{\int_{x=0}^{1} \int_{y=0}^{1-x}}_{\substack{\text{Integral} \\ \text{over triangle}}} \int_{z=x+y}^{3x+5y} z \, dz \, dy \, dx$$

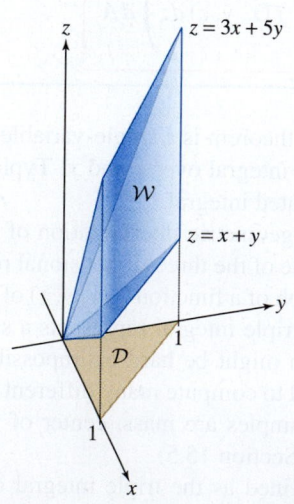

DF **FIGURE 4** Region $\mathcal{W}$ between the planes $z = x + y$ and $z = 3x + 5y$ lying over the triangle $\mathcal{D}$.

We computed the inner integral in the previous example [see Eq. (3)]:

$$\int_{z=x+y}^{3x+5y} z \, dz = \frac{1}{2} z^2 \Big|_{x+y}^{3x+5y} = 4x^2 + 14xy + 12y^2$$

Next, we integrate with respect to y (omitting some intermediate steps):

$$\int_{y=0}^{1-x} (4x^2 + 14xy + 12y^2)\, dy = 4x^2 y + 7xy^2 + 4y^3 \Big|_{y=0}^{1-x} = 4 - 5x + 2x^2 - x^3$$

And finally,

$$\iiint_{\mathcal{W}} z\, dV = \int_{x=0}^{1} (4 - 5x + 2x^2 - x^3)\, dx = \frac{23}{12} \qquad ■$$

EXAMPLE 4 **Region Between Intersecting Surfaces** Integrate $f(x, y, z) = x$ over the region $\mathcal{W}$ bounded above by $z = 4 - x^2 - y^2$ and below by $z = x^2 + 3y^2$ in the octant $x \geq 0, y \geq 0, z \geq 0$. [*Note:* The region bounded between the paraboloids is shown in Figure 5(A). The part of that region that is $\mathcal{W}$ is shown in Figure 5(B).]

Solution The region $\mathcal{W}$ is z-simple, so

$$\iiint_{\mathcal{W}} x\, dV = \iint_{\mathcal{D}} \int_{z=x^2+3y^2}^{4-x^2-y^2} x\, dz\, dA$$

where $\mathcal{D}$ is the projection of $\mathcal{W}$ onto the xy-plane. To evaluate the integral over $\mathcal{D}$, we must find the equation of the curved part of the boundary of $\mathcal{D}$.

Step 1. **Find the boundary of $\mathcal{D}$.**
The upper and lower surfaces intersect at points (x, y, z) where both

$$z = x^2 + 3y^2 \quad \text{and} \quad z = 4 - x^2 - y^2$$

are satisfied. Thus,

$$4 - x^2 - y^2 = z = x^2 + 3y^2 \qquad \text{or} \qquad x^2 + 2y^2 = 2$$

Therefore, as we see in Figure 5(B), $\mathcal{W}$ projects onto the domain $\mathcal{D}$ consisting of the quarter of the inside of the ellipse $x^2 + 2y^2 = 2$ in the first quadrant. This ellipse hits the axes at $(\sqrt{2}, 0)$ and $(0, 1)$.

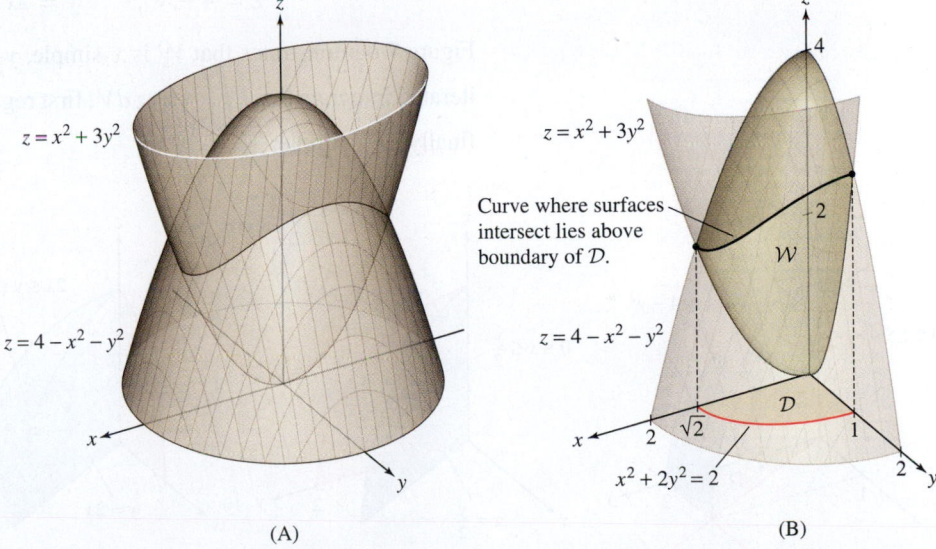

FIGURE 5 The region between the paraboloids $z = x^2 + 3y^2$ and $z = 4 - x^2 - y^2$ is shown in (A). The region we are integrating over is shown in (B).

(A)

(B)

Step 2. **Express $\mathcal{D}$ as a simple domain.**
Since $\mathcal{D}$ is both vertically and horizontally simple, we can integrate in either $dy\, dx$ order or $dx\, dy$ order. If we choose $dx\, dy$, then y varies from 0 to 1 and the domain is described by

$$\mathcal{D} : 0 \leq y \leq 1, \quad 0 \leq x \leq \sqrt{2 - 2y^2}$$

Step 3. **Write the triple integral as an iterated integral.**

$$\iiint_{\mathcal{W}} x\, dV = \int_{y=0}^{1} \int_{x=0}^{\sqrt{2-2y^2}} \int_{z=x^2+3y^2}^{4-x^2-y^2} x\, dz\, dx\, dy$$

Step 4. **Evaluate.**

Here are the results of evaluating the integrals in order:

Inner integral: $\displaystyle \int_{z=x^2+3y^2}^{4-x^2-y^2} x\, dz = xz \Big|_{z=x^2+3y^2}^{4-x^2-y^2} = 4x - 2x^3 - 4y^2 x$

Middle integral:

$$\int_{x=0}^{\sqrt{2-2y^2}} (4x - 2x^3 - 4y^2 x)\, dx = \left(2x^2 - \frac{1}{2}x^4 - 2x^2 y^2\right)\Big|_{x=0}^{\sqrt{2-2y^2}}$$

$$= 2 - 4y^2 + 2y^4$$

Triple integral: $\displaystyle \iiint_{\mathcal{W}} x\, dV = \int_0^1 (2 - 4y^2 + 2y^4)\, dy = 2 - \frac{4}{3} + \frac{2}{5} = \frac{16}{15}$ ∎

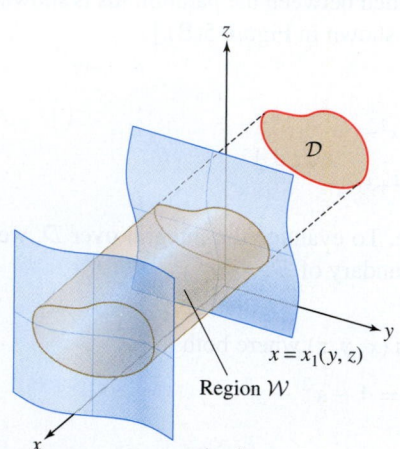

FIGURE 6 $\mathcal{D}$ is the projection of $\mathcal{W}$ onto the yz-plane.

So far, we have evaluated triple integrals over regions $\mathcal{W}$ that were identified as z-simple and having a projection that is a domain in the xy-plane. We can integrate equally well over x-simple and y-simple regions. For example, if $\mathcal{W}$ is the x-simple region between the graphs of $x = x_1(y, z)$ and $x = x_2(y, z)$ lying over a domain $\mathcal{D}$ in the yz-plane (Figure 6), then

$$\iiint_{\mathcal{W}} f(x, y, z)\, dV = \iint_{\mathcal{D}} \left(\int_{x=x_1(y,z)}^{x_2(y,z)} f(x, y, z)\, dx \right) dA$$

EXAMPLE 5 **Writing a Triple Integral in Three Ways** A region $\mathcal{W}$ is bounded by

$$z = 4 - y^2, \qquad y = 2x, \qquad z = 0, \qquad x = 0$$

Figure 7 demonstrates that $\mathcal{W}$ is x-simple, y-simple, and z-simple. Set up three separate iterated integrals for $\displaystyle \iiint_{\mathcal{W}} xyz\, dV$, first regarding $\mathcal{W}$ as z-simple, then as x-simple, and finally as y-simple.

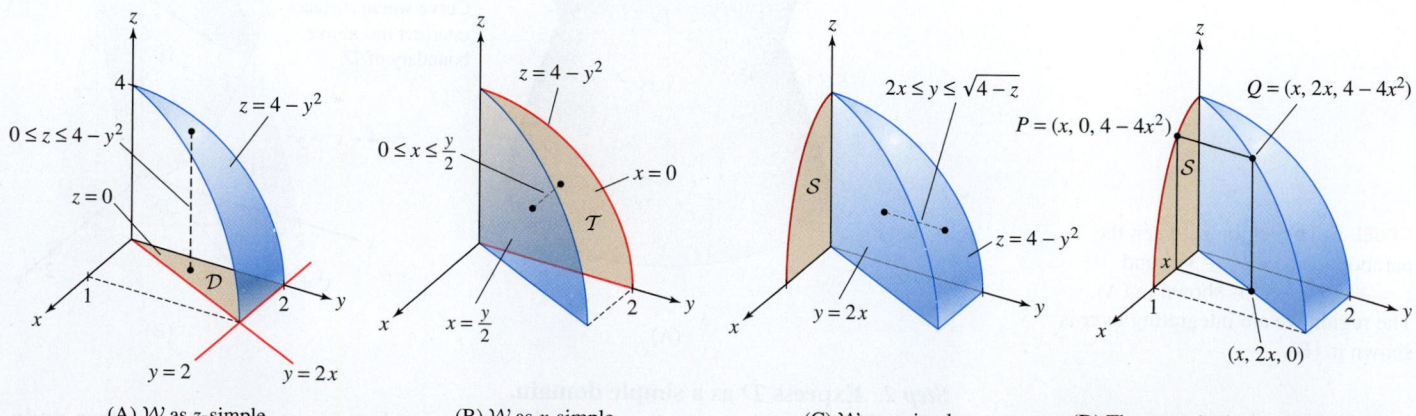

(A) $\mathcal{W}$ as z-simple

(B) $\mathcal{W}$ as x-simple

(C) $\mathcal{W}$ as y-simple

(D) The curve in the boundary of $\mathcal{S}$ consists of points $P = (x, 0, 4 - 4x^2)$

FIGURE 7

Solution We consider each case separately.

Consider W as z-simple. In Figure 7(A), we can see that W is z-simple because the region is enclosed between the plane $z = 0$ and the surface $z = 4 - y^2$. Therefore, the inner integral is an integral with respect to z with $0 \leq z \leq 4 - y^2$. To set up the outer two integrals, we need to determine the domain in the xy-plane for these two integrals. Still examining Figure 7(A), we see that the projection of W onto the xy-plane is a triangle $\mathcal{D}$ with $0 \leq x \leq 1$, $2x \leq y \leq 2$. We then have

$$W : 0 \leq x \leq 1, \quad 2x \leq y \leq 2, \quad 0 \leq z \leq 4 - y^2$$

$$\iiint_W xyz \, dV = \int_{x=0}^{1} \int_{y=2x}^{2} \int_{z=0}^{4-y^2} xyz \, dz \, dy \, dx \qquad \boxed{4}$$

Consider W as x-simple. In Figure 7(B), we can see that W is x-simple because it lies between the yz-plane ($x = 0$) and the plane $x = \frac{y}{2}$. Thus, the inner integral is an integral with respect to x with $0 \leq x \leq \frac{y}{2}$. We need to determine the domain in the yz-plane for the outer two integrals. The projection of W onto the yz-plane is the domain $\mathcal{T}$ in Figure 7(B) with $0 \leq y \leq 2$, $0 \leq z \leq 4 - y^2$. Therefore, we have

$$W : 0 \leq y \leq 2, \quad 0 \leq z \leq 4 - y^2, \quad 0 \leq x \leq \frac{1}{2}y$$

$$\iiint_W xyz \, dV = \int_{y=0}^{2} \int_{z=0}^{4-y^2} \int_{x=0}^{y/2} xyz \, dx \, dz \, dy$$

Consider W as y-simple. Observe that Figure 7(C) shows that W is enclosed between the plane $y = 2x$ and the surface $z = 4 - y^2$, and therefore the inner integral is an integral with respect to y with $2x \leq y \leq \sqrt{4 - z}$. We need to determine the domain in the xz-plane for the outer two integrals. The challenge is to describe the projection of W onto the xz-plane; that is, the region $\mathcal{S}$ in Figure 7(C). We need the equation of the boundary curve of $\mathcal{S}$. A point P on this curve is the projection of a point $Q = (x, y, z)$ on the boundary of the left face [Figure 7(D)]. Since Q lies on both the plane $y = 2x$ and the surface $z = 4 - y^2$, $Q = (x, 2x, 4 - 4x^2)$. The projection of Q is $P = (x, 0, 4 - 4x^2)$. We see that the projection of W onto the xz-plane is the domain $\mathcal{S}$ with $0 \leq x \leq 1$, $0 \leq z \leq 4 - 4x^2$. Thus,

$$W : 0 \leq x \leq 1, \quad 0 \leq z \leq 4 - 4x^2, \quad 2x \leq y \leq \sqrt{4 - z}$$

$$\iiint_W xyz \, dV = \int_{x=0}^{1} \int_{z=0}^{4-4x^2} \int_{y=2x}^{\sqrt{4-z}} xyz \, dy \, dz \, dx \qquad \boxed{5}$$

The **average value** of a function of three variables is defined as in the case of two variables:

$$\overline{f} = \frac{1}{\text{volume}(W)} \iiint_W f(x, y, z) \, dV \qquad \boxed{6}$$

where $\text{volume}(W) = \iiint_W 1 \, dV$. Also, as in the case of two variables, $\overline{f}$ lies between the minimum and maximum values of f on W. And, furthermore, the Mean Value Theorem holds: If W is connected and f is continuous on W, then there exists a point $P \in W$ such that $f(P) = \overline{f}$.

EXAMPLE 6 A solid piece of crystal W in the first octant of space is bounded by the five planes given by $z = 0$, $y = 0$, $x = 0$, $x + z = 1$, and $x + y + z = 3$. The temperature at every point in the crystal is given by $T(x, y, z) = x$ in degrees centigrade. Find the average temperature for all points in the crystal.

You can check that all three ways of writing the triple integral in Example 5 yield the same answer:

$$\iiint_W xyz \, dV = \frac{2}{3}$$

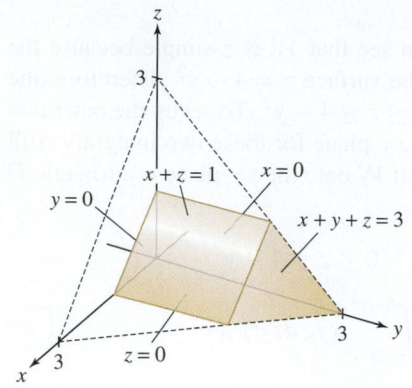

FIGURE 8 Finding the average temperature of a crystal.

Solution To find the average temperature, we first find the volume. The region $\mathcal{W}$, which appears in Figure 8, is best handled as y-simple, lying between the planes $x + y + z = 3$ and $y = 0$. Furthermore, it projects onto the triangle in the xz-plane given by $0 \le z \le 1 - x$ for $0 \le x \le 1$. Thus, $\mathcal{W}$ is described by

$$0 \le x \le 1, \quad 0 \le z \le 1 - x, \quad 0 \le y \le 3 - x - z$$

$$\text{volume}(\mathcal{W}) = \iiint_{\mathcal{W}} dV = \int_{x=0}^{1} \int_{z=0}^{1-x} \int_{y=0}^{3-x-z} dy \, dz \, dx$$

$$= \int_{x=0}^{1} \int_{z=0}^{1-x} (3 - x - z) \, dz \, dx$$

$$= \int_{x=0}^{1} \left(3(1-x) - x(1-x) - \frac{(1-x)^2}{2} \right) dx = \frac{7}{6}$$

To obtain the average temperature, we must next compute $\iiint_{\mathcal{W}} T(x, y, z) \, dV$. The only difference from our previous calculation is that we are now integrating $T(x, y, z) = x$ on $\mathcal{W}$:

$$\iiint_{\mathcal{W}} T(x, y, z) \, dV = \int_{x=0}^{1} \int_{z=0}^{1-x} \int_{y=0}^{3-x-z} x \, dy \, dz \, dx$$

$$= \int_{x=0}^{1} \int_{z=0}^{1-x} \left(3x - x^2 - zx \right) dz \, dx$$

$$= \int_{x=0}^{1} \left(3x(1-x) - x^2(1-x) - \frac{x(1-x)^2}{2} \right) dx = \frac{3}{8}$$

Therefore,

$$\overline{T} = \frac{1}{\text{volume}(\mathcal{W})} \iiint_{\mathcal{W}} T(x, y, z) \, dV = \left(\frac{6}{7} \right) \left(\frac{3}{8} \right) = \frac{9}{28} {}^{\circ}\text{C}$$

Note that the Mean Value Theorem implies that there is some point in the crystal where the temperature is exactly $\frac{9}{28}{}^{\circ}\text{C}$. ∎

The Volume of the Sphere in Higher Dimensions

Archimedes (287–212 BCE) proved the beautiful formula $V = \frac{4}{3}\pi r^3$ for the volume of a sphere nearly 2000 years before calculus was invented, by means of a brilliant geometric argument showing that the volume of a sphere is equal to two-thirds the volume of the circumscribed cylinder. According to Plutarch (ca. 45–120 CE), Archimedes valued this achievement so highly that he requested that a sphere with a circumscribed cylinder be engraved on his tomb.

We can use integration to generalize Archimedes's formula to n dimensions. The ball of radius r in $\mathbf{R}^n$, denoted $B_n(r)$, is the set of points $(x_1, \ldots, x_n)$ in $\mathbf{R}^n$ such that

$$x_1^2 + x_2^2 + \cdots + x_n^2 \le r^2$$

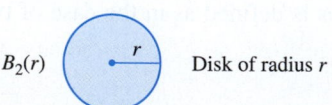

FIGURE 9 Balls of radius r in dimensions $n = 1, 2, 3$.

The balls $B_n(r)$ in dimensions 1, 2, and 3 are the interval, disk, and ball shown in Figure 9. In dimensions $n \ge 4$, the ball $B_n(r)$ is difficult, if not impossible, to visualize, but we can compute its volume. Denote this volume by $V_n(r)$. For $n = 1$, the "volume" $V_1(r)$ is the length of the interval $B_1(r)$, and for $n = 2$, $V_2(r)$ is the area of the disk $B_2(r)$. We know that

$$V_1(r) = 2r, \qquad V_2(r) = \pi r^2, \qquad V_3(r) = \frac{4}{3}\pi r^3$$

For $n \ge 4$, $V_n(r)$ is sometimes called the **hypervolume**.

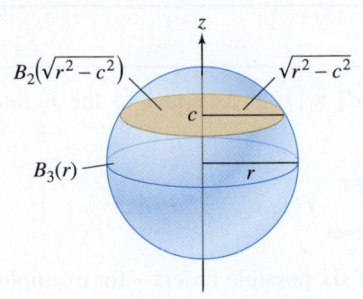

FIGURE 10 The volume $V_3(r)$ is the integral of cross-sectional area $V_2(\sqrt{r^2 - c^2})$.

The key idea is to determine $V_n(r)$ from the formula for $V_{n-1}(r)$ by integrating cross-sectional volume. Consider the case $n = 3$, where the horizontal slice at height $z = c$ is a two-dimensional ball (a disk) of radius $\sqrt{r^2 - c^2}$ (Figure 10). The volume $V_3(r)$ is equal to the integral of the areas of these horizontal slices:

$$V_3(r) = \int_{z=-r}^{r} V_2\left(\sqrt{r^2 - z^2}\right) dz = \int_{z=-r}^{r} \pi(r^2 - z^2)\, dz = \frac{4}{3}\pi r^3$$

Next, we show by induction that for all $n \geq 1$, there is a constant A_n such that

$$\boxed{V_n(r) = A_n r^n} \qquad \boxed{7}$$

The slice of $B_n(r)$ at height $x_n = c$ has equation

$$x_1^2 + x_2^2 + \cdots + x_{n-1}^2 + c^2 = r^2$$

This slice is the ball $B_{n-1}\left(\sqrt{r^2 - c^2}\right)$ of radius $\sqrt{r^2 - c^2}$, and $V_n(r)$ is obtained by integrating the volume of these slices:

$$V_n(r) = \int_{x_n=-r}^{r} V_{n-1}\left(\sqrt{r^2 - x_n^2}\right) dx_n = A_{n-1}\int_{x_n=-r}^{r} \left(\sqrt{r^2 - x_n^2}\right)^{n-1} dx_n$$

Using the substitution $x_n = r\sin\theta$ and $dx_n = r\cos\theta\, d\theta$, we have

$$V_n(r) = A_{n-1} r^n \int_{-\pi/2}^{\pi/2} \cos^n\theta\, d\theta = A_{n-1} C_n r^n$$

where $C_n = \int_{\theta=-\pi/2}^{\pi/2} \cos^n\theta\, d\theta$. This proves Eq. (7) with

$$\boxed{A_n = A_{n-1}C_n} \qquad \boxed{8}$$

In Exercise 45, you are asked to use Integration by Parts to verify the relation

$$C_n = \left(\frac{n-1}{n}\right) C_{n-2} \qquad \boxed{9}$$

It is easy to check directly that $C_0 = \pi$ and $C_1 = 2$. By Eq. (9), $C_2 = \frac{1}{2}C_0 = \frac{\pi}{2}$, $C_3 = \frac{2}{3}(2) = \frac{4}{3}$, and so on. Here are the first few values of C_n:

n	0	1	2	3	4	5	6	7
C_n	π	2	$\dfrac{\pi}{2}$	$\dfrac{4}{3}$	$\dfrac{3\pi}{8}$	$\dfrac{16}{15}$	$\dfrac{5\pi}{16}$	$\dfrac{32}{35}$

Note that $A_n = V_n(1)$ by Eq. (7), and therefore A_n is the volume of the n-dimensional unit ball (i.e., ball of radius 1). We know that $A_1 = 2$ and $A_2 = \pi$, so we can use the values of C_n together with Eq. (8) to obtain the values of A_n in Table 1. We see, for example, that the ball of radius r in four dimensions has volume $V_4(r) = \frac{1}{2}\pi^2 r^4$. The general formula depends on whether n is even or odd. Using induction and formulas (8) and (9), we can prove that

$$\boxed{A_{2m} = \frac{\pi^m}{m}, \qquad A_{2m+1} = \frac{2^{m+1}\pi^m}{1 \cdot 3 \cdot 5 \cdots (2m+1)}}$$

This sequence of numbers A_n has a curious property. Setting $r = 1$ in Eq. (7), we see that A_n is the volume of the unit ball in n dimensions. From Table 1, it appears that the volumes increase up to dimension 5 and then begin to decrease. In Exercise 46, you are asked to verify that the five-dimensional unit ball has the largest volume. Furthermore, the volumes A_n tend to 0 as $n \to \infty$.

TABLE 1   Volume of an n-dimensional unit ball is $V_n(1) = A_n$

n	A_n
1	2
2	$\pi \approx 3.14$
3	$\dfrac{4}{3}\pi \approx 4.19$
4	$\dfrac{\pi^2}{2} \approx 4.93$
5	$\dfrac{8\pi^2}{15} \approx 5.26$
6	$\dfrac{\pi^3}{6} \approx 5.17$
7	$\dfrac{16\pi^3}{105} \approx 4.72$

15.3 SUMMARY

- The triple integral over a box $\mathcal{B} = [a, b] \times [c, d] \times [p, q]$ is equal to the iterated integral

$$\iiint_{\mathcal{B}} f(x, y, z)\, dV = \int_{x=a}^{b} \int_{y=c}^{d} \int_{z=p}^{q} f(x, y, z)\, dz\, dy\, dx$$

The iterated integral may be written in any one of six possible orders—for example,

$$\int_{z=p}^{q} \int_{y=c}^{d} \int_{x=a}^{b} f(x, y, z)\, dx\, dy\, dz$$

- A *z-simple region* $\mathcal{W}$ in $\mathbf{R}^3$ is a region consisting of the points (x, y, z) between two surfaces $z = z_1(x, y)$ and $z = z_2(x, y)$, where $z_1(x, y) \leq z_2(x, y)$, lying over a domain $\mathcal{D}$ in the xy-plane. In other words, $\mathcal{W}$ is defined by

$$(x, y) \in \mathcal{D}, \qquad z_1(x, y) \leq z \leq z_2(x, y)$$

Similarly, we have x-simple regions and y-simple regions.

- The triple integral over a z-simple region $\mathcal{W}$ is equal to an iterated integral:

$$\iiint_{\mathcal{W}} f(x, y, z)\, dV = \iint_{\mathcal{D}} \left(\int_{z=z_1(x,y)}^{z_2(x,y)} f(x, y, z)\, dz \right) dA$$

- The volume of a region $\mathcal{W}$ is

$$V = \iiint_{\mathcal{W}} 1\, dV$$

- The *average value* of $f(x, y, z)$ on a region $\mathcal{W}$ of volume V is the quantity

$$\overline{f} = \frac{1}{V} \iiint_{\mathcal{W}} f(x, y, z)\, dV$$

15.3 EXERCISES

Preliminary Questions

1. Which of (a)–(c) is *not* equal to $\int_0^1 \int_3^4 \int_6^7 f(x, y, z)\, dz\, dy\, dx$?

(a) $\int_6^7 \int_0^1 \int_3^4 f(x, y, z)\, dy\, dx\, dz$

(b) $\int_3^4 \int_0^1 \int_6^7 f(x, y, z)\, dz\, dx\, dy$

(c) $\int_0^1 \int_3^4 \int_6^7 f(x, y, z)\, dx\, dz\, dy$

2. Which of the following is *not* a meaningful triple integral?

(a) $\int_0^1 \int_0^x \int_{x+y}^{2x+y} e^{x+y+z}\, dz\, dy\, dx$

(b) $\int_0^1 \int_0^z \int_{x+y}^{2x+y} e^{x+y+z}\, dz\, dy\, dx$

3. Describe the projection of the region of integration $\mathcal{W}$ onto the xy-plane:

(a) $\int_0^1 \int_0^x \int_0^{x^2+y^2} f(x, y, z)\, dz\, dy\, dx$

(b) $\int_0^1 \int_0^{\sqrt{1-x^2}} \int_2^4 f(x, y, z)\, dz\, dy\, dx$

Exercises

In Exercises 1–8, evaluate $\iiint_{\mathcal{B}} f(x, y, z)\, dV$ for the specified function f and box $\mathcal{B}$.

1. $f(x, y, z) = xz + yz^2$; $\quad 0 \leq x \leq 2, \quad 2 \leq y \leq 4, \quad 0 \leq z \leq 4$

2. $f(x, y, z) = xy + z^2$; $\quad [-2, 2] \times [0, 1] \times [0, 2]$

3. $f(x, y, z) = xe^{y-2z}$; $\quad 0 \leq x \leq 2, \quad 0 \leq y \leq 1, \quad 0 \leq z \leq 1$

4. $f(x, y, z) = \dfrac{x}{(y+z)^2}$; $\quad [0, 2] \times [2, 4] \times [-1, 1]$

5. $f(x, y, z) = (x - y)(y - z)$; $\quad [0, 1] \times [0, 3] \times [0, 3]$

6. $f(x, y, z) = \dfrac{z}{x}$; $\quad 1 \le x \le 3, \quad 0 \le y \le 2, \quad 0 \le z \le 4$

7. $f(x, y, z) = (x + z)^3$; $\quad [0, a] \times [0, b] \times [0, c]$

8. $f(x, y, z) = (x + y - z)^2$; $\quad [0, a] \times [0, b] \times [0, c]$

In Exercises 9–14, evaluate $\iiint_{\mathcal{W}} f(x, y, z)\, dV$ *for the function* f *and region* $\mathcal{W}$ *specified.*

9. $f(x, y, z) = x + y$; $\quad \mathcal{W} : y \le z \le x, \ 0 \le y \le x, \ 0 \le x \le 1$

10. $f(x, y, z) = e^{x+y+z}$; $\quad \mathcal{W} : 0 \le z \le 1, \ 0 \le y \le x, \ 0 \le x \le 1$

11. $f(x, y, z) = xyz$; $\quad \mathcal{W} : 0 \le z \le 1, \ 0 \le y \le \sqrt{1 - x^2}, \ 0 \le x \le 1$

12. $f(x, y, z) = x$; $\quad \mathcal{W} : x^2 + y^2 \le z \le 4$

13. $f(x, y, z) = e^z$; $\quad \mathcal{W} : x + y + z \le 1, \ x \ge 0, \quad y \ge 0, \ z \ge 0$

14. $f(x, y, z) = z$; $\quad \mathcal{W} : 0 \le x \le 1, \ x^2 \le y \le 2,$ $x - y \le z \le x + y$

15. Calculate the integral of $f(x, y, z) = z$ over the region $\mathcal{W}$ in Figure 11, below the hemisphere of radius 3 and lying over the triangle $\mathcal{D}$ in the xy-plane bounded by $x = 1$, $y = 0$, and $x = y$.

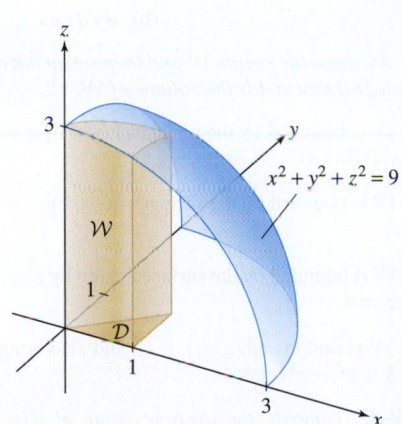

FIGURE 11

16. Calculate the integral of $f(x, y, z) = e^z$ over the tetrahedron $\mathcal{W}$ in Figure 12 (the region in the first octant under the plane shown).

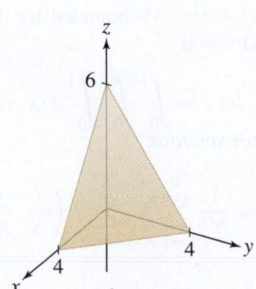

FIGURE 12

17. Integrate $f(x, y, z) = x$ over the region in the first octant bounded above by $z = 8 - 2x^2 - y^2$ and below by $z = y^2$.

18. Compute the integral of $f(x, y, z) = y^2$ over the region within the cylinder $x^2 + y^2 = 4$, where $0 \le z \le y$.

19. Find the triple integral of the function $F(x, y, z) = z$ over the region in Figure 13.

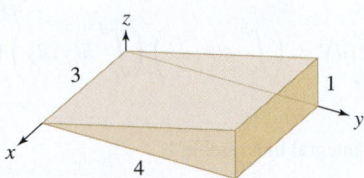

FIGURE 13

20. Find the volume of the solid in $\mathbf{R}^3$ bounded by $y = x^2$, $x = y^2$, $z = x + y + 5$, and $z = 0$.

21. Find the volume of the solid in the first octant bounded between the planes $x + y + z = 1$ and $x + y + 2z = 1$.

22. Calculate $\iiint_{\mathcal{W}} y\, dV$, where $\mathcal{W}$ is the region above $z = x^2 + y^2$ and below $z = 5$, and bounded by $y = 0$ and $y = 1$.

23. Evaluate $\iiint_{\mathcal{W}} xz\, dV$, where $\mathcal{W}$ is the domain bounded by the elliptic cylinder $\dfrac{x^2}{4} + \dfrac{y^2}{9} = 1$ and the sphere $x^2 + y^2 + z^2 = 16$ in the first octant (Figure 14).

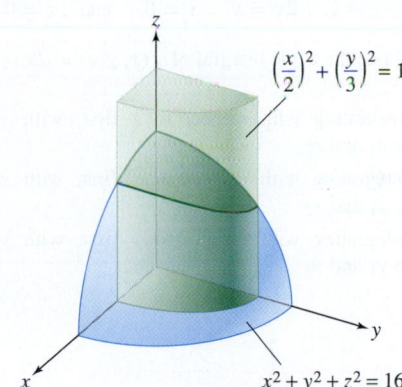

FIGURE 14

24. Describe the domain of integration and evaluate:

$$\int_0^3 \int_0^{\sqrt{9-x^2}} \int_0^{\sqrt{9-x^2-y^2}} xy\, dz\, dy\, dx$$

25. Describe the domain of integration of the following integral:

$$\int_{-2}^2 \int_{-\sqrt{4-z^2}}^{\sqrt{4-z^2}} \int_1^{\sqrt{5-x^2-z^2}} f(x, y, z)\, dy\, dx\, dz$$

26. Let $\mathcal{W}$ be the region below the paraboloid

$$x^2 + y^2 = z - 2$$

that lies above the part of the plane $x + y + z = 1$ in the first octant. Express

$$\iiint_{\mathcal{W}} f(x, y, z)\, dV$$

as an iterated integral (for an arbitrary function f).

27. Assume $f(x, y, z)$ can be expressed as a product, $f(x, y, z) = g(x)h(y)k(z)$. Show that the integral of f over a box $\mathcal{B} = [a, b] \times [c, d] \times [p, q]$ can be expressed as a product of integrals as follows:

$$\iiint_{\mathcal{B}} f(x, y, z)\, dV = \left(\int_a^b g(x)\, dx \right) \left(\int_c^d h(y)\, dy \right) \left(\int_p^q k(z)\, dz \right)$$

28. Consider the integral in Example 1:

$$\int_1^4 \int_0^3 \int_2^6 x^2 e^{y+3z}\, dz\, dy\, dx$$

Show that the integrand can be expressed as a product $g(x)h(y)k(z)$. Then verify the equation in Exercise 27 by computing the product of integrals on the right-hand side and showing it equals the result obtained in the example.

29. In Example 5, we expressed a triple integral as an iterated integral in the three orders

$$dz\, dy\, dx, \quad dx\, dz\, dy, \quad \text{and} \quad dy\, dz\, dx$$

Write this integral in the three other orders:

$$dz\, dx\, dy, \quad dx\, dy\, dz, \quad \text{and} \quad dy\, dx\, dz$$

30. Let $\mathcal{W}$ be the region shown in Figure 15, bounded by

$$y + z = 2, \quad 2x = y, \quad x = 0, \quad \text{and} \quad z = 0$$

Express and evaluate the triple integral of $f(x, y, z) = 2x - 4y + 6z$ considering $\mathcal{W}$ as:

(a) z-simple, integrating with respect to z first, with $z_1(x, y) \leq z \leq z_2(x, y)$ for some z_1 and z_2.

(b) x-simple, integrating with respect to x first, with $x_1(y, z) \leq x \leq x_2(y, z)$ for some x_1 and x_2.

(c) y-simple, integrating with respect to y first, with $y_1(x, z) \leq y \leq y_2(x, z)$ for some y_1 and y_2.

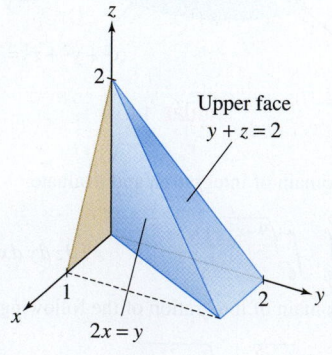

FIGURE 15

31. Let

$$\mathcal{W} = \left\{ (x, y, z) : \sqrt{x^2 + y^2} \leq z \leq 1 \right\}$$

(see Figure 16). Express $\iiint_{\mathcal{W}} f(x, y, z)\, dV$ as an iterated integral in the order $dz\, dy\, dx$ (for an arbitrary function f).

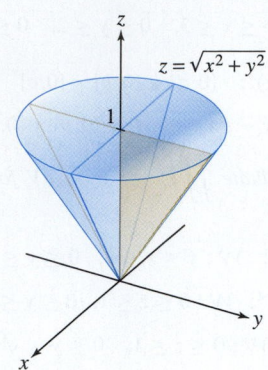

FIGURE 16

32. Repeat Exercise 31 for the order $dx\, dy\, dz$.

33. Let $\mathcal{W}$ be the region bounded by $z = 1 - y^2$, $y = x^2$, and the plane $z = 0$. Calculate the volume of $\mathcal{W}$ as a triple integral in the order $dz\, dy\, dx$.

34. Calculate the volume of the region $\mathcal{W}$ in Exercise 33 as a triple integral in the following orders:

(a) $dx\, dz\, dy$ **(b)** $dy\, dz\, dx$

In Exercises 35–38, draw the region $\mathcal{W}$ and then set up but do not compute a single triple integral that yields the volume of $\mathcal{W}$.

35. The region $\mathcal{W}$ is bounded by the surfaces given by $z = 1 - y^2$, $x = 0$ and $z = 0$, $z + x = 3$.

36. The region $\mathcal{W}$ is bounded by the surfaces given by $z = x^2$, $z + y = 1$, and $z - y = 1$.

37. The region $\mathcal{W}$ is bounded by the surfaces given by $z = y^2$, $y = z^2$ and $x = 0$, $x + y + z = 4$.

38. The region $\mathcal{W}$ is underneath $z = 1 - x^2$ and also bounded by $y = 0$, $z = 0$, and $y = 3 - x^2 - z^2$.

In Exercises 39–42, compute the average value of $f(x, y, z)$ over the region $\mathcal{W}$.

39. $f(x, y, z) = xy \sin(\pi z); \quad \mathcal{W} = [0, 1] \times [0, 1] \times [0, 1]$

40. $f(x, y, z) = xyz; \quad \mathcal{W} : 0 \leq z \leq y \leq x \leq 1$

41. $f(x, y, z) = e^y; \quad \mathcal{W} : 0 \leq y \leq 1 - x^2, \quad 0 \leq z \leq x$

42. $f(x, y, z) = x^2 + y^2 + z^2; \quad \mathcal{W}$ bounded by the planes $2y + z = 1$, $x = 0$, $x = 1$, $z = 0$, and $y = 0$

In Exercises 43 and 44, let $I = \int_0^1 \int_0^1 \int_0^1 f(x, y, z)\, dV$ and let $S_{N,N,N}$ be the Riemann sum approximation

$$S_{N,N,N} = \frac{1}{N^3} \sum_{i=1}^N \sum_{j=1}^N \sum_{k=1}^N f\left(\frac{i}{N}, \frac{j}{N}, \frac{k}{N} \right)$$

43. [CAS] Calculate $S_{N,N,N}$ for $f(x, y, z) = e^{x^2 - y - z}$ for $N = 10, 20, 30$. Then evaluate I and find an N such that $S_{N,N,N}$ approximates I to two decimal places.

44. [CAS] Calculate $S_{N,N,N}$ for $f(x, y, z) = \sin(xyz)$ for $N = 10, 20, 30$. Then use a computer algebra system to calculate I numerically and estimate the error $|I - S_{N,N,N}|$.

Further Insights and Challenges

45. Use Integration by Parts to verify Eq. (9).

46. Using Eq. (8), compute the volume A_n of the unit ball in $\mathbf{R}^n$ for $n = 8$, 9, 10. Show that $C_n \leq 1$ for $n \geq 6$ and use this to prove that of all unit balls, the five-dimensional ball has the largest volume. Can you explain why A_n tends to 0 as $n \to \infty$?

15.4 Integration in Polar, Cylindrical, and Spherical Coordinates

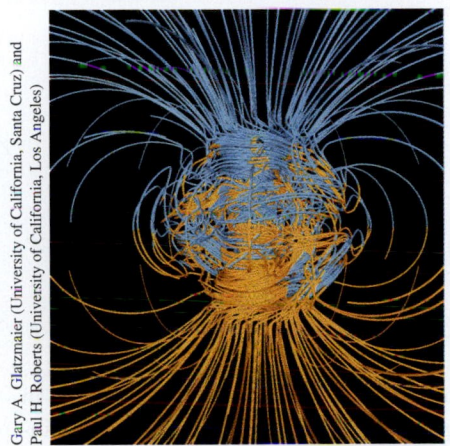

FIGURE 1 Spherical coordinates are used in mathematical models of Earth's magnetic field. This computer simulation, based on the Glatzmaier–Roberts model, shows the magnetic lines of force, representing inward- and outward-directed field lines in blue and yellow, respectively.

In single-variable calculus, a well-chosen substitution (also called a change of variables) often transforms a complicated integral into a simpler one. Change of variables is also useful in multivariable calculus, but the emphasis is different. In the multivariable case, we are usually interested in simplifying not just the integrand, but also the representation of the domain of integration.

This section treats three of the most useful changes of variables, in which an integral is expressed in polar, cylindrical, or spherical coordinates. As in Figure 1, certain physical systems are much more easily modeled with the right coordinate system. The general Change of Variables Formula is discussed in Section 15.6.

Double Integrals in Polar Coordinates

Polar coordinates are convenient when the domain of integration is an angular sector or a **polar rectangle** (Figure 2):

$$\mathcal{R} : \theta_1 \leq \theta \leq \theta_2, \quad r_1 \leq r \leq r_2 \qquad \boxed{1}$$

We assume throughout that $r_1 \geq 0$ and that all radial coordinates are nonnegative. Recall that rectangular and polar coordinates are related by

$$x = r\cos\theta, \qquad y = r\sin\theta$$

Thus, we write a function $f(x, y)$ in polar coordinates as $f(r\cos\theta, r\sin\theta)$. The Change of Variables Formula for a polar rectangle $\mathcal{R}$ is

$$\iint_{\mathcal{R}} f(x, y)\, dA = \int_{\theta_1}^{\theta_2} \int_{r_1}^{r_2} f(r\cos\theta, r\sin\theta)\, r\, dr\, d\theta \qquad \boxed{2}$$

Notice the extra factor r in the integrand on the right. It will become clear why it is included when we derive the Change of Variables Formula next.

Equation (2) expresses the integral of $f(x, y)$ over the polar rectangle in Figure 2 as the integral of a new function $r f(r\cos\theta, r\sin\theta)$ over the ordinary rectangle $[\theta_1, \theta_2] \times [r_1, r_2]$. In this sense, the change of variables simplifies the domain of integration.

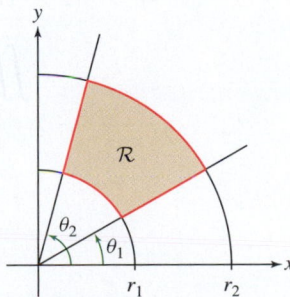

FIGURE 2 Polar rectangle.

To derive Eq. (2), the key step is to estimate the area ΔA of the small polar rectangle shown in Figure 3. If Δr and $\Delta\theta$ are small, then this polar rectangle is very nearly an ordinary rectangle of sides Δr and $r \Delta \theta$, as in the Reminder in the margin, and therefore $\Delta A \approx r \, \Delta r \, \Delta \theta$. In fact, ΔA is the difference of areas of two sectors:

$$\Delta A = \frac{1}{2}(r + \Delta r)^2 \, \Delta \theta - \frac{1}{2}r^2 \, \Delta \theta = r(\Delta r \, \Delta \theta) + \frac{1}{2}(\Delta r)^2 \Delta \theta \approx r \, \Delta r \, \Delta \theta$$

The error in our approximation is the term $\frac{1}{2}(\Delta r)^2 \Delta \theta$, which has a smaller order of magnitude than $\Delta r \, \Delta \theta$ when Δr and $\Delta \theta$ are both small.

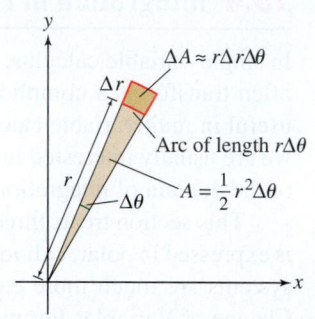

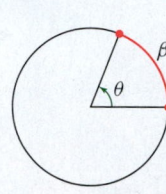

REMINDER In Figure 4, the length β of the arc subtended by the angle θ is the fraction of the entire circumference that θ is of the entire angle 2π. Hence, $\beta = \frac{\theta}{2\pi} 2\pi r = r\theta$. Similarly, the area of the sector subtended by θ is $\frac{1}{2}r^2\theta$.

DF **FIGURE 3** Small polar rectangle.

FIGURE 4

Now, decompose $\mathcal{R}$ into an $N \times M$ grid of small polar subrectangles $\mathcal{R}_{ij}$ as in Figure 5, and choose a sample point P_{ij} in $\mathcal{R}_{ij}$. If $\mathcal{R}_{ij}$ is small and $f(x, y)$ is continuous, then by approximation (11) in Section 15.2, we have

$$\iint_{\mathcal{R}_{ij}} f(x, y) \, dx \, dy \approx f(P_{ij}) \, \text{area}(\mathcal{R}_{ij}) \approx f(P_{ij}) \, r_{ij} \, \Delta r \, \Delta \theta$$

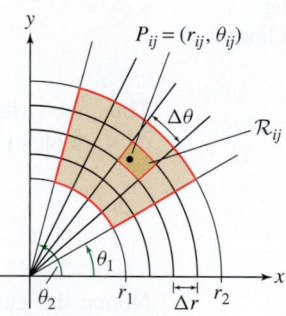

FIGURE 5 Decomposition of a polar rectangle into subrectangles.

Note that each polar rectangle $\mathcal{R}_{ij}$ has angular width $\Delta \theta = (\theta_2 - \theta_1)/N$ and radial width $\Delta r = (r_2 - r_1)/M$. The integral over $\mathcal{R}$ is the sum

$$\iint_{\mathcal{R}} f(x, y) \, dx \, dy = \sum_{i=1}^{N} \sum_{j=1}^{M} \iint_{\mathcal{R}_{ij}} f(x, y) \, dx \, dy$$

$$\approx \sum_{i=1}^{N} \sum_{j=1}^{M} f(P_{ij}) \, \text{Area}(\mathcal{R}_{ij})$$

$$\approx \sum_{i=1}^{N} \sum_{j=1}^{M} f(r_{ij} \cos \theta_{ij}, r_{ij} \sin \theta_{ij}) \, r_{ij} \, \Delta r \, \Delta \theta$$

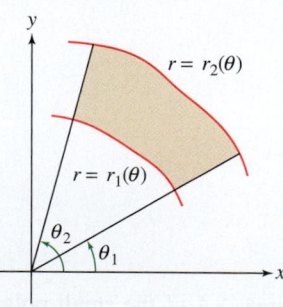

FIGURE 6 A region between two polar curves.

This is a Riemann sum for the double integral of $r f(r \cos \theta, r \sin \theta)$ over the region where $r_1 \le r \le r_2$, $\theta_1 \le \theta \le \theta_2$, and we can prove that it approaches the double integral as $N, M \to \infty$. A similar derivation is valid for domains (Figure 6) that can be described as the region between two polar curves $r = r_1(\theta)$ and $r = r_2(\theta)$ for θ between the values θ_1 and θ_2. This gives us the following Theorem 1.

> **THEOREM 1 Double Integral in Polar Coordinates** For a continuous function f on the domain
>
> $$\mathcal{D} : \theta_1 \leq \theta \leq \theta_2, \quad r_1(\theta) \leq r \leq r_2(\theta)$$
>
> $$\iint_{\mathcal{D}} f(x, y) \, dA = \int_{\theta_1}^{\theta_2} \int_{r=r_1(\theta)}^{r_2(\theta)} f(r \cos \theta, r \sin \theta) r \, dr \, d\theta \qquad \boxed{3}$$

Equation (3) is summarized in the expression for the area element dA in polar coordinates:

$$\boxed{dA = r \, dr \, d\theta}$$

We call such a region $\mathcal{D}$ **radially simple**. It has the property that every ray from the origin intersects the region in a single point or in a line segment that begins on $r = r_1(\theta)$ and ends on $r = r_2(\theta)$.

EXAMPLE 1 Compute $\iint_{\mathcal{D}} (x + y) \, dA$, where $\mathcal{D}$ is the quarter annulus in Figure 7.

Solution The quarter annulus is an example of a domain that is radially simple.

Step 1. Describe $\mathcal{D}$ and f in polar coordinates.
The quarter annulus $\mathcal{D}$ is defined by the inequalities (Figure 7)

$$\mathcal{D} : 0 \leq \theta \leq \frac{\pi}{2}, \quad 2 \leq r \leq 4$$

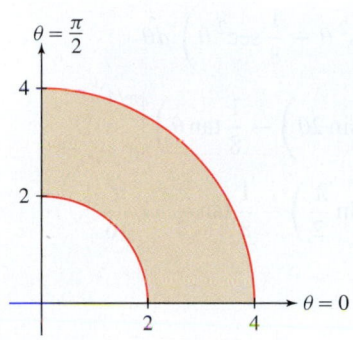

FIGURE 7 Quarter annulus $0 \leq \theta \leq \frac{\pi}{2}$, $2 \leq r \leq 4$.

In polar coordinates,

$$f(x, y) = x + y = r \cos \theta + r \sin \theta = r(\cos \theta + \sin \theta)$$

Step 2. Change variables and evaluate.
To write the integral in polar coordinates, we replace dA by $r \, dr \, d\theta$:

$$\iint_{\mathcal{D}} (x + y) \, dA = \int_0^{\pi/2} \int_{r=2}^{4} r(\cos \theta + \sin \theta) r \, dr \, d\theta$$

The inner integral is

$$\int_{r=2}^{4} (\cos \theta + \sin \theta) r^2 \, dr = (\cos \theta + \sin \theta) \left(\frac{4^3}{3} - \frac{2^3}{3} \right) = \frac{56}{3} (\cos \theta + \sin \theta)$$

and

$$\iint_{\mathcal{D}} (x + y) \, dA = \frac{56}{3} \int_0^{\pi/2} (\cos \theta + \sin \theta) \, d\theta = \frac{56}{3} (\sin \theta - \cos \theta) \Big|_0^{\pi/2} = \frac{112}{3} \quad \blacksquare$$

EXAMPLE 2 Calculate $\iint_{\mathcal{D}} (x^2 + y^2)^{-2} \, dA$ for the shaded domain $\mathcal{D}$ in Figure 8.

Solution

Step 1. Describe $\mathcal{D}$ and f in polar coordinates.
The quarter circle lies in the angular sector $0 \leq \theta \leq \frac{\pi}{4}$ because the line through $P = (1, 1)$ makes an angle of $\frac{\pi}{4}$ with the x-axis (Figure 8).
To determine the limits on r, recall from Section 11.3 (Examples 5 and 8) that

- The vertical line $x = 1$ has polar equation $r \cos \theta = 1$ or $r = \sec \theta$.
- The circle of radius 1 and center $(1, 0)$ has polar equation $r = 2 \cos \theta$.

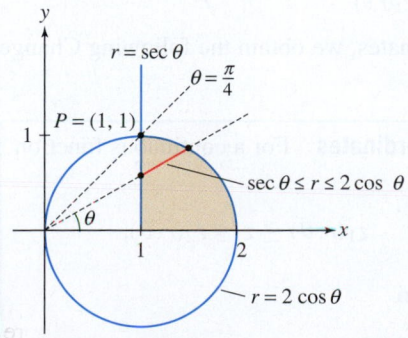

FIGURE 8

Therefore, a ray of angle θ intersects $\mathcal{D}$ in the segment where r ranges from $\sec \theta$ to $2 \cos \theta$. In other words, our domain is radially simple with polar description

$$\mathcal{D} : 0 \leq \theta \leq \frac{\pi}{4}, \quad \sec \theta \leq r \leq 2 \cos \theta$$

The function in polar coordinates is

$$f(x, y) = (x^2 + y^2)^{-2} = (r^2)^{-2} = r^{-4}$$

Step 2. Change variables and evaluate.

$$\iint_{\mathcal{D}} (x^2 + y^2)^{-2} \, dA = \int_0^{\pi/4} \int_{r=\sec\theta}^{2\cos\theta} r^{-4} r \, dr \, d\theta = \int_0^{\pi/4} \int_{r=\sec\theta}^{2\cos\theta} r^{-3} \, dr \, d\theta$$

The inner integral is

REMINDER

$$\int \cos^2\theta \, d\theta = \frac{1}{2}\left(\theta + \frac{1}{2}\sin 2\theta\right) + C$$

$$\int \sec^2\theta \, d\theta = \tan\theta + C$$

$$\int_{r=\sec\theta}^{2\cos\theta} r^{-3} \, dr = -\frac{1}{2}r^{-2}\bigg|_{r=\sec\theta}^{2\cos\theta} = -\frac{1}{8}\sec^2\theta + \frac{1}{2}\cos^2\theta$$

Therefore,

$$\iint_{\mathcal{D}} (x^2 + y^2)^{-2} \, dA = \int_0^{\pi/4} \left(\frac{1}{2}\cos^2\theta - \frac{1}{8}\sec^2\theta\right) d\theta$$

$$= \left(\frac{1}{4}\left(\theta + \frac{1}{2}\sin 2\theta\right) - \frac{1}{8}\tan\theta\right)\bigg|_0^{\pi/4}$$

$$= \frac{1}{4}\left(\frac{\pi}{4} + \frac{1}{2}\sin\frac{\pi}{2}\right) - \frac{1}{8}\tan\frac{\pi}{4} = \frac{\pi}{16} \qquad \blacksquare$$

Triple Integrals in Cylindrical Coordinates

Cylindrical coordinates, introduced in Section 12.7, are useful when the domain has **axial symmetry**—that is, symmetry with respect to an axis. In cylindrical coordinates (r, θ, z), the axis of symmetry is the z-axis. Recall the relations (Figure 9)

$$x = r\cos\theta, \qquad y = r\sin\theta, \qquad z = z$$

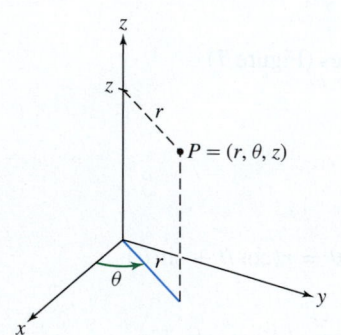

FIGURE 9 Cylindrical coordinates.

To set up a triple integral in cylindrical coordinates, we assume that the domain of integration $\mathcal{W}$ can be described as the region between two surfaces (Figure 10)

$$z_1(r, \theta) \leq z \leq z_2(r, \theta)$$

lying over a radially simple domain $\mathcal{D}$ in the xy-plane with polar description

$$\mathcal{D}: \theta_1 \leq \theta \leq \theta_2, \quad r_1(\theta) \leq r \leq r_2(\theta)$$

A triple integral over $\mathcal{W}$ can be written as an iterated integral (see Theorem 2 of Section 15.3):

$$\iiint_{\mathcal{W}} f(x, y, z) \, dV = \iint_{\mathcal{D}} \left(\int_{z=z_1(r,\theta)}^{z_2(r,\theta)} f(x, y, z) \, dz\right) dA$$

By expressing the integral over $\mathcal{D}$ in polar coordinates, we obtain the following Change of Variables Formula.

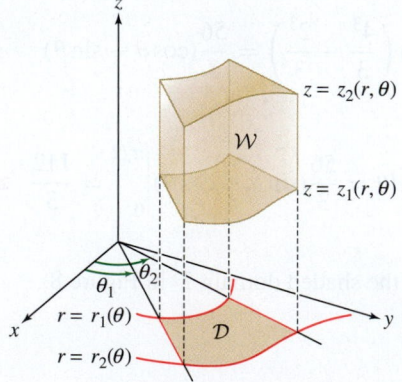

FIGURE 10 Region described in cylindrical coordinates.

Equation (4) is summarized in the expression for the volume element dV in cylindrical coordinates:

$$\boxed{dV = r \, dz \, dr \, d\theta}$$

> **THEOREM 2 Triple Integrals in Cylindrical Coordinates** For a continuous function f on the region
> $$\theta_1 \leq \theta \leq \theta_2, \qquad r_1(\theta) \leq r \leq r_2(\theta), \qquad z_1(r, \theta) \leq z \leq z_2(r, \theta),$$
> the triple integral $\displaystyle\iiint_{\mathcal{W}} f(x, y, z) \, dV$ is equal to
> $$\boxed{\int_{\theta_1}^{\theta_2} \int_{r=r_1(\theta)}^{r_2(\theta)} \int_{z=z_1(r,\theta)}^{z_2(r,\theta)} f(r\cos\theta, r\sin\theta, z) \, r \, dz \, dr \, d\theta}$$

4

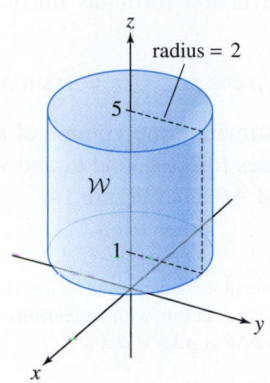

FIGURE 11 The region $x^2 + y^2 \le 4$, $1 \le z \le 5$.

EXAMPLE 3 Integrate $f(x, y, z) = z\sqrt{x^2 + y^2}$ over the cylindrical region $\mathcal{W}$ where $x^2 + y^2 \le 4$ for $1 \le z \le 5$ (Figure 11).

Solution The domain of integration $\mathcal{W}$ lies above the disk of radius 2 centered at the origin, so in cylindrical coordinates,

$$\mathcal{W} : 0 \le \theta \le 2\pi, \quad 0 \le r \le 2, \quad 1 \le z \le 5$$

We write the function in cylindrical coordinates:

$$f(x, y, z) = z\sqrt{x^2 + y^2} = zr$$

and integrate using $dV = r \, dz \, dr \, d\theta$.

$$\iiint_{\mathcal{W}} z\sqrt{x^2 + y^2} \, dV = \int_0^{2\pi} \int_{r=0}^2 \int_{z=1}^5 (zr) r \, dz \, dr \, d\theta$$

$$= \int_0^{2\pi} \int_{r=0}^2 12r^2 \, dr \, d\theta$$

$$= \int_0^{2\pi} 32 \, d\theta = 64\pi \qquad \blacksquare$$

EXAMPLE 4 Compute the integral of $f(x, y, z) = z$ over the region $\mathcal{W}$ within the cylinder $x^2 + y^2 \le 4$, where $0 \le z \le y$.

Solution

Step 1. **Express $\mathcal{W}$ in cylindrical coordinates.**

The condition $0 \le z \le y$ tells us that $y \ge 0$, so $\mathcal{W}$ projects onto the semicircle $\mathcal{D}$ in the xy-plane shown in Figure 12. The semicircle has radius 2 and corresponds to $y \ge 0$. In polar coordinates we have

$$\mathcal{D} : 0 \le \theta \le \pi, \quad 0 \le r \le 2$$

The z-coordinate in $\mathcal{W}$ varies from $z = 0$ to $z = y$, and in polar coordinates $y = r \sin\theta$, so the region has the description

$$\mathcal{W} : 0 \le \theta \le \pi, \quad 0 \le r \le 2, \quad 0 \le z \le r \sin\theta$$

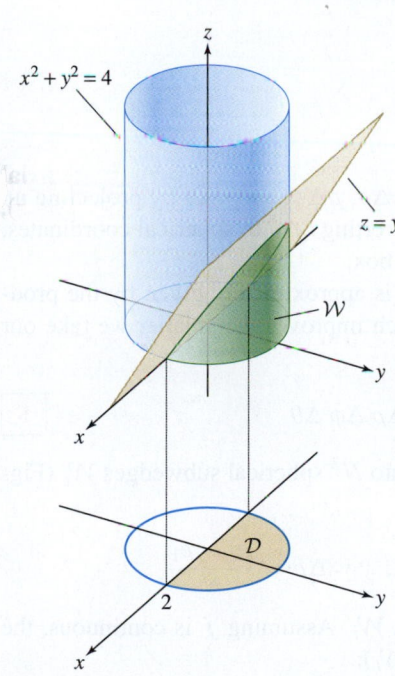

FIGURE 12

Step 2. **Change variables and evaluate.**

$$\iiint_{\mathcal{W}} f(x, y, z) \, dV = \int_0^\pi \int_{r=0}^2 \int_{z=0}^{r\sin\theta} zr \, dz \, dr \, d\theta$$

$$= \int_0^\pi \int_{r=0}^2 \frac{1}{2}(r\sin\theta)^2 r \, dr \, d\theta$$

$$= \int_0^\pi 2\sin^2\theta \, d\theta = \pi \qquad \blacksquare$$

← **REMINDER**

$$\int \sin^2\theta \, d\theta = \frac{1}{2}\left(\theta - \frac{1}{2}\sin 2\theta\right) + C$$

$$\int_0^\pi \sin^2\theta \, d\theta = \frac{\pi}{2}$$

Triple Integrals in Spherical Coordinates

We noted that the Change of Variables Formula in cylindrical coordinates has a volume element expressed as $dV = r \, dr \, d\theta \, dz$. In spherical coordinates (ρ, θ, ϕ) (introduced in Section 12.7), the analog is the formula

$$\boxed{dV = \rho^2 \sin\phi \, d\rho \, d\phi \, d\theta}$$

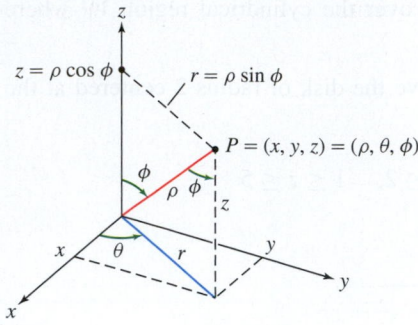

$z = \rho \cos \phi$ $r = \rho \sin \phi$

$P = (x, y, z) = (\rho, \theta, \phi)$

DF **FIGURE 13** Spherical coordinates.

FIGURE 14 A spherical wedge with volume approximately $\rho^2 \sin \phi \, \Delta \rho \, \Delta \phi \, \Delta \theta$.

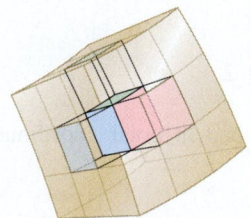

FIGURE 15 Decomposition of a spherical wedge into subwedges.

To begin the derivation of this formula, recall the conversion formulas illustrated in Figure 13:

$$x = \rho \sin \phi \cos \theta, \qquad y = \rho \sin \phi \sin \theta, \qquad z = \rho \cos \phi, \qquad r = \rho \sin \phi$$

A key step in the derivation of the formula for dV is estimating the volume of a small **spherical wedge** $\mathcal{W}$. Suppose it is defined by fixing values for ρ, ϕ, and θ, and varying each coordinate by a small amount given by $\Delta\rho$, $\Delta\phi$, and $\Delta\theta$ as in Figure 14.

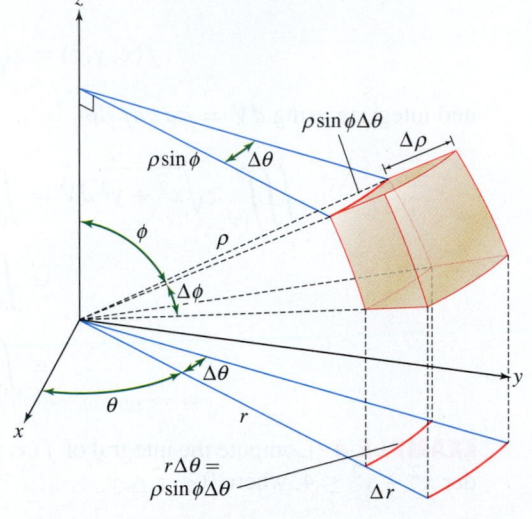

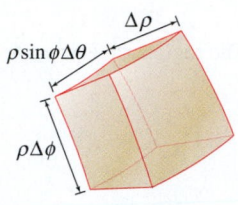

For small increments, the wedge is nearly a rectangular box with dimensions $\rho \sin \phi \Delta\theta \times \rho \Delta\phi \times \Delta\rho$.

The spherical wedge is nearly a box with sides $\Delta\rho$, $\rho\Delta\phi$, and $r\Delta\theta$ by projecting up from the corresponding length in the xy-plane. Converting $r\Delta\theta$ to spherical coordinates, we obtain $\rho \sin \phi \Delta\theta$ for this third dimension of the box.

Therefore, the volume of the spherical wedge is approximately given by the product of these three dimensions, the accuracy of which improves the smaller we take our changes in the variables:

$$\text{volume}(\mathcal{W}) \approx \rho^2 \sin \phi \, \Delta\rho \, \Delta\phi \, \Delta\theta \qquad \boxed{5}$$

Following the usual steps, we decompose $\mathcal{W}$ into N^3 spherical subwedges $\mathcal{W}_i$ (Figure 15) with increments

$$\Delta\theta = \frac{\theta_2 - \theta_1}{N}, \qquad \Delta\phi = \frac{\phi_2 - \phi_1}{N}, \qquad \Delta\rho = \frac{\rho_2 - \rho_1}{N}$$

and choose a sample point $P_i = (\rho_i, \theta_i, \phi_i)$ in each $\mathcal{W}_i$. Assuming f is continuous, the following approximation holds for large N (small $\mathcal{W}_i$):

$$\iiint_{\mathcal{W}_i} f(x, y, z) \, dV \approx f(P_i)\text{volume}(\mathcal{W}_i)$$

$$\approx f(P_i)\rho_i^2 \sin \phi_i \, \Delta\rho \, \Delta\phi \, \Delta\theta$$

Taking the sum over i, we obtain

$$\iiint_{\mathcal{W}} f(x, y, z) \, dV \approx \sum_i f(P_i)\rho_i^2 \sin \phi_i \, \Delta\rho \, \Delta\phi \, \Delta\theta \qquad \boxed{6}$$

The sum on the right is a Riemann sum for the function

$$f(\rho \cos \theta \sin \phi, \rho \sin \theta \sin \phi, \rho \cos \phi) \rho^2 \sin \phi$$

on the domain $\mathcal{W}$. Equation (7) follows by passing to the limit as $N \to \infty$ [and showing that the error in Eq. (6) tends to zero]. This argument applies generally to regions defined by an inequality $\rho_1(\theta, \phi) \leq \rho \leq \rho_2(\theta, \phi)$ with $\theta_1 \leq \theta \leq \theta_2$ and $\phi_1 \leq \phi \leq \phi_2$.

Equation (7) is summarized in the expression for the volume element dV in spherical coordinates:

$$dV = \rho^2 \sin\phi \, d\rho \, d\phi \, d\theta$$

THEOREM 3 Triple Integrals in Spherical Coordinates For a region $\mathcal{W}$ defined by

$$\theta_1 \leq \theta \leq \theta_2, \qquad \phi_1 \leq \phi \leq \phi_2, \qquad \rho_1(\theta,\phi) \leq \rho \leq \rho_2(\theta,\phi)$$

the triple integral $\iiint_{\mathcal{W}} f(x,y,z)\,dV$ is equal to

$$\int_{\theta_1}^{\theta_2} \int_{\phi=\phi_1}^{\phi_2} \int_{\rho=\rho_1(\theta,\phi)}^{\rho_2(\theta,\phi)} f(\rho\sin\phi\cos\theta, \rho\sin\phi\sin\theta, \rho\cos\phi)\,\rho^2 \sin\phi \, d\rho \, d\phi \, d\theta \qquad \boxed{7}$$

We call such a region $\mathcal{W}$ **centrally simple** in that every ray from the origin intersects the solid in a point or a single line segment such that the first endpoint of the segment lies on the surface $\rho = \rho_1(\theta,\phi)$ and the second endpoint lies on the surface $\rho = \rho_2(\theta,\phi)$.

EXAMPLE 5 Compute the integral of $f(x,y,z) = x^2 + y^2$ over the sphere $\mathcal{S}$ of radius 4 centered at the origin (Figure 16).

Solution First, write $f(x,y,z)$ in spherical coordinates:

$$f(x,y,z) = x^2 + y^2 = (\rho\sin\phi\cos\theta)^2 + (\rho\sin\phi\sin\theta)^2$$

$$= \rho^2 \sin^2\phi(\cos^2\theta + \sin^2\theta) = \rho^2 \sin^2\phi$$

Since we are integrating over the entire sphere $\mathcal{S}$ of radius 4, this is a centrally simple region, with each ray beginning at the origin and ending on the sphere. So, ρ varies from 0 to 4, θ from 0 to 2π, and ϕ from 0 to π. In the following computation, we integrate first with respect to θ. This is justified because the result of the inner two integrals is independent of θ, and therefore the outer integral can be treated as the integral of a constant with respect to θ:

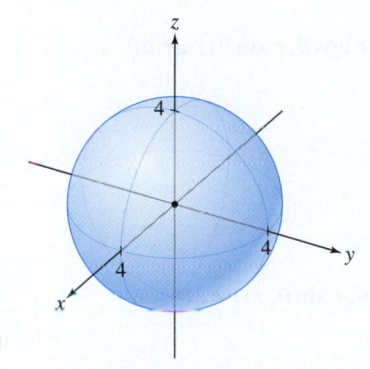

FIGURE 16 Sphere of radius 4.

$$\iiint_{\mathcal{S}} (x^2 + y^2)\,dV = \int_0^{2\pi} \int_{\phi=0}^{\pi} \int_{\rho=0}^4 (\rho^2 \sin^2\phi)\,\rho^2 \sin\phi \, d\rho \, d\phi \, d\theta$$

$$= 2\pi \int_{\phi=0}^{\pi} \int_{\rho=0}^4 \rho^4 \sin^3\phi \, d\rho \, d\phi = 2\pi \int_0^{\pi} \left(\frac{\rho^5}{5}\bigg|_0^4\right) \sin^3\phi \, d\phi$$

$$= \frac{2048\pi}{5} \int_0^{\pi} \sin^3\phi \, d\phi$$

$$= \frac{2048\pi}{5} \left(\frac{1}{3}\cos^3\phi - \cos\phi\right)\bigg|_0^{\pi} = \frac{8192\pi}{15} \qquad \blacksquare$$

← **REMINDER** We can integrate $\sin^3\phi$ by writing $\sin^3\phi = (1 - \cos^2\phi)\sin\phi$ and using a substitution $u = \cos\phi$. This yields

$$\int \sin^3\phi \, d\phi = \frac{1}{3}\cos^3\phi - \cos\phi + C$$

EXAMPLE 6 Integrate $f(x,y,z) = z$ over the ice cream cone–shaped region $\mathcal{W}$ in Figure 17, lying above the cone and below the sphere.

Solution The cone has equation $x^2 + y^2 = z^2$, which in spherical coordinates is

$$(\rho\sin\phi\cos\theta)^2 + (\rho\sin\phi\sin\theta)^2 = (\rho\cos\phi)^2$$

$$\rho^2 \sin^2\phi(\cos^2\theta + \sin^2\theta) = \rho^2 \cos^2\phi$$

$$\sin^2\phi = \cos^2\phi$$

$$\sin\phi = \pm\cos\phi \quad \Rightarrow \quad \phi = \frac{\pi}{4}, \frac{3\pi}{4}$$

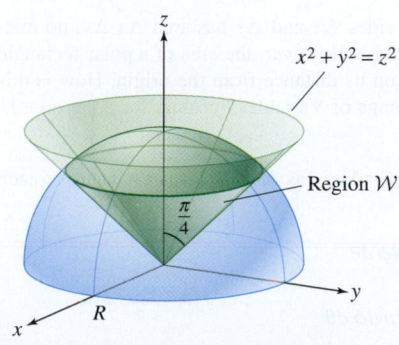

FIGURE 17 Ice cream cone defined by $0 \leq \rho \leq R, 0 \leq \phi \leq \pi/4$.

The half of the cone above the xy-plane has the equation $\phi = \frac{\pi}{4}$. On the other hand, the sphere has equation $\rho = R$, so the ice cream cone has the description

$$\mathcal{W}: 0 \leq \theta \leq 2\pi, \quad 0 \leq \phi \leq \frac{\pi}{4}, \quad 0 \leq \rho \leq R$$

We have the following integral that, as in the previous example, we integrate first with respect to θ because the result of the inner two integrals is independent of θ.

$$\iiint_{\mathcal{W}} z\, dV = \int_0^{2\pi} \int_{\phi=0}^{\pi/4} \int_{\rho=0}^{R} (\rho\cos\phi)\rho^2 \sin\phi\, d\rho\, d\phi\, d\theta$$

$$= 2\pi \int_{\phi=0}^{\pi/4} \int_{\rho=0}^{R} \rho^3 \cos\phi \sin\phi\, d\rho\, d\phi = \frac{\pi R^4}{2} \int_0^{\pi/4} \sin\phi\cos\phi\, d\phi = \frac{\pi R^4}{8}$$

∎

15.4 SUMMARY

The area and volume elements:

$$\boxed{dA = r\, dr\, d\theta}$$

$$\boxed{dV = r\, dz\, dr\, d\theta}$$

$$\boxed{dV = \rho^2 \sin\phi\, d\rho\, d\phi\, d\theta}$$

• Double integral in *polar coordinates:*

$$\iint_{\mathcal{D}} f(x,y)\, dA = \int_{\theta_1}^{\theta_2} \int_{r=r_1(\theta)}^{r_2(\theta)} f(r\cos\theta, r\sin\theta)\, r\, dr\, d\theta$$

• Triple integral $\iiint_{\mathcal{W}} f(x,y,z)\, dV$

   − In *cylindrical coordinates:*

$$\int_{\theta_1}^{\theta_2} \int_{r=r_1(\theta)}^{r_2(\theta)} \int_{z=z_1(r,\theta)}^{z_2(r,\theta)} f(r\cos\theta, r\sin\theta, z)\, r\, dz\, dr\, d\theta$$

   − In *spherical coordinates:*

$$\int_{\theta_1}^{\theta_2} \int_{\phi=\phi_1}^{\phi_2} \int_{\rho=\rho_1(\theta,\phi)}^{\rho_2(\theta,\phi)} f(\rho\sin\phi\cos\theta, \rho\sin\phi\sin\theta, \rho\cos\phi)\, \rho^2 \sin\phi\, d\rho\, d\phi\, d\theta$$

15.4 EXERCISES

Preliminary Questions

1. Which of the following represent the integral of $f(x,y) = x^2 + y^2$ over the unit circle?

(a) $\displaystyle\int_0^1 \int_0^{2\pi} r^2\, dr\, d\theta$

(b) $\displaystyle\int_0^{2\pi} \int_0^1 r^2\, dr\, d\theta$

(c) $\displaystyle\int_0^1 \int_0^{2\pi} r^3\, dr\, d\theta$

(d) $\displaystyle\int_0^{2\pi} \int_0^1 r^3\, dr\, d\theta$

2. What are the limits of integration in $\iiint f(r,\theta,z)\, r\, dr\, d\theta\, dz$ if the integration extends over the following regions?

(a) $x^2 + y^2 \le 4,\quad -1 \le z \le 2$

(b) Lower hemisphere of the sphere of radius 2, center at origin

3. What are the limits of integration in

$$\iiint f(\rho,\phi,\theta)\, \rho^2 \sin\phi\, d\rho\, d\phi\, d\theta$$

if the integration extends over the following spherical regions centered at the origin?

(a) Sphere of radius 4

(b) Region between the spheres of radii 4 and 5

(c) Lower hemisphere of the sphere of radius 2

4. An ordinary rectangle of sides Δx and Δy has area $\Delta x\, \Delta y$, no matter where it is located in the plane. However, the area of a polar rectangle of sides Δr and $\Delta\theta$ depends on its distance from the origin. How is this difference reflected in the Change of Variables Formula for polar coordinates?

5. The volume of a sphere of radius 3 is 36π. What is the value of each of the following integrals?

(a) $\displaystyle\int_0^{\pi} \int_0^{\pi} \int_0^3 \rho^2 \sin\phi\, d\rho\, d\phi\, d\theta$

(b) $\displaystyle\int_0^{2\pi} \int_0^{\pi/2} \int_0^3 \rho^2 \sin\phi\, d\rho\, d\phi\, d\theta$

(c) $\displaystyle\int_0^{\pi/4} \int_0^{\pi/2} \int_0^3 \rho^2 \sin\phi\, d\rho\, d\phi\, d\theta$

(d) $\displaystyle\int_{-\pi/4}^{\pi/4} \int_0^{\pi} \int_0^3 \rho^2 \sin\phi\, d\rho\, d\phi\, d\theta$

Exercises

In Exercises 1–6, sketch the region $\mathcal{D}$ indicated and integrate $f(x, y)$ over $\mathcal{D}$ using polar coordinates.

1. $f(x, y) = \sqrt{x^2 + y^2}, \quad x^2 + y^2 \leq 2$

2. $f(x, y) = x^2 + y^2; \quad 1 \leq x^2 + y^2 \leq 4$

3. $f(x, y) = xy; \quad x \geq 0, \quad y \geq 0, \quad x^2 + y^2 \leq 4$

4. $f(x, y) = y(x^2 + y^2)^3; \quad y \geq 0, \quad x^2 + y^2 \leq 1$

5. $f(x, y) = y(x^2 + y^2)^{-1}; \quad y \geq \frac{1}{2}, \quad x^2 + y^2 \leq 1$

6. $f(x, y) = e^{x^2 + y^2}; \quad x^2 + y^2 \leq R$

In Exercises 7–14, sketch the region of integration and evaluate by changing to polar coordinates.

7. $\displaystyle\int_{-2}^{2} \int_{0}^{\sqrt{4-x^2}} (x^2 + y^2) \, dy \, dx$

8. $\displaystyle\int_{0}^{3} \int_{0}^{\sqrt{9-y^2}} \sqrt{x^2 + y^2} \, dx \, dy$

9. $\displaystyle\int_{0}^{1/2} \int_{\sqrt{3}x}^{\sqrt{1-x^2}} x \, dy \, dx$

10. $\displaystyle\int_{0}^{4} \int_{0}^{\sqrt{16-x^2}} \tan^{-1}\frac{y}{x} \, dy \, dx$

11. $\displaystyle\int_{0}^{5} \int_{0}^{y} x \, dx \, dy$

12. $\displaystyle\int_{0}^{2} \int_{x}^{\sqrt{3}x} y \, dy \, dx$

13. $\displaystyle\int_{-1}^{2} \int_{0}^{\sqrt{4-x^2}} (x^2 + y^2) \, dy \, dx$

14. $\displaystyle\int_{1}^{2} \int_{0}^{\sqrt{2x-x^2}} \frac{1}{\sqrt{x^2 + y^2}} \, dy \, dx$

In Exercises 15–20, calculate the integral over the given region by changing to polar coordinates.

15. $f(x, y) = (x^2 + y^2)^{-2}; \quad x^2 + y^2 \leq 2, \quad x \geq 1$

16. $f(x, y) = y; \quad 2 \leq x^2 + y^2 \leq 9$

17. $f(x, y) = |xy|; \quad x^2 + y^2 \leq 1$

18. $f(x, y) = (x^2 + y^2)^{-3/2}; \quad x^2 + y^2 \leq 1, \quad x + y \geq 1$

19. $f(x, y) = x - y; \quad x^2 + y^2 \leq 1, \quad x + y \geq 1$

20. $f(x, y) = y; \quad x^2 + y^2 \leq 1, \quad (x-1)^2 + y^2 \leq 1$

21. Find the volume of the wedge-shaped region (Figure 18) contained in the cylinder $x^2 + y^2 = 9$, bounded above by the plane $z = x$ and below by the xy-plane.

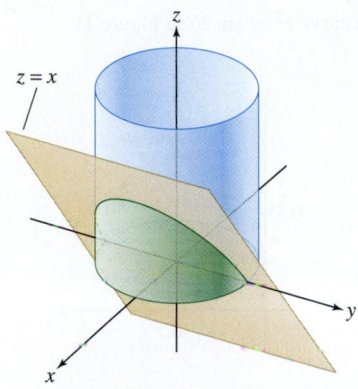

FIGURE 18

22. Let $\mathcal{W}$ be the region above the sphere $x^2 + y^2 + z^2 = 6$ and below the paraboloid $z = 4 - x^2 - y^2$.

(a) Show that the projection of $\mathcal{W}$ on the xy-plane is the disk $x^2 + y^2 \leq 2$ (Figure 19).

(b) Compute the volume of $\mathcal{W}$ using polar coordinates.

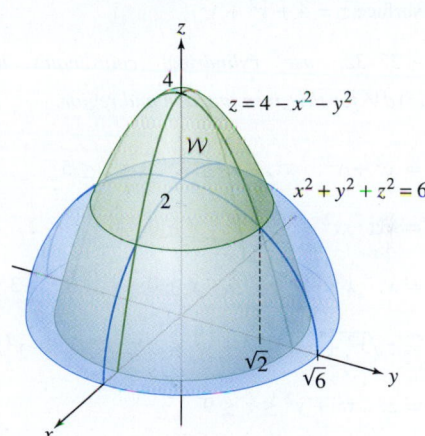

FIGURE 19

23. Evaluate $\displaystyle\iint_{\mathcal{D}} \sqrt{x^2 + y^2} \, dA$, where $\mathcal{D}$ is the domain in Figure 20. *Hint:* Find the equation of the inner circle in polar coordinates and treat the right and left parts of the region separately.

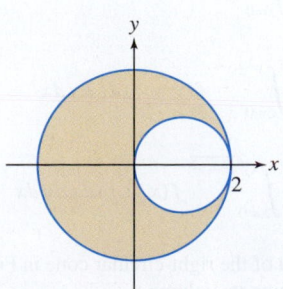

FIGURE 20

24. Evaluate $\iint_{\mathcal{D}} x\sqrt{x^2 + y^2}\, dA$, where $\mathcal{D}$ is the shaded region enclosed by the lemniscate curve $r^2 = \sin 2\theta$ in Figure 21.

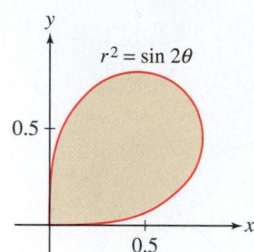

$r^2 = \sin 2\theta$

FIGURE 21

25. Let $\mathcal{W}$ be the region above the plane $z = 2$ and below the paraboloid $z = 6 - (x^2 + y^2)$.

(a) Describe $\mathcal{W}$ in cylindrical coordinates.

(b) Use cylindrical coordinates to compute the volume of $\mathcal{W}$.

26. Use cylindrical coordinates to calculate the integral of the function $f(x, y, z) = z$ over the region above the disk $x^2 + y^2 \leq 1$ in the xy-plane and below the surface $z = 4 + x^2 + y^2$.

In Exercises 27–32, use cylindrical coordinates to calculate $\iiint_{\mathcal{W}} f(x, y, z)\, dV$ *for the given function and region.*

27. $f(x, y, z) = x^2 + y^2;\quad x^2 + y^2 \leq 9,\quad 0 \leq z \leq 5$

28. $f(x, y, z) = xz;\quad x^2 + y^2 \leq 1,\quad x \geq 0,\quad 0 \leq z \leq 2$

29. $f(x, y, z) = x;\quad x^2 + y^2 \leq 16,\quad x \geq 0,\quad y \geq 0,\quad -3 \leq z \leq 3$

30. $f(x, y, z) = z\sqrt{x^2 + y^2};\quad x^2 + y^2 \leq z \leq 8 - (x^2 + y^2)$

31. $f(x, y, z) = z;\quad x^2 + y^2 \leq z \leq 9$

32. $f(x, y, z) = z;\quad 0 \leq z \leq x^2 + y^2 \leq 9$

In Exercises 33–36, express the triple integral in cylindrical coordinates.

33. $\displaystyle\int_{-1}^{1} \int_{y=-\sqrt{1-x^2}}^{y=\sqrt{1-x^2}} \int_{z=0}^{4} f(x, y, z)\, dz\, dy\, dx$

34. $\displaystyle\int_{-1}^{0} \int_{y=0}^{y=\sqrt{1-x^2}} \int_{z=0}^{2} f(x, y, z)\, dz\, dy\, dx$

35. $\displaystyle\int_{-1}^{1} \int_{y=0}^{y=\sqrt{1-x^2}} \int_{z=0}^{x^2+y^2} f(x, y, z)\, dz\, dy\, dx$

36. $\displaystyle\int_{0}^{2} \int_{y=0}^{y=\sqrt{2x-x^2}} \int_{z=0}^{\sqrt{x^2+y^2}} f(x, y, z)\, dz\, dy\, dx$

37. Find the equation of the right-circular cone in Figure 22 in cylindrical coordinates and compute its volume.

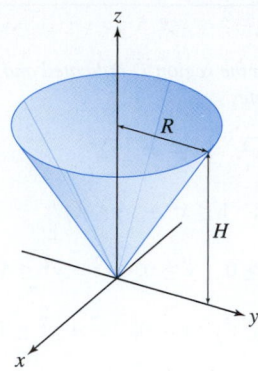

FIGURE 22

38. Use cylindrical coordinates to integrate $f(x, y, z) = z$ over the intersection of the solid hemisphere $x^2 + y^2 + z^2 \leq 4$, $z \geq 0$, and the cylinder $x^2 + y^2 \leq 1$.

39. Find the volume of the region appearing between the two surfaces in Figure 23.

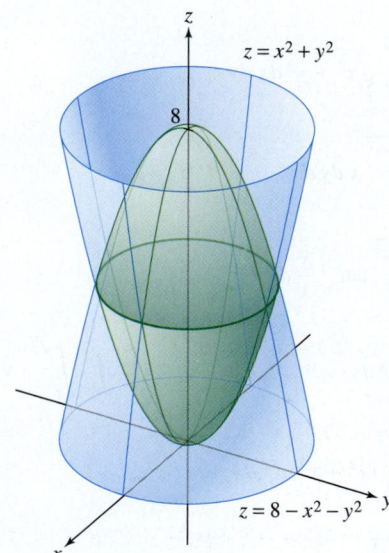

$z = x^2 + y^2$

$z = 8 - x^2 - y^2$

FIGURE 23

40. Use cylindrical coordinates to find the volume of a sphere of radius $2a$ from which a central cylinder of radius a has been removed.

41. Use cylindrical coordinates to show that the volume of a sphere of radius a from which a central cylinder of radius b has been removed, where $0 < b < a$, only depends on the height of the band that results. In particular, this implies that such a band of radius 2 m and height 1 m has the same volume as such a band of radius 6,400 km (the radius of the earth) and height 1 m.

42. Use cylindrical coordinates to find the volume of the region bounded below by the plane $z = 1$ and above by the sphere $x^2 + y^2 + z^2 = 4$.

43. Use spherical coordinates to find the volume of the region bounded below by the plane $z = 1$ and above by the sphere $x^2 + y^2 + z^2 = 4$.

44. Use spherical coordinates to find the volume of a sphere of radius 2 from which a central cylinder of radius 1 has been removed.

In Exercises 45–50, use spherical coordinates to calculate the triple integral of $f(x, y, z)$ over the given region.

45. $f(x, y, z) = y;$ $x^2 + y^2 + z^2 \le 1,$ $x, y, z \le 0$

46. $f(x, y, z) = \dfrac{1}{x^2 + y^2 + z^2};$ $5 \le x^2 + y^2 + z^2 \le 25$

47. $f(x, y, z) = x^2 + y^2;$ $\rho \le 1$

48. $f(x, y, z) = 1;$ $x^2 + y^2 + z^2 \le 4z,$ $z \ge \sqrt{x^2 + y^2}$

49. $f(x, y, z) = \sqrt{x^2 + y^2 + z^2};$ $x^2 + y^2 + z^2 \le 2z$

50. $f(x, y, z) = \rho;$ $x^2 + y^2 + z^2 \le 4,$ $z \le 1,$ $x \ge 0$

51. Use spherical coordinates to evaluate the triple integral of $f(x, y, z) = z$ over the region

$$0 \le \theta \le \frac{\pi}{3}, \qquad 0 \le \phi \le \frac{\pi}{2}, \qquad 1 \le \rho \le 2$$

52. Find the volume of the region lying above the cone $\phi = \phi_0$ and below the sphere $\rho = R$.

53. Calculate the integral of

$$f(x, y, z) = z(x^2 + y^2 + z^2)^{-3/2}$$

over the part of the ball $x^2 + y^2 + z^2 \le 16$ defined by $z \ge 2$.

54. Calculate the volume of the cone in Figure 22, using spherical coordinates.

55. Calculate the volume of the sphere $x^2 + y^2 + z^2 = a^2$, using both spherical and cylindrical coordinates.

56. Let $\mathcal{W}$ be the region within the cylinder $x^2 + y^2 = 2$ between $z = 0$ and the cone $z = \sqrt{x^2 + y^2}$. Calculate the integral of $f(x, y, z) = x^2 + y^2$ over $\mathcal{W}$, using both spherical and cylindrical coordinates.

57. Bell-Shaped Curve   One of the key results in calculus is the computation of the area under the bell-shaped curve (Figure 24):

$$I = \int_{-\infty}^{\infty} e^{-x^2}\, dx$$

This integral appears throughout engineering, physics, and statistics, and although e^{-x^2} does not have an elementary antiderivative, we can compute I using multiple integration.

(a) Show that $I^2 = J$, where J is the improper double integral

$$J = \int_{-\infty}^{\infty} \int_{-\infty}^{\infty} e^{-x^2 - y^2}\, dx\, dy$$

Hint: Use Fubini's Theorem and $e^{-x^2 - y^2} = e^{-x^2} e^{-y^2}$.

(b) Evaluate J in polar coordinates.

(c) Prove that $I = \sqrt{\pi}$.

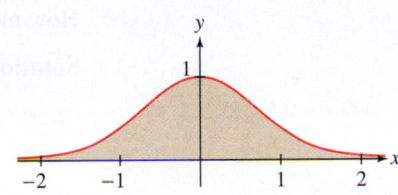

FIGURE 24 The bell-shaped curve $y = e^{-x^2}$.

Further Insights and Challenges

58. An Improper Multiple Integral   Show that a triple integral of $(x^2 + y^2 + z^2 + 1)^{-2}$ over all of $\mathbf{R}^3$ is equal to π^2. This is an improper integral, so integrate first over $\rho \le R$ and let $R \to \infty$.

59. Prove the formula

$$\iint_{\mathcal{D}} \ln r\, dA = -\frac{\pi}{2}$$

where $r = \sqrt{x^2 + y^2}$ and $\mathcal{D}$ is the unit disk $x^2 + y^2 \le 1$. This is an improper integral since $\ln r$ is not defined at $(0, 0)$, so integrate first over the annulus $a \le r \le 1$, where $0 < a < 1$, and let $a \to 0$.

60. Recall that the improper integral $\displaystyle\int_0^1 x^{-a}\, dx$ converges if and only if $a < 1$. For which values of a does $\displaystyle\iint_{\mathcal{D}} r^{-a}\, dA$ converge, where $r = \sqrt{x^2 + y^2}$ and $\mathcal{D}$ is the unit disk $x^2 + y^2 \le 1$?

15.5 **Applications of Multiple Integrals**

Previously we used the variable ρ to represent density in applications. We have also used ρ as a spherical-coordinates variable. To avoid confusion between these two uses, we will also use δ to represent density.

This section discusses some applications of multiple integrals. First, we consider quantities (such as mass, charge, and population) that are distributed with a given density δ in $\mathbf{R}^2$ or $\mathbf{R}^3$. In single-variable calculus, we saw that the "total amount" is defined as the integral of density. Similarly, the total amount of a quantity distributed in $\mathbf{R}^2$ or $\mathbf{R}^3$ is defined as the double or triple integral:

$$\text{total amount} = \iint_{\mathcal{D}} \delta(x, y)\, dA \quad \text{or} \quad \iiint_{\mathcal{W}} \delta(x, y, z)\, dV \qquad \boxed{1}$$

The density function δ has units of amount per unit area (or per unit volume).

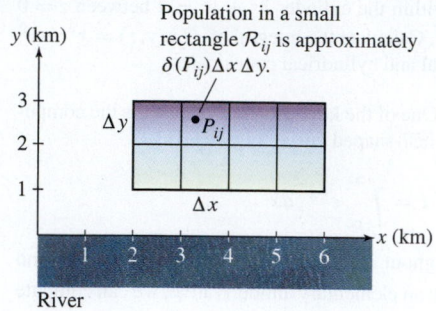

FIGURE 1 An aerial view of a region near the river.

The intuition behind Eq. (1) is similar to that of the single-variable case. Suppose, for example, that $\delta(x, y)$ is population density (Figure 1). When density is constant, the total population is simply density times area:

$$\text{population} = \text{density (people/km}^2\text{)} \times \text{area (km}^2\text{)}$$

To treat variable density in the case, say, of a rectangle $\mathcal{R}$, we divide $\mathcal{R}$ into smaller rectangles $\mathcal{R}_{ij}$ of area $\Delta x \, \Delta y$ on which δ is nearly constant (assuming that δ is continuous on $\mathcal{R}$). The population in $\mathcal{R}_{ij}$ is approximately $\delta(P_{ij}) \, \Delta x \, \Delta y$ for any sample point P_{ij} in $\mathcal{R}_{ij}$, and the sum of these approximations is a Riemann sum that converges to the double integral:

$$\int_{\mathcal{R}} \delta(x, y) \, dA \approx \sum_i \sum_j \delta(P_{ij}) \Delta x \, \Delta y$$

EXAMPLE 1   Population Density   The population in a rural area near a river has density

$$\delta(x, y) = 40xe^{0.1y} \text{ people per km}^2$$

How many people live in the region $\mathcal{R}$: $2 \le x \le 6$, $1 \le y \le 3$ (Figure 1)?

Solution   The total population is the integral of population density:

$$\iint_{\mathcal{R}} 40xe^{0.1y} \, dA = \int_1^3 \int_2^6 40xe^{0.1y} \, dx \, dy$$

$$= \int_1^3 \left(20x^2 e^{0.1y} \Big|_{x=2}^6 \right) dy = \int_1^3 640 e^{0.1y} \, dy$$

$$= 6400 e^{0.1y} \Big|_{y=1}^3 \approx 1566 \text{ people} \qquad \blacksquare$$

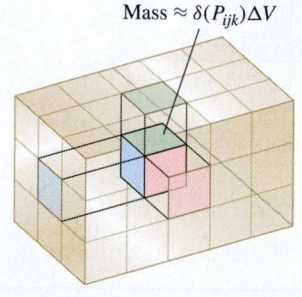

FIGURE 2 The mass of a small box is approximately $\delta(P_{ijk}) \, \Delta V$.

In the next example, we compute the mass of an object as the integral of mass density. In three dimensions, we justify this computation by dividing $\mathcal{W}$ into boxes $\mathcal{B}_{ijk}$ of volume ΔV that are so small that the mass density is nearly constant on $\mathcal{B}_{ijk}$ (Figure 2). The mass of $\mathcal{B}_{ijk}$ is approximately $\delta(P_{ijk}) \, \Delta V$, where P_{ijk} is any sample point in $\mathcal{B}_{ijk}$, and the sum of these approximations is a Riemann sum that converges to the triple integral:

$$\iiint_{\mathcal{W}} \delta(x, y, z) \, dV \approx \sum_i \sum_j \sum_k \underbrace{\delta(P_{ijk}) \Delta V}_{\substack{\text{Approximate mass} \\ \text{of } \mathcal{B}_{ijk}}}$$

When δ is constant, we say that the solid has a **uniform** mass density. In this case, the triple integral has the value δV and the mass is simply $M = \delta V$.

EXAMPLE 2   Let $a > 0$. Find the mass of the solid $\mathcal{W}$ enclosed between the paraboloid $z = a(x^2 + y^2)$ and the plane $z = H$ (Figure 3). Assume a mass density of $\delta(x, y, z) = z$.

Solution   Because the solid is symmetric with respect to the z-axis, we use cylindrical coordinates (r, θ, z). Recall that $r^2 = x^2 + y^2$, so the cylindrical equation of the paraboloid is $z = ar^2$. A point (r, θ, z) lies *above* the paraboloid if $z \ge ar^2$, so it lies *in* the solid if $ar^2 \le z \le H$. In other words, the solid is described by

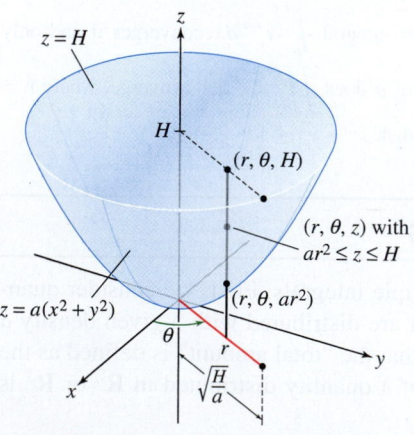

FIGURE 3 A solid enclosed between the paraboloid $z = a(x^2 + y^2)$ and the plane $z = H$.

$$0 \le \theta \le 2\pi, \qquad 0 \le r \le \sqrt{\frac{H}{a}}, \qquad ar^2 \le z \le H$$

The mass of the solid is the integral of mass density:

$$M = \iiint_{\mathcal{W}} \delta(x, y, z) \, dV = \int_{\theta=0}^{2\pi} \int_{r=0}^{\sqrt{H/a}} \int_{z=ar^2}^{H} (z) r \, dz \, dr \, d\theta$$

$$= \int_{\theta=0}^{2\pi} \int_{r=0}^{\sqrt{H/a}} \left(\frac{1}{2} H^2 - \frac{1}{2} a^2 r^4 \right) r \, dr \, d\theta$$

$$= \int_{\theta=0}^{2\pi} \left(\left(\frac{H^2 r^2}{4} - \frac{a^2 r^6}{12} \right) \Big|_{r=0}^{\sqrt{H/a}} \right) d\theta$$

$$= \int_{\theta=0}^{2\pi} \frac{H^3}{6a} \, d\theta = \frac{\pi H^3}{3a} \qquad \blacksquare$$

Next, we compute centers of mass. In Section 8.4, we computed centers of mass of laminas (thin plates in the plane) that had constant mass density. Multiple integration enables us to treat variable mass density. We define the moments of a lamina $\mathcal{D}$ with respect to the coordinate axes:

In $\mathbf{R}^2$, we denote the integral of $x\delta(x, y)$ by M_y and call it the moment with respect to the y-axis because x is the signed distance from (x, y) to the y-axis.

$$M_y = \iint_{\mathcal{D}} x\delta(x, y) \, dA, \qquad M_x = \iint_{\mathcal{D}} y\delta(x, y) \, dA$$

The **center of mass** (COM) is the point $P_{CM} = (x_{CM}, y_{CM})$, where

$$\boxed{x_{CM} = \frac{M_y}{M}, \qquad y_{CM} = \frac{M_x}{M}}$$ **2**

You can think of the coordinates x_{CM} and y_{CM} as **weighted averages**—they are the averages of x and y in which the factor δ assigns a larger coefficient to points with larger mass density.

If $\mathcal{D}$ has uniform mass density (δ constant), then the factors of δ in the numerator and denominator in Eq. (2) cancel, and the center of mass coincides with the **centroid**, defined as the point whose coordinates are the averages of the coordinates over the domain:

$$\boxed{\bar{x} = \frac{1}{A} \iint_{\mathcal{D}} x \, dA, \qquad \bar{y} = \frac{1}{A} \iint_{\mathcal{D}} y \, dA}$$

Here, $A = \iint_{\mathcal{D}} 1 \, dA$ is the area of $\mathcal{D}$.

In $\mathbf{R}^3$, the moments of a solid region $\mathcal{W}$ are defined not with respect to the axes as in $\mathbf{R}^2$, but with respect to the coordinate planes:

In $\mathbf{R}^3$, we denote the integral of $x\delta(x, y, z)$ by M_{yz} and call it the moment with respect to the yz-plane because x is the signed distance from (x, y, z) to the yz-plane.

$$M_{yz} = \iiint_{\mathcal{W}} x\delta(x, y, z) \, dV$$

$$M_{xz} = \iiint_{\mathcal{W}} y\delta(x, y, z) \, dV$$

$$M_{xy} = \iiint_{\mathcal{W}} z\delta(x, y, z) \, dV$$

The center of mass is the point $P_{CM} = (x_{CM}, y_{CM}, z_{CM})$ with coordinates

$$x_{CM} = \frac{M_{yz}}{M}, \qquad y_{CM} = \frac{M_{xz}}{M}, \qquad z_{CM} = \frac{M_{xy}}{M}$$

The centroid of $\mathcal{W}$ is the point $P = (\overline{x}, \overline{y}, \overline{z})$, which, as before, coincides with the center of mass when δ is constant:

$$\overline{x} = \frac{1}{V} \iiint_{\mathcal{W}} x \, dV, \qquad \overline{y} = \frac{1}{V} \iiint_{\mathcal{W}} y \, dV, \qquad \overline{z} = \frac{1}{V} \iiint_{\mathcal{W}} z \, dV$$

where $V = \iiint_{\mathcal{W}} 1 \, dV$ is the volume of $\mathcal{W}$.

Symmetry can often be used to simplify COM calculations. We say that a region $\mathcal{W}$ in $\mathbf{R}^3$ is symmetric with respect to the xy-plane if $(x, y, -z)$ lies in $\mathcal{W}$ whenever (x, y, z) lies in $\mathcal{W}$. The density δ is symmetric with respect to the xy-plane if

$$\delta(x, y, -z) = \delta(x, y, z)$$

In other words, the mass density is the same at points located symmetrically with respect to the xy-plane. If both $\mathcal{W}$ and δ have this symmetry, then $M_{xy} = 0$ and the COM lies on the xy-plane—that is, $z_{\text{CM}} = 0$. Similar remarks apply to the other coordinate axes and to domains in the plane.

EXAMPLE 3   Center of Mass   Find the center of mass of the domain $\mathcal{D}$ bounded by $y = 1 - x^2$ and the x-axis, assuming a mass density of $\delta(x, y) = y$ (Figure 4).

Solution   The domain $\mathcal{D}$ is symmetric with respect to the y-axis, and so too is the mass density because $\delta(x, y) = \delta(-x, y) = y$. Therefore, $x_{\text{CM}} = 0$. We need only compute y_{CM}:

$$M_x = \iint_{\mathcal{D}} y\delta(x, y) \, dA = \int_{x=-1}^{1} \int_{y=0}^{1-x^2} y^2 \, dy \, dx = \int_{x=-1}^{1} \left(\frac{1}{3} y^3 \Big|_{y=0}^{1-x^2} \right) dx$$

$$= \frac{1}{3} \int_{x=-1}^{1} \left(1 - 3x^2 + 3x^4 - x^6 \right) dx = \frac{1}{3} \left(2 - 2 + \frac{6}{5} - \frac{2}{7} \right) = \frac{32}{105}$$

$$M = \iint_{\mathcal{D}} \delta(x, y) \, dA = \int_{x=-1}^{1} \int_{y=0}^{1-x^2} y \, dy \, dx = \int_{x=-1}^{1} \left(\frac{1}{2} y^2 \Big|_{y=0}^{1-x^2} \right) dx$$

$$= \frac{1}{2} \int_{x=-1}^{1} \left(1 - 2x^2 + x^4 \right) dx = \frac{1}{2} \left(2 - \frac{4}{3} + \frac{2}{5} \right) = \frac{8}{15}$$

Therefore, $y_{\text{CM}} = \dfrac{M_x}{M} = \dfrac{32}{105} \left(\dfrac{8}{15} \right)^{-1} = \dfrac{4}{7}$.   ∎

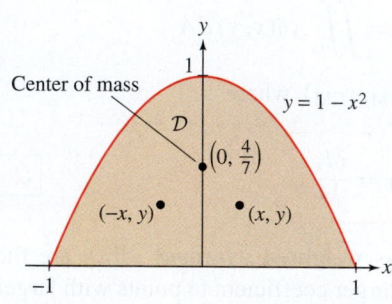

FIGURE 4 Since $\delta(x, y) = \delta(-x, y)$ mass density is symmetric with respect to the y-axis.

EXAMPLE 4   Find the center of mass of the solid $\mathcal{W}$ in Example 2, enclosed between the paraboloid $z = a(x^2 + y^2)$ and the plane $z = H$, assuming a mass density of $\delta(x, y, z) = z$.

Solution   The domain is shown in Figure 3.

Step 1. Use symmetry.
The solid $\mathcal{W}$ and the mass density are both symmetric with respect to the z-axis, so we can expect the COM to lie on the z-axis. In fact, the density satisfies both $\delta(-x, y, z) = \delta(x, y, z)$ and $\delta(x, -y, z) = \delta(x, y, z)$, and thus we have $M_{xz} = M_{yz} = 0$. It remains to compute the moment M_{xy}.

Step 2. Compute the moment.
In Example 2, we described the solid in cylindrical coordinates as

$$0 \leq \theta \leq 2\pi, \qquad 0 \leq r \leq \sqrt{\frac{H}{a}}, \qquad ar^2 \leq z \leq H$$

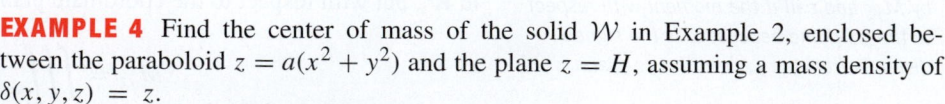

and we computed the solid's mass as $M = \dfrac{\pi H^3}{3a}$. The moment is

$$M_{xy} = \iiint_{\mathcal{W}} z\,\delta(x, y, z)\,dV = \iiint_{\mathcal{W}} z^2\,dV$$

$$= \int_{\theta=0}^{2\pi} \int_{r=0}^{\sqrt{H/a}} \int_{z=ar^2}^{H} z^2 r\,dz\,dr\,d\theta$$

$$= \int_{\theta=0}^{2\pi} \int_{r=0}^{\sqrt{H/a}} \left(\frac{1}{3}H^3 - \frac{1}{3}a^3 r^6\right) r\,dr\,d\theta$$

$$= \int_{\theta=0}^{2\pi} \left(\frac{1}{6}H^3 r^2 - \frac{1}{24}a^3 r^8\right)\Bigg|_{r=0}^{\sqrt{H/a}} d\theta$$

$$= \int_{\theta=0}^{2\pi} \frac{H^4}{8a}\,d\theta = \frac{\pi H^4}{4a}$$

The z-coordinate of the center of mass is

$$z_{CM} = \frac{M_{xy}}{M} = \frac{\pi H^4/(4a)}{\pi H^3/(3a)} = \frac{3}{4}H$$

and the center of mass itself is $\left(0, 0, \frac{3}{4}H\right)$. ∎

Moments of inertia are used to analyze rotation about an axis. For example, the spinning yo-yo in Figure 5 rotates about its center as it falls downward, and according to physics, it has a rotational kinetic energy equal to

$$\text{rotational KE} = \frac{1}{2}I\omega^2$$

Here, ω is the angular velocity (in radians per second) about this axis and I is the **moment of inertia** with respect to the **axis of rotation**. The quantity I is a rotational analog of the mass m, which appears in the expression $\frac{1}{2}mv^2$ for translational kinetic energy.

By definition, the moment of inertia with respect to an axis L is the integral of the square of the distance from the axis, weighted by mass density. We confine our attention to the coordinate axes. Thus, for a lamina in the plane $\mathbf{R}^2$, we define the moments of inertia

$$I_x = \iint_{\mathcal{D}} y^2 \delta(x, y)\,dA$$

$$I_y = \iint_{\mathcal{D}} x^2 \delta(x, y)\,dA$$ 3

$$I_0 = \iint_{\mathcal{D}} (x^2 + y^2)\delta(x, y)\,dA$$

The quantity I_0 is called the **polar moment of inertia**. It is the moment of inertia relative to the z-axis because $x^2 + y^2$ is the square of the distance from a point in the xy-plane to the z-axis. Notice that $I_0 = I_x + I_y$.

For a solid object occupying the region $\mathcal{W}$ in $\mathbf{R}^3$,

$$I_x = \iiint_{\mathcal{W}} (y^2 + z^2)\delta(x, y, z)\,dV$$

$$I_y = \iiint_{\mathcal{W}} (x^2 + z^2)\delta(x, y, z)\,dV$$

$$I_z = \iiint_{\mathcal{W}} (x^2 + y^2)\delta(x, y, z)\,dV$$

Moments of inertia have units of mass times length squared.

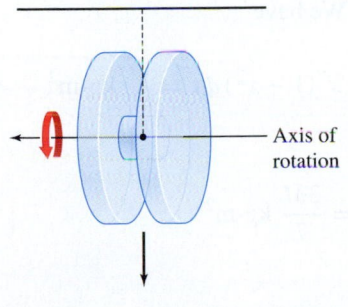

— Axis of rotation

FIGURE 5 A spinning yo-yo has rotational kinetic energy $\frac{1}{2}I\omega^2$, where I is the moment of inertia and ω is the angular velocity. See Exercise 49.

EXAMPLE 5 A lamina $\mathcal{D}$ of uniform mass density and total mass M kilograms occupies the region between $y = 1 - x^2$ and the x-axis (with distance measured in meters). Calculate the rotational kinetic energy if $\mathcal{D}$ rotates with angular velocity $\omega = 4$ radians per second about:

(a) the x-axis.   **(b)** the z-axis.

Solution The lamina is shown in Figure 6. To find the rotational kinetic energy about the x- and z-axes, we need to compute I_x and I_0, respectively.

Step 1. **Find the mass density.**
The mass density is uniform (i.e., δ is constant), but this does not mean that $\delta = 1$. In fact, the area of $\mathcal{D}$ is $\int_{-1}^{1} (1 - x^2)\, dx = \frac{4}{3}$, so the mass density (mass per unit area) is

$$\delta = \frac{\text{mass}}{\text{area}} = \frac{M}{\frac{4}{3}} = \frac{3M}{4} \text{ kg/m}^2$$

Step 2. **Calculate the moments.**

$$I_x = \int_{-1}^{1} \int_{y=0}^{1-x^2} y^2 \delta\, dy\, dx = \int_{-1}^{1} \frac{1}{3}(1 - x^2)^3 \left(\frac{3M}{4}\right) dx$$

$$= \frac{M}{4} \int_{-1}^{1} (1 - 3x^2 + 3x^4 - x^6)\, dx = \frac{8M}{35} \text{ kg-m}^2$$

To calculate I_0, we use the relation $I_0 = I_x + I_y$. We have

$$I_y = \int_{-1}^{1} \int_{y=0}^{1-x^2} x^2 \delta\, dy\, dx = \left(\frac{3M}{4}\right) \int_{-1}^{1} x^2(1 - x^2)\, dx = \frac{M}{5} \text{ kg-m}^2$$

and thus

$$I_0 = I_x + I_y = \frac{8M}{35} + \frac{M}{5} = \frac{3M}{7} \text{ kg-m}^2 \qquad \boxed{4}$$

Step 3. **Calculate kinetic energy.**
Assuming an angular velocity of $\omega = 4$ rad/s,

$$\text{rotational KE about } x\text{-axis} = \frac{1}{2} I_x \omega^2 = \frac{1}{2} \left(\frac{8M}{35}\right) 4^2 \approx 1.8M \text{ joules}$$

$$\text{rotational KE about } z\text{-axis} = \frac{1}{2} I_0 \omega^2 = \frac{1}{2} \left(\frac{3M}{7}\right) 4^2 \approx 3.4M \text{ joules}$$

The unit of energy is the joule (J), equal to 1 kg-m^2/s^2. ∎

A point mass m located a distance r from an axis has moment of inertia $I = mr^2$ with respect to that axis. Given an extended object of total mass M (not necessarily a point mass) whose moment of inertia with respect to the axis is I, we define the **radius of gyration** by $r_g = (I/M)^{1/2}$. With this definition, the moment of inertia would not change if all of the mass of the object were concentrated at a point located a distance r_g from the axis.

EXAMPLE 6   Radius of Gyration of a Hemisphere Find the radius of gyration about the z-axis of the solid hemisphere $\mathcal{W}$ above the xy-plane and inside the sphere $x^2 + y^2 + z^2 = R^2$, assuming a mass density of $\delta(x, y, z) = z$ kg/m^3.

Solution To compute the radius of gyration about the z-axis, we must compute I_z and the total mass M. We use spherical coordinates, and we compute the outer integral with respect to θ first since the inner two integrals have no dependence on θ.

$$x^2 + y^2 = (\rho \cos\theta \sin\phi)^2 + (\rho \sin\theta \sin\phi)^2 = \rho^2 \sin^2\phi, \qquad z = \rho \cos\phi$$

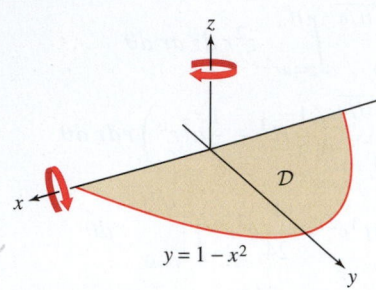

FIGURE 6 Rotating about the z-axis, the plate remains in the xy-plane. About the x-axis, it rotates out of the xy-plane.

CAUTION The relation

$$I_0 = I_x + I_y$$

is valid for a lamina in the xy-plane. However, there is no relation of this type for solid objects in $\mathbf{R}^3$.

$$I_z = \iiint_{\mathcal{W}} (x^2 + y^2)z \, dV = \int_{\theta=0}^{2\pi} \int_{\phi=0}^{\pi/2} \int_{\rho=0}^{R} (\rho^2 \sin^2 \phi)(\rho \cos \phi)\rho^2 \sin \phi \, d\rho \, d\phi \, d\theta$$

$$= 2\pi \int_{\phi=0}^{\pi/2} \int_{\rho=0}^{R} \rho^5 \sin^3 \phi \cos \phi \, d\rho \, d\phi$$

$$= \frac{\pi R^6}{3} \int_{\phi=0}^{\pi/2} \sin^3 \phi \cos \phi \, d\phi$$

$$= \frac{\pi R^6}{3} \left(\frac{\sin^4 \phi}{4} \Big|_0^{\pi/2} \right) = \frac{\pi R^6}{12} \text{ kg-m}^2$$

$$M = \iiint_{\mathcal{W}} z \, dV = \int_{\theta=0}^{2\pi} \int_{\phi=0}^{\pi/2} \int_{\rho=0}^{R} (\rho \cos \phi)\rho^2 \sin \phi \, d\rho \, d\phi \, d\theta$$

The integral computation for M is similar to that for I_z and results in $M = \pi R^4/4$ kg. Therefore, the radius of gyration is $r_g = (I_z/M)^{1/2} = (R^2/3)^{1/2} = R/\sqrt{3}$ m. ■

Probability Theory

In Section 8.1, we discussed how probabilities can be represented as areas under curves (Figure 7). Recall that a *random variable X* is defined as the outcome of an experiment or measurement whose value is not known in advance. The probability that the value of X lies between a and b is denoted $P(a \leq X \leq b)$. Furthermore, X is a *continuous random variable* if there is a continuous function p of one variable, called the *probability density function*, such that (Figure 7)

$$P(a \leq X \leq b) = \int_a^b p(x) \, dx$$

Double integration enters the picture when we compute "joint probabilities" of two random variables X and Y. We let

$$P(a \leq X \leq b; \ c \leq Y \leq d)$$

denote the probability that X and Y satisfy

$$a \leq X \leq b, \qquad c \leq Y \leq d$$

For example, if X is the height (in centimeters) and Y is the weight (in kilograms) in a certain population, then

$$P(160 \leq X \leq 170; \ 52 \leq Y \leq 63)$$

is the probability that a person chosen at random has height between 160 and 170 cm and weight between 52 and 63 kg.

We say that X and Y are jointly continuous if there is a continuous function $p(x, y)$, called the **joint probability density function** (or simply the joint density), such that for all intervals $[a, b]$ and $[c, d]$ (Figure 8),

$$P(a \leq X \leq b; c \leq Y \leq d) = \int_{x=a}^{b} \int_{y=c}^{d} p(x, y) \, dy \, dx$$

In the margin, we recall two conditions that a probability density function must satisfy. Joint density functions must satisfy similar conditions: First, $p(x, y) \geq 0$ for all x and y (because probabilities cannot be negative), and second,

$$\int_{-\infty}^{\infty} \int_{-\infty}^{\infty} p(x, y) \, dy \, dx = 1 \qquad \boxed{5}$$

This is often called the **normalization condition**. It holds because it is certain (the probability is 1) that X and Y each take on some value between $-\infty$ and ∞.

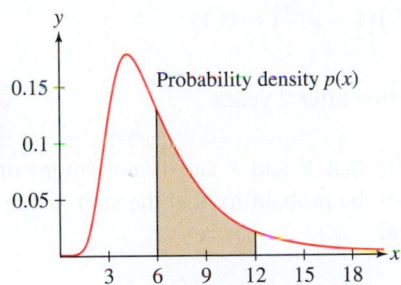

DF **FIGURE 7** The shaded area is the probability that X lies between 6 and 12.

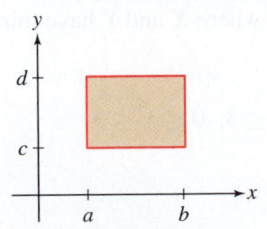

FIGURE 8 The probability $P(a \leq X \leq b; c \leq Y \leq d)$ is equal to the integral of $p(x, y)$ over the rectangle.

← REMINDER *Conditions on a probability density function:*

- $p(x) \geq 0$,
- $p(x)$ satisfies $\int_J p(x) \, dx = 1$ where J is the domain of the density function.

EXAMPLE 7 Without proper maintenance, the time to failure (in months) of two sensors in an aircraft are random variables X and Y with joint density

$$p(x, y) = \begin{cases} \dfrac{1}{864} e^{-x/24 - y/36} & \text{for } x \geq 0, y \geq 0 \\[2mm] 0 & \text{otherwise} \end{cases}$$

What is the probability that neither sensor functions after 2 years?

Solution The problem asks for the probability $P(0 \leq X \leq 24; 0 \leq Y \leq 24)$. For simplicity in integration, we rewrite $e^{-x/24 - y/36}$ as $e^{-x/24} e^{-y/36}$:

$$\int_{x=0}^{24} \int_{y=0}^{24} p(x, y)\, dy\, dx = \frac{1}{864} \int_{x=0}^{24} \int_{y=0}^{24} e^{-x/24} e^{-y/36}\, dy\, dx$$

$$= \frac{1}{864} \int_{x=0}^{24} e^{-x/24} \left(-36 e^{-y/36} \Big|_0^{24} \right) dx$$

$$= \frac{1}{24} \left(1 - e^{-24/36} \right) \left(-24 e^{-x/24} \Big|_0^{24} \right)$$

$$= \left(1 - e^{-24/36} \right) \left(1 - e^{-1} \right) \approx 0.31$$

There is a 31% chance that neither sensor will function after 2 years.　■

More generally, we can compute the probability that X and Y satisfy conditions of various types. For example, $P(X + Y \leq M)$ denotes the probability that the sum $X + Y$ is at most M. This probability is equal to the integral

$$P(X + Y \leq M) = \iint_{\mathcal{D}} p(x, y)\, dy\, dx$$

where $\mathcal{D} = \{(x, y) : x + y \leq M\}$.

EXAMPLE 8 Calculate the probability that $X + Y \leq 3$, where X and Y have joint probability density

$$p(x, y) = \begin{cases} \dfrac{1}{81}(2xy + 2x + y) & 0 \leq x \leq 3,\ 0 \leq y \leq 3 \\[2mm] 0 & \text{otherwise} \end{cases}$$

Solution The probability density function $p(x, y)$ is nonzero only on the square in Figure 9. Within that square, the inequality $x + y \leq 3$ holds only on the shaded triangle, so the probability that $X + Y \leq 3$ is equal to the integral of $p(x, y)$ over the triangle:

$$\int_{x=0}^{3} \int_{y=0}^{3-x} p(x, y)\, dy\, dx = \frac{1}{81} \int_{x=0}^{3} \left(xy^2 + \frac{1}{2} y^2 + 2xy \right) \Big|_{y=0}^{3-x} dx$$

$$= \frac{1}{81} \int_{x=0}^{3} \left(x^3 - \frac{15}{2} x^2 + 12x + \frac{9}{2} \right) dx$$

$$= \frac{1}{81} \left(\frac{1}{4} 3^4 - \frac{5}{2} 3^3 + 6(3^2) + \frac{9}{2}(3) \right) = \frac{1}{4}$$　■

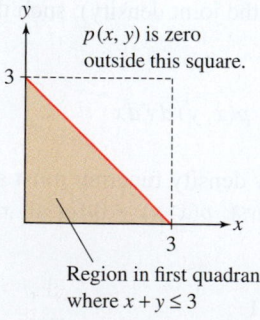

$p(x, y)$ is zero outside this square.

Region in first quadrant where $x + y \leq 3$

FIGURE 9

15.5 SUMMARY

- If the mass density is constant, then the center of mass coincides with the *centroid*, whose coordinates $\overline{x}$, $\overline{y}$ (and $\overline{z}$ in three dimensions) are the average values of x, y, and z over the domain. For a domain in $\mathbf{R}^2$,

$$\overline{x} = \frac{1}{A} \iint_{\mathcal{D}} x \, dA, \qquad \overline{y} = \frac{1}{A} \iint_{\mathcal{D}} y \, dA, \qquad A = \iint_{\mathcal{D}} 1 \, dA$$

	In $\mathbf{R}^2$	In $\mathbf{R}^3$
Total mass	$M = \iint_{\mathcal{D}} \delta(x, y) \, dA$	$M = \iiint_{\mathcal{W}} \delta(x, y, z) \, dV$
Moments	$M_x = \iint_{\mathcal{D}} y\delta(x, y) \, dA$ $M_y = \iint_{\mathcal{D}} x\delta(x, y) \, dA$	$M_{yz} = \iiint_{\mathcal{W}} x\delta(x, y, z) \, dV$ $M_{xz} = \iiint_{\mathcal{W}} y\delta(x, y, z) \, dV$ $M_{xy} = \iiint_{\mathcal{W}} z\delta(x, y, z) \, dV$
Center of mass	$x_{CM} = \dfrac{M_y}{M}, \quad y_{CM} = \dfrac{M_x}{M}$	$x_{CM} = \dfrac{M_{yz}}{M}, \quad y_{CM} = \dfrac{M_{xz}}{M}, \quad z_{CM} = \dfrac{M_{xy}}{M}$
Moments of inertia	$I_x = \iint_{\mathcal{D}} y^2\delta(x, y) \, dA$ $I_y = \iint_{\mathcal{D}} x^2\delta(x, y) \, dA$ $I_0 = \iint_{\mathcal{D}} (x^2 + y^2)\delta(x, y) \, dA$ $(I_0 = I_x + I_y)$	$I_x = \iiint_{\mathcal{W}} (y^2 + z^2)\delta(x, y, z) \, dV$ $I_y = \iiint_{\mathcal{W}} (x^2 + z^2)\delta(x, y, z) \, dV$ $I_z = \iiint_{\mathcal{W}} (x^2 + y^2)\delta(x, y, z) \, dV$

- Radius of gyration: $r_g = (I/M)^{1/2}$
- Random variables X and Y have joint probability density function $p(x, y)$ if

$$P(a \le X \le b; c \le Y \le d) = \int_{x=a}^{b} \int_{y=c}^{d} p(x, y) \, dy \, dx$$

- A joint probability density function must satisfy $p(x, y) \ge 0$ and

$$\int_{x=-\infty}^{\infty} \int_{y=-\infty}^{\infty} p(x, y) \, dy \, dx = 1$$

15.5 EXERCISES

Preliminary Questions

1. What is the mass density $\delta(x, y, z)$ of a solid of volume 5 m³ with uniform mass density and total mass 25 kg?

2. A domain $\mathcal{D}$ in $\mathbf{R}^2$ with uniform mass density is symmetric with respect to the y-axis. Which of the following are true?

(a) $x_{CM} = 0$ (b) $y_{CM} = 0$ (c) $I_x = 0$ (d) $I_y = 0$

3. If $p(x, y)$ is the joint probability density function of random variables X and Y, what does the double integral of $p(x, y)$ over $[0, 1] \times [0, 1]$ represent? What does the integral of $p(x, y)$ over the triangle bounded by $x = 0$, $y = 0$, and $x + y = 1$ represent?

Exercises

1. Find the total mass of the rectangle $0 \le x \le 1, 0 \le y \le 2$ assuming a mass density of

$$\delta(x, y) = 2x^2 + y^2$$

2. Calculate the total mass of a plate bounded by $y = 0$ and $y = x^{-1}$ for $1 \le x \le 4$ (in meters) assuming a mass density of $\delta(x, y) = y/x$ kg/m^2.

3. Find the total charge in the region under the graph of $y = 4e^{-x^2/2}$ for $0 \le x \le 10$ (in centimeters) assuming a charge density of $\delta(x, y) = 10^{-6}xy$ coulombs per square centimeter (C/cm^2).

4. Find the total population within a 4-km radius of the city center (located at the origin) assuming a population density of $\delta(x, y) = 2000(x^2 + y^2)^{-0.2}$ people per square kilometer.

5. Find the total population within the sector $2|x| \le y \le 8$ assuming a population density of $\delta(x, y) = 100e^{-0.1y}$ people per square kilometer.

6. Find the total mass of the solid region $\mathcal{W}$ defined by $x \ge 0$, $y \ge 0$, $x^2 + y^2 \le 4$, and $x \le z \le 32 - x$ (in centimeters) assuming a mass density of $\delta(x, y, z) = 6y$ g/cm^3.

7. Calculate the total charge of the solid ball $x^2 + y^2 + z^2 \le 5$ (in centimeters) assuming a charge density (in coulombs per cubic centimeter) of

$$\delta(x, y, z) = (3 \cdot 10^{-8})(x^2 + y^2 + z^2)^{1/2}$$

8. Compute the total mass of the plate in Figure 10 assuming a mass density of $f(x, y) = x^2/(x^2 + y^2)$ g/cm^2.

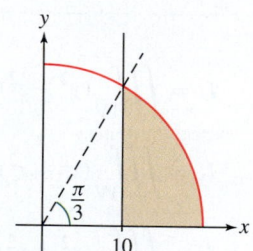

FIGURE 10

9. Assume that the density of the atmosphere as a function of altitude h (in kilometers) above sea level is $\delta(h) = ae^{-bh}$ kg/km^3, where $a = 1.225 \times 10^9$ and $b = 0.13$. Calculate the total mass of the atmosphere contained in the cone-shaped region $\sqrt{x^2 + y^2} \le h \le 3$.

10. Calculate the total charge on a plate $\mathcal{D}$ in the shape of the ellipse with the polar equation

$$r^2 = \left(\frac{1}{6} \sin^2 \theta + \frac{1}{9} \cos^2 \theta \right)^{-1}$$

with the disk $x^2 + y^2 \le 1$ removed (Figure 11) assuming a charge density of $\rho(r, \theta) = 3r^{-4}$ C/cm^2.

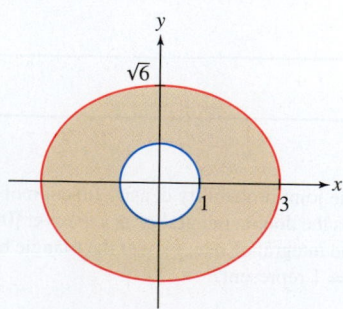

FIGURE 11

In Exercises 11–16, find the centroid of the given region assuming the density $\delta(x, y) = 1$.

11. Region bounded by $y = 1 - x^2$ and $y = 0$

12. Region bounded by $y^2 = x + 4$ and $x = 4$

13. Quarter circle $x^2 + y^2 \le R^2$, $x \ge 0$, $y \ge 0$

14. Quarter circle $x^2 + y^2 \le R^2$, $y \ge |x|$

15. Lamina bounded by the x- and y-axes, the line $x = M$, and the graph of $y = e^{-x}$

16. Infinite lamina bounded by the x- and y-axes and the graph of $y = e^{-x}$

17. CAS Use a computer algebra system to compute numerically the centroid of the shaded region in Figure 12 bounded by $r^2 = \cos 2\theta$ for $x \ge 0$.

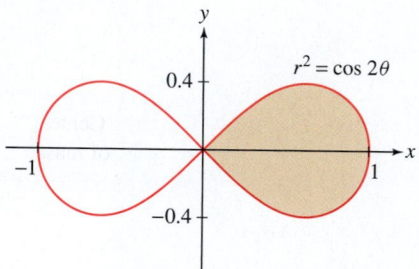

FIGURE 12

18. Show that the centroid of the sector in Figure 13 has y-coordinate

$$\bar{y} = \left(\frac{2R}{3} \right) \left(\frac{\sin \alpha}{\alpha} \right)$$

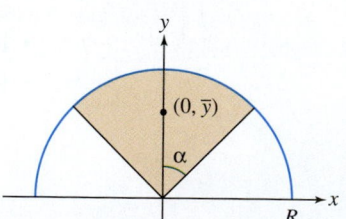

FIGURE 13

In Exercises 19–21, find the centroid of the given solid region assuming a density of $\delta(x, y) = 1$.

19. Hemisphere $x^2 + y^2 + z^2 \le R^2$, $z \ge 0$

20. Region bounded by the xy-plane, the cylinder $x^2 + y^2 = R^2$, and the plane $x/R + z/H = 1$, where $R > 0$ and $H > 0$

21. The "ice cream cone" region $\mathcal{W}$ bounded, in spherical coordinates, by the cone $\phi = \pi/3$ and the sphere $\rho = 2$

22. Show that the z-coordinate of the centroid of the tetrahedron bounded by the coordinate planes and the plane

$$\frac{x}{a} + \frac{y}{b} + \frac{z}{c} = 1$$

in Figure 14 is $\bar{z} = c/4$. Conclude by symmetry that the centroid is $(a/4, b/4, c/4)$.

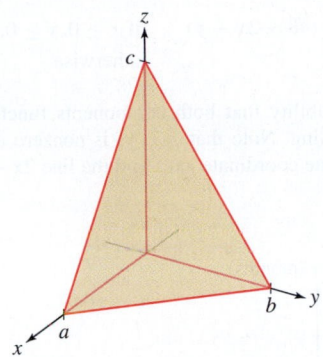

FIGURE 14

23. Find the centroid of the region $\mathcal{W}$ in Figure 15, lying above the sphere $x^2 + y^2 + z^2 = 6$ and below the paraboloid $z = 4 - x^2 - y^2$.

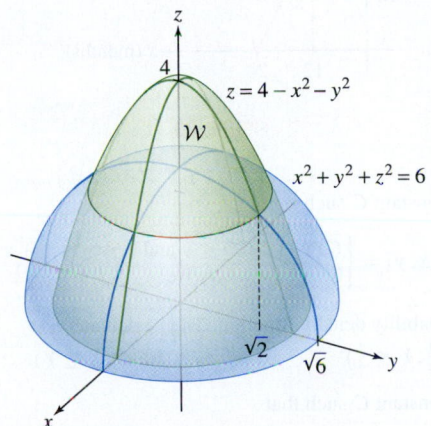

FIGURE 15

24. Let $R > 0$ and $H > 0$, and let $\mathcal{W}$ be the upper half of the ellipsoid $x^2 + y^2 + (Rz/H)^2 = R^2$, where $z \geq 0$ (Figure 16). Find the centroid of $\mathcal{W}$ and show that it depends on the height H but not on the radius R.

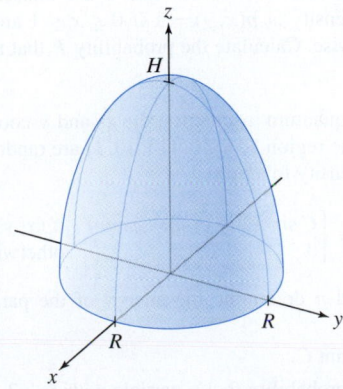

FIGURE 16 Upper half of ellipsoid $x^2 + y^2 + (Rz/H)^2 = R^2$, $z \geq 0$.

In Exercises 25–28, find the center of mass of the region with the given mass density δ.

25. Region bounded by $y = 6 - x$, $x = 0$, $y = 0$; $\quad \delta(x, y) = x^2$

26. Region bounded by $y^2 = x + 4$ and $x = 0$; $\quad \delta(x, y) = |y|$

27. Region $|x| + |y| \leq 1$; $\quad \delta(x, y) = (x + 1)(y + 1)$

28. Semicircle $x^2 + y^2 \leq R^2$, $y \geq 0$; $\quad \delta(x, y) = y$

29. Find the z-coordinate of the center of mass of the first octant of the unit sphere with mass density $\delta(x, y, z) = y$ (Figure 17).

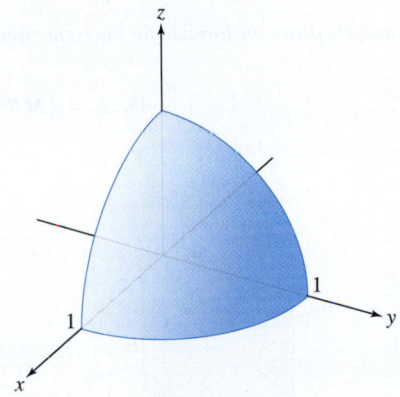

FIGURE 17

30. Find the center of mass of a cylinder of radius 2 and height 4 and mass density e^{-z}, where z is the height above the base.

31. Let $\mathcal{R}$ be the rectangle $[-a, a] \times [b, -b]$ with uniform density and total mass M. Calculate:

(a) The mass density δ of $\mathcal{R}$

(b) I_x and I_0

(c) The radius of gyration about the x-axis

32. Calculate I_x and I_0 for the rectangle in Exercise 31 assuming a mass density of $\delta(x, y) = x$.

33. Calculate I_0 and I_x for the disk $\mathcal{D}$ defined by $x^2 + y^2 \leq 16$ (in meters), with total mass 1000 kg and uniform mass density. *Hint:* Calculate I_0 first and observe that $I_0 = 2I_x$. Express your answer in the correct units.

34. Calculate I_x and I_y for the half-disk $x^2 + y^2 \leq R^2$, $x \geq 0$ (in meters), with total mass M kilograms and uniform mass density.

In Exercises 35–38, let $\mathcal{D}$ be the triangular domain bounded by the coordinate axes and the line $y = 3 - x$, with mass density $\delta(x, y) = y$. Compute the given quantities.

35. Total mass

36. Center of mass

37. I_x

38. I_0

In Exercises 39–42, let $\mathcal{D}$ be the domain between the line $y = bx/a$ and the parabola $y = bx^2/a^2$, where $a, b > 0$. Assume the mass density is $\delta(x, y) = 1$ for Exercise 39 and $\delta(x, y) = xy$ for Exercises 40–42. Compute the given quantities.

39. Centroid

40. Center of mass

41. I_x

42. I_0

43. Calculate the moment of inertia I_x of the disk $\mathcal{D}$ defined by $x^2 + y^2 \leq R^2$ (in meters), with total mass M kilograms. How much kinetic energy (in joules) is required to rotate the disk about the x-axis with angular velocity 10 radians per second?

44. Calculate the moment of inertia I_z of the box $\mathcal{W} = [-a, a] \times [-a, a] \times [0, H]$ assuming that $\mathcal{W}$ has total mass M.

45. Show that the moment of inertia of a sphere of radius R of total mass M with uniform mass density about any axis passing through the center of the sphere is $\frac{2}{5}MR^2$. Note that the mass density of the sphere is $\delta = M/(\frac{4}{3}\pi R^3)$.

46. Use the result of Exercise 45 to calculate the radius of gyration of a uniform sphere of radius R about any axis through the center of the sphere.

In Exercises 47 and 48, prove the formula for the right circular cylinder in Figure 18.

47. $I_z = \frac{1}{2}MR^2$

48. $I_x = \frac{1}{4}MR^2 + \frac{1}{12}MH^2$

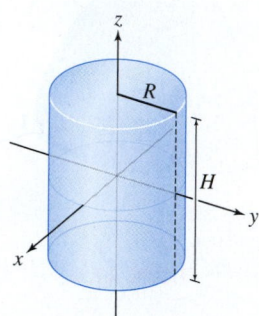

FIGURE 18

49. The yo-yo in Figure 19 is made up of two disks of radius $r = 3$ cm and an axle of radius $b = 1$ cm. Each disk has mass $M_1 = 20$ g, and the axle has mass $M_2 = 5$ g.

(a) Use the result of Exercise 47 to calculate the moment of inertia I of the yo-yo with respect to the axis of symmetry. Note that I is the sum of the moments of the three components of the yo-yo.

(b) The yo-yo is released and falls to the end of a 100-cm string, where it spins with angular velocity ω. The total mass of the yo-yo is $m = 45$ g, so the potential energy lost is PE $= mgh = (45)(980)100$ g-cm^2/s^2. Find ω using the fact that the potential energy is the sum of the rotational kinetic energy and the translational kinetic energy and that the velocity $v = b\omega$ because the string unravels at this rate.

Axle of radius b

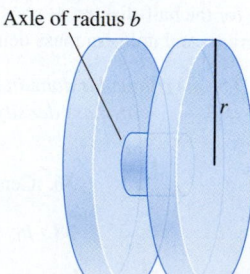

FIGURE 19

50. Calculate I_z for the solid region $\mathcal{W}$ inside the hyperboloid $x^2 + y^2 = z^2 + 1$ between $z = 0$ and $z = 1$.

51. Calculate $P(0 \le X \le 2; 1 \le Y \le 2)$, where X and Y have joint probability density function

$$p(x, y) = \begin{cases} \frac{1}{72}(2xy + 2x + y) & \text{if } 0 \le x \le 4 \text{ and } 0 \le y \le 2 \\ 0 & \text{otherwise} \end{cases}$$

52. Calculate the probability that $X + Y \le 2$ for random variables with joint probability density function as in Exercise 51.

53. The lifetime (in months) of two components in a certain device are random variables X and Y that have joint probability density function

$$p(x, y) = \begin{cases} \frac{1}{9216}(48 - 2x - y) & \text{if } x \ge 0, y \ge 0, 2x + y \le 48 \\ 0 & \text{otherwise} \end{cases}$$

Calculate the probability that both components function for at least 12 months without failing. Note that $p(x, y)$ is nonzero only within the triangle bounded by the coordinate axes and the line $2x + y = 48$ shown in Figure 20.

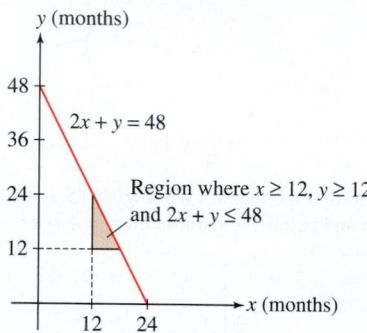

FIGURE 20

54. Find a constant C such that

$$p(x, y) = \begin{cases} Cxy & \text{if } 0 \le x \text{ and } 0 \le y \le 1 - x \\ 0 & \text{otherwise} \end{cases}$$

is a joint probability density function. Then calculate:

(a) $P\left(X \le \frac{1}{2}; Y \le \frac{1}{4}\right)$ **(b)** $P(X \ge Y)$

55. Find a constant C such that

$$p(x, y) = \begin{cases} Cy & \text{if } 0 \le x \le 1 \text{ and } x^2 \le y \le x \\ 0 & \text{otherwise} \end{cases}$$

is a joint probability density function. Then calculate the probability that $Y \ge X^{3/2}$.

56. Numbers X and Y between 0 and 1 are chosen randomly. The joint probability density is $p(x, y) = 1$ if $0 \le x \le 1$ and $0 \le y \le 1$, and $p(x, y) = 0$ otherwise. Calculate the probability P that the product XY is at least $\frac{1}{2}$.

57. According to quantum mechanics, the x- and y-coordinates of a particle confined to the region $\mathcal{R} = [0, 1] \times [0, 1]$ are random variables with joint probability density function

$$p(x, y) = \begin{cases} C \sin^2(2\pi \ell x) \sin^2(2\pi n y) & \text{if } (x, y) \in \mathcal{R} \\ 0 & \text{otherwise} \end{cases}$$

The integers ℓ and n determine the energy of the particle, and C is a constant.

(a) Find the constant C.

(b) Calculate the probability that a particle with $\ell = 2$, $n = 3$ lies in the region $\left[0, \frac{1}{4}\right] \times \left[0, \frac{1}{8}\right]$.

58. The wave function for the 1s state of an electron in the hydrogen atom is

$$\psi_{1s}(\rho) = \frac{1}{\sqrt{\pi a_0^3}} e^{-\rho/a_0}$$

where a_0 is the Bohr radius. The probability of finding the electron in a region $\mathcal{W}$ of $\mathbf{R}^3$ is equal to

$$\iiint_{\mathcal{W}} p(x, y, z)\, dV$$

where, in spherical coordinates,

$$p(\rho) = |\psi_{1s}(\rho)|^2$$

Use integration in spherical coordinates to show that the probability of finding the electron at a distance greater than the Bohr radius is equal to $5/e^2 \approx 0.677$. (The Bohr radius is $a_0 = 5.3 \times 10^{-11}$ m, but this value is not needed.)

59. According to Coulomb's Law, the attractive force between two electric charges of magnitude q_1 and q_2 separated by a distance r is kq_1q_2/r^2 (k is a constant). Let F be the net force on a charged particle P of charge Q coulombs located d centimeters above the center of a circular disk of radius R, with a uniform charge distribution of density ρ coulombs per square meter (Figure 21). By symmetry, F acts in the vertical direction.

(a) Let $\mathcal{R}$ be a small polar rectangle of size $\Delta r \times \Delta \theta$ located at distance r. Show that $\mathcal{R}$ exerts a force on P whose vertical component is

$$\left(\frac{k\rho Q d}{(r^2 + d^2)^{3/2}}\right) r\, \Delta r\, \Delta \theta$$

(b) Explain why F is equal to the following double integral, and evaluate:

$$F = k\rho Q d \int_0^{2\pi} \int_0^R \frac{r\, dr\, d\theta}{(r^2 + d^2)^{3/2}}$$

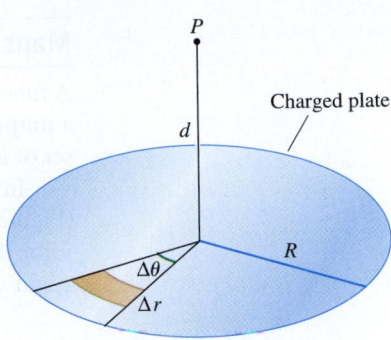

FIGURE 21

60. Let $\mathcal{D}$ be the annular region

$$-\frac{\pi}{2} \le \theta \le \frac{\pi}{2}, \quad a \le r \le b$$

where $b > a > 0$. Assume that $\mathcal{D}$ has a uniform charge distribution of ρ coulombs per square meter. Let F be the net force on a charged particle of charge Q coulombs located at the origin (by symmetry, F acts along the x-axis).

(a) Argue as in Exercise 59 to show that

$$F = k\rho Q \int_{\theta = -\pi/2}^{\pi/2} \int_{r=a}^{b} \left(\frac{\cos\theta}{r^2}\right) r\, dr\, d\theta$$

(b) Compute F.

Further Insights and Challenges

61. Let $\mathcal{D}$ be the domain in Figure 22. Assume that $\mathcal{D}$ is symmetric with respect to the y-axis; that is, both $g_1(x)$ and $g_2(x)$ are even functions.

(a) Prove that the centroid lies on the y-axis—that is, that $\overline{x} = 0$.

(b) Show that if the mass density satisfies $\delta(-x, y) = \delta(x, y)$, then $M_y = 0$ and $x_{CM} = 0$.

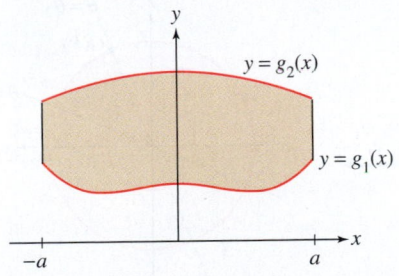

FIGURE 22

62. Pappus's Theorem Let A be the area of the region $\mathcal{D}$ between two graphs $y = g_1(x)$ and $y = g_2(x)$ over the interval $[a, b]$, where $g_2(x) \ge g_1(x) \ge 0$. Prove Pappus's Theorem: The volume of the solid obtained by revolving $\mathcal{D}$ about the x-axis is $V = 2\pi A\overline{y}$, where $\overline{y}$ is the y-coordinate of the centroid of $\mathcal{D}$ (the average of the y-coordinate). *Hint:* Show that

$$A\overline{y} = \int_{x=a}^{b} \int_{y=g_1(x)}^{g_2(x)} y\, dy\, dx$$

63. Use Pappus's Theorem in Exercise 62 to show that the torus obtained by revolving a circle of radius b centered at $(0, a)$ about the x-axis (where $b < a$) has volume $V = 2\pi^2 ab^2$.

64. Use Pappus's Theorem to compute $\overline{y}$ for the upper half of the disk $x^2 + y^2 \le a^2$, $y \ge 0$. *Hint:* The disk revolved about the x-axis is a sphere.

65. Parallel-Axis Theorem Let $\mathcal{W}$ be a region in $\mathbf{R}^3$ with center of mass at the origin. Let I_z be the moment of inertia of $\mathcal{W}$ about the z-axis, and let I_h be the moment of inertia about the vertical axis through a point $P = (a, b, 0)$, where $h = \sqrt{a^2 + b^2}$. By definition,

$$I_h = \iiint_{\mathcal{W}} ((x - a)^2 + (y - b)^2)\delta(x, y, z)\, dV$$

Prove the Parallel-Axis Theorem: $I_h = I_z + Mh^2$.

66. Let $\mathcal{W}$ be a cylinder of radius 10 cm and height 20 cm, with total mass $M = 500$ g. Use the Parallel-Axis Theorem (Exercise 65) and the result of Exercise 47 to calculate the moment of inertia of $\mathcal{W}$ about an axis that is parallel to and at a distance of 30 cm from the cylinder's axis of symmetry.

15.6 Change of Variables

The formulas for integration in polar, cylindrical, and spherical coordinates are important special cases of the general Change of Variables Formula for multiple integrals. In this section, we discuss the general formula.

Maps from $\mathbf{R}^2$ to $\mathbf{R}^2$

A function $G : X \to Y$ from a set X (the domain) to another set Y is often called a **map** or a **mapping**. For $x \in X$, the element $G(x)$ belongs to Y and is called the **image** of x. The set of all images $G(x)$ is called the image or **range** of G. We denote the image by $G(X)$.

In this section, we consider maps $G : \mathcal{D} \to \mathbf{R}^2$ defined on a domain $\mathcal{D}$ in $\mathbf{R}^2$ (Figure 1). To prevent confusion, we'll often use u, v as our domain variables and x, y for the range. Thus, we will write $G(u, v) = (x(u, v), y(u, v))$, where the components x and y are functions of u and v:

$$x = x(u, v), \qquad y = y(u, v)$$

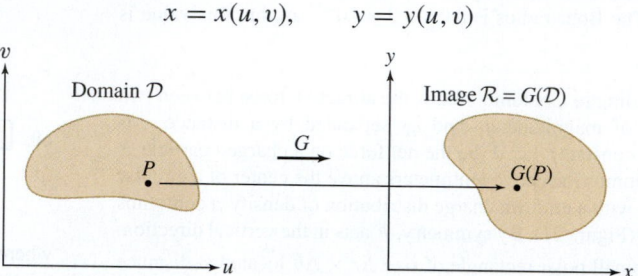

FIGURE 1 G maps $\mathcal{D}$ to $\mathcal{R}$.

One map we are familiar with is the map defining polar coordinates. For this map, we use variables r, θ instead of u, v. The **polar coordinates map** $G : \mathbf{R}^2 \to \mathbf{R}^2$ is defined by

$$G(r, \theta) = (r \cos \theta, r \sin \theta)$$

EXAMPLE 1 **Polar Coordinates Map** Describe the image of a polar rectangle $\mathcal{R} = [r_1, r_2] \times [\theta_1, \theta_2]$ under the polar coordinates map.

Solution Referring to Figure 2, we see that

- A vertical line $r = r_1$ (shown in red) is mapped to the set of points with radial coordinate r_1 and arbitrary angle. This is the circle of radius r_1.
- A horizontal line $\theta = \theta_1$ (dashed line in the figure) is mapped to the set of points with polar angle θ_1 and arbitrary r-coordinate. This is the line through the origin of angle θ_1.

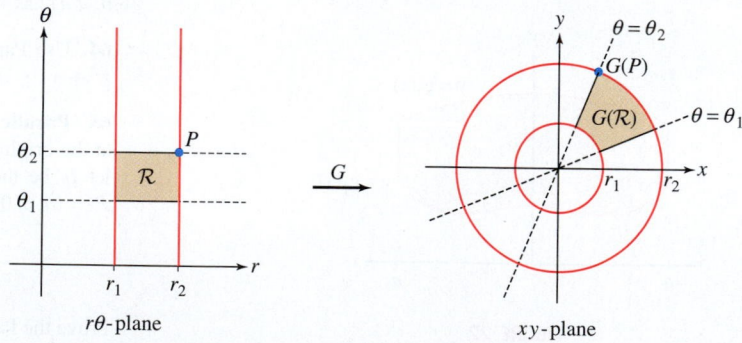

FIGURE 2 The polar coordinates map $G(r, \theta) = (r \cos \theta, r \sin \theta)$.

The image of $\mathcal{R} = [r_1, r_2] \times [\theta_1, \theta_2]$ under the polar coordinates map $G(r, \theta)$ is the polar rectangle in the xy-plane defined by $r_1 \leq r \leq r_2, \theta_1 \leq \theta \leq \theta_2$. ∎

General mappings can be quite complicated, so it is useful to study the simplest case—linear maps—in detail. A map $G(u, v)$ is **linear** if it has the form

$$G(u, v) = (Au + Cv, Bu + Dv) \quad (A, B, C, D \text{ are constants})$$

We can get a clear picture of this linear map by thinking of G as a map from vectors in the uv-plane to vectors in the xy-plane. Then G has the following linearity properties (see Exercise 46):

$$G(u_1 + u_2, v_1 + v_2) = G(u_1, v_1) + G(u_2, v_2) \qquad \boxed{1}$$

$$G(cu, cv) = cG(u, v) \quad (c \text{ any constant}) \qquad \boxed{2}$$

A consequence of these properties is that G maps the parallelogram spanned by any two vectors **a** and **b** in the uv-plane to the parallelogram spanned by the images $G(\mathbf{a})$ and $G(\mathbf{b})$, as shown in Figure 3.

More generally, *G maps the segment joining any two points P and Q to the segment joining G(P) and G(Q)* (see Exercise 47). The grid generated by basis vectors $\mathbf{i} = \langle 1, 0 \rangle$ and $\mathbf{j} = \langle 0, 1 \rangle$ is mapped to the grid generated by the image vectors (Figure 3)

$$\mathbf{r} = G(1,0) = \langle A, B \rangle$$

$$\mathbf{s} = G(0,1) = \langle C, D \rangle$$

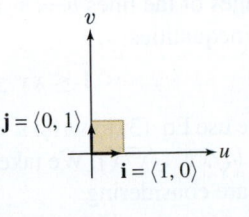
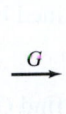
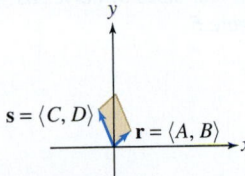
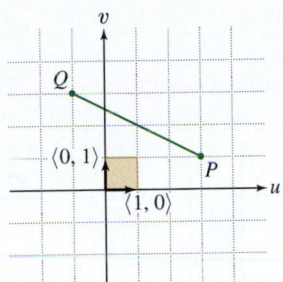
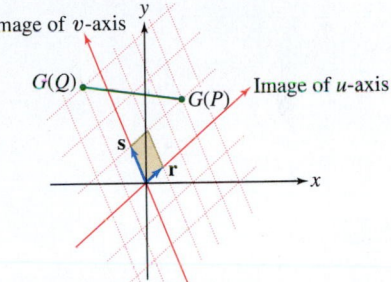

DF **FIGURE 3** A linear mapping G maps a parallelogram to a parallelogram.

EXAMPLE 2   **Image of a Triangle**   Find the image of the triangle $\mathcal{T}$ with vertices $(1, 2)$, $(2, 1)$, $(3, 4)$ under the linear map $G(u, v) = (2u - v, u + v)$.

Solution   Because G is linear, it maps the segment joining two vertices of $\mathcal{T}$ to the segment joining the images of the two vertices. Therefore, the image of $\mathcal{T}$ is the triangle whose vertices are the images (Figure 4)

$$G(1, 2) = (0, 3), \qquad G(2, 1) = (3, 3), \qquad G(3, 4) = (2, 7) \qquad \blacksquare$$

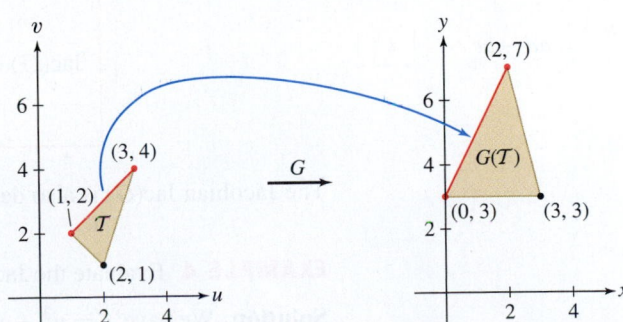

FIGURE 4 The map
$G(u, v) = (2u - v, u + v)$.

To understand a nonlinear map, it is usually helpful to determine the images of horizontal and vertical lines, as we did for the polar coordinate mapping.

EXAMPLE 3   Let $G(u, v) = (uv^{-1}, uv)$ for $u > 0$, $v > 0$. Determine the images of

(a) The lines $u = c$ and $v = c$ **(b)** $[1, 2] \times [1, 2]$

Find the inverse map G^{-1}.

Solution In this map, we have $x = uv^{-1}$ and $y = uv$. Thus,

$$xy = u^2, \qquad \frac{y}{x} = v^2 \qquad \boxed{3}$$

(a) By the first part of Eq. (3), G maps a point (c, v) to a point in the xy-plane with $xy = c^2$. In other words, G maps the vertical line $u = c$ to the hyperbola $xy = c^2$. Similarly, by the second part of Eq. (3), the horizontal line $v = c$ is mapped to the set of points where $y/x = c^2$, or $y = c^2 x$, which is the line through the origin of slope c^2. See Figure 5.

The term "curvilinear rectangle" refers to a region bounded on four sides by curves as on the right in Figure 5.

(b) The image of $[1, 2] \times [1, 2]$ is the *curvilinear* rectangle bounded by the four curves that are the images of the lines $u = 1$, $u = 2$, and $v = 1$, $v = 2$. By Eq. (3), this region is defined by the inequalities

$$1 \le xy \le 4, \qquad 1 \le \frac{y}{x} \le 4$$

To find G^{-1}, we use Eq. (3) to write $u = \sqrt{xy}$ and $v = \sqrt{y/x}$. Therefore, the inverse map is $G^{-1}(x, y) = \left(\sqrt{xy}, \sqrt{y/x}\right)$. We take positive square roots because $u > 0$ and $v > 0$ on the domain we are considering. ∎

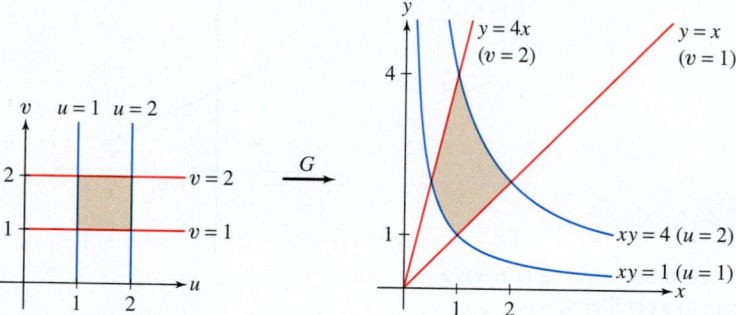

FIGURE 5 The mapping $G(u, v) = (uv^{-1}, uv)$.

How Area Changes Under a Mapping: The Jacobian Determinant

The **Jacobian determinant** (or simply the Jacobian) of a map

$$G(u, v) = (x(u, v), y(u, v))$$

is the determinant

← REMINDER The definition of a 2 × 2 determinant is

$$\begin{vmatrix} a & b \\ c & d \end{vmatrix} = ad - bc \qquad \boxed{4}$$

$$\mathrm{Jac}(G) = \begin{vmatrix} \dfrac{\partial x}{\partial u} & \dfrac{\partial x}{\partial v} \\[2mm] \dfrac{\partial y}{\partial u} & \dfrac{\partial y}{\partial v} \end{vmatrix} = \frac{\partial x}{\partial u}\frac{\partial y}{\partial v} - \frac{\partial x}{\partial v}\frac{\partial y}{\partial u}$$

The Jacobian $\mathrm{Jac}(G)$ is also denoted $\dfrac{\partial(x, y)}{\partial(u, v)}$. Note that $\mathrm{Jac}(G)$ is a function of u and v.

EXAMPLE 4 Evaluate the Jacobian of $G(u, v) = (u^3 + v, uv)$ at $(u, v) = (2, 1)$.

Solution We have $x = u^3 + v$ and $y = uv$, so

$$\mathrm{Jac}(G) = \frac{\partial(x, y)}{\partial(u, v)} = \begin{vmatrix} \dfrac{\partial x}{\partial u} & \dfrac{\partial x}{\partial v} \\[2mm] \dfrac{\partial y}{\partial u} & \dfrac{\partial y}{\partial v} \end{vmatrix}$$

$$= \begin{vmatrix} 3u^2 & 1 \\ v & u \end{vmatrix} = 3u^3 - v$$

The value of the Jacobian at $(2, 1)$ is $\mathrm{Jac}(G)(2, 1) = 3(2)^3 - 1 = 23$. ∎

The Jacobian tells us how area changes under a map G. We can see this most directly in the case of a linear map $G(u, v) = (Au + Cv, Bu + Dv)$.

THEOREM 1 Jacobian of a Linear Map The Jacobian of a linear map

$$G(u, v) = (Au + Cv, Bu + Dv)$$

is *constant* with value

$$\text{Jac}(G) = \begin{vmatrix} A & C \\ B & D \end{vmatrix} = AD - BC \tag{5}$$

Under G, the area of a region $\mathcal{D}$ is multiplied by the factor $|\text{Jac}(G)|$; that is,

$$\boxed{\text{area}(G(\mathcal{D})) = |\text{Jac}(G)|\,\text{area}(\mathcal{D})} \tag{6}$$

Proof Equation (5) is verified by direct calculation: Because

$$x = Au + Cv \quad \text{and} \quad y = Bu + Dv$$

the partial derivatives in the Jacobian are the constants A, B, C, D.

We sketch a proof of Eq. (6). It certainly holds for the unit square $\mathcal{D} = [1, 0] \times [0, 1]$ because $G(\mathcal{D})$ is the parallelogram spanned by the vectors $\langle A, B \rangle$ and $\langle C, D \rangle$ (Figure 6) and this parallelogram has area

$$|\text{Jac}(G)| = |AD - BC|$$

by Eq. (10) in Section 12.4. Similarly, we can check directly that Eq. (6) holds for arbitrary parallelograms (see Exercise 48). To verify Eq. (6) for a general domain $\mathcal{D}$, we use the fact that $\mathcal{D}$ can be approximated as closely as desired by a union of rectangles in a fine grid of lines parallel to the u- and v-axes. ■

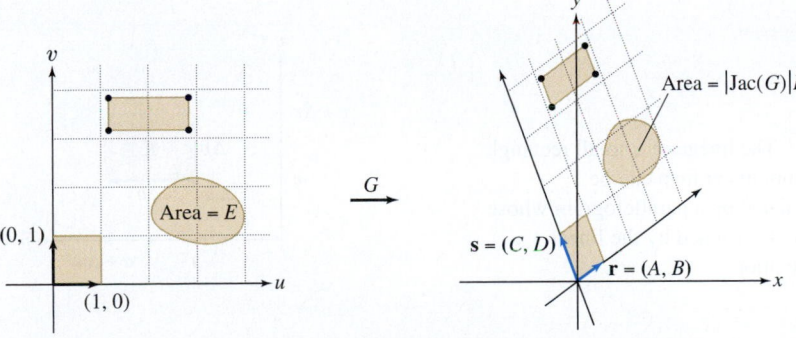

FIGURE 6 A linear map G expands (or shrinks) area by the factor $|\text{Jac}(G)|$.

We cannot expect Eq. (6) to hold for a nonlinear map. In fact, it would not make sense as stated because the value $\text{Jac}(G)(P)$ may vary from point to point. However, it is *approximately* true if the domain $\mathcal{D}$ is small and P is a sample point in $\mathcal{D}$:

$$\boxed{\text{area}(G(\mathcal{D})) \approx |\text{Jac}(G)(P)|\,\text{area}(\mathcal{D})} \tag{7}$$

This result may be stated more precisely as the limit relation:

$$|\text{Jac}(G)(P)| = \lim_{|\mathcal{D}| \to 0} \frac{\text{area}(G(\mathcal{D}))}{\text{area}(\mathcal{D})} \tag{8}$$

Here, we write $|\mathcal{D}| \to 0$ to indicate the limit as the diameter of $\mathcal{D}$ (the maximum distance between two points in $\mathcal{D}$) tends to zero.

CONCEPTUAL INSIGHT Although a rigorous proof of Eq. (8) is too technical to include here, we can understand Eq. (7) as an application of linear approximation. Consider a rectangle $\mathcal{R}$ with vertex at $P = (u, v)$ and sides of lengths Δu and Δv, assumed to be small as in Figure 7. The image $G(\mathcal{R})$ is not a parallelogram, but it is approximated well by the parallelogram spanned by the vectors **A** and **B** in the figure:

$$\mathbf{A} = G(u + \Delta u, v) - G(u, v)$$

$$= \langle x(u + \Delta u, v) - x(u, v), y(u + \Delta u, v) - y(u, v) \rangle$$

$$\mathbf{B} = G(u, v + \Delta v) - G(u, v)$$

$$= \langle x(u, v + \Delta v) - x(u, v), y(u, v + \Delta v) - y(u, v) \rangle$$

REMINDER Eqs. (9) and (10) use the linear approximations

$$x(u + \Delta u, v) - x(u, v) \approx \frac{\partial x}{\partial u} \Delta u$$

$$y(u + \Delta u, v) - y(u, v) \approx \frac{\partial y}{\partial u} \Delta u$$

and

$$x(u, v + \Delta v) - x(u, v) \approx \frac{\partial x}{\partial v} \Delta v$$

$$y(u, v + \Delta v) - y(u, v) \approx \frac{\partial y}{\partial v} \Delta v$$

The linear approximation applied to the components of G yields

$$\mathbf{A} \approx \left\langle \frac{\partial x}{\partial u} \Delta u, \frac{\partial y}{\partial u} \Delta u \right\rangle \qquad \boxed{9}$$

$$\mathbf{B} \approx \left\langle \frac{\partial x}{\partial v} \Delta v, \frac{\partial y}{\partial v} \Delta v \right\rangle \qquad \boxed{10}$$

Using Eq. (10) from Section 12.4 for the area of a parallelogram spanned by vectors **A** and **B**, we obtain the desired approximation:

$$\text{area}(G(\mathcal{R})) \approx \left| \det \begin{pmatrix} \mathbf{A} \\ \mathbf{B} \end{pmatrix} \right| = \left| \det \begin{pmatrix} \frac{\partial x}{\partial u} \Delta u & \frac{\partial y}{\partial u} \Delta u \\ \frac{\partial x}{\partial v} \Delta v & \frac{\partial y}{\partial v} \Delta v \end{pmatrix} \right|$$

$$= \left| \frac{\partial x}{\partial u} \frac{\partial y}{\partial v} - \frac{\partial y}{\partial u} \frac{\partial x}{\partial v} \right| \Delta u \, \Delta v$$

$$= |\text{Jac}(G)(P)| \, \text{area}(\mathcal{R})$$

The last equation holds since the area of $\mathcal{R}$ is $\Delta u \, \Delta v$.

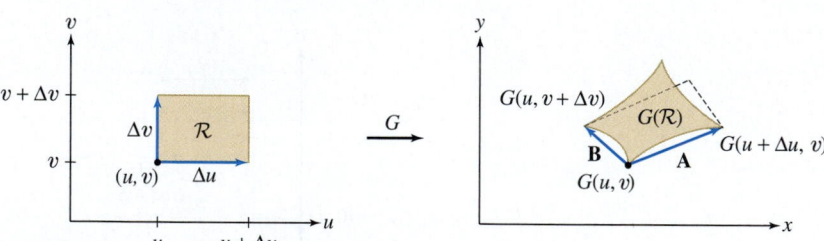

FIGURE 7 The image of a small rectangle under a nonlinear map can be approximated by a parallelogram whose sides are determined by the linear approximation.

The Change of Variables Formula

Recall the formula for integration in polar coordinates:

$$\iint_{\mathcal{D}} f(x, y) \, dx \, dy = \int_{\theta_1}^{\theta_2} \int_{r_1}^{r_2} f(r \cos \theta, r \sin \theta) \, r \, dr \, d\theta \qquad \boxed{11}$$

Here, $\mathcal{D}$ is the polar rectangle consisting of points $(x, y) = (r \cos \theta, r \sin \theta)$ in the xy-plane (see Figure 2). On the right the rectangle $\mathcal{R} = [r_1, r_2] \times [\theta_1, \theta_2]$ in the $r\theta$-plane is the domain of integration. Thus, $\mathcal{D}$ is the image of the domain on the right under the polar coordinates map.

The general Change of Variables Formula has a similar form. Given a map

$$G : \quad \underset{\text{in } uv\text{-plane}}{\mathcal{D}_0} \quad \rightarrow \quad \underset{\text{in } xy\text{-plane}}{\mathcal{D}}$$

from a domain in the uv-plane to a domain in the xy-plane (Figure 8), our formula expresses an integral over $\mathcal{D}$ as an integral over $\mathcal{D}_0$. The Jacobian plays the role of the factor r on the right-hand side of Eq. (11).

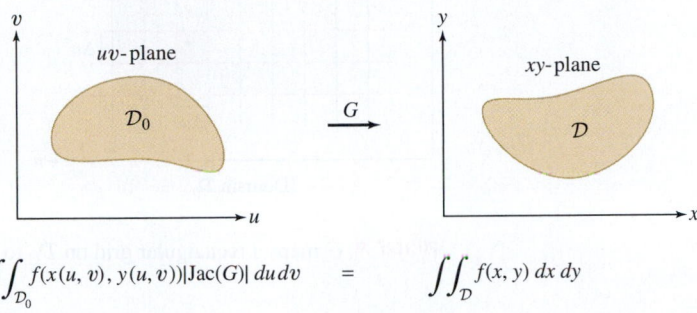

$$\iint_{\mathcal{D}_0} f(x(u, v), y(u, v))|\text{Jac}(G)|\, du\, dv \quad = \quad \iint_{\mathcal{D}} f(x, y)\, dx\, dy$$

FIGURE 8 The Change of Variables Formula expresses a double integral over $\mathcal{D}$ as a double integral over $\mathcal{D}_0$.

← **REMINDER** G is called "one-to-one" if $G(P) = G(Q)$ only for $P = Q$.

A few technical assumptions are necessary. First, we assume that G is one-to-one, at least on the interior of $\mathcal{D}_0$, because we want G to cover the target domain $\mathcal{D}$ just once. We also assume that G is a C^1 **map**, by which we mean that the component functions x and y have continuous partial derivatives. Under these conditions, if $f(x, y)$ is continuous, we have the following result.

THEOREM 2 Change of Variables Formula Let $G : \mathcal{D}_0 \to \mathcal{D}$ be a C^1 mapping that is one-to-one on the interior of $\mathcal{D}_0$. If $f(x, y)$ is continuous, then

$$\iint_{\mathcal{D}} f(x, y)\, dx\, dy = \iint_{\mathcal{D}_0} f(x(u, v), y(u, v)) \left| \frac{\partial(x, y)}{\partial(u, v)} \right| du\, dv \qquad \boxed{12}$$

Equation (12) is summarized by the equality

$$dx\, dy = \left| \frac{\partial(x, y)}{\partial(u, v)} \right| du\, dv$$

Recall that $\dfrac{\partial(x, y)}{\partial(u, v)}$ denotes the Jacobian Jac(G).

← **REMINDER** If $\mathcal{D}$ is a domain of small diameter, $P \in \mathcal{D}$ is a sample point, and $f(x, y)$ is continuous, then (see Section 15.2)

$$\iint_{\mathcal{D}} f(x, y)\, dx\, dy \approx f(P)\text{area}(\mathcal{D})$$

Proof We sketch the proof. Observe first that Eq. (12) is *approximately* true if the domains $\mathcal{D}_0$ and $\mathcal{D}$ are small. Let $P = G(P_0)$, where P_0 is any sample point in $\mathcal{D}_0$. Since $f(x, y)$ is continuous, the approximation recalled in the margin together with Eq. (7) yields

$$\iint_{\mathcal{D}} f(x, y)\, dx\, dy \approx f(P)\text{area}(\mathcal{D})$$

$$\approx f(G(P_0))\, |\text{Jac}(G)(P_0)|\, \text{area}(\mathcal{D}_0)$$

$$\approx \iint_{\mathcal{D}_0} f(G(u, v))\, |\text{Jac}(G)(u, v)|\, du\, dv$$

If $\mathcal{D}$ is not small, divide it into small subdomains $D_j = G(\mathcal{D}_{0j})$ (Figure 9 shows a rectangle divided into smaller rectangles), apply the approximation to each subdomain, and sum:

$$\iint_{\mathcal{D}} f(x, y)\, dx\, dy = \sum_j \iint_{\mathcal{D}_j} f(x, y)\, dx\, dy$$

$$\approx \sum_j \iint_{\mathcal{D}_{0j}} f(G(u, v))\, |\text{Jac}(G)(u, v)|\, du\, dv$$

$$= \iint_{\mathcal{D}_0} f(G(u, v))\, |\text{Jac}(G)(u, v)|\, du\, dv$$

Careful estimates show that the error tends to zero as the maximum of the diameters of the subdomains $\mathcal{D}_j$ tends to zero. This yields the Change of Variables Formula. ∎

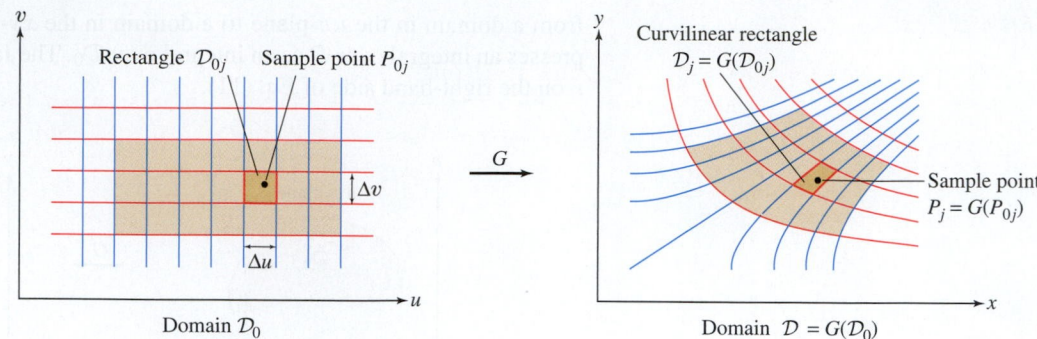

FIGURE 9 G maps a rectangular grid on $\mathcal{D}_0$ to a curvilinear grid on $\mathcal{D}$.

EXAMPLE 5 **Polar Coordinates Revisited** Use the Change of Variables Formula to derive the formula for integration in polar coordinates.

Solution The Jacobian of the polar coordinate map $G(r, \theta) = (r \cos \theta, r \sin \theta)$ is

$$\mathrm{Jac}(G) = \begin{vmatrix} \dfrac{\partial x}{\partial r} & \dfrac{\partial x}{\partial \theta} \\[2mm] \dfrac{\partial y}{\partial r} & \dfrac{\partial y}{\partial \theta} \end{vmatrix} = \begin{vmatrix} \cos \theta & -r \sin \theta \\ \sin \theta & r \cos \theta \end{vmatrix} = r(\cos^2 \theta + \sin^2 \theta) = r$$

Let $\mathcal{D} = G(\mathcal{R})$ be the image under the polar coordinates map G of the rectangle $\mathcal{R}$ defined by $r_0 \leq r \leq r_1$, $\theta_0 \leq \theta \leq \theta_1$ (see Figure 2). Then Eq. (12) yields the expected formula for polar coordinates:

$$\iint_{\mathcal{D}} f(x, y)\, dx\, dy = \int_{\theta_0}^{\theta_1} \int_{r_0}^{r_1} f(r \cos \theta, r \sin \theta)\, r\, dr\, d\theta \qquad \boxed{13}$$

■

Assumptions Matter In the Change of Variables Formula, we assume that G is one-to-one on the interior but not necessarily on the boundary of the domain. Thus, we can apply Eq. (12) to the polar coordinates map G on the rectangle $\mathcal{D}_0 = [0, 1] \times [0, 2\pi]$. In this case, G is one-to-one on the interior but not on the boundary of $\mathcal{D}_0$ since $G(0, \theta) = (0, 0)$ for all θ and $G(r, 0) = G(r, 2\pi)$ for all r. On the other hand, Eq. (12) cannot be applied to G on the rectangle $[0, 1] \times [0, 4\pi]$ because it is not one-to-one on the interior.

EXAMPLE 6 Use the Change of Variables Formula to calculate $\displaystyle\iint_{\mathcal{P}} e^{4x-y}\, dx\, dy$, where $\mathcal{P}$ is the parallelogram spanned by the vectors $\langle 4, 1 \rangle$, $\langle 3, 3 \rangle$ in Figure 10.

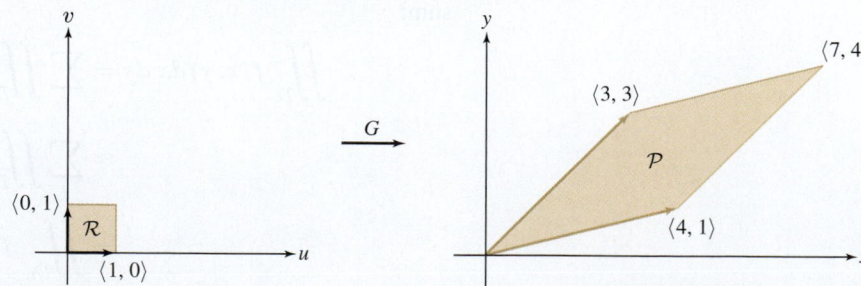

FIGURE 10 The map
$G(u, v) = (4u + 3v, u + 3v)$.

Recall that the linear map

$$G(u, v) = (Au + Cv, Bu + Dv)$$

satisfies

$$G(1, 0) = (A, B), \quad G(0, 1) = (C, D)$$

Solution

Step 1. **Define a map.**

We can convert our double integral to an integral over the unit square $\mathcal{R} = [0, 1] \times [0, 1]$ if we can find a map that sends $\mathcal{R}$ to $\mathcal{P}$. The following linear map does the job:

$$G(u, v) = (4u + 3v, u + 3v)$$

Indeed, $G(1, 0) = (4, 1)$ and $G(0, 1) = (3, 3)$, so it maps $\mathcal{R}$ to $\mathcal{P}$ because linear maps map parallelograms to parallelograms.

Step 2. **Compute the Jacobian.**

$$\text{Jac}(G) = \begin{vmatrix} \dfrac{\partial x}{\partial u} & \dfrac{\partial x}{\partial v} \\ \dfrac{\partial y}{\partial u} & \dfrac{\partial y}{\partial v} \end{vmatrix} = \begin{vmatrix} 4 & 3 \\ 1 & 3 \end{vmatrix} = 9$$

Step 3. **Express $f(x, y)$ in terms of the new variables.**

Since $x = 4u + 3v$ and $y = u + 3v$, we have

$$e^{4x - y} = e^{4(4u + 3v) - (u + 3v)} = e^{15u + 9v}$$

Step 4. **Apply the Change of Variables Formula.**

The Change of Variables Formula tells us that $dx \, dy = 9 \, du \, dv$:

$$\iint_{\mathcal{P}} e^{4x - y} \, dx \, dy = \iint_{\mathcal{R}} e^{15u + 9v} |\text{Jac}(G)| \, du \, dv = \int_0^1 \int_0^1 e^{15u} e^{9v} \, 9 \, du \, dv$$

$$= \frac{3}{5}(e^{15} - 1) \int_0^1 e^{9v} \, dv = \frac{1}{15}(e^{15} - 1)(e^9 - 1) \qquad \blacksquare$$

EXAMPLE 7 Use the Change of Variables Formula to compute

$$\iint_{\mathcal{D}} (x^2 + y^2) \, dx \, dy$$

where $\mathcal{D}$ is the domain $1 \le xy \le 4$, $1 \le y/x \le 4$ (Figure 11).

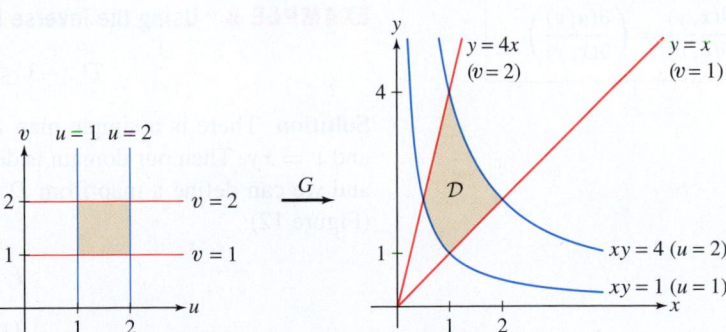

FIGURE 11

Solution In Example 3, we studied the map $G(u, v) = (uv^{-1}, uv)$, which can be written

$$x = uv^{-1}, \qquad y = uv$$

We showed (Figure 11) that G maps the rectangle $\mathcal{D}_0 = [1, 2] \times [1, 2]$ to our domain $\mathcal{D}$. Indeed, because $xy = u^2$ and $yx^{-1} = v^2$, the two conditions $1 \le xy \le 4$ and $1 \le y/x \le 4$ that define $\mathcal{D}$ become $1 \le u \le 2$ and $1 \le v \le 2$.

The Jacobian is

$$\text{Jac}(G) = \frac{\partial(x, y)}{\partial(u, v)} = \begin{vmatrix} \dfrac{\partial x}{\partial u} & \dfrac{\partial x}{\partial v} \\[2mm] \dfrac{\partial y}{\partial u} & \dfrac{\partial y}{\partial v} \end{vmatrix} = \begin{vmatrix} v^{-1} & -uv^{-2} \\ v & u \end{vmatrix} = \frac{2u}{v}$$

To apply the Change of Variables Formula, we write $f(x, y)$ in terms of u and v:

$$f(x, y) = x^2 + y^2 = \left(\frac{u}{v}\right)^2 + (uv)^2 = u^2(v^{-2} + v^2)$$

By the Change of Variables Formula,

$$\iint_{\mathcal{D}} (x^2 + y^2)\, dx\, dy = \iint_{\mathcal{D}_0} u^2(v^{-2} + v^2) \left|\frac{2u}{v}\right| du\, dv$$

$$= 2 \int_{v=1}^{2} \int_{u=1}^{2} u^3(v^{-3} + v)\, du\, dv$$

$$= \frac{15}{2} \int_{v=1}^{2} (v^{-3} + v)\, dv$$

$$= \frac{15}{2} \left(-\frac{1}{2}v^{-2} + \frac{1}{2}v^2 \Big|_1^2 \right) = \frac{225}{16} \qquad \blacksquare$$

Keep in mind that the Change of Variables Formula turns an xy-integral into a uv-integral, but the map G goes from the uv-domain to the xy-domain. Sometimes, it is easier to find a map F going in the *wrong direction*, from the xy-domain to the uv-domain. The desired map G is then the inverse $G = F^{-1}$. The next example shows that in some cases, we can evaluate the integral without solving for G. The key fact is that the Jacobian of G is the reciprocal of the Jacobian of F (see Exercises 49–51):

The relationship between Jacobians in Equation (14) can be written in the suggestive form

$$\frac{\partial(x, y)}{\partial(u, v)} = \left(\frac{\partial(u, v)}{\partial(x, y)} \right)^{-1}$$

> If $G = F^{-1}$ and $\text{Jac}(F) \neq 0$, then $\text{Jac}(G) = \text{Jac}(F)^{-1}$ **14**

EXAMPLE 8   Using the Inverse Map   Integrate $f(x, y) = xy(x^2 + y^2)$ over

$$\mathcal{D}: -3 \le x^2 - y^2 \le 3, \quad 1 \le xy \le 4$$

Solution   There is a simple map F that goes in the *wrong* direction. Let $u = x^2 - y^2$ and $v = xy$. Then our domain is defined by the inequalities $-3 \le u \le 3$ and $1 \le v \le 4$, and we can define a map from $\mathcal{D}$ to the rectangle $\mathcal{R} = [-3, 3] \times [1, 4]$ in the uv-plane (Figure 12):

$$F : \mathcal{D} \to \mathcal{R}$$

$$(x, y) \to (x^2 - y^2, xy)$$

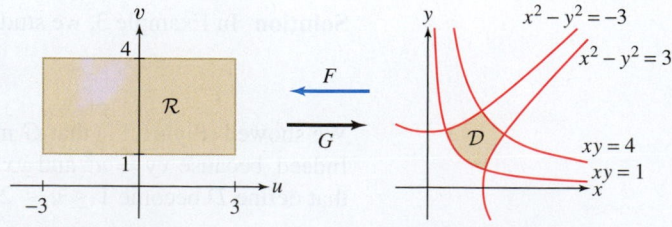

DF **FIGURE 12** The map F goes in the "wrong" direction.

To convert the integral over $\mathcal{D}$ into an integral over the rectangle $\mathcal{R}$, we have to apply the Change of Variables Formula to the inverse mapping:

$$G = F^{-1} : \mathcal{R} \to \mathcal{D}$$

We will see that it is not necessary to find G explicitly. Since $u = x^2 - y^2$ and $v = xy$, the Jacobian of F is

$$\mathrm{Jac}(F) = \begin{vmatrix} \dfrac{\partial u}{\partial x} & \dfrac{\partial u}{\partial y} \\[2mm] \dfrac{\partial v}{\partial x} & \dfrac{\partial v}{\partial y} \end{vmatrix} = \begin{vmatrix} 2x & -2y \\ y & x \end{vmatrix} = 2(x^2 + y^2)$$

By Eq. (14),

$$\mathrm{Jac}(G) = \mathrm{Jac}(F)^{-1} = \frac{1}{2(x^2 + y^2)}$$

Normally, the next step would be to express $f(x, y)$ in terms of u and v. We can avoid doing this in our case by observing that the Jacobian cancels with one factor of $f(x, y)$:

$$\iint_{\mathcal{D}} xy(x^2 + y^2)\, dx\, dy = \iint_{\mathcal{R}} f(x(u, v), y(u, v))\, |\mathrm{Jac}(G)|\, du\, dv$$

$$= \iint_{\mathcal{R}} xy(x^2 + y^2) \frac{1}{2(x^2 + y^2)}\, du\, dv$$

$$= \frac{1}{2} \iint_{\mathcal{R}} xy\, du\, dv$$

$$= \frac{1}{2} \iint_{\mathcal{R}} v\, du\, dv \qquad (\text{because } v = xy)$$

$$= \frac{1}{2} \int_{-3}^{3} \int_{1}^{4} v\, dv\, du = \frac{1}{2}(6)\left(\frac{1}{2}4^2 - \frac{1}{2}1^2\right) = \frac{45}{2} \qquad \blacksquare$$

Change of Variables in Three Variables

The Change of Variables Formula has the same form in three (or more) variables as in two variables. Let

$$G : \mathcal{W}_0 \to \mathcal{W}$$

be a mapping from a three-dimensional region $\mathcal{W}_0$ in (u, v, w)-space to a region $\mathcal{W}$ in (x, y, z)-space, say,

$$x = x(u, v, w), \qquad y = y(u, v, w), \qquad z = z(u, v, w)$$

← REMINDER 3×3 determinants are defined in Eq. (2) of Section 12.4.

The Jacobian $\mathrm{Jac}(G)$ is the 3×3 determinant:

$$\mathrm{Jac}(G) = \frac{\partial(x, y, z)}{\partial(u, v, w)} = \begin{vmatrix} \dfrac{\partial x}{\partial u} & \dfrac{\partial x}{\partial v} & \dfrac{\partial x}{\partial w} \\[2mm] \dfrac{\partial y}{\partial u} & \dfrac{\partial y}{\partial v} & \dfrac{\partial y}{\partial w} \\[2mm] \dfrac{\partial z}{\partial u} & \dfrac{\partial z}{\partial v} & \dfrac{\partial z}{\partial w} \end{vmatrix}$$ **15**

The Change of Variables Formula states

$$\boxed{\; dx\, dy\, dz = \left| \frac{\partial(x, y, z)}{\partial(u, v, w)} \right|\, du\, dv\, dw \;}$$

More precisely, if G is C^1 and one-to-one on the interior of $\mathcal{W}_0$, and if f is continuous, then

$$\iiint_{\mathcal{W}} f(x, y, z)\,dx\,dy\,dz$$

$$= \iiint_{\mathcal{W}_0} f(x(u, v, w), y(u, v, w), z(u, v, w)) \left| \frac{\partial(x, y, z)}{\partial(u, v, w)} \right| du\,dv\,dw \qquad \boxed{16}$$

In Exercises 42 and 43, you are asked to use the general Change of Variables Formula to derive the formulas for integration in cylindrical and spherical coordinates developed in Section 15.4.

15.6 SUMMARY

- Let $G(u, v) = (x(u, v), y(u, v))$ be a mapping. The Jacobian of G is the determinant

$$\mathrm{Jac}(G) = \frac{\partial(x, y)}{\partial(u, v)} = \begin{vmatrix} \dfrac{\partial x}{\partial u} & \dfrac{\partial x}{\partial v} \\[2mm] \dfrac{\partial y}{\partial u} & \dfrac{\partial y}{\partial v} \end{vmatrix}$$

- If $G = F^{-1}$ and $\mathrm{Jac}(F) \neq 0$, then $\mathrm{Jac}(G) = \mathrm{Jac}(F)^{-1}$.
- Change of Variables Formula: If $G : \mathcal{D}_0 \to \mathcal{D}$ has component functions with continuous partial derivatives and is one-to-one on the interior of $\mathcal{D}_0$, and if f is continuous, then

$$\iint_{\mathcal{D}} f(x, y)\,dx\,dy = \iint_{\mathcal{D}_0} f(x(u, v), y(u, v)) \left| \frac{\partial(x, y)}{\partial(u, v)} \right| du\,dv$$

- The Change of Variables Formula is written in two and three variables as

$$dx\,dy = \left| \frac{\partial(x, y)}{\partial(u, v)} \right| du\,dv, \qquad dx\,dy\,dz = \left| \frac{\partial(x, y, z)}{\partial(u, v, w)} \right| du\,dv\,dw$$

15.6 EXERCISES

Preliminary Questions

1. Which of the following maps is linear?

(a) (uv, v) (b) $(u + v, u)$ (c) $(3, e^u)$

2. Suppose that G is a linear map such that $G(2, 0) = (4, 0)$ and $G(0, 3) = (-3, 9)$. Find the images of:

(a) $G(1, 0)$ (b) $G(1, 1)$ (c) $G(2, 1)$

3. What is the area of $G(\mathcal{R})$ if $\mathcal{R}$ is a rectangle of area 9 and G is a mapping whose Jacobian has constant value 4?

4. Estimate the area of $G(\mathcal{R})$, where $\mathcal{R} = [1, 1.2] \times [3, 3.1]$ and G is a mapping such that $\mathrm{Jac}(G)(1, 3) = 3$.

Exercises

1. Determine the image under $G(u, v) = (2u, u + v)$ of the following sets:

(a) The u- and v-axes

(b) The rectangle $\mathcal{R} = [0, 5] \times [0, 7]$

(c) The line segment joining $(1, 2)$ and $(5, 3)$

(d) The triangle with vertices $(0, 1)$, $(1, 0)$, and $(1, 1)$

2. Describe [in the form $y = f(x)$] the images of the lines $u = c$ and $v = c$ under the mapping $G(u, v) = (u/v, u^2 - v^2)$.

3. Let $G(u, v) = (u^2, v)$. Is G one-to-one? If not, determine a domain on which G is one-to-one. Find the image under G of:

(a) The u- and v-axes

(b) The rectangle $\mathcal{R} = [-1, 1] \times [-1, 1]$

(c) The line segment joining $(0, 0)$ and $(1, 1)$

(d) The triangle with vertices $(0, 0)$, $(0, 1)$, and $(1, 1)$

4. Let $G(u, v) = (e^u, e^{u+v})$.

(a) Is G one-to-one? What is the image of G?

(b) Describe the images of the vertical lines $u = c$ and the horizontal lines $v = c$.

In Exercises 5–12, let $G(u, v) = (2u + v, 5u + 3v)$ be a map from the uv-plane to the xy-plane.

5. Show that the image of the horizontal line $v = c$ is the line with equation $y = \frac{5}{2}x + \frac{1}{2}c$. What is the image (in slope-intercept form) of the vertical line $u = c$?

6. Describe the image of the line through the points $(u, v) = (1, 1)$ and $(u, v) = (1, -1)$ under G in slope-intercept form.

7. Describe the image of the line $v = 4u$ under G in slope-intercept form.

8. Show that G maps the line $v = mu$ to the line of slope $(5 + 3m)/(2 + m)$ through the origin in the xy-plane.

9. Show that the inverse of G is

$$G^{-1}(x, y) = (3x - y, -5x + 2y)$$

Hint: Show that $G(G^{-1}(x, y)) = (x, y)$ and $G^{-1}(G(u, v)) = (u, v)$.

10. Use the inverse in Exercise 9 to find:

(a) A point in the uv-plane mapping to $(2, 1)$

(b) A segment in the uv-plane mapping to the segment joining $(-2, 1)$ and $(3, 4)$

11. Calculate $\text{Jac}(G) = \dfrac{\partial(x, y)}{\partial(u, v)}$.

12. Calculate $\text{Jac}(G^{-1}) = \dfrac{\partial(u, v)}{\partial(x, y)}$.

In Exercises 13–18, compute the Jacobian (at the point, if indicated).

13. $G(u, v) = (3u + 4v, u - 2v)$

14. $G(r, s) = (rs, r + s)$

15. $G(r, t) = (r \sin t, r - \cos t), \quad (r, t) = (1, \pi)$

16. $G(u, v) = (v \ln u, u^2 v^{-1}), \quad (u, v) = (1, 2)$

17. $G(r, \theta) = (r \cos \theta, r \sin \theta), \quad (r, \theta) = \left(4, \frac{\pi}{6}\right)$

18. $G(u, v) = (ue^v, e^u)$

19. Find a linear mapping G that maps $[0, 1] \times [0, 1]$ to the parallelogram in the xy-plane spanned by the vectors $\langle 2, 3 \rangle$ and $\langle 4, 1 \rangle$.

20. Find a linear mapping G that maps $[0, 1] \times [0, 1]$ to the parallelogram in the xy-plane spanned by the vectors $\langle -2, 5 \rangle$ and $\langle 1, 7 \rangle$.

21. Let $\mathcal{D}$ be the parallelogram in Figure 13. Apply the Change of Variables Formula to the map $G(u, v) = (5u + 3v, u + 4v)$ to evaluate $\displaystyle\iint_{\mathcal{D}} xy \, dx \, dy$ as an integral over $\mathcal{D}_0 = [0, 1] \times [0, 1]$.

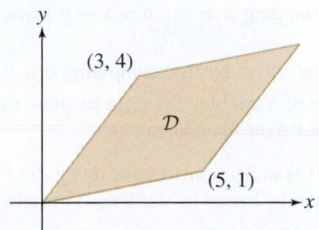

FIGURE 13

22. Let $G(u, v) = (u - uv, uv)$.

(a) Show that the image of the horizontal line $v = c$ is $y = \dfrac{c}{1 - c}x$ if $c \neq 1$, and is the y-axis if $c = 1$.

(b) Determine the images of vertical lines in the uv-plane.

(c) Compute the Jacobian of G.

(d) Observe that by the formula for the area of a triangle, the region $\mathcal{D}$ in Figure 14 has area $\frac{1}{2}(b^2 - a^2)$. Compute this area again, using the Change of Variables Formula applied to G.

(e) Calculate $\displaystyle\iint_{\mathcal{D}} xy \, dx \, dy$.

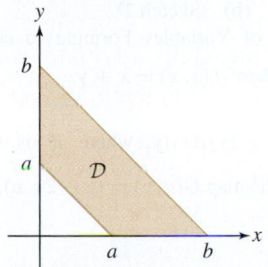

FIGURE 14

23. Let $G(u, v) = (3u + v, u - 2v)$. Use the Jacobian to determine the area of $G(\mathcal{R})$ for:

(a) $\mathcal{R} = [0, 3] \times [0, 5]$ **(b)** $\mathcal{R} = [2, 5] \times [1, 7]$

24. Find a linear map T that maps $[0, 1] \times [0, 1]$ to the parallelogram $\mathcal{P}$ in the xy-plane with vertices $(0, 0), (2, 2), (1, 4), (3, 6)$. Then calculate the double integral of e^{2x-y} over $\mathcal{P}$ via change of variables.

25. With G as in Example 3, use the Change of Variables Formula to compute the area of the image of $[1, 4] \times [1, 4]$.

In Exercises 26–28, let $\mathcal{R}_0 = [0, 1] \times [0, 1]$ be the unit square. The translate of a map $G_0(u, v) = (\phi(u, v), \psi(u, v))$ is a map

$$G(u, v) = (a + \phi(u, v), b + \psi(u, v))$$

where a, b are constants. Observe that the map G_0 in Figure 15 maps $\mathcal{R}_0$ to the parallelogram $\mathcal{P}_0$ and that the translate

$$G_1(u, v) = (2 + 4u + 2v, 1 + u + 3v)$$

maps $\mathcal{R}_0$ to $\mathcal{P}_1$.

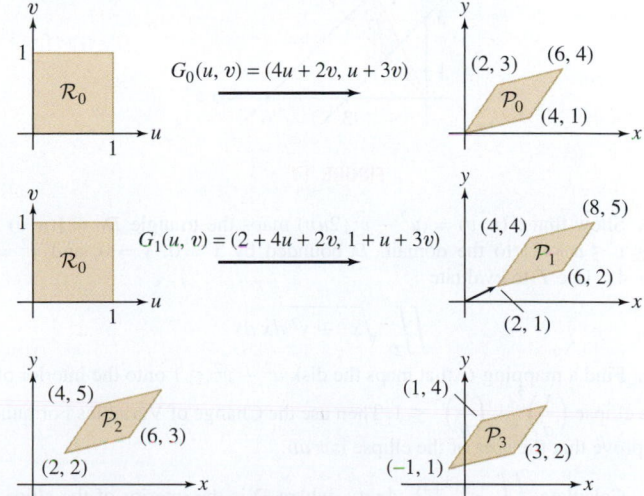

FIGURE 15

26. Find translates G_2 and G_3 of the mapping G_0 in Figure 15 that map the unit square $\mathcal{R}_0$ to the parallelograms $\mathcal{P}_2$ and $\mathcal{P}_3$.

27. Sketch the parallelogram $\mathcal{P}$ with vertices $(1, 1), (2, 4), (3, 6), (4, 9)$ and find the translate of a linear mapping that maps $\mathcal{R}_0$ to $\mathcal{P}$.

28. Find the translate of a linear mapping that maps $\mathcal{R}_0$ to the parallelogram spanned by the vectors $\langle 3, 9 \rangle$ and $\langle -4, 6 \rangle$ based at $(4, 2)$.

29. Let $\mathcal{D} = G(\mathcal{R})$, where $G(u, v) = (u^2, u + v)$ and $\mathcal{R} = [1, 2] \times [0, 6]$. Calculate $\iint_{\mathcal{D}} y \, dx \, dy$. *Note:* It is not necessary to describe $\mathcal{D}$.

30. Let $\mathcal{D}$ be the image of $\mathcal{R} = [1, 4] \times [1, 4]$ under the map $G(u, v) = (u^2/v, v^2/u)$.
(a) Compute $\text{Jac}(G)$. **(b)** Sketch $\mathcal{D}$.
(c) Use the Change of Variables Formula to compute $\text{Area}(\mathcal{D})$ and $\iint_{\mathcal{D}} f(x, y) \, dx \, dy$, where $f(x, y) = x + y$.

31. Compute $\iint_{\mathcal{D}} (x + 3y) \, dx \, dy$, where $\mathcal{D}$ is the shaded region in Figure 16. *Hint:* Use the map $G(u, v) = (u - 2v, v)$.

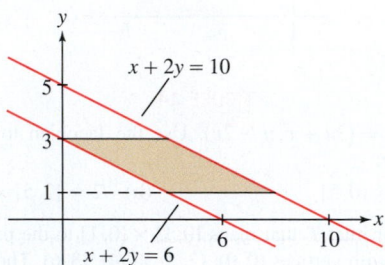

FIGURE 16

32. Use the map $G(u, v) = \left(\dfrac{u}{v + 1}, \dfrac{uv}{v + 1} \right)$ to compute

$$\iint_{\mathcal{D}} (x + y) \, dx \, dy$$

where $\mathcal{D}$ is the shaded region in Figure 17.

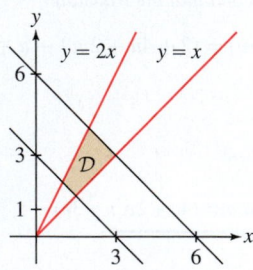

FIGURE 17

33. Show that $T(u, v) = (u^2 - v^2, 2uv)$ maps the triangle $\mathcal{D}_0 = \{(u, v) : 0 \leq v \leq u \leq 1\}$ to the domain $\mathcal{D}$ bounded by $x = 0$, $y = 0$, and $y^2 = 4 - 4x$. Use T to evaluate

$$\iint_{\mathcal{D}} \sqrt{x^2 + y^2} \, dx \, dy$$

34. Find a mapping G that maps the disk $u^2 + v^2 \leq 1$ onto the interior of the ellipse $\left(\dfrac{x}{a} \right)^2 + \left(\dfrac{y}{b} \right)^2 \leq 1$. Then use the Change of Variables Formula to prove that the area of the ellipse is πab.

35. Calculate $\iint_{\mathcal{D}} e^{9x^2 + 4y^2} \, dx \, dy$, where $\mathcal{D}$ is the interior of the ellipse $\left(\dfrac{x}{2} \right)^2 + \left(\dfrac{y}{3} \right)^2 \leq 1$.

36. Let $\mathcal{D}$ be the region inside the ellipse $x^2 + 2xy + 2y^2 - 4y = 8$. Compute the area of $\mathcal{D}$ as an integral in the variables $u = x + y$, $v = y - 2$.

37. Sketch the domain $\mathcal{D}$ bounded by $y = x^2$, $y = \frac{1}{2}x^2$, and $y = x$. Use a change of variables with the map $x = uv$, $y = u^2$ to calculate

$$\iint_{\mathcal{D}} y^{-1} \, dx \, dy$$

This is an improper integral since $f(x, y) = y^{-1}$ is undefined at $(0, 0)$, but it becomes proper after changing variables.

38. Find an appropriate change of variables to evaluate

$$\iint_{\mathcal{R}} (x + y)^2 e^{x^2 - y^2} \, dx \, dy$$

where $\mathcal{R}$ is the square with vertices $(1, 0)$, $(0, 1)$, $(-1, 0)$, $(0, -1)$.

39. Let G be the inverse of the map $F(x, y) = (xy, x^2 y)$ from the xy-plane to the uv-plane. Let $\mathcal{D}$ be the domain in Figure 18. Show, by applying the Change of Variables Formula to the inverse $G = F^{-1}$, that

$$\iint_{\mathcal{D}} e^{xy} \, dx \, dy = \int_{10}^{20} \int_{20}^{40} e^u v^{-1} \, dv \, du$$

and evaluate this result. *Hint:* See Example 8.

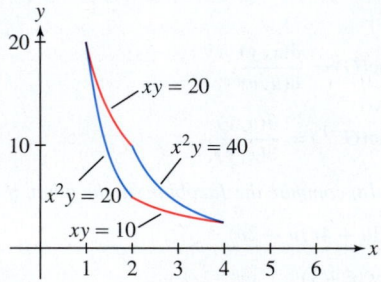

FIGURE 18

40. Sketch the domain

$$\mathcal{D} = \{(x, y) : 1 \leq x + y \leq 4, \, -4 \leq y - 2x \leq 1\}$$

(a) Let F be the map $u = x + y$, $v = y - 2x$ from the xy-plane to the uv-plane, and let G be its inverse. Use Eq. (14) to compute $\text{Jac}(G)$.
(b) Compute $\iint_{\mathcal{D}} e^{x+y} \, dx \, dy$ using the Change of Variables Formula with the map G. *Hint:* It is not necessary to solve for G explicitly.

41. Let $I = \iint_{\mathcal{D}} (x^2 - y^2) \, dx \, dy$, where

$$\mathcal{D} = \{(x, y) : 2 \leq xy \leq 4, 0 \leq x - y \leq 3, x \geq 0, y \geq 0\}$$

(a) Show that the mapping $u = xy$, $v = x - y$ maps $\mathcal{D}$ to the rectangle $\mathcal{R} = [2, 4] \times [0, 3]$.
(b) Compute $\partial(x, y)/\partial(u, v)$ by first computing $\partial(u, v)/\partial(x, y)$.
(c) Use the Change of Variables Formula to show that I is equal to the integral of $f(u, v) = v$ over $\mathcal{R}$ and evaluate.

42. Derive formula (4) in Section 15.4 for integration in cylindrical coordinates from the general Change of Variables Formula.

43. Derive formula (7) in Section 15.4 for integration in spherical coordinates from the general Change of Variables Formula.

44. Use the Change of Variables Formula in three variables to prove that the volume of the ellipsoid $\left(\dfrac{x}{a} \right)^2 + \left(\dfrac{y}{b} \right)^2 + \left(\dfrac{z}{c} \right)^2 = 1$ is equal to $abc \times$ the volume of the unit sphere.

Further Insights and Challenges

45. Use the map

$$x = \frac{\sin u}{\cos v}, \qquad y = \frac{\sin v}{\cos u}$$

to evaluate the integral

$$\int_0^1 \int_0^1 \frac{dx\,dy}{1 - x^2 y^2}$$

This integral is an improper integral since the integrand is infinite if $x = \pm 1$ and $y = \pm 1$, but applying the Change of Variables Formula shows that the result is finite.

46. Verify properties (1) and (2) for linear functions and show that any map satisfying these two properties is linear.

47. Let P and Q be points in $\mathbf{R}^2$. Show that a linear map $G(u, v) = (Au + Cv, Bu + Dv)$ maps the segment joining P and Q to the segment joining $G(P)$ to $G(Q)$. *Hint:* The segment joining P and Q has parametrization

$$(1 - t)\overrightarrow{OP} + t\overrightarrow{OQ} \quad \text{for} \quad 0 \le t \le 1$$

48. [✎] Let G be a linear map. Prove Eq. (6) in the following steps.

(a) For any set $\mathcal{D}$ in the uv-plane and any vector $\mathbf{u}$, let $\mathcal{D} + \mathbf{u}$ be the set obtained by translating all points in $\mathcal{D}$ by $\mathbf{u}$. By linearity, G maps $\mathcal{D} + \mathbf{u}$ to the translate $G(\mathcal{D}) + G(\mathbf{u})$ [Figure 19(C)]. Therefore, if Eq. (6) holds for $\mathcal{D}$, it also holds for $\mathcal{D} + \mathbf{u}$.

(b) In the text, we verified Eq. (6) for the unit rectangle. Use linearity to show that Eq. (6) also holds for all rectangles with vertex at the origin and sides parallel to the axes. Then argue that it also holds for each triangular half of such a rectangle, as in Figure 19(A).

(c) Figure 19(B) shows that the area of a parallelogram is a difference of the areas of rectangles and triangles covered by steps (a) and (b). Use this to prove Eq. (6) for arbitrary parallelograms.

49. The product of 2×2 matrices A and B is the matrix AB defined by

$$\underbrace{\begin{pmatrix} a & b \\ c & d \end{pmatrix}}_{A} \underbrace{\begin{pmatrix} a' & b' \\ c' & d' \end{pmatrix}}_{B} = \underbrace{\begin{pmatrix} aa' + bc' & ab' + bd' \\ ca' + dc' & cb' + dd' \end{pmatrix}}_{AB}$$

The (i, j)-entry of A is the **dot product** of the ith row of A and the jth column of B. Prove that $\det(AB) = \det(A)\det(B)$.

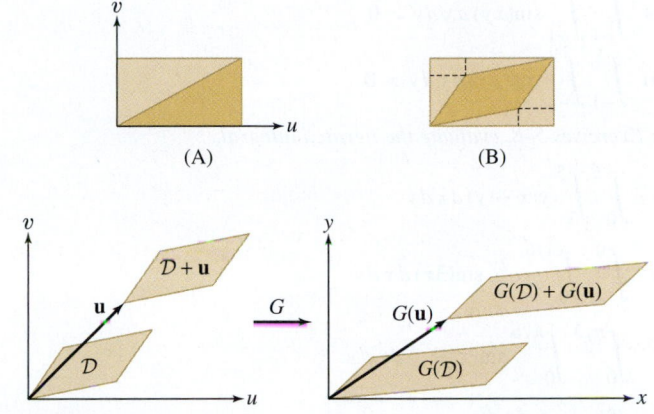

FIGURE 19

50. Let $G_1 : \mathcal{D}_1 \to \mathcal{D}_2$ and $G_2 : \mathcal{D}_2 \to \mathcal{D}_3$ be C^1 maps, and let $G_2 \circ G_1 : \mathcal{D}_1 \to \mathcal{D}_3$ be the composite map. Use the Multivariable Chain Rule and Exercise 49 to show that

$$\mathrm{Jac}(G_2 \circ G_1) = \mathrm{Jac}(G_2)\mathrm{Jac}(G_1)$$

51. Use Exercise 50 to prove that

$$\mathrm{Jac}(G^{-1}) = \mathrm{Jac}(G)^{-1}$$

Hint: Verify that $\mathrm{Jac}(I) = 1$, where I is the identity map $I(u, v) = (u, v)$.

52. Let $(\overline{x}, \overline{y})$ be the centroid of a domain $\mathcal{D}$. For $\lambda > 0$, let $\lambda\mathcal{D}$ be the **dilation** of $\mathcal{D}$, defined by

$$\lambda\mathcal{D} = \{(\lambda x, \lambda y) : (x, y) \in \mathcal{D}\}$$

Use the Change of Variables Formula to prove that the centroid of $\lambda\mathcal{D}$ is $(\lambda\overline{x}, \lambda\overline{y})$.

CHAPTER REVIEW EXERCISES

1. Calculate the Riemann sum $S_{2,3}$ for $\displaystyle\int_1^4 \int_2^6 x^2 y\,dx\,dy$ using two choices of sample points:

(a) Lower-left vertex

(b) Midpoint of rectangle

Then calculate the exact value of the double integral.

2. Let $S_{N,N}$ be the Riemann sum for $\displaystyle\int_0^1 \int_0^1 \cos(xy)\,dx\,dy$ using midpoints as sample points.

(a) Calculate $S_{4,4}$.

(b) [CAS] Use a computer algebra system to calculate $S_{N,N}$ for $N = 10, 50, 100$.

3. Let $\mathcal{D}$ be the shaded domain in Figure 1.

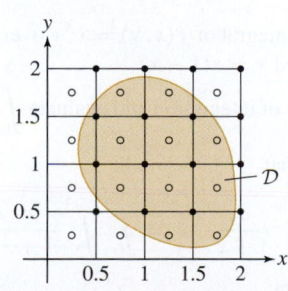

FIGURE 1

Estimate $\displaystyle\iint_{\mathcal{D}} xy\,dA$ by the Riemann sum whose sample points are the midpoints of the squares in the grid.

4. Explain the following:

(a) $\int_{-1}^{1}\int_{-1}^{1}\sin(xy)\,dx\,dy = 0$

(b) $\int_{-1}^{1}\int_{-1}^{1}\cos(xy)\,dx\,dy > 0$

In Exercises 5–8, evaluate the iterated integral.

5. $\int_{0}^{2}\int_{3}^{5}y(x-y)\,dx\,dy$

6. $\int_{1/2}^{0}\int_{0}^{\pi/6}e^{2y}\sin(3x)\,dx\,dy$

7. $\int_{0}^{\pi/3}\int_{0}^{\pi/6}\sin(x+y)\,dx\,dy$

8. $\int_{1}^{2}\int_{1}^{2}\dfrac{y\,dx\,dy}{x+y^2}$

In Exercises 9–14, sketch the domain $\mathcal{D}$ and calculate $\iint_{\mathcal{D}}f(x,y)\,dA$.

9. $\mathcal{D} = \{0 \le x \le 4,\ 0 \le y \le x\},\quad f(x,y) = \cos y$

10. $\mathcal{D} = \{0 \le x \le 2,\ 0 \le y \le 2x - x^2\},\quad f(x,y) = \sqrt{xy}$

11. $\mathcal{D} = \{0 \le x \le 1,\ 1 - x \le y \le 2 - x\},\quad f(x,y) = e^{x+2y}$

12. $\mathcal{D} = \{1 \le x \le 2,\ 0 \le y \le 1/x\},\quad f(x,y) = \cos(xy)$

13. $\mathcal{D} = \{0 \le y \le 1,\ 0.5y^2 \le x \le y^2\},\quad f(x,y) = ye^{1+x}$

14. $\mathcal{D} = \{1 \le y \le e,\ y \le x \le 2y\},\quad f(x,y) = \ln(x+y)$

15. Express $\int_{-3}^{3}\int_{0}^{9-x^2}f(x,y)\,dy\,dx$ as an iterated integral in the order $dx\,dy$.

16. Let $\mathcal{W}$ be the region bounded by the planes $y = z,\ 2y + z = 3$, and $z = 0$ for $0 \le x \le 4$.

(a) Express the triple integral $\iiint_{\mathcal{W}}f(x,y,z)\,dV$ as an iterated integral in the order $dy\,dx\,dz$ (project $\mathcal{W}$ onto the xz-plane).

(b) Evaluate the triple integral for $f(x,y,z) = 1$.

(c) Compute the volume of $\mathcal{W}$ using geometry and check that the result coincides with the answer to (b).

17. Let $\mathcal{D}$ be the domain between $y = x$ and $y = \sqrt{x}$. Calculate $\iint_{\mathcal{D}}xy\,dA$ as an iterated integral in the order $dx\,dy$ and $dy\,dx$.

18. Find the double integral of $f(x,y) = x^3y$ over the region between the curves $y = x^2$ and $y = x(1-x)$.

19. Change the order of integration and evaluate $\int_{0}^{9}\int_{0}^{\sqrt{y}}\dfrac{x\,dx\,dy}{(x^2+y)^{1/2}}$.

20. Verify directly that

$$\int_{2}^{3}\int_{0}^{2}\frac{dy\,dx}{1+x-y} = \int_{0}^{2}\int_{2}^{3}\frac{dx\,dy}{1+x-y}$$

21. Prove the formula

$$\int_{0}^{1}\int_{0}^{y}f(x)\,dx\,dy = \int_{0}^{1}(1-x)f(x)\,dx$$

Then use it to calculate $\int_{0}^{1}\int_{0}^{y}\dfrac{\sin x}{1-x}\,dx\,dy$.

22. Rewrite $\int_{0}^{1}\int_{-\sqrt{1-y^2}}^{\sqrt{1-y^2}}\dfrac{y\,dx\,dy}{(1+x^2+y^2)^2}$ by interchanging the order of integration, and evaluate.

23. Use cylindrical coordinates to compute the volume of the region defined by $4 - x^2 - y^2 \le z \le 10 - 4x^2 - 4y^2$.

24. Evaluate $\iint_{\mathcal{D}}x\,dA$, where $\mathcal{D}$ is the shaded domain in Figure 2.

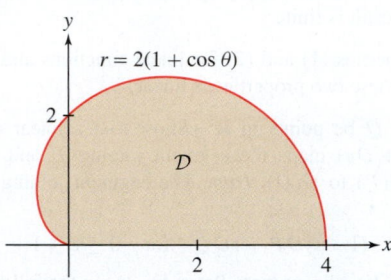

$r = 2(1 + \cos\theta)$

$\mathcal{D}$

FIGURE 2

25. Find the volume of the region between the graph of the function $f(x,y) = 1 - (x^2 + y^2)$ and the xy-plane.

26. Evaluate $\int_{0}^{3}\int_{1}^{4}\int_{2}^{4}(x^3 + y^2 + z)\,dx\,dy\,dz$.

27. Calculate $\iiint_{\mathcal{B}}(xy + z)\,dV$, where

$$\mathcal{B} = \{0 \le x \le 2,\ 0 \le y \le 1,\ 1 \le z \le 3\}$$

as an iterated integral in two different ways.

28. Calculate $\iiint_{\mathcal{W}}xyz\,dV$, where

$$\mathcal{W} = \{0 \le x \le 1,\ x \le y \le 1,\ x \le z \le x + y\}$$

29. Evaluate $I = \int_{-1}^{1}\int_{0}^{\sqrt{1-x^2}}\int_{0}^{1}(x + y + z)\,dz\,dy\,dx$.

30. Describe a region whose volume is equal to:

(a) $\int_{0}^{2\pi}\int_{0}^{\pi/2}\int_{4}^{9}\rho^2\sin\phi\,d\rho\,d\phi\,d\theta$

(b) $\int_{-2}^{1}\int_{\pi/3}^{\pi/4}\int_{0}^{2}r\,dr\,d\theta\,dz$

(c) $\int_{0}^{2\pi}\int_{0}^{3}\int_{-\sqrt{9-r^2}}^{0}r\,dz\,dr\,d\theta$

31. Find the volume of the solid contained in the cylinder $x^2 + y^2 = 1$ below the surface $z = (x + y)^2$ and above the surface $z = -(x - y)^2$.

32. Use polar coordinates to evaluate $\iint_{\mathcal{D}} x \, dA$, where $\mathcal{D}$ is the shaded region between the two circles of radius 1 in Figure 3.

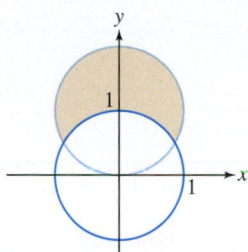

FIGURE 3

33. Use polar coordinates to calculate $\iint_{\mathcal{D}} \sqrt{x^2 + y^2} \, dA$, where $\mathcal{D}$ is the region in the first quadrant bounded by the spiral $r = \theta$, the circle $r = 1$, and the x-axis.

34. Calculate $\iint_{\mathcal{D}} \sin(x^2 + y^2) \, dA$, where

$$\mathcal{D} = \left\{ \frac{\pi}{2} \le x^2 + y^2 \le \pi \right\}$$

35. Express in cylindrical coordinates and evaluate:

$$\int_0^1 \int_0^{\sqrt{1-x^2}} \int_0^{\sqrt{x^2+y^2}} z \, dz \, dy \, dx$$

36. Use spherical coordinates to calculate the triple integral of $f(x, y, z) = x^2 + y^2 + z^2$ over the region

$$1 \le x^2 + y^2 + z^2 \le 4$$

37. Convert to spherical coordinates and evaluate:

$$\int_{-2}^{2} \int_{-\sqrt{4-x^2}}^{\sqrt{4-x^2}} \int_0^{\sqrt{4-x^2-y^2}} e^{-(x^2+y^2+z^2)^{3/2}} \, dz \, dy \, dx$$

38. Find the average value of $f(x, y, z) = xy^2 z^3$ on the box $[0, 1] \times [0, 2] \times [0, 3]$.

39. Let $\mathcal{W}$ be the ball of radius R in $\mathbf{R}^3$ centered at the origin, and let $P = (0, 0, R)$ be the North Pole. Let $d_P(x, y, z)$ be the distance from P to (x, y, z). Show that the average value of d_P over the ball $\mathcal{W}$ is equal to $\bar{d} = 6R/5$. *Hint:* Show that

$$\bar{d} = \frac{1}{\frac{4}{3}\pi R^3} \int_{\theta=0}^{2\pi} \int_{\rho=0}^{R} \int_{\phi=0}^{\pi} \rho^2 \sin\phi \sqrt{R^2 + \rho^2 - 2\rho R \cos\phi} \, d\phi \, d\rho \, d\theta$$

and evaluate.

40. ⊡CAS Express the average value of $f(x, y) = e^{xy}$ over the ellipse $\dfrac{x^2}{2} + y^2 = 1$ as an iterated integral, and evaluate numerically using a computer algebra system.

41. Use cylindrical coordinates to find the mass of the solid bounded by $z = 8 - x^2 - y^2$ and $z = x^2 + y^2$, assuming a mass density of $f(x, y, z) = (x^2 + y^2)^{1/2}$.

42. Let $\mathcal{W}$ be the portion of the half-cylinder $x^2 + y^2 \le 4, x \ge 0$ such that $0 \le z \le 3y$. Use cylindrical coordinates to compute the mass of $\mathcal{W}$ if the mass density is $\rho(x, y, z) = z^2$.

43. Use cylindrical coordinates to find the mass of a cylinder of radius 4 and height 10 if the mass density at a point is equal to the square of the distance from the cylinder's central axis.

44. Find the centroid of the region $\mathcal{W}$ bounded, in spherical coordinates, by $\phi = \phi_0$ and the sphere $\rho = R$.

45. Find the centroid of the solid bounded by the xy-plane, the cylinder $x^2 + y^2 = R^2$, and the plane $x/R + z/H = 1$.

46. Using cylindrical coordinates, prove that the centroid of a right circular cone of height h and radius R is located at height $\frac{h}{4}$ on the central axis.

47. Find the centroid of solid (A) in Figure 4 defined by $x^2 + y^2 \le R^2$, $0 \le z \le H$, and $\frac{\pi}{6} \le \theta \le 2\pi$, where θ is the polar angle of (x, y).

48. Calculate the coordinate y_{CM} of the centroid of solid (B) in Figure 4 defined by $x^2 + y^2 \le 1$ and $0 \le z \le \frac{1}{2}y + \frac{3}{2}$.

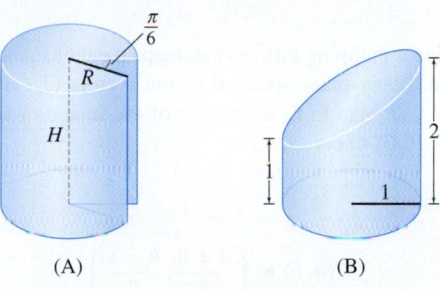

(A)   (B)

FIGURE 4

49. Find the center of mass of the cylinder $x^2 + y^2 \le 1$ for $0 \le z \le 1$, assuming a mass density of $\delta(x, y, z) = z$.

50. Find the center of mass of the sector of central angle $2\theta_0$ (symmetric with respect to the y-axis) in Figure 5, assuming that the mass density is $\delta(x, y) = x^2$.

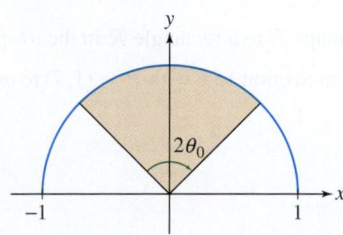

FIGURE 5

51. Find the center of mass of the part of the ball $x^2 + y^2 + z^2 \le 1$, in the first octant assuming a mass density of $\delta(x, y, z) = x$.

52. Find a constant C such that

$$p(x, y) = \begin{cases} C(4x - y + 3) & \text{if } 0 \le x \le 2 \text{ and } 0 \le y \le 3 \\ 0 & \text{otherwise} \end{cases}$$

is a joint probability density function and calculate $P(X \le 1; Y \le 2)$.

53. Calculate $P(3X + 2Y \ge 6)$ for the probability density in Exercise 52.

54. The lifetimes X and Y (in years) of two machine components have joint probability density

$$p(x, y) = \begin{cases} \frac{6}{125}(5 - x - y) & \text{if } 0 \le x \le 5 - y \text{ and } 0 \le y \le 5 \\ 0 & \text{otherwise} \end{cases}$$

What is the probability that both components are still functioning after 2 years?

55. An insurance company issues two kinds of policies: A and B. Let X be the time until the next claim of type A is filed, and let Y be the time (in days) until the next claim of type B is filed. The random variables have joint probability density

$$p(x, y) = 12e^{-4x - 3y}$$

Find the probability that $X \le Y$.

56. Compute the Jacobian of the map

$$G(r, s) = \left(e^r \cosh(s), e^r \sinh(s)\right)$$

57. Find a linear mapping $G(u, v)$ that maps the unit square to the parallelogram in the xy-plane spanned by the vectors $\langle 3, -1 \rangle$ and $\langle 1, 4 \rangle$. Then use the Jacobian to find the area of the image of the rectangle $\mathcal{R} = [0, 4] \times [0, 3]$ under G.

58. Use the map

$$G(u, v) = \left(\frac{u + v}{2}, \frac{u - v}{2}\right)$$

to compute $\iint_{\mathcal{R}} \left((x - y)\sin(x + y)\right)^2 dx\, dy$, where $\mathcal{R}$ is the square with vertices $(\pi, 0)$, $(2\pi, \pi)$, $(\pi, 2\pi)$, and $(0, \pi)$.

59. Let $\mathcal{D}$ be the shaded region in Figure 6, and let F be the map

$$u = y + x^2, \qquad v = y - x^3$$

(a) Show that F maps $\mathcal{D}$ to a rectangle $\mathcal{R}$ in the uv-plane.

(b) Apply Eq. (7) in Section 15.6 with $P = (1, 7)$ to estimate Area($\mathcal{D}$).

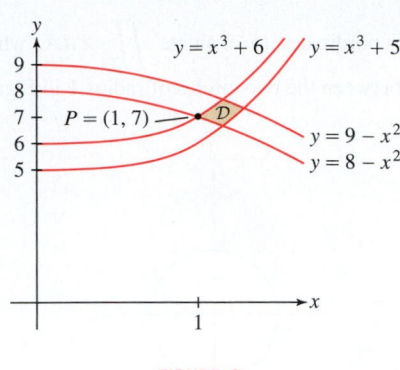

FIGURE 6

60. Calculate the integral of $f(x, y) = e^{3x - 2y}$ over the parallelogram in Figure 7.

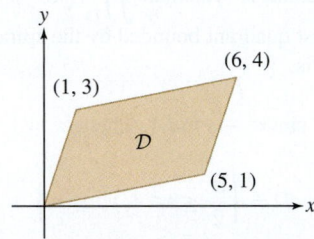

FIGURE 7

61. Sketch the region $\mathcal{D}$ bounded by the curves $y = 2/x$, $y = 1/(2x)$, $y = 2x$, $y = x/2$ in the first quadrant. Let F be the map $u = xy$, $v = y/x$ from the xy-plane to the uv-plane.

(a) Find the image of $\mathcal{D}$ under F.

(b) Let $G = F^{-1}$. Show that $|\text{Jac}(G)| = \dfrac{1}{2|v|}$.

(c) Apply the Change of Variables Formula to prove the formula

$$\iint_{\mathcal{D}} f\left(\frac{y}{x}\right) dx\, dy = \frac{3}{4} \int_{1/2}^{2} \frac{f(v)\, dv}{v}$$

(d) Apply (c) to evaluate $\iint_{\mathcal{D}} \dfrac{ye^{y/x}}{x} dx\, dy$.

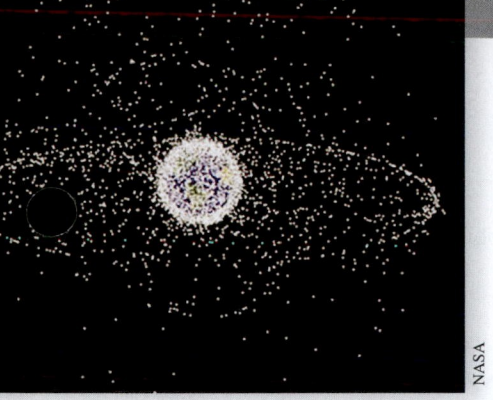

NASA

There are over 20,000 objects larger than a softball orbiting the earth. Some are functioning satellites; most are space "debris" of some sort. The primary force on these objects, holding them in orbit, is Earth's gravity, which is modeled as a field of force vectors everywhere directed inward toward the center of the earth.

16 LINE AND SURFACE INTEGRALS

In the previous chapter, we generalized integration from one variable to several variables. In this chapter, we generalize still further to include integration over curves and surfaces, and we will integrate not just functions but also vector fields. Integrals of vector fields are used in the study of phenomena such as electromagnetism, fluid dynamics, and heat transfer. To lay the groundwork, the chapter begins with a discussion of vector fields.

16.1 Vector Fields

How can we describe a physical object such as the wind, which consists of a large number of molecules moving in a region of space? What we need is a new type of function called a **vector field**. In this case, a vector field **F** assigns to each point $P = (x, y, z)$ a vector $\mathbf{F}(x, y, z)$ that represents the velocity (speed and direction) of the wind at that point (Figure 1). Vector fields describe many other physical phenomena that have magnitude and direction such as force fields, electric fields, and magnetic fields.

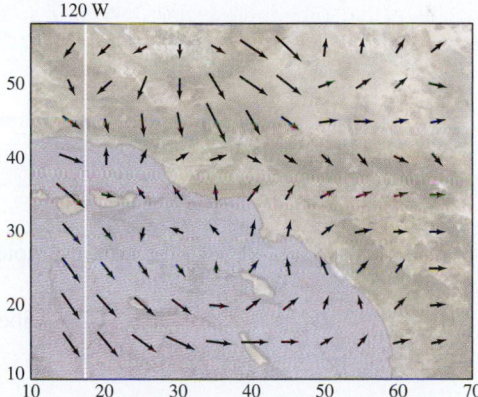

FIGURE 1 Vector field of wind velocity off the coast at Los Angeles.

Mathematically, a vector field in $\mathbf{R}^3$ is represented by a vector whose components are functions:

$$\mathbf{F}(x, y, z) = \langle F_1(x, y, z), F_2(x, y, z), F_3(x, y, z) \rangle$$

To each point $P = (a, b, c)$ is associated the vector $\mathbf{F}(a, b, c)$, which we also denote by $\mathbf{F}(P)$. Alternatively,

$$\mathbf{F} = F_1\mathbf{i} + F_2\mathbf{j} + F_3\mathbf{k}$$

When drawing a vector field, we draw $\mathbf{F}(P)$ as a vector based at P. The **domain** of **F** is the set of points P for which $\mathbf{F}(P)$ is defined. Vector fields in the plane are written in a similar fashion:

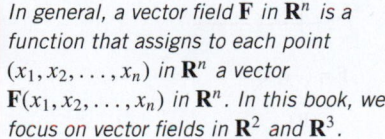

*In general, a vector field **F** in **R**^n^ is a function that assigns to each point $(x_1, x_2, \ldots, x_n)$ in **R**^n^ a vector $\mathbf{F}(x_1, x_2, \ldots, x_n)$ in **R**^n^. In this book, we focus on vector fields in **R**^2^ and **R**^3^.*

$$\mathbf{F}(x, y) = \langle F_1(x, y), F_2(x, y) \rangle = F_1\mathbf{i} + F_2\mathbf{j}$$

Throughout this chapter, we assume that the component functions F_j are smooth—that is, they have partial derivatives of all orders on their domains.

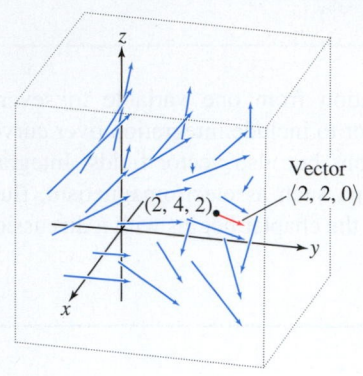

FIGURE 2

EXAMPLE 1 Which vector corresponds to the point $P = (2, 4, 2)$ for the vector field $\mathbf{F}(x, y, z) = \langle y - z, x, z - \sqrt{y} \rangle$?

Solution The vector attached to P is

$$\mathbf{F}(2, 4, 2) = \langle 4 - 2, 2, 2 - \sqrt{4} \rangle = \langle 2, 2, 0 \rangle$$

Some vectors from the vector field are shown in Figure 2, and $\mathbf{F}(2, 4, 2)$ is in red. ■

Although it is not practical to sketch complicated vector fields in three dimensions by hand, computer graphing tools can produce useful visual representations (Figure 3). The vector field in Figure 3(B) is an example of a **constant vector field**. It assigns the same vector $\langle 1, -1, 3 \rangle$ to every point in $\mathbf{R}^3$.

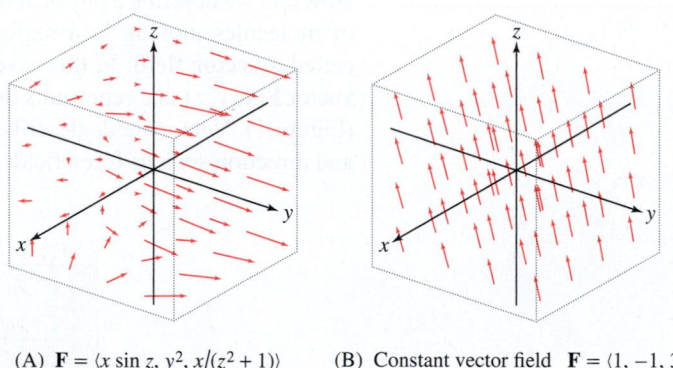

(A) $\mathbf{F} = \langle x \sin z, y^2, x/(z^2 + 1) \rangle$ (B) Constant vector field $\mathbf{F} = \langle 1, -1, 3 \rangle$

FIGURE 3

In the next example, we analyze two vector fields in the plane qualitatively.

EXAMPLE 2 Describe the following vector fields in $\mathbf{R}^2$:

(a) $\mathbf{G} = \mathbf{i} + x\mathbf{j}$ **(b)** $\mathbf{F} = \langle -y, x \rangle$

Solution **(a)** The vector field $\mathbf{G} = \mathbf{i} + x\mathbf{j}$ assigns the vector $\langle 1, a \rangle$ to the point (a, b). In particular, it assigns the same vector to all points with the same x-coordinate [Figure 4(A)]. Notice that $\langle 1, a \rangle$ has slope a and length $\sqrt{1 + a^2}$. We may describe $\mathbf{G}$ as follows: $\mathbf{G}$ assigns a vector of slope a and length $\sqrt{1 + a^2}$ to all points with $x = a$.
(b) To visualize $\mathbf{F}$, observe that $\mathbf{F}(a, b) = \langle -b, a \rangle$ has length $r = \sqrt{a^2 + b^2}$. It is perpendicular to the radial vector $\langle a, b \rangle$ and points counterclockwise. Thus, $\mathbf{F}$ has the following description: The vectors along the circle of radius r all have length r and they are tangent to the circle, pointing counterclockwise [Figure 4(B)]. ■

The English physicist and Nobel laureate Paul Dirac (1902–1984) introduced a generalization of vectors called "spinors" to unify the special theory of relativity with quantum mechanics. This led to the discovery of the positron, an elementary particle used today in PET-scan imaging.

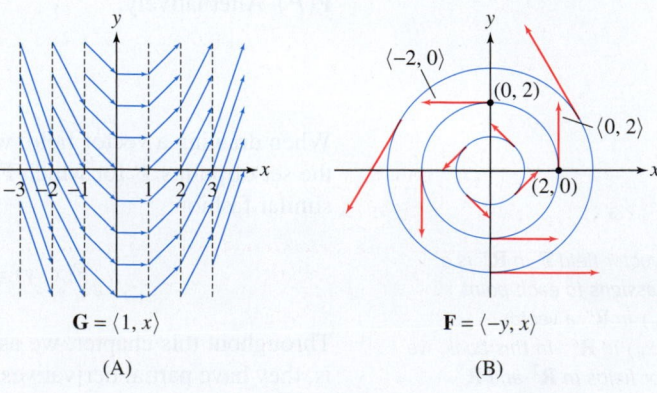

$\mathbf{G} = \langle 1, x \rangle$

(A)

$\mathbf{F} = \langle -y, x \rangle$

(B)

DF **FIGURE 4**

A **unit vector field** is a vector field $\mathbf{F}$ such that $\|\mathbf{F}(P)\| = 1$ for all points P. A vector field $\mathbf{F}$ is called a **radial vector field** if $\mathbf{F}(P)$ is parallel to $\overrightarrow{OP}$ and $\|\mathbf{F}(P)\|$ depends only on the distance r from P to the origin. Here, we use the notation $r = (x^2 + y^2)^{1/2}$ for $\mathbf{R}^2$ and $r = (x^2 + y^2 + z^2)^{1/2}$ for $\mathbf{R}^3$. Two important vector fields are the unit radial vector fields in two and three dimensions [Figures 5(A) and (B)]:

$$\mathbf{e}_r = \left\langle \frac{x}{r}, \frac{y}{r} \right\rangle = \left\langle \frac{x}{\sqrt{x^2 + y^2}}, \frac{y}{\sqrt{x^2 + y^2}} \right\rangle \qquad \boxed{1}$$

$$\mathbf{e}_r = \left\langle \frac{x}{r}, \frac{y}{r}, \frac{z}{r} \right\rangle = \left\langle \frac{x}{\sqrt{x^2 + y^2 + z^2}}, \frac{y}{\sqrt{x^2 + y^2 + z^2}}, \frac{z}{\sqrt{x^2 + y^2 + z^2}} \right\rangle \qquad \boxed{2}$$

A gravitational vector field for a point mass and an electrostatic vector field for a point charge are radial vector fields. They can be conveniently expressed in the form $f(r)\mathbf{e}_r$ with a scalar function f. We work with such fields often in the remainder of the text.

Observe that $\mathbf{e}_r(P)$ is a unit vector pointing away from the origin at P. Note, however, that $\mathbf{e}_r$ is not defined at the origin where $r = 0$.

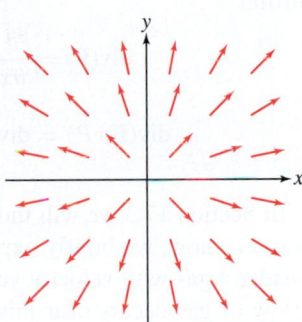

(A) Unit radial vector field in the plane
$\mathbf{e}_r = \langle x/r, y/r \rangle$

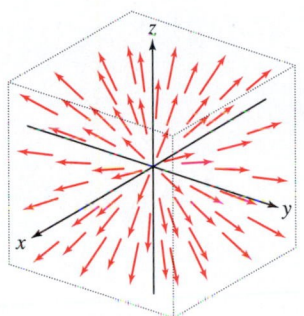

(B) Unit radial vector field in 3-space
$\mathbf{e}_r = \langle x/r, y/r, z/r \rangle$

FIGURE 5

Operations on Vector Fields

Three important derivative operations in multivariable calculus are the gradient, divergence, and curl. Gradient was defined in Section 14.5. Divergence and curl are operations on vector fields that we introduce here. Each of the three operations is defined using the **del operator** ∇, which is a vector of derivative operators:

$$\nabla = \left\langle \frac{\partial}{\partial x}, \frac{\partial}{\partial y}, \frac{\partial}{\partial z} \right\rangle$$

The operation of ∇ on a scalar function f produces the gradient of f. Notationally, we treat this operation like multiplication of a vector by a scalar, but the resulting vector components are derivative operations on functions rather than products:

$$\nabla f = \left\langle \frac{\partial}{\partial x}, \frac{\partial}{\partial y}, \frac{\partial}{\partial z} \right\rangle f = \left\langle \frac{\partial}{\partial x} f, \frac{\partial}{\partial y} f, \frac{\partial}{\partial z} f \right\rangle = \left\langle \frac{\partial f}{\partial x}, \frac{\partial f}{\partial y}, \frac{\partial f}{\partial z} \right\rangle$$

Operations of ∇ on a vector field $\mathbf{F}$ are expressed via dot product (producing divergence) and cross product (producing curl). We introduce divergence and curl briefly here. They will play a significant role in the next chapter. For a vector field $\mathbf{F} = \langle F_1, F_2, F_3 \rangle$, we define the **divergence** of $\mathbf{F}$, denoted $\mathrm{div}(\mathbf{F})$, by

$$\mathrm{div}(\mathbf{F}) = \nabla \cdot \mathbf{F} = \left\langle \frac{\partial}{\partial x}, \frac{\partial}{\partial y}, \frac{\partial}{\partial z} \right\rangle \cdot \langle F_1, F_2, F_3 \rangle = \frac{\partial F_1}{\partial x} + \frac{\partial F_2}{\partial y} + \frac{\partial F_3}{\partial z}$$

That is,

$$\text{div}(\mathbf{F}) = \frac{\partial F_1}{\partial x} + \frac{\partial F_2}{\partial y} + \frac{\partial F_3}{\partial z}$$ **3**

Note that the divergence of a vector field $\mathbf{F}(x, y, z)$ is a scalar function of x, y, and z. Divergence obeys the **linearity** rules:

$$\text{div}(\mathbf{F} + \mathbf{G}) = \text{div}(\mathbf{F}) + \text{div}(\mathbf{G})$$

$$\text{div}(c\mathbf{F}) = c\,\text{div}(\mathbf{F}) \qquad (c \text{ any constant})$$

EXAMPLE 3 Evaluate the divergence of $\mathbf{F} = \langle e^{xy}, xy, z^4 \rangle$ at $P = (1, 0, 2)$.

Solution

$$\text{div}(\mathbf{F}) = \frac{\partial}{\partial x} e^{xy} + \frac{\partial}{\partial y} xy + \frac{\partial}{\partial z} z^4 = ye^{xy} + x + 4z^3$$

$$\text{div}(\mathbf{F})(P) = \text{div}(\mathbf{F})(1, 0, 2) = 0 \cdot e^0 + 1 + 4 \cdot 2^3 = 33 \qquad \blacksquare$$

In Section 17.3 we will thoroughly investigate divergence and its physical interpretation. For now, we briefly explore its meaning in the context of a physics application. Consider a gas with velocity vector field given by $\mathbf{F}$. When $\text{div}(\mathbf{F}) > 0$ at a point P, an outflow of gas occurs near this point. In other words, the gas is expanding around th point, as might occur when the gas is heated. When $\text{div}(\mathbf{F}) < 0$ at a point P, the gas i compressing toward P, as might occur when the gas is cooled. When $\text{div}(\mathbf{F}) = 0$, the gas is neither compressing nor expanding near P.

For example, the vector field $\mathbf{F} = \langle x, y, z \rangle$, appearing in Figure 6(A), has $\text{div}(\mathbf{F}) = 3$ everywhere. Thinking of this as the velocity vector field for a gas, at every point, the gas is expanding. This is most obvious at the origin, but even at other points, the gas is expanding in the sense that more gas atoms are moving away from the point than are moving toward it. We say that each of these points is a **source**.

For the vector field $\mathbf{F} = \langle -x, -y, -z \rangle$ appearing in Figure 6(B), $\text{div}(\mathbf{F}) = -3$ for all points P, and the gas is compressing at every point. We say that every point is a **sink**.

For the vector field $\mathbf{F} = \langle 0, 1, 0 \rangle$, appearing in Figure 6(C), $\text{div}(\mathbf{F}) = 0$. At each point, the gas is neither expanding nor compressing. Rather, it is simply shifting in the positive y-direction. In this case, no points are sources or sinks and we say that the vector field is **incompressible**. For other vector fields, there can be points that are sources, points that are sinks, and points that are neither.

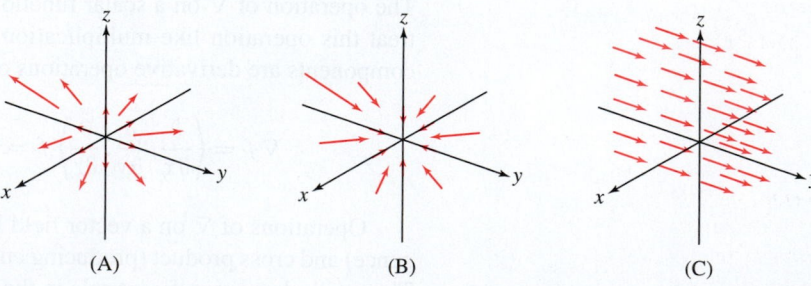

(A)   (B)   (C)

FIGURE 6

The other operation involving ∇ and vector fields $\mathbf{F} = \langle F_1, F_2, F_3 \rangle$ is the **curl** of $\mathbf{F}$, denoted $\mathrm{curl}(\mathbf{F})$ and defined using the cross product as follows:

$$
\mathrm{curl}(\mathbf{F}) = \nabla \times \mathbf{F} =
\begin{vmatrix}
\mathbf{i} & \mathbf{j} & \mathbf{k} \\
\dfrac{\partial}{\partial x} & \dfrac{\partial}{\partial y} & \dfrac{\partial}{\partial z} \\
F_1 & F_2 & F_3
\end{vmatrix}
$$

$$
= \left(\frac{\partial F_3}{\partial y} - \frac{\partial F_2}{\partial z} \right) \mathbf{i} - \left(\frac{\partial F_3}{\partial x} - \frac{\partial F_1}{\partial z} \right) \mathbf{j} + \left(\frac{\partial F_2}{\partial x} - \frac{\partial F_1}{\partial y} \right) \mathbf{k}
$$

That is,

$$
\mathrm{curl}(\mathbf{F}) = \left\langle \frac{\partial F_3}{\partial y} - \frac{\partial F_2}{\partial z}, \; \frac{\partial F_1}{\partial z} - \frac{\partial F_3}{\partial x}, \; \frac{\partial F_2}{\partial x} - \frac{\partial F_1}{\partial y} \right\rangle
$$

Note that, in contrast to divergence, the curl of a vector field is itself a vector field. It is straightforward to check that curl obeys the **linearity** rules:

$$
\mathrm{curl}(\mathbf{F} + \mathbf{G}) = \mathrm{curl}(\mathbf{F}) + \mathrm{curl}(\mathbf{G})
$$

$$
\mathrm{curl}(c\mathbf{F}) = c \, \mathrm{curl}(\mathbf{F}) \qquad (c \text{ any constant})
$$

EXAMPLE 4   **Calculating the Curl**   Calculate the curl of $\mathbf{F} = \langle xy, e^x, y + z \rangle$.

Solution   We compute the curl as the determinant:

$$
\mathrm{curl}(\mathbf{F}) =
\begin{vmatrix}
\mathbf{i} & \mathbf{j} & \mathbf{k} \\
\dfrac{\partial}{\partial x} & \dfrac{\partial}{\partial y} & \dfrac{\partial}{\partial z} \\
xy & e^x & y + z
\end{vmatrix}
$$

$$
= \left(\frac{\partial}{\partial y}(y + z) - \frac{\partial}{\partial z}e^x \right) \mathbf{i} - \left(\frac{\partial}{\partial x}(y + z) - \frac{\partial}{\partial z}xy \right) \mathbf{j} + \left(\frac{\partial}{\partial x}e^x - \frac{\partial}{\partial y}xy \right) \mathbf{k}
$$

$$
= \mathbf{i} + (e^x - x)\mathbf{k} \qquad\qquad \blacksquare
$$

The magnitude of the vector $\mathrm{curl}\,(\mathbf{F})(P)$ is a measure of how fast the vector field $\mathbf{F}$, when considered as the velocity vector field of a fluid flow, would turn a paddle wheel inserted into the fluid as in Figure 7. The direction of $\mathrm{curl}\,(\mathbf{F})(P)$ is the direction of the paddle-wheel axis at P that results in a maximal rate of rotation of the paddle wheel. The magnitude of $\mathrm{curl}\,(\mathbf{F})(P)$ is that maximum rate of rotation. If $\mathrm{curl}(\mathbf{F}) = \mathbf{0}$, then the vector field $\mathbf{F}$ is said to be **irrotational**. We examine these interpretations of $\mathrm{curl}(\mathbf{F})$ further in Sections 17.1 and 17.2, where we investigate the physical significance of the curl.

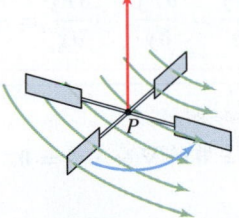

curl $\mathbf{F}(P)$

DF **FIGURE 7**   curl $\mathbf{F}(P)$ tells us about the rotation of the fluid.

Conservative Vector Fields

Vector fields that can be expressed as the gradient of a scalar function are important in multivariable calculus and its applications. A vector field $\mathbf{F}$ is called **conservative** if there is a differentiable function $f(x, y, z)$ such that

- The term "conservative" comes from physics and the Law of Conservation of Energy (see Section 16.3).
- Any letter can be used to denote a potential function. We use f. Some textbooks use V, which suggests "volt," the unit of electric potential. Others use $\phi(x, y, z)$ or $U(x, y, z)$.

$$
\mathbf{F} = \nabla f = \left\langle \frac{\partial f}{\partial x}, \frac{\partial f}{\partial y}, \frac{\partial f}{\partial z} \right\rangle
$$

The function f is called a **potential function** (or scalar potential function) for $\mathbf{F}$.

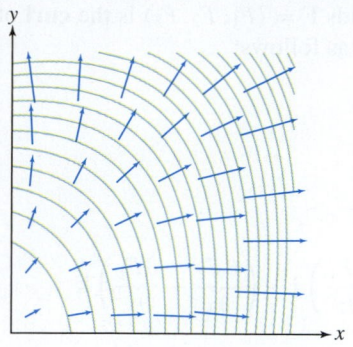

FIGURE 8 A conservative vector field is the gradient of a potential function and therefore is orthogonal to the potential function's level curves.

The same terms apply in two variables and, more generally, in n variables. Recall that the gradient vectors are orthogonal to the level curves, and thus in a conservative vector field, the vector at every point P is orthogonal to the level curve of a potential function through P (Figure 8). Conservative vector fields have critically important properties. For instance, in Section 16.3, we will see that the work done by a conservative vector field, as a particle travels from one point to another, is independent of the path taken. In physics, conservative vector fields appear naturally as force fields corresponding to physical systems in which energy is conserved.

EXAMPLE 5 Verify that $f(x, y, z) = xy + yz^2$ is a potential function for the vector field $\mathbf{F} = \langle y, x + z^2, 2yz \rangle$.

Solution We compute the gradient of f:

$$\frac{\partial f}{\partial x} = y, \qquad \frac{\partial f}{\partial y} = x + z^2, \qquad \frac{\partial f}{\partial z} = 2yz$$

Thus, $\nabla f = \langle y, x + z^2, 2yz \rangle = \mathbf{F}$ as claimed. ∎

Next, we show the important fact that the curl of a conservative vector field is the trivial vector field $\mathbf{0}$.

THEOREM 1   Curl of a Conservative Vector Field

1. In $\mathbf{R}^2$, if the vector field $\mathbf{F} = \langle F_1, F_2 \rangle$ is conservative, then

$$\frac{\partial F_1}{\partial y} = \frac{\partial F_2}{\partial x}$$

2. In $\mathbf{R}^3$, if the vector field $\mathbf{F} = \langle F_1, F_2, F_3 \rangle$ is conservative, then

$$\text{curl}(\mathbf{F}) = \mathbf{0}, \quad \text{or equivalently,} \quad \frac{\partial F_1}{\partial y} = \frac{\partial F_2}{\partial x}, \quad \frac{\partial F_2}{\partial z} = \frac{\partial F_3}{\partial y}, \quad \frac{\partial F_3}{\partial x} = \frac{\partial F_1}{\partial z}$$

We refer to the partial derivatives of the vector-field component functions appearing in Theorem 1 as **cross-partial derivatives**. Also, we refer to the individual equations in the theorem as the **cross-partials equations** and the highlighted necessary conditions for the vector field to be conservative as the **cross-partials conditions**.

Note that we could also write this result as $\text{curl}(\nabla f) = \mathbf{0}$ or $\nabla \times \nabla f = \mathbf{0}$.

Proof We provide the proof for a vector field in $\mathbf{R}^3$, but the same idea works for a vector field in $\mathbf{R}^2$. If $\mathbf{F} = \nabla f$, then

$$F_1 = \frac{\partial f}{\partial x}, \qquad F_2 = \frac{\partial f}{\partial y}, \qquad F_3 = \frac{\partial f}{\partial z}$$

Now compare the cross-partial derivatives:

$$\frac{\partial F_1}{\partial y} = \frac{\partial}{\partial y}\left(\frac{\partial f}{\partial x}\right) = \frac{\partial^2 f}{\partial y \partial x}$$

$$\frac{\partial F_2}{\partial x} = \frac{\partial}{\partial x}\left(\frac{\partial f}{\partial y}\right) = \frac{\partial^2 f}{\partial x \partial y}$$

Clairaut's Theorem (Section 14.3) tells us that $\dfrac{\partial^2 f}{\partial y \, \partial x} = \dfrac{\partial^2 f}{\partial x \, \partial y}$, and thus

$$\frac{\partial F_1}{\partial y} = \frac{\partial F_2}{\partial x}$$

Similarly, $\dfrac{\partial F_2}{\partial z} = \dfrac{\partial F_3}{\partial y}$ and $\dfrac{\partial F_3}{\partial x} = \dfrac{\partial F_1}{\partial z}$. It follows that $\operatorname{curl}(\mathbf{F}) = \mathbf{0}$. ∎

From Theorem 1, we can see that most vector fields are *not* conservative. Indeed, an arbitrary triple of functions $\langle F_1, F_2, F_3 \rangle$ does not satisfy the cross-partials condition. Here is an example.

EXAMPLE 6 Show that $\mathbf{F} = \left\langle xy, \dfrac{x^2}{2}, zy \right\rangle$ is not conservative.

Solution Since $\mathbf{F}$ is a vector field in $\mathbf{R}^3$, we must show that at least one of the three cross-partials equations in the second part of Theorem 1 is *not* satisfied. The first one is satisfied because $\dfrac{\partial F_1}{\partial y} = \dfrac{\partial F_2}{\partial x} = x$. Checking the second:

$$\frac{\partial F_2}{\partial z} = \frac{\partial}{\partial z}\left(\frac{x^2}{2}\right) = 0, \qquad \frac{\partial F_3}{\partial y} = \frac{\partial}{\partial y}(zy) = z$$

Thus, $\dfrac{\partial F_2}{\partial z} \neq \dfrac{\partial F_3}{\partial y}$, and Theorem 1 implies that $\mathbf{F}$ is not conservative. ∎

In this example, all we needed to do to prove that $\mathbf{F}$ is not conservative was show that just one of the cross-partials equations was not satisfied. While two of the cross-partials equations *are* satisfied for this $\mathbf{F}$, showing $\dfrac{\partial F_2}{\partial z} \neq \dfrac{\partial F_3}{\partial y}$ was enough to establish that $\mathbf{F}$ is not conservative.

Potential functions, like antiderivatives in one variable, are unique to within an additive constant. To state this precisely, we must assume that the domain $\mathcal{D}$ of the vector field is open and connected.

> **THEOREM 2 Uniqueness of Potential Functions** If $\mathbf{F}$ is conservative on an open connected domain, then any two potential functions of $\mathbf{F}$ differ by a constant.

Proof If both f_1 and f_2 are potential functions of $\mathbf{F}$, then

$$\nabla(f_1 - f_2) = \nabla f_1 - \nabla f_2 = \mathbf{F} - \mathbf{F} = \mathbf{0}$$

However, a function whose gradient is zero on an open connected domain is a constant function (this generalizes the fact from single-variable calculus that a function on an interval with zero derivative is a constant function—see Exercise 57). Thus, $f_1 - f_2 = C$ for some constant C, and hence $f_1 = f_2 + C$. ∎

The next two examples consider two important radial vector fields.

EXAMPLE 7 **Unit Radial Vector Fields** Show that

$$f(x, y, z) = r = \sqrt{x^2 + y^2 + z^2}$$

is a potential function for the unit radial vector field $\mathbf{e}_r = \left\langle \dfrac{x}{r}, \dfrac{y}{r}, \dfrac{z}{r} \right\rangle$. That is, $\mathbf{e}_r = \nabla r$.

Solution We have

$$\frac{\partial r}{\partial x} = \frac{\partial}{\partial x}\sqrt{x^2 + y^2 + z^2} = \frac{x}{\sqrt{x^2 + y^2 + z^2}} = \frac{x}{r}$$

Similarly, $\frac{\partial r}{\partial y} = \frac{y}{r}$ and $\frac{\partial r}{\partial z} = \frac{z}{r}$. Therefore, $\nabla r = \left\langle \frac{x}{r}, \frac{y}{r}, \frac{z}{r} \right\rangle = \mathbf{e}_r$.

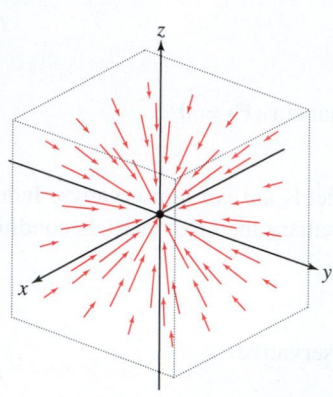

FIGURE 9 The vector field $-\dfrac{Gm\mathbf{e}_r}{r^2}$ represents the force of gravitational attraction due to a point mass located at the origin.

The gravitational force exerted by a point mass m is described by an inverse-square force field (Figure 9) whose magnitude is inversely proportional to the square of the distance from the mass. A point mass located at the origin exerts a gravitational force $\mathbf{F}$ on a unit mass located at (x, y, z) equal to

$$\mathbf{F} = -\frac{Gm}{r^2}\mathbf{e}_r = -Gm\left\langle \frac{x}{r^3}, \frac{y}{r^3}, \frac{z}{r^3} \right\rangle$$

where G is the universal gravitation constant. The negative sign indicates that the force is attractive (it pulls in the direction of the point mass at the origin). The electrostatic force field due to a charged particle is also an inverse-square vector field. The next example shows that these vector fields are conservative.

EXAMPLE 8 **Inverse-Square Vector Field** Show that

$$\frac{\mathbf{e}_r}{r^2} = \nabla\left(\frac{-1}{r}\right)$$

← **REMINDER** The Chain Rule for Gradients:

$$\nabla(f(r(x, y, z))) = f'(r(x, y, z))\nabla r$$

Solution Use the Chain Rule for Gradients (Theorem 1 in Section 14.5) and Example 7:

$$\nabla(-r^{-1}) = r^{-2}\nabla r = r^{-2}\mathbf{e}_r$$

16.1 SUMMARY

- A *vector field* assigns a vector to each point in a domain. A vector field in $\mathbf{R}^3$ is represented by a triple of functions $\mathbf{F} = \langle F_1, F_2, F_3 \rangle$. A vector field in $\mathbf{R}^2$ is represented by a pair of functions $\mathbf{F} = \langle F_1, F_2 \rangle$. We always assume that the components F_j are smooth functions on their domains.

- The del operator $\nabla = \left\langle \dfrac{\partial}{\partial x}, \dfrac{\partial}{\partial y}, \dfrac{\partial}{\partial z} \right\rangle$ is used to define gradient (∇f), divergence ($\nabla \cdot \mathbf{F}$), and curl ($\nabla \times \mathbf{F}$).

- The *divergence* of a vector field $\mathbf{F} = \langle F_1, F_2, F_3 \rangle$ is the scalar function given by

$$\text{div}(\mathbf{F}) = \nabla \cdot \mathbf{F} = \frac{\partial F_1}{\partial x} + \frac{\partial F_2}{\partial y} + \frac{\partial F_3}{\partial z}$$

- The *curl* of a vector field $\mathbf{F} = \langle F_1, F_2, F_3 \rangle$ is the vector field given by

$$\text{curl}(\mathbf{F}) = \nabla \times \mathbf{F} = \left(\frac{\partial F_3}{\partial y} - \frac{\partial F_2}{\partial z}\right)\mathbf{i} - \left(\frac{\partial F_3}{\partial x} - \frac{\partial F_1}{\partial z}\right)\mathbf{j} + \left(\frac{\partial F_2}{\partial x} - \frac{\partial F_1}{\partial y}\right)\mathbf{k}$$

- If $\mathbf{F} = \nabla f$, then $\mathbf{F}$ is called *conservative* and f is called a *potential function* for $\mathbf{F}$.
- Any two potential functions for a conservative vector field differ by a constant (on an open, connected domain).

- A conservative vector field $\mathbf{F} = \langle F_1, F_2, F_3 \rangle$ satisfies the condition

$$\operatorname{curl}(\mathbf{F}) = \mathbf{0}, \quad \text{or equivalently,} \quad \frac{\partial F_1}{\partial y} = \frac{\partial F_2}{\partial x}, \quad \frac{\partial F_2}{\partial z} = \frac{\partial F_3}{\partial y}, \quad \frac{\partial F_3}{\partial x} = \frac{\partial F_1}{\partial z}$$

- The radial unit vector field and the inverse-square vector field are conservative:

$$\mathbf{e}_r = \left\langle \frac{x}{r}, \frac{y}{r}, \frac{z}{r} \right\rangle = \nabla r, \quad \frac{\mathbf{e}_r}{r^2} = \left\langle \frac{x}{r^3}, \frac{y}{r^3}, \frac{z}{r^3} \right\rangle = \nabla(-r^{-1}), \quad \text{where} \quad r = \sqrt{x^2 + y^2 + z^2}$$

16.1 EXERCISES

Preliminary Questions

1. Which of the following is a unit vector field in the plane?

(a) $\mathbf{F} = \langle y, x \rangle$

(b) $\mathbf{F} = \left\langle \dfrac{y}{\sqrt{x^2 + y^2}}, \dfrac{x}{\sqrt{x^2 + y^2}} \right\rangle$

(c) $\mathbf{F} = \left\langle \dfrac{y}{x^2 + y^2}, \dfrac{x}{x^2 + y^2} \right\rangle$

2. Sketch an example of a nonconstant vector field in the plane in which each vector is parallel to $\langle 1, 1 \rangle$.

3. Show that the vector field $\mathbf{F} = \langle -z, 0, x \rangle$ is orthogonal to the position vector $\overrightarrow{OP}$ at each point P. Give an example of another vector field with this property.

4. Show that $f(x, y, z) = xyz$ is a potential function for $\langle yz, xz, xy \rangle$ and give an example of a potential function other than f.

Exercises

1. Compute and sketch the vector assigned to the points $P = (1, 2)$ and $Q = (-1, -1)$ by the vector field $\mathbf{F} = \langle x^2, x \rangle$.

2. Compute and sketch the vector assigned to the points $P = (1, 2)$ and $Q = (-1, -1)$ by the vector field $\mathbf{F} = \langle -y, x \rangle$.

3. Compute and sketch the vector assigned to the points $P = (0, 1, 1)$ and $Q = (2, 1, 0)$ by the vector field $\mathbf{F} = \langle xy, z^2, x \rangle$.

4. Compute the vector assigned to the points $P = (1, 1, 0)$ and $Q = (2, 1, 2)$ by the vector fields $\mathbf{e}_r$, $\dfrac{\mathbf{e}_r}{r}$, and $\dfrac{\mathbf{e}_r}{r^2}$.

In Exercises 5–12, sketch the following planar vector fields by drawing the vectors attached to points with integer coordinates in the rectangle $-3 \le x \le 3$, $-3 \le y \le 3$. Instead of drawing the vectors with their true lengths, scale them if necessary to avoid overlap.

5. $\mathbf{F} = \langle 1, 0 \rangle$

6. $\mathbf{F} = \langle 1, 1 \rangle$

7. $\mathbf{F} = x\mathbf{i}$

8. $\mathbf{F} = y\mathbf{i}$

9. $\mathbf{F} = \langle 0, x \rangle$

10. $\mathbf{F} = x^2\mathbf{i} + y\mathbf{j}$

11. $\mathbf{F} = \left\langle \dfrac{x}{x^2 + y^2}, \dfrac{y}{x^2 + y^2} \right\rangle$

12. $\mathbf{F} = \left\langle \dfrac{-y}{\sqrt{x^2 + y^2}}, \dfrac{x}{\sqrt{x^2 + y^2}} \right\rangle$

In Exercises 13–16, match each of the following planar vector fields with the corresponding plot in Figure 10.

13. $\mathbf{F} = \langle 2, x \rangle$

14. $\mathbf{F} = \langle 2x + 2, y \rangle$

15. $\mathbf{F} = \langle y, \cos x \rangle$

16. $\mathbf{F} = \langle x + y, x - y \rangle$

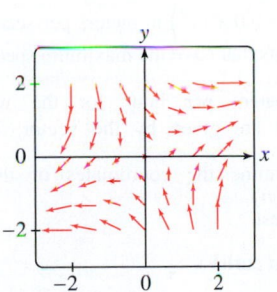
(A)

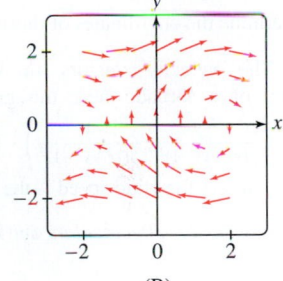
(B)

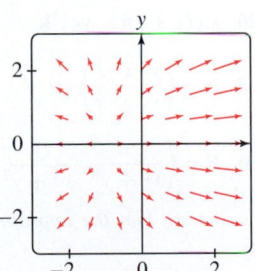
(C)

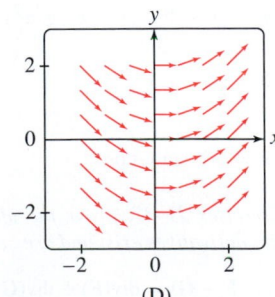
(D)

FIGURE 10

In Exercises 17–20, match each three-dimensional vector field with the corresponding plot in Figure 11.

17. $\mathbf{F} = \langle 1, 1, 1 \rangle$

18. $\mathbf{F} = \langle x, 0, z \rangle$

19. $\mathbf{F} = \langle x, y, z \rangle$

20. $\mathbf{F} = \mathbf{e}_r$

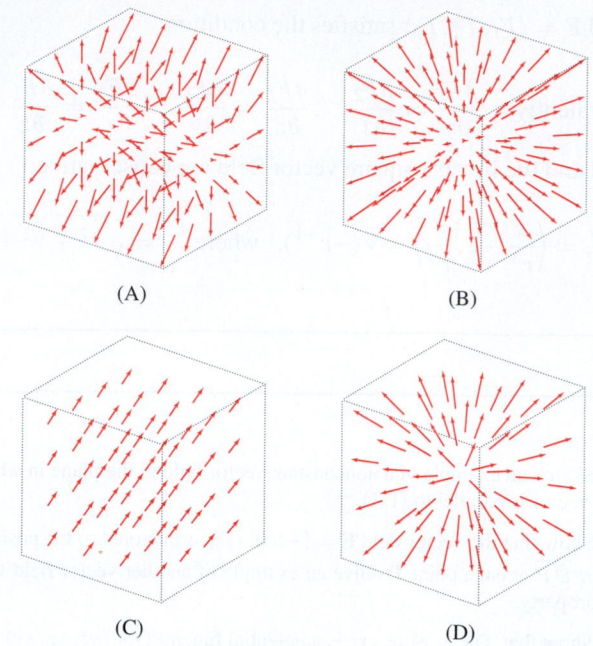

(A)

(B)

(C)

(D)

FIGURE 11

21. A river 200 meters wide is modeled by the region in the xy-plane given by $-100 \le x \le 100$. The velocity vector field on the surface of the river is given by $\mathbf{F} = \langle -0.05x, 20 - 0.0001x^2 \rangle$ in meters per second. Determine the coordinates of those points that have the maximum speed.

22. The velocity vectors in kilometers per hour for the wind speed of a tornado near the ground are given by the vector field $\mathbf{F} = \left\langle \dfrac{-y}{e^{(x^2+y^2-1)^2}}, \dfrac{x}{e^{(x^2+y^2-1)^2}} \right\rangle$. Determine the coordinates of those points where the wind speed is the highest.

In Exercises 23–30, calculate div(**F**) *and* curl(**F**).

23. $\mathbf{F} = \langle x, y, z \rangle$

24. $\mathbf{F} = \langle y, z, x \rangle$

25. $\mathbf{F} = \langle x - 2zx^2, z - xy, z^2x^2 \rangle$

26. $\sin(x+z)\mathbf{i} - ye^{xz}\mathbf{k}$

27. $\mathbf{F} = \langle yz, xz, xy \rangle$

28. $\mathbf{F} = \left\langle \dfrac{y}{x}, \dfrac{y}{z}, \dfrac{z}{x} \right\rangle$

29. $\mathbf{F} = \langle e^y, \sin x, \cos x \rangle$

30. $\mathbf{F} = \left\langle \dfrac{x}{x^2+y^2}, \dfrac{y}{x^2+y^2}, 0 \right\rangle$

In Exercises 31–37, prove the identities assuming that the appropriate partial derivatives exist and are continuous.

31. $\operatorname{div}(\mathbf{F} + \mathbf{G}) = \operatorname{div}(\mathbf{F}) + \operatorname{div}(\mathbf{G})$

32. $\operatorname{curl}(\mathbf{F} + \mathbf{G}) = \operatorname{curl}(\mathbf{F}) + \operatorname{curl}(\mathbf{G})$

33. $\operatorname{div}\operatorname{curl}(\mathbf{F}) = 0$

34. $\operatorname{div}(\mathbf{F} \times \mathbf{G}) = \mathbf{G} \cdot \operatorname{curl}(\mathbf{F}) - \mathbf{F} \cdot \operatorname{curl}(\mathbf{G})$

35. If f is a scalar function, then $\operatorname{div}(f\mathbf{F}) = f\operatorname{div}(\mathbf{F}) + \mathbf{F} \cdot \nabla f$.

36. $\operatorname{curl}(f\mathbf{F}) = f\operatorname{curl}(\mathbf{F}) + (\nabla f) \times \mathbf{F}$

37. $\operatorname{div}(\nabla f \times \nabla g) = 0$

38. Find (by inspection) a potential function for $\mathbf{F} = \langle x, 0 \rangle$ and prove that $\mathbf{G} = \langle y, 0 \rangle$ is not conservative.

In Exercises 39–47, find a potential function for the vector field $\mathbf{F}$ by inspection or show that one does not exist.

39. $\mathbf{F} = \langle x, y \rangle$

40. $\mathbf{F} = \langle y, x \rangle$

41. $\mathbf{F} = \langle y^2z, 1 + 2xyz, xy^2 \rangle$

42. $\mathbf{F} = \langle yz, xz, y \rangle$

43. $\mathbf{F} = \langle ye^{xy}, xe^{xy} \rangle$

44. $\mathbf{F} = \langle 2xyz, x^2z, x^2yz \rangle$

45. $\mathbf{F} = \langle yz^2, xz^2, 2xyz \rangle$

46. $\mathbf{F} = \langle 2xze^{x^2}, 0, e^{x^2} \rangle$

47. $\mathbf{F} = \langle yz\cos(xyz), xz\cos(xyz), xy\cos(xyz) \rangle$.

48. Find potential functions for $\mathbf{F} = \dfrac{\mathbf{e}_r}{r^3}$ and $\mathbf{G} = \dfrac{\mathbf{e}_r}{r^4}$ in $\mathbf{R}^3$. *Hint:* See Example 8.

49. Show that $\mathbf{F} = \langle 3, 1, 2 \rangle$ is conservative. Then prove more generally that any constant vector field $\mathbf{F} = \langle a, b, c \rangle$ is conservative.

50. Let $\varphi = \ln r$, where $r = \sqrt{x^2 + y^2}$. Express $\nabla\varphi$ in terms of the unit radial vector $\mathbf{e}_r$ in $\mathbf{R}^2$.

51. For $P = (a, b)$, we define the unit radial vector field based at P:

$$\mathbf{e}_P = \frac{\langle x - a, y - b \rangle}{\sqrt{(x-a)^2 + (y-b)^2}}$$

(a) Verify that $\mathbf{e}_P$ is a unit vector field.

(b) Calculate $\mathbf{e}_P(1, 1)$ for $P = (3, 2)$.

(c) Find a potential function for $\mathbf{e}_P$.

52. Which of (A) or (B) in Figure 12 is the contour plot of a potential function for the vector field $\mathbf{F}$? Recall that the gradient vectors are perpendicular to the level curves.

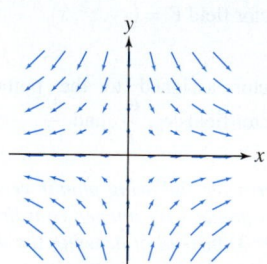

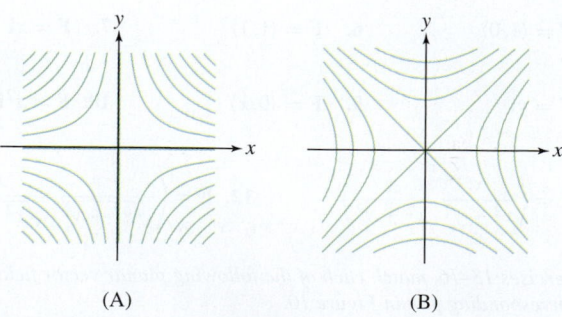

(A)

(B)

FIGURE 12

53. Which of (A) or (B) in Figure 13 is the contour plot of a potential function for the vector field $\mathbf{F}$?

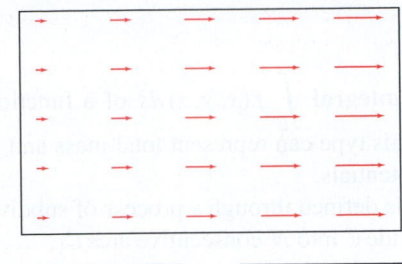

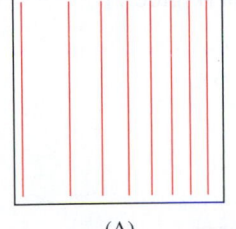 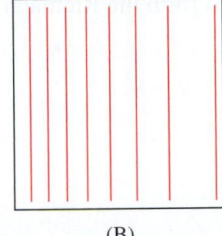

(A)   (B)

FIGURE 13

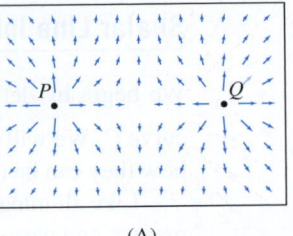 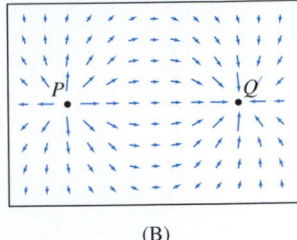

(A)   (B)

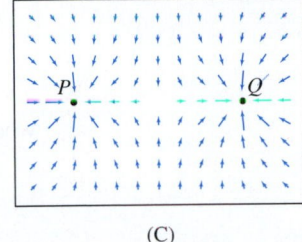

(C)

FIGURE 14

54. Match each of these descriptions with a vector field in Figure 14.

(a) The gravitational field created by two planets of equal mass located at P and Q

(b) The electrostatic field created by two equal and opposite charges located at P and Q (representing the force on a negative test charge; opposite charges attract and like charges repel)

55. In this exercise, we show that the vector field $\mathbf{F}$ in Figure 15 is not conservative. Explain the following statements:

(a) If a potential function f for $\mathbf{F}$ exists, then the level curves of f must be vertical lines.

(b) If a potential function f for $\mathbf{F}$ exists, then the level curves of f must grow farther apart as y increases.

(c) Explain why (a) and (b) are incompatible, and hence f cannot exist.

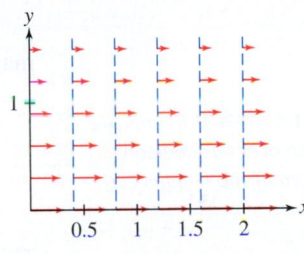

FIGURE 15

Further Insights and Challenges

56. Show that any vector field of the form

$$\mathbf{F} = \langle f(x), g(y), h(z) \rangle$$

has a potential function. Assume that f, g, and h are continuous.

57. Let $\mathcal{D}$ be a disk in $\mathbf{R}^2$. This exercise shows that if

$$\nabla f(x, y) = \mathbf{0}$$

for all (x, y) in $\mathcal{D}$, then f is constant. Consider points $P = (a, b)$, $Q = (c, d)$, and $R = (c, b)$ as in Figure 16.

(a) Use single-variable calculus to show that f is constant along the segments $\overline{PR}$ and $\overline{RQ}$.

(b) Conclude that $f(P) = f(Q)$ for any two points $P, Q \in \mathcal{D}$.

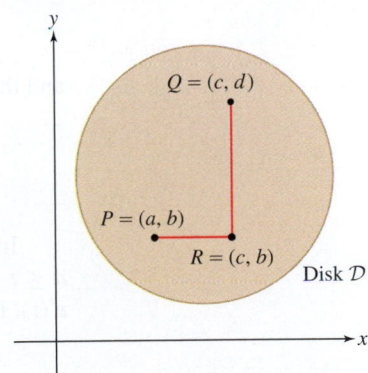

FIGURE 16

16.2 Line Integrals

In this section, we introduce two types of integrals over curves: integrals of functions and integrals of vector fields. These are traditionally called **line integrals**, although it would be more appropriate to call them curve integrals or path integrals.

Scalar Line Integrals

We begin by defining the **scalar line integral** $\int_{\mathcal{C}} f(x, y, z)\, ds$ of a function f over a curve $\mathcal{C}$. We will see how integrals of this type can represent total mass and charge, and how they can be used to find electric potentials.

Like all integrals, this line integral is defined through a process of subdivision, summation, and passage to the limit. We divide $\mathcal{C}$ into N consecutive arcs $\mathcal{C}_1, \ldots, \mathcal{C}_N$, choose a sample point P_i in each arc $\mathcal{C}_i$, and form the Riemann sum (Figure 1)

$$\sum_{i=1}^{N} f(P_i)\, \text{length}(\mathcal{C}_i) = \sum_{i=1}^{N} f(P_i)\, \Delta s_i$$

where Δs_i is the length of $\mathcal{C}_i$.

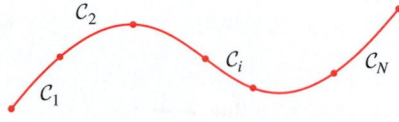
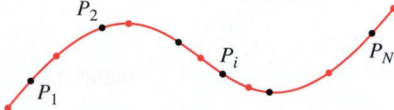

Partition of $\mathcal{C}$ into N small arcs · Choice of sample points P_i in each arc

FIGURE 1 The curve $\mathcal{C}$ is divided into N small arcs.

The line integral of f over $\mathcal{C}$ is the limit (if it exists) of these Riemann sums as the maximum of the lengths Δs_i approaches zero:

In Eq. (1), we write $\{\Delta s_i\} \to 0$ to indicate that the limit is taken over all Riemann sums as the maximum of the lengths Δs_i tends to zero.

$$\boxed{\int_{\mathcal{C}} f(x, y, z)\, ds = \lim_{\{\Delta s_i\} \to 0} \sum_{i=1}^{N} f(P_i)\, \Delta s_i}$$

1

This definition also applies to functions $f(x, y)$ of two variables over a curve in $\mathbf{R}^2$.

The scalar line integral of the function $f(x, y, z) = 1$ is simply the length of $\mathcal{C}$. In this case, all the Riemann sums have the same value:

$$\sum_{i=1}^{N} 1\, \Delta s_i = \sum_{i=1}^{N} \text{length}(\mathcal{C}_i) = \text{length}(\mathcal{C})$$

and thus

$$\boxed{\int_{\mathcal{C}} 1\, ds = \text{length}(\mathcal{C})}$$

In practice, line integrals are computed using parametrizations. Suppose $\mathbf{r}(t)$, for $a \leq t \leq b$, is a parametrization that directly traverses $\mathcal{C}$ and has a continuous derivative $\mathbf{r}'(t)$. Recall that the derivative is the tangent vector

$$\mathbf{r}'(t) = \langle x'(t), y'(t), z'(t) \rangle$$

We divide $\mathcal{C}$ into N consecutive arcs $\mathcal{C}_1, \ldots, \mathcal{C}_N$ corresponding to a partition of the interval $[a, b]$,

$$a = t_0 < t_1 < \cdots < t_{N-1} < t_N = b$$

where each $\mathcal{C}_i$ is parametrized by $\mathbf{r}(t)$ for $t_{i-1} \leq t \leq t_i$ (Figure 2), then we choose sample points $P_i = \mathbf{r}(t_i^*)$ with t_i^* in $[t_{i-1}, t_i]$. Now according to the arc length formula (Section 13.3),

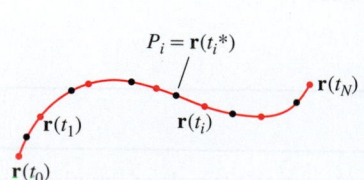

FIGURE 2 Partition of parametrized curve $\mathbf{r}(t)$.

$$\text{length}(\mathcal{C}_i) = \Delta s_i = \int_{t_{i-1}}^{t_i} \|\mathbf{r}'(t)\|\, dt$$

Because $\mathbf{r}'(t)$ is continuous, the function $\|\mathbf{r}'(t)\|$ is nearly constant on $[t_{i-1}, t_i]$ if the length $\Delta t_i = t_i - t_{i-1}$ is small, and thus $\int_{t_{i-1}}^{t_i} \|\mathbf{r}'(t)\| \, dt \approx \|\mathbf{r}'(t_i^*)\| \Delta t_i$. This gives us the approximation

$$\sum_{i=1}^{N} f(P_i) \, \Delta s_i \approx \sum_{i=1}^{N} f(\mathbf{r}(t_i^*)) \|\mathbf{r}'(t_i^*)\| \, \Delta t_i \qquad \boxed{2}$$

The sum on the right is a Riemann sum that converges to the integral

$$\int_a^b f(\mathbf{r}(t)) \|\mathbf{r}'(t)\| \, dt \qquad \boxed{3}$$

as the maximum of the lengths Δt_i tends to zero. By estimating the errors in this approximation, we can show that the sums on the left-hand side of (2) also approach (3). This gives us the following formula for the scalar line integral.

> **THEOREM 1** **Computing a Scalar Line Integral** Let $\mathbf{r}(t)$ be a parametrization that directly traverses $\mathcal{C}$ for $a \leq t \leq b$. If $f(x, y, z)$ and $\mathbf{r}'(t)$ are continuous, then
>
> $$\int_{\mathcal{C}} f(x, y, z) \, ds = \int_a^b f(\mathbf{r}(t)) \|\mathbf{r}'(t)\| \, dt \qquad \boxed{4}$$

The symbol ds is intended to suggest arc length s and is often referred to as the **line element** or **arc length differential**. The arc length differential is related to the parameter differential dt via

$$ds = \|\mathbf{r}'(t)\| \, dt \quad \text{with} \quad \|\mathbf{r}'(t)\| = \sqrt{x'(t)^2 + y'(t)^2 + z'(t)^2}$$

Since arc length along a curve is given by $s(t) = \int_a^t \|\mathbf{r}'(t)\| \, dt$, the Fundamental Theorem of Calculus says that $\dfrac{ds}{dt} = \|\mathbf{r}'(t)\|$. Hence, it makes sense to call $ds = \dfrac{ds}{dt} dt = \|\mathbf{r}'(t)\| \, dt$ the arc length differential.

Note that the integral on the right side in Eq. (4) is a single-variable calculus integral, one that we can attempt to compute using the tools and techniques from earlier in the text.

EXAMPLE 1 **Integrating Along a Helix** Calculate

$$\int_{\mathcal{C}} (x + y + z) \, ds$$

where $\mathcal{C}$ is the helix $\mathbf{r}(t) = \langle \cos t, \sin t, t \rangle$ for $0 \leq t \leq 3\pi$ (Figure 3).

Solution

***Step 1.* Compute ds.**

$$\mathbf{r}'(t) = \langle -\sin t, \cos t, 1 \rangle$$

$$\|\mathbf{r}'(t)\| = \sqrt{(-\sin t)^2 + \cos^2 t + 1} = \sqrt{2}$$

$$ds = \|\mathbf{r}'(t)\| dt = \sqrt{2} \, dt$$

***Step 2.* Write out the integrand and evaluate.**
We have $f(x, y, z) = x + y + z$, and so

$$f(\mathbf{r}(t)) = f(\cos t, \sin t, t) = \cos t + \sin t + t$$

$$f(x, y, z) \, ds = f(\mathbf{r}(t)) \|\mathbf{r}'(t)\| \, dt = (\cos t + \sin t + t) \sqrt{2} \, dt$$

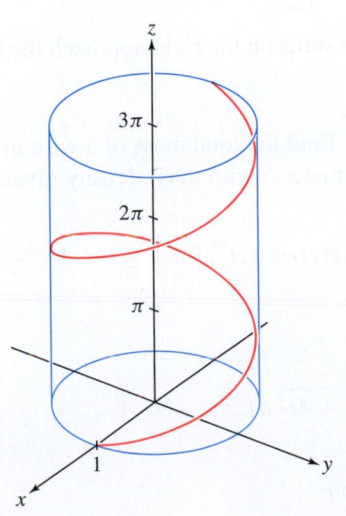

FIGURE 3 The helix $\mathbf{r}(t) = \langle \cos t, \sin t, t \rangle$.

By Eq. (4),

$$\int_{\mathcal{C}} f(x, y, z)\, ds = \int_0^{3\pi} f(\mathbf{r}(t))\, \|\mathbf{r}'(t)\|\, dt = \int_0^{3\pi} (\cos t + \sin t + t)\sqrt{2}\, dt$$

$$= \sqrt{2}\left(\sin t - \cos t + \frac{1}{2}t^2\right)\Bigg|_0^{3\pi}$$

$$= \sqrt{2}\left(0 + 1 + \frac{1}{2}(3\pi)^2\right) - \sqrt{2}\,(0 - 1 + 0) = 2\sqrt{2} + \frac{9\sqrt{2}}{2}\pi^2 \quad \blacksquare$$

EXAMPLE 2 Calculate $\int_{\mathcal{C}} 1\, ds$ for the helix $\mathbf{r}(t) = (\cos t, \sin t, t)$ in the previous example, defined for $0 \le t \le 3\pi$. What does this integral represent?

Solution In the previous example, we showed that $ds = \sqrt{2}\, dt$, and thus

$$\int_{\mathcal{C}} 1\, ds = \int_0^{3\pi} \sqrt{2}\, dt = 3\pi\sqrt{2}$$

This is the length of the helix for $0 \le t \le 3\pi$. $\quad\blacksquare$

Applications of the Scalar Line Integral

In Section 15.5, we discussed the general principle that the integral of a density is the total quantity. This applies to scalar line integrals. For example, we can view the curve $\mathcal{C}$ as a wire with continuous **mass density** $\rho(x, y, z)$, given in units of mass per unit length. The total mass is defined as the integral of mass density:

$$\boxed{\text{total mass of } \mathcal{C} = \int_{\mathcal{C}} \rho(x, y, z)\, ds} \qquad \boxed{5}$$

A similar formula for total charge is valid if $\rho(x, y, z)$ is the charge density along the curve. As in Section 15.5, we justify this interpretation by dividing $\mathcal{C}$ into N arcs $\mathcal{C}_i$ of length Δs_i with N large. The mass density is nearly constant on $\mathcal{C}_i$, and therefore the mass of $\mathcal{C}_i$ is approximately $\rho(P_i)\Delta s_i$, where P_i is any sample point on $\mathcal{C}_i$ (Figure 4). The total mass is the sum

$$\text{total mass of } \mathcal{C} = \sum_{i=1}^{N} \text{mass of } \mathcal{C}_i \approx \sum_{i=1}^{N} \rho(P_i)\,\Delta s_i$$

As the maximum of the lengths Δs_i tends to zero, the sums on the right approach the line integral in Eq. (5).

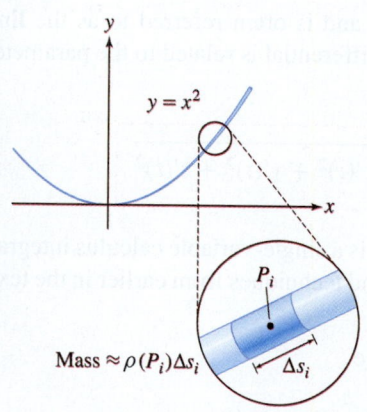

$y = x^2$

Mass $\approx \rho(P_i)\Delta s_i$ P_i Δs_i

FIGURE 4

EXAMPLE 3 **Scalar Line Integral as Total Mass** Find the total mass of a wire in the shape of the parabola $y = x^2$ for $1 \le x \le 4$ (in centimeters) with mass density given by $\rho(x, y) = y/x$ g/cm.

Solution The arc of the parabola is parametrized by $\mathbf{r}(t) = \langle t, t^2 \rangle$ for $1 \le t \le 4$.

Step 1. Compute ds.

$$\mathbf{r}'(t) = \langle 1, 2t \rangle$$

$$ds = \|\mathbf{r}'(t)\|\, dt = \sqrt{1 + 4t^2}\, dt$$

Step 2. Write out the integrand and evaluate.
We have $\rho(\mathbf{r}(t)) = \rho(t, t^2) = t^2/t = t$, and thus

$$\rho(x, y)\, ds = \rho(\mathbf{r}(t))\sqrt{1 + 4t^2}\, dt = t\sqrt{1 + 4t^2}\, dt$$

We evaluate the line integral of mass density using the substitution $u = 1 + 4t^2$, $du = 8t \, dt$, and the limits of integration changing from 1 and 4 to $u(1) = 5$ and $u(4) = 65$, respectively:

$$\int_{\mathcal{C}} \rho(x, y) \, ds = \int_1^4 \rho(\mathbf{r}(t)) \|\mathbf{r}'(t)\| \, dt = \int_1^4 t\sqrt{1 + 4t^2} \, dt$$

$$= \frac{1}{8} \int_5^{65} \sqrt{u} \, du = \frac{1}{12} u^{3/2} \Big|_5^{65}$$

$$= \frac{1}{12}(65^{3/2} - 5^{3/2}) \approx 42.74$$

The total mass of the wire is approximately 42.74 g. ∎

Scalar line integrals are also used to compute electric potentials. When an electric charge is distributed continuously along a curve $\mathcal{C}$ in $\mathbf{R}^3$, with charge density $\rho(x, y, z)$, the charge distribution sets up an electrostatic field $\mathbf{E}$ that is a conservative vector field. Coulomb's Law tells us that $\mathbf{E} = -\nabla V$, where

By definition, $\mathbf{E}$ is the vector field with the property that the electrostatic force on a point charge q placed at location $P = (x, y, z)$ is the vector $q\mathbf{E}(x, y, z)$.

$$\boxed{V(P) = k \int_{\mathcal{C}} \frac{\rho(x, y, z)}{D_P(x, y, z)} \, ds} \qquad \boxed{6}$$

In this integral, $D_P(x, y, z)$ denotes the distance from (x, y, z) to P. The constant k has the value $k = 8.99 \times 10^9$ N-m^2/C^2. In a situation like this, we use V to denote the function and call it the **electric potential**. It is defined for all points P that do not lie on $\mathcal{C}$ and has units of volts (1 volt is 1 N-m/C).

The constant k is usually written as $\dfrac{1}{4\pi\epsilon_0}$, where ϵ_0 is the vacuum permittivity.

EXAMPLE 4   Electric Potential   A charged semicircle of radius R centered at the origin in the xy-plane (Figure 5) has charge density

$$\rho(x, y, 0) = 10^{-8} \left(2 - \frac{x}{R}\right) \text{ C/m}$$

Find the electric potential at a point $P = (0, 0, a)$ if $R = 0.1$ m.

Solution   To compute the integral, we use $\mathbf{r}(t) = \langle R \cos t, R \sin t, 0 \rangle$ to parametrize the semicircle, with t such that $-\pi/2 \leq t \leq \pi/2$:

$$\|\mathbf{r}'(t)\| = \|\langle -R \sin t, R \cos t, 0 \rangle\| = \sqrt{R^2 \sin^2 t + R^2 \cos^2 t + 0} = R$$

$$ds = \|\mathbf{r}'(t)\| \, dt = R \, dt$$

$$\rho(\mathbf{r}(t)) = \rho(R \cos t, R \sin t, 0) = 10^{-8}\left(2 - \frac{R \cos t}{R}\right) = 10^{-8}(2 - \cos t)$$

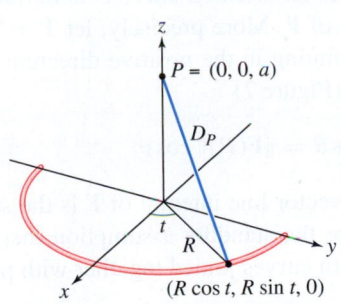

FIGURE 5

In our case, the distance D_P from P to a point $(x, y, 0)$ on the semicircle has the constant value $D_P = \sqrt{R^2 + a^2}$ (Figure 5). Thus,

$$V(P) = k \int_{\mathcal{C}} \frac{\rho(x, y, z) \, ds}{D_P} = k \int_{\mathcal{C}} \frac{10^{-8}(2 - \cos t) R \, dt}{\sqrt{R^2 + a^2}}$$

$$= \frac{10^{-8} k R}{\sqrt{R^2 + a^2}} \int_{-\pi/2}^{\pi/2} (2 - \cos t) \, dt = \frac{10^{-8} k R}{\sqrt{R^2 + a^2}}(2\pi - 2)$$

With $R = 0.1$ m and $k = 8.99 \times 10^9$, we then obtain $10^{-8} k R(2\pi - 2) \approx 38.5$ and

$$V(P) \approx \frac{38.5}{\sqrt{0.01 + a^2}} \text{ volts.} \qquad ∎$$

Vector Line Integrals

When you carry a backpack up a mountain, you do work against the earth's gravitational field. The work, or energy expended, is one example of a quantity represented by a vector line integral.

An important difference between vector and scalar line integrals is that vector line integrals depend on the direction along the curve. This is reasonable if you think of the vector line integral as work, because the work performed going down the mountain is the negative of the work performed going up.

A specified direction along a curve C is called an **orientation** (Figure 6), and with an orientation, C is called an **oriented curve**. We refer to the specified direction as the **positive** direction along C and the opposite direction as the **negative** direction. In Figure 6(A), if we reversed the orientation, the positive direction would become the direction from Q to P.

*The unit tangent vector **T** varies from point to point along the curve. When it is necessary to stress this dependence, we write $\mathbf{T}(P)$.*

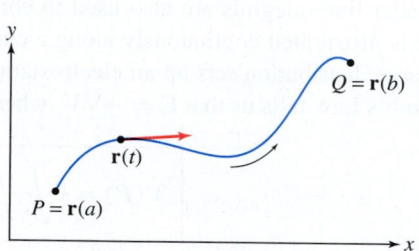

(A) An oriented curve from P to Q

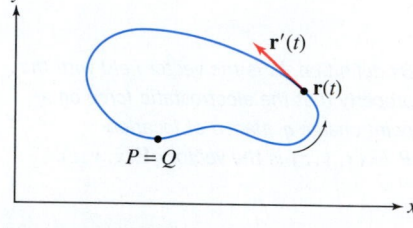

(B) A closed oriented curve

FIGURE 6

The vector line integral of a vector field $\mathbf{F}$ over an oriented curve C is defined as the scalar line integral of the tangential component of $\mathbf{F}$. More precisely, let $\mathbf{T} = \mathbf{T}(P)$ denote the unit tangent vector at a point P on C pointing in the positive direction. The **tangential component** of $\mathbf{F}$ at P is the dot product (Figure 7)

$$\mathbf{F}(P) \cdot \mathbf{T}(P) = \|\mathbf{F}(P)\| \, \|\mathbf{T}(P)\| \cos\theta = \|\mathbf{F}(P)\| \cos\theta$$

where θ is the angle between $\mathbf{F}(P)$ and $\mathbf{T}(P)$. The vector line integral of $\mathbf{F}$ is the scalar line integral of the scalar function $\mathbf{F} \cdot \mathbf{T}$. We make the standing assumption that C is piecewise smooth (it consists of finitely many smooth curves joined together with possible corners).

FIGURE 7 The vector line integral is the integral of the tangential component of $\mathbf{F}$ along C.

> **DEFINITION** **Vector Line Integral** The line integral of a vector field $\mathbf{F}$ along an oriented curve C is the integral of the tangential component of $\mathbf{F}$:
>
> $$\int_C (\mathbf{F} \cdot \mathbf{T}) \, ds$$

Another notation for the vector line integral is obtained by expressing the product of the unit tangent vector $\mathbf{T}$ and the arc length differential ds as the **vector differential** $d\mathbf{r} = \mathbf{T} \, ds$. Thus,

$$\int_C (\mathbf{F} \cdot \mathbf{T}) \, ds = \int_C \mathbf{F} \cdot d\mathbf{r}$$

We use parametrizations to evaluate vector line integrals, but there is one important difference with the scalar case: The parametrization $\mathbf{r}(t)$ must be *positively oriented*; that is, $\mathbf{r}(t)$ must directly traverse C in the positive direction. We assume also that $\mathbf{r}(t)$ is regular (see Section 13.4); that is, $\mathbf{r}'(t) \neq \mathbf{0}$ for $a \leq t \leq b$. Then $\mathbf{r}'(t)$ is a nonzero tangent vector pointing in the positive direction, and

$$\mathbf{T} = \frac{\mathbf{r}'(t)}{\|\mathbf{r}'(t)\|}$$

In terms of the arc length differential $ds = \|\mathbf{r}'(t)\|\, dt$, we have

$$(\mathbf{F} \cdot \mathbf{T})\, ds = \left(\mathbf{F}(\mathbf{r}(t)) \cdot \frac{\mathbf{r}'(t)}{\|\mathbf{r}'(t)\|} \right) \|\mathbf{r}'(t)\|\, dt = \mathbf{F}(\mathbf{r}(t)) \cdot \mathbf{r}'(t)\, dt$$

Therefore, we obtain the following theorem.

THEOREM 2 Computing a Vector Line Integral   If $\mathbf{r}(t)$ is a positively oriented regular parametrization of an oriented curve $\mathcal{C}$ for $a \leq t \leq b$, then

$$\int_{\mathcal{C}} \mathbf{F} \cdot d\mathbf{r} = \int_{\mathcal{C}} \mathbf{F} \cdot \mathbf{T}\, ds = \int_a^b \mathbf{F}(\mathbf{r}(t)) \cdot \mathbf{r}'(t)\, dt \qquad \boxed{7}$$

The vector differential $d\mathbf{r}$ is related to the parameter differential dt via the equation

$$\boxed{d\mathbf{r} = \mathbf{r}'(t)\, dt = \langle x'(t), y'(t), z'(t) \rangle\, dt}$$

Vector line integrals are usually easier to calculate than scalar line integrals, because the length $\|\mathbf{r}'(t)\|$, which involves a square root, does not appear in the integrand.

Equation (7) tells us that to evaluate a vector line integral, we replace the integrand $\mathbf{F} \cdot d\mathbf{r}$ with $\mathbf{F}(\mathbf{r}(t)) \cdot \mathbf{r}'(t)\, dt$ and integrate over the parameter interval $a \leq t \leq b$. Like scalar line integrals, we are converting a vector line integral to a simple single-variable definite integral.

EXAMPLE 5   Evaluate $\displaystyle\int_{\mathcal{C}} \mathbf{F} \cdot d\mathbf{r}$, where $\mathbf{F} = \langle z, y^2, x \rangle$ and $\mathcal{C}$ is parametrized (in the positive direction) by $\mathbf{r}(t) = \langle t + 1, e^t, t^2 \rangle$ for $0 \leq t \leq 2$.

Solution   There are two steps in evaluating a line integral.

Step 1. **Calculate the integrand.**

$$\mathbf{r}(t) = \langle t + 1, e^t, t^2 \rangle$$

$$\mathbf{F}(\mathbf{r}(t)) = \langle z, y^2, x \rangle = \langle t^2, e^{2t}, t + 1 \rangle$$

$$\mathbf{r}'(t) = \langle 1, e^t, 2t \rangle$$

The integrand (as a differential) is the dot product:

$$\mathbf{F}(\mathbf{r}(t)) \cdot \mathbf{r}'(t)\, dt = \langle t^2, e^{2t}, t + 1 \rangle \cdot \langle 1, e^t, 2t \rangle\, dt = (e^{3t} + 3t^2 + 2t)\, dt$$

Step 2. **Evaluate the line integral.**

$$\int_{\mathcal{C}} \mathbf{F} \cdot d\mathbf{r} = \int_0^2 \mathbf{F}(\mathbf{r}(t)) \cdot \mathbf{r}'(t)\, dt$$

$$= \int_0^2 (e^{3t} + 3t^2 + 2t)\, dt = \left(\frac{1}{3}e^{3t} + t^3 + t^2 \right) \Big|_0^2$$

$$= \left(\frac{1}{3}e^6 + 8 + 4 \right) - \frac{1}{3} = \frac{1}{3}\left(e^6 + 35 \right) \qquad \blacksquare$$

Another standard notation for the line integral $\int_{\mathcal{C}} \mathbf{F} \cdot d\mathbf{r}$ is

$$\int_{\mathcal{C}} F_1\, dx + F_2\, dy + F_3\, dz$$

In this notation, we write $d\mathbf{r}$ as a vector differential:

$$d\mathbf{r} = \langle dx, dy, dz \rangle$$

so that

$$\mathbf{F} \cdot d\mathbf{r} = \langle F_1, F_2, F_3 \rangle \cdot \langle dx, dy, dz \rangle = F_1\, dx + F_2\, dy + F_3\, dz$$

In terms of a parametrization $\mathbf{r}(t) = \langle x(t), y(t), z(t) \rangle$,

$$d\mathbf{r} = \left\langle \frac{dx}{dt}, \frac{dy}{dt}, \frac{dz}{dt} \right\rangle dt$$

$$\mathbf{F} \cdot d\mathbf{r} = \left(F_1(\mathbf{r}(t))\frac{dx}{dt} + F_2(\mathbf{r}(t))\frac{dy}{dt} + F_3(\mathbf{r}(t))\frac{dz}{dt} \right) dt$$

So, we have the following formula:

$$\int_{\mathcal{C}} F_1\, dx + F_2\, dy + F_3\, dz = \int_a^b \left(F_1(\mathbf{r}(t))\frac{dx}{dt} + F_2(\mathbf{r}(t))\frac{dy}{dt} + F_3(\mathbf{r}(t))\frac{dz}{dt} \right) dt$$

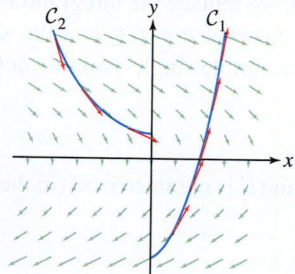

FIGURE 8

GRAPHICAL INSIGHT   The magnitude of a vector line integral (or even whether it is positive or negative) depends on the angles between $\mathbf{F}$ and $\mathbf{T}$ along the curve. Consider the line integral of the vector field $\mathbf{F}$ along the curves $\mathcal{C}_1$ and $\mathcal{C}_2$ illustrated in Figure 8.

- Along $\mathcal{C}_1$, the angles θ between $\mathbf{F}$ and $\mathbf{T}$ appear to be mostly obtuse. Consequently, $\mathbf{F} \cdot \mathbf{T} \leq 0$ and the line integral is negative. We are primarily going against the vector field as we travel along the curve.
- Along $\mathcal{C}_2$, the angles θ appear to be mostly acute. Consequently, $\mathbf{F} \cdot \mathbf{T} \geq 0$ and the line integral is positive. We are going with the vector field as we travel along the curve.

A vector field that has a number of interesting properties is known as the **vortex field** (Figure 9) and is given by

$$\mathbf{F} = \left\langle \frac{-y}{x^2 + y^2}, \frac{x}{x^2 + y^2} \right\rangle$$

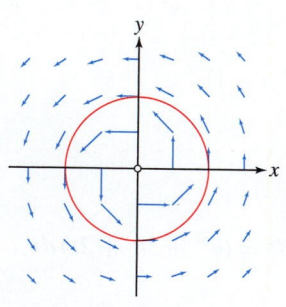

FIGURE 9   The vortex field.

We will examine some of its properties in this chapter and the next. To begin, we show that the integral of this vector field along any circle centered at the origin and oriented in the counterclockwise direction is 2π.

EXAMPLE 6   Show that if $\mathcal{C}$ is the circle of radius R centered at the origin, oriented counterclockwise, then

$$\int_{\mathcal{C}} \mathbf{F} \cdot d\mathbf{r} = \int_{\mathcal{C}} \frac{-y}{x^2 + y^2}\, dx + \frac{x}{x^2 + y^2}\, dy = 2\pi$$

Solution   The circle is parameterized by $\mathbf{r}(t) = \langle R\cos t, R\sin t \rangle$ for $0 \leq t \leq 2\pi$. We have

$$\frac{dx}{dt} = -R\sin t, \qquad \frac{dy}{dt} = R\cos t$$

The integrand of the line integral is

$$\frac{-y}{x^2 + y^2}\,dx + \frac{x}{x^2 + y^2}\,dy = \left(\frac{-y}{x^2 + y^2}\right)\left(\frac{dx}{dt}\right)dt + \left(\frac{x}{x^2 + y^2}\right)\left(\frac{dy}{dt}\right)dt$$

$$= \left(\frac{-R\sin t}{R^2}(-R\sin t) + \frac{R\cos t}{R^2}(R\cos t)\right)dt$$

$$= (\sin^2 t + \cos^2 t)dt$$

$$= dt$$

Therefore,

$$\int_C \frac{-y}{x^2 + y^2}\,dx + \frac{x}{x^2 + y^2}\,dy = \int_0^{2\pi} dt = 2\pi \qquad \blacksquare$$

We now state some basic properties of vector line integrals. First, given an oriented curve C, we write $-C$ to denote the curve C with the opposite orientation (Figure 10). The unit tangent vector changes sign from $\mathbf{T}$ to $-\mathbf{T}$ when we change orientation, so the tangential component of $\mathbf{F}$ and the line integral also change sign:

$$\int_{-C} \mathbf{F} \cdot d\mathbf{r} = -\int_C \mathbf{F} \cdot d\mathbf{r}$$

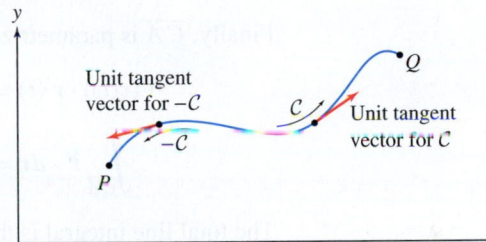

DF **FIGURE 10** The curve between P and Q has two possible orientations.

Next, if we are given n oriented curves $C_1, \ldots, C_n$, we write

$$C = C_1 + \cdots + C_n$$

to indicate the union of the curves, and we define the line integral over C as the sum

$$\int_C \mathbf{F} \cdot d\mathbf{r} = \int_{C_1} \mathbf{F} \cdot d\mathbf{r} + \cdots + \int_{C_n} \mathbf{F} \cdot d\mathbf{r}$$

We use this formula to define the line integral when C is **piecewise smooth**, meaning that C is a union of smooth curves $C_1, \ldots, C_n$. For example, the triangle in Figure 11 is piecewise smooth but not smooth. The next theorem summarizes the main properties of vector line integrals.

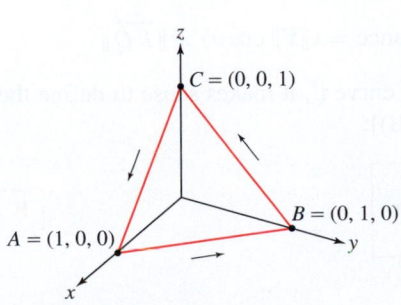

FIGURE 11 The triangle is piecewise smooth: It is the union of its three edges, each of which is smooth.

THEOREM 3 Properties of Vector Line Integrals   Let C be a smooth oriented curve, and let $\mathbf{F}$ and $\mathbf{G}$ be vector fields.

(i) Linearity: $\displaystyle \int_C (\mathbf{F} + \mathbf{G}) \cdot d\mathbf{r} = \int_C \mathbf{F} \cdot d\mathbf{r} + \int_C \mathbf{G} \cdot d\mathbf{r}$

$$\int_C k\mathbf{F} \cdot d\mathbf{r} = k \int_C \mathbf{F} \cdot d\mathbf{r} \quad (k \text{ a constant})$$

(ii) Reversing orientation: $\displaystyle \int_{-C} \mathbf{F} \cdot d\mathbf{r} = -\int_C \mathbf{F} \cdot d\mathbf{r}$

(iii) Additivity: If C is a union of n smooth curves $C_1 + \cdots + C_n$, then

$$\int_C \mathbf{F} \cdot d\mathbf{r} = \int_{C_1} \mathbf{F} \cdot d\mathbf{r} + \cdots + \int_{C_n} \mathbf{F} \cdot d\mathbf{r}$$

EXAMPLE 7 Compute $\int_{\mathcal{C}} \mathbf{F} \cdot d\mathbf{r}$, where $\mathbf{F} = \langle e^z, e^y, x + y \rangle$ and $\mathcal{C}$ is the triangle joining $(1, 0, 0)$, $(0, 1, 0)$, and $(0, 0, 1)$ oriented in the counterclockwise direction when viewed from above (Figure 11).

Solution The line integral is the sum of the line integrals over the edges of the triangle:

$$\int_{\mathcal{C}} \mathbf{F} \cdot d\mathbf{r} = \int_{\overline{AB}} \mathbf{F} \cdot d\mathbf{r} + \int_{\overline{BC}} \mathbf{F} \cdot d\mathbf{r} + \int_{\overline{CA}} \mathbf{F} \cdot d\mathbf{r}$$

Segment $\overline{AB}$ is parametrized by $\mathbf{r}(t) = \langle 1 - t, t, 0 \rangle$ for $0 \le t \le 1$. We have

$$\mathbf{F}(\mathbf{r}(t)) \cdot \mathbf{r}'(t) = \mathbf{F}(1 - t, t, 0) \cdot \langle -1, 1, 0 \rangle = \langle e^0, e^t, 1 \rangle \cdot \langle -1, 1, 0 \rangle = -1 + e^t$$

$$\int_{\overline{AB}} \mathbf{F} \cdot d\mathbf{r} = \int_0^1 (e^t - 1)\, dt = (e^t - t)\Big|_0^1 = (e - 1) - 1 = e - 2$$

Similarly, $\overline{BC}$ is parametrized by $\mathbf{r}(t) = \langle 0, 1 - t, t \rangle$ for $0 \le t \le 1$, and

$$\mathbf{F}(\mathbf{r}(t)) \cdot \mathbf{r}'(t) = \langle e^t, e^{1-t}, 1 - t \rangle \cdot \langle 0, -1, 1 \rangle = -e^{1-t} + 1 - t$$

$$\int_{\overline{BC}} \mathbf{F} \cdot d\mathbf{r} = \int_0^1 (-e^{1-t} + 1 - t)\, dt = \left(e^{1-t} + t - \frac{1}{2}t^2 \right)\Big|_0^1 = \frac{3}{2} - e$$

Finally, $\overline{CA}$ is parametrized by $\mathbf{r}(t) = \langle t, 0, 1 - t \rangle$ for $0 \le t \le 1$, and

$$\mathbf{F}(\mathbf{r}(t)) \cdot \mathbf{r}'(t) = \langle e^{1-t}, 1, t \rangle \cdot \langle 1, 0, -1 \rangle = e^{1-t} - t$$

$$\int_{\overline{CA}} \mathbf{F} \cdot d\mathbf{r} = \int_0^1 (e^{1-t} - t)\, dt = \left(-e^{1-t} - \frac{1}{2}t^2 \right)\Big|_0^1 = -\frac{3}{2} + e$$

The total line integral is the sum

$$\int_{\mathcal{C}} \mathbf{F} \cdot d\mathbf{r} = (e - 2) + \left(\frac{3}{2} - e \right) + \left(-\frac{3}{2} + e \right) = e - 2 \qquad \blacksquare$$

Applications of the Vector Line Integral

Recall that in physics, "work" refers to the energy expended when a force is applied to an object as it moves along a path. By definition, the work W performed along the straight segment from P to Q by applying a constant force $\mathbf{F}$ at an angle θ [Figure 12(A)] is

$$W = (\text{tangential component of } \mathbf{F}) \times \text{distance} = (\|\mathbf{F}\| \cos \theta) \times \|\overrightarrow{PQ}\|$$

When the force acts on the object moving along a curve $\mathcal{C}$, it makes sense to define the work W performed as the line integral [Figure 12(B)]:

$$\boxed{W = \int_{\mathcal{C}} \mathbf{F} \cdot d\mathbf{r}} \qquad \boxed{8}$$

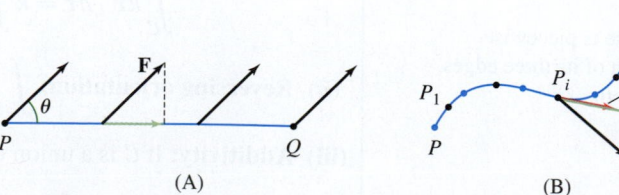

(A)   (B)

FIGURE 12

This is the work performed by the field $\mathbf{F}$. The idea is that we can divide $\mathcal{C}$ into a large number of short consecutive arcs $\mathcal{C}_1, \ldots, \mathcal{C}_N$, where $\mathcal{C}_i$ has length Δs_i. The work W_i performed along $\mathcal{C}_i$ is approximately equal to the tangential component $\mathbf{F}(P_i) \cdot \mathbf{T}(P_i)$ times the length Δs_i, where P_i is a sample point in $\mathcal{C}_i$. Thus, we have

$$W = \sum_{i=1}^{N} W_i \approx \sum_{i=1}^{N} (\mathbf{F}(P_i) \cdot \mathbf{T}(P_i)) \Delta s_i$$

The right-hand side approaches $\displaystyle\int_{\mathcal{C}} \mathbf{F} \cdot d\mathbf{r}$ as the lengths Δs_i tend to zero.

Often, we are interested in calculating the work required to move an object along a path in the presence of a force field $\mathbf{F}$ (such as an electrical or gravitational field). In this case, $\mathbf{F}$ acts on the object and we must work *against* the force field to move the object. The work required is the negative of the line integral in Eq. (8):

$$\text{work performed against } \mathbf{F} = -\int_{\mathcal{C}} \mathbf{F} \cdot d\mathbf{r}$$

EXAMPLE 8 **Calculating Work** Calculate the work performed against $\mathbf{F}$ in moving a particle from $P = (1, 1, 1)$ to $Q = (4, 8, 2)$ along the path

$$\mathbf{r}(t) = \left\langle t^2, t^3, t \right\rangle \quad \text{(in meters)} \qquad \text{for } 1 \le t \le 2$$

in the presence of a force field $\mathbf{F} = \left\langle x^2, -z, -yz^{-1} \right\rangle$ in newtons.

Solution We have

$$\mathbf{F}(\mathbf{r}(t)) = \mathbf{F}(t^2, t^3, t) = \left\langle t^4, -t, -t^2 \right\rangle$$

$$\mathbf{r}'(t) = \left\langle 2t, 3t^2, 1 \right\rangle$$

$$\mathbf{F} \cdot d\mathbf{r} = \mathbf{F}(\mathbf{r}(t)) \cdot \mathbf{r}'(t)\, dt = \left\langle t^4, -t, -t^2 \right\rangle \cdot \left\langle 2t, 3t^2, 1 \right\rangle dt = (2t^5 - 3t^3 - t^2)\, dt$$

The work performed against the force field in joules is

$$W = -\int_{\mathcal{C}} \mathbf{F} \cdot d\mathbf{r} = -\int_{1}^{2} (2t^5 - 3t^3 - t^2)\, dt = \frac{89}{12} \qquad \blacksquare$$

Line integrals are also used to define what is known as the flux of a vector field across a plane curve. Instead of integrating the tangential component of the vector field, the flux across a plane curve is defined as the integral of the normal component of the vector field. Given an oriented curve $\mathcal{C}$ in the plane, we define the positive direction *across* $\mathcal{C}$ to be the direction going from left to right relative to the positive direction *along* $\mathcal{C}$ given by the orientation. Note that this makes sense for a curve in the plane, but in $\mathbf{R}^3$, there is no natural choice of a positive direction across a curve (in $\mathbf{R}^3$, flux is computed across surfaces). We let $\mathbf{n}$ represent a unit normal vector in the positive direction across $\mathcal{C}$ and define the **flux of $\mathbf{F}$ across** $\mathcal{C}$ as the integral $\displaystyle\int_{\mathcal{C}} (\mathbf{F} \cdot \mathbf{n})\, ds$ (Figure 13).

To compute the flux, let $\mathbf{r}(t)$, for $a \le t \le b$, be a positively oriented parametrization of an oriented curve $\mathcal{C}$. The derivative vector $\mathbf{r}'(t) = \left\langle x'(t), y'(t) \right\rangle$ is tangent to the curve, pointing in the positive direction along $\mathcal{C}$. The vector $\mathbf{N}(t) = \left\langle y'(t), -x'(t) \right\rangle$ is orthogonal to $\mathbf{r}'(t)$ and points to the right. Let $\mathbf{n}(t)$ be a unit vector in the direction of $\mathbf{N}(t)$. These normal vectors point in the positive direction across $\mathcal{C}$.

Now, note that since $\mathbf{N}(t)$ and $\mathbf{r}'(t)$ have the same magnitude, it follows that

$$\mathbf{n}(t) = \frac{\mathbf{N}(t)}{\|\mathbf{N}(t)\|} = \frac{\mathbf{N}(t)}{\|\mathbf{r}'(t)\|}$$

◄── REMINDER *Work has units of energy. The SI unit of force is the newton, and the unit of energy is the joule, defined as 1 newton-meter. The British unit is the foot-pound.*

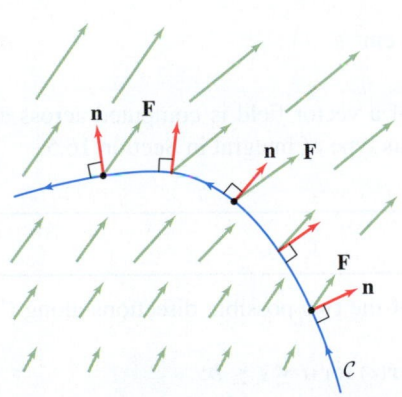

FIGURE 13 The flux of $\mathbf{F}$ across $\mathcal{C}$ is the integral of $\mathbf{F} \cdot \mathbf{n}$, the normal component of $\mathbf{F}$, over $\mathcal{C}$.

Given any nonzero vector $\mathbf{v} = \langle p, q \rangle$, the vectors $\langle q, -p \rangle$ and $\langle -q, p \rangle$ are both orthogonal to $\mathbf{v}$, the former pointing to the right of $\mathbf{v}$, the latter to the left.

*CAUTION In Sections 13.4 and 13.5, **N** was the principal unit normal vector to a curve in space. Here, it represents a normal vector, not necessarily a unit vector, to a curve in the plane, and **n** represents the corresponding unit normal vector. This conforms to common usage.*

The flux across $\mathcal{C}$ is then computed via

$$\int_{\mathcal{C}} (\mathbf{F} \cdot \mathbf{n}) \, ds = \int_a^b \mathbf{F}(\mathbf{r}(t)) \cdot \frac{\mathbf{N}(t)}{\|\mathbf{r}'(t)\|} \|\mathbf{r}'(t)\| \, dt = \int_a^b \mathbf{F}(\mathbf{r}(t)) \cdot \mathbf{N}(t) \, dt \qquad \boxed{9}$$

If $\mathbf{F}$ is the velocity field of a fluid (modeled as a two-dimensional fluid), then the flux is the quantity of fluid flowing across the curve per unit time.

EXAMPLE 9 Flux Across a Curve Calculate the flux of the velocity vector field $\mathbf{v} = \langle 3 + 2y - y^2/3, 0 \rangle$ (in centimeters per second) across the quarter-ellipse $\mathbf{r}(t) = \langle 3 \cos t, 6 \sin t \rangle$ for $0 \le t \le \frac{\pi}{2}$ (Figure 14).

Solution Note that along the curve the vector field crosses left to right relative to the orientation. Thus, we expect the resulting flux to be positive. The vector field along the path is

$$\mathbf{v}(\mathbf{r}(t)) = \left\langle 3 + 2(6 \sin t) - (6 \sin t)^2/3, 0 \right\rangle = \left\langle 3 + 12 \sin t - 12 \sin^2 t, 0 \right\rangle$$

The tangent vector is $\mathbf{r}'(t) = \langle -3 \sin t, 6 \cos t \rangle$, and thus $\mathbf{N}(t) = \langle 6 \cos t, 3 \sin t \rangle$. We integrate the dot product

$$\mathbf{v}(\mathbf{r}(t)) \cdot \mathbf{N}(t) = \left\langle 3 + 12 \sin t - 12 \sin^2 t, 0 \right\rangle \cdot \langle 6 \cos t, 3 \sin t \rangle$$

$$= (3 + 12 \sin t - 12 \sin^2 t)(6 \cos t)$$

$$= 18 \cos t + 72 \sin t \cos t - 72 \sin^2 t \cos t$$

to obtain the flux:

$$\int_a^b \mathbf{v}(\mathbf{r}(t)) \cdot \mathbf{N}(t) \, dt = \int_0^{\pi/2} (18 \cos t + 72 \sin t \cos t - 72 \sin^2 t \cos t) \, dt$$

$$= 18 + 36 - 24 = 30 \text{ cm}^2/\text{s} \qquad \blacksquare$$

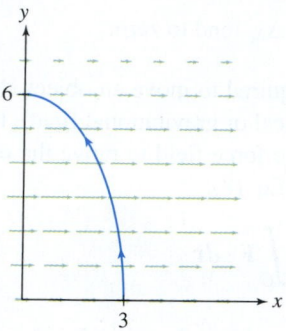

FIGURE 14

As we indicated previously, in $\mathbf{R}^3$, the flux of a vector field is computed across a surface, rather than across a line. We will define this type of integral in Section 16.5.

16.2 SUMMARY

- An *oriented curve* $\mathcal{C}$ is a curve in which one of the two possible directions along $\mathcal{C}$ (called the *positive direction*) is chosen.
- Line integral over a curve with parametrization $\mathbf{r}(t)$ for $a \le t \le b$:
  - Arc length differential: $ds = \|\mathbf{r}'(t)\| \, dt$. Scalar line integral:

$$\int_{\mathcal{C}} f(x, y, z) \, ds = \int_a^b f(\mathbf{r}(t)) \, \|\mathbf{r}'(t)\| \, dt$$

  - Vector differential: $d\mathbf{r} = \mathbf{T} ds = \mathbf{r}'(t) \, dt$. Vector line integral

$$\int_{\mathcal{C}} \mathbf{F} \cdot d\mathbf{r} = \int_{\mathcal{C}} (\mathbf{F} \cdot \mathbf{T}) \, ds = \int_a^b \mathbf{F}(\mathbf{r}(t)) \cdot \mathbf{r}'(t) \, dt$$

$$= \int_{\mathcal{C}} F_1 \, dx + F_2 \, dy + F_3 \, dz \text{ (in three dimensions)}$$

- The scalar line integral and the vector line integral depend on the orientation of the curve $\mathcal{C}$. The parametrization $\mathbf{r}(t)$ must be regular (i.e., $\mathbf{r}'(t) \ne \mathbf{0}$), and it must trace $\mathcal{C}$ in the positive direction.

- We write $-\mathcal{C}$ for the curve $\mathcal{C}$ with the opposite orientation. Then

$$\int_{-\mathcal{C}} \mathbf{F} \cdot d\mathbf{r} = - \int_{\mathcal{C}} \mathbf{F} \cdot d\mathbf{r}$$

- If $\rho(x, y, z)$ is the mass or charge density along $\mathcal{C}$, then the total mass or charge is equal to the scalar line integral $\int_{\mathcal{C}} \rho(x, y, z) \, ds$.

- The vector line integral is used to compute the work W exerted on an object along a curve $\mathcal{C}$:

$$W = \int_{\mathcal{C}} \mathbf{F} \cdot d\mathbf{r}$$

The work performed *against* $\mathbf{F}$ is the quantity $- \int_{\mathcal{C}} \mathbf{F} \cdot d\mathbf{r}$.

- For a curve $\mathcal{C}$ in $\mathbf{R}^2$, flux across $\mathcal{C} = \int_{\mathcal{C}} (\mathbf{F} \cdot \mathbf{n}) \, ds = \int_a^b \mathbf{F}(\mathbf{r}(t)) \cdot \mathbf{N}(t) \, dt$, where $\mathbf{N}(t) = \langle y'(t), -x'(t) \rangle$.

16.2 EXERCISES

Preliminary Questions

1. What is the line integral of the constant function $f(x, y, z) = 10$ over a curve $\mathcal{C}$ of length 5?

2. Which of the following have a zero line integral over the vertical segment from $(0, 0)$ to $(0, 1)$?

(a) $f(x, y) = x$ (b) $f(x, y) = y$
(c) $\mathbf{F} = \langle x, 0 \rangle$ (d) $\mathbf{F} = \langle y, 0 \rangle$
(e) $\mathbf{F} = \langle 0, x \rangle$ (f) $\mathbf{F} = \langle 0, y \rangle$

3. State whether each statement is true or false. If the statement is false, give the correct statement.

(a) The scalar line integral does not depend on how you parametrize the curve.

(b) If you reverse the orientation of the curve, neither the vector line integral nor the scalar line integral changes sign.

4. Suppose that $\mathcal{C}$ has length 5. What is the value of $\int_{\mathcal{C}} \mathbf{F} \cdot d\mathbf{r}$ if:

(a) $\mathbf{F}(P)$ is normal to $\mathcal{C}$ at all points P on $\mathcal{C}$?

(b) $\mathbf{F}(P)$ is a unit vector pointing in the negative direction along the curve?

Exercises

1. Let $f(x, y, z) = x + yz$, and let $\mathcal{C}$ be the line segment from $P = (0, 0, 0)$ to $(6, 2, 2)$.

(a) Calculate $f(\mathbf{r}(t))$ and $ds = \|\mathbf{r}'(t)\| \, dt$ for the parametrization $\mathbf{r}(t) = \langle 6t, 2t, 2t \rangle$ for $0 \le t \le 1$.

(b) Evaluate $\int_{\mathcal{C}} f(x, y, z) \, ds$.

2. Repeat Exercise 1 with the parametrization $\mathbf{r}(t) = \langle 3t^2, t^2, t^2 \rangle$ for $0 \le t \le \sqrt{2}$.

3. Let $\mathbf{F} = \langle y^2, x^2 \rangle$, and let $\mathcal{C}$ be the curve $y = x^{-1}$ for $1 \le x \le 2$, oriented from left to right.

(a) Calculate $\mathbf{F}(\mathbf{r}(t))$ and $d\mathbf{r} = \mathbf{r}'(t) \, dt$ for the parametrization of $\mathcal{C}$ given by $\mathbf{r}(t) = \langle t, t^{-1} \rangle$.

(b) Calculate the dot product $\mathbf{F}(\mathbf{r}(t)) \cdot \mathbf{r}'(t) \, dt$ and evaluate $\int_{\mathcal{C}} \mathbf{F} \cdot d\mathbf{r}$.

4. Let $\mathbf{F}(x, y, z) = \langle z^2, x, y \rangle$, and let $\mathcal{C}$ be the curve that is given by $\mathbf{r}(t) = \langle 3 + 5t^2, 3 - t^2, t \rangle$ for $0 \le t \le 2$.

(a) Calculate $\mathbf{F}(\mathbf{r}(t))$ and $d\mathbf{r} = \mathbf{r}'(t) \, dt$.

(b) Calculate the dot product $\mathbf{F}(\mathbf{r}(t)) \cdot \mathbf{r}'(t) \, dt$ and evaluate $\int_{\mathcal{C}} \mathbf{F} \cdot d\mathbf{r}$.

In Exercises 5–8, compute the integral of the scalar function or vector field over $\mathbf{r}(t) = \langle \cos t, \sin t, t \rangle$ for $0 \le t \le \pi$.

5. $f(x, y, z) = x^2 + y^2 + z^2$ **6.** $f(x, y, z) = xy + z$

7. $\mathbf{F}(x, y, z) = \langle x, y, z^2 \rangle$ **8.** $\mathbf{F}(x, y, z) = \langle xy, 2, z^3 \rangle$

In Exercises 9–16, compute $\int_{\mathcal{C}} f \, ds$ for the curve specified.

9. $f(x, y) = \sqrt{1 + 9xy}, \quad y = x^3$ for $0 \le x \le 2$

10. $f(x, y) = \dfrac{y^3}{x^7}, \quad y = \frac{1}{4}x^4$ for $1 \le x \le 2$

11. $f(x, y, z) = z^2, \quad \mathbf{r}(t) = \langle 2t, 3t, 4t \rangle$ for $0 \le t \le 2$

12. $f(x, y, z) = 3x - 2y + z, \quad \mathbf{r}(t) = \langle 2 + t, 2 - t, 2t \rangle$ for $-2 \le t \le 1$

13. $f(x, y, z) = xe^{z^2}$, piecewise linear path from $(0, 0, 1)$ to $(0, 2, 0)$ to $(1, 1, 1)$

14. $f(x, y, z) = x^2 z, \quad \mathbf{r}(t) = \langle e^t, \sqrt{2}t, e^{-t} \rangle$ for $0 \le t \le 1$

15. $f(x, y, z) = 2x^2 + 8z, \quad \mathbf{r}(t) = \langle e^t, t^2, t \rangle, \quad 0 \le t \le 1$

16. $f(x,y,z) = 6xz - 2y^2$, $\mathbf{r}(t) = \left\langle t, \dfrac{t^2}{\sqrt{2}}, \dfrac{t^3}{3} \right\rangle$, $0 \le t \le 2$

17. Calculate $\displaystyle\int_C 1\,ds$, where the curve C is parametrized by $\mathbf{r}(t) = \langle 4t, -3t, 12t \rangle$ for $2 \le t \le 5$. What does this integral represent?

18. Calculate $\displaystyle\int_C 1\,ds$, where the curve C is parametrized by $\mathbf{r}(t) = \langle e^t, \sqrt{2}t, e^{-t} \rangle$ for $0 \le t \le 2$.

In Exercises 19–26, compute $\displaystyle\int_C \mathbf{F} \cdot d\mathbf{r}$ for the oriented curve specified.

19. $\mathbf{F}(x,y) = \langle 1 + x^2, xy^2 \rangle$, line segment from $(0,0)$ to $(1,3)$

20. $\mathbf{F}(x,y) = \langle -2, y \rangle$, half-circle $x^2 + y^2 = 1$ with $y \ge 0$, oriented counterclockwise

21. $\mathbf{F}(x,y) = \langle x^2, xy \rangle$, part of circle $x^2 + y^2 = 9$ with $x \le 0$, $y \ge 0$, oriented clockwise

22. $\mathbf{F}(x,y) = \langle e^{y-x}, e^{2x} \rangle$, piecewise linear path from $(1,1)$ to $(2,2)$ to $(0,2)$

23. $\mathbf{F}(x,y) = \langle 3zy^{-1}, 4x, -y \rangle$, $\mathbf{r}(t) = \langle e^t, e^t, t \rangle$ for $-1 \le t \le 1$

24. $\mathbf{F}(x,y) = \left\langle \dfrac{-y}{(x^2+y^2)^2}, \dfrac{x}{(x^2+y^2)^2} \right\rangle$, circle of radius R with center at the origin oriented counterclockwise

25. $\mathbf{F}(x,y,z) = \left\langle \dfrac{1}{y^3+1}, \dfrac{1}{z+1}, 1 \right\rangle$, $\mathbf{r}(t) = \langle t^3, 2, t^2 \rangle$ for $0 \le t \le 1$

26. $\mathbf{F}(x,y,z) = \langle z^3, yz, x \rangle$, quarter of the circle of radius 2 in the yz-plane with center at the origin where $y \ge 0$ and $z \ge 0$, oriented clockwise when viewed from the positive x-axis

In Exercises 27–34, evaluate the line integral.

27. $\displaystyle\int_C x\,dx$, over $y = x^3$ for $0 \le x \le 3$

28. $\displaystyle\int_C y\,dy$, over $y = x^3$ for $0 \le x \le 3$

29. $\displaystyle\int_C y\,dx - x\,dy$, parabola $y = x^2$ for $0 \le x \le 2$

30. $\displaystyle\int_C y\,dx + z\,dy + x\,dz$, $\mathbf{r}(t) = \langle 2 + t^{-1}, t^3, t^2 \rangle$ for $0 \le t \le 1$

31. $\displaystyle\int_C (x-y)\,dx + (y-z)\,dy + z\,dz$, line segment from $(0,0,0)$ to $(1,4,4)$

32. $\displaystyle\int_C z\,dx + x^2\,dy + y\,dz$, $\mathbf{r}(t) = \langle \cos t, \tan t, t \rangle$ for $0 \le t \le \dfrac{\pi}{4}$

33. $\displaystyle\int_C \dfrac{-y\,dx + x\,dy}{x^2 + y^2}$, segment from $(1,0)$ to $(0,1)$

34. $\displaystyle\int_C y^2\,dx + z^2\,dy + (1-x^2)\,dz$, quarter of the circle of radius 1 in the xz-plane with center at the origin in the quadrant $x \ge 0$, $z \le 0$, oriented counterclockwise when viewed from the positive y-axis

35. [CAS] Let $f(x,y,z) = x^{-1}yz$, and let C be the curve parametrized by $\mathbf{r}(t) = \langle \ln t, t, t^2 \rangle$ for $2 \le t \le 4$. Use a computer algebra system to calculate $\displaystyle\int_C f(x,y,z)\,ds$ to four decimal places.

36. [CAS] Use a CAS to calculate $\displaystyle\int_C \langle e^{x-y}, e^{x+y} \rangle \cdot d\mathbf{r}$ to four decimal places, where C is the curve $y = \sin x$ for $0 \le x \le \pi$, oriented from left to right.

In Exercises 37 and 38, calculate the line integral of $\mathbf{F}(x,y,z) = \langle e^z, e^{x-y}, e^y \rangle$ over the given path.

37. The blue path from P to Q in Figure 15

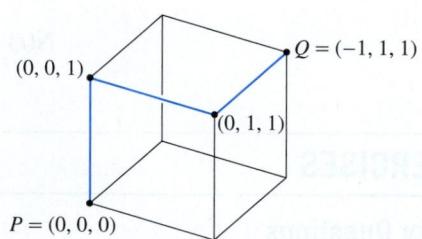

FIGURE 15

38. The closed path $ABCA$ in Figure 16

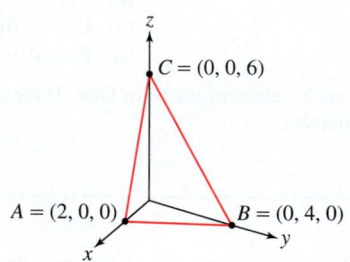

FIGURE 16

In Exercises 39 and 40, C is the path from P to Q in Figure 17 that traces C_1, C_2, and C_3 in the orientation indicated, and $\mathbf{F}$ is a vector field such that

$$\int_C \mathbf{F} \cdot d\mathbf{r} = 5, \qquad \int_{C_1} \mathbf{F} \cdot d\mathbf{r} = 8, \qquad \int_{C_3} \mathbf{F} \cdot d\mathbf{r} = 8$$

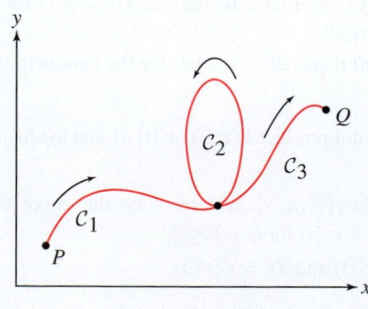

FIGURE 17

39. Determine:

(a) $\displaystyle\int_{-C_3} \mathbf{F}\cdot d\mathbf{r}$ (b) $\displaystyle\int_{C_2} \mathbf{F}\cdot d\mathbf{r}$ (c) $\displaystyle\int_{-C_1-C_3} \mathbf{F}\cdot d\mathbf{r}$

40. Find the value of $\displaystyle\int_{C'} \mathbf{F}\cdot d\mathbf{r}$, where C' is the path that traverses the loop C_2 four times in the clockwise direction.

41. The values of a function $f(x, y, z)$ and vector field $\mathbf{F}(x, y, z)$ are given at six sample points along the path ABC in Figure 18. Estimate the line integrals of f and $\mathbf{F}$ along ABC.

Point	$f(x, y, z)$	$\mathbf{F}(x, y, z)$
$\left(1, \frac{1}{6}, 0\right)$	3	$\langle 1, 0, 2\rangle$
$\left(1, \frac{1}{2}, 0\right)$	3.3	$\langle 1, 1, 3\rangle$
$\left(1, \frac{5}{6}, 0\right)$	3.6	$\langle 2, 1, 5\rangle$
$\left(1, 1, \frac{1}{6}\right)$	4.2	$\langle 3, 2, 4\rangle$
$\left(1, 1, \frac{1}{2}\right)$	4.5	$\langle 3, 3, 3\rangle$
$\left(1, 1, \frac{5}{6}\right)$	4.2	$\langle 5, 3, 3\rangle$

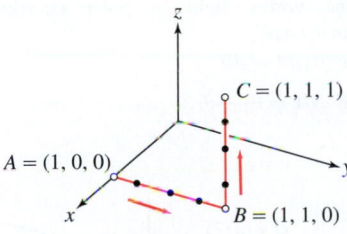

FIGURE 18

42. Estimate the line integrals of $f(x, y)$ and $\mathbf{F}(x, y)$ along the quarter-circle (oriented counterclockwise) in Figure 19 using the values at the three sample points along each path.

Point	$f(x, y)$	$\mathbf{F}(x, y)$
A	1	$\langle 1, 2\rangle$
B	-2	$\langle 1, 3\rangle$
C	4	$\langle -2, 4\rangle$

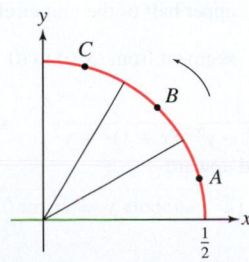

FIGURE 19

43. Determine whether the line integrals of the vector fields around the circle (oriented counterclockwise) in Figure 20 are positive, negative, or zero.

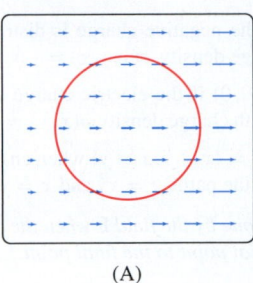 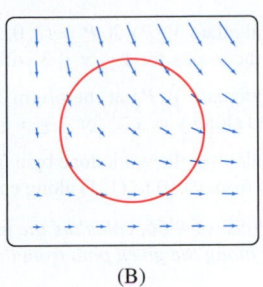

(A) (B)

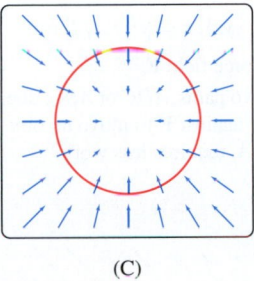

(C)

FIGURE 20

44. Determine whether the line integrals of the vector fields along the oriented curves in Figure 21 are positive or negative.

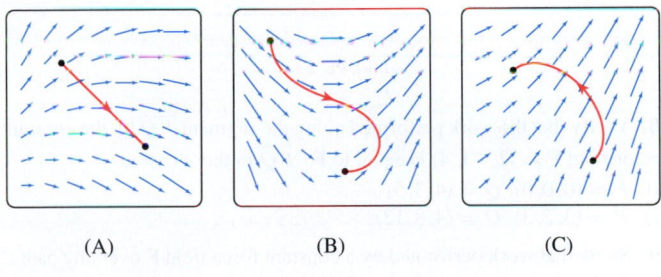

(A) (B) (C)

FIGURE 21

45. Calculate the total mass of a circular piece of wire of radius 4 cm centered at the origin whose mass density is $\rho(x, y) = x^2$ g/cm.

46. Calculate the total mass of a metal tube in the helical shape $\mathbf{r}(t) = (\cos t, \sin t, t^2)$ (distance in centimeters) for $0 \le t \le 2\pi$ if the mass density is $\rho(x, y, z) = \sqrt{z}$ g/cm.

47. Find the total charge on the curve $y = x^{4/3}$ for $1 \le x \le 8$ (in centimeters) assuming a charge density of $\rho(x, y) = x/y$ (in units of 10^{-6} C/cm).

48. Find the total charge on the curve $\mathbf{r}(t) = (\sin t, \cos t, \sin^2 t)$ in centimeters for $0 \le t \le \frac{\pi}{8}$ assuming a charge density of $\rho(x, y, z) = xy(y^2 - z)$ (in units of 10^{-6} C/cm).

In Exercises 49–52, use Eq. (6) to compute the electric potential $V(P)$ at the point P for the given charge density (in units of 10^{-6} C).

49. Calculate $V(P)$ at $P = (0, 0, 12)$ if the electric charge is distributed along the quarter circle of radius 4 centered at the origin with charge density $\rho(x, y, z) = xy$.

50. Calculate $V(P)$ at the origin $P = (0, 0)$ if the negative charge is distributed along $y = x^2$ for $1 \le x \le 2$ with charge density $\rho(x, y) = -y\sqrt{x^2 + 1}$.

51. Calculate $V(P)$ at $P = (2, 0, 2)$ if the negative charge is distributed along the y-axis for $1 \le y \le 3$ with charge density $\rho(x, y, z) = -y$.

52. Calculate $V(P)$ at the origin $P = (0, 0)$ if the electric charge is distributed along $y = x^{-1}$ for $\frac{1}{2} \le x \le 2$ with charge density $\rho(x, y) = x^3 y$.

53. Calculate the work done by a field $\mathbf{F} = \langle x + y, x - y \rangle$ when an object moves from $(0, 0)$ to $(1, 1)$ along each of the paths $y = x^2$ and $x = y^2$.

In Exercises 54–56, calculate the work done by the field $\mathbf{F}$ when the object moves along the given path from the initial point to the final point.

54. $\mathbf{F}(x, y, z) = \langle x, y, z \rangle$, $\mathbf{r} = \langle \cos t, \sin t, t \rangle$ for $0 \le t \le 3\pi$

55. $\mathbf{F}(x, y, z) = \langle xy, yz, xz \rangle$, $\mathbf{r} = \langle t, t^2, t^3 \rangle$ for $0 \le t \le 1$

56. $\mathbf{F}(x, y, z) = \langle e^x, e^y, xyz \rangle$, $\mathbf{r} = \langle t^2, t, t/2 \rangle$ for $0 \le t \le 1$

57. Figure 22 shows a force field $\mathbf{F}$.
(a) Over which of the two paths, ADC or ABC, does $\mathbf{F}$ perform less work?
(b) If you have to work against $\mathbf{F}$ to move an object from C to A, which of the paths, CBA or CDA, requires less work?

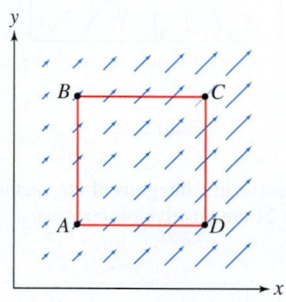

FIGURE 22

58. Verify that the work performed along the segment $\overline{PQ}$ by the constant vector field $\mathbf{F} = \langle 2, -1, 4 \rangle$ is equal to $\mathbf{F} \cdot \overrightarrow{PQ}$ in these cases:
(a) $P = (0, 0, 0)$, $Q = (4, 3, 5)$
(b) $P = (3, 2, 3)$, $Q = (4, 8, 12)$

59. Show that work performed by a constant force field $\mathbf{F}$ over any path $\mathcal{C}$ from P to Q is equal to $\mathbf{F} \cdot \overrightarrow{PQ}$.

60. Note that a curve $\mathcal{C}$ in polar form $r = f(\theta)$ is parametrized by $\mathbf{r}(\theta) = (f(\theta)\cos\theta, f(\theta)\sin\theta))$ because the x- and y-coordinates are given by $x = r\cos\theta$ and $y = r\sin\theta$.
(a) Show that $\|\mathbf{r}'(\theta)\| = \sqrt{f(\theta)^2 + f'(\theta)^2}$.
(b) Evaluate $\displaystyle\int_{\mathcal{C}} (x - y)^2 \, ds$, where $\mathcal{C}$ is the semicircle in Figure 23 with polar equation $r = 2\cos\theta$, $0 \le \theta \le \frac{\pi}{2}$.

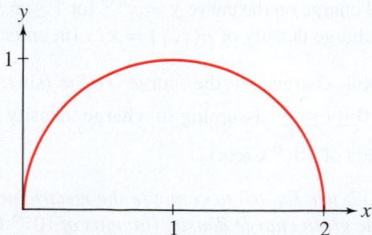

FIGURE 23 Semicircle $r = 2\cos\theta$.

61. Charge is distributed along the spiral with polar equation $r = \theta$ for $0 \le \theta \le 2\pi$. The charge density is $\rho(r, \theta) = r$ (assume distance is in centimeters and charge in units of 10^{-6} C/cm). Use the result of Exercise 60(a) to compute the total charge.

In Exercises 62 and 63, let $\mathbf{F}$ be the vortex field (Figure 24):

$$\mathbf{F}(x, y) = \left\langle \frac{-y}{x^2 + y^2}, \frac{x}{x^2 + y^2} \right\rangle$$

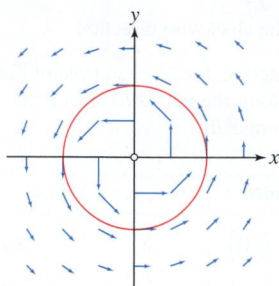

FIGURE 24

62. Let $a > 0$, $b < c$. Show that the integral of $\mathbf{F}$ along the segment [Figure 25(A)] from $P = (a, b)$ to $Q = (a, c)$ is equal to the angle $\angle POQ$.

63. Let $\mathcal{C}$ be a curve in polar form $r = f(\theta)$ for $\theta_1 \le \theta \le \theta_2$ [Figure 25(B)], parametrized by $\mathbf{r}(\theta) = (f(\theta)\cos\theta, f(\theta)\sin\theta))$ as in Exercise 60.
(a) Show that the vortex field in polar coordinates is written $\mathbf{F}(r, \theta) = r^{-1}\langle -\sin\theta, \cos\theta \rangle$.
(b) Show that $\mathbf{F} \cdot \mathbf{r}'(\theta)\, d\theta = d\theta$.
(c) Show that $\displaystyle\int_{\mathcal{C}} \mathbf{F} \cdot d\mathbf{r} = \theta_2 - \theta_1$.

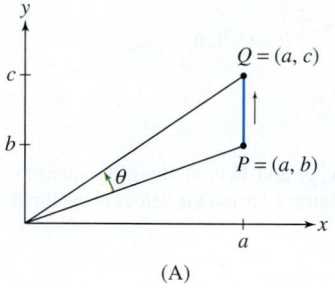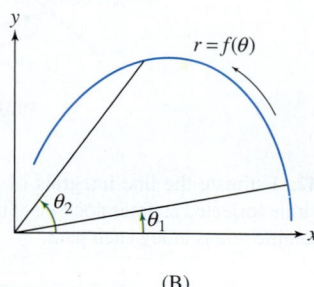

(A)   (B)

FIGURE 25

In Exercises 64–67, use Eq. (9) to calculate the flux of the vector field across the curve specified.

64. $\mathbf{F}(x, y) = \langle -y, x \rangle$; upper half of the unit circle, oriented clockwise

65. $\mathbf{F}(x, y) = \langle x^2, y^2 \rangle$; segment from $(3, 0)$ to $(0, 3)$, oriented upward

66. $\mathbf{F}(x, y) = \left\langle \dfrac{x + 1}{(x + 1)^2 + y^2}, \dfrac{y}{(x + 1)^2 + y^2} \right\rangle$; segment $1 \le y \le 4$ along the y-axis, oriented upward

67. $\mathbf{F}(x, y) = \langle e^y, 2x - 1 \rangle$; parabola $y = x^2$ for $0 \le x \le 1$, oriented left to right

68. Let $I = \displaystyle\int_{\mathcal{C}} f(x, y, z) \, ds$. Assume that $f(x, y, z) \ge m$ for some number m and all points (x, y, z) on $\mathcal{C}$. Which of the following conclusions is correct? Explain.
(a) $I \ge m$
(b) $I \ge mL$, where L is the length of $\mathcal{C}$

Further Insights and Challenges

69. Let $\mathbf{F}(x, y) = \langle x, 0 \rangle$. Prove that if C is any path from (a, b) to (c, d), then

$$\int_C \mathbf{F} \cdot d\mathbf{r} = \frac{1}{2}(c^2 - a^2)$$

70. Let $\mathbf{F}(x, y) = \langle y, x \rangle$. Prove that if C is any path from (a, b) to (c, d), then

$$\int_C \mathbf{F} \cdot d\mathbf{r} = cd - ab$$

71. We wish to define the **average value** $\mathrm{Av}(f)$ of a continuous function f along a curve C of length L. Divide C into N consecutive arcs $C_1, \ldots, C_N$, each of length L/N, and let P_i be a sample point in C_i (Figure 26). The sum

$$\frac{1}{N} \sum_{i=1}^{N} f(P_i)$$

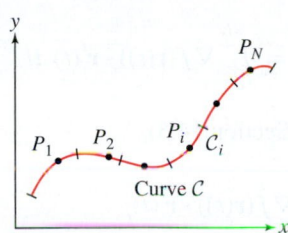

FIGURE 26

may be considered an approximation to $\mathrm{Av}(f)$, so we define

$$\mathrm{Av}(f) = \lim_{N \to \infty} \frac{1}{N} \sum_{i=1}^{N} f(P_i)$$

Prove that

$$\mathrm{Av}(f) = \frac{1}{L} \int_C f(x, y, z)\, ds \qquad \boxed{10}$$

Hint: Show that $\dfrac{L}{N} \sum_{i=1}^{N} f(P_i)$ is a Riemann sum approximation to the line integral of f along C.

72. Use Eq. (10) to calculate the average value of $f(x, y) = x - y$ along the segment from $P = (2, 1)$ to $Q = (5, 5)$.

73. Use Eq. (10) to calculate the average value of $f(x, y) = x$ along the curve $y = x^2$ for $0 \le x \le 1$.

74. The temperature (in degrees centigrade) at a point P on a circular wire of radius 2 cm centered at the origin is equal to the square of the distance from P to $P_0 = (2, 0)$. Compute the average temperature along the wire.

75. The value of a scalar line integral does not depend on the choice of parametrization (because it is defined without reference to a parametrization). Prove this directly. That is, suppose that $\mathbf{r}_1(t)$ and $\mathbf{r}(t)$ are two parametrizations such that $\mathbf{r}_1(t) = \mathbf{r}(\varphi(t))$, where $\varphi(t)$ is an increasing function. Use the Change of Variables Formula to verify that

$$\int_c^d f(\mathbf{r}_1(t)) \|\mathbf{r}_1'(t)\|\, dt = \int_a^b f(\mathbf{r}(t)) \|\mathbf{r}'(t)\|\, dt$$

where $a = \varphi(c)$ and $b = \varphi(d)$.

16.3 Conservative Vector Fields

← **REMINDER**

- A vector field $\mathbf{F}$ is conservative if $\mathbf{F} = \nabla f$ for some function $f(x, y, z)$.
- f is called a potential function.

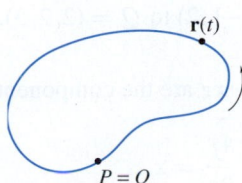

FIGURE 1 The circulation around a closed path is denoted $\oint_C \mathbf{F} \cdot d\mathbf{r}$.

In this section, we study conservative vector fields in greater depth. One important property we will see is that the vector line integral of a conservative vector field around a closed curve is zero.

When a curve C is closed, we often refer to the line integral of any vector field $\mathbf{F}$ around C as the **circulation** of $\mathbf{F}$ around C (Figure 1) and denote it with the symbol $\oint$:

$$\oint_C \mathbf{F} \cdot d\mathbf{r}$$

It actually does not matter what point we take as the starting point when we have a closed curve.

Suppose A and B are two points on the closed curve. If we start at A, the circulation as we travel around the curve back to A is the sum of the line integral from A to B and the line integral from B the rest of the way along the curve back to A. Switching the order of these two line integrals yields the circulation as we start from B and then return to B, thereby obtaining the same result.

Our first result establishes the fundamental **path independence** of conservative vector fields. This theorem indicates that the line integral of $\mathbf{F}$ along a path from P to Q depends only on the endpoints P and Q, and not on the particular path followed from P to Q (Figure 2).

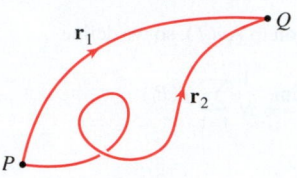

FIGURE 2 Path independence: If **F** is conservative, then the line integrals over $\mathbf{r}_1$ and $\mathbf{r}_2$ are equal.

THEOREM 1 Fundamental Theorem for Conservative Vector Fields Assume that $\mathbf{F} = \nabla f$ on a domain $\mathcal{D}$.

1. If **r** is a path along a curve $\mathcal{C}$ from P to Q in $\mathcal{D}$, then

$$\int_{\mathcal{C}} \mathbf{F} \cdot d\mathbf{r} = f(Q) - f(P)$$
$$\boxed{1}$$

In particular, **F** is path independent.

2. The circulation around a closed curve $\mathcal{C}$ (i.e., $P = Q$) is zero:

$$\oint_{\mathcal{C}} \mathbf{F} \cdot d\mathbf{r} = 0$$

Proof Let $\mathbf{r}(t)$ be a path along the curve $\mathcal{C}$ in $\mathcal{D}$ for $a \leq t \leq b$ with $\mathbf{r}(a) = P$ and $\mathbf{r}(b) = Q$. Then

$$\int_{\mathcal{C}} \mathbf{F} \cdot d\mathbf{r} = \int_{\mathcal{C}} \nabla f \cdot d\mathbf{r} = \int_a^b \nabla f(\mathbf{r}(t)) \cdot \mathbf{r}'(t) \, dt$$

By the Chain Rule for Paths (Theorem 2 in Section 14.5),

$$\frac{d}{dt} f(\mathbf{r}(t)) = \nabla f(\mathbf{r}(t)) \cdot \mathbf{r}'(t)$$

Thus, we can apply the Fundamental Theorem of Calculus:

$$\int_{\mathcal{C}} \mathbf{F} \cdot d\mathbf{r} = \int_a^b \frac{d}{dt} f(\mathbf{r}(t)) \, dt = f(\mathbf{r}(t))\Big|_a^b = f(\mathbf{r}(b)) - f(\mathbf{r}(a)) = f(Q) - f(P)$$

This proves Eq. (1). It also proves path independence, because the quantity $f(Q) - f(P)$ depends on the endpoints but not on the path **r**. If **r** is a closed path, then $P = Q$ and $f(Q) - f(P) = 0$. ∎

EXAMPLE 1 Let $\mathbf{F}(x, y, z) = \langle 2xy + z, x^2, x \rangle$.

(a) Verify that $f(x, y, z) = x^2 y + xz$ is a potential function for **F**.

(b) Evaluate $\displaystyle\int_{\mathcal{C}} \mathbf{F} \cdot d\mathbf{r}$, where $\mathcal{C}$ is a curve from $P = (1, -1, 2)$ to $Q = (2, 2, 3)$.

Solution (a) The partial derivatives of $f(x, y, z) = x^2 y + xz$ are the components of **F**:

$$\frac{\partial f}{\partial x} = 2xy + z, \qquad \frac{\partial f}{\partial y} = x^2, \qquad \frac{\partial f}{\partial z} = x$$

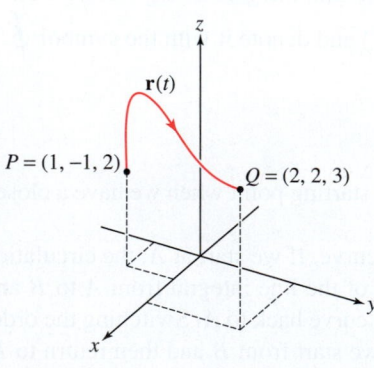

FIGURE 3 An arbitrary path from $(1, -1, 2)$ to $(2, 2, 3)$.

Therefore, $\nabla f = \langle 2xy + z, x^2, x \rangle = \mathbf{F}$, implying that f is a potential function for **F**.

(b) By Theorem 1, the line integral over any path $\mathbf{r}(t)$ from $P = (1, -1, 2)$ to $Q = (2, 2, 3)$ (Figure 3) has the value

$$\int_{\mathcal{C}} \mathbf{F} \cdot d\mathbf{r} = f(Q) - f(P) = f(2, 2, 3) - f(1, -1, 2)$$

$$= \left(2^2(2) + 2(3)\right) - \left(1^2(-1) + 1(2)\right) = 13$$

∎

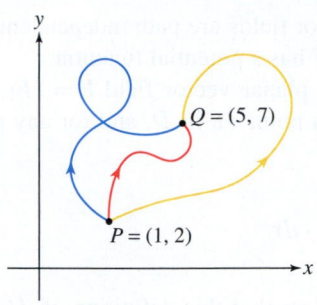

FIGURE 4 Paths from $(1, 2)$ to $(5, 7)$.

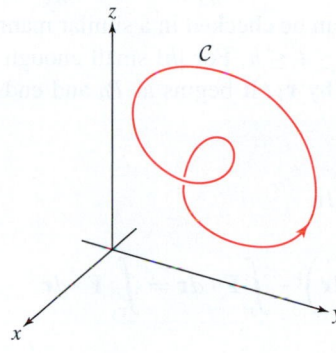

FIGURE 5 The line integral of a conservative vector field around a closed curve is zero.

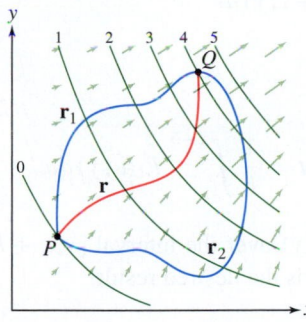

FIGURE 6 Vector field $\mathbf{F} = \nabla f$ with the contour lines of f.

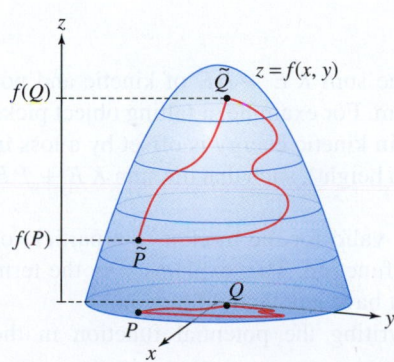

FIGURE 7 The potential surface $z = f(x, y)$.

EXAMPLE 2 Find a potential function for $\mathbf{F} = \langle 2x + y, x \rangle$ and use it to evaluate $\int_C \mathbf{F} \cdot d\mathbf{r}$, where $\mathbf{r}$ is any path (Figure 4) from $(1, 2)$ to $(5, 7)$.

Solution Later in this section, we will develop a general method for finding potential functions. At this point, we can see that $f(x, y) = x^2 + xy$ satisfies $\nabla f = \mathbf{F}$:

$$\frac{\partial f}{\partial x} = \frac{\partial}{\partial x}(x^2 + xy) = 2x + y, \qquad \frac{\partial f}{\partial y} = \frac{\partial}{\partial y}(x^2 + xy) = x$$

Therefore, for any path $\mathbf{r}$ from $(1, 2)$ to $(5, 7)$,

$$\int_C \mathbf{F} \cdot d\mathbf{r} = f(5, 7) - f(1, 2) = (5^2 + 5(7)) - (1^2 + 1(2)) = 57 \qquad \blacksquare$$

EXAMPLE 3 **Integral Around a Closed Path** Let $f(x, y, z) = xy \sin(yz)$. Evaluate $\oint_C \nabla f \cdot d\mathbf{r}$, where C is the closed curve in Figure 5.

Solution By Theorem 1, the integral of a gradient vector field around any closed path is zero. In other words, $\oint_C \nabla f \cdot d\mathbf{r} = 0$. $\qquad \blacksquare$

CONCEPTUAL INSIGHT A good way to think about path independence is in terms of the contour map of the potential function. Consider a vector field $\mathbf{F} = \nabla f$ in the plane (Figure 6). The level curves of f are called **equipotential curves**, and the value $f(P)$ is called the potential at P.

When we integrate $\mathbf{F}$ along a path $\mathbf{r}(t)$ from P to Q, the integrand is

$$\mathbf{F}(\mathbf{r}(t)) \cdot \mathbf{r}'(t) = \nabla f(\mathbf{r}(t)) \cdot \mathbf{r}'(t)$$

Now recall that by the Chain Rule for Paths,

$$\nabla f(\mathbf{r}(t)) \cdot \mathbf{r}'(t) = \frac{d}{dt} f(\mathbf{r}(t))$$

In other words, the integrand is the rate at which the potential changes along the path, and thus the integral itself is the net change in potential:

$$\int \mathbf{F} \cdot d\mathbf{r} = \underbrace{f(Q) - f(P)}_{\text{Net change in potential}}$$

Note that the change in potential depends on only the equipotential curves at the beginning and the end of the path. It does not matter how we get from one end to the other, whether we cross equipotential curves only once along the way, or double back and cross them multiple times.

We can also interpret the line integral in terms of the graph of the potential function $z = f(x, y)$. The line integral computes the change in height as we move on the surface (Figure 7). Again, this change in height does not depend on the path from P to Q. Of course, these interpretations apply only to conservative vector fields—otherwise, there is no potential function.

You might wonder whether there exist any path-independent vector fields other than the conservative ones. The answer is no. By the next theorem, a path-independent vector field is necessarily conservative.

> **THEOREM 2** A vector field $\mathbf{F}$ on an open connected domain $\mathcal{D}$ is path independent if and only if it is conservative.

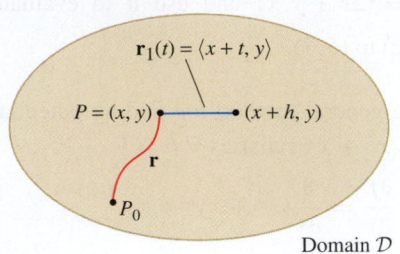

Domain $\mathcal{D}$

FIGURE 8

Proof We have already shown that conservative vector fields are path independent. So, we assume that **F** is path independent and prove that **F** has a potential function.

To simplify the notation, we treat the case of a planar vector field $\mathbf{F} = \langle F_1, F_2 \rangle$. The proof for vector fields in $\mathbf{R}^3$ is similar. Choose a point P_0 in $\mathcal{D}$, and for any point $P = (x, y) \in \mathcal{D}$, define

$$f(P) = f(x, y) = \int_{\mathcal{C}} \mathbf{F} \cdot d\mathbf{r}$$

where **r** is any path in $\mathcal{D}$ from P_0 to P (Figure 8). Note that this definition of $f(P)$ is meaningful only because we assume that the line integral does not depend on the path **r**.

We will prove that $\mathbf{F} = \nabla f$, which involves showing that $\dfrac{\partial f}{\partial x} = F_1$ and $\dfrac{\partial f}{\partial y} = F_2$. We will verify only the first equation, as the second can be checked in a similar manner. Let $\mathbf{r}_1$ be the horizontal path $\mathbf{r}_1(t) = (x + t, y)$ for $0 \leq t \leq h$. For $|h|$ small enough, $\mathbf{r}_1$ lies inside $\mathcal{D}$. Let $\mathbf{r} + \mathbf{r}_1$ denote the path **r** followed by $\mathbf{r}_1$. It begins at P_0 and ends at $(x + h, y)$, so

$$f(x + h, y) - f(x, y) = \int_{\mathbf{r}+\mathbf{r}_1} \mathbf{F} \cdot d\mathbf{r} - \int_{\mathbf{r}} \mathbf{F} \cdot d\mathbf{r}$$

$$= \left(\int_{\mathbf{r}} \mathbf{F} \cdot d\mathbf{r} + \int_{\mathbf{r}_1} \mathbf{F} \cdot d\mathbf{r} \right) - \int_{\mathbf{r}} \mathbf{F} \cdot d\mathbf{r} = \int_{\mathbf{r}_1} \mathbf{F} \cdot d\mathbf{r}$$

The path $\mathbf{r}_1$ has tangent vector $\mathbf{r}_1'(t) = \langle 1, 0 \rangle$, so

$$\mathbf{F}(\mathbf{r}_1(t)) \cdot \mathbf{r}_1'(t) = \langle F_1(x + t, y), F_2(x + t, y) \rangle \cdot \langle 1, 0 \rangle = F_1(x + t, y)$$

$$f(x + h, y) - f(x, y) = \int_{\mathbf{r}_1} \mathbf{F} \cdot d\mathbf{r} = \int_0^h F_1(x + t, y)\, dt$$

Using the substitution $u = x + t$, we have

$$\frac{f(x + h, y) - f(x, y)}{h} = \frac{1}{h} \int_0^h F_1(x + t, y)\, dt = \frac{1}{h} \int_x^{x+h} F_1(u, y)\, du$$

The integral on the right is the average value of $F_1(u, y)$ over the interval $[x, x + h]$. It converges to the value $F_1(x, y)$ as $h \to 0$, and this yields the desired result:

$$\frac{\partial f}{\partial x} = \lim_{h \to 0} \frac{f(x + h, y) - f(x, y)}{h} = \lim_{h \to 0} \frac{1}{h} \int_x^{x+h} F_1(u, y)\, du = F_1(x, y) \qquad \blacksquare$$

Conservative Fields in Physics

The Conservation of Energy principle says that the sum $KE + PE$ of kinetic and potential energy remains constant in an isolated system. For example, a falling object picks up kinetic energy as it falls to Earth, but this gain in kinetic energy is offset by a loss in gravitational potential energy (g times the change in height), such that the sum $KE + PE$ remains unchanged.

We show now that conservation of energy is valid for the motion of a particle of mass m under a force field **F** if **F** has a potential function. This explains why the term "conservative" is used to describe vector fields that have a potential function.

We follow the convention in physics of writing the potential function in the form $-V$. Thus,

*In a conservative force field, the work W against **F** required to move the particle from P to Q is equal to the change in potential energy:*

$$W = -\int_{\mathcal{C}} \mathbf{F} \cdot d\mathbf{r} = V(Q) - V(P)$$

$$\mathbf{F} = -\nabla V$$

When the particle is located at $P = (x, y, z)$, it is said to have **potential energy** $V(P)$. Suppose that the particle moves along a path $\mathbf{r}(t)$. The particle's velocity is $\mathbf{v} = \mathbf{r}'(t)$, and its **kinetic energy** is $KE = \frac{1}{2}m\|\mathbf{v}\|^2 = \frac{1}{2}m\mathbf{v} \cdot \mathbf{v}$. By definition, the **total energy** at time t is the sum

$$E = KE + PE = \frac{1}{2}m\mathbf{v} \cdot \mathbf{v} + V(\mathbf{r}(t))$$

> **THEOREM 3 Conservation of Energy** The total energy E of a particle moving under the influence of a conservative force field $\mathbf{F} = -\nabla V$ is constant in time. That is,
> $$\frac{dE}{dt} = 0.$$

Proof Let $\mathbf{a} = \mathbf{v}'(t)$ be the particle's acceleration and m its mass. According to Newton's Second Law of Motion, $\mathbf{F}(\mathbf{r}(t)) = m\mathbf{a}(t)$, and thus

$$\frac{dE}{dt} = \frac{d}{dt}\left(\frac{1}{2}m\mathbf{v} \cdot \mathbf{v} + V(\mathbf{r}(t))\right)$$

$$= \frac{1}{2}m\left(\frac{d\mathbf{v}}{dt} \cdot \mathbf{v} + \mathbf{v} \cdot \frac{d\mathbf{v}}{dt}\right) + \nabla V(\mathbf{r}(t)) \cdot \mathbf{r}'(t) \qquad \text{(Product and Chain Rules)}$$

$$= m\mathbf{v} \cdot \mathbf{a} + \nabla V(\mathbf{r}(t)) \cdot \mathbf{r}'(t)$$

$$= \mathbf{v} \cdot m\mathbf{a} - \mathbf{F} \cdot \mathbf{v} \qquad \text{(since } \mathbf{F} = -\nabla V \text{ and } \mathbf{r}'(t) = \mathbf{v}\text{)}$$

$$= \mathbf{v} \cdot (m\mathbf{a} - \mathbf{F}) = 0 \qquad \text{(since } \mathbf{F} = m\mathbf{a}\text{)} \qquad \blacksquare$$

In Example 8 of Section 16.1, we verified that inverse-square vector fields are conservative:

$$\mathbf{F} = k\frac{\mathbf{e}_r}{r^2} = -\nabla f \quad \text{with} \quad f = \frac{k}{r}$$

Basic examples of inverse-square vector fields are the gravitational and electrostatic forces due to a point mass or charge. By convention, these fields have units of force *per unit mass or unit charge*. Thus, if $\mathbf{F}$ is a gravitational field, the force on a particle of mass m is $m\mathbf{F}$ and its potential energy is mf, where $\mathbf{F} = -\nabla f$.

EXAMPLE 4 **Work Against Gravity** Compute the work W against the earth's gravitational field required to move a satellite of mass $m = 600$ kg along any path from an orbit of altitude 2000 km to an orbit of altitude 4000 km.

Solution The earth's gravitational field is the inverse-square field

$$\mathbf{F} = -k\frac{\mathbf{e}_r}{r^2} = -\nabla f, \qquad f = -\frac{k}{r}$$

where r is the distance from the center of the earth and $k = 4 \cdot 10^{14}$ (see marginal note). The radius of the earth is approximately $6.4 \cdot 10^6$ m, so the satellite must be moved from $r = 8.4 \cdot 10^6$ meters to $r = 10.4 \cdot 10^6$ m. The force on the satellite is $m\mathbf{F} = 600\mathbf{F}$, and the work W required to move the satellite along a path $\mathbf{r}$ is

$$W = -\int_{\mathbf{r}} m\mathbf{F} \cdot d\mathbf{r} = 600\int_{\mathbf{r}} \nabla f \cdot d\mathbf{r}$$

$$= -\frac{600k}{r}\Bigg|_{8.4\cdot10^6}^{10.4\cdot10^6}$$

$$\approx -\frac{2.4 \cdot 10^{17}}{10.4 \cdot 10^6} + \frac{2.4 \cdot 10^{17}}{8.4 \cdot 10^6} \approx 5.5 \cdot 10^9 \text{ joules} \qquad \blacksquare$$

Potential functions first appeared in 1774 in the writings of Joseph-Louis Lagrange (1736–1813). One of the greatest mathematicians of his time, Lagrange made fundamental contributions to physics, analysis, algebra, and number theory. He was born in Turin, Italy, to a family of French origin but spent most of his career first in Berlin and then in Paris. After the French Revolution, Lagrange was required to teach courses in elementary mathematics, but apparently he spoke above the heads of his audience. A contemporary wrote, "Whatever this great man says deserves the highest degree of consideration, but he is too abstract for youth."

Example 8 of Section 16.1 showed that

$$\frac{\mathbf{e}_r}{r^2} = -\nabla\left(\frac{1}{r}\right)$$

The constant k is equal to GM_e, where $G \approx 6.67 \cdot 10^{-11}$ m^3 kg^{-1} s^{-2} and the mass of the earth is $M_e \approx 5.98 \cdot 10^{24}$ kg:

$$k = GM_e \approx 4 \cdot 10^{14} \text{ } m^3 s^{-2}$$

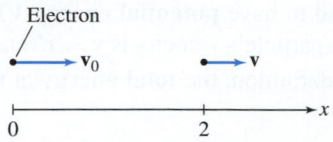

FIGURE 9 An electron moving in an electric field.

EXAMPLE 5 An electron is traveling in the positive x-direction with speed $v_0 = 10^7$ m/s. When it passes $x = 0$, a horizontal electric field $\mathbf{E} = 100x\mathbf{i}$ (in newtons per coulomb) is turned on. Find the electron's velocity after it has traveled 2 m (see Figure 9). Assume that $q_e/m_e = -1.76 \cdot 10^{11}$ C/kg, where m_e and q_e are the mass and charge of the electron, respectively.

Solution We have $\mathbf{E} = -\nabla V$, where $V(x, y, z) = -50x^2$, so the electric field is conservative. Since V depends only on x, we write $V(x)$ for $V(x, y, z)$. By the Law of Conservation of Energy, the electron's total energy E is constant, and therefore is the same when the electron is at $x = 2$ as it is when the electron is at $x = 0$. That is,

$$E = \frac{1}{2}m_e v_0^2 + q_e V(0) = \frac{1}{2}m_e v^2 + q_e V(2)$$

Since $V(0) = 0$, we obtain

$$\frac{1}{2}m_e v_0^2 = \frac{1}{2}m_e v^2 + q_e V(2) \quad \Rightarrow \quad v = \sqrt{v_0^2 - 2(q_e/m_e)V(2)}$$

Using the numerical value of q_e/m_e, we have

$$v \approx \sqrt{10^{14} - 2(-1.76 \cdot 10^{11})(-50(2)^2)} \approx \sqrt{2.96 \cdot 10^{13}} \approx 5.4 \cdot 10^6 \text{ m/s}$$

Note that the velocity has decreased. This is because $\mathbf{E}$ exerts a force in the negative x-direction on a negative charge. ∎

Finding Potential Functions

We do not yet have an effective way of telling whether a given vector field is conservative. By Theorem 1 in Section 16.1, every conservative vector field in $\mathbf{R}^3$ satisfies the condition

$$\boxed{\text{curl}(\mathbf{F}) = \mathbf{0}, \quad \text{or equivalently,} \quad \frac{\partial F_1}{\partial y} = \frac{\partial F_2}{\partial x}, \quad \frac{\partial F_2}{\partial z} = \frac{\partial F_3}{\partial y}, \quad \frac{\partial F_3}{\partial x} = \frac{\partial F_1}{\partial z}}$$

$\boxed{2}$

But does this condition guarantee that $\mathbf{F}$ is conservative? The answer is a qualified yes; the cross-partials condition does guarantee that $\mathbf{F}$ is conservative, but only on domains $\mathcal{D}$ with a property called simple connectedness.

Roughly speaking, a domain $\mathcal{D}$ in the plane is **simply connected** if it is connected and it does not have any "holes" (Figure 10). More precisely, $\mathcal{D}$ is simply connected if every loop in $\mathcal{D}$ can be shrunk to a point *while staying within* $\mathcal{D}$ as in Figure 11(A). Examples of simply connected regions in $\mathbf{R}^2$ are disks, rectangles, and the entire plane $\mathbf{R}^2$. By contrast, the disk with a point removed in Figure 11(B) is not simply connected: The loop cannot be drawn down to a point without passing through the point that was removed. In $\mathbf{R}^3$, the interiors of balls and boxes are simply connected, as is the entire space $\mathbf{R}^3$.

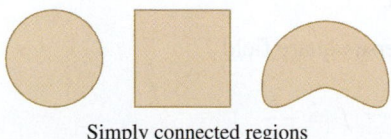

Simply connected regions

Nonsimply connected regions

FIGURE 10 Simply connected regions have no holes.

> **THEOREM 4 Existence of a Potential Function**   Let $\mathbf{F}$ be a vector field on a simply connected domain $\mathcal{D}$. If $\mathbf{F}$ satisfies the cross-partials condition, then $\mathbf{F}$ is conservative.

Rather than prove Theorem 4, we illustrate a practical procedure for finding a potential function when the cross-partials condition is satisfied. The proof itself involves Stokes' Theorem and is somewhat technical because of the role played by the simply connected property of the domain.

(A) Simply connected region:
Any loop can be drawn down to
a point within the region.

(B) Nonsimply connected region:
A loop around the hole cannot
be drawn tight without
passing through the hole.

FIGURE 11

EXAMPLE 6 **Finding a Potential Function** Show that

$$\mathbf{F} = \langle 2xy + y^3, x^2 + 3xy^2 + 2y \rangle$$

is conservative and find a potential function.

Solution First we observe that the cross-partial derivatives are equal:

$$\frac{\partial F_1}{\partial y} = \frac{\partial}{\partial y}(2xy + y^3) \qquad = 2x + 3y^2$$

$$\frac{\partial F_2}{\partial x} = \frac{\partial}{\partial x}(x^2 + 3xy^2 + 2y) = 2x + 3y^2$$

Furthermore, $\mathbf{F}$ is defined on all of $\mathbf{R}^2$, which is a simply connected domain. Therefore, a potential function exists by Theorem 4.

Now, the potential function f satisfies

$$\frac{\partial f}{\partial x} = F_1(x, y) = 2xy + y^3$$

This tells us that f is an antiderivative of $F_1(x, y)$, regarded as a function of x alone:

$$f(x, y) = \int F_1(x, y)\, dx$$

$$= \int \left(2xy + y^3\right) dx$$

$$= x^2 y + xy^3 + g(y)$$

As usual, when we antidifferentiate, we include a constant of integration, a term whose derivative with respect to the integration variable is zero. When we antidifferentiate with respect to x, the constant of integration could depend on the other variables present. In this case, the constant of integration depends on y.

where $g(y)$ is a constant (with respect to x) of integration. Similarly, we have

$$f(x, y) = \int F_2(x, y)\, dy$$

$$= \int \left(x^2 + 3xy^2 + 2y\right) dy$$

$$= x^2 y + xy^3 + y^2 + h(x)$$

The two expressions for $f(x, y)$ must be equal:

$$x^2 y + xy^3 + g(y) = x^2 y + xy^3 + y^2 + h(x)$$

As always, it is good to check your result. Compute ∇f and make sure it equals $\mathbf{F}$.

From this it follows that $g(y) = y^2$ and $h(x) = 0$, up to the addition of an arbitrary numerical constant C. Thus, we obtain the general potential function

$$f(x, y) = x^2 y + xy^3 + y^2 + C \qquad \blacksquare$$

In the next example, we show that the approach used in Example 6 can be carried over to find a potential function for vector fields in $\mathbf{R}^3$.

EXAMPLE 7 Find a potential function for

$$\mathbf{F} = \left\langle 2xyz^{-1}, z + x^2z^{-1}, y - x^2yz^{-2} \right\rangle$$

Solution If a potential function f exists, then it satisfies

$$f(x, y, z) = \int 2xyz^{-1}\, dx \qquad = x^2yz^{-1} + f(y, z)$$

$$f(x, y, z) = \int \left(z + x^2z^{-1}\right) dy \quad = zy + x^2z^{-1}y + g(x, z)$$

$$f(x, y, z) = \int \left(y - x^2yz^{-2}\right) dz = yz + x^2yz^{-1} + h(x, y)$$

These three ways of writing $f(x, y, z)$ must be equal:

$$x^2yz^{-1} + f(y, z) = zy + x^2z^{-1}y + g(x, z) = yz + x^2yz^{-1} + h(x, y)$$

These equalities hold if $f(y, z) = yz$, $g(x, z) = 0$, and $h(x, y) = 0$. Thus, $\mathbf{F}$ is conservative and, for any constant C, a potential function is

$$f(x, y, z) = x^2yz^{-1} + yz + C \qquad \blacksquare$$

In Example 7, $\mathbf{F}$ is defined only for $z \neq 0$, so the domain has two halves: $z > 0$ and $z < 0$. We are free to choose different constants C on the two halves, if desired.

Assumptions Matter We cannot expect the method for finding a potential function to work if $\mathbf{F}$ does not satisfy the cross-partials condition (because in this case, no potential function exists). What goes wrong? Consider $\mathbf{F} = \langle y, 0 \rangle$. If we attempted to find a potential function, we would calculate

$$f(x, y) = \int y\, dx = xy + g(y)$$

$$f(x, y) = \int 0\, dy = 0 + h(x)$$

However, there is no choice of $g(y)$ and $h(x)$ for which $xy + g(y) = h(x)$. If there were, and we differentiated this equation twice, once with respect to x and once with respect to y, we would obtain the contradiction $1 = 0$. The method fails in this case because $\mathbf{F}$ does not satisfy the cross-partials condition, and thus is not conservative.

The Vortex Field

Why does Theorem 4 require that the domain is simply connected? This is an interesting question that we can answer by examining the vortex field that we introduced in the previous section,

$$\mathbf{F} = \left\langle \frac{-y}{x^2 + y^2}, \frac{x}{x^2 + y^2} \right\rangle$$

EXAMPLE 8 Show that the vortex field satisfies the cross-partials condition but is not conservative. Does this contradict Theorem 4?

Solution We check the cross-partials condition directly:

$$\frac{\partial}{\partial x}\left(\frac{x}{x^2 + y^2}\right) = \frac{(x^2 + y^2) - x(\partial/\partial x)(x^2 + y^2)}{(x^2 + y^2)^2} = \frac{y^2 - x^2}{(x^2 + y^2)^2}$$

$$\frac{\partial}{\partial y}\left(\frac{-y}{(x^2 + y^2)}\right) = \frac{-(x^2 + y^2) + y(\partial/\partial y)(x^2 + y^2)}{(x^2 + y^2)^2} = \frac{y^2 - x^2}{(x^2 + y^2)^2}$$

In Example 6 in the previous section, we showed that $\oint_{\mathcal{C}} \mathbf{F} \cdot d\mathbf{r} = 2\pi \neq 0$ for any circle $\mathcal{C}$ centered at the origin. If $\mathbf{F}$ were conservative, its circulation around every closed

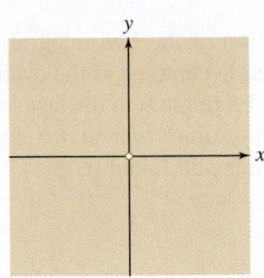

FIGURE 12 The domain $\mathcal{D}$ of the vortex field **F** is the plane with the origin removed. This domain is not simply connected.

$\longleftarrow$ **REMINDER** $\dfrac{d}{dt}\tan^{-1}t = \dfrac{1}{1+t^2}$

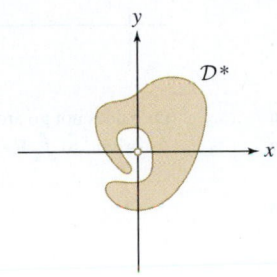

FIGURE 13 There is a potential function for **F** on $\mathcal{D}^*$.

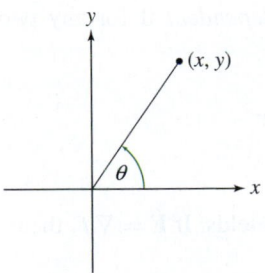

FIGURE 14 The angle θ is the inverse tangent of y/x.

curve would be zero by Theorem 1. Thus, **F** cannot be conservative, even though it satisfies the cross-partials condition.

This result does not contradict Theorem 4 because the domain of **F** does not satisfy the simply connected condition of the theorem. Because **F** is not defined at $(x, y) = (0, 0)$, its domain is $\mathcal{D} = \{(x, y) \neq (0, 0)\}$, and this domain is not simply connected (Figure 12). $\blacksquare$

CONCEPTUAL INSIGHT Although the vortex field **F** is not conservative on its domain $\mathcal{D} = \{(x, y) \neq (0, 0)\}$, it is conservative on any simply connected domain contained in $\mathcal{D}$. For example, on the right half-plane $\{(x, y) : x > 0\}$ and on the left half-plane $\{(x, y) : x < 0\}$, **F** is conservative with potential function $f(x, y) = \tan^{-1}\frac{y}{x}$. We can verify that f is a potential function for **F** by directly computing the partial derivatives:

$$\frac{\partial f}{\partial x} = \frac{\partial}{\partial x}\tan^{-1}\frac{y}{x} = \frac{-y/x^2}{1+(y/x)^2} = \frac{-y}{x^2+y^2} \qquad (x \neq 0)$$

$$\frac{\partial f}{\partial y} = \frac{\partial}{\partial y}\tan^{-1}\frac{y}{x} = \frac{1/x}{1+(y/x)^2} = \frac{x}{x^2+y^2} \qquad (x \neq 0)$$

Furthermore, on the upper half-plane and on the lower half-plane, **F** is conservative with potential function $g(x, y) = -\tan^{-1}\frac{x}{y}$. (See Exercise 32.)

Even if a simply connected domain $\mathcal{D}^*$ is irregularly shaped, like the domain in Figure 13, we can specify a potential function for **F**, although the function may not be expressed as simply as f or g. Nevertheless, we can define a potential function as follows: Fix a point $(x_0, y_0) \in \mathcal{D}^*$, and for every $(x, y) \in \mathcal{D}^*$, choose a path $\mathcal{C}_{(x,y)}$ in $\mathcal{D}^*$ from (x_0, y_0) to (x, y). It can be shown that the function

$$h(x, y) = \int_{\mathcal{C}_{(x,y)}} \mathbf{F} \cdot d\mathbf{r}$$

is defined, independent of the path chosen, and is a potential function for **F** on $\mathcal{D}^*$.

GRAPHICAL INSIGHT There is an interesting geometric interpretation of the integral of the vortex field over a curve. First, we saw that in the right half-plane the function $f(x, y) = \tan^{-1}\frac{y}{x}$ is a potential function for **F**. But $\tan^{-1}\frac{y}{x}$ is just the angle θ illustrated in Figure 14. Thus, by the Fundamental Theorem for Conservative Vector Fields, the integral of **F** along a curve in the right half plane is the difference between the angles at the end and the beginning of the curve; that is, the change in θ along the curve [Figure 15(A)].

We can show that this relationship is true for any curve $\mathcal{C}$ in $\mathcal{D} = \{(x, y) \neq (0, 0)\}$, not just those in the right half plane. Assume we have a parametrization $\mathbf{r}(t) = \langle x(t), y(t) \rangle$ of a curve $\mathcal{C}$ in $\mathcal{D}$. Consider the equation $\tan \theta = \frac{y}{x}$ and differentiate implicitly with respect to t. We obtain

$$\left(\sec^2 \theta\right)\frac{d\theta}{dt} = \left(\frac{-y}{x^2}\right)\frac{dx}{dt} + \left(\frac{1}{x}\right)\frac{dy}{dt}$$

With θ as in Figure 14, it can be shown that $\sec^2 \theta = \frac{x^2+y^2}{x^2}$. Substituting this for $\sec^2 \theta$ and simplifying, we have

$$\frac{d\theta}{dt} = \left(\frac{-y}{x^2+y^2}\right)\frac{dx}{dt} + \left(\frac{x}{x^2+y^2}\right)\frac{dy}{dt}$$

Now, if we integrate both sides of this equation with respect to t along $\mathcal{C}$, on the left side we obtain the net change in the angle θ along $\mathcal{C}$. On the right side, we obtain an expression for the line integral of **F** along $\mathcal{C}$. Thus, we have

$$\text{net change in } \theta \text{ along } \mathcal{C} = \int_{\mathcal{C}} \mathbf{F} \cdot d\mathbf{r}$$

as illustrated in Figures 15(A) and (B). This interpretation of the line integral of the vortex field **F** explains why the integral of **F** around a circle centered at the origin and oriented counterclockwise is 2π, a result we obtained in Example 6 in the last section.

In general, if a closed path **r** winds around the origin n times (where n is negative if the curve winds in the clockwise direction), then [Figures 15(C) and (D)]

$$\oint_C \mathbf{F} \cdot d\mathbf{r} = 2\pi n$$

The number n is called the *winding number* of the path. It plays an important role in the mathematical field of topology.

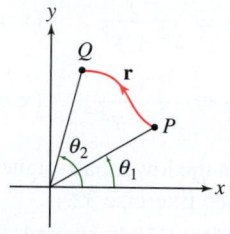

(A) $\int_C \mathbf{F} \cdot d\mathbf{r} = \theta_2 - \theta_1$

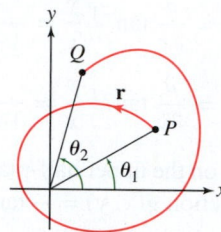

(B) $\int_C \mathbf{F} \cdot d\mathbf{r} = \theta_2 + 2\pi - \theta_1$

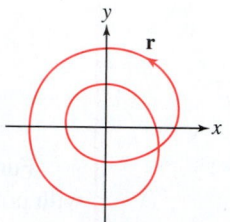

(C) **r** goes around the origin twice, so $\int_C \mathbf{F} \cdot d\mathbf{r} = 4\pi$.

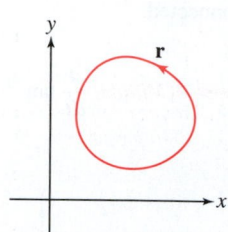

(D) **r** does not go around the origin, so $\int_C \mathbf{F} \cdot d\mathbf{r} = 0$.

FIGURE 15 The line integral of the vortex field **F** is equal to the change in θ along the path.

16.3 SUMMARY

- A vector field **F** on a domain $\mathcal{D}$ is conservative if there exists a function f such that $\nabla f = \mathbf{F}$ on $\mathcal{D}$. The function f is called a *potential function* of **F**.
- A vector field **F** on a domain $\mathcal{D}$ is called *path independent* if for any two points $P, Q \in \mathcal{D}$, we have

$$\int_{C_1} \mathbf{F} \cdot d\mathbf{r} = \int_{C_2} \mathbf{F} \cdot d\mathbf{r}$$

for any two curves C_1 and C_2 in $\mathcal{D}$ from P to Q.
- The Fundamental Theorem for Conservative Vector Fields: If $\mathbf{F} = \nabla f$, then

$$\int_C \mathbf{F} \cdot d\mathbf{r} = f(Q) - f(P)$$

for any path **r** from P to Q in the domain of **F**. This shows that conservative vector fields are path independent. In particular, if **r** is a *closed path* ($P = Q$), then

$$\oint_C \mathbf{F} \cdot d\mathbf{r} = 0$$

- The converse is also true: On an open, connected domain, a path independent vector field is conservative.
- Conservative vector fields satisfy the cross-partials condition

$$\frac{\partial F_1}{\partial y} = \frac{\partial F_2}{\partial x}, \qquad \frac{\partial F_2}{\partial z} = \frac{\partial F_3}{\partial y}, \qquad \frac{\partial F_3}{\partial x} = \frac{\partial F_1}{\partial z}$$

- Equality of the cross partial derivatives guarantees that **F** is conservative if the domain $\mathcal{D}$ is simply connected—that is, if any loop in $\mathcal{D}$ can be drawn down to a point within $\mathcal{D}$.

16.3 EXERCISES

Preliminary Questions

1. The following statement is false. *If* **F** *is a gradient vector field, then the line integral of* **F** *along every curve is zero.* Which single word must be added to make it true?

2. Which of the following statements are true for all vector fields, and which are true only for conservative vector fields?

(a) The line integral along a path from P to Q does not depend on which path is chosen.

(b) The line integral over an oriented curve C does not depend on how C is parametrized.

(c) The line integral around a closed curve is zero.

(d) The line integral changes sign if the orientation is reversed.

(e) The line integral is equal to the difference of a potential function at the two endpoints.

(f) The line integral is equal to the integral of the tangential component along the curve.

(g) The cross partial derivatives of the components are equal.

3. Let **F** be a vector field on an open, connected domain $\mathcal{D}$ with continuous second partial derivatives. Which of the following statements are always true, and which are true under additional hypotheses on $\mathcal{D}$?

(a) If **F** has a potential function, then **F** is conservative.

(b) If **F** is conservative, then the cross partial derivatives of **F** are equal.

(c) If the cross partial derivatives of **F** are equal, then **F** is conservative.

4. Let $\mathcal{C}$, $\mathcal{D}$, and $\mathcal{E}$ be the oriented curves in Figure 16, and let $\mathbf{F} = \nabla f$ be a gradient vector field such that $\int_{\mathcal{C}} \mathbf{F} \cdot d\mathbf{r} = 4$. What are the values of the following integrals?

(a) $\displaystyle\int_{\mathcal{D}} \mathbf{F} \cdot d\mathbf{r}$

(b) $\displaystyle\int_{\mathcal{E}} \mathbf{F} \cdot d\mathbf{r}$

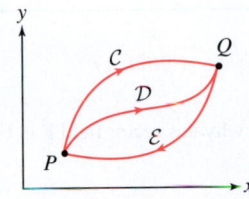

FIGURE 16

Exercises

1. Let $f(x, y, z) = xy \sin(yz)$ and $\mathbf{F} = \nabla f$. Evaluate $\displaystyle\int_{\mathcal{C}} \mathbf{F} \cdot d\mathbf{r}$, where $\mathcal{C}$ is any path from $(0, 0, 0)$ to $(1, 1, \pi)$.

2. Let $\mathbf{F}(x, y, z) = \langle x^{-1}z, y^{-1}z, \ln(xy)\rangle$.

(a) Verify that $\mathbf{F} = \nabla f$, where $f(x, y, z) = z \ln(xy)$.

(b) Evaluate $\displaystyle\int_{\mathcal{C}} \mathbf{F} \cdot d\mathbf{r}$, where $\mathbf{r}(t) = \langle e^t, e^{2t}, t^2 \rangle$ for $1 \le t \le 3$.

(c) Evaluate $\displaystyle\int_{\mathcal{C}} \mathbf{F} \cdot d\mathbf{r}$ for any path $\mathcal{C}$ from $P = (\frac{1}{2}, 4, 2)$ to $Q = (2, 2, 3)$ contained in the region $x > 0$, $y > 0$.

(d) In part (c), why is it necessary to specify that the path lies in the region where x and y are positive?

In Exercises 3–6, verify that $\mathbf{F} = \nabla f$ *and evaluate the line integral of* **F** *over the given path.*

3. $\mathbf{F}(x, y) = \langle 3, 6y \rangle$, $\quad f(x, y) = 3x + 3y^2$; $\quad \mathbf{r}(t) = \langle t, 2t^{-1} \rangle$ on the interval $1 \le t \le 4$

4. $\mathbf{F}(x, y) = \langle \cos y, -x \sin y \rangle$, $\ f(x, y) = x \cos y$; upper half of the unit circle centered at the origin, oriented counterclockwise

5. $\mathbf{F}(x, y, z) = ye^z \mathbf{i} + xe^z \mathbf{j} + xye^z \mathbf{k}$, $\quad f(x, y, z) = xye^z$; $\mathbf{r}(t) = \langle t^2, t^3, t - 1 \rangle$ for $1 \le t \le 2$

6. $\mathbf{F}(x, y, z) = \dfrac{z}{x}\mathbf{i} + \mathbf{j} + \ln x\,\mathbf{k}$, $\quad f(x, y, z) = y + z \ln x$; circle $(x - 4)^2 + y^2 = 1$ in the clockwise direction

In Exercises 7–18, find a potential function for **F** *or determine that* **F** *is not conservative.*

7. $\mathbf{F} = \langle x, y, z \rangle$

8. $\mathbf{F} = \langle y, x, z \rangle$

9. $\mathbf{F} = \langle z, x, y \rangle$

10. $\mathbf{F} = x\mathbf{j} + y\mathbf{k}$

11. $\mathbf{F} = y^2 \mathbf{i} + (2xy + e^z)\mathbf{j} + ye^z \mathbf{k}$

12. $\mathbf{F} = \langle y, x, z^3 \rangle$

13. $\mathbf{F} = \langle \cos(xz), \sin(yz), xy \sin z \rangle$

14. $\mathbf{F} = \langle \cos z, 2y, -x \sin z \rangle$

15. $\mathbf{F} = \langle z \sec^2 x, z, y + \tan x \rangle$

16. $\mathbf{F} = \langle e^x(z + 1), -\cos y, e^x \rangle$

17. $\mathbf{F} = \langle 2xy + 5, x^2 - 4z, -4y \rangle$

18. $\mathbf{F} = \langle yze^{xy}, xze^{xy} - z, e^{xy} - y \rangle$

19. Evaluate

$$\int_{\mathcal{C}} 2xyz\,dx + x^2z\,dy + x^2y\,dz$$

over the path $\mathbf{r}(t) = (t^2, \sin(\pi t/4), e^{t^2 - 2t})$ for $0 \le t \le 2$.

20. Evaluate

$$\oint_{\mathcal{C}} \sin x\,dx + z\cos y\,dy + \sin y\,dz$$

where $\mathcal{C}$ is the ellipse $4x^2 + 9y^2 = 36$, oriented clockwise.

In Exercises 21–22, let $\mathbf{F} = \nabla f$, *and determine directly* $\displaystyle\int_{\mathcal{C}} \mathbf{F} \cdot d\mathbf{r}$ *for each of the two paths given, showing that they both give the same answer, which is* $f(Q) - f(P)$.

21. $f = x^2y - z, \mathbf{r}_1 = \langle t, t, 0 \rangle$ for $0 \le t \le 1$, and $\mathbf{r}_2 = \langle t, t^2, 0 \rangle$ for $0 \le t \le 1$

22. $f = zy + xy + xz, \mathbf{r}_1 = \langle t, t, t \rangle$ for $0 \le t \le 1$, $\mathbf{r}_2 = \langle t, t^2, t^3 \rangle$ for $0 \le t \le 1$

23. A vector field $\mathbf{F}$ and contour lines of a potential function for $\mathbf{F}$ are shown in Figure 17. Calculate the common value of $\int_C \mathbf{F} \cdot d\mathbf{r}$ for the curves shown in Figure 17 oriented in the direction from P to Q.

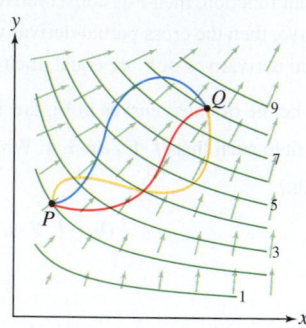

FIGURE 17

24. ✏️ Give a reason why the vector field $\mathbf{F}$ in Figure 18 is not conservative.

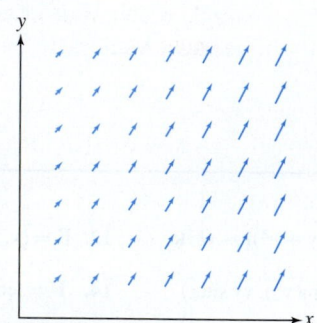

FIGURE 18

25. Calculate the work expended when a particle is moved from O to Q along segments $\overline{OP}$ and $\overline{PQ}$ in Figure 19 in the presence of the force field $\mathbf{F} = \langle x^2, y^2 \rangle$. How much work is expended moving in a complete circuit around the square?

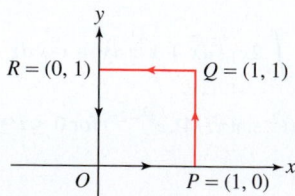

FIGURE 19

26. Let $\mathbf{F}(x, y) = \left\langle \dfrac{1}{x}, \dfrac{-1}{y} \right\rangle$. Calculate the work against F required to move an object from $(1, 1)$ to $(3, 4)$ along any path in the first quadrant.

27. Compute the work W against the earth's gravitational field required to move a satellite of mass $m = 1000$ kg along any path from an orbit of altitude 4000 km to an orbit of altitude 6000 km.

28. An electric dipole with dipole moment $p = 4 \times 10^{-5}$ C-m sets up an electric field (in newtons per coulomb)

$$\mathbf{F}(x, y, z) = \frac{kp}{r^5} \left\langle 3xz, 3yz, 2z^2 - x^2 - y^2 \right\rangle$$

where $r = (x^2 + y^2 + z^2)^{1/2}$ with distance in meters and $k = 8.99 \times 10$ with units N-m^2/C^2. Calculate the work against $\mathbf{F}$ required to move a particle of charge $q = 0.01$ C from $(1, -5, 0)$ to $(3, 4, 4)$. *Note:* The force on q is $q\mathbf{F}$ newtons.

29. On the surface of the earth, the gravitational field (with z as vertical coordinate measured in meters) is $\mathbf{F} = \langle 0, 0, -g \rangle$.

(a) Find a potential function for $\mathbf{F}$.

(b) Beginning at rest, a ball of mass $m = 2$ kg moves under the influence of gravity (without friction) along a path from $P = (3, 2, 400)$ to $Q = (-21, 40, 50)$. Find the ball's velocity when it reaches Q.

30. An electron at rest at $P = (5, 3, 7)$ moves along a path ending at $Q = (1, 1, 1)$ under the influence of the electric field (in newtons per coulomb)

$$\mathbf{F}(x, y, z) = 400(x^2 + z^2)^{-1} \langle x, 0, z \rangle$$

(a) Find a potential function for $\mathbf{F}$.

(b) What is the electron's speed at point Q? Use Conservation of Energy and the value $q_e/m_e = -1.76 \times 10^{11}$ C/kg, where q_e and m_e are the charge and mass on the electron, respectively.

31. Let $\mathbf{F} = \left\langle \dfrac{-y}{x^2 + y^2}, \dfrac{x}{x^2 + y^2} \right\rangle$ be the vortex field. Determine $\int_C \mathbf{F} \cdot d\mathbf{r}$ for each of the paths in Figure 20.

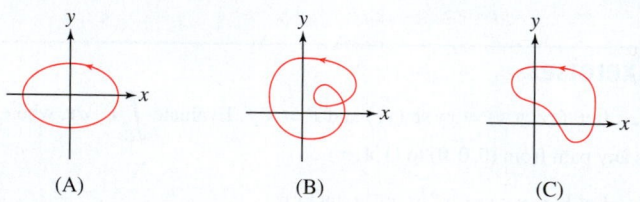

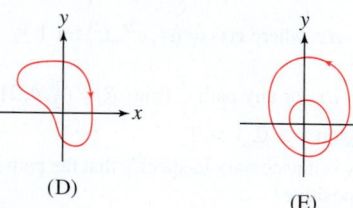

FIGURE 20

32. Show that $g(x, y) = -\tan^{-1}\dfrac{x}{y}$ is a potential function for the vortex field.

33. Determine whether or not the vector field $\mathbf{F}(x, y) = \left\langle \dfrac{x^2}{x^2 + y^2}, \dfrac{y^2}{x^2 + y^2} \right\rangle$ has a potential function.

34. The vector field $\mathbf{F}(x, y) = \left\langle \dfrac{x}{x^2 + y^2}, \dfrac{y}{x^2 + y^2} \right\rangle$ is defined on the domain $\mathcal{D} = \{(x, y) \neq (0, 0)\}$.

(a) Is $\mathcal{D}$ simply connected?

(b) Show that $\mathbf{F}$ satisfies the cross-partials condition. Does this guarantee that $\mathbf{F}$ is conservative?

(c) Show that $\mathbf{F}$ is conservative on $\mathcal{D}$ by finding a potential function.

(d) Do these results contradict Theorem 4?

Further Insights and Challenges

35. Suppose that **F** is defined on $\mathbf{R}^3$ and that $\oint_C \mathbf{F} \cdot d\mathbf{r} = 0$ for all closed paths C in $\mathbf{R}^3$. Prove:

(a) **F** is path independent; that is, for any two paths C_1 and C_2 in $\mathcal{D}$ with the same initial and terminal points,

$$\int_{C_1} \mathbf{F} \cdot d\mathbf{r} = \int_{C_2} \mathbf{F} \cdot d\mathbf{r}$$

(b) **F** is conservative.

16.4 Parametrized Surfaces and Surface Integrals

The basic idea of an integral appears in several guises. So far, we have defined single, double, and triple integrals and, in the previous section, line integrals over curves. Now, we consider one last type of integral: integrals over surfaces. We treat scalar surface integrals in this section and vector surface integrals in the following section.

Just as parametrized curves are a key ingredient in the discussion of line integrals, surface integrals require the notion of a **parametrized surface**—that is, a surface $\mathcal{S}$, in $\mathbf{R}^3$, whose points are described in the form

$$G(u, v) = (x(u, v), y(u, v), z(u, v))$$

The variables u, v (called parameters) vary in a region $\mathcal{D}$, in the uv-plane, called the **parameter domain**. Two parameters u and v are needed to parametrize a surface because the surface is two-dimensional.

Figure 1 shows a surface $\mathcal{S}$ in $\mathbf{R}^3$ with parametrization $G(u, v)$ defined for (u, v) in $\mathcal{D}$ in the uv-plane.

> For $G(u, v)$ in $\mathbf{R}^3$, we allow both interpretations as a point and as a vector. The intent should be clear from the context or the notation used. Typically for a parametrization, we regard $G(u, v)$ as points on the surface and $\dfrac{\partial G}{\partial u}(u, v)$ and $\dfrac{\partial G}{\partial v}(u, v)$ as vectors tangent to the surface.

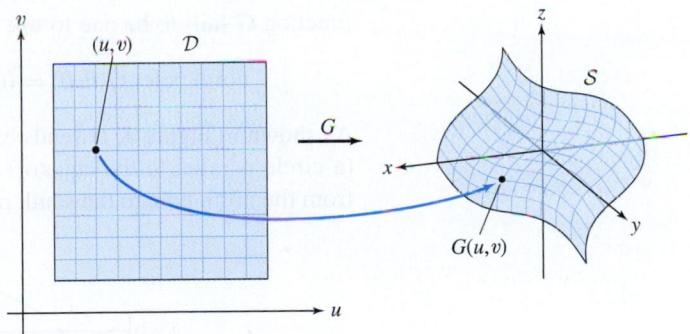

FIGURE 1 A parametrized surface $\mathcal{S}$.

EXAMPLE 1 Find a parametrization for the cylinder $x^2 + y^2 = 1$.

Solution The cylinder of radius 1 with equation $x^2 + y^2 = 1$ is conveniently parametrized in cylindrical coordinates (Figure 2). Points on the cylinder have cylindrical coordinates $(1, \theta, z)$, so we use θ and z as parameters.

We obtain

$$G(\theta, z) = (\cos\theta, \sin\theta, z), \qquad 0 \le \theta < 2\pi, \quad -\infty < z < \infty \qquad \blacksquare$$

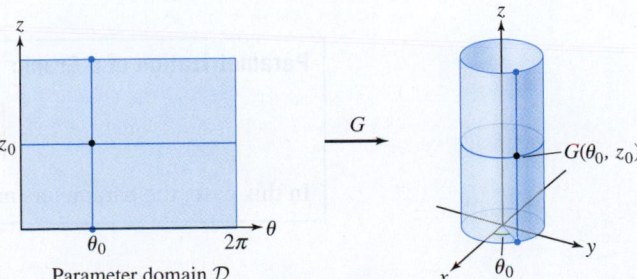

Parameter domain $\mathcal{D}$

FIGURE 2 The parametrization of a cylinder by cylindrical coordinates amounts to wrapping the rectangle around the cylinder.

Similarly, we obtain the parametrization for any vertical cylinder of radius R, given by $x^2 + y^2 = R^2$:

> **Parametrization of a Cylinder:**
>
> $$G(\theta, z) = (R\cos\theta, R\sin\theta, z), \qquad 0 \le \theta < 2\pi, \quad -\infty < z < \infty$$

EXAMPLE 2 Find a parametrization for the sphere of radius 2.

Solution The sphere of radius 2 with its center at the origin is parametrized conveniently using spherical coordinates (ρ, θ, ϕ) with $\rho = 2$ and each of the x, y, and z coordinates expressed by their spherical-coordinates representation (Figure 3).

$$G(\theta, \phi) = (2\cos\theta\sin\phi, 2\sin\theta\sin\phi, 2\cos\phi), \quad 0 \le \theta < 2\pi, \quad 0 \le \phi \le \pi \qquad \blacksquare$$

More generally, we can parametrize a sphere of radius R as follows:

> **Parametrization of a Sphere:**
>
> $$G(\theta, \phi) = (R\cos\theta\sin\phi, R\sin\theta\sin\phi, R\cos\phi), \quad 0 \le \theta < 2\pi, \quad 0 \le \phi \le \pi$$

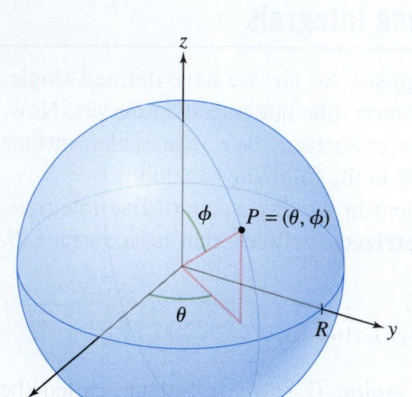

FIGURE 3 Spherical coordinates on a sphere of radius R.

The north and south poles correspond to $\phi = 0$ and $\phi = \pi$ with any value of θ (the function G fails to be one-to-one at the poles):

north pole: $G(\theta, 0) = (0, 0, R)$, south pole: $G(\theta, \pi) = (0, 0, -R)$

As shown in Figure 4, G sends each horizontal segment $\phi = c$ ($0 < c < \pi$) to a latitude (a circle parallel to the equator) and each vertical segment $\theta = c$ to a longitudinal arc from the north pole to the south pole.

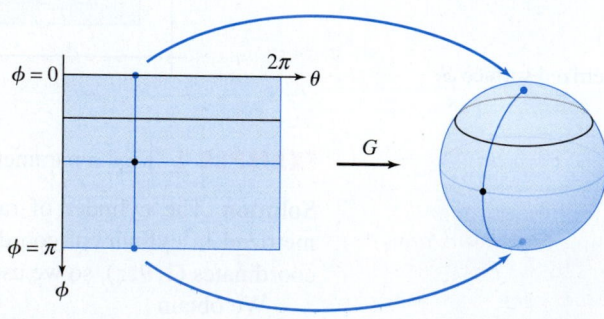

FIGURE 4 The parametrization by spherical coordinates amounts to wrapping the rectangle around the sphere. The top and bottom edges of the rectangle are collapsed to the north and south poles.

A simple situation for generating a parametrization of a surface occurs when the surface is the **graph of a function** $z = f(x, y)$, as in Figure 5.

> **Parametrization of a Graph:**
>
> $$G(x, y) = (x, y, f(x, y))$$

In this case, the parameters are x and y.

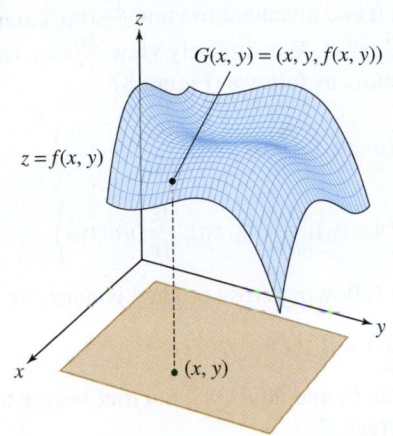

FIGURE 5 Parametrizing the graph of a function.

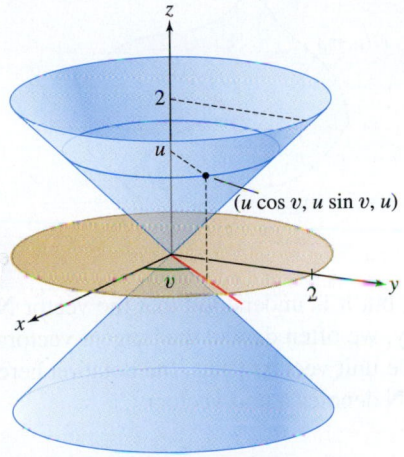

FIGURE 6 The cone $x^2 + y^2 = z^2$.

In essence, a parametrization labels each point P on $\mathcal{S}$ by a unique pair (u_0, v_0) in the parameter domain. We can think of (u_0, v_0) as coordinates of P determined by the parametrization. They are sometimes called **curvilinear coordinates**.

EXAMPLE 3 Find a parametrization of the paraboloid given by $f(x, y) = x^2 + y^2$.

Solution We can immediately define $G(x, y) = (x, y, x^2 + y^2)$. Then G sends the xy-plane onto the paraboloid. ∎

Most surfaces in which we are interested do not appear as graphs of functions. In this case, we need to find some other parametrization.

EXAMPLE 4 **Parametrization of a Cone** Find a parametrization of the portion $\mathcal{S}$ of the cone with equation $x^2 + y^2 = z^2$ lying above and below the disk $x^2 + y^2 \leq 4$. Specify the domain $\mathcal{D}$ of the parametrization.

Solution Notice as in Figure 6 that this portion of the cone is not the graph of a function since it includes that part of the cone that is above the xy-plane with that part that lies below the xy-plane. However, each point on the cone is uniquely determined by its cylindrical coordinates. Because the cone satisfies $r^2 = z^2$, the r-coordinate of a point can be expressed in terms of the z-coordinate, and therefore we can parametrize the surface via the z- and θ-coordinates. Letting the parameter u correspond to the z-coordinate, and v correspond to the θ-coordinate, a point on the cone at height u has coordinates $(u \cos v, u \sin v, u)$ for some angle v. Thus, the cone has the parametrization

$$G(u, v) = (u \cos v, u \sin v, u)$$

Since we are interested in the portion of the cone where $x^2 + y^2 = u^2 \leq 4$, the height variable u satisfies $-2 \leq u \leq 2$. The angular variable v varies in the interval $[0, 2\pi)$, and therefore the parameter domain is $\mathcal{D} = [-2, 2] \times [0, 2\pi)$. ∎

Grid Curves, Normal Vectors, and the Tangent Plane

Suppose that a surface $\mathcal{S}$ has a parametrization

$$G(u, v) = (x(u, v), y(u, v), z(u, v))$$

that is one-to-one on a domain $\mathcal{D}$. We shall always assume that G is **continuously differentiable**, meaning that the functions $x(u, v)$, $y(u, v)$, and $z(u, v)$ have continuous partial derivatives.

In the uv-plane, we can form a grid of lines parallel to the coordinates axes. These grid lines correspond under G to a system of **grid curves** on the surface (Figure 7). More precisely, the horizontal and vertical lines through (u_0, v_0) in the domain correspond to the grid curves $G(u, v_0)$ and $G(u_0, v)$ that intersect at the point $P = G(u_0, v_0)$.

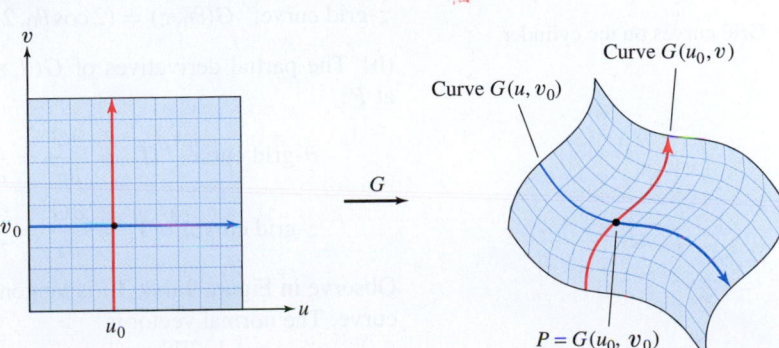

FIGURE 7 Grid curves.

With $G(u, v_0)$ representing a curve through P, it is convenient to view $\frac{\partial G}{\partial u}(u_0, v_0)$ a a vector tangent to that curve (and to the surface $\mathcal{S}$) at P. We similarly view $\frac{\partial G}{\partial v}(u_0, v_0)$ as a tangent vector at P. Thus, we have tangent vectors as follows (Figure 8):

$$\text{For } G(u, v_0): \quad \mathbf{T}_u(P) = \frac{\partial G}{\partial u}(u_0, v_0) = \left\langle \frac{\partial x}{\partial u}(u_0, v_0), \frac{\partial y}{\partial u}(u_0, v_0), \frac{\partial z}{\partial u}(u_0, v_0) \right\rangle$$

$$\text{For } G(u_0, v): \quad \mathbf{T}_v(P) = \frac{\partial G}{\partial v}(u_0, v_0) = \left\langle \frac{\partial x}{\partial v}(u_0, v_0), \frac{\partial y}{\partial v}(u_0, v_0), \frac{\partial z}{\partial v}(u_0, v_0) \right\rangle$$

The parametrization G is called **regular** at P if the following cross product is nonzero:

$$\mathbf{N}(P) = \mathbf{N}(u_0, v_0) = \mathbf{T}_u(P) \times \mathbf{T}_v(P)$$

In this case, $\mathbf{T}_u$ and $\mathbf{T}_v$ span the tangent plane to $\mathcal{S}$ at P, and $\mathbf{N}(P)$ is a **normal vector** to the tangent plane. We call $\mathbf{N}(P)$ a normal to the surface $\mathcal{S}$.

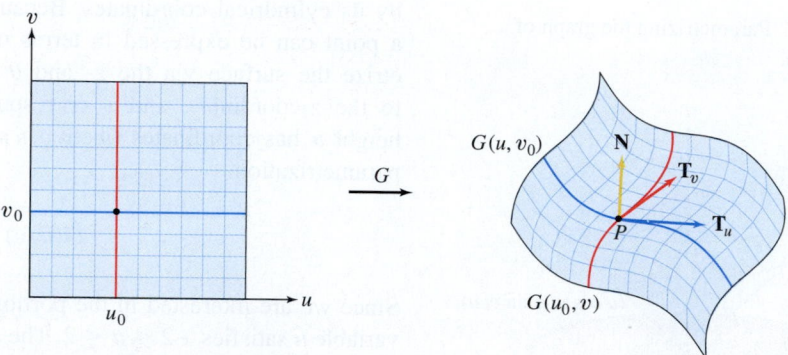

FIGURE 8 The vectors $\mathbf{T}_u$ and $\mathbf{T}_v$ are tangent to the grid curves through $P = G(u_0, v_0)$.

At each point on a surface, the normal vector points in one of two opposite directions. If we change the parametrization, the length of $\mathbf{N}$ may change and its direction may be reversed.

We often write $\mathbf{N}$ instead of $\mathbf{N}(P)$ or $\mathbf{N}(u, v)$, but it is understood that the vector $\mathbf{N}$ varies from point to point on the surface. Similarly, we often denote the tangent vectors by $\mathbf{T}_u$ and $\mathbf{T}_v$. Note that $\mathbf{T}_u$, $\mathbf{T}_v$, and $\mathbf{N}$ need not be unit vectors (thus, the notation here differs from that in Sections 13.4 and 13.5, where $\mathbf{N}$ denotes a unit vector).

EXAMPLE 5 Consider the parametrization $G(\theta, z) = (2\cos\theta, 2\sin\theta, z)$ of the cylinder $x^2 + y^2 = 4$:

(a) Describe the grid curves.

(b) Compute $\mathbf{T}_\theta$, $\mathbf{T}_z$, and $\mathbf{N}(\theta, z)$.

(c) Find an equation of the tangent plane at $P = G(\frac{\pi}{4}, 5)$.

Solution

(a) The grid curves on the cylinder through $P = (\theta_0, z_0)$ are (Figure 9)

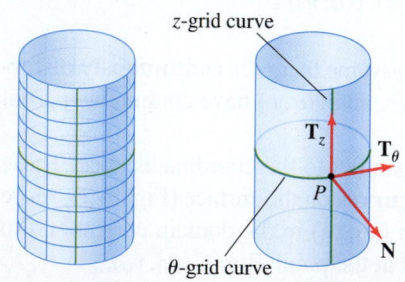

FIGURE 9 Grid curves on the cylinder.

θ-grid curve: $G(\theta, z_0) = (2\cos\theta, 2\sin\theta, z_0)$ (circle of radius 2 at height $z = z_0$)

z-grid curve: $G(\theta_0, z) = (2\cos\theta_0, 2\sin\theta_0, z)$ (vertical line through P with $\theta = \theta_0$)

(b) The partial derivatives of $G(\theta, z) = (2\cos\theta, 2\sin\theta, z)$ give us the tangent vectors at P:

$$\theta\text{-grid curve:} \quad \mathbf{T}_\theta = \frac{\partial G}{\partial \theta} = \frac{\partial}{\partial \theta}(2\cos\theta, 2\sin\theta, z) = \langle -2\sin\theta, 2\cos\theta, 0 \rangle$$

$$z\text{-grid curve:} \quad \mathbf{T}_z = \frac{\partial G}{\partial z} = \frac{\partial}{\partial z}(2\cos\theta, 2\sin\theta, z) = \langle 0, 0, 1 \rangle$$

Observe in Figure 9 that $\mathbf{T}_\theta$ is tangent to the θ-grid curve and $\mathbf{T}_z$ is tangent to the z-grid curve. The normal vector is

$$\mathbf{N}(\theta, z) = \mathbf{T}_\theta \times \mathbf{T}_z = \begin{vmatrix} \mathbf{i} & \mathbf{j} & \mathbf{k} \\ -2\sin\theta & 2\cos\theta & 0 \\ 0 & 0 & 1 \end{vmatrix} = 2\cos\theta\,\mathbf{i} + 2\sin\theta\,\mathbf{j}$$

The coefficient of $\mathbf{k}$ is zero, and $\mathbf{N}$ points horizontally out of the cylinder.

(c) For $\theta = \frac{\pi}{4}$, $z = 5$,

$$P = G\left(\frac{\pi}{4}, 5\right) = \left(\sqrt{2}, \sqrt{2}, 5\right), \qquad \mathbf{N} = \mathbf{N}\left(\frac{\pi}{4}, 5\right) = \langle \sqrt{2}, \sqrt{2}, 0 \rangle$$

The tangent plane through P has normal vector $\mathbf{N}$ and thus has equation

$$\langle x - \sqrt{2}, y - \sqrt{2}, z - 5 \rangle \cdot \langle \sqrt{2}, \sqrt{2}, 0 \rangle = 0$$

This can be written

$$\sqrt{2}(x - \sqrt{2}) + \sqrt{2}(y - \sqrt{2}) = 0 \qquad \text{or} \qquad x + y = 2\sqrt{2}$$

The tangent plane is vertical (because z does not appear in the equation). ■

> ◀ **REMINDER** An equation of the plane through $P = (x_0, y_0, z_0)$ with normal vector $\mathbf{N}$ is
>
> $$\langle x - x_0, y - y_0, z - z_0 \rangle \cdot \mathbf{N} = 0$$

EXAMPLE 6 **Helicoid Surface** The surface S with parametrization

$$G(u, v) = (u \cos v, u \sin v, v), \qquad -1 \le u \le 1, \quad 0 \le v < 2\pi$$

is called a helicoid (Figure 10).

(a) Describe the grid curves on S associated with $G(u, v)$.

(b) Compute $\mathbf{N}(u, v)$ at $u = \frac{1}{2}$, $v = \frac{\pi}{2}$.

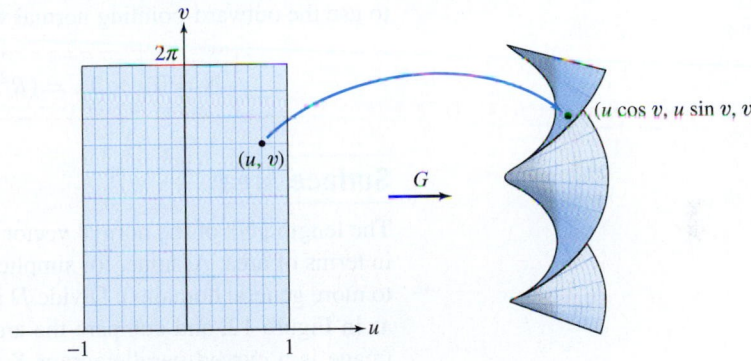

DF **FIGURE 10** The helicoid parametrized by $G(u, v)$.

Solution

(a) First, for each fixed value $u = a$, the grid curve $G(a, v) = (a \cos v, a \sin v, v)$ is a helix of radius a. Therefore, as u varies from -1 to 1, $G(u, v)$ describes a family of helices of radius u.

For fixed $v = b$, the grid curve $G(u, b) = (u \cos b, u \sin b, b)$ is a line segment in the plane $z = b$, at an angle of $\theta = b$ from the xz-plane. As v varies from 0 to 2π, the segments rise in the z-direction (because the z-coordinate is increasing) and rotate around the z-axis (because the angle that the segment makes with the xz-plane is increasing).

(b) The tangent and normal vectors are

$$\mathbf{T}_u = \frac{\partial G}{\partial u} = \langle \cos v, \sin v, 0 \rangle$$

$$\mathbf{T}_v = \frac{\partial G}{\partial v} = \langle -u \sin v, u \cos v, 1 \rangle$$

$$\mathbf{N}(u, v) = \mathbf{T}_u \times \mathbf{T}_v = \begin{vmatrix} \mathbf{i} & \mathbf{j} & \mathbf{k} \\ \cos v & \sin v & 0 \\ -u \sin v & u \cos v & 1 \end{vmatrix} = (\sin v)\mathbf{i} - (\cos v)\mathbf{j} + u\mathbf{k}$$

At $u = \frac{1}{2}$, $v = \frac{\pi}{2}$, we have $\mathbf{N} = \mathbf{i} + \frac{1}{2}\mathbf{k}$. ■

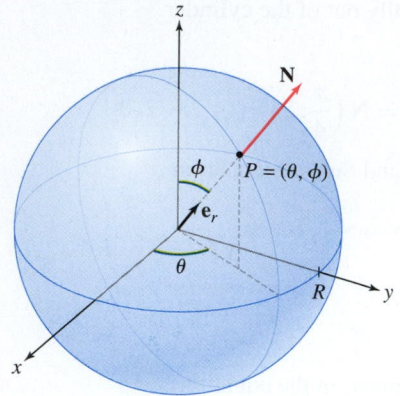

FIGURE 11 The normal vector $\mathbf{N}$ points in the radial direction $\mathbf{e}_r$.

For future reference, we compute the outward-pointing normal vector in the standard parametrization of the sphere of radius R centered at the origin (Figure 11):

$$G(\theta, \phi) = (R \cos \theta \sin \phi, R \sin \theta \sin \phi, R \cos \phi)$$

Note first that since the distance from $G(\theta, \phi)$ to the origin is R, the *unit* radial vector at $G(\theta, \phi)$ is obtained by dividing by R:

$$\mathbf{e}_r = \langle \cos \theta \sin \phi, \sin \theta \sin \phi, \cos \phi \rangle$$

Furthermore,

$$\mathbf{T}_\theta = \langle -R \sin \theta \sin \phi, R \cos \theta \sin \phi, 0 \rangle$$

$$\mathbf{T}_\phi = \langle R \cos \theta \cos \phi, R \sin \theta \cos \phi, -R \sin \phi \rangle$$

$$\mathbf{N} = \mathbf{T}_\theta \times \mathbf{T}_\phi = \begin{vmatrix} \mathbf{i} & \mathbf{j} & \mathbf{k} \\ -R \sin \theta \sin \phi & R \cos \theta \sin \phi & 0 \\ R \cos \theta \cos \phi & R \sin \theta \cos \phi & -R \sin \phi \end{vmatrix}$$

$$= -R^2 \cos \theta \sin^2 \phi \, \mathbf{i} - R^2 \sin \theta \sin^2 \phi \, \mathbf{j} - R^2 \cos \phi \sin \phi \, \mathbf{k}$$

$$= -R^2 \sin \phi \, \langle \cos \theta \sin \phi, \sin \theta \sin \phi, \cos \phi \rangle \qquad \boxed{1}$$

$$= -(R^2 \sin \phi) \, \mathbf{e}_r$$

This is an inward-pointing normal vector. However, in most computations, it is standard to use the outward-pointing normal vector:

$$\boxed{\mathbf{N} = \mathbf{T}_\phi \times \mathbf{T}_\theta = (R^2 \sin \phi) \, \mathbf{e}_r, \qquad \|\mathbf{N}\| = R^2 \sin \phi} \qquad \boxed{2}$$

Surface Area

The length $\|\mathbf{N}\|$ of the normal vector in a parametrization has an important interpretation in terms of area. Assume, for simplicity, that $\mathcal{D}$ is a rectangle (the argument also applies to more general domains). Divide $\mathcal{D}$ into a grid of small rectangles $\mathcal{R}_{ij}$ of size $\Delta u \times \Delta v$, as in Figure 12, and compare the area of $\mathcal{R}_{ij}$ with the area of its image under G. This image is a curved parallelogram $\mathcal{S}_{ij} = G(\mathcal{R}_{ij})$. Assume points P_0, Q_0, and S_0 are at corners of $\mathcal{R}_{ij}$, shown in the figure, and that P, R, and S are the corresponding points on $\mathcal{S}_{ij}$.

First, we note that if Δu and Δv in Figure 12 are small, then the curved parallelogram $\mathcal{S}_{ij}$ has approximately the same area as the parallelogram with sides $\overrightarrow{PQ}$ and $\overrightarrow{PS}$. Recall from Section 12.4 that the area of the parallelogram spanned by two vectors is the length of their cross product, so

$$\text{area}(\mathcal{S}_{ij}) \approx \|\overrightarrow{PQ} \times \overrightarrow{PS}\|$$

Next, we use the linear approximation to estimate the vectors $\overrightarrow{PQ}$ and $\overrightarrow{PS}$:

$$\overrightarrow{PQ} = G(u_{ij} + \Delta u, v_{ij}) - G(u_{ij}, v_{ij}) \approx \frac{\partial G}{\partial u}(u_{ij}, v_{ij}) \Delta u = \mathbf{T}_u \Delta u$$

$$\overrightarrow{PS} = G(u_{ij}, v_{ij} + \Delta v) - G(u_{ij}, v_{ij}) \approx \frac{\partial G}{\partial v}(u_{ij}, v_{ij}) \Delta v = \mathbf{T}_v \Delta v$$

Thus, we have

$$\text{area}(\mathcal{S}_{ij}) \approx \|\mathbf{T}_u \Delta u \times \mathbf{T}_v \Delta v\| = \|\mathbf{T}_u \times \mathbf{T}_v\| \, \Delta u \, \Delta v$$

Since $\mathbf{N}(u_{ij}, v_{ij}) = \mathbf{T}_u \times \mathbf{T}_v$ and $\text{area}(\mathcal{R}_{ij}) = \Delta u \Delta v$, we obtain

$$\boxed{\text{area}(\mathcal{S}_{ij}) \approx \|\mathbf{N}(u_{ij}, v_{ij})\| \text{area}(\mathcal{R}_{ij})} \qquad \boxed{3}$$

The approximation (3) is valid for any small region $\mathcal{R}$ in the uv-plane:

$$\text{area}(\mathcal{S}) \approx \|\mathbf{N}(u_0, v_0))\| \text{area}(\mathcal{R})$$

where $\mathcal{S} = G(\mathcal{R})$ and (u_0, v_0) is any sample point in $\mathcal{R}$. Here, "small" means contained in a small disk. We do not allow $\mathcal{R}$ to be very thin and wide.

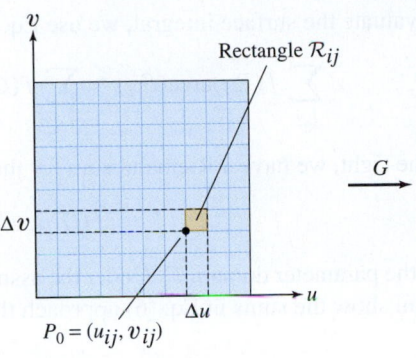
Rectangle $\mathcal{R}_{ij}$

Δv

Δu

$P_0 = (u_{ij}, v_{ij})$

G

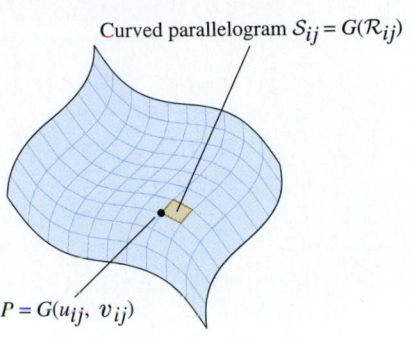
Curved parallelogram $\mathcal{S}_{ij} = G(\mathcal{R}_{ij})$

$P = G(u_{ij}, v_{ij})$

S_0

Δv 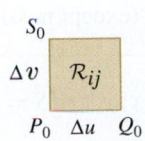 $\mathcal{R}_{ij}$

P_0 Δu Q_0

G

$T_v \Delta v$

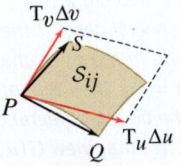 S

P $\mathcal{S}_{ij}$

Q $T_u \Delta u$

FIGURE 12

Our conclusion: $\|\mathbf{N}\|$ *is a scaling factor that measures how the area of a small rectangle* $\mathcal{R}_{ij}$ *is altered under the map* G.

To compute the surface area of $\mathcal{S}$, we assume that G is one-to-one and regular, except possibly on the boundary of $\mathcal{D}$. Recall that "regular" means $\mathbf{N}(u, v)$ is nonzero.

The entire surface $\mathcal{S}$ is the union of the small patches $\mathcal{S}_{ij}$, so we can apply the approximation on each patch to obtain

Note: We require only that G be one-to-one on the interior of $\mathcal{D}$. Many common parametrizations (such as the parametrizations by cylindrical and spherical coordinates) fail to be one-to-one on the boundary of their domains.

$$\text{area}(\mathcal{S}) = \sum_{i,j} \text{area}(\mathcal{S}_{ij}) \approx \sum_{i,j} \|\mathbf{N}(u_{ij}, v_{ij})\| \Delta u \, \Delta v \qquad \boxed{4}$$

The sum on the right is a Riemann sum for the double integral of $\|\mathbf{N}(u, v)\|$ over the parameter domain $\mathcal{D}$. As Δu and Δv tend to zero, these Riemann sums converge to a double integral, which we take as the definition of surface area:

$$\boxed{\text{area}(\mathcal{S}) = \iint_{\mathcal{D}} \|\mathbf{N}(u, v)\| \, du \, dv}$$

Surface Integral

Now, we can define the surface integral of a function $f(x, y, z)$ on a surface $\mathcal{S}$:

$$\iint_{\mathcal{S}} f(x, y, z) \, dS$$

This is similar to the definition of the line integral of a function along a curve. Choose a sample point $P_{ij} = G(u_{ij}, v_{ij})$ in each small patch $\mathcal{S}_{ij}$ and form the sum:

$$\sum_{i,j} f(P_{ij}) \text{area}(\mathcal{S}_{ij}) \qquad \boxed{5}$$

The limit of these sums as Δu and Δv tend to zero (if it exists) is the **surface integral**:

$$\iint_{\mathcal{S}} f(x, y, z) \, dS = \lim_{\Delta u, \Delta v \to 0} \sum_{i,j} f(P_{ij}) \text{area}(\mathcal{S}_{ij})$$

To evaluate the surface integral, we use Eq. (3) to write

$$\sum_{i,j} f(P_{ij})\text{area}(\mathcal{S}_{ij}) \approx \sum_{i,j} f(G(u_{ij},v_{ij}))\|\mathbf{N}(u_{ij},v_{ij})\| \, \Delta u \, \Delta v \qquad \boxed{6}$$

On the right, we have a Riemann sum for the double integral of

$$f(G(u,v))\|\mathbf{N}(u,v)\|$$

over the parameter domain $\mathcal{D}$. Under the assumption that G is continuously differentiable, we can show the sums in Eq. (6) approach the same limit. This yields the next theorem.

> **THEOREM 1 Surface Integrals and Surface Area**   Let $G(u,v)$ be a parametrization of a surface $\mathcal{S}$ with parameter domain $\mathcal{D}$. Assume that G is continuously differentiable, one-to-one, and regular (except possibly at the boundary of $\mathcal{D}$). Then
>
> $$\iint_{\mathcal{S}} f(x,y,z)\,dS = \iint_{\mathcal{D}} f(G(u,v))\|\mathbf{N}(u,v)\| \, du\, dv \qquad \boxed{7}$$
>
> For $f(x,y,z) = 1$, we obtain the surface area of $\mathcal{S}$:
>
> $$\text{area}(\mathcal{S}) = \iint_{\mathcal{D}} \|\mathbf{N}(u,v)\| \, du\, dv$$

It is interesting to note that Eq. (7) includes the Change of Variables Formula for double integrals (Theorem 2 in Section 15.6) as a special case. If the surface $\mathcal{S}$ is a domain in the xy-plane [in other words, $z(u,v) = 0$], then the integral over $\mathcal{S}$ reduces to the double integral of the function $f(x,y,0)$. We may view $G(u,v)$ as a mapping from the uv-plane to the xy-plane, and we find that $\|\mathbf{N}(u,v)\|$ is the Jacobian of this mapping.

Equation (7) yields the following important relationship between the **surface area differential** dS and the parameter differentials du and dv, enabling us to compute surface integrals as iterated integrals:

$$\boxed{dS = \|\mathbf{N}(u,v)\| \, du\, dv}$$

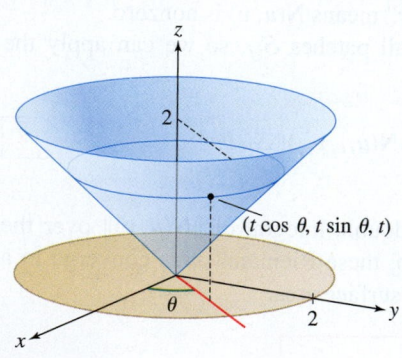

FIGURE 13 Portion $\mathcal{S}$ of the cone $x^2 + y^2 = z^2$ lying over the disk $x^2 + y^2 \le 4$.

EXAMPLE 7   Calculate the surface area of the portion $\mathcal{S}$ of the cone $x^2 + y^2 = z^2$ lying above the disk $x^2 + y^2 \le 4$ (Figure 13). Then calculate $\displaystyle\iint_{\mathcal{S}} x^2 z \, dS$.

Solution   Similar to Example 4, but here using variables θ and t, we parametrize $\mathcal{S}$ by

$$G(\theta, t) = (t\cos\theta, t\sin\theta, t), \qquad 0 \le t \le 2, \quad 0 \le \theta < 2\pi$$

Step 1. Compute the tangent and normal vectors.

$$\mathbf{T}_\theta = \frac{\partial G}{\partial \theta} = \langle -t\sin\theta, t\cos\theta, 0 \rangle, \qquad \mathbf{T}_t = \frac{\partial G}{\partial t} = \langle \cos\theta, \sin\theta, 1 \rangle$$

$$\mathbf{N} = \mathbf{T}_\theta \times \mathbf{T}_t = \begin{vmatrix} \mathbf{i} & \mathbf{j} & \mathbf{k} \\ -t\sin\theta & t\cos\theta & 0 \\ \cos\theta & \sin\theta & 1 \end{vmatrix} = t\cos\theta\,\mathbf{i} + t\sin\theta\,\mathbf{j} - t\mathbf{k}$$

The normal vector has length

$$\|\mathbf{N}\| = \sqrt{t^2\cos^2\theta + t^2\sin^2\theta + (-t)^2} = \sqrt{2t^2} = \sqrt{2}\,|t|$$

Thus, $dS = \sqrt{2}\,|t|\, d\theta\, dt$. Since $t \ge 0$ on our domain, we drop the absolute value.

Step 2. Calculate the surface area.

$$\text{area}(\mathcal{S}) = \iint_{\mathcal{D}} \|\mathbf{N}\| \, d\theta\, dt = \int_0^2 \int_0^{2\pi} \sqrt{2}\,t \, d\theta\, dt = \sqrt{2}\pi t^2 \Big|_0^2 = 4\sqrt{2}\pi$$

Step 3. **Calculate the surface integral.**

We express $f(x, y, z) = x^2 z$ in terms of the parameters t and θ and evaluate:

$$f(G(\theta, t)) = f(t \cos \theta, t \sin \theta, t) = (t \cos \theta)^2 t = t^3 \cos^2 \theta$$

$$\iint_{S} f(x, y, z) \, dS = \int_{t=0}^{2} \int_{\theta=0}^{2\pi} f(G(\theta, t)) \, \|\mathbf{N}(\theta, t)\| \, d\theta \, dt$$

$$= \int_{t=0}^{2} \int_{\theta=0}^{2\pi} (t^3 \cos^2 \theta)(\sqrt{2} t) \, d\theta \, dt$$

◀── REMINDER

$$\int_{0}^{2\pi} \cos^2 \theta \, d\theta = \int_{0}^{2\pi} \frac{1 + \cos 2\theta}{2} \, d\theta = \pi$$

$$= \sqrt{2} \int_{0}^{2} \int_{0}^{2\pi} t^4 \cos^2 \theta \, d\theta \, dt = \sqrt{2} \int_{0}^{2} \pi t^4 \, dt$$

$$= \sqrt{2}\pi \left(\frac{32}{5} \right) = \frac{32\sqrt{2}\pi}{5} \qquad \blacksquare$$

In previous discussions of multiple and line integrals, we applied the principle that the integral of a density is the total quantity. This applies to surface integrals as well. For example, a surface with mass density $\delta(x, y, z)$ (in units of mass per area) is the surface integral of the mass density:

$$\text{mass of } S = \iint_{S} \delta(x, y, z) \, dS$$

Similarly, if an electric charge is distributed over S with charge density $\delta(x, y, z)$, then the surface integral of $\delta(x, y, z)$ is the total charge on S.

EXAMPLE 8 **Total Charge on a Surface** Find the total charge (in coulombs) on a sphere S of radius 5 cm whose charge density in spherical coordinates is $\delta(\theta, \phi) = 0.003 \cos^2 \phi$ C/cm^2.

Solution We parametrize S in spherical coordinates:

$$G(\theta, \phi) = (5 \cos \theta \sin \phi, 5 \sin \theta \sin \phi, 5 \cos \phi)$$

By Eq. (2), $\|\mathbf{N}\| = 5^2 \sin \phi$ and

$$\text{total charge} = \iint_{S} \delta(\theta, \phi) \, dS = \int_{\theta=0}^{2\pi} \int_{\phi=0}^{\pi} \delta(\theta, \phi) \|\mathbf{N}\| \, d\phi \, d\theta$$

$$= \int_{\theta=0}^{2\pi} \int_{\phi=0}^{\pi} (0.003 \cos^2 \phi)(25 \sin \phi) \, d\phi \, d\theta$$

$$= (0.075)(2\pi) \int_{\phi=0}^{\pi} \cos^2 \phi \sin \phi \, d\phi$$

$$= 0.15\pi \left(-\frac{\cos^3 \phi}{3} \right) \Big|_{0}^{\pi} = 0.15\pi \left(\frac{2}{3} \right) \approx 0.1\pi \text{ coulombs} \qquad \blacksquare$$

When a graph $z = g(x, y)$ is parametrized by $G(x, y) = (x, y, g(x, y))$, the tangent and normal vectors are

$$\mathbf{T}_x = (1, 0, g_x), \qquad \mathbf{T}_y = (0, 1, g_y)$$

$$\mathbf{N} = \mathbf{T}_x \times \mathbf{T}_y = \begin{vmatrix} \mathbf{i} & \mathbf{j} & \mathbf{k} \\ 1 & 0 & g_x \\ 0 & 1 & g_y \end{vmatrix} = -g_x \mathbf{i} - g_y \mathbf{j} + \mathbf{k}, \qquad \|\mathbf{N}\| = \sqrt{1 + g_x^2 + g_y^2} \qquad \boxed{8}$$

The surface integral of $f(x, y, z)$ over the portion of a graph lying over a domain $\mathcal{D}$ in the xy-plane is

$$\text{surface integral over a graph} = \iint_{\mathcal{D}} f(x, y, g(x, y))\sqrt{1 + g_x^2 + g_y^2}\, dx\, dy \qquad \boxed{9}$$

EXAMPLE 9 Calculate $\iint_{\mathcal{S}} (z - x)\, dS$, where $\mathcal{S}$ is the portion of the graph of $z = x + y^2$, where $0 \le x \le y$, $0 \le y \le 1$ (Figure 14).

Solution Let $z = g(x, y) = x + y^2$. Then $g_x = 1$ and $g_y = 2y$, and

$$dS = \sqrt{1 + g_x^2 + g_y^2}\, dx\, dy = \sqrt{1 + 1 + 4y^2}\, dx\, dy = \sqrt{2 + 4y^2}\, dx\, dy$$

On the surface $\mathcal{S}$, we have $z = x + y^2$, and thus

$$f(x, y, z) = z - x = (x + y^2) - x = y^2$$

By Eq. (9),

$$\iint_{\mathcal{S}} f(x, y, z)\, dS = \int_{y=0}^{1} \int_{x=0}^{y} y^2\sqrt{2 + 4y^2}\, dx\, dy$$

$$= \int_{y=0}^{1} \left(y^2\sqrt{2 + 4y^2}\right) x \Big|_{x=0}^{y}\, dy = \int_{0}^{1} y^3\sqrt{2 + 4y^2}\, dy$$

Now use the substitution $u = 2 + 4y^2$, $du = 8y\, dy$. Then $y^2 = \frac{1}{4}(u - 2)$, and

$$\int_{0}^{1} y^3\sqrt{2 + 4y^2}\, dy = \frac{1}{8}\int_{2}^{6} \frac{1}{4}(u - 2)\sqrt{u}\, du = \frac{1}{32}\int_{2}^{6} (u^{3/2} - 2u^{1/2})\, du$$

$$= \frac{1}{32}\left(\frac{2}{5}u^{5/2} - \frac{4}{3}u^{3/2}\right)\Big|_{2}^{6} = \frac{1}{30}(6\sqrt{6} + \sqrt{2}) \approx 0.54 \qquad \blacksquare$$

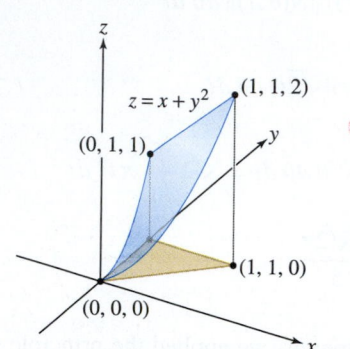

FIGURE 14 The surface $z = x + y^2$ over $0 \le x \le y$, $0 \le y \le 1$.

Gravitational Potential of a Sphere

In physics, it is an important fact that the gravitational field $\mathbf{F}$ corresponding to any arrangement of masses is conservative; that is, $\mathbf{F} = -\nabla V$ for some function V (recall that the negative sign is a convention of physics). The field at a point P due to a mass m located at point Q is $\mathbf{F} = -\dfrac{Gm}{r^2}\mathbf{e}_r$, where G is the universal gravitational constant ($\approx 6.67408 \times 10^{-11}\,\mathrm{m}^3\mathrm{kg}^{-1}\mathrm{s}^{-2}$), $\mathbf{e}_r$ is the unit vector pointing from Q to P, and r is the distance from P to Q, which we denote by $|P - Q|$. It follows by Example 8 in Section 16.1 that $\mathbf{F} = -\nabla V$ for

$$V(P) = -\frac{Gm}{r} = -\frac{Gm}{|P - Q|}$$

If, instead of a single mass, we have K point masses $m_1, \ldots, m_K$ located at $Q_1, \ldots, Q_K$, then the gravitational potential is the sum

$$V(P) = -G\sum_{i=1}^{K} \frac{m_i}{|P - Q_i|} \qquad \boxed{10}$$

If mass is distributed continuously over a thin surface $\mathcal{S}$ with mass density function $\delta(x, y, z)$, we replace the sum by the surface integral

$$V(P) = -G\iint_{\mathcal{S}} \frac{\delta(x, y, z)\, dS}{|P - Q|} = -G\iint_{\mathcal{S}} \frac{\delta(x, y, z)\, dS}{\sqrt{(x - a)^2 + (y - b)^2 + (z - c)^2}} \qquad \boxed{11}$$

where $P = (a, b, c)$. However, this surface integral cannot usually be evaluated explicitly unless the surface and mass distribution are sufficiently symmetric, as in the case of a hollow sphere of uniform mass density (Figure 15).

The French mathematician Pierre Simon, Marquis de Laplace (1749–1827), showed that the gravitational potential satisfies the Laplace equation

$$\frac{\partial^2 V}{\partial x^2} + \frac{\partial^2 V}{\partial y^2} + \frac{\partial^2 V}{\partial z^2} = 0$$

This equation plays an important role in more advanced branches of math and physics.

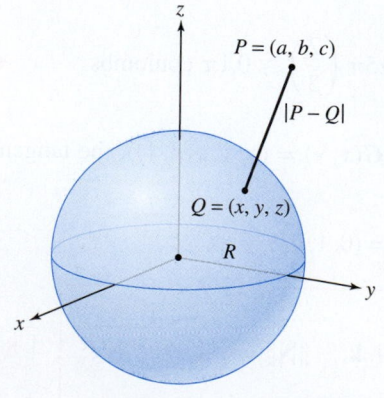

FIGURE 15

> **THEOREM 2** **Gravitational Potential of a Uniform Hollow Sphere** The gravitational potential V due to a hollow sphere of radius R with uniform mass distribution of total mass m at a point P located at a distance r from the center of the sphere is
>
> $$V(P) = \begin{cases} \dfrac{-Gm}{r} & \text{if } r > R \quad (P \text{ outside the sphere}) \\[2mm] \dfrac{-Gm}{R} & \text{if } r < R \quad (P \text{ inside the sphere}) \end{cases}$$
>
> **12**

We leave this calculation as an exercise (Exercise 48), because we will derive it again with much less effort using Gauss's Law in Section 17.3.

In his magnum opus, *Principia Mathematica*, Isaac Newton proved that a sphere of uniform mass density (whether hollow or solid) attracts a particle outside the sphere as if the entire mass were concentrated at the center. In other words, a uniform sphere behaves like a point mass as far as gravity is concerned. Furthermore, if the sphere is hollow, then the sphere exerts no gravitational force on a particle inside it. Newton's result follows from Eq. (12). Outside the sphere, V has the same formula as the potential due to a point mass. Inside the sphere, the potential is *constant* with value $-Gm/R$. But constant potential means zero force because the force is the (negative) gradient of the potential. This discussion applies equally well to the electrostatic force. In particular, a uniformly charged sphere behaves like a point charge (when viewed from outside the sphere).

16.4 SUMMARY

- A *parametrized surface* is a surface $\mathcal{S}$ whose points are described in the form

$$G(u, v) = (x(u, v), y(u, v), z(u, v))$$

where the *parameters* u and v vary in a domain $\mathcal{D}$ in the uv-plane.
- Tangent and normal vectors:

$$\mathbf{T}_u = \frac{\partial G}{\partial u} = \left\langle \frac{\partial x}{\partial u}, \frac{\partial y}{\partial u}, \frac{\partial z}{\partial u} \right\rangle, \qquad \mathbf{T}_v = \frac{\partial G}{\partial v} = \left\langle \frac{\partial x}{\partial v}, \frac{\partial y}{\partial v}, \frac{\partial z}{\partial v} \right\rangle$$

$$\mathbf{N} = \mathbf{N}(u, v) = \mathbf{T}_u \times \mathbf{T}_v$$

The parametrization is *regular* at (u, v) if $\mathbf{N}(u, v) \neq \mathbf{0}$.
- The quantity $\|\mathbf{N}\|$ is an area scaling factor. If $\mathcal{D}$ is a small region in the uv-plane and $\mathcal{S} = G(\mathcal{D})$, then

$$\text{area}(\mathcal{S}) \approx \|\mathbf{N}(u_0, v_0)\| \text{area}(\mathcal{D})$$

where (u_0, v_0) is any sample point in $\mathcal{D}$.
- Surface integrals and surface area:

$$\iint_{\mathcal{S}} f(x, y, z)\, dS = \iint_{\mathcal{D}} f(G(u, v))\, \|\mathbf{N}(u, v)\|\, du\, dv$$

$$\text{area}(\mathcal{S}) = \iint_{\mathcal{D}} \|\mathbf{N}(u, v)\|\, du\, dv$$

- Some standard parametrizations:

  - Cylinder of radius R (z-axis as central axis):

$$G(\theta, z) = (R \cos \theta, R \sin \theta, z)$$

Outward normal: $\mathbf{N} = \mathbf{T}_\theta \times \mathbf{T}_z = R \langle \cos \theta, \sin \theta, 0 \rangle$

$$dS = \|\mathbf{N}\|\, d\theta\, dz = R\, d\theta\, dz$$

– Sphere of radius R, centered at the origin:

$$G(\theta, \phi) = (R \cos\theta \sin\phi, R \sin\theta \sin\phi, R \cos\phi)$$

Unit radial vector: $\mathbf{e}_r = \langle \cos\theta \sin\phi, \sin\theta \sin\phi, \cos\phi \rangle$

Outward normal: $\mathbf{N} = \mathbf{T}_\phi \times \mathbf{T}_\theta = (R^2 \sin\phi)\,\mathbf{e}_r$

$$dS = \|\mathbf{N}\|\, d\phi\, d\theta = R^2 \sin\phi\, d\phi\, d\theta$$

– Graph of $z = g(x, y)$:

$$G(x, y) = (x, y, g(x, y))$$

$$\mathbf{N} = \mathbf{T}_x \times \mathbf{T}_y = \langle -g_x, -g_y, 1 \rangle$$

$$dS = \|\mathbf{N}\|\, dx\, dy = \sqrt{1 + g_x^2 + g_y^2}\, dx\, dy$$

16.4 EXERCISES

Preliminary Questions

1. What is the surface integral of the function $f(x, y, z) = 10$ over a surface of total area 5?

2. What interpretation can we give to the length $\|\mathbf{N}\|$ of the normal vector for a parametrization $G(u, v)$?

3. A parametrization maps a rectangle of size 0.01×0.02 in the uv-plane onto a small patch S of a surface. Estimate area(S) if $\mathbf{T}_u \times \mathbf{T}_v = \langle 1, 2, 2 \rangle$ at a sample point in the rectangle.

4. A small surface S is divided into three small pieces, each of area 0.2. Estimate $\iint_S f(x, y, z)\, dS$ if $f(x, y, z)$ takes the values 0.9, 1, and 1.1 at sample points in these three pieces.

5. A surface S has a parametrization whose domain is the square $0 \le u$, $v \le 2$ such that $\|\mathbf{N}(u, v)\| = 5$ for all (u, v). What is area(S)?

6. What is the outward-pointing unit normal to the sphere of radius 3 centered at the origin at $P = (2, 2, 1)$?

Exercises

1. Match each parametrization with the corresponding surface in Figure 16.

(a) $(u, \cos v, \sin v)$

(b) $(u, u + v, v)$

(c) (u, v^3, v)

(d) $(\cos u \sin v, 3 \cos u \sin v, \cos v)$

(e) $(u, u(2 + \cos v), u(2 + \sin v))$

2. Show that $G(r, \theta) = (r \cos\theta, r \sin\theta, 1 - r^2)$ parametrizes the paraboloid $z = 1 - x^2 - y^2$. Describe the grid curves of this parametrization.

3. Show that $G(u, v) = (2u + 1, u - v, 3u + v)$ parametrizes the plane $2x - y - z = 2$. Then

(a) Calculate $\mathbf{T}_u$, $\mathbf{T}_v$, and $\mathbf{N}(u, v)$.

(b) Find the area of $S = G(\mathcal{D})$, where $\mathcal{D} = \{(u, v) : 0 \le u \le 2, 0 \le v \le 1\}$.

(c) Express $f(x, y, z) = yz$ in terms of u and v, and evaluate $\iint_S f(x, y, z)\, dS$.

4. Let $S = G(\mathcal{D})$, where $\mathcal{D} = \{(u, v) : u^2 + v^2 \le 1, u \ge 0, v \ge 0\}$ and G is as defined in Exercise 3.

(a) Calculate the surface area of S.

(b) Evaluate $\iint_S (x - y)\, dS$. *Hint:* Use polar coordinates.

5. Let $G(x, y) = (x, y, xy)$.

(a) Calculate $\mathbf{T}_x$, $\mathbf{T}_y$, and $\mathbf{N}(x, y)$.

(b) Let S be the part of the surface with parameter domain $\mathcal{D} = \{(x, y) : x^2 + y^2 \le 1, x \ge 0, y \ge 0\}$. Verify the following formula and evaluate using polar coordinates:

$$\iint_S 1\, dS = \iint_{\mathcal{D}} \sqrt{1 + x^2 + y^2}\, dx\, dy$$

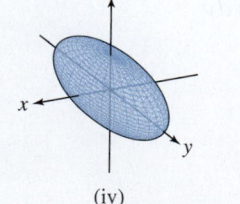

(i)   (ii)   (iii)

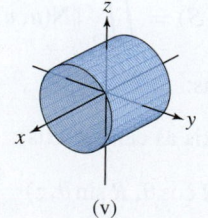

(iv)   (v)

FIGURE 16

(c) Verify the following formula and evaluate:

$$\iint_S z\,dS = \int_0^{\pi/2} \int_0^1 (\sin\theta\cos\theta)r^3\sqrt{1+r^2}\,dr\,d\theta$$

6. A surface S has a parametrization $G(u,v)$ whose domain $\mathcal{D}$ is the square in Figure 17. Suppose that G has the following normal vectors:

$$\mathbf{N}(A) = \langle 2,1,0 \rangle, \quad \mathbf{N}(B) = \langle 1,3,0 \rangle$$

$$\mathbf{N}(C) = \langle 3,0,1 \rangle, \quad \mathbf{N}(D) = \langle 2,0,1 \rangle$$

Estimate $\iint_S f(x,y,z)\,dS$, where f is a function such that $f(G(u,v)) = u + v$.

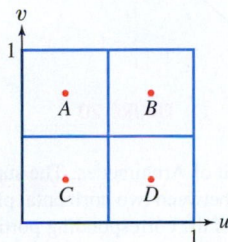

FIGURE 17

In Exercises 7–10, calculate $\mathbf{T}_u$, $\mathbf{T}_v$, and $\mathbf{N}(u,v)$ for the parametrized surface at the given point. Then find the equation of the tangent plane to the surface at that point.

7. $G(u,v) = (2u+v, u-4v, 3u); \quad u=1, \quad v=4$

8. $G(u,v) = (u^2-v^2, u+v, u-v); \quad u=2, \quad v=3$

9. $G(\theta,\phi) = (\cos\theta\sin\phi, \sin\theta\sin\phi, \cos\phi); \quad \theta = \frac{\pi}{2}, \quad \phi = \frac{\pi}{4}$

10. $G(r,\theta) = (r\cos\theta, r\sin\theta, 1-r^2); \quad r = \frac{1}{2}, \quad \theta = \frac{\pi}{4}$

11. Use the normal vector computed in Exercise 8 to estimate the area of the small patch of the surface $G(u,v) = (u^2-v^2, u+v, u-v)$ defined by

$$2 \le u \le 2.1, \qquad 3 \le v \le 3.2$$

12. Sketch the small patch of the sphere whose spherical coordinates satisfy

$$\frac{\pi}{2} - 0.15 \le \theta \le \frac{\pi}{2} + 0.15, \qquad \frac{\pi}{4} - 0.1 \le \phi \le \frac{\pi}{4} + 0.1$$

Use the normal vector computed in Exercise 9 to estimate its area.

In Exercises 13–26, calculate $\iint_S f(x,y,z)\,dS$ for the given surface and function.

13. $G(u,v) = (u\cos v, u\sin v, u), \quad 0 \le u \le 1, \quad 0 \le v \le 1;$
$f(x,y,z) = z(x^2+y^2)$

14. $G(r,\theta) = (r\cos\theta, r\sin\theta, \theta), \quad 0 \le r \le 1, \quad 0 \le \theta \le 2\pi;$
$f(x,y,z) = \sqrt{x^2+y^2}$

15. $y = 4 - z^2, \quad 0 \le x \le 2, 0 \le z \le 2; \quad f(x,y,z) = 3z$

16. $y = 4 - z^2, \quad 0 \le x \le z \le 2; \quad f(x,y,z) = 3$

17. $x^2+y^2+z^2 = 1, \quad x,y,z \ge 0; \quad f(x,y,z) = x^2$

18. $z = 4 - x^2 - y^2, \quad 0 \le z \le 3; \quad f(x,y,z) = x^2/(4-z)$

19. $x^2+y^2 = 4, \quad 0 \le z \le 4; \quad f(x,y,z) = e^{-z}$

20. $G(u,v) = (u, v^3, u+v), \quad 0 \le u \le 1, 0 \le v \le 1;$
$f(x,y,z) = y$

21. Part of the plane $x+y+z = 1$, where $x,y,z \ge 0;$
$f(x,y,z) = z$

22. Part of the plane $x+y+z = 0$ contained in the cylinder $x^2+y^2 = 1;$
$f(x,y,z) = z^2$

23. $x^2+y^2+z^2 = 4, 1 \le z \le 2; \quad f(x,y,z) = z^2(x^2+y^2+z^2)^{-1}$

24. $x^2+y^2+z^2 = 4, 0 \le y \le 1; \quad f(x,y,z) = y$

25. Part of the surface $z = x^3$, where $0 \le x \le 1, \quad 0 \le y \le 1;$
$f(x,y,z) = z$

26. Part of the unit sphere centered at the origin, where $x \ge 0$ and $|y| \le x;$
$f(x,y,z) = x$

27. A surface S has a parametrization $G(u,v)$ with rectangular domain $0 \le u \le 2, 0 \le v \le 4$ such that the following partial derivatives are constant:

$$\frac{\partial G}{\partial u} = \langle 2,0,1 \rangle, \qquad \frac{\partial G}{\partial v} = \langle 4,0,3 \rangle$$

What is the surface area of S?

28. Let S be the sphere of radius R centered at the origin. Explain using symmetry:

$$\iint_S x^2\,dS = \iint_S y^2\,dS = \iint_S z^2\,dS$$

Then show that $\iint_S x^2\,dS = \frac{4}{3}\pi R^4$ by adding the integrals.

29. Calculate $\iint_S (xy+e^z)\,dS$, where S is the triangle in Figure 18 with vertices $(0,0,3)$, $(1,0,2)$, and $(0,4,1)$.

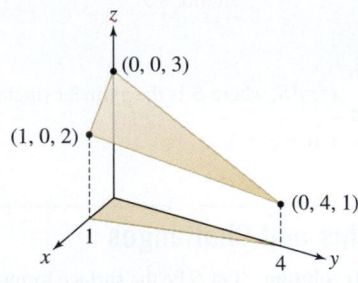

FIGURE 18

30. Use spherical coordinates to compute the surface area of a sphere of radius R.

31. Use cylindrical coordinates to compute the surface area of a sphere of radius R.

32. CAS Let S be the surface with parametrization

$$G(u,v) = \big((3+\sin v)\cos u, (3+\sin v)\sin u, v\big)$$

for $0 \le u \le 2\pi, 0 \le v \le 2\pi$. Using a computer algebra system:

(a) plot S from several different viewpoints. Is S best described as a "vase that holds water" or a "bottomless vase"?

(b) calculate the normal vector $\mathbf{N}(u,v)$.

(c) calculate the surface area of S to four decimal places.

33. CAS Let S be the surface $z = \ln(5 - x^2 - y^2)$ for $0 \le x \le 1$, $0 \le y \le 1$. Using a computer algebra system:

(a) calculate the surface area of S to four decimal places.

(b) calculate $\iint_S x^2 y^3 \, dS$ to four decimal places.

34. Find the area of the portion of the plane $2x + 3y + 4z = 28$ lying above the rectangle $1 \le x \le 3$, $2 \le y \le 5$ in the xy-plane.

35. Use a surface integral to compute the area of that part of the plane $ax + by + cz = d$ corresponding to $0 \le x, y \le 1$.

36. Find the surface area of the part of the cone $x^2 + y^2 = z^2$ between the planes $z = 2$ and $z = 5$.

37. Find the surface area of the portion S of the cone $z^2 = x^2 + y^2$, where $z \ge 0$, contained within the cylinder $y^2 + z^2 \le 1$.

38. Calculate the integral of ze^{2x+y} over the surface of the box in Figure 19.

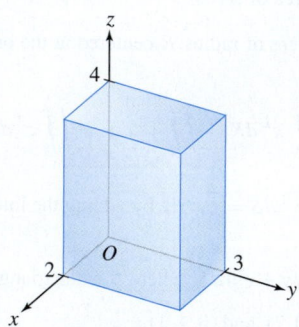

FIGURE 19

39. Calculate $\iint_S x^2 z \, dS$, where S is the cylinder (including the top and bottom) $x^2 + y^2 = 4$, $0 \le z \le 3$.

40. Let S be the portion of the sphere $x^2 + y^2 + z^2 = 9$, where $1 \le x^2 + y^2 \le 4$ and $z \ge 0$ (Figure 20). Find a parametrization of S in polar coordinates and use it to compute:

(a) The area of S **(b)** $\iint_S z^{-1} \, dS$

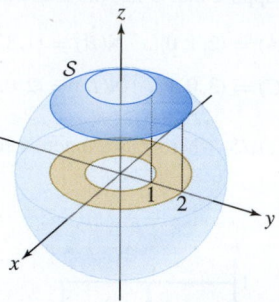

FIGURE 20

41. Prove a famous result of Archimedes: The surface area of the portion of the sphere of radius R between two horizontal planes $z = a$ and $z = b$ is equal to the surface area of the corresponding portion of the circumscribed cylinder (Figure 21).

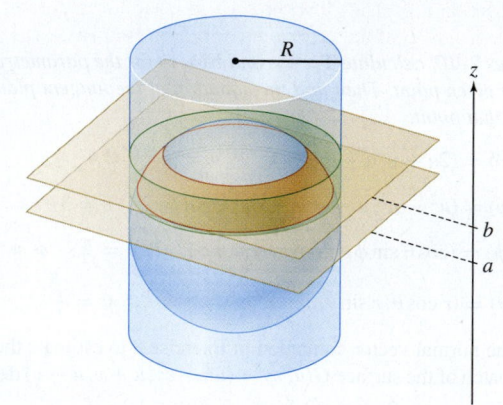

FIGURE 21

Further Insights and Challenges

42. Surfaces of Revolution Let S be the surface formed by rotating the region under the graph $z = g(y)$ in the yz-plane for $c \le y \le d$ about the z-axis, where $c \ge 0$ (Figure 22).

(a) Show that the circle generated by rotating a point $(0, a, b)$ about the z-axis is parametrized by

$$(a \cos \theta, a \sin \theta, b), \quad 0 \le \theta \le 2\pi$$

(b) Show that S is parametrized by

$$G(y, \theta) = (y \cos \theta, y \sin \theta, g(y)) \qquad \boxed{13}$$

for $c \le y \le d$, $0 \le \theta \le 2\pi$.

(c) Use Eq. (13) to prove the formula

$$\text{area}(S) = 2\pi \int_c^d y \sqrt{1 + g'(y)^2} \, dy \qquad \boxed{14}$$

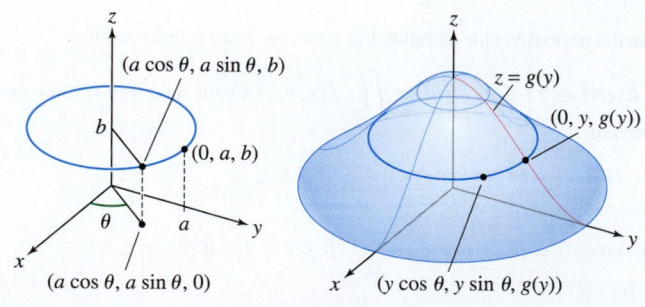

FIGURE 22

43. Use Eq. (14) to compute the surface area of $z = 4 - y^2$ for $0 \le y \le 2$ rotated about the z-axis.

44. Describe the upper half of the cone $x^2 + y^2 = z^2$ for $0 \le z \le d$ as a surface of revolution (Figure 6) and use Eq. (14) to compute its surface area.

45. Area of a Torus Let $\mathcal{T}$ be the torus obtained by rotating the circle in the yz-plane of radius a centered at $(0, b, 0)$ about the z-axis (Figure 23). We assume that $b > a > 0$.

(a) Use Eq. (14) to show that

$$\text{area}(\mathcal{T}) = 4\pi \int_{b-a}^{b+a} \frac{ay}{\sqrt{a^2 - (b-y)^2}} \, dy$$

(b) Show that $\text{area}(\mathcal{T}) = 4\pi^2 ab$.

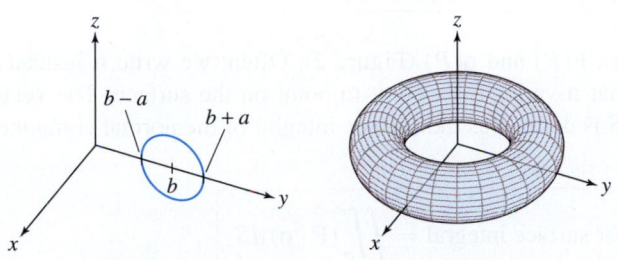

FIGURE 23 The torus obtained by rotating a circle of radius a.

46. Pappus's Theorem (also called **Guldin's Rule**), which we introduced in Section 8.4, states that the area of a surface of revolution $\mathcal{S}$ is equal to the length L of the generating curve times the distance traversed by the center of mass. Use Eq. (14) to prove Pappus's Theorem. If $\mathcal{C}$ is the graph $z = g(y)$ for $c \leq y \leq d$, then the center of mass is defined as the point $(\bar{y}, \bar{z})$ with

$$\bar{y} = \frac{1}{L} \int_{\mathcal{C}} y \, ds, \qquad \bar{z} = \frac{1}{L} \int_{\mathcal{C}} z \, ds$$

47. Compute the surface area of the torus in Exercise 45 using Pappus's Theorem.

48. ✍ **Potential Due to a Uniform Sphere** Let $\mathcal{S}$ be a hollow sphere of radius R with its center at the origin with a uniform mass distribution of total mass m [since $\mathcal{S}$ has surface area $4\pi R^2$, the mass density is $\delta = m/(4\pi R^2)$]. With G representing the universal gravitational constant, the gravitational potential $V(P)$ due to $\mathcal{S}$ at a point $P = (a, b, c)$ is equal to

$$-G \iint_{\mathcal{S}} \frac{\delta \, dS}{\sqrt{(x-a)^2 + (y-b)^2 + (z-c)^2}}$$

(a) Use symmetry to conclude that the potential depends only on the distance r from P to the center of the sphere. Therefore, it suffices to compute $V(P)$ for a point $P = (0, 0, r)$ on the z-axis (with $r \neq R$).

(b) Use spherical coordinates to show that $V(0, 0, r)$ is equal to

$$\frac{-Gm}{4\pi} \int_0^\pi \int_0^{2\pi} \frac{\sin\phi \, d\theta \, d\phi}{\sqrt{R^2 + r^2 - 2Rr\cos\phi}}$$

(c) Use the substitution $u = R^2 + r^2 - 2Rr\cos\phi$ to show that

$$V(0, 0, r) = \frac{-mG}{2Rr}\big(|R + r| - |R - r|\big)$$

(d) Verify Eq. (12) for V.

49. Calculate the gravitational potential V for a hemisphere of radius R with uniform mass distribution.

50. The surface of a cylinder of radius R and length L has a uniform mass distribution δ (the top and bottom of the cylinder are excluded). Use Eq. (11) to find the gravitational potential at a point P located along the axis of the cylinder.

51. Let S be the part of the graph $z = g(x, y)$ lying over a domain $\mathcal{D}$ in the xy-plane. Let $\phi = \phi(x, y)$ be the angle between the normal to S and the vertical. Prove the formula

$$\text{area}(S) = \iint_{\mathcal{D}} \frac{dA}{|\cos\phi|}$$

16.5 Surface Integrals of Vector Fields

The word "flux" is derived from the Latin word "fluere," which means "to flow."

The last integrals we will consider are surface integrals of vector fields. These integrals represent flux or rates of flow through a surface. One example is the flux of molecules across a cell membrane (number of molecules per unit time).

Because flux through a surface S goes from one side of the surface to the other, we need to specify a *positive direction* of flow. This is done by means of an **orientation**, which is a choice of unit normal vector $\mathbf{n}(P)$ at each point P of S, chosen in a contin-

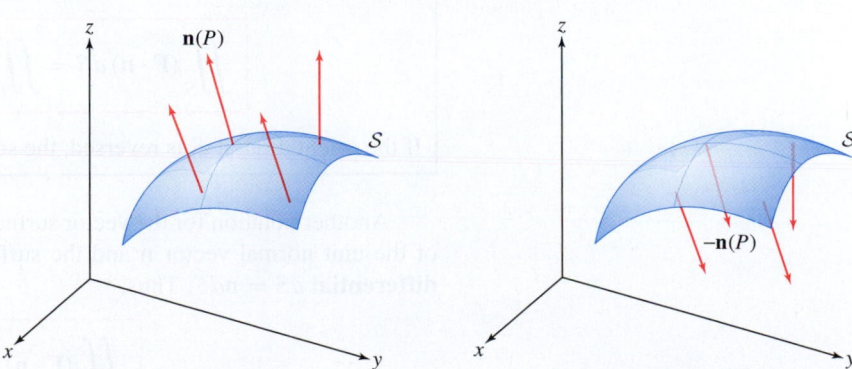

FIGURE 1 The surface S has two possible orientations.

(A) One possible orientation of S

(B) The opposite orientation

uously varying manner (Figure 1). There are two normal directions at each point, so the orientation serves to specify one of the two sides of the surface in a consistent manner. The unit vectors $-\mathbf{n}(P)$ define the *opposite orientation*. For example, if the $\mathbf{n}$ vectors are outward-pointing unit normal vectors on a sphere, then a flow from the inside of the sphere to the outside has a positive flux.

The **normal component** of a vector field $\mathbf{F}$ at a point P on an oriented surface $\mathcal{S}$ is the dot product

$$\text{normal component at } P = \mathbf{F}(P) \cdot \mathbf{n}(P) = \|\mathbf{F}(P)\| \cos\theta$$

where θ is the angle between $\mathbf{F}(P)$ and $\mathbf{n}(P)$ (Figure 2). Often, we write $\mathbf{n}$ instead of $\mathbf{n}(P)$, but it is understood that $\mathbf{n}$ varies from point to point on the surface. The **vector surface integral** of $\mathbf{F}$ over $\mathcal{S}$ is defined as the surface integral of the normal component of $\mathbf{F}$:

$$\text{vector surface integral} = \iint_{\mathcal{S}} (\mathbf{F} \cdot \mathbf{n}) \, dS$$

This quantity is also called the **flux** of $\mathbf{F}$ across or through $\mathcal{S}$.

An oriented parametrization $G(u, v)$ is a regular parametrization [meaning that $\mathbf{N}(u, v)$ is nonzero for all u, v] whose unit normal vector defines the orientation:

$$\mathbf{n} = \mathbf{n}(u, v) = \frac{\mathbf{N}(u, v)}{\|\mathbf{N}(u, v)\|}$$

Applying Eq. (1) in the margin to $\mathbf{F} \cdot \mathbf{n}$, we obtain

$$\iint_{\mathcal{S}} (\mathbf{F} \cdot \mathbf{n}) \, dS = \iint_{\mathcal{D}} (\mathbf{F} \cdot \mathbf{n}) \|\mathbf{N}(u, v)\| \, du \, dv$$

$$= \iint_{\mathcal{D}} \mathbf{F}(G(u, v)) \cdot \left(\frac{\mathbf{N}(u, v)}{\|\mathbf{N}(u, v)\|} \right) \|\mathbf{N}(u, v)\| \, du \, dv$$

$$= \iint_{\mathcal{D}} \mathbf{F}(G(u, v)) \cdot \mathbf{N}(u, v) \, du \, dv \qquad \boxed{2}$$

This formula remains valid even if $\mathbf{N}(u, v)$ is zero at points on the boundary of the parameter domain $\mathcal{D}$. If we reverse the orientation of $\mathcal{S}$ in a vector surface integral, $\mathbf{N}(u, v)$ is replaced by $-\mathbf{N}(u, v)$ and the integral changes sign.

Thus, we obtain the following theorem.

> **THEOREM 1** **Vector Surface Integral** Let $G(u, v)$ be an oriented parametrization of a surface $\mathcal{S}$ with parameter domain $\mathcal{D}$. Assume that G is one-to-one and regular, except possibly at points on the boundary of $\mathcal{D}$. Then
>
> $$\iint_{\mathcal{S}} (\mathbf{F} \cdot \mathbf{n}) \, dS = \iint_{\mathcal{D}} \mathbf{F}(G(u, v)) \cdot \mathbf{N}(u, v) \, du \, dv \qquad \boxed{3}$$
>
> If the orientation of $\mathcal{S}$ is reversed, the surface integral changes sign.

Another notation for the vector surface integral is obtained by expressing the product of the unit normal vector $\mathbf{n}$ and the surface area differential dS as the **vector surface differential** $d\mathbf{S} = \mathbf{n} \, dS$. Thus,

$$\iint_{\mathcal{S}} (\mathbf{F} \cdot \mathbf{n}) \, dS = \iint_{\mathcal{S}} \mathbf{F} \cdot d\mathbf{S}$$

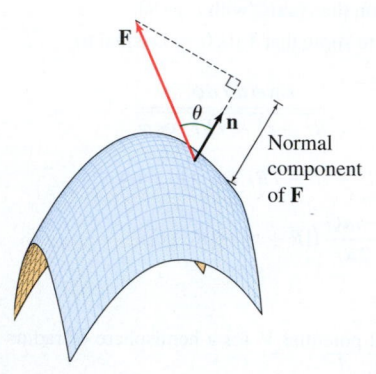

DF FIGURE 2 The normal component of a vector to a surface.

⬅ **REMINDER** *Formula for a scalar surface integral in terms of a parametrization:*

$$\iint_{\mathcal{S}} f(x, y, z) \, dS$$
$$= \iint f(G(u, v)) \|\mathbf{N}(u, v)\| \, du \, dv \qquad \boxed{1}$$

EXAMPLE 1 Calculate $\iint_{\mathcal{S}} \mathbf{F} \cdot d\mathbf{S}$, where $\mathbf{F} = \langle 0, 0, x \rangle$ and $\mathcal{S}$ is the surface with parametrization $G(u, v) = (u^2, v, u^3 - v^2)$ for $0 \leq u \leq 1$, $0 \leq v \leq 1$ and oriented by upward-pointing normal vectors.

Solution

Step 1. **Compute the tangent and normal vectors.**

$$\mathbf{T}_u = \langle 2u, 0, 3u^2 \rangle, \qquad \mathbf{T}_v = \langle 0, 1, -2v \rangle$$

$$\mathbf{N}(u, v) = \mathbf{T}_u \times \mathbf{T}_v = \begin{vmatrix} \mathbf{i} & \mathbf{j} & \mathbf{k} \\ 2u & 0 & 3u^2 \\ 0 & 1 & -2v \end{vmatrix}$$

$$= -3u^2\mathbf{i} + 4uv\mathbf{j} + 2u\mathbf{k} = \langle -3u^2, 4uv, 2u \rangle$$

The z-component of $\mathbf{N}$ is positive on the domain $0 \leq u \leq 1$, so $\mathbf{N}$ is the upward-pointing normal (Figure 3).

Step 2. **Evaluate $\mathbf{F} \cdot \mathbf{N}$.**

Write $\mathbf{F}$ in terms of the parameters u and v. Since $x = u^2$,

$$\mathbf{F}(G(u, v)) = \langle 0, 0, x \rangle = \langle 0, 0, u^2 \rangle$$

and

$$\mathbf{F}(G(u, v)) \cdot \mathbf{N}(u, v) = \langle 0, 0, u^2 \rangle \cdot \langle -3u^2, 4uv, 2u \rangle = 2u^3$$

Step 3. **Evaluate the surface integral.**

The parameter domain is $0 \leq u \leq 1$, $0 \leq v \leq 1$, so

$$\iint_{\mathcal{S}} \mathbf{F} \cdot d\mathbf{S} = \int_{u=0}^{1} \int_{v=0}^{1} \mathbf{F}(G(u, v)) \cdot \mathbf{N}(u, v) \, dv \, du$$

$$= \int_{u=0}^{1} \int_{v=0}^{1} 2u^3 \, dv \, du = \int_{u=0}^{1} 2u^3 \, du = \frac{1}{2} \qquad \blacksquare$$

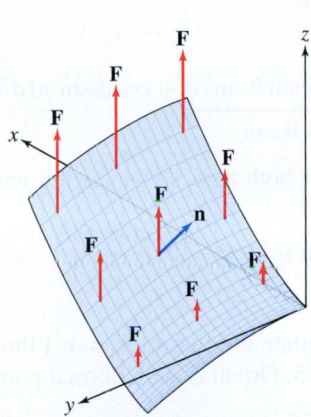

DF FIGURE 3 The surface $G(u, v) = (u^2, v, u^3 - v^2)$ with an upward-pointing normal. The vector field $\mathbf{F} = \langle 0, 0, x \rangle$ points in the vertical direction.

EXAMPLE 2 **Integral over a Hemisphere** Calculate the flux of $\mathbf{F} = \langle z, x, 1 \rangle$ across the upper hemisphere $\mathcal{S}$ of the sphere $x^2 + y^2 + z^2 = 1$, oriented with outward-pointing normal vectors (Figure 4).

Solution Parametrize the hemisphere by spherical coordinates:

$$G(\theta, \phi) = (\cos \theta \sin \phi, \sin \theta \sin \phi, \cos \phi), \qquad 0 \leq \phi \leq \frac{\pi}{2}, \quad 0 \leq \theta < 2\pi$$

Step 1. **Compute the normal vector.**

According to Eq. (2) in Section 16.4, the outward-pointing normal vector is

$$\mathbf{N} = \mathbf{T}_\phi \times \mathbf{T}_\theta = \sin \phi \langle \cos \theta \sin \phi, \sin \theta \sin \phi, \cos \phi \rangle$$

Step 2. **Evaluate $\mathbf{F} \cdot \mathbf{N}$.**

$$\mathbf{F}(G(\theta, \phi)) = \langle z, x, 1 \rangle = \langle \cos \phi, \cos \theta \sin \phi, 1 \rangle$$

$$\mathbf{F}(G(\theta, \phi)) \cdot \mathbf{N}(\theta, \phi) = \langle \cos \phi, \cos \theta \sin \phi, 1 \rangle \cdot \langle \cos \theta \sin^2 \phi, \sin \theta \sin^2 \phi, \cos \phi \sin \phi \rangle$$

$$= \cos \theta \sin^2 \phi \cos \phi + \cos \theta \sin \theta \sin^3 \phi + \cos \phi \sin \phi$$

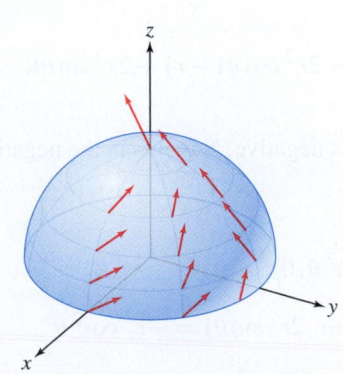

FIGURE 4 The vector field $\mathbf{F} = \langle z, x, 1 \rangle$.

Step 3. **Evaluate the surface integral.**

$$\iint_{\mathcal{S}} \mathbf{F} \cdot d\mathbf{S} = \int_{\phi=0}^{\pi/2} \int_{\theta=0}^{2\pi} \mathbf{F}(G(\theta,\phi)) \cdot \mathbf{N}(\theta,\phi)\, d\theta\, d\phi$$

$$= \int_{\phi=0}^{\pi/2} \int_{\theta=0}^{2\pi} \underbrace{(\cos\theta \sin^2\phi \cos\phi + \cos\theta \sin\theta \sin^3\phi}_{\text{Integral over } 0 \le \theta \le 2\pi \text{ is zero}} + \cos\phi \sin\phi)\, d\theta\, d\phi$$

The integrals of $\cos\theta$ and $\cos\theta \sin\theta$ over $[0, 2\pi]$ are both zero, so we are left with

$$\int_{\phi=0}^{\pi/2} \int_{\theta=0}^{2\pi} \cos\phi \sin\phi\, d\theta\, d\phi = 2\pi \int_{\phi=0}^{\pi/2} \cos\phi \sin\phi\, d\phi = -2\pi \frac{\cos^2\phi}{2}\bigg|_0^{\pi/2} = \pi \quad\blacksquare$$

EXAMPLE 3 **Surface Integral over a Graph**   Calculate the flux of $\mathbf{F} = x^2\mathbf{j}$ through the surface $\mathcal{S}$ defined by $y = 1 + x^2 + z^2$ for $1 \le y \le 5$. Orient $\mathcal{S}$ with normal pointing in the negative y-direction.

Solution   This surface is the graph of the function $y = 1 + x^2 + z^2$, where x and z are the independent variables (Figure 5).

Step 1. **Find a parametrization.**
It is convenient to use x and z because y is given explicitly as a function of x and z. Thus, we define

$$G(x,z) = (x, 1 + x^2 + z^2, z)$$

What is the parameter domain? Since $y = 1 + x^2 + z^2$, the condition $1 \le y \le 5$ is equivalent to $1 \le 1 + x^2 + z^2 \le 5$ or $0 \le x^2 + z^2 \le 4$. Therefore, the parameter domain is the disk of radius 2 in the xz-plane—that is, $\mathcal{D} = \{(x,z) : x^2 + z^2 \le 4\}$.

Because the parameter domain is a disk, it makes sense to use the polar variables r and θ in the xz-plane. In other words, we write $x = r\cos\theta$, $z = r\sin\theta$. Then

$$y = 1 + x^2 + z^2 = 1 + r^2$$

$$G(r,\theta) = (r\cos\theta, 1 + r^2, r\sin\theta), \quad 0 \le \theta \le 2\pi, \quad 0 \le r \le 2$$

Step 2. **Compute the tangent and normal vectors.**

$$\mathbf{T}_r = \langle \cos\theta, 2r, \sin\theta \rangle, \qquad \mathbf{T}_\theta = \langle -r\sin\theta, 0, r\cos\theta \rangle$$

$$\mathbf{N} = \mathbf{T}_r \times \mathbf{T}_\theta = \begin{vmatrix} \mathbf{i} & \mathbf{j} & \mathbf{k} \\ \cos\theta & 2r & \sin\theta \\ -r\sin\theta & 0 & r\cos\theta \end{vmatrix} = 2r^2\cos\theta\,\mathbf{i} - r\mathbf{j} + 2r^2\sin\theta\,\mathbf{k}$$

The coefficient of $\mathbf{j}$ is $-r$. Because this coefficient is negative, $\mathbf{N}$ points in the negative y-direction, as required.

Step 3. **Evaluate $\mathbf{F} \cdot \mathbf{N}$.**

$$\mathbf{F}(G(r,\theta)) = x^2\mathbf{j} = r^2\cos^2\theta\,\mathbf{j} = \langle 0, r^2\cos^2\theta, 0 \rangle$$

$$\mathbf{F}(G(r,\theta)) \cdot \mathbf{N} = \langle 0, r^2\cos^2\theta, 0 \rangle \cdot \langle 2r^2\cos\theta, -r, 2r^2\sin\theta \rangle = -r^3\cos^2\theta$$

$$\iint_{\mathcal{S}} \mathbf{F} \cdot d\mathbf{S} = \iint_{\mathcal{D}} \mathbf{F}(G(r,\theta)) \cdot \mathbf{N}\, dr\, d\theta = \int_0^{2\pi} \int_0^2 (-r^3\cos^2\theta)\, dr\, d\theta$$

$$= -\left(\int_0^{2\pi} \cos^2\theta\, d\theta \right)\left(\int_0^2 r^3\, dr \right)$$

$$= -(\pi)\left(\frac{2^4}{4} \right) = -4\pi$$

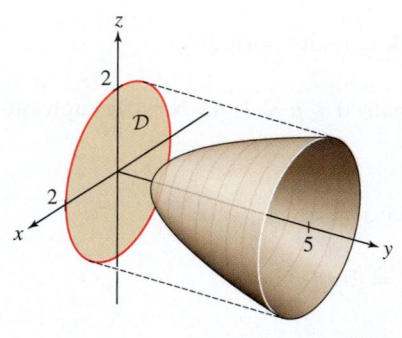

FIGURE 5

CAUTION *In Step 3, we integrate $\mathbf{F} \cdot \mathbf{N}$ with respect to $dr\, d\theta$, and not $r\, dr\, d\theta$. The factor of r in $r\, dr\, d\theta$ is a Jacobian factor that we add only when changing variables in a double integral. In surface integrals, the Jacobian factor is incorporated into the magnitude of $\mathbf{N}$ (recall that $\|\mathbf{N}\|$ is an area scaling factor).*

It is not surprising that the flux is negative since the positive normal direction was chosen to be the negative y-direction but the vector field **F** points in the positive y-direction. ∎

CONCEPTUAL INSIGHT Since a vector surface integral depends on the orientation of the surface, this integral is defined only for surfaces that have two sides. However, some surfaces, such as the Möbius strip (discovered in 1858 independently by August Möbius and Johann Listing), cannot be oriented because they are one-sided. You can construct a Möbius strip M with a rectangular strip of paper: Join the two ends of the strip together with a 180° twist. Unlike an ordinary two-sided strip, the Möbius strip M has only one side, and it is impossible to specify a positive normal direction in a consistent manner (Figure 6). If you choose a unit normal vector at a point P and carry that unit vector continuously around M, when you return to P, the vector will point in the opposite direction. Therefore, we cannot integrate a vector field over a Möbius strip, and it is not meaningful to speak of the flux across M. On the other hand, it is possible to integrate a scalar function. For example, the integral of mass density would equal the total mass of the Möbius strip.

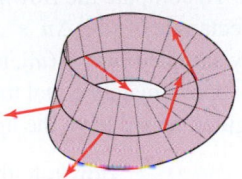
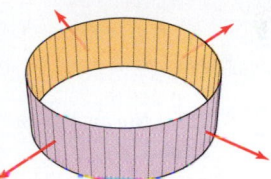

Möbius strip   Ordinary (untwisted) band

FIGURE 6 It is not possible to choose a continuously varying unit normal vector on a Möbius strip.

Fluid Flux

Imagine dipping a net into a stream of flowing water (Figure 7). The **flow rate** is the volume of water that flows through the net per unit time.

To compute the flow rate, let **v** be the velocity vector field. At each point P, $\mathbf{v}(P)$ is the velocity vector of the fluid particle located at the point P. We claim that *the flow rate through a surface S is equal to the surface integral of* **v** *over* S.

To explain why, suppose first that S is a rectangle of area A and that **v** is a constant vector field with value $\mathbf{v}_0$ perpendicular to the rectangle. The particles travel at speed $\|\mathbf{v}_0\|$, say, in meters per second, so a given particle flows through S within a 1-second time interval if its distance to S is at most $\|\mathbf{v}_0\|$ meters—in other words, if its velocity vector passes through S (see Figure 8). Thus, the block of fluid passing through S in a 1-s interval is a box of volume $\|\mathbf{v}_0\|A$ (Figure 9), and

$$\text{flow rate} = (\text{velocity})(\text{area}) = \|\mathbf{v}_0\|A$$

FIGURE 7 Velocity field of a fluid flow.

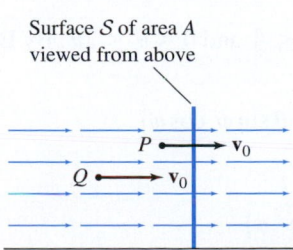

Surface S of area A viewed from above

FIGURE 8 The particle P flows through S within a 1-second interval, but Q does not.

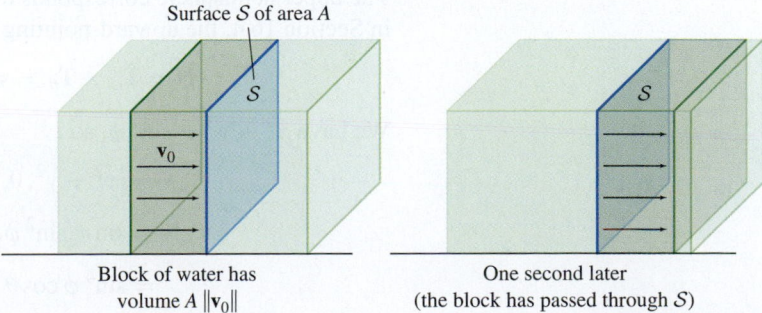

Surface S of area A

Block of water has volume $A\|\mathbf{v}_0\|$

One second later (the block has passed through S)

FIGURE 9

If the fluid flows at an angle θ relative to $\mathcal{S}$, then the block of water is a parallelepiped (rather than a box) of volume $A\|\mathbf{v}_0\|\cos\theta$ (Figure 10). If $\mathbf{N}$ is a vector normal to $\mathcal{S}$ of length equal to the area A, then we can write the flow rate as a dot product:

$$\text{flow rate} = A\|\mathbf{v}_0\|\cos\theta = \mathbf{v}_0 \cdot \mathbf{N}$$

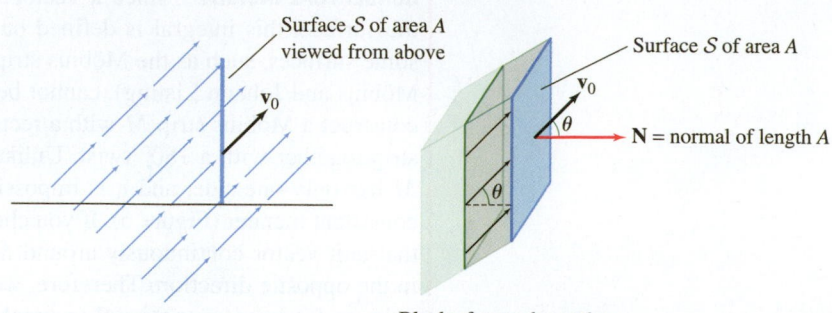

Surface $\mathcal{S}$ of area A viewed from above

Surface $\mathcal{S}$ of area A

$\mathbf{v}_0$

$\mathbf{N}$ = normal of length A

Block of water has volume $A\|\mathbf{v}_0\|\cos\theta = \mathbf{N}\cdot\mathbf{v}_0$

FIGURE 10 Water flowing at constant velocity $\mathbf{v}_0$, making an angle θ with a rectangular surface.

In the general case, the velocity field $\mathbf{v}$ is not constant, and the surface $\mathcal{S}$ may be curved. To compute the flow rate, we choose a parametrization $G(u, v)$ and we consider a small rectangle of size $\Delta u \times \Delta v$ that is mapped by G to a small patch $\mathcal{S}_0$ of $\mathcal{S}$ (Figure 11). For any sample point $G(u_0, v_0)$ in $\mathcal{S}_0$, the vector $\mathbf{N}(u_0, v_0)\,\Delta u\,\Delta v$ is a normal vector of length approximately equal to the area of $\mathcal{S}_0$ [Eq. (3) in Section 16.4]. This patch is nearly rectangular, so we have the approximation

$$\text{flow rate through } \mathcal{S}_0 \approx \mathbf{v}(u_0, v_0) \cdot \mathbf{N}(u_0, v_0)\,\Delta u\,\Delta v$$

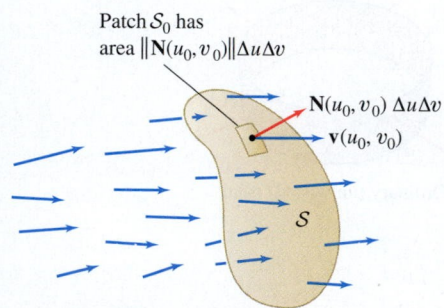

Patch $\mathcal{S}_0$ has area $\|\mathbf{N}(u_0,v_0)\|\Delta u \Delta v$

$\mathbf{N}(u_0,v_0)\,\Delta u \Delta v$

$\mathbf{v}(u_0,v_0)$

$\mathcal{S}$

The total flow per second is the sum of the flows through all of the small patches covering $\mathcal{S}$. As usual, the limit of the sums as Δu and Δv tend to zero is the integral of $\mathbf{v}(u, v) \cdot \mathbf{N}(u, v)$, which is the surface integral of $\mathbf{v}$ over $\mathcal{S}$.

FIGURE 11 The flow rate across the small patch $\mathcal{S}_0$ is approximately $\mathbf{v}(u_0, v_0) \cdot \mathbf{N}(u_0, v_0)\,\Delta u\,\Delta v$.

Flow Rate Through a Surface   For a fluid with velocity vector field $\mathbf{v}$,

$$\text{flow rate across the } \mathcal{S} \text{ (volume per unit time)} = \iint_{\mathcal{S}} \mathbf{v} \cdot d\mathbf{S} \qquad \boxed{4}$$

EXAMPLE 4 Let $\mathbf{v} = \langle x^2 + y^2, 0, z^2 \rangle$ be the velocity field (in centimeters per second) of a fluid in $\mathbf{R}^3$. Compute the flow rate upward through the upper hemisphere $\mathcal{S}$ of the unit sphere centered at the origin.

Solution We use spherical coordinates:

$$x = \cos\theta\sin\phi, \qquad y = \sin\theta\sin\phi, \qquad z = \cos\phi$$

The upper hemisphere corresponds to the ranges $0 \le \phi \le \frac{\pi}{2}$ and $0 \le \theta \le 2\pi$. By Eq. (2) in Section 16.4, the upward-pointing normal is

$$\mathbf{N} = \mathbf{T}_\phi \times \mathbf{T}_\theta = \sin\phi\langle\cos\theta\sin\phi, \sin\theta\sin\phi, \cos\phi\rangle$$

We have $x^2 + y^2 = \sin^2\phi$, so

$$\mathbf{v} = \langle x^2 + y^2, 0, z^2 \rangle = \langle \sin^2\phi, 0, \cos^2\phi \rangle$$

$$\mathbf{v} \cdot \mathbf{N} = \sin\phi\langle\sin^2\phi, 0, \cos^2\phi\rangle \cdot \langle\cos\theta\sin\phi, \sin\theta\sin\phi, \cos\phi\rangle$$

$$= \sin^4\phi\cos\theta + \sin\phi\cos^3\phi$$

$$\iint_{\mathcal{S}} \mathbf{v} \cdot d\mathbf{S} = \int_{\phi=0}^{\pi/2} \int_{\theta=0}^{2\pi} (\sin^4\phi\cos\theta + \sin\phi\cos^3\phi)\,d\theta\,d\phi$$

The integral of $\sin^4 \phi \cos \theta$ with respect to θ is zero, so we are left with

$$\int_{\phi=0}^{\pi/2} \int_{\theta=0}^{2\pi} \sin \phi \cos^3 \phi \, d\theta \, d\phi = 2\pi \int_{\phi=0}^{\pi/2} \cos^3 \phi \sin \phi \, d\phi$$

$$= 2\pi \left(-\frac{\cos^4 \phi}{4} \right) \Big|_{\phi=0}^{\pi/2} = \frac{\pi}{2} \text{ cm}^3/\text{s}$$

Since **N** is an upward-pointing normal, this is the rate at which fluid flows across the hemisphere from below to above. ■

Electric and Magnetic Fields

The laws of electricity and magnetism are expressed in terms of two vector fields, the electric field **E** and the magnetic field **B**, whose properties are summarized in Maxwell's four equations. One of these equations is **Faraday's Law of Induction**, which can be formulated either as a partial differential equation or in the following integral form:

$$\boxed{\int_{\mathcal{C}} \mathbf{E} \cdot d\mathbf{r} = -\frac{d}{dt} \iint_{S} \mathbf{B} \cdot d\mathbf{S}} \qquad \boxed{5}$$

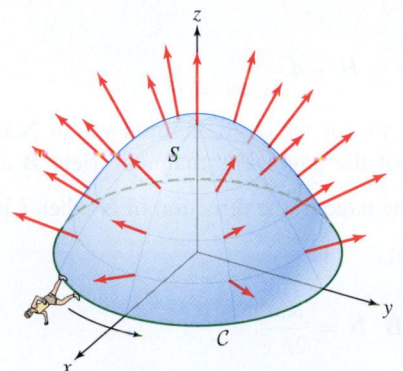

FIGURE 12 The positive direction along the boundary curve $\mathcal{C}$ is defined so that if a pedestrian walks in the positive direction with the surface to her left, then her head points in the positive (normal) direction.

In this equation, S is an oriented surface with boundary curve $\mathcal{C}$, oriented as indicated in Figure 12. The line integral of **E** is equal to the voltage drop around the boundary curve (the work performed by **E** moving a positive unit charge around $\mathcal{C}$).

To illustrate Faraday's Law, consider an electric current of i amperes flowing through a straight wire. According to the Biot-Savart Law, this current produces a magnetic field **B** of magnitude $B(r) = \dfrac{\mu_0 |i|}{2\pi r}$ T, where r is the distance (in meters) from the wire and $\mu_0 = 4\pi \cdot 10^{-7}$ T-m/A. At each point P, **B** is tangent to the circle through P perpendicular to the wire as in Figure 13(A), with the direction determined by the right-hand rule: If the thumb of your right hand points in the direction of the current, then your fingers curl in the direction of **B**.

*The **tesla** (T) is the SI unit of magnetic field strength. A 1-coulomb point charge passing through a magnetic field of 1 tesla at 1 m/s experiences a force of 1 newton.*

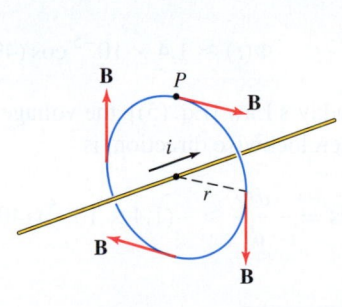

(A) Magnetic field **B** due to a current in a wire

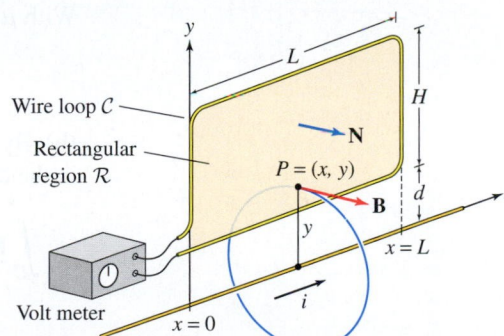

(B) The magnetic field **B** points in the direction **N** normal to $\mathcal{R}$

FIGURE 13

*The electric field **E** is conservative when the charges are stationary or, more generally, when the magnetic field **B** is constant. When **B** varies in time, the integral on the right in Eq. (5) is nonzero for some surface, and hence the circulation of **E** around the boundary curve $\mathcal{C}$ is also nonzero. This shows that **E** is not conservative when **B** varies in time.*

EXAMPLE 5 A varying current of magnitude (t in seconds)

$$i = 28 \cos(400t) \text{ amperes}$$

flows through a straight wire [Figure 13(B)]. A rectangular wire loop $\mathcal{C}$ of length $L = 1.2$ m and width $H = 0.7$ m is located a distance $d = 0.1$ m from the wire as in the figure. The loop encloses a rectangular surface $\mathcal{R}$, which is oriented by normal vectors $\mathbf{N}$ pointing out of the page.

(a) Calculate the flux $\Phi(t)$ of $\mathbf{B}$ through $\mathcal{R}$.

(b) Use Faraday's Law to determine the voltage drop (in volts) around the loop $\mathcal{C}$.

Magnetic flux as a function of time is often denoted by the Greek letter Φ:

$$\Phi(t) = \iint_{\mathcal{S}} \mathbf{B} \cdot d\mathbf{S}$$

Solution We choose coordinates (x, y) on rectangle $\mathcal{R}$ as in Figure 13, so that y is the distance from the wire and $\mathcal{R}$ is the region

$$0 \leq x \leq L, \qquad d \leq y \leq H + d$$

Our parametrization of $\mathcal{R}$ is simply $G(x, y) = (x, y)$, for which the normal vector $\mathbf{N}$ is the unit vector perpendicular to $\mathcal{R}$, pointing out of the page. The magnetic field $\mathbf{B}$ at $P = (x, y)$ has magnitude $\dfrac{\mu_0 |i|}{2\pi y}$. It points out of the page in the direction of $\mathbf{N}$ when i is positive and into the page when i is negative. Thus,

$$\mathbf{B} = \frac{\mu_0 i}{2\pi y} \mathbf{N} \qquad \text{and} \qquad \mathbf{B} \cdot \mathbf{N} = \frac{\mu_0 i}{2\pi y}$$

(a) The flux $\Phi(t)$ of $\mathbf{B}$ through $\mathcal{R}$ at time t is

$$\Phi(t) = \iint_{\mathcal{R}} \mathbf{B} \cdot d\mathbf{S} = \int_{x=0}^{L} \int_{y=d}^{H+d} \mathbf{B} \cdot \mathbf{N} \, dy \, dx$$

$$= \int_{x=0}^{L} \int_{y=d}^{H+d} \frac{\mu_0 i}{2\pi y} \, dy \, dx = \frac{\mu_0 L i}{2\pi} \int_{y=d}^{H+d} \frac{dy}{y}$$

$$= \frac{\mu_0 L}{2\pi} \left(\ln \frac{H+d}{d} \right) i$$

$$= \frac{\mu_0 (1.2)}{2\pi} \left(\ln \frac{0.8}{0.1} \right) 28 \cos (400t)$$

With $\mu_0 = 4\pi \cdot 10^{-7}$, we obtain

$$\Phi(t) \approx 1.4 \times 10^{-5} \cos (400t) \text{ T-m}^2$$

(b) By Faraday's Law [Eq. (5)], the voltage drop around the rectangular loop $\mathcal{C}$, oriented in the counterclockwise direction, is

$$\int_{\mathcal{C}} \mathbf{E} \cdot d\mathbf{s} = -\frac{d\Phi}{dt} \approx -(1.4 \times 10^{-5})(400) \sin (400t) = -0.0056 \sin (400t) \text{ volts} \quad \blacksquare$$

Types of Integrals

We end with a list of the types of integrals we have introduced in this chapter.

1. Scalar line integral along a curve $\mathcal{C}$ given by $\mathbf{r}(t)$ for $a \leq t \leq b$ (can be used to compute arc length, mass, electric potential):

$$\int_{\mathcal{C}} f(x, y, z) \, ds = \int_{a}^{b} f(\mathbf{r}(t)) \| \mathbf{r}'(t) \| \, dt$$

2. Vector line integral to calculate work along a curve $\mathcal{C}$ given by $\mathbf{r}(t)$ for $a \leq t \leq b$:

$$\int_{\mathcal{C}} \mathbf{F} \cdot d\mathbf{r} = \int_{a}^{b} \mathbf{F}(\mathbf{r}(t)) \cdot \mathbf{r}'(t) \, dt = \int_{\mathcal{C}} F_1 \, dx + F_2 \, dy + F_3 \, dz$$

3. **Vector line integral** to calculate flux across a curve $\mathcal{C}$ given by $\mathbf{r}(t)$ for $a \leq t \leq b$:

$$\int_{\mathcal{C}} \mathbf{F} \cdot \mathbf{n} \, ds = \int_a^b \mathbf{F}(\mathbf{r}(t)) \cdot \mathbf{N}(t) \, dt$$

4. **Surface integral** over a surface with parametrization $G(u, v)$ and parameter domain $\mathcal{D}$ (can be used to calculate surface area, total charge, gravitational potential):

$$\iint_{\mathcal{S}} f(x, y, z) \, dS = \iint_{\mathcal{D}} f(G(u, v)) \|\mathbf{N}(u, v)\| \, du \, dv$$

5. **Vector surface integral** to calculate flux of a vector field $\mathbf{F}$ across a surface $\mathcal{S}$ with parametrization $G(u, v)$ and parameter domain $\mathcal{D}$:

$$\iint_{\mathcal{S}} (\mathbf{F} \cdot \mathbf{n}) \, dS = \iint_{\mathcal{S}} \mathbf{F} \cdot d\mathbf{S} = \iint_{\mathcal{D}} \mathbf{F}(G(u, v)) \cdot \mathbf{N}(u, v) \, du \, dv$$

16.5 SUMMARY

- A surface $\mathcal{S}$ is *oriented* if a continuously varying unit normal vector $\mathbf{n}(P)$ is specified at each point on $\mathcal{S}$. This distinguishes a positive direction across the surface.
- The integral of a vector field $\mathbf{F}$ over an oriented surface $\mathcal{S}$ is defined as the surface integral of the normal component $\mathbf{F} \cdot \mathbf{n}$ over $\mathcal{S}$.
- Vector surface integrals are computed using the formula

$$\iint_{\mathcal{S}} (\mathbf{F} \cdot \mathbf{n}) \, dS = \iint_{\mathcal{S}} \mathbf{F} \cdot d\mathbf{S} = \iint_{\mathcal{D}} \mathbf{F}(G(u, v)) \cdot \mathbf{N}(u, v) \, du \, dv$$

Here, $G(u, v)$ is a parametrization of $\mathcal{S}$ such that $\mathbf{N}(u, v) = \mathbf{T}_u \times \mathbf{T}_v$ points in the direction of the unit normal vector specified by the orientation.
- The surface integral of a vector field $\mathbf{F}$ over $\mathcal{S}$ is also called the *flux* of $\mathbf{F}$ through $\mathcal{S}$. If $\mathbf{F}$ is the velocity field of a fluid, then $\iint_{\mathcal{S}} \mathbf{F} \cdot d\mathbf{S}$ is the rate at which fluid flows through $\mathcal{S}$ per unit time.

16.5 EXERCISES

Preliminary Questions

1. Let $\mathbf{F}$ be a vector field and $G(u, v)$ a parametrization of a surface $\mathcal{S}$, and set $\mathbf{N} = \mathbf{T}_u \times \mathbf{T}_v$. Which of the following is the normal component of $\mathbf{F}$?

(a) $\mathbf{F} \cdot \mathbf{N}$ (b) $\mathbf{F} \cdot \mathbf{n}$

2. The vector surface integral $\iint_{\mathcal{S}} \mathbf{F} \cdot d\mathbf{S}$ is equal to the scalar surface integral of the function (choose the correct answer):

(a) $\|\mathbf{F}\|$.

(b) $\mathbf{F} \cdot \mathbf{N}$, where $\mathbf{N}$ is a normal vector.

(c) $\mathbf{F} \cdot \mathbf{n}$, where $\mathbf{n}$ is the unit normal vector.

3. $\iint_{\mathcal{S}} \mathbf{F} \cdot d\mathbf{S}$ is zero if (choose the correct answer):

(a) $\mathbf{F}$ is tangent to $\mathcal{S}$ at every point.

(b) $\mathbf{F}$ is perpendicular to $\mathcal{S}$ at every point.

4. If $\mathbf{F}(P) = \mathbf{n}(P)$ at each point on $\mathcal{S}$, then $\iint_{\mathcal{S}} \mathbf{F} \cdot d\mathbf{S}$ is equal to which of the following?

(a) Zero   (b) Area($\mathcal{S}$)   (c) Neither

5. Let $\mathcal{S}$ be the disk $x^2 + y^2 \leq 1$ in the xy-plane oriented with normal in the positive z-direction. Determine $\iint_{\mathcal{S}} \mathbf{F} \cdot d\mathbf{S}$ for each of the following vector constant fields:

(a) $\mathbf{F} = \langle 1, 0, 0 \rangle$ (b) $\mathbf{F} = \langle 0, 0, 1 \rangle$ (c) $\mathbf{F} = \langle 1, 1, 1 \rangle$

6. Estimate $\iint_{\mathcal{S}} \mathbf{F} \cdot d\mathbf{S}$, where $\mathcal{S}$ is a tiny oriented surface of area 0.05 and the value of $\mathbf{F}$ at a sample point in $\mathcal{S}$ is a vector of length 2 making an angle $\frac{\pi}{4}$ with the normal to the surface.

7. A small surface $\mathcal{S}$ is divided into three pieces of area 0.2. Estimate $\iint_{\mathcal{S}} \mathbf{F} \cdot d\mathbf{S}$ if $\mathbf{F}$ is a unit vector field making angles of 85°, 90°, and 95° with the normal at sample points in these three pieces.

Exercises

1. Let $\mathbf{F} = \langle z, 0, y \rangle$, and let $\mathcal{S}$ be the oriented surface parametrized by $G(u, v) = (u^2 - v, u, v^2)$ for $0 \leq u \leq 2, -1 \leq v \leq 4$. Calculate:

(a) $\mathbf{N}$ and $\mathbf{F} \cdot \mathbf{N}$ as functions of u and v

(b) The normal component of $\mathbf{F}$ to the surface at $P = (3, 2, 1) = G(2, 1)$

(c) $\iint_{\mathcal{S}} \mathbf{F} \cdot d\mathbf{S}$

2. Let $\mathbf{F} = \langle y, -x, x^2 + y^2 \rangle$, and let $\mathcal{S}$ be the portion of the paraboloid $z = x^2 + y^2$ where $x^2 + y^2 \leq 3$.

(a) Show that if $\mathcal{S}$ is parametrized in polar variables $x = r \cos \theta$, $y = r \sin \theta$, then $\mathbf{F} \cdot \mathbf{N} = r^3$.

(b) Show that $\iint_{\mathcal{S}} \mathbf{F} \cdot d\mathbf{S} = \int_0^{2\pi} \int_0^3 r^3 \, dr \, d\theta$ and evaluate.

3. Let $\mathcal{S}$ be the unit square in the xy-plane shown in Figure 14, oriented with the normal pointing in the positive z-direction. Estimate

$$\iint_{\mathcal{S}} \mathbf{F} \cdot d\mathbf{S}$$

where $\mathbf{F}$ is a vector field whose values at the labeled points are

$$\mathbf{F}(A) = \langle 2, 6, 4 \rangle, \qquad \mathbf{F}(B) = \langle 1, 1, 7 \rangle$$
$$\mathbf{F}(C) = \langle 3, 3, -3 \rangle, \qquad \mathbf{F}(D) = \langle 0, 1, 8 \rangle$$

4. Suppose that $\mathcal{S}$ is a surface in $\mathbf{R}^3$ with a parametrization G whose domain $\mathcal{D}$ is the square in Figure 14. The values of a function f, a vector field $\mathbf{F}$, and the normal vector $\mathbf{N} = \mathbf{T}_u \times \mathbf{T}_v$ at $G(P)$ are given for the four sample points in $\mathcal{D}$ in the following table. Estimate the surface integrals of f and $\mathbf{F}$ over $\mathcal{S}$.

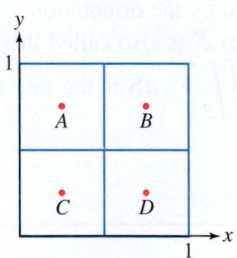

FIGURE 14

Point P in $\mathcal{D}$	f	$\mathbf{F}$	$\mathbf{N}$
A	3	$\langle 2, 6, 4 \rangle$	$\langle 1, 1, 1 \rangle$
B	1	$\langle 1, 1, 7 \rangle$	$\langle 1, 1, 0 \rangle$
C	2	$\langle 3, 3, -3 \rangle$	$\langle 1, 0, -1 \rangle$
D	5	$\langle 0, 1, 8 \rangle$	$\langle 2, 1, 0 \rangle$

In Exercises 5–17, compute $\iint_{\mathcal{S}} \mathbf{F} \cdot d\mathbf{S}$ *for the given oriented surface.*

5. $\mathbf{F} = \langle y, z, x \rangle$, plane $3x - 4y + z = 1$, $0 \leq x \leq 1$, $0 \leq y \leq 1$, upward-pointing normal

6. $\mathbf{F} = \langle e^z, z, x \rangle$, $G(r, s) = (rs, r + s, r)$, $0 \leq r \leq 1, 0 \leq s \leq 1$, oriented by $\mathbf{T}_r \times \mathbf{T}_s$

7. $\mathbf{F} = \langle 0, 3, x \rangle$, part of sphere $x^2 + y^2 + z^2 = 9$, where $x \geq 0$, $y \geq 0$, $z \geq 0$, outward-pointing normal

8. $\mathbf{F} = \langle x, y, z \rangle$, part of sphere $x^2 + y^2 + z^2 = 1$, where $\dfrac{1}{2} \leq z \leq \dfrac{\sqrt{3}}{2}$, inward-pointing normal

9. $\mathbf{F} = \langle z, z, x \rangle$, $z = 9 - x^2 - y^2$, $x \geq 0$, $y \geq 0$, $z \geq 0$, upward-pointing normal

10. $\mathbf{F} = \langle \sin y, \sin z, yz \rangle$, rectangle $0 \leq y \leq 2$, $0 \leq z \leq 3$ in the (y, z)-plane, normal pointing in negative x-direction

11. $\mathbf{F} = y^2 \mathbf{i} + 2\mathbf{j} - x\mathbf{k}$, portion of the plane $x + y + z = 1$ in the octant $x, y, z \geq 0$, upward-pointing normal

12. $\mathbf{F} = \langle x, y, e^z \rangle$, cylinder $x^2 + y^2 = 4, 1 \leq z \leq 5$, outward-pointing normal

13. $\mathbf{F} = \langle xz, yz, z^{-1} \rangle$, disk of radius 3 at height 4 parallel to the xy-plane, upward-pointing normal

14. $\mathbf{F} = \langle xy, y, 0 \rangle$, cone $z^2 = x^2 + y^2$, $x^2 + y^2 \leq 4$, $z \geq 0$, downward-pointing normal

15. $\mathbf{F} = \langle 0, 0, e^{y+z} \rangle$, boundary of unit cube $0 \leq x \leq 1$, $0 \leq y \leq 1$, $0 \leq z \leq 1$, outward-pointing normal

16. $\mathbf{F} = \langle 0, 0, z^2 \rangle$, $G(u, v) = (u \cos v, u \sin v, v)$, $0 \leq u \leq 1$, $0 \leq v \leq 2\pi$, upward-pointing normal

17. $\mathbf{F} = \langle y, z, 0 \rangle$, $G(u, v) = (u^3 - v, u + v, v^2)$, $0 \leq u \leq 2$, $0 \leq v \leq 3$, downward-pointing normal

18. [✎] Let $\mathcal{S}$ be the oriented half-cylinder in Figure 15. In (a)–(f), determine whether $\iint_{\mathcal{S}} \mathbf{F} \cdot d\mathbf{S}$ is positive, negative, or zero. Explain your reasoning.

(a) $\mathbf{F} = \mathbf{i}$ **(b)** $\mathbf{F} = \mathbf{j}$ **(c)** $\mathbf{F} = \mathbf{k}$

(d) $\mathbf{F} = y\mathbf{i}$ **(e)** $\mathbf{F} = -y\mathbf{j}$ **(f)** $\mathbf{F} = x\mathbf{j}$

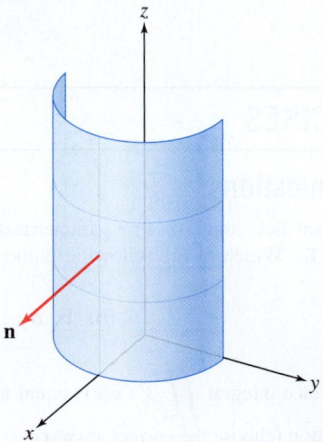

FIGURE 15

19. Let $\mathbf{e_r} = \langle x/r, y/r, z/r \rangle$ be the unit radial vector, where $r = \sqrt{x^2 + y^2 + z^2}$. Calculate the integral of $\mathbf{F} = e^{-r} \mathbf{e_r}$ over:

(a) the upper hemisphere of $x^2 + y^2 + z^2 = 9$, outward-pointing normal.

(b) the octant $x \geq 0, y \geq 0, z \geq 0$ of the unit sphere centered at the origin.

20. Show that the flux of $\mathbf{F} = \dfrac{\mathbf{e}_r}{r^2}$ through a sphere centered at the origin does not depend on the radius of the sphere.

21. The electric field due to a point charge located at the origin in $\mathbf{R}^3$ is $\mathbf{E} = k\dfrac{\mathbf{e}_r}{r^2}$, where $r = \sqrt{x^2 + y^2 + z^2}$ and k is a constant. Calculate the flux of $\mathbf{E}$ through the disk D of radius 2 parallel to the xy-plane with center $(0, 0, 3)$.

22. Let $\mathcal{S}$ be the ellipsoid $\left(\dfrac{x}{4}\right)^2 + \left(\dfrac{y}{3}\right)^2 + \left(\dfrac{z}{2}\right)^2 = 1$. Calculate the flux of $\mathbf{F} = z\mathbf{i}$ over the portion of $\mathcal{S}$ where $x, y, z \leq 0$ with upward-pointing normal. *Hint: Parametrize $\mathcal{S}$ using a modified form of spherical coordinates (θ, ϕ).*

23. Let $\mathbf{v} = z\mathbf{k}$ be the velocity field (in meters per second) of a fluid in $\mathbf{R}^3$. Calculate the flow rate (in cubic meters per second) through the upper hemisphere $(z \geq 0)$ of the sphere $x^2 + y^2 + z^2 = 1$.

24. Calculate the flow rate of a fluid with velocity field $\mathbf{v} = \langle x, y, x^2y \rangle$ (in meters per second) through the portion of the ellipse $\left(\dfrac{x}{2}\right)^2 + \left(\dfrac{y}{3}\right)^2 = 1$ in the xy-plane, where $x, y \geq 0$, oriented with the normal in the positive z-direction.

In Exercises 25–28, a net is dipped in a river. Determine the flow rate of water across the net if the velocity vector field for the river is given by $\mathbf{v}$ and the net is described by the given equations.

25. $\mathbf{v} = \langle x - y, z + y + 4, z^2 \rangle$, net given by $x^2 + z^2 \leq 1, y = 0$, oriented in the positive y-direction

26. $\mathbf{v} = \langle x - y, z + y + 4, z^2 \rangle$, net given by $y = 1 - x^2 - z^2, y \geq 0$, oriented in the positive y-direction

27. $\mathbf{v} = \langle x - y, z + y + 4, z^2 \rangle$, net given by $y = \sqrt{1 - x^2 - z^2}, y \geq 0$, oriented in the positive y-direction

28. $\mathbf{v} = \langle zy, xz, xy \rangle$, net given by $y = 1 - x - z$, for $x, y, z \geq 0$ oriented in the positive y-direction

In Exercises 29–30, let $\mathcal{T}$ be the triangular region with vertices $(1, 0, 0)$, $(0, 1, 0)$, and $(0, 0, 1)$ oriented with upward-pointing normal vector (Figure 16). Assume distances are in meters.

29. A fluid flows with constant velocity field $\mathbf{v} = 2\mathbf{k}$ (meters per second). Calculate:
(a) the flow rate through $\mathcal{T}$.
(b) the flow rate through the projection of $\mathcal{T}$ onto the xy-plane [the triangle with vertices $(0, 0, 0)$, $(1, 0, 0)$, and $(0, 1, 0)$].

30. Calculate the flow rate through $\mathcal{T}$ if $\mathbf{v} = -\mathbf{j}$ m/s.

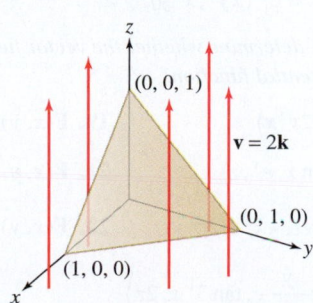

FIGURE 16

31. Prove that if $\mathcal{S}$ is the part of a graph $z = g(x, y)$ lying over a domain $\mathcal{D}$ in the xy-plane, then

$$\iint_{\mathcal{S}} \mathbf{F} \cdot d\mathbf{S} = \iint_{\mathcal{D}} \left(-F_1 \frac{\partial g}{\partial x} - F_2 \frac{\partial g}{\partial y} + F_3\right) dx\, dy$$

In Exercises 32–33, a varying current $i(t)$ flows through a long straight wire in the xy-plane as in Example 5. The current produces a magnetic field $\mathbf{B}$ whose magnitude at a distance r from the wire is $B = \dfrac{\mu_0 i}{2\pi r}$ T, where $\mu_0 = 4\pi \cdot 10^{-7}$ T-m/A. Furthermore, $\mathbf{B}$ points into the page at points P in the xy-plane.

32. Assume that $i(t) = t(12 - t)$ A (t in seconds). Calculate the flux $\Phi(t)$, at time t, of $\mathbf{B}$ through a rectangle of dimensions $L \times H = 3 \times 2$ m whose top and bottom edges are parallel to the wire and whose bottom edge is located $d = 0.5$ m above the wire, similar to Figure 13(B). Then use Faraday's Law to determine the voltage drop around the rectangular loop (the boundary of the rectangle) at time t.

33. Assume that $i = 10e^{-0.1t}$ A (t in seconds). Calculate the flux $\Phi(t)$, at time t, of $\mathbf{B}$ through the isosceles triangle of base 12 cm and height 6 cm whose bottom edge is 3 cm from the wire, as in Figure 17. Assume the triangle is oriented with normal vector pointing out of the page. Use Faraday's Law to determine the voltage drop around the triangular loop (the boundary of the triangle) at time t.

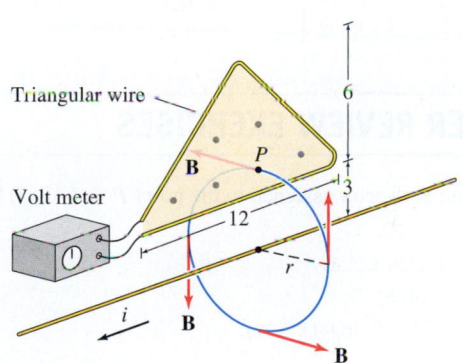

FIGURE 17

In Exercises 34–35, a solid material that has thermal conductivity K in kilowatts per meter-kelvin and temperature given at each point by $w(x, y, z)$ has heat flow given by the vector field $\mathbf{F} = -K\nabla w$ and rate of heat flow across a surface $\mathcal{S}$ within the solid given by $-K\iint_{\mathcal{S}} \nabla w\, d\mathbf{S}$.

34. Find the rate of heat flow out of a sphere of radius 1 m inside a large cube of copper $(K = 400$ kW/m-k) with temperature function given by $w(x, y, z) = 20 - 5(x^2 + y^2 + z^2)$°C.

35. An insulated cylinder of solid gold $(K = 310$ kW/m-k) of radius $\sqrt{2}$ m and height 5 m is heated at one end until the temperature at each point in the cylinder is given by $w(x, y, z) = (30 - z^2)(2 - (x^2 + y^2))$. Determine the rate of heat flow across each horizontal disk given by $z = 1$, $z = 2$, and $z = 3$, identifying which has the greatest rate of heat flow across it.

Further Insights and Challenges

36. A point mass m is located at the origin. Let Q be the flux of the gravitational field $\mathbf{F} = -Gm\dfrac{\mathbf{e}_r}{r^2}$ through the cylinder $x^2 + y^2 = R^2$ for $a \leq z \leq b$, including the top and bottom (Figure 18). Show that $Q = -4\pi Gm$ if $a < 0 < b$ (m lies inside the cylinder) and $Q = 0$ if $0 < a < b$ (m lies outside the cylinder).

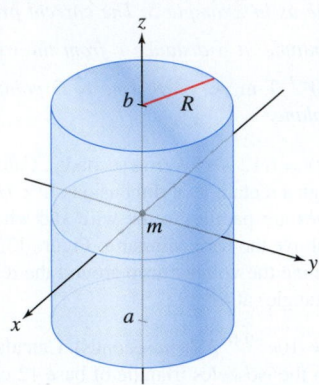

FIGURE 18

In Exercises 37 and 38, let S be the surface with parametrization

$$G(u, v) = \left(\left(1 + v\cos\frac{u}{2}\right)\cos u, \left(1 + v\cos\frac{u}{2}\right)\sin u, v\sin\frac{u}{2}\right)$$

for $0 \leq u \leq 2\pi$, $-\frac{1}{2} \leq v \leq \frac{1}{2}$.

37. [CAS] Use a computer algebra system.

(a) Plot S and confirm visually that S is a Möbius strip.

(b) The intersection of S with the xy-plane is the unit circle $G(u, 0) = (\cos u, \sin u, 0)$. Verify that the normal vector along this circle is

$$\mathbf{N}(u, 0) = \left\langle \cos u \sin\frac{u}{2}, \sin u \sin\frac{u}{2}, -\cos\frac{u}{2} \right\rangle$$

(c) As u varies from 0 to 2π, the point $G(u, 0)$ moves once around the unit circle, beginning and ending at $G(0,0) = G(2\pi, 0) = (1, 0, 0)$. Verify that $\mathbf{N}(u, 0)$ is a unit vector that varies continuously but that $\mathbf{N}(2\pi, 0) = -\mathbf{N}(0, 0)$. This shows that S is not orientable—that is, it is not possible to choose a nonzero normal vector at each point on S in a continuously varying manner (if it were possible, the unit normal vector would return to itself rather than to its negative when carried around the circle).

38. [CAS] We cannot integrate vector fields over S because S is not orientable, but it is possible to integrate functions over S. Using a computer algebra system:

(a) Verify that

$$\|\mathbf{N}(u, v)\|^2 = 1 + \frac{3}{4}v^2 + 2v\cos\frac{u}{2} + \frac{1}{2}v^2\cos u$$

(b) Compute the surface area of S to four decimal places.

(c) Compute $\displaystyle\iint_S (x^2 + y^2 + z^2)\, dS$ to four decimal places.

CHAPTER REVIEW EXERCISES

1. Compute the vector assigned to the point $P = (-3, 5)$ by the vector field:

(a) $\mathbf{F}(x, y) = \langle xy, y - x \rangle$

(b) $\mathbf{F}(x, y) = \langle 4, 8 \rangle$

(c) $\mathbf{F}(x, y) = \langle 3^{x+y}, \log_2(x + y) \rangle$

2. Find a vector field $\mathbf{F}$ in the plane such that $\|\mathbf{F}(x, y)\| = 1$ and $\mathbf{F}(x, y)$ is orthogonal to $\mathbf{G}(x, y) = \langle x, y \rangle$ for all x, y.

In Exercises 3–6, sketch the vector field.

3. $\mathbf{F}(x, y) = \langle y, 1 \rangle$ **4.** $\mathbf{F}(x, y) = \langle 4, 1 \rangle$

5. ∇f, where $f(x, y) = x^2 - y$

6. $\mathbf{F}(x, y) = \left\langle \dfrac{4y}{\sqrt{x^2 + 4y^2}}, \dfrac{-x}{\sqrt{x^2 + 16y^2}} \right\rangle$

Hint: Show that $\mathbf{F}$ is a unit vector field tangent to the family of ellipses $x^2 + 4y^2 = c^2$.

In Exercises 7–14, calculate div($\mathbf{F}$) and curl($\mathbf{F}$).

7. $\mathbf{F} = \langle x^2, y^2, z^2 \rangle$ **8.** $\mathbf{F} = \langle yz, xz, xy \rangle$

9. $\mathbf{F} = \langle x^3 y, xz^2, y^2 z \rangle$ **10.** $\mathbf{F} = \langle \sin xy, \cos yz, \sin xz \rangle$

11. $\mathbf{F} = y\mathbf{i} - z\mathbf{k}$ **12.** $\mathbf{F} = \langle e^{x+y}, e^{y+z}, xyz \rangle$

13. $\mathbf{F} = \nabla(e^{-x^2-y^2-z^2})$

14. $\mathbf{e}_r = r^{-1}\langle x, y, z \rangle$ $(r = \sqrt{x^2 + y^2 + z^2})$

15. Show that if F_1, F_2, and F_3 are differentiable functions of one variable, then

$$\text{curl}\left((\langle F_1(x), F_2(y), F_3(z)\rangle\right) = \mathbf{0}$$

Use this to calculate the curl of

$$\mathbf{F}(x, y, z) = \left\langle x^2 + y^2, \ln y + z^2, z^3\sin(z^2)e^{z^3} \right\rangle$$

16. Give an example of a nonzero vector field $\mathbf{F}$ such that curl($\mathbf{F}$) = $\mathbf{0}$ and div($\mathbf{F}$) = 0.

17. Verify the identity div(curl($\mathbf{F}$)) = 0 for the vector fields $\mathbf{F} = \langle xz, ye^x, yz \rangle$ and $\mathbf{G} = \langle z^2, xy^3, x^2 y \rangle$.

In Exercises 18–26, determine whether the vector field is conservative, and if so, find a potential function.

18. $\mathbf{F}(x, y) = \langle x^2 y, y^2 x \rangle$ **19.** $\mathbf{F}(x, y) = \langle 4x^3 y^5, 5x^4 y^4 \rangle$

20. $\mathbf{F}(x, y, z) = \langle \sin x, e^y, z \rangle$ **21.** $\mathbf{F}(x, y, z) = \langle 2, 4, e^z \rangle$

22. $\mathbf{F}(x, y, z) = \langle xyz, \frac{1}{2}x^2 z, 2z^2 y \rangle$ **23.** $\mathbf{F}(x, y) = \langle y^4 x^3, x^4 y^3 \rangle$

24. $\mathbf{F}(x, y, z) = \left\langle \dfrac{y}{1 + x^2}, \tan^{-1} x, 2z \right\rangle$

25. $\mathbf{F}(x, y, z) = \left\langle \dfrac{2xy}{x^2 + z}, \ln(x^2 + z), \dfrac{y}{x^2 + z} \right\rangle$

26. $\mathbf{F}(x, y, z) = \langle xe^{2x}, ye^{2z}, ze^{2y} \rangle$

27. Find a conservative vector field of the form $\mathbf{F} = \langle g(y), h(x) \rangle$ such that $\mathbf{F}(0,0) = \langle 1, 1 \rangle$, where $g(y)$ and $h(x)$ are differentiable functions. Determine all such vector fields.

In Exercises 28–31, compute the line integral $\int_C f(x, y)\, ds$ for the given function and path or curve.

28. $f(x, y) = xy$, the path $\mathbf{r}(t) = \langle t, 2t - 1 \rangle$ for $0 \le t \le 1$

29. $f(x, y) = x - y$, the unit semicircle $x^2 + y^2 = 1$, $y \ge 0$

30. $f(x, y, z) = e^x - \dfrac{y}{2\sqrt{2}z}$, the path $\mathbf{r}(t) = \langle \ln t, \sqrt{2}t, \frac{1}{2}t^2 \rangle$ for $1 \le t \le 2$

31. $f(x, y, z) = x + 2y + z$, the helix $\mathbf{r}(t) = \langle \cos t, \sin t, t \rangle$ for $0 \le t \le \pi/2$

32. Find the total mass of an L-shaped rod consisting of the segments $(2t, 2)$ and $(2, 2 - 2t)$ for $0 \le t \le 1$ (length in centimeters) with mass density $\delta(x, y) = x^2 y$ g/cm.

33. Calculate $\mathbf{F} = \nabla f$, where $f(x, y, z) = xye^z$, and compute $\int_C \mathbf{F} \cdot d\mathbf{r}$, where:

(a) C is any curve from $(1, 1, 0)$ to $(3, e, -1)$.

(b) C is the boundary of the square $0 \le x \le 1$, $0 \le y \le 1$ oriented counterclockwise.

34. Calculate $\int_{C_1} y\, dx + x^2 y\, dy$, where C_1 is the oriented curve in Figure 1(A).

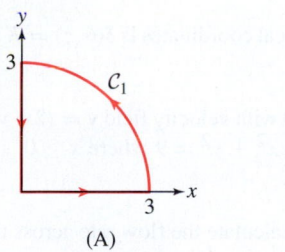 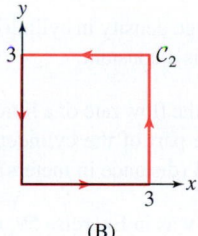

FIGURE 1

35. Let $\mathbf{F}(x, y) = \langle 9y - y^3, e^{\sqrt{y}}(x^2 - 3x) \rangle$, and let C_2 be the oriented curve in Figure 1(B).

(a) Show that $\mathbf{F}$ is not conservative.

(b) Show that $\int_{C_2} \mathbf{F} \cdot d\mathbf{r} = 0$ without explicitly computing the integral. *Hint:* Show that $\mathbf{F}$ is orthogonal to the edges along the square.

In Exercises 36–39, compute the line integral $\int_C \mathbf{F} \cdot d\mathbf{r}$ for the given vector field and path.

36. $\mathbf{F}(x, y) = \left\langle \dfrac{2y}{x^2 + 4y^2}, \dfrac{x}{x^2 + 4y^2} \right\rangle$, the path $\mathbf{r}(t) = \langle \cos t, \frac{1}{2}\sin t \rangle$ for $0 \le t \le 2\pi$

37. $\mathbf{F}(x, y) = \langle 2xy, x^2 + y^2 \rangle$, the part of the unit circle in the first quadrant oriented counterclockwise

38. $\mathbf{F}(x, y) = \langle x^2 y, y^2 z, z^2 x \rangle$, the path $\mathbf{r}(t) = \langle e^{-t}, e^{-2t}, e^{-3t} \rangle$ for $0 \le t < \infty$

39. $\mathbf{F} = \nabla f$, where $f(x, y, z) = 4x^2 \ln(1 + y^4 + z^2)$, the path $\mathbf{r}(t) = \langle t^3, \ln(1 + t^2), e^t \rangle$ for $0 \le t \le 1$

40. Consider the line integrals $\int_C \mathbf{F} \cdot d\mathbf{r}$ for the vector fields $\mathbf{F}$ and paths $\mathbf{r}$ in Figure 2. Which two of the line integrals appear to have a value of zero? Which of the other two appears to have a negative value?

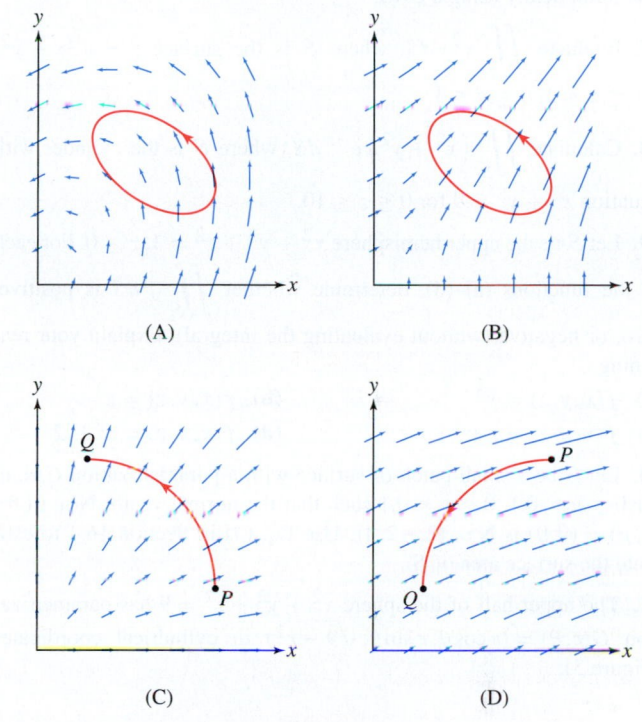

FIGURE 2

41. Calculate the work required to move an object from $P = (1, 1, 1)$ to $Q = (3, -4, -2)$ against the force field $\mathbf{F}(x, y, z) = -12r^{-4}\langle x, y, z \rangle$ (distance in meters, force in newtons), where $r = \sqrt{x^2 + y^2 + z^2}$. *Hint:* Find a potential function for $\mathbf{F}$.

42. Find constants a, b, c such that

$$G(u, v) = (u + av, bu + v, 2u - c)$$

parametrizes the plane $3x - 4y + z = 5$. Calculate $\mathbf{T}_u$, $\mathbf{T}_v$, and $\mathbf{N}(u, v)$.

43. Calculate the integral of $f(x, y, z) = e^z$ over the portion of the plane $x + 2y + 2z = 3$, where $x, y, z \ge 0$.

44. Let S be the surface parametrized by

$$G(u, v) = \left(2u \sin \frac{v}{2}, 2u \cos \frac{v}{2}, 3v \right)$$

for $0 \le u \le 1$ and $0 \le v \le 2\pi$.

(a) Calculate the tangent vectors $\mathbf{T}_u$ and $\mathbf{T}_v$ and the normal vector $\mathbf{N}(u, v)$ at $P = G(1, \frac{\pi}{3})$.

(b) Find the equation of the tangent plane at P.

(c) Compute the surface area of S.

45. (CAS) Plot the surface with parametrization

$$G(u, v) = (u + 4v, 2u - v, 5uv)$$

for $-1 \leq v \leq 1$, $-1 \leq u \leq 1$. Express the surface area as a double integral and use a computer algebra system to compute the area numerically.

46. (CAS) Express the surface area of the surface $z = 10 - x^2 - y^2$ for $-1 \leq x \leq 1$, $-3 \leq y \leq 3$ as a double integral. Evaluate the integral numerically using a CAS.

47. Evaluate $\iint_S x^2 y \, dS$, where S is the surface $z = \sqrt{3}x + y^2$, $-1 \leq x \leq 1, 0 \leq y \leq 1$.

48. Calculate $\iint_S \left(x^2 + y^2\right) e^{-z} \, dS$, where S is the cylinder with equation $x^2 + y^2 = 9$ for $0 \leq z \leq 10$.

49. Let S be the upper hemisphere $x^2 + y^2 + z^2 = 1$, $z \geq 0$. For each of the functions (a)–(d), determine whether $\iint_S f \, dS$ is positive, zero, or negative (without evaluating the integral). Explain your reasoning.

(a) $f(x, y, z) = y^3$ **(b)** $f(x, y, z) = z^3$

(c) $f(x, y, z) = xyz$ **(d)** $f(x, y, z) = z^2 - 2$

50. Let S be a small patch of surface with a parametrization $G(u, v)$ for $0 \leq u \leq 0.1$, $0 \leq v \leq 0.1$ such that the normal vector $\mathbf{N}(u, v)$ for $(u, v) = (0, 0)$ is $\mathbf{N} = \langle 2, -2, 4 \rangle$. Use Eq. (3) in Section 16.4 to estimate the surface area of S.

51. The upper half of the sphere $x^2 + y^2 + z^2 = 9$ has parametrization $G(r, \theta) = (r \cos \theta, r \sin \theta, \sqrt{9 - r^2})$ in cylindrical coordinates (Figure 3).

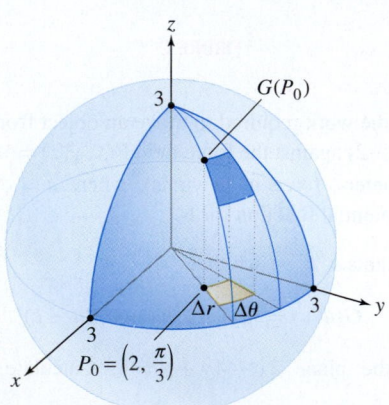

FIGURE 3

(a) Calculate the normal vector $\mathbf{N} = \mathbf{T}_r \times \mathbf{T}_\theta$ at the point $G(2, \frac{\pi}{3})$.

(b) Use Eq. (3) in Section 16.4 to estimate the surface area of $G(\mathcal{R})$, where $\mathcal{R}$ is the small domain defined by

$$2 \leq r \leq 2.1, \qquad \frac{\pi}{3} \leq \theta \leq \frac{\pi}{3} + 0.05$$

In Exercises 52–57, compute $\iint_S \mathbf{F} \cdot d\mathbf{S}$ for the given oriented surface or parametrized surface.

52. $\mathbf{F}(x, y, z) = \langle y, x, e^{xz} \rangle$, $x^2 + y^2 = 9$, $x \geq 0$, $y \geq 0$, $-3 \leq z \leq 3$,   outward-pointing normal

53. $\mathbf{F}(x, y, z) = \langle -y, z, -x \rangle$, $G(u, v) = (u + 3v, v - 2u, 2v + 5)$, $0 \leq u \leq 1, 0 \leq v \leq 1$,   upward-pointing normal

54. $\mathbf{F}(x, y, z) = \langle 0, 0, x^2 + y^2 \rangle$, $x^2 + y^2 + z^2 = 4$, $z \geq 0$, outward-pointing normal

55. $\mathbf{F}(x, y, z) = \langle z, 0, z^2 \rangle$, $G(u, v) = (v \cosh u, v \sinh u, v)$, $0 \leq u \leq 1, 0 \leq v \leq 1$,   upward-pointing normal

56. $\mathbf{F}(x, y, z) = \langle 0, 0, xze^{xy} \rangle$, $z = xy$, $0 \leq x \leq 1, 0 \leq y \leq 1$, upward-pointing normal

57. $\mathbf{F}(x, y, z) = \langle 0, 0, z \rangle$, $3x^2 + 2y^2 + z^2 = 1$, $z \geq 0$, upward-pointing normal

58. Calculate the total charge on the cylinder

$$x^2 + y^2 = R^2, \qquad 0 \leq z \leq H$$

if the charge density in cylindrical coordinates is $\delta(\theta, z) = Kz^2 \cos^2 \theta$, where K is a constant.

59. Find the flow rate of a fluid with velocity field $\mathbf{v} = \langle 2x, y, xy \rangle$ m/s across the part of the cylinder $x^2 + y^2 = 9$ where $x \geq 0, y \geq 0$, and $0 \leq z \leq 4$ (distance in meters).

60. With $\mathbf{v}$ as in Exercise 59, calculate the flow rate across the part of the elliptic cylinder $\frac{x^2}{4} + y^2 = 1$, where $x \geq 0, y \geq 0$, and $0 \leq z \leq 4$.

61. Calculate the flux of the vector field $\mathbf{E}(x, y, z) = \langle 0, 0, x \rangle$ through the part of the ellipsoid

$$4x^2 + 9y^2 + z^2 = 36$$

where $z \geq 3$, $x \geq 0$, $y \geq 0$. *Hint:* Use the parametrization

$$G(r, \theta) = \left(3r \cos \theta, 2r \sin \theta, 6\sqrt{1 - r^2}\right)$$

Niday Picture Library/Alamy

17 FUNDAMENTAL THEOREMS OF VECTOR ANALYSIS

Adding up the local swirling (curl) over Van Gogh's *Starry Night* sky nets the overall circulation around the boundary of the region of sky in the painting. In this chapter, with vector fields, curl, and surface and line integrals, we paint the mathematically formal version of this relationship between curl and circulation as Stokes' Theorem, $\iint_{\mathcal{S}} (\nabla \times \mathbf{F}) \cdot d\mathbf{S} = \oint_{\partial \mathcal{S}} \mathbf{F} \cdot d\mathbf{r}$.

I n this final chapter, we study three generalizations of the Fundamental Theorem of Calculus, Part I, which we have seen indicates that $\int_a^b F'(x)\,dx = F(b) - F(a)$. If we think of the boundary of the interval $[a, b]$ as being given by the two points $\{a, b\}$, then FTC I says that we can find the integral of the derivative of a function over an interval just by evaluating that function on the boundary of the interval. The first of these new theorems, Green's Theorem, says that we can find a double integral of a certain type of derivative over a region in the xy-plane by finding a line integral around the boundary of the region. The second theorem, Stokes' Theorem, allows us to find a surface integral of a certain derivative (involving curl) over a surface with boundary curves in space by evaluating a line integral on the boundary curves. The third theorem, the Divergence Theorem, allows us to find the triple integral of another kind of derivative (involving divergence) over a solid in space by evaluating a surface integral over the boundary surface of the solid.

This is a culmination of our efforts to extend the ideas of single-variable calculus to the multivariable setting. However, vector analysis is not so much an endpoint as a gateway to the more advanced mathematical theory of differential forms and manifolds and to a host of applications in many fields, including physics, engineering, biology, and environmental science.

17.1 Green's Theorem

In Section 16.3, we showed that the circulation of a conservative vector field $\mathbf{F}$ around every closed path is zero. For vector fields in the plane, Green's Theorem tells us what happens when $\mathbf{F}$ is not conservative.

To formulate Green's Theorem, we need some notation. Consider a domain $\mathcal{D}$ in the plane whose boundary $\mathcal{C}$ is a **simple closed curve**—that is, a closed curve that does not intersect itself (Figure 1). We follow standard usage and denote the boundary curve $\mathcal{C}$ by $\partial \mathcal{D}$. The **boundary orientation** of $\partial \mathcal{D}$ is the direction to traverse the boundary such that the region is always to your left, as in Figure 1. When there is a single boundary curve, the boundary orientation is the counterclockwise orientation.

Recall the following two notations for the line integral of $\mathbf{F} = \langle F_1, F_2 \rangle$:

$$\int_{\mathcal{C}} \mathbf{F} \cdot d\mathbf{r} \qquad \text{and} \qquad \int_{\mathcal{C}} F_1\,dx + F_2\,dy$$

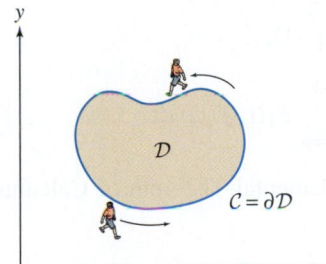

FIGURE 1 The boundary of $\mathcal{D}$ is a simple closed curve $\mathcal{C}$ that is denoted $\partial \mathcal{D}$. The boundary is oriented in the counterclockwise direction.

If $\mathcal{C}$ is parametrized by $\mathbf{r}(t) = \langle x(t), y(t) \rangle$ for $a \le t \le b$, then

$$dx = x'(t)\,dt, \qquad dy = y'(t)\,dt$$

$$\int_{\mathcal{C}} F_1\,dx + F_2\,dy = \int_a^b \left(F_1(x(t), y(t))x'(t) + F_2(x(t), y(t))y'(t) \right) dt \qquad \boxed{1}$$

Throughout this chapter, we assume that the components of all vector fields have continuous second-order partial derivatives, and also that $\mathcal{C}$ is smooth ($\mathcal{C}$ has a parametrization with derivatives of all orders) or piecewise smooth (a finite union of smooth curves joined together at endpoints).

← **REMINDER** The line integral of a vector field over a closed curve is called the circulation and is often denoted by the symbol $\oint$.

Green's Theorem can also be written

$$\oint_{\partial \mathcal{D}} \mathbf{F} \cdot d\mathbf{r} = \iint_{\mathcal{D}} \left(\frac{\partial F_2}{\partial x} - \frac{\partial F_1}{\partial y} \right) dA$$

THEOREM 1 Green's Theorem Let $\mathcal{D}$ be a domain whose boundary $\partial \mathcal{D}$ is a simple closed curve, oriented counterclockwise. If F_1 and F_2 have continuous partial derivatives in an open region containing $\mathcal{D}$, then

$$\oint_{\partial \mathcal{D}} F_1\,dx + F_2\,dy = \iint_{\mathcal{D}} \left(\frac{\partial F_2}{\partial x} - \frac{\partial F_1}{\partial y} \right) dA \qquad \boxed{2}$$

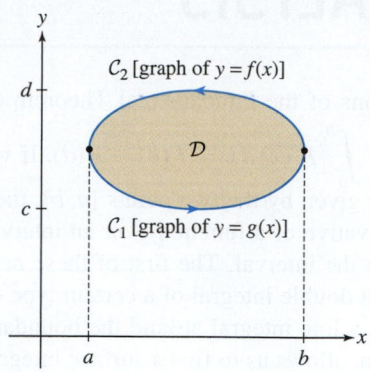

FIGURE 2 The boundary curve $\partial \mathcal{D}$ is the union of the graphs of $y = g(x)$ and $y = f(x)$ oriented counterclockwise.

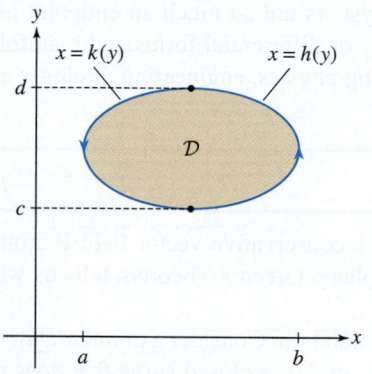

FIGURE 3 The boundary curve $\partial \mathcal{D}$ is also the union of the graphs of $x = h(y)$ and $x = k(y)$ oriented counterclockwise.

Proof Because a complete proof is quite technical, we shall make the simplifying assumption that the boundary of $\mathcal{D}$ can be described as the union of two graphs $y = g(x)$ and $y = f(x)$ with $g(x) \le f(x)$ as in Figure 2 and also as the union of two graphs $x = k(y)$ and $x = h(y)$, with $k(y) \le h(y)$ as in Figure 3.

From the terms in Eq. (2), we construct two separate equations to prove, one for F_1 and one for F_2:

$$\oint_{\partial \mathcal{D}} F_1 \, dx = - \iint_{\mathcal{D}} \frac{\partial F_1}{\partial y} \, dA \qquad \boxed{3}$$

$$\oint_{\partial \mathcal{D}} F_2 \, dy = \iint_{\mathcal{D}} \frac{\partial F_2}{\partial x} \, dA \qquad \boxed{4}$$

If we can show that both these equations hold, then we obtain a proof of Green's Theorem by adding them together. To prove Eq. (3), we write

$$\oint_{\partial \mathcal{D}} F_1 \, dx = \int_{\mathcal{C}_1} F_1 \, dx + \int_{\mathcal{C}_2} F_1 \, dx$$

where $\mathcal{C}_1$ is the graph of $y = g(x)$ and $\mathcal{C}_2$ is the graph of $y = f(x)$, oriented as in Figure 2. To compute these line integrals, we parametrize the graphs from left to right using $t = x$ as the parameter:

Graph of $y = g(x)$: $\qquad \mathbf{r}_1(t) = \langle t, g(t) \rangle, \qquad a \le t \le b$

Graph of $y = f(x)$: $\qquad \mathbf{r}_2(t) = \langle t, f(t) \rangle, \qquad a \le t \le b$

Since $\mathcal{C}_2$ is oriented from right to left, the line integral over $\partial \mathcal{D}$ is the difference

$$\oint_{\partial \mathcal{D}} F_1 \, dx = \int_{\mathcal{C}_1} F_1 \, dx - \int_{\mathcal{C}_2} F_1 \, dx$$

In both parametrizations, $x = t$, so $dx = dt$, and by Eq. (1),

$$\oint_{\partial \mathcal{D}} F_1 \, dx = \int_{t=a}^{b} F_1(t, g(t)) \, dt - \int_{t=a}^{b} F_1(t, f(t)) \, dt \qquad \boxed{5}$$

Now, the key step is to apply Part I of the Fundamental Theorem of Calculus to $\dfrac{\partial F_1}{\partial y}(t, y)$ as a function of y with t held constant:

$$F_1(t, f(t)) - F_1(t, g(t)) = \int_{y=g(t)}^{f(t)} \frac{\partial F_1}{\partial y}(t, y) \, dy$$

Substituting the integral on the right in Eq. (5), we obtain Eq. (3):

$$\oint_{\partial \mathcal{D}} F_1 \, dx = - \int_{t=a}^{b} \int_{y=g(t)}^{f(t)} \frac{\partial F_1}{\partial y}(t, y) \, dy \, dt = - \iint_{\mathcal{D}} \frac{\partial F_1}{\partial y} \, dA$$

Eq. (4) is proved in a similar fashion, by expressing $\partial \mathcal{D}$ as the union of the graphs of $x = h(y)$ and $x = k(y)$ as in Figure 3. ∎

Recall from Section 16.1 that if $\mathbf{F}$ is a conservative vector field, that is, if $\mathbf{F} = \nabla f$, then the cross-partial condition is satisfied:

$$\frac{\partial F_2}{\partial x} - \frac{\partial F_1}{\partial y} = 0$$

In this case, Green's Theorem merely confirms what we already know: The line integral of a conservative vector field around any closed curve is zero.

EXAMPLE 1 **Verifying Green's Theorem** Verify Green's Theorem for the line integral along the unit circle $\mathcal{C}$, oriented counterclockwise (Figure 4):

$$\oint_{\mathcal{C}} xy^2 \, dx + x \, dy$$

Solution

Step 1. **Evaluate the line integral directly.**

We use the standard parametrization of the unit circle:

$$x = \cos\theta, \qquad\qquad y = \sin\theta$$

$$dx = -\sin\theta \, d\theta, \qquad dy = \cos\theta \, d\theta$$

The integrand in the line integral is

$$xy^2 \, dx + x \, dy = \cos\theta \sin^2\theta(-\sin\theta \, d\theta) + \cos\theta(\cos\theta \, d\theta)$$

$$= \left(-\cos\theta \sin^3\theta + \cos^2\theta\right) d\theta$$

and

$$\oint_{\mathcal{C}} xy^2 \, dx + x \, dy = \int_0^{2\pi} \left(-\cos\theta \sin^3\theta + \cos^2\theta\right) d\theta$$

$$= -\frac{\sin^4\theta}{4}\bigg|_0^{2\pi} + \frac{1}{2}\left(\theta + \frac{1}{2}\sin 2\theta\right)\bigg|_0^{2\pi}$$

$$= 0 + \frac{1}{2}(2\pi + 0) = \boxed{\pi}$$

Step 2. **Evaluate the line integral using Green's Theorem.**

In this example, $F_1 = xy^2$ and $F_2 = x$, so

$$\frac{\partial F_2}{\partial x} - \frac{\partial F_1}{\partial y} = \frac{\partial}{\partial x}x - \frac{\partial}{\partial y}xy^2 = 1 - 2xy$$

According to Green's Theorem,

$$\oint_{\mathcal{C}} xy^2 \, dx + x \, dy = \iint_{\mathcal{D}} \left(\frac{\partial F_2}{\partial x} - \frac{\partial F_1}{\partial y}\right) dA = \iint_{\mathcal{D}} (1 - 2xy) \, dA$$

where $\mathcal{D}$ is the disk $x^2 + y^2 \leq 1$ enclosed by $\mathcal{C}$. The integral of $2xy$ over $\mathcal{D}$ is zero by symmetry—the contributions for positive and negative x cancel. We can check this directly:

$$\iint_{\mathcal{D}} (-2xy) \, dA = -2\int_{x=-1}^{1} \int_{y=-\sqrt{1-x^2}}^{\sqrt{1-x^2}} xy \, dy \, dx = -\int_{x=-1}^{1} xy^2\bigg|_{y=-\sqrt{1-x^2}}^{\sqrt{1-x^2}} dx = 0$$

Therefore,

$$\iint_{\mathcal{D}} \left(\frac{\partial F_2}{\partial x} - \frac{\partial F_1}{\partial y}\right) dA = \iint_{\mathcal{D}} 1 \, dA = \text{area}(\mathcal{D}) = \boxed{\pi}$$

This agrees with the value in Step 1. So Green's Theorem is verified in this case. ∎

EXAMPLE 2 **Computing a Line Integral Using Green's Theorem** Compute the circulation of $\mathbf{F}(x, y) = \left\langle \sin x, x^2 y^3 \right\rangle$ around the triangular curve $\mathcal{C}$ with the counterclockwise orientation shown in Figure 5.

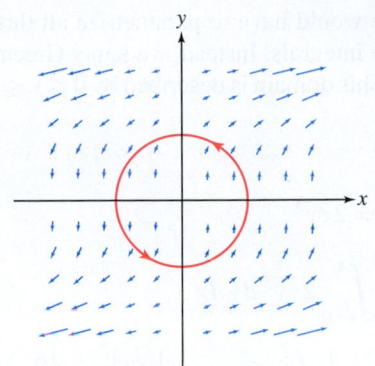

FIGURE 4 The vector field $\mathbf{F}(x, y) = \langle xy^2, x \rangle$.

REMINDER To integrate $\cos^2\theta$, use the identity $\cos^2\theta = \frac{1}{2}(1 + \cos 2\theta)$.

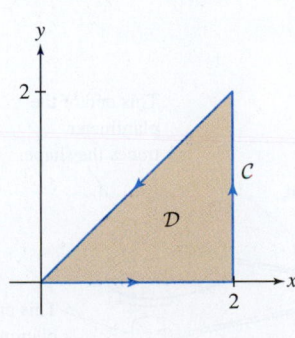

FIGURE 5 The region $\mathcal{D}$ is described by $0 \leq x \leq 2, 0 \leq y \leq x$.

Solution To compute the line integral directly, we would have to parametrize all three sides of the triangle and compute three separate line integrals. Instead, we apply Green's Theorem to the domain $\mathcal{D}$ enclosed by the triangle. This domain is described by $0 \le x \le 2$, $0 \le y \le x$.

Applying Green's Theorem, we obtain

$$\frac{\partial F_2}{\partial x} - \frac{\partial F_1}{\partial y} = \frac{\partial}{\partial x} x^2 y^3 - \frac{\partial}{\partial y} \sin x = 2xy^3$$

$$\oint_{\mathcal{C}} \sin x \, dx + x^2 y^3 \, dy = \iint_{\mathcal{D}} 2xy^3 \, dA = \int_0^2 \int_{y=0}^x 2xy^3 \, dy \, dx$$

$$= \int_0^2 \left(\frac{1}{2} xy^4 \Big|_{y=0}^x \right) dx = \frac{1}{2} \int_0^2 x^5 \, dx = \frac{1}{12} x^6 \Big|_0^2 = \frac{16}{3} \qquad \blacksquare$$

Area via Green's Theorem

We can use Green's Theorem to obtain formulas for the area of the domain $\mathcal{D}$ enclosed by a simple closed curve $\mathcal{C}$ (Figure 6). The trick is to choose a vector field $\mathbf{F} = \langle F_1, F_2 \rangle$ such that $\dfrac{\partial F_2}{\partial x} - \dfrac{\partial F_1}{\partial y} = 1$. Here are a few possibilities:

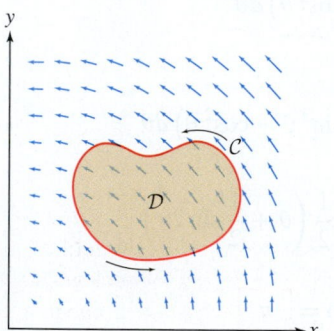

FIGURE 6 The line integral of the vector field $\langle -y/2, x/2 \rangle$ around $\mathcal{C}$ is equal to the area of the region $\mathcal{D}$ enclosed by $\mathcal{C}$.

If we choose $\mathbf{F}(x, y) = \langle 0, x \rangle$, then $\dfrac{\partial F_2}{\partial x} - \dfrac{\partial F_1}{\partial y} = \dfrac{\partial}{\partial x} x - \dfrac{\partial}{\partial y} 0 = 1$

If we choose $\mathbf{F}(x, y) = \langle -y, 0 \rangle$, then $\dfrac{\partial F_2}{\partial x} - \dfrac{\partial F_1}{\partial y} = \dfrac{\partial}{\partial x} 0 - \dfrac{\partial}{\partial y} (-y) = 1$

If we choose $\mathbf{F}(x, y) = \langle -y/2, x/2 \rangle$, then $\dfrac{\partial F_2}{\partial x} - \dfrac{\partial F_1}{\partial y} = \dfrac{\partial}{\partial x} \left(\dfrac{x}{2} \right) - \dfrac{\partial}{\partial y} \left(\dfrac{-y}{2} \right)$

$$= \frac{1}{2} + \frac{1}{2} = 1$$

By Green's Theorem, in all three cases, we have

$$\oint_{\mathcal{C}} F_1 \, dx + F_2 \, dy = \iint_{\mathcal{D}} \left(\frac{\partial F_2}{\partial x} - \frac{\partial F_1}{\partial y} \right) dA = \iint_{\mathcal{D}} 1 \, dA = \text{area}(\mathcal{D})$$

Plugging in F_1 and F_2 for each of these three cases, we obtain the following three formulas for the area of the domain $\mathcal{D}$ enclosed by $\mathcal{C}$.

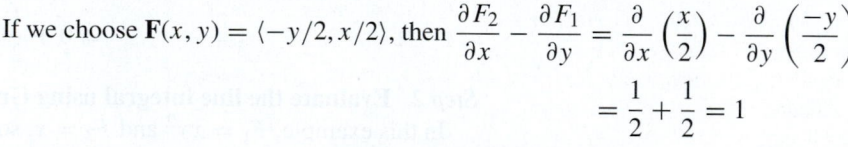

$$\text{area enclosed by } \mathcal{C} = \oint_{\mathcal{C}} x \, dy = \oint_{\mathcal{C}} -y \, dx = \frac{1}{2} \oint_{\mathcal{C}} x \, dy - y \, dx \qquad \boxed{6}$$

These remarkable formulas tell us how to compute an enclosed area by making measurements only along the boundary. It is the mathematical basis of the **planimeter**, a device that computes the area of an irregular shape when you trace the boundary with a pointer at the end of a movable arm (Figure 7).

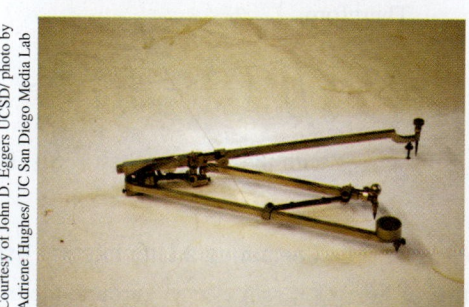

FIGURE 7 A planimeter is a mechanical device used for measuring the areas of irregular shapes.

Courtesy of John D. Eggers UCSD/ photo by Adriene Hughes/ UC San Diego Media Lab

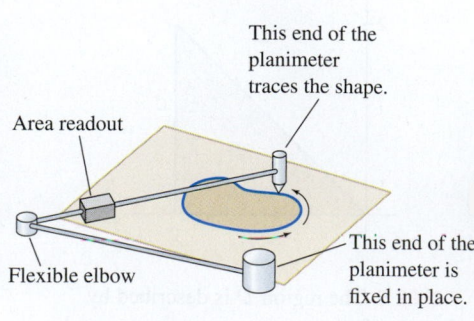

This end of the planimeter traces the shape.

Area readout

Flexible elbow

This end of the planimeter is fixed in place.

"Fortunately (for me), I was the only one in the local organization who had even heard of Green's Theorem... although I was not able to make constructive contributions, I could listen, nod my head and exclaim in admiration at the right places."
John M. Crawford, geophysicist and director of research at Conoco Oil, 1951–1971, writing about his first job interview in 1943, when a scientist visiting the company began speaking about applications of mathematics to oil exploration

◀ **REMINDER** We use the fact that

$$\int_0^{2\pi} \cos^2 \theta \, d\theta = \pi$$

which follows immediately from the identity

$$\cos^2 \theta = \frac{1 + \cos 2\theta}{2}$$

as in Example 1.

Stokes' Theorem in the next section generalizes Green's Theorem to three dimensions, relating the circulation of a vector field around a simple closed curve in 3-space to the integral of the curl over a surface that the curve bounds.

EXAMPLE 3 **Computing Area via Green's Theorem** Compute the area of the ellipse $\left(\frac{x}{a}\right)^2 + \left(\frac{y}{b}\right)^2 = 1$ using a line integral.

Solution We parametrize the boundary of the ellipse by

$$x = a\cos\theta, \qquad y = b\sin\theta, \qquad 0 \le \theta < 2\pi$$

We can use any of the three formulas in Eq. (6). We will use the first. See Exercises 16 and 17 for the computation using the other two.

$$\text{enclosed area} = \oint_C x \, dy = \int_0^{2\pi} (a\cos\theta)(b\cos\theta) \, d\theta$$

$$= ab \int_0^{2\pi} \cos^2\theta \, d\theta = \pi ab$$

Thus, the area of an ellipse $\left(\frac{x}{a}\right)^2 + \left(\frac{y}{b}\right)^2 = 1$ is πab. ∎

The Circulation Form of Green's Theorem

Green's Theorem can be written in a form that relates the circulation of a vector field around a simple closed curve to the integral of the curl of the vector field over the domain enclosed by the curve. To show this, think of a two-dimensional vector field $\mathbf{F} = \langle F_1, F_2\rangle$ as a three-dimensional vector field with a third component 0. So, $\mathbf{F} = \langle F_1, F_2, 0\rangle$. Then when we take the curl, keeping in mind that F_1 and F_2 depend only on x and y, we find

$$\text{curl}(\mathbf{F}) = \begin{vmatrix} \mathbf{i} & \mathbf{j} & \mathbf{k} \\ \frac{\partial}{\partial x} & \frac{\partial}{\partial y} & \frac{\partial}{\partial z} \\ F_1 & F_2 & 0 \end{vmatrix}$$

$$= 0\mathbf{i} + 0\mathbf{j} + \left(\frac{\partial F_2}{\partial x} - \frac{\partial F_1}{\partial y}\right)\mathbf{k}$$

The z-component of the result is $\frac{\partial F_2}{\partial x} - \frac{\partial F_1}{\partial y}$, which is the integrand that appears in Green's Theorem. Thus, we define

$$\text{curl}_z(\mathbf{F}) = \text{curl}(\mathbf{F}) \cdot \mathbf{k} = \frac{\partial F_2}{\partial x} - \frac{\partial F_1}{\partial y}$$

We can interpret this scalar quantity as the curl of the two-dimensional vector field $\mathbf{F}$. Then Green's Theorem becomes

$$\oint_C \mathbf{F} \cdot d\mathbf{r} = \iint_D \text{curl}_z(\mathbf{F}) \, dA \qquad \boxed{7}$$

We refer to this form as the **Circulation Form of Green's Theorem.**

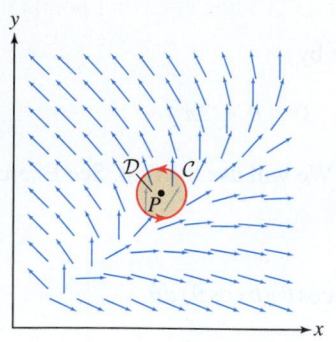

FIGURE 8 The circulation of **F** around $\mathcal{C}$ is approximately $\text{curl}_z(\mathbf{F})(P) \cdot \text{area}(\mathcal{D})$.

CONCEPTUAL INSIGHT Interpretation of curl$_z$ The Circulation Form of Green's Theorem says a circulation integral and an integral of $\text{curl}_z(\mathbf{F})$ are equal. Consequently, it provides us with an interpretation of $\text{curl}_z(\mathbf{F})$ in terms of circulation.

Let $\mathcal{D}$ be a small domain whose boundary is a circle $\mathcal{C}$ centered at P. If $\mathcal{D}$ is small enough we can approximate $\text{curl}_z(\mathbf{F})$ by the constant value $\text{curl}_z(\mathbf{F})(P)$ over $\mathcal{D}$. The Circulation Form of Green's Theorem yields the following approximation (Figure 8):

$$\oint_{\mathcal{C}} \mathbf{F} \cdot d\mathbf{r} = \iint_{\mathcal{D}} \text{curl}_z(\mathbf{F}) \, dA \approx \text{curl}_z(\mathbf{F})(P) \iint_{\mathcal{D}} dA$$

$$\approx \text{curl}_z(\mathbf{F})(P) \cdot \text{area}(\mathcal{D})$$

Thus,

$$\text{curl}_z(\mathbf{F})(P) \approx \frac{1}{\text{area}(\mathcal{D})} \oint_{\mathcal{C}} \mathbf{F} \cdot d\mathbf{r} \qquad \boxed{8}$$

In other words, the curl is approximately the circulation around a small circle divided by the area of the circle. The approximation improves as the circle shrinks, and thus we can think of $\text{curl}_z(\mathbf{F})(P)$ as the *circulation of* **F** *per unit area near* P.

GRAPHICAL INSIGHT   If we think of **F** as the velocity field of a fluid, then we can measure the curl by placing a small paddle wheel in the stream at a point P and observing how fast it rotates (Figure 9). Because the fluid pushes each paddle to move with a velocity equal to the tangential component of **F**, we can assume that the wheel itself rotates with a velocity v_a equal to the *average tangential component* of **F**. If the paddle wheel is a circle $\mathcal{C}_r$ of radius r (and hence length $2\pi r$), then the average tangential component of velocity is

$$v_a = \frac{1}{2\pi r} \oint_{\mathcal{C}_r} \mathbf{F} \cdot d\mathbf{r}$$

Angular Velocity   *An arc of ℓ meters on a circle of radius r meters has radian measure ℓ/r. Therefore, an object moving along the circle with a speed of v meters per second travels v/r radians per second. In other words, the object has angular velocity v/r.*

On the other hand, the paddle encloses an area of πr^2, and for small r, we can apply the approximation formula (8)

$$v_a \approx \frac{1}{2\pi r}(\pi r^2)\text{curl}_z(\mathbf{F})(P) = \left(\frac{1}{2}r\right)\text{curl}_z(\mathbf{F})(P)$$

Now, if an object moves along a circle of radius r with speed v_a, then its angular velocity (in radians per unit time) is $v_a/r \approx \frac{1}{2}\text{curl}_z(\mathbf{F})(P)$. Therefore, *the angular velocity of the paddle wheel is approximately one-half the curl.*

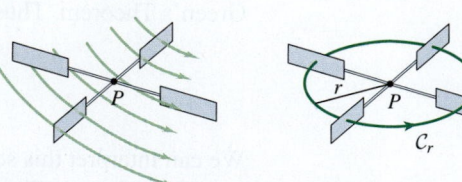

FIGURE 9 The curl is approximately equal to twice the angular velocity of a small paddle wheel placed at P.

Figure 10 shows vector fields such that $\text{curl}_z(\mathbf{F})$ is constant. Field (A) describes a fluid rotating counterclockwise around the origin, and field (B) describes a fluid that spirals into the origin. In both these cases, a small paddle wheel placed anywhere in the fluid rotates counterclockwise (corresponding to positive curl). On the other hand, a nonzero curl does not mean that the fluid itself is necessarily rotating. It means only that a small paddle wheel rotates if placed in the fluid. For example, field (C) is a **shear flow** (also known as a Couette flow). It has nonzero curl, but unlike cases (A) and (B), the fluid does not rotate about any point. However, the paddle wheel rotates clockwise (corresponding to negative curl) wherever it is placed.

In contrast to fields (A)–(C), the fields in cases (D) and (E) have zero curl. In either case, a paddle wheel placed anywhere in the vector field does not rotate. Unlike these examples, for most vector fields $\mathbf{F}$, $\mathrm{curl}_z(\mathbf{F})$ varies over the plane, having points where the paddle wheel turns counterclockwise, points where it turns clockwise, and points where it does not turn at all.

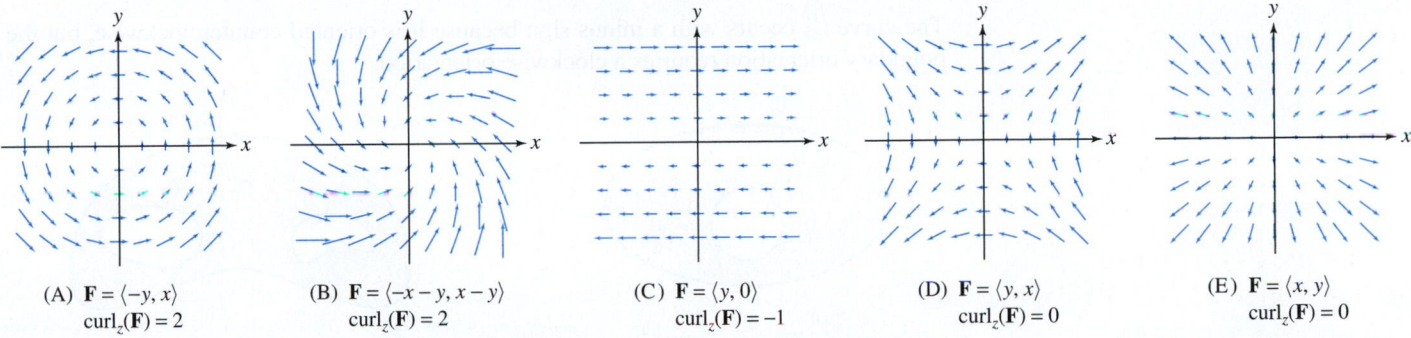

(A) $\mathbf{F} = \langle -y, x \rangle$
$\mathrm{curl}_z(\mathbf{F}) = 2$

(B) $\mathbf{F} = \langle -x - y, x - y \rangle$
$\mathrm{curl}_z(\mathbf{F}) = 2$

(C) $\mathbf{F} = \langle y, 0 \rangle$
$\mathrm{curl}_z(\mathbf{F}) = -1$

(D) $\mathbf{F} = \langle y, x \rangle$
$\mathrm{curl}_z(\mathbf{F}) = 0$

(E) $\mathbf{F} = \langle x, y \rangle$
$\mathrm{curl}_z(\mathbf{F}) = 0$

FIGURE 10 Examples of vector fields $\mathbf{F}$ and the corresponding $\mathrm{curl}_z(\mathbf{F})$.

Additivity of Circulation

Circulation around a closed curve has an important additivity property: If we decompose a domain $\mathcal{D}$ into two (or more) nonoverlapping domains $\mathcal{D}_1$ and $\mathcal{D}_2$ that intersect only on part of their boundaries as in Figure 11, then

$$\oint_{\partial \mathcal{D}} \mathbf{F} \cdot d\mathbf{r} = \oint_{\partial \mathcal{D}_1} \mathbf{F} \cdot d\mathbf{r} + \oint_{\partial \mathcal{D}_2} \mathbf{F} \cdot d\mathbf{r}$$

9

To verify this equation, note first that

$$\oint_{\partial \mathcal{D}} \mathbf{F} \cdot d\mathbf{r} = \int_{\mathcal{C}_{\text{top}}} \mathbf{F} \cdot d\mathbf{r} + \int_{\mathcal{C}_{\text{bot}}} \mathbf{F} \cdot d\mathbf{r}$$

with $\mathcal{C}_{\text{top}}$ and $\mathcal{C}_{\text{bot}}$ as in Figure 11, with the orientations shown. Then observe that the dashed segment $\mathcal{C}_{\text{middle}}$ occurs in both $\partial \mathcal{D}_1$ and $\partial \mathcal{D}_2$ but with opposite orientations. If $\mathcal{C}_{\text{middle}}$ is oriented right to left, then

$$\oint_{\partial \mathcal{D}_1} \mathbf{F} \cdot d\mathbf{r} = \int_{\mathcal{C}_{\text{top}}} \mathbf{F} \cdot d\mathbf{r} - \int_{\mathcal{C}_{\text{middle}}} \mathbf{F} \cdot d\mathbf{r}$$

$$\oint_{\partial \mathcal{D}_2} \mathbf{F} \cdot d\mathbf{r} = \int_{\mathcal{C}_{\text{bot}}} \mathbf{F} \cdot d\mathbf{r} + \int_{\mathcal{C}_{\text{middle}}} \mathbf{F} \cdot d\mathbf{r}$$

We obtain Eq. (9) by adding these two equations:

$$\oint_{\partial \mathcal{D}_1} \mathbf{F} \cdot d\mathbf{r} + \oint_{\partial \mathcal{D}_2} \mathbf{F} \cdot d\mathbf{r} = \int_{\mathcal{C}_{\text{top}}} \mathbf{F} \cdot d\mathbf{r} + \int_{\mathcal{C}_{\text{bot}}} \mathbf{F} \cdot d\mathbf{r} = \oint_{\partial \mathcal{D}} \mathbf{F} \cdot d\mathbf{r}$$

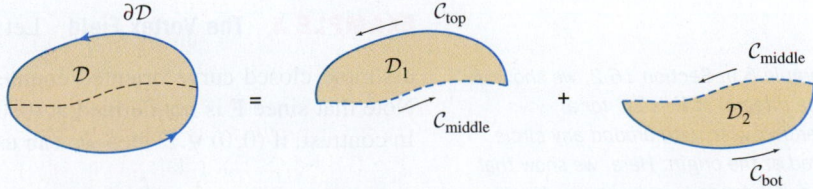

FIGURE 11 The domain $\mathcal{D}$ is the union of $\mathcal{D}_1$ and $\mathcal{D}_2$.

More General Form of Green's Theorem

← REMINDER In the boundary orientation, the region lies to the left as the curve is traversed in the orientation direction.

Consider a domain $\mathcal{D}$ whose boundary consists of more than one simple closed curve as in Figure 12. As before, $\partial\mathcal{D}$ denotes the boundary of $\mathcal{D}$ with its boundary orientation. For the domains in Figure 12,

$$\partial\mathcal{D}_1 = \mathcal{C}_1 + \mathcal{C}_2, \qquad \partial\mathcal{D}_2 = \mathcal{C}_3 + \mathcal{C}_4 - \mathcal{C}_5$$

The curve $\mathcal{C}_5$ occurs with a minus sign because it is oriented counterclockwise, but the boundary orientation requires a clockwise orientation.

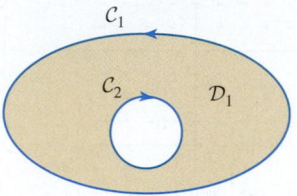
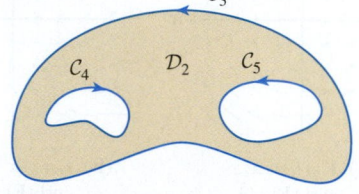

(A) Oriented boundary of $\mathcal{D}_1$ is $\mathcal{C}_1 + \mathcal{C}_2$.   (B) Oriented boundary of $\mathcal{D}_2$ is $\mathcal{C}_3 + \mathcal{C}_4 - \mathcal{C}_5$.

FIGURE 12

Green's Theorem holds for more general domains of this type:

$$\oint_{\partial\mathcal{D}} \mathbf{F}\cdot d\mathbf{r} = \iint_{\mathcal{D}} \left(\frac{\partial F_2}{\partial x} - \frac{\partial F_1}{\partial y} \right) dA$$

10

This equality is proved by decomposing $\mathcal{D}$ into smaller domains, each of which is bounded by a simple closed curve. To illustrate, consider the region $\mathcal{D}$ in Figure 13. We decompose $\mathcal{D}$ into domains $\mathcal{D}_1$ and $\mathcal{D}_2$. Then

$$\partial\mathcal{D} = \partial\mathcal{D}_1 + \partial\mathcal{D}_2$$

because the edges common to $\partial\mathcal{D}_1$ and $\partial\mathcal{D}_2$ occur with opposite orientation, and therefore cancel. By Eq. 9 and Green's Theorem applied to both $\mathcal{D}_1$ and $\mathcal{D}_2$, we have

$$\oint_{\partial\mathcal{D}} \mathbf{F}\cdot d\mathbf{r} = \int_{\partial\mathcal{D}_1} \mathbf{F}\cdot d\mathbf{r} + \int_{\partial\mathcal{D}_2} \mathbf{F}\cdot d\mathbf{r}$$

$$= \iint_{\mathcal{D}_1} \left(\frac{\partial F_2}{\partial x} - \frac{\partial F_1}{\partial y} \right) dA + \iint_{\mathcal{D}_2} \left(\frac{\partial F_2}{\partial x} - \frac{\partial F_1}{\partial y} \right) dA$$

$$= \iint_{\mathcal{D}} \left(\frac{\partial F_2}{\partial x} - \frac{\partial F_1}{\partial y} \right) dA$$

FIGURE 13 Here, $\partial\mathcal{D}$ is the sum $\partial\mathcal{D}_1 + \partial\mathcal{D}_2$. Note that the dashed edges cancel.

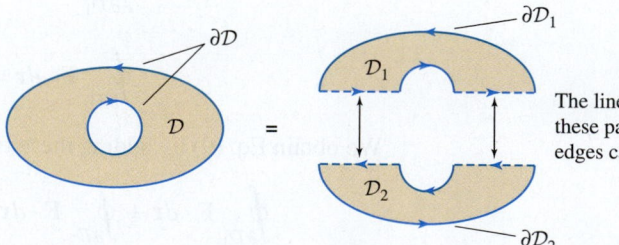

The line integrals over these pairs of dashed edges cancel.

In Example 6 in Section 16.2, we showed that the integral of $\mathbf{F}$ is 2π for a counterclockwise path around any circle centered at the origin. Here, we show that this result extends to any simple closed curve enclosing the origin.

EXAMPLE 4   The Vortex Field   Let $\mathbf{F}(x,y) = \left\langle \dfrac{-y}{x^2+y^2}, \dfrac{x}{x^2+y^2} \right\rangle$. Assume that $\mathcal{C}$ is a simple closed curve oriented counterclockwise and that $\mathcal{D}$ is the region it encloses. Note that since $\mathbf{F}$ is not defined at $(0,0)$, Green's Theorem does not apply if $(0,0) \in \mathcal{D}$. In contrast, if $(0,0) \notin \mathcal{D}$ then we can use Green's Theorem. Show that

$$\oint_{\mathcal{C}} F_1\, dx + F_2\, dy = \begin{cases} 0 & \text{if } (0,0) \notin \mathcal{D} \\ 2\pi & \text{if } (0,0) \in \mathcal{D} \end{cases}$$

Solution Applying the Quotient Rule to compute $\dfrac{\partial F_2}{\partial x}$ and $\dfrac{\partial F_1}{\partial y}$, we obtain

$$\frac{\partial F_2}{\partial x} = \frac{\partial F_1}{\partial y} = \frac{y^2 - x^2}{(x^2 + y^2)^2}$$

If $(0,0)$ is not in $\mathcal{D}$, then we may apply Green's Theorem, and it follows that

$$\oint_{\mathcal{C}} F_1 \, dx + F_2 \, dy = \iint_{\mathcal{D}} \left(\frac{\partial F_2}{\partial x} - \frac{\partial F_1}{\partial y} \right) dA = \iint_{\mathcal{D}} 0 \, dA = 0$$

as we wished to show.

Now, assume $(0,0) \in \mathcal{D}$. To compute the integral using Green's Theorem, we need to modify the enclosed region so that it satisfies the theorem. The idea is to cut out the "bad" part at the origin. Thus, we choose a small enough R so that the circle $\mathcal{C}^*$ of radius R centered at the origin is contained in $\mathcal{D}$, and we let $\mathcal{D}^*$ be the region between $\mathcal{C}$ and $\mathcal{C}^*$ (Figure 14). If we assume that $\mathcal{C}$ and $\mathcal{C}^*$ are both oriented counterclockwise, then the oriented boundary of $\mathcal{D}^*$ is $\partial \mathcal{D}^* = \mathcal{C} - \mathcal{C}^*$. Since $(0,0)$ is not in $\mathcal{D}^*$, we may apply Green's Theorem:

$$\oint_{\partial \mathcal{D}^*} F_1 \, dx + F_2 \, dy = \iint_{\mathcal{D}^*} \left(\frac{\partial F_2}{\partial x} - \frac{\partial F_1}{\partial y} \right) dA = \iint_{\mathcal{D}*} 0 \, dA = 0$$

Now, with $\partial \mathcal{D}^* = \mathcal{C} - \mathcal{C}^*$, we have

$$\oint_{\mathcal{C}} F_1 \, dx + F_2 \, dy - \oint_{\mathcal{C}^*} F_1 \, dx + F_2 \, dy = \oint_{\partial \mathcal{D}^*} F_1 \, dx + F_2 \, dy = 0$$

From this, we conclude that

$$\oint_{\mathcal{C}} F_1 \, dx + F_2 \, dy = \oint_{\mathcal{C}^*} F_1 \, dx + F_2 \, dy$$

By Example 6 in Section 16.2, $\displaystyle\oint_{\mathcal{C}^*} F_1 \, dx + F_2 \, dy = 2\pi$, and therefore it follows that

$\displaystyle\oint_{\mathcal{C}} F_1 \, dx + F_2 \, dy = 2\pi$ in the case $(0,0) \in \mathcal{D}$ as we wanted to show. ∎

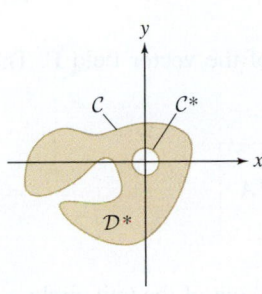

FIGURE 14

The Circulation Form of Green's Theorem relates a circulation integral to an integral of curl. Next, we introduce a form of Green's Theorem that relates a flux integral to an integral of divergence.

Flux Form of Green's Theorem

Recall from Section 16.2 that the flux of a vector field $\mathbf{F}$ across a curve $\mathcal{C}$ is the integral of the normal component of $\mathbf{F}$ along $\mathcal{C}$, as in Figure 15. For a simple closed curve $\mathcal{C}$, we are interested in the flux across the curve in the direction out of the enclosed region. We refer to this as the **outward flux** or **flux out of** $\mathcal{C}$. It is given by the integral $\displaystyle\oint_{\mathcal{C}} \mathbf{F} \cdot \mathbf{n} \, ds$, where $\mathbf{n}$ points away from the enclosed region. We assume that $\mathcal{C}$ is parametrized by $\mathbf{r}(t) = \langle x(t), y(t) \rangle$ for $a \le t \le b$, such that $\mathbf{r}'(t) \ne \mathbf{0}$. Then the unit tangent vector is given by $\mathbf{T} = \dfrac{\mathbf{r}'(t)}{\|\mathbf{r}'(t)\|} = \left\langle \dfrac{x'(t)}{\|\mathbf{r}'(t)\|}, \dfrac{y'(t)}{\|\mathbf{r}'(t)\|} \right\rangle$ and the outward unit normal vector is given by $\mathbf{n}(t) = \left\langle \dfrac{y'(t)}{\|\mathbf{r}'(t)\|}, \dfrac{-x'(t)}{\|\mathbf{r}'(t)\|} \right\rangle$ since its dot product with $\mathbf{T}$ is 0 and $\mathbf{n}$ points to the right as we travel around the curve.

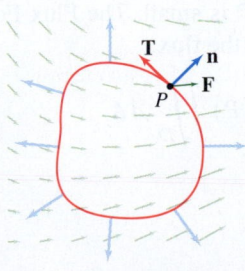

FIGURE 15 The flux of $\mathbf{F}$ is the integral of the normal component $\mathbf{F} \cdot \mathbf{n}$ around the curve.

Thus, the flux of $\mathbf{F}$ out of $\mathcal{C}$ is

$$\oint_{\mathcal{C}} \mathbf{F} \cdot \mathbf{n} \, ds = \int_a^b (\mathbf{F} \cdot \mathbf{n})(t) \|\mathbf{r}'(t)\| \, dt = \int_a^b \left[\frac{F_1 \, y'(t)}{\|\mathbf{r}(t)\|} - \frac{F_2 \, x'(t)}{\|\mathbf{r}(t)\|} \right] \|\mathbf{r}(t)\| \, dt$$

$$= \int_a^b F_1 y'(t) \, dt - F_2 x'(t) \, dt = \oint_{\mathcal{C}} F_1 \, dy - F_2 \, dx$$

We can apply Green's Theorem to the last integral, but we have to realize that the roles of F_1 and F_2 are switched and there is a negative sign with the second term. Since $\mathcal{D}$ is the region enclosed by $\mathcal{C}$, and $\mathcal{C} = \partial \mathcal{D}$, Green's Theorem gives us

$$\int_{\partial \mathcal{D}} F_1 \, dy - F_2 \, dx = \iint_{\mathcal{D}} \left(\frac{\partial F_1}{\partial x} + \frac{\partial F_2}{\partial y} \right) dA$$

Now, the integrand $\dfrac{\partial F_1}{\partial x} + \dfrac{\partial F_2}{\partial y}$ is the divergence of the vector field $\mathbf{F}$. Thus, we obtain the **Flux Form of Green's Theorem**:

$$\boxed{\oint_{\partial \mathcal{D}} \mathbf{F} \cdot \mathbf{n} \, ds = \iint_{\mathcal{D}} \operatorname{div}(\mathbf{F}) \, dA} \qquad \boxed{11}$$

> *The Divergence Theorem in Section 17.3 generalizes the Flux Form of Green's Theorem to three dimensions, relating the outward flux of a vector field across a closed surface in 3-space to the integral of the divergence over the volume enclosed by the surface.*

EXAMPLE 5 Calculate the flux of $\mathbf{F}(x, y) = \langle x^3, y^3 + y \rangle$ out of the unit circle.

Solution We find $\operatorname{div} \mathbf{F} = \dfrac{\partial F_1}{\partial x} + \dfrac{\partial F_2}{\partial y} = 3x^2 + 3y^2 + 1$. Therefore, the flux of $\mathbf{F}$ out of the unit circle is given by

$$\text{flux} = \iint_{\mathcal{D}} \operatorname{div}(\mathbf{F}) \, dA = \iint_{\mathcal{D}} (3x^2 + 3y^2 + 1) \, dA$$

Converting to polar coordinates, we have

$$\text{flux} = \int_0^{2\pi} \int_0^1 (3r^2 + 1) r \, dr \, d\theta = \int_0^{2\pi} \int_0^1 (3r^3 + r) \, dr \, d\theta$$

$$= 2\pi \left(\frac{3r^4}{4} + \frac{r^2}{2} \right) \Big|_0^1 = \frac{5\pi}{2} \qquad \blacksquare$$

> **CONCEPTUAL INSIGHT Interpretation of Divergence** The Flux Form of Green's Theorem relates a flux integral to an integral of divergence, and therefore it gives us an interpretation of divergence in terms of flux out of a simple closed curve.
>
> Let $\mathcal{D}$ be a small domain bounded by a circle $\mathcal{C}$ centered at P. Over $\mathcal{D}$, we can approximate $\operatorname{div}(\mathbf{F})$ by the constant value $\operatorname{div}(\mathbf{F})(P)$ if $\mathcal{D}$ is small. The Flux Form of Green's Theorem yields the following approximation for the flux:
>
> $$\oint_{\mathcal{C}} \mathbf{F} \cdot \mathbf{n} \, ds = \iint_{\mathcal{D}} \operatorname{div}(\mathbf{F}) \, dA \approx \operatorname{div}(\mathbf{F})(P) \iint_{\mathcal{D}} dA$$
>
> $$\approx \operatorname{div}(\mathbf{F})(P) \cdot \operatorname{area}(\mathcal{D})$$
>
> Thus,
>
> $$\operatorname{div}(\mathbf{F})(P) \approx \frac{1}{\operatorname{area}(\mathcal{D})} \oint_{\mathcal{C}} \mathbf{F} \cdot d\mathbf{r}$$
>
> This indicates that the divergence is approximately the flux out of a small circle, divided by the area of the circle. The approximation improves as the circle shrinks, and thus we can think of $\operatorname{div}(\mathbf{F})(P)$ as the *outward flux of* $\mathbf{F}$ *per unit area near* P.

17.1 SUMMARY

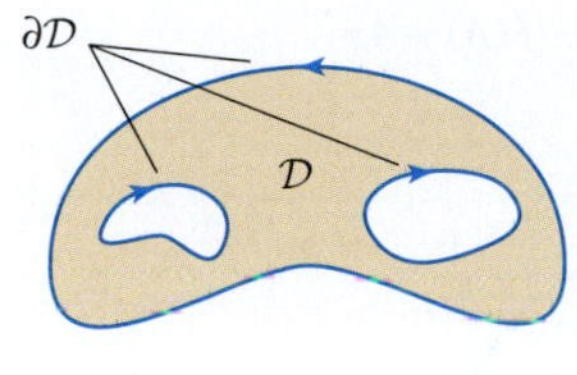

- We have two notations for the line integral of a vector field on the plane:

$$\int_C \mathbf{F} \cdot d\mathbf{r} \qquad \text{and} \qquad \int_C F_1 \, dx + F_2 \, dy$$

- $\partial \mathcal{D}$ denotes the boundary of $\mathcal{D}$ with its boundary orientation (Figure 16).
- Green's Theorem:

$$\oint_{\partial \mathcal{D}} F_1 \, dx + F_2 \, dy = \iint_{\mathcal{D}} \left(\frac{\partial F_2}{\partial x} - \frac{\partial F_1}{\partial y} \right) dA$$

- Formulas for the area of the region $\mathcal{D}$ enclosed by $\mathcal{C}$:

$$\text{area}(\mathcal{D}) = \oint_C x \, dy = \oint_C -y \, dx = \frac{1}{2} \oint_C x \, dy - y \, dx$$

- Circulation Form of Green's Theorem:

$$\oint_{\partial \mathcal{D}} \mathbf{F} \cdot d\mathbf{r} = \iint_{\mathcal{D}} \text{curl}_z(\mathbf{F}) \, dA$$

where $\text{curl}_z(\mathbf{F}) = \dfrac{\partial F_2}{\partial x} - \dfrac{\partial F_1}{\partial y}$.

- For a two-dimensional vector field $\mathbf{F}$, the quantity $\text{curl}_z(\mathbf{F})$ is interpreted as *circulation per unit area*. If $\mathcal{C}$ is a small circle centered at P, enclosing domain $\mathcal{D}$, then

$$\text{curl}_z(\mathbf{F})(P) \approx \frac{1}{\text{area}(\mathcal{D})} \oint_C \mathbf{F} \cdot d\mathbf{r}$$

- Flux Form of Green's Theorem:

$$\oint_{\partial \mathcal{D}} \mathbf{F} \cdot \mathbf{n} \, ds = \iint_{\mathcal{D}} \text{div}(\mathbf{F}) \, dA$$

- For a two-dimensional vector field $\mathbf{F}$, the quantity $\text{div}(\mathbf{F})$ is interpreted as *outward flux per unit area*. If $\mathcal{C}$ is a small circle centered at P, enclosing domain $\mathcal{D}$, then

$$\text{div}(\mathbf{F})(P) \approx \frac{1}{\text{area}(\mathcal{D})} \oint_C \mathbf{F} \cdot \mathbf{n} \, ds$$

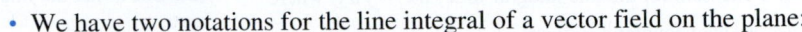

FIGURE 16 The boundary orientation is chosen so that the region lies to your left as you walk along the curve.

17.1 EXERCISES

Preliminary Questions

1. Which vector field $\mathbf{F}$ is being integrated in the line integral $\oint x^2 \, dy - e^y \, dx$?

2. Draw a domain in the shape of an ellipse and indicate with an arrow the boundary orientation of the boundary curve. Do the same for the annulus (the region between two concentric circles).

3. The circulation of a conservative vector field around a closed curve is zero. Is this fact consistent with Green's Theorem? Explain.

4. Indicate which of the following vector fields possess this property: For every simple closed curve $\mathcal{C}$, $\oint_C \mathbf{F} \cdot d\mathbf{r}$ is equal to the area enclosed by $\mathcal{C}$.

    (a) $\mathbf{F}(x, y) = \langle -y, 0 \rangle$

    (b) $\mathbf{F}(x, y) = \langle x, y \rangle$

    (c) $\mathbf{F}(x, y) = \langle \sin(x^2), x + e^{y^2} \rangle$

5. Let A be the area enclosed by a simple closed curve $\mathcal{C}$, and assume that $\mathcal{C}$ is oriented counterclockwise. Indicate whether the value of each integral is 0, $-A$, or A.

    (a) $\oint_C x \, dx$

    (b) $\oint_C y \, dx$

    (c) $\oint_C y \, dy$

    (d) $\oint_C x \, dy$

Exercises

1. Verify Green's Theorem for the line integral $\oint_C xy\,dx + y\,dy$, where C is the unit circle, oriented counterclockwise.

2. Let $I = \oint_C \mathbf{F} \cdot d\mathbf{r}$, where $\mathbf{F}(x, y) = \langle y + \sin x^2, x^2 + e^{y^2} \rangle$ and C is the circle of radius 4 centered at the origin.
(a) Which is easier, evaluating I directly or using Green's Theorem?
(b) Evaluate I using the easier method.

In Exercises 3–12, use Green's Theorem to evaluate the line integral. Orient the curve counterclockwise unless otherwise indicated.

3. $\oint_C y^2\,dx + x^2\,dy$, where C is the boundary of the square that is given by $0 \le x \le 1, 0 \le y \le 1$

4. $\oint_C y^2\,dx + x^2\,dy$, where C is the boundary of the square $-1 \le x \le 1$, $-1 \le y \le 1$

5. $\oint_C 5y\,dx + 2x\,dy$, where C is the triangle with vertices $(-1, 0)$, $(1, 0)$, and $(0, 1)$

6. $\oint_C e^{2x+y}\,dx + e^{-y}\,dy$, where C is the triangle with vertices $(0, 0)$, $(1, 0)$, and $(1, 1)$

7. $\oint_C x^2 y\,dx$, where C is the unit circle centered at the origin

8. $\oint_C \mathbf{F} \cdot d\mathbf{r}$, where $\mathbf{F}(x, y) = \langle x + y, x^2 - y \rangle$ and C is the boundary of the region enclosed by $y = x^2$ and $y = \sqrt{x}$ for $0 \le x \le 1$

9. $\oint_C \mathbf{F} \cdot d\mathbf{r}$, where $\mathbf{F}(x, y) = \langle x^2, x^2 \rangle$ and C consists of the arcs $y = x^2$ and $y = x$ for $0 \le x \le 1$

10. $\oint_C (\ln x + y)\,dx - x^2\,dy$, where C is the rectangle with vertices $(1, 1)$, $(3, 1)$, $(1, 4)$, and $(3, 4)$

11. The line integral of $\mathbf{F}(x, y) = \langle e^{x+y}, e^{x-y} \rangle$ along the curve (oriented clockwise) consisting of the line segments by joining the points $(0, 0)$, $(2, 2)$, $(4, 2)$, $(2, 0)$, and back to $(0, 0)$ (Note the orientation.)

12. $\int_C xy\,dx + (x^2 + x)\,dy$, where C is the path in Figure 17

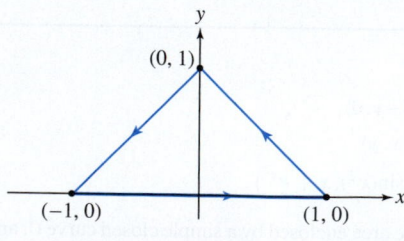

FIGURE 17

13. Let $\mathbf{F}(x, y) = \langle 2xe^y, x + x^2 e^y \rangle$ and let C be the quarter-circle path from A to B in Figure 18. Evaluate $I = \int_C \mathbf{F} \cdot d\mathbf{r}$ as follows:
(a) Find a function $f(x, y)$ such that $\mathbf{F} = \mathbf{G} + \nabla f$, where $\mathbf{G} = \langle 0, x \rangle$.

(b) Show that the line integrals of $\mathbf{G}$ along the segments $\overline{OA}$ and $\overline{OB}$ are zero.
(c) Evaluate I. *Hint:* Use Green's Theorem to show that
$$I = f(B) - f(A) + 4\pi$$

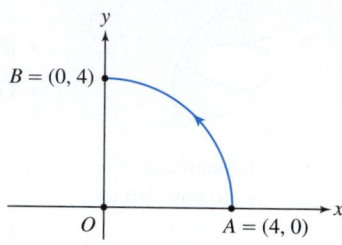

FIGURE 18

14. Compute the line integral of $\mathbf{F}(x, y) = \langle x^3, 4x \rangle$ along the path from A to B in Figure 19. To save work, use Green's Theorem to relate this line integral to the line integral along the vertical path from B to A.

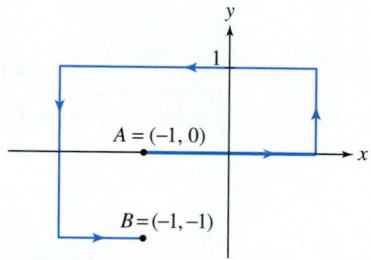

FIGURE 19

15. Evaluate $I = \int_C (\sin x + y)\,dx + (3x + y)\,dy$ for the nonclosed path $ABCD$ in Figure 20. Use the method of Exercise 14.

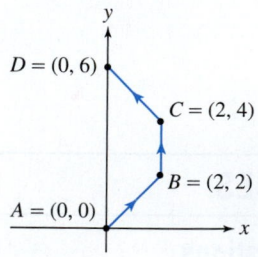

FIGURE 20

16. Use $\oint_C y\,dx$ to compute the area of the ellipse $\left(\frac{x}{a}\right)^2 + \left(\frac{y}{b}\right)^2 = 1$.

17. Use $\frac{1}{2}\oint_C x\,dy - y\,dx$ to compute the area of the ellipse $\left(\frac{x}{a}\right)^2 + \left(\frac{y}{b}\right)^2 = 1$.

In Exercises 18–21, use one of the formulas in Eq. (6) to calculate the area of the given region.

18. The circle of radius 3 centered at the origin

19. The triangle with vertices $(0, 0)$, $(1, 0)$, and $(1, 1)$

20. The region between the x-axis and the cycloid parametrized by $\mathbf{r}(t) = \langle t - \sin t, 1 - \cos t \rangle$ for $0 \le t \le 2\pi$ (Figure 21)

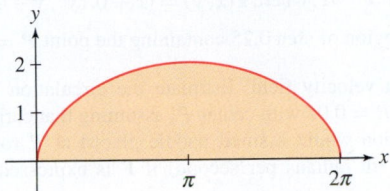

FIGURE 21 Cycloid.

21. The region between the graph of $y = x^2$ and the x-axis for $0 \le x \le 2$

22. A square with vertices $(1, 1), (-1, 1), (-1, -1)$, and $(1, -1)$ has area 4. Calculate this area three times using the formulas in Eq. (6).

23. Let $x^3 + y^3 = 3xy$ be the **folium of Descartes** (Figure 22).

(a) Show that the folium has a parametrization in terms of $t = y/x$ given by

$$x = \frac{3t}{1 + t^3}, \qquad y = \frac{3t^2}{1 + t^3} \qquad (-\infty < t < \infty) \quad (t \ne -1)$$

(b) Show that

$$x \, dy - y \, dx = \frac{9t^2}{(1 + t^3)^2} \, dt$$

Hint: By the Quotient Rule,

$$x^2 \, d\left(\frac{y}{x}\right) = x \, dy - y \, dx$$

(c) Find the area of the loop of the folium. *Hint:* The limits of integration are 0 and ∞.

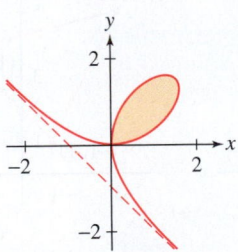

FIGURE 22 Folium of Descartes.

24. Find a parametrization of the lemniscate $(x^2 + y^2)^2 = xy$ (see Figure 23) by using $t = y/x$ as a parameter (see Exercise 23). Then use Eq. (6) to find the area of one loop of the lemniscate.

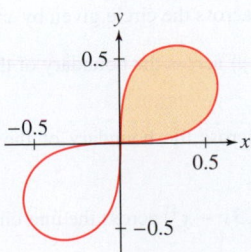

FIGURE 23 Lemniscate.

25. The Centroid via Boundary Measurements The centroid (see Section 15.5) of a domain $\mathcal{D}$ enclosed by a simple closed curve $\mathcal{C}$ is the point with coordinates $(\overline{x}, \overline{y}) = (M_y/M, M_x/M)$, where M is the area of $\mathcal{D}$ and the moments are defined by

$$M_x = \iint_{\mathcal{D}} y \, dA, \qquad M_y = \iint_{\mathcal{D}} x \, dA$$

Show that $M_x = \oint_{\mathcal{C}} xy \, dy$. Find a similar expression for M_y.

26. Use the result of Exercise 25 to compute the moments of the semicircle $x^2 + y^2 = R^2$, $y \ge 0$ as line integrals. Verify that the centroid is $(0, 4R/(3\pi))$.

27. Let $\mathcal{C}_R$ be the circle of radius R centered at the origin. Use the general form of Green's Theorem to determine $\oint_{\mathcal{C}_2} \mathbf{F} \cdot d\mathbf{r}$, where $\mathbf{F}$ is a vector field such that $\oint_{\mathcal{C}_1} \mathbf{F} \cdot d\mathbf{r} = 9$ and $\dfrac{\partial F_2}{\partial x} - \dfrac{\partial F_1}{\partial y} = x^2 + y^2$ for (x, y) in the annulus $1 \le x^2 + y^2 \le 4$.

28. Referring to Figure 24, suppose that $\oint_{\mathcal{C}_2} \mathbf{F} \cdot d\mathbf{r} = 12$. Use Green's Theorem to determine $\oint_{\mathcal{C}_1} \mathbf{F} \cdot d\mathbf{r}$, assuming that $\dfrac{\partial F_2}{\partial x} - \dfrac{\partial F_1}{\partial y} = -3$ in $\mathcal{D}$.

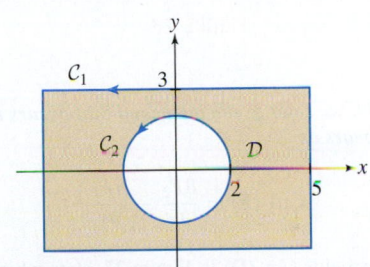

FIGURE 24

29. Referring to Figure 25, suppose that

$$\oint_{\mathcal{C}_2} \mathbf{F} \cdot d\mathbf{r} = 3\pi, \qquad \oint_{\mathcal{C}_3} \mathbf{F} \cdot d\mathbf{r} = 4\pi$$

Use Green's Theorem to determine the circulation of $\mathbf{F}$ around $\mathcal{C}_1$, assuming that $\dfrac{\partial F_2}{\partial x} - \dfrac{\partial F_1}{\partial y} = 9$ on the shaded region.

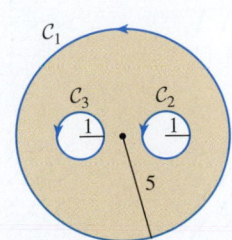

FIGURE 25

30. Let $\mathbf{F}$ be the vector field

$$\mathbf{F}(x, y) = \left\langle \frac{x}{x^2 + y^2}, \frac{y}{x^2 + y^2} \right\rangle$$

and assume that C_R is the circle of radius R centered at the origin and oriented counterclockwise.

(a) Show that $\dfrac{\partial F_2}{\partial x} - \dfrac{\partial F_1}{\partial y} = 0$.

(b) Explain why we cannot use Green's Theorem to argue that $\displaystyle\int_{C_R} \mathbf{F} \cdot d\mathbf{r} = 0$.

(c) By direct computation of the line integral, show that $\displaystyle\int_{C_R} \mathbf{F} \cdot d\mathbf{r} = 0$.

(d) Let $\mathcal{C}$ be the curve shown in Figure 26. Explain why we *can* use Green's Theorem, along with the result of (c), to conclude that $\displaystyle\int_{\mathcal{C}} \mathbf{F} \cdot d\mathbf{r} = 0$.

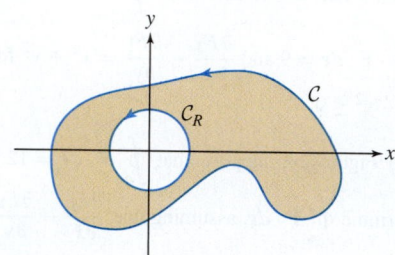

FIGURE 26

In Exercises 31–34, we refer to the integrand that occurs in Green's Theorem and that appears as

$$\operatorname{curl}_z(\mathbf{F}) = \frac{\partial F_2}{\partial x} - \frac{\partial F_1}{\partial y}$$

31. For the vector fields (A)–(D) in Figure 27, state whether curl_z at the origin appears to be positive, negative, or zero.

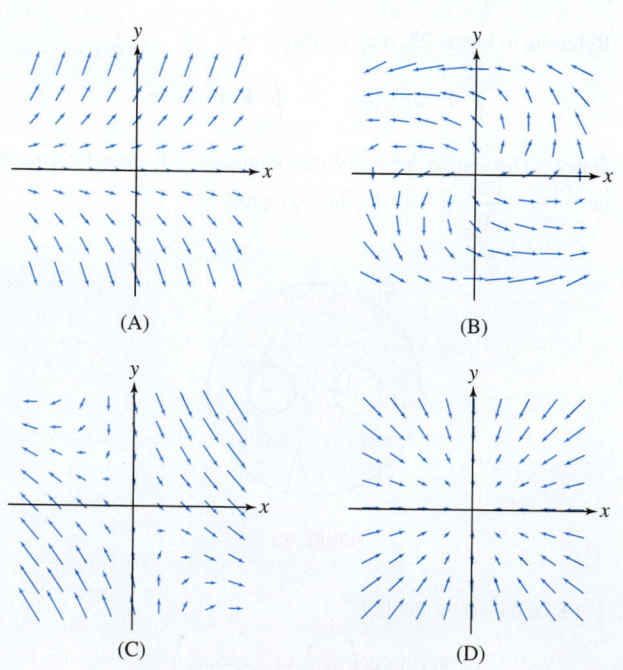

FIGURE 27

32. Estimate the circulation of a vector field $\mathbf{F}$ around a circle of radius $R = 0.1$, assuming that $\operatorname{curl}_z(\mathbf{F})$ takes the value 4 at the center of the circle.

33. Estimate $\displaystyle\oint_{\mathcal{C}} \mathbf{F} \cdot d\mathbf{r}$, where $\mathbf{F}(x, y) = \langle x + 0.1y^2, y - 0.1x^2 \rangle$ and $\mathcal{C}$ encloses a small region of area 0.25 containing the point $P = (1, 1)$.

34. Let $\mathbf{F}$ be a velocity field. Estimate the circulation of $\mathbf{F}$ around a circle of radius $R = 0.05$ with center P, assuming that $\operatorname{curl}_z(\mathbf{F})(P) = -3$. In which direction would a small paddle placed at P rotate? How fast would it rotate (in radians per second) if $\mathbf{F}$ is expressed in meters per second?

35. Let C_R be the circle of radius R centered at the origin. Use Green's Theorem to find the value of R that maximizes $\displaystyle\oint_{C_R} y^3 \, dx + x \, dy$.

36. Area of a Polygon Green's Theorem leads to a convenient formula for the area of a polygon.

(a) Let $\mathcal{C}$ be the line segment joining (x_1, y_1) to (x_2, y_2). Show that

$$\frac{1}{2}\int_{\mathcal{C}} -y \, dx + x \, dy = \frac{1}{2}(x_1 y_2 - x_2 y_1)$$

(b) Prove that the area of the polygon with vertices $(x_1, y_1), (x_2, y_2), \ldots, (x_n, y_n)$ is equal [where we set $(x_{n+1}, y_{n+1}) = (x_1, y_1)$] to

$$\frac{1}{2} \sum_{i=1}^{n} (x_i y_{i+1} - x_{i+1} y_i)$$

37. Use the result of Exercise 36 to compute the areas of the polygons in Figure 28. Check your result for the area of the triangle in (A) using geometry.

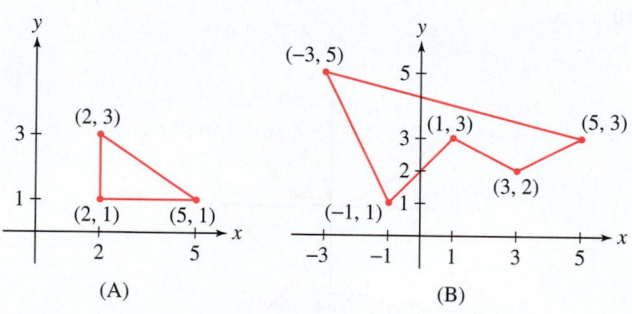

FIGURE 28

In Exercises 38–43, compute the flux $\displaystyle\oint_{\mathcal{C}} \mathbf{F} \cdot \mathbf{n} \, ds$ of $\mathbf{F}$ across the curve $\mathcal{C}$ for the given vector field and curve using the Flux Form of Green's Theorem.

38. $\mathbf{F}(x, y) = \langle 3x, 2y \rangle$ across the circle given by $x^2 + y^2 = 9$

39. $\mathbf{F}(x, y) = \langle xy, x - y \rangle$ across the boundary of the square $-1 \le x \le 1$, $-1 \le y \le 1$

40. $\mathbf{F}(x, y) = \langle x^2, y^2 \rangle$ across the boundary of the triangle with vertices $(0, 0)$, $(1, 0)$, and $(0, 1)$

41. $\mathbf{F}(x, y) = \langle 2x + y^3, 3y - x^4 \rangle$ across the unit circle

42. $\mathbf{F}(x, y) = \langle \cos y, \sin y \rangle$ across the boundary of the square $0 \le x \le 2$, $0 \le y \le \dfrac{\pi}{2}$

43. $\mathbf{F}(x, y) = \langle xy^2 + 2x, x^2y - 2y \rangle$ across the simple closed curve that is the boundary of the half-disk given by $x^2 + y^2 \leq 3, y \geq 0$

44. If $\mathbf{v}$ is the velocity field of a fluid, the flux of $\mathbf{v}$ across $\mathcal{C}$ is equal to the flow rate (amount of fluid flowing across $\mathcal{C}$ in square meters per second). Find the flow rate across the circle of radius 2 centered at the origin if $\text{div}(\mathbf{v}) = x^2$.

45. A buffalo stampede (Figure 29) is described by a velocity vector field $\mathbf{F} = \langle xy - y^3, x^2 + y \rangle$ kilometers per hour in the region $\mathcal{D}$ defined by $2 \leq x \leq 3, 2 \leq y \leq 3$ in units of kilometers (Figure 30). Assuming a density of $\rho = 500$ buffalo per square kilometer, use the Flux Form of Green's Theorem to determine the net number of buffalo leaving or entering $\mathcal{D}$ per minute (equal to ρ times the flux of $\mathbf{F}$ across the boundary of $\mathcal{D}$).

C. K. Lorenz/Science Source

FIGURE 29 Buffalo stampede.

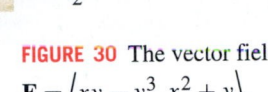

FIGURE 30 The vector field
$\mathbf{F} = \langle xy - y^3, x^2 + y \rangle$.

Further Insights and Challenges

In Exercises 46–49, the **Laplace operator** Δ *is defined by*

$$\Delta \varphi = \frac{\partial^2 \varphi}{\partial x^2} + \frac{\partial^2 \varphi}{\partial y^2} \qquad \boxed{12}$$

For any vector field $\mathbf{F} = \langle F_1, F_2 \rangle$, *define the conjugate vector field* $\mathbf{F}^* = \langle -F_2, F_1 \rangle$.

46. Show that if $\mathbf{F} = \nabla \varphi$, then $\text{curl}_z(\mathbf{F}^*) = \Delta \varphi$.

47. Let $\mathbf{n}$ be the outward-pointing unit normal vector to a simple closed curve $\mathcal{C}$. The **normal derivative** of a function φ, denoted $\dfrac{\partial \varphi}{\partial \mathbf{n}}$, is the directional derivative $D_\mathbf{n}(\varphi) = \nabla \varphi \cdot \mathbf{n}$. Prove that

$$\oint_{\mathcal{C}} \frac{\partial \varphi}{\partial \mathbf{n}} \, ds = \iint_{\mathcal{D}} \Delta \varphi \, dA$$

where $\mathcal{D}$ is the domain enclosed by a simple closed curve $\mathcal{C}$. *Hint:* Let $\mathbf{F} = \nabla \varphi$. Show that $\dfrac{\partial \varphi}{\partial \mathbf{n}} = \mathbf{F}^* \cdot \mathbf{T}$, where $\mathbf{T}$ is the unit tangent vector, and apply Green's Theorem.

48. Let $P = (a, b)$ and let $\mathcal{C}_r$ be the circle of radius r centered at P. The average value of a continuous function φ on $\mathcal{C}_r$ is defined as the integral

$$I_\varphi(r) = \frac{1}{2\pi} \int_0^{2\pi} \varphi(a + r\cos\theta, b + r\sin\theta) \, d\theta$$

(a) Show that

$$\frac{\partial \varphi}{\partial \mathbf{n}}(a + r\cos\theta, b + r\sin\theta)$$
$$= \frac{\partial \varphi}{\partial r}(a + r\cos\theta, b + r\sin\theta)$$

(b) Use differentiation under the integral sign to prove that

$$\frac{d}{dr} I_\varphi(r) = \frac{1}{2\pi r} \int_{\mathcal{C}_r} \frac{\partial \varphi}{\partial \mathbf{n}} \, ds$$

(c) Use Exercise 47 to conclude that

$$\frac{d}{dr} I_\varphi(r) = \frac{1}{2\pi r} \iint_{\mathcal{D}(r)} \Delta \varphi \, dA$$

where $\mathcal{D}(r)$ is the interior of $\mathcal{C}_r$.

49. Prove that $m(r) \leq I_\varphi(r) \leq M(r)$, where $m(r)$ and $M(r)$ are the minimum and maximum values of φ on $\mathcal{C}_r$. Then use the continuity of φ to prove that $\lim\limits_{r \to 0} I_\varphi(r) = \varphi(P)$.

In Exercises 50 and 51, let $\mathcal{D}$ be the region bounded by a simple closed curve $\mathcal{C}$. A function $\varphi(x, y)$ on $\mathcal{D}$ (whose second-order partial derivatives exist and are continuous) is called **harmonic** *if $\Delta \varphi = 0$, where $\Delta \varphi$ is the Laplace operator defined in Eq. (12).*

50. Use the results of Exercises 48 and 49 to prove the **mean-value property** of harmonic functions: If φ is harmonic, then $I_\varphi(r) = \varphi(P)$ for all r.

51. Show that $f(x, y) = x^2 - y^2$ is harmonic. Verify the mean-value property for $f(x, y)$ directly [expand $f(a + r\cos\theta, b + r\sin\theta)$ as a function of θ and compute $I_\varphi(r)$]. Show that $x^2 + y^2$ is not harmonic and does not satisfy the mean-value property.

17.2 Stokes' Theorem

Stokes' Theorem is an extension of Green's Theorem to three dimensions in which circulation is related to a surface integral over a surface in $\mathbf{R}^3$ (rather than to a double integral over a region in the plane). In order to state it, we introduce some definitions and terminology.

Figure 1 shows three surfaces with different types of boundaries. The boundary of a surface $\mathcal{S}$ is denoted $\partial \mathcal{S}$. Observe that the boundary in (A) is a single, simple closed curve and the boundary in (B) consists of three simple closed curves. The surface in (C) is called a **closed surface** because its boundary is empty. In this case, we write $\partial \mathcal{S} = \varnothing$.

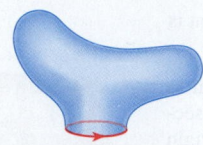
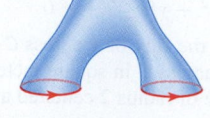

FIGURE 1 Surfaces and their boundaries.

(A) The boundary consists of a single closed curve.

(B) The boundary consists of three closed curves.

(C) The boundary is empty. This is a closed surface.

Recall from Section 16.5 that an orientation of a surface S is a continuously varying choice of unit normal vector at each point of S. When S is oriented, we can specify an orientation of ∂S, called the **boundary orientation**. Imagine that you are a unit normal vector walking along the boundary curve with your head at the head end of the vector and your feet at the tail end. The boundary orientation is the direction for which the surface is on your left as you walk. For example, the boundary of the surface in Figure 2 consists of two curves, C_1 and C_2. In (A), the normal vector points to the outside. The woman (representing the normal vector) is walking along C_1 and has the surface to her left, so she is walking in the positive direction. The curve C_2 is oriented in the opposite direction because she would have to walk along C_2 in that direction to keep the surface on her left. The boundary orientations in (B) are reversed because the opposite normal has been selected to orient the surface.

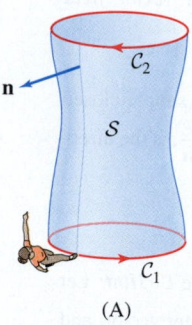
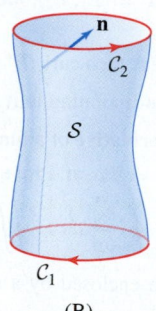

FIGURE 2 The orientation of the boundary ∂S for each of the two possible orientations of the surface S.

(A)

(B)

In the next theorem, we assume that S is an oriented surface with parametrization $G : \mathcal{D} \to S$, where $\mathcal{D}$ is a domain in the plane bounded by smooth, simple closed curves, and G is one-to-one and regular, except possibly on the boundary of $\mathcal{D}$. More generally, S may be a finite union of surfaces of this type. The surfaces in applications we consider, such as spheres, cubes, and graphs of functions, satisfy these conditions.

THEOREM 1 Stokes' Theorem   Let S be a surface as described earlier, and let $\mathbf{F}$ be a vector field whose components have continuous partial derivatives on an open region containing S.

$$\oint_{\partial S} \mathbf{F} \cdot d\mathbf{r} = \iint_S \operatorname{curl}(\mathbf{F}) \cdot d\mathbf{S} \qquad \boxed{1}$$

The integral on the left is defined relative to the boundary orientation of ∂S.

If S is a closed surface, then

$$\iint_S \operatorname{curl}(\mathbf{F}) \cdot d\mathbf{S} = 0$$

The curl measures the extent to which $\mathbf{F}$ fails to be conservative. If $\mathbf{F}$ is conservative, then $\operatorname{curl}(\mathbf{F}) = 0$ and Stokes' Theorem merely confirms what we already know: The circulation of a conservative vector field around a closed path is zero.

With the notation $\nabla \times \mathbf{F} = \operatorname{curl}(\mathbf{F})$, Stokes' Theorem is also written in the form

$$\oint_{\partial S} \mathbf{F} \cdot d\mathbf{r} = \iint_S (\nabla \times \mathbf{F}) \cdot d\mathbf{S}$$

Again, we see the analogy with the Fundamental Theorem of Calculus–Part I. A double integral over a surface of a derivative, in this case the curl, yields a single integral over the boundary of the surface.

Proof The left side of Eq. (1) is equal to a sum over the components of **F**:

$$\oint_C \mathbf{F} \cdot d\mathbf{r} = \oint_C F_1 \, dx + F_2 \, dy + F_3 \, dz \qquad \boxed{2}$$

Considering $\mathbf{F} = F_1\mathbf{i} + F_2\mathbf{j} + F_3\mathbf{k}$, and using the additivity property of the curl operator, we have $\text{curl}(\mathbf{F}) = \text{curl}(F_1\mathbf{i}) + \text{curl}(F_2\mathbf{j}) + \text{curl}(F_3\mathbf{k})$ and therefore:

$$\iint_S \text{curl}(\mathbf{F}) \cdot d\mathbf{S} = \iint_S \text{curl}(F_1\mathbf{i}) \cdot d\mathbf{S} + \iint_S \text{curl}(F_2\mathbf{j}) \cdot d\mathbf{S} + \iint_S \text{curl}(F_3\mathbf{k}) \cdot d\mathbf{S} \qquad \boxed{3}$$

The proof consists of showing that the F_1, F_2, and F_3 terms in Eqs. (2) and (3) are separately equal.

Because a complete proof is quite technical, we will prove it under the simplifying assumption that S is the graph of a function $z = f(x, y)$ lying over a domain $\mathcal{D}$ in the xy-plane. Furthermore, we will carry the details only for the F_1 terms. The calculation for the F_2 terms is similar, and we leave as an exercise the equality of the F_3 terms (Exercise 37). Thus, we shall prove that

$$\oint_C F_1 \, dx = \iint_S \text{curl}(F_1(x, y, z)\mathbf{i}) \cdot d\mathbf{S} \qquad \boxed{4}$$

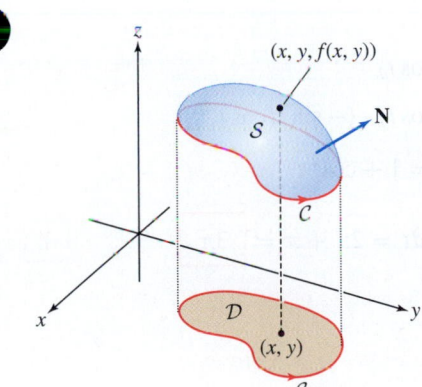

(x, y, f(x, y))

N

$\mathcal{S}$

$\mathcal{C}$

$\mathcal{D}$

(x, y)

$\mathcal{C}_0$

Orient S with an upward-pointing normal as in Figure 3, and let $C = \partial S$ be the boundary curve with orientation determined by the orientation of S. Let C_0 be the boundary of $\mathcal{D}$ in the xy-plane, and let $\mathbf{r}_0(t) = \langle x(t), y(t) \rangle$ (for $a \le t \le b$) be a counterclockwise parametrization of C_0 as in Figure 3. The boundary curve C projects onto C_0, so C has parametrization

$$\mathbf{r}(t) = \langle x(t), y(t), f(x(t), y(t)) \rangle$$

and thus

$$\oint_C F_1(x, y, z) \, dx = \int_a^b F_1\big(x(t), y(t), f(x(t), y(t))\big) \frac{dx}{dt} \, dt$$

The integral on the right-hand side of this equation is precisely the integral we obtain by integrating $F_1\big(x, y, f(x, y)\big) \, dx$ over the curve C_0 in the plane $\mathbf{R}^2$ with parametrization $\mathbf{r}_0(t)$. In other words,

$$\oint_C F_1(x, y, z) \, dx = \oint_{C_0} F_1\big(x, y, f(x, y)\big) \, dx$$

By Green's Theorem applied to the integral on the right, we get

$$\oint_C F_1(x, y, z) \, dx = - \iint_\mathcal{D} \frac{\partial}{\partial y} F_1(x, y, f(x, y)) \, dA$$

By the Chain Rule,

$$\frac{\partial}{\partial y} F_1\big(x, y, f(x, y)\big) = F_{1y}\big(x, y, f(x, y)\big) + F_{1z}\big(x, y, f(x, y)\big) f_y(x, y)$$

so finally we obtain

$$\oint_C F_1 \, dx = - \iint_\mathcal{D} \Big(F_{1y}\big(x, y, f(x, y)\big) + F_{1z}\big(x, y, f(x, y)\big) f_y(x, y) \Big) \, dA \qquad \boxed{5}$$

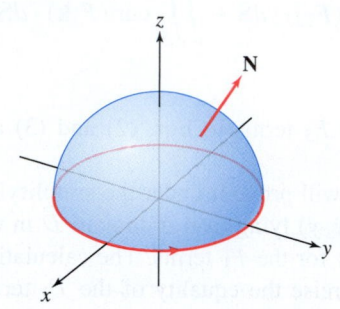

FIGURE 4 Upper hemisphere with oriented boundary.

To finish the proof, we compute the surface integral of $\operatorname{curl}(F_1\mathbf{i})$ using the parametrization $G(x, y) = (x, y, f(x, y))$ of $\mathcal{S}$:

$$\mathbf{N} = \langle -f_x(x, y), -f_y(x, y), 1 \rangle \qquad \text{(upward-pointing normal)}$$

$$\operatorname{curl}(F_1\mathbf{i}) \cdot \mathbf{N} = \langle 0, F_{1z}, -F_{1y} \rangle \cdot \langle -f_x(x, y), -f_y(x, y), 1 \rangle$$

$$= -F_{1z}(x, y, f(x, y)) f_y(x, y) - F_{1y}(x, y, f(x, y))$$

$$\iint_{\mathcal{S}} \operatorname{curl}(F_1\mathbf{i}) \cdot d\mathbf{S} = -\iint_{\mathcal{D}} \Big(F_{1z}(x, y, z) f_y(x, y) + F_{1y}(x, y, f(x, y)) \Big) \, dA \qquad \boxed{6}$$

The right-hand sides of Eq. (5) and Eq. (6) are equal. This proves Eq. (4). ∎

EXAMPLE 1 **Verifying Stokes' Theorem** Verify Stokes' Theorem for

$$\mathbf{F}(x, y, z) = \langle -y, 2x, x + z \rangle$$

and the upper hemisphere with outward-pointing normal vectors (Figure 4):

$$\mathcal{S} = \{(x, y, z) : x^2 + y^2 + z^2 = 1, z \geq 0\}$$

Solution We will show that both the line integral and the surface integral in Stokes' Theorem are equal to 3π.

Step 1. **Compute the line integral around the boundary curve.**
The boundary of $\mathcal{S}$ is the unit circle oriented in the counterclockwise direction with parametrization $\mathbf{r}(t) = \langle \cos t, \sin t, 0 \rangle$. Thus,

$$\mathbf{r}'(t) = \langle -\sin t, \cos t, 0 \rangle$$

$$\mathbf{F}(\mathbf{r}(t)) = \langle -\sin t, 2\cos t, \cos t \rangle$$

$$\mathbf{F}(\mathbf{r}(t)) \cdot \mathbf{r}'(t) = \langle -\sin t, 2\cos t, \cos t \rangle \cdot \langle -\sin t, \cos t, 0 \rangle$$

$$= \sin^2 t + 2\cos^2 t = 1 + \cos^2 t$$

$$\oint_{\partial \mathcal{S}} \mathbf{F} \cdot d\mathbf{r} = \int_0^{2\pi} (1 + \cos^2 t) \, dt = 2\pi + \pi = \boxed{3\pi} \qquad \boxed{7}$$

Step 2. **Compute the curl.**

$$\operatorname{curl}(\mathbf{F}) = \begin{vmatrix} \mathbf{i} & \mathbf{j} & \mathbf{k} \\ \dfrac{\partial}{\partial x} & \dfrac{\partial}{\partial y} & \dfrac{\partial}{\partial z} \\ -y & 2x & x + z \end{vmatrix}$$

$$= \left(\frac{\partial}{\partial y}(x + z) - \frac{\partial}{\partial z} 2x \right) \mathbf{i} - \left(\frac{\partial}{\partial x}(x + z) - \frac{\partial}{\partial z}(-y) \right) \mathbf{j}$$

$$+ \left(\frac{\partial}{\partial x} 2x - \frac{\partial}{\partial y}(-y) \right) \mathbf{k}$$

$$= \langle 0, -1, 3 \rangle$$

Step 3. **Compute the surface integral of the curl.**
We parametrize the hemisphere using spherical coordinates:

$$G(\theta, \phi) = (\cos\theta \sin\phi, \sin\theta \sin\phi, \cos\phi)$$

By Eq. (1) of Section 16.4, the outward-pointing normal vector is

$$\mathbf{N} = \sin\phi \, \langle \cos\theta \sin\phi, \sin\theta \sin\phi, \cos\phi \rangle$$

Therefore,

$$\text{curl}(\mathbf{F}) \cdot \mathbf{N} = \sin\phi \, \langle 0, -1, 3\rangle \cdot \langle \cos\theta\sin\phi, \sin\theta\sin\phi, \cos\phi\rangle$$

$$= -\sin\theta\sin^2\phi + 3\cos\phi\sin\phi$$

The upper hemisphere S corresponds to $0 \le \phi \le \frac{\pi}{2}$ and $0 \le \theta \le 2\pi$, so

$$\iint_S \text{curl}(\mathbf{F}) \cdot d\mathbf{S} = \int_{\phi=0}^{\pi/2} \int_{\theta=0}^{2\pi} (-\sin\theta\sin^2\phi + 3\cos\phi\sin\phi) \, d\theta \, d\phi$$

$$= 0 + 2\pi \int_{\phi=0}^{\pi/2} 3\cos\phi\sin\phi \, d\phi = 2\pi \left(\frac{3}{2}\sin^2\phi \right)\Bigg|_{\phi=0}^{\pi/2}$$

$$= \boxed{3\pi} \qquad \blacksquare$$

EXAMPLE 2 Use Stokes' Theorem to show that $\oint_{\mathcal{C}} \mathbf{F} \cdot d\mathbf{r} = 0$, where

$$\mathbf{F}(x, y, z) = \left\langle \sin(x^2), e^{y^2} + x^2, z^4 + 2x^2 \right\rangle$$

and $\mathcal{C}$ is the boundary of the triangle in Figure 5 with the indicated orientation.

Solution Note that if we wanted to evaluate $\oint \mathbf{F} \cdot d\mathbf{r}$ directly, we would need to parame-trize each of the three edges of $\mathcal{C}$, and do three integrals. Instead, we let S be the triangular surface bounded by $\mathcal{C}$ and apply Stokes' Theorem:

$$\oint_{\mathcal{C}} \mathbf{F} \cdot d\mathbf{r} = \iint_S \text{curl}(\mathbf{F}) \cdot d\mathbf{S}$$

and show that the integral on the right is zero. We first compute the curl:

$$\text{curl}\left(\left\langle \sin x^2, e^{y^2} + x^2, z^4 + 2x^2 \right\rangle\right) = \begin{vmatrix} \mathbf{i} & \mathbf{j} & \mathbf{k} \\ \dfrac{\partial}{\partial x} & \dfrac{\partial}{\partial y} & \dfrac{\partial}{\partial z} \\ \sin x^2 & e^{y^2} + x^2 & z^4 + 2x^2 \end{vmatrix} = \langle 0, -4x, 2x \rangle$$

Now, in this particular case, it turns out that we can show the surface integral is zero without actually computing it. Note that the triangular surface S lies in the plane through $(3, 0, 0)$, $(0, 2, 0)$, and $(0, 0, 1)$. That plane has equation

$$\frac{x}{3} + \frac{y}{2} + z = 1$$

Therefore, $\mathbf{N} = \left\langle \frac{1}{3}, \frac{1}{2}, 1 \right\rangle$ is a normal vector to this plane (Figure 5). But $\mathbf{N}$ and $\text{curl}(\mathbf{F})$ are orthogonal:

$$\text{curl}(\mathbf{F}) \cdot \mathbf{N} = \langle 0, -4x, 2x \rangle \cdot \left\langle \frac{1}{3}, \frac{1}{2}, 1 \right\rangle = -2x + 2x = 0$$

Thus, if $\mathbf{n}$ is a unit vector in the direction of $\mathbf{N}$, then $\text{curl}(\mathbf{F}) \cdot \mathbf{n} = 0$. Furthermore, the given orientation of $\mathcal{C}$ is the boundary orientation associated with $\mathbf{n}$, and therefore

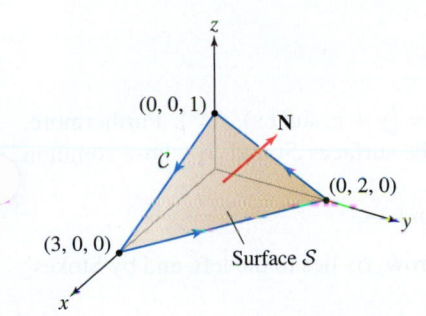

FIGURE 5

← REMINDER For a vector field $\mathbf{G}$,

$$\iint_S \mathbf{G} \cdot d\mathbf{S} = \iint_S \mathbf{G} \cdot \mathbf{n} \, dS \text{ by definition}$$

of the vector surface integral.

$$\oint_{\mathcal{C}} \mathbf{F} \cdot d\mathbf{r} = \iint_S \text{curl}(\mathbf{F}) \cdot d\mathbf{S} = \iint_S \text{curl}(\mathbf{F}) \cdot \mathbf{n} \, dS = \iint_S 0 \, dS = 0 \qquad \blacksquare$$

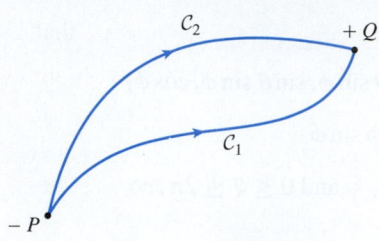

FIGURE 6 Two paths with the same boundary $Q - P$.

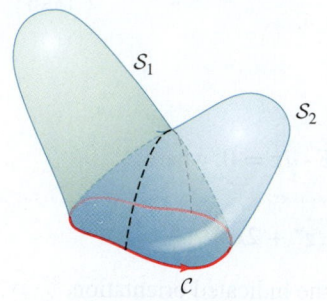

FIGURE 7 Surfaces $\mathcal{S}_1$ and $\mathcal{S}_2$ have the same oriented boundary.

Vector potentials are not unique: If $\mathbf{F} = \text{curl}(\mathbf{A})$, then $\mathbf{F} = \text{curl}(\mathbf{A} + \mathbf{B})$ for any vector field $\mathbf{B}$ such that $\text{curl}(\mathbf{B}) = \mathbf{0}$.

⬅ **REMINDER** By the **flux** of a vector field through a surface, we mean the surface integral of the vector field.

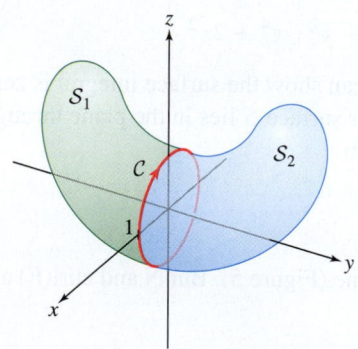

FIGURE 8

CONCEPTUAL INSIGHT Recall that if $\mathbf{F}$ is conservative—that is, $\mathbf{F} = \nabla f$—then for any two paths $\mathcal{C}_1$ and $\mathcal{C}_2$ from P to Q (Figure 6),

$$\int_{\mathcal{C}_1} \mathbf{F} \cdot d\mathbf{r} = \int_{\mathcal{C}_2} \mathbf{F} \cdot d\mathbf{r} = f(Q) - f(P)$$

Thus, the line integral of $\mathbf{F}$ is path independent, and, in particular, $\oint_{\mathcal{C}} \mathbf{F} \cdot d\mathbf{r}$ is zero if $\mathcal{C}$ is closed.

Analogous facts are true for surface integrals of a vector field $\mathbf{F}$ when $\mathbf{F} = \text{curl}(\mathbf{A})$. The vector field $\mathbf{A}$ is called a **vector potential** for $\mathbf{F}$. Stokes' Theorem tells us that for any two surfaces $\mathcal{S}_1$ and $\mathcal{S}_2$ with the same oriented boundary $\mathcal{C}$ (Figure 7),

$$\iint_{\mathcal{S}_1} \mathbf{F} \cdot d\mathbf{S} = \iint_{\mathcal{S}_2} \mathbf{F} \cdot d\mathbf{S} = \oint_{\mathcal{C}} \mathbf{A} \cdot d\mathbf{r}$$

In other words, *the surface integral of a vector field with vector potential $\mathbf{A}$ is surface independent*. Furthermore, if the surface is closed, then the surface integral is zero:

$$\iint_{\mathcal{S}} \mathbf{F} \cdot d\mathbf{S} = 0 \qquad \text{if} \quad \mathbf{F} = \text{curl}(\mathbf{A}) \text{ and } \mathcal{S} \text{ is closed}$$

EXAMPLE 3 Let $\mathbf{F} = \text{curl}(\mathbf{A})$, where $\mathbf{A}(x, y, z) = \langle y + z, \sin(xy), e^{xyz} \rangle$. Furthermore, let $\mathcal{S}$ be the closed surface in Figure 8 made up of the surfaces $\mathcal{S}_1$ and $\mathcal{S}_2$ whose common boundary $\mathcal{C}$ is the unit circle in the xz-plane.

Find the outward flux of $\mathbf{F}$ across each of $\mathcal{S}_1$ and $\mathcal{S}_2$.

Solution With $\mathcal{C}$ oriented in the direction of the arrow, $\mathcal{S}_1$ lies to the left, and by Stokes' Theorem,

$$\iint_{\mathcal{S}_1} \mathbf{F} \cdot d\mathbf{S} = \iint_{\mathcal{S}_1} \text{curl}(\mathbf{A}) \cdot d\mathbf{S} = \oint_{\mathcal{C}} \mathbf{A} \cdot d\mathbf{r}$$

We shall compute the line integral on the right. The parametrization $\mathbf{r}(t) = \langle \cos t, 0, \sin t \rangle$ traces $\mathcal{C}$ in the direction of the arrow because it begins at $\mathbf{r}(0) = \langle 1, 0, 0 \rangle$ and moves in the direction of $\mathbf{r}\left(\frac{\pi}{2}\right) = \langle 0, 0, 1 \rangle$. We have

$$\mathbf{A}(\mathbf{r}(t)) = \langle 0 + \sin t, \sin(0), e^0 \rangle = \langle \sin t, 0, 1 \rangle$$

$$\mathbf{A}(\mathbf{r}(t)) \cdot \mathbf{r}'(t) = \langle \sin t, 0, 1 \rangle \cdot \langle -\sin t, 0, \cos t \rangle = -\sin^2 t + \cos t$$

$$\oint_{\mathcal{C}} \mathbf{A} \cdot d\mathbf{r} = \int_0^{2\pi} (-\sin^2 t + \cos t) \, dt = -\pi$$

We conclude that $\iint_{\mathcal{S}_1} \mathbf{F} \cdot d\mathbf{S} = -\pi$.

Since $\mathcal{S}$ is closed and $\mathbf{F}$ is the curl of a vector field $\mathbf{A}$, Stokes' Theorem implies that the outward flux of $\mathbf{F}$ across $\mathcal{S}$ is zero; that is, $\iint_{\mathcal{S}} \mathbf{F} \cdot d\mathbf{S} = 0$. Furthermore, $\iint_{\mathcal{S}} \mathbf{F} \cdot d\mathbf{S} = \iint_{\mathcal{S}_1} \mathbf{F} \cdot d\mathbf{S} + \iint_{\mathcal{S}_2} \mathbf{F} \cdot d\mathbf{S}$; therefore,

$$\iint_{\mathcal{S}_2} \mathbf{F} \cdot d\mathbf{S} = -\iint_{\mathcal{S}_1} \mathbf{F} \cdot d\mathbf{S} = \pi$$

Thus, the outward flux of $\mathbf{F}$ across $\mathcal{S}_1$ is $-\pi$ and across $\mathcal{S}_2$ is π. ∎

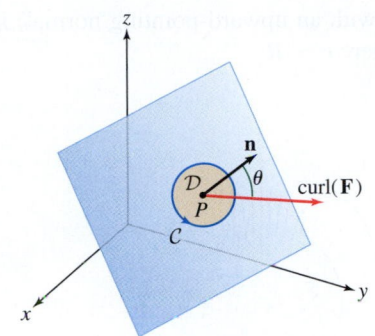

FIGURE 9 The circle C is centered at P and lies in the plane through P with normal vector $\mathbf{n}$.

CONCEPTUAL INSIGHT **Interpretation of the Curl** In Section 17.1, we showed that the quantity $\dfrac{\partial F_2}{\partial x} - \dfrac{\partial F_1}{\partial y}$ in Green's Theorem is the circulation per unit area. A similar interpretation is valid in $\mathbf{R}^3$.

Consider a plane through a point P with unit normal vector $\mathbf{n}$. Let C be a small circle in the plane, centered at P, and enclosing region $\mathcal{D}$ (Figure 9). By Stokes' Theorem,

$$\oint_C \mathbf{F} \cdot d\mathbf{r} \approx \iint_{\mathcal{D}} \mathrm{curl}(\mathbf{F}) \cdot \mathbf{n}\, dS \qquad \boxed{8}$$

The vector field $\mathrm{curl}(\mathbf{F})$ is continuous, and if C is sufficiently small, we can approximate $\mathrm{curl}(\mathbf{F})$ by the constant value $\mathrm{curl}(\mathbf{F})(P)$, and thus

$$\iint_{\mathcal{D}} (\mathrm{curl}\mathbf{F}) \cdot \mathbf{n}\, dS \approx \iint_{\mathcal{D}} \mathrm{curl}(\mathbf{F})(P) \cdot \mathbf{n}\, dS \qquad \boxed{9}$$
$$\approx (\mathrm{curl}(\mathbf{F})(P) \cdot \mathbf{n})\, \mathrm{area}(\mathcal{D})$$

Since $\mathrm{curl}(\mathbf{F})(P) \cdot \mathbf{n} = \|\mathrm{curl}(\mathbf{F})(P)\| \cos\theta$, where θ is the angle between $\mathrm{curl}(\mathbf{F})$ and $\mathbf{n}$, Eqs. (8) and (9) give us the approximations

$$\mathrm{curl}(\mathbf{F})(P) \cdot \mathbf{n} \approx \frac{1}{\mathrm{area}(\mathcal{D})} \oint_C \mathbf{F} \cdot d\mathbf{r} \quad \text{and} \quad \|\mathrm{curl}(\mathbf{F})(P)\|(\cos\theta) \approx \frac{1}{\mathrm{area}(\mathcal{D})} \oint_C \mathbf{F} \cdot d\mathbf{r}$$

This is a remarkable result. It tells us that $\mathrm{curl}(\mathbf{F})$ encodes the circulation per unit area in every plane through P in a simple way—namely, as the dot product $\mathrm{curl}(\mathbf{F})(P) \cdot \mathbf{n}$. In particular, the circulation rate is directly related to the cosine of the angle θ between $\mathrm{curl}(\mathbf{F})(P)$ and $\mathbf{n}$.

We can also argue (as in Section 17.1 for vector fields in the plane) that if $\mathbf{F}$ is the velocity field of a fluid, then a small paddle wheel with normal $\mathbf{n}$ will rotate with an angular velocity of approximately $\frac{1}{2}\mathrm{curl}(\mathbf{F})(P) \cdot \mathbf{n}$ (see Figure 10). At any given point, the angular velocity is maximized when the normal vector to the paddle wheel $\mathbf{n}$ points in the direction of $\mathrm{curl}(\mathbf{F})$.

FIGURE 10 The paddle wheel can be oriented in different ways, as specified by the normal vector $\mathbf{n}$. In the direction of $\mathrm{curl}(\mathbf{F})$ it rotates the fastest.

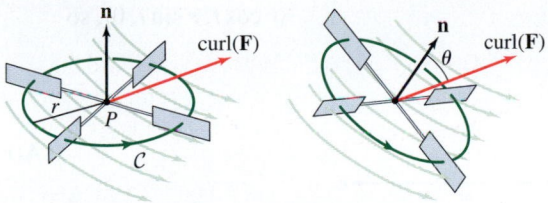

EXAMPLE 4 **Vector Potential for a Solenoid** An electric current I flowing through a solenoid (a tightly wound spiral of wire; see Figure 11) creates a magnetic field $\mathbf{B}$. If we assume that the solenoid is infinitely long, with radius R and the z-axis as the central axis, then

$$\mathbf{B}(r) = \begin{cases} \mathbf{0} & \text{if } r > R \\ B\mathbf{k} & \text{if } r < R \end{cases}$$

where $r = (x^2 + y^2)^{1/2}$ is the distance to the z-axis, and B is a constant that depends on the current strength I and the spacing of the turns of wire.

(a) Show that a vector potential for $\mathbf{B}$ is

$$\mathbf{A}(r) = \begin{cases} \dfrac{1}{2} R^2 B \left\langle -\dfrac{y}{r^2}, \dfrac{x}{r^2}, 0 \right\rangle & \text{if } r > R \\[2mm] \dfrac{1}{2} B \left\langle -y, x, 0 \right\rangle & \text{if } r < R \end{cases}$$

(b) Calculate the flux of **B** through the surface $\mathcal{S}$ (with an upward-pointing normal) in Figure 11 whose boundary is a circle of radius r, where $r > R$.

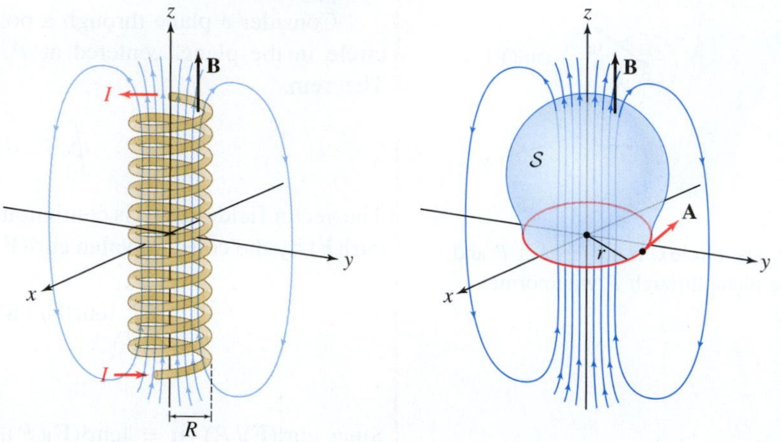

FIGURE 11 The magnetic field of a long solenoid is nearly uniform inside and weak outside. In practice, we treat the solenoid as infinitely long if it is very long in comparison with its radius.

Solution

(a) For any functions f and g,

$$\text{curl}(\langle f, g, 0 \rangle) = \langle -g_z, f_z, g_x - f_y \rangle$$

Applying this to **A** for $r < R$, we obtain

$$\text{curl}(\mathbf{A}) = \frac{1}{2} B \left\langle 0, 0, \frac{\partial}{\partial x} x - \frac{\partial}{\partial y}(-y) \right\rangle = \langle 0, 0, B \rangle = B\mathbf{k} = \mathbf{B}$$

We leave it as an exercise (Exercise 35) to show that $\text{curl}(\mathbf{A}) = \mathbf{B} = \mathbf{0}$ for $r > R$.

(b) The boundary of $\mathcal{S}$ is a circle with counterclockwise parametrization $\mathbf{r}(t) = \langle r \cos t, r \sin t, 0 \rangle$, so

$$\mathbf{r}'(t) = \langle -r \sin t, r \cos t, 0 \rangle$$

$$\mathbf{A}(\mathbf{r}(t)) = \frac{1}{2} R^2 B r^{-1} \langle -\sin t, \cos t, 0 \rangle$$

$$\mathbf{A}(\mathbf{r}(t)) \cdot \mathbf{r}'(t) = \frac{1}{2} R^2 B \left((-\sin t)^2 + \cos^2 t \right) = \frac{1}{2} R^2 B$$

By Stokes' Theorem, the flux of **B** through $\mathcal{S}$ is equal to

$$\iint_{\mathcal{S}} \mathbf{B} \cdot d\mathbf{S} = \oint_{\partial \mathcal{S}} \mathbf{A} \cdot d\mathbf{r} = \int_0^{2\pi} \mathbf{A}(\mathbf{r}(t)) \cdot \mathbf{r}'(t)\, dt = \frac{1}{2} R^2 B \int_0^{2\pi} dt = \pi R^2 B \qquad \blacksquare$$

*The vector potential **A** is continuous but not differentiable on the cylinder $r = R$, that is, on the solenoid itself (Figure 12). The magnetic field $\mathbf{B} = \text{curl}(\mathbf{A})$ has a jump discontinuity where $r = R$. We take for granted the fact that Stokes' Theorem remains valid in this setting.*

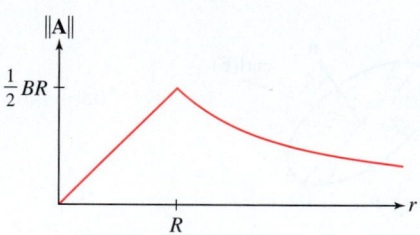

FIGURE 12 The magnitude $\|\mathbf{A}\|$ of the vector potential as a function of distance r to the z-axis.

CONCEPTUAL INSIGHT   There is an interesting difference between scalar and vector potentials. If $\mathbf{F} = \nabla f$, then the scalar potential f is constant in regions where the field **F** is zero (since a function with zero gradient is constant). This is not true for vector potentials. As we saw in Example 4, the magnetic field **B** produced by a solenoid is zero everywhere outside the solenoid, but the vector potential **A** is not constant outside the solenoid. In fact, **A** is proportional to $\left\langle -\frac{y}{r^2}, \frac{x}{r^2}, 0 \right\rangle$. This is related to an intriguing phenomenon in physics called the *Aharonov–Bohm (AB) effect*, first proposed on theoretical grounds in the 1940s.

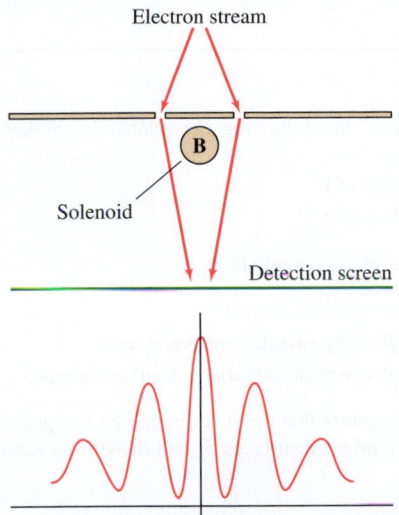

Electron stream

B

Solenoid

Detection screen

FIGURE 13 A stream of electrons passing through a double slit produces an interference pattern on the detection screen. The pattern shifts slightly when an electric current flows through the solenoid.

According to electromagnetic theory, a magnetic field **B** exerts a force on a moving electron, causing a deflection in the electron's path. We do not expect any deflection when an electron moves past a solenoid because **B** is zero outside the solenoid (in practice, the field is not actually zero for a solenoid of finite length, but it is very small—we ignore this difficulty). However, according to quantum mechanics, electrons have both particle and wave properties. In a double-slit experiment, a stream of electrons passing through two small slits creates a wavelike interference pattern on a detection screen (Figure 13).

The AB effect predicts that if we place a small solenoid between the slits as in the figure (the solenoid is so small that the electrons never pass through it), then the interference pattern will shift slightly. It is as if the electrons are "aware" of the magnetic field inside the solenoid, even though they never encounter the field directly.

The AB effect was hotly debated until it was confirmed definitively in 1985, in experiments carried out by a team of Japanese physicists led by Akira Tonomura. The AB effect appeared to contradict classical electromagnetic theory, according to which the trajectory of an electron is determined by **B** alone. There is no such contradiction in quantum mechanics, because the behavior of the electrons is governed not by **B** but by a "wave function" derived from the nonconstant vector potential **A**.

17.2 SUMMARY

- The *boundary* of a surface S is denoted ∂S. We say that S is *closed* if ∂S is empty.
- Suppose that S is oriented (a continuously varying unit normal is specified at each point of S). The *boundary orientation* of ∂S is defined as follows: If you walk along the boundary in the positive direction with your head pointing in the normal direction, then the surface is on your left.
- Stokes' Theorem relates the circulation around the boundary to the surface integral of the curl:

$$\oint_{\partial S} \mathbf{F} \cdot d\mathbf{r} = \iint_{S} \operatorname{curl}(\mathbf{F}) \cdot d\mathbf{S}$$

- Surface independence: If $\mathbf{F} = \operatorname{curl}(\mathbf{A})$, then the flux of $\mathbf{F}$ through a surface S depends only on the oriented boundary ∂S and not on the surface itself:

$$\iint_{S} \mathbf{F} \cdot d\mathbf{S} = \iint_{S} \operatorname{curl}(\mathbf{A}) \cdot d\mathbf{S} = \oint_{\partial S} \mathbf{A} \cdot d\mathbf{r}$$

In particular, if S is *closed* (i.e., ∂S is empty) and $\mathbf{F} = \operatorname{curl}(\mathbf{A})$, then $\iint_{S} \mathbf{F} \cdot d\mathbf{S} = 0$.

If S_1 and S_2 are oriented surfaces that share an oriented boundary and $\mathbf{F} = \operatorname{curl}(\mathbf{A})$, then

$$\iint_{S_1} \mathbf{F} \cdot d\mathbf{S} = \iint_{S_2} \mathbf{F} \cdot d\mathbf{S}$$

- The curl is interpreted as a vector that encodes circulation per unit area: If P is any point and **n** is a unit normal vector, then

$$\operatorname{curl}(\mathbf{F})(P) \cdot \mathbf{n} = \|\operatorname{curl}(\mathbf{F})(P)\|(\cos\theta) \approx \frac{1}{\operatorname{area}(\mathcal{D})} \oint_{C} \mathbf{F} \cdot d\mathbf{r}$$

where C is a small circle centered at P in the plane through P with normal vector **n**, $\mathcal{D}$ is the region enclosed by C, and θ is the angle between $\operatorname{curl}(\mathbf{F})(P)$ and **n**.

17.2 EXERCISES

Preliminary Questions

1. Indicate with an arrow the boundary orientation of the boundary curves of the surfaces in Figure 14, oriented by the outward-pointing normal vectors.

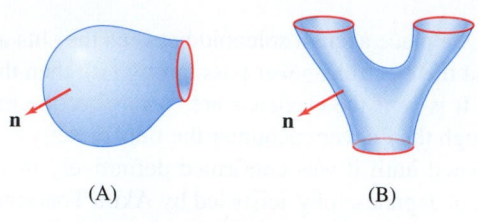

(A)   (B)

FIGURE 14

2. Let $\mathbf{F} = \operatorname{curl}(\mathbf{A})$. Which of the following are related by Stokes' Theorem?

(a) The circulation of $\mathbf{A}$ and flux of $\mathbf{F}$

(b) The circulation of $\mathbf{F}$ and flux of $\mathbf{A}$

3. What is the definition of a vector potential?

4. Which of the following statements is correct?

(a) The flux of $\operatorname{curl}(\mathbf{A})$ through every oriented surface is zero.

(b) The flux of $\operatorname{curl}(\mathbf{A})$ through every closed, oriented surface is zero.

5. Which condition on $\mathbf{F}$ guarantees that the flux through $\mathcal{S}_1$ is equal to the flux through $\mathcal{S}_2$ for any two oriented surfaces $\mathcal{S}_1$ and $\mathcal{S}_2$ with the same oriented boundary?

Exercises

In Exercises 1–4, verify Stokes' Theorem for the given vector field and surface, oriented with an upward-pointing normal.

1. $\mathbf{F} = \langle 2xy, x, y+z \rangle$, the surface $z = 1 - x^2 - y^2$ for $x^2 + y^2 \le 1$

2. $\mathbf{F} = \langle yz, 0, x \rangle$, the portion of the plane $\dfrac{x}{2} + \dfrac{y}{3} + z = 1$, where $x, y, z \ge 0$

3. $\mathbf{F} = \langle e^{y-z}, 0, 0 \rangle$, the square with vertices $(1,0,1)$, $(1,1,1)$, $(0,1,1)$, and $(0,0,1)$

4. $\mathbf{F} = \langle y, x, x^2 + y^2 \rangle$, the upper hemisphere $x^2 + y^2 + z^2 = 1$, $z \ge 0$

In Exercises 5–10, calculate $\operatorname{curl}(\mathbf{F})$ and then apply Stokes' Theorem to compute the flux of $\operatorname{curl}(\mathbf{F})$ through the given surface using a line integral.

5. $\mathbf{F} = \left\langle e^{z^2} - y, e^{z^3} + x, \cos(xz) \right\rangle$, the upper half of the unit sphere $x^2 + y^2 + z^2 = 1$, $z \ge 0$ with outward-pointing normal

6. $\mathbf{F} = \left\langle x+y, z^2 - 4, x\sqrt{y^2 + 1} \right\rangle$, surface of the wedge-shaped box in Figure 15 (bottom included, top excluded) with outward-pointing normal

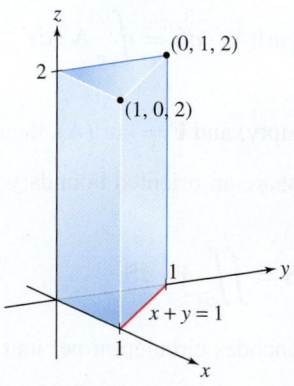

FIGURE 15

7. $\mathbf{F} = \langle 3z, 5x, -2y \rangle$, that part of the paraboloid $z = x^2 + y^2$ that lies below the plane $z = 4$ with upward-pointing unit normal vector

8. $\mathbf{F} = \langle yz, -xz, z^3 \rangle$, that part of the cone $z = \sqrt{x^2 + y^2}$ that lies between the two planes $z = 1$ and $z = 3$ with upward-pointing unit normal vector

9. $\mathbf{F} = \langle yz, xz, xy \rangle$, that part of the cylinder $x^2 + y^2 = 1$ that lies between the two planes $z = 1$ and $z = 4$ with outward-pointing unit normal vector

10. $\mathbf{F} = \langle 2y, e^z, -\arctan x \rangle$, that part of the paraboloid $z = 4 - x^2 - y^2$ cut off by the xy-plane with upward-pointing unit normal vector

In Exercises 11–16, apply Stokes' Theorem to evaluate $\displaystyle\oint_C \mathbf{F} \cdot d\mathbf{r}$ by finding the flux of $\operatorname{curl}(\mathbf{F})$ across an appropriate surface.

11. $\mathbf{F} = \langle 3y, -2x, 3y \rangle$, $\ C$ is the circle $x^2 + y^2 = 9, z = 2$, oriented counterclockwise as viewed from above.

12. $\mathbf{F} = \langle yz, xy, xz \rangle$, $\ C$ is the square with vertices $(0,0,2)$, $(1,0,2)$, $(1,1,2)$, and $(0,1,2)$, oriented counterclockwise as viewed from above.

13. $\mathbf{F} = \langle xz, xy, yz \rangle$, $\ C$ is the rectangle with vertices $(0,0,0)$, $(0,0,2)$, $(3,0,2)$, and $(3,0,0)$, oriented counterclockwise as viewed from the positive y-axis.

14. $\mathbf{F} = \langle y + 2x, 2x + 5z, 7y + 8x \rangle$, $\ C$ is the circle with radius 5, center at $(2,0,0)$, in the plane $x = 2$, and oriented counterclockwise as viewed from the origin $(0,0,0)$.

15. $\mathbf{F} = \langle y, z, x \rangle$, $\ C$ is the triangle with vertices $(0,0,0)$, $(3,0,0)$, and $(0,3,3)$, oriented counterclockwise as viewed from above.

16. $\mathbf{F} = \langle y, -2z, 4x \rangle$, $\ C$ is the boundary of that portion of the plane $x + 2y + 3z = 1$ that is in the first octant of space, oriented counterclockwise as viewed from above.

17. Let $\mathcal{S}$ be the surface of the cylinder (not including the top and bottom) of radius 2 for $1 \le z \le 6$, oriented with outward-pointing normal (Figure 16).

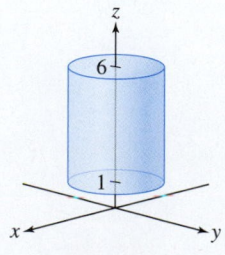

FIGURE 16

(a) Indicate with an arrow the orientation of ∂S (the top and bottom circles).

(b) Verify Stokes' Theorem for S and $\mathbf{F} = \langle yz^2, 0, 0 \rangle$.

18. Let S be the portion of the plane $z = x$ contained in the half-cylinder of radius R depicted in Figure 17. Use Stokes' Theorem to calculate the circulation of $\mathbf{F} = \langle z, x, y + 2z \rangle$ around the boundary of S (a half-ellipse) in the counterclockwise direction when viewed from above. *Hint:* Show that curl($\mathbf{F}$) is orthogonal to the normal vector to the plane.

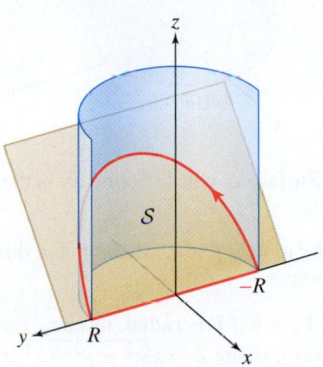

FIGURE 17

19. Let I be the flux of $\mathbf{F} = \langle e^y, 2xe^{x^2}, z^2 \rangle$ through the upper hemisphere S of the unit sphere.

(a) Let $\mathbf{G} = \langle e^y, 2xe^{x^2}, 0 \rangle$. Find a vector field $\mathbf{A}$ such that curl($\mathbf{A}$) = $\mathbf{G}$.

(b) Use Stokes' Theorem to show that the flux of $\mathbf{G}$ through S is zero. *Hint:* Calculate the circulation of $\mathbf{A}$ around ∂S.

(c) Calculate I. *Hint:* Use (b) to show that I is equal to the flux of $\langle 0, 0, z^2 \rangle$ through S.

20. Let $\mathbf{F} = \langle 0, -z, 1 \rangle$. Let S be the spherical cap $x^2 + y^2 + z^2 \leq 1$, where $z \geq \frac{1}{2}$. Evaluate $\iint_S \mathbf{F} \cdot d\mathbf{S}$ directly as a surface integral. Then verify that $\mathbf{F} = $ curl($\mathbf{A}$), where $\mathbf{A} = \langle 0, x, xz \rangle$ and evaluate the surface integral again using Stokes' Theorem.

21. Let $\mathbf{A}$ be the vector potential and $\mathbf{B}$ the magnetic field of the infinite solenoid of radius R in Example 4. Use Stokes' Theorem to compute:

(a) The flux of $\mathbf{B}$ through a surface whose boundary is a circle in the xy-plane of radius $r < R$

(b) The circulation of $\mathbf{A}$ around the boundary $\mathcal{C}$ of a surface lying outside the solenoid

22. The magnetic field $\mathbf{B}$ due to a small current loop (which we place at the origin) is called a **magnetic dipole** (Figure 18). For ρ large, $\mathbf{B} = $ curl($\mathbf{A}$), where

$$\mathbf{A} = \left\langle -\frac{y}{\rho^3}, \frac{x}{\rho^3}, 0 \right\rangle \text{ and } \rho = \sqrt{x^2 + y^2 + z^2}$$

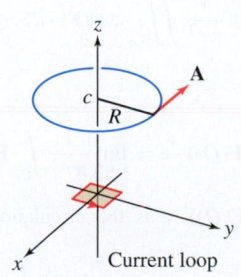

FIGURE 18

(a) Let $\mathcal{C}$ be a horizontal circle of radius R with center $(0, 0, c)$, where c is large. Show that $\mathbf{A}$ is tangent to $\mathcal{C}$.

(b) Use Stokes' Theorem to calculate the flux of $\mathbf{B}$ through $\mathcal{C}$.

23. A uniform magnetic field $\mathbf{B}$ has constant strength b in the z-direction [i.e., $\mathbf{B} = \langle 0, 0, b \rangle$].

(a) Verify that $\mathbf{A} = \frac{1}{2}\mathbf{B} \times \mathbf{r}$ is a vector potential for $\mathbf{B}$, where $\mathbf{r} = \langle x, y, 0 \rangle$.

(b) Calculate the flux of $\mathbf{B}$ through the rectangle with vertices A, B, C, and D in Figure 19.

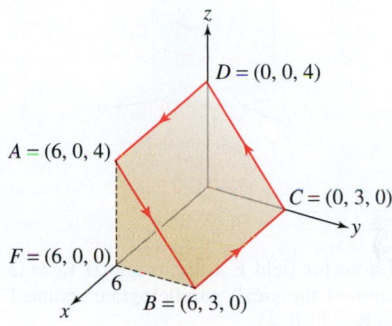

FIGURE 19

24. Let $\mathbf{F} = \langle -x^2y, x, 0 \rangle$. Referring to Figure 19, let $\mathcal{C}$ be the closed path $ABCD$. Use Stokes' Theorem to evaluate $\int_{\mathcal{C}} \mathbf{F} \cdot d\mathbf{r}$ in two ways. First, regard $\mathcal{C}$ as the boundary of the rectangle with vertices A, B, C, and D. Then treat $\mathcal{C}$ as the boundary of the wedge-shaped box with an open top.

25. Let $\mathbf{F} = \langle y^2, 2z + x, 2y^2 \rangle$. Use Stokes' Theorem to find a plane with equation $ax + by + cz = 0$ (where a, b, c are not all zero) such that $\oint_{\mathcal{C}} \mathbf{F} \cdot d\mathbf{r} = 0$ for every closed $\mathcal{C}$ lying in the plane. *Hint:* Choose a, b, c so that curl($\mathbf{F}$) lies in the plane.

26. Let $\mathbf{F} = \langle -z^2, 2zx, 4y - x^2 \rangle$, and let $\mathcal{C}$ be a simple closed curve in the plane $x + y + z = 4$ that encloses a region of area 16 (Figure 20). Calculate $\oint_{\mathcal{C}} \mathbf{F} \cdot d\mathbf{r}$, where $\mathcal{C}$ is oriented in the counterclockwise direction (when viewed from above the plane).

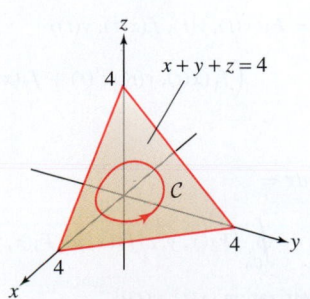

FIGURE 20

27. Let $\mathbf{F} = \langle y^2, x^2, z^2 \rangle$. Show that

$$\int_{\mathcal{C}_1} \mathbf{F} \cdot d\mathbf{r} = \int_{\mathcal{C}_2} \mathbf{F} \cdot d\mathbf{r}$$

for any two closed curves going around a cylinder whose central axis is the z-axis as shown in Figure 21.

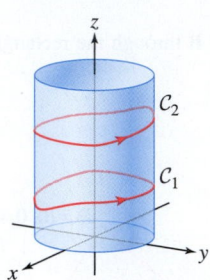

FIGURE 21

28. The curl of a vector field $\mathbf{F}$ at the origin is $\mathbf{v}_0 = \langle 3, 1, 4 \rangle$. Estimate the circulation around the small parallelogram spanned by the vectors $\mathbf{A} = \langle 0, \frac{1}{2}, \frac{1}{2} \rangle$ and $\mathbf{B} = \langle 0, 0, \frac{1}{3} \rangle$.

29. You know two things about a vector field $\mathbf{F}$:

(i) $\mathbf{F}$ has a vector potential $\mathbf{A}$ (but $\mathbf{A}$ is unknown).

(ii) The circulation of $\mathbf{A}$ around the unit circle (oriented counterclockwise) is 25.

Determine the flux of $\mathbf{F}$ through the surface $\mathcal{S}$ in Figure 22, oriented with an upward-pointing normal.

30. Suppose that $\mathbf{F}$ has a vector potential and that $\mathbf{F}(x, y, 0) = \mathbf{k}$. Find the flux of $\mathbf{F}$ through the surface $\mathcal{S}$ in Figure 22, oriented with an upward-pointing normal.

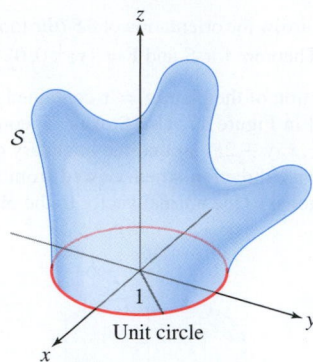

FIGURE 22 Surface $\mathcal{S}$ whose boundary is the unit circle.

31. Prove that $\operatorname{curl}(f\mathbf{a}) = \nabla f \times \mathbf{a}$, where f is a differentiable function and $\mathbf{a}$ is a constant vector.

32. Show that $\operatorname{curl}(\mathbf{F}) = \mathbf{0}$ if $\mathbf{F}$ is **radial**, meaning that $\mathbf{F} = f(\rho) \langle x, y, z \rangle$ for some function $f(\rho)$, where $\rho = \sqrt{x^2 + y^2 + z^2}$. *Hint:* It is enough to show that one component of $\operatorname{curl}(\mathbf{F})$ is zero, because it will then follow for the other two components by symmetry.

33. Prove the following Product Rule:

$$\operatorname{curl}(f\mathbf{F}) = f\operatorname{curl}(\mathbf{F}) + \nabla f \times \mathbf{F}$$

34. Assume that f and g have continuous partial derivatives of order 2. Prove that

$$\oint_{\partial \mathcal{S}} f\nabla g \cdot d\mathbf{r} = \iint_{\mathcal{S}} \nabla f \times \nabla g \cdot d\mathbf{S}$$

35. Verify that $\mathbf{B} = \operatorname{curl}(\mathbf{A})$ for $r > R$ in the setting of Example 4.

36. ⬜ Explain carefully why Green's Theorem is a special case of Stokes' Theorem.

Further Insights and Challenges

37. In this exercise, we use the notation of the proof of Theorem 1 and prove

$$\oint_{\mathcal{C}} F_3(x, y, z)\mathbf{k} \cdot d\mathbf{r} = \iint_{\mathcal{S}} \operatorname{curl}(F_3(x, y, z)\mathbf{k}) \cdot d\mathbf{S} \qquad \boxed{10}$$

In particular, $\mathcal{S}$ is the graph of $z = f(x, y)$ over a domain $\mathcal{D}$, and $\mathcal{C}$ is the boundary of $\mathcal{S}$ with parametrization $(x(t), y(t), f(x(t), y(t)))$.

(a) Use the Chain Rule to show that

$$F_3(x, y, z)\mathbf{k} \cdot d\mathbf{r} = F_3(x(t), y(t), f(x(t), y(t)))$$
$$\left(f_x(x(t), y(t))x'(t) + f_y(x(t), y(t))y'(t) \right) dt$$

and verify that

$$\oint_{\mathcal{C}} F_3(x, y, z)\mathbf{k} \cdot d\mathbf{r} =$$
$$\oint_{\mathcal{C}_0} \left\langle F_3(x, y, z)f_x(x, y), F_3(x, y, z)f_y(x, y) \right\rangle \cdot d\mathbf{r}$$

where $\mathcal{C}_0$ has parametrization $(x(t), y(t))$.

(b) Apply Green's Theorem to the line integral over $\mathcal{C}_0$ and show that the result is equal to the right-hand side of Eq. (10).

38. Let $\mathbf{F}$ be a continuously differentiable vector field in $\mathbf{R}^3$, Q a point, and $\mathcal{S}$ a plane containing Q with unit normal vector $\mathbf{e}$. Let $\mathcal{C}_r$ be a circle of radius r centered at Q in $\mathcal{S}$, and let $\mathcal{S}_r$ be the disk enclosed by $\mathcal{C}_r$. Assume $\mathcal{S}_r$ is oriented with unit normal vector $\mathbf{e}$.

(a) Let $m(r)$ and $M(r)$ be the minimum and maximum values of $\operatorname{curl}(\mathbf{F}(P)) \cdot \mathbf{e}$ for $P \in \mathcal{S}_r$. Prove that

$$m(r) \le \frac{1}{\pi r^2} \iint_{\mathcal{S}_r} \operatorname{curl}(\mathbf{F}) \cdot d\mathbf{S} \le M(r)$$

(b) Prove that

$$\operatorname{curl}(\mathbf{F}(Q)) \cdot \mathbf{e} = \lim_{r \to 0} \frac{1}{\pi r^2} \int_{\mathcal{C}_r} \mathbf{F} \cdot d\mathbf{r}$$

This proves that $\operatorname{curl}(\mathbf{F}(Q)) \cdot \mathbf{e}$ is the circulation per unit area in the plane $\mathcal{S}$.

17.3 Divergence Theorem

We have studied several Fundamental Theorems involving integrals and derivatives. Each of these is a relation of the type:

$$\begin{array}{ccc} \text{Integral of a derivative} & = & \text{Integral over the oriented} \\ \text{on an oriented domain} & & \text{boundary of the domain} \end{array}$$

Here are the examples we have seen so far:

* In single-variable calculus, the Fundamental Theorem of Calculus, Part I (FTC I) relates the integral of $f'(x)$ over an interval $[a, b]$ to the "integral" of $f(x)$ over the boundary of $[a, b]$ consisting of two points a and b:

$$\underbrace{\int_a^b f'(x)\,dx}_{\text{Integral of derivative over } [a,b]} = \underbrace{f(b) - f(a)}_{\text{"Integral" over the boundary of } [a,b]}$$

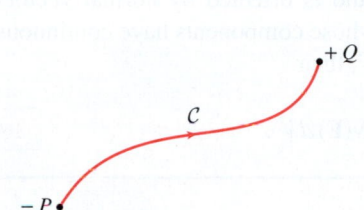

FIGURE 1 The oriented boundary of $\mathcal{C}$ is $\partial\mathcal{C} = Q - P$.

The boundary of $[a, b]$ is oriented by assigning a plus sign to b and a minus sign to a.

* The Fundamental Theorem for Conservative Vector Fields generalizes FTC I: Instead of taking an integral over an interval $[a, b]$ (a path from a to b along the x-axis), we take an integral along any path from points P to Q in $\mathbf{R}^3$ (Figure 1), and instead of $f'(x)$, we use the gradient:

$$\underbrace{\int_{\mathcal{C}} \nabla f \cdot d\mathbf{r}}_{\text{Integral of derivative over a curve}} = \underbrace{f(Q) - f(P)}_{\substack{\text{"Integral" over the} \\ \text{boundary } \partial\mathcal{C} = Q - P}}$$

* Green's Theorem is a two-dimensional version of FTC I that relates the integral of a certain derivative over a domain $\mathcal{D}$ in the plane to an integral over its boundary curve $\mathcal{C} = \partial\mathcal{D}$ (Figure 2):

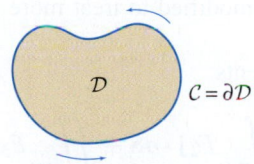

FIGURE 2 Domain $\mathcal{D}$ in $\mathbf{R}^2$ with boundary curve $\mathcal{C} = \partial\mathcal{D}$.

$$\underbrace{\iint_{\mathcal{D}} \left(\frac{\partial F_2}{\partial x} - \frac{\partial F_1}{\partial y} \right) dA}_{\text{Integral of derivative over domain}} = \underbrace{\int_{\mathcal{C}} \mathbf{F} \cdot d\mathbf{r}}_{\text{Integral over boundary curve}}$$

* Stokes' Theorem extends Green's Theorem: Instead of a domain in the plane (a flat surface), we allow any surface in $\mathbf{R}^3$ (Figure 3). The appropriate derivative is the curl:

$$\underbrace{\iint_{\mathcal{S}} \text{curl}(\mathbf{F}) \cdot d\mathbf{S}}_{\text{Integral of derivative over surface}} = \underbrace{\int_{\mathcal{C}} \mathbf{F} \cdot d\mathbf{r}}_{\text{Integral over boundary curve}}$$

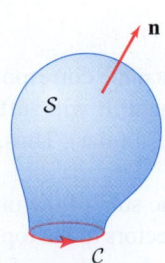

FIGURE 3 The oriented boundary of $\mathcal{S}$ is $\mathcal{C} = \partial\mathcal{S}$.

Our last theorem—the Divergence Theorem—also follows this pattern:

$$\underbrace{\iiint_{\mathcal{W}} \text{div}(\mathbf{F})\,dV}_{\text{Integral of derivative over three-dimensional region}} = \underbrace{\iint_{\mathcal{S}} \mathbf{F} \cdot d\mathbf{S}}_{\text{Integral over boundary surface}}$$

Here, $\mathcal{S}$ is a closed surface that encloses a three-dimensional region $\mathcal{W}$. In other words, $\mathcal{S}$ is the boundary of $\mathcal{W}$, so $\mathcal{S} = \partial\mathcal{W}$. Figure 4 shows two examples of regions and boundary surfaces that we will consider.

We consider a piecewise smooth closed surface $\mathcal{S}$, which means $\mathcal{S}$ consists of one smooth surface or at most finitely many smooth surfaces that have been glued together along their boundaries, as in the example of the cube.

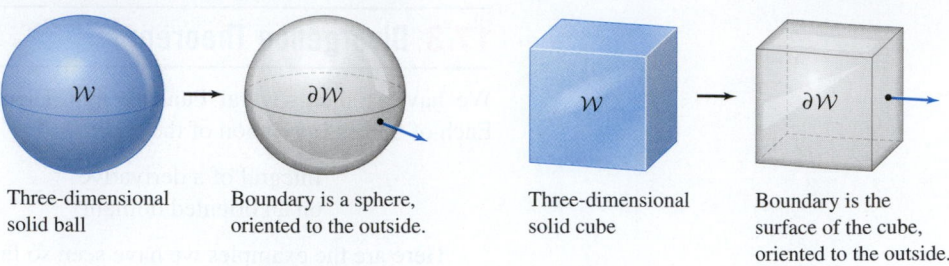

Three-dimensional solid ball

Boundary is a sphere, oriented to the outside.

Three-dimensional solid cube

Boundary is the surface of the cube, oriented to the outside.

FIGURE 4

THEOREM 1 Divergence Theorem   Let $\mathcal{S}$ be a closed surface that encloses a region $\mathcal{W}$ in $\mathbf{R}^3$. Assume that $\mathcal{S}$ is piecewise smooth and is oriented by normal vectors pointing to the outside of $\mathcal{W}$. If $\mathbf{F}$ is a vector field whose components have continuous partial derivatives in an open domain containing $\mathcal{W}$, then

$$\iint_{\mathcal{S}} \mathbf{F} \cdot d\mathbf{S} = \iiint_{\mathcal{W}} \operatorname{div}(\mathbf{F})\, dV \qquad \boxed{1}$$

With the notation $\nabla \cdot \mathbf{F} = \operatorname{div}(\mathbf{F})$, the Divergence Theorem is also written in the form

$$\iint_{\mathcal{S}} \mathbf{F} \cdot d\mathbf{S} = \iiint_{\mathcal{W}} \nabla \cdot \mathbf{F}\, dV$$

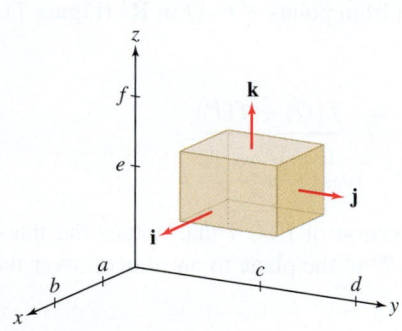

FIGURE 5 A box $\mathcal{W} = [a,b] \times [c,d] \times [e,f]$.

Proof We prove the Divergence Theorem in the special case that $\mathcal{W}$ is a rectangular box $[a,b] \times [c,d] \times [e,f]$ as in Figure 5. The proof can be modified to treat more general regions such as the interiors of spheres and cylinders.

We write each side of Eq. (1) as a sum over components:

$$\iint_{\partial\mathcal{W}} (F_1\mathbf{i} + F_2\mathbf{j} + F_3\mathbf{k}) \cdot d\mathbf{S} = \iint_{\partial\mathcal{W}} F_1\mathbf{i} \cdot d\mathbf{S} + \iint_{\partial\mathcal{W}} F_2\mathbf{j} \cdot d\mathbf{S} + \iint_{\partial\mathcal{W}} F_3\mathbf{k} \cdot d\mathbf{S}$$

$$\iiint_{\mathcal{W}} \operatorname{div}(F_1\mathbf{i} + F_2\mathbf{j} + F_3\mathbf{k})\, dV = \iiint_{\mathcal{W}} \operatorname{div}(F_1\mathbf{i})\, dV + \iiint_{\mathcal{W}} \operatorname{div}(F_2\mathbf{j})\, dV$$
$$+ \iiint_{\mathcal{W}} \operatorname{div}(F_3\mathbf{k})\, dV$$

As in the proofs of Green's and Stokes' Theorems, we show that the corresponding terms in the sums on the right-hand sides of these equations are equal. It will suffice to carry out the argument for the $\mathbf{i}$ terms (the other two components are similar). Thus, we assume that $\mathbf{F} = F_1\mathbf{i}$.

The surface integral over the boundary $\mathcal{S}$ of the box is the sum of the integrals over the six faces. However, $\mathbf{F} = F_1\mathbf{i}$ is orthogonal to the normal vectors to the top and bottom as well as the two side faces because $\mathbf{F} \cdot \mathbf{j} = \mathbf{F} \cdot \mathbf{k} = 0$. Therefore, the surface integrals over these faces are zero. Nonzero contributions come only from the front and rear faces, which we denote $\mathcal{S}_f$ and $\mathcal{S}_r$ (Figure 6):

$$\iint_{\mathcal{S}} \mathbf{F} \cdot d\mathbf{S} = \iint_{\mathcal{S}_f} \mathbf{F} \cdot d\mathbf{S} + \iint_{\mathcal{S}_r} \mathbf{F} \cdot d\mathbf{S}$$

To evaluate these integrals, we parametrize $\mathcal{S}_f$ and $\mathcal{S}_r$ by

$$G_f(y,z) = (b,y,z), \qquad c \le y \le d,\ e \le z \le f$$
$$G_r(y,z) = (a,y,z), \qquad c \le y \le d,\ e \le z \le f$$

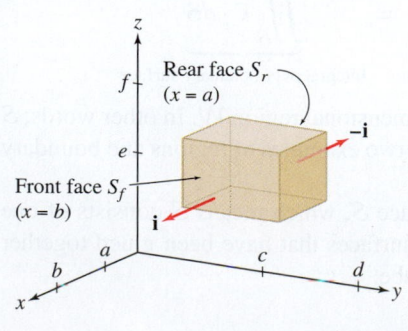

Rear face $\mathcal{S}_r$ $(x=a)$

Front face $\mathcal{S}_f$ $(x=b)$

FIGURE 6

The normal vectors for these parametrizations are

$$\frac{\partial G_f}{\partial y} \times \frac{\partial G_f}{\partial z} = \mathbf{j} \times \mathbf{k} = \mathbf{i}$$

$$\frac{\partial G_r}{\partial y} \times \frac{\partial G_r}{\partial z} = \mathbf{j} \times \mathbf{k} = \mathbf{i}$$

On $\mathcal{S}_r$, we take $\mathbf{n} = -\mathbf{i}$ as the outward-pointing unit normal, and therefore $\mathbf{F} \cdot d\mathbf{S} = (F_1 \mathbf{i}) \cdot (-\mathbf{i}) dS = -F_1 dy\, dz$. Thus,

$$\iint_{\mathcal{S}_f} \mathbf{F} \cdot d\mathbf{S} + \iint_{\mathcal{S}_r} \mathbf{F} \cdot d\mathbf{S} = \int_e^f \int_c^d F_1(b, y, z)\, dy\, dz - \int_e^f \int_c^d F_1(a, y, z)\, dy\, dz$$

$$= \int_e^f \int_c^d \Big(F_1(b, y, z) - F_1(a, y, z) \Big)\, dy\, dz$$

By the Fundamental Theorem of Calculus, Part I,

$$F_1(b, y, z) - F_1(a, y, z) = \int_a^b \frac{\partial F_1}{\partial x}(x, y, z)\, dx$$

Since $\mathrm{div}(\mathbf{F}) = \mathrm{div}(F_1 \mathbf{i}) = \dfrac{\partial F_1}{\partial x}$, we obtain the desired result:

$$\iint_{\mathcal{S}} \mathbf{F} \cdot d\mathbf{S} = \int_e^f \int_c^d \int_a^b \frac{\partial F_1}{\partial x}(x, y, z)\, dx\, dy\, dz = \iiint_{\mathcal{W}} \mathrm{div}(\mathbf{F})\, dV \qquad \blacksquare$$

The names attached to mathematical theorems often conceal a more complex historical development. What we call Green's Theorem was stated by Augustin Cauchy in 1846, but it was never stated by George Green himself (he published a result that implies Green's Theorem in 1828). Stokes' Theorem first appeared as a problem on a competitive exam written by George Stokes at Cambridge University, but William Thomson (Lord Kelvin) had previously stated the theorem in a letter to Stokes. Gauss published special cases of the Divergence Theorem in 1813 and later in 1833 and 1839, while the general theorem was stated and proved by the Russian mathematician Michael Ostrogradsky in 1826. For this reason, the Divergence Theorem is also referred to as Gauss's Theorem or the Gauss–Ostrogradsky Theorem.

EXAMPLE 1 **Verifying the Divergence Theorem** Verify Theorem 1 for $\mathbf{F}(x, y, z) = \langle y, yz, z^2 \rangle$ and the closed cylinder in Figure 7.

Solution We must verify that the flux $\displaystyle\iint_{\mathcal{S}} \mathbf{F} \cdot d\mathbf{S}$, where $\mathcal{S}$ is the surface of the cylinder, is equal to the integral of $\mathrm{div}(\mathbf{F})$ over the cylinder. We compute the flux through $\mathcal{S}$ first: It is the sum of three surface integrals over the side, the top, and the bottom.

Step 1. **Integrate over the side of the cylinder.**
We use the standard parametrization of the cylinder:

$$G(\theta, z) = (2\cos\theta, 2\sin\theta, z), \qquad 0 \le \theta < 2\pi, \quad 0 \le z \le 5$$

The normal vector is

$$\mathbf{N} = \mathbf{T}_\theta \times \mathbf{T}_z = \langle -2\sin\theta, 2\cos\theta, 0 \rangle \times \langle 0, 0, 1 \rangle = \langle 2\cos\theta, 2\sin\theta, 0 \rangle$$

and $\mathbf{F}(G(\theta, z)) = \langle y, yz, z^2 \rangle = \langle 2\sin\theta, 2z\sin\theta, z^2 \rangle$. Thus,

$$\mathbf{F} \cdot d\mathbf{S} = \langle 2\sin\theta, 2z\sin\theta, z^2 \rangle \cdot \langle 2\cos\theta, 2\sin\theta, 0 \rangle\, d\theta\, dz$$

$$= (4\cos\theta\sin\theta + 4z\sin^2\theta)\, d\theta\, dz$$

$$\iint_{\text{side}} \mathbf{F} \cdot d\mathbf{S} = \int_0^5 \int_0^{2\pi} (4\cos\theta\sin\theta + 4z\sin^2\theta)\, d\theta\, dz$$

REMINDER In Eq. (2), we use

$$\int_0^{2\pi} \cos\theta\sin\theta\, d\theta = 0$$

$$\int_0^{2\pi} \sin^2\theta\, d\theta = \pi$$

$$= 0 + 4\pi \int_0^5 z\, dz = 4\pi \left(\frac{25}{2} \right) = 50\pi \qquad \boxed{2}$$

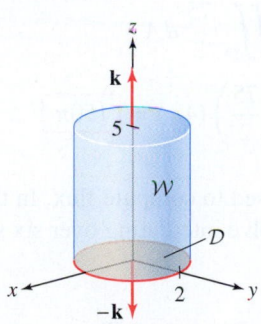

DF **FIGURE 7** Closed cylinder of radius 2 and height 5.

Step 2. **Integrate over the top and bottom of the cylinder.**
The top of the cylinder is at height $z = 5$, so we can parametrize the top by $G(x, y) = (x, y, 5)$ for (x, y) in the disk $\mathcal{D}$ of radius 2:

$$\mathcal{D} = \{(x, y) : x^2 + y^2 \le 4\}$$

Then

$$\mathbf{N} = \mathbf{T}_x \times \mathbf{T}_y = \langle 1, 0, 0 \rangle \times \langle 0, 1, 0 \rangle = \langle 0, 0, 1 \rangle$$

and since $\mathbf{F}(G(x, y)) = \mathbf{F}(x, y, 5) = \langle y, 5y, 5^2 \rangle$, we have

$$\mathbf{F}(G(x, y)) \cdot \mathbf{N} = \langle y, 5y, 5^2 \rangle \cdot \langle 0, 0, 1 \rangle = 25$$

$$\iint_{\text{top}} \mathbf{F} \cdot d\mathbf{S} = \iint_{\mathcal{D}} 25 \, dA = 25 \, \text{area}(\mathcal{D}) = 25(4\pi) = 100\pi$$

Along the bottom disk of the cylinder, we have $z = 0$ and $\mathbf{F}(x, y, 0) = \langle y, 0, 0 \rangle$. It follows that $\mathbf{F}$ is orthogonal to the vector $-\mathbf{k}$ that is normal to the bottom disk, and the integral along the bottom is zero.

Step 3. Find the total flux.

$$\iint_{\mathcal{S}} \mathbf{F} \cdot d\mathbf{S} = \text{side} + \text{top} + \text{bottom} = 50\pi + 100\pi + 0 = \boxed{150\pi}$$

Step 4. Compare with the integral of divergence.

$$\text{div}(\mathbf{F}) = \text{div}\left(\langle y, yz, z^2 \rangle\right) = \frac{\partial}{\partial x} y + \frac{\partial}{\partial y} (yz) + \frac{\partial}{\partial z} z^2 = 0 + z + 2z = 3z$$

The cylinder $\mathcal{W}$ consists of all points (x, y, z) for $0 \le z \le 5$ and (x, y) in the disk $\mathcal{D}$. We see that the integral of the divergence is equal to the total flux as required:

$$\iiint_{\mathcal{W}} \text{div}(\mathbf{F}) \, dV = \iint_{\mathcal{D}} \int_{z=0}^{5} 3z \, dV = \iint_{\mathcal{D}} \frac{75}{2} \, dA$$

$$= \left(\frac{75}{2}\right) (\text{area}(\mathcal{D})) = \left(\frac{75}{2}\right) (4\pi) = \boxed{150\pi} \quad \blacksquare$$

In many applications, the Divergence Theorem is used to compute flux. In the next example, we reduce a flux computation (that would involve integrating over six sides of a box) to a more simple triple integral.

EXAMPLE 2 Using the Divergence Theorem Use the Divergence Theorem to evaluate $\iint_{\mathcal{S}} \langle x^2, z^4, e^z \rangle \cdot d\mathbf{S}$, where $\mathcal{S}$ is the box in Figure 8 that encloses the region $\mathcal{W}$.

Solution First, compute the divergence:

$$\text{div}\left(\langle x^2, z^4, e^z \rangle\right) = \frac{\partial}{\partial x} x^2 + \frac{\partial}{\partial y} z^4 + \frac{\partial}{\partial z} e^z = 2x + e^z$$

Then apply the Divergence Theorem and integrate:

$$\iint_{\mathcal{S}} \langle x^2, z^4, e^z \rangle \cdot d\mathbf{S} = \iiint_{\mathcal{W}} (2x + e^z) \, dV = \int_0^2 \int_0^3 \int_0^1 (2x + e^z) \, dz \, dy \, dx$$

$$= \int_0^2 \int_0^3 (2x + e - 1) \, dy \, dx = \int_0^2 (6x + 3e - 3) \, dx = 6e + 6 \quad \blacksquare$$

EXAMPLE 3 A Vector Field with Zero Divergence Compute the flux of

$$\mathbf{F} = \langle z^2 + xy^2, \cos(x + z), e^{-y} - zy^2 \rangle$$

outward through the boundary of the surface $\mathcal{S}$ in Figure 9.

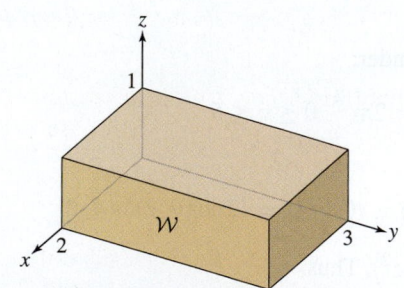

FIGURE 8

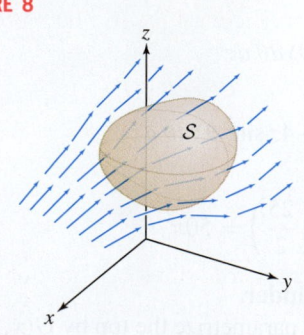

FIGURE 9

Solution Although **F** is rather complicated, its divergence is zero:

$$\text{div}(\mathbf{F}) = \frac{\partial}{\partial x}(z^2 + xy^2) + \frac{\partial}{\partial y}\cos(x + z) + \frac{\partial}{\partial z}(e^{-y} - zy^2) = y^2 - y^2 = 0$$

The Divergence Theorem shows that the flux is zero. Letting $\mathcal{W}$ be the region enclosed by $\mathcal{S}$, we have

$$\iint_{\mathcal{S}} \mathbf{F} \cdot d\mathbf{S} = \iiint_{\mathcal{W}} \text{div}(\mathbf{F})\, dV = \iiint_{\mathcal{W}} 0\, dV = 0 \qquad \blacksquare$$

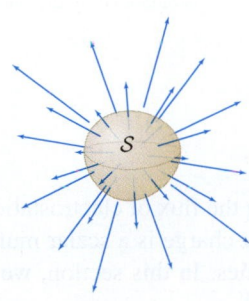

FIGURE 10 For a velocity field, the flux through a surface is the flow rate (in volume per unit time) of fluid across the surface.

GRAPHICAL INSIGHT **Interpretation of Divergence** Let's assume again that **F** is the velocity field of a fluid (Figure 10). Then the flux of **F** through a surface $\mathcal{S}$ is the flow rate (volume of fluid passing through $\mathcal{S}$ per unit time). If $\mathcal{S}$ encloses a region $\mathcal{W}$, then by the Divergence Theorem,

$$\text{flow rate across } \mathcal{S} = \iint_{\mathcal{S}} \mathbf{F} \cdot d\mathbf{S} = \iiint_{\mathcal{W}} \text{div}(\mathbf{F})\, dV \qquad \boxed{3}$$

Now, assume that $\mathcal{S}$ is a small sphere centered at a point P. Because $\text{div}(\mathbf{F})$ is continuous, we can approximate $\text{div}(\mathbf{F})$ on $\mathcal{W}$ by the constant value $\text{div}(\mathbf{F})(P)$. This gives us the approximation

$$\iint_{\mathcal{S}} \mathbf{F} \cdot d\mathbf{S} = \iiint_{\mathcal{W}} \text{div}(\mathbf{F})\, dV \approx \text{div}(\mathbf{F})(P) \cdot \text{Vol}(\mathcal{W}) \qquad \boxed{4}$$

Therefore,

$$\text{div}(\mathbf{F})(P) \approx \frac{1}{\text{Vol}(\mathcal{W})} \iint_{\mathcal{S}} \mathbf{F} \cdot d\mathbf{S}$$

Thus, the divergence is approximately the flow rate across a small sphere, divided by the volume of the sphere. The approximation becomes more exact as the sphere shrinks, and therefore $\text{div}(\mathbf{F})(P)$ has an interpretation as *outward flow rate (or flux) of* **F** *per unit volume near* P.

- If $\text{div}(\mathbf{F})(P) > 0$, there is a net outflow of fluid across any small closed surface enclosing P, or, in other words, a net "creation" of fluid near P. In this case, we call P a *source*.

  Because of this, $\text{div}(\mathbf{F})$ is sometimes called the *source density* of the field.

- If $\text{div}(\mathbf{F})(P) < 0$, there is a net inflow of fluid across any small closed surface enclosing P, or, in other words, a net "destruction" of fluid near P. In this case, we call P a *sink*.

- If $\text{div}(\mathbf{F})(P) = 0$, then the net flow across any small closed surface enclosing P is approximately zero. A vector field such that $\text{div}(\mathbf{F}) = 0$ everywhere is called incompressible.

To visualize these cases, consider the two-dimensional situation, where

$$\text{div}(\langle F_1, F_2 \rangle) = \frac{\partial F_1}{\partial x} + \frac{\partial F_2}{\partial y}$$

Do the units match up in Eq. (4)? The flow rate has units of volume per unit time. On the other hand, the divergence is a sum of derivatives of velocity with respect to distance. Therefore, the divergence has units of distance per unit time per unit distance, or units time^{-1}, and the right-hand side of Eq. (4) also has units of volume per unit time.

In Figure 11, field (A) has positive divergence. There is a positive net flow of fluid across every circle per unit time. Similarly, field (B) has negative divergence. By contrast, field (C) is incompressible. The fluid flowing into every circle is balanced by the fluid flowing out.

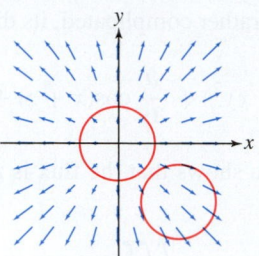

(A) The field $\mathbf{F} = \langle x, y \rangle$ with div($\mathbf{F}$) = 2. There is a net outflow through every circle.

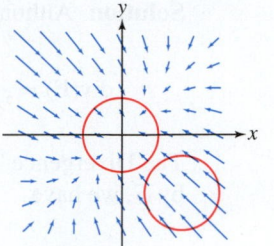

(B) The field $\mathbf{F} = \langle y - 2x, x - 2y \rangle$ with div($\mathbf{F}$) = −4. There is a net inflow into every circle.

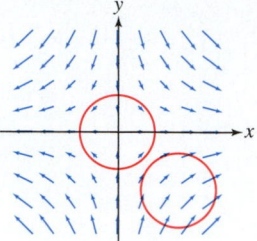

(C) The field $\mathbf{F} = \langle x, -y \rangle$ with div($\mathbf{F}$) = 0. The inflow is balanced by the outflow.

FIGURE 11

Applications to Electrostatics

The Divergence Theorem is a powerful tool for computing the flux of electrostatic fields. This is due to the fact that the electrostatic field of a point charge is a scalar multiple of the inverse-square vector field, which has special properties. In this section, we denote the inverse-square vector field by $\mathbf{F}_{IS}$:

←**REMINDER**

$$r = \sqrt{x^2 + y^2 + z^2}$$

For $r \neq 0$,

$$\mathbf{e}_r = \frac{\langle x, y, z \rangle}{r} = \frac{\langle x, y, z \rangle}{\sqrt{x^2 + y^2 + z^2}}$$

$$\boxed{\mathbf{F}_{IS} = \frac{\mathbf{e}_r}{r^2} = \frac{\mathbf{r}}{r^3}}$$

The unit radial vector field $\mathbf{e}_r$ appears in Figure 12. Note that $\mathbf{F}_{IS}$ is defined only for $r \neq 0$. The next example verifies the key property that div($\mathbf{F}_{IS}$) = 0.

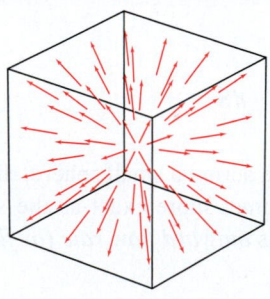

FIGURE 12 Unit radial vector field $\mathbf{e}_r$.

EXAMPLE 4   The Inverse-Square Vector Field   Verify that $\mathbf{F}_{IS} = \dfrac{\mathbf{e}_r}{r^2}$ has zero divergence:

$$\text{div}\left(\frac{\mathbf{e}_r}{r^2}\right) = 0$$

Solution   Write the field as

$$\mathbf{F}_{IS} = \langle F_1, F_2, F_3 \rangle = \frac{1}{r^2}\left\langle \frac{x}{r}, \frac{y}{r}, \frac{z}{r} \right\rangle = \langle xr^{-3}, yr^{-3}, zr^{-3} \rangle$$

We have

$$\frac{\partial r}{\partial x} = \frac{\partial}{\partial x}(x^2 + y^2 + z^2)^{1/2} = \frac{1}{2}(x^2 + y^2 + z^2)^{-1/2}(2x) = \frac{x}{r}$$

$$\frac{\partial F_1}{\partial x} = \frac{\partial}{\partial x}(xr^{-3}) = r^{-3} - 3xr^{-4}\frac{\partial r}{\partial x} = r^{-3} - (3xr^{-4})\frac{x}{r} = \frac{r^2 - 3x^2}{r^5}$$

The derivatives $\dfrac{\partial F_2}{\partial y}$ and $\dfrac{\partial F_3}{\partial z}$ are similar, so

$$\text{div}(\mathbf{F}_{IS}) = \frac{r^2 - 3x^2}{r^5} + \frac{r^2 - 3y^2}{r^5} + \frac{r^2 - 3z^2}{r^5} = \frac{3r^2 - 3(x^2 + y^2 + z^2)}{r^5} = 0 \quad ∎$$

The next theorem shows that the flux of $\mathbf{F}_{IS}$ through a closed surface $\mathcal{S}$ depends only on whether $\mathcal{S}$ contains the origin.

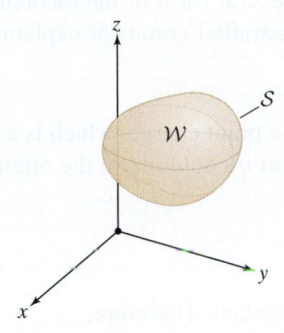

FIGURE 13 $\mathcal{W}$ is contained in the domain of $\mathbf{F}_{IS}$ (away from the origin).

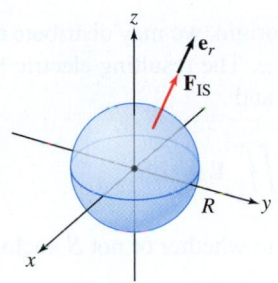

FIGURE 14

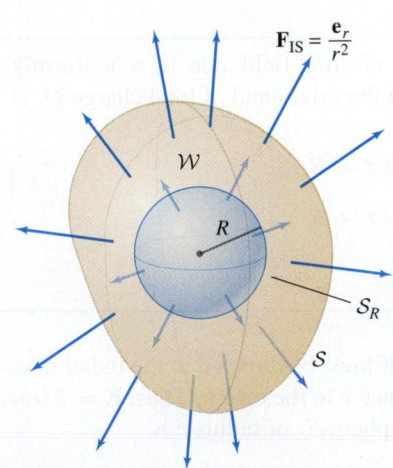

FIGURE 15 $\mathcal{W}$ is the region between $\mathcal{S}$ and the sphere $\mathcal{S}_R$.

THEOREM 2 Flux of the Inverse-Square Field The flux of $\mathbf{F}_{IS} = \dfrac{\mathbf{e}_r}{r^2}$ through closed surfaces has the following remarkable description:

$$\iint_{\mathcal{S}} \left(\frac{\mathbf{e}_r}{r^2}\right) \cdot d\mathbf{S} = \begin{cases} 4\pi & \text{if } \mathcal{S} \text{ encloses the origin} \\ 0 & \text{if } \mathcal{S} \text{ does not enclose the origin} \end{cases}$$

Proof First, assume that $\mathcal{S}$ does not enclose the origin (Figure 13). Then the region $\mathcal{W}$ enclosed by $\mathcal{S}$ is contained in the domain of $\mathbf{F}_{IS}$ and we can apply the Divergence Theorem. By Example 4, $\text{div}(\mathbf{F}_{IS}) = 0$, and therefore

$$\iint_{\mathcal{S}} \left(\frac{\mathbf{e}_r}{r^2}\right) \cdot d\mathbf{S} = \iiint_{\mathcal{W}} \text{div}(\mathbf{F}_{IS})\, dV = \iiint_{\mathcal{W}} 0\, dV = 0$$

Next, let us prove the theorem for $\mathcal{S} = \mathcal{S}_R$, the sphere of radius R centered at the origin (Figure 14). We cannot use the Divergence Theorem because $\mathcal{S}_R$ encloses a point (the origin) where $\mathbf{F}_{IS}$ is not defined. However, we can directly compute the flux of $\mathbf{F}_{IS}$ through $\mathcal{S}_R$ via a surface integral using spherical coordinates. Recall from Section 16.4 [Eq. (2)] that the outward-pointing normal vector to the sphere in spherical coordinates is

$$\mathbf{N} = \mathbf{T}_\phi \times \mathbf{T}_\theta = (R^2 \sin\phi)\mathbf{e}_r$$

The inverse-square field on $\mathcal{S}_R$ is simply $\mathbf{F}_{IS} = R^{-2}\mathbf{e}_r$, and thus

$$\mathbf{F}_{IS} \cdot \mathbf{N} = (R^{-2}\mathbf{e}_r) \cdot (R^2 \sin\phi\, \mathbf{e}_r) = \sin\phi(\mathbf{e}_r \cdot \mathbf{e}_r) = \sin\psi$$

$$\iint_{\mathcal{S}_R} \mathbf{F}_{IS} \cdot d\mathbf{S} = \int_0^{2\pi} \int_0^{\pi} \mathbf{F}_{IS} \cdot \mathbf{N}\, d\phi\, d\theta$$

$$= \int_0^{2\pi} \int_0^{\pi} \sin\phi\, d\phi\, d\theta$$

$$= 2\pi \int_0^{\pi} \sin\phi\, d\phi = 4\pi$$

To extend this result to *any* surface $\mathcal{S}$ enclosing the origin, choose a sphere $\mathcal{S}_R$ whose radius $R > 0$ is so small that $\mathcal{S}_R$ is enclosed inside $\mathcal{S}$. Let $\mathcal{W}$ be the region *between* $\mathcal{S}_R$ and $\mathcal{S}$ (Figure 15). The oriented boundary of $\mathcal{W}$ is the difference:

$$\partial\mathcal{W} = \mathcal{S} - \mathcal{S}_R$$

This means that $\mathcal{S}$ is oriented by outward-pointing normals and $\mathcal{S}_R$ by inward-pointing normals. We have

$$\iint_{\mathcal{S}} \mathbf{F}_{IS} \cdot d\mathbf{S} - \iint_{\mathcal{S}_R} \mathbf{F}_{IS} \cdot d\mathbf{S} = \iint_{\partial\mathcal{W}} \mathbf{F}_{IS} \cdot d\mathbf{S}$$

$$= \iiint_{\mathcal{W}} \text{div}(\mathbf{F}_{IS})\, dV \qquad \text{(Divergence Theorem)}$$

$$= \iiint_{\mathcal{W}} 0\, dV = 0 \qquad [\text{Because div}(\mathbf{F}_{IS}) = 0]$$

This proves that the fluxes through $\mathcal{S}$ and $\mathcal{S}_R$ are equal, and hence both equal 4π.

To prove that the Divergence Theorem is valid for regions between two surfaces, such as the region $\mathcal{W}$ in Figure 15, we cut $\mathcal{W}$ down the middle. Each half is a region enclosed by a surface, so the Divergence Theorem as we have stated it applies. By adding the results for the two halves, we obtain the Divergence Theorem for $\mathcal{W}$. This uses the fact that the fluxes through the common face of the two halves cancel since the common faces have opposite orientations.

Notice that we just applied the Divergence Theorem to a region $\mathcal{W}$ that lies *between two surfaces, one enclosing the other*. This is a more general form of the theorem than the one we stated formally in Theorem 1 above. The marginal comment explains why this is justified. ∎

This result applies directly to the electric field **E** of a point charge, which is a scalar multiple of the inverse-square vector field. For a charge of q coulombs at the origin,

$$\mathbf{E} = \left(\frac{q}{4\pi\epsilon_0} \right) \frac{\mathbf{e}_r}{r^2}$$

where $\epsilon_0 = 8.85 \times 10^{-12}$ C^2/N-m^2 is the permittivity constant. Therefore,

$$\text{flux of } \mathbf{E} \text{ through } \mathcal{S} = \begin{cases} \dfrac{q}{\epsilon_0} & \text{if } q \text{ is inside } \mathcal{S} \\ 0 & \text{if } q \text{ is outside } \mathcal{S} \end{cases}$$

Now, instead of placing just one point charge at the origin, we may distribute a finite number M of point charges q_i at different points in space. The resulting electric field **E** is the sum of the fields $\mathbf{E}_i$ due to the individual charges, and

$$\iint_{\mathcal{S}} \mathbf{E} \cdot d\mathbf{S} = \iint_{\mathcal{S}} \mathbf{E}_1 \cdot d\mathbf{S} + \cdots + \iint_{\mathcal{S}} \mathbf{E}_M \cdot d\mathbf{S}$$

Each integral on the right is either 0 or q_i/ϵ_0, according to whether or not $\mathcal{S}$ encloses q_i, so we conclude that

$$\boxed{\iint_{\mathcal{S}} \mathbf{E} \cdot d\mathbf{S} = \frac{\text{total charge enclosed by } \mathcal{S}}{\epsilon_0}} \qquad \boxed{5}$$

This fundamental relation is called **Gauss's Law**. A limiting argument shows that Eq. (5) remains valid for the electric field due to a *continuous* distribution of charge.

The next theorem, describing the electric field due to a uniformly charged sphere, is a classic application of Gauss's Law.

We proved Theorem 3 in the analogous case of a gravitational field (also a radial inverse-square field) by a laborious calculation in Exercise 48 of Section 16.4. Here, we have derived it from Gauss's Law and a simple appeal to symmetry.

THEOREM 3 Uniformly Charged Sphere The electric field due to a uniformly charged hollow sphere $\mathcal{S}_R$ of radius R, centered at the origin and of total charge Q, is

$$\mathbf{E} = \begin{cases} \left(\dfrac{Q}{4\pi\epsilon_0} \right) \dfrac{\mathbf{e}_r}{r^2} & \text{if } r > R \\ \mathbf{0} & \text{if } r < R \end{cases} \qquad \boxed{6}$$

where $\epsilon_0 = 8.85 \times 10^{-12}$ C^2/N-m^2.

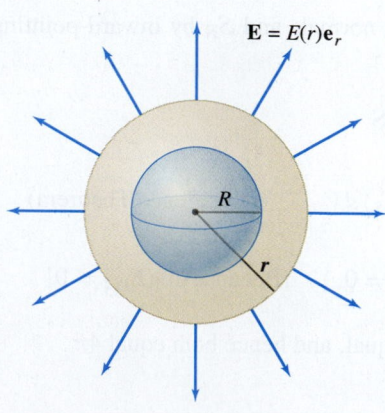

$\mathbf{E} = E(r)\mathbf{e}_r$

FIGURE 16 The electric field due to a uniformly charged sphere.

Proof By symmetry (Figure 16), the electric field **E** must be directed in the radial direction $\mathbf{e}_r$ with magnitude depending only on the distance r to the origin. Thus, $\mathbf{E} = E(r)\mathbf{e}_r$ for some function $E(r)$. The flux of **E** through the sphere $\mathcal{S}_r$ of radius r is

$$\iint_{\mathcal{S}_r} \mathbf{E} \cdot d\mathbf{S} = E(r) \underbrace{\iint_{\mathcal{S}_r} \mathbf{e}_r \cdot d\mathbf{S}}_{\text{Surface area of sphere}} = 4\pi r^2 E(r)$$

By Gauss's Law, this flux is equal to C/ϵ_0, where C is the charge enclosed by $\mathcal{S}_r$. If $r < R$, then $C = 0$ and $\mathbf{E} = \mathbf{0}$. If $r > R$, then $C = Q$ and $4\pi r^2 E(r) = Q/\epsilon_0$, or $E(r) = Q/(\epsilon_0 4\pi r^2)$. This proves Eq. (6). ∎

CONCEPTUAL INSIGHT Here is a summary of the basic operations on functions and vector fields:

$$
f \xrightarrow[\text{gradient}]{\nabla} \mathbf{F} \xrightarrow[\text{curl}]{\nabla \times} \mathbf{G} \xrightarrow[\text{div}]{\nabla \cdot} g
$$

function   vector field   vector field   function

One basic fact is that the result of two consecutive operations in this diagram is zero:

$$\text{curl}(\text{gradient}(f)) = \mathbf{0}, \qquad \text{div}(\text{curl}(\mathbf{F})) = 0$$

$$\nabla \times (\nabla f) = \mathbf{0}, \qquad \nabla \cdot (\nabla \times \mathbf{F}) = 0$$

The first identity follows from Theorem 1 of Section 16.1. The second identity appeared as Exercise 33 in Section 16.1. An interesting question is whether every vector field satisfying curl(**F**) = **0** is necessarily conservative—that is, **F** = ∇f for some function f. The answer is yes, but only if the domain $\mathcal{D}$ is simply connected. For example, in $\mathbf{R}^2$ the vortex field satisfies $\text{curl}_z(\mathbf{F}) = 0$ and yet cannot be conservative because its circulation around the unit circle is nonzero (which is not possible for conservative vector fields since their circulation around a closed path must be zero). However, the domain of the vortex vector field is $\mathbf{R}^2$ with the origin removed, and this domain is not simply connected.

The situation for vector potentials is similar. Can every vector field **G** satisfying div(**G**) = 0 be written in the form **G** = curl(**A**) for some vector potential **A**? Again, the answer is yes—provided that the domain is a region $\mathcal{W}$ in $\mathbf{R}^3$ that has no "holes," a region like a ball, a solid cube, or all of $\mathbf{R}^3$. The inverse-square field $\mathbf{F}_{\text{IS}} = \mathbf{e}_r/r^2$ plays the role of the vortex field in this setting: Although $\text{div}(\mathbf{F}_{\text{IS}}) = 0$, $\mathbf{F}_{\text{IS}}$ cannot have a vector potential over its whole domain because, as shown in Theorem 2, its flux through the unit sphere is nonzero (which is not possible for a vector field with a vector potential since Stokes' Theorem implies that the flux of such a vector field over a closed surface must be zero). In this case, the domain of $\mathbf{F}_{\text{IS}} = \mathbf{e}_r/r^2$ is $\mathbf{R}^3$ with the origin removed, which has a hole.

These properties of the vortex and inverse-square vector fields are significant because they relate line and surface integrals to topological properties of the domain, such as whether the domain is simply connected or has holes. They are a first hint of the important and fascinating connections between vector analysis and the area of mathematics called topology.

17.3 SUMMARY

- The Divergence Theorem: If $\mathcal{W}$ is a region in $\mathbf{R}^3$ whose boundary $\partial \mathcal{W}$ is a surface, oriented by normal vectors pointing outside $\mathcal{W}$, then

$$\iint_{\partial \mathcal{W}} \mathbf{F} \cdot d\mathbf{S} = \iiint_{\mathcal{W}} \text{div}(\mathbf{F}) \, dV$$

- Corollary: If div(**F**) = 0, then **F** has zero flux through the boundary $\partial \mathcal{W}$ of any $\mathcal{W}$ contained in the domain of **F**.
- The divergence div(**F**) is interpreted as flux per unit volume, which means that the flux through a small closed surface containing a point P is approximately equal to div(**F**)(P) times the enclosed volume.
- Basic operations on functions and vector fields:

$$
f \xrightarrow{\nabla} \mathbf{F} \xrightarrow{\text{curl}} \mathbf{G} \xrightarrow{\text{div}} g
$$

function   vector field   vector field   function

- In these cases, the result of two consecutive operations is zero:

$$\mathrm{curl}(\nabla f) = \mathbf{0}, \qquad \mathrm{div}(\mathrm{curl}(\mathbf{F})) = 0$$

- The inverse-square field $\mathbf{F}_{\mathrm{IS}} = \mathbf{e}_r/r^2$, defined for $r \neq 0$, satisfies $\mathrm{div}(\mathbf{F}_{\mathrm{IS}}) = 0$. The flux of $\mathbf{F}_{\mathrm{IS}}$ through a closed surface $\mathcal{S}$ is 4π if $\mathcal{S}$ contains the origin and is zero otherwise.

HISTORICAL PERSPECTIVE

SSPL/The Image Works

—James Clerk Maxwell
(1831–1879)

Vector analysis was developed in the nineteenth century, in large part, to express the laws of electricity and magnetism. Electromagnetism was studied intensively in the period 1750–1890, culminating in the famous Maxwell Equations, which provide a unified understanding in terms of two vector fields: the electric field $\mathbf{E}$ and the magnetic field $\mathbf{B}$. In a region of empty space (where there are no charged particles), the Maxwell Equations are

$$\mathrm{div}(\mathbf{E}) = 0, \qquad\qquad \mathrm{div}(\mathbf{B}) = 0$$
$$\mathrm{curl}(\mathbf{E}) = -\frac{\partial \mathbf{B}}{\partial t}, \qquad \mathrm{curl}(\mathbf{B}) = \mu_0 \epsilon_0 \frac{\partial \mathbf{E}}{\partial t}$$

where μ_0 and ϵ_0 are experimentally determined constants. In SI units,

$$\mu_0 = 4\pi \times 10^{-7} \text{ henries/m}$$
$$\epsilon_0 \approx 8.85 \times 10^{-12} \text{ farads/m}$$

These equations led Maxwell to make two predictions of fundamental importance: (1) that electromagnetic waves exist (this was confirmed by H. Hertz in 1887), and (2) that light is an electromagnetic wave.

How do the Maxwell Equations suggest that electromagnetic waves exist? And why did Maxwell conclude that light is an electromagnetic wave? It was known to mathematicians in the eighteenth century that waves traveling with velocity c may be described by functions $\varphi(x, y, z, t)$ that satisfy the *wave equation*

$$\Delta\varphi = \frac{1}{c^2}\frac{\partial^2\varphi}{\partial t^2} \qquad \boxed{7}$$

where Δ is the Laplace operator (also known as the Laplacian)

$$\Delta\varphi = \frac{\partial^2\varphi}{\partial x^2} + \frac{\partial^2\varphi}{\partial y^2} + \frac{\partial^2\varphi}{\partial z^2}$$

We will show that the components of $\mathbf{E}$ satisfy this wave equation. Take the curl of both sides of Maxwell's third equation:

$$\mathrm{curl}(\mathrm{curl}(\mathbf{E})) = \mathrm{curl}\left(-\frac{\partial \mathbf{B}}{\partial t}\right) = -\frac{\partial}{\partial t}\mathrm{curl}(\mathbf{B})$$

Then apply Maxwell's fourth equation to obtain

$$\mathrm{curl}(\mathrm{curl}(\mathbf{E})) = -\frac{\partial}{\partial t}\left(\mu_0\epsilon_0\frac{\partial \mathbf{E}}{\partial t}\right)$$
$$= -\mu_0\epsilon_0\frac{\partial^2\mathbf{E}}{\partial t^2} \qquad \boxed{8}$$

Finally, let us define the Laplacian of a vector field

$$\mathbf{F} = \langle F_1, F_2, F_3 \rangle$$

by applying the Laplacian Δ to each component, $\Delta\mathbf{F} = \langle \Delta F_1, \Delta F_2, \Delta F_3 \rangle$. Then the following identity holds (see Exercise 36):

$$\mathrm{curl}(\mathrm{curl}(\mathbf{F})) = \nabla(\mathrm{div}(\mathbf{F})) - \Delta\mathbf{F}$$

Applying this identity to $\mathbf{E}$, we obtain $\mathrm{curl}(\mathrm{curl}(\mathbf{E})) = -\Delta\mathbf{E}$ because $\mathrm{div}(\mathbf{E}) = 0$ by Maxwell's first equation. Thus, Eq. (8) yields

$$\boxed{\Delta\mathbf{E} = \mu_0\epsilon_0\frac{\partial^2\mathbf{E}}{\partial t^2}}$$

In other words, each component of the electric field satisfies the wave equation (7), with $c = (\mu_0\epsilon_0)^{-1/2}$. This tells us that the $\mathbf{E}$-field (and similarly the $\mathbf{B}$-field) can propagate through space like a wave, giving rise to electromagnetic radiation (Figure 17).

Maxwell computed the velocity c of an electromagnetic wave:

$$c = (\mu_0\epsilon_0)^{-1/2} \approx 3 \times 10^8 \text{ m/s}$$

and observed that the value is suspiciously close to the velocity of light (first measured by Olaf Römer in 1676). This had to be more than a coincidence, as Maxwell wrote in 1862: "We can scarcely avoid the conclusion that light consists in the transverse undulations of the same medium which is the cause of electric and magnetic phenomena." Needless to say, the wireless technologies that drive our modern society rely on the unseen electromagnetic radiation whose existence Maxwell first predicted on mathematical grounds.

> This is not just mathematical elegance... but beauty.
> It is so simple and yet it describes something so complex.
>
> Francis Collins (1950–), leading geneticist and former director of the Human Genome Project, speaking of the Maxwell Equations.

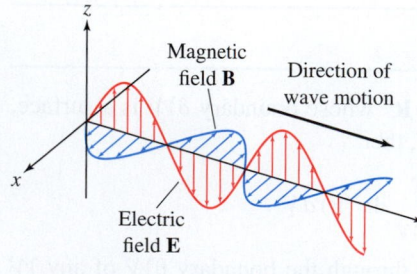

FIGURE 17 The $\mathbf{E}$ and $\mathbf{B}$ fields of an electromagnetic wave along an axis of motion.

Magnetic field $\mathbf{B}$ · Direction of wave motion · Electric field $\mathbf{E}$

17.3 EXERCISES

Preliminary Questions

1. What is the flux of $\mathbf{F} = \langle 1, 0, 0 \rangle$ through a closed surface?

2. Justify the following statement: The flux of $\mathbf{F} = \langle x^3, y^3, z^3 \rangle$ through every closed surface is positive.

3. Which of the following expressions are meaningful (where $\mathbf{F}$ is a vector field and f is a function)? Of those that are meaningful, which are automatically zero?

(a) $\operatorname{div}(\nabla f)$ **(b)** $\operatorname{curl}(\nabla f)$ **(c)** $\nabla \operatorname{curl}(f)$
(d) $\operatorname{div}(\operatorname{curl}(\mathbf{F}))$ **(e)** $\operatorname{curl}(\operatorname{div}(\mathbf{F}))$ **(f)** $\nabla(\operatorname{div}(\mathbf{F}))$

4. Which of the following statements is correct (where $\mathbf{F}$ is a continuously differentiable vector field defined everywhere)?

(a) The flux of $\operatorname{curl}(\mathbf{F})$ through all surfaces is zero.

(b) If $\mathbf{F} = \nabla \varphi$, then the flux of $\mathbf{F}$ through all surfaces is zero.

(c) The flux of $\operatorname{curl}(\mathbf{F})$ through all closed surfaces is zero.

5. How does the Divergence Theorem imply that the flux of the vector field $\mathbf{F} = \langle x^2, y - e^z, y - 2zx \rangle$ through a closed surface is equal to the enclosed volume?

Exercises

In Exercises 1–4, verify the Divergence Theorem for the vector field and region.

1. $\mathbf{F}(x, y, z) = \langle z, x, y \rangle$,   the box $[0, 4] \times [0, 2] \times [0, 3]$

2. $\mathbf{F}(x, y, z) = \langle y, x, z \rangle$,   the region $x^2 + y^2 + z^2 \le 4$

3. $\mathbf{F}(x, y, z) = \langle 2x, 3z, 3y \rangle$,   the region $x^2 + y^2 \le 1, 0 \le z \le 2$

4. $\mathbf{F}(x, y, z) = \langle x, 0, 0 \rangle$,   the region $x^2 + y^2 \le z \le 4$

In Exercises 5–16, use the Divergence Theorem to evaluate the flux $\iint_{\mathcal{S}} \mathbf{F} \cdot d\mathbf{S}$.

5. $\mathbf{F}(x, y, z) = \langle 0, 0, z^3/3 \rangle$, $\mathcal{S}$ is the sphere $x^2 + y^2 + z^2 = 1$.

6. $\mathbf{F}(x, y, z) = \langle y, z, x \rangle$, $\mathcal{S}$ is the sphere $x^2 + y^2 + z^2 = 1$.

7. $\mathbf{F}(x, y, z) = \langle xy^2, yz^2, zx^2 \rangle$, $\mathcal{S}$ is the boundary of the cylinder given by $x^2 + y^2 \le 4, 0 \le z \le 3$.

8. $\mathbf{F}(x, y, z) = \langle x^2z, yx, xyz \rangle$, $\mathcal{S}$ is the boundary of the tetrahedron given by $x + y + z \le 1, 0 \le x, 0 \le y, 0 \le z$.

9. $\mathbf{F}(x, y, z) = \langle x + z^2, xz + y^2, zx - y \rangle$, $\mathcal{S}$ is the surface that bounds the solid region with boundary given by the parabolic cylinder $z = 1 - x^2$, and the planes $z = 0$, $y = 0$, and $z + y = 5$.

10. $\mathbf{F}(x, y, z) = \langle zx, yx^3, x^2z \rangle$, $\mathcal{S}$ is the surface that bounds the solid region with boundary given by $y = 4 - x^2 - z^2$, $y = 0$.

11. $\mathbf{F}(x, y, z) = \langle x^3, 0, z^3 \rangle$, $\mathcal{S}$ is the boundary of the region in the first octant of space given by $x^2 + y^2 + z^2 \le 4$, $x \ge 0$, $y \ge 0$, $z \ge 0$.

12. $\mathbf{F}(x, y, z) = \langle e^{x+y}, e^{x+z}, e^{x+y} \rangle$, $\mathcal{S}$ is the boundary of the unit cube $0 \le x \le 1, 0 \le y \le 1, 0 \le z \le 1$.

13. $\mathbf{F}(x, y, z) = \langle x, y^2, z + y \rangle$, $\mathcal{S}$ is the boundary of the region contained in the cylinder $x^2 + y^2 = 4$ between the planes $z = x$ and $z = 8$.

14. $\mathbf{F}(x, y, z) = \langle x^2 - z^2, e^{z^2} - \cos x, y^3 \rangle$, $\mathcal{S}$ is the boundary of the region bounded by $x + 2y + 4z = 12$ and the coordinate planes in the first octant.

15. $\mathbf{F}(x, y, z) = \langle x + y, z, z - x \rangle$, $\mathcal{S}$ is the boundary of the region between the paraboloid $z = 9 - x^2 - y^2$ and the xy-plane.

16. $\mathbf{F}(x, y, z) = \langle e^{z^2}, 2y + \sin(x^2z), 4z + \sqrt{x^2 + 9y^2} \rangle$, $\mathcal{S}$ is the region $x^2 + y^2 \le z \le 8 - x^2 - y^2$.

17. Calculate the flux of the vector field $\mathbf{F} = 2xy\mathbf{i} - y^2\mathbf{j} + \mathbf{k}$ through the surface $\mathcal{S}$ in Figure 18. *Hint:* Apply the Divergence Theorem to the closed surface consisting of $\mathcal{S}$ and the unit disk.

18. Let $\mathcal{S}_1$ be the closed surface consisting of $\mathcal{S}$ in Figure 18 together with the unit disk. Find the volume enclosed by $\mathcal{S}_1$, assuming that

$$\iint_{\mathcal{S}_1} \langle x, 2y, 3z \rangle \cdot d\mathbf{S} = 72$$

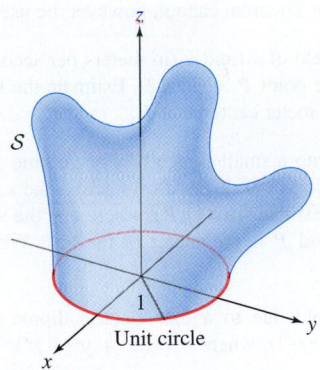

$\mathcal{S}$

Unit circle

FIGURE 18 Surface $\mathcal{S}$ whose boundary is the unit circle.

19. Let $\mathcal{S}$ be the half-cylinder $x^2 + y^2 = 1$, $x \ge 0$, $0 \le z \le 1$. Assume that $\mathbf{F}$ is a horizontal vector field (the z-component is zero) such that $\mathbf{F}(0, y, z) = zy^2\mathbf{i}$. Let $\mathcal{W}$ be the solid region enclosed by $\mathcal{S}$, and assume that

$$\iiint_{\mathcal{W}} \operatorname{div}(\mathbf{F}) \, dV = 4$$

Find the flux of $\mathbf{F}$ through the curved side of $\mathcal{S}$.

20. Volume as a Surface Integral Let $\mathbf{F}(x, y, z) = \langle x, y, z \rangle$. Prove that if $\mathcal{W}$ is a region in $\mathbf{R}^3$ with a smooth boundary $\mathcal{S}$, then

$$\operatorname{volume}(\mathcal{W}) = \frac{1}{3} \iint_{\mathcal{S}} \mathbf{F} \cdot d\mathbf{S} \qquad \boxed{9}$$

21. Use Eq. (9) to calculate the volume of the unit ball as a surface integral over the unit sphere.

22. Verify that Eq. (9) applied to the box $[0, a] \times [0, b] \times [0, c]$ yields the volume $V = abc$.

23. Let $\mathcal{W}$ be the region in Figure 19 bounded by the cylinder $x^2 + y^2 = 4$, the plane $z = x + 1$, and the xy-plane. Use the Divergence Theorem to compute the flux of $\mathbf{F}(x, y, z) = \langle z, x, y + z^2 \rangle$ through the boundary of $\mathcal{W}$.

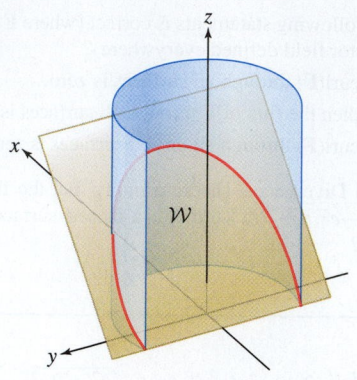

FIGURE 19

24. Let $I = \displaystyle\iint_{\mathcal{S}} \mathbf{F} \cdot d\mathbf{S}$, where

$$\mathbf{F}(x, y, z) = \left\langle \frac{2yz}{r^2}, -\frac{xz}{r^2}, -\frac{xy}{r^2} \right\rangle$$

$(r = \sqrt{x^2 + y^2 + z^2})$ and $\mathcal{S}$ is the boundary of a region $\mathcal{W}$.

(a) Check that $\mathbf{F}$ is divergence free.

(b) [icon] Show that $I = 0$ if $\mathcal{S}$ is a sphere centered at the origin. Explain why the Divergence Theorem cannot, however, be used to prove this.

25. The velocity field of a fluid $\mathbf{v}$ (in meters per second) has divergence $\mathrm{div}(\mathbf{v})(P) = 3$ at the point $P = (2, 2, 2)$. Estimate the flow rate out of the sphere of radius 0.5 meter centered at P.

26. A hose feeds into a small screen box of volume 10 cm³ that is suspended in a swimming pool. Water flows across the surface of the box at a rate of 12 cm³/s. Estimate $\mathrm{div}(\mathbf{v})(P)$, where $\mathbf{v}$ is the velocity field of the water in the pool and P is the center of the box. What are the units of $\mathrm{div}(\mathbf{v})(P)$?

27. The electric field due to a unit electric dipole oriented in the $\mathbf{k}$-direction is $\mathbf{E} = \nabla(z/r^3)$, where $r = (x^2 + y^2 + z^2)^{1/2}$ (Figure 20). Let $\mathbf{e}_r = r^{-1} \langle x, y, z \rangle$.

(a) Show that $\mathbf{E} = r^{-3}\mathbf{k} - 3zr^{-4}\mathbf{e}_r$.

(b) Calculate the flux of $\mathbf{E}$ through a sphere centered at the origin.

(c) Calculate $\mathrm{div}(\mathbf{E})$.

(d) [icon] Can we use the Divergence Theorem to compute the flux of $\mathbf{E}$ through a sphere centered at the origin?

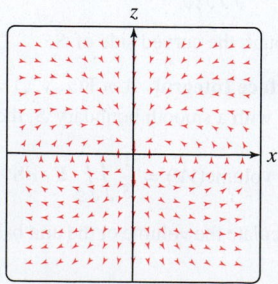

FIGURE 20 The dipole vector field restricted to the xz-plane.

28. Let $\mathbf{E}$ be the electric field due to a long, uniformly charged rod of radius R with charge density δ per unit length (Figure 21). By symmetry, we may assume that $\mathbf{E}$ is everywhere perpendicular to the rod and its magnitude $E(d)$ depends only on the distance d to the rod (strictly speaking, this would hold only if the rod were infinite, but it is nearly true if the rod is long enough). Show that $E(d) = \delta/2\pi\epsilon_0 d$ for $d > R$. *Hint:* Apply Gauss's Law to a cylinder of radius R and of unit length with its axis along the rod.

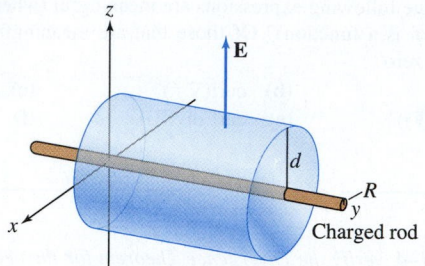

FIGURE 21

29. Let $\mathcal{W}$ be the region between the sphere of radius 4 and the cube of side 1, both centered at the origin. What is the flux through the boundary $\mathcal{S} = \partial\mathcal{W}$ of a vector field $\mathbf{F}$ whose divergence has the constant value $\mathrm{div}(\mathbf{F}) = -4$?

30. Let $\mathcal{W}$ be the region between the sphere of radius 3 and the sphere of radius 2, both centered at the origin. Use the Divergence Theorem to calculate the flux of $\mathbf{F} = x\mathbf{i}$ through the boundary $\mathcal{S} = \partial\mathcal{W}$.

31. Let f be a scalar function and $\mathbf{F}$ be a vector field. Prove the following Product Rule for Divergence:

$$\mathrm{div}(f\mathbf{F}) = f\,\mathrm{div}(\mathbf{F}) + \nabla f \cdot \mathbf{F}$$

32. Let $\mathbf{F}$ and $\mathbf{G}$ be vector fields. Prove the following Product Rule for Divergence:

$$\mathrm{div}(\mathbf{F} \times \mathbf{G}) = \mathrm{curl}(\mathbf{F}) \cdot \mathbf{G} - \mathbf{F} \cdot \mathrm{curl}(\mathbf{G})$$

In Exercises 33 and 34, use the product rules in Exercises 31 and 32. A vector field $\mathbf{F}$ is incompressible *if $\mathrm{div}(\mathbf{F}) = 0$ and is* **irrotational** *if $\mathrm{curl}(\mathbf{F}) = \mathbf{0}$.*

33. Let $\mathbf{F}$ be an incompressible vector field that is everywhere tangent to level surfaces of f. Prove that $f\mathbf{F}$ is incompressible.

34. Prove that the cross product of two irrotational vector fields is incompressible, and explain why this implies that the cross product of two conservative vector fields is incompressible.

In Exercises 35–38, Δ denotes the Laplace operator defined by

$$\Delta\varphi = \frac{\partial^2\varphi}{\partial x^2} + \frac{\partial^2\varphi}{\partial y^2} + \frac{\partial^2\varphi}{\partial z^2}$$

35. Prove the identity

$$\mathrm{div}(\nabla\varphi) = \Delta\varphi$$

36. Prove the identity

$$\mathrm{curl}(\mathrm{curl}(\mathbf{F})) = \nabla(\mathrm{div}(\mathbf{F})) - \Delta\mathbf{F}$$

where $\Delta\mathbf{F}$ denotes $\langle \Delta F_1, \Delta F_2, \Delta F_3 \rangle$.

37. A function φ satisfying $\Delta\varphi = 0$ is called **harmonic**.

(a) Show that $\Delta\varphi = \mathrm{div}(\nabla\varphi)$ for any function φ.

(b) Show that φ is harmonic if and only if $\mathrm{div}(\nabla\varphi) = 0$.

(c) Show that if $\mathbf{F}$ is the gradient of a harmonic function, then $\mathrm{curl}(F) = 0$ and $\mathrm{div}(F) = 0$.

(d) Show that $\mathbf{F}(x, y, z) = \left\langle xz, -yz, \frac{1}{2}(x^2 - y^2) \right\rangle$ is the gradient of a harmonic function. What is the flux of $\mathbf{F}$ through a closed surface?

38. Let $\mathbf{F} = r^n \mathbf{e}_r$, where n is any number, $r = (x^2 + y^2 + z^2)^{1/2}$, and $\mathbf{e}_r = r^{-1} \langle x, y, z \rangle$ is the unit radial vector.

(a) Calculate $\mathrm{div}(\mathbf{F})$.

(b) Calculate the flux of $\mathbf{F}$ through the surface of a sphere of radius R centered at the origin. For which values of n is this flux independent of R?

(c) Prove that $\nabla(r^n) = n \, r^{n-1} \mathbf{e}_r$.

(d) Use (c) to show that $\mathbf{F}$ is conservative for $n \neq -1$. Then show that $\mathbf{F} = r^{-1} \mathbf{e}_r$ is also conservative by computing the gradient of $\ln r$.

(e) What is the value of $\int_C \mathbf{F} \cdot d\mathbf{s}$, where C is a closed curve that does not pass through the origin?

(f) Find the values of n for which the function $\varphi = r^n$ is harmonic.

Further Insights and Challenges

39. Let S be the boundary surface of a region $\mathcal{W}$ in $\mathbf{R}^3$, and let $D_\mathbf{n}\varphi$ denote the directional derivative of φ, where $\mathbf{n}$ is the outward unit normal vector. Let Δ be the Laplace operator defined earlier.

(a) Use the Divergence Theorem to prove that

$$\iint_S D_\mathbf{n}\varphi \, dS = \iiint_\mathcal{W} \Delta\varphi \, dV$$

(b) Show that if φ is a harmonic function (defined in Exercise 37), then

$$\iint_S D_\mathbf{n}\varphi \, dS = 0$$

40. Assume that φ is harmonic. Show that $\mathrm{div}(\varphi\nabla\varphi) = \|\nabla\varphi\|^2$ and conclude that

$$\iint_S \varphi D_\mathbf{n}\varphi \, dS = \iiint_\mathcal{W} \|\nabla\varphi\|^2 \, dV$$

41. Let $\mathbf{F} = \langle P, Q, R \rangle$ be a vector field defined on $\mathbf{R}^3$ such that $\mathrm{div}(\mathbf{F}) = 0$. Use the following steps to show that $\mathbf{F}$ has a vector potential.

(a) Let $\mathbf{A} = \langle f, 0, g \rangle$. Show that

$$\mathrm{curl}(\mathbf{A}) = \left\langle \frac{\partial g}{\partial y}, \frac{\partial f}{\partial z} - \frac{\partial g}{\partial x}, -\frac{\partial f}{\partial y} \right\rangle$$

(b) Fix any value y_0 and show that if we define

$$f(x, y, z) = -\int_{y_0}^y R(x, t, z) \, dt + \alpha(x, z)$$

$$g(x, y, z) = \int_{y_0}^y P(x, t, z) \, dt + \beta(x, z)$$

where α and β are any functions of x and z, then $\partial g/\partial y = P$ and $-\partial f/\partial y = R$.

(c) It remains for us to show that α and β can be chosen so $Q = \partial f/\partial z - \partial g/\partial x$. Verify that the following choice works (for any choice of z_0):

$$\alpha(x, z) = \int_{z_0}^z Q(x, y_0, t) \, dt, \qquad \beta(x, z) = 0$$

Hint: You will need to use the relation $\mathrm{div}(\mathbf{F}) = 0$.

42. Show that

$$\mathbf{F}(x, y, z) = \langle 2y - 1, 3z^2, 2xy \rangle$$

has a vector potential and find one.

43. Show that

$$\mathbf{F}(x, y, z) = \langle 2ye^z - xy, y, yz - z \rangle$$

has a vector potential and find one.

44. In the text, we observed that although the inverse-square radial vector field $\mathbf{F} = \dfrac{\mathbf{e}_r}{r^2}$ satisfies $\mathrm{div}(\mathbf{F}) = 0$, $\mathbf{F}$ cannot have a vector potential on its domain $\{(x, y, z) \neq (0, 0, 0)\}$ because the flux of $\mathbf{F}$ through a sphere containing the origin is nonzero.

(a) Show that the method of Exercise 41 produces a vector potential $\mathbf{A}$ such that $\mathbf{F} = \mathrm{curl}(\mathbf{A})$ on the restricted domain $\mathcal{D}$ consisting of $\mathbf{R}^3$ with the y-axis removed.

(b) Show that $\mathbf{F}$ also has a vector potential on the domains obtained by removing either the x-axis or the z-axis from $\mathbf{R}^3$.

(c) Does the existence of a vector potential on these restricted domains contradict the fact that the flux of $\mathbf{F}$ through a sphere containing the origin is nonzero?

CHAPTER REVIEW EXERCISES

1. Let $\mathbf{F}(x, y) = \langle x + y^2, x^2 - y \rangle$, and let C be the unit circle, oriented counterclockwise. Evaluate $\oint_C \mathbf{F} \cdot d\mathbf{r}$ directly as a line integral and using Green's Theorem.

2. Let $\partial\mathcal{R}$ be the boundary of the rectangle in Figure 1, and let $\partial\mathcal{R}_1$ and $\partial\mathcal{R}_2$ be the boundaries of the two triangles, all oriented counterclockwise.

(a) Determine $\oint_{\partial\mathcal{R}_1} \mathbf{F} \cdot d\mathbf{r}$ if $\oint_{\partial\mathcal{R}} \mathbf{F} \cdot d\mathbf{r} = 4$ and $\oint_{\partial\mathcal{R}_2} \mathbf{F} \cdot d\mathbf{r} = -2$.

(b) What is the value of $\oint_{\partial\mathcal{R}} \mathbf{F} \, d\mathbf{r}$ if $\partial\mathcal{R}$ is oriented clockwise?

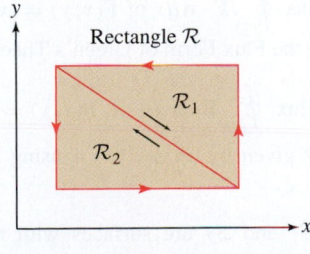

FIGURE 1

In Exercises 3–6, use Green's Theorem to evaluate the line integral around the given closed curve.

3. $\oint_C xy^3\,dx + x^3y\,dy$, where C is the boundary of the rectangle $-1 \le x \le 2, -2 \le y \le 3$, oriented counterclockwise

4. $\oint_C (3x + 5y - \cos y)\,dx + x\sin y\,dy$, where C is any closed curve enclosing a region with area 4, oriented counterclockwise

5. $\oint_C y^2\,dx - x^2\,dy$, where C consists of the arcs $y = x^2$ and $y = \sqrt{x}, 0 \le x \le 1$, oriented clockwise

6. $\oint_C ye^x\,dx + xe^y\,dy$, where C is the triangle with vertices $(-1, 0)$, $(0, 4)$, and $(0, 1)$, oriented counterclockwise

7. Let $\mathbf{r}(t) = \langle t^2(1 - t), t(t - 1)^2 \rangle$.

(a) [GU] Plot the path $\mathbf{r}(t)$ for $0 \le t \le 1$.

(b) Calculate the area A of the region enclosed by $\mathbf{r}(t)$ for $0 \le t \le 1$ using the formula $A = \dfrac{1}{2}\oint_C (x\,dy - y\,dx)$.

8. Calculate the area of the region bounded by the two curves $y = x^2$ and $y = 4$ using the formula $A = \oint_C x\,dy$.

9. Calculate the area of the region bounded by the two curves $y = x^2$ and $y = \sqrt{x}$ for $x \ge 0$ using the formula $A = \oint_C x\,dy$.

10. Calculate the area of the region bounded by the two curves $y = x^2$ and $y = 4$ using the formula $A = \oint_C -y\,dx$.

11. Calculate the area of the region bounded by the two curves $y = x^2$ and $y = \sqrt{x}$ for $x \ge 0$ using the formula $A = \oint_C -y\,dx$.

12. In (a)–(d), state whether the equation is an identity (valid for all $\mathbf{F}$ or f). If it is not, provide an example in which the equation does not hold.

(a) $\text{curl}(\nabla f) = \mathbf{0}$ **(b)** $\text{div}(\nabla f) = 0$

(c) $\text{div}(\text{curl}(\mathbf{F})) = 0$ **(d)** $\nabla(\text{div}(\mathbf{F})) = \mathbf{0}$

13. Let $\mathbf{F}(x, y) = \langle x^2y, xy^2 \rangle$ be the velocity vector field for a fluid in the plane. Find all points where the angular velocity of a small paddle wheel inserted into the fluid would be 0.

14. Compute the flux $\oint_{\partial D} \mathbf{F} \cdot \mathbf{n}\,ds$ of $\mathbf{F}(x, y) = \langle x^3, yx^2 \rangle$ across the unit square D using the Flux Form of Green's Theorem.

15. Compute the flux $\oint_{\partial D} \mathbf{F} \cdot \mathbf{n}\,ds$ of $\mathbf{F}(x, y) = \langle x^3 + 2x, y^3 + y \rangle$ across the circle D given by $x^2 + y^2 = 4$ using the Flux Form of Green's Theorem.

16. Suppose that S_1 and S_2 are surfaces with the same oriented boundary curve C. In each case, does the condition guarantee that the flux of $\mathbf{F}$ through S_1 is equal to the flux of $\mathbf{F}$ through S_2?

(a) $\mathbf{F} = \nabla f$ for some function f

(b) $\mathbf{F} = \text{curl}(\mathbf{G})$ for some vector field $\mathbf{G}$

17. Prove that if $\mathbf{F}$ is a gradient vector field, then the flux of $\text{curl}(\mathbf{F})$ through a smooth surface S (whether closed or not) is equal to zero.

18. Verify Stokes' Theorem for $\mathbf{F}(x, y, z) = \langle y, z - x, 0 \rangle$ and the surface $z = 4 - x^2 - y^2, z \ge 0$, oriented by outward-pointing normals.

19. Let $\mathbf{F}(x, y, z) = \langle z^2, x + z, y^2 \rangle$, and let S be the upper half of the ellipsoid

$$\frac{x^2}{4} + y^2 + z^2 = 1$$

oriented by outward-pointing normals. Use Stokes' Theorem to compute $\iint_S \text{curl}(\mathbf{F}) \cdot d\mathbf{S}$.

20. Use Stokes' Theorem to evaluate $\oint_C \langle y, z, x \rangle \cdot d\mathbf{r}$, where C is the curve in Figure 2.

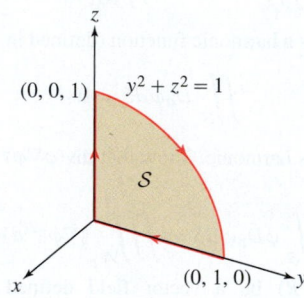

$(0, 0, 1)$ $y^2 + z^2 = 1$

$\mathcal{S}$

$(0, 1, 0)$

FIGURE 2

21. Let S be the side of the cylinder $x^2 + y^2 = 4, 0 \le z \le 2$ (not including the top and bottom of the cylinder). Use Stokes' Theorem to compute the flux of $\mathbf{F}(x, y, z) = \langle 0, y, -z \rangle$ through S (with outward-pointing normal) by finding a vector potential $\mathbf{A}$ such that $\text{curl}(\mathbf{A}) = \mathbf{F}$.

22. Verify the Divergence Theorem for $\mathbf{F}(x, y, z) = \langle 0, 0, z \rangle$ and the region $x^2 + y^2 + z^2 = 1$.

In Exercises 23–26, use the Divergence Theorem to calculate $\iint_S \mathbf{F} \cdot d\mathbf{S}$ for the given vector field and surface.

23. $\mathbf{F}(x, y, z) = \langle xy, yz, x^2z + z^2 \rangle$, S is the boundary of the box $[0, 1] \times [2, 4] \times [1, 5]$.

24. $\mathbf{F}(x, y, z) = \langle xy, yz, x^2z + z^2 \rangle$, S is the boundary of the unit sphere.

25. $\mathbf{F}(x, y, z) = \langle xyz + xy, \frac{1}{2}y^2(1 - z) + e^x, e^{x^2 + y^2} \rangle$, S is the boundary of the solid bounded by the cylinder $x^2 + y^2 = 16$ and the planes $z = 0$ and $z = y - 4$.

26. $\mathbf{F}(x, y, z) = \langle \sin(yz), \sqrt{x^2 + z^4}, x\cos(x - y) \rangle$, S is any smooth closed surface that is the boundary of a region in $\mathbf{R}^3$.

27. Find the volume of a region $\mathcal{W}$ if

$$\iint_{\partial \mathcal{W}} \left\langle x + xy + z, x + 3y - \frac{1}{2}y^2, 4z \right\rangle \cdot d\mathbf{S} = 16$$

28. Show that the circulation of $\mathbf{F}(x, y, z) = \langle x^2, y^2, z(x^2 + y^2) \rangle$ around any curve C on the surface of the cone $z^2 = x^2 + y^2$ is equal to zero (Figure 3).

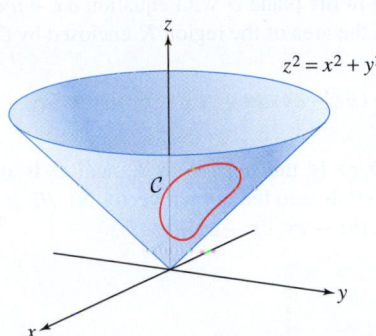

FIGURE 3

In Exercises 29–32, let $\mathbf{F}$ be a vector field whose curl and divergence at the origin are

$$\text{curl}(\mathbf{F})(0, 0, 0) = \langle 2, -1, 4 \rangle, \qquad \text{div}(\mathbf{F})(0, 0, 0) = -2$$

29. Estimate $\oint_C \mathbf{F} \cdot d\mathbf{r}$, where C is the circle of radius 0.03 in the xy-plane centered at the origin.

30. Estimate $\oint_C \mathbf{F} \cdot d\mathbf{r}$, where C is the boundary of the square of side 0.03 in the yz-plane centered at the origin. Does the estimate depend on how the square is oriented within the yz-plane? Might the actual circulation depend on how it is oriented?

31. Suppose that $\mathbf{F}$ is the velocity field of a fluid and imagine placing a small paddle wheel at the origin. Find the equation of the plane in which the paddle wheel should be placed to make it rotate as quickly as possible.

32. Estimate the flux of $\mathbf{F}$ through the box of side 0.5 in Figure 4. Does the result depend on how the box is oriented relative to the coordinate axes?

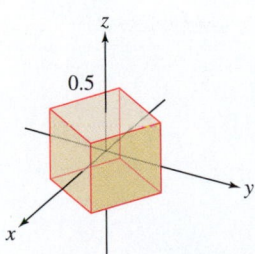

FIGURE 4

33. The velocity vector field of a fluid (in meters per second) is

$$\mathbf{F}(x, y, z) = \langle x^2 + y^2, 0, z^2 \rangle$$

Let $\mathcal{W}$ be the region between the hemisphere

$$\mathcal{S} = \{(x, y, z) : x^2 + y^2 + z^2 = 1, \quad z \geq 0\}$$

and the disk $\mathcal{D} = \{(x, y, 0) : x^2 + y^2 \leq 1\}$ in the xy-plane. Recall that the flow rate of a fluid across a surface is equal to the flux of $\mathbf{F}$ through the surface.

(a) Show that the flow rate across $\mathcal{D}$ is zero.

(b) Use the Divergence Theorem to show that the flow rate across $\mathcal{S}$, oriented with an outward-pointing normal, is equal to

$$\iiint_{\mathcal{W}} \text{div}(\mathbf{F}) \, dV.$$ Then compute this triple integral.

34. The velocity field of a fluid (in meters per second) is

$$\mathbf{F} = (3y - 4)\mathbf{i} + e^{-y(z+1)}\mathbf{j} + (x^2 + y^2)\mathbf{k}$$

(a) Estimate the flow rate (in cubic meters per second) through a small surface $\mathcal{S}$ around the origin if $\mathcal{S}$ encloses a region of volume 0.01 m³.

(b) Estimate the circulation of $\mathbf{F}$ about a circle in the xy-plane of radius $r = 0.1$ m centered at the origin (oriented counterclockwise when viewed from above).

(c) Estimate the circulation of $\mathbf{F}$ about a circle in the yz-plane of radius $r = 0.1$ m centered at the origin (oriented counterclockwise when viewed from the positive x-axis).

35. Let $f(x, y) = x + \dfrac{x}{x^2 + y^2}$. The vector field $\mathbf{F} = \nabla f$ (Figure 5) provides a model in the plane of the velocity field of an incompressible, irrotational fluid flowing past a cylindrical obstacle (in this case, the obstacle is the unit circle $x^2 + y^2 = 1$).

(a) Verify that $\mathbf{F}$ is irrotational [by definition, $\mathbf{F}$ is irrotational if $\text{curl}(\mathbf{F}) = \mathbf{0}$].

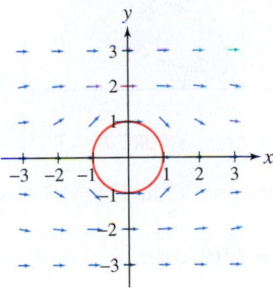

FIGURE 5 The vector field ∇f for $f(x, y) = x + \dfrac{x}{x^2 + y^2}$.

(b) Verify that $\mathbf{F}$ is tangent to the unit circle at each point along the unit circle except $(1, 0)$ and $(-1, 0)$ (where $\mathbf{F} = \mathbf{0}$).

(c) What is the circulation of $\mathbf{F}$ around the unit circle?

(d) Calculate the line integral of $\mathbf{F}$ along the upper and lower halves of the unit circle separately.

36. Figure 6 shows the vector field $\mathbf{F} = \nabla f$, where

$$f(x, y) = \ln\left(x^2 + (y - 1)^2\right) + \ln\left(x^2 + (y + 1)^2\right)$$

which is the velocity field for the flow of a fluid with sources of equal strength at $(0, \pm 1)$ (note that f is undefined at these two points). Show that $\mathbf{F}$ is both irrotational and incompressible—that is, $\text{curl}_z(\mathbf{F}) = 0$ and $\text{div}(\mathbf{F}) = 0$ [in computing $\text{div}(\mathbf{F})$, treat $\mathbf{F}$ as a vector field in $\mathbf{R}^3$ with a zero z-component]. Is it necessary to compute $\text{curl}_z(\mathbf{F})$ to conclude that it is zero?

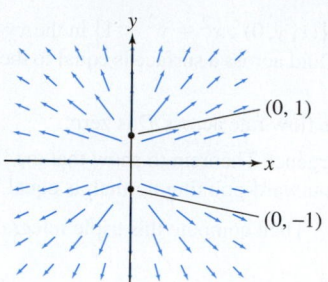

FIGURE 6 The vector field ∇f for $f(x, y) = \ln(x^2 + (y - 1)^2) + \ln(x^2 + (y + 1)^2)$.

37. In Section 17.1, we showed that if C is a simple closed curve, oriented counterclockwise, then the area enclosed by C is given by

$$\text{area enclosed by } C = \frac{1}{2} \oint_C x \, dy - y \, dx \qquad \boxed{1}$$

Suppose that C is a path from P to Q that is not closed but has the property that every line through the origin intersects C in at most one point, as in Figure 7. Let $\mathcal{R}$ be the region enclosed by C and the two radial segments joining P and Q to the origin. Show that the line integral in Eq. (1) is equal to the area of $\mathcal{R}$. *Hint:* Show that the line integral of $\mathbf{F} = \langle -y, x \rangle$ along the two radial segments is zero and apply Green's Theorem.

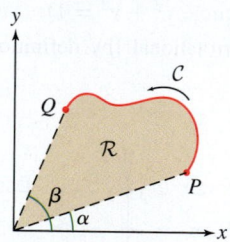

FIGURE 7

38. Suppose that the curve C in Figure 7 has the polar equation $r = f(\theta)$.

(a) Show that $\mathbf{r}(\theta) = \langle f(\theta) \cos \theta, f(\theta) \sin \theta \rangle$ is a counterclockwise parametrization of C.

(b) In Section 11.4, we showed that the area of the region $\mathcal{R}$ is given by the formula

$$\text{area of } \mathcal{R} = \frac{1}{2} \int_\alpha^\beta f(\theta)^2 \, d\theta$$

Use the result of Exercise 37 to give a new proof of this formula. *Hint:* Evaluate the line integral in Eq. (1) using $\mathbf{r}(\theta)$.

39. Prove the following generalization of Eq. (1). Let C be a simple closed curve in the plane $\mathcal{S}$ with equation $ax + by + cz + d = 0$ (Figure 8). Then the area of the region $\mathcal{R}$ enclosed by C is equal to

$$\frac{1}{2\|\mathbf{N}\|} \oint_C (bz - cy) \, dx + (cx - az) \, dy + (ay - bx) \, dz$$

where $\mathbf{N} = \langle a, b, c \rangle$ is the normal to $\mathcal{S}$, and C is oriented as the boundary of $\mathcal{R}$ (relative to the normal vector $\mathbf{N}$). *Hint:* Apply Stokes' Theorem to $\mathbf{F} = \langle bz - cy, cx - az, ay - bx \rangle$.

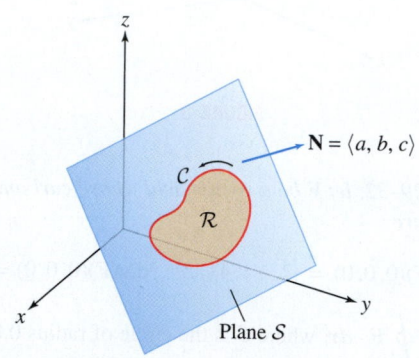

FIGURE 8

40. Use the result of Exercise 39 to calculate the area of the triangle with vertices $(1, 0, 0)$, $(0, 1, 0)$, and $(0, 0, 1)$ as a line integral. Verify your result using geometry.

41. Show that $G(\theta, \phi) = (a \cos \theta \sin \phi, b \sin \theta \sin \phi, c \cos \phi)$ is a parametrization of the ellipsoid

$$\left(\frac{x}{a}\right)^2 + \left(\frac{y}{b}\right)^2 + \left(\frac{z}{c}\right)^2 = 1$$

Then calculate the volume of the ellipsoid as the surface integral of $\mathbf{F} = \frac{1}{3} \langle x, y, z \rangle$ (this surface integral is equal to the volume by the Divergence Theorem).

A THE LANGUAGE OF MATHEMATICS

One of the challenges in learning calculus is growing accustomed to its precise language and terminology, especially in the statements of theorems. In this section, we analyze a few details of logic that are helpful, and indeed essential, in understanding and applying theorems properly.

Many theorems in mathematics involve an **implication**. If A and B are statements, then the implication $A \Rightarrow B$ is the assertion that A implies B:

$$A \Rightarrow B: \qquad \textit{If A is true, then B is true.}$$

Statement A is called the **hypothesis** (or premise) and statement B the **conclusion** of the implication. Here is an example: *If m and n are even integers, then m + n is an even integer*. This statement may be divided into a hypothesis and conclusion:

$$\underbrace{m \text{ and } n \text{ are even integers}}_{A} \quad \Rightarrow \quad \underbrace{m + n \text{ is an even integer}}_{B}$$

In everyday speech, implications are often used in a less precise way. An example is: *If you work hard, then you will succeed*. Furthermore, some statements that do not initially have the form $A \Rightarrow B$ may be restated as implications. For example, the statement, "Cats are mammals," can be rephrased as follows:

$$\text{Let } X \text{ be an animal.} \quad \underbrace{X \text{ is a cat}}_{A} \quad \Rightarrow \quad \underbrace{X \text{ is a mammal}}_{B}$$

When we say that an implication $A \Rightarrow B$ is true, we do not claim that A or B is necessarily true. Rather, we are making the conditional statement that *if* A happens to be true, *then* B is also true. In the above, if X does not happen to be a cat, the implication tells us nothing.

The **negation** of a statement A is the assertion that A is false and is denoted $\neg A$.

Statement A	Negation $\neg A$
X lives in California.	X does not live in California.
$\triangle ABC$ is a right triangle.	$\triangle ABC$ is not a right triangle.

The negation of the negation is the original statement: $\neg(\neg A) = A$. To say that X does *not not live in California* is the same as saying that *X lives in California*.

EXAMPLE 1 State the negation of each statement.

(a) The door is open and the dog is barking.

(b) The door is open or the dog is barking (or both).

Solution

(a) The first statement is true if two conditions are satisfied (door open and dog barking), and it is false if at least one of these conditions is not satisfied. So the negation is

Either the door is not open *OR* the dog is not barking *(or both)*.

(b) The second statement is true if at least one of the conditions (door open or dog barking) is satisfied, and it is false if neither condition is satisfied. So the negation is

The door is not open *AND* the dog is not barking. ■

Contrapositive and Converse

Two important operations are the formation of the contrapositive and the formation of the converse of a statement. The **contrapositive** of $A \Rightarrow B$ is the statement "If B is false, then A is false":

Keep in mind that when we form the contrapositive, we reverse the order of A and B. The contrapositive of $A \Rightarrow B$ is NOT $\neg A \Rightarrow \neg B$.

> The contrapositive of $\ \ A \Rightarrow B \ \ $ is $\ \ \neg B \Rightarrow \neg A$.

Here are some examples:

Statement	Contrapositive
If X is a cat, then X is a mammal.	If X is not a mammal, then X is not a cat.
If you work hard, then you will succeed.	If you did not succeed, then you did not work hard.
If m and n are both even, then $m + n$ is even.	If $m + n$ is not even, then m and n are not both even.

A key observation is this:

The contrapositive and the original implication are equivalent.

The fact that $A \Rightarrow B$ is equivalent to its contrapositive $\neg B \Rightarrow \neg A$ is a general rule of logic that does not depend on what A and B happen to mean. This rule belongs to the subject of "formal logic," which deals with logical relations between statements without concern for the actual content of these statements.

In other words, if an implication is true, then its contrapositive is automatically true, and vice versa. In essence, an implication and its contrapositive are two ways of saying the same thing. For example, the contrapositive, "If X is not a mammal, then X is not a cat," is a roundabout way of saying that cats are mammals.

The **converse** of $A \Rightarrow B$ is the *reverse* implication $B \Rightarrow A$:

Implication: $\ A \Rightarrow B$	Converse $B \Rightarrow A$
If A is true, then B is true.	If B is true, then A is true.

The converse plays a very different role than the contrapositive because *the converse is NOT equivalent to the original implication*. The converse may be true or false, even if the original implication is true. Here are some examples:

True Statement	Converse	Converse True or False?
If X is a cat, then X is a mammal.	If X is a mammal, then X is a cat.	False
If m is even, then m^2 is even.	If m^2 is even, then m is even.	True

A counterexample is an example that satisfies the hypothesis but not the conclusion of a statement. If a single counterexample exists, then the statement is false. However, we cannot prove that a statement is true merely by giving an example.

EXAMPLE 2 **An Example Where the Converse Is False** Show that the converse of, "If m and n are even, then $m + n$ is even," is false.

Solution The converse is, "If $m + n$ is even, then m and n are even." To show that the converse is false, we display a counterexample. Take $m = 1$ and $n = 3$ (or any other pair of odd numbers). The sum is even (since $1 + 3 = 4$) but neither 1 nor 3 is even. Therefore, the converse is false. ∎

EXAMPLE 3 **An Example Where the Converse Is True** State the contrapositive and converse of the Pythagorean Theorem. Are either or both of these true?

Solution Consider a triangle with sides a, b, and c, and let θ be the angle opposite the side of length c, as in Figure 1. The Pythagorean Theorem states that if $\theta = 90°$, then $a^2 + b^2 = c^2$. Here are the contrapositive and converse:

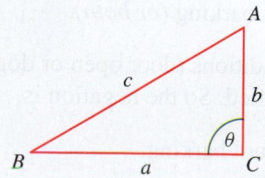

FIGURE 1

Pythagorean Theorem	$\theta = 90° \Rightarrow a^2 + b^2 = c^2$	True
Contrapositive	$a^2 + b^2 \neq c^2 \Rightarrow \theta \neq 90°$	Automatically true
Converse	$a^2 + b^2 = c^2 \Rightarrow \theta = 90°$	True (but not automatic)

The contrapositive is automatically true because it is just another way of stating the original theorem. The converse is not automatically true since there could conceivably exist a nonright triangle that satisfies $a^2 + b^2 = c^2$. However, the converse of the Pythagorean Theorem is, in fact, true. This follows from the Law of Cosines (see Exercise 38). ∎

When both a statement $A \Rightarrow B$ and its converse $B \Rightarrow A$ are true, we write $A \Longleftrightarrow B$. In this case, A and B are **equivalent**. We often express this with the phrase

$$A \Longleftrightarrow B \qquad A \text{ is true } \textit{if and only if } B \text{ is true.}$$

For example,

$$a^2 + b^2 = c^2 \qquad \text{if and only if} \qquad \theta = 90°$$

$$\text{It is morning} \qquad \text{if and only if} \qquad \text{the sun is rising.}$$

We mention the following variations of terminology involving implications that you may come across:

Statement	Is Another Way of Saying
A is true <u>if</u> B is true.	$B \Rightarrow A$
A is true <u>only if</u> B is true.	$A \Rightarrow B$ (A cannot be true unless B is also true.)
For A to be true, it is <u>necessary</u> that B be true.	$A \Rightarrow B$ (A cannot be true unless B is also true.)
For A to be true, <u>it is sufficient</u> that B be true.	$B \Rightarrow A$
For A to be true, it is <u>necessary and sufficient</u> that B be true.	$B \Longleftrightarrow A$

Analyzing a Theorem

To see how these rules of logic arise in calculus, consider the following result from Section 4.2:

> **THEOREM 1 Existence of Extrema on a Closed Interval**   A continuous function f on a closed (bounded) interval $I = [a, b]$ takes on both a minimum and a maximum value on I (Figure 2).

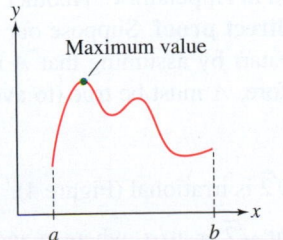

FIGURE 2 A continuous function on a closed interval $I = [a, b]$ has a maximum value.

To analyze this theorem, let's write out the hypotheses and conclusion separately:

Hypotheses A: f is continuous and I is closed.

Conclusion B: f takes on a minimum and a maximum value on I.

A first question to ask is: "Are the hypotheses necessary?" Is the conclusion still true if we drop one or both assumptions? To show that both hypotheses are necessary, we provide counterexamples:

- **The continuity of f is a necessary hypothesis.** Figure 3(A) shows the graph of a function on a closed interval $[a, b]$ that is not continuous. This function has no maximum value on $[a, b]$, which shows that the conclusion may fail if the continuity hypothesis is not satisfied.
- **The hypothesis that I is closed is necessary.** Figure 3(B) shows the graph of a continuous function on an *open interval* (a, b). This function has no maximum value, which shows that the conclusion may fail if the interval is not closed.

We see that both hypotheses in Theorem 1 are necessary. In stating this, we do not claim that the conclusion *always* fails when one or both of the hypotheses are not satisfied. We claim only that the conclusion *may* fail when the hypotheses are not satisfied. Next, let's analyze the contrapositive and converse:

- **Contrapositive** $\neg B \Rightarrow \neg A$ **(automatically true):** If f does not have a minimum and a maximum value on I, then either f is not continuous or I is not closed (or both).
- **Converse** $B \Rightarrow A$ **(in this case, false):** If f has a minimum and a maximum value on I, then f is continuous and I is closed. We prove this statement false with a counterexample [Figure 3(C)].

As we know, the contrapositive is merely a way of restating the theorem, so it is automatically true. The converse is not automatically true, and in fact, in this case it is false. The function in Figure 3(C) provides a counterexample to the converse: f has a maximum value on $I = (a, b)$, but f is not continuous and I is not closed.

> The technique of proof by contradiction *is also known by its Latin name* reductio ad absurdum *or "reduction to the absurd." The ancient Greek mathematicians used proof by contradiction as early as the fifth century* BCE, *and Euclid (325–265* BCE) *employed it in his classic treatise on geometry entitled* The Elements. *A famous example is the proof that* $\sqrt{2}$ *is irrational in Example 4. The philosopher Plato (427–347* BCE) *wrote: "He is unworthy of the name of man who is ignorant of the fact that the diagonal of a square is incommensurable with its side."*

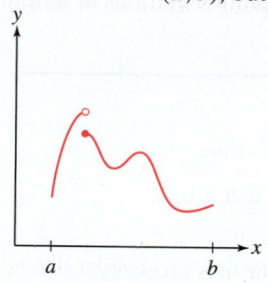

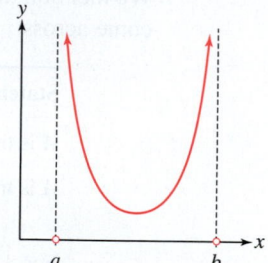

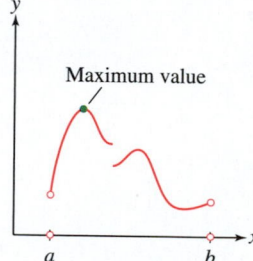

(A) The interval is closed but the function is not continuous. The function has no maximum value.

(B) The function is continuous but the interval is open. The function has no maximum value.

(C) This function is not continuous and the interval is not closed, but the function does have a maximum value.

FIGURE 3

Mathematicians have devised various general strategies and methods for proving theorems. The method of proof by induction is discussed in Appendix C. Another important method is **proof by contradiction**, also called **indirect proof**. Suppose our goal is to prove statement A. In a proof by contradiction, we start by assuming that A is false and then show that this leads to a contradiction. Therefore, A must be true (to avoid the contradiction).

EXAMPLE 4 **Proof by Contradiction** The number $\sqrt{2}$ is irrational (Figure 4).

Solution Assume that the theorem is false, namely that $\sqrt{2} = p/q$, where p and q are whole numbers. We may assume that p/q is in lowest terms, and therefore, at most one of p and q is even. Note that if the square m^2 of a whole number is even, then m itself must be even.

The relation $\sqrt{2} = p/q$ implies that $2 = p^2/q^2$ or $p^2 = 2q^2$. This shows that p must be even. But if p is even, then $p = 2m$ for some whole number m, and $p^2 = 4m^2$. Because $p^2 = 2q^2$, we obtain $4m^2 = 2q^2$, or $q^2 = 2m^2$. This shows that q is also even. But we chose p and q so that at most one of them is even. This contradiction shows that our original assumption, that $\sqrt{2} = p/q$, must be false. Therefore, $\sqrt{2}$ is irrational. ■

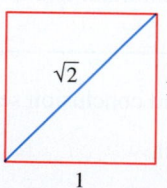

FIGURE 4 The diagonal of the unit square has length $\sqrt{2}$.

One of the most famous problems in mathematics is known as Fermat's Last Theorem. It states that the equation

$$x^n + y^n = z^n$$

has no solutions in positive integers if $n \geq 3$. In a marginal note written around 1630, Fermat claimed to have a proof, and over the centuries, that assertion was verified for many values of the exponent n. However, only in 1994 did the British-American mathematician Andrew Wiles, working at Princeton University, find a complete proof.

CONCEPTUAL INSIGHT   The hallmark of mathematics is precision and rigor. A theorem is established, not through observation or experimentation, but by a proof that consists of a chain of reasoning with no gaps.

This approach to mathematics comes down to us from the ancient Greek mathematicians, especially Euclid, and it remains the standard in contemporary research. In recent decades, the computer has become a powerful tool for mathematical experimentation and data analysis. Researchers may use experimental data to discover potential new mathematical facts, but the title "theorem" is not bestowed until someone writes down a proof.

This insistence on theorems and proofs distinguishes mathematics from the other sciences. In the natural sciences, facts are established through experiment and are subject to change or modification as more knowledge is acquired. In mathematics, theories are also developed and expanded, but previous results are not invalidated. The Pythagorean Theorem was discovered in antiquity and is a cornerstone of plane geometry. In the nineteenth century, mathematicians began to study more general types of geometry (of the type that eventually led to Einstein's four-dimensional space-time geometry in the Theory of Relativity). The Pythagorean Theorem does not hold in these more general geometries, but its status in plane geometry is unchanged.

A. SUMMARY

- The implication $A \Rightarrow B$ is the assertion, "If A is true, then B is true."
- The *contrapositive* of $A \Rightarrow B$ is the implication $\neg B \Rightarrow \neg A$, which says, "If B is false, then A is false." An implication and its contrapositive are equivalent (one is true if and only if the other is true).
- The *converse* of $A \Rightarrow B$ is $B \Rightarrow A$. An implication and its converse are not necessarily equivalent. One may be true and the other false.
- A and B are *equivalent* if $A \Rightarrow B$ and $B \Rightarrow A$ are both true.
- In a proof by contradiction (in which the goal is to prove statement A), we start by assuming that A is false and show that this assumption leads to a contradiction.

A. EXERCISES

Preliminary Questions

1. Which is the contrapositive of $A \Rightarrow B$?

(a) $B \Rightarrow A$ **(b)** $\neg B \Rightarrow A$

(c) $\neg B \Rightarrow \neg A$ **(d)** $\neg A \Rightarrow \neg B$

2. Which of the choices in Question 1 is the converse of $A \Rightarrow B$?

3. Suppose that $A \Rightarrow B$ is true. Which is then automatically true, the converse or the contrapositive?

4. Restate as an implication: "A triangle is a polygon."

Exercises

1. Which is the negation of the statement, "The car and the shirt are both blue"?

(a) Neither the car nor the shirt is blue.

(b) The car is not blue and/or the shirt is not blue.

2. Which is the contrapositive of the implication, "If the car has gas, then it will run"?

(a) If the car has no gas, then it will not run.

(b) If the car will not run, then it has no gas.

In Exercises 3–8, state the negation.

3. The time is 4 o'clock.

4. $\triangle ABC$ is an isosceles triangle.

5. m and n are odd integers.

6. Either m is odd or n is odd.

7. x is a real number and y is an integer.

8. f is a linear function.

In Exercises 9–14, state the contrapositive and converse.

9. If m and n are odd integers, then mn is odd.

10. If today is Tuesday, then we are in Belgium.

11. If today is Tuesday, then we are not in Belgium.

12. If $x > 4$, then $x^2 > 16$.

13. If m^2 is divisible by 3, then m is divisible by 3.

14. If $x^2 = 2$, then x is irrational.

In Exercise 15–18, give a counterexample to show that the converse of the statement is false.

15. If m is odd, then $2m + 1$ is also odd.

16. If $\triangle ABC$ is equilateral, then it is an isosceles triangle.

17. If m is divisible by 9 and 4, then m is divisible by 12.

18. If m is odd, then $m^3 - m$ is divisible by 3.

In Exercise 19–22, determine whether the converse of the statement is false.

19. If $x > 4$ and $y > 4$, then $x + y > 8$.

20. If $x > 4$, then $x^2 > 16$.

21. If $|x| > 4$, then $x^2 > 16$.

22. If m and n are even, then mn is even.

In Exercises 23 and 24, state the contrapositive and converse (it is not necessary to know what these statements mean).

23. If f and g are differentiable, then fg is differentiable.

24. If the force field is radial and decreases as the inverse square of the distance, then all closed orbits are ellipses.

*In Exercises 25–28, the **inverse** of $A \Rightarrow B$ is the implication $\neg A \Rightarrow \neg B$.*

25. Which of the following is the inverse of the implication, "If she jumped in the lake, then she got wet"?

(a) If she did not get wet, then she did not jump in the lake.

(b) If she did not jump in the lake, then she did not get wet.

Is the inverse true?

26. State the inverses of these implications:

(a) If X is a mouse, then X is a rodent.

(b) If you sleep late, you will miss class.

(c) If a star revolves around the sun, then it's a planet.

27. 🖉 Explain why the inverse is equivalent to the converse.

28. 🖉 State the inverse of the Pythagorean Theorem. Is it true?

29. Theorem 1 in Section 2.4 states the following: "If f and g are continuous functions, then $f + g$ is continuous." Does it follow logically that if f and g are not continuous, then $f + g$ is not continuous?

30. Write out a proof by contradiction for this fact: There is no smallest positive rational number. Base your proof on the fact that if $r > 0$, then $0 < r/2 < r$.

31. Use proof by contradiction to prove that if $x + y > 2$, then $x > 1$ or $y > 1$ (or both).

In Exercises 32–35, use proof by contradiction to show that the number is irrational.

32. $\sqrt{\frac{1}{2}}$ **33.** $\sqrt{3}$ **34.** $\sqrt[3]{2}$ **35.** $\sqrt[4]{11}$

36. An isosceles triangle is a triangle with two equal sides. The following theorem holds: If $\triangle$ is a triangle with two equal angles, then $\triangle$ is an isosceles triangle.

(a) What is the hypothesis?

(b) Show by providing a counterexample that the hypothesis is necessary.

(c) What is the contrapositive?

(d) What is the converse? Is it true?

37. Consider the following theorem: Let f be a quadratic polynomial with a positive leading coefficient. Then f has a minimum value.

(a) What are the hypotheses?

(b) What is the contrapositive?

(c) What is the converse? Is it true?

Further Insights and Challenges

38. Let a, b, and c be the sides of a triangle and let θ be the angle opposite c. Use the Law of Cosines (Theorem 1 in Section 1.4) to prove the converse of the Pythagorean Theorem.

39. Carry out the details of the following proof by contradiction that $\sqrt{2}$ is irrational (this proof is due to R. Palais). If $\sqrt{2}$ is rational, then $n\sqrt{2}$ is a whole number for some whole number n. Let n be the smallest such whole number and let $m = n\sqrt{2} - n$.

(a) Prove that $m < n$.

(b) Prove that $m\sqrt{2}$ is a whole number.

Explain why (a) and (b) imply that $\sqrt{2}$ is irrational.

40. Generalize the argument of Exercise 39 to prove that $\sqrt{A}$ is irrational if A is a whole number but not a perfect square. *Hint:* Choose n as before and let $m = n\sqrt{A} - n\lfloor \sqrt{A} \rfloor$, where $\lfloor x \rfloor$ is the greatest integer function.

41. Generalize further and show that for any whole number r, the rth root $\sqrt[r]{A}$ is irrational unless A is an rth power. *Hint:* Let $x = \sqrt[r]{A}$. Show that if x is rational, then we may choose a smallest whole number n such that nx^j is a whole number for $j = 1, \ldots, r - 1$. Then consider $m = nx - n[x]$ as before.

42. 🖉 Given a finite list of prime numbers $p_1, \ldots, p_N$, let $M = p_1 \cdot p_2 \cdots p_N + 1$. Show that M is not divisible by any of the primes $p_1, \ldots, p_N$. Use this and the fact that every number has a prime factorization to prove that there exist infinitely many prime numbers. This argument was advanced by Euclid in *The Elements*.

B PROPERTIES OF REAL NUMBERS

The ingenious method of expressing every possible number using a set of 10 symbols (each symbol having a place value and an absolute value) emerged in India. The idea seems so simple nowadays that its significance and profound importance is no longer appreciated. Its simplicity lies in the way it facilitated calculation and placed arithmetic foremost amongst useful inventions. The importance of this invention is more readily appreciated when one considers that it was beyond the two greatest men of Antiquity, Archimedes and Apollonius.

—Pierre-Simon Laplace, one of the great French mathematicians of the eighteenth century

In this appendix, we discuss the basic properties of real numbers. First, let us recall that a real number is a number that may be represented by a finite or infinite decimal (also called a decimal expansion). The set of all real numbers is denoted **R** and is often visualized as the "number line" (Figure 1).

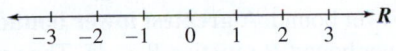

FIGURE 1 The real number line.

Thus, a real number a is represented as

$$a = \pm n.a_1a_2a_3a_4\ldots,$$

where n is any whole number and each digit a_j is a whole number between 0 and 9. For example, $10\pi = 31.41592\ldots$. Recall that a is rational if its expansion is finite or repeating and is irrational if its expansion is nonrepeating. Furthermore, the decimal expansion is unique apart from the following exception: Every finite expansion is equal to an expansion in which the digit 9 repeats. For example, $0.5 = 0.4999\cdots = 0.4\bar{9}$.

We shall take for granted that the operations of addition and multiplication are defined on **R**—that is, on the set of all decimals. Roughly speaking, addition and multiplication of infinite decimals are defined in terms of finite decimals. For $d \geq 1$, define the dth truncation of $a = n.a_1a_2a_3a_4\ldots$ to be the finite decimal $a(d) = a.a_1a_2\ldots a_d$ obtained by truncating at the dth place. To form the sum $a + b$, assume that both a and b are infinite (possibly ending with repeated nines). This eliminates any possible ambiguity in the expansion. Then the nth digit of $a + b$ is equal to the nth digit of $a(d) + b(d)$ for d sufficiently large [from a certain point onward, the nth digit of $a(d) + b(d)$ no longer changes, and this value is the nth digit of $a + b$]. Multiplication is defined similarly. Furthermore, the Commutative, Associative, and Distributive Laws hold (Table 1).

TABLE 1 Algebraic Laws

Commutative Laws:	$a + b = b + a, \quad ab = ba$
Associative Laws:	$(a + b) + c = a + (b + c), \quad (ab)c = a(bc)$
Distributive Law:	$a(b + c) = ab + ac$

Every real number x has an additive inverse $-x$ such that $x + (-x) = 0$, and every nonzero real number x has a multiplicative inverse x^{-1} such that $x(x^{-1}) = 1$. We do not regard subtraction and division as separate algebraic operations because they are defined in terms of inverses. By definition, the difference $x - y$ is equal to $x + (-y)$, and the quotient x/y is equal to $x(y^{-1})$ for $y \neq 0$.

In addition to the algebraic operations, there is an **order relation** on **R**: For any two real numbers a and b, precisely one of the following is true:

$$\text{Either} \quad a = b, \quad \text{or} \quad a < b, \quad \text{or} \quad a > b$$

To distinguish between the conditions $a \leq b$ and $a < b$, we often refer to $a < b$ as a **strict inequality**. Similar conventions hold for $>$ and $\geq$. The rules given in Table 2 allow us to manipulate inequalities. The last order property says that an inequality reverses direction when multiplied by a negative number c. For example,

$$-2 < 5 \quad \text{but} \quad (-3)(-2) > (-3)5$$

TABLE 2 Order Properties

If $a < b$ and $b < c$,	then $a < c$.
If $a < b$ and $c < d$,	then $a + c < b + d$.
If $a < b$ and $c > 0$,	then $ac < bc$.
If $a < b$ and $c < 0$,	then $ac > bc$.

The algebraic and order properties of real numbers are certainly familiar. We now discuss the less familiar **Least Upper Bound (LUB) Property** of the real numbers. This property is one way of expressing the so-called **completeness** of the real numbers. There are other ways of formulating completeness (such as the so-called nested interval property discussed in any book on analysis) that are equivalent to the LUB Property and serve the same purpose. Completeness is used in calculus to construct rigorous proofs of basic theorems about continuous functions, such as the Intermediate Value Theorem, (IVT) or the existence of extreme values on a closed interval. The underlying idea is that the real number line "has no holes." We elaborate on this idea below. First, we introduce the necessary definitions.

Suppose that S is a nonempty set of real numbers. A number M is called an **upper bound** for S if

$$x \leq M \qquad \text{for all } x \in S$$

If S has an upper bound, we say that S is **bounded above**. A **least upper bound** L is an upper bound for S such that every other upper bound M satisfies $M \geq L$. For example (Figure 2),

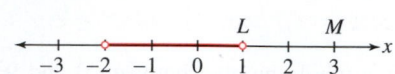

FIGURE 2 $M = 3$ is an upper bound for the set $S = (-2, 1)$. The LUB is $L = 1$.

- $M = 3$ is an upper bound for the open interval $S = (-2, 1)$.
- $L = 1$ is the LUB for $S = (-2, 1)$.

We now state the LUB Property of the real numbers.

> **THEOREM 1 Existence of a Least Upper Bound** Let S be a nonempty set of real numbers that is bounded above. Then S has an LUB.

In a similar fashion, we say that a number B is a **lower bound** for S if $x \geq B$ for all $x \in S$. We say that S is **bounded below** if S has a lower bound. A **greatest lower bound** (GLB) is a lower bound M such that every other lower bound B satisfies $B \leq M$. The set of real numbers also has the GLB Property: If S is a nonempty set of real numbers that is bounded below, then S has a GLB. This may be deduced immediately from Theorem 1. For any nonempty set of real numbers S, let $-S$ be the set of numbers of the form $-x$ for $x \in S$. Then $-S$ has an upper bound if S has a lower bound. Consequently, $-S$ has an LUB L by Theorem 1, and $-L$ is a GLB for S.

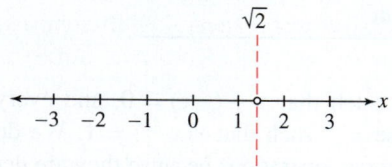

FIGURE 3 The rational numbers have a "hole" at the location $\sqrt{2}$.

CONCEPTUAL INSIGHT   Theorem 1 may appear quite reasonable, but perhaps it is not clear why it is useful. We suggested above that the LUB Property expresses the idea that $\mathbf{R}$ is "complete" or "has no holes." To illustrate this idea, let's compare $\mathbf{R}$ to the set of rational numbers, denoted $\mathbf{Q}$. Intuitively, $\mathbf{Q}$ is not complete because the irrational numbers are missing. For example, $\mathbf{Q}$ has a "hole" where the irrational number $\sqrt{2}$ should be located (Figure 3). This hole divides $\mathbf{Q}$ into two halves that are not connected to each other (the half to the left and the half to the right of $\sqrt{2}$). Furthermore, the half on the left is bounded above but no rational number is an LUB, and the half on the right is bounded below but no rational number is a GLB. The LUB and GLB are both equal to the irrational number $\sqrt{2}$, which exists in only $\mathbf{R}$ but not $\mathbf{Q}$. So unlike $\mathbf{R}$, the rational numbers $\mathbf{Q}$ do not have the LUB property.

EXAMPLE 1   Show that 2 has a square root by applying the LUB Property to the set

$$S = \{x : x^2 < 2\}$$

Solution   First, we note that S is bounded with the upper bound $M = 2$. Indeed, if $x > 2$, then x satisfies $x^2 > 4$, and hence x does not belong to S. By the LUB Property, S has a least upper bound. Call it L. We claim that $L = \sqrt{2}$, or, equivalently, that $L^2 = 2$. We prove this by showing that $L^2 \geq 2$ and $L^2 \leq 2$.

If $L^2 < 2$, let $b = L + h$, where $h > 0$. Then

$$b^2 = L^2 + 2Lh + h^2 = L^2 + h(2L + h) \qquad \boxed{1}$$

We can make the quantity $h(2L + h)$ as small as desired by choosing $h > 0$ small enough. In particular, we may choose a positive h so that $h(2L + h) < 2 - L^2$. For this choice, $b^2 < L^2 + (2 - L^2) = 2$ by Eq. (1). Therefore, $b \in S$. But $b > L$ since $h > 0$, and thus

L is not an upper bound for S, in contradiction to our hypothesis on L. We conclude that $L^2 \geq 2$.

If $L^2 > 2$, let $b = L - h$, where $h > 0$. Then

$$b^2 = L^2 - 2Lh + h^2 = L^2 - h(2L - h)$$

Now choose h positive but small enough so that $0 < h(2L - h) < L^2 - 2$. Then $b^2 > L^2 - (L^2 - 2) = 2$. But $b < L$, so b is a smaller lower bound for S. Indeed, if $x \geq b$, then $x^2 \geq b^2 > 2$, and x does not belong to S. This contradicts our hypothesis that L is the LUB. We conclude that $L^2 \leq 2$, and since we have already shown that $L^2 \geq 2$, we have $L^2 = 2$ as claimed. ■

We now prove three important theorems, the third of which is used in the proof of the LUB Property below.

THEOREM 2 Bolzano–Weierstrass Theorem   Let S be a bounded, infinite set of real numbers. Then there exists a sequence of distinct elements $\{a_n\}$ in S such that the limit $L = \lim_{n \to \infty} a_n$ exists.

Proof For simplicity of notation, we assume that S is contained in the unit interval $[0, 1]$ (a similar proof works in general). If $k_1, k_2, \ldots, k_n$ is a sequence of n digits (i.e., each k_j is a whole number and $0 \leq k_j \leq 9$), let

$$S(k_1, k_2, \ldots, k_n)$$

be the set of $x \in S$ whose decimal expansion begins $0.k_1 k_2 \ldots k_n$. The set S is the union of the subsets $S(0), S(1), \ldots, S(9)$, and since S is infinite, at least one of these subsets must be infinite. Therefore, we may choose k_1 so that $S(k_1)$ is infinite. In a similar fashion, at least one of the set $S(k_1, 0), S(k_2, 1), \ldots, S(k_1, 9)$ must be infinite, so we may choose k_2 so that $S(k_1, k_2)$ is infinite. Continuing in this way, we obtain an infinite sequence $\{k_n\}$ such that $S(k_1, k_2, \ldots, k_n)$ is infinite for all n. We may choose a sequence of elements $a_n \in S(k_1, k_2, \ldots, k_n)$ with the property that a_n differs from $a_1, \ldots, a_{n-1}$ for all n. Let L be the infinite decimal $0.k_1 k_2 k_3 \ldots$. Then $\lim_{n \to \infty} a_n = L$ since $|L - a_n| < 10^{-n}$ for all n. ■

We use the Bolzano–Weierstrass Theorem to prove two important results about sequences $\{a_n\}$. Recall that an upper bound for $\{a_n\}$ is a number M such that $a_j \leq M$ for all j. If an upper bound exists, $\{a_n\}$ is said to be bounded from above. Lower bounds are defined similarly and $\{a_n\}$ is said to be bounded from below if a lower bound exists. A sequence is bounded if it is bounded from above and below. A **subsequence** of $\{a_n\}$ is a sequence of elements $a_{n_1}, a_{n_2}, a_{n_3}, \ldots$, where $n_1 < n_2 < n_3 < \cdots$.

Now consider a bounded sequence $\{a_n\}$. If infinitely many of the a_n are distinct, the Bolzano–Weierstrass Theorem implies that there exists a subsequence $\{a_{n_1}, a_{n_2}, \ldots\}$ such that $\lim_{n \to \infty} a_{n_k}$ exists. Otherwise, infinitely many of the a_n must coincide, and these terms form a convergent subsequence. This proves the next result.

| Section 10.1

THEOREM 3   Every bounded sequence has a convergent subsequence.

THEOREM 4 Bounded Monotonic Sequences Converge

- If $\{a_n\}$ is increasing and $a_n \leq M$ for all n, then $\{a_n\}$ converges and $\lim_{n \to \infty} a_n \leq M$.
- If $\{a_n\}$ is decreasing and $a_n \geq m$ for all n, then $\{a_n\}$ converges and $\lim_{n \to \infty} a_n \geq m$.

Proof   Suppose that $\{a_n\}$ is increasing and bounded above by M. Then $\{a_n\}$ is automatically bounded below by $m = a_1$ since $a_1 \le a_2 \le a_3 \cdots$. Hence, $\{a_n\}$ is bounded, and by Theorem 3, we may choose a convergent subsequence $a_{n_1}, a_{n_2}, \ldots$. Let

$$L = \lim_{k \to \infty} a_{n_k}$$

Observe that $a_n \le L$ for all n. For if not, then $a_n > L$ for some n and then $a_{n_k} \ge a_n > L$ for all k such that $n_k \ge n$. But this contradicts that $a_{n_k} \to L$. Now, by definition, for any $\epsilon > 0$, there exists $N_\epsilon > 0$ such that

$$|a_{n_k} - L| < \epsilon \qquad \text{if } n_k > N_\epsilon$$

Choose m such that $n_m > N_\epsilon$. If $n \ge n_m$, then $a_{n_m} \le a_n \le L$, and therefore,

$$|a_n - L| \le |a_{n_m} - L| < \epsilon \qquad \text{for all } n \ge n_m$$

This proves that $\lim_{n \to \infty} a_n = L$, as desired. It remains to prove that $L \le M$. If $L > M$, let $\epsilon = (L - M)/2$ and choose N so that

$$|a_n - L| < \epsilon \qquad \text{if } k > N$$

Then $a_n > L - \epsilon = M + \epsilon$. This contradicts our assumption that M is an upper bound for $\{a_n\}$. Therefore, $L \le M$ as claimed.   ∎

Proof of Theorem 1   We now use Theorem 4 to prove the LUB Property (Theorem 1). As above, if x is a real number, let $x(d)$ be the truncation of x of length d. For example,

$$\text{If } x = 1.41569, \text{ then } x(3) = 1.415$$

We say that x is a *decimal of length d* if $x = x(d)$. Any two distinct decimals of length d differ by at least 10^{-d}. It follows that for any two real numbers $A < B$, there are at most finitely many decimals of length d between A and B.

Now let S be a nonempty set of real numbers with an upper bound M. We shall prove that S has an LUB. Let $S(d)$ be the set of truncations of length d:

$$S(d) = \{x(d) : x \in S\}$$

We claim that $S(d)$ has a maximum element. To verify this, choose any $a \in S$. If $x \in S$ and $x(d) > a(d)$, then

$$a(d) \le x(d) \le M$$

Thus, by the remark of the previous paragraph, there are at most finitely many values of $x(d)$ in $S(d)$ larger than $a(d)$. The largest of these is the maximum element in $S(d)$.

For $d = 1, 2, \ldots$, choose an element x_d such that $x_d(d)$ is the maximum element in $S(d)$. By construction, $\{x_d(d)\}$ is an increasing sequence (since the largest dth truncation cannot get smaller as d increases). Furthermore, $x_d(d) \le M$ for all d. We now apply Theorem 4 to conclude that $\{x_d(d)\}$ converges to a limit L. We claim that L is the LUB of S. Observe first that L is an upper bound for S. Indeed, if $x \in S$, then $x(d) \le L$ for all d and thus $x \le L$. To show that L is the LUB, suppose that M is an upper bound such that $M < L$. Then $x_d \le M$ for all d and hence $x_d(d) \le M$ for all d. But then

$$L = \lim_{d \to \infty} x_d(d) \le M$$

This is a contradiction since $M < L$. Therefore, L is the LUB of S.   ∎

As mentioned above, the LUB Property is used in calculus to establish certain basic theorems about continuous functions. As an example, we prove the IVT. Another example is the theorem on the existence of extrema on a closed interval (see Appendix D).

> **THEOREM 5** **Intermediate Value Theorem**   If f is continuous on a closed interval $[a, b]$ then for every value M, strictly between $f(a)$ and $f(b)$, there exists at least one value $c \in (a, b)$ such that $f(c) = M$.

Proof   Assume first that $M = 0$. Replacing $f(x)$ by $-f(x)$ if necessary, we may assume that $f(a) < 0$ and $f(b) > 0$. Now let

$$S = \{x \in [a, b] : f(x) < 0\}$$

Then $a \in S$ since $f(a) < 0$ and thus S is nonempty. Clearly, b is an upper bound for S. Therefore, by the LUB Property, S has an LUB L. We claim that $f(L) = 0$. If not, set $r = f(L)$. Assume first that $r > 0$.

Since f is continuous, there exists a number $\delta > 0$ such that

$$\text{if } |x - L| < \delta, \text{ then } |f(x) - f(L)| = |f(x) - r| < \frac{1}{2}r$$

Equivalently,

$$\text{if } |x - L| < \delta, \text{ then } \frac{1}{2}r < f(x) < \frac{3}{2}r$$

The number $\frac{1}{2}r$ is positive, so we conclude that

$$\text{if } L - \delta < x < L + \delta, \text{ then } f(x) > 0$$

By definition of L, $f(x) \geq 0$ for all $x \in [a, b]$ such that $x > L$, and thus $f(x) \geq 0$ for all $x \in [a, b]$ such that $x > L - \delta$. Thus, $L - \delta$ is an upper bound for S. This is a contradiction since L is the LUB of S, and it follows that $r = f(L)$ cannot satisfy $r > 0$. Similarly, r cannot satisfy $r < 0$. We conclude that $f(L) = 0$ as claimed.

Now, if M is nonzero, let $g(x) = f(x) - M$. Then 0 lies between $g(a)$ and $g(b)$, and by what we have proved, there exists $c \in (a, b)$ such that $g(c) = 0$. But then $f(c) = g(c) + M = M$, as desired. ∎

C INDUCTION AND THE BINOMIAL THEOREM

The Principle of Induction is a method of proof that is widely used to prove that a given statement $P(n)$ is valid for all natural numbers $n = 1, 2, 3, \ldots$. Here are two statements of this kind:

- $P(n)$: The sum of the first n odd numbers is equal to n^2.
- $P(n)$: $\dfrac{d}{dx}x^n = nx^{n-1}$.

The first statement claims that for all natural numbers n,

$$\underbrace{1 + 3 + \cdots + (2n - 1)}_{\text{Sum of first } n \text{ odd numbers}} = n^2$$

1

We can check directly that $P(n)$ is true for the first few values of n:

$$P(1) \text{ is the equality:} \qquad 1 = 1^2 \quad \text{(true)}$$
$$P(2) \text{ is the equality:} \qquad 1 + 3 = 2^2 \quad \text{(true)}$$
$$P(3) \text{ is the equality:} \qquad 1 + 3 + 5 = 3^2 \quad \text{(true)}$$

The Principle of Induction may be used to establish $P(n)$ for all n.

The Principle of Induction applies if $P(n)$ is an assertion defined for $n \geq n_0$, where n_0 is a fixed integer. Assume that

(i) **Initial step:** $P(n_0)$ is true.

(ii) **Induction step:** If $P(n)$ is true for $n = k$, then $P(n)$ is also true for $n = k + 1$.

Then $P(n)$ is true for all $n \geq n_0$.

THEOREM 1 Principle of Induction Let $P(n)$ be an assertion that depends on a natural number n. Assume that

(i) **Initial step:** $P(1)$ is true.
(ii) **Induction step:** If $P(n)$ is true for $n = k$, then $P(n)$ is also true for $n = k + 1$.

Then $P(n)$ is true for all natural numbers $n = 1, 2, 3, \ldots$.

EXAMPLE 1 Prove that $1 + 3 + \cdots + (2n - 1) = n^2$ for all natural numbers n.

Solution As above, we let $P(n)$ denote the equality

$$P(n): \qquad 1 + 3 + \cdots + (2n - 1) = n^2$$

Step 1. **Initial step: Show that $P(1)$ is true.**
We checked this above. $P(1)$ is the equality $1 = 1^2$.

Step 2. **Induction step: Show that if $P(n)$ is true for $n = k$, then $P(n)$ is also true for $n = k + 1$.**
Assume that $P(k)$ is true. Then

$$1 + 3 + \cdots + (2k - 1) = k^2$$

Add $2k + 1$ to both sides:

$$\left[1 + 3 + \cdots + (2k - 1)\right] + (2k + 1) = k^2 + 2k + 1 = (k + 1)^2$$
$$1 + 3 + \cdots + (2k + 1) = (k + 1)^2$$

This is precisely the statement $P(k + 1)$. Thus, $P(k + 1)$ is true whenever $P(k)$ is true. By the Principle of Induction, $P(k)$ is true for all k. ∎

The intuition behind the Principle of Induction is the following. If $P(n)$ were not true for all n, then there would exist a smallest natural number k such that $P(k)$ is false.

Furthermore, $k > 1$ since $P(1)$ is true. Thus, $P(k-1)$ is true [otherwise, $P(k)$ would not be the smallest "counterexample"]. On the other hand, if $P(k-1)$ is true, then $P(k)$ is also true by the induction step. This is a contradiction. So $P(k)$ must be true for all k.

EXAMPLE 2   Use Induction and the Product Rule to prove that for all whole numbers n,

$$\frac{d}{dx}x^n = nx^{n-1}$$

Solution   Let $P(n)$ be the formula $\dfrac{d}{dx}x^n = nx^{n-1}$.

Step 1. **Initial step: Show that $P(1)$ is true.**
We use the limit definition to verify $P(1)$:

$$\frac{d}{dx}x = \lim_{h \to 0}\frac{(x+h)-x}{h} = \lim_{h \to 0}\frac{h}{h} = \lim_{h \to 0}1 = 1$$

Step 2. **Induction step: Show that if $P(n)$ is true for $n = k$, then $P(n)$ is also true for $n = k + 1$.**
To carry out the induction step, assume that $\dfrac{d}{dx}x^k = kx^{k-1}$, where $k \geq 1$. Then, by the Product Rule,

$$\frac{d}{dx}x^{k+1} = \frac{d}{dx}(x \cdot x^k) = x\frac{d}{dx}x^k + x^k\frac{d}{dx}x = x(kx^{k-1}) + x^k$$

$$= kx^k + x^k = (k+1)x^k$$

This shows that $P(k+1)$ is true.

By the Principle of Induction, $P(n)$ is true for all $n \geq 1$.   ■

As another application of induction, we prove the Binomial Theorem, which describes the expansion of the binomial $(a+b)^n$. The first few expansions are familiar:

$$(a+b)^1 = a+b$$

$$(a+b)^2 = a^2 + 2ab + b^2$$

$$(a+b)^3 = a^3 + 3a^2b + 3ab^2 + b^3$$

In general, we have an expansion

$$(a+b)^n = a^n + \binom{n}{1}a^{n-1}b + \binom{n}{2}a^{n-2}b^2 + \binom{n}{3}a^{n-3}b^3$$

$$+ \cdots + \binom{n}{n-1}ab^{n-1} + b^n$$

2

where the coefficient of $a^{n-k}b^k$, denoted $\binom{n}{k}$, is called the **binomial coefficient**. Note that the first term in Eq. (2) corresponds to $k = 0$ and the last term to $k = n$; thus, $\binom{n}{0} = \binom{n}{n} = 1$. In summation notation,

$$(a+b)^n = \sum_{k=0}^{n}\binom{n}{k}a^{n-k}b^k$$

In Pascal's Triangle, the nth row displays the coefficients in the expansion of $(a+b)^n$:

n														
0							1							
1						1		1						
2					1		2		1					
3				1		3		3		1				
4			1		4		6		4		1			
5		1		5		10		10		5		1		
6	1		6		15		20		15		6		1	

The triangle is constructed as follows: Each entry is the sum of the two entries above it in the previous line. For example, the entry 15 in line $n = 6$ is the sum $10 + 5$ of the entries above it in line $n = 5$. The recursion relation guarantees that the entries in the triangle are the binomial coefficients.

Pascal's Triangle (described in the marginal note on page A13) can be used to compute binomial coefficients if n and k are not too large. The Binomial Theorem provides the following general formula:

$$\binom{n}{k} = \frac{n!}{k!\,(n-k)!} = \frac{n(n-1)(n-2)\cdots(n-k+1)}{k(k-1)(k-2)\cdots 2\cdot 1} \qquad \boxed{3}$$

Before proving this formula, we prove a recursion relation for binomial coefficients. Note, however, that Eq. (3) is certainly correct for $k=0$ and $k=n$ (recall that by convention, $0!=1$):

$$\binom{n}{0} = \frac{n!}{(n-0)!\,0!} = \frac{n!}{n!} = 1, \qquad \binom{n}{n} = \frac{n!}{(n-n)!\,n!} = \frac{n!}{n!} = 1$$

THEOREM 2 Recursion Relation for Binomial Coefficients

$$\binom{n}{k} = \binom{n-1}{k} + \binom{n-1}{k-1} \qquad \text{for } 1 \le k \le n-1$$

Proof We write $(a+b)^n$ as $(a+b)(a+b)^{n-1}$ and expand in terms of binomial coefficients:

$$(a+b)^n = (a+b)(a+b)^{n-1}$$

$$\sum_{k=0}^{n} \binom{n}{k} a^{n-k}b^k = (a+b)\sum_{k=0}^{n-1}\binom{n-1}{k}a^{n-1-k}b^k$$

$$= a\sum_{k=0}^{n-1}\binom{n-1}{k}a^{n-1-k}b^k + b\sum_{k=0}^{n-1}\binom{n-1}{k}a^{n-1-k}b^k$$

$$= \sum_{k=0}^{n-1}\binom{n-1}{k}a^{n-k}b^k + \sum_{k=0}^{n-1}\binom{n-1}{k}a^{n-(k+1)}b^{k+1}$$

Replacing k by $k-1$ in the second sum, we obtain

$$\sum_{k=0}^{n}\binom{n}{k}a^{n-k}b^k = \sum_{k=0}^{n-1}\binom{n-1}{k}a^{n-k}b^k + \sum_{k=1}^{n}\binom{n-1}{k-1}a^{n-k}b^k$$

On the right-hand side, the first term in the first sum is a^n and the last term in the second sum is b^n. Thus, we have

$$\sum_{k=0}^{n}\binom{n}{k}a^{n-k}b^k = a^n + \left(\sum_{k=1}^{n-1}\left(\binom{n-1}{k}+\binom{n-1}{k-1}\right)a^{n-k}b^k\right) + b^n$$

The recursion relation follows because the coefficients of $a^{n-k}b^k$ on the two sides of the equation must be equal. ∎

We now use induction to prove Eq. (3). Let $P(n)$ be the claim

$$\binom{n}{k} = \frac{n!}{k!\,(n-k)!} \qquad \text{for } 0 \le k \le n$$

We have $\binom{1}{0} = \binom{1}{1} = 1$ since $(a+b)^1 = a+b$, so $P(1)$ is true. Furthermore,

$$\binom{n}{n} = \binom{n}{0} = 1 \text{ as observed above, since } a^n \text{ and } b^n \text{ have coefficient 1 in the expansion}$$

of $(a + b)^n$. For the inductive step, assume that $P(n)$ is true. By the recursion relation, for $1 \le k \le n$, we have

$$\binom{n+1}{k} = \binom{n}{k} + \binom{n}{k-1} = \frac{n!}{k!\,(n-k)!} + \frac{n!}{(k-1)!\,(n-k+1)!}$$

$$= n!\left(\frac{n+1-k}{k!\,(n+1-k)!} + \frac{k}{k!\,(n+1-k)!}\right) = n!\left(\frac{n+1}{k!\,(n+1-k)!}\right)$$

$$= \frac{(n+1)!}{k!\,(n+1-k)!}$$

Thus, $P(n + 1)$ is also true and the Binomial Theorem follows by induction.

EXAMPLE 3 Use the Binomial Theorem to expand $(x + y)^5$ and $(x + 2)^3$.

Solution The fifth row in Pascal's Triangle yields

$$(x + y)^5 = x^5 + 5x^4 y + 10x^3 y^2 + 10x^2 y^3 + 5xy^4 + y^5$$

The third row in Pascal's Triangle yields

$$(x + 2)^3 = x^3 + 3x^2(2) + 3x(2)^2 + 2^3 = x^3 + 6x^2 + 12x + 8 \qquad \blacksquare$$

C. EXERCISES

In Exercises 1–4, use the Principle of Induction to prove the formula for all natural numbers n.

1. $1 + 2 + 3 + \cdots + n = \dfrac{n(n+1)}{2}$

2. $1^3 + 2^3 + 3^3 + \cdots + n^3 = \dfrac{n^2(n+1)^2}{4}$

3. $\dfrac{1}{1\cdot 2} + \dfrac{1}{2\cdot 3} + \cdots + \dfrac{1}{n(n+1)} = \dfrac{n}{n+1}$

4. $1 + x + x^2 + \cdots + x^n = \dfrac{1 - x^{n+1}}{1 - x}$ for any $x \ne 1$

5. Let $P(n)$ be the statement $2^n > n$.
(a) Show that $P(1)$ is true.
(b) Observe that if $2^n > n$, then $2^n + 2^n > 2n$. Use this to show that if $P(n)$ is true for $n = k$, then $P(n)$ is true for $n = k+1$. Conclude that $P(n)$ is true for all n.

6. Use induction to prove that $n! > 2^n$ for $n \ge 4$.

Let $\{F_n\}$ be the Fibonacci sequence, defined by the recursion formula

$$F_n = F_{n-1} + F_{n-2}, \qquad F_1 = F_2 = 1$$

The first few terms are $1, 1, 2, 3, 5, 8, 13, \ldots$. In Exercises 7–10, use induction to prove the identity.

7. $F_1 + F_2 + \cdots + F_n = F_{n+2} - 1$

8. $F_1^2 + F_2^2 + \cdots + F_n^2 = F_{n+1} F_n$

9. $F_n = \dfrac{R_+^n - R_-^n}{\sqrt{5}}$, where $R_\pm = \dfrac{1 \pm \sqrt{5}}{2}$

10. $F_{n+1} F_{n-1} = F_n^2 + (-1)^n$. *Hint:* For the induction step, show that

$$F_{n+2} F_n = F_{n+1} F_n + F_n^2$$

$$F_{n+1}^2 = F_{n+1} F_n + F_{n+1} F_{n-1}$$

11. Use induction to prove that $f(n) = 8^n - 1$ is divisible by 7 for all natural numbers n. *Hint:* For the induction step, show that

$$8^{k+1} - 1 = 7 \cdot 8^k + (8^k - 1)$$

12. Use induction to prove that $n^3 - n$ is divisible by 3 for all natural numbers n.

13. Use induction to prove that $5^{2n} - 4^n$ is divisible by 7 for all natural numbers n.

14. Use Pascal's Triangle to write out the expansions of $(a + b)^6$ and $(a - b)^4$.

15. Expand $(x + x^{-1})^4$.

16. What is the coefficient of x^9 in $(x^3 + x)^5$?

17. Let $S(n) = \displaystyle\sum_{k=0}^{n} \binom{n}{k}$.
(a) Use Pascal's Triangle to compute $S(n)$ for $n = 1, 2, 3, 4$.
(b) Prove that $S(n) = 2^n$ for all $n \ge 1$. *Hint:* Expand $(a + b)^n$ and evaluate at $a = b = 1$.

18. Let $T(n) = \displaystyle\sum_{k=0}^{n} (-1)^k \binom{n}{k}$.
(a) Use Pascal's Triangle to compute $T(n)$ for $n = 1, 2, 3, 4$.
(b) Prove that $T(n) = 0$ for all $n \ge 1$. *Hint:* Expand $(a + b)^n$ and evaluate at $a = 1, b = -1$.

D ADDITIONAL PROOFS

In this appendix, we provide proofs of several theorems that were stated or used in the text.

I Section 2.3

> **THEOREM 1 Basic Limit Laws** Assume that $\lim_{x \to c} f(x)$ and $\lim_{x \to c} g(x)$ exist. Then:
>
> **(i)** $\lim_{x \to c} \big(f(x) + g(x)\big) = \lim_{x \to c} f(x) + \lim_{x \to c} g(x)$
>
> **(ii)** For any number k, $\lim_{x \to c} k f(x) = k \lim_{x \to c} f(x)$
>
> **(iii)** $\lim_{x \to c} f(x)g(x) = \Big(\lim_{x \to c} f(x) \Big) \Big(\lim_{x \to c} g(x) \Big)$
>
> **(iv)** If $\lim_{x \to c} g(x) \neq 0$, then
>
> $$\lim_{x \to c} \frac{f(x)}{g(x)} = \frac{\lim_{x \to c} f(x)}{\lim_{x \to c} g(x)}$$

Proof Let $L = \lim_{x \to c} f(x)$ and $M = \lim_{x \to c} g(x)$. The Sum Law (i) was proved in Section 2.9. Observe that (ii) is a special case of (iii), where $g(x) = k$ is a constant function. Thus, it will suffice to prove the Product Law (iii). We write

$$f(x)g(x) - LM = f(x)(g(x) - M) + M(f(x) - L)$$

and apply the Triangle Inequality to obtain

$$|f(x)g(x) - LM| \leq |f(x)(g(x) - M)| + |M(f(x) - L)| \qquad \boxed{1}$$

By the limit definition, we may choose $\delta > 0$ so that

$$\text{if } 0 < |x - c| < \delta, \text{ then } |f(x) - L| < 1$$

If follows that $|f(x)| < |L| + 1$ for $0 < |x - c| < \delta$. Now choose any number $\epsilon > 0$. Applying the limit definition again, we see that by choosing a smaller δ if necessary, we may also ensure that if $0 < |x - c| < \delta$, then

$$|f(x) - L| \leq \frac{\epsilon}{2(|M| + 1)} \qquad \text{and} \qquad |g(x) - M| \leq \frac{\epsilon}{2(|L| + 1)}$$

Using Eq. (1), we see that if $0 < |x - c| < \delta$, then

$$|f(x)g(x) - LM| \leq |f(x)|\,|g(x) - M| + |M|\,|f(x) - L|$$

$$\leq (|L| + 1)\frac{\epsilon}{2(|L| + 1)} + |M|\frac{\epsilon}{2(|M| + 1)}$$

$$\leq \frac{\epsilon}{2} + \frac{\epsilon}{2} = \epsilon$$

Since ϵ is arbitrary, this proves that $\lim_{x \to c} f(x)g(x) = LM$. To prove the Quotient Law (iv), it suffices to verify that if $M \neq 0$, then

$$\lim_{x \to c} \frac{1}{g(x)} = \frac{1}{M} \qquad \boxed{2}$$

For if Eq. (2) holds, then we may apply the Product Law to $f(x)$ and $g(x)^{-1}$ to obtain the Quotient Law:

$$\lim_{x\to c}\frac{f(x)}{g(x)} = \lim_{x\to c} f(x)\frac{1}{g(x)} = \left(\lim_{x\to c} f(x)\right)\left(\lim_{x\to c}\frac{1}{g(x)}\right)$$

$$= L\left(\frac{1}{M}\right) = \frac{L}{M}$$

We now verify Eq. (2). Since $g(x)$ approaches M and $M \neq 0$, we may choose $\delta > 0$ so that if $0 < |x - c| < \delta$, then $|g(x)| \geq |M|/2$. Now choose any number $\epsilon > 0$. By choosing a smaller δ if necessary, we may also ensure that

$$\text{for } 0 < |x - c| < \delta, \text{ then } |M - g(x)| < \epsilon|M|\left(\frac{|M|}{2}\right)$$

Then

$$\left|\frac{1}{g(x)} - \frac{1}{M}\right| = \left|\frac{M - g(x)}{Mg(x)}\right| \leq \left|\frac{M - g(x)}{M(M/2)}\right| \leq \frac{\epsilon|M|(|M|/2)}{|M|(|M|/2)} = \epsilon$$

Since ϵ is arbitrary, the limit in Eq. (2) is proved.   ∎

The following result was used in the text.

> **THEOREM 2 Limits Preserve Inequalities**   Let (a, b) be an open interval and let $c \in (a, b)$. Suppose that $f(x)$ and $g(x)$ are defined on (a, b), except possibly at c. Assume that
>
> $$f(x) \leq g(x) \qquad \text{for } x \in (a, b), \quad x \neq c$$
>
> and that the limits $\lim_{x\to c} f(x)$ and $\lim_{x\to c} g(x)$ exist. Then
>
> $$\lim_{x\to c} f(x) \leq \lim_{x\to c} g(x)$$

Proof Let $L = \lim_{x\to c} f(x)$ and $M = \lim_{x\to c} g(x)$. To show that $L \leq M$, we use proof by contradiction. If $L > M$, let $\epsilon = \frac{1}{2}(L - M)$. By the formal definition of limits, we may choose $\delta > 0$ so that the following two conditions are satisfied:

$$\text{If } |x - c| < \delta, \text{ then } |M - g(x)| < \epsilon.$$

$$\text{If } |x - c| < \delta, \text{ then } |L - f(x)| < \epsilon.$$

But then

$$f(x) > L - \epsilon = M + \epsilon > g(x)$$

This is a contradiction since $f(x) \leq g(x)$. We conclude that $L \leq M$.   ∎

> **THEOREM 3 Limit of a Composite Function**   Assume that the following limits exist:
>
> $$L = \lim_{x\to c} g(x) \qquad \text{and} \qquad M = \lim_{x\to L} f(x)$$
>
> Then $\lim_{x\to c} f(g(x)) = M$.

Proof Let $\epsilon > 0$ be given. By the limit definition, there exists $\delta_1 > 0$ such that

$$\text{if } 0 < |x - L| < \delta_1, \text{ then } |f(x) - M| < \epsilon.$$

Similarly, there exists $\delta > 0$ such that

$$\text{if } 0 < |x - c| < \delta, \text{ then } |g(x) - L| < \delta_1.$$

4

We replace x by $g(x)$ in Eq. (3) and apply Eq. (4) to obtain:

$$\text{If } 0 < |x - c| < \delta, \text{ then } |f(g(x)) - M| < \epsilon.$$

Since ϵ is arbitrary, this proves that $\lim_{x \to c} f(g(x)) = M$.   ∎

| Section 2.4

> **THEOREM 4 Continuity of Composite Functions**   Let $F(x) = f(g(x))$ be a composite function. If g is continuous at $x = c$ and f is continuous at $x = g(c)$, then F is continuous at $x = c$.

Proof By definition of continuity,

$$\lim_{x \to c} g(x) = g(c) \qquad \text{and} \qquad \lim_{x \to g(c)} f(x) = f(g(c))$$

Therefore, we may apply Theorem 3 to obtain

$$\lim_{x \to c} f(g(x)) = f(g(c))$$

This proves that $F(x) = f(g(x))$ is continuous at $x = c$.   ∎

| Section 2.6

> **THEOREM 5 Squeeze Theorem**   Assume that for $x \neq c$ (in some open interval containing c),
>
> $$l(x) \leq f(x) \leq u(x) \qquad \text{and} \qquad \lim_{x \to c} l(x) = \lim_{x \to c} u(x) = L$$
>
> Then $\lim_{x \to c} f(x)$ exists and $\lim_{x \to c} f(x) = L$.

Proof Let $\epsilon > 0$ be given. We may choose $\delta > 0$ such that

$$\text{if } 0 < |x - c| < \delta, \text{ then } |l(x) - L| < \epsilon \text{ and } |u(x) - L| < \epsilon.$$

In principle, a different δ may be required to obtain the two inequalities for $l(x)$ and $u(x)$, but we may choose the smaller of the two deltas. Thus, if $0 < |x - c| < \delta$, we have

$$L - \epsilon < l(x) < L + \epsilon$$

and

$$L - \epsilon < u(x) < L + \epsilon$$

Since $f(x)$ lies between $l(x)$ and $u(x)$, it follows that

$$L - \epsilon < l(x) \leq f(x) \leq u(x) < L + \epsilon$$

and therefore $|f(x) - L| < \epsilon$ if $0 < |x - c| < \delta$. Since ϵ is arbitrary, this proves that $\lim_{x \to c} f(x) = L$, as desired.   ∎

| Section 4.2

> **THEOREM 6 Existence of Extrema on a Closed Interval**   A continuous function f on a closed (bounded) interval $I = [a, b]$ takes on both a minimum and a maximum value on I.

Proof We prove that f takes on a maximum value in two steps (the case of a minimum is similar).

Step 1. **Prove that f is bounded from above.**

We use proof by contradiction. If f is not bounded from above, then there exist points $a_n \in [a, b]$ such that $f(a_n) \geq n$ for $n = 1, 2, \ldots$. By Theorem 3 in Appendix B, we may choose a subsequence of elements $a_{n_1}, a_{n_2}, \ldots$ that converges to a limit in $[a, b]$—say, $\lim_{k \to \infty} a_{n_k} = L$. Since f is continuous, there exists $\delta > 0$ such that

$$\text{if } x \in [a, b] \text{ and } |x - L| < \delta, \text{ then } |f(x) - f(L)| < 1.$$

Therefore,

$$\text{if } x \in [a, b] \text{ and } x \in (L - \delta, L + \delta), \text{ then } f(x) < f(L) + 1. \qquad \boxed{5}$$

For k sufficiently large, a_{n_k} lies in $(L - \delta, L + \delta)$ because $\lim_{k \to \infty} a_{n_k} = L$. By Eq. (5), $f(a_{n_k})$ is bounded by $f(L) + 1$. However, $f(a_{n_k}) = n_k$ tends to infinity as $k \to \infty$. This is a contradiction. Hence, our assumption that f is not bounded from above is false.

Step 2. **Prove that f takes on a maximum value.**

The range of f on $I = [a, b]$ is the set

$$S = \{f(x) : x \in [a, b]\}$$

By the previous step, S is bounded from above and therefore has a least upper bound M by the LUB Property. Thus, $f(x) \leq M$ for all $x \in [a, b]$. To complete the proof, we show that $f(c) = M$ for some $c \in [a, b]$. This will show that f attains the maximum value M on $[a, b]$.

By definition, $M - 1/n$ is not an upper bound for $n \geq 1$, and therefore, we may choose a point b_n in $[a, b]$ such that

$$M - \frac{1}{n} \leq f(b_n) \leq M$$

Again by Theorem 3 in Appendix B, there exists a subsequence of elements $\{b_{n_1}, b_{n_2}, \ldots\}$ in $\{b_1, b_2, \ldots\}$ that converges to a limit—say,

$$\lim_{k \to \infty} b_{n_k} = c$$

Furthermore, this limit c belongs to $[a, b]$ because $[a, b]$ is closed. Let $\epsilon > 0$. Since f is continuous, we may choose k so large that the following two conditions are satisfied: $|f(c) - f(b_{n_k})| < \epsilon/2$ and $n_k > 2/\epsilon$. Then

$$|f(c) - M| \leq |f(c) - f(b_{n_k})| + |f(b_{n_k}) - M| \leq \frac{\epsilon}{2} + \frac{1}{n_k} \leq \frac{\epsilon}{2} + \frac{\epsilon}{2} = \epsilon$$

Thus, $|f(c) - M|$ is smaller than ϵ for all positive numbers ϵ. But this is not possible unless $|f(c) - M| = 0$. Thus, $f(c) = M$, as desired.   ∎

> **THEOREM 7 Continuous Functions Are Integrable**   If f is continuous on $[a, b]$, or if f is continuous except at finitely many jump discontinuities in $[a, b]$, then f is integrable over $[a, b]$.

| *Section 5.2*

Proof We shall make the simplifying assumption that f is differentiable and that its derivative f' is bounded. In other words, we assume that $|f'(x)| \leq K$ for some constant K. This assumption is used to show that f cannot vary too much in a small interval. More precisely, let us prove that if $[a_0, b_0]$ is any closed interval contained in $[a, b]$ and if m and M are the minimum and maximum values of f on $[a_0, b_0]$, then

$$|M - m| \leq K|b_0 - a_0| \qquad \boxed{6}$$

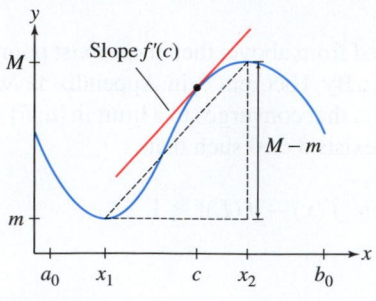

FIGURE 1 Since $M - m = f'(c)(x_2 - x_1)$, we conclude that $M - m \leq K(b_0 - a_0)$.

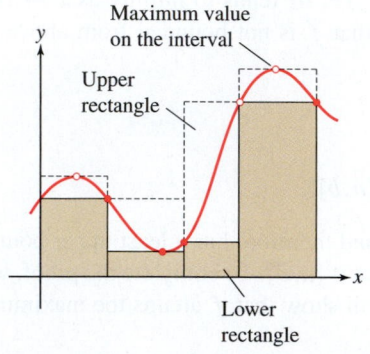

FIGURE 2 Lower and upper rectangles for a partition of length $N = 4$.

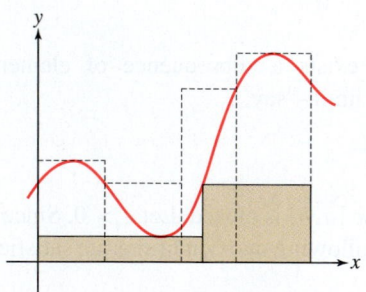

FIGURE 3 The lower rectangles always lie below the upper rectangles, even when the partitions are different.

Figure 1 illustrates the idea behind this inequality. Suppose that $f(x_1) = m$ and $f(x_2) = M$, where x_1 and x_2 lie in $[a_0, b_0]$. If $x_1 \neq x_2$, then by the Mean Value Theorem (MVT), there is a point c between x_1 and x_2 such that

$$\frac{M - m}{x_2 - x_1} = \frac{f(x_2) - f(x_1)}{x_2 - x_1} = f'(c)$$

Since x_1, x_2 lie in $[a_0, b_0]$, we have $|x_2 - x_1| \leq |b_0 - a_0|$, and thus,

$$|M - m| = |f'(c)| \, |x_2 - x_1| \leq K|b_0 - a_0|$$

This proves Eq. (6).

We divide the rest of the proof into two steps. Consider a partition P:

$$P: \qquad x_0 = a < x_1 < \quad \cdots \quad < x_{N-1} < x_N = b$$

Let m_i be the minimum value of f on $[x_{i-1}, x_i]$ and M_i the maximum on $[x_{i-1}, x_i]$. We define the *lower* and *upper* Riemann sums

$$L(f, P) = \sum_{i=1}^{N} m_i \, \Delta x_i, \qquad U(f, P) = \sum_{i=1}^{N} M_i \, \Delta x_i$$

These are the particular Riemann sums in which the intermediate point in $[x_{i-1}, x_i]$ is the point where f takes on its minimum or maximum on $[x_{i-1}, x_i]$. Figure 2 illustrates the case $N = 4$.

Step 1. **Prove that the lower and upper sums approach a limit.**

We observe that

$$L(f, P_1) \leq U(f, P_2) \quad \text{for any two partitions } P_1 \text{ and } P_2 \qquad \boxed{7}$$

Indeed, if a subinterval I_1 of P_1 overlaps with a subinterval I_2 of P_2, then the minimum of f on I_1 is less than or equal to the maximum of f on I_2 (Figure 3). In particular, the lower sums are bounded above by $U(f, P)$ for all partitions P. Let L be the least upper bound of the lower sums. Then for all partitions P,

$$L(f, P) \leq L \leq U(f, P) \qquad \boxed{8}$$

According to Eq. (6), $|M_i - m_i| \leq K \Delta x_i$ for all i. Since $\|P\|$ is the largest of the widths Δx_i, we see that $|M_i - m_i| \leq K\|P\|$ and

$$|U(f, P) - L(f, P)| \leq \sum_{i=1}^{N} |M_i - m_i| \, \Delta x_i$$

$$\leq K\|P\| \sum_{i=1}^{N} \Delta x_i = K\|P\| \, |b - a| \qquad \boxed{9}$$

Let $c = K|b - a|$. Using Eq. (8) and Eq. (9), we obtain

$$|L - U(f, P)| \leq |U(f, P) - L(f, P)| \leq c\|P\|$$

We conclude that $\lim_{\|P\| \to 0} |L - U(f, P)| = 0$. Similarly,

$$|L - L(f, P)| \leq c\|P\|$$

and

$$\lim_{\|P\| \to 0} |L - L(f, P)| = 0$$

Thus, we have

$$\lim_{\|P\| \to 0} U(f, P) = \lim_{\|P\| \to 0} L(f, P) = L$$

Step 2. **Prove that** $\int_a^b f(x)\,dx$ **exists and has value** *L.*

Recall that for any choice C of intermediate points $c_i \in [x_{i-1}, x_i]$, we define the Riemann sum

$$R(f, P, C) = \sum_{i=1}^{N} f(c_i)\Delta x_i$$

We have

$$L(f, P) \le R(f, P, C) \le U(f, P)$$

Indeed, since $c_i \in [x_{i-1}, x_i]$, we have $m_i \le f(c_i) \le M_i$ for all i and

$$\sum_{i=1}^{N} m_i\,\Delta x_i \le \sum_{i=1}^{N} f(c_i)\,\Delta x_i \le \sum_{i=1}^{N} M_i\,\Delta x_i$$

It follows that

$$|L - R(f, P, C)| \le |U(f, P) - L(f, P)| \le c\|P\|$$

This shows that $R(f, P, C)$ converges to L as $\|P\| \to 0$.   ■

| Section 10.1

> **THEOREM 8**   If f is continuous and $\{a_n\}$ is a sequence such that the limit $\lim_{n\to\infty} a_n = L$ exists, then
>
> $$\lim_{n\to\infty} f(a_n) = f(L)$$

Proof Choose any $\epsilon > 0$. Since f is continuous, there exists $\delta > 0$ such that

$$\text{if } 0 < |x - L| < \delta, \text{ then } |f(x) - f(L)| < \epsilon.$$

Since $\lim_{n\to\infty} a_n = L$, there exists $N > 0$ such that $|a_n - L| < \delta$ for $n > N$. Thus,

$$|f(a_n) - f(L)| < \epsilon \qquad \text{for } n > N$$

It follows that $\lim_{n\to\infty} f(a_n) = f(L)$.   ■

| Section 14.3

> **THEOREM 9 Clairaut's Theorem**   If f_{xy} and f_{yx} both exist and are continuous on a disk D, then $f_{xy}(a, b) = f_{yx}(a, b)$ for all $(a, b) \in D$.

Proof We prove that both $f_{xy}(a, b)$ and $f_{yx}(a, b)$ are equal to the limit

$$L = \lim_{h\to 0} \frac{f(a+h, b+h) - f(a+h, b) - f(a, b+h) + f(a, b)}{h^2}$$

Let $F(x) = f(x, b+h) - f(x, b)$. The numerator in the limit is equal to

$$F(a+h) - F(a)$$

and $F'(x) = f_x(x, b+h) - f_x(x, b)$. By the MVT, there exists a_1 between a and $a+h$ such that

$$F(a+h) - F(a) = hF'(a_1) = h(f_x(a_1, b+h) - f_x(a_1, b))$$

By the MVT applied to f_x, there exists b_1 between b and $b+h$ such that

$$f_x(a_1, b+h) - f_x(a_1, b) = hf_{xy}(a_1, b_1)$$

Thus,

$$F(a+h) - F(a) = h^2 f_{xy}(a_1, b_1)$$

and

$$L = \lim_{h \to 0} \frac{h^2 f_{xy}(a_1, b_1)}{h^2} = \lim_{h \to 0} f_{xy}(a_1, b_1) = f_{xy}(a, b)$$

The last equality follows from the continuity of f_{xy} since (a_1, b_1) approaches (a, b) as $h \to 0$. To prove that $L = f_{yx}(a, b)$, repeat the argument using the function $F(y) = f(a+h, y) - f(a, y)$, with the roles of x and y reversed. ∎

❚ Section 14.4

> **THEOREM 10 Criterion for Differentiability** If $f_x(x, y)$ and $f_y(x, y)$ exist and are continuous on an open disk D, then $f(x, y)$ is differentiable on D.

Proof Let $(a, b) \in D$ and set

$$L(x, y) = f(a, b) + f_x(a, b)(x - a) + f_y(a, b)(y - b)$$

It is convenient to switch to the variables h and k, where $x = a + h$ and $y = b + k$. Set

$$\Delta f = f(a+h, b+k) - f(a, b)$$

Then

$$L(x, y) = f(a, b) + f_x(a, b)h + f_y(a, b)k$$

and we may define the function

$$e(h, k) = f(x, y) - L(x, y) = \Delta f - (f_x(a, b)h + f_y(a, b)k)$$

To prove that $f(x, y)$ is differentiable, we must show that

$$\lim_{(h,k) \to (0,0)} \frac{e(h, k)}{\sqrt{h^2 + k^2}} = 0$$

To do this, we write Δf as a sum of two terms:

$$\Delta f = (f(a+h, b+k) - f(a, b+k)) + (f(a, b+k) - f(a, b))$$

and apply the MVT to each term separately. We find that there exist a_1 between a and $a+h$, and b_1 between b and $b+k$, such that

$$f(a+h, b+k) - f(a, b+k) = h f_x(a_1, b+k)$$

$$f(a, b+k) - f(a, b) = k f_y(a, b_1)$$

Therefore,

$$e(h, k) = h(f_x(a_1, b+k) - f_x(a, b)) + k(f_y(a, b_1) - f_y(a, b))$$

and for $(h, k) \neq (0, 0)$,

$$\left| \frac{e(h, k)}{\sqrt{h^2 + k^2}} \right| = \left| \frac{h(f_x(a_1, b+k) - f_x(a, b)) + k(f_y(a, b_1) - f_y(a, b))}{\sqrt{h^2 + k^2}} \right|$$

$$\leq \left| \frac{h(f_x(a_1, b+k) - f_x(a, b))}{\sqrt{h^2 + k^2}} \right| + \left| \frac{k(f_y(a, b_1) - f_y(a, b))}{\sqrt{h^2 + k^2}} \right|$$

$$= |f_x(a_1, b+k) - f_x(a, b)| + |f_y(a, b_1) - f_y(a, b)|$$

In the second line, we use the Triangle Inequality [see Eq. (1) in Section 1.1], and we may pass to the third line because $\left| h/\sqrt{h^2 + k^2} \right|$ and $\left| k/\sqrt{h^2 + k^2} \right|$ are both less than 1. Both terms in the last line tend to zero as $(h, k) \to (0, 0)$ because f_x and f_y are assumed to be continuous. This completes the proof that $f(x, y)$ is differentiable. ∎

ANSWERS TO ODD-NUMBERED EXERCISES

Chapter 1

Section 1.1 Preliminary Questions

1. $a = -3$ and $b = 1$

2. The numbers $a \geq 0$ satisfy $|a| = a$ and $|-a| = a$. The numbers $a \leq 0$ satisfy $|a| = -a$.

3. $a = -3$ and $b = 1$ **4.** No   **5.** $(9, -4)$

6. **(a)** First quadrant   **(b)** Second quadrant
(c) Fourth quadrant   **(d)** Third quadrant

7. 3   **8.** **(b)**   **9.** Symmetry with respect to the origin

10. The only function that is both even and odd is the constant function $f(x) = 0$.

Section 1.1 Exercises

1. **(a)** Correct   **(b)** Correct   **(c)** Incorrect   **(d)** Correct

3. $128 - 448x + 672x^2 - 560x^3 + 280x^4 - 84x^5 + 14x^6 - x^7$

5. a, c   **7.** $|x| \leq 2$ **9.** $|x - 2| < 2$ **11.** $|x - 3.5| \leq 4.5$

13. $-8 < x < 8$ **15.** $-3 < x < 2$ **17.** $(-4, 4)$ **19.** $(2, 6)$

21. $\left[-\frac{7}{4}, \frac{9}{4}\right]$ **23.** $(-\infty, 2) \cup (6, \infty)$ **25.** $(-\infty, -\sqrt{3}) \cup (\sqrt{3}, \infty)$

27. **(a)** (i)   **(b)** (iii)   **(c)** (v)   **(d)** (vi)   **(e)** (ii)   **(f)** (iv)

29. $(-3, 1)$

31. If a and b are both positive, then $a > b \Rightarrow 1 > \frac{b}{a} \Rightarrow \frac{1}{b} > \frac{1}{a}$. If a and b are both negative, then $a > b \Rightarrow 1 < \frac{b}{a} \Rightarrow \frac{1}{b} > \frac{1}{a}$. If $a > 0$ and $b < 0$, then $\frac{1}{a} > 0$ and $\frac{1}{b} < 0$ so $\frac{1}{b} < \frac{1}{a}$.

33.
$|a + b - 13| = |(a - 5) + (b - 8)| \leq |a - 5| + |b - 8| < \frac{1}{2} + \frac{1}{2} = 1$

35. **(a)** 11   **(b)** 1   **37.** $r_1 = \frac{3}{11}$ and $r_2 = \frac{4}{15}$

39. **(a)** $d = \sqrt{(3-1)^2 + (2-4)^2} = \sqrt{2^2 + (-2)^2} = \sqrt{8} = 2\sqrt{2}$

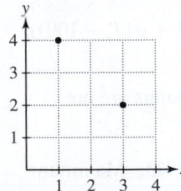

(b) $d = \sqrt{(2-2)^2 + (4-1)^2} = \sqrt{9} = 3$

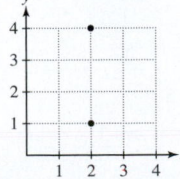

41. **(a)** $(x - 2)^2 + (y - 4)^2 = 9$ **(b)** $(x - 2)^2 + (y - 4)^2 = 26$

43. $D = \{r, s, t, u\}$; $R = \{A, B, E\}$ **45.** D: all reals; R: all reals

47. D: all reals; R: all reals   **49.** D: all reals; R: $\{y : y \geq 0\}$

51. D: $\{x : x \neq 0\}$; R: $\{y : y > 0\}$ **53.** On the interval $(-1, \infty)$

55. On the interval $(0, \infty)$

57. Zeros: ± 2; Increasing: $x > 0$; Decreasing: $x < 0$; Symmetry: $f(-x) = f(x)$, so y-axis symmetry

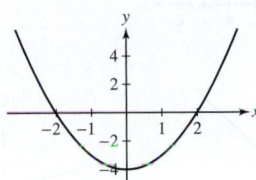

59. Zeros: $0, \pm 2$; Symmetry: $f(-x) = -f(x)$, so origin symmetry

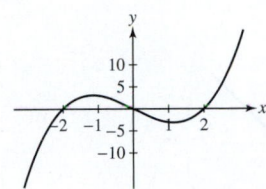

61. This is an x-axis reflection of $y = x^3$ translated up 2 units. There is one zero at $x = \sqrt[3]{2}$.

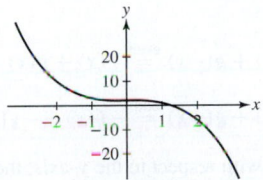

63. B, E, F   **65.** **(a)** Odd   **(b)** Neither even nor odd   **(c)** Even

67. $f(x) = g(x) + h(x)$, where $g(x) = 2x^4 + 12x^2 + 4$ and $h(x) = -5x^3 - 3x$

69. We have
$$f(-x) = p\left(\frac{2 - (-x)}{2 + (-x)}\right) = p\left(\frac{2 + x}{2 - x}\right) = p(2 - x) - p(2 + x) =$$
$$-(p(2 + x) - p(2 - x)) = -p\left(\frac{2 - x}{2 + x}\right) = -f(x). \text{ Since}$$
$f(-x) = -f(x)$ it follows that f is an odd function.

71. D: $[0, 4]$; R: $[0, 4]$

73.

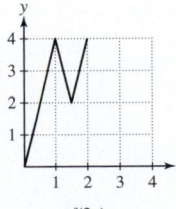

$f(2x)$

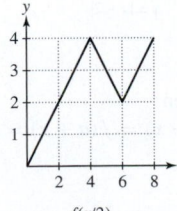

$f(x/2)$

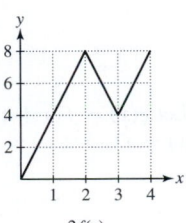

$2f(x)$

75.

77. **(a)** D: $[4, 8]$, R: $[5, 9]$ **(b)** D: $[1, 5]$, R: $[2, 6]$
(c) D: $\left[\frac{4}{3}, \frac{8}{3}\right]$, R: $[2, 6]$ **(d)** D: $[4, 8]$, R: $[6, 18]$

ANS1

79. (a) $f(x) = (2(x-5))^4 - (2(x-5))^2$
(b) $f(x) = (2x-5)^4 - (2x-5)^2$
(c)

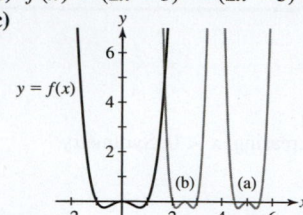

81.

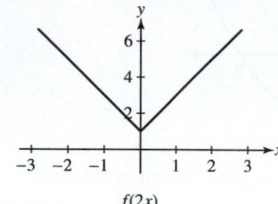 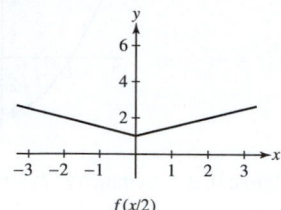

$f(2x)$ $f(x/2)$

83.

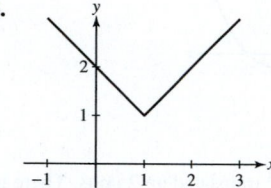

D: all reals; R: $\{y \mid y \geq 1\}$; $f(x) = |x-1| + 1$

85. Even:
$(f+g)(-x) = f(-x) + g(-x) \overset{\text{even}}{=} f(x) + g(x) = (f+g)(x)$
Odd:
$(f+g)(-x) = f(-x) + g(-x) \overset{\text{odd}}{=} -f(x) + -g(x) = -(f+g)(x)$

87. If f is symmetric with respect to the y-axis, then $f(-x) = f(x)$. If f is also symmetric with respect to the origin, then $f(-x) = -f(x)$. Thus $f(x) = -f(x)$ or $2f(x) = 0$ or $f(x) = 0$.

91. (a) There are many possibilities, one of which is

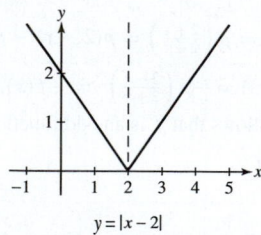

$y = |x-2|$

(b) Let $g(x) = f(x+a)$. Then
$g(-x) = f(-x+a) = f(a-x) = f(a+x) = g(x)$

Section 1.2 Preliminary Questions

1. -4 **2.** No

3. Parallel to the y-axis when $b = 0$; parallel to the x-axis when $a = 0$

4. $\Delta y = 9$ **5.** -4 **6.** $(x-0)^2 + 1$

7. For $a \neq 0$, vertex at $(0, -1)$. Large $a < 0$: opens down, narrow. Small $a < 0$: opens down, wider. At $a = 0$: a horizontal line. Small $a > 0$: opens up, wide. Large $a > 0$: opens up, more narrow.

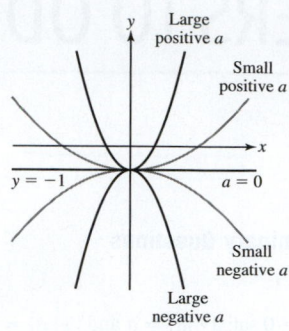

8. Open up with intercepts at 0 and $-b$. Vertex at $\left(-\frac{b}{2}, -\frac{b^2}{4}\right)$. For $b < 0$, vertex in fourth quadrant. For $b = 0$, vertex at the origin. For $b > 0$, vertex in third quadrant.

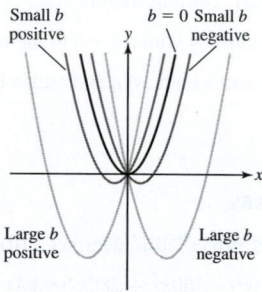

Section 1.2 Exercises

1. $m = 3$; $y = 12$; $x = -4$ **3.** $m = -\frac{4}{9}$; $y = \frac{1}{3}$; $x = \frac{3}{4}$

5. $m = 3$ **7.** $m = -\frac{3}{4}$ **9.** $y = 3x + 8$ **11.** $y = 3x - 12$

13. $y = -2$ **15.** $y = 3x - 2$ **17.** $5x - 3y = 1$ **19.** $y = 4$

21. $y = -2x + 9$ **23.** $3x + 4y = 12$

25. (a) $c = -\frac{1}{4}$ **(b)** $c = -2$ **(c)** No value for c that will make this slope equal to 0 **(d)** $c = 0$

27. (a) $N(P) = -5P + 15{,}000$; **(b)** Slope $= -5$ computers/ dollar. For every dollar increase in price, five fewer computers are sold. **(c)** $\Delta N = -500$

29. (a) Slope $= -70$ students/week. Enrollment dropped by 70 students per week during Fall 2017.

(b) Slope $= 3.5$ dollars/person. The rental cost increases by 3.5 dollars for each person attending.

31. (a) 40.0248 cm **(b)** 64.9597 cm **(c)** $L = 65(1 + \alpha(T - 100))$

33. $b = 4$

35. No, because the slopes between consecutive data points are not equal. **37. (a)** 1 or $-\frac{1}{4}$ **(b)** $1 \pm \sqrt{2}$

39. Minimum value is 0 **41.** Minimum value is -7 **43.** Maximum value is $\frac{137}{16}$ **45.** Maximum value is $\frac{1}{3}$

47.

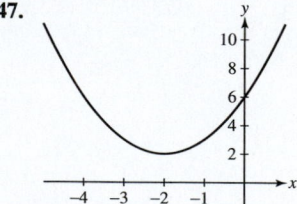

49. A double root occurs when $c = \pm 2$. There are no real roots when $-2 < c < 2$.

51. (a) With $y = 3x - 25$, there are no intersection points. With $y = 6x - 25$, there is a single intersection point at $(2, -13)$. With $y = 9x - 25$, there are intersection points at $\left(\frac{7-\sqrt{33}}{2}, \frac{13-9\sqrt{33}}{22}\right)$ and $\left(\frac{7+\sqrt{33}}{2}, \frac{13+9\sqrt{33}}{2}\right)$.

(b)

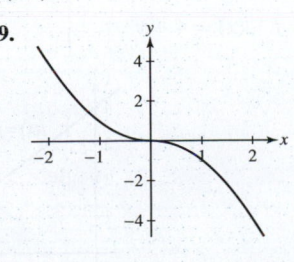

(c) All have y-intercept -25. At $c = 0$, horizontal. For $0 \le c < 6$, no intersection with parabola. For $c = 6$, intersection at one point, $(2, -13)$. For $c > 6$, intersection in two points.

53. x **55.** $4 \pm \sqrt{8}$

57. If $f(x) = mx + b$ and $g(x) = nx + d$, then $f(x) + g(x) = mx + b + nx + d = (m + n)x + (b + d)$, which is linear. fg is not generally linear.

59. For x^2, $\frac{\Delta y}{\Delta x} = \frac{x_2^2 - x_1^2}{x_2 - x_1} = x_2 + x_1$

63. $(x - \alpha)(x - \beta) = x^2 - \alpha x - \beta x + \alpha\beta = x^2 + (-\alpha - \beta)x + \alpha\beta$

Section 1.3 Preliminary Questions

1. A rational function is a quotient of polynomials. Both $h(x) = 1$ and $k(x) = x^3 + 1$ are polynomials, and $f(x) = \frac{k(x)}{h(x)}$, while $g(x) = \frac{h(x)}{k(x)}$. Thus, $f(x)$ and $g(x)$ are quotients of polynomials and therefore are rational functions.

2. $y = |x|$ is not a polynomial; $y = |x^2 + 1|$ is a polynomial.

3. The domain of $f \circ g$ is the empty set.

4. Decreasing

5. The numerator and the denominator of both $f(x)$ and $g(x)$ are roots of polynomials (in three cases, polynomials themselves). Since a quotient of roots of polynomials is an algebraic function, it follows that both f and g are algebraic functions.

6. **(a)** h **(b)** k **(c)** g **(d)** f

Section 1.3 Exercises

1. $x \ge 0$ **3.** All reals   **5.** $t \neq -2$ **7.** $u \neq \pm 2$ **9.** $x \neq 0, 1$

11. $y > 0$ **13.** Polynomial   **15.** Algebraic   **17.** Transcendental

19. Rational   **21.** Transcendental   **23.** Rational   **25.** Yes

27. $f(g(x)) = \sqrt{x + 1}$; D: $x \ge -1$, $g(f(x)) = \sqrt{x} + 1$; D: $x \ge 0$

29. $f(g(x)) = 2^{x^2}$; D: $\mathbf{R}$, $g(f(x)) = (2^x)^2 = 2^{2x}$; D: $\mathbf{R}$

31. $f(g(x)) = \cos(x^3 + x^2)$; D: $\mathbf{R}$, $g(f(\theta)) = \cos^3 \theta + \cos^2 \theta$; D: $\mathbf{R}$

33. $f(g(t)) = \frac{1}{\sqrt{-t^2}}$; D: Not valid for any t,

$g(f(t)) = -\left(\frac{1}{\sqrt{t}}\right)^2 = -\frac{1}{t}$; D: $t > 0$

35. $r(V) = \left(\frac{3V}{4\pi}\right)^{\frac{1}{3}}$ and $S(V) = \sqrt[3]{\pi}(6V)^{\frac{2}{3}}$

37.

39.

41. **(a)** Domain $= \{x : x \neq 0\}$; range $= \{-1, 1\}$

(b)

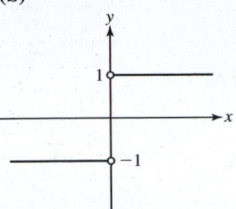

(c) $f(x) = \begin{cases} -1 & \text{when } x < 0 \\ 1 & \text{when } x > 0 \end{cases}$

43. $P(t + 10) = 30 \cdot 2^{0.1(t+10)} = 30 \cdot 2^{0.1t+1} = 2(30 \cdot 2^{0.1t}) = 2P(t)$; $g\left(t + \frac{1}{k}\right) = a2^{k(t+1/k)} = a2^{kt+1} = 2a2^{kt} = 2g(t)$

45. $f(x) = x^2$: $\delta f(x) = f(x + 1) - f(x) = (x + 1)^2 - x^2 = 2x + 1$
$f(x) = x$: $\delta f(x) = x + 1 - x = 1$
$f(x) = x^3$: $\delta f(x) = (x + 1)^3 - x^3 = 3x^2 + 3x + 1$

47.

$\delta(f + g) = (f(x + 1) + g(x + 1)) - (f(x) + g(x))$

$= (f(x + 1) - f(x)) + (g(x + 1) - g(x)) = \delta f(x) + \delta g(x)$

$\delta(cf) = cf(x + 1) - cf(x) = c(f(x + 1) - f(x)) = c\delta f(x)$.

Section 1.4 Preliminary Questions

1. It is possible if the rotations differ by an integer multiple of 2π.

2. $\frac{9\pi}{4}$ and $\frac{41\pi}{4}$ **3.** $-\frac{5\pi}{3}$ **4.** **(a)**

5. Let O denote the center of the unit circle, and let P be a point on the unit circle such that the radius $\overline{OP}$ makes an angle θ with the positive x-axis. Then, $\sin \theta$ is the y-coordinate of the point P.

6. Let O denote the center of the unit circle, and let P be a point on the unit circle such that the radius $\overline{OP}$ makes an angle θ with the positive x-axis. The angle $\theta + 2\pi$ is obtained from the angle θ by making one full revolution around the circle. The angle $\theta + 2\pi$ will therefore have the radius $\overline{OP}$ as its terminal side.

Section 1.4 Exercises

1. $5\pi/4$

3. **(a)** $\frac{180°}{\pi} \approx 57.3°$ **(b)** $60°$ **(c)** $\frac{75°}{\pi} \approx 23.87°$ **(d)** $225°$

5. $s = r\theta = 3.6$; $s = r\phi = 8$

7.

θ	$(\cos\theta, \sin\theta)$	θ	$(\cos\theta, \sin\theta)$
$\frac{\pi}{2}$	$(0, 1)$	$\frac{5\pi}{4}$	$\left(\frac{-\sqrt{2}}{2}, \frac{-\sqrt{2}}{2}\right)$
$\frac{2\pi}{3}$	$\left(\frac{-1}{2}, \frac{\sqrt{3}}{2}\right)$	$\frac{4\pi}{3}$	$\left(\frac{-1}{2}, \frac{-\sqrt{3}}{2}\right)$
$\frac{3\pi}{4}$	$\left(\frac{-\sqrt{2}}{2}, \frac{\sqrt{2}}{2}\right)$	$\frac{3\pi}{2}$	$(0, -1)$
$\frac{5\pi}{6}$	$\left(\frac{-\sqrt{3}}{2}, \frac{1}{2}\right)$	$\frac{5\pi}{3}$	$\left(\frac{1}{2}, \frac{-\sqrt{3}}{2}\right)$
π	$(-1, 0)$	$\frac{7\pi}{4}$	$\left(\frac{\sqrt{2}}{2}, \frac{-\sqrt{2}}{2}\right)$
$\frac{7\pi}{6}$	$\left(\frac{-\sqrt{3}}{2}, \frac{-1}{2}\right)$	$\frac{11\pi}{6}$	$\left(\frac{\sqrt{3}}{2}, \frac{-1}{2}\right)$

9. $\theta = \frac{\pi}{3}, \frac{5\pi}{3}$ **11.** $\theta = \frac{3\pi}{4}, \frac{7\pi}{4}$ **13.** $x = \frac{\pi}{3}, \frac{2\pi}{3}$

15.

θ	$\frac{\pi}{6}$	$\frac{\pi}{4}$	$\frac{\pi}{3}$	$\frac{\pi}{2}$	$\frac{2\pi}{3}$	$\frac{3\pi}{4}$	$\frac{5\pi}{6}$
$\tan\theta$	$\frac{1}{\sqrt{3}}$	1	$\sqrt{3}$	und	$-\sqrt{3}$	-1	$-\frac{1}{\sqrt{3}}$
$\sec\theta$	$\frac{2}{\sqrt{3}}$	$\sqrt{2}$	2	und	-2	$-\sqrt{2}$	$-\frac{2}{\sqrt{3}}$

17. The hypotenuse of the triangle will have length $\sqrt{1 + c^2}$.

19. $\sin\theta = \frac{12}{13}$ and $\tan\theta = \frac{12}{5}$

21. $\sin\theta = \frac{2}{\sqrt{53}}$, $\sec\theta = \frac{\sqrt{53}}{7}$, and $\cot\theta = \frac{7}{2}$ **23.** $23/25$

25. $\cos\theta = -\frac{\sqrt{21}}{5}$ and $\tan\theta = -\frac{2\sqrt{21}}{21}$

27. $\cos\theta = -\frac{4}{5}$

29. Let's start with the four points in Figure 22(A).

- The point in the first quadrant:

$$\sin\theta = 0.918, \quad \cos\theta = 0.3965, \quad \text{and}$$

$$\tan\theta = \frac{0.918}{0.3965} = 2.3153$$

- The point in the second quadrant:

$$\sin\theta = 0.3965, \quad \cos\theta = -0.918, \quad \text{and}$$

$$\tan\theta = \frac{0.3965}{-0.918} = -0.4319$$

- The point in the third quadrant:

$$\sin\theta = -0.918, \quad \cos\theta = -0.3965, \quad \text{and}$$

$$\tan\theta = \frac{-0.918}{-0.3965} = 2.3153$$

- The point in the fourth quadrant:

$$\sin\theta = -0.3965, \quad \cos\theta = 0.918, \quad \text{and}$$

$$\tan\theta = \frac{-0.3965}{0.918} = -0.4319$$

Now consider the four points in Figure 22(B).

- The point in the first quadrant:

$$\sin\theta = 0.918, \quad \cos\theta = 0.3965, \quad \text{and}$$

$$\tan\theta = \frac{0.918}{0.3965} = 2.3153$$

- The point in the second quadrant:

$$\sin\theta = 0.918, \quad \cos\theta = -0.3965, \quad \text{and}$$

$$\tan\theta = \frac{0.918}{-0.3965} = -2.3153$$

- The point in the third quadrant:

$$\sin\theta = -0.918, \quad \cos\theta = -0.3965, \quad \text{and}$$

$$\tan\theta = \frac{-0.918}{-0.3965} = 2.3153$$

- The point in the fourth quadrant:

$$\sin\theta = -0.918, \quad \cos\theta = 0.3965, \quad \text{and}$$

$$\tan\theta = \frac{-0.918}{0.3965} = -2.3153$$

31. $\cos\psi = 0.3$, $\sin\psi = \sqrt{0.91}$, $\cot\psi = \frac{0.3}{\sqrt{0.91}}$, and $\csc\psi = \frac{1}{\sqrt{0.91}}$

33. $\sin\frac{7\pi}{12} = \frac{\sqrt{2}+\sqrt{6}}{4}$ and $\cos\frac{7\pi}{12} = \frac{\sqrt{2}-\sqrt{6}}{4}$

35.

37.

39. $3\cos(\theta/2)$; period 4π; amplitude 3

41. Wolf Point: $L(t) = 12 + 3.9\sin(\frac{2\pi}{365}t)$; on April 1, $L(11) \approx 12.73$ hours; on July 15, $L(116) \approx 15.55$ hours; on November 1, $L(225) \approx 9.39$ hours

   Mexico City: $L(t) = 12 + 1.3\sin(\frac{2\pi}{365}t)$; on April 1, $L(11) \approx 12.24$ hours; on July 15, $L(116) \approx 13.18$ hours; on November 1, $L(225) \approx 11.13$ hours

43. If $|c| > 1$, no points of intersection; if $|c| = 1$, one point of intersection; if $|c| < 1$, two points of intersection.

45. $\theta = 0, \frac{\pi}{3}, \pi, \frac{5\pi}{3}$

47. $\cos 2\theta = \cos(\theta + \theta) = \cos\theta\cos\theta - \sin\theta\sin\theta = \cos^2\theta - \sin^2\theta = 2\cos^2\theta - 1$

49. By the corresponding double-angle formula,
$\sin^2\frac{\theta}{2} = \frac{1}{2}(1 - \cos 2(\frac{\theta}{2})) = \frac{1-\cos\theta}{2}$.

51. $\cos(\theta + \pi) = \cos\theta\cos\pi - \sin\theta\sin\pi = \cos\theta(-1) = -\cos\theta$

53. Using Exercises 50 and 51,
$$\tan(\pi - \theta) = \frac{\sin(\pi - \theta)}{\cos(\pi - \theta)} = \frac{-\sin(-\theta)}{-\cos(-\theta)} = \frac{\sin\theta}{-\cos\theta} = -\tan\theta.$$

55. $\dfrac{\sin 2x}{1 + \cos 2x} = \dfrac{2\sin x\cos x}{1 + 2\cos^2 x - 1} = \dfrac{2\sin x\cos x}{2\cos^2 x} = \dfrac{\sin x}{\cos x} = \tan x$

57. $\tan(\theta + \pi) = \dfrac{\sin(\theta + \pi)}{\cos(\theta + \pi)} = \dfrac{-\sin\theta}{-\cos\theta} = \tan\theta$, and
$\cot(\theta + \pi) = \dfrac{\cos(\theta + \pi)}{\sin(\theta + \pi)} = \dfrac{-\cos\theta}{-\sin\theta} = \cot\theta$. Thus, both $\tan\theta$ and $\cot\theta$ are periodic with period π.

59. $\cos^2\dfrac{\theta}{2} = \dfrac{1 + \cos\theta}{2} \Rightarrow$

$\cos^2\dfrac{\pi}{8} = \dfrac{1 + \cos\frac{\pi}{4}}{2} = \dfrac{1 + \frac{\sqrt{2}}{2}}{2} = \dfrac{1}{2} + \dfrac{\sqrt{2}}{4}$

$\cos\dfrac{\pi}{8} > 0$, so $\cos\dfrac{\pi}{8} = \sqrt{\dfrac{1}{2} + \dfrac{\sqrt{2}}{4}}$

61. 16.928

65. Using the distances labeled in Figure 28(A), we see that the slope of the line is given by the ratio r/s. The tangent of the angle θ is given by the same ratio. Therefore, $m = \tan\theta$.

Section 1.5 Preliminary Questions

1. (a), (b), (f)

2. Many different teenagers will have the same last name, so this function will not be one-to-one.

3. This function is one-to-one, and $f^{-1}(6{:}27) =$ Hamilton Township.

4. The graph of the inverse function is the reflection of the graph of $y = f(x)$ through the line $y = x$.

5. (b) and (c) **6.** $\theta = 3\pi$; no

Section 1.5 Exercises

1. $f^{-1}(x) = \frac{x+4}{7}$ **3.** $[-\pi/2, \pi/2]$

5. $\cdot$ $f(g(x)) = ((x - 3)^{1/3})^3 + 3 = x - 3 + 3 = x$
   $\cdot$ $g(f(x)) = (x^3 + 3 - 3)^{1/3} = (x^3)^{1/3} = x$

7. $R(v) = \frac{2GM}{v^2}$ **9.** $r(V) = (\frac{3V}{\pi})^{\frac{1}{2}}$

11. $f^{-1}(x) = 4 - x$

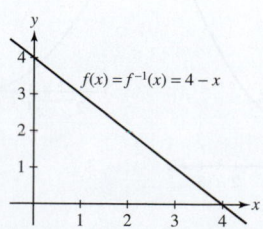

13. $f^{-1}(x) = \frac{1}{7x} + \frac{3}{7}$

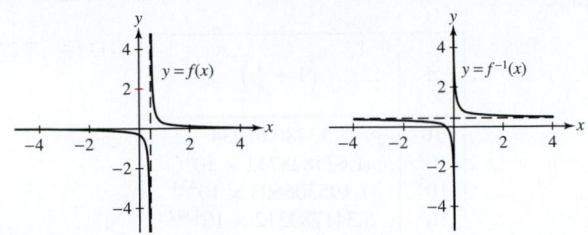

15. Domain $\{x : x \geq 0\}$: $f^{-1}(x) = \frac{\sqrt{1-x^2}}{x}$; domain $\{x : x \leq 0\}$: $f^{-1}(x) = -\frac{\sqrt{1-x^2}}{x}$

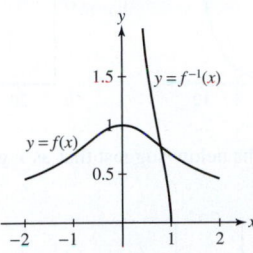

17. $f^{-1}(x) = (x^2 - 9)^{1/3}$

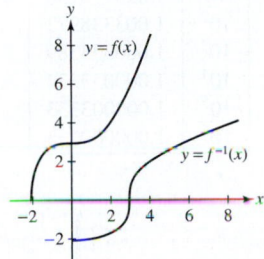

19. Figures (B) and (C)

21. (a)

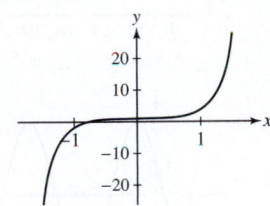

(b) $(-\infty, \infty)$ **(c)** $f^{-1}(3) = 1$

23. Domain $x \leq 1$: $f^{-1}(x) = 1 - \sqrt{x+1}$; domain $x \geq 1$: $f^{-1}(x) = 1 + \sqrt{x+1}$

25. $f^{-1}(x) = \begin{cases} x & \text{when } x < 0 \\ \frac{1}{2}x & \text{when } x \geq 0 \end{cases}$

27. f is not one-to-one.   **29.** 0   **31.** $\frac{\pi}{4}$ **33.** $\frac{\pi}{3}$ **35.** $\frac{\pi}{3}$ **37.** $\frac{\pi}{2}$

39. $-\frac{\pi}{4}$ **41.** π **43.** Not defined   **45.** $\frac{\sqrt{1-x^2}}{x}$ **47.** $\frac{1}{\sqrt{x^2-1}}$

49. $\frac{\sqrt{5}}{3}$ **51.** $\frac{4}{3}$ **53.** $\sqrt{3}$ **55.** $\frac{1}{20}$

57. **(a)** From the graph of f, the Horizontal Line Test implies that f is not one-to-one and therefore is not invertible.

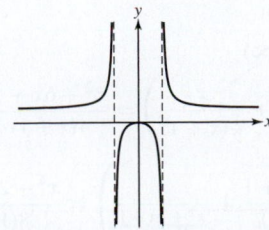

If we restrict the domain to $[0, \infty)$, the Horizontal Line Test implies that f is one-to-one and therefore is invertible.

(b) The function $g(x) = \sqrt{\frac{x}{x-1}}$ is the inverse of f restricted to $[0, \infty)$.

Section 1.6 Preliminary Questions

1. For $0 < x < 1$ **2.** $\ln(-3)$ is not defined.

3. This phrase is a verbal description of the general property of logarithms that states $\log(ab) = \log a + \log b$.

4. $D: x > 0$; R: real numbers

5. Note that $f(1) = 1$, $f(2) = 8$, and $f(3) = 27$. As x increases by 1, from 1 to 2, f increases by 700%, and as x increases by 1, from 2 to 3, f increases by 237.5%. Since these percent increases in f are not the same, while the increases in x are, f does not grow exponentially.

6. $\log_{b^2}(b^4) = 2$ **7.** $f(x) = \cosh x$ and $f(x) = \operatorname{sech} x$

8. $f(x) = \sinh x$ and $f(x) = \tanh x$

9. Both types of functions have the same parity, they share similar identities, and values of trigonometric functions lie on a circle while those of hyperbolic functions lie on a hyperbola.

Section 1.6 Exercises

1. $x = 1$ **3.** $x = -1/2$ **5.** $x = -1/3$ **7.** $k = 9$ **9.** 3   **11.** 0

13. $\frac{5}{3}$ **15.** $\frac{1}{3}$ **17.** $\frac{5}{6}$ **19.** 1   **21.** 7   **23.** 29

25. **(a)** $\ln 1600$ **(b)** $\ln(9x^{7/2})$ **27.** $t = \frac{1}{5}\ln\left(\frac{100}{7}\right)$

29. $x = -1$ or $x = 3$ **31.** $x = e$ **33.** $y = (3 + \ln x)/2$

35.

x	-3	0	5
$\sinh x = \dfrac{e^x - e^{-x}}{2}$	-10.0179	0	74.203
$\cosh x = \dfrac{e^x + e^{-x}}{2}$	10.0677	1	74.210

37. $\ln(2 \cdot 1) \neq (\ln 2)(\ln 1)$

39. $\tanh(-x) = \dfrac{\sinh(-x)}{\cosh(-x)} = -\dfrac{\sinh(x)}{\cosh(x)} = -\tanh(x)$

41. **(a)** $I(D) = 10^{\frac{D-120}{10}}$

(b) Assume D increases by 20 going from D_1 to D_2, and $I_1 = I(D_1)$, $I_2 = I(D_2)$. Then $I_2 = 10^{\frac{D_2 - 120}{10}} = 10^{\frac{D_1 + 20 - 120}{10}} = 10^{2 + \frac{D_1 - 120}{10}} = 10^2 10^{\frac{D_1 - 120}{10}} = 100 I(D_1) = 100 I_1$. Since $I_2 = 100 I_1$, I increases by a factor of 100. Thus, when D increases by 20, I increases by a factor of 100.

47. **(a)** By Galileo's Law, $w = 500 + 10 = 510$ m/s. Using Einstein's Law, $w = c \cdot \tanh(1.7 \times 10^{-6}) \approx 510$ m/s.

(b) By Galileo's Law, $u + v = 10^7 + 10^6 = 1.1 \times 10^7$ m/s. By Einstein's Law, $w \approx c \cdot \tanh(0.036679) \approx 1.09988 \times 10^7$ m/s.

49. Let $y = \log_b x$. Then $x = b^y$ and $\log_a x = \log_a b^y = y \log_a b$. Thus, $y = \dfrac{\log_a x}{\log_a b}$.

53. $13 \cosh x - 3 \sinh x$

Section 1.7 Preliminary Questions

1. It is best to experiment.

2. **(a)** The screen will display nothing. **(b)** The screen will display the portion of the parabola between the points $(0, 3)$ and $(1, 4)$.

3. No

4. Experiment with the viewing window to zoom in on the lowest point on the graph of the function. The y-coordinate of the lowest point on the graph is the minimum value of the function.

Section 1.7 Exercises

1.

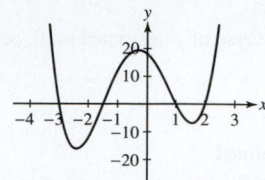

$x = -3, x = -1.5, x = 1,$ and $x = 2$

3. Two positive solutions   **5.** There are no solutions.

7. Nothing; an appropriate viewing window: [50, 150] by [1000, 2000]

9.

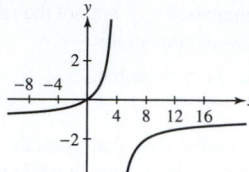

11.

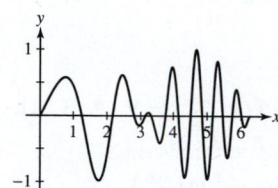
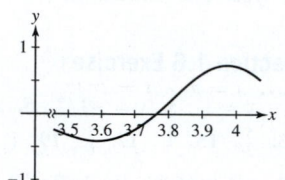

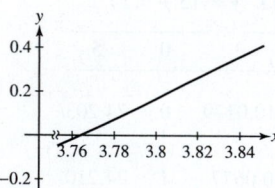

13. The maximum is approximately 0.604, occurring at $x \approx -0.716$.

15. $N \approx 161$

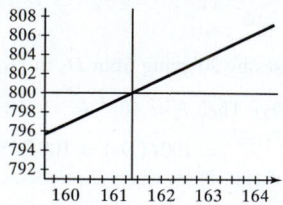

17. The table and graphs below suggest that as n gets large, $n^{1/n}$ approaches 1.

n	$n^{1/n}$
10	1.258925412
10^2	1.047128548
10^3	1.006931669
10^4	1.000921458
10^5	1.000115136
10^6	1.000013816

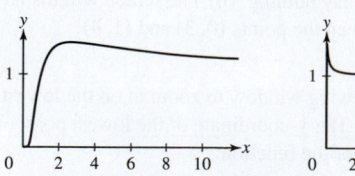

19. The table and graphs below suggest that as n gets large, $f(n)$ tends toward ∞.

n	$\left(1 + \frac{1}{n}\right)^{n^2}$
10	13780.61234
10^2	$1.635828711 \times 10^{43}$
10^3	$1.195306603 \times 10^{434}$
10^4	$5.341783312 \times 10^{4342}$
10^5	$1.702333054 \times 10^{43429}$
10^6	$1.839738749 \times 10^{434294}$

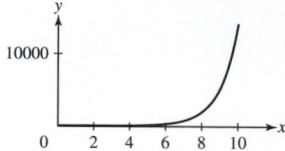
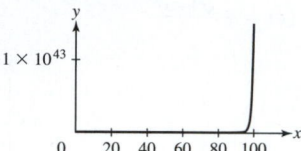

21. The table and graphs below suggest that as x gets large, $f(x)$ approaches 1.

x	$\left(x \tan \frac{1}{x}\right)^x$
10	1.033975758
10^2	1.003338973
10^3	1.000333389
10^4	1.000033334
10^5	1.000003333
10^6	1.000000333

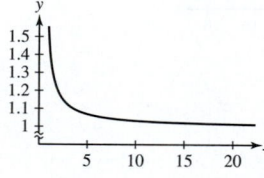
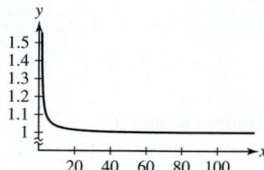

23.

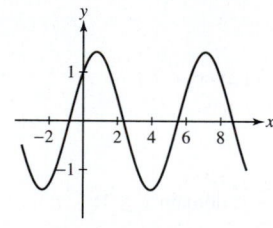
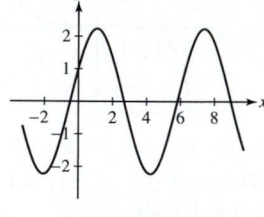

$(A, B) = (1, 1)$ $(A, B) = (1, 2)$

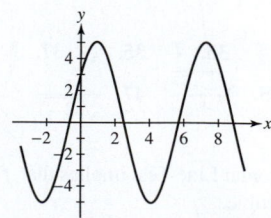

$(A, B) = (3, 4)$

25. $x \in (-2, 0) \cup (3, \infty)$

27.
$$f_3(x) = \frac{1}{2}\left(\frac{1}{2}(x+1) + \frac{x}{\frac{1}{2}(x+1)}\right) = \frac{x^2 + 6x + 1}{4(x+1)}$$

$$f_4(x) = \frac{1}{2}\left(\frac{x^2 + 6x + 1}{4(x+1)} + \frac{x}{\frac{x^2 + 6x + 1}{4(x+1)}}\right) = \frac{x^4 + 28x^3 + 70x^2 + 28x + 1}{8(1+x)(1 + 6x + x^2)}$$

and

$$f_5(x) = \frac{1 + 120x + 1820x^2 + 8008x^3 + 12870x^4 + 8008x^5 + 1820x^6 + 120x^7 + x^8}{16(1+x)(1+6x+x^2)(1+28x+70x^2+28x^3+x^4)}$$

It appears as if the f_n are asymptotic to $\sqrt{x}$.

Chapter 1 Review

1. **(a)** No match **(b)** No match **(c)** **(i)** **(d)** **(iii)**

3. $\{x : |x - 7| < 3\}$ **5.** $[-5, -1] \cup [3, 7]$

7. $(x, 0)$ with $x \geq 0$; $(0, y)$ with $y < 0$

9.

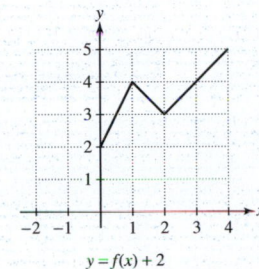

$y = f(x) + 2$

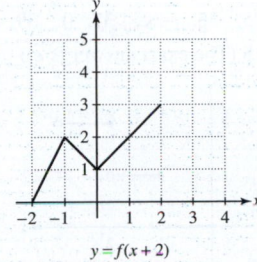

$y = f(x + 2)$

11.

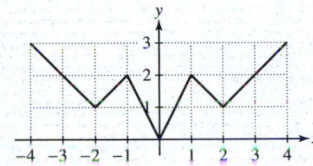

13. $D: \{x : x \geq -1\}$; $R: \{y : y \geq 0\}$

15. $D: \{x : x \neq 3\}$; $R: \{y : y \neq 0\}$

17. **(a)** Decreasing **(b)** Neither **(c)** Neither **(d)** Increasing

19. $2x - 3y = -14$ **21.** $6x - y = 53$

23. $x + 3y = 5$ **25.** $x + y = 5$ **27.** Yes

29. **(a)** $C(P) - 2250P + 1{,}225{,}000$

(b) Slope $= -2250$ customers/dollar. For every dollar increase in the monthly price, there are 2250 fewer customers.
(c) 225,000 customers.

31. Roots: $x = -2$, $x = 0$ and $x = 2$; decreasing: $x < -1.4$ and $0 < x < 1.4$

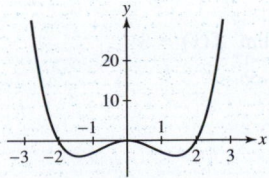

33. $f(x) = 10x^2 + 2x + 5$; minimum value is $\frac{49}{10}$

35.

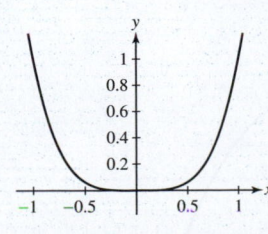

37.

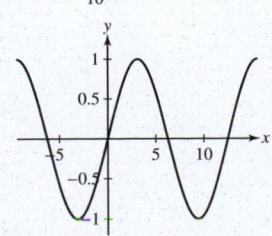

39.

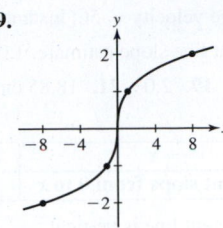

41. Let $g(x) = f\left(\frac{1}{3}x\right)$. Then
$g(x - 3b) = f\left(\frac{1}{3}(x - 3b)\right) = f\left(\frac{1}{3}x - b\right)$. The graph of $y = \left|\frac{1}{3}x - 4\right|$:

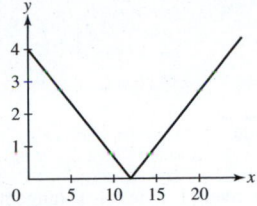

43. $f(t) = t^4$ and $g(t) = 12t + 9$ **45.** **(a)** π **(b)** 4π **(c)** 4π

47. $A = 1.5$; $B = \pi/12$; $C = 16.5$

49. **(a)** $a = b = \pi/2$ **(b)** $a = \pi$

51. $x = \pi/2$, $x = 7\pi/6$, $x = 3\pi/2$ and $x = 11\pi/6$

53. There are no solutions

55. **(a)** **(ii)** **(b)** No match **(c)** **(iii)** **(d)** No match

57. $10 \log_{10} 5000 \approx 36.99$ decibels greater

59. $f^{-1}(x) = \sqrt[3]{x^2 + 8}$; $D: \{x : x \geq 0\}$; $R: \{y : y \geq 2\}$

61. For $\{t : t \leq 3\}$, $h^{-1}(t) = 3 - \sqrt{t}$. For $t \geq 3$, $h^{-1}(t) = 3 + \sqrt{t}$.

63. **(a)** Yes **(b)** Yes $f^{-1}(x) = \begin{cases} -\sqrt{-x} & \text{when } x < 0 \\ x & \text{when } x \geq 0 \end{cases}$

65. **(a)** **(iii)** **(b)** **(iv)** **(c)** **(ii)** **(d)** **(i)**

Chapter 2

Section 2.1 Preliminary Questions

1. The graph of position as a function of time

2. No. Instantaneous velocity is defined as the limit of average velocity as time elapsed shrinks to zero.

3. **(a)** 63 mi/h **(b)** 42 mi/h **(c)** 0 mi/h

4. The slope of the line tangent to the graph of position as a function of time at $t = t_0$

Section 2.1 Exercises

1. **(a)** 11.025 m **(b)** 22.05 m/s
(c)

Time interval	[2, 2.01]	[2, 2.005]	[2, 2.001]	[2, 2.00001]
Average velocity	19.649	19.6245	19.6049	19.600049

The instantaneous velocity at $t = 2$ is 19.6 m/s.

3. Average velocity $= 22$ km/h; instantaneous velocity $= 22$ km/h

5. Average velocity $= 17.1$ m/s; instantaneous velocity estimate: 11.4 m/s

7. 0.3 m/s

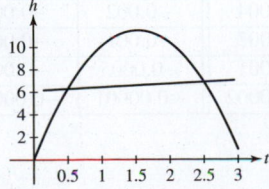

9. Average velocity $= 56$; instantaneous velocity estimate: 24.0

11. Tangent line slope estimate: 1.0   **13.** 12   **15.** 0.75

17. 15.15   **19.** 2.0   **21.** 18.85 cm/s

23. **(a)**

x	1	0.1	0.01	0.001	0.0001
Secant slope from 0 to x	1	3.16	10	31.62	100

(b) The tangent line is vertical.

(c)

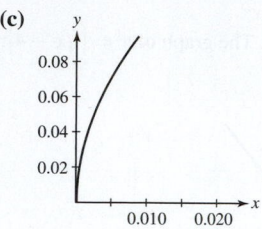

27. (B)

29. Slope of secant line over $[1, t]$ is $t + 1$; tangent line slope estimate: 2; slope of secant line over $[2, t]$ is $t + 2$; tangent line slope estimate: 4

31. The secant line over $[-3, x]$ has slope
$\frac{x^3 + 27}{x + 3} = \frac{(x+3)(x^2 - 3x + 9)}{x + 3} = x^2 - 3x + 9$. The tangent line slope estimate is 27.

33. **(a)** With 2 subintervals, the total rectangle area $= 9/16$. With 3 subintervals, the total rectangle area $= 4/9$. With 5 subintervals, the total rectangle area $= 9/25$. With 10 subintervals, the total rectangle area $= 0.3025$. **(b)** $A(2) = 9/16$; $A(3) = 4/9$; $A(5) = 9/25$; $A(10) = 0.3025$ **(c)** $A(100) = 0.255025$; $A(1000) = 0.25050025$; $A(10,000) = 0.2500500025$; conjecture $A = 1/4$

Section 2.2 Preliminary Questions

1. 1   **2.** π **3.** 20   **4.** Yes; $f(x) = \frac{x^2 - 9}{x - 3}$ at $c = 3$

5. $\lim\limits_{x \to 1^-} f(x) = \infty$ and $\lim\limits_{x \to 1^+} f(x) = 3$

6. No because $\lim\limits_{x \to 5^-} f(x)$ may not be equal to $\lim\limits_{x \to 5^+} f(x)$

7. Yes

Section 2.2 Exercises

1.

x	0.998	0.999	0.9995	0.99999
$f(x)$	1.498501	1.499250	1.499625	1.499993

x	1.00001	1.0005	1.001	1.002
$f(x)$	1.500008	1.500375	1.500750	1.501500

The limit as $x \to 1$ is $\frac{3}{2}$.

3.

y	1.998	1.999	1.9999
$f(y)$	0.59984	0.59992	0.599992

y	2.0001	2.001	2.002
$f(y)$	0.600008	0.60008	0.60016

The limit as $y \to 2$ is $\frac{3}{5}$.

5.

t	$f(t)$	t	$f(t)$
0.002	0.004	-0.002	-0.004
0.001	0.002	-0.001	-0.002
0.0005	0.001	-0.0005	-0.001
0.00001	0.00002	-0.00001	-0.00002

Limit guess: 0

7. 1.5   **9.** 21

11.

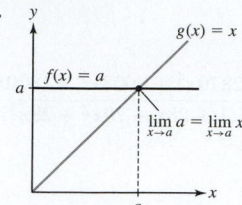

13. $|3x - 12| = 3|x - 4|$

15. $|(5x + 2) - 17| = |5x - 15| = 5|x - 3|$

17. Suppose $|x| < 1$ so that $|x^2 - 0| = |x + 0||x - 0| = |x||x| < |x|$.

19. If $|x| < 1$, $|4x + 2|$ can be no larger than 6, so $|4x^2 + 2x + 5 - 5| = |4x^2 + 2x| = |x||4x + 2| < 6|x|$.

21. $\frac{1}{2}$ **23.** $\frac{5}{3}$ **25.** 2   **27.** 1   **29.** 0

31. As $x \to 4^-$, $f(x) \to -\infty$; similarly, as $x \to 4^+$, $f(x) \to \infty$.

33. $-1/5$ **35.** $-\infty$ **37.** 0   **39.** 1

41. 2.718 (The exact answer is e.)   **43.** ∞

45.

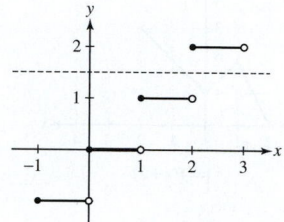

(a) $c - 1$ **(b)** c **(c)** 2

47. $\lim\limits_{x \to 0^-} f(x) = -1$, $\lim\limits_{x \to 0^+} f(x) = 1$

49. $\lim\limits_{x \to 0^-} f(x) = \infty$, $\lim\limits_{x \to 0^+} f(x) = \frac{1}{6}$

51. $\lim\limits_{x \to -2^-} \frac{4x^2 + 7}{x^3 + 8} = -\infty$, $\lim\limits_{x \to -2^+} \frac{4x^2 + 7}{x^3 + 8} = \infty$

53. $\lim\limits_{x \to 1^{\pm}} \frac{x^5 + x - 2}{x^2 + x - 2} = 2$

55. • $\lim\limits_{x \to 2^-} f(x) = \infty$ and $\lim\limits_{x \to 2^+} f(x) = \infty$.
• $\lim\limits_{x \to 4^-} f(x) = -\infty$ and $\lim\limits_{x \to 4^+} f(x) = 10$.
The vertical asymptotes are the vertical lines $x = 2$ and $x = 4$.

57.

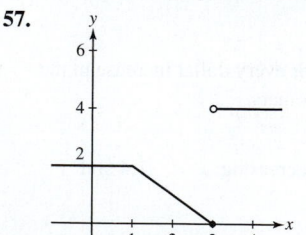

59.

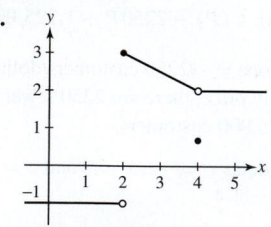

61. • $\lim\limits_{x \to 1^-} f(x) = \lim\limits_{x \to 1^+} f(x) = 3$
• $\lim\limits_{x \to 3^-} f(x) = -\infty$
• $\lim\limits_{x \to 3^+} f(x) = 4$
• $\lim\limits_{x \to 5^-} f(x) = 2$
• $\lim\limits_{x \to 5^+} f(x) = -3$
• $\lim\limits_{x \to 6^-} f(x) = \lim\limits_{x \to 6^+} f(x) = \infty$

63. $\frac{5}{2}$

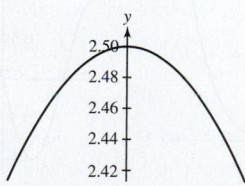

65. 0.693 (The exact answer is ln 2.)

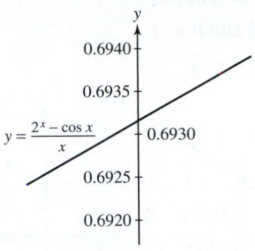

$$y = \frac{2^x - \cos x}{x}$$

67. -12

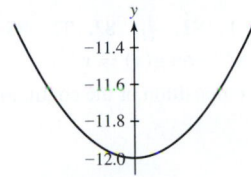

69. For n even

71. (a) No   (b) $f\left(\frac{1}{2n}\right) = 1$ for all integers n

(c) At $x = 1, \frac{1}{3}, \frac{1}{5}, \ldots$, the value of $f(x)$ is always -1.

73. $\lim\limits_{\theta \to 0} \dfrac{\sin n\theta}{\theta} = n$ **75.** $\frac{1}{2}, 2, \frac{3}{2}, \frac{2}{3}$; $\lim\limits_{x \to 1} \dfrac{x^n - 1}{x^m - 1} = \dfrac{n}{m}$

77. (a)   (b) $L = 5.545$

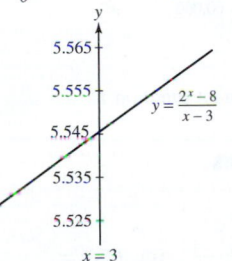

$$y = \frac{2^x - 8}{x - 3}$$

$x = 3$

Section 2.3 Preliminary Questions

1. Suppose $\lim\limits_{x \to c} f(x)$ and $\lim\limits_{x \to c} g(x)$ both exist. The Sum Law states that

$$\lim_{x \to c}(f(x) + g(x)) = \lim_{x \to c} f(x) + \lim_{x \to c} g(x)$$

Provided $\lim\limits_{x \to c} g(x) \neq 0$, the Quotient Law states that

$$\lim_{x \to c} \frac{f(x)}{g(x)} = \frac{\lim_{x \to c} f(x)}{\lim_{x \to c} g(x)}$$

2. (b)   **3.** (a)

Section 2.3 Exercises

1. 9   **3.** $\frac{1}{16}$ **5.** $\frac{1}{2}$ **7.** 4.6   **9.** 1   **11.** 9   **13.** 1/8   **15.** $-\frac{2}{5}$
17. 10   **19.** $\frac{1}{5}$ **21.** $\frac{1}{5}$ **23.** $\frac{2}{5}$ **25.** 64

27. $\lim\limits_{x \to c}\left(\dfrac{1}{f(x)}\right) = \dfrac{\left(\lim\limits_{x \to c} 1\right)}{\left(\lim\limits_{x \to c}(f(x))\right)} = \dfrac{1}{\lim\limits_{x \to c} f(x)}$ **29.** 3   **31.** $\frac{1}{16}$ **33.** No

35. (a) 0   (b) $2/\pi$ (c) The limit does not exist.   (d) 0
37. $f(x) = 1/x$ and $g(x) = -1/x$ **39.** $f(x) = 1/x$ and $g(x) = -1/x$
41. Write $g(t) = \dfrac{tg(t)}{t}$.   **43.** (b)

Section 2.4 Preliminary Questions

1. Continuity   **2.** $f(3) = \frac{1}{2}$ **3.** No   **4.** No; yes
5. (a) False. The correct statement is "f is continuous at $x = a$ if the left- and right-hand limits of $f(x)$ as $x \to a$ exist and equal $f(a)$."
(b) True
(c) False. The correct statement is "If the left- and right-hand limits of $f(x)$ as $x \to a$ are equal but not equal to $f(a)$, then f has a removable discontinuity at $x = a$."

(d) True
(e) False. The correct statement is "If f and g are continuous at $x = a$ and $g(a) \neq 0$, then f/g is continuous at $x = a$."

Section 2.4 Exercises

1. • The function f is discontinuous at $x = 1$; it is left-continuous there.
• The function f is discontinuous at $x = 3$; it is neither left-continuous nor right-continuous there.
• The function f is discontinuous at $x = 5$; it is left-continuous there.
None of these discontinuities is removable.

3. $x = 3$; redefine $g(3) = 4$

5. The function f is discontinuous at $x = 0$, at which $\lim\limits_{x \to 0^-} f(x) = \infty$ and $\lim\limits_{x \to 0^+} f(x) = 2$. The function f is also discontinuous at $x = 2$, at which $\lim\limits_{x \to 2^-} f(x) = 6$ and $\lim\limits_{x \to 2^+} f(x) = 6$. The discontinuity at $x = 2$ is removable. Assigning $f(2) = 6$ makes f continuous at $x = 2$.

7. $y = x$ and $y = \sin x$ are continuous; so is $f(x) = x + \sin x$ by Continuity Law (i).

9. Since $y = x$ and $y = \sin x$ are continuous, so are $y = 3x$ and $y = 4 \sin x$ by Continuity Law (ii). Thus, $f(x) = 3x + 4 \sin x$ is continuous by Continuity Law (i).

11. Since $y = x$ is continuous, so is $y = x^2$ by Continuity Law (iii). Recall that constant functions, such as 1, are continuous. Thus, $y = x^2 + 1$ is continuous by Continuity Law (i). Finally, $f(x) = \dfrac{1}{x^2 + 1}$ is continuous by Continuity Law (iv) because $x^2 + 1$ is never 0.

13. The function f is a composite of two continuous functions: $y = \cos x$ and $y = x^2$, so f is continuous by Theorem 5.

15. The functions $g(x) = 3^x$ and $h(x) = \cos 3x$ are continuous (the latter by Theorem 5 since it is a composition of continuous functions). Because f is the product of g and h, Theorem 1 (iii) implies that f is continuous.

17. Discontinuous at $x = 0$, at which there is an infinite discontinuity. The function is neither left- nor right-continuous at $x = 0$.

19. Discontinuous at $x = 1$, at which there is an infinite discontinuity. The function is neither left- nor right-continuous at $x = 1$.

21. Discontinuous at even integers, at which there are jump discontinuities. The function is right-continuous at the even integers but not left-continuous.

23. Infinite discontinuities at $x = \pm 2$. h is neither left- nor right-continuous at both of these points.

25. Discontinuous at $x = \frac{1}{2}$, at which there is an infinite discontinuity. The function is neither left- nor right-continuous at $x = \frac{1}{2}$.

27. Continuous for all x

29. Jump discontinuity at $x = 2$. The function is left-continuous at $x = 2$ but not right-continuous.

31. Removable discontinuity at $x = -5$. f is neither left- nor right-continuous at $x = -5$.

33. Discontinuous whenever $t = \dfrac{(2n+1)\pi}{4}$, where n is an integer. At every such value of t, there is an infinite discontinuity. The function is neither left- nor right-continuous at any of these points of discontinuity.

35. Continuous everywhere

37. Discontinuous at $x = 0$, at which there is an infinite discontinuity. The function is neither left- nor right-continuous at $x = 0$.

39. The domain is all real numbers. Both $y = \sin x$ and $y = \cos x$ are continuous on this domain, so $f(x) = 2 \sin x + 3 \cos x$ is continuous by Continuity Laws (i) and (ii).

41. Domain is $x \geq 0$. Since $y = \sqrt{x}$ and $y = \sin x$ are continuous, so is $f(x) = \sqrt{x} \sin x$ by Continuity Law (iii).

43. Domain is all real numbers. Both $y = x^{2/3}$ and $y = 2^x$ are continuous on this domain, so $f(x) = x^{2/3} 2^x$ is continuous by Continuity Law (iii).

45. Domain is $x \neq 0$. Because the function $y = x^{4/3}$ is continuous and not equal to zero for $x \neq 0$, $f(x) = x^{-4/3}$ is continuous for $x \neq 0$ by Continuity Law (iv).

47. Domain is all $x \neq \pm(2n-1)\pi/2$, where n is a positive integer. Because $y = \tan x$ is continuous on this domain, it follows from Continuity Law (iii) that $f(x) = \tan^2 x$ is also continuous on this domain.

49. Domain of $f(x) = (x^4 + 1)^{3/2}$ is all real numbers. Because $y = x^{3/2}$ and the polynomial $y = x^4 + 1$ are both continuous, so is the composite function $f(x) = (x^4 + 1)^{3/2}$.

51. Domain is all $x \neq \pm 1$. Because the functions $y = \cos x$ and $y = x^2$ are continuous on this domain, so is the composite function $y = \cos(x^2)$. Finally, because the polynomial $y = x^2 - 1$ is continuous and not equal to zero for $x \neq \pm 1$, the function $f(x) = \frac{\cos(x^2)}{x^2 - 1}$ is continuous by Continuity Law (iv).

53. Right-hand limit at $x = 1$: 9; left-hand limit at $x = 1$: 4; both right- and left-hand limits at $x = 2$: 8; f is right-continuous at $x = 1$; f is continuous at $x = 2$.

55.

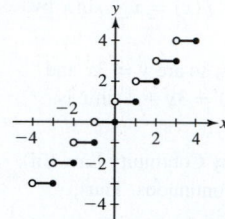

f has a jump discontinuity at $x = n$ for all integers n. At each discontinuity, f is left-continuous.

57. The function f is continuous everywhere.

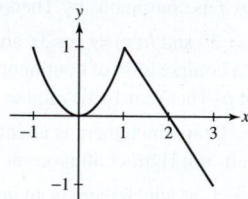

59. The function f is neither left- nor right-continuous at $x = 2$.

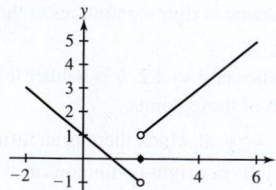

61. $\lim\limits_{x \to 4} \frac{x^2 - 16}{x - 4} = \lim\limits_{x \to 4}(x + 4) = 8 \neq 10 = f(4)$

63.

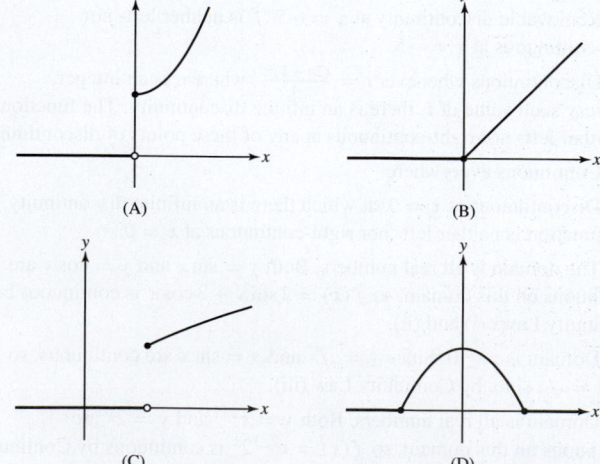

(A) (B)

(C) (D)

f is continuous in (b) and (d). f has a discontinuity at $x = 0$ in (a). f has a discontinuity at $x = 2$ in (c).

65. $c = \frac{5}{3}$ **67.** $a = 2$ and $b = 1$

69. **(a)** No **(b)** $g(1) = -\frac{\pi}{2}$

71. **73.**

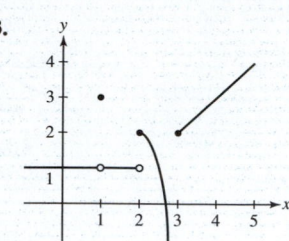

75. -6 **77.** $\frac{1}{3}$ **79.** -1 **81.** $\frac{1}{32}$ **83.** 27 **85.** 1000 **87.** $\frac{\pi}{2}$

89. No. Take $f(x) = -x^{-1}$ and $g(x) = x^{-1}$.

91. $f(x) = |g(x)|$ is a composition of the continuous functions g and $y = |x|$.

93. No

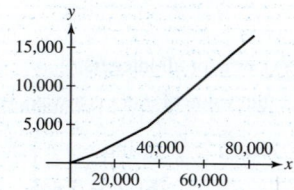

95. $f(x) = 3$ and $g(x) = \lfloor x \rfloor$

97. In this case, $y = f(x)^2$ is the constant function 1.

Section 2.5 Preliminary Questions

1. $\frac{x^2 - 1}{\sqrt{x + 3} - 2}$

2. **(a)** $f(x) = \frac{x^2 - 1}{x - 1}$ **(b)** $f(x) = \frac{x^2 - 1}{x - 1}$ **(c)** $f(x) = \frac{1}{x}$

3. The simplify and plug in strategy is based on simplifying a function that is indeterminate to a continuous function. Once the simplification has been made, the limit of the remaining continuous function is obtained by evaluation.

Section 2.5 Exercises

1. $\lim\limits_{x \to 6} \frac{x^2 - 36}{x - 6} = \lim\limits_{x \to 6} \frac{(x - 6)(x + 6)}{x - 6} = \lim\limits_{x \to 6}(x + 6) = 12$

3. 0 **5.** $\frac{1}{14}$ **7.** -1 **9.** $\frac{11}{10}$ **11.** 2 **13.** 1 **15.** 2 **17.** $\frac{1}{8}$

19. $-\frac{1}{4}$

21. Limit does not exist

- As $h \to 0+$, $\dfrac{\sqrt{h + 2} - 2}{h} \to -\infty$.

- As $h \to 0-$, $\dfrac{\sqrt{h + 2} - 2}{h} \to \infty$.

23. 2 **25.** $\frac{1}{4}$ **27.** 1 **29.** $-\frac{1}{2}$ **31.** 9 **33.** $\frac{1}{2}$

35. -1, does not exist; 0

37. Maximum height estimate: 183.67 m.

39. $\lim\limits_{x \to 4} f(x) \approx 2.00$; to two decimal places, this matches the value of 2 obtained in Exercise 23.

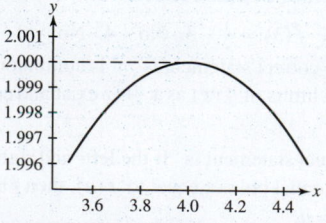

43. 12   **45.** -1 **47.** $\frac{4}{3}$ **49.** $2a$ **51.** $-4+5a$ **53.** $\frac{1}{2\sqrt{a}}$

55. $3a^2$ **57.** $\frac{1}{4}$ **59.** $c=-1$ and $c=6$ **61.** $c=3$ **63.** +

Section 2.6 Preliminary Questions

1. $\lim\limits_{x\to 0} f(x) = 0$; no

2. Assume that for $x \neq c$ (in some open interval containing c),

$$l(x) \leq f(x) \leq u(x)$$

and that $\lim\limits_{x\to c} l(x) = \lim\limits_{x\to c} u(x) = L$. Then $\lim\limits_{x\to c} f(x)$ exists and

$$\lim\limits_{x\to c} f(x) = L$$

3. **(a)**

Section 2.6 Exercises

1. $\lim\limits_{x\to 0} x^2 \cos \frac{1}{x} = 0$ **3.** $\lim\limits_{x\to 1} (x-1) \sin \frac{\pi}{x-1} = 0$

5. $\lim\limits_{t\to 0} (2^t - 1) \cos \frac{1}{t} = 0$ **7.** $\lim\limits_{t\to 2} (t^2 - 4) \cos \frac{1}{t-2} = 0$

9. $\lim\limits_{\theta\to \frac{\pi}{2}} \cos \theta \cos(\tan \theta) = 0$

11. For all $x \neq 1$ on the open interval $(0, 2)$ containing $x = 1$, $\ell(x) \leq f(x) \leq u(x)$. Moreover,

$$\lim\limits_{x\to 1} \ell(x) = \lim\limits_{x\to 1} u(x) = 2$$

Therefore, by the Squeeze Theorem,

$$\lim\limits_{x\to 1} f(x) = 2$$

13. $\lim\limits_{x\to 7} f(x) = 6$; no

15. **(a)** *Not* sufficient information   **(b)** $\lim\limits_{x\to 1} f(x) = 1$
(c) $\lim\limits_{x\to 1} f(x) = 3$

17. 1   **19.** 3   **21.** 1   **23.** 0   **25.** $\frac{2\sqrt{2}}{\pi}$ **27.** 11   **29.** 9   **31.** $\frac{1}{5}$
33. $\frac{7}{3}$ **35.** $\frac{1}{25}$ **37.** 6   **39.** $-\frac{3}{4}$ **41.** $\frac{1}{2}$ **43.** $\frac{6}{5}$ **45.** 0   **47.** 0
49. 0   **55.** $-\frac{9}{2}$

57. $\lim\limits_{t\to 0^+} \dfrac{\sqrt{1-\cos t}}{t} = \dfrac{\sqrt{2}}{2}$; $\lim\limits_{t\to 0^-} \dfrac{\sqrt{1-\cos t}}{t} = -\dfrac{\sqrt{2}}{2}$;
does not exist

61. **(a)**

x	$c-0.01$	$c-0.001$	$c+0.001$	$c+0.01$
$\dfrac{\sin x - \sin c}{x - c}$	0.999983	0.99999983	0.99999983	0.999983

Here, $c = 0$ and $\cos c = 1$

x	$c-0.01$	$c-0.001$	$c+0.001$	$c+0.01$
$\dfrac{\sin x - \sin c}{x - c}$	0.868511	0.866275	0.865775	0.863511

Here, $c = \frac{\pi}{6}$ and $\cos c = \frac{\sqrt{3}}{2} \approx 0.866025$

x	$c-0.01$	$c-0.001$	$c+0.001$	$c+0.01$
$\dfrac{\sin x - \sin c}{x - c}$	0.504322	0.500433	0.499567	0.495662

Here, $c = \frac{\pi}{3}$ and $\cos c = \frac{1}{2}$

x	$c-0.01$	$c-0.001$	$c+0.001$	$c+0.01$
$\dfrac{\sin x - \sin c}{x - c}$	0.710631	0.707460	0.706753	0.703559

Here, $c = \frac{\pi}{4}$ and $\cos c = \frac{\sqrt{2}}{2} \approx 0.707107$

x	$c-.01$	$c-.001$	$c+.001$	$c+.01$
$\dfrac{\sin x - \sin c}{x - c}$	0.005000	0.000500	-0.000500	-0.005000

Here, $c = \frac{\pi}{2}$ and $\cos c = 0$

(b) $\lim\limits_{x\to c} \dfrac{\sin x - \sin c}{x - c} = \cos c$

(c)

x	$c-.01$	$c-.001$	$c+.001$	$c+.01$
$\dfrac{\sin x - \sin c}{x - c}$	-0.411593	-0.415692	-0.416601	-0.420686

Here, $c = 2$ and $\cos c = \cos 2 \approx -0.416147$

x	$c-0.01$	$c-0.001$	$c+0.001$	$c+0.01$
$\dfrac{\sin x - \sin c}{x - c}$	0.863511	0.865775	0.866275	0.868511

Here, $c = -\frac{\pi}{6}$ and $\cos c = \frac{\sqrt{3}}{2} \approx 0.866025$

Section 2.7 Preliminary Questions

1. **(a)** Correct   **(b)** Not correct   **(c)** Not correct   **(d)** Correct

2. **(a)** $\lim\limits_{x\to\infty} x^3 = \infty$ **(b)** $\lim\limits_{x\to-\infty} x^3 = -\infty$
(c) $\lim\limits_{x\to-\infty} x^4 = \infty$

3.

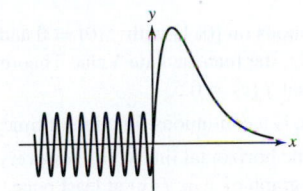

4. Negative   **5.** Negative

6. As $x \to \infty$, $\frac{1}{x} \to 0$, so

$$\lim\limits_{x\to\infty} \sin \frac{1}{x} = \sin 0 = 0$$

On the other hand, $\frac{1}{x} \to \pm\infty$ as $x \to 0$, and as $\frac{1}{x} \to \pm\infty$, $\sin \frac{1}{x}$ oscillates infinitely often.

Section 2.7 Exercises

1. $y = 1$ and $y = 2$

3.

5. **(a)** From the table below, it appears that

$$\lim\limits_{x\to\pm\infty} \frac{x^2}{x^2 + 1} = 1$$

x	± 50	± 100	± 500	± 1000
$f(x)$	0.999600	0.999900	0.999996	0.999999

(b) From the graph below, it also appears that

$$\lim_{x \to \pm\infty} \frac{x^2}{x^2 + 1} = 1$$

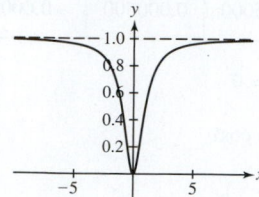

(c) The horizontal asymptote of f is $y = 1$.

7. 1 **9.** 0 **11.** $\frac{7}{4}$ **13.** $-\infty$ **15.** ∞ **17.** $y = \frac{1}{4}$ **19.** $y = \frac{2}{3}$ and $y = -\frac{2}{3}$ **21.** $y = 0$ **23.** $y = 0$ and $y = 10$

25. $\lim_{x \to \infty} f(x) = 3$. The expression "never touches it" is incorrect. Here the graph of f coincides with the horizontal asymptote $y = 3$ for all $x > 0$. **27.** 0 **29.** 2 **31.** $\frac{1}{16}$ **33.** 0

35. $\frac{\pi}{2}$; the graph of $y = \tan^{-1} x$ has a horizontal asymptote at $y = \frac{\pi}{2}$.

37. **(a)** $M \approx 1863.3$; $A \approx 9.808$; $k \approx 0.576$

(b) Approximately 1863.3 million **(c)** Approximately 2020

39. 0 **41.** ∞ **43.** $\ln \frac{3}{2}$ **45.** $-\frac{\pi}{2}$

49. **(a)** $\lim_{s \to \infty} R(s) = \lim_{s \to \infty} \frac{As}{K + s} = \lim_{s \to \infty} \frac{A}{1 + \frac{K}{s}} = A$.

(b) $R(K) = \dfrac{AK}{K + K} = \dfrac{AK}{2K} = \dfrac{A}{2}$ half of the limiting value.

(c) 3.75 mM

Section 2.8 Preliminary Questions

1. Observe that $f(x) = x^2$ is continuous on $[0, 1]$ with $f(0) = 0$ and $f(1) = 1$. Because $f(0) < 0.5 < f(1)$, the Intermediate Value Theorem guarantees there is a $c \in [0, 1]$ such that $f(c) = 0.5$.

2. We must assume that temperature is a continuous function of time.

3. If f is continuous on $[a, b]$, then the horizontal line $y = k$ for every k between $f(a)$ and $f(b)$ intersects the graph of $y = f(x)$ at least once.

4.

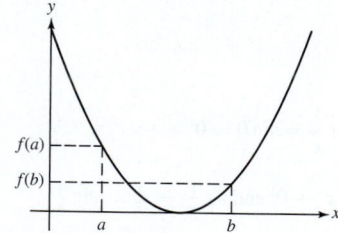

5. **(a)** Sometimes true **(b)** Always true **(c)** Never true
(d) Sometimes true

Section 2.8 Exercises

1. Observe that $f(1) = 2$ and $f(2) = 10$. Since f is a polynomial, it is continuous everywhere, in particular on $[1, 2]$. Therefore, by the IVT, there is a $c \in [1, 2]$ such that $f(c) = 9$.

3. $g(0) = 0$ and $g(\frac{\pi}{4}) = \frac{\pi^2}{16}$. g is continuous for all t between 0 and $\frac{\pi}{4}$, and $0 < \frac{1}{2} < \frac{\pi^2}{16}$; therefore, by the IVT, there is a $c \in [0, \frac{\pi}{4}]$ such that $g(c) = \frac{1}{2}$.

5. Let $f(x) = x - \cos x$. Observe that f is continuous with $f(0) = -1$ and $f(1) = 1 - \cos 1 \approx 0.46$. Therefore, by the IVT, there is a $c \in [0, 1]$ such that $f(c) = c - \cos c = 0$.

7. Let $f(x) = \sqrt{x} + \sqrt{x + 2} - 3$. Note that f is continuous on $\left[\frac{1}{4}, 2\right]$ with $f(\frac{1}{4}) = -1$ and $f(2) = \sqrt{2} - 1 \approx 0.41$. Therefore, by the IVT, there is a $c \in \left[\frac{1}{4}, 2\right]$ such that $f(c) = \sqrt{c} + \sqrt{c + 2} - 3 = 0$.

9. Let $f(x) = x^2$. Observe that f is continuous with $f(1) = 1$ and $f(2) = 4$. Therefore, by the IVT, there is a $c \in [1, 2]$ such that $f(c) = c^2 = 2$.

11. For each positive integer k, let $f(x) = x^k - \cos x$. Observe that f is continuous on $\left[0, \frac{\pi}{2}\right]$ with $f(0) = -1$ and $f(\frac{\pi}{2}) = \left(\frac{\pi}{2}\right)^k > 0$. Therefore, by the IVT, there is a $c \in \left[0, \frac{\pi}{2}\right]$ such that $f(c) = c^k - \cos(c) = 0$.

13. Let $f(x) = 2^x + 3^x - 4^x$. Observe that f is continuous on $[0, 2]$ with $f(0) = 1 > 0$ and $f(2) = -3 < 0$. Therefore, there is a $c \in (0, 2)$ such that $f(c) = 2^c + 3^c - 4^c = 0$.

15. Let $f(x) = e^x + \ln x$. Observe that f is continuous on $[e^{-2}, 1]$ with $f(e^{-2}) = e^{e^{-2}} - 2 < 0$ and $f(1) = e > 0$. Therefore, by the IVT, there is a $c \in (e^{-2}, 1) \subset (0, 1)$ such that $f(c) = e^c + \ln c = 0$.

17. Apply Corollary 2 to the Intermediate Value Theorem. f is a polynomial and continuous everywhere. $f(-3) = 170$, $f(-2) = -25$, $f(-1) = 2$, $f(0) = -1$, $f(1) = 2$, $f(2) = -25$, and $f(3) = 170$. Thus, f has a zero in each of these intervals: $(-3, -2)$, $(-2, -1)$, $(-1, 0)$, $(0, 1)$, $(1, 2)$, $(2, 3)$, and f must have six distinct solutions.

19. The IVT does not apply since g is not continuous on the interval $[-1, 1]$. However, this function does take on all values between $g(-1) = 1$ and $g(1) = 2$.

21. **(a)** $f(1) = 1$; $f(1.5) = 2^{1.5} - (1.5)^3 < 3 - 3.375 < 0$. Hence, $f(x) = 0$ for some x between 1 and 1.5.

(b) $f(1.25) \approx 0.4253 > 0$ and $f(1.5) < 0$. Hence, $f(x) = 0$ for some x between 1.25 and 1.5.

(c) $f(1.375) \approx -0.0059$. Hence, $f(x) = 0$ for some x between 1.25 and 1.375.

23. $[1.25, 1.5]$

25. 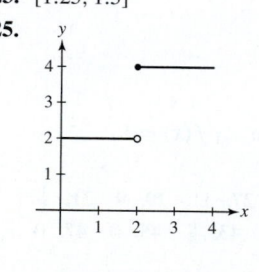 **27.**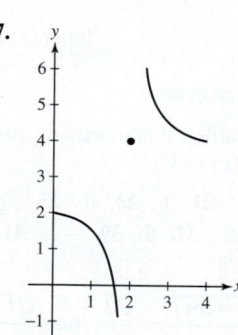

29. No; no **31.** At time c

Section 2.9 Preliminary Questions

1. **(c)** **2.** **(b)** and **(d)** are true.

Section 2.9 Exercises

1. $L = 4$, $\epsilon = 0.8$, and $\delta = 0.1$

3.

With $\delta = \epsilon$, the gap is within ϵ of a.

5. **(a)**
$$|f(x) - 35| = |8x + 3 - 35| = |8x - 32| = |8(x - 4)| = 8\,|x - 4|$$
(b) Let $\epsilon > 0$. Let $\delta = \epsilon/8$ and suppose $|x - 4| < \delta$. By part **(a)**, $|f(x) - 35| = 8|x - 4| < 8\delta$. Substituting $\delta = \epsilon/8$, we see $|f(x) - 35| < 8\epsilon/8 = \epsilon$.

7. **(a)** If $0 < |x - 2| < \delta = .01$, then $|x| < 3$ and $|x^2 - 4| = |x - 2||x + 2| \leq |x - 2|\,(|x| + 2) < 5|x - 2| < 0.05$.

(b) If $0 < |x - 2| < \delta = .0002$, then $|x| < 2.0002$ and

$$\left|x^2 - 4\right| = |x - 2||x + 2| \le |x - 2|\,(|x| + 2) < 4.0002|x - 2|$$

$$< 0.00080004 < 0.0009.$$

(c) $\delta = 10^{-5}$

9. $\delta = 6 \times 10^{-4}$

11. $\delta = 0.25$

13. $\delta = 0.198$

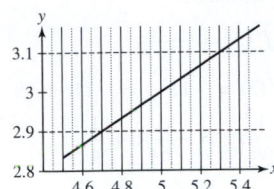
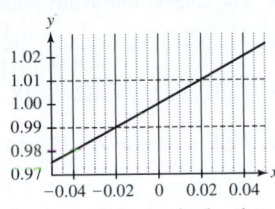

15. (a) Since $|x - 2| < 1$, it follows that $1 < x < 3$, in particular that $x > 1$. Because $x > 1$, then $\dfrac{1}{x} < 1$ and

$$\left|\frac{1}{x} - \frac{1}{2}\right| = \left|\frac{2 - x}{2x}\right| = \frac{|x - 2|}{2x} < \frac{1}{2}|x - 2|$$

(b) Choose $\delta = .02$.

(c) Let $\delta = \min\{1, 2\epsilon\}$ and suppose that $|x - 2| < \delta$. Then by part (a) we have

$$\left|\frac{1}{x} - \frac{1}{2}\right| < \frac{1}{2}|x - 2| < \frac{1}{2}\delta < \frac{1}{2} \cdot 2\epsilon = \epsilon$$

Let $\epsilon > 0$ be given. Then whenever $0 < |x - 2| < \delta = \min\{1, 2\epsilon\}$, we have

$$\left|\frac{1}{x} - \frac{1}{2}\right| < \frac{1}{2}\delta \le \epsilon$$

17.

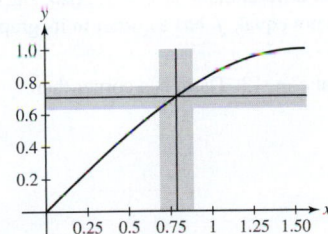

19. Given $\epsilon > 0$, we let

$$\delta = \min\left\{|c|, \frac{\epsilon}{3|c|}\right\}$$

Then, for $|x - c| < \delta$, we have

$$|x^2 - c^2| = |x - c|\,|x + c| < 3|c|\delta < 3|c|\frac{\epsilon}{3|c|} = \epsilon$$

21. Let $\epsilon > 0$ be given. Let $\delta = \min(1, 3\epsilon)$. If $|x - 4| < \delta$,

$$|\sqrt{x} - 2| = |x - 4|\left|\frac{1}{\sqrt{x} + 2}\right| < |x - 4|\frac{1}{3} < \delta\frac{1}{3} < 3\epsilon\frac{1}{3} = \epsilon$$

23. Let $\epsilon > 0$ be given. Let $\delta = \min(1, \frac{\epsilon}{7})$, and assume $|x - 1| < \delta$. Since $\delta < 1, 0 < x < 2$. Since $x^2 + x + 1$ increases as x increases for $x > 0$, $x^2 + x + 1 < 7$ for $0 < x < 2$, so

$$\left|x^3 - 1\right| = |x - 1|\left|x^2 + x + 1\right| < 7|x - 1| < 7\frac{\epsilon}{7} = \epsilon$$

25. Let $\epsilon > 0$ be given. Let $\delta = \min(1, \frac{4}{5}\epsilon)$, and suppose $|x - 2| < \delta$. Since $\delta < 1, |x - 2| < 1$, so $1 < x < 3$. This means that $4x^2 > 4$ and $|2 + x| < 5$ so that $\frac{2 + x}{4x^2} < \frac{5}{4}$. We get

$$\left|x^{-2} - \frac{1}{4}\right| = |2 - x|\left|\frac{2 + x}{4x^2}\right| < \frac{5}{4}|x - 2| < \frac{5}{4} \cdot \frac{4}{5}\epsilon = \epsilon$$

27. Let L be any real number. Let $\delta > 0$ be any small positive number. Let $x = \frac{\delta}{2}$, which satisfies $|x| < \delta$, and $f(x) = 1$. We consider two cases:

- $(|f(x) - L| \ge \frac{1}{2})$: We are done.

- $(|f(x) - L| < \frac{1}{2})$: This means $\frac{1}{2} < L < \frac{3}{2}$. In this case, let $x = -\frac{\delta}{2}$. $f(x) = -1$ so $\frac{3}{2} < L - f(x)$.

In either case, there exists an x such that $|x| < \delta$, but $|f(x) - L| \ge \frac{1}{2}$.

29. Let $\epsilon > 0$ and let $\delta = \min(1, \frac{\epsilon}{2})$. Then, whenever $|x - 1| < \delta$, it follows that $0 < x < 2$. If $1 < x < 2$, then $\min(x, x^2) = x$ and

$$|f(x) - 1| = |x - 1| < \delta < \frac{\epsilon}{2} < \epsilon$$

On the other hand, if $0 < x < 1$, then $\min(x, x^2) = x^2$, $|x + 1| < 2$ and

$$|f(x) - 1| = |x^2 - 1| = |x - 1|\,|x + 1| < 2\delta < \epsilon$$

Thus, whenever $|x - 1| < \delta$, $|f(x) - 1| < \epsilon$.

33. Suppose that $\lim\limits_{x \to c} f(x) = L$. Let $\epsilon > 0$ be given. Since $\lim\limits_{x \to c} f(x) = L$, we know there is a $\delta > 0$ such that $|x - c| < \delta$ forces $|f(x) - L| < \epsilon/|a|$. Suppose $|x - c| < \delta$. Then $|af(x) - aL| = |a|\,|f(x) - L| < |a|(\epsilon/|a|) = \epsilon$.

Chapter 2 Review

1. Average velocity: approximately 0.954 m/s; instantaneous velocity: approximately 0.894 m/s

3.

x	Slope of secant from 16 to x	x	Slope of secant from 16 to x
15.99	0.176804	16.01	0.176749
15.999	0.176779	16.001	0.176774
15.9999	0.176777	16.0001	0.176776

Tangent line slope estimate: 0.1768

5. 1.50 **7.** 1.69 **9.** 2.00 **11.** 5 **13.** $-\frac{1}{2}$ **15.** $\frac{1}{6}$ **17.** 2

19. Does not exist;

$$\lim_{t \to 9^-} \frac{t - 6}{\sqrt{t} - 3} = -\infty \quad \text{and} \quad \lim_{t \to 9^+} \frac{t - 6}{\sqrt{t} - 3} = \infty$$

21. ∞

23. Does not exist;

$$\lim_{x \to 1^-} \frac{x^3 - 2x}{x - 1} = \infty \quad \text{and} \quad \lim_{x \to 1^+} \frac{x^3 - 2x}{x - 1} = -\infty$$

25. 2 **27.** 0 **29.** $-\frac{1}{2}$ **31.** $3b^2$ **33.** $\frac{1}{9}$ **35.** ∞

37. Does not exist;

$$\lim_{\theta \to \frac{\pi}{2}^-} \theta \sec \theta = \infty \quad \text{and} \quad \lim_{\theta \to \frac{\pi}{2}^+} \theta \sec \theta = -\infty$$

39. Does not exist;

$$\lim_{\theta \to 0^-} \frac{\cos \theta - 2}{\theta} = \infty \quad \text{and} \quad \lim_{\theta \to 0^+} \frac{\cos \theta - 2}{\theta} = -\infty$$

41. ∞ **43.** ∞

45. Does not exist;

$$\lim_{x \to \frac{\pi}{2}^-} \tan x = \infty \quad \text{and} \quad \lim_{x \to \frac{\pi}{2}^+} \tan x = -\infty$$

47. 0 **49.** 0

51. According to the graph of f,

$$\lim_{x \to 0^-} f(x) = \lim_{x \to 0^+} f(x) = 1$$

$$\lim_{x \to 2^-} f(x) = \lim_{x \to 2^+} f(x) = \infty$$

$$\lim_{x \to 4^-} f(x) = -\infty$$

$$\lim_{x \to 4^+} f(x) = \infty$$

The function is both left- and right-continuous at $x = 0$ and neither left- nor right-continuous at $x = 2$ and $x = 4$.

53. At $x = 0$, the function has an infinite discontinuity but is left-continuous.

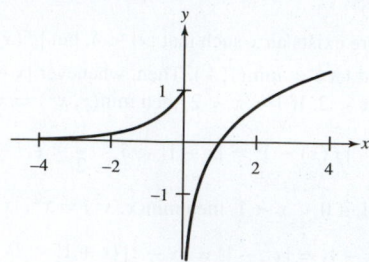

55. g has a jump discontinuity at $x = -1$; g is left-continuous at $x = -1$.

57. $b = 7$; h has a jump discontinuity at $x = -2$.

59. Does not have any horizontal asymptotes

61. $y = 2$ **63.** $y = 1$ **65.** $y = 0$ and $y = 2n$

67. $M \approx 62.78$; $A \approx 5.278$; $k \approx 0.4737$; horizontal asymptotes at $y = 0$ and $y = 62.78$ (approximately)

69. $B = B \cdot 1 = B \cdot L =$

$$\lim_{x \to a} g(x) \cdot \lim_{x \to a} \frac{f(x)}{g(x)} = \lim_{x \to a} g(x) \frac{f(x)}{g(x)} = \lim_{x \to a} f(x) = A$$

71. $f(x) = \dfrac{1}{(x - a)^3}$ and $g(x) = \dfrac{1}{(x - a)^5}$

75. Let $f(x) = x^2 - \cos x$. Now, f is continuous over the interval $[0, \frac{\pi}{2}]$, $f(0) = -1 < 0$, and $f(\frac{\pi}{2}) = \frac{\pi^2}{4} > 0$. Therefore, by the Intermediate Value Theorem, there exists a $c \in (0, \frac{\pi}{2})$ such that $f(c) = 0$; consequently, the curves $y = x^2$ and $y = \cos x$ intersect.

77. Let $f(x) = e^{-x^2} - x$. Observe that f is continuous on $[0, 1]$ with $f(0) = e^0 - 0 = 1 > 0$ and $f(1) = e^{-1} - 1 < 0$. Therefore, the IVT guarantees there exists a $c \in (0, 1)$ such that $f(c) = e^{-c^2} - c = 0$.

79. $g(x) = \lfloor x \rfloor$; on the interval

$$x \in \left[\frac{a}{2 + 2\pi a}, \frac{a}{2} \right] \subset [-a, a]$$

$\frac{1}{x}$ runs from $\frac{2}{a}$ to $\frac{2}{a} + 2\pi$, so the sine function covers one full period and clearly takes on every value from $-\sin a$ through $\sin a$.

81. $\delta = 0.55$;

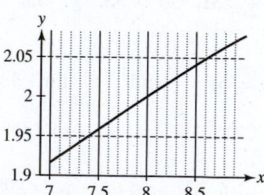

83. Let $\epsilon > 0$ and take $\delta = \epsilon/8$. Then, whenever $|x - (-1)| = |x + 1| < \delta$,

$$|f(x) - (-4)| = |4 + 8x + 4| = 8|x + 1| < 8\delta = \epsilon$$

Chapter 3

Section 3.1 Preliminary Questions

1. B and D **2.** $\dfrac{f(x) - f(a)}{x - a}$ and $\dfrac{f(a + h) - f(a)}{h}$

3. $a = 3$ and $h = 2$

4. Derivative of the function $f(x) = \tan x$ at $x = \frac{\pi}{4}$

5. **(a)** The difference in height between the points $(0.9, \sin 0.9)$ and $(1.3, \sin 1.3)$ **(b)** The slope of the secant line between the points $(0.9, \sin 0.9)$ and $(1.3, \sin 1.3)$ **(c)** The slope of the tangent line to the graph at $x = 0.9$

6. **(a)** Horizontal   **7.** **(b)** Vertical

Section 3.1 Exercises

1. $f'(3) = 30$ **3.** $f'(0) = 9$ **5.** $f'(-1) = -2$ **7.** $f'(1) = 5$

9. Slope of the secant line $= 1$; the secant line through $(2, f(2))$ and $(2.5, f(2.5))$ has a larger slope than the tangent line at $x = 2$.

11. $f'(1) \approx 0$; $f'(2) \approx 0.8$

13. $f'(1) = f'(2) = 0$; $f'(4) = \frac{1}{2}$; $f'(7) = 0$

15. $f'(5.5)$ **17.** $f'(x) = 7$ **19.** $g'(t) = -3$ **21.** $y = 2x - 1$

23. The tangent line at any point is the line itself.

25. $f(-2 + h) = \dfrac{1}{-2 + h}$; $-\dfrac{1}{3}$ **27.** $f'(5) = -\dfrac{1}{10\sqrt{5}}$

29. $f'(3) = 22$; $y = 22x - 18$ **31.** $f'(3) = -11$; $y = -11t + 18$

33. $f'(0) = 1$; $y = x$ **35.** $f'(8) = -\dfrac{1}{64}$; $y = -\dfrac{1}{64}x + \dfrac{1}{4}$

37. $f'(-2) = -1$; $y = -x - 1$

39. $f'(1) = \dfrac{1}{2\sqrt{5}}$; $y = \dfrac{1}{2\sqrt{5}}x + \dfrac{9}{2\sqrt{5}}$

41. $f'(4) = -\dfrac{1}{16}$; $y = -\dfrac{1}{16}x + \dfrac{3}{4}$

43. $f'(3) = \dfrac{3}{\sqrt{10}}$; $y = \dfrac{3}{\sqrt{10}}t + \dfrac{1}{\sqrt{10}}$ **45.** $f'(0) = 0$; $y = 1$

47. $\displaystyle\lim_{h \to 0^+} \frac{f(1 + h) - f(1)}{h} = \lim_{h \to 0^+} \frac{(1 + h)^2 - 1}{h}$

$\displaystyle = \lim_{h \to 0^+} \frac{2h + h^2}{h} = \lim_{h \to 0^+} (2 + h) = 2$

$\displaystyle\lim_{h \to 0^-} \frac{f(1 + h) - f(1)}{h} = \lim_{h \to 0^-} \frac{1 - 1}{h} = 0$

Since these one-sided limits exist but are not equal, $\displaystyle\lim_{h \to 0} \frac{f(1+h) - f(1)}{h}$ does not exist and therefore f is not differentiable at $x = 1$. Also, since these one-sided limits exist but are not equal, f has a corner in its graph at $x = 1$.

49. The derivative does not exist at $c = -3$. There is a corner there.

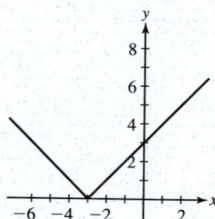

51. The derivative does not exist at $c = \pm 2$. There are corners at both points.

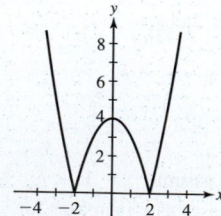

53. $f'(0) \approx -0.69$

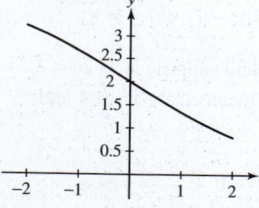 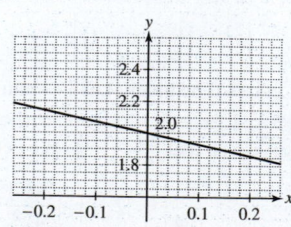

55. For $1 < x < 2.5$ and for $x > 3.5$ **57.** $f(x) = x^3$ and $a = 5$

59. $f(x) = \sin x$ and $a = \frac{\pi}{6}$ **61.** $f(x) = 5^x$ and $a = 2$

63. $f'\left(\dfrac{\pi}{4}\right) \approx 0.7071$

65. • On curve (A), $f'(1)$ is larger than

$$\frac{f(1+h) - f(1)}{h}$$

The curve is bending downward, so the secant line to the right is at a lower angle than the tangent line.

• On curve (B), $f'(1)$ is smaller than

$$\frac{f(1+h) - f(1)}{h}$$

The curve is bending upward, so the secant line to the right is at a steeper angle than the tangent line.

67. (b) $f'(4) \approx 20.0000$

(c) $y = 20x - 48$

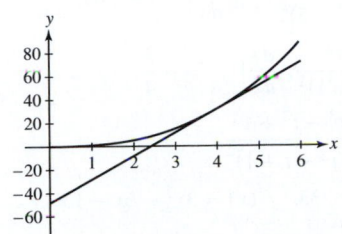

69. $c \approx 0.37$

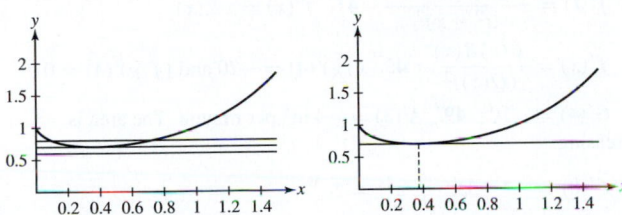

71. $P'(293) \approx 0.00204$; $P'(313) \approx 0.00503$; $P'(333) \approx 0.01106$ with units of atm/K

73. $P'(1997) \approx 0.10$; $P'(2001) \approx 0.35$; $P'(2005) \approx 0.89$; $P'(2009) \approx 2.36$ with units of billion gallons per year

75. $P'(303) \approx 0.00265$; $P'(313) \approx 0.004145$; $P'(333) \approx 0.00931$; $P'(343) \approx 0.013435$ with units of atm/K

77. -0.375 kph·km/car **79.** $i(3) = 0.06$ amperes

81. $v'(4) \approx 160$; $C \approx 0.2$ farads

83. (a) Secant slope $= \dfrac{f(x+h) - f(x-h)}{(x+h) - (x-h)} = \dfrac{f(x+h) - f(x-h)}{2h}$

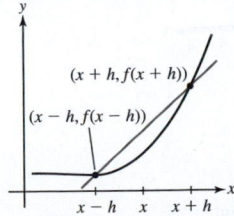

(b) The secant slope from x to $x + h$ is $\dfrac{f(x+h) - f(x)}{h}$. The secant slope from $x - h$ to x is $\dfrac{f(x) - f(x-h)}{h}$. The average of these two slopes is then

$$\frac{1}{2}\left(\frac{f(x+h) - f(x)}{h} + \frac{f(x) - f(x-h)}{h}\right) = \frac{f(x+h) - f(x-h)}{2h}$$

85. Let $f(x) = px^2 + qx + r$; SDQ: $\big(f(a+h) - f(a-h)\big)/2h = 2h(2pa + q)/2h = 2pa + q$; $f'(a) = \lim\limits_{h\to 0}\big(f(a+h) - f(a)\big)/h = 2pa + q$

Section 3.2 Preliminary Questions

1. 8 **2.** $(f - g)'(1) = -2$ and $(3f + 2g)'(1) = 19$

3. (a), (b), (c), and (f)

4. (a) False. For example, $f(x) = |x|$ is continuous at $x = 0$ but is not differentiable there. **(b)** True

Section 3.2 Exercises

1. $f'(x) = 3$ **3.** $f'(x) = 3x^2$ **5.** $f'(x) = 1 - \dfrac{1}{2\sqrt{x}}$

7. $\dfrac{d}{dx}x^4\Big|_{x=-2} = 4(-2)^3 = -32$

9. $\dfrac{d}{dt}t^{2/3}\Big|_{t=8} = \dfrac{2}{3}(8)^{-1/3} = \dfrac{1}{3}$

11. $0.35x^{-0.65}$ **13.** $\sqrt{17}t^{\sqrt{17}-1}$

15. $f'(x) = 4x^3$; $y = 32x - 48$

17. $f'(x) = 5 - 16x^{-1/2}$; $y = -3x - 32$

19. (a) $\dfrac{d}{dx}12e^x = 12e^x$ **(b)** $\dfrac{d}{dt}(25t - 8e^t) = 25 - 8e^t$

(c) $\dfrac{d}{dt}e^{t-3} = e^{t-3}$

21. $f'(x) = 6x^2 - 6x$ **23.** $f'(x) = \dfrac{20}{3}x^{2/3} + 6x^{-3}$

25. $g'(z) = -\dfrac{5}{2}z^{-19/14} - 5z^{-6}$ **27.** $f'(s) = \dfrac{1}{4}s^{-3/4} + \dfrac{1}{3}s^{-2/3}$

29. $g'(x) = 0$ **31.** $h'(t) = 5e^{t-3}$ **33.** $P'(s) = 32s - 24$

35. $f'(x) = -2x$ **37.** $g'(x) = -6x^{-5/2}$ **39.** 1 **41.** -60

43. $1 - e^4$

45. • The graph in (A) matches the derivative in (III).
• The graph in (B) matches the derivative in (I).
• The graph in (C) matches the derivative in (II).
• The graph in (D) matches the derivative in (III).
(A) and (D) have the same derivative because the graph in (D) is just a vertical translation of the graph in (A).

47. Label the graph in (A) as f, the graph in (B) as h, and the graph in (C) as g.

49. Let $f(x) = mx + b$. Then,

$$
\begin{aligned}
f'(x) &= (mx)' + (b)' && \text{(Sum Rule)}\\
&= m(x)' + 0 && \text{(Constant Multiple Rule and Constant Rule)}\\
&= m && \text{(First-Power Rule)}
\end{aligned}
$$

51. (B) might be the graph of the derivative of f.

53. (a) $\dfrac{d}{dt}ct^3 = 3ct^2$ **(b)** $\dfrac{d}{dz}(5z + 4cz^2) = 5 + 8cz$

(c) $\dfrac{d}{dy}(9c^2y^3 - 24c) = 27c^2y^2$

55. $x = \dfrac{1}{2}$ **57.** $a = 2$ and $b = -3$

59. • $f'(x) = 3x^2 - 3 \geq -3$ since $3x^2$ is nonnegative. Also $f'(x) = m$ for $x \pm \sqrt{\dfrac{m+3}{3}}$. There are two values of x if $x > -3$, there is one if $x = -3$, and there are none if $x < -3$.
• The two parallel tangent lines with slope 2 are shown with the graph of f here.

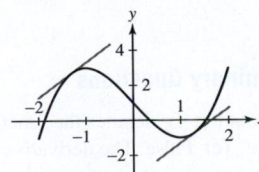

61. $f'(x) = \dfrac{3}{2}x^{1/2}$

63. (a) $f'(0) = 1$; $y = x$

(b) $f'(0) = \lim\limits_{h\to 0}\dfrac{f(0+h) - f(0)}{h} = \lim\limits_{h\to 0}\dfrac{h^2 e^h}{h} = \lim\limits_{h\to 0}he^h = 0$. The tangent line has slope 0 and passes through $(0, f(0)) = (0, 0)$. Therefore $y = 0$ is an equation for the tangent line.

65. Approximately -1.8×10^{-8} W/m^2 per meter.

67. $P'(303) \approx 0.00265$; $P'(313) \approx 0.004145$; $P'(323) \approx 0.006295$;
$P'(333) \approx 0.00931$; $P'(343) \approx 0.013435$ with units of atm/K
$\frac{T^2}{P}\frac{dP}{dT}$ is roughly constant, suggesting that the Clausius–Clapeyron law is
valid, and that $k \approx 5000$.

71.

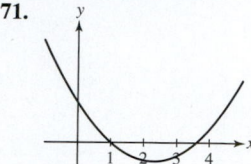

73. For $x < 0$, $f(x) = -x^2$, and $f'(x) = -2x$. For $x > 0$, $f(x) = x^2$,
and $f'(x) = 2x$. Thus, $f'(0) = 0$.

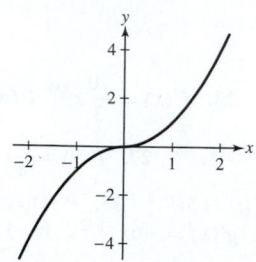

75. It appears that f is not differentiable at $a = 0$. Moreover, the tangent
line does not exist at this point.

77. It appears that f is not differentiable at $a = 3$. Moreover, the tangent
line appears to be vertical.

79. It appears that f is not differentiable at $a = 0$. Moreover, the tangent
line does not exist at this point.

81. $(1, 8)$ **83.** $\dfrac{10}{7}$

85. The normal line intersects the x-axis at the point T with coordinates
$(x + f(x)f'(x), 0)$. The point R has coordinates $(x, 0)$, so the subnormal
is $|x + f(x)f'(x) - x| = |f(x)f'(x)|$.

87. The tangent line to f at $x = a$ is $y = 2ax - a^2$. The x-intercept of
this line is $\frac{a}{2}$, so the subtangent is $a - a/2 = a/2$.

89. The subtangent is $\dfrac{1}{n}a$. **91.** $r \le \dfrac{1}{2}$

97. Note that

$$\lim_{x \to c^+} (f(x) - f(c)) = \lim_{x \to c^+} \left((x - c)\frac{f(x) - f(c)}{x - c} \right)$$

$$= \left(\lim_{x \to c^+} (x - c) \right) \left(\lim_{x \to c^+} \frac{f(x) - f(c)}{x - c} \right)$$

Now, $\lim_{x \to c^+} (x - c) = 0$, and by assumption $\lim_{x \to c^+} \frac{f(x) - f(c)}{x - c}$ exists. It
follows that $\lim_{x \to c^+} (f(x) - f(c)) = 0$. Similarly,
$\lim_{x \to c^-} (f(x) - f(c)) = 0$. This implies $\lim_{x \to c} f(x) = f(c)$ and therefore f
is continuous at c.

Section 3.3 Preliminary Questions

1. **(a)** False. The notation fg denotes the function whose value at x is
$f(x)g(x)$. **(b)** True **(c)** False. The derivative of a product fg is
$f'(x)g(x) + f(x)g'(x)$. **(d)** False. $\dfrac{d}{dx}(fg)\Big|_{x=4} = f(4)g'(4) + g(4)f'(4)$. **(e)** True

2. -1 **3.** 5

Section 3.3 Exercises

1. $f'(x) = 10x^4 + 3x^2$ **3.** $f'(x) = e^x(x^2 + 2x)$

5. $\dfrac{dh}{ds} = -\dfrac{7}{2}s^{-3/2} + \dfrac{3}{2}s^{-5/2} + 14$; $\dfrac{dh}{ds}\Big|_{s=4} = \dfrac{871}{64}$

7. $f'(x) = \dfrac{-2}{(x - 2)^2}$ **9.** $\dfrac{dg}{dt} = -\dfrac{4t}{(t^2 - 1)^2}$; $\dfrac{dg}{dt}\Big|_{t=-2} = \dfrac{8}{9}$

11. $g'(x) = -\dfrac{e^x}{(1 + e^x)^2}$

13. $f'(x) = 3x^2x^{-3} + x^3(-3x^{-4}) = 3x^{-1} - 3x^{-1} = 0$. Alternatively,
$f(x) = 1$, and therefore $f'(x) = 0$.

15. $f'(t) = 6t^2 + 2t - 4$ **17.** $h'(t) = 1$ for $t \ne 1$

19. $f'(x) = 6x^5 + 4x^3 + 18x^2 + 5$

21. $\dfrac{dy}{dx} = -\dfrac{1}{(x + 10)^2}$; $\dfrac{dy}{dx}\Big|_{x=3} = -\dfrac{1}{169}$

23. $f'(x) = 1$ for $x \ge 0$

25. $\dfrac{dy}{dx} = \dfrac{2x^5 - 20x^3 + 8x}{(x^2 - 5)^2}$; $\dfrac{dy}{dx}\Big|_{x=2} = -80$

27. $\dfrac{dz}{dx} = -\dfrac{3x^2}{(x^3 + 1)^2}$; $\dfrac{dz}{dx}\Big|_{x=1} = -\dfrac{3}{4}$

29. $h'(t) = \dfrac{-2t^3 - t^2 + 1}{(t^3 + t^2 + t + 1)^2}$

31. $f'(x) = 2e^2x$ **33.** $f'(x) = 3x^2 - 6x - 13$

35. $f'(x) = \dfrac{xe^x}{(x + 1)^2}$ **37.** For $z \ne -2$ and $z \ne 1$, $g'(z) = 2z - 1$

39. $f'(t) = \dfrac{-xt^2 + 8t - x^2}{(t^2 - x)^2}$ **41.** $f'(x) = xP(x)$

43. $f'(x) = \dfrac{P(x)R(x)}{(Q(x))^2}$ **45.** $(fg)'(4) = -20$ and $(f/g)'(4) = 0$

47. $G'(4) = -10$ **49.** $A'(3) = -4$ in^2 per minute. The area is
decreasing.

51. $F'(0) = -7$ **53.** $\dfrac{d}{dx}e^{2x} = 2e^{2x}$

55. From the plot of f shown, we see that f is decreasing on its domain
$\{x : x \ne \pm 1\}$. Consequently, $f'(x)$ must be negative. Using the quotient
rule, we find

$$f'(x) = \frac{(x^2 - 1)(1) - x(2x)}{(x^2 - 1)^2} = -\frac{x^2 + 1}{(x^2 - 1)^2}$$

which is negative for all $x \ne \pm 1$.

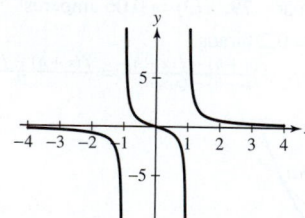

57. $a = 1$

59. **(a)** Given $R(t) = N(t)S(t)$, it follows that

$$\frac{dR}{dt} = N(t)S'(t) + S(t)N'(t)$$

(b) $\dfrac{dR}{dt}\Big|_{t=0} = 1{,}250{,}000$ dollars per month

(c) The term $5S(0)$ is larger than the term $10{,}000N(0)$. Thus, if only one
leg of the campaign can be implemented, it should be part A: increase the
number of stores by five per month.

61. • At $x = -1$, the tangent line is $y = \dfrac{1}{2}x + 1$.

• At $x = 1$, the tangent line is $y = -\dfrac{1}{2}x + 1$.

63. Let $g = f^2 = ff$. Then $g' = \left(f^2\right)' = (ff)' = ff' + ff' = 2ff'$.

65. Let $p = fgh$. Then
$p' = (fgh)' = f\left(gh' + hg'\right) + ghf' = f'gh + fg'h + fgh'$.

69. $\dfrac{d}{dx}(xf(x)) = \lim\limits_{h\to 0} \dfrac{(x+h)f(x+h) - xf(x)}{h}$

$$= \lim\limits_{h\to 0}\left(x\,\dfrac{f(x+h) - f(x)}{h} + f(x+h)\right)$$

$$= x\lim\limits_{h\to 0} \dfrac{f(x+h) - f(x)}{h} + \lim\limits_{h\to 0} f(x+h)$$

$$= xf'(x) + f(x)$$

73. **(a)** Is a multiple root **(b)** Not a multiple root

Section 3.4 Preliminary Questions

1. **(a)** atmospheres/meter **(b)** moles/(liter · hour) **2.** 90 mph

3. If the velocity of an object in motion is negative and decreasing, then the magnitude of the velocity is increasing, and therefore the object is speeding up.

4.

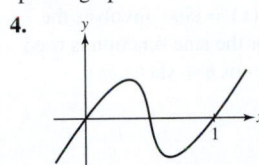

Section 3.4 Exercises

1. 10 square units per unit increase

3.

c	ROC of $f(x)$ with respect to x at $x = c$
1	$f'(1) = \frac{1}{3}$
8	$f'(8) = \frac{1}{12}$
27	$f'(27) = \frac{1}{27}$

5. $d' = 2$ **7.** $dV/dr = 3\pi r^2$

9. **(a)** 100 km/h **(b)** 100 km/h **(c)** 0 km/h **(d)** −50 km/h

11. **(a)** (i) **(b)** (ii) **(c)** (iii)

13. $t = 180$ or 200, $dT/dt \approx -0.1°\text{C/min}$

15. **(a)** C **(b)** A **(c)** B

17.

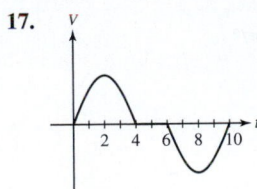

19.

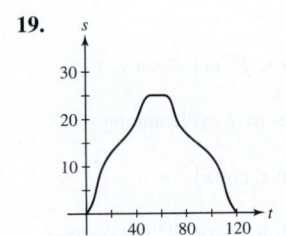

21. -8×10^{-6} 1/s

23. $\left.\dfrac{dT}{dh}\right|_{h=30} \approx 2.3°\text{C/km}$; $\left.\dfrac{dT}{dh}\right|_{h=70} \approx -2.8°\text{C/km}$; $\dfrac{dT}{dh} = 0$ over the interval $[13, 23]$, and near the points $h = 50$ and $h = 90$

25. $v'_{\text{esc}}(r) = -1.41 \times 10^7 r^{-3/2}$

27. The particle passes through the origin when $t = 0$ s and when $t = 3\sqrt{2} \approx 4.24$ s. The particle is instantaneously motionless when $t = 0$ s and when $t = 3$ s.

29. $\dfrac{9000}{49} \approx 183.7$ m

31. Initial velocity: $v_0 = 19.6$ m/s; maximum height: 19.6 m

35. **(a)** $\dfrac{dV}{dv} = -1$ **(b)** −4

37. $S'(3)$ is likely larger because he is probably learning at a faster rate after 3 hours of tutoring than after 30 hours.

39. Rate of change of BSA with respect to mass: $\dfrac{\sqrt{5}}{20\sqrt{m}}$; $m = 70$ kg, rate of change is $\approx 0.0134 \frac{m^2}{kg}$; $m = 80$ kg, rate of change is $\frac{1}{80} \frac{m^2}{kg}$; BSA increases more rapidly at lower body mass.

41. 2

43. The cost of producing 2000 bagels is \$796. The cost of the 2001st bagel is approximately \$0.244, which is indistinguishable from the estimated cost.

47. **(a)** The average income among households in the bottom rth part is

$$\dfrac{F(r)T}{rN} = \dfrac{F(r)}{r} \cdot \dfrac{T}{N} = \dfrac{F(r)}{r}A$$

(b) The average income of households belonging to an interval $[r, r + \Delta r]$ is equal to

$$\dfrac{F(r + \Delta r)T - F(r)T}{\Delta r N} = \dfrac{F(r + \Delta r) - F(r)}{\Delta r} \cdot \dfrac{T}{N}$$

$$= \dfrac{F(r + \Delta r) - F(r)}{\Delta r}A$$

(c) Take the result from part (b) and let $\Delta r \to 0$. Because

$$\lim\limits_{\Delta r \to 0} \dfrac{F(r + \Delta r) - F(r)}{\Delta r} = F'(r)$$

we find that a household in the $100r$th percentile has income $F'(r)A$.

(d) The point P in Figure 15(B) has an r-coordinate of 0.6, while the point Q has an r-coordinate of roughly 0.75. Thus, on curve L_1, 40% of households have $F'(r) > 1$ and therefore have above-average income. On curve L_2, roughly 25% of households have above-average income.

51. By definition, the slope of the line through $(0, 0)$ and $(x, C(x))$ is

$$\dfrac{C(x) - 0}{x - 0} = \dfrac{C(x)}{x} = C_{\text{avg}}(x)$$

- At point A, average cost is greater than marginal cost.
- At point B, average cost is greater than marginal cost.
- At point C, average cost and marginal cost are nearly the same.
- At point D, average cost is less than marginal cost.

Section 3.5 Preliminary Questions

1. The first derivative of stock prices must be positive, while the second derivative must be negative. The first derivative of the reservoir level is negative, while the second derivative is positive. The first derivative of the distance between the asteroid and Earth is negative, and the second derivative is also negative.

2. **3.**

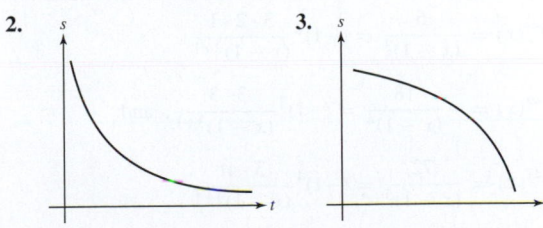

4. True **5.** All quadratic polynomials **6.** e^x

7. $f^{(7)}(x) = 5040$, $f^{(8)}(x) = 0$

Section 3.5 Exercises

1. $y'' = 28$ and $y''' = 0$ **3.** $y'' = 12x^2 - 50$ and $y''' = 24x$

5. $y'' = 8\pi r$ and $y''' = 8\pi$

7. $y'' = -\dfrac{16}{5}t^{-6/5} + \dfrac{4}{3}t^{-4/3}$ and $y''' = \dfrac{96}{25}t^{-11/5} - \dfrac{16}{9}t^{-7/3}$

9. $y'' = -8z^{-3}$ and $y''' = 24z^{-4}$

11. $y'' = 12\theta + 14$ and $y''' = 12$

13. $y'' = -8x^{-3}$ and $y''' = 24x^{-4}$

15. $y'' = (x^5 + 10x^4 + 20x^3)e^x$ and $y''' = (x^5 + 15x^4 + 60x^3 + 60x^2)e^x$

17. $f^{(4)}(1) = 24$ **19.** $\left.\dfrac{d^2 y}{dt^2}\right|_{t=1} = 54$

21. $\left.\dfrac{d^4 x}{dt^4}\right|_{t=16} = \dfrac{3465}{134217728}$ **23.** $f'''(-3) = 4e^{-3} - 6$

25. $h''(1) = \dfrac{7}{4}e$

27. $y^{(0)}(0) = d,\ y^{(1)}(0) = c,\ y^{(2)}(0) = 2b,\ y^{(3)}(0) = 6a,\ y^{(4)}(0) = 24,$ and $y^{(5)}(0) = 0$

29. $\dfrac{d^6}{dx^6}x^{-1} = 720x^{-7}$ **31.** $f^{(n)}(x) = (-1)^n(n+1)!\,x^{-(n+2)}$

33. $f^{(n)}(x) = (-1)^n\dfrac{(2n-1)\cdot(2n-3)\cdot\ldots\cdot 1}{2^n}x^{-(2n+1)/2}$

35. $f^{(n)}(x) = (-1)^n(x-n)e^{-x}$

37. (a) $a(5) = -120\ \text{m/min}^2$

(b) The acceleration of the helicopter for $0 \le t \le 6$ is shown in the figure below. As the acceleration of the helicopter is negative, the velocity of the helicopter must be decreasing. Because the velocity is positive for $0^- \le t < 5$, the helicopter is slowing down between 0 and 5 min and speeding up between 5 and 6 min.

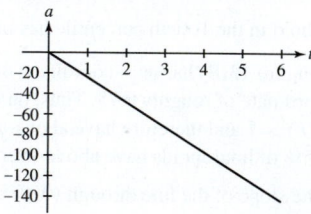

39. (A) f'' **(B)** f' **(C)** f

41. Roughly from time 10 to time 20 and from time 30 to time 40

43. $n = 4, -1$ **45. (a)** $v(t) = -39.1\ \text{m/s}$ **(b)** $v(t) = -55.3\ \text{m/s}$

47.

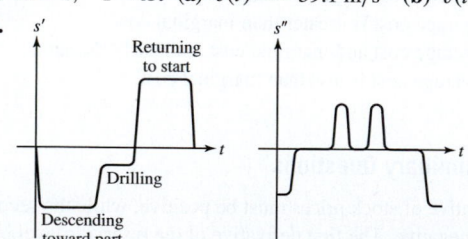

49.
$$f'(x) = -\dfrac{3}{(x-1)^2} = (-1)^1\dfrac{3\cdot 1}{(x-1)^{1+1}};$$
$$f''(x) = \dfrac{6}{(x-1)^3} = (-1)^2\dfrac{3\cdot 2\cdot 1}{(x-1)^{2+1}};$$
$$f'''(x) = -\dfrac{18}{(x-1)^4} = (-1)^3\dfrac{3\cdot 3!}{(x-1)^{3+1}};\ \text{and}$$
$$f^{(4)}(x) = \dfrac{72}{(x-1)^5} = (-1)^4\dfrac{3\cdot 4!}{(x-1)^{4+1}}$$

From the pattern observed, we conjecture

$$f^{(k)}(x) = (-1)^k\dfrac{3\cdot k!}{(x-1)^{k+1}}$$

51. $99!$

53. $(fg)''' = f'''g + 3f''g' + 3f'g'' + fg''';$
$$(fg)^{(n)} = \sum_{k=0}^{n}\binom{n}{k}f^{(n-k)}g^{(k)}$$

Section 3.6 Preliminary Questions

1. (a) $\dfrac{d}{dx}(\sin x + \cos x) = -\sin x + \cos x$

(b) $\dfrac{d}{dx}\sec x = \sec x \tan x$ **(c)** $\dfrac{d}{dx}\cot x = -\csc^2 x$

2. (a) This function can be differentiated using the Product Rule.
(b) We have not yet discussed how to differentiate a function like this.
(c) This function can be differentiated using the Product Rule.

3. (a) $y = x$ **(b)** $y = 1$

4. The difference quotient for the function $f(x) = \sin x$ involves the expression $\sin(x + h)$. The addition formula for the sine function is used to expand this expression as $\sin(x + h) = \sin x \cos h + \sin h \cos x$.

Section 3.6 Exercises

1. $y = \dfrac{\sqrt{2}}{2}x + \dfrac{\sqrt{2}}{2}\left(1 - \dfrac{\pi}{4}\right)$ **3.** $y = 2x + 1 - \dfrac{\pi}{2}$

5. $f'(x) = \cos^2 x - \sin^2 x$ **7.** $f'(x) = \sin x + x \cos x$

9. $H'(t) = \sec t + 2\sin t \sec^2 t \tan t$

11. $f'(\theta) = \sec\theta\left(\sec^2\theta + \tan^2\theta\right)$

13. $f'(x) = \left(8x^3 + 4x^{-2}\right)\sec x + \left(2x^4 - 4x^{-1}\right)\sec x \tan x$

15. $y' = \dfrac{\theta \sec\theta \tan\theta - \sec\theta}{\theta^2}$ **17.** $R'(y) = \dfrac{4\cos y - 3}{\sin^2 y}$

19. $f'(x) = \dfrac{2\sec^2 x}{(1 - \tan x)^2}$ **21.** $f'(x) = e^x(\cos x + \sin x)$

23. $f'(\theta) = e^\theta\left(5\sin\theta + 5\cos\theta - 4\tan\theta - 4\sec^2\theta\right)$

25. $y = 1$ **27.** $y = \dfrac{2}{3}t + \left(3\sqrt{3} - 2\pi\right)/9$

29. $y = \left(1 - \sqrt{3}\right)\left(\theta - \dfrac{\pi}{3}\right) + 1 + \sqrt{3}$

31. $y = x + 1$ **33.** $y = 2e^{\pi/2}\left(t - \dfrac{\pi}{2}\right) + e^{\pi/2}$

35. $\cot x = \dfrac{\cos x}{\sin x}$; use the Quotient Rule.

37. $\csc x = \dfrac{1}{\sin x}$; use the Quotient Rule.

39. $f''(\theta) = 2\cos\theta - \theta\sin\theta$

41. $y'' = 2\sec^2 x \tan x$
$y''' = 2\sec^4 x + 4\sec^2 x \tan^2 x$

43. • Then $f'(x) = -\sin x$, $f''(x) = -\cos x$, $f'''(x) = \sin x$, $f^{(4)}(x) = \cos x$, and $f^{(5)}(x) = -\sin x$.
• Accordingly, the successive derivatives of f cycle among

$$\{-\sin x,\ -\cos x,\ \sin x,\ \cos x\}$$

in that order. Since 8 is a multiple of 4, we have $f^{(8)}(x) = \cos x$.
• Since 36 is a multiple of 4, we have $f^{(36)}(x) = \cos x$. Therefore, $f^{(37)}(x) = -\sin x$.

45. If $r = 0$, then $f^{(n)}(x) = \sin x$. If $r = 1$, then $f^{(n)}(x) = \cos x$. If $r = 2$, then $f^{(n)}(x) = -\sin x$. If $r = 3$, then $f^{(n)}(x) = -\cos x$.

47. (a) From $\sin^2 x + \cos^2 x = 1$, we have $f(x) + g(x) = 1$. Take the derivative of both sides of this equation to obtain $f'(x) + g'(x) = 0$. This implies $f'(x) = -g'(x)$.

(b) $f'(x) = 2\sin x \cos x$, and $g'(x) = 2(\cos x)(-\sin x) = -2\sin x \cos x$. So, $f'(x) = -g'(x)$.

49. $x = \dfrac{\pi}{4}, \dfrac{3\pi}{4}, \dfrac{5\pi}{4}, \dfrac{7\pi}{4}$

51. (a)

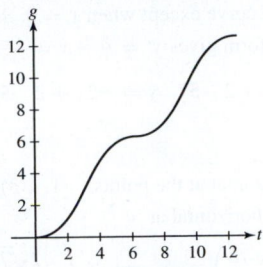

(b) Since $g'(t) = 1 - \cos t \geq 0$ for all t, the slope of the tangent line to g is always nonnegative.   **(c)** $t = 0, 2\pi, 4\pi$

53. $f'(x) = \sec^2 x = \frac{1}{\cos^2 x}$. Note that $f'(x) = \frac{1}{\cos^2 x}$ has numerator 1; the equation $f'(x) = 0$ therefore has no solution. The least slope for a tangent line to $\tan x$ is 1. Here is a graph of f'.

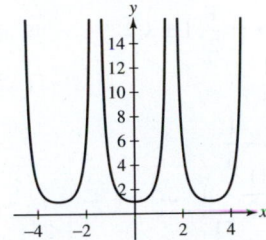

55. $\frac{dR}{d\theta} = \left(\frac{2v_0^2}{g}\right)(\cos^2\theta - \sin^2\theta) = \left(\frac{2v_0^2}{g}\right)\cos(2\theta)$. $\frac{dR}{d\theta} = 0$ when $\cos(2\theta) = 0$, and for $0 \leq \theta \leq \pi/2$ that occurs only when $\theta = \pi/4$. The corresponding range is

$$\left(\frac{2v_0^2}{g}\right)\sin(\pi/4)\cos(\pi/4) = \left(\frac{2v_0^2}{g}\right)\left(\frac{\sqrt{2}}{2}\right)\left(\frac{\sqrt{2}}{2}\right) = \frac{v_0^2}{g}$$

57.
$$f'(x) = \lim_{h \to 0} \frac{\cos(x+h) - \cos x}{h}$$
$$= \lim_{h \to 0} \frac{\cos x \cos h - \sin x \sin h - \cos x}{h}$$
$$= \lim_{h \to 0} \left((-\sin x)\frac{\sin h}{h} + (\cos x)\frac{\cos h - 1}{h}\right)$$
$$= (-\sin x) \cdot 1 + (\cos x) \cdot 0 = -\sin x$$

Section 3.7 Preliminary Questions

1. **(a)** The outer function is $\sqrt{x}$, and the inner function is $4x + 9x^2$.
(b) The outer function is $\tan x$, and the inner function is $x^2 + 1$.
(c) The outer function is x^5, and the inner function is $\sec x$.
(d) The outer function is x^4, and the inner function is $1 + e^x$.

2. The function $\frac{x}{x+1}$ can be differentiated using the Quotient Rule, and the functions $\sqrt{x} \cdot \sec x$ and xe^x can be differentiated using the Product Rule. The functions $\tan(7x^2 + 2)$, $x\sqrt{\sec x}$, and $\sin(e^x)$ require the Chain Rule.

3. **(b)**

4. We do not have enough information to compute $F'(4)$. We are missing the value of $f'(1)$.

Section 3.7 Exercises

1.

$f(g(x))$	$f'(u)$	$f'(g(x))$	$g'(x)$	$(f \circ g)'$
$(x^4 + 1)^{3/2}$	$\frac{3}{2}u^{1/2}$	$\frac{3}{2}(x^4 + 1)^{1/2}$	$4x^3$	$6x^3(x^4 + 1)^{1/2}$

3.

$f(g(x))$	$f'(u)$	$f'(g(x))$	$g'(x)$	$(f \circ g)'$
$\tan(x^4)$	$\sec^2 u$	$\sec^2(x^4)$	$4x^3$	$4x^3 \sec^2(x^4)$

5. $4(x + \sin x)^3(1 + \cos x)$

7. **(a)** $2x\sin(9 - x^2)$ **(b)** $\frac{\sin(x^{-1})}{x^2}$ **(c)** $-\sec^2 x \sin(\tan x)$

9. 12

11. Multiplying out, $f(x) = 4x^4 - 20x^2 + 25$, so $f'(x) = 16x^3 - 40x = 8x(2x^2 - 5)$. To use the Product Rule, write $f(x) = (2x^2 - 5)(2x^2 - 5)$. We have $f'(x) = (4x)(2x^2 - 5) + (2x^2 - 5)(4x) = 8x(2x^2 - 5)$. Using the Chain Rule, $f'(x) = 2(2x^2 - 5)(4x) = 8x(2x^2 - 5)$.

13. $12x^3(x^4 + 5)^2$ **15.** $\frac{7}{2\sqrt{7x - 3}}$

17. $-2(x^2 + 9x)^{-3}(2x + 9)$ **19.** $-4\cos^3\theta \sin\theta$

21. $9(2\cos\theta + 5\sin\theta)^8(5\cos\theta - 2\sin\theta)$

23. e^{x-12} **25.** $2\cos(2x + 1)$ **27.** $e^{x+x^{-1}}\left(1 - x^{-2}\right)$

29.
$$\frac{d}{dx}f(g(x)) = -\sin(x^2 + 1)(2x) = -2x\sin(x^2 + 1)$$
$$\frac{d}{du}g(f(u)) = -2\sin u \cos u$$

31. $2x\cos(x^2)$ **33.** $\frac{t}{\sqrt{t^2 + 9}}$

35. $\frac{2}{3}\left(x^4 - x^3 - 1\right)^{-1/3}\left(4x^3 - 3x^2\right)$ **37.** $\frac{8(1+x)^3}{(1-x)^5}$

39. $-\frac{\sec(1/x)\tan(1/x)}{x^2}$ **41.** $(1 - \sin\theta)\sec^2(\theta + \cos\theta)$

43. $-18te^{2-9t^2}$ **45.** $(2x + 4)\sec^2(x^2 + 4x)$

47. $\cos(1 - 3x) + 3x\sin(1 - 3x)$

49. $2(4t + 9)^{-1/2}$ **51.** $4(\sin x - 3x^2)(x^3 + \cos x)^{-5}$

53. $\frac{\cos 2x}{\sqrt{2\sin 2x}}$ **55.** $\frac{x\cos(x^2) - 3\sin 6x}{\sqrt{\cos 6x + \sin(x^2)}}$

57. $3(\tan^2 x \sec^2 x + x^2 \sec^2(x^3))$ **59.** $\frac{-1}{\sqrt{z+1}(z-1)^{3/2}}$

61. $\frac{\sin(-1) - \sin(1 + x)}{(1 + \cos x)^2}$ **63.** $-35x^4\cot^6(x^5)\csc^2(x^5)$

65. $-180x^3\cot^4(x^4 + 1)\csc^2(x^4 + 1)\left(1 + \cot^5(x^4 + 1)\right)^8$

67. $24(2e^{3x} + 3e^{-2x})^3(e^{3x} - e^{-2x})$

69. $4(x + 1)(x^2 + 2x + 3)e^{(x^2+2x+3)^2}$

71. $\frac{1}{8\sqrt{x}\sqrt{1 + \sqrt{x}}\sqrt{1 + \sqrt{1 + \sqrt{x}}}}$

73. $-\frac{k}{3}(kx + b)^{-4/3}$ **75.** $2\cos(x^2) - 4x^2\sin(x^2)$

77. $-336(9 - x)^5$ **79.** $\left.\frac{dv}{dP}\right|_{P=1.5} = \frac{290\sqrt{3}}{3}\frac{m}{s \cdot atm}$

81. **(a)** When $r = 3$, $\frac{dV}{dt} = 1.6\pi(3)^2 \approx 45.24$ cm³/s.

(b) When $t = 3$, we have $r = 1.2$. Hence, $\frac{dV}{dt} = 1.6\pi(1.2)^2 \approx 7.24$ cm³/s.

83. $L'(t) = \frac{6.8\pi}{365}\cos\left(\frac{2\pi}{365}t\right)$. December 1: $L'(255) \approx -0.019$ h/day ≈ -1.1 min/day. January 1: $L'(286) \approx 0.012$ h/day ≈ 0.7 min/day. February 1: $L'(317) \approx 0.04$ h/day ≈ 2.4 min/day. The lengths of the days are decreasing in late fall, but then are increasing once the winter starts. As the winter progresses, the rate of increase of the lengths of the days is increasing.

85. $\pm\frac{1}{\sqrt{2k}}$ **87.** $W'(10) \approx 0.3566$ kg/year   **89.** **(a)** -9 **(b)** $-3/2$ **(c)** 18   **91.** $5\sqrt{3}$ **93.** 12   **95.** $\frac{1}{16}$ **97.** $\left.\frac{dP}{dt}\right|_{t=3} = -0.727$ $\frac{dollars}{year}$

99. $\frac{dP}{dh} = -4.08569 \times 10^{-13}(288.14 - 0.00649h)^{4.256}$

101. 0.0973 kelvins/year   **103.** $f''(g(x))(g'(x))^2 + f'(g(x))g''(x)$

105. Let $u = h(x)$, $v = g(u)$, and $w = f(v)$. Then

$$\frac{dw}{dx} = \frac{df}{dv}\frac{dv}{dx} = \frac{df}{dv}\frac{dv}{du}\frac{du}{dx} = f'(g(h(x)))g'(h(x))h'(x)$$

109. For $n = 1$, we find

$$\frac{d}{dx}\sin x = \cos x = \sin\left(x + \frac{\pi}{2}\right)$$

as required. Now, suppose that for some positive integer k,

$$\frac{d^k}{dx^k}\sin x = \sin\left(x + \frac{k\pi}{2}\right)$$

Then

$$\frac{d^{k+1}}{dx^{k+1}}\sin x = \frac{d}{dx}\sin\left(x + \frac{k\pi}{2}\right)$$

$$= \cos\left(x + \frac{k\pi}{2}\right) = \sin\left(x + \frac{(k+1)\pi}{2}\right)$$

Section 3.8 Preliminary Questions

1. The Chain Rule

2. (a) This is correct.   (b) This is correct.

(c) This is incorrect. Because the differentiation is with respect to the variable x, the Chain Rule is needed to obtain

$$\frac{d}{dx}\sin(y^2) = 2y\cos(y^2)\frac{dy}{dx}$$

3. There are two mistakes in Jason's answer. First, Jason should have applied the Product Rule to the second term to obtain

$$\frac{d}{dx}(2xy) = 2x\frac{dy}{dx} + 2y$$

Second, he should have applied the General Power Rule to the third term to obtain

$$\frac{d}{dx}y^3 = 3y^2\frac{dy}{dx}$$

4. (b)   **5.** $g(x) = \tan^{-1}x$

6. The derivatives of $\sin^{-1}x$ and $\cos^{-1}x$ are negatives of each other.

7. $\frac{d}{dx}a^2 = 0$, $\frac{d}{dx}x^2 = 2x$, $\frac{d}{dx}y^2 = 2y\frac{dy}{dx}$

Section 3.8 Exercises

1. $(2, 1)$, $\frac{dy}{dx} = -\frac{2}{3}$ **3.** $\frac{d}{dx}\left(x^2y^3\right) = 3x^2y^2y' + 2xy^3$

5. $\frac{d}{dx}\left(\left(x^2 + y^2\right)^{3/2}\right) = 3\left(x + yy'\right)\sqrt{x^2 + y^2}$

7. $\sqrt[3]{y} + \frac{1}{3}xy^{-2/3}\frac{dy}{dx}$ **9.** $\frac{d}{dx}\frac{y}{y+1} = \frac{y'}{(y+1)^2}$ **11.** $y' = -\frac{2x}{9y^2}$

13. $y' = \frac{2xy + 6x^2y - 1}{1 - x^2 - 2x^3}$ **15.** $R' = -\frac{3R}{5x}$

17. $y' = \frac{y(y^2 - x^2)}{x(y^2 - x^2 - 2xy^2)}$ **19.** $y' = \frac{9}{4}x^{1/2}y^{5/3}$

21. $y' = \frac{(2x + 1)y^2}{y^2 - 1}$ **23.** $y' = \frac{1 - \cos(x + y)}{\cos(x + y) + \sin y}$

25. $y' = \frac{e^y - 2y}{2x + 3y^2 - xe^y}$ **27.** $\frac{dy}{dx} = \frac{1 - e^x}{1 + e^y}$ **29.** $5/4$ **31.** $\frac{1}{4\sqrt{15}}$

33. $\frac{7}{\sqrt{1 - 49x^2}}$ **35.** $\frac{-2x}{\sqrt{1 - x^4}}$ **37.** $\tan^{-1}x + \frac{x}{x^2 + 1}$

39. $\frac{e^x}{\sqrt{1 - e^{2x}}}$ **41.** $\frac{1 - t}{\sqrt{1 - t^2}}$ **43.** $\frac{3(\tan^{-1}x)^2}{x^2 + 1}$ **45.** 0

47. $\cos y = x$, so implicit differentiation gives $-\sin(y)y' = 1$, or
$$\frac{dy}{dx} = -\csc y = \frac{-1}{\sqrt{1 - x^2}}.$$

49. Since $1 + \tan^2 y = \sec^2 y$, and $\sec y = x$, $\Rightarrow \tan y = \pm\sqrt{x^2 - 1}$. If $x \geq 1$, $\text{arcsec } x = y \in \left[0, \frac{\pi}{2}\right) \Rightarrow \tan y \geq 0$. If $x \leq -1$, $\text{arcsec } x = y \in \left(\pi/2, \pi\right] \Rightarrow \tan y = -\sqrt{x^2 - 1} \leq 0$.

51. Multiplying both sides of $x + yx^{-1} = 1$ by x gives $x^2 + y = x$, so the two define the same curve except when $x = 0$. Since $y = x - x^2$, differentiating the first form gives $y' = \frac{y}{x} - x = \frac{x - x^2}{x} - x = 1 - 2x$.

53. $\frac{1}{4}$ **55.** $y = -\frac{1}{2}x + 2$ **57.** $y = -2x + 2$ **59.** $y = -\frac{12}{5}x + \frac{32}{5}$

61. $y = \frac{4}{3}x + \frac{4}{3}$

63. The tangent is horizontal at the points $(-1, \sqrt{3})$ and $(-1, -\sqrt{3})$.

65. The tangent line is horizontal at

$$\left(\frac{2\sqrt{78}}{13}, -\frac{4\sqrt{78}}{13}\right) \quad \text{and} \quad \left(-\frac{2\sqrt{78}}{13}, \frac{4\sqrt{78}}{13}\right)$$

67. At $(0, 2^{1/4})$, $\frac{dy}{dx} = \frac{-2^{1/4} - 1}{2^{11/4}}$, and the tangent line has equation $y = \left(\frac{-2^{1/4} - 1}{2^{11/4}}\right)x + 2^{1/4}$. At $(0, -2^{1/4})$, $\frac{dy}{dx} = \frac{1 - 2^{1/4}}{2^{11/4}}$, and the tangent line has equation $y = \left(\frac{1 - 2^{1/4}}{2^{11/4}}\right)x - 2^{1/4}$

69. $(2^{1/3}, 2^{2/3})$ **71.** $x = \frac{1}{2}, 1 \pm \sqrt{2}$

73. • At $(1, 2)$, $y' = \frac{1}{3}$

• At $(1, -2)$, $y' = -\frac{1}{3}$

• At $(1, \frac{1}{2})$, $y' = \frac{11}{12}$

• At $(1, -\frac{1}{2})$, $y' = -\frac{11}{12}$

75. There are vertical tangent lines at six points $(\pm 1, 0)$ and $\left(\pm\frac{\sqrt{3}}{2}, \pm\frac{\sqrt{2}}{2}\right)$.

77. $\frac{dx}{dy} = \frac{2y}{3x^2 - 4}$; it follows that $\frac{dx}{dy} = 0$ when $y = 0$, so the tangent line to this curve is vertical at the points where the curve intersects the x-axis.

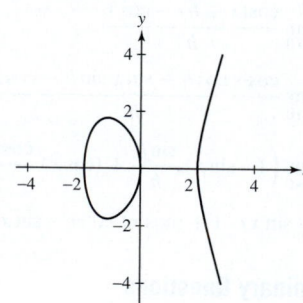

79. By implicit differentiation, $y' = \frac{y}{1-x}$ and $y'' = \frac{2y}{(1-x)^2}$. Solving the original equation for y, we obtain $y = \frac{2}{1-x}$. Differentiating, we obtain $y' = \frac{2}{(1-x)^2}$ and $y'' = \frac{4}{(1-x)^3}$. If we substitute $y = \frac{2}{1-x}$ into the expressions for y' and y'' found by implicit differentiation, we obtain the expressions obtained by direct differentiation.

81. $y'' = (y^2 - 2xyy')/y^4 = (y^2 - 2xyx/y^2)/y^4 = (y^3 - 2x^2)/y^5$

83. (a) $y' = -y^2/(2xy + 1)$; $y'|(1, 1) = -1/3$

(b) $y'' = (2y^3 - 2xy^2y' - 2yy')/(2xy + 1)^2$; $y''|(1, 1) = 10/27$

85. (a) $r = \frac{(16y^2 + x^2)^{3/2}}{64}$

(b) At $(4, 0)$, $r = 1$. At $(2, \sqrt{3})$, $r = \frac{13^{3/2}}{8} \approx 5.86$. At $(0, 2)$, $r = 8$.

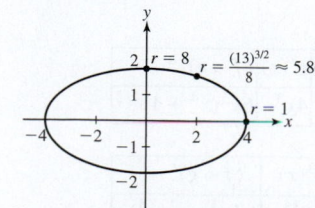

87. $\dfrac{dy}{dt} = -6x^{-3}\dfrac{dx}{dt}$ **89.** $\dfrac{dy}{dt} = \dfrac{dx}{dt}\left(\dfrac{y-4x}{4y^3-x}\right)$

91. The derivatives of these two curves are $y' = x/y$ and $y' = -y/x$, respectively. So at any point (x, y) satisfying both equations, the slopes of one tangent line will be the negative reciprocal of the other slope. That is, the tangent lines will be perpendicular.

93. • Upper branch:

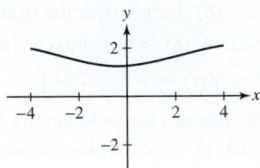

• Lower part of lower left curve:

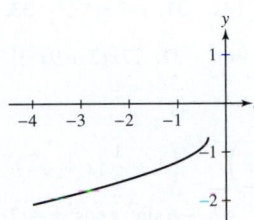

• Upper part of lower left curve:

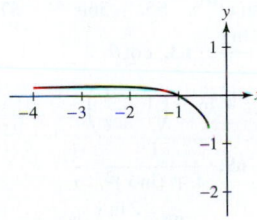

• Upper part of lower right curve:

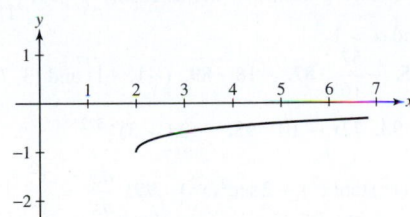

• Lower part of lower right curve:

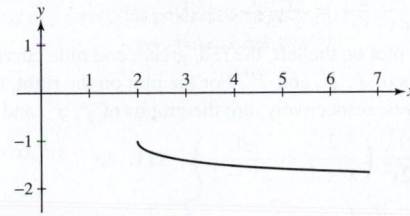

Section 3.9 Preliminary Questions

1. $\ln 4$ **2.** $\dfrac{1}{10}$ **3.** e^2 **4.** e^3

5. $y^{(100)} = \cosh x$ and $y^{(101)} = \sinh x$

Section 3.9 Exercises

1. $\dfrac{d}{dx}x\ln x = \ln x + 1$ **3.** $\dfrac{d}{dx}2^{x^3} = 2^{x^3}\cdot 3x^2 \cdot \ln 2$

5. $\dfrac{d}{dx}\ln(9x^2-8) = \dfrac{18x}{9x^2-8}$ **7.** $\dfrac{d}{dx}(\ln x)^2 = \dfrac{2}{x}\ln x$

9. $\dfrac{d}{dx}e^{(\ln x)^2} = \dfrac{2}{x}\ln x \cdot e^{(\ln x)^2}$ **11.** $\dfrac{d}{dx}\ln(\ln x) = \dfrac{1}{x\ln x}$

13. $\dfrac{d}{dx}(\ln(\ln x))^3 = \dfrac{3(\ln(\ln x))^2}{x\ln x}$

15. $\dfrac{d}{dx}\ln((x+1)(2x+9)) = \dfrac{4x+11}{(x+1)(2x+9)}$

17. $\dfrac{d}{dx}11^x = \ln 11 \cdot 11^x$

19. $\dfrac{d}{dx}\dfrac{2^x - 3^{-x}}{x} = \dfrac{x(2^x\ln 2 + 3^{-x}\ln 3) - (2^x - 3^{-x})}{x^2}$

21. $f'(x) = \dfrac{1}{x}\cdot\dfrac{1}{\ln 2}$ **23.** $\dfrac{d}{dt}\log_3(\sin t) = \dfrac{\cot t}{\ln 3}$

25. $y = 36\ln 6(x-2) + 36$ **27.** $y = 3^{20}\ln 3(t-2) + 3^{18}$

29. $y = 5^{-1}$ **31.** $y = -1(t-1) + \ln 4$

33. $y = \dfrac{12}{25\ln 5}(z-3) + 2$ **35.** $y = \dfrac{8}{\ln 2}\left(w - \dfrac{1}{8}\right) - 3$

37. $y' = 2x + 14$ **39.** $y' = 3x^2 - 12x - 79$

41. $y' = \dfrac{x(x^2+1)}{\sqrt{x+1}}\left(\dfrac{1}{x} + \dfrac{2x}{x^2+1} - \dfrac{1}{2(x+1)}\right)$

43. $y' = \dfrac{1}{2}\sqrt{\dfrac{x(x+2)}{(2x+1)(3x+2)}}\cdot\left(\dfrac{1}{x} + \dfrac{1}{x+2} - \dfrac{2}{2x+1} - \dfrac{3}{3x+2}\right)$

45. $\dfrac{d}{dx}x^{3x} = x^{3x}(3+3\ln x)$ **47.** $\dfrac{d}{dx}x^{e^x} = x^{e^x}\left(\dfrac{e^x}{x} + e^x\ln x\right)$

49. $y' = x^{\cos x}((\cos x)/x - \sin x\ln x)$ **51.** $\dfrac{d}{dx}\sinh(9x) = 9\cosh(9x)$

53. $\dfrac{d}{dt}\cosh^2(9-3t) = -6\cosh(9-3t)\sinh(9-3t)$

55. $\dfrac{d}{dx}\sqrt{\cosh x + 1} = \dfrac{1}{2}(\cosh x + 1)^{-1/2}\sinh x$

57. $\dfrac{dy}{dt} = -\dfrac{\operatorname{csch} t(\operatorname{csch} t + 2\operatorname{sech} t)}{(1+\tanh t)^2}$ **59.** $\dfrac{d}{dx}\sinh(\ln x) = \dfrac{\cosh(\ln x)}{x}$

61. $\dfrac{d}{dx}\tanh(e^x) = e^x\operatorname{sech}^2(e^x)$

63. $\dfrac{d}{dx}\operatorname{sech}(\sqrt{x}) = -\dfrac{1}{2}x^{-1/2}\operatorname{sech}\sqrt{x}\tanh\sqrt{x}$

65. $\dfrac{d}{dx}\operatorname{sech} x\coth x = -\operatorname{csch} x\coth x$

67. $\dfrac{d}{dx}\cosh^{-1}(3x) = \dfrac{3}{\sqrt{9x^2-1}}$

69. $\dfrac{d}{dx}(\sinh^{-1}(x^2))^3 = 3(\sinh^{-1}(x^2))^2\dfrac{2x}{\sqrt{x^4+1}}$

71. $\dfrac{d}{dx}e^{\cosh^{-1}x} = e^{\cosh^{-1}x}\left(\dfrac{1}{\sqrt{x^2-1}}\right)$

73. $\dfrac{d}{dt}\tanh^{-1}(\ln t) = \dfrac{1}{t(1-(\ln t)^2)}$

75. $\dfrac{d}{dx}\operatorname{sech} x = \dfrac{d}{dx}\dfrac{1}{\cosh x} = \dfrac{-\sinh x}{\cosh^2 x} = -\left(\dfrac{1}{\cosh x}\right)\left(\dfrac{\sinh x}{\cosh x}\right) = -\operatorname{sech} x\tanh x$

79. At $h = 60$ m, $v \approx 15.57$ m/s and $dv/dh \approx 0.052$ m/s per meter.

81.
$$\lim_{h\to 0}\frac{f(x+h)-f(x)}{h} = \lim_{h\to 0}\frac{a^{x+h}-a^x}{h} = \lim_{h\to 0}\frac{a^x a^h - a^x}{h} = \left(\lim_{h\to 0}\frac{a^h-1}{h}\right)a^x$$
Since this limit equals $f'(x)$ and since $f'(x) = (\ln a)a^x$, it follows that $\lim_{h\to 0}\dfrac{a^h-1}{h} = \ln a$.

85. (a) $\dfrac{dP}{dT} = -\dfrac{1}{T\ln 10}$ (b) $\Delta P \approx -0.054$

87. $\dfrac{d}{dx}e^{n\ln x} = (e^{n\ln x})(n)\left(\dfrac{1}{x}\right) = n(x^n)(x^{-1}) = nx^{n-1}$

Section 3.10 Preliminary Questions

1. When $x = -3$, $\frac{dy}{dt} = -18$. When $x = 2$, $\frac{dy}{dt} = 12$. When $x = 5$, $\frac{dy}{dt} = 30$.

2. When $x = -4$, $\frac{dy}{dt} = 96$. When $x = 2$, $\frac{dy}{dt} = 24$. When $x = 6$, $\frac{dy}{dt} = 216$.

3. Let s and V denote the length of the side and the corresponding volume of a cube, respectively. Determine $\frac{dV}{dt}$ if $\frac{ds}{dt} = 0.5$ cm/s.

4. $\frac{dV}{dt} = 4\pi r^2 \frac{dr}{dt}$ **5.** Determine $\frac{dh}{dt}$ if $\frac{dV}{dt} = 2$ cm^3/min.

6. Determine $\frac{dV}{dt}$ if $\frac{dh}{dt} = 1$ cm/min.

Section 3.10 Exercises

1. 0.039 ft/min

3. **(a)** $100\pi \approx 314.16$ m^2/min   **(b)** $24\pi \approx 75.40$ m^2/min

5. 27000π cm^3/min   **7.** 9600π cm^2/min

9. When $h = 1$, $\frac{dh}{dt} = \frac{9}{8\pi} \approx 0.36$ m/min. When $h = 2$, $\frac{dh}{dt} = \frac{9}{2\pi} \approx 1.43$ m/min.

11. -0.632 m/s   **13.** $x \approx 4.737$ m; $\frac{dx}{dt} \approx 0.405$ m/s

15. $\frac{1000\pi}{3} \approx 1047.20$ cm^3/s   **17.** 0.675 m/s

19. **(a)** 594.6 km/h   **(b)** 0 km/h

21. 1.22 km/min   **23.** $\frac{1200}{241} \approx 4.98$ rad/h

25. **(a)** $\frac{100\sqrt{13}}{13} \approx 27.735$ km/h   **(b)** 112.963 km/h

27. $\sqrt{16.2} \approx 4.025$ m   **29.** $\frac{5}{3}$ m/s   **31.** -1.92 kPa/min

33. $-\frac{1}{8}$ rad/s

35. **(b)** when $x = 1$, $L'(t) = 0$; when $x = 2$, $L'(t) = \frac{16}{3}$

37. $-4\sqrt{5} \approx -8.94$ ft/s   **39.** -0.79 m/min

41. Let the equation $y = f(x)$ describe the shape of the roller coaster track. Taking $\frac{d}{dt}$ of both sides of this equation yields $\frac{dy}{dt} = f'(x)\frac{dx}{dt}$.

43. **(a)** The distance formula gives

$$L = \sqrt{(x - r\cos\theta)^2 + (-r\sin\theta)^2}$$

Thus,

$$L^2 = (x - r\cos\theta)^2 + r^2\sin^2\theta$$

(b) From (a), we have

$$0 = 2(x - r\cos\theta)\left(\frac{dx}{dt} + r\sin\theta\frac{d\theta}{dt}\right) + 2r^2\sin\theta\cos\theta\frac{d\theta}{dt}$$

(c) $-80\pi \approx -251.33$ cm/min

45. **(c)** $\frac{3\sqrt{5}}{250} \approx 0.027$ m/min

Chapter 3 Review

1. 3; the slope of the secant line through the points $(2, 7)$ and $(0, 1)$ on the graph of $f(x)$

3. $\approx \frac{7}{3}$; the value of the difference quotient is larger than the value of the derivative

5. $f'(1) = 1$; $y = x - 1$ **7.** $f'(4) = -\frac{1}{16}$; $y = -\frac{1}{16}x + \frac{1}{2}$

9. $-2x$ **11.** $\frac{1}{(2 - x)^2}$ **13.** $f'(1)$, where $f(x) = \sqrt{x}$

15. $f'(\pi)$, where $f(t) = \sin t \cos t$ **17.** $f(4) = -2$; $f'(4) = 3$

19. (C) is the graph of $f'(x)$.

21.

23. **(a)** 8.05 cm/year   **(b)** Larger over the first half
(c) $h'(3) \approx 7.8$ cm/year; $h'(8) \approx 6.0$ cm/year $\left(\text{using the difference quotient approximation } h'(t) \approx \frac{h(t+1)-h(t)}{1}\right)$

25. $A'(t)$ measures the rate of change in automobile production in the United States; $A'(1971) \approx 0.25$ million automobiles/year; $\left(\text{using the difference quotient approximation } A'(t) \approx \frac{A(t+1)-A(t)}{1}\right)$ $A'(1974)$ would be negative.

27. **(b)**   **29.** $15x^4 - 14x$ **31.** $-7.3t^{-8.3}$ **33.** $\frac{1 - 2x - x^2}{(x^2 + 1)^2}$

35. $6(4x^3 - 9)(x^4 - 9x)^5$ **37.** $27x(2 + 9x^2)^{1/2}$

39. $\frac{2 - z}{2(1 - z)^{3/2}}$ **41.** $2x - \frac{3}{2}x^{-5/2}$

43. $\frac{1}{2}\left(x + \sqrt{x + \sqrt{x}}\right)^{-1/2}\left(1 + \frac{1}{2}(x + \sqrt{x})^{-1/2}\left(1 + \frac{1}{2}x^{-1/2}\right)\right)$

45. $-3t^{-4}\sec^2(t^{-3})$ **47.** $-6\sin^2 x \cos^2 x + 2\cos^4 x$

49. $\frac{1 + \sec t - t \sec t \tan t}{(1 + \sec t)^2}$ **51.** $\frac{8\csc^2\theta}{(1 + \cot\theta)^2}$

53. $y' = -100x^{99}\sin(x^{100})$ **55.** $-36e^{-4x}$ **57.** $(4 - 2t)e^{4t - t^2}$

59. $\frac{8x}{4x^2 + 1}$ **61.** $\frac{2\ln s}{s}$ **63.** $\cot\theta$

65. $\sec(z + \ln z)\tan(z + \ln z)\left(1 + \frac{1}{z}\right)$

67. $-2(\ln 7)(7^{-2x})$ **69.** $\frac{1}{1 + (\ln x)^2} \cdot \frac{1}{x}$

71. $-\frac{1}{|x|\sqrt{x^2 - 1}\csc^{-1} x}$ **73.** $\frac{2\ln s}{s}s^{\ln s}$

75. $2(\sin^2 t)^t(t\cot t + \ln\sin t)$ **77.** $2t\cosh(t^2)$ **79.** $\frac{e^x}{1 - e^{2x}}$

81. $\alpha = 0$ and $\alpha > 1$

83. -27 **85.** $-\frac{57}{16}$ **87.** -18 **89.** $(-1, -1)$ and $(3, 7)$

91. $a = \frac{1}{6}$ **93.** $72x - 10$ **95.** $-(2x + 3)^{-3/2}$

97. $8x^2\sec^2(x^2)\tan(x^2) + 2\sec^2(x^2)$ **99.** $\frac{dy}{dx} = \frac{x^2}{y^2}$

101. $\frac{dy}{dx} = \frac{y^2 + 4x}{1 - 2xy}$ **103.** $\frac{dy}{dx} = \frac{\cos(x + y)}{1 - \cos(x + y)}$

105. $\frac{dy}{dx} = \frac{x}{4y}$, $\frac{d^2y}{dx^2} = \frac{4y^2 - x^2}{16y^3}$

107. For the plot on the left, the red, green, and blue curves, respectively, are the graphs of f, f', and f''. For the plot on the right, the green, red, and blue curves, respectively, are the graphs of f, f', and f''.

109. $\frac{(x + 1)^3}{(4x - 2)^2}\left(\frac{3}{x + 1} - \frac{4}{2x - 1}\right)$ **111.** $4e^{(x-1)^2}e^{(x-3)^2}(x - 2)$

113. $\frac{e^{3x}(x - 2)^2}{(x + 1)^2}\left(3 + \frac{2}{x - 2} - \frac{2}{x + 1}\right)$

115. $\frac{dh}{dt} = \frac{20}{240 + 15(4)} = \frac{1}{15}$ m/min

117. $\frac{ds}{dt} = \frac{476}{6\sqrt{5536}} \approx 1.066$ km/min

119. **(a)** $\frac{d\theta}{dt} = -\frac{5}{4\sqrt{3}} \approx -0.72$ rad/s

(b) $\frac{dD}{dt} = \frac{16 + 24 - \frac{20}{\sqrt{3}} - 12\sqrt{3} - 8\sqrt{3}}{8\sqrt{2 - \sqrt{3}}} \approx -1.49$ cm/s

$\frac{dD}{dt} = \frac{15 - 10\sqrt{3}}{3\sqrt{2 - \sqrt{3}}}$

Chapter 4

Section 4.1 Preliminary Questions

1. True   2. $g(1.2) - g(1) \approx 0.8$ 3. $f(2.1) \approx 1.3$

4. The Linear Approximation tells us that up to a small error, the change in output Δf is directly proportional to the change in input Δx when Δx is small.

Section 4.1 Exercises

1. $\Delta f \approx 0.12$ 3. $\Delta f \approx -0.00222$ 5. $\Delta f \approx 0.003333$

7. $\Delta f \approx 0.0074074$

9. $\Delta f \approx 0.05$; error is 0.000610; the percentage error is 1.24%.

11. $\Delta f \approx -0.03$; error is 0.0054717; the percentage error is 22.31%.

13. $\Delta f \approx 0.1$; $f(26) \approx 5.1$; error ≈ 0.00098

15. $\Delta f \approx -0.0005$; $f(101) \approx 0.0995$; error ≈ 0.000004

17. $\Delta f \approx \dfrac{1}{12} \approx 0.08333$; $f(9) \approx 2.08333$; error ≈ 0.00325

19. $\Delta f \approx -0.1$; $f(-0.1) \approx 0.9$; error ≈ 0.0048

21. $L(x) = 4x - 3$; $f(0.96) \approx 0.84$

23. $L(x) = x - \dfrac{\pi}{4} + \dfrac{1}{2}$; $f\left(\dfrac{1.1\pi}{4}\right) \approx 0.5785$

25. $L(x) = -\dfrac{1}{2}x + 1$; $f(0.08) \approx 0.96$

27. $L(x) = \dfrac{1}{2}e(x + 1)$; $f(0.85) \approx 2.5144$

29. $\Delta y \approx -0.007$ 31. $\Delta y \approx -0.026667$ 33. $f(4.03) \approx 2.01$

35. $\sqrt{2.1} - \sqrt{2}$ is larger than $\sqrt{9.1} - \sqrt{9}$.

37. $R(9) = 25110$ euros; if p is raised by 0.5 euros, then $\Delta R \approx 585$ euros; on the other hand, if p is lowered by 0.5 euros, then $\Delta R \approx -585$ euros.

39. $\Delta L \approx -0.00171$ cm

41. (a) $\Delta P \approx -0.434906$ kilopascals (kPa)   (b) The actual change in pressure is -0.418274 kPa; the percentage error is 3.98%.

43. (a) $\Delta W \approx W'(R)\Delta x = -\dfrac{2wR^2}{R^3}h = -\dfrac{2wh}{R} \approx -0.0005wh$

(b) $\Delta W \approx -0.7$ pounds (lb)

45. (a) $\Delta h \approx 0.71$ cm   (b) $\Delta h \approx 1.02$ cm   (c) There is a bigger effect at higher velocities.

47. (a) If $\theta = 34°$ (i.e., $t = \frac{17}{90}\pi$), then

$$\Delta s \approx s'(t)\Delta t = \dfrac{625}{16}\cos\left(\dfrac{17}{45}\pi\right)\Delta t$$

$$= \dfrac{625}{16}\cos\left(\dfrac{17}{45}\pi\right)\Delta\theta \cdot \dfrac{\pi}{180} \approx 0.255\Delta\theta$$

(b) If $\Delta\theta = 2°$, this gives $\Delta s \approx 0.51$ ft, in which case the shot would not have been successful, having been off half a foot.

(c) $\Delta s \approx 2.897$ ft

49. $\Delta V \approx 4\pi(25)^2(0.5) \approx 3927$ cm^3; $\Delta S \approx 8\pi(25)(0.5) \approx 314.2$ cm^2

51. $P = 6$ atm; $\Delta P \approx \pm 0.45$ atm   53. $f(2) = 8$

55. $\sqrt{16.2} \approx L(16.2) = 4.025$. Graphs of f and L are shown below. Because the graph of L lies above the graph of f, we expect that the estimate from the Linear Approximation is too large.

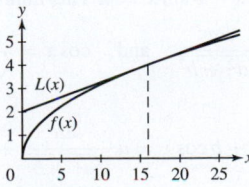

57. $\dfrac{1}{\sqrt{17}} \approx L(17) \approx 0.24219$; the percentage error is 0.14%.

59. $\dfrac{1}{(10.03)^2} \approx L(10.03) = 0.00994$; the percentage error is 0.0027%.

61. $(64.1)^{1/3} \approx L(64.1) \approx 4.002083$; the percentage error is 0.000019%.

63. $\cos^{-1}(0.52) \approx L(0.02) = 1.024104$; the percentage error is 0.015%.

65. $e^{-0.012} \approx L(-0.012) = 0.988$; the percentage error is 0.0073%.

67. Let $f(x) = \sqrt{x}$. Then $f(9) = 3$, $f'(x) = \frac{1}{2}x^{-1/2}$ and $f'(9) = \frac{1}{6}$. Therefore, by the Linear Approximation,

$$f(9 + h) - f(9) = \sqrt{9 + h} - 3 \approx \dfrac{1}{6}h$$

Moreover, $f''(x) = -\frac{1}{4}x^{-3/2}$, so $|f''(x)| = \frac{1}{4}x^{-3/2}$. Because this is a decreasing function, it follows that for $x \geq 9$,

$$K = \max|f''(x)| \leq |f''(9)| = \dfrac{1}{108} < 0.01$$

From the following table, we see that for $h = 10^{-n}$, $1 \leq n \leq 4$, $E \leq \frac{1}{2}Kh^2$.

h	$E = \left|\sqrt{9 + h} - 3 - \frac{1}{6}h\right|$	$\frac{1}{2}Kh^2$
10^{-1}	4.604×10^{-5}	5.00×10^{-5}
10^{-2}	4.627×10^{-7}	5.00×10^{-7}
10^{-3}	4.629×10^{-9}	5.00×10^{-9}
10^{-4}	4.627×10^{-11}	5.00×10^{-11}

69. $\left.\dfrac{dy}{dx}\right|_{(2,1)} = -\dfrac{1}{3}$; $y \approx L(2.1) = 0.967$

71. $L(x) = -\dfrac{14}{25}x + \dfrac{36}{25}$; $y \approx L(-1.1) = 2.056$

73. Let $f(x) = x^2$. Then

$$\Delta f = f(5 + h) - f(5) = (5 + h)^2 - 5^2 = h^2 + 10h$$

and

$$E = |\Delta f - f'(5)h| = |h^2 + 10h - 10h| = h^2 = \dfrac{1}{2}(2)h^2 = \dfrac{1}{2}Kh^2$$

Section 4.2 Preliminary Questions

1. A critical point is a value of the independent variable x in the domain of a function f at which either $f'(x) = 0$ or $f'(x)$ does not exist.

2. (b)   3. (b)

4. (a) False. For example, $x = 0$ is a critical point of $f(x) = x^3$ but is neither a local minimum nor a local maximum.

(b) False. For example, the maximum of $f(x) = x^2 + 1$ on $[1, 2]$ is 5, occurring at $x = 2$, but $f'(2) \neq 0$.   (c) True   (d) False. For example, the function $f(x) = 2x^2 - x^3$ has a single local minimum on $[-1, 3]$, at $x = 0$, but the absolute minimum on $[-1, 3]$ occurs at the endpoint $x = 3$.

Section 4.2 Exercises

1. (a) 3, 5, 7   (b) Maximum: 6, minimum: 1   (c) Local maximum 5 at $x = 5$, local minima: 3 at $x = 3$, and 1 at $x = 7$ (d) $[2, 6]$ is an example.   (e) $(0, 2)$ is an example.   (f) $(4, 6)$ is an example.

3. $x = 1$ 5. $x = -3$ and $x = 6$ 7. $x = 2$ 9. $x = \pm 1$

11. $t = 3$ and $t = -1$ 13. $x = -\frac{1}{2}$ 15. $\theta = \dfrac{n\pi}{2}$

17. $x = \dfrac{1}{e}$ 19. $x = \pm\dfrac{\sqrt{3}}{2}, \pm 1$

21. (a) Critical point at $x = 2$; $f(2) = -1$ (b) Minimum: -1, maximum: 17   (c) Minimum: 1, maximum: 71

23. $x = \dfrac{\pi}{4}$; maximum value: $\sqrt{2}$; minimum value: 1

25. Maximum = 5, minimum = 3.

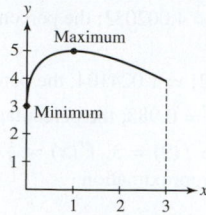

27. Minimum: $f(-1) = 3$, maximum: $f(2) = 21$

29. Minimum: $f(0) = 0$, maximum: $f(3) = 9$

31. Minimum: $f(4) = -24$, maximum: $f(6) = 8$

33. Minimum $= -19$, maximum $= 3$.

35. Minimum: $f(1) = 5$, maximum: $f(2) = 28$

37. Minimum: $f(2) = -128$, maximum: $f(-2) = 128$

39. Minimum: $f(6) = 18.5$, maximum: $f(5) = 26$

41. Minimum: $f(1) = -1$, maximum: $f(0) = f(3) = 0$

43. Minimum: $f(0) = 2\sqrt{6} \approx 4.9$, maximum: $f(2) = 4\sqrt{2} \approx 5.66$

45. Minimum: $f\left(\dfrac{\sqrt{3}}{2}\right) \approx -0.589980$, maximum: $f(4) \approx 0.472136$

47. Minimum: $f(0) = f\left(\dfrac{\pi}{2}\right) = 0$, maximum: $f\left(\dfrac{\pi}{4}\right) = \dfrac{1}{2}$

49. Minimum: $f(0) = -1$, maximum:
$f\left(\dfrac{\pi}{4}\right) = \sqrt{2}\left(\dfrac{\pi}{4} - 1\right) \approx -0.303493$

51. $x = \dfrac{1}{3}$ and $x = -1$ are critical points. Minimum: -2, maximum: 10.

53. Minimum: $g\left(\dfrac{\pi}{3}\right) = \dfrac{\pi}{3} - \sqrt{3} \approx -0.685$, maximum:
$g\left(\dfrac{5}{3}\pi\right) = \dfrac{5}{3}\pi + \sqrt{3} \approx 6.968$

55. Minimum: $f\left(\dfrac{\pi}{4}\right) = 1 - \dfrac{\pi}{2} \approx -0.570796$, maximum: $f(0) = 0$

57. Minimum: $f(1) = 0$, maximum: $f(e) = e^{-1} \approx 0.367879$

59. Minimum: $f(1) = 3e - e^2 \approx 0.765789$, maximum: $f\left(\ln\left(\dfrac{3}{2}\right)\right) = \dfrac{9}{4}$

61.

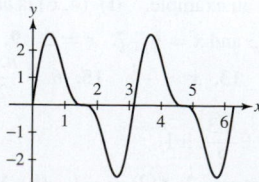

There appears to be a minimum between $x = 1$ and $x = 2$. Since $f(x) \to \infty$ as x approaches 0 from the right, there is no maximum value. Using calculus, the minimum is $2\sqrt{2}$ occurring at $x = \sqrt{2}$.

63. (d) $\dfrac{\pi}{6}, \dfrac{\pi}{2}, \dfrac{5\pi}{6}, \dfrac{7\pi}{6}, \dfrac{3\pi}{2}$, and $\dfrac{11\pi}{6}$; the maximum value is
$f(\tfrac{\pi}{6}) = f(\tfrac{7\pi}{6}) = \dfrac{3\sqrt{3}}{2}$ and the minimum value is
$f(\tfrac{5\pi}{6}) = f(\tfrac{11\pi}{6}) = -\dfrac{3\sqrt{3}}{2}$.
(e) We can see that there are six points where the graph has a horizontal tangent line on the graph between 0 and 2π, as predicted. There are four local extrema, and two points at $(\tfrac{\pi}{2}, 0)$ and $(\tfrac{3\pi}{2}, 0)$ where the graph has neither a local maximum nor a local minimum.

65. Critical point: $x = 2$; minimum value: $f(2) = 0$, maximum:
$f(0) = f(4) = 2$

67. Critical point: $x = 2$; minimum value: $f(2) = 0$, maximum:
$f(4) = 20$

69. $c = 1$ **71.** $c = \dfrac{15}{4}$

73. $f(0) < 0$ and $f(2) > 0$, so there is at least one root by the Intermediate Value Theorem; there cannot be another root because $f'(x) \geq 4$ for all x.

75. There cannot be a root $c > 0$ because $f'(x) > 4$ for all $x > 0$.

79. $b \approx 2.86$

81. (a) $F = \dfrac{1}{2}\left(1 - \dfrac{v_2^2}{v_1^2}\right)\left(1 + \dfrac{v_2}{v_1}\right)$ **(b)** $F(r)$ achieves its maximum value when $r = 1/3$.   **(c)** If v_2 were 0, then no air would be passing through the turbine, which is not realistic.

85. • The maximum value of f on $[0, 1]$ is
$$f\left(\left(\dfrac{a}{b}\right)^{1/(b-a)}\right) = \left(\dfrac{a}{b}\right)^{a/(b-a)} - \left(\dfrac{a}{b}\right)^{b/(b-a)}$$
• $\dfrac{1}{4}$

87. Critical points: $x = 1$, $x = 4$, and $x = \dfrac{5}{2}$; maximum value: $f(1) = f(4) = \dfrac{5}{4}$, minimum value: $f(-5) = \dfrac{17}{70}$

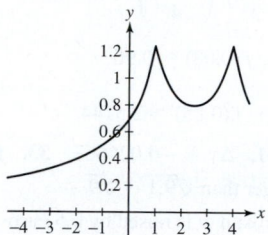

89. (a) There are therefore four points at which the derivative is zero:
$$(-1, -\sqrt{2}), (-1, \sqrt{2}), (1, -\sqrt{2}), (1, \sqrt{2})$$
There are also critical points where the derivative does not exist:
$$(0, 0), (\pm\sqrt[4]{27}, 0)$$
(b) The curve $27x^2 = (x^2 + y^2)^3$ and its horizontal tangents are plotted here.

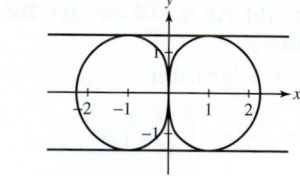

91.

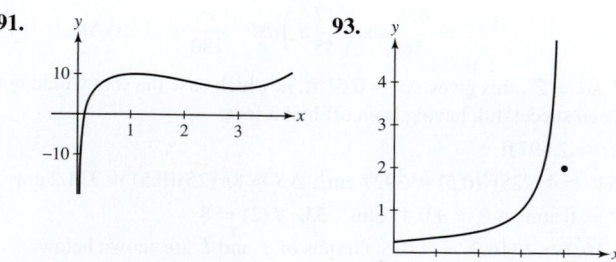

93.

95. If $f(x) = a \sin x + b \cos x$, then $f'(x) = a \cos x - b \sin x$, so $f'(x) = 0$ implies $a \cos x - b \sin x = 0$. This implies $\tan x = \dfrac{a}{b}$. Then
$$\sin x = \dfrac{\pm a}{\sqrt{a^2 + b^2}} \quad \text{and} \quad \cos x = \dfrac{\pm b}{\sqrt{a^2 + b^2}}$$
Therefore,
$$f(x) = a \sin x + b \cos x = a \dfrac{\pm a}{\sqrt{a^2 + b^2}} + b \dfrac{\pm b}{\sqrt{a^2 + b^2}}$$
$$= \pm \dfrac{a^2 + b^2}{\sqrt{a^2 + b^2}} = \pm\sqrt{a^2 + b^2}$$

97. Let $f(x) = x^2 + rx + s$ and suppose that $f(x)$ takes on both positive and negative values. This will guarantee that f has two real roots. By the quadratic formula, the roots of f are

$$x = \frac{-r \pm \sqrt{r^2 - 4s}}{2}$$

Observe that the midpoint between these roots is

$$\frac{1}{2}\left(\frac{-r + \sqrt{r^2 - 4s}}{2} + \frac{-r - \sqrt{r^2 - 4s}}{2}\right) = -\frac{r}{2}$$

Next, $f'(x) = 2x + r = 0$ when $x = -\frac{r}{2}$ and, because the graph of f is an upward-opening parabola, it follows that $f(-\frac{r}{2})$ is a minimum.

99. $b > \frac{1}{4}a^2$

101. • Let f be a continuous function with $f(a)$ and $f(b)$ local minima on the interval $[a, b]$. By Theorem 1, $f(x)$ must take on both a minimum and a maximum on $[a, b]$. Since local minima occur at $f(a)$ and $f(b)$, the maximum must occur at some other point in the interval, call it c, where $f(c)$ is a local maximum.
• The function graphed here is discontinuous at $x = 0$.

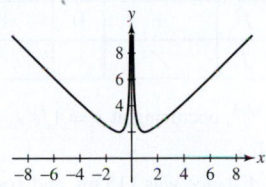

Section 4.3 Preliminary Questions

1. $m = 3$ **2.** (c)
. Yes. The figure below displays a function that takes on only negative values but has a positive derivative.

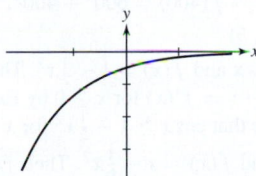

4. (a) $f(c)$ must be a local maximum.   (b) No

5. $f(x) = \sec x$ and $f(x) = \csc x$

6. $f(x) = |\sin x|$ is an example.

Section 4.3 Exercises

1. $c = 4$ **3.** $c = \dfrac{3\pi}{4}$ or $\dfrac{7\pi}{4}$ **5.** $c = \pm\sqrt{7}$

7. $c = -\dfrac{1}{2}\ln\left(\dfrac{1 - e^{-6}}{6}\right)$

9. The slope of the secant line between $x = 0$ and $x = 1$ is

$$\frac{f(1) - f(0)}{1 - 0} = 1$$

Since $f'(x) = 2x$, solving $2c = 1$ gives $c = \dfrac{1}{2}$. A graph of f and the tangent line appears here:

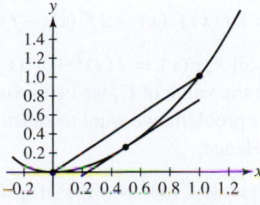

11. The slope of the secant line between $x = 0$ and $x = 1$ is

$$\frac{f(1) - f(0)}{1 - 0} = e - 1$$

Since $f'(x) = e^x$, solving $e^c = e - 1$ gives $c = \ln(e - 1)$. A graph of f and the tangent line is given here:

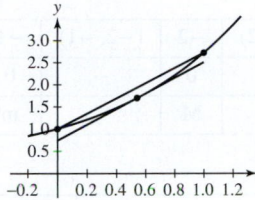

13. The slope of the secant line between $x = 0$ and $x = 1$ is

$$\frac{f(1) - f(0)}{1 - 0} = \frac{2 - 0}{1} = 2$$

It appears that the x-coordinate of the point of tangency is approximately 0.62.

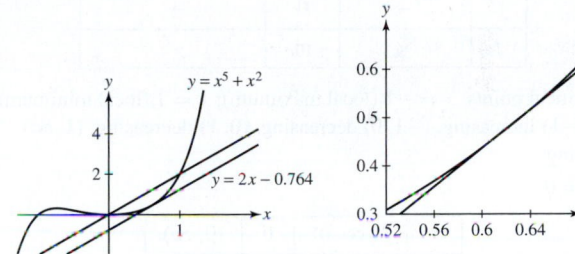

15. The derivative is positive on the intervals $(-\infty, 1) \cup (3, 5)$ and negative on the intervals $(1, 3) \cup (5, 6)$.

17. $f(2)$ is a local maximum; $f(4)$ is a local minimum.

19.   **21.**

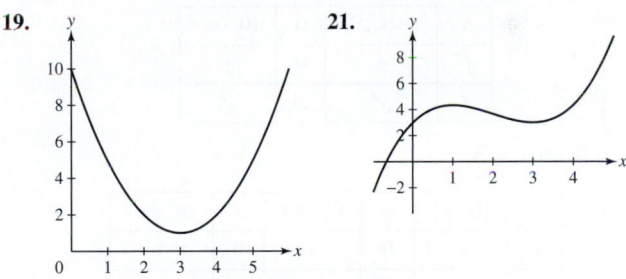

23. Critical point: $c = 3$; since the derivative changes sign from $+$ to $-$, this is a point of local maximum.

25. Critical points: $c = -2$ and $c = 0$. Since $f'(x)$ changes sign from $+$ to $-$ at $c = -2$, this is a point of local maximum. Since $f'(x)$ changes sign from $-$ to $+$ at $c = 0$, this is a point of local minimum.

27. $c = \dfrac{7}{2}$

x	$\left(-\infty, \frac{7}{2}\right)$	$7/2$	$\left(\frac{7}{2}, \infty\right)$
f'	$+$	0	$-$
f	↗	M	↘

29. $c = 0, 8$

x	$(-\infty, 0)$	0	$(0, 8)$	8	$(8, \infty)$
f'	$+$	0	$-$	0	$+$
f	↗	M	↘	m	↗

31. $c = -2, -1, 1$

x	$(-\infty, -2)$	-2	$(-2, -1)$	-1	$(-1, 1)$	1	$(1, \infty)$
f'	$-$	0	$+$	0	$-$	0	$+$
f	↘	m	↗	M	↘	m	↗

33. $c = -2, -1$

x	$(-\infty, -2)$	-2	$(-2, -1)$	-1	$(-1, \infty)$
f'	$+$	0	$-$	0	$+$
f	↗	M	↘	m	↗

35. $c = 0$

x	$(-\infty, 0)$	0	$(0, \infty)$
f'	$+$	0	$+$
f	↗	$\neq$	↗

37. $c = \left(\frac{3}{2}\right)^{2/5}$

x	$\left(0, \left(\frac{3}{2}\right)^{2/5}\right)$	$\frac{3}{2}^{2/5}$	$\left(\left(\frac{3}{2}\right)^{2/5}, \infty\right)$
f'	$-$	0	$+$
f	↘	m	↗

39. Critical points: $x = -1$ (local maximum); $x = 1$ (local minimum); $(-\infty, -1)$ increasing, $(-1, 0)$ decreasing, $(0, 1)$ decreasing, $(1, \infty)$ increasing

41. $c = 0$

x	$(-\infty, 0)$	0	$(0, \infty)$
f'	$+$	0	$-$
f	↗	M	↘

43. $c = 0$

x	$(-\infty, 0)$	0	$(0, \infty)$
f'	$+$	0	$+$
f	↗	¬	↗

45. $c = \frac{\pi}{2}$ and $c = \pi$

x	$\left(0, \frac{\pi}{2}\right)$	$\frac{\pi}{2}$	$\left(\frac{\pi}{2}, \pi\right)$	π	$(\pi, 2\pi)$
f'	$+$	0	$-$	0	$+$
f	↗	M	↘	m	↗

47. $c = \frac{\pi}{2}, \frac{7\pi}{6}, \frac{3\pi}{2}$, and $\frac{11\pi}{6}$

x	$\left(0, \frac{\pi}{2}\right)$	$\frac{\pi}{2}$	$\left(\frac{\pi}{2}, \frac{7\pi}{6}\right)$	$\frac{7\pi}{6}$	$\left(\frac{7\pi}{6}, \frac{3\pi}{2}\right)$
f'	$+$	0	$-$	0	$+$
f	↗	M	↘	m	↗

x	$\frac{3\pi}{2}$	$\left(\frac{3\pi}{2}, \frac{11\pi}{6}\right)$	$\frac{11\pi}{6}$	$\left(\frac{11\pi}{6}, 2\pi\right)$
f'	0	$-$	0	$+$
f	M	↘	m	↗

49. $c = 0$

x	$(-\infty, 0)$	0	$(0, \infty)$
f'	$-$	0	$+$
f	↘	m	↗

51. $c = -\frac{\pi}{4}$

x	$\left[-\frac{\pi}{2}, -\frac{\pi}{4}\right)$	$-\frac{\pi}{4}$	$\left(-\frac{\pi}{4}, \frac{\pi}{2}\right]$
f'	$+$	0	$-$
f	↗	M	↘

53. $c = \pm 1$

x	$(-\infty, -1)$	-1	$(-1, 1)$	1	$(1, \infty)$
f'	$-$	0	$+$	0	$-$
f	↘	m	↗	M	↘

55. Critical point: $x = 2$ (local minimum); $(-\infty, 0)$ increasing, $(0, 2)$ decreasing, $(2, \infty)$ increasing

57. $c = 0$; f' is positive on $(-\infty, 0)$ and on $(0, \infty)$ and is undefined at 0; f is increasing on $(-\infty, 0)$ and $(0, \infty)$; $x = 0$ is not a local minimum or maximum.

x	$(-\infty, 0)$	0	$(0, \infty)$
f'	$+$	0	$+$
f	↗	$-$	↗

59. Maximum $= e^{(1/e)}$, occurring at $x = 1/e$

61. $f'(x) > 0$ for all x

63. Your change in distance was 115 mi, and your change in time was 95 min. So your average velocity was 115/95 mi/min, and that is approximately 72.63 mi/h. The Mean Value Theorem implies that if your average velocity was 72.63 mi/h, then at some time you must have had an instantaneous velocity of 72.63 mi/h. At that time, your speed exceeded 70 mi/h.

65. $f'(x) < 0$ as long as $x < 500$, so $800^2 + 200^2 = f(200) > f(400) = 600^2 + 400^2$.

67. every point $c \in (a, b)$

75. **(a)** Let $g(x) = \cos x$ and $f(x) = 1 - \frac{1}{2}x^2$. Then $f(0) = g(0) = 1$ and $g'(x) = -\sin x \geq -x = f'(x)$ for $x \geq 0$ by Exercise 73. Now apply Exercise 73 to conclude that $\cos x \geq 1 - \frac{1}{2}x^2$ for $x \geq 0$.

(b) Let $g(x) = \sin x$ and $f(x) = x - \frac{1}{6}x^3$. Then $f(0) = g(0) = 0$ and $g'(x) = \cos x \geq 1 - \frac{1}{2}x^2 = f'(x)$ for $x \geq 0$ by part (a). Now apply Exercise 73 to conclude that $\sin x \geq x - \frac{1}{6}x^3$ for $x \geq 0$.

(c) Let $g(x) = 1 - \frac{1}{2}x^2 + \frac{1}{24}x^4$ and $f(x) = \cos x$. Then $f(0) = g(0) = 1$ and $g'(x) = -x + \frac{1}{6}x^3 \geq -\sin x = f'(x)$ for $x \geq 0$ by part (b). Now apply Exercise 73 to conclude that $\cos x \leq 1 - \frac{1}{2}x^2 + \frac{1}{24}x^4$ for $x \geq 0$.

(d) The next inequality in the series is $\sin x \leq x - \frac{1}{6}x^3 + \frac{1}{120}x^5$, valid for $x \geq 0$.

77. • Let $f''(x) = 0$ for all x. Then $f'(x) = $ constant for all x. Since $f'(0) = m$, we conclude that $f'(x) = m$ for all x.
• Let $g(x) = f(x) - mx$. Then $g'(x) = f'(x) - m = m - m = 0$, which implies that $g(x) = $ constant for all x and, consequently, $f(x) - mx = $ constant for all x. Rearranging the statement, $f(x) = mx + $ constant. Since $f(0) = b$, we conclude that $f(x) = mx + b$ for all x.

79. **(a)** Let $g(x) = f(x)^2 + f'(x)^2$. Then

$$g'(x) = 2f(x)f'(x) + 2f'(x)f''(x)$$

$$= 2f(x)f'(x) + 2f'(x)(-f(x)) = 0$$

Because $g'(0) = 0$ for all x, $g(x) = f(x)^2 + f'(x)^2$ must be a constant function. To determine the value of C, we can substitute any number for x. In particular, for this problem, we want to substitute $x = 0$ and find $C = f(0)^2 + f'(0)^2$. Hence,

$$f(x)^2 + f'(x)^2 = f(0)^2 + f'(0)^2$$

(b) Let $f(x) = \sin x$. Then $f'(x) = \cos x$ and $f''(x) = -\sin x$, so $f''(x) = -f(x)$. Finally, if we take $f(x) = \sin x$, the result from part (a) guarantees that

$$\sin^2 x + \cos^2 x = \sin^2 0 + \cos^2 0 = 0 + 1 = 1$$

Section 4.4 Preliminary Questions

1. **(a)** increasing   **2.** $f(c)$ is a local maximum.

3. False   **4.** False. For example, with $f(x) = x^4$, we have $f''(0) = 0$, but there is not an inflection point at $x = 0$ since the concavity does not change there.

5. No. An inflection point is a point on the graph of a function where the concavity changes. Since f is not defined at $x = 0$, there is no point on the graph of f at $x = 0$ and therefore no inflection point corresponding to the change in concavity that occurs going from negative to positive values of x.

6. Yes. For example, for $f(x) = x^3$, there is a critical point at $x = 0$, and $(0, 0)$ is an inflection point.

Section 4.4 Exercises

1. **(a)** In C, we have $f''(x) < 0$ for all x.   **(b)** In A, $f''(x)$ goes from $+$ to $-$.   **(c)** In B, we have $f''(x) > 0$ for all x.   **(d)** In D, $f''(x)$ goes from $-$ to $+$.

3.

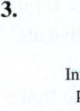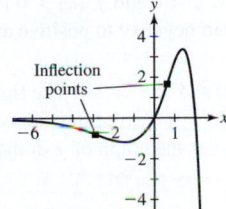

Inflection points $\left(-1 - \sqrt{3}, (-6 - 4\sqrt{3})e^{-1-\sqrt{3}}\right)$ and $\left(-1 + \sqrt{3}, (-6 + 4\sqrt{3})e^{-1+\sqrt{3}}\right)$

5. Concave up everywhere; no points of inflection

7. Concave up for $x < -\sqrt{3}$ and for $0 < x < \sqrt{3}$; concave down for $-\sqrt{3} < x < 0$ and for $x > \sqrt{3}$; point of inflection at $x = 0$ and at $x = \pm\sqrt{3}$

9. Concave up for $0 < \theta < \pi$; concave down for $\pi < \theta < 2\pi$; point of inflection at $\theta = \pi$

11. Concave down for $0 < x < 9$; concave up for $x > 9$; point of inflection when $x = 9$

13. Concave up on $(0, 1)$; concave down on $(-\infty, 0) \cup (1, \infty)$; point of inflection at both $x = 0$ and $x = 1$

15. Concave up for $|x| > 1$; concave down for $|x| < 1$; point of inflection at both $x = -1$ and $x = 1$

17. $(-\infty, -1)$ concave up, $(-1, 0)$ concave down, $(0, \infty)$ concave up; inflection point: $(0, 0)$

19. Concave down for $x < \frac{2}{3}$; concave up for $x > \frac{2}{3}$; point of inflection at $x = \frac{2}{3}$

21. Concave down for $x < \frac{1}{2}$; concave up for $x > \frac{1}{2}$; point of inflection at $x = \frac{1}{2}$

23. $(-\infty, -\sqrt{3/2})$ concave down, $(-\sqrt{3/2}, 0)$ concave up, $(0, \sqrt{3/2})$ concave down, $(\sqrt{3/2}, \infty)$ concave up; inflection points: $(-\sqrt{3/2}, (-\sqrt{3/2})e^{-3/2})$, $(0, 0)$, $(\sqrt{3/2}, (\sqrt{3/2})e^{-3/2})$

25. **(a)** Starts 100 km away traveling toward us at slower and slower speeds, stops when it gets to us (after 2 h), then turns around and goes back to 100 km away, traveling at increasing speeds

(b) The velocity is always increasing. When the ambulance is moving toward us (negative velocity), that means the speed is decreasing; when it is moving away, it means the speed is increasing.

27. Near the point of inflection, the curve is roughly a straight line going through $(55, 200)$ and $(35, 100)$, so the rate of change is roughly $\frac{200 - 100}{55 - 35} = 5$ cm/day. So when the growth rate starts to slow down, the height is growing at about 5 cm/day. Plots of the first and second derivatives are

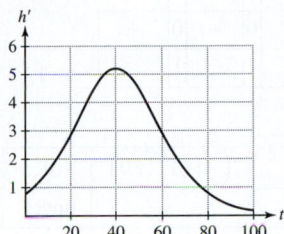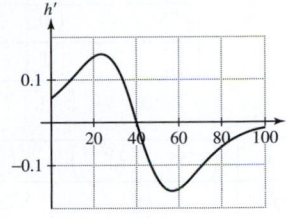

29. Points of inflection are a, d, and f. The function is concave down on $[0, a) \cup (d, f)$.

31. **(a)** f is increasing on $(0, 0.4)$.   **(b)** f is decreasing on $(-\infty, 0) \cup (0.4, 1) \cup (1, 1.2)$.   **(c)** f is concave up on $(0, 0.17) \cup (0.64, 1)$.   **(d)** f is concave down on $(0.17, 0.64) \cup (1, 1.2)$.

33. Critical points are $x = 3$ and $x = 5$; $f(3) = 54$ is a local maximum, and $f(5) = 50$ is a local minimum.

35. Critical points are $x = 0$ and $x = 1$; $f(0) = 0$ is a local minimum; and the Second Derivative Test is inconclusive at $x = 1$.

37. Critical points are $x = -4$ and $x = 2$; $f(-4) = -16$ is a local maximum, and $f(2) = -4$ is a local minimum.

39. Critical points are $x = 0$ and $x = \frac{2}{9}$; $f\left(\frac{2}{9}\right)$ is a local minimum; $f''(x)$ is undefined at $x = 0$, so the Second Derivative Test cannot be applied there.

41. Critical points are $x = 0$, $x = \frac{\pi}{3}$, and $x = \pi$; $f(0)$ is a local minimum, $f(\frac{\pi}{3})$ is a local maximum, and $f(\pi)$ is a local minimum.

43. Critical points are $x = \pm\frac{\sqrt{2}}{2}$; $f\left(\frac{\sqrt{2}}{2}\right)$ is a local maximum and $f\left(-\frac{\sqrt{2}}{2}\right)$ is a local minimum.

45. The critical point is $x = e^{-1/3}$; $f\left(e^{-1/3}\right)$ is a local minimum.

47.

x	$\left(-\infty, \frac{1}{3}\right)$	$\frac{1}{3}$	$\left(\frac{1}{3}, 1\right)$	1	$(1, \infty)$
f'	$+$	0	$-$	0	$+$
f	↗	M	↘	m	↗

x	$\left(-\infty, \frac{2}{3}\right)$	$\frac{2}{3}$	$\left(\frac{2}{3}, \infty\right)$
f''	$-$	0	$+$
f	⌢	I	⌣

49.

t	$(-\infty, 0)$	0	$\left(0, \frac{2}{3}\right)$	$\frac{2}{3}$	$\left(\frac{2}{3}, \infty\right)$
f'	$-$	0	$+$	0	$-$
f	↘	m	↗	M	↘

t	$\left(-\infty, \frac{1}{3}\right)$	$\frac{1}{3}$	$\left(\frac{1}{3}, \infty\right)$
f''	$+$	0	$-$
f	⌣	I	⌢

51. $f''(x) > 0$ for all $x \geq 0$, which means there are no inflection points.

x	0	$(0, (2)^{2/3})$	$(2)^{2/3}$	$((2)^{2/3}, \infty)$
f'	U	$-$	0	$+$
f	M	↘	m	↗

53.

x	$(-\infty, -3\sqrt{3})$	$-3\sqrt{3}$	$(-3\sqrt{3}, 3\sqrt{3})$	$3\sqrt{3}$	$(3\sqrt{3}, \infty)$
f'	$-$	0	$+$	0	$-$
f	↘	m	↗	M	↘

x	$(-\infty, -9)$	-9	$(-9, 0)$	0	$(0, 9)$	9	$(9, \infty)$
f''	$-$	0	$+$	0	$-$	0	$+$
f	⌢	I	⌣	I	⌢	I	⌣

55.

x	$\left(-\infty, -\left(\frac{3}{5}\right)^{3/2}\right)$	$-\left(\frac{3}{5}\right)^{3/2}$	$\left(-\left(\frac{3}{5}\right)^{3/2}, 0\right)$	0
f''	$-$	$-$	$-$	undef

$\left(0, \left(\frac{3}{5}\right)^{3/2}\right)$	$\left(\frac{3}{5}\right)^{3/2}$	$\left(\left(\frac{3}{5}\right)^{3/2}, \infty\right)$
$+$	$+$	$+$

57.

θ	$(0, \pi)$	π	$(\pi, 2\pi)$
f'	$+$	0	$+$
f	↗	⌐	↗

θ	0	$(0, \pi)$	π	$(\pi, 2\pi)$	2π
f''	0	$-$	0	$+$	0
f	⌐	⌢	I	⌣	⌐

59.

x	$\left(-\frac{\pi}{2}, \frac{\pi}{2}\right)$
f'	$+$
f	↗

x	$\left(-\frac{\pi}{2}, 0\right)$	0	$\left(0, \frac{\pi}{2}\right)$
f''	$-$	0	$+$
f	⌢	I	⌣

61.

x	$\left(0, 1+\sqrt{3}\right)$	$1+\sqrt{3}$	$\left(1+\sqrt{3}, \infty\right)$
f'	$+$	0	$-$
f	↗	M	↘

x	$(0, 4)$	4	$(4, \infty)$
f''	$-$	0	$+$
f	⌢	I	⌣

63.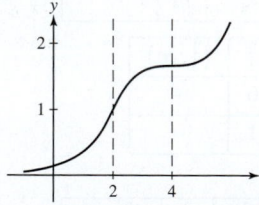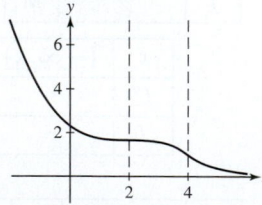

65.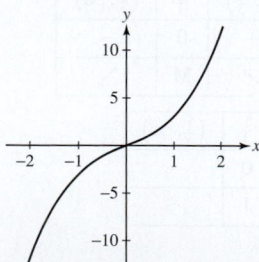

67. (a) Near the beginning of the epidemic, the graph of R is concave up. Near the epidemic's end, R is concave down.

(b) "Epidemic subsiding: number of new cases declining."

69. The point of inflection should occur when the water level is equal to the radius of the sphere. A possible graph of V is shown here.

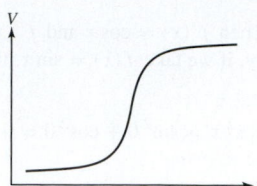

71. (a) $f'(u) = \dfrac{be^{b(a-u)}}{(1+e^{b(a-u)})^2} > 0$ **(b)** $u = a$

73. (a) From the definition of the derivative, we have
$$f''(c) = \lim_{h \to 0} \frac{f'(c+h) - f'(c)}{h} = \lim_{h \to 0} \frac{f'(c+h)}{h}$$

(b) We are given that $f''(c) > 0$. By part (a), it follows that
$$\lim_{h \to 0} \frac{f'(c+h)}{h} > 0$$

In other words, for sufficiently small h,
$$\frac{f'(c+h)}{h} > 0$$

Now, if h is sufficiently small but negative, then $f'(c+h)$ must also be negative [so that the ratio $f'(c+h)/h$ will be positive] and $c + h < c$. On the other hand, if h is sufficiently small but positive, then $f'(c+h)$ must also be positive and $c + h > c$. Thus, there exists an open interval (a, b) containing c such that $f'(x) < 0$ for $a < x < c$ and $f'(c) > 0$ for $c < x < b$. Finally, because $f'(x)$ changes from negative to positive at $x = c$, $f(c)$ must be a local minimum.

75. (b) $f(x)$ has a point of inflection at $x = 0$ and at $x = \pm 1$. The figure shows the graph of $y = f(x)$ and its tangent lines at each of the points of inflection. It is clear that each tangent line crosses the graph of f at the inflection point.

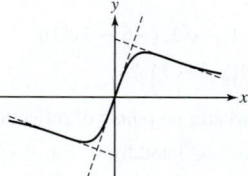

77. Let $f(x) = a_n x^n + a_{n-1} x^{n-1} + \cdots + a_1 x + a_0$ be a polynomial of degree n. Then
$$f'(x) = na_n x^{n-1} + (n-1)a_{n-1}x^{n-2} + \cdots + 2a_2 x + a_1$$
and
$$f''(x) = n(n-1)a_n x^{n-2}$$
$$+ (n-1)(n-2)a_{n-1}x^{n-3} + \cdots + 6a_3 x + 2a_2$$

If $n \geq 3$ and is odd, then $n - 2$ is also odd and f'' is a polynomial of odd degree. Therefore, f'' must take on both positive and negative values. It follows that $f''(x)$ has at least one root c such that $f''(x)$ changes sign at c. The function f will then have a point of inflection at $x = c$. On the other hand, the functions $f(x) = x^2$, x^4, and x^8 are polynomials of even degree that do not have any points of inflection.

Section 4.5 Preliminary Questions

1. Not of the form $\frac{0}{0}$ or $\frac{\infty}{\infty}$ **2.** No

3. You do not apply the quotient rule on $\frac{\ln(1-x)}{x}$. You separately differentiate the numerator and denominator, and work with the new rational expression obtained this way.

4. The function does not have an indeterminate form at $x = 0$. it has the form 0^∞, corresponding to a limit that equals 0.

5. It is a continuous function.

Section 4.5 Exercises

1. L'Hôpital's Rule does not apply.

3. L'Hôpital's Rule does not apply.

5. L'Hôpital's Rule does not apply.

7. L'Hôpital's Rule does not apply.

9. 0 **11.** Quotient of the form $\frac{\infty}{\infty}$; $-\frac{9}{2}$

13. Quotient of the form $\frac{\infty}{\infty}$; 0 **15.** Quotient of the form $\frac{\infty}{\infty}$; 0

17. $\frac{5}{6}$ **19.** $-\frac{3}{5}$ **21.** $-\frac{7}{3}$ **23.** $\frac{9}{7}$ **25.** $\frac{2}{7}$ **27.** 1 **29.** 2 **31.** -1

33. $\frac{1}{2}$ **35.** 0 **37.** $-\frac{2}{\pi}$ **39.** 1 **41.** Does not exist **43.** 0

45. $\ln a$ **47.** e **49.** $e^{-3/2}$ **51.** 1 **53.** $\frac{1}{\pi}$

55.

$$\lim_{x \to \pi/2} \frac{\cos mx}{\cos nx} = \begin{cases} (-1)^{(m-n)/2}, & m, n \text{ even} \\ \text{does not exist}, & m \text{ even}, n \text{ odd} \\ 0 & m \text{ odd}, n \text{ even} \\ (-1)^{(m-n)/2} \frac{m}{n}, & m, n \text{ odd} \end{cases}$$

57. (a) ∞ (b) 1 **59.** $\frac{9000}{49} \approx 183.7$ m

61. (a) Here, $f(x) \to 1$, so $\ln f(x) \to 0$, and $g(x) \to \infty$; therefore, $g(x) \ln f(x)$ has the indeterminate form $\infty \cdot 0$.

(b) Here, $f(x) \to \infty$, so $\ln f(x) \to \infty$, and $g(x) \to 0$; therefore, $g(x) \ln f(x)$ has the form $0 \cdot \infty$.

63. (a) $\lim_{x \to 0+} f(x) = 0$; $\lim_{x \to \infty} f(x) = e^0 = 1$

(b) f is increasing for $0 < x < e$, is decreasing for $x > e$, and has a maximum at $x = e$. The maximum value is $f(e) = e^{1/e} \approx 1.444668$.

65. Neither **67.** $\lim_{x \to \infty} \frac{\ln x}{x^a} = \lim_{x \to \infty} \frac{x^{-1}}{ax^{a-1}} = \lim_{x \to \infty} \frac{1}{a} x^{-a} = 0$

71. (a) $1 \leq 2 + \sin x \leq 3$, so

$$\frac{x}{x^2 + 1} \leq \frac{x(2 + \sin x)}{x^2 + 1} \leq \frac{3x}{x^2 + 1}$$

It follows by the Squeeze Theorem that

$$\lim_{x \to \infty} \frac{x(2 + \sin x)}{x^2 + 1} = 0$$

(b) $\lim_{x \to \infty} f(x) = \lim_{x \to \infty} x(2 + \sin x) \geq \lim_{x \to \infty} x = \infty$ and $\lim_{x \to \infty} g(x) = \lim_{x \to \infty} (x^2 + 1) = \infty$, but

$$\lim_{x \to \infty} \frac{f'(x)}{g'(x)} = \lim_{x \to \infty} \frac{x(\cos x) + (2 + \sin x)}{2x}$$

does not exist since $\cos x$ oscillates. This does not violate L'Hôpital's Rule since the theorem clearly states

$$\lim_{x \to \infty} \frac{f(x)}{g(x)} = \lim_{x \to \infty} \frac{f'(x)}{g'(x)}$$

"provided the limit on the right exists."

73. (a) Using Exercise 70, we see that $G(b) = e^{H(b)}$. Thus, $G(b) = 1$ if $0 \leq b \leq 1$ and $G(b) = b$ if $b > 1$.

(b)

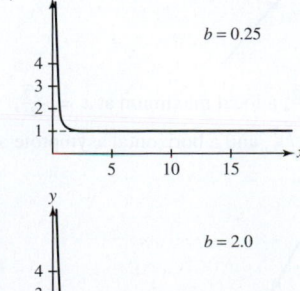

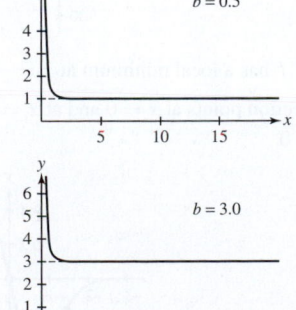

75. $\lim_{x \to 0} \frac{f(x)}{x^k} = \lim_{x \to 0} \frac{1}{x^k e^{1/x^2}}$. Let $t = 1/x$. As $x \to 0$, $t \to \infty$. Thus,

$$\lim_{x \to 0} \frac{1}{x^k e^{1/x^2}} = \lim_{t \to \infty} \frac{t^k}{e^{t^2}} = 0$$

by Exercise 72.

77. For $x \neq 0$, $f'(x) = e^{-1/x^2} \left(\frac{2}{x^3} \right)$. Here, $P(x) = 2$ and $r = 3$.

Assume $f^{(k)}(x) = \frac{P(x)e^{-1/x^2}}{x^r}$. Then

$$f^{(k+1)}(x) = e^{-1/x^2} \left(\frac{x^3 P'(x) + (2 - rx^2)P(x)}{x^{r+3}} \right)$$

which is of the form desired.

Moreover, from Exercise 74, $f'(0) = 0$. Suppose $f^{(k)}(0) = 0$. Then

$$f^{(k+1)}(0) = \lim_{x \to 0} \frac{f^{(k)}(x) - f^{(k)}(0)}{x - 0} = \lim_{x \to 0} \frac{P(x)e^{-1/x^2}}{x^{r+1}}$$

$$= P(0) \lim_{x \to 0} \frac{f(x)}{x^{r+1}} = 0$$

81. $\lim_{x \to 0} \frac{\sin x}{x} = \lim_{x \to 0} \frac{\cos x}{1} = 1$. To use L'Hôpital's Rule to evaluate $\lim_{x \to 0} \frac{\sin x}{x}$, we must know that the derivative of $\sin x$ is $\cos x$, but to determine the derivative of $\sin x$, we must be able to evaluate $\lim_{x \to 0} \frac{\sin x}{x}$.

83. (a) $e^{-1/6} \approx 0.846481724$

x	1	0.1	0.01
$\left(\frac{\sin x}{x} \right)^{1/x^2}$	0.841471	0.846435	0.846481

(b) $1/3$

x	± 1	± 0.1	± 0.01
$\frac{1}{\sin^2 x} - \frac{1}{x^2}$	0.412283	0.334001	0.333340

Section 4.6 Preliminary Questions

1. An arc with the sign combination $++$ (increasing, concave up) is shown here at the left. An arc with the sign combination $-+$ (decreasing, concave up) is shown here at the right.

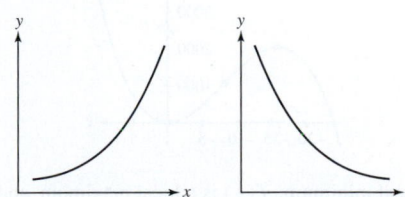

2. (c) **3.** $x = 4$ is not in the domain of f.

Section 4.6 Exercises

1. • In A, f is decreasing and concave up, so $f' < 0$ and $f'' > 0$.
• In B, f is increasing and concave up, so $f' > 0$ and $f'' > 0$.
• In C, f is increasing and concave down, so $f' > 0$ and $f'' < 0$.
• In D, f is decreasing and concave down, so $f' < 0$ and $f'' < 0$.
• In E, f is decreasing and concave up, so $f' < 0$ and $f'' > 0$.
• In F, f is increasing and concave up, so $f' > 0$ and $f'' > 0$.
• In G, f is increasing and concave down, so $f' > 0$ and $f'' < 0$.

3. This function changes from concave up to concave down at $x = -1$ and from increasing to decreasing at $x = 0$.

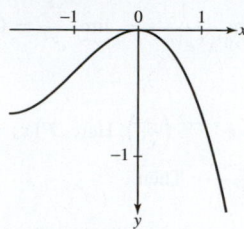

5. The function is decreasing everywhere and changes from concave up to concave down at $x = -1$ and from concave down to concave up at $x = -\frac{1}{2}$.

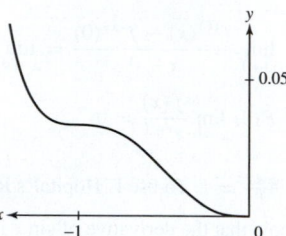

7.

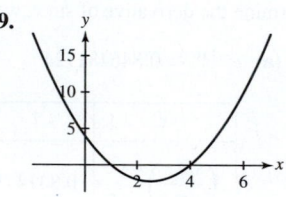

9.

11.

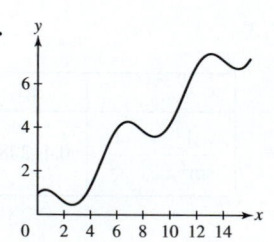

13.

15. Local maximum at $x = -16$, a local minimum at $x = 0$, and an inflection point at $x = -8$

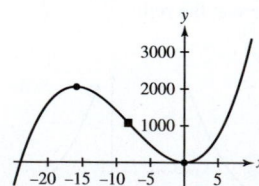

17. $f(0)$ is a local minimum, $f(\frac{1}{6})$ is a local maximum, and there is a point of inflection at $x = \frac{1}{12}$.

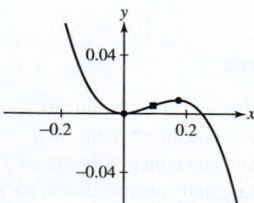

19. f has local minima at $x = \pm\sqrt{6}$, a local maximum at $x = 0$, and inflection points at $x = \pm\sqrt{2}$.

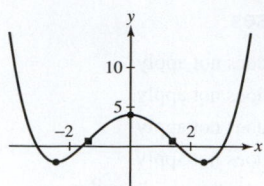

21. The graph has no critical points and is always increasing, with inflection point at $(0, 0)$.

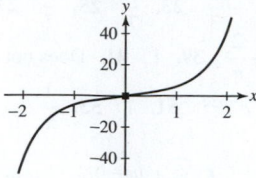

23. $f\left(\frac{1-\sqrt{33}}{8}\right)$ and $f(2)$ are local minima, and $f\left(\frac{1+\sqrt{33}}{8}\right)$ is a local maximum; points of inflection both at $x = 0$ and $x = \frac{3}{2}$.

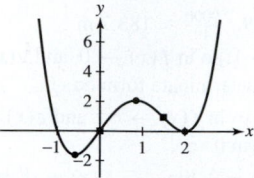

25. $f(0)$ is a local maximum, $f(12)$ is a local minimum, and there is a point of inflection at $x = 10$.

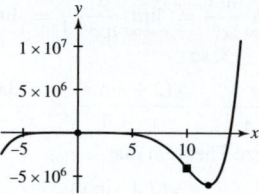

27. $f(4)$ is a local minimum, and the graph is always concave up.

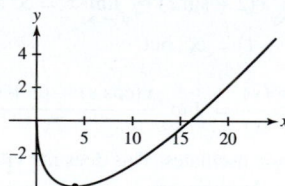

29. f has a local maximum at $x = 6$ and inflection points at $x = 8$ and $x = 12$.

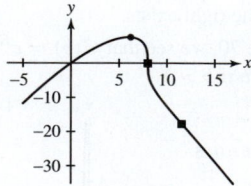

31. f has a local minimum at $x = -\frac{\sqrt{2}}{2}$, a local maximum at $x = \frac{\sqrt{2}}{2}$, inflection points at $x = 0$ and at $x = \pm\sqrt{\frac{3}{2}}$, and a horizontal asymptote at $y = 0$.

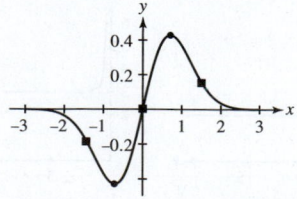

33. $f(2)$ is a local minimum and the graph is always concave up.

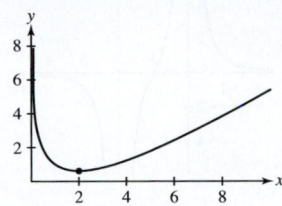

35. f has a local minimum at $x = 1$ and no inflection points. It is concave up everywhere. It has a vertical asymptote at $x = 0$.

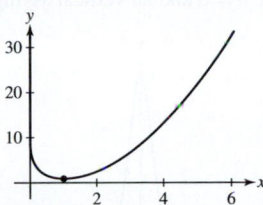

37. The graph has an inflection point at $x = \frac{3}{5}$, a local maximum at $x = 1$ (at which the graph has a cusp), and a local minimum at $x = \frac{9}{5}$.

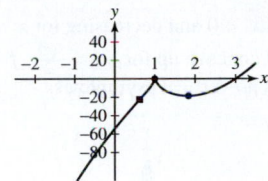

39. f has a local maximum at $x = 0$, local minima at $x = \pm 3$, and points of inflection at $x = \pm\sqrt{-6 + 3\sqrt{5}}$.

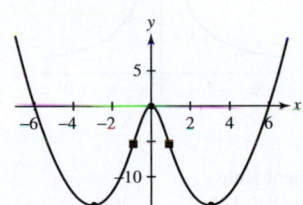

41. f has local minima at $x = -1.473$ and $x = 1.347$, a local maximum at $x = 0.126$, and points of inflection at $x = \pm\sqrt{\frac{2}{3}}$.

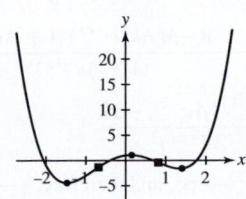

43. The graph has an inflection point at $x = \pi$, and no local maxima or minima.

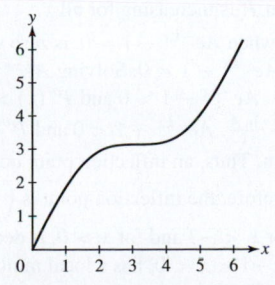

45. Local maximum at $x = \frac{\pi}{2}$, a local minimum at $x = \frac{3\pi}{2}$, and inflection points at $x = \frac{\pi}{6}$ and $x = \frac{5\pi}{6}$.

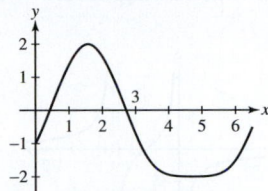

47. Local maximum at $x = \frac{\pi}{6}$ and a point of inflection at $x = \frac{2\pi}{3}$.

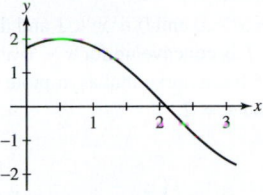

49. In both cases, there is a point where f is not differentiable at the transition from increasing to decreasing or decreasing to increasing.

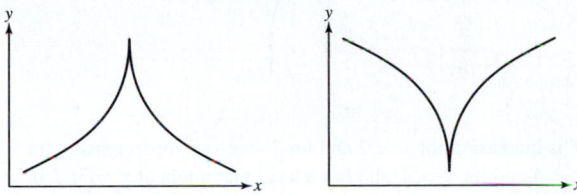

51.

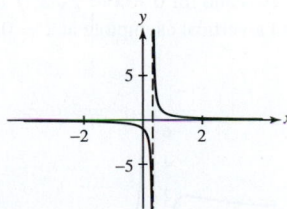

53. (B) is the graph of $f(x) = \dfrac{3x^2}{x^2 - 1}$; (A) is the graph of $f(x) = \dfrac{3x}{x^2 - 1}$.

55. f is decreasing for all $x \neq \frac{1}{3}$, is concave up for $x > \frac{1}{3}$, is concave down for $x < \frac{1}{3}$, and has a horizontal asymptote at $y = 0$ and a vertical asymptote at $x = \frac{1}{3}$.

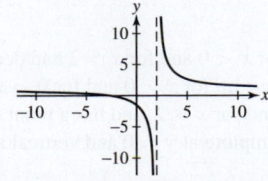

57. f is decreasing for all $x \neq 2$, is concave up for $x > 2$, is concave down for $x < 2$, and has a horizontal asymptote at $y = 1$ and a vertical asymptote at $x = 2$.

59. f is decreasing for all $x \neq 0, 1$, is concave up for $0 < x < \frac{1}{2}$ and $x > 1$, is concave down for $x < 0$ and $\frac{1}{2} < x < 1$, and has a horizontal asymptote at $y = 0$ and vertical asymptotes at $x = 0$ and $x = 1$.

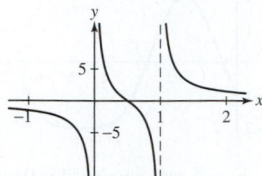

61. f is increasing for $x < 0$ and $0 < x < 1$ and decreasing for $1 < x < 2$ and $x > 2$; f is concave up for $x < 0$ and $x > 2$ and concave down for $0 < x < 2$; f has a horizontal asymptote at $y = 0$ and vertical asymptotes at $x = 0$ and $x = 2$.

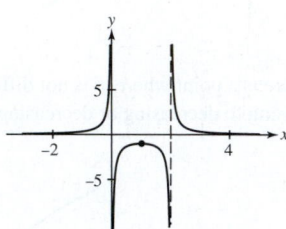

63. f is increasing for $x < 2$ and for $2 < x < 3$, is decreasing for $3 < x < 4$ and for $x > 4$, and has a local maximum at $x = 3$; f is concave up for $x < 2$ and for $x > 4$ and is concave down for $2 < x < 4$; f has a horizontal asymptote at $y = 0$ and vertical asymptotes at $x = 2$ and $x = 4$.

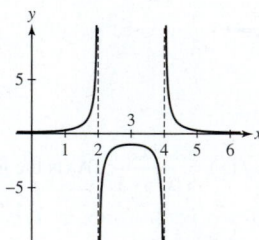

65. f is increasing for $|x| > 2$ and decreasing for $-2 < x < 0$ and for $0 < x < 2$; f is concave down for $-2\sqrt{2} < x < 0$ and for $x > 2\sqrt{2}$ and concave up for $x < -2\sqrt{2}$ and for $0 < x < 2\sqrt{2}$; f has a horizontal asymptote at $y = 1$ and a vertical asymptote at $x = 0$.

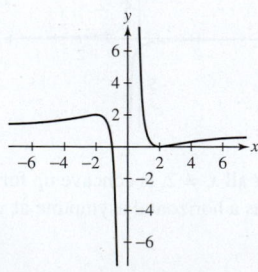

67. f is increasing for $x < 0$ and for $x > 2$ and decreasing for $0 < x < 2$; f is concave up for $x < 0$ and for $0 < x < 1$, is concave down for $1 < x < 2$ and for $x > 2$, and has a point of inflection at $x = 1$; f has a horizontal asymptote at $y = 0$ and vertical asymptotes at $x = 0$ and $x = 2$.

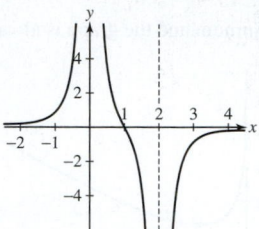

69. f is increasing for $x < 0$, is decreasing for $x > 0$, and has a local maximum at $x = 0$; f is concave up for $|x| > 1/\sqrt{5}$, is concave down for $|x| < 1/\sqrt{5}$, and has points of inflection at $x = \pm 1/\sqrt{5}$; f has a horizontal asymptote at $y = 0$ and no vertical asymptotes.

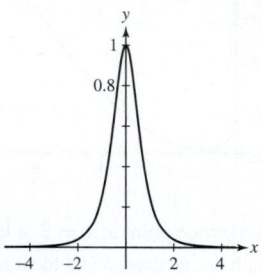

71. f is increasing for $x < 0$ and decreasing for $x > 0$; f is concave down for $|x| < \frac{\sqrt{2}}{2}$ and concave up for $|x| > \frac{\sqrt{2}}{2}$; f has a horizontal asymptote at $y = 0$ and no vertical asymptotes.

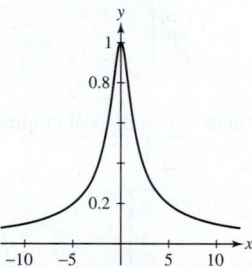

73. **(a)** By the Quotient Rule,
$$P'(x) = \frac{(1+Ae^{-kx})(0) - (M)(-k)Ae^{-kx}}{(1+Ae^{-kx})^2} = \frac{MAke^{-kx}}{(1+Ae^{-kx})^2}.$$ By the Quotient Rule,

$$P''(x) = \frac{(1 + Ae^{-kx})^2(-MAk^2 e^{-kx}) - (MAke^{-kx})(2)(1 + Ae^{-kx})(-k)(Ae^{-kx})}{(1 + Ae^{-kx})^4}$$

$$= \frac{(1 + Ae^{-kx})(-MAk^2 e^{-kx})(1 + Ae^{-kx} - 2Ae^{-kx})}{(1 + Ae^{-kx})^4}$$

$$= \frac{MAk^2 e^{-kx}(Ae^{-kx} - 1)}{(1 + Ae^{-kx})^3}$$

(b) As $x \to -\infty$, $Ae^{-kx} \to \infty$, and therefore, $\frac{M}{1+Ae^{-kx}} \to 0$. As $x \to \infty$, $Ae^{-kx} \to 0$, and therefore, $\frac{M}{1+Ae^{-kx}} \to M$

(c) Since A, M, and k are all positive, all factors in the numerator of P' are positive. The denominator is also positive, and therefore, $P'(x) > 0$ for all x. It follows that P is increasing for all x.

(d) $P''(x)$ is positive when $Ae^{-kx} - 1 > 0$, is zero when $Ae^{-kx} - 1 = 0$, and is negative when $Ae^{-kx} - 1 < 0$. Solving $Ae^{-kx} - 1 = 0$, we obtain $x = \frac{\ln A}{k}$. For $x < \frac{\ln A}{k}$, $Ae^{-kx} - 1 > 0$ and $P''(x) > 0$, implying that P is concave up. For $x > \frac{\ln A}{k}$, $Ae^{-kx} - 1 < 0$ and $P''(x) < 0$, implying that P is concave down. Thus, an inflection point occurs at $x = \frac{\ln A}{k}$. $P(\frac{\ln A}{k}) = \frac{M}{2}$ and therefore, the inflection point is $\left(\frac{\ln A}{k}, \frac{M}{2}\right)$.

77. f is increasing for $x < -2$ and for $x > 0$, is decreasing for $-2 < x < -1$ and for $-1 < x < 0$, has a local minimum at $x = 0$, has a local maximum at $x = -2$, is concave down on $(-\infty, -1)$ and concave up on $(-1, \infty)$; f has a vertical asymptote at $x = -1$; by polynomial division, $f(x) = x - 1 + \frac{1}{x+1}$ and

$$\lim_{x \to \pm\infty} \left(x - 1 + \frac{1}{x+1} - (x-1) \right) = 0$$

which implies that the slant asymptote is $y = x - 1$.

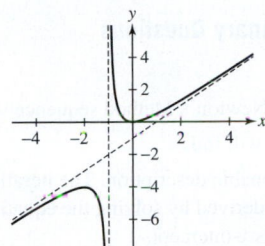

79. $y = x + 2$ is the slant asymptote of f; local minimum at $x = 2 + \sqrt{3}$, a local maximum at $x = 2 - \sqrt{3}$, and f is concave down on $(-\infty, 2)$ and concave up on $(2, \infty)$; vertical asymptote at $x = 2$.

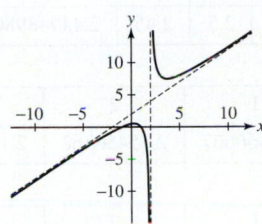

Section 4.7 Preliminary Questions

1. $b + h + \sqrt{b^2 + h^2} = 10$

2. If the function tends to infinity at the endpoints of the interval, then the function must take on a minimum value at a critical point.

3. No

Section 4.7 Exercises

1. (a) $y = \frac{3}{2} - x$ (b) $A = x(\frac{3}{2} - x) = \frac{3}{2}x - x^2$

(c) Closed interval $[0, \frac{3}{2}]$ (d) The maximum area 0.5625 m² is

achieved with $x = y = \frac{3}{4}$ m.

3. One side of length 6 and two of length 3   **5.** 4 and 32

7. Allot approximately 5.28 m of the wire to the circle.

9. 20 and 20   **11.** $x = 40$; $y = 20$

13. (a) The box should be a cube with side length $12^{1/3}$

(b) The box should be a cube with side length $\dfrac{\sqrt{30}}{3}$.

15. The corral of maximum area has dimensions

$$x = \frac{300}{1 + \pi/4} \text{ m} \quad \text{and} \quad y = \frac{150}{1 + \pi/4} \text{ m}$$

where x is the width of the corral and therefore the diameter of the semicircle and y is the height of the rectangular section.

17. Square of side length $4\sqrt{2}$

19. (a) We have $T(x) = \frac{\sqrt{900 + x^2}}{r} + \frac{50 - x}{h}$. Therefore,

$T'(x) = \frac{x}{r\sqrt{900 + x^2}} - \frac{1}{h}$.

Setting $T'(x) = 0$ and solving, we obtain

$$0 = \frac{x}{r\sqrt{900 + x^2}} - \frac{1}{h}$$

$$\frac{x}{r\sqrt{900 + x^2}} = \frac{1}{h}$$

$$\frac{xh}{r} = \sqrt{900 + x^2}$$

$$\frac{x^2 h^2}{r^2} = 900 + x^2$$

$$x^2 \left(\frac{h^2}{r^2} - 1 \right) = 900$$

$$x = \frac{30}{\sqrt{(h/r)^2 - 1}}$$

If $r \geq h$, then $(h/r)^2 - 1 < 0$, implying that there is no solution to $T'(x) = 0$ and therefore no critical point.

(b) Since the numerator in the expression for the critical point is nonzero, the critical point must be nonzero as well. Note that for fixed r, as $h \to \infty$, $\frac{30}{\sqrt{(h/r)^2 - 1}} \to 0$ and therefore, by making h large enough, we can have the critical point arbitrarily close to 0.

21. About 1.43 m   **23.** $\left(\frac{1}{2}, \frac{1}{2} \right)$ **25.** $(0.632784, -1.090410)$

27. $\theta = \dfrac{\pi}{2}$ **29.** $\dfrac{3\sqrt{3}}{4}r^2$

31. 60 cm wide by 100 cm high for the full poster (48 cm by 80 cm for the printed matter)

33. Radius: $\sqrt{\frac{2}{3}}R$; half-height: $\frac{R}{\sqrt{3}}$

35. $x = 10\sqrt{5} \approx 22.36$ m and $y = 20\sqrt{5} \approx 44.72$ m, where x is the length of the brick wall and y is the length of an adjacent side

37. 1.0718   **39.** $LH + \frac{1}{2}(L^2 + H^2)$ **41.** $y = -3x + 24$

45. $s = 3\sqrt[3]{4}$ m and $h = 2\sqrt[3]{4}$ m, where s is the length of the side of the square bottom of the box and h is the height of the box.

47. (a) Each compartment has length of 600 m and width of 400 m.

(b) 240,000 m²

49. $N \approx 58.14$ lb and $P \approx 77.33$ lb   **51.** $990

53. 1.2 million euros in equipment and 600,000 euros in labor

55. Brandon swims diagonally to a point located 20.2 m downstream and then runs the rest of the way.

59. $A = B = 30$ cm   **61.** $x = \dfrac{x_1 + x_2 + \cdots + x_n}{n}$

65. (a) 900 m² when $x = -10$ (b) $[0, 20]$; 800 m² when $x = 0$

67. $\theta = \arctan(0.4) \approx 21.8°$ or 0.3805 rad

69. (b) $\left(\dfrac{17}{0.003} \right)^{1/4} \approx 8.676247$ (d) $v_d \approx 11.418583$;

$D(v_d) \approx 191.741$ km

71. $s = \left(\dfrac{b^{2/3}}{2^{2/3}} + h^{2/3} \right)^{3/2}$ **73.** $\left(a^{2/3} + b^{2/3} \right)^{3/2}$

75. (a) $\alpha = 0$ corresponds to shooting the ball directly at the basket, while $\alpha = \pi/2$ corresponds to shooting the ball directly upward. In neither case is it possible for the ball to go into the basket. If the angle α is extremely close to 0, the ball is shot almost directly at the basket; on the other hand, if the angle α is extremely close to $\pi/2$, the ball is launched almost vertically. In either one of these cases, the ball has to travel at an enormous speed.

(b) The minimum clearly occurs where $\theta = \pi/3$.

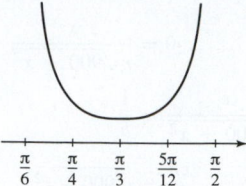

(c) $v^2 = \dfrac{16d}{F(\theta)}$; hence, v^2 is smallest whenever $F(\theta)$ is greatest.

(d) A critical point of F occurs where $\cos(\alpha - 2\theta) = 0$ so that $\alpha - 2\theta = -\frac{\pi}{2}$ (negative because $2\theta > \theta > \alpha$), and this gives us $\theta = \alpha/2 + \pi/4$. The minimum value $F(\theta_0)$ takes place at $\theta_0 = \alpha/2 + \pi/4$.

(e) Plug in $\theta_0 = \alpha/2 + \pi/4$. From Figure 38, we see that

$$\cos \alpha = \frac{d}{\sqrt{d^2 + h^2}} \qquad \text{and} \qquad \sin \alpha = \frac{h}{\sqrt{d^2 + h^2}}$$

(f) This shows that the minimum velocity required to launch the ball to the basket drops as shooter height increases. This shows one of the ways height is an advantage in free throws; a taller shooter need not shoot the ball as hard to reach the basket.

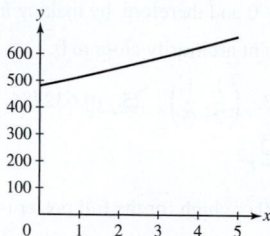

77. (a) From the figure, we see that

$$\theta(x) = \tan^{-1} \frac{c - f(x)}{x} - \tan^{-1} \frac{b - f(x)}{x}$$

Then

$\theta'(x)$

$= \dfrac{b - (f(x) - xf'(x))}{x^2 + (b - f(x))^2} - \dfrac{c - (f(x) - xf'(x))}{x^2 + (c - f(x))^2}$

$= (b - c)\dfrac{x^2 - bc + (b + c)(f(x) - xf'(x)) - (f(x))^2 + 2xf(x)f'(x)}{(x^2 + (b - f(x))^2)(x^2 + (c - f(x))^2)}$

$= (b - c)\dfrac{(x^2 + (xf'(x))^2) - (bc - (b + c)(f(x) - xf'(x)) + (f(x) - xf'(x))^2)}{(x^2 + (b - f(x))^2)(x^2 + (c - f(x))^2)}$

$= (b - c)\dfrac{(x^2 + (xf'(x))^2) - (b - (f(x) - xf'(x)))(c - (f(x) - xf'(x)))}{(x^2 + (b - f(x))^2)(x^2 + (c - f(x))^2)}$

(b) The point Q is the y-intercept of the line tangent to the graph of f at point P. The equation of this tangent line is

$$Y - f(x) = f'(x)(X - x)$$

The y-coordinate of Q is then $f(x) - xf'(x)$.

(c) From the figure, we see that

$$BQ = b - (f(x) - xf'(x)),$$
$$CQ = c - (f(x) - xf'(x))$$

and

$$PQ = \sqrt{x^2 + (f(x) - (f(x) - xf'(x)))^2} = \sqrt{x^2 + (xf'(x))^2}$$

Comparing these expressions with the numerator of $d\theta/dx$, it follows that $\dfrac{d\theta}{dx} = 0$ is equivalent to

$$PQ^2 = BQ \cdot CQ$$

(d) The equation $PQ^2 = BQ \cdot CQ$ is equivalent to

$$\frac{PQ}{BQ} = \frac{CQ}{PQ}$$

In other words, the sides CQ and PQ from the triangle $\triangle QCP$ are proportional in length to the sides PQ and BQ from the triangle $\triangle QPB$. As $\angle PQB = \angle CQP$, it follows that triangles $\triangle QCP$ and $\triangle QPB$ are similar.

Section 4.8 Preliminary Questions

1. One

2. Every term in the Newton's Method sequence will remain x_0.

3. Newton's Method will fail.

4. Yes, that is a reasonable description. The iteration formula for Newton's Method was derived by solving the equation of the tangent line to $y = f(x)$ at x_0 for its x-intercept.

Section 4.8 Exercises

1.

n	1	2	3
x_n	2.5	2.45	2.44948980

3.

n	1	2	3
x_n	2.16666667	2.15450362	2.15443469

5.

n	1	2	3
x_n	0.28540361	0.24288009	0.24267469

7. We take $x_0 = -1.4$, based on the figure, and then calculate

n	1	2	3
x_n	-1.330964467	-1.328272820	-1.328268856

9. $r_1 \approx 0.259$ and $r_2 \approx 2.543$

11. $\sqrt{11} \approx 3.317$; a calculator yields 3.31662479.

13. $2^{7/3} \approx 5.040$; a calculator yields 5.0396842.

15. 2.093064358 **17.** -2.225 **19.** $x \approx 2.331$ **21.** 1.749

23. $x = 4.49341$, which is approximately 1.4303π

25. $(2.7984, -0.941684)$

27. (a) $P \approx \$156.69$ **(b)** $b \approx 1.02121$; the interest rate is around 25.45%.

29. (a) The sector SAB is the slice OAB with the triangle OBS removed. OAB is a central sector with arc θ and radius $\overline{OA} = a$, and therefore has area $\frac{a^2\theta}{2}$. OBS is a triangle with height $a \sin \theta$ and base length $\overline{OS} = ea$. Hence, the area of the sector is

$$\frac{a^2}{2}\theta - \frac{1}{2}ea^2 \sin \theta = \frac{a^2}{2}(\theta - e \sin \theta)$$

(b) Since Kepler's Second Law indicates that the area of the sector is proportional to the time t since the planet passed point A, we get

$$\pi a^2 (t/T) = a^2/2 (\theta - e \sin \theta)$$

$$2\pi \frac{t}{T} = \theta - e \sin \theta$$

(c) From the point of view of the Sun, Mercury has traversed an angle of approximately 1.76696 radians $= 101.24°$. Mercury has therefore traveled more than one fourth of the way around (from the point of view of central angle) during this time.

31. (a)
$$x_{n+1} = x_n - \frac{x_n e^{-x_n}}{-x_n e^{-x_n} + e^{-x_n}} = x_n - \frac{x_n}{1 - x_n}$$
$$= \frac{x_n - x_n^2 - x_n}{1 - x_n} = \frac{x_n^2}{x_n - 1}$$

(b)

0	0.8	0	5
1	−3.2	1	6.25
2	−2.4381	2	7.440476
3	−1.72895	3	8.595744
4	−1.09539	4	9.727397
5	−0.57263	5	10.84198
6	−0.20851	6	11.94358
7	−0.03597	7	13.03496
8	−0.00125	8	14.11805
9	-1.6×10^{-6}	9	15.19428
10	-2.4×10^{-12}	10	16.26473

It appears that with $x_0 = 0.8$, the sequence converges to the root at 0, but with $x_0 = 5$, the sequence diverges.

33. The sequence of iterates diverges spectacularly, since $x_n = (-2)^n x_0$.

35. (a) Let $f(x) = \frac{1}{x} - c$. Then
$$x - \frac{f(x)}{f'(x)} = x - \frac{\frac{1}{x} - c}{-x^{-2}} = 2x - cx^2$$

(b) For $c = 10.3$, we have $f(x) = \frac{1}{x} - 10.3$ and thus $x_{n+1} = 2x_n - 10.3x_n^2$.

• Take $x_0 = 0.1$.

n	1	2	3
x_n	0.097	0.0970873	0.09708738

• Take $x_0 = 0.5$.

n	1	2	3
x_n	−1.575	−28.7004375	−8541.66654

(c) The graph is disconnected. If $x_0 = 0.5$, $(x_1, f(x_1))$ is on the other portion of the graph, which will never converge to any point under Newton's Method.

37. $\theta \approx 1.2757$; hence, $h = L\dfrac{1 - \cos\theta}{2\sin\theta} \approx 1.11181$

39. (a) $a = 46.95$ **(b)** $s = 29.24$

41. (a) $a \approx 28.46$ **(b)** $\Delta L = 1$ ft yields $\Delta s \approx 0.61$; $\Delta L = 5$ yields $\Delta s \approx 3.05$. **(c)** $s(161) - s(160) = 0.62$, very close to the approximation obtained from the Linear Approximation; $s(165) - s(160) = 3.02$, again very close to the approximation obtained from the Linear Approximation

Chapter 4 Review

1. $8.1^{1/3} - 2 \approx 0.00833333$; the error is 3.445×10^{-5}.

3. $625^{1/4} - 624^{1/4} \approx 0.002$; the error is 1.201×10^{-6}.

5. $\frac{1}{1.02} \approx 0.98$; the error is 3.922×10^{-4}.

7. $L(x) = 5 + \dfrac{1}{10}(x - 25)$ **9.** $L(r) = 36\pi(r - 2)$

11. $L(x) = \dfrac{1}{\sqrt{e}}(2 - x)$ **13.** $\Delta s \approx 0.632$

15. (a) An increase of $1500 in revenue.   **(b)** A small increase in price would result in a decrease in revenue.

19. $c = \dfrac{3}{\ln 4} \approx 2.164 \in (1, 4)$

21. Let $x > 0$. Because f is continuous on $[0, x]$ and differentiable on $(0, x)$, the Mean Value Theorem guarantees there exists a $c \in (0, x)$ such that
$$f'(c) = \frac{f(x) - f(0)}{x - 0} \quad \text{or} \quad f(x) = f(0) + xf'(c)$$

Now, we are given $f(0) = 4$ and $f'(x) \le 2$ for $x > 0$. Therefore, for all $x \ge 0$,
$$f(x) \le 4 + x(2) = 2x + 4$$

23. $x = \frac{2}{3}$ and $x = 2$ are critical points; $f\left(\frac{2}{3}\right)$ is a local maximum, while $f(2)$ is a local minimum.

25. $x = 0$, $x = -2$, and $x = -\frac{4}{5}$ are critical points; $f(-2)$ is neither a local maximum nor a local minimum, $f\left(-\frac{4}{5}\right)$ is a local maximum, and $f(0)$ is a local minimum.

27. $\theta = \dfrac{3\pi}{4} + n\pi$ is a critical point for all integers n; $g\left(\dfrac{3\pi}{4} + n\pi\right)$ is neither a local maximum nor a local minimum for any integer n.

29. Maximum value is 21; minimum value is −11.

31. Minimum value is −1; maximum value is $\dfrac{5}{4}$.

33. Minimum value is −1; maximum value is 3.

35. Minimum value is $12 - 12\ln 12 \approx -17.818880$; maximum value is $40 - 12\ln 40 \approx -4.266553$.

37. Critical points are $x = 1$ and $x = 3$. The minimum value is 2, occurring at $x = 3$, and the maximum value is 17, occurring at the right endpoint $x = 8$.

39. $x = \dfrac{4}{3}$ **41.** $x = \pm\dfrac{2}{\sqrt{3}}$ **43.** $x = 1$ and $x = 4$

45. No horizontal asymptotes; no vertical asymptotes

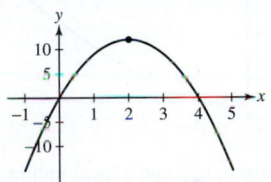

47. No horizontal asymptotes; no vertical asymptotes

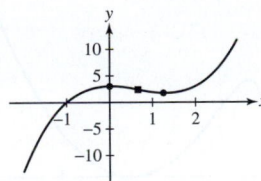

49. $y = 0$ is a horizontal asymptote; $x = -1$ is a vertical asymptote.

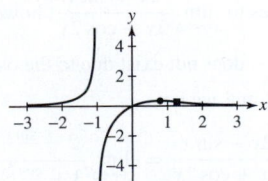

51. Horizontal asymptote of $y = 0$; no vertical asymptotes

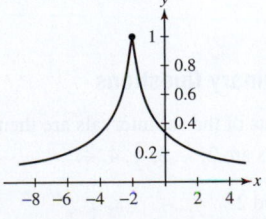

53.

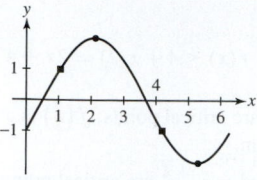

55.

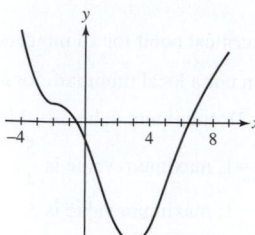

57. $b = \sqrt[3]{12}$ m and $h = \frac{1}{3}\sqrt[3]{12}$ m   **61.** $\dfrac{16}{9}\pi$ **67.** $\sqrt[3]{25} = 2.9240$

69. $(0, \frac{2}{e})$ is a local minimum.

71. Local minimum at $x = e^{-1}$; no points of inflection; $\lim\limits_{x\to 0+} x \ln x = 0$; $\lim\limits_{x\to\infty} x \ln x = \infty$

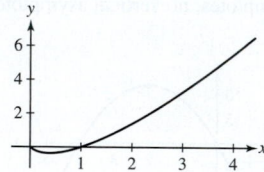

73. Local maximum at $x = e^{-2}$ and a local minimum at $x = 1$; point of inflection at $x = e^{-1}$; $\lim\limits_{x\to 0+} x(\ln x)^2 = 0$; $\lim\limits_{x\to\infty} x(\ln x)^2 = \infty$

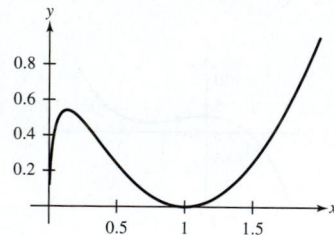

75. As $x \to \infty$, both $2x - \sin x$ and $3x + \cos 2x$ tend toward infinity, so L'Hôpital's Rule applies to $\lim\limits_{x\to\infty} \dfrac{2x - \sin x}{3x + \cos 2x}$; however, the resulting limit, $\lim\limits_{x\to\infty} \dfrac{2 - \cos x}{3 - 2\sin 2x}$, does not exist due to the oscillation of $\sin x$ and $\cos x$. To evaluate the limit, we note

$$\lim_{x\to\infty} \frac{2x - \sin x}{3x + \cos 2x} = \lim_{x\to\infty} \frac{2 - \frac{\sin x}{x}}{3 + \frac{\cos 2x}{x}} = \frac{2}{3}$$

77. 4   **79.** 0   **81.** 3   **83.** $\ln 2$ **85.** $\dfrac{1}{6}$ **87.** 2

Chapter 5

Section 5.1 Preliminary Questions

1. The right endpoints of the subintervals are then $\frac{5}{2}, 3, \frac{7}{2}, 4, \frac{9}{2}, 5$, while the left endpoints are $2, \frac{5}{2}, 3, \frac{7}{2}, 4, \frac{9}{2}$.

2. (a) $\dfrac{9}{2}$ (b) $\dfrac{3}{2}$ and 2

3. (a) *Are* the same   (b) *Not* the same   (c) *Are* the same   (d) *Are* the same

4. The first term in the sum $\sum_{j=0}^{100} j$ is equal to zero, so it may be dropped; on the other hand, the first term in $\sum_{j=0}^{100} 1$ is not zero.

5. On $[3, 7]$, the function $f(x) = x^{-2}$ is a decreasing function.

Section 5.1 Exercises

1. Over the interval $[0, 3]$: 0.96 km; over the interval $[1, 2.5]$: 0.5 km

3. 28.5 cm; The figure below is a graph of the rainfall as a function of time. The area of the shaded region represents the total rainfall.

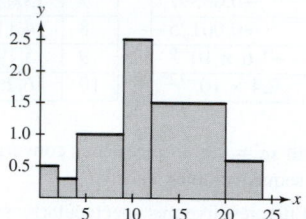

5. $L_5 = 46$; $R_5 = 44$

7. (a) $L_6 = 16.5$; $R_6 = 19.5$

(b) Via geometry (see figure), the exact area is $A = 18$. Thus, L_6 underestimates the true area ($L_6 - A = -1.5$), while R_6 overestimates the true area ($R_6 - A = +1.5$).

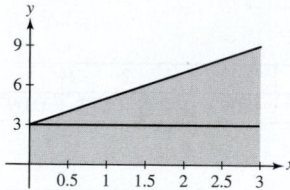

9. $R_3 = 32$; $L_3 = 20$; the area under the graph is larger than L_3 but smaller than R_3.

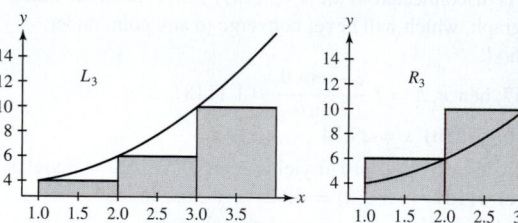

11. $R_3 = 2.5$; $M_3 = 2.875$; $L_6 = 3.4375$

13. (a) $L_4 = 1.75$ (b) $R_4 = 3.75$ (c) The actual area A under the curve $f(x) = x^2$ over the interval $[0, 2]$ satisfies $L_4 < A < R_4$.

15. $L_4 \approx 2.1730$ **17.** $R_6 \approx 1.2963$ **19.** $M_5 \approx 1.30$

21. $L_4 \approx 0.410236$ **23.** $\sum\limits_{k=4}^{8} k^7$ **25.** $\sum\limits_{k=2}^{5} (2^k + 2)$

27. $\sum\limits_{i=1}^{n} \dfrac{i}{(i+1)(i+2)}$

29. (a) 45   (b) 24   (c) 99

31. (a) -1 (b) 13   (c) 12

33. 15,050   **35.** 352,800   **37.** 1,093,350   **39.** 41,650

41. $-123,165$ **43.** $\dfrac{1}{2}$ **45.** $\dfrac{1}{3}$

47. 18; the region under the graph is a triangle with base 2 and height 18.

49. 12; the region under the curve is a trapezoid with base width 4 and heights 2 and 4.

51. 2; the region under the curve over $[0, 2]$ is a triangle with base and height 2.

53. $\lim\limits_{N\to\infty} R_N = 16$ **55.** $R_N = \dfrac{26}{3} + \dfrac{8}{n} + \dfrac{4}{3n^2}$; area $= \dfrac{26}{3}$

57. $R_N = 222 + \dfrac{189}{N} + \dfrac{27}{N^2}$; 222   **59.** $R_N = 2 + \dfrac{6}{N} + \dfrac{8}{N^2}$; 2

61. $R_N = (b-a)(2a+1) + (b-a)^2 + \dfrac{(b-a)^2}{N}$; $(b^2+b) - (a^2+a)$

63. The area between the graph of $f(x) = x^4$ and the x-axis over the interval $[0, 1]$

65. The area between the graph of $y = e^x$ and the x-axis over the interval $[-2, 3]$

67. $\lim\limits_{N\to\infty} R_N = \lim\limits_{N\to\infty} \dfrac{\pi}{N} \sum\limits_{k=1}^{N} \sin\left(\dfrac{k\pi}{N}\right)$

69. $\lim\limits_{N\to\infty} L_N = \lim\limits_{N\to\infty} \dfrac{4}{N} \sum\limits_{j=0}^{N-1} \sqrt{15 + \dfrac{8j}{N}}$

71. $\lim\limits_{N\to\infty} M_N = \lim\limits_{N\to\infty} \dfrac{1}{2N} \sum\limits_{j=1}^{N} \tan\left(\dfrac{1}{2} + \dfrac{1}{2N}\left(j - \dfrac{1}{2}\right)\right)$

73. Represents the area between the graph of $y = f(x) = \sqrt{1 - x^2}$ and the x-axis over the interval $[0, 1]$. This is the portion of the circular disk $x^2 + y^2 \le 1$ that lies in the first quadrant. Accordingly, its area is $\dfrac{\pi}{4}$.

75. Of the three approximations, R_N is the least accurate, and then L_N and finally M_N are the most accurate.

77. The area A under the curve is somewhere between $L_4 \approx 0.518$ and $R_4 \approx 0.768$.

79. f is increasing over the interval $[0, \pi/2]$, so $0.79 \approx L_4 \le A \le R_4 \approx 1.18$.

81. $L_{100} = 0.793988$; $R_{100} = 0.80399$; $L_{200} = 0.797074$; $R_{200} = 0.802075$; thus, $A = 0.80$ to two decimal places.

83. $R_{100} \approx 1.4142$ **85.** $R_{100} \approx 0.9946$; guess: area $= 1$

87. (a) Let $f(x) = e^x$ on $[0, 1]$. With $n = N$, $\Delta x = (1 - 0)/N = 1/N$ and

$$x_j = a + j\Delta x = \dfrac{j}{N}$$

for $j = 0, 1, 2, \ldots, N$. Therefore,

$$L_N = \Delta x \sum\limits_{j=0}^{N-1} f(x_j) = \dfrac{1}{N} \sum\limits_{j=0}^{N-1} e^{j/N}$$

(b) Applying Eq. (8) with $r = e^{1/N}$, we have

$$L_N = \dfrac{1}{N} \dfrac{(e^{1/N})^N - 1}{e^{1/N} - 1} = \dfrac{e - 1}{N(e^{1/N} - 1)}$$

(c) $A = e - 1$

89.

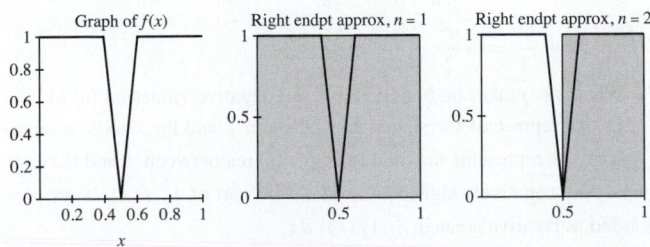

Graph of $f(x)$ | Right endpt approx, $n = 1$ | Right endpt approx, $n = 2$

91. When f' is large, the graph of f is steeper and hence there is more gap between f and L_N or R_N.

95. $N > 30,000$

Section 5.2 Preliminary Questions

1. 2

2. (a) False. $\int_a^b f(x)\, dx$ is the *signed* area between the graph and the x-axis.   **(b)** True   **(c)** True

3. Because $\cos(\pi - x) = -\cos x$, the "negative" area between the graph of $y = \cos x$ and the x-axis over $[\frac{\pi}{2}, \pi]$ exactly cancels the "positive" area between the graph and the x-axis over $[0, \frac{\pi}{2}]$.

4. $\displaystyle\int_{-1}^{-5} 8\, dx$

Section 5.2 Exercises

1. The region bounded by the graph of $y = 2x$ and the x-axis over the interval $[-3, 3]$ consists of two right triangles. One has area $\frac{1}{2}(3)(6) = 9$ below the axis, and the other has area $\frac{1}{2}(3)(6) = 9$ above the axis. Hence,

$$\int_{-3}^{3} 2x\, dx = 9 - 9 = 0$$

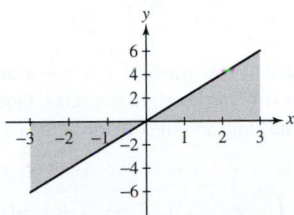

3. The region bounded by the graph of $y = 3x + 4$ and the x-axis over the interval $[-2, 1]$ consists of two right triangles. One has area $\frac{1}{2}(\frac{2}{3})(2) = \frac{2}{3}$ below the axis, and the other has area $\frac{1}{2}(\frac{7}{3})(7) = \frac{49}{6}$ above the axis. Hence,

$$\int_{-2}^{1} (3x + 4)\, dx = \dfrac{49}{6} - \dfrac{2}{3} = \dfrac{15}{2}$$

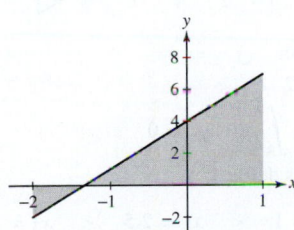

5. The region bounded by the graph of $y = 7 - x$ and the x-axis over the interval $[6, 8]$ consists of two right triangles. One triangle has area $\frac{1}{2}(1)(1) = \frac{1}{2}$ above the axis, and the other has area $\frac{1}{2}(1)(1) = \frac{1}{2}$ below the axis. Hence,

$$\int_{6}^{8} (7 - x)\, dx = \dfrac{1}{2} - \dfrac{1}{2} = 0$$

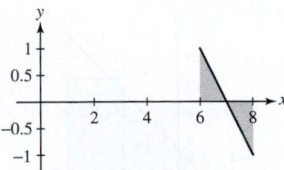

7. The region bounded by the graph of $y = \sqrt{25 - x^2}$ and the x-axis over the interval $[0, 5]$ is one-quarter of a circle of radius 5. Hence,

$$\int_{0}^{5} \sqrt{25 - x^2}\, dx = \dfrac{1}{4}\pi(5)^2 = \dfrac{25\pi}{4}$$

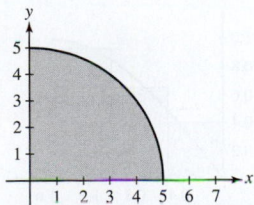

9. The region bounded by the graph of $y = 2 - |x|$ and the x-axis over the interval $[-2, 2]$ is a triangle above the axis with base 4 and height 2. Consequently,

$$\int_{-2}^{2} (2 - |x|)\,dx = \frac{1}{2}(2)(4) = 4$$

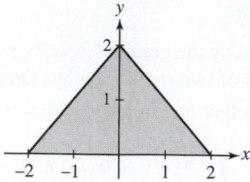

11. (a) $\displaystyle \lim_{N \to \infty} R_N = \lim_{N \to \infty} \left(30 - \frac{50}{N}\right) = 30$

(b) The region bounded by the graph of $y = 8 - x$ and the x-axis over the interval $[0, 10]$ consists of two right triangles. One triangle has area $\frac{1}{2}(8)(8) = 32$ above the axis, and the other has area $\frac{1}{2}(2)(2) = 2$ below the axis. Hence,

$$\int_{0}^{10} (8 - x)\,dx = 32 - 2 = 30$$

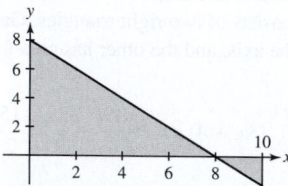

13. (a) $-\dfrac{\pi}{2}$ **(b)** $\dfrac{3\pi}{2}$

15. $\displaystyle \int_{0}^{3} g(t)\,dt = \frac{3}{2}; \int_{3}^{5} g(t)\,dt = 0$

17. The partition P is defined by

$$x_0 = 0 \;\;<\;\; x_1 = 1 \;\;<\;\; x_2 = 2.5 \;\;<\;\; x_3 = 3.2 \;\;<\;\; x_4 = 5$$

The set of sample points is given by $C = \{c_1 = 0.5, c_2 = 2, c_3 = 3, c_4 = 4.5\}$. Finally, the value of the Riemann sum is

$$34.25(1 - 0) + 20(2.5 - 1) + 8(3.2 - 2.5) + 15(5 - 3.2) = 96.85$$

19. $R(f, P, C) = 70$; here is a sketch of the graph of f and the rectangles.

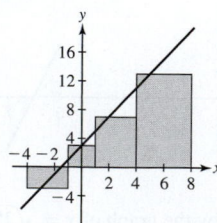

21. $R(f, P, C) = 1.029225$; here is a sketch of the graph of f and the rectangles.

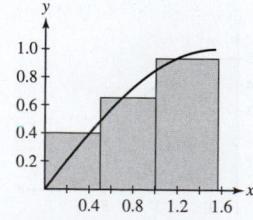

23. Using a left-hand Riemann sum approximation with

$$\Delta t = 1, \int_{0}^{12} E(t)\,dt \approx E(0)\Delta t + E(1)\Delta t + \cdots + E(11)\Delta t \approx$$

$(1.1)(1) + (1)(1) + (1)(1) + (1.1)(1) + (1.3)(1) + (1.8)(1) +$
$(2.5)(1) + (2)(1) + (0.3)(1) + (0.5)(1) + (0.4)(1) + (-0.8)(1) \approx 12.2$
kilowatt-hours

25. 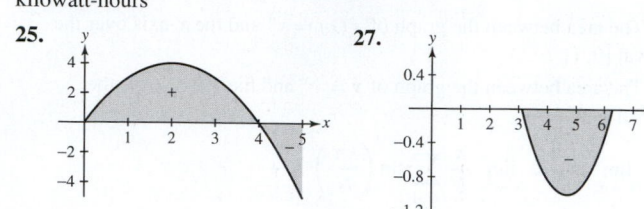 **27.**

29.

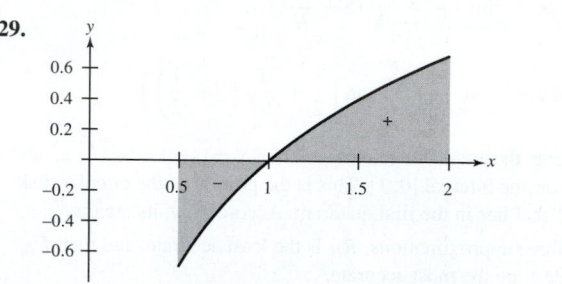

31. The integrand is always positive. The integral must therefore be positive, since the signed area has only a positive part.

33. Since $y = x$ is an increasing function on $[0, 2\pi]$, and $\sin x \geq 0$ on $[0, \pi]$ but $\sin x \leq 0$ on $[\pi, 2\pi]$, it follows that the area below the x-axis is enclosed by $y = x \sin x$ on the interval $[0, \pi]$. Hence, the total area on $[0, 2\pi]$ will be negative, so the definite integral will be negative as well.

35. 36 **37.** 243 **39.** $-\dfrac{2}{3}$ **41.** $\dfrac{196}{3}$ **43.** $\dfrac{1}{3}a^3 - \dfrac{1}{2}a^2 + \dfrac{5}{6}$

45. 17 **47.** -12 **49.** $\displaystyle \int_{a}^{b} H(x)dx = \begin{cases} 0 & \text{if } b \leq 0 \\ b & \text{if } a \leq 0 \text{ and } b > 0 \\ b - a & \text{if } a > 0 \end{cases}$

51. The integral formula holds for $b > 0$ by Exercise 50, and it holds for $b = 0$ by definition of a definite integral on an interval of length zero. Now, consider the situation where $b < 0$. By symmetry,

$$\int_{b}^{0} x^3 dx = -\int_{0}^{|b|} x^3 dx.$$ By Exercise 50,

$$-\int_{0}^{|b|} x^3 dx = -\frac{|b|^4}{4} = -\frac{b^4}{4}.$$ Therefore, $\displaystyle \int_{0}^{b} x^3 dx = \frac{b^4}{4}.$

53. $-\dfrac{63}{4}$ **55.** 7 **57.** 8 **59.** -7 **61.** $\displaystyle \int_{0}^{7} f(x)\,dx$

63. $\displaystyle \int_{5}^{9} f(x)\,dx$ **65.** $\displaystyle \int_{a}^{b} x\,dx = \int_{a}^{0} x\,dx + \int_{0}^{b} x\,dx =$

$$-\int_{0}^{a} x\,dx + \frac{b^2}{2} = -\frac{a^2}{2} + \frac{b^2}{2} = \frac{b^2 - a^2}{2}$$

67. When $f(x)$ takes on both positive and negative values on $[a, b]$, $\int_{a}^{b} f(x)\,dx$ represents the signed area between f and the x-axis, whereas $\int_{a}^{b} |f(x)|\,dx$ represents the total (unsigned) area between f and the x-axis. Any negatively signed areas that were part of $\int_{a}^{b} f(x)\,dx$ are regarded as positive areas in $\int_{a}^{b} |f(x)|\,dx$.

69. $[-1, \sqrt{2}]$ or $[-\sqrt{2}, 1]$ **71.** 9 **73.** $\dfrac{1}{2}$

75. On the interval $[0, 1]$, $x^5 \leq x^4$; on the other hand, $x^4 \leq x^5$ for $x \in [1, 2]$.

77. $y = \sin x$ is increasing on $[0.2, 0.3]$. Accordingly, for $0.2 \leq x \leq 0.3$, we have

$$m = 0.198 \leq 0.19867 \approx \sin 0.2 \leq \sin x \leq \sin 0.3$$

$$\approx 0.29552 \leq 0.296 = M$$

Therefore, by the Comparison Theorem, we have

$$0.0198 = m(0.3 - 0.2) = \int_{0.2}^{0.3} m\,dx \leq \int_{0.2}^{0.3} \sin x\,dx \leq \int_{0.2}^{0.3} M\,dx$$

$$= M(0.3 - 0.2) = 0.0296$$

79. f is decreasing and nonnegative on the interval $[\pi/4, \pi/2]$. Therefore, $0 \leq f(x) \leq f(\pi/4) = \frac{2\sqrt{2}}{\pi}$ for all x in $[\pi/4, \pi/2]$.

81. The assertion $f'(x) \leq g'(x)$ is false. Consider $a = 0, b = 1$, $f(x) = x, g(x) = 2$. $f(x) \leq g(x)$ for all x in the interval $[0, 1]$, but $f'(x) = 1$, while $g'(x) = 0$ for all x.

83. If f is an odd function, then $f(-x) = -f(x)$ for all x. Accordingly, for every positively signed area in the right half-plane where f is above the x-axis, there is a corresponding negatively signed area in the left half-plane where f is below the x-axis. Similarly, for every negatively signed area in the right half-plane where f is below the x-axis, there is a corresponding positively signed area in the left half-plane where f is above the x-axis.

Section 5.3 Preliminary Questions

1. Any constant function is an antiderivative for the function $f(x) = 0$.

2. No difference   **3.** No

4. **(a)** False. Even if $f(x) = g(x)$, the antiderivatives F and G may differ by an additive constant.

(b) True. This follows from the fact that the derivative of any constant is 0.

(c) False. If the functions f and g are different, then the antiderivatives F and G differ by a linear function: $F(x) - G(x) = ax + b$ for some constants a and b.

5. No

Section 5.3 Exercises

1. $6x^3 + C$ **3.** $\frac{2}{5}x^5 - 8x^3 + 12\ln|x| + C$

5. $2\sin x + 9\cos x + C$ **7.** $12e^x + 5x^{-1} + C$

9. **(a)** (ii)   **(b)** (iii)   **(c)** (i)   **(d)** (iv)

11. $4x - 9x^2 + C$ **13.** $\frac{11}{5}t^{5/11} + C$ **15.** $3t^6 - 2t^5 - 14t^2 + C$

17. $5z^{1/5} - \frac{3}{5}z^{5/3} + \frac{4}{9}z^{9/4} + C$ **19.** $\frac{3}{2}x^{2/3} + C$ **21.** $-\frac{18}{t^2} + C$

23. $\frac{2}{5}t^{5/2} + \frac{1}{2}t^2 + \frac{2}{3}t^{3/2} + t + C$ **25.** $\frac{1}{2}x^2 + 3\ln|x| + 4x^{-1} + C$

27. $12\sec x + C$ **29.** $-\csc t + C$ **31.** $\frac{1}{3}x^3 - \tan x + C$

33. $\sec\theta + \tan\theta + C$ **35.** $\frac{3}{5}e^{5x} + C$ **37.** $4x^2 + 2e^{5-2x} + C$

39. Graph (B) does not have the same local extrema as indicated by $y = f(x)$ and therefore is *not* an antiderivative of $y = f(x)$.

41. $\frac{d}{dx}\left(\frac{1}{7}(x + 13)^7 + C\right) = (x + 13)^6$

43. $\frac{d}{dx}\left(\frac{1}{12}(4x + 13)^3 + C\right) = \frac{1}{4}(4x + 13)^2(4) = (4x + 13)^2$

45. $G'(x) = 2xe^x + x^2e^x$, and this is not equal to $f(x)$. However, $H'(x) = 2e^x + 2xe^x - 2e^x = 2xe^x = f(x)$.

47. $y = \frac{1}{4}x^4 + 4$ **49.** $y = t^2 + 3t^3 - 2$ **51.** $y = \frac{2}{3}t^{3/2} + \frac{1}{3}$

53. $y = \frac{1}{12}(3x + 2)^4 - \frac{1}{3}$ **55.** $y = 1 - \cos x$

57. $y = e^x - e^2$ **59.** $y = -3e^{12-3t} + 10$

61. $f'(x) = 6x^2 + 1; f(x) = 2x^3 + x + 2$

63. $f'(x) = \frac{1}{4}x^4 - x^2 + x + 1; f(x) = \frac{1}{20}x^5 - \frac{1}{3}x^3 + \frac{1}{2}x^2 + x$

65. $f'(t) = -2t^{-1/2} + 2; f(t) = -4t^{1/2} + 2t + 4$

67. $f'(t) = \frac{1}{2}t^2 - \sin t + 2; f(t) = \frac{1}{6}t^3 + \cos t + 2t - 3$

69. The differential equation satisfied by $s(t)$ is

$$\frac{ds}{dt} = v(t) = 6t^2 - t$$

and the associated initial condition is $s(1) = 0$; $s(t) = 2t^3 - \frac{1}{2}t^2 - \frac{3}{2}$.

71. $v_y = -49$ m/s

73. $\frac{ds}{dt} = \sin t, s(0) = 0$; solution: $s(t) = 1 - \cos t$

75. 6.25 s; 78.125 m   **77.** 300 m/s   **81.** $c_1 = c_2 = -3$

83. **(a)** By the Chain Rule, we have

$$\frac{d}{dx}\left(\frac{1}{2}F(2x)\right) = \frac{1}{2}F'(2x) \cdot 2 = F'(2x) = f(2x)$$

Thus, $y = \frac{1}{2}F(2x)$ is an antiderivative of $y = f(2x)$.

(b) $\frac{1}{k}F(kx) + C$

Section 5.4 Preliminary Questions

1. **(a)** 4   **(b)** The signed area between $y = f(x)$ and the x-axis

2. 3

3. **(a)** False. The FTC I is valid for continuous functions.

(b) False. The FTC I works for any antiderivative of the integrand.

(c) False. If you cannot find an antiderivative of the integrand, you cannot use the FTC I to evaluate the definite integral, but the definite integral may still exist.

4. 0

Section 5.4 Exercises

1. $A = \frac{1}{3}$

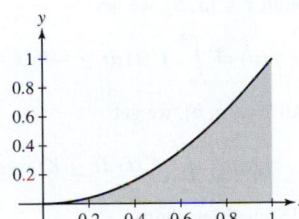

3. $A = \frac{1}{2}$

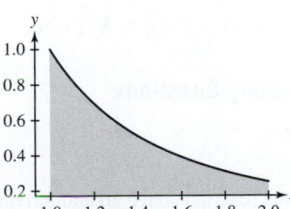

5. $\frac{27}{2}$ **7.** -1 **9.** 128   **11.** $\frac{27}{2}$ **13.** $\frac{16}{3}$ **15.** $\frac{31}{40}$ **17.** $\frac{2}{3}$

19. 12   **21.** $\frac{11}{6}$ **23.** $60\sqrt{3} - \frac{8}{3}$ **25.** $1 + \frac{1}{\sqrt{2}}$ **27.** $\frac{\sqrt{3}}{2}$

29. $\pi - 2$ **31.** $e - 1$ **33.** $\frac{1}{6}(e - e^{-17})$ **35.** $\ln 5$ **37.** e

39. $3e^{-6} - 9$ **41.** Lasts for more than 100 h: probability ≈ 0.905; lasts for more than 1000 h: probability ≈ 0.368

43. $\int_{-2}^{0} -x\,dx + \int_{0}^{1} x\,dx = \frac{5}{2}$ **45.** $\int_{-2}^{0} -x^3\,dx + \int_{0}^{3} x^3\,dx = \frac{97}{4}$

47. $\int_{0}^{\pi/2} \cos x\,dx + \int_{\pi/2}^{\pi} -\cos x\,dx = 2$ **49.** $\frac{1}{4}\left(b^4 - 1\right)$

51. $\frac{1}{6}(b^6 - 1)$ **53.** $\ln 5$ **55.** $\frac{707}{12}$

57. Graphically speaking, for an odd function, the positively signed area from $x = 0$ to $x = 1$ cancels the negatively signed area from $x = -1$ to $x = 0$.

59. 24

61. $\int_0^1 x^n \, dx$ represents the area between the positive curve $f(x) = x^n$ and the x-axis over the interval $[0, 1]$. This area gets smaller as n gets larger, as is readily evident in the following graph, which shows curves for several values of n.

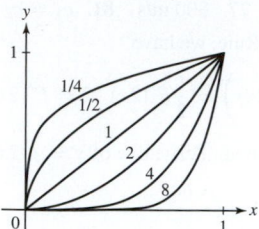

63. First, if $a = b$, then by definition, $\int_a^b f(x) \, dx = 0$. Also $F(b) - F(a) = 0$, so in this case, $\int_a^b f(x) \, dx = F(b) - F(a)$ holds.

Now, assume that $b < a$. Then

$$\int_a^b f(x) \, dx = -\int_b^a f(x) \, dx = -(F(a) - F(b)) = F(b) - F(a)$$

Therefore, $\int_a^b f(x) \, dx = F(b) - F(a)$ holds when $b < a$ too.

69. Let $a > b$ be real numbers, and let $f(x)$ be such that $|f'(x)| \leq K$ for $x \in [a, b]$. By FTC,

$$\int_a^x f'(t) \, dt = f(x) - f(a)$$

Since $f'(x) \geq -K$ for all $x \in [a, b]$, we get

$$f(x) - f(a) = \int_a^x f'(t) \, dt \geq -K(x - a)$$

Since $f'(x) \leq K$ for all $x \in [a, b]$, we get

$$f(x) - f(a) = \int_a^x f'(t) \, dt \leq K(x - a)$$

Combining these two inequalities yields

$$-K(x - a) \leq f(x) - f(a) \leq K(x - a)$$

so that, by definition,

$$|f(x) - f(a)| \leq K|x - a|$$

Section 5.5 Preliminary Questions

1. **(a)** No **(b)** Yes

2. **(c)**

3. Yes. All continuous functions have an antiderivative, namely, $\int_a^x f(t) \, dt$.

4. **(b)**, **(e)**, and **(f)**

Section 5.5 Exercises

1. $A(x) = 4x - x^2$; $A'(x) = 4 - 2x$

3. $A(x) = 2x^2 + 2x^3$; $A'(x) = 4x + 6x^2$

5. $A(x) = \frac{1}{3}x^3 + \cos x - 1$; $A'(x) = x^2 - \sin x$

7. $A(x) = \frac{1}{2}e^{2x} - \frac{1}{2}$; $A'(x) = e^{2x}$

9. $F(0) = 0$; $F(3) \approx 5.72$; $F'(0) = 0$; $F'(3) = 2\sqrt{3}$

11. $F(-2) = 0$; $F(2) \approx 2.21$; $F'(0) = 1$; $F'(2) = \frac{1}{5}$

13. $\frac{1}{5}x^5 - \frac{32}{5}$ **15.** $1 - \cos x$ **17.** $\frac{1}{3}e^{3x} - \frac{1}{3}e^{12}$ **19.** $\frac{1}{2}x^4 - \frac{1}{2}$

21. $-e^{-9x-2} + e^{-3x}$ **25.** $x^5 - 9x^3$ **27.** $\sec(5t - 9)$

29. **(a)** $A(2) = 4$; $A(3) = 6.5$; $A'(2) = 2$ and $A'(3) = 3$

(b)

$$A(x) = \begin{cases} 2x, & 0 \leq x < 2 \\ \frac{1}{2}x^2 + 2, & 2 \leq x \leq 4 \end{cases}$$

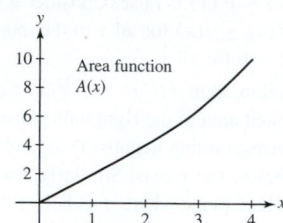

31.

x	0	1	2	3	4	5	6	7	8	9	10
$A(x)$	0	3.3	6.4	6.5	2.8	−1	−3.6	−4.6	−3.3	−1.5	0.1
$A'(x)$		3.2	1.6	−1.8	−3.75	−3.2	−1.8	0.15	1.55	1.7	

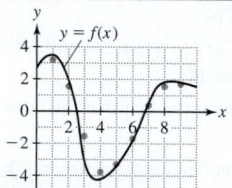

33. $\frac{2x^3}{x^2 + 1}$ **35.** $-\cos^4 s \sin s$ **37.** $2x \tan(x^2) - \frac{\tan(\sqrt{x})}{2\sqrt{x}}$

39. The minimum value of $A(x)$ is $A(1.5) = -1.25$; the maximum value of A is $A(4.5) = 1.25$.

41. $A(x) = (x - 2) - 1$ and $B(x) = (x - 2)$

43. **(a)** A does not have a local maximum at P. **(b)** A has a local minimum at R. **(c)** A has a local maximum at S. **(d)** True

45. **(a)** If $x = c$ is an inflection point of A, then $A''(c) = f'(c) = 0$.

(b) If A is concave up, then $A''(x) > 0$. Since A is the area function associated with f, $A'(x) = f(x)$ by FTC II, so $A''(x) = f'(x)$. Therefore, $f'(x) > 0$, so f is increasing.

(c) If A is concave down, then $A''(x) < 0$. Since A is the area function associated with $f(x)$, $A'(x) = f(x)$ by FTC II, so $A''(x) = f'(x)$. Therefore, $f'(x) < 0$, so f is decreasing.

47. **(a)** A is increasing on the intervals $(0, 4)$ and $(8, 12)$ and is decreasing on the intervals $(4, 8)$ and $(12, \infty)$.

(b) Local minimum: $x = 8$; local maximum: $x = 4$ and $x = 12$

(c) A has inflection points at $x = 2$, $x = 6$, and $x = 10$.

(d) A is concave up on the intervals $(0, 2)$ and $(6, 10)$ and is concave down on the intervals $(2, 6)$ and $(10, \infty)$.

49. The graph of one such function is

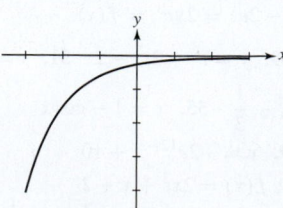

51. Smallest positive critical point: $x = (\pi/2)^{2/3}$ corresponds to a local maximum; smallest positive inflection point: $x = \pi^{2/3}$, $y = F(x)$ changes from concave down to concave up

53. **(a)** $\dfrac{d}{dx}\mathrm{erf}(x) = \dfrac{2}{\sqrt{\pi}}e^{-x^2}$, that is, positive for all x, and therefore $\mathrm{erf}(x)$ is an increasing function.

(b) $\mathrm{erf}(-x) = \dfrac{2}{\sqrt{\pi}}\displaystyle\int_0^{-x} e^{-t^2}\,dt = -\dfrac{2}{\sqrt{\pi}}\displaystyle\int_{-x}^0 e^{-t^2}\,dt$. Since $f(t) = e^{-t^2}$ is an even function, $\displaystyle\int_{-x}^0 e^{-t^2}\,dt = \int_0^x e^{-t^2}\,dt$. It follows that

$\mathrm{erf}(-x) = -\dfrac{2}{\sqrt{\pi}}\displaystyle\int_{-x}^0 e^{-t^2}\,dt = -\dfrac{2}{\sqrt{\pi}}\displaystyle\int_0^x e^{-t^2}\,dt = -\mathrm{erf}(x)$. Therefore, $\mathrm{erf}(x)$ is an odd function.

(c) $\mathrm{erf}(1/2) \approx 0.5205$; $\mathrm{erf}(1) \approx 0.8427$; $\mathrm{erf}(3/2) \approx 0.9661$; $\mathrm{erf}(2) \approx 0.9953$; $\mathrm{erf}(5/2) \approx 0.9996$

(d) $y = -1$ and $y = 1$

(e)

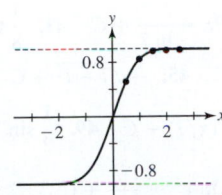

55. **(a)** Then by the FTC, Part II, $A'(x) = f(x)$ and thus $y = A(x)$ and $y = F(x)$ are both antiderivatives of $y = f(x)$. Hence, $F(x) = A(x) + C$ for some constant C.

(b)

$$F(b) - F(a) = (A(b) + C) - (A(a) + C) = A(b) - A(a)$$

$$= \int_a^b f(t)\,dt - \int_a^a f(t)\,dt$$

$$= \int_a^b f(t)\,dt - 0 = \int_a^b f(t)\,dt$$

which proves the FTC, Part I.

57. Write

$$\int_{u(x)}^{v(x)} f(x)\,dx = \int_{u(x)}^0 f(x)\,dx + \int_0^{v(x)} f(x)\,dx$$

$$= \int_0^{v(x)} f(x)\,dx - \int_0^{u(x)} f(x)\,dx$$

Then, by the Chain Rule and the FTC,

$$\frac{d}{dx}\int_{u(x)}^{v(x)} f(x)\,dx = \frac{d}{dx}\int_0^{v(x)} f(x)\,dx - \frac{d}{dx}\int_0^{u(x)} f(x)\,dx$$

$$= f(v(x))v'(x) - f(u(x))u'(x)$$

Section 5.6 Preliminary Questions

1. The total drop in temperature of the metal object in the first T minutes after being submerged in the cold water

2. 560 km

3. Quantities **(a)** and **(c)** would naturally be represented as derivatives; quantities **(b)** and **(d)** would naturally be represented as integrals.

Section 5.6 Exercises

1. 15,250 gal **3.** 3,660,000 **5.** 33 m **7.** 3.675 m

9. Displacement: 10 m; distance: 26 m

11. Displacement: 0 m; distance: 1 m

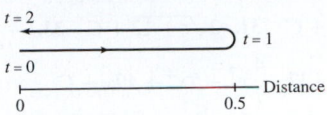

13. 39 m/s **15.** 9200 cars

17. Total cost: \$650; average cost of first 10: \$37.50; average cost of last 10: \$27.50

19. 112.5 ft

21. The integral represents the total snowfall over the 24-h period. It was approximately 35.8 in.

23. **(a)** 2.916×10^{10} **(b)** Approximately 240,526 asteroids of diameter 50 km **25.** $\displaystyle\int_0^{365} R(t)\,dt \approx 605.05$ billion ft³

27. $100 \le t \le 150$: 404.968 families; $350 \le t \le 400$: 245.812 families

29. The particle's velocity is $v(t) = s'(t) = t^{-2}$, an antiderivative for which is $F(t) = -t^{-1}$. Hence, the particle's position at time t is

$$s(t) = \int_1^t s'(u)\,du = F(u)\Big|_1^t = F(t) - F(1) = 1 - \frac{1}{t} < 1$$

for all $t \ge 1$. Thus, the particle will never pass $x = 1$, which implies it will never pass $x = 2$ either.

31. **(a)** $\mathrm{CS} = \displaystyle\int_0^{q^*} [D(q) - p^*]\,dq$ **(b)** $\mathrm{PS} = \displaystyle\int_0^{q^*} [p^* - S(q)]\,dq$

Section 5.7 Preliminary Questions

1. **(a)** and **(b)**

2. **(a)** $u(x) = x^2 + 9$ **(b)** $u(x) = x^3$ **(c)** $u(x) = \cos x$

3. **(c)**

Section 5.7 Exercises

1. $du = (3x^2 - 2x)\,dx$ **3.** $du = -2x\sin(x^2)\,dx$

5. $du = 4e^{4x+1}\,dx$

7. $\displaystyle\int (x+8)^4\,dx = \int u^4\,du = \frac{1}{5}u^5 + C = \frac{1}{5}(x+8)^5 + C$

9. $\displaystyle\int (3t-4)^5\,dt = \int \frac{1}{3}u^5\,du = \frac{1}{18}u^6 + C = \frac{1}{18}(3t-5)^6 + C$

11. $\displaystyle\int t\sqrt{t^2+1}\,dt = \frac{1}{2}\int u^{1/2}\,du = \frac{1}{3}u^{3/2} + C = \frac{1}{3}(t^2+1)^{3/2} + C$

13. $\displaystyle\int \frac{t^3}{(4-2t^4)^{11}}\,dt = -\frac{1}{8}\int u^{-11}\,du = \frac{1}{80}u^{-10} + C$

$= \dfrac{1}{80}(4-2t^4)^{-10} + C$

15. $\displaystyle\int x(x+1)^9\,dx = \int (u-1)u^9\,du = \int (u^{10} - u^9)\,du$

$= \dfrac{1}{11}u^{11} - \dfrac{1}{10}u^{10} + C = \dfrac{1}{11}(x+1)^{11} - \dfrac{1}{10}(x+1)^{10} + C$

17.
$\displaystyle\int x^2\sqrt{4-x}\,dx = \int -(4-u)^2\sqrt{u}\,du = \int \left(-u^{5/2} + 8u^{3/2} - 16u^{1/2}\right)du$

$= -\dfrac{2}{7}u^{7/2} + \dfrac{16}{5}u^{5/2} - \dfrac{32}{3}u^{3/2} + C =$

$- \dfrac{2}{7}(4-x)^{7/2} + \dfrac{16}{5}(4-x)^{5/2} - \dfrac{32}{3}(4-x)^{3/2} + C$

19. $\displaystyle\int \sin\theta\cos^3\theta\,d\theta = \int -u^3\,du = -\frac{1}{4}u^4 + C = -\frac{1}{4}\cos^4\theta + C$

21. $\displaystyle\int xe^{-x^2}\,dx = -\frac{1}{2}\int e^u\,du = -\frac{1}{2}e^u + C = -\frac{1}{2}e^{-x^2} + C$

23. $\displaystyle\int \frac{(\ln x)^2}{x}\,dx = \int u^2\,du = \frac{1}{3}u^3 + C = \frac{1}{3}(\ln x)^3 + C$

25. $u = x^4$; $\frac{1}{4}\sin(x^4) + C$ **27.** $u = x^{3/2}$; $\frac{2}{3}\sin(x^{3/2}) + C$

29. $\frac{1}{40}(4x+5)^{10} + C$ **31.** $2\sqrt{t+12} + C$ **33.** $-\frac{1}{4(x^2+2x)^2} + C$

35. $\sqrt{x^2+9} + C$ **37.** $\frac{4}{7}x^7 - 7x^4 + 49x + C$

39. $\frac{1}{24}(2x^3-7)^4 + C$ **41.** $\frac{1}{36}(3x+8)^{12} + C$

43. $\frac{2}{9}(x^3+1)^{3/2} + C$ **45.** $-\frac{1}{2}(x+5)^{-2} + C$

47. $\frac{1}{39}(z^3+1)^{13} + C$ **49.** $\frac{4}{9}(x+1)^{9/4} + \frac{4}{5}(x+1)^{5/4} + C$

51. $\frac{1}{3}\cos(8-3\theta) + C$ **53.** $2\sin\sqrt{t} + C$

55. $\frac{1}{4}\ln|\sec(4\theta+9)| + C$ **57.** $\ln|\sin x| + C$

59. $\frac{1}{4}\tan(4x+9) + C$ **61.** $2\tan(\sqrt{x}) + C$

63. $-\frac{1}{6}(\cos 4x+1)^{3/2} + C$ **65.** $\frac{1}{2}(\sec\theta - 1)^2 + C$

67. $\frac{1}{14}e^{14x-7} + C$ **69.** $-\frac{1}{3(e^x+1)^3} + C$ **71.** $-\frac{1}{e^t+1} + C$

73. $\frac{1}{5}(\ln x)^5 + C$ **75.** $-\ln|\cos(\ln x)| + C$

77. $-\frac{2}{1+\sqrt{x}} + \frac{1}{(1+\sqrt{x})^2} + C$

79. With $u = \sin x$, $\frac{1}{2}\sin^2 x + C_1$; with $u = \cos x$, $-\frac{1}{2}\cos^2 x + C_2$; the two results differ by a constant.

81. $u = \pi$ and $u = 4\pi$ **83.** 78 **85.** $3 - \sqrt{5}$ **87.** $\frac{3}{16}$ **89.** $\frac{98}{3}$

91. $\frac{243}{4}$ **93.** $\frac{1}{2}$ **95.** $\frac{1}{2}\ln(\sec 1)$ **97.** $\frac{1}{4}$ **99.** $\frac{20}{3}\sqrt{5} - \frac{32}{5}\sqrt{3}$

101. (a) The probability that $v \in [0, b]$ is

$$\int_0^b \frac{1}{32}ve^{-v^2/64}\,dv$$

Let $u = -v^2/64$. Then $du = -v/32\,dv$ and

$$\int_0^b \frac{1}{32}ve^{-v^2/64}\,dv = -\int_0^{-b^2/64} e^u\,du$$

$$= -e^u\Big|_0^{-b^2/64} = -e^{-b^2/64} + 1$$

(b) $e^{-1/16} - e^{-25/64}$

103. $\frac{1}{4}f(x)^4 + C$ **105.** $\ln|f(x)| + C$

107. Let $u = \sin\theta$. Then $u(\pi/6) = 1/2$ and $u(0) = 0$, as required. Furthermore, $du = \cos\theta\,d\theta$, so

$$d\theta = \frac{du}{\cos\theta}$$

If $\sin\theta = u$, then $u^2 + \cos^2\theta = 1$, so $\cos\theta = \sqrt{1-u^2}$. Therefore, $d\theta = du/\sqrt{1-u^2}$. This gives

$$\int_0^{\pi/6} f(\sin\theta)\,d\theta = \int_0^{1/2} f(u)\frac{1}{\sqrt{1-u^2}}\,du$$

109. $I = \pi/4$

Section 5.8 Preliminary Questions

1. (a) $b = 3$ (b) $b = e^3$ **2.** $b = \sqrt{3}$ **3.** (b) **4.** $x = 4u$

Section 5.8 Exercises

1. $\ln 9$ **3.** 3 **5.** $\frac{1}{3}\ln 4$ **7.** $\frac{\pi}{12}$ **9.** $\frac{\pi}{6}$ **11.** $\frac{7}{\ln 2}$

13. Let $u = x/3$. Then $x = 3u$, $dx = 3\,du$, $9 + x^2 = 9(1 + u^2)$, and

$$\int \frac{dx}{9+x^2} = \int \frac{3\,du}{9(1+u^2)} = \frac{1}{3}\int \frac{du}{1+u^2}$$

$$= \frac{1}{3}\tan^{-1}u + C = \frac{1}{3}\tan^{-1}\frac{x}{3} + C$$

15. $2\tan^{-1}2x + C$ **17.** $\frac{\pi}{3\sqrt{3}}$ **19.** $\frac{1}{4}\sin^{-1}(4t) + C$

21. $\frac{1}{\sqrt{3}}\sin^{-1}\sqrt{\frac{3}{5}}t + C$ **23.** $\frac{1}{\sqrt{3}}\sec^{-1}(2x) + C$

25. $\frac{1}{2}\sec^{-1}x^2 + C$ **27.** $\frac{\pi}{4} - \tan^{-1}(1/2)$ **29.** $\frac{(\tan^{-1}x)^2}{2} + C$

31. $\frac{1}{\ln 2}$ **33.** $-\frac{1}{\ln 9}\cos(9^x) + C$ **35.** $\frac{1}{2}e^{y^2} + C$

37. $\frac{1}{4}\sqrt{4x^2+9} + C$ **39.** $-\frac{7^{-x}}{\ln 7} + C$ **41.** $\frac{1}{8}\tan^8\theta + C$

43. $-\frac{2}{3}(1-\ln w)^{3/2} + C$ **45.** $-\sqrt{7-t^2} + C$

47. $\frac{3}{2}\ln(x^2+4) + \tan^{-1}(x/2) + C$ **49.** $\frac{1}{2}\sin^{-1}\left(\frac{2}{3}x\right) + C$

51. $-e^{-x} - 2x^2 + C$ **53.** $e^x - \frac{e^{3x}}{3} + C$

55. $-\sqrt{4-x^2} + 5\sin^{-1}(x/2) + C$ **57.** $\sin(e^x) + C$

59. $\frac{1}{4}\sin^{-1}\left(\frac{4x}{3}\right) + C$ **61.** $\frac{e^{7x}}{7} + \frac{3e^{5x}}{5} + e^{3x} + e^x + C$

63. $\frac{1}{3}\ln|x^3+2| + C$ **65.** $\ln|\sin x| + C$ **67.** $\frac{1}{8}(4\ln x + 5)^2 + C$

69. $\frac{3^{x^2}}{2\ln 3} + C$ **71.** $\frac{(\ln(\sin x))^2}{2} + C$

73. $\frac{2}{7}(t-3)^{7/2} + \frac{12}{5}(t-3)^{5/2} + 6(t-3)^{3/2} + C$

75. The definite integral $\displaystyle\int_0^x \sqrt{1-t^2}\,dt$ represents the area of the region under the upper half of the unit circle from 0 to x. The region consists of a sector of the circle and a right triangle. The sector has a central angle of $\frac{\pi}{2} - \theta$, where $\cos\theta = x$, and the right triangle has a base of length x and a height of $\sqrt{1-x^2}$.

77. Show that $\dfrac{d}{dt}\left(\sqrt{1-t^2} + t\sin^{-1}t\right) = \sin^{-1}t$.

79. Integrating both sides of the inequality $e^t \geq 1$ yields

$$\int_0^x e^t\,dt = e^x - 1 \geq x \quad\text{or}\quad e^x \geq 1 + x$$

Integrating both sides of this new inequality then gives

$$\int_0^x e^t\,dt = e^x - 1 \geq x + x^2/2 \quad\text{or}\quad e^x \geq 1 + x + x^2/2$$

Finally, integrating both sides again gives

$$\int_0^x e^t\,dt = e^x - 1 \geq x + x^2/2 + x^3/6$$

or

$$e^x \geq 1 + x + x^2/2 + x^3/6$$

as requested.

81. By Exercise 79, $e^x \geq 1 + x + \frac{x^2}{2} + \frac{x^3}{6}$. Thus,

$$\frac{e^x}{x^2} \geq \frac{1}{x^2} + \frac{1}{x} + \frac{1}{2} + \frac{x}{6} \geq \frac{x}{6}$$

Since $\lim_{x\to\infty} x/6 = \infty$, $\lim_{x\to\infty} e^x/x^2 = \infty$. More generally, by Exercise 80,

$$e^x \geq 1 + \frac{x^2}{2} + \cdots + \frac{x^{n+1}}{(n+1)!}$$

Thus,

$$\frac{e^x}{x^n} \geq \frac{1}{x^n} + \cdots + \frac{x}{(n+1)!} \geq \frac{x}{(n+1)!}$$

Since $\lim_{x\to\infty} \frac{x}{(n+1)!} = \infty$, $\lim_{x\to\infty} \frac{e^x}{x^n} = \infty$.

83. (a) The domain of G is $x > 0$ and, by part (i) of the previous exercise, the range of G is **R**. Now,

$$G'(x) = \frac{1}{x} > 0$$

for all $x > 0$. Thus, G is increasing on its domain, which implies that G has an inverse. The domain of the inverse is **R** and the range is $\{x : x > 0\}$. Let F denote the inverse of G.

(b) Let x and y be real numbers and suppose that $x = G(w)$ and $y = G(z)$ for some positive real numbers w and z. Then, using part (b) of the previous exercise,

$$F(x + y) = F(G(w) + G(z)) = F(G(wz)) = wz = F(x) + F(y)$$

(c) Let r be any real number. By part (k) of the previous exercise, $G(E^r) = r$. By definition of an inverse function, it then follows that $F(r) = E^r$.

(d) By the formula for the derivative of an inverse function,

$$F'(x) = \frac{1}{G'(F(x))} = \frac{1}{1/F(x)} = F(x)$$

85.

$$\lim_{n \to -1} \int_1^x t^n \, dt = \lim_{n \to -1} \frac{t^{n+1}}{n+1}\bigg|_1^x = \lim_{n \to -1}\left(\frac{x^{n+1}}{n+1} - \frac{1^{n+1}}{n+1}\right)$$

$$= \lim_{n \to -1} \frac{x^{n+1} - 1}{n+1} = \lim_{n \to -1} (x^{n+1}) \ln x$$

$$= \ln x = \int_1^x t^{-1} \, dt$$

87. (a) Interpreting the graph with y as the independent variable, we see that the function is $x = e^y$. Integrating in y then gives the area of the shaded region as $\int_0^{\ln a} e^y \, dy$.

(b) We can obtain the area under the graph of $y = \ln x$ from $x = 1$ to $x = a$ by computing the area of the rectangle extending from $x = 0$ to $x = a$ horizontally and from $y = 0$ to $y = \ln a$ vertically and then subtracting the area of the shaded region. This yields

$$\int_1^a \ln x \, dx = a \ln a - \int_0^{\ln a} e^y \, dy$$

(c) By direct calculation,

$$\int_0^{\ln a} e^y \, dy = e^y \bigg|_0^{\ln a} = a - 1$$

Thus,

$$\int_1^a \ln x \, dx = a \ln a - (a - 1) = a \ln a - a + 1$$

(d) Based on these results, it appears that

$$\int \ln x \, dx = x \ln x - x + C$$

Chapter 5 Review

1. $L_4 = \frac{23}{4}$; $M_4 = 7$

3. In general, R_N is larger than $\int_a^b f(x) \, dx$ on any interval $[a, b]$ over which $f(x)$ is increasing. Given the graph of f, we may take $[a, b] = [0, 2]$. In order for L_4 to be larger than $\int_a^b f(x) \, dx$, f must be decreasing over the interval $[a, b]$. We may therefore take $[a, b] = [2, 3]$.

5. $R_6 = \frac{625}{8}$

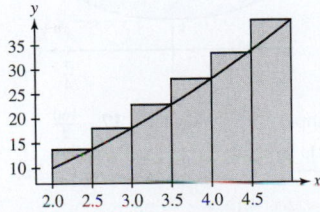

$M_6 = \frac{1127}{16}$

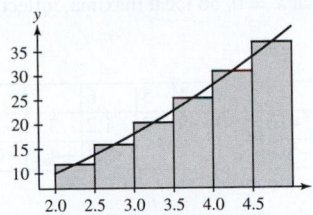

$L_6 = \frac{505}{8}$; The rectangles corresponding to this approximation are shown below.

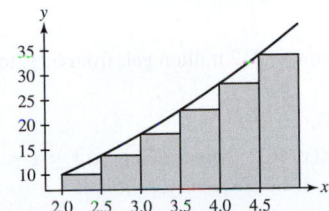

7. $R_N = \frac{141}{2} + \frac{45}{N} + \frac{9}{2N^2}$; $\frac{141}{2}$

9. $R_5 \approx 0.733732$; $M_5 \approx 0.786231$; $L_5 \approx 0.833732$

11. The area represented by the shaded rectangles is R_5; $R_5 = 90$; $L_5 = 90$.

13. $\lim_{N \to \infty} \frac{\pi}{6N} \sum_{j=1}^{N} \sin\left(\frac{\pi}{3} + \frac{\pi j}{6N}\right) = \int_{\pi/3}^{\pi/2} \sin x \, dx = \frac{1}{2}$

15. $\lim_{N \to \infty} \frac{5}{N} \sum_{j=1}^{N} \sqrt{4 + 5j/N} = \int_4^9 \sqrt{x} \, dx = \frac{38}{3}$

17. $x^4 - \frac{2}{3}x^3 + C$ **19.** $-\cos(\theta - 8) + C$

21. $-2t^{-2} + 4t^{-3} + C$ **23.** $\tan x + C$ **25.** $\frac{1}{5}(y+2)^5 + C$

27. $e^x - \frac{x^2}{2} + C$ **29.** $4 \ln |x| + C$ **31.** $y(x) = x^4 + 3$

33. $y(x) = 2\sqrt{x} - 1$ **35.** $y(x) = -e^{-x} + 4$

37. $f(t) = \frac{t^2}{2} - \frac{t^3}{3} - t + 2$

39. $\frac{1}{4} \ln \frac{5}{3}$ **41.** $\frac{1}{5}\left(1 - \frac{9\sqrt{3}}{32}\right)$

43. $4x^5 - \frac{9}{4}x^4 - x^2 + C$ **45.** $\frac{4}{5}x^5 - 3x^4 + 3x^3 + C$

47. $\frac{1}{4}x^4 + x^3 + C$ **49.** $\frac{46}{3}$ **51.** $27/2$

53. $\frac{1}{150}(10t - 7)^{15} + C$ **55.** $-\frac{1}{24}(3x^4 + 9x^2)^{-4} + C$ **57.** 506

59. $-\frac{3\sqrt{3}}{2\pi}$ **61.** $\frac{1}{27}\tan(9t^3 + 1) + C$ **63.** $\frac{1}{2}\cot(9 - 2\theta) + C$

65. $3 - \frac{3\sqrt[3]{4}}{2}$ **67.** $-\frac{1}{2}e^{9-2x} + C$ **69.** $\frac{1}{3}e^{x^3} + C$

71. $\frac{10^x e^x}{\ln 10 + 1} + C$ **73.** $\frac{1}{2(e^{-x} + 2)^2} + C$ **75.** $\frac{1}{2}\ln 2$

77. $\tan^{-1}(\ln t) + C$ **79.** $\frac{1}{2}$ **81.** $\frac{1}{6}\tan^{-1}\left(\frac{2x}{3}\right) + C$

83. $\sec^{-1} 12 - \sec^{-1} 4$ **85.** $\frac{\pi}{12}$ **87.** $\frac{1}{2}\sin^{-1}(x^2) + C$

89. $\frac{1}{\sqrt{2}}\tan^{-1}(4\sqrt{2})$ **91.** $\frac{\pi^4}{1024}$ **93.** $\int_{-2}^{6} f(x)\,dx$

95. Local minimum at $x = 0$, no local maxima, inflection points at $x = \pm 1$

97.

x	0	1	2	3	4	5	6	7	8	9	10
$A(x)$	0	−6.8	−10	−10	−5	1.2	1.2	−3.6	−7.2	−5.4	−1.4
$A'(x)$		−5.2	−1.8	2.7	5.8	3.1	−2.4	−4.2	−0.9	2.9	

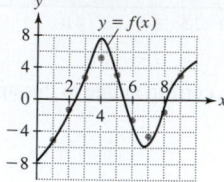

99. Daily consumption: 9.312 million gal; from 6 PM to midnight: 1.68 million gal

101. \$208,245 **103.** 0

107. The function $f(x) = 2^x$ is increasing, so $1 \le x \le 2$ implies that $2 = 2^1 \le 2^x \le 2^2 = 4$. Consequently,

$$2 = \int_1^2 2\,dx \le \int_1^2 2^x\,dx \le \int_1^2 4\,dx = 4$$

On the other hand, the function $f(x) = 3^{-x}$ is decreasing, so $1 \le x \le 2$ implies that

$$\frac{1}{9} = 3^{-2} \le 3^{-x} \le 3^{-1} = \frac{1}{3}$$

It then follows that

$$\frac{1}{9} = \int_1^2 \frac{1}{9}\,dx \le \int_1^2 3^{-x}\,dx \le \int_1^2 \frac{1}{3}\,dx = \frac{1}{3}$$

109. $\frac{4}{3} \le \int_0^1 f(x)\,dx \le \frac{5}{3}$ **111.** $-\frac{1}{1+\pi}$

113. $\sin^3 x \cos x$ **115.** -2

117. Consider the figure below, which displays a portion of the graph of a linear function.

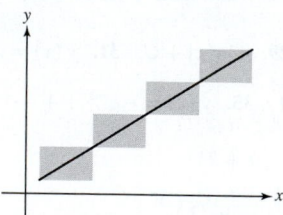

The shaded rectangles represent the differences between the right-endpoint approximation R_N and the left-endpoint approximation L_N. Because the graph of $y = f(x)$ is a line, the lower portion of each shaded rectangle is exactly the same size as the upper portion. Therefore, if we average L_N and R_N, the error in the two approximations will exactly cancel, leaving

$$\frac{1}{2}(R_N + L_N) = \int_a^b f(x)\,dx$$

119. Let

$$F(x) = x\sqrt{x^2 - 1} - 2\int_1^x \sqrt{t^2 - 1}\,dt$$

Then

$$\frac{dF}{dx} = \sqrt{x^2 - 1} + \frac{x^2}{\sqrt{x^2 - 1}} - 2\sqrt{x^2 - 1}$$

$$= \frac{x^2}{\sqrt{x^2 - 1}} - \sqrt{x^2 - 1} = \frac{1}{\sqrt{x^2 - 1}}$$

Also, $\frac{d}{dx}(\cosh^{-1} x) = \frac{1}{\sqrt{x^2 - 1}}$; therefore, $y = F(x)$ and $y = \cosh^{-1} x$ have the same derivative. We conclude that $y = F(x)$ and $y = \cosh^{-1} x$ differ by a constant:

$$F(x) = \cosh^{-1} x + C$$

Now, let $x = 1$. Because $F(1) = 0$ and $\cosh^{-1} 1 = 0$, it follows that $C = 0$. Therefore,

$$F(x) = \cosh^{-1} x$$

Chapter 6

Section 6.1 Preliminary Questions

1. Area of the region between the graphs of $y = f(x)$ and $y = g(x)$, bounded on the left by the vertical line $x = a$ and on the right by the vertical line $x = b$

2. Yes **3.** $\int_0^3 (f(x) - g(x))\,dx + \int_3^5 (g(x) - f(x))\,dx$

4. Negative

5. The area of the region bounded by the graphs of the functions f and g, and the vertical lines $x = a$ and $x = b$

6.

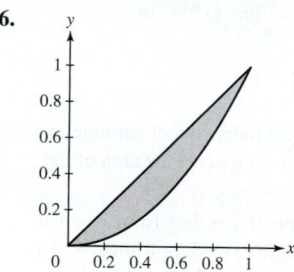

Section 6.1 Exercises

1. 102 **3.** $\frac{32}{3}$

5. $\sqrt{2} - 1$

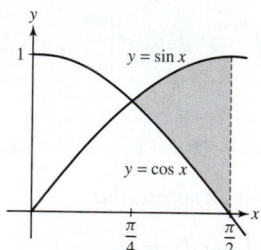

7. $\frac{343}{3}$

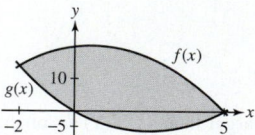

9. $\frac{1}{2}e^2 - e + \frac{1}{2}$

11. $\pi - 2$

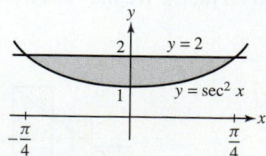

13. Horizontally simple **15.** Neither **17.** $\frac{160}{3}$

19. $\frac{12\sqrt{3} - 12 + (\sqrt{3} - 2)\pi}{24}$ **21.** $2 - \frac{\pi}{2}$ **23.** $\frac{1331}{6}$ **25.** 256 **27.** $\frac{32}{3}$

29. $\frac{64}{3}$

31. Area $= \frac{1000}{3}$

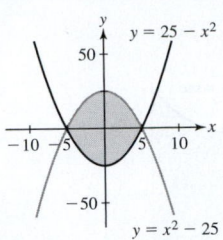

33. 2

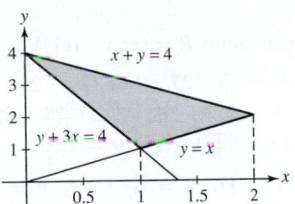

35. Area $= 125$

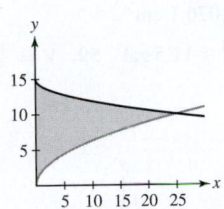

37. $\frac{1}{2}$

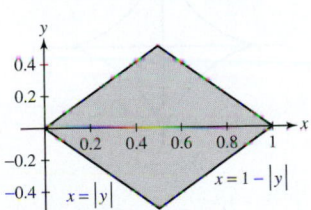

39. $\frac{1225}{8}$

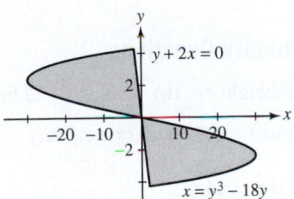

41. $\frac{32}{3}$

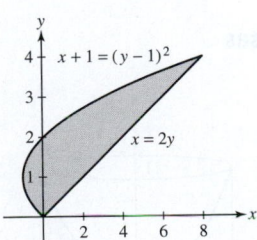

43. $\frac{3\sqrt{3}}{4}$

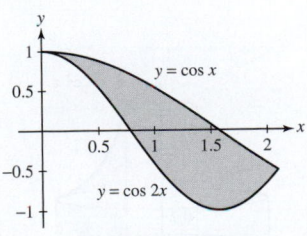

45. $\frac{2-\sqrt{2}}{2}$

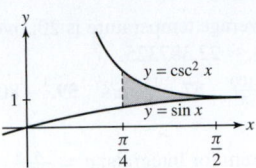

47. $4 \ln 2 - 2 \approx 0.77259$

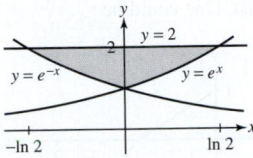

49. Area $= \frac{1}{2}$

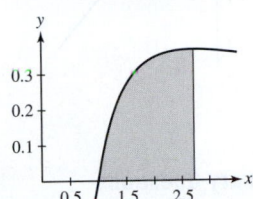

51. ≈ 0.7567130951

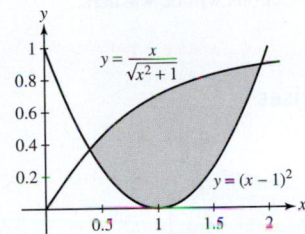

53. (a) How far Athlete 1 is ahead of Athlete 2 (b) ii
(c) $t = 5, 10, 15$, Athlete 1; $t = 20$, approximately tied;
$t = 25, 30$, Athlete 2
55. $\frac{8}{3}c^{3/2}$; $c = \frac{9^{1/3}}{4} \approx 0.520021$
57. $\int_{-\sqrt{(-1+\sqrt{5})/2}}^{\sqrt{(-1+\sqrt{5})/2}} \left[(1+x^2)^{-1} - x^2\right] dx$
59. 0.8009772242
61. Left endpoint approximation $= 101{,}750$ ft^2; right endpoint
approximation $= 96{,}250$ ft^2
63. (b) $\frac{1}{3}$ (c) 0 (d) 1
65. $m = 1 - \left(\frac{1}{2}\right)^{1/3} \approx 0.206299$

Section 6.2 Preliminary Questions

1. 3 **2.** 15
3. Flow rate is the volume of fluid that passes through a cross-sectional
area at a given point per unit time.
4. The fluid velocity depended only on the radial distance from the
center of the tube.
5. 15

Section 6.2 Exercises

1. (a) $\frac{4}{25}(20 - y)^2$ (b) $\frac{1280}{3}$
3. $\frac{\pi R^2 h}{3}$ **5.** $\pi\left(Rh^2 - \frac{h^3}{3}\right)$ **7.** $\frac{1}{6}abc$ **9.** $\frac{8}{3}$ **11.** 36 **13.** 18
15. $\frac{\pi}{3}$ **17.** 96π
21. (a) $2\sqrt{r^2 - y^2}$ (b) $4(r^2 - y^2)$ (c) $\frac{16}{3}r^3$
23. 160π **25.** $12 \ln \frac{3}{2} \approx 4.87$ kg **27.** 0.36 g
29. $P \approx 4423.59$ thousand **31.** $L_6 = 233.86$, $R_6 = 290.56$,
Average $= 262.21$ **33.** $P \approx 61$ deer **35.** $Q = 128\pi$ cm^3/s
37. $Q = \frac{8\pi}{3}$ cm^3/s **39.** 16 **41.** $\frac{3}{\pi}$ **43.** $\frac{1}{10}$ **45.** -4 **47.** $\frac{2n}{\pi}$
49. $\frac{a^n}{n+1}$

51. Over $[0,24]$, the average temperature is 20; over $[2,6]$, the average temperature is $20 + \frac{15}{2\pi} \approx 22.387325$.

53. $\approx 79.65°F$ **55.** $\frac{100}{\pi}$ **57.** $\frac{17}{2}$ m/s   **59.** -80m/s^2; 159.033 m/s

61. $\frac{3}{5^{1/4}} \approx 2.006221$

63. Mean Value Theorem for Integrals; $c = \frac{A}{\sqrt[3]{4}}$

65. Over $[0, 1]$, $f(x)$; over $[1, 2]$, g

67. Many solutions exist. One could be

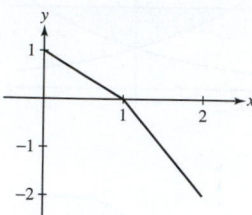

69. $v_0/2$

Section 6.3 Preliminary Questions

1. (a), (c)   **2.** True

3. False. The cross sections will be washers.

4. (b)

Section 6.3 Exercises

1. (a)

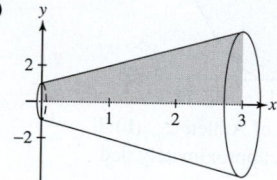

(b) Disk with radius $x + 1$

(c) $V = 21\pi$

3. (a)

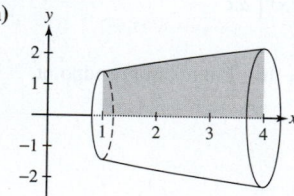

(b) Disk with radius $\sqrt{x + 1}$

(c) $V = \frac{21\pi}{2}$

5. $V = \frac{81\pi}{10}$ **7.** $V = \frac{24,573\pi}{13}$ **9.** $V = \pi$

11. $V = \frac{\pi}{2}\left(e^2 - 1\right)$ **13.** $\frac{3\pi}{\sqrt{2}}$ **15.** (ii)

17. (a)

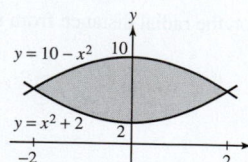

(b) A washer with outer radius $R = 10 - x^2$ and inner radius $r = x^2 + 2$

(c) $V = 256\pi$

19. (a)

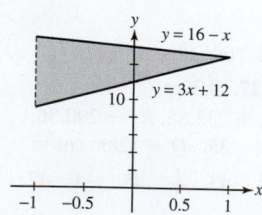

(b) A washer with outer radius $R = 16 - x$ and inner radius $r = 3x + 12$

(c) $V = \frac{656\pi}{3}$

21. (a)

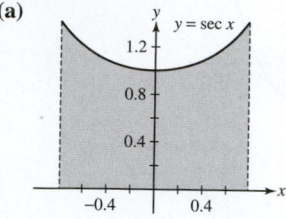

(b) A circular disk with radius $R = \sec x$ (c) $V = 2\pi$

23. $V = \frac{15\pi}{2}$ **25.** $V = \frac{3\pi}{10}$ **27.** $V = 32\pi$ **29.** $V = \frac{704\pi}{15}$

31. $V = \frac{128\pi}{5}$ **33.** $V = 40\pi$ **35.** $V = \frac{376\pi}{15}$ **37.** $V = \frac{824\pi}{15}$

39. $V = \frac{32\pi}{3}$ **41.** $V = \frac{1872\pi}{5}$ **43.** $V = \frac{1400\pi}{3}$

45. $V = \pi\left(\frac{7\pi}{9} - \sqrt{3}\right)$ **47.** $V = \frac{96\pi}{5}$ **49.** $V = \frac{16\pi}{35}$

51. $V = \frac{1184\pi}{15}$ **53.** $V = 7\pi\,(1 - \ln 2)$

55. $V \approx 12,120\pi \approx 38,076.1$ cm^3

57. Volume ≈ 2650 in.$^3 \approx 11.5$ gal   **59.** $V = \frac{1}{3}\pi r^2 h$

61. $V = \frac{32\pi}{105}$

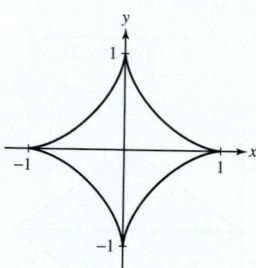

63. $V = 4\pi\sqrt{3}$ **65.** $V = \frac{4}{3}\pi a^2 b$

Section 6.4 Preliminary Questions

1. (a) Radius h and height r (b) Radius r and height h

2. (a) With respect to x (b) With respect to y

3. $V = 2\pi \displaystyle\int_0^8 y \cdot 1 \, dy = 64\pi$

Section 6.4 Exercises

1. $V = \frac{2}{5}\pi$

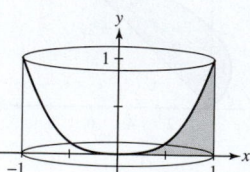

3. $V = 4\pi$

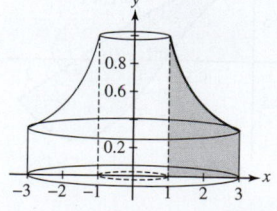

5. $V = 18\pi \left(2\sqrt{2} - 1\right)$

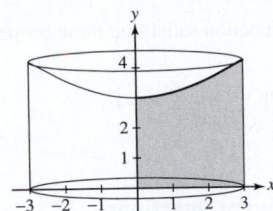

7. $V = \frac{32\pi}{3}$ **9.** $V = 16\pi$ **11.** $V = \frac{32\pi}{5}$

13. $\pi(e^4 - e) \approx 162.99$

15. The point of intersection is $x = 1.376769504$; $V = 1.321975576$.

17. $V = \frac{3\pi}{5}$

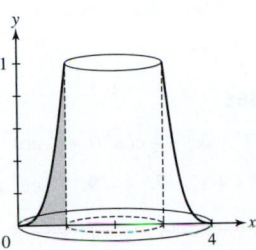

19. $V = \frac{280\pi}{81}$

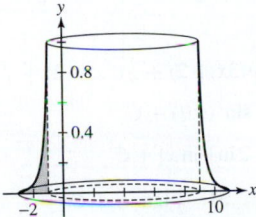

21. $V = \frac{1}{3}\pi a^3 + \pi a^2$

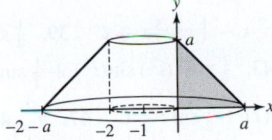

23. $V = \frac{\pi}{3}$

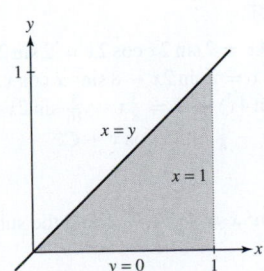

25. $V = \frac{128\pi}{3}$

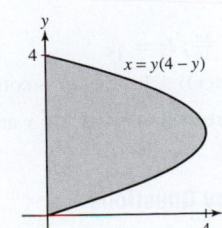

27. $V = \frac{256\pi}{15}$

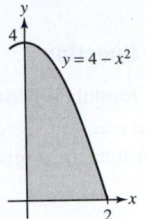

29. (b)

31. (a) $V = \frac{576\pi}{7}$ **(b)** $V = \frac{96\pi}{5}$

33. (a) $\overline{AB}$ generates a disk with radius $R = h(y)$; $\overline{CB}$ generates a shell with radius x and height $f(x)$.

(b) Shell, $V = 2\pi \int_0^2 x f(x)\, dx$; disk, $V = \pi \int_0^{1.3} (h(y))^2\, dy$

35. $V = \frac{602\pi}{5}$ **37.** $V = 8\pi$ **39.** $V = \frac{40\pi}{3}$ **41.** $V = \frac{1024\pi}{15}$

43. $V = 16\pi$ **45.** $V = \frac{32\pi}{3}$ **47.** $V = \frac{776\pi}{15}$ **49.** $V = \frac{625\pi}{6}$

51. $\frac{3\pi}{10}$ **53.** $V = \frac{121\pi}{525}$ **55.** $V = \frac{563\pi}{30}$

57. $V = \frac{4}{3}\pi r^3$ **59.** $V = 2\pi^2 ab^2$

61. $\pi(1 + 3e^4) \approx 517.72$

63. $x^2 + y^2 = 1$ over $[0, 1]$; $y = 1 - x$ over $[0, 1]$

Section 6.5 Preliminary Questions

1. Because the required force is not constant through the stretching process

2. The weight of the water and the distance it travels is not a constant. However, the weight of the tank and the distance it travels is a constant.

3. $\frac{1}{2}kx^2$

4. When the force applied is in the opposite direction to the motion

Section 6.5 Exercises

1. $W = 627.2$ J **3.** $W = 5.76$ J **5.** $W = 8$ J

7. (a) Progressive **(b)** Degressive

9. Degressive; work $= 960 - 240\sqrt{2} \approx 621$ N-cm

11. $W = 105,840$ J

13. $W = \frac{56,448\pi}{5}$ J $\approx 3.547 \times 10^4$ J **15.** $W \approx 1.842 \times 10^{12}$ J

17. $W = 3.92 \times 10^6$ J **19.** $W \approx 1.18 \times 10^8$ J **21.** $W = 9800\pi \ell r^3$ J

23. $W = 2.94 \times 10^6$ J **25.** $W \approx 3.79 \times 10^6$ J **27.** $W = 3920$ J

29. $W = 529.2$ J

31. $W = 1470$ J **33.** $W = 374.85$ J

37. $W \approx 5.16 \times 10^9$ J **41.** $\sqrt{2GM_e \left(\frac{1}{R_e} - \frac{1}{r+R_e}\right)}$ m/s

43. $v_{\text{esc}} = \sqrt{\frac{2GM_e}{R_e}}$ m/s

Chapter 6 Review

1. $\frac{32}{3}$ **3.** $\frac{1}{2}$ **5.** 24 **7.** $\frac{1}{2}$ **9.** $3\sqrt{2} - 1$ **11.** $e - \frac{3}{2}$

13. Intersection points $x = 0$, $x = 0.7145563847$; area $= 0.08235024596$

15. $V = 4\pi$ **17.** 2.7552 kg **19.** $\frac{9}{4}$ **21.** $\frac{1}{2}\sinh 1$ **23.** $\frac{3\pi}{4}$ **25.** 27

27. $\frac{2\pi m^5}{15}$ **29.** $V = \frac{162\pi}{5}$ **31.** $V = 64\pi$ **33.** $V = 8\pi$

35. $V = \frac{56\pi}{15}$ **37.** $V = \frac{128\pi}{15}$ **39.** $V = 4\pi \left(1 - \frac{1}{\sqrt{e}}\right)$

41. $V = 2\pi \left(c + \frac{c^3}{3}\right)$ **43.** $V = c\pi$

45. (a) $\displaystyle\int_0^1 \left(\sqrt{1 - (x-1)^2} - \left(1 - \sqrt{1 - x^2}\right)\right) dx$

(b) $\pi \displaystyle\int_0^1 \left[(1 - (x-1)^2) - (1 - \sqrt{1 - x^2})^2\right] dx$

47. $V = \pi \displaystyle\int_0^{1/a} \left(a\sqrt{x} - ax^2\right)^2 dx = \frac{\pi}{6}$ **49.** $W = 1.6$ J

51. $W = 1.93 \times 10^{10}$ J **53.** $9800\pi \left(h^3 + 2h^2 - \frac{1}{4}h^4\right)$ J

Chapter 7

Section 7.1 Preliminary Questions

1. The Integration by Parts formula is derived from the Product Rule.

3. Transforming $v' = x$ into $v = \frac{1}{2}x^2$ increases the power of x and makes the new integral harder than the original.

Section 7.1 Exercises

1. $-x\cos x + \sin x + C$ **3.** $e^x(2x+7) + C$

5. $\frac{x^4}{16}(4\ln x - 1) + C$ **7.** $-e^{-x}(4x+1) + C$

9. $\frac{1}{25}(5x-1)e^{5x+2} + C$ **11.** $\frac{1}{2}x\sin 2x + \frac{1}{4}\cos 2x + C$

13. $-x^2\cos x + 2x\sin x + 2\cos x + C$

15. $-\frac{1}{2}e^{-x}(\sin x + \cos x) + C$

17. $-\frac{1}{26}e^{-5x}(\cos(x) + 5\sin(x)) + C$

19. $\frac{1}{4}x^2(2\ln x - 1) + C$ **21.** $\frac{x^3}{3}\left(\ln x - \frac{1}{3}\right) + C$

23. $x\left[(\ln x)^2 - 2\ln x + 2\right] + C$ **25.** $x\cos^{-1}x - \sqrt{1-x^2} + C$

27. $x\sec^{-1}x - \ln|x + \sqrt{x^2-1}| + C$

29. $\dfrac{3^x(\sin x + \ln 3\cos x)}{1 + (\ln 3)^2} + C$

31. $(x^2+2)\sinh x - 2x\cosh x + C$

33. $x\tanh^{-1}4x + \frac{1}{8}\ln|1 - 16x^2| + C$ **35.** $2e^{\sqrt{x}}(\sqrt{x} - 1) + C$

37. With $u = x$, $dv = \tan x\,dx$, Integration by Parts gives $\int x\tan x\,dx = -x\ln|\cos x| + \int \ln|\cos x|dx$. This does not simplify the integral for us because it leaves us with determining $\int \ln|\cos x|dx$, an integral that we have no simple result for.

With $u = \tan x$, $dv = x\,dx$, Integration by Parts gives $\int x\tan x\,dx = \frac{1}{2}x^2\tan x - \int \frac{1}{2}x^2\sec^2 x\,dx$. This leaves us with determining $\int x^2\sec^2 x\,dx$, an integral that is more complicated than what we started with.

39. $\frac{1}{4}x\sin 4x + \frac{1}{16}\cos 4x + C$ **41.** $\frac{2}{3}(x+1)^{3/2} - 2(x+1)^{1/2} + C$

43. $\sin x\ln(\sin x) - \sin x + C$

45. $2xe^{\sqrt{x}} - 4\sqrt{x}e^{\sqrt{x}} + 4e^{\sqrt{x}} + C$

47. $\frac{1}{4}(\ln x)^2[2\ln(\ln x) - 1] + C$ **49.** $\frac{1}{16}(11e^{12} + 1)$

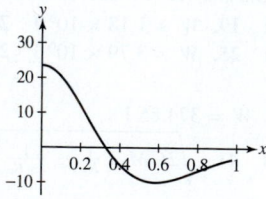

51. $2\ln 2 - \frac{3}{4}$ **53.** $1 - \frac{2}{e}$ **55.** $\frac{3\ln 3 - 2}{\ln^2 3}$

57. $\frac{e^\pi + 1}{2}$ **59.** ≈ 13.4 cm

61. $-\frac{1}{5}\sin^4 x\cos x - \frac{4}{15}\sin^2 x\cos x - \frac{8}{15}\cos x + C$

63. $\frac{1}{3}\cos^2 x\sin x + \frac{2}{3}\sin x + C$

65. $e^x(x^3 - 3x^2 + 6x - 6) + C$

67. $\int x^n e^{-x}\,dx = -x^n e^{-x} + n\int x^{n-1}e^{-x}\,dx$

69. $\pi(\pi^2 - 4)$ **71.** $6\pi^2/59.22$

73. Use Integration by Parts, with $u = \ln x$ and $dv = \sqrt{x}\,dx$.

75. Use substitution, followed by algebraic manipulation, with $u = 4 - x^2$ and $du = -2x\,dx$.

77. Use substitution with $u = x^2 + 4x + 3$, $\frac{du}{2} = (x+2)\,dx$.

79. Use Integration by Parts, with $u = x$ and $dv = \sin(3x+4)\,dx$.

81. $x(\sin^{-1}x)^2 + 2\sqrt{1-x^2}\sin^{-1}x - 2x + C$

83. $\frac{1}{4}x^4\sin(x^4) + \frac{1}{4}\cos(x^4) + C$ **85.** Area ≈ 6.12 **87.** \$42,995.10

89. For $k = 2$: $x(\ln x)^2 - 2x\ln x + 2x + C$; for $k = 3$: $x(\ln x)^3 - 3x(\ln x)^2 + 6x\ln x - 6x + C$

91. Use Integration by Parts with $u = x$ and $v' = b^x$.

93. (b) $V(x) = \frac{1}{2}x^2 + \frac{1}{2}$ is simpler and yields $\frac{1}{2}(x^2\tan^{-1}x - x + \tan^{-1}x) + C$

95. An example of a function satisfying these properties for some λ is $f(x) = \sin\pi x$.

97. (a) $I_n = \frac{1}{2}x^{n-1}\sin(x^2) - \frac{n-1}{2}J_{n-2}$;
    (c) $\frac{1}{2}x^2\sin(x^2) + \frac{1}{2}\cos(x^2) + C$

Section 7.2 Preliminary Questions

1. Rewrite $\sin^5 x = \sin x\sin^4 x = \sin x(1 - \cos^2 x)^2$ and then substitute $u = \cos x$.

3. No, a reduction formula is not needed because the sine function is raised to an odd power.

5. The second integral requires the use of reduction formulas and therefore more work.

Section 7.2 Exercises

1. $\sin x - \frac{1}{3}\sin^3 x + C$ **3.** $-\frac{1}{3}\cos^3\theta + \frac{1}{5}\cos^5\theta + C$

5. $-\frac{1}{4}\cos^4 t + \frac{1}{6}\cos^6 t + C$ **7.** $\frac{2}{3}$ **9.** $\frac{1}{3}\sec^3 x - \sec x + C$

11. $\frac{1}{5}\tan x\sec^4 x - \frac{1}{15}\tan(x)\sec^2 x - \frac{2}{15}\tan x + C$

13. $-\frac{1}{2}\cot^2 x + \ln|\csc x| + C$ **15.** $-\frac{1}{6}\cot^6 x + C$ **17.** $1 - \frac{\pi}{4}$

19. $\frac{5}{16}x - \frac{5}{16}\sin x\cos x - \frac{5}{24}\sin^3 x\cos x - \frac{1}{6}\sin^5 x\cos x + C$

21. $-\frac{1}{6}\cos^6 x + C$

23. $\frac{1}{12}\cos^3(3x+2)\sin(3x+2) + \frac{1}{8}(3x+2) + \frac{1}{16}\sin(6x+4) + C$

25. $\frac{1}{5\pi}\sin^5(\pi\theta) - \frac{1}{7\pi}\sin^7(\pi\theta) + C$

27. $\frac{1}{2}\sin^2 x - \frac{1}{2\sin^2 x} - 2\ln|\sin x| + C$

29. $\frac{1}{2}\cot(3 - 2x) + C$ **31.** $\frac{1}{2}\tan^2 x + C$

33. $\frac{1}{8}\sec^8 x - \frac{1}{3}\sec^6 x + \frac{1}{4}\sec^4 x + C$

35. $\frac{1}{9}\tan^9 x + \frac{1}{7}\tan^7 x + C$

37. $-\frac{1}{9}\csc^9 x + \frac{2}{7}\csc^7 x - \frac{1}{5}\csc^5 x + C$ **39.** $\frac{1}{4}\sin^2 2x + C$

41. $-\frac{2}{5}\cos^5 x + C$ **43.** $\frac{1}{6}\cos^2(t^2)\sin(t^2) + \frac{1}{3}\sin(t^2) + C$

45. $\frac{1}{2}\cos(\sin t)\sin(\sin t) + \frac{1}{2}\sin t + C$ **47.** π **49.** $\frac{8}{15}$

51. $\ln\left(\sqrt{2}+1\right)$ **53.** $\ln 2$ **55.** $\frac{8}{3}$ **57.** $-\frac{6}{7}$ **59.** $\frac{1}{24}$

61. Area $= \dfrac{2}{(n+1)(n+3)}$

63. First, observe $\sin 4x = 2\sin 2x\cos 2x = 2\sin 2x(1 - 2\sin^2 x) = 2\sin 2x - 4\sin 2x\sin^2 x = 2\sin 2x - 8\sin^3 x\cos x$. Then $\frac{1}{32}(12x - 8\sin 2x + \sin 4x) + C = \frac{3}{8}x - \frac{3}{16}\sin 2x - \frac{1}{4}\sin^3 x\cos x + C = \frac{3}{8}x - \frac{3}{8}\sin x\cos x - \frac{1}{4}\sin^3 x\cos x + C$.

65. $\frac{\pi^2}{2}$

67. Use the identity $\tan^2 x = \sec^2 x - 1$ and the substitution $u = \tan x$, $du = \sec^2 x\,dx$.

69. (a) $I_0 = \int_0^{\pi/2}\sin^0 x\,dx = \frac{\pi}{2}$; $I_1 = \int_0^{\pi/2}\sin x\,dx = 1$

(b) $\frac{m-1}{m}\int_0^{\pi/2}\sin^{m-2}x\,dx$

(c) $I_2 = \frac{\pi}{4}$; $I_3 = \frac{2}{3}$; $I_4 = \frac{3\pi}{16}$; $I_5 = \frac{8}{15}$

71. $\cos(x) - \cos(x)\ln(\sin(x)) + \ln|\csc(x) - \cot(x)| + C$

75. Use Integration by Parts with $u = \sec^{m-2}x$ and $v' = \sec^2 x$.

Section 7.3 Preliminary Questions

1. (a) $x = 3\sin\theta$ (b) $x = 4\sec\theta$ (c) $x = 4\tan\theta$
(d) $x = \sqrt{5}\sec\theta$ **3.** $2x\sqrt{1-x^2}$

Section 7.3 Exercises

1. (a) $\theta + C$ (b) $\sin^{-1}\left(\frac{x}{3}\right) + C$

3. (a) $\int \frac{dx}{\sqrt{4x^2+9}} = \frac{1}{2}\int \sec\theta\, d\theta$

(b) $\frac{1}{2}\ln|\sec\theta + \tan\theta| + C$ (c) $\frac{1}{2}\ln\left|\sqrt{4x^2+9}+2x\right| + C$

5. $\frac{8}{\sqrt{5}}\arcsin(\frac{x\sqrt{5}}{4}) + \frac{1}{2}x\sqrt{16-5x^2} + C$ **7.** $\frac{1}{3}\sec^{-1}\left(\frac{x}{3}\right) + C$

9. $\frac{-x}{4\sqrt{x^2-4}} + C$ **11.** $\sqrt{x^2-4} + C$

13. (a) $-\sqrt{1-x^2}$ (b) $\frac{1}{8}(\arcsin x - x\sqrt{1-x^2}(1-2x^2))$

(c) $-\frac{1}{3}(1-x^2)^{\frac{3}{2}} + \frac{1}{5}(1-x^2)^{\frac{5}{2}}$

(d) $\sqrt{1-x^2}(-\frac{x^3}{4} - \frac{3x}{8}) + \frac{3}{8}\arcsin(x)$

15. $\frac{9}{2}\sin^{-1}\left(\frac{x}{3}\right) - \frac{1}{2}x\sqrt{9-x^2} + C$

17. $\frac{1}{4}\ln\left|\frac{\sqrt{x^2+16}-4}{x}\right| + C$ **19.** $\ln\left|x+\sqrt{x^2-9}\right| + C$

21. $-\frac{\sqrt{5-y^2}}{5y} + C$ **23.** $\frac{1}{5}\ln\left|5x+\sqrt{25x^2+2}\right| + C$

25. $\frac{1}{16}\sec^{-1}\left(\frac{z}{2}\right) + \frac{\sqrt{z^2-4}}{8z^2} + C$

27. $\frac{1}{12}x\sqrt{6x^2-49} + \frac{49\ln\left(x\sqrt{6}+\sqrt{6x^2-49}\right)}{12\sqrt{6}} + C$

29. $\frac{1}{54}\tan^{-1}\left(\frac{1}{3}\right) + \frac{1}{180}$ **31.** $-\frac{x}{\sqrt{x^2-1}} + \ln\left|x+\sqrt{x^2+1}\right| + C$

33. ≈ 79.06

35. (a) With $x = a\tan\theta$ and $dx = a\sec^2\theta\, d\theta$, we have $\int \frac{dx}{x^2+a^2} =$

$\int \frac{a\sec^2\theta\, d\theta}{a^2\tan^2\theta + a^2} = \int \frac{a\sec^2\theta\, d\theta}{a^2\sec^2\theta} = \frac{1}{a}\int d\theta = \frac{1}{a}\theta + C = \frac{1}{a}\tan^{-1}\frac{x}{a} + C$

(b) $\frac{d}{dx}\left(\frac{1}{a}\tan^{-1}\frac{x}{a}\right) = \frac{1}{a}\left(\frac{1}{1+(\frac{x}{a})^2}\right)\frac{1}{a} = \frac{1}{a^2+x^2}$

37. (a) $x^2 - 4x + 8 = x^2 - 4x + 4 + 4 = (x-2)^2 + 4$

(b) $\ln\left|\sqrt{u^2+4}+u\right| + C$ (c) $\ln\left|\sqrt{(x-2)^2+4}+x-2\right| + C$

39. $\ln\left|\sqrt{x^2+4x+13}+x+2\right| + C$

41. $\frac{1}{\sqrt{6}}\ln\left|12x+1+2\sqrt{6}\sqrt{x+6x^2}\right| + C$

43. $\frac{1}{2}(x-2)\sqrt{x^2-4x+3} - \frac{1}{2}\ln\left|x-2+\sqrt{x^2-4x+3}\right| + C$

45. $x\sec^{-1}x - \ln\left|x+\sqrt{x^2-1}\right| + C$

47. $x(\ln(x^2+1)-2) + 2\tan^{-1}x + C$

49. $\frac{\pi}{4}$ **51.** $4\pi\left[\sqrt{3} - \ln\left|2+\sqrt{3}\right|\right]$ **53.** $\frac{1}{2}\ln\left(\frac{x-1}{x+1}\right) + C$

55. (a) $1.789 \times 10^6\ \frac{V}{m}$ (b) $3.526 \times 10^6\ \frac{V}{m}$

Section 7.4 Preliminary Questions

1. (a) $x = \sinh t$ (b) $x = 3\sinh t$ (c) $3x = \sinh t$

3. $\frac{1}{2}\ln\left|\frac{1+x}{1-x}\right|$

Section 7.4 Exercises

1. $\frac{1}{3}\sinh(3x) + C$ **3.** $x\cosh x - \sinh x + C$

5. $-\frac{1}{2}\tanh(1-2x) + C$ **7.** $\frac{\tanh^2 x}{2} + C$ **9.** $\ln\cosh x + C$

11. $\ln|\sinh x| + C$ **13.** $\frac{1}{16}\sinh(8x-18) - \frac{1}{2}x + C$

15. $\frac{1}{32}\sinh 4x - \frac{1}{8}x + C$ **17.** $\cosh^{-1}x + C$ **19.** $\sinh^{-1}\left(\frac{x}{2}\right) + C$

21. $\frac{1}{2}x\sqrt{x^2-1} - \frac{1}{2}\cosh^{-1}x + C$ **23.** $2\tanh^{-1}\left(\frac{1}{2}\right)$ **25.** $\sinh^{-1}1$

27. $\frac{1}{4}\left(\operatorname{csch}^{-1}\left(-\frac{1}{4}\right) - \operatorname{csch}^{-1}\left(-\frac{3}{4}\right)\right)$ **29.** $\cosh^{-1}x - \frac{\sqrt{x^2-1}}{x} + C$

31. Let $x = \sinh t$ for the first formula and $x = \cosh t$ for the second.

33. $\frac{1}{2}x\sqrt{x^2+16} + 8\ln\left|\frac{x}{4} + \sqrt{\left(\frac{x}{4}\right)^2 + 1}\right| + C$

35. Using Integration by Parts with $u = \cosh^{n-1}x$ and $v' = \cosh x$ to begin proof

37. $-\frac{1}{2}\left(\tanh^{-1}x\right)^2 + C$ **39.** $x\tanh^{-1}x + \frac{1}{2}\ln|1-x^2| + C$

41. (a) Area $= \cosh(5) - 1 \approx 73.21$

(b) Area $= 5\sinh(5) - \cosh(5) + 1 \approx 297.81$

(c) $\cosh(5) - 1 + 5\sinh(5) - \cosh(5) + 1 = 5\sinh 5$

(d) The area in (a) is the area of the shaded region under the graph of $y = \sinh x$. Areas under the graph of $y = \sinh^{-1}$ correspond to areas to the left of the graph of $y = \sinh x$ in the figure. In particular, the area in (b) is the area of the shaded region to the left of the graph. The sum of the areas in (a) and (b) is the sum of the two shaded areas in the figure. That is the area of the rectangle, which is $5\sinh 5$.

43. $u = \sqrt{\frac{\cosh x - 1}{\cosh x + 1}}$. From this, it follows that $\cosh x = \frac{1+u^2}{1-u^2}$, $\sinh x = \frac{2u}{1-u^2}$, and $dx = \frac{2du}{1-u^2}$.

45. $\int du = u + C = \tanh\frac{x}{2} + C$

47. Let $gd(y) = \tan^{-1}(\sinh y)$. Then

$$\frac{d}{dy}gd(y) = \frac{1}{1+\sinh^2 y}\cosh y = \frac{1}{\cosh y} = \operatorname{sech} y$$

where we have used the identity $1 + \sinh^2 y = \cosh^2 y$.

49. Let $x = gd(y) = \tan^{-1}(\sinh y)$. Solving for y yields $y = \sinh^{-1}(\tan x)$. Therefore, $gd^{-1}(y) = \sinh^{-1}(\tan y)$.

Section 7.5 Preliminary Questions

1. No, f cannot be a rational function because the integral of a rational function cannot contain a term with a noninteger exponent such as $\sqrt{x+1}$.

3. (a) Square is already completed; irreducible

(b) Square is already completed; factors as $(x - \sqrt{5})(x + \sqrt{5})$

(c) $x^2 + 4x + 6 = (x+2)^2 + 2$; irreducible

(d) $x^2 + 4x + 2 = (x+2)^2 - 2$; factors as $(x + 2 - \sqrt{2})(x + 2 + \sqrt{2})$

Section 7.5 Exercises

1. (a) $\frac{x^2+4x+12}{(x+2)(x^2+4)} = \frac{1}{x+2} + \frac{4}{x^2+4}$

(b) $\frac{2x^2+8x+24}{(x+2)^2(x^2+4)} = \frac{1}{x+2} + \frac{2}{(x+2)^2} + \frac{-x+2}{x^2+4}$

(c) $\frac{x^2-4x+8}{(x-1)^2(x-2)^2} = \frac{-8}{x-2} + \frac{4}{(x-2)^2} + \frac{8}{x-1} + \frac{5}{(x-1)^2}$

(d) $\frac{x^4-4x+8}{(x+2)(x^2+4)} = x - 2 + \frac{4}{x+2} - \frac{4x-4}{x^2+4}$

3. -2 **5.** $\frac{1}{9}(3x + 4\ln|3x-4|) + C$

7. $\frac{x^3}{3} + \ln|x+2| + C$ **9.** $-\frac{1}{2}\ln|x-2| + \frac{1}{2}\ln|x-4| + C$

11. $\ln x - \ln(3x+1) + C$ **13.** $x - 3\arctan\left(\frac{x}{3}\right) + C$

15. $2\ln|x+3| - \ln|x+5| - \frac{2}{3}\ln|3x-2| + C$

17. $3\ln|x-1| - 2\ln|x+1| - \frac{5}{x+1} + C$

19. $2\ln|x-1| - \frac{1}{x-1} - 2\ln|x-2| - \frac{1}{x-2} + C$

21. $\ln|x| - \ln|x+2| + \frac{2}{x+2} + \frac{2}{(x+2)^2} + C$

23. $\frac{1}{2\sqrt{6}}\ln\left|\sqrt{2}x - \sqrt{3}\right| - \frac{1}{2\sqrt{6}}\ln\left|\sqrt{2}x + \sqrt{3}\right| + C$

25. $\frac{1}{2(x+1)} + \frac{1}{4}\ln|x-1| - \frac{1}{4}\ln|x+1| + C$

27. $\frac{5}{2x+5} - \frac{5}{4(2x+5)^2} + \frac{1}{2}\ln|2x+5| + C$

29. $-\ln|x| + \ln|x-1| + \frac{1}{x-1} - \frac{1}{2(x-1)^2} + C$

31. $x + \ln|x| - 3\ln|x+1| + C$

33. $2\ln|x-1| + \frac{1}{2}\ln|x^2+1| - 3\tan^{-1}x + C$

35. $\frac{1}{25}\ln|x| - \frac{1}{50}\ln|x^2+25| + C$

37. $6x - 14\ln|x+3| + 2\ln|x-1| + C$

39. $-\frac{1}{5}\ln|x-1| - \frac{1}{x-1} + \frac{1}{10}\ln|x^2+9| - \frac{4}{15}\tan^{-1}\left(\frac{x}{3}\right) + C$

41. $\frac{1}{64}\ln|x| - \frac{1}{128}\ln|x^2+8| + \frac{1}{16(x^2+8)} + C$

43. $\frac{1}{6}\ln|x+2| - \frac{1}{12}\ln|x^2+4x+10| + C$

45. $\ln|x| - \frac{1}{2}\ln|x^2+2x+5| + \frac{15-5x}{8(x^2+2x+5)} - \frac{13}{16}\tan^{-1}\left(\frac{x+1}{2}\right) + C$

47. $\frac{1}{2}\arctan(x^2) + C$ **49.** $\frac{1}{2}\ln(e^x-1) - \frac{1}{2}\ln(e^x+1) + C$

51. $2\sqrt{x} + \ln|\sqrt{x}-1| - \ln|\sqrt{x}+1| + C$

53. $\ln|x^{1/4}-2| - \ln|x^{1/4}+2| + C$

55. $\ln\left|\frac{x}{\sqrt{x^2-1}} - \frac{1}{\sqrt{x^2-1}}\right| + C = \ln\left|\frac{x-1}{\sqrt{x^2-1}}\right| + C$

57. If $\theta = 2\tan^{-1}t$, then $d\theta = 2\,dt/(1+t^2)$. We also have $\cos(\frac{\theta}{2}) = 1/\sqrt{1+t^2}$ and $\sin(\frac{\theta}{2}) = t/\sqrt{1+t^2}$. To find $\cos\theta$, we use the double angle identity $\cos\theta = 1 - 2\sin^2(\frac{\theta}{2})$. This gives us $\cos\theta = \frac{1-t^2}{1+t^2}$. To find $\sin\theta$, we use the double angle identity $\sin\theta = 2\sin(\frac{\theta}{2})\cos(\frac{\theta}{2})$. This gives us $\sin\theta = \frac{2t}{1+t^2}$. It follows then that

$$\int \frac{d\theta}{\cos\theta + \frac{3}{4}\sin\theta} = -\frac{4}{5}\ln\left|2 - \tan\left(\frac{\theta}{2}\right)\right| +$$
$$\frac{4}{5}\ln\left|1 + 2\tan\left(\frac{\theta}{2}\right)\right| + C.$$

59. Partial fraction decomposition shows $\frac{1}{(x-a)(x-b)} = \frac{\frac{1}{a-b}}{x-a} + \frac{\frac{1}{b-a}}{x-b}$. This can be used to show $\int \frac{dx}{(x-a)(x-b)} = \frac{1}{a-b}\ln\left|\frac{x-a}{x-b}\right| + C$.

61. $\frac{2}{x-6} + \frac{1}{x+2}$

Section 7.6 Preliminary Questions

1. Integration by parts with $u = x$, $dv = \sin x\,dx$

2. Trigonometric substitution $x = \tan\theta$, $dx = \sec^2\theta\,d\theta$

3. Partial fractions

4. Substitute $u = \cos x$, $du = -\sin x\,dx$.

5. Integration by parts with $u = \ln x$, $dv = x\,dx$

6. Trigonometric substitution $x = \sin\theta$, $dx = \cos\theta\,d\theta$

7. **Trig Method.** Rewrite as $\int (1 - \sin^2 x)\cos^2 x \sin x\,dx$. Then use u-substitution with $u = \cos x$ and $du = -\sin x\,dx$.

8.
$$\int \frac{u^2\,du}{a+bu} = \frac{1}{2b^3}\left[(a+bu)^2 - 4a(a+bu) + 2a^2\ln|a+bu|\right] + C$$

9. Trig substitution with $x = \frac{5}{4}\tan\theta$ and $dx = \frac{5}{4}\sec^2\theta\,d\theta$

10. $\int \sec^3 u\,du = \frac{1}{2}\sec u\tan u + \frac{1}{2}\ln|\sec u + \tan u| + C$

11. Complete the square, followed by u-substitution with $u = x+1$ and $du = dx$. Then use trig substitution with $u = 2\tan\theta$ and $du = 2\sec^2\theta\,d\theta$.

Section 7.6 Exercises

1. Complete the square to get $\int \frac{x\,dx}{\sqrt{21-(x+3)^2}}$ and then use the substitution $x+3 = \sqrt{21}\sin u$ so that $dx = \sqrt{21}\cos u\,du$.

3. **Trig Method.** Rewrite as $\int \sin^3 x(1-\sin^2 x)\cos x\,dx$ followed by u-substitution with $u = \sin x$ and $du = \cos x\,dx$.

5. Trig substitution $x = 3\sin\theta$ and $dx = 3\cos\theta\,d\theta$

7. None   **9.** Partial fractions   **11.** $-\frac{\sqrt{4-x^2}}{4x} + C$

13. $\frac{x}{2} + \frac{1}{8}\sin 4x\cos 4x + C$ **15.** $x\sin x + \cos x + C$

17. $\frac{x}{18(9+x^2)} + \frac{1}{54}\tan^{-1}\left(\frac{x}{3}\right) + C$

19. $\sec x - \frac{2}{3}\sec^3 x + \frac{1}{5}\sec^5 x + C$

21. $x\ln|x^4+1| - 4x + 2\tan^{-1}x - \ln|x-1| + \ln|x+1| + C$

23. $\ln\left|\sqrt{x^2-1}+x\right| - \frac{x}{\sqrt{x^2-1}} + C$

25. $-\frac{x+6}{8(x^2+4x+8)} - \frac{1}{16}\tan^{-1}\left(\frac{x+2}{2}\right) + C$

27. $6\tan^{-1}\left(x^{1/6}\right) - 6x^{1/6} + 2\sqrt{x} - \frac{6}{5}x^{5/6} + \frac{6}{7}x^{7/6} + C$

29. $\ln(1+e^x) + C$ **31.** $\frac{1}{3}x^3\ln x - \frac{x^3}{9} + C$ **33.** $\frac{1}{2}\tan^{-1}\left(\frac{x+1}{2}\right) + C$

35. $\frac{2}{9}(1+x^3)^{3/2} + C$ **37.** $x - \ln(1+e^x) + C$

39. $2\sqrt{x} + x + \frac{2}{3}x^{3/2} + 2\ln|\sqrt{x}-1| + C$

41. $\frac{2}{5}(x+1)^{5/2} - \frac{4}{3}(x+1)^{3/2} + 2\sqrt{x+1} + C$

43. $x + \cos x - \frac{1}{4}\sin 2x - \frac{1}{3}\cos 3x + \frac{1}{8}\sin 4x + C$

45. $\frac{1}{2}\cos^2 x - \ln|\cos x| + C$

47. $x\ln(x^2+9) + 6\tan^{-1}\left(\frac{x}{3}\right) - 2x + C$

49. $-\frac{1}{3}\cos^3 x + \frac{2}{5}\cos^5 x - \frac{1}{7}\cos^7 x + C$

51. $\sin x - \sin^3 x + \frac{3}{5}\sin^5 x - \frac{1}{7}\sin^7 x + C$

53. $\frac{1}{2}x^2 - \frac{1}{4}\ln(x^2+1) + \frac{1}{4}\ln|x-1| + \frac{1}{4}\ln|x+1| + C$

55. $\frac{1}{4}\sin 2x + 9\tan x - \frac{11}{2}x + C$

57. $\frac{1}{3}(1+\ln x)^3 + C$ **59.** $\cosh^{-1}\left(\frac{x}{6}\right) + C$

Section 7.7 Preliminary Questions

1. **(a)** The integral converges.   **(b)** The integral diverges.
(c) The integral diverges.   **(d)** The integral converges.

3. Any value of b satisfying $|b| \geq 2$ will make this an improper integral.

5. Knowing that an integral is smaller than a divergent integral does not allow us to draw any conclusions using the comparison test.

Section 7.7 Exercises

1. **(a)** Improper. The function $y = x^{-1/3}$ is infinite at 0.
(b) Improper; infinite interval of integration
(c) Improper; infinite interval of integration
(d) Proper. The function $y = e^{-x}$ is continuous on the finite interval $[0, 1]$.
(e) Improper. The function $y = \sec x$ is infinite at $\frac{\pi}{2}$.
(f) Improper; infinite interval of integration
(g) Proper. The function $y = \sin x$ is continuous on the finite interval $[0, 1]$.
(h) Proper. The function $y = 1/\sqrt{3-x^2}$ is continuous on the finite interval $[0, 1]$.
(i) Improper; infinite interval of integration
(j) Improper. The function $y = \ln x$ is infinite at 0.

3.
$\int_1^\infty x^{-2/3}\,dx = \lim_{R\to\infty}\int_1^R x^{-2/3}\,dx = \lim_{R\to\infty} 3\left(R^{1/3}-1\right) = \infty$

5. The integral does not converge.

7. The integral converges; $I = 10{,}000e^{0.0004}$.

9. The integral does not converge.

11. The integral converges; $I = 4$.

13. The integral converges; $I = \frac{1}{8}$.

15. The integral converges; $I = 2$.

17. The integral converges; $I = 0$.

19. The integral converges; $I = \frac{1}{3e^{12}}$.

21. The integral converges; $I = \frac{1}{3}$.

23. The integral converges; $I = 2\sqrt{2}$.

25. The integral does not converge.

27. The integral converges; $I = \frac{1}{2}$.   **29.** Converges to $\frac{1}{9}$

31. The integral converges; $I = \frac{\pi}{2}$.

33. The integral does not converge.

35. The integral does not converge.

37. The integral converges; $I = -1$.

39. The integral does not converge.

41. **(a)** Partial fractions yield $\frac{dx}{(x-2)(x-3)} = \frac{dx}{x-3} - \frac{dx}{x-2}$. This yields $\int_4^R \frac{dx}{(x-2)(x-3)} = \ln\left|\frac{R-3}{R-2}\right| - \ln\frac{1}{2}$.

(b) $I = \lim_{R \to \infty} \left(\ln \left| \frac{R-3}{R-2} \right| - \ln \frac{1}{2} \right) = \ln 1 - \ln \frac{1}{2} = \ln 2$

43. The integral does not converge.

45. The integral does not converge.

47. The integral converges; $I = 0$.

49. $\int_{-1}^{1} \frac{dx}{x^{1/3}} = \int_{-1}^{0} \frac{dx}{x^{1/3}} + \int_{0}^{1} \frac{dx}{x^{1/3}} = 0$

51. The integral converges for $a < 0$.   **53.** $\int_{-\infty}^{\infty} \frac{dx}{1+x^2} = \pi$

55. $\frac{1}{x^3+4} \le \frac{1}{x^3}$. Therefore, by the Comparison Test, the integral converges.

57. For $x \ge 1$, $x^2 \ge x$, so $-x^2 \le -x$ and $e^{-x^2} \le e^{-x}$. Now $\int_{1}^{\infty} e^{-x} \, dx$ converges, so $\int_{1}^{\infty} e^{-x^2} \, dx$ converges by the Comparison Test. We conclude that our integral converges by writing it as a sum: $\int_{0}^{\infty} e^{-x^2} \, dx = \int_{0}^{1} e^{-x^2} \, dx + \int_{1}^{\infty} e^{-x^2} \, dx$.

59. Let $f(x) = \frac{1 - \sin x}{x^2}$. Since $f(x) \le \frac{2}{x^2}$ and $\int_{1}^{\infty} 2x^{-2} \, dx = 2$, it follows that $\int_{1}^{\infty} \frac{1 - \sin x}{x^2} \, dx$ converges by the Comparison Test.

61. The integral converges.   **63.** The integral does not converge.

65. The integral converges.   **67.** The integral does not converge.

69. The integral converges.   **71.** The integral converges.

73. The integral converges.   **75.** The integral does not converge.

77. $\int_{0}^{1} \frac{dx}{x^{1/2}(x+1)}$ and $\int_{1}^{\infty} \frac{dx}{x^{1/2}(x+1)}$ both converge; therefore, J converges.

79. $\$\frac{250}{0.07} = \$3571.43$ **81.** $\$2,000,000$

83. $W = \lim_{T \to \infty} CV^2 \left(\frac{1}{2} - e^{-T/RC} + \frac{1}{2} e^{-2T/RC} \right) = CV^2 \left(\frac{1}{2} - 0 + 0 \right) = \frac{1}{2} CV^2$

85. 2π

87. The integrand is infinite at the upper limit of integration, $x = \sqrt{2E/k}$, so the integral is improper; $T = \lim_{R \to \sqrt{2E/k}} T(R) = 4\sqrt{\frac{m}{k}} \sin^{-1}(1) = 2\pi\sqrt{\frac{m}{k}}$

89. $Lf(s) = \frac{-1}{s^2 + \alpha^2} \lim_{t \to \infty} e^{-st}(s \sin(\alpha t) + \alpha \cos(\alpha t)) - \alpha$

91. $\frac{s}{s^2 + \alpha^2}$ **93.** $J_n = \frac{n}{\alpha} J_{n-1} = \frac{n}{\alpha} \cdot \frac{(n-1)}{\alpha^n} = \frac{n}{\alpha^{n+1}}$

95. $E = \frac{8\pi h}{c^3} \int_{0}^{\infty} \frac{v^3}{e^{\alpha v} - 1} \, dv$. Because $\alpha > 0$ and $8\pi h/c^3$ is a constant, we know E is finite by Exercise 92.

97. Because $t > \ln t$ for $t > 2$, $F(x) = \int_{2}^{x} \frac{dt}{\ln t} > \int_{2}^{x} \frac{dt}{t} > \ln x$. Thus, $F(x) \to \infty$ as $x \to \infty$. Moreover, $\lim_{x \to \infty} G(x) = \lim_{x \to \infty} \frac{1}{1/x} = \lim_{x \to \infty} x = \infty$. Thus, $\lim_{x \to \infty} \frac{F(x)}{G(x)}$ is of the form ∞/∞, and L'Hôpital's Rule applies. Finally, $L = \lim_{x \to \infty} \frac{F(x)}{G(x)} = \lim_{x \to \infty} \frac{\frac{1}{\ln x}}{\frac{\ln x - 1}{(\ln x)^2}} = \lim_{x \to \infty} \frac{\ln x}{\ln x - 1} = 1$.

99. The integral is absolutely convergent. Use the Comparison Test with $\frac{1}{x^2}$.

Section 7.8 Preliminary Questions

1. $T_1 = 6$; $T_2 = 7$

3. The Trapezoidal Rule integrates linear functions exactly, so the error will be zero.

5. The two graphical interpretations of the Midpoint Rule are the sum of the areas of the midpoint rectangles and the sum of the areas of the tangential trapezoids.

Section 7.8 Exercises

1. $M_N = 8.625$; $T_N = 8.75$; $\int_{1}^{3} x^2 dx = \frac{26}{3} \approx 8.6667$

3. $M_N = 19.97$; $T_N = 20.81$; $\int_{0}^{3} x^3 dx = 20.25$

5. $T_6 \approx 1.4054$; $M_6 \approx 1.3769$ **7.** $T_6 \approx 1.1703$; $M_6 \approx 1.2063$

9. $T_5 \approx 0.3846$; $M_5 \approx 0.3871$ **11.** $T_5 \approx 0.7444$; $M_5 \approx 0.7481$

13. $S_N \approx 8.6667$; $\int_{1}^{3} x^2 dx = \frac{26}{3} \approx 8.6667$

15. $S_N \approx 0.95023$; $\int_{0}^{3} e^{-x} dx = 1 - e^{-3} \approx 0.95021$ **17.** $S_6 \approx 1.1090$

19. $S_4 \approx 0.7469$ **21.** $S_8 \approx 2.5450$ **23.** $S_{10} \approx 0.3466$

25. ≈ 2.4674 **27.** ≈ 1.8769 **29.** 214.75 in.^2

31. (a) Assuming the speed of the tsunami is a continuous function, at x miles from the shore, the speed is $\sqrt{15 f(x)}$. Covering an infinitesimally small distance, dx, the time T required for the tsunami to cover that distance becomes $\frac{dx}{\sqrt{15 f(x)}}$. It follows from this that $T = \int_{0}^{M} \frac{dx}{\sqrt{15 f(x)}}$.

(b) ≈ 3.347 h

33. $T_6 = 4.1111$

(a) Since x^3 is concave up on $[0, 2]$, T_6 is too large.

(b) We have $f'(x) = 3x^2$ and $f''(x) = 6x$. Since $|f''(x)| = |6x|$ is *increasing* on $[0, 2]$, its maximum value occurs at $x = 2$ and we may take $K_2 = |f''(2)| = 12$. Thus, error$(T_6) \le \frac{2}{9}$.

(c) Error$(T_6) \approx 0.1111 < \frac{2}{9}$

35. T_{10} will overestimate the integral. Error$(T_{10}) \le 0.045$.

37. M_{10} will overestimate the integral. Error$(M_{10}) \le 0.0113$

39. $N \ge 10^3$; error $\approx 3.333 \times 10^{-7}$

41. $N \ge 750$; error $\approx 2.805 \times 10^{-7}$

43. Error$(T_{10}) \le 0.0225$; error$(M_{10}) \le 0.01125$

45. $S_8 \approx 4.0467$; error $(S_8) \le 0.00833$; $N \ge 78$

47. Error$(S_{40}) \le 1.017 \times 10^{-4}$ **49.** $N = 306$ **51.** $N = 186$

53. (a) The maximum value of $|f^{(4)}(x)|$ on the interval $[0, 1]$ is 24.

(b) $N = 20$; $S_{20} \approx 0.785398$; $\left| 0.785398 - \frac{\pi}{4} \right| \approx 1.55 \times 10^{-10}$

55. (a) Notice $|f''(x)| = |2 \cos(x^2) - 4x^2 \sin(x^2)|$; the proof follows.

(b) When $K_2 = 6$, error$(M_N) \le \frac{1}{4N^2}$.   **(c)** $N \ge 16$

57. Error$(T_4) \approx 0.1039$; error$(T_8) \approx 0.0258$; error$(T_{16}) \approx 0.0064$; error$(T_{32}) \approx 0.0016$; error$(T_{64}) \approx 0.0004$. These are about twice as large as the error in M_N.

59. $S_2 = \frac{1}{4}$. This is the exact value of the integral.

61. $T_N = \frac{r(b^2 - a^2)}{2} + s(b - a) = \int_{a}^{b} f(x) \, dx$

63. (a) This result follows because the even-numbered interior endpoints overlap:

$$\sum_{i=0}^{(N-2)/2} S_2^{2j} = \frac{b-a}{6} \left[(y_0 + 4y_1 + y_2) + (y_2 + 4y_3 + y_4) + \cdots \right]$$

$$= \frac{b-a}{6} \left[y_0 + 4y_1 + 2y_2 + 4y_3 + 2y_4 + \cdots + 4y_{N-1} + y_N \right] = S_N$$

(b) If $f(x)$ is a quadratic polynomial, then by part (a), we have

$$S_N = S_2^0 + S_2^2 + \cdots + S_2^{N-2} = \int_{a}^{b} f(x) \, dx$$

65. Let $f(x) = ax^3 + bx^2 + cx + d$, with $a \ne 0$, be any cubic polynomial. Then $f^{(4)}(x) = 0$, so we can take $K_4 = 0$. This yields error$(S_N) \le \frac{0}{180N^4} = 0$. In other words, S_N is exact for all cubic polynomials for all N.

Chapter 7 Review

1. (a) (v)   **(b)** (iv)   **(c)** (iii)   **(d)** (i)   **(e)** (ii)

3. $\frac{\sin^9 \theta}{9} - \frac{\sin^{11} \theta}{11} + C$

5. $\frac{\tan \theta \sec^5 \theta}{6} - \frac{7 \tan \theta \sec^3 \theta}{24} + \frac{\tan \theta \sec \theta}{16} + \frac{1}{16} \ln |\sec \theta + \tan \theta| + C$

7. $-\frac{1}{\sqrt{x^2 - 1}} - \sec^{-1} x + C$ **9.** $2 \tan^{-1} \sqrt{x} + C$

11. $-\frac{\tan^{-1} x}{x} + \ln |x| - \frac{1}{2} \ln (1 + x^2) + C$

13. $\frac{5}{32} e^4 - \frac{1}{32} \approx 8.50$ **15.** $\frac{\cos^{12} 6\theta}{72} - \frac{\cos^{10} 6\theta}{60} + C$

17. $5\ln|x-1| + \ln|x+1| + C$ **19.** $\frac{\tan^3\theta}{3} + \tan\theta + C$ **21.** $\ln 8 - 1$

23. $-\frac{\cos^5\theta}{5} + \frac{2\cos^3\theta}{3} - \cos\theta + C$ **25.** $-\frac{1}{4}$

27. $\frac{2}{3}(\tan x)^{3/2} + C$ **29.** $\frac{\sin^6\theta}{6} - \frac{\sin^8\theta}{8} + C$

31. $-\frac{1}{3}u^3 + C = -\frac{1}{3}\cot^3 x + C$

33. $\int_{\pi/4}^{\pi/3} \cot^2(x)\csc^3(x)\,dx = \frac{1}{16}\ln 3 + \frac{1}{8}\ln\left(\sqrt{2}-1\right) + \frac{3}{8}\sqrt{2} - \frac{5}{56}$

35. $\frac{1}{49}\ln\left|\frac{t+4}{t-3}\right| - \frac{1}{7}\cdot\frac{1}{t-3} + C$ **37.** $\frac{1}{2}\sec^{-1}\frac{x}{2} + C$

39. $\displaystyle\int \frac{dx}{x^{3/2} + ax^{1/2}} = \begin{cases} \frac{2}{\sqrt{a}}\tan^{-1}\sqrt{\frac{x}{a}} + C & a > 0 \\ \frac{1}{\sqrt{-a}}\ln\left|\frac{\sqrt{x}-\sqrt{-a}}{\sqrt{x}+\sqrt{-a}}\right| + C & a < 0 \\ -\frac{2}{\sqrt{x}} + C & a = 0 \end{cases}$

41. $\ln|x+2| + 5/(x+2) - 3/(x+2)^2 + C$

43. $-\ln|x-2| - 2\frac{1}{x-2} + \frac{1}{2}\ln\left(x^2+4\right) + C$

45. $\frac{1}{3}\tan^{-1}\left(\frac{x+4}{3}\right) + C$

47. $\frac{1}{2}\ln|x+1| - \frac{5}{2}\ln|x-1| + 2\ln|x-2| + C$

49. $-\frac{(x^2+4)^{3/2}}{48x^3} + \frac{\sqrt{x^2+4}}{16x} + C$ **51.** $-\frac{1}{9}e^{4-3x}(3x+4) + C$

53. $\frac{1}{2}x^2\sin x^2 + \frac{1}{2}\cos x^2 + C$

55. $\frac{x^2}{2}\tanh^{-1}x + \frac{x}{2} - \frac{1}{4}\ln\left|\frac{1+x}{1-x}\right| + C$

57. $x\ln\left(x^2+9\right) - 2x + 6\tan^{-1}\left(\frac{x}{3}\right) + C$

59. $\frac{1}{2}\sinh 2$ **61.** $t + \frac{1}{4}\coth(1-4t) + C$ **63.** $\frac{\pi}{3}$

65. $\tan^{-1}(\tanh x) + C$

67. (a) $I_n = \displaystyle\int \frac{x^n}{x^2+1}\,dx = \int \frac{x^{n-2}(x^2+1-1)}{x^2+1}\,dx =$

$\displaystyle\int x^{n-2}\,dx - \int \frac{x^{n-2}}{x^2+1}\,dx = \frac{x^{n-1}}{n-1} - I_{n-2}$

(b) $I_0 = \tan^{-1}x + C$; $I_1 = \frac{1}{2}\ln\left(x^2+1\right) + C$; $I_2 = x - \tan^{-1}x + C$;

$I_3 = \frac{x^2}{2} - \frac{1}{2}\ln\left(x^2+1\right) + C$; $I_4 = \frac{x^3}{3} - x + \tan^{-1}x + C$;

$I_5 = \frac{x^4}{4} - \frac{x^2}{2} + \frac{1}{2}\ln\left(x^2+1\right) + C$

(c) Prove by induction; show it works for $n = 1$, then assume it works for $n = k$ and use that to show it works for $n = k+1$.

69. Integral converges; $I = \frac{1}{2}$ **71.** Integral converges; $I = 3\sqrt[3]{4}$

73. Integral converges; $I = \frac{\pi}{2}$ **75.** The integral does not converge.

77. The integral does not converge. **79.** The integral converges.

81. The integral converges. **83.** The integral converges.

85. π **89.** $\frac{2}{(s-\alpha)^3}$

91. (a) T_N is smaller and M_N is larger than the integral.

(b) M_N is smaller and T_N is larger than the integral.

(c) M_N is smaller and T_N is larger than the integral.

(d) T_N is smaller and M_N is larger than the integral.

93. $M_5 \approx 0.7481$ **95.** $M_4 \approx 0.7450$ **97.** $S_4 \approx 0.7469$

99. $V \approx T_9 \approx 20$ acre-ft $= 871{,}200\text{ ft}^3$

101. Error $\leq \frac{3}{128}$ **103.** $N \geq 29$

Chapter 8

Section 8.1 Preliminary Questions

1. No, $p(x) \geq 0$ fails. **3.** $p(x) = 4e^{-4x}$

Section 8.1 Exercises

1. $C = 2$; $P(0 \leq X \leq 1) = \frac{3}{4}$

3. $C = \frac{1}{\pi}$; $P\left(-\frac{1}{2} \leq X \leq \frac{1}{2}\right) = \frac{1}{3}$

5. $C = \frac{1}{2}$; $P(\frac{\pi}{4} \leq X \leq \frac{3\pi}{4}) = \frac{\sqrt{2}}{2} \approx 0.71$

7. $C = \frac{2}{\pi}$; $P\left(-\frac{1}{2} \leq X \leq 1\right) = \frac{2}{3} + \frac{\sqrt{3}}{4\pi}$

9. $\int_1^\infty 3x^{-4} = 1$; $\mu = \frac{3}{2}$

11. Integration confirms $\int_0^\infty \frac{1}{50}e^{-t/50} = 1$.

13. $e^{-\frac{3}{2}} \approx 0.2231$ **15.** $\frac{1}{2}\left(2 - 10e^{-2}\right) \approx 0.32$

17. $F(-\frac{2}{3}) - F(-\frac{13}{6}) \approx 0.2374$

19. (a) ≈ 0.8849 **(b)** ≈ 0.6554

21. $1 - F(z)$ and $F(-z)$ are the same area on opposite tails of the distribution function. Simple algebra with the standard normal cumulative distribution function shows $P(\mu - r\sigma \leq X \leq \mu + r\sigma) = 2F(r) - 1$

23. ≈ 0.0062 **27.** $\mu = \frac{5}{3}$; $\sigma = \frac{2\sqrt{5}}{3}$ **29.** $\mu = 3$; $\sigma = 3$

31. (a) $f(t)$ is the fraction of initial atoms present at time t. Therefore, the fraction of atoms that decay is going to be the rate of change of the total number of atoms. Over a small interval, this is simply $-f'(t)\Delta t$.

(b) The fraction of atoms that decay over an arbitrarily small interval is equivalent to the probability that an individual atom will decay over that same interval. Thus, the probability density function becomes $-f'(t)$.

(c) $\int_0^\infty -tf'(t)\,dt = \frac{1}{k}$

Section 8.2 Preliminary Questions

1. $\int_0^\pi \sqrt{1 + \sin^2 x}\,dx$ **2.** $\int_0^h 2\pi r\,dx = 2\pi rh$ **3.** Yes

4. The graph of $y = f(x) + C$ is a vertical translation of the graph of $y = f(x)$; hence, the two graphs should have the same arc length. We can explicitly establish this as follows:

$$\text{length of } y = f(x) + C = \int_a^b \sqrt{1 + \left[\frac{d}{dx}(f(x) + C)\right]^2}\,dx$$

$$= \int_a^b \sqrt{1 + [f'(x)]^2}\,dx$$

$$= \text{length of } y = f(x)$$

5. Since $\sqrt{1 + f'(x)^2} \geq 1$ for any function f, we have

$$\text{length of graph of } y = f(x) \text{ over } [1, 4] = \int_1^4 \sqrt{1 + f'(x)^2}\,dx$$

$$\geq \int_1^4 1\,dx = 3$$

Section 8.2 Exercises

1. $L = \int_2^6 \sqrt{1 + 16x^6}\,dx$ **3.** $\frac{13}{12}$ **5.** $3\sqrt{10}$

7. $\frac{1}{27}(22\sqrt{22} - 13\sqrt{13})$ **9.** $e^2 + \frac{\ln 2}{2} + \frac{1}{4}$

11. $\int_1^2 \sqrt{1 + x^6}\,dx \approx 3.957736$ **13.** $\int_1^2 \sqrt{1 + \frac{1}{x^4}}\,dx \approx 1.132123$

15. $\int_1^3 \sqrt{1 + \frac{1}{x^2}}\,dx = 2.29896$ **17.** ≈ 320.0 **19.** 6 **21.** 7.6337

23. $a = \sinh^{-1}(5) = \ln(5 + \sqrt{26})$

27. Let s denote the arc length. Then
$s = \frac{a}{2}\sqrt{1 + 4a^2} + \frac{1}{4}\ln|\sqrt{1 + 4a^2} + 2a|$. Thus, when $a = 1$,
$s = \frac{1}{2}\sqrt{5} + \frac{1}{4}\ln(\sqrt{5} + 2) \approx 1.478943$.

29. $\sqrt{1 + e^{2a}} + \frac{1}{2}\ln\frac{\sqrt{1+e^{2a}}-1}{\sqrt{1+e^{2a}}+1} - \sqrt{2} + \frac{1}{2}\ln\frac{1+\sqrt{2}}{\sqrt{2}-1}$ **31.** $\ln(1 + \sqrt{2})$

35. 1.552248 **37.** $16\pi\sqrt{2}$ **39.** $\frac{\pi}{27}(145^{3/2} - 1)$

41. $\frac{\pi}{32}\left(132\sqrt{17} - \ln\left(4 + \sqrt{17}\right)\right) \approx 53.23$

43. $\frac{384\pi}{5}$ **45.** $\frac{\pi}{16}(e^4 - 9)$ **47.** $2\pi\int_1^3 x^{-1}\sqrt{1 + x^{-4}}\,dx \approx 7.60306$

49. $2\pi\int_0^2 e^{-x^2/2}\sqrt{1 + x^2 e^{-x^2}}\,dx \approx 8.222696$ **51.** $2\pi\ln 2 + \frac{15\pi}{8}$

53. $4\pi^2 br$ **55.** $\frac{\pi}{6}\left(\ln(\sqrt{145} + 12) + 12\sqrt{145}\right) \approx 77.33$

57. ≈ 261.13 **59.** $2\pi b^2 + \frac{2\pi ba^2}{\sqrt{a^2 - b^2}}\sin^{-1}\left(\frac{\sqrt{a^2 - b^2}}{a}\right)$

Section 8.3 Preliminary Questions

1. Pressure is defined as force per unit area.
2. The factor of proportionality is the weight density of the fluid, $w = \rho g$.
3. Fluid force acts in the direction perpendicular to the side of the submerged object.
4. Pressure depends only on depth and does not change horizontally at a given depth.
5. When a plate is submerged vertically, the pressure is not constant along the plate, so the fluid force is not equal to the pressure times the area.

Section 8.3 Exercises

1. (a) Top: $F = 176,400$ newtons; bottom: $F = 705,600$ newtons
(b) $F \approx \sum_{j=1}^{N} \rho g 3 y_j \, \Delta y$ (c) $F = \int_2^8 \rho g 3 y \, dy$
(d) $F = 882,000$ newtons
3. Difference = force when top edge is 1 meter below the water – force when top edge is level with water = 19,600 newtons
5. (a) The width of the triangle varies linearly from 0 at a depth of $y = 3$ m to 1 at a depth of $y = 5$ m. Thus, $f(y) = \frac{1}{2}(y - 3)$.
(b) The area of the strip at depth y is $\frac{1}{2}(y - 3)\Delta y$, and the pressure at depth y is $\rho g y$, where $\rho = 10^3$ kg/m^3 and $g = 9.8$. Thus, the fluid force acting on the strip at depth y is approximately equal to $\rho g \frac{1}{2} y (y - 3) \Delta y$.
(c) $F \approx \sum_{j=1}^{N} \rho g \frac{1}{2} y_j (y_j - 3) \, \Delta y \to \int_3^5 \rho g \frac{1}{2} y (y - 3) \, dy$
(d) $F = \frac{127,400}{3}$ newtons
7. (b) $F = \frac{19,600}{3} r^3$ newtons
9. $F = \frac{19,600}{3} r^3 + 4900\pi m r^2$ newtons 11. $F \approx 328,224,000$ lb
13. 333,200 newtons 15. $F = \frac{815,360}{3}$ newtons
17. $F \approx 6153.18$ newtons 19. $F \approx 5652.4$ newtons
21. $F = 940,800$ newtons 23. 5.4604×10^{11} newtons
25. $F = (15b + 30a)h^2$ lb
27. Front and back: $F = \frac{62.5\sqrt{3}}{9} H^3$; slanted sides: $F = \frac{62.5\sqrt{3}}{3} \ell H^2$

Section 8.4 Preliminary Questions

1. $M_x = M_y = 0$ 2. $M_x = 21$ 3. $M_x = 5$; $M_y = 10$
4. Because a rectangle is symmetric with respect to both the vertical line and the horizontal line through the center of the rectangle, the Symmetry Principle guarantees that the centroid of the rectangle must lie along both these lines. The only point in common to both lines of symmetry is the center of the rectangle, so the centroid of the rectangle must be the center of the rectangle.
5. If the plate looks like a ring, then the center of mass doesn't occur at any point on the plate.

Section 8.4 Exercises

1. (a) $-\frac{7}{8}$ (b) $\frac{19}{5}$
3. (a) $M_x = 4m$; $M_y = 9m$; center of mass: $\left(\frac{9}{4}, 1\right)$
(b) $\left(\frac{46}{17}, \frac{14}{17}\right)$
7. A sketch of the lamina is shown here.

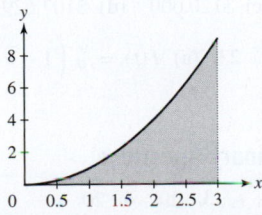

(a) $M_x = \frac{729}{10}$; $M_y = \frac{243}{4}$
(b) Area = 9 cm^2; center of mass: $\left(\frac{9}{4}, \frac{27}{10}\right)$
9. $M_x = \frac{64\delta}{7}$; $M_y = \frac{32\delta}{5}$; center of mass: $\left(\frac{8}{5}, \frac{16}{7}\right)$
11. (a) $M_x = 24$
(b) $M = 12$, so $y_{cm} = 2$; center of mass: $(0, 2)$
13. $\left(\frac{2}{3}, \frac{4}{3}\right)$ 15. $\left(\frac{93}{35}, \frac{45}{56}\right)$ 17. $\left(\frac{9}{8}, \frac{18}{5}\right)$
19. $\left(\frac{1 - 5e^{-4}}{1 - e^{-4}}, \frac{1 - e^{-8}}{4(1 - e^{-4})}\right)$ 21. $\left(\frac{\pi}{2}, \frac{\pi}{8}\right)$
23. A sketch of the region is shown here.

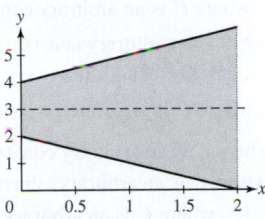

The region is clearly symmetric about the line $y = 3$, so we expect the centroid of the region to lie along this line. We find $M_x = 24$, $M_y = \frac{28}{3}$, centroid: $\left(\frac{7}{6}, 3\right)$.
25. $\left(\frac{9}{20}, \frac{9}{20}\right)$ 27. $\left(\frac{1}{2(e - 2)}, \frac{e^2 - 3}{4(e - 2)}\right)$
29. $\left(\frac{\pi\sqrt{2} - 4}{4(\sqrt{2} - 1)}, \frac{1}{4(\sqrt{2} - 1)}\right)$
31. A sketch of the region is shown here. Centroid: $\left(0, \frac{2}{7}\right)$

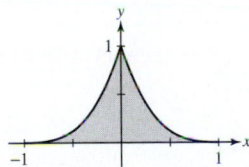

33. $\left(0, \frac{4b}{3\pi}\right)$ 35. $\left(\frac{4}{3\pi}, \frac{4}{3\pi}\right)$
37. $\left(0, \frac{\frac{2}{3}(r^2 - h^2)^{3/2}}{r^2 \sin^{-1}\sqrt{1 - h^2/r^2} - h\sqrt{r^2 - h^2}}\right)$; with $r = 1$ and $h = \frac{1}{2}$:
$\left(0, \frac{3\sqrt{3}}{4\pi - 3\sqrt{3}}\right) \approx (0, 0.71)$
39. $V = \frac{\pi r^2 H}{3}$ 41. $\left(0, \frac{49}{24}\right)$ 43. $\left(-\frac{4}{9\pi}, \frac{4}{9\pi}\right)$
45. For the square on the left: $(4, 4)$; for the square on the right: $\left(4, \frac{25}{7}\right)$

Chapter 8 Review

1. $\frac{3}{4}$ 3. $C = 2$; $p(0 \le X \le 1) = 1 - \frac{2}{e}$
5. (a) 0.1587 (b) 0.49997
7. $\frac{779}{240}$ 9. $4\sqrt{17}$ 13. $24\pi\sqrt{2}$ 15. $\frac{67\pi}{36}$
17. $12\pi + 4\pi^2$ 19. 176,400 newtons
21. Fluid force on a triangular face: $183,750\sqrt{3} + 306,250$ newtons; fluid force on a slanted rectangular edge: $122,500\sqrt{3} + 294,000$ newtons
23. $M_x = 20,480$; $M_y = 25,600$; center of mass: $\left(2, \frac{8}{5}\right)$
25. $\left(0, \frac{2}{\pi}\right)$ 27. $\frac{3\pi}{10}$

Chapter 9

Section 9.1 Preliminary Questions

1. (a) First order (b) First order (c) Order 3 (d) Order 2
2. b, d, e 3. a, c, d, f 4. b, c, d, e

Section 9.1 Exercises

1. Let $y = 4x^2$. Then $y' = 8x$ and $y' - 8x = 8x - 8x = 0$.

3. Let $y = 25e^{-2x^2}$. Then $y' = -100xe^{-2x^2}$ and
$y' + 4xy = -100xe^{-2x^2} + 4x(25e^{-2x^2}) = 0$.

5. Let $y = 4x^4 - 12x^2 + 3$. Then
$$y'' - 2xy' + 8y = (48x^2 - 24) - 2x(16x^3 - 24x) + 8(4x^4 - 12x^2 + 3)$$
$$= 48x^2 - 24 - 32x^4 + 48x^2 + 32x^4 - 96x^2 + 24 = 0$$

7. $y(t) = -1$ and $y(t) = 0.001x - 1.001$

9. (d) $y = \ln\left|\frac{1}{x} - \frac{1}{2} + e^4\right|$

11. $y = (8x^3 + C)^{1/4}$, where C is an arbitrary constant.

13. $y = Ce^{-x^3/3}$, where C is an arbitrary constant.

15. $y = \ln(4t^5 + C)$, where C is an arbitrary constant.

17. $y = Ce^{-5x/2} + \frac{4}{5}$, where C is an arbitrary constant.

19. $y = Ce^{-\sqrt{1-x^2}}$, where C is an arbitrary constant.

21. $y = \pm\sqrt{x^2 + C}$, where C is an arbitrary constant.

23. $x = \tan(\frac{1}{2}t^2 + t + C)$, where C is an arbitrary constant.

25. $y = \sin^{-1}\left(\frac{1}{2}x^2 + C\right)$, where C is an arbitrary constant.

27. $y = C \sec t$, where C is an arbitrary constant.

29. $y = 75e^{-2x}$ **31.** $y = -\sqrt{\ln(x^2 + e^4)}$ **33.** $y = 2 + 2e^{x(x-2)/2}$

35. $y = \tan(x^2/2)$ **37.** $y = e^{1-e^{-t}}$ **39.** $y = \frac{et}{e^{1/t}} - 1$

41. $y = \sin^{-1}\left(\frac{1}{2}e^x\right)$

43. (a) With an approximation of the numerical values involved in the differential equation and solution, $T(x) = 0.091x$.

(b)
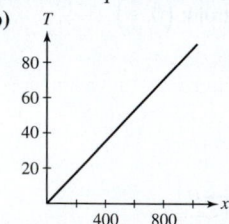

45. (a) $\approx$1145 s or 19.1 min (b) $\approx$3910 s or 65.2 min

47. $y = 8 - (8 + 0.0002215t)^{2/3}$; $t_e \approx 66000$ s or 18 hr, 20 min

51. After 5 h, amount $\approx$ 184 g; After 10 h, amount $\approx$ 68 g

53. (a) $\frac{dN}{dt} = kN$, $N(0) = 11.3$, $N(10) = 11.7$; solution:
$N(t) = 11.3e^{0.0035t}$. (b) Approximately 2.3 million

55. (a) $t = \frac{\ln p}{k}$ (b) Doubles in approximately 0.46 h; triples in approximately 0.73 h; increases ten-fold in approximately 1.54 h

57. (a) $q(t) = CV\left(1 - e^{-t/RC}\right)$
(c) $\lim_{t\to\infty} q(t) = \lim_{t\to\infty} CV\left(1 - e^{-t/RC}\right) = \lim_{t\to\infty} CV\left(1 - 0\right) = CV$
(d) $q(RC) = CV\left(1 - e^{-1}\right) \approx (0.63)CV$

59. cubic; $V = (kt/3 + C)^3$, V increases roughly with the cube of time.

61. (a)
$\frac{d}{dt}(ye^{-kt}) = \frac{dy}{dt}(e^{-kt}) + y(-ke^{-kt}) = ky(e^{-kt}) - ky(e^{-kt}) = 0$
(b) Since $\frac{d}{dt}(ye^{-kt}) = 0$ [by part (a)] the corollary implies that $ye^{-kt} = D$ for some constant D. Therefore, $y = De^{kt}$.

63. $y = Cx^3$ and $y = \pm\sqrt{A - \frac{x^2}{3}}$

65. (b) $v(t) = -9.8t + 100(\ln(50) - \ln(50 - 4.75t))$;
$v(10) = -98 + 100(\ln(50) - \ln(2.5)) \approx 201.573$ m/s

71. (c) $C = \frac{14\pi}{15B\sqrt{2g}} \cdot R^{5/2}$

Section 9.2 Preliminary Questions

1. $y(t) = 5 - ce^{4t}$ for any positive constant c

2. No **3.** True

4. The difference in temperature between a cooling object and the ambient temperature is decreasing. Hence, the rate of cooling, which is proportional to this difference, is also decreasing in magnitude.

Section 9.2 Exercises

1. General solution: $y(t) = 10 + ce^{2t}$; solution satisfying $y(0) = 25$: $y(t) = 10 + 15e^{2t}$; solution satisfying $y(0) = 5$: $y(t) = 10 - 5e^{2t}$

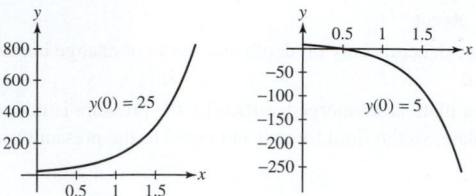

3. $y = -6 + 11e^{4x}$

5. (a) $y' = -0.02(y - 10)$ (b) $y = 10 + 90e^{-\frac{1}{50}t}$

(c) $100\ln 3 \text{ s} \approx 109.8$ s

7. $\approx$ 5:50 AM **9.** $\approx$ 0.77 min = 46.6 s

11. $500\ln\frac{3}{2}$ s ≈ 203 s $= 3$ min 23 s

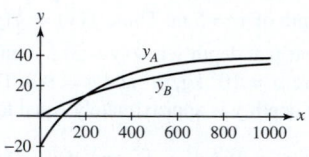

13. -58.8 m/s **15.** -11.8 m/s

17. (a)
$$0 = -\frac{9.8}{k} + \left(30 + \frac{9.8}{k}\right)e^{-kt^*}$$
$$9.8 = (30k + 9.8)e^{-kt^*}$$
$$e^{kt^*} = \frac{30k}{9.8} + 1$$
$$kt^* = \ln\left(\frac{30k}{9.8} + 1\right)$$
$$t^* = \frac{1}{k}\ln\left(\frac{30k}{9.8} + 1\right)$$

(b)
$$y(t^*) = -\frac{9.8}{k^2}\ln\left(\frac{30k}{9.8} + 1\right) + \frac{1}{k}\left(30 + \frac{9.8}{k}\right)\left(1 - e^{-\ln\left(\frac{30k}{9.8}+1\right)}\right)$$
$$= -\frac{9.8}{k^2}\ln\left(\frac{30k}{9.8} + 1\right) + \frac{1}{k}\left(30 + \frac{9.8}{k}\right)\left(1 - \frac{9.8}{30k + 9.8}\right)$$
$$= -\frac{9.8}{k^2}\ln\left(\frac{30k}{9.8} + 1\right) + \frac{1}{k}\left(\frac{30k + 9.8}{k}\right)\left(\frac{30k}{30k + 9.8}\right)$$
$$= -\frac{9.8}{k^2}\ln\left(\frac{30k}{9.8} + 1\right) + \frac{30}{k}$$
$$= \frac{30k - 9.8\ln(\frac{150k}{49} + 1)}{k^2}$$

19. (a) i. $17,563.94 ii. approximately 13.86 years or about 13 years 10 months (c) $120,000 (d) $107,629.00 (e) 8%

21. $4068.73 per year **23.** (a) $I(t) = \frac{V}{R}\left(1 - e^{-\left(\frac{R}{L}\right)t}\right)$

Section 9.3 Preliminary Questions

1. 7 **2.** $y = \pm\sqrt{1 + t}$ **3.** (b) **4.** 20

Section 9.3 Exercises

1.
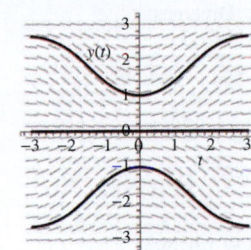
3.
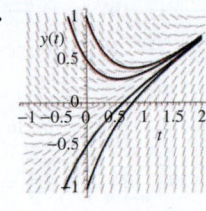

5. (a)
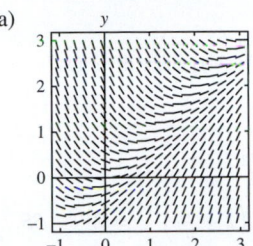
7.
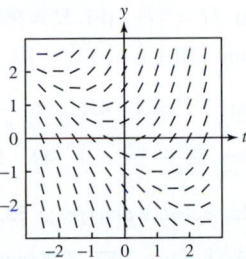

9. For $y' = t$, y' depends only on t. The isoclines of any slope c will be the vertical lines $t = c$.

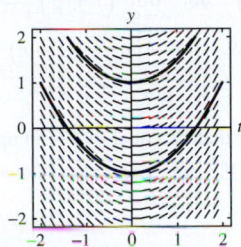

11. (i) C (ii) B (iii) F (iv) D (v) A (vi) E

13. (a)

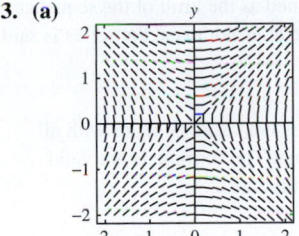

15. (a) $y_1 = 3.1$ (b) $y_2 = 3.231$

(c) $y_3 = 3.3919$, $y_4 = 3.58171$, $y_5 = 3.799539$, $y_6 = 4.0445851$

(d) $y(2.2) \approx 3.231$, $y(2.5) \approx 3.799539$

17. $y(0.5) \approx 1.7210$ **19.** $y(3.3) \approx 3.3364$ **21.** $y(2) \approx 2.8838$

25. $y(0.5) \approx 1.794894$ **27.** $y(0.25) \approx 1.094871$

Section 9.4 Preliminary Questions

1. (a) No (b) Yes (c) No (d) Yes

2. $y(t) = 0$ and $y(t) = A$ **3.** Yes

Section 9.4 Exercises

1. $y = \dfrac{5}{1 - e^{-3t}/C}$ and $y = \dfrac{5}{1 + (3/2)e^{-3t}}$

3. (a) $y(t) = 6$ (b) $y(t) = \dfrac{6}{1 + 0.5e^{-18t}}$ (c) $y(t) = 0$

5. (a) $P(t) = \dfrac{2000}{1 + 3e^{-0.6t}}$ (b) $t = \dfrac{1}{0.6}\ln 3 \approx 1.83$ years

7. $k = \ln\dfrac{81}{31} \approx 0.96$ year^{-1}; $t = \dfrac{\ln 9}{2\ln 9 - \ln 31} \approx 2.29$ years

9. After $t = 7.6$ h, or at 3:36 PM

11. (a) $y_1(t) = \dfrac{10}{10 - 9e^{-t}}$ and $y_2(t) = \dfrac{1}{1 - 2e^{-t}}$

(b) $t = \ln\dfrac{9}{8}$ (c) $t = \ln 2$

13. (a) $A(t) = 16(1 - \tfrac{5}{3}e^{t/40})^2/(1 + \tfrac{5}{3}e^{t/40})^2$

(b) $A(10) \approx 2.1$

(c)
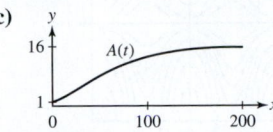

15. ≈ 943 million   **17.** (d) $t = -\tfrac{1}{k}(\ln y_0 - \ln(A - y_0))$

Section 9.5 Preliminary Questions

1. (a) Yes (b) No (c) Yes (d) No

2. (b)   **3.** $P(x) = x^{-1}$ **4.** $P(x) = 1$

Section 9.5 Exercises

1. (c) $y = \dfrac{x^4}{5} + \dfrac{C}{x}$ (d) $y = \dfrac{x^4}{5} - \dfrac{1}{5x}$

5. $y = \tfrac{1}{2}x + \dfrac{C}{x}$ **7.** $y = -\tfrac{1}{4}x^{-1} + Cx^{1/3}$

9. $y = \tfrac{1}{5}x^2 + \tfrac{1}{3} + Cx^{-3}$ **11.** $y = -x\ln x + Cx$

13. $y = \tfrac{1}{2}e^x + Ce^{-x}$ **15.** $y = x\cos x + C\cos x$

17. $y = Ce^{2x} - \tfrac{1}{3}e^{-x}$ **19.** $y = x^x + Cx^x e^{-x}$

21. $y = \tfrac{1}{5}e^{2x} - \tfrac{6}{5}e^{-3x}$ **23.** $y = \dfrac{\ln|x|}{x+1} - \dfrac{1}{x(x+1)} + \dfrac{5}{x+1}$

25. $y = -\cos x + \sin x$ **27.** $y = \tanh x + 3\operatorname{sech} x$

29. The differential equation is first-order linear with $p(x) = 0$ and $Q(x) = x$. Constant functions are antiderivatives of $p(x)$, and therefore we can take $\alpha(x) = e^c$, for any C, as an integrating factor. By Theorem 1 then, the solution to the differential equation is $y = \dfrac{1}{e^C}\int e^C x\,dx = \int x\,dx$. From here, obtaining the solution amounts to directly integrating the right side of the differential equation.

31. For $m \neq -n$: $y = \dfrac{1}{m+n}e^{mx} + Ce^{-nx}$; for $m = -n$: $y = (x + C)e^{-nx}$

33. (a) $y' = 4000 - \dfrac{40y}{500 + 40t}$; $y = 1000\dfrac{4t^2 + 100t + 125}{2t + 25}$

(b) 40 g/L

35. 50 g/L

37. (a) $\dfrac{dV}{dt} = \dfrac{20}{1+t} - 5$ and $V(t) = 20\ln(1 + t) - 5t + 100$

(b) The maximum value is $V(3) = 20\ln 4 - 15 + 100 \approx 112.726$.

(c) Estimate empty at ≈ 34 min

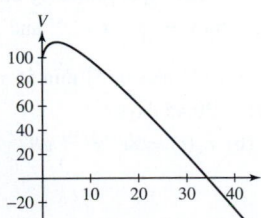

39. $I(t) = \dfrac{1}{10}\left(1 - e^{-20t}\right)$

41. (a) $I(t) = \dfrac{V}{R} - \dfrac{V}{R}e^{-(R/L)t}$ (c) Approximately 0.0184 s

43. (b) $c_1(t) = 10e^{-t/6}$

Chapter 9 Review

1. (a) No, first order (b) Yes, first order (c) No, order 3
(d) Yes, second order

3. $y = \pm\left(\tfrac{4}{3}t^3 + C\right)^{1/4}$, where C is an arbitrary constant.

5. $y = Cx^2 - \tfrac{3}{2}$, where C is an arbitrary constant.

7. $y = \tfrac{1}{2}\left(x + \tfrac{1}{2}\sin 2x\right) + \tfrac{\pi}{4}$ **9.** $y = \dfrac{4}{13 - 12x^2}$

11. 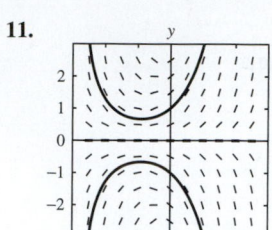   **13.**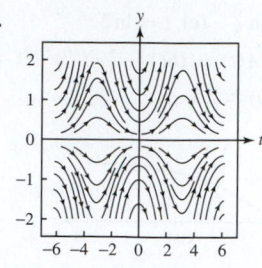

15. $y(t) = \tan t$

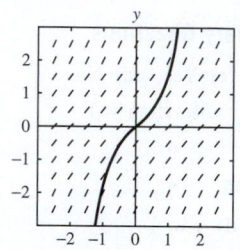

17. $y(0.1) \approx 1.1$; $y(0.2) \approx 1.209890$; $y(0.3) \approx 1.329919$

19. $y = \frac{5}{3}\sqrt{x} - \frac{2}{3}x^2$ **21.** $y = \frac{1}{2} + e^{-x} - \frac{11}{2}e^{-2x}$

23. $y = \frac{1}{2}\sin 2x - 2\cos x$ **25.** $y = 1 - \sqrt{t^2 + 15}$

27. $w = \tan\left(k\ln x + \frac{\pi}{4}\right)$

29. $y = -\cos x + \frac{\sin x}{x} + \frac{C}{x}$, where C is an arbitrary constant.

31. Solution satisfying $y(0) = 3$: $y(t) = 4 - e^{-2t}$; solution satisfying $y(0) = 4$: $y(t) = 4$

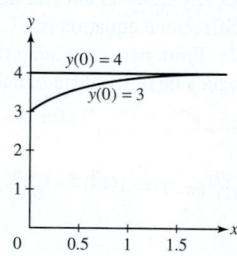

33. (a) 12   (b) ∞, if $y(0) > 12$; 12, if $y(0) = 12$; $-\infty$, if $y(0) < 12$
(c) -3

35. $400{,}000 - 200{,}000e^{0.25} \approx \$143{,}194.91$ **37.** \$400,000

39. (a) $P(t) = 500e^{0.262t}$ (b) Approximately 2.65 h

41. Solutions are of the form $y = \frac{B}{A} + Ce^{-At}$ and $\lim\limits_{t\to\infty} y = \frac{B}{A}$.

43. $\frac{dy}{dt} = \frac{-7\sqrt{10y}}{(30y + 8100)}$; $t = 3225.88$ s or 51 min 56 s

45. 2   **47.** $t = 5\ln 441 \approx 30.45$ days

51. (a) $\frac{dc_1}{dt} = -\frac{2}{5}c_1$ (b) $c_1(t) = 8e^{(-2/5)t}$ g/L

Chapter 10

Section 10.1 Preliminary Questions

1. $a_4 = 12$ **2.** (c)   **3.** $\lim\limits_{n\to\infty} a_n = \sqrt{2}$ **4.** (b)

5. (a) False; counterexample: $a_n = \cos\pi n$
(b) True   (c) False; counterexample: $a_n = (-1)^n$

Section 10.1 Exercises

1. (a) (iv)   (b) (i)   (c) (iii)   (d) (ii)

3. $c_1 = 3$; $c_2 = \frac{9}{2}$; $c_3 = \frac{9}{2}$; $c_4 = \frac{27}{8}$

5. $a_1 = 2$; $a_2 = 5$; $a_3 = 47$; $a_4 = 4415$

7. $b_1 = 4$; $b_2 = 6$; $b_3 = 4$; $b_4 = 6$

9. $c_1 = 1$; $c_2 = \frac{3}{2}$; $c_3 = \frac{11}{6}$; $c_4 = \frac{25}{12}$

11. $b_1 = 2$; $b_2 = 3$; $b_3 = 8$; $b_4 = 19$

13. (a) $a_n = \frac{(-1)^{n+1}}{n^3}$ (b) $a_n = \frac{n+1}{n+5}$ **15.** Diverges

17. $\lim\limits_{n\to\infty} \frac{5n-1}{12n+9} = \frac{5}{12}$ **19.** Diverges

21. The sequence diverges.

23. $\lim\limits_{n\to\infty} \frac{n}{\sqrt{n^2+1}} = 1$ **25.** $\lim\limits_{n\to\infty} \ln\left(\frac{12n+2}{-9+4n}\right) = \ln 3$

27. Limit $= 0$ **29.** $\lim\limits_{n\to\infty} \sqrt{4 + \frac{1}{n}} = 2$

31. $\lim\limits_{n\to\infty} \cos^{-1}\left(\frac{n^3}{2n^3+1}\right) = \frac{\pi}{3}$ **33.** Limit ≈ 1.61803

35. (a) $M = 999$ (b) $M = 99{,}999$

39. $\lim\limits_{n\to\infty}\left(10 + \left(-\frac{1}{9}\right)^n\right) = 10$ **41.** The sequence diverges.

43. $\lim\limits_{n\to\infty} 2^{1/n} = 1$ **45.** $\lim\limits_{n\to\infty} \frac{9^n}{n!} = 0$

47. $\lim\limits_{n\to\infty} \frac{3n^2+n+2}{2n^2-3} = \frac{3}{2}$ **49.** $\lim\limits_{n\to\infty} \frac{\cos n}{n} = 0$

51. The sequence diverges.   **53.** $\lim\limits_{n\to\infty}\left(2 + \frac{4}{n^2}\right)^{1/3} = 2^{1/3}$ **55.** $\lim\limits_{n\to\infty} \ln\left(\frac{2n+1}{3n+4}\right) = \ln\frac{2}{3}$ **57.** The sequence diverges.   **59.** $\lim\limits_{n\to\infty} \frac{e^n + (-3)^n}{5^n} = 0$ **61.** $\lim\limits_{n\to\infty} n\sin\frac{\pi}{n} = \pi$

63. $\lim\limits_{n\to\infty} \frac{3-4^n}{2+7\cdot4^n} = -\frac{1}{7}$ **65.** $\lim\limits_{n\to\infty}\left(1 + \frac{1}{n}\right)^n = e$

67. $\lim\limits_{n\to\infty} \frac{(\ln n)^2}{n} = 0$ **69.** $\lim\limits_{n\to\infty} n\left(\sqrt{n^2+1} - n\right) = \frac{1}{2}$

71. $\lim\limits_{n\to\infty} \frac{1}{\sqrt{n^4+n^8}} = 0$ **73.** $\lim\limits_{n\to\infty}(2^n + 3^n)^{1/n} = 3$ **75.** (b)

77. Any number greater than or equal to 3 is an upper bound.

79. Example: $a_n = (-1)^n$ **83.** Example: $f(x) = \sin\pi x$

91. (e) $AGM\left(1, \sqrt{2}\right) \approx 1.198$

Section 10.2 Preliminary Questions

1. The sum of an infinite series is defined as the limit of the sequence of partial sums. If the limit of this sequence does not exist, the series is said to diverge.

2. $S = \frac{1}{2}$

3. The result is negative, so the result is not valid: A series with all positive terms cannot have a negative sum. The formula is not valid because a geometric series with $|r| \geq 1$ diverges.

4. No   **5.** No   **6.** $N = 13$

7. No, S_N is increasing and converges to 1, so $S_N \leq 1$ for all N.

8. Example: $\displaystyle\sum_{n=1}^{\infty} \frac{1}{n^{9/10}}$

Section 10.2 Exercises

1. (a) $a_n = \frac{1}{3^n}$ (b) $a_n = \left(\frac{5}{2}\right)^{n-1}$

(c) $a_n = (-1)^{n+1}\frac{n^n}{n!}$ (d) $a_n = \frac{1 + \frac{(-1)^{n+1}+1}{2}}{n^2+1}$

3. $S_2 = \frac{5}{4}$; $S_4 = \frac{205}{144}$; $S_6 = \frac{5369}{3600}$

5. $S_2 = \frac{2}{3}$; $S_4 = \frac{4}{5}$; $S_6 = \frac{6}{7}$ **7.** $S_6 = 1.24992$

9. $S_{10} = 0.03535167962$; $S_{100} = 0.03539810274$; $S_{500} = 0.03539816290$; $S_{1000} = 0.03539816334$; yes.

11. $S_3 = \frac{3}{10}$; $S_4 = \frac{1}{3}$; $S_5 = \frac{5}{14}$; $\displaystyle\sum_{n=1}^{\infty}\left(\frac{1}{n+1} - \frac{1}{n+2}\right) = \frac{1}{2}$

13. $S_3 = \frac{3}{7}$; $S_4 = \frac{4}{9}$; $S_5 = \frac{5}{11}$; $\displaystyle\sum_{n=1}^{\infty} \frac{1}{4n^2-1} = \frac{1}{2}$

15. $S = \frac{1}{2}$ **17.** $\lim\limits_{n\to\infty} \frac{n}{10n+12} = \frac{1}{10} \neq 0$

19. $\lim\limits_{n\to\infty}(-1)^{n+1}\left(\frac{n-1}{n}\right)$ does not exist.

21. $\lim\limits_{n\to\infty} a_n = \lim\limits_{n\to\infty} \cos\frac{1}{n+1} = 1 \neq 0$ **23.** $\frac{6}{5}$

25. $\frac{7}{2}$ **27.** The series diverges.   **29.** $S = \frac{59,049}{3328}$

31. $S = \frac{1}{e-1}$ **33.** $S = \frac{35}{3}$ **35.** $S = 4$ **37.** $S = \frac{7}{15}$ **39.** $\frac{2}{9}$

41. $\frac{31}{99}$ **43.** $\frac{37}{300}$

45. $0.999999\ldots = \frac{9}{10} + \frac{9}{100} + \frac{9}{1000} + \cdots = \frac{\frac{9}{10}}{1-\frac{1}{10}} = 1$

47. (b) and (c)   **51.** (a) Counterexample: $\sum\limits_{n=1}^{\infty} \left(\frac{1}{2}\right)^n = 1$

(b) Counterexample: If $a_n = 1$, then $S_N = N$

(c) Counterexample: $\sum\limits_{n=1}^{\infty} \frac{1}{n}$ diverges

(d) Counterexample: $\sum\limits_{n=1}^{\infty} \cos 2\pi n \neq 1$

53. (a) $(.55)^n (.48)^{n-1} (.52)$

(b) $\sum_{n=1}^{\infty} (.55)^n (.48)^{n-1} (.52) = \sum_{n=1}^{\infty} (.55)(.52)((.55)(.48))^{n-1} = \frac{0.286}{1-0.264} \approx 0.39$

(c) $\sum_{n=1}^{\infty} (.52)((.48)(.55))^{n-1} = \frac{0.52}{1-0.264} \approx 0.71$

55. The total area is $\frac{1}{4}$.

57. (a) $De^{-k} + De^{-2k} + De^{-3k} + \cdots = \frac{De^{-k}}{1-e^{-k}}$

(b) $De^{-kt} + De^{-2kt} + De^{-3kt} + \cdots = \frac{De^{-kt}}{1-e^{-kt}}$

(c) $t \geq -\frac{1}{k} \ln\left(1 - \frac{D}{S}\right)$

59. The total length of the path is $2 + \sqrt{2}$.

63. (a) As $x \to \infty$ the diameter of the horn goes to zero. If we can fill the horn all of the way to the end, then the paint must be capable of being spread thin enough to fit all of the way into the horn.   (b) Use a volume of $\frac{1}{2^n}$ milliliters of paint to paint the portion of the horn between $x = n$ and $x = n+1$. Overall we use a volume of $\frac{1}{2} + \frac{1}{4} + \frac{1}{8} + \frac{1}{16} + \cdots = 1$ milliliter of paint to paint the surface.

Section 10.3 Preliminary Questions

1. (b)

2. A function f such that $a_n = f(n)$ must be positive, decreasing, and continuous for $x \geq 1$.

3. Convergence of p-series or integral test

4. Comparison Test

5. No; $\sum\limits_{n=1}^{\infty} \frac{1}{n}$ diverges, but since $\frac{e^{-n}}{n} < \frac{1}{n}$ for $n \geq 1$, the Comparison Test tells us nothing about the convergence of $\sum\limits_{n=1}^{\infty} \frac{e^{-n}}{n}$.

Section 10.3 Exercises

1. $\int_{x=1}^{\infty} \frac{1}{(x+1)^4}\, dx$ converges, so the series converges.

3. $\int_{1}^{\infty} x^{-1/3}\, dx = \infty$, so the series diverges.

5. $\int_{25}^{\infty} \frac{x^2}{(x^3+9)^{5/2}}\, dx$ converges, so the series converges.

7. $\int_{1}^{\infty} \frac{dx}{x^2+1}$ converges, so the series converges.

9. $\int_{x=1}^{\infty} \frac{1}{x(x+5)}\, dx$ converges, so the series converges.

11. $\int_{2}^{\infty} \frac{1}{x(\ln x)^2}\, dx$ converges, so the series converges.

13. $\frac{1}{n^3+8n} \leq \frac{1}{n^3}$, so the series converges.

17. $\frac{1}{n2^n} \leq \left(\frac{1}{2}\right)^n$, so the series converges.

19. $\frac{1}{n^{1/3}+2^n} \leq \left(\frac{1}{2}\right)^n$, so the series converges.

21. $\frac{4}{m!+4^m} \leq 4\left(\frac{1}{4}\right)^m$, so the series converges.

23. $0 \leq \frac{\sin^2 k}{k^2} \leq \frac{1}{k^2}$, so the series converges.

25. $\frac{2}{3^n+3^{-n}} \leq 2\left(\frac{1}{3}\right)^n$, so the series converges.

27. $\frac{1}{(n+1)!} \leq \frac{1}{n^2}$, so the series converges.

29. $\frac{\ln n}{n^3} \leq \frac{1}{n^2}$ for $n \geq 1$, so the series converges.

31. $\frac{(\ln n)^{100}}{n^{1.1}} \leq \frac{1}{n^{1.09}}$ for n sufficiently large, so the series converges.

33. $\frac{n}{3^n} \leq \left(\frac{2}{3}\right)^n$ for $n \geq 1$, so the series converges.

37. The series converges.   **39.** The series diverges.

41. The series converges.   **43.** The series diverges.

45. The series converges.   **47.** The series converges.

49. The series diverges.   **51.** The series converges.

53. The series diverges.   **55.** The series converges.

57. The series diverges.   **59.** The series diverges.

61. The series diverges.   **63.** The series converges.

65. The series diverges.   **67.** The series diverges.

69. The series converges.   **71.** The series converges.

73. The series diverges.   **75.** The series converges.

77. The series converges for $a > 1$ and diverges for $a \leq 1$.

79. The series converges for $p > 1$ and diverges for $p \leq 1$.

87. $\sum\limits_{n=1}^{\infty} n^{-5} \approx 1.0369277551$

91. $\sum\limits_{n=1}^{1000} \frac{1}{n^2} = 1.6439345667$ and $1 + \sum\limits_{n=1}^{100} \frac{1}{n^2(n+1)} = 1.6448848903$. The second sum is a better approximation to $\frac{\pi^2}{6} \approx 1.6449340668$.

Section 10.4 Preliminary Questions

1. Example: $\sum \frac{(-1)^n}{\sqrt[3]{n}}$ **2.** (b)   **3.** No

4. $|S - S_{100}| < 10^{-3}$, and S is larger than S_{100}.

Section 10.4 Exercises

3. Converges conditionally   **5.** Converges absolutely

7. Converges absolutely   **9.** Converges conditionally

11.

n	S_n	n	S_n
1	1	6	0.899782407
2	0.875	7	0.902697859
3	0.912037037	8	0.900744734
4	0.896412037	9	0.902116476
5	0.904412037	10	0.901116476

13. $S_5 = 0.947$ **15.** $S_{44} = 0.06567457397$

17. Converges (by geometric series)

19. Converges (by Limit Comparison Test)

21. Converges (by Limit Comparison Test)

23. Diverges (by Limit Comparison Test)

25. Converges (by geometric series and linearity)

27. Converges absolutely (by Integral Test)

29. Converges (by Alternating Series Test)

31. Converges (by Integral Test)   **33.** Converges conditionally

Section 10.5 Preliminary Questions

1. (a) $|r|$ (b) $|c|^{1/n}|r|$ **2.** (a) $\left(\frac{n}{n+1}\right)^p$ (b) Nothing

3. Yes in the first case, no in the second.

4. Yes in the first case, no in the second.

Section 10.5 Exercises

1. Converges absolutely   **3.** Converges absolutely

5. The Ratio Test is inconclusive   **7.** Diverges

9. Converges absolutely   **11.** Converges absolutely

13. Diverges   **15.** The Ratio Test is inconclusive

17. Converges absolutely   **19.** Converges absolutely

21. $\rho = \frac{1}{3} < 1$ **23.** $\rho = 2|x|$ **25.** $\rho = |r|$

27. Converges   **29.** Converges absolutely

31. The Ratio Test is inconclusive, so the series may converge or diverge.   **33.** Converges absolutely

35.
$$\lim_{n\to\infty} \sqrt[n]{n^{-p}} = \lim_{n\to\infty} n^{-p/n} = \lim_{n\to\infty} e^{-p\left(\frac{\ln n}{n}\right)} = e^{-p\left(\lim_{n\to\infty} \frac{\ln n}{n}\right)} = e^0 = 1.$$
Therefore the root test is inconclusive.

37. Converges absolutely   **39.** Converges absolutely

41. Converges absolutely

43. Converges (by geometric series and linearity)

45. Diverges (by the Divergence Test)

47. Converges (by the Direct Comparison Test)

49. Diverges (by the Direct Comparison Test)

51. Converges (by the Ratio Test)

53. Converges (by the Limit Comparison Test)

55. Diverges (by p-series)   **57.** Converges (by geometric series)

59. Converges (by Limit Comparison Test)

61. Diverges (by Divergence Test)

65. (b) $\sqrt{2\pi} \approx 2.50663$

n	$\frac{e^n n!}{n^{n+1/2}}$
1000	2.506837
1500	2.506768
2000	2.506733
2500	2.506712
3000	2.506698

Section 10.6 Preliminary Questions

1. Yes. The series must converge for both $x = 4$ and $x = -3$.

2. (a), (c)   **3.** $R = 4$

4. $F'(x) = \sum_{n=1}^{\infty} n^2 x^{n-1}$; $R = 1$

Section 10.6 Exercises

1. $R = 2$. It does not converge at the endpoints.

3. $R = 3$ for all three series.

9. $(-1, 1)$ **11.** $[-\sqrt{2}, \sqrt{2}]$ **13.** $[-1, 1]$ **15.** $(-\infty, \infty)$

17. $(-\infty, \infty)$ **19.** $(-1, 1]$ **21.** $(-1, 1)$ **23.** $[-1, 1)$ **25.** $(2, 4)$

27. $(6, 8)$ **29.** $\left[-\frac{7}{2}, -\frac{5}{2}\right)$ **31.** $(-\infty, \infty)$

33. $\left(2 - \frac{1}{e}, 2 + \frac{1}{e}\right)$ **35.** $\sum_{n=0}^{\infty} 3^n x^n$ on the interval $\left(-\frac{1}{3}, \frac{1}{3}\right)$

37. $\sum_{n=0}^{\infty} \frac{x^n}{3^{n+1}}$ on the interval $(-3, 3)$

39. $\sum_{n=0}^{\infty} x^{3n}$ on the interval $(-1, 1)$

41. $g(x) = \sum_{n=1}^{\infty} 3n x^{3n-1}$

43. (a) $h(x) = \sum_{n=1}^{\infty} n x^{3n-3} \cdot \lim_{n\to\infty} \left|\frac{a_{n+1}}{a_n}\right| =$

$\lim_{n\to\infty} \frac{(n+1)|x|^{3(n+1)-3}}{n|x|^{3n-3}} = |x|^3$. It follows that the radius of convergence is 1.   **(b)** Multiplying term-by-term, we find that
$(f(x))^2 = 1 + x^3 + x^6 + x^9 + x^3 + x^6 + x^9 + x^6 + x^9 + x^9 + \cdots = 1 + 2x^3 + 3x^6 + 4x^9 + \cdots$. Expanding out $h(x)$ up to $n = 4$, we have
$h(x) = 1 + 2x^3 + 3x^6 + 4x^9 + \cdots$.

47. $\sum_{n=0}^{\infty} (-1)^{n+1}(x - 5)^n$ on the interval $(4, 6)$

51. (c) $S_4 = \frac{69}{640}$ and $|S - S_4| \approx 0.000386 < a_5 = \frac{1}{1920}$

53. $R = 1$ **55.** $\sum_{n=1}^{\infty} \frac{n}{2^n} = 2$ **57.** $F(x) = \frac{1 - x - x^2}{1 - x^3}$ **59.** $-1 \le x \le 1$

61. $P(x) = \sum_{n=0}^{\infty} (-1)^n \frac{x^n}{n!}$

63. N must be at least 5; $S_5 = 0.3680555556$

65. $P(x) = 1 - \frac{1}{2}x^2 - \sum_{n=2}^{\infty} \frac{1 \cdot 3 \cdot 5 \cdots (2n-3)}{(2n)!} x^{2n}$; $R = \infty$

Section 10.7 Preliminary Questions

1. $T_3(x) = 9 + 8(x - 3) + 2(x - 3)^2 + 2(x - 3)^3$

2. The polynomial graphed on the right is a Maclaurin polynomial.

3. A Maclaurin polynomial gives the value of $f(0)$ exactly.

4. The correct statement is **(b)**: $|T_3(2) - f(2)| \le \frac{2}{3}$.

Section 10.7 Exercises

1. $T_2(x) = x$; $T_3(x) = x - \frac{x^3}{6}$

3. $T_2(x) = \frac{1}{3} - \frac{1}{9}(x - 2) + \frac{1}{27}(x - 2)^2$;
$T_3(x) = \frac{1}{3} - \frac{1}{9}(x - 2) + \frac{1}{27}(x - 2)^2 - \frac{1}{81}(x - 2)^3$

5. $T_2(x) = 75 + 106(x - 3) + 54(x - 3)^2$;
$T_3(x) = 75 + 106(x - 3) + 54(x - 3)^2 + 12(x - 3)^3$

7. $T_2(x) = 1 + \frac{1}{2}(x - 1) - \frac{1}{8}(x - 1)^2$, $T_3(x) =$
$1 + \frac{1}{2}(x - 1) - \frac{1}{8}(x - 1)^2 + \frac{1}{16}(x - 1)^3$

9. $T_2(x) = x$; $T_3(x) = x + \frac{x^3}{3}$

11. $T_2(x) = 2 - 3x + \frac{5x^2}{2}$; $T_3(x) = 2 - 3x + \frac{5x^2}{2} - \frac{3x^3}{2}$

13. $T_2(x) = \frac{1}{e} + \frac{1}{e}(x - 1) - \frac{1}{2e}(x - 1)^2$;
$T_3(x) = \frac{1}{e} + \frac{1}{e}(x - 1) - \frac{1}{2e}(x - 1)^2 - \frac{1}{6e}(x - 1)^3$

15. $T_2(x) = (x - 1) - \frac{3(x-1)^2}{2}$; $T_3(x) = (x - 1) - \frac{3(x-1)^2}{2} + \frac{11(x-1)^3}{6}$

17. $f(1) = p + q + r$, $f'(1) = 2p + q$, $f''(1) = 2p$, therefore
$T_2(x) = (p + q + r) + (2p + q)(x - 1) + \frac{2p}{2}(x - 1)^2 = (p + q + r - 2p - q + p) + (2p + q - 2p)x + px^2 = px^2 + qx + r = f(x)$.

19. Let $f(x) = e^x$. Then, for all n,

$$f^{(n)}(x) = e^x \quad \text{and} \quad f^{(n)}(0) = 1$$

It follows that

$$T_n(x) = 1 + \frac{x}{1!} + \frac{x^2}{2!} + \cdots + \frac{x^n}{n!}$$

23. $T_n(x) = 1 - x + x^2 - x^3 + \cdots + (-1)^n x^n$

25. $T_n(x) = e + e(x - 1) + \frac{e(x-1)^2}{2!} + \cdots + \frac{e(x-1)^n}{n!}$

27. $T_n(x) =$
$1 - 2(x - 1) + 3(x - 1)^2 - 4(x - 1)^3 + \cdots + (-1)^n(n + 1)(x - 1)^n$

29. $T_n(x) = \frac{1}{\sqrt{2}} - \frac{1}{\sqrt{2}}\left(x - \frac{\pi}{4}\right) - \frac{1}{2\sqrt{2}}\left(x - \frac{\pi}{4}\right)^2 + \frac{1}{6\sqrt{2}}\left(x - \frac{\pi}{4}\right)^3 \cdots$
In general, the coefficient of $(x - \pi/4)^n$ is

$$\pm \frac{1}{(\sqrt{2})n!}$$

with the pattern of signs $+, -, -, +, +, -, -, \ldots$.

31. $T_2(x) = 1 + x + \frac{x^2}{2}$; $|f(-0.5) - T_2(-0.5)| \approx 0.018469$

33. $T_2(x) = 1 - \frac{2}{3}(x-1) + \frac{5}{9}(x-1)^2$;
$|f(1.2) - T_2(1.2)| \approx 0.00334008$

35. $T_3(x) = 1 + \frac{1}{2}(x-1) - \frac{1}{8}(x-1)^2 + \frac{1}{16}(x-1)^3$; $1 \le c \le 2.9$

37. $\frac{e^{1.1}|1.1|^4}{4!}$

39. $T_5(x) = 1 - \frac{x^2}{2} + \frac{x^4}{24}$; maximum error $= \frac{(0.25)^6}{6!}$

41. $T_3(x) = \frac{1}{2} - \frac{1}{16}(x-4) + \frac{3}{256}(x-4)^2 - \frac{5}{2048}(x-4)^3$; maximum error $= \frac{35(0.3)^4}{65,536}$

43. $T_3(x) = x - \frac{x^3}{3}$; $T_3\left(\frac{1}{2}\right) = \frac{11}{24}$; with $K = 5$,

$$\left| T_3\left(\frac{1}{2}\right) - \tan^{-1}\frac{1}{2} \right| \le \frac{5\left(\frac{1}{2}\right)^4}{4!} = \frac{5}{384}$$

45. $K = 6.25$ is acceptable.   **47.** $n = 4$ **49.** $n = 6$

53. $n = 4$ **57.** $T_{4n}(x) = 1 - \frac{x^4}{2} + \frac{x^8}{4!} + \cdots + (-1)^n \frac{x^{4n}}{(2n)!}$

59. At $a = 0$,

$$T_1(x) = -4 - x$$
$$T_2(x) = -4 - x + 2x^2$$
$$T_3(x) = -4 - x + 2x^2 + 3x^3 = f(x)$$
$$T_4(x) = T_3(x)$$
$$T_5(x) = T_3(x)$$

At $a = 1$,

$$T_1(x) = 12(x-1)$$
$$T_2(x) = 12(x-1) + 11(x-1)^2$$
$$T_3(x) = 12(x-1) + 11(x-1)^2 + 3(x-1)^3$$
$$= -4 - x + 2x^2 + 3x^3 = f(x)$$
$$T_4(x) = T_3(x)$$
$$T_5(x) = T_3(x)$$

61. $T_2(t) = 60 + 24t - \frac{3}{2}t^2$; truck's distance from the intersection after 4 s is ≈ 132 m.

63. (a) $T_3(x) = -\frac{k}{R^3}x + \frac{3k}{2R^5}x^3$

71. $T_4(x) = 1 - x^2 + \frac{1}{2}x^4$; the error is approximately $|0.461458 - 0.461281| = 0.000177$.

73. (b) $\int_0^{1/2} T_4(x)\, dx = \frac{1841}{3840}$; error bound:

$$\left| \int_0^{1/2} \cos x\, dx - \int_0^{1/2} T_4(x)\, dx \right| < \frac{\left(\frac{1}{2}\right)^7}{6!}$$

75. (a) $T_6(x) = x^2 - \frac{1}{6}x^6$

Section 10.8 Preliminary Questions

1. $f(0) = 3$ and $f'''(0) = 30$

2. $f(-2) = 0$ and $f^{(4)}(-2) = 48$

3. Substitute x^2 for x in the Maclaurin series for $\sin x$.

4. $f(x) = 4 + \sum_{n=1}^{\infty} \frac{(x-3)^{n+1}}{n(n+1)}$ **5.** (c)

Section 10.8 Exercises

1. $f(x) = 2 + 3x + 2x^2 + 2x^3 + \cdots$

3. $\frac{1}{(1+10x)} = \sum_{n=0}^{\infty} (-10)^n x^n$ on the interval $\left(-\frac{1}{10}, \frac{1}{10}\right)$

5. $\cos 3x = \sum_{n=0}^{\infty} (-1)^n \frac{9^n x^{2n}}{(2n)!}$ on the interval $(-\infty, \infty)$

7. $\sin(x^2) = \sum_{n=0}^{\infty} (-1)^n \frac{x^{4n+2}}{(2n+1)!}$ on the interval $(-\infty, \infty)$

9. $\ln(1-x^2) = -\sum_{n=1}^{\infty} \frac{x^{2n}}{n}$ on the interval $(-1, 1)$

11. $\tan^{-1}(x^2) = \sum_{n=0}^{\infty} (-1)^n \frac{x^{4n+2}}{2n+1}$ on the interval $[-1, 1]$

13. $e^{x-2} = \sum_{n=0}^{\infty} \frac{x^n}{e^2 n!}$ on the interval $(-\infty, \infty)$

15. $\ln(1-5x) = -\sum_{n=1}^{\infty} \frac{5^n x^n}{n}$ on the interval $\left[-\frac{1}{5}, \frac{1}{5}\right)$

17. $\sinh x = \sum_{k=0}^{\infty} \frac{x^{2k+1}}{(2k+1)!}$ on the interval $(-\infty, \infty)$

19. $e^x \sin x = x + x^2 + \frac{x^3}{3} - \frac{x^5}{30} + \cdots$

21. $\frac{\sin x}{1-x} = x + x^2 + \frac{5x^3}{6} + \frac{5x^4}{6} + \cdots$

23. $(1+x)^{1/4} = 1 + \frac{1}{4}x - \frac{3}{32}x^2 + \frac{7}{128}x^3 + \cdots$

25. $e^x \tan^{-1} x = x + x^2 + \frac{1}{6}x^3 - \frac{1}{6}x^4 + \cdots$

27. $e^{\sin x} = 1 + x + \frac{1}{2}x^2 - \frac{1}{8}x^4 + \cdots$ **29.** $1 + \frac{x^4}{2}$

31. $\frac{1}{x} = \sum_{n=0}^{\infty} (-1)^n (x-1)^n$ on the interval $(0, 2)$

33. $\frac{1}{1-x} = \sum_{n=0}^{\infty} (-1)^{n+1} \frac{(x-5)^n}{4^{n+1}}$ on the interval $(1, 9)$

35. $21 + 35(x-2) + 24(x-2)^2 + 8(x-2)^3 + (x-2)^4$ on the interval $(-\infty, \infty)$

37. $\frac{1}{x^2} = \sum_{n=0}^{\infty} (-1)^n (n+1) \frac{(x-4)^n}{4^{n+2}}$ on the interval $(0, 8)$

39. $\frac{1}{1-x^2} = \sum_{n=0}^{\infty} \frac{(-1)^{n+1}(2^{n+1}-1)}{2^{2n+3}}(x-3)^n$ on the interval $(1, 5)$

41. $\cos^2 x = \frac{1}{2} + \frac{1}{2}\sum_{n=0}^{\infty} (-1)^n \frac{(4)^n x^{2n}}{(2n)!}$ **47.** $S_4 = 0.1822666667$

49. (a) 5   (b) $S_4 = 0.7474867725$

51. $\int_0^1 \cos(x^2)\, dx = \sum_{n=0}^{\infty} \frac{(-1)^n}{(2n)!(4n+1)}$; $S_3 = 0.9045227920$

53. $\int_0^1 e^{-x^3}\, dx = \sum_{n=0}^{\infty} \frac{(-1)^n}{n!(3n+1)}$; $S_5 = 0.8074461996$

55. $\int_0^x \frac{1-\cos(t)}{t}\, dt = \sum_{n=1}^{\infty} (-1)^{n+1} \frac{x^{2n}}{(2n)!2n}$

57. $\int_0^x \ln(1+t^2)\, dt = \sum_{n=1}^{\infty} (-1)^{n-1} \frac{x^{2n+1}}{n(2n+1)}$

59. $\frac{1}{1+2x}$ **61.** $\cos \pi = -1$ **67.** e^{x^3} **69.** $1 - 5x + \sin 5x$

71. $\frac{1}{(1-2x)(1-x)} = \sum_{n=0}^{\infty} \left(2^{n+1} - 1\right) x^n$

73. $I(t) = \frac{V}{R}\sum_{n=1}^{\infty} \frac{(-1)^{n+1}}{n!}\left(\frac{Rt}{L}\right)^n$

75. $f(x) = \sum_{n=0}^{\infty} \frac{(-1)^n x^{6n}}{(2n)!}$ and $f^{(6)}(0) = -360$

77. $e^{x^{20}} = 1 + x^{20} + \frac{x^{40}}{2} + \cdots$

79. No

n	Series value when $x = 2$
5	2.54297
10	−0.239933
15	41.9276
20	−764.272
25	16,595.8

85. $\lim\limits_{x \to 0} \dfrac{\sin x - x + \frac{x^3}{6}}{x^5} = \dfrac{1}{120}$ **87.** $\lim\limits_{x \to 0} \left(\dfrac{\sin(x^2)}{x^4} - \dfrac{\cos x}{x^2} \right) = \dfrac{1}{2}$

89. (a) $\frac{1}{\sqrt{2}} - \frac{1}{\sqrt{2}}i$ (b) 1 (c) $\frac{3\sqrt{3}}{2} + \frac{3}{2}i$

91. $\dfrac{e^{iz} - e^{-iz}}{2i} = \dfrac{\cos z + i \sin z - (\cos(-z) + i \sin(-z))}{2i} =$
$\dfrac{\cos z + i \sin z - \cos z + i \sin z}{2i} = \dfrac{2i \sin z}{2i} = \sin z$

93. (c) $S = \frac{\pi}{4} - \frac{1}{2} \ln 2$ **97.** $L \approx 28.369$

Chapter 10 Review

1. (a) $a_1^2 = 4, a_2^2 = \frac{1}{4}, a_3^2 = 0$

(b) $b_1 = \frac{1}{24}, b_2 = \frac{1}{60}, b_3 = \frac{1}{240}$

(c) $a_1 b_1 = -\frac{1}{12}, a_2 b_2 = -\frac{1}{120}, a_3 b_3 = 0$

(d) $2a_2 - 3a_1 = 5, 2a_3 - 3a_2 = \frac{3}{2}, 2a_4 - 3a_3 = \frac{1}{12}$

3. $\lim\limits_{n \to \infty} (5a_n - 2a_n^2) = 2$ **5.** $\lim\limits_{n \to \infty} e^{a_n} = e^2$

7. $\lim\limits_{n \to \infty} (-1)^n a_n$ does not exist.

9. $\lim\limits_{n \to \infty} \left(\sqrt{n+5} - \sqrt{n+2} \right) = 0$ **11.** $\lim\limits_{n \to \infty} 2^{1/n^2} = 1$

13. The sequence diverges. **15.** $\lim\limits_{n \to \infty} \tan^{-1} \left(\frac{n+2}{n+5} \right) = \frac{\pi}{4}$

17. $\lim\limits_{n \to \infty} \left(\sqrt{n^2 + n} - \sqrt{n^2 + 1} \right) = \frac{1}{2}$

19. $\lim\limits_{m \to \infty} \left(1 + \frac{1}{m} \right)^{3m} = e^3$

21. $\lim\limits_{n \to \infty} \left(n \left(\ln(n+1) - \ln n \right) \right) = 1$

25. $\lim\limits_{n \to \infty} \dfrac{a_{n+1}}{a_n} = 3$ **27.** $S_4 = -\frac{11}{60}, S_7 = \frac{41}{630}$

29. $\sum\limits_{n=2}^{\infty} \left(\frac{2}{3} \right)^n = \frac{4}{3}$ **31.** $S = \frac{4}{37}$ **33.** $\sum\limits_{n=-1}^{\infty} \dfrac{2^{n+3}}{3^n} = 36$

35. $a_n = \left(\frac{1}{2} \right)^n + 1 - 2^n, b_n = 2^n - 1$

37. $S = \frac{47}{180}$ **39.** The series diverges.

41. $\int_1^{\infty} \dfrac{1}{(x+2)(\ln(x+2))^3} \, dx = \dfrac{1}{2(\ln(3))^2}$, so the series converges.

43. $\dfrac{1}{(n+1)^2} < \dfrac{1}{n^2}$, so the series converges.

45. $\sum\limits_{n=1}^{\infty} \dfrac{1}{n^{1.5}}$ converges, so the series converges.

47. $\dfrac{n}{\sqrt{n^5 + 5}} < \dfrac{1}{n^{3/2}}$, so the series converges.

49. $\sum\limits_{n=0}^{\infty} \left(\frac{10}{11} \right)^n$ converges, so the series converges.

51. Converges

55. (b) $0.3971162690 \le S \le 0.3971172688$, so the maximum size of the error is 10^{-6}.

57. Converges absolutely **59.** Diverges

61. (a) 500 (b) $K \approx \sum\limits_{k=0}^{499} \dfrac{(-1)^k}{(2k+1)^2} = 0.9159650942$

63. (a) Converges (b) Converges (c) Diverges
(d) Converges

65. Converges **67.** Converges **69.** Diverges

71. Diverges **73.** Converges **75.** Converges

77. Converges (by geometric series)

79. Converges (by geometric series)

81. Converges (by the Alternating Series Test)

83. Converges (by the Alternating Series Test)

85. Diverges (by the Divergence Test)

87. Converges (absolutely, by a Direct Comparison with the p-series $\sum\limits_{n=1}^{\infty} \frac{1}{n^{3/2}}$

89. Converges (by the Root Test)

91. Converges (by the Direct Comparison Test)

93. Converges using partial sums (the series is telescoping)

95. Diverges (by the Direct Comparison Test)

97. Converges (by the Direct Comparison Test)

99. Converges (by the Limit Comparison Test)

101. Converges on the interval $(-\infty, \infty)$

103. Converges on the interval $[2, 4]$ **105.** Converges at $x = 0$

107. $\dfrac{2}{4-3x} = \dfrac{1}{2} \sum\limits_{n=0}^{\infty} \left(\frac{3}{4} \right)^n x^n$. The series converges on the interval $\left(-\frac{4}{3}, \frac{4}{3} \right)$.

109. (c)

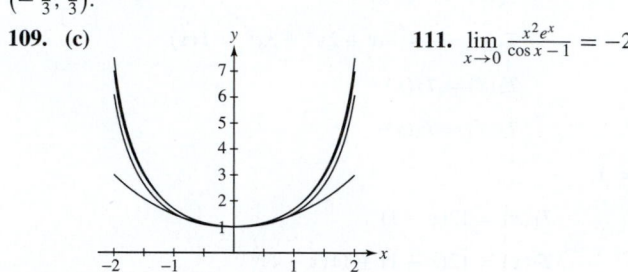

111. $\lim\limits_{x \to 0} \dfrac{x^2 e^x}{\cos x - 1} = -2$

113. $T_3(x) = 1 + 3(x - 1) + 3(x - 1)^2 + (x - 1)^3$

115. $T_4(x) = (x - 1) + \frac{1}{2}(x - 1)^2 - \frac{1}{6}(x - 1)^3 + \frac{1}{12}(x - 1)^4$

117. $T_4(x) = x - x^3$

119. $T_n(x) = 1 + 3x + \frac{1}{2!}(3x)^2 + \frac{1}{3!}(3x)^3 + \cdots + \frac{1}{n!}(3x)^n$

121. $T_3(1.1) = 0.832981496; \left| T_3(1.1) - \tan^{-1} 1.1 \right| = 2.301 \times 10^{-7}$

123. $n = 11$ is sufficient.

125. The nth Maclaurin polynomial for $g(x) = \dfrac{1}{1+x}$ is $T_n(x) = 1 - x + x^2 - x^3 + \cdots + (-x)^n$.

127. $e^{4x} = \sum\limits_{n=0}^{\infty} \dfrac{4^n}{n!} x^n$

129. $x^4 = 16 + 32(x - 2) + 24(x - 2)^2 + 8(x - 2)^3 + (x - 2)^4$

131. $\sin x = \sum\limits_{n=0}^{\infty} \dfrac{(-1)^{n+1}(x - \pi)^{2n+1}}{(2n+1)!}$

133. $\dfrac{1}{1 - 2x} = \sum\limits_{n=0}^{\infty} \dfrac{2^n}{5^{n+1}}(x + 2)^n$ **135.** $\ln \frac{x}{2} = \sum\limits_{n=1}^{\infty} \dfrac{(-1)^{n+1}(x - 2)^n}{n 2^n}$

137. $(x^2 - x)e^{x^2} = \sum\limits_{n=0}^{\infty} \left(\dfrac{x^{2n+2} - x^{2n+1}}{n!} \right)$ so $f^{(3)}(0) = -6$

139. $\dfrac{1}{1 + \tan x} = 1 - x + x^2 - \frac{4}{3} x^3 + \cdots$, so $f^{(3)}(0) = -8$

141. $\frac{\pi}{2} - \dfrac{\pi^3}{2^3 3!} + \dfrac{\pi^5}{2^5 5!} - \dfrac{\pi^7}{2^7 7!} + \cdots = \sin \frac{\pi}{2} = 1$

Chapter 11

Section 11.1 Preliminary Questions

1. A circle of radius 3 centered at the origin.

2. The center is at $(4, 5)$. **3.** Maximum height: 4

4. Yes; no

5. (a) The line $y = x$ traversed left to right.
(b) The same path traversed left to right twice as fast.

6. Possible answer: $c_1(t) = (t, t^3); c_2(t) = (\sqrt[3]{t}, t); c_3(t) = (t^3, t^9)$

Section 11.1 Exercises

1. $(t = 0)(1, 9); (t = 2)(9, -3); (t = 4)(65, -39)$

3. $y = 2.5x - 0.000766x^2$

5. **(a)** 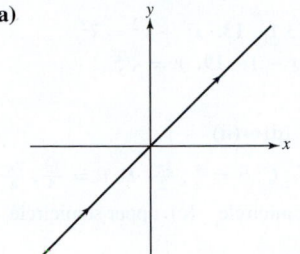   **(b)**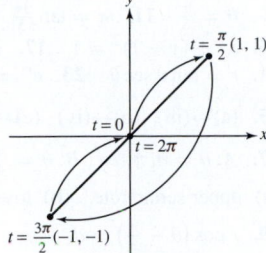

(c) 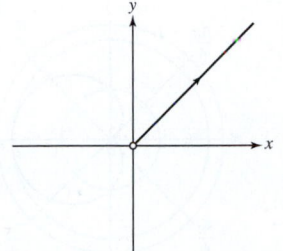   **(d)**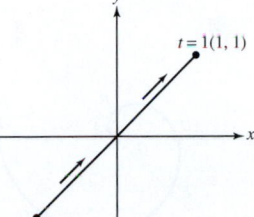

7. $y = 4x - 12$ **9.** $y = (x + 1)^{2/3} + 1$

11. $y = \frac{6}{x^2}$ (where $x > 0$)   **13.** $y = 2 - e^x$

15. 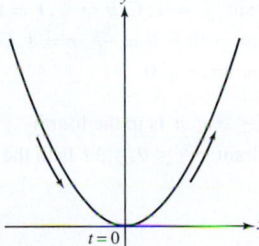   **17.**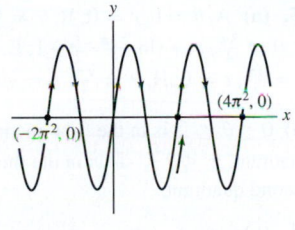

19. (a) ↔ (iv) (b) ↔ (ii) (c) ↔ (iii) (d) ↔ (i)

21. **(a)** $y_{\max} = 100$ cm   **(b)** lands at $x = 2040$ cm from the origin, when $t = 20$ s

23. $c(t) = (t, 9 - 4t)$ **25.** $c(t) = \left(\frac{5+t^2}{4}, t\right)$

27. $c(t) = (-9 + 7\cos t, 4 + 7\sin t)$ **29.** $c(t) = (-4 + t, 9 + 8t)$

31. $c(t) = (3 - 8t, 1 + 3t)$ **33.** $c(t) = (1 + t, 1 + 2t)$ $(0 \le t \le 1)$

35. $c(t) = (3 + 4\cos t, 9 + 4\sin t)$ **37.** $c(t) = \left(-4 + t, -8 + t^2\right)$

39. $c(t) = (2 + t, 2 + 3t)$ **41.** $c(t) = \left(3 + t, (3 + t)^2\right)$

43. $x = -5\cos t; y = 2\sqrt{5}\sin t$, for $0 \le t \le \pi$ **45.** $y = \sqrt{x^2 - 1}$ $(1 \le x < \infty)$ **47.** Plot III

49. **(a)** At point C, the prey species is near a minimum. Since the resources for the predator are low we expect that the predator population is decreasing. At point D on the curve, the predator population is close to its minimum value. Circumstances are then favorable for the prey, and we expect its population to be increasing. Both of these situations support the conclusion that the curve is traced in the counterclockwise direction.

(b) p, q

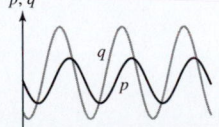

(c) The predator has its peaks shortly after the prey does. This makes sense because when the predator population is large, the prey population should be increasing due to the abundance of prey. After the prey reaches its peak and begins to decline, the predator will soon reach its peak and begin to decline in response to the decrease in available prey.

51. $\left.\frac{dy}{dx}\right|_{t=-4} = -\frac{1}{6}$ **53.** $\left.\frac{dy}{dx}\right|_{s=-1} = -\frac{3}{4}$ **55.** $-\frac{2}{3}$ **57.** -4

59. $y = -\frac{9}{2}x + \frac{11}{2}; \frac{dy}{dx} = -\frac{9}{2}$

61. $y = x^2 + x^{-1}; \frac{dy}{dx} = 2x - \frac{1}{x^2}$

63. $y = \ln(1 - x) - 1$, so $\frac{dy}{dx} = -\frac{1}{1-x}$. By Eq. (6), $\frac{dy}{dx} = -\frac{1}{e^t}$

65. $(0, 0), (96, 180)$

67.

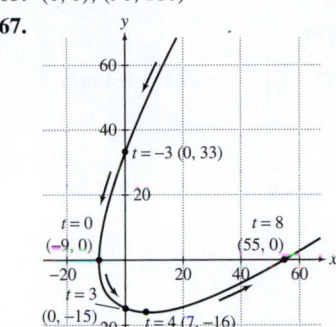

The graph is in quadrant i for $t < -3$ or $t > 8$, quadrant ii for $-3 < t < 0$, quadrant iii for $0 < t < 3$, and quadrant iv for $3 < t < 8$.

69. $(55, 0)$

71. The coordinates of P, $(R\cos\theta, r\sin\theta)$, describe an ellipse for $0 \le \theta \le 2\pi$.

75. $c(t) = (3 - 9t + 24t^2 - 16t^3, 2 + 6t^2 - 4t^3), 0 \le t \le 1$

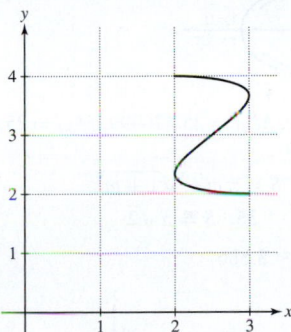

79. $y = -\sqrt{3}x + \frac{\sqrt{3}}{2}$

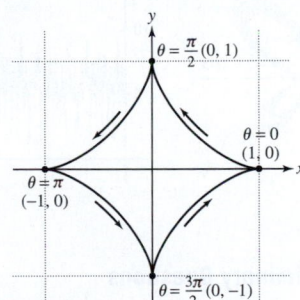

81. $((2k - 1)\pi, 2)$, $k = 0, \pm 1, \pm 2, \ldots$ **91.** $\left.\frac{d^2 y}{dx^2}\right|_{t=2} = -\frac{21}{512}$

93. $\left.\frac{d^2 y}{dx^2}\right|_{t=-3} = 0$ **95.** Concave up: $t > 0$ **97.** $\frac{1}{3}$ **99.** $\frac{2}{5}$ **101.** $\frac{2}{3}$

Section 11.2 Preliminary Questions

1. $S = \int_a^b \sqrt{x'(t)^2 + y'(t)^2}\, dt$

2. No. They are equal when the curve traced by $c(t)$ is a line segment from the initial to the final point, and $c(t)$ is a one-to-one function.

3. The speed at time t **4.** Displacement: 5; no   **5.** $L = 180$ cm

6. 4π

Section 11.2 Exercises

1. $\int_0^5 \sqrt{3^2 + (-2)^2}\,dt = \int_0^5 \sqrt{13}\,dt = 5\sqrt{13}$. The path is a line segment from $(-1, 2)$ to $(14, -8)$, and the length of the segment is $\sqrt{(14 - (-1))^2 + ((-8) - 2)^2} = \sqrt{325} = 5\sqrt{13}$. **3.** $S = 16\sqrt{13}$

5. $S = \frac{1}{2}(65^{3/2} - 5^{3/2}) \approx 256.43$ **7.** $S = 3\pi$

9. $S = -8\left(\frac{\sqrt{2}}{2} - 1\right) \approx 2.34$

13. $S = \frac{1}{2}\sqrt{5} + \frac{1}{4}\ln(2 + \sqrt{5}) \approx 1.479$

15. $\left.\frac{ds}{dt}\right|_{t=2} = 4\sqrt{10} \approx 12.65$ m/s **17.** $\left.\frac{ds}{dt}\right|_{t=9} = \sqrt{41} \approx 6.4$ m/s

19. $\left.\frac{ds}{dt}\right|_{t=0} = 1$ m/s

21. $\left(\frac{ds}{dt}\right)_{min} \approx \sqrt{4.89} \approx 2.21$ **23.** $\frac{ds}{dt} = 8$

25.

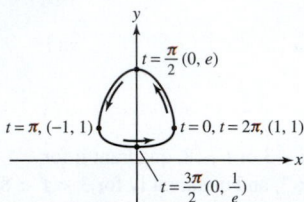

$M_{10} = 6.903734$, $M_{20} = 6.915035$, $M_{30} = 6.914949$, $M_{50} = 6.914951$

27.

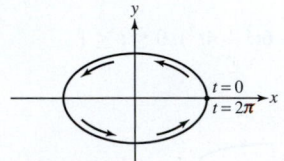

$M_{10} = 25.528309$, $M_{20} = 25.526999$, $M_{30} = 25.526999$, $M_{50} = 25.526999$

29. $s = 2\pi^2 R$ **31.** $S = \pi m A^2 \sqrt{1 + m^2}$

33. $S = \frac{\pi}{6}(5\sqrt{5} - 1)$ **35.** $S = \pi\sqrt{2}$

37. $S = 64\frac{\pi}{3}$ **39.** ≈ 8.886

41. (a)

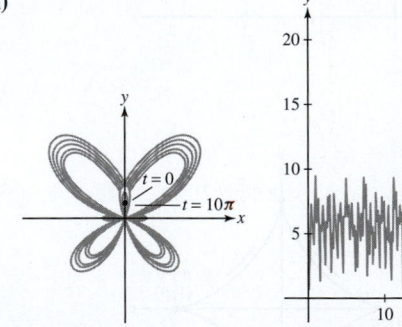

(b) $L \approx 212.096$

Section 11.3 Preliminary Questions

1. (a) **2.** Positive: $(r, \theta) = \left(1, \frac{\pi}{2}\right)$ negative: $(r, \theta) = \left(-1, \frac{3\pi}{2}\right)$

3. (a) Equation of the circle of radius 2 centered at the origin
(b) Equation of the circle of radius $\sqrt{2}$ centered at the origin
(c) Equation of the vertical line through the point $(2, 0)$ **4.** (a)

Section 11.3 Exercises

1. (A): $\left(3\sqrt{2}, \frac{3\pi}{4}\right)$ (B): $(3, \pi)$ (C): $\left(\sqrt{5}, \pi + 0.46\right) \approx \left(\sqrt{5}, 3.60\right)$

(D): $\left(\sqrt{2}, \frac{5\pi}{4}\right)$ (E): $\left(\sqrt{2}, \frac{\pi}{4}\right)$

(F): $\left(4, \frac{\pi}{6}\right)$ (G): $\left(4, \frac{11\pi}{6}\right)$

3. (a) $(1, 0)$ (b) $\left(\sqrt{12}, \frac{\pi}{6}\right)$ (c) $\left(\sqrt{8}, \frac{3\pi}{4}\right)$ (d) $\left(2, \frac{2\pi}{3}\right)$

5. (a) $\left(\frac{3\sqrt{3}}{2}, \frac{3}{2}\right)$ (b) $\left(-\frac{6}{\sqrt{2}}, \frac{6}{\sqrt{2}}\right)$ (c) $(0, 0)$ (d) $(0, -5)$

7. (A): $0 \le r \le 3$, $\pi \le \theta \le 2\pi$, (B): $0 \le r \le 3$, $\frac{\pi}{4} \le \theta \le \frac{\pi}{2}$ (C): $3 \le r \le 5$, $\frac{3\pi}{4} \le \theta \le \pi$

9. $\theta = \frac{\pi}{6}$ **11.** $m = \tan\frac{3\pi}{5} \approx -3.1$ **13.** $x^2 + y^2 = 7^2$

15. $x^2 + (y - 1)^2 = 1$ **17.** $y = x - 1$ **19.** $r = \sqrt{5}$

21. $r = \tan\theta\sec\theta$ **23.** $e^r = 1$

25. (a)↔(iii) (b)↔(iv) (c)↔(i) (d)↔(ii)

27. A: $\theta = 0, \pi, 2\pi$; B: $\theta = \frac{\pi}{4}, \frac{5\pi}{4}$; C: $\theta = \frac{\pi}{2}, \frac{3\pi}{2}$; D: $\theta = \frac{3\pi}{4}, \frac{7\pi}{4}$
(a) upper semicircle (b) lower semicircle (c) upper semicircle

29. $r\cos\left(\theta - \frac{\pi}{3}\right) = d$

31.

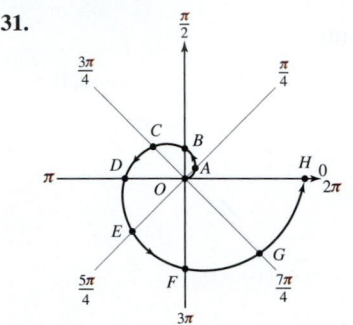

33.

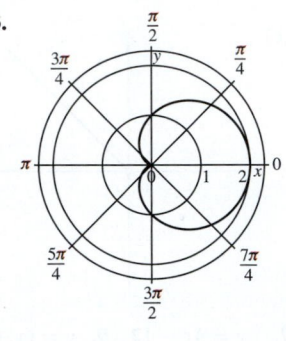

35. (a) A, $\theta = 0, r = 0$; B, $\theta = \frac{\pi}{4}, r = \sin\frac{2\pi}{4} = 1$; C, $\theta = \frac{\pi}{2}, r = 0$; D, $\theta = \frac{3\pi}{4}, r = \sin\frac{2\cdot3\pi}{4} = -1$; E, $\theta = \pi, r = 0$; F, $\theta = \frac{5\pi}{4}, r = 1$; G, $\theta = \frac{3\pi}{2}, r = 0$; H, $\theta = \frac{7\pi}{4}, r = -1$; I, $\theta = 2\pi, r = 0$

(b) $0 \le \theta \le \frac{\pi}{2}$ is in the first quadrant. $\frac{\pi}{2} \le \theta \le \pi$ is in the fourth quadrant. $\pi \le \theta \le \frac{3\pi}{2}$ is in the third quadrant. $\frac{3\pi}{2} \le \theta \le 2\pi$ is in the second quadrant.

37. (a)

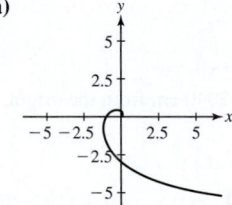

(b) $\lim_{\theta \to 2\pi^-} r\cos\theta = \infty$ and $\lim_{\theta \to 2\pi^-} r\sin\theta = -2\pi$

(c) As θ approaches 2π from the left, x approaches ∞ and y approaches -2π. Therefore, $y = -2\pi$ is a horizontal asymptote.

39.

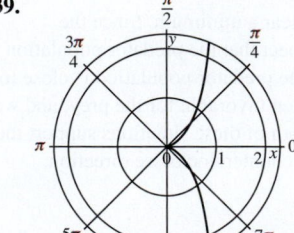

41. $\left(x - \frac{a}{2}\right)^2 + \left(y - \frac{b}{2}\right)^2 = \frac{a^2 + b^2}{4}$; $r = \frac{\sqrt{x^2 + y^2}}{2}$, centered at the point $\left(\frac{a}{2}, \frac{b}{2}\right)$

43. $r^2 = \sec 2\theta$ **45.** $(x^2 + y^2)^2 = x^3 - 3y^2 x$

47. $r = 2\sec\left(\theta - \frac{\pi}{9}\right)$ **49.** $r = 2\sqrt{10}\sec(\theta - 4.39)$

53. $r^2 = 2a^2 \cos 2\theta$

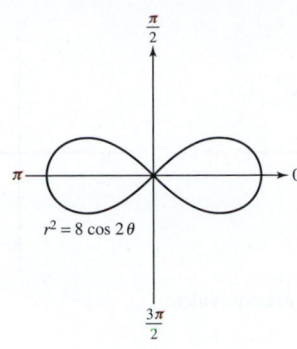

$r^2 = 8 \cos 2\theta$

57. $\theta = \frac{\pi}{2}, m = -\frac{2}{\pi}; \theta = \pi, m = \pi$

59. $\left(\frac{\sqrt{2}}{2}, \frac{\pi}{6}\right), \left(\frac{\sqrt{2}}{2}, \frac{5\pi}{6}\right), \left(\frac{\sqrt{2}}{2}, \frac{7\pi}{6}\right), \left(\frac{\sqrt{2}}{2}, \frac{11\pi}{6}\right)$

61. A: $m = 1$, B: $m = -1$, C: $m = 1$

Section 11.4 Preliminary Questions

1. (b) **2.** Yes **3.** (c)

Section 11.4 Exercises

1. $A = \frac{1}{2} \int_{\pi/2}^{\pi} r^2 \, d\theta = \frac{25\pi}{4}$

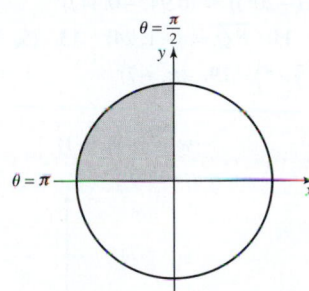

3. $A = \frac{1}{2} \int_0^{\pi} r^2 \, d\theta = 4\pi$ **5.** $A = \frac{\pi^5}{320} + \frac{\pi^4}{16} + \frac{\pi^3}{3}$

7. $A = \frac{3\pi}{2}$ **9.** $A = \frac{\pi}{8} \approx 0.39$

11. Area $= \frac{1}{2}(\pi - 2) \approx 0.571$

13. $A = \frac{\sqrt{15}}{2} + 7 \cos^{-1}\left(\frac{1}{4}\right) \approx 11.163$

15. $A = \pi - \frac{3\sqrt{3}}{2} \approx 0.54$ **17.** $A = \frac{\pi}{8} - \frac{1}{4} \approx 0.14$ **19.** $A = 4\pi$

21. $A = \frac{9\pi}{2} - 4\sqrt{2}$

23. (a)

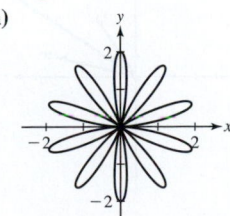

(b) Area $= \frac{3\pi}{2}$

(c) $r_n(0) = 0$. As θ goes from 0 to $\frac{\pi}{n}$ to $\frac{2\pi}{n}$, $r_n(\theta)$ increases from 0 to 2 and then decreases back to 0. Thus, a petal is traced out as θ goes from 0 to $\frac{2\pi}{n}$. This repeats over each successive interval of length $\frac{2\pi}{n}$, thus tracing out n petals as θ goes from 0 to 2π.

(d) Area $= \frac{1}{2} \int_0^{2\pi} (1 - \cos(n\theta))^2 \, d\theta =$
$\frac{1}{2} \left(\frac{3}{2} x + \frac{2 \sin(nx)}{n} + \frac{\sin(nx) \cos(nx)}{2n} \right) \Big|_0^{2\pi} = \frac{3\pi}{2}$. *Note*: this equals $\frac{3}{8}(4\pi)$ and 4π is the area of a circle of radius 2.

25. $S = 4\pi$ **27.** $L = \frac{1}{3}\left((\pi^2 + 4)^{3/2} - 8\right) \approx 14.55$

29. $L = \pi$ **31.** $L = \sqrt{2}\pi/4 \approx 1.11$

33. $L = \int_0^{2\pi} \sqrt{\cos^4\theta + 4\cos^2\theta \sin^2\theta} \, d\theta \approx 5.52$

35. $\int_0^{\pi/2} \sqrt{2e^{2\theta} + 2e^{\theta} + 1} \, d\theta$ **37.** $\int_0^{2\pi} \sin^2\theta \sqrt{1 + 8\cos^2\theta} \, d\theta$

39. $L \approx 6.682$ **41.** $L \approx 79.56$

Section 11.5 Preliminary Questions

1. (a) Hyperbola (b) Parabola (c) Ellipse
(d) Not a conic section

2. Hyperbolas **3.** The points $(0, c)$ and $(0, -c)$

4. $\pm\frac{b}{a}$ are the slopes of the two asymptotes of the hyperbola.

Section 11.5 Exercises

1. $F_1 = \left(-\sqrt{65}, 0\right), F_2 = \left(\sqrt{65}, 0\right)$. The vertices are $(9, 0)$, $(-9, 0)$, $(0, 4)$, and $(0, -4)$.

3. $F_1 = \left(\sqrt{97}, 0\right), F_2 = \left(\sqrt{97}, 0\right)$. The vertices are $(4, 0)$ and $(-4, 0)$.

5. Vertices: $(-2, -3)$, $(10, -3)$; foci: $(4 - \sqrt{61}, -3)$, $(4 + \sqrt{61}, -3)$

7. $\frac{x^2}{6^2} + \frac{y^2}{3^2} = 1$ **9.** $\frac{(x-14)^2}{6^2} + \frac{(y+4)^2}{3^2} = 1$

11. $\frac{x^2}{9} + \frac{y^2}{25} = 1$ **13.** $\frac{x^2}{(40/3)^2} + \frac{y^2}{(50/3)^2} = 1$

15. $\left(\frac{x}{3}\right)^2 - \left(\frac{y}{4}\right)^2 = 1$ **17.** $x^2 - \frac{y^2}{8} = 1$

19. $\left(\frac{x-2}{5}\right)^2 - \left(\frac{y}{10\sqrt{2}}\right)^2 = 1$ **21.** $y = 3x^2$

23. $y = \frac{1}{20}x^2$ **25.** $y = \frac{1}{16}x^2$ **27.** $x = \frac{1}{8}y^2$

29. Vertices: $(\pm 4, 0)$, $(0, \pm 2)$; foci: $\left(\pm\sqrt{12}, 0\right)$; centered at the origin

31. Vertices: $(7, -5)$, $(-1, -5)$; foci: $\left(\sqrt{65} + 3, -5\right)$, $\left(-\sqrt{65} + 3, -5\right)$; center: $(3, -5)$; asymptotes $y = \frac{7}{4}x - \frac{41}{4}$ and $y = -\frac{7}{4}x + \frac{1}{4}$

33. Vertices: $(5, 5)$, $(-7, 5)$; foci: $\left(\sqrt{84} - 1, 5\right)$, $\left(-\sqrt{84} - 1, 5\right)$; center: $(-1, 5)$; asymptotes: $y = \frac{\sqrt{48}}{6}(x + 1) + 5 \approx 1.15x + 6.15$ and $y = -\frac{\sqrt{48}}{6}(x + 1) + 5 \approx -1.15x + 3.85$

35. Vertex: $(4, 0)$; focus: $\left(4, \frac{1}{16}\right)$

37. Vertices: $\left(1 \pm \frac{5}{2}, \frac{1}{5}\right), \left(1, \frac{1}{5} \pm 1\right)$; foci: $\left(-\frac{\sqrt{21}}{2} + 1, \frac{1}{5}\right)$, $\left(\frac{\sqrt{21}}{2} + 1, \frac{1}{5}\right)$; centered at $\left(1, \frac{1}{5}\right)$

39. $D = -87$; ellipse **41.** $D = 40$; hyperbola

47. Focus: $(0, c)$; directrix: $y = -c$ **49.** $A = \frac{8}{3}c^2$ **51.** $r = \frac{3}{2 + \cos\theta}$

53. $r = \frac{4}{1 + \cos\theta}$

55. Hyperbola, $e = 4$; directrix, $x = 2$

57. Ellipse, $e = \frac{3}{4}$; directrix, $x = \frac{8}{3}$ **59.** $r = \frac{-12}{5 + 6\cos\theta}$

61. $\left(\frac{x+3}{5}\right)^2 + \left(\frac{y}{4}\right)^2 = 1$ **63.** 4.5 billion miles

65. To begin, we note that the relationships between a, b, c, and e implies $e^2 = 1 - \frac{b^2}{a^2}$. Assume $P = (x, y)$ lies on the ellipse. Now,
$(x - c)^2 + y^2 = x^2 - 2cx + c^2 - \frac{b^2 x^2}{a^2} + b^2 = \left(1 - \frac{b^2}{a^2}\right)x^2 - 2cx + c^2 + b^2 = e^2 x^2 - 2aex + a^2 = (ex - a)^2 = e^2\left(x - \frac{a}{e}\right)^2$. It follows that $\sqrt{(x - c)^2 + y^2} = e\sqrt{x - \frac{a}{e}}$, and therefore $PF = ePD$.

Chapter 11 Review

1. (a), (c)

3. $c(t) = (1 + 2\cos t, 1 + 2\sin t)$. The intersection points with the y-axis are $\left(0, 1 \pm \sqrt{3}\right)$. The intersection points with the x-axis are $\left(1 \pm \sqrt{3}, 0\right)$.

5. $c(\theta) = (\cos(\theta + \pi), \sin(\theta + \pi))$ **7.** $c(t) = (1 + 2t, 3 + 4t)$

9. $y = -\frac{x}{4} + \frac{37}{4}$ **11.** $y = -\frac{8}{(x-3)^3} + \frac{3-x}{2}$

13. $\left.\frac{dy}{dx}\right|_{t=3} = \frac{3}{14}$ **15.** $\left.\frac{dy}{dx}\right|_{t=20} = \frac{\cos 20}{e^{20}}$ **17.** $(1.41, 1.60)$

19. $c(t) = \left(-1 + 6t^2 - 4t^3, -1 + 6t - 6t^2\right)$

21. $\frac{ds}{dt} = \sqrt{3 + 2(\cos t - \sin t)}$; maximal speed: $\sqrt{3 + 2\sqrt{2}}$

23. $s = \sqrt{2}$

25.

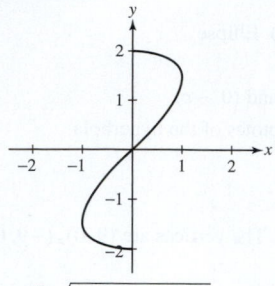

$s = 2\int_0^\pi \sqrt{\cos^2 2t + \sin^2 t}\, dt \approx 6.0972$

27. $\left(1, \frac{\pi}{6}\right)$ and $\left(3, \frac{5\pi}{4}\right)$ have rectangular coordinates $\left(\frac{\sqrt{3}}{2}, \frac{1}{2}\right)$ and $\left(-\frac{3\sqrt{2}}{2}, -\frac{3\sqrt{2}}{2}\right)$.

29. $\sqrt{x^2 + y^2} = \frac{2x}{x-y}$ **31.** $r = 3 + 2\sin\theta$

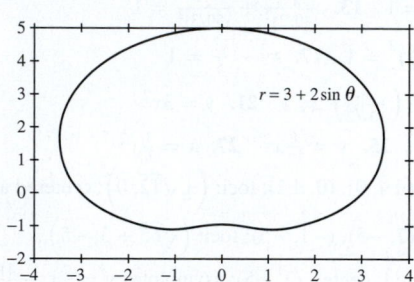

33. $A = \frac{\pi}{16}$ **35.** $e - \frac{1}{e}$
  Note: One needs to double the integral from $-\frac{\pi}{2}$ to $\frac{\pi}{2}$ in order to account for both sides of the graph.

37. $A = \frac{3\pi a^2}{2}$

39. Outer: $L \approx 36.121$; inner: $L \approx 7.5087$; difference: 28.6123

41. Ellipse. Vertices: $(\pm 3, 0)$, $(0, \pm 2)$; foci: $(\pm\sqrt{5}, 0)$

43. Ellipse. Vertices: $\left(\pm\frac{2}{\sqrt{5}}, 0\right)$, $\left(0, \pm\frac{4}{\sqrt{5}}\right)$; foci: $\left(0, \pm\sqrt{\frac{12}{5}}\right)$

45. $\left(\frac{x}{8}\right)^2 + \left(\frac{y}{\sqrt{61}}\right)^2 = 1$ **47.** $\left(\frac{x}{8}\right)^2 - \left(\frac{y}{6}\right)^2 = 1$ **49.** $x = \frac{1}{32}y^2$

51. $y = \sqrt{3}x + \left(\sqrt{3} - 5\right)$ and $y = -\sqrt{3}x + \left(-\sqrt{3} - 5\right)$

Chapter 12

Section 12.1 Preliminary Questions

1. (a) True (b) False (c) True (d) True
2. $\|-3\mathbf{a}\| = 15$ **3.** The components are not changed. **4.** $\langle 0, 0\rangle$
5. (a) True (b) False

Section 12.1 Exercises

1. $\mathbf{v}_1 = \langle 2, 0\rangle$, $\|\mathbf{v}_1\| = 2$ $\mathbf{v}_2 = \langle 2, 0\rangle$, $\|\mathbf{v}_2\| = 2$

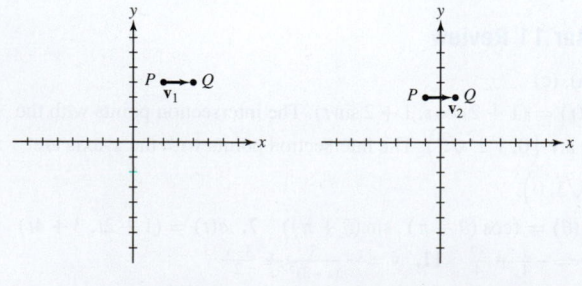

$\mathbf{v}_3 = \langle 3, 1\rangle$, $\|\mathbf{v}_3\| = \sqrt{10}$ $\mathbf{v}_4 = \langle 2, 2\rangle$, $\|\mathbf{v}_4\| = 2\sqrt{2}$

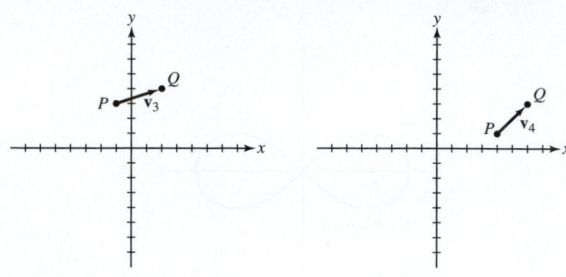

Vectors $\mathbf{v}_1$ and $\mathbf{v}_2$ are equivalent.

3. $(3, 5)$

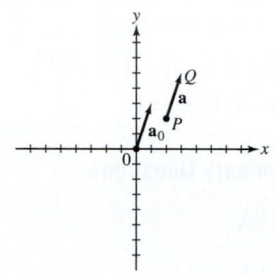

5. $\left\langle\frac{\sqrt{2}}{2}, \frac{\sqrt{2}}{2}\right\rangle \approx \langle 0.707, 0.707\rangle$

7. $\langle\cos(-20°), \sin(-20°)\rangle \approx \langle 0.94, -0.342\rangle$

9. $\overrightarrow{PQ} = \langle -1, 5\rangle$ **11.** $\overrightarrow{PQ} = \langle -1, 24\rangle$ **13.** $\langle 5, 5\rangle$

15. $\langle 30, 10\rangle$ **17.** $\left(\frac{5}{2}, 5\right)$ **19.** $\langle\pi, -7\rangle$

21. Vector (B)

23. $2\mathbf{v} = \langle 4, 6\rangle$ $-\mathbf{w} = \langle -4, -1\rangle$

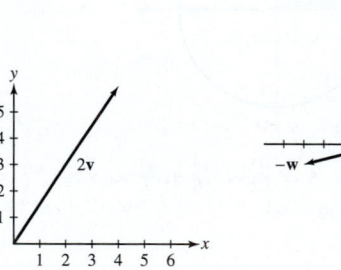

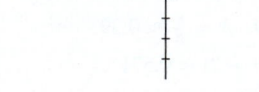

$2\mathbf{v} - \mathbf{w} = \langle 0, 5\rangle$ $\mathbf{v} + \mathbf{w} = \langle 6, 4\rangle$

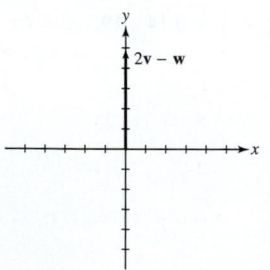

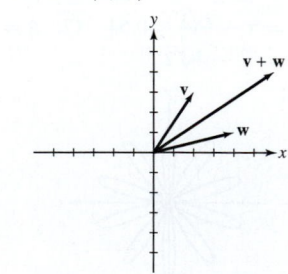

25. $3\mathbf{v} + \mathbf{w} = \langle -2, 10\rangle$, $2\mathbf{v} - 2\mathbf{w} = \langle 4, -4\rangle$

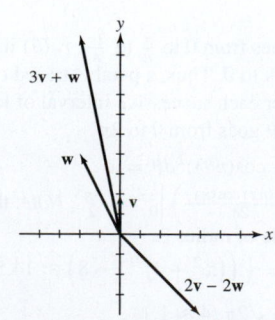

27.

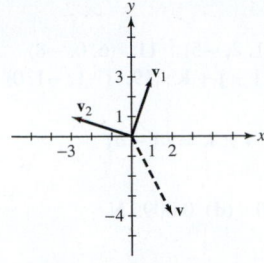

29. (b) and (c)   **31.** $\overrightarrow{AB} = \langle 2, 6 \rangle$ and $\overrightarrow{PQ} = \langle 2, 6 \rangle$; equivalent

33. $\overrightarrow{AB} = \langle 3, -2 \rangle$ and $\overrightarrow{PQ} = \langle 3, -2 \rangle$; equivalent

35. $\overrightarrow{AB} = \langle 2, 3 \rangle$ and $\overrightarrow{PQ} = \langle 6, 9 \rangle$; parallel and point in the same direction   **37.** $\overrightarrow{AB} = \langle -8, 1 \rangle$ and $\overrightarrow{PQ} = \langle 8, -1 \rangle$; parallel and point in opposite directions   **39.** $\left\| \overrightarrow{OR} \right\| = \sqrt{53}$ **41.** $P = (0, 0)$ **43.** $\left\langle \frac{3}{5}, \frac{4}{5} \right\rangle$

45. $4\mathbf{e_u} = \left\langle -2\sqrt{2}, -2\sqrt{2} \right\rangle$

47. $2\mathbf{e_{-v}} = -\sqrt{2}\mathbf{i} + \sqrt{2}\mathbf{j}$

49. $\mathbf{e} = \left\langle \cos \frac{4\pi}{7}, \sin \frac{4\pi}{7} \right\rangle \approx \langle -0.22, 0.97 \rangle$

51. $\frac{1}{\sqrt{5}} \langle -1, 2 \rangle$ **53.** $\lambda = \pm \frac{1}{\sqrt{13}}$ **55.** $P = (4, 6)$

57. (a) → (ii), (b) → (iv), (c) → (iii), (d) → (i)   **59.** $9\mathbf{i} + 7\mathbf{j}$

61. $-5\mathbf{i} - 3\mathbf{j}$

63.

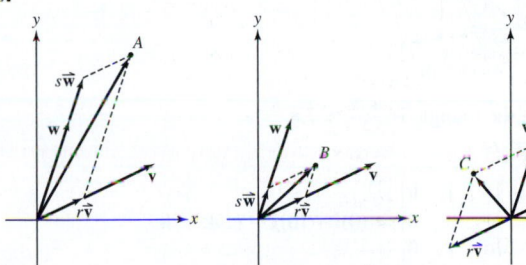

65. $\mathbf{u} = 2\mathbf{v} - \mathbf{w}$

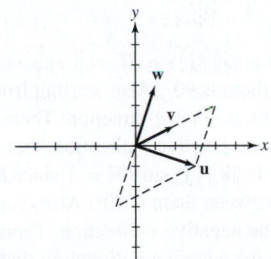

67. The force on cable 1 is $\approx$ 444 newtons, and force on cable 2 is $\approx$ 207 newtons

69. 230 km/h   **71.** $\mathbf{r} \approx \langle 6.45, 0.38 \rangle$

Section 12.2 Preliminary Questions

1. $(4, 3, 2)$ **2.** $\langle 3, 2, 1 \rangle$ **3.** (a)   **4.** (c)

5. Infinitely many direction vectors   **6.** True   **7.** $\sqrt{5}$

8. $(0, 0, 0), (0, 0, -1), (1, -1, 1), (1, 1, 0)$

Section 12.2 Exercises

1. $\|\mathbf{v}\| = \sqrt{14}$

3. The head of $\mathbf{v} = \overrightarrow{PQ}$ is $Q = (1, 2, 1)$.

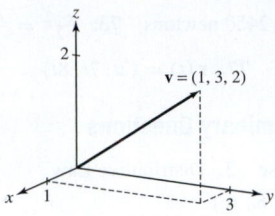

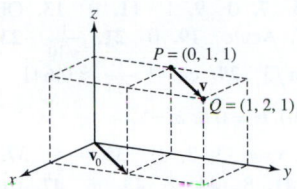

5. $\overrightarrow{PQ} = \langle 1, 1, -1 \rangle$ **7.** $\overrightarrow{PQ} = \left\langle -\frac{9}{2}, -\frac{3}{2}, 1 \right\rangle$

9. $\left\| \overrightarrow{OR} \right\| = \sqrt{26} \approx 5.1$ **11.** $P = (-2, 6, 0)$

13. (a) Parallel and same direction   (b) Not parallel
(c) Parallel and opposite directions   (d) Not parallel

15. Not equivalent   **17.** Not equivalent

19. $\langle -8, -18, -2 \rangle$ **21.** $\langle -2, -2, 3 \rangle$

23. $\langle 16, -1, 9 \rangle$ **25.** Not parallel   **27.** Not parallel

29. $\mathbf{e_w} = \left\langle \frac{4}{\sqrt{21}}, \frac{-2}{\sqrt{21}}, \frac{-1}{\sqrt{21}} \right\rangle$ **31.** $-\mathbf{e_v} = \left\langle \frac{2}{3}, -\frac{2}{3}, -\frac{1}{3} \right\rangle$

33. The top half of a sphere of radius 2, centered at $(0, 0, 2)$

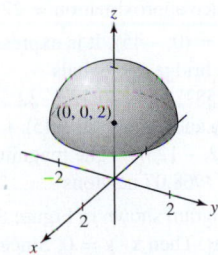

35. The part of the cylinder of radius $\sqrt{7}$, with the z-axis as central axis, lying between $z = -7$ and $z = 7$

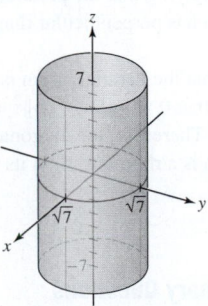

37. $x^2 + y^2 + (z + 3)^2 = 9$

39. $(x - 6)^2 + (y + 3)^2 + (z - 11)^2 = 277$

41. $(x - 1)^2 + (y + 1)^2 = 2$

43. $\mathbf{r}(t) = \langle 1 + 2t, 2 + t, -8 + 3t \rangle$ **45.** $\mathbf{r}(t) = \langle 4 + 7t, 0, 8 + 4t \rangle$

47. $\mathbf{r}(t) = \langle 1 + 2t, 1 - 6t, 1 + t \rangle$ **49.** $\mathbf{r}(t) = \langle 4t, t, t \rangle$

51. $\mathbf{r}(t) = \langle 0, 0, t \rangle$ **53.** $\mathbf{r}(t) = \langle -t, -2t, 4 - 2t \rangle$ **55.** (c)

59. $\mathbf{r_1}(t) = \langle 5, 5, 2 \rangle + t \langle 0, -2, 1 \rangle$;
$\mathbf{r_2}(t) = \langle 5, 5, 2 \rangle + t \langle 0, -20, 10 \rangle$

63. The lines intersect at $(3, 4, 7)$.

65. 4 min   **67.** $(-1, -3, 2)$ and $\left(\frac{23}{17}, -\frac{11}{17}, -\frac{26}{17} \right)$

69. $\left\langle 0, \frac{1}{2}, -\frac{1}{2} \right\rangle$ **71.** 2450 newtons **73.** $\frac{x+2}{2} = \frac{y-3}{4} = \frac{z-3}{3}$

75. $\frac{x-3}{2} = \frac{y-4}{-9} = \frac{z}{12}$ **77.** $\mathbf{r}(t) = \langle 2t, 7t, 8t \rangle$

Section 12.3 Preliminary Questions

1. Scalar **2.** Obtuse **3.** Distributive Law
4. (a) **5.** (b); (c) **6.** (c)

Section 12.3 Exercises

1. 15 **3.** 2 **5.** 5 **7.** 0 **9.** 1 **11.** 0 **13.** Obtuse
15. Orthogonal **17.** Acute **19.** 0 **21.** $\frac{1}{\sqrt{10}}$ **23.** $\pi/4$
25. ≈ 0.615 **27.** $2\pi/3$ **29.** $\cos^{-1}\frac{3}{\sqrt{14}} \approx 0.641$
31. (a) $b = -\frac{1}{2}$ (b) $b = 0$ or $b = \frac{1}{2}$
33. $\mathbf{v}_1 = \langle 0, 1, 0 \rangle$, $\mathbf{v}_2 = \langle 3, 2, 2 \rangle$ **35.** $-\frac{3}{2}$ **37.** $\|\mathbf{v}\|^2$
39. $\|\mathbf{v}\|^2 - \|\mathbf{w}\|^2$ **41.** 8 **43.** 2 **45.** π **47.** (b) 7 **51.** $51.91°$
53. (a) (b) $\mathbf{u}_{\|\mathbf{v}}$

55. $\left\langle \frac{7}{2}, \frac{7}{2} \right\rangle$ **57.** $\left\langle -\frac{4}{5}, 0, -\frac{2}{5} \right\rangle$ **59.** $-4\mathbf{k}$ **61.** $a\mathbf{i}$ **63.** $2\sqrt{2}$
65. $\sqrt{17}$ **67.** $\mathbf{a} = \left\langle \frac{1}{2}, \frac{1}{2} \right\rangle + \left\langle \frac{1}{2}, -\frac{1}{2} \right\rangle$
69. $\mathbf{a} = \left\langle 0, -\frac{1}{2}, -\frac{1}{2} \right\rangle + \left\langle 4, -\frac{1}{2}, \frac{1}{2} \right\rangle$
71. $\left\langle \frac{x-y}{2}, \frac{y-x}{2} \right\rangle + \left\langle \frac{x+y}{2}, \frac{y+x}{2} \right\rangle$
75. $\approx 35°$ **77.** $\overrightarrow{AD}$ **79.** Diagonal length $= \sqrt{52,359.625}$
≈ 228.82 in.; split difference approximation $= 228\frac{3}{4}$ in. **81.** $\approx 109.5°$
85. The wind vector is $\mathbf{w} = \langle 0, -45 \rangle$. It is expressed as $\mathbf{w} = \mathbf{w}_{\|} + \mathbf{w}_{\perp}$, where $\mathbf{w}_{\|}$ is parallel to the bridge and equals
$-45 \sin 58° \langle \cos 58°, \sin 58° \rangle \approx \langle -20.22, -32.36 \rangle$, while $\mathbf{w}_{\perp}$ is perpendicular to the bridge and equals $\langle 0, -45 \rangle + 45 \sin 58°$
$\langle \cos 58°, \sin 58° \rangle \approx \langle 20.22, -12.64 \rangle$. The magnitude of the perpendicular term ≈ 23.85 km/h. **87.** ≈ 68.07 newtons
93. Consider the parallelogram shown in Figure 8. Assume that the diagonals are perpendicular. Then $\mathbf{x} \cdot \mathbf{y} = 0$. Since $\mathbf{x} = \mathbf{a} + \mathbf{b}$ and $\mathbf{y} = \mathbf{a} - \mathbf{b}$, it follows that

$$0 = (\mathbf{a} + \mathbf{b}) \cdot (\mathbf{a} - \mathbf{b}) = \mathbf{a} \cdot \mathbf{a} - \mathbf{b} \cdot \mathbf{b} = \|\mathbf{a}\|^2 - \|\mathbf{b}\|^2$$

Therefore, $\|\mathbf{a}\| = \|\mathbf{b}\|$, implying that the parallelogram is a rhombus. Thus, if the parallelogram has perpendicular diagonals, then it is a rhombus.
Conversely, assume that the parallelogram is a rhombus. Then $\|\mathbf{a}\| = \|\mathbf{b}\|$. This implies that $0 = \|\mathbf{a}\|^2 - \|\mathbf{b}\|^2 = \mathbf{a} \cdot \mathbf{a} - \mathbf{b} \cdot \mathbf{b} = (\mathbf{a} + \mathbf{b}) \cdot (\mathbf{a} - \mathbf{b}) = \mathbf{x} \cdot \mathbf{y}$. Therefore, the diagonals are perpendicular. Thus, if the parallelogram is a rhombus, then its diagonals are perpendicular.
105. $2x + 2y - 2z = 1$

Section 12.4 Preliminary Questions

1. $\begin{vmatrix} -5 & -1 \\ 4 & 0 \end{vmatrix}$ **2.** $\|\mathbf{e} \times \mathbf{f}\| = \frac{1}{2}$ **3.** $\mathbf{u} \times \mathbf{w} = \langle -2, -2, -1 \rangle$
4. (a) 0 (b) 0 **5.** $\mathbf{i} \times \mathbf{j} = \mathbf{k}$ and $\mathbf{i} \times \mathbf{k} = -\mathbf{j}$
6. $\mathbf{v} \times \mathbf{w} = \mathbf{0}$ if either $\mathbf{v}$ or $\mathbf{w}$ (or both) is the zero vector or $\mathbf{v}$ and $\mathbf{w}$ are parallel vectors.
7. (a) Not meaningful because you cannot find the cross product of a scalar and a vector
(b) Meaningful because it represents the dot product of two vectors
(c) Meaningful because it is the product of two scalars
(d) Meaningful because it is a scalar multiple of a vector
8. (b)

Section 12.4 Exercises

1. -5 **3.** -15 **5.** -8 **7.** 0 **9.** $\langle 1, 2, -5 \rangle$ **11.** $\langle 6, 0, -8 \rangle$
13. $\langle 0.02, -0.01, 0 \rangle$ **15.** $-\mathbf{j} + \mathbf{i}$ **17.** $\mathbf{i} + \mathbf{j} + \mathbf{k}$ **19.** $\langle -1, -1, 0 \rangle$
21. $\langle -2, -2, -2 \rangle$ **23.** $\langle 4, 4, 0 \rangle$
25. $\mathbf{v} \times \mathbf{i} = c\mathbf{j} - b\mathbf{k}$; $\mathbf{v} \times \mathbf{j} = -c\mathbf{i} + a\mathbf{k}$; $\mathbf{v} \times \mathbf{k} = b\mathbf{i} - a\mathbf{j}$
27. $-\mathbf{u}$ **29.** $\langle 0, 3, 3 \rangle$ **33.** $\mathbf{e}'$
35. (a) 0.0067 N (b) 0.0067 N (c) 0 (d) 0.0095 N
37. $\mathbf{F}_1$ **41.** $2\sqrt{138}$
43. The volume is 4.

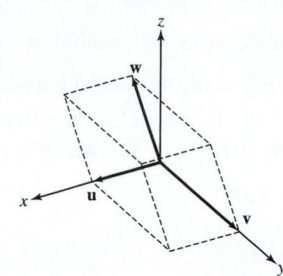

45. $\sqrt{35} \approx 5.92$
47.

The area of the triangle is $\frac{9\sqrt{3}}{2} \approx 7.8$.
49. 3 **51.** $\frac{33}{2}\sqrt{3}$
57. (a) $\mathbf{i} \times \mathbf{j} = \begin{vmatrix} \mathbf{i} & \mathbf{j} & \mathbf{k} \\ 1 & 0 & 0 \\ 0 & 1 & 0 \end{vmatrix} = (0)\mathbf{i} + (0)\mathbf{j} + (1)\mathbf{k} = \mathbf{k}$

$$\mathbf{k} \times \mathbf{j} = \begin{vmatrix} \mathbf{i} & \mathbf{j} & \mathbf{k} \\ 0 & 0 & 1 \\ 0 & 1 & 0 \end{vmatrix} = (-1)\mathbf{i} + (0)\mathbf{j} + (0)\mathbf{k} = -\mathbf{i}$$

(b) The length of $\mathbf{i} \times \mathbf{j}$ is $\|\mathbf{i}\| \|\mathbf{j}\| \sin 90° = 1$ since $\mathbf{i}$ and $\mathbf{j}$ are unit vectors and the angle between them is $90°$. Also, curling from $\mathbf{i}$ to $\mathbf{j}$, the right-hand rule yields the positive z-direction. Therefore, $\mathbf{i} \times \mathbf{j}$ is a vector of length 1, pointing in the positive z-direction; that is, $\mathbf{i} \times \mathbf{j} = \mathbf{k}$.
The length of $\mathbf{k} \times \mathbf{j}$ is $\|\mathbf{k}\| \|\mathbf{j}\| \sin 90° = 1$ since $\mathbf{k}$ and $\mathbf{j}$ are unit vectors and the angle between them is $90°$. Also, curling from $\mathbf{k}$ to $\mathbf{j}$, the right-hand rule yields the negative x-direction. Therefore $\mathbf{k} \times \mathbf{j}$ is a vector of length 1, pointing in the negative x-direction; that is, $\mathbf{k} \times \mathbf{j} = -\mathbf{i}$.
63. $\mathbf{X} = \langle a, a, a+1 \rangle$ **67.** $\tau = 250 \sin 125° \, \mathbf{k} \approx 204.79 \, \mathbf{k}$ N-m

Section 12.5 Preliminary Questions

1. $3x + 4y - z = 0$ **2.** (c): $z = 1$ **3.** Plane (c) **4.** xz-plane
5. (c): $x + y = 0$ **6.** Statement (a)

Section 12.5 Exercises

1. $x + 3y + 2z = 3$ **3.** $-x + 2y + z = 3$ **5.** $x = 3$
7. $z = 2$ **9.** $x = 0$ **11.** Statements (b) and (d)
13. $\langle 9, -4, -11 \rangle$ **15.** $\langle 3, -8, 11 \rangle$ **17.** $4x - 9y + z = 0$
19. $x = 4$ **21.** $6x + 9y + 4z = 19$ **23.** $x + 2y - z = 1$
25. (a) Let the points be P, Q, and R. Then $\mathbf{n} = \overrightarrow{PQ} \times \overrightarrow{PR}$ is a normal vector to the plane. (b) Let $\mathbf{v}_1$ and $\mathbf{v}_2$ be direction vectors for the lines. Then $\mathbf{n} = \mathbf{v}_1 \times \mathbf{v}_2$ is a normal vector to the plane.

27. (a) Do not intersect **(b)** Have a single point of intersection; an equation for the plane containing the lines is $x - y - z = 4$.

29. (a) The point is on the line. **(b)** The point is not on the line. An equation for the plane containing the point and the line is $x - y = 6$.

31. (a) They are distinct parallel lines. An equation for the plane containing the lines is $4x - 5y + 6z = -13$.

(b) The lines are not parallel.

33.
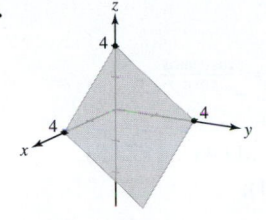

35.
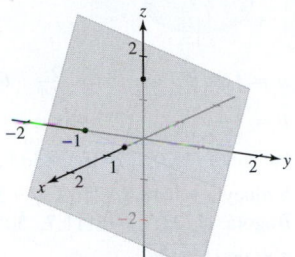

37.
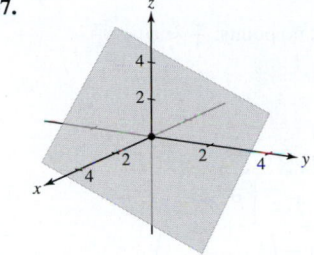

39. $10x + 15y + 6z = 30$ **41.** $(1, 5, 8)$ **43.** $(-2, 3, 12)$
45. $-9y + 4z = 5$ **47.** $x = -\frac{2}{3}$ **49.** $x = -4$

51. The two planes have no common points.

53. $y - 4z = 0$
$x + y - 4z = 0$

55. $(3\lambda)x + by + (2\lambda)z = 5\lambda$, $\lambda \neq 0$ **57.** $\theta = \pi/2$
59. $\theta \approx 1.143$ radians or $\theta \approx 65.49°$ **61.** $\theta \approx 55.0°$
63. $x + y + z = 1$ **65.** $x - y - z = d/a$
67. $x = \frac{9}{5} + 2t$, $y = -\frac{6}{5} - 3t$, $z = 2 + 5t$ **69.** $\pm 24 \langle 1, 2, -2 \rangle$
75. $\left(\frac{2}{3}, -\frac{1}{3}, \frac{2}{3}\right)$ **77.** $\frac{6}{\sqrt{30}} \approx 1.095$ **79.** $|a|$

Section 12.6 Preliminary Questions

1. True, mostly, except at $x = \pm a$, $y = \pm b$, or $z = \pm c$

2. False **3.** Hyperbolic paraboloid **4.** No **5.** Ellipsoid

6. All vertical lines passing through a parabola c in the xy-plane

Section 12.6 Exercises

1. Ellipsoid **3.** Ellipsoid **5.** Hyperboloid of one sheet

7. Ellipsoid **9.** Elliptic paraboloid **11.** Hyperbolic paraboloid

13. Hyperbolic paraboloid **15.** Elliptic cone

17. Ellipsoid; the trace is a circle on the xz-plane.

19. Ellipsoid; the trace is an ellipse parallel to the xy-plane.

21. Hyperboloid of one sheet; the trace is a hyperbola.

23. Parabolic cylinder; the trace is the parabola $y = 3x^2$.

25. (a) ↔ Figure b; (b) ↔Figure c; (c) ↔ Figure a

27. $y = \left(\frac{x}{2}\right)^2 + \left(\frac{z}{4}\right)^2$

29.
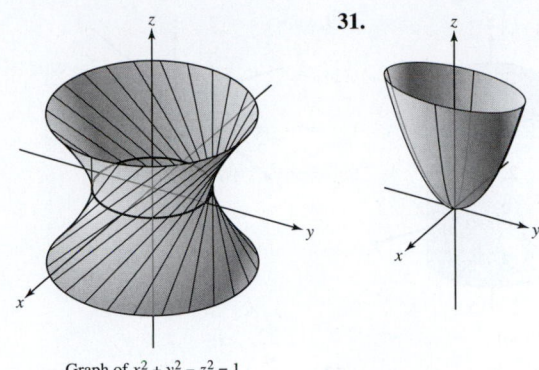

Graph of $x^2 + y^2 - z^2 = 1$

31.
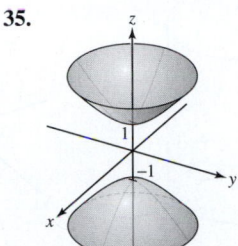

33.
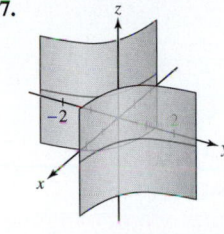

35.
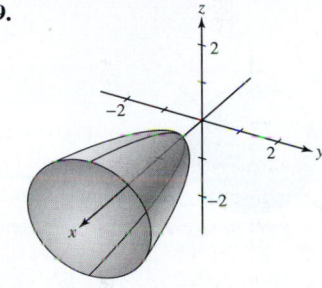

37.
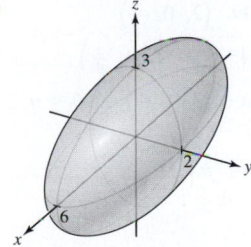

39.

41.

43. $\left(\frac{x}{2}\right)^2 + \left(\frac{y}{4}\right)^2 + \left(\frac{z}{6}\right)^2 = 1$ **45.** $\left(\frac{x}{4}\right)^2 + \left(\frac{y}{6}\right)^2 - \left(\frac{z}{3\sqrt{3}}\right)^2 = 1$

47. One or two vertical lines, or an empty set **49.** An elliptic cone

Section 12.7 Preliminary Questions

1. Cylinder of radius R whose axis is the z-axis, sphere of radius R centered at the origin

2. (b) **3.** (a) **4.** $\phi = 0$, π **5.** $\phi = \frac{\pi}{2}$; the xy-plane

Section 12.7 Exercises

1. $(-4, 0, 4)$ **3.** $\left(0, 0, \frac{1}{2}\right)$ **5.** $\left(\sqrt{2}, \frac{7\pi}{4}, 1\right)$ **7.** $\left(2, \frac{\pi}{3}, 7\right)$

9. $\left(5, \frac{\pi}{4}, 2\right)$ **11.** $r^2 \leq 3$ **13.** $r^2 + z^2 \leq 4$, $\theta = \frac{\pi}{2}$ or $\theta = \frac{3\pi}{2}$

15. $r^2 \leq 9$, $\frac{5\pi}{4} \leq \theta \leq 2\pi$ and $0 \leq \theta \leq \frac{\pi}{4}$

17.
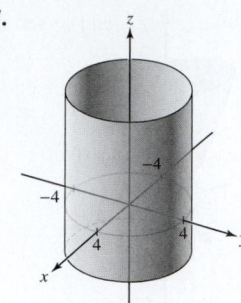

19.
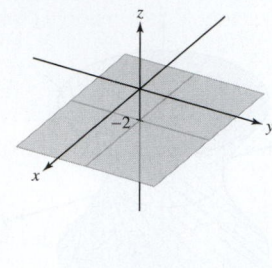

21.
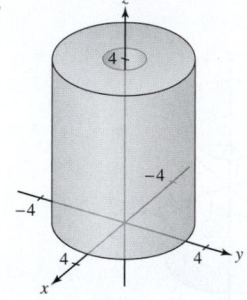

23.
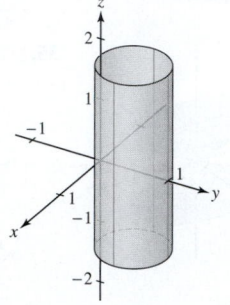

25.
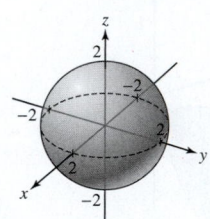

63.
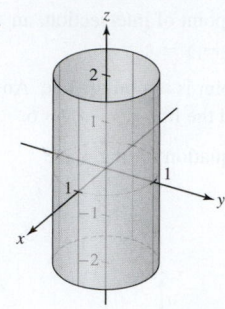

65. $\rho = 3 \csc \phi$ **67.** $\rho = \frac{2}{\cos \phi}$ **69.** $\rho = \frac{\cos \theta \tan \phi}{\cos \phi}$

71. $\rho = \frac{2}{\sin \phi \sqrt{\cos 2\theta}}$ **73.** (b)

75. Helsinki: $(25.0°, \ 29.9°)$; São Paulo: $(313.48°, \ 113.52°)$

77. Sydney: $(-4618.8, \ 2560.3, \ -3562.1)$;
Bogotá: $(1723.7, \ -6111.7, \ 503.5)$

79. $z = \pm r \sqrt{\cos 2\theta}$

81. $\left\{ (r, \theta, z) : -\sqrt{4 - r^2} \le z \le \sqrt{4 - r^2}, 1 \le r \le 2, 0 \le \theta \le 2\pi \right\}$

85. $r = \sqrt{z^2 + 1}$ and $\rho = \sqrt{-\frac{1}{\cos 2\phi}}$; no points; $\frac{\pi}{4} < \phi < \frac{3\pi}{4}$

Chapter 12 Review

1. $\langle 21, \ -25 \rangle$ and $\langle -19, \ 31 \rangle$ **3.** $\left\langle \frac{-2}{\sqrt{29}}, \ \frac{5}{\sqrt{29}} \right\rangle$

5. $\mathbf{i} = \frac{2}{11}\mathbf{v} + \frac{5}{11}\mathbf{w}$ **7.** $\overrightarrow{PQ} = \langle -4, \ 1 \rangle$; $\left\| \overrightarrow{PQ} \right\| = \sqrt{17}$

9. $\left\langle \frac{3}{\sqrt{2}}, \ -\frac{3}{\sqrt{2}} \right\rangle$ **11.** $\beta = \frac{3}{2}$ **13.** $\mathbf{u} = \left\langle \frac{1}{3}, \ -\frac{11}{6}, \ \frac{7}{6} \right\rangle$

15. $\mathbf{r}_1(t) = \langle 1 + 3t, \ 4 + t, \ 5 + 6t \rangle$; $\mathbf{r}_2(t) = \langle 1 + 3t, \ t, \ 6t \rangle$

17. $a = -2, \ b = 2$

19.
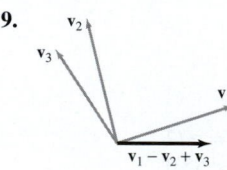

27. $r = \frac{z}{\cos \theta + \sin \theta}$ **29.** $r = \frac{z \tan \theta}{\cos \theta}$ **31.** $r = 2$ **33.** $(3, 0, 0)$

35. $(0, 0, 3)$ **37.** $\left(\frac{3\sqrt{3}}{2}, \ \frac{3}{2}, \ -3\sqrt{3} \right)$ **39.** $\left(2, 0, \frac{\pi}{3} \right)$

41. $\left(\sqrt{3}, \ \frac{\pi}{4}, \ 0.955 \right)$ **43.** $\left(2, \ \frac{\pi}{3}, \ \frac{\pi}{6} \right)$ **45.** $\left(2\sqrt{2}, \ 0, \ \frac{\pi}{4} \right)$

47. $\left(2\sqrt{2}, 0, 2\sqrt{2} \right)$ **49.** $\rho \le 10$ **51.** $\rho = \sqrt{10}, \ 0 \le \phi, \theta \le \pi/2$

53. $\left\{ (\rho, \ \theta, \ \phi) : 0 \le \rho \le 2, \ \theta = \frac{\pi}{2} \text{ or } \theta = \frac{3\pi}{2} \right\}$

21. $\mathbf{v} \cdot \mathbf{w} = -9$ **23.** $\mathbf{v} \times \mathbf{w} = \langle 10, \ -8, \ -7 \rangle$ **25.** $V = 48$ **29.** $\frac{5}{3}$

31. $\|\mathbf{F}_1\| = \frac{2\|\mathbf{F}_2\|}{\sqrt{3}}$; $\|\mathbf{F}_1\| = 980$ N

33. $\mathbf{v} \times \mathbf{w} = \langle -6, \ 7, \ -2 \rangle$ **35.** -47 **37.** $5\sqrt{2}$

41. $\|\mathbf{e} - 4\mathbf{f}\| = \sqrt{13}$ **47.** $(x - 0) + 4(y - 1) - 3(z + 1) = 0$

49. $17x - 21y - 13z = -28$ **51.** $3x - 2y = 4$ **53.** Ellipsoid

55. Elliptic paraboloid   **57.** Elliptic cone

59. (a) Empty set   (b) Hyperboloid of one sheet
(c) Hyperboloid of two sheets

61. $(r, \ \theta, \ z) = \left(5, \ \tan^{-1} \frac{4}{3}, \ -1 \right)$,
$(\rho, \ \theta, \ \phi) = \left(\sqrt{26}, \ \tan^{-1} \frac{4}{3}, \ \cos^{-1} \left(\frac{-1}{\sqrt{26}} \right) \right)$

63. $(r, \ \theta, \ z) = \left(\frac{3\sqrt{3}}{2}, \ \frac{\pi}{6}, \ \frac{3}{2} \right)$

65. $z = 2x$

55.
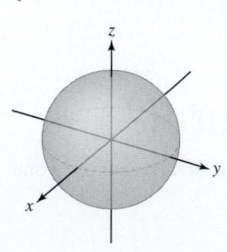

57.
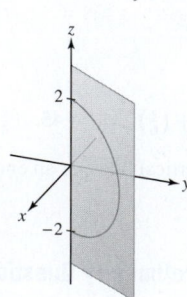

59.
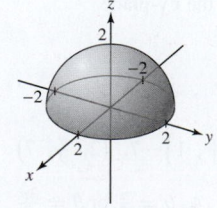

61.
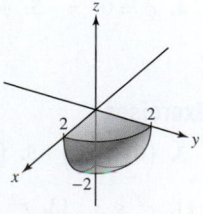

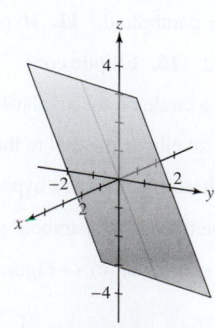

69. $A < -1$: Hyperboloid of one sheet
$A = -1$: Cylinder with the z-axis as its central axis
$A > -1$: Ellipsoid
$A = 0$: Sphere

Chapter 13

Section 13.1 Preliminary Questions

1. (c)   **2.** The curve $z = e^x$

3. The projection onto the xz-plane   **4.** The point $(-2, 2, 3)$

5. As t increases from 0 to 2π, a point on $\sin t\mathbf{i} + \cos t\mathbf{j}$ moves clockwise and a point on $\cos t\mathbf{i} + \sin t\mathbf{j}$ moves counterclockwise.

6. (a), (c), and (d)

Section 13.1 Exercises

1. $D = \{t \in \mathbf{R}, \ t \neq 0, \ t \neq -1\}$

3. $\mathbf{r}(2) = \left\langle 0, 4, \frac{1}{5} \right\rangle$; $\mathbf{r}(-1) = \left\langle -1, 1, \frac{1}{2} \right\rangle$

5. $\mathbf{r}(t) = (3 + 3t)\mathbf{i} - 5\mathbf{j} + (7 + t)\mathbf{k}$

7. Yes, when $t = (2n - 1)\pi$, where n is an integer, at the points $\left(0, 0, (2n - 1)\pi\right)$

9. No intersection

11. The path given by $\mathbf{r}(t)$ intersects the yz-plane when $1 - \cos(2t) = 0$, so for $t = n\pi$ for all integers n. The corresponding points are $(0, n\pi, (n\pi)^2)$. Since $1 - \cos(2t) \geq 0$ for all t, the x-coordinate of every point on the curve is nonnegative, and therefore no point on the curve is on the negative x side of the yz-plane. Thus, the curve doesn't cross through the yz-plane.

13. A ↔ ii;   B ↔ i;   C ↔ iii

15. (a) = (v); (b) = (i); (c) = (ii); (d) = (vi); (e) = (iv); (f) = (iii)

17. C ↔ i;   A ↔ ii;   B ↔ iii

19. This is a circle of radius 9 centered at the origin lying in the xy-plane.

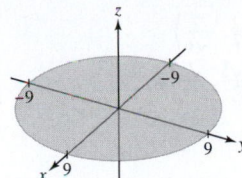

21. Radius 1, center $(0, 0, 4)$, xz-plane

23. (a) $x^2 + y^2 + z^2 = \cos^2(2t) \sin^2 t + \sin^2(2t) + \cos^2(2t)$
$\cos^2 t = \cos^2(2t)(\cos^2 t + \sin^2 t) + \sin^2(2t) = \cos^2(2t) + \sin^2(2t) = 1$. Since $x^2 + y^2 + z^2 = 1$, the curve $\mathcal{C}$ lies on the sphere of radius 1 centered at the origin.

(b) $\mathbf{r}(3\pi/2) = \langle 1, 0, 0 \rangle$; $\mathbf{r}(\pi/4) = \langle 0, 1, 0 \rangle$; $\mathbf{r}(0) = \langle 0, 0, 1 \rangle$

25.

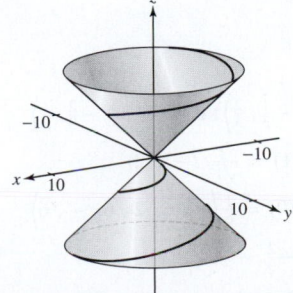

27. $(0, 1, 0), (0, -1, 0), \left(\frac{1}{\sqrt{2}}, \frac{1}{\sqrt{2}}, 0\right), \left(\frac{1}{\sqrt{2}}, -\frac{1}{\sqrt{2}}, 0\right),$ $\left(-\frac{1}{\sqrt{2}}, -\frac{1}{\sqrt{2}}, 0\right), \left(-\frac{1}{\sqrt{2}}, \frac{1}{\sqrt{2}}, 0\right)$

29. $x = 2t^2 - 7$; $y = t$; $z = \sqrt{(9 - t^2)}$, with $-3 \leq t \leq 3$

31. (a) $\mathbf{r}(t) = \left\langle \pm t\sqrt{1 - t^2}, \ t^2, \ t \right\rangle$ for $-1 \leq t \leq 1$

(b) The projection is a circle in the xy-plane with radius $\frac{1}{2}$ and centered at the xy-point $\left(0, \frac{1}{2}\right)$.

33. $x = \cos(t)$; $y = \sin(t)$; $z = 1 - \cos(t) - \sin(t)$, with $0 \leq t < 2\pi$.

35. $\mathbf{r}(t) = \langle \cos t, \ \sin t, \ 4\cos^2 t \rangle, 0 \leq t \leq 2\pi$

37. Collide at the point $(12, 4, 2)$ and intersect at the points $(4, 0, -6)$ and $(12, 4, 2)$

39. $\mathbf{r}(t) = \langle 3, 2, t \rangle, \ -\infty < t < \infty$

41. $\mathbf{r}(t) = \langle t, 3t, 15t \rangle, \ -\infty < t < \infty$

43. $\mathbf{r}(t) = \langle 1, \ 2 + 2\cos t, \ 5 + 2\sin t \rangle, \ 0 \leq t \leq 2\pi$

45. $\mathbf{r}(t) = \left\langle \frac{\sqrt{3}}{2} \cos t, \ \frac{1}{2}, \ \frac{\sqrt{3}}{2} \sin t \right\rangle, \ 0 \leq t \leq 2\pi$

47. $\mathbf{r}(t) = \langle 3 + 2\cos t, \ 1, \ 5 + 3\sin t \rangle, \ 0 \leq t \leq 2\pi$

49.

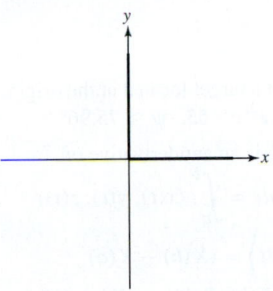

$\mathbf{r}(t) = \langle |t| + t, \ |t| - t \rangle$

Section 13.2 Preliminary Questions

1. $\frac{d}{dt} (f(t)\mathbf{r}(t)) = f(t)\mathbf{r}'(t) + f'(t)\mathbf{r}(t)$

$\frac{d}{dt} (\mathbf{r}_1(t) \cdot \mathbf{r}_2(t)) = \mathbf{r}_1(t) \cdot \mathbf{r}'_2(t) + \mathbf{r}'_1(t) \cdot \mathbf{r}_2(t)$

$\frac{d}{dt} (\mathbf{r}_1(t) \times \mathbf{r}_2(t)) = \mathbf{r}_1(t) \times \mathbf{r}'_2(t) + \mathbf{r}'_1(t) \times \mathbf{r}_2(t)$

2. True **3.** True **4.** True **5.** False **6.** False

7. (a) Vector (b) Scalar (c) Vector

Section 13.2 Exercises

1. $\lim\limits_{t \to 3} \left\langle t^2, \ 4t, \ \frac{1}{t} \right\rangle = \left\langle 9, \ 12, \ \frac{1}{3} \right\rangle$

3. $\lim\limits_{t \to 0} (e^{2t}\mathbf{i} + \ln(t + 1)\mathbf{j} + 4\mathbf{k}) = \mathbf{i} + 4\mathbf{k}$

5. $\lim\limits_{h \to 0} \frac{\mathbf{r}(t+h) - \mathbf{r}(t)}{h} = \left\langle -\frac{1}{t^2}, \ \cos t, \ 0 \right\rangle$ **7.** $\frac{d\mathbf{r}}{dt} = \langle 1, \ 2t, \ 3t^2 \rangle$

9. $\mathbf{r}'(s) = \langle -e^{1-s}, -1, 1/(s - 1) \rangle$ **11.** $\mathbf{c}'(t) = -t^{-2}\mathbf{i} - 2e^{2t}\mathbf{k}$

13. $\mathbf{r}'(t) = \langle 1, \ 2t, \ 3t^2 \rangle$; $\mathbf{r}''(t) = \langle 0, \ 2, \ 6t \rangle$

15.

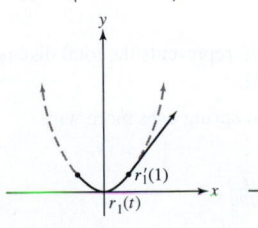
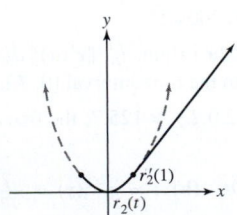

17. $t = 2\pi/3$

19. $\frac{d}{dt} (\mathbf{r}_1(t) \cdot \mathbf{r}_2(t)) =$
$\qquad 2t^3e^{2t} + 3t^2e^{3t} + 2te^{3t} + 3t^2e^{2t} + te^t + e^t$

21. $\frac{d}{dt} (\mathbf{r}_1(t) \times \mathbf{r}_2(t)) =$

$\left\langle \begin{array}{l} 3t^2e^t - 2te^{2t} - e^{2t} + t^3e^t, \ e^{3t} + 3te^{3t} - t^2e^t - 2te^t, \\ 2te^{2t} + 2t^2e^{2t} - 3t^2e^{3t} - 3t^3e^{3t} \end{array} \right\rangle$

23. $2 + 4e$ **25.** $\frac{d}{dt}\mathbf{r}(g(t)) = \langle 2e^{2t}, \ -e^t \rangle$

27. $\frac{d}{dt}\mathbf{r}(g(t)) = \langle 4e^{4t+9}, \ 8e^{8t+18}, \ 0 \rangle$

29. $\frac{d}{dt} (\mathbf{r}(t) \cdot \mathbf{a}(t))|_{t=2} = 13$ **31.** $\mathbf{L}(t) = \langle 4 - 4t, \ 16 - 32t \rangle$

33. $\mathbf{L}(t) = \langle -3 - 4t, \ 10 + 5t, \ 16 + 24t \rangle$

35. $\mathbf{L}(s) = \left\langle 2 - s, \ 0, \ -\frac{1}{3} + \frac{1}{2}s \right\rangle$

37. $\frac{d}{dt}(\mathbf{r} \times \mathbf{r}') = \langle (t^2 - 2)e^t, \ -te^t, \ 2t \rangle$ **41.** $\langle \frac{16}{3}, 0 \rangle$ **43.** $\langle 0, 0 \rangle$

45. $\langle -2, 3\pi^2, \pi^2 \rangle$ **47.** $(\ln 4)\mathbf{i} + \frac{56}{3}\mathbf{j} - \frac{496}{5}\mathbf{k}$

49. $\mathbf{r}(t) = \langle -t^2 + t + c_1, 2t^2 + c_2 \rangle$; with initial conditions
$\mathbf{r}(t) = \langle -t^2 + t + 3, 2t^2 + 1 \rangle$

51. $\mathbf{r}(t) = \left(\frac{1}{3}t^3\right)\mathbf{i} + \left(\frac{5t^2}{2}\right)\mathbf{j} + t\mathbf{k} + \mathbf{c}$; with initial conditions
$\mathbf{r}(t) = \left(\frac{1}{3}t^3 - \frac{1}{3}\right)\mathbf{i} + \left(\frac{5}{2}t^2 - \frac{3}{2}\right)\mathbf{j} + (t+1)\mathbf{k}$

53. $\mathbf{r}(t) = (8t^2)\mathbf{k} + \mathbf{c}_1 t + \mathbf{c}_2$; with initial conditions
$\mathbf{r}(t) = \mathbf{i} + t\mathbf{j} + (8t^2)\mathbf{k}$

55. $\mathbf{r}(t) = \langle 0, \ t^2, \ 0 \rangle + \mathbf{c}_1 t + \mathbf{c}_2$; with initial conditions
$\mathbf{r}(t) = \langle 1, \ t^2 - 6t + 10, \ t - 3 \rangle$

57. $\mathbf{r}(3) = \langle \frac{45}{4}, \ 5 \rangle$

59. Only at time $t = 3$ can the pilot hit a target located at the origin.
61. $\mathbf{r}(t) = (t-1)\mathbf{v} + \mathbf{w}$ **63.** $\mathbf{r}(t) = e^{2t}\mathbf{c}$ **65.** $\psi \approx 75.96°$

69. Assume $\mathbf{R}(t) = \langle X(t), Y(t), Z(t) \rangle$ is an antiderivative of
$\mathbf{r}(t) = \langle x(t), y(t), z(t) \rangle$. Then $\int_a^b \mathbf{r}(t)\,dt = \int_a^b \langle x(t), y(t), z(t) \rangle$
$dt = \left\langle \int_a^b x(t)\,dt, \int_a^b y(t)\,dt, \int_a^b z(t)\,dt \right\rangle = \langle X(b) - X(a),$
$Y(b) - Y(a), Z(b) - Z(a) \rangle = \langle X(b), Y(b), Z(b) \rangle - \langle X(a), Y(a),$
$Z(a) \rangle = \mathbf{R}(b) - \mathbf{R}(a)$

Section 13.3 Preliminary Questions

1. $2\mathbf{r}' = \langle 50, -70, 20 \rangle$; $-\mathbf{r}' = \langle -25, 35, -10 \rangle$

2. Statement (b) is true.

3. (a) $L'(2) = 4$

(b) $L(t)$ is the distance along the path traveled, which is usually different from the distance from the origin.

4. 6

Section 13.3 Exercises

1. $L = 3\sqrt{61}$ **3.** $L = 15 + \ln 4$ **5.** $L = \frac{544\sqrt{34}-2}{135} \approx 23.48$

7. $L = \pi\sqrt{4\pi^2 + 10} + 5\ln\frac{2\pi + \sqrt{4\pi^2 + 10}}{\sqrt{10}} \approx 29.3$

9. Arc length ≈ 44.87 **11.** $s(t) = \frac{1}{27}\left((20 + 9t^2)^{3/2} - 20^{3/2}\right)$

13. $v(4) = \sqrt{21}$ **15.** $v(1) = \sqrt{2}$ **17.** $v\left(\frac{\pi}{2}\right) = 5$

19. $\mathbf{r}' = \langle 100\sqrt{5}, 200\sqrt{5} \rangle$

21. The bee is at the origin. $\int_0^T \|\mathbf{r}'(u)\|\,du$ represents the total distance the bee traveled on the time interval $[0, T]$.

23. (c) $L_1 \approx 132.0, L_2 \approx 125.7$; the first spring uses more wire.

25. (a) $t = \pi$

27. (a) $s = \sqrt{29}t$ (b) $t = g^{-1}(s) = \frac{s}{\sqrt{29}}$

29. $\left\langle 1 + \frac{3s}{\sqrt{50}}, \ 2 + \frac{4s}{\sqrt{50}}, \ 3 + \frac{5s}{\sqrt{50}} \right\rangle$

31. $\mathbf{r}(s) = \langle 2 + 4\cos(2s), 10, -3 + 4\sin(2s) \rangle$

33. $\mathbf{r}_1(s) = \left\langle \cos\left[\left(\frac{3}{2}s + 1\right)^{2/3} - 1\right], \sin\left[\left(\frac{3}{2}s + 1\right)^{2/3} - 1\right], \right.$
$\left. \frac{2}{3}\left[\left(\frac{3}{2}s + 1\right)^{2/3} - 1\right]^{3/2} \right\rangle, s \geq 0$

35. $\mathbf{r}_1(s) = \left\langle \frac{1}{9}(27s + 8)^{2/3} - \frac{4}{9}, \ \pm\frac{1}{27}\left((27s + 8)^{2/3} - 4\right)^{3/2} \right\rangle$

37. $\left\langle \frac{s}{\sqrt{1+m^2}}, \ \frac{sm}{\sqrt{1+m^2}} \right\rangle$

39. (a) $\sqrt{17}e^t$ (b) $\frac{s}{\sqrt{17}}\left\langle \cos\left(4\ln\frac{s}{\sqrt{17}}\right), \ \sin\left(4\ln\frac{s}{\sqrt{17}}\right) \right\rangle$

41. $L = \int_{-\infty}^{\infty} \|\mathbf{r}'(t)\|\,dt = 2\int_{-\infty}^{\infty} \frac{dt}{1+t^2} = 2\pi$

Section 13.4 Preliminary Questions

1. $\left\langle -\frac{2}{3}, -\frac{1}{3}, \frac{2}{3} \right\rangle$ **2.** $\frac{1}{4}$

3. The curvature of a circle of radius 2   **4.** Zero curvature

5. $\kappa = \sqrt{14}$ **6.** 4   **7.** $\frac{1}{9}$

Section 13.4 Exercises

1. $\mathbf{r}'(t) = \langle 8t, 9 \rangle$; $\mathbf{T}(t) = \frac{1}{\sqrt{64t^2+81}}\langle 8t, \ 9 \rangle$; $\mathbf{T}(1) = \left\langle \frac{8}{\sqrt{145}}, \ \frac{9}{\sqrt{145}} \right\rangle$

3. $\mathbf{r}'(t) = \langle 4, -5, 9 \rangle$; $\mathbf{T}(t) = \left\langle \frac{4}{\sqrt{122}}, \ -\frac{5}{\sqrt{122}}, \ \frac{9}{\sqrt{122}} \right\rangle$; $\mathbf{T}(1) = \mathbf{T}(t)$

5. $\mathbf{r}'(t) = \langle -\pi\sin\pi t, \pi\cos\pi t, 1 \rangle$;
$\mathbf{T}(t) = \frac{1}{\sqrt{\pi^2+1}}\langle -\pi\sin\pi t, \ \pi\cos\pi t, \ 1 \rangle$;
$\mathbf{T}(1) = \left\langle 0, \ -\frac{\pi}{\sqrt{\pi^2+1}}, \ \frac{1}{\sqrt{\pi^2+1}} \right\rangle$

7. $\kappa(t) = \frac{e^t}{(1+e^{2t})^{3/2}}$ **9.** $\kappa(t) = 0$ **11.** $\kappa = \frac{2\sqrt{74}}{27}$

13. $\kappa = \frac{\sqrt{t^2+5}}{(t^2+1)^{3/2}} \approx 0.108$ **15.** $\kappa(3) = \frac{e^3}{(e^6+1)^{3/2}} \approx 0.0025$

17. $\kappa(2) = \frac{48\sqrt{41}}{210,125} \approx 0.0015$

19. $\kappa\left(\frac{\pi}{3}\right) = \frac{\sqrt{330}}{4} \approx 4.54$; $\kappa\left(\frac{\pi}{2}\right) = \frac{1}{5} = 0.2$ **23.** $\alpha = \pm\sqrt{2}$

29. $\kappa(2) = \frac{3\sqrt{10}}{800} \approx 0.012$ **31.** $\kappa(\pi) = \frac{\pi\sqrt{2}}{4} \approx 1.11$ **35.** $\kappa(t) = t^2$

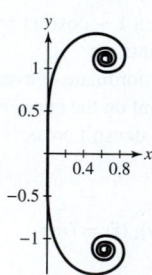

37. $\mathbf{N}(t) = \langle 0, -\sin 2t, -\cos 2t \rangle$

39. $\mathbf{N}\left(\frac{\pi}{4}\right) = \left\langle -\frac{\sqrt{2}}{\sqrt{6}}, -\frac{2}{\sqrt{6}} \right\rangle$; $\mathbf{N}\left(\frac{3\pi}{4}\right) = \left\langle \frac{\sqrt{2}}{\sqrt{6}}, \frac{2}{\sqrt{6}} \right\rangle$

41. $\mathbf{T}(1) = \left\langle 0, \frac{\sqrt{5}}{5}, \frac{2\sqrt{5}}{5} \right\rangle$; $\mathbf{N}(1) = \left\langle 0, -\frac{2\sqrt{5}}{5}, \frac{\sqrt{5}}{5} \right\rangle$;
$\mathbf{B}(1) = \langle 1, 0, 0 \rangle$

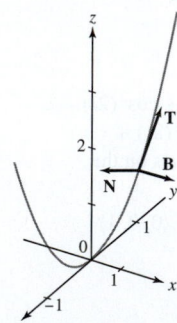

43. $\mathbf{T}(1) = \left\langle \frac{1}{3}, \frac{2}{3}, \frac{2}{3} \right\rangle$; $\mathbf{N}(1) = \left\langle -\frac{2}{3}, -\frac{1}{3}, \frac{2}{3} \right\rangle$; $\mathbf{B}(1) = \left\langle \frac{2}{3}, -\frac{2}{3}, \frac{1}{3} \right\rangle$

45. $\mathbf{N}(\pi^{1/3}) = \left\langle \frac{1}{2}, -\frac{\sqrt{3}}{2} \right\rangle$ **47.** $\mathbf{N}(1) = \frac{1}{\sqrt{13}}\langle -3, 2 \rangle$

49. $\mathbf{N}(1) = \frac{1}{\sqrt{2}}\langle 0, 1, -1 \rangle$ **51.** $\mathbf{N}(0) = \frac{1}{6}\left\langle -\sqrt{6}, 2\sqrt{6}, -\sqrt{6} \right\rangle$

53. (a) $\mathbf{T}(1) = \left\langle \frac{1}{3}, \frac{2}{3}, \frac{2}{3} \right\rangle$; $\mathbf{N}(1) = \left\langle -\frac{2}{3}, -\frac{1}{3}, \frac{2}{3} \right\rangle$;
$\mathbf{B}(1) = \left\langle \frac{2}{3}, -\frac{2}{3}, \frac{1}{3} \right\rangle$ (b) $6x - 6y + 3z = 1$

55. (a) $\mathbf{T}(t) = \left\langle \frac{1}{\sqrt{2+4t^2}}, \frac{-1}{\sqrt{2+4t^2}}, \frac{2t}{\sqrt{2+4t^2}} \right\rangle$;
$\mathbf{N}(t) = \left\langle \frac{-t\sqrt{2}}{\sqrt{2+4t^2}}, \frac{t\sqrt{2}}{\sqrt{2+4t^2}}, \frac{\sqrt{2}}{\sqrt{2+4t^2}} \right\rangle$

(b) $\mathbf{B}(t) = \left\langle -\frac{1}{\sqrt{2}}, -\frac{1}{\sqrt{2}}, 0 \right\rangle$

(c) The osculating planes are all parallel to each other, with equation $x + y = c$ for some c.

59. $(x + 4)^2 + \left(y - \frac{7}{2}\right)^2 = \frac{125}{4}$ **61.** $\left(x - \frac{\pi}{2}\right)^2 + y^2 = 1$

63. $(x + 2)^2 + (y - 3)^2 = 8$ **65.** $x^2 + y^2 = 1$

67. $(x - 1)^2 + \left(y - \frac{5}{2}\right)^2 = \frac{1}{4}$ **75.** $\kappa(\theta) = 1$ **77.** $\kappa(\theta) = \frac{1}{\sqrt{2}}e^{-\theta}$

85. Either $-\mathbf{k}$ or $\mathbf{k}$. Since $\mathbf{T}(t) = \mathbf{r}'(t)/\|\mathbf{r}'(t)\|$, it follows that $\mathbf{T}(t)$ is in the form $\mathbf{T}(t) = \langle f(t), g(t), 0\rangle$ for some functions $f(t)$ and $g(t)$. Similarly, since $\mathbf{N}(t) = \mathbf{T}'(t)/\|\mathbf{T}'(t)\|$, it follows that $\mathbf{N}(t)$ is in the form $\mathbf{N}(t) = \langle p(t), q(t), 0\rangle$ for some functions $p(t)$ and $q(t)$. Now, $\mathbf{B}(t) = \mathbf{T}(t) \times \mathbf{N}(t)$, and taking the cross product, we find that $\mathbf{B}(t)$ is in the form $\mathbf{B}(t) = \langle 0, 0, w(t)\rangle$ for some function $w(t)$. Since $\mathbf{B}(t)$ is a unit vector, we must have $w(t) = \pm 1$, and therefore either $\mathbf{B}(t) = -\mathbf{k}$ or $\mathbf{B}(t) = \mathbf{k}$.

93. (a) $\mathbf{N}(0) = \left\langle -\frac{1}{\sqrt{5}}, 0, \frac{2}{\sqrt{5}}\right\rangle$ **(b)** $\mathbf{N}(1) = \frac{1}{\sqrt{66}}\langle 4, 7, -1\rangle$

Section 13.5 Preliminary Questions

1. No, since the particle may change its direction. **2.** $\mathbf{a}(t)$

3. All three statements are false.

4. If the speed is constant, the tangential component of acceleration is zero. The acceleration consists only of the normal component, which is orthogonal to the velocity.

5. Description (b), parallel **6.** $\|\mathbf{a}(t)\| = 8$ cm/s^2 **7.** $a_\mathbf{N}$

Section 13.5 Exercises

1.
$h = -0.2$: $\langle -0.085, 1.91, 2.635\rangle$
$h = -0.1$: $\langle -0.19, 2.07, 2.97\rangle$
$h = 0.1$: $\langle -0.41, 2.37, 4.08\rangle$
$h = 0.2$: $\langle -0.525, 2.505, 5.075\rangle$
$\mathbf{v}(1) \approx \langle -0.3, 2.2, 3.5\rangle$; $v(1) \approx 4.1$

3. $\mathbf{v}(1) = \langle 3, -1, 8\rangle$; $\mathbf{a}(1) = \langle 6, 0, 8\rangle$; $v(1) = \sqrt{74}$

5. $\mathbf{v}\left(\frac{\pi}{3}\right) = \left\langle \frac{1}{2}, -\frac{\sqrt{3}}{2}, 0\right\rangle$; $\mathbf{a}\left(\frac{\pi}{3}\right) = \left\langle -\frac{\sqrt{3}}{2}, -\frac{1}{2}, 9\right\rangle$; $v\left(\frac{\pi}{3}\right) = 1$

7. $\mathbf{a}(t) = -2\left\langle \cos\frac{t}{2}, \sin\frac{t}{2}\right\rangle$; $\mathbf{a}\left(\frac{\pi}{4}\right) \approx \langle -1.85, -0.77\rangle$; $\mathbf{r}(t) = 8\left\langle \cos\frac{t}{2}, \sin\frac{t}{2}\right\rangle$

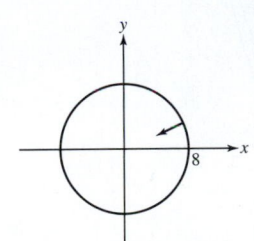

9.

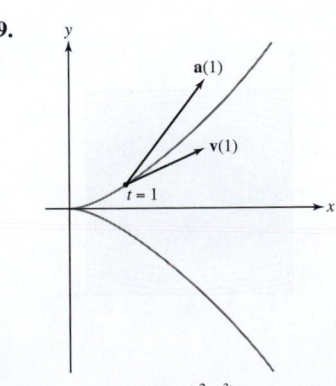

$\mathbf{r}(t) = \langle t^2, t^3\rangle$

11. $\mathbf{v}(t) = \left\langle \frac{3t^2 + 2}{6}, 4t - 2\right\rangle$ **13.** $\mathbf{v}(t) = \mathbf{i} + t\mathbf{k}$

15. $\mathbf{v}(t) = \left\langle \frac{t^2}{2} + 3, 4t - 2\right\rangle$; $\mathbf{r}(t) = \left\langle \frac{t^3}{6} + 3t, 2t^2 - 2t\right\rangle$

17. $\mathbf{v}(t) = \mathbf{i} + \frac{t^2}{2}\mathbf{k}$, $\mathbf{r}(t) = t\mathbf{i} + \mathbf{j} + \frac{t^3}{6}\mathbf{k}$

19. $v_0 = \sqrt{5292} \approx 72.746$ m/s **21.** Approximately 663.1 m

25. (a) Assume that $r(0) = \langle 150, 75, 5\rangle$

$$a(t) = \langle 0, 0, -32\rangle; v(t) = \langle 40, 35, -32t + 32\rangle$$
$$r(t) = \left\langle 40t + 150, 35t + 75, -16t^2 + 32t + 5\right\rangle$$

(b) $z = 5$ when $t = 0$ or 2. At $t = 2, r(2) = \langle 230, 145, 5\rangle$, so the player is in bounds, since $(300, 150, z)$ is the maximum possible point to be in bounds.

27. $\mathbf{r}(10) = \langle 45, -20\rangle$

29. (a) At its original position **(b)** No

31. The speed is decreasing.

33. $a_\mathbf{T} = 0$; $a_\mathbf{N} = 1$ **35.** $a_\mathbf{T} = \frac{7}{\sqrt{6}}$; $a_\mathbf{N} = \sqrt{\frac{53}{6}}$

37. $\mathbf{a}(-1) = -\frac{2}{\sqrt{10}}\mathbf{T} + \frac{6}{\sqrt{10}}\mathbf{N}$ with $\mathbf{T} = \frac{1}{\sqrt{10}}\langle 1, -3\rangle$ and $\mathbf{N} = \frac{1}{\sqrt{10}}\langle -3, -1\rangle$

39. $a_\mathbf{T}(4) = 4$; $a_\mathbf{N}(4) = 1$, so $\mathbf{a} = 4\mathbf{T} + \mathbf{N}$, with $\mathbf{T} = \left\langle \frac{1}{9}, \frac{4}{9}, \frac{8}{9}\right\rangle$ and $\mathbf{N} = \left\langle -\frac{4}{9}, -\frac{7}{9}, \frac{4}{9}\right\rangle$

41. $\mathbf{a}(0) = \sqrt{3}\mathbf{T} + \sqrt{2}\mathbf{N}$, with $\mathbf{T} = \frac{1}{\sqrt{3}}\langle 1, 1, 1\rangle$ and $\mathbf{N} = \frac{1}{\sqrt{2}}\langle -1, 0, 1\rangle$

43. $\mathbf{a}\left(\frac{\pi}{2}\right) = -\frac{\pi}{2\sqrt{3}}\mathbf{T} + \frac{\pi}{\sqrt{6}}\mathbf{N}$, with $\mathbf{T} = \frac{1}{\sqrt{3}}\langle 1, -1, 1\rangle$ and $\mathbf{N} = \frac{1}{\sqrt{6}}\langle 1, -1, -2\rangle$

45. $a_\mathbf{T} = 0$; $a_\mathbf{N} = 0.25$ cm/s^2

47. The tangential acceleration is $\frac{50}{\sqrt{2}} \approx 35.36$ m/min^2, $v \approx \sqrt{35.36(30)} \approx 32.57$ m/min.

49. $\|\mathbf{a}\| = 1.157 \times 10^5$ km/h^2 **51.** $\mathbf{a} = \left\langle -\frac{1}{2\sqrt{2}}, -\frac{9}{2}, \frac{1}{2\sqrt{2}}\right\rangle$

53. (A) Slowing down (B) Speeding up (C) Slowing down

59. After 139.91 s, the car will begin to skid. **61.** $R \approx 105$ m

Section 13.6 Preliminary Questions

1. $\frac{dA}{dt} = \frac{1}{2}\|\mathbf{J}\|$ **3.** The period is increased eight-fold.

Section 13.6 Exercises

1. The data support Kepler's prediction; $T \approx \sqrt{a^3 \cdot 3 \cdot 10^{-4}} \approx 11.9$ years

3. $M \approx 1.897 \times 10^{27}$ kg **5.** $M \approx 2.6225 \times 10^{41}$ kg

11. The satellite's orbit is in the plane $20x - 29y + 9z = 0$.

Chapter 13 Review

1. (a) $-1 < t < 0$ or $0 < t \le 1$ (b) $0 < t \le 2$

3. $\mathbf{r}(t) = \left\langle t^2, t, \sqrt[3]{3 - t^4}\right\rangle$, $-\infty < t < \infty$

5. $\mathbf{r}'(t) = \left\langle -1, -2t^{-3}, \frac{1}{t}\right\rangle$ **7.** $\mathbf{r}'(0) = \langle 2, 0, 6\rangle$

9. $\frac{d}{dt}e^t\langle 1, t, t^2\rangle = e^t\langle 1, 1 + t, 2t + t^2\rangle$

11. $\frac{d}{dt}(6\mathbf{r}_1(t) - 4\mathbf{r}_2(t))|_{t=3} = \langle 0, -8, -10\rangle$

13. $\frac{d}{dt}(\mathbf{r}_1(t) \cdot \mathbf{r}_2(t))|_{t=3} = 2$

15. $\int_0^3 \langle 4t + 3, t^2, -4t^3\rangle\, dt = \langle 27, 9, -81\rangle$

17. $\left(3, 3, \frac{16}{3}\right)$ **19.** $\mathbf{r}(t) = \left\langle 2t^2 - \frac{8}{3}t^3 + t, t^4 - \frac{1}{6}t^3 + 1\right\rangle$

21. $L = 2\sqrt{13}$ **23.** $\left\langle 5\cos\frac{2\pi s}{5\sqrt{1+4\pi^2}}, 5\sin\frac{2\pi s}{5\sqrt{1+4\pi^2}}, \frac{s}{\sqrt{1+4\pi^2}}\right\rangle$

25. $v_0 \approx 67.279$ m/s **27.** $\left(0, \frac{11}{2}, 38\right)$ **29.** $\mathbf{T}(\pi) = \left\langle \frac{-1}{\sqrt{2}}, \frac{1}{\sqrt{2}}, 0\right\rangle$

31. $\kappa(1) = \frac{1}{2^{3/2}}$

33. $\mathbf{a} = \frac{1}{\sqrt{2}}\mathbf{T} + 4\mathbf{N}$, where $\mathbf{T} = \langle -1, 0\rangle$ and $\mathbf{N} = \langle 0, -1\rangle$

35. $\kappa = \frac{13}{16}$ **37.** $\left(x - \frac{25}{2}\right)^2 + (y + 32)^2 = \frac{4913}{4}$

39. $2x - 4y + 2z = -3$

Chapter 14

Section 14.1 Preliminary Questions

1. Same shape, but located in parallel planes

2. The parabola $z = x^2$ in the xz-plane   **3.** Not possible

4. The vertical lines $x = c$ with distance of 1 unit between adjacent lines

5. In the contour map of $g(x, y) = 2x$, the distance between two adjacent vertical lines is $\frac{1}{2}$.

Section 14.1 Exercises

1. $f(1, 2) = 3$, $f(-1, 6) = -7$, $f(e, \pi) = e + \pi e^3$

3. $h(20, 2) = \frac{2}{9}$; $h(1, -2)$ and $h(1, 1)$ are not defined.

5. $h(3, 7, -2) = \frac{21}{4}$, $h(3, 2, 1/4) = 96$; $h(4, -4, 0)$ is not defined.

7. The domain is the entire xy-plane.

9. 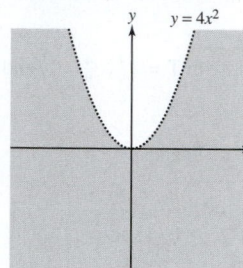   **11.** $\mathcal{D} = \{(y, z) : z \neq -y^2\}$

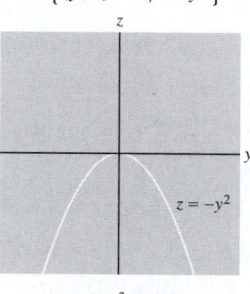

13.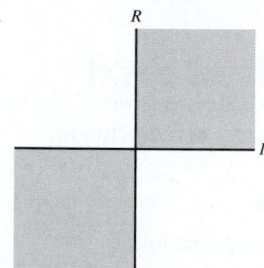

15. Domain: entire (x, y, z)−space; range: entire real line

17. Domain: $\{(r, s, t) : |rst| \leq 4\}$; range: $\{P : 0 \leq P \leq 4\}$

19. $f \leftrightarrow$ (B), $g \leftrightarrow$ (A)

21. **(a)** D   **(b)** C   **(c)** E   **(d)** B   **(e)** A   **(f)** F

23.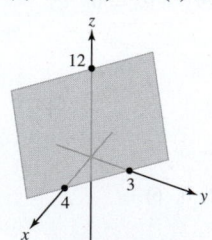

Horizontal trace: $3x + 4y = 12 - c$ in the plane $z = c$

Vertical trace: $z = (12 - 3a) - 4y$ and $z = -3x + (12 - 4b)$ in the planes $x = a$, and $y = b$, respectively

25.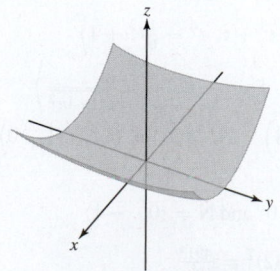

The horizontal traces are ellipses for $c > 0$.

The vertical trace in the plane $x = a$ is the parabola $z = a^2 + 4y^2$.

The vertical trace in the plane $y = b$ is the parabola $z = x^2 + 4b^2$.

27.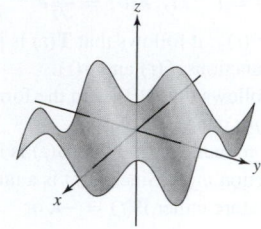

The horizontal traces in the plane $z = c$, $|c| \leq 1$, are the lines $x - y = \sin^{-1} c + 2k\pi$ and $x - y = \pi - \sin^{-1} c + 2k\pi$, for integer k.

The vertical trace in the plane $x = a$ is $z = \sin(a - y)$.

The vertical trace in the plane $y = b$ is $z = \sin(x - b)$.

29. $m = 1 : m = 2 :$

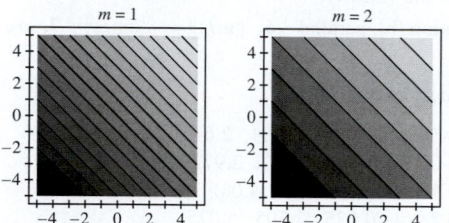

31.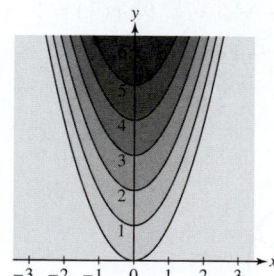

33. 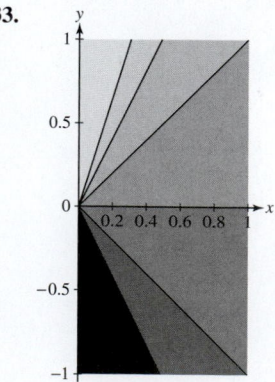   **35.**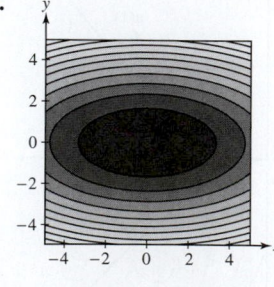

37.

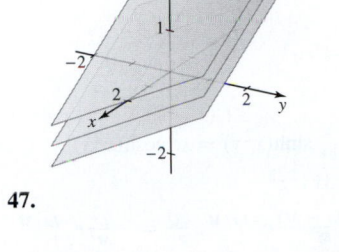

39. $m = 6:$ $f(x, y) = 2x + 6y + 6$
$m = 3:$ $f(x, y) = x + 3y + 3$

41. **(a)** A and B   **(b)** B

43. Colorado, North Dakota, Arkansas, Wisconsin

45.

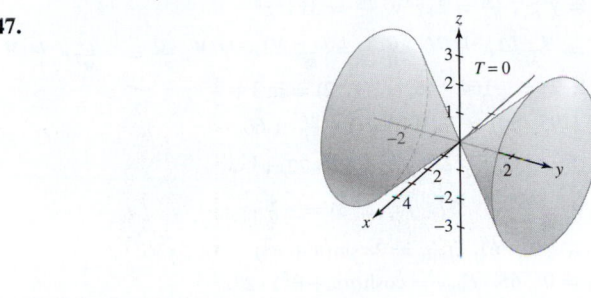

47.

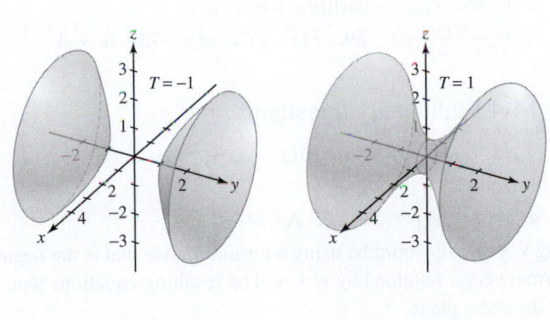

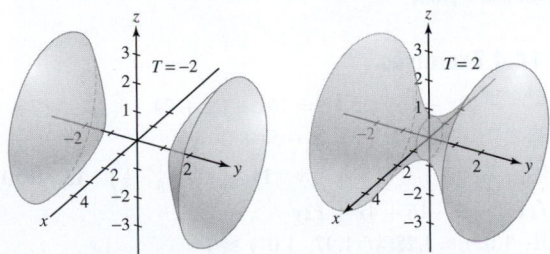

49. Average ROC from B to $C = 0.000625$ kg/m^3 · ppt

51. At point A

53. Average ROC from A to $B \approx 0.0737$, average ROC from A to $C \approx 0.0457$

55.

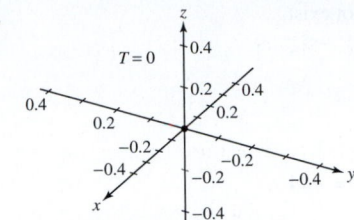

Contour interval = 20 m   0   1   2 km

57.

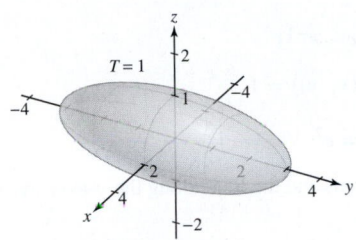

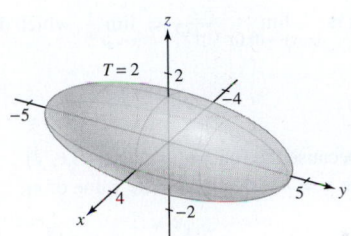

59.

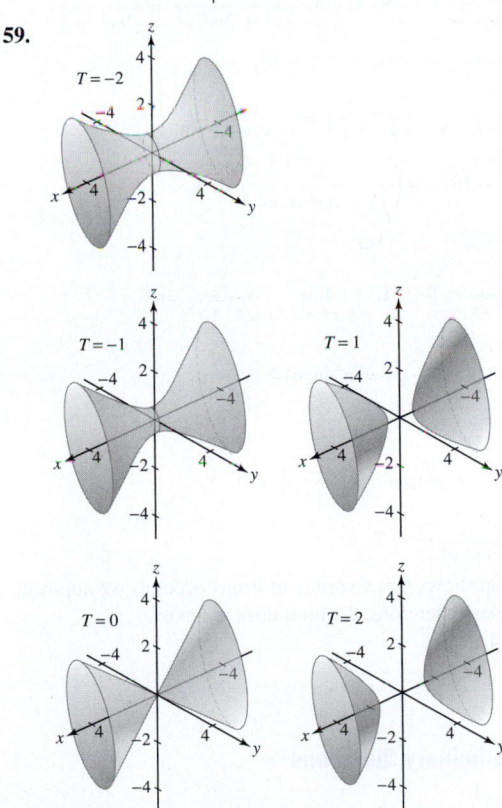

61. $f(r, \theta) = \cos\theta$; the level curves are
$\theta = \pm\cos^{-1}(c)$ for $|c| < 1$, $c \neq 0$;
the y-axis for $c = 0$;
the positive x-axis for $c = 1$;
the negative x-axis for $c = -1$.

Section 14.2 Preliminary Questions

1. $D^*(P, r)$ consists of all points in $D(P, r)$ other than P itself.

2. $f(2, 3) = 27$ **3.** All three statements are true.

4. $\lim\limits_{(x,\,y)\to(0,\,0)} f(x,\,y)$ does not exist.

Section 14.2 Exercises

1. $\lim\limits_{(x,y)\to(1,2)} (x^2+y) = 3$

3. $\lim\limits_{(x,y)\to(-2,1)} (x^2 y - 3x^4 y^3) = -44$

5. $\lim\limits_{(x,y)\to(\frac{\pi}{4},0)} \tan x \cos y = 1$

7. $\lim\limits_{(x,y)\to(1,1)} \frac{e^{x^2}-e^{-y^2}}{x+y} = \frac{1}{2}(e - e^{-1})$

9. $\lim\limits_{(x,y)\to(2,5)} (g(x,\,y) - 2f(x,\,y)) = 1$

11. $\lim\limits_{(x,y)\to(2,5)} e^{f(x,y)^2 - g(x,y)} = e^2$

13. No; the limit along the x-axis and the limit along the y-axis are different.

15. The limit is $\frac{1+m^3}{m^2}$ for all $m \neq 0$.

17. The limit along the x-axis is $\lim\limits_{(x,y)\to(0,0)} \frac{x}{x^2+y^2} = \lim\limits_{x\to 0} \frac{1}{x}$, which does not exist.

19. $\lim\limits_{(x,y)\to(0,0)} \frac{x^2-y^2}{\sqrt{x^2+y^2}} = 0$

21. The limit does not exist, because the function values as $(x,\,y)$ approaches $(0,0)$ along the line $y = mx$ depend on the value of m.

$$\lim\limits_{(x,y)\to(0,0)y=mx} \frac{xy}{3x^2+2y^2} = \lim\limits_{x\to 0} \frac{mx^2}{3x^2+2m^2x^2} = \frac{m}{2m^2+3}$$

23. Along the x-coordinate axis $(y = z = 0)$,

$$\lim\limits_{(x,y,z)\to(0,0,0)} \frac{x+y+z}{x^2+y^2+z^2} = \lim\limits_{(x,y,z)\to(0,0,0)} \frac{1}{x} = \infty$$

25. $\lim\limits_{(x,y)\to(4,0)} (x^2-16)\cos\left(\frac{1}{(x-4)^2+y^2}\right) = 0$

27. $\lim\limits_{(z,w)\to(-2,1)} \frac{z^4 \cos(\pi w)}{e^{z+w}} = -16e$

29. $\lim\limits_{(x,y)\to(4,2)} \frac{y-2}{\sqrt{x^2-4}} = 0$ **31.** $\lim\limits_{(x,y)\to(3,4)} \frac{1}{\sqrt{x^2+y^2}} = \frac{1}{5}$

33. Does not exist.

35. $\lim\limits_{(x,y)\to(1,-3)} e^{x-y}\ln(x-y) = e^4 \ln(4)$

37. $\lim\limits_{(x,y)\to(-3,-2)} (x^2 y^3 + 4xy) = -48$

39. $\lim\limits_{(x,y)\to(0,0)} \tan(x^2+y^2)\tan^{-1}\left(\frac{1}{x^2+y^2}\right) = 0$

41. $\lim\limits_{(x,y)\to(0,0)} \frac{x^2+y^2}{\sqrt{x^2+y^2+1}-1} = 2$

45. The contour map shows that a variety of limits occur as we approach P along different lines. Therefore, the limit does not exist.
$$\lim\limits_{(x,y)\to Q} g(x,\,y) = 4$$

47. Yes

Section 14.3 Preliminary Questions

1. $\frac{\partial}{\partial x}(x^2 y^2) = 2xy^2$

2. In this case, the Constant Multiple Rule can be used. In the second part, since y appears in both the numerator and the denominator, the Quotient Rule is preferred.

3. (a), (c) **4.** $f_x = 0$ **5.** (a), (d)

Section 14.3 Exercises

3.
$$f_x = (2x)(x - y^2) + (x^2 - y)(1) = 3x^2 - 2xy^2 - y$$
$$f_y = (-1)(x - y^2) + (x^2 - y)(-2y) = 3y^2 - 2x^2 y - x$$

5. $\frac{\partial}{\partial y}\frac{y}{x+y} = \frac{x}{(x+y)^2}$ **7.** $f_z(2,\,3,\,1) = 6$

9. $m = 10$ **11.** $f_x(A) \approx 10$, $f_y(A) \approx -20$ **13.** NW

15. $\frac{\partial}{\partial x}(x^2+y^2) = 2x$, $\frac{\partial}{\partial y}(x^2+y^2) = 2y$

17. $\frac{\partial}{\partial x}(x^4 y + xy^{-2}) = 4x^3 y + y^{-2}$, $\frac{\partial}{\partial y}(x^4 y + xy^{-2}) = x^4 - 2xy^{-3}$

19. $\frac{\partial}{\partial x}\left(\frac{x}{y}\right) = \frac{1}{y}$, $\frac{\partial}{\partial y}\left(\frac{x}{y}\right) = \frac{-x}{y^2}$

21.
$$\frac{\partial}{\partial x}\left(\sqrt{9-x^2-y^2}\right) = \frac{-x}{\sqrt{9-x^2-y^2}}, \ \frac{\partial}{\partial y}\left(\sqrt{9-x^2-y^2}\right) = \frac{-y}{\sqrt{9-x^2-y^2}}$$

23. $\frac{\partial z}{\partial x} = (\cos x)(\cos y)$, $\frac{\partial z}{\partial y} = -(\sin x)(\sin y)$

25. $\frac{\partial z}{\partial x} = \frac{1}{y}\sin\left(\frac{1-x}{y}\right)$, $\frac{\partial z}{\partial y} = \frac{1-x}{y^2}\sin\left(\frac{1-x}{y}\right)$

27. $\frac{\partial w}{\partial x} = \frac{2x}{x^2-y^2}$, $\frac{\partial w}{\partial y} = \frac{-2y}{x^2-y^2}$

29. $\frac{\partial}{\partial r}e^{r+s} = e^{r+s}$, $\frac{\partial}{\partial s}e^{r+s} = e^{r+s}$

31. $\frac{\partial}{\partial x}e^{xy} = ye^{xy}$, $\frac{\partial}{\partial y}e^{xy} = xe^{xy}$

33. $\frac{\partial z}{\partial y} = -2xe^{-x^2-y^2}$, $\frac{\partial z}{\partial y} = -2ye^{-x^2-y^2}$

35. $\frac{\partial U}{\partial t} = -e^{-rt}$, $\frac{\partial U}{\partial r} = \frac{-e^{-rt}(rt+1)}{r^2}$

37. $\frac{\partial}{\partial x}\sinh(x^2 y) = 2xy\cosh(x^2 y)$, $\frac{\partial}{\partial y}\sinh(x^2 y) = x^2 \cosh(x^2 y)$

39. $\frac{\partial w}{\partial x} = y^2 z^3$, $\frac{\partial w}{\partial y} = 2xz^3 y$, $\frac{\partial w}{\partial z} = 3xy^2 z^2$

41. $\frac{\partial Q}{\partial L} = \frac{M-Lt}{M^2}e^{-Lt/M}$, $\frac{\partial Q}{\partial M} = \frac{L(Lt-M)}{M^3}e^{-Lt/M}$, $\frac{\partial Q}{\partial t} = -\frac{L^2}{M^2}e^{-Lt/M}$

43. $f_x(1,\,2) = -164$ **45.** $g_u(1,\,2) = \ln 3 + \frac{1}{3}$

47. (a) $I(95,\,50) \approx 73.1913$ (b) $\frac{\partial I}{\partial T}$; 1.66

51. (a), (b) **53.** $\frac{\partial^2 f}{\partial x^2} = 6y$, $\frac{\partial^2 f}{\partial y^2} = -72xy^2$

55. $h_{vv} = \frac{32u}{(u+4v)^3}$ **57.** $f_{yy}(2,\,3) = -\frac{4}{9}$

59. $f_{xyxzy} = 0$ **61.** $f_{uuv} = 2v\sin(u+v^2)$

63. $F_{rst} = 0$ **65.** $F_{uu\theta} = \cosh(uv + \theta^2)\cdot 2\theta v^2$

67. $g_{xyz} = \frac{3xyz}{(x^2+y^2+z^2)^{5/2}}$ **69.** $f(x,\,y) = x^2 y$ **73.** $B = A^2$

Section 14.4 Preliminary Questions

1. $L(x,\,y) = f(a,\,b) + f_x(a,\,b)(x-a) + f_y(a,\,b)(y-b)$

2. It is horizontal.

3. (b) **4.** $f(2,\,3.1) \approx 8.7$ **5.** $\Delta f \approx -0.1$

6. Using $\mathbf{v} \times \mathbf{w}$, we would be using a normal vector that is the negative of the normal vector obtained by $\mathbf{w} \times \mathbf{v}$. The resulting equations would represent the same plane.

Section 14.4 Exercises

1. $z = -34 - 20x + 16y$ **3.** $z = 16x - 8y + 24$

5. $z = 8x - 2y - 13$ **7.** $z = 4r - 5s + 2$

9. $z = \left(\frac{4}{5} + \frac{12}{25}\ln 2\right) - \frac{12}{25}x + \frac{12}{25}y$ **11.** $\left(-\frac{1}{4},\,\frac{1}{8},\,\frac{1}{8}\right)$ **13.** $(0,\,0)$

15. (a) $f(x,\,y) = -16 + 4x + 12y$

(b) $f(2.01,\,1.02) \approx 4.28$; $f(1.97,\,1.01) \approx 4$

17. $\Delta f \approx 3.56$ **19.** $f(0.01,\,-0.02) \approx 0.98$

21. $L(x,y,z) = \frac{5}{12}\sqrt{3}x + \frac{5}{12}\sqrt{3}y + 2\sqrt{3}z - 5\sqrt{3}$

23. 5.07 **25.** 8.44 **27.** 4.998 **29.** 3.945

31. $f(-2.1, 3.1) \approx 4.2$ **33.** $\Delta I \approx 0.5644$

35. (b) $\Delta H \approx 0.022$ m **37.** (b) 6% (c) 1% error in r

39. (a) \$7.10 (b) \$28.85, \$57.69 (c) $-$\$74.24

41. The maximum error in V is about 8.948 m.

Section 14.5 Preliminary Questions

1. (b) $\langle 3,\,4\rangle$ **2.** False

3. ∇f points in the direction of the maximum rate of increase of f and is normal to the level curve of f.

4. (b) NW and (c) SE **5.** $3\sqrt{2}$

Section 14.5 Exercises

1. (a) $\nabla f = \langle y^2,\ 2xy \rangle$, $\mathbf{r}'(t) = \langle t,\ 3t^2 \rangle$
(b) $\frac{d}{dt}\left(f(\mathbf{r}(t))\right)\big|_{t=1} = 4$; $\frac{d}{dt}\left(f(\mathbf{r}(t))\right)\big|_{t=-1} = -4$

3. A: zero; B: negative; C: positive; D: zero

5. $\nabla f = -\sin(x^2 + y)\langle 2x,\ 1 \rangle$

7. $\nabla h = \langle yz^{-3},\ xz^{-3},\ -3xyz^{-4} \rangle$

9. $\frac{d}{dt}\left(f(\mathbf{r}(t))\right)\big|_{t=0} = -7$ **11.** $\frac{d}{dt}\left(f(\mathbf{r}(t))\right)\big|_{t=0} = -3$

13. $\frac{d}{dt}\left(f(\mathbf{r}(t))\right)\big|_{t=0} = 5\cos 1 \approx 2.702$

15. $\frac{d}{dt}\left(f(\mathbf{r}(t))\right)\big|_{t=4} = -56$

17. $\frac{d}{dt}\left(f(\mathbf{r}(t))\right)\big|_{t=\pi/4} = -1 + \frac{8}{\pi} \approx 1.546$ **19.** $\frac{d}{dt}\left(g(\mathbf{r}(t))\right)\big|_{t=1} = 0$

21. $D_{\mathbf{u}} f(1,\ 2) = \frac{44}{5}$ **23.** $D_{\mathbf{u}} f\left(\frac{1}{6},\ 3\right) = \frac{39}{4\sqrt{2}}$

25. $D_{\mathbf{u}} f(3,\ 4) = \frac{7\sqrt{2}}{290}$ **27.** $D_{\mathbf{u}} f(1,\ 0) = \frac{6}{\sqrt{13}}$

29. $D_{\mathbf{u}} f(1,\ 2,\ 0) = -\frac{1}{\sqrt{3}}$ **31.** $D_{\mathbf{u}} f(3,\ 2) = \frac{-50}{\sqrt{13}}$

33. Direction: $\langle 1,\ -2 \rangle$; rate: $\sqrt{5}$

35. Direction: $\left\langle -\frac{1}{3},\ \frac{1}{3},\ \frac{1}{9} \right\rangle$; rate: $\frac{\sqrt{19}}{9}$

37. f is increasing at P in the direction of v.

39. $D_{\mathbf{u}} f(P) = \frac{\sqrt{6}}{2}$ **41.** $\langle 6,\ 2,\ -4 \rangle$

43. $\left(\frac{4}{\sqrt{17}},\ \frac{9}{\sqrt{17}},\ -\frac{2}{\sqrt{17}} \right)$ and $\left(-\frac{4}{\sqrt{17}},\ -\frac{9}{\sqrt{17}},\ \frac{2}{\sqrt{17}} \right)$

45. From Eq. 5, we have

$$ \mathbf{w} = \langle w_1,\ w_2 \rangle = C \left\langle -\frac{\partial p}{\partial y},\ \frac{\partial p}{\partial x} \right\rangle $$

for a constant C. Note that $\sin L$ is negative since $L < 0$ in the Southern Hemisphere. Since m, ω, and V are all positive, it then follows that $C < 0$. The vector $\left\langle -\frac{\partial p}{\partial y},\ \frac{\partial p}{\partial x} \right\rangle$ is ∇p rotated counterclockwise by $90°$, and $C\left\langle -\frac{\partial p}{\partial y},\ \frac{\partial p}{\partial x} \right\rangle$ points opposite to that, that is, in the direction of ∇p rotated clockwise by $90°$. So the parcel velocity vector (i.e., the wind vector) is proportional in magnitude to the pressure gradient and points in the direction $90°$ clockwise to it. Thus, the wind blows with low pressure to the right, and the closer together the isobars, the stronger the winds. In particular, winds blow clockwise around low pressure systems and counterclockwise around high pressure systems.

47. $9x + 10y + 5z = 33$

49. $0.5217x + 0.7826y - 1.2375z = -5.309$

51.

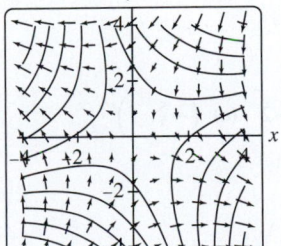

53. $f(x,\ y,\ z) = x^2 + y + 2z$

55. $f(x,\ y,\ z) = xz + y^2$ **59.** $\Delta f \approx 0.08$

61. (a) $\langle 34,\ 18,\ 0 \rangle$

(b) $\left\langle 2 + \frac{32}{\sqrt{21}}t,\ 2 + \frac{16}{\sqrt{21}}t,\ 8 - \frac{8}{\sqrt{21}}t \right\rangle$; ≈ 4.58 s

65. $x = 1 - 4t$, $y = 2 + 26t$, $z = 1 - 25t$ **77.** $y = \sqrt{1 - \ln(\cos^2 x)}$

Section 14.6 Preliminary Questions

1. (a) $\frac{\partial f}{\partial x}$ and $\frac{\partial f}{\partial y}$ (b) u and v

2. (a) **3.** $f(u,\ v)|_{(r,s)=(1,1)} = e^2$ **4.** (b) **5.** (c) **6.** No

Section 14.6 Exercises

1. (a) $\frac{\partial f}{\partial x} = 2xy^3$, $\frac{\partial f}{\partial y} = 3x^2y^2$, $\frac{\partial f}{\partial z} = 4z^3$

(b) $\frac{\partial x}{\partial s} = 2s$, $\frac{\partial y}{\partial s} = t^2$, $\frac{\partial z}{\partial s} = 2st$

(c) $\frac{\partial f}{\partial s} = 7s^6t^6 + 8s^7t^4$

3. $\frac{\partial f}{\partial s} = 6rs^2$, $\frac{\partial f}{\partial r} = 2s^3 + 4r^3$

5. $\frac{\partial g}{\partial x} = (y+1)\sec^2(xy + x + y)$,
$\frac{\partial g}{\partial y} = (x+1)\sec^2(xy + x + y)$

7. $\frac{\partial F}{\partial y} = xe^{x^2 + xy}$ **9.** $\frac{\partial h}{\partial t_2} = 0$

11. $\frac{\partial f}{\partial u}\Big|_{(u,v)=(-1,-1)} = 1$, $\frac{\partial f}{\partial v}\Big|_{(u,v)=(-1,-1)} = -2$

13. $\frac{\partial g}{\partial \theta}\Big|_{(r,\theta)=(2\sqrt{2},\ \pi/4)} = -\frac{1}{6}$ **15.** $\frac{\partial g}{\partial u}\Big|_{(u,v)=(0,1)} = 2\cos 2$

17. (a) $g'(x) = \frac{dg}{dx} = \frac{\partial f}{\partial x}\frac{dx}{dx} + \frac{\partial f}{\partial y}\frac{dy}{dx} = \frac{\partial f}{\partial x} + \frac{\partial f}{\partial y}y'(x)$
(b)
$g'(x) = 3x^2 - y^2 - 2xy(-1) = 3x^2 - (1-x)^2 + 2x(1-x) = 4x - 1$
(c) $g(x) = x^3 - x(1-x)^2 = 2x^2 - x$, so $g'(x) = 4x - 1$

19. -26.8 ft/s **21.** $4\pi^3 - 3\pi^2 - 1$

27. (a) $F_x = z^2 + y$, $F_y = 2yz + x$, $F_z = 2xz + y^2$

(b) $\frac{\partial z}{\partial x} = -\frac{z^2 + y}{2xz + y^2}$, $\frac{\partial z}{\partial y} = -\frac{2yz + x}{2xz + y^2}$

29. $\frac{\partial z}{\partial x} = -\frac{2xy + z^2}{2xz + y^2}$ **31.** $\frac{\partial z}{\partial y} = -\frac{xe^{xy} + 1}{x\cos(xz)}$

33. $\frac{\partial w}{\partial y} = \frac{-y(w^2 + x^2)^2}{w\left((w^2 + y^2)^2 + (w^2 + x^2)^2\right)}$; at $(1, 1, 1)$, $\frac{\partial w}{\partial y} = -\frac{1}{2}$

37. $\nabla\left(\frac{1}{r}\right) = -\frac{1}{r^3}\mathbf{r}$ **39.** (c) $\frac{\partial z}{\partial x} = \frac{x-6}{z+4}$

41. $\frac{\partial P}{\partial T} = -\frac{F_T}{F_P} = -\frac{-nR}{V-nb} = \frac{nR}{V-nb}$; $\frac{\partial V}{\partial P} = -\frac{F_P}{F_V} = \frac{nb-V}{P - \frac{an^2}{V^2} + \frac{2an^3b}{V^3}}$

Section 14.7 Preliminary Questions

1. f has a local (and global) min at $(0,\ 0)$; g has a saddle point at $(0,\ 0)$.

2.

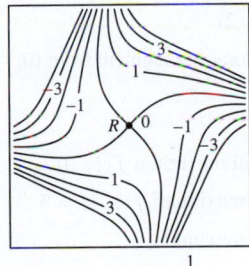

Point R is a saddle point.

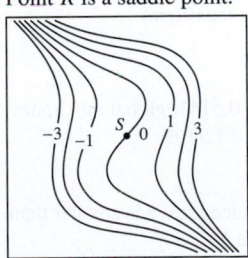

Point S is neither a local extremum nor a saddle point.

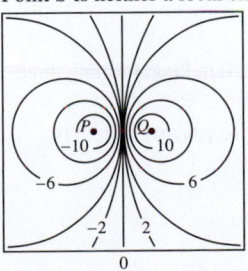

Point P is a local minimum and point Q is a local maximum.

3. Statement (a)

Section 14.7 Exercises

1. **(b)** $P_1 = (0, 0)$ is a saddle point, $P_2 = \left(2\sqrt{2}, \sqrt{2}\right)$ and $P_3 = \left(-2\sqrt{2}, -\sqrt{2}\right)$ are local minima; the absolute minimum value of f is -4.

3. $(0, 0)$ saddle point, $\left(\frac{13}{64}, -\frac{13}{32}\right)$ and $\left(-\frac{1}{4}, \frac{1}{2}\right)$ local minima

5. **(c)** $(0, 0)$, $(1, 0)$, and $(0, -1)$ saddle points; $\left(\frac{1}{3}, -\frac{1}{3}\right)$ local minimum

7. $\left(-\frac{2}{3}, -\frac{1}{3}\right)$ local minimum

9. $(-2, -1)$ local maximum, $\left(\frac{5}{3}, \frac{5}{6}\right)$ saddle point

11. $\left(0, \pm\sqrt{2}\right)$ saddle points, $\left(\frac{2}{3}, 0\right)$ local maximum, $\left(-\frac{2}{3}, 0\right)$ local minimum

13. $(0, 0)$ saddle point, $(1, 1)$ and $(-1, -1)$ local minima

15. $(0, 0)$ saddle point, $\left(\frac{1}{\sqrt{2}}, \frac{1}{\sqrt{2}}\right)$ and $\left(-\frac{1}{\sqrt{2}}, -\frac{1}{\sqrt{2}}\right)$ local maximum, $\left(\frac{1}{\sqrt{2}}, -\frac{1}{\sqrt{2}}\right)$ and $\left(-\frac{1}{\sqrt{2}}, \frac{1}{\sqrt{2}}\right)$ local minimum

17. Critical points are $\left(j\pi, \ k\pi + \frac{\pi}{2}\right)$, for
j, k even: saddle points
j, k odd: local maxima
j even, k odd: local minima
j odd, k even: saddle points

19. $\left(1, \frac{1}{2}\right)$ local maximum   **21.** $\left(\frac{3}{2}, -\frac{1}{2}\right)$ saddle point

23. $\left(-\frac{1}{6}, -\frac{17}{18}\right)$ local minimum

27. $x = y = 0.27788$ local minimum

29. Global maximum 2, global minimum 0

31. Global maximum 1, global minimum $\frac{1}{35}$

33. Max $= 28$ at $(0, -4)$; min $= -8$ at $(0, 2)$

35. Max $= 72$ at $(-3, 0)$; min $= -8$ all along the segment from $(0, -4)$ to $(1, 0)$

39. Maximum value $\frac{1}{3}$

41. Global minimum $f(0, 1) = -2$, global maximum $f(1, 0) = 1$

43. Global minimum, $f(0, 0) = 0$, global maximum $f(1, 1) = 3$

45. Global minimum, $f(0, 0) = 0$, global maximum $f(1, 0) = f(0, 1) = 1$

47. Global minimum, $f(1, 2) = -5$, global maximum $f(0, 0) = f(3, 3) = 0$

49. Global minimum, $f(-0.4343, 0.9) = f(-0.4343, -0.9) \approx -0.5161$, global maximum $f(0.7676, 0.6409) = f(0.7676, -0.6409) \approx 1.2199$

51. Maximum volume $\frac{3}{4}$

55. **(a)** No. In the box B with minimal surface area, z is smaller than $\sqrt[3]{V}$, which is the side of a cube with volume V.
(b) Width: $x = (2V)^{1/3}$; length: $y = (2V)^{1/3}$; height: $z = \left(\frac{V}{4}\right)^{1/3}$

57. The fence should be cut into 12 pieces 10 m long, forming three 10×10 squares.

59. $V = \frac{64,000}{\pi} \approx 20,372 \text{ cm}^3$ **61.** $f(x) = 1.9629x - 1.5519$

Section 14.8 Preliminary Questions

1. Statement (b)

2. f had a local maximum 2, under the constraint, at A; $f(B)$ is neither a local minimum nor a local maximum of f.

3. **(a)**

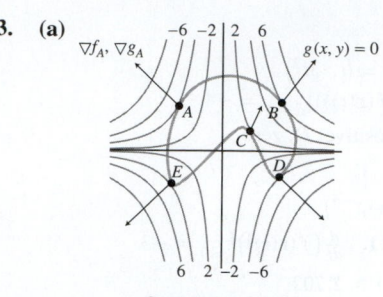

Contour plot of $f(x, y)$
(contour interval 2)

(b) Global minimum -4, global maximum 6

Section 14.8 Exercises

1. **(c)** Critical points $(-1, -2)$ and $(1, 2)$
(d) Maximum 10, minimum -10

3. Maximum $4\sqrt{2}$, minimum $-4\sqrt{2}$

5. Minimum $\frac{36}{13}$, no maximum value

7. Maximum $\frac{8}{3}$, minimum $-\frac{8}{3}$

9. Maximum $\sqrt{2}$, minimum 1

11. Maximum 3.7, minimum -3.7

13. Maxima at $f(\pm 4, \pm 4, 2) = 20$, and minima at $f(\pm 4, \mp 4, -2) = -20$

15. Maximum $2\sqrt{2}$, minimum $-2\sqrt{2}$ **17.** $(-1, \ e^{-1})$

19. **(a)** $\frac{h}{r} = \sqrt{2}$ **(b)** $\frac{h}{r} = \sqrt{2}$
(c) There is no cone of fixed V with maximal S.

21. $(8, \ -2)$ **23.** $\left(\frac{48}{97}, \frac{108}{97}\right)$ **25.** $\frac{a^a b^b}{(a+b)^{a+b}}$ **27.** $\sqrt{\frac{a^a b^b}{(a+b)^{a+b}}}$

33. $r = 3$, $h = 6$ **35.** $x + y + z = 3$

41. $\frac{25}{3}$ **43.** $\left(\frac{-6}{\sqrt{105}}, \frac{-3}{\sqrt{105}}, \frac{30}{\sqrt{105}}\right)$ **45.** $(-1, 0, 2)$

47. Minimum $\frac{138}{11} \approx 12.545$, no maximum value

51. **(b)** $\lambda = \frac{c}{2p_1 p_2}$

Chapter 14 Review

1. **(a)**

(b) $f(3, 1) = \frac{\sqrt{2}}{3}$, $f(-5, \ -3) = -2$ **(c)** $\left(-\frac{5}{3}, 1\right)$

3.

Vertical and horizontal traces: the line $z = (c^2 + 1) - y$ in the plane $x = c$, the parabola $z = x^2 - c + 1$ in the plane $y = c$

5. **(a)** Graph (B) **(b)** Graph (C) **(c)** Graph (D) **(d)** Graph (A)

7. **(a)** Parallel lines $4x - y = \ln c$, $c > 0$, in the xy-plane
(b) Parallel lines $4x - y = e^c$ in the xy-plane
(c) Hyperbolas $3x^2 - 4y^2 = c$ in the xy-plane
(d) Parabolas $x = c - y^2$ in the xy-plane

9. $\lim\limits_{(x,y)\to(1,-3)}(xy+y^2)=6$

11. The limit does not exist.

13. $\lim\limits_{(x,y)\to(1,-3)}(2x+y)e^{-x+y}=-e^{-4}$

17. $f_x=2,\ f_y=2y$

19. $f_x=e^{-x-y}(y\cos(xy)-\sin(xy))$
$f_y=e^{-x-y}(x\cos(yx)-\sin(yx))$

21. $f_{xxyz}=-\cos(x+z)$ **23.** $z=33x+8y-42$

25. Estimate, 12.146; calculator value to three places, 11.996

27. Statements (ii) and (iv) are true.

29. $\frac{d}{dt}\big(f(\mathbf{c}(t))\big)\big|_{t=2}=3+4e^4\approx221.4$

31. $\frac{d}{dt}\big(f(\mathbf{c}(t))\big)\big|_{t=1}=4e-e^{3e}\approx-3469.3$

33. $D_\mathbf{u}f(3,-1)=-\frac{54}{\sqrt5}$

35. $D_\mathbf{u}f(P)=-\frac{\sqrt2 e}{5}$ **37.** $\left\langle\frac{1}{\sqrt2},\frac{1}{\sqrt2},0\right\rangle$

41. $\frac{\partial f}{\partial s}=3s^2t+4st^2+t^3-2st^3+6s^2t^2$
$\frac{\partial f}{\partial t}=4s^2t+3st^2+s^3+4s^3t-3s^2t^2$

45. $\frac{\partial z}{\partial x}=-\frac{e^z-1}{xe^z+e^y}$

47. $(0,0)$ saddle point, $(1,1)$ and $(-1,-1)$ local minima

49. $\left(\frac12,\frac12\right)$ saddle point

53. Global maximum $f(2,4)=10$, global minimum $f(-2,4)=-18$

55. Maximum $\frac{26}{\sqrt{13}}$, minimum $-\frac{26}{\sqrt{13}}$

57. Maximum $\frac{12}{\sqrt3}$, minimum $-\frac{12}{\sqrt3}$

59. Minimum $=f\left(\frac{1+\sqrt3}{3},\frac13,\frac{1-\sqrt3}{3}\right)=\frac{6-2\sqrt3}{3}$
Maximum $=f\left(\frac{1-\sqrt3}{3},\frac13,\frac{1+\sqrt3}{3}\right)=\frac{6+2\sqrt3}{3}$

61. $r=h=\sqrt[3]{\frac{V}{\pi}},\ S=3\pi\left(\frac{V}{\pi}\right)^{2/3}$

Chapter 15

Section 15.1 Preliminary Questions

1. $\Delta A=1$; the number of subrectangles is 32.

2. $\iint_R f\,dA\approx S_{1,1}=0.16$ **3.** $\iint_R 5\,dA=50$

4. The signed volume between the graph $z=f(x,y)$ and the xy-plane. The region below the xy-plane is treated as negative volume.

5. (b) **6.** (b), (c); volume above xy-plane $=$ volume below xy-plane

Section 15.1 Exercises

1. $S_{4,3}=13.5$ **3.** (A) $S_{3,2}=42$ (B) $S_{3,2}=43.5$

5. (A) $S_{3,2}=60$ (B) $S_{3,2}=62$

7. Two possible solutions are $S_{3,2}=\frac{77}{72}$ and $S_{3,2}=\frac{79}{72}$.

9. $\frac{225}{2}$

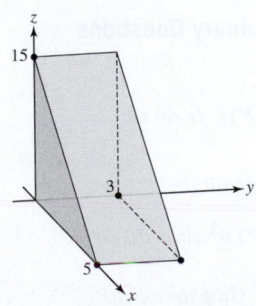

11. 0.19375 **13.** 1.0731; 1.0783; 1.0809 **15.** 0 **17.** 0 **19.** 40

21. 55 **23.** $\frac43$ **25.** 84 **27.** 4 **29.** $\frac{1858}{15}$

31. $6\ln6-2\ln2-5\ln5\approx1.317$ **33.** $\frac43\left(19-5\sqrt5\right)\approx10.426$

35. $\frac12(\ln3)(-2+\ln48)\approx1.028$ **37.** $6\ln3\approx6.592$

39. 1 **41.** $\left(e^2-1\right)\left(1-\frac{\sqrt2}{2}\right)\approx1.871$ **43.** 9/2 **45.** $m=\frac34$

47. (a) With respect to x since the u-substitution doesn't involve a product rule (b) $\frac{2}{15}(8\sqrt2-7)\approx0.575161$

49. (a) With respect to x since the u-substitution doesn't involve a quotient rule (b) $\ln(4)-1\approx0.386$ **53.** $\frac{e^3}{3}-\frac13-e+1\approx4.644$

Section 15.2 Preliminary Questions

1. (b), (c)

2.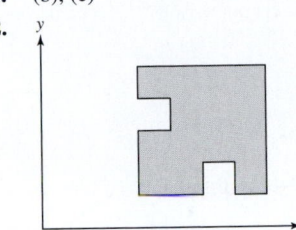

3.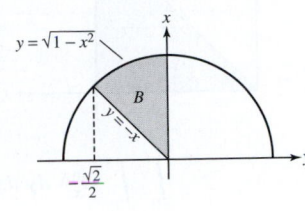

4. (b)

Section 15.2 Exercises

1. (a) Sample points •, $S_{3,4}=-3$

(b) Sample points ∘, $S_{3,4}=-4$

3. As a vertically simple region: $0\le x\le1,\ 0\le y\le1-x^2$; as a horizontally simple region: $0\le y\le1,\ 0\le x\le\sqrt{1-y}$,
$\int_0^1\left(\int_0^{1-x^2}(xy)\,dy\right)dx=\frac{1}{12}$

5. $\frac{192}{5}=38.4$ **7.** $\frac{608}{15}\approx40.53$ **9.** $2\frac14$ **11.** $-\frac34+\ln4$

13. $\frac{16}{3}\approx5.33$ **15.** $\frac{11}{60}$ **17.** 66875/8 **19.** $\frac{e-2}{2}\approx0.359$ **21.** 29/70

23. $2e^{12}-\frac12e^9+\frac12e^5\approx321,532.2$

25.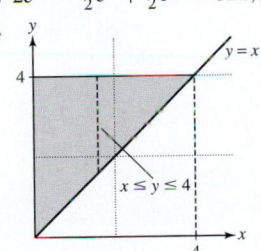
$$\int_0^4\int_x^4 f(x,y)\,dy\,dx=\int_0^4\int_0^y f(x,y)\,dx\,dy$$

27.
$$\int_4^9\int_2^{\sqrt y} f(x,y)\,dx\,dy=\int_2^3\int_{x^2}^9 f(x,y)\,dy\,dx$$

29.
$$\int_0^2\int_0^{x^2}\sqrt{4x^2+5y}\,dy\,dx=\frac{152}{15}$$

31. $\int_1^e \int_{\ln^2 y}^{\ln y} (\ln y)^{-1} \, dx \, dy = e - 2 \approx 0.718$

33.

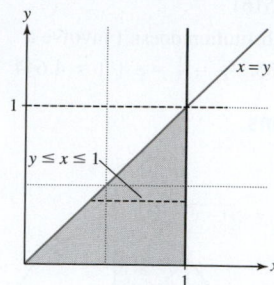

$$\int_0^1 \int_0^x \frac{\sin x}{x} \, dy \, dx = 1 - \cos 1 \approx 0.460$$

35.

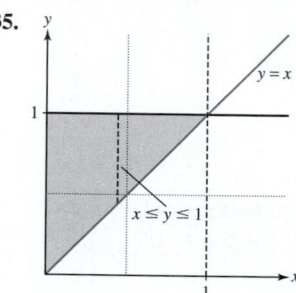

$$\int_0^1 \int_0^y xe^{y^3} \, dx \, dy = \frac{e-1}{6} \approx 0.286$$

37.

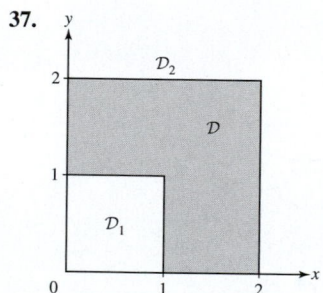

$$\iint_D e^{x+y} \, dA = (e^2 - 1)^2 - (e - 1)^2 \approx 37.878$$

39. $\int_0^4 \int_{x/4}^{3x/4} e^{x^2} \, dy \, dx = \frac{1}{4}\left(e^{16} - 1\right)$

41. $\int_2^4 \int_{y-1}^{7-y} \frac{x}{y^2} \, dx \, dy = 6 - 6\ln 2 \approx 1.841$

43. $\iint_D \frac{\sin y}{y} \, dA = \cos 1 - \cos 2 \approx 0.956$

45. $\int_{-2}^2 \int_0^{4-x^2} (40 - 10y) \, dy \, dx = 256$

47. $\int_{-2}^2 \int_{x^2}^{8-x^2} (16 - y - y) \, dy \, dx = \frac{512}{3}$

49. $\int_{-2}^2 \left(\int_{-\sqrt{4-x^2}}^{\sqrt{4-x^2}} [(8 - x^2 - y^2) - (x^2 + y^2)] \, dy \right) dx$

51. ≈ 7.541 billion cubic meters

53. $\int_0^1 \int_0^1 e^{x+y} \, dx \, dy = e^2 - 2e + 1 \approx 2.952$

55. $\frac{1}{\pi} \int_0^1 \int_0^\pi y^2 \sin x \, dx \, dy = \frac{2}{3\pi}$ **57.** $\bar{f} = p$

63. One possible solution is $P = \left(\frac{2}{3}, 2\right)$.

65. $\iint_D f(x, y) \, dA \approx 57.01$

Section 15.3 Preliminary Questions

1. (c) **2.** (b)

3. (a) $D = \{(x, y) : 0 \le x \le 1, \, 0 \le y \le x\}$

(b) $D = \left\{(x, y) : 0 \le x \le 1, \, 0 \le y \le \sqrt{1 - x^2}\right\}$

Section 15.3 Exercises

1. 288 **3.** $(e - 1)(1 - e^{-2})$ **5.** $-\frac{27}{4} = -6.75$

7. $\frac{b}{20}\left[(a+c)^5 - a^5 - c^5\right]$ **9.** $\frac{1}{6}$ **11.** $\frac{1}{16}$ **13.** $e - \frac{5}{2}$ **15.** $2\frac{1}{12}$

17. $\frac{128}{15}$ **19.** $\int_0^3 \int_0^4 \int_0^{y/4} dz \, dy \, dx = 6$ **21.** $\frac{1}{12}$

23. $\frac{126}{5}$

25. Region enclosed by sphere $x^2 + y^2 + z^2 = 5$ to right of plane $y = 1$.

27. $\iiint_B f(x, y, z) \, dV = \int_a^b \int_c^d \int_p^q g(x)h(y)k(z) \, dz \, dy \, dx = \int_a^b g(x) \, dx \int_c^d h(y) \, dy \int_p^q k(z) \, dz$

29. $\int_0^2 \int_0^{y/2} \int_0^{4-y^2} xyz \, dz \, dx \, dy$, $\int_0^4 \int_0^{\sqrt{4-z}} \int_0^{y/2} xyz \, dx \, dy \, dz$, and $\int_0^4 \int_0^{\sqrt{1-(z/4)}} \int_{2x}^{\sqrt{4-z}} xyz \, dy \, dx \, dz$

31. $\int_{-1}^1 \int_{-\sqrt{1-x^2}}^{\sqrt{1-x^2}} \int_{\sqrt{x^2+y^2}}^1 f(x, y, z) \, dz \, dy \, dx$ **33.** $\frac{16}{21}$

35. $\int_{-1}^1 \left(\int_0^{1-y^2} \left(\int_0^{3-z} dx \right) dz \right) dy$

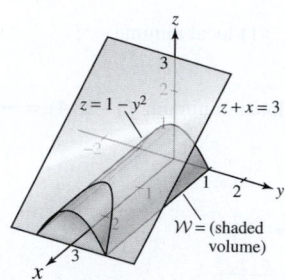

$W =$ (shaded volume)

37. $\int_0^1 \left(\int_{\sqrt{y}}^{y^2} \left(\int_0^{4-y-z} dx \right) dz \right) dy$

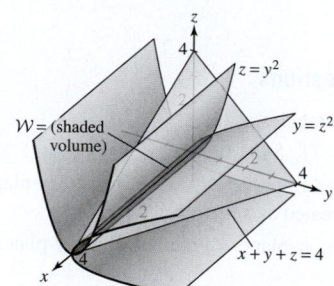

$W =$ (shaded volume)

39. $\frac{1}{2\pi}$ **41.** $2e - 4 \approx 1.437$

43. $S_{N,N,N} \approx 0.561, \, 0.572, \, 0.576; \, I \approx 0.584; \, N = 100$

Section 15.4 Preliminary Questions

1. (d)

2. (a) $\int_{-1}^2 \int_0^{2\pi} \int_0^2 f(P) \, r \, dr \, d\theta \, dz$

(b) $\int_{-2}^0 \int_0^{2\pi} \int_0^{\sqrt{4-z^2}} r \, dr \, d\theta \, dz$

3. (a) $\int_0^{2\pi} \int_0^\pi \int_0^4 f(P) \, \rho^2 \sin\phi \, d\rho \, d\phi \, d\theta$

(b) $\int_0^{2\pi} \int_0^\pi \int_4^5 f(P) \, \rho^2 \sin\phi \, d\rho \, d\phi \, d\theta$

(c) $\int_0^{2\pi} \int_{\pi/2}^\pi \int_0^2 f(P) \, \rho^2 \sin\phi \, d\rho \, d\phi \, d\theta$

4. $\Delta A \approx r(\Delta r \, \Delta\theta)$, and the factor r appears in $dA = r \, dr \, d\theta$ in the Change of Variables formula.

5. (a) 18π (b) 18π (c) $\frac{9\pi}{4}$ (d) 9π

Section 15.4 Exercises

1.

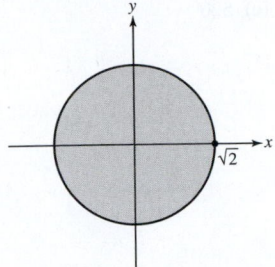

$$\iint_D \sqrt{x^2 + y^2}\, dA = \frac{4\sqrt{2}\pi}{3}$$

3.

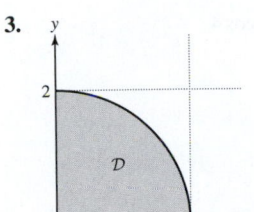

$$\iint_D xy\, dA = 2$$

5.

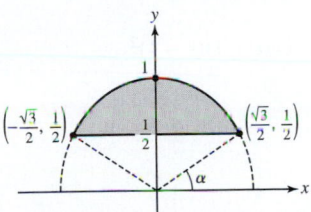

$$\iint_D y\left(x^2 + y^2\right)^{-1} dA = \sqrt{3} - \frac{\pi}{3} \approx 0.685$$

7.

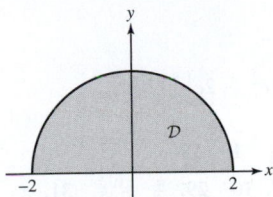

$$\int_{-2}^{2} \int_{0}^{\sqrt{4-x^2}} (x^2 + y^2)\, dy\, dx = 4\pi$$

9.

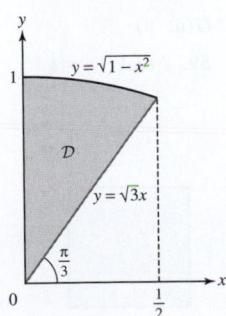

$$\int_{0}^{1/2} \int_{\sqrt{3}x}^{\sqrt{1-x^2}} x\, dy\, dx = \frac{1}{3}\left(1 - \frac{\sqrt{3}}{2}\right) \approx 0.045$$

11.

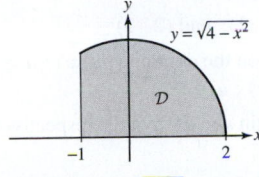

$$\int_{0}^{\pi/4} \left(\int_{0}^{5/\cos(\theta)} \left(r^2 \cos(\theta)\right) dr \right) d\theta = \frac{125}{3}$$

13.

$$\int_{-1}^{2} \int_{0}^{\sqrt{4-x^2}} (x^2 + y^2)\, dy\, dx = \frac{\sqrt{3}}{2} + \frac{8\pi}{3} \approx 9.244$$

15. $\frac{1}{4}$ **17.** $\frac{1}{2}$ **19.** 0   **21.** 18   **23.** $\frac{48\pi - 32}{9} \approx 13.2$

25. (a) $W: 0 \le \theta \le 2\pi,\ 0 \le r \le 2,\ 2 \le z \le 6 - r^2$ **(b)** 8π

27. $\frac{405\pi}{2} \approx 636.17$ **29.** 128   **31.** 243π

33. $\int_{0}^{2\pi} \int_{0}^{1} \int_{0}^{4} f(r\cos\theta,\ r\sin\theta,\ z)\, r\, dz\, dr\, d\theta$

35. $\int_{0}^{\pi} \int_{0}^{1} \int_{0}^{r^2} f(r\cos\theta,\ r\sin\theta,\ z)\, r\, dz\, dr\, d\theta$

37. $z = \frac{H}{R}r;\ V = \frac{\pi R^2 H}{3}$ **39.** 16π

41.

$$V = 2\int_{0}^{2\pi} \left(\int_{b}^{a} \left(\int_{0}^{\sqrt{a^2-r^2}} (r)\, dz \right) dr \right) d\theta = \frac{4}{3}\pi\left(a^2 - b^2\right)^{3/2} = \frac{4}{3}\pi r h^3$$

43. $V = \int_{0}^{2\pi} \left(\int_{0}^{\pi/3} \left(\int_{\sec(\phi)}^{2} \left(\rho^2 \sin(\phi)\right) d\rho \right) d\phi \right) d\theta = \frac{5\pi}{3}$

45. $-\frac{\pi}{16}$ **47.** $\frac{8\pi}{15}$ **49.** $\frac{8\pi}{5}$ **51.** $\frac{5\pi}{8}$ **53.** π **55.** $\frac{4\pi a^3}{3}$

57. (b) $J = \int_{0}^{2\pi} \left(\int_{0}^{\infty} \left(e^{-r^2} r\right) dr \right) d\theta = \pi$

Section 15.5 Preliminary Questions

1. 5 kg/m^3 **2. (a)**

3. The probability that $0 \le X \le 1$ and $0 \le Y \le 1$; the probability that $0 \le X + Y \le 1$

Section 15.5 Exercises

1. 4   **3.** $4\left(1 - e^{-100}\right) \times 10^{-6}$ C $\approx 4 \times 10^{-6}$ C

5. $10{,}000 - 18{,}000e^{-4/5} \approx 1912$

7. $25\pi\left(3 \times 10^{-8}\ \text{C}\right) \approx 2.356 \times 10^{-6}$ C

9. $\approx 2.593 \times 10^{10}$ kg   **11.** $\left(0, \frac{2}{5}\right)$ **13.** $\left(\frac{4R}{3\pi}, \frac{4R}{3\pi}\right)$

15. $\left(1 - \frac{M}{e^M - 1}, \frac{1 - e^{-2M}}{4(1 - e^{-M})}\right)$ **17.** $(0.555, 0)$ **19.** $\left(0, 0, \frac{3R}{8}\right)$

21. $\left(0, 0, \frac{9}{8}\right)$ **23.** $\left(0, 0, \frac{13}{2(17 - 6\sqrt{6})}\right)$ **25.** $\left(\frac{18}{5}, \frac{6}{5}\right)$ **27.** $\left(\frac{1}{6}, \frac{1}{6}\right)$

29. $\frac{16}{15\pi}$

31. (a) $\frac{M}{4ab}$ **(b)** $I_x = \frac{Mb^2}{3};\ I_0 = \frac{M(a^2 + b^2)}{3}$ **(c)** $\frac{b}{\sqrt{3}}$

33. $I_0 = 8{,}000$ kg $\cdot$ m^2; $I_x = 4{,}000$ kg $\cdot$ m^2

35. $\frac{9}{2}$ **37.** $\frac{243}{20}$ **39.** $\left(\frac{a}{2}, \frac{2b}{5}\right)$ **41.** $\frac{a^2 b^4}{60}$

43. $I_x = \frac{MR^2}{4}$; kinetic energy required is $\frac{25MR^2}{2}$ J

49. (a) $I = 182.5$ g $\cdot$ cm^2 **(b)** $\omega \approx 196.90$ rad/s

51. $\frac{13}{72}$ **53.** $\frac{1}{64}$ **55.** $C = 15$; probability is $\frac{5}{8}$

57. (a) $C = 4$ **(b)** $\frac{1}{48\pi} + \frac{1}{32} \approx 0.038$

Section 15.6 Preliminary Questions

1. (b)

2. (a) $G(1, 0) = (2, 0)$ (b) $G(1, 1) = (1, 3)$

(c) $G(2, 1) = (3, 3)$

3. Area $(G(R)) = 36$ **4.** Area $(G(R)) \approx 0.06$

Section 15.6 Exercises

1. (a) Image of the u-axis is the line $y = \frac{1}{2}x$; image of the v-axis is the y-axis

(b) The parallelogram with vertices $(0, 0)$, $(10, 5)$ $(10, 12)$, $(0, 7)$

(c) The segment joining the points $(2, 3)$ and $(10, 8)$

(d) The triangle with vertices $(0, 1)$, $(2, 1)$, and $(2, 2)$

3. G is not one-to-one; G is one-to-one on the domain $\{(u, v) : u \geq 0\}$, and G is one-to-one on the domain $\{(u, v) : u \leq 0\}$.

(a) The positive x-axis including the origin and the y-axis, respectively

(b) The rectangle $[0, 1] \times [-1, 1]$

(c) The curve $y = \sqrt{x}$ for $0 \leq x \leq 1$

(d)

5. $y = 3x - c$ **7.** $y = \frac{17}{6}x$ **11.** Jac$(G) = 1$ **13.** Jac$(G) = -10$
15. Jac$(G) = 1$ **17.** Jac$(G) = 4$

19. $G(u, v) = (4u + 2v, u + 3v)$ **21.** $\frac{2329}{12} \approx 194.08$

23. (a) Area $(G(R)) = 105$ (b) Area $(G(R)) = 126$

25. Jac$(G) = \frac{2u}{v}$; for $R = [1, 4] \times [1, 4]$, area $(G(R)) = 15 \ln 4$

27.

$$G(u, v) = (1 + 2u + v, 1 + 5u + 3v)$$

29. 82   **31.** 80   **33.** $\frac{56}{45}$ **35.** $\frac{\pi(e^{36} - 1)}{6}$

37.

$$\iint_D y^{-1} \, dx \, dy = 1$$

39. $\iint_D e^{xy} \, dA = (e^{20} - e^{10}) \ln 2$

41. (b) $-\frac{1}{x+y}$ (c) $I = 9$ **45.** $\frac{\pi^2}{8}$

Chapter 15 Review

1. (a) $S_{2,3} = 240$ (b) $S_{2,3} = 510$ (c) 520

3. $S_{4,4} = 2.9375$ **5.** $\frac{32}{3}$ **7.** $\frac{\sqrt{3} - 1}{2}$

9.

$$\iint_D \cos y \, dA = 1 - \cos 4$$

11.

$$\iint_D e^{x+2y} \, dA = \frac{1}{2}e(e + 1)(e - 1)^2$$

13.

$$\iint_D y e^{1+x} \, dA = 0.5(e^2 - 2e^{1.5} + e)$$

15. $\int_0^9 \int_{-\sqrt{9-y}}^{\sqrt{9-y}} f(x, y) \, dx \, dy$ **17.** $\frac{1}{24}$ **19.** $18(\sqrt{2} - 1)$

21. $1 - \cos 1$ **23.** 6π **25.** $\pi/2$ **27.** 10   **29.** $\frac{\pi}{4} + \frac{2}{3}$ **31.** π

33. $\frac{1}{4}$ **35.** $\int_0^{\pi/2} \int_0^1 \int_0^r zr \, dz \, dr \, d\theta = \pi/16$ **37.** $\frac{2\pi(-1 + e^8)}{3e^8}$

41. $\frac{256\pi}{15} \approx 53.62$ **43.** 1280π **45.** $\left(-\frac{1}{4}R, 0, \frac{5}{8}H\right)$

47. $\left(-\frac{2}{11\pi}R, -\frac{2}{11\pi}R(2 - \sqrt{3}), \frac{1}{2}H\right)$.   **49.** $\left(0, 0, \frac{2}{3}\right)$

51. $\left(\frac{8}{15}, \frac{16}{15\pi}, \frac{16}{15\pi}\right)$ **53.** $\frac{19}{33}$ **55.** $\frac{4}{7}$ **57.** $G(u, v)$

$= (3u + v, -u + 4v)$; area $(G(R)) = 156$ **59.** Area$(D) \approx \frac{1}{5}$

61. (a)

(d) $\frac{3}{4}(e^2 - \sqrt{e})$

Chapter 16

Section 16.1 Preliminary Questions

1. (b)   **2.**

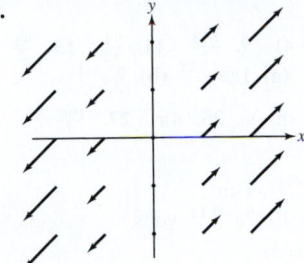

3. $\mathbf{F} = \langle 0, -z, y \rangle$ **4.** $f_1(x, y, z) = xyz + 1$

Section 16.1 Exercises

1. $\mathbf{F}(1, 2) = \langle 1, 1 \rangle$, $\mathbf{F}(-1, -1) = \langle 1, -1 \rangle$

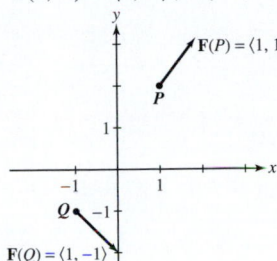

3. $\mathbf{F}(P) = \langle 0, 1, 0 \rangle$; $\mathbf{F}(Q) = \langle 2, 0, 2 \rangle$

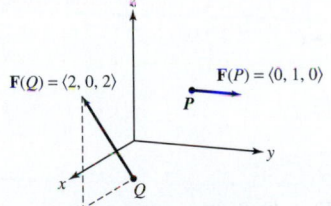

5. $\mathbf{F} = \langle 1, 0 \rangle$

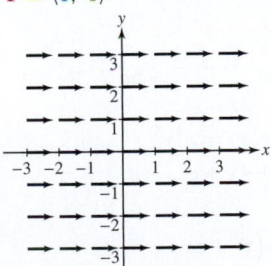

7. $\mathbf{F} = x\mathbf{i}$

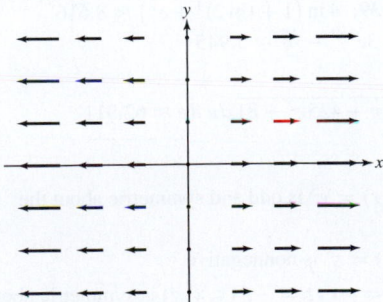

9. $\mathbf{F}(x, y) = \langle 0, x \rangle$

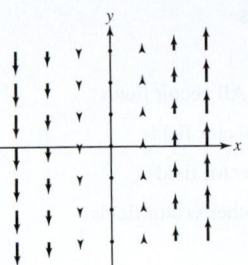

11. $\mathbf{F} = \left\langle \frac{x}{x^2 + y^2}, \ \frac{y}{x^2 + y^2} \right\rangle$

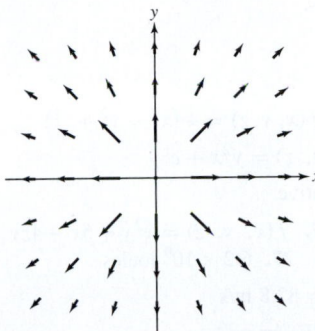

13. Plot (D)   **15.** Plot (B)   **17.** Plot (C)   **19.** Plot (B)
21. $(0, y)$ **23.** $\operatorname{div}(\mathbf{F}) = 3$; $\operatorname{curl}(\mathbf{F}) = 0$
25. $\operatorname{div}(\mathbf{F}) = 1 - 4xz - x + 2x^2z$,
$\operatorname{curl}(\mathbf{F}) = \langle -1, 2x^2 - 2xz^2, -y \rangle$
27. $\operatorname{div}(\mathbf{F}) = 0$; $\operatorname{curl}(\mathbf{F}) = 0$
29. $\operatorname{div}(\mathbf{F}) = 0$, $\operatorname{curl}(\mathbf{F}) = \langle 0, \sin x, \cos x - e^y \rangle$
39. $f(x, y) = \frac{1}{2}x^2 + \frac{1}{2}y^2 + K$
41. $f(x, y, z) = \langle xy^2z, y + xy^2z, xy^2z \rangle$
43. $f(x, y) = e^{xy} + K$
45. $f(x, y, z) = xyz^2 + K$ **47.** $f(x, y, z) = \sin(xyz) + K$
49. $f(x, y, z) = ax + by + cz + K$
51. (b) $e_P(1, 1) = \langle -2, -1 \rangle / \sqrt{5}$;
　　(c) $f(x, y, z) = \sqrt{(x - a)^2 + (y - b)^2}$
53. (A)

Section 16.2 Preliminary Questions

1. 50   **2.** (a), (c), (d), (e)
3. (a) True   (b) False. Reversing the orientation of the curve changes the sign of the vector line integral.
4. (a) 0   (b) -5

Section 16.2 Exercises

1. (a) $f(\mathbf{r}(t)) = 6t + 4t^2$; $ds = 2\sqrt{11}\, dt$
(b) $\int_0^1 (6t + 4t^2) 2\sqrt{11}\, dt = \frac{26\sqrt{11}}{3}$
3. (a) $\mathbf{F}(\mathbf{r}(t)) = \langle t^{-2}, t^2 \rangle$; $d\mathbf{r} = \langle 1, -t^{-2} \rangle dt$
(b) $\int_1^2 (t^{-1} - 1) dt = -\frac{1}{2}$
5. $\sqrt{2}\left(\pi + \frac{\pi^3}{3}\right)$ **7.** $\frac{\pi^3}{3}$
9. 298/5   **11.** $\frac{128\sqrt{29}}{3} \approx 229.8$
13. $\frac{\sqrt{3}}{2}(e - 1) \approx 1.488$ **15.** $\frac{2}{3}\left((e^2 + 5)^{3/2} - 2^{3/2}\right)$
17. 39; the distance between $(8, -6, 24)$ and $(20, -15, 60)$
19. 97/12   **21.** 0   **23.** $2(e^2 - e^{-2}) - (e - e^{-1}) \approx 12.157$ **25.** $\frac{10}{9}$
27. 9/4   **29.** $-\frac{8}{3}$ **31.** $\frac{13}{2}$ **33.** $\frac{\pi}{2}$ **35.** 339.5587   **37.** $2 - e - \frac{1}{e}$
39. (a) -8 (b) -11 (c) -16
41. ≈ 7.6; ≈ 4 **43.** (A) Zero (B) Negative (C) Zero   **45.** 64π g
47. $\approx 10.4 \times 10^{-6}$ C   **49.** $\approx 22{,}743.10$ volts   **51.** $\approx -10{,}097$ volts
53. 1 joule   **55.** $\frac{27}{28}$ joule   **57.** (a) ABC (b) CBA
61. $\frac{1}{3}\left((4\pi^2 + 1)^{3/2} - 1\right) \approx 85.5 \times 10^{-6}$ C   **67.** 18   **73.** ≈ 0.574

Section 16.3 Preliminary Questions

1. Closed

2. (a) Conservative vector fields (b) All vector fields

(c) Conservative vector fields (d) All vector fields

(e) Conservative vector fields (f) All vector fields

(g) Conservative vector fields and *some* other vector fields

3. (a) Always true (b) Always true

(c) True under additional hypotheses on D

4. (a) 4 (b) -4

Section 16.3 Exercises

1. 0 **3.** $-\frac{9}{4}$ **5.** $32e - 1$ **7.** $f(x, y, z) = \frac{1}{2}\left(x^2 + y^2 + z^2\right)$

9. Not conservative **11.** $f(x, y, z) = y^2x + e^z y$

13. The vector field is not conservative.

15. $f(x, y, z) = z \tan x + zy$ **17.** $f(x, y, z) = x^2y + 5x - 4zy$
19. 16 **21.** 1 **23.** 6 **25.** $\frac{2}{3}$; 0 **27.** 6.2×10^9 joules

29. (a) $f(x, y, z) = -gz$ (b) ≈ 82.8 m/s

31. (A) 2π (B) 2π (C) 0 (D) -2π (E) 4π

33. Not conservative

Section 16.4 Preliminary Questions

1. 50

2. A distortion factor that indicates how much the area of R_{ij} is altered under the map G **3.** area$(S) \approx 0.0006$ **4.** $\iint_S f(x, y, z)\,dS \approx 0.6$
5. area$(S) = 20$ **6.** $\left\langle \frac{2}{3}, \frac{2}{3}, \frac{1}{3} \right\rangle$

Section 16.4 Exercises

1. (a) v (b) iii (c) i (d) iv (e) ii

3. (a) $\mathbf{T}_u = \langle 2, 1, 3 \rangle$, $\mathbf{T}_v = \langle 0, -1, 1 \rangle$, $\mathbf{N}(u, v) = \langle 4, -2, -2 \rangle$

(b) area$(S) = 4\sqrt{6}$ (c) $\iint_S f(x, y, z)\,dS = \frac{32\sqrt{6}}{3}$

5. (a) $\mathbf{T}_x = \langle 1, 0, y \rangle$, $\mathbf{T}_y = \langle 0, 1, x \rangle$,

$\mathbf{N}(x, y) = \langle -y, -x, 1 \rangle$ (b) $\frac{(2\sqrt{2} - 1)\pi}{6}$ (c) $\frac{\sqrt{2}+1}{15}$

7. $\mathbf{T}_u = \langle 2, 1, 3 \rangle$; $\mathbf{T}_v = \langle 1, -4, 0 \rangle$;
$\mathbf{N}(u, v) = 3\langle 4, 1, -3 \rangle$, $4x + y - 3z = 0$

9. $\mathbf{T}_\theta = \langle -\sin\theta \sin\phi, \cos\theta \sin\phi, 0 \rangle$;
$\mathbf{T}_\phi = \langle \cos\theta \cos\phi, \sin\theta \cos\phi, -\sin\phi \rangle$;
$\mathbf{N}(u, v) = -\cos\theta \sin^2\phi \mathbf{i} - \sin\theta \sin^2\phi \mathbf{j} - \sin\phi \cos\phi \mathbf{k}$,
$y + z = \sqrt{2}$

11. area$(S) \approx 0.2078$ **13.** $\frac{\sqrt{2}}{5}$ **15.** $\frac{1}{2}\left(17\sqrt{17} - 1\right) \approx 34.546$

17. $\frac{\pi}{6}$ **19.** $4\pi(1 - e^{-4})$ **21.** $\frac{\sqrt{3}}{2}$ **23.** $\frac{7\pi}{3}$ **25.** $\frac{5\sqrt{10}}{27} - \frac{1}{54}$
27. area$(S) = 16$ **29.** $3e^3 - 6e^2 + 3e + 1 \approx 25.08$
31. area$(S) = 4\pi R^2$

33. (a) area$(S) \approx 1.0780$ (b) ≈ 0.09814

35. $\sqrt{1 + \left(\frac{a}{c}\right)^2 + \left(\frac{b}{d}\right)^2}$ **37.** area$(S) = \pi$ **39.** 48π

43. area$(S) = \frac{\pi}{6}\left(17\sqrt{17} - 1\right) \approx 36.18$ **47.** $4\pi^2 ab$

49. $f(r) = -\frac{Gm}{2Rr}\left(\sqrt{R^2 + r^2} - |R - r|\right)$

Section 16.5 Preliminary Questions

1. (b) **2.** (c) **3.** (a) **4.** (b) **5.** (a) 0 (b) π (c) π
6. $\approx 0.05\sqrt{2} \approx 0.0707$ **7.** 0

Section 16.5 Exercises

1. (a) $\mathbf{N} = \langle 2v, -4uv, 1 \rangle$; $\mathbf{F} \cdot \mathbf{N} = 2v^3 + u$

(b) $\frac{4}{\sqrt{69}}$ (c) 265

3. 4 **5.** -4 **7.** $\frac{27}{12}(3\pi + 4)$ **9.** $\frac{693}{5}$ **11.** $\frac{11}{12}$ **13.** $\frac{9\pi}{4}$
15. $(e - 1)^2$ **17.** 270 **19.** (a) $18\pi e^{-3}$ (b) $\frac{\pi}{2}e^{-1}$

21. $\left(2 - \frac{6}{\sqrt{13}}\right)\pi k$ **23.** $\frac{2\pi}{3}$ m^3/s **25.** 4π **27.** $\frac{16\pi}{3}$

29. (a) 1 (b) 1

33. $\Phi(t) = -1.56 \times 10^{-5}e^{-0.1t}$ T-m^2;
voltage drop $= -1.56 \times 10^{-6}e^{-0.1t}$ volts

35. The flow is greatest at $z = 3$.

Chapter 16 Review

1. (a) $\langle -15, 8 \rangle$ (b) $\langle 4, 8 \rangle$ (c) $\langle 9, 1 \rangle$

3.

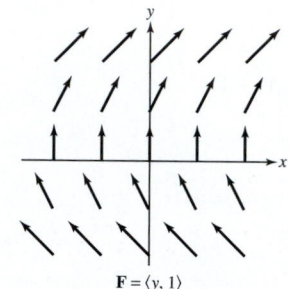

$$\mathbf{F} = \langle y, 1 \rangle$$

5. $\mathbf{F}(x, y) = \langle 2x, -1 \rangle$

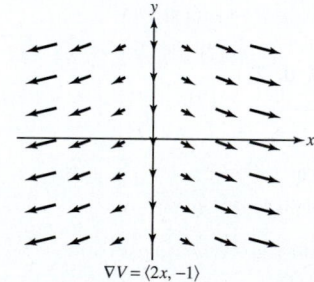

$$\nabla V = \langle 2x, -1 \rangle$$

7. div$(\mathbf{F}) = 2x + 2y + 2z$; curl$(\mathbf{F}) = \langle 0, 0, 0 \rangle$

9. div$(\mathbf{F}) = 3x^2y + y^2$; curl$(\mathbf{F}) = \langle 2yz - 2xz, 0, z^2 - x^3 \rangle$

11. div$(\mathbf{F}) = -1$; curl$(\mathbf{F}) = \langle 0, 0, -1 \rangle$

13. div$(\mathbf{F}) =$
$4x^2e^{-x^2-y^2-z^2} + 4y^2e^{-x^2-y^2-z^2} + 4z^2e^{-x^2-y^2-z^2} - 6e^{-x^2-y^2-z^2}$;
curl$(\mathbf{F}) = \langle 0, 0, 0 \rangle$

15. curl$(\mathbf{F}) = \langle -2z, 0, -2y \rangle$ **19.** $f(x, y) = x^4y^5$

21. $\mathbf{F}$ is conservative; $f(x, y, z) = 2x + 4y + e^z$

23. $f(x, y) = \frac{x^4y^4}{4}$

25. $\mathbf{F}$ is conservative; $f(x, y, z) = y \ln\left(x^2 + z\right)$

27. $\mathbf{F} = \langle 1 + by, 1 + bx \rangle$ **29.** -2

31. $\sqrt{2}(\sin 3 - 2\cos 3 + \sin 1 + 2\cos 1 + 4) \approx 11.375$

33. (a) 2 (b) 0 **37.** $\frac{1}{3}$ **39.** $4\ln\left(1 + (\ln 2)^4 + e^2\right) \approx 8.616$
41. $\frac{52}{29} \approx 1.79$ joules **43.** $3e^{3/2} - \frac{15}{2} \approx 5.945$

45. area $(S) =$
$\int_{-1}^{1}\int_{-1}^{1}\sqrt{125u^2 - 100uv + 425v^2 + 81}\,du\,dv \approx 62.911$

47. $\frac{4}{9}\left(2\sqrt{2} - 1\right)$

49. (a) Zero since $f(x, y, z) = y^3$ is odd and symmetric about the xz-plane

(b) Positive since $f(x, y, z) = z^3$ is nonnegative

(c) Zero since $f(-x, y, z) = -xyz = -f(x, y, z)$ is symmetric about the yz-plane

(d) Negative since $f(x, y, z) = z^2 - 2$ is negative

51. (a) $\mathbf{N} = \left\langle \frac{2}{\sqrt{5}}, \frac{2\sqrt{3}}{\sqrt{5}}, 2 \right\rangle$

(b) area $(G(R)) = \frac{6}{\sqrt{5}} \cdot 0.1 \cdot 0.05 \approx 0.0134$

53. $\int \int_S \mathbf{F} \cdot d\mathbf{S} =$

$\int_0^1 \left(\int_0^1 (-4(2u - v) - 2(2v + 5) + 7(-u - 3v)) \, dv \right) du = -28$

55. $\frac{1}{4} - \frac{\sinh 1}{3}$ **57.** $\frac{\sqrt{6}\pi}{9}$ **59.** 27π **61.** $\frac{9\sqrt{3}}{4}$

Chapter 17

Section 17.1 Preliminary Questions

1. $\mathbf{F} = \left\langle -e^y, \ x^2 \right\rangle$

2.

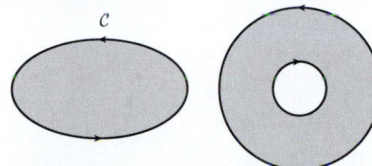

3. Yes   **4.** (a), (c)

5. (a) 0 (b) $-A$ (c) 0 (d) A

Section 17.1 Exercises

1. $\oint_C xy \, dx + y \, dy =$

$\int_0^{2\pi} (\cos\theta \sin\theta (-\sin\theta) + \sin\theta \cos\theta) \, d\theta = 0 =$

$\int_{-1}^1 \int_{-\sqrt{1-x^2}}^{\sqrt{1-x^2}} (0 - x) \, dy \, dx = \int \int_D \left(\frac{\partial}{\partial x} y - \frac{\partial}{\partial y} (xy) \right) dA$

3. 0   **5.** -3 **7.** $-\frac{\pi}{4}$ **9.** $\frac{1}{6}$ **11.** $\frac{(e^2 - 1)(e^4 - 5)}{2}$

13. (a) $f(x, y) = x^2 e^y$ (c) $4\pi - 16$ **15.** $I = 34$ **17.** πab

19. $\frac{1}{2}$ **21.** $\frac{8}{3}$ **23.** (c) $A = \frac{3}{2}$ **27.** $9 + \frac{15\pi}{2}$ **29.** 214π

31. (A) Zero (B) Positive (C) Negative (D) Zero

33. -0.10 **35.** $R = \sqrt{\frac{2}{3}}$ **37.** Triangle (A), 3; polygon (B), 12

39. -4 **41.** 5π **43.** $\frac{9\pi}{4}$ **45.** 29.2 buffalo leaving per minute

Section 17.2 Preliminary Questions

1.

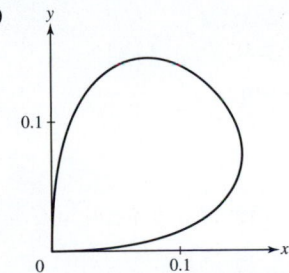

(A)   (B)

2. (a)

3. A vector field $\mathbf{A}$ such that $\mathbf{F} = \text{curl}(\mathbf{A})$ is a vector potential for $\mathbf{F}$.

4. (b)   **5.** $\mathbf{F}$ must be the curl of some other vector field $\mathbf{A}$.

Section 17.2 Exercises

1. $\oint_{\partial S} \mathbf{F} \cdot d\mathbf{r} = \int \int_S \text{curl}(\mathbf{F}) \cdot d\mathbf{S} = \pi$

3. $\oint_{\partial S} \mathbf{F} \cdot d\mathbf{r} = \int \int_S \text{curl}(\mathbf{F}) \cdot d\mathbf{S} = e^{-1} - 1$

5. $\left\langle -3z^2 e^{z^3}, \ 2ze^{z^2} + z\sin(xz), \ 2 \right\rangle$; 2π

7. $\langle -2, 3, 5 \rangle$, 20π **9.** 0   **11.** -45π **13.** $\frac{1}{2}$ **15.** 0

17. (a)

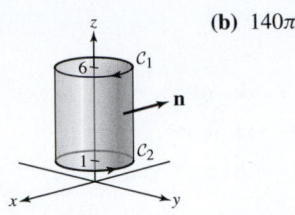

(b) 140π

19. (a) $\mathbf{A} = \left\langle 0, \ 0, \ e^y - e^{x^2} \right\rangle$ (c) $\int \int_S \mathbf{F} \cdot d\mathbf{S} = \frac{\pi}{2}$

21. (a) $\int \int_S \mathbf{B} \cdot d\mathbf{S} = r^2 B\pi$ (b) $\oint_C \mathbf{A} \cdot d\mathbf{r} = 0$

23. $\int \int_S B \, dS = 18b$ **25.** $c = 2a$ and b is arbitrary.

29. $\int \int_S \mathbf{F} \cdot d\mathbf{S} = 25$

Section 17.3 Preliminary Questions

1. $\int \int_S \mathbf{F} \cdot d\mathbf{S} = 0$

2. $\text{div}(\mathbf{F}) = 3x^2 + 3y^2 + 3z^2$; Since the integrand is positive for all $(x, y, z) \neq (0, 0, 0)$, the triple integral, hence also the flux, is positive.

3. (a), (b), (d), (f) are meaningful; (b) and (d) are automatically zero.

4. (c)   **5.** $\text{div}(\mathbf{F}) = 1$ and flux $= \int \int \int \text{div}(\mathbf{F}) \, dV = $ volume

Section 17.3 Exercises

1. $\int \int_S \mathbf{F} \cdot d\mathbf{S} = \int \int \int_W \text{div}(\mathbf{F}) \, dV = \int \int \int_W 0 \, dV = 0$

3. $\int \int_S \mathbf{F} \cdot d\mathbf{S} = \int \int \int_W \text{div}(\mathbf{F}) \, dV = 4\pi$ **5.** $\frac{4\pi}{15}$ **7.** 60π

9. $\frac{3,616}{105}$ **11.** $\frac{32\pi}{5}$ **13.** 64π **15.** 81π **17.** π **19.** $\frac{13}{3}$ **21.** $\frac{4\pi}{3}$

23. $\frac{16\pi}{3} + \frac{9\sqrt{3}}{2} \approx 24.549$ **25.** $\approx 1.57 \ \text{m}^3/\text{s}$

27. (b) 0   (c) 0   (d) Since $\mathbf{E}$ is not defined at the origin, which is inside the ball W, we cannot use the Divergence Theorem.

29. $(-4) \cdot \left[\frac{256\pi}{3} - 1 \right] \approx -1068.33$

37. (d) Flux through a closed surface is 0.

Chapter 17 Review

1. 0   **3.** -30 **5.** $\frac{3}{5}$

7. (a)

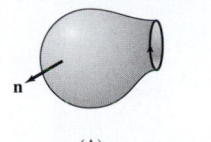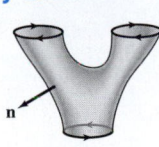

(b) $A = \frac{1}{60}$

9. $\frac{1}{3}$ **11.** $\frac{1}{3}$ **13.** $\{(x, y) \mid y = x \text{ or } y = -x\}$ **15.** 36π **19.** 2π

21. $\mathbf{A} = \langle yz, 0, 0 \rangle$ and the flux is 8π.   **23.** $\frac{296}{3}$ **25.** -128π

27. volume$(W) = 2$ **29.** $4 \cdot 0.0009\pi \approx 0.0113$

31. $2x - y + 4z = 0$ **33.** (b) $\frac{\pi}{2}$

35. (c) 0   (d) $\int_{C_1} \mathbf{F} \cdot d\mathbf{r} = -4$, $\int_{C_2} \mathbf{F} \cdot d\mathbf{r} = 4$

41. $V = \frac{4\pi}{3} abc$

REFERENCES

The online source MacTutor History of Mathematics Archive **www-history.mcs .st-and.ac.uk** has been a valuable source of historical information.

Section 1.1

(EX 83) Adapted from *Calculus Problems for a New Century*, Robert Fraga, ed., Mathematical Association of America, Washington, DC, 1993, p. 9.

Section 1.2

(EX 25) Adapted from *Calculus Problems for a New Century*, Robert Fraga, ed., Mathematical Association of America, Washington, DC, 1993, p. 9.

Section 1.7

(EXMP 4) Adapted from B. Waits and F. Demana, "The Calculator and Computer Pre-Calculus Project," in *The Impact of Calculators on Mathematics Instruction*, University of Houston, 1994.

(EX 16) Adapted from B. Waits and F. Demana, "The Calculator and Computer Pre-Calculus Project," in *The Impact of Calculators on Mathematics Instruction*, University of Houston, 1994.

Section 2.2

(EX 69) Adapted from *Calculus Problems for a New Century*, Robert Fraga, ed., Mathematical Association of America, Washington, DC, 1993, Note 28.

Section 2.3

(EX 44) Adapted from *Calculus Problems for a New Century*, Robert Fraga, ed., Mathematical Association of America, Washington, DC, 1993, Note 28.

Chapter 2 Review

(EX 72) Adapted from *Calculus Problems for a New Century*, Robert Fraga, ed., Mathematical Association of America, Washington, DC, 1993, Note 28.

Section 3.1

(EX 85) Problem suggested by Dennis DeTurck, University of Pennsylvania.

Section 3.2

(EX 90) Problem suggested by Chris Bishop, SUNY Stony Brook.

(EX 91) Problem suggested by Chris Bishop, SUNY Stony Brook.

Section 3.4

(PQ 2) Adapted from *Calculus Problems for a New Century*, Robert Fraga, ed., Mathematical Association of America, Washington, DC, 1993, p. 25.

(EX 46) Karl J. Niklas and Brian J. Enquist, "Invariant Scaling Relationships for Interspecific Plant Biomass Production Rates and Body Size," *Proc. Natl. Acad. Sci.* 98, no. 5:2922-2927 (February 27, 2001).

Section 3.5

(EX 45) Adapted from Walter Meyer, "Falling Raindrops," in *Applications of Calculus*, P. Straffin, ed., Mathematical Association of America, Washington, DC, 1993.

(EX 50, 54) Problems suggested by Chris Bishop, SUNY Stony Brook.

Section 3.10

(EX 32) Adapted from *Calculus Problems for a New Century*, Robert Fraga, ed., Mathematical Association of America, Washington, DC, 1993.

(EX 34) Problem suggested by Kay Dundas.

(EX 38, 44) Adapted from *Calculus Problems for a New Century*, Robert Fraga, ed., Mathematical Association of America, Washington, DC, 1993.

Chapter 3 Review

(EX 81, 92, 116) Problems suggested by Chris Bishop, SUNY Stony Brook.

Section 4.5

(EX 48, 81) Adapted from *Calculus Problems for a New Century*, Robert Fraga, ed., Mathematical Association of America, Washington, DC, 1993.

Section 4.6

(EX 28–29) Adapted from *Calculus Problems for a New Century*, Robert Fraga, ed., Mathematical Association of America, Washington, DC, 1993.

(EX 34) From Michael Helfgott, "Thomas Simpson and Maxima and Minima," *Convergence Magazine*, published online by the Mathematical Association of America.

(EX 42) Problem suggested by John Haverhals, Bradley University. *Source:* Illinois Agrinews.

Section 4.7

(EX 44) Adapted from *Calculus Problems for a New Century*, Robert Fraga, ed., Mathematical Association of America, Washington, DC, 1993.

(EX 68–70) Adapted from B. Noble, *Applications of Undergraduate Mathematics in Engineering*, Macmillan, New York, 1967.

(EX 72) Adapted from Roger Johnson, "A Problem in Maxima and Minima," *American Mathematical Monthly*, 35:187-188 (1928).

Section 4.7

(EX 21) Inspired by "Do Dogs Know Calculus?" Timothy Pennings, *The College Mathematic Journal*, Vol. 34, No. 3 (May, 2003), pp. 178–182.

(EX 40) Adapted from *Calculus for a Real and Complex World* by Frank Wattenberg, PWS Publishing, Boston, 1995.

(EX 79) Adapted from Robert J. Bumcrot, "Some Subtleties in L' Hôpital's Rule," in *A Century of Calculus*, Part II, Mathematical Association of America, Washington, DC, 1992.

Section 4.8

(EX 22) Adapted from *Calculus Problems for a New Century*, Robert Fraga, ed., Mathematical Association of America, Washington, DC, 1993, p. 52.

(EX 36–37) Adapted from E. Packel and S. Wagon, *Animating Calculus*, Springer-Verlag, New York, 1997, p. 79.

Chapter 4 Review

(EX 66) Adapted from *Calculus Problems for a New Century*, Robert Fraga, ed., Mathematical Association of America, Washington, DC, 1993.

Section 5.1

(EX 3) Problem suggested by John Polhill, Bloomsburg University.

(EX 90) Problem suggested by Chris Bishop, SUNY Stony Brook.

Section 5.6

(EX 25–26) M. Newman and G. Eble, "Decline in Extinction Rates and Scale Invariance in the Fossil Record," *Paleobiology* 25:434-439 (1999).

(EX 28) From H. Flanders, R. Korfhage, and J. Price, *Calculus,* Academic Press, New York, 1970.

Section 5.7

(EX 78) Adapted from *Calculus Problems for a New Century*, Robert Fraga, ed., Mathematical Association of America, Washington, DC, 1993, p. 121.

Section 6.1

(EX 54) Adapted from Tom Farmer and Fred Gass, "Miami University: An Alternative Calculus" in *Priming the Calculus Pump*, Thomas Tucker, ed., Mathematical Association of America, Washington, DC, 1990, Note 17.

(EX 67) Adapted from *Calculus Problems for a New Century*, Robert Fraga, ed., Mathematical Association of America, Washington, DC, 1993.

Section 6.3

(EX 64, 66) Adapted from G. Alexanderson and L. Klosinski, "Some Surprising Volumes of Revolution," *Two-Year College Mathematics Journal* 6, 3:13-15 (1975).

Section 7.1

(EX 66–68, 73, 74–76, 79) Problems suggested by Brian Bradie, Christopher Newport University.

(EX 93) Adapted from J. L. Borman, "A Remark on Integration by Parts," *American Mathematical Monthly*, 51:32-33 (1944).

Section 7.3

(EX 54) Adapted from *Calculus Problems for a New Century*, Robert Fraga, ed., Mathematical Association of America, Washington, DC, 1993, p. 118.

Section 7.6

(EX 1–5, 9) Problems suggested by Brian Bradie, Christopher Newport University.

Section 7.8

See R. Courant and F. John, *Introduction to Calculus and Analysis, Vol. 1*, Springer-Verlag, New York, 1989.

Section 8.2

(EX 62) Adapted from G. Klambauer, *Aspects of Calculus*, Springer-Verlag, New York, 1986, Ch 6.

Section 9.1

(EX 59) Adapted from E. Batschelet, *Introduction to Mathematics for Life Scientists*, Springer-Verlag, New York, 1979.

(EX 62, 67) Adapted from M. Tenenbaum and H. Pollard, *Ordinary Differential Equations*, Dover, New York, 1985.

Section 10.1

(EX 72) Adapted from G. Klambauer, *Aspects of Calculus*, Springer-Verlag, New York, 1986, p. 393.

Section 10.2

(EX 52) Adapted from *Calculus Problems for a New Century*, Robert Fraga, ed., Mathematical Association of America, Washington, DC, 1993, p. 137.

(EX 55) Adapted from *Calculus Problems for a New Century*, Robert Fraga, ed., Mathematical Association of America, Washington, DC, 1993, p. 138.

(EX 67) Adapted from George Andrews, "The Geometric Series in Calculus," *American Mathematical Monthly* 105, 1:36-40 (1998).

(EX 70) Adapted from Larry E. Knop, "Cantor's Disappearing Table," *The College Mathematics Journal* 16, 5:398-399 (1985).

Section 10.4

(EX 33) Adapted from *Calculus Problems for a New Century*, Robert Fraga, ed., Mathematical Association of America, Washington, DC, 1993, p. 145.

Section 11.2

(EX 43) Adapted from Richard Courant and Fritz John, *Differential and Integral Calculus*, Wiley-Interscience, New York, 1965.

Section 11.3

(EX 62) Adapted from *Calculus Problems for a New Century*, Robert Fraga, ed., Mathematical Association of America, Washington, DC, 1993.

Section 12.4

(EX 69) Adapted from Ethan Berkove and Rich Marchand, "The Long Arm of Calculus," *The College Mathematics Journal* 29, 5:376-386 (November 1998).

Section 13.3

(EX 23) Adapted from *Calculus Problems for a New Century*, Robert Fraga, ed., Mathematical Association of America, Washington, DC, 1993.

Section 13.4

(EX 70) Damien Gatinel, Thanh Hoang-Xuan, and Dimitri T. Azar, "Determination of Corneal Asphericity After Myopia Surgery with the Excimer Laser: A Mathematical Model," *Investigative Opthalmology and Visual Science* 42: 1736-1742 (2001).

Section 13.5

(EX 57, 60) Adapted from notes to the course "Dynamics and Vibrations" at Brown University, http://www.engin.brown.edu/courses/en4/.

Section 14.8

(EX 46) Adapted from C. Henry Edwards, "Ladders, Moats, and Lagrange Multipliers," *Mathematica Journal* 4, Issue 1 (Winter 1994).

Section 15.3

(FIGURE 10 COMPUTATION) The computation is based on Jeffrey Nunemacher, "The Largest Unit Ball in Any Euclidean Space," in *A Century of Calculus*, Part II, Mathematical Association of America, Washington DC, 1992.

Section 15.6

(CONCEPTUAL INSIGHT) See R. Courant and F. John, *Introduction to Calculus and Analysis*, Springer-Verlag, New York, 1989, p. 534.

Section 16.2

(FIGURE 10) Inspired by Tevian Dray and Corinne A. Manogue, "The Murder Mystery Method for Determining Whether a Vector Field Is Conservative," *The College Mathematics Journal*, May 2003.

Section 16.3

(EX 23) Adapted from *Calculus Problems for a New Century*, Robert Fraga, ed., Mathematical Association of America, Washington, DC, 1993.

Appendix D

(PROOF OF THEOREM 6) A proof without this simplifying assumption can be found in R. Courant and F. John, *Introduction to Calculus and Analysis, Vol. 1*, Springer-Verlag, New York, 1989.

INDEX

ALGEBRA

Lines

Slope of the line through $P_1 = (x_1, y_1)$ and $P_2 = (x_2, y_2)$:

$$m = \frac{y_2 - y_1}{x_2 - x_1}$$

Slope-intercept equation of line with slope m and y-intercept b:

$$y = mx + b$$

Point-slope equation of line through $P_1 = (x_1, y_1)$ with slope m:

$$y - y_1 = m(x - x_1)$$

Point-point equation of line through $P_1 = (x_1, y_1)$ and $P_2 = (x_2, y_2)$:

$$y - y_1 = m(x - x_1) \quad \text{where } m = \frac{y_2 - y_1}{x_2 - x_1}$$

Lines of slope m_1 and m_2 are parallel if and only if $m_1 = m_2$.
Lines of slope m_1 and m_2 are perpendicular if and only if $m_1 = -\frac{1}{m_2}$.

Circles

Equation of the circle with center (a, b) and radius r:

$$(x - a)^2 + (y - b)^2 = r^2$$

Distance and Midpoint Formulas

Distance between $P_1 = (x_1, y_1)$ and $P_2 = (x_2, y_2)$:

$$d = \sqrt{(x_2 - x_1)^2 + (y_2 - y_1)^2}$$

Midpoint of $\overline{P_1 P_2}$: $\left(\dfrac{x_1 + x_2}{2}, \dfrac{y_1 + y_2}{2} \right)$

Laws of Exponents

$$x^m x^n = x^{m+n} \qquad \frac{x^m}{x^n} = x^{m-n} \qquad (x^m)^n = x^{mn}$$

$$x^{-n} = \frac{1}{x^n} \qquad (xy)^n = x^n y^n \qquad \left(\frac{x}{y}\right)^n = \frac{x^n}{y^n}$$

$$x^{1/n} = \sqrt[n]{x} \qquad \sqrt[n]{xy} = \sqrt[n]{x}\,\sqrt[n]{y} \qquad \sqrt[n]{\frac{x}{y}} = \frac{\sqrt[n]{x}}{\sqrt[n]{y}}$$

$$x^{m/n} = \sqrt[n]{x^m} = \left(\sqrt[n]{x}\right)^m$$

Special Factorizations

$$x^2 - y^2 = (x + y)(x - y)$$
$$x^3 + y^3 = (x + y)(x^2 - xy + y^2)$$
$$x^3 - y^3 = (x - y)(x^2 + xy + y^2)$$

Binomial Theorem

$$(x + y)^2 = x^2 + 2xy + y^2$$
$$(x - y)^2 = x^2 - 2xy + y^2$$
$$(x + y)^3 = x^3 + 3x^2y + 3xy^2 + y^3$$
$$(x - y)^3 = x^3 - 3x^2y + 3xy^2 - y^3$$

$$(x + y)^n = x^n + nx^{n-1}y + \frac{n(n-1)}{2}x^{n-2}y^2$$
$$+ \cdots + \binom{n}{k}x^{n-k}y^k + \cdots + nxy^{n-1} + y^n$$

where $\dbinom{n}{k} = \dfrac{n(n-1)\cdots(n-k+1)}{1\cdot 2\cdot 3\cdot\ \cdots\ \cdot k}$

Quadratic Formula

If $ax^2 + bx + c = 0$, then $x = \dfrac{-b \pm \sqrt{b^2 - 4ac}}{2a}$.

Inequalities and Absolute Value

If $a < b$ and $b < c$, then $a < c$.

If $a < b$, then $a + c < b + c$.

If $a < b$ and $c > 0$, then $ca < cb$.

If $a < b$ and $c < 0$, then $ca > cb$.

$|x| = x \quad$ if $x \geq 0$

$|x| = -x \quad$ if $x \leq 0$

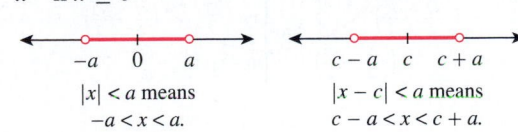

$|x| < a$ means $-a < x < a$. $\qquad$ $|x - c| < a$ means $c - a < x < c + a$.

GEOMETRY

Formulas for area A, circumference C, and volume V

Triangle	Circle	Sector of Circle	Sphere	Cylinder	Cone	Cone with arbitrary base
$A = \frac{1}{2}bh$	$A = \pi r^2$	$A = \frac{1}{2}r^2\theta$	$V = \frac{4}{3}\pi r^3$	$V = \pi r^2 h$	$V = \frac{1}{3}\pi r^2 h$	$V = \frac{1}{3}Ah$
$= \frac{1}{2}ab\sin\theta$	$C = 2\pi r$	$s = r\theta$	$A = 4\pi r^2$		$A = \pi r\sqrt{r^2 + h^2}$	where A is the area of the base
		(θ in radians)				

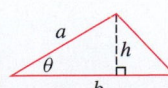 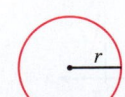 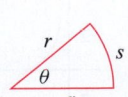 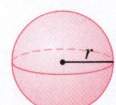 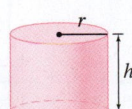 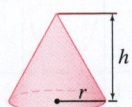 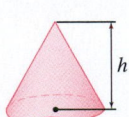

Pythagorean Theorem: For a right triangle with hypotenuse of length c and legs of lengths a and b, $c^2 = a^2 + b^2$.

TRIGONOMETRY

Angle Measurement

π radians $= 180°$

$1° = \dfrac{\pi}{180}$ rad 1 rad $= \dfrac{180°}{\pi}$

$s = r\theta$ (θ in radians)

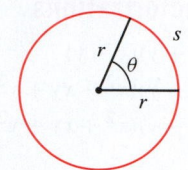

Right Triangle Definitions

$\sin\theta = \dfrac{\text{opp}}{\text{hyp}}$ $\cos\theta = \dfrac{\text{adj}}{\text{hyp}}$

$\tan\theta = \dfrac{\sin\theta}{\cos\theta} = \dfrac{\text{opp}}{\text{adj}}$ $\cot\theta = \dfrac{\cos\theta}{\sin\theta} = \dfrac{\text{adj}}{\text{opp}}$

$\sec\theta = \dfrac{1}{\cos\theta} = \dfrac{\text{hyp}}{\text{adj}}$ $\csc\theta = \dfrac{1}{\sin\theta} = \dfrac{\text{hyp}}{\text{opp}}$

Trigonometric Functions

$\sin\theta = \dfrac{y}{r}$ $\csc\theta = \dfrac{r}{y}$

$\cos\theta = \dfrac{x}{r}$ $\sec\theta = \dfrac{r}{x}$

$\tan\theta = \dfrac{y}{x}$ $\cot\theta = \dfrac{x}{y}$

$\displaystyle\lim_{\theta\to 0}\dfrac{\sin\theta}{\theta} = 1$ $\displaystyle\lim_{\theta\to 0}\dfrac{1-\cos\theta}{\theta} = 0$

$P = (r\cos\theta, r\sin\theta)$

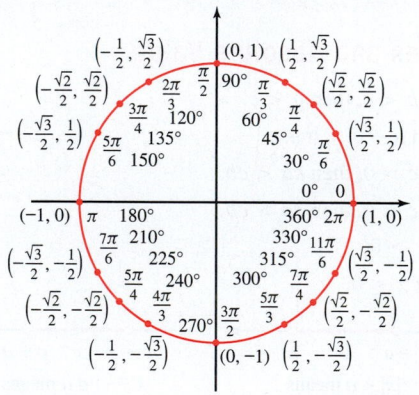

Fundamental Identities

$\sin^2\theta + \cos^2\theta = 1$ $\sin(-\theta) = -\sin\theta$

$1 + \tan^2\theta = \sec^2\theta$ $\cos(-\theta) = \cos\theta$

$1 + \cot^2\theta = \csc^2\theta$ $\tan(-\theta) = -\tan\theta$

$\sin\left(\dfrac{\pi}{2} - \theta\right) = \cos\theta$ $\sin(\theta + 2\pi) = \sin\theta$

$\cos\left(\dfrac{\pi}{2} - \theta\right) = \sin\theta$ $\cos(\theta + 2\pi) = \cos\theta$

$\tan\left(\dfrac{\pi}{2} - \theta\right) = \cot\theta$ $\tan(\theta + \pi) = \tan\theta$

The Law of Sines

$\dfrac{\sin A}{a} = \dfrac{\sin B}{b} = \dfrac{\sin C}{c}$

The Law of Cosines

$a^2 = b^2 + c^2 - 2bc\cos A$

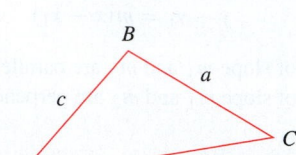

Addition and Subtraction Formulas

$\sin(x + y) = \sin x\cos y + \cos x\sin y$

$\sin(x - y) = \sin x\cos y - \cos x\sin y$

$\cos(x + y) = \cos x\cos y - \sin x\sin y$

$\cos(x - y) = \cos x\cos y + \sin x\sin y$

$\tan(x + y) = \dfrac{\tan x + \tan y}{1 - \tan x\tan y}$

$\tan(x - y) = \dfrac{\tan x - \tan y}{1 + \tan x\tan y}$

Double-Angle Formulas

$\sin 2x = 2\sin x\cos x$

$\cos 2x = \cos^2 x - \sin^2 x = 2\cos^2 x - 1 = 1 - 2\sin^2 x$

$\tan 2x = \dfrac{2\tan x}{1 - \tan^2 x}$

$\sin^2 x = \dfrac{1 - \cos 2x}{2}$ $\cos^2 x = \dfrac{1 + \cos 2x}{2}$

Graphs of Trigonometric Functions

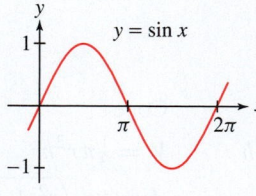

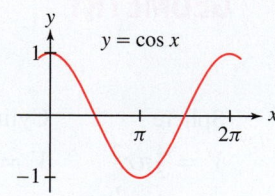

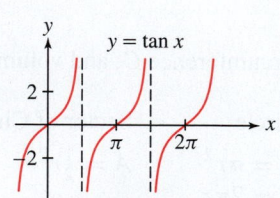

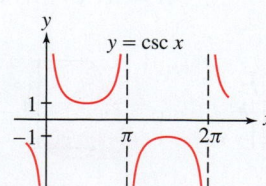

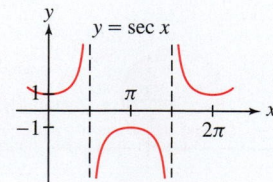

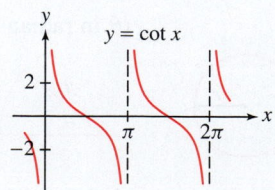

ELEMENTARY FUNCTIONS

Power Functions $f(x) = x^a$

$f(x) = x^n$, n a positive integer

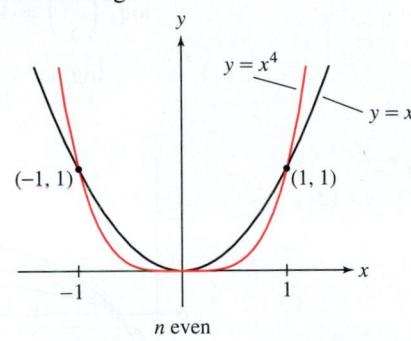

n even

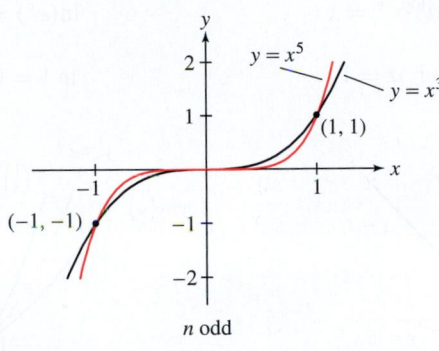

n odd

Asymptotic behavior of a polynomial function of even degree and positive leading coefficient

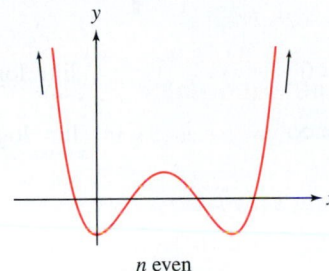

n even

Asymptotic behavior of a polynomial function of odd degree and positive leading coefficient

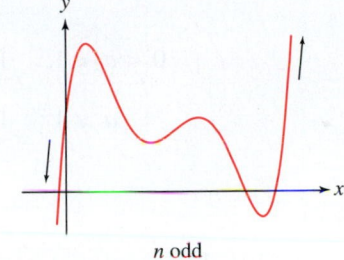

n odd

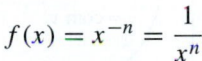
$$f(x) = x^{-n} = \frac{1}{x^n}$$

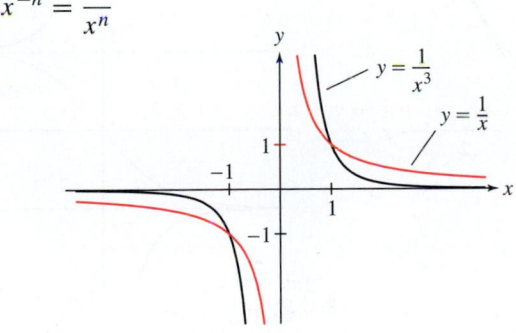

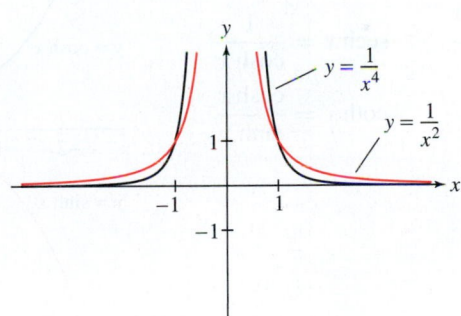

Inverse Trigonometric Functions

$\arcsin x = \sin^{-1} x = \theta$

$\Leftrightarrow \quad \sin\theta = x, \quad -\dfrac{\pi}{2} \le \theta \le \dfrac{\pi}{2}$

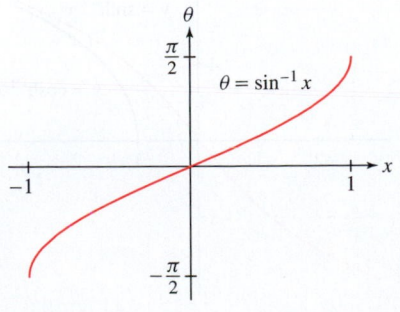

$\arccos x = \cos^{-1} x = \theta$

$\Leftrightarrow \quad \cos\theta = x, \quad 0 \le \theta \le \pi$

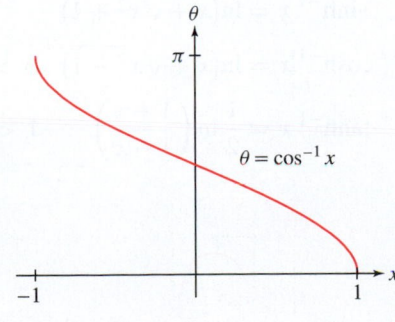

$\arctan x = \tan^{-1} x = \theta$

$\Leftrightarrow \quad \tan\theta = x, \quad -\dfrac{\pi}{2} < \theta < \dfrac{\pi}{2}$

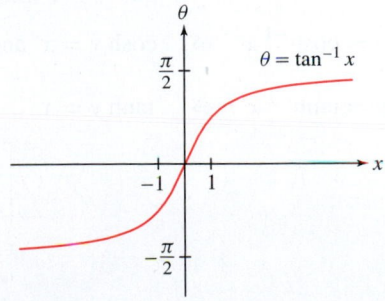

Exponential and Logarithmic Functions

$$\log_a x = y \quad \Leftrightarrow \quad a^y = x$$

$$\ln x = y \quad \Leftrightarrow \quad e^y = x$$

$$\log_a(xy) = \log_a x + \log_a y$$

$$\log_a(a^x) = x \qquad a^{\log_a x} = x$$

$$\ln(e^x) = x \qquad e^{\ln x} = x$$

$$\log_a\left(\frac{x}{y}\right) = \log_a x - \log_a y$$

$$\log_a 1 = 0 \qquad \log_a a = 1$$

$$\ln 1 = 0 \qquad \ln e = 1$$

$$\log_a(x^r) = r \log_a x$$

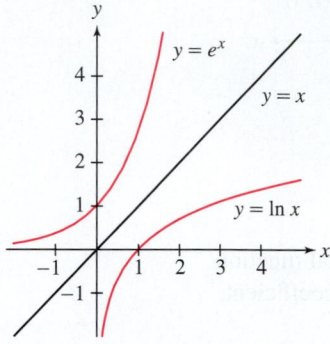

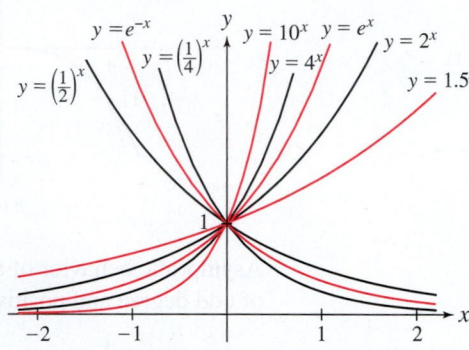

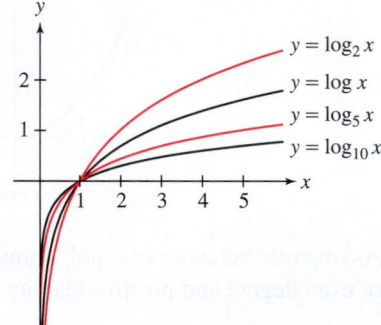

$$0 < a < 1: \quad \lim_{x \to -\infty} a^x = \infty, \ \lim_{x \to \infty} a^x = 0$$

$$\lim_{x \to 0^+} \log_a x = -\infty$$

$$a > 1: \quad \lim_{x \to -\infty} a^x = 0, \quad \lim_{x \to \infty} a^x = \infty$$

$$\lim_{x \to \infty} \log_a x = \infty$$

Hyperbolic Functions

$$\sinh x = \frac{e^x - e^{-x}}{2} \qquad \operatorname{csch} x = \frac{1}{\sinh x}$$

$$\cosh x = \frac{e^x + e^{-x}}{2} \qquad \operatorname{sech} x = \frac{1}{\cosh x}$$

$$\tanh x = \frac{\sinh x}{\cosh x} \qquad \coth x = \frac{\cosh x}{\sinh x}$$

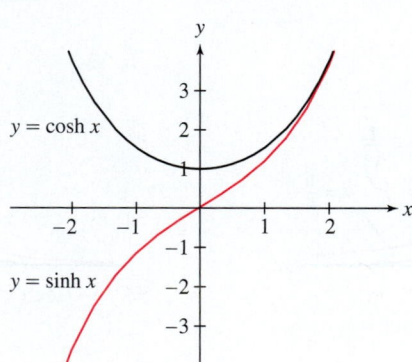

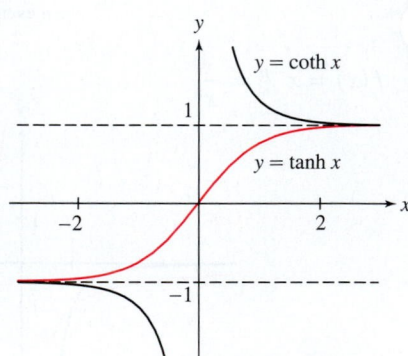

$$\sinh(x + y) = \sinh x \cosh y + \cosh x \sinh y$$

$$\sinh 2x = 2 \sinh x \cosh x$$

$$\cosh(x + y) = \cosh x \cosh y + \sinh x \sinh y$$

$$\cosh 2x = \cosh^2 x + \sinh^2 x$$

Inverse Hyperbolic Functions

$$y = \sinh^{-1} x \quad \Leftrightarrow \quad \sinh y = x$$

$$\sinh^{-1} x = \ln\left(x + \sqrt{x^2 + 1}\right)$$

$$y = \cosh^{-1} x \quad \Leftrightarrow \quad \cosh y = x \text{ and } y \geq 0$$

$$\cosh^{-1} x = \ln\left(x + \sqrt{x^2 - 1}\right) \quad x > 1$$

$$y = \tanh^{-1} x \quad \Leftrightarrow \quad \tanh y = x$$

$$\tanh^{-1} x = \frac{1}{2} \ln\left(\frac{1 + x}{1 - x}\right) \quad -1 < x < 1$$

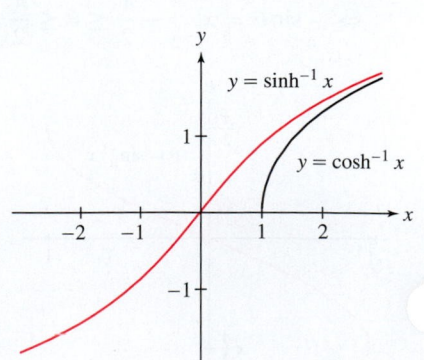

DIFFERENTIATION

Differentiation Rules

1. $\dfrac{d}{dx}(c) = 0$

2. $\dfrac{d}{dx}x = 1$

3. $\dfrac{d}{dx}(x^n) = nx^{n-1}$ (Power Rule)

4. $\dfrac{d}{dx}[cf(x)] = cf'(x)$

5. $\dfrac{d}{dx}[f(x) + g(x)] = f'(x) + g'(x)$

6. $\dfrac{d}{dx}[f(x)g(x)] = f(x)g'(x) + g(x)f'(x)$ (Product Rule)

7. $\dfrac{d}{dx}\left[\dfrac{f(x)}{g(x)}\right] = \dfrac{g(x)f'(x) - f(x)g'(x)}{[g(x)]^2}$ (Quotient Rule)

8. $\dfrac{d}{dx}f(g(x)) = f'(g(x))g'(x)$ (Chain Rule)

9. $\dfrac{d}{dx}f(x)^n = nf(x)^{n-1}f'(x)$ (General Power Rule)

10. $\dfrac{d}{dx}f(kx + b) = kf'(kx + b)$

11. $\dfrac{d}{dx}g(x) = \dfrac{1}{f'(g(x))}$ where g is the inverse f^{-1}

12. $\dfrac{d}{dx}\ln f(x) = \dfrac{f'(x)}{f(x)}$

Trigonometric Functions

13. $\dfrac{d}{dx}\sin x = \cos x$

14. $\dfrac{d}{dx}\cos x = -\sin x$

15. $\dfrac{d}{dx}\tan x = \sec^2 x$

16. $\dfrac{d}{dx}\csc x = -\csc x \cot x$

17. $\dfrac{d}{dx}\sec x = \sec x \tan x$

18. $\dfrac{d}{dx}\cot x = -\csc^2 x$

Inverse Trigonometric Functions

19. $\dfrac{d}{dx}(\sin^{-1} x) = \dfrac{1}{\sqrt{1 - x^2}}$

20. $\dfrac{d}{dx}(\cos^{-1} x) = -\dfrac{1}{\sqrt{1 - x^2}}$

21. $\dfrac{d}{dx}(\tan^{-1} x) = \dfrac{1}{1 + x^2}$

22. $\dfrac{d}{dx}(\csc^{-1} x) = -\dfrac{1}{|x|\sqrt{x^2 - 1}}$

23. $\dfrac{d}{dx}(\sec^{-1} x) = \dfrac{1}{|x|\sqrt{x^2 - 1}}$

24. $\dfrac{d}{dx}(\cot^{-1} x) = -\dfrac{1}{1 + x^2}$

Exponential and Logarithmic Functions

25. $\dfrac{d}{dx}(e^x) = e^x$

26. $\dfrac{d}{dx}(a^x) = (\ln a)a^x$

27. $\dfrac{d}{dx}\ln|x| = \dfrac{1}{x}$

28. $\dfrac{d}{dx}(\log_a x) = \dfrac{1}{(\ln a)x}$

Hyperbolic Functions

29. $\dfrac{d}{dx}(\sinh x) = \cosh x$

30. $\dfrac{d}{dx}(\cosh x) = \sinh x$

31. $\dfrac{d}{dx}(\tanh x) = \operatorname{sech}^2 x$

32. $\dfrac{d}{dx}(\operatorname{csch} x) = -\operatorname{csch} x \coth x$

33. $\dfrac{d}{dx}(\operatorname{sech} x) = -\operatorname{sech} x \tanh x$

34. $\dfrac{d}{dx}(\coth x) = -\operatorname{csch}^2 x$

Inverse Hyperbolic Functions

35. $\dfrac{d}{dx}(\sinh^{-1} x) = \dfrac{1}{\sqrt{1 + x^2}}$

36. $\dfrac{d}{dx}(\cosh^{-1} x) = \dfrac{1}{\sqrt{x^2 - 1}}$

37. $\dfrac{d}{dx}(\tanh^{-1} x) = \dfrac{1}{1 - x^2}$

38. $\dfrac{d}{dx}(\operatorname{csch}^{-1} x) = -\dfrac{1}{|x|\sqrt{x^2 + 1}}$

39. $\dfrac{d}{dx}(\operatorname{sech}^{-1} x) = -\dfrac{1}{x\sqrt{1 - x^2}}$

40. $\dfrac{d}{dx}(\coth^{-1} x) = \dfrac{1}{1 - x^2}$

INTEGRATION

Substitution

If an integrand has the form $f(u(x))u'(x)$, then rewrite the entire integral in terms of u and its differential $du = u'(x)\,dx$:

$$\int f(u(x))u'(x)\,dx = \int f(u)\,du$$

Integration by Parts Formula

$$\int uv'\,dx = uv - \int u'v\,dx$$

TABLE OF INTEGRALS

Basic Forms

1. $\displaystyle\int u^n\,du = \frac{u^{n+1}}{n+1} + C, \quad n \neq -1$

2. $\displaystyle\int \frac{du}{u} = \ln|u| + C$

3. $\displaystyle\int e^u\,du = e^u + C$

4. $\displaystyle\int a^u\,du = \frac{a^u}{\ln a} + C$

5. $\displaystyle\int \sin u\,du = -\cos u + C$

6. $\displaystyle\int \cos u\,du = \sin u + C$

7. $\displaystyle\int \sec^2 u\,du = \tan u + C$

8. $\displaystyle\int \csc^2 u\,du = -\cot u + C$

9. $\displaystyle\int \sec u \tan u\,du = \sec u + C$

10. $\displaystyle\int \csc u \cot u\,du = -\csc u + C$

11. $\displaystyle\int \tan u\,du = \ln|\sec u| + C$

12. $\displaystyle\int \cot u\,du = \ln|\sin u| + C$

13. $\displaystyle\int \sec u\,du = \ln|\sec u + \tan u| + C$

14. $\displaystyle\int \csc u\,du = \ln|\csc u - \cot u| + C$

15. $\displaystyle\int \frac{du}{\sqrt{a^2 - u^2}} = \sin^{-1}\frac{u}{a} + C$

16. $\displaystyle\int \frac{du}{a^2 + u^2} = \frac{1}{a}\tan^{-1}\frac{u}{a} + C$

Exponential and Logarithmic Forms

17. $\displaystyle\int ue^{au}\,du = \frac{1}{a^2}(au - 1)e^{au} + C$

18. $\displaystyle\int u^n e^{au}\,du = \frac{1}{a}u^n e^{au} - \frac{n}{a}\int u^{n-1}e^{au}\,du$

19. $\displaystyle\int e^{au}\sin bu\,du = \frac{e^{au}}{a^2 + b^2}(a\sin bu - b\cos bu) + C$

20. $\displaystyle\int e^{au}\cos bu\,du = \frac{e^{au}}{a^2 + b^2}(a\cos bu + b\sin bu) + C$

21. $\displaystyle\int \ln u\,du = u\ln u - u + C$

22. $\displaystyle\int u^n \ln u\,du = \frac{u^{n+1}}{(n+1)^2}[(n+1)\ln u - 1] + C$

23. $\displaystyle\int \frac{1}{u\ln u}\,du = \ln|\ln u| + C$

Hyperbolic Forms

24. $\displaystyle\int \sinh u\,du = \cosh u + C$

25. $\displaystyle\int \cosh u\,du = \sinh u + C$

26. $\displaystyle\int \tanh u\,du = \ln\cosh u + C$

27. $\displaystyle\int \coth u\,du = \ln|\sinh u| + C$

28. $\displaystyle\int \operatorname{sech} u\,du = \tan^{-1}|\sinh u| + C$

29. $\displaystyle\int \operatorname{csch} u\,du = \ln\left|\tanh \frac{1}{2}u\right| + C$

30. $\displaystyle\int \operatorname{sech}^2 u\,du = \tanh u + C$

31. $\displaystyle\int \operatorname{csch}^2 u\,du = -\coth u + C$

32. $\displaystyle\int \operatorname{sech} u \tanh u\,du = -\operatorname{sech} u + C$

33. $\displaystyle\int \operatorname{csch} u \coth u\,du = -\operatorname{csch} u + C$

Trigonometric Forms

34. $\displaystyle\int \sin^2 u\,du = \frac{1}{2}u - \frac{1}{4}\sin 2u + C$

35. $\displaystyle\int \cos^2 u\,du = \frac{1}{2}u + \frac{1}{4}\sin 2u + C$

36. $\displaystyle\int \tan^2 u\,du = \tan u - u + C$

37. $\displaystyle\int \cot^2 u\,du = -\cot u - u + C$

38. $\displaystyle\int \sin^3 u\,du = -\frac{1}{3}(2 + \sin^2 u)\cos u + C$

39. $\displaystyle\int \cos^3 u\,du = \frac{1}{3}(2 + \cos^2 u)\sin u + C$

40. $\displaystyle\int \tan^3 u\,du = \frac{1}{2}\tan^2 u + \ln|\cos u| + C$

41. $\int \cot^3 u \, du = -\dfrac{1}{2} \cot^2 u - \ln|\sin u| + C$

42. $\int \sec^3 u \, du = \dfrac{1}{2} \sec u \tan u + \dfrac{1}{2} \ln|\sec u + \tan u| + C$

43. $\int \csc^3 u \, du = -\dfrac{1}{n} \csc u \cot u + \dfrac{1}{n} \ln|\csc u - \cot u| + C$

44. $\int \sin^n u \, du = -\dfrac{1}{n} \sin^{n-1} u \cos u + \dfrac{n-1}{n} \int \sin^{n-2} u \, du$

45. $\int \cos^n u \, du = \dfrac{1}{n} \cos^{n-1} u \sin u + \dfrac{n-1}{n} \int \cos^{n-2} u \, du$

46. $\int \tan^n u \, du = \dfrac{1}{n-1} \tan^{n-1} u - \int \tan^{n-2} u \, du$

47. $\int \cot^n u \, du = \dfrac{-1}{n-1} \cot^{n-1} u - \int \cot^{n-2} u \, du$

48. $\int \sec^n u \, du = \dfrac{1}{n-1} \tan u \sec^{n-2} u + \dfrac{n-2}{n-1} \int \sec^{n-2} u \, du$

49. $\int \csc^n u \, du = \dfrac{-1}{n-1} \cot u \csc^{n-2} u + \dfrac{n-2}{n-1} \int \csc^{n-2} u \, du$

50. $\int \sin au \sin bu \, du = \dfrac{\sin(a-b)u}{2(a-b)} - \dfrac{\sin(a+b)u}{2(a+b)} + C$

51. $\int \cos au \cos bu \, du = \dfrac{\sin(a-b)u}{2(a-b)} + \dfrac{\sin(a+b)u}{2(a+b)} + C$

52. $\int \sin au \cos bu \, du = -\dfrac{\cos(a-b)u}{2(a-b)} - \dfrac{\cos(a+b)u}{2(a+b)} + C$

53. $\int u \sin u \, du = \sin u - u \cos u + C$

54. $\int u \cos u \, du = \cos u + u \sin u + C$

55. $\int u^n \sin u \, du = -u^n \cos u + n \int u^{n-1} \cos u \, du$

56. $\int u^n \cos u \, du = u^n \sin u - n \int u^{n-1} \sin u \, du$

57. $\int \sin^n u \cos^m u \, du$

$\quad = -\dfrac{\sin^{n-1} u \cos^{m+1} u}{n+m} + \dfrac{n-1}{n+m} \int \sin^{n-2} u \cos^m u \, du$

$\quad = \dfrac{\sin^{n+1} u \cos^{m-1} u}{n+m} + \dfrac{m-1}{n+m} \int \sin^n u \cos^{m-2} u \, du$

Inverse Trigonometric Forms

58. $\int \sin^{-1} u \, du = u \sin^{-1} u + \sqrt{1-u^2} + C$

59. $\int \cos^{-1} u \, du = u \cos^{-1} u - \sqrt{1-u^2} + C$

60. $\int \tan^{-1} u \, du = u \tan^{-1} u - \dfrac{1}{2} \ln(1+u^2) + C$

61. $\int u \sin^{-1} u \, du = \dfrac{2u^2-1}{4} \sin^{-1} u + \dfrac{u\sqrt{1-u^2}}{4} + C$

62. $\int u \cos^{-1} u \, du = \dfrac{2u^2-1}{4} \cos^{-1} u - \dfrac{u\sqrt{1-u^2}}{4} + C$

63. $\int u \tan^{-1} u \, du = \dfrac{u^2+1}{2} \tan^{-1} u - \dfrac{u}{2} + C$

64. $\int u^n \sin^{-1} u \, du = \dfrac{1}{n+1} \left[u^{n+1} \sin^{-1} u - \int \dfrac{u^{n+1} \, du}{\sqrt{1-u^2}} \right], \quad n \neq -1$

65. $\int u^n \cos^{-1} u \, du = \dfrac{1}{n+1} \left[u^{n+1} \cos^{-1} u + \int \dfrac{u^{n+1} \, du}{\sqrt{1-u^2}} \right], \quad n \neq -1$

66. $\int u^n \tan^{-1} u \, du = \dfrac{1}{n+1} \left[u^{n+1} \tan^{-1} u - \int \dfrac{u^{n+1} \, du}{1+u^2} \right], \quad n \neq -1$

Forms Involving $\sqrt{a^2 - u^2}$, $a > 0$

67. $\int \sqrt{a^2 - u^2} \, du = \dfrac{u}{2} \sqrt{a^2 - u^2} + \dfrac{a^2}{2} \sin^{-1} \dfrac{u}{a} + C$

68. $\int u^2 \sqrt{a^2 - u^2} \, du = \dfrac{u}{8}(2u^2 - a^2)\sqrt{a^2 - u^2} + \dfrac{a^4}{8} \sin^{-1} \dfrac{u}{a} + C$

69. $\int \dfrac{\sqrt{a^2 - u^2}}{u} \, du = \sqrt{a^2 - u^2} - a \ln\left| \dfrac{a + \sqrt{a^2 - u^2}}{u} \right| + C$

70. $\int \dfrac{\sqrt{a^2 - u^2}}{u^2} \, du = -\dfrac{1}{u}\sqrt{a^2 - u^2} - \sin^{-1} \dfrac{u}{a} + C$

71. $\int \dfrac{u^2 \, du}{\sqrt{a^2 - u^2}} = -\dfrac{u}{2}\sqrt{a^2 - u^2} + \dfrac{a^2}{2} \sin^{-1} \dfrac{u}{a} + C$

72. $\int \dfrac{du}{u\sqrt{a^2 - u^2}} = -\dfrac{1}{a} \ln\left| \dfrac{a + \sqrt{a^2 - u^2}}{u} \right| + C$

73. $\int \dfrac{du}{u^2\sqrt{a^2 - u^2}} = -\dfrac{1}{a^2 u} \sqrt{a^2 - u^2} + C$

74. $\int (a^2 - u^2)^{3/2} \, du = -\dfrac{u}{8}(2u^2 - 5a^2)\sqrt{a^2 - u^2} + \dfrac{3a^4}{8} \sin^{-1} \dfrac{u}{a} + C$

75. $\int \dfrac{du}{(a^2 - u^2)^{3/2}} = \dfrac{u}{a^2\sqrt{a^2 - u^2}} + C$

Forms Involving $\sqrt{u^2 - a^2}$, $a > 0$

76. $\int \sqrt{u^2 - a^2} \, du = \dfrac{u}{2}\sqrt{u^2 - a^2} - \dfrac{a^2}{2} \ln|u + \sqrt{u^2 - a^2}| + C$

77. $\int u^2 \sqrt{u^2 - a^2} \, du$

$\quad = \dfrac{u}{8}(2u^2 - a^2)\sqrt{u^2 - a^2} - \dfrac{a^4}{8} \ln|u + \sqrt{u^2 - a^2}| + C$

78. $\int \dfrac{\sqrt{u^2 - a^2}}{u} \, du = \sqrt{u^2 - a^2} - a \cos^{-1} \dfrac{a}{|u|} + C$

79. $\int \dfrac{\sqrt{u^2 - a^2}}{u^2} \, du = -\dfrac{\sqrt{u^2 - a^2}}{u} + \ln|u + \sqrt{u^2 - a^2}| + C$

80. $\int \dfrac{du}{\sqrt{u^2 - a^2}} = \ln|u + \sqrt{u^2 - a^2}| + C$

81. $\int \dfrac{u^2 \, du}{\sqrt{u^2 - a^2}} = \dfrac{u}{2}\sqrt{u^2 - a^2} + \dfrac{a^2}{2} \ln|u + \sqrt{u^2 - a^2}| + C$

82. $\int \dfrac{du}{u^2\sqrt{u^2 - a^2}} = \dfrac{\sqrt{u^2 - a^2}}{a^2 u} + C$

83. $\int \dfrac{du}{(u^2 - a^2)^{3/2}} = -\dfrac{u}{a^2\sqrt{u^2 - a^2}} + C$

Forms Involving $\sqrt{a^2 + u^2}$, $a > 0$

84. $\int \sqrt{a^2 + u^2} \, du = \dfrac{u}{2}\sqrt{a^2 + u^2} + \dfrac{a^2}{2} \ln(u + \sqrt{a^2 + u^2}) + C$

85. $\int u^2 \sqrt{a^2 + u^2} \, du$

$\quad = \dfrac{u}{8}(a^2 + 2u^2)\sqrt{a^2 + u^2} - \dfrac{a^4}{8} \ln(u + \sqrt{a^2 + u^2}) + C$

86. $\displaystyle\int \frac{\sqrt{a^2+u^2}}{u}\,du = \sqrt{a^2+u^2} - a\ln\left|\frac{a+\sqrt{a^2+u^2}}{u}\right| + C$

87. $\displaystyle\int \frac{\sqrt{a^2+u^2}}{u^2}\,du = -\frac{\sqrt{a^2+u^2}}{u} + \ln\left(u+\sqrt{a^2+u^2}\right) + C$

88. $\displaystyle\int \frac{du}{\sqrt{a^2+u^2}} = \ln\left(u+\sqrt{a^2+u^2}\right) + C$

89. $\displaystyle\int \frac{u^2\,du}{\sqrt{a^2+u^2}} = \frac{u}{2}\sqrt{a^2+u^2} - \frac{a^2}{2}\ln\left(u+\sqrt{a^2+u^2}\right) + C$

90. $\displaystyle\int \frac{du}{u\sqrt{a^2+u^2}} = -\frac{1}{a}\ln\left|\frac{\sqrt{a^2+u^2}+a}{u}\right| + C$

91. $\displaystyle\int \frac{du}{u^2\sqrt{a^2+u^2}} = -\frac{\sqrt{a^2+u^2}}{a^2 u} + C$

92. $\displaystyle\int \frac{du}{(a^2+u^2)^{3/2}} = \frac{u}{a^2\sqrt{a^2+u^2}} + C$

Forms Involving $a+bu$

93. $\displaystyle\int \frac{u\,du}{a+bu} = \frac{1}{b^2}\left(a+bu - a\ln|a+bu|\right) + C$

94. $\displaystyle\int \frac{u^2\,du}{a+bu} = \frac{1}{2b^3}\left[(a+bu)^2 - 4a(a+bu) + 2a^2\ln|a+bu|\right] + C$

95. $\displaystyle\int \frac{du}{u(a+bu)} = \frac{1}{a}\ln\left|\frac{u}{a+bu}\right| + C$

96. $\displaystyle\int \frac{du}{u^2(a+bu)} = -\frac{1}{au} + \frac{b}{a^2}\ln\left|\frac{a+bu}{u}\right| + C$

97. $\displaystyle\int \frac{u\,du}{(a+bu)^2} = \frac{a}{b^2(a+bu)} + \frac{1}{b^2}\ln|a+bu| + C$

98. $\displaystyle\int \frac{du}{u(a+bu)^2} = \frac{1}{a(a+bu)} - \frac{1}{a^2}\ln\left|\frac{a+bu}{u}\right| + C$

99. $\displaystyle\int \frac{u^2\,du}{(a+bu)^2} = \frac{1}{b^3}\left(a+bu - \frac{a^2}{a+bu} - 2a\ln|a+bu|\right) + C$

100. $\displaystyle\int u\sqrt{a+bu}\,du = \frac{2}{15b^2}(3bu-2a)(a+bu)^{3/2} + C$

101. $\displaystyle\int u^n\sqrt{a+bu}\,du$

$\displaystyle\qquad = \frac{2}{b(2n+3)}\left[u^n(a+bu)^{3/2} - na\int u^{n-1}\sqrt{a+bu}\,du\right]$

102. $\displaystyle\int \frac{u\,du}{\sqrt{a+bu}} = \frac{2}{3b^2}(bu-2a)\sqrt{a+bu} + C$

103. $\displaystyle\int \frac{u^n\,du}{\sqrt{a+bu}} = \frac{2u^n\sqrt{a+bu}}{b(2n+1)} - \frac{2na}{b(2n+1)}\int \frac{u^{n-1}\,du}{\sqrt{a+bu}}$

104. $\displaystyle\int \frac{du}{u\sqrt{a+bu}} = \frac{1}{\sqrt{a}}\ln\left|\frac{\sqrt{a+bu}-\sqrt{a}}{\sqrt{a+bu}+\sqrt{a}}\right| + C, \quad \text{if } a > 0$

$\displaystyle\qquad\quad = \frac{2}{\sqrt{-a}}\tan^{-1}\sqrt{\frac{a+bu}{-a}} + C, \qquad \text{if } a < 0$

105. $\displaystyle\int \frac{du}{u^n\sqrt{a+bu}} = -\frac{\sqrt{a+bu}}{a(n-1)u^{n-1}} - \frac{b(2n-3)}{2a(n-1)}\int \frac{du}{u^{n-1}\sqrt{a+bu}}$

106. $\displaystyle\int \frac{\sqrt{a+bu}}{u}\,du = 2\sqrt{a+bu} + a\int \frac{du}{u\sqrt{a+bu}}$

107. $\displaystyle\int \frac{\sqrt{a+bu}}{u^2}\,du = -\frac{\sqrt{a+bu}}{u} + \frac{b}{2}\int \frac{du}{u\sqrt{a+bu}}$

Forms Involving $\sqrt{2au-u^2}$, $a>0$

108. $\displaystyle\int \sqrt{2au-u^2}\,du = \frac{u-a}{2}\sqrt{2au-u^2} + \frac{a^2}{2}\cos^{-1}\left(\frac{a-u}{a}\right) + C$

109. $\displaystyle\int u\sqrt{2au-u^2}\,du$

$\displaystyle\qquad = \frac{2u^2-au-3a^2}{6}\sqrt{2au-u^2} + \frac{a^3}{2}\cos^{-1}\left(\frac{a-u}{a}\right) + C$

110. $\displaystyle\int \frac{du}{\sqrt{2au-u^2}} = \cos^{-1}\left(\frac{a-u}{a}\right) + C$

111. $\displaystyle\int \frac{du}{u\sqrt{2au-u^2}} = -\frac{\sqrt{2au-u^2}}{au} + C$

ESSENTIAL THEOREMS

Intermediate Value Theorem

If f is continuous on a closed interval $[a, b]$ and $f(a) \neq f(b)$, then for every value M between $f(a)$ and $f(b)$, there exists at least one value $c \in (a, b)$ such that $f(c) = M$.

Mean Value Theorem

If f is continuous on a closed interval $[a, b]$ and differentiable on (a, b), then there exists at least one value $c \in (a, b)$ such that

$$f'(c) = \frac{f(b) - f(a)}{b - a}$$

Extreme Values on a Closed Interval

If f is continuous on a closed interval $[a, b]$, then f attains both a minimum and a maximum value on $[a, b]$. Furthermore, if $c \in [a, b]$ and $f(c)$ is an extreme value (min or max), then c is either a critical point of f in (a, b) or one of the endpoints a or b.

The Fundamental Theorem of Calculus, Part I

Assume that f is continuous on $[a, b]$ and let F be an antiderivative of f on $[a, b]$. Then

$$\int_a^b f(x)\,dx = F(b) - F(a)$$

Fundamental Theorem of Calculus, Part II

Assume that f is a continuous function on $[a, b]$. Then the area function $A(x) = \displaystyle\int_a^x f(t)\,dt$ is an antiderivative of f, that is,

$$A'(x) = f(x) \quad \text{or equivalently} \quad \frac{d}{dx}\int_a^x f(t)\,dt = f(x)$$

Furthermore, $A(x)$ satisfies the initial condition $A(a) = 0$.